Chapter 9
REFRIGERANTS 335

Chapter 10
REFRIGERANT RECOVERY/RECYCLING/RECLAIMING 365

Robinair Division, SPX Corporation

Chapter 11
DOMESTIC REFRIGERATORS AND FREEZERS 381

Chapter 12
SERVICING AND INSTALLING SMALL HERMETIC SYSTEMS 417

Chapter 13
COMMERCIAL SYSTEMS 473

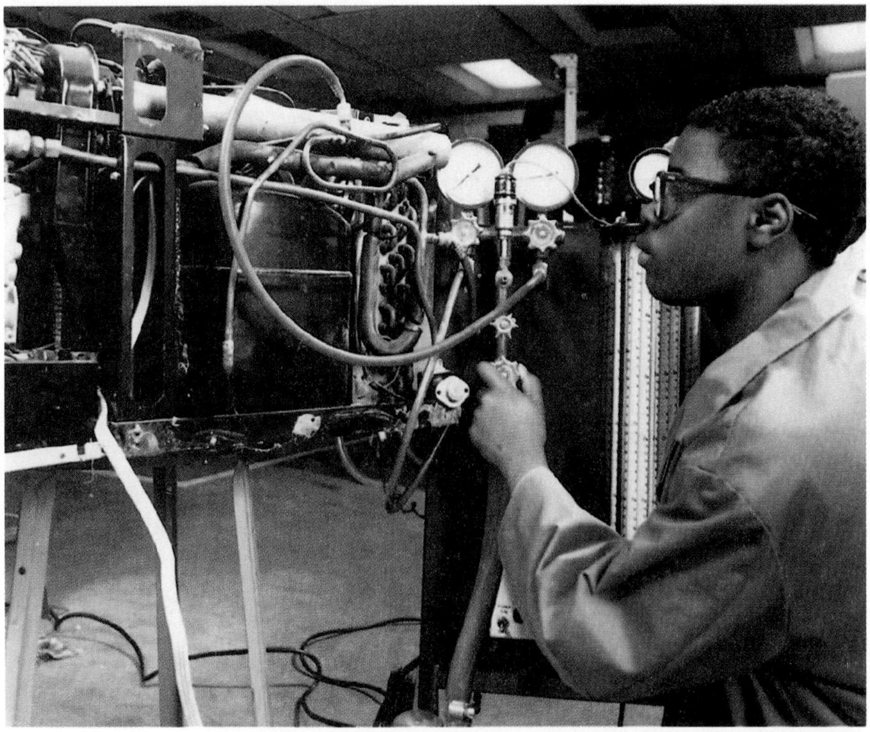

Southeast Oakland Vocational Education Center

Chapter 14
COMMERCIAL SYSTEMS—APPLICATIONS 547

Chapter 15
SERVICING AND INSTALLING COMMERCIAL SYSTEMS 571

Leybold Inficon, Inc.

Chapter 16
COMMERCIAL SYSTEMS—HEAT LOADS AND PIPING 633

Chapter 17
ABSORPTION SYSTEMS—PRINCIPLES AND APPLICATIONS 685

Chapter 18
SPECIAL REFRIGERATION SYSTEMS AND APPLICATIONS 709

Recycling Specialists International

Chapter 19
FUNDAMENTALS OF AIR CONDITIONING 727

Chapter 20
BASIC HEATING AND AIR CONDITIONING SYSTEMS 761

Chapter 21
HEATING AND HUMIDIFICATION SYSTEMS 789

Chapter 22
COOLING AND DEHUMIDIFYING SYSTEMS 853

Chapter 23
AIR DISTRIBUTION, MEASUREMENT, AND CLEANING 873

Knauf Fiber Glass GmbH

Chapter 24
HEAT PUMPS AND COMPLETE AIR CONDITIONING SYSTEMS 915

Chapter 25
SOLAR ENERGY 969

Chapter 26
AIR CONDITIONING AND HEATING CONTROL SYSTEMS 983

Chapter 27
AIR CONDITIONING SYSTEMS—HEAT LOADS 1033

Chapter 28
AUTOMOTIVE AIR CONDITIONING 1057

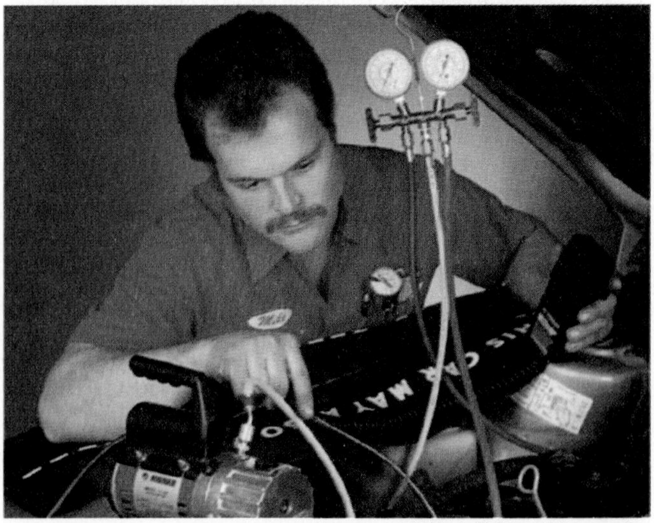

Robinair Division, SPX Corporation

Chapter 29
SERVICING AND TROUBLESHOOTING SIMPLIFIED 1085

Chapter 30
PASSING EPA EXAMS 1105

Knauf Fiber Glass GmbH

Knauf Fiber Glass GmbH

Chapter 31
TECHNICAL CHARACTERISTICS 1113

CAREER OPPORTUNITIES IN REFRIGERATION AND AIR CONDITIONING

In the past fifty years, the heating, ventilating and air conditioning (HVAC) field has experienced massive technological change. It has gone from the era of the iceman to that of the educated and highly trained technician. See **Figures A** and **B.**

The most rapid advances have occurred in the last ten years. Today's technician needs more than a small box of tools and a cylinder of refrigerant. You must now have a broad background in working with computerized, automated electronic HVAC equipment.

Many of the recent changes in the HVAC field are due to rapid changes in technology and a growing concern for the environment. Scientists have warned that continued release of refrigerants to the atmosphere will destroy the earth's ozone layer. This layer, located approximately 35 miles above the ground, protects the earth from the damaging ultraviolet rays of the sun. Destruction of the ozone layer would affect humans, animal, plants, and sea life.

To prevent continuing damage to the ozone layer, laws have been passed governing the types of refrigerants manufactured and how they can be used. New equipment has been developed that requires skill and training for proper operation.

A technician today must be familiar with the complex electronic devices used in refrigeration systems. It is common to see fully automated heating and cooling systems in homes. These systems can be set for varying temperatures and humidity levels for each individual room. Commercial buildings are using computerized systems that are even more sophisticated.

Figure A. *Only a half-century ago, the iceman was a familiar figure in most neighborhoods, delivering blocks of ice to keep food cold. (Edward Hulyk Studio)*

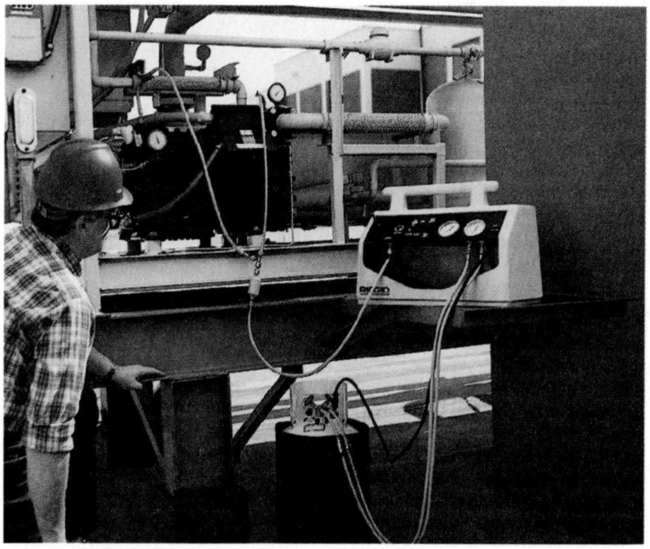

Figure B. *Today's HVAC technician must be able to work with computer-controlled electronic systems and use sophisticated diagnostic and charging equipment. (The Ridge Tool Company)*

Importance of Refrigeration and Air Conditioning

There are few phases of modern living untouched by refrigeration and air conditioning. Business operations, manufacturing processes, storage, and shipping are almost always carried out today under controlled-temperature conditions. Skilled specialists are required to design, install, and maintain controlled environments in enclosed areas that range from homes to space satellites. The use of computerized equipment has increased the need for facilities that are totally energy-controlled.

The refrigeration and air conditioning industry helps make possible this system of living. Air conditioning has improved business and industrial efficiency, while adding to human comfort. More and more factories and heavy industries are being air conditioned. The present scale of farming is made possible, to a great extent, by the use of air-conditioned tractor cabs and refrigerated harvesting equipment. Many fruits and vegetables are refrigerated immediately upon being harvested. The quality of such products is much better for this reason.

Cooling and freezing of meat and meat products makes possible their handling in a much more sanitary way than would be possible without mechanical refrigeration. Beverages, desserts, and even staple foods are all at least partially processed by refrigeration equipment.

Designing, manufacturing, selling, installing, and maintaining this equipment provides for many, many jobs that did not exist less than a generation ago. Opportunities for employment in writing specifications for refrigeration and air conditioning equipment and selling this equipment have naturally grown with the industry. Since refrigeration is used in so many enterprises, it follows that anyone who has to work in these industries must be familiar with the basic air conditioning and refrigeration processes. All the careers in this field are available to anyone interested, regardless of race, creed, color, or sex.

The Air Conditioning and Refrigeration Industry

The air conditioning and refrigeration industry is usually divided into three areas:

- Domestic.
- Commercial.
- Industrial.

The *domestic field* covers home refrigerators, freezers, and window air conditioners.

The *commercial field* includes all small automatic systems. Such systems are used for stores, supermarkets, domestic central air conditioning, water coolers, beverage coolers, marine refrigeration and air conditioning, automotive air conditioning, and truck refrigeration and air conditioning systems. See **Figures C and D.**

The ***industrial field*** includes large processing and air conditioning systems, packing plants, cold storage, and ice rinks. These systems require the attention of a refrigeration operating engineer. **See Figure E.**

Career Opportunities

Some of the opportunities for employment in refrigeration and air conditioning include:

Jobs at various levels:

- Senior skilled.
- Skilled.
- Technicians.
- Supervisors.
- Professional personnel.

Various specialties (partial list):

- Engineers.
- Technicians.
- Test technicians.
- Sales engineers.
- Application engineers.
- Installers.
- Testers.
- Maintenance technicians.
- Service persons.
- Repair specialists.
  - Helpers.
- Wholesalers.
  - Sales engineers.
  - Salespersons.
  - Counter persons.
  - Parts persons.
  - Shipping and receiving persons.
- Operating engineers.
  - Refrigeration.
  - Industrial.
- Sheet metal experts.
  - Helpers.

Figure C. *Two service technicians repairing a rooftop air conditioning unit. (Superior Contract Services, Inc.)*

Figure D. *Installation and servicing of home air conditioners is one of the many careers open to a person with knowledge and hands-on training in refrigeration and air conditioning. (Lennox International, Inc.)*

Figure E. *An air movement system being installed. A sales engineer was required to select and specify the proper equipment, and skilled workers were needed to install it. A service technician will be needed to periodically service and maintain the system. (Knauf Fiber Glass GmbH)*

The Work They Do

As you might expect, the responsibilities of persons working in the air conditioning and heating industry will vary greatly. So will the kind of work that is done.

Consider air conditioning, heating, and refrigeration technicians, for example. Working under the supervision of engineers, they help design, manufacture, sell, and service equipment. Often, a technician will specialize in one area, such as research and development.

Those working in manufacturing may design and test or supervise production of equipment. They may also work as manufacturer's representatives or field salespersons. In such cases, responsibilities would typically include supplying contractors and engineering firms with data on installation, maintenance, operating costs, and performance specifications of equipment.

Some technicians are employed by contractors to help design and prepare installation instructions. Others work in customer relations and may be responsible for supervising the installation and maintenance of equipment.

Another group employed by the industry works on installation and service. They travel about in service trucks, servicing units in homes, offices, schools, and other buildings. This group includes:

Air conditioning and refrigeration technicians. These workers install and repair units ranging in size from small window air conditioners up to large central systems. They must follow blueprints and specifications to install motors, compressors, evaporators, and other components. They connect ducts, refrigerant lines, and piping, as well as make power hookups. In event of breakdown, they find the cause and make repairs.

Furnace installers or ***heating equipment installers.*** They read blueprints and specifications and install oil, gas, and electric heating. Installation work includes placement of fuel supply lines, air ducts, pumps, and other parts of a heating system. After connecting the electrical wiring and controls, they check units for proper operation.

Oil heating system technicians. These technicians keep oil-fueled heating systems in good working order. Their work varies with the season. In fall and winter, they service and adjust burners. During the summer, they service the heating unit, replace oil and air filters, and vacuum vents, ducts, and other parts of the system.

Gas heating system technicians or ***gas appliance service persons.*** Their duties are similar to those of an oil burner technician. They determine why a burner will not work and adjust or repair it.

Cooling and heating systems are sometimes installed or repaired by other types of technicians or tradespeople. For example, ductwork on a large heating or air conditioning job may be done by sheet metalworkers; electrical work by an electrician; and piping by pipe fitters. This is often the case on large installations where union members of the building trades are involved.

Educational Requirements

To qualify for employment, you should have strong communication skills, a good grasp of practical mathematics, and some physics and chemistry knowledge. A minimum requirement is at least a one-year training program in refrigeration, heating, and air conditioning systems. The program should include both theory and hands-on laboratory work. **Figure F** provides a graphic representation of employment opportunities in the refrigeration and air conditioning fields.

Specific employment information is available from the nearest branch of the United States Employment Service and the local State Employment Service. Local school or public libraries also offer reference materials such as the *Dictionary of Occupational Titles* and the *Occupational Outlook Handbook.*

EMPLOYMENT OPPORTUNITIES

EDUCATIONAL REQUIREMENTS

- **A** GRADE 12 GRADUATE
- **APPRENTICESHIP B** VOCATIONAL/ TECHNICAL SCHOOL
- **C** COMMUNITY COLLEGE
- **D** UNIVERSITY

EDUCATION	CODE	JOB TITLES
D	1	• Outside Sales Engineer
AC	2	• Outside Sales Tech. or Rep.
ACD	3	• Inside Sales Representative
AC	4	• Inside Sales Order Desk/Counter
ABCD	5	• Estimator
CD	6	• Sales Manager
CD	7	• Administration
CD	8	• Application Engineer/Tech.
AC	9	• Draftsman—Layout
D	10	• Consulting/Design Engineer
CD	11	• Project Manager
ACD	12	• Purchasing
ABC	13	• Production
BCD	14	• Inspector
BC	15	• Job Foreman or Supervisor
ABCD	16	• Specification Writer
BC	17	• Field Service & Installation
ABC	18	• Shop Service and Repair
B	19	• Journeyman-Technician
AB	20	• Apprentice Technician
ABCD	21	• Service Manager
C	22	• Lab. Technician/Technologist
CD	23	• Research & Development
ABC	24	• Service Salesperson
AC	25	• Inventory Control

AREAS OF EMPLOYMENT

Area	Codes	Description
CONSULTING	5, 7, 9, 10, 11, 14, 16	**Consulting** Companies devoted to designing and engineering the heating, cooling, ventilating, plumbing, and electrical systems for buildings. Includes preparation of drawings, specifications, estimate of cost, supervision of installation, and final approval.
CONTRACTING	5, 7, 8, 9, 10, 11, 12, 15, 17, 19, 20	**Contracting** Companies that sell and install mechanical systems. Includes installation and fabrication of system components, preparation of drawings, estimation of costs, and supervision of installation according to specifications.
SERVICING	1, 2, 3, 4, 5, 6, 7, 8, 15, 17, 18, 19, 20, 21, 24	**Servicing** Companies that repair and maintain mechanical systems. Includes repair and maintaining of system components, sale and installation of replacement components for economic and efficient system operation.
MANUFACTURING	1, 2, 3, 4, 5, 6, 7, 8, 9, 10, 12, 13, 14, 17, 18, 19, 20, 21, 22, 23, 24, 25	**Manufacturing** Companies that purchase raw material or components and fabricate or assemble into equipment for sale. Includes sales and marketing, production, design, and research development.
MERCHANDISING & SALES	1, 2, 3, 4, 5, 6, 7, 8, 12, 25	**Merchandising and Sales** Companies which promote and sell equipment and components which have been manufactured by others. Includes sales promotion, advertising, warehousing, and technical assistance.
GOVERNMENT & UTILITIES	5, 7, 8, 9, 10, 11, 12, 14, 16, 19, 20, 22	**Government and Utilities** Governments, utilities, and other agencies provide their own consulting and servicing functions, set standards, test equipment, and approve installations.

Figure F. *Employment opportunities in the refrigeration and air conditioning field. (American Society of Heating, Refrigerating, and Air-Conditioning Engineers)*

As the technology in the heating, refrigerating, and air conditioning field has changed, so has the leading textbook in the field, ***Modern Refrigerating and Air Conditioning.*** *This edition of the highly acclaimed text contains the most recent information and advances in the field, including technical changes, EPA rulings, and recovery, recycling, and reclaiming of refrigerants.*

Chapter 1

FUNDAMENTALS OF REFRIGERATION

Modules:

Key Terms:

Btu
condenser
evaporator
horsepower
latent heat
power
pressure
refrigerant
sensible heat
temperature

Learning Objectives:

By studying this chapter, you will be able to:

- Describe the early development of refrigeration.
- Discuss the basic physical, chemical, and engineering principles which apply to refrigeration.
- Explain how cold preserves food.
- Define basic refrigeration terms.
- Explain principles of heat transfer.
- Compare Fahrenheit, Celsius, Kelvin, and Rankine temperature scales.
- Use temperature conversion formulas to convert from one temperature scale to another.
- Determine area and volume of cabinets.
- Explain the difference between psia (absolute pressure) and psig (gauge pressure).
- Describe the basic operation of a refrigerator.
- Discuss the differences between sensible heat, specific heat, and latent heat. Describe their applications.
- Explain physical laws which apply to refrigeration.
- Demonstrate and explain the relationship between SI metric and U.S. conventional measurement.
- Recognize and use various symbols for SI metric units of measure.
- Make conversions between U.S. conventional and SI metric systems of measurement.
- Calculate the enthalpy of water at a variety of temperatures.
- **Follow approved safety procedures.**

Users of this text who are unfamiliar with SI metrics have no cause for concern. Conventional measurements are carried alongside the metric. Reference is made to Chapter 31 as problems arise affecting metric usage.

HISTORY AND FUNDAMENTALS OF REFRIGERATION MODULE

1.1 Development of Refrigeration

Modern refrigeration has many applications. The first, and probably still the most important, is the preservation of food.

Most foods kept at room temperature spoil rapidly. This is due to the rapid growth of bacteria. At common refrigeration temperatures of about 40°F (4°C), bacteria grow quite slowly. Food at this temperature will keep much longer. Refrigeration preserves food by keeping it cold.

Other important uses of refrigeration include air conditioning, beverage cooling, and humidity control. Many manufacturing processes also use refrigeration.

The refrigeration industry became important commercially during the 18th century. Early refrigeration was obtained by use of ice. Ice from lakes and ponds was cut and stored in the winter in insulated storerooms for summer use.

The use of natural ice required building insulated containers or iceboxes for stores, restaurants, and homes. These units appeared on a large scale during the 19th century.

Ice was first made artificially about 1820 as an experiment. Not until 1834 did artificial ice manufacturing become practical. Jacob Perkins, an American engineer, invented the machine which led to our modern compression systems. Michael Faraday discovered the principles for the absorption type of refrigeration as early as 1824. It was not actually built until 1855 by a German engineer.

Little artificial ice was produced until shortly after 1890. During 1890, a warm winter resulted in a shortage of natural ice. This helped start the mechanical ice-making industry.

Mechanical domestic refrigeration first appeared about 1910. J.M. Larsen produced a manually operated household machine in 1913. By 1918 Kelvinator produced the first automatic refrigerator for the American market. They sold 67 machines that year. Now millions of units are sold each year.

The first of the sealed or "hermetic" automatic refrigeration units was introduced by General Electric in 1928. It was named the Monitor Top.

Beginning with 1920, domestic refrigeration became an important industry. The Electrolux, which was an automatic domestic absorption unit, appeared in 1927.

Fast freezing to preserve food for extended periods was developed about 1923. This marked the beginning of the modern frozen foods industry. Automatic refrigeration units, for the comfort cooling part of air conditioning, appeared in 1927.

Mechanical refrigeration systems were first connected to heating plants to provide summer cooling in the late 1920s. By 1940, practically all domestic units were of the hermetic type. Commercial units had also been successfully made and used. These units were capable of refrigerating large commercial food storage systems. They could provide comfort cooling of large auditoriums. They could also produce low temperatures used in many commercial operations.

In 1935, Frederick McKinley Jones produced an automatic refrigeration system for long-haul trucks. From a small, slow start in the late 1930s, air conditioning of automobiles has also grown rapidly.

Starting in the 1960s, the home air conditioning market experienced tremendous growth. Energy was inexpensive, and therefore, simple air conditioning became common in many homes. Solar energy and other alternative energy sources became additional sources for powering heating and cooling systems.

Due to a tremendous growth in technology, by 1990 all areas of refrigeration and air conditioning were using microprocessor control systems. The purpose of these systems is to increase reliability and efficiency of the heating and cooling units. By 1990, the automobile air conditioner became as standard as the automatic transmission.

1.2 How a Mechanical Refrigerator Operates

There are four basic parts in a mechanical refrigeration system. The ***compressor*** pumps refrigerant vapor. The ***condenser*** releases heat from the refrigerant, similar to a vehicle's radiator releasing heat from the cooling system. The ***refrigerant control*** releases vapor refrigerant when it is needed. Finally, the ***evaporator*** is the area that absorbs heat.

Removing heat from inside a refrigerator is somewhat like removing water from a leaking canoe. A sponge may be used to soak up the water in the canoe. The sponge is held over the side, squeezed, and the water is released overboard. The operation may be repeated as often as necessary. This transfers the water from the canoe into the lake.

In a refrigerator, heat instead of water is transferred. Inside the refrigerating mechanism, heat is absorbed. It is "soaked up" by evaporating the liquid refrigerant in the evaporator (cooling unit). This occurs as the refrigerant changes from a liquid to a vapor (gas), **Figure 1-1.**

The refrigerant, which has absorbed heat, has now turned into a vapor. It is pumped into the condensing unit located outside the refrigerated space. The condenser works the opposite of the evaporator. In the evaporator, the refrigerant enters as a liquid, absorbs heat, and flows out the other end as a vapor. By the time it reaches the end of the evaporator, it is all a vapor. Now this vapor flows into the condenser under a high pressure and high temperature. The vapor gives up its heat to the surrounding air. As it reaches the end of the condenser, the refrigerant is now cooled. It has become a liquid again. We say that, in the condenser, the heat is "squeezed out." This cycle repeats until the desired temperature is reached.

Heat enters a refrigerator in many ways. It leaks through the insulated walls or enters when the door is opened. Still more heat is introduced when warm substances are placed in the refrigerator.

Heat is not destroyed to make the refrigerator cold. It is simply removed from the refrigerated space and released outside.

1.3 Color Coding System in this Text

When presenting refrigeration systems illustrations like **Figure 1-1,** a common color coding system will be followed in this text:

Dark Red—High-Pressure Liquid ■
Light Red—High-Pressure Vapor ■
Dark Blue—Low-Pressure Liquid ■
Light Blue—Low-Pressure Vapor ■

The following paragraphs will provide the technical foundation needed to understand the heat removal operation. This background is important for service and repair.

Service managers of refrigerating and air conditioning companies prefer service and installation technicians who are good mechanics. Knowledge of the principles of mathematics and physics as these apply to refrigeration is also important.

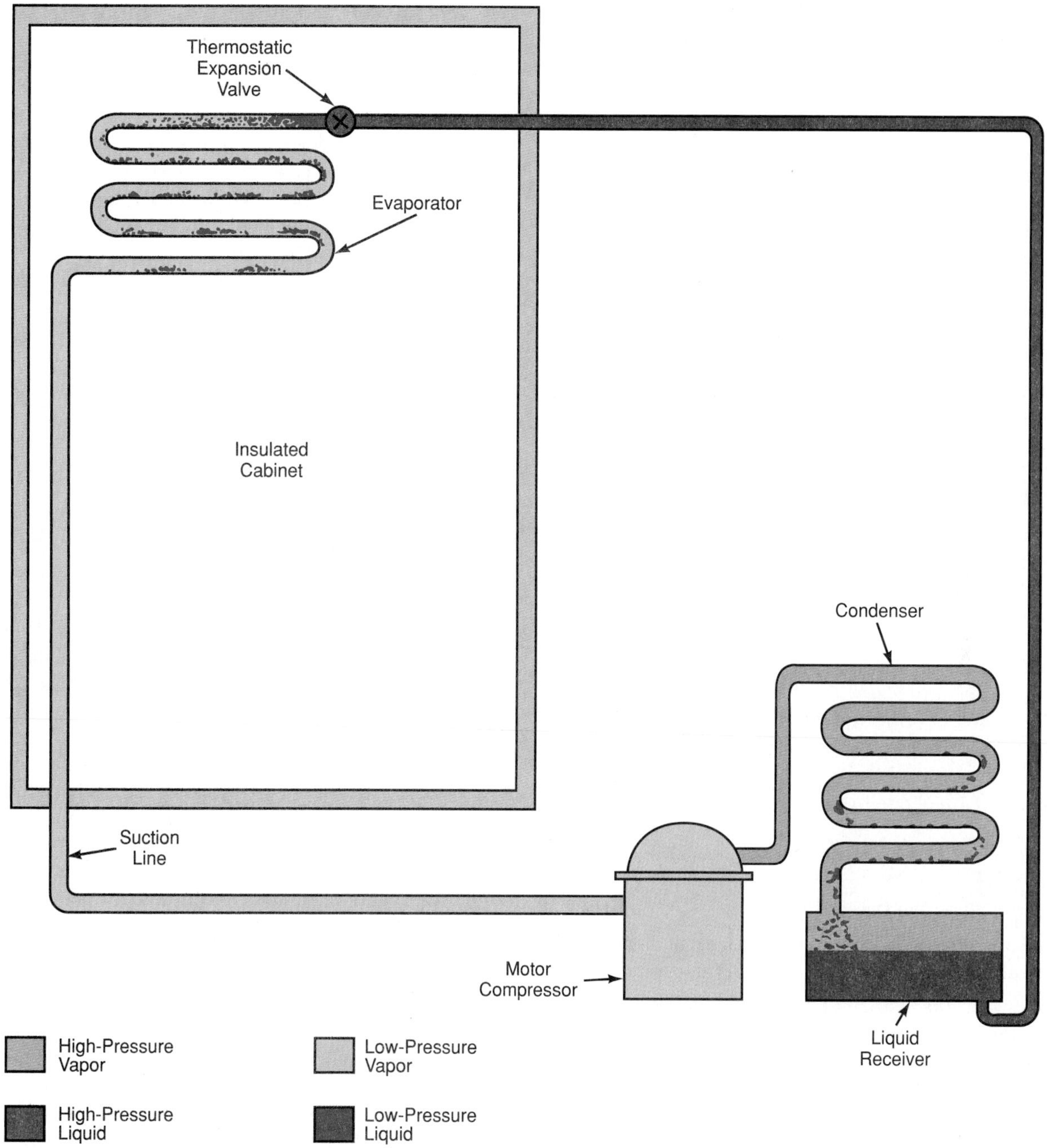

Figure 1-1. *Elementary mechanical refrigerator. In operation, liquid refrigerant under high pressure (dark red) flows from liquid receiver to pressure reducing valve (refrigerant control) and into evaporator. Here pressure is greatly reduced (dark blue). Liquid refrigerant boils and absorbs heat from evaporator. Now a vapor, refrigerant (light blue) flows back to compressor and is compressed to high pressure (light red). Its temperature is greatly increased. In the condenser, heat is transferred to the surrounding air and the refrigerant cools, becoming liquid again. It flows back into the liquid receiver and the cooling cycle is repeated.*

1.4 Heat

Heat is a form of energy. It has a relationship to the atom, the smallest indivisible part of an element. ("Indivisible" means if an atom was broken down into more pieces it would no longer be that element.) All substances are made up of tiny atoms, which combine to make molecules. All the atoms are in a state of rapid motion.

As the temperature of a substance increases, the atoms move more rapidly. As the temperature drops, they slow down. If all heat is removed from a substance (absolute zero), all molecular motion stops.

The U.S. conventional unit of heat is the ***British thermal unit (Btu).*** The metric unit of heat is the ***joule (J).*** If a substance is warmed, heat is added; if cooled, heat is removed.

The amount of heat in a substance equals the mass of the substance multiplied by its temperature. The amount of heat in a substance may greatly affect the nature of the substance. Adding heat causes most substances to expand. Removing heat causes them to contract.

Most substances change their physical state with the addition or removal of heat. For instance, water ice is a solid (under atmospheric pressure at a temperature below 0°C). If heat is added to the ice, it will melt and become water (a liquid). Further, addition of heat will cause the water to turn into a vapor (steam). The compression-type refrigeration cycle makes use of this principle in its operation.

1.4.1 Heat Flow

Heat always flows from a warmer to a cooler substance. The faster moving atoms give up some of their energy to slower moving atoms. Therefore, each fast atom slows down a little and the slower one moves a little faster.

Heat causes some solids to become liquids or gases, or liquids to become gases. Cooling will reverse the process. The atoms making up the molecules of these substances act in a different way to temperature. Instead of moving faster or slower, one or more of the atoms in the molecule shift their positions.

1.5 Cold

Cold means low temperature or lack of heat. Cold is the result of removing heat. A refrigerator produces "cold" by drawing heat from the inside of the refrigerator cabinet.

The refrigerator does not destroy the heat. It pumps heat from the inside of the cabinet to the outside. Heat always travels from a substance at a higher temperature to a substance at a lower temperature (second law of thermodynamics—see Chapter 31). Heat cannot travel spontaneously from a cold body to a hot body.

1.5.1 Cold Preserves Food

Spoiling of food is actually the growth of bacteria in it. As the molecules move slowly, they have an important effect on the bacteria present in most foods. Slowing movement by cooling the molecules makes all organisms more sluggish. Cold, or low temperature, slows down the growth of these bacteria. Foods, thus, do not spoil as fast. If the bacteria can be kept from increasing, the food will be edible longer. Often, a small change in temperature (just a few degrees) can make a large difference in the growth rate of bacteria.

Most foods contain a considerable amount of water. Food, therefore, must be kept slightly above freezing temperatures (32°F, 0°C).

If food is frozen slowly at or near the freezing point of water, the ice crystals formed are large. Their growth breaks down the food tissues. When defrosted, it spoils rapidly; appearance and taste are ruined.

Fast freezing at very low temperatures, 0 to −15°F (−18 to −26°C), forms small crystals which do not injure the food tissues. Food freezers are maintained at or below 0°F (−18°C). Food placed in freezers will freeze quickly. Keep in mind the difference between refrigerating and freezing. The correct refrigerating temperature for fresh food is 35°F (1.7°C) to 45°F (7.3°C). To make ice, a temperature lower than 32°F (0°C) is needed.

TEMPERATURE, PRESSURE, AND MEASUREMENTS MODULE

1.6 Temperature and Temperature Measurement

Temperature measures the heat intensity or heat level of a substance. Temperature alone does not give the amount of heat in a substance. It indicates the degree of warmth, or how hot or cold the substance or body is. In the molecular theory of heat, temperature indicates the speed of motion of the molecules. It is important not to use the words "heat" and "temperature" carelessly.

Temperature measures the speed of motion of the atom. ***Heat*** is the thermal energy of the atom multiplied by the number of atoms (mass) so affected.

For example, a small copper dish weighing a few grams, heated to 1340°F (727°C) does not contain as much heat as 5 kilograms of copper heated to 284°F (140°C). However, its heat level is higher. Its intensity of heat is greater.

The U.S. conventional unit of temperature is the degree Fahrenheit. The SI unit of temperature is the kelvin (K). The temperature intervals (space between degrees) on the Kelvin scale are the same as Celsius.

Temperature is measured with a thermometer. This is usually through uniform expansion of a liquid in a sealed glass tube. There is a bulb at the bottom of the tube and a quantity of liquid (mercury or alcohol) inside.

The glass does not expand or contract as much as the liquid during a temperature change. The liquid will rise and fall in the tube as the temperature changes. The tube is "calibrated" or marked off in degrees using the desired temperature scale. **Figure 1-2** shows glass stem thermometers used in refrigeration and air conditioning work.

Another type of temperature measurement instrument is shown in **Figure 1-2C.** This is known as a thermometer-pyrometer. The term "pyrometer" means high temperature. This instrument has a digital scale. It has the capability of measuring from −40°F (−40°C) to 1999°F (1100°C). It is used when accurate readings at

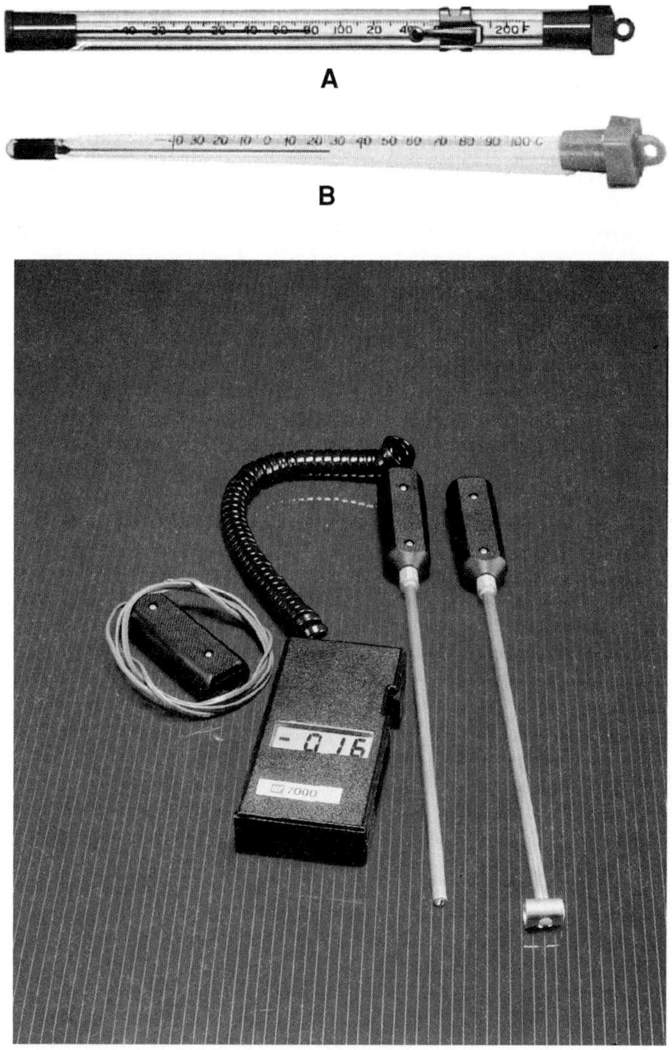

Figure 1-2. *Thermometers used in refrigeration work. A—Glass stem Fahrenheit thermometers has range from −40°F to 210°F. B—Glass stem Celsius thermometer has range from −40°C to 100°C. (Marsh Instrument Co.) C—Digital thermometer-pyrometer measures temperatures in either Fahrenheit or Celsius. (TIF Instruments, Inc.)*

various temperatures are needed. It will indicate the temperature in about 2–10 seconds.

Some thermometers use metal to measure temperature. Metal will expand and contract as temperature rises and falls. This moves an indicator up and down the scale.

Other thermometers indicate temperature by measurement of a very small electric voltage generated in a thermocouple. (See Chapters 6 and 19.) They are especially useful in the measurement of high temperatures.

A radiometer is a thermometer which detects infrared rays produced by a substance. This thermometer is very easy to use. No contact is needed with the substance whose temperature is to be measured. (See **Figure 31-1.**)

Thermistors may also be used to measure temperatures. A thermistor is a type of thermometer, which is operated by electrical current. (See Chapter 6.)

1.6.1 Thermometer Scales—Fahrenheit and Celsius

The two most common thermometer scales are the Fahrenheit and the Celsius scales. Celsius is sometimes called the Centigrade scale. The Celsius scale is named in honor of Anders Celsius, the Swedish astronomer who recommended the new system.

Two temperatures determine the calibration of a thermometer:

- The temperature of melting ice.
- The temperature of boiling water.

(Both must be at a pressure of 1 atmosphere or at sea level.)

On the Fahrenheit thermometer, the temperature of melting ice is 32°F. The temperature of boiling water is 212°F. This provides 180 spaces or degrees between the freezing and boiling temperatures.

On the Celsius thermometer, the temperature of melting ice is 0°C. The temperature of boiling water is 100°C. There are 100 spaces or degrees on the scale between freezing and boiling. For a comparison of the Fahrenheit and Celsius scales, see **Figure 1-3.** (Also, see Chapter 31.)

The freezing point and boiling point are based on freezing and boiling temperatures of water at standard atmospheric pressure. Effects of pressure on these temperatures is explained in Section 1.19 and Section 1.20.

1.6.2 Absolute Temperature Scales, Kelvin and Rankine

Absolute zero is that temperature where molecular motion stops. It is the lowest temperature possible. There is no more heat in the substance at this point.

Two absolute temperature scales are used in cryogenics (very low temperature work). (See Section 1.32.) These two scales are the Rankine (Fahrenheit Absolute) scale and the Kelvin (Celsius Absolute) scale.

The Rankine scale uses the same divisions as the Fahrenheit scale. Zero on this scale (0°R) is located 460 degrees below 0°F.

The Kelvin scale uses the same divisions as the Celsius scale. However, zero on the Kelvin scale (0 K) is 273 degrees below 0°C. (Physical scientists often omit the degree symbol (°) when writing Kelvin temperatures.) Therefore, Kelvin temperatures are often expressed with just the K, not °K.)

These temperature scales are not used by the technician in normal service work. The absolute temperature scales are used by engineers. They use them in designing and manufacturing various parts of heating and air conditioning systems. The absolute temperature scales

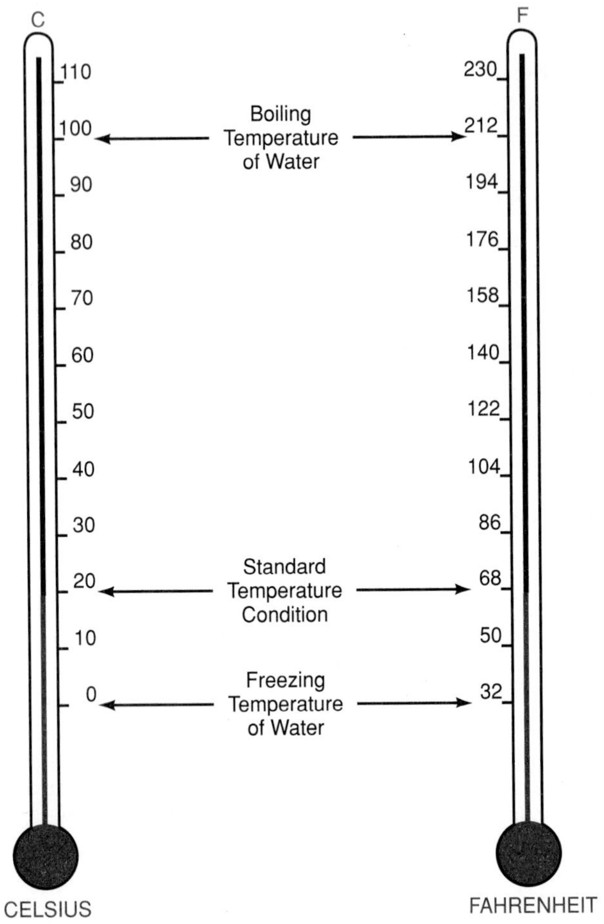

Figure 1-3. *A comparison of Celsius and Fahrenheit thermometer scales.*

are also used to identify the operational performance of a product. These ratings can then be used by the technician to compare one manufacturer's products with those of another manufacturer.

Figure 1-4 compares the Celsius, Kelvin, Fahrenheit, and Rankine thermometer scales.

Problem:

What is the temperature at which water freezes and boils using the Kelvin scale?

Solution, Freezing Point:

Water freezes at 0°C. The Kelvin scale zero is 273 degrees below 0°C. The freezing temperature of water is, therefore, 273 degrees above zero kelvin (K), or 273 kelvin. The freezing temperature is 273 K.

Solution, Boiling Point:

Water boils at 100 degrees above 0°C. The boiling point of water on the Kelvin scale will be: 100 + 273 = 373 K. Therefore, the boiling point is 373 K.

1.7 Basic Arithmetic

Basic mathematics plays an important role in a technician's day-to-day operations. Being able to quickly and accurately compute basic mathematic formulas is an asset for the technician. The following paragraphs provide you with some background in the types of mathematic operations a technician may be expected to perform on the job.

+ means plus or add.
= means equal to or of the same value.
Example: $4 + 4 = 8$

– means minus, subtract, or take away.
Example: $4 - 3 = 1$

× means multiply by, or times.
Example: $4 \times 5 = 20$

÷ means divide by.
Example: $12 \div 2 = 6$

· means multiply by, or times.
( ) are parentheses; do the arithmetic inside the parentheses first.
Example: $(7 - 3) + 2 = (4) + 2 = 6$

Some calculations use parentheses instead of a multiplication sign.
Example: $(4)(5) = 20$

$( )^2$ means that the number inside the parentheses is to be multiplied by itself, or squared.
Example: $(4)^2 = 4 \times 4 = 16$;
in this example, multiply together two 4s

$( )^3$ means that the number inside the parentheses is to be multiplied by itself three times, or cubed.
Example: $(4)^3 = 4 \times 4 \times 4 = 64$;
in this example, multiply together three 4s

$\frac{a}{b}$ means that the top number, "a," is to be divided by the bottom number, "b."
Example: If "a" = 6, and "b" = 2

$$\frac{a}{b} = \frac{6}{2} = 6 \div 2 = 3$$

Δ (delta) means a difference
Example: ΔT = temperature difference, for instance, 0°C to 40°C.

1.7.1 Basic Unit or Digits

Most calculations include the use of basic units. Basic units are expressed in digits. In the statement, $7 \times 8 = 56$, 7 and 8 are digits; 56 is made up of two digits, 5 and 6. In the metric system, multiples of digits are on the basis of 10. For example: the digit 1, if divided by 10, would be 0.1; each subsequent division of 10 would result in 0.01, 0.001, and the like. The prefix (name) for these follow. The digit 1, if multiplied by 10,

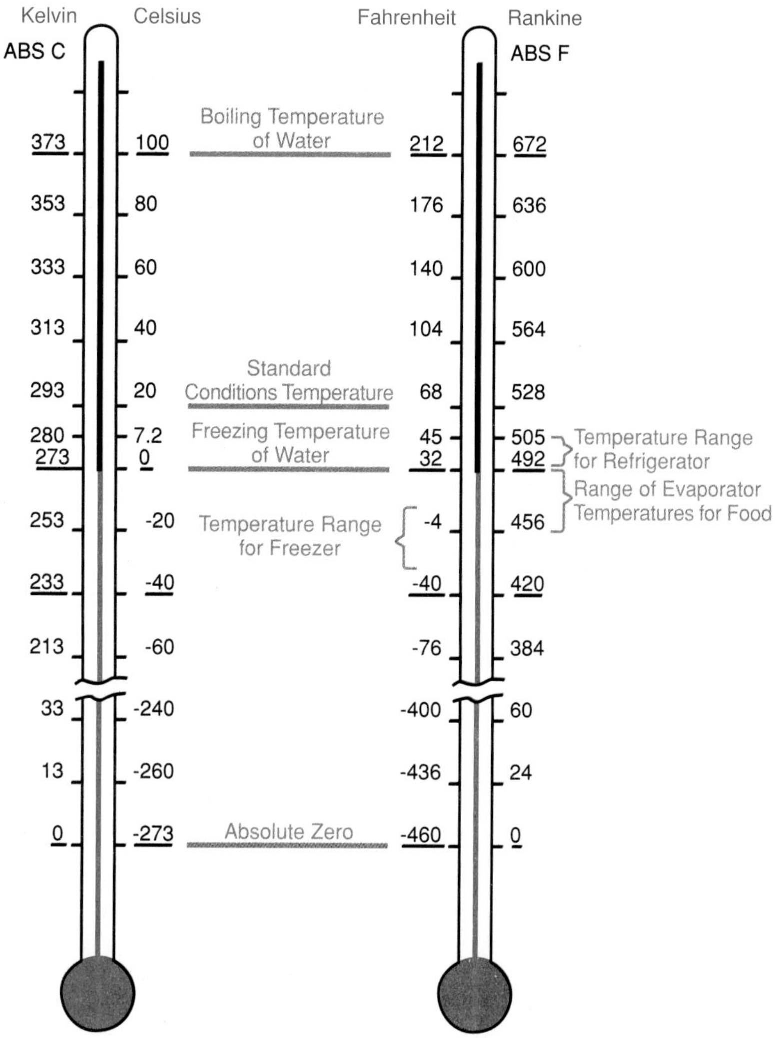

Figure 1-4. *Kelvin, Celsius, Fahrenheit, and Rankine thermometer scales are compared.*

would be 10; each subsequent multiplication by 10 would result in 100, 1000, 10,000, 100,000, and the like. Each level of multiplication or division has a name:

Symbol	Prefix	Quantity	Pronunciation
M	mega	= 1,000,000	like megaphone
k	kilo	= 1000	kill'-oh
h	hecto	= 100	heck'-toe
da	deka	= 10	deck'-uh
basic unit		= 1	
d	deci	= 0.1	dess'-ih
c	centi	= 0.01	sen'-tih
m	milli	= 0.001	like military
μ	micro	= 0.000 001	my'-crow

In many calculations, it is difficult to work with numbers using many zeros either ahead of or behind the decimal point. A special number, called a "powers of ten," may be used to express these types of numbers.

"Power of 10" means that the number 10 is multiplied by itself the desired number of times to obtain the required number of zeros. The small number above and to the right of the number 10 is called the "exponent." It works as follows:

For numbers larger than one:

$1000 = 10^3$ means multiply together three 10s $(10 \times 10 \times 10)$

$100 = 10^2$ means multiply together two 10s (10×10)

$10 = 10^1$ or (10)

For numbers less than one, a minus sign is placed before the exponent. This means that 1/10 or 0.1 is to be multiplied together, instead of 10:

$0.1 = 10^{-1}$ or (0.10)

$0.01 = 10^{-2}$ or (0.10×0.10)

$0.001 = 10^{-3}$ or $(0.10 \times 0.10 \times 0.10)$

1.7.2 Rounding Numbers

In refrigeration calculations, it is not usually necessary to use fractions or decimals of a unit. When the decimal is less than 5, round to the number and ignore the decimal. When the decimal is 5 or over, round to the next larger number. For instance, 35.5 becomes 36. If a problem has been carried two or more decimal places and less accuracy is required, it is acceptable to round such numbers to a single decimal. For instance: 3.52 may be rounded to 4.

1.8 Temperature Conversion

It is often necessary to convert a temperature from one scale to another. Formulas have been developed for this purpose. It is not necessary to memorize these formulas, only to refer to them when needed.

°C means temperature in degrees Celsius
°F means temperature in degrees Fahrenheit
K means temperature in degrees Kelvin
°R means temperature in degrees Rankine

To convert from one of these scales to another, follow the procedures outlined in the following examples.

1.8.1 Degrees Celsius to Degrees Fahrenheit

Formula:

$$\text{Temperature in °F} = \left(\frac{180}{100} \times \text{Temperature °C}\right) + 32$$

or

$$\text{Temperature in °F} = \left(\frac{9}{5} \times \text{°C}\right) + 32$$

Example:

Convert 75°C to Fahrenheit.

Solution:

$$\text{°F} = \left(\frac{9}{5} \times 75\right) + 32$$

$$\text{°F} = (1.8 \times 75) + 32$$

$$\text{°F} = 135 + 32$$

$$\text{°F} = 167\text{°F}$$

1.8.2 Degrees Fahrenheit to Degrees Celsius

Formula:

$$\text{Temperature in °C} = \frac{100}{180} \times (\text{Temperature °F} - 32)$$

or

$$\text{Temperature in °C} = \frac{5}{9} \times (\text{°F} - 32)$$

Example:

Convert 212°F to °C.

Solution:

$$\text{°C} = \frac{5}{9} \times (212 - 32)$$

$$\text{°C} = \frac{5}{9} \times 180$$

$$\text{°C} = .56 \times 180$$

$$\text{°C} = 100\text{°C}$$

1.8.3 Degrees Fahrenheit to Degrees Rankine (Fahrenheit Absolute)

Formula:

Temperature in °R (F_A) = °F + 460

Example:

Convert 40°F to °R (F_A).

Solution:

°R (F_A) = 40 + 460
°R (F_A) = 500°R (F_A)

1.8.4 Degrees Rankine to Degrees Fahrenheit

Formula:

Temperature in °F = °R − 460

Example:

Convert 180°R to °F.

Solution:

°F = 180 − 460
°F = −280°F

1.8.5 Degrees Celsius to Kelvin

Formula:

K = °C + 273

Example:

Convert −10°C to K.

Solution:

K = −10 + 273
K = 263 K

1.8.6 Kelvin to Degrees Celsius

Formula:

Temperature in °C = K − 273

Example:

Convert 400 K to °C.

Solution:

°C = 400 − 273
°C = 127°C

1.8.7 Degrees Rankine to Kelvin

Formula:

$$\text{Temperature in K} = \frac{5}{9}\,\text{°R}$$

Example:

Convert 180°R to K.

Solution:

$$\text{K} = \frac{5}{9} \times 180$$

$$\text{K} = 101\ \text{K}$$

1.8.8 Kelvin to Degrees Rankine

Formula:

$$\text{Temperature °R} = \frac{9}{5}\,\text{K}$$

Example:

Convert 263 K to °R.

Solution:

$$°R = \frac{9}{5} \times 263$$

$$°R = 473°R$$

1.9 Temperature Difference Calculations

Calculations which require converting Fahrenheit temperature difference to Celsius temperature difference and Celsius temperature difference to Fahrenheit temperature difference may be computed as follows:

Formula:

$$°C \text{ temperature difference} = \frac{5}{9}(°F \text{ temperature difference})$$

Example—Fahrenheit to Celsius:

When the outside temperature is 10°F and the inside temperature is 75°F, the temperature difference is 65°F. What is the temperature difference in °C?

Solution:

$$°C \text{ temperature difference} = \frac{5}{9} \times 65 = 36°C$$

Formula:

$$°F \text{ temperature difference} = \frac{9}{5}(°C \text{ temperature difference})$$

Example—Celsius to Fahrenheit:

When the outside temperature is 10°C and the inside temperature is 26°C, the temperature difference is 16°C. What is the temperature difference in Fahrenheit?

Solution:

$$°F \text{ temperature difference} = \frac{9}{5} \times 16 = 28.8 \text{ or } 29°F$$

Throughout this text, temperatures are given in both Fahrenheit and Celsius. Most of the Fahrenheit temperature values are rounded to whole numbers. The equivalent Celsius temperature is shown rounded to the nearest possible whole number. Where the Celsius temperature ends in 0.5 or more, the next higher temperature is used. If the Celsius temperature ends in a 0.4 or less, the next lower Celsius temperature is chosen. For example, 40°F is equivalent to 4.4°C. This number is rounded off to 4°C. Another example is 0°F. Carried to one decimal place, it equals −17.8°C. This is rounded to −18°C.

1.10 Dimensions

Dimensions, as used in this text, are measurements which are necessary in determining lengths, areas, and volumes. The following paragraphs discuss these measurements.

1.10.1 Linear Measurement (Le

Linear measurement considers only
Finding the length of a piece of copper tu
ample of linear measurement.

U.S. Conventional Units
Decimals and Fractions of an Inch

Measurement	How to Express the Measurement
0.001 in.	one-thousandth of an inch
0.01 in.	one-hundredth of an inch
0.1 in.	one-tenth of an inch
1/64 in.	one sixty-fourth of an inch
1/32 in.	one thirty-second of an inch
1/16 in.	one-sixteenth of an inch
1/8 in.	one-eighth of an inch
1/4 in.	one-fourth of an inch
1/2 in.	one-half of an inch

Sometimes the symbol (″) indicates inches; for example, 6″. Occasionally the symbol (′) indicates feet. The following is an example: 6′.

Units of Conventional Linear Measurement

12 inches = 1 foot
3 feet = 1 yard
5280 feet = 1 statute mile
6080 feet = 1 nautical mile

Metric Units and U.S. Conventional Unit Equivalents

1 millimeter (mm)	= 0.039 in.
10 mm	= 1 centimeter (cm) = 0.394 in.
10 cm	= 1 decimeter (dm) = 3.937 in.
10 dm	= 1 meter (m) = 100 cm = 39.37 in. = 3.28 ft.
1000 m	= 1 kilometer (km) = 3280.8 ft.
2.54 cm	= 1 in.

Some linear metric units useful to a service technician are shown in **Figure 1-5.** In measuring very tiny particles, the micron (μ) unit has been most used. The micron is one-thousandth of a millimeter (mm).

1.10.2 Area Measurement

The measurement of area involves the measurement of two-dimensional space. The area of an object is found by multiplying its length by its width.

Formula:

Area (A) = Length (L) × Width (W)

Example:

The width of a tabletop is 2′ and the length of the table is 4′. Determine the area of the tabletop.

Solution:

Area (A) = Length (L) × Width (W)

A = 2′ × 4′

A = 8 sq. ft. The area of the tabletop is 8 sq. ft.

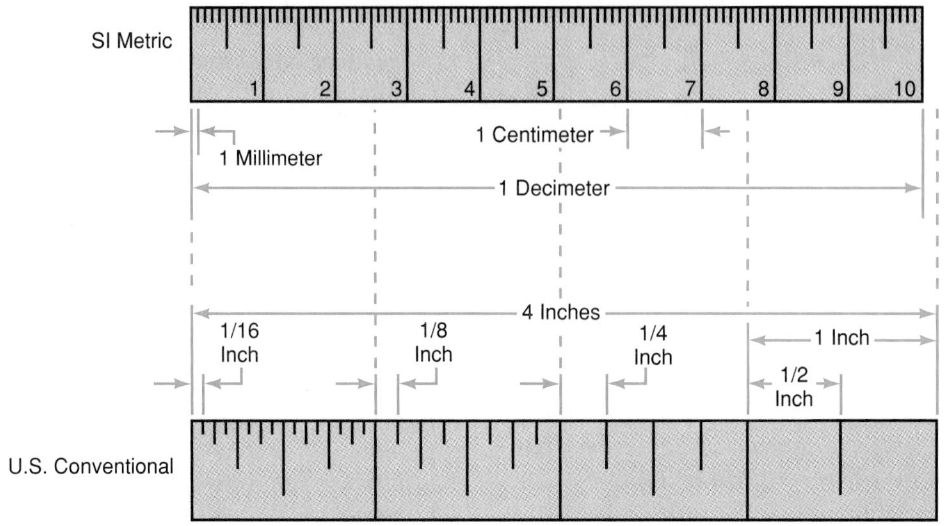

Figure 1-5. *Comparison of U.S. conventional and SI metric units of linear measurement.*

Some special formulas must be used when finding the area of certain objects. For example, the area of a circle is found by using the formula $A = \pi r^2$. The symbol, π, is always 3.1416, and r is the radius of a circle. It is equal to one-half the diameter. Therefore, this formula may also be expressed as:

$$A = \pi \frac{D^2}{4}$$

U.S. Conventional Units

square inches (sq. in.)

144 sq. in. = 1 square foot
9 sq. ft. = 1 square yard (sq. yd.)

These units are shown in **Figure 1-6A.** The area of a circle is shown in **Figure 1-6B.**

Metric Units

1 square centimeter (cm^2 or sq. cm)
= 0.155 square inch
1 square decimeter (dm^2 or sq. dm)
= 10 cm × 10 cm = 100 cm^2 = 15.5 sq. in.
1 square meter (m^2 or sq. m) = 1550 sq. in.
= 10 dm × 10 dm = 100 square decimeters (dm^2)
= 10.76 sq. ft.

These units are shown in **Figure 1-7.** Refer back to **Figure 1-6** for the area of a circle.

1.10.3 Volume Measurement

The measurement of volume involves the measurement of three-dimensional space (cubic). The volume of an object is determined by multiplying the width by the length by the height. An example is finding the volume of a cube (width × length × height, or W × L × H).

Some special formulas must be used when finding the volume of certain objects. To determine the volume

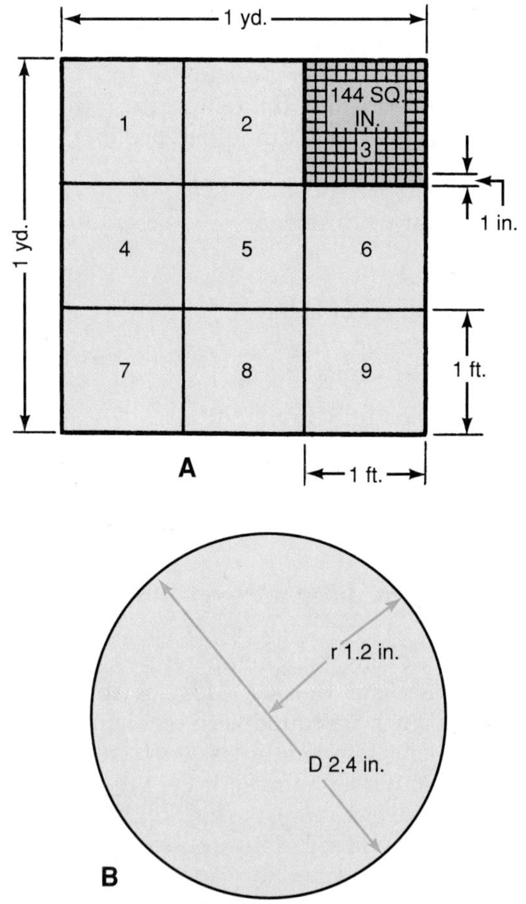

Figure 1-6. *Calculating standard areas using U.S. customary units. A—Area of rectangle is calculated by multiplying width by length. Remember, 144 sq. in. equal 1 sq. ft. and 9 sq. ft. equal 1 sq. yd. B—Area of a circle is calculated using the formula πr^2. Value of π is 3.1416. If diameter (D) of circle is 2.4 in., the radius (r), which is half the diameter, is 1.2 in.: $r^2 = r \times r = 1.2 \times 1.2 = 1.44$. Area of circle is $3.1416 \times 1.44 = 4.52$ in.^{2}*

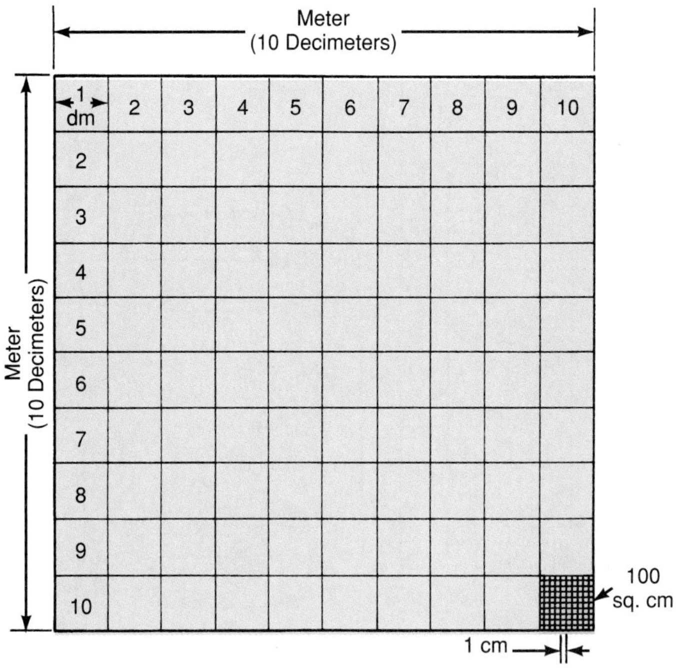

Figure 1-7. *Calculations of standard areas using SI metric units. Area of a rectangle is calculated by multiplying width by length. Area of circle is calculated with same formula as in Figure 1-6B.*

of a cylinder, for example, multiply the area of one end (πr^2) by the length (L) of the cylinder.

U.S. Conventional Units

cubic inches (in.^{3} or cu. in.)
cubic feet (ft^3 or cu. ft.)
cubic yards (yd.^{3} or cu. yd.)

1728 cu. in. = 1 cu. ft.
27 cu. ft. = 1 cu. yd.
1 cu. ft. = 7.48 gal.

These units are shown in **Figure 1-8A.** The volume of a cylinder is shown in **Figure 1-8B.**

SI Metric Units

1 liter (L) = 1000 cubic centimeters (cm^3)
= 1.05 quarts (qt.)
= 61 cu. in. = 0.035 cu. ft.

1000 cubic centimeters (cm^3) = 1 cubic decimeter (dm^3)
1 cubic meter (m^3) = 1.3 cu. yd.

These are shown in **Figure 1-9A.** The volume of a cylinder is shown in **Figure 1-9B.**

1.10.4 Angular Measurement

Circles and arcs of a circle are measured in ***degrees.*** A complete circle has 360°. See **Figure 1-10.** A degree is further divided into minutes. Sixty (60) minutes equal one degree.

Minutes are further divided into seconds. Sixty (60) seconds equal one minute. These are the same names—minutes and seconds—that are used in measuring time. In angular measurement, they have a different meaning.

The angle does not depend on the size of the circle. It does not depend on the length of the diameter or radius. A part of a circle is called an ***arc.*** It is formed by two lines going out from the center of the circle. They cut across the circumference. For a given central angle, the larger the circle, the longer the arc. The arc of the

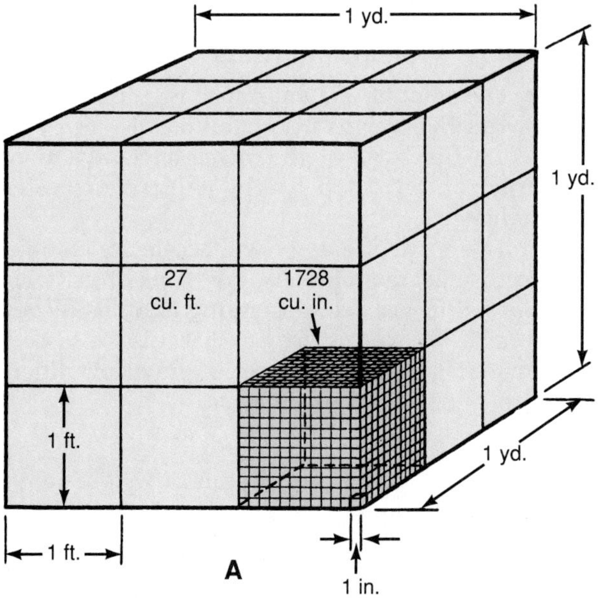

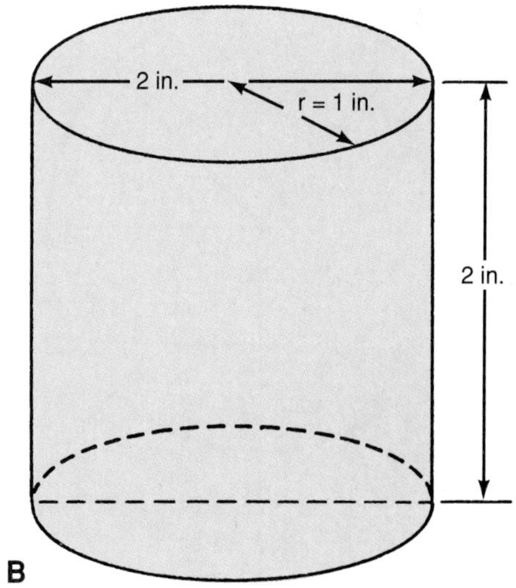

Figure 1-8. *Calculation of standard volumes using U.S. customary units. A—Volume is calculated by multiplying width by length by height. B—Volume of cylinder is determined by multiplying the area of the end by the length. (Cylinder dimensions are usually expressed in inches and decimals of an inch.)*

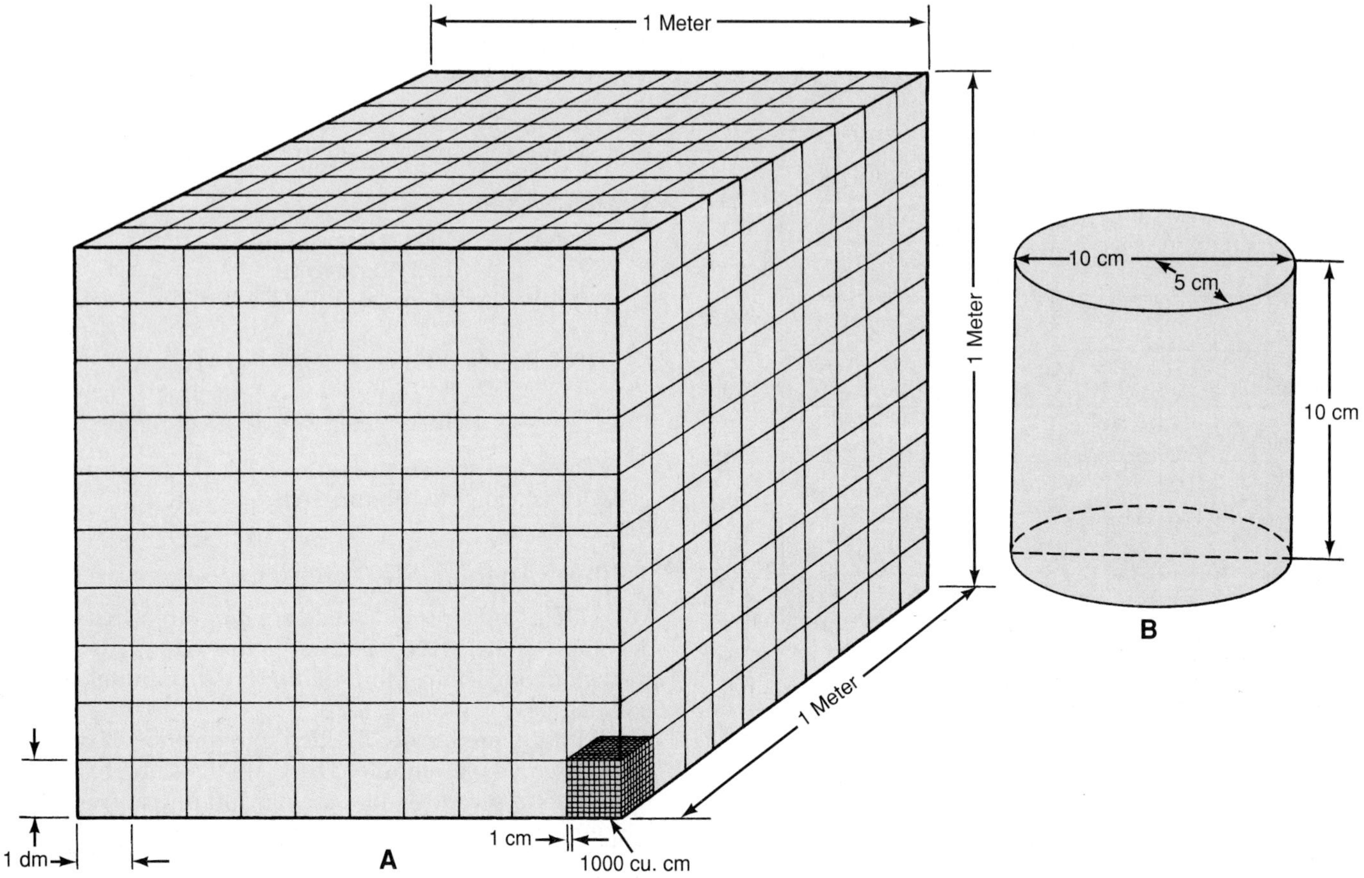

Figure 1-9. *Calculation of standard volumes using SI metric units. A—Volume is calculated by multiplying width by length by height. B—Volume of cylinder is calculated by multiplying the area of the end by the length.*

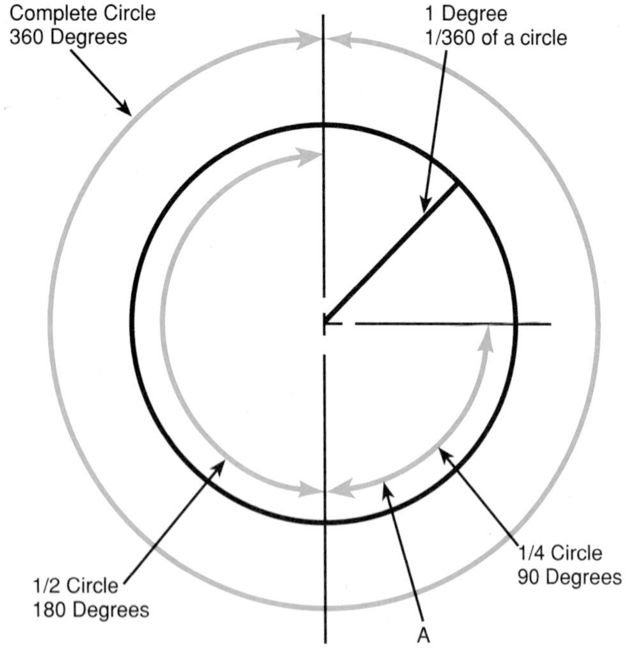

Figure 1-10. *Angular measurement. A complete circle consists of 360 degrees. One-half of a circle equals 180° and one-quarter of a circle equals 90°. Arc "A" is a 90° arc.*

circle which includes a 90° degree central angle is called a 90° arc.

1.10.5 Weight and Mass

The amount of a substance is commonly related to how much it weighs. Food and metals, for example, are sold on the basis of their weight. Gravitational force exerted on an object by the earth is expressed as its weight.

This force of gravity will accelerate an object if the object is released and falls. The same object will accelerate at a different rate depending upon its distance from the earth. To express the fact that it is the same quantity of material, even if the force of gravity is different, this quantity is defined as its ***mass.***

U.S. Conventional Units

The U.S. conventional units of weight are the grain, ounce, pound, and ton. This weight is often expressed as the force an object exerts on a scale.

If an object weighs one pound, then at the earth's surface—where the gravitational force will accelerate it at 32.2 ft./sec.^{2}—it is said to have a mass of one pound. Under these conditions, $lb._f$ represents the weight and $lb._m$ represents the mass of the object.

SI Metric Units

In SI metric measurement units, the mass is measured in kilograms (kg). A kilogram mass is equivalent to 2.2 $lb._m$. In SI units, the unit of force is called the newton (N). A newton equals the force exerted on an object having a mass of 1 kilogram, where the gravitational acceleration is 1 m/sec.^{2}. A 1-kilogram mass at the earth's surface will exert a force of 9.8 newtons. This is because in SI metric units the gravitational acceleration is 9.8 m/sec.^{2}. A 1-pound mass will have a mass of $\frac{1}{2.2}$ or 0.455 kilograms. However, it will exert a force of $\frac{9.8}{2.2}$ or 4.455 newtons on a weighing scale.

1.11 Pressure

Pressure is the force per unit area. It is expressed in pounds per square inch (psi). It is also expressed in pascals (Pa) or kilopascals (kPa) in the metric system.

The normal pressure of the atmosphere at sea level is 14.7 pounds per square inch (psi) or 101.3 kPa. In technical practice, this is usually rounded to 15 psi or 100 kPa.

Operation of a refrigerating system depends mainly on pressure differences in the system. Substances always push on the surfaces supporting or containing them. A block of ice (a solid) exerts a pressure on its support. If the support were removed, the block would fall to another supporting level.

A liquid exerts a pressure on the sides and bottom of its container. A gas exerts a pressure on all surfaces of its container. **Figure 1-11** illustrates these types of pressures.

A solid weight of 1 pound with a bottom surface area of 1 inch square would exert a pressure of 1 pound (lb.) per square inch (1 psi) upon a flat surface.

Liquid in a container maintains an increasing pressure on the sides and bottom as the liquid depth increases. The pressure of gas in the container will depend on the quantity of the gas and the temperature.

1.11.1 Pascal's Law

To honor the scientist Pascal, the SI metric system uses the term "pascal" as a unit of pressure. A ***pascal*** is a newton per square meter (N/m^2).

A ***newton*** is the metric unit of force. One newton is equal to the mass of 1 kilogram being accelerated at a rate of 1 meter per second per second.

Pascal's Law states that pressure applied upon a confined fluid is transmitted equally in all directions. It is the basis of operation of most hydraulic and pneumatic systems.

Figure 1-12 illustrates Pascal's Law. It shows a fluid-filled cylinder. A piston having a cross-sectional area of 645 mm^2 (one square inch) is fitted into a small cylinder connected to the larger cylinder. A force of 85 psig or 100 psia (690 kPa) is applied to the piston in the small cylinder. The pressure gauges show the pressure being transmitted equally in all directions.

One psia equals 6894.8 pascals (Pa). This can be rounded to 6.9 kPa. One psig equals 15.7 psia. Then 15.7 psia is converted to the metric unit as 108 kPa.

The refrigeration technician must deal with pressures both above and below atmospheric pressure. Metric gauges are calibrated so that zero on the gauge means that there is no pressure at all. This means not even atmospheric pressure is present. No negative pressures are used. A pressure of 5 kPa, then, is the same as 0.75

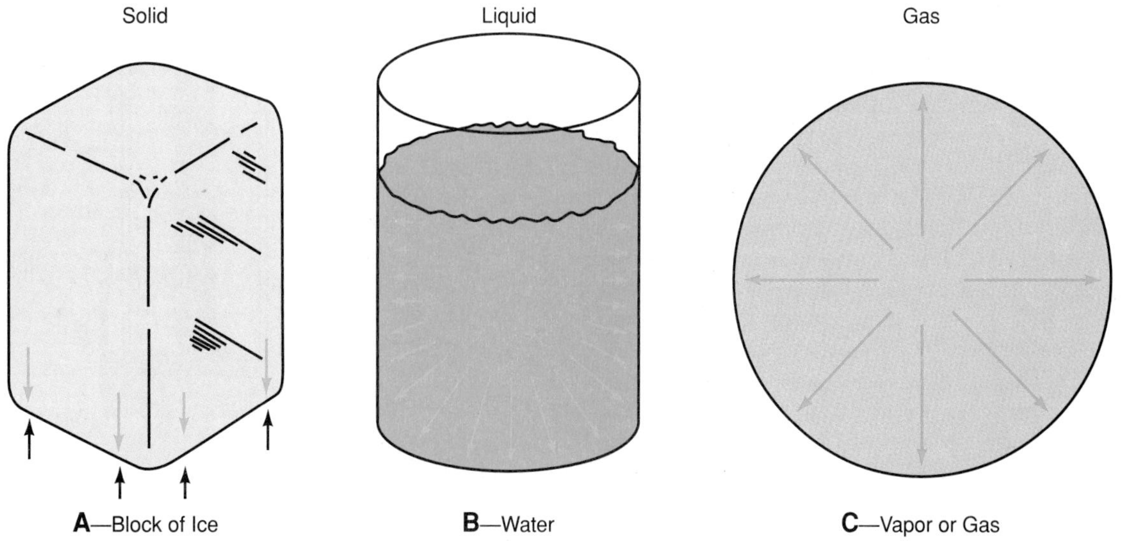

Figure 1-11. *Three states of a substance, such as water. A—Solid state. A mass of ice exerts downward force only. B—Liquid state. Water exerts pressure on container both vertically and horizontally. C—Gaseous state. Vapor or gas in rubber balloon exerts pressure uniformly in all directions.*

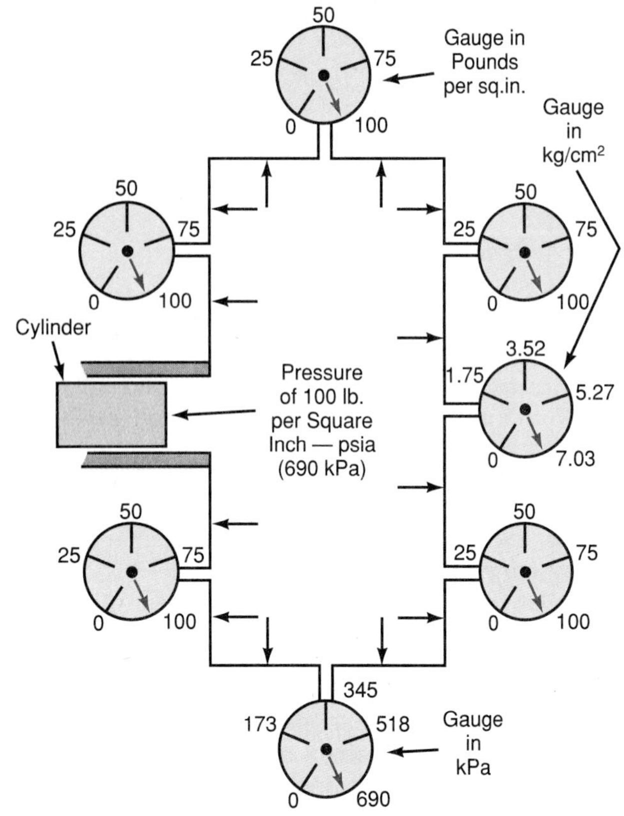

Figure 1-12. *Illustration of Pascal's Law. Pressure of 100 psia (690 kPa) is pressing against piston having head area of 1 sq. in. All gauges have same reading. Bottom gauge, calibrated in kilopascals, reads 690 kPa (100 psia).* ***Figure 1-17*** *is a reference which summarizes data about pressure scales.*

pounds per square inch absolute (psia). Expressed in inches of mercury vacuum, it would be about the same as 28.5 inches of Hg vacuum.

U.S. Conventional Units

In the U.S. conventional units, pressures above atmospheric pressure are measured in pounds per square inch (psi). Pressures below atmospheric are measured in inches of mercury (in. Hg) column. See Section 1.11.2.

Some older instruments are calibrated in atmospheres, bars, and torr. A ***bar*** is equal to one atmosphere; a millibar (mb) is equal to 0.001 bar. An ***atmosphere*** is approximately 14.7 pounds per square inch absolute (psia). Gauges calibrated in atmospheres are marked 1, 2, 3, 4, and up. The numbers represent the number of atmospheres. Arbitrarily, the atmospheric gauges have been calibrated in atmospheres of 15 pounds per square inch absolute (psia). One torr is 1/760 of an atmosphere.

1.11.2 Pressures—Atmospheric, Gauge, and Absolute

Atmospheric pressures are expressed in pounds per unit of area or in inches of liquid column height. The most popular gauges are those that register in pounds per square inch above atmospheric pressure (psig or psi).

A reading of 0 psi on the gauge is equal to the atmospheric pressure, which is about 14.7 psia. (Fifteen pounds per square inch is often used for working out problems). The absolute pressure scale registers zero at a pressure which cannot be further reduced. A perfect vacuum is 0 pounds per square inch absolute (0 psia).

Pressure may also be indicated in other ways: 1. Inches of mercury (in. Hg). 2. Feet or inches of water column. These gauges may be calibrated either above atmospheric pressure or absolute pressure. This is dependent upon the construction. A mercury gauge is often used for measuring below atmospheric pressure.

The barometer in **Figure 1-13** is a mercury gauge. With a vacuum at the closed top of the tube, the atmospheric pressure will support a mercury column 29.92″ high at sea level under standard conditions.

A unit of measure which has been used for reading high vacuums (pressure close to an absolute vacuum) is the ***torr.*** One torr equals a pressure of 1 mm of mercury (mm Hg, 0°C). It is named after the man who invented the mercury barometer. The unit torr may be expressed in fractions of an atmosphere. One torr = 1/760 of an atmosphere. A pressure of one torr is almost a perfect vacuum.

In solving most pressure and volume problems, it is necessary to use absolute pressures (psia). Absolute pressure is gauge pressure plus atmospheric pressure.

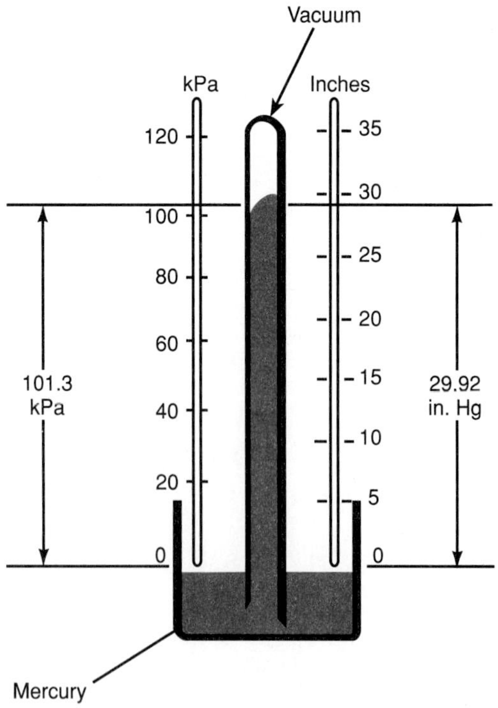

Figure 1-13. *Mercury barometer used for measuring atmospheric pressure. It consists of glass tube closed at one end and open at other end. Fill tube with mercury. Then, sealing open end, invert it in container of mercury. When seal is removed, mercury will drop to level corresponding with atmospheric pressure. Glass tube should be about 34 in. (86 cm) long.*

Example:

Calculate absolute pressure when the pressure gauge reading is 21 psi (psi always indicates gauge pressure, and psia indicates absolute pressure).

Solution:

Absolute pressure equals gauge pressure plus atmospheric pressure.

psi + 15 = psia
21 + 15 = 36 psia

Air pressure or a vacuum can be measured with a column of water. To equal 29.92″ Hg, it would be about 34′ high. The height is greater because water is so much lighter than mercury.

The service technician must often test both pressures and vacuums in the same system. Therefore, pressure gauges are made which will measure both. They are called compound gauges. Compound gauges have two or more pressure scales. One measures pressures below atmospheric pressure. The other measures pressures above atmospheric pressure. **Figure 1-14** illustrates such a gauge.

It is often necessary to convert inches of mercury into pounds per square inch absolute (psia). You can also convert pounds per square inch absolute into inches of mercury. Formulas are available for making an accurate conversion. Roughly 2″ Hg equals 1 psia. The chart shown in **Figure 1-15** makes converting easy.

Water columns are usually designed for measuring small pressures above or below atmospheric pressure. They can be used for pressures in air ducts, gas lines, and the like. A water column 2.3′ high (or about 28″) equals 1 psi.

These pressure measuring devices are called manometers. They are calibrated in inches of water column. **Figure 1-16** shows types of water manometers.

In some high-pressure refrigerating machines, pressure gauges are calibrated in atmospheres. One atmosphere is about 15 pounds per square inch (psi). Two atmospheres equal 30 psi. Three atmospheres equal 45 psi.

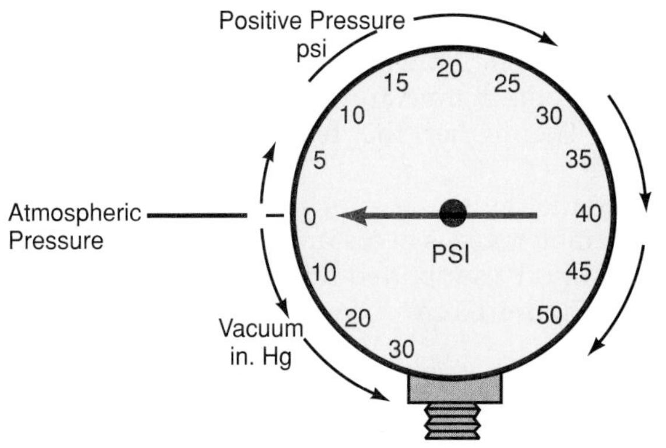

Figure 1-14. *Compound gauge measures both pressure above atmospheric in psi and pressure below atmospheric using units of in. Hg vacuum. Zero on this scale is atmospheric pressure.*

Inches of Hg	mm of Hg	psia	Ft. of Water
30		15	
(29.92)	760	(14.7)	33.40
29		14.5	
28	711	14	32.2
27		13.5	
26	660	13	29.9
25		12.5	
24	610	12	27.6
23		11.5	
22	559	11	25.3
21		10.5	
20	508	10	23.0
19		9.5	
18	457	9	20.7
17		8.5	
16	408	8	18.4
15		7.5	
14	356	7	16.1
13		6.5	
12	305	6	13.8
11		5.5	
10	254	5	11.5
9		4.5	
8	203	4	9.2
7		3.5	
6	152	3	6.9
5		2.5	
4	102	2	4.6
3		1.5	
2	51	1	2.3
1		0.5	
0	0	0	0

Figure 1-15. *Chart converts inches of mercury (in. Hg) into pounds per square inch absolute (psia).*

SI Metric Units

Figure 1-17 uses SI units. Compare the scales. In SI units, atmospheric pressures are expressed in kPa (kilopascals). Normal atmospheric pressure is 101.3 kPa. For practical purposes, gauges are often calibrated at 100 kPa for atmospheric pressure.

Pressures lower than atmospheric are called partial vacuums. Zero on the absolute pressure scale is at a pressure which cannot be further reduced. Thus, a perfect vacuum is 0 Pa (pascals). The pascal, rather than the kilopascal, is used for measuring high vacuums (pressures close to an absolute vacuum).

Figure 1-18 illustrates a pressure gauge used in refrigeration work. It is calibrated in kilopascals rather than in psi.

1.12 The Three Physical States

Substances exist in three states, depending on their temperature, pressure, and heat content. For example, water at atmospheric pressure is a solid at temperatures below 32°F (0°C). It is a liquid from 32°F (0°C) to 212°F (100°C). At 212°F (100°C) and above it is a vapor (gas).

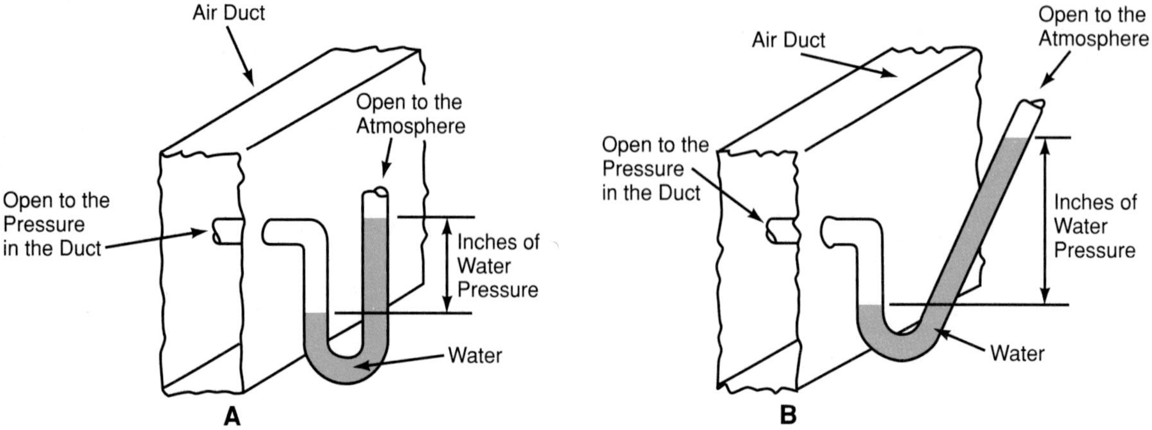

Figure 1-16. *Types of water manometers. A—This manometer is used to measure low pressure in air ducts and gas lines. Pressure is indicated in inches of water. It is measured by difference in level between surface of water in two branches of tube. B—For easier reading, the end open to atmosphere is often placed at a low angle. Red dye in water makes gauges easier to read.*

	Pounds per Square Inch		Inches Mercury Vacuum	cm Hg	kPa
	Absolute psia	Gauge psig or psi	in. Hg		
Positive Pressure	105	90			725
	90	75			621
	75	60			518
	60	45			414
	45	30			311
	30	15			207
Atmospheric Pressure	14.7	0	0	0	101.3
Negative Pressure or Vacuum	10	−5	10	25.4	69
	5	−10	20	50.8	35
	0	−15	29.92	76.0	0

Figure 1-17. *Table compares various pressure scales.*

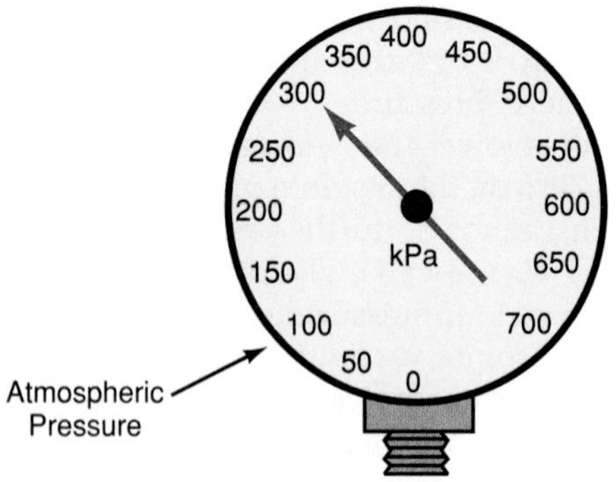

Figure 1-18. *Pressure gauge calibrated in kilopascals. Pressures from 0 to 100 are partial vacuums. Atmospheric pressure is set at 100 kPa.*

Water is shown in its three states in **Figure 1-11.** In this example, the physical state is controlled both by temperature and pressure. The temperatures just given apply only when atmospheric pressure is at 14.7 pounds per square inch (psi) or 101.3 kilopascals (kPa).

1.12.1 Solids

A ***solid*** is any physical substance which keeps its shape even when not contained. It consists of billions of molecules, all exactly the same size, mass, and shape. These stay in the same relative position to each other. Yet, they are in a condition of rapid motion or vibration. The rate of vibration will depend upon the temperature. The lower the temperature, the slower the molecules vibrate. The higher the temperature, the faster the vibration.

The molecules are strongly attracted to each other. Considerable force is necessary to separate them. A solid must always be supported by an upward force or it will fall. See **Figure 1-11A.**

1.12.2 Liquids

A ***liquid*** is any physical substance which will freely take on the shape of its container **(Figure 1-11B).** However, its molecules strongly attract each other.

Think of the molecules as swimming among their fellow molecules without ever leaving them. The higher the temperature, the faster the molecules swim. Warmer molecules will move upwards toward the top of the container. This is because they take up more space by their rapid movement. They become lighter (less dense) than colder molecules.

1.12.3 Gases

A ***gas*** is any physical substance which must be enclosed in a sealed container to prevent its escape into the atmosphere.

The molecules, having little or no attraction for each other, travel (fly) in a straight line. They bounce off each other, off molecules of other substances, or off the container walls. They have little or no attraction for any other substance. The pressures shown in the gas-filled balloon in **Figure 1-11C** illustrate how gases behave.

Almost any substance can be made to exist as a solid, a liquid, or a gas. Any molecule can be made to vibrate, swim, or fly. It depends on two things: temperature and pressure. To understand this change of state, one must study temperature and pressure relationships.

1.13 Density

Density is a substance's mass per unit of volume. Some substances are heavier than others. Comparative weights of gases, liquids, and solids may be shown by either density or specific gravity (see Section 1.13). Density is expressed as pounds per cubic foot (lb./ft.^{3}) or kilograms per cubic meter (kg/m^3).

1.13.1 Specific Volume

When comparing densities of gases, it is common to express the densities in specific volumes. ***Specific volume*** is the volume of one pound of a gas at standard conditions.

Standard conditions are 68°F at 29.92 in. of mercury column pressure. The volume of 1 lb. of dry, clean air at standard atmospheric conditions is 13.454 cu. ft. By comparison, 1 lb. of hydrogen occupies 178.9 cu. ft. One pound of the refrigerant, ammonia (R-717), occupies 21 cu. ft. Carbon dioxide (R-744) only occupies 8.15 cu. ft.

In SI terms, specific volume is the volume of one kilogram of a gas at standard conditions. Standard conditions are 20°C at 101.3 kPa pressure. The volume of 1 kg of dry, clean air at standard atmospheric conditions is 0.840 m^3. By comparison, 1 kg of hydrogen occupies 11.17 m^3. One kilogram of the refrigerant ammonia (R-717) occupies 1.311 m^3. Carbon dioxide (R-744) only occupies 0.509 m^3.

If 1 kg of gas occupies a greater space than air, the gas is called a light gas. If it occupies less space than air, it is classified as a heavy gas. The specific volume is $\frac{1}{\text{density}}$.

Equivalents

1 lb./ft.^{3} = 16 kg/m^3
1 kg/m^3 = 0.0625 lb./ft.^{3}

1.13.2 Specific Gravity (Relative Density)

Specific gravity (sp. gr) is the ratio of the mass of a certain volume of a liquid or a solid as compared to the mass of an equal volume of water. Specific gravity is sometimes called relative density.

Water is given a specific gravity of one. Objects which float on water have a specific gravity of less than one. Objects which sink in water have a specific gravity greater than one.

Mixtures of salt and water (brine) have a specific gravity greater than one. A calcium chloride brine adjusted to freeze at 0°F (−18°C) will have a specific gravity of 1.18. See Chapter 31 for a table of brine densities and freezing temperatures.

The ***relative density*** of gases is defined as the ratio of the mass of a certain volume of a gas as compared to the mass of an equal volume of hydrogen. The readings are taken at 68°F and 29.92" Hg pressure.

1.14 Force

Force applied to a body at rest causes it to move. The unit of force is the pound force (lb.$_f$). In SI metric units, it is the newton (N). The pound force is that force which, applied to a one-pound mass, will result in an acceleration of 32.173 ft./sec.^{2}.

At the surface of the earth, where the acceleration of gravity is 32.173 ft./sec.^{2}, a 1-lb. mass weighs 1-lb. force (it exerts 1-lb. force on the surface upon which it rests). If the object of 1-lb. mass were on the moon where the gravity is about 1/6 that on earth, the weight would be 1/6 lb.$_f$.

Example:

Determine the force on the head of a piston 10 square inches (sq. in.) in area and under a pressure of 25 psi.

Formula:

$F = A \times P$

in which:

F = force
A = area of the piston head (10 sq. in.)
P = pressure (25 psi)

Solution:

$F = A \times P$
$F = 10 \times 25$
F = 250 pounds (lb.)

1.14.1 SI Metric Units

In SI units, the newton is that force which, when applied to a body having a mass of one kilogram, gives it an acceleration of one meter per second per second. Force may also be called accumulated pressure.

One newton equals one kilogram times one meter divided by seconds squared: 1 kg × 1 m/sec.^{2} or 1 kg m/sec.^{2} = 1 N. (This unit of force is similar to the pound force in the U.S. conventional system.)

Example:

Determine the force on the head of a piston 645 mm^2 in area and under a pressure of 0.172 Mpa.

Formula:

$F = A \times P$

in which:

F = force
A = area of the piston head (645 mm^2)
P = pressure (0.172 MPa)

Solution:

$F = A \times P$
$F = 645 \times 0.172$
F = 111 newtons (N)

A pascal is the unit of pressure. The newton is the total force, which equals the unit pressure times the area. Equivalents:

1N = 0.225 lb. force
1 lb. force = 4.45 N

1.15 Work

Work (W) is force (F) multiplied by the distance (D) through which it travels.

1.15.1 U.S. Conventional Units

The unit of work is called the foot-pound. One *foot-pound* is the amount of work done in lifting a 1 lb. weight a vertical distance of 1 ft. Work is sometimes expressed in inch-pounds. At such times, the distance through which the force acts is measured in inches.

Example:

Calculate the work when lifting a weight of 2000 lb. a vertical distance of 10 ft.

Formula:

Work = Force × Distance

Solution:

W = F × D (F = 2000 lb.; D = 10 ft.)
W = 2000 × 10
W = 20,000 foot-pounds (ft.-lb.)

or, expressed in inch units,

W = 2000 × 10 × 12
W = 240,000 inch-pounds (in.-lb.)

1.15.2 SI Metric Units

The unit of work is called the joule (J) in SI metric units. The *joule* is the amount of work done by a force of one newton moving its point of application a distance of one meter.

Example:

The propeller on a boat pushes the boat through the water with a force of 200 newtons. If the boat travels 10 km, how much work is done?

Solution:

10 km = 10,000 m
Work = Force × Distance
$W = F \times D$
$W = 200 \times 10{,}000$
$W = 2 \times 10^2 \times 1 \times 10^4$
$W = 2 \times 10^6\ N \cdot m = 2\ MJ$

1.16 Energy

Energy is the capacity or ability to do work. The electric motor supplies the energy to drive the refrigerator compressor. There are three kinds of energy:

- *Potential energy* is stored energy. Examples are water behind a dam, electrical energy in a battery, and weight which can fall or drop.
- *Kinetic energy* is energy doing work. Examples are water flowing over a dam, a battery lighting a bulb, and a falling weight.
- *Heat energy* (see Section 1.4).

The work formula is expressed as $W = F \times D$, or newtons times meters ($N \cdot m$). $1\ J = 1\ N \times 1\ m = 1\ N \cdot m$.

Equivalents

1 ft.-lb. force = 1.356 J = 1.356 N · m
1 J = 1 N · m = 0.737 ft.-lb.

1.17 Power

Power is the time rate of doing work.

1.17.1 U.S. Conventional Units

The unit of mechanical power is the ***horsepower.*** One horsepower (hp) is the equivalent of 33,000 foot-pounds (ft.-lb.) of work per minute. If a 2000-lb. weight is lifted 10 ft. in two minutes, the power required would be:

$$\text{Horsepower} = \text{weight in pounds} \times \frac{\text{distance in feet}}{\text{time in minutes}} \times 33{,}000$$

$$\text{Horsepower} = \frac{2000 \times 10}{2 \times 33{,}000} = \frac{20{,}000}{66{,}000} = 0.3\ \text{hp}$$

1.17.2 SI Metric Units

In SI units, power is expressed in watts. A *watt* is the force of one newton moving through a distance of one meter in one second.

The common unit of mechanical power is the kilowatt (kW). A kilowatt is equal to 1000 watts. The formula for power is force times distance divided by time. It is expressed in watts (W). 1 W = 1 joule per second = 1 J/sec. (See Sections 1.15 and 1.16.)

Example:

What is the power required to lift a mass of 100 kilograms at the rate of 10 meters per second?

Solution:

$$\text{Power} = \frac{\text{Force} \times \text{Distance}}{\text{Time}} = \frac{\text{newtons} \times \text{meters}}{\text{seconds}}$$

Force = 100 × 9.8 newtons
Distance = 10 meters
Time = 1 second

$$\text{Power} = \frac{100 \times 9.8 \times 10}{1}$$

Power = 9800 W = 9.8 kW

Equivalents

1 hp = 746 W
1 W = 0.0013 hp

1.18 Unit of Heat

The unit of heat is the British thermal unit (Btu). The ***Btu*** is the amount of heat required to raise the temperature of one pound of water one degree Fahrenheit. (See **Figure 1-19A.**)

Whether a substance such as water is cooled or heated, the heat calculation is made in the same way. The temperature difference multiplied by the number of pounds of water gives the number of Btu. Where large heat loads are involved, the unit therm, which equals 100,000 Btu, is often used.

Example:

Calculate the amount of heat required to raise the temperature of 62.4 lb. (1 cu. ft.) of water from 40°F to 80°F.

Solution:

Btu = wt. in lb. × temperature change in °F
Btu = 62.4 lb. × (80 − 40)
Btu = 62.4 × 40
Btu = 2496 Btu
If a substance is cooled, heat is removed.

Example:

Determine the amount of heat removed to cool 50 lb. of water from 80°F to 35°F.

Solution:

Btu = wt. in lb. × temperature change in °F
Btu = 50 lb. × (80 − 35)
Btu = 50 × 45
Btu = 2250 Btu

SI Metric Units

In SI metric, the unit of heat is the joule (J). A ***joule*** is a very small unit of heat. For refrigeration work, the kilojoule (kJ), 1000 joules, is used. The amount of heat required to raise the temperature of 1 kg of water 1°C is equal to 4.187 kJ. (See **Figure 1-19B.**) Conversely, the amount of heat removed to lower the temperature of 1 kg of water 1°C is also equal to 4.187 kJ.

The mass in kilograms multiplied by the degrees Celsius temperature difference multiplied by 4.187 kJ equals the amount of heat added or subtracted.

Example:

Find the amount of heat required to raise the temperature of 1 kg (approximately 1 liter) of water from 4°C to 27°C.

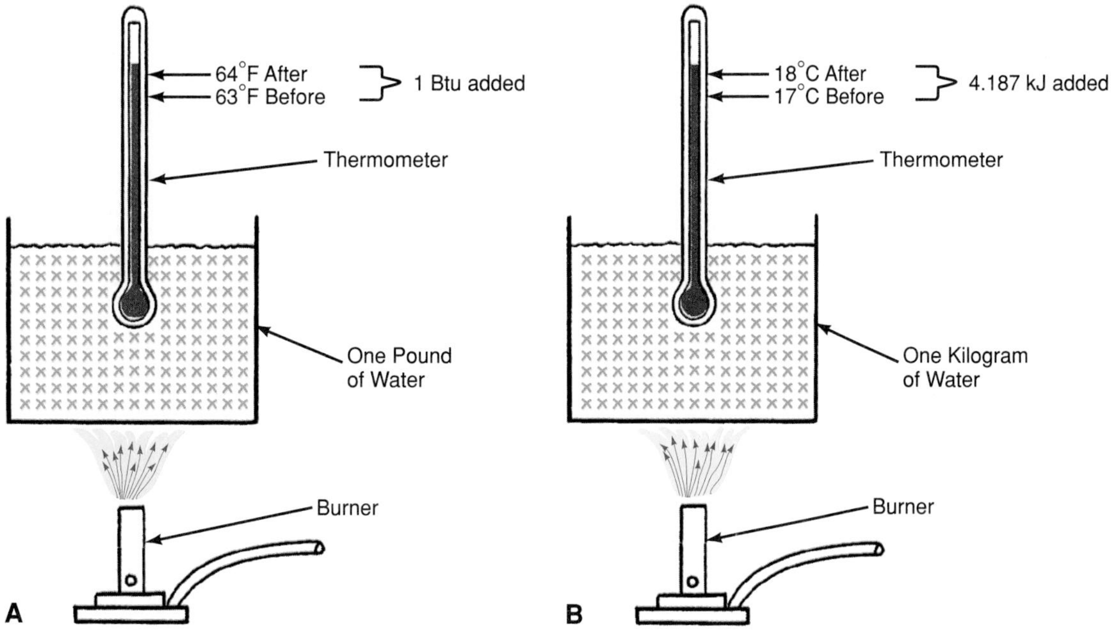

Figure 1-19. *Experiment in heating. A—Raising temperature of one pound of water from 63°F to 64°F requires one British thermal unit of heat. B—It takes 4.187 kJ of heat to raise the temperature of 1 kg of water from 17°C to 18°C. Note: An accurate definition of the calorie uses 4.1840, not 4.187.*

Solution:

kJ = 4.187 × mass in kilograms × temperature change in °C
kJ = 4.187 × 1 × (27 − 4)
kJ = 4.187 × 1 × 23
kJ = 96.301 kJ

Conversely, if a substance is cooled, heat is removed.

Example:

Determine the amount of heat removed to cool 19 kg of water from 27°C to 1°C.

kJ = 4.187 × mass in kilograms × temperature change in °C
kJ = 4.187 × 19 × (27 − 1)
kJ = 4.187 × 19 × 26
kJ = 2068.378 = 2068 kJ

Another metric unit, the calorie, is the amount of heat required to raise the temperature of one gram of water one degree Celsius. However, the calorie is such a small unit that it is no longer used in refrigeration work. Most calculations in engineering science are made using the kilocalorie which equals 1000 calories.

Example:

Find the amount of heat required to raise the temperature of 150 grams of water from 10°C to 90°C.

Solution:

Calorie = mass in grams × temperature change in °C
Calorie = 150 grams × (90 − 10)
Calorie = 150 × 80
Calorie = 12,000 calories = 12 kilocalories

Equivalents

1 kJ = 239 cal = 0.948 Btu
1 cal = 0.004 kJ = 0.004 Btu
1 Btu = 1.055 kJ = 252 cal = 0.252 kilocalories

1.18.1 First Law of Thermodynamics

The *First Law of Thermodynamics* states that "heat and mechanical energy are mutually convertible."

U.S. Conventional Units

In Section 1.4, the Btu is defined as the unit of heat. In Section 1.15, the unit of work is defined as the ft.-lb. Since work is convertible to heat, the conversion factor from Btu to ft.-lbs. is used:

1 Btu = 778 ft.-lbs.

Example:

Change 10 Btu to ft.-lbs.

Solution:

10 Btu = 10 × 778 = 7780 ft.-lbs.

SI Metric Units

Similarly, in SI units, heat and mechanical energy are mutually convertible. In Section 1.4, the joule is defined as the unit of heat. In Section 1.15, the unit of work is also the joule, or newton-meter. For example, 100 joules of work equal 100 joules of heat, or 100 newton-meters.

1.18.2 Sensible Heat

Heat which causes a change in temperature in a substance is called ***sensible heat.*** When a substance is heated (heat added) and the temperature rises as the heat is added the increase in heat is called sensible heat. Likewise, heat may be removed from a substance (heat subtracted). If the temperature falls, the heat removed is, again, sensible heat.

1.18.3 Specific Heat Capacity

The ***specific heat capacity*** of a substance is the amount of heat added or released to change the temperature of one pound of the substance one degree Fahrenheit.

The sensible heat required to cause a temperature change in substances varies with the kind and amount of substance. The specific heat capacity of water is 1.0 Btu/lb.°F. Different substances require different amounts of heat per unit mass to cause these changes of temperature. The specific heat capacity of several common substances is shown in **Figure 1-20.**

Section 1.18 shows how to calculate the heat (Btu) added or removed from water. The amount of heat necessary to cause a desired change of temperature is calculated in the following manner. Multiply the mass of the substance by the specific heat capacity and by the temperature change (provided there is no change of state). The formula for sensible heat added or removed is: Heat in Btu = weight × sp. ht. capacity × °F change.

Example:

Determine the amount of Btu which must be removed to cool 40 lb. of 20% salt brine (see **Figure 1-20**) from 60°F to 20°F.

Material	Specific Heat Capacity	
	Btu/lb./F	kJ/kg·K
Wood	0.327	1.369
Water	1.0	4.187
Ice	0.504	2.110
Iron	0.129	0.540
Mercury	0.0333	0.139
Alcohol	0.615	2.575
Copper	0.095	0.398
Sulphur	0.177	0.741
Glass	0.187	0.783
Graphite	0.200	0.837
Brick	0.200	0.837
Glycerine	0.576	2.412
R-717 (Liquid ammonia at 40 °F)	1.1	4.606
R-744 (Carbon dioxide at 40 °F)	0.6	2.512
R-502	0.255	1.068
Salt Brine 20%	0.85	3.559
R-12	0.213	0.892
R-22	0.26	1.089
The above values may be used for computations which involve no "change of state." If a change of state is involved, the specific heat for each state of the substance must be used.		

Figure 1-20. *Table of specific heat capacity values for some substances.*

Solution:

Btu = wt. × sp. ht. capacity × °F change
Btu = 40 lb. × 0.85 sp. ht. capacity × (60°F − 20°F)
Btu = 40 × 0.85 × 40
Btu = 1360 Btu

SI Metric Units

In SI metric units, the specific heat capacity of a substance is the amount of heat that must be added or released to change the temperature of one kilogram of the substance one degree kelvin (K). The specific heat capacity of water is 4.187 kJ/kg · K. See **Figure 1-20** for the specific heat capacity of several common substances. Section 1.18 shows how to calculate the heat in kilojoules (kJ) added or removed from water.

The specific heat capacity unit is expressed as joules per kilogram Kelvin. The expression is J/kg · K. The formula for sensible heat added or removed is: Heat in kJ = mass × sp. ht. capacity × K change.

Example:

Find the amount of heat, in kJ, which must be removed to cool 15 kg of 20% salt brine (see **Figure 1-20**) from 16°C to 7°C.

Solution:

kJ = mass × sp. ht. capacity × K change
kJ = 15 kg × 3.559 sp. ht. capacity × (16°C − 7°C)
kJ = 15 × 3.559 × 9
kJ = 480.5

Specific Heat Capacity Equivalents

1 cal/g · °C = 4.187 J/g · K
1 Btu/lb. · °F = 4.187 kJ/kg · K (kilojoule per kilogram Kelvin)
1 kJ//kg · K = 0.2388 Btu/ lb. · °F

1.18.4 Latent Heat

Heat which brings about a change of state with no change in temperature is called ***latent*** (hidden) ***heat.***

All pure substances are able to change their state. Solids become liquids, liquids become gas. These changes of state occur at the same temperature and pressure combinations for any given substance. It takes the addition of heat or the removal of heat to produce these changes.

In **Figure 1-21,** note that considerable heat (144 Btu/lb., 335 kJ/kg) was added between points B and C. Even so, the temperature did not change. This heat was required to change the ice to water. This heat is called "latent heat of melting" or "latent heat of fusion," which means the same thing.

Likewise, between points D and E, 970 Btu/lb. (2257 kJ/kg) were added and the temperature did not change. This heat was required to change the water to steam. This heat is called "latent heat of vaporization." When cooling the steam to water, the latent heat removed is called the "latent heat of condensation."

There are two latent heats for each substance, solid to liquid (melting and freezing) and liquid to gaseous (vaporizing and condensing). **Figure 1-22** shows the latent heat for water and several common refrigerants.

The latent heat of fusion of ice is 144 Btu/lb. (335 kJ/kg). The latent heat of vaporization for water (at 212°F, 100°C) = 970 Btu/lb. (2257 kJ/kg).

The addition of heat to a solid increases the vibration of the molecules. This continues until they separate at the change-of-state point. In the liquid form, the molecules are only weakly attracted to other molecules. Thus, they are free to move around. At the transition

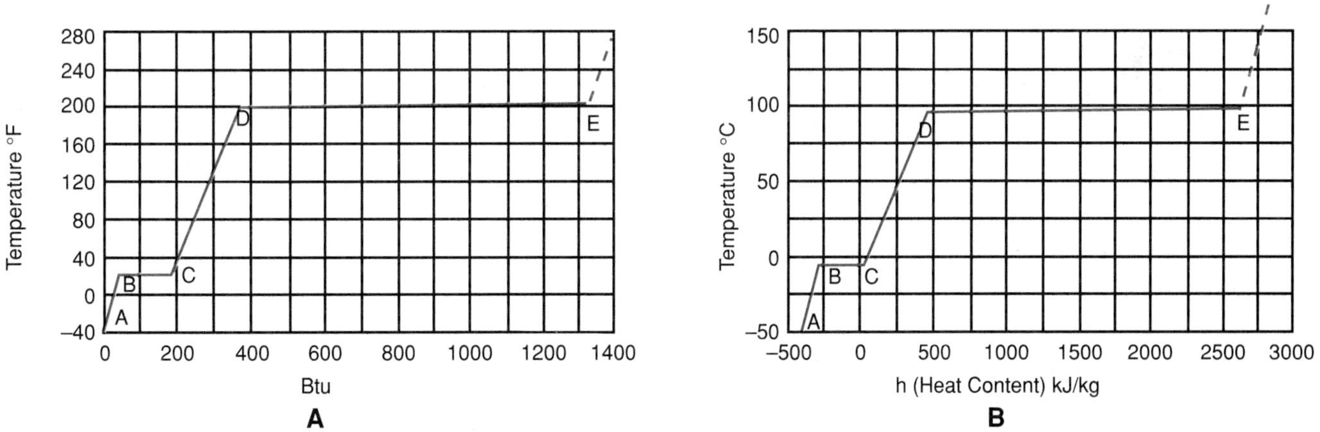

Figure 1-21. *A—Temperature-heat diagram for one pound of water at atmospheric pressure, heated from −40°F through complete vaporization. From A to B, 36.3 Btu were added to heat ice from −40°F to 32°F (−40°F to 32°F = 72°F temperature change). (Btu = 1 lb. × 0.504 × 72 = 36.288). From B to C, 144 Btu were added to melt the ice. The temperature did not change from B to C. From C to D, 180 Btu were added to heat the water from 32°F to 212°F. From D to E, 970 Btu were added to vaporize water. Note that temperature did not change from D to E. B—Temperature-heat diagram for one kilogram of water at atmospheric pressure (100 kPa) heated from −50°C through complete vaporization. From A to B, 100 kJ of heat are added to increase ice temperature from −50°C to 0°C. This is 2 kJ/kg °C × 50°C = 100 kJ/kg. From B to C, 335 kJ are added to melt the ice without changing its temperature. From C to D, 420 kJ were added to heat the water to its boiling point (4.2 kJ/kg °C × 100°C = 420 kJ). From D to E, 2260 kJ were added to convert the water to steam without changing its temperature. More heat increases the temperature of steam as shown in dotted line.*

Material	Freezing or Melting Btu/lb.	Latent Heat of Vaporization or Condensation Btu/lb.	Freezing or Melting kJ/kg	Latent Heat of Vaporization or Condensation kJ/kg
Water	144	970.4 at 212°F	335	2257 at 100°C
R-717 (Ammonia)	–	565.0 at 5°F	–	1314 at −15°C
R-502	–	68.96 at 5°F	–	160 at −15°C
R-40 (Methyl chloride)	–	178.5 at 5°F	–	415 at −15°C
R-12	–	68.2 at 5°F	–	159 at −15°C
R-22	–	93.2 at 5°F	–	217 at −15°C

Figure 1-22. *Table of latent heat of vaporization of water and some common refrigerants. Latent heat of fusion is only given for water, as refrigerants do not freeze at temperatures commonly handled by refrigeration service engineers.*

from a solid to a liquid, some molecules are attached in a solid form. Others are weakly attracted in a liquid form. When all the solid attachments are broken, further heating causes the molecules in liquid form to move around faster.

It requires as much energy to change the molecular attachment of a block of ice as it does to raise the temperature of the same amount of liquid from 32°F to 176°F. In SI metric, it requires as much heat to change 1 kg of ice to 1 kg of water as it does to raise the temperature of that same amount of water from 0°C to 80°C.

All of the basic operations of the compression refrigeration cycle are based upon these two heats—*sensible* and *latent.*

Equivalents:

1 kJ/kg = 0.4299 Btu/lb.
1 Btu/lb. = 2.326 kJ/kg
1 kJ/kg = 0.2388 kcal/kg
1 kcal/kg = 4.187 kJ/kg
1 kJ/kg = 0.2388 cal/g
1 cal/g = 4.187 kJ/kg

1.18.5 Application of Latent Heat

In refrigeration work, the physics of latent heat is especially important. Applications of this principle give the cold or freezing temperature desired.

As ice melts, its temperature remains constant. Nevertheless, it absorbs a considerable amount of heat in changing from ice to water. To melt 1 lb. of ice, 144 Btu are required. (To melt 1 ton of ice, 288,000 Btu are required.) To melt one kg of ice, 335 kJ of heat are required.

When a substance passes from a liquid to a vapor, its ability to absorb heat is very high. This principle is useful in the operation of the mechanical refrigerator.

The temperature at which a substance changes its state depends on the pressure. The higher the pressure, the higher the temperature needed to bring about the change. The opposite is also true. If the pressure is lowered, the temperature at which the change of state will take place is also lowered. This principle is shown in **Figure 1-23.**

A liquid under low pressure will boil at a lower temperature. If the vapor resulting from this boiling is then compressed, it will condense back into a liquid at a higher temperature.

Every substance has a different latent heat value. This is because each substance has a different molecular structure. Latent heat temperatures for water and the more common refrigerants are shown in **Figure 1-22.** See Chapter 9 for more information concerning refrigerants.

In a modern refrigerator, freezer, or air conditioner, liquid refrigerant is piped under pressure to the evaporator. In the evaporator, the pressure is greatly reduced. The refrigerant boils (vaporizes), absorbing heat from the evaporator. This produces a low temperature and cools the evaporator.

The compressor pumps this vaporized refrigerant out of the evaporator. It compresses (squeezes) the refrigerant into the condenser. Here the heat that was absorbed in the evaporator is "squeezed out." It is released to the surrounding atmosphere. Having lost this heat of vaporization, the refrigerant becomes a liquid again. The cycle is then repeated.

1.19 Effect of Pressure on Evaporating Temperatures

The evaporating (boiling) temperature for any liquid is controlled by the pressure placed upon it. Water at atmospheric pressure (15 psia or 100 kPa) normally boils at 212°F (100°C). If the pressure is increased to 45 psia (311 kPa), the boiling temperature is raised to 271°F (133°C). If the pressure is lowered to 3 psia (20 kPa), the boiling temperature will be lowered to 142°F (62°C), as shown in **Figure 1-23.**

Mechanical and absorption refrigerators use the effect of reduced pressure to lower the boiling temperature. (See Chapter 3.) Consider the refrigerant, R-12. It boils under atmospheric pressure (15 psia or 100 kPa) at −20°F (−29°C). If the pressure is lowered to 9 psia (62 kPa), the boiling temperature is −42°F (−41°C). A refrigerator can then cool to −42°F (−41°C), if the pressure on the evaporator is lowered to 9 psia (62 kPa). However, if the evaporator were to operate at atmospheric pressure, the lowest temperature possible with R-12 would be −20°F (−29°C). **Figure 1-24** shows the effect

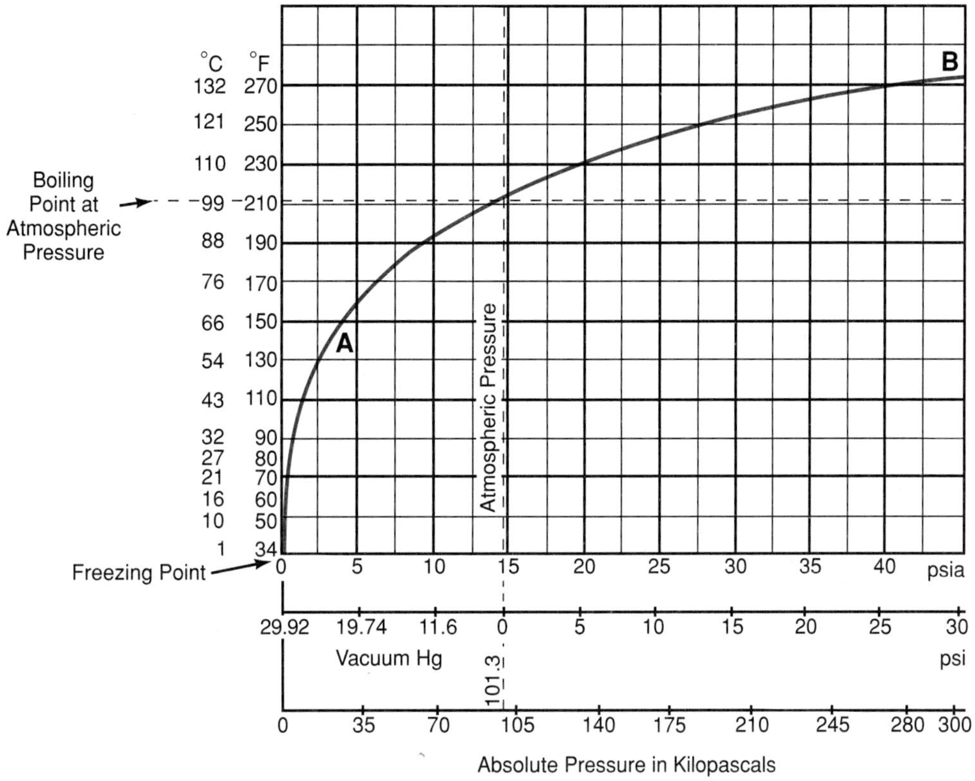

Figure 1-23. *Temperature-pressure curve for water. At atmospheric pressure, water boils at 212°F (100°C). At point A, with vacuum of 24" Hg (20 kPa), water boils at 142°F (62°C). Increasing pressure above atmospheric raises boiling temperature. At B, which is at a pressure of 45 psia (311 kPa), boiling temperature is 271°F (133°C).*

Substance	Evaporating Temperature in °C at:		
	60 kPa Pressure	Atmospheric Pressure 100 kPa	200 kPa Pressure
Water	89	100	122
R-12	–41	–29	–10
R-717 (Ammonia)	–40	–38	–18

Figure 1-24. *Effect of pressure is shown on evaporating temperatures of three substances.*

of pressure change on the boiling temperature of three substances used in refrigeration work.

1.20 Effect of Pressure on Freezing Temperature of Water

The temperature at which water freezes is affected by the pressure on the surface of the water. Increasing the pressure lowers the freezing temperature. Decreasing the pressure raises the freezing temperature. **Figure 1-25** shows this relationship.

This relationship goes the opposite way from the general rule given in Section 1.18.4. This is because water expands when it freezes. Most substances expand when they melt, and obey the rule in Section 1.18.4. For them, the higher the pressure, the higher the melting temperature.

1.21 Refrigerating Effect of Ice

Ice is still important to the refrigeration industry. As stated before, ice changes to water at 32°F (0°C) and atmospheric pressure. Heat absorption to produce this change is 144 Btu/lb. (335 kJ/kg).

The specific heat equation for changing ice to water is: Heat = wt. of ice × sp. ht. capacity of ice × temperature change. Heat will be in Btu. The weight (Wt.) will be in pounds. Specific heat (Sp. ht.) will be given in Btu/lb. (The specific heat of water is 1 Btu/lb. The specific heat of ice is 0.50 Btu/lb.) The latent heat of fusion of ice is 144 Btu/lb.

Example:

How many Btu will be absorbed in changing 25 lb. of ice at 5°F to water at 40°F?

Solution (in steps):

1. Raise the temperature of ice from 5°F to 32°F:

   Btu = wt. of ice × sp. ht. capacity of ice × temperature change
   $\text{Btu} = 25 \times 0.50 \times (32 - 5)$
   $\text{Btu} = 25 \times 0.50 \times 27$
   Btu = 337.5 Btu

2. To melt the ice at 32°F:

   Btu = wt. of ice × latent heat of fusion of ice
   $\text{Btu} = 25 \times 144$
   Btu = 3600 Btu

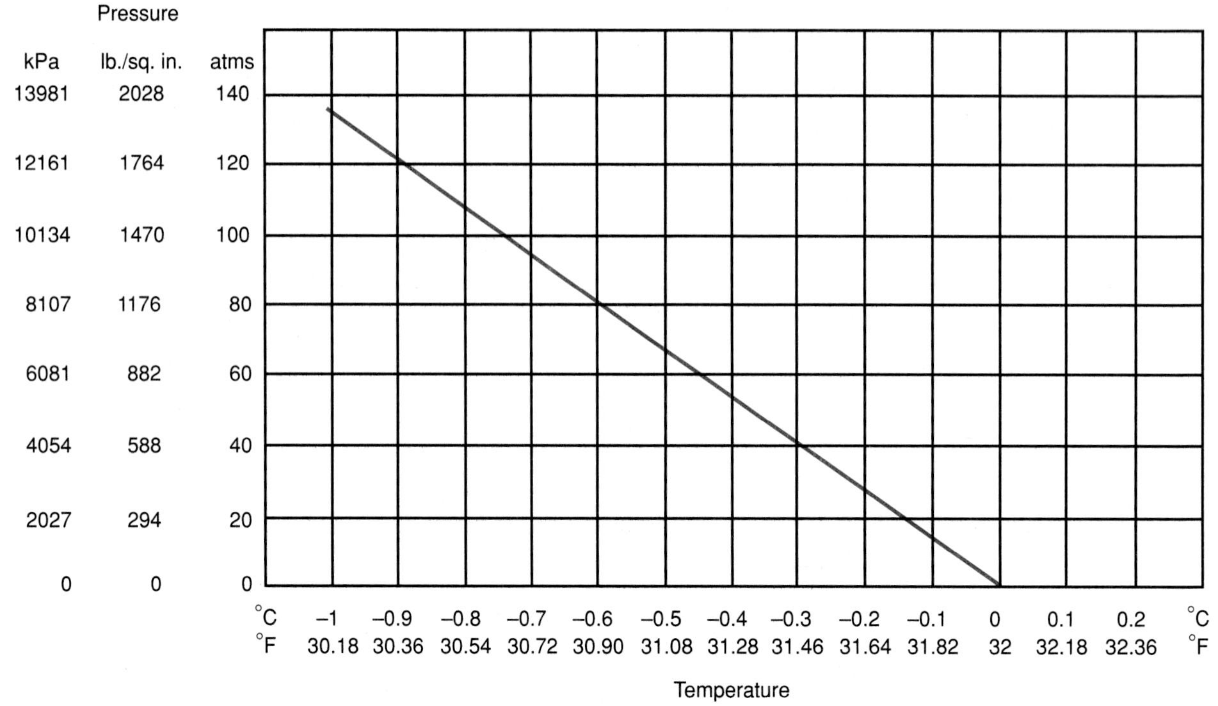

Figure 1-25. *Chart shows effect of pressure on freezing temperature of water.*

3. To warm the water from 32°F to 40°F:

   Btu = wt. of water × sp. ht. capacity of water × temperature change
   Btu = 25 × 1 × (40 − 32)
   Btu = 25 × 1 × 8
   Btu = 200 Btu

4. Total heat required to change 25 lb. of ice at 5°F to water at 40°F:

   Btu = 337.5 + 3600 + 200 = 4137.5 Btu

SI Metric Units

In SI metric, the specific heat capacity of ice = 2.11 kJ/kg · K. Its heat absorption ability, when changing from a temperature below 0°C to 0°C, is 2.11 kJ/kg per degree change. The latent heat of fusion (melting) of ice = 335 kJ/kg. The specific heat of water = 4.19 kJ/kg · K.

Example:

How many kJ will be absorbed in changing 93 kg of ice at −20°C to water at 4°C?

Solution (in three steps):

1. To find heat needed to bring ice from −20°C to 0°C:

   kJ = mass of ice × sp. ht. capacity of ice × temperature change
   kJ = 93 × 2.11 × 20
   kJ = 3924.5 kJ

2. To find heat needed to melt ice at 0°C:

   kJ = mass of ice × latent heat of fusion of ice
   kJ = 93 × 335
   kJ = 31 155.0 kJ

3. To find heat needed to raise water temperature from 0°C to 4°C:

   kJ = mass of water × sp. ht. capacity of water × temperature change
   kJ = 93 × 4.19 × (4 − 0)
   kJ = 93 × 4.19 × 4 = 1558.7 kJ

4. Total heat required to change 93 kg of ice at −20°C to water at 4°C

   kJ = 3924.6 kJ + 31 155.0 kJ + 1558.7 kJ
   kJ = 36 638.3 kJ

Equivalents

1 kJ/kg · K = 0.239 Btu/lb.°F
1 Btu/lb.°F = 4.187 kJ/kg · K

1.21.1 Ice and Salt Mixtures

Refrigerating by ice alone will not provide temperatures below 32°F (0°C). Therefore, to get the lower temperatures required in some instances, ice and salt mixtures are used. These mixtures, ice and salt (sodium chloride or NaCl), and ice and calcium chloride ($CaCl_2$), lower the melting temperature of ice. An ice and salt mixture may be made which will melt at 0°F (−18°C).

A solution of water and salt freezes at a lower temperature because more energy must be removed from the solution before ice will start to form. This phenomenon is called "freezing point depression."

1.21.2 Ton of Refrigeration Effect

The cooling capacity of older refrigeration units is often indicated in "tons of refrigeration." A ***ton of refrigeration*** represents the heat energy absorbed when a ton (2000 lb.) of ice melts during one 24-hour day. The ice is assumed to be a solid at 32°F (0°C) initially and becomes water at 32°F (0°C). The energy absorbed by the ice is the latent heat of ice times the total weight.

Today, refrigeration units are often rated in Btu/hr. instead of tons. The Btu equivalent of one ton of refrigeration is easy to calculate. Multiply the weight of one ton of ice (2000 lb.) by the latent heat of fusion (melting) of ice (144 Btu/lb.). Then, divide by 24 hours to obtain Btu/hr.

One ton of refrigeration effect = 2000 × 144/24
One ton of refrigeration effect = 288,000 Btu/24 hours
One ton of refrigeration effect = 12,000 Btu/hr.

A 12,000 Btu/hr. cooling capacity is equivalent to one ton of refrigeration.

A refrigerating or air conditioning mechanism capable of absorbing heat can be rated in tons per 24 hours by its heat-absorbing ability (HA) in Btu divided by 288,000.

T = tons of refrigeration effect
A = heat-absorbing ability

$$T = \frac{HA}{288{,}000} = \text{tons of refrigeration effect}$$

Example:

The heat-absorbing ability of a refrigerator unit is 1,440,000 Btu per 24 hours. What is its ton rating?

Solution:

$$T = \frac{1{,}440{,}000}{288{,}000}$$

T = 5 tons of refrigeration effect

Example:

What will be the "ton" rating of a refrigerating mechanism capable of absorbing 1,728,000 Btu in 24 hours?

Solution:

$$T = \frac{1{,}728{,}000}{288{,}000}$$

T = 6 tons of refrigeration effect

Example:

What is the Btu heat absorbing capacity of a 1/2-ton refrigerating system?

Solution:

1/2 × 288,000 = 144,000 Btu per day, or 6000 Btu per hour

Most room air conditioners are rated on their heat-absorbing ability in Btu per hour. A 1-ton machine is rated:

$$\frac{288{,}000}{24} = 12{,}000 \text{ Btu per hour}$$

SI Metric Units

The SI metric system has no unit which can be compared with the "ton of refrigeration."

1 ton = approximately 907 kg
latent heat = 335 kJ/kg
energy absorbed = latent heat × weight
energy absorbed = 335 kJ/kg × 907 kg
energy absorbed = 303 845 kJ

The melting of this ice in one day has a cooling or refrigeration capacity of 303 845 kJ.

To convert the rating to kilowatts:

1kW = 1 kJ/sec.
1 ton refrigeration capacity = 303 845/(24 × 3600 sec)
1 ton refrigeration capacity = 303 845/86 400 sec.
1 ton refrigeration capacity = 3.52 kJ/sec.
1 ton refrigeration capacity = 3.52 kW

A refrigerator which produces an equivalent cooling rate of this ice melting will be rated as a 1-ton unit.

Equivalents

1 kW = 3415 Btu/hr.
1 Btu/hr = 0.29W
1 "ton" = 12,000 Btu/hr.

1.22 Ambient Temperature

Ambient temperature is the temperature of the air surrounding a motor, a control mechanism, or any other device. For example, a motor operated at full power may be guaranteed not to get hotter than 72°F (40°C) above the ambient temperature. Then, if the room temperature (ambient temperature) is 86°F (30°C), the temperature of the motor could get as high as 158°F (70°C) when working at full power.

Ambient temperature is not usually constant. It may change day by day and hour by hour, depending on usage of the space, sunshine, and many other factors.

1.23 Heat of Compression

As a gas is compressed, its temperature rises. See **Figure 1-1.** This is due to the work (energy) added to the gas by the compressor. The energy added is often termed "heat of compression."

The temperature of the vaporized (gas) refrigerant returning to the compressor from the evaporator will probably be at, or slightly below, the room temperature. This same vapor, leaving the compressor and entering the condenser, will be at a much higher temperature.

The compression of vaporized refrigerant in a refrigerator compressor is considered to be a nearly adiabatic process. Adiabatic compression is a process in which a gas is compressed without losing heat to the surroundings. The refrigerant passes through the compressor very quickly. Therefore, it only loses a small amount of its heat of compression there.

The compressed vapor in the condenser is now much warmer than the temperature of the surrounding air. Heat of compression will be rapidly transferred through the condenser walls to the surrounding air or condenser cooling water. (See **Second Law of Thermodynamics,** Chapter 31.)

1.24 Energy Units

In refrigeration work, three common, related forms of energy must be considered: mechanical, electrical, and heat energy. The study of refrigeration deals mainly with heat energy. However, a refrigerator must make use of electrical and mechanical energy to move heat energy from one place to another.

In a compression refrigerating unit, electrical energy flows into an electric motor. There this electrical energy is turned into mechanical energy. The mechanical energy is used to turn a compressor. The compressor, in turn, compresses the vapor to a high pressure and high temperature. This process transforms mechanical energy into heat energy.

Various units are used for measuring mechanical, heat, and electrical energy. Energy conversion units are expressed as follows:

Mechanical to heat	1 hp = 2546 Btu/hr
	778 ft.-lb. = 1 Btu
Mechanical to electrical	1 hp = 746 watts (W)
Electrical to mechanical	746 watts = 1 hp
Electrical to heat	1 watt (1 joule/sec.) = 3.412 Btu/hr
	1 kilowatt (kW) = 3412 Btu/hr
Heat to mechanical	1 Btu/hr = 0.000393 hp
Heat to electrical	1 Btu/hr = 0.293 watts

These conversion units are used in calculating loads and determining the capacity of equipment required for specific refrigeration applications.

SI Metric Units

In SI metric, the unit for measuring energy in all three of these forms is the joule. The kilowatt hour is widely used, however, as a measure of electric energy.

Equivalents

1 J = 0.7376 ft.-lb.
1 ft.-lb. = 1.3558 J
1 W = 0.7376 ft.-lb./sec.
1 ft.-lb./sec. = 1.3558 W
1 kW = 1.34 hp = 3412 Btu/h
1 hp = 0.746 kW

REFRIGERATION SYSTEMS AND TERMS MODULE

1.25 Refrigerant

In refrigerating systems, fluids which absorb heat inside the cabinet and release it outside are called ***refrigerants.***

In the evaporator, under a reduced pressure, the fluid changes from a liquid to a vapor, thus absorbing heat. In vapor form, the fluid is taken into the compressor. There the temperature and pressure are increased. This allows the heat that was absorbed in the evaporator to be released (squeezed out) in the condenser. The refrigerant is then returned to a liquid form for another cycle.

Chapter 9 describes the refrigerants most used. Their technical characteristics are also explained.

1.26 Heat Transfer

Heat may be transferred or moved from one body to another by one of three methods: radiation, conduction, or convection. Some systems of heat transfer use a combination of these three methods.

1.26.1 Radiation

Radiation is the transfer of heat by heat rays. The earth receives heat from the sun by radiation. Light rays from the sun turn into heat as they strike opaque or translucent materials. These materials will absorb some or all of the rays. (Opaque means light cannot shine through. Translucent means light can go through but one cannot see through. See Chapter 25.)

Air is heated very little as light rays pass through it. Likewise, a glass pane absorbs little heat as rays pass through it.

Sunlight generates more heat when striking dark-colored objects than when striking light-colored or polished surfaces. This is because light-colored and polished objects reflect the rays. Reflected rays are not absorbed and changed into heat. Rough, dark-colored surfaces will get hotter than light-colored or polished surfaces, so they will radiate more heat.

Any heated surface loses heat to cooler surrounding space or surfaces through radiation.

Likewise, a cold surface will absorb radiated heat. Some space heating systems depend on radiant heating sources to the room. These may be located in the ceilings, walls, or floors.

1.26.2 Conduction

Conduction is the flow of heat between parts of a substance by molecular vibrations. The flow can also be from one substance to another substance in direct contact.

A piece of iron with one end in a fire will soon become warm from end to end. This is an example of the transfer of heat by conduction. The heat travels through the iron, using the metal as the conducting medium.

Substances differ in their ability to conduct heat. In general, substances which are good conductors of electricity are also good conductors of heat (Wiedmann-Frank Law).

Substances which conduct heat poorly are called insulators. Such substances are used to insulate refrigerators and homes. Any structure that is to be maintained at a temperature difference from its surroundings may use insulators.

1.26.3 Convection

Convection is the movement of heat from one place to another by way of fluid or air. For example, heated air moves from a furnace into the rooms of a house. It releases its heat to the rooms. Then cooled air returns through cold air ducts to receive another supply of heat.

The same method may be used to cool a space. Unwanted heat is collected and discharged outside the space.

1.26.4 Control of Heat Flow

The flow of heat by radiation, conduction, and convection can be controlled. The transfer of heat by each can be increased or cut back according to need.

Use of materials that are good radiators of heat improves transfer of heat by radiation. A color known to be a good radiator can also be used. Radiation may also be improved by the type of receiving surfaces used. Materials or colors that are good absorbers (poor reflectors) of radiated heat should be used. Radiation may be reduced by reversing this application.

Dark materials or colors absorb and radiate readily. Light-colored or shiny materials have the opposite properties.

Conduction may be improved by providing large conducting surfaces. Good conducting materials, such as copper, aluminum, and iron also improve conduction. Cork, foam plastics, mineral wool, and many other similar materials are poor conductors of heat. Poor conductors of heat are commonly referred to as ***heat insulators*** (insulation).

Convection may be improved by increasing the flow of the conveying medium. Forced-air circulation heating systems are an example. A blower speeds up airflow. Conversely, convection can be slowed by retarding the circulation of air.

Heat transfer is also controlled (affected) by the temperature difference (T. D.). The greater the temperature difference, the greater the heat flow.

1.27 Brine and Sweet Water

Some refrigeration and air conditioning applications require that water be kept from freezing at temperatures considerably below the normal freezing temperature of 32°F (0°C). Other applications require that water at atmospheric pressure be kept from boiling at temperatures above 212°F (100°C).

Salt, sodium chloride (NaCl), or calcium chloride ($CaCl_2$), added to water, raises the temperature at which the water will boil. It also lowers the temperature at which it will freeze. See Chapter 31 for tables of brine solutions with a freezing point and specific gravity for each.

Some refrigerating and air conditioning installations use tap water without adding any salt or other substance. This is referred to as "sweet water."

1.28 Dry Ice

Solid carbon dioxide (CO_2) is sometimes used for refrigeration. Solid CO_2 is a white crystalline (like crystals) substance. It is formed when liquid carbon dioxide is allowed to escape into a snow chamber (heat-insulated box).

The heat for vaporizing the liquid is drawn from the interior of the chamber. A very low temperature (−108°F, −78°C) is formed. As a result, quantities of the carbon dioxide solidify.

This solid is pressed into various shapes and sizes and sold for refrigeration purposes. It is given such names as dry ice, zero ice, and so forth. It remains at a temperature of −108°F (−78°C) while in a solid state at atmospheric pressure.

Dry ice does not melt into a liquid. It goes directly from the solid to the vapor state. This is called "sublimation." Dry ice has some desirable characteristics. It does not wet the surfaces that it touches. The vapor given off is a preservative. The low temperature maintained permits handling frozen foods without using a heavily insulated container.

The latent heat of sublimation is 248 Btu/lb. (577 kJ/kg). The heat absorbed by the vapor in passing from −108°F (−78°C) to 32°F (0°C) is approximately 27 Btu/lb. (63 kJ/kg). This, added to the latent heat of sublimation, makes a total heat-absorbing capacity of 275 Btu/lb. (640 kJ/kg).

Dry ice has a greater heat-absorbing capability than does water ice. It is generally more expensive than water ice.

Equivalents

1 kJ/kg = 0.4299 Btu/lb.
1 Btu/lb. = 2.326 kJ/kg

Never place dry ice in a sealed container! At ordinary temperatures, the dry ice will sublime (turn into a vapor). The resulting pressure may cause the container to explode.

Avoid touching dry ice. It will instantly freeze the skin.

1.29 Critical Temperature

The *critical temperature* of a substance is the highest temperature at which the substance may be liquefied, regardless of the pressure applied upon it. Chapter 31 lists critical temperatures for common refrigerants.

The condensing temperature for a refrigerant must be kept below its critical temperature. Otherwise, the refrigerator will not operate. Carbon dioxide (R-744) has a critical temperature of 87.8°F (31°C). This refrigerant is not used in air-cooled compression systems. This is because the condensing temperature would usually be above this temperature.

1.30 Critical Pressure

The *critical pressure* of a substance is the minimum pressure necessary to liquefy a gas that is at its critical temperature. Less pressure will not liquefy it.

1.31 Enthalpy

Enthalpy is the measure of the heat content of a substance. The amount of enthalpy is determined by both the temperature and the pressure of the substance.

U.S. Conventional Units

Enthalpy is all the heat in one pound of a substance calculated from an accepted reference temperature, 32°F. This reference temperature can be used for water and water vapor calculations. For refrigerant calculations, the accepted reference temperature is −40°F. See **Figure 1-21A.**

Example:

What is the enthalpy of 1 lb. of water at 212°F, assuming 0 enthalpy at 32°F?

Solution:

Enthalpy at 32°F = 0

The specific heat of water is C_p = 1 Btu/lb./°F

Heat to raise the temperature of 1 lb. of water from 32°F to 212°F:

212 − 32 = 180°F

Enthalpy = Mass of water × Specific heat of water × Change in temperature (°F)

$H = M \times C_p \times \Delta\ °F$

$H = 1 \times 1 \times 180$

H = 180 Btu

Total enthalpy at 212°F = 180 Btu

SI Metric Units

In SI, the zero enthalpies for water, refrigerants, and air are also taken at some convenient temperature (reference temperature or T_r) and pressure, as shown:

1. For water, zero enthalpy is at 0°C and 100 kPa.
2. For refrigerants, −40°C and 100 kPa.
3. For air, 25°C and 100 kPa.

The enthalpy is measured in joules (J) or kilojoules (kJ).

Example:

What is the total enthalpy of 5 kilograms of water at 80°C?

Solution:

The enthalpy (H) of 5 kilograms of water at 80°C is calculated using the following formula:

Enthalpy = Mass × Specific heat × Temperature change ($T - T_r$)

$H = M \times C_p \times (T - T_r)$

M = 5 kg

$C_p = 4.19\ kJ/kg \cdot K$

T = 80°C temperature

T_r = 0°C, the reference temperature for water

$H = 5 \times 4.19 \times (80 - 0)$

$H = 5 \times 4.19 \times 80$

H = 1 676 kJ

Figure 1-21B shows the relationship of enthalpy to temperature.

1.31.1 Specific Enthalpy

Specific enthalpy is enthalpy per unit mass. It is measured in Btu per pound (J/kg). Tables of the enthalpy of substances and pressure-enthalpy diagrams, such as **Figure 1-21,** use specific enthalpy.

Example:

If 100 lb. of a substance absorbs 2000 Btu of energy when heated from the reference state of 0 Btu/lb., what is the specific enthalpy?

Solution:

Specific enthalpy = enthalpy absorbed ÷ mass

$$h = \frac{H}{M}$$

$$h = \frac{2000\ \text{Btu}}{100\ \text{lbs.}}$$

h = 20 Btu/lb.

1.32 Cryogenics

Cryogenics refers to creating and using temperatures in the range of 116 K down to 0 K (−251°F down to −460°F, or −157°C down to −273°C).

The term cryogenics has increased in common usage. This is due to frequent use of liquid helium, nitrogen, and liquid hydrogen in refrigeration. The term is also applied to the low-temperature liquefaction of gases and their handling and storage. It includes insulation of containers, instrumentation, and techniques used in such work.

Figure 1-26 shows evaporating temperatures at atmospheric pressure of some common cryogenic fluids. It also indicates the cryogenic range of temperatures. These temperatures are in Kelvin (K).

Boiling Temperature at Atmospheric Pressure				
FLUID	Fahrenheit °F	Rankine °R (Absolute °F Scale)	Celsius °C	Kelvin K (Absolute °C Scale)
Water	212	672	100	373
R-12 Refrigerant	–22	438	–30	243
R-22 Refrigerant	–41	419	–41	230
R-744 Refrigerant Carbon Dioxide	–109	351	–78	195
R-1150 Refrigerant Ethylene	–135	325	–93	180
Beginning of the Cryogenic Range	–250	210	–157	116
Methane	–258	202	–161	112
Oxygen	–297	163	–183	90
Air	–313	147	–192	81
Nitrogen	–320	140	–196	77
Neon	–411	49	–246	27
Hydrogen	–423	37	–253	20
Helium	–452	8	–270	4
Absolute Zero	–460	0	–273	0

Figure 1-26. *Boiling temperatures of some common refrigerants and some other fluids at atmospheric pressure. Note difference between boiling points of some commonly used refrigerants and boiling points of fluids in the cryogenic range.*

1.33 Perfect Gas Equation

If a quantity of gas is enclosed in a tight container, the relationship between pressure, temperature, and volume may be expressed by the formula: PV = MRT.

P = Pressure in pounds per square foot absolute
V = Volume in cubic feet
M = Mass of gas in pounds
R = Gas constant (R will differ for different gases). (**Figure 1-27** gives the value of R for some common substances.)
T = Absolute temperature in °R

Material	Gas Constant (R) J/kg·K	Gas Constant (R) $\frac{\text{ft.-lb.}}{\text{lb. R}}$	Specific Heat kJ/kg·K C_p	Specific Heat kJ/kg·K C_v
Air	288.68	53.34	1.00	0.71
R-717 (Ammonia)	666.98	123.24	2.13	1.46
R-744 (Carbon Dioxide)	210.10	38.82	0.92	0.71
Ether	125.08	23.11	2.01	1.88
Oxygen	262.76	48.55	0.92	0.67
Alcohol	224.87	41.55	1.88	1.67
Water Vapor	450.45	83.23	2.03	1.55

Figure 1-27. *Table lists gas values (constants) for some substances used in refrigeration work.*

Example:

What will be the volume of 2 lb. of carbon dioxide at 240°F when the pressure is 185 psi?

Formula:

PV = MRT
P = (185 + 15) = 200 psia = 200 × 144 = 28,800 lb. per sq. ft. absolute
M = 2 lb.
R = 38.82 ft.-lb./lb.°R*
T = (240 + 460) = 700°R

$$V = \frac{MRT}{P}$$

$$V = \frac{2 \times 38.82 \times 700}{28{,}800}$$

$$V = \frac{54{,}348}{28{,}800} = 1.89$$

V = 1.89 cu. ft.

*The unit "ft.-lb." is used instead of Btu. Btu is not compatible with feet, pounds, and °R, just as cal are not compatible with N · m = Joule, kg, and °K.

1.33.1 SI Metric Units

The equation in SI may be expressed: PV = MRT.

P = Pressure in pascals (Pa)
V = Container volume in cubic meters (m^3)
M = Mass of the gas in the container in kilograms (kg)
R = Gas constant which has a value depending on the gas properties. **Figure 1-27** gives values of R for some substances used in refrigeration work.
T = Absolute temperature, kelvin (K) which is: (273 + T°C)

The equation shows that if the container of gas is heated so that the temperature increases, then the pressure will also rise. Cooling a container will reduce both the temperature and the pressure.

Example:

If 0.2 kg of air at a pressure of 1.00 kPa is contained in a volume of 50 m^3, what is the temperature?

Solution:

Solving for T (absolute temperature):

$$T = \frac{PV}{MR}$$

P = 1.00 kPa = 1000 newtons/m^2
V = 50 m^3
R = 288.68 J/kg · K (from **Figure 1-27**)
M = 0.2 kg

$$T = \frac{1000 \times 50}{0.2 \times 288.68}$$

T = 866 K

The temperature in Celsius is then:

$T\ °C = T\ K - 273$
$T\ °C = 866 - 273$
$T\ °C = 593°C$

Note: The temperature used in the perfect gas equation must always be in Kelvin.

Example:
If the air in the container is heated until the temperature reaches 1000°C, find the new pressure:

Solution:

$$T = 1000 + 273 = 1\,273\ K$$

$$P = \frac{MRT}{V}$$

$$P = \frac{0.2 \times 288.68 \times 1\,273}{50}$$

$$P = 1\,470\ Pa = 1.47\ kPa$$

1.34 Dalton's Law

Dalton's Law of partial pressures is the foundation of the principle of operation of one of the absorption type refrigerating systems. The law states:

- The total pressure of a confined mixture of gases is the sum of the pressures of each of the gases in the mixture.
- The total pressure of the air in a compressed air cylinder is the sum of the oxygen, nitrogen, and the carbon dioxide gases, and the water vapor pressure.

The law further explains that each gas behaves as if it occupies the space alone. To illustrate, the absorption refrigerator uses two gases, ammonia and hydrogen. The ammonia, at room temperature, is absorbed by the water in the closed system.

Heating this solution drives out the ammonia. (The hydrogen is not absorbed by the water and remains as a gas.) Due to the pressure it is under, the ammonia condenses into a liquid in the condenser. The pressure is uniform throughout the system. Total pressure in the system is the sum of the vapor pressure of the ammonia plus the hydrogen pressure. When the pressure of the ammonia vapor is below the pressure corresponding to the vapor pressure for ammonia alone, the ammonia continues to evaporate. It tries to reach a vapor pressure corresponding to the temperature in the absorber.

1.35 Evaporator

In this text, the word *evaporator* is used to indicate a part of the refrigerator mechanism. This is where the liquid refrigerant boils or evaporates and absorbs heat. In some trade literature, the term "cooling coil" is used to indicate the part in which such cooling takes place. The correct technical term, however, is evaporator. Coils cooled by brine or any fluid which does not evaporate in the coil may properly be called *cooling coils.*

1.36 Vapor—Gas

The word *vapor* in this text indicates refrigerant which has become heated, usually in the evaporator, and has changed to a vapor or gaseous state. In some trade and service literature, refrigerant in this state is called a gas. The correct technical term is vapor.

1.36.1 Saturated Vapor

The term *saturated vapor* identifies a condition of balance on an enclosed quantity of a vaporized fluid. The balance is such that some condensate (liquid) will be produced if there is even the slightest lowering of the temperature or increase in pressure.

There is usually some of the substance present in liquid form when the vapor is saturated. In a saturated condition, all of the substance has been vaporized that can be vaporized under the existing conditions of pressure and temperature.

1.37 Humidity—Relative Humidity

The word *humidity,* as used in connection with refrigeration, air conditioning, and weather information, refers to water vapor or moisture in the air.

Air absorbs moisture. The amount depends on the temperature of the air. The higher the temperature of the air, the more moisture it will absorb.

Relative humidity is the amount of moisture carried in a sample of air, compared to the total amount which it can absorb at the stated pressure and temperature. Relative humidity is covered in Chapter 19.

A relative humidity of 50% indicates that the air has 50% as much moisture as it will hold at that particular temperature and pressure. See Chapter 31 for tables of moisture-holding ability for air at various temperatures, with volume and heat data.

1.38 Elementary Refrigerator

In **Figure 1-28,** a refrigerant cylinder, A, is shown with the valve closed. The pressure inside is 72.7 psig (87.4 psia, 603 kPa) and the temperature is 71.5°F. All conditions inside the cylinder are balanced. The number of molecules leaving the vapor state by diving back into the liquid, and the liquid molecules flying out of the liquid into the vapor state are equal.

In cylinder B, the valve has been opened slightly. Some of the vapor is escaping. The results are twofold:

1. The pressure over the liquid refrigerant in the cylinder is reduced to 47.4 psig (62.1 psia). This causes change. There is now more liquid changing to a vapor than there is vapor changing back into a liquid.

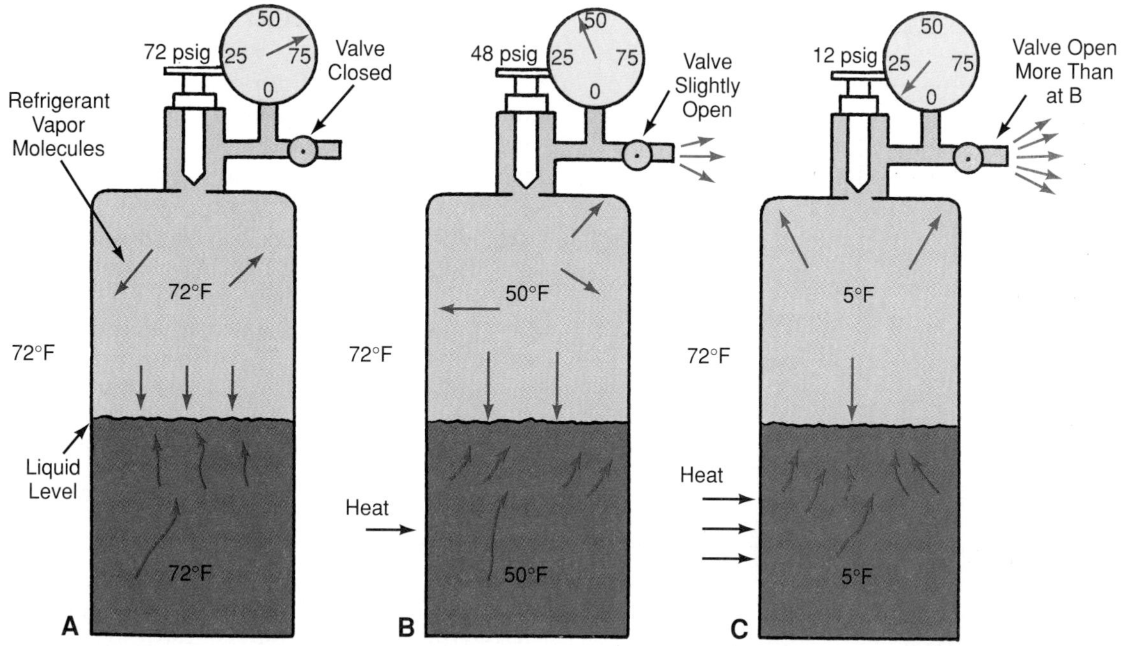

Figure 1-28. *Cooling effect of different pressures operating on surface of liquid refrigerant R-12. A—Cylinder of refrigerant with valve closed. Refrigerant is in state of equilibrium. B—Cylinder of refrigerant with valve slightly opened and refrigerant vapor beginning to leave cylinder. C—Cylinder of refrigerant with valve open wider. Large amount of refrigerant vapor is flowing from cylinder.*

2. With more liquid turning into a vapor than vapor returning to a liquid, heat is absorbed. The liquid refrigerant and cylinder will be cooled. The temperature of the refrigerant and cylinder is now 50°F. Some heat from the surrounding area, which is at 72°F, will now flow into the cylinder and the refrigerant.

In cylinder C, the valve has been opened more than in B. Refrigerant vapor now flows out more rapidly. This further lowers the pressure on the liquid refrigerant, which evaporates even faster.

The increase in the rate of evaporation lowers the temperature of the refrigerant and the cylinder. Now heat will flow even more rapidly from the 72°F air into the cold cylinder.

In cylinder A, there is a state of equilibrium (balance), with all temperatures and pressures in balance.

In cylinder B, there is a slight imbalance due to the vapor escaping through the valve. If this condition were to continue for a considerable time, a condition of balance would again exist. In this new condition, its balance would not be static as in A. Instead, it would be a balance between the rate of heat flow into the cylinder, the evaporation of refrigerant, and the flow of refrigerant vapor out of the cylinder valve. In this condition of balance, the refrigerant is cooling the cylinder and its surroundings.

As long as the valve is open and vapor can escape, the temperature will stay low. More molecules are escaping from the liquid into the vapor. Only a few vapor molecules are diving back into the liquid.

This vapor bombardment is called vapor pressure. If pressure can be reduced, the temperature of the liquid can be reduced since evaporation will be increased.

If the vapor molecules can be removed fast enough, the vapor pressure may be low enough to create refrigerant boiling temperatures in the refrigerating range (low temperature). Vapor molecules are usually removed with a compressor or by using chemicals which absorb the molecules.

The operation of the mechanical refrigerator relies on the heat absorption property of a fluid passing from the liquid to the vapor state. **Figure 1-29** displays an elementary refrigerator. A cylinder of refrigerant is placed in a box. Its vapor is vented to the outside. The components inside the box act as a heat absorber. This is the same method that is used in mechanical refrigeration, as explained in Section 1.39. Thus, the liquid can boil only at or above its evaporation temperature. The liquid will remain at this temperature until is has completely evaporated.

Additional heat would only cause more rapid evaporation of the liquid. **Figure 1-29** is different from **Figure 1-28.** In **Figure 1-29,** the refrigerant evaporates in the evaporator, C, not in the cylinder, B. Pressure in C remains constant, at 14.7 psia.

Since the liquid is at this low temperature, there is a transfer of heat to it from the surrounding objects. This heat increases the evaporation. The heat itself is carried away as the vapor passes off. Thus, the evaporating gets the heat energy for doing this from the objects surrounding it. That same heat is removed with the vapor to the outside of the box.

This type of refrigerating system works nicely. However, it is expensive because the refrigerant fluid is

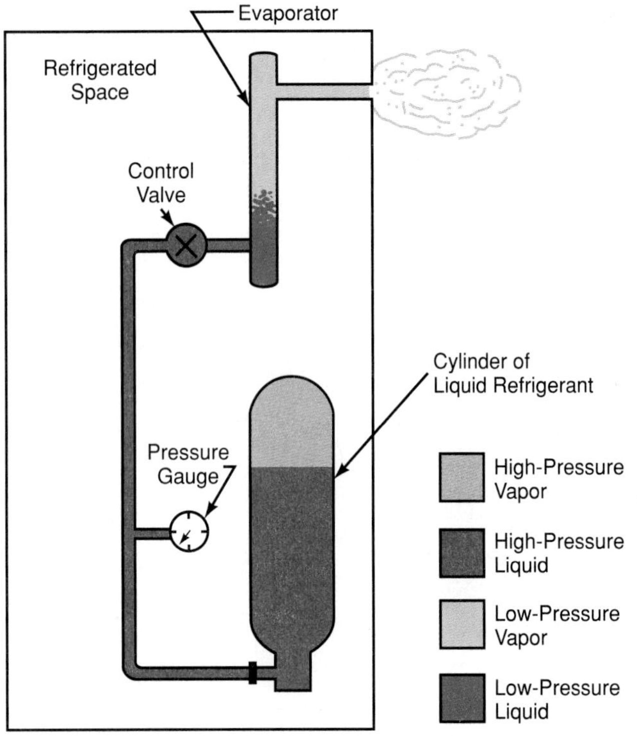

Figure 1-29. *Components of an elementary refrigerator.*

lost. Some mobile refrigerating units (trucks) use this method. The refrigerant is usually a liquid nitrogen, which is relatively inexpensive. It is called "expendable refrigerant" refrigeration.

1.39 Mechanical Refrigerating System

In a mechanical refrigerator, the vaporized refrigerant is not discarded. Instead, it is captured, compressed, and cooled to a liquid state again. This is done so that it can be reused, as shown in **Figure 1-30.**

To follow the refrigerant cycle, begin with the refrigerant in the receiver B. The refrigerant is R-134a.

The refrigerant at B is under a pressure corresponding to the room temperature of 72°F (22°C). For R-134a, this pressure will be approximately 73 psig (606 kPa) when the unit is idle. The pressure will be higher when the unit is running.

At the refrigerant control, C, this pressure is reduced to provide low-pressure, low-temperature evaporation in the evaporator (cooling coil). Since the box or inside temperature is to be held at 35°F (1.5°C), the pressure in the evaporator, D, must be held at or below 30.4 psig (311 kPa).

The purpose of the refrigerant control is to allow refrigerant to flow from the liquid receiver (high side) into the evaporator (low side). It must maintain the pressure difference between the high-pressure side (high side) and the low-pressure side (low side).

In the evaporator, D, the liquid refrigerant is now under a much reduced pressure. It will evaporate or boil very rapidly. This, in turn, cools the evaporator. The compressor, E, creates a low pressure by drawing (sucking) the vapor from the evaporator. Then it compresses the vapor back into the high side.

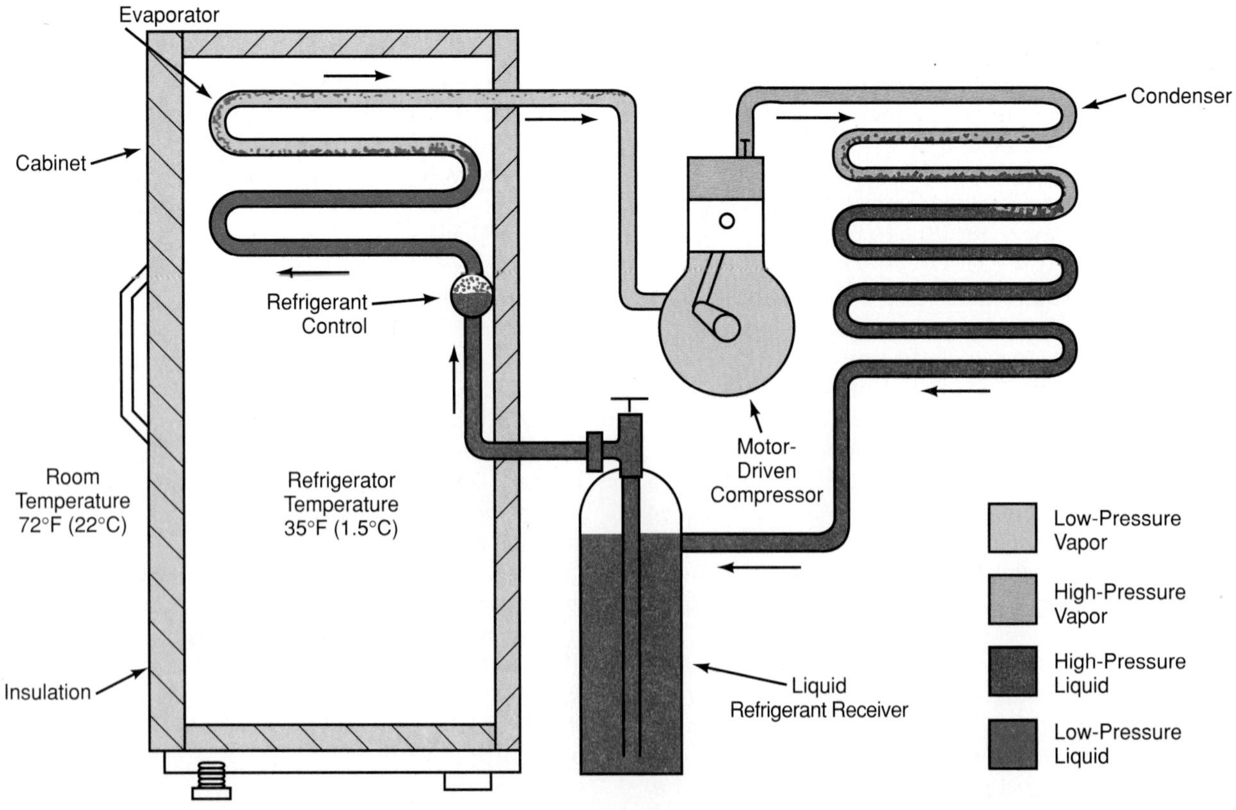

Figure 1-30. *Components of an elementary mechanical refrigerator.*

From the compressor, the warm (see heat of compression, Section 1.23), high-pressure vaporized refrigerant flows into the condenser, F. The temperature of the vapor entering the condenser will be several degrees warmer than the room temperature. This difference causes a very rapid transfer of the heat from the condenser to the surrounding air.

The vapor, as it flows through the condenser, cools and loses its heat of vaporization. It returns to the liquid state (condensed). As a liquid, it flows from the condenser back into the liquid receiver, B.

This refrigeration cycle is repeated over and over until the desired temperature is reached. A thermostat then opens the electrical circuit to the driving motor. The compressor stops.

The temperature of the evaporator determines the pressure at which the refrigerant is evaporated. The amount of heat removed depends on the amount (mass) of refrigerant vaporized.

1.40 Review of Abbreviations and Symbols

The following is a review of the various abbreviations and symbols studied so far in this chapter. This review includes both the U.S. conventional units and the SI metric units.

1.40.1 U.S. Conventional Units

Btu = British thermal unit
Btu/h = British thermal units per hour
°F = degrees Fahrenheit
F_A = degrees Fahrenheit absolute
°R = degrees Rankine = degrees absolute F
lb. = pound
psi = pounds per square inch = lb. per sq. in.
psia = pounds per square inch absolute = psi + atmospheric pressure
in. = inches = i = ″
ft. = foot or feet = f = ′
sq. in. = square inch = in^2
sq. ft. = square foot = ft^2
cu. in. = cubic inch = in^3
cu. ft. = cubic foot = ft^3
ft.-lb. = foot-pound
ton = ton of refrigeration effect
lb./cft. = pounds per cubic foot
in. Hg = inches of mercury vacuum = ″Hg
hp = horsepower
qt. = quart
gr = grain

1.40.2 SI Metric Units

°C = degrees Celsius
K = kelvin
mm = millimeter
cm = centimeter
cm^2 = centimeter squared
cm^3 = centimeter cubed
dm = decimeter
dm^2 = decimeter squared
dm^3 = decimeter cubed
m = meter
m^2 = meter squared
m^3 = meter cubed
L = liter
g = gram
cal = calorie
kg = kilogram
J = joule
kJ = kilojoule
N = newton
Pa = pascal
kPa = kilopascal
W = watt
kW = kilowatt
MW = megawatt

1.40.3 Miscellaneous Abbreviations

P = pressure
hr. or h = hours
sec. = seconds
T_r = reference temperature
Δ = difference
C_p = specific heat (sp. ht.)
h = enthalpy per unit mass
H = total enthalpy
D = diameter
r = radius of circle
A = area
π = 3.1416 (a constant used in determining the area of a circle)
V = volume
R = gas constant
∞ = infinity
M = mass
Hz = hertz

1.41 Review of Safety

The term "safety," as applied to any refrigeration or air conditioning activity, may have three different applications. It may apply to:

1. ***Safety of the operator.*** **When refrigeration and air conditioning equipment is properly handled, there is relatively little danger to the operator.**

   Always pull on a wrench (instead of pushing), to prevent possible slippage of the wrench. Slippage could cause rounded corners on nuts and bolts and possible injury to hands.

   Always use leg muscles when lifting objects, never the back muscles. A hoist is recommended for

lifting anything weighing over 35 lb. (16 kg). Make certain there is no oil or water on the floor.

Always wear safety goggles when working with refrigerants.

Most refrigerating mechanisms are electrically driven and controlled. When working on electrical circuits, make sure the circuit is disconnected from the power source. This can usually be accomplished by opening the switch at the power panel. Never work on "hot" electrical circuits.

2. ***Safety of the equipment.*** **Many parts of refrigeration and air conditioning equipment are quite fragile. Parts may be ruined by overtightening nuts and bolts, not tightening them in the correct order, or using the wrong size wrench. Make certain that all connections are tightened before operating a compressor. Before operating open compressors, be sure that the flywheel and pulley are aligned and that guards are in place.**
3. ***Safety of the contents.*** **Safety of the contents of the refrigerated space depends entirely on the accuracy and care given the installation and adjustment of the various parts of the system.**

    Tables listing proper operating temperatures for various types of refrigerated space are found throughout this text. These operating temperatures must be observed if the unit is to provide safe conditions for the refrigerated or air conditioned space.

It is advisable to observe these three points during the service work explained in the chapters following. Each chapter has a "Review of Safety." These serve as a reminder of potential hazards which may be present when working with the equipment and supplies.

There is no exception to the rule that "The safe way is the right way."

1.42 Test Your Knowledge

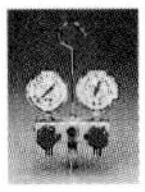

Please do not write in this text. Place your answers on a separate sheet of paper.

HISTORY AND FUNDAMENTALS OF REFRIGERATION MODULE

1. Normal refrigerator temperature is approximately _______°F.
    A. 30
    B. 35
    C. 40
    D. 42
2. What was used in the 18th century to obtain refrigeration?
    A. Water.
    B. Cold water.
    C. Ice.
    D. Salt and water solution.
3. A mechanical refrigerator removes heat from its contents by _______.
    A. evaporating the liquid refrigerant in the evaporator
    B. changing the refrigerant in the evaporator into a liquid
    C. allowing the refrigerant to evaporate into the air
    D. All of the above.
4. Heat is absorbed or soaked up by evaporating a liquid refrigerant in the _______.
    A. compressor
    B. condenser
    C. evaporator
    D. All of the above.
5. What color designates a high-pressure liquid?
    A. Dark red.
    B. Light red.
    C. Dark blue.
    D. Light blue.
6. What color designates a low-pressure liquid?
    A. Dark red.
    B. Light red.
    C. Dark blue.
    D. Light blue.
7. Heat flows in what direction?
    A. Cold to warm.
    B. Warm to cold.
    C. Liquid to gas.
    D. All of the above.
8. Temperature is an indication of the degree of _______.
    A. warmth
    B. cold
    C. Both A and B.
    D. None of the above.
9. Most substances change their physical state by _______.
    A. becoming invisible
    B. the addition or removal of heat
    C. becoming molecules
    D. the addition of heat only
10. Cold or low temperatures _______.
    A. cause the molecules to move more quickly
    B. slow bacteria growth
    C. cause increased bacteria growth
    D. cause food to become spoiled

TEMPERATURE, PRESSURE, AND MEASUREMENTS MODULE

11. How many centimeters are equal to one inch?
    A. 25.4 cm.
    B. 0.254 cm.
    C. 2.54 cm.
    D. 1.5 cm.

12. What is the piston displacement of a compressor with 2″ bore and 3″ strokes?
    A. 9.42 in^3.
    B. 12.00 in^3.
    C. 14.14 in^3.
    D. 37.70 in^3.
13. How many 90° arcs fit in a complete circle?
    A. 8.
    B. 4.
    C. 2.5.
    D. 6.
14. What determines the temperature at which a refrigerant will vaporize?
    A. Pressure.
    B. Temperature.
    C. Volume.
    D. Enthalpy.
15. Express standard atmospheric pressure in pounds per square foot.
    A. 176 lbs./ft^2.
    B. 2117 lbs./ft^2.
    C. 2593 lbs./ft^2.
    D. 14,590 lbs./ft^2.
16. A substance has a temperature of 20°C. What is the temperature in °F?
    A. 52°F.
    B. 68°F.
    C. 64°F.
    D. None of the above.
17. A substance has a temperature of 78°F. What is the temperature in °C?
    A. 32°C.
    B. −5°C.
    C. 46°C.
    D. 26°C.
18. In SI units, the joule is a unit of ______.
    A. temperature
    B. heat
    C. work
    D. Both heat and work.
19. How much heat will be required to change 5 lbs. of ice at 32°F into water at 82°F?
    A. 194 Btu.
    B. 720 Btu.
    C. 970 Btu.
    D. 1925 Btu.
20. ______ has the largest specific heat capacity.
    A. Water
    B. Iron
    C. Glass
    D. Brick

REFRIGERATION SYSTEMS AND TERMS MODULE

21. What is dry ice?
    A. A liquid form of ice.
    B. Solid carbon dioxide.
    C. Sulfur dioxide.
    D. A vapor.
22. Given two objects of the same size—one chrome-plated and one painted black—which one will absorb more radiant heat?
    A. Chrome-painted object.
    B. Black-painted object.
    C. Both absorb equally.
    D. Neither absorb.
23. Which of the following illustrates the principle of convection?
    A. Movement of heat by fluid or air.
    B. Heat waves from the sun.
    C. Heat traveling through a solid object from end to end.
    D. Radiation from the sun.
24. Will glass or copper conduct heat the most rapidly?
    A. Glass.
    B. Both conduct heat at an equal rate.
    C. Neither are conductors of heat.
    D. Copper.
25. Should refrigerants be operated at temperatures above or below their critical temperature?
    A. Above.
    B. Below.
    C. Neither above or below. They should be at their critical temperature.
    D. Either above or below.
26. The desired temperature in a domestic cabinet is ______°F.
    A. 15
    B. 0
    C. 50
    D. 35
27. The desired temperature in a domestic cabinet is ______°C.
    A. 35
    B. 2.5
    C. 1.5
    D. None of the above.
28. Refrigerant fluids are ______.
    A. fluids that absorb heat inside a cabinet
    B. fluids that release heat outside of a cabinet
    C. fluids that absorb heat outside of a cabinet
    D. Both A and B.
29. Heat may be transferred from one body to another by ______.
    A. radiation
    B. conduction
    C. convection
    D. All of the above.
30. Liquid can boil only ______ temperature.
    A. at its evaporation
    B. below its evaporation
    C. at or above its evaporation
    D. at or below its freezing

A technician brazing lines on a compressor at a manufacturing plant. Note the safety equipment being worn. (Bryant Air Conditioning/Heating)

Chapter 2

REFRIGERATION TOOLS AND MATERIALS

Modules:

Key Terms:

ACR tubing
annealed
brazing
double-thickness flare
flare
flux
one-way service valve
refrigerant oil
swaging
two-way service valve

Learning Objectives:

By studying this chapter, you will be able to:

- List and discuss the various types of tubing used in refrigeration work.
- Cut and fit tubing using approved methods.
- Demonstrate soldering and brazing techniques.
- Repair cracks and leaks in evaporators.
- Select the proper tools for servicing and maintaining domestic refrigerators.
- Explain how to use various hand tools.
- Discuss the procedures for threading steel pipe.
- Identify thread types.
- Identify different types of threaded fasteners.
- Demonstrate standard procedures for basic mechanical service and repair operations.
- Explain how to maintain and calibrate gauges.
- Compare cleaning methods and use of solvents.
- Explain the use of vacuum and compound gauges.
- Define various types of service valves.
- Discuss the importance of oil in refrigerating systems.
- Define purging and explain how it is done.
- Discuss the evacuation of a system.
- **Follow approved safety procedures.**

TUBING AND FITTINGS MODULE

2.1 Tubing

Most tubing used in refrigeration and air conditioning is made of copper. However, some aluminum, steel, stainless steel, and plastic tubing is also used.

Instructions in this chapter will deal mainly with copper tubing. All tubing used in air conditioning and refrigeration work is carefully processed to be sure that it is clean and dry inside. The ends must be kept sealed until it is used.

2.1.1 Copper Tubing

Most copper tubing used in air conditioning and refrigeration work is known as ***Air Conditioning and Refrigeration (ACR) tubing.*** The tubing is intended for air conditioning and refrigeration. It has been processed to give the desired characteristics.

ACR tubing is usually charged with gaseous nitrogen. This keeps the tubing clean and dry until it is used. Nitrogen should be fed through the tubing during brazing and soldering operations. Take care; it is dangerous to use. See Chapter 12.

The nitrogen will eliminate the danger of oxidation inside the tube. All tubing ends should be plugged immediately after cutting a length from the piece.

Copper tubing is available in soft and hard types. Both are available in two wall thicknesses—K and L. Type K is a heavy wall. Type L is a medium thick wall. Most ACR tubing currently being used is the Type L. Soft copper tubing is supplied in 25′, 50′, and 100′ rolls.

Another type of copper tubing used in heating and plumbing is called "nominal size." This type of copper tubing is used on water lines, drains, and in other applications. It is never used with refrigerants.

Soft Copper Tubing

Soft copper tubing is used in domestic work and in some commercial refrigeration and air conditioning work. It is ***annealed*** (heated and then allowed to cool). This makes it flexible and easy to bend and flare. (A ***flare*** is an enlargement at the end of a piece of flexible tubing by which the tubing is connected to a fitting or another piece of tubing.) Being easily bent, this tubing must be supported by clamps or brackets. Soft copper tubing is most often used in connection with flared fittings (Society of Automotive Engineers (SAE) standards) and soft soldered fittings. Soft copper tubing is sold in 25′+, 50′+, and 100′ rolls. Sizes most commonly used are 3/16″, 1/4″, 5/16″, 3/8″, 7/16″, 1/2″, 9/16″, 5/8″, and 3/4″ outside diameter (OD). Wall thickness is usually specified in thousandths of an inch. **Figure 2-1** is a table of common copper tube diameters and thicknesses.

Outside Diameter	Wall Thickness
1/4	.030
3/8	.032
1/2	.032
5/8	.035
3/4	.035
7/8	.045
1 1/8	.050
1 3/8	.055

Figure 2-1. *Copper tube sizes used in refrigeration work. Both soft- and hard-drawn sizes are the same as the measurements listed in the table. Outside diameter size for this tubing is the actual outside diameter of tube.*

Nominal Size Inches	Type	OD Inches		Wall Thickness Inches
1/4	K	0.375	3/8	0.035
	L	0.375	3/8	0.030
3/8	K	0.500	1/2	0.049
	L	0.500	1/2	0.035
1/2	K	0.625	5/8	0.049
	L	0.625	5/8	0.040
5/8	K	0.750	3/4	0.049
	L	0.750	3/4	0.042
3/4	K	0.875	7/8	0.065
	L	0.875	7/8	0.045
1	K	1.125	1 1/8	0.065
	L	1.125	1 1/8	0.050

Figure 2-2. *Nominal-size copper tubing. Type K—heavy wall is available in hard and soft temper. Type L—medium wall is available in hard and soft temper. Type K is used where corrosion conditions are severe. Type L is used where conditions may be considered normal. Outside diameter sizes indicated by dimension are 1/8″ (.125″) larger than nominal size.*

Soft copper tubing may be worked to give it certain properties. It can be hardened by repeated bending or hammering. This is called ***work hardening.*** It can be softened by annealing, as explained earlier.

Tubing must be installed so that there is no strain on it when the job is completed. Horizontal loops may be used to keep vibration from crystallizing the copper, making it crack or break.

Hard-Drawn Copper Tubing

Hard-drawn copper tubing is used in commercial refrigeration and air conditioning applications. Being hard and stiff, it needs few clamps or supports, particularly in larger diameters.

Hard tubing should not be bent. Use straight lengths and fittings to form necessary tubing connections.

Hard-drawn refrigeration tubing joints should be brazed. Soft solder should be used only on water lines. Hard-drawn tubing is supplied in 20′ lengths. It is available in the same diameters and thicknesses as soft copper tubing.

Nominal-Size Copper Tubing

Nominal-size copper tubing is a type used on water lines, drains, and in other applications. Nominal-size copper tubing is never used with refrigerants. It is available in both soft- and hard-drawn grades. The table, **Figure 2-2,** shows commonly used sizes and wall thicknesses of this tubing. The wall thickness (type) is indicated by use of a letter after the nominal size.

Copper tubing used for such applications is often referred to by its nominal size. If you measure the OD, you will notice that the OD is 1/8″ larger than that listed under nominal size.

When purchasing fittings for this tubing, it is important that the fitting size is the same as the size tubing purchased. You should order all the tubing, valves, and fittings by nominal size or order all by OD, to avoid problems.

2.1.2 Steel Tubing

Some thin-wall steel tubing is used in refrigeration and air conditioning work. These sizes are practically the same as for copper tubing. Connections may be made on steel tubing by using either flared joints or brazed joints.

Copper or brass tubing must not be used with refrigerant R-717 (ammonia). Use steel tubing instead. There is a chance of chemical reaction (corrosion) between ammonia and copper.

Two types of steel tubing are in common use. One type has a double lap brazed construction using SAE 1008 mild steel. The other is butt welded, using the same type steel.

2.1.3 Stainless Steel Tubing

Stainless steel tubing comes in the typical refrigeration tube sizes. The most common sizes are listed in the table, **Figure 2-3.** Stainless steel is strong and very resistant to corrosion. It may be easily connected to fittings by flare fittings or brazing.

Stainless steel tubing No. 304 is commonly used. This is a low-carbon iron alloy containing 18% chromium and 8% nickel. It is often used in food processing, ice cream manufacture, milk handling systems, and the like. Type 304 stainless steel is not magnetic.

2.1.4 Metric Tube Sizes

Metric-sized tubing is used in some refrigeration and air conditioning systems. The standard sizes are 6, 8, 10, 12, 14, and 15 millimeters (mm) OD.

2.1.5 Plastic Tubing

Polyethylene is one of the most common substances used in the manufacture of plastic tubing. Sizes and suggestions for its use are shown in **Figure 2-4.**

The usual safe temperature range for polyethylene tubing is from −100 to +175°F (−73 to 79°C). Therefore, never use this tubing in installations where fluid temperature goes beyond these limits.

In general, polyethylene tubing is not used in the refrigerating cycle mechanisms. It is used for cold water lines, water-cooled condensers, and the like. It is very easy to use and may be cut with a knife. It may also be easily bent.

Special fittings are available for connecting polyethylene tubing to refrigeration and air conditioning mechanisms.

2.1.6 Flexible Tubing (Hose)

In many refrigeration and air conditioning applications, the liquid lines and suction lines must be flexible, **Figure 2-5.** This is particularly true in many commercial and industrial refrigeration and air conditioning applications.

Air conditioning equipment on motor vehicles requires flexible tubing. This type of hose is made from a variety of special materials. Such materials do not age and remain flexible. These materials do not allow fluid to leak through the hose wall. They are easy to attach to fittings.

Outside Diameter	Wall Thickness	Burst Pressure psi	Minimum Bend Radius
1/8"	.020	500	1/2"
3/16"	.030	500	1/2"
1/4"	.040	400	1"
5/16"	.062	600	1 1/8"
3/8"	.062	350	1 1/4"
1/2"	.062	250	2 1/2"

Figure 2-4. *Plastic tubing specifications. Note that there are four different thicknesses used in this size range of plastic tubing. (Imperial Eastman, Imperial Division)*

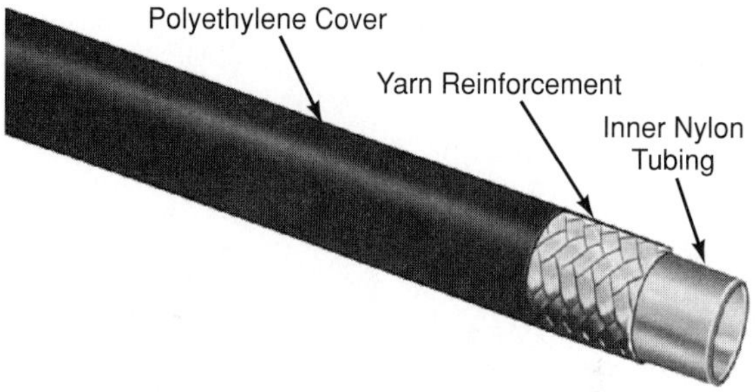

Figure 2-5. *Thermoplastic hose used in refrigeration systems.*

Flexible Hose Fittings

There are various types of flexible hose fittings available. See the descriptions listed in **Figure 2-6.**

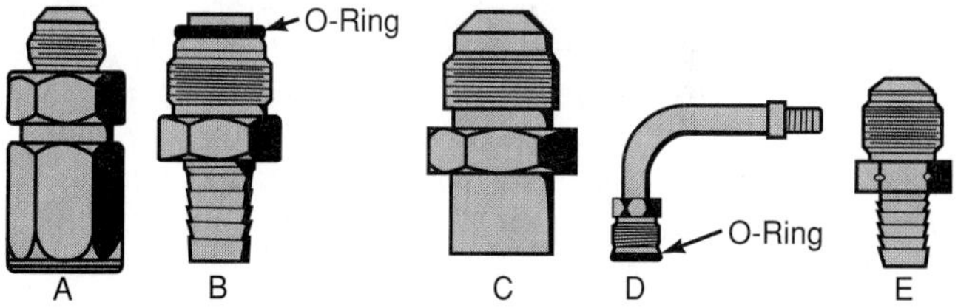

Figure 2-6. *Assorted nylon fittings suitable for use with refrigerant hose. A—Coupling, straight male 45° flare, threaded reusable. B—Coupling, straight male push-on barb-type reusable, with O-ring seal. C—Coupling, straight male 45° flare permanent (crimped-on and not reusable). D—Coupling, 90° male push-on barb-type reusable, with O-ring seal. E—Coupling, straight male 45° flare push-on.*

	Outside Diameter							Wall Thickness
Fractions	1/4	3/8	1/2	5/8	3/4	1	1 1/4	(All of the stainless steel tubing is available in various wall thicknesses (BWG)* – 31 to 20 gage.)
Decimals	.250	.375	.500	.625	.750	1.000	1.250	
Millimeters	6.35	9.52	12.7	15.87	19.05	25.40	31.75	

*– Birmingham Wire Gage

Figure 2-3. *Stainless steel tubing sizes are given in fractions and decimals of an inch, and in millimeters.*

Flexible hose fittings are usually made of brass. Nylon fittings are sometimes used. Synthetic rubber O-rings are put on some of these fittings to provide a better seal. The attachment end of these fittings conforms to the standard SAE fittings specifications.

2.2 Cutting and Bending Tools

There are numerous types of cutting and bending tools used on copper and plastic tubing. See **Figure 2-7.** These tools and their uses are described in the following paragraphs.

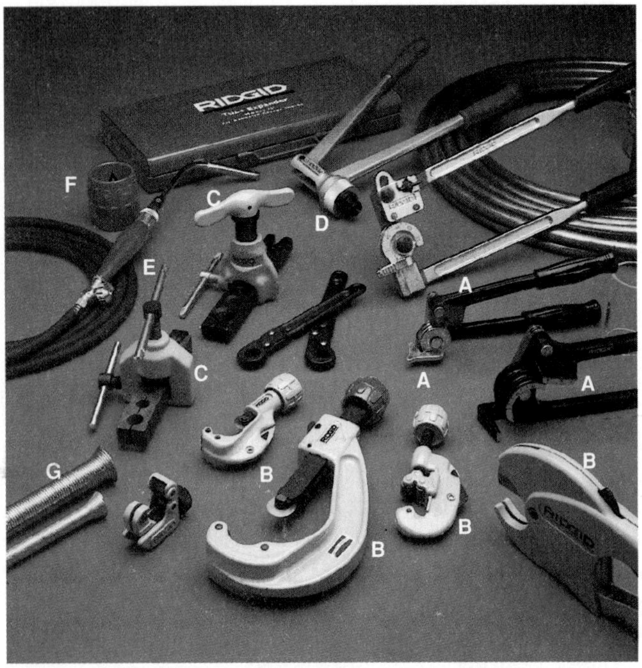

Figure 2-7. *An assortment of cutting, bending, and flaring tools. A—Lever-type tube bender. B—Tube cutter. C—Flaring tools. D—Tube expander. E—Propane torch. F—Inner/outer reamer. G—Spring-type tube bender. (Ridge Tools Company)*

2.2.1 Cutting Tubing

To cut tubing, use either a hacksaw or a tube cutter. The tube cutter is usually used on smaller, annealed (soft) copper tubing. The hacksaw is preferred for cutting larger, hard copper tubing. If a saw is used, a wave set blade of 32 teeth per inch will do the best job. (See Section 2.9.12.) The tubing should be straight and cut squarely (90°) to eliminate an off-center flare. The cutter usually leaves some sharp burrs on the cut ends. Burrs must be removed by reaming (scraping with a pointed tool). Most tube cutters have a reamer.

Figure 2-8 illustrates a wheel-type cutter. For confined areas, the technician can use a mini-tubing cutter, **Figure 2-9.** Its operation is similar to the standard tube cutter. This type of cutter is available for cutting copper tubing from 1/8″ to 1 1/8″.

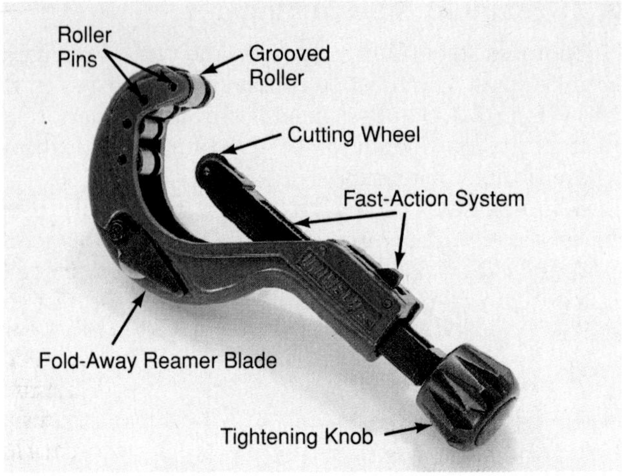

Figure 2-8. *A tube cutter. Note attached reamer which is used to remove burrs from inside of tube after cutting. Grooves in the rollers allow cutter to be used to remove flare from tube with little tubing waste. (Uniweld Products, Inc.)*

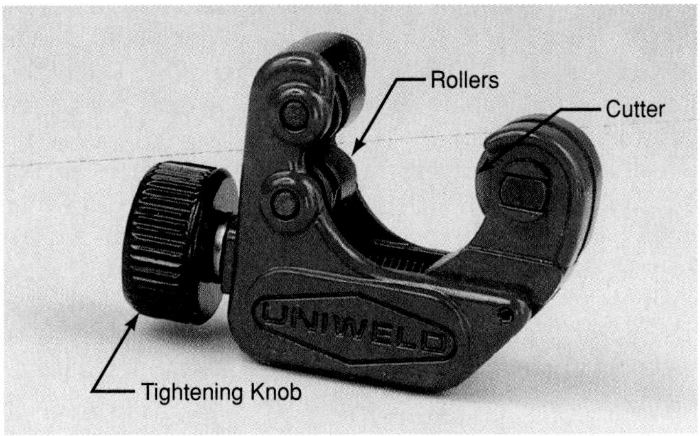

Figure 2-9. *A mini-tubing cutter is used in compact areas. (Uniweld Products, Inc.)*

It is important that no filings or chips of any kind enter the tubing. When cutting tubing with a hacksaw, do not allow the chips to fall into the section that is to be used. **Figure 2-10** illustrates a sawing fixture.

If soft tubing is used, pinch the end of the tube on the unused side of the cut. This eliminates the danger of chips entering the tubing. It also seals the tubing against moisture and protects it until used. In hard copper tubing installation, cap or plug the ends of the unused section.

To provide a full-wall thickness at the end of the tubing, many service technicians file the end of the tubing using a smooth or medium cut mill file. (See Section 2.9.11.) Again, do not allow any filings or other materials to fall into the tubing.

2.2.2 Bending Tubing

It takes practice to become good at bending tubing. Special bending tools are not needed for smaller sizes

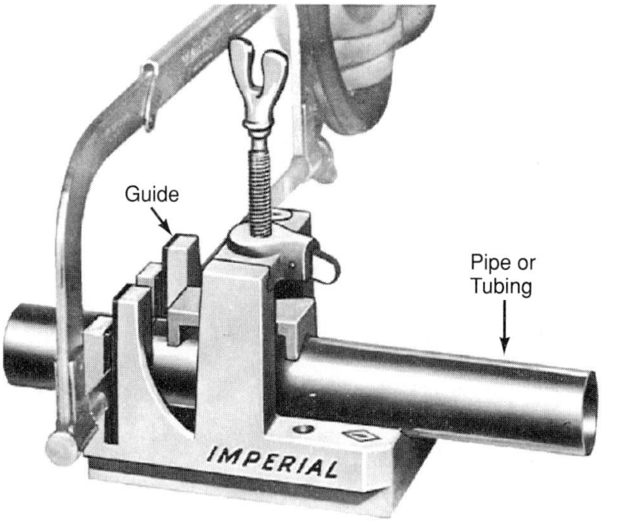

Figure 2-10. *Sawing fixture to ensure square and accurate cuts when using a hacksaw to cut tubing. This method is recommended for cutting hard-drawn copper and steel tubing. (Imperial Eastman, Imperial Division)*

used in domestic appliances. However, a much neater and more satisfactory job is possible with such tools.

Tubing should be bent so that it does not place any strain on the fittings after it is installed. Be very careful when bending the tubing to keep it round. Do not allow the tubing to kink, flatten, or buckle. The minimum radius for a tubing bend is between 5 and 10 times the diameter of the tubing as shown in **Figure 2-11.**

Tubes should be bent quite slowly and carefully. It is always wise to use as large a radius as possible. This reduces the amount of flattening. It is also easier to bend a large radius. Do not try to make the complete bend in one operation; rather, bend the tubing gradually. There is less danger that the sudden stress will break or buckle the tubing.

An inexpensive tool called a bending spring is shown in **Figure 2-12.** It may be easily carried in a tool kit. These are available in a variety of sizes. They can be used both inside and outside the tubing. Bending springs can be used internally for making bends near the end of the tubing.

An internal bending spring for 1/2″ OD tubing may be used as an external bending spring for 1/4″ OD tubing.

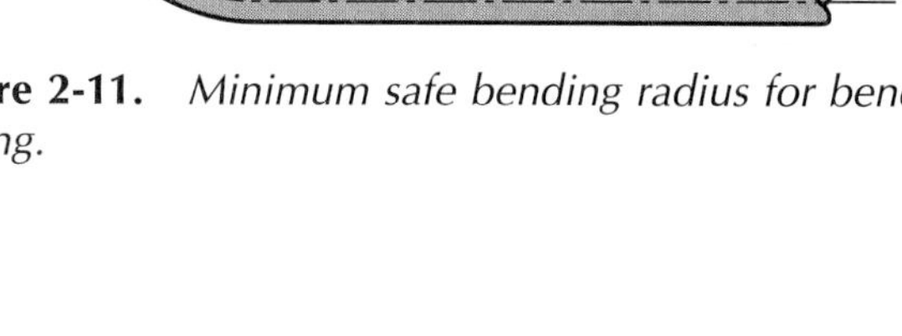

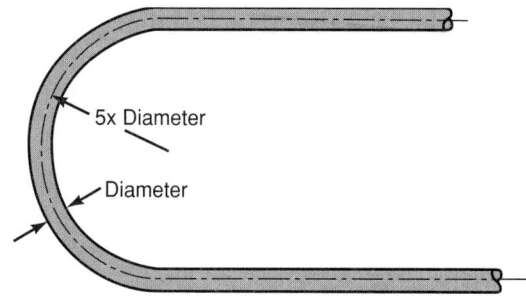

Figure 2-11. *Minimum safe bending radius for bending tubing.*

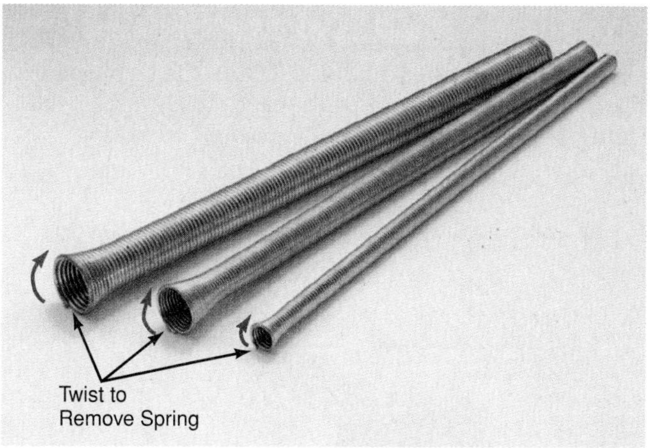

Figure 2-12. *Three spring benders used for 1/4″ through 3/4″ tubing. Tube bending spring may be fitted either outside or inside copper tube while bending tube. Bending spring reduces danger of flattening tube while it is being bent. (Uniweld Products, Inc.)*

Use the spring bender externally in the middle of long lengths of tubing.

Bending springs tend to bind on the tubing after the bend. They may be easily removed by twisting the spring. This changes the spring diameter slightly so the grip on the tubing is released.

If a bend is to be made near the flare and an external spring is to be used, bend the tubing before making the flare. An internal spring can be used either before or after the flaring operation.

Other tools are available for bending operations. A gear-type bending tool is shown in **Figure 2-13.**

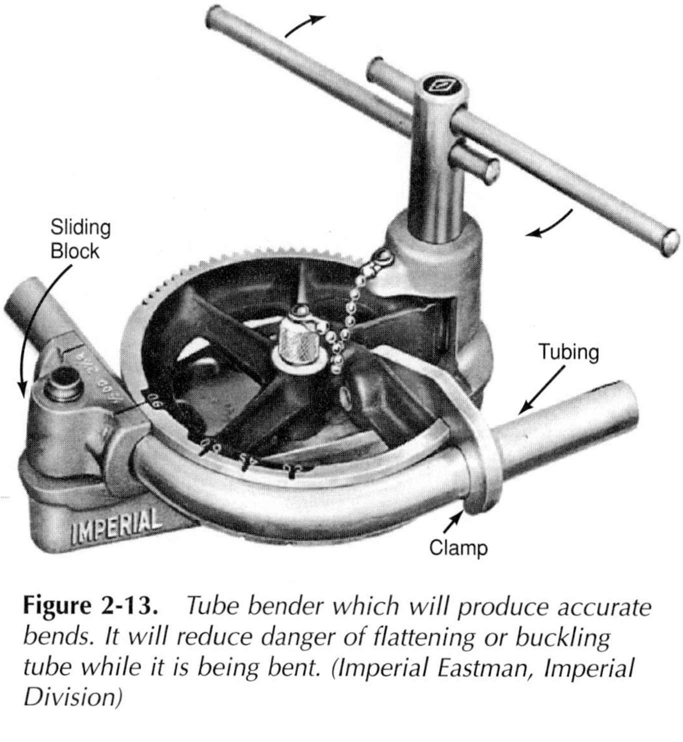

Figure 2-13. *Tube bender which will produce accurate bends. It will reduce danger of flattening or buckling tube while it is being bent. (Imperial Eastman, Imperial Division)*

A triple-size tube bender is shown in **Figure 2-14.** This type of bender is used for 1/4″, 5/16″, and 3/8″ OD tubing. The calibrated markings allow the technician to make accurate left-hand, right-hand, and offset bends. **Figure 2-15** shows some practice bends on tubing.

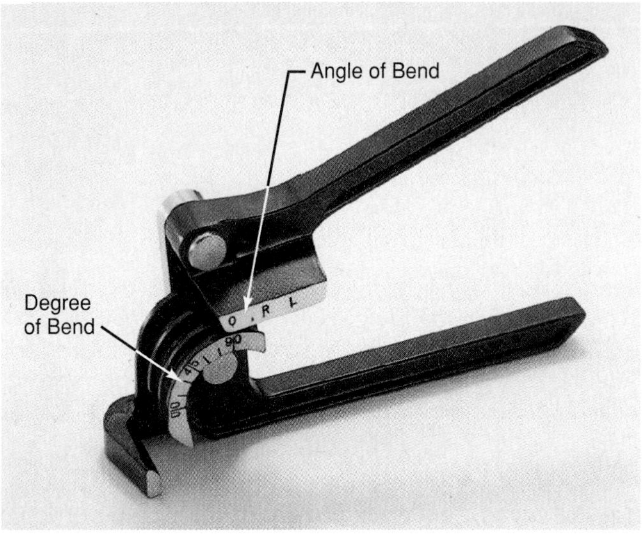

Figure 2-14. *Triple-size tube bender. As shown, tool is making 90° offset bend. (Uniweld Products, Inc.)*

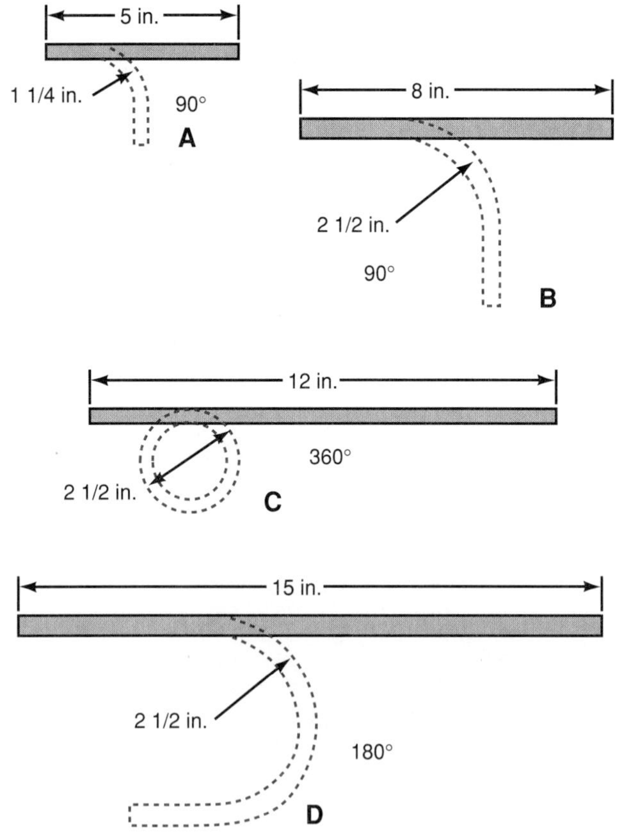

Figure 2-15. *Some practice bends on tubing. A—90° bend on 1/4″ tubing. B—90° bend on 1/2″ tubing. C—360° bend on 1/4″ tubing. D—180° bend on 1/2″ tubing.*

2.3 Connecting Tubing

Tubing walls are too thin for threading. Therefore, other methods of joining tubing to tubing and tubing to fittings must be used. The three common methods are:

- Flared connections.
- Soldered connections.
- Brazed connections.

2.3.1 Flared Connections

When connecting tubing to fittings, it is common practice to flare the end of the tube. Fittings designed to grip the tube are then used. Special tools are used for making flares.

Figure 2-16 illustrates how a tubing flare is used to form a leakproof joint between a tube and fitting. It also shows some flares which were incorrectly made. A correctly formed flare is squeezed tightly between the flare nut and the coupling. A vapor-tight seal results.

Some flares are made from a single thickness of the tube. Other flares are made with a double thickness of metal in the flare surface. These ***double flares*** are stronger and usually cause few problems if properly made.

Most flares are made at a 45° angle to the tube. Flares on steel tubing, however, are usually made at a 37° angle. This is because steel tubing is harder to flare than copper tubing.

Single-Thickness Flare

To make a flare of the correct size using a flaring block, do the following:

1. Carefully prepare the end of the tube for flaring. The end must be straight and square with the tube. The burr from the cutting operation must be removed by reaming. **Figure 2-17** shows the steps necessary to prepare a tube for flaring.
2. Use a 10″ smooth mill file to square the end of the tube. Use great care that no filings enter the tubing. Next, use a burring reamer to remove the slight burr remaining after the cut-off operation.
3. A flaring tool which may be used to make a single-thickness flare is shown in **Figure 2-18A.** A flaring tool suited for flaring either fractional size or millimeter size tubing is shown in **Figure 2-18B.**
4. Place the flare nut on the tubing with the open end toward the end of the tubing. Insert the tube in the flaring tool so that it extends above the surface of the block as shown in **Figure 2-19A.** This allows enough metal to form a full flare. Tighten the clamp so the tube cannot move.
5. If the tube extends above the block more than the amount shown, the flare will be too large in diameter and the flare nut will not fit over it. If the tube does not extend above the block, the flare will be too small. It may be squeezed out of the fitting as the flare nut is tightened. **Figure 2-19B** shows appearance of completed flare.

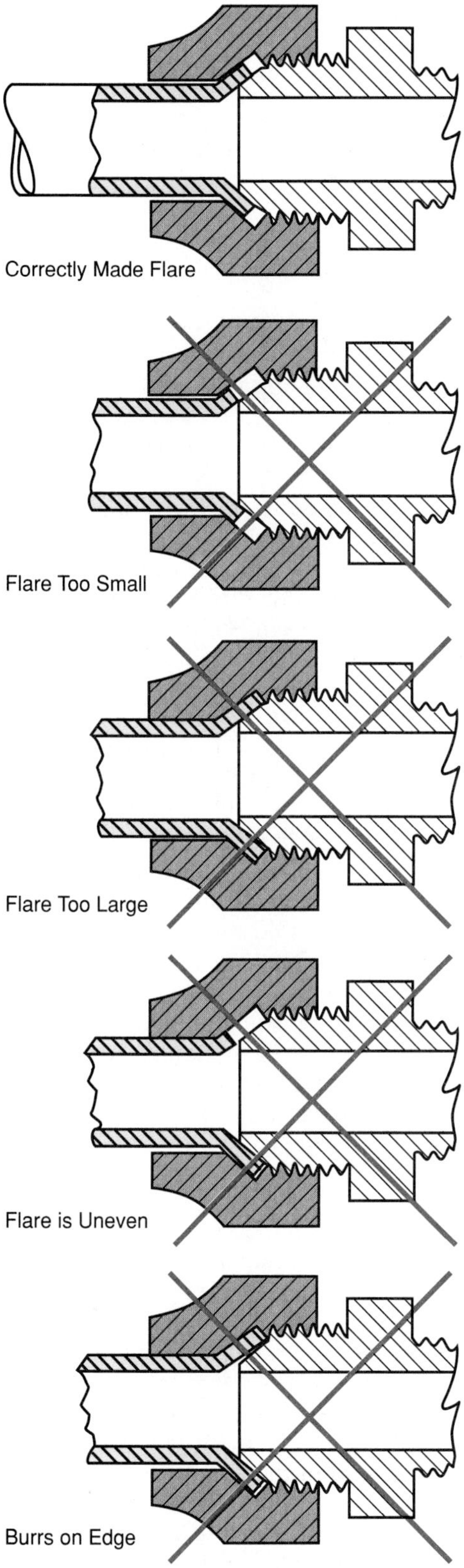

Figure 2-16. *Flared fittings.*

6. To form the flare, first put a drop of refrigerant oil on the flaring tool spinner where it will contact the tubing. Tighten the spinner against the tube end one-half turn and back it off one-quarter turn. Advance it three-quarters of a turn and again back it off one-quarter turn. Repeat the forward movement and backing off until the flare is formed.

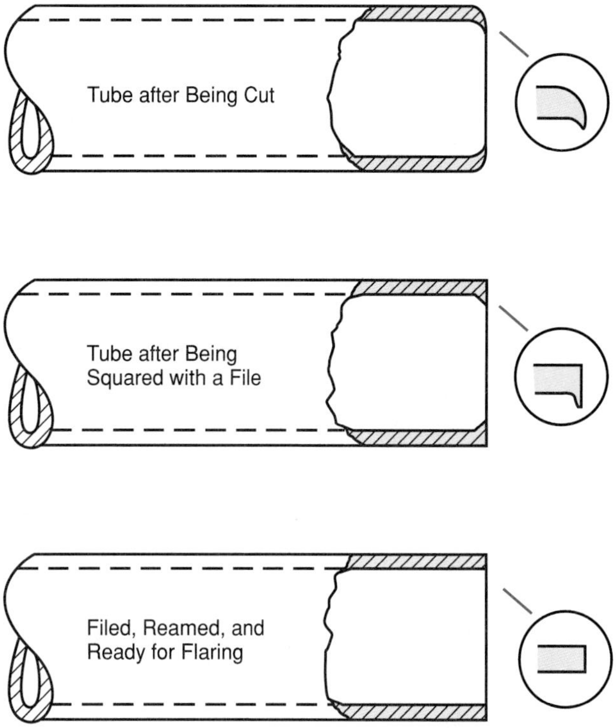

Figure 2-17. *End of tube must be carefully prepared before flaring.*

Some technicians make the flare using one continuous motion of the flaring tool. That is, they do not use a back-and-forth motion. It is believed by some that the constant turning of the tool, without back turning, may work harden the tubing. It would then be more likely to split.

Other technicians like to use a flare which is not completely formed—about seven-eighths complete. They depend on the tightening of the flare nut on the flare to complete it.

Do not tighten the spinning tool too much. This would thin the wall of the tubing at the flare and weaken it.

Always place the flare nut on the tube in the proper position before the flare is made. It cannot, in most cases, be installed on the tube after it has been flared.

Double-Thickness Flare

Double-thickness flares are formed with special tools. **Figure 2-20** illustrates a cross section through a simple block-and-punch type of tool used to make a double flare. The correct shape of the double flare is shown in the final operation, **Figure 2-20D.** Some flaring tools have double flare adaptors, **Figure 2-21.** These make it possible to form either single or double flares.

Figure 2-22 shows the steps for making a double flare using the tool shown in **Figure 2-21.** Double-thickness flares are recommended only for larger size

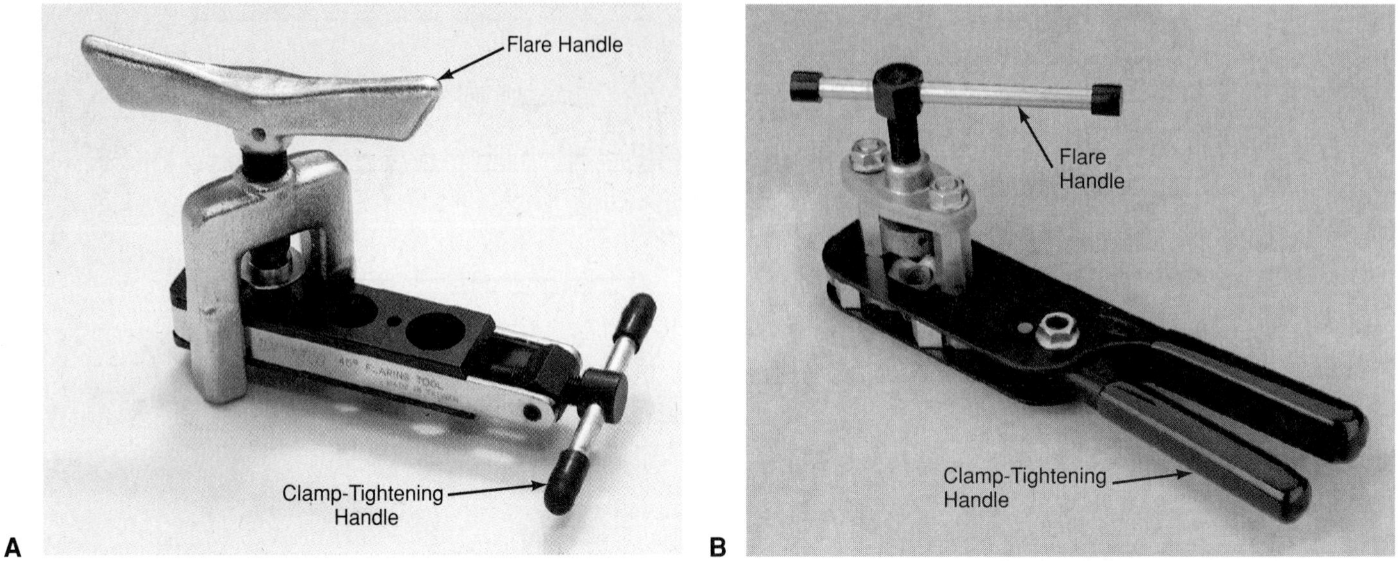

Figure 2-18. *Flaring tools. A—Popular style used for making single-thickness flares on refrigeration tubing. Flaring block is split, making it easy to insert the clamp tubing in place for flaring. Note 45° chamfer in block, which gives the flare its correct shape. (Uniweld Products, Inc.) B—Flaring tool having an adjustable tube-holding mechanism which permits flaring tubing 3/16" to 5/8" OD and 5 mm to 16 mm OD. (Reed Manufacturing Co.)*

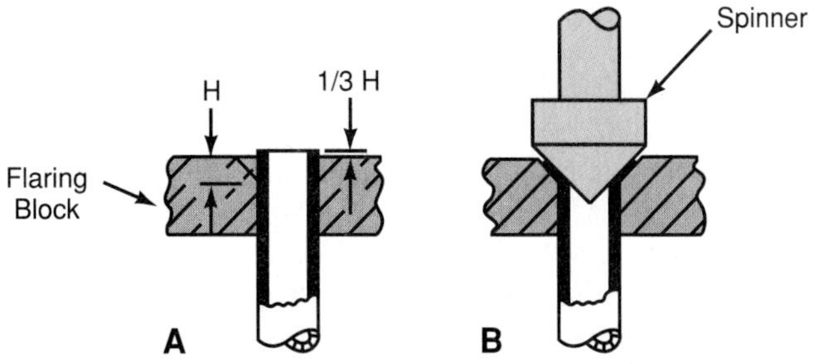

Figure 2-19. *Tubing to be flared should extend slightly above flaring block to allow enough metal to form a satisfactory flare. Amount to allow is about one-third the height of the flare. A—Proper position of tube in flaring tool before flaring. B—Completed flare.*

tubing, 5/16" and over. Such flares are not easily formed on smaller tubing. The double flare makes a stronger joint than a single flare.

Annealing Tubing

If a tube splits while being flared, it may be due to the age of the tubing. Old tubing becomes brittle after a period of use and cannot be flared satisfactorily.

To remedy this brittle condition, anneal (soften) the tubing by heating to a dull, cherry red or blue color and allowing it to cool slowly. Rough handling (such as pounding) or bending the tubing tends to work harden it. Hard-drawn tubing cannot be bent or flared unless annealed.

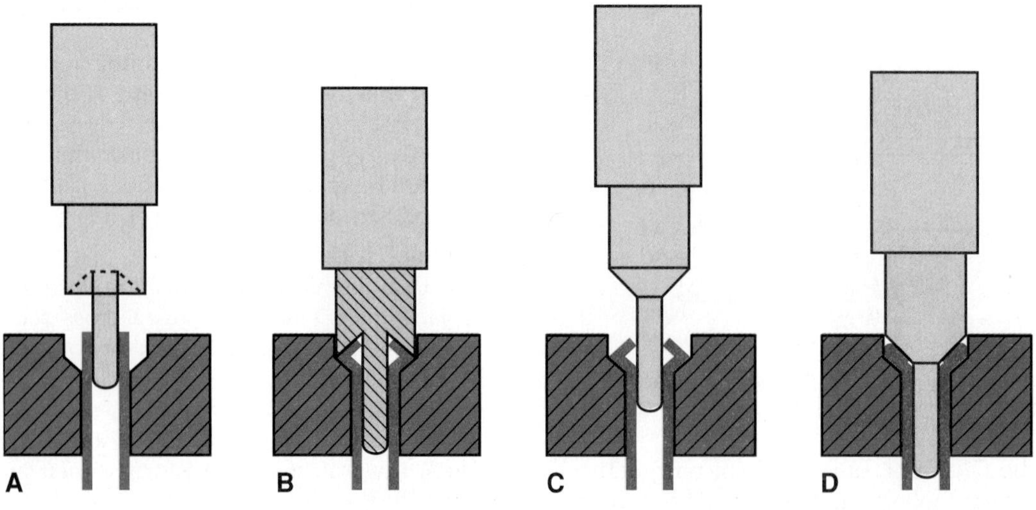

Figure 2-20. *Simple block and punch tool for forming double flares on copper tubing. A—Tube is clamped in body of flaring block. B—Female punch bends end of tube inward. C—Male punch is inserted in partially flared tube. D—Male punch folds end of tube downward to form double thickness and expand flare into final form.*

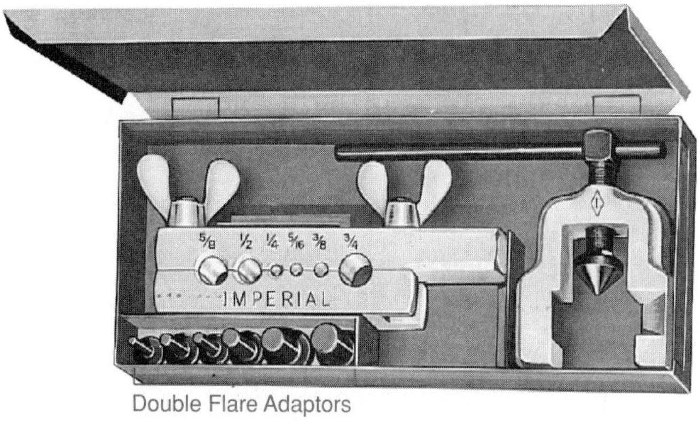

Figure 2-21. *A flaring tool which, with necessary adaptors, is capable of producing either single or double flares. (Imperial Eastman, Imperial Division)*

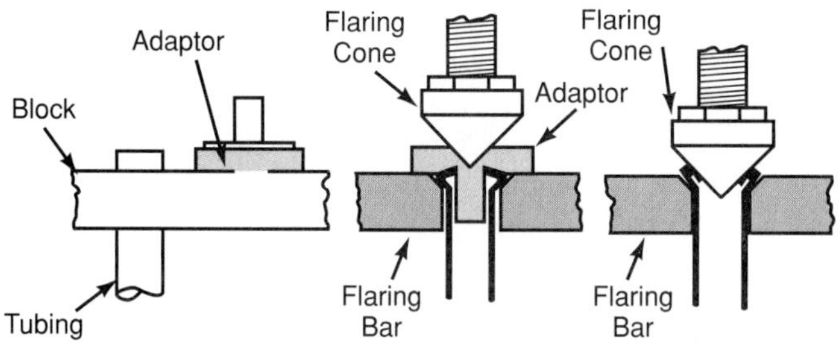

Figure 2-22. *Correct procedure for forming a double flare using the tool shown in* ***Figure 2-21.***

2.3.2 Flared Tube Fittings

To attach a fitting to soft copper tubing, a flared-type connection is generally used. There are many different fitting designs on the market. The accepted standard for refrigeration is a forged fitting. Some of these have National Pipe (NP) threads. Some have Society of Automotive Engineers (SAE) National Fine (NF) threads. See **Figure 2-23.**

The fittings are usually made of drop-forged brass. They are accurately machined to form the threads, the hexagonal shapes for wrench attachment, and the 45° flare for fitting against the tubing flare. These threaded fittings must be carefully handled to prevent damage to them.

All fitting sizes are based on the tubing size. For example, a 1/4″ flare nut attaches 1/4″ tubing to a flared fitting even though it has 7/16″ NF internal threads and uses a 3/4″ wrench to turn it.

Figure 2-24 is a table of common flared fitting sizes. Catalog listings of tube fittings usually provide a code number to indicate the size. The code number 3 indicates that the fitting fits 3/16″ tubing. Code number 4 indicates that it fits 1/4″ (4/16). Code number 8 fits 1/2″ tubing (8/16).

Some tubing fittings have pipe threads on one end. Pipe threads taper 1/16″ in diameter for every inch in length.

With more plastic tubing being used, special fittings for plastic are now available. These fittings are made of

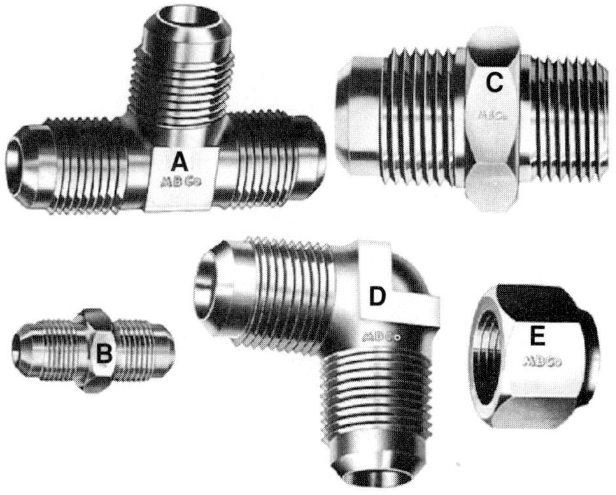

Figure 2-23. *Some of the more common flare-type fittings used in refrigeration and air conditioning work. A—Flared tee fitting, male flare × male flare. B—Flared union coupling, male flare × male flare. C—Flared half union coupling, male flare × male pipe. D—Flared 90° elbow, male flare × male flare. E—Flare nut. (Mueller Refrigeration Products Co., Division of Mueller Industries, Inc.)*

brass, aluminum, and polyethylene materials. Plastic tubing is not flared like copper. Rather, a compression-type fitting is used, as shown in **Figure 2-25.**

2.3.3 Metric-Size Tube Fittings

Metric-size tubing, as described in Section 2.1.4, requires metric-size fittings. These are very similar to U.S. conventional-size fittings and are used in the same way. The technician must be careful not to mix U.S. conventional-size fittings with metric-size fittings.

2.4 Soldered or Brazed Tubing Fittings

Most tube and fitting connections are made by either soldering or brazing. Soldered joints are used for water pipes and drains. Brazed joints are used for refrigerant pipes and tubing. The terms "soft soldering" and "brazing" are often misused. The difference between soldering and brazing is the lower temperature at which solder flows. If the temperature required to melt the alloy used to join copper tubing is below 800°F (427°C), it is considered ***soldering.*** If the temperature required to flow the alloy is above 800°F (427°C), it is referred to as ***brazing.***

Soldered joints use a capillary action to draw molten solder into the area between the fitting and the tube. The selection of a solder is based upon two factors: operating pressure and temperature of the line. A tin-antimony solder is appropriate for moderate pressures and temperatures. It melts at 360°F (182°C) and is fluid at 415°F (213°C). For higher pressures or greater joint strength, a 95/5 tin-antimony solder is used. This mixture contains 95% tin and 5% antimony. A 95/5 solder melts at 450°F (232°C) and is fully liquid at 465°F (241°C).

Refrigeration Fittings (Flared Type) Sizes Are Based on the outside Diameter of the Tubing						
Name and Description	1/4	5/16	3/8	7/16	1/2	5/8
Nut Forged	X	X	X	X	X	X
Union (Threads same size).	X	X	X	X	X	X
Half Union (1/8 Pipe)	X	X				
Half Union (1/4 Pipe)	X	X	X	X		
Half Union (3/8 Pipe)					X	
Half Union (1/2 Pipe)						X
Elbow	X	X	X	X	X	X
Elbow (One 1/8 Pipe).	X	X				
Elbow (One 1/4 Pipe).	X	X	X	X		
Elbow (One 3/8 Pipe).					X	
Elbow (One 1/2 Pipe).						X
Tee (Threads same size)	X	X	X	X	X	X
Tee (One 1/8 Pipe)	X	X				
Tee (One 1/4 Pipe)			X	X		
Tee (One 3/8 Pipe)					X	
Tee (One 1/2 Pipe)						X
Cross	X	X	X	X	X	X
Flared Tube Sealing Plug	X	X	X	X	X	X
Flared Tube Sealing Cap.		X	X	X	X	X
Flared Tube Copper Seal Cap	X	X	X	X	X	X
Union (Reducing).	5/16-1/4	3/8-1/4	1/2-1/4	1/2-3/8		
Elbow (Reducing).	5/16-1/4	3/8-1/4	1/2-1/4	1/2-3/8	5/8-1/2	
Tee (Reducing)	5/16-1/4	3/8-1/4	1/2-1/4	1/2-3/8	5/8-1/2	

Figure 2-24. *Some of the more popular flared copper tube fittings used in refrigeration and air conditioning. The reducing fittings are used for connecting tubing of different sizes.*

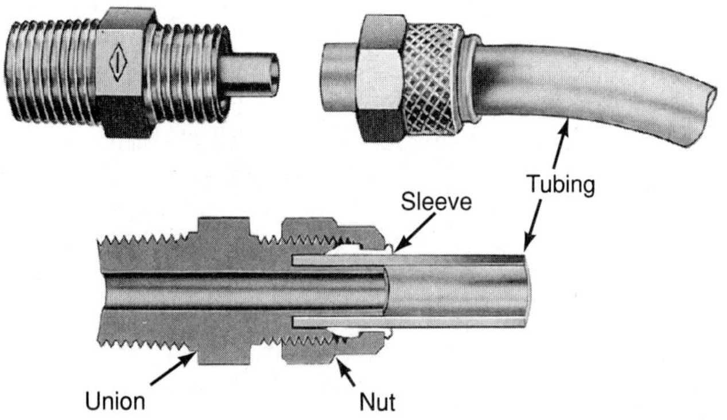

Figure 2-25. *A compression-type fitting used with polyethylene tubing. Caution: Polyethylene is a soft substance and very little tightening is needed. While most fittings are made with flats for wrench tightening, most polyethylene installations require little more than "finger tightness." (Imperial Eastman, Imperial Division)*

Brazing produces a stronger bond than soldering. Brazing filler metals can join similar and dissimilar metals at brazing temperature. Brazing filler metals melt at temperatures in the range of 1000°F to 1500°F (538°C to 816°C). Some filler metals used for brazing copper tubing are of two categories: alloys containing 30% to 60% silver. Others are copper alloys which contain some phosphorus. These two classes vary in melting, flowing, and fluxing characteristics. Strong joints can be made with either class of filler metal. Strength of a brazed copper joint depends more upon proper clearance between the tube and the socket of the fitting.

2.4.1 Soldering

Soldering is a process of applying molten (melted) metal to metals that are heated but are not molten. It is an adhesion process. (In adhesion processes, one part is bonded to or is stuck to another by a third material.) The molten solder flows into the pores of the surface of the metals being joined. As the solder solidifies (hardens), a good bond is obtained.

A fitting soldered to a tube is shown in **Figure 2-26.** A good sweat joint begins with first cleaning the parts to be joined, followed by fluxing and assembling them. The assembly is then heated. As soon as the joint reaches

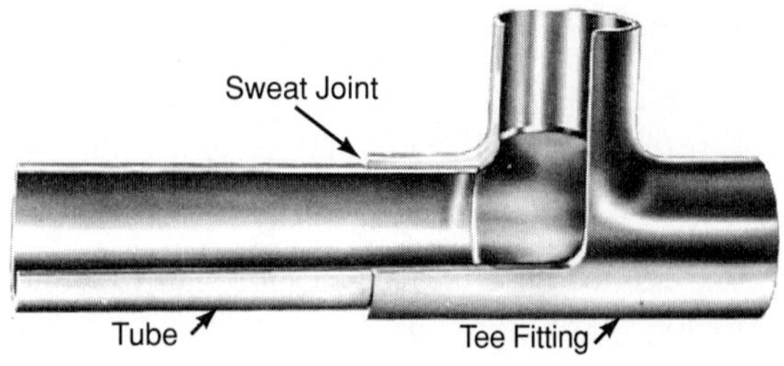

Figure 2-26. *A cross section of tee fitting soldered to hard ACR tube.*

the flowing temperature of the solder, solder is added to the joint and flows into it. After the solder cools, it will seal and connect the surfaces. The step-by-step procedure for making a sweat joint is shown in **Figure 2-27.**

When assembling swaged (shaped) tube-to-tube joints, or when connecting tubing to a fitting, thoroughly clean the mating parts. Next, apply flux to the outside of the tube. (***Flux*** is a substance that does not actually clean the metal. Rather, it keeps the metal clean once soil has been removed.) Insert the tube into the fitting 1/16″ to 1/8″. Rotate one of the pieces to spread the flux evenly over both the internal and external surface. See **Figure 2-28.** Using this rotation technique will eliminate any possibility of flux entering the system. When this is done, apply the necessary heat for soldering or brazing.

Avoid ***swaging*** steel tubing (shaping by hammering). It is harder (less ductile) and may crack or split. Sometimes the process tube of a hermetic motor compressor is made of steel. Many technicians clean the

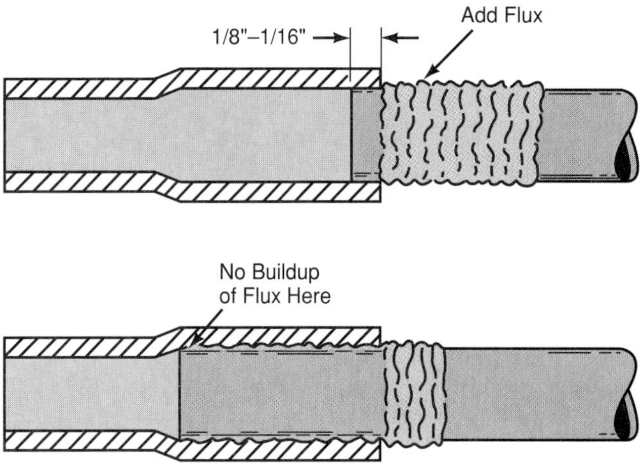

Figure 2-28. *Brazing or soldering flux may be a source of corrosion in a system. Apply flux to joints as above so that it will not get into system.*

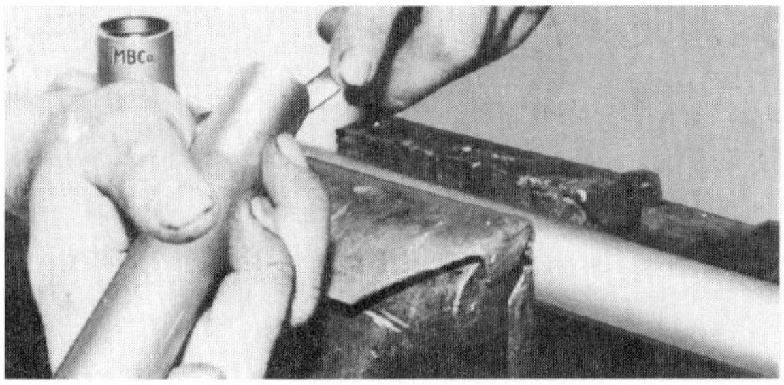

Step 1. Cut tube to length and remove burr with file or scraper.

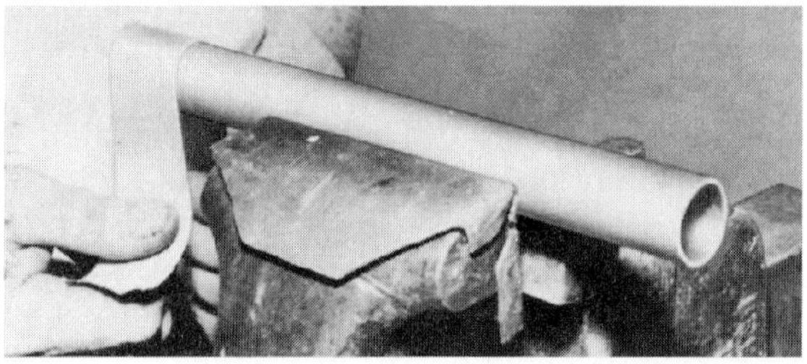

Step 2. Clean outside of tube with clean abrasive paper or cloth.

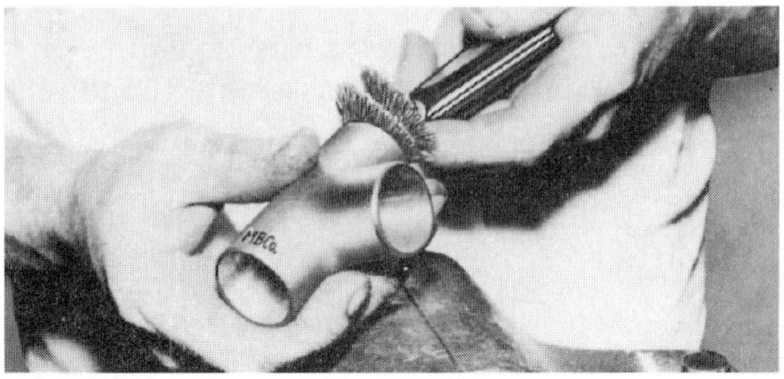

Step 3. Clean inside of fitting with a clean wire brush, or abrasive paper or cloth. Do not use emery cloth.

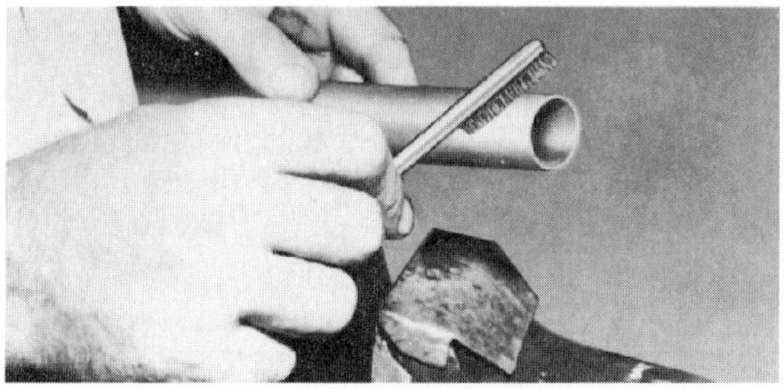

Step 4. Apply flux thoroughly to outside of tube—assemble tube and fitting.

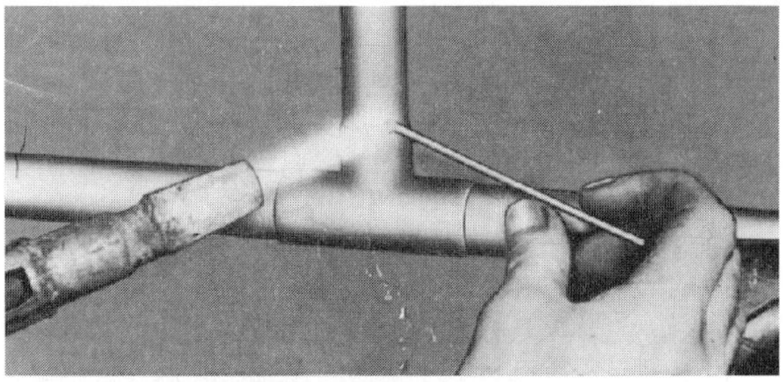

Step 5. Apply heat with torch. When solder melts upon contact with heated fitting, the proper temperature for soldering has been reached. Remove flame and feed solder to the joint at one or two points until a ring of solder appears at the end of the fitting.

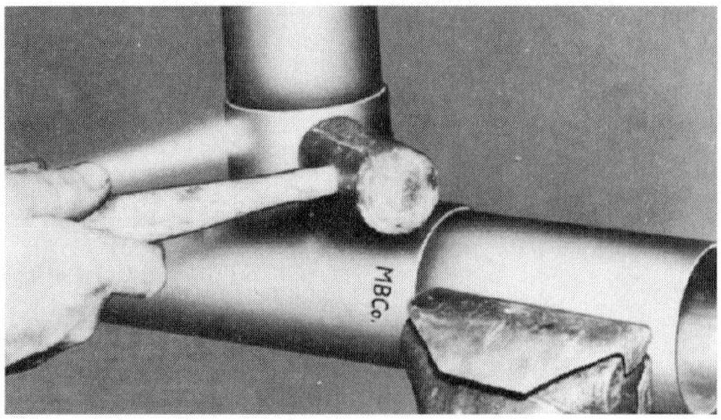

Step 6. Tap larger sized fittings with mallet while soldering, to break surface tension and to distribute solder evenly in joint.

Figure 2-27. *Recommended step-by-step procedures to follow when soldering tubing. (Mueller Refrigeration Products Co., Division of Mueller Industries, Inc.)*

tubing before cutting it. This helps to keep dirt out of the system.

Sometimes a small tube is soldered directly into a larger tube. One tubing should extend into the other the same distance as the diameter of the larger tubing. For example, if a 1/4″ tube is placed into 5/16″ tubing, they should overlap 5/16″.

Figure 2-29 illustrates some common fittings, which may be either soldered or brazed to tubing. All brass and copper parts may be easily soldered. To solder:

1. The surfaces to be soldered must be very clean.
2. A good clean flux must be used.
3. A good source of clean heat must be on hand.
4. The parts being soldered must be firmly supported during the soldering operation.

Surfaces being soldered must be free of grease, dirt, and oxides. Flux does not clean the metal. It keeps the metal clean once soil has been removed. Before soldering, thoroughly clean the surfaces to be soldered. Surfaces can be cleaned by filing, scraping, sanding, or by using steel wool and wire brushes.

Apply flux thoroughly to outside of tube. Flux for this type of work should have no corrosive properties. Acid flux should not be used. It tends to corrode fittings, making them unsightly and difficult to work on later.

Solder in usually used in wire form. Hard-to-reach surfaces can be easily supplied with solder just by bending the wire to the needed shape.

A tin-antimony alloy is usually satisfactory for soft soldering. Solders containing as much as 95% tin are now being recommended for soldered joints subjected to very low temperatures. Do not solder with 100% tin. Pure tin may slowly disintegrate when exposed to cold.

A portable air-acetylene torch is shown in **Figure 2-30.** The air-acetylene mixture provides maximum temperature of 2700°F (1482°C). Therefore, this type of system is used mainly in residential and small commercial systems. These types of systems are not under heavy movement as large, commercial systems might be.

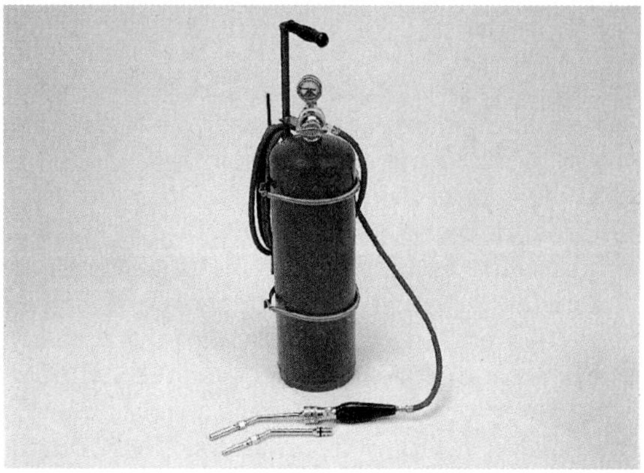

Figure 2-30. *Portable acetylene-air torch. Unit shown has two different types of flame tips for soldering and brazing. (Uniweld Products, Inc.)*

For good soldering, the metal being joined must be hot enough to melt the solder. This is the only way the solder will go into the pores of the metal. When the metal is hot enough, touch the solder to the metal. Do not overheat. Keep testing the metal with the solder wire. Heat the metal only until the solder flows.

While soldering, it helps to "wipe" the surfaces after putting some solder on. Use a clean cloth, a brush, or the solder wire itself. This action will remove any dirt and will help coat the surface.

If the parts are at the correct temperature and have been cleaned and fluxed, the solder will flow over the surface quickly. Remember not to heat the solder directly with the torch.

Figure 2-31 illustrates a tube soldering practice assembly. For leak-testing this assembly, connect it to a cylinder of the type of refrigerant used and nitrogen.

Never use oxygen when testing for leaks. Any oil in contact with oxygen under pressure will form an explosive mixture.

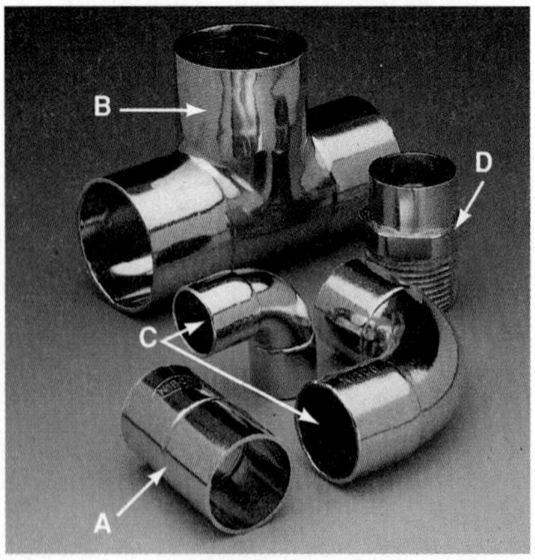

Figure 2-29. *Common fittings which may be either soldered or brazed to tubing. A—Coupling with rolled stop, sweat × sweat. B—Tee, sweat × sweat × sweat. C—90° elbow, sweat × sweat. D—Adaptor, sweat × male pipe thread (mpt). (NIBCO Inc.)*

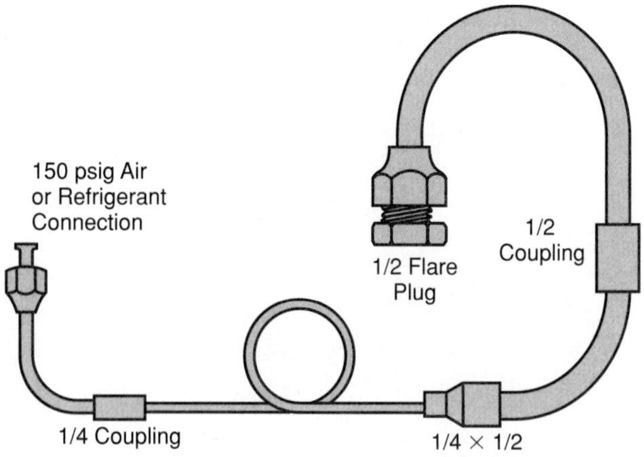

Figure 2-31. *Practice piece used to develop skill in copper tube soldering. Completed assembly should be pressurized as indicated and all of the joints tested for leaks using soapsuds.*

2.4.2 Brazing

Brazing is one of the best methods of making leakproof connections. These joints are very strong and will stand up under the most extreme temperature conditions.

Oxyacetylene brazing equipment is used to achieve maximum strength and a leakproof joint. A small, portable system is shown in **Figure 2-32.** "Oxyacetylene" means the addition of pure oxygen to burning acetylene. The combined mixture produces a maximum temperature of 6000°F (3316°C). This allows a minimal amount of heat to be transferred down the copper tubing when brazing in a compressor, line drier, filter, etc. The technician is able to braze replacement items in place without damaging them through heat transfer. See **Figure 2-33.**

Correct use of oxyacetylene requires metering of the flow of oxygen and acetylene. The oxygen tank and the acetylene tank have pressure regulators and a set of gauges. One gauge registers tank pressure; the other displays the pressure at the torch.

Acetylene is a highly flammable gas, especially when mixed with oxygen. Therefore, safety glasses should always be worn when brazing. Never point the torch (lit or unlit) toward an open flame or source of sparks. Light the torch only with a sparker—do not use matches.

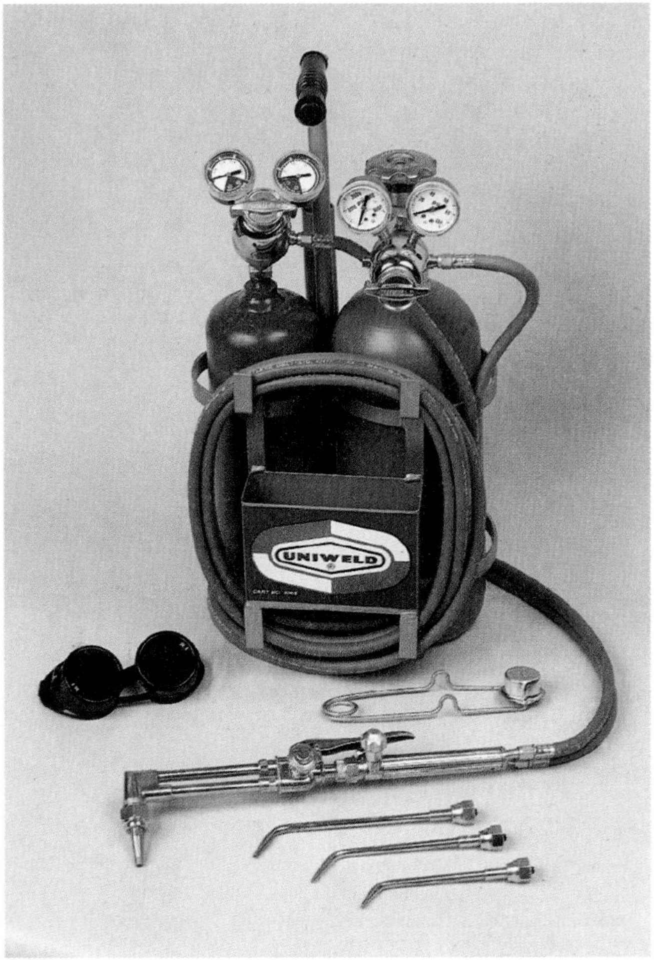

Figure 2-32. *Portable oxyacetylene outfit with tanks and different types of tips. (Uniweld Products, Inc.)*

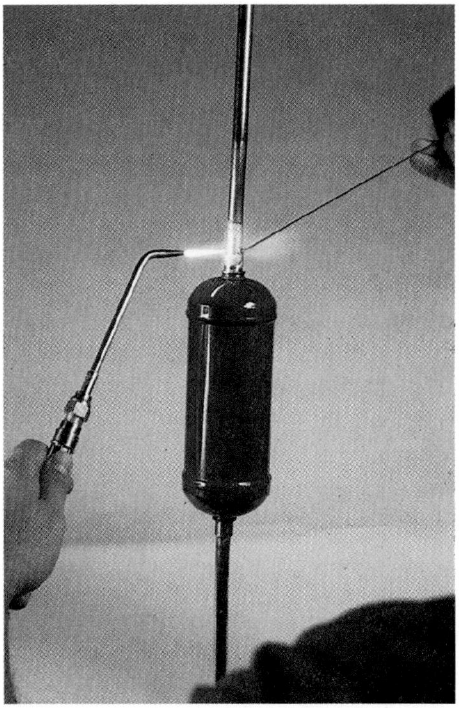

Figure 2-33. *Oxyacetylene brazing in a drier. (Uniweld Products, Inc.)*

The acetylene valve adjusts the flame size. Slowly turn the oxygen valve to obtain type of flame required. A "neutral flame" is most efficient in brazing. It has a blue cone with a bit of reddish purple at the tip.

Brazing can be done easily if the correct procedures are followed:

1. Degrease parts and clean the joints thoroughly.
2. Fit the joints closely and support all parts.
3. Apply the clean flux recommended for the brazing alloy. Follow the manufacturer's instructions.
4. Heat evenly to recommended temperature. Keep the torch moving constantly in a "figure-8" motion.
5. Apply brazing alloy to the heated parts. Do not heat (melt) the brazing alloy with the torch.
6. Cool the joint.
7. Clean the joint thoroughly, using warm water and a brush. Be sure all flux has been removed.

An oxyacetylene torch is an excellent heat source for brazing. However, you must have training in its safe use. Be sure to use flashback arrestors at both the acetylene and oxygen regulators.

There are various brazing alloys on the market. Most have a 35% to 45% silver content. This material usually starts melting at 1120°F (604°C) and flows at 1145°F (618°C). Contact a local welding supply house or air conditioning and refrigeration wholesaler for brazing supplies.

Caution: Carefully check the specifications of the brazing alloy used. If it contains any amount of cadmium (Cd), be sure that the work space is well ventilated. Do

not breathe any of the fumes. Keep fumes away from your eyes and skin. Cadmium fumes are very poisonous.

The parts to be brazed must be carefully cleaned and fitted accurately. Dirt must be removed from any external surface. Use a fine grade of stainless steel wool for cleaning the exterior. Internal surfaces can be cleaned with stainless steel wire brushes or stainless steel wool rolled on a rod.

The parts must have contacting surfaces of sufficient area, such as a tube sliding into a fitting (not a drive fit), **Figure 2-34.** The contacting surfaces need not be very large (three times the smallest section).

If the parts are dented or are out of round, these faults must be corrected before brazing. It is important to support all the parts securely so they will not move during brazing.

Make sure that no flux enters the system during brazing, as it cannot be easily removed. Avoid overfluxing by applying the flux to the surface that is to slide into the part, **Figure 2-28.** The excess flux will then stay on the outside.

All air must be removed from the tubing being brazed. This can be best done by purging the tubing with either carbon dioxide or nitrogen, as shown in Figure 2-35. Any oil inside the tubing or part may be vaporized by the heat of the torch. Oil vapor mixed with air will explode if ignited. Using a nonflammable gas such as carbon dioxide or nitrogen will eliminate this hazard.

Caution: Never use a refrigerant, oxygen, or compressed air when brazing.

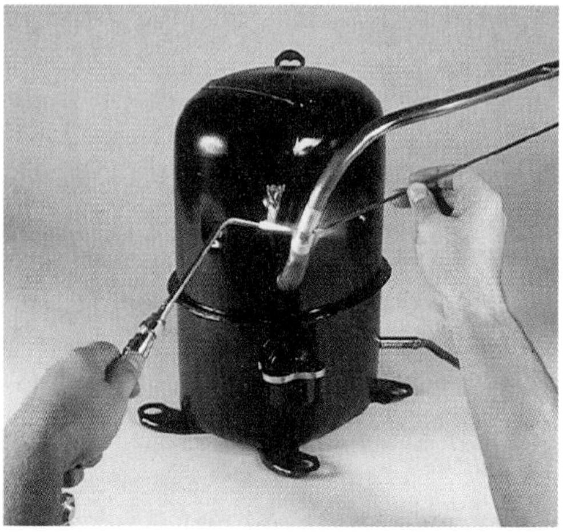

Figure 2-35. *A line being brazed to compressor. Low-pressure carbon dioxide or nitrogen is purged through the line into the compressor during brazing to prevent fire or explosion. (Uniweld Products, Inc.)*

Heating of the joint must be done carefully. The flux behavior is a good indication of the temperature of the joint as the heating progresses.

1. Keep the joint covered with the flame all during the operation to prevent air getting to it.
2. The moisture (water) will boil off. At 212°F (100°C), the flux will turn milky in color.
3. Next, it will bubble at about 600°F (316°C).
4. At 800°F (427°C), the flux lies on the surface and has a milky appearance.

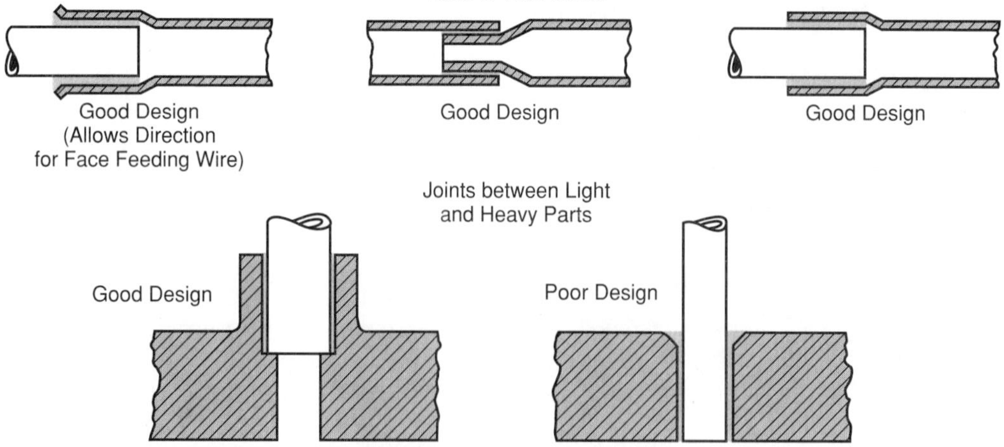

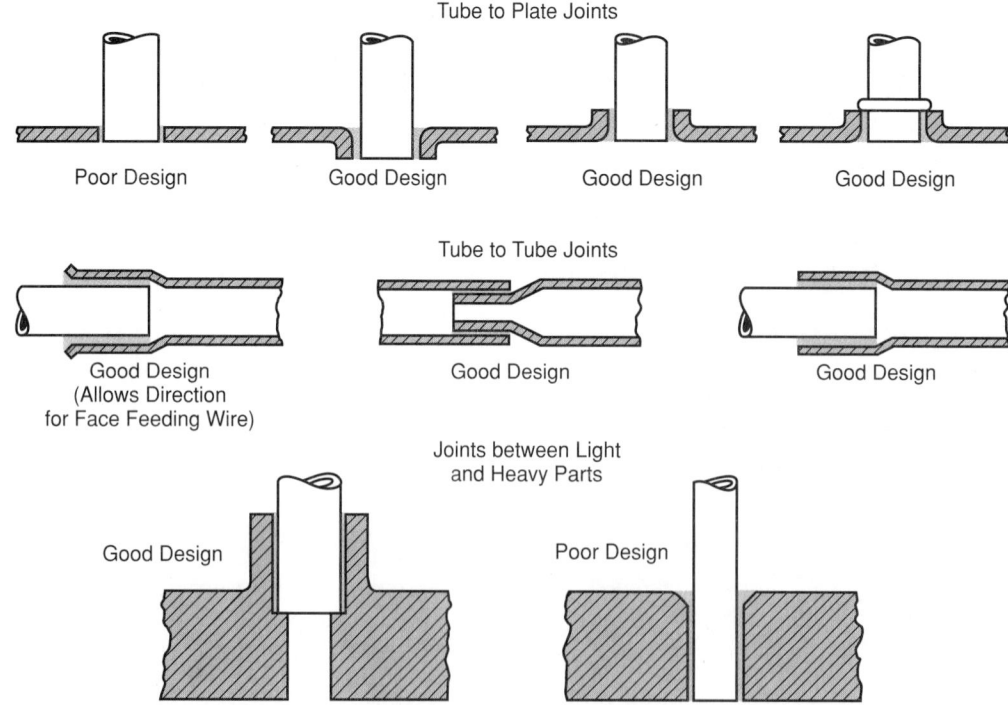

Figure 2-34. *Suggestions for making joints to be brazed. Actual thickness of brazing is exaggerated to show its application. (Lucas-Milhaupt, Inc., A Handy & Harman Company)*

5. Following this, it will turn into a clear liquid at about 1100°F (593°C). This point is just short of the brazing temperature.

The alloy itself begins to melt at 1120°F (604°C) and flows at 1145°F (618°C). A torch tip several sizes larger than the one used for soldering should be used. Be sure to heat both pieces which are to have the alloy adhered to them.

The proper brazing temperature will be indicated by the color of the secondary flame. The flame will start to show a green shade as the brazing temperature is reached. For silver brazing, a clear flux and/or a green flame show the proper temperature.

When heating a copper-to-steel joint, heat the copper first. It takes more heat because it carries it away faster. Put some flux on the brazing rod to help it flow quicker.

When cutting capillary tubing, notch all around it with a triangular file. Break the tubing by bending back and forth (small bends). The tubing ID will then remain full size. A tube cutter would reduce inside diameter.

When brazing a capillary tube, do not let brazing material run to the end of the tube. It might partially close the hole (ID) of the capillary. Leaving the end of the tube uncleaned will help prevent this from happening.

When brazing, the torch is never held in one spot. It should be moved around the entire area to be brazed. Many technicians prefer to move the torch in a "figure-8" motion. Larger torch tip sizes are recommended for brazing. This allows a soft flame and a large quantity of heat without excess pressure or "blow." A slight feather on the inner cone of the flame is good. See **Figure 2-36.**

Cleaning the Brazed Joint

Thoroughly wash with water and scrub the outside of the completed brazed joint. This is always necessary. Flux left on the metals will tend to corrode them or may temporarily stop a leak which will only show up later.

The joint may be cooled quickly or slowly. Cooling with water is allowable. The same water may be used to wash the joint.

Visual inspection will quickly reveal any places where the alloy did not adhere. It is best to watch for poor adhesion (dark cup-shaped areas). Then, any corrections can be made during the brazing operation immediately while the parts are still hot.

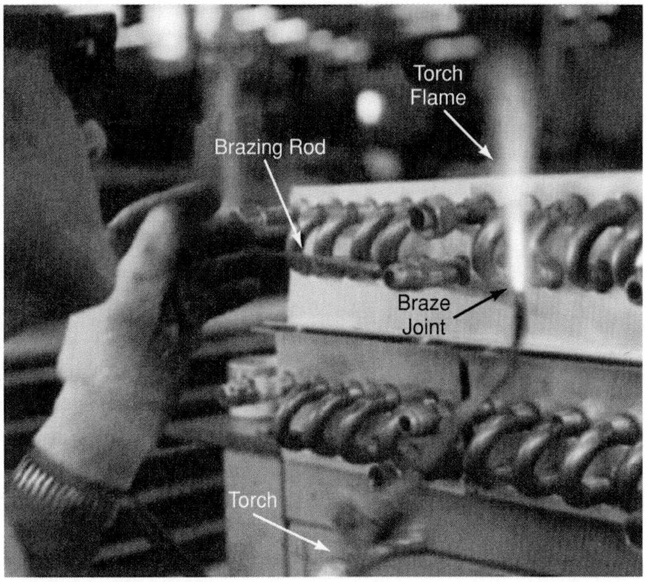

Figure 2-36. *Brazing copper tubing connection. See text for suggestions on flame movement. (Kramer Trenton, Co.,* Brazing Book*)*

2.5 Tube Couplings

Tube couplings may be used to join aluminum tubes to copper tubes. This requires a process different than joining copper to copper. There are a variety of methods available for joining aluminum to copper. These include threaded mechanical fittings, flared and compression fittings, and epoxy resin and adhesive kits. **Figure 2-37** shows a threaded mechanical fitting. **Figure 2-38** shows an adhesive kit being applied. It uses a tube coupling that shrinks when heated with a propane torch. It forms its copper-aluminum joint in about twenty seconds of heating.

2.6 Swaging Copper Tubing

Two pieces of soft copper tubing of the same diameter can be joined together without fittings. One piece is ***swaged*** (enlarged to receive another piece of tubing of the same diameter), as shown in **Figure 2-39.**

Swaging of copper tubing is often done. It is more convenient to solder one joint than to make two flared connections. The length of the overlap of the two

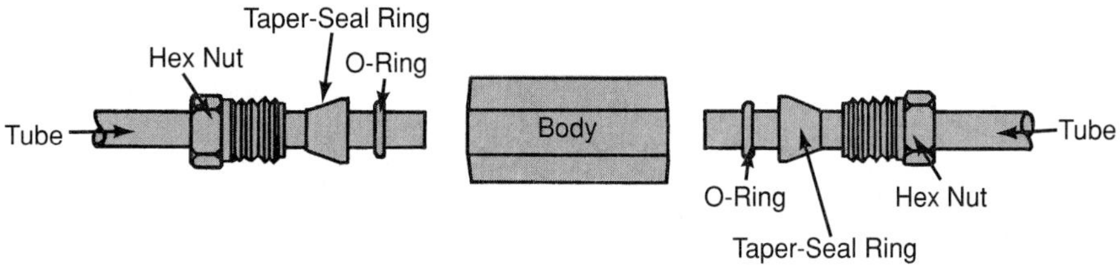

Figure 2-37. *Threaded mechanical fitting. Note the hex nut, tapered seal ring, and O-ring. (Watsco Components, Inc.)*

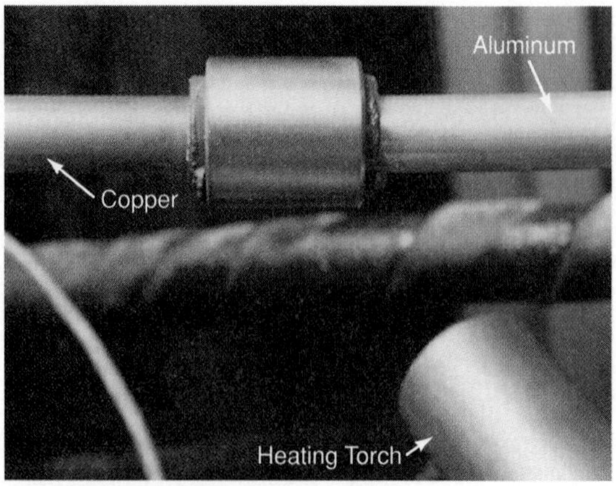

Figure 2-38. *Copper and aluminum tubing being joined with adhesive kit and heat from torch.*

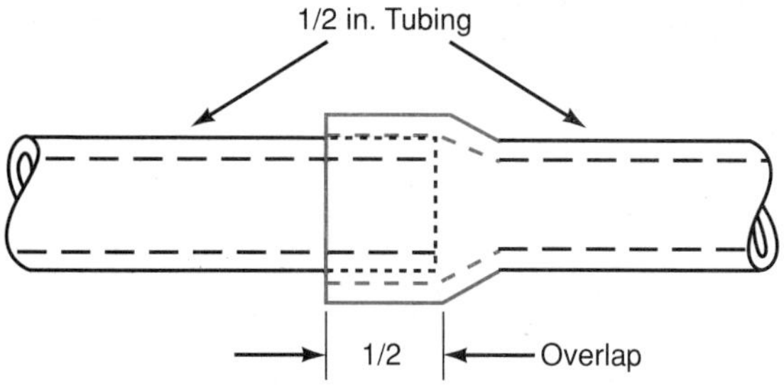

Figure 2-39. *Two pieces of soft copper tubing assembled and ready for soldering or brazing to make joint. Note that pieces were of same diameter before one was swaged.*

pieces of tubing is important. As a rule, the length of overlap should equal the outside diameter (OD) of the tubing.

Two types of swaging tools are commonly used—the punch-type and the lever-type. In both cases, different tool sizes are available for the many sizes of copper tubing. The punch-type swaging tool has an anvil block with several holes. See **Figure 2-40.** The copper tubing is inserted into the correct hole size in the anvil block. The tube is clamped in place. Then, a punch is hammered into the end of the tubing the desired distance.

A combination flaring and swaging tool is shown in **Figure 2-40.** This type of tool can be used as a flaring tool using Block A or B. The block used will depend upon the diameter of the tubing. To use it as a swaging tool, swaging adaptors (C) are used. Turning the lever expands the tube to the proper size.

2.7 Tube Constrictor

Often, two tubes which fit together rather loosely must be soldered or brazed together. Good practice demands that the tubes be sized as close as .003″ to each other. **Figure 2-41** shows a special tool used to constrict the outer tube until it fits the OD of the inner tube. With

A

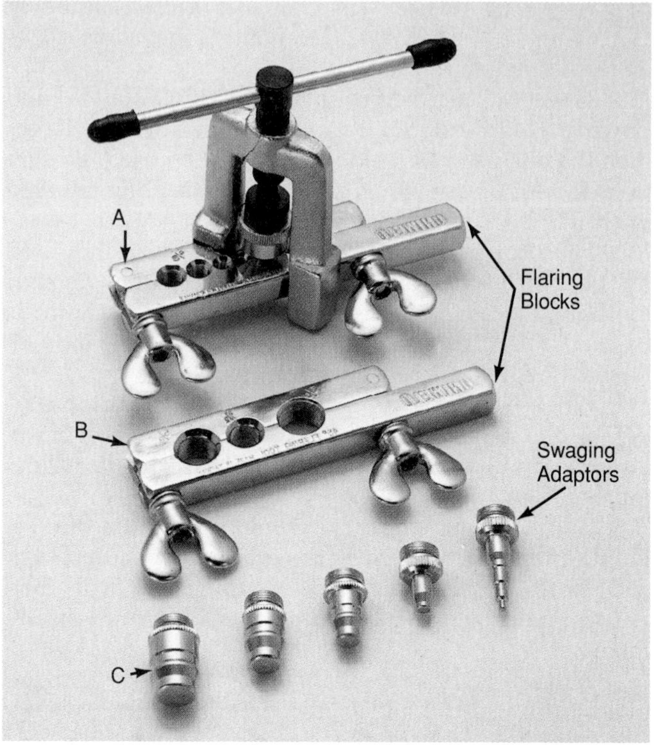

B

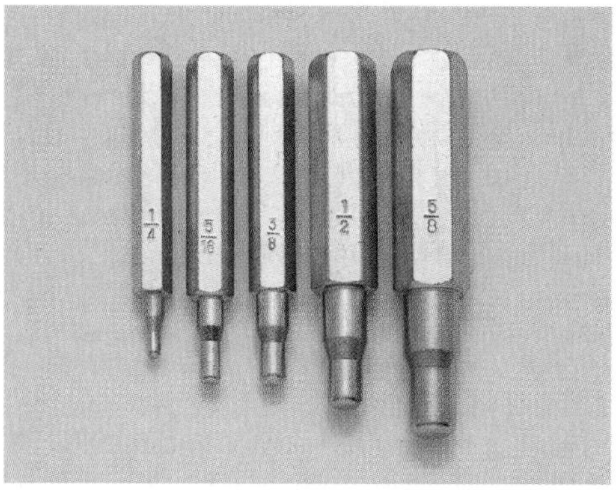

Figure 2-40. *A—Combination flaring and swaging tool. B—Punch-type swaging tool. (Uniweld Products, Inc.)*

this tool, the joint can be easily soldered or brazed without leaks, while keeping flux out of the system.

2.8 Pipe Fittings and Sizes

Air conditioning and refrigeration installations make wide use of pipe fittings and pipe threads (National Pipe Threads or NP). Taper pipe threads are specially formed V-threads made on a conical spiral. This taper causes the threads to seal as the fitting is tightened. Pipe threads taper 1/16″ in diameter for every inch of length. Untapered threads can be made leakproof by use of a gasket or an American Standards Association (ASA) machined shoulder.

Besides being tapered (or in a conical spiral), pipe threads are different from the National Fine (NF) series

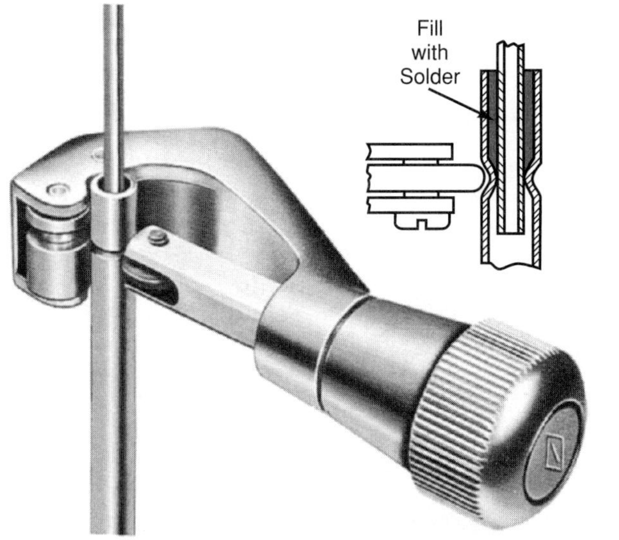

Figure 2-41. *Combination tube cutter and constrictor. Different wheels are used for cutting and constricting. Inset shows two tubes joined by soldering or brazing. Outer tube has been constricted before soldering. (Imperial Eastman, Imperial Division)*

and the National Coarse (NC) series. NF and NC sizes are based on outside diameter. Pipe thread sizes are based on flow diameter, or roughly the diameter of the hole in the pipe (inside diameter or ID).

Figure 2-42 illustrates a male thread on a 1/2″ pipe. The external threads are cut with a pipe die. The die is turned by a standard die stock, ratchet die stock, or power-driven die stock.

Figure 2-43 shows the pipe thread tap. It is used with a tap wrench to cut female, or internal, threads. Special thread-cutting compound should be used when cutting pipe threads. The taps and dies must be kept clean and sharp.

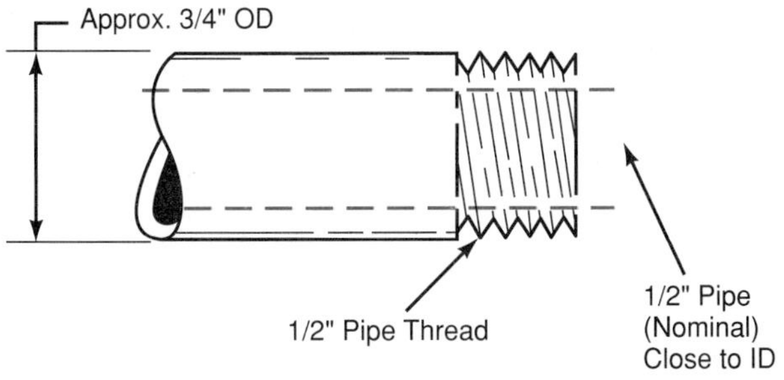

Figure 2-42. *A male thread on a 1/2″ pipe.*

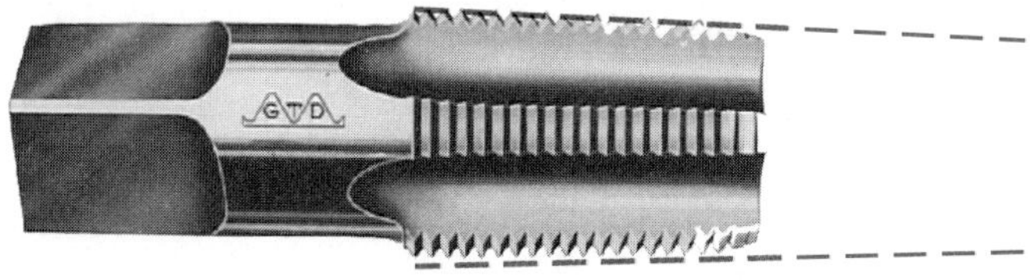

Figure 2-43. *Pipe tap. Note taper of threads. (TRW Greenfield Tap & Die Div.)*

Pipe fittings are supplied with the threads already cut. The most common fittings are the coupling, reducing coupling, union, nipple, 90° elbow, reducing elbow, 45° elbow, and street ell. The street ell is usually a 90° fitting with a male thread on one end and a female thread on the other end. A male-threaded pipe should be turned into the female fitting for a distance of five threads for a good seal.

The threads are made self-sealing by the pressing together of the sharp V-threads as they are assembled. Various commercial compounds are available to help seal these threads. When brushed on pipe threads before assembly, the compound will make a strong, leakproof joint.

REFRIGERATION TOOLS MODULE

2.9 Hand Tools

The refrigeration and air conditioning service technician performs work chiefly with hand tools. To be successful, the technician must pick good tools, take good care of them, and be skilled in their use. Most service failures can be traced to poor hand tool skills.

The refrigerating mechanism, in comparison to an automobile engine, is relatively light. It can easily be damaged by abuse. Great care is necessary to avoid damaging the units. **Figure 2-44** shows an assortment of basic hand tools needed by the service technician. The following paragraphs provide useful suggestions for the selection, care, and use of hand tools.

2.9.1 Wrenches

Most refrigeration and air conditioning installation and servicing requires the use of various types of wrenches. Many fasteners and parts are copper or brass, and therefore, are rather soft. Never use pliers on parts designed to be handled with wrenches.

A service technician needs several types and sizes of wrenches. Wrenches should be made of good alloy steel, and should be properly heat treated. They should be accurately machined and ground to fit the nut or bolt head. The wrench must fit the nut or bolt head accurately and it must fit as much of the hexagon as possible. For these reasons, the wrench types are listed in the order preferred.

1. Socket wrenches.
2. Box wrenches.
3. Open end wrenches.
4. Adjustable wrenches.

Use wrenches properly so that they fit completely on the nut or bolt. Sockets should be inserted all the way on the nut or bolt head. A loose or worn wrench may slip and spoil the corners on nuts. Proper servicing then becomes impossible without replacing the part.

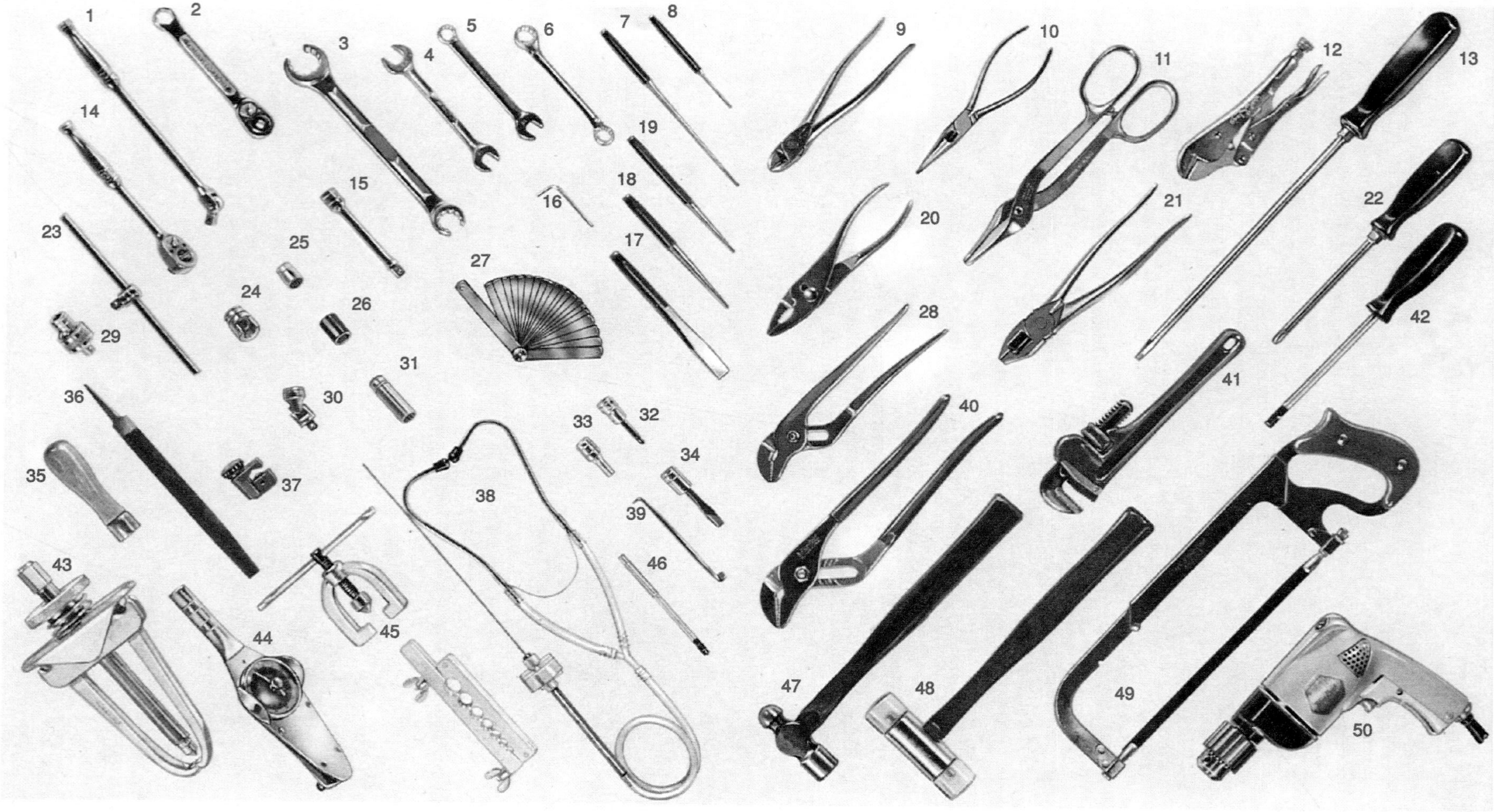

Figure 2-44. *Basic hand tool assortment for refrigeration service. 1—Nut spinner. 2—Refrigeration ratchet wrench. 3—Flare nut wrench. 4—Open end wrench. 5—Combination wrench, double hex. 6—Box socket, double hex offset wrench. 7—Taper punch. 8—Pin punch. 9—Diagonal cutting pliers. 10—Needle-nose pliers. 11—Tinner's snips. 12—Pinch-off pliers. 13—Standard-tip screwdriver. 14—Ratchet handle. 15—4" extension. 16—Allen (hex head) wrench. 17—Cold chisel. 18—Center punch. 19—Starter punch. 20—Slip-joint pliers. 21—Lineman pliers. 22—Phillips screwdriver. 23—Sliding bar handle. 24—Weatherhead socket. 25—Double hex socket. 26—Magnetic socket. 27—Feeler gage. 28—Interlocking-joint pliers. 29—Adaptor. 30—Universal joint. 31—Double hex socket, deep. 32—Phillips screwdriver socket. 33—Clutch screwdriver socket. 34—Standard screwdriver socket. 35—File handle. 36—Half-round file. 37—Tube cutter. 38—Technician's stethoscope. 39—Offset screwdriver. 40—Large, interlocking-joint pliers. 41—Pipe wrench. 42—Clutch screwdriver. 43—Two-jaw puller. 44—Torque wrench, U.S. conventional/SI metric. 45—Flaring tool. 46—Screw starter. 47—Ball peen hammer, plastic tip. 49—Hacksaw. 50—Electric drill. (Snap-on Tools Corp.)*

Always pull on a wrench rather than push on it. Otherwise, sudden loosening of the nut or bolt may cause a serious hand injury. Figure 2-52 shows the proper direction to pull on an adjustable wrench.

Avoid pounding on a wrench to obtain greater turning force or torque. Avoid using a length of pipe or another wrench for more turning force or torque.

A tight bolt or nut may be loosened safely by soaking the threads with a penetrating oil. Heating the nut or bolt may also help. Some service technicians tap a nut or bolt lightly with a hammer. Any of these methods can be used to loosen corroded threads.

Socket Wrenches

If the nut or bolt head has enough room around it, the six- or twelve-point socket is the best wrench to use. These sockets are usually made of chromium-vanadium steel. They are turned by handles that have a 1/4″, 3/8″, or 1/2″ square drive. The handles come in a variety of designs, as shown in **Figure 2-45.**

A variation of the socket wrench is the nut driver. A nut driver is a small direct-drive socket wrench. It has a plastic handle that can be used with assorted drive sockets.

Sockets are now available which will hold the nut or cap screw securely. This is designed to prevent the nut or screw from falling out during alignment and initial threading. This feature is very useful, since a dropped nut or screw can cause problems.

Box end and socket wrenches are more usable if they are double broached (12-point). **Figure 2-46** illustrates both the 6-point and the 12-point box end wrench.

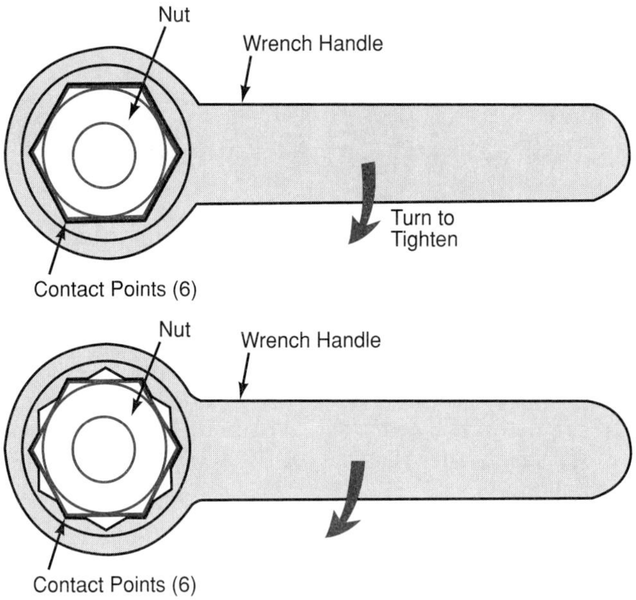

Figure 2-46. *Box end socket wrenches as they appear fitting over hex nuts. Upper wrench is a 6-point box end. Lower is a 12-point box end.*

The 12-point wrench is easier to use if the handle must be operated in a small or restricted space. The 6-point socket is best for worn hex nuts or bolts.

Metric-size nuts and bolts require metric-size wrenches. **Figure 2-47** shows a set of metric 6-point sockets commonly used when working with metric-size nuts and bolts. The size marked on the socket corresponds to

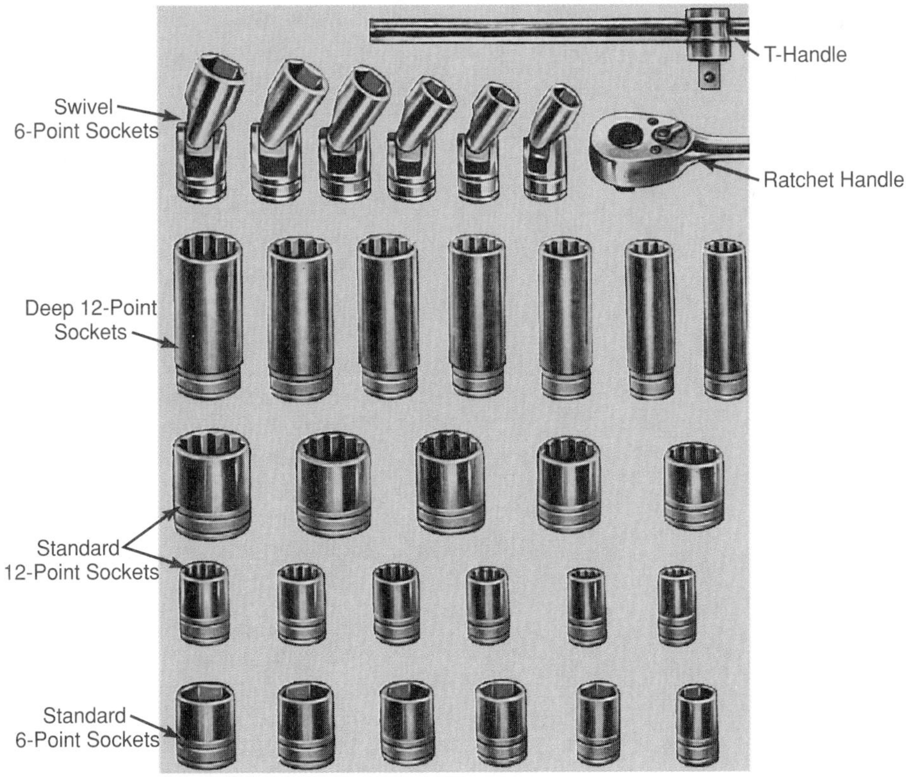

Figure 2-45. *Typical socket wrenches and handles.*

Figure 2-47. *A set of metric-size sockets. Note that the size marked on each socket corresponds to the diameter in millimeters (mm) of the bolt or capscrew. Black rings indicate metric sockets. (Snap-on Tools Corp.)*

the diameter of the cap screw or bolt. It is not the distance across the flats as it is with fractional-inch wrenches.

Box Wrenches

Often, a box wrench can be used in a tight space where a socket wrench cannot go. Box wrenches are usually 12-point and provide a powerful nondamaging grip on the nut or bolt, **Figure 2-48.**

Box wrenches may be either straight, offset, or double offset. Most box wrenches are double-ended. Both ends may be of the same size with one end offset, or they may be of the same pattern and different sizes.

The table in **Figure 2-49** shows what size wrench will fit the most common bolts and nuts. Below 1/2″ bolt size, the wrench size is 3/16″ larger than the bolt size. A 1/4″ bolt uses a 7/16″ wrench size (4/16 + 3/16 = 7/16).

At 1/2″ bolt size and larger, the wrench size is 1/4″ larger than the bolt size. For example, on a 5/8″ bolt, a 7/8″ wrench size is needed (5/8 + 1/4 = 7/8).

The size of the wrench opening (across the flats) is marked on the wrench. Box wrenches having both flat and 15° handles are necessary for a complete tool kit.

Figure 2-48. *An alloy steel box wrench with 12-point or double hex ends. Ends are offset (double offset) to provide gripping or swinging clearance above mechanism. Socket wrenches are safest; box wrenches are next safest. Box wrenches are less likely to slip than open end wrenches. (Duro/Indestro, Duro Metal Products Co.)*

Nominal Bolt Size	Wrench Size: Head and Nut Width Across Flats	
1/4	7/16	Wrench is 3/16" Larger than the Bolt
5/16	1/2	
3/8	9/16	
7/16	5/8	
1/2	3/4	Wrench is 1/4" Larger than the Bolt
9/16	13/16	
5/8	7/8	
3/4	1	

Figure 2-49. *Table of wrench openings for standard bolt heads and nuts.*

Flare Nut Wrenches

A flare nut wrench used with SAE fittings is shown in **Figure 2-50.** Its opening allows the wrench to slip over the tubing to reach the flare nut. A box wrench cannot do this. An open end wrench could be used, but the flare nut wrench grips the nut better.

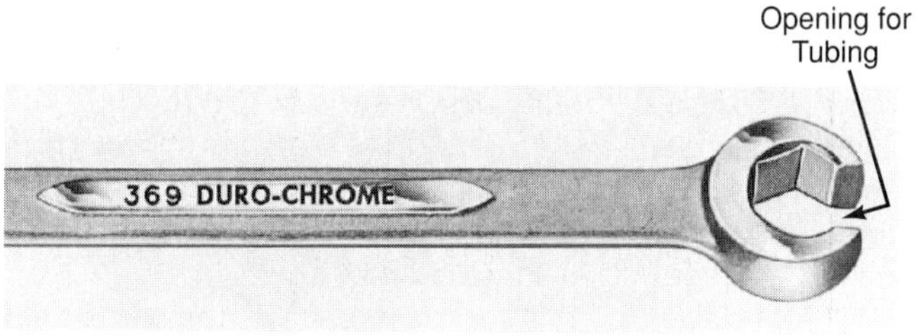

Figure 2-50. *Flare nut wrench used when turning SAE flare nuts. (Duro/Indestro, Duro Metal Products Co.)*

Other types of flare nut wrenches have been devised. **Figure 2-51** shows a strong, easy-to-operate, opening-type flare nut wrench.

Forged flare nut sizes are an SAE standard used in automotive, marine, and refrigeration service. See Chapter 31 for a table of flare nut wrench sizes.

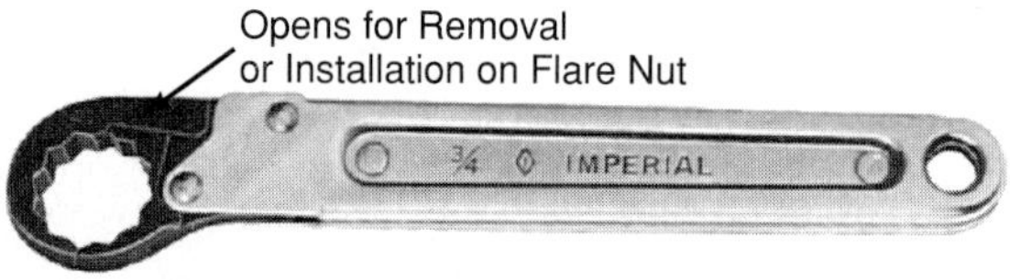

Figure 2-51. *Special type of flare nut wrench opens to pass over tubing and closes on flare nut to give positive contact. (Imperial Eastman, Imperial Division)*

Open End Wrenches

Open end wrenches can slide on the nut or bolt head from the side. They are used in close spaces on unions and other places where the socket wrench and box wrench cannot be used.

An open end wrench should not be used for refrigeration work if its jaws are spread or have burrs. Open end wrenches used in servicing work should have a thick jaw. Thin wrench jaws have a tendency to bite into soft brass and copper parts.

Popular sizes for open end wrenches are:

- The 7/16″ across flats, often needed for 1/4″ screws and bolts.
- The 1/2″ across flats for 5/16″ NC and NF cap screws, commonly used on compressors and expansion valves.
- The 5/8″ across flats for 1/4″ flare nuts.
- The 15/16″ across flats for the 1/2″ flare nuts.

- The 1″ across flats which fit the 1/2″ flare nuts. A typical open end wrench, No. 4, is shown in the assortment making up **Figure 2-44.**

Another popular wrench used in refrigeration work is the combination open end and box socket. Both ends are the same size. This wrench is illustrated in **Figure 2-44, No. 5.**

Adjustable Wrenches

Often, odd-size nuts and bolts are found in refrigeration work. Therefore, wrenches with adjustable jaws, **Figure 2-52,** are necessary in the tool kit. Adjustable wrenches must be kept in good repair. If the wrench does not fit tightly, it may slip and result in a ruined wrench, bruised hand, and a ruined nut or bolt head.

The forces on the jaws of the wrench should be in the right direction, **Figure 2-52.** This will give solid support against both the nut and the body of the wrench.

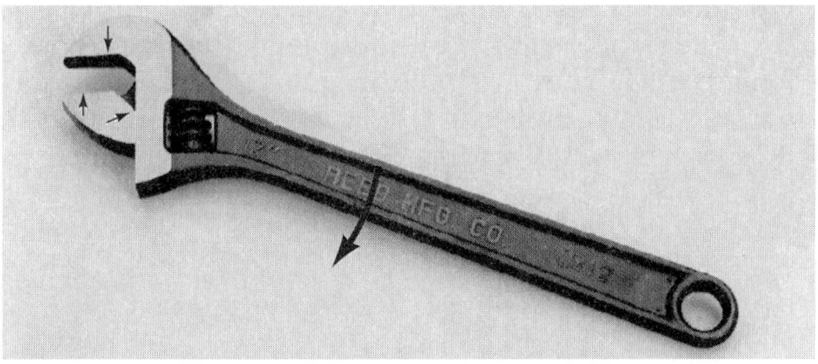

Figure 2-52. *A popular type of adjustable wrench. Handle should be pulled as shown by direction of arrow on handle. Note that wrench is adjusted to fit nut tightly. The red arrows show the pressure of the wrench against the corners of the nut. Turning wrench in the direction shown tends to press movable jaw against wrench body, thus tightening the grip. (Reed Manufacturing Co.)*

Pipe Wrenches

The pipe wrench is designed to grip pipes, studs, and other cylindrical (round) surfaces. The greater the torque on the wrench handle, the tighter the wrench will grip the object. Pipe wrenches should not be used on nuts or bolt heads. The typical pipe wrench is pictured in **Figure 2-44, No. 41.**

An internal-type pipe wrench, **Figure 2-53,** may be used for installing pipe, nipples, or fittings.

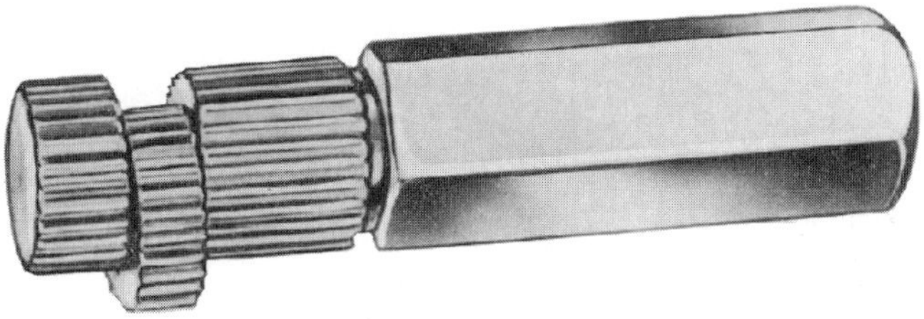

Figure 2-53. *Internal-type pipe wrench. It grips the pipe from the inside. (Snap-on Tools Corp.)*

A chain wrench, **Figure 2-54,** is another type of adjustable pipe wrench. The chain wrench can be used on square, round, or irregular shapes, and also used in confined areas.

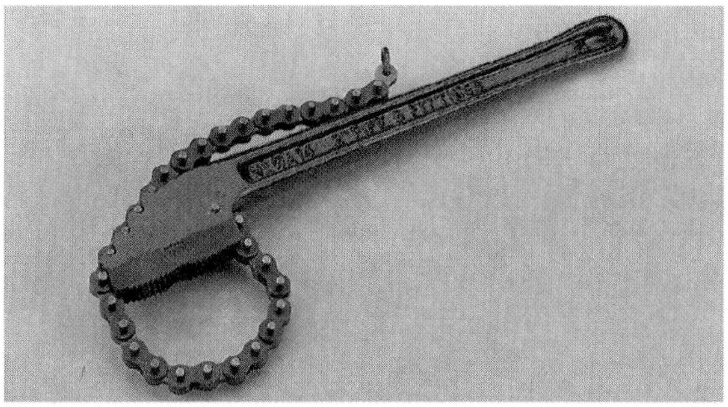

Figure 2-54. *A light-duty chain wrench for use in close quarters. (Reed Manufacturing Co.)*

Hex Key Wrenches

Hex key wrenches are constructed of alloy steel with a hexagonal (six-point) tip. A common type of hex key is the fold-up tool with many key sizes in one handle, **Figure 2-55A.** Individual L-keys and T-handle hex keys, **Figure 2-55B,** are frequently used for long-reach operations, such as set screws on pulleys.

Another type of wrench similar to the hex wrench is the Torx®, which is star-like in appearance, **Figure 2-55C.** This allows better metal-to-metal contact. It is less likely to damage the socket or itself.

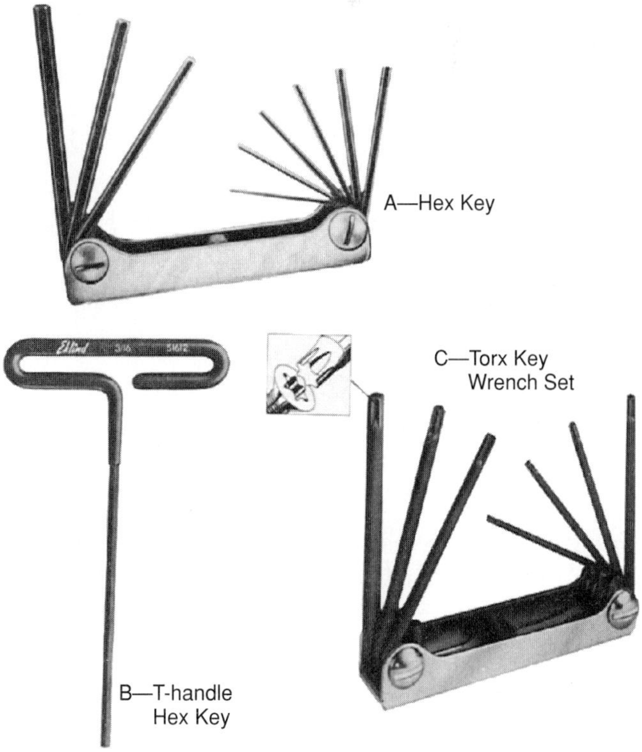

Figure 2-55. *Key wrenches. (Eklind Tool Co.)*

Service Valve Wrenches

Service valve stems usually have a square end milled on the valve shaft. A special service valve wrench is needed to turn them, **Figure 2-56.** This tool usually has a ratchet and a fixed end.

When "cracking" valves, the fixed end only should be used. ***Cracking*** is the slight opening required to cause the valve needle or plunger to leave its seat, but allow only a very slow flow of refrigerant.

The fixed end of the wrench provides good control of the slight opening and closing of a valve. For rapid opening and closing of valves, the ratchet end may be used.

Some service valve wrenches have a reversible ratchet. The operator can reverse the direction of turn ing without removing this wrench from the stem. A reversible ratchet wrench is shown in **Figure 2-57.** This wrench is often used to open or close a compressor access valve. It may also be used to tighten or loosen a nut or bolt by changing the reversible ratchet.

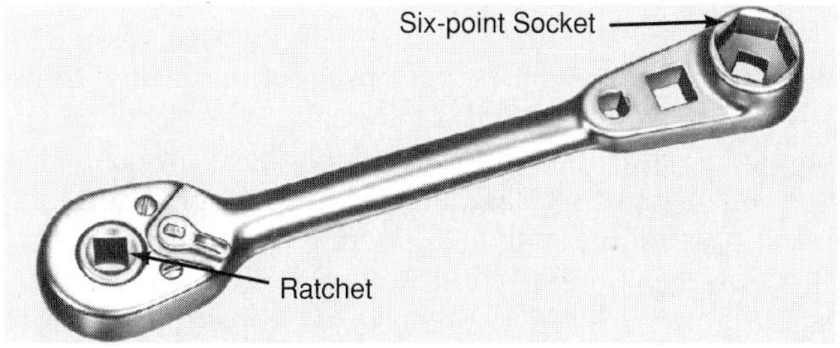

Figure 2-56. *Refrigeration service valve wrench. Fixed end is for "cracking" valves. Ratchet end is for rapid valve stem operation. Left end has 1/4" square drive for use with valve stem and packing gland nut sockets. Other openings are 3/16", 1/4", and 5/16" square. The 6-point socket fits 3/8" nuts. (Duro/Indestro, Duro Metal Products Co.)*

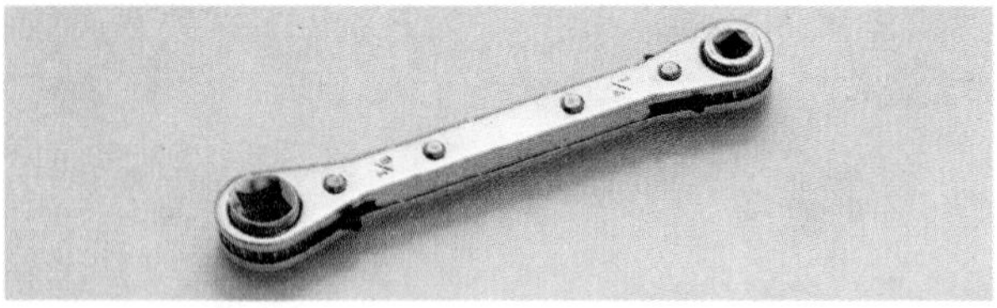

Figure 2-57. *Reversible ratchet wrench. Square openings with 1/4" and 3/16" at one end, and 3/8" and 5/16" at the other end. (Uniweld Products, Inc.)*

Service Valve Wrench Adaptors

Many manufacturers use valve stems other than the 1/4" square. Some valve stems are made so that the milled end is inside the valve body. This requires a good socket wrench to turn it.

To accommodate these valves, adaptors are available in various sizes. The male or drive part of the socket is usually 1/4" square. There are a few which use a larger drive (9/32"). The socket which fits the valve stem comes in five sizes: 3/16", 7/32", 1/4", 5/16", and 3/8". These sockets usually have eight points to simplify their use.

Most valve stems have internal packing gland nuts. Special sockets must be used on these valves. It is best to use sockets with ball bearing grippers. There is less chance of losing tools when working in difficult positions. **Figure 2-58** illustrates a set of these special tools. Also included are sockets for packing gland fittings.

Torque Wrenches

All materials are elastic (will stretch, compress, and twist). Even cast iron and hardened steels used in the construction of compressors are elastic up to a point. When tightening bolts, nuts, and other attachments on compressor parts and assemblies, it is important to measure the amount of tightness. Otherwise, warpage or other part damage may occur. To measure the amount of tightness, a torque wrench is used, **Figure 2-59.**

Torque wrenches are usually wrench handles only. They are made to be used with sockets of different sizes. The handle is equipped with a dial or pointer, which measures the foot-pounds or inch-pounds of torque.

The torque is found by multiplying the length of the handle (in feet) by the pull (in pounds) applied to the handle (foot-pounds). A 1′ long wrench handle pulled by a spring scale reading 50 pounds will produce a

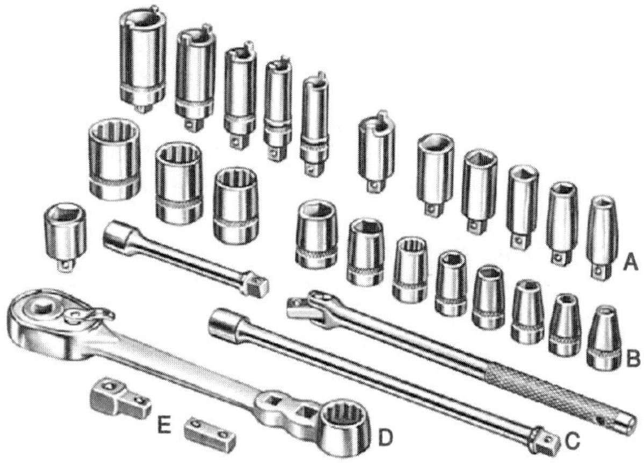

Figure 2-58. *Special service valve socket set. A—Packing gland sockets and valve stem sockets. B—Variety of 6-point and 12-point sockets. C—Handles and extenders. D—Ratchet wrench. E—Adaptors.*

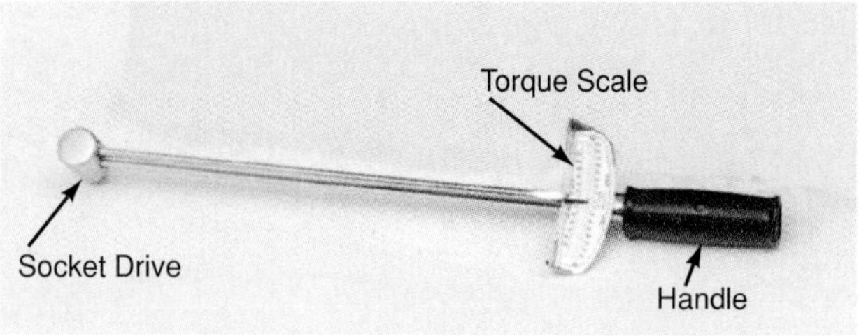

Figure 2-59. *Torque wrench used to measure the amount of tightness of nuts and screws. This wrench is made to be used with standard sockets. (Reed Manufacturing Co.)*

torque of 50 foot-pounds. (Technically, foot-pounds is the wrong term. The correct term should be pounds-feet. The foot-pound is a unit of work. However, popular usage has made the term foot-pound acceptable for the measurement of torque.)

To calculate inch-pounds, multiply the length of the handle (in inches) by the pull on the handle (in pounds).

The manufacturers of equipment (automobiles, airplanes, refrigerating equipment, etc.) are able to determine the proper torque that should be applied to the fasteners on their various mechanisms. The recommended torque for the many parts of refrigerating mechanisms are specified in manufacturers' service manuals.

To use a torque wrench, the operator fits the proper size of socket onto the wrench. The socket is then applied to the nut, and the handle of the wrench is pulled until the indicator shows that the required torque has been applied. At that torque, the nut is at the tightness recommended by the manufacturer.

2.9.2 Hammers

A hammer is a necessity in the refrigeration shop. The 12- or 16-ounce ball peen hammer is a useful tool. See **Figure 2-44, No. 47.** A carpenter's claw hammer may also be needed for mounting pipe supports and fastening sheet metal to wood. It is important that the hammer head be firmly fastened to the handle. The handle must also be in good condition.

Grasp the handle about two-thirds of the way back from the head. For light, accurate blows, hold the hammer with the index finger on the top of the handle and use wrist action. For heavy blows, hold the hammer with fingers around the handle and use elbow muscles.

2.9.3 Mallets

In service work, a mallet is often needed to drive parts into place or to separate them without injury to their surfaces. For such work, a 1 1/2-1b. to 2-1b. mallet is desirable, made of rawhide, rubber, wood, plastic, or lead. A mallet is shown in **Figure 2-44, No. 48.**

2.9.4 Pliers

Pliers are universal tools. Pliers are made of alloy steel, usually with manganese, although some are chrome-vanadium steel. Top-quality pliers are usually drop forged. Many different types are available. Use only pliers with insulated handles when working on electrical parts.

- Gas pliers are slip joint combination pliers, which are handy for general use. However, they should not be used on nuts, bolts, or fittings. They could slip and injure the surface. See **Figure 2-44, No. 20.**
- Cutting pliers are mostly used when working on refrigerator wiring. One type, called the lineman's pliers, is a powerful cutting and gripping tool. Another type, called the diagonal pliers, is used to cut in close quarters. Refer to **Figure 2-44, No. 9.**
- Nut pliers are used to good advantage on some jobs. The jaws always stay parallel. Some have an adjustable cam action that locks the jaws on the nut or bolt. In general, it is not good practice to use nut pliers on bolts or nuts. However, on a job such as holding a bolt head while turning the nut with a wrench, the use of nut pliers is permissible.
- Slim-nose pliers, needle-nose pliers, and duckbill pliers are frequently used in hard-to-reach places. See **Figure 2-44, No. 10.**
- Round-nose pliers are used to shape wire into loops and to bend sheet metal edges.

2.9.5 Screwdrivers

A complete set of screwdrivers is very necessary both for installation and for shop work. The length of a screwdriver is measured from the blade tip to the handle. Handles are not measured. The recommended average sizes are 2 1/2″, 4″, 6″, and 8″.

The types of screwdrivers are named for the shape of the end of the blade or bit. See **Figure 2-60.** Most popular is the straight blade, slot blade, or regular screwdriver. The screwdriver bit should fit the screw slot snugly. The blade should be wide enough to fill the screw slot end-to-end. Also see **Figure 2-44, No. 13.**

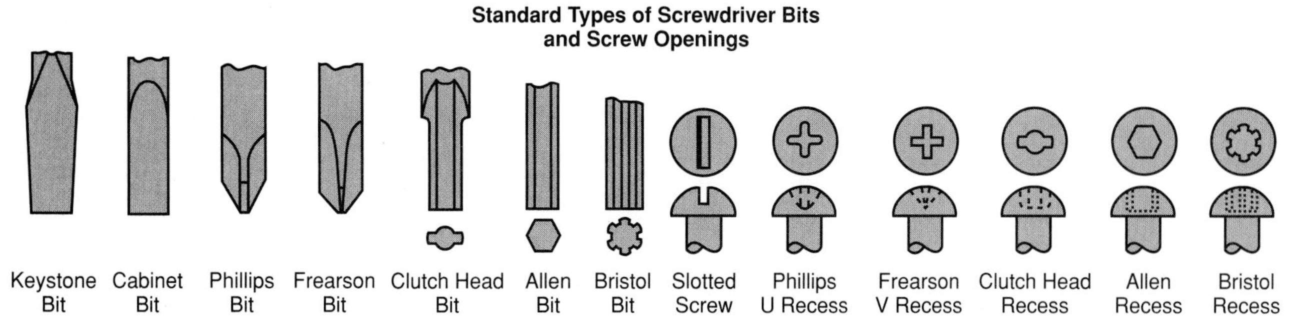

Figure 2-60. *Several types of screwdrivers. Flat-bladed Keystone and Cabinet bits and the Phillips bit are most popular. (Klein Tools, Inc.)*

The Phillips screwdriver has a tip which fits a recessed cross in the head of the screw. Phillips screwdrivers are available in the 3″ size for No. 4 and smaller screws; the 4″ size for No. 5 to No. 9 screws; the 5″ size for No. 10 to No. 16 screws; and the 8″ size for No. 18 screws and larger.

Stubby (short) screwdrivers are available for working in small spaces. Some screwdrivers may be equipped with a clip that holds screws while starting them. Better quality screwdrivers have strong handles firmly bonded to the blade. Plastic handles are popular.

An offset screwdriver is necessary in refrigeration work. There are many places where it alone can be used.

Never use a hammer to pound on a screwdriver. If a screwdriver is needed for heavy service, use one with a solid steel handle.

2.9.6 Vises

Sturdy machinist's vises are necessary in the shop. They are particularly convenient for holding parts while drilling, filing, or assembling.

One vise should be large enough to hold most compressor bodies. A special pipe vise, which has a hacksaw blade slot, is useful for a large service shop. This blade slot allows accurate cutting of piping and tubing.

Always use soft jaws made of sheet copper or aluminum when clamping a part which must not be marred. These are available as inserts which fit over regular vise jaws.

2.9.7 Twist Drills

Twist drills are frequently used for installation and repair work. Drill designs are available for working metal, wood, plastic, and masonry. Twist drills may be turned by drill presses, portable electric drills, or hand braces.

Most commonly, twist drills have straight shanks. This means that the section gripped by a three-jaw chuck is straight and perfectly cylindrical in shape. See **Figure 2-61.** Split-joint twist drills are often used with portable electric drills because they penetrate many metals easily.

The shank of a twist drill carries a stamped identification giving the kind and size of the drill. Twist drills may be made from high carbon steel, or from alloy steel (HSS) for high-speed use.

Drills are sized by bit diameter. Those intended for working metal come in three different set sizes. Identification systems for sizes include fractional numbers, whole numbers, and letters.

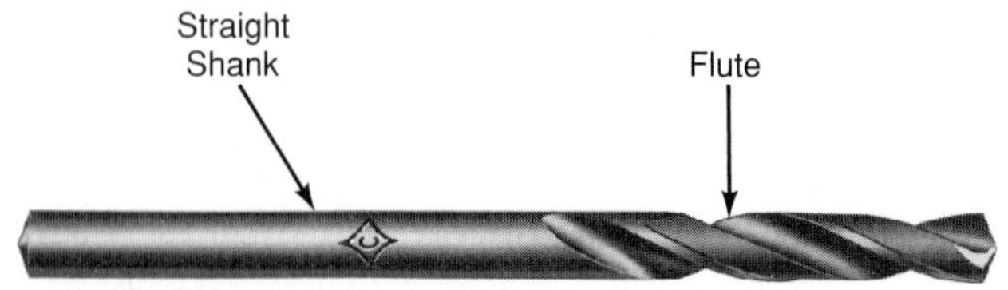

Figure 2-61. *Straight-shank twist drill for use on metal. (Cleveland Twist Drill Co.)*

Fractional sizes come in sets usually beginning with 1/16″ and going up to 1/2″ in steps of 1/64″. Larger fractional sizes are also available.

Numbered sets begin with No. 1 and range through No. 80 (.228″–.0135″). The higher the number, the smaller the drill. Most commonly used sizes are No. 1 through No. 60.

Letter size twist drills are larger than 1/4″ in diameter and vary from .234″ for the "A" size to .413″ for the "Z" drill.

Note that the numbered twist drills cover a range of sizes—approximately .013″ through 1/4″. Letter sizes range from approximately 1/4″ to nearly 1/2″. These two twist drill sets are often used as tap drills in making holes for inside threads. They provide a greater range of sizes than the fractional-inch twist drills. For a table of various drill diameters, see Chapter 31.

Speed of drilling depends upon the type of material being drilled and the diameter of the hole. In general, the smaller the twist drill, the faster it should be turned.

Most twist drills have two cutting edges or "lips." These edges must be sharp and equal in length. They must also have clearance and rake angles. See **Figure 2-62.**

Twist drills have flutes which remove chips from the hole. Most flutes are spiraled at an angle which automatically provides a rake angle for the cutting edges.

Always be sure the drill is cutting when it is being used. If the cutting edges are just rubbing against the stock, they will quickly heat up. Overheating will destroy the hardness of the drill.

To ensure that the drill forms the correct size hole, both cutting lips must be exactly the same length and

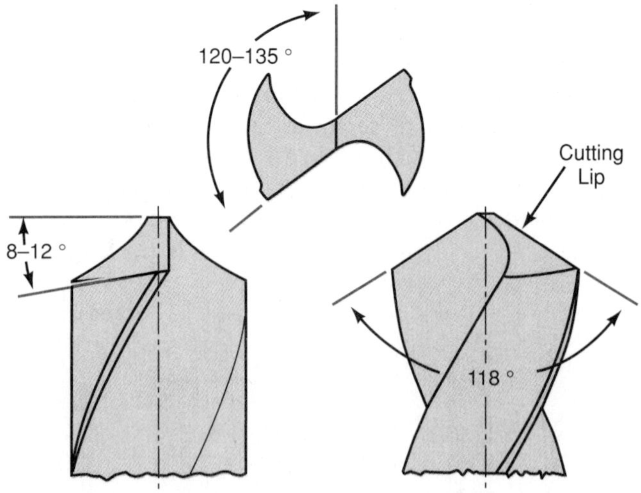

Figure 2-62. *Correctly ground twist drill point for steel. Clearance angle shown, 8°–12°, is used for mild steel and cast iron for drills in 1/2″ range. As diameters are reduced, clearance angles increase. A 1/16″ diameter twist drill should have a clearance angle of about 20°. (Cleveland Twist Drill Co.)*

angle. See **Figure 2-63.** If one lip is longer, the hole being drilled will be oversize. If one lip has a smaller angle, it will do all the cutting and soon grow dull.

Always wear safety glasses to protect eyes from flying chips when using either a drill press or portable drill.

Electric drills should be grounded for safety. The metal frame of the drill should be electrically connected to a good ground (water pipe or a ground rod). Most electric drills are equipped with a three-prong grounded plug. If the circuit to which the drill is connected is not provided with a three-prong grounded socket, a grounded adaptor should be used. Some hand drills have the electric motor insulated from the case, and do not need grounding. Grounding is covered further in Chapter 6.

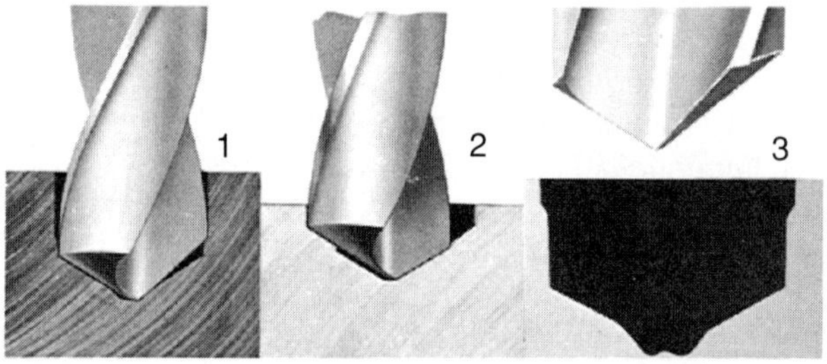

Figure 2-63. *An illustration of what happens when a drill is incorrectly sharpened. 1—Lips are equal in lengths but at different angles. 2—Lips are at equal angles but are of different lengths. 3—Lips are at different angles and at different lengths.*

2.9.8 Taps

Many assemblies of metal parts are fastened with machine screws or tap screws threaded into tapped holes. A ***tap*** is used for making threads inside a hole. Taps are accurately made from hard alloy steel. Clearance pockets are provided for chips. The threads are made with a very small clearance angle.

There are taps for every size or diameter thread and also for each kind of thread—National Fine (NF), National Coarse (NC), American Standard Taper Pipe Thread (ASA), or metric. Taps are of three types—taper, plug, and bottoming. Most common is the plug tap. See **Figure 2-64.**

Taper taps are used for starting a cut. They are also used for tapping thin pieces in which the tapped hole goes all the way through. Plug taps are used to do most of the cutting in blind holes (holes not going all of the way through the piece). Bottoming taps are used to cut full threads to the bottom of a blind hole.

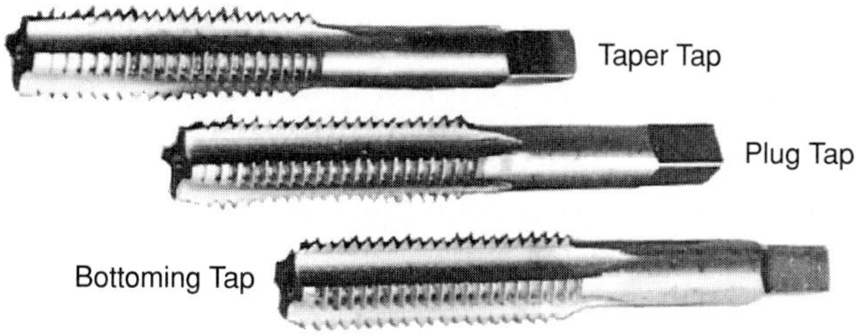

Figure 2-64. *Set of taps for cutting 1/2" NC threads.*

The shank of the tap is ground to a square at the end. A tap wrench is used to turn it. Power tools may also be used for driving. However, a special tap-driving accessory must also be used.

Since tapping is basically a cutting operation, the general rules for cutting metals apply. Most taps have four cutting edges for each thread. These edges must be kept sharp. They must have a ground cutting face and cutting clearance.

Always use a special thread-cutting lubricant when doing any kind of threading. The single exception is gray cast iron. It contains enough graphite to provide necessary lubrication. Thread-cutting lubricant, if applied generously, also serves as a coolant.

Tap-Drill Sizes

It is very important that the hole to be tapped is first drilled to the correct size. If the hole is oversize, threads will not be full size. If the hole is undersize, the tap must remove too much metal and will probably break. See **Figure 2-65.**

The tap drill should be slightly larger than the root diameter of the threads for which the hole is being drilled. Threads which are 75% of full size are generally considered satisfactory. Always refer to tap-drill size tables for the correct size drill. For most refrigeration and

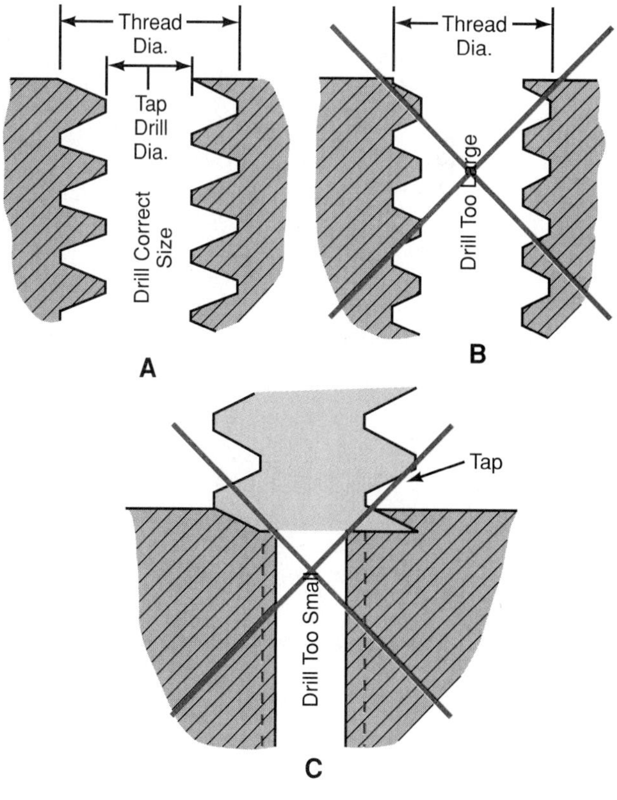

Figure 2-65. *Hole size is important when tapping threads in metal. A—Tap drill correct size, correct thread depth. B—Tap drill too large, threads not full depth. C—Tap drill too small, tap likely to break.*

air conditioning work, the tap-drill table, **Figure 2-66,** will be satisfactory.

2.9.9 Dies

Dies cut external threads on round stock. The threads match those cut by the tap. Since a tap is non-adjustable, dies can usually be adjusted to permit careful matching of the threads. Taps and dies may also be used to clean threads that are corroded or damaged. Dies are held and turned by a diestock. A diestock is shown in **Figure 2-67.**

Tap	Tap Drill	Tap	Tap Drill
4/36	No. 43	14-20	No. 9
	No. 44		No. 10
	No. 45		No. 11
4-40	3/32	14-24	No. 6
	No. 43		No. 7
	No. 44		No. 8
4-48	No. 41	1/4-20	No. 5
	No. 42		No. 6
5-40	No. 37		13/64
	No. 38		No. 7
	No. 39		No. 8
5-44	No. 36	1/4-28	7/32
	No. 37		No. 3
	No. 38	5/16-18	17/64
6-32	No. 33		G
	No. 34		F
	7/64	5/64-24	J
	No. 36		I
6-40	No. 32	3/8-16	O
	No. 33		5/16
8-32	No. 29	3/8-24	R
8-36	No. 28		Q
	No. 29	7/16-14	3/8
10-24	No. 24		U
	No. 25	7/16-20	25/64
	No. 26		W
10-32	No. 19	1/2-13	27/64
	No. 20	1/2-20	29/64
	No. 21	9/16-12	31/64
	No. 22	9/16-18	33/64
12-24	No. 15	5/8-11	17/32
	No. 16	5/8-18	37/64
	No. 17	3/4-10	21/32
12-28	3/16	3/4-16	11/16
	No. 13		
	No. 14	A	B
	No. 15		

Figure 2-66. *Tap drill sizes recommended for common tapping operations. Note that for certain sizes, the tap drill may be a fractional-inch size, a number size, or a letter-size drill bit. A—Outside diameter. B—Number of threads per inch.*

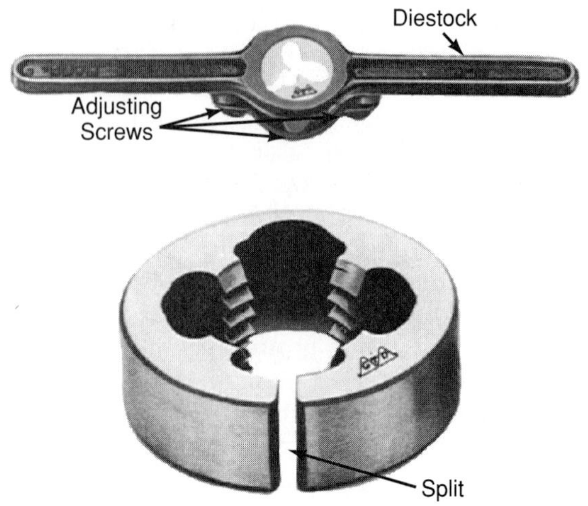

Figure 2-67. *Adjustable diestock has three screws. Two of the screws hold the die and apply contracting pressure. The third screw expands the die at the split. (TRW Greenfield Tap & Die Div.)*

As with taps, there are dies for each type of thread and size. Dies must be made of tool steel because they are also cutting tools. They, too, must be carefully shaped to cut correctly.

Special precautions should be taken to start threading very straight. Guides are available for this purpose. The small guide is located inside the threading die.

Round stock must also be accurately sized. Stock only a few thousandths of an inch oversize might break the die.

First adjust the die to full open position to make the initial cut. Then adjust the die to cut deeper until the threads match the tapped thread. Always advance the die (or tap) one-quarter to one-half turn, back off by reversing the direction of the diestock, and then repeat.

2.9.10 Cold Chisels and Punches

A cold chisel is used on various jobs. As an example, one may find corroded fasteners on evaporators. A chisel is needed to remove the nut or screw.

A 3/4" flat cold chisel is a popular size. Be sure to keep the head (hammering end) of the chisel free from "mushrooming." Flying pieces from a mushroomed head may cause injuries. **Always wear goggles or head shield when working with a chisel. See Figure 2-44, No. 17.**

Punches are available in various lengths and are usually made of carefully heat-treated chrome-alloy steel. The cutting edge or point is hard, while the head is tough and shatterproof. Always grind away any mushroom head that forms. A fairly heavy 6" punch will be the most satisfactory. **Figure 2-68** illustrates the shapes of various punches.

Four common types of punches are the center punch, drift punch, pin punch, and prick punch. See **Figure 2-44, Nos. 7, 8, 18,** and **19.**

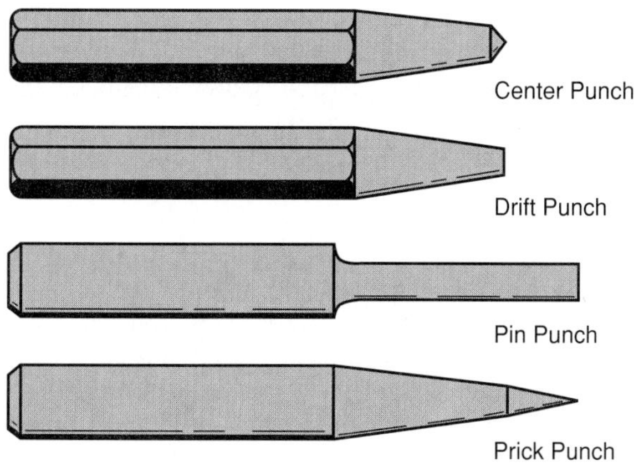

Figure 2-68. *Punches most used in refrigeration and air conditioning work.*

- The ***center punch*** has a 60° to 90° point. It is used for center punching the location of a hole to be drilled. A heavy blow on the punch makes a depression in which the drill may be started. Center punches may also be used to make alignment marks on refrigeration parts before dismantling.
- A ***drift punch*** is used to drive out keys and to line up holes in mating surfaces. The punch tapers from its flat point to the stock diameter.
- The ***pin punch*** is used for driving retainer pins in or out. The blunt end is called the bill. Pin punches are measured in overall length, by diameter of the stock, and by diameter of the bill. Bill diameters are available from 3/32″ to 5/16″.
- The ***prick punch*** has a long, sharp point and is used only for layout work. Be careful not to damage this sharp point.

2.9.11 Files

Various sizes and types of files are needed for cleaning metal surfaces and shaping metal parts. They are classified according to tooth size, shape, and the number of directions the teeth are cut on the file.

Files are either single or double cut, **Figure 2-69.** The single cut is used for finishing surfaces and double cut for fast metal removal.

Files come in different lengths: 4″, 6″, 8″, 10″, 12″, and so on. The size of the teeth varies from dead smooth, smooth, second cut, bastard, to coarse. The longer the file of a given type, the coarser the teeth. Thus, a second cut 12″ file has coarser teeth than a second cut 6″ file.

Many file shapes are available. They include rectangular, half round, round, triangular, square, wedge shape, and so on. See **Figure 2-44, Nos. 35** and **36.**

One oddity in file shapes is that there are three types of rectangular cross-section files—mill, hand, and flat. The mill file has only single-cut teeth. It is uniform in thickness but tapers slightly in width. The hand and flat files have double-cut teeth. The edges are parallel but the thickness varies slightly. The hand file has one edge that has no teeth, called a safe edge. The flat file has teeth on all four surfaces.

Use file brushes and file cards to clean file teeth which quickly become filled with metal. If clogging material is not removed, the files become useless. Do not use a file card for any other purpose than file cleaning. The bristles may become clogged with dirt.

2.9.12 Hacksaws

Hacksaws are used for cutting tubing and for other work requiring metal cutting. **Figure 2-70** illustrates a popular saw with a rigid frame and a 12″ blade. Blades are available with different numbers of teeth per inch. Blades with 14 teeth per inch are used for soft metal and wide cuts; 18 teeth per inch for medium soft metals; 24 teeth per inch for general work; and 32 teeth per inch for thin metal, tubing, or hard metal. A thinner and/or harder metal will require a blade with more teeth per inch.

A hacksaw blade should not be stroked faster than 60 strokes per minute. Most blades are made of high carbon steel, and their cutting edges (points) are very sharp and very small. Too rapid use will cause these points to overheat and lose their temper.

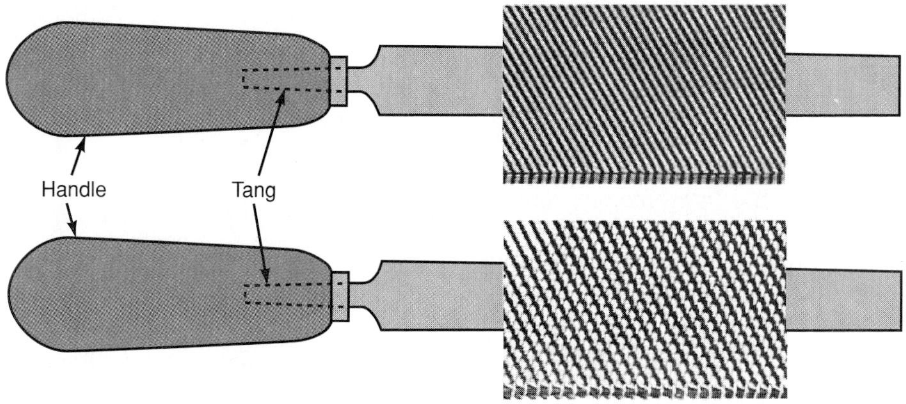

Figure 2-69. *Hand files may be either single-cut (upper) or double-cut (lower). Files should always have handles to avoid hand injuries. (Cooper Tools, Nicholson)*

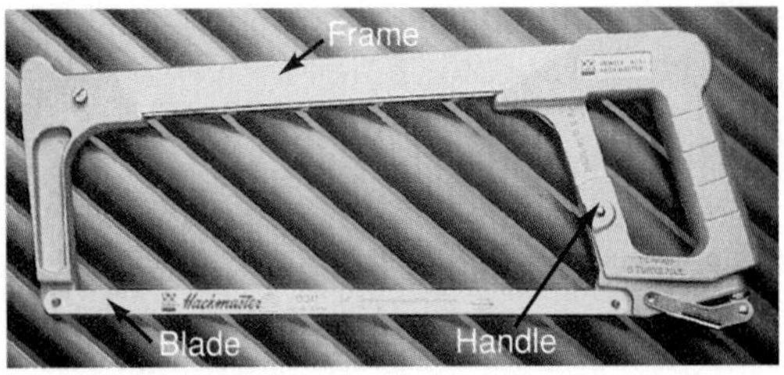

Figure 2-70. *Hacksaw with a rigid frame to hold the blade in proper tension. (American Saw and Mfg. Company)*

Always lift the blade slightly on the back stroke. This helps to keep the cutting edges sharp. If the blade is not lifted, chips may roll between the work and the cutting edge of the blade, dulling the teeth.

With most blades, the teeth are hardened while the back of the blade is soft and flexible. Some higher quality blades may be made of tungsten or molybdenum steel alloy. **Figure 2-71** illustrates a bimetal hand hacksaw blade. It has a high-speed steel cutting edge and a die steel flexible backing.

Special hacksaw frames are available for working in small holes. There is also a stub hacksaw blade and an adaptor drive to fit electric drills.

Figure 2-71. *Hacksaw blade. Three basic types are: standard high-speed blades for general use; high-speed flexible blades made of molybdenum steel; and bimetal blades with high-speed steel teeth and flexible die steel backing. The type of blade, its length, and number of teeth per inch are indicated on the blade. (Ridge Tool Company)*

INSTRUMENTS AND GAUGES MODULE

2.10 Instruments and Gauges

The technician uses instruments and gauges to determine conditions (pressure and temperature) inside the operating mechanism. The most common instruments are thermometers, pressure gauges, and vacuum gauges. Later chapters will cover special instruments such as recording thermometers, hygrometers, ammeters, voltmeters, ohmmeters, and others.

Instruments must be carefully handled and kept in good condition if they are to remain accurate. If accuracy is doubtful, the instrument should be sent to a repair company for testing and calibration.

2.10.1 Thermometers

The thermometer displays the temperature of the evaporator, refrigerator cabinet, liquid line, suction line, or condensing unit. An ice and water bath may be used to determine a thermometer's accuracy. When in this solution, the thermometer should check within 1°F of 32°F (within 1°C of 0°C).

Many sizes and types of thermometers have been developed for the technician's use. Figure 1-2 shows a popular pocket-type thermometer. It has a glass stem mounted in a case.

Glass-stem thermometers usually read from −30°F to 120°F (−35°C to 49°C) in 2 degree increments (marked spaces). The tube may contain either mercury or a red liquid. The mercury-filled thermometer is faster but more difficult to read. Some thermometers have a special magnifying front built into the glass. This enlarges the liquid line for easier reading.

There are numerous other thermometers that are popular and easy to use. The dial-stem thermometer shown in **Figure 2-72** may be operated either by a bimetal strip or by a bellows charged with a volatile (vaporizes readily) fluid. Its temperature range varies, but it is usually from −40°F to 120°F (−40°C to 49°C) in 2 degree increments. Never expose any kind of thermometer to temperatures beyond the limits of the scale. Doing so may ruin the instrument.

A battery-powered digital thermometer with dial/head swivel is shown in **Figure 2-73.** The temperature range for this instrument also varies, but may be from −40°F to 230°F (−40°C to 110°C).

Dial thermometers are very useful for taking pipe temperatures. **Figure 2-74** shows how these instruments may be clamped to pipes or tubes to check temperatures.

The thermocouple and the thermistor are two types of electrical thermometers. The fundamentals of these are described in Chapter 6. Additional information regarding thermometers and their applications may be found in Chapter 19.

Figure 2-75 illustrates a hand-held digital thermometer used for troubleshooting systems where the knowledge of the specific temperature of the evaporator or condenser is necessary.

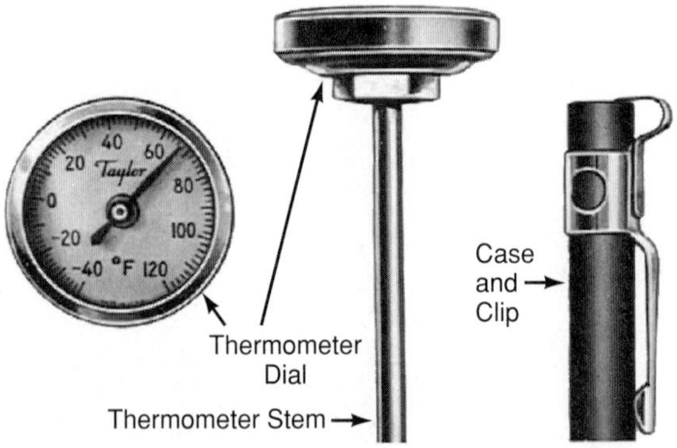

Figure 2-72. *Dial-stem thermometer is calibrated in 2-degree increments from −40°F to 120°F (−40°C to 49°C). This is the temperature range most used by technicians in refrigeration and air conditioning work. (Taylor Environmental Instruments)*

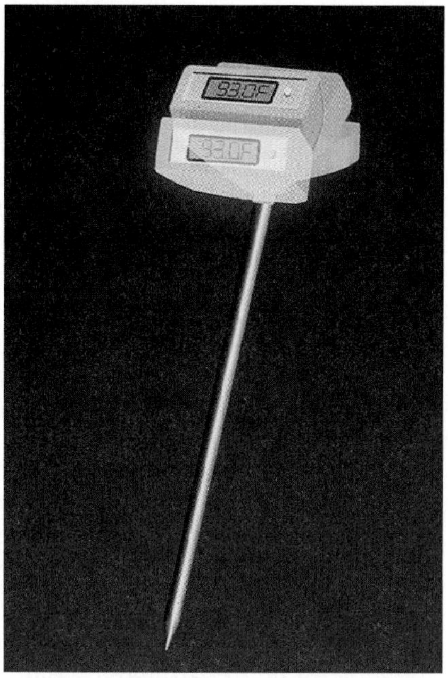

Figure 2-73. *A digital-type thermometer with swivel head for greater flexibility. (PSG Industries, Inc.)*

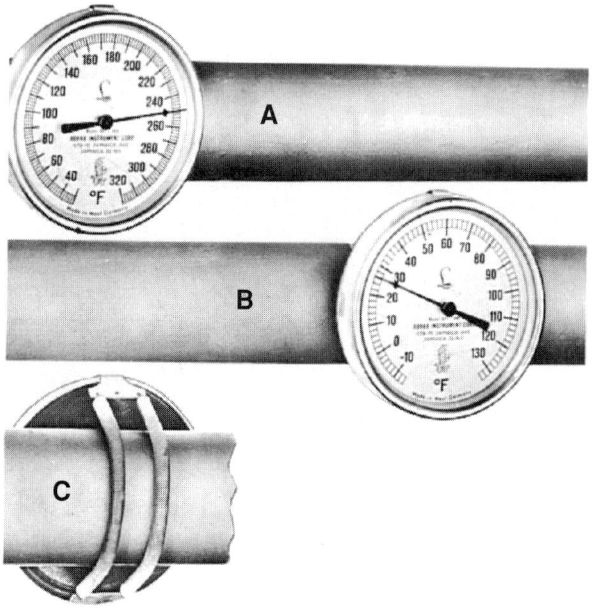

Figure 2-74. *Dial thermometers are easily clamped to pipes to indicate temperature of pipe. A—Shows calibration from 30°F to 320°F by 2-degree increments. Temperature of 250°F indicates that this thermometer is attached to steam or hot water pipe. B—Calibration is from −10°F to 130°F by 2-degree increments. Temperature (26°F) indicates thermometer is attached to evaporator suction line. C—Spring arrangement for attaching thermometer to pipe.*

Figure 2-76 shows a maximum and minimum thermometer. This type is useful when attached to a system which is unattended for a few cycles.

Figure 2-75. *A hand-held thermometer indicating the condensing temperature. (Reproduced with permission of Fluke Corporation)*

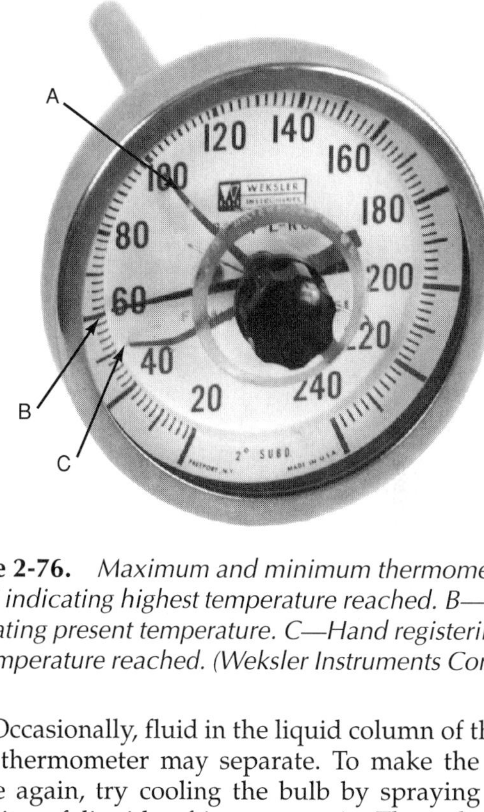

Figure 2-76. *Maximum and minimum thermometer. A—Hand indicating highest temperature reached. B—Hand indicating present temperature. C—Hand registering lowest temperature reached. (Weksler Instruments Corp.)*

Occasionally, fluid in the liquid column of the glass-stem thermometer may separate. To make the column whole again, try cooling the bulb by spraying a small quantity of liquid refrigerant on it. The column will shrink into the bulb; and, when it re-expands, the break should have disappeared.

Caution: If the mercury is frozen into a solid, the thermometer will break.

Another way to reconnect the column is to heat the thermometer as in Figure 2-77. Heat the thermometer very, very slowly. Do not allow fluid to reach top end of stem. If overheated, thermometer will burst. Wear goggles.

2.10.2 Pressure Gauges

Pressure gauges are used by the technician to help determine what is happening inside the system. Gauges

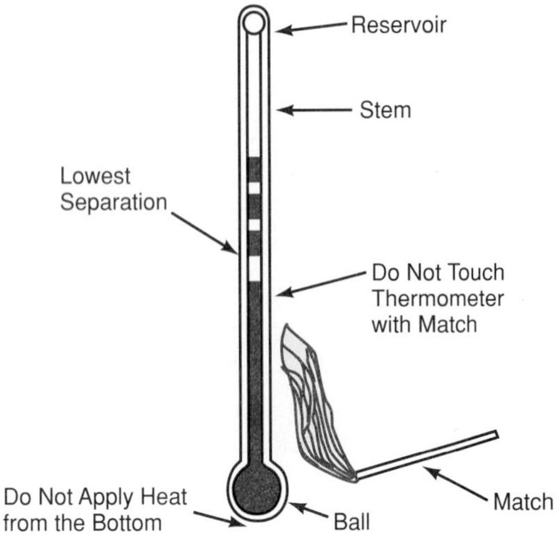

Figure 2-77. *Using heat to connect a break in the liquid column of a glass-stem thermometer. Be careful. Wear goggles. Avoid putting match below bulb. Keep flame moving to avoid hot spots. (White-Rodgers Div., Emerson Electric Co.)*

use a Bourdon tube as the operating element. The ***Bourdon tube*** is a flattened metal tube (usually copper alloy) sealed at one end, curved and soldered to the gauge fitting at the other end. **Figure 2-78** shows the typical construction of a pressure gauge.

A pressure rise in a Bourdon tube makes it straighten. This movement will pull on the link, which will turn the gear sector counterclockwise. The pointer shaft will then turn clockwise to move the needle.

Some gauges have a retarder to permit accurate readings in the usual operating range. The retarder uses an extra spring at pressures above normal. These gauges are easily recognized by the change in graduations at the higher readings of the positive pressure scale.

Most popular gauges have a 2 1/2″ dial and are connected into the refrigerating system with a 1/8″ male pipe thread. Some gauges, however, have a 1/8″ female pipe thread. Continual installation and removal quickly wears the threads. It is advisable to use a 1/8″ pipe nipple on any heavily used gauge.

Three types of pressure gauges are used in refrigeration service work: high-pressure, vacuum, and compound gauge.

Maximum pressure at which a gauge should be continuously used should be no greater than 75% of the full-scale range.

Gauge Manifolds

A gauge manifold includes both a high-side gauge and a low-side (vacuum) gauge. It allows the service technician to check operating pressures, add or recover refrigerant, add oil, and perform other necessary operations.

The gauge manifold illustrated in **Figure 2-79** is color coded. Manufacturers often color code the exterior of the gauge—low-side blue and high-side red. Low-side hoses are color coded blue, and the high-side hoses are red.

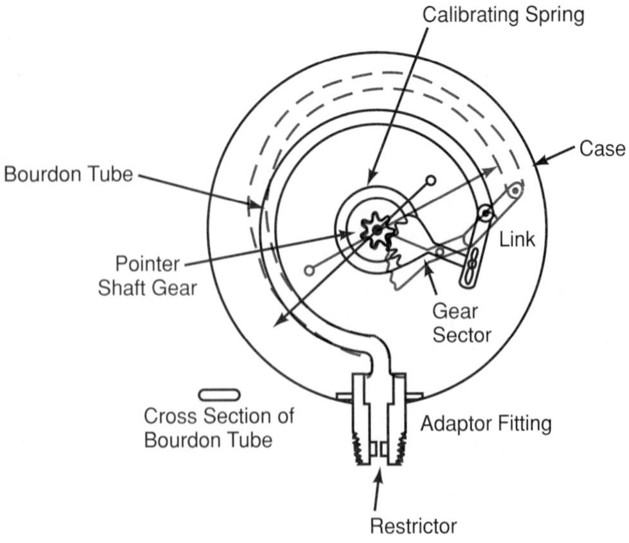

Figure 2-78. *Internal construction of pressure gauge. Red outline at top indicates how pressure in Bourdon tube causes it to straighten and operate gauge.*

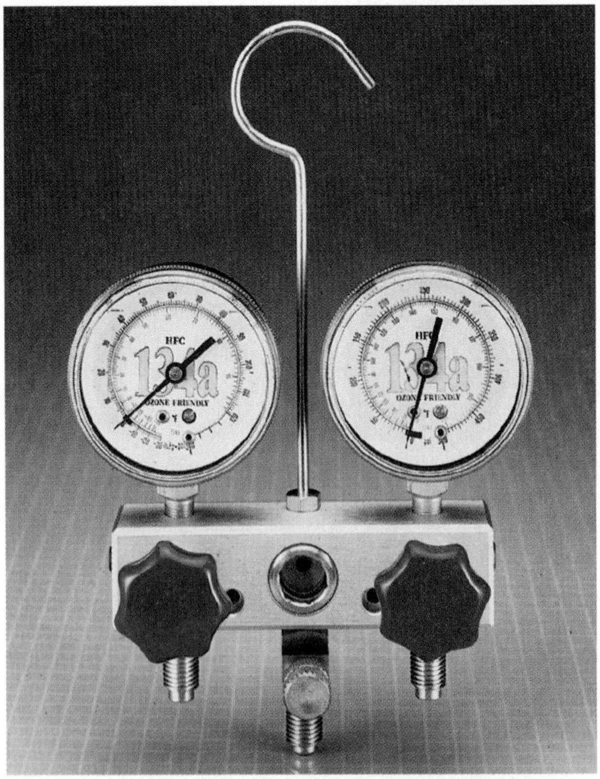

Figure 2-79. *A testing manifold for R-134a. The compound (suction) gauge (shown in blue) is mounted on the left. Its hose leads to the equipment suction service valve. The high-pressure (discharge) gauge (shown in red) is mounted on the right. Its hose leads to the discharge service valve or liquid line. (TIF Instruments, Inc.)*

High-Pressure Gauges

The high-pressure gauge has a single continuous scale, usually marked off to read from 0 to 500 psi. The scale usually has either 2-1b. or 5-1b. increments and usually is connected into the high-pressure side of the refrigerating mechanism. **Figure 2-80A** shows a manifold gauge set that can be used with all types of refrigerants, including substitutes and blends. The gauges show pressure only. The technician looks at the pressure-temperature chart on the back of the gauge manifold to read the corresponding temperature for the specific refrigerant.

Figure 2-80B shows a manifold with two battery-operated gauges and a liquid crystal display (LCD). The low-side gauge indicates either a vacuum or pressure reading from 0 to 29.9 inches of mercury or 0 to 99.9 psi pressure. The high-side gauge reads up through a maximum of 999 psi.

Vacuum Gauges

The vacuum gauge measures lower-than-atmosphere pressure. It will have one of four calibrations—inches of mercury (in. Hg), pounds per square inch absolute (psia), millimeters of mercury (mm Hg), or torrs or microns for very high vacuum. The micron is explained in Chapter 12.

A mercury barometer, **Figure 1-13,** measures vacuum in the normal ranges of refrigeration work. For measurement of very high vacuums, a special instrument, called the ***McLeod gauge,*** **Figure 2-81,** is usually used. Such instruments are calibrated in millimeters of mercury (torr). See Section 1.11.1 for the definition of a torr. The vacuum calibration on the compound gauge (in. of Hg) is most used in refrigeration work for measuring pressures below atmospheric.

For very high vacuums, the thermocouple gauge (shown in Chapter 12) should be used. It is accurate between 1 and 1000 microns. A mercury manometer is accurate for 1000 microns and above.

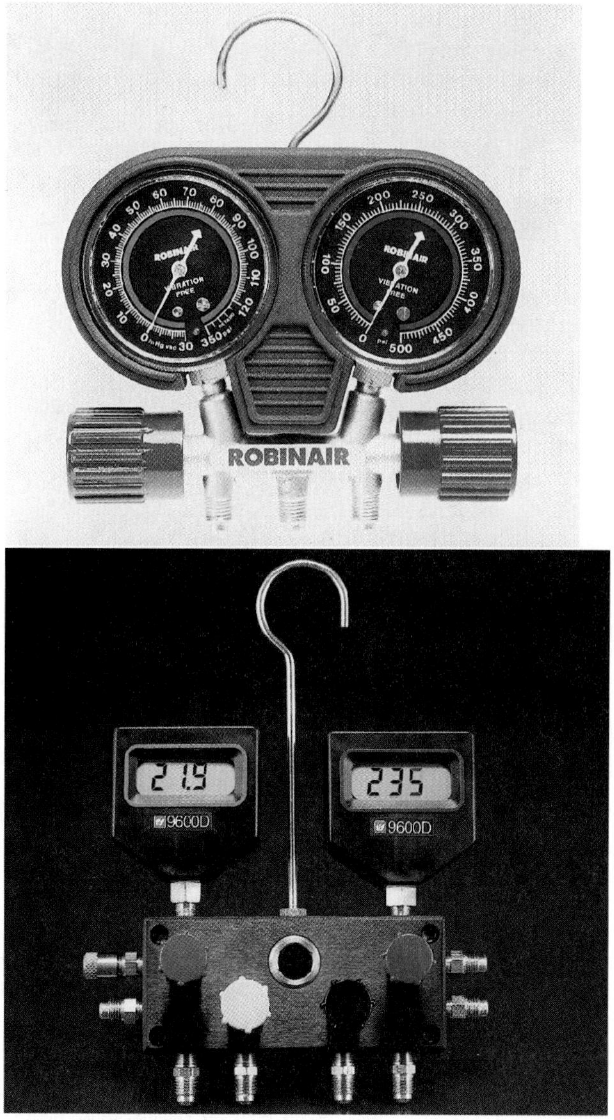

Figure 2-80. *High-pressure gauges. A—Manifold gauge set used for all existing refrigerants and for new substitutes and blends. (Robinair Division, SPX Corporation) B—Two-way digital manifold gauge set with LCD. (TIF Instruments, Inc.)*

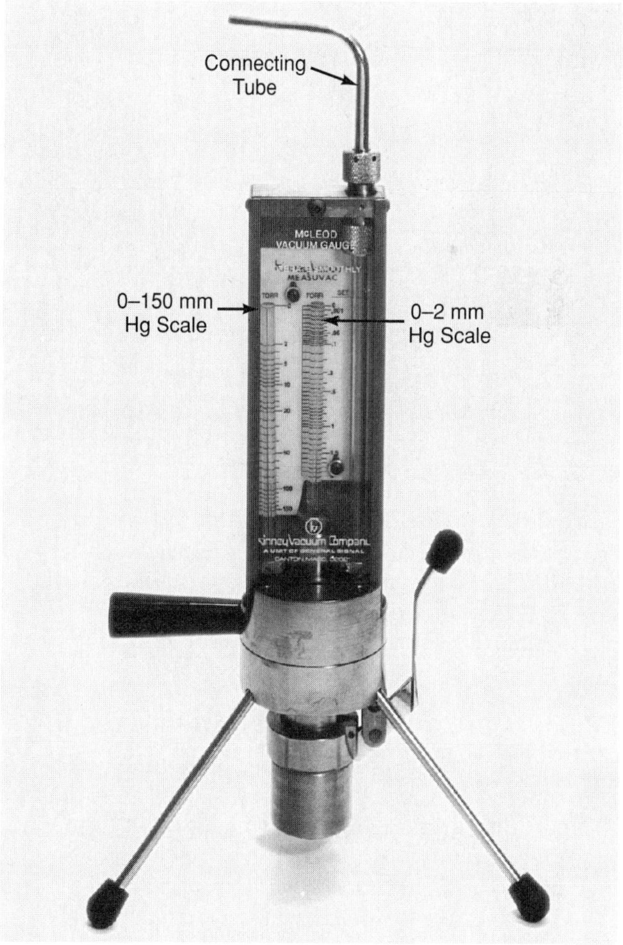

Figure 2-81. *A very high vacuum gauge using the McLeod principle. It can measure from 150 torr (150 mm) to 1 millitorr.*

Compound Gauges

The compound gauge, **Figure 2-82,** measures both pressure and vacuum. It is usually calibrated from 0 to 30″ Hg and from 0 psi to 240 psi.

Some compound gauges have scales calibrated according to the evaporating temperature of various common refrigerants. With these extra scales, it is unnecessary to refer to pressure-temperature tables and curves in order to check for correct pressure-temperature relationships.

A 30″ Hg, 0 psi–240 psi gauge should be used when connecting to a refrigerating mechanism in which the high pressure may back up through the compressor or balance through the refrigerant control while the compressor is stopped. Never use compound gauges continuously on the high pressure side of the system.

Figure 2-82. *Compound gauge with scale of 0 psi to 240 psi. Note the scale is shortened between 100 psi and 240 psi. This is accomplished by use of retarder spring. A—1/8″ npt connection. B—Calibration adjustment. (TIF Instruments, Inc.)*

2.10.3 Care and Calibration of Gauges

Rapidly changing fluctuating pressures quickly destroy gauge accuracy. A sudden release of high pressure (300 psi) into a gauge also may damage it. Sometimes it is necessary to connect a gauge into a rapidly fluctuating pressure condition. If so, it should be attached through a connector having a very tiny bore. This will help to dampen (choke) the pressure fluctuations entering the gauge. Some gauges are filled with liquid, which tends to prevent rapid fluctuations in the instrument.

Gauges that are used in refrigeration work must be accurate. When it is time for a periodic recalibration of gauges, instruments are available to do this accurately. Any shop that uses a large number of gauges or is remotely situated should have one. Such instruments usually use "dead weights" for calibrations above atmospheric pressure, and a mercury column for pressures below atmospheric or vacuum. Gauges should be checked over their full operating range or scale.

When checking gauge accuracy, remember that calibrating equipment is made to show a "0" reading at sea level. A gauge calibrated on equipment so adjusted will not be accurate at either above or below sea level.

Figure 2-83 shows change in atmospheric pressure with altitude. Any gauge on which a "0" reading means atmospheric pressure must be adjusted for elevation of use. This includes pressure, vacuum, and compound gauges. To make the adjustment, disconnect the gauge so that it is open to the air. Then, set the needle at 0. The pressure recorded by the gauge will then be accurate enough for the technician to use. Any gauge that shows absolute pressure (psia, etc.) should not be adjusted for elevation.

A compound gauge is accurate to about 1 psi or 2″ Hg. A mercury manometer is accurate to 1 mm of mercury (1000 microns).

There are many dial scales in use; some common ones are shown in **Figure 2-84.** You must be careful when reading a pressure gauge to use the correct scale spacings and values.

Figure 2-85 shows a compound gauge that can be used for various types of refrigerants. To speed installation of gauges, a quick coupler system may be used.

Elevation	Pressure
Sea Level	14.7 psi
2000 ft.	13.7 psi
4000 ft.	12.9 psi
5000 ft.	12.2 psi

Figure 2-83. *Atmospheric pressure change with altitude.*

2.11 Measuring Tools

Measuring tools used by technicians include rules, tapes, and micrometers. Even though rules and tapes can assure accuracy up to 1/32″ increments, micrometers can provide more accurate readings.

2.11.1 Rules and Tapes

A 9″ or 12″ stainless steel rule is frequently needed when overhauling or installing units. The rule should be graduated in increments of 1/32″. Numerals and graduations should be clearly visible. Installation workers will find a 6′ flexible steel tape useful when laying out a job.

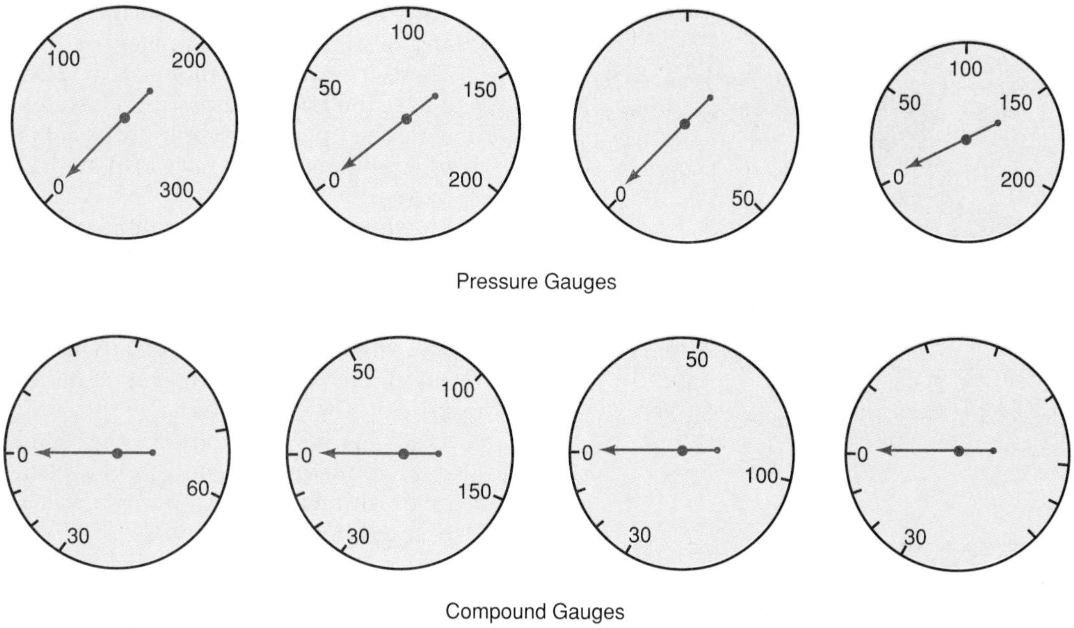

Figure 2-84. *Some common pressure gauge dials. Pressure gauges are shown at the top. Compound gauges are shown at the bottom.*

Figure 2-85. *Compound refrigeration gauge shows evaporating temperature for R-12, R-22, R-502, or R-134a. Gauge reads from 0 to 500 or 160 to 210, depending on the type of refrigerant. (Robinair Division, SPX Corporation)*

2.11.2 Micrometers

Often, the technician must check sizes and make accurate measurements to a few thousandths of an inch or hundredths of a millimeter. The common micrometer is commonly used for this purpose.

Micrometers are available in both U.S. conventional and SI metric units. The U.S. conventional micrometer is calibrated in inches and decimals of an inch. The metric micrometer is calibrated in millimeters and decimals of a millimeter.

English Micrometers

Figure 2-86 illustrates a 1-inch micrometer caliper. When reading it, be careful to observe the micrometer calibration range. Note that a 1-inch micrometer, **Figure 2-86,** has a range of 0 to 1″. A 2-inch micrometer has a range of 1″ to 2″, and so on. If a 3-inch micrometer is used, 2″ must be added to the micrometer reading. This is because when the gap is set to 2″, the reading is 0.

A micrometer is not a heavy-duty tool, so do not tighten it like a vise. Simply close the gap on the piece being measured until the slightest resistance is felt. Then, read the dimension.

The following points may help the beginner learn to read the micrometer:

1. The divisions on the sleeve are tenths of an inch (.1″).
2. The four divisions between the tenths markers are 1/40″ or .025 (twenty-five thousandths of an inch).
3. The thimble (right end of the micrometer) carries the spindle. It is threaded into the micrometer body on a 40 thread-per-inch screw. One turn of the thimble moves the spindle twenty-five thousandths of a inch (.025″).
4. There are 25 graduations on the thimble. Turning the thimble one graduation moves the spindle .001 (1/1000) of an inch.
5. The micrometer, **Figure 2-86,** reads .200 inch. The inserts (portions of a micrometer) read as noted in the caption.

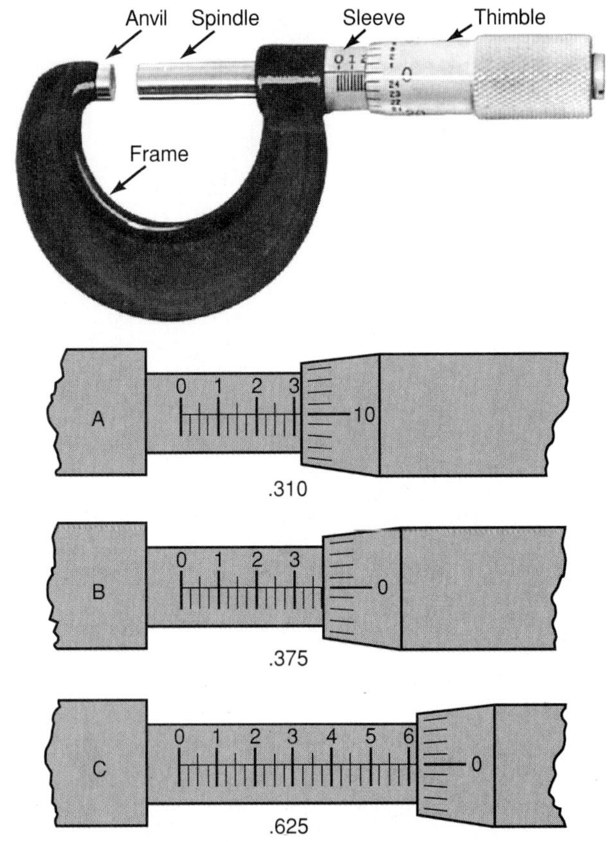

Figure 2-86. *Reading a micrometer caliper. Can you tell why the reading in the photograph is .200? Note drawings A, B, and C to see how readings are made. Answers are in thousandths of an inch.*

Metric Micrometers

Figure 2-87 shows a metric micrometer. The calibration range is 0 to 25 millimeters. Twenty-five millimeters are almost one inch (1 inch = 25.4 mm).

The following points may help the beginner learn to read the metric scale:

- The metric micrometer is calibrated in millimeters and decimals of a millimeter.
- There are two calibrations on the sleeve (see **Figure 2-87**). The lower calibration (A scale) is in millimeters. The upper calibration (B scale) is one-half (0.5) of a millimeter. The lines of B are halfway between those of A.
- The thimble (C scale) is calibrated in hundredths of a millimeter. There are 50 graduations on the thimble.
- The thread pitch on the metric micrometer spindle is 2 threads per millimeter. This means that the spindle moves a one-half (0.5) millimeter for each turn.
- There are 50 divisions on the thimble. Since the thimble moves a .5 mm in one turn, one division on the thimble (1/50) moves the spindle 1/2 of 1/50 of a millimeter. This equals 1/100 (0.01) of a millimeter.
- When reading this type of metric micrometer, it is not necessary to observe the micrometer calibration range. The lower calibration on the sleeve starts with a millimeter reading corresponding to the gap between the anvil and the spindle. As an example, the 0 mm to 25 mm range starts with 0 on the sleeve; the 25 mm to 50 mm range starts with 25 on the sleeve; and the 50 mm to 75 mm range starts with the number 50 on the sleeve.

SUPPLIES AND USE MODULE

2.12 Fastening Devices

Many items are made in one piece today that were considered impossible to fabricate a short time ago. However, some mechanisms must still be made of

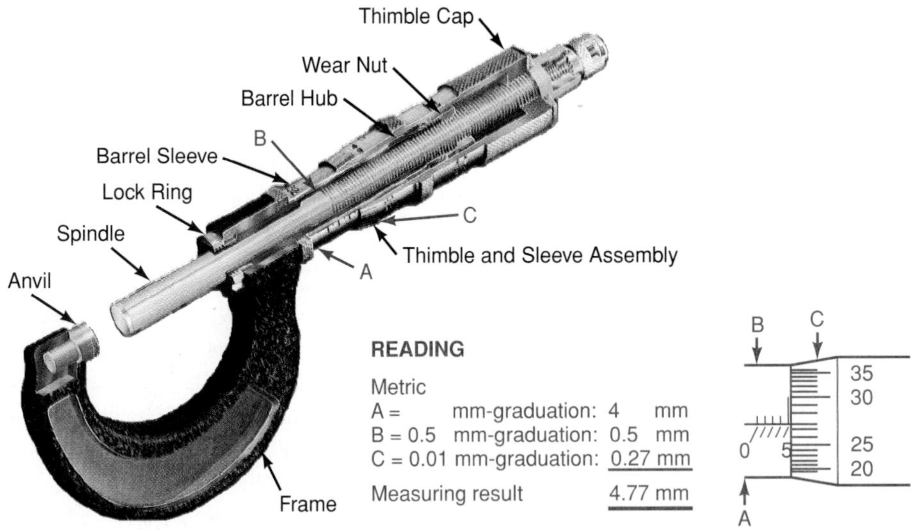

Figure 2-87. *A 0–25 mm micrometer. The reading as shown in the insert is 4.77 mm. (S-T Industries, Inc.)*

several pieces and then assembled. If there is motion in the mechanism—such as a piston in a cylinder—the apparatus must be made of two or more pieces.

Various techniques have been developed to fasten pieces together. In metal work, soldering, brazing, welding, crimping, rivets, bolts, machine screws, pins, spring fasteners, and force fits have been used with success.

The type of assembly device used depends first on the kind and condition of the metal, and secondly on how frequently the pieces must be dismantled. If the parts are to be put together permanently, riveting, welding, soldering, and brazing are popular assembly methods.

If the parts must be dismantled for frequent repair or service, assembly devices must be used that can be easily removed without injuring the parts. Nuts and bolts, cap screws, machine screws, and set screws are then used. **Figure 2-88** shows an assortment of fastening devices. In the metric system, the diameter of fastening sizes is specified in millimeters.

2.12.1 Machine Screws, Bolts, and Cap Screws

Many small parts are assembled using specially threaded devices called ***machine screws.*** Machine screws are made of steel, stainless steel, brass, Monel metal, or other materials. They are available in a variety of head shapes. Various methods are used to turn these screws. See Figures 2-55, 2-60, and 2-88. Some less commonly used screw heads are shown in **Figure 2-89.**

Machine screws come in various diameters. Eight are in the numbered sizes, while three are in the fractional-inch sizes. Each size may have either fine or coarse threads; the larger the number, the larger the diameter. A table of machine screw sizes and threads is given in **Figure 2-90.**

In general, bolts and cap screws are used in sizes 1/4″ and larger. The length of threads on a bolt is usually two times the bolt diameter. Most bolts are provided with nuts.

The threading on a cap screw is usually longer than the threading on a bolt. Threads sometimes extend up to the head of the cap screw. Cap screws are threaded into a part of the mechanism and do not require a nut.

Metric screws can have four thread types, known as coarse, average, fine, and extra fine threads. The popular diameters and thread pitch sizes are shown in **Figure 2-91.**

Diameters of metric bolts, nuts, and screws, as well as the thread pitches, are in millimeters. For example: 10 mm diameter with a 1.50 mm pitch. The wrench size is the same as the size given for the diameter of the bolt. A 10 mm wrench fits a 10 mm bolt.

2.12.2 Gaskets

Most mating surfaces are somewhat rough. To make a leakproof joint, ***gaskets*** are often used between surfaces. Gaskets, being soft, seal joints between assembled

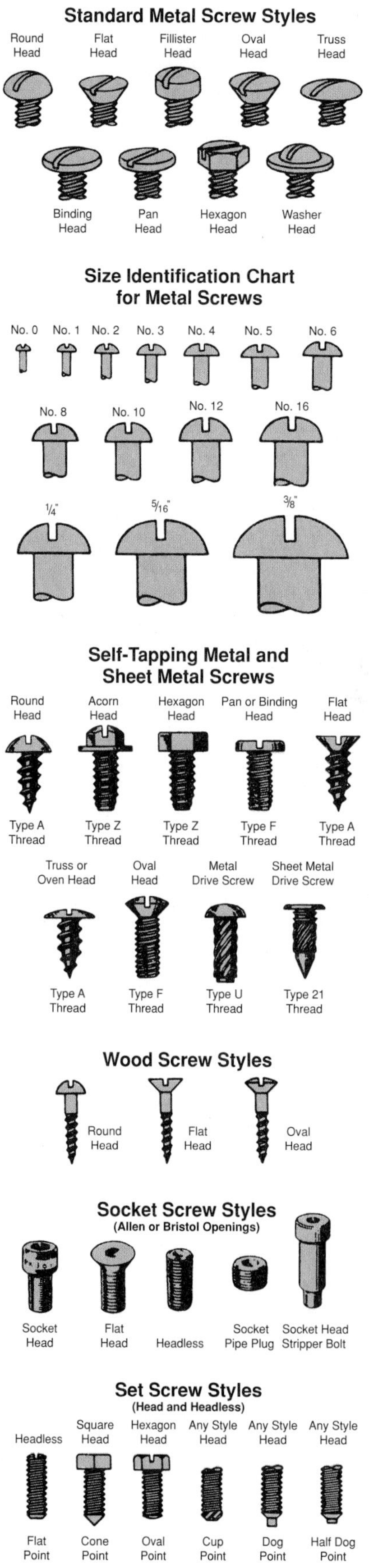

Figure 2-88. *Fasteners must be carefully used and driven with proper tools. (Klein Tools, Inc.)*

Figure 2-89. *Less commonly used screw heads.*

Machine Screw Number	Diameter (in.)	Threads per Inch	
		Coarse	Fine
2	.086	56	64
3	.099	48	56
4	.112	40	48
5	.125	40	44
6	.138	32	40
8	.164	32	36
10	.190	24	32
12	.216	24	28
1/4	.250	20	28
5/16	.3125	18	24
3/8	.375	16	24

Figure 2-90. *Common machine screw sizes.*

Diameter (mm)	Pitch (mm)	Wrench Size (mm)
3	0.60	3
4	0.70	4
4	0.75	4
5	0.80	5
5	0.90	5
6	1.00	6
7	1.00	7
8	1.00	8
8	1.25	8
9	1.00	9
9	1.25	9
10	1.25	
10	1.50	
11	1.50	
12	1.25	
12	1.50	
12	1.75	

Figure 2-91. *Table of metric screw information.*

parts. They keep the refrigerant from leaking out, prevent oil leakage, and keep air out of the system. They are used between the valve plate and the compressor body, between the service valve and the compressor body, and between the valve plate and the compressor head. Gaskets are also used on the crankcase and at the crankshaft seal on open or external drive units.

Metal is the most common gasket material. Lead is popular, being soft and noncorrosive. Aluminum has also been used. Composition gaskets made of plastic-impregnated paper are also popular.

Gaskets must not restrict the openings. They must not lose their compressibility. Replacement gaskets must not be thicker than the original gaskets. The surfaces of parts that contact the gasket must be free of burrs, bruises, and foreign matter.

2.13 Refrigeration Supplies

Space prevents giving specifications for all refrigeration supplies. However, a few basic items will be named and some information given about their specifications, handling, and use.

2.13.1 Abrasives

Surfaces may be cleaned, smoothed, or made to accurate size with abrasives. Abrasives are sand-like grinding particles often attached to paper or cloth by glue or other adhesives.

Sandpaper was the only abrasive for many years. It is still excellent for wood finishing or where a dry surface is wanted. Today, emery, aluminum oxide, and silicon carbide are also commonly used.

Each abrasive has several grades or variations in coarseness.

- **Emery cloth:** 0000 (finest), 000 (extra fine), 00 (very fine), 0 (fine), 1/2 (medium fine), and 1 (medium).
- **Silicon carbide:** 500 (finest), 360 (very fine), 320 (fine), 220 (medium fine), and 180 (medium).
- **Aluminum oxides:** 320 (extra fine), 240 (fine), 150 (medium fine), and 100 (medium).

These abrasives come in 9″ × 11″ sheets or in rolls usually 1″ wide.

Sheet abrasives, whether paper or cloth, should be backed by a block of wood, metal, felt, or rubber. Special sanding blocks may also be used. Always use clean abrasive paper.

2.13.2 Brushes

A clean steel wire brush is an excellent tool to prepare copper and steel surfaces for welding or brazing. These brushes can be bought in a variety of shapes and sizes. They should have fine steel wire bristles, which are thickly set. The handle should be comfortable. Special cylindrical brushes are good for cleaning outside and inside surfaces of tubing and fittings. **Figure 2-92** shows a cylindrically shaped wire brush. They are available in all sizes.

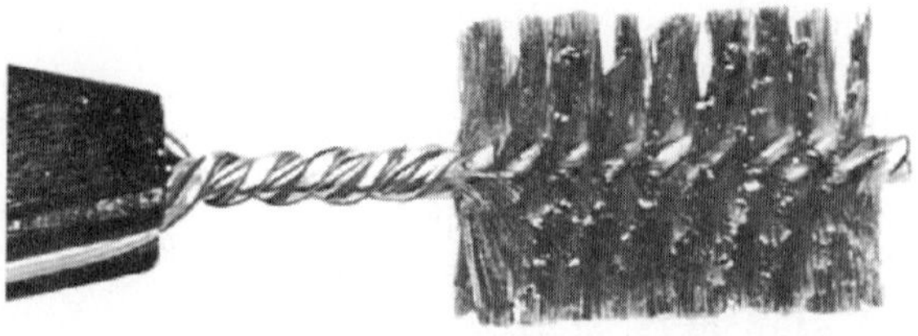

Figure 2-92. *Wire brush used for cleaning inside surfaces of fittings before soldering or brazing. (Schaefer Brush Mfg. Co., Inc.)*

2.13.3 Cleaning Solvents

Any refrigeration mechanism must be thoroughly cleaned before and after repair. Many methods have been used. Some do not do a thorough job; some are dangerous.

Any method must remove oil, grease, and sludge. In refrigeration and air conditioning, the cleaning method must remove moisture, or at least it should not add moisture. It must not damage the parts nor harm the user. There are several cleaning methods:

- **Steam Cleaning.** If the parts are exposed to hot water or steam, the grease will usually become fluid and float off the surface. Steam and hot water may burn the user if they are carelessly used.
- **Caustic Solution Cleaning.** An alkaline cleaner dissolved in hot water will remove grease and oil. This solution must be carefully used or the user may suffer burns or eye injury.
- **Oleum or Mineral Spirits.** This petroleum product is popular for cleaning. It has a flash point of approximately 140°F (60°C). Kerosene has a flash point of 130°F (54°C). It cleans well and leaves a smudge-free surface. However, it presents a fire hazard and should always be used in small amounts. It should be contained in self-closing tanks. The area should be exhaust-ventilated (have a hood and an explosion-proof exhaust fan).
- **Refrigerant R-11.** This refrigerant is a good cleaning solvent for flushing systems contaminated by hermetic motor burnout. R-11 has a high boiling point, 74.8°F (24°C). It is nontoxic and nonflammable. It leaves no noncondensable residue and has no reaction with the electrical insulation. R-11 refrigerant can be filtered and reused.
- **Alcohol.** Alcohol is also a good cleaning fluid. However, it is both flammable and toxic. Special precautions must be taken: excellent ventilation, no open flames, use in small amounts.
- **Vapor degreasing.** This is a system using a cleaning fluid contained in a tank. The fluid is warmed, filling the upper part of the tank with vapors of the cleaner. Any parts suspended in this cleaning vapor are quickly and thoroughly cleaned. Such a tank must be specially vented.
- **Patented cleaning fluids.** When these are used, one should read and carefully follow the manufacturer's instructions.

Carbon tetrachloride should never be used in cleaning refrigeration or air conditioning mechanisms. It is toxic, and can be absorbed through the respiratory system or the skin.

Never use gasoline for cleaning. It has a low flash point. Gasoline fumes are heavy and may travel far to ignition sources causing an explosion or flash fire. Do not use a propane or LP cylinder to clean parts. The liquid or gas will quickly evaporate and may burn or explode.

Refer to Chapter 31 for more information on cleaning.

2.13.4 Refrigerants

The refrigeration service technician is required to handle refrigerants and charge them into refrigerating mechanisms. Several different refrigerants are carefully described in Chapter 9 along with safe handling methods.

Refrigerants must be kept *dry* and *clean*. Remember, all exposed surfaces absorb moisture if left in the open. If a compressor is torn down, overhauled, and reassembled, it must first be completely dried before it can be charged with refrigerant. See Chapter 12 for detailed instructions.

Refrigerants are stored and handled by the service technician in portable refrigerant cylinders. Refrigerants are identified by a cylinder color code. See Chapter 9 for a table of color codes for common refrigerants.

Cylinders for different refrigerants must not be interchanged. Refrigerants should always be stored in the cylinder specified (color code).

Never fill refrigerant cylinders over 80% of capacity. With a temperature increase, hydrostatic pressure may burst the overfilled cylinder.

Do not vent refrigerant directly into the atmosphere.

2.13.5 Refrigerant Oil

In the mechanical refrigerating system, moving parts must be lubricated with oil for long life and efficient performance. There are a variety of refrigerant oils, including mineral oils, polyol ester-based oils, alkyl benzene, and polyalkylene glycol (PAG). The type of oil used must match the type of refrigerant used. The new azeotropic mixtures and refrigerant 134a use polyol ester-based oils. The traditional refrigerants require mineral oil. Never mix different types of oil within a system.

Use oils that have a low ***pour point*** (temperature at which the oil begins to flow). This will avoid wax separation at the lowest temperature in the system. Wax could clog the refrigerant control orifice.

Due to the low temperatures at which they operate, food freezer and frozen food units need oils with extra-low pour points and a very low wax content. The oil cannot have any hydrocarbons of the type that may collect on the compressor valves or other parts.

The viscosity of the oil must be accurately determined for the temperature ranges to which the refrigerating system may be exposed. See Chapter 9 for further information. When possible, follow the manufacturer's recommendations. See Chapter 31.

Refrigerant oil must be kept in sealed containers. It must be transferred in chemically cleaned containers and lines, and it must not be exposed to air where it will absorb moisture. When refilling a refrigerant, always use new oil. To retrofit a system and change oils, see Chapter 10 for step-by-step procedures.

Refrigeration oil comes in one- or five-gallon cans and in barrels. It is advisable to purchase it in small sealed containers, holding only enough for each separate service operation. Unused oil allowed to remain in the container or oil transferred from one container to another may pick up some moisture, and perhaps even dirt.

Special pressure pumps may be used to pump oil into the low side of the system.

The ***pour point*** of any oil is the temperature at which it starts to flow. The price of refrigeration oil varies with the grade. A low pour point oil is more expensive.

Domestic machines with refrigerant temperatures as low as 0°F to 5°F (−18°C to −15°C) need oil with a pour point of −20°F (−29°C). For food freezers with refrigerant temperatures as low as −50°F (−46°C), a pour point of −60°F (−51°C) is desirable. Always seal an oil container after drawing oil from it.

2.13.6 Epoxy Resin

Epoxy resin may be used to repair cracks and leaks in evaporators and joints. Such resins have good adhesion (sticking) qualities when used with steel, copper, wood, and many plastics. Epoxy adhesives are available from many manufacturers or refrigeration supply wholesalers.

The most desirable type of epoxy resin is the two-part system. This consists of an epoxy resin and a hardener, in two jars or tubes.

These two, paste-like substances will harden at room temperature when mixed together. The one-part adhesive must be heated in order to harden it.

When repairing tubing, first determine the size of the leak. Small leaks or holes up to 1/16″ in diameter can often be successfully sealed by placing the mixed epoxy over the leak and allowing it to cure. The same procedure is recommended for small tubing cracks. For larger holes, a patch of the same type of tubing material is recommended.

The shelf life of most epoxy resins is about six months. Epoxy compounds should be purchased from a refrigeration wholesaler because some epoxies available elsewhere are not compatible with refrigerants R-12 and R-22.

Care must be taken when using epoxy compounds because they contain chemicals which may irritate the skin. Long contact with the skin should be avoided. In case of contact, remove the epoxy and clean the skin with rubbing alcohol. Then wash thoroughly with soap and water.

The procedure for using epoxy compound is as follows:

1. Obtain a two-part epoxy kit, **Figure 2-93.**
2. Clean the surface or surfaces to be bonded. Use clean, coarse sandpaper, or scrub the surface with clean steel wool.
3. Clean the surface with a suitable recommended solvent such as methyl ethyl ketone, toluene, acetone, or a similar industrial solvent. **Obey all solvent safety rules.**
4. Mix together equal parts of resin and hardener on a clean surface such as a piece of cardboard. Keep mixing until the mixture has a uniform color.
5. Apply the epoxy mixture to the surface if it has only a small hole. Apply to mating surfaces if a patch is to be used as shown in **Figure 2-94.** Epoxy compounds should be used immediately after mixing since chemical hardening or setting starts immediately.

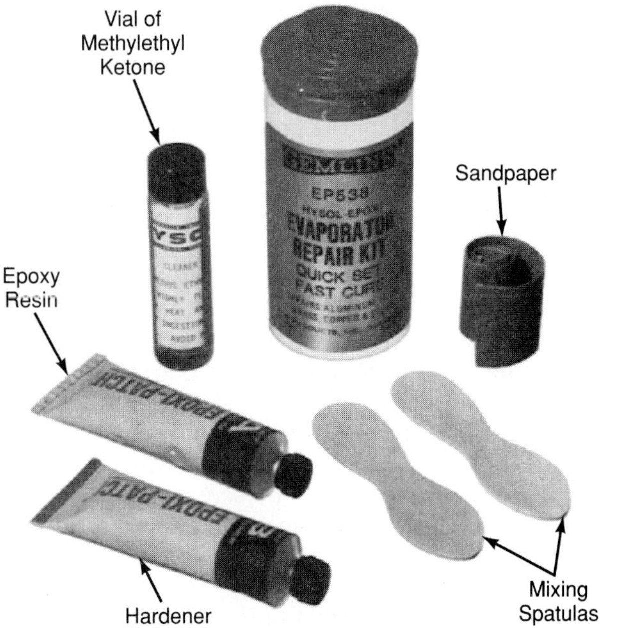

Figure 2-93. *Evaporator repair kit. (Gem Products, Inc.)*

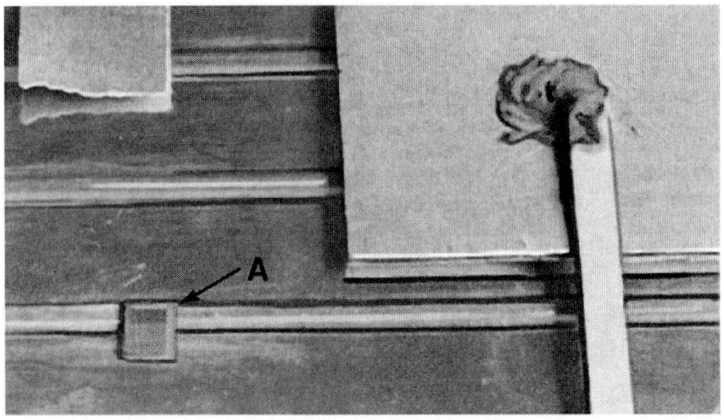

Figure 2-94. *Epoxy compound is used to join the patch material to the damaged part. A—Completed patch.*

2.14 Service Valves

Service technicians must be familiar with manual valves in refrigerating systems. These valves enable them to seal off parts of the system while installing gauges, recharging, or discharging the system.

Several kinds of manual or hand valves are used. Such valves may have handwheels on their stems, but most are made so that a valve wrench is needed to turn them. Valve stems are made of steel or brass. The body of the valve is usually made of drop-forged brass. Packing is installed around the valve stem. A packing adjusting nut keeps the joint from leaking.

One-way and two-way service valves are commonly used. ***One-way service valves*** have only one opening which can be either opened or closed. The ***two-way service valve*** has two openings. One may be open while the other is closed, or both may be open.

The two-way valve usually closes or shuts off the refrigerant flow in the system when the stem is turned all the way in (clockwise). It shuts off the charging, discharging, or gauge opening when the valve stem is turned all the way out (counterclockwise). When the valve stem is turned part way, both of the openings allow the fluid (refrigerant) to flow through, **Figure 2-95.**

The tubing or pipe is fastened to a valve by flare connection or by brazing. The valve may also be attached to the refrigerating mechanism either by pipe threads or by bolted flanges.

It is good practice to open a valve by first "cracking" it (opening it 1/16 or 1/8 turn). This prevents a shock pressure rush which may damage mechanisms, gauges, flush oil in abnormal amounts, or injure the technician.

Be sure the valve stem is clean before turning it in (clockwise). A scarred or dirty valve stem will ruin the valve packing. See Chapter 15.

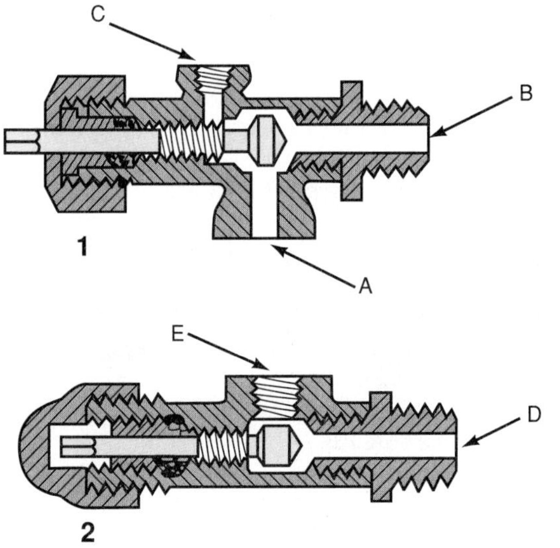

Figure 2-95. *Two typical service valve designs. No. 1 is a two-way valve. A—Opening to compressor. B—Opening to the refrigerant line. C—Service opening. No. 2 is a one-way valve. D—Opening to the liquid line. E—Opening to the liquid receiver. Valve No. 1 has an open valve stem. Valve No. 2 has a cap over the valve stem.*

2.15 Purging

Purging means the process of removing unwanted air, vapors, dirt, or moisture from the system. A neutral gas or the recommended refrigerant flows through the parts, forcing unwanted air and vapors out. **Never purge or vent a refrigerant directly to the atmosphere.**

2.16 Evacuating

A refrigerating system must contain only the refrigerant in either liquid or vapor state along with dry oil. All other vapors, gases, and fluids must be removed.

These substances can be removed best by connecting the system to a vacuum pump, **Figure 2-96,** and allowing the pump to run continuously for some time while a deep vacuum is drawn on the system. There are numerous types of vacuum pumps, depending upon the application. Residential units will utilize a smaller system than those used for commercial units.

It is sometimes necessary to warm the parts to 120°F (49°C) while under a high vacuum in order to remove all unwanted moisture. Heat the parts using warm air, heat lamps, or warm water. **Never use a torch!**

Figure 2-96. *Heavy-duty portable vacuum pump. Note offset rotary vanes used as a means of compression. Unit is capable of producing a vacuum of 20 microns, and is usable in air conditioning and refrigeration systems using CFC, HCFC, or HFC refrigerants, in conjunction with mineral oil, polyol ester oil, PAG oil, or alkylbenzene oil as lubricants. (Robinair Division, SPX Corporation)*

2.17 Review of Safety

Before working with tools, review the safety steps.

1. **Tubing should be bent in as large a radius as possible.**
2. **"Mushroom" heads should be ground off from chisels and punches. Fragments may fly off when struck with the hammer, causing serious injury.**
3. **Files should never be used without handles. The tang may injure the hands.**
4. **Wear goggles when drilling. Chips may fly.**

5. Emery cloth should not be used to clean tubing to prepare for soldering. It may leave an oily deposit. Also, the hard grit could cause considerable damage if allowed to enter the refrigerating mechanism.
6. When pressure testing for leaks in tubing circuits, use low or medium pressure carbon dioxide or nitrogen. Never use oxygen. Caution: see Chapter 12.
7. Carbon tetrachloride should not be used for any purpose as it is toxic and can be absorbed through the skin.
8. Brazing materials sometimes contain cadmium. Fumes from heated cadmium are poisonous. Be sure that the work space is well ventilated. If possible, use brazing alloys which *do not* contain cadmium.
9. It is recommended that refrigerant cylinders never be filled above 80% of their capacity. If overfilled, hydrostatic pressure may cause them to burst.
10. Wrenches used on refrigeration line fittings should always fit snugly. Poorly fitted wrenches will ruin nuts and bolt heads. Also, the wrench may slip and cause an injury to the technician.
11. Always "crack" service valves and cylinder valves before opening. This gives quick control of the flow of gases if there is any danger.
12. Moisture is always a hazard to refrigerating mechanisms. Keep everything connected with a refrigerating mechanism thoroughly dry.
13. Never purge refrigerant directly to the atmosphere.
14. Epoxy bonding materials may irritate the skin or many membranes of the user.

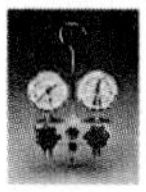

2.18 Test Your Knowledge

Please do not write in this text. Place your answers on a separate sheet of paper.

TUBING AND FITTINGS MODULE

1. A ______ fitting is the best type of fitting for use with polyethylene tubing.
   A. push-on-type barb
   B. flare
   C. compression-type
   D. solar-type brass
2. When cutting copper tubing with a hacksaw, what precautions must be taken?
   A. Cut squarely at 90° and remove all sharp burrs.
   B. Do not let chips or filings get into the tubing.
   C. Close off ends of unused part of the tubing.
   D. All of the above.
3. What is the OD of 1/4" refrigeration tubing?
   A. 0.250"
   B. 0.280"
   C. 0.310"
   D. 0.375"
4. Why do some technicians file the ends of copper tubing?
   A. To remove burrs.
   B. To restore full wall thickness.
   C. Both A and B.
   D. Only steel tubing should be filed.
5. How can an internal bending spring be removed from a flared tube after making a bend?
   A. Bend the tube before making the flare.
   B. Grease the spring before insertion.
   C. Twist the spring.
   D. Only an internal spring should be used.
6. Before soldering metal, are both cleaning and fluxing necessary?
   A. No. Only fluxing is necessary.
   B. No. If the metal is clean, it does not need flux.
   C. Yes. First clean the surface and then add flux.
   D. Yes, except for aluminum. It can be soldered without flux.
7. Soldered metal must be ______ when soldering.
   A. about 1150°F
   B. hot enough to anneal the metal
   C. hot enough to make the solder fully liquid
   D. hot enough to drive the water out of the flux
8. Why must a joint be cleaned after brazing?
   A. It only needs cleaning before brazing.
   B. It only needs cleaning if acid flux is used.
   C. To remove flux, which may corrode tubing.
   D. To remove excess brazing alloy.
9. What indicates that the correct brazing temperature has been reached?
   A. The metal surface turns green.
   B. A green color appears in the flame.
   C. The brazing alloy melts when held in the flame.
   D. The flux bubbles.
10. When mating a steel pipe with a male NP thread to a fitting that has an NF internal thread, what precautions are necessary?
    A. Use pipe compound, and insert for a distance of five threads.
    B. Make sure the sizes are equal i.e.: 3/4" NP and 3/4" NF.
    C. Both A and B.
    D. Do not mate NP to NF.

REFRIGERATION TOOLS MODULE

11. A(n) ______ wrench should be used on a valve stem.
    A. Allen wrench
    B. miniature pipe
    C. service valve
    D. Torx® key
12. When should you push and pull on a wrench?
    A. Push to loosen, pull to tighten.
    B. Pull to loosen, push to tighten.
    C. Always push.
    D. Always pull.

13. What tool should *not* be used on a slotted hexagon head bolt?
    A. Straight screwdriver.
    B. Nut pliers.
    C. Hex key wrench.
    D. Twelve-point socket.
14. Which tool would be easiest to use when operating in a small or restricted space?
    A. Six-point socket wrench.
    B. Twelve-point box and wrench.
    C. Metric wrenches only.
    D. Nut driver.
15. Which of the following would *not* describe a box wrench?
    A. Straight.
    B. Offset.
    C. Double offset.
    D. Open ended.
16. Which of the following is *not* a popular size for open ended wrenches?
    A. 7/16″.
    B. 1/2″.
    C. 3/16″.
    D. 5/8″.
17. How much torque will be produced if a 2′ long wrench handle is pulled by a spring scale reading 40 lbs.?
    A. 80 ft.lbs.
    B. 80 in.lbs.
    C. 960 ft.lbs.
    D. 960 in.lbs.
18. A double cut file is a file ______.
    A. with teeth on both sides
    B. that has no safe edge
    C. that cuts on both the forward and back strokes
    D. with teeth cut in two directions
19. Which of the following is *not* a common type of punch?
    A. Drift punch.
    B. Pin punch.
    C. Prick punch.
    D. Mushroom punch.
20. A tight nut or bolt may be loosened safely by ______.
    A. soaking the threads with penetrating oil
    B. heating it
    C. tapping it lightly with a hammer
    D. Any of the above.

INSTRUMENTS AND GAUGES MODULE

21. What is the most common way to check a thermometer for accuracy?
    A. Spray on R-134a; it should read −26°C.
    B. Put in ice and salt mix; it should read −21°C.
    C. Put in ice and water mix; it should read within 1°C of 0°C.
    D. Put in mouth; it should read 37°C.
22. A compound gauge is used to ______.
    A. show the highest and lowest temperatures reached
    B. show pressure-temperature relationships
    C. measure pressure (psig) and vacuum (in. Hg)
    D. give accurate readings in the usual range and approximate readings over a wider range
23. Which of the following is the most common tool to use for determining pressure and temperature?
    A. Thermometer.
    B. Pressure gauge.
    C. Vacuum gauge.
    D. None of the above.
24. The thermometer is frequently used to determine the temperature in a(n) ______.
    A. evaporator and condensing unit
    B. refrigerator
    C. liquid liner or suction line
    D. All of the above.
25. The vacuum gauge measures pressures that are lower than atmospheric pressure. Which of the following is *not* a common calibration?
    A. Inches of mercury (in. Hg).
    B. Pounds per square inch (psi).
    C. Pounds per square inch absolute (psia).
    D. Pounds per square foot (psf).
26. A gauge manifold includes a ______.
    A. center gauge
    B. high-side gauge
    C. low-side gauge
    D. high- and low-side gauge
27. A compound gauge shows a change in atmospheric pressure with altitude. At sea level the gauge should be set for ______.
    A. 0
    B. 14.7
    C. 13.7
    D. 12.9
28. Micrometers are used to measure ______.
    A. hundredths of an inch
    B. thousandths of an inch or hundredths of a millimeter
    C. ten-thousandths of an inch or hundredths of a millimeter
    D. Any of the above.
29. Calibrating equipment used to check the accuracy of a gauge is made to show a ______ reading at sea level.
    A. 0
    B. 14.7
    C. 10.5
    D. 15
30. A gauge that has a 0 reading must be reset to ______ when used at high elevations.
    A. 0
    B. 14.7
    C. 10
    D. None of the above.

SUPPLIES AND USE MODULE

31. Why must refrigerant oil be practically wax free?
    A. To prevent foaming.
    B. Wax can clog the refrigerant control orifice.
    C. Wax can cause copper tubing to corrode.
    D. All of the above.
32. What is a very important precaution one must take when filling refrigerant cylinders?
    A. Obey refrigerant color code.
    B. Never fill cylinders over 80% of capacity.
    C. Both A and B.
    D. Never reuse refrigerant cylinders.
33. What could cause a one-way service valve to leak at the valve stem?
    A. A dirty valve stem.
    B. A loose packing adjusting nut.
    C. Valve was "cracked" too much before opening.
    D. Either A or B.
34. When a system is left connected to a vacuum pump for a time to clean it, this procedure is referred to as ______.
    A. evacuating
    B. flushing
    C. purging
    D. vapor degreasing
35. Never fill refrigerant cylinders over ______% of capacity.
    A. 80
    B. 90
    C. 50
    D. 100
36. Purging is a term which describes ______.
    A. leakage of refrigerant into the atmosphere
    B. removing unwanted air, vapors, dirt or moisture from a system
    C. removing excess dirt from the atmosphere
    D. adding refrigerant to a system
37. Which one of the following should *not* be used as a cleaning solvent?
    A. Oleum of mineral spirits.
    B. Alchohol.
    C. Refrigerant R-11.
    D. Carbon tetrachloride.
38. What type of lubricant does R-134a use?
    A. Mineral oil.
    B. Polyol ester-based.
    C. Both A and B.
    D. Neither A nor B.
39. In a two-way service valve ______.
    A. only one opening can be opened
    B. only one opening can be closed
    C. both openings can be opened.
    D. None of the above.
40. Which of the following is used to repair a 1/8" hole in a copper evaporator tube?
    A. Sandpaper and cleaning solvent.
    B. Epoxy resin and hardener.
    C. Copper patch made by cutting a scrap of tubing in half.
    D. All of the above.

Chapter 3

BASIC REFRIGERATION SYSTEMS

Key Terms:

cascade refrigerating system
chemical refrigeration
compound system
dry ice
expendable refrigerant system
humidity
hermetic
sight glass
thermoelectric refrigeration
vapor

Learning Objectives:

By studying this chapter, you will be able to:

- Explain the operation of a simple ice refrigerator.
- Explain how evaporation provides a cooling effect.
- Name the basic mechanical refrigeration systems.
- Explain various applications for mechanical refrigeration systems.
- Describe the operation of various mechanical refrigeration systems.
- Compare compression- and absorption-type systems.
- Discuss refrigeration systems using icemakers and water coolers.
- Explain how a system using an expendable-type of refrigerant works.
- Discuss and compare domestic and commercial refrigeration systems.
- Explain the operation of thermoelectric refrigeration.
- Compare the differences between hot-gas and electric defrost systems.

3.1 Ice Refrigeration

For years, ice (frozen water) was the only refrigerating means available. It is still used in many refrigerating applications. The typical ice refrigerator, **Figure 3-1,** is an insulated cabinet equipped with a tray or tank at the top for holding blocks or pieces of ice (aqua).

Shelves for food are located below the ice compartment. Cold air (green arrows) flows downward from

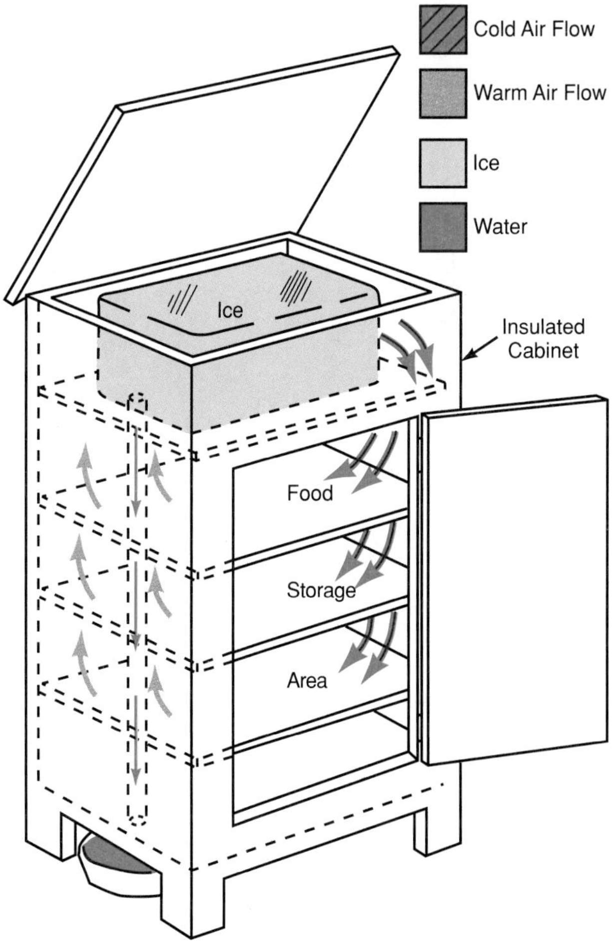

Figure 3-1. *Basic design and operation of an ice refrigerator.*

the ice compartment. It cools the food on the shelves below. The air becomes warmer and rises from the bottom of the cabinet (light red arrows). It travels up the sides and back of the cabinet. Flowing over the ice, it cools and again flows down over the shelves.

Ice refrigeration has the advantage of maintaining the interior of the cabinet at a fairly high ***humidity*** (moisture level). Food stored in this type of refrigerator does not dry out rapidly.

Until the development of the mechanical refrigerator, natural ice refrigeration was quite widely used. Since then, artificial ice has been manufactured for refrigeration. Temperatures inside an ice refrigerator are controlled by air flow. The air flows over the ice and through the cabinet. Temperatures will usually range between 40°F and 50°F (4.4°C and 10°C).

When it is necessary to use ice for cooling temperatures below 32°F (0°C), ice and salt mixtures may be used. Temperatures down to 0°F (−18°C) may be obtained with ice and salt mixtures. See Chapter 31 for a table of ice and salt mixtures.

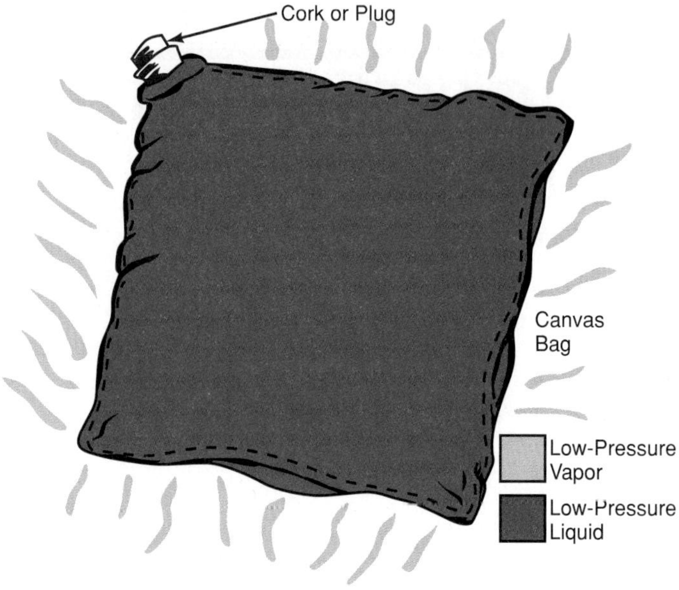

Figure 3-2. *The desert bag is an example of cooling by evaporative refrigeration.*

3.2 Evaporative Refrigeration (Desert Bag)

When a fluid evaporates, heat is absorbed. Evaporation of water is an example. This is why humans and animals perspire. Evaporation of moisture from the skin surface helps to keep a person cool.

Another example of the evaporative principle is the ***desert bag*** used to keep drinking water cool. This bag, **Figure 3-2,** made of a tightly woven fabric, is filled with drinking water. Since the bag is not waterproof, some water seeps through. Thus, the outside surface of the bag remains moist. Desert conditions are usually both hot and dry. Moisture on the surface of the bag evaporates rapidly.

Much of the heat which causes this evaporation comes from the bag and its water. This heat removal cools the drinking water inside the canvas. The water temperature is now several degrees below the temperature of the surrounding air.

3.3 Evaporative Refrigeration (Snow Making)

Another common application of water evaporation refrigeration is the method of making artificial snow for ski slopes. A snow machine, **Figure 3-3,** consists of a water nozzle into which a high-pressure jet of air

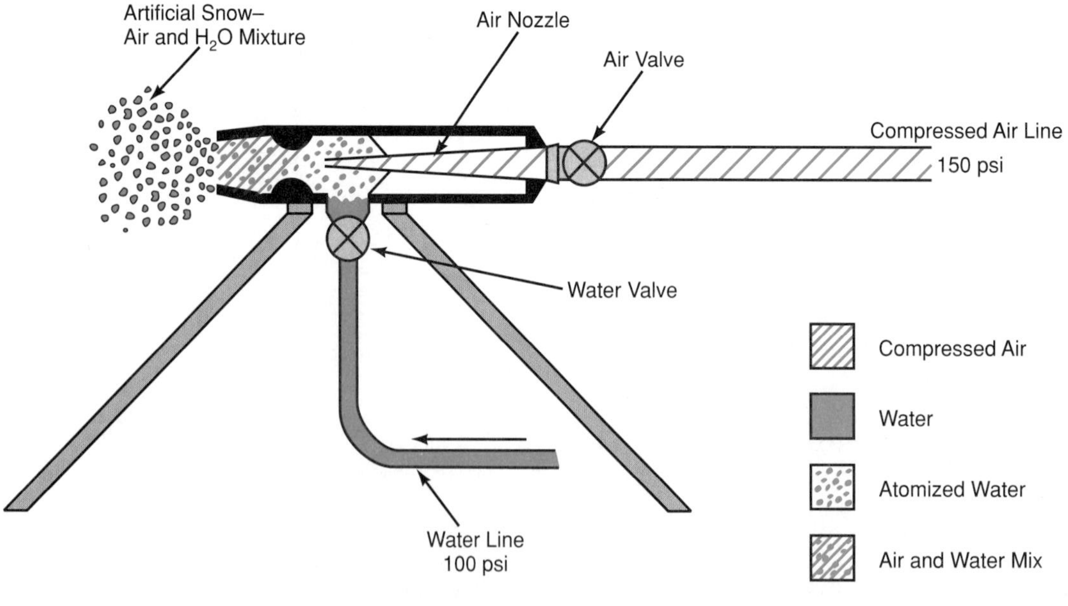

Figure 3-3. *A water/compressed air nozzle is used for making artificial snow.*

is inserted. Water (dark green) flows from the nozzle. The air (green stripe) under high pressure causes the water to break up into tiny droplets. The droplets are similar to a fog. The surrounding air temperature must be near freezing or below freezing for snow to form. The droplets of water will tend to evaporate and rapidly cool. At this point, tiny drops of ice are formed.

Using this method, artificial snow can be made when the temperature of the surrounding air temperature is 32°F (0°C) or lower. In low humidity, artificial snow can be made when the temperature is as high as 34°F (1°C). This is possible because of the rapid evaporation and evaporative cooling caused by the low humidity.

Another example of evaporative cooling is an evaporative condenser. Evaporative condensers are often used in connection with air conditioners. See Chapter 13. The evaporation of water helps cool the condenser.

3.4 Compression System Using Low-Side Float Refrigerant Control

The ***low-side float refrigerant control system*** was often used in early refrigerating mechanisms. It is also known as a ***flooded system.***

Figure 3-4 is a schematic diagram of this system. The liquid refrigerant flows from the liquid receiver through the liquid line. It continues to flow up to the low-side float needle. The evaporator in this system consists of a finned tank (evaporator). The tank contains a float and needle control. These maintain a constant level of liquid refrigerant under a low-side pressure.

This refrigerant, since it is a liquid on the low side, is at a low temperature. The cold liquid refrigerant will absorb much heat in both the on and the off cycles.

Vaporized refrigerant moves through the suction (vapor) line to the compressor. There it is compressed to a high pressure and discharged into the condenser. It is cooled by the condenser, returns to a liquid and flows into the liquid receiver. The operation continues until the desired low temperature is reached.

The pressure on the low side in a flooded system such as this will vary with the temperature. The higher the temperature, the higher the low-side pressure.

The system shown in **Figure 3-4** uses a pressure motor control. A spring-loaded pressure-sensitive device is located on the suction line or on the evaporator. It activates a motor control switch. As the motor drives the compressor, the pressure and temperature in the evaporator will be reduced. At a given pressure setting, the motor compressor will stop.

When the pressure in the evaporator rises to a level corresponding to a preset refrigerant temperature, the cycle will repeat. The motor compressor will then start again.

The cabinet temperature may be controlled by the temperature control switch. In this case, the temperature-sensitive element may be clamped to the fins on the evaporator.

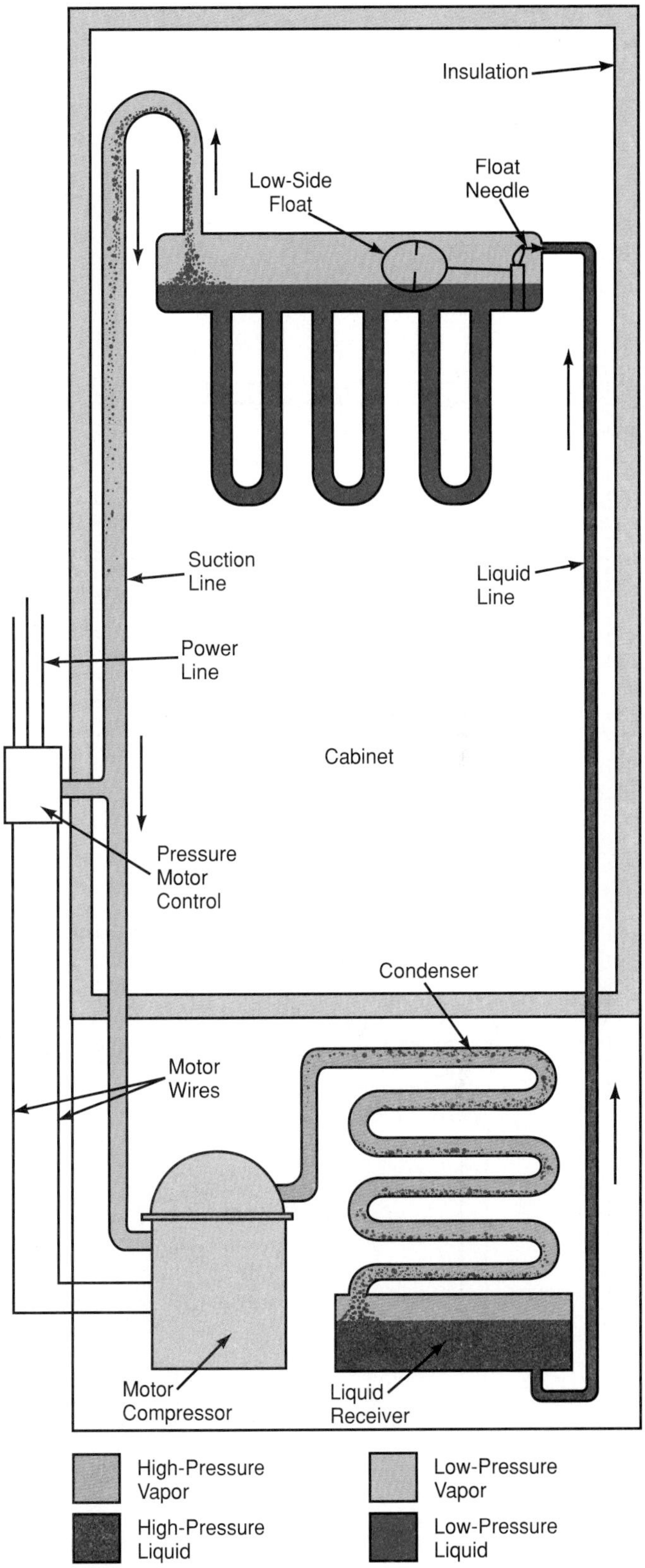

Figure 3-4. *Compression system using low-side float refrigerant control.*

This refrigerating cycle is useful when a constant temperature is desired. It is often used on drinking fountains and other installations requiring a constant temperature.

The pressures do not balance on the off cycle. Therefore, it is necessary to use a motor which will start under a load. Such a system requires a rather large refrigerant charge. This is because there is liquid refrigerant in both the liquid receiver and in the evaporator.

All flooded systems are quite efficient. Cold, liquid refrigerant wets the evaporator surfaces, providing excellent heat transfer. These systems are easy to service. The float needle and seat must be kept in good condition to avoid possible flooding of the low side.

3.5 External-Drive (Open) Refrigerating System

In the ***external-drive system,*** the compressor is usually belt-driven from an electric motor. The speed of the compressor is usually considerably less than the speed of the motor. A small pulley is used on the motor shaft. A larger pulley (flywheel) is used on the compressor shaft. Most early refrigerating systems were like this. **Figure 3-5** illustrates an open system.

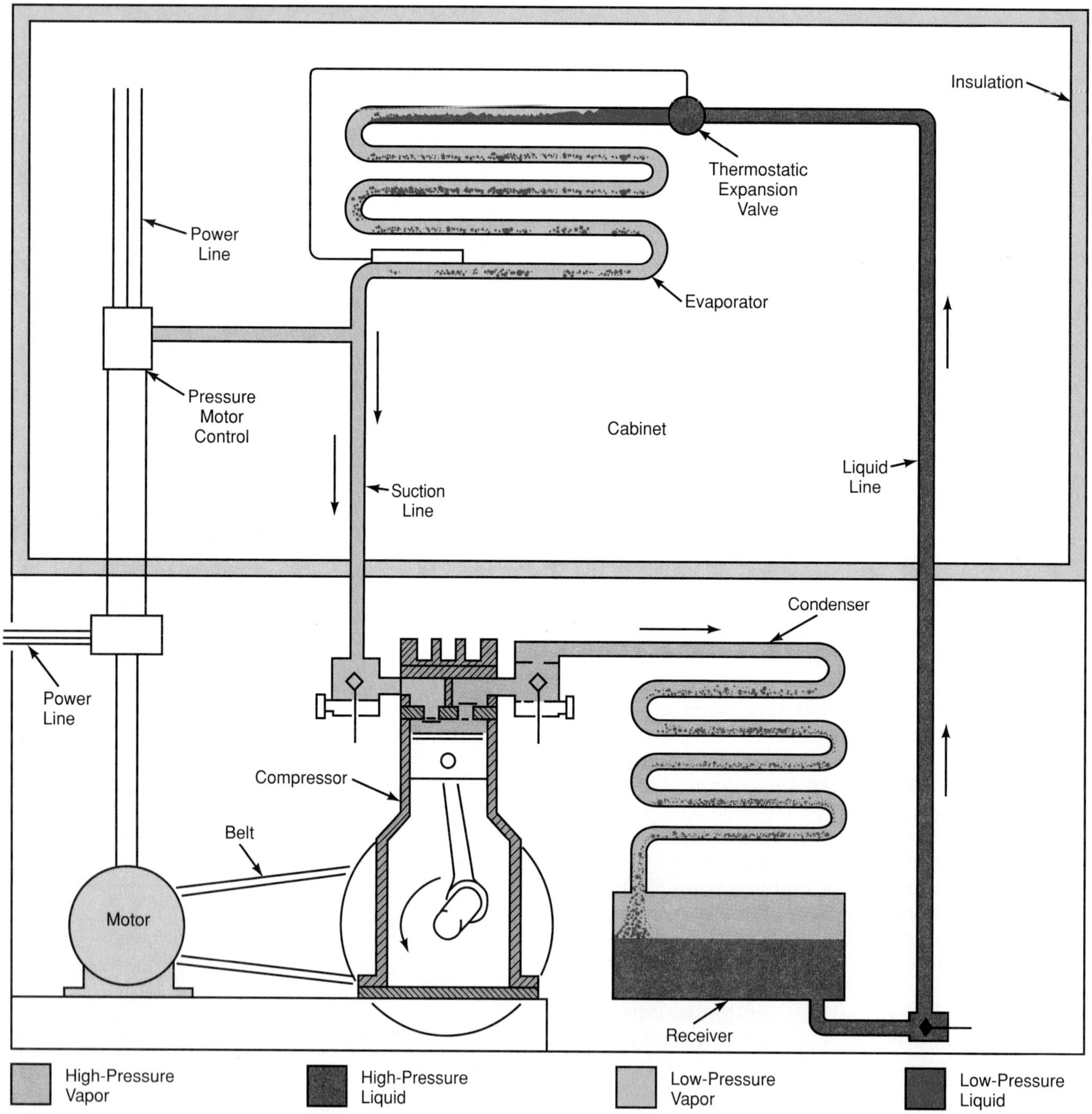

Figure 3-5. *Compression system using external-drive (open) compressor. A crankshaft seal is required at the place where crankshaft extends through crankcase of the compressor.*

The liquid refrigerant (dark red), under high pressure, flows through the thermostatic expansion valve. It then enters the evaporator where it is under low pressure. It boils, vaporizes, and absorbs heat in the evaporator.

When the compressor is running, the vaporized refrigerant (light blue) is drawn through the suction line and into the compressor. The refrigerant is compressed to a high pressure (light red) before being discharged into the condenser.

In the condenser, the ***vapor*** (vaporized refrigerant) gives up its latent heat of vaporization. It is cooled, and returns to a liquid (dark red). From here, the cycle is repeated.

A thermostatic bulb motor control is shown. The starting mechanism on external-drive (open) system motors is usually built into the motor.

An external-drive system requires a crankshaft seal on the compressor. The motor and the compressor drive are at atmospheric pressure. The pressure inside the crankcase will vary depending on the refrigerant used and the temperature. Sometimes it may be considerably above atmospheric pressure; at other times, it may be below. Refrigerant vapor cannot be allowed to flow out or air to flow into the crankcase. Either would quickly ruin the operation.

3.6 Compression System Using High-Side Float Refrigerant Control

The high-side float system is a flooded system. The evaporator is always filled with liquid refrigerant.

Figure 3-6 is a schematic diagram of a high-side float refrigerant control system. As the compressor runs, refrigerant from the condenser flows into the high-side float mechanism.

When enough liquid refrigerant has entered the high-side float mechanism, it raises the float ball. The refrigerant will then begin to flow through the control to the evaporator. The evaporator is under low pressure. Therefore, the tubing connecting the high-side float and the evaporator should be insulated. A capillary tube refrigerant line is frequently used.

If a different size line is used, it should have a weight valve at the evaporator. This prevents the refrigerant from evaporating in the connecting line. **Figure 3-6** shows a weight valve in the connecting line.

Refrigerant entering the evaporator is under low pressure (dark blue). It will rapidly evaporate (boil) and absorb heat from the evaporator.

The vapor (light blue) then flows through the suction line to the compressor. There it is compressed (squeezed) to the high-side pressure (light red). In the condenser, the heat absorbed in the evaporator is removed. The refrigerant is returned to the liquid state (dark red). It flows into the high-side mechanism where the cycle is repeated.

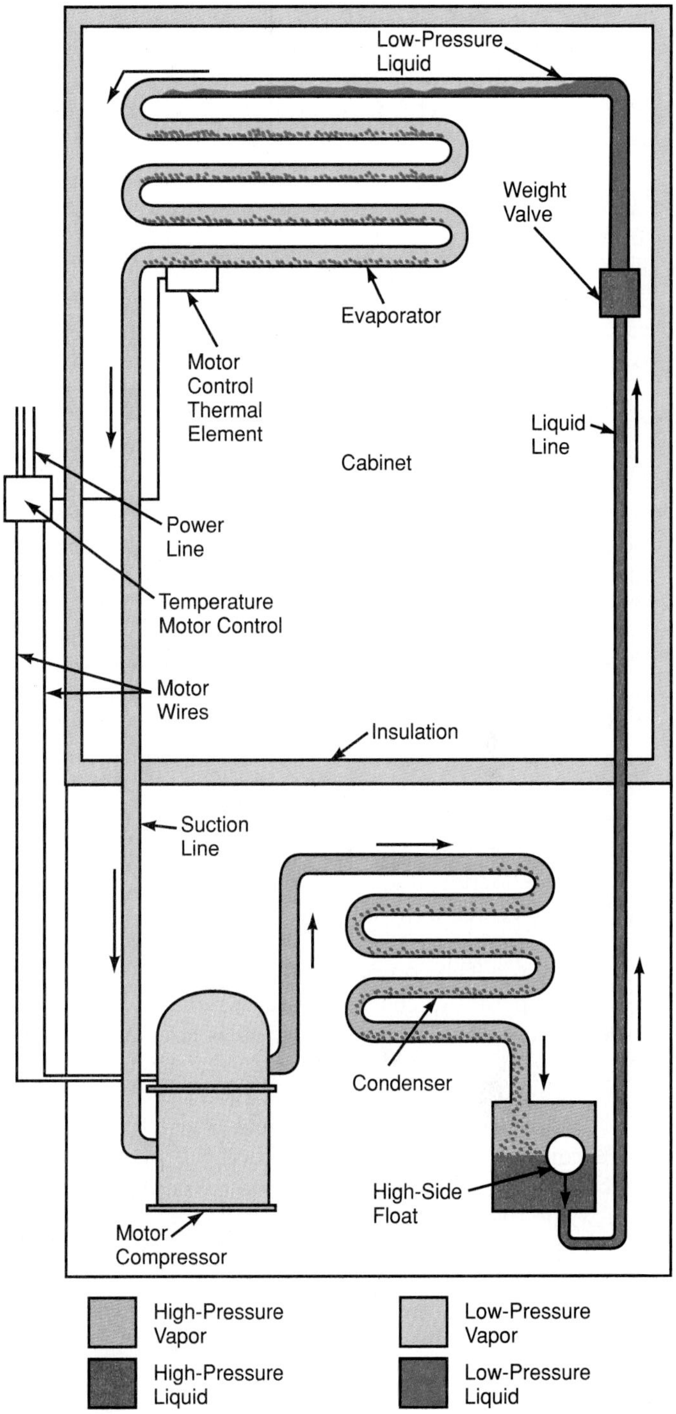

Figure 3-6. *Compression system using high-side float refrigerant control.*

Either a temperature or a pressure motor control may be used on this refrigeration cycle. **Figure 3-6** shows a temperature motor control located in the refrigerated space.

This system is most used in commercial applications where high operating efficiency is desired. It is easy to service. However, the amount of refrigerant charged into the system must be very accurately measured.

3.7 Compression System Using Automatic Expansion Valve (AEV) Refrigerant Control

The operation of an automatic expansion valve (AEV) refrigerant control refrigerating mechanism is shown in **Figure 3-7.** The compressor, motor, and condenser (condensing unit) are in the base of the cabinet. Liquid refrigerant (dark red) flows from the liquid receiver through the liquid line. It flows through the filter to the automatic expansion valve.

Refrigerant can flow through an AEV only if the evaporator pressure is reduced by the compressor running. As the compressor runs, liquid refrigerant flows through the automatic expansion valve. It is sprayed into the evaporator (dark blue). Here, due to low pressure, the refrigerant boils rapidly and absorbs heat. This vaporized refrigerant (light blue) moves back to the compressor through the suction line.

In the compressor, the refrigerant is compressed to the high-side pressure as vapor (light red). While flowing through the condenser, it is cooled. The refrigerant gives up the heat that it absorbed in the evaporator and returns to a liquid (dark red). It then flows into the liquid receiver ready to repeat the cycle.

The motor control thermal element is clamped to the end of the evaporator. This is located at the beginning of the suction line. After the evaporator is cooled to its proper temperature, the control bulb pressure causes the motor control to turn off the current to the driving motor. The compressor is stopped.

The operating characteristics of this system are quite satisfactory. The refrigerant oil is circulated without trouble. The temperature control limits can also be kept quite close.

This type of refrigeration cycle is used widely in small commercial applications. Since the pressures do not balance on the off cycle, the motor compressor must start while under load.

A faulty needle or seat in the expansion valve will allow refrigerant to leak in the off cycle. Liquid refrigerant may flow into the suction line. When the compressor starts, this will be indicated by frosting of the suction line. Such a problem may result in liquid refrigerant entering the compressor through the suction line. This may cause the compressor to knock severely.

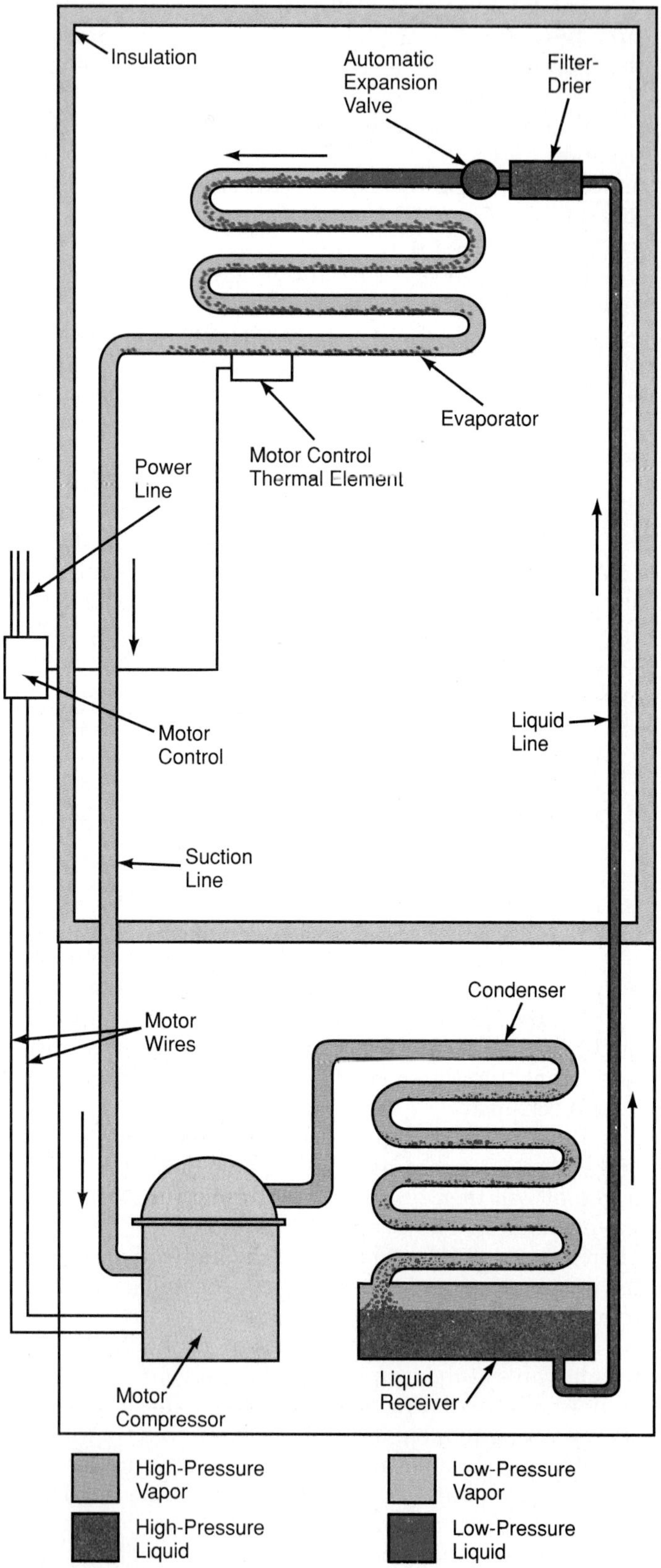

Figure 3-7. *Compression system using automatic expansion valve (AEV) refrigerant control.*

3.8 Compression System Using Thermostatically Controlled Expansion Valve (TEV)

A schematic diagram of a thermostatically controlled expansion valve (TEV) refrigeration cycle is shown in **Figure 3-8.** This system is used on large commercial refrigerators as well as on many air conditioning applications.

The liquid refrigerant (dark red) flows from the liquid receiver through the liquid line. It flows to the filter-drier and to the thermostatic expansion valve.

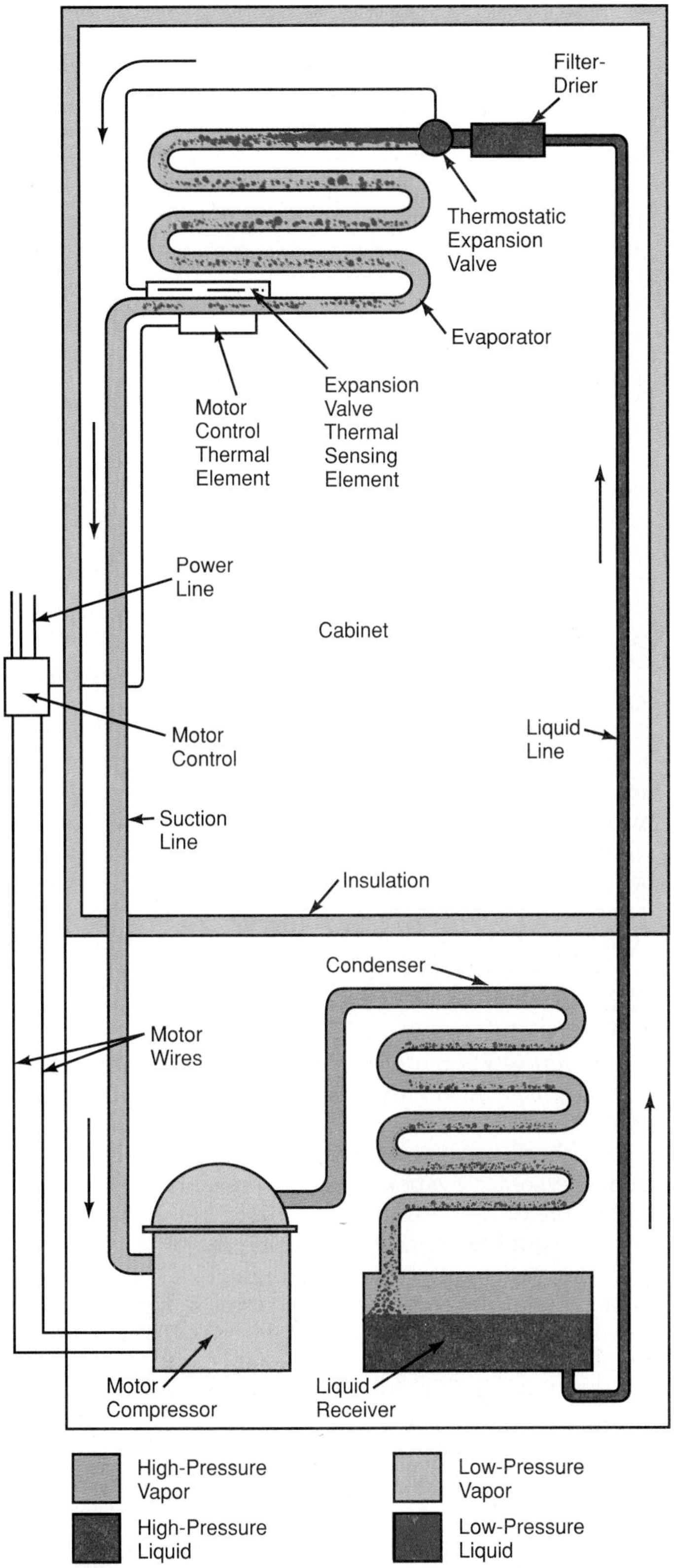

Figure 3-8. *Compression system using thermostatically controlled expansion valve (TEV).*

The operation of the thermostatic expansion valve is controlled by two conditions. These are the temperature of the TEV control bulb and the pressure in the evaporator. The temperature of the TEV control bulb must be higher than the evaporator refrigerant temperature before the valve will open. The amount of opening will be governed by the temperature of the evaporator. If the evaporator is quite warm, the needle will open quite wide. This allows a rapid flow of liquid (dark blue) into the evaporator. In this manner, cooling is speeded up. As the temperature of the evaporator drops, the TEV needle valve will decrease the refrigerant flow.

Low-pressure (vaporized) refrigerant (light blue) from the evaporator moves back into the compressor. There it is compressed back to the high-side pressure (light red). As it flows through the condenser, the refrigerant gives up the heat absorbed in the evaporator. Now cooled, the refrigerant is condensed to a liquid (dark red). It flows back into the liquid receiver. The refrigerating cycle is then repeated.

When the evaporator reaches the desired temperature, the motor control will turn off current to the motor and stop the compressor. When this happens, the TEV needle valve will close. No more refrigerant will flow through it until the compressor again lowers the pressure in the evaporator.

Pressures do not balance on the off cycle. Therefore, it is necessary to provide a motor compressor which will start under load.

The TEV control remains closed unless the evaporator is under reduced pressure and the temperature is above normal. A leaking valve will usually be indicated by a frosted or sweating suction line.

3.9 Compression System Using Capillary Tube Refrigerant Control

The ***capillary tube system,*** shown in **Figure 3-9,** is one of the most popular compression-type systems. This system is commonly used in household refrigerators, freezers, air conditioners, dehumidifiers, and small commercial applications.

Liquid refrigerant (dark red) flows from the condenser up through the liquid line. It then flows through the filter (which may also be a drier). From the filter, refrigerant flows through the capillary tube refrigerant control into the evaporator. The pressure of the liquid refrigerant, entering the capillary tube at the filter end, is at a high pressure (dark red). This is the high-pressure side. The pressure in the evaporator is low.

The capillary tube is designed so that it maintains a pressure difference while the compressor is operating. The compressor maintains a low pressure in the evaporator. The refrigerant boils, rapidly absorbing heat. The vaporized refrigerant (light blue) moves through the suction line back to the compressor. Here it is compressed to a high pressure and discharged into the condenser (light red). The vaporized refrigerant is cooled in the condenser and returns to a liquid (dark red). It again flows into the liquid line.

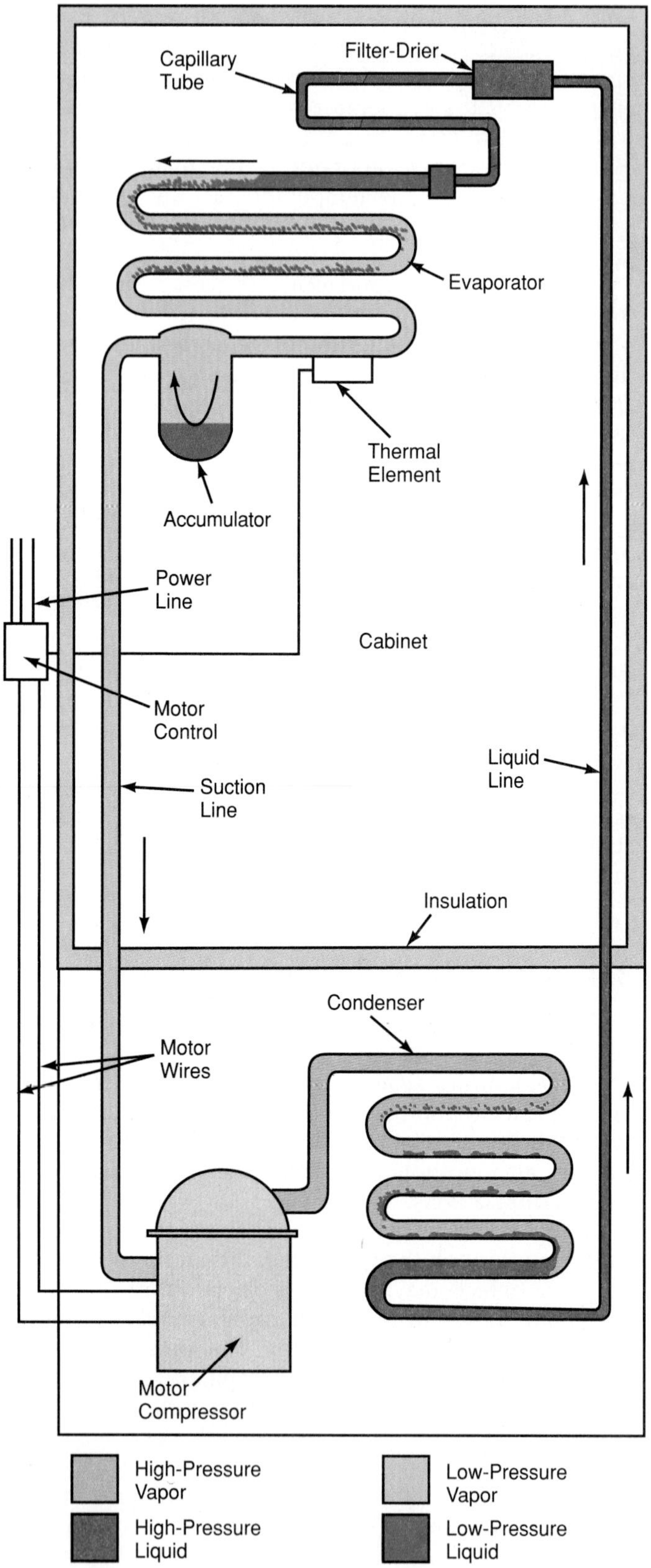

Figure 3-9. *Compression system using capillary tube refrigerant control.*

This operation continues until the thermal element has been cooled to a preset low temperature. When that temperature is reached, the thermal element operates the motor control mechanism. It turns off power to the motor. The refrigeration cycle stops. It will remain stopped until the thermal element warms up. The thermal bulb pressure will close the motor control contacts to again operate the compressor. This type of cycle is quite satisfactory for most refrigerating applications.

In the off cycle, the capillary tube allows the pressures to balance between the high and low sides. It is not usually necessary, then, to use a motor with a high starting torque.

3.10 Multiple Evaporator System

Some commercial refrigerating systems have one condensing unit connected to two or more evaporators, **Figure 3-10.** Multiple (two or more) evaporator refrigeration systems are commonly used in commercial refrigeration applications.

Liquid refrigerant (dark red) flows through the thermostatic expansion valves to the evaporators. The evaporators may have identical or different evaporator temperatures.

If the evaporator temperatures are identical, the system uses only a low-side float or the TEV to control the refrigerant. If two or more evaporating temperatures are desired (a frozen foods temperature and a water cooling temperature, for example), a device must be used to keep one of the evaporators at a higher low-side pressure. Look at the schematic shown in **Figure 3-10.** A two-temperature valve in the suction line (upper-left) keeps the low-side pressure refrigerant liquid (dark blue) and vapor (light blue) in evaporator B at a higher pressure than at evaporator A. The evaporator temperature is governed by the evaporating pressure. The lower the pressure, the lower the temperature.

A check valve is located in the suction line coming from the colder evaporator, A. It prevents the warmer, higher pressure low-side vapor (light blue) from entering the colder evaporator, A, during the off cycle.

The vaporized refrigerant (light blue) is returned to the motor compressor. It becomes a high-pressure and high-temperature vapor (light red). This vapor is cooled in the condenser, becoming a high-pressure liquid (dark red) to be stored in the receiver until needed.

Note the filter-drier on the liquid line. It keeps the refrigerant clean and dry.

A ***sight glass*** (liquid indicator) is often included in the liquid line. The technician may use it to see if there is enough refrigerant in the system. Bubbles will indicate a refrigerant shortage. This system, as shown, uses a pressure motor control. The operating pressure is taken from the low side of the system.

A line from the high-pressure side also enters the motor control. This operates a safety device which stops the motor if the condensing pressure (high side) goes too high.

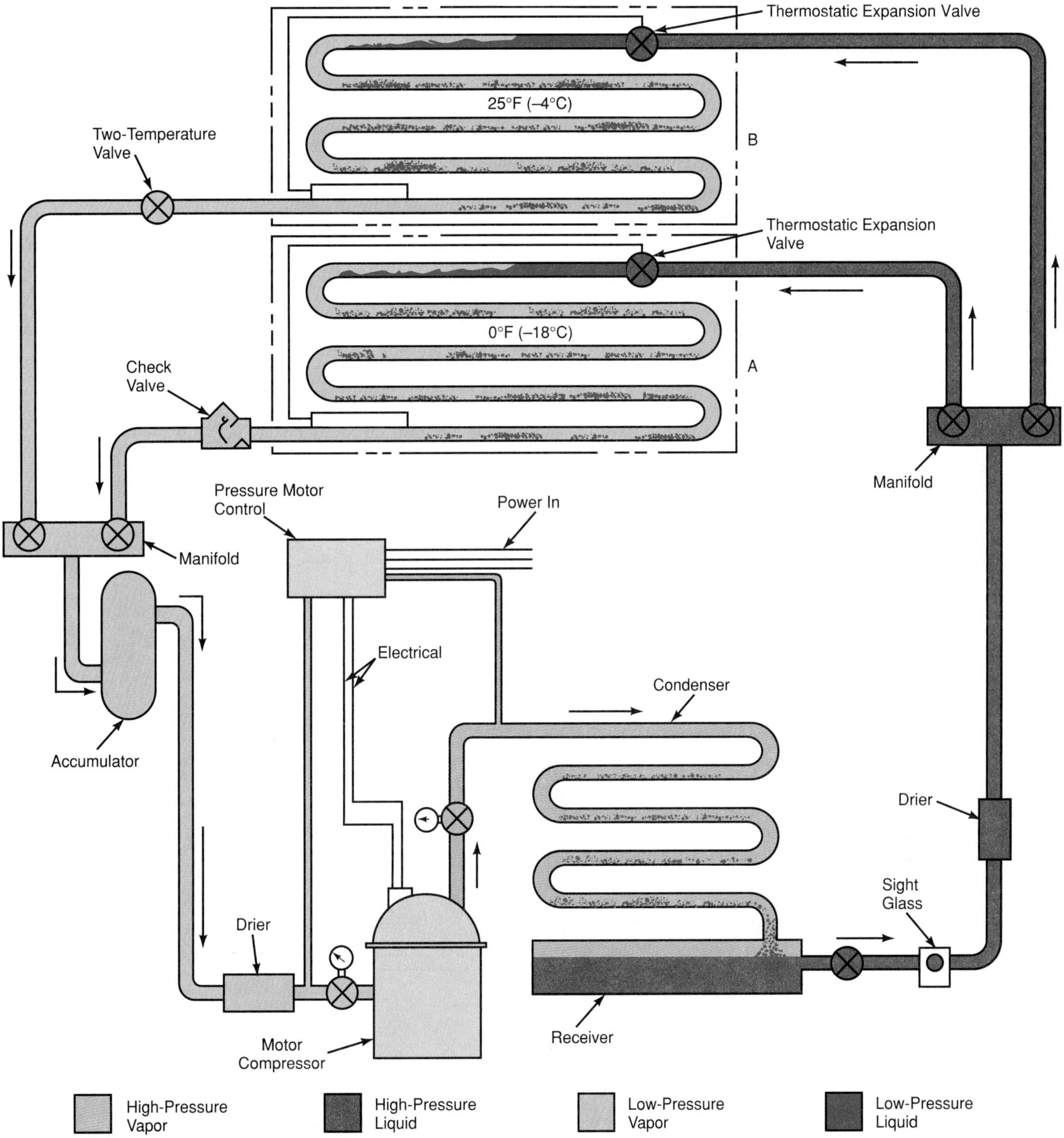

Figure 3-10. *A multiple evaporator system. Evaporator A operates at 0°F (−18°C). Evaporator B operates at 25°F (−4°C).*

3.11 Compound Refrigerating Systems

In compound refrigerating systems, two or more compressors are connected in series, **Figure 3-11.** In this illustration, compressor No. 1 discharges into the intake side of compressor No. 2. Compressor No. 2 then discharges into the condenser (light red). Here the vapor condenses. The liquid refrigerant (dark red) flows into the liquid receiver. Refrigerant vapor is not condensed between compressors. An intercooler lowers the vapor temperature. This type of installation usually requires an oil separator for each compressor.

From the liquid receiver the liquid refrigerant (dark red) flows up to the thermostatic expansion valve. It then

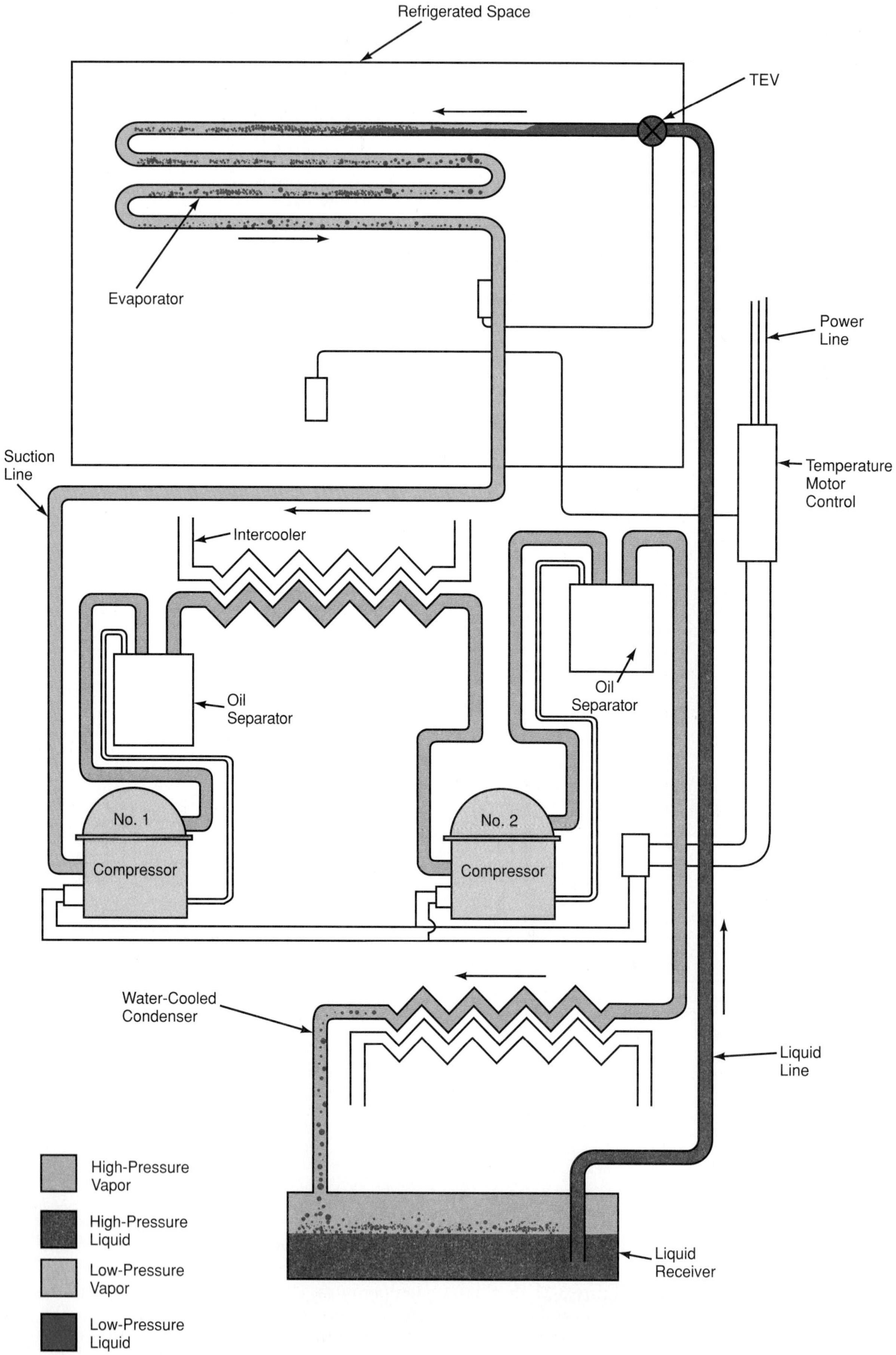

Figure 3-11. *Compound refrigerating system.*

enters the evaporator. In the evaporator (dark blue) the refrigerant boils and absorbs heat (light blue). From the evaporator, the vaporized refrigerant flows back to compressor No. 1. From here the cycle is repeated.

A ***compound system*** increases capacity when pulling down to extremely low pressures (low temperatures). A single compressor would have difficulty reaching these pressures/temperatures.

A single-temperature motor control operates all motors. A thermostatic expansion valve controls the liquid refrigerant flow into the evaporator.

The pressures do not balance on the off cycle. Therefore, motors capable of starting under load are required.

Compound installations usually operate under rather heavy service requirements. Condensers and refrigerant must be kept clean. Compressor valves must be kept in good condition.

3.12 Cascade Refrigerating Systems

In a cascade refrigerating system, two or more refrigerating systems are connected as shown in **Figure 3-12.** Cascade systems are often used in industrial processes where objects must be cooled to temperatures below −50°F (−46°C).

Both systems operate at the same time. System A (on the right) has its evaporator, A, (heat-absorbing part) arranged to cool the condenser B for the system B. The evaporator for system B supplies the cooling effect desired. Each system has a thermostatic expansion valve (TEV) for refrigerant control.

The low-pressure liquid (dark blue) of system A cools the high-pressure vapor (light red) of system B.

One motor control is used for both motors. It is connected to a temperature-sensing bulb on evaporator B.

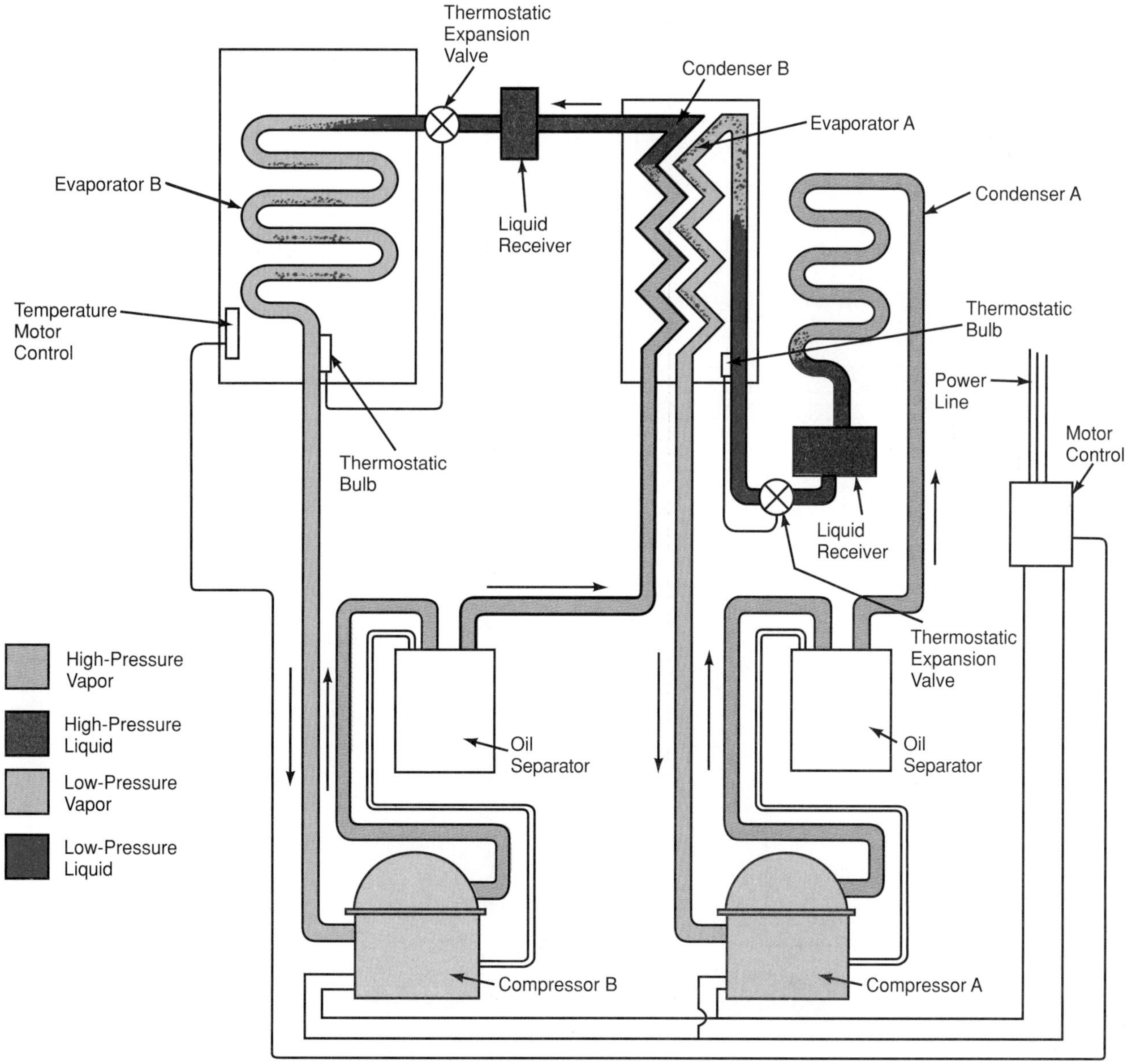

Figure 3-12. *Cascade refrigerating system.*

Motors used on cascade systems must be capable of starting under load. With the use of thermostatic expansion valves, the pressures do not balance on the off cycle.

The condenser-evaporator is usually of the shell-and-tube flooded evaporator type.

Since these systems operate at very low temperatures, the refrigerant must be very dry. Any moisture would condense at the needle seat of the TEV and stop the refrigerant flow. System B must have special refrigerant oil (wax-free, moisture-free, and the ability to flow at extra low temperatures).

Oil separators should be installed in the compressor-to-condenser lines on both of these condensing units. This will help keep the oil in the compressors.

3.13 Modulating Refrigeration Cycle

Most refrigeration installations have enough cooling or refrigerating capacity to maintain the desired temperature under the heaviest load. This temperature is maintained by the motor control. It starts the motor compressor when cooling (heat removal) is required. It shuts off as soon as the desired temperature is reached.

However, if the heat load is light, this single system may be overcapacity for the job. The operating expense is greater than if the machine capacity more easily matched the needed load. The system tends to cool too fast and turn on and off too quickly.

A modulating (varying capacity) system has been developed to fit the machine capacity more closely to the needed heat load. This is sometimes done by using two or more compressors connected in parallel. Each compressor is operated by a motor control.

If the heat load increases and the temperature starts to rise, one compressor will still run. However, if the temperature keeps rising, the second compressor will start to operate. Additional compressors may cut in until enough capacity is obtained.

Figure 3-13 illustrates a typical cycle diagram for a modulated installation. This installation has three compressors. A pressure control connected to the suction lines operates the motors. The control contains a special

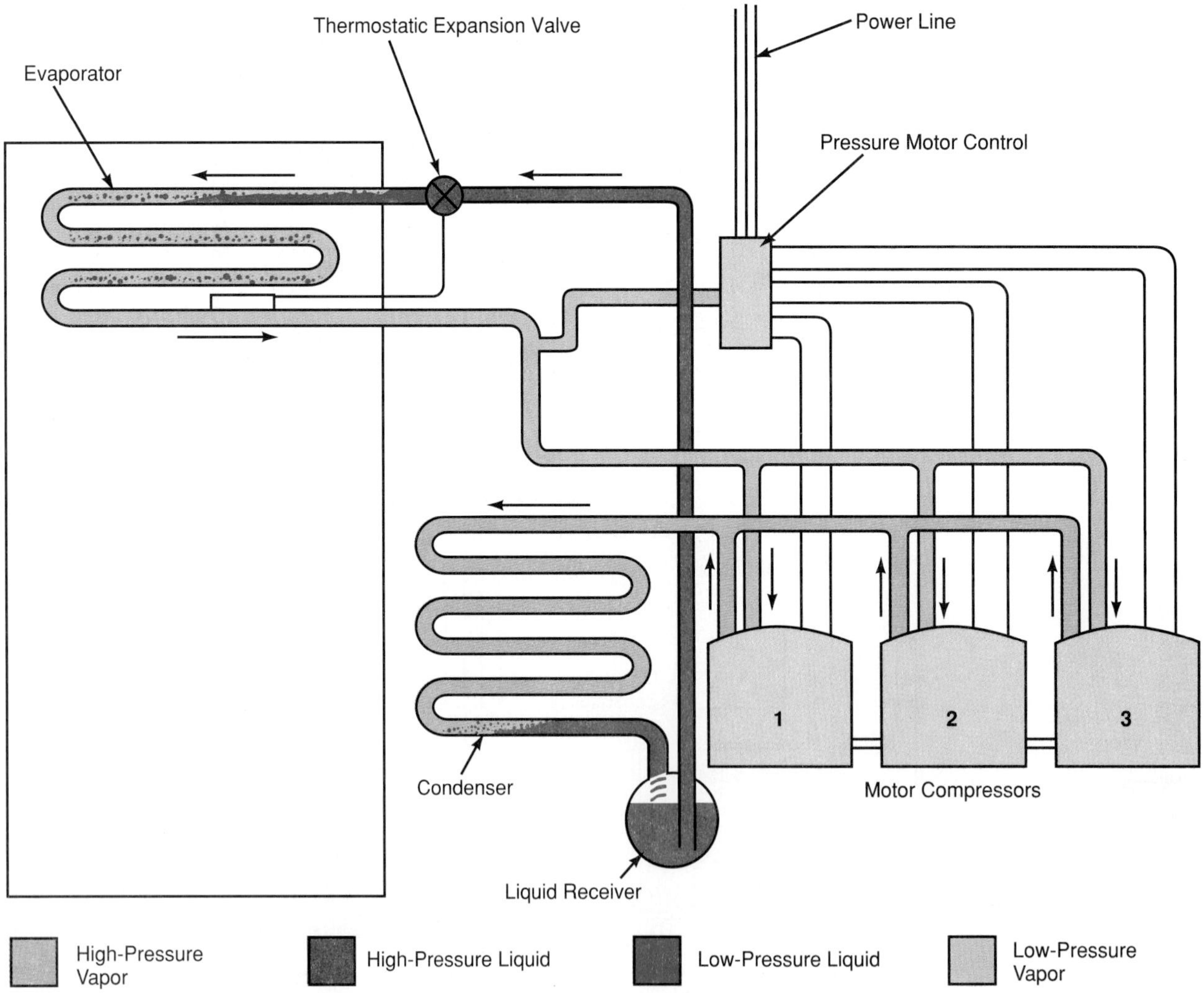

Figure 3-13. *Modulating refrigeration cycle mechanism, which uses three motor compressors. Pressure motor control is arranged to operate one or more compressors as needed.*

switching device. This rotates the service of the various compressors. Each compressor will be used about the same amount of time. The modulating cycle maintains uniform temperatures and operates economically.

Any conventional refrigerant control can be used. However, the thermostatic expansion valve is most common type.

The same condenser and liquid receiver may be used by all the compressors, or each may have its own condenser and receiver. The same evaporator is connected to all the compressors.

A modulating system may use a multiple cylinder compressor. Each cylinder is equipped with an unloader device. Variable-speed motors are also used to provide a modulated refrigeration capacity.

3.14 Ice Maker

Ice makers use various types of refrigerating systems. Note in the simple ice-making unit, **Figure 3-14,** the motor compressor and condenser are usually located in the bottom of the cabinet. Liquid refrigerant (dark red) flows from the bottom of the condenser up through a filter-drier. It enters the evaporator through a capillary tube. The evaporator surrounds the inverted (upside-down) ice cube molds.

From the evaporator, the refrigerant vapor (light blue) flows into an accumulator. This container has a coil from the liquid refrigerant line in it or around it. Such an arrangement serves as a heat exchanger. The refrigerant vapor (light blue) is drawn from the accumulator back to the compressor. Here it is compressed up to the high-side pressure (light red). It is forced into the condenser. From here the cycle is repeated.

The mechanism which makes and handles the ice is also shown. Cold water is sprayed into the inverted ice cube molds. The temperature of the molds is very low. Water striking the molds freezes to the mold surface. It gradually builds up until complete ice cubes are formed. Next, an electric heating unit heats the ice cube molds until the cubes fall out. They slide down a chute into the ice cube bin. Then the refrigerating cycle is stopped. Most surfaces in contact with water and ice are stainless steel for cleanliness.

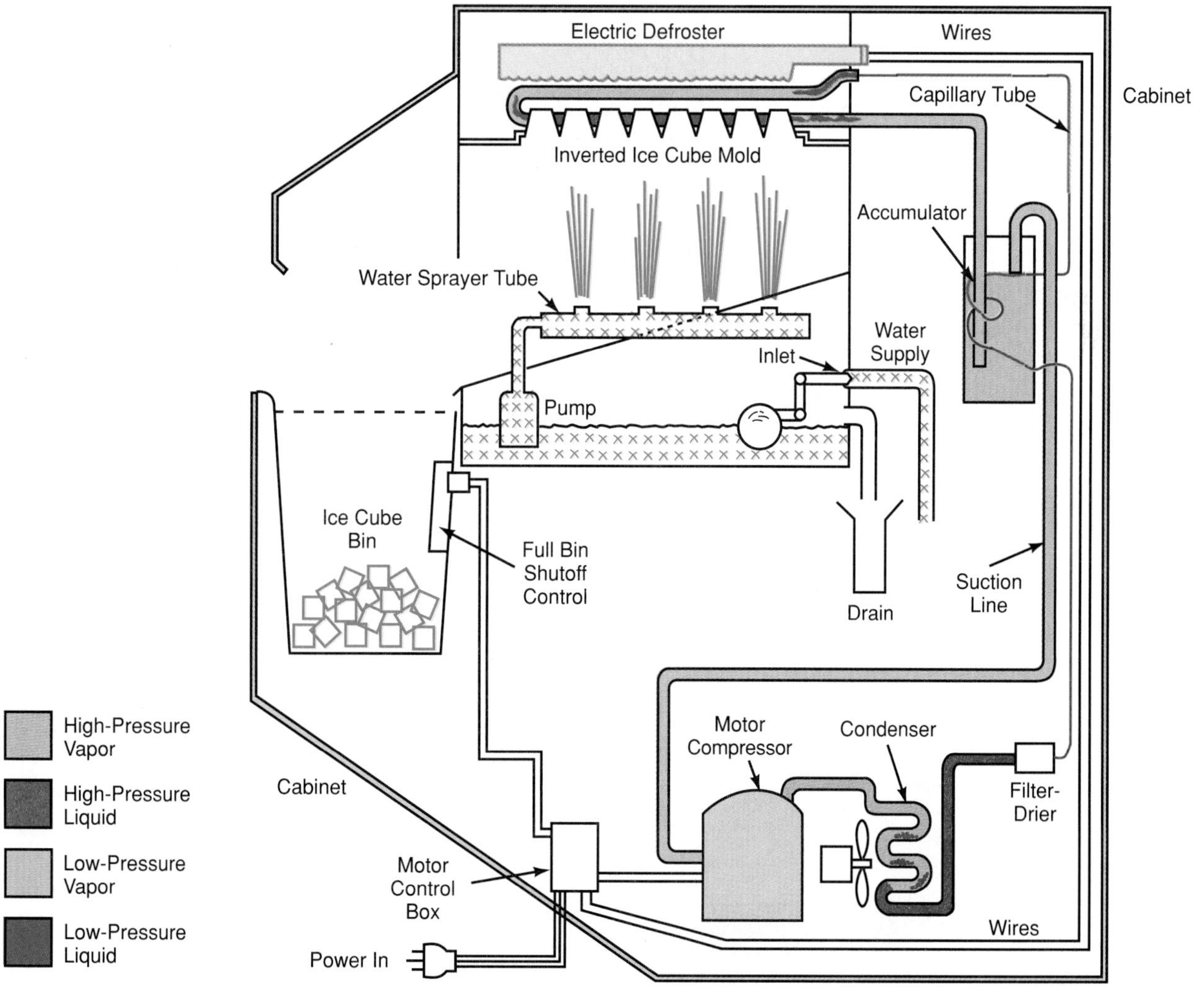

Figure 3-14. *In an ice maker, water is sprayed into ice cube molds to produce clear ice cubes.*

3.15 Drinking Water Cooler

The water cooler is a special use of a refrigerating mechanism. It is used to cool water "on tap" at a drinking fountain. The usual ***hermetic*** (airtight) compression refrigerating system is used. The refrigerant control is a capillary tube. A schematic of a drinking water system is shown in **Figure 3-15.**

Liquid refrigerant flows from the bottom of the condenser through the liquid line. It flows through a filter-drier (dark red) and into the capillary tube. As it flows into the evaporator, it vaporizes and absorbs heat from the evaporator surface (light blue). The evaporator is next to or surrounds the drinking water coil or water cooling tank.

From the evaporator, the refrigerant vapor goes into an accumulator in the suction line. The accumulator stops any liquid refrigerant from flowing into the suction line and/or into the motor compressor.

From the accumulator, the vapor is drawn into the motor compressor where it is pumped into the condenser (light red). Here the heat picked up in the evaporator is released. Meanwhile, the refrigerant returns to a liquid and collects in the bottom of the condenser. From here, the cycle is repeated.

Since the demand on a drinking fountain is very irregular, it must have some hold-over capacity. Still it must not overcool the water. The necessary capacity is provided by using either an insulated storage tank or large cooling surfaces in the evaporator.

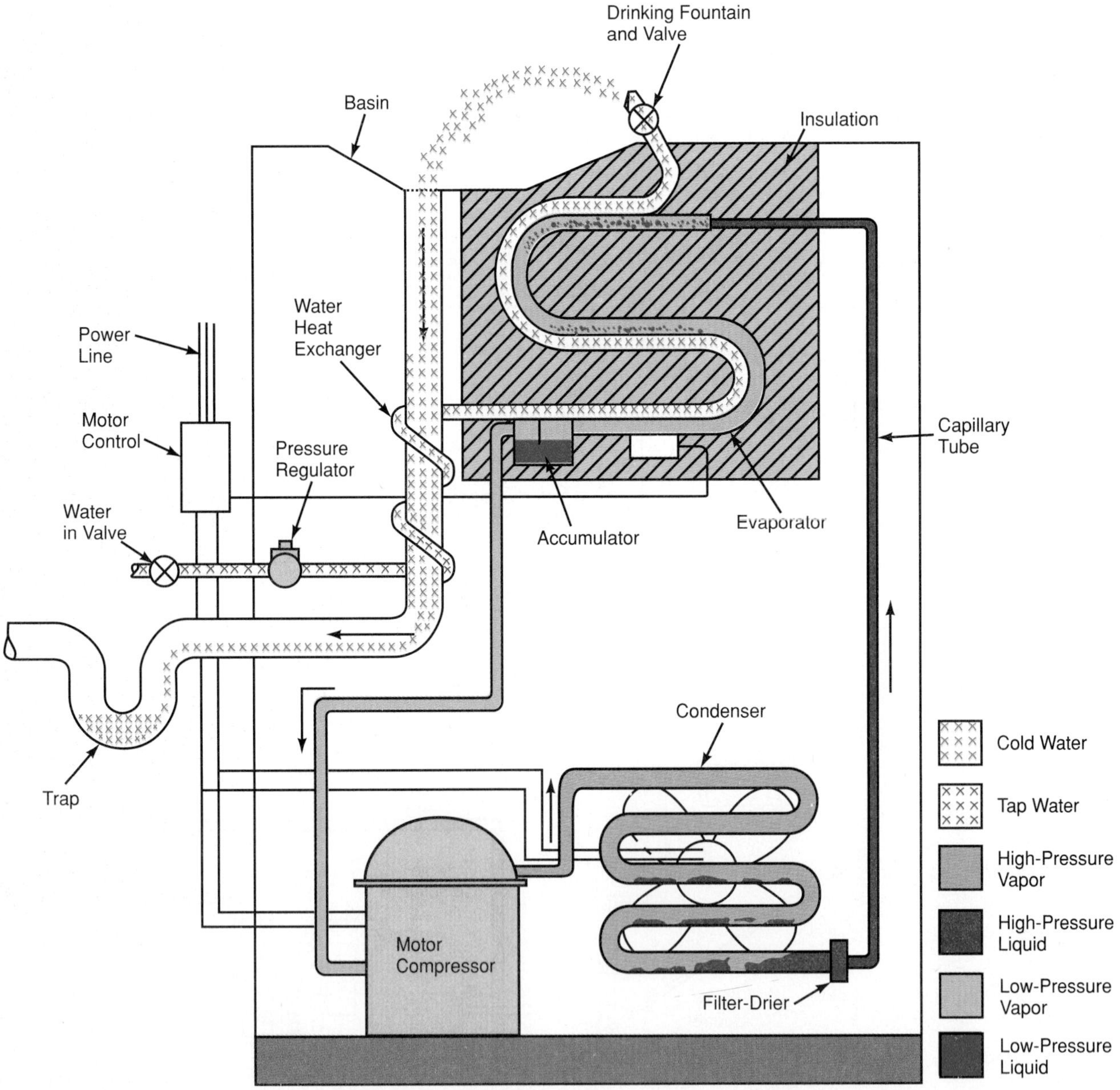

Figure 3-15. *A drinking fountain cooled by a compression system refrigerating mechanism.*

To increase the mechanism's efficiency, the waste water flows down a tube alongside or attached to the fresh water inlet. In this way, the warmer fresh water (water-in) is cooled by the cooler waste water leaving the fountain.

A water pressure regulator adjusts the water flow. The condensing unit is air-cooled. This ensures that the fountain can deliver enough cold water under heavy demand. A condenser fan is used to increase the condenser capacity. The fan is connected into the electrical circuit. It runs whenever the condensing unit is running.

A thermostat with a control bulb is attached to the water-dispensing tube. It maintains the desired drinking water temperature in the fountain. Water leaving the fountain should be at approximately 50°F (10°C).

3.16 Expendable Refrigerant Refrigeration System

This simple system, sometimes called ***chemical refrigeration*** or ***open-cycle refrigeration,*** is becoming increasingly popular. It is used on trucks and other vehicles in the transportation industry and in the storage of refrigerated or frozen foods. Basically, an expendable refrigerant refrigerating system is a heavily insulated space. It may be cooled by being surrounded by tubes carrying evaporating liquid nitrogen. Another method of cooling consists of spraying liquid nitrogen directly into the space to be cooled. In any event, an ***expendable refrigerant system*** is one in which the system discards the refrigerant after it has evaporated.

Figure 3-16 illustrates the spray system. The liquid nitrogen (dark red), supplied from a cylinder inside the refrigerated space, is kept under pressure (200 psi). Dark blue indicates low-pressure liquid refrigerant.

The pressurized cylinder is insulated. However, an automatic pressure relief valve will open as a safety measure, if necessary. It would allow the nitrogen vapor to escape should pressure exceed the relief valve setting. Heat surrounding the cylinder may cause the vapor pressure to rise above the automatic pressure release setting. Cold nitrogen vapor is then released by the automatic pressure release valve. It is discharged into the refrigerated space or into the refrigerating tubes, depending on the system being used.

A temperature-sensing element, control box, and liquid control valve, control the flow of liquid nitrogen

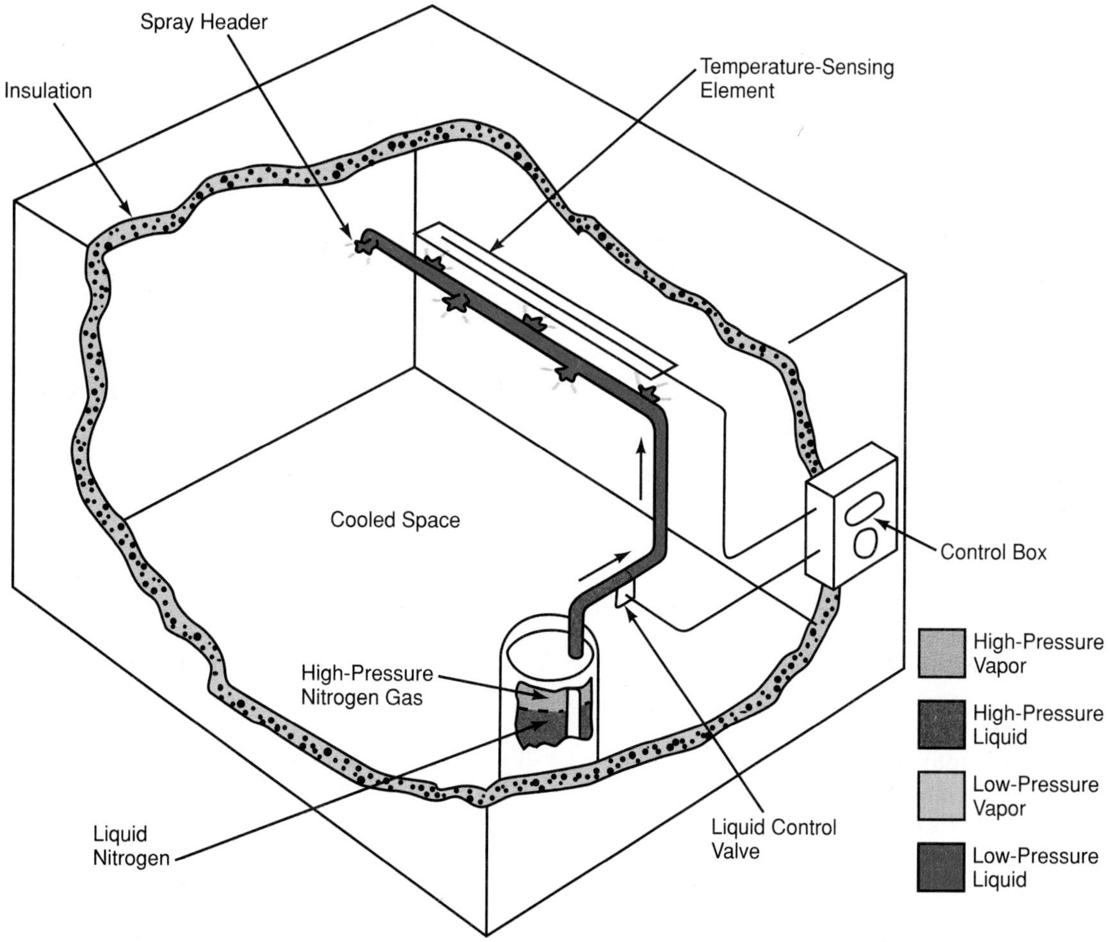

Figure 3-16. *Expendable refrigerant refrigeration system.*

from the nozzles. They maintain the desired temperatures inside the refrigerated space.

Liquid nitrogen (dark red) vaporizes (boils and turns into a gas) at a temperature of −320°F (−196°C) at atmospheric pressure **(Figure 1-26).** This type of system is excellent for shipping frozen foods. Temperatures may be kept as low as desired—usually about −20°F (−29°C).

Simple construction such as this demands little attention. Occasionally, it may be necessary to replace or recharge the nitrogen storage cylinder. Another advantage to this system is its ability to operate without a power source. Safety devices in spaces refrigerated by liquid nitrogen shut off the flow of nitrogen when a person opens a door to the space. See Chapter 18.

Another form of expendable refrigerant refrigeration system is natural gas shipped in liquid form in large tanker ships. Natural gas, which is a liquid under pressure, will evaporate. Some of the gas is allowed to evaporate. This evaporative cooling maintains the remaining natural gas in liquid form. The evaporated natural gas is then ducted to the tanker engines. There it is burned to provide power to drive the tanker.

A thermostat inside the refrigerated space controls the current flow through the transformer-rectifier. The transformer-rectifier supplies a controlled dc current to the modules. In this manner, the temperature inside the refrigerator is controlled.

There are no moving parts in thermoelectric refrigeration. Aside from the construction of the modules, it is quite simple. Thermal efficiency is low. The amount of refrigerating effect obtained for the electrical energy spent is less than with a conventional compressor-type refrigeration system.

Reversing the direction of the flow of current through a thermoelectric device reverses the hot and cold surfaces. Thus, the same device can be used for both heating and cooling an insulated space.

A thermoelectric device has been used in the air conditioning and heating of nuclear submarines. The thermoelectric device is also often used to control temperatures in electronic equipment (computers, aerospace devices, etc.).

Refer to Chapter 18 for further technical information concerning thermoelectric refrigeration and air conditioning devices.

3.17 Thermoelectric Refrigeration

The physical principle (Peltier effect), upon which thermoelectric refrigeration is based, has been known since 1834. ***Thermoelectric refrigeration*** transfers heat energy from one place to another using electrons rather than refrigerants.

Figure 3-17A represents a simple thermoelectric couple. The couple moves heat from the inside of an insulated space to a heat exchanger. The heat exchanger is located on the outside. Electrons, rather than refrigerants, carry away the heat.

Fins on the evaporator (dark blue) increase the heat flow. Fins on the outside of the heat exchanger (dark red) help give off the heat to the surrounding air.

Semiconductors are materials that conduct electricity, but not as well as typical metals. They may be made from elements such as silicon, germanium, or a combination of elements. Semiconductors may be processed so that ***N-type semiconductors*** conduct electricity by the flow of negatively charged particles (usually electrons). Others, called ***P-type semiconductors,*** conduct electricity by the flow of positively charged particles (often called "holes" or electron holes).

Figure 3-17A shows current being forced to flow from a P-type material into an N-type material. The junction where N and P are connected absorbs heat. The opposite ends become hot and give off heat. This is the ***Peltier effect.*** A single junction produces only a small cooling effect. Therefore, several N-P paired junctions are connected in series to produce significant cooling. See **Figure 3-17B.** Groups of modules may be connected together in parallel to increase the capacity still further. (Chapter 6 for series-parallel connections.)

3.18 Dry Ice Refrigeration

Dry ice is solid carbon dioxide. It may be pressed into various sizes and shapes, blocks, or slabs. As it absorbs heat, it changes directly from a solid to a vapor. It does not go through the liquid state. This change from solid to vapor is called ***sublimation.*** At atmospheric pressure, solid carbon dioxide vaporizes at −109°F (−78°C).

Figure 3-18A illustrates a common method of using dry ice as a frozen food refrigerating device. Dry ice (aqua) is usually packed either beside or on top of the food packages. Carbon dioxide, as it changes to a vapor, keeps the food frozen. The dry vapor tends to replace the air in the container or cabinet. This helps to preserve the food.

A device has been developed which uses dry ice for refrigerating materials carried on aircraft. See **Figure 3-18B.** A closed refrigerating circuit is connected to an evaporator in the space to be refrigerated. It also connects to a condenser located in an insulated bin. The bin holds dry ice pellets. The circuit contains a common refrigerant.

The condenser operates at a very low temperature (−109°F or −78°C). This causes refrigerant vapor entering the condenser to condense quickly to a liquid. The liquid refrigerant flows by gravity into the evaporator. There, it absorbs heat as it vaporizes and flows upward into the condenser. From here the cycle is repeated.

A thermostatically operated control valve controls the flow of refrigerant into the evaporator. It is located in the liquid line. This device is illustrated in **Figure 3-18B.**

Dry ice is usually stored in heavily insulated cabinets. **Never handle it with bare hands. It will cause instant freeze burns. Always wear heavy gloves.**

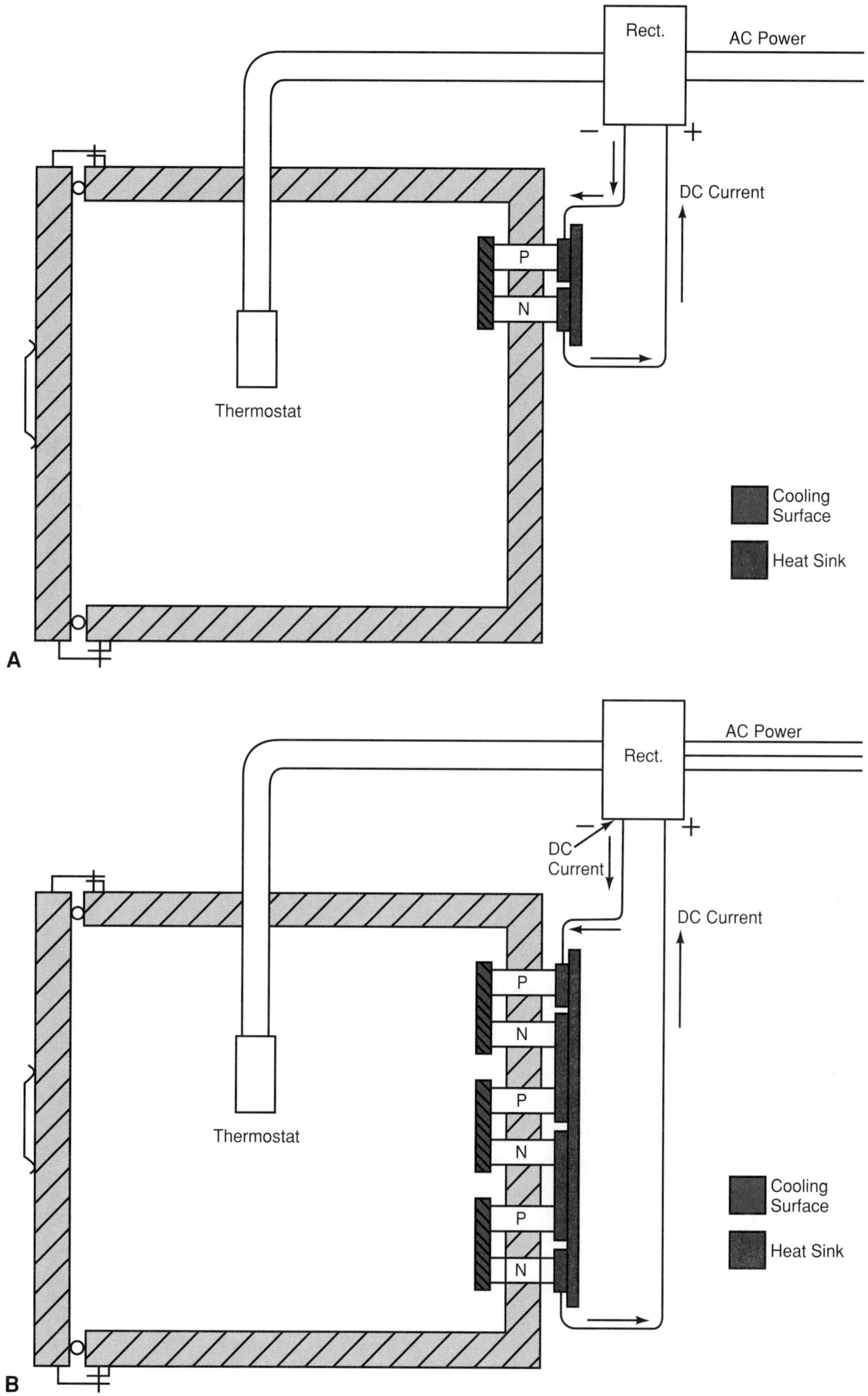

Figure 3-17. *A—Diagram of simple thermoelectric couple, used for refrigerating an insulated space. Heat absorbed by thermoelectric couple is released to outside by fins attached to heat-radiating surface (heat sink). B—Thermoelectric module cooling device. Three couples are connected in series to increase heat absorbing effect. Electrons flow into N-type section. See Chapter 6.*

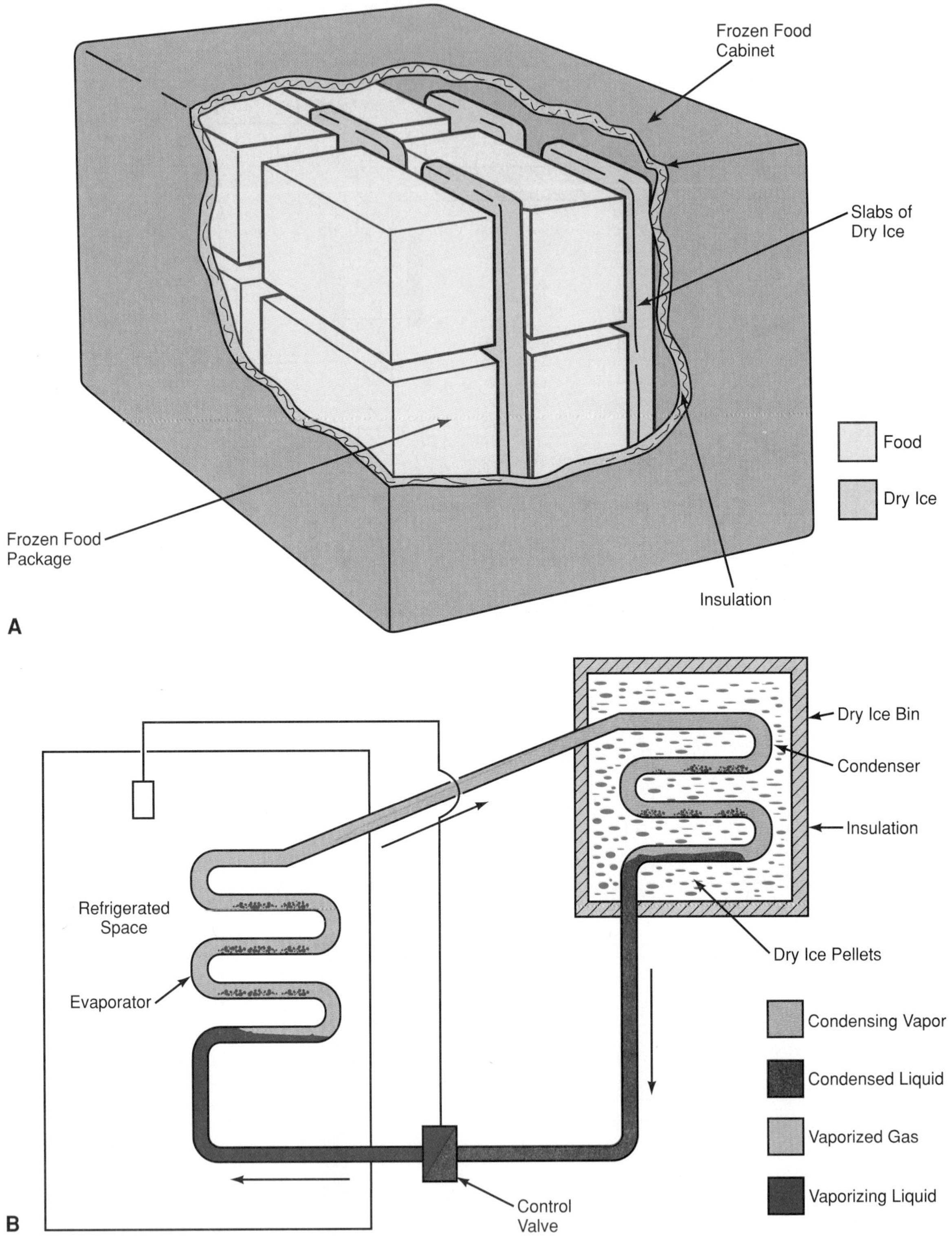

Figure 3-18. *A—Dry ice frozen food container. B—A dry ice refrigerator does not require a compressor.*

3.19 Intermittent Absorption System

The intermittent absorption system uses a generator charged with water and ammonia. A heat source, usually a kerosene flame, heats this solution in the generator. The ammonia is vaporized and is driven off.

A condenser, at the top of the system, condenses the ammonia vapor into a liquid. The liquid flows by gravity into the liquid receiver and then into the evaporator. During the generating cycle, little or no refrigerating effect is taking place. As the system cools, the pressure drops, causing the liquid ammonia in the evaporator to boil and absorb heat. The cycle is completed when vaporized ammonia is reabsorbed in the generator.

Figure 3-19A illustrates the generating cycle. In operation, the kerosene burner tank is filled with just enough kerosene for one cycle. This cycle is usually once a day. The burner is filled and lighted. It heats the water and ammonia mixture (reddish-brown) in the generator. The ammonia vapor (light reddish-brown) is driven off through the tube, A, up to the condenser, C. There the ammonia gas is cooled and condensed to liquid ammonia (red). The liquid flows into the receiver.

When the kerosene has all been burned (usually from 20 to 40 minutes), the generating cycle ends. The refrigeration cycle now begins. See **Figure 3-19B.**

The pressure in the system drops as the water cools and absorbs ammonia vapor. Liquid ammonia (dark blue) flows into the evaporator, begins to evaporate, and cools it. Evaporated ammonia (light blue) flows back through the tube, B. It is again absorbed by the water in the generator. Refrigeration continues, usually until the next firing of the kerosene burner.

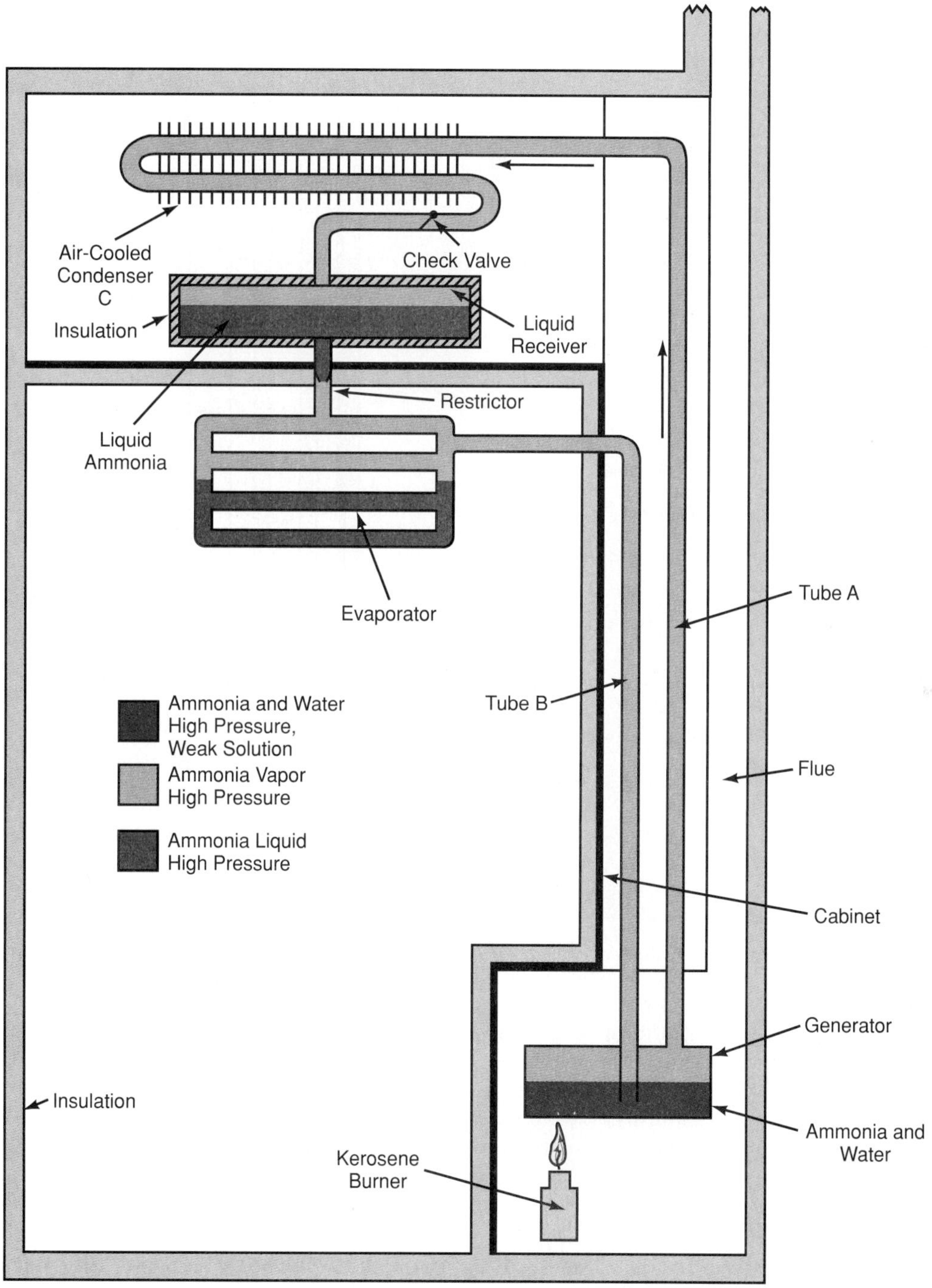

Figure 3-19A. *Intermittent absorption system during its generating cycle. The system is under high, or condensing, pressure.*

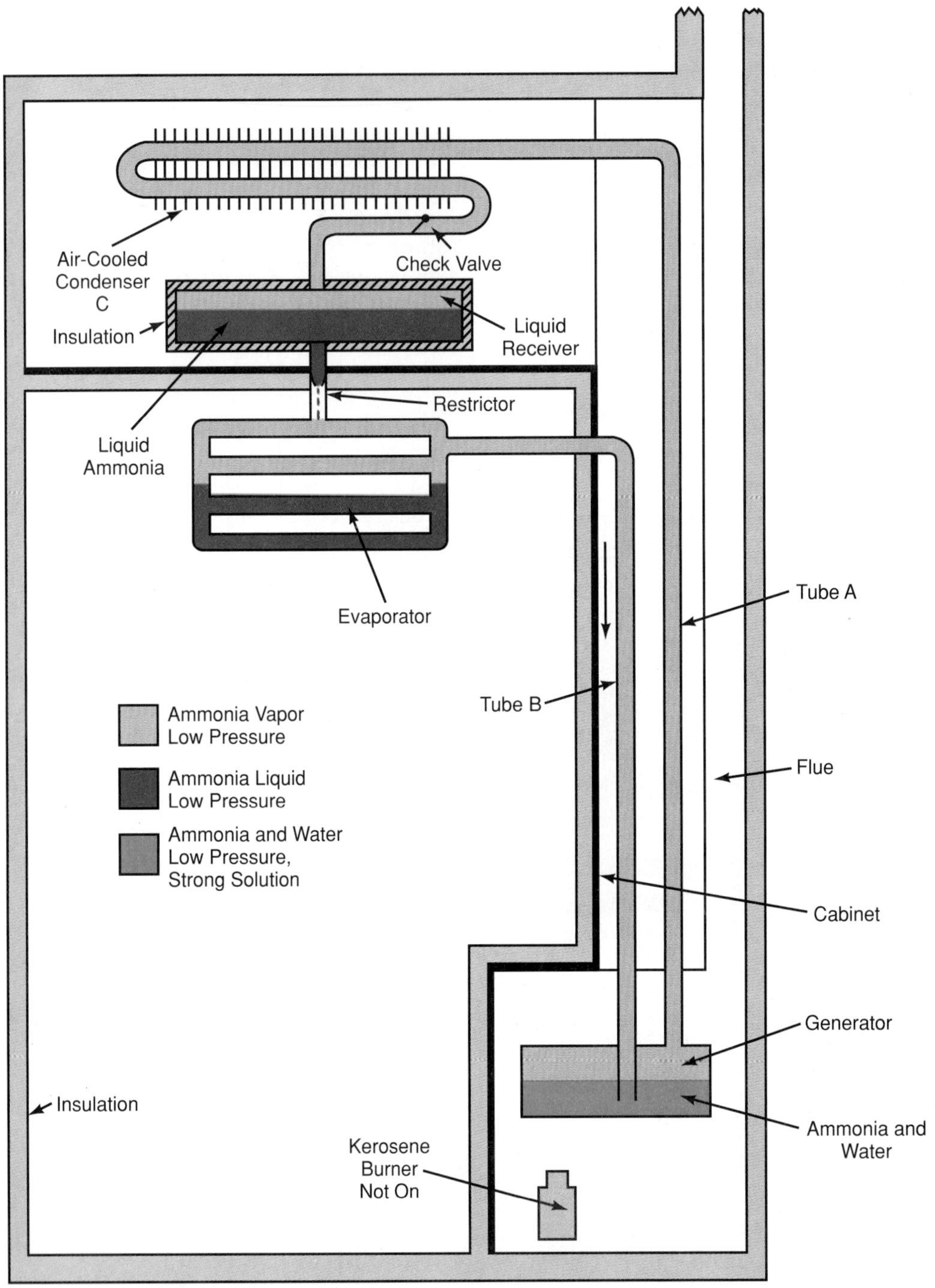

Figure 3-19B. *Intermittent absorption system during its refrigerating cycle. The system is under low, or refrigerating, pressure.*

This type of refrigerating system is quite simple. The piping is welded steel because the pressures on the generating cycle are quite high. The refrigerating ability is quite good. Kerosene flame heated absorption refrigerators are popular in areas where electric power is not available.

3.20 Continuous-Cycle Absorption System

The continuous-cycle absorption cooling unit is operated by the application of a limited amount of heat. This heat is furnished by gas, electricity, or kerosene. No moving parts are employed. The operation of the refrigerating mechanism is based on Dalton's Law. See Section 1.34. This refrigerating device is widely used in domestic refrigerators and recreation vehicles. It is also used in year-around air conditioning of both homes and larger buildings. Modern absorption systems are illustrated in Chapter 17.

The unit consists of four main parts—the boiler, condenser, evaporator, and absorber. See **Figure 3-20.**

When the unit operates on kerosene or gas, the heat is supplied by a burner. This element is fitted

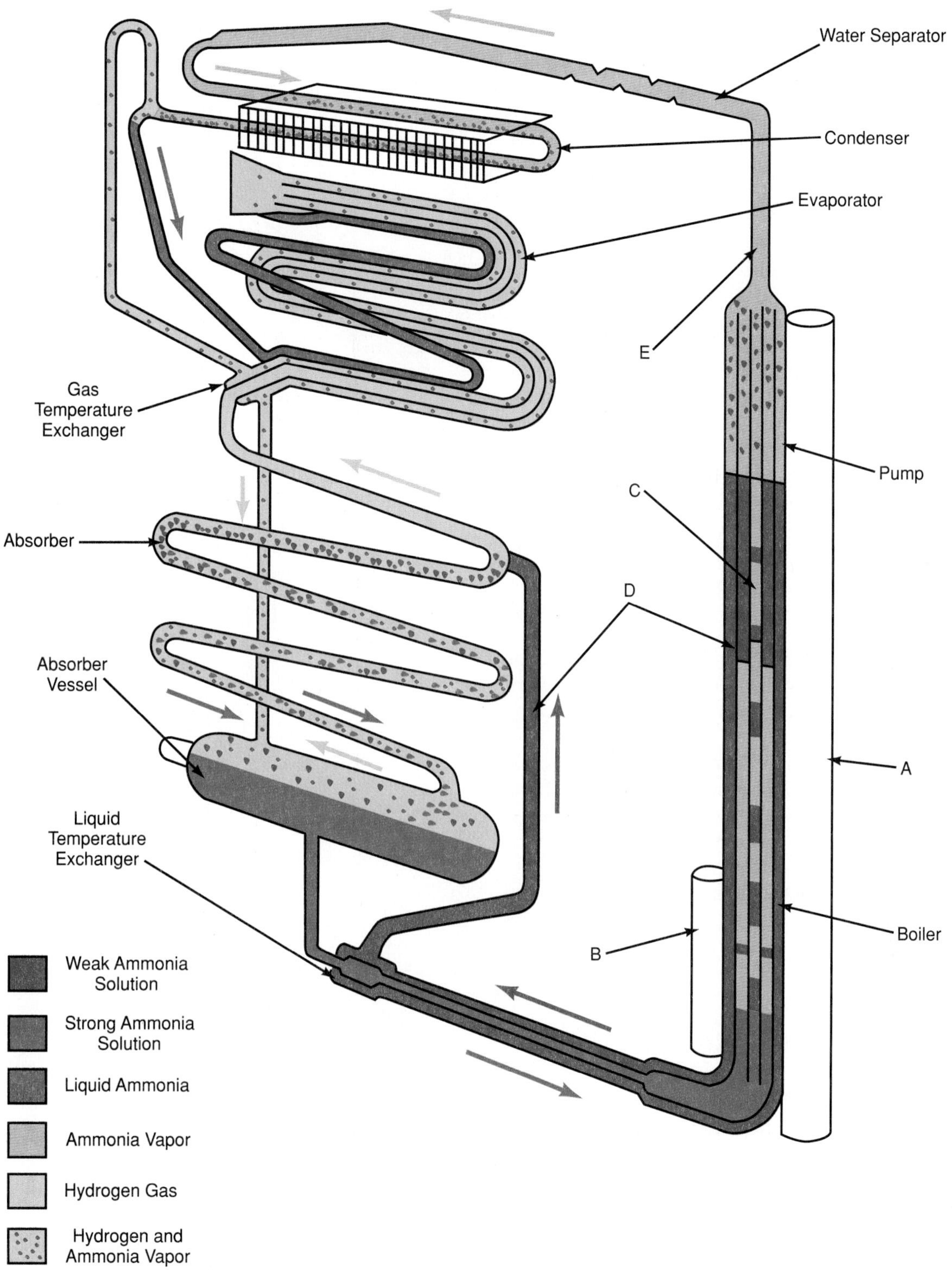

Figure 3-20. *A continuous-cycle absorption system. (Electrolux AB)*

underneath the central tube (A). When operating on electricity, the heat is supplied by an element inserted in the pocket (B).

The unit charge consists of a quantity of ammonia, water, and hydrogen. These are at a sufficient pressure to condense ammonia at room temperature. When heat is supplied to the boiler system, bubbles of ammonia gas are produced. They rise and carry with them quantities of weak ammonia solution through the siphon pump (C). This weak solution passes into tube (D), while the ammonia vapor passes into the vapor pipe (E) and on to the water separator. Here any water vapor is condensed and runs back into the boiler system, leaving the dry ammonia vapor to pass to the condenser.

Air circulating over the fins of the condenser removes heat from the ammonia vapor. It condenses into liquid ammonia and then flows into the evaporator.

The evaporator is supplied with hydrogen. The hydrogen passes across the surface of the ammonia. It lowers the ammonia vapor pressure enough to allow the liquid ammonia to evaporate. The evaporation of the ammonia extracts heat from the evaporator. This, in turn, extracts heat from the food storage space, lowering the temperature inside the refrigerator.

The mixture of ammonia and hydrogen vapor passes from the evaporator to the absorber. A continuous trickle of weak ammonia solution enters the upper portion of the absorber. It is fed by gravity from the tube (D). This weak solution flows down through the absorber. It comes into contact with the mixed ammonia and hydrogen gases. This readily absorbs the ammonia from the mixture. The hydrogen is free to rise through the absorber coil and to return to the evaporator. The hydrogen circulates continuously between the absorber and the evaporator.

The strong ammonia solution produced in the absorber flows down to the absorber vessel. It passes on to the boiler system, thus completing the full cycle of operation.

This cycle operates continuously as long as the boiler is heated. A thermostat which controls the heat source regulates the temperature of the refrigerated space.

Since the refrigerant is ammonia, it can produce quite low temperatures. Most systems require electrical devices, so both gas and electricity must be supplied. Except for the thermostatic controls and (in some cases) fans, there are no moving parts.

Service is usually quite simple. The burner and stack must be kept clean. The refrigerator should be carefully leveled before being placed in operation.

3.21 Solid Absorbent Refrigeration

Various kinds of solid absorbent refrigerators have been developed. All have depended on the original Faraday experiment.

In 1824, Michael Faraday tried to liquefy certain "fixed" gases. These were gases which certain scientists believed could exist only in vapor form. Among them was ammonia, then regarded as a "fixed" gas.

Faraday knew that silver chloride, a white powder, could absorb large amounts of ammonia vapor. He exposed silver chloride to dry ammonia vapor. He allowed the powder to absorb all of the vapor it would take. Then he sealed the ammonia-silver chloride compound in a test tube which was shaped like an inverted "V."

Faraday then heated the end of the tube containing the powder (dark red). See **Figure 3-21A.** At the same time, he cooled the opposite end with water. The heat released ammonia vapor. Drops of a colorless liquid

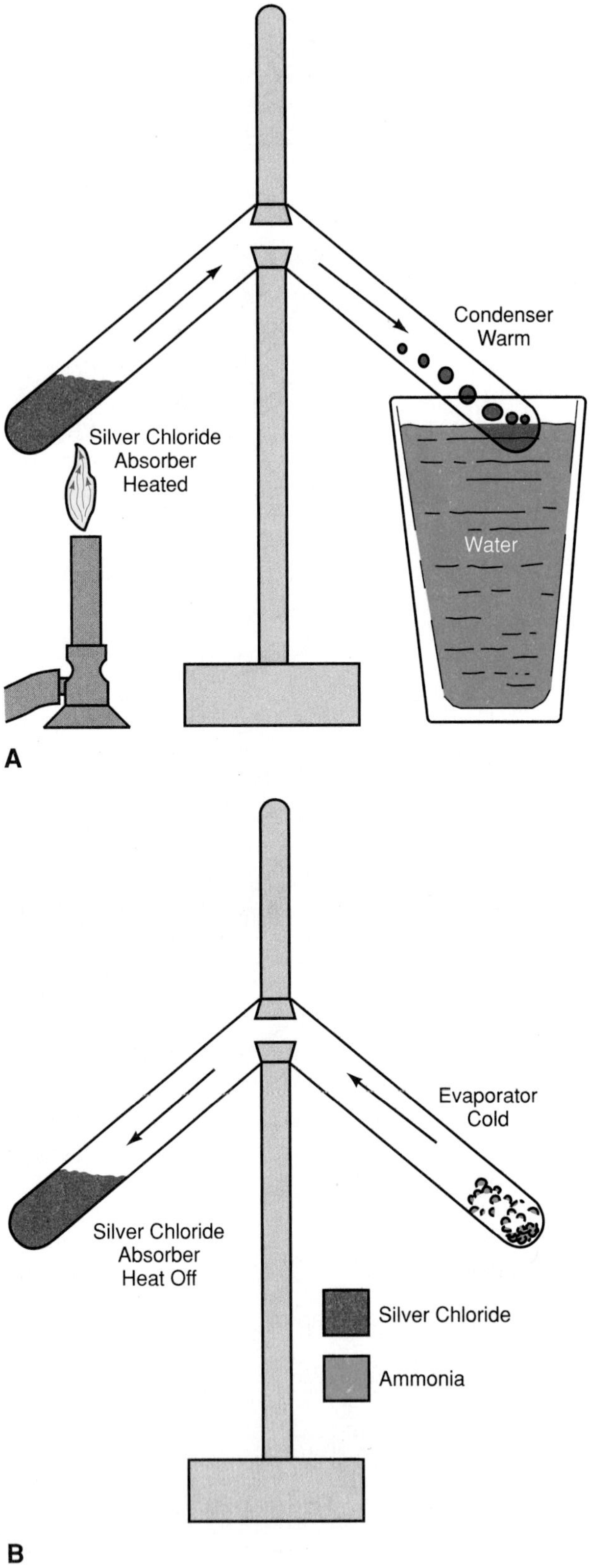

Figure 3-21. *Solid absorbent refrigerator principle in experiment done by Michael Faraday. A—Faraday heated the silver chloride compound while cooling the opposite end of the test tube with water. B—When the heat and water were removed, the liquid ammonia rapidly changed to a vapor and was reabsorbed by the silver chloride powder.*

soon began to appear in the cool end of the tube. It was liquid ammonia.

Faraday continued the heating process until he had enough liquid ammonia for his purpose. Then, he took away the heat, removed the cooling water, and watched the newly discovered substance.

Moments later, Faraday saw something unusual. The liquid ammonia did not remain quietly in the sealed test tube. It began to bubble and then boil, **Figure 3-21B.** The liquid was rapidly changing back into a vapor. The vapor was being reabsorbed by the powder.

When Faraday touched the end of the tube containing the boiling liquid, he found it intensely cold. Ammonia, in changing from liquid to vapor form, had removed heat. It took this heat from the nearest thing at hand—the test tube itself. At one time, many refrigeration cycles used this principle. These are not commonly used anymore. However, cooling mechanisms have been developed on this principle. They use water as the refrigerant and lithium bromide or lithium chloride as the absorbent. See Chapter 17.

3.22 Sophisticated Commercial Systems

Up to now, the compression-type refrigerating systems described have been quite simple. Different conditions and refrigeration requirements require accessory (add-on) devices. Pressure regulators, vibration dampeners, crankcase heaters, and such make the refrigerating systems more efficient and safer. **Figure 3-22** illustrates a small commercial-type refrigerating system using a variety of accessories.

Beginning with the evaporator in the top (warmer) cabinet, refrigerant evaporates (light blue). It flows back through the suction line toward the motor compressor. The suction line then enters an evaporator pressure regulator. From there it leads to the suction line accumulator. Any liquid refrigerant which may come from the evaporator will stay here and evaporate. This prevents it from slugging into or entering the compressor.

The vapor then goes through a suction line filter-drier. The filter-drier traps any moisture or solid impurities. A compressor pressure regulator protects the compressor from excessive low-side pressures.

The suction line vapor then enters a vibration dampener. This is a flexible connection on the low side between the motor compressor and the suction line. The sensing element for the motor compressor control is also attached to the suction line.

A suction service valve (SSV) is located at the entrance to the low side of the motor compressor. Along with the service valve on the high side, this valve makes servicing the motor compressor easy.

A crankcase heater keeps refrigerant from liquefying in the motor compressor during the off cycle when the unit is operating in a cold space.

From the high-side service valve (HSV), the compressed vapor (light red) enters an oil separator. The oil which the separator has removed from the high-pressure refrigerant is returned to the compressor crankcase.

From the oil separator, the high-pressure vapor enters the condenser. The condenser has a head pressure control. When head pressure gets too high, it shuts off the system.

A service valve is placed in the line between the condenser and the liquid receiver. The liquid receiver is a reservoir for liquid refrigerant (dark red).

Another service valve is located at the outlet of the liquid receiver (LRSV). This makes it possible to remove the receiver from the system or store the refrigerant in the receiver during service operations.

The line leaving the liquid receiver is also fitted with a vibration dampener. This flexible tube stops carry-over of vibration to other parts of the system.

As the liquid moves on, it passes through a filter-drier which helps keep the refrigerant clean and dry. A moisture and liquid indicator allows visual inspection. It indicates whether enough refrigerant is flowing. At the same time, its color indicates presence of any moisture in the refrigerant.

Sometimes the motor compressor overheats. This can occur when the suction gas temperature is too high. The liquid desuperheater valve permits light injections of refrigerant to flow into the low side of the system, so that the suction gas is cooled immediately.

A manifold with hand valves lets the liquid pass into one of the two evaporators. The refrigerant flows through a solenoid valve to the top evaporator. This makes it possible to automatically control the flow into the evaporator.

From here, the liquid refrigerant flows into a thermostatic expansion valve. The thermostatic expansion valve regulates the rate of flow (dark blue) of low-pressure liquid. Flow depends on the temperature and pressure of the refrigerant as it leaves the evaporator.

The second evaporator is fitted with an electrically operated defrost control device. When opened, it allows hot compressed vapor to enter the evaporator. The vapor flows back to the compressor without going through the expansion valve. This heats the evaporator quickly. Any frost accumulation on it will be quickly melted.

A hot-gas bypass solenoid valve is used to allow hot refrigerant vapor to enter the suction line. This occurs if the suction line gets too cold and may allow liquid refrigerant to enter the compressor.

This brief description of some of the accessories and their purpose may help in understanding the commercial systems as described later in the text.

3.23 Hot-Gas Defrost

In the hot-gas defrost system, a timing mechanism directs hot high-pressure vapor (light red) through the

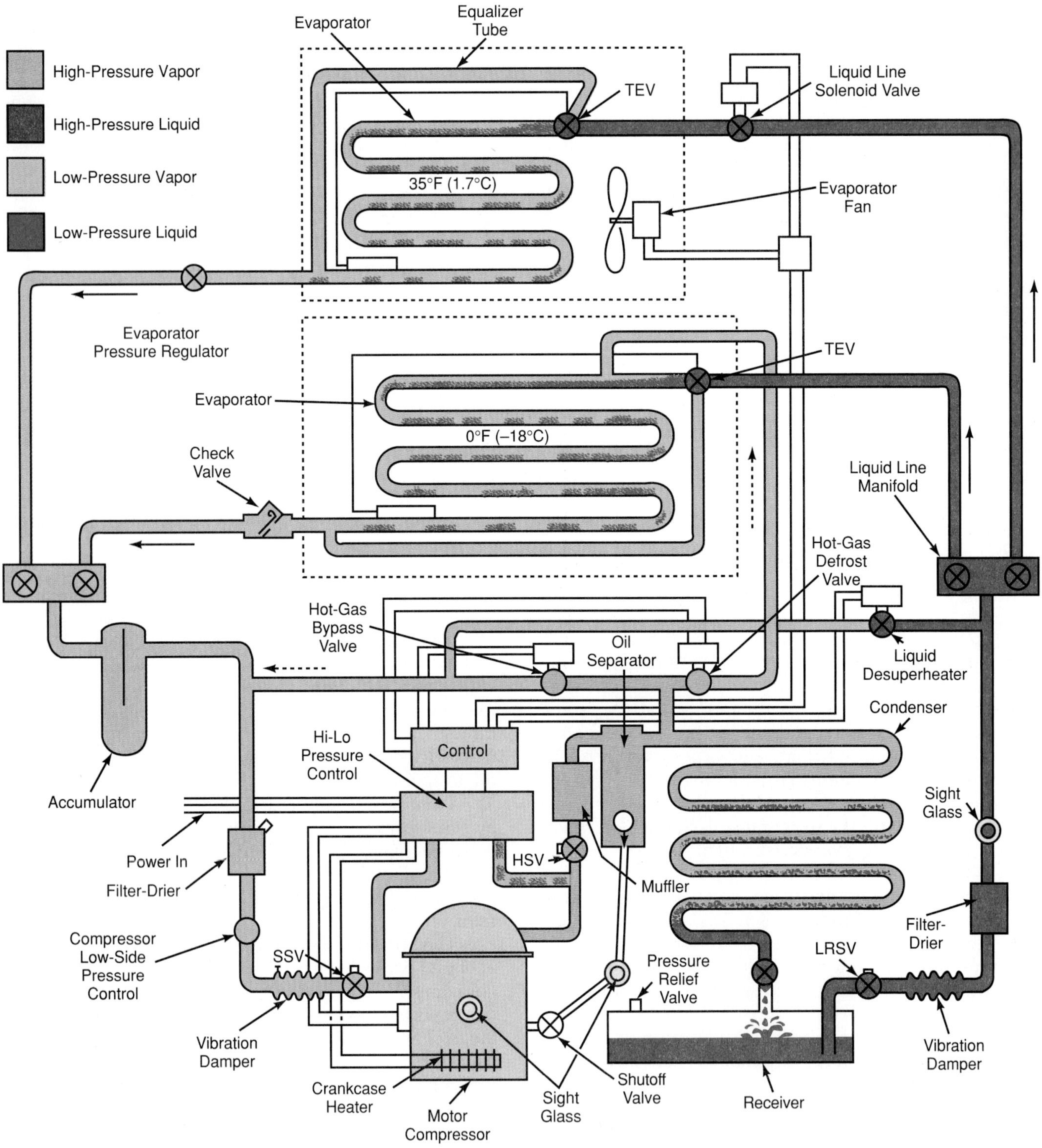

Figure 3-22. *Accessory parts on this commercial system make the unit work better. It is also easier to service.*

evaporator. Its job is to remove frost and ice. **Figure 3-23A** shows how it operates during the refrigerating cycle; **Figure 3-23B** illustrates defrost cycle operation.

Two solenoid valves in the refrigerant circuit control the system. They determine whether to provide either the refrigerating cycle or the defrost cycle. During the refrigerating cycle, as in **Figure 3-23A,** solenoid valve No. 1 is open. The refrigerator is operating normally for a refrigerator using a thermostatically controlled expansion valve.

The liquid refrigerant (dark red) flows from the liquid receiver up through the liquid line. It travels through solenoid valve No. 1, through the thermostatic expansion valve, and into the evaporator. It evaporates under low pressure and absorbs heat. The refrigerant returns as a vapor (light blue) first through the accumulator and then

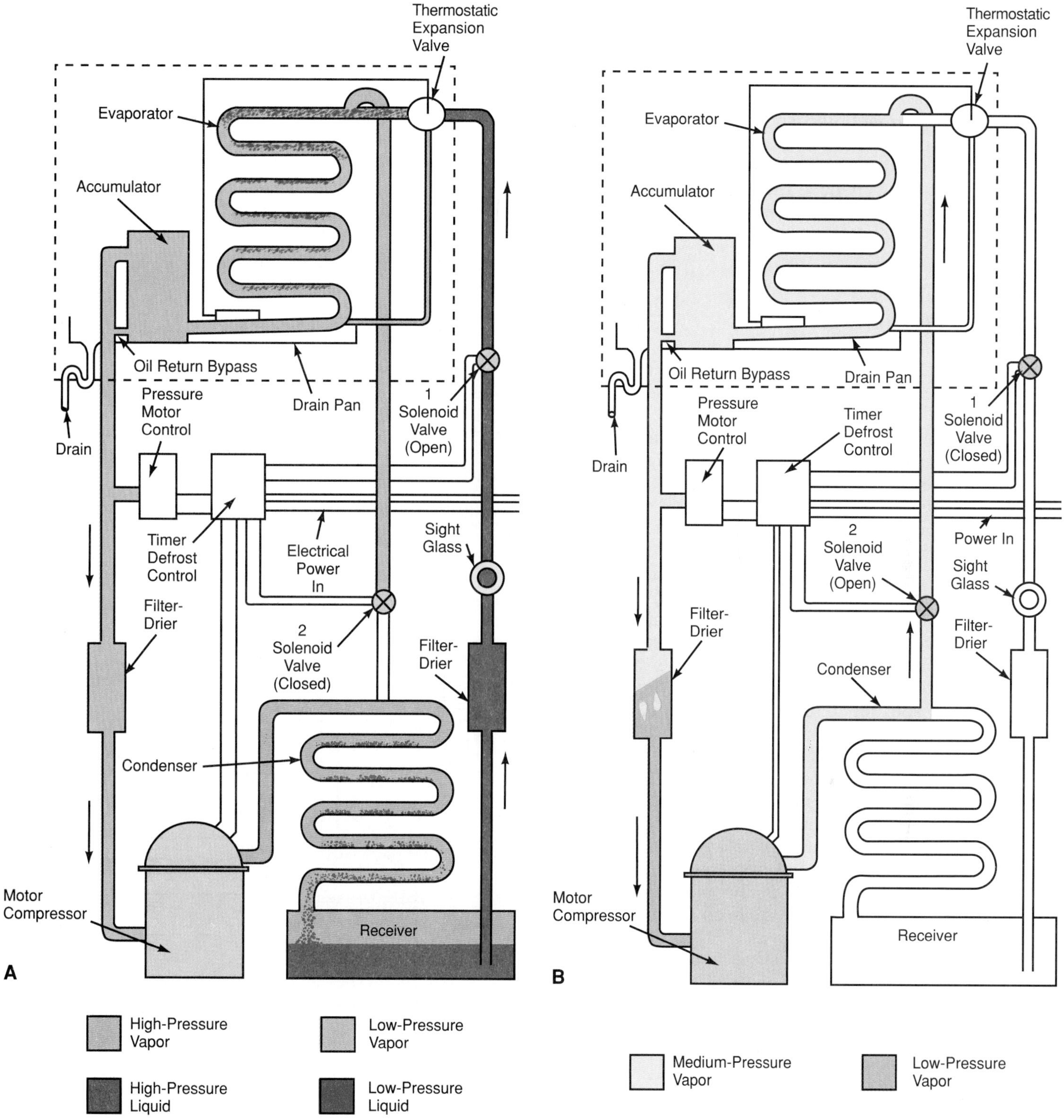

Figure 3-23. *A—hot-gas defrost system during the refrigerating cycle. B—The defrost cycle of a hot-gas defrost system. There is a pressure drop in the evaporator due to ice temperature.*

through the suction line. It then arrives back at the compressor. From the compressor, the hot, compressed vapor (light red) is forced into the condenser. The vapor's heat is removed, and it is condensed back into a liquid.

During the defrost cycle, **Figure 3-23B,** solenoid valve No. 1 is closed. Solenoid valve No. 2 is open. Since solenoid valve No. 1 is closed, no liquid refrigerant is flowing through the thermostatic expansion valve into the evaporator. Since solenoid valve No. 2 is open, the hot compressed refrigerant vapor (light red) flows through it directly into the evaporator. It passes through the evaporator, through the accumulator, and back through the suction line to the suction side of the hermetic (airtight) compressor. As it passes through the evaporator, the hot vapor (light red) melts the ice from the evaporator surface.

The vaporized refrigerant (light blue) picks up some heat passing through the suction line. As it passes through the compressor, the heat of compression raises the temperature to such a level that it continues to heat the evaporator and remove the frost. Little or no condensation of vaporized refrigerant takes place during this cycle.

3.24 Electric Defrost

Electric heating elements placed alongside the evaporator surfaces get warm. They melt the frost and ice buildup from the evaporator. A timer or control mechanism operates the heater during the time that the refrigerating mechanism is in the off cycle. **Figure 3-24A** shows the refrigerating cycle; **Figure 3-24B** shows the defrost cycle.

In **Figure 3-24A,** the electric heating mechanism is in the refrigerating part of the cycle. Liquid refrigerant is vaporized in the evaporator. It absorbs heat and becomes a vapor. While in the evaporator, it flows through an accumulator and passes on to the suction line back to the compressor. In the compressor, it is compressed to a high pressure and high temperature, and then flows into the condenser. Here the heat of vaporization is removed and the refrigerant returns to a liquid flowing into the liquid receiver. The cycle is then repeated.

In **Figure 3-24B,** the same system is in the defrost cycle. The compressor is stopped and the defrost control mechanism lets electric current flow through the resistance heating elements (red) alongside the evaporator surface. Heat warms the evaporator surfaces until the frost and ice are melted. The moisture empties into a drain pan. The drain tubes go to the building drain.

The operation of the resistance units is usually timed. They control both the frequency and the duration of the electric heating. This timing provides for adequate frost removal. The system operates more efficiently with little or no frost on the evaporator surfaces.

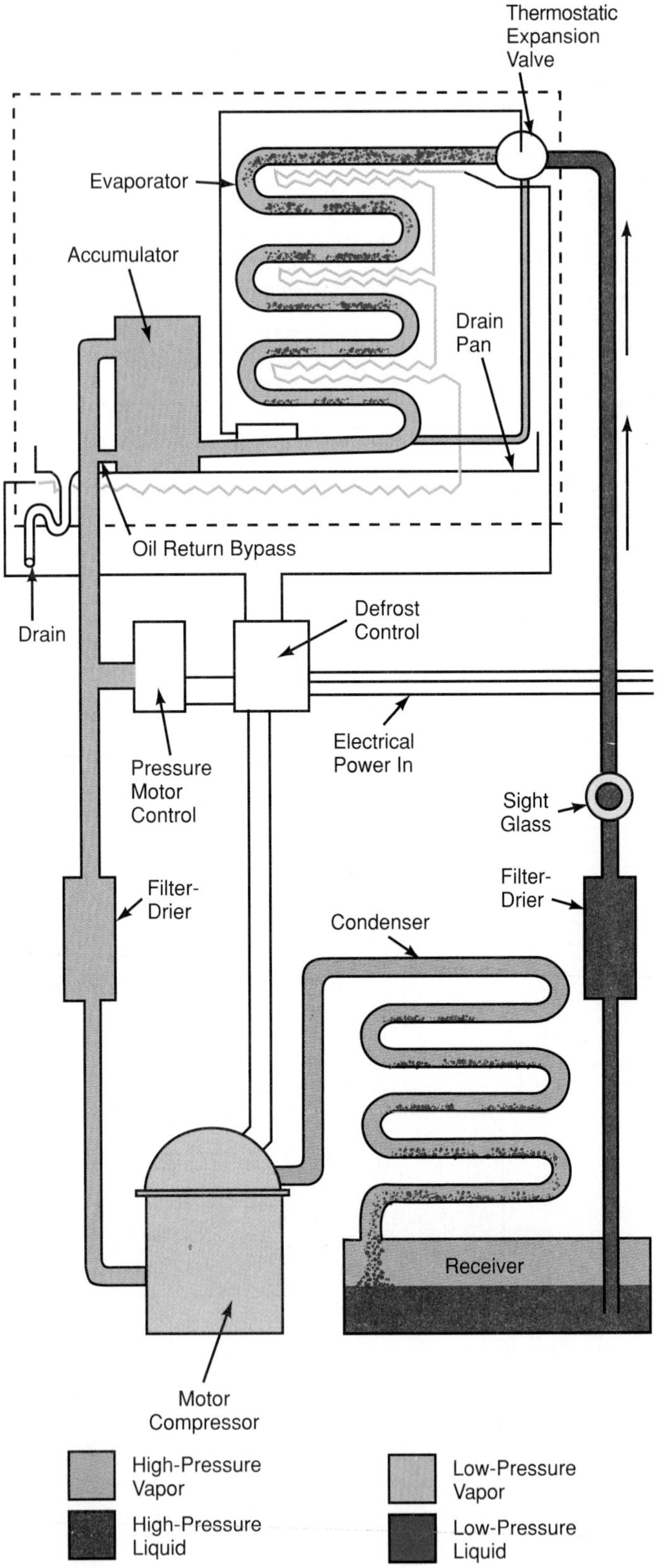

Figure 3-24A. *An electric defrost system during the refrigerating cycle.*

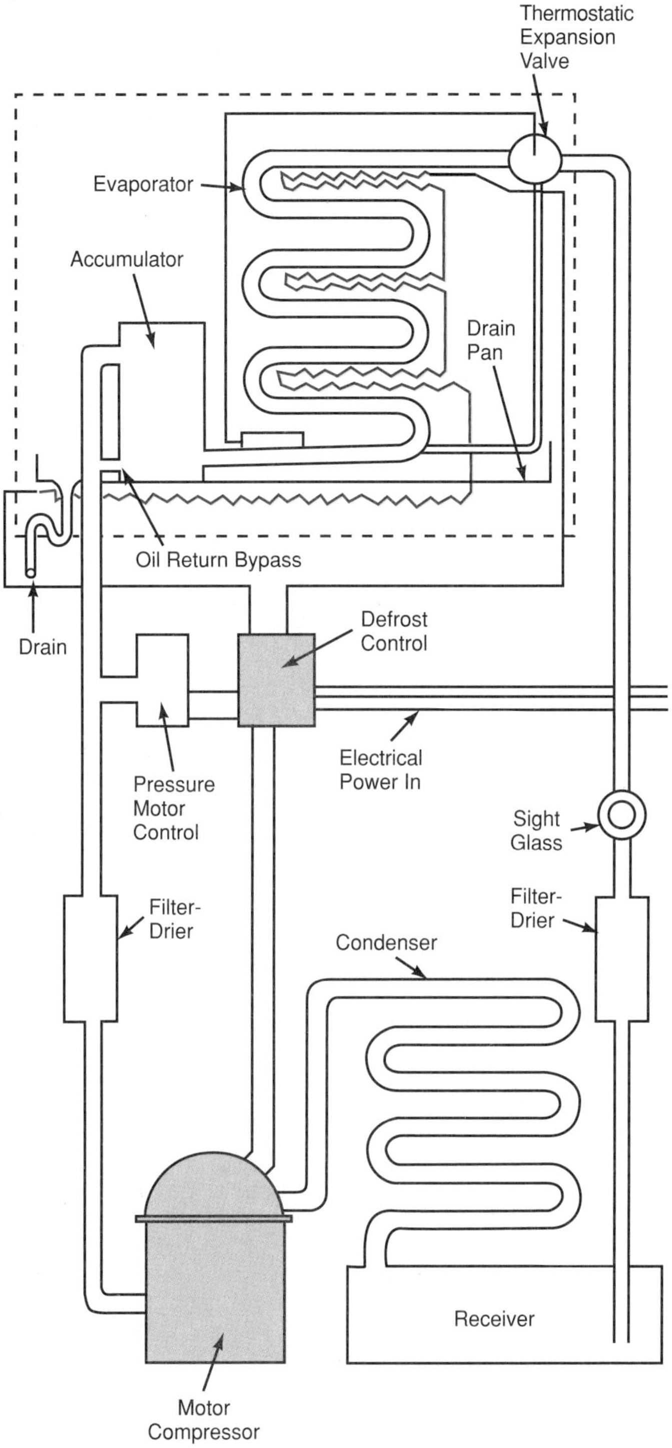

Figure 3-24B. *An electric defrost system during the defrost cycle. Note that the refrigerating unit is not running; therefore, there is no refrigerant flow.*

3.25 Test Your Knowledge

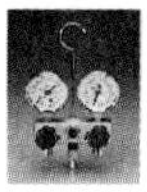

Please do not write in this text. Place your answers on a separate sheet of paper.

1. Which system does *not* get its cooling effect by turning a substance into a vapor?
   A. Desert bag.
   B. Ice box.
   C. Snow machine.
   D. Dry ice cabinet.
2. In the low-side float system of refrigerant control, when does the needle valve open?
   A. This system has no needle valve.
   B. When the liquid level in the evaporator falls.
   C. When the compressor reduces pressure in the evaporator.
   D. When the evaporator pressure is low and the temperature is above normal.
3. In the capillary tube system of refrigerant control, when does the needle valve open?
   A. This system has no needle valve.
   B. When the liquid level in the evaporator falls.
   C. When the compressor reduces pressure in the evaporator.
   D. When the evaporator pressure is low and the temperature is above normal.
4. In the AEV system, when does the needle valve open?
   A. This system has no needle valve.
   B. When the liquid level in the evaporator falls.
   C. When the compressor reduces pressure in the evaporator.
   D. When the evaporator pressure is low and the temperature is above normal.
5. In the TEV system, when does the needle valve open?
   A. This system has no needle valve.
   B. When the liquid level in the evaporator falls.
   C. When the compressor reduces pressure in the evaporator.
   D. When the evaporator pressure is low and the temperature is above normal.
6. _______ are most often used inside an intermittent absorption refrigerator?
   A. Ammonia and water
   B. Hydrogen and water
   C. Lithium bromide (LiBr) and water
   D. Silver chloride (AgCl) and ammonia
7. A _______ system has two compressors with an intercooler between them.
   A. cascade refrigerating
   B. compound refrigerating
   C. modulating refrigerating
   D. multiple evaporator

8. A _______ system has two complete systems linked by a combined evaporator/condenser.
   A. cascade refrigerating
   B. compound refrigerating
   C. modulating refrigerating
   D. multiple evaporator
9. A _______ system would probably be used for a supermarket installation that needs different temperature cabinets.
   A. cascade refrigerating
   B. compound refrigerating
   C. modulating refrigerating
   D. multiple evaporator
10. A _______ system sometimes uses only part of its compressor capacity.
   A. cascade refrigerating
   B. compound refrigerating
   C. modulating refrigerating
   D. multiple evaporator
11. Which of the following characteristics applies to the low-side float-type of refrigerant control?
   A. System is complicated and hard to service.
   B. System is not very efficient.
   C. System requires a large charge of refrigerant.
   D. Refrigerant charge must be carefully measured.
12. Which of the following characteristics applies to thermoelectric refrigeration?
   A. System is complicated and hard to service.
   B. System is not very efficient.
   C. System requires a large charge of refrigerant.
   D. Refrigerant charge must be carefully measured.
13. What principle is used in the continuous-type absorption refrigerator?
   A. Dalton's Law.
   B. The Peltier effect.
   C. The Faraday effect.
   D. The Faraday ice bucket principle.
14. Which of these names does *not* apply to a food truck installation that uses liquid nitrogen?
   A. Chemical refrigeration.
   B. Expendable refrigerant system.
   C. Open-cycle refrigeration.
   D. Open refrigerating system.
15. Dry ice is better than water ice for food shipment in many ways. Which of these "advantages" is *not* true?
   A. Dry ice can absorb more heat than water ice.
   B. Dry ice humidifies food.
   C. Dry ice keeps air away from the food.
   D. Dry ice produces no melt water.
16. In an ice box, the refrigeration temperature will be about _______.
   A. 50°F (10°C)
   B. 32°F (0°C)
   C. −20°F (−29°C)
   D. −50°F (−46°C) and below
17. In a water cooler, the refrigeration temperature will be about _______.
   A. 50°F (10°C)
   B. 32°F (0°C)
   C. −20°F (−29°C)
   D. −50°F (−46°C) and below
18. In a cascade system, the refrigeration temperature will be about _______.
   A. 50°F (10°C)
   B. 32°F (0°C)
   C. −20°F (−29°C)
   D. −50°F (−46°C) and below
19. In a frozen food truck using liquid nitrogen, the refrigeration temperature will be about _______.
   A. 50°F (10°C)
   B. 32°F (0°C)
   C. −20°F (−29°C)
   D. −50°F (−46°C) and below
20. Why do most kitchen refrigerators have a defrost cycle?
   A. Frost buildup corrodes the evaporator tubing.
   B. Frost buildup can squeeze the tubing closed.
   C. Frost buildup on evaporator reduces efficiency.
   D. All of the above.

Chapter 4

COMPRESSION SYSTEMS AND COMPRESSORS

Modules:

Key Words:

flash gas
centrifugal compressor
heat absorber
heat dissipator
reciprocating
Scotch yoke
screw compressor
scroll compressor
swash plate compressor
volumetric efficiency

Learning Objectives:

After studying this chapter, you will be able to:

- State the five thermal laws relating to refrigeration.
- Explain the compression cycle for a domestic refrigerator.
- List the components of a refrigeration compression system.
- Explain the operation of each component of a compression system.
- Trace the flow of refrigerant through a complete refrigeration system.
- Name the two types of motor controls and discuss their operation and purpose.
- Describe the five principal types of refrigerant controls and their operation.
- Name four different types of compressors.
- Explain how compressors operate.
- Identify the internal parts of a compressor.
- **Follow approved safety procedures.**

COMPRESSION SYSTEMS MODULE

4.1 Laws of Refrigeration

All refrigerating systems depend on five thermal laws:

1. Fluids absorb heat while changing from a liquid state to a vapor state. Fluids give up heat in changing from a vapor to a liquid.
2. The temperature at which a change of state occurs is constant during the change, provided the pressure remains constant.
3. Heat flows only from a body which is at a higher temperature to a body which is at a lower temperature (hot to cold).
4. Metallic parts of the evaporating and condensing units use metals that have a high heat conductivity (copper, brass, aluminum).
5. Heat energy and other forms of energy are interchangeable. For example, electricity may be converted to heat; heat to electrical energy; and heat to mechanical energy.

4.2 Compression Cycle

The compressor changes the refrigerant vapor from low pressure to high pressure during the compression cycle. This pumping transfers heat from the inside of the cabinet to the outside. The compressor transfers heat from one place to another, similar to the heat pump.

A refrigerating system consists of a high-pressure side and a low-pressure side, **Figure 4-1.** A refrigeration cycle follows these steps: From the liquid receiver, liquid refrigerant (at high pressure) flows through the refrigerant control. The refrigerant control is a pressure reducer. The refrigerant moves into the evaporator. The evaporator is under a low pressure. Here the liquid refrigerant vaporizes (boils) and absorbs heat.

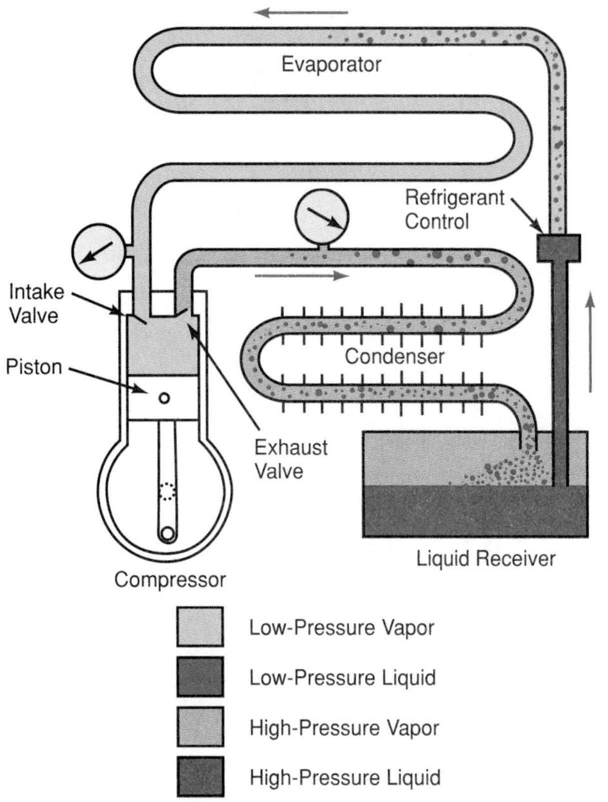

Figure 4-1. *Compression cycle showing the two pressure conditions. Low-pressure side extends from refrigerant control, through evaporator, to the compressor intake valve. High-pressure side begins in the cylinder above the piston, on the compression stroke. It extends from exhaust valve, through condenser, liquid receiver, and liquid line, to refrigerant control.*

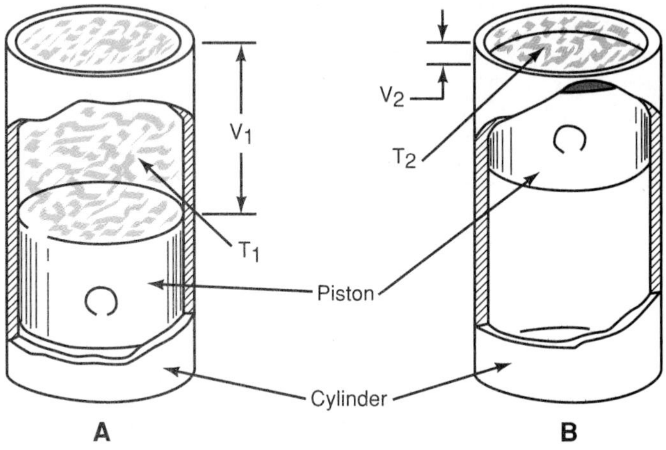

Figure 4-2. *Heat of the vapor compressed into a small space raises vapor temperature greatly. A—Volume V_1 = volume of vapor at the end of the intake stroke = 8 in^3 (131 cm^3). T_1 = temperature of vapor at end of intake stroke = 50°F (10°C). B—Volume V_2 = the volume of the vapor at the end of compression stroke = 1/2 in^3 (8.2 cm^3). T_2 = the temperature of the vapor at the end of the compression stroke = 250°F (121°C).*

The vapor then flows into the compressor through the intake valve and back into the compressor cylinder. On the compression stroke, the piston squeezes the vapor into a small space. This increases the vapor temperature. **Figure 4-2** illustrates this principle.

The compressed high-temperature vapor is pushed through the exhaust valve and into the condenser. (See **Figure 4-1**.) In the condenser, heat from the refrigerant is passed onto the surrounding air. In giving up this heat, it returns to a liquid. This liquid is stored in the receiver. From here the cycle is repeated.

In operation, the system transfers heat from one place to another place. It takes heat from the inside of a refrigerator to the outside air. A water cooler takes heat from the inside water cooler to the outside air. This is similar to using a sponge. Water is picked up in one place and released in another place by squeezing the sponge.

To have a transfer of heat, there must be a temperature difference. To get the temperature difference, there must be a low-pressure side ***(heat absorber)*** and a high-pressure side ***(heat dissipator).*** Various compression cycles are illustrated in Chapter 3.

4.2.1 Operation of Compression Cycle

Figure 4-3 illustrates a typical compression cycle as used in a domestic refrigerator. It has certain necessary parts which will be explained.

In any compression refrigeration system, there are two different pressure conditions. One is called the ***low side*** and the other side is called the ***high side.*** The evaporator is in the low side. Heat is absorbed in the low side. The accumulator, suction line, and entrance to the compressor suction valve are also on the low side.

The condenser is in the high side. This is where the heat is released from the refrigerant. The compressor exhaust valve and liquid receiver (if used) are on the high side. The liquid line filter-drier, liquid line, and the refrigerant control are also on the high side.

A thermostat maintains correct operating temperature by controlling the motor electrical circuit.

4.2.2 Temperature and Pressure Conditions in the Compression Cycle

Upon starting the compressor, it moves refrigerant molecules from the low-pressure side to the high-pressure side. This is done without much difficulty. The molecules are not moving about much faster. See **Figure 4-4A.** These molecules of refrigerant enter the condenser from the compressor through the opening at 1. The temperatures are the same (70°F [21°C]) inside and out. In this case, the *pressure* is the sum of the bombarding molecules. The *temperature* is the speed of molecular motion (how fast the molecules move to and from).

It is necessary therefore to speed up the molecules. Their temperature must be increased to a point where they give up heat to surrounding cooling surfaces (air and water).

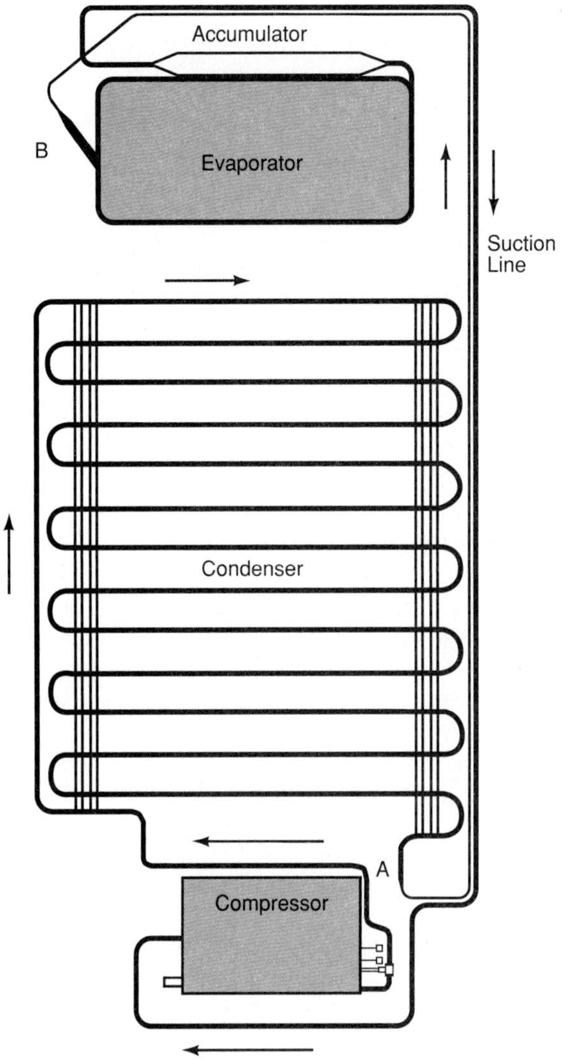

Figure 4-3. *Compression cycle showing the flow of refrigerant. Note that the capillary tube extends from A to B. (Hotpoint Div., General Electric Co.)*

The longer the compressor runs, the more vapor molecules it squeezes into the condenser. With each stroke, the pressure and temperature increase. This is due to more molecules hitting the sides of the container. The compressor piston, pushing the vapor molecules against the higher pressure, hits them harder. This speeds up the molecules and increases their temperature.

During compression, the pressure increases (due to Boyle's Law). At the same time, the temperature increases (Charles' Law). This continues until the vapor temperature is higher than the temperature of the condenser cooling medium. Boyle's Law and Charles' Law are discussed in Chapter 31.

The higher temperature **(Figure 4-4B)** causes a flow of heat to the surrounding metal and air. Heat now moves from the vaporized refrigerant, 2, to the cooling medium. This cooling continues until enough heat loss makes some vapor molecules become liquid molecules

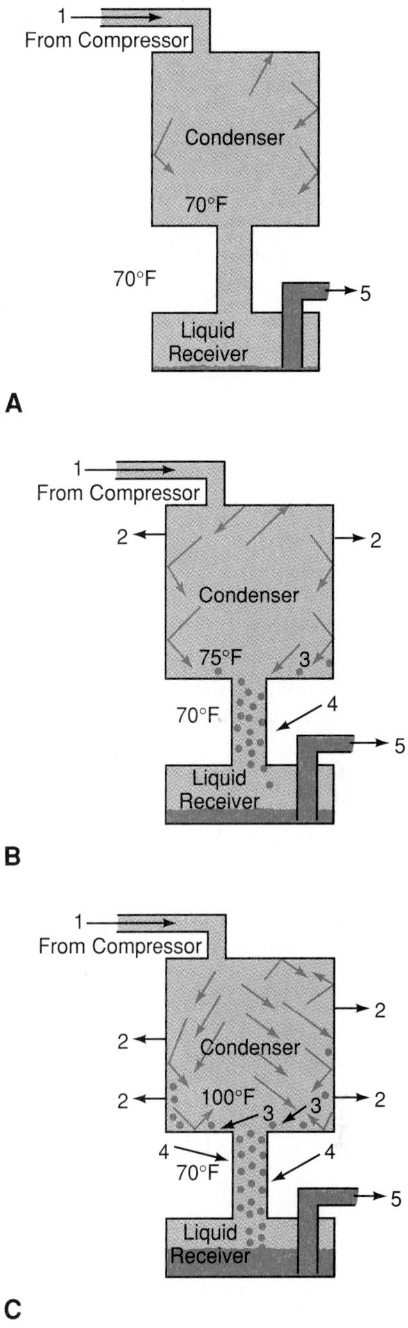

Figure 4-4. *Refrigerant condition as it changes from vapor to liquid in condenser. At A, condensing unit is just starting. At B, the unit has been operating long enough to condense some refrigerant vapor. At C, unit is in state of equilibrium (balance)—heat is being removed and vaporized refrigerant is being condensed at same rate it is being pumped into condenser. A, B, C—1. Vapor enters under pressure. B—2. Heat moving from condenser (small amount). B—3. Vapor losing heat and condensing to liquid (small amount). B—4. Condensed refrigerant enters liquid receiver (small amount). C—2. Heat moving from condenser (large amount). C—3. Condensed refrigerant droplets (large amount). C—4. Condensed refrigerant flowing into liquid receiver (larger amount). Refrigerant liquid line to refrigerant control is shown at 5.*

(see B3). As these collect, they flow into the liquid receiver (see B4).

The temperature and pressure will continue to rise until a balance is reached. Just as many vapor molecules condense into a liquid as the compressor pumps into the condenser (see **Figure 4-4C**).

If anything changes this balance, the condensing pressure and temperature will adjust accordingly. For example, if the room gets warmer, the pressure and temperature will rise again. This continues until just as many vapor molecules are condensing as are being pumped into the condenser.

After condensing (liquefying), the refrigerant is stored in the liquid receiver until needed. When needed, it passes through the high-pressure liquid line, 5, to the refrigerant control. Here refrigerant pressure is reduced to allow evaporation of the liquid at a low temperature.

The evaporating liquid absorbs much heat, thereby supplying refrigeration. The increase in volume, as it evaporates, pushes the vaporized refrigerant through the "suction" line. Finally, the refrigerant goes to the compressor intake, where pressure is greatly reduced. It then passes through the intake valve of the compressor. It travels into the cylinder where a new cycle begins.

4.3 Evaporator

Liquid refrigerant entering the evaporator from the refrigerant flow control is suddenly under low pressure. This makes it vaporize (boil) and absorb heat. The vapors move on into the suction line. The accumulator holds any refrigerant which has not vaporized. This prevents liquid refrigerant from flowing into the suction line.

There are two main types of evaporators—a dry system and the flooded system. ***Dry system evaporators*** are fed refrigerant as quickly as is needed to maintain the desired temperature. In the ***flooded system,*** the evaporator is always filled with liquid refrigerant. The type of refrigerant control used determines the type of evaporator to be used.

Evaporators are made in four different styles:

- Shell-type, **Figure 4-5.**
- Shelf-type, **Figure 4-6.**
- Wall-type, used in chest-type freezer, **Figure 4-7.**
- Fin tube-type with forced circulation. This type of evaporator is most used with frost-free construction. (See **Figure 4-8A and B.**)

Frost-free refrigerators usually need a fan or fans. They circulate the air over the evaporator and distribute cold air throughout the cabinet.

There are many types of commercial system evaporators. Air cooling and liquid cooling are two basic designs. They are constructed of plain or finned tubing or have a flat plate design. Details of commercial system evaporators are explained in Chapter 13.

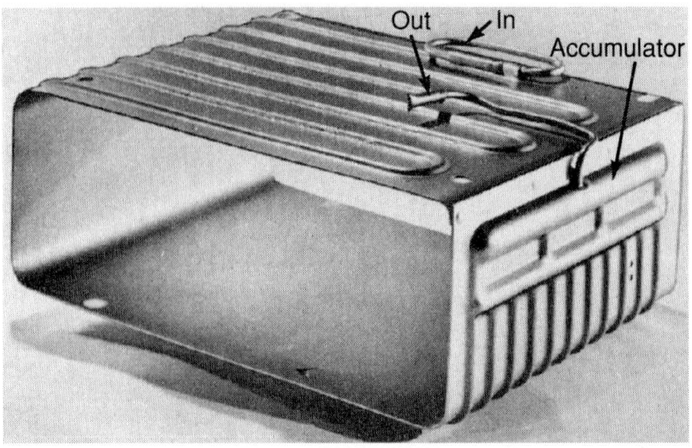

Figure 4-5. *Shell-type evaporator used with capillary tube or high-side float-type refrigerant control.*

4.4 Accumulator

The ***accumulator*** is a safety device. It prevents liquid refrigerant from flowing into the suction line and into the compressor. Liquid refrigerant, flowing into the compressor, may cause considerable knocking and damage to the compressor.

A typical accumulator, **Figures 4-5** and **4-6,** has the outlet at the top. Any liquid refrigerant that flows into the accumulator will be evaporated. Then vapor only will flow into the suction line. Since the accumulator is located inside the cabinet, it also provides some refrigeration.

Not all refrigerating systems have accumulators. Commercial system accumulators are explained in Chapter 13.

4.5 Suction Line

The ***suction line*** carries the refrigerant vapor from the evaporator to the compressor. The line must be large enough to carry the vaporized refrigerant with minimal flow resistance. It should slope from the evaporator or accumulator down to the compressor. If it does not slope, pockets of oil collect.

The liquid line may be in contact with all or part of the suction line length. This cools the liquid refrigerant, helping to reduce flash gas in the evaporator. (***Flash gas*** is the instantaneous evaporation of some of the liquid refrigerant in an evaporator, which cools the remaining liquid refrigerant to the desired evaporation temperature.) It also adds some superheat to the refrigerant vapor entering the compressor. See Section 5.1.2 for an explanation of superheat.

4.5.1 Low-Side Filter-Drier

Some systems include a low-side filter-drier at the compressor end of the suction line. These may be a part of the original system. They may also be placed in the system for a short time to clean it. **Figure 4-9** shows a typical suction line filter-drier. The filter-drier used in the suction

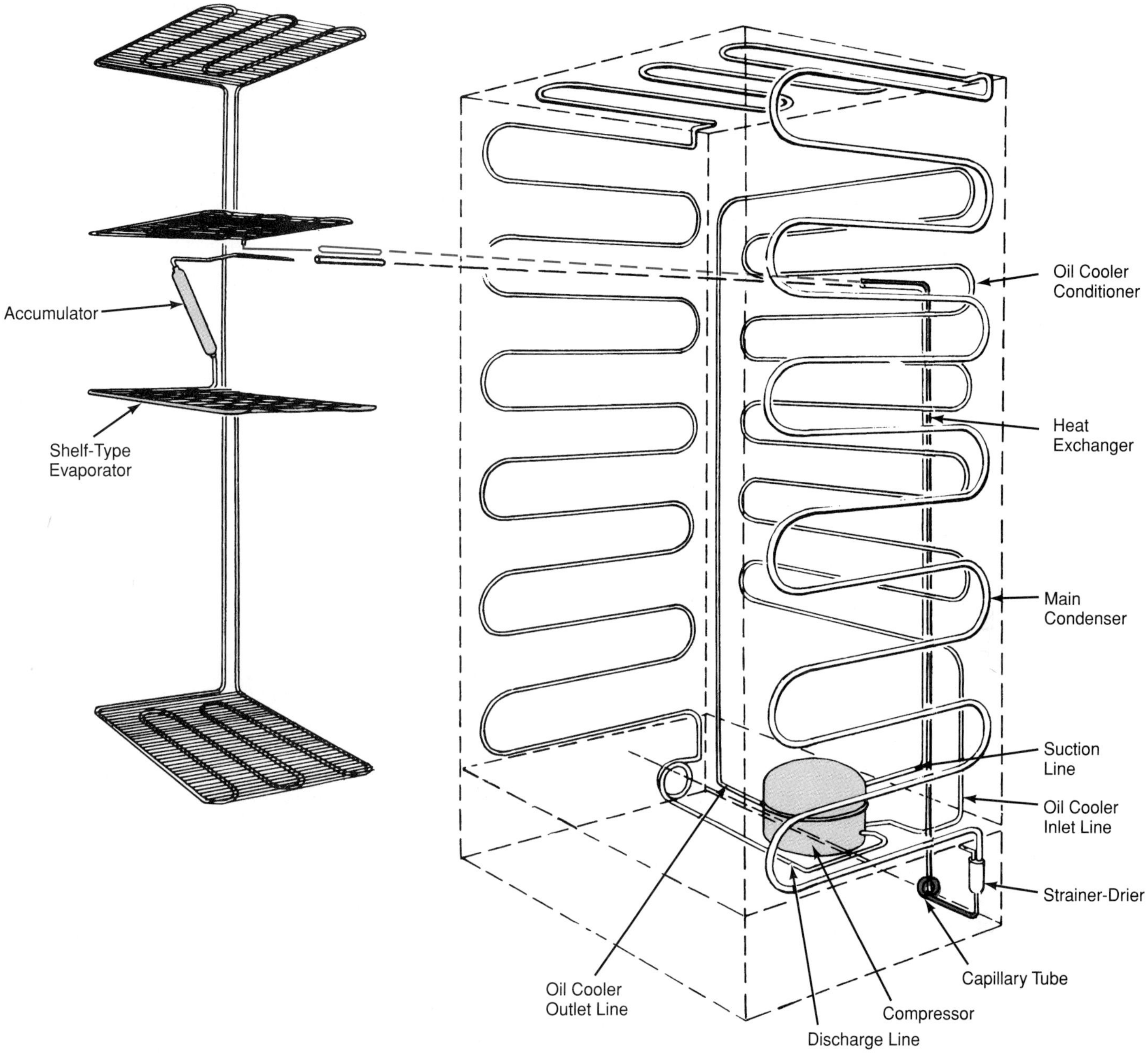

Figure 4-6. *This illustration shows a shelf-type evaporator as it forms the shelf in upright freezer. An accumulator is located at outlet of evaporator. This is a small reservoir to catch refrigerant not needed in evaporator.*

line should offer little resistance to vaporized refrigerant flow. The pressure difference between the evaporator and the inlet to the compressor should be small.

4.5.2 Compressor Low-Side or Suction Service Valve

Many systems have ***service valves*** that allow the technician to connect gauges to the system. These valves will also allow checking pressures, and adding or removing refrigerant or oil.

A typical compressor suction service valve is pictured in **Figure 4-10.** This valve is connected to the compressor at the compressor inlet union. The suction line from the evaporator is attached at the low-side inlet. Sealing caps protect the charging and gauge opening port and valve stem when the valve is not in use.

More recent domestic models do not have service valves. The service technician must use a saddle valve (covered in Chapter 12).

4.6 Compressor

The refrigeration ***compressor*** is a motor-driven device, which removes the heat-laden vapor refrigerant from the evaporator. The compressor compresses

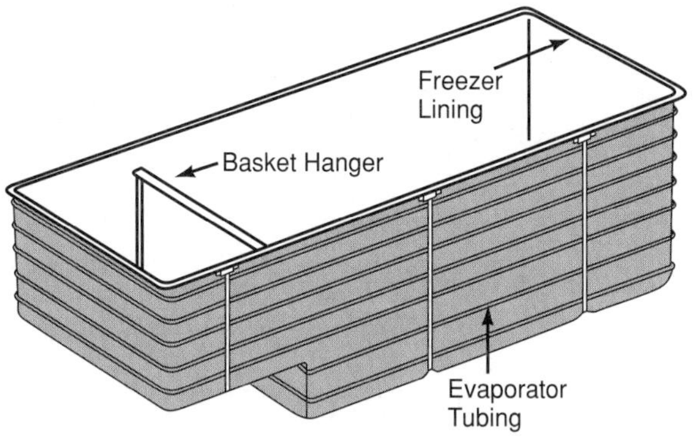

Figure 4-7. *Wall-type evaporator. Note that evaporator tubing is attached to lining of freezing cabinet. This arrangement provides smooth inside surface with uniform cooling throughout cabinet.*

(squeezes) the vapor into a small volume at a high temperature. The various types of pumping mechanisms (compressors) used are explained later in this chapter.

4.6.1 Compressor High-Side Service Valve

The ***compressor high-side service valve*** provides a shutoff between the compressor and the condenser. It also provides an opening for a high-pressure gauge or a gauge manifold.

With the valve closed, the compressor may be disconnected from the condenser without refrigerant leakage. When the valve stem is all the way out, the opening for the gauge is closed.

Figure 4-11 illustrates a cross section of the service valve. It is not used on all refrigerating systems.

4.7 Oil Separator

Refrigeration compressors get their lubrication from a small amount of special lubricating oil. This oil is placed inside the compressor crankcase or housing. It is circulated to various compressor parts. In a hermetic (airtight) system, this oil also lubricates the motor bearings.

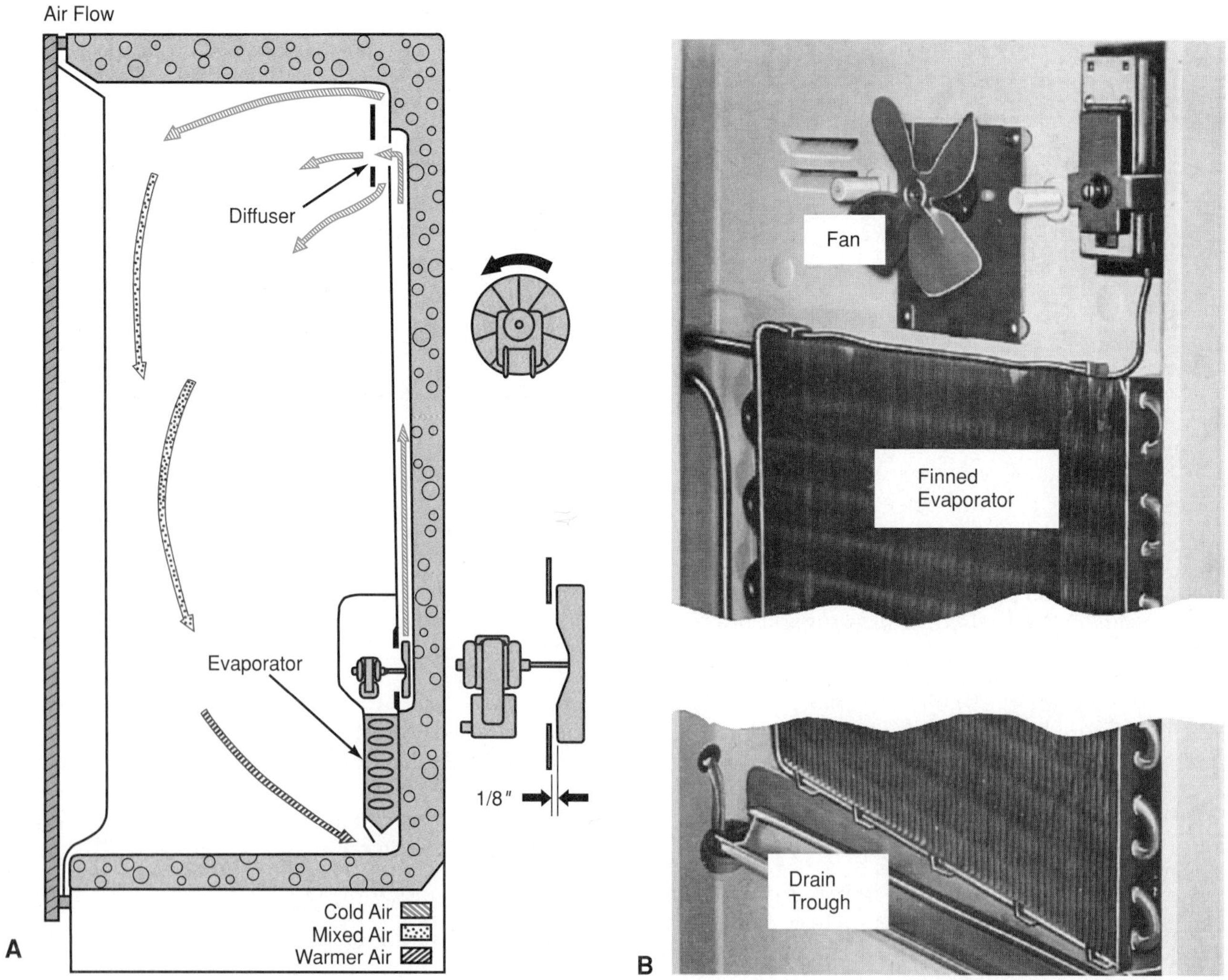

Figure 4-8. *Forced convection evaporator. A—A forced circulation evaporator is used in this upright freezer cabinet. Door switch stops fan when cabinet door is opened. B—Fin-type evaporator. Note the trough to collect and carry away the defrost moisture which drains from the evaporator during the defrost part of the cycle. (Frigidaire Company)*

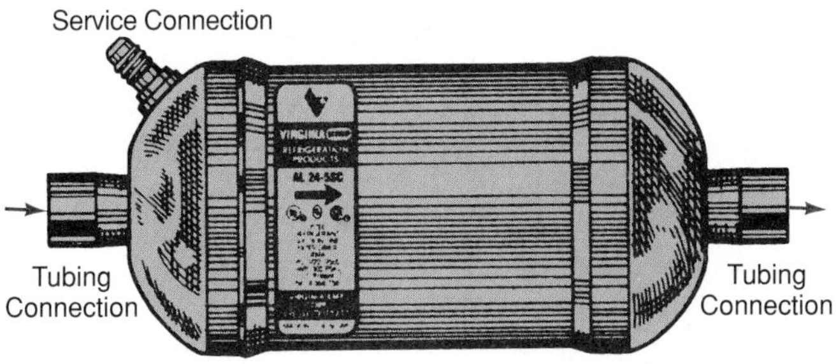

Figure 4-9. *Suction line filter-drier. Direction of refrigerant vapor flow is indicated. (Virginia KMP Corp.)*

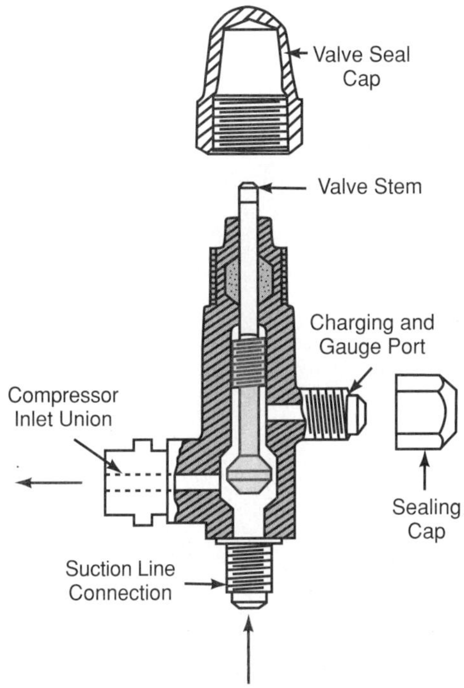

Figure 4-10. *Compressor low-side or suction service valve. If valve stem is turned all the way in, it closes off the connection from the compressor to the suction line. In this position, if valve is removed from compressor, suction line remains sealed. If valve stem is turned out as far as possible, it closes off the opening called the "charging and gauge port." This makes it possible to install a compound gauge or charging line.*

When the compressor operates, small amounts of oil are pumped out with the hot, compressed vapor. A small amount of oil throughout the system does no harm. Too much oil entering the condenser, refrigerant controls, evaporator, and filters interferes with their operation.

It is possible to separate the oil from the hot, compressed vapor. This involves placing an oil separator between the compressor exhaust and the condenser. The location and operation of such a separator is shown in **Figure 4-12.** The separator is enlarged in the illustration to help show details.

The oil separator is a tank or cylinder. It contains a series of baffles or screens which collect the oil. The oil,

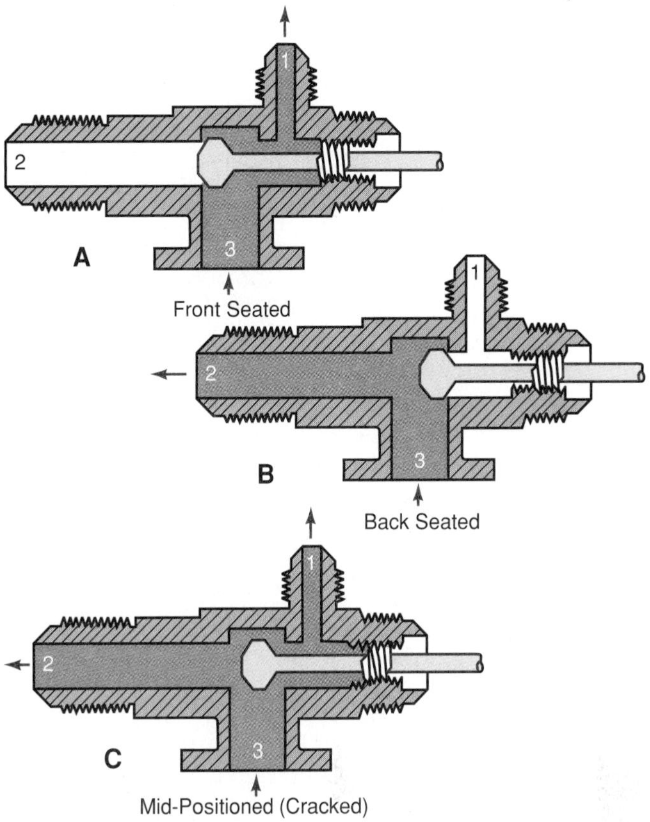

Figure 4-11. *A compressor high-side service valve. A—If valve is turned all the way in, it shuts off connection between compressor, 3, and condenser, 2. B—If valve is turned all the way out, it closes off connection to gauge port, 1. C—At mid-position, all passages are open. (Murray Corp.)*

separated from the hot, compressed vapors, drops to the bottom of the separator.

A float arrangement controls a needle valve. This opens an oil return line to the compressor crankcase. When the oil level is high enough, the float rises and opens the needle valve. The pressure in the separator is considerably higher than the pressure in the compressor crankcase. This causes the oil to return quickly to the compressor crankcase.

Oil separators are quite efficient. Very little oil passes on into the system. They are most commonly used in large commercial installations.

4.8 Condenser

The ***condenser*** in the refrigeration cycle removes the condensation heat from the refrigerant vapor. This heat is picked up in the evaporator. Domestic refrigerators commonly use the following types of condensers, as shown in **Figure 4-13:**

- Finned–static (natural convection).
- Finned–forced convection.
- Wire–static.
- Plate–static.

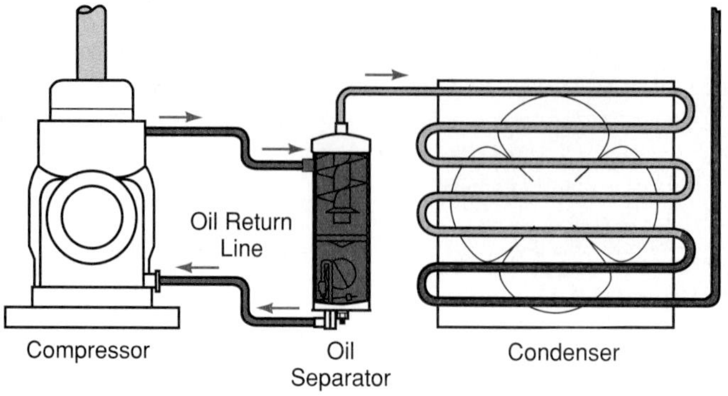

Figure 4-12. *An oil separator located in the discharge line. Note flow of refrigerant and oil. (AC&R Components, Inc.)*

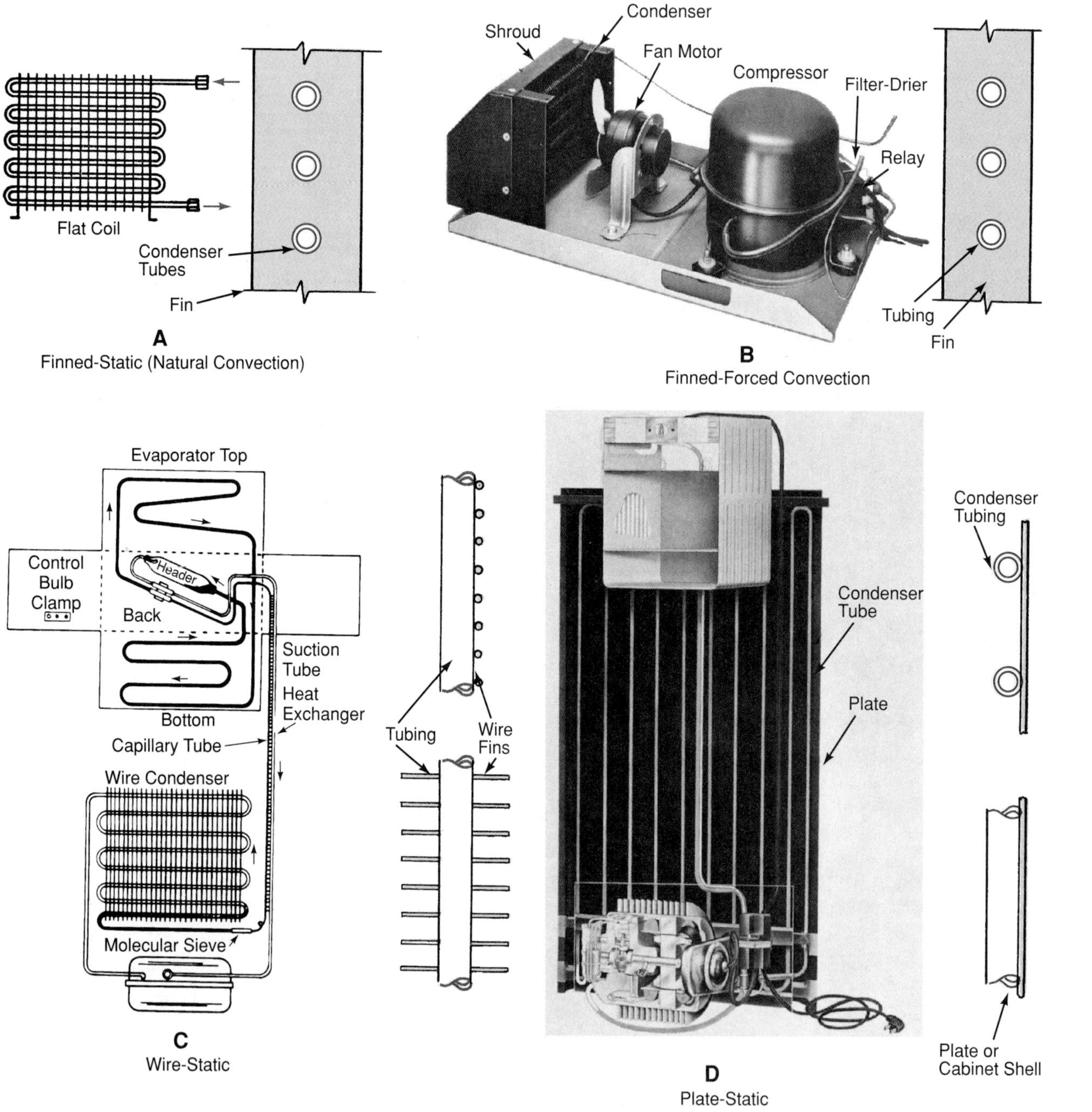

Figure 4-13. *Common domestic condensers.*

Figure 4-13A is a common finned-type static condenser. ***Static*** means that air circulation through the condenser tubing and fins is by natural convection. This is because warm air tends to rise. As the air in contact with the fins and tubes becomes heated, it rises. Cooler air takes its place. The tubes and fins are usually made of copper or steel.

Figure 4-13B is a forced convection finned-type condenser. Whenever the compressor is operating, the motor-driven fan forces air through the condenser.

Figure 4-13C shows a wire-type condenser, which uses small metal wires brazed or spotwelded to the condenser tubing. It is usually a static condenser.

The plate-type condenser is shown in **Figure 4-13D.** In this type of condenser, the tubes are soldered or brazed to a flat metal surface. This is a very common type of condenser construction. It is used on many chest-type freezers. The condenser tubes are attached to the inside (insulation side) of the freezer's outer shell. This type of condenser is very easy to keep clean. It is only necessary to wipe off the surface of the cabinet shell. For proper removal of heat from the refrigerant vapor, always keep the condenser area clean.

Commercial systems use three types of condensers:

- Finned–static, air-cooled.
- Finned–forced convection, air-cooled.
- Water-cooled, tube-in-a-tube, and shell.

Finned–static and finned–forced convection condensers are built much the same as the domestic condensers. However, finned–static and finned–forced convection condensers are larger. Finned–forced convection condensers are used on many commercial refrigeration installations. Some applications of these are shown in Chapter 13.

Water-cooled condensers usually consist of two tubes, one within the other. Water circulates through the inside tube. Hot, compressed vapor circulates through the space between the tubes. These condensers are usually called ***tube-within-a-tube condensers.*** They are very efficient.

Commercial refrigeration and air conditioning have developed rapidly. Many communities now have difficulty in supplying enough water for water-cooled condensers. Forced convection air-cooled condensers and "cooling towers" are increasingly being used in connection with water-cooled condensers.

See Chapter 13 for illustrations of cooling tower applications and water-cooled condensers. **Figure 4-14** shows a large, six-fan, roof-mounted, condensing unit. Diagram of a water-cooled condenser is shown in the bottom view.

4.9 Liquid Receiver

The ***liquid receiver*** is a storage tank for liquid refrigerant. Refrigerant is pumped out of various parts and stored in the liquid receiver during servicing. Its use makes the quantity of refrigerant in a system less critical.

Occasionally, a liquid receiver is built into the bottom of the condenser. Most receivers have service valves. See **Figure 4-15.** A fine copper mesh in the outlet prevents dirt from entering the refrigerant control valves.

Liquid receivers are often found on systems which use the low-side float or the expansion valve-type refrigerant control. Capillary tube systems do not use liquid receivers. (All liquid refrigerant is stored in the evaporator during the Off part of their cycle.) There has been a greater use of hermetic systems and capillary tube refrigerant controls. This has reduced the need for liquid receivers in domestic systems and many small commercial units.

On larger commercial systems, the receiver provides reserve liquid refrigerant. This ensures that the liquid line refrigerant is subcooled and free of flash gas. The receiver must provide enough room for refrigerant during automatic pumpdowns (for defrost purposes and when some of the evaporators are not in use).

Some systems, which have an outdoor air-cooled condenser, need room in the receiver for extra refrigerant. Without extra room, liquid partly fills the condenser when the head pressure is too low. The liquid will not move through the condenser.

4.10 Liquid Line

Copper tubing is commonly used to carry the liquid refrigerant from the condenser to the evaporator. However, domestic units often use steel. These lines are mounted in back of the refrigerator cabinet. They may also be hidden behind the breaker strip at the refrigerator door jamb (frame).

The lines are soldered or brazed to fittings. It is important to avoid pinching or buckling these lines. Support them to prevent wear or breakage due to vibration. Refrigerant lines in commercial units may be connected by soldering, brazing, or by flared fittings.

Often, the liquid line runs parallel to and in contact with the suction line. The reason for this is explained in Section 4.5.

4.10.1 Liquid Line Filter-Drier

It is common practice to install a filter-drier in the liquid line. This tank-like accessory keeps moisture, dirt, metal, and chips from entering the refrigerant flow control. The drying element in the filter removes moisture. This moisture might otherwise freeze in the refrigerant flow control. Moisture is also harmful when mixed with oil in a system since it forms sludges and acids. Moisture is especially harmful to hermetic units. A liquid line filter-drier is shown in **Figure 4-16.**

Some filter-driers are equipped with a sight glass which will indicate refrigerant level. Many sight glasses also have a chemical which will change color. The color change indicates that the system has moisture in it.

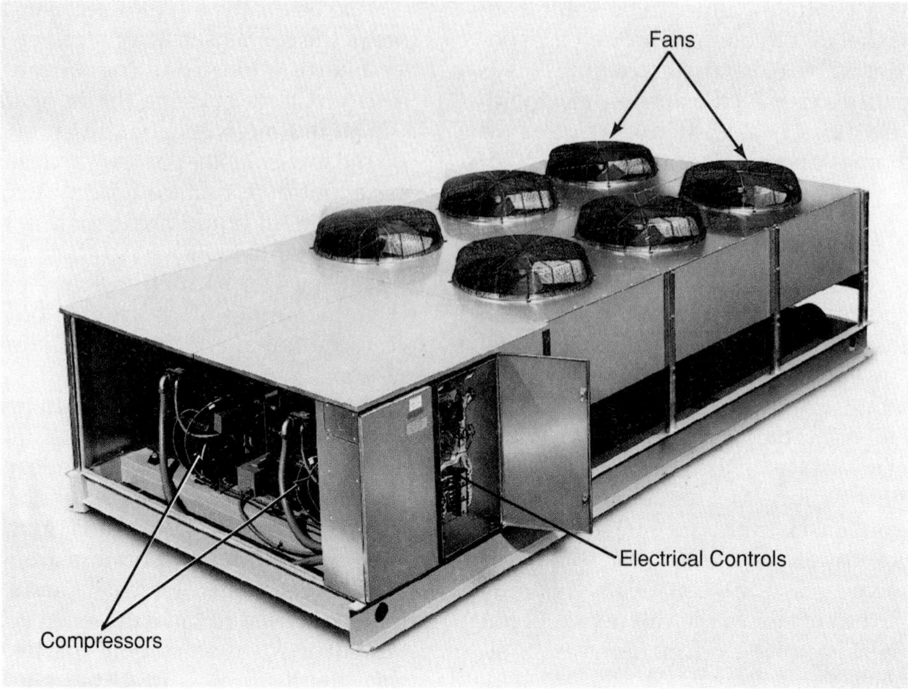

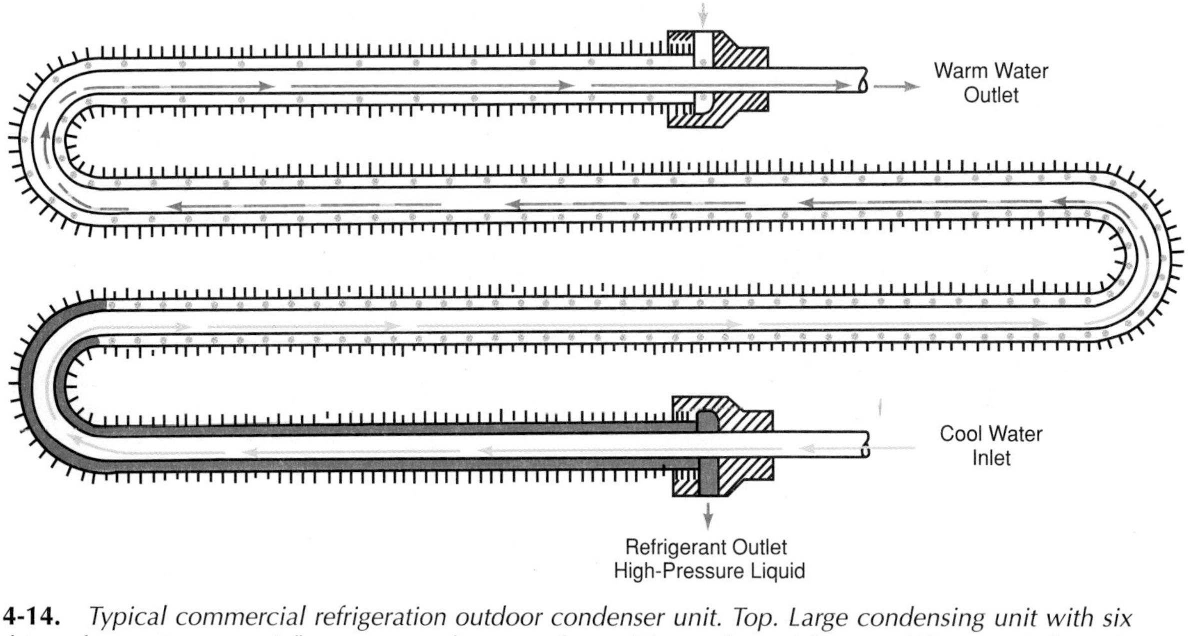

Figure 4-14. *Typical commercial refrigeration outdoor condenser unit. Top. Large condensing unit with six motor-driven fans to increase airflow over condenser surfaces. (Heatcraft, Inc.) Bottom. Water-cooled condenser flow diagram. Refrigerant flows through condenser in opposite direction of water.*

4.11 Types of Refrigerant Flow Control

The ***refrigerant flow control*** has two jobs. It allows liquid refrigerant to enter the evaporator. At the same time, it maintains the required evaporating pressure in the evaporator.

Several types of refrigerant flow controls are used in modern refrigerating mechanisms. These controls and their characteristics are explained in Chapter 5.

There are five principal types of refrigerant flow controls:

- Capillary Tube (CAP)—(dry system).
- Automatic Expansion Valve (AEV)—(dry system).
- Thermostatic Expansion Valve (TEV)—(dry system).
- Low-Side Float (LSF)—(flooded system).
- High-Side Float (HSF)—(flooded system).

4.11.1 Capillary (CAP.) Tube

Domestic refrigerators, freezers, room air conditioners, and small commercial installations commonly use a capillary tube. A typical installation is diagrammed in **Figure 4-17.** The capillary tube is a long length of small diameter tubing. It reduces pressure by reducing the flow of refrigerant through its length.

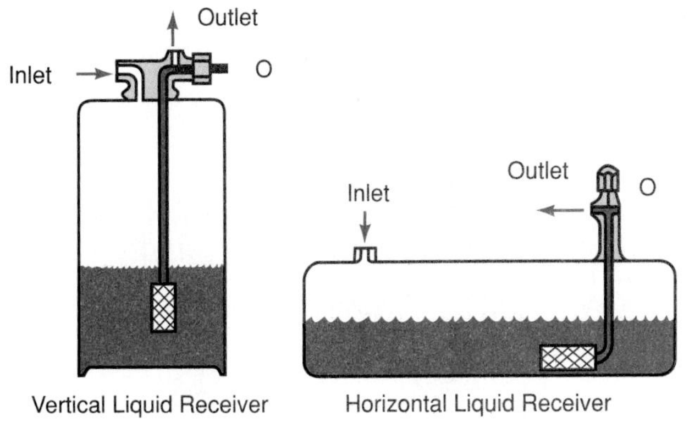

Figure 4-15. *Two common-types of liquid receivers. Note the liquid line service valve at O. It allows easy charging of refrigerant.*

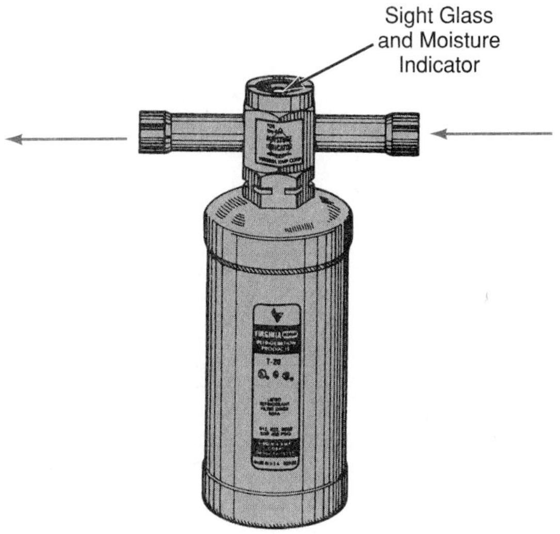

Figure 4-16. *A liquid line filter-drier with sight glass indicator. (Virginia KMP Corp.)*

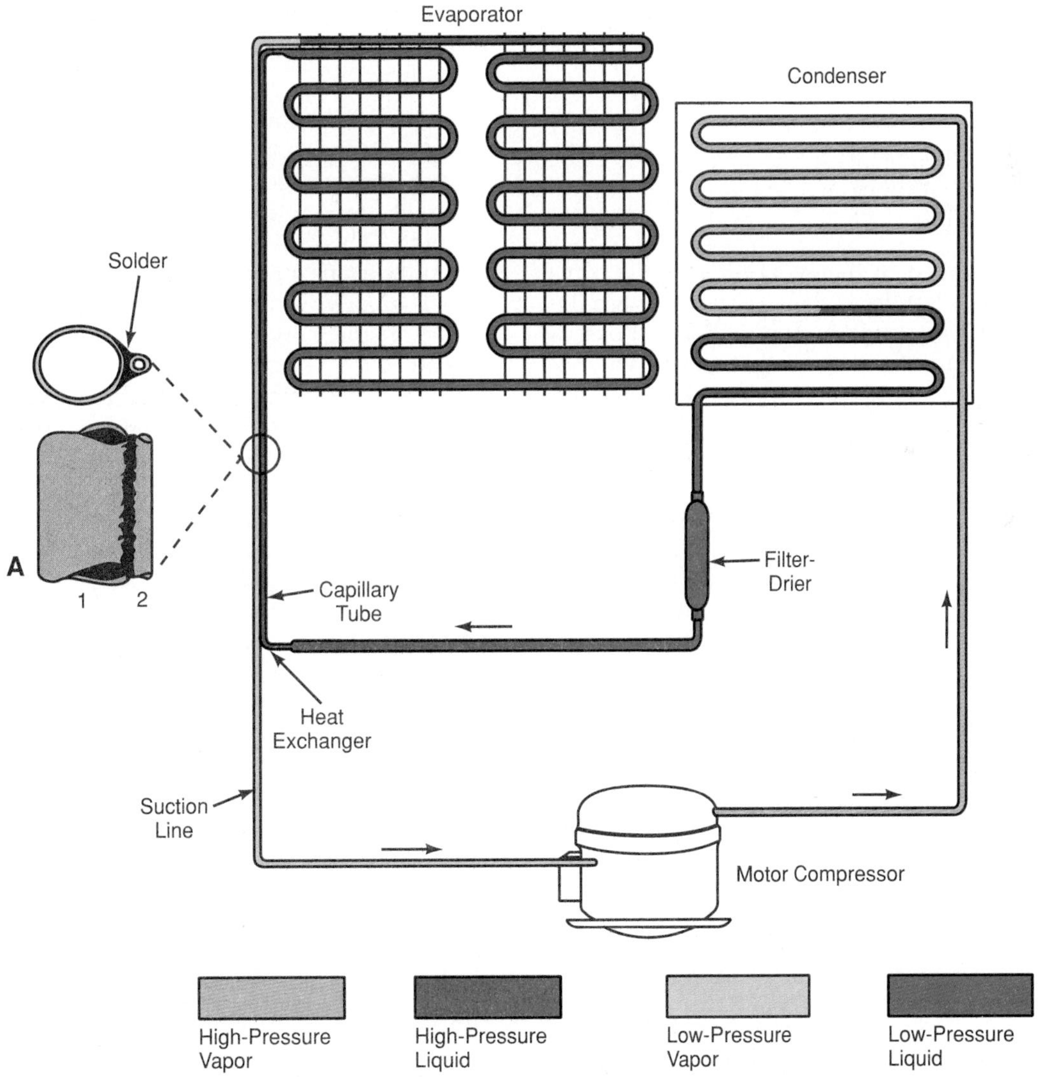

Figure 4-17. *Refrigerating system using capillary tube refrigerant control. Filter-drier is located in liquid line ahead of connection to capillary tube. Most of the capillary tube is fastened to suction line, which provides heat exchange. A—Enlarged cross section of suction line, 1, and capillary tube, 2, showing how they are soldered or brazed together. (Frigidaire Company)*

Figure 4-18 shows another type of capillary tube refrigerant control. The tube's inside diameter may vary. The diameter depends upon the refrigerant, the capacity of the unit, and the length of the line.

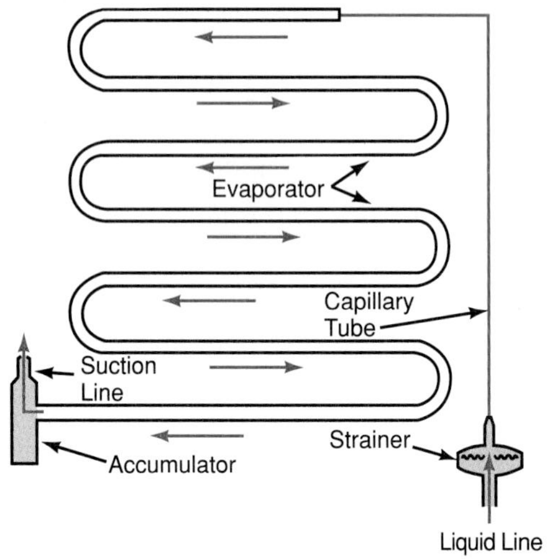

Figure 4-18. *Capillary tube refrigerant control. A strainer is located in liquid line at entrance to capillary tube. Note the accumulator, which serves as receiver for any liquid refrigerant overflow.*

Liquid is vaporized in the evaporator as the compressor operates. The capillary tube is placed between the liquid line and the evaporator. Just enough liquid passes through to make up for the amount that was vaporized.

The capillary tube reduces the liquid refrigerant from its high pressure to its evaporating pressure. The pressure of the liquid drops slightly in the first two-thirds of the length of the capillary tube.

Then some of the liquid starts to change to vapor. When the refrigerant reaches the end of the tube, 10% to 20% of it has vaporized. There is an increased volume of the vapor. This causes most of the pressure drop to occur at the end of the tube nearest the liquid line.

A recent development in capillary tube design uses a larger and longer tube (20′ to 30′). Being larger in diameter, it is less likely to become plugged.

The capillary tube refrigerant control does not use a check valve or a direction control valve. The high and low pressures equalize during the Off part of the cycle. This allows for easier starting. The compressor starts with equal pressures on the high and low side. The system must not have an overcharge of refrigerant. Extra refrigerant would tend to fill the evaporator too full. Severe frosting of the suction line when starting the motor indicates an overcharge.

4.11.2 Automatic Expansion Valve (AEV)

One of the dry systems uses an automatic expansion valve (AEV) as refrigerant flow control, **Figure 4-19.** This valve may be used only with the temperature-operated motor control. As pressure drops on the low side, the expansion valve opens and liquid refrigerant flows into the evaporator. It absorbs heat then while evaporating under low pressure. The valve maintains constant pressure in the evaporator when the system is running. This system operates independently of the amount of refrigerant in the system.

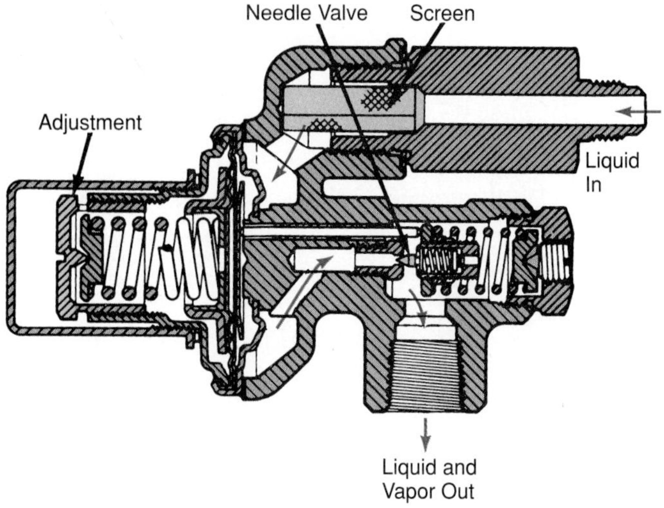

Figure 4-19. *Cross section of an automatic expansion valve shows flow of refrigerant through valve. This valve is designed to control the flow of liquid into the evaporator. It also maintains constant low pressure in the evaporator while compressor is running.*

The AEV is a division point between the high-pressure and low-pressure sides of the system. See Chapter 5 for a detailed explanation of expansion valves.

The automatic expansion valve, **Figure 4-20,** may be adjusted to the correct evaporator pressure. Turning the adjustment clockwise increases the rate of flow, thereby increasing the low-side pressure.

The rate of refrigerant flow through the AEV is controlled by the evaporator. Refrigerant will not flow through the valve unless the compressor is running. The evaporator must also be under a low pressure for the refrigerant to flow. Remember, lowering the pressure in the evaporator lowers the temperature at which the refrigerant evaporates.

4.11.3 Thermostatic Expansion Valve (TEV)

Many units, especially commercial ones, are equipped with a temperature-controlled expansion valve. This is called a ***thermostatic expansion valve (TEV).*** **Figure 4-21** shows a cross section of a thermostatic expansion valve.

This valve has a temperature-sensing bulb mounted on the outlet of the evaporator. The bulb temperature controls the opening of the thermostat valve needle.

Addition of this thermal element to the valve lets the evaporator fill more quickly and permits more efficient cooling. The thermostatic expansion valve keeps the evaporator full of liquid refrigerant when the system is running. See Chapter 5 for a more detailed explanation of its operation.

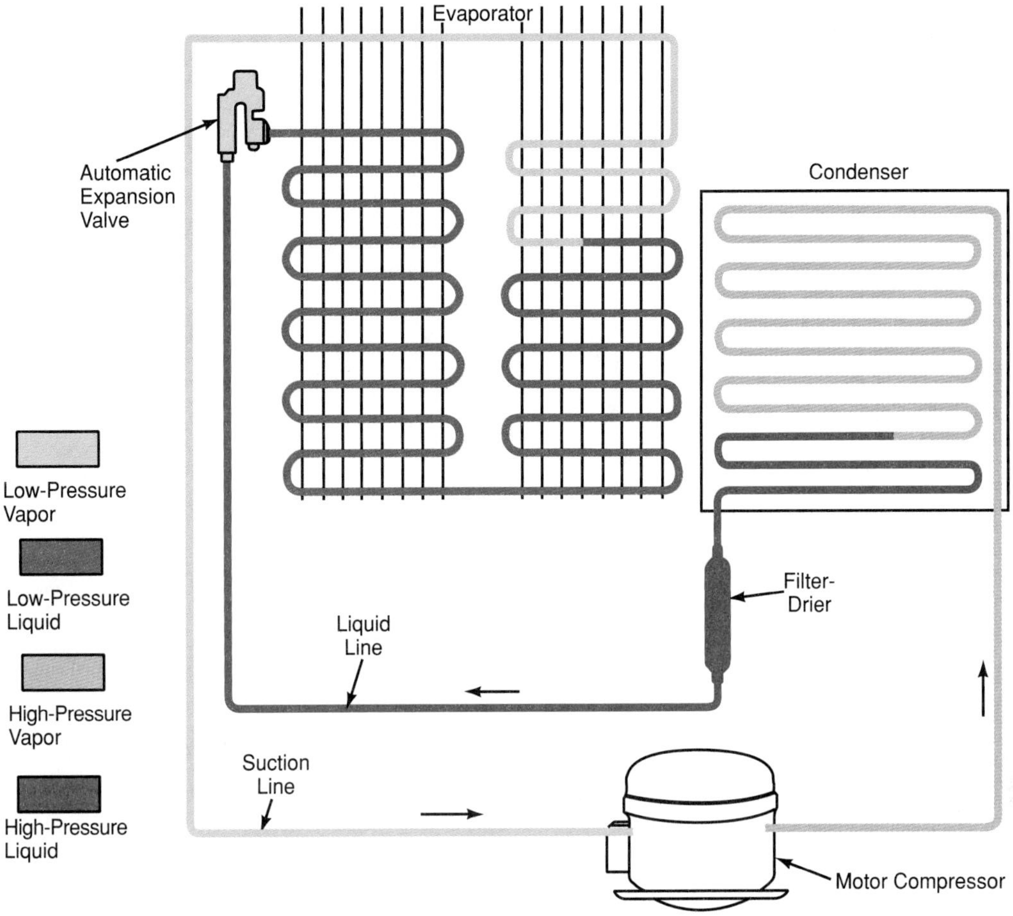

Figure 4-20. *A refrigerating system using an automatic expansion valve refrigerant control. A filter-drier is located in the liquid line ahead of the automatic expansion valve.*

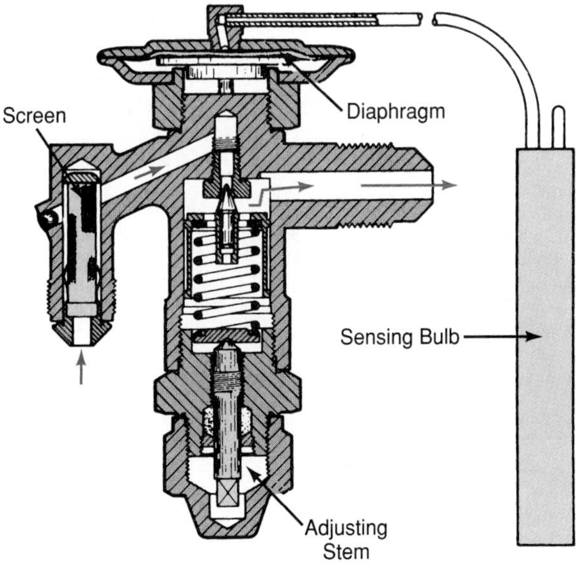

Figure 4-21. *Diaphragm-type thermostatic expansion valve. Sensing bulb pressure operates on upper surface of diaphragm. As the sensing bulb temperature increases, pressure on top of the diaphragm increases and tends to open valve, allowing liquid refrigerant to enter evaporator. Note screen at liquid line connection and direction of refrigerant flow through valve. (Sporlan Valve Co.)*

As the evaporator becomes colder, the TEV reduces the rate of flow of refrigerant into the evaporator. There is no flow at all unless the compressor is running.

Figure 4-22 demonstrates how a thermostatic expansion valve is connected into a refrigerating system. The valve may be used with either a pressure- or temperature-operated motor control. A thermostatic expansion valve can also be used in a multiple evaporator system.

A thermoelectric element may replace the temperature-sensing element in the thermostatic expansion valve. More details regarding this valve are found in Section 5.1.2.

4.11.4 Low-Side Float (LSF)

A ***low-side float*** is used on flooded systems where the evaporator is flooded with refrigerant and the refrigerant level is controlled by a float valve. A float is used for control. As the refrigerant evaporates, the liquid level falls. This lowers a float and, in turn, opens the needle valve connected to it. More liquid enters from the high-pressure liquid line, taking the place of the evaporated liquid.

Figure 4-23 suggests the exterior and the interior construction of a flooded evaporator using a low-side

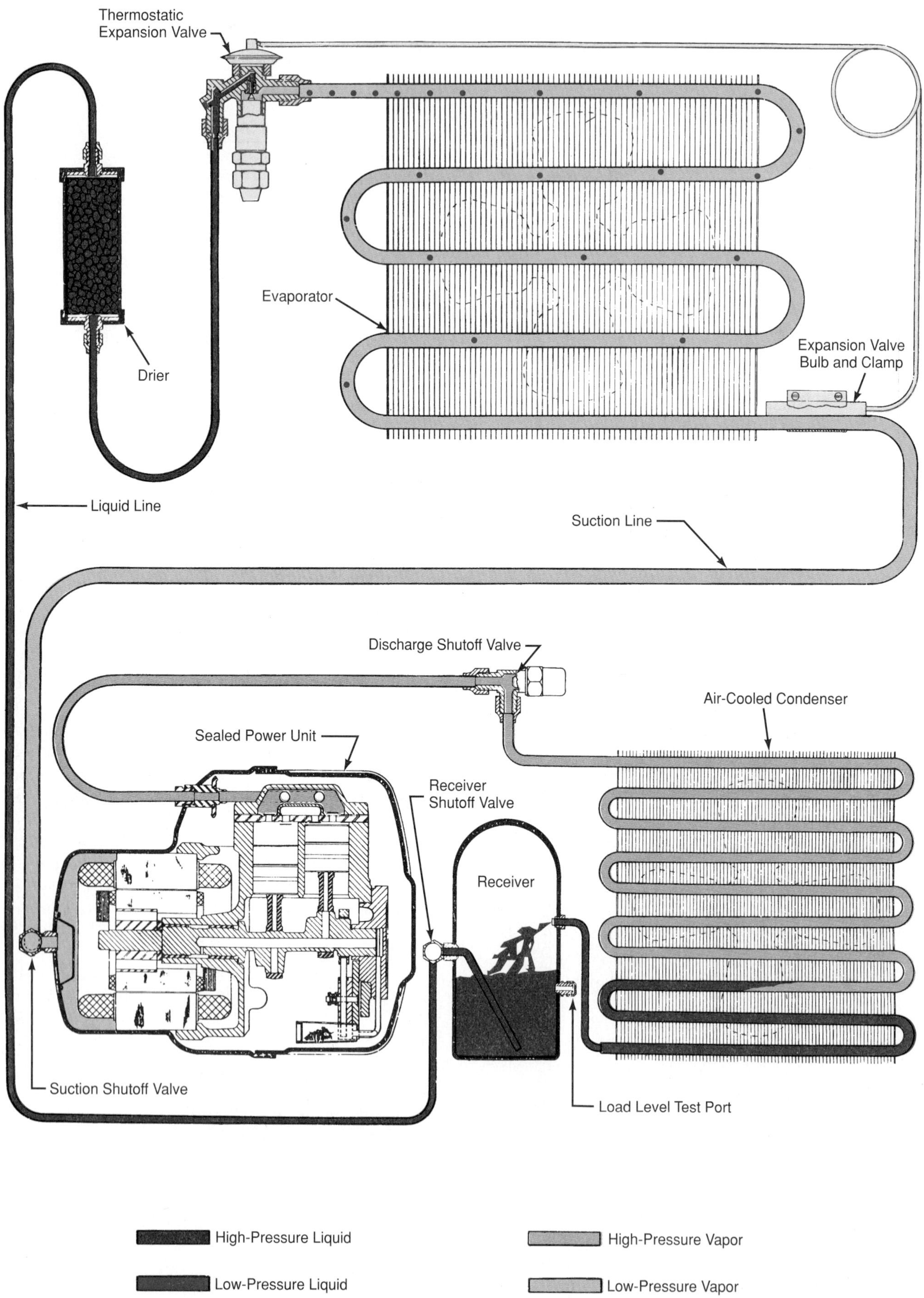

Figure 4-22. *Typical thermostatic expansion valve system. Note the two-cylinder hermetic compressor.*

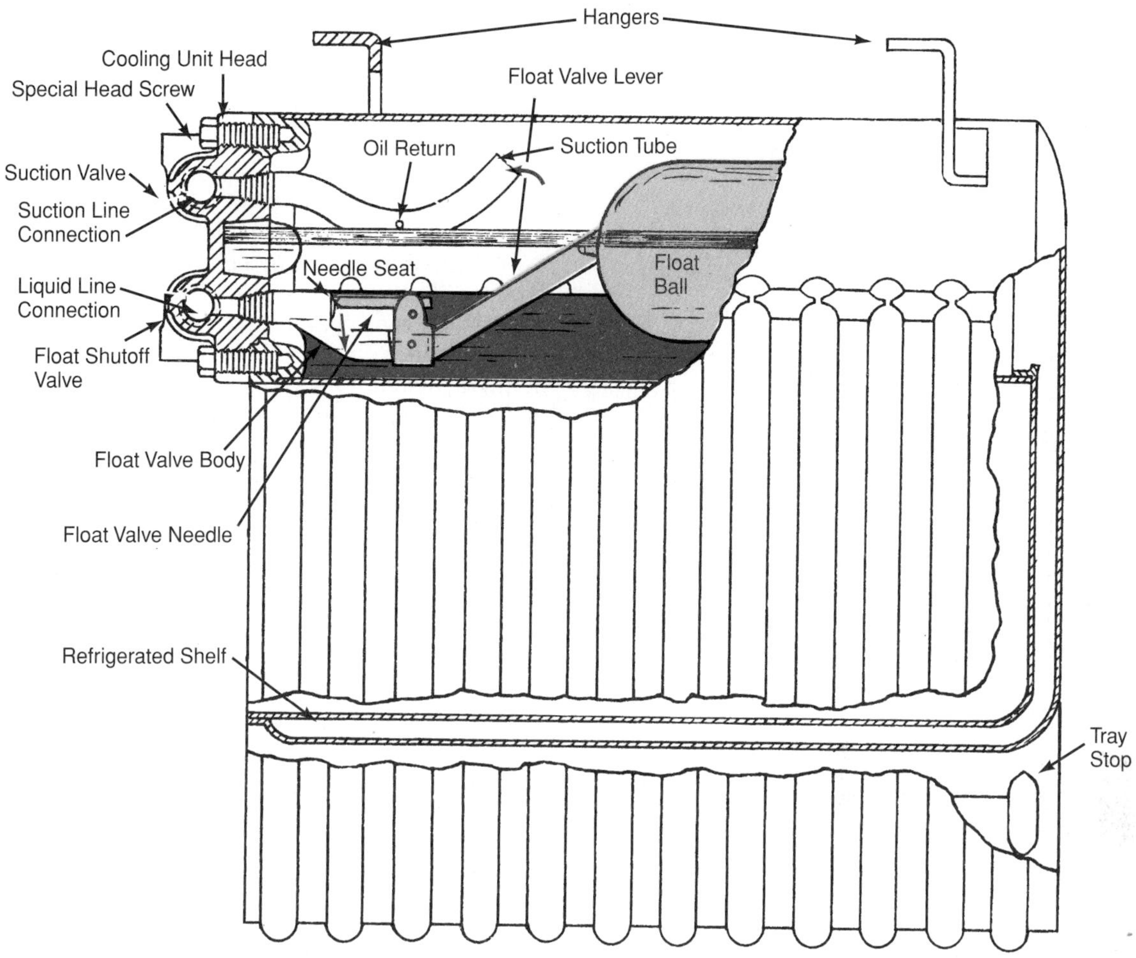

Figure 4-23. *Low-side float refrigerant control. Note the suction line and the liquid line connections. Float and needle mechanism maintain constant level of liquid refrigerant in evaporator.*

float. Either a temperature- or pressure-operated motor control may be used.

A low-side float system usually has a large liquid receiver. The receiver must be large enough to store all the refrigerant in the system.

Oil picked up by the vapor is normally returned through a small opening at a predetermined level in the suction return tubing. Since the diameter of the hole is small, if the unit is not level, the oil will not return to the compressor and "oil binding" may result. When this occurs, the oil forms a layer on the surface of the liquid refrigerant. It prevents the refrigerant from evaporating at a rapid rate or at the temperature corresponding to the pressure.

The low-side float refrigerant control can be used in multiple evaporator systems.

4.11.5 High-Side Float (HSF)

To make a high-side float system operate, a float is located in the liquid receiver tank or in a chamber in the high-pressure side. Liquefied refrigerant collects in the float chamber. When enough refrigerant has collected, the float will rise enough to open the needle valve. Liquid flows into the low-pressure side or evaporator. The float controls the level of liquid refrigerant on the high-pressure side.

The amount of refrigerant in a system must be carefully measured. The evaporator must receive the correct amount for the system to operate correctly. Extra refrigerant will overcharge the evaporator and cause frosting of the suction line.

Refer to **Figure 4-24** for an illustration of a high-side float mechanism. A more detailed description is given in Chapter 5. This refrigerant flow control can be used with either a pressure- or temperature-operated motor control.

4.12 Motor Control

Most electric refrigerators are designed with more cooling capacity than needed. Therefore, under normal use, they do not run all of the time. To get correct refrigeration temperature, the motor must be turned off upon reaching the desired low temperature. It is turned on again when the evaporator has warmed to a certain

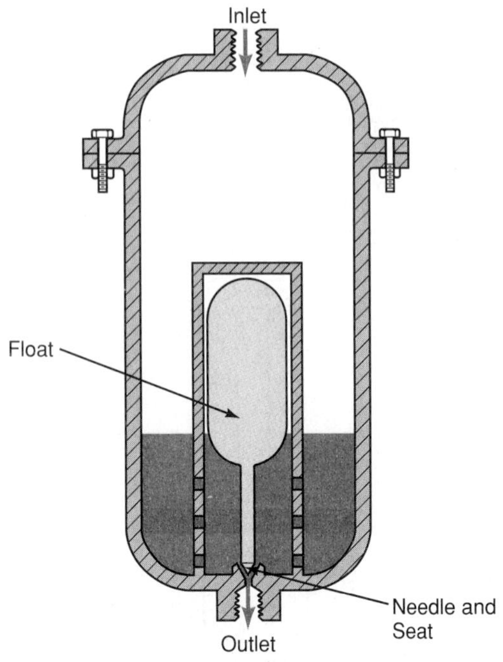

Figure 4-24. *A high-side float allows liquid refrigerant to flow to the evaporator only when enough refrigerant collects to raise the float and open the needle valve.*

temperature. Two principal types of motor controls are used to turn the motors on and off:

- Temperature-operated motor control (thermostatic).
- Pressure-operated motor control (low-side pressure).

Thermostatic control is the most popular, especially on small installations. The thermostatic temperature control, **Figure 4-25,** has a sensing bulb. The sensing bulb is connected by a capillary tube to a diaphragm or bellows. This element is charged with a volatile fluid. The fluid expands to increase the pressure as the bulb becomes warmer. It will contract again to decrease the pressure as the bulb cools.

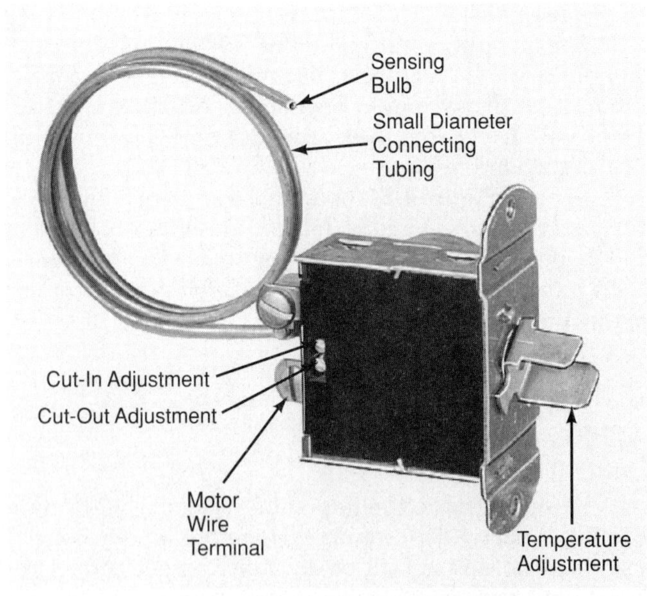

Figure 4-25. *A temperature-operated motor control. Note the temperature adjustment. (Eaton Corp.)*

As bulb pressure increases, the diaphragm moves. Since it is connected to a toggle or snap-action switch, it will turn on this switch (close the circuit). As the bulb cools, the diaphragm or bellows moves the other way. The toggle switch will move (to open the circuit).

These controls have adjustments that permit differences in operating temperatures. Many controls have a manual switch. This switch permits shutting off or turning on of the system as desired. They also may include an overload protector. It will open the switch if the unit draws too much current.

Thermostats may also be electrically connected to timers for automatic defrosting of the evaporator.

Many commercial units use a pressure-operated motor control. It opens the circuit when the pressure drops enough. It closes the circuit when the pressure has risen enough. A pressure-operated motor control may be used with the TEV. It may also be used with high- or low-side refrigerant control systems.

The pressure of the vapor in the low-pressure side varies with the temperature. Therefore, the pressure may indicate temperature. This permits the use of pressures to control the stopping and starting of the motor. Therefore, pressure controls the temperature of the cabinet. The details concerning the operation of these controls are explained fully in Chapter 8.

COMPRESSORS MODULE

4.13 External-Drive Compressors

The purpose of a compressor is described in Section 4.6. An external-drive (open) compressor is bolted together. Its crankshaft extends through the crankcase. The crankshaft is driven by a flywheel (pulley) and belt. It may also be driven directly by an electric motor. A crankshaft seal is required where the crankshaft comes through the crankcase.

Figure 4-26 illustrates a cross section through an open compressor. This is a four-cylinder V-type compressor with an eccentric crankshaft. (An ***eccentric*** is a shaft section which is larger and has a different center than the shaft.) The pistons are fitted with rings.

A master connecting rod is mounted on each eccentric. It is connected to a piston in one bank of the "V." The connecting rod attached to the piston in the other bank is connected with a pin. This pin is connected through a flange on the master connecting rod. These connecting rods are somewhat shorter than the master connecting rod. The shorter rods are called ***articulated connecting rods.***

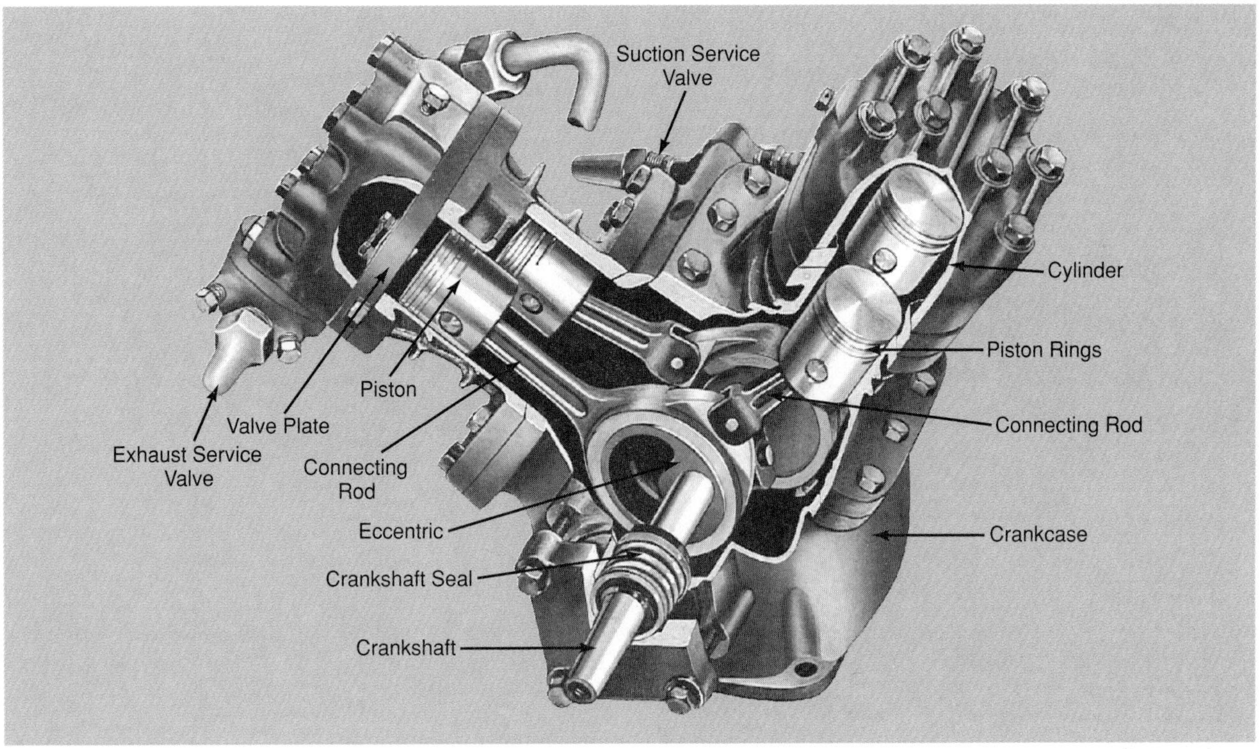

Figure 4-26. *A four-cylinder, external-drive, V-type, air-cooled compressor. (Frick Co.)*

4.14 Hermetic Compressors

The motor in a hermetic compressor is sealed inside a dome or housing with the compressor. It is directly connected to the compressor. A crankshaft seal is not needed.

A motor rotor is usually press fit onto the compressor crankshaft. Some motor compressors are made with the motor at the top. Others have the motor at the bottom and the compressor at the top.

A hermetic unit is usually spring-mounted inside the hermetic dome. This prevents most of the compressor vibration from being felt outside of the dome.

The exhaust and suction lines inside the dome are made flexible. A connection through the dome allows fastening the compressor lines to the remaining system. The electrical connections to the motor pass through the dome by means of an insulated leakproof seal.

A hermetic compressor is lubricated in the following manner. The return suction gas is fed into a hollow disk. This disk is mounted on the motor compressor shaft. Centrifugal force throws the oil and a liquid refrigerant to the outer rim of the disk. (Centrifugal force action rotates things to pull spinning particles away from the center of the rotation.) The oil and refrigerant flow over the motor windings. Only the vapor refrigerant remains at the center and is drawn into the cylinders of the compressor.

Two types of hermetically sealed compressors are shown in **Figure 4-27. Figure 4-27A** is a conventional twin-piston hermetic compressor. **Figure 4-27B** is a conventional twin-piston compressor with a valve design routing the refrigerant vapor through a discharge muffler to increase efficiency and tolerate liquid slugging.

Some motor compressors are two-speed. These are popular in large systems and in air conditioning where heat loads change.

Figure 4-28 illustrates a hermetically sealed compressor. It has an internal and external steel shell and is combined into a single housing. The suction gas goes through the motor area, cooling the motor. The unit has an internal accumulator. This prevents liquid from returning to the cylinder area. The discharge line is coiled to the unit. It keeps the oil warm enough to evaporate any liquid refrigerant that may have returned. The piston head is sculptured and has a circular valve plate design. This provides a balanced inlet and outlet flow of the vapor. The unit has rotolock connections which allow for servicing. This unit contains an internal discharge muffler which prevents excessive vapor pulsation and vibration. See Section 4.15.7 for additional information. The system also has an internal overload that senses temperature and amperage.

4.15 Types of Compressors

There are five basic types of compressors in use in the refrigeration industry:

- Reciprocating (piston-cylinder).
- Rotary.
- Scroll.
- Screw.
- Centrifugal.

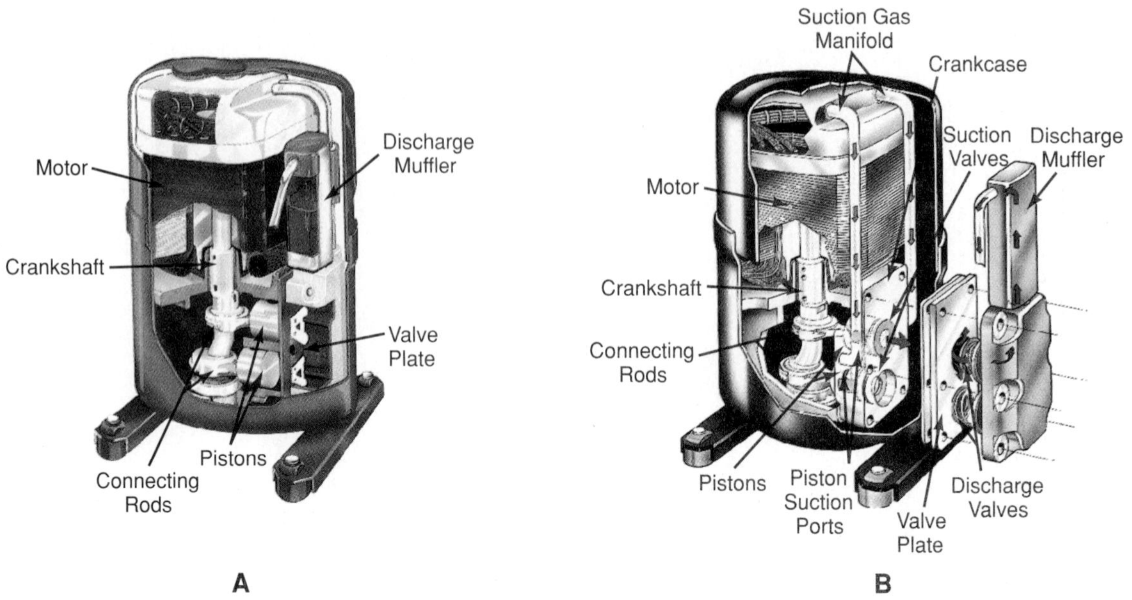

Figure 4-27. *Two types of hermetically sealed compressors. A—Conventional hermetic compressor. B—Compressor with valve design that routes refrigerant vapor to increase efficiency and to tolerate liquid slugging. (Bristol Compressors)*

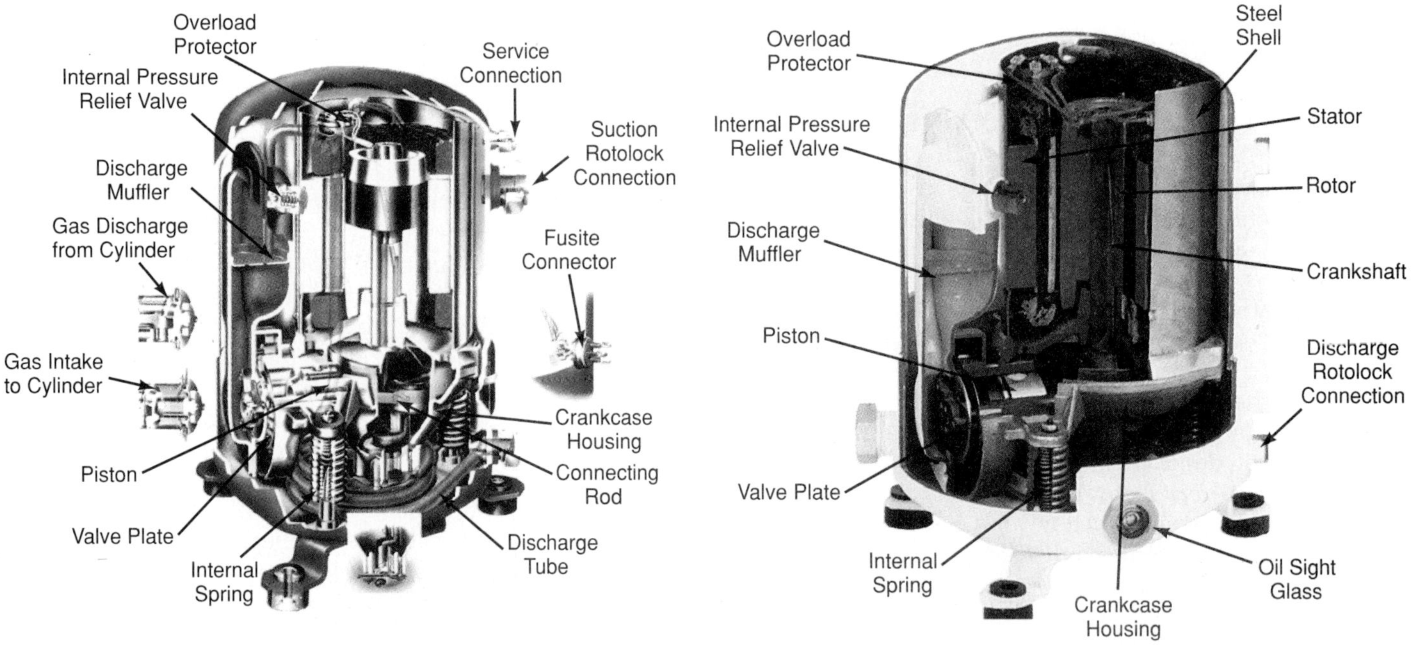

Figure 4-28. *Steel-shell hermetic compressor with a winding that is sealed and running gear enclosed, allowing the compressor to take continuous liquid slugging. (Maneurop Inc.)*

4.15.1 Reciprocating Compressors

The original energy source for reciprocating compressors is usually an electric motor. Its rotary motion must be changed to ***reciprocating*** (back-and-forth action in a straight line) motion. This change is usually made by a crank and a rod. The ***rod*** connects the crank to the piston. The complete mechanism is housed in a leakproof container called a ***crankcase.*** It is very efficient. Its construction resembles that of the automobile engine.

Basically, a reciprocating compressor is a cylinder and a piston. **Figure 4-29** shows the principle of operation of a reciprocating compressor. In **Figure 4-29A,** the piston has moved downward in the cylinder, A. It has moved refrigerant vapor from the suction line through the intake valve. From there the refrigerant vapor has moved into the cylinder space. In **Figure 4-29B,** the piston has moved upward. It has compressed the vaporized refrigerant into a much smaller space (clearance

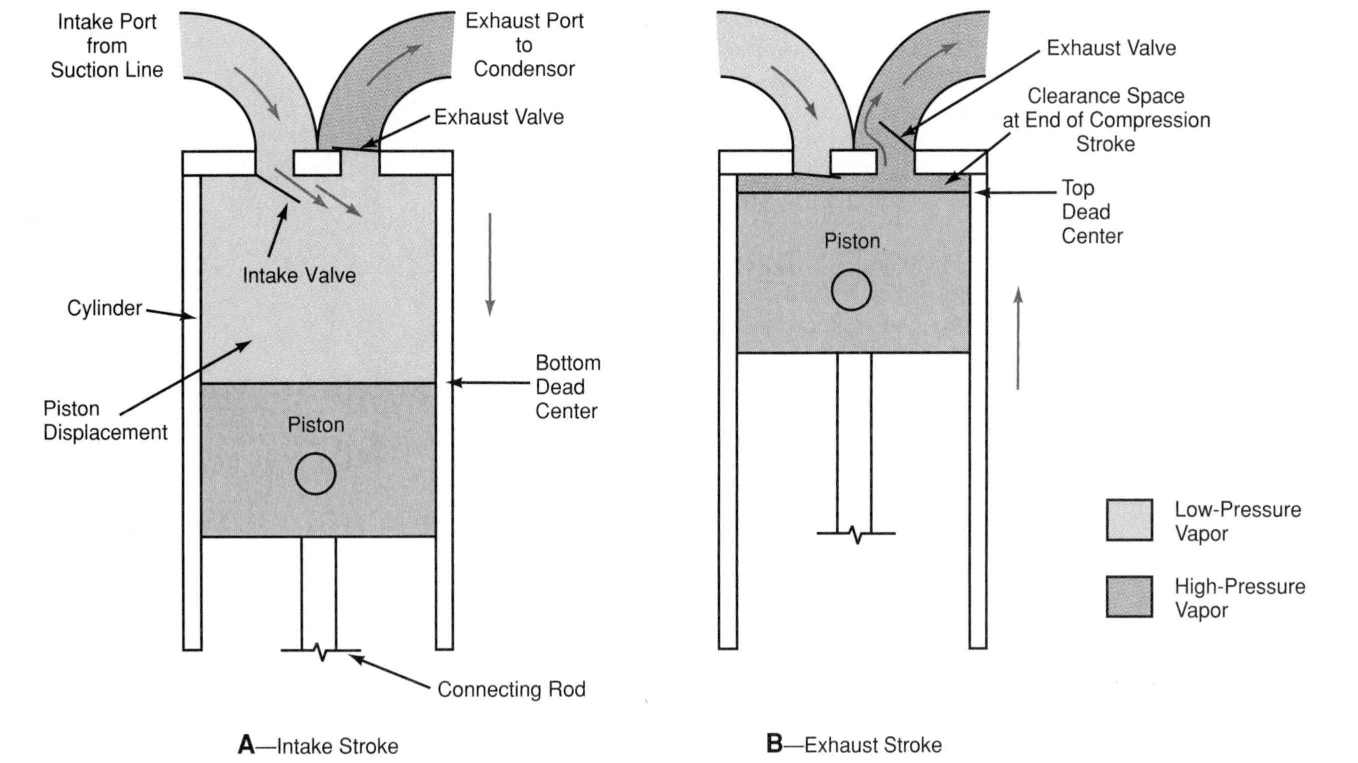

Figure 4-29. *Basic construction of reciprocating compressor. A—Intake stroke. B—Exhaust stroke.*

space). The compressed vapor has been pushed through the exhaust valve into the condenser.

Piston Cylinder Crank Arrangements

Compressors may have more than one cylinder (multicylinder). In these compressors, the crankshaft and cylinders are arranged to make the compressor as compact as possible. **Figure 4-30** illustrates some common cylinder and crankshaft arrangements. More pumping capacity should be provided for each revolution of the crankshaft. Most two-cylinder compressors use a side-by-side arrangement of the cylinders and a 180° crankshaft. While one piston is at the top of the stroke, the other piston is at the bottom. Other two-cylinder compressors have two cylinders at a 90° V. See **Figure 4-26.** With this cylinder arrangement, a single-throw crank is used.

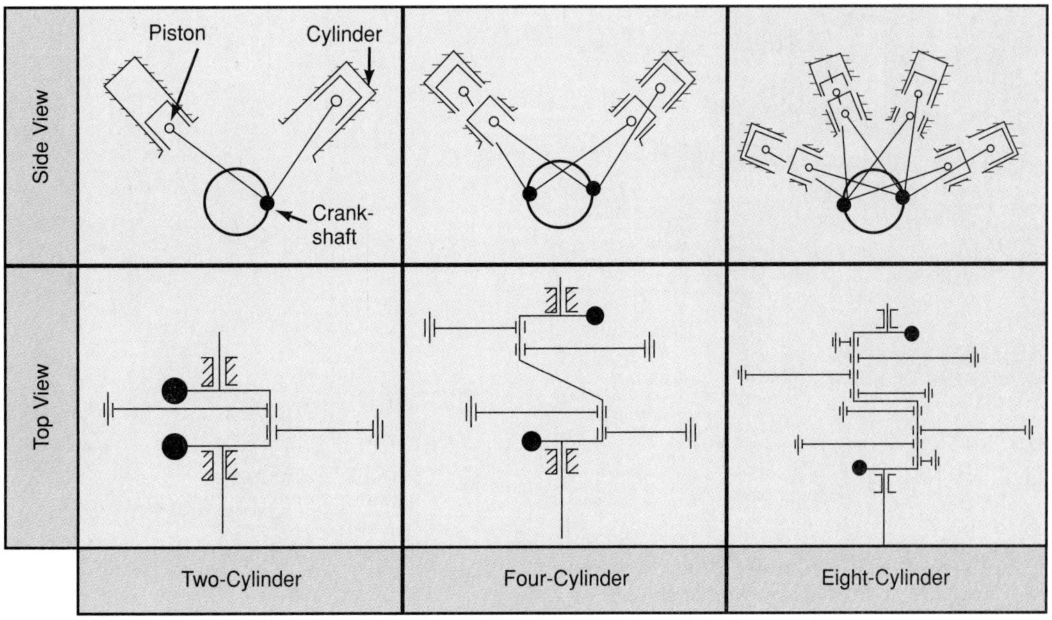

Figure 4-30. *Piston, cylinder, and crankshaft arrangements for two-, four-, and eight-cylinder compressors.*

Cylinders

Compressor cylinders for external-drive compressors are usually made of cast iron. The cast iron must be dense enough to prevent seepage. Some nickel is usually added to give the casting the desired density.

Small compressors have fins cast with the cylinders to provide better air cooling. Larger compressors may have water jackets surrounding the cylinders for cooling. Some compressors are built with cylinder liners or sleeves, which may be replaced when worn.

Usually, the crankcase is part of the same casting as the cylinder. This practice cuts down the number of joints that might leak. It also permits close alignment between crankshaft main bearings and cylinder. The main bearings are ball-type. Construction is shown in **Figure 4-31.** The side-by-side cylinder arrangement is commonly used on open compressors. This compressor is designed for use on vehicle air conditioning. However, the same design may be used in other air conditioning applications.

Hermetic compressors usually have cast-iron cylinders. Some may be made of aluminum or other materials. A typical hermetic compressor cylinder is shown in **Figure 4-27.** Another type of hermetic compressor is pictured in **Figure 4-32.** This is a bolted hermetic, which can be dismantled easily for servicing.

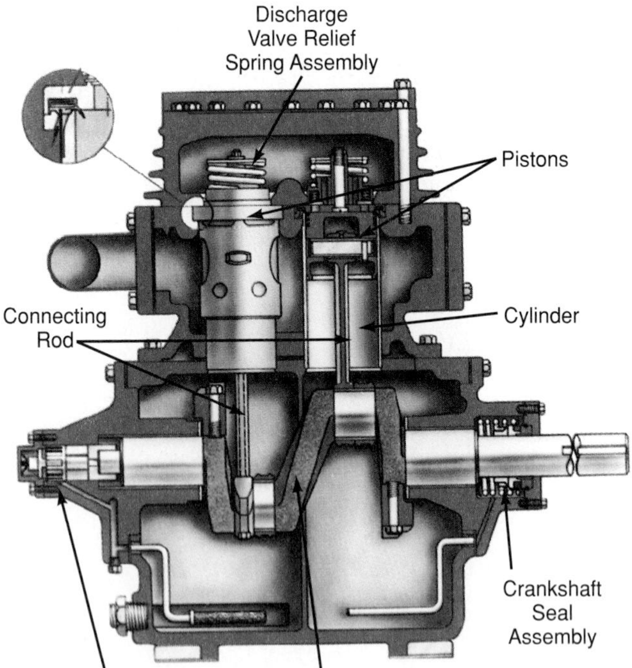

Figure 4-31. *Cutaway view of small, external-drive, two-cylinder reciprocating compressor. The body is a lightweight alloy casting. Cast-iron cylinder liners are permanently cast into crankcase body. (Grasso, Inc.)*

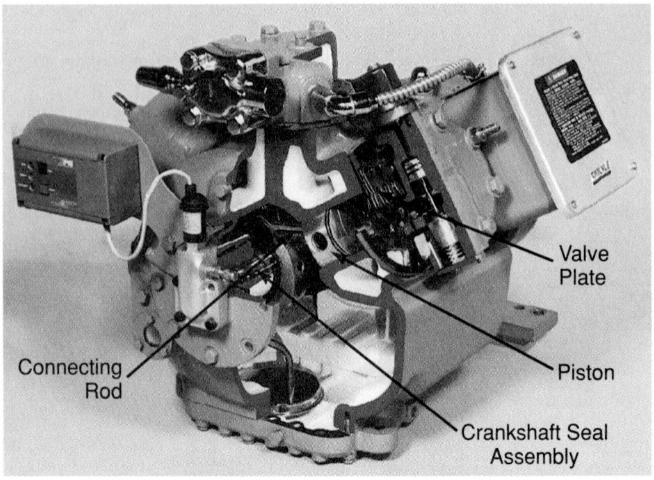

Figure 4-32. *A serviceable four-cylinder compressor used for air conditioning and refrigeration. (Carlyle Compressor Company, Division of Carrier Corporation)*

Pistons and Piston Rings

Pistons used in external-drive compressors are usually made of cast iron. In small, high-speed hermetic compressors they are of die-cast aluminum. Smaller sizes do not have piston rings.

The temperature of pistons seldom goes higher than 250°F (121°C). Thus, there is not much expansion of either piston or cylinder. Pistons may be fitted with as little as .0002″ (.0051 mm) clearance for each inch diameter.

The smaller pistons have oil grooves cut in them. **Figure 4-33** illustrates a common piston/connecting rod assembly. **Figure 4-34** illustrates a commercial-type piston and connecting rod assembly. This one is fitted with piston rings.

There are two types of piston rings. The upper ring or rings are known as ***compression rings.*** The lower is designed to control the oil "flow" past the piston. It is an ***oil ring.***

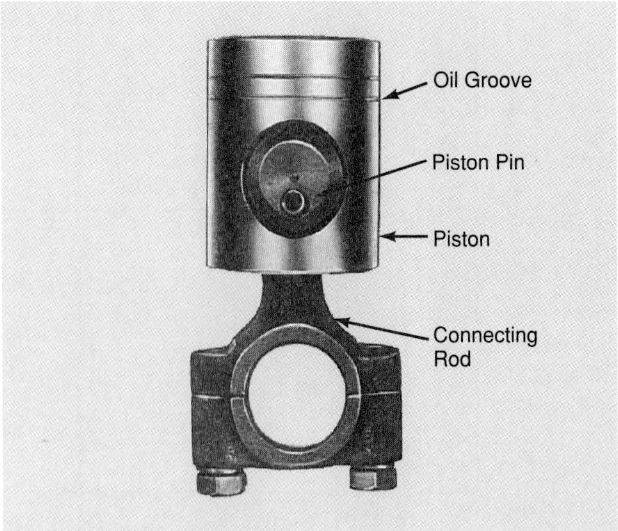

Figure 4-33. *Piston and connecting rod assembly. Note the oil grooves cut into the piston. (General Electric Co.)*

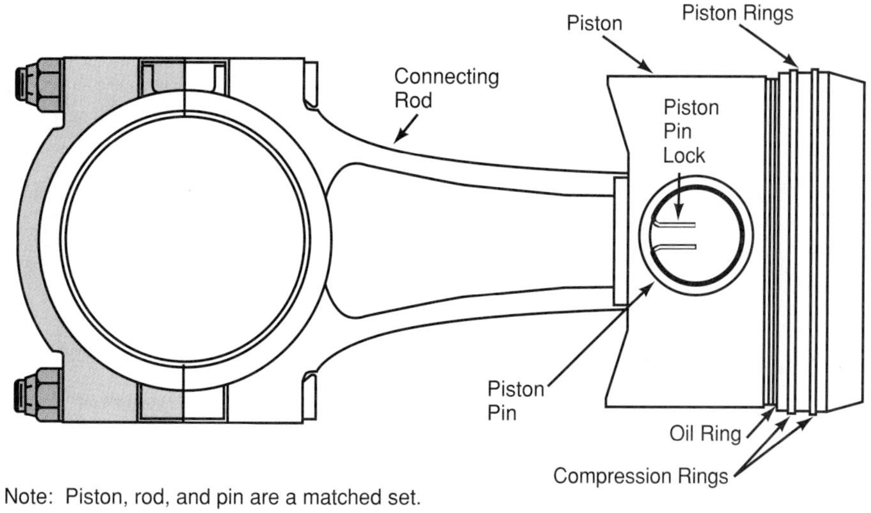

Figure 4-34. *Compressor piston and connecting rod assembly. Note how the connecting rod's lower (left) end is split and then bolted together to provide bearing for the crankshaft journal.*

Piston rings are usually made of cast iron. Some bronze rings have been used, however. Rings should be fitted to the groove as closely as possible and still allow movement. A ring is a complete circle with a gap in it. A 45° tapered or angled ring gap permits the ring to push out against the cylinder wall. The gap should be about .001″ (.0254 mm) for each inch of piston diameter.

Piston pins are made of case-hardened, high carbon steel accurately ground to size. They are hollow to reduce weight.

Piston pins are usually of the full floating type. This means the pin is free to turn in both the connecting rod bushing and piston bushings.

The piston is designed to come as close as possible to the cylinder head without touching it. This presses as much of the vapor into the high-pressure side as possible.

When the piston is at top dead center (TDC) of its stroke, there is a very small clearance. The clearance between the piston and cylinder head is approximately .010″ to .020″ (.254 mm to .508 mm). The volume of space created is called ***clearance space.*** (Refer back to **Figure 4-29B.**)

There is a valve plate under the cylinder head. It has both the intake and exhaust valve located in it, **Figure 4-35.**

In hermetic systems, the pistons and rings, if used, are constructed much the same as those used in external-drive compressors. However, the hermetic compressors usually run at a higher speed than external-drive compressors. Therefore, the pistons are smaller in diameter and are made as light as possible.

Figure 4-36 illustrates a hermetic compressor. It uses a Scotch yoke piston arrangement. **Figure 4-37** shows a hermetic motor compressor with four opposing pistons on a flat plane.

Connecting Rods

The ***connecting rod*** attaches the piston to the crankshaft. A conventional connecting rod is shown in **Figure 4-34.**

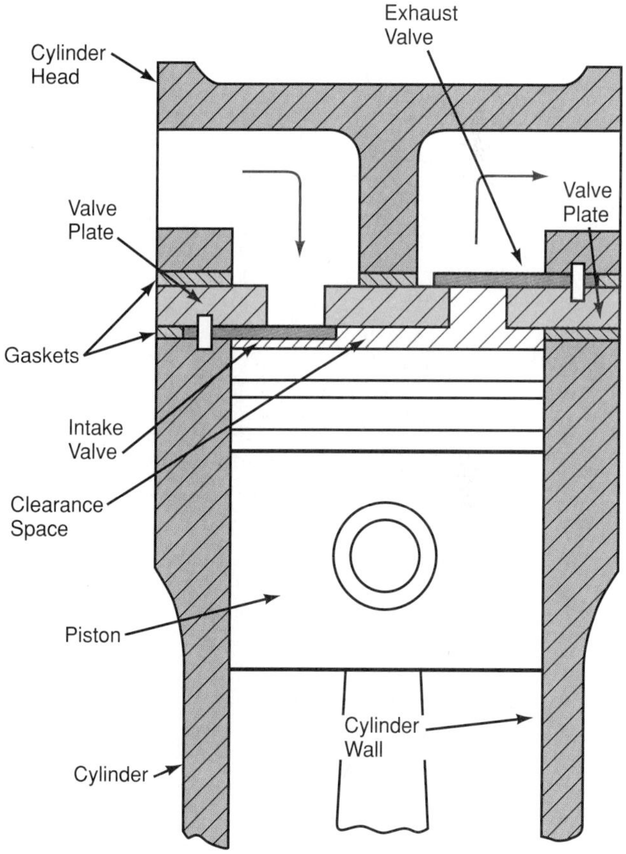

Figure 4-35. *A cross section through a compressor cylinder.*

Connecting rods for external-drive compressors are usually made of drop forged steel. Crankshafts with a throw use a connecting rod having a split lower end. This end clamps around the crankshaft journal. The rod bearing must be fitted to a clearance of about .001″ (.0254 mm). It is important, therefore, that the bolts be carefully tightened (torqued).

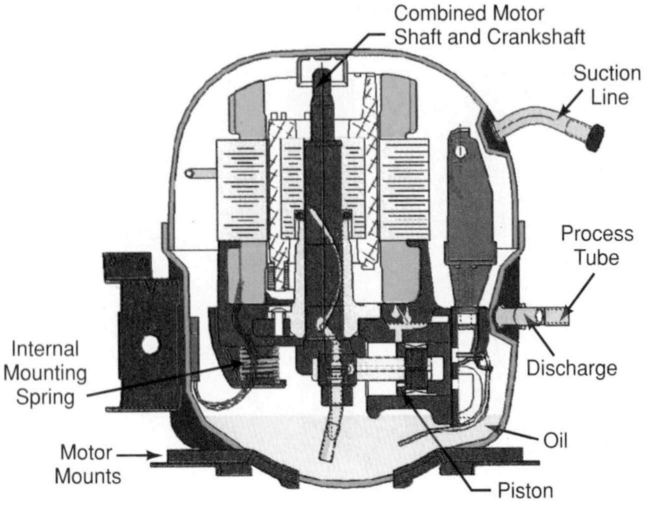

Figure 4-36. *Single-cylinder hermetic compressor with Scotch yoke piston crankcase mechanism. (Americold Compressor Division of White Consolidated Industries, Inc.)*

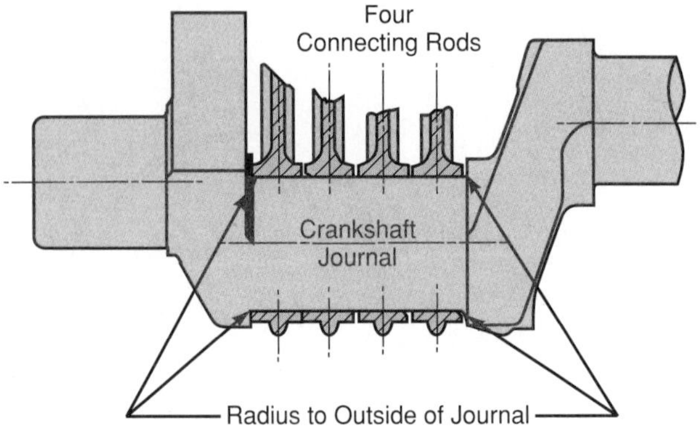

Figure 4-38. *Four connecting rods mounted on one crankshaft journal. This arrangement could be used on radial-type compressor.*

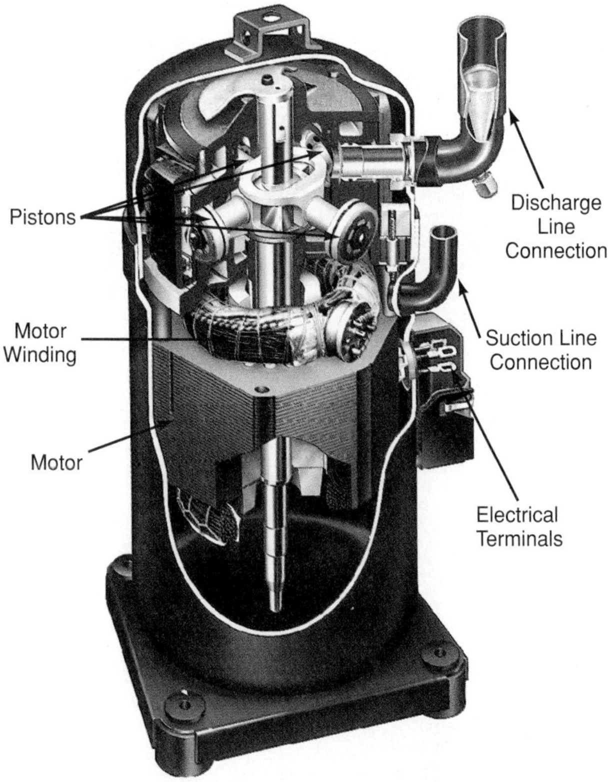

Figure 4-37. *Four opposing pistons—"Quadro-Flex." The four opposing pistons on a flat plane reduce the vibration of the unit and, therefore, it can be used for rooftop heat pumps. (Tecumseh Products Company)*

Figure 4-38 shows four connecting rods mounted on one crankshaft journal. This arrangement could be used in a radial or a V-type compressor.

The eccentric-type connecting rod usually has a cast iron bearing surface. The crankthrow end is a solid ring. It must be mounted on the eccentric before the crankshaft is assembled to the eccentric. The construction is shown in **Figure 4-39.**

A small piston and connecting rod assembly used in a hermetic unit is shown in **Figure 4-33.** The connecting rod is attached rigidly to a large piston pin. It is attached by means of a locking pin and spring. The dismantled unit is illustrated in **Figure 4-40.**

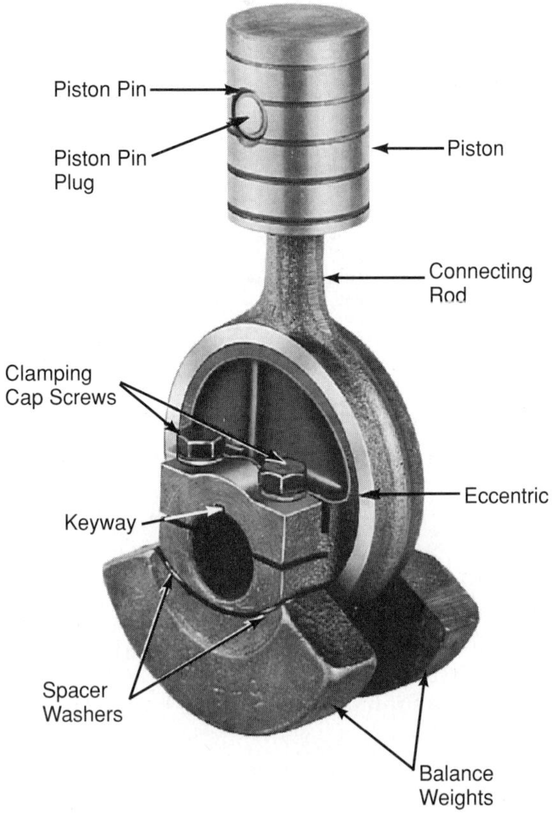

Figure 4-39. *Eccentric crankshaft assembly. Note the clamping cap screws and balance weights.*

Cylinder Head

Cylinder heads for both external-drive and hermetic compressors are usually made of cast iron. The

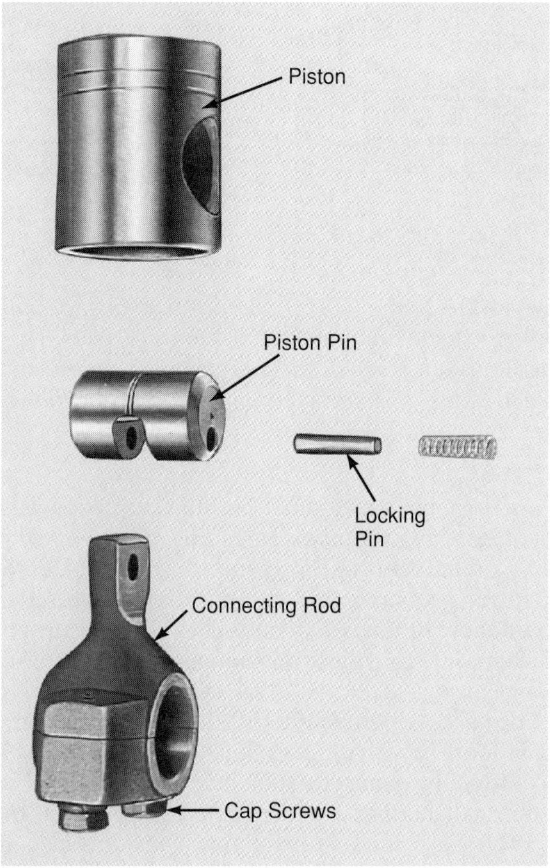

Figure 4-40. *A piston and connecting rod assembly dismantled. This is the same assembly shown in* ***Figure 4-33.*** *Note the small pin which locks the connecting rod to a large piston pin. Also note the two cap screws used to attach the cap to the crankshaft end of the connecting rod.*

head serves as a pressure plate. It supports and holds the valves and valve plate in position. It also provides the vapor passages into and out of the compressor. The pressures of compression may amount to as much as 300 psi (2170 kPa). These pressures depend upon the kind of refrigerant used. The valve plate must, therefore, have good support. There must be no leakage at the gaskets on either side of the valve.

In some hermetic systems, the entire compressor housing is inside a dome. The entire space within the dome is open to the suction line. Consequently, the whole dome is under low-side pressure. In such systems, no intake manifold is required. Only an opening into the intake valve or valves is required.

The cylinder head is usually attached to the cylinder with cap screws. **Figure 4-41** is a cutaway view of a commercial multicylinder reciprocating hermetic compressor.

The suction line connects to the shutoff valve on the right end. The exhaust, which connects to the condenser, connects to the discharge shutoff valve (left end).

Note the location of the crankcase heater (see Section 4.27). Also, note the discharge header safety spring. This relieves the pressure in the cylinder if it exceeds a safe working limit. (This excessive pressure may be due to slugging of liquid refrigerant or refrigerant oil.)

Valves and Valve Plates

Valve assemblies usually consist of a valve plate, an intake valve, an exhaust valve, and valve retainers. Refer to **Figure 4-42.**

Valve plates are sometimes made of cast iron. Hardened steel is also used, as plates can be thinner with longer wearing valve seats.

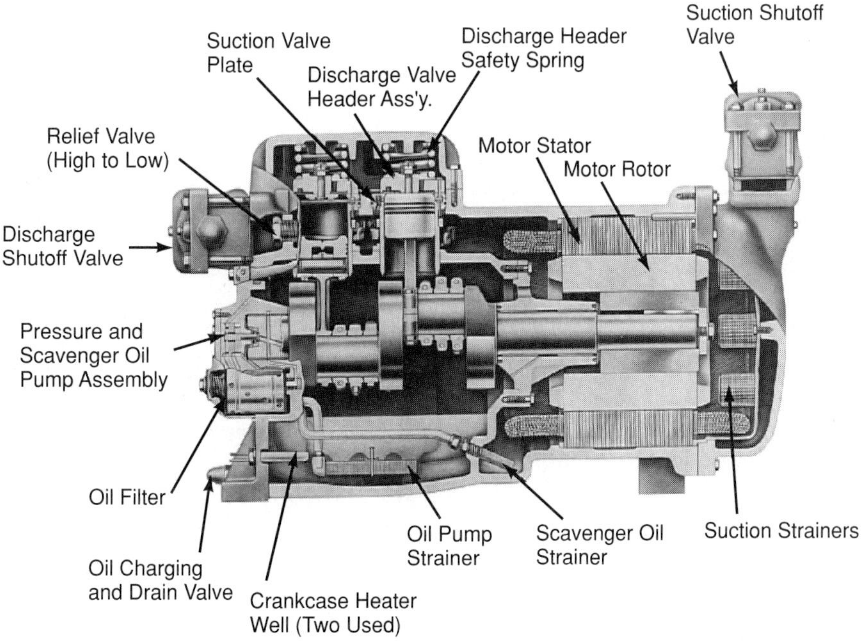

Figure 4-41. *A commercial, hermetic reciprocating compressor. It has four banks of two cylinders each (four connecting rods on each crankthrow) and is bolted for ease in servicing.*

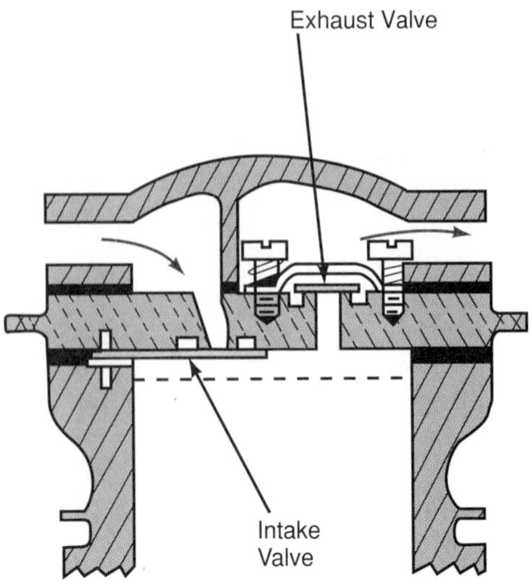

Figure 4-42. *Typical compressor valve plate. Heavy springs on the exhaust valve cage permit a greater valve lift to protect the compressor in case of severe liquid refrigerant or oil pumping.*

Compressor valves are usually made of high carbon alloy steel. They are heat treated to give them the properties of spring steel and ground to a perfectly flat surface.

The intake valve is usually kept in place by small pins. It may also be kept in place by the clamping action between the compressor head and valve plate. Exhaust valves may be clamped the same way.

Some different valve designs are displayed in **Figure 4-43. Figure 4-44** shows a typical valve plate assembly.

The valve disks or reeds must be perfectly flat. A defect of only .0001″ (.00254 mm) will cause the valve to leak.

Of the two valves—the intake and the exhaust—the intake valve presents fewer problems. This is because it is constantly lubricated by oil circulating with the cool refrigerant vapors. Also, it operates at a relatively cool temperature.

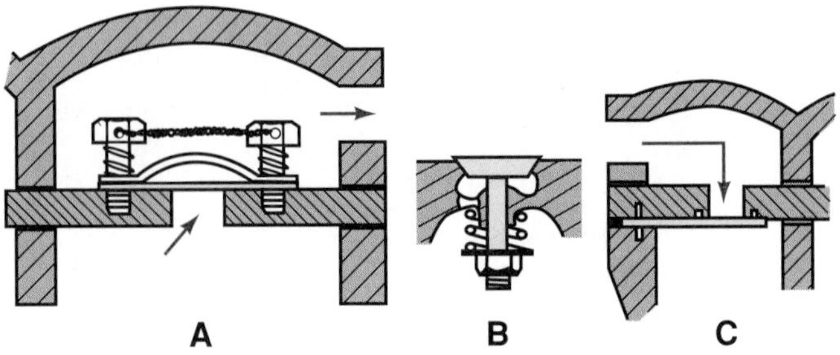

Figure 4-43. *Some typical compressor valve designs. A—Reed valve, spring-closed. B—Poppet valve, spring-closed. It is used on some large compressors. C—Reed valve. The pressure difference keeps the valve closed.*

The exhaust valve must be fitted with special care. It operates at high temperatures and must be leakproof against a relatively high pressure difference. Due to the high vapor pressures and the high temperatures, there is a tendency for the heavy ends (heavy molecules of hydrocarbon oils) to collect on the valve and valve seat as carbon.

The valves open about .010″ (.254 mm). If the movement is more, a valve noise develops. If the movement is too little, not enough vapor can move past the valve.

In small high-speed hermetic compressors, the intake valves are made very light. They are also made as large as possible. The cylinder intake valve is only open a fraction of a second. The valve design allows a greater amount of refrigerant vapor to enter during that time.

Crankshaft Seal

Some refrigerating systems use an external motor (open-type) to drive the compressor. These systems need a leakproof joint where the crankshaft comes out of the compressor crankcase. This is absolutely necessary, as the pressures vary greatly in the crankcase. **Figure 4-45** illustrates some popular crankshaft seals.

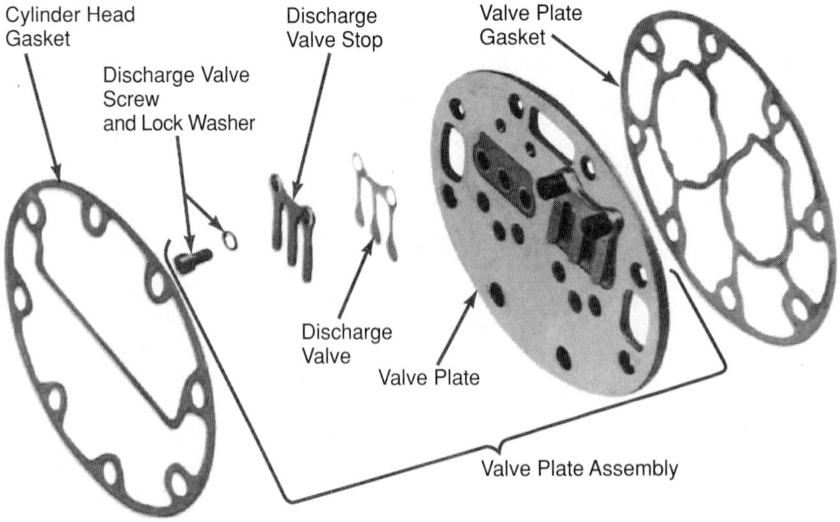

Figure 4-44. *Reciprocating compressor valve plate assembly. (Carrier Corp., Subsidiary of United Technologies Corp.)*

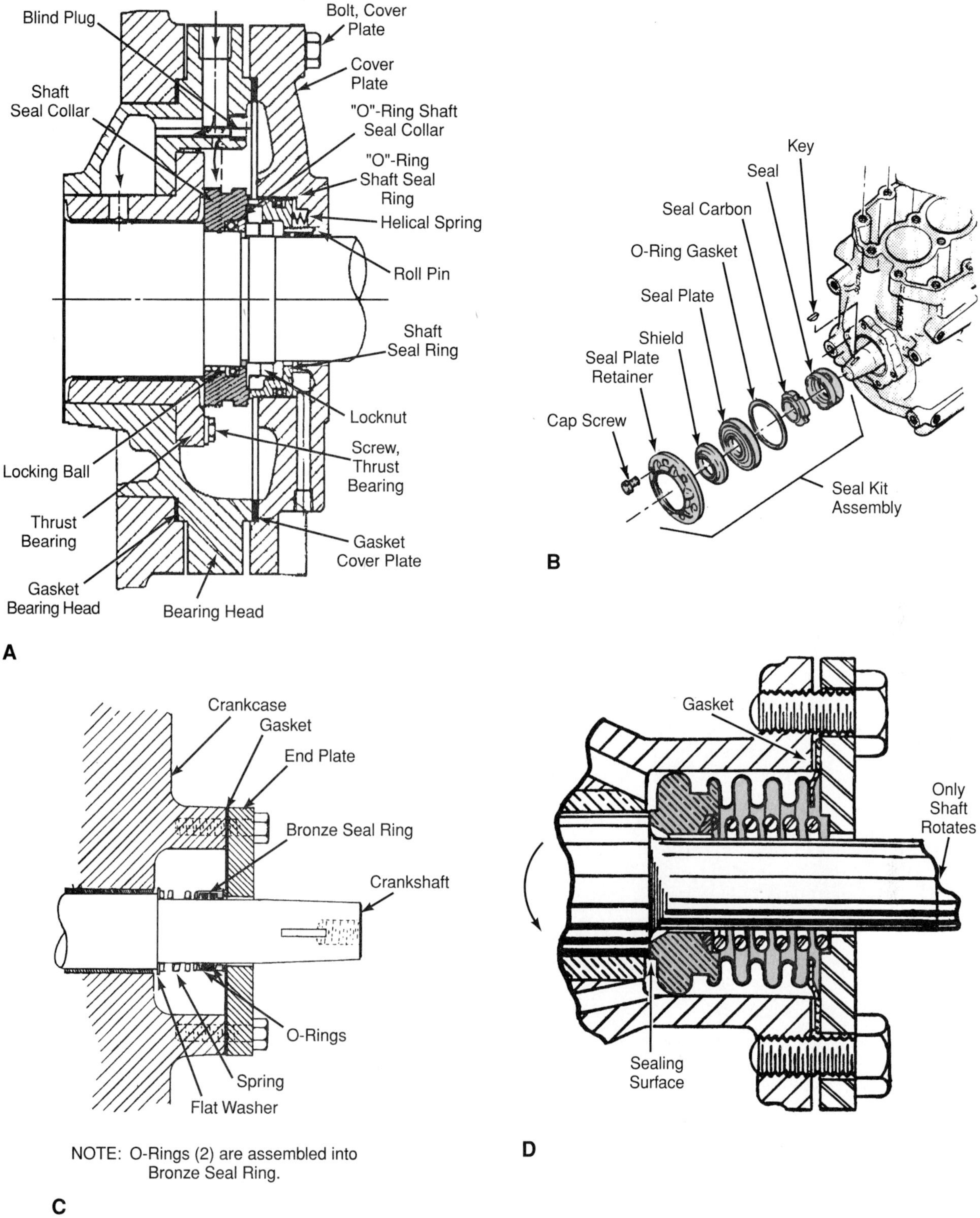

Figure 4-45. *Crankshaft seal construction for external-drive compressors. A—Seal used in commercial compressors. (Mycom Corp.) B—Seal used with an automobile air conditioning compressor. (Ford Motors) C—Replacement seal. (Chicago Valve Plate & Seal Co.) D—Bellows-type seal.*

The seal is the place where the shaft rotates part of the time and rests part of the time. Therefore, the seal must be carefully designed and installed.

All seals use two rubbing surfaces. One surface turns with the crankshaft. It is sealed to the shaft with an O-ring of synthetic material. The other surface is stationary and mounted on the housing with leakproof gaskets. The surface materials are accurate to .000001″ (.0000254 mm) and are optically flat. They are made of either hardened steel and bronze, or ceramics and carbon. The two rubbing surfaces must be lubricated or they will wear and start to leak.

Teflon™ is often used as a gasket material on automobile air conditioning compressors. The crankshaft seal must operate at a high temperature. It is usually made of ceramics and carbon.

Compressor Drive (External-Drive)

External-drive compressors are usually driven by a V-belt. The V-belt provides a quiet, efficient drive. Most V-belts are driven at less than the motor speed. This means that the motor belt pulley will be smaller than the compressor pulley. The diameter of the motor drive pulley and the compressor flywheel governs the compressor speed.

In large capacity installations, more than one belt may be used. This is necessary in order to transmit the required horsepower. **Figure 4-46** illustrates a two-belt, external-drive system.

Pulleys must be in perfect alignment and proper tension must be provided on the belt. Pulley shafts (motor and compressor) must be exactly parallel to each other. Most V-belt pulleys are made of cast iron. Some are built up from stamped steel parts.

Figure 4-46. *External-drive compressor with two-belt drive. The service technician is turning over the belt to check for cracks and excessive wear. (The Gates Rubber Co.)*

Crankshaft

Reciprocating compressors must change the rotary motion of the motor into reciprocating motion in the compressor. The crankthrow-connecting rod-piston combination is most frequently used. The crankshaft in these designs is usually made of forged or cast steel. Refer to **Figure 4-47.**

Some compressors use an eccentric fastened to a straight shaft. This is used in place of the conventional eccentric crankshaft. This construction is used to reduce vibration. It also removes the need for connecting rod caps and bolts. See **Figures 4-48** and **4-39.**

The crankshaft main bearings support the crank. They also must carry any end load. Crankshaft and connecting rod bearings are fitted with great accuracy. Clearance for lubrication is usually .001" (.0254 mm). In external-drive compressors, various methods are used to attach the drive pulley to the crankshaft. These include a standard taper, a key, and a nut-lock washer combination.

Many hermetic systems use the crankthrow/crankshaft connecting rod/piston arrangement. See **Figure 4-27.**

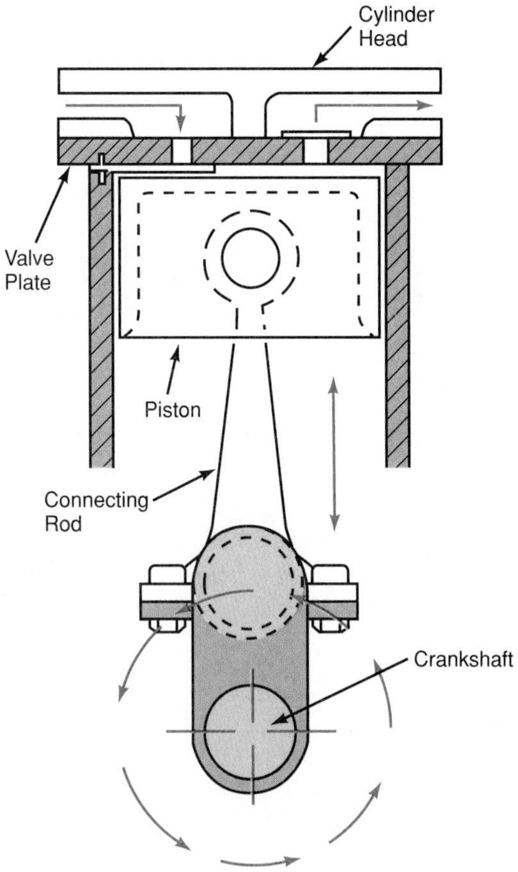

Figure 4-47. *Crankthrow-type crankshaft. As the crankshaft revolves, a piston reciprocates (moves up and down). The piston pin oscillates (swings back and forth) as it reciprocates with the piston. The lower end of the connecting rod rotates with the crankshaft.*

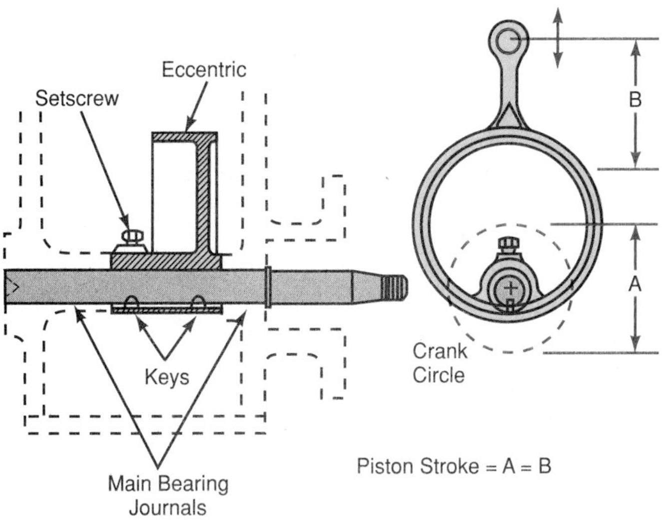

Figure 4-48. *Eccentric-type crankshaft mechanism. Note that the eccentric is attached to the crankshaft with keys and a setscrew.*

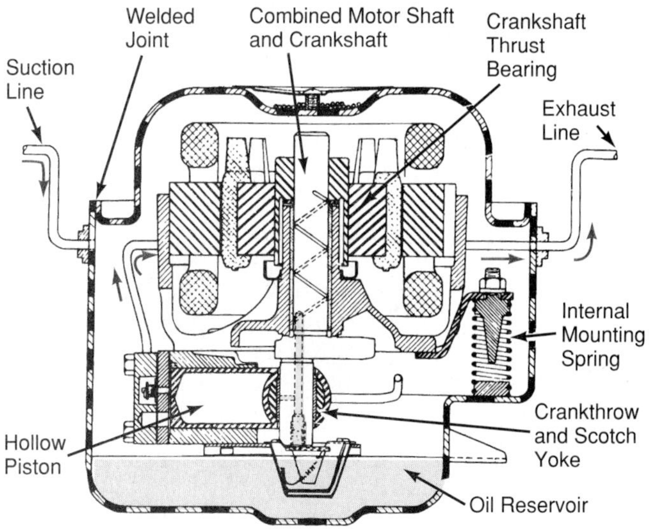

Figure 4-50. *Hermetic compressor using Scotch yoke mechanism.*

Scotch Yoke

The Scotch yoke mechanism is shown in **Figures 4-49** and **4-50.** There is no connecting rod. The cylinder and piston are both quite long. Even at the lower end of the stroke, the piston is guided by the cylinder wall. The crankshaft pin, also called the ***crankthrow,*** connects to the lower end of the piston. It is connected by means of a floating bearing. The Scotch yoke is popular in small high-speed compressors.

Swash Plate

The reciprocating compressor used on many automobile air conditioning systems is known as a ***swash plate compressor.*** It is also known as the ***wobble plate***

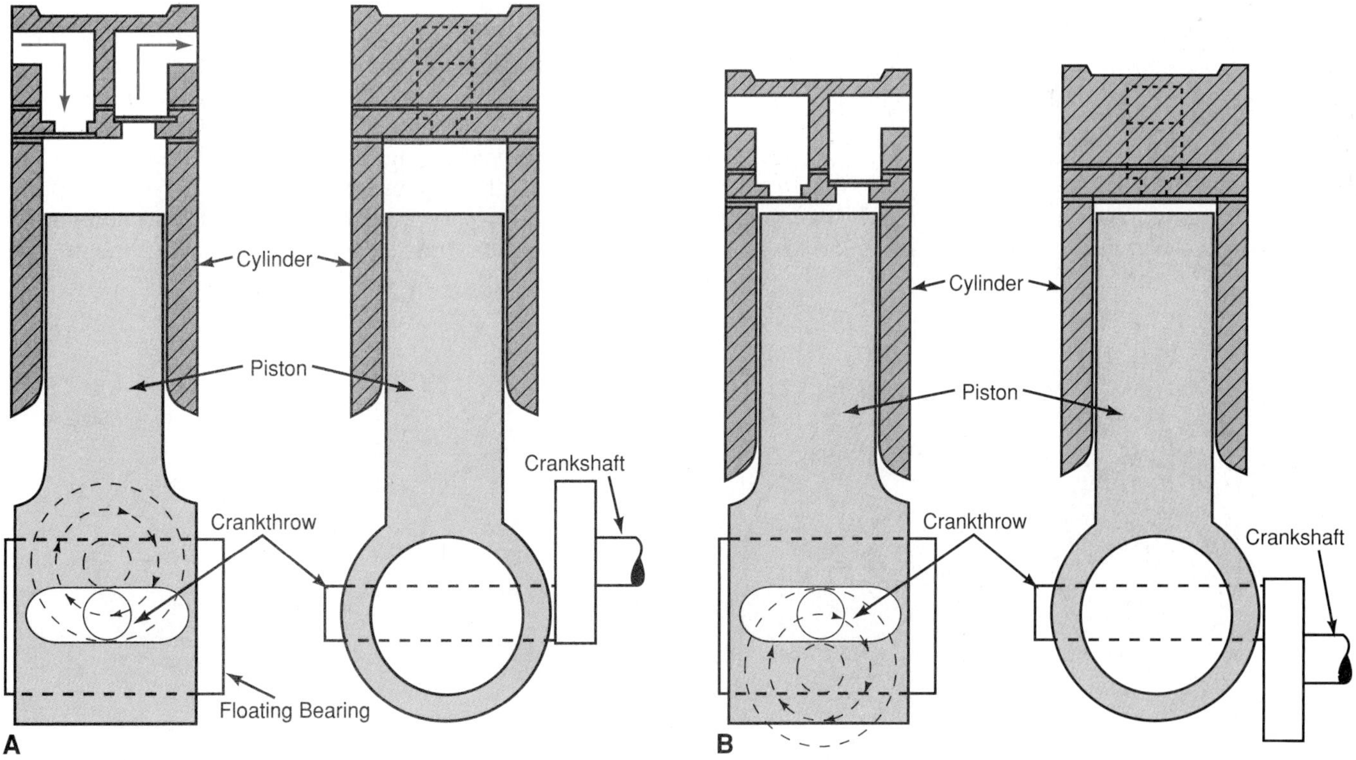

Figure 4-49. *Scotch yoke mechanism used to connect piston to crankshaft. No connecting rod is used. The piston extends to the yoke mechanism and the compressor cylinder serves as a guide. A—Shows piston at bottom of stroke (end of intake stroke). B—Shows piston at top of stroke (end of exhaust stroke).*

compressor. No connecting rod is used in this type of compressor. The cylinder and pistons are mounted as in **Figure 4-51.**

As the shaft revolves, the swash plate causes the pistons to reciprocate in the cylinders. Usually the swash plate compressor has three or more cylinders. These cylinders are arranged in a circle around the drive shaft.

The compressor is double acting—that is, compression takes place at each end of the stroke. Therefore, a three-cylinder compressor gives a pumping action like a six-cylinder conventional compressor of the same cylinder and stroke dimensions. This is an external-drive compressor. It requires a seal where the drive shaft passes through the compressor housing.

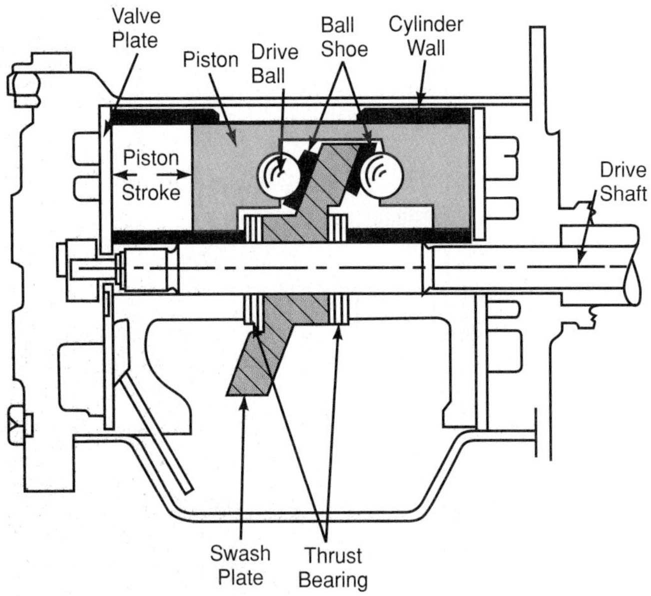

Figure 4-51. *Cross section through a swash plate reciprocating compressor. As drive shaft and swash plate revolve, double-end piston is moved back and forth in cylinder.*

Compressor Housing–Crankcase

In both the external-drive and hermetic compressor, the compressor housing gives support. It supports the cylinders, crankshaft, valves, oil pump, lubrication lines, and refrigerant inlet and exhaust openings.

In hermetic systems, the housing also supports and aligns the driving motor. Typical compressor housing designs are shown in **Figures 4-32** and **4-50.**

Some hermetic designs are bolted together and are provided with service valves. These are called ***serviceable hermetics,*** **Figure 4-32.** Many hermetic housings, particularly in the smaller sizes, are welded together. Refer to **Figures 4-27** and **4-36.**

Intake and Exhaust Ports

Conventional external-drive compressors provide inlet and exhaust ports as part of the cylinder head. These ports are usually fitted with service valves. See **Figure 4-26.**

Some hermetic compressors also have service valves. In small hermetic compressors, the motor compressor mechanism is enclosed in a welded dome. The inlet and exhaust lines go directly from the compressor inlet and exhaust port through the compressor dome. They are not generally supplied with service valves. See the line connections in **Figure 4-50.**

4.15.2 Rotary Compressors

There are two basic types of rotary compressors. One has blades (vanes) that rotate with the shaft. The other has a blade which remains stationary and is part of the housing assembly. In both types, the blade itself slides, providing a continuous seal for the refrigerant vapor. **Figure 4-52** shows a typical rotating two-blade compressor. The low-pressure vapor from the suction line is drawn into the opening. The vapor fills the space behind the blade as it revolves. As the blades revolve, trapped vapor in the space ahead of the blade is compressed until it can be pushed into the exhaust line to the condenser.

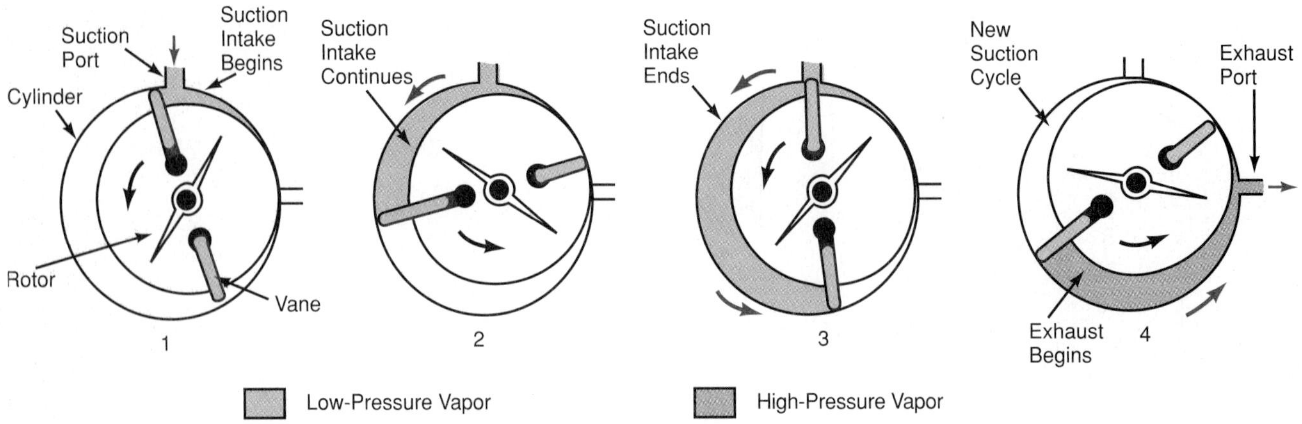

Figure 4-52. *A rotary blade compressor. Black arrows indicate direction of rotation of rotor. Red arrows indicate refrigerant vapor flow.*

A commercial rotary blade compressor, using eight blades, is pictured in **Figure 4-53.** The basic operation of the eight-blade compressor is the same as the two-blade.

Figure 4-54 illustrates a section through an eight-blade rotary compressor. This is an external-drive compressor. The shaft seal is shown at the right end.

Figure 4-55A shows the cylinder with the intake and exhaust ports. The relative positions of the rotor and cylinder are shown in **Figure 4-55B.**

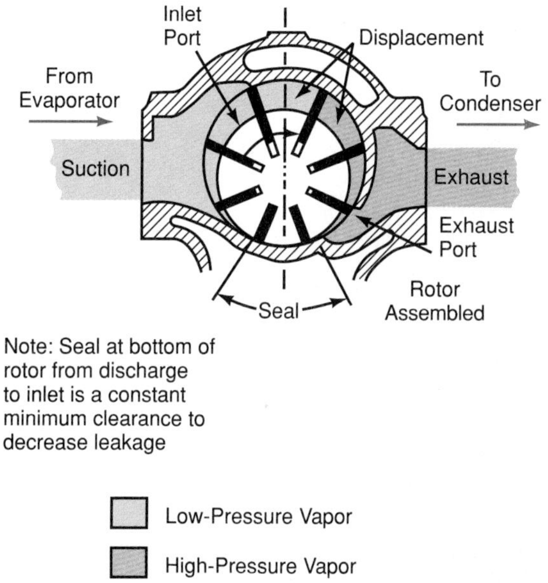

Figure 4-53. *Eight-blade rotary compressor. Black arrow indicates direction of rotation. Red arrows show direction of vapor flow. Inlet port is much larger than exhaust port. Large inlet port is needed to collect enough refrigerant vapor from the sparse low-pressure side (light blue).*

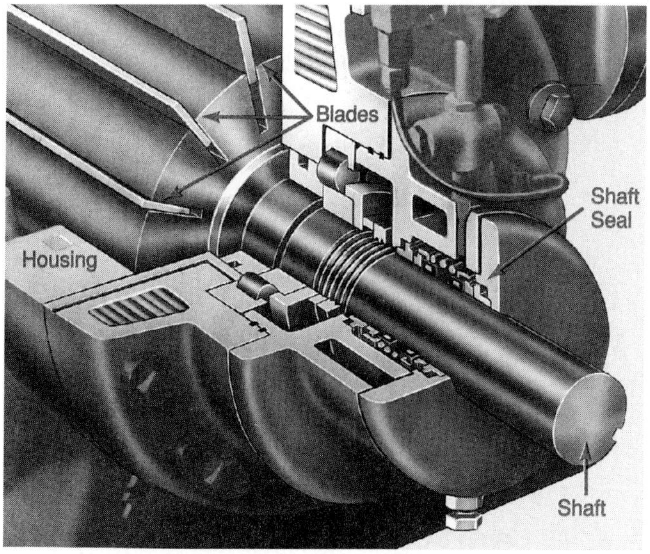

Figure 4-54. *Section through rotating vane compressor. This is an external-drive compressor. Note the shaft seal at the right end.*

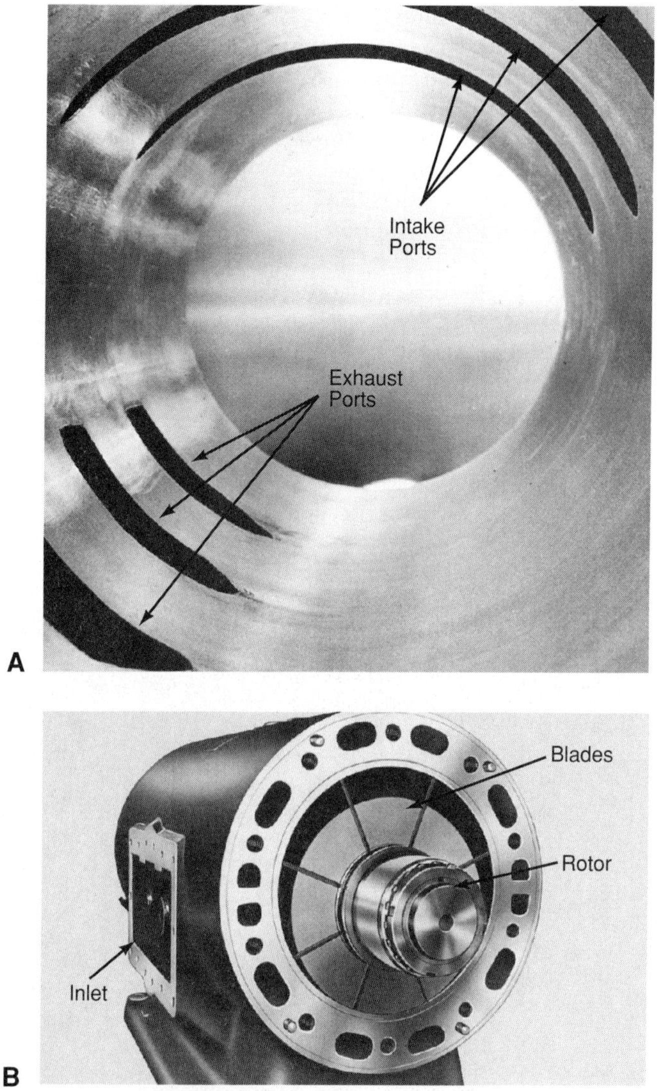

Figure 4-55. *Detail of cylinder and rotor shown in* ***Figure 4-54.*** *A—The inside of the cylinder showing port openings. Note that the intake ports are longer than the exhaust ports. B—Relative position of rotor and blades inside cylinder.*

Rotating vane compressors are frequently used as the "booster" compressor in cascade systems. This is the name commonly given to the first compressor in a cascade system.

These compressors have three advantages:

- They provide a large size opening into the suction line.
- They provide large inlet port openings.
- They have a very small clearance volume.

The low-side pressure may be quite low. The low-side vapor will be drawn into the compressor under a very small pressure difference. These compressors provide a large opening into the compressor from the low side. Thus, more vapor will be drawn in on the intake stroke. The clearance space provided in these compressors is small. Therefore, all the vapor drawn in on the intake stroke is pushed out on the exhaust stroke. This increases the compressor efficiency. The cascade system is explained in Chapter 18.

Figure 4-56 represents a stationary blade (often called a divider block) rotary compressor. An eccentric shaft rotates an impeller in a cylinder. This impeller constantly rubs against the outer wall of the cylinder.

As the impeller (or roller) revolves, the blade traps quantities of vapor. The vapor is compressed into a smaller and smaller space. The pressure and temperature build up. Finally the vapor is forced through the exhaust port. It enters the high-pressure side of the system (condenser).

The compression action on one quantity of vapor takes place at the same time another quantity of vapor

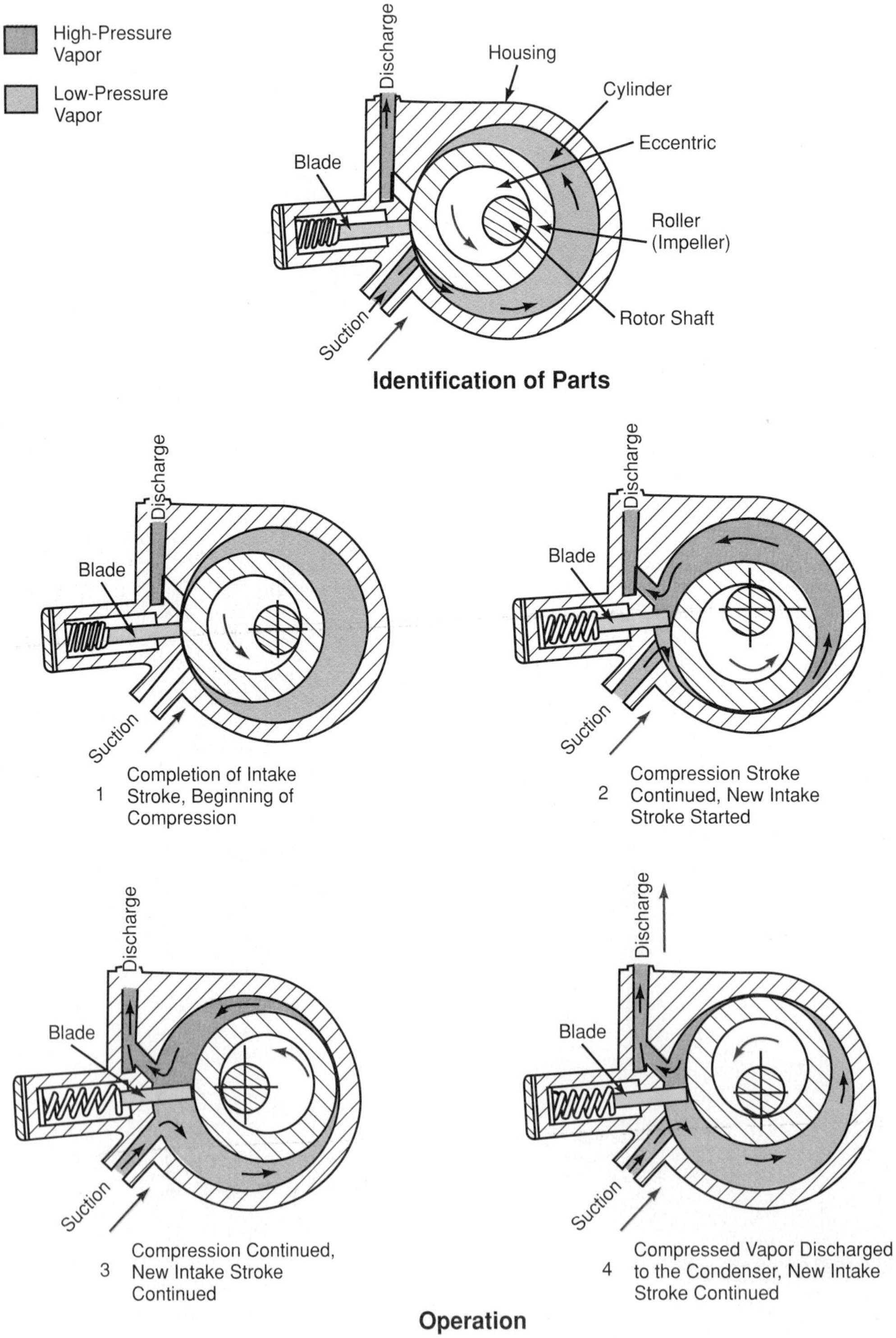

Figure 4-56. *Rotary compressor. Stationary blade or divider block is in contact with an impeller.*

is filling the cylinder on the intake stroke. All of the parts must be fitted to extremely close tolerances and clearances. The dimensions are very accurate and the surfaces quite smooth. Therefore, no gaskets are needed in the compressor assembly.

Figure 4-57 shows a hermetic rotary compressor using a stationary blade (dividing block).

In rotary compressors, check valves are usually used in the suction line. They prevent the high-pressure vapor and compressor oil from flowing back into the evaporator.

Rotary Cylinder Construction

Rotary cylinders are usually made of cast iron. Each is accurately machined, honed, and lapped (finished) on the inner surface and on the ends.

All cylinders have intake and exhaust ports. Some models have oil passages for lubrication. Cylinders are usually mounted on an end plate, which is part of the main compressor crankcase. Refrigerant passages continue into the end plate.

The exhaust valve reed is mounted on the exhaust port outlet of the compressor. It is mounted as close to the compression chamber as possible. Four or more bolts hold the cylinder to the main part of the compressor.

One or more steel dowel pins help align the cylinder on the back plate. Another accurately finished plate seals the other end of the cylinder.

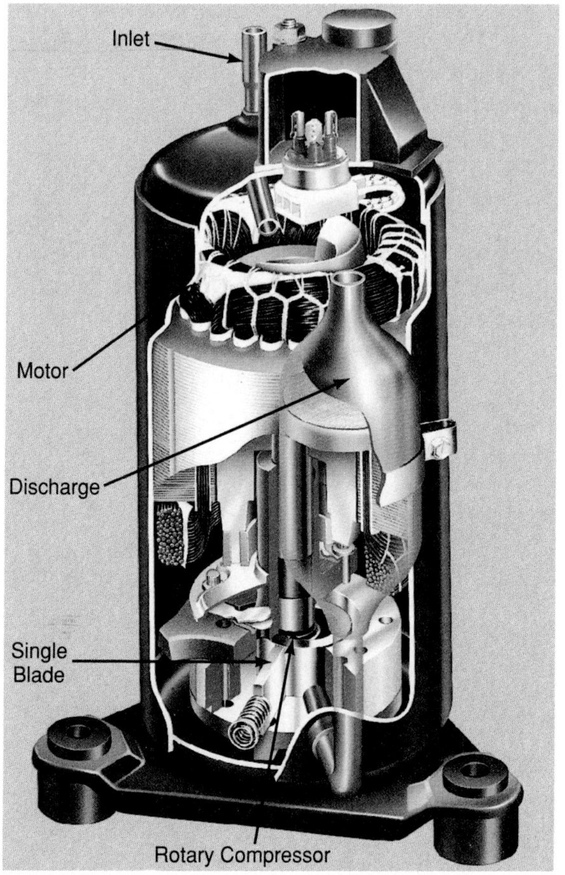

Figure 4-57. *Hermetic, single stationary-blade rotary compressor. (Tecumseh Products Company)*

Rotor–Compressor Construction

In the rotating blade compressor, the rotor is a fixed part of the shaft. The rotor length must be accurate to .0005″ (.0127 mm). Usually the slots for the blades are on a radius to the center of the shaft. To lower the starting load, one design puts the slots at an angle. This prevents the blades from touching the cylinder until the compressor nears its operating speed.

In the stationary blade compressor, the rotor (sometimes called the ***impeller***) accurately fits the eccentric. The eccentric is a fixed part of the shaft.

Figure 4-54 illustrates a popular type of rotor construction used on external-drive commercial compressors. **Figure 4-58** illustrates a stationary-blade rotary compressor. This is a hermetic compressor used both in refrigeration and air conditioning units.

Blade (Vane) Construction

Rotating blade compressors use two or more blades. **Figure 4-55,** Part B, shows a rotor with eight of them. These blades may be made of cast iron, steel, aluminum, or carbon.

The compressor efficiency depends greatly on the condition of the blade edge where it rubs on the cylinder. The blade must be very accurately ground. It must be ground to fit the slots, the ends of the cylinder, and other contact surfaces with the cylinder.

4.15.3 Scroll Compressors

A ***scroll compressor*** generates a series of crescent-shaped gas pockets between two scrolls, **Figure 4-59.** One scroll—the fixed scroll—remains stationary. The

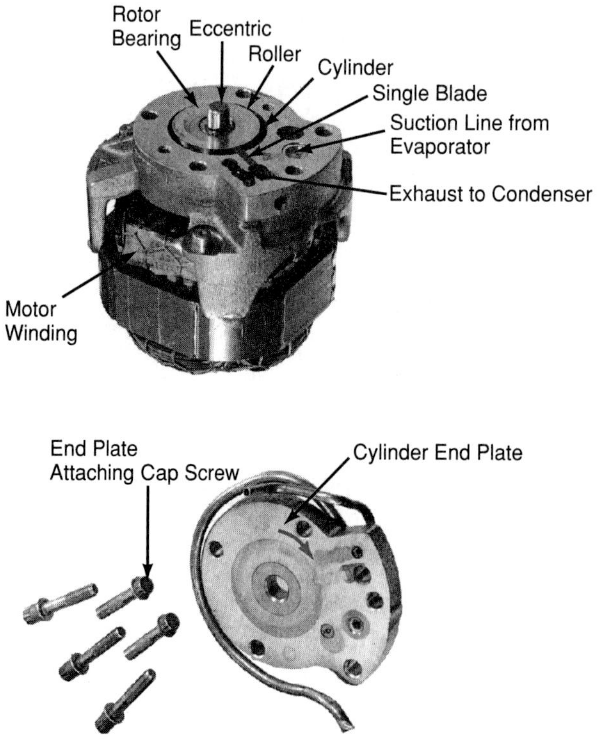

Figure 4-58. *A hermetic stationary-blade rotary compressor.*

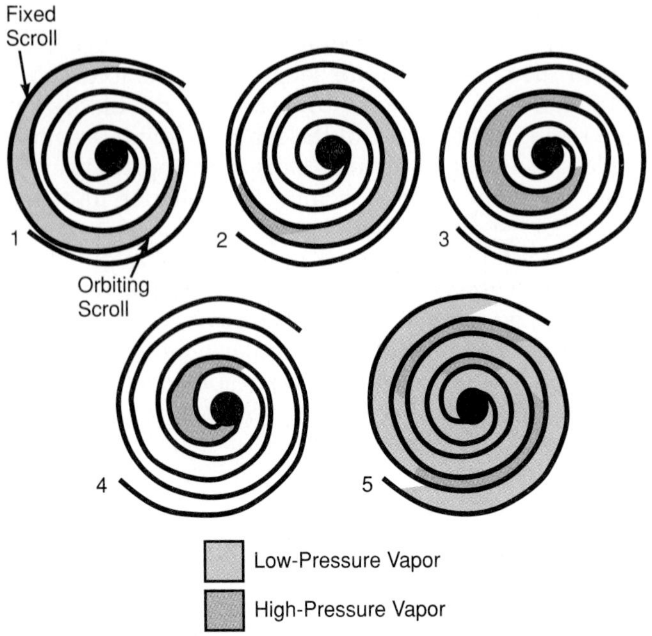

Figure 4-59. *Compression in the scroll is caused by the interaction of an orbiting scroll mated within a stationary scroll. 1—Gas is drawn into an outer opening as one of the scrolls orbits. 2—As the orbiting motion continues, the open passage is sealed off and the gas is forced to the center of the scroll. 3—The pocket becomes progressively small in volume. This creates increasingly higher gas pressures. 4—Discharge pressure is reached at the center of the pocket. Gas is released from the port of the stationary scroll member. 5—In actual operation, six gas passages are in various stages of compression at all times. This creates nearly continuous suction and discharge. (Lennox Industries, Inc.)*

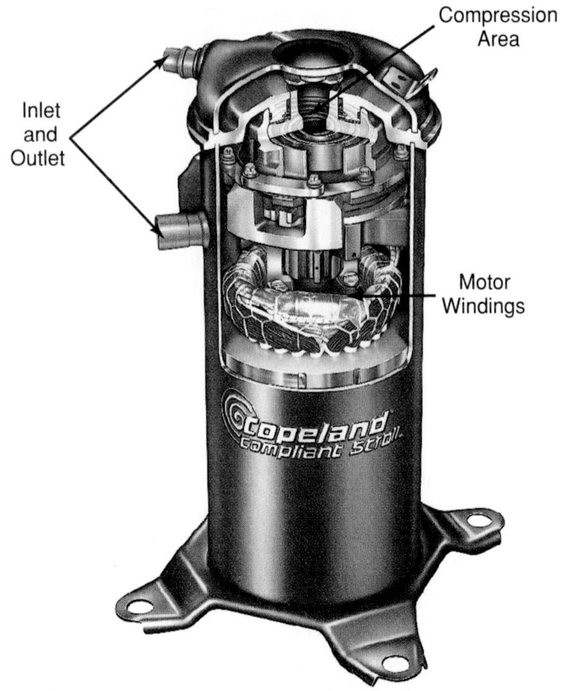

Figure 4-61. *Scroll compressor used in light commercial installations. Note the compression area. (Copeland Corporation)*

other scroll—the orbiting scroll—rotates through the use of the swing link. As the motion occurs, the pockets between the two forms are slowly pushed to the center of the two scrolls. This reduces the gas volume. When the pocket reaches the center of the scroll, the gas is at a high pressure. It is discharged out of the center port. During this compression process, several pockets are being formed at the same time. The suction process from the outer portion of the scroll and the discharge from the inner portion are continuous. This continuous process gives the compressor very smooth action.

Scroll compressor design is shown in **Figure 4-60.** A scroll compressor used on domestic room air conditioners is shown in **Figure 4-61.** A scroll compressor designed for commercial installations is shown in **Figure 4-62.** It has a low sound level. Note the compressor comes equipped with service valves on the inlet and outlet sides.

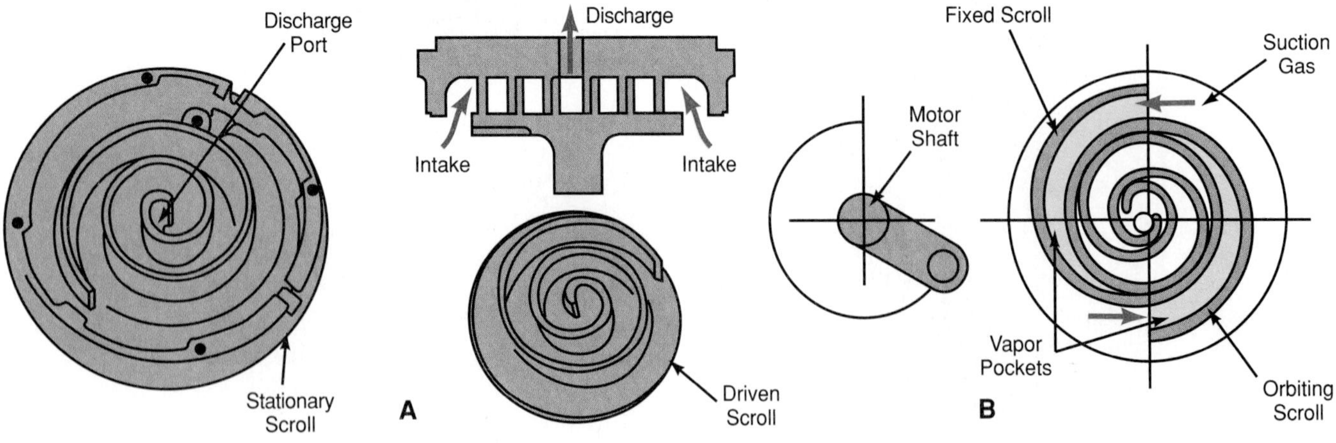

Figure 4-60. *Scroll compressor design. A—Two scrolls are used to produce a vapor compression. The upper scroll is stationary and the lower scroll is driven. Note intake and discharge ports. B—Note how the rotation of the motor shaft causes the orbiting scroll to* orbit*—not rotate about the shaft center. (The Trane Co.)*

Figure 4-62. *Hermetically sealed scroll compressor. Note the location of the service valves for ease of access. (Copeland Corporation)*

The scroll compressor has fewer moving parts and less torque variation than reciprocating compressors. This results in very smooth and quiet operation.

4.15.4 Screw Compressors

The screw compressor uses a pair of special helical rotors. They trap and compress air as they revolve in an accurately machined compressor cylinder. Screw compressors are available in either external-drive or hermetic construction. They are used in large systems (20 ton capacity and up).

Figure 4-63 illustrates a cross section of a screw compressor. The two rotors are not the same shape. One is male, the other female. The male rotor, A, is driven by the motor. It has four lobes. The female rotor, B, meshes with and is driven by the male rotor. It has six interlobe spaces. The cylinder, C, encloses both rotors.

In operation, the refrigerant vapor is drawn in as shown in **Figure 4-64.** The intake (low-pressure vapor) enters at one end of the compressor and is discharged (compressed vapor) at the opposite end.

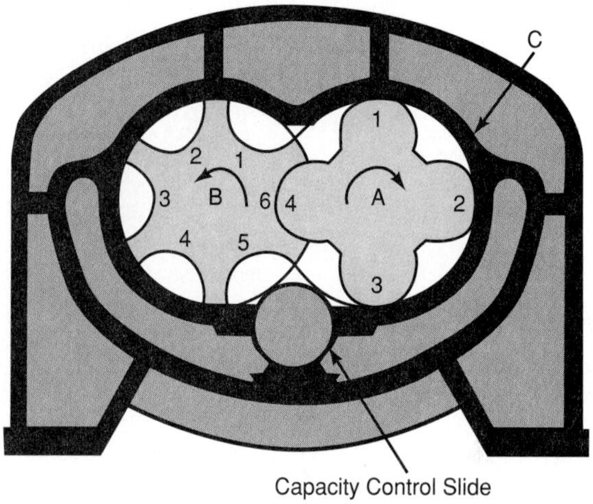

Figure 4-63. *Cross section of screw compressor. A—Male rotor. B—Female rotor. C—Cylinder. Vaporized refrigerant enters at one end and exhausts at other end. (ABB Stal Refrigeration Corporation)*

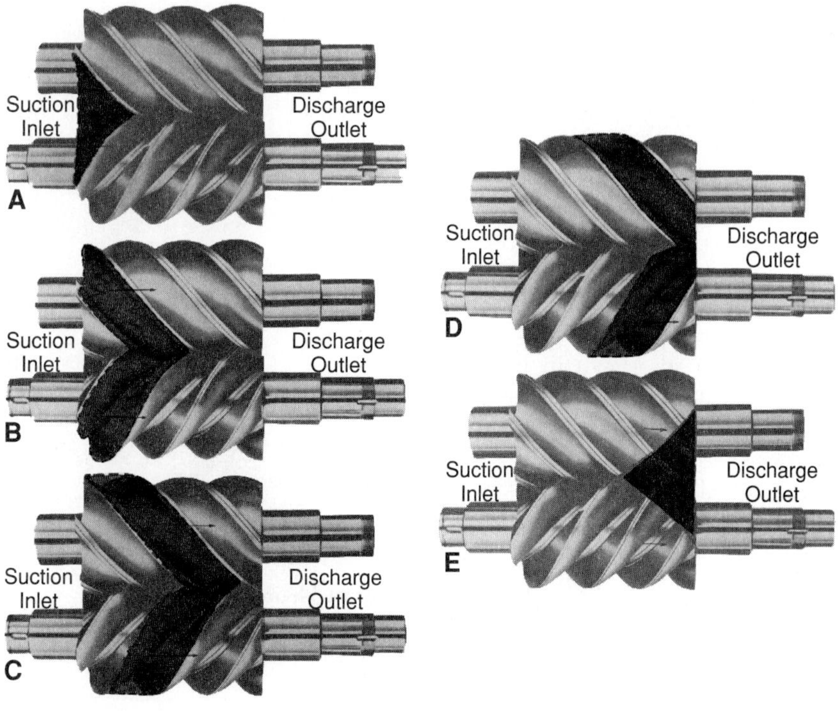

Figure 4-64. *Basic operation of screw compressor. Revolving rotor compresses vapor. A—Compressor interlobe spaces being filled. B—Beginning of compression. C—Full compression of trapped vapor. D—Beginning of discharge of compressed vapor. E—Compressed vapor fully discharged from interlobe spaces. (Dunham-Bush, Inc.)*

Figure 4-65. *Screw compressor which uses a matched set of helical rotors. (Dunham-Bush, Inc.)*

The male rotor revolves more rapidly than the female rotor. (There are four lobes on the male rotor and six on the female rotor.) The rotors are helixes. They provide a continuous pumping action rather than pulsating as with a reciprocating compressor. With this pumping action, there is very little vibration during operation.

Figure 4-65 illustrates a cutaway view of an external-drive screw compressor. It is powered by an external-electric motor, which drives the drive shaft. As previously mentioned, the motor drives the male rotor. Two matched helical rotors—male and female—turn together. This action traps and compresses the refrigerant. The male rotor in the illustration is an extension of the drive shaft. The other rotor is made to turn by the action of the male rotor. A control device mounted outside the housing regulates the capacity of the unit.

A hermetic screw compressor is shown in **Figure 4-66.** Its capacity control device is mounted inside the housing. Its inlet port is located at right angles to the rotors. Outlet port is through the motor housing.

Figure 4-67 illustrates a 3600-rpm, single screw compressor. It utilizes one main rotor that meshes with two diametrically opposed star-shaped gate rotors. The main rotor contains six grooves. It has straight roller bearings at the shaft ends. Two capacity control slide valves, one on each side, help to determine the capacity control. **Figure 4-68** is a complete single screw compressor unit using a microprocessor control system.

Many screw compressors operate with oil injection. This seals the clearance between the rotors and between the rotors and the cylinder. It also helps cool the compressor. The efficiency of these compressors is quite high. They may be used with most of the common refrigerants.

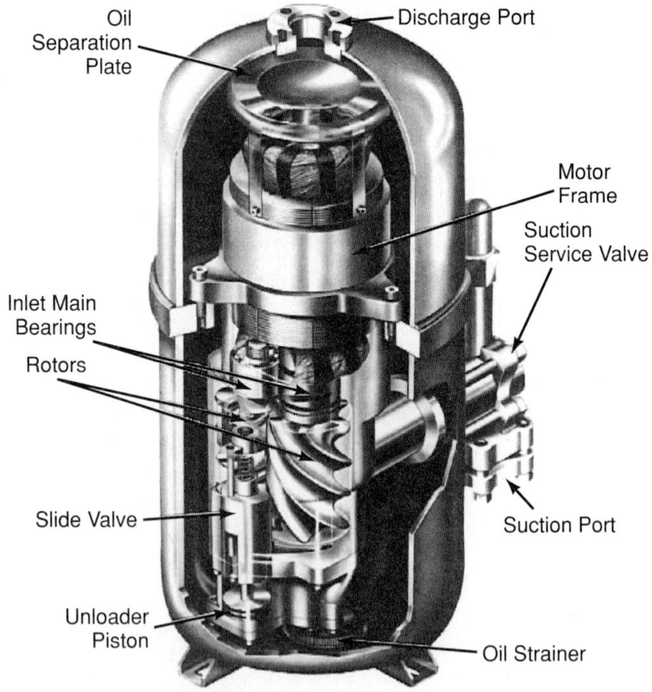

Figure 4-66. *Cutaway view of hermetic screw compressor. Note the helical rotors and capacity control slide valve. (Dunham-Bush, Inc.)*

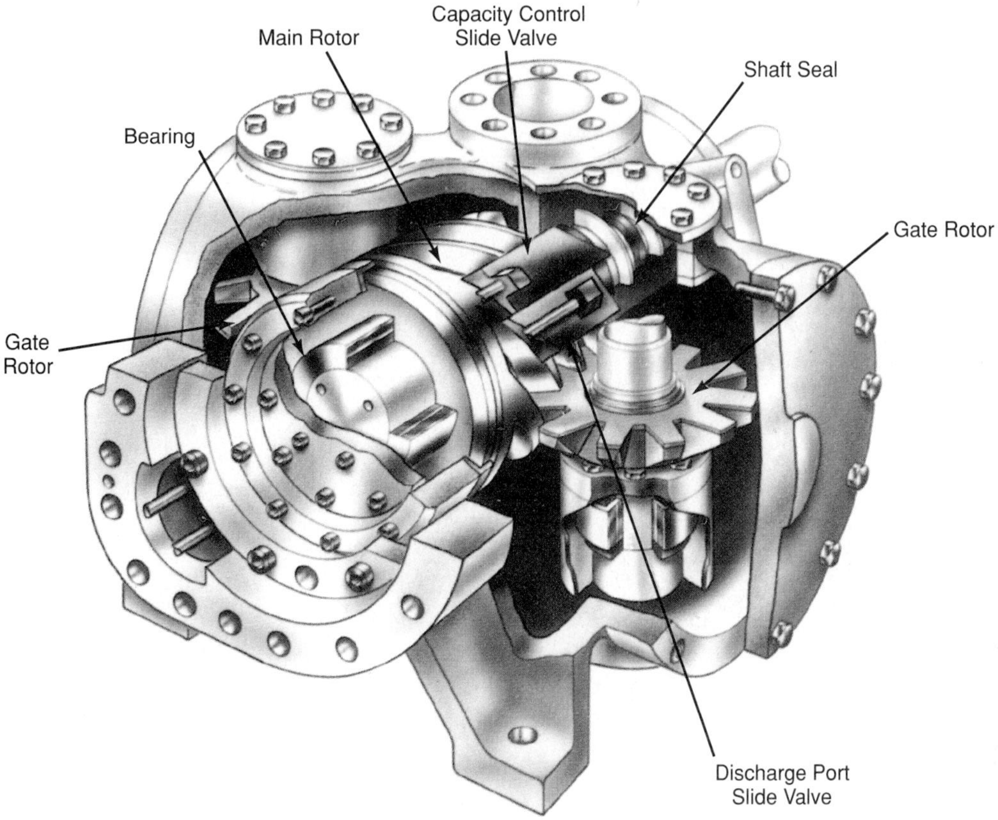

Figure 4-67. *Single screw compressor. Note the location of the main rotor in relation to the two gate rotors. (Vilter Manufacturing Corporation)*

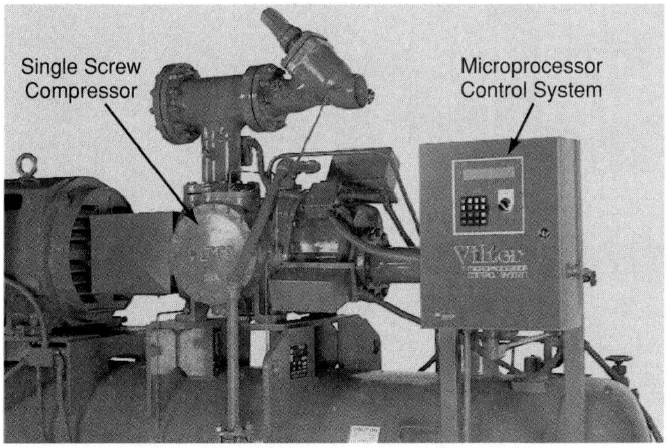

Figure 4-68. *A commercial single screw compressor system. (Vilter Manufacturing Corporation)*

Figure 4-69 illustrates a pair of screw-type rotors in operating position. Since screw compressors operate at fairly high speed, adequate bearings are needed for good rotor bearing life.

4.15.5 Centrifugal Compressors

Centrifugal compressors are used successfully in large refrigerating systems. In this type of compressor, vapor moves outward as it is moved rapidly in a circular path. This action is called ***centrifugal force.*** (However, the correct term is "centripetal force.")

The vapor is fed into a housing near the center of the compressor. A disk with radial blades (impellers) spins rapidly in this housing. This forces vapor against the outer diameter.

The pressure gained is small, so several of these compressor wheels or impellers are put in series. This creates greater pressure difference and pumps a sufficient volume of vapor. A centrifugal compressor looks like a steam turbine or axial flow air compressor for a gas turbine engine.

The centrifugal compressor has the advantage of simplicity. There are no valves or pistons and cylinders. The only wearing parts are the main bearings. Pumping efficiency increases with speed, so the compressors are designed to operate at high speeds.

Figure 4-70 is a cross section through a two-stage centrifugal compressor. The driving motor is mounted between stages. The inlet is at the left on the illustration. The discharge is in the back at the right end of the illustration and is not shown.

Figure 4-71 pictures a centrifugal compressor mounted in a refrigerating system. A cutaway view showing the refrigerant as it passes through the system is illustrated in **Figure 4-71B.** The compressor continuously draws refrigerant vapor from the cooler. As the compressor suction reduces the pressure in the cooler, the remaining refrigerant boils off. Energy required for

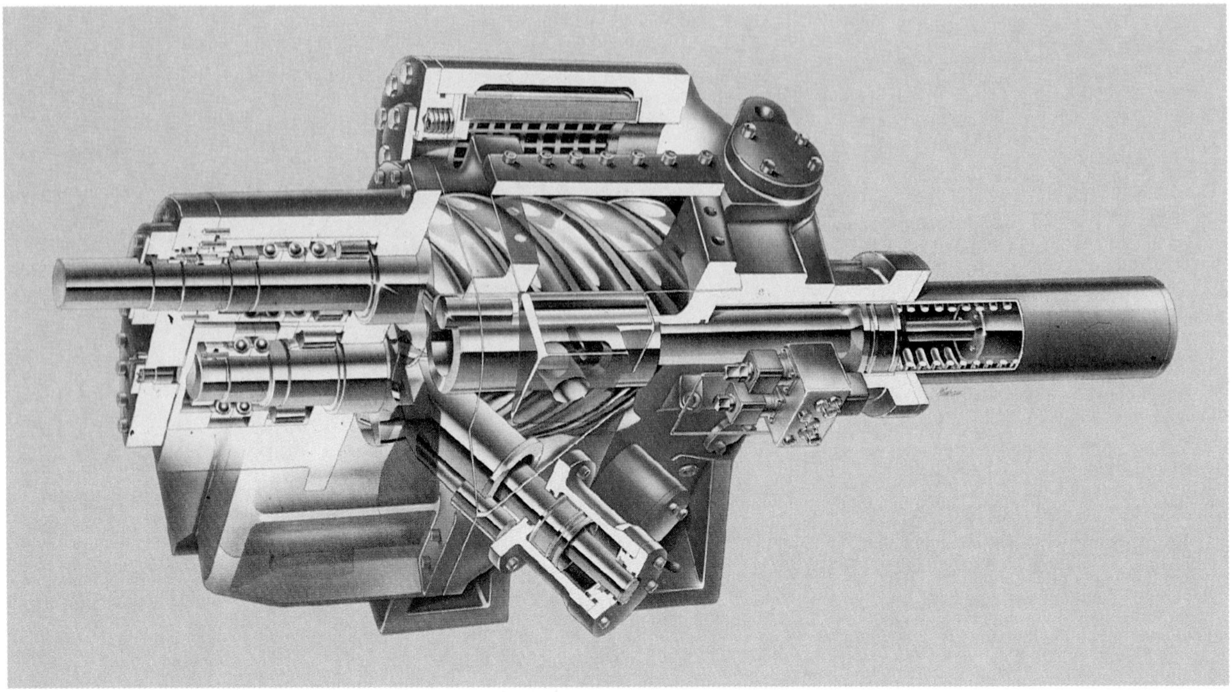

Figure 4-69. *Screw compressor with matched set of helical rotors. Designed to operate with ammonia, R-134a, or some other types of refrigerants. (ABB Stal Refrigeration Corporation)*

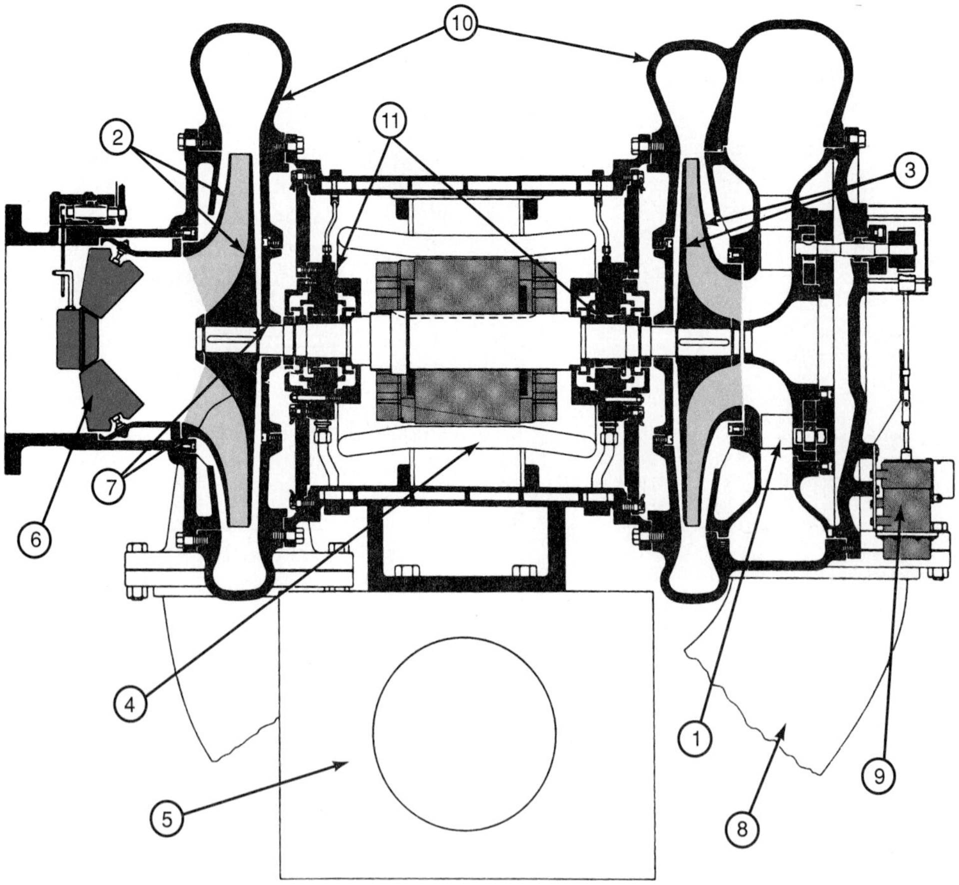

Figure 4-70. *Two-stage centrifugal compressor. 1—Second-stage variable inlet guide vane. 2—First-stage impeller. 3—Second-stage impeller. 4—Water-cooled motor. 5—Base, oil tank, and lubricating oil pump assembly. 6—First-stage guide vanes and capacity control. 7—Labyrinth seal. 8—Cross-over connection. 9—Guide vane actuator. 10—Volute casing. 11—Pressure-lubricated sleeve bearing. Note that discharge opening is not shown.*

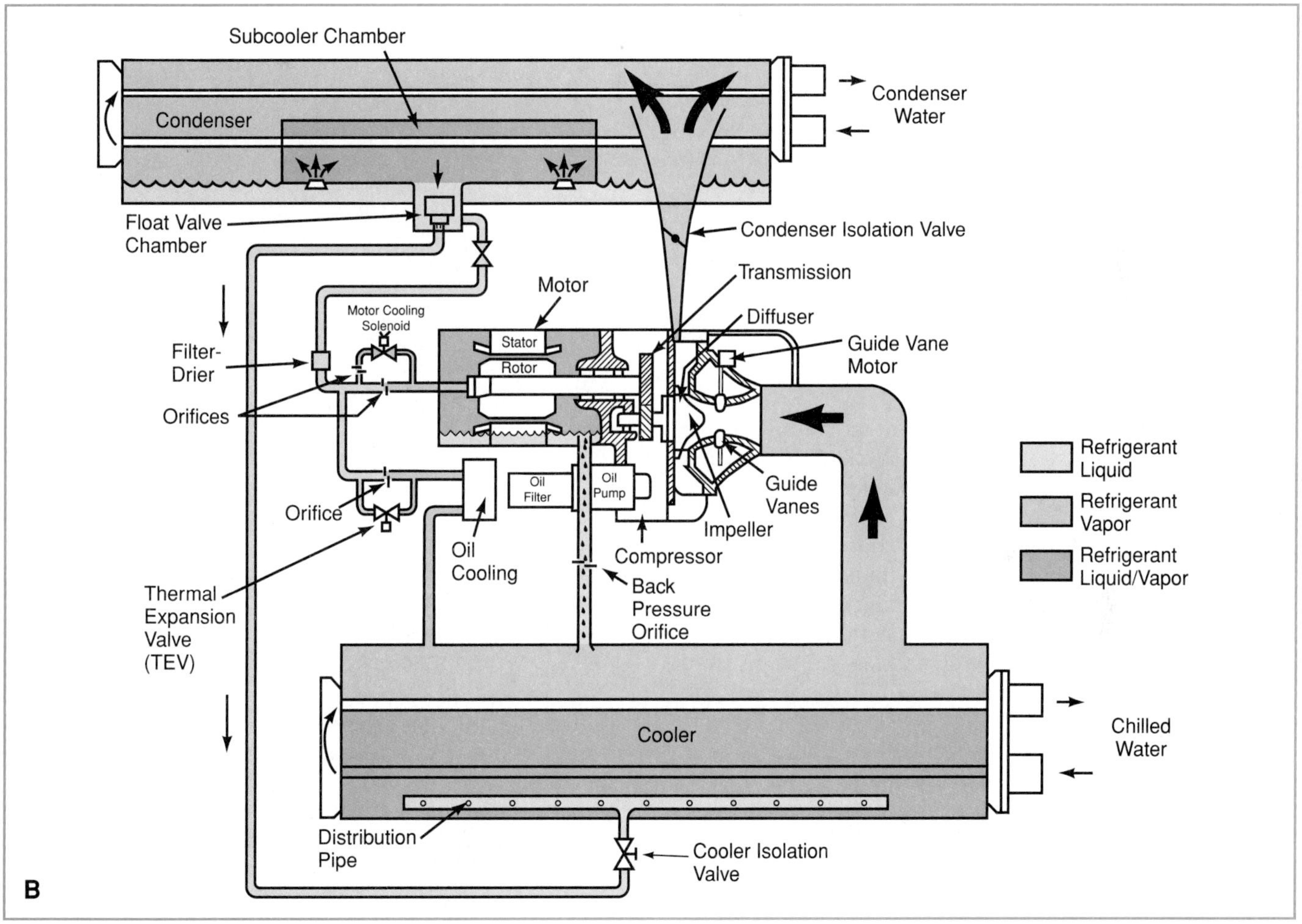

Figure 4-71. *A hermetic centrifugal liquid chiller, single-stage compressor. Using HCFC-22, 300 to 600 nominal tons; using HFC-134a, 200 through 530 nominal tons. The system can utilize either R-22 or R-134a, which allows for conversion from R-22 to R-134a if needed. The unit has a microprocessor to control the system. A—Parts identified. B—Cutaway view showing the refrigeration cycle. (Carrier Corporation, Subsidiary of United Technologies Corporation)*

boiling is obtained from the water flowing through the cooler tubes. With the heat energy removed, the water becomes cold enough for use in an air conditioning circuit. After removing heat from the water, the refrigerant vapor is compressed. It is then discharged from the compressor into the condenser. Cooled water flowing into the condenser tubes removes heat from the refrigerant. The vapor condenses to a liquid. The liquid refrigerant passes through the orifices into the flash subcooler chamber. Part of the liquid flashes to vapor,

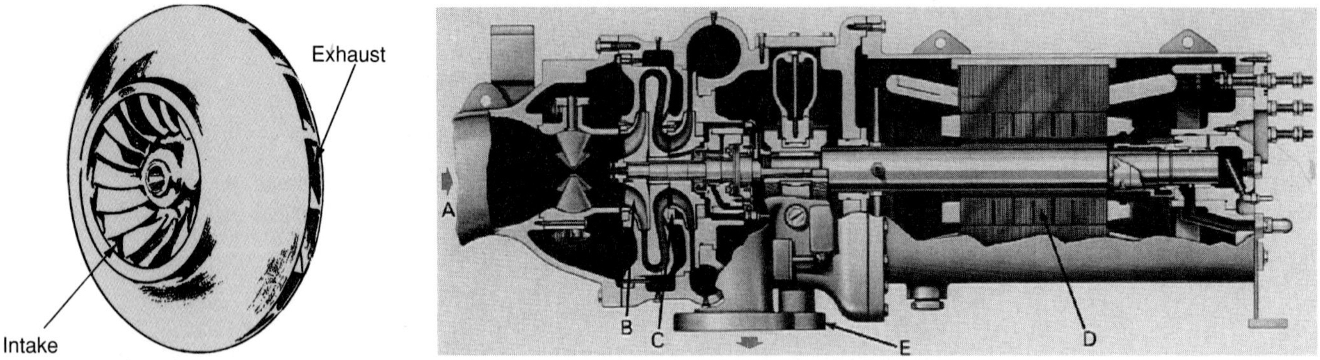

Figure 4-72. *Hermetic centrifugal compressor. The impeller is shown at left above. Major components of the compressor are: A—Intake. B—First-stage impeller. C—Second-stage impeller. D—Hermetic motor. E—Exhaust. (Carrier Corporation, Subsidiary of United Technologies Corporation)*

cooling the remaining liquid. The flash vapor is recondensed on the tubes, which are cooled by entering condenser water. The liquid drains into a float valve chamber by the flash chamber and cooler. The float valve prevents any of the flash chamber vapor from entering the cooler. Refrigerant is now at a temperature and pressure at which the cycle began. The cycle is completed.

Figure 4-72, right view, shows a section through a hermetic centrifugal compressor. These compressors operate at a high speed. They are usually driven by an electric motor or steam turbine.

Stator and Rotor Construction

The stator or casing of a centrifugal compressor is usually made of cast iron. It has a changing radius inside. The radius adapts itself to the vapor pickup by the impellers.

The casing (cylinder) also holds the main bearings, oil pressure pump, and the vapor intake and exhaust ports. When an external motor is used, the cylinder also holds the shaft seal where the shaft extends out for the power drive. Both the first stage and second stage have adjustable inlet vanes to control the capacity of the pump.

The rotor or impeller in a centrifugal compressor is keyed to the compressor shaft. It is made of cast iron or steel. It is specially designed to move the vapors without going above gas velocity limits. It is designed so there will be no vapor-trapping pockets. A typical rotor is shown in **Figure 4-72,** left view.

4.16 Motors

Two different types of motors are used for driving refrigeration compressors. External-drive compressors use conventional motors. The compressor is driven with one or more V-belts or by direct drive. **Figure 4-73** shows a sectional view of a typical motor used on small external-drive compressors.

In hermetic compressors, the motor is mounted under the same dome as the compressor. These motors are lubricated by the oil carried in the refrigerant. Hermetic motors do not use brushes or open points inside the dome. Arcing would cause pollution in both the oil and the refrigerant. This would lead to an electrical burnout. A special electrical starting device is located outside the dome. **Figure 4-74** shows the three main parts of a motor used in a hermetic refrigeration compressor.

Figure 4-73. *Motor used on external-drive compressor. A—Motor shaft. B—Motor rotor. C—Motor stator. D—Starting mechanism. E—Motor bearing. F—Motor frame. (A.O. Smith Electrical Products Co.)*

See Chapter 7 for information concerning motors. Chapter 8 has more information about starting devices and motor controls.

4.17 Service Valves

Service valves are shown in **Figures 4-10** and **4-11.** Many small hermetic systems do not have service valves of any type. To attach gauges and service manifolds, refrigerant lines must be tapped. Special tapping valves are available. These are clamped to a tube so that the valves pierce the tube. At the same time, the necessary gauge and service connections are provided.

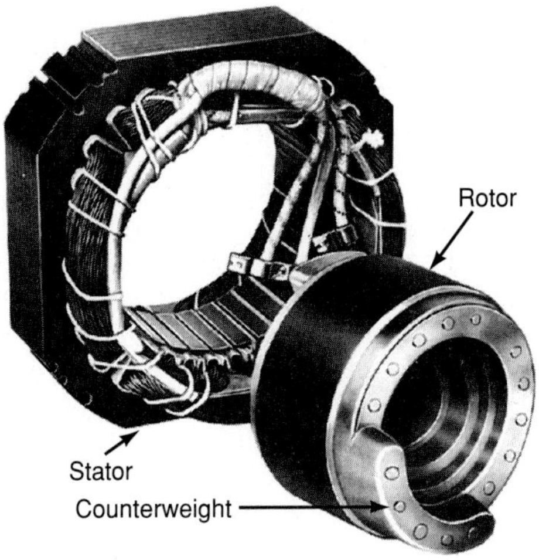

Figure 4-74. *Hermetic motor stator and rotor. Rotor is mounted directly on compressor crankshaft. Note counterweight, which balances weight of crank, connecting rod, and piston.*

4.18 Mufflers

Most hermetic units and many external-drive systems have noise-reducing devices. These are called mufflers. They are located on both the intake and the exhaust openings of compressors. Mufflers reduce the sharp gasping sound on the intake stroke. They also reduce the even sharper puff of the exhaust. These mufflers are brazed cylinders with baffle plates mounted inside. **Figure 4-75** shows a unit equipped with a suction, or "intake" muffler.

Some suction and exhaust mufflers are connected directly to the compressor cylinder head with lines. See **Figure 4-76.**

4.19 Compressor Cooling

The temperature of the compressor is greatly affected by the heat of compression. As vapor is

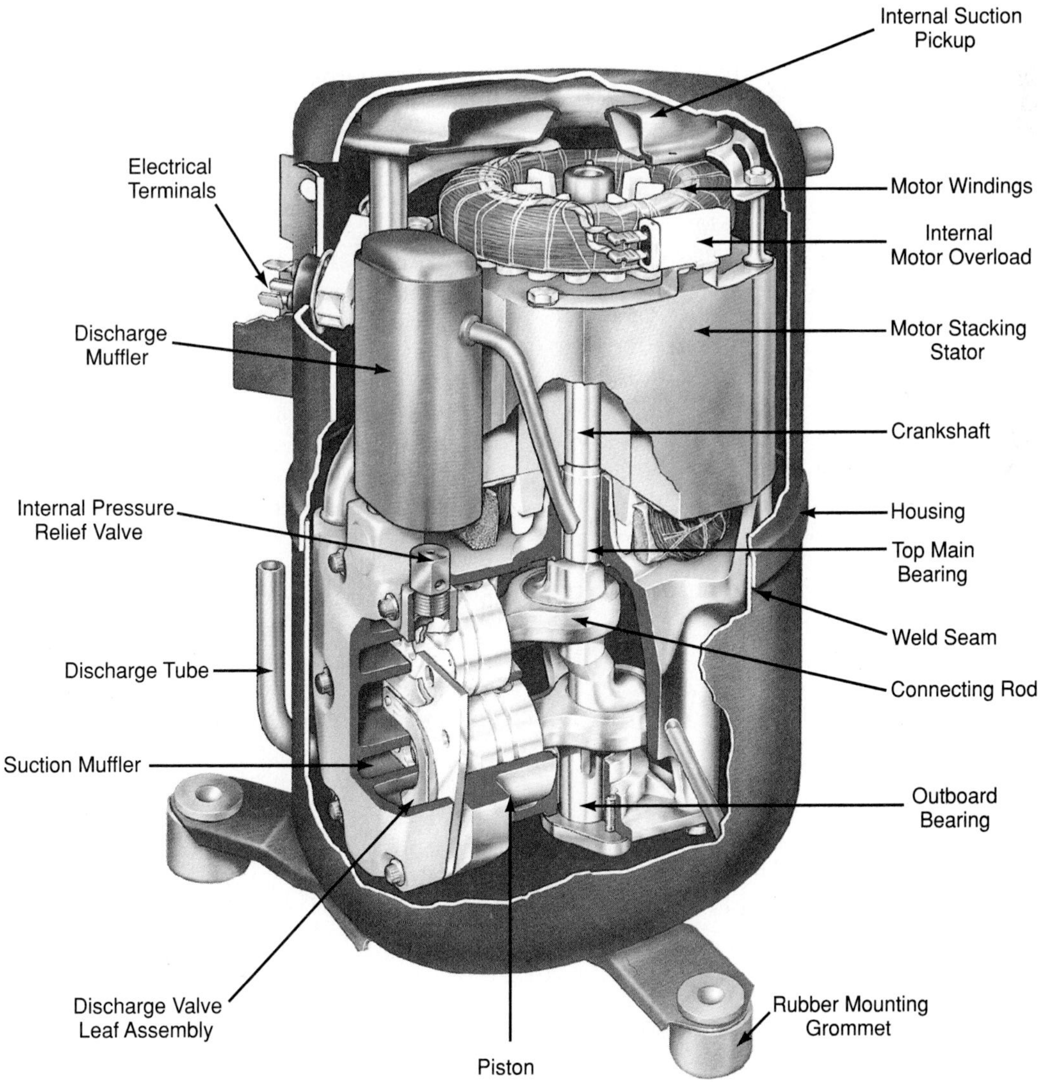

Figure 4-75. *A two-cylinder hermetic compressor suitable for use either in a commercial application or as part of a residence air conditioning system. Note suction and discharge mufflers. (Tecumseh Products Company)*

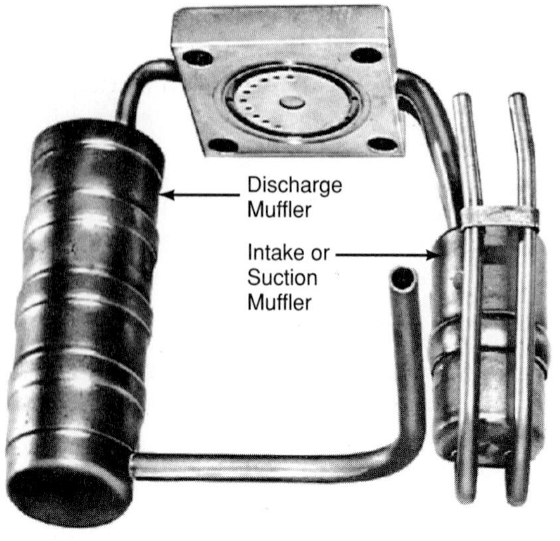

Figure 4-76. *Hermetic compressor cylinder head. Mufflers are attached to suction and exhaust openings.*

"squeezed" and forced into the condenser, the vapor temperature is raised. Friction (rubbing) between moving parts also adds to compressor temperature. This heat must be removed to prevent loss of efficiency of the pump. This heat must also be removed in order to maintain the lubricating qualities of the oil.

The oil that circulates in the compressor removes much of the heat from motor and compressor. As it flows over the heated surfaces, it picks up this heat. It carries the heat to cooler surfaces.

Many compressors and many domes have metal fins on their outer surfaces. These fins help carry the heat away. Some units even use a motor-driven fan to force cooling air over the compressor.

With a water-cooled condenser, the water is often used to cool the compressor or dome.

Motors are often cooled by passing the suction vapors and return oil over the windings. Some units circulate the crankcase oil through an air-cooled coil. The cooled oil then helps to cool the motor compressor. Some larger hermetic units are water cooled.

4.20 Lubrication

Lubricating oils have been developed especially for reciprocating and rotary refrigeration compressors. Usually, these are mineral oils, which are completely dehydrated, wax-free, and nonfoaming. They have a viscosity that is best for the refrigerant and for the refrigeration temperatures. (***Viscosity*** indicates the ability to flow at given temperatures).

Some refrigerating oil contains additives to improve lubricating qualities. The additives may also improve the oil's viscosity properties.

Reciprocating compressors may be lubricated either by a splash or a pressure (force feed) system. In the splash system, the crankcase is filled with the correct oil up to the bottom of the main bearings or to the middle of the crankshaft main bearings. At each crankshaft revolution, the crankthrow, or the eccentric, dips into the oil. It splashes the oil around the inside of the compressor. Oil is thrown onto cylinder walls and piston pin bushings. It is also thrown into small openings where it can drain into the main bearings.This is an excellent system for normal use in small compressors.

Some compressor connecting rods have little dips or scoops attached to the lower ends. These scoop up small amounts of oil and sling it around to other parts.

Clearances between the moving parts must be less in this type system. Noisy bearings will occur at smaller clearances than in the pressure system. This is because there is no oil under pressure to cushion the bearing surface.

The force feed or pressure system uses a small oil pump. The pump forces oil to the main bearings, lower connecting rod bearings, and in some cases, piston pins. It is a more expensive system. A pump is required. Also, the crankshaft and connecting rod must have oil passages drilled in them. With the pressure system, the compressor gets better protection from the oil. It will also run quieter, even though it has greater bearing clearances.

The oil pump is usually mounted on one end of the crankshaft. Whenever used, an overload relief valve must be built into the pump. This will protect it and the system against oil pressures that are too high.

Figure 4-77 shows the oil path in a pressure-type lubrication system. The oil pump delivers oil, under pressure, to all bearing surfaces.

Larger, pressure-lubricated compressors sometimes use pressure-controlled electric switches. These will stop the unit if the oil pressure drops too low. It is sometimes necessary to use some kind of an unloading device. This enables the compressor to start easily with no vapor pressure load in the cylinder.

In a rotary compressor, a constant film of oil is needed on the cylinder, roller, and blade surfaces. When the compressor operates, the oil feeds through the main bearings into the cylinder. The cylinder is located so the oil level half covers the main bearings.

Larger units, and some smaller units, use a force-feed lubrication system. Some units use a separate oil pump. Some use the pumping action of the blades moving in and out of their slots.

4.21 Compressor Volumetric Efficiency

Compressor ***volumetric efficiency*** is the actual volume of refrigerant gas pumped divided by the calculated volume of the compressor cylinder. This value is then multiplied by 100 to obtain a percentage.

Example:

A compressor is designed to pump 10 in^3 of vapor each revolution or stroke (this is called the ***piston displacement***). If it pumps only 6 in^3 each revolution, what is the volumetric efficiency of the pump?

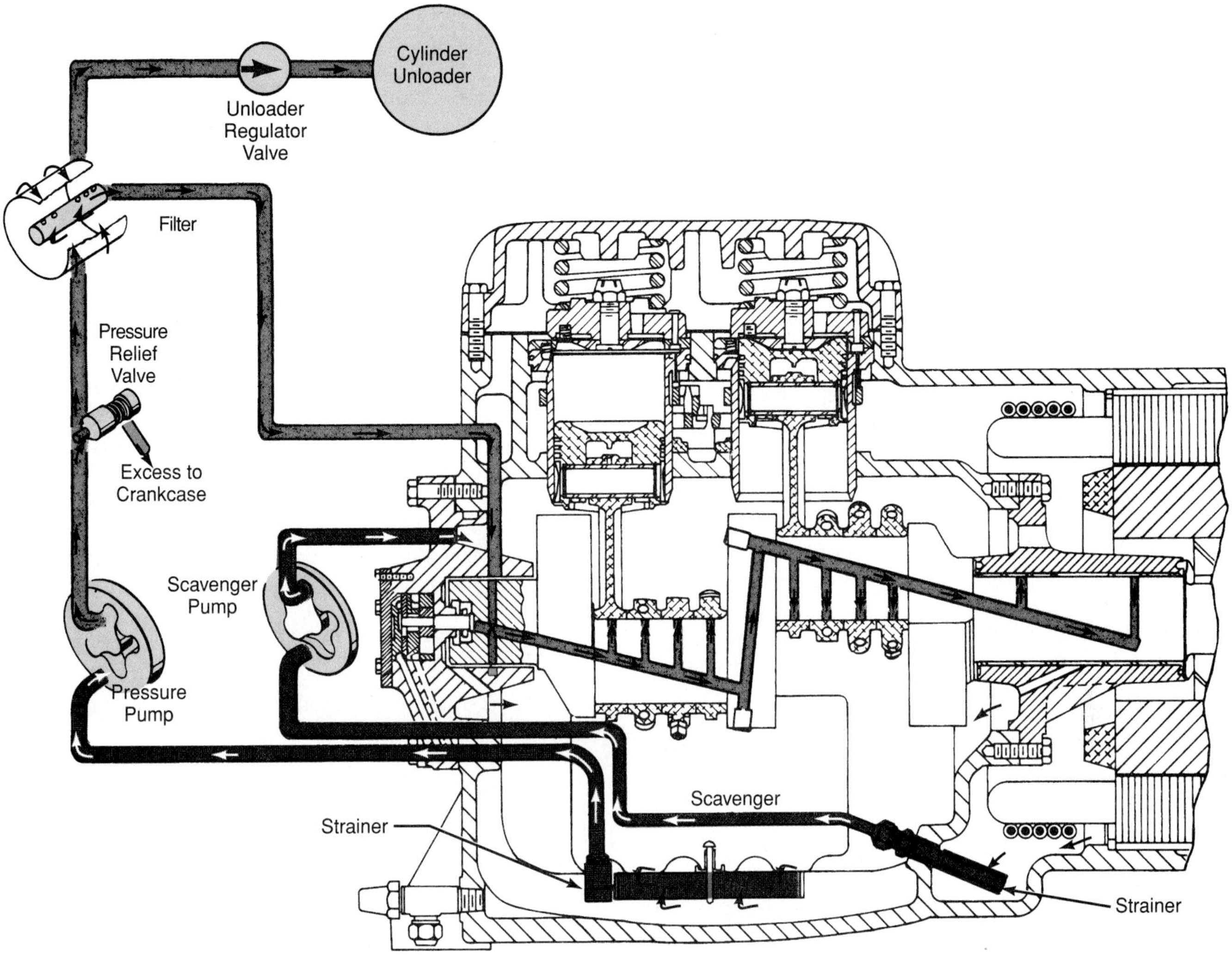

Figure 4-77. *A pressure-lubricated, hermetic, radial, multicylinder reciprocating refrigerator compressor. Scavenger pump returns oil from motor end of the motor compressor back to the lower portion of the crankcase at the cylinder, which serves as an oil sump to store the oil charge. Note pressure relief valve in pressure line. Cylinder unloader is operated by the oil pressure.*

Formula:

$$\text{Volumetric efficiency} = \left(\frac{\text{Actual volume}}{\text{Calculated volume}}\right) \times 100$$

Solution (U.S. conventional):

$$\text{Volumetric efficiency} = \left(\frac{\text{Actual volume}}{\text{Calculated volume}}\right) \times 100$$

Volumetric efficiency = $(6 \text{ in}^3 / 10 \text{ in}^3) \times 100$

Volumetric efficiency = $.6 \times 100$

Volumetric efficiency = 60%

Solution (SI metric):

$$\text{Volumetric efficiency} = \left(\frac{\text{Actual volume}}{\text{Calculated volume}}\right) \times 100$$

Volumetric efficiency = $(98 \text{ cm}^3 / 164 \text{ cm}^3) \times 100$

Volumetric efficiency = $.6 \times 100$

Volumetric efficiency = 60%

For efficient operation, the volumetric efficiency must be as high as possible. Several things affect this efficiency.

The ***head pressure*** is the pressure the compressor must pump against. If the head pressure increases, the amount pumped per stroke will decrease. This is because the compressed vapor in the clearance space will expand on the intake stroke. Fresh vapor cannot move into the cylinder until the cylinder pressure is lower than the suction line pressure. The higher the compressed pressure, the more the compressed vapor in the clearance space expands.

Second, if the low-side pressure decreases, it is harder for the vapors to fill the cylinder. The amount of vapor pumped per stroke will decrease.

Third, if the clearance space is enlarged, the amount pumped per stroke will decrease. ***Clearance space*** is the space left in the cylinder when the roller or piston ends its compression stroke. This is shown in **Figure 4-35.**

The efficiency of a compressor also depends on the size and condition of the valve openings. If the intake valve reduces low-side vapor flow into the cylinder, the cylinder will not be filled. The efficiency of the compressor will be lowered. If the exhaust valves stick, this extra pressure in the cylinder reduces compressor pumping efficiency. This is also true if the line from the compressor to the condenser is pinched.

4.22 Compression Ratio

Compression ratio is the ratio of the total volume of the cylinder to the clearance space. In a refrigeration system, this is the relationship of the high-side absolute pressure to the low-side absolute pressure. Ratios vary up to 10:1 (10 to 1) for single-stage compressors. If the ratio is higher, two-stage compressors must be used. This factor is discussed in Chapter 31.

4.23 Check Valves

Check valves allow fluid flow in only one direction in a system. See **Figure 4-78.** Sometimes they are put in refrigerant suction lines. They prevent refrigerant vapor, oil, or even liquid refrigerant from backing up into the evaporator or other devices. If backup occurs, these substances may condense or lodge in the evaporator or other devices. Check valves are often used in multiple evaporator installations and in heat pump installations.

4.24 Unloader

To make it easier to start the compressor, some installations provide an unloader. *Unloaders* temporarily reduce the high-side pressure at the cylinder head during compressor start-up. Unloaders may be operated mechanically, electrically, hydraulically, or by a solenoid valve. A system using an unloader is shown in **Figure 4-77.**

In small systems using a capillary tube refrigerant control, the pressures balance when the compressor is off. This balancing of the low- and high-side pressures serves as an unloader. Unloader mechanisms may also be used to vary the pumping capacity. This is necessary when there is a changing heat load, such as in an air conditioner.

4.25 Gaskets

On external-drive compressors, joining surfaces are usually sealed with ***gaskets.*** Gaskets are needed between bolted parts, such as cylinder heads, valve plates, and crankcase openings.

The gaskets may be made of either special paper, synthetic material, or lead. Some are made of a plastic substance. Gaskets must be completely free from moisture before use.

4.26 O-Rings

An *O-ring* is commonly used as a sealing device. They are needed particularly where there may be some motion between the assembled parts. **Figure 4-79** illustrates three typical O-ring installations.

The materials used for O-rings depend on various factors. These factors include temperature, pressure, fluids to be controlled, and useful life required. They are usually made of fluid-resistant elastomer compounds.

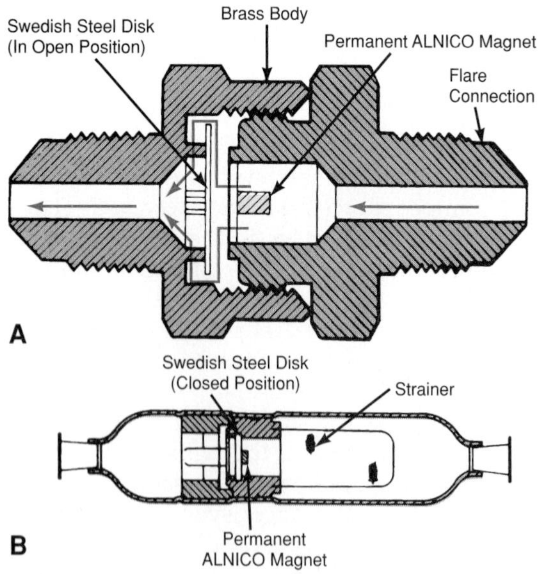

Figure 4-78. *Check valves in which a permanent magnet provides a slight pull. This tends to keep the valves closed. A—Flare-type connector in open position. B—Solder-type connector in closed position. (Watsco Components, Inc.)*

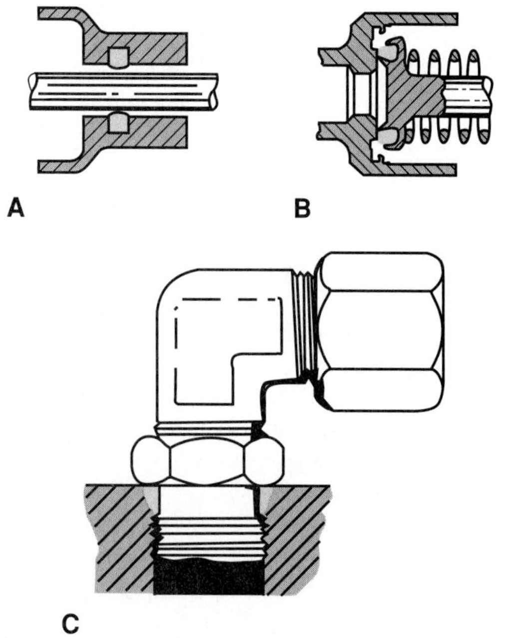

Figure 4-79. *Typical O-ring installations. A—An O-ring installed as seal between a shaft and its housing. B—O-ring installed to serve as a seat seal. C—An O-ring installed as a pipe fitting assembly seal. (Parker Hannifin Corporation, Seal Group)*

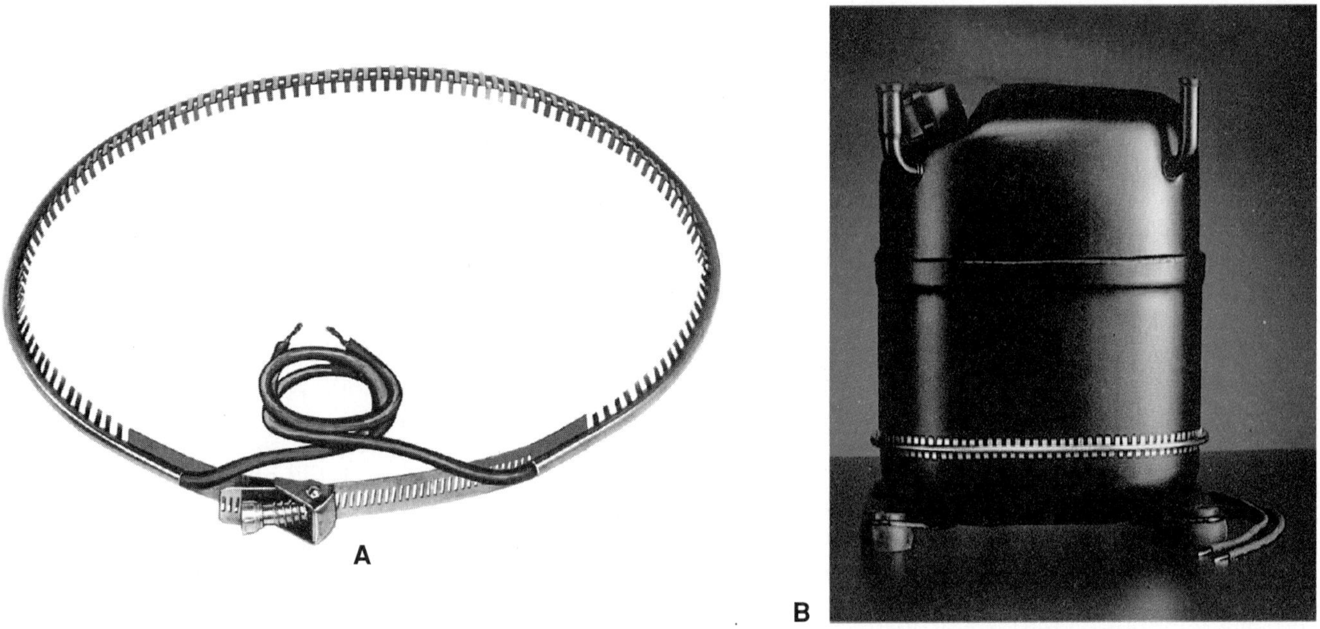

Figure 4-80. *Crankcase heater. A—Accessory electric crankcase heater. B—Crankcase heater installed on a hermetic compressor. (Tutco)*

4.27 Crankcase Heater

In many condensing unit applications, it is necessary to heat the compressor crankcase to evaporate liquid refrigerant trapped in the oil. Most large compressors for commercial applications are fitted with a crankcase heater when the compressor is manufactured. This is especially true if the compressor may be exposed to cold temperatures (outdoor units). These heaters may be operated during the Off cycle or they may be thermostatically controlled.

Smaller installations do not usually require a built-in crankcase heater. An accessory heater may be purchased and attached to the crankcase. Crankcase heaters are required on remote installations which may operate in ambient temperatures lower than evaporator temperatures. Refrigerant oil mixtures in the compressor begin to foam when the compressor starts. This causes the oil charge to pump out of the compressor. This is called ***oil slugging.*** Slugging can cause broken valves, damaged pistons, and broken head gaskets. Flooded starts occur when the compressor picks up refrigerant mixed with oil and feeds it to the cylinders and bearings. There may be severe damage as a result. The heating coil keeps the crankcase coil up to 30°F (17°C) warmer than system temperature. This forces the refrigerant to remain in the condenser, evaporator, or accumulator.

A compressor with a built-in crankcase heater can be seen in **Figure 4-41. Figure 4-80** shows a crankcase heater which may be attached to the hermetic compressor housing.

4.28 Review of Safety

The service technician should always be alert to avoid service procedures which may:

1. Be a hazard to the technician.
2. Be injurious to the equipment.
3. Cause refrigeration to fail.

Normal servicing of refrigerator mechanisms is not considered hazardous. There are, however, recommended procedures which should be followed. This will ensure that the service operations are performed under the safest possible conditions. To service a system safely, the technician must understand the construction and operation of all the parts.

The technician should always wear goggles when working on refrigerating systems.

When installing or servicing a unit, the parts must be handled with care. Every dismantled part must be clean and dry.

It is the little things that count most in servicing refrigerator mechanisms. They include taking care in tightening tube connections, installing gaskets, replacing electrical terminals, and soldering fittings. These may determine whether the entire job will be safe and satisfactory.

Even a slight amount of moisture allowed to enter a refrigerating mechanism may cause system failure. This may happen through the formation of ice in the refrigerant control. Moisture may also cause the formation of a "sludge" in combination with the refrigerant oil.

One important purpose of refrigeration is to preserve food. If the refrigerator does not maintain the desired temperatures, food may spoil. The technician must make the necessary settings and adjustments correctly.

4.29 Test Your Knowledge

Please do not write in this text. Place your answers on a separate sheet of paper.

COMPRESSION SYSTEMS MODULE

1. In a compression refrigerator, oil buildup can cause a problem in the ______.
   A. crankcase
   B. evaporator
   C. motor
   D. oil separator
2. In a compression refrigerator, where can moisture cause a problem?
   A. It can form ice in the refrigerant control device.
   B. It can make the refrigerant oil acidic.
   C. It can turn the refrigerant oil into sludge.
   D. All of the above.
3. In a compression refrigerator, liquid refrigerant can be a problem in the ______.
   A. compressor
   B. expansion valve
   C. filter-drier
   D. sight glass
4. What control determines the refrigerator cabinet temperature?
   A. Motor control.
   B. Refrigerant control.
   C. Scotch yoke.
   D. Shutoff valve.
5. In a low-side float system, ______.
   A. a flooded evaporator cannot be used
   B. a float inside the evaporator controls refrigerant level
   C. the control float is located below the bottom of the cabinet
   D. the low-side temperature is allowed to vary up and down
6. Automatic expansion valves usually adjust the ______.
   A. temperature in the cabinet
   B. temperature in the evaporator
   C. pressure in the evaporator
   D. Both B and C.
7. What does a high-side float do?
   A. It turns the motor on and off.
   B. It controls the temperature in the cabinet.
   C. It controls the liquid level in the receiver.
   D. Both A and B.
8. When does the condensing pressure stop rising?
   A. When the pressure equals the vapor pressure of refrigerant oil.
   B. When oil binding begins to occur.
   C. When liquid starts flowing through the liquid line.
   D. When just as many vapor molecules condense into liquid as the compressor pumps into the condenser.
9. What basic conditions are necessary to produce refrigeration?
   A. The thermostat must be calling for cooling action.
   B. The compressor must be operating.
   C. The condenser pressure must be high enough.
   D. Refrigerant must be vaporizing within the evaporator.
10. What is the purpose of a compressor?
    A. It removes heat-laden vapor from the evaporator.
    B. It compresses the vapor to a high temperature so that heat can flow out from the vapor in the condenser.
    C. It compresses the vapor to a pressure high enough that it will condense into liquid.
    D. All of the above.

COMPRESSORS MODULE

11. In a compression refrigerator, the ______ is usually found in the low-pressure side.
    A. condenser
    B. liquid line
    C. liquid receiver
    D. motor
12. A(n) ______ compressor does not use a crankshaft seal.
    A. hermetic
    B. open-type
    C. scroll
    D. swash plate
13. How does a swash plate compressor get twice the pumping rate from the same number of cylinders?
    A. The swash plate compressor uses a 2-stroke cycle.
    B. The swash plate doubles as a valve plate.
    C. Both ends of the piston pump vapor.
    D. The piston reciprocates twice during each shaft revolution.
14. A full floating piston pin is ______.
    A. lubricated by the pressure (force flood) system
    B. hollow to reduce weight
    C. made of a lightweight carbon-boron alloy
    D. free to turn both in the connecting rod bushing and in the piston bushings
15. Compressor pistons are made of ______.
    A. cast iron or die-cast aluminum
    B. forged stainless steel or copper
    C. high carbon steel, or bronze
    D. copper-nickel alloys
16. In a rotary compressor, a ______ is sometimes used in place of an intake valve.
    A. labyrinth seal
    B. reed valve
    C. rotolock connection
    D. None of the above.

17. A compressor has a piston displacement of 8 $in.^3$, but measurements show that it is actually pumping only 5 $in.^3$ per crankshaft revolution. What is probably wrong with this refrigerator?
    A. The compression rate is incorrect.
    B. The condenser is too large.
    C. There is oil in the refrigerant.
    D. There isn't anything wrong.
18. A centrifugal compressor may have a minimum of _______ impeller(s).
    A. No impellers are needed.
    B. one
    C. two
    D. three
19. Why does a compression system sometimes use two mufflers?
    A. It will need two if it has double-acting pistons.
    B. One for the crankcase exhaust and one in the exhaust line.
    C. One in the intake line and one in the exhaust line.
    D. One for the refrigeration cycle and one for the defrost cycle.
20. A compressor service valve has _______ opening(s).
    A. one
    B. two
    C. three
    D. four

Refrigeration systems use many types of control valves and other necessary equipment. A—Replaceable catch-all filter-drier. B—Ammonia solenoid valve. C—Sealed model suction filter. D—Replaceable suction filter. E—Sealed model suction line filter-drier. F—Sealed model catch-all filter-drier. G—Evaporator pressure regulating valve. H—Discharge bypass valve. I—Ammonia thermostatic expansion valve with level control. J—Three-way heat reclaim valve. K—Solenoid valve. L—Normally open solenoid valve. M—Ammonia thermostatic expansion valve. N—Evaporator pressure regulating valve. O—Balanced port TEV. P—Thermostatic expansion valve. Q—Nonadjustable solenoid valve. R—Pressure differential valve. S—Solenoid valve with manual lift stem. T—Acid test kit. U—Thermostatic expansion valve. V—Refrigerant distributors. W—Moisture and liquid indicators. (Sporlan Valve Co.)

Chapter 5

REFRIGERANT CONTROLS

Key Words:

adsorption
check valve
cross-charged
dual-pressure regulator
equalizer
flash gas
frost back
hunting
oil binding
starve
superheat
thermistor
vapor lock

Learning Objectives:

After studying this chapter, you will be able to:

- Explain the purpose and operation of refrigerant control devices.
- Name the six main types of controls and explain their operation.
- Define terms related to refrigerant control operations.
- Compare the various charging elements used on refrigerant controls.
- Explain the fast evaporation of liquid into a vapor.
- Determine the proper size capillary tube to be used for specific applications.
- Explain the operation of special refrigerant controls.
- Define the purpose and function of three types of solenoid valves.
- **Follow approved safety standards.**

A refrigerant control is the device used in a refrigeration system to change the pressure of the refrigerant. Refrigerant in the evaporator must be at a low pressure to evaporate at a low temperature. The liquid refrigerant in the condensing unit is at a relatively high pressure. So that the unit may operate automatically, an automatic refrigerant flow control must be used. This control is placed in the circuit between the liquid line and the evaporator. It reduces the high pressure in the liquid line to low pressure in the evaporator.

There are six main types of automatic refrigerant flow controls:

- Automatic Expansion Valve (AEV or AXV).
- Thermostatic Expansion Valve (TEV or TXV).
- Thermal-Electric Expansion Valve (THEXV).
- Low-Pressure Side Float (LSF).
- High-Pressure Side Float (HSF).
- Capillary Tube (Cap. Tube).

The basic systems using these controls are described in Chapter 3 and Chapter 4.

5.1 Compression System Refrigerant Controls

Modern refrigeration systems are automatic in operation. Devices have been developed that control refrigerant flow into the evaporator. They also control the electric motor, which drives the mechanism.

The refrigerant controls may be divided into three principal classes. A combination of these controls may also be used:

- Control based on pressure changes.
- Control based on temperature changes.
- Control based on volume or quantity changes.

Automatic refrigerant and motor controls are needed to maintain the temperature in the refrigerated space within specific limits. Automatic motor controls are explained in Chapter 8.

Popular uses and temperature ranges include:

- For fresh food storage, temperatures usually are maintained at 35°F to 45°F (2°C to 7°C).
- For frozen food storage, temperatures usually are between −10°F and 0°F (−23°C and −18°C).
- For comfort cooling, temperatures are usually maintained no more than 10°F to 12°F (5.6°C to 6.6°C) below the ambient temperature.

5.1.1 Automatic Expansion Valve Principles

An ***automatic expansion valve (AEV*** or ***AXV)*** or pressure-controlled expansion valve is a refrigerant control operated by low-side pressure. The valve throttles the liquid refrigerant in the liquid line down to a constant pressure. While the compressor is running, the liquid refrigerant is sprayed into the evaporator (low-pressure side). A system using an automatic expansion valve is sometimes called a dry system. The evaporator is never filled with liquid refrigerant, but with a mist or fog.

Automatic Expansion Valve Design

Figure 5-1 shows flexible bellows linked up to a needle valve. The evaporator pressure, P_2, is on the inside and atmospheric or confined gas pressure, P_1, is on the outside. Spring force, F_1, tends to open the valve; spring force, F_2, tends to close the valve.

As the evaporator pressure decreases, the pressure difference will force the bellows toward the valve body.

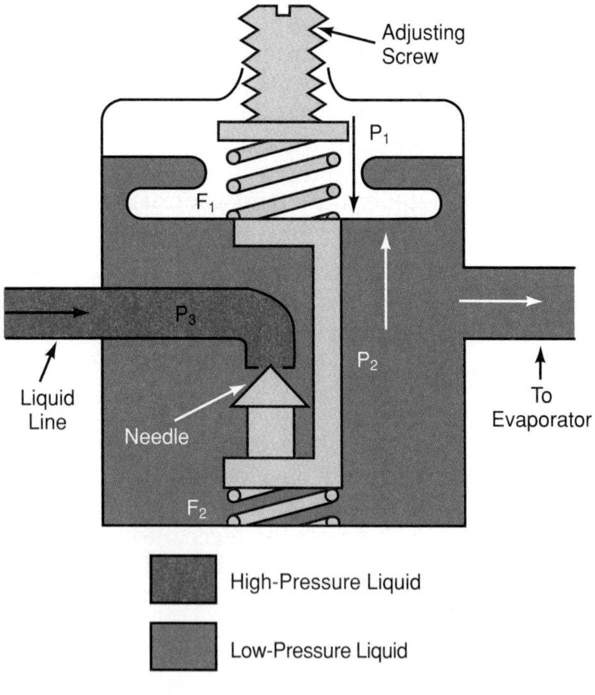

Figure 5-1. *Automatic expansion valve shows various pressures inside valve which control its operation. P_1—Atmospheric pressure. P_2—Suction or evaporator pressure. P_3—Liquid line pressure. F_1—Adjustable spring. F_2—Nonadjustable pressure spring.*

Being attached to the needle, it will open the needle valve. Some liquid refrigerant will spray into the evaporator. Because the refrigerant evaporates at a constant low pressure, the evaporator and cabinet temperature stay within design limits.

The expansion valve opens only when the evaporator pressure drops. Pressure drops occur only when the compressor is running. The expansion valve, however, will not flood the low side when the compressor is running. When the evaporating refrigerant, liquid, spray, and vapor reach the end of the evaporator, the motor control will cool. This motor control (sensing bulb) is attached to the suction line. The switch will open and stop the motor. The low-side pressure will then immediately build up enough to close the expansion valve.

These valves are adjustable. They permit the opening of the needle valve over a wide range of pressures. Expansion valves must be adjusted for atmospheric pressure, P_1, which affects their operation. High altitudes will cause a decrease in atmospheric pressure. The adjusting screw must be turned in to make up for the lower atmospheric pressure. Refrigerants with different evaporating pressures have different expansion valve settings.

There are many different designs of automatic expansion valves. The flexible part is either a diaphragm, **Figure 5-2,** or a bellows. Usually, it is made of phosphor bronze and soldered or brazed to the valve body. Flexible elements must move in and out time after time without losing flexibility. The valve body is usually drop-forged brass, but sometimes it is cast. It must be seepage (leak) proof. The liquid inlet has either a soldered connection, standard flange, flared connection, or pipe thread. It usually has a screen designed for easy removal. The screen is made of brass or stainless steel wire 60 to 100 mesh. (This means 60 to 100 openings per square inch.)

It is important to remember that valve capacity should equal pump capacity. The liquid refrigerant flowing past the expansion valve needle is the same weight as the vapor (gas) pumped by the compressor.

Use a one-ton valve with a one-ton capacity condensing unit. An under-capacity valve tends to starve an evaporator (too little refrigerant gets through). An over-capacity valve will tend to allow too much refrigerant into the evaporator when the valve opens. This may cause sweat backs or ***frost backs*** on the suction line.

Bellows Automatic Expansion Valve

A bellows automatic expansion valve, **Figure 5-3,** usually has valve seats softer than the needles. This reduces the wearing of a shoulder on the needle. Usually, needles are Stellite and seats are Monel metal. The spring, C, is attached at both ends. It can be adjusted for either pressure or tension. This eliminates the need to have a spring inside the refrigerant space. The needle, L, is mounted in a ball and socket to ensure full seating. It allows the needle to align with the seat, M. The

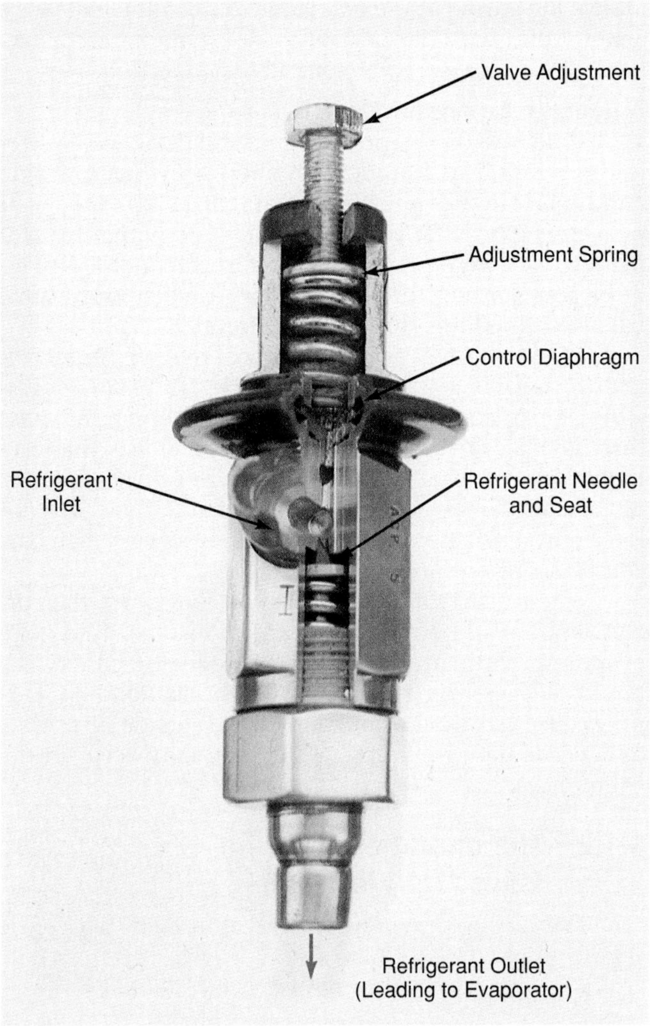

Figure 5-2. *Diaphragm automatic expansion valve. (Alco Controls Division, Emerson Electric Company)*

bellows, D, made of special brass, is flexible. It is soldered to both the body and the disk. The bellows is made of metal .005″ to .010″ thick.

The liquid line, H, usually has an outside diameter (OD) of 1/4″. It is fastened to the valve by a special nut, which permits easy removal of the screen. Note how the valve is sealed at O and B so moisture will not enter. If present, moisture could freeze on the bellows and interfere with the accurate operation of the valve.

Valves are attached to the evaporator by threaded fittings or a two-bolt flange. Flanges sealed with lead gaskets are usually preferred.

The bellows automatic expansion valve is mainly used on domestic air conditioning units. It is also used on vending machines and as a replacement for capillary tubes.

Diaphragm Automatic Expansion Valve

Diaphragm automatic expansion valves have stops to prevent too great a movement of the diaphragm. See **Figure 5-4.** Note that the diaphragm has concentric

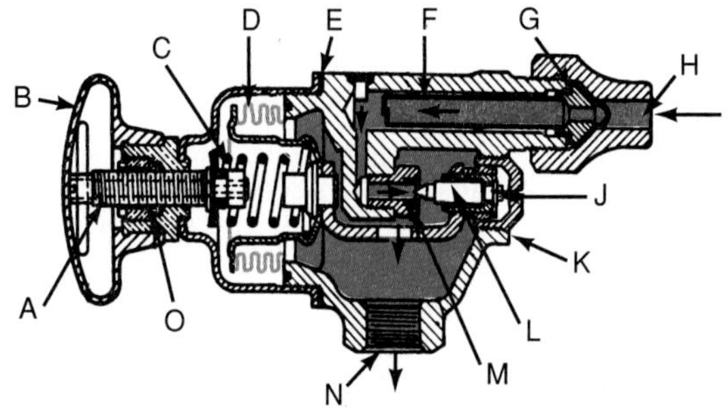

Figure 5-3. *Bellows expansion valve. A—Adjusting screw. B—Rubber cap. C—Adjustment spring. D—Bellows. E—Gasket. F—Screen. G—Shipping plug. H—Liquid refrigerant inlet. J—Needle shoulder. K—Seating plug. L—Needle. M—Needle valve seat. N—Refrigerant outlet. O—Packing gland. Arrows show direction of refrigerant flow through valve.*

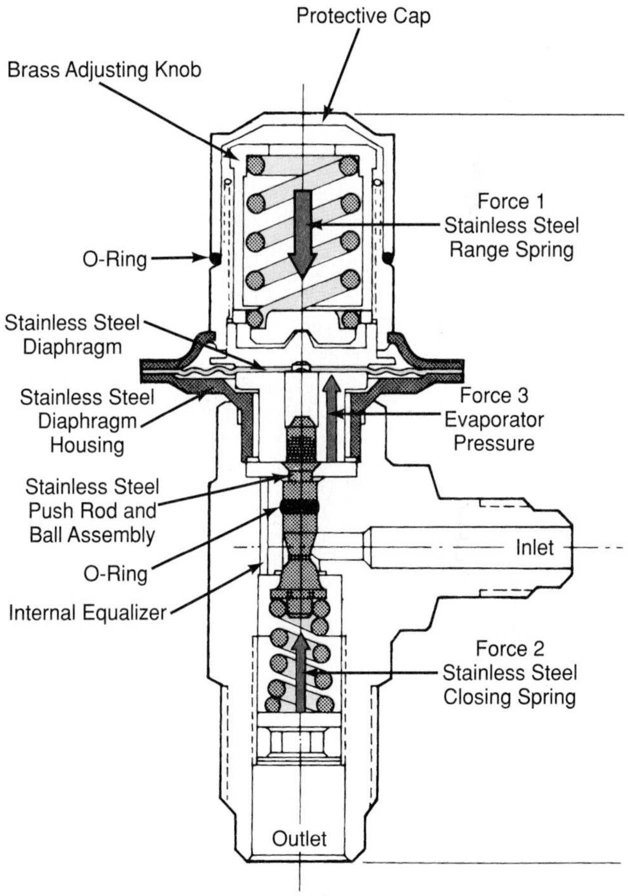

Figure 5-4. *Cross section of a diaphragm automatic expansion valve, showing three forces that control operation of valve. Force 1, adjustable range opening, moves diaphragm. Force 2 moves push rod and ball assembly. Force 3 is evaporator pressure. Valve is designed to control flow of refrigerant to evaporator, maintaining a constant evaporator pressure. (Refrigeration & Air Conditioning Div., Parker Hannifin Corp.)*

corrugations (ripples) to improve its flexibility. The diaphragm separates the atmospheric pressure and the system pressure. Three basic forces control the operation of the valve. Force 1 is the adjustable spring. This moves the diaphragm down, opening the valve. Force 2 is a spring beneath the diaphragm. It moves the push rod and ball assembly up, closing the valve. Force 3 is the outlet pressure underneath the diaphragm. This pressure is controlled when the spring force, F_1, is equal to sum of the forces F_2 and F_3.

Figure 5-5 is an external view of an automatic expansion valve. The arrow on the inlet port ensures proper installation. Its capacity is often marked on the cap.

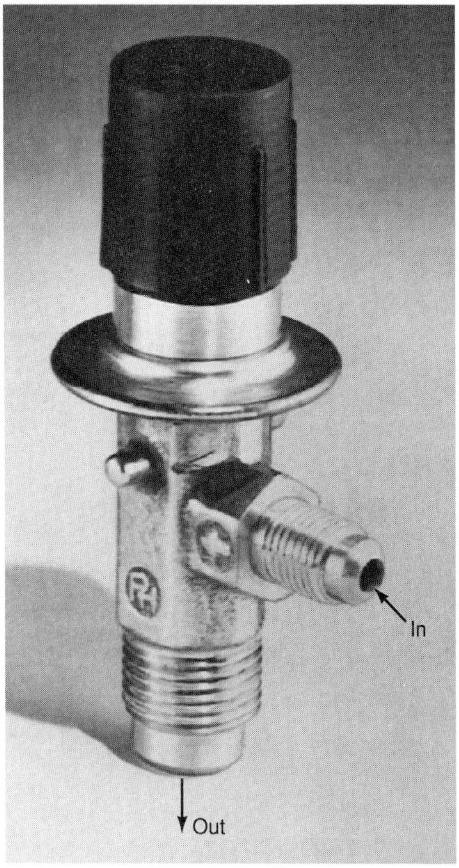

Figure 5-5. *Diaphragm type automatic expansion valve. Note refrigerant flow direction. (Refrigeration & Air Conditioning Div., Parker Hannifin Corp.)*

Another diaphragm expansion valve design is shown in **Figure 5-6.** The diaphragm movement is limited by the body and the cap. Threads fasten the diaphragm assembly to the body of the valve. The tightly-fitted cap or cover plate, which protects the pressure adjustment, can be removed to adjust the valve.

The diaphragm has a disk on its low-pressure side. This disk presses on a pin that moves the ball valve away from the seat. When the low-side pressure increases, the diaphragm moves against the adjustment spring. This allows the spring at the ball valve to push the ball valve against the seat. The inlet is a 1/4″ male flare connection. The outlet is a 3/8″ male flare connection.

Automatic Expansion Valve Bypass

Many motor compressor units are designed to start under a low load (torque) condition, as when low-side and high-side pressures are equal (balanced). The equal pressures allow the compressor to start without pushing against a high pressure. Therefore, the motor will require less starting torque. Capillary tube systems have balanced pressures during the Off cycle.

Automatic expansion valves seal the refrigerant orifice during the Off part of the cycle. To balance pressures, an opening is designed so refrigerant can pass through the valve. Typically, V-shaped slot is made in the valve seat. This is shown in **Figure 5-7.** The bypass, or bleeder, openings are quite small. They do not interfere with the operation of the valve when the compressor is running.

When using this type of expansion valve, the correct amount of refrigerant charge must be used. The evaporator outlet must have an accumulator. Otherwise, liquid refrigerant may travel down the suction line. This may cause sweating or frosting on the suction line. Also, dangerous liquid refrigerant slugging may occur in the compressor.

5.1.2 Thermostatic Expansion Valve (TEV) Principles

Thermostatic expansion valves are of two basic types:

- Sensing bulb.
- Thermal-electric.

The sensing bulb is further divided into four types:

- Liquid-charged.
- Gas-charged.
- Liquid cross-charged.
- Gas cross-charged.

In the liquid-charged and the gas-charged elements, the same refrigerant is used in the system and the bulb. ***Cross-charged*** means that the sensing bulb fluid is different than the system refrigerant.

In the automatic expansion valve, the refrigerant flows through the valve and into the evaporator. Flow is controlled by the pressure in the evaporator.

In the thermostatic expansion valve, the flow is also through the valve and into the evaporator. However, it is controlled by both the low-side pressure and the temperature of the evaporator outlet. The valve provides a high flow rate as the evaporator empties (warms). It reduces the flow as the evaporator fills (cools) with refrigerant.

The sensing bulb valve is operated by accumulated pressure or force difference between the sensing bulb bellows and valve low-side pressure. See **Figure 5-8.**

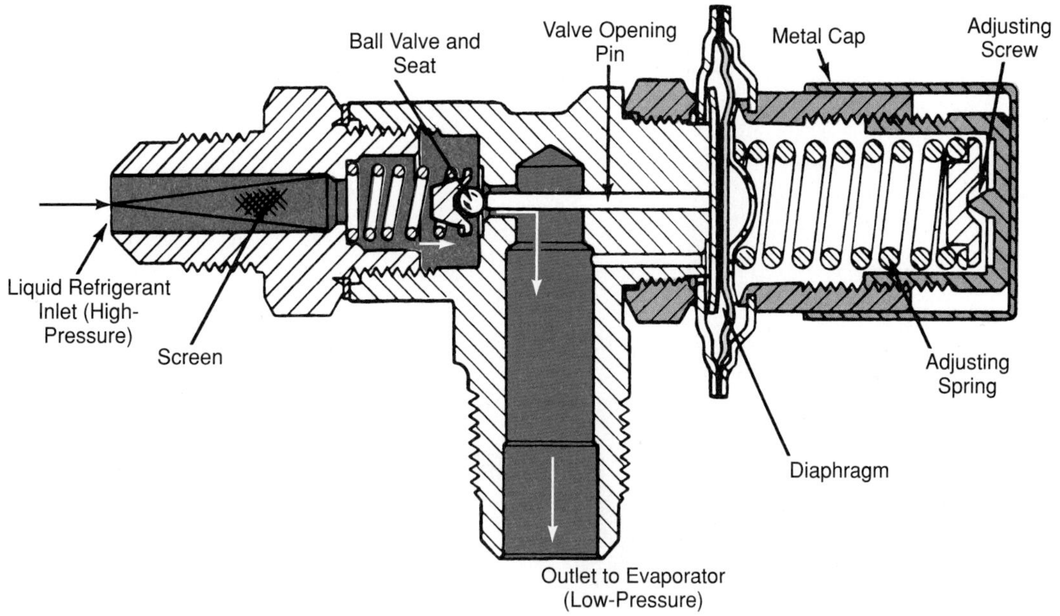

Figure 5-6. *Parts and operation of an automatic expansion valve. Directional arrows indicate flow of refrigerant through expansion valve.*

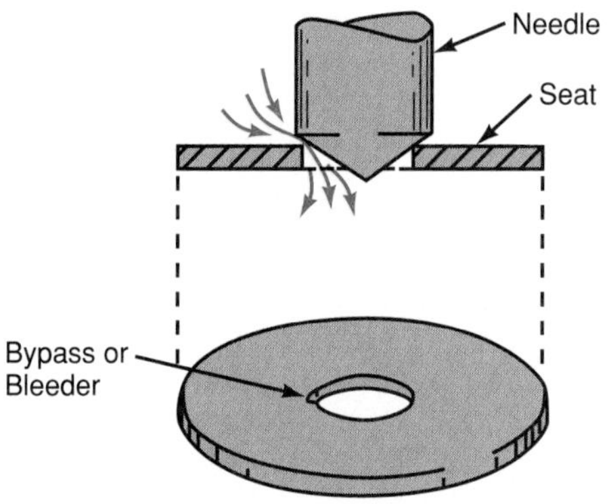

Figure 5-7. *Needle valve and seat used on bypass or bleeder automatic expansion valve. Bypass permits pressures to balance during Off part of cycle.*

With the unit running, the refrigerant temperature in the sensing bulb, T_1, is usually about 10°F (5.6°C) warmer than the refrigerant temperature in the evaporator, T_2. This temperature difference produces the different pressures and therefore the different forces. This temperature difference is often described as the superheat of the bulb over the refrigerant temperature inside the evaporator.

The pressure in the sensing bulb, P_1, is greater than the pressure in the evaporator, P_2. Note that as the temperatures increase or decrease, the pressure will also increase or decrease.

When the compressor stops, the low-side pressure and the sensing bulb pressure tend to equalize. The total expansion valve internal force, $F_2 + F_3$, overpowers the sensing bulb force, F_1. The needle is forced firmly into its seat. Refrigerant flow stops. The needle will stay closed until the sensing bulb force overcomes the low-side force.

This valve-opening action should only happen when the unit is running. If the valve is adjusted correctly, it closes when the compressor is idle. This will prevent flooding of the low side with liquid refrigerant.

The thermostatic expansion valve does not regulate the low-side pressure. It controls the filling of the evaporator with refrigerant. The pumping action of the compressor establishes the low-side pressure. **Figure 5-9** illustrates a bellows thermostatic expansion valve.

The valve can be adjusted so the needle seats itself while the unit is running. The needle will then close even though there is a greater temperature difference (about 15°F or 8.3°C) between the refrigerant in the evaporator and that in the sensing bulb. This is shown in **Figure 5-10.** The evaporator liquid refrigerant will not reach the sensing bulb location in this case. Only the low-pressure vapor will be cold enough to reduce the sensing bulb temperature (and therefore the pressure) to the closing point. The needle will close before the evaporator becomes full of liquid refrigerant droplets. The evaporator will be starved.

If the valve is adjusted the other way, the needle moves away from the seat. The temperature of the sensing bulb refrigerant will become nearer to that of the evaporator refrigerant, about 5°F to 7°F (2.8°C to 3.9°C). See **Figure 5-11.**

The evaporator would then become full of liquid refrigerant droplets to bring the temperature (pressure) difference down to this value. The evaporator would be

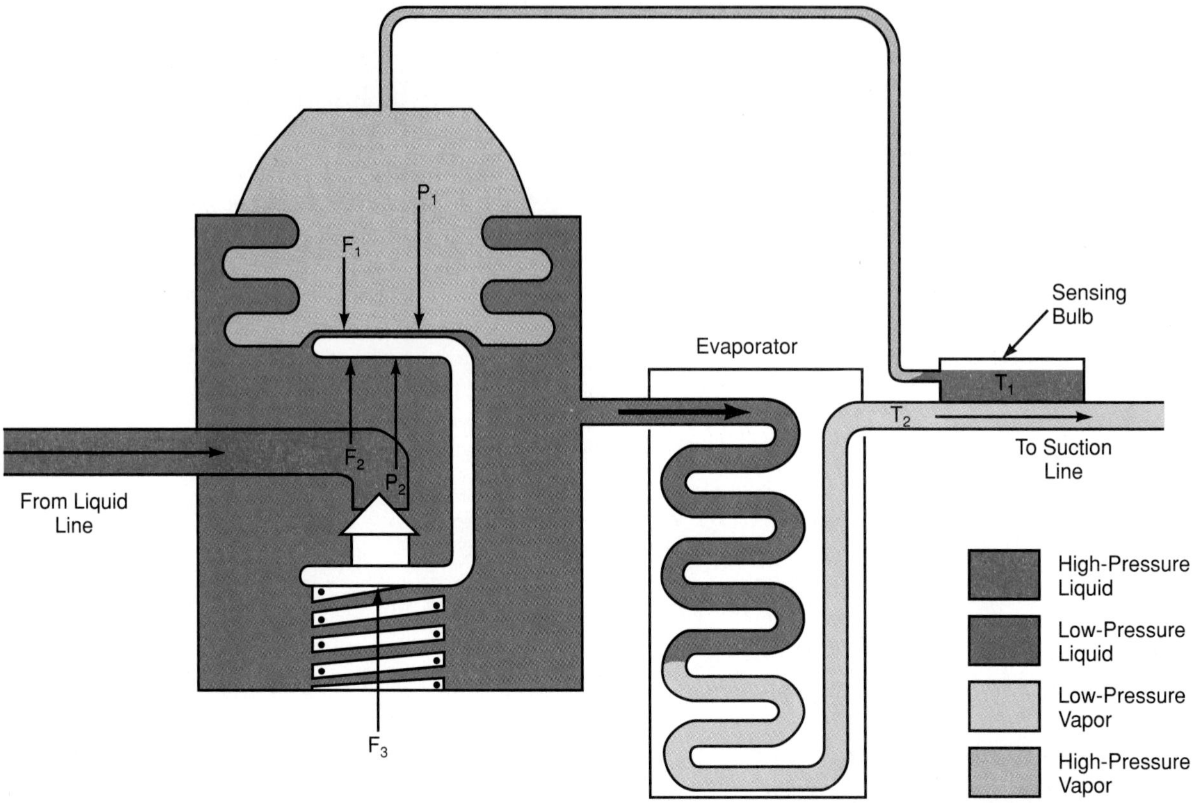

Figure 5-8. *Thermostatic expansion valve showing various pressures and temperatures within valve that operate it. F_1—Sensing bulb pressure (force) tending to open valve. F_2—Low-side pressure (force) tending to close valve. F_3—Spring force tending to close valve. P_1—Sensing bulb pressure tending to open valve. P_2—Suction pressure (low side) tending to close valve. T_1—Sensing bulb temperature. T_2—Evaporator refrigerant temperature (low side). Valve opens when F_1 is greater than combined force of F_2 and F_3. Valve closes when combined F_2 and F_3 forces are greater than F_1.*

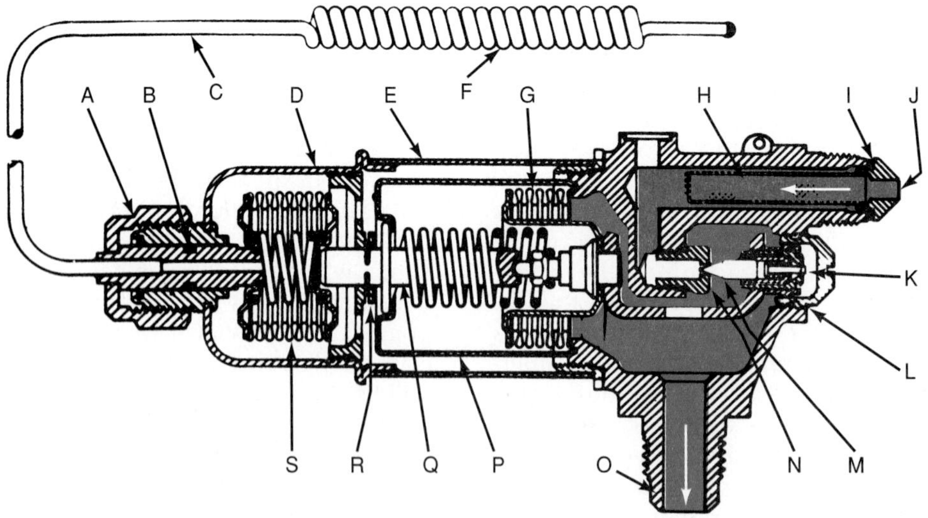

Figure 5-9. *Thermostatic expansion valve. A—Adjusting nut. B—Seal ring. C—Capillary tube. D—Bellows housing. E—Housing spacer. F—Temperature sensing bulb. G—Body bellows. H—Screen. I—Gasket. J—Refrigerant inlet. K—Needle pin. L—Sealed fitting. M—Needle. N—Seat. O—Evaporator connection. P—Inner spacer. Q—Spacer rod. R—Snap ring. S—Thermal bellows.*

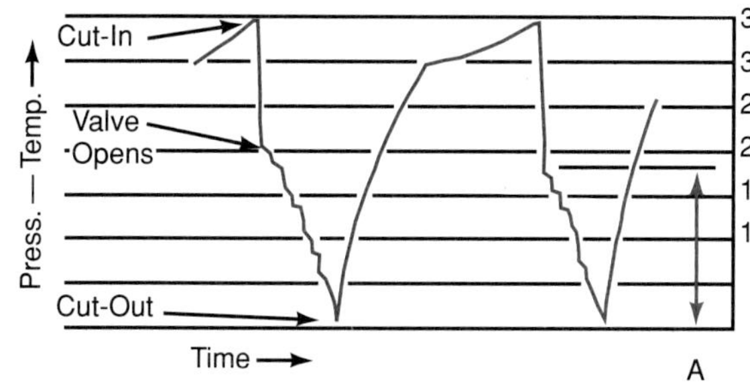

Figure 5-10. *A thermostatic expansion valve low-side pressure-time cycle diagram using a high superheat adjustment. A—Pressure drop on low side between opening of valve and cut-out point.*

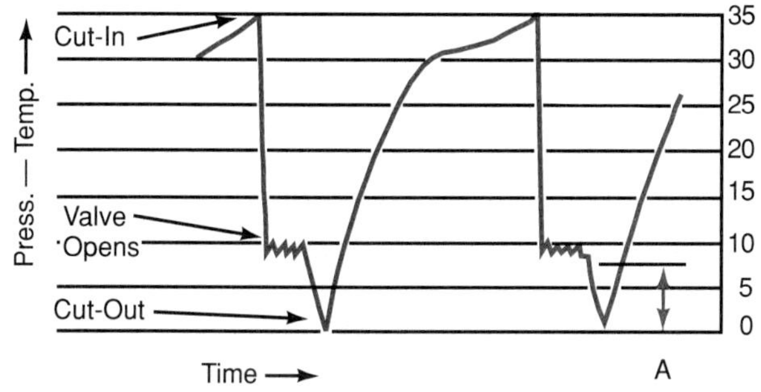

Figure 5-11. *A thermostatic expansion valve low-side pressure-time cycle diagram using a low superheat adjustment. A—Pressure drop on low side between opening of valve and cut-out point. Note: Compare with A, in **Figure 5-10,** to understand how superheat adjustment controls evaporator (low-side) pressure.*

flooded. Some liquid droplets may even go into the suction line. This would cause a sweating or frosting of the suction line and may harm the compressor (slugging).

The adjusting nut or screw is sensitive. It should not be turned more than one-quarter revolution while the unit is operating. The system should be given 10 to 15 minutes before another adjustment is made.

Some thermostatic expansion valves use diaphragms instead of bellows. In either design, the valve is closed when the unit is not running.

Chapter 3 describes the basic principles of a TEV system.

See Chapter 15 for instructions on installing and servicing thermostatic expansion valves.

Some large refrigeration installations (50 tons and over) may use a pilot-controlled thermostatic expansion valve. In these installations, a conventional thermostatic expansion valve is mounted on a large auxiliary valve body. The auxiliary valve provides a larger pressure-operated needle and orifice. The conventional thermostatic refrigerant control (pilot) regulates the pressure that operates the larger valve. **Figure 5-12** illustrates one type of pilot-controlled thermostatic expansion valve.

Some large air conditioning systems may use as many as six thermostatic expansion valves on one evaporator. In this way it is possible to:

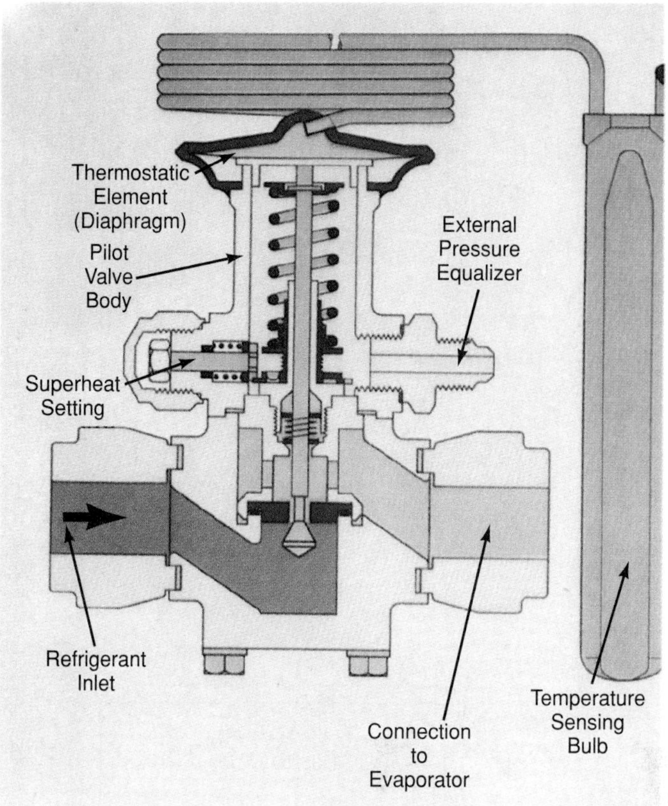

Figure 5-12. *A pilot-controlled thermostatic expansion valve. (Danfoss Automatic Controls, Division of Danfoss, Inc.)*

- Maintain constant pressures and temperatures.
- Make sure that the evaporator has a full charge of refrigerant.
- Reduce pressure drop through the evaporator.

Thermostatic Expansion Valve Design

Thermostatic expansion valves are usually used on multiple-evaporator systems. However, the low-side float may also be used on multiple systems. A multiple system using thermostatic expansion valves can provide different temperatures in the various cabinets. This valve is also commonly used on air conditioning systems.

The correct sensing element must be chosen for each installation. The correct size valve is also needed.

The thermostatic expansion valve has a brass body into which the liquid line and evaporator line are connected. The needle and seat are inside the body. The needle is joined to a flexible metal bellows or diaphragm. The bellows is moved by a rod connected at the other end to a sealed bellows or diaphragm (power element). This is joined to the sensing bulb by means of a capillary tube. **Figure 5-13** shows the refrigerant behavior in the sensing element.

Each manufacturer has a code for identifying the fluid that charges the sensing element. Some use letters; others use colors or numbers. Some valves are marked with the refrigerant number. Some valves

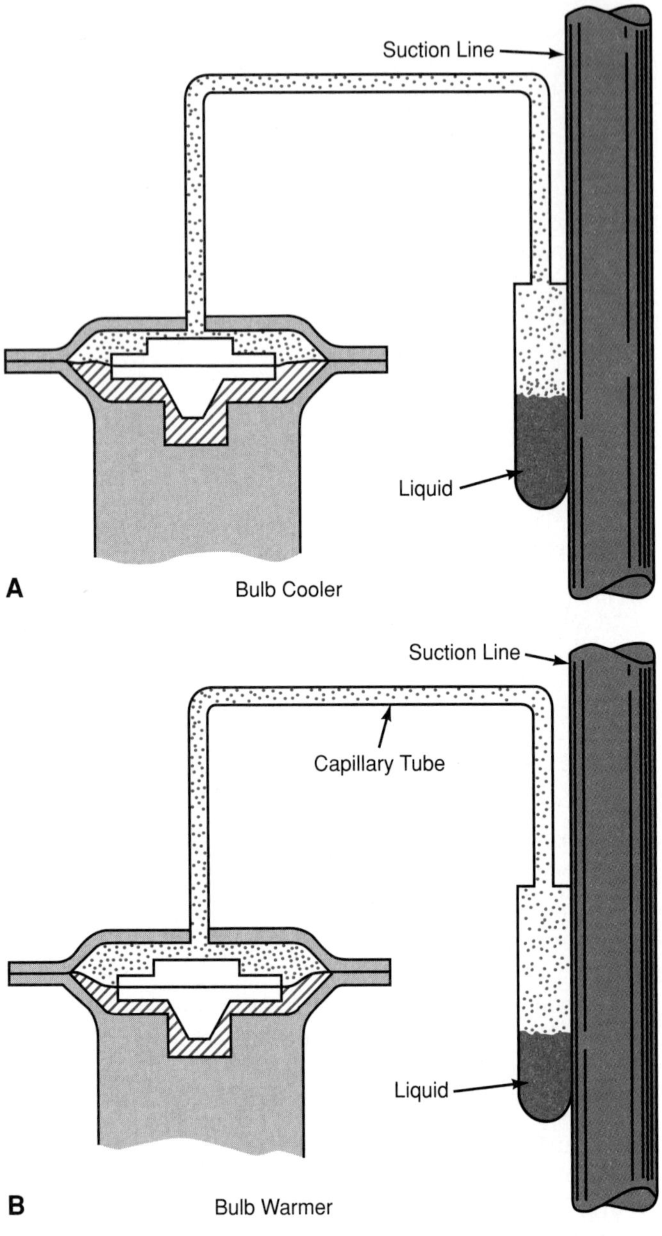

Figure 5-13. *Effect of temperature on sensing bulb of thermostatic expansion valve. A—Sensing bulb is cold, pressure is low, and a considerable quantity of control fluid is shown as a condensed liquid in bulb (solid red). B—Sensing bulb is warmer and some of control fluid has evaporated (red dots) and is exerting pressure on expansion valve diaphragm, which will cause it to admit more refrigerant into evaporator. For accurate control, there must be enough control fluid to ensure that it never completely evaporates. Red crosshatching indicates suction line pressure.*

intended for use with refrigerant R-12 are color-coded yellow.

The valve is sealed to prevent moisture seeping into it. A strainer (screen) is located between the liquid line connection and the orifice. It keeps dirt away from the needle and seat. See **Figure 5-14.**

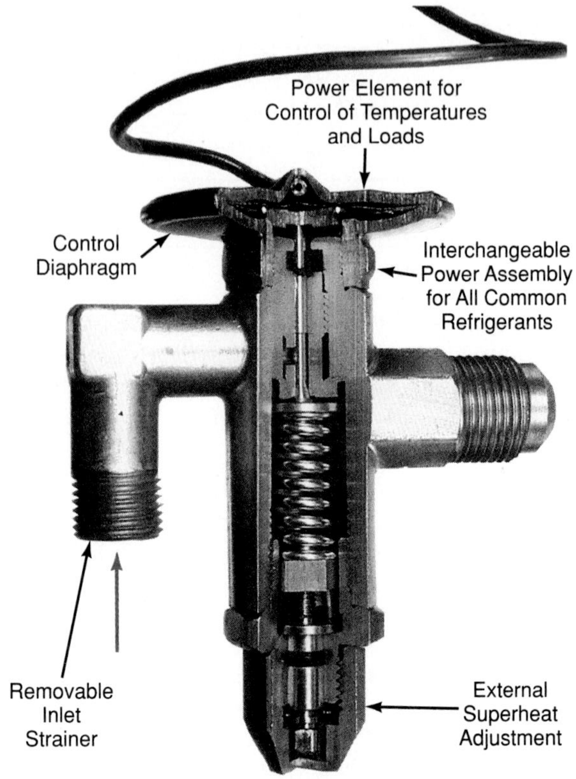

Figure 5-14. *Diaphragm thermostatic expansion valve with flared inlet and outlet connections. Turning adjustment screw in will starve evaporator; turning it out will flood evaporator. (Alco Controls Division, Emerson Electric Company)*

Figure 5-15 is an exploded view of a thermostatic expansion valve. Note that the inlet flare surface is mounted on the strainer. The needle is usually made of Stellite, Hastelloy, or stainless steel. The needles are usually pointed cones. Spherical (ball) valves and flat orifice closers may also be used. The needle is popular for small capacity valves. The ball or the flat orifice is used in larger capacity valves.

Figure 5-16 illustrates a large capacity flat valve seat. A filter-drier should be placed in the liquid line immediately ahead of the thermostatic expansion valve.

Figure 5-17 shows a cut-away view of a diaphragm type TEV. **Figure 5-18** shows a cutaway of a ball valve.

Another type thermostatic expansion valve is shown in **Figure 5-19.** The single diaphragm valve is designed for service in air conditioning applications. It is equipped with a bleed valve for rapid pressure balancing (RPB). The bleed mechanism only works when the compressor is not running. When the compressor is operating, the secondary bleed port closes. Then the valve operates in a normal manner. Rods carry the diaphragm action to the needle. The liquid inlet is on the left; the evaporator connection on the right.

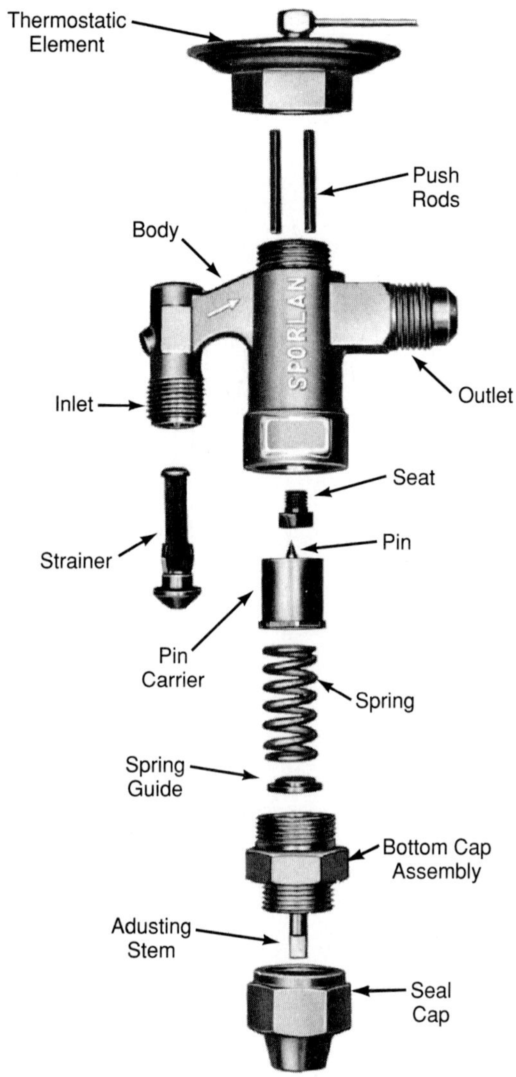

Figure 5-15. *Parts of a thermostatic expansion valve. In this valve, thermostatic element is threaded on body of valve. (Sporlan Valve Co.)*

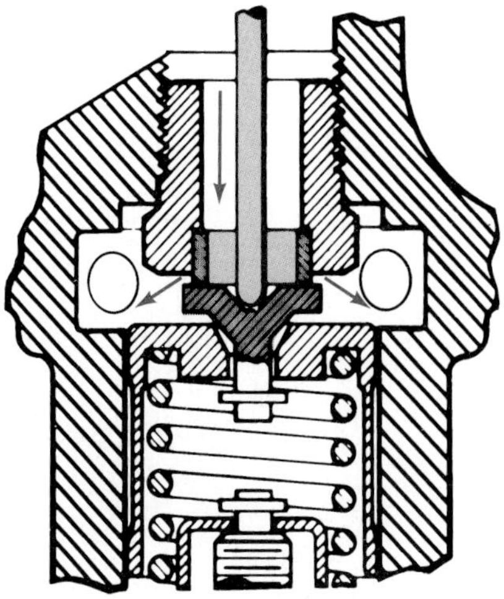

Figure 5-16. *This cutaway shows a thermostatic valve orifice that has been designed for large capacity. Note that both valve and valve seat are formed by flat surfaces. Arrows indicate direction of flow of refrigerant.*

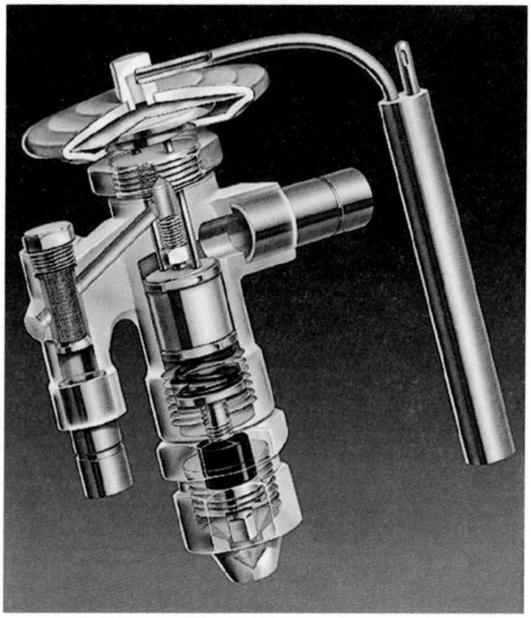

Figure 5-17. *Cutaway view of a thermostatic expansion valve with soldered connections. Note replaceable cartridge allowing technician to select a valve orifice or cartridge to match the system capacity. (Sporlan Valve Company)*

Flash Gas

The term ***flash gas*** indicates that portion of the refrigerant which evaporates instantly (flashes) and turns into a vapor as it passes through the refrigerant control orifice. The instant vaporizing of some of the liquid refrigerant (flash gas) cools the remaining liquid to the evaporating temperature.

The amount of flash gas depends on the temperature of the refrigerant in the liquid line and the pressure inside the evaporator. Flash gas reduces the valve capacity. See Chapter 16.

One method of reducing flash gas is to clamp the liquid line to the suction line. The liquid coming from the condenser is warm and the vapor coming from the evaporator is cold. Clamping the two lines together causes a heat transfer from the liquid line to the suction line.

Cooling the liquid in the liquid line decreases the flash gas. Raising the temperature of the suction line vapor decreases the possibility of liquid refrigerant entering the compressor.

Superheat

The term ***superheat,*** when used with thermostatic expansion valves, refers to the difference in temperature between the vapor in the low side and in the sensing bulb. A system adjusted to operate at a normal 10°F (5.6°C) superheat is shown in **Figure 5-20.**

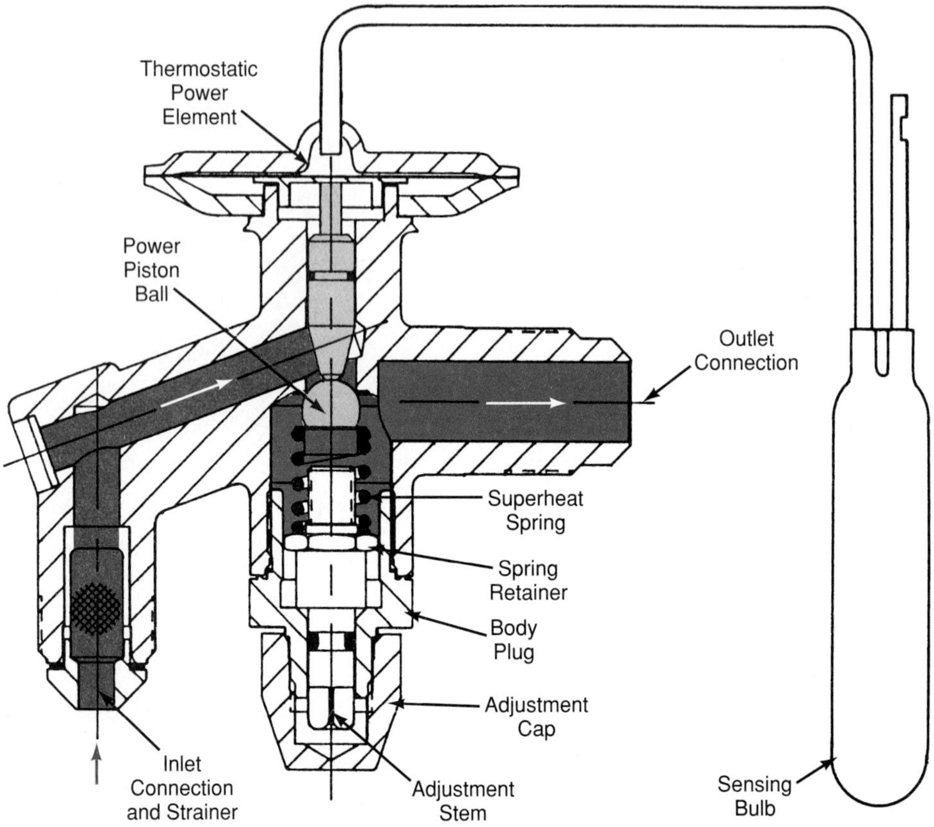

Figure 5-18. *Diaphragm thermostatic expansion valve. Note that a ball valve is used, replacing the needle. Direction of refrigerant flow is marked by arrows. (Refrigeration & Air Conditioning Div., Parker Hannifin Corp.)*

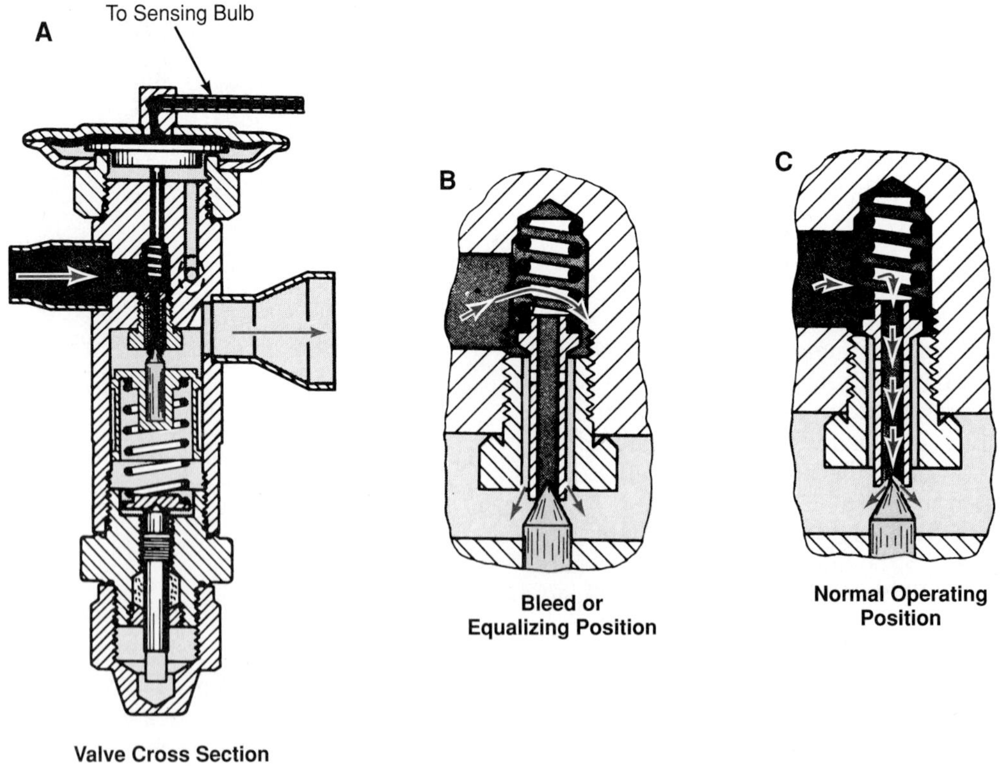

Figure 5-19. *Diaphragm thermostatic expansion valve. Sensing bulb pressure operates on top surface of diaphragm. As sensing bulb temperature increases, pressure on top of diaphragm tends to open valve, allowing refrigerant to enter evaporator. Note direction of refrigerant flow through valve. A—Cross section through entire valve. B—Bleed valve shown in normal operation. C—Needle valve in normal operating position. (Sporlan Valve Co.)*

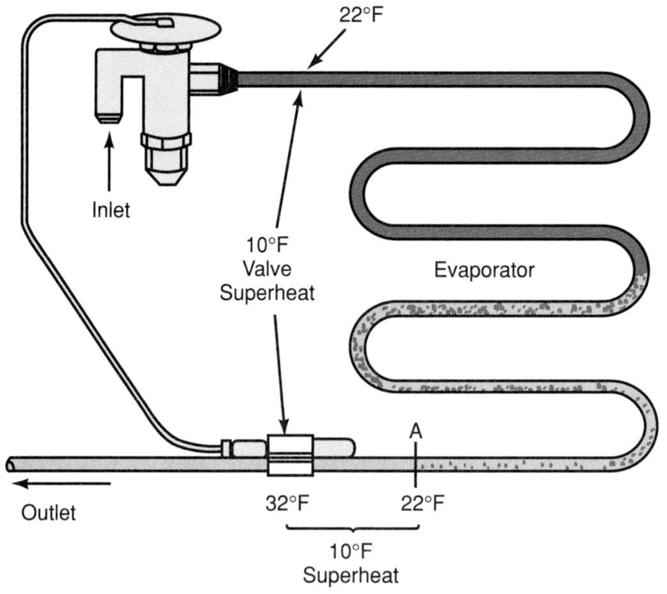

Figure 5-20. *A thermostatic expansion valve adjusted to give a normal 10°F (5.6°C) superheat. Liquid refrigerant will reach point A before valve controlling refrigerant closes. (Alco Controls Division, Emerson Electric Co.)*

Increasing the superheat tends to ***starve*** the evaporator. Starving the evaporator means only part of the evaporator is filled with liquid refrigerant. **Figure 5-21** illustrates a superheat setting of 15°F (8.3°C). At this setting, the evaporator is said to be starved.

As the temperature of the evaporator drops, the amount of superheat increases. The amount of superheat in the suction line is largely determined by the

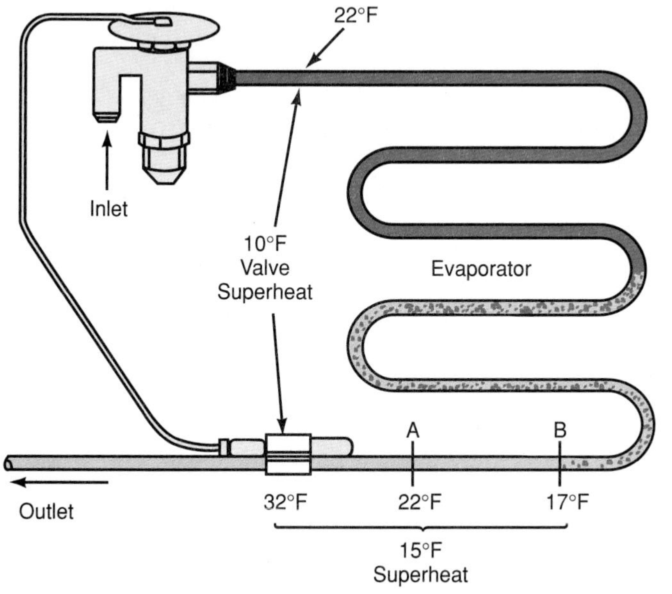

Figure 5-21. *A starved evaporator results when there is a pressure drop. Liquid reaches point B at 17°F (−8°C) and vapor warms to 22°F (−6°C). Valve closes at this point with too little liquid in evaporator. (Alco Controls Division, Emerson Electric Co.)*

refrigerant control. The best superheat setting for an evaporator is when the thermal bulb temperature changes the least while the system is running. This setting is called the Minimum Stable Signal (MSS) point or setting.

For example, if a valve and evaporator combination behaves as follows:

- At 12°F superheat, bulb temperature changes from 14°F to 10°F.
- At 10°F superheat, bulb temperature changes from 11°F to 9°F.
- At 8°F superheat, bulb temperature changes from 8.5°F to 7.5°F.
- At 6°F superheat, bulb temperature changes from 8°F to 4°F.

The least change (most stable condition) is at 8°F superheat setting, because it produces the smallest variation in the bulb temperature.

Liquid-Charged Sensing Element

The liquid-charged sensing bulb is charged with the same refrigerant as the system. Thus, the valve maintains a constant superheat setting even though low-side pressures and temperatures change, as shown in **Figure 5-22.** The quantity of fluid in the liquid-charged sensing bulb is sufficient so there is always some liquid in the bulb regardless of its temperature.

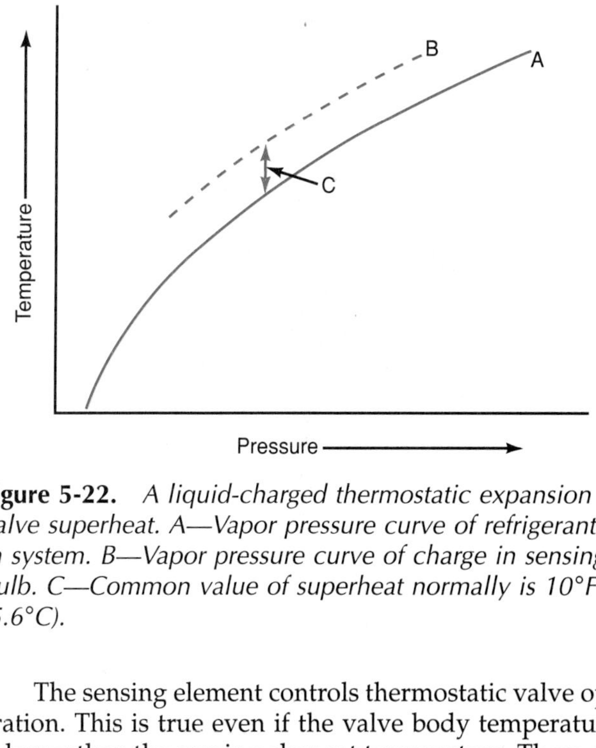

Figure 5-22. *A liquid-charged thermostatic expansion valve superheat. A—Vapor pressure curve of refrigerant in system. B—Vapor pressure curve of charge in sensing bulb. C—Common value of superheat normally is 10°F (5.6°C).*

The sensing element controls thermostatic valve operation. This is true even if the valve body temperature is lower than the sensing element temperature. These elements are designed for a temperature range of from approximately −20°F to 40°F (−28.9°C to 4.4°C).

This valve may cause some evaporator flooding when pulling down from normal ambient temperatures. The valve is used in air conditioners and some refrigeration systems.

Liquid Cross-Charged Sensing Element

The liquid cross-charged sensing bulb uses a liquid different from the refrigerant in the system. It may use a mixture of fluids to give the desired operating characteristics.

Liquid cross-charged elements share many similarities with liquid-charged elements. Some liquid is always present in the element, regardless of temperature. Also, the sensing element continues to control the valve even if the valve body temperature dips below that of the element.

The valve closes quickly when the compressor stops. This is because the evaporator pressure increases faster than sensing bulb pressure as the evaporator warms.

The load on the compressor is reduced at startup. As the suction pressure is reduced, the superheat is reduced, thus using maximum evaporator surface.

Hunting is reduced because the pressure-temperature curve of the sensing element is flatter. The valve is more responsive to changes in suction pressure than to changes in sensing bulb temperature. See Section 5.1.3 for an explanation of hunting.

Liquid cross-charged elements are designed for a temperature range from −40°F to 40°F (−40°C to 4.4°C). These valves are usually used for either commercial low-temperature applications or with extremely low-temperature systems.

Figure 5-23 shows the operating characteristics of the liquid cross-charged element in graph form. **Figure 5-24** shows the difference in the superheat curve of the cross-charged element as compared to a charged element.

Gas-Charged Sensing Element

The gas-charged sensing bulb uses the same refrigerant as the system. The amount of charge is such that all the liquid is vaporized at a predetermined temperature. Increasing the temperature above this point does not cause an increase in element pressure. The expansion valve does not open more with an increase in the cabinet temperature.

However, if the valve body becomes colder than the sensing bulb, the vaporized control fluid condenses in the valve body. Control is lost, and the valve closes.

For example: Just enough control fluid is put in the element to produce a maximum pressure of 40 psig (377 kPa). The element pressure will not exceed this pressure, no matter how warm the bulb is. When the low side exceeds this pressure, the valve will not open. See **Figure 5-25.** Thus, low-side pressure will not have to operate above 40 psi.

Gas-charged elements are designed for a temperature range from 30°F to 60°F (−1.1°C to 15.6°C).

Gas Cross-Charged Sensing Element—Adsorption

The gas cross-charged sensing bulb is charged with a liquid different from the system refrigerant. The amount of charge is such that, at the desired temperature, all the liquid has vaporized. Increasing the

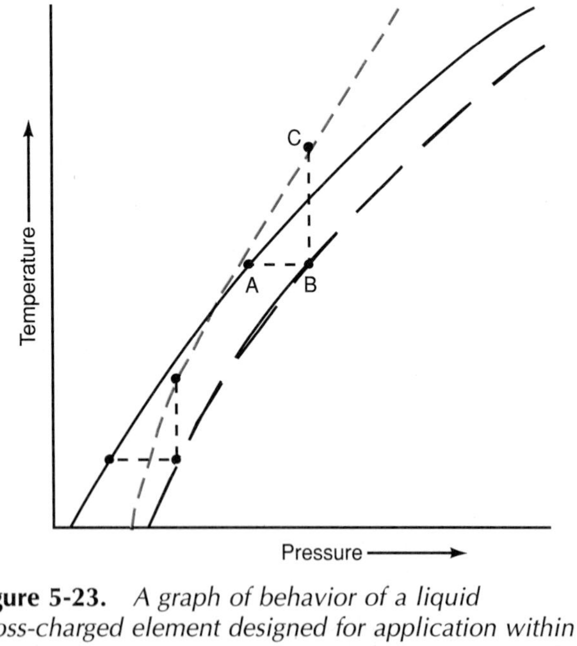

Figure 5-23. *A graph of behavior of a liquid cross-charged element designed for application within a specific temperature range. It provides a rapid pull-down and is used for normal refrigeration. Superheat setting at top end is wide to prevent flooding of unit. A—Given evaporator pressure and corresponding saturation temperature. B—Evaporator pressure plus equivalent superheat spring pressure. C—Corresponding sensing bulb and power assembly pressure. B-C is superheat setting for this evaporator pressure and spring setting.*

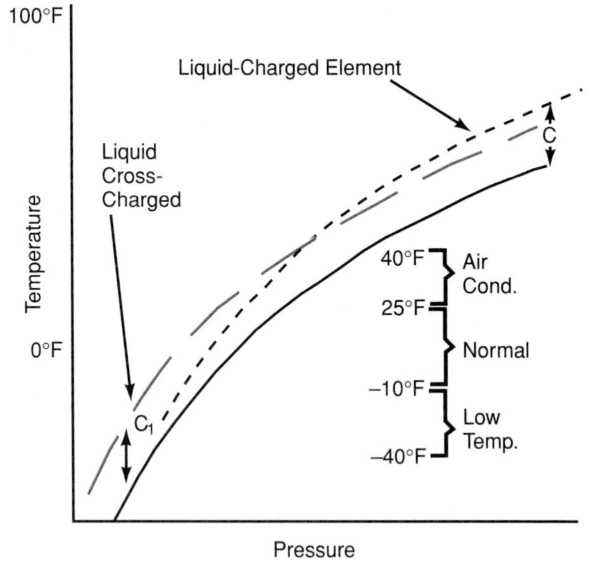

Figure 5-24. *Graph showing constant superheat of liquid cross-charged element designed to be used for all three applications—low temperature, normal, and air conditioning—as compared to a liquid-charged element in wide temperature ranges. Red dashed line shows degrees of superheat for a liquid cross-charged thermostatic power element. Black dashed line shows degrees of superheat for liquid charge. The liquid charge, C, changes as temperature drops, but the C, liquid cross charge remains constant as temperature drops.*

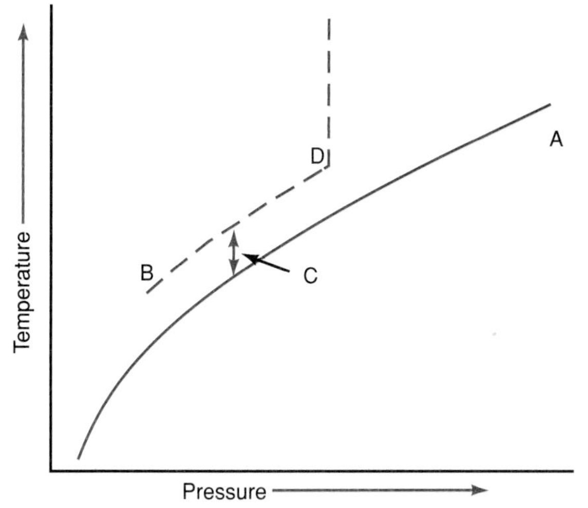

Figure 5-25. *A gas-charged thermostatic expansion valve superheat. A—Vapor pressure curve of refrigerant in system. B—Vapor pressure curve of charge in sensing bulb. C—Superheat. D—Point at which all fluid in sensing bulb becomes vaporized.*

temperature above this point does not cause an increase in element pressure. The superheat characteristics are like the liquid cross-charged element.

Some types of gas cross-charged sensing elements depend upon a different principle. In these thermostatic expansion valves, the sensing element contains two substances. One is a noncondensing gas, such as carbon dioxide, which provides the pressure in the element. The other is a solid such as carbon, silica gel, or charcoal. These substances have the ability to adsorb gas.

Adsorption is the adhesion of a layer of gas one molecule thick over the surface of a solid substance. There is no chemical combination between the gas and the solid substance (adsorber).

The ability of a substance to adsorb gas depends upon the temperature. Substances more readily adsorb gas at low temperatures. As the sensing element warms, the pressure in the element will increase. This is due to release of the adsorbed gas. As the sensing element cools, its pressure will decrease due to the adsorption of gas back to the solid substance. The pressure change controls the refrigerant needle valve opening in the thermostatic expansion valve.

These thermostatic expansion valves have the advantage of a pressure-temperature lag in their operation. They have very wide temperature applications and may be used on any refrigerating or air conditioning system.

Thermal-Electric (Solid State) Expansion Valve

The thermal-electric controlled expansion valve depends upon the use of ***thermistors*** (see Chapter 6), directly exposed to the refrigerant in the suction line, to control the expansion valve needle opening. It does not use a pressure element, as in the thermostatic expansion valve.

The resistance to electrical flow in the thermistor changes with temperature. Increasing temperature reduces resistance. Therefore, with a given voltage, increasing the temperature also increases the rate of current flow. This increased current flow heats and bends the bimetal in the valve body, opening the valve.

Figure 5-26 illustrates a typical thermal-electric expansion valve installation. The thermistor, C, is placed in immediate contact with the refrigerant vapor inside the suction line from the evaporator.

A low-voltage transformer is the power source. This is connected to the expansion valve control mechanism at B. The transformer is in series with the thermistor and electric device at B. Increasing current flow through the thermistor increases the opening of the expansion valve. This opening allows an increased the rate of flow of the refrigerant into the evaporator.

The refrigerant flow is controlled by the temperature in the suction line. The control mechanism is not dependent on the pressure in the evaporator. The thermal-electric expansion valve controls the suction line superheat in order to prevent flooding of the compressor.

Figure 5-27 shows a cross section of a thermal-electric expansion valve. This illustrates, in some detail, the electrical connections and the mechanisms that control the operation of this expansion valve. The bleed valves have a small slot in the valve seat. This causes the system pressure to balance during the Off cycle. When the next running cycle begins, the motor starts under practically no load. This allows the use of low-starting-torque compressor motors. Another purpose is to prevent complete close-off at the end of the machine's On cycle. Refrigerant is permitted to flow at a reduced rate.

A complete thermal-electric expansion valve and thermistor ready for installation is shown in **Figure 5-28,** view A. The thermistor, electrically connected to the thermal expansion valve, is shown in view B.

Off cycle operation is possible in two ways. The thermal-electric expansion valve may be electrically connected in parallel to the operating system. During the Off cycle, the thermistor becomes warmed and the thermal-electric expansion valve remains open. This unloading is similar to the pressure-balancing effect when a capillary tube refrigerant control is used.

In the second case, the valve may be electrically connected (interlocked) into the motor circuit. It is only energized when the compressor is running. With this type of connection, the valve will be closed on the Off cycle.

Pressure Limiters

Sometimes, a pressure-limiting expansion valve is used to prevent overloading the condensing unit. Pressure limiters are designed for systems in which the evaporator pressure must not exceed a safe operating limit.

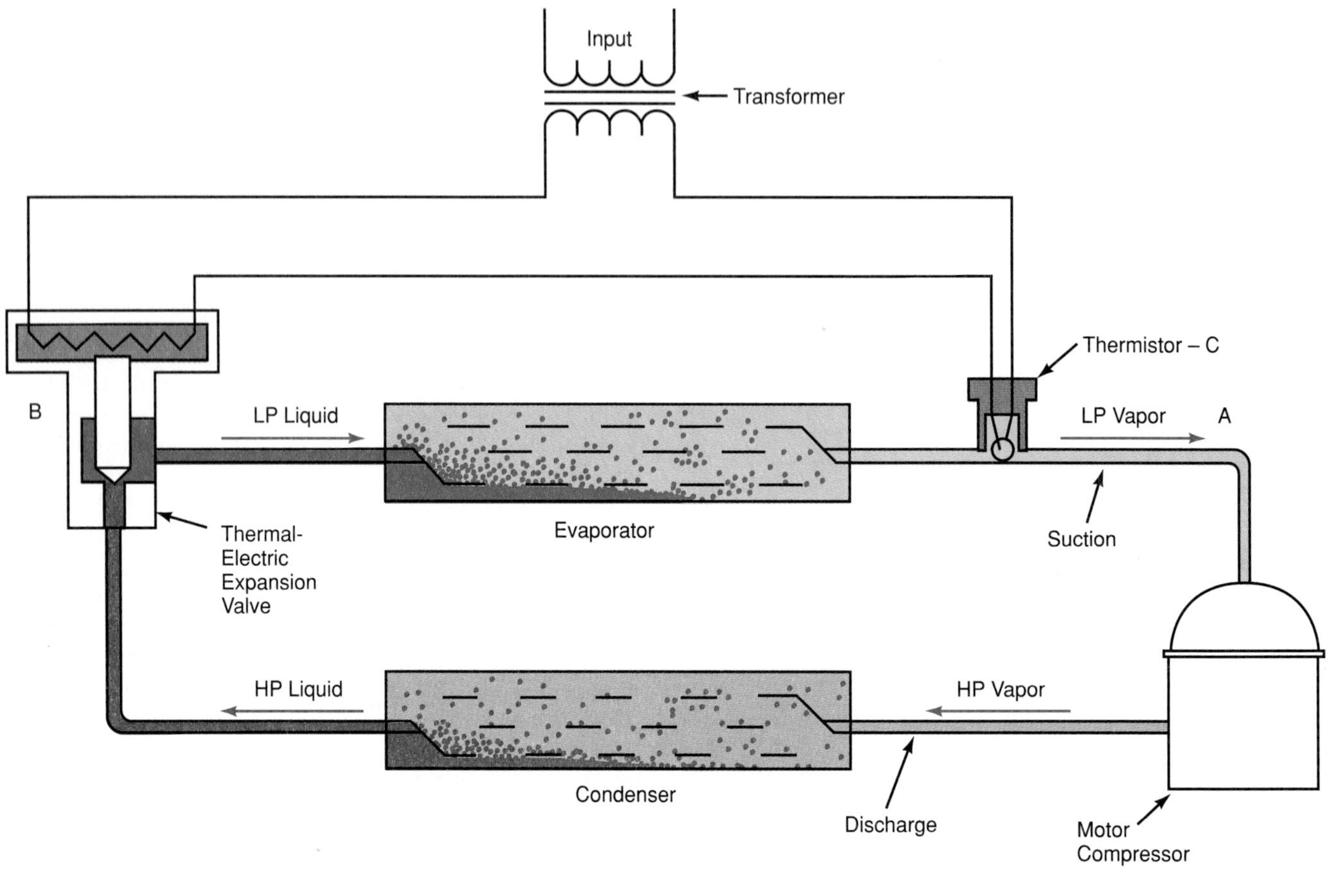

Figure 5-26. *A thermal-electric expansion valve installation.*

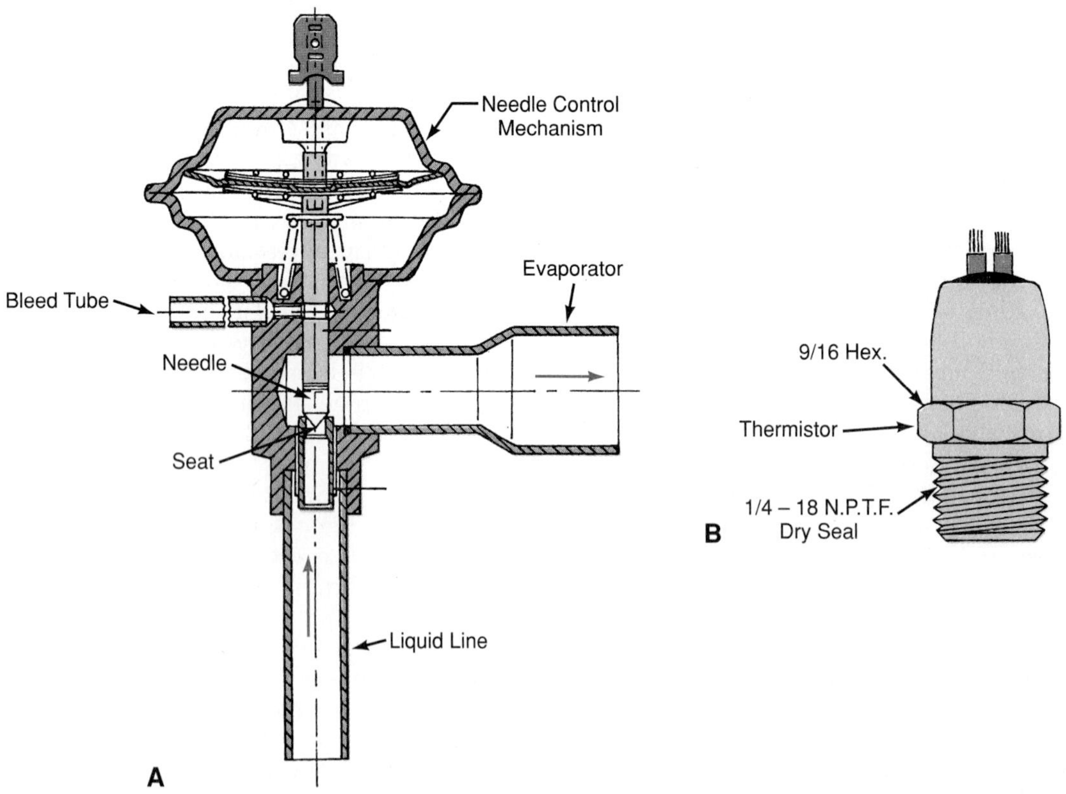

Figure 5-27. *A thermal-electric expansion valve. A—Cross section of valve. (Eaton Corporation) B—Thermistor.*

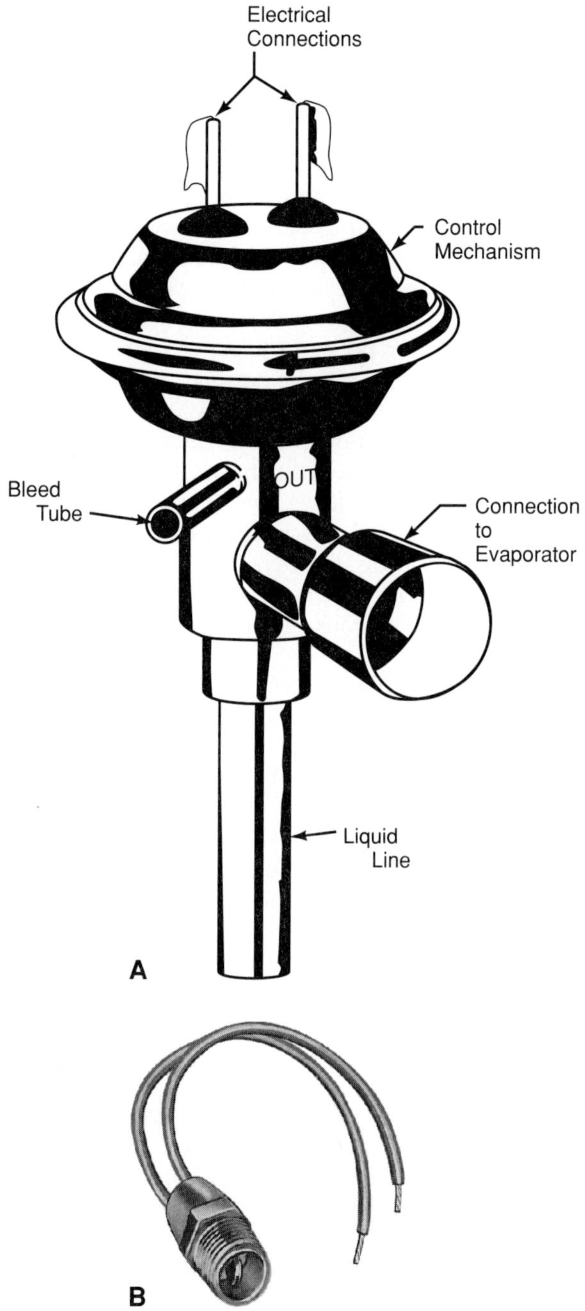

Figure 5-28. *Components of a thermal-electric expansion valve system. A—Valve. B—Thermistor.*

A pressure limiter is placed between the sensing element and the needle valve. Composed of a diaphragm and a spring, it is designed to collapse at a certain force. Thus, if the element is designed to collapse at 40 psig (377 kPa), the needle will close if the low-side pressure exceeds this amount, regardless of the evaporator temperature-pressure. These valves offer rapid pull-down on start-up. See **Figure 5-29.**

A gas-charged collapsible element can provide a limit to the pressure that will open the valve. When the low-side pressure exceeds a certain set value, the diaphragm will collapse. The gas used is noncondensable and obeys Charles' and Boyle's Laws.

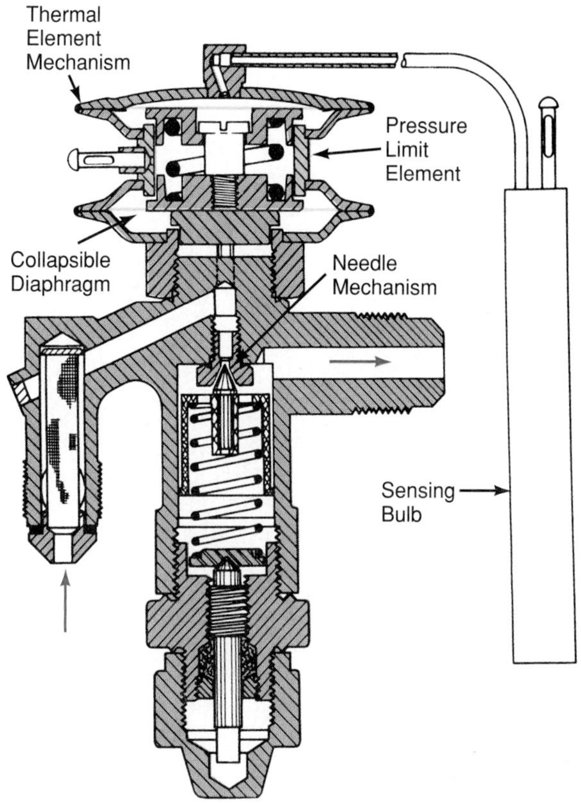

Figure 5-29. *Thermostatic expansion valve with mechanical pressure limiter. Element limits pressure by means of two diaphragms and a spring. Whenever suction pressure gets near motor overload point, spring between two diaphragms compresses and valve reduces flow of refrigerant to evaporator.*

When the refrigerator is warm, the pressure limiter prevents a long running time at excessive low-side pressures. An example is shown in **Figure 5-30.** The cycle record shows a pressure drop from over 50 psig to 10 psig (446 kPa to 170 kPa) in just a few minutes. The unit then runs for two hours before it shuts off.

Still another type of pressure-limiting thermostatic expansion valve is shown in **Figure 5-31.** This valve has an adjustable pressure limiter. Above a certain pressure setting, the spring above the diaphragm compresses instead of the valve needle being opened.

Frequently, it is necessary to have two different pressure levels controlling a given valve. A control valve with two pilot pressure regulators is called a ***dual-pressure regulator.*** This system uses a switching mechanism for the selection of either high- or low-pressure control.

Maximum operating pressure (MOP) or pressure limit may be achieved by a specially designed thermostatic charge. The MOP thermostatic charge is a modification of the conventional limited liquid thermostatic charge. It allows a predetermined valve bulb temperature and corresponding bulb pressure (opening pressure) to be reached. The valve throttles or closes. Increase

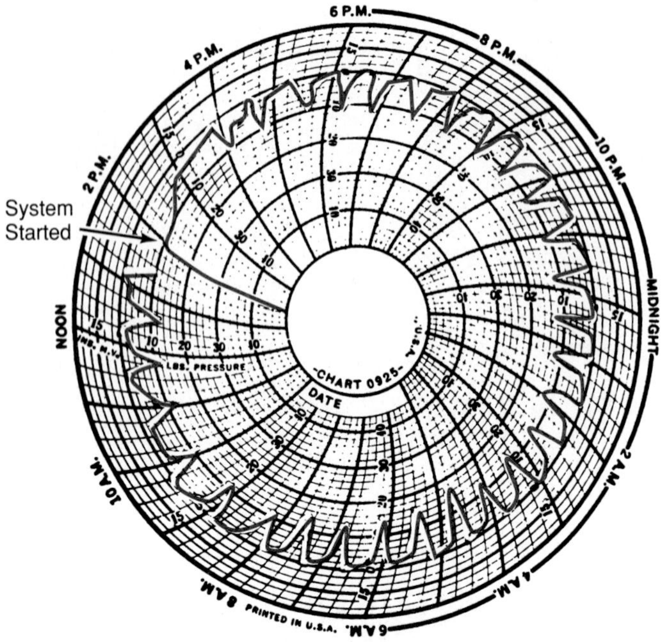

Figure 5-30. *A 24-hour record of pressure-time relationship for a food freezer as it is cycled, beginning with a warm condition at 2 p.m. Freezer is equipped with pressure-limiting thermostatic expansion valve and thermostatic motor control. Note quick reduction of pressure until valve opened at 2 p.m. (Sporlan Valve Co.)*

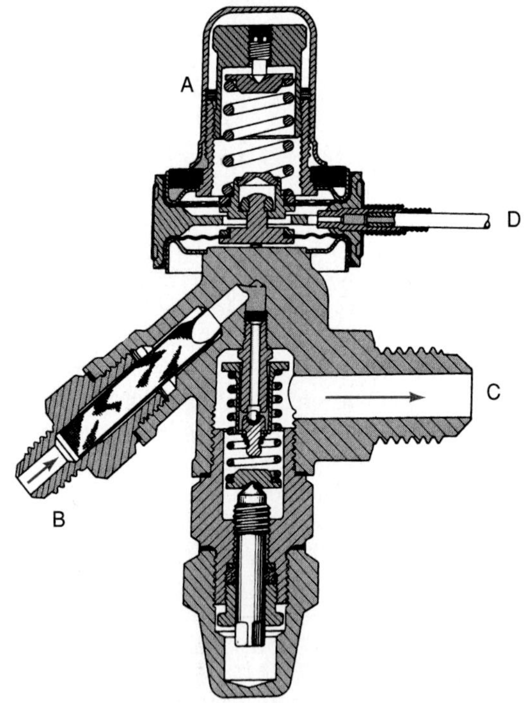

Figure 5-31. *Thermostatic expansion valve with adjustable pressure limiter. A—Pressure limiter adjustment. B—Liquid refrigerant inlet. C—Evaporator connection. D—Capillary tube sensing bulb connection.*

in bulb temperature above the predetermined value causes little or no increase in bulb pressure. The MOP setting is comparable to the pressure-limiting setting of the mechanical pressure limiter.

There are two differences between MOP thermostatic charges and conventional thermostatic charges. The first is that MOP thermostatic expansion valves close tightly during Off cycle. As the evaporator warms up in the Off cycle, the point of maximum bulb pressure is reached. Increase in bulb temperature results in no increase in bulb pressure (opening pressure). Therefore, the evaporator pressure (closing pressure) continues to rise. Assisted by the spring pressure (closing pressure), the valve closes tightly.

The second difference is that thermostatic expansion valves remain closed during pull-down. Although temperatures and pressures are relatively high in the evaporator during pull-down, the valve remains closed until the evaporating temperature is reduced below MOP of the thermostatic charge. This permits rapid pull-down, avoiding floodback and overloading of the compressor motor.

Thermostatic charges using a maximum operating pressure (MOP) require the diaphragm and capillary tubing to be kept at a temperature warmer than the bulb during the operating cycle. This is necessary so the valve will be controlled by the bulb. If the diaphragm case becomes colder than the bulb, the thermostatic charge may move to the diaphragm case. Control from the bulb will be lost. The valve will then close or throttle.

Maximum operating pressure thermostatic charges are used with comfort cooling systems. They are also used with indoor and outdoor coils of heat pumps.

Sensing Bulb Mounting

The location and the actual mounting of the sensing bulb is very important. It must be in good thermal contact with the evaporator outlet, **Figure 5-32.** The bulb should be mounted on the top of the suction line. Liquid in the bulb is close to the suction line, as shown in **Figure 5-32A.** To mount the bulb on a vertical suction line, the capillary tube of the bulb should always enter from the top of the bulb. It should never enter from the bottom.

The bulb must not be affected by the air or liquid being cooled. It should be wrapped in insulation, as in C. Special insulation forms are available so that only suction line temperature affects the bulb. Plastic tape can also be used for this purpose.

Copper straps and non-rusting machine screws and nuts should be used to fasten the bulb to the suction line. The bulb must have excellent thermal contact with the suction line. The connection must be clean and tight. Both the suction line and the bulb should be cleaned with steel wool before assembling.

The suction line carries chiefly vaporized refrigerant. However, there will be some droplets of liquid

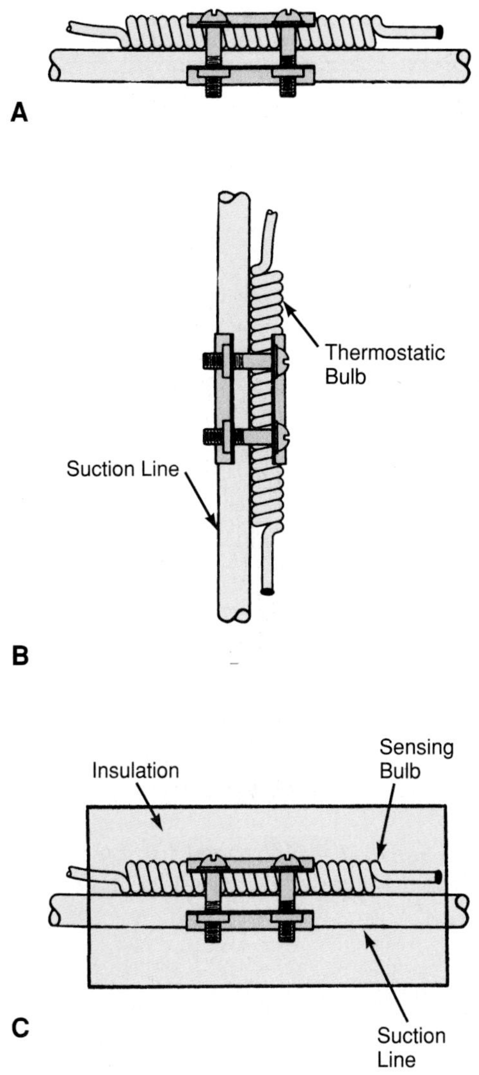

Figure 5-32. *Correct way to attach thermostatic expansion valve sensing bulb to suction line. A—Thermostatic sensing bulb mounted in horizontal position. B—Thermostatic sensing bulb mounted in vertical position. C—Insulated sensing bulb installation. Note that thermal bulb is best mounted in horizontal position and on top of suction line.*

refrigerant and some oil. **Figure 5-33** illustrates conditions inside the suction line, particularly on installations requiring a large diameter line. In **Figure 5-33A,** refrigerant vapor and some droplets of liquid refrigerant are flowing through a rather large diameter suction line. Due to the large diameter, the velocity of the vaporized refrigerant at times will be quite slow. The droplets of liquid refrigerant and oil will settle on the bottom of the line. The suction area in **Figure 5-33B** is smaller. As a result, the velocity of the vaporized refrigerant will be higher than in **Figure 5-33A.** This means less separating of the oil and liquid refrigerant from the flowing vapor. The inside of the tube will be rather uniformly coated with oil. The suction line in **Figure 5-33C** is shown in vertical position. In this position, there will be no separation of the droplets of refrigerant from the vapor. However, the oil will uniformly coat the inside of the suction line.

The temperature of the vaporized refrigerant and the droplets of liquid refrigerant will be a few degrees colder than the suction line surface. This is due to the insulating quality of the oil that coats the inside of the suction line.

The sensing bulb on large suction lines should be located near the underside of the line rather than on top because droplets of liquid refrigerant tend to separate from the flowing vapors. Also, oil has an insulating effect. The recommended bulb position (on large suction lines) is shown in **Figure 5-33D.** Some sensing bulbs are crimped or creased lengthwise to provide double contact and help align the sensing bulb with the surface of the suction line.

Thermostatic Expansion Valve Capacities

The capacity of a thermostatic expansion valve (TEV) varies according to:

- Orifice size.
- Pressure difference between the high side and the low side.
- The temperature and condition of the refrigerant in the liquid line.

The capacity of most thermostatic expansion valves is selected from the size of the orifice and needle assembly. The same body may be used for many capacities. The larger the orifice, the more liquid refrigerant can be fed into the evaporator in a given time.

Valves are rated in tons of refrigeration. However, the same orifice usually has three different tonnage capacities. This capacity range depends on the pressure difference between the high side and the low side. Increasing this pressure difference will increase the rate of refrigerant flow.

For example, if a valve is used on an R-12 refrigerant system, the following will occur: A valve that has a 1/2 ton (.455 metric ton) rating at 13 psi (193 kPa) pressure on the low side will have a 3/4 ton to 1 ton (.72 to .98 metric ton) capacity at 5″ (12.7 cm) Hg vacuum on a frozen food unit. The same valve has only a 1/3 ton (.3 metric ton) capacity at a low-side pressure of 40 psi (380 kPa) on an air conditioner.

In the first case, there is a 130 − 13=117 psi (807 kPa) pressure difference, assuming a 130 psi (1 000 kPa) head pressure. In the second case, it is a matter of 130 plus 2 1/2 psi (5″ of vacuum=2 1/2 psi=84 kPa) = 132 1/2 psi (914 kPa) pressure difference. In the last case, it is 130 − 40=90 psi (621 kPa) pressure difference.

It is important to use a valve of the correct capacity. With an undersize valve orifice, the evaporator will be starved regardless of the superheat setting. The full capacity of the evaporator cannot be reached.

If the orifice is oversize, the valve will hunt, or surge. When the valve opens, too much refrigerant will pass into the evaporator. The suction line will sweat or

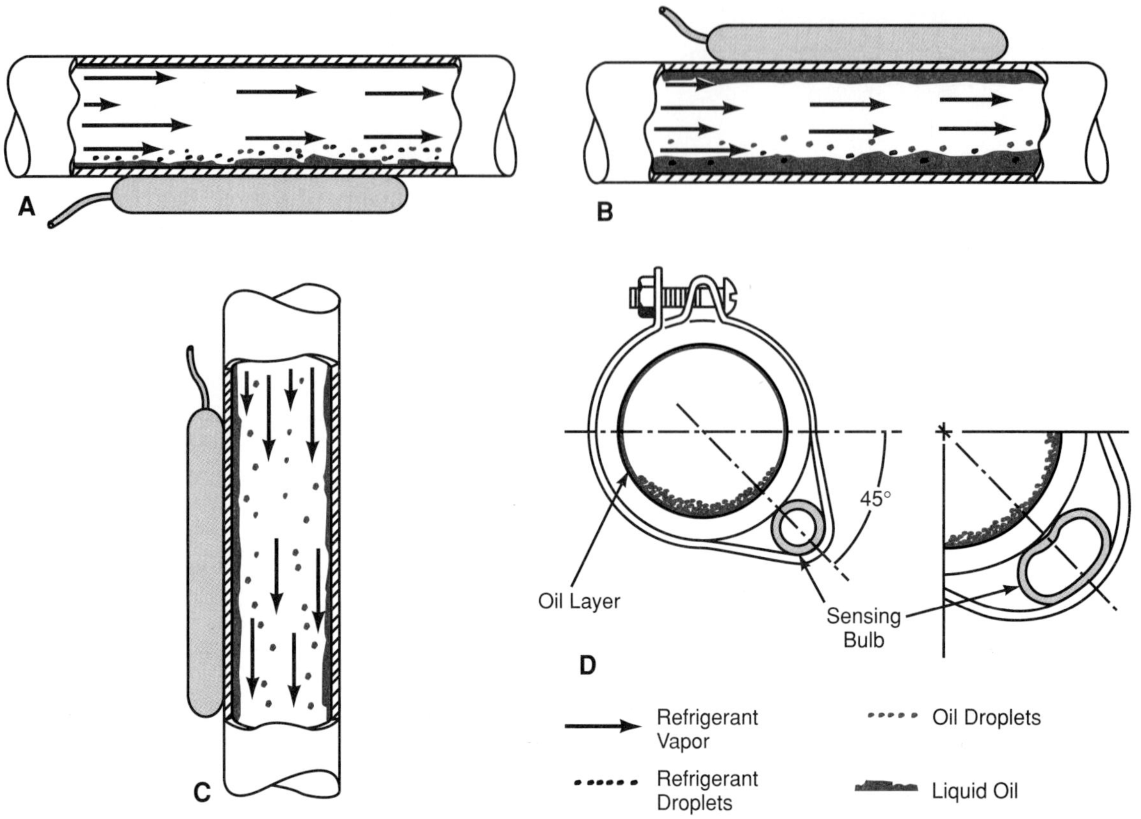

Figure 5-33. *Flow conditions in suction line. A—Horizontal suction line (large). B—Horizontal suction line (small). C—Vertical suction line. D—Sensing bulb on large diameter suction line. Oil is shown in red; refrigerant vapor by black arrows; refrigerant droplets by black dots.*

frost before the thermal element can close the valve. Increasing the superheat setting to correct this condition results in the evaporator being starved much of the time.

Special Thermostatic Expansion Valves

Many different thermostatic expansion valve designs are available. One type uses a separate six-circuit refrigerant distributor connected to the expansion valve. See **Figure 5-34.**

This design is used to reduce the pressure drop in a large evaporator. It provides several parallel refrigerant paths through the evaporator. It is popular for air conditioning applications. Careful engineering is needed, as each tube must receive an equal amount of refrigerant.

A diagram of a special thermostatic expansion valve is shown in **Figure 5-35.** It provides multiple connections to the evaporator. **Figure 5-36** shows a distributor used on large capacity evaporators.

5.1.3 Solenoid Valve Principles

A solenoid valve is used in many refrigerating applications. It will automatically close or open refrigerant circuits to get the desired refrigerating effect. It is easily installed and uses only simple electrical control circuits.

A solenoid valve is simply an electromagnet with a movable core or center. The basic construction includes

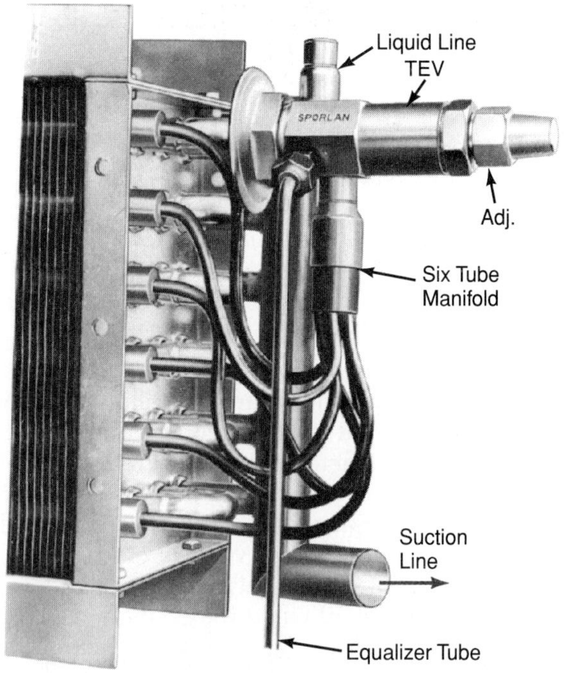

Figure 5-34. *Special thermostatic expansion valve installation for air conditioning applications. Refrigerant distributor is brazed onto outlet connection of valve. This valve uses an equalizer tube which connects to suction line and enters under valve diaphragm. (Sporlan Valve Co.)*

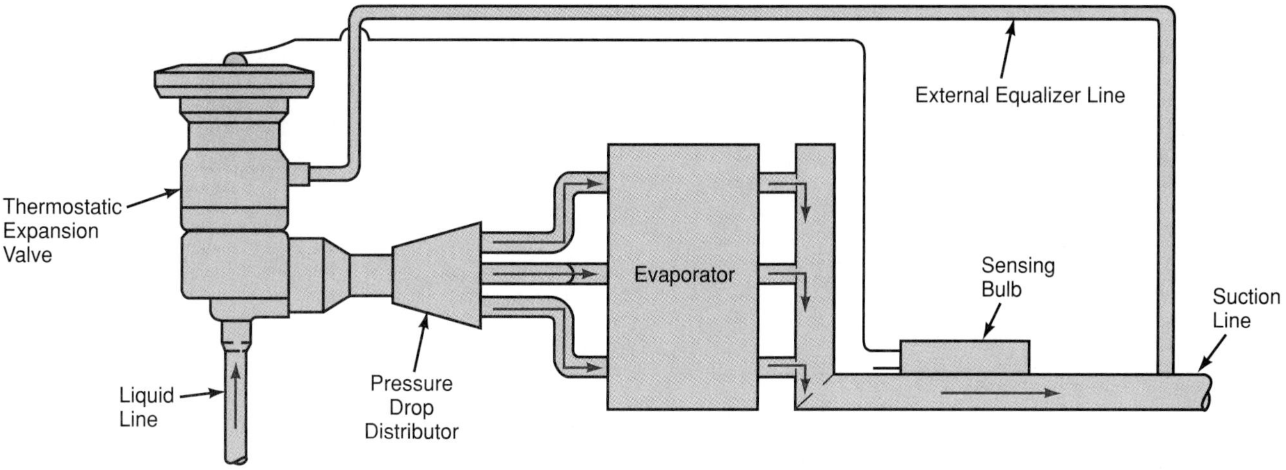

Figure 5-35. *A thermostatic expansion valve that provides multiple connections to evaporator. (Alco Controls Division Emerson Electric Company)*

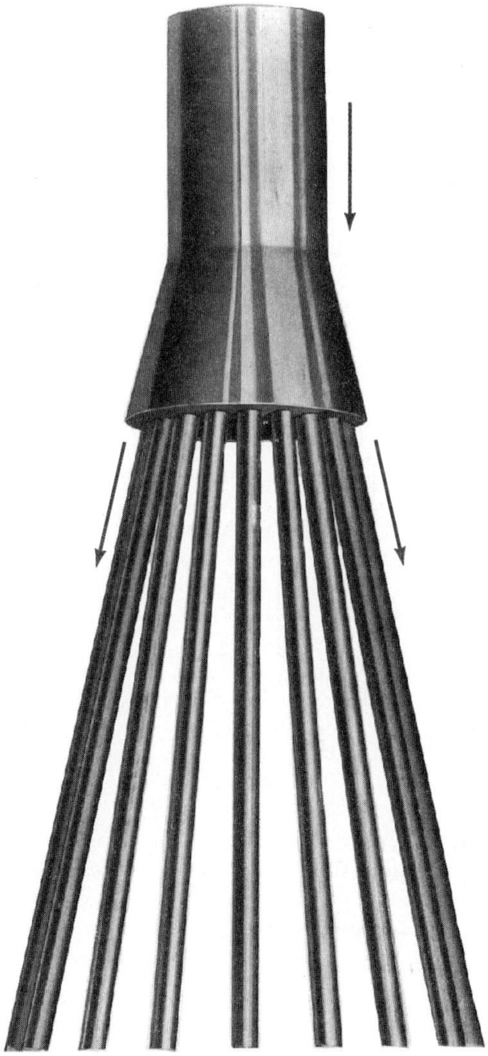

Figure 5-36. *A distributor tube. Thermostatic expansion valve supplies refrigerant at top. In turn, distributor tube supplies equal quantities of refrigerant to parallel paths through evaporator (bottom). (Alco Controls Division, Emerson Electric Company)*

a movable armature made of an iron alloy and attached to the valve needle. This element is sealed into the valve body. The armature can raise and lower the valve needle. A coil is wound around the valve housing that contains the armature.

The basic construction of a solenoid valve is shown in **Figure 5-37.** As the coil is energized, the magnetic armature moves upward toward the center of the coil. It opens the valve. When the circuit is opened, the coil is de-energized and, therefore, demagnetized. The spring and weight of the armature force the valve against the valve seat.

Types of Solenoid Valves

Three types of solenoid valves are in common use:

- The two-way valve that controls the flow of refrigerant through a single line. See **Figure 5-38.**
- The three-way valve with an inlet that is common to two opposite openings. It controls refrigerant flow in two different lines. See **Figure 5-39.**
- The four-way reversing valve used on heat pumps. These are illustrated in Chapter 24.

Solenoid valves may be activated by a thermostat. These valves are used to control the temperature of a refrigerator or a room.

Three-way solenoid valves are used mainly on commercial refrigerating units. They may be used to control two separate refrigerant circuits for defrosting, two-temperature evaporators, etc.

In the three-way solenoid valve shown in **Figure 5-39,** opening A is the common opening and is never closed. As the electromagnet is de-energized, the following occurs: The weight of the solenoid plunger assembly holds the valve against the lower seat. The force of the upper spring also holds the valve firmly against the lower seat. This action closes port B to the common port and opens port C.

When the electrical circuit is closed, the solenoid becomes energized. The piston with the two valves attached to it will move in the opposite direction. This

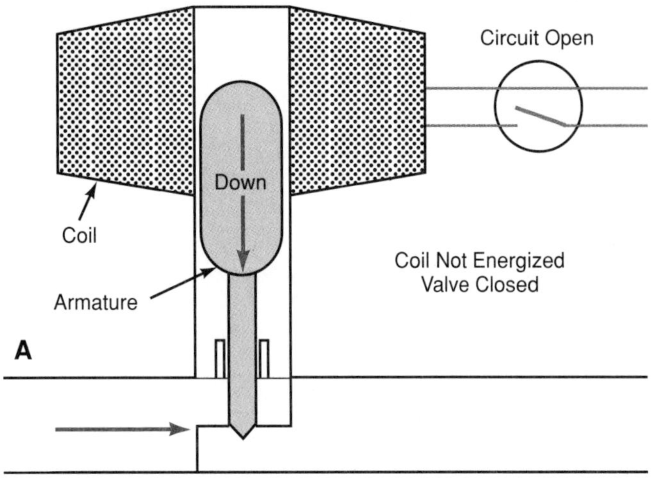

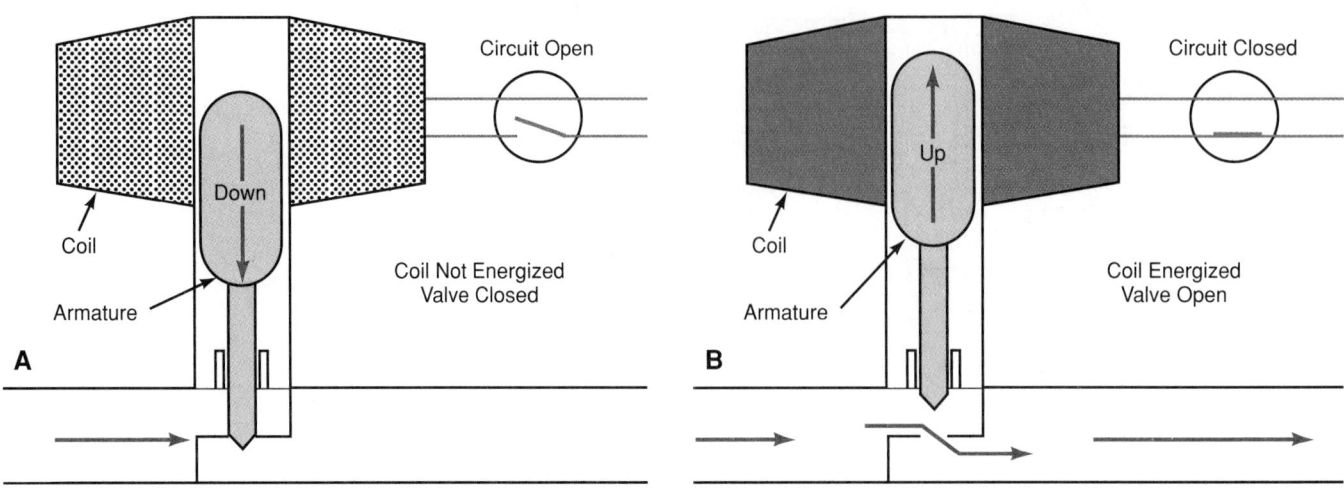

Figure 5-37. *Solenoid valve. A—Circuit is open, coil is not magnetized and armature, due to its weight, drops and closes valve. B—Circuit is closed, coil is magnetized armature, due to magnetic field, is pulled up and opens valve.*

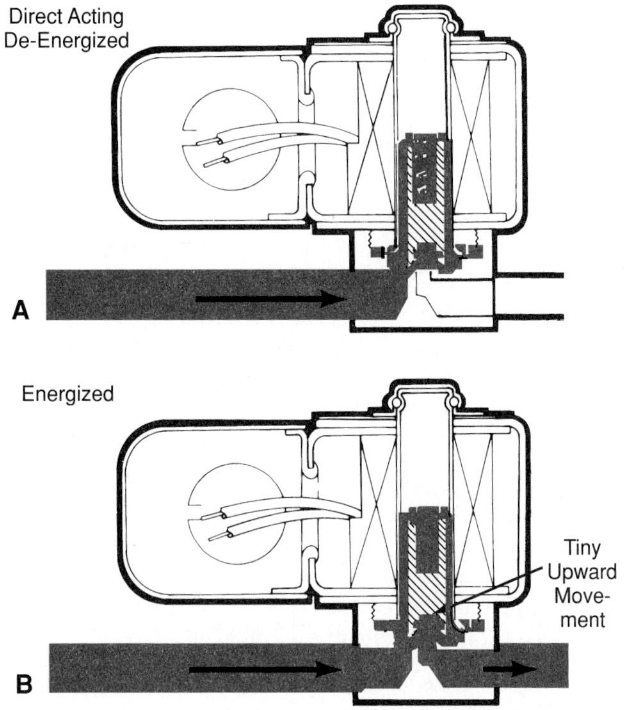

Figure 5-38. *Two-way solenoid valve. A—De-energized, closed position. B—Energized, open position. (Alco Control Division, Emerson Electric Company)*

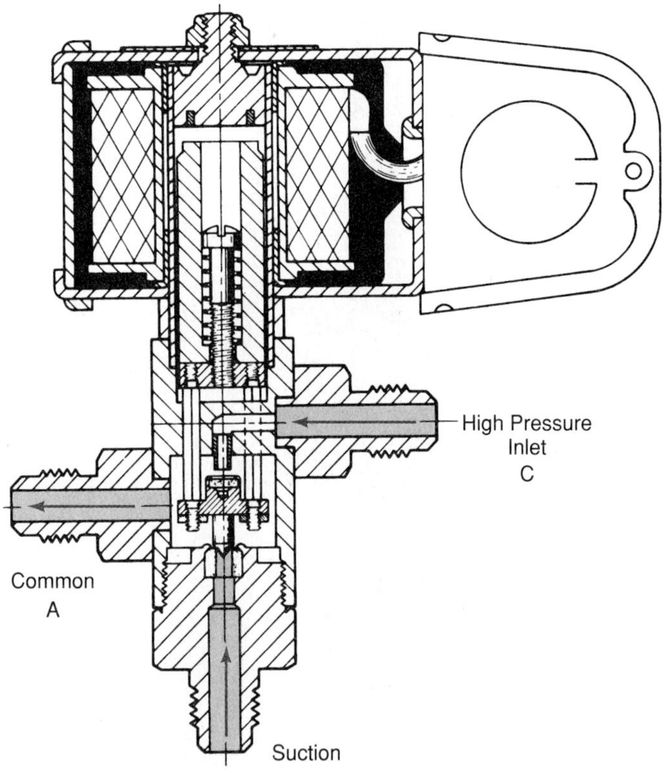

Figure 5-39. *Three-way solenoid valve used to close thermostatic expansion valve during Off part of cycle. When compressor is running, openings marked suction and common are connected. When compressor is off, high pressure inlet and common are connected. (Sporlan Valve Co.)*

action opens port B to common port A. It closes port C to the common port.

Four-way solenoid valves are often called reversing valves. See Chapter 24. Four-way solenoid valves are used chiefly on heat pumps. They control the cycle for either heating or cooling, as needed. When the solenoid is de-energized, the valve stem closes several ports. It opens others to reverse the flow of the refrigerant to the condenser and evaporator. The heat pump then becomes a cooling system. When the solenoid valve becomes energized, the four-way valve stem is drawn upward. The heat pump system becomes a heating system again.

For large commercial applications, it is desirable to use a pilot-operated solenoid valve. In these valves, the solenoid operates a pilot mechanism. The pressures in the lines operate in a piston arrangement. This causes the opening and closing of the main control valve.

Figure 5-40 is a cutaway of a pilot-operated solenoid valve. When the solenoid, A, is energized, the plunger, B, will be pulled from its seat. The pressure in the diaphragm spring area, D, will leave to cylinder and the diaphragm, E, will move up. The movement of the diaphragm controls will open the pilot port, F.

When the solenoid valve is de-energized, plunger, B, returns to its seat. Pressure from G goes through a small opening and builds up pressure in diaphragm spring area, D. The spring then closes the pilot port, F.

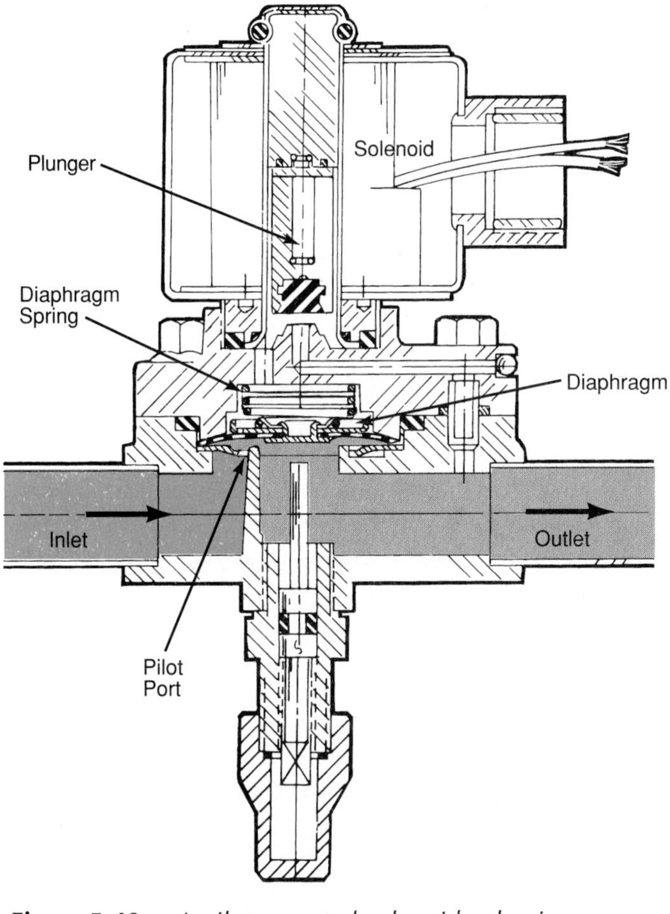

Figure 5-40. *A pilot-operated solenoid valve in open position. (Alco Controls Division, Emerson Electric Company)*

Equalizers

An ***equalizer*** is a small tube—usually 1/4″ OD. It joins the suction line at the outlet of the evaporator. The other end opens beneath the expansion valve diaphragm. The equalizer compensates for any pressure drop through the evaporator while the compressor is running. An equalizer tube is shown in **Figure 5-41.**

There is always some pressure drop through the evaporator. An equalizer should be used if the pressure drop between the inlet of the evaporator and the outlet is more than 4 psi (28 kPa). The equalizing tube provides

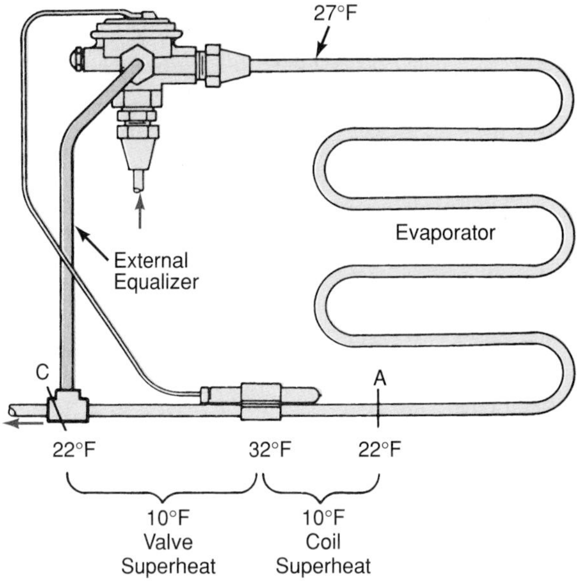

Figure 5-41. *A thermostatic expansion valve fitted with an equalizer tube. Equalizer connects suction line pressure at sensing bulb to underside of valve bellows or diaphragm (low-pressure side). This enables low-side pressure operating valve to be same as pressure at sensing bulb. This compensates for any pressure drop through evaporator while compressor is running.*

the same pressure as is in the suction line at the sensing bulb location. This equalizing of pressure will permit accurate superheating adjustments. Pressure drop in the evaporator tends to increase the superheat effect and to starve the evaporator.

The thermostatic expansion valve may open intermittently (opening and closing frequently) during the Off cycle. This may be caused by temperature fluctuations (changes) that occur during cabinet opening and closing. To prevent this, the high-side pressure may be used on the valve, forcing it closed during the Off cycle. A solenoid valve may be used to control this pressure.

A special solenoid valve connected into the equalizer tube is shown in **Figure 5-42.** Note valve construction in **Figure 5-39.**

The electrical circuit to the solenoid valve is opened when the motor compressor circuit opens. On opening the circuit, the solenoid core falls. It closes the equalizer tube to the suction line. The high-pressure refrigerant enters the top of the solenoid valve. It passes upward through the equalizer tube. This forces the thermostatic diaphragm up, closing the valve.

The thermostatic expansion valve, **Figure 5-43,** uses an unusual adjustment. The equalizer tube connects into the valve at connection K.

Figure 5-44 is a large capacity thermostatic expansion valve. It has flanged refrigerant line connections bolted together and gasketed. Instead of a needle and seat, it has a flat (disk) valve surface and seat. An equal izer tube connection is shown in the upper right part of the body.

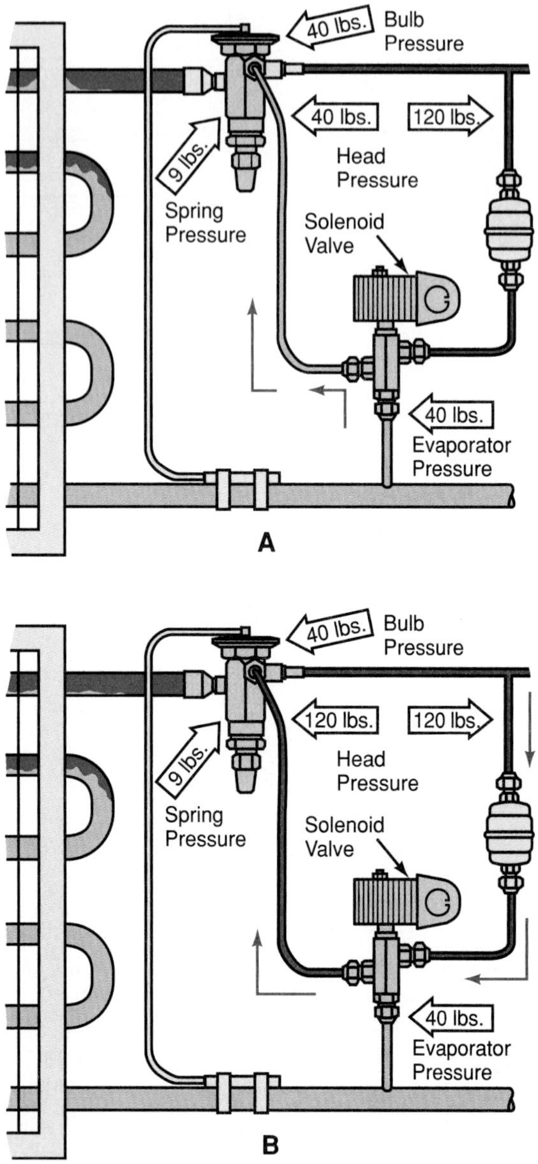

Figure 5-42. *A three-way solenoid valve is used to keep thermostatic expansion valve tightly closed during Off cycle. A—When motor compressor is on the On part of cycle, solenoid valve is energized. Pressure in suction line is transmitted through equalizer tube to thermostatic expansion valve. B—With motor compressor on the Off part of cycle, solenoid valve is de-energized so high pressure can flow up equalizer tube to close expansion valve. (Sporlan Valve Co.)*

Hunting

The term ***hunting,*** applied to any type of mechanism, means that the mechanism is first going too far in one direction. Then it returns too far in the other direction. Sometimes, this is also called surging.

Hunting, in refrigerating systems, identifies the changes in refrigerant flow through the control operation. There is always some hunting while the system is operating. If the valves hunt, they will alternately open up too wide. This allows too much refrigerant to flow into the evaporator. Then they close down too far, not

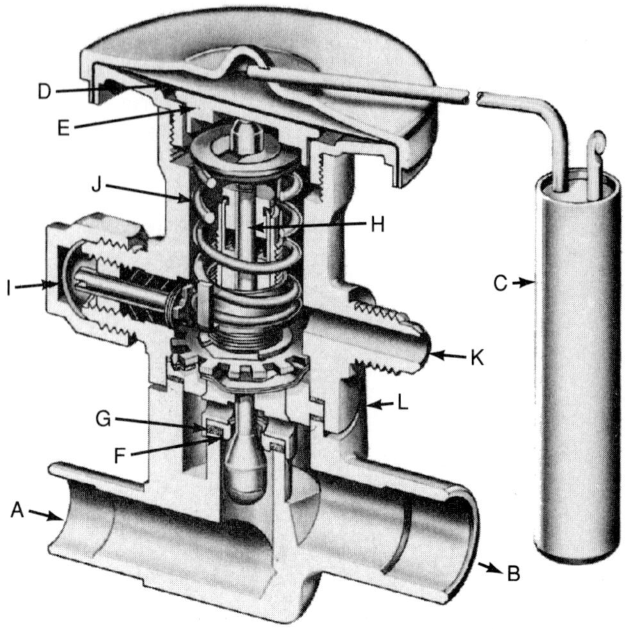

Figure 5-43. *A thermostatic expansion valve that uses an equalizer tube. A—Liquid line connection. B—Suction line connection. C—Sensing bulb. D—Diaphragm. E—Buffer plate. F—Valve pin. G—Valve seat. H—Push rod. I—Superheat adjustment. J—Superheat spring. K—Equalizer tube connection. L—Valve body. (Alco Controls Division, Emerson Electric Company)*

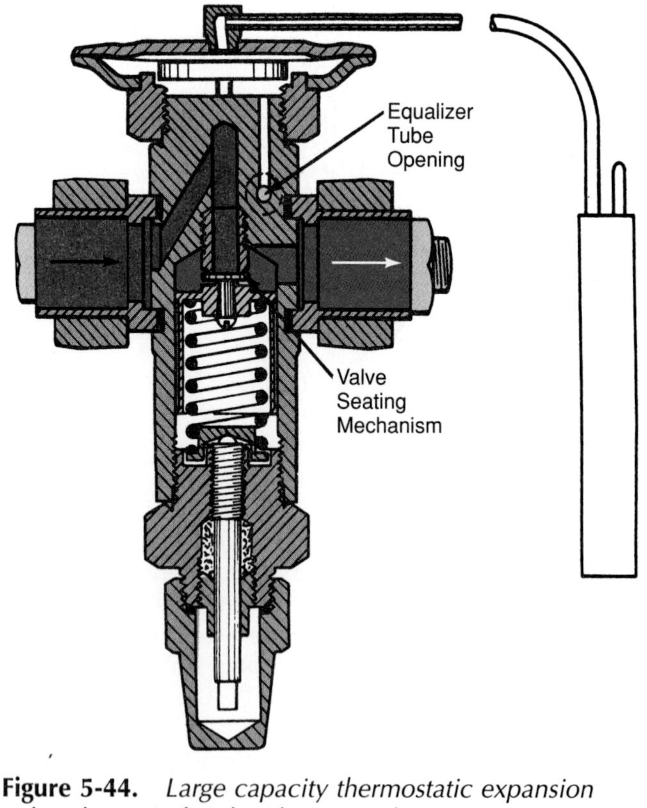

Figure 5-44. *Large capacity thermostatic expansion valve shown is fitted with an equalizer connection. (Sporlan Valve Co.)*

allowing enough refrigerant into the evaporator. This effect is suggested by dashed lines in **Figure 5-45.**

The less hunting, the more effective the system will be. When the valve is hunting too much, a uniform amount of refrigerant is not provided to the evaporator. It may even allow liquid refrigerant to reach the compressor and cause compressor damage. In some cases, this condition may be caused by a valve that is too large for the system.

Each thermostatic expansion valve and evaporator combination has to be adjusted. The superheat adjustment should be used to reduce surging and hunting to a minimum. However, it should still permit full evaporator use.

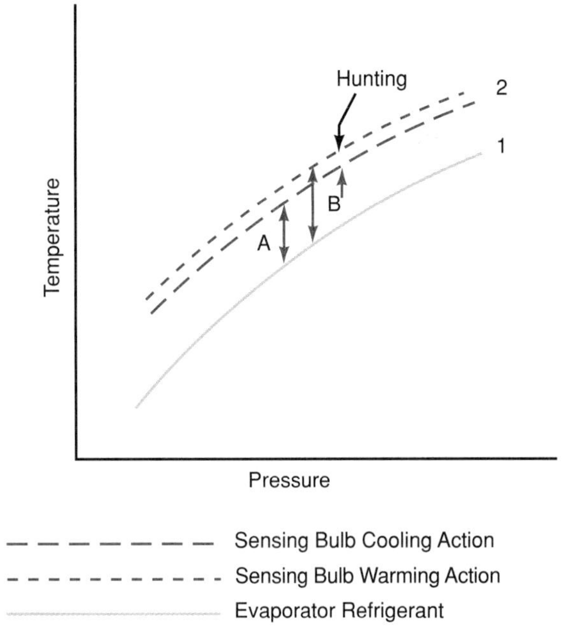

Figure 5-45. *Thermostatic expansion valve superheat. 1—Vapor pressure curve of refrigerant in system. 2—Pressure-temperature curve of liquid in sensing bulb. A—Superheat during warming of bulb. B—Superheat during cooling of bulb. The difference between the two lengths for A and B is called hunting. It is mainly the result of weight of valve parts and bending resistance of the bellows or diaphragm.*

5.1.4 Low-Side Float

The low-side float is an efficient, yet simple, refrigerant control. Its job is to maintain a constant level of liquid refrigerant in the evaporator. The system is an efficient heat transfer device. Heat moves easily from the evaporator to the liquid refrigerant. The basic principles of the low-side float system are explained in Chapter 3.

The float itself may be a sealed ball, a cylinder, or an open pan. It is connected by levers to a needle or ball valve. This valve closes when the liquid level reaches the correct height. The valve opens when some of the refrigerant evaporates and the liquid level drops.

The low-side float has the possible disadvantage of ***oil binding.*** The mechanism will float in either oil or liquid refrigerant. Special provisions must be made to return any excess oil to the compressor. This is done by using a wick or a small bypass at the liquid refrigerant surface. Extra refrigerant is stored in a liquid receiver.

The suction tube on these evaporators extends to the float chamber. With pan floats, it extends to the bottom of the pan. See **Figure 5-46.** This design ensures a more positive oil return than the ball float.

Low-side float systems are used in large industrial systems and in some water cooling systems. Pressure-operated motor controls or thermostatic motor controls may be used with this system.

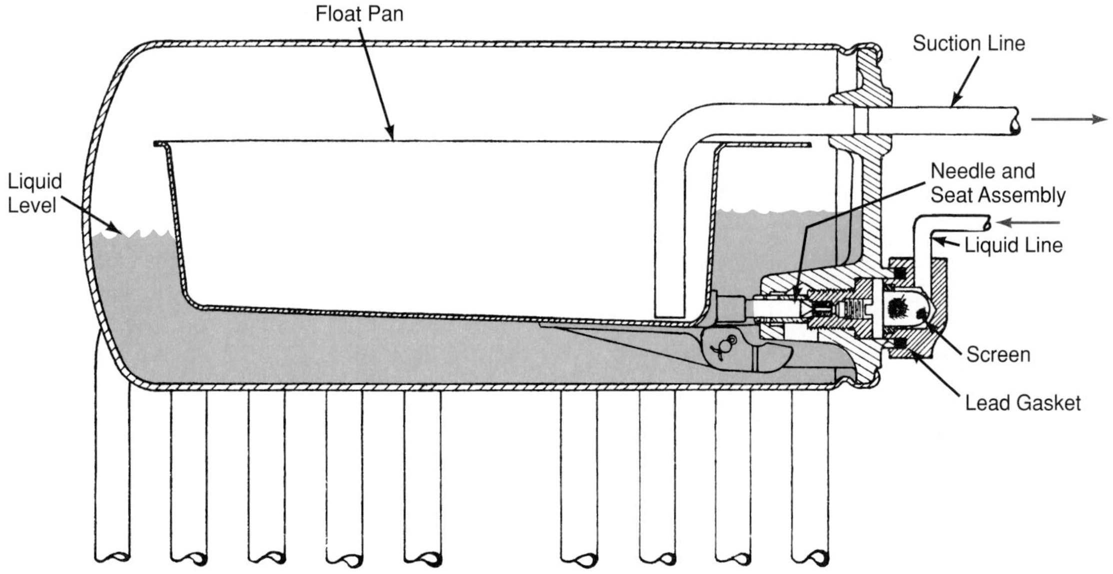

Figure 5-46. *Low-pressure side float refrigerant control. A bucket or pan float is used in this refrigerant level control. Suction line dips to bottom of open float in order to remove oil that might otherwise accumulate in open pan.*

The thermostatic expansion valve (TEV) or the low-side float may be used in multiple evaporator systems. The low-side float control also controls water level in cooling towers, evaporative condensers, and humidifiers.

5.1.5 High-Side Float

A high-side float refrigerant control is like the low-side float mechanism, but it is located in the high-pressure side.

When the compressor is running, condensed refrigerant from the condenser flows directly to the high-side float chamber. No liquid receiver is used. As the liquid refrigerant level rises, the float inside the chamber opens a valve, allowing liquid refrigerant to flow into the evaporator, where it is stored.

As the liquid level falls in the float chamber, the float moves down. It closes the valve opening into the evaporator. In this way the pressure difference between the high side and the low side is maintained. The high-side float maintains a constant level of liquid refrigerant on the high-pressure side.

Floats are made of either copper or steel. In hermetic units, steel is usually used. Such units cannot use a liquid receiver unless the float is located within it.

High-side float controls do not have as much trouble with oil binding as low-side float controls. At higher pressure, the oil dissolves and circulates more readily in the liquid refrigerant. The evaporator used with a high-side float control must be equipped with a special oil return, or oil binding will occur.

The float may be connected directly to the needle, or may operate the needle with a lever. **Figure 5-47** shows a float operated with a simple lever. Needles and seats are made of long-wearing alloys, such as stainless steel or hard-surfaced alloys.

Some high-side float systems use a weight check valve. This valve is used when the float chamber is a long distance away from the evaporator. Chapter 3 explains how this type system operates.

Either a thermostatic or pressure motor control may be used with a high-side float refrigerant control.

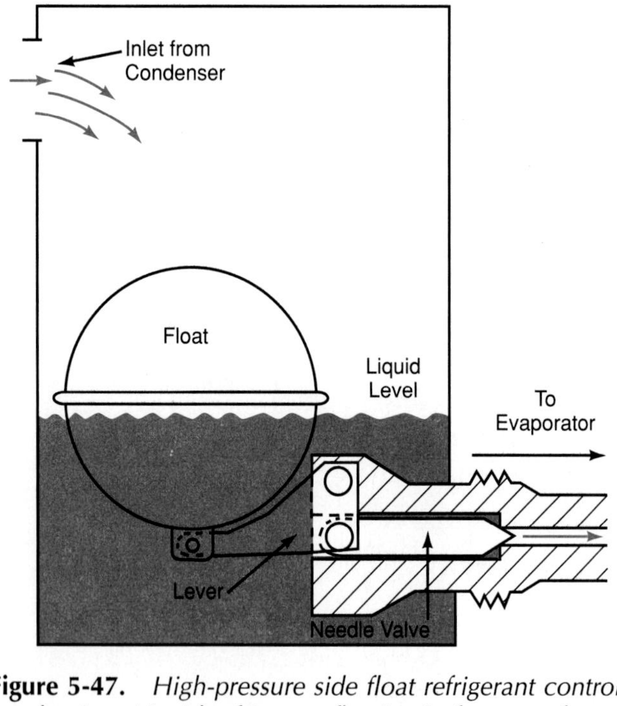

Figure 5-47. *High-pressure side float refrigerant control mechanism. Liquid refrigerant flowing in from condenser will cause float to rise and open needle valve. Then liquid refrigerant flows into evaporator.*

5.1.6 Capillary Tube

The capillary tube type of refrigerant controls simply a length of seamless tubing with a small and accurate inside diameter. It acts as a constant throttle on the refrigerant. It usually is equipped with a fine filter or filter-drier, which removes any moisture or dirt from the refrigerant at the tube inlet.

The amount of refrigerant in the system must be carefully calibrated. All the liquid refrigerant moves into the low side during the Off cycle as the pressures balance. Too much refrigerant will cause the unit to frost back on the low side. This control must be used with a thermostatic motor control.

Pressure decreases as the small liquid flow moves through the tube, which restricts fluid flow. The liquid starts to evaporate in the tube. This vapor formation provides a sudden pressure and temperature drop in approximately the last quarter of the tube length. The refrigerant is finally cooled to evaporator temperature, and its pressure is reduced to evaporator pressure.

This vapor formation in the capillary tube is called ***vapor lock.*** The design of the capillary tube depends on four variables:

- Tube length.
- Inside diameter.
- Tightness of tube windings.
- Temperature of tubing.

A typical capillary tube refrigerant control is shown in **Figure 5-48.** Most domestic refrigerators use this type of control.

The capillary tube refrigerant control has no moving parts. Therefore, it has several advantages. First, there are no parts to wear or stick. Second, the pressures balance in the system when the unit stops. This condition places a minimum starting load on the motor.

The capillary tube may be coiled for part of its length. It is usually attached to the suction line. This permits suction line vapor to cool the liquid refrigerant in the capillary tube.

Recent designs of capillary tubes use a larger diameter and a longer tube. The larger diameter tube is less likely to become plugged with dirt, ice, or wax. These larger diameter capillary tubes are normally used on air conditioning applications.

Capillary Tube Capacities

Popular capillary tube sizes are shown in **Figure 5-49.** Tubing .114″ OD by .049″ ID is often used for refrigerant R-12 in domestic units. This is suitable for

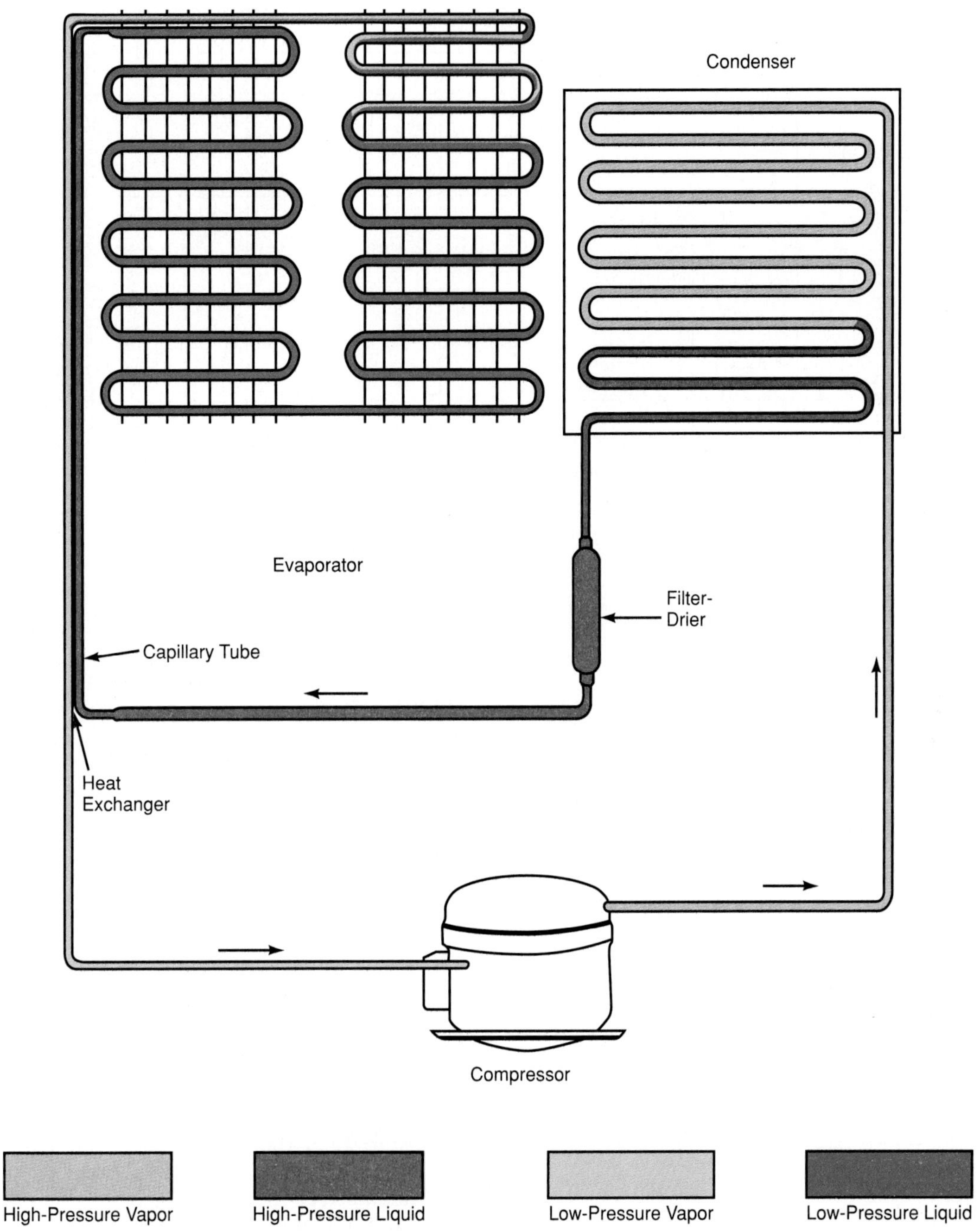

Figure 5-48. *A capillary tube refrigerant control. Note heat exchanger, which cools capillary tube by transferring heat from capillary tube to suction line. (Frigidaire Company)*

Capillary Tube Diameters	
Outside Diameter	Inside Diameter
.083	.031
.094	.036
.109	.042
.114	.049
.120	.055
.130	.065

Figure 5-49. *Some common capillary tube diameters.*

average temperatures and frozen food temperatures. Approximate sizes for capillary tube installations for R-12 are shown in **Figure 5-50.**

Capillary Tube Fittings

The capillary tube connections are often brazed at both the condenser and evaporator end. In other applications, the capillary tube is attached to the evaporator and condenser or drier. Fittings must be leakproof and able to withstand vibration.

Figure 5-51 illustrates ways to make these connections. Detail 1 shows the use of a special nut that squeezes against both the capillary tube and the fitting.

Horsepower	Temp.	Required Length of Capillary Tubing in Feet						
		.031 ID	.036 ID	.040 ID	.042 ID	.049 ID	.055 ID	.065 ID
1/8	H	1.1	2.2	3.5	4.5	9	15	
	M	4	8	13	16	32	56	
	L	9	18	29	36	72	126	
1/5	H						10	
	M	2.2	4.4	7.0	9	18	31	
	L	5.2	10.5	17	21	42	73	
1/4	H						5	
	M	1.1	2.2	3.5	4.5	9	15	
	L							7.5
1/3	M						9.5	
	L	1.75	3.5	5.6	7	14	2.5	

Figure 5-50. *Capillary tube sizes for use with R-12 refrigerant. Temperatures are shown as: high, H; medium, M; or low, L. Required length of tube in feet is indicated for each tube inside diameter (ID).*

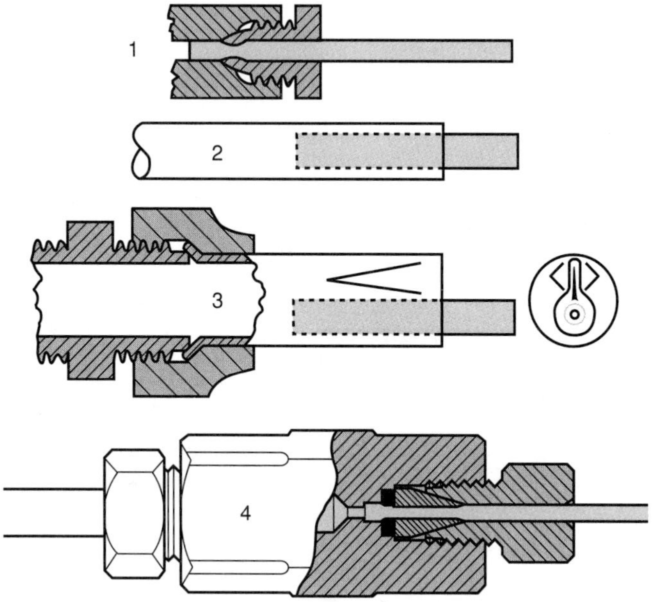

Figure 5-51. *Some typical capillary tube connections. A standard flared fitting connected to a special capillary tube fitting is shown in View 4. (Watsco Components, Inc.)*

The nose section is deformed as the nut is tightened. The nut should always be replaced when the capillary tube is serviced.

Detail 2 shows the capillary tube brazed to a 1/4″ OD soft copper tubing. The larger tube may then be connected to the system by the usual flared fitting.

Detail 3 shows a larger tube brazed to the capillary tube. The larger tube connection is then made with a flared fitting. Mechanical connections are often substituted during overhaul using special capillary tube fittings.

Detail 4 shows a standard flared fitting connected to a special capillary tube fitting. The diameter of the capillary tube determines the size of the fitting.

5.2 Comparing Refrigerant Controls

The pressure-time diagram for one operating cycle for any refrigerant control will show its operating characteristics for one running cycle. The pressure-time diagram will vary with the refrigerant used. **Figure 5-52** shows the pressure-time characteristics for six different refrigerant controls using refrigerant R-12.

5.3 Check Valve

Check valves allow fluid to flow through them in only one direction. Rotary and gear compressors have check valves in the suction line. This prevents the high-pressure vapor and refrigerant oil from backing up into the evaporator during the Off cycle.

A typical check valve is shown in Chapter 4, **Figure 4-78.** A check valve may use either a disk or a solid ball in its construction. Some use a spring or magnet to keep the valve against the seat. Others are mounted so that the weight of the valve keeps it against its seat.

Multiple systems have evaporators that operate at different temperatures. They use check valves. These valves keep refrigerant vapors in warmer evaporators from backing up into the colder evaporators. This is explained in Chapter 13.

5.4 Suction Pressure Valves

Many systems use bellows or diaphragm-operated suction pressure-regulating valves on the low side of the system. These valves are required on multiple systems in which the evaporators operate at different temperatures.

Some of these valves are used to keep the compressor suction pressure at a safe level. This prevents

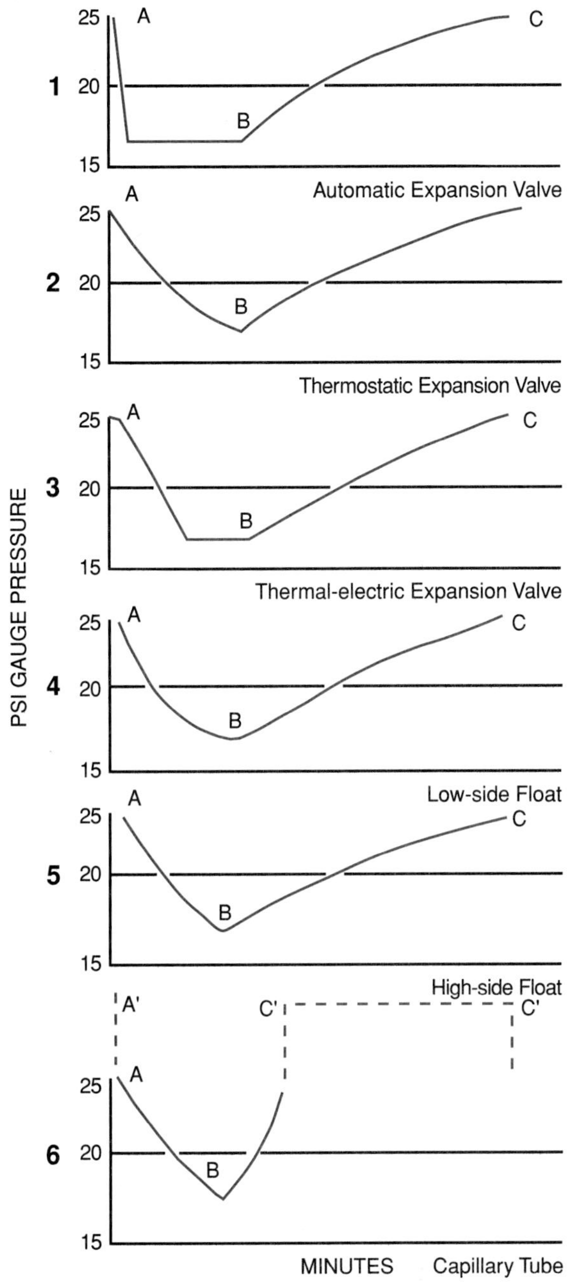

Figure 5-52. *A comparison of low-side pressure-time cycles for various refrigerant controls using R-12 refrigerant. 1—Automatic expansion valve. 2—Thermostatic expansion valve. 3—Thermal-electric expansion valve. 4—Low-side float. 5—High-side float. 6—Capillary tube. Note that pressure goes up until it balances with high-side pressure. Point A—Cut-in pressure. Point B—Compressor stops. Point C—Cycle repeats.*

overloading the compressor. A modified suction pressure valve is required on most automobile air conditioning systems. On these systems, the compressor operates at various speeds. The valve is needed to maintain a fairly constant temperature in the evaporator.

Suction pressure valves may be controlled by the evaporator pressure. These valves control the evaporator temperature. Some operate from crankcase pressure, working to keep the compressor from being overloaded.

As a crankcase pressure regulator, the valve modulates (adjusts) suction pressure to the compressor intake. This provides overload protection for the condensing unit motor. **Figure 5-53** illustrates a low-side pressure regulator installation.

Additional information concerning suction pressure valves is given in Chapter 13.

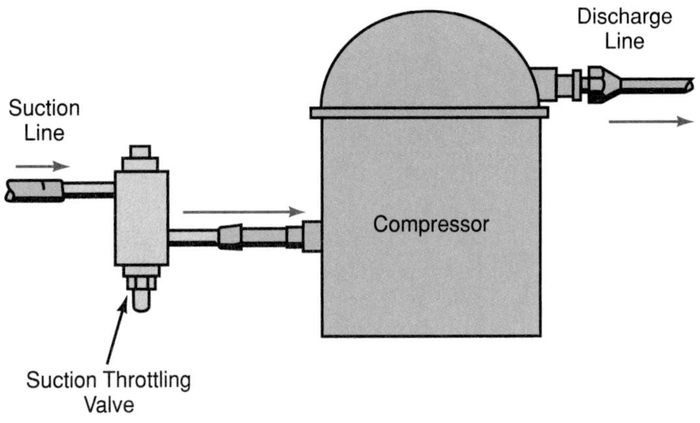

Figure 5-53. *A low-side pressure control valve located in suction line.*

5.5 Review of Safety

The most damaging condition of refrigeration operation is allowing liquid refrigerant to enter the compressor. This is called refrigerant slugging. Liquid refrigerant is not compressible. The piston striking the liquid refrigerant in the cylinder will cause knocking. There is considerable overloading of the piston, connecting rod, bearings, and compressor valves. Most compressors provide some kind of a safety overload device in the valve arrangement. This device attempts to protect the compressor under severe slugging conditions.

A 24-hour pressure time recorder attached to any new installation is an excellent safety device. It ensures that the system is operating within safe pressure limits. A 24-hour temperature recorder may also be used for this purpose.

Keep the floor clear of debris—oil, water, or other slippery material.

Wear safety goggles when working on refrigerating systems.

Avoid lifting objects which weigh over 35 pounds (15.9 kg). Use leg muscles when lifting.

Always have good ventilation and good lighting when working on a refrigerating system.

In order to avoid the possibility of electrical shock, all electrical circuits must be well insulated.

All metal parts of refrigerating mechanisms should be grounded.

When removing a valve from a system, always use two wrenches. This will avoid twisting the valve or the tubing.

5.6 Test Your Knowledge

Please do not write in this text. Place your answers on a separate sheet of paper.

1. How many pressures or forces influence the needle movement of an AEV needle?
   A. AEVs are designed so that they are not affected by forces.
   B. One or two, depending on type of valve.
   C. Three or four, depending on type of valve.
   D. Five or more, depending on type of valve.
2. Why are expansion valves adjustable?
   A. To adjust for different altitudes.
   B. To adjust for different types of refrigerant.
   C. To adjust for different sizes of liquid receiver.
   D. Both A and B.
3. What is thermostatic expansion valve superheat?
   A. If an AEV is replaced by a TEV, the system will run warmer. This difference is called superheat.
   B. Superheat is the difference between the sensing bulb temperature and the temperature of the evaporating refrigerant.
   C. Superheat is the difference between the cabinet temperature and the temperature of the evaporating refrigerant.
   D. None of the above.
4. The following measurements were made on a TEV installation. Which is the best superheat setting for this unit?
   A. 14°F superheat; bulb temperature varies from 11.5°F to 16.5°F.
   B. 12°F superheat; bulb temperature varies from 10.5°F to 13.5°F.
   C. 10°F superheat; bulb temperature varies from 9°F to 11°F.
   D. 8°F superheat; bulb temperature varies from 5°F to 11°F.
5. How is a capillary tube usually fastened to copper tubing?
   A. Flatten tubing around capillary.
   B. Use an epoxy patch kit with sleeve.
   C. Use special capillary tube fittings.
   D. None of the above.
6. Does an AEV system have a dry or a flooded evaporator?
   A. Dry.
   B. Flooded.
   C. Dry during refrigeration cycle; flooded during Off cycle.
   D. Flooded during refrigeration cycle; dry during Off cycle.
7. What is the most common expansion valve body material?
   A. Aluminum alloys.
   B. Brass.
   C. Phosphor bronze.
   D. Stainless steel.
8. What is the operating principle of a sensing bulb that contains carbon dioxide and charcoal?
   A. Absorption.
   B. Adsorption.
   C. Dalton's Law.
   D. Both A and C.
9. Why does the suction line extend down into the pan low-side float?
   A. To aid in the removal of oil that collects there.
   B. To open the needle in case of suction line reverse flow.
   C. To prevent siphoning.
   D. None of the above.
10. High-side floats are usually made of ______.
    A. aluminum alloys
    B. copper
    C. steel
    D. Either B or C.
11. If the bulb temperature of a maximum operating pressure thermostatic charge continues to increase, what happens to the bulb pressure?
    A. Pressure will stop rising and then decrease, due to adsorption.
    B. Pressure will stop rising and then decrease, due to vapor saturation.
    C. Pressure will stop rising rapidly, and then will increase very slowly.
    D. The bulb fills completely with liquid. Pressure then becomes extremely high, and usually bursts the bulb.
12. A capillary tube, .042″ ID and 9′ long, must be replaced. The technician has run out of .042″ capillary and wishes to use .049″ instead. How long should the .049″ capillary be cut?
    A. Shorter than 9′.
    B. 9′.
    C. Longer than 9′.
    D. .049″ capillary cannot be used.
13. Which of these systems probably has a suction line check valve?
    A. Scotch yoke compressor system.
    B. Stationary blade rotary compressor system.
    C. Swash plate compressor system.
    D. None of the above.
14. Which of these design features is usually found on a capillary tube system?
    A. Heat exchanger.
    B. High starting torque motor.
    C. Liquid receiver.
    D. Pressure motor control.

15. _______ causes a liquid cross-charged TEV to open.
    A. Vapor pressure
    B. Gravity
    C. An electromagnet
    D. A bimetal strip bends when heated by a resistor
16. What force causes a thermal-electric expansion valve to open?
    A. Vapor pressure.
    B. Gravity.
    C. An electromagnet.
    D. A bimetal strip bends when heated by a resistor.
17. What force causes a solenoid valve to open?
    A. Vapor pressure.
    B. Gravity.
    C. An electromagnet.
    D. A bimetal strip bends when heated by a resistor.
18. In a TEV system, what can an equalizer tube do?
    A. It can compensate for the pressure drop through the evaporator.
    B. It can hold the valve tightly closed during the Off cycle.
    C. It can supply equal amounts of refrigerant to each section of the evaporator.
    D. It can do both A and B, if a solenoid valve is used.
19. What may be the trouble if a capillary tube unit frosts down the suction tube?
    A. Not enough refrigerant in the system.
    B. Too much refrigerant in the system.
    C. Wrong size capillary tube.
    D. Either A or B.
20. What type of valve is used in a reversible heat pump to switch from cooling to heating?
    A. Suction pressure valve.
    B. Three check valves.
    C. 3-way solenoid valve.
    D. 4-way solenoid valve.

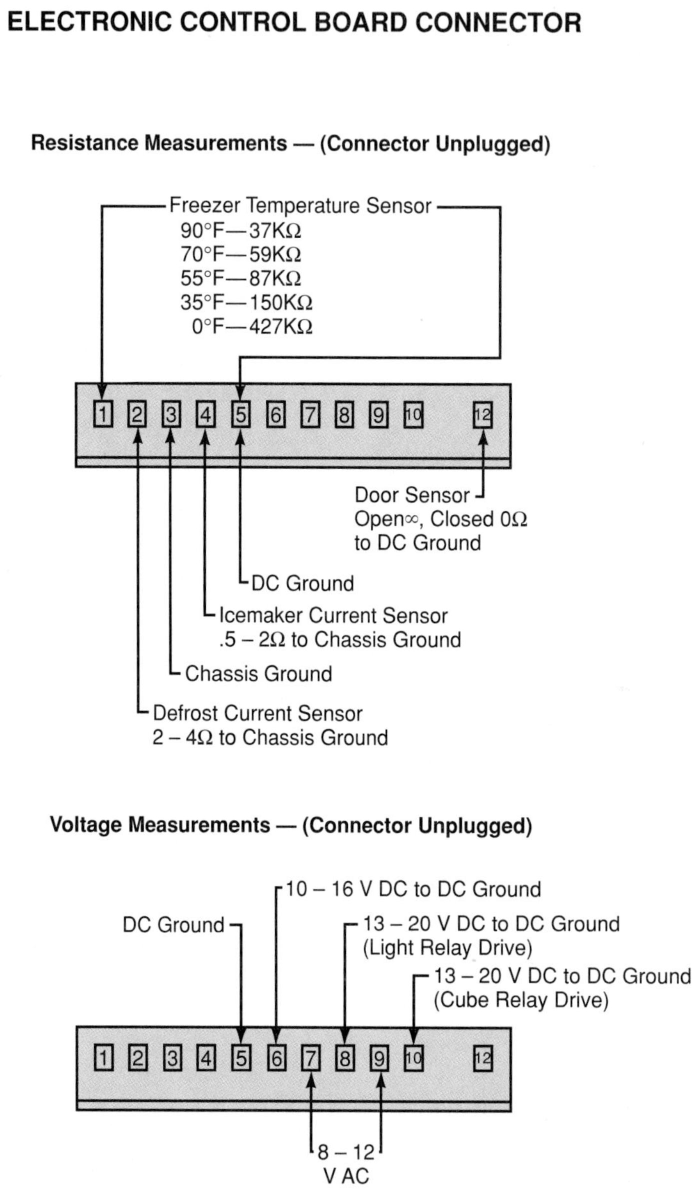

Electronic control board connector. This is part of an electronic control console monitor on a domestic refrigerator. It is used to diagnose various voltage and resistance measurements of electronic systems, such as those for an ice maker or for defrost. (General Electric Co.)

Chapter 6

ELECTRICAL-MAGNETIC FUNDAMENTALS

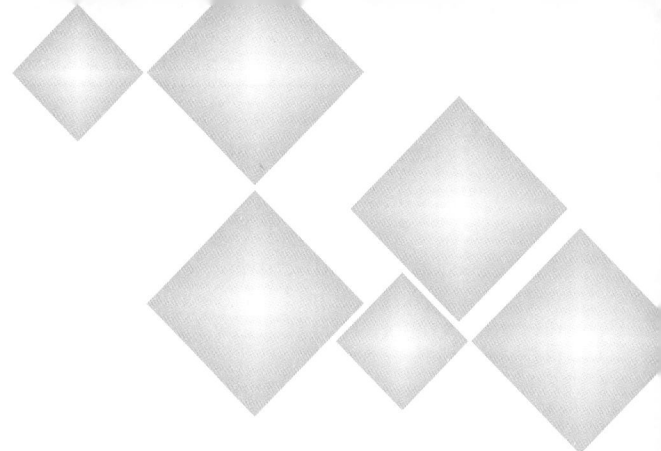

Modules:

Key Words:

alternating current (ac)
ammeter
ampere
direct current (dc)
electricity
electronics
inductance
magnetism
voltage
wattmeter
watts

Learning Objectives:

After studying this chapter, you will be able to:

- Define the terms *electricity* and *electronics.*
- Distinguish between types of electricity—static and current.
- Explain the difference between direct and alternating current.
- Define electrical and electronic terms.
- Describe the difference between parallel circuits and series circuits.
- Discuss the basic theory of electric motors and related devices.
- Use various electrical formulas to solve problems.
- Explain the use of computers in refrigeration controls.
- Explain the components of an electrical circuit.
- Identify and use the proper electrical symbols.
- **Follow approved safety procedures.**

ELECTRICAL FUNDAMENTALS MODULE

There may be some confusion in the understanding of the terms *electricity* and *electronics.* ***Electronics*** is a branch of physics covering the behavior and effects of electron flow in vacuum tubes, gases, and semiconductors. ***Electricity*** is a branch of physics concerning a natural phenomenon. Electricity is known only by its effects: electric charge, electric current, electric field, and electromagnetism. Electricity usually refers to power generation, transmission, and use. Electronics usually involves the use of low power to control high-power circuits.

6.1 Generating Electricity

Electricity used in refrigeration and air conditioning is usually generated by electromechanical equipment (generator). This may be either direct or alternating current.

Electricity also may be generated chemically, such as dry cells used in flashlights. The voltage of a dry cell is approximately 1.5 volts (V). Chemically-generated electricity is always direct current. The chemicals react to generate electricity and are consumed in the process. When the chemicals are gone, the battery is "dead" and the cell is discarded.

The automobile storage battery does not generate electricity, but merely stores it. Electricity from the vehicle's generator flows into the storage battery. This causes a reversible chemical action between the electrolyte (acid solution) and the battery plates (lead or lead oxide). The electrical energy has been stored and can be used later.

Electricity can be caused to flow from other energy forms. These include heat energy, friction, mechanical energy, light, chemistry, and magnetism. Any method that produces a movement of free electrons causes an electrical potential (pressure). Free electrons will flow if a conductor (wire) is present.

6.2 Types of Electricity

Refrigeration and air conditioning is concerned with two common types of electricity. These are static electricity and current electricity. ***Static electricity*** is often defined as electricity at rest. ***Current electricity*** is electricity flowing through conductors (wires).

Lightning is created through the discharge of static electricity. Under certain conditions, materials such as paper and clothing may become charged with static electricity. That is why they sometimes cling together. Static electricity is often produced by friction.

Current electricity is commonly used in homes and in industries. It is used to drive motors, weld, and start engines. It is usually produced by coiled wires moving in an electromagnetic field.

Electricity is usually supplied to homes and industries by electric utility companies. The supply from the utility company to the home ends in a control panel. This panel contains fuses or circuit breakers for each of the circuits.

The electricity supplied to most homes is at 120 or 240 V alternating current. Most lighting and appliances use 120 V. Electric stoves, water heaters, air conditioners, and larger size refrigerating systems use 240 V.

Service technicians should understand the fundamentals of electricity. This will help them service the electrical system of a refrigerator or air conditioner. See **Figure 6-1.**

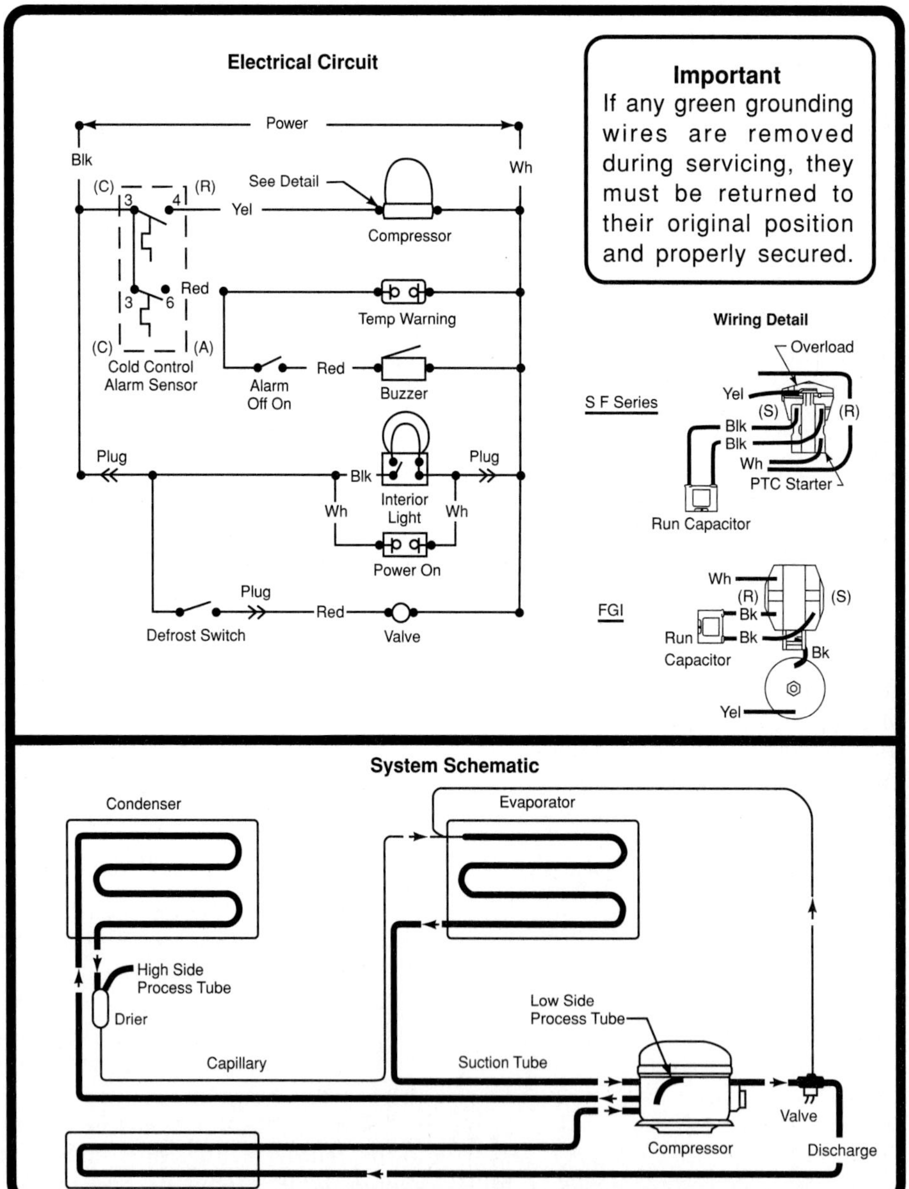

Figure 6-1. *Typical electrical circuit diagram and system schematic for a freezer. Note the temperature warning system. (Frigidaire Company)*

6.2.1 Electrostatic Electricity

There are two kinds of electrostatic charges—positive and negative. Objects with the same kind of charge repel (push apart) each other. Two substances with different charges tend to attract one another.

Two devices used to indicate electrostatic charge are the electrometer and the electroscope. A simple electroscope consists of two small aluminum foil strips attached to a metal rod. The rod is extended through the top of a stoppered glass bottle. When this rod contacts a body having an electrostatic charge, the two strips of aluminum move away from each other. They have both become charged with the same kind of electrical charges.

Placing like poles of two magnets close to each other demonstrates the same reaction. The magnets act as though a strong invisible spring had come between them, preventing contact.

A common example of electrostatic generation occurs when a person walks across a carpet. This charges the entire body with static electricity. In touching a filing cabinet, faucet, or doorknob, this charge will be quickly dissipated (spent). There may be a visible and easily heard spark discharge.

Silk rubbed over glass produces a positive charge of static electricity on the glass surface. Plastic rubbed with wool or fur will generate a negative charge.

The refrigeration service technician is concerned with static electricity in two applications. It applies to condensers used with capacitor type electric motors. The term ***capacitor*** means practically the same as condenser. Capacitance, measured in farads, is the unit of measurement that describes the capacity of a capacitor for storing electrostatic charge. See Section 6.5.7. Electrostatic electricity is also used in electrostatic air cleaners.

6.2.2 Current Electricity

Current electricity is the movement of electrons along an electrical conductor. For example, pushing the button on a doorbell closes (completes) the circuit, causing electrons (current) to flow through the circuit. This electrical flow rings the bell.

There are two common types of electric current:

- Direct current (dc).
- Alternating current (ac).

Direct current is the continuous flow of electrons in the same direction. It is the type of current used in automobile starting, lighting, and ignition. Direct current is used in most solid-state circuits.

Direct current is used on cordless electric appliances such as toothbrushes, electric shavers, and drills. It is also used extensively in electronics. Direct current is needed for battery charging since batteries produce only direct current.

Alternating current is the flow of electrons along a conductor in one direction, then in another. It is the type of current used for most power and light applications.

Direct Current

Direct current (dc) is electron flow along a conductor in one direction. It is the type of current produced by batteries. Most vehicles operate on a 12 V direct current circuit supplied by the storage battery.

Some power companies still produce and sell direct current. However, its use for power and lighting is fast disappearing. At present, its chief uses are in electronics, elevator service, electric welding, and automobiles.

Generally, in both elevator operation and electric welding, direct current is generated at the site. It is made using an alternating current rectifier, driving a dc generator with an ac motor, or driving a dc generator with a gasoline or diesel engine.

Lightweight storage cells and rectifiers have been developed. Many small appliances are using cordless dc power. Motors in some appliances operate on either ac or dc. These units are called universal motors.

Alternating Current

Alternating current (ac) is electron flow along a conductor, first in one direction, then in the other. Because the current alternates its direction of flow, it is called alternating current.

A cycle diagram for an alternating current is shown in **Figure 6-2.** In this illustration, the voltage starts at zero. It reaches its maximum (V_{max}) at a quarter of the cycle. At half cycle, it is again at zero. The voltage reaches its maximum in the opposite direction at three-quarters of the cycle. It is again zero at the end. Positive voltage flows in one direction, negative voltage in the other.

This diagram pictures the ideal alternating current voltage cycle. Most alternating current is 60 Hz. Therefore, this pattern is repeated in the alternating current circuit 60 times per second. The peak voltage (V_{max}) depends upon the voltage supplied by the power source.

Due to this alternating nature of a typical ac circuit, most references to voltage, current, and power refer to the root mean square (rms) values. The rms value is an average that is equal to the maximum value times a constant. For standard ac this constant is .707.

Typical household voltage:

$$V_{rms} = V_{max} \times \text{Constant}$$
$$120\text{ V} = 170\text{ V} \times .707$$

Many ac meters measure ac voltage, current, and power, and indicate the root mean square (rms) value.

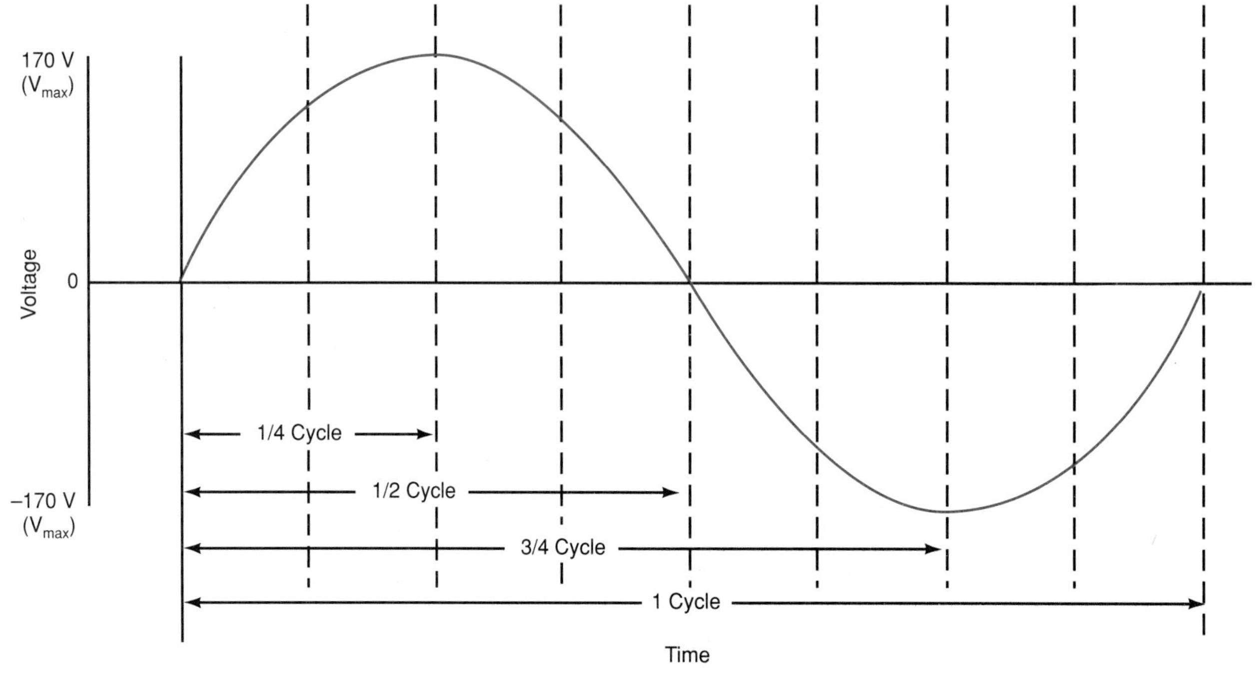

Figure 6-2. *Red line represents one complete cycle of alternating current flow in a typical household (120 V) circuit. Most ac is at a rate of 60 cycles per second (60 Hz). Note that voltage varies from 0 to V_{max} twice each cycle.*

Chapter 31 provides additional information on root mean square values.

Hertz (Hz) is the accepted unit of measurement for electrical cycles per second. Household electricity in most countries is 60 Hz. Some hand tools operate on 180 Hz current. These high-cycle tools do not draw heavy currents and do not overheat if stalled.

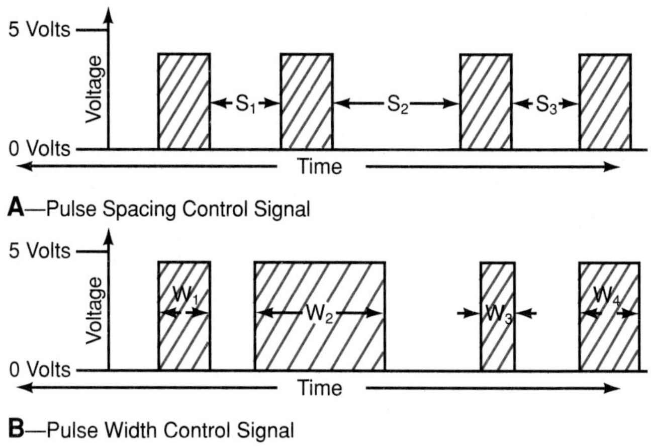

Figure 6-3. *Digital control signals used in pulse wave applications. A—Control using variable spacing between pulses, (S). B—Control using variable pulse width, (W).*

Pulse Wave and Digital Control Signals

In computer or digital control applications, a second type of alternating current is used—pulse wave electronics. The signals in these applications are electrical pulses, as shown in **Figure 6-3.** The control is obtained by the spacing of the pulses and the width of the pulses. Most control systems using computers have 5-volt pulses. If they are used in motor control, the voltage is amplified to the voltage required by the motor.

6.3 Circuit Fundamentals

In order to understand an electrical circuit's operation, compare it to a water system. Just like water in pipes, wires must be large enough to carry the current (amount).

There is always a pressure loss (volts) in a wire when electricity flows along it. This action is due to resistance in the wire. To relate this to water pressure, see **Figure 6-4.** The upper portion shows a water pipe with the inlet valve open and the outlet valve closed. Pressure produced by the pumps is the same throughout when the water is not flowing. When the outlet valve is opened, water flows and the pressure drops.

Notice in the lower portion that there is a small pressure drop for increased distance from the pump. This pressure drop indicates energy lost pushing the water that distance. The pressure drop between gauges C and D is greatest because additional effort is needed to push water around the three sharp bends.

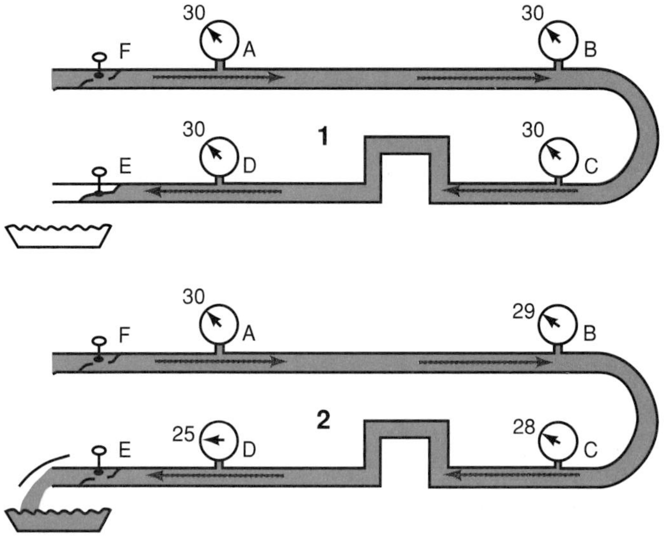

Figure 6-4. *Water flow. 1—Valve E closed, valve F open. No water flow, all gauges read the same. 2—Valve E open, valve F open, water flowing. All gauges show a pressure drop. Most of drop is between C and D because of bend in pipe.*

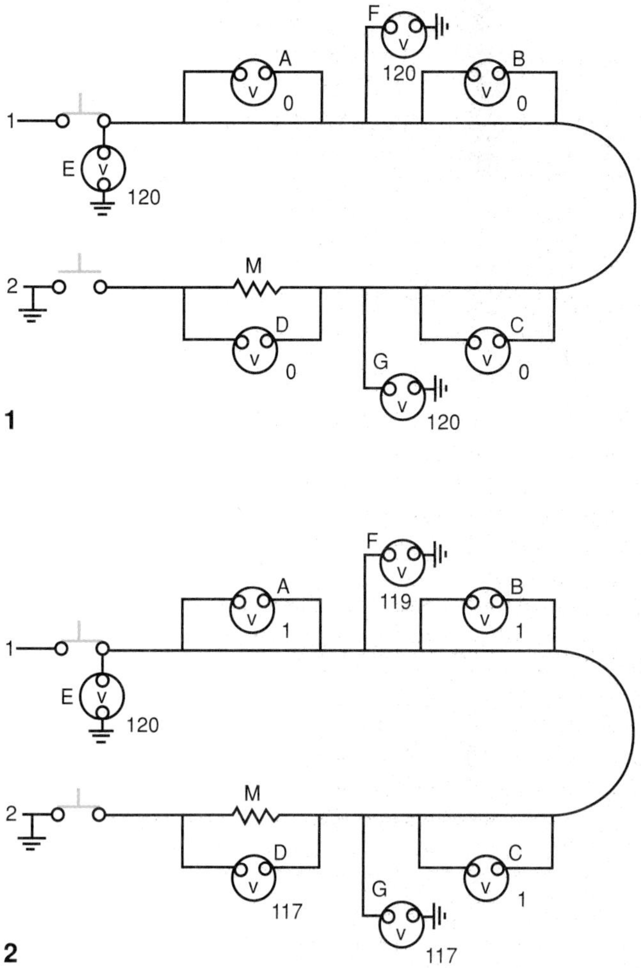

Figure 6-5. *Voltmeter tests of electricity flow show how potential (pressure), resistance, and potential drop operate in a system. 1—Electricity not flowing, with switch No. 1 closed, switch No. 2 open. Pressure to ground is 120 V, but other pressure drops are zero. 2—Electricity is flowing, with switch No. 1 and No. 2 closed. Voltage drop at A, B, and C is 1 V each, and resistance at D has a 117 V drop.*

In **Figure 6-5** a similar electrical system setup is shown (drawing is in symbols). In both views, a wire extends from switch No. 1 to the resistance, lightbulb, or motor at M. In the upper portion of the figure, switch No. 1 is closed, but switch No. 2 is open. The potential (voltage) is 120 V up to switch No. 2. However, no electricity is flowing (open circuit). Therefore, there is no voltage drop along the line. Voltmeters A, B, C, and D show no voltage (no pressure difference) between their lead connections to the main line. However, the voltmeters at E, F, and G show the full 120 V, because there *is* an electric pressure difference between the line and the ground.

The lower part of **Figure 6-5** shows switch No. 2 closed. Current is flowing through the circuit (closed circuit). Note that now there is a small voltage drop at A. This drop takes place in any line in which current is flowing. If the line is large enough to carry the current flow, voltage drop will be very small (.001 to .0001 V). If the line is too small, voltage drop will be greater. Usually, an undersize wire will become warmer than usual while current is flowing.

Note in **Figure 6-5** that the voltage difference between the line and the ground is less at points F and G in the second part. This is caused by voltage drop in the line up to these measuring points. Also note that at D the voltage drop is 117 V. This is the remaining pressure difference between the hot wire and ground.

If voltage drop were greater than shown at A, B, and C, the voltage at D would be less. This is one cause of motor trouble because voltage drop reduces motor voltage. A motor designed to operate at 120 V will lose speed if there is a large amount of voltage drop in the circuit. The rotor will start slipping relative to the magnetic field in the stator (stationary windings). The rotor will slow down too far below its synchronous speed. This causes the magnetic fields to grow large at the wrong time. The motor will heat up and it may even burn up.

6.3.1 Circuits and Circuit Symbols

An ***electrical circuit*** is a complete path (or paths) for the electrons to follow. (See Chapter 31.) It might consist of a battery, switch, lamp, and conductors (wires). See **Figure 6-6.** When a conductor is used to connect the battery to the lamp, the lamp to the switch, and the switch to the battery, an electrical circuit is made. If the switch is closed, the electron path is complete and the lamp will light.

An ***open circuit*** means that an electrical switch or other device is open or disconnected. Current cannot flow. A ***closed circuit*** means that an electrical switch or

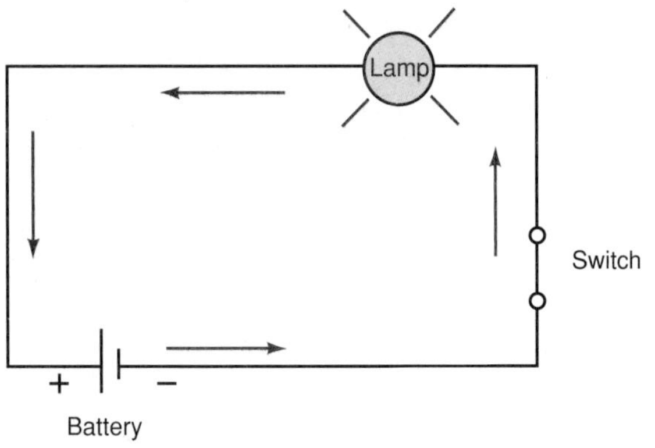

Figure 6-6. *Simple electrical circuit using a one-cell battery, three conductors, a switch, and a lamp. Arrows show direction of electron (current) travel, from negative (−) to positive (+).*

other device is closed or connected. The electrons are allowed to leave from the original source and return to it. This creates a closed path for electron flow. A closed circuit may also be called a continuous circuit.

Figure 6-7 illustrates three common circuit troubles. No. 1 shows an open (disconnected) circuit. It has an open switch or a broken conductor.

No. 2 shows a short circuit. The electrons have taken a shortcut back to their source. For example: A small conductor is placed across the terminals of a lamp. Most of the electrons will flow along the path of least resistance (the new conductor). The light will go out. Because of low resistance, too many electrons may flow, and the wires may overheat. Amperage flow will greatly increase (because of a decrease in resistance). Trouble usually results. Another more common example of a "short" is the touching of two adjacent conductors.

No. 3 shows a ground condition. This can occur when a conductor touches the metal structure of a device. For example, a bare conductor may touch the metal frame of the lamp. The metal frame becomes "hot."

Electrical wiring diagrams use symbols for many of the electrical parts. **Figure 6-8** illustrates standard electrical symbols.

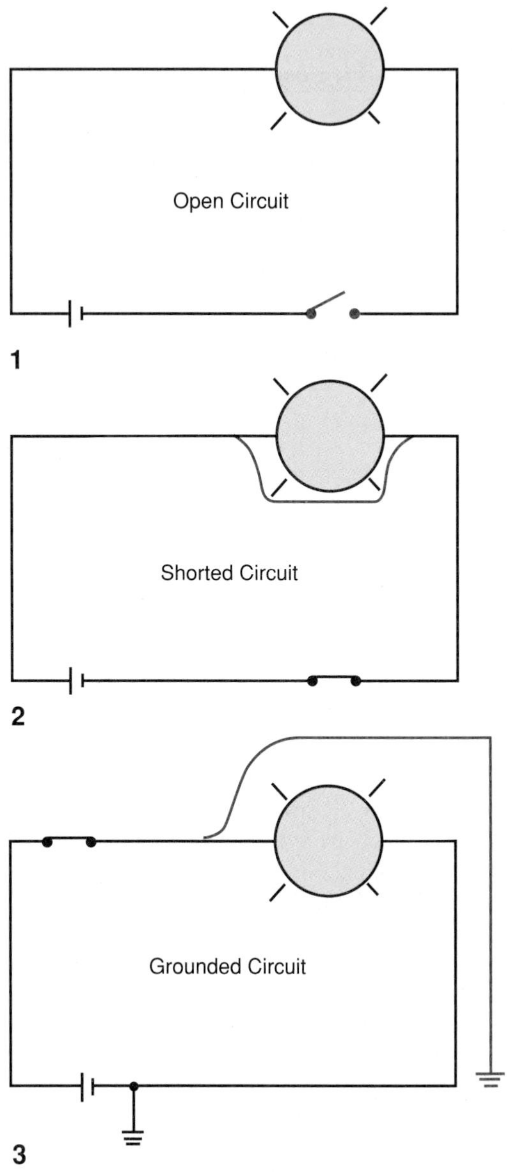

Figure 6-7. *Three common electrical circuit troubles. 1—Open circuit (open switch shown here or a broken conductor). 2—Shorted circuit. 3—Grounded circuit.*

6.3.2 Electromotive Force (EMF)—Volt

Electromotive force, or emf, is used to indicate electrical pressure or voltage that causes current to flow. One volt is the electromotive force required to send one ampere through a resistance of one ohm. The abbreviation for volt (emf) is E.

A ***volt*** is the unit of electrical pressure. It is similar to pressure used to make gases or liquids flow. Water pressure causes the water to flow through pipes, hoses, nozzles, and the like. Increasing the water pressure increases the water flow, **Figure 6-9.**

In an electrical circuit, increasing the voltage increases the current flow. **Figure 6-10** shows the effect of increasing the electrical pressure—voltage (emf)—on an electrical circuit. Note the increased light coming from the lamp in the circuit with a two-cell battery.

Voltmeter

Instruments called voltmeters have been developed to measure the electromotive force. Voltage in a direct current circuit is measured with the use of a direct current voltmeter. An alternating current voltmeter must be used to measure the voltage in an alternating current circuit (such as a home lighting circuit).

The force (electromotive force) needed to move one ampere through a one-ohm resistance is called a volt. One kilovolt (kV) is 1000 volts. A millivolt (mV) is 1/1000 (.001) of a volt. A microvolt (μV) is 1/1,000,000 (.000 001) volt.

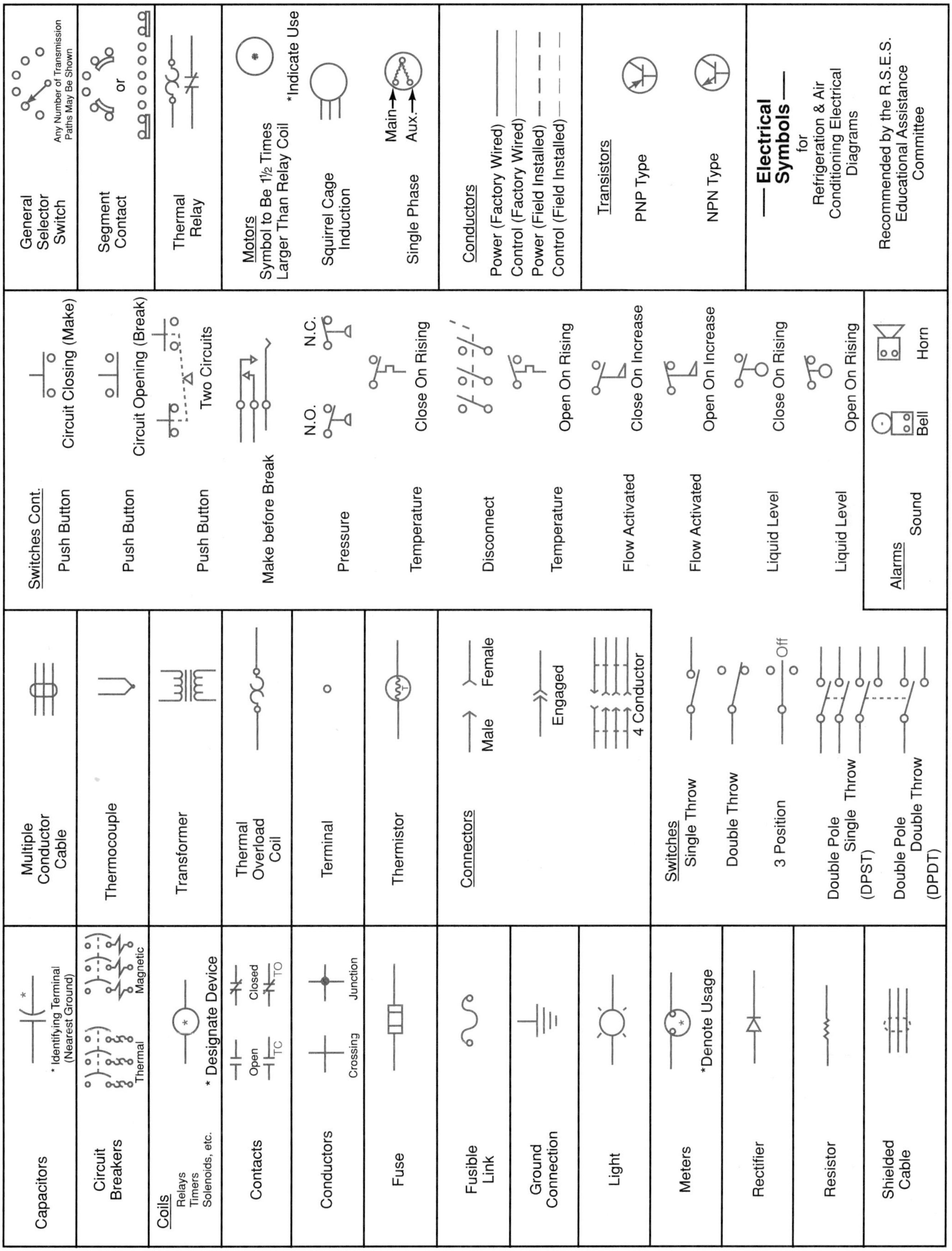

Capacitors
* Identifying Terminal (Nearest Ground)
Circuit Breakers
Thermal
Magnetic
Coils
Relays
Timers
Solenoids, etc.
* Designate Device
Contacts
Open
Closed
TC
TO
Conductors
Crossing
Junction
Fuse
Fusible Link
Ground Connection
Light
Meters
*Denote Usage
Rectifier
Resistor
Shielded Cable
Multiple Conductor Cable
Thermocouple
Transformer
Thermal Overload Coil
Terminal
Thermistor
T
Connectors
Male
Female
Engaged
4 Conductor
Switches
Single Throw
Double Throw
3 Position
Off
Double Pole Single Throw (DPST)
Double Pole Double Throw (DPDT)
Switches Cont.
Push Button
Circuit Closing (Make)
Push Button
Circuit Opening (Break)
Push Button
Two Circuits
Make before Break
Pressure
N.O.
N.C.
Temperature
Close On Rising
Disconnect
Temperature
Open On Rising
Flow Activated
Close On Increase
Flow Activated
Open On Increase
Liquid Level
Close On Rising
Liquid Level
Open On Rising
Alarms
Sound
Bell
Horn
General Selector Switch
Any Number of Transmission Paths May Be Shown
Segment Contact
or
Thermal Relay
Motors
Symbol to Be 1½ Times Larger Than Relay Coil
*Indicate Use
Squirrel Cage Induction
Single Phase
Main
Aux.
Conductors
Power (Factory Wired)
Control (Factory Wired)
Power (Field Installed)
Control (Field Installed)
Transistors
PNP Type
NPN Type
Electrical Symbols
for
Refrigeration & Air Conditioning Electrical Diagrams
Recommended by the R.S.E.S. Educational Assistance Committee

Figure 6-8. *Electrical symbols commonly used in wiring diagrams.*

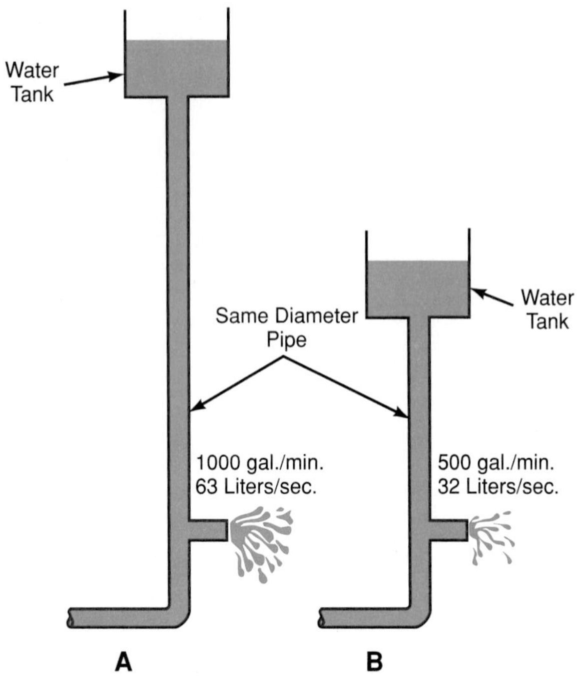

Figure 6-9. *Rate of flow depends upon pressure. A—Higher tank of water provides greater pressure, greater flow. B—Lower tank provides lower pressure, lesser flow.*

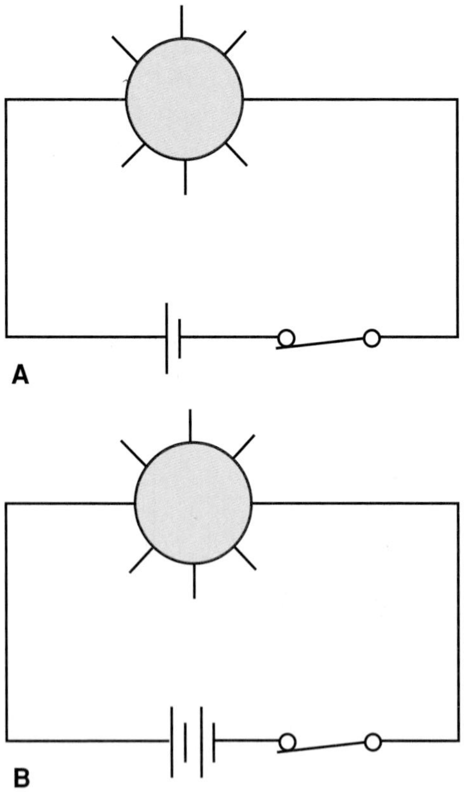

Figure 6-10. *A—Circuit with one-cell battery (1 1/2 V). Bulb lights, but is dim. B—Same circuit with two one-cell batteries (3 V). Light is much brighter (more current flowing).*

There are five types of electromechanical voltmeters for measuring electromotive force. The first four use forces made by current flow and magnetism. The fifth uses electrostatic forces:

- Permanent magnet-moving coil (D'Arsonval), **Figure 6-11.**
- Electrodynamic (dynamotor movement), **Figure 6-12.**
- Moving vane (iron), **Figure 6-13.**
- Moving magnet (polarized iron or iron vane), **Figure 6-14.**
- Moving plate (the electrostatic type), **Figure 6-15.**

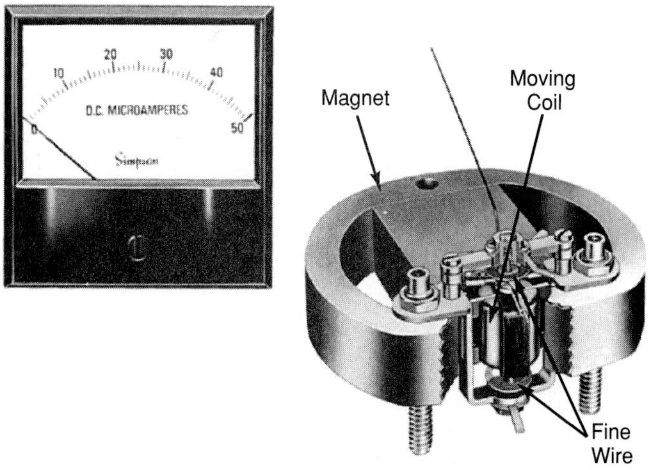

Figure 6-11. *Left—A microampere instrument used for testing dc circuits. Right—The permanent magnet and moving coil movement. (Simpson Electric Company)*

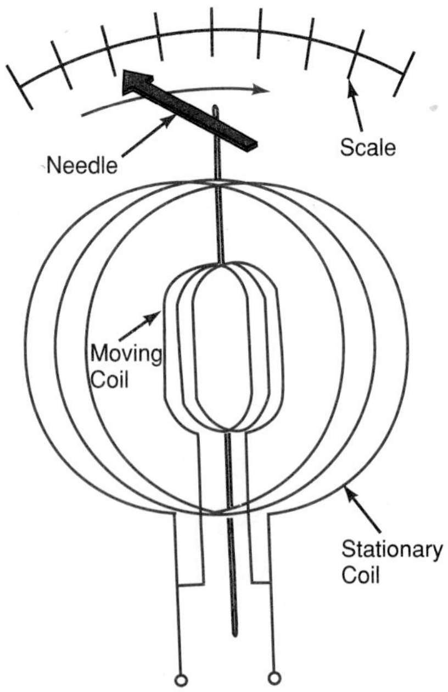

Figure 6-12. *Electrodynamic instrument. Two coils create magnetism, and inner coil movement is related to electron flow. Used with dc, or ac to 200 Hz. Single coil is used for voltage, current, or power. Double coil is used for power, motor phase angle, frequency, and capacitance.*

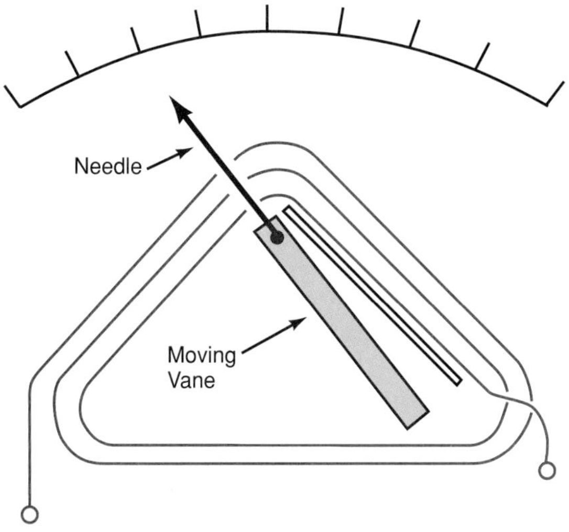

Figure 6-13. *This moving iron or moving vane instrument is used for dc, or ac to 125 Hz.*

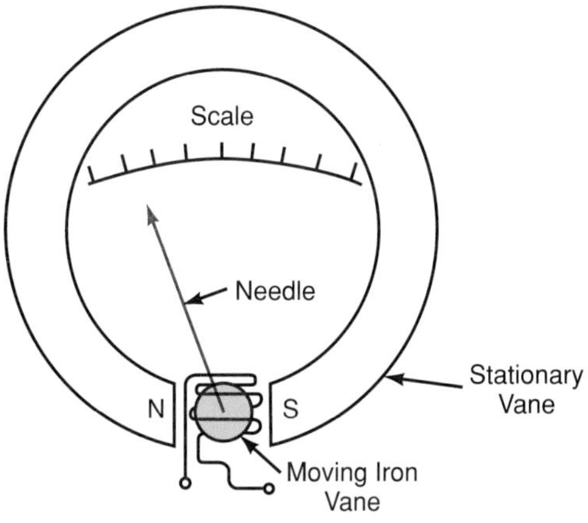

Figure 6-14. *A moving magnet voltmeter is designed for use with dc.*

A sixth type of voltmeter is a digital voltmeter. Most new meters use electronic circuitry instead of electromagnetic effects. See **Figure 6-16.** They have several advantages over the electromagnetic effect devices:

- No moving mechanical parts.
- Easy readability.
- Smaller size.

These devices use solid-state semiconductors. They withstand shock and vibration better than delicate meter movements. They require no lubrication or compensation for the position of the instrument. The display is a number (digit) instead of a pointer, which allows rapid reading.

Some designs offer auto-ranging. The voltmeter automatically senses the voltage level and selects the proper scale. Electrical power requirements for electronic meters are lower than those for electromagnetic meters.

Voltmeters are always connected in parallel with the circuit. The method of connecting them is shown in **Figure 6-17.**

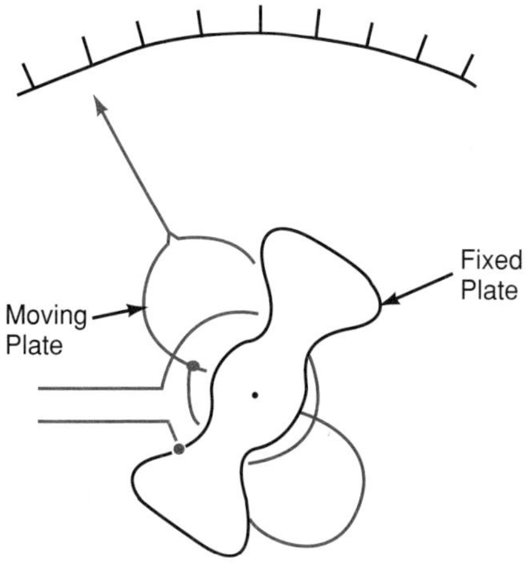

Figure 6-15. *Electrostatic meter. Potential difference exists between the moving plate and the fixed plate. Used for testing voltage in dc or ac circuits over 10 V.*

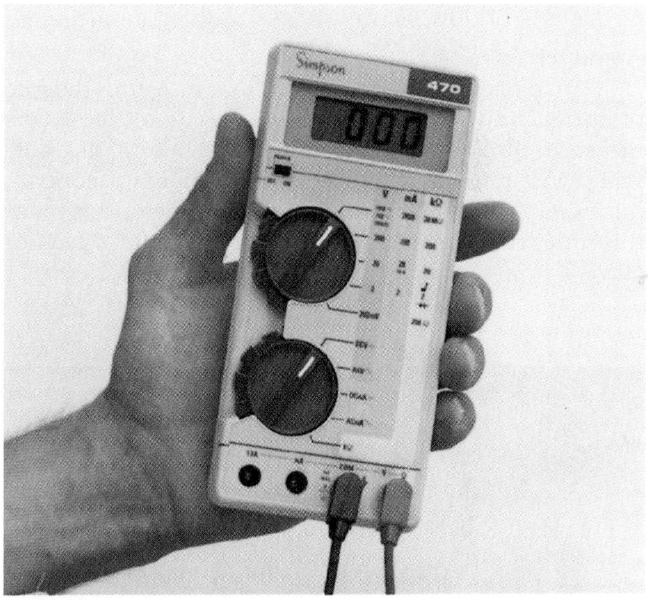

Figure 6-16. *Hand-held digital multimeter. Using electronic circuitry, this device measures current, voltage, and resistance. (Simpson Electric Company)*

6.3.3 Coulombs

A coulomb is a count of the number of electrons passing a given point on a conductor. It is the amount of electricity passing a given point in a conductor in one second, while the rate of current flow is constant at one ampere.

The number of electrons in a coulomb equals 6.24×10^{18}, or 6,240,000,000,000,000,000. A one-coulomb flow per second equals one ampere. This is like the rate of water flow in gallons per minute (gpm).

6.3.4 Ampere

The ***ampere*** measures the rate of flow of current. It does not measure electrons. It has a one-to-one relationship with coulombs. Ten amperes flowing past a point

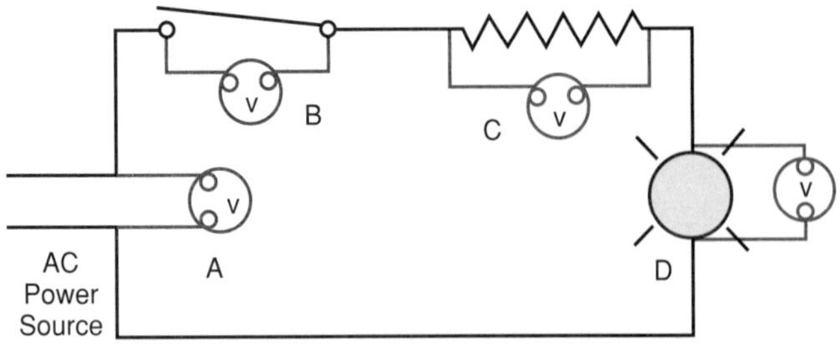

Figure 6-17. *Correct connections for testing with voltmeter. Meter will show: A—115 to 120 V if power is on. B—Whether switch is open or closed (zero reading if closed, 120 V if open). C—Voltage drop across resistor. D—Voltage drop across lamp (zero reading if shorted, 120 V if broken or open). Voltages of B + C + D should equal A when switch is closed.*

in one second equals 10 coulombs. "Current" usually means rate of flow. It is measured with an ammeter.

Ammeter

An ***ammeter*** is an instrument that measures the rate of current flow in amperes. There are two types. One measures direct current flow; the other, alternating current. Most refrigeration and air conditioning work requires an ac ammeter. See **Figure 6-18.** The operation of this ammeter depends upon the magnetic effect of current in a conductor. The magnetism causes current flow at the base of the tongs (clamp) surrounding the conductor.

The tongs can be opened and slipped around a conductor. Electricity is generated in the tongs by the magnetism. It moves the indicator to give the ampere flow reading.

Only one wire of the circuit should be placed in the jaws of the meter. This type of ammeter can only be used on an alternating current circuit. See **Figure 6-19.**

The reading from a "clip on" (tong-type) ammeter may be too low to read accurately. Wrapping the wire once around one of the jaws doubles the reading. (The wire goes between jaws twice.) Divide this reading by two to get the correct reading. Some of these meters have multipliers available as accessories.

The method of connecting the two types of ammeters is shown in **Figure 6-20.** *Caution: Ammeters should always be connected in series with a circuit.* (For series circuit, see Section 6.3.10.) *If an in-line ammeter is accidentally connected in parallel, it will be burned up.*

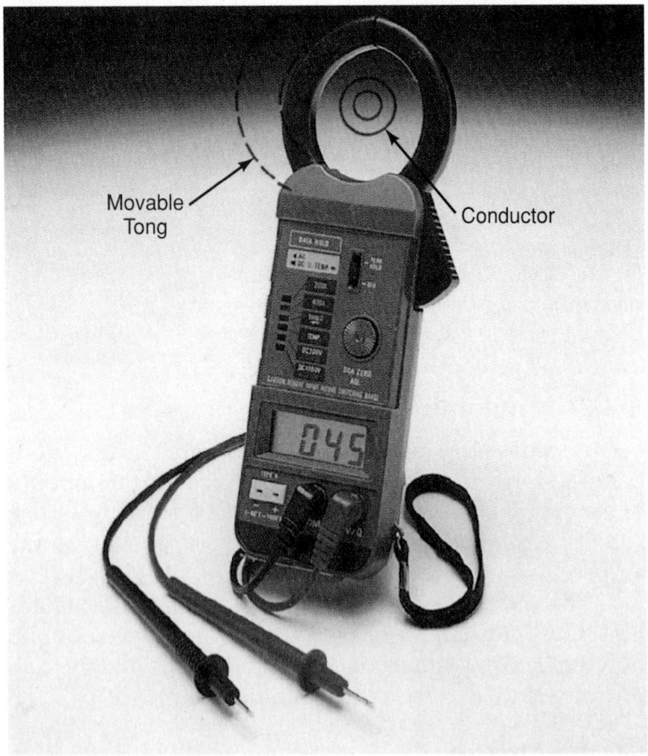

Figure 6-18. *Digital clampmeter with six functions and red and black test leads. Meter measures ac and dc voltage and temperature.*

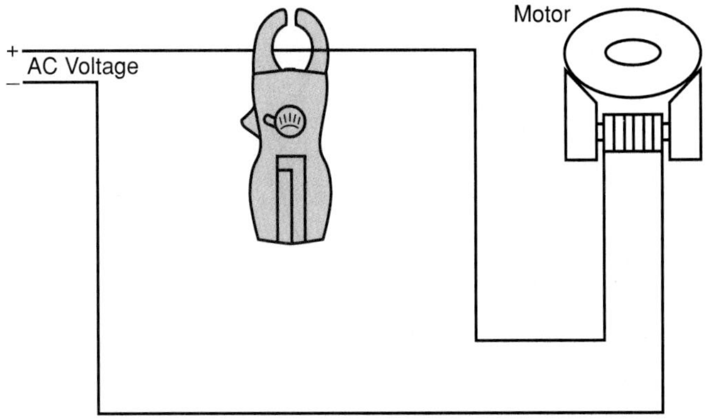

Figure 6-19. *Tong ammeter usage.*

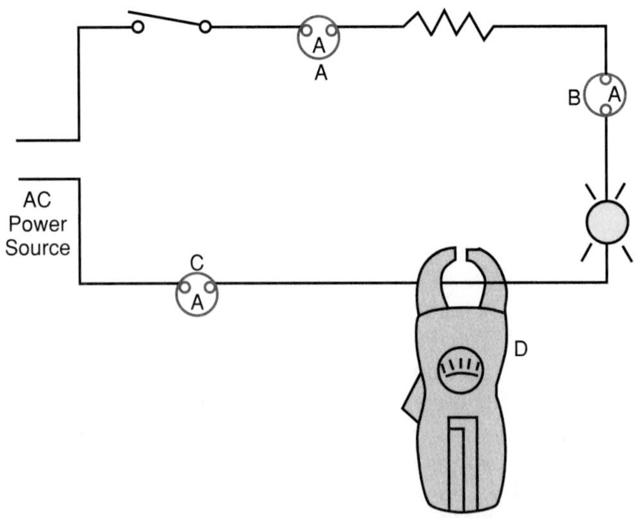

Figure 6-20. *Ammeter connections. Ammeters A, B, C, and D should all read the same: Zero when switch is open, 5 A when switch is closed. D is special induction type ammeter described in* ***Figure 6-18.***

6.3.5 Watts

Power is the time rate of doing work. When amperes flow (coulombs per second) at a certain pressure (emf or volts), this is power.

Electrical power is measured in watts (W) and kilowatts (kW). See Section 6.7. When computing electrical horsepower (hp), 746 W or 0.746 kW equal 1 hp.

In direct current circuits, the wattage may be obtained in the following way. The current (amperes) is multiplied by the emf (volts), $I \times E = W$.

In alternating current circuits, the wattage may be less than the amount obtained by multiplying the amperes and volts together. This is because current in an ac circuit may not be exactly in phase with voltage.

Wattmeter

A ***wattmeter*** is an instrument used to measure the wattage consumed by an electric motor or an electrical device. A wattmeter is connected in series with the circuit being measured. (See Section 6.3.10.) **Figure 6-21** illustrates a portable wattmeter. It shows how the wattmeter is connected to measure wattage drawn by an electric drill. This is a combination instrument and may be used for measuring volts or watts.

A wattmeter indicates the true wattage in a circuit. It automatically adjusts for the power factor (see Section 6.3.6). This is done by two coils in the wattmeter—a voltage coil and a current coil. The influence of the two coils drives the wattmeter. If the voltage and current are out of phase, this influence is reduced.

6.3.6 Power Factor

The ***power factor*** represents that fraction of the total possible power that can be generated in a circuit. See Section 6.7.1 for additional information.

Current usually lags behind the emf, depending on the type of circuit. When the emf is at 50% of its maximum, the current flow may be 35% or 40% of its maximum. Therefore, the wattage or power of an ac circuit must be obtained as follows. Multiply the voltage times the current times the power factor ($E \times I \times PF$ = watts).

Figure 6-22A shows an alternating current cycle. The voltage and the current are in phase. The current and the voltage are at their maximum at the same time. The voltmeter reads 120 V. An ammeter connected into the circuit reads 10 A. The product of the voltage times the current equals 1200 W. A wattmeter connected to this circuit will also read 1200 W. This indicates a power factor of 100%.

Figure 6-22B shows an alternating current cycle out of phase. The current is lagging behind the voltage one-eighth of a cycle. A voltmeter connected to the circuit reads 120 V. An ammeter connected to the circuit reads 10 A. The product of the voltmeter reading multiplied by the ammeter reading will be 1200. However, a wattmeter applied to this circuit will read 1000 W. The power factor is the ratio of the wattmeter reading over the calculated wattage, which is the current multiplied by the voltage.

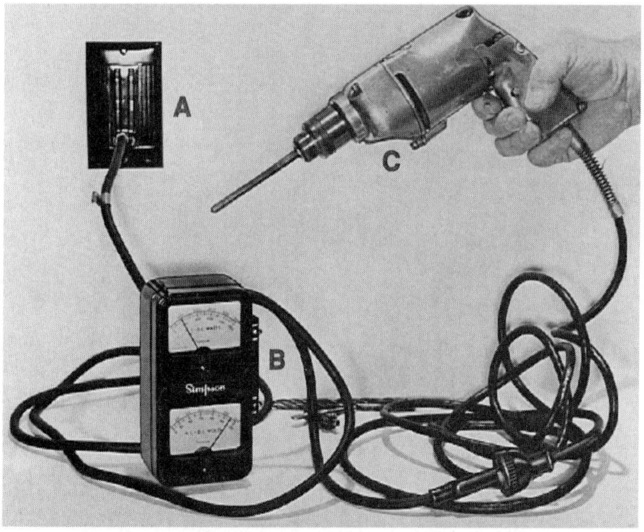

Figure 6-21. *Top. Combination voltmeter and wattmeter. Bottom. Volt-wattmeter being used to test an electric portable drill. A—Power source. B—Volt-wattmeter. C—Electric load. (Simpson Electric Company)*

Formula:

Power Factor = PF

$$= \frac{\text{wattmeter reading}}{(\text{ammeter reading} \times \text{voltmeter reading})}$$

Example:

wattmeter reading = 1000
voltmeter reading = 120
ammeter reading = 10

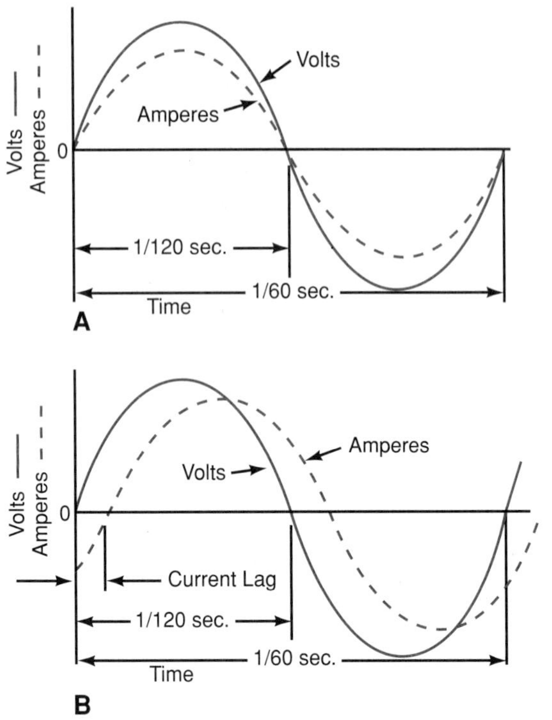

Figure 6-22. *Alternating current cycles. A—Current and voltage in phase. B—Current and voltage not in phase.*

Solution:

$$PF = \frac{1000}{(120 \times 10)} = \frac{1000}{1200} = .833$$

To change to percentage: .833 × 100 = 83.3%. To calculate the wattage, multiply the voltmeter reading by the ammeter reading. The answer must be multiplied by .83 (the power factor), to give the correct wattage:

$$120 \times 10 \times .83 = 1000.$$

Various things in the electrical circuit affect the power factor. If the electrical load is entirely a resistance load, such as electric heating, this is called a ***noninductive load.*** It does not seriously affect the power factor.

Most electric motors, transformer welders, and similar devices affect the power factor. These are called ***inductive loads.*** It takes time for current to build up in any coil. This is due to the counter emf generated in the coil. This lag affects the power factor. See Section 6.5.9, which discusses the effects of counter electromotive force.

Power Factor Meter

A power factor meter is an instrument used to provide a direct reading of the power factor in an electrical circuit. A direct-reading meter saves time.

This meter first measures the total available wattage in a circuit. It finds the product of the measured voltage times the measured current. It then determines the true wattage using the same method described in Section 6.3.5. The ratio of the true wattage to the available wattage is determined and displayed. This value is the power factor. See **Figure 6-23.**

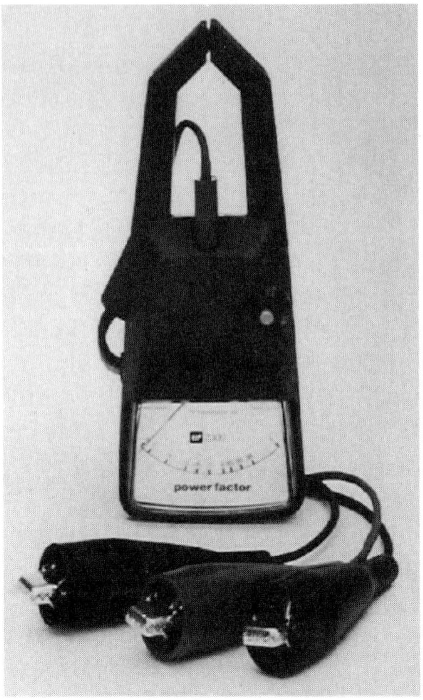

Figure 6-23. *Power factor meters determine the effect of individual loads and systems. (TIF Instruments, Inc.)*

Power factor meters can be used to assist in increasing the effectiveness of a circuit. For example, assume a circuit has a measured power factor of .80 (80% of the total possible wattage is being used). This value is quite common in an inductive (motor, transformer) circuit. The proper addition of a capacitive load would result in an increase in true wattage. This increase would then bring the power factor closer to 1.00.

6.3.7 Resistance—Resistors

Most electrical conductors are made of metal. No material is a perfect conductor of electricity. However, some metals are better conductors than others. Silver, copper, and aluminum are very good. Iron, steel, and carbon will also conduct electricity, but their resistance is quite high. Carbon is sometimes used in electrical circuits, but it is not a very good conductor.

Extremely poor conductors are called ***resistors*** or resistances. They have no free electrons or, at best, very few free electrons in the atom. It is difficult for these free electrons to travel through and around the other atoms. The word ***impedance*** is often used when a complete ac circuit, or part of one, resists the flow of free electrons.

The harder it is for the free electrons to move, the greater the heat generated in the conductor. This heating action is shown best in iron, steel, and steel alloys used for electric heating purposes. The resistance values of conductors, semiconductors, and nonconductors are discussed in Chapter 31.

All materials have some resistance to the flow of electricity. The resistance of electrical conductors usually increases with an increase in temperature. The resistance also increases with an increase in the conductor length. Also, the resistance increases as the diameter (thickness) of the conductor decreases. In the case of some semiconductors, increasing the temperature increases their ability to conduct electricity.

Components designed to offer specific levels of resistance in a circuit are called resistors. See **Figure 6-24.** These resistors usually have a series of color bands. The color bands represent both the amount and accuracy of resistance. See Chapter 31 for a brief summary of these color bands.

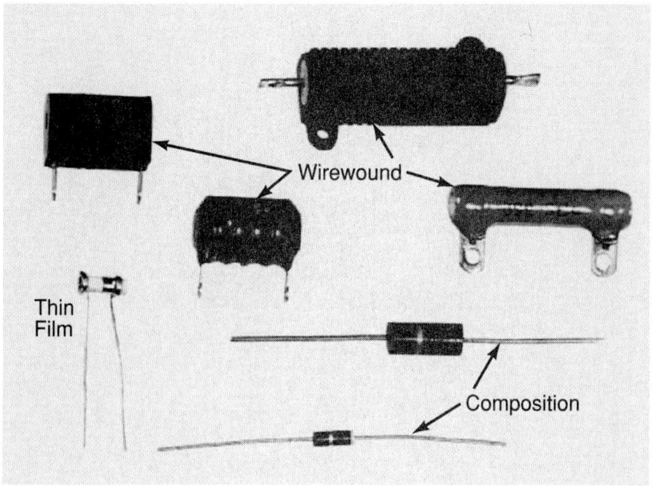

Figure 6-24. *Typical resistors.*

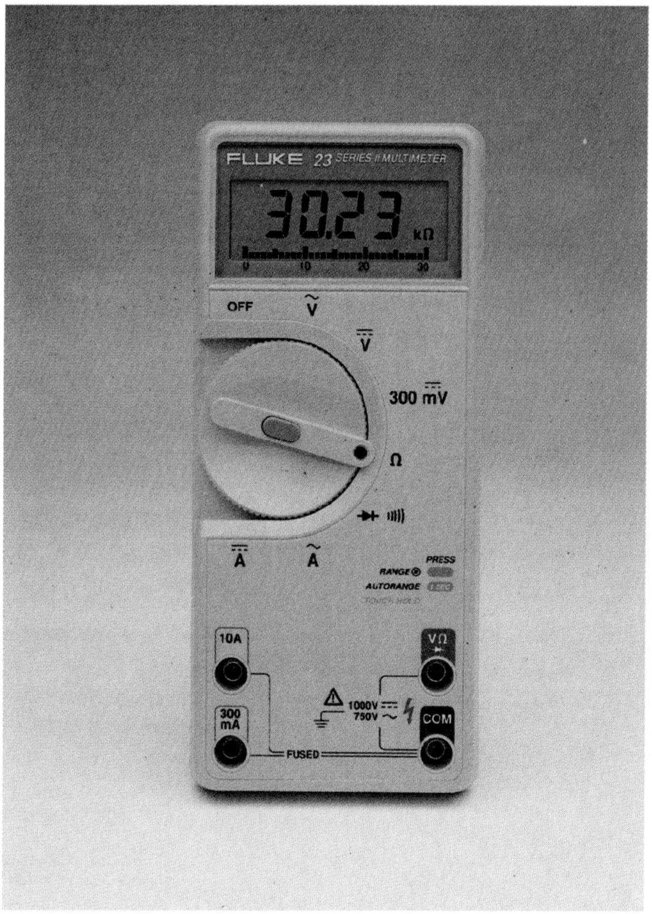

Figure 6-25. *Multimeter that reads voltage (ac and dc) and ohms. (Reproduced with permission of Fluke Corp.)*

6.3.8 Ohms

Electrical resistance is measured in ohms. An ohm is the amount of resistance in an electrical circuit. It allows an emf of 1 volt to cause 1 ampere to flow through the circuit. The symbol for ohm is the Greek letter omega (Ω).

The resistance in a conductor depends upon four things:

- Material used.
- Diameter or size (thickness) of conductor.
- Length of conductor.
- Temperature of conductor.

Ohmmeter

Ohmmeters are used to check circuits for resistance, open circuits, and grounds. **Power must be off when the ohmmeter is being used.** The instrument may be ruined otherwise. A universal meter that measures ohms is shown in **Figure 6-25.** This meter also measures voltage (ac and dc) and a small amount of amperage. **Figure 6-26** shows the correct method of connecting an ohmmeter into a circuit to measure the resistance.

6.3.9 Ohm's Law

The relationship between the volt, the ampere, and the ohm is known as Ohm's Law.

If a conductor 1′ long will allow 1 A to flow with an emf of 1 V, this same conductor will allow 2 A to flow if the emf is 2 V. In addition, if the conductor is 2′ long and the emf is 1 V, only 1/2 A will flow. Therefore, it is evident that emf is the product of the current intensity (amperes) multiplied by the resistance (ohms). In developing a formula, begin with these symbols:

E = Electromotive force (emf) in volts.
I = Intensity of current in amperes.
R = Resistance in ohms.

Emf (Volts) = Intensity (Amperes) × Resistance (Ohms)

$$E = IR$$

$$\text{and therefore } I = \frac{E}{R}$$

$$\text{or } R = \frac{E}{I}$$

Figure 6-27 shows the relationship of I, E, and R.

This basic law indicates that, if the resistance stays constant, the current can only be increased by increasing the emf. Also, if the resistance in a circuit becomes low, the amperage will become high.

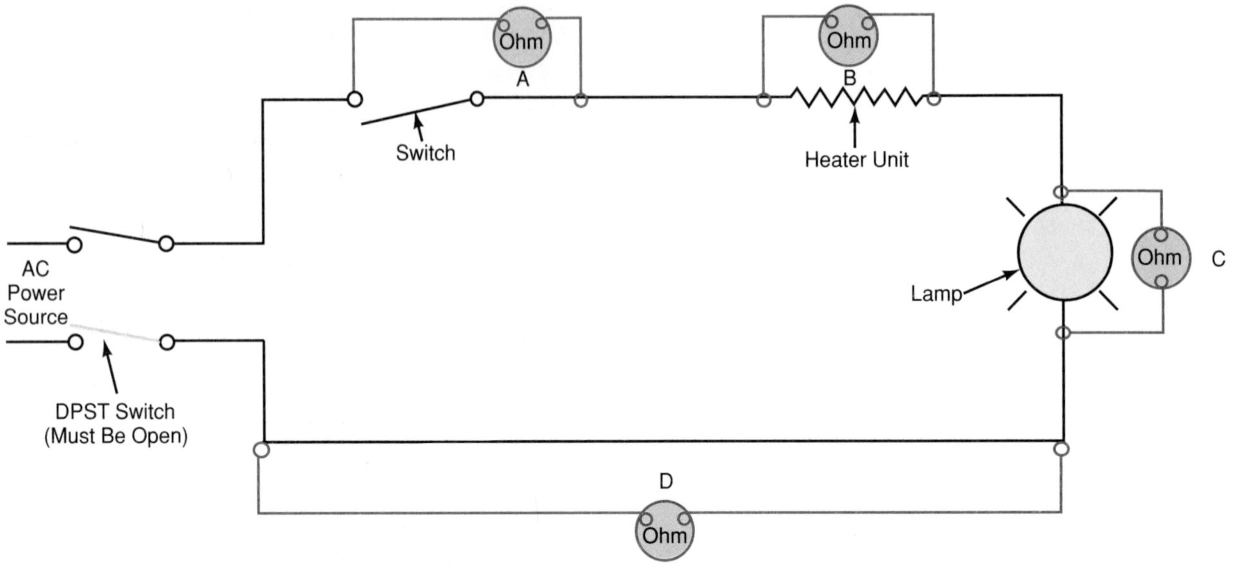

Figure 6-26. *Ohmmeter connections. There must be no power in circuits. A—Testing resistance across a switch in open, then closed position. With switch closed, resistance indicated on meter indicates a poor connection somewhere in switch. B—Testing resistance (ohms) across heater unit. C—Testing resistance across a lamp. D—Measuring continuity of conductor. If meter reads zero, conductor is not broken and connections are good. If meter reads infinity (maximum resistance), conductor or a connection is broken.*

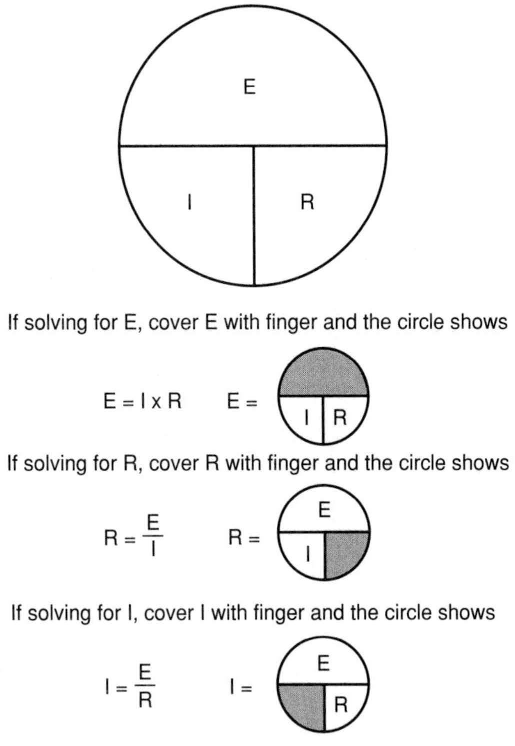

E
I
R
If solving for E, cover E with finger and the circle shows
E = I x R
E =
I R
If solving for R, cover R with finger and the circle shows
R = E/I
R =
E
I
If solving for I, cover I with finger and the circle shows
I = E/R
I =
E
R

Figure 6-27. *Ohm's Law equations shown in graphic form.*

Formula:

$$E = IR \rightarrow R = \frac{E}{I}$$

Example:

A 240 W lamp draws 2 A at 120 V; what is its resistance?

Solution:

$$R = \frac{E}{I}$$

$$R = \frac{120V}{2A}$$

$$R = 60 \text{ ohm } (\Omega)$$

The formula $E = I \times R$ also means that doubling the emf in a circuit (the resistance remaining the same), doubles the current flow. This may cause trouble. An example would be to connect a 120 V motor into a 240 V power outlet. If the fuse does not "blow," the motor windings will become too hot. This is due to their carrying this excessive current. The insulation on these wires may be destroyed and the motor may be ruined.

The resistance is increased when a smaller diameter conductor is used. It is also increased when there is a dirty or loose connection. When this resistance increase occurs, and the emf stays the same, the current will decrease. This causes a loss of power. The wire will heat or the poor connection will become very warm and may even cause a fire.

6.3.10 Series Circuits

A circuit having only a single path for current flow is called a ***series circuit.***

In the series circuit in **Figure 6-28,** each resistance has the same current flowing through it. In a series circuit, all the resistances are added together to determine the total resistance. Here, the total voltage at the power supply equals the sum of the voltages across the three resistances.

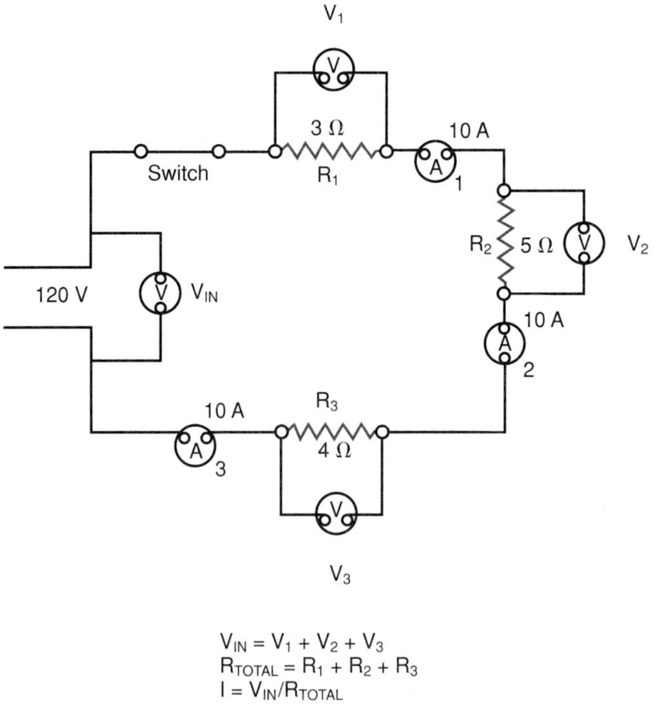

Figure 6-28. *A series circuit with three resistances. R_1 = 3 ohms, R_2 = 5 ohms, and R_3 = 4 ohms. Total resistance is 12 ohms. All three ammeters will read the same, 10 A. V_1 = 30 V, V_2 = 50 V, and V_3 = 40 V. $V_{IN} = V_1 + V_2 + V_3$ = 120 V.*

One object in a series circuit—switch, lightbulb, resistor, etc.—may not allow current flow. In this case, the circuit will not operate.

6.3.11 Circuits, Parallel

A circuit that allows the current to flow along either one of two or more conductors or electrical paths at the same time is called a parallel circuit. See **Figure 6-29.** Here, the sum of the current flowing through the

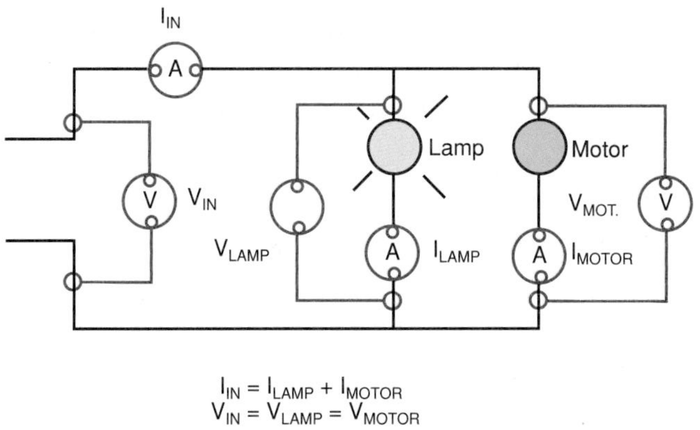

Figure 6-29. *A parallel circuit showing the incoming current and the current draw. I_{IN} represents the incoming current. I_{LAMP} equals the current through the lamp. I_{MOTOR} is the current through the motor. $I_{IN} = I_{LAMP} + I_{MOTOR}$. $V_{IN} = V_{LAMP} = V_{MOTOR}$.*

lamp and motor equals total input current. The current flow is based on the resistance of the conductors. If the lamp has 1/4 of the resistance of the motor, 4/5 of the current will flow through the lamp and 1/5 through the motor. The voltage is the same across each of the parallel paths in the circuit.

6.3.12 Circuits, Series—Parallel

A combination of series and parallel circuits is sometimes used. In some automatic defrost systems, two or more heating elements are connected in parallel. These heating elements in turn connect into the system when the compressor motor is stopped.

A timed switch controls the heating elements. This switch is in series with the heating elements. See **Figure 6-30.**

6.3.13 Voltage Drop (IR)

The sum of the voltage drops in a circuit is always equal to the voltage applied. The voltage drop across any part of a circuit is equal to the current (amperes) multiplied by the resistance (ohms) across that part of the circuit.

Figure 6-31 shows a typical refrigeration circuit. The voltmeter indicates an applied voltage of 120 V. The alternating current ammeter indicates a current draw of 5 A.

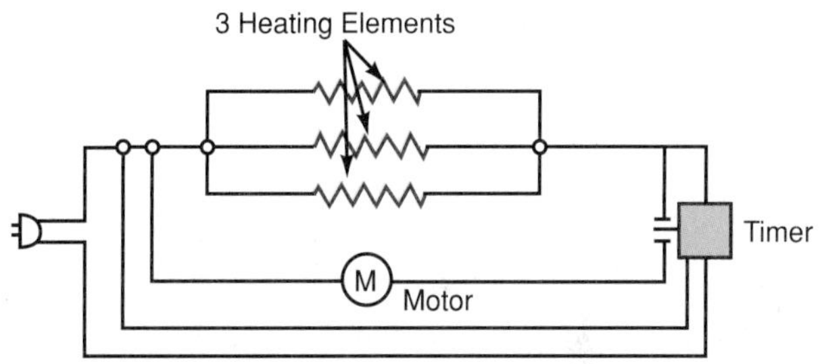

Figure 6-30. *Series parallel circuit. Timer controls three heating elements (resistances). Current through heating element section divides three ways. Some current goes through each of the three heating elements.*

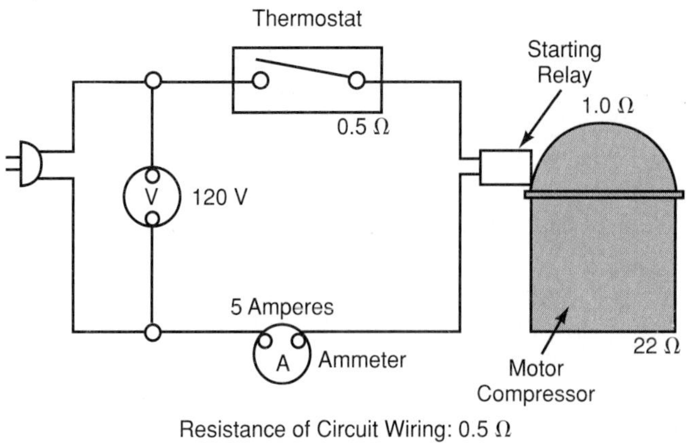

Figure 6-31. *Voltage drop in an electrical circuit is equal to voltage applied.*

To determine the voltage drop, use the formula E = IR.

Equivalent resistance of circuit wiring is 0.5 Ω	5 × .5 =	2.5
Equivalent resistance of thermostat is 0.5 Ω	5 × .5 =	2.5
Equivalent resistance of starting relay is 1.0 Ω	5 × 1 =	5.0
Equivalent resistance of motor compressor is 22.0 Ω	5 × 22 =	110.0
2.5 + 2.5 + 5.0 + 110.0 = 120.0		
Therefore, total voltage drop is		120.0 V

There is always some electrical resistance across any electrical control, relay, or circuit wiring. A very sensitive instrument is needed to measure this low resistance.

6.3.14 Power Loss (I^2R)

Power loss in a circuit due to resistance is equal to the square of the current multiplied by the resistance. This is usually expressed as I^2R loss = $I^2 \times R$ = Power Loss.

Referring again to **Figure 6-31,** the power applied is at 120 V. The total current in the circuit is 5 A; the total resistance in the circuit is 24 Ω.

$5^2 = 25$
25 × 24 = 600 W power loss.

Power applied is equal to voltage times amperage:

5 A × 120 V = 600 W.

Power loss results in the generation of heat. One watt converted into heat equals:

3.4144 Btu/hr. = 860 calories/hr. (1 Btu = 252 calories = 0.252 kg-calories).

6.3.15 Instrument Connecting and Handling

Basic electrical instruments may be used to measure either volts, amperes, or ohms. The difference is in the external wiring of the instrument. The scale and use of instruments vary depending upon the wiring inside of the instrument. Any instrument needs only a very few electrons to activate the pointer mechanism.

The voltmeter has a high internal resistance. The ammeter is designed to bypass (shunt around) most of the current outside the instrument. The ohmmeter allows only a few electrons in the circuit.

The mechanisms for each type of meter vary widely. These instruments must be delicate enough to read accurately. Therefore, they should not be dropped. They should not be used above their maximum reading, or carelessly handled or stored.

The voltmeter is connected in parallel with the circuit. A high value resistor in the instrument coil circuit will allow a maximum of 0.000005 A, or 5 microamperes, to flow into the moving coil through the springs.

Be careful! A 120 V voltmeter used to measure a 240 V circuit will ruin the meter. (Double the milliamperes will flow through the meter.)

A low-resistance shunt is placed across the terminals of an ammeter. The instrument is placed in series with the circuit when current is being measured.

An ohmmeter is connected in parallel with the circuit while a known voltage is applied. (The voltage source is usually a low-voltage battery inside the meter housing.)

6.3.16 Shunt

As indicated in Section 6.3.15, internal construction of many electrical meters is very much alike. There are various methods of connecting into the circuit. Various accessory devices may be used. This enables the same basic instrument to be used as either an ammeter, voltmeter, or ohmmeter. See **Figure 6-32.**

Voltmeters are connected across or in parallel with circuits in which voltage is being measured. Ammeters are connected in series with the circuit. This would indicate that all the current being measured by the ammeter must go into the meter. This is not true. A shunt (a second circuit path) is placed in parallel with the instrument. The greater portion of the current goes through the shunt rather than through the meter.

Several different shunts can be used with the same meter. Shunts are calibrated for the current range in which each is to be used. Sometimes shunts are built into the meter. The electrical connection to the meter should be made to terminals within the meter's range.

When there is doubt of the circuit's voltage or amperage, connect to the upper range of the instrument. If the reading is within the lower range, a more accurate reading may be made. This is done by making a connection that uses almost the full scale of the instrument. See **Figure 6-33.**

6.4 Electrical Materials

There are three physical materials that are used in electrical and electronic systems:

- Conductors (metals, such as silver, copper, and aluminum).
- Semiconductors (metal oxides or metal compounds).
- Nonconductors or insulators (nonmetals such as glass, wood, paper, and mica).

Any of these three may exist in any of the three forms of matter. However, most are solids rather than liquids or gases.

6.4.1 Conductors

A conductor has atoms with free electrons in its structure. Any electromotive force (pressure) will cause these electrons to travel from one atom to another. This moves the electrical energy through the material. In a

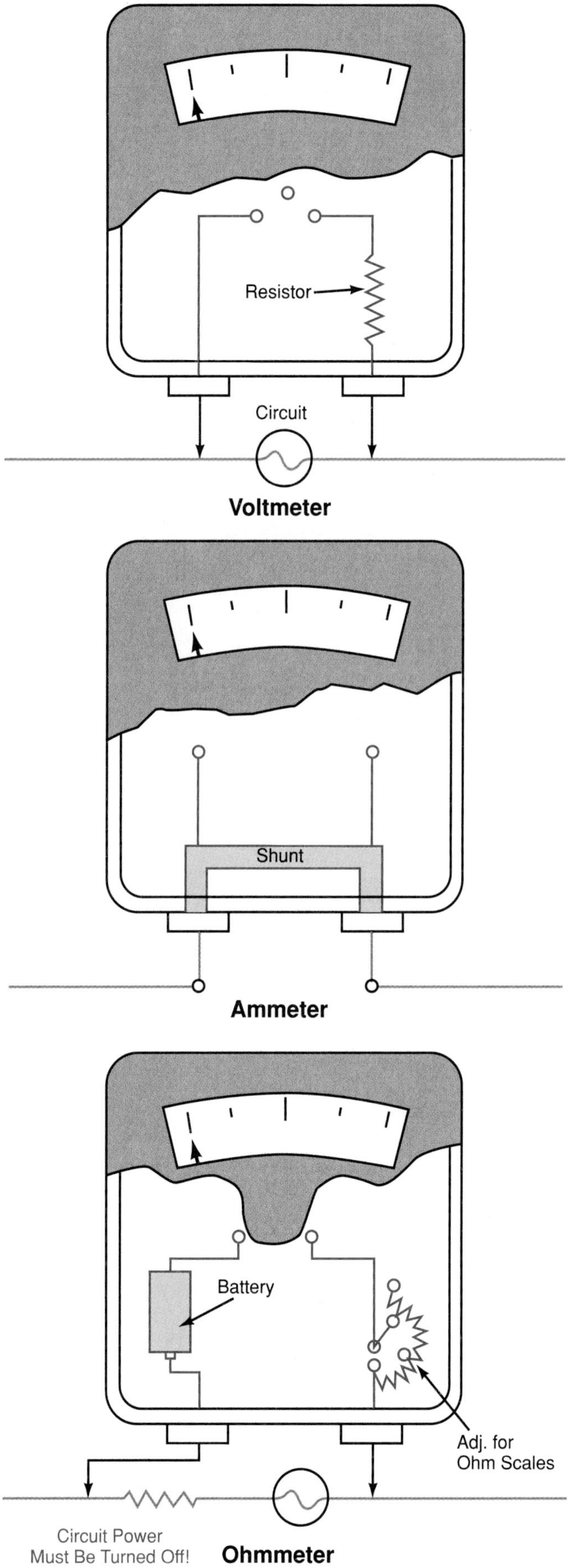

Figure 6-32. *How the basic meter movement is usually wired inside the voltmeter, ammeter, and ohmmeter.*

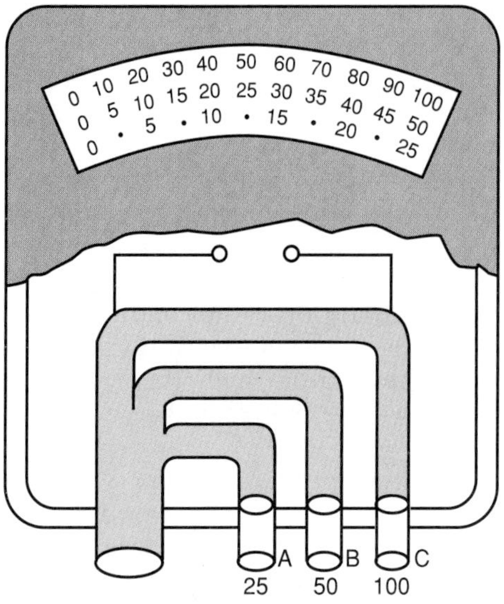

Figure 6-33. *Ammeter having built-in shunts. A—0 to 25 A. B—0 to 50 A. C—0 to 100 A.*

wire, for example, energy moves from one end to the other, **Figure 6-34.** Note that electron movement is from negative (−) to positive (+).

Each atom allows the free electrons to move with varying ease. In this respect, gold and silver atoms are excellent. Copper, mercury, and aluminum atoms are very good, also. Most good conductors are metals, yet free electrons movement is somewhat more difficult in iron.

Conductivity of metal is usually expressed in ohms per circular mil foot at standard temperature. Wire with 1 circular mil cross-sectional area has a diameter of 0.001″. The standard temperature used when measuring the conductivity is 68°F (20°C).

Wires (solid conductors) are popular for carrying electricity from one electrical device to another. Most electricians call wires "electrical leads."

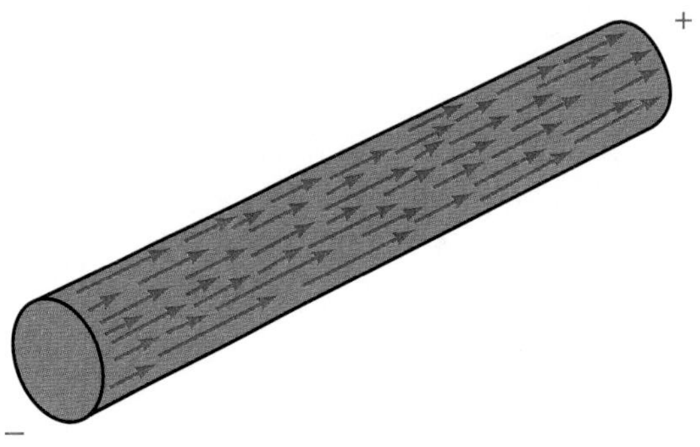

Figure 6-34. *A wire with free electrons traveling from negative (−) to positive (+).*

6.4.2 Semiconductors

Conductors such as metals conduct heat and electricity readily. Insulators conduct heat and electricity very poorly. In between these materials are semiconductors. Semiconductors are ordinarily insulators. However, under certain conditions, they can be made to conduct electricity easily. These materials are the basis of the present electronics industry. The term "solid-state electronics" refers to electronic devices which are made up of semiconductor elements.

Transistors, diodes, and photocells are semiconductors. Many modern motor controls consist of silicon controlled rectifiers (SCRs). These are semi-conductive switching devices.

The conductivity of a semiconductor can be controlled in various ways: an electrical signal, light intensity, pressure, temperature, and other signaling devices. Semiconductors can, therefore, serve as relays and switches.

The ac current produced by an alternator is converted to dc current by semiconductor diodes. These are the sensitive devices that can be damaged if a car is started with the wrong jumper connection between batteries. Photocells used on automatic door openers are semiconductor switching devices activated by light. See Section 6.6.1 for some semiconductor applications.

6.4.3 Nonconductors (Insulators)

Nonconductors resist electron flow. The atoms have virtually no free electrons. A perfect vacuum is also a nonconductor.

Nonconductors are as useful in electrical systems as conductors or semiconductors. There are many parts of the system in which electrical flow must be stopped. Insulation is needed for these.

Some common nonconductors are quartz, ceramics, mica, glass, and organic substances (rubber, wood, paper, and plastics). The resistance values for nonconductors (insulators) range between 10^9 to 10^{18} ohms.

6.4.4 Insulation Testers

Poor insulation can cause equipment breakdowns. It may also create a dangerous shock hazard. An insulation tester is a meter used to detect leaks or possible areas of failure along nonconductors or insulators. See **Figure 6-35.**

Insulation testers may be used in two ways. They can be applied to a live circuit. One lead is connected to ground and the probe is run along the insulation to see if any voltage reading (leak) occurs. This would indicate a break in the insulation.

They can also be used on a circuit where power has been disconnected. In this case, one lead is connected to the conductor. The probe is run along the insulation. This reading will indicate the high resistances that exist. A quick drop in this reading would identify a break in the insulation. It might also indicate an area of heavy wear where failure is likely to occur.

6.5 Magnetism

Operation of the magnetic compass depends on the fact that the earth is a magnet. The north magnetic pole

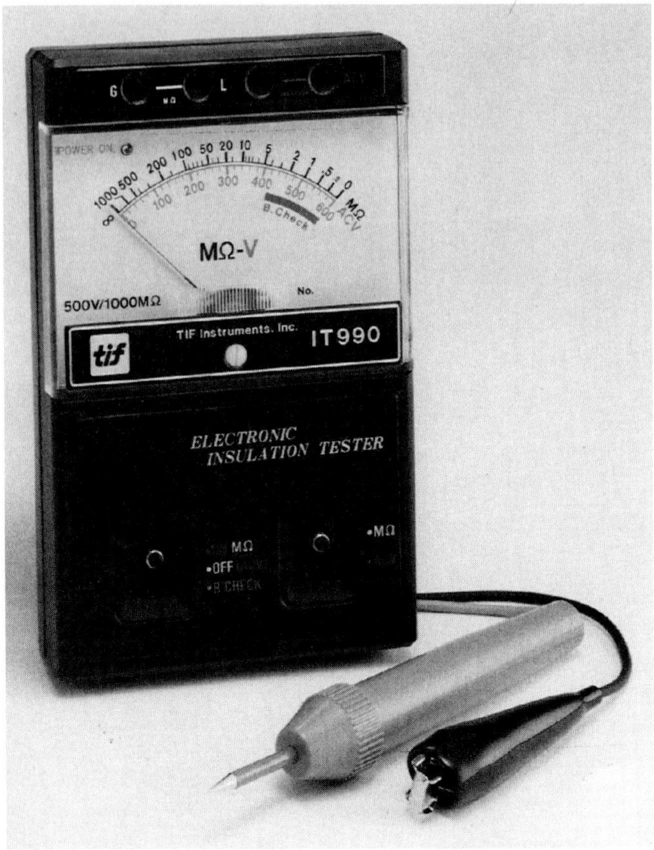

Figure 6-35. *Insulation tester, used for testing of electrical insulation and resistance. Unit can measure leak up to 1000 megohms. (TIF Instruments, Inc.)*

is near the north geographic pole. The compass needle is free to turn and "line up" with the earth's magnetic field. See **Figure 6-36.** It is more correct to say there is a south magnetic pole at the earth's north geographic pole. The north pole of the compass is attracted to the south magnetic pole. Therefore, it points north.

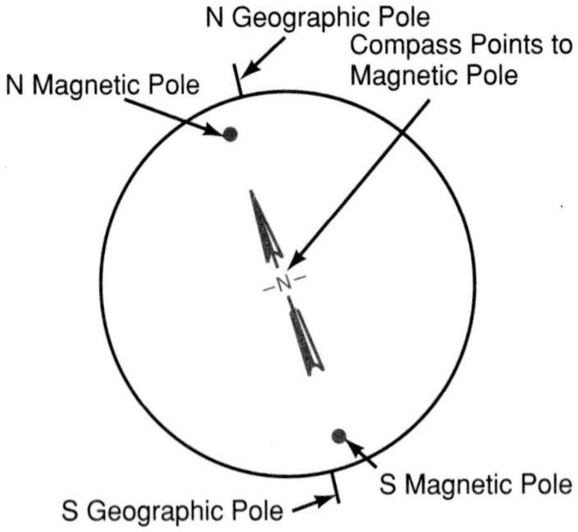

Figure 6-36. *A magnetic compass is a bar of magnetized steel mounted to turn freely on a vertical axis. It points in direction of earth's magnetic lines of force. It is more correct to say there is a south magnetic pole at the earth's north geographic pole which attracts the north pole of compass magnet.*

All magnets have a north pole and a south pole. Like poles repel each other (try to move apart). Unlike poles attract (pull toward each other). **Figure 6-37** illustrates several shapes of magnets. The attraction and repulsion of magnetic poles is shown in **Figure 6-38.**

There are lines of magnetic force connecting the north and south poles of a magnet. These lines of force are called flux. The space in which a magnetic force is operating is called a magnetic field. Magnetic flux will flow through any substance. It is not stopped by glass, mica, wood, air, or any usual electrical insulating substances. Some substances, particularly soft iron, are better conductors of magnetic flux than other substances. This is why certain parts of electric motors and generators are made of soft iron.

Instruments may be shielded from a magnetic field by using a soft iron case. The soft iron is a good conductor of magnetic flux. Therefore, the magnetic field will pass around the instrument. Some nonmagnetic watches use this method.

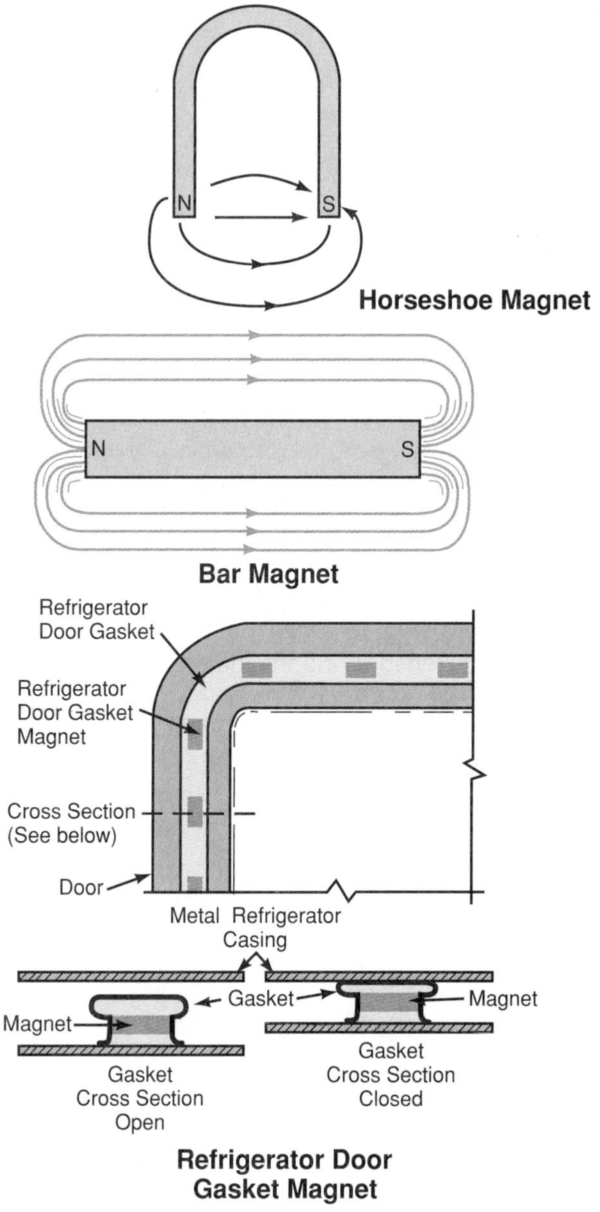

Figure 6-37. *Various shapes of magnets.*

6.5.1 Permanent Magnetism

Permanent magnets are usually made of hardened steel. Once magnetized, they remain magnetized. Some patented alloys of iron, aluminum, nickel, and cobalt make strong and super-permanent magnets.

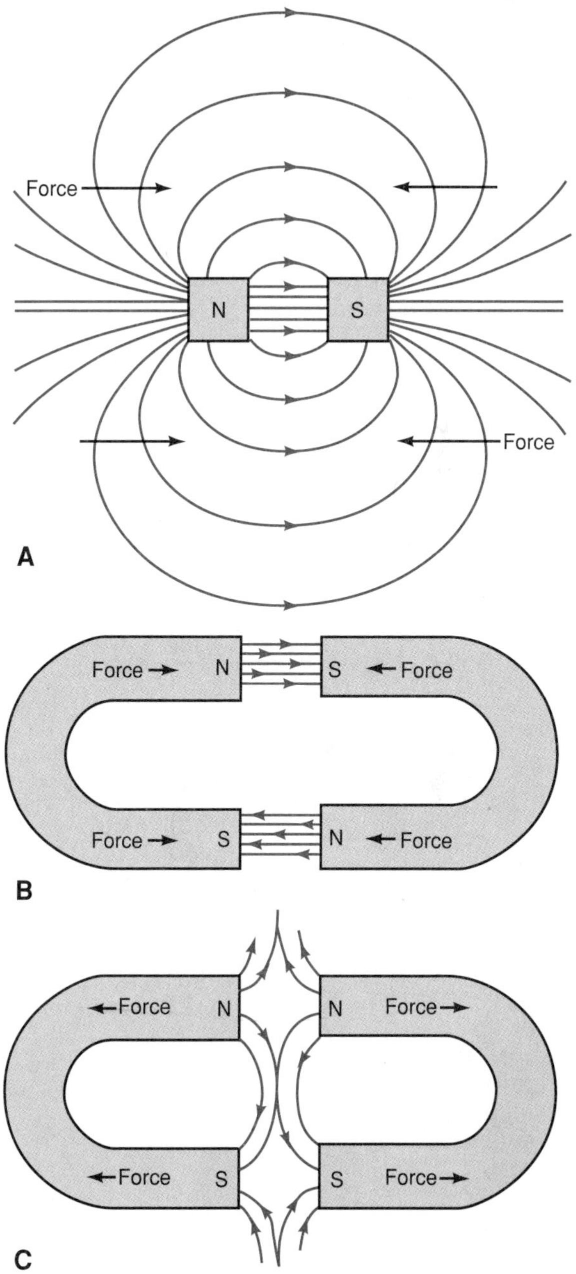

Figure 6-38. *Attraction and repulsion of magnetic poles. A—Magnetic flux around poles of a horseshoe magnet, shown end on. B—Magnetic flux provides a force which tends to pull unlike poles of magnets together (attract). C—Magnetic flux provides a force which tends to push like poles away from each other (repel).*

Magnetic lines (flux) are like stretched rubber bands. They tend to become as short as possible. This shortening force has many industrial applications.

Permanent magnets are used in some controls to provide a snap action for electrical contacts. They are also used in small control motors. In these motors, either the stator or rotor has permanent magnets. These provide precision movement of a motor rotor (dc, servo motors).

6.5.2 Induced Magnetism

A property of magnetism is that it may produce magnetism in some other metals nearby.

In **Figure 6-39,** the permanent magnet is surrounded by its magnetic field. The soft (low carbon), mild steel piece has no magnetism. The mild steel piece, placed near the permanent magnet or touching it, also becomes a magnet.

Any material capable of being magnetized becomes a magnet if placed in a magnetic field. This is called induced magnetism. Since the bar is soft iron, it loses its magnetism when removed from the magnetic field.

6.5.3 Electromagnetism

If a current of electricity is passed through a conductor, the conductor becomes surrounded by a magnetic field. See **Figure 6-40.** If the current is turned off, the magnetic field will disappear.

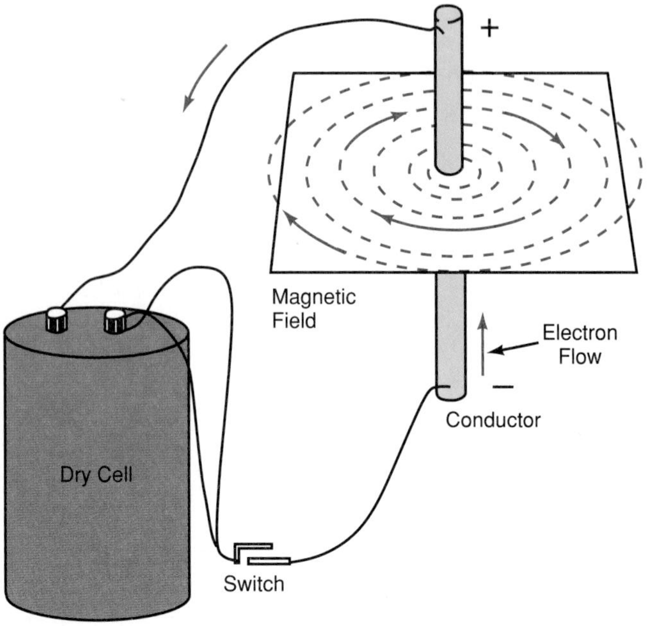

Figure 6-40. *Magnetic field encircles current flowing in a straight conductor. To demonstrate, conductor is passed vertically through the center of a sheet of cardboard. Iron filings are sprinkled over surface of cardboard. When ends of vertical conductor are reconnected to a dry cell, iron filings form pattern of circles around conductor. Changing the placement of cardboard will not change pattern of magnetic field.*

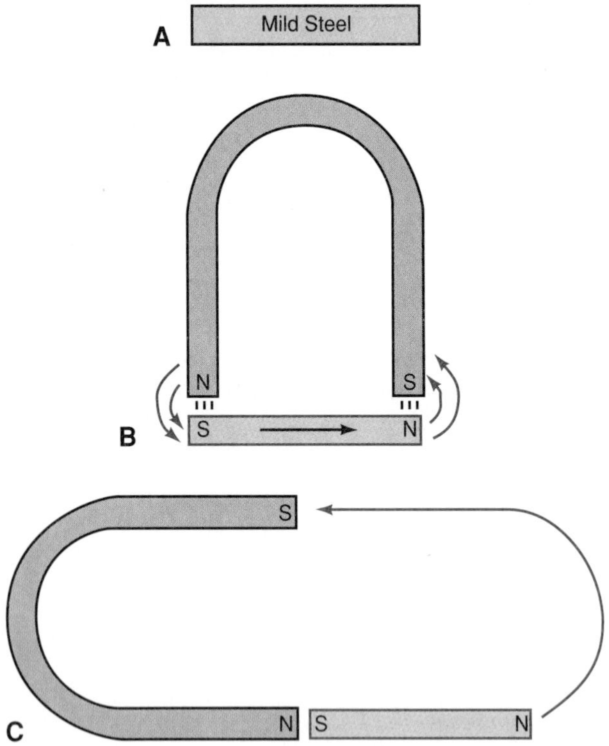

Figure 6-39. *Induced magnetism. A—Nonmagnetized mild steel bar. B—Bar has become a magnet by being placed in a magnetic field. C—Bar is again magnetized by being placed against one pole of permanent magnet.*

If the conductor is wound around a piece of soft iron (a good magnetic conductor), current will pass through it. Then the soft iron becomes a magnet. Turning off the current (opening the circuit) stops the magnetic effect. The magnetic effect is called electromagnetism. Magnets formed in this manner are called electromagnets, **Figure 6-41.** The iron part is called the core. The current-carrying conductor is called the winding.

The characteristics of a magnetic field surrounding a current-carrying conductor (wire) are shown in **Figure 6-42.** One very important characteristic is that the magnetic field travels around the wire in a specific direction.

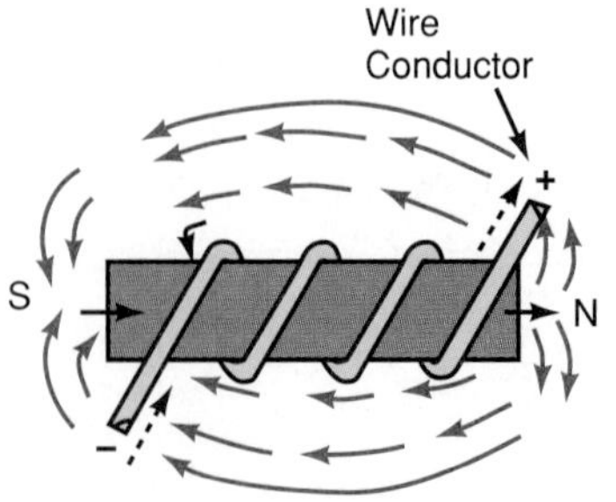

Figure 6-41. *Simple electromagnet has several turns of conductor (wire) placed on soft iron core. Individual circular magnetic fields are combined in core to form one magnet.*

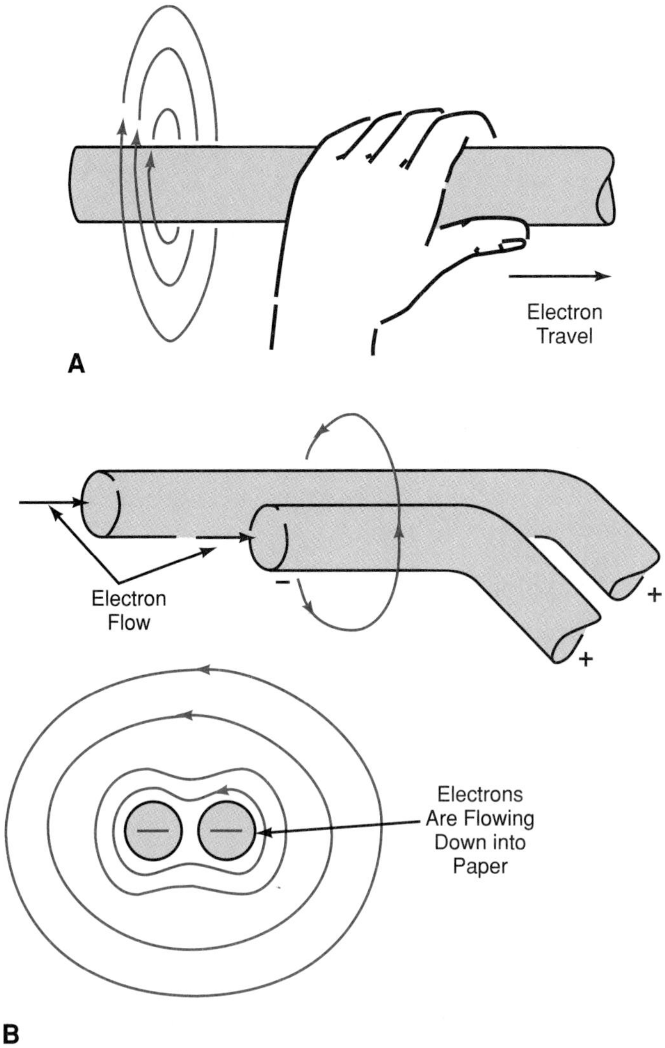

Figure 6-42. *Magnetic fields. A—Around a single wire. B—Around groups of wires. Note that left-hand rule establishes direction of magnetic field arrows.*

The direction depends on the direction of electron flow in the wire. This, in turn, determines which end of the electromagnet is the north pole. To determine the direction of magnetic field travel around a wire, use the left-hand rule. This is illustrated in **Figure 6-42.**

Left-hand rule: If one wraps the left hand fingers around a wire with the thumb pointing in the direction of electron flow (− to +), the fingers will point in the direction of the magnetic field travel.

The strength of an electromagnet is determined by two factors. The first factor is the number of turns in the winding around the core. The second factor is the current flowing through the winding. This strength is indicated by the term "ampere-turns." To determine this, multiply the number of turns in the winding by the amperes flowing through the winding.

Electromagnets are used in motors, relays, and solenoids, and in many other electromagnetic applications. **Figure 6-43** shows magnetic flow through the field poles and rotor of a four-pole motor. Note the outer part

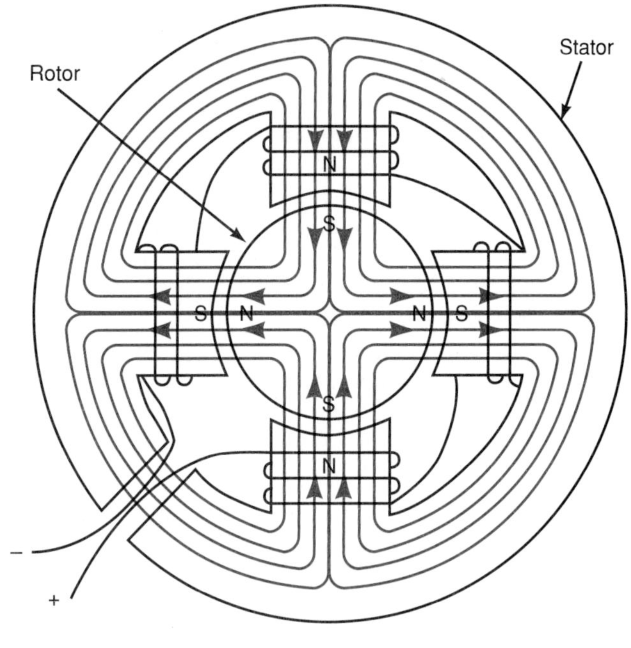

Figure 6-43. *Magnetic lines of force flowing through rotor and stator on a four-pole motor.*

of the stator to which the pole shoes are attached. It carries the same magnetic field as the pole shoes. This is why the iron in the motor stator must be so thick and heavy.

Polarity of Electromagnets

The polarity of an electromagnet is determined by the direction of flow of current (or electrons). The flow is through the winding on the core of the magnet. If the electromagnet is held in the left hand with the fingers grasping the magnet in the direction of electric current flow, the thumb will point to the north pole of the magnet. This is known as the left-hand rule. See **Figure 6-44.**

Electromagnetic Induction

A magnet may cause another substance to become a magnet. This was shown in **Figure 6-39. Figure 6-41** illustrates how the flow of electricity may produce magnetism. Electromagnetic induction uses both of these

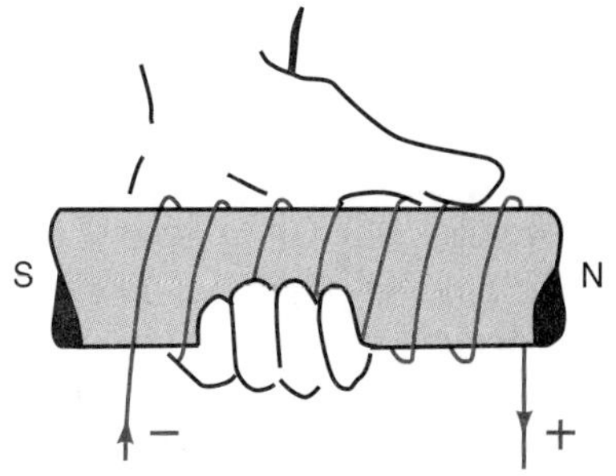

Figure 6-44. *Electromagnetic polarity. Use left-hand rule to determine which end of magnet is north pole.*

principles. It is the basis of operation of the induction-motor, the motor most used for driving refrigeration and air conditioning compressors.

The principle of electromagnetic induction is illustrated in **Figure 6-45.** Coils A and B are separate circuits. The leads to Coil A are connected to an alternating current supply. When switch S is closed, both lamps will light. This is because the alternating field around Coil A induces a magnetic field around Coil B. The magnetic flux builds up, then reverses. The coils of wire around Coil B have an emf generated in them. The building up and collapsing of the magnetic field produces the effect of wires cutting across a magnetic field. This is the basic principle of the electric generator explained in Section 6.5.9. This also explains the basic operation of the electric transformer, as described in Section 6.7.5.

Magnetic induction is used in electric motors. The rotor is made of mild steel. Magnetism in the rotor is induced by magnetism created in the stator magnets or field windings.

If a current travels in a coil of wire, it will create a magnetic field. An emf will be created in a wire or another coil close to the first coil. It is created by the magnetic field cutting across these wires.

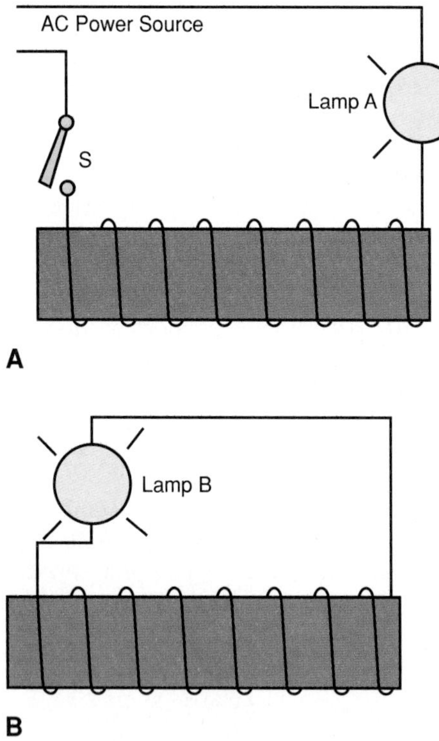

Figure 6-45. *Electromagnetic induction. Coils A and B are wound on soft iron cores. Coil A is connected through a switch to an alternating current source. Lamp in circuit controls amount of current flowing through Coil A. Coil B is wound on a core adjacent to Coil A, but they are not connected electrically. A low voltage lamp is connected to terminals of Coil B. When switch to Coil A is closed, both lamps will light, indicating that a current is being generated in Coil B by electromagnetic induction.*

Remember: In order to create magnetic induction, either one or a combination of the following must occur:

- The magnetic field must be changing.
- The magnet must be moving.
- The wires must be moving.

6.5.4 Magnetic Field Strength

Magnetic field strength depends upon the density of magnetic lines of force. The lines of force become less dense away from the magnet. The strength of the magnetic field rapidly decreases. To illustrate this effect: Magnets placed 1/2″ from each other may be pulled together with a 20 lbs. force. At a distance of 1″, the pull would be only 5 lbs. The magnetic field density is measured in units of Gauss.

6.5.5 Solenoid

A coil, wound on a nonmagnetic substance (paper or plastic) that is made to carry a current of electricity, will become a magnet. See **Figure 6-46A.** However, the magnetic effect will not be as great as if an iron core were placed inside the coil. This is illustrated in **Figure 6-46B.**

If this electromagnet is fitted with a movable iron core and the coil is made to carry a current of electricity, the soft iron core will be drawn into the coil. This is shown in **Figure 6-46C.** Coils like these are known as solenoids.

Solenoids are used in many places in refrigeration and air conditioning work. They are used to open and close valves. They operate dampers and make other desired movements in a mechanism.

A solenoid may be used on either alternating or direct current. The lines of magnetic force tend to become as short as possible in either case.

6.5.6 Permeability—Reluctance

Some substances, particularly soft iron, are better conductors of magnetic flux than other substances. The property of the material which determines its flux density under a magnetic field is called its ***magnetic permeability.*** It indicates the ease with which a material may be magnetized. Air is considered to have a permeability of one. Soft iron has a much greater permeability than air.

Reluctance is the resistance offered to the passage of magnetic lines of force through a substance.

Figure 6-47 shows magnetic behavior when glass or iron is placed near a magnet.

6.5.7 Capacitance—Capacitors

Capacitance may be defined as a system of conductors and insulators that permit the storage of electricity (free electrons). This ability is indicated by the letter C. Capacities of a series or group of capacitors are usually designated as C_1, C_2, C_3, and so on.

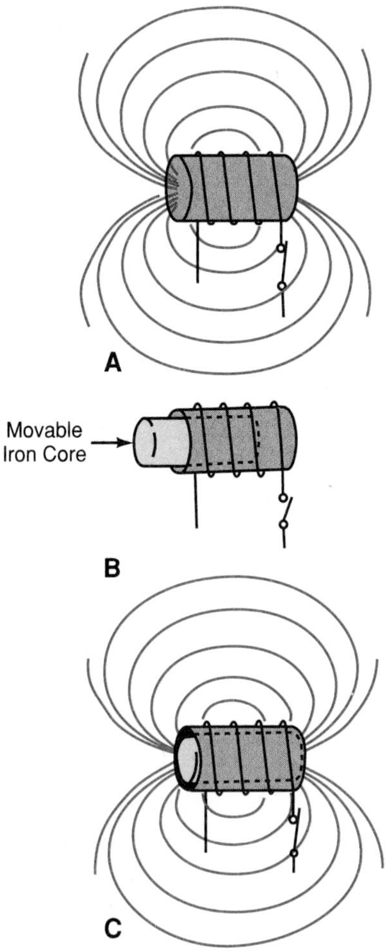

Figure 6-46. *A—Magnetic field is created if a solenoid is nonmagnetic material and current is flowing through windings. B—Solenoid fitted with an iron core with no electric current flowing. C—With current flowing, iron core is drawn into solenoid winding.*

The unit of capacity is the farad. The symbol for the farad is the capital letter F. A farad may be defined as a charge of one coulomb on the capacitor surface with a potential difference of one volt between the plates.

A farad is a rather large unit of capacity. Most capacitors (condensers) used in the refrigeration industry are rated in microfarads. A microfarad is one-millionth (.000 001) of a farad, or 10^{-6}. The symbol for the microfarad is (μF).

Any device that has capacitance (stores free electrons) is a capacitor. Large capacitors are metal surfaces such as aluminum foil separated by insulating material (dielectric). See **Figure 6-48.**

Capacitors are classified by the materials used for the dielectric. They include air, mica, paper, oil-filled, ceramic, and electrolytic. **Figure 6-49** shows several types of capacitors.

The capacitor, in series with the load, changes the sine wave and makes the current wave (lead) the voltage curve of an ac circuit. Capacitors are used to help start motors, increase their efficiency, and improve the power factor.

The capacity value of capacitors in series may be expressed by the formula:

$$\frac{1}{C_n} = \frac{1}{C_1} + \frac{1}{C_2}$$

C_n = net capacitance (effective value)
C_1 = capacity of capacitor No. 1
C_2 = capacity of capacitor No. 2

The capacity of capacitors in parallel may be expressed by the formula:

$$C_n = C_1 + C_2$$

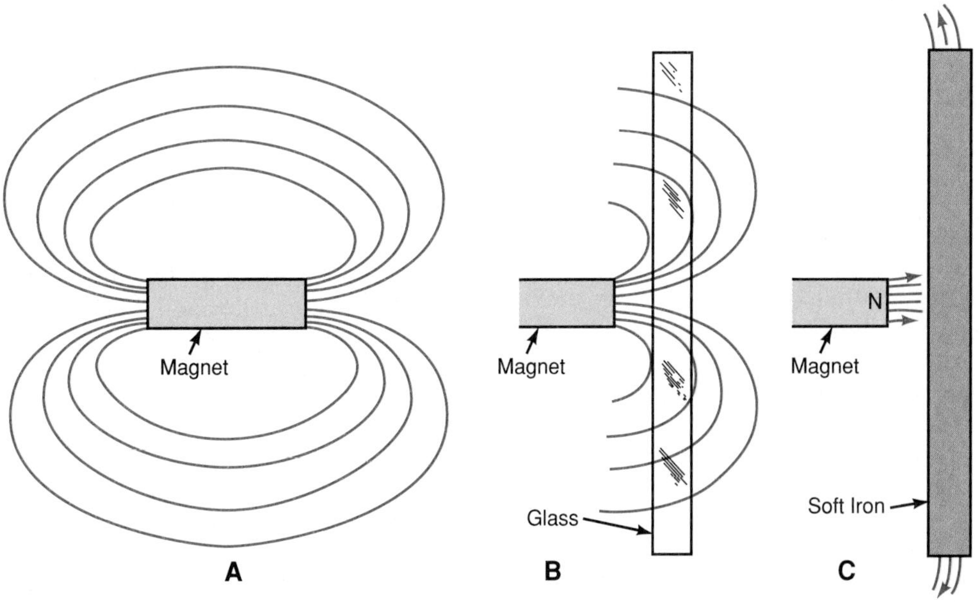

Figure 6-47. *Fundamentals of magnetic behavior. A—Typical magnetic field surrounding a magnet. B—Plate of glass has no effect on shape of magnetic field. C—Effect of a soft iron plate placed in a magnetic field.*

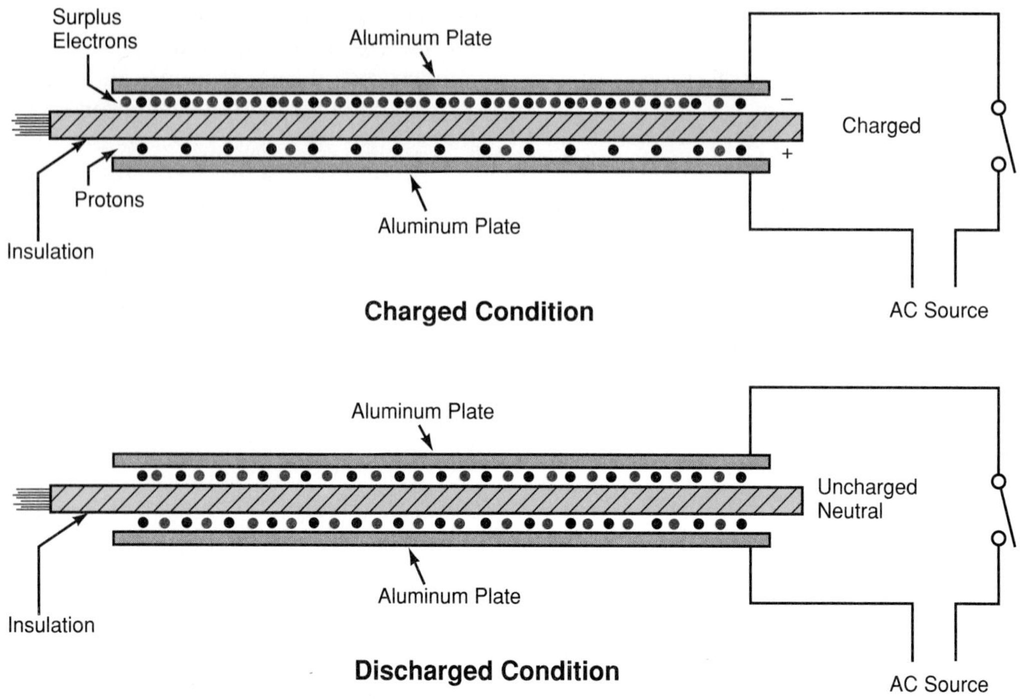

Figure 6-48. *Capacitor construction.*

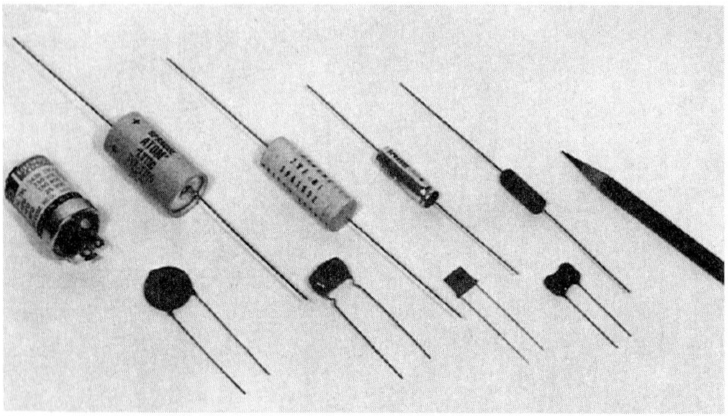

Figure 6-49. *Photo of some typical capacitors, including mica, paper, and electrolytic.*

or by simply adding together the values of all the capacitors connected in parallel.

Large capacitors in a machine can store dangerously high voltages. Before handling or replacing them, drain off the charge. A 20,000 ohm (20KΩ), 2 watt resistor may be used. A high voltage capacitor may store electricity at 600 V in a three-phase electrical motor circuit. This is dangerous.

6.5.8 Reactance

Reactance is the opposition to an alternating current flow in a circuit.

There are two kinds of reactance: capacitive reactance and inductive reactance. Capacitive reactance is caused by the capacity or condenser effect in the circuit. Inductive reactance is caused by the generation of counter-emf in a circuit. It is usually produced by a wire coil or an electromagnet.

6.5.9 Electrical Generator

If a conductor is moved across a magnetic field, an electrical potential (emf) will be generated in the conductor. This is illustrated in **Figure 6-50.**

The loop in **Figure 6-50A** is not cutting across the lines of magnetic force. It is running parallel to them. No emf is generated in the loop during the interval it is in this position. **Figure 6-50B** shows the loop positioned so that the two sides cut across the magnetic field. In this position, an emf will be generated in both sides of the loop.

If the two ends of the loop are connected into an electrical circuit, no current will flow when the loop is vertical. As the loop starts to revolve, it will cut across the magnetic field. Current will begin to flow. When the loop has revolved 90° and is horizontal, current flow will reach its maximum. Once past this point, current flow will diminish. After another 90°, no current will be generated. Then, the loop continues to revolve. It cuts through the magnetic field at opposite poles of the magnet. This will cause the current generated in the loop to flow in the opposite direction.

An electrical generator has a revolving conductor (armature). The wires of the generator first cut the magnetic field in one direction, then in the other. This generates an alternate (or opposite) flow. This happens each revolution of the armature as the conductors pass the magnetic poles and generate alternating

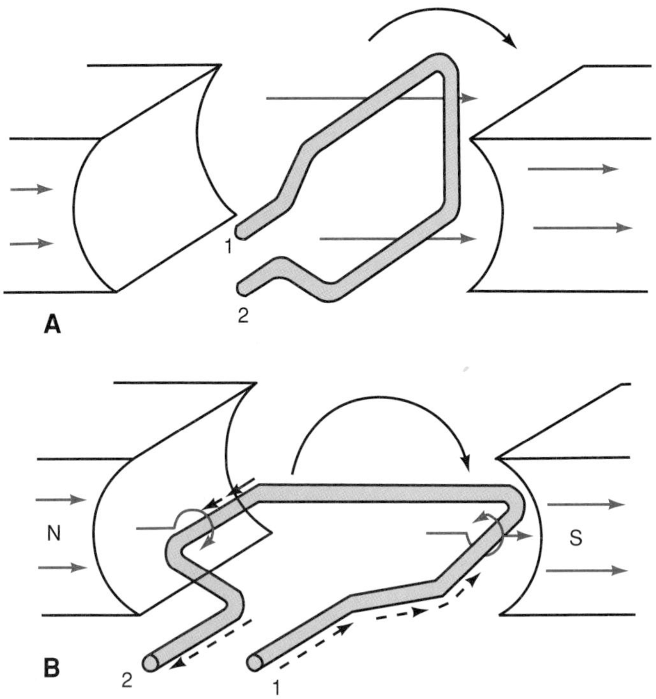

Figure 6-50. *Generation of current. Conductor loop has two sides, 1 and 2, and it is revolving clockwise. A—In this position, wires are traveling nearly parallel to lines of force and no emf is being generated. B—Loop has revolved 90° and current is being generated.*

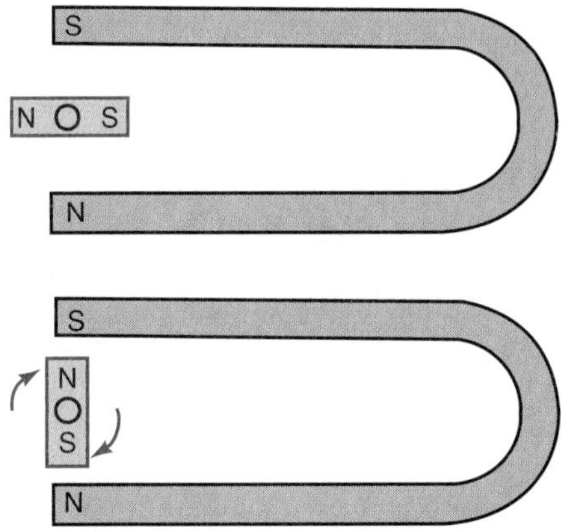

Figure 6-51. *Magnet mounted on an axis will turn when put in another magnetic field. Note direction of movement of pivoting magnet.*

current (ac). However, commutators and brushes on direct current generators rectify (correct) the current leaving the generator. The current flows in one direction only.

As current flows from the generator, a magnetic field surrounds the conductor in which the current is generated. This field opposes the movement of the conductor across the generator field. The greater the current produced, the greater the power required to drive the generator. It takes a great deal of power to drive a generator while it is producing a current.

This principle is stated in Lenz's Law: "The magnetic effect surrounding the conductor in which a current is induced opposes the movement by which the current is induced."

The Elementary Electric Motor

Electrical energy is changed to mechanical energy in an electric motor. First, electrical energy becomes magnetism. Magnetism may then be used to cause motion.

Like poles repel (N repels N and S repels S). Unlike poles attract (N attracts S and S attracts N). These actions can be used to produce motion. One magnet is placed on a shaft and another is mounted in a fixed position. This is shown in **Figure 6-51.**

The bar magnet on the shaft will turn until its S pole is near the fixed N pole. This movement places the shaft bar magnet's N pole near the fixed S pole. The fixed magnet is called the stator. The other magnet, which rotates, is called the rotor or armature.

The rotor will stay in the vertical position until the magnetism of the stator or rotor is reversed. Then the rotor will rotate another half turn.

The magnetism may be reversed by using electromagnets instead of permanent magnets, **Figure 6-52.** When the rotor reaches the vertical position, the alternating current reverses (due to its cycling). Stator polarity reverses. The rotor S pole is now near the S pole of the stator. They will repel each other. The rotor will now revolve a half revolution, or 180°.

The direction of movement of the armature (or rotor) depends on polarity. This is shown in **Figure 6-53A.** The direction of movement of a current carrying conductor in a magnetic field may be determined by the "left-hand motor rule." The thumb and first two fingers of the left hand are placed at right angles. The

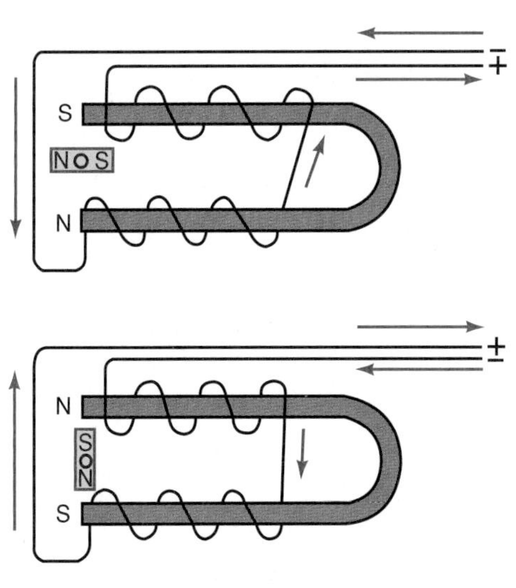

Figure 6-52. *Elementary electric motor. By reversing flow of current through field (stator) winding, its magnetic polarity will be reversed. This causes rotor magnet to turn on its axis.*

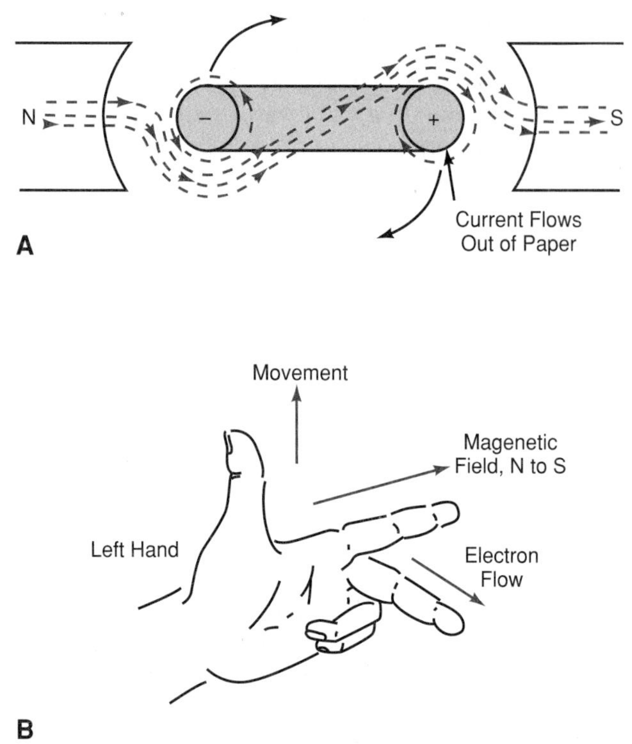

Figure 6-53. *Left-hand motor rule. A—Magnetic lines of force from north pole to south pole merge with magnetic field around conductor. The combined fields cause movement of rotor. B—Position fingers of left hand as shown, and thumb will point in direction of movement.*

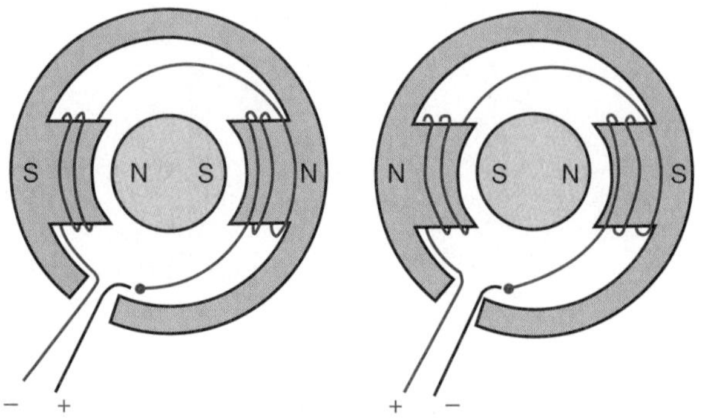

Figure 6-54. *Two-pole stator motor. Rotor will make a half turn with each half of current cycle.*

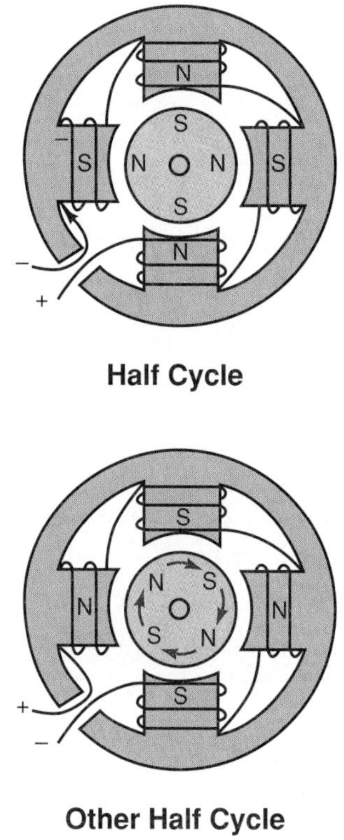

Figure 6-55. *Four-pole stator. Armature makes one-half turn as current completes one cycle in field windings.*

index finger is pointed in the direction of the magnetic force. The middle finger is pointed in the direction of the current flow in the conductor. The thumb is pointed in the direction of the force or movement of the conductor.

If the rotor turns a half revolution during one half of 60 Hz (frequency), then it turns a half turn during 1/120 of a second. It will then turn 60 revolutions per second or 3600 revolutions in one minute. The two-pole motor (3600 rpm) is a popular hermetic motor in refrigerating and air conditioning. Its construction is shown in **Figure 6-54.**

If four poles are used in the stator, the motor will turn only 1800 rpm. That is, the rotor will only turn one-fourth revolution in 1/120 of a second. See **Figure 6-55.** Most open motors and some hermetic motors are of this design.

Speeds of 3600 rpm and 1800 rpm are called synchronous speeds. Under actual conditions, the 3600 rpm motor usually operates at approximately 3450 rpm. The 1800 rpm motor operates at approximately 1750 rpm. This reduction in speed is due to slight magnetic slippage, depending on the load. An overloaded motor will slow down slightly.

The stator of an induction motor produces a rotating magnetic field. The rotor cannot keep up with the field.

The electric motor shown in **Figure 6-56** is an open capacitor-start motor. It can be used on external drive refrigeration compressors, pumps, and the like. See Chapter 7 for further information on various types of motors used on refrigeration and air conditioning machines.

Commutators

A conductor moving across a magnetic field will have an emf generated in it. In **Figure 6-57,** the generated current in the wires of an armature will flow in one direction. Then the conductors move from one field pole to the other. This causes the current to flow in the other direction. Alternating current is being produced.

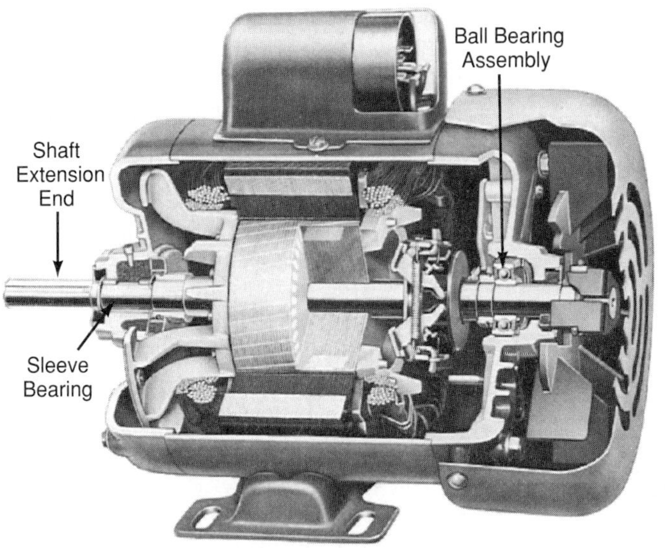

Figure 6-56. *Section through motor suitable for use on external drive refrigeration and air conditioning installations. (MagneTek)*

To produce direct current, a commutator and brushes are used in the generator. **Figure 6-58** illustrates an elementary type of armature fitted with a commutator and brushes.

The commutator and brushes provide a movable electric contact between the rotating armature wires and the stationary electrical device. The movable contact surface on the rotating armature is called the commutator. The parts that contact the rotating parts of the commutator are called the brushes.

As the armature revolves, the commutator contacts are made so that one brush always carries current into the commutator in a negative to positive direction. The other brush always carries current from the commutator to the charging circuit. Therefore, direct current generators are always fitted with commutators and brushes.

The starting mechanism on some ac motors with wound motor armatures also use commutators and brushes. These commutators and brushes are used only when the motor is starting. Universal motors (both ac and dc operation) use commutators and brushes whenever the motor is in use. Motors of this type are used in electric drills and mixers.

Counter emf

A running motor develops an emf in the rotor bars or windings. This is called a ***counter emf.*** It opposes the applied emf, which is driving the motor.

Counter emf depends upon the speed of the rotor. Under no-load, synchronous speed, the counter emf practically balances the applied emf. As the load increases and the speed decreases, the counter emf drops. As a result, the applied emf sends more current through the windings. This tends to maintain the speed constant.

If the motor is slowed considerably by a heavy load, the current supplied increases greatly. This is due to the reduced counter emf. The motor then will overheat. Under continuous overload, it is likely to burn out.

If the rotor is locked so it cannot turn and current is applied, the current will be very high. A motor will quickly burn out under this "locked rotor condition." Counter emf also occurs in all coils and electromagnets.

6.5.10 Inductance

A magnet may be formed first by winding a coil of wire (conductor) on a soft iron core. An electrical current is then passed through the conductor. The entire coil is surrounded and saturated by magnetic lines of force (flux).

As current is switched on in this coil, the magnetism is not built up instantly. There is a delay of perhaps

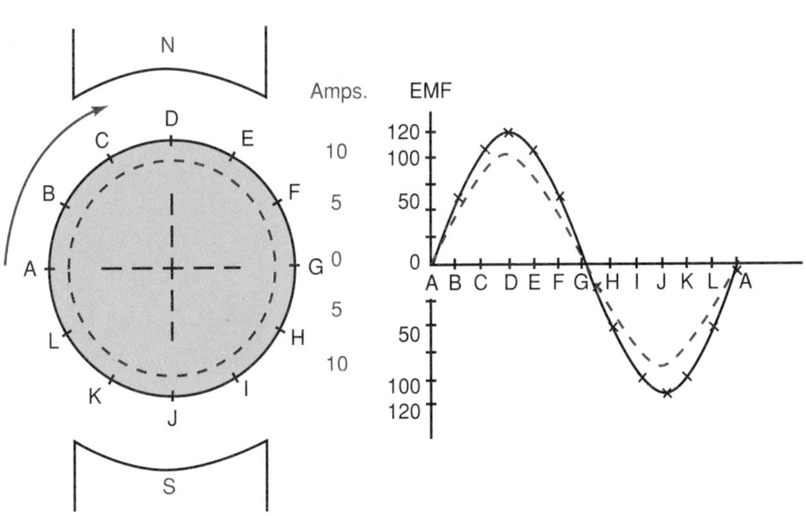

Figure 6-57. *Current (amps) and voltage (emf) changes in an alternating current circuit are graphed for one revolution of armature.*

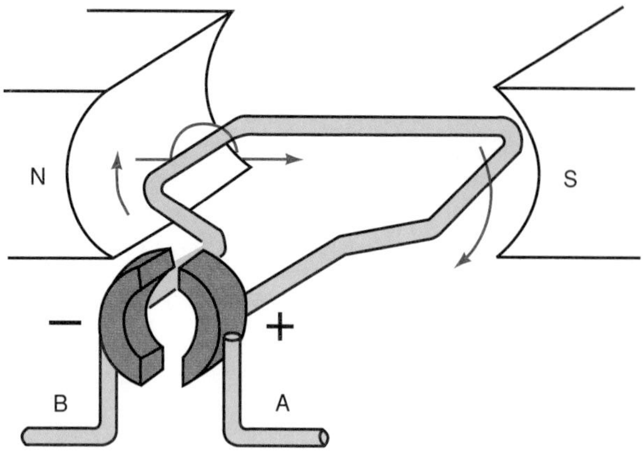

Figure 6-58. *A generator commutator. Brushes A and B contact ends of generating loop. Current from brushes always flows in same direction.*

a few hundredths of a second. During this time, the current continues to increase until it reaches its full value. This value depends on the resistance and emf of the circuit. Likewise, when the switch is opened and current turned off, it does not stop flowing instantly. The magnetic lines build up as the switch is turned on. They collapse as the switch is turned off.

There is a tendency to generate an electromotive force within the coil. This emf counteracts the change in the current flow. This counteracting force is the counter electromotive force or counter emf.

The principle of inducing a voltage in a coil due to the change in current flow rate is called inductance. It acts much like a flywheel on a piece of machinery. The flywheel requires power to give it a rotating motion. Likewise, it gives up power if it is forced to stop.

Inductors

Induced magnetism is useful in electric motors. Electrical windings are placed on the field poles. Magnetism can then be created or induced in the rotor. There is a slight time delay in this induction. Therefore, if the rotor has started to turn, it will continue because the field pole (stator) always has the opposite polarity.

When the field pole changes its polarity, it induces an opposite pole in the rotor. This repels the rotor pole toward the next stator pole, which has become an opposite pole in the meantime. The rotor must turn the full distance between the two poles before the induced magnetism can change. If it does not, the motor will not keep running. It takes only 1/120 of a second to change the magnetism from a full strength N pole to a full strength S pole.

This principle is used in the design of the split-phase motor. There are two windings in this motor. One is a starting winding and the other a running winding. The starting winding is a smaller diameter wire than the running winding. However, it has a greater number of turns. As a result, its magnetic inductance will be greater than that of the running winding. The starting winding is always behind the running winding in both building up and stopping its magnetic field. This type of inductance is generally known as self inductance.

Mutual inductance is the flow of electricity in a conductor produced by the magnetic field of another conductor. The principle of mutual induction is used in all induction motors.

In induction motors, the magnetic effect of the current flowing in the field windings induces the current in conductors on the rotor. The magnetic effect of this induced rotor current causes the rotor to revolve. This characteristic of the flow of current in an electromagnet—to resist the flow when the current is turned on and to resist the stopping of the flow when the current is turned off—depends on Lenz's Law. The law states the polarity of an induced voltage is such that it opposes the motion of the flux inducing it.

APPLIED ELECTRONICS AND ELECTRICITY MODULE

6.6 Electronics

Electronics concerns electron flow through gases, vacuums, and semiconductors. Electronics developed with the discovery of vacuum tubes and gas-filled tubes.

It was found that electrons would flow from a heated element in a tube to another element. This occurs only if a potential difference exists between the two elements.

Figure 6-59 shows a simple circuit of this type. The first radios used the vacuum tube to control and amplify the radio signals. These tubes were fragile and relatively large. The invention of solid state semiconductors has greatly expanded the application of electronics.

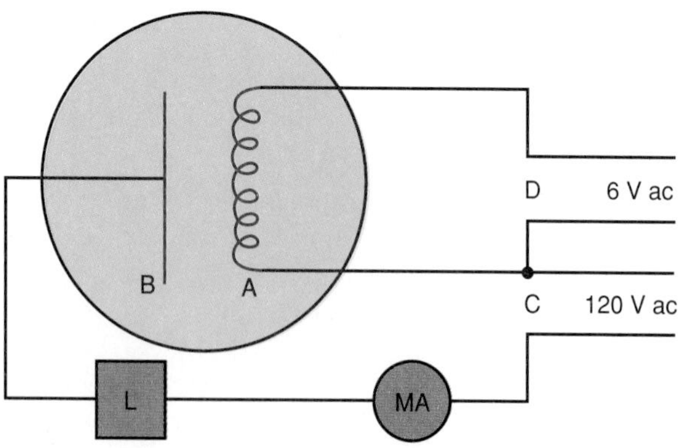

Figure 6-59. *Diagram of vacuum tube rectifier. A—Heating element from which electrons will flow. B—Plate to which electrons flow from A. C—120 V ac power source. D—6 V ac to operate heating element. L—Direct current load. This might be a storage cell being charged. MA—Milliammeter, which shows rate of dc flow.*

In the following paragraphs, some electronics will be explained, including the following:

- Semiconductors.
- Diodes—diacs.
- Rectifiers.
- Silicon controlled rectifiers—triacs.
- Transistors.
- Sensors.
- Thermistors.
- Amplifiers.
- Transducers.
- Thermocouples and thermoelectric devices.
- Photoelectric devices.
- Integrated circuits.
- Integrated circuit boards.

With the development of rather complicated automatic controls on refrigerating and air conditioning appliances, more and more electronic devices are being used. It is necessary to understand these devices in order to understand the circuits in which they are used.

6.6.1 Semiconductor Applications

Solid-state electronic devices make up most of the modern electrical control system. They are important in computer control systems and in devices which must withstand hard use.

Semiconductors are of two general types:

- ***Intrinsic semiconductors.*** These are pure substances like silicon and germanium or combined substances like lead sulfide. These are particularly useful as thermometers and as other temperature sensing devices.
- ***Extrinsic semiconductors.*** These are combinations of intrinsic semiconductors with very small impurities. These semiconductors are very sensitive to electrical forces. They are the basic materials used in electronics. Thermoelectric refrigerators (discussed in Section 3.17) use them to produce cooling.

6.6.2 Diodes, Diacs

A solid-state diode is a solid wafer or capsule composed of two materials that allow the electrons to flow through in one direction only. See **Figure 6-60.** The diode acts as an electron flow check valve.

Vacuum tubes can also serve as diodes, **Figure 6-61.** Electrons will flow from the pointed filament to the plate on half the cycle. During the other half cycle, the electrons will not flow from the plate to the filament.

A diac (ac diode) is similar to a diode. However, it allows current to flow in both directions. The diac will not conduct current until a preset voltage is exceeded. See **Figure 6-62.** Schematically, it operates similar to two diodes in parallel. See **Figure 6-63.** The diac is used in ac circuits where both halves of the sinusoidal voltage are required. They are often used as part of the switching circuit in motor controls (see Chapter 8).

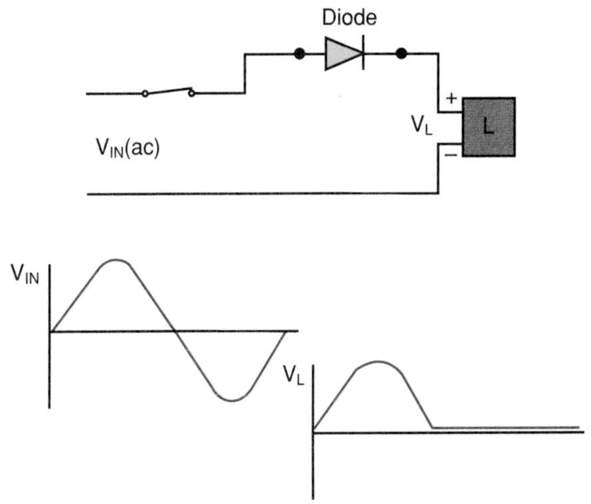

Figure 6-60. *When an alternating voltage is applied to a diode (V_{IN}), the diode allows current flow only when the voltage is positive. As a result, voltage at load (V_L) reflects only the positive component of input voltage.*

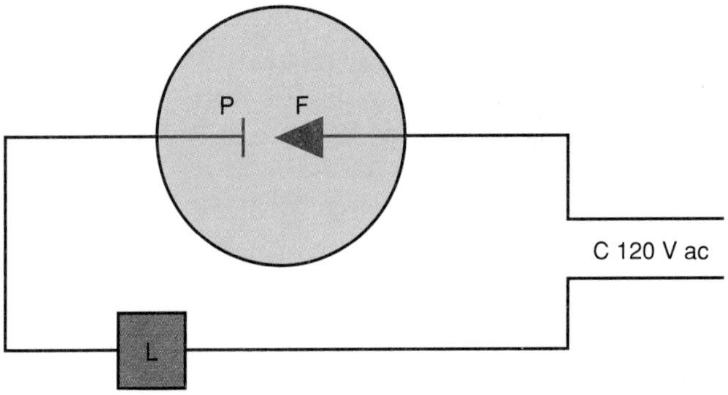

Figure 6-61. *Diagram of a vacuum tube serving as an alternating current rectifier or diode. Note that electrons will only flow from pointed filament (F) to plate (P). L is direct current load, which may be a storage battery being charged.*

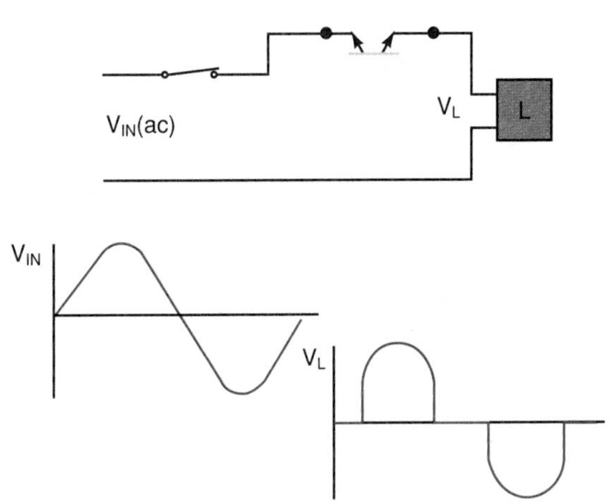

Figure 6-62. *When an alternating voltage (V_{IN}) is applied to a diac, the diac allows current to flow after a preset voltage level. Result is voltage at load (V_L) as shown.*

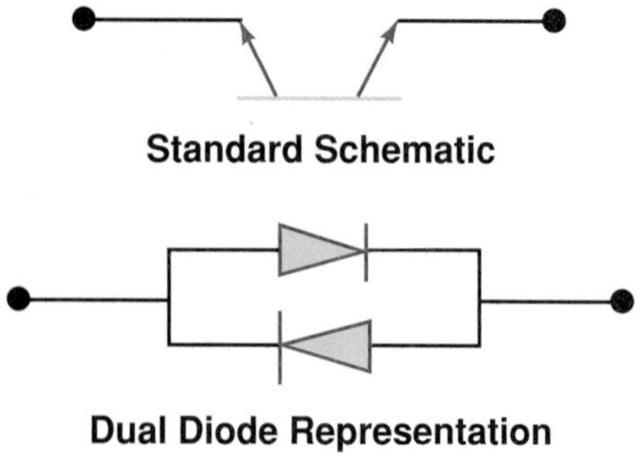

Figure 6-63. *Schematic representation of a diac.*

6.6.3 Rectifiers

Rectifiers are electronic valves which permit the flow of current in one direction only. These devices change alternating current to direct current. A simple rectifier uses only one half of the sine wave. Four diodes are needed to use both halves of the sine wave and still produce only direct current. See **Figure 6-64.**

Silicon Controlled Rectifiers, Triacs

Among silicon semiconductors are diodes and silicon controlled rectifiers (SCRs). The SCR has three connections as shown in **Figure 6-65A.** It conducts current from A to C. This occurs when both:

- The voltage at A is greater than at C, and
- A preset voltage has been applied at B.

If these conditions are not met, the device is "off" and not conducting current.

SCRs are used extensively in electrical motor controls. They are also used to convert ac voltage to dc voltage in inverter devices. SCRs are also used in many applications where relays were formerly used for switching.

A triac, **Figure 6-65B,** is similar to an SCR. However, it can conduct current in both directions— from A to C and C to A—when a preset voltage is applied at B.

6.6.4 Inverter

Electrical energy stored in a battery is available as direct current (dc) energy. The voltage supplied by the battery is a steady voltage. It gradually decreases with time as the charge is drained from the battery. An electric motor powered by a battery must be a dc motor.

Dc motors are heavier and more expensive than ac motors. It is often advantageous to change the battery voltage so an ac motor can be used. The device used to do this is called an inverter. The device does the opposite of the rectifier discussed in Section 6.6.3. (The rectifier converts ac power to dc power.) **Figure 6-66** indicates the inverter function, converting direct current (dc) to alternating current (ac).

Older electrical systems used a dc motor connected to an ac generator to do this inverting. Newer solid-state electronic devices do this without any mechanically

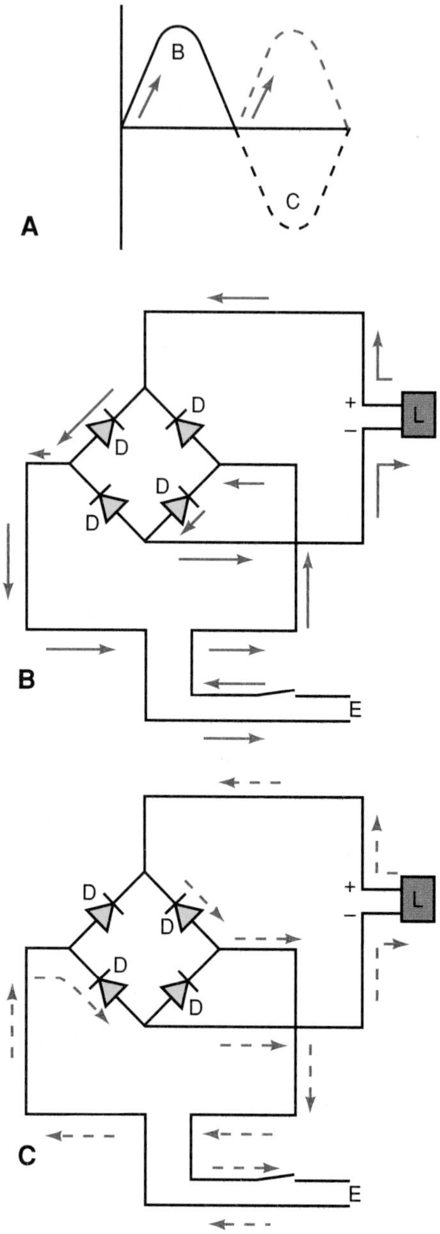

Figure 6-64. *Circuit for full wave rectifier. A—Note how two halves of an ac cycle are made to provide dc during all parts of cycle. B—Solid red arrows show current flow for one-half wave. C—Dashed red arrows show current flow for other half wave. D—Four diodes are needed. E—60 cycle ac power supply. L—The dc load.*

moving parts. The basic elements used in a solid-state inverter are:

- A crystal that oscillates at the frequency of the ac power required.
- A switching circuit using silicon controlled rectifiers (SCRs) to switch dc power on and off.

A simple inverter, using a set of standard diodes, produces a square wave output.

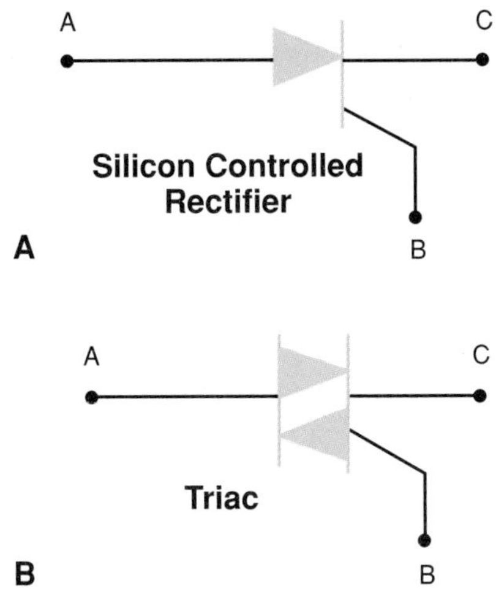

Figure 6-65. *Schematics of semiconductors.*

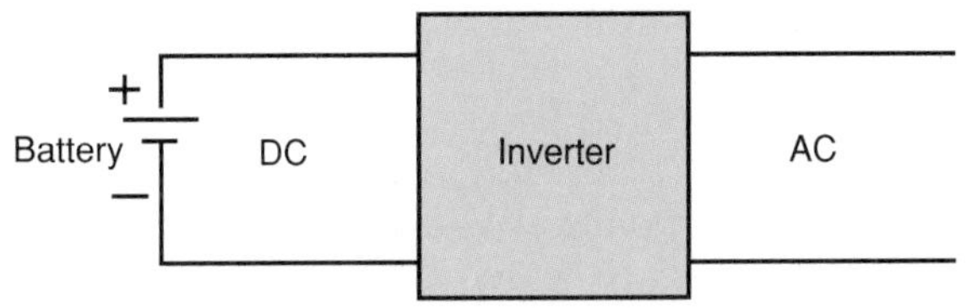

Figure 6-66. *Inverter converts direct current (dc) supplied by battery to alternating current (ac).*

Most ac motors and controls are designed to operate only with alternating (ac) power, similar to that provided by the power company. These devices will operate with a square wave. However, they will not operate as efficiently. Their lifetimes will usually be reduced.

An inverter is usually required in solar electric energy systems. The output of solar cells is dc power.

6.6.5 Transistors

A ***transistor*** is a three-layer sandwich of two different components that consist chiefly of silicon semiconductor material. Electrically, the three wafers are connected as shown in **Figure 6-67.**

The materials are labeled for their properties. P is for positive, meaning a lack of electrons. (It has "holes" ready to receive electrons.) N is for negative, meaning the material has a surplus of electrons.

Three conductors are connected to the transistor. One attaches to the base or middle wafer. One connects to one of the outer wafers, called the emitter. The third connects to the collector. The outer two wafers are of the same material. The base material (middle wafer) is different.

A small electron flow from the base to the emitter will control a large electron flow. This flow is from the emitter to the collector. The device, therefore, acts as a valve and as a relay. An electron signal circuit from base to emitter may control a collector electron flow as much as 1000 times larger than base to emitter.

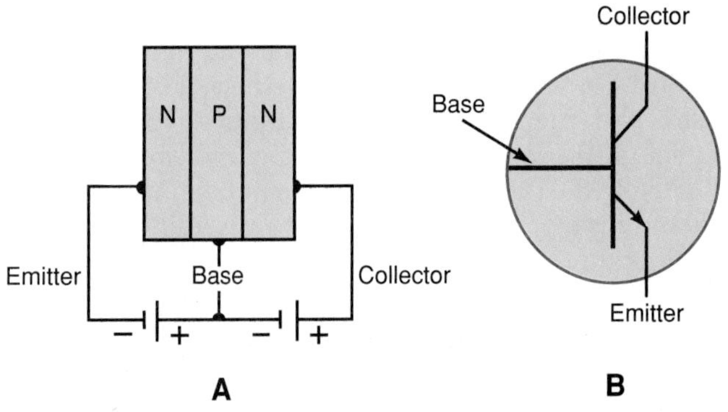

Figure 6-67. *A—Transistor in a circuit. B—Symbol for a transistor.*

All three parts are semiconductors having added substances to give the desired characteristics. **Figure 6-67A** is a schematic showing the basic construction of a transistor. The two basic types of transistors are shown in **Figure 6-68.**

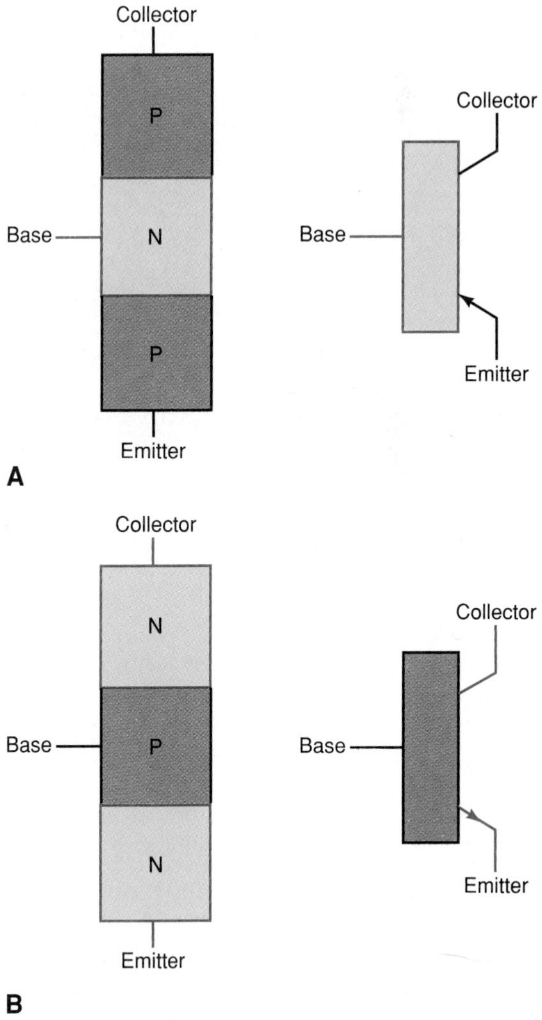

Figure 6-68. *Two basic types of transistors. A—PNP transistor. B—NPN transistor. In A, current flow is controlled by negative charge carried in base. In B, current flow is controlled by positive charge in base.*

A transistor connected as an amplifier is shown in **Figure 6-69.** The low energy signal enters the circuit at the left. The signal is amplified by the action of the transistor and energy supplied by batteries. The amplified signal (at the load) is shown leaving the circuit at right.

Note the transistors in the circuit board, **Figure 6-70.** Each transistor amplifies one signal. A unit may control parts of a refrigeration system.

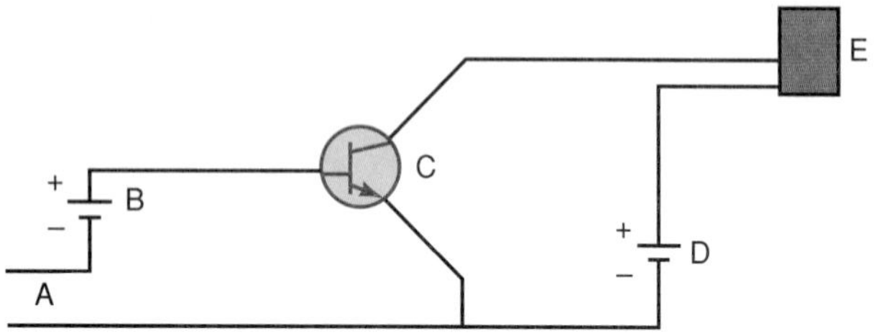

Figure 6-69. *Circuit diagram showing transistor used in amplifier circuit. A—Signal (current) to be amplified enters here. B—Battery. C—Transistor. D—Battery. E—Load, which receives amplified current.*

Figure 6-70. *Circuit board with transistors and other solid-state devices. (Acme Electric Corp.)*

6.6.6 Sensors

In many electronic circuits, the control signal is triggered by a sensor. The sensor is a solid-state semiconductor material. It controls electron flow as its temperature or pressure changes. Pressure sensors can respond to pressure changes in the refrigerating system.

A typical, completely automatic automotive air conditioner uses three sensors in its circuitry:

- One for outside (ambient) temperature.
- One for in-the-car temperature.
- One for the air discharge duct temperature.

6.6.7 Thermistors

A thermistor is a solid-state semiconductor that allows fewer electrons to flow through as its temperature increases. Most thermistors are made of lithium chloride or doped barium titanate.

The resistance changes about 3% for each 1°F change (6% for 1°C). In some circuits, thermistors are used instead of bimetal strips or temperature-sensitive power elements.

The thermistor is used in three ways:

- A temperature-operated electric circuit control.
- To measure temperatures.
- To stop electric power flow to a motor if the windings' temperature increases to the danger point.

A typical thermistor circuit is shown in **Figure 6-71.** The thermistor temperature sensor is used to control the temperature of a room or conditioned space which is heated by a 120 V ac electric resistance heater.

The temperature control, A, is set to the desired room temperature. If the room is below this desired temperature, the sensor, C, will change current flow to transistors D and E. Here the changes are amplified. The thermal relay heater, I, will cause the contact points at J to close. This action brings the electric space heater, K, into operation.

When the desired temperature is reached, the sensor, C, causes the current to the heater coil to be switched off. Then contact points at J will open and stop the flow of current to the space heater. These devices are very sensitive. They will maintain the space temperature within a fraction of a degree.

There is a special thermistor that increases its resistance as the temperature rises. However, it changes from low resistance to high resistance within two degrees. It can, therefore, be used as a switch. At present, these operate between 150°F (65°C) and 356°F (180°C).

The thermistor may be used to control crankcase heaters. (It shuts off current when oil temperature

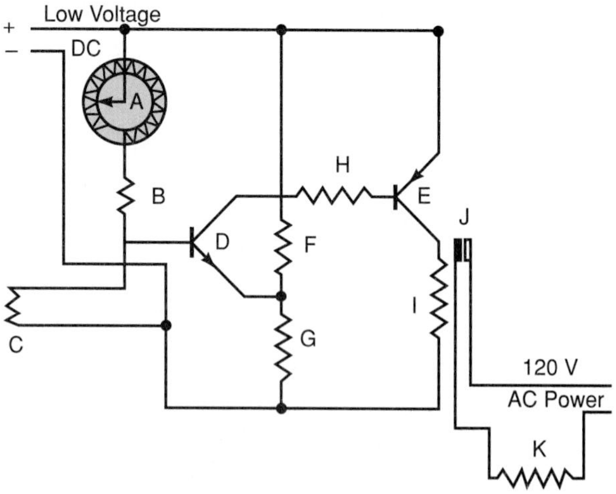

Figure 6-71. *Typical thermistor circuit. A—Temperature control knob, which controls variable resistance. B—Fixed resistance. C—Thermistor temperature sensor. D and E—Transistors. F and G—Bias resistors. H—Current-limiting resistor. I—Thermal relay heater. J—Thermal relay. K—120 V ac electric space heater. Sensor will maintain temperature of heated space within close limits.*

reaches design conditions.) It also may be used to sequence heating systems. The number of heaters in operation would increase or decrease according to heating need.

The thermistor can do the same for a cooling system. It may be used to control a defrosting system on an ice cube maker release circuit. It is small enough to place in motor windings to protect them from too much heat.

6.6.8 Amplifiers

An ***amplifier*** is an electronic device. Upon receiving a small input signal, it will increase it to produce a larger output signal. For example, the signal from a phonograph record is not large enough to be heard. It must be amplified.

Amplifiers are often used in control systems. They increase a small signal from a sensor to a level high enough to control another device. See **Figure 6-72.**

Differential amplifiers are used to determine the difference between a changing input voltage and a constant base voltage.

6.6.9 Transducers

Transducers include a variety of devices that are sensitive to changes in intensity of some form of energy. The transducer responds by controlling the intensity of some other form of energy.

Transducers may be operated by pressure, temperature, fluid flow, vibration, electrical potential, and other means.

A small varying current flow through a transducer may be amplified by an amplifier. The amplified current may then operate a control circuit.

An application of a transducer is shown in **Figure 6-73.** In this application, a pressure transducer is connected to a pipe carrying fluid under pressure. The transducer will change pressure variation into electric current variation.

In the amplifier, electric current variations are amplified and connected to a relay. This relaying device

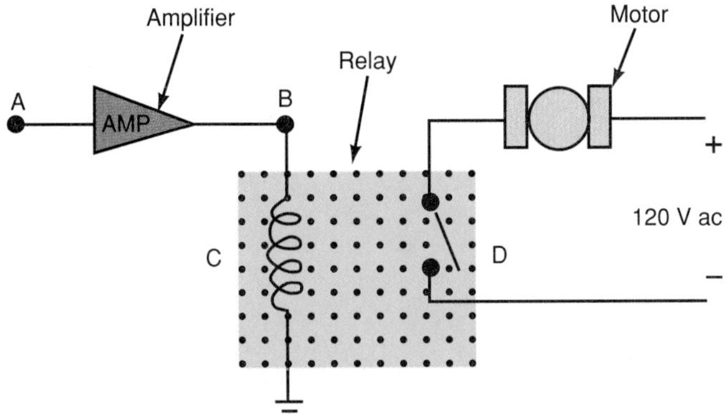

Figure 6-72. *Typical amplifier circuit. Small signal received at A is amplified to produce a voltage at B. Voltage at B is strong enough to energize Coil C and close relay switch D. Relay then turns motor on.*

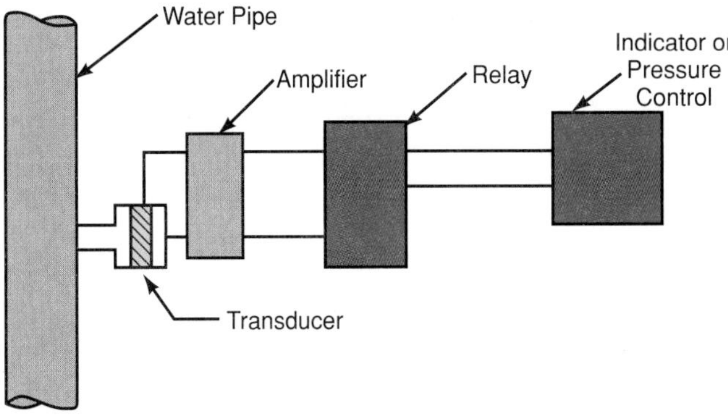

Figure 6-73. *This transducer application indicates pressure in a pipe.*

translates what was a weak pipe pressure variation into an indicator or pressure control. This may be a recorder, pressure gauge, signal light, oscilloscope, or other signal indicating device.

6.6.10 Thermocouple and Thermoelectric

Thermocouples may be used to measure temperatures or to operate controls. The principle of operation depends on many facts. If two dissimilar (different) metals are connected together and the point of connection heated, an emf (voltage) difference will occur. This difference will be across the other ends of the two metals. Copper and iron may be used as the two different metals. However, other thermocouples have been developed using tungsten and rhenium, as well as other materials.

If a thermocouple is connected to a sensitive voltmeter, the voltage indicated varies with the junction temperature. This principle is the basis of operation of the thermocouple thermometer.

Two or more such junctions may be connected in a series. A sensitive voltmeter will indicate the voltage varying directly with the number of the junctions. A multiple thermocouple installation can generate as much as 500 mV. It can, therefore, operate special solenoids. These will operate valves on a gas furnace. It may also be used as a safety device to shut off a gas supply if a pilot light is extinguished.

The electrical efficiency of a thermocouple is quite low. It is not an efficient way to generate electricity.

In 1820, the German physicist Thomas J. Seebeck discovered that when a closed circuit is made through two different metals in contact with each other, an electric current will flow in the circuit when heat is applied to one of the junctions.

In 1834, Jean Peltier discovered that if direct current is passed through a junction of two dissimilar metals, the junction becomes either hot or cold. This is dependent upon the direction of the current flow.

In 1837, Emil Lenz showed the importance of both Peltier's and Seebeck's discoveries. He placed a drop of water on the junction of two dissimilar metals. When current passed in one direction, the drop of water froze. When the current was reversed, the ice melted and the water was warmed.

The principle of the thermocouple, therefore, can be used in various ways. It can be used as a temperature measuring instrument. It can serve as a current generator for some sensitive controls. It may also be used for its refrigerating effect. **Figure 6-74** shows these three applications of the Seebek, Peltier, and Lenz discoveries.

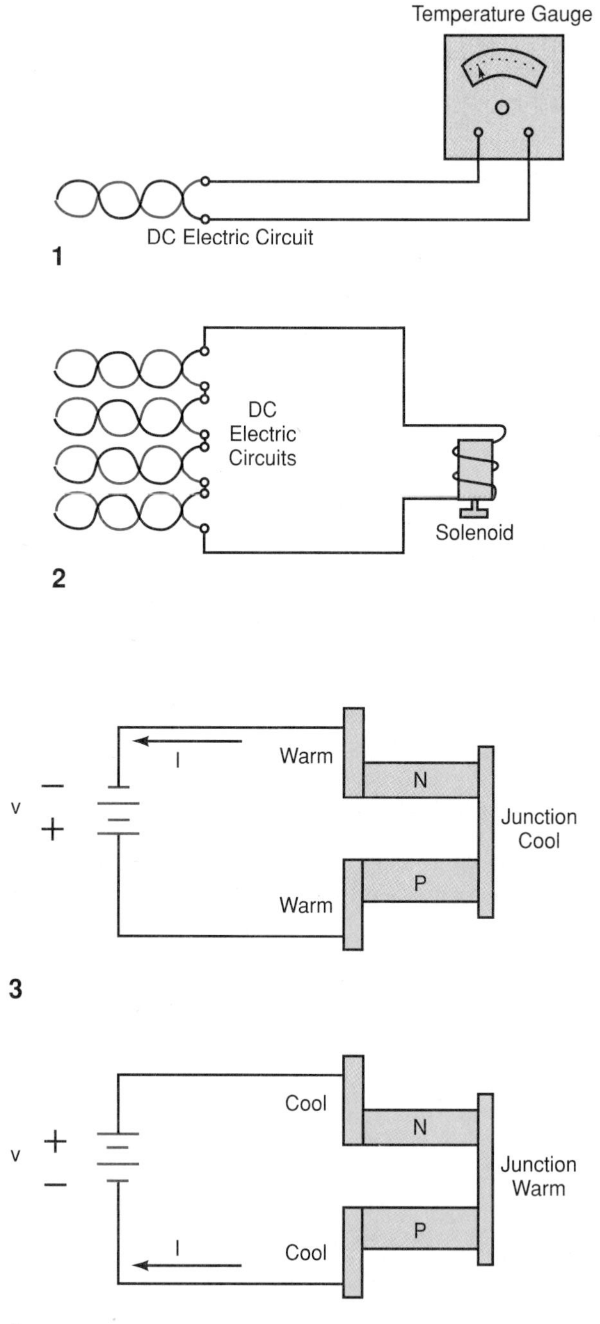

Figure 6-74. *Three basic uses of thermocouple principle. 1—Temperature measurement (used for −300°F to 1700°F temperatures). 2—Generating dc electricity to operate solenoid. Solenoid then operates electric circuit for gas valve. 3—When current enters N-type material, junction cools. 4—When current enters P-type material, junction warms.*

View 1 of **Figure 6-74** shows the thermocouple being used as a temperature measuring instrument. View 2 shows the current-generating effect. This current is used to control a solenoid valve. View 3 illustrates the effect when the thermocouple junction is used for cooling. View 4 illustrates the effect when the thermocouple is used for heating.

6.6.11 Photoelectricity

Photoelectric devices are of three particular types:

- Photoconductor.
- Photovoltaic.
- Photoemissive.

Photoconductors are semiconductor devices that increase their conductivity when illuminated. They are used in electric eye devices and in infrared camera devices.

Photovoltaic devices include solar cells. These are semiconductor devices that produce electrical energy when they absorb light. They are used in solar energy conversion and in light meters for photography.

Photoemissive devices give off light when electrical energy is added. The semiconductor in a light-emitting diode (LED) gives off the light that shows the numbers in a calculator or wristwatch. Other light emitting devices include fluorescent lights and lasers.

6.6.12 Integrated Circuits

Methods have been developed to integrate separate semiconductor circuit components into small self-contained devices. The device that incorporates multiple transistors and other semiconductor devices in a single small component is known as an integrated circuit chip. **Figure 6-75** shows typical chips.

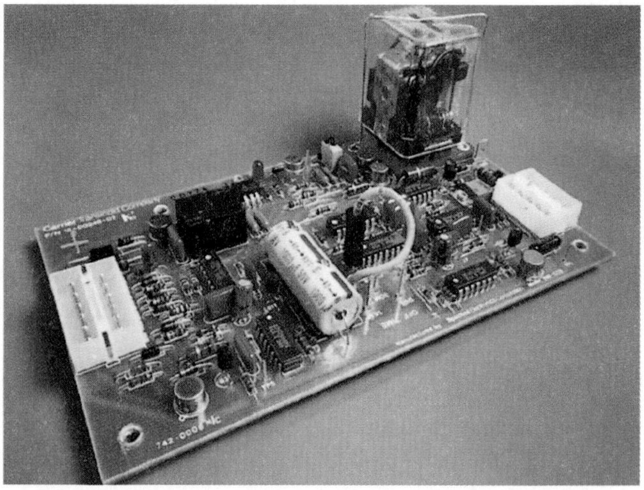

Figure 6-75. *Circuit board using seven integrated chips. Board controls automatic start/stop system for trailer refrigeration. (Carrier Transicold Division, Carrier Corporation)*

An integrated circuit chip is usually constructed as follows:

1. The proper base material is selected. This material is in the form of semiconductor layers similar to those found in a transistor.
2. A circuit is designed and laid out on the chip material.
3. The circuit is "burned" into the material, usually using lasers or acid.
4. Input and output locations are identified and attached to metal connectors (legs) on the chip.
5. The chip is then tested and packaged.

A single component containing many circuits can be designed using the integrated circuit technique. This device is commonly known as a microprocessor. Microprocessors are capable of accepting information, storing it, and reacting in some preset way. See **Figure 6-76.**

The microprocessor is the "brains" of many electronic devices used in modern HVAC (heating, ventilating, air conditioning) systems. Typically, programmable thermostats or the electronic controls on a modern refrigeration system use a microprocessor.

6.6.13 Printed Circuit Boards

A printed circuit board is a support for electronic circuits. Electronic components are combined with resistors, capacitors, and other electrical devices. These form the various circuits on the board. A given circuit board is usually related to a specific function in a device. A series of individual circuit boards are then used like building blocks. The electrical system for that device is constructed with these building blocks.

Circuit boards help to simplify the servicing of electrical systems. See **Figure 6-77.** When a specific electrical function does not work in a device, replacing the circuit board is usually recommended. The failed circuit board can then be repaired at a later time or discarded.

Figure 6-76. *Microprocessor control panel used for large commercial trailer refrigeration units. (Carrier Transicold Division, Carrier Corporation)*

6.6.14 Computers

Computers have assumed an important role in the HVAC industry. They are being used as an integral part of many electronic controls. The increasingly popular programmable home thermostat is shown in **Figure 6-78.** A more complicated computerized industrial control system is shown in **Figure 6-79.** In most cases, a computer supported HVAC system can even provide a diagnostic analysis for that system.

A computer is any device constructed to provide specific outputs based on input data. The electronic components reviewed in this book can be assembled to do that. Most modern computers have microprocessors as their "thinking" component. These microprocessors are then combined with input devices (keyboard, mouse, etc.), output devices (monitor, LEDs, etc.), and data storage to make up the total computer system.

Computers use the on-off characteristics of integrated circuitry. That is, there is one-way current flow of a diode or the relay action of a transistor. This information can be programmed as a series of 1s (on) and 0s (off). All computer devices—from large computers used

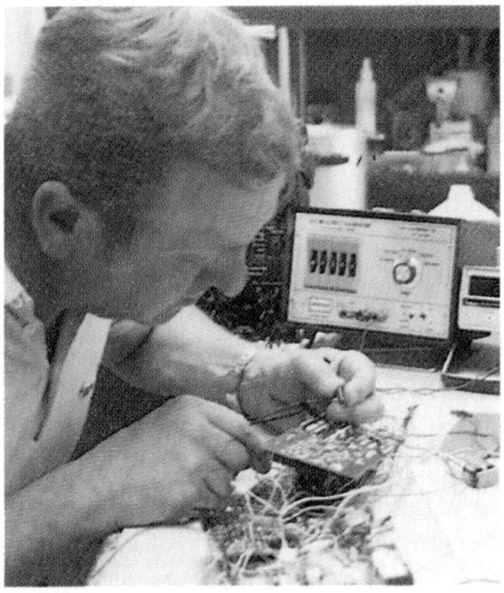

Figure 6-77. *Checking a printed circuit board. (Simpson Electric Company)*

Figure 6-78. *Programmable home thermostat. Unit is designed to control a heat pump.*

Figure 6-79. *Computerized industrial control system. (McQuay, SnyderGeneral Corp.)*

to control corporate HVAC systems to handheld calculators—use the on-off principle of storing and reacting to information.

Using 1s and 0s for all programming of a computer can be very cumbersome. Additional computer languages have been developed to make this process easier. These languages try to make programming more closely related to actual commands or instructions. Some of the more common languages include BASIC, FORTRAN, COBOL, and C++.

6.7 Electrical Power

Electrical power is measured in watts (W), kilowatts (kW), and megawatts (MW). A watt is the rate at which energy is produced by a current of one ampere flowing under an electrical potential of one volt. A simple example is the rate at which heat is given off by a wire connected to a 1-volt battery when the current in the wire is 1 ampere.

This can be expressed mathematically:

Power (watts) = current (amps) × electrical potential (volts)

Formula:

$P = I \times V$

Example:

What is the power used by an electric motor that draws a current of 20 A (amperes) from a 120-V (volt) power source?

Solution:

$P = 20 \times 120 = 2400\ W = 2.4\ kW$

The electrical potential is sometimes called the electromotive force (emf).

In Section 6.3.14, the power loss was indicated as I^2R. If the circuit load is only a resistance load, then the power is lost as heat. The power loss can then be calculated as either I^2R or $I \times V$, where $V = I \times R$, from Section 6.3.13.

A simple electrical circuit represents a dc fan motor speed control in **Figure 6-80.** When the switch is at A, the fan is off, and no current flows. When the switch is at C, the fan is on high speed and the power used is the square current (I_c) times the motor resistance (R_M).

$$P_C = I^2_C R_M$$

$$\text{The current } I_C = \frac{V}{R_M}$$

$$\text{Then } P_C = VI_C = \frac{V^2}{R_M}$$

If the switch is moved to B, the fan is on low speed. The power is then:

$$P_B = I^2_B (R_1 + R_M)$$

$$\text{The current is } I_B = \frac{V}{(R_1 + R_M)}$$

$$\text{Then, } P_B = VI_B$$

$$P_B = \frac{V^2}{(R_1 + R_M)}$$

The power (P_B) used with the resistance R_1 in the circuit is less than the power (P_c) used without the resistance. With the switch on low speed, the power used by the fan motor is lower.

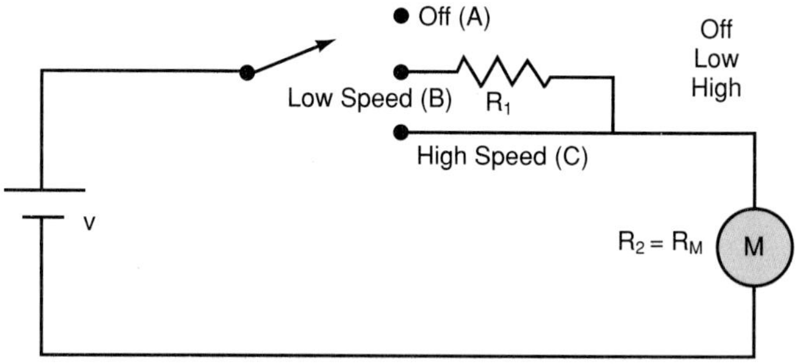

Figure 6-80. *Example of two-speed fan motor circuit.*

6.7.1 Electrical Efficiency—Power Factor

If voltage and current vary within the cycle, as shown in **Figure 6-81,** the power must be calculated differently. The current and voltage are not varying together. Therefore, the product of the voltage and current varies with time in the cycle. An average voltage times current product must then be used to calculate the power.

An average voltage for a voltage that varies as shown is called the root mean square (rms) voltage (V_{rms}). It is equal to the maximum voltage (V_{max}) times a constant.

$$V_{rms} = V_{max} \times 0.707$$

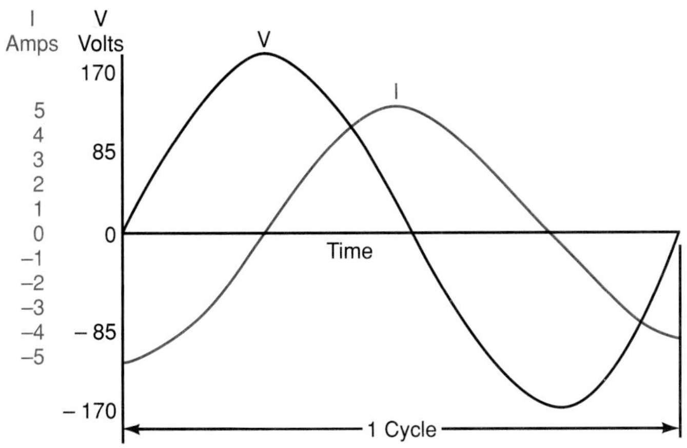

Figure 6-81. *Combined voltage and current flow curves for an inductive load. Current is at a maximum when voltage is zero. This is the situation in a low resistance motor or coil. Power factor is at a minimum.*

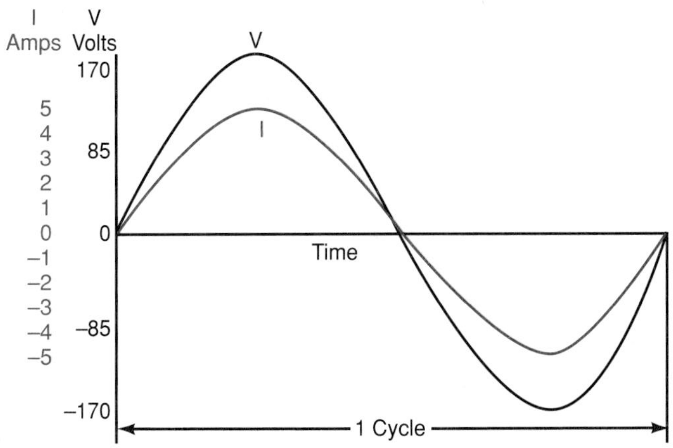

Figure 6-82. *Combined voltage and current flow curves for a resistive load (lamp, heater, or high resistance motor). Current is at a maximum when voltage is at a maximum. Power factor is one.*

Usually, ac circuits are designated by the rms voltage, such as 120, etc. In 1/2 cycle:

$$V_{average} = V_{max} \times 0.637$$

The rms current is also:

$$I_{rms} = I_{max} \times 0.707$$

Voltage and current may vary so that the maximum voltage and current occur at the same time. This is shown in **Figure 6-82.** The power will be:

$$P = V_{rms} \times I_{rms}$$

If the voltage and current maximum do not occur at the same time, the power is multiplied by the power factor, PF:

$$P = V_{rms} \times I_{rms} \times PF$$

The voltage and current are then said to be "out of phase." If the electrical load in a circuit contains inductor or capacitor elements, then the voltage and current are out of phase. Inductive loads include electrical motor windings, transformers, solenoids, relays, and electrical coils. Capacitor loads include condensers and crystals. Elements that produce out-of-phase voltage and current store electrical energy during part of the cycle and then release it later. This produces the out-of-phase behavior.

Inductive load efficiencies, like motors, can be improved by increasing the power factor. This can be done by connecting the proper value capacitor across the motor terminals. The increase in the power factor reduces the current flow (I_{rms}) through the motor resistance. The product

$$P = V_{rms} \times I_{rms} \times PF$$

is smaller.

Electrical utilities usually limit the power factor allowable in industrial and commercial loads. Normally a power factor of at least 0.85 is required.

To increase motor efficiency, the power factor should be brought as close as possible to 1.0. To do this, a wattmeter should be connected to the motor. Capacitors should then be connected across the motor terminals until a minimum wattage reading is obtained. Power draw by the motor will then be at a minimum. The power factor will be as high as possible.

6.7.2 Grounding

Most soil (ground) is a fairly good conductor of electricity. Moist ground is a better conductor than dry ground. In early telephone power distribution systems, the ground was frequently used for the return circuit. A wire for the return circuit was merely extended into the ground. Electricity flowed through the ground to the end of the circuit. The symbol for the ground became as shown in **Figure 6-83.**

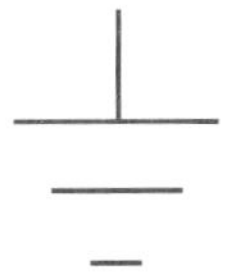

Figure 6-83. *Ground symbol indicates that conductor is attached to frame, shell, or other structural part of a device.*

Now, however, the wire ground refers to an electrical circuit that is attached to the frame, shell, or other structural part of a mechanism. It is common practice to electrically connect the ground connection to a convenient water pipe, but the best ground is a metal stake driven about 8′ into the ground.

Since the 1970s, grounding has been required by Underwriter's Laboratories on all major household appliances. Therefore, all units produced after September 2, 1969, have a three-wire grounded service cord.

Under no conditions is the grounding prong to be cut off or removed. Sometimes a grounded appliance must be installed where there is no three-wire grounded receptacle. The customer is responsible to contact a qualified electrician. A properly grounded three-prong wall receptacle should be installed in accordance with the appropriate electrical code.

If a two-prong adaptor plug is required temporarily, the customer is responsible to have it replaced. It should be replaced with a properly grounded three-prong receptacle. The two-prong adaptor may also be properly grounded by a qualified electrician in accordance with the appropriate electrical code.

The standard accepted color coding for ground wires is green or green with yellow stripe. These ground leads are not to be used as current-carrying conductors.

Electrical components must be grounded. This includes compressors, condensers, evaporator fan motors, defrost timers, temperature controls, and ice makers. They are grounded by using an individual wire attached to the electrical component and to another part of the appliance. Ground wires should not be removed from individual components while servicing. The exception to this rule is if the component is to be removed and replaced.

Grounded components may require servicing that necessitates the removal of the ground wire. It is extremely important that the service technician replace any and all grounds prior to completing the service call. Under no conditions should a ground wire be left off. It is a potential hazard to the service technician and the customer.

Figure 6-84 illustrates a properly grounded receptacle. A grounding adaptor, which may be used temporarily, is illustrated in **Figure 6-85.** A properly grounded receptacle is shown in **Figure 6-86A.** A method used to ground an old style ungrounded receptacle is illustrated in **Figure 6-86B.**

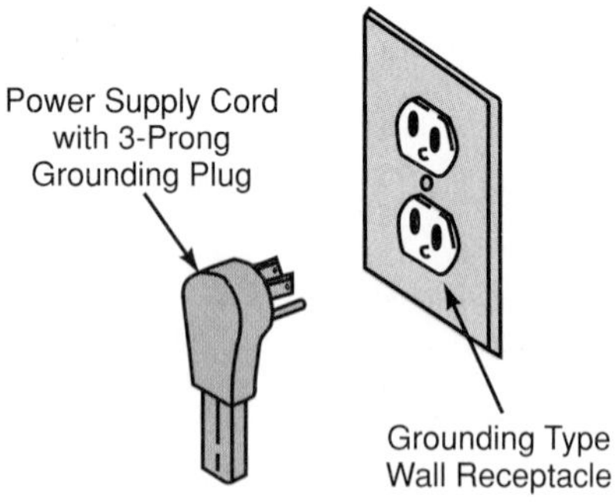

Figure 6-84. *An approved appliance cord receptacle. Slotted openings are for hot and neutral wire. Round opening is ground connection.*

A ground fault protector or circuit interrupter (GFCI) is recommended for use on certain circuits. It is a circuit breaker. It will open electrical supply circuits if as little as 5 milliamperes (.005 amperes) leak out of the circuit and into the ground.

A GFCI is required on outdoor outlets, outdoor lighting, and swimming pools. It should also be used by the technician when using portable tools on extension cords outdoors. These devices are small and may be plugged into an outlet. This device gives the best protection against shock.

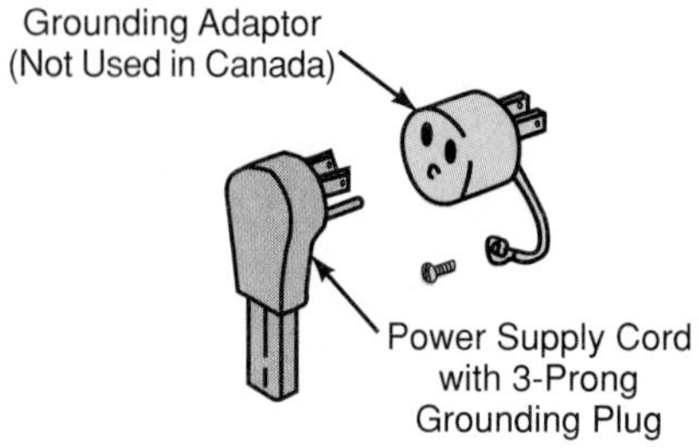

Figure 6-85. *Typical grounding adaptor.*

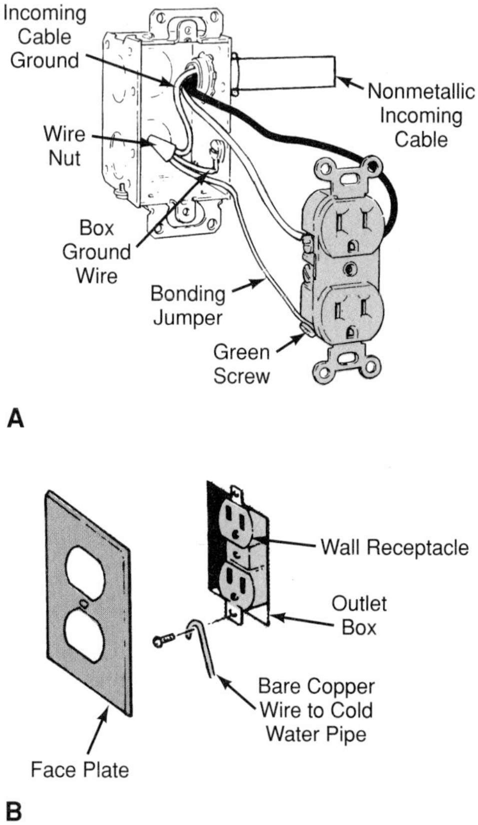

Figure 6-86. *Properly grounded wall receptacles. A—Jumper runs from green screw on receptacle to grounding wire of incoming cable. If outlet box is plastic, grounding wire from cable goes directly to green screw on receptacle. B—How to ground an old style ungrounded receptacle. A grounding adaptor is required.*

6.7.3 Single-Phase—Three-Phase

A single-phase cycle is explained and illustrated in Section 6.2.2. As shown in that section, the voltage and current start at 0. They rise to a maximum, and fall to 0 again as the cycle repeats. There is no power produced during the instant that the voltage and current are 0.

Other cycles may be imposed on the above cycle so that voltage and current flow through the circuit at all times. Such an arrangement is called polyphase. The two most common phases in use are single-phase and three-phase.

The voltage and current characteristics of the three-phase cycle are shown in **Figure 6-87.** Note that, in the three-phase system, there is always a considerable voltage applied.

Three-phase motors are generally more efficient than single-phase motors. Three-phase motor sizes begin at about 1/2 hp and extend upward. Single-phase motors are not commonly used above 1 hp.

6.7.4 Power Circuits

Electric motors used in a system must be designed to be used with the electric utility's power.

These motor properties must match the power source in:

- Emf (volts).
- Cycle (Hertz).
- Phase.

Wires must be large enough to carry the full or maximum current that the motor will use. The voltage may be:

- 110 V.
- 115 V.
- 120 V.
- 208 V.
- 220 V.
- 230 V.
- 240 V.
- 277 V.

The number of cycles per second (Hertz) may be:

- 25.
- 50.
- 60.

The phase may be:

- Single-phase.
- Two-phase.
- Three-phase.
- Four-phase.

Some popular power electrical sources are:

- 115 V, 60 cycle, single-phase.
- 120 V, 60 cycle, single-phase.
- 208 V, 60 cycle, single-phase.
- 230 V, 60 cycle, single-phase.
- 240 V, 60 cycle, single-phase.
- 230 V, 60 cycle, three-phase.
- 240 V, 60 cycle, three-phase.
- 480 V, 60 cycle, three-phase.

Carefully check the power source before purchasing or installing equipment. Check with the electrical utility before installing equipment of any sizable horsepower.

One concern of the electrical utility is the flicker in lights, television sets, and radios when a motor compressor starts. It is especially critical if the motor compressor starts more than four times an hour. If wires are too small or too long, electrical flow to the motor will not be strong enough and the flicker will be worsened.

6.7.5 Transformer Principles

Several types of transformers are used by electrical utilities. They transform high voltage to electrical power satisfactory to the user. The systems used depend on where the power is to be used. The uses may be mainly for residential, for industrial power, or for commercial lighting.

At the generating station, the power is stepped up. It is increased to a voltage considerably above that used by appliances and motors for either domestic service or

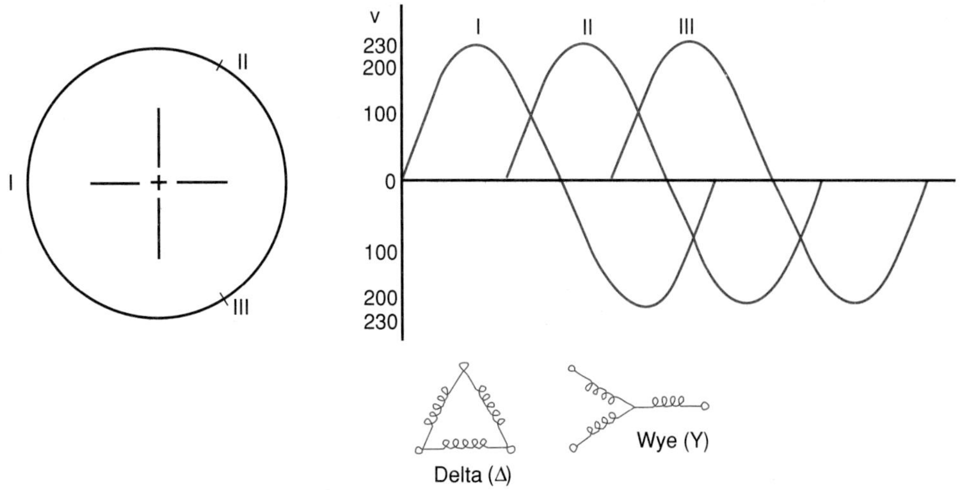

Figure 6-87. *Three sine curves of three-phase circuit. System can be wired for either Delta design or Wye (Star) design.*

by industry. Current flow is usually sent across country over a high voltage transmission line at 120,000 V. A schematic diagram, **Figure 6-88,** shows the basics of a power distributing system.

Step-down transformer stations are located along the high voltage transmission line in areas receiving power. At these region stations, there are step-down transformers. These reduce the high voltage transmission line electricity to 40,000 V.

This is carried to the communities to be served, where it is again stepped down to 13,200 V or 4800 V. Circuits of this voltage must be handled by certified utility line workers.

Electricity is carried to the customers by the primary distributing lines. At the customer's business or home, the electricity is again stepped down to 120 V, 240 V, or 480 V, depending on the customer's needs.

Transformer and Motor Circuits (Characteristics)

Transformers are required to step down high voltage to the final voltage used by the consumer. The power coming into the transformer windings is called the primary. The power going out of the transformer windings is called the secondary. This is true regardless of which of the two are of the higher voltage.

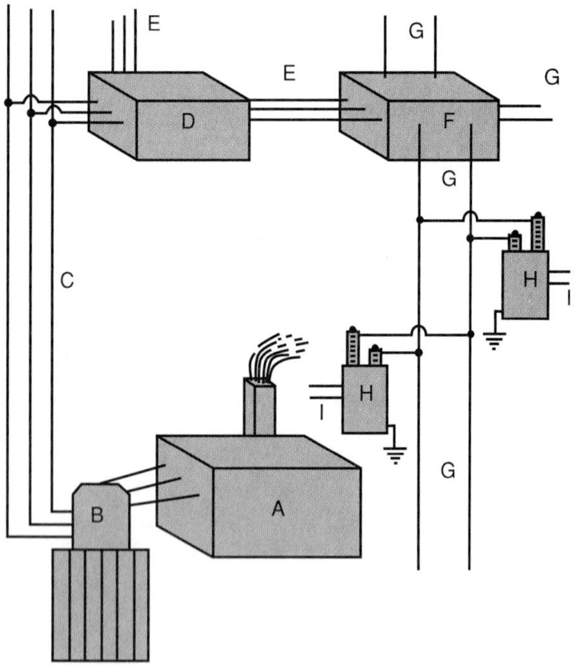

Figure 6-88. *Schematic diagram of power generating and distribution system. A—Steam power generating plant. B—Step-up transformer. Generated power is stepped up to 120,000 V. C—120,000 V transfer lines. D—Regional transformer station. Power is stepped down to 40,000 V. E—Subtransmission line 40,000 V. F—Area transformer station. Voltage is stepped down to 13,200 or 4800 V. G—Primary distribution circuits. H—House or neighborhood transformers. I—Secondary circuit to homes, businesses, and industries. Secondary circuit is 120 V, 240 V, or 480 V.*

The output voltage of a transformer is determined by the ratio of primary to secondary turns. See **Figure 6-89.** For example: If the primary has 100 turns and the secondary has 10 turns, the turn ratio is 10:1. Therefore, if 200 V are applied to the primary, the secondary will put out 20 V.

Common types of transformers are:

- Delta (Δ).
  A. Open Delta.
  B. Closed Delta.
- Wye (Star).

Figure 6-90 shows a wiring diagram for the Open Delta transformer. The Closed Delta transformer will be explained later.

The input voltage from the power station is usually 4200 V. The circuit is three-phase (three hot wires). The Open Delta system is different from the Closed Delta system in that two connected transformers are used (1 and 2) instead of three.

The voltage of each secondary outlet is designed to be 240 V (A to B or B to D). A ground wire is connected to the middle of the secondary winding of 1. Therefore, the emf between D and C is 120 V and B and C is 120 V. The emf between A and C (a geometric change of angle) is 208 V, because the secondary winding is all of 2 and then angles halfway down the secondary winding of 1.

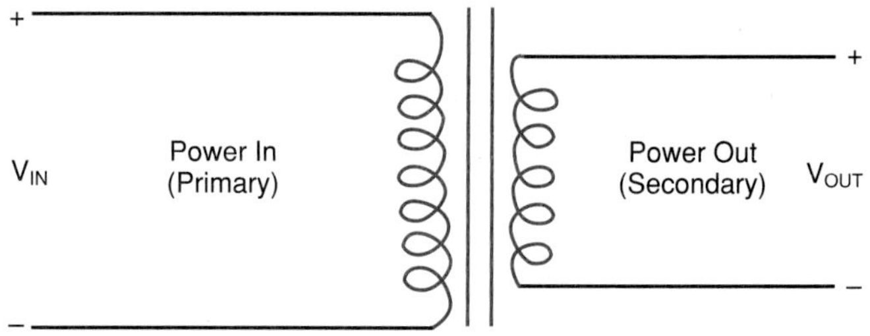

Figure 6-89. *Standard transformer design. Polarity during one part of cycle is shown.*

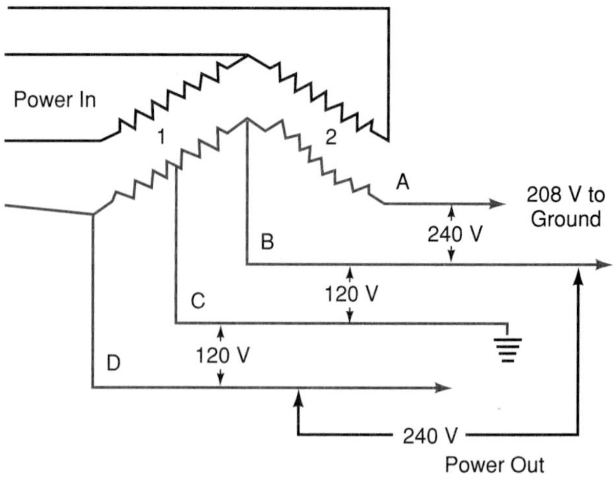

Figure 6-90. *Circuit diagram of Open Delta transformer. Power may be obtained from four taps, A, B, C, and D.*

This 208 V is popular where the building's main electrical load is the lighting load. It is not a good motor voltage, but many motor compressors are connected to this type of circuit. The motor must be designed to operate at this voltage. If it is not, a correction line voltage transformer must be used. (See Section 6.7.5.)

A dangerous condition may develop if there is a voltage drop of over 5% at the motor compressor. (This would be less than 95% of the desired voltage.) For example: If a 208 V circuit, due to its length, conductor size, and ampere flow load, has only 197.6 V (208 V × 95% = 208 V × .95 = 197.6 V), it is at the very lowest usable voltage. The voltage must not go below this value. If it does, the motor may work poorly or the windings may burn out.

A serious situation may occur if a 240 V motor compressor is connected to a 208 V line. (240 × 95% = 240 × .95 = 228 V.) A 228 V voltage is the lowest voltage on which a 240 V motor will work satisfactorily. The 208 V circuit is far below the 228 V circuit desired. The 240 V motor will operate poorly, the overload protection will operate, or the motor may fail.

However, motors can use a voltage slightly over their rating. In fact, a 208 V motor connected to a 220 V line will operate very well. It will start faster, and will give more power to the compressor. A 10% over-normal voltage will let a motor carry a 20% overload. However, a 208 V motor on a 240 V line will be much noisier, a condition normally undesirable in air conditioning.

A 220 V motor operates very well on a 240 V line. A 220 V motor can operate on a 208 V line. However, it will have lower torque. These are single-phase motors. Occasionally, only single-phase is available and more than 3 horsepower is needed. In this event, two or more motor compressors should be used in the system.

Figure 6-91 is a table of changes in a motor's operating characteristics as the input voltage changes.

A 240 V motor compressor works with the same efficiency as a 120 V motor compressor. It is not true that 240 V motors use less kilowatt hours to do the same amount of work. For example, a 120 V motor uses 5 A to create 600 W (120 V × 5 A = 600 W) or .600 kW. To provide .600 kW, a 240 V circuit must carry 2.5 A, but the electrical cost is the same. The only advantage is that the 240 V unit may use smaller conductors from the meter box to the unit. There will be less voltage drop between the meter box and the motor.

Some motors are labeled 208–220 V. This indicates that they may be used with either voltage. However, these motors are sensitive to voltages below 208 V. Voltages of 195 V or lower must not be used.

For example: In a home air conditioner, the power circuit is usually 240 V, one-phase, three wire. This means there is 120 V between the hot conductors and the ground conductor. This is shown in **Figure 6-92.**

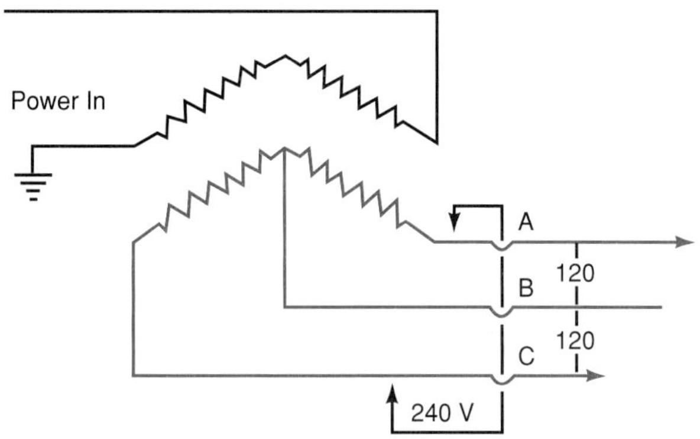

Figure 6-92. *Schematic circuit diagram of transformer used to serve average home. Note that 120 V are available between A and B, and between B and C. B is a ground. Also, 240 V are available between lines A and C.*

These voltages should be checked (verified) with a voltmeter. Do not trust a test light and guess at its bulb brilliance.

When studying a building wiring system to decide on the circuit necessary for a motor compressor, the wire sizes should be checked for capacity. All electrical appliances should be on at the same time. Also, the average load per day should be checked. A recording ammeter may be used for this purpose if the utility does not already have the data. A demand meter will usually be put in by the electrical utility if requested.

Line Voltage Transformer

Improper line voltage may cause refrigeration and air conditioning motors to burn out. Line voltages may vary as much as 5% to 10%. Equipment designed to operate on 120 V (plus or minus 10 V) may not operate well if voltage drops below 100 V.

When equipment designed for 120 V is used on 240 V lines, a line voltage transformer is used. This overcomes the problem of voltage fluctuation. The line voltage transformer increases and decreases the input voltage to the motor as needed.

Figure 6-93 shows a basic wiring diagram of a "boost-and-buck" voltage transformer being used to

Input Voltage Variations versus Motor Characteristics					
Input Voltage	Current	Torque	Temperature	Cycle Slip	Efficiency
+10% −10%	− 7% +11%	+21% −15%	−3°C = (5.5°F) +7°C = (12.5°F)	17% decrease 23% increase	1% increase 2% decrease

Figure 6-91. *Effect of input voltage on operating characteristics of motor.*

correct the line voltage to 120 V. **Figure 6-94** shows a wiring diagram of a transformer that changes a 208 V circuit to 240 V. Some line voltage transformers have one coil. Part of the secondary coil is shared with the primary coil. This is usually called an ***autotransformer,* Figure 6-94B.**

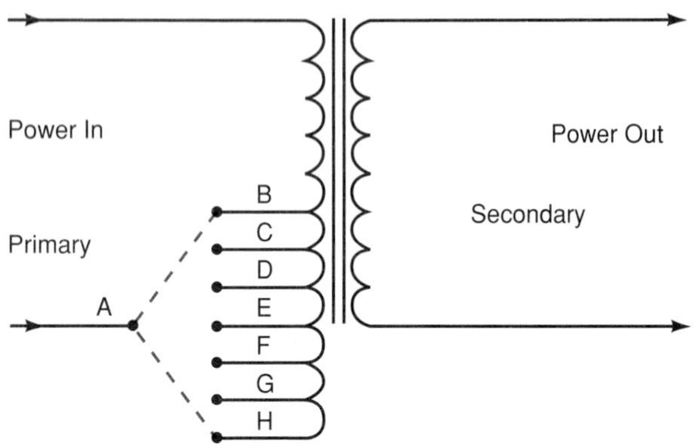

Figure 6-93. *Wiring diagram of transformer designed to step up or step down voltages to provide correct voltage for driving motor. Connecting terminal A to B, C, or D will step up power-out voltage. Connecting A to F, G, or H will step down power-out voltage. Connecting A to E will not change voltage, but will match load to source and will limit power surges. A transformer like this used as an isolation transformer can improve safety against electrical shock.*

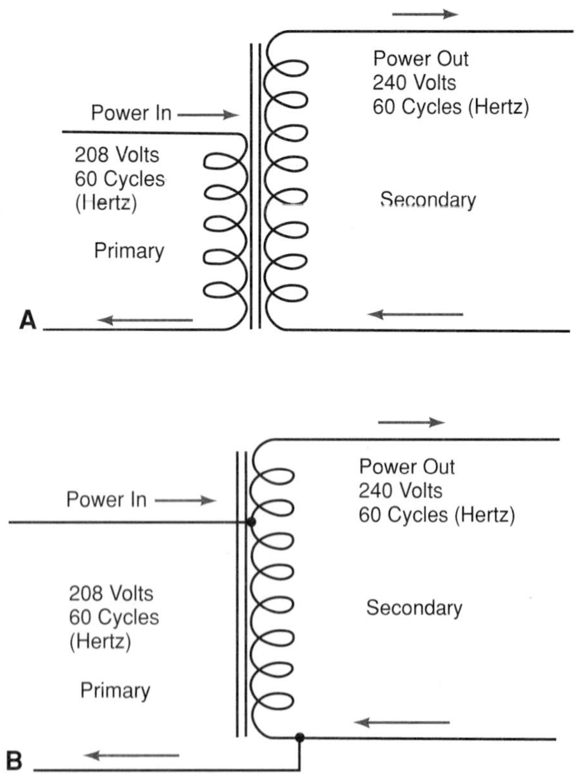

Figure 6-94. *A—Wiring diagram of voltage transformer used to step up 208 V to 240 V. B—An autotransformer used to step up 208 V to 240 V. Note that part of the secondary coil is shared with the primary coil.*

A voltage of 208 V is quite common when the main electrical load is lighting. A line transformer will be required if 120 V or 240 V motors are to be used on the 208 V current supply. This also is called an autotransformer. See **Figure 6-95.**

A 120 V transformer output can be produced from an input ranging from 95 to 260 V. The technician must first determine the required voltage from the motor identification plate and specifications.

A voltmeter is used to measure the voltage and ampere flow at the motor compressor. It is measured when the unit is running. A transformer is then selected that will carry the current. It is adjusted to raise or lower the voltage as needed.

Transformers are rated in kVA or VA output. The abbreviation, kVA, means "kilovolt amperes." VA means "volt amperes." To determine kVA, the output voltage is multiplied by the amperes.

$$\text{kVA} = \frac{\text{Volts} \times \text{Amperes}}{1000}$$

There is often confusion between volt amperes and watts. The difference is explained as follows. Suppose a 10 kVA transformer is supplying power to a load that is partially inductive. The load voltage multiplied by the load current is 10 kVA. Because of the inductance of the load, the power factor is 0.8. The quantity of kilowatts supplied by the transformer is calculated as follows:

$$\text{kW} = \text{kVA} \times \text{Power Factor}$$
$$\text{kW} = 10 \text{ kVA} \times 0.8 = 8 \text{ kW}$$

Only 8 kW are being supplied to the load for producing heat or useful mechanical motion. Although the transformer has a 10 kVA rating, it supplies a real power of 8 kW. The transformer will supply 10 kW to a purely

Figure 6-95. *Line transformer used to transform 208 V ac to 240 V ac. (Acme Electric Corp.)*

resistive load, such as an electric space heater. Transformers that drive inductive or capacitive loads must always be larger and heavier than transformers that drive resistive loads.

The total load of the circuits fed by the transformer must be less than the transformer output. The transformer is connected into the circuit between the electrical service and the motor. The technician should check the completed installation for both ampere flow and voltage. If these values are different than the motor ratings, further adjustments will be necessary.

Three-Phase, Four-Wire Transformer

Another transformer design used by utilities is the Wye type shown in **Figure 6-96.** This system uses one transformer, connected as shown. Note than a 208 V circuit is also possible from this system. In fact, any two of the three conductors produces 208 V. This is made possible by the angle of the two secondary windings.

Some utilities recommend a 277–480 V system. It is the least critical and therefore is good for commercial/industrial use. This system is shown in **Figure 6-97.**

Utility circuits may show as much as 212 V at the meter on a 208 V system, or they may show as much as 250 V on a 240 V system. A voltage increase is as much a disadvantage to lightbulb life as it is an advantage for a motor compressor. Therefore, the utilities keep their circuit voltages as close to the required value as possible.

A 277 V circuit is used for some refrigeration or air conditioning units of 15,000 to 35,000 Btu/hour capacity. It may be obtained by an angular tap from a Wye type transformer. One lead is from any of the three legs. The other lead is from the center of the Wye. The secondary voltage—between A and B and between A and C—is 480 V.

The Closed Delta system, **Figure 6-98,** is sometimes found in large industries. The transformer is not grounded. The voltage between any two terminals in **Figure 6-98** is 480 V.

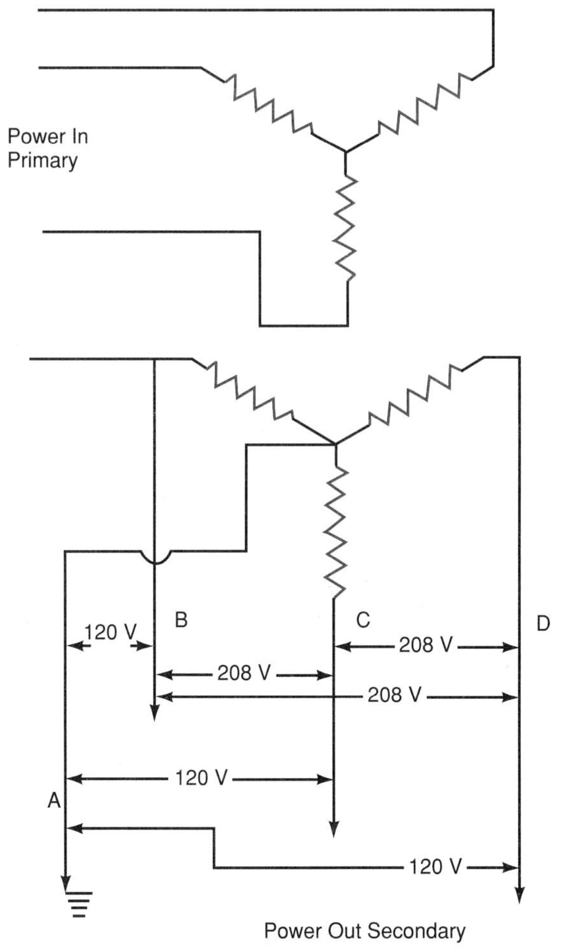

Figure 6-96. *Wye transformer. Note that center of Wye is grounded in secondary circuit. Choice of emf may be obtained from this transformer by taking power from pairs of terminals as follows: A and B—120 V. A and C—120 V. A and D—120 V. B and C—208 V. C and D—208 V. B and D—208 V.*

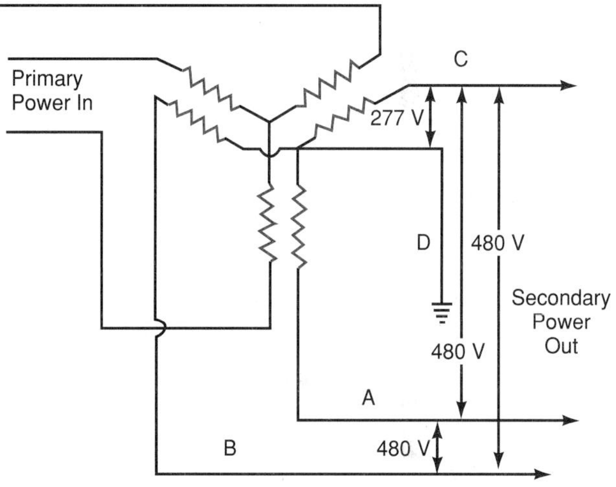

Figure 6-97. *Schematic of Wye transformer used in serving industrial and commercial applications. The following voltages are available from this transformer: Between C and D, 277 V; between A and B, 480 V; between A and C, 480 V; between B and C, 480 V.*

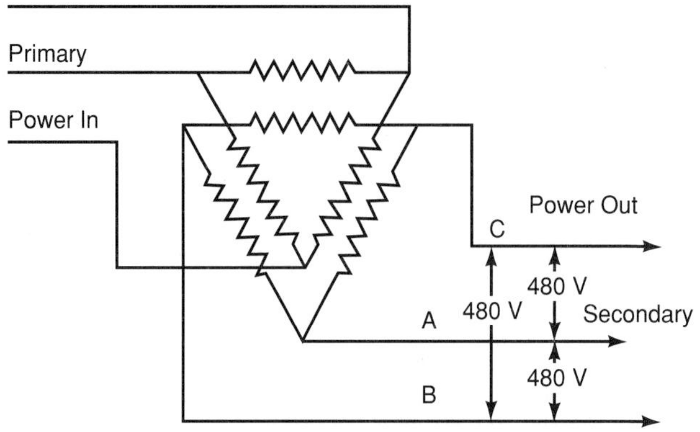

Figure 6-98. *Diagram of Closed Delta type transformer that will deliver 480 V between either A and B or A and C.*

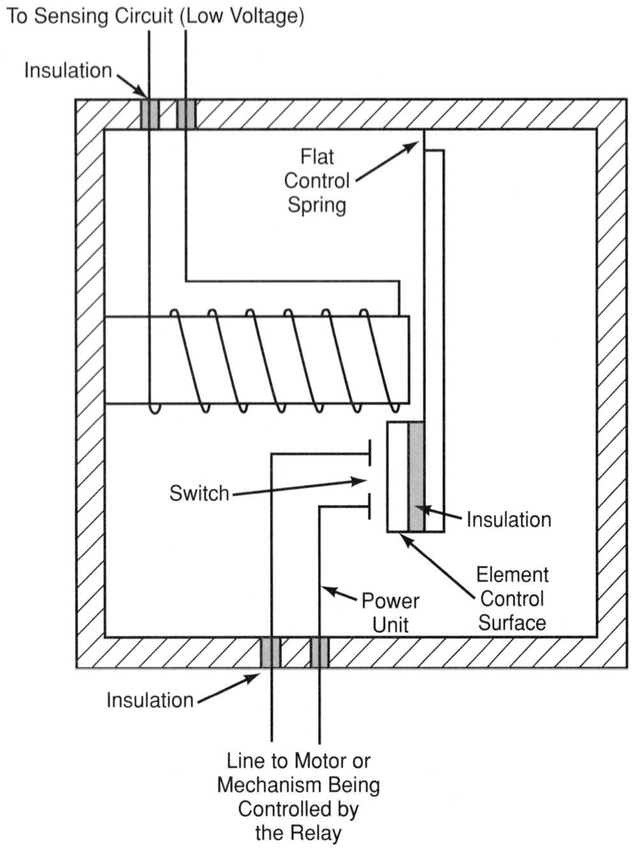

Figure 6-99. *Electrical relay. Low-voltage thermostat or sensing instrument controls operation of relay. Relay in turn switches power circuit on and off. Power circuit is normally off (called normally open, NO). Power circuit is on only when relay low-voltage circuit is energized. Some relays are normally closed, NC.*

6.7.6 Relays

Various types of relays are used in electrical circuits on refrigerating and air conditioning mechanisms. Popular relays include fan relays, motor starting relays, and main power line relays.

These relays usually have an enclosed switch operated by an electromagnet. Relays are designed to protect motors. They disconnect the motor from the line when it becomes overheated or overloaded. The amount of current used to operate a relay is very small.

The contact points in the relay then operate circuits that can carry quite heavy currents. **Figure 6-99** shows the electrical circuit for a typical relay.

Electronic relays are also now in use. These relays are usually self-enclosed devices. They have connections for *normally open* (NO), *normally closed* (NC), and input/controlled voltage.

6.8 Electrical Codes

The National Electrical Code (NEC) establishes rules and regulations covering materials and methods used in installing electrical systems. In addition, many cities and communities have supplementary local codes. All electrical installations should be made in conformity with national and local codes.

The National Electrical Code provides for a type of wiring called Class 2. These circuits are usually used for controlling relays, bells, signal systems, and communications. The current is usually supplied by a small transformer having two windings. These transformers have a capacity of about 100 volt amperes.

Refrigeration and air conditioning service technicians are permitted to make Class 2 connections and installations. Under this Code, the current from the secondary winding of the transformer is limited according to the voltage:

Below 15 V—up to 5 A
15 to 30 V—up to 3 A
30 to 60 V—up to 1.5 A
above 60 V—up to 1 A.

Underwriters Laboratories tests and approves electrical components such as switches, extension cords, and relays.

6.9 Circuit Protection

Previous paragraphs explain the heating and magnetic effect of current flowing through electrical circuits. Appliances may be seriously damaged or ruined by overheating. Instruments may be damaged by too much magnetism.

Safety devices are used to avoid the possibility of an accidental current surge. There are four types of these safety devices in common use.

The most common protection is the fuse. See **Figure 6-100.** This is a soft metal conductor placed in series with the circuit with either a plug or a cartridge arrangement. The size, length, and material in the fuse are such that its resistance will cause enough heat to melt the fuse. Therefore, the fuse will open the circuit if the current exceeds the fuse rating.

Fuses are rated in amperes. Those used in refrigeration and air conditioning circuits are usually designed

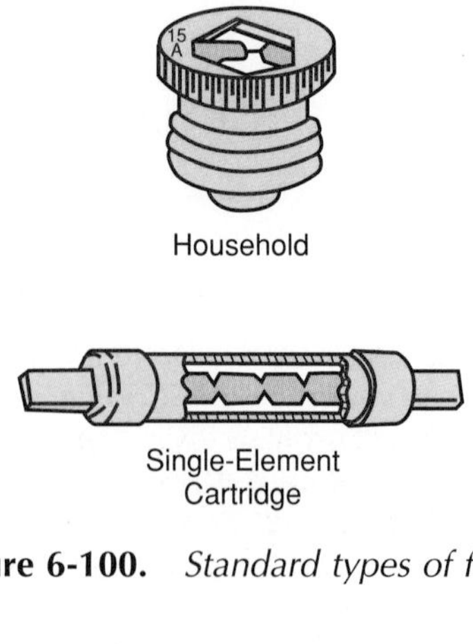

Figure 6-100. *Standard types of fuses.*

to carry 5, 10, 15, 20, or 30 A. Fuses used in electronic circuits are available from 1/500 A to 2 A. Some electronic fuses, however, are available with higher ampere ratings.

The circuit breaker is another protective device. See **Figure 6-101.** Current flowing through the circuit being protected passes through a solenoid in the breaker. The magnetic effect of the solenoid current may cause the solenoid to trip a spring-loaded switch. This occurs if the current in the circuit exceeds a predetermined level (amperage).

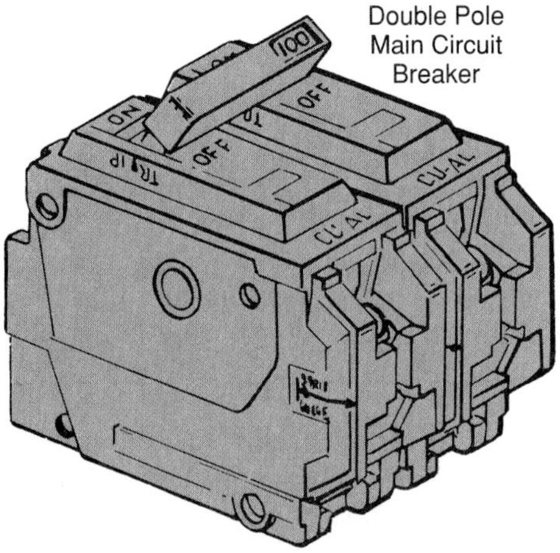

Figure 6-101. *A circuit breaker will automatically switch to "off" during an overload. (General Electric Co., Wiring Devices Dept.)*

Figure 6-102 shows the exterior and a cut-away of a circuit breaker. The function of this switch is to open the circuit when overload occurs. The red portion is the trip indicator. When the current exceeds the set limit, the circuit is broken. The red indicator will show in the small window.

Another common circuit breaker device has a bimetal breaker. It is usually connected in series with the breaker point and a resistance heater. The current flowing in the circuit will cause the resistance heater to heat the bimetal strip causing it to bend. The heating effect will depend upon the current flow. The more current, the more heat. The instrument will open the points when the current flow is greater than safely allowed.

The fourth type of circuit protection makes use of the thermistor. Thermistors may regulate (modulate) the current flow. In some cases, they may cause the current flow to be reduced to a safe value.

Caution: Never connect alternating current appliances or instruments into direct current circuits. Never connect direct current appliances or instruments into alternating circuits.

6.9.1 Wire Sizes

The current-carrying capacity of a solid conductor depends on its diameter. Larger wires may carry a heavier current than smaller wires. The following information is based on the use of copper conductors (wires).

A

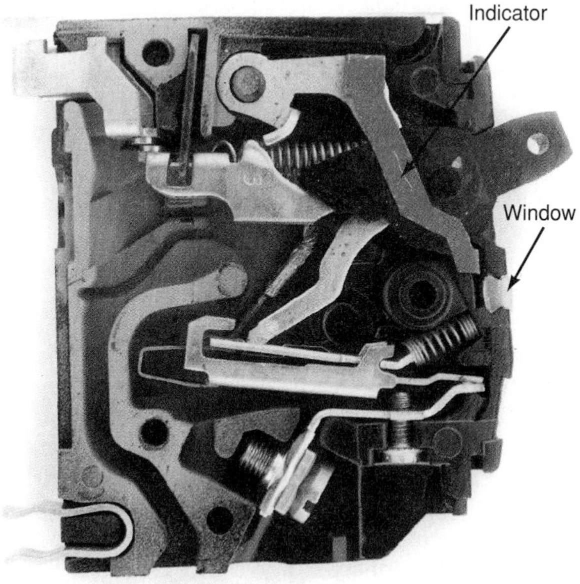

B

Figure 6-102. *Circuit breaker. A—Exterior view; note location of window. B—Cutaway. (Square D Company)*

The electrical supply wire to outlet receptacles must be of adequate size. Fifteen- and twenty-ampere outlets should be supplied with No. 12 wire. No. 10 wire should be used for 30 A outlets. No. 8 stranded wire should be used for 40 A outlets.

Wire sizes are measured in circular mils. A mil is the area of a circle 1/1000″ (.001″) in diameter. **Figure 6-103** is a table of the more common conductor sizes.

The cross-sectional area of a No. 10 wire is about 10,000 circular mils. Its resistance is about 1.0 ohm per 1000′. A No. 7 wire has approximately double the circular mil area of a No. 10 wire. It has half the resistance (approximately 0.5 ohm per 1000′). A No. 4 wire has a circular mil area of approximately 40,000. Its resistance is approximately .25 ohms per 1000′. The circular mil area doubles every third size as the wire becomes larger and the resistance is cut in half. Using this data, the wire size and resistance for commonly used conductors can be calculated.

Wire Size AWG*	Conductor Diameter in Inch Decimals	Circular Mils	Ohms per 1000 ft. at 68 °F
14	.0641	4,110	2.52
13	.0720	5,180	2.00
12	.0808	6,530	1.59
11	.0907	8,230	1.26
10	.1019	10,380	1.00
9	.1144	13,090	.7925
8	.1285	16,510	.6281
7	.1443	20,820	.4981
6	.1620	26,240	.3952
5	.1819	33,090	.3134
4	.2043	41,740	.2485
3	.2294	52,620	.1971
2	.2576	66,360	.1563
1	.2893	83,690	.1239
0	.3249	105,600	.09825

* American Wire Gauge

Figure 6-103. *Wire size data for rubber or thermoplastic covered conductors.*

6.9.2 Attachment Plug Configurations (Terminals)

It is sometimes necessary to connect electrical devices using flexible leads and attachment plugs. Most electrical devices are designed for a particular power supply specification. It is important that connections to the power supply conform to the electrical specifications for the equipment.

For instance, an appliance designed for 120 V should not be connected into a 240 V circuit. The appliance would very quickly "burn out." Likewise, an appliance with protection up to 15 A should not be connected into a circuit of 30 A capacity. It could be burned out, or the safety device could be ruined.

The National Electrical Manufacturers Association (NEMA) has established attachment plug configurations (terminals). **Figure 6-104** shows receptacle and plug configurations commonly used with refrigeration and air conditioning equipment.

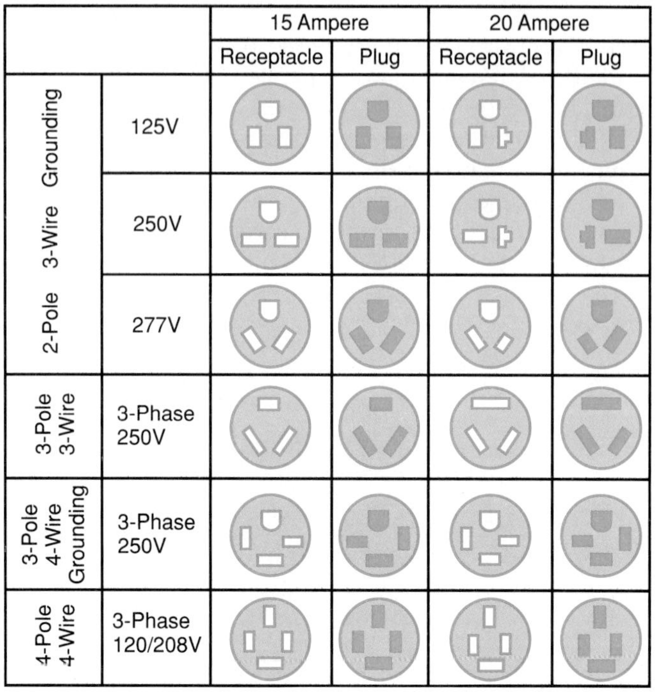

Figure 6-104. *NEMA configurations are shown for general purpose, nonlocking plugs and receptacles. Blade spacing and configuration vary with specified amperage.*

6.9.3 Switches

Any device used to open or close an electrical circuit is called a switch. Switches are made of many materials and there are numerous designs. Some are manually operated, while some are operated automatically.

Figure 6-105 shows the basic types of switches. All are open contact point. Other types, such as knife blade contact and mercury-in-a-tube contact, use the same basic arrangements. Normally closed (NC) switches have the contacts touching when there is no power in the circuit. Normally open (NO) switches have the contacts separated when there is no power in the circuit.

6.9.4 Circuit Testing Instruments

A test light, **Figure 6-106,** makes it easy to test electrical circuits without causing damage. The test light with the two prongs may be used when the device being tested is connected to electrical power.

For example: A refrigerator, air conditioner, or heating system is plugged in and will not start. This test light (about 25 to 50 W) is used to determine if there is power to the wall outlet. The lighting of the light will indicate power is available up to the wire probes. If the wall outlet has power, then the open circuit is in the refrigerator wiring.

The test light shown in **Figure 6-106B** is used to check electrical devices not connected to power. This test light should be connected to a power source. If the bulb lights when the probes are touched together, the test light is functioning.

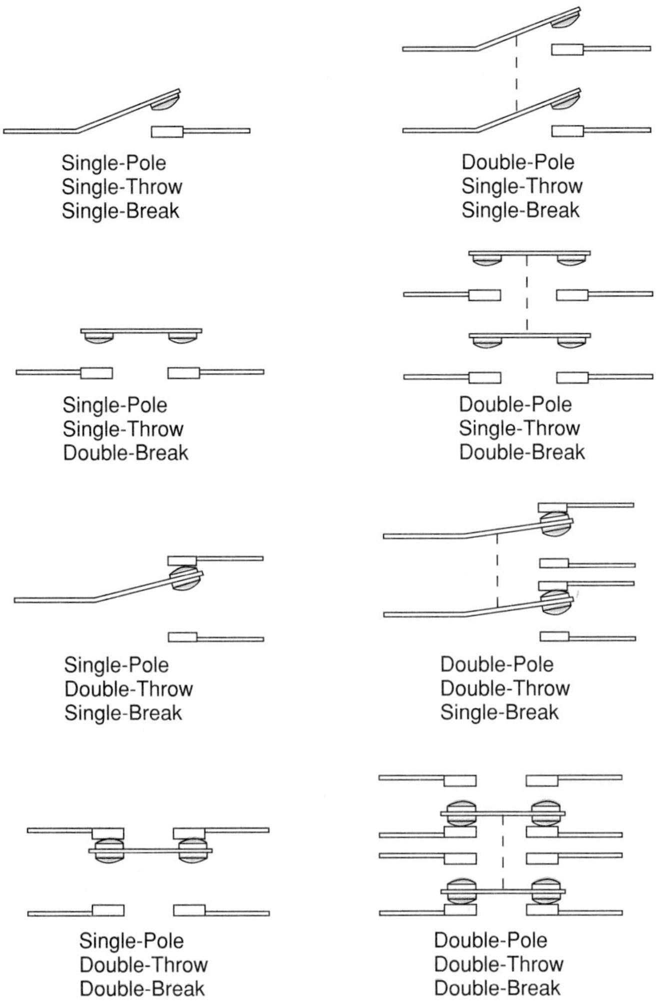

Figure 6-105. *Types of switches. Orange contacts indicate movable breaker points. (Micro Switch, Div. of Honeywell, Inc.)*

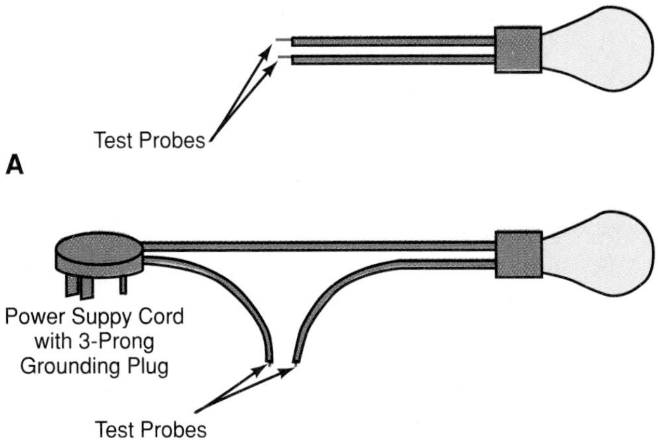

Figure 6-106. *Two types of test lights. A—Light bulb with two probes is used to test circuits that have power on. B—Test light connected to electric power source is used to check circuits not connected to power source.*

To use this light, one probe is placed on one end of a wire. The other probe is placed on whatever that wire is supposed to connect to. If the bulb lights, the circuit is continuous (there is continuity).

Figure 6-107 shows a special test light used to test a three-phase circuit. It may be used to determine which leads of a three-phase circuit connect to the three-phase power lines. If the tester glows, the rotation is 1, 2, 3. Reverse any of the two leads and the tester should not glow. If the tester glows when connected in any order, one of the three-phase circuits is open.

These test lights should not be used on solid-state circuits. They may damage diodes, transistors, and other parts.

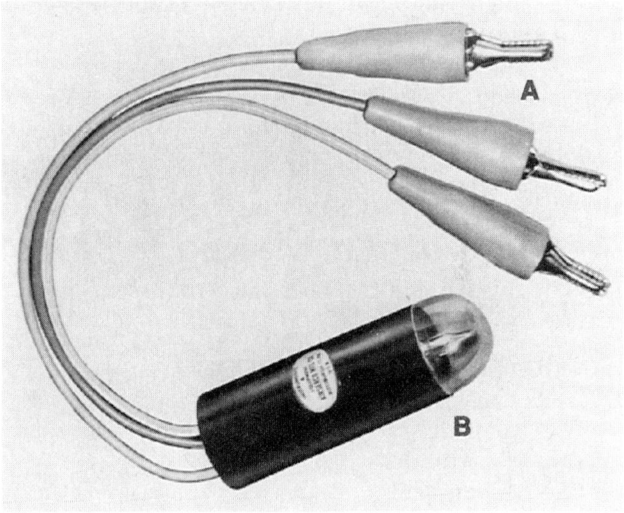

Figure 6-107. *Test light for testing a three-phase circuit. Connect three leads to three-phase terminals. If tester bulb glows, rotation is 1, 2, 3. If it does not glow, reverse pairs of leads until it does glow. If tester glows regardless of connection order, one circuit is open. A—Terminals. B—Bulb.*

6.10 Review of Safety

There are two different factors to be considered in reviewing safety in this chapter.

- ***Safety to the operator.***
- ***Safety to the equipment.***

One must have great respect for electricity. It only requires 0.25 mA to seriously injure or kill a person. Voltages as low as 30 V can force this much current through vital organs. Never handle live circuits when in contact with pipes, other wires, damp floors, damp ground, or water.

Remember: Electricity cannot be seen. Only its effect is seen in light, heat, and magnetism. It is too late to learn that a circuit is live after touching it. Use instruments to determine if the wires or equipment are safe before starting to work.

Always use insulated tools (or dry rubber gloves) when working on circuits of 30 V or higher. Even a large battery can injure one if a tool, wrist watch, or ring short circuits it. The metal will become hot enough to cause severe burns.

Before working on household, commercial, or industrial electrical circuits, disconnect the power. Make sure no one can turn the power on while someone is working on the circuit. Use signs, locks, or other devices.

Equipment and instruments used in refrigeration and air conditioning, while durable, are quite sensitive to abuse. When connecting an electrical instrument into a circuit, make sure that the instrument and its setting are within the voltage and current range of the instrument. An instrument adjusted to measure a voltage within a 150 V range will be ruined if connected into a 440 V circuit.

When capacitor testing, the capacitor should not be left in the circuit any length of time. Usually, a second or two is sufficient. A longer period may destroy the capacitor.

6.11 Test Your Knowledge

Please do not write in this text. Place your answers on a separate piece of paper.

ELECTRICAL FUNDAMENTALS MODULE

1. _______ is a unit of electromotive force.
   A. Kilowatt
   B. Microfarad
   C. Milliampere
   D. Volt
2. _______ is a unit of electric current.
   A. Kilowatt
   B. Microfarad
   C. Ampere
   D. Volt
3. _______ is a unit of capacitance.
   A. Kilowatt
   B. Microfarad
   C. Milliampere
   D. Volt
4. _______ is a unit of electrical power.
   A. Kilowatt
   B. Milliampere
   C. Ohm
   D. Volt
5. Why is it incorrect to say "ac current?"
   A. "Ac current" would mean "alternating current current."
   B. Ac is actually voltage.
   C. Current applies only to dc.
   D. Both B and C.
6. What measurement of electricity is similar to "gallons per hour" of water?
   A. Electric charge.
   B. Electric current.
   C. Kilowatt hours.
   D. Power loss.
7. Which of these formulas is *not* Ohm's Law?
   A. $E = IR$.
   B. $I = E/R$.
   C. $R = E/I$.
   D. All three formulas are Ohm's Law.
8. What is the difference between an "open" circuit and an "open" AEV?
   A. The open circuit is ac.
   B. The open circuit prevents flow. The open AEV allows flow to occur.
   C. The open circuit produces heat.
   D. None of the above.
9. How long does current flow in one direction when 60 Hertz (60 cycle) current is used?
   A. 1/120 second.
   B. 1/60 second.
   C. 0.030 second.
   D. 0.060 second.
10. What is the formula for power loss?
    A. $P = ER^2$.
    B. $P = E^2R$.
    C. $P = IR^2$.
    D. $P = I^2R$.

APPLIED ELECTRONICS AND ELECTRICITY MODULE

11. Thermistors are used in refrigeration work _______.
    A. to measure temperatures
    B. to protect an overheated motor from burn-out
    C. as a temperature-operated electric circuit control
    D. All of the above.
12. Which operating condition is most likely to overheat a motor?
    A. Full speed, full load.
    B. Full speed, no load.
    C. Half speed, full load.
    D. Locked rotor.
13. How is a transistor similar to a pilot-controlled valve?
    A. They both contain an electromagnet.
    B. They are both made of copper.
    C. In both, a small flow is used to turn on or off a larger flow.
    D. None of the above. They are totally different.
14. Which is the most popular polyphase circuit?
    A. 2-phase.
    B. 3-phase.
    C. 4-phase.
    D. 6-phase.
15. What is the integrated circuit device used as the "brains" in many HVAC programmable controls?
    A. Farad.
    B. Microprocessor.
    C. Amplifier.
    D. Voltage.

16. What does an amplifier do to a small input signal?
    A. Increases signal.
    B. Decrease signal.
    C. Signal remains constant.
    D. None of the above.
17. When the temperature is increased in a thermistor, what happens to the electrical resistance?
    A. Increase.
    B. Decrease.
    C. Remains the same.
18. Rectifiers are electronic valves that allow the flow of current through them in how many directions?
    A. One.
    B. Two.
    C. Three.
    D. Four.
19. Computers use what characteristics of integrated circuits as the basis for their performance?
    A. On.
    B. Off.
    C. On-off.
    D. Two-way electric flow.
20. Electricity is carried through a primary distributing line to a business or home at what voltage?
    A. 120 V, 240 VB, and 480 V.
    B. 110 V—120 V.
    C. 120 V—240 V.
    D. Any of the above.

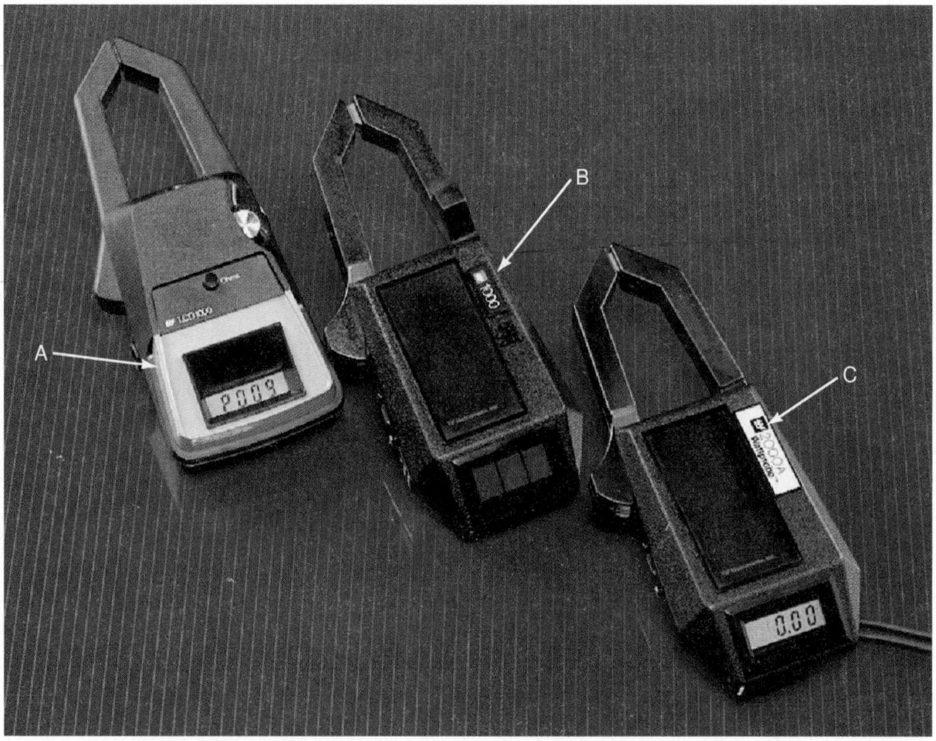

Three types of electrical analyzers, all with digital readouts. A—Digital clamp-on ammeter. B—Digital clamp-on meter, used for amperage, voltage, and resistance (ohm) readings. C—Wattprobe, used to calculate power factor. Provides readings from 0.1 kW to 199.9 kW. (TIF Instruments, Inc.)

Chapter 7

ELECTRIC MOTORS

Modules:

Key Words:

capacitor-start, induction-run motor
dielectric
end bells
field winding
hermetic unit
permanent split capacitor motor
repulsion-start, inductance-run motor
serpentine belt
shaded-pole motor
split-phase induction motor
stator
synchronous speed

Learning Objectives:

After studying this chapter, you will be able to:

- List several types of electric motors.
- Explain the operating principles of various types of electric motors.
- List and describe devices that protect motors from overloads and overheating.
- Demonstrate service procedures required for several types of motors.
- **List safety procedures for servicing electric motors.**
- Use various electrical testing instruments to check motor windings, shorts, and grounds.
- **Follow approved safety procedures.**

ELECTRIC MOTORS MODULE

7.1 Electric Motor Applications

Refrigerating systems operate either with external-drive or hermetic motors. The external-drive motor drives the compressor directly off the shaft or by means of a belt. External-drive electric motors are also often used to drive many accessory devices. The hermetic motor is built inside the compressor dome. It usually drives the compressor directly.

Motors may be grouped into four general classifications according to use:

1. To drive compressors.
   - External-drive (open belt-drive).
   - Hermetic direct-drive.
2. To drive fans for:
   - Condensers.
   - Evaporators.
   - Air circulation.
   - Induced draft.
   - Forced draft.
3. To drive pumps.
   - Condensate pumps.
   - Chilled water pumps.
   - Condenser water pumps.
   - Ice-making machine water pumps.
   - Oil pumps.
4. To drive miscellaneous devices.
   - Vending machines.
   - Automatic ice cube makers.

7.2 The Motor Structure

All motors have a similar basic construction. See **Figure 7-1.** Each has two main parts:

- The stator.
- The rotor.

Figure 7-1. *Internal view of a common air conditioning motor. It uses a fan at one end of the rotating shaft to keep the operating temperatures low. (The Lincoln Electric Company)*

The ***stator*** is also known as the ***frame.*** This frame is usually cylindrical in shape. The field poles with ***field windings*** on them are part of the stator. The identification plate is also mounted on the stator. The rotor is mounted on a shaft, which has two journal bearings (one at each end).

The stator has end bells or plates attached to it. The ***end bells*** hold the bearings. When the rotor shaft journals are mounted in the bearings, the bells support the rotor.

The bearings are accurately machined to provide the proper amount of end play for the rotor. There is a clearance of .001″ to .002″ between the motor shaft and the bearing. In hermetic units built into the compressor dome, the compressor bearings may also serve as rotor bearings.

The windings are insulated copper wire. This insulation is usually a polyester material. It is resistant to moisture and has considerable ***dielectric*** and mechanical strength.

7.3 Types of Electric Motors

Both alternating current (ac) and direct current (dc) may be used to operate electric motors. Alternating current is most commonly used. Direct current motors are found in areas supplied with direct current only.

The following alternating current motors are used to drive compressors. Those listed in italics are most common and will be explained in greater detail.

1. Basic types:
   - Single-phase.
   - Two-, three-, and four-phase (polyphase).
2. The external-drive motor. These are used on open, belt-driven compressors. Types include:
   - *Repulsion-start, induction-run.*
   - *Capacitor-start, induction-run.*
   - Capacitor-start, capacitor-run.
   - Capacitor-run.
   - Permanent split capacitor.
   - Induction polyphase.
3. Hermetic motors. These are used on sealed systems in which the motor and compressor are enclosed in a common housing or dome:
   - *Capacitor-start, induction-run.*
   - Capacitor-start, capacitor-run.
   - Capacitor-run.
   - Permanent split capacitor.
   - Induction two-phase and polyphase.
4. The condenser and evaporator fan motor-type:
   - Split-phase.
   - *Shaded-pole.*
   - *Capacitor.*
   - Permanent split capacitor.

Today, the capacitor motor is used for most applications. It is one of the most popular motors for single-phase hermetic units.

7.3.1 External-Drive Motors

Four main types of motors are used for external-drive compressors:

- Repulsion-start, induction-run motor, **Figure 7-2.**
- Capacitor-start, induction-run motor, **Figure 7-3.**
- Permanent split capacitor (PCS) motor. See Section 7.8.3.
- Three-phase motors.

Figure 7-2. *Section view of a repulsion-start, induction-run 1/2 hp motor. Note that this motor has a wound rotor and centrifugal mechanism. The centrifugal mechanism shorts the rotor winding and raises the brushes off the commutator when the motor attains about 75% of running speed. A—Centrifugal mechanism. B—Commutator shorting segments. C—Brushes and brush holder. Wick oilers are used at the bearings.*

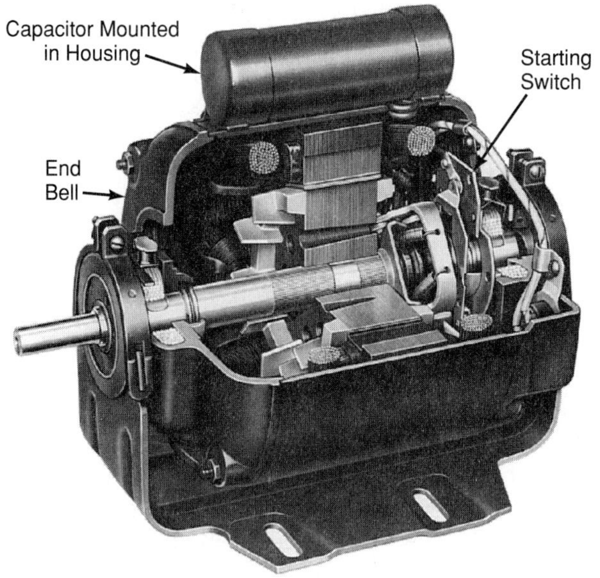

Figure 7-3. *Capacitor-start, induction-run motor. Motor has rubber mounts at each end which provide flexibility. Wick oiling is used.*

A V-belt connects the motor to the compressor. A direct connection may also be made with a coupling. The speed reduction for belt drives is usually about 3:1 (three to one). This means that the compressor flywheel diameter is three times larger than the motor pulley diameter. Section 7.3.2 through 7.3.5 explain features common to most external-drive motors. **Figure 7-4** shows the parts of a capacitor type motor.

7.3.2 Induction Motors

Induction motors have no windings on the armature (rotor). Instead, the rotor has copper bars or other conducting material. These are in its outer surface and lie about parallel to the motor shaft. When current is

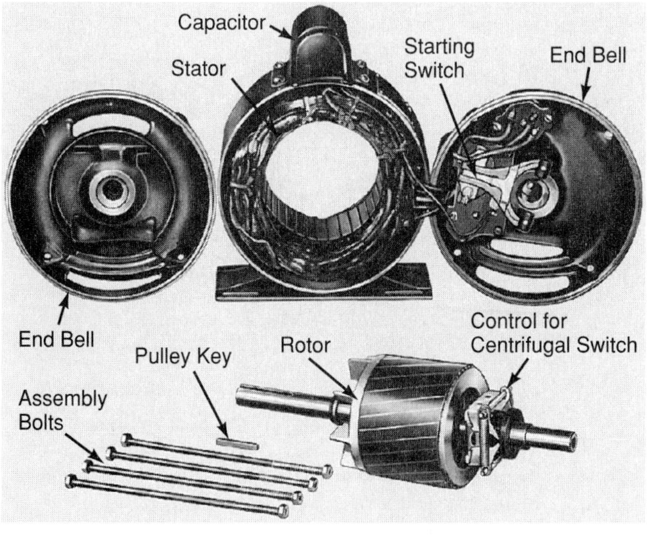

Figure 7-4. *Parts of a capacitor-type motor used on external-drive compressor. (Emerson Electric Co.)*

induced (made to flow) in them, turning power (torque) is produced by the magnetism created.

One or more field windings are mounted in the stator. Alternating current, passing through these windings, creates a changing magnetic field. This magnetism passes through the rotor, inducing (building up) current in the rotor bars.

Induced current creates an opposite magnetic field in the rotor. The opposing magnetism causes the rotor to revolve as it tries to keep up with the changing field polarities in the field windings.

All hermetic machines use induction motors. Most of these motors use two field windings:

- A starting winding.
- A running winding.

Many hermetic motors need a starting relay. It is always located outside of the motor compressor dome. Electrical contacts would quickly fail from the oil and refrigerant mist inside the dome. Section 8.12 describes the various types of starting relays and how they are used.

The ***starting relay*** temporarily connects the starting winding of the motor to the power circuit. When the motor reaches about 75% of operating speed, the relay opens the circuit. The starting winding is then disconnected from the power line.

7.3.3 Split-Phase Induction Motor

The split-phase induction motor is used in most fractional horsepower appliances. Two stator windings are used, one for starting and one for running.

The running windings are made into two or four coils in series depending on the motor speed desired. These windings are ***energized*** (have current flowing in them) during the whole time the motor is running.

All of the starting windings have the same number of coils. They are usually rotated several degrees from the running windings. See Section 1.10.4 for information about degrees of arc measurement.

A smaller gauge of wire is used in the starting winding than in the running winding. However, it has more turns than the running winding. Counter emf (electromotive force) in the greater number of turns causes the current to build up more slowly in this winding. Therefore, the magnetic effect will be several electrical degrees behind the running winding. This will create torque on the rotor to make it start turning in the correct direction. The starting windings are disconnected when the motor reaches approximately 75% of its running speed.

There are no windings on the rotor, but magnetism builds up around the rotor bars. Some rotors have heavy copper bars fitted into slots in the laminated iron. The ends of these copper bars are braze-welded to heavy copper rings at each end. This completes the induced (by magnetism) electrical circuit. Such an arrangement is often called a ***squirrel-cage winding.*** (The term ***electrical degrees*** means that the maximum magnetic effect on the rotor is a few degrees away from (behind) the magnetic effect in the stator.)

7.3.4 Repulsion-Start, Induction-Run Motor

A repulsion-start, induction-run motor has an electrical winding on the rotor for starting purposes. Typical applications are external-drive refrigerators and air compressors. In many cases, they have been replaced by capacitor-type motors.

A special winding in the armature gives this motor a high starting torque. The motor starts as a repulsion motor, using brushes against a commutator in the armature winding circuit. This increases the induced electrical flow in the armature and produces more magnetic power. As soon as it reaches a certain speed, the armature windings are shorted. Then the brushes are usually lifted from the commutator. The motor then operates as an induction motor, **Figure 7-5.** Section 6.5.9 explains the operation of commutators and brushes.

7.3.5 Capacitor-Start, Induction-Run Motor

The capacitor-start, induction-run motor uses a capacitor in the starting winding. (Recall that a capacitor is a device for storing electrical energy.) One type, **Figure 7-6,** becomes a two-phase motor when starting. More torque is provided that way. Then it becomes a single-phase motor at about 75% of its full speed. The capacitor-start, induction-run motor is a popular type.

During starting, the capacitor changes the phase angle of the current in the starting winding. This produces two-phase electrical characteristics.

The capacitor is usually placed on top of the motor in a metal or plastic cylinder. The capacitor is connected to a centrifugal switch built into the motor and to a starting winding in the stator. See **Figure 7-7.** (A ***centrifugal switch*** is one which whirls around with the shaft. Weights are moved by the whirling motion to open the contacts.)

Operation of a capacitor-start, induction-run motor is quite simple. The centrifugal switch is closed when the motor is not running. Starting up the motor causes the current to pass through both the starting and the running winding. The starting winding is connected in series with the capacitor. This capacitor puts the electrical surges in the starting winding out-of-step or out-of-phase with those of the running winding. The motor then acts as a temporary two-phase motor. It has a very high starting torque.

At about 75% of the motor's rated speed, the centrifugal switch opens. This disconnects the starting winding. The unit, however, continues to run as an induction motor.

The capacitor has two terminals. One terminal connects to one leg of the power line. The other connects to the starting winding terminal.

The simplest method for producing greater torque is to change the single-phase motor into a two-phase motor during starting and/or running. A capacitor is used.

There are two types of capacitors:

- Starting capacitor (electrolytic–dry).
- Running capacitor (oil-filled).

The starting capacitor is usually a electrolytic type. It is constructed of two sheets of conductor metal

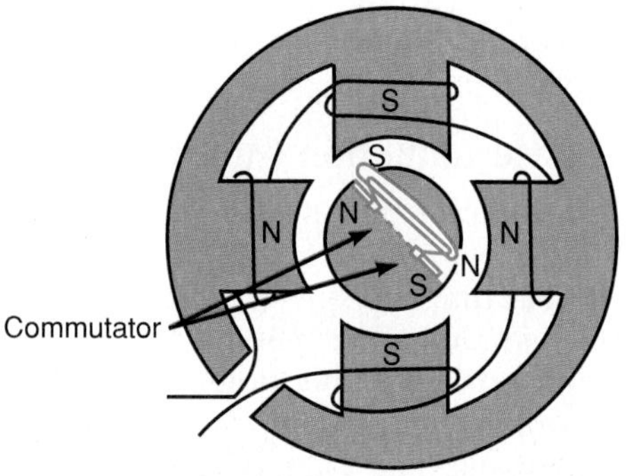

Figure 7-5. *Schematic sketch of repulsion-start, induction-run motor. Brushes that contact commutator are grounded and complete circuit between two commutator bars.*

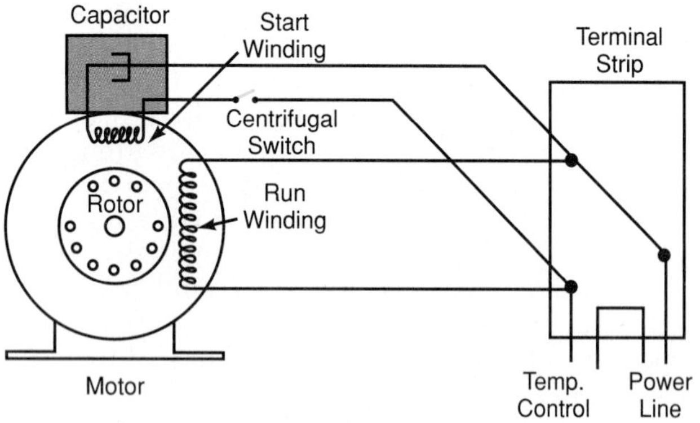

Figure 7-6. *Wiring diagram of capacitor-start, induction-run motor. Note that capacitor is in series with starting field winding as motor starts. Rotor-operated centrifugal switch opens this winding as soon as motor reaches about 75% of full speed.*

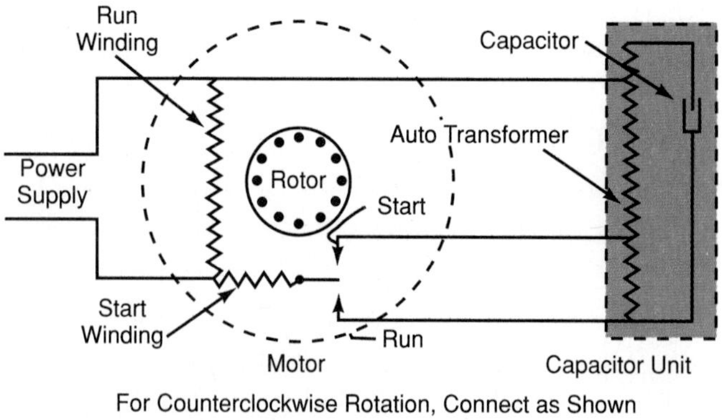

Figure 7-7. *Wiring diagram of combination capacitor and autotransformer motor. Note double-throw single-pole switch for changing transformer output when centrifugal switch moves from start to run. (Fedders North America)*

separated by an insulator. This is shown in **Figure 7-8.** A typical capacitor is shown at the top of **Figure 7-3.**

A capacitor placed in an alternating current line is charged during the buildup of the voltage and current. The surge of power in the line causes this buildup. Then, the current flow decreases in the power line, as the ac reverses. The capacitor discharges. This causes another power surge in the motor starting windings, **Figure 7-9.**

The running capacitor operates in the same way, except that it keeps on operating when the motor is running. This capacitor is usually larger and is made of metal to provide better heat removal. The insulation, plates, and terminals of capacitors are designed for durability.

7.4 Motor Speeds

Induction motors—including split-phase, repulsion-start, and capacitor-start—are made for use on 25, 50, or 60 Hz current. The speed of a motor depends on the frequency (cycles per second, measured in Hz) and number of field poles. Motor speeds are calculated from the synchronous speed.

Synchronous speed is related to rotating magnetic fields. The stator windings of a motor produce a rotating magnetic field. (The field advances one pole for every one-half cycle of current.) If the rotor keeps up with the rotating field, the motor runs at synchronous speed.

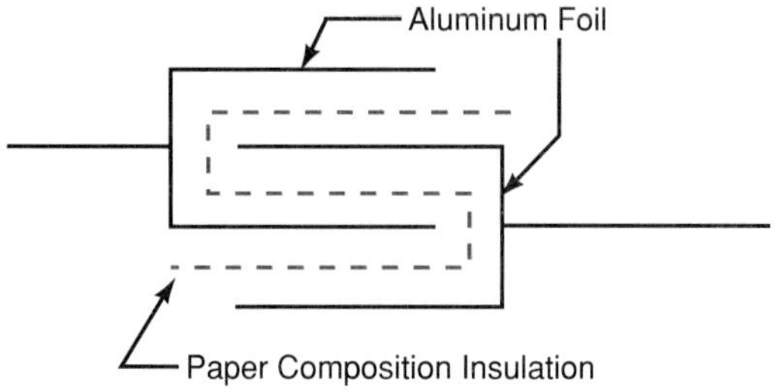

Figure 7-8. *Capacitor construction. Two layers of metal foil are separated by special insulating paper. The two sheets of foil are connected to the two terminals of the unit. Capacitors used on motors are rolled or folded into compact package and installed in metal housing.*

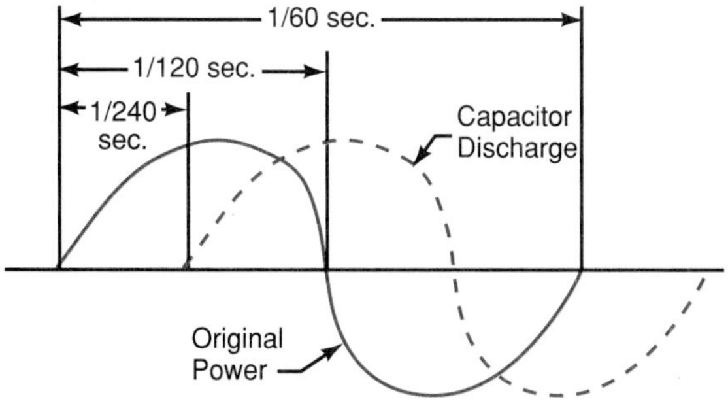

Figure 7-9. *Cycle diagram showing effect of capacitor in series with power flow in single-phase circuit.*

The synchronous speed for two- and four-pole motors is shown in **Figure 7-10.**

Motors do not operate exactly at synchronous speed. They are usually operating under a load and there is some magnetic slippage. Thus, motors are not rated at the synchronous speed. Rather, they are rated at a speed which corresponds with a normal load. **Figure 7-10** shows operating speed for two- and four-pole motors operating at 60 Hz, 50 Hz, and 25 Hz.

Hermetic motor compressors may operate as either two-pole or four-pole motors. This arrangement makes it possible to operate the motor compressor either at 3600 rpm or 1800 rpm. As you can see from the chart, the operating speed is lower than the synchronous speed.

A residential hermetic motor compressor, **Figure 7-11,** may be wired to operate at either low or high speed. **Figure 7-12** is the electrical wiring diagram. In **Figure 7-12A,** the run windings are connected in series. Leads 1 and 2 (L1 and L2) are connected to terminals 1

Motor Speed – RPM						
No. of Poles	60 Hz Syn.	60 Hz Op.	50 Hz Syn.	50 Hz Op.	25 Hz Syn.	25 Hz Op.
2	3600	3450	3000	2850	1500	1450
4	1800	1750	1500	1450	750	700

Figure 7-10. *Synchronous and operational speed for two- and four-pole motors. Note that operational (Op.) speed is approximate.*

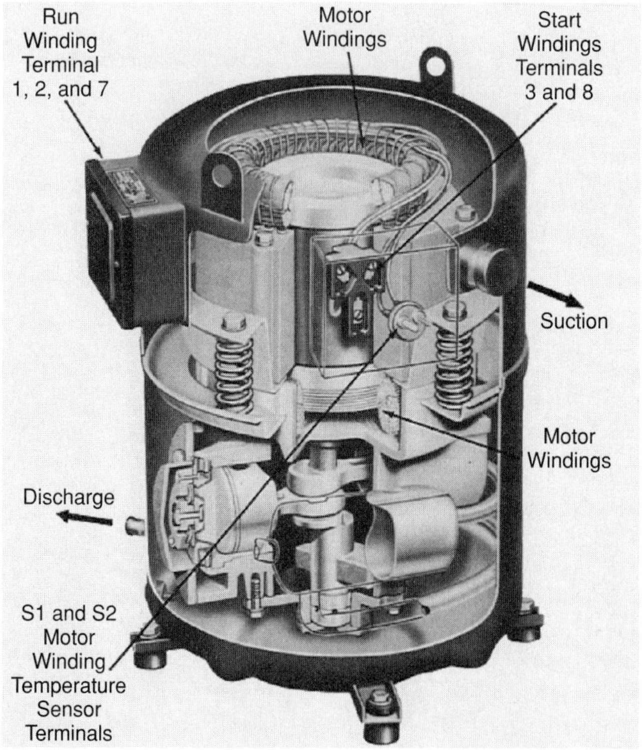

Figure 7-11. *Residential, two-speed hermetic motor compressor. It uses single-phase electrical power. (Lennox International, Inc.)*

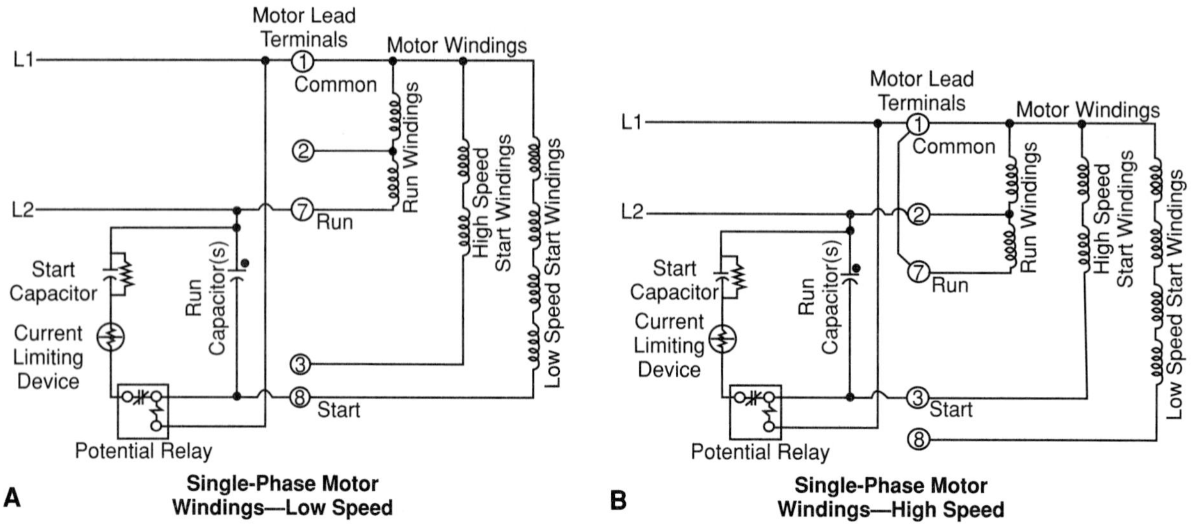

Figure 7-12. *Residential hermetic motor compressor winding connections. A—Connections for low speed. B—Connections for high speed. (Lennox International, Inc.)*

and 7. These windings form a four-pole motor operating at a low speed of 1800 rpm. In **Figure 7-12B,** the unit operates at a high speed when the run windings are connected in parallel (L1 and L2 to terminals 1 and 2). This winding forms a two-pole motor operating at 3600 rpm.

A two-speed, three-phase motor compressor is shown in **Figure 7-13.** It operates as a two-pole motor at high speed and as a four-pole motor at low speed. The circuit diagram for a two-speed motor compressor is shown in **Figure 7-14.** Solid-state circuits control the compressor speeds.

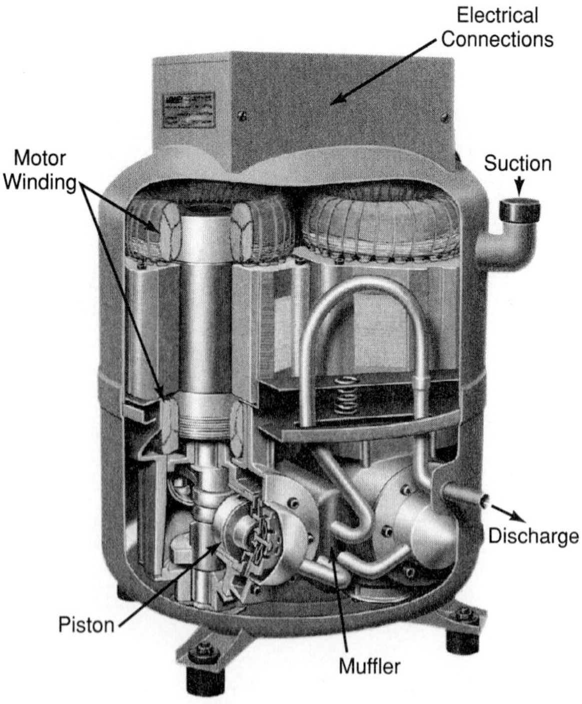

Figure 7-13. *Commercial two-speed hermetic motor compressor. It uses three-phase electrical power. Note muffler in discharge line. Muffler reduces noise and stabilizes discharge pressure. (Lennox International, Inc.)*

Motors designed to operate on a certain frequency will not operate at a different frequency. This is because the number of turns of wire on the field poles and the amount of iron in the magnetic circuit is different for each frequency. Some motors have field connections in either series or parallel. This allows either 120 V or 240 V operation. On 120 V, the field winding would be connected in parallel. Operating on 240 V, the field winding would be connected in series.

It is possible to build motors having six, eight, or more poles. However, such motors are not in common use.

7.5 Starting and Running Windings

As mentioned in previous paragraphs, most motors have a running winding and a starting winding. These windings are mounted on the stator.

During starting, current goes through both windings. When the motor reaches 60% to 75% of its running speed, the starting winding circuit is opened. The motor operates on the running winding only. The 60% value is for 6- and 12-pole motors. A starting winding is found on split-phase motors and on all types of capacitor motors.

The starting winding has the same number of coils as the running winding. However, it has smaller diameter wire and a greater number of turns. Its induction action, therefore, splits the phase, creating a rotating magnetic field.

Electricity passes through the running winding the entire time the motor is in operation. Heavier wire is used than in the starting winding. It is installed on the field poles on both the two- and four-pole motors.

Figure 7-15 shows the electrical circuits of a split-

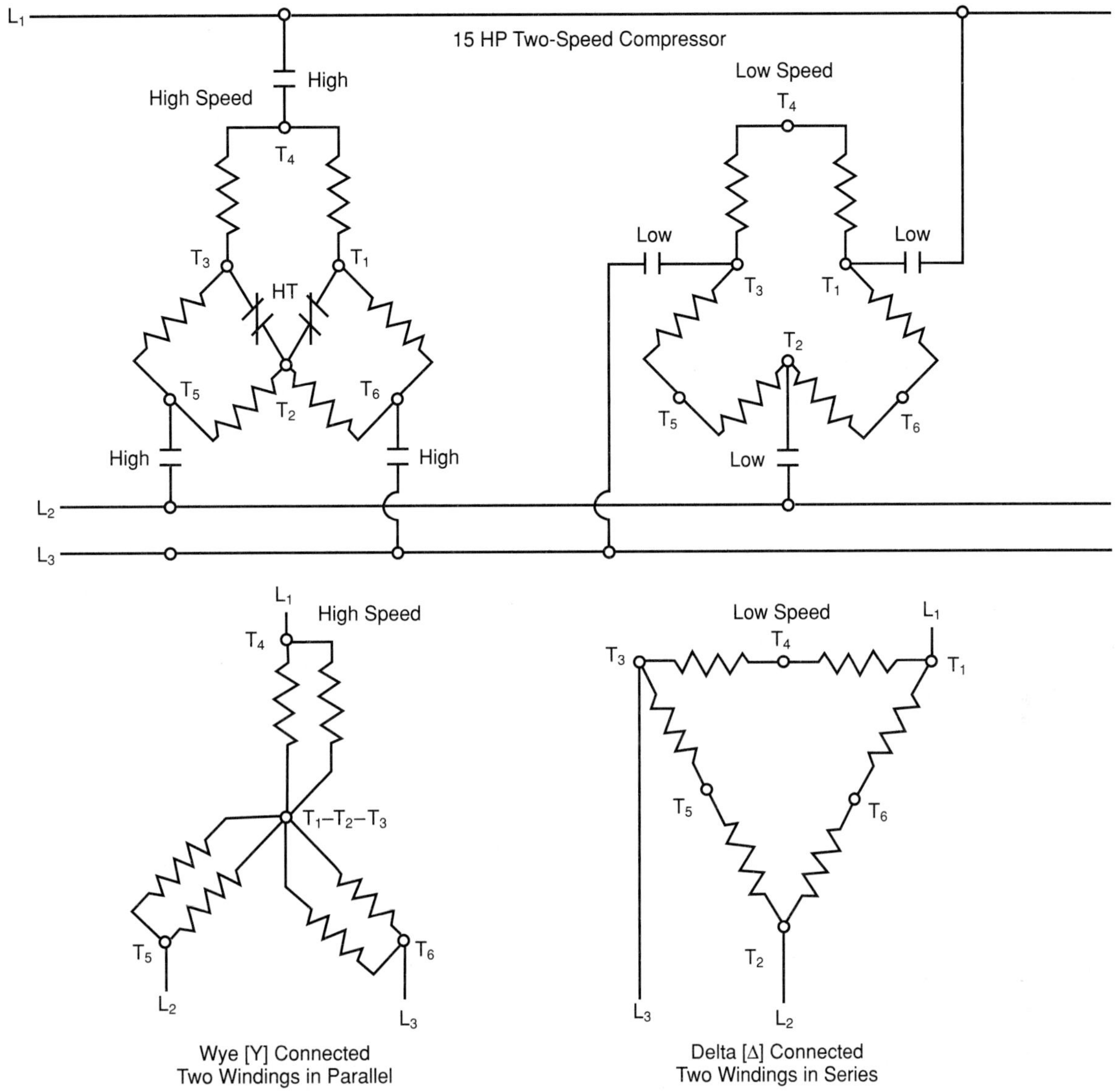

Figure 7-14. *Circuit diagram of two-speed, three-phase hermetic motor compressor. (Lennox International, Inc.)*

phase electric motor while the motor is starting. When the current goes into the R terminal (during one-half of the cycle), the electrons separate. Most will go through the running winding but some will go through the starting winding. Magnetism builds up faster in the running winding. Thus, when the magnetism is created a few thousandths of a second later in the starting winding, a turning force is exerted upon the rotor.

When current enters terminal R during the other half cycle, most will go through the running winding. Some will go through the starting winding. This action reverses the polarity of the electromagnets. There is again a delay of magnetic build up in the starting winding. The rotor is attracted and repulsed in the same direction of rotation as in the first half of the cycle.

To change the direction of rotation on some of these types of motors, disconnect the starting winding leads from the two terminals and reverse them. That is, put the old S connection (lead) on terminal R. Put the old R connection (lead) on terminal S. Reversing the two main leads will not reverse the rotation of the motor. (With starting windings disconnected, one can start the motor manually in either direction.)

If the starting winding is left in the circuit, it may overheat. Therefore, a switch is mounted in the starting winding's circuit. See **Figure 7-16.** After a motor reaches approximately 75% of its rated speed, the switch disconnects the winding. It is open while the motor runs.

7.6 Starting Current

The instant the starting circuit is closed on a motor, there is little or no counter emf. Therefore, the amount of current starting to flow will be determined by the resistance in the circuit. At the instant of starting, the current flow will be quite high (two to four times the running current).

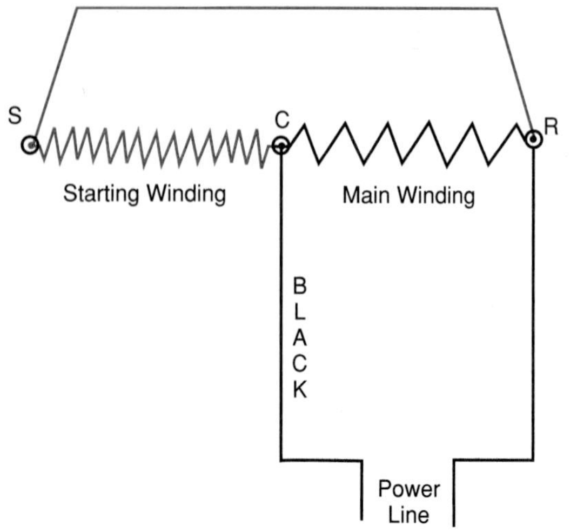

Figure 7-15. *Circuit diagram of split-phase motor. C—Common terminal for both windings. C to R—Main or running winding. C to S—Starting winding. R—Running winding terminal. S—Starting winding terminal. (Copeland Corporation)*

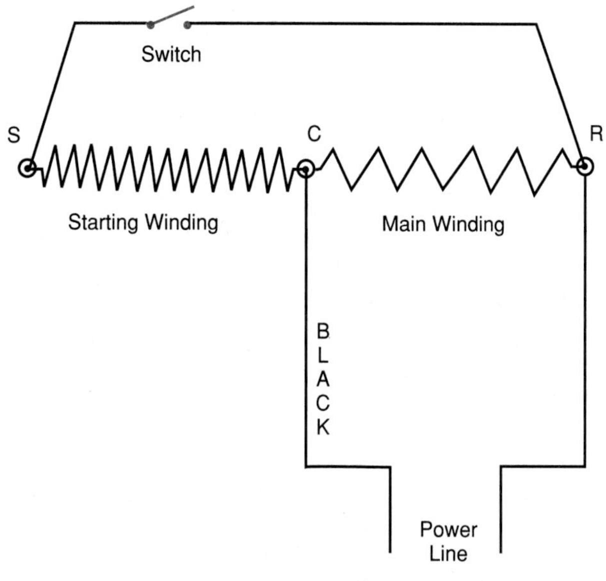

Figure 7-16. *Circuit diagram of split-phase motor using switch in starting winding circuit. Switch is operated by rotor speed or relay and is closed as motor starts. Switch will open when motor reaches nearly 75% of running speed. (Copeland Corporation)*

This initial high flow of amperage is called ***locked rotor amperage.*** During this high amperage flow, the voltage drops to its lowest reading.

Frequent starting of a motor causes these high current flows to happen too often. The motor will overheat—especially those of 1/2 hp or more.

Fuses and circuit breakers must be designed with this fact in mind. If protective devices are set to allow only running current, they will open at the instant of start. The protective device must have delayed action—usually a heating strip and a bimetal switch.

7.7 Motor Connections

Except for those that are reversible or specially designed, motors usually come with two or four leads. The leads connect the motor to one of two different voltages.

Provided the correct voltage is maintained at the motor terminals, 120 V, 240 V, or other corresponding voltages are equally good. These motors may be used on circuits having voltages of 105 V to 125 V or 210 V to 250 V.

Most electrical power companies are now providing 120 V and 240 V electrical power. These voltages are safe to use on motors labeled 110 V to 220 V or 115 V to 230 V.

It is recommended that, when possible, all motors be connected to 240 V service. Wires of adequate size are needed to supply adequate voltage at the motor terminals. This is true especially if the motor is heavily loaded.

Most repulsion-start induction-run motors use four motor leads. These leads are connected to field windings, two leads to a winding. The windings are connected in parallel if the voltage to be used is 120 V. The leads are connected in series if 240 V is used. Power is the product of the emf in volts multiplied by the amperage. Therefore, the amperage flow for any specified power using 240 V will be half the current flow if using 120 V.

Be sure to avoid connecting a 240 V motor to a 120 V circuit. If this is done, the motor will overheat and "burn out." The maintenance of proper voltage at the motor terminals is absolutely necessary.

Fan motors and similar units are usually wired either for 120 V or 240 V only. Only two leads are connected to the field windings. Air conditioning evaporator fans often have two or three speeds. These motors have three to four leads.

The power that an alternating current motor develops varies directly to the square of the voltage at the motor terminals. Low voltage will also result in a rapid drop in horsepower capacity of any motor.

See Chapter 31 for information on the wire size to be used. Loose connections cause excessive voltage drop and are a fire hazard. All connections should be made with solder or with Underwriters Laboratories-approved pressure connectors.

A 120 V motor supplied with current at 100 V will develop only about 75% as much power as it would if supplied with 120 V current. Therefore, to do the job and develop its full horsepower, it must draw more current. This increase in current may overheat the electrical power leads and/or the motor windings. Insulation may fail.

Figure 7-17 shows a method of making connections for single-phase motors. Note how the direction of rotation may be changed. Special motors requiring other connections are shipped with diagrams showing proper connections.

A 120 V to 240 V motor is connected to a 120 V line through an across-the-line switch. This switch is usually operated by the cabinet temperature. However, low-side pressure may also be used. The same motor control may be used to operate a 120 V to 240 V motor installed on either the 120 V or 240 V circuit.

The field winding leads come out of the motor frame through an insulated opening. The wire is normally protected with rubber, fabric, or plastic grommet. A small metal box is usually mounted over the leads to protect the electrical connections.

7.8 Hermetic System Motors

When the motor and the compressor are placed inside a dome or housing, it is called a ***hermetic unit.*** In such systems, the motor must drive the compressor directly. This requires good, leakproof electrical connections. The motor must have adequate power characteristics and must be of an induction type.

Motors using rotor windings requiring either brushes or slip rings cannot be used. Development of the split-phase motor and the capacitor motor made the hermetic motor possible. The original hermetic motors were four-pole, operating at approximately 1750 rpm.

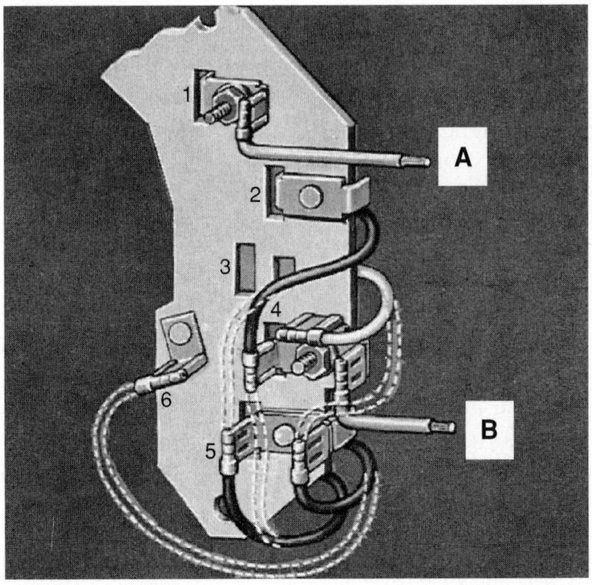

Figure 7-17. *How to make motor leads give desired direction of rotation and connect motor to voltage supplied. Leads for power input are shown at A and B. Wiring diagram inside motor cover plate indicates correct connection for each voltage and direction of rotation. Dotted lines show method of reversing. (MagneTek)*

To calculate the motor speed, the formula is:

$$N = 120 \times \frac{f}{P}$$

where N = rpm
f = frequency (cycles per second)
P = number of poles

In the formula, the 120 represents the 60 seconds per minute × 2. (The magnetism or polarity changes two times per cycle.)

Formula:

$$N = 120 \times \frac{f}{P}$$

Example:

What is the speed of a two-pole motor?

Solution:

$$N = 120 \times \frac{f}{P}$$

$$N = 120 \times \frac{60}{2}$$

$$N = 120 \times 30$$

N = 3600 rpm synchronous speed or about 3400 rpm actual speed

The motor is almost twice as fast because the rotor has to travel one-half revolution during one-half cycle (instead of one-fourth revolution as with the four-pole motor). These two-pole motors are about two-thirds the size of four-pole motors of the same power.

Hermetic motors present some problems not found with external-drive motors:

- Special cooling provisions must be made.
- Wiring insulation must be resistant to oil and chemicals in the refrigerant. (This is particularly true in the presence of moisture and/or high temperatures.)
- Manufacturing standards must provide exact alignment of the stator, rotor, and compressor.
- Electrical connections through the dome must be electrically perfect and leakproof.

The motors may be cooled by several methods. One way is pressing the stator into the dome, then putting cooling fins on the dome. Another method uses a water coil to cool motor windings while the unit is running. Some designs pass partly cooled condenser refrigerant around the motor housing, cooling the electric motor. Most systems pull the cool return suction line vapor and oil over the motor windings.

Most manufacturers have developed special synthetic wire coatings (usually synthetic enamels) which have good insulating qualities. They are also safe to use with most of the popular refrigerants.

The air gap between the rotor and the stator is only a few thousandths of an inch. It must be equal on all sides; otherwise, the motor may hum. Note, in particular, that many motor compressors are mounted on springs inside the dome.

Figure 7-18 shows a hermetic motor compressor unit. The stator and rotor of a hermetic motor are shown in **Figure 7-19.**

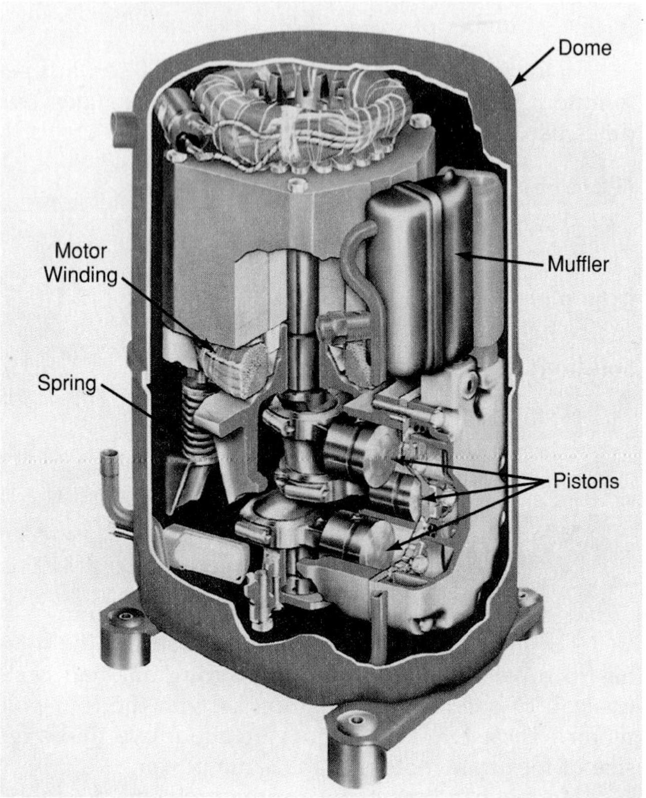

Figure 7-18. *Hermetic motor compressor unit. Note that compressor is mounted on springs inside dome. Muffler helps prevent liquid refrigerant from filling tubing on top of cylinder valve. (Tecumseh Products Co.)*

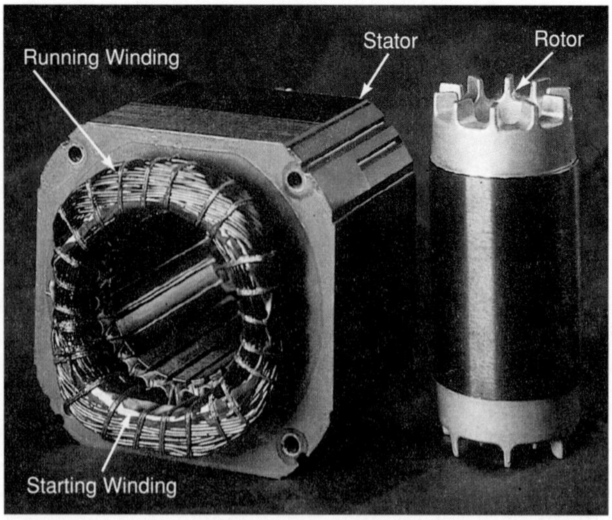

Figure 7-19. *The hermetic frame, motor, stator and rotor for a 2 hp reciprocating compressor used in residential units. (A.O. Smith Electrical Products Co.)*

7.8.1 Hermetic Motors Electrical Characteristics

The design characteristics of electric motors used in hermetic compressors will depend on whether the unit starts under load, under no-load, or under a balanced pressure condition.

The basic operation of the motor is the same as for external-drive motors. All torque is developed by induction only.

Units which start under load need a higher starting torque (turning effort). They require the use of larger conductors in the starting circuit. Usually, manufacturers try to provide starting power equal to twice the running power. In other words, a 1/6 hp motor is designed to produce 1/3 hp during starting. Various methods are used to shut off the special starting devices after the motor reaches full speed.

Figure 7-20 shows the external circuit of a hermetic motor. Note the solid state ***positive temperature coefficient (PTC)*** resistor. This increases in resistance as its temperature is increased. Also shown is a thermally-operated overload protection and a run capacitor.

7.8.2 Hermetic Motor Types

Hermetic motors are either single-phase or polyphase. Four types of single-phase induction motors are used:

- Split-phase (SP).
- Capacitor-start, induction-run (CSIR).
- Capacitor-start, capacitor-run (CSCR).
- Permanent split capacitor (PSC).

Hermetic Split-phase (SP) Induction Motor

Split-phase induction is the basic type motor for small hermetic condensing units. The principle of operation is simple. There are two windings—one for starting and one for running. Since starting torque is low, such motors must be used on systems with low starting load.

The split-phase (SP) motor is very popular on systems that use the capillary tube refrigerant control. Pressures in these systems balance in the Off cycle. Thus, the compressor is not required to start under a load.

These motors may also be used where the system has an electrical, mechanical, or hydraulic pressure unloading device. In these installations, any type of refrigerant control may be used.

A split-phase motor used in hermetic motor compressor systems must have some type of outside starting relay. This may be a thermal, current, or potential relay. See Chapter 8 for information on these relays.

A schematic wiring diagram for a hermetic split-phase induction motor with a current starting relay is shown in **Figure 7-21. Figure 7-22** shows the same motor using a potential relay. These motors are mainly used on small hermetic condensing units of 1/10 hp, 1/6 hp, to 1/3 hp.

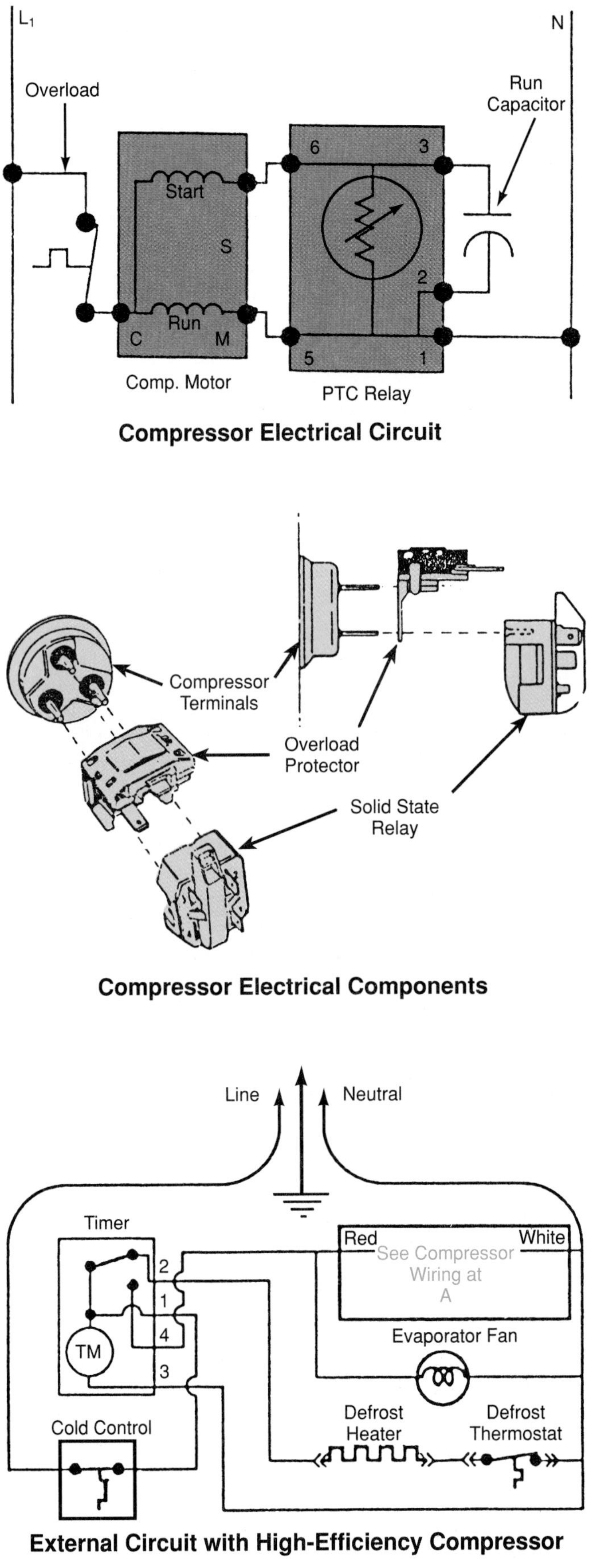

Figure 7-20. *External circuit of a hermetic motor. (Frigidaire Company)*

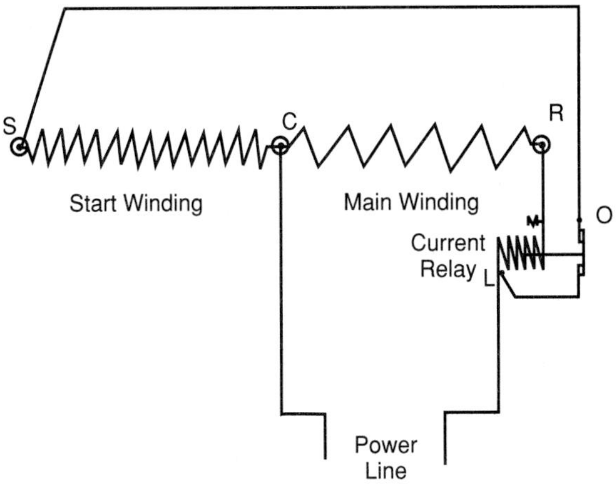

Figure 7-21. *Wiring diagram of hermetic split-phase induction motor. With first burst of current from power line, current relay control switch O is closed and start winding is cut in. As motor begins to start, current flowing through relay control winding and main winding keeps switch O closed. As motor reaches about 75% of running speed, current through main winding and relay coil is reduced. This opens switch O to cut out start winding, and motor operates on main winding. C—Common terminal. R—Running terminal. S—Starting terminal. (Copeland Corporation)*

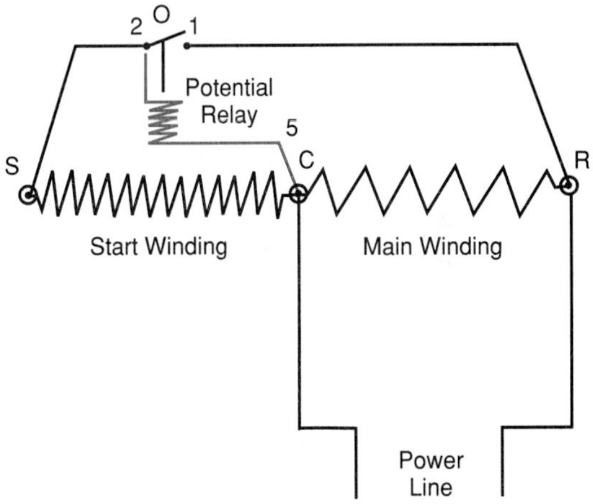

Figure 7-22. *Wiring diagram of hermetic split-phase induction motor with a potential relay in running position. Relay control winding develops current from generation of electricity by starting winding. When enough magnetic effect is produced, it will open switch O just after motor starts. C—Common terminal. R—Running terminal. S—Starting terminal. O—Relay control switch. (Copeland Corporation)*

7.8.3 Motor Capacitors

The capacitor in an alternating current circuit basically changes electrical flow. It is changed from a single-phase electrical flow to a two-phase electrical flow. Capacitors are used both in starting windings and running windings.

When capacitors are used in the starting winding only, they are called ***start capacitors.*** Start capacitors are only used for a fraction of a second. They have no overheating problems. However, the start capacitor must not remain in the start winding circuit too long. If it remains too long, damage to the windings will occur. The start relay must drop the start capacitor out of the starting winding circuit.

A ***run capacitor*** is designed to dissipate heat generated during the running of the motor. Never use a starting capacitor in the run winding circuit.

Always use the correct microfarad (mfd) and voltage rating when replacing a capacitor.

Hermetic Capacitor-start, Induction-run (CSIR) Motor

Capacitor-start, induction-run (CSIR) motor is a very popular hermetic motor for refrigerating units. It has a good starting torque, which is obtained by placing a capacitor in series with the starting winding. It can use any one of several starting relays—current relay, potential relay, or a hot wire (thermal) relay.

Figure 7-23 is a schematic wiring diagram for such a motor with current starting relay. A similar installation using a potential relay is shown in **Figure 7-24.** The starting capacitor circuit is kept open during the running cycle by the induced emf generated in the starting winding (across terminals S and C). Induced voltage pushes enough current through the potential relay coil to produce a magnetic effect. The magnetic force keeps the relay points open.

See Chapter 8 for detailed information concerning construction and operation of the different types of motor controls.

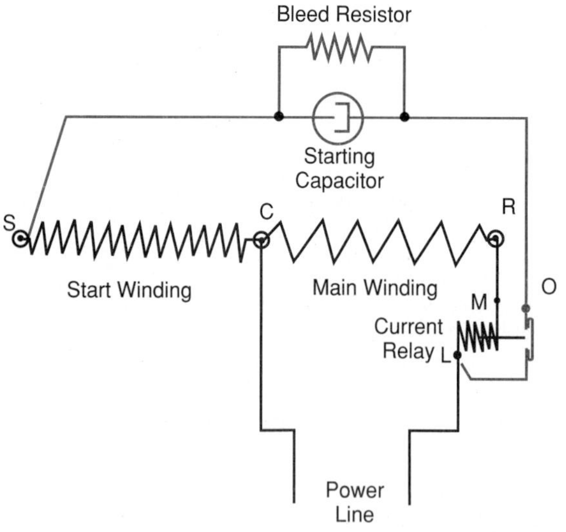

Figure 7-23. *Current relay is shown in starting position (points closed). C—Common terminal. R—Running terminal. S—Starting terminal. O—Relay control switch. (Copeland Corporation)*

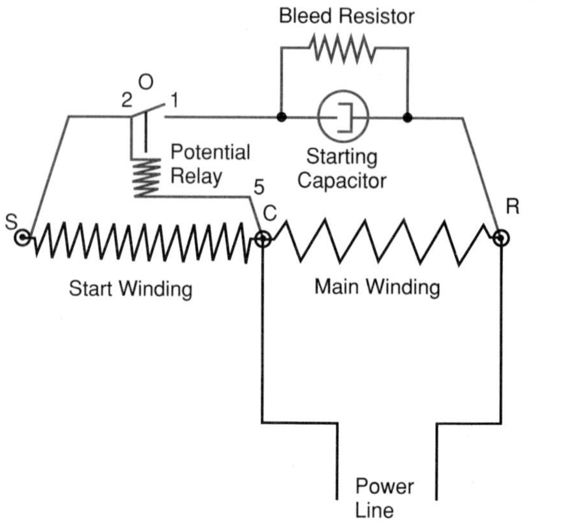

Figure 7-24. *Capacitor-start, induction-run motor with potential relay in running position (points open). C—Common terminal. R—Main or running terminal. S—Starting terminal. O—Relay control switch. (Copeland Corporation)*

Hermetic Capacitor-start, Capacitor-run (CSCR) Motor

The capacitor-start, capacitor-run motor generally uses two capacitors. Both are in the starting winding circuit but only the start capacitor is controlled by the relay switch.

When the motor is started, the capacitors turn the motor power surges into two-phase power and produce a high starting torque. After the motor reaches 60% to 75% of its rated speed, the relay opens the circuit to the starting capacitor. The running capacitor is left in the circuit.

This action produces a two-phase motor that is very efficient. The power factor is improved. Larger hermetic units in commercial systems use this type of motor.

Figure 7-25 shows the wiring diagram. Note that the running capacitor is in series with the starting winding. In **Figure 7-26,** the motor wiring circuit shows two starting capacitors and two running capacitors. These are connected in series to increase the voltage capacity. (Two 120 V capacitors in series can be used in a 240 V circuit.)

Permanent Split Capacitor (PSC) Motor

The permanent split capacitor motor is popular for air conditioning systems. It does not use a relay. Current flows through both the running winding and the starting winding when power is on. (See **Figure 7-27A.**) A running capacitor is connected between the running (R) and starting (S) terminals and is in series with the starting winding.

Such motors are sensitive to line voltage. A 5% to 10% drop will cause starting difficulty and overheating. To prevent damage, thermal protection will open the circuit.

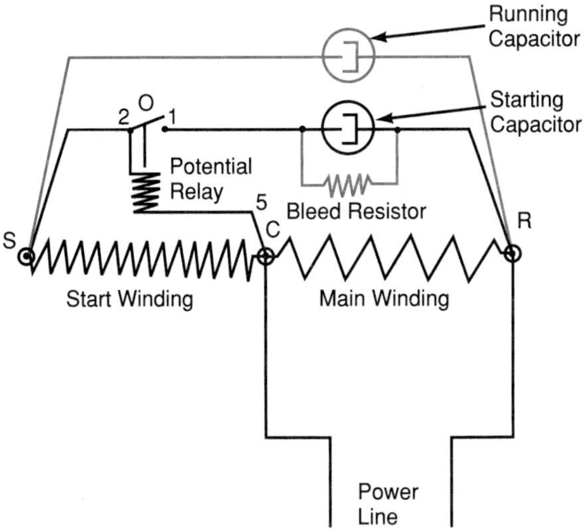

Figure 7-25. *Schematic diagram of capacitor-start, capacitor-run motor using potential relay. Relay is in running position (points open). C—Common terminal. R—Main or running terminal. S—Starting terminal. O—Relay control switch. (Copeland Corporation)*

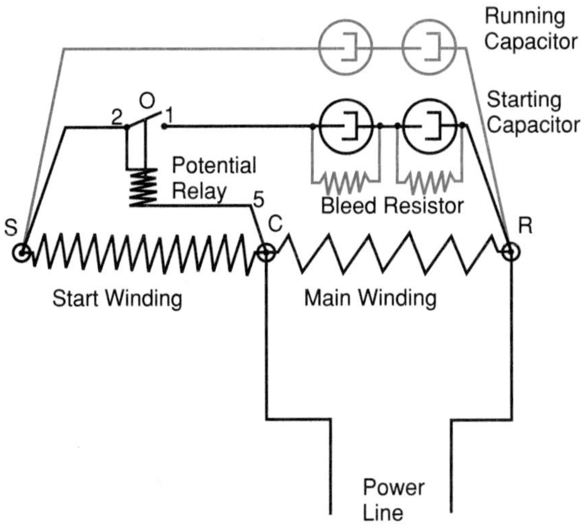

Figure 7-26. *Schematic wiring diagram of hermetic motor. This circuit has potential relay in running position. C—Common terminal. R—Running terminal. S—Starting terminal. O—Relay control switch. (Copeland Corporation)*

Starting torque is low. Thus, if the motor tries to start when the system's pressures are not balanced, the motor will overheat. Thermal protectors will open the circuit.

Figure 7-27B shows an open, vented PCS motor. **Figure 7-27C** shows a closed-type motor. A cutaway view of a PCS external-drive motor is shown in **Figure 7-27D.** This type of motor is used on furnaces and air conditioning fans.

Chapter 22 explains how to change a PSC into a capacitor-start, capacitor-run motor in a window air conditioner.

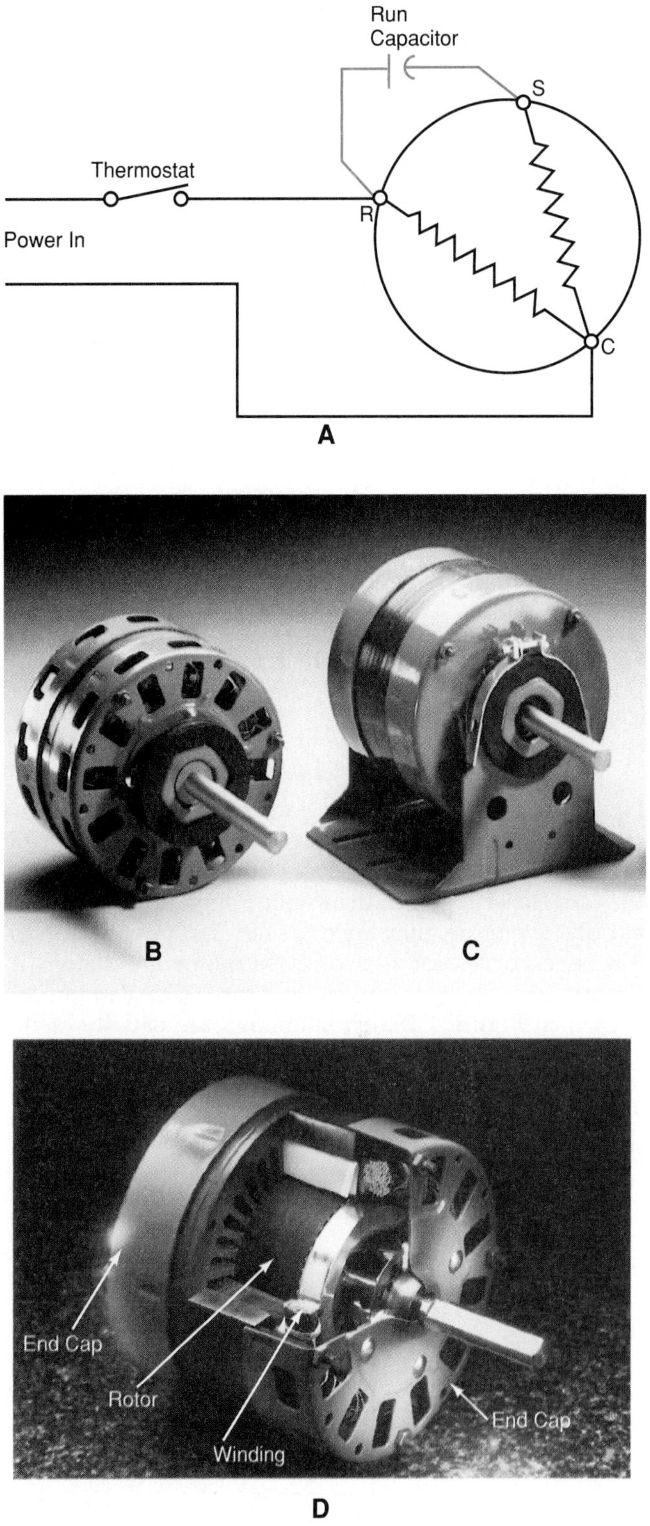

Figure 7-27. *Permanent split capacitor (PSC) motors. A—Diagram of a PSC motor. Note the run capacitor. B—Open, vented motor used in open areas. C—Closed-type motor with mounting brackets. Used in enclosed low air movement areas. D—Cutaway view of PSC external-drive motor. Note the two endcaps. (Fasco Motors Group)*

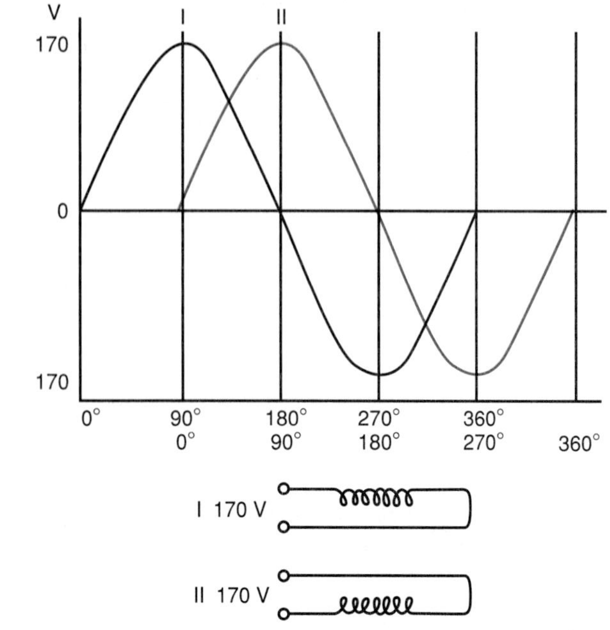

Figure 7-28. *Two voltage sine curves for two-phase power circuit. Note that there are two separate circuits. They are 90° out-of-phase.*

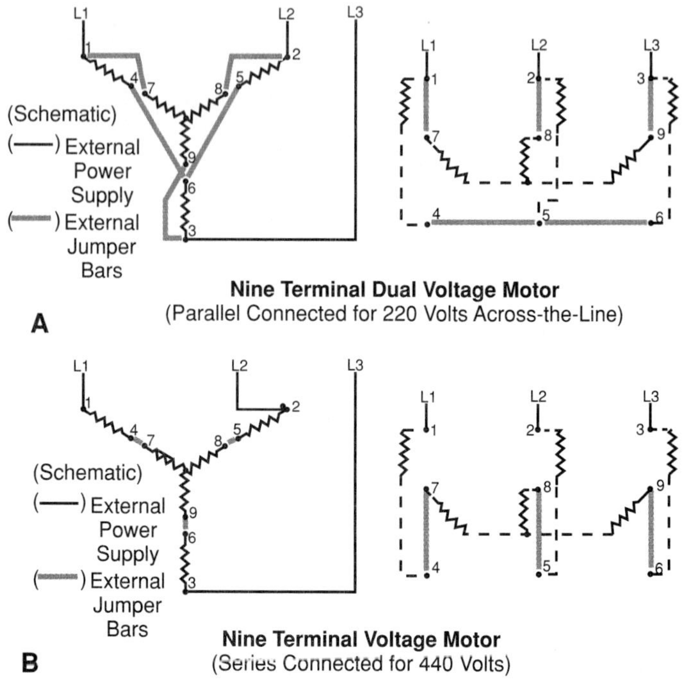

Figure 7-30. *Schematic wiring diagram showing circuits and connections for three-phase hermetic motor. A—Circuit as connected for 220 V. B—Circuit as connected for 440 V. Note that L1, L2, and L3 are the three-phase line connections. Numbers 1-2-3, 4-5-6, 7-8-9 are the connections to the motor windings. Each motor winding coil is designed for 220 V; for example, the coil between terminals 1 and 4 is rated for 220 V.*

Hermetic Polyphase Motor

Large hermetic compressors usually are driven by three-phase motors. The surges of current in these motors are closer together than with a single-phase current supply. Therefore, they are more efficient power sources. The sine curves for a two-phase motor are shown in **Figure 7-28.** The sine curves for a three-phase motor are shown in **Figure 7-29.** These motors are usually of the 220 V or 440 V type. The dome terminal block has nine terminals. The technician may wire the motor for either 220 V or 440 V. See **Figure 7-30.** Some of these motors use a 550 V supply.

Three-phase motors are available from 1/2 hp size and up. The building in which the unit is to be placed must be wired for three-phase service. Very few residences have three-phase electrical power. However, most industries and some commercial buildings are wired for three-phase.

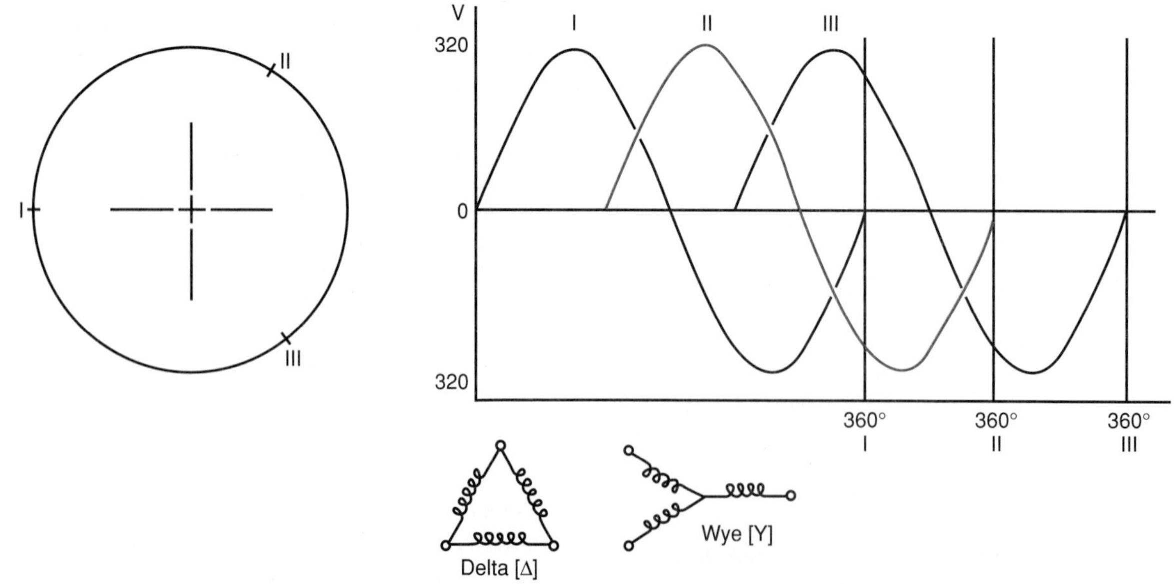

Figure 7-29. *Three voltage sine curves of three-phase circuit. System can be wired for either Delta or Wye design.*

The higher voltages are very dangerous. Disconnect the power and lock the switch open (use an actual lock) before starting to service the machines.

Three-phase motors use contactors or motor starters. They do not have the usual starting relays.

Figure 7-31 shows a three-phase motor wiring circuit with its starting and protection circuit. It is always best to have an electrical journeyman do the electrical work on these units. Since each unit may have certain differences, it is important to use the manufacturer's wiring diagram when servicing the system.

Operating characteristics of two types of polyphase motors are shown in **Figure 7-32.**

Occasionally, a three-phase motor may blow a fuse or open a circuit breaker on one phase only. The motor will attempt to operate on the remaining two phases. The motor will quickly overheat and may burn out if there is too much load on it. This is because the remaining two windings must carry all the load. Each one will need 1 1/2 times the current to compensate for the lost phase.

Phase loss monitors are sometimes used to shut down a motor to prevent it from damage. Each phase of a three-phase motor must be tested individually using a voltmeter. There will be about 50 volts difference between the open line and one of the other lines. The circuit having the "blown" fuse will indicate below normal voltage.

The direction of rotation of a three-phase motor may be reversed. This is done by changing any two of the power leads to the motor.

7.8.4 Hermetic Motor Terminals

The electrical terminals that carry the current through the dome must be electrically insulated from the dome or housing. They must also be leakproof.

Most motor terminals are fused to glass. The glass, in turn, is fused to a metal disk. See **Figure 7-33.** This assembly may be welded to the hermetic dome or housing. The terminal must be leakproof after thousands of heating and cooling and expansion and contraction cycles. Furthermore, it must have a high insulating value.

A fused glass multiple terminal installation is shown in **Figure 7-34.** The wire terminals are spring clips that tightly grip the hermetic terminals.

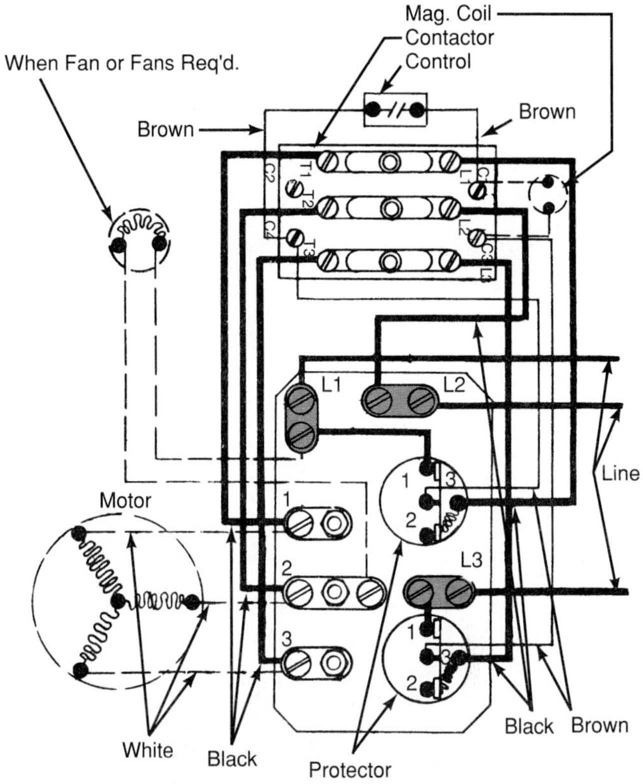

Figure 7-31. *Three-phase hermetic motor circuit. L1, L2, and L3 are three-phase line connections. Overload protection is shown in L1 and L3 circuits and operates magnetic coil of starter switch (top part of drawing). Special magnetic starter is required. Automatic temperature control would be connected to control at top of illustration. (Copeland Corporation)*

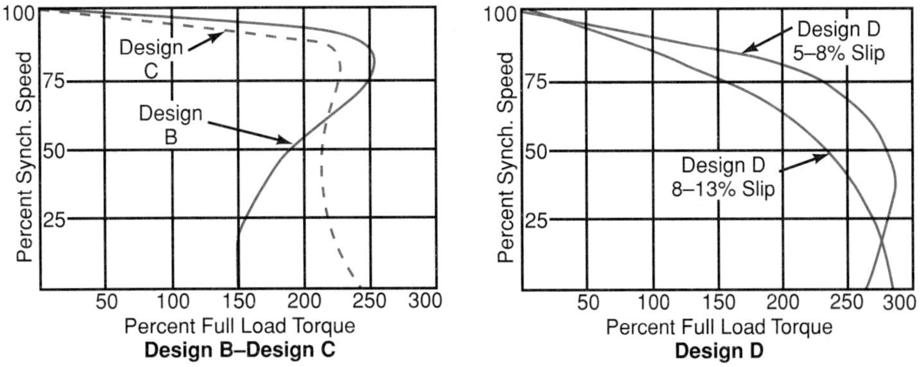

Figure 7-32. *Speed/torque curves compared for two polyphase motor designs. Note that Design B and C motors provide starting torque of 150% and 250% of full load torque. Design D motors provide starting torque of 260% to 280%. However, speed of Design D motors will fluctuate more than Design B and C motors as load changes. (MagneTek)*

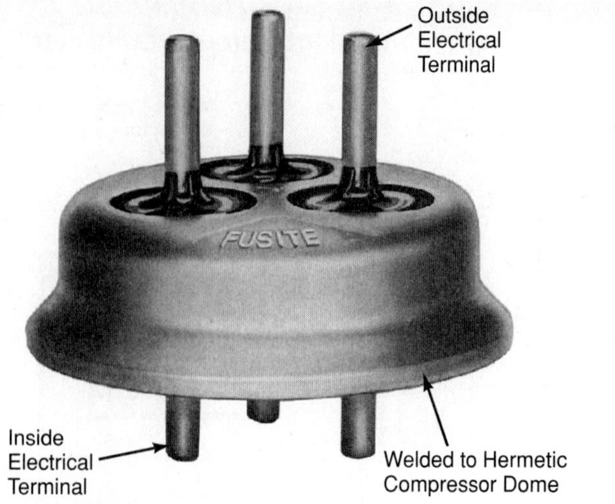

Figure 7-33. *Metal electric terminals which may be welded to hermetic compressor dome. (Fusite Division, Emerson Electric Co.)*

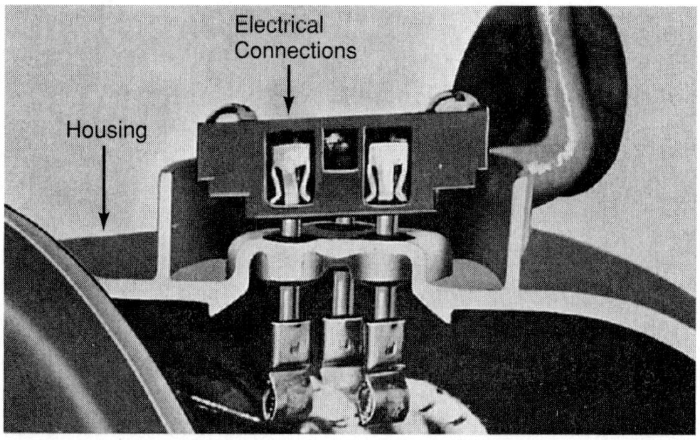

Figure 7-34. *Cross-sectional view of metal-glass fused hermetic electrical terminal installed. Wires to relay are connected to terminals with spring-loaded clips. Note metal structure around terminals to protect them from abuse.*

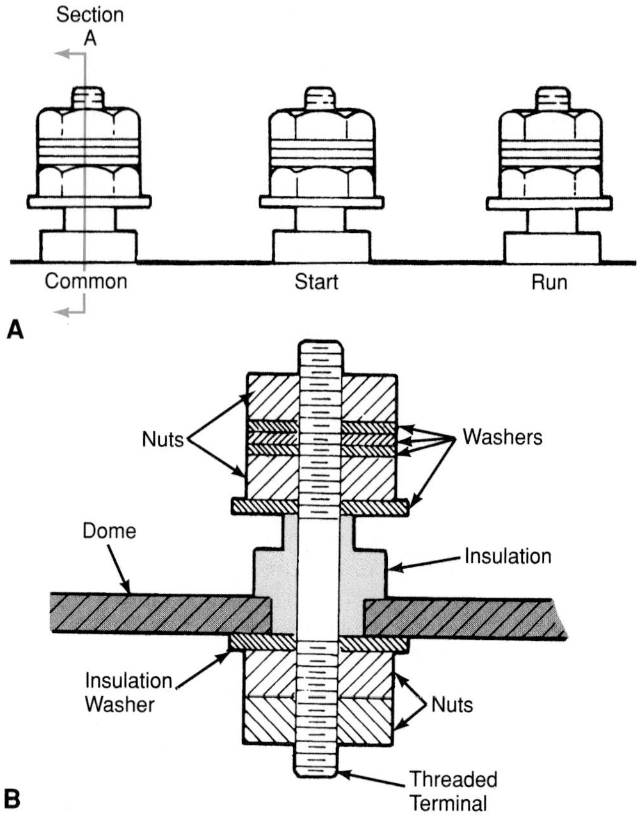

Figure 7-35. *A—Built-up hermetic motor terminals. B—Cross section of one of the terminals, showing insulation.*

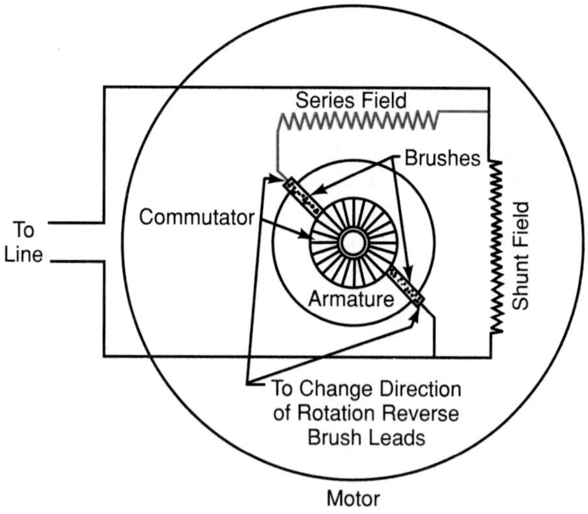

Figure 7-36. *Wiring diagram of compound wound direct current motor. Note that current going through armature must pass through the series field. Shunt field winding is in parallel with armature/series field winding circuit. Series winding gives motor high starting torque. (Fedders North America)*

Some compressors use what is known as ***built-up terminals.*** These are attached to the compressor dome, as shown in **Figure 7-35.**

Replacement terminals are used by many service technicians. Gaskets of synthetic material are used to make a leakproof joint.

7.9 Direct Current and Universal Motors

Areas with dc power must use direct current motors in refrigerators. These motors are ***compound wound.*** Direct current may be used only on external-drive systems.

Direct current motors have a mechanical likeness to both capacitor- and repulsion-start induction motors. However, the electrical circuits are quite different. A circuit diagram is shown in **Figure 7-36.**

Compound wound motors have two types of field windings. One is in parallel, while the other is in series

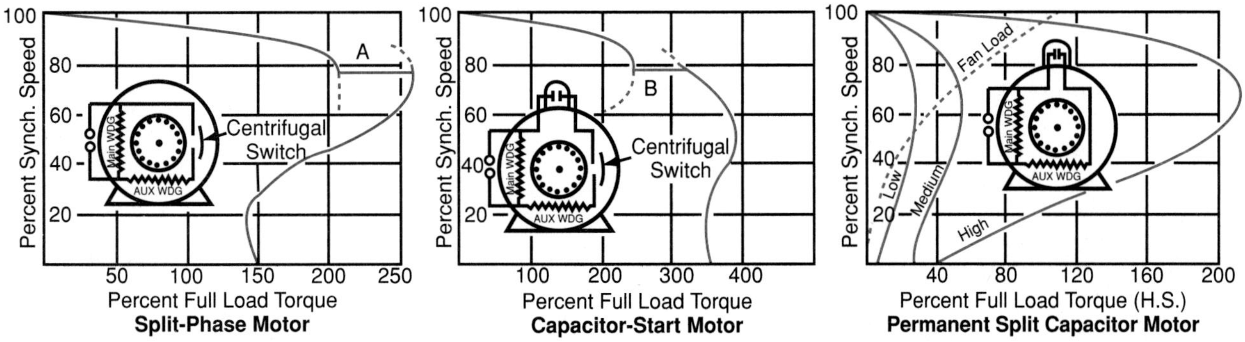

Figure 7-37. *Torque and speed characteristics of three types of single-phase fractional horsepower motors. Left—Split-phase. Starting winding is disconnected by rotor-operated centrifugal switch at 75% of synchronous speed as shown at A. This type gives fair starting torque. It is available in horsepower ratings up to 3/4 hp and for either 120 V or 240 V. Center—Capacitor-start. Starting winding is disconnected by rotor-operated centrifugal switch at about 75% of synchronous speed as shown at B. Higher starting torque is obtained by adding capacitor in starting circuit. Available up to 3 hp and for either 120 V or 240 V. Right—Permanent split capacitor. Capacitor stays in auxiliary winding whenever motor is running. Its use is limited to easy-to-start loads. PSC type may be used as multispeed motor using simple and inexpensive controls (see high, medium, low, fan in chart). Used in low-horsepower applications. Available for either 120 V or 240 V. (MagneTek)*

with the armature winding. Since dc is used, the field poles always have the same magnetic polarity. Also, because dc current is going through the armature coil, the magnetic polarity of the armature will remain constant. It is positioned to cause a turning effect or torque in the armature. The series field helps to keep the motor speed constant.

The series winding strength builds up to increase the speed if the motor tends to slow down. It weakens to reduce the speed if the motor tends to speed up. Reversing the brush leads will reverse the direction of rotation of these motors. (It reverses the armature magnetic polarity.)

The possibilities for wear are slightly greater in direct current motors than in the others because of the armature design. The brushes are in constant contact with the commutator. This presents problems such as a dirty commutator, worn brushes, high mica insulation between the bars, shorted armature, and squeaky brushes.

7.10 Motor Horsepower and Motor Characteristics

Energy, work, and power are explained in Chapter 1. Motors are rated by horsepower. One horsepower is equal to lifting 33,000 ft. lb. per minute or 550 ft. lb. per second. At 100% efficiency, 746 W equals one horsepower. Refrigeration motors of 1/100 hp to several hundred hp are in use. Motors as small as 1/20 hp are used to drive compressors.

In refrigeration, a motor's torque is as important as its horsepower. ***Torque*** is the force of a twisting or turning action (such as turning the crankshaft of a compressor).

It is important to know the properties of single-phase motors. They are the most popular of all motors. **Figure 7-37** shows the operating curves of the various types based on motor speed and percent of full load torque.

The average full load amperage and locked rotor amperage for various size of ac motors is shown in **Figure 7-38.**

	Amperage			
	120 Volts		240 Volts	
HP	Full Load	Locked Rotor	Full Load	Locked Rotor
1/6	4.4	26.4	2.2	13.2
1/4	5.8	34.8	2.9	17.4
1/3	7.2	43.2	3.6	21.6
1/2	9.8	58.8	4.9	29.4
3/4	13.8	82.8	6.9	41.4
1	16.0	96.0	8.0	48.0
1 1/2	20.0	120.0	10.0	60.0
2	24.0	144.0	12.0	72.0
3	34.0	204.0	17.0	102.0

Figure 7-38. *Full load and locked rotor amperage for single-phase ac motors.*

7.11 Electric Motor Grounding

Electric motors used on refrigeration and air conditioning systems must be grounded. Grounding of electric motors and other refrigerator parts and appliances is explained in Section 6.7.2.

In all refrigeration and electrical circuits, the ground wire is green. This is never used as a current-carrying conductor. Its main purpose is to provide protection to the operator in the event of an accidental ground in one of the mechanisms.

7.12 Motor Protection

The most common causes of motor failure are overloads and overheating. An overload condition may result in melted conductors or burned insulation on the motor's conductors. Considerable damage may also result if the compressor and/or motor overheat. This overheating may occur without the current draw becoming excessive.

It is important to protect a motor from both current overloads and overheating. Therefore, it is necessary to use both current- and heat-sensitive devices. These will open the circuit before there is damage to the motor. The following devices are used for motor protection:

- Fuses and circuit breakers.
- Bimetal switches.
- Electronic thermistors.

Each of these types of motor protection devices is described in detail in the following paragraphs.

7.12.1 Fuses and Circuit Breakers

Fuses and circuit breakers are often used to protect motors from burning out due to current overloads. Fuses and circuit breakers are usually located outside the motor for easy access.

Fuses conduct electrical current normally when operating below their maximum rating. (**Figure 7-39** shows the full-load current draw in amperes for various size ac and dc motors. It also shows the fuse ratings.)

When an overcurrent condition exists that exceeds the fuse's maximum rating, heat builds up inside the fuse. This causes the conduction element inside the fuse to melt. The element melts and opens the circuit. Electrical current can no longer flow to the motor. A fuse with a melted element is called a ***blown*** fuse. A circuit using this type of protection is shown in **Figure 7-40.**

There are four basic types of fuses:

- Fast-acting.
- Time-delay.
- Multipurpose.
- Current-limiting.

HP	AC Motors Single-Phase Split-Phase or Capacitor		DC Motors Compound Wound	
	120V	240V	120V	240V
1/6	4.4	2.2	- - -	- - -
1/4	5.8	2.9	2.9	1.5
1/3	7.2	3.6	3.6	1.8
1/2	9.8	4.9	5.2	2.6
3/4	13.8	6.9	7.4	3.7
1	16.0	8.0	9.4	4.7
1 1/2	20.0	10.0	13.2	6.6

Figure 7-39. *Maximum fuse ratings for motor running protection.*

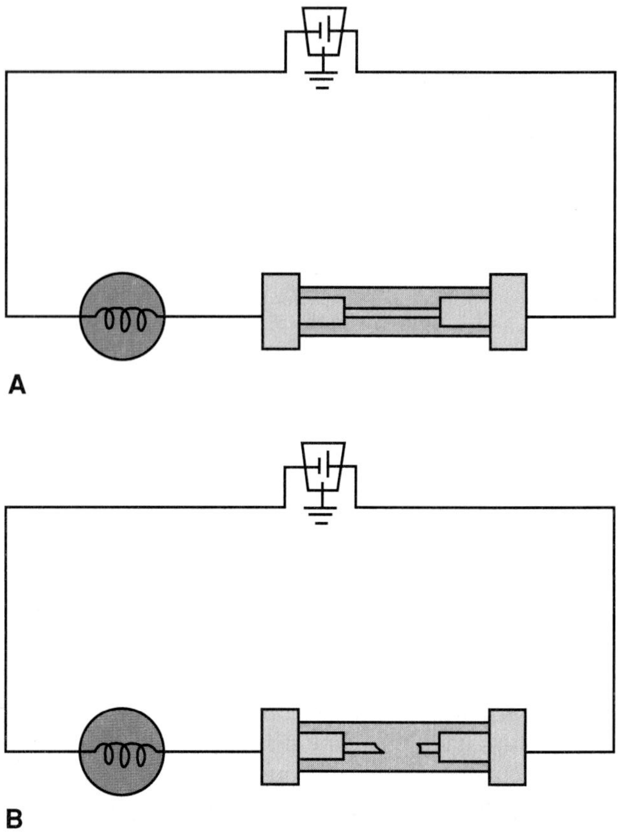

Figure 7-40. *One way to get motor protection. A—During normal conditions, fuse conducts current and motor is On. B—After overload condition, fuse can no longer conduct current, due to burnout, and motor is Off.*

The starting current of a motor can be from two to six times the running current of the motor. ***Fast-acting fuses*** blow immediately after the maximum rating of the fuse is exceeded. A fast-acting fuse used on a motor with a high starting current will blow before the motor can start running.

Time-delay fuses will not blow unless an overload condition exists for an extended period of time, typically 10 seconds. The time delay is usually required when a motor has high starting currents.

The time-delay fuse has a disadvantage to the fast-acting fuse if an extremely high current overload occurs. The motor could be damaged from the high current before the time delay is over. The fast-acting fuse does not have the time delay. It can shut the motor off before damage may occur.

The ***multipurpose fuse*** has the advantages of both the fast-acting and time-delay fuses. The multipurpose fuse will not blow during small overloads lasting only short periods of time, such as when the motor is starting. If an extremely high overload occurs—over 500% maximum current rating—the fuse will blow immediately. The multipurpose fuse provides good motor protection from both long-term small overloads and short-term large overloads.

The *current-limiting fuse* will never blow regardless of conditions. It prevents the electrical current to the motor from exceeding the rated current.

If a fuse continues to blow, check that it is the proper size rating for the application. If the fuse is the correct size, there could be another cause.

Many homes and businesses use circuit breakers rather than fuses. A ***circuit breaker*** is an automatic switch which will open a circuit if the current draw is too great.

Circuit breakers are usually rated the same as fuses. An "opened" circuit breaker must be manually reset. As with fuses, a circuit which is continually opening the breaker should be carefully examined. If the breaker has sufficient capacity, there may be a short or other trouble in the circuit.

Fuses and circuit breakers are not necessarily interchangeable. The UL (Underwriters Laboratories) nameplate on an HVAC device may indicate the type of overcurrent protection required by the National Electrical Code®.

7.12.2 Bimetal Switches

Bimetal overload devices are now the most common safety device. They are located at important places in the unit. If these parts overheat, the circuit will be opened by the bimetal snap switches.

All these devices, however, will only stop the mechanism if the current load is too high. If the motor should overheat from other troubles, the unit may still run and cause damage. Other sources of excessive heat may be high exhaust temperatures, poor air circulation, poor refrigerant circulation, and friction.

Present refrigerating systems have safety devices installed which open the electrical circuit when necessary. They give protection if the motor draws too much current, overheats for any reason, or if the compressor becomes too hot.

Figure 7-41 illustrates the action of a bimetal device. The device opens the circuit if the bimetal disk reaches a temperature that causes it to snap in the other direction. **Figure 7-42** shows the construction of a motor protection.

Overheating could occur without the current draw becoming excessive. It is, therefore, necessary to use heat-sensitive overload devices. These will open the circuit before heat can cause damage. They are temperature-operated. Their heat comes from the motor, compressor, and the current draw of the motor. **Figure 7-43** shows overload protection located on the motor compressor dome or housing. An external overload protection used on polyphase motors is shown in **Figure 7-44.**

The action of a bimetal snap-action, current- and heat-actuated overload protection as used on an external-drive motor is shown in **Figure 7-45.**

Some of the temperature-sensitive motor protection is quite compact. They may be easily installed inside the motor winding. Such protection is described in the following section (Section 7.12.3).

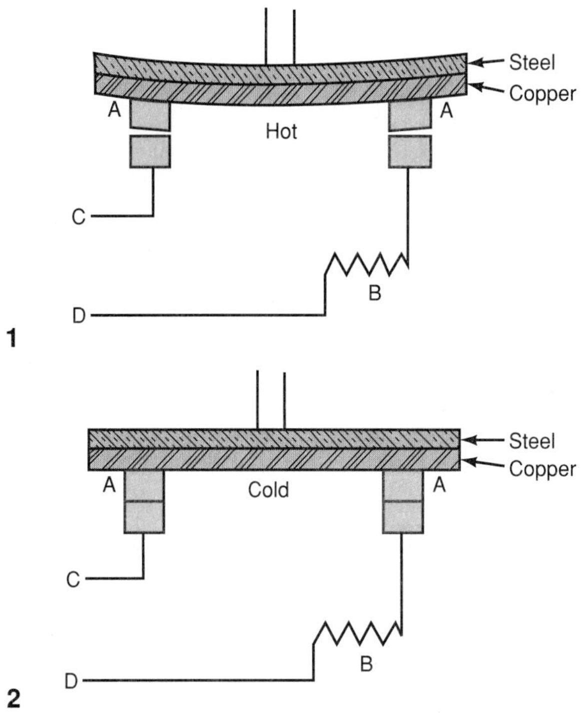

Figure 7-41. *Snap-action bimetal motor overload protection. High speed of snap-action motion avoids burning of points due to arcing. A—Contact points. B—Heater coil. C—Motor winding connection. D—Power source connection. 1—Circuit open because of increased temperature or voltage. 2—Points closed in normal operation.*

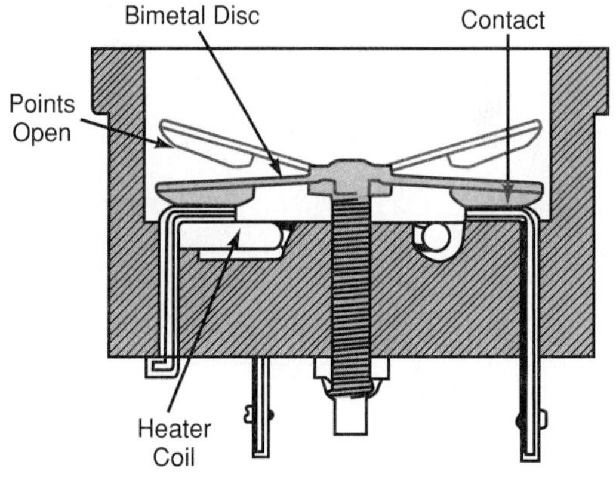

Figure 7-42. *Motor overload protection. Excess heat will cause bimetal disc to snap and open contact points. (Texas Instruments, Inc.)*

7.12.3 Motor Internal Overload and Overheating Protection

Internal overload motor protection is mainly used on hermetic motors. Along with the two-pole motor, they were developed for large units. These are another example of a bimetal protection device.

A motor may overheat if there is too little refrigerant flow. (The refrigerant vapor cools the motor.) It may also overheat if the unit has to start again too soon after

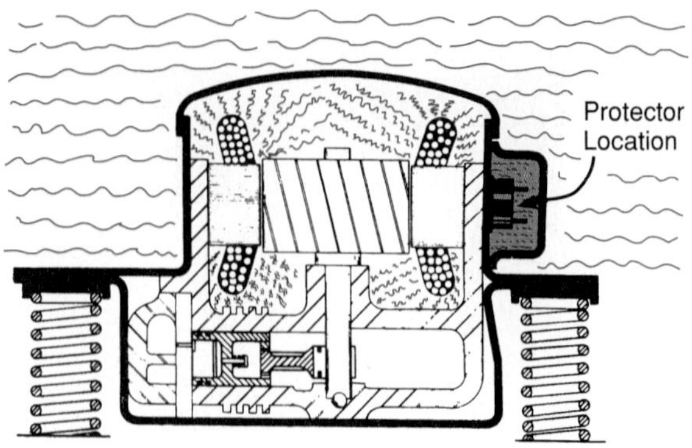

Figure 7-43. *Suitable location for motor compressor overload or excess temperature protection.*

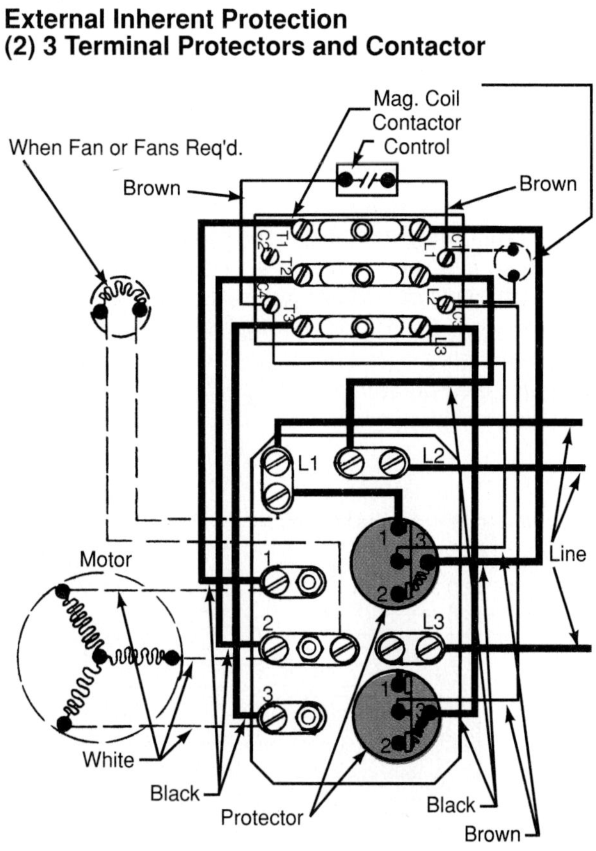

Figure 7-44. *Wiring diagram for three-phase hermetic motor which uses two external motor protectors. (Copeland Corporation)*

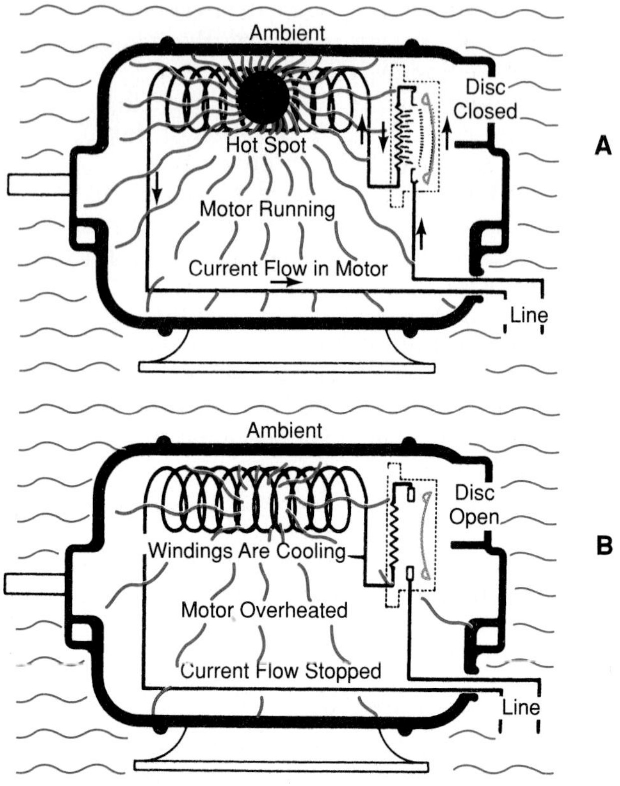

Figure 7-45. *Bimetal overload protection installed inside external-drive electric motor. A—Current flowing, motor running. B—Current cutoff, motor stopped.*

The motors normally operate at 125°F (52°C). When the temperature reaches 200°F to 250°F (93°C to 121°C), the protection device will open the circuit and stop the motor. It will then close at about 150°F to 175°F (66°C to 79°C).

One type of internal motor overload protection is the bimetal disc. The contact points are on a bimetal strip. They are normally in a closed position as in **Figure 7-46.** When an excessive temperature is reached, the points will open the circuit. When the temperature in the disc decreases enough, the strip returns to its normal position and the contact points will close.

The internal overload protection is placed inside the hermetically sealed compressors, directly on or in the windings. The internal overload protection will open if shutting off. Too much current draw may be caused by a stuck or locked rotor or compressor.

In most refrigeration and air conditioning equipment, the motor compressor unit is designed to start under a condition of balanced pressures. There is danger of overheating the motor if it must start against a high head pressure. If there is an increased starting load, the internal overload protection will open the motor circuit. This will protect the motor from such abuse.

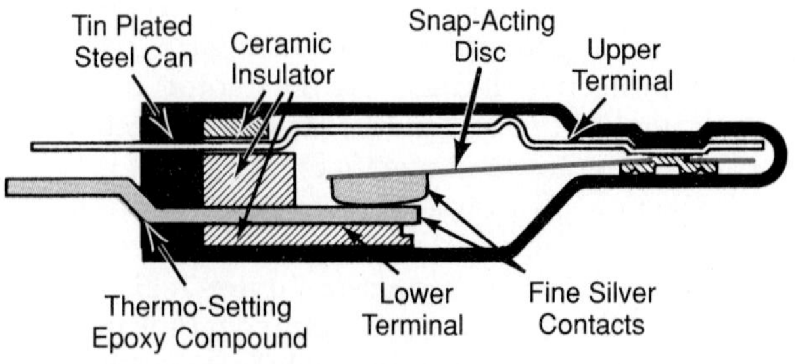

Figure 7-46. *Compact motor protection designed to fit into motor windings.*

there is excessive current draw, excessive temperature, or both. Loss of refrigerant, a restriction in the system, or low suction pressure could lead to a motor burnout if the overload protection were not installed.

It may be an hour to two hours after the protection opens the circuit before it will close. This depends on the ambient temperature conditions. Use forced air, dry ice, or carbon dioxide spray to speed up dome cooling. Do not tap on the controls in an attempt to operate the points. The points may vibrate and arc, causing them to burn out quickly.

The leads to this overload protection must never be shorted. Even a few moments of operating a unit without this protection may burn out the motor.This protection cannot be taken out of the circuit. They are the best possible protection for a hermetic compressor. Motors having this protection are usually labeled "Internal Overload Protected."

In Wye-type three-phase motors, the internal overload is at the common terminal of the three windings. It will open all three circuits when its points open, **Figure 7-47.** This internal protection is very reliable. Cases of failure are almost unknown. **Figure 7-48** shows an instrument used to check the three circuits of a three-phase system. An instrument is used to check if one lead of a three-phase system is open.

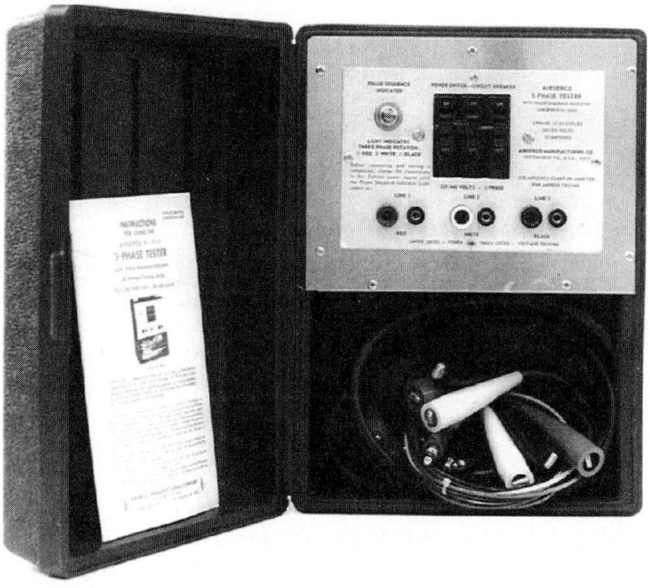

Figure 7-48. *Instrument used to test all three circuits of a three-phase system. (Airserco Mfg. Co.)*

7.12.4 Electronic Thermistor

Thermistors that have a positive temperature coefficient (PTC) are used for motor protection. As the temperature increases, the resistance of the PTC thermistor also increases. The PTC thermistor is connected in series with the copper windings of a motor. It prevents current from conducting when the temperature of the motor increases beyond a safe value. After the motor and thermistor cool down to a safe temperature, current can begin to flow again. The motor will then start up.

Another type of electronic thermistor has a ***negative temperature coefficient (NTC).*** The thermistor is placed in a capsule within the motor. As the temperature increases, the resistance of the NTC thermistor decreases. If the temperature rises to about 200°F (93°C), the increased current flow through the thermistor will operate a relay circuit and open the circuit. This shuts the motor off. When the temperature falls back to a safe value, the current decreases below the amount required to hold the relay open. The relay will then close so the motor can run again.

7.13 Motor Temperature

The temperature of the hottest part of the motor should not be more than 72°F (40°C) over the room temperature. This means an average maximum temperature of approximately 150°F (66°C).

Such temperature is difficult to measure except in a laboratory. Therefore, it is better to check the ***ambient***

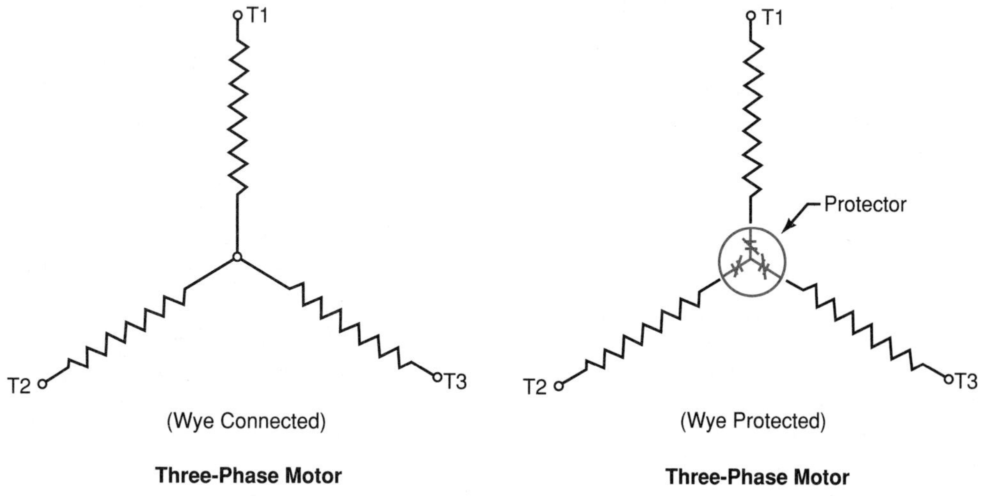

Figure 7-47. *Method of connecting motor protection in Wye-type three-phase motor. (Copeland Corporation)*

(surrounding) temperature to be sure it is not too high. Check the motor for cleanliness and airflow. Motors depend on rather cool ambient air for cooling. If this air is too warm or if the airflow is restricted, the motor will overheat.

Then check the current draw of the motor. If the draw exceeds the rating on the motor identification plate or in the motor manual specifications, the motor will overheat. Current overdraw may be due to overloading the motor. It may also be caused by a shorting in the motor windings.

Always measure motor temperature with a thermometer. A motor with too high of a frame or stator temperature may have high motor winding temperatures. These may become so high that the insulation on the wires will fail.

The thermal overload protection for motors usually opens the circuit when the temperature reaches 200°F (93°C). It closes the circuit when the temperature drops to about 150°F (66°C).

7.14 Standard Motor Data

The data on the motor identification plate usually gives the following information:

- Required voltage (emf) supply.
- Hertz (cycles per second).
- Running current (amperes).
- Locked current draw. This indicates internal circuit condition when the rotor is locked so it cannot turn. The locked current draw is also the starting current draw.
- Temperature rise. The temperature rise is usually specified in degrees Celsius.
- Seasonal energy efficiency ratio (SEER) (detailed in Chapter 16).

Compressor speed on external-drive systems can be controlled by using two- or four-pole motors. It can also be controlled by changing pulley sizes. Direct-connected compressors must operate at motor speed. If a four-pole motor is used to replace a two-pole motor, a compressor of greater displacement must be used with the slower rpm motor. The unit will be running at half its former speed. Therefore, the compressor displacement per stroke must be double.

The size of the conductor used in the refrigeration mechanism is very important. If the conductor is too small or too long, it will heat up and eventually cause a fire. Long circuits also add an unnecessary resistance to the flow of electricity. This causes excessive voltage drop. The voltage to the motor should not be less than 90% of the rating of the motor. If it is less, there is danger of the motor being overheated and ruined. A table of wire sizes recommended for 120 V circuits is given in **Figure 7-49.**

The efficiency of small motors is only 50% to 60% because of clearances and efficiencies of the winding.

Wire No.	Diameter of Wire in Inches	Ampere Capacity Plastic Insulation
18	.040	5
16	.051	10
14	.064	20
12	.081	25
10	.102	30
8	.128	50
6	.162	70
4	.204	90
2	.258	125

Figure 7-49. *Recommended wire sizes for various ampere capacity circuits. This table is calculated on the basis that wire is used for 120 V circuit.*

Therefore, they consume nearly twice as much current as they should, compared to larger motors such as 1/2 hp and over. A 1/6 hp motor, which should theoretically use only 124 W or 1 1/4 A will need approximately 2 1/2 A to 3 A (280 W to 400 W) to develop the 1/6 horsepower.

When making electrical connections to a domestic refrigerator, the thermostat should be connected into the ***hot wire*** of the circuit. This wire has black insulation. The other wire is called the ***common wire*** of the circuit. It has white insulation. It should be run directly to the motor. The common wire carries the same amount of current as is carried in the black wire.

A system of green grounding wires grounds all mechanisms in a refrigerator or air conditioner. This ground is not a current-carrying wire. It is for safety only. It is used to avoid an electric shock should a short circuit or a ground occur in the electrical system.

7.15 Fan Motors

Many hermetic units use motor-driven fans to:

- Force condenser cooling air through the ducts and over the condenser and condensing units.
- Circulate air in refrigerated parts of domestic and commercial units. See **Figure 7-50.**

To create efficient air movement, the fan and condenser are housed in sheet metal or plastic. The fans are carefully balanced and run almost noiselessly. They are usually attached to the motor shaft with Allen setscrews.

Some of these motors have sealed bearings (bushings) and require no oiling. Others need oiling (SAE 10 or 20) amounting to one drop per bearing each six months. A few motors on the market have only one bearing. The motors are usually attached to brackets and are mounted in rubber. **Figure 7-51** shows a replacement condenser fan.

Generally, the condenser fan motor leads are connected to the common terminal and the running wind-

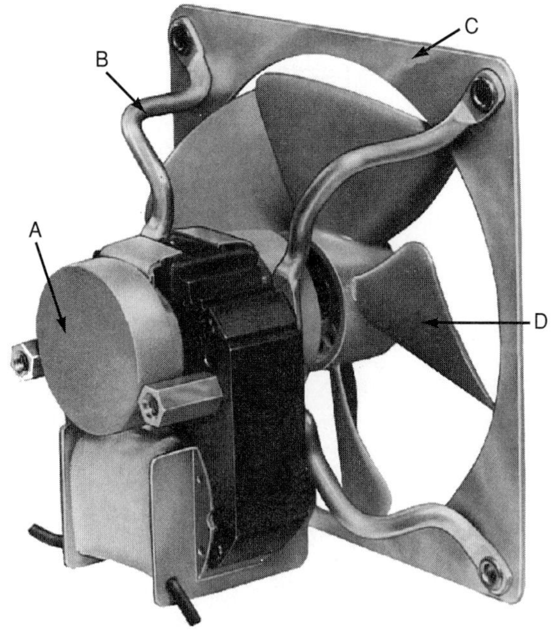

Figure 7-50. *Shaded-pole motor and fan. A—Motor. B—Motor support brackets. C—Mounting plate. D—Molded fan. (General Electric Co.)*

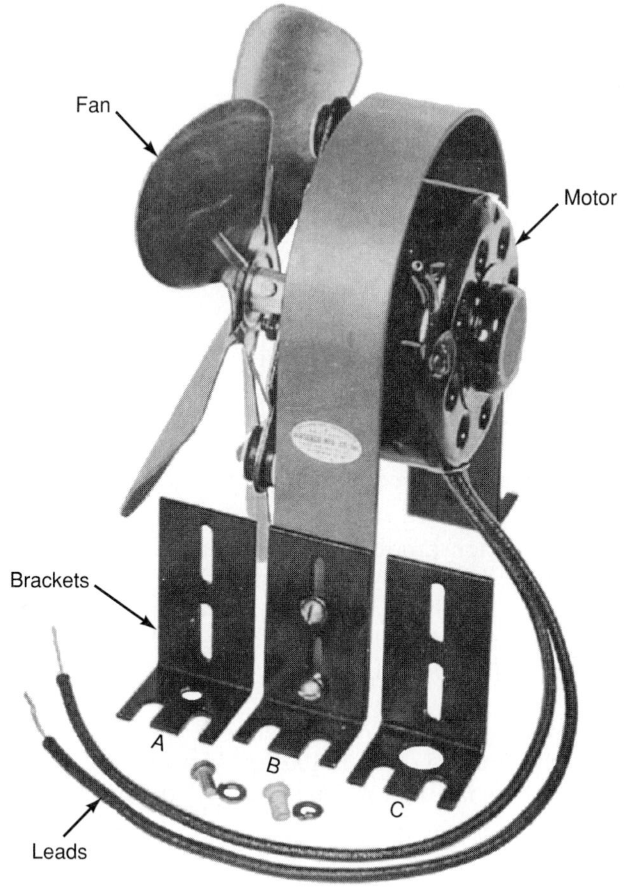

Figure 7-51. *Replacement motor and fan for refrigerant condenser. Motor unit has universal mounting brackets (A, B, C) which may be used with a variety of condensing unit designs.*

ing terminal of the compressor motor. This connection puts the fan motor in parallel with the compressor motor and allows it to be controlled by the thermostat. The safety overload cutout is also put in the circuit ahead of the fan. It will also cut out the fan motor.

Some fan motors have their own thermal safety controls. Many are of the two- or three-speed type. The variation in speed may be obtained by using extra poles in the stator or by using a solid state control.

The speed of a fan motor is quite sensitive to the applied voltage. As the voltage drops, so will the fan speed. **Figure 7-52** shows the relationship between the voltage and fan speed.

Figure 7-53 is a schematic of some of the common fan motor circuits. One-, two-, and three-speed motor circuits are shown. Refer to Section 7.18.6 for information on voltage drop tests, troubleshooting, and servicing fan motors.

7.16 Shaded-Pole Motors

Shaded-pole motor construction is different than that of the motors previously described. See **Figure 7-54.** The shaded-pole produces a moving magnetic field perpendicular to the field pole.

Approximately half of each pole face has a small copper plate insert, A, with a small winding. This insert slows down the build-up of the magnetic field through the copper plate. It is slowed down just enough to cause a magnetic motion toward the copper plate.

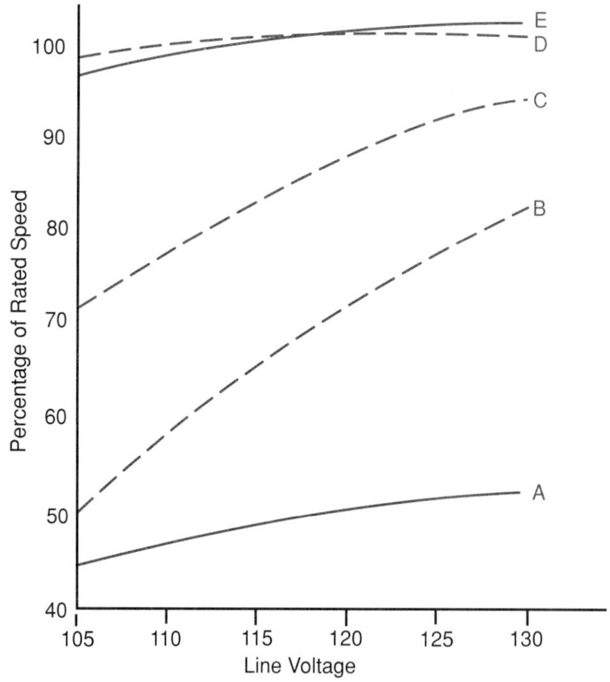

Figure 7-52. *How voltage affects fan speed for six- and twelve-pole motors. A—Twelve-pole. B—Low speed for a high-efficiency twelve-pole motor. C—Medium speed, twelve-pole. D—High speed, twelve-pole. E—Six-pole.*

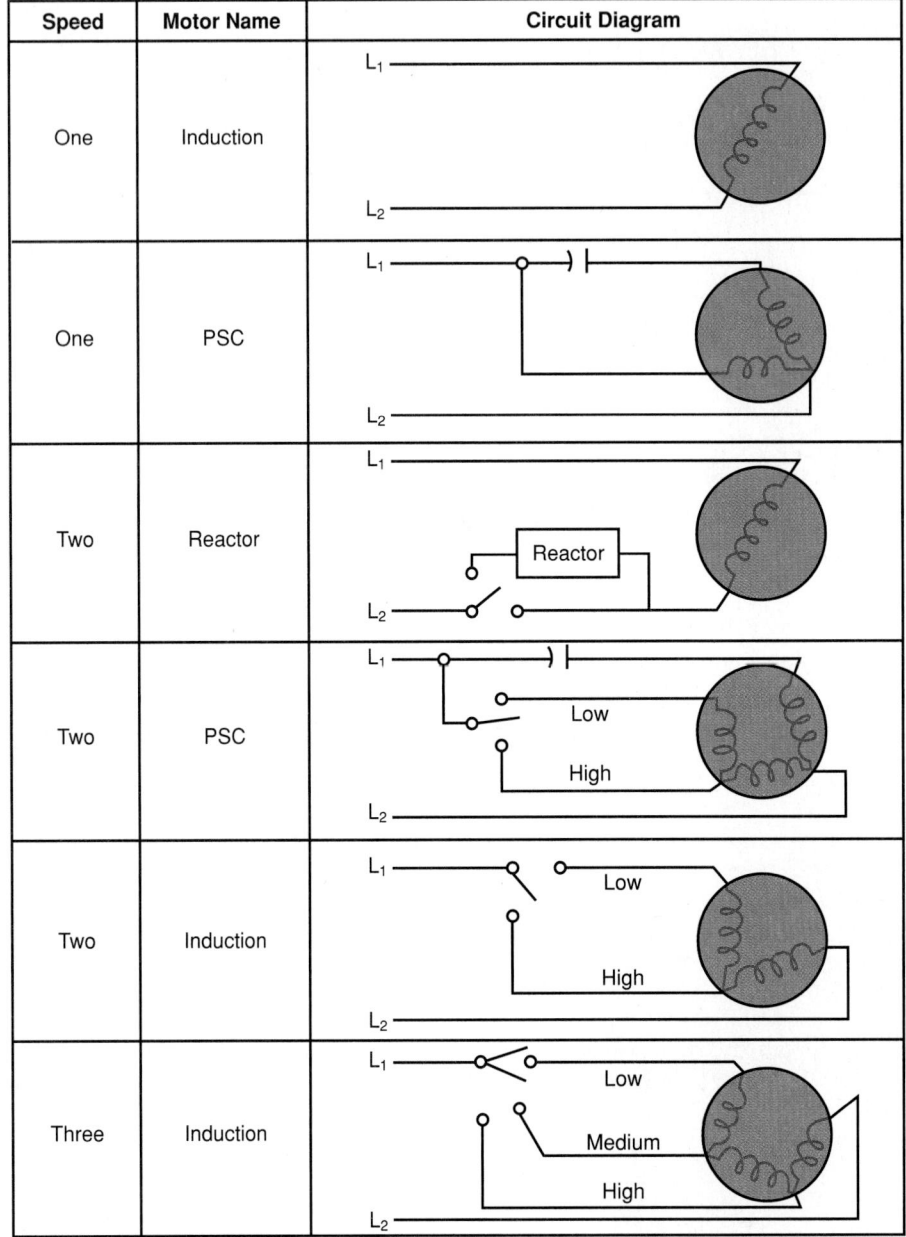

Speed	Motor Name	Circuit Diagram
One	Induction	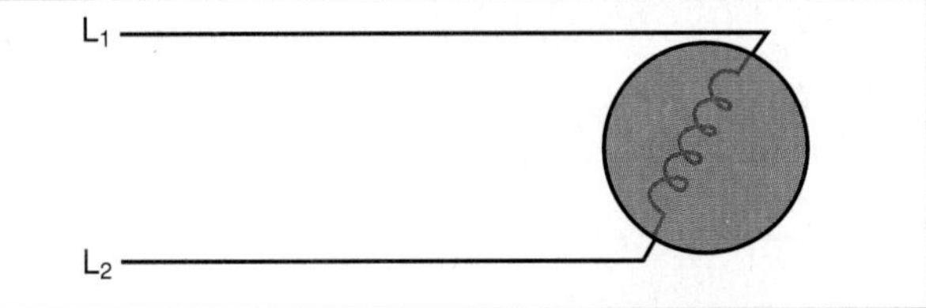
One	PSC	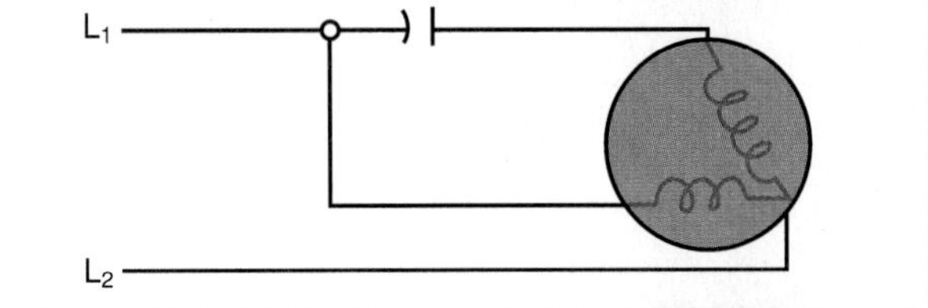
Two	Reactor	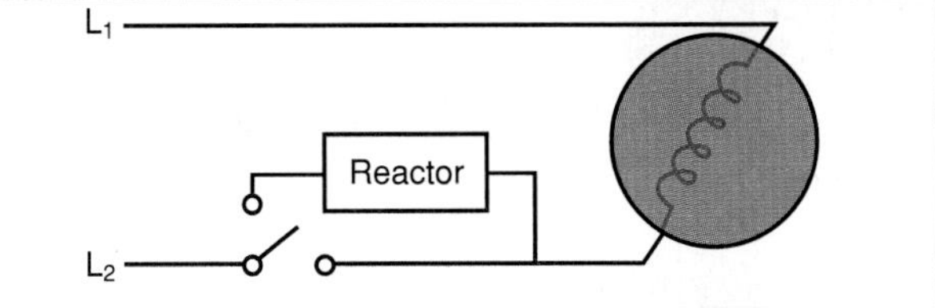
Two	PSC	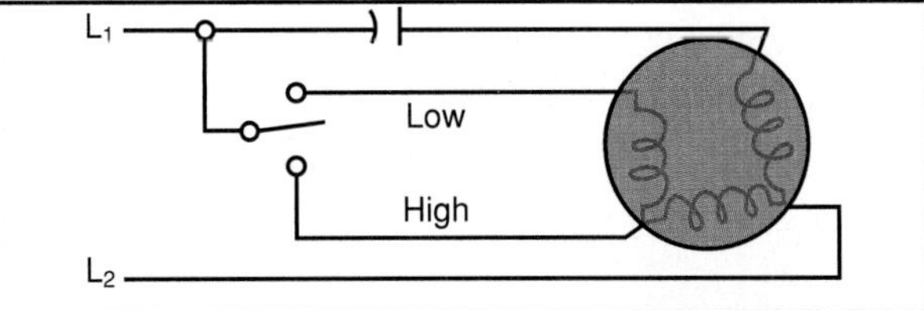
Two	Induction	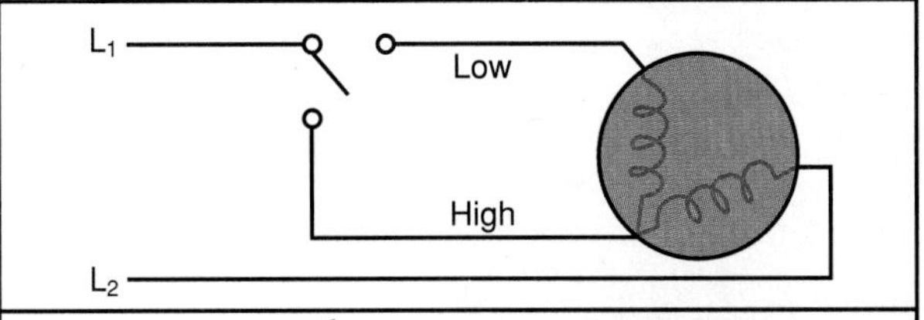
Three	Induction	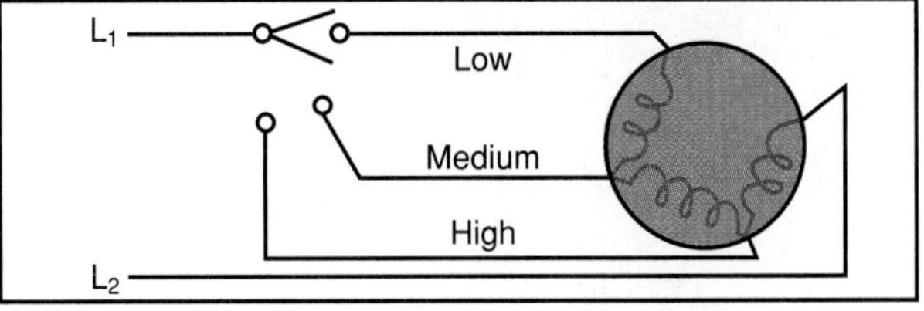

Figure 7-53. *Schematic of some common fan motor circuits.*

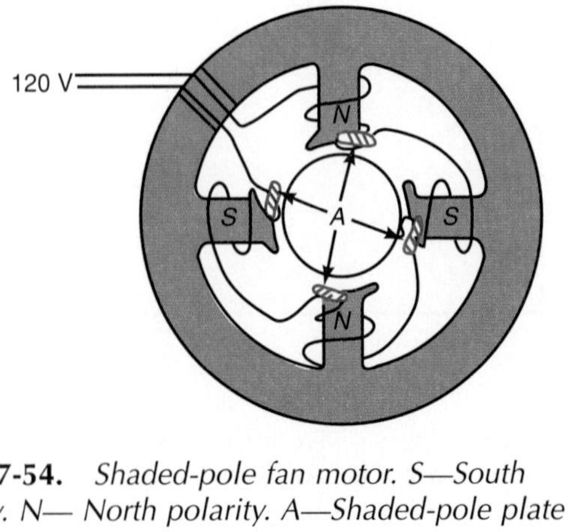

Figure 7-54. *Shaded-pole fan motor. S—South polarity. N— North polarity. A—Shaded-pole plate (copper).*

This action produces a lag for induced magnetism in the rotor (opposite magnetism). The rotor turns as it is attracted by the magnetism. Movement of the rotor will continue as the alternating current changes the polarity of the poles and the rotor.

Shaded-pole motors have less starting torque than other types of motors. Nevertheless, it is very successful in small motors 1/6 to 1/100 hp. **Figure 7-55** shows a double-shaft fan motor. The end bell of this motor is shown in **Figure 7-56.**

7.17 Electronic Variable Speed Motors

A method employing transistor switching instead of the brush or commutator has been developed for

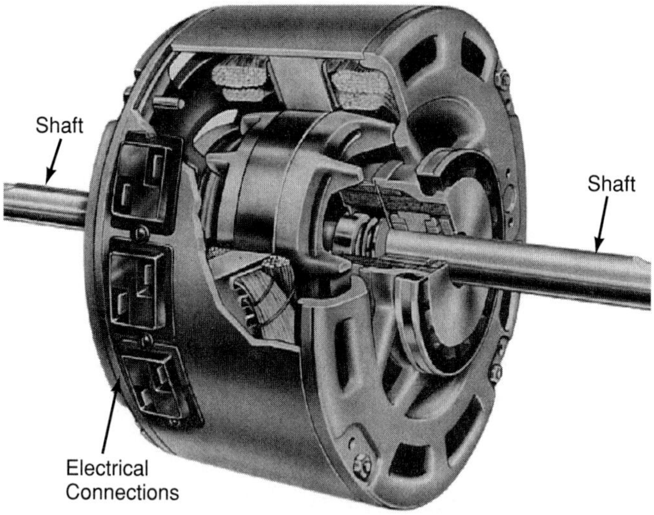

Figure 7-55. *Low starting torque shaded-pole, double-shaft fan motor. Three terminals for electrical connections provide a variety of fan speeds. (General Electric Co.)*

Figure 7-56. *Aluminum end bell for fan motor. Note assembly screws and nuts.*

low-horsepower motors. Brushless motors operate with silicon rectifiers, transistors, and special circuitry. There are a number of advantages of the transistorized motor. These include high speed, compactness, performance, durability, and variable speed. The transistorized motor also eliminates sparking and brush noise.

Speed is changed by adding a small variable resistance ***(potentiometer).*** This will vary the resistance within the circuitry of the control. Motor rotation direction can be quickly reversed by manipulating the motor control's switching devices.

Figure 7-57 is the wiring diagram for a solid state motor speed control. The control reacts to signals from the unit-mounted sensor. **Figure 7-58** illustrates the same unit complete with case and controls.

Figure 7-59 shows a schematic diagram for yet another control commonly used to regulate motor speed. The internal schematic diagram of the control is shown in **Figure 7-60.** The circuit operates as follows:

1. Resistances R_2, R_3, R_4, and capacitor C_2 form an R-C (resistance-capacitance) charging network. The

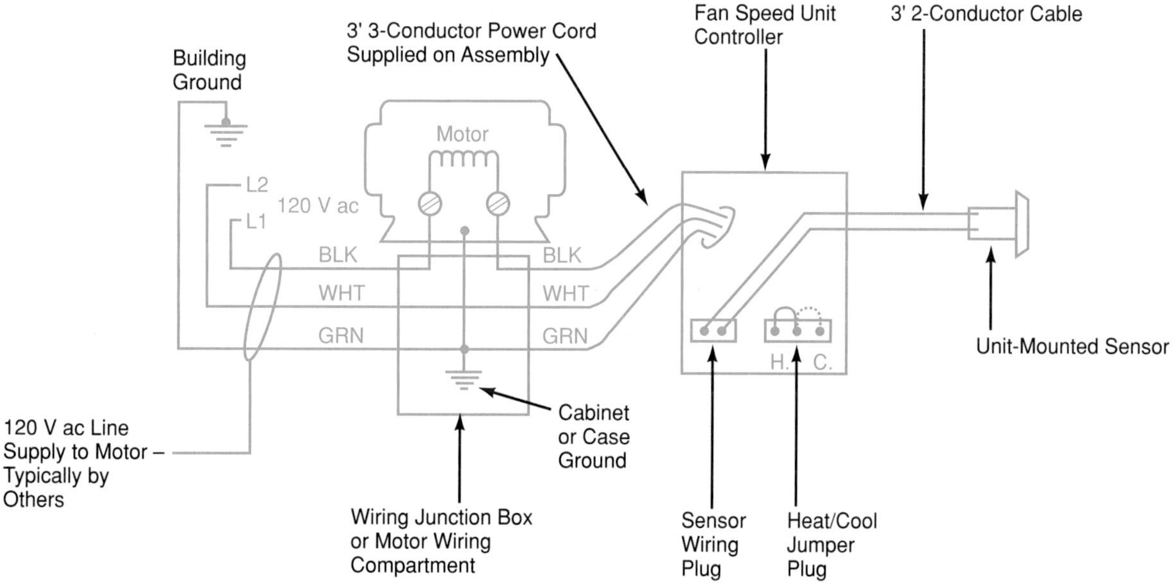

Figure 7-57. *Installation wiring diagram for solid state motor control. (Barber-Colman Co.)*

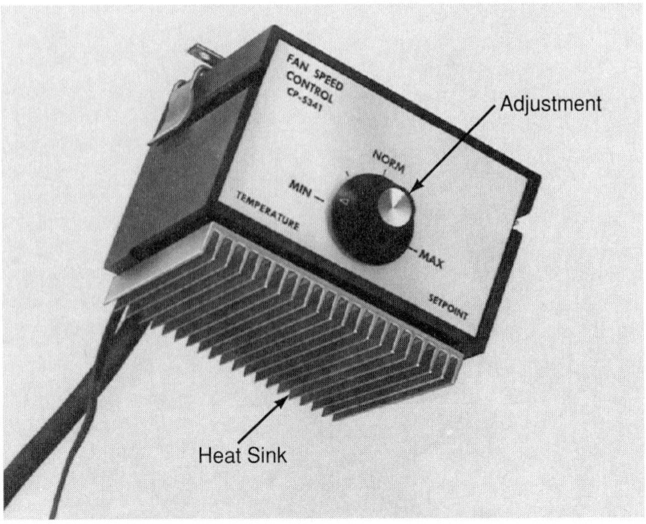

Figure 7-58. *Solid state fan speed control. (Siebe Environmental Controls)*

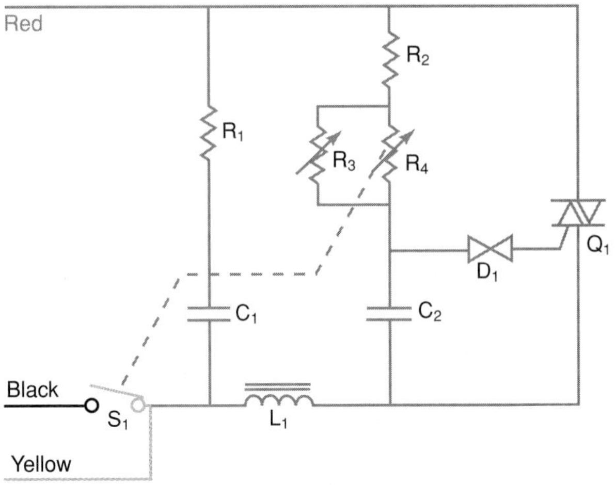

Figure 7-60. *The electronic circuit for a variable speed control used with permanent split capacitor (PSC) motors, shaded-pole motors, and universal motors. Refer to **Figure 7-59.** (Lutron Electronics Co., Inc.)*

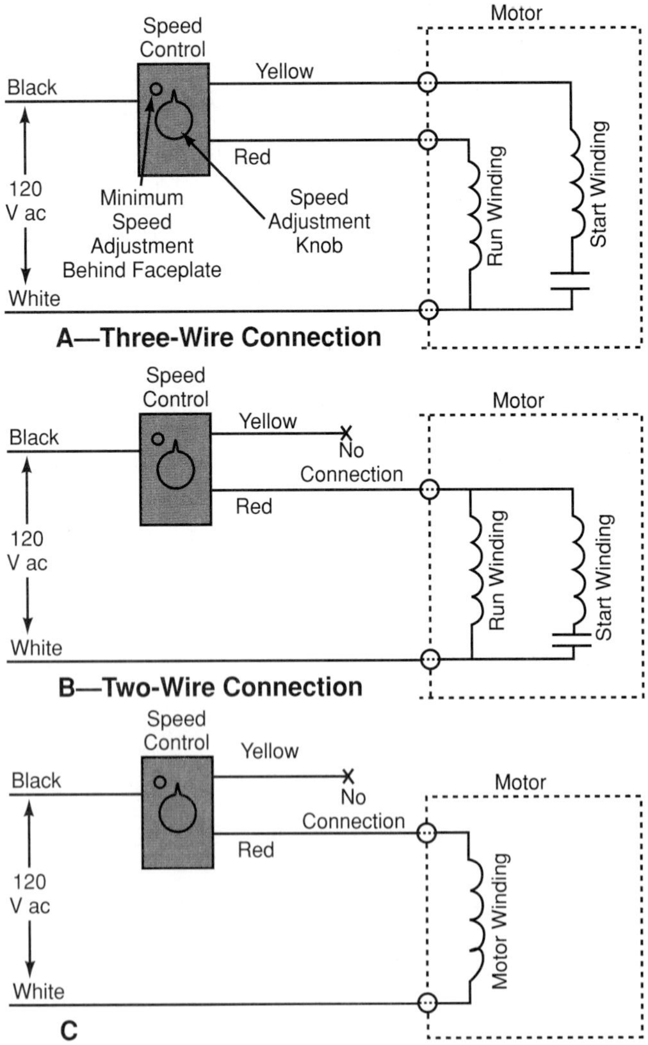

Figure 7-59. *A and B—Connecting an electronic variable speed control to a permanent split capacitor motor. C—Same control connected to a shaded-pole motor. (Lutron Electronics Co., Inc.)*

effective resistance of R_2, R_3, and R_4, charges the capacitor, C_2. It is charged up to a voltage that causes diac, D_1, to conduct an electric current. This fires triac, Q_1.

2. When Q_1 is fired, it conducts until the current through it drops below the holding current (25 milliamperes) of the triac. At that point, the triac turns off. The control's charge-up, fire, conduct, and turn-off process repeats every half cycle of alternating current power.
3. The speed of the motor is controlled by varying the resistance of R_4. The variable resistor is operated by the knob on the front of the control.
4. The lower the resistance of R_4, the faster the motor will turn. The minimum speed of the motor can be set by adjusting the resistance of trimmer, R_3.
5. R_1, C_1, and L_1 provide suppression of radio frequency interference caused by the fast switching of Q_1. This circuit can be used with permanent split capacitor, shaded-pole, and universal motors.

SERVICING ELECTRIC MOTORS MODULE

7.18 Servicing Electric Motors

The maintenance, troubleshooting, removal, repair, and installation of electric motors as well as their accessories is a major portion of a service technician's job. You must know how to use instruments and be knowledgeable about electricity to accurately determine the trouble. The following paragraphs describe these operations:

- General service.
- External-drive motors.
- Hermetic motors.
- Fan motors.

It is important that a solid base be provided for the installation of any motor. The motor must be bolted down securely. The armature shaft should be level for a horizontal motor. It should be exactly vertical (plumb) for a vertical motor.

If a motor is severely damaged and must be replaced, the following information must be obtained:

- Type of motor.
- Operating line voltage.
- Maximum current draw.
- Direction of rotation and speed.
- Mounting hardware.

Use the manufacturer's nameplate to get all the required motor specifications. Sometimes a replacement motor from another manufacturer can be found by using cross-reference charts.

Make certain that the replacement motor has all the same specifications as the original motor. The replacement motor may have a maximum current draw 10% larger. It must never have a current draw less than the motor being replaced.

7.18.1 Watt Reading to Determine Motor Troubles

One way to learn the condition of a motor compressor unit is to observe the wattage consumption (watts drawn) of the unit. The wattmeter will provide two different wattage readings:

- Combined starting and running winding reading (only 1 to 1 1/2 seconds long).
- The running winding reading during the time the unit is running.

Note: The wattmeter needle will overswing slightly at the instant that the motor compressor starts. The combined starting and running winding watt reading is for only a fraction of a second. The technician must allow for the overswing. Approximate watt readings for small hermetic motors are shown in **Figure 7-61.**

When the thermostat contacts close, the wattmeter indicator will swing to the right. Then it will quickly move back to the combined reading. In a few seconds, the pointer will fall to the running winding reading only.

Motor HP	Watts at 120 Volts		
	Running		Starting or Locked
	At 70°F	At 110°F	
1/16	66	100	375
1/9	117	160	740
1/8	108	163	743
1/7	160	218	970
1/5	242	295	1450
1/4	235	320	1250

Figure 7-61. *Approximate watt readings for small hermetic motor compressors. Indicated temperatures are ambient.*

If the starting winding circuit is open, the wattmeter pointer will swing to the right. It will then move back to the running winding value only. This action indicates a bad relay or starting winding. The overload safety cutout should open the circuit in two or three seconds.

7.18.2 Radio and TV Interference

Some commercial refrigeration units cause a slight amount of radio or TV interference. This interference will usually amount to a slight snap or click in the radio or TV at the instant the unit stops or starts. It should be no more noticeable than turning off a light.

This interference may be reduced by grounding the frame of the motor to a water pipe. Another means of reducing the interference is to place a capacitor between the frame of the motor and a ground. In general, the interference is not a problem and may usually be disregarded.

There may be excessive radio interference of a continuing nature when the unit motor operates, or when it starts. This indicates a loose electrical connection or some fault in the mechanism of the motor. These include worn brushes, a badly pitted commutator, or loose connections. The particular trouble can be easily determined by a careful examination of the motor.

Occasionally a static charge will be built upon the belt of a belt-driven compressor. The discharge of this buildup will cause radio interference. If the motor and compressor are grounded together, this noise will be eliminated.

7.18.3 Testing Capacitors

When a motor or motor compressor does not start or run properly, there is a good possibility that the trouble is in the capacitor. Most motors have only one—a starting capacitor—but some have two or more. In such cases, one might be a starting capacitor and the other a running capacitor. Run and start capacitors are not interchangeable. Some motors have two or more capacitors connected in parallel. This provides for additional capacitance. Some capacitors may be connected in series for additional voltage.

The starting capacitor is connected in series with the starting winding. It is usually wired into the circuit between the relay and the starting winding terminal of the motor (Off-and-On, or ***intermittent*** operation).

The running capacitor is also in the starting winding circuit. However, it stays in operation while the unit runs ***(continuous operation).***

There are two types of capacitors:

- Starting capacitor (electrolytic—dry; for intermittent operation).
- Running capacitor (oil-filled; for continuous operation).

Both types may be tested in the same way. (The dry capacitor is described in Section 7.3.5.)

The simplest capacitor test is to substitute a good capacitor for the one being tested. If the motor operates, the old capacitor is faulty. The replacement capacitor should be the same capacity as the old. If one of a different capacity must be used, it should be 10% over capacity rather than under.

Figure 7-62 illustrates a capacitor troubleshooting instrument. This type of tester is used for checking capacitor shorts, leaks, grounds, and open circuits. It indicates whether the capacitor is the problem in the circuit and should be replaced.

Note: Never place your fingers across the terminals of a capacitor. When discharging a capacitor, place it in a protective case, then discharge it through a resistor connected between the terminals.

7.18.4 Servicing Capacitors

Some capacitors have resistors connected across the terminals. This resistor, visible at the capacitor terminal, slowly bleeds the capacitor of its charge to lessen the arcing of the motor control points. Such arcing may occur if the unit cycles frequently.

When testing such a capacitor, remove the resistor from one terminal. To discharge a capacitor, use a 20,000 ohm, 2 W resistor in the circuit. Avoid shorting the terminals. The sudden discharge may rupture the thin metal foil in the capacitor.

Capacitor size must be accurately suited to the motor and the motor load. It is general practice to permit up to 10% over capacity. For example, a 110 μf can be used for a 100 μf capacitor. An undersized capacitor should never be used. If at all possible, use an exact replacement. The make, model, and model number are usually placed on each capacitor. If this information is unavailable, or if an emergency capacitor must be used temporarily, there are several ways to determine the proper size. One way is to look at the capacitor size used in a similar refrigerating unit by the same manufacturer. Another way to determine the proper size is to look in a service manual for a similar unit by any manufacturer.

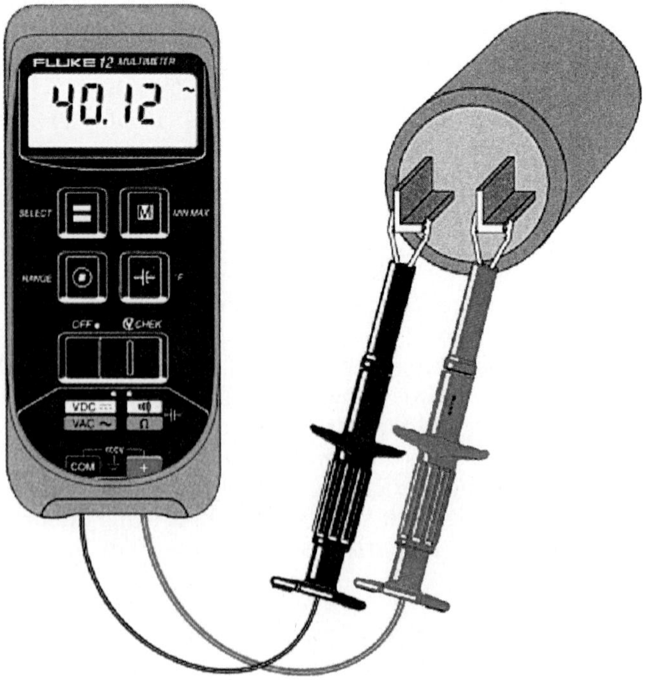

Figure 7-62. *Capacitor tester and analyzer being used to measure capacitance. The capacitor is discharged using a resistor. (Reproduced with permission of Fluke Corporation)*

A better method to determine capacitor size is to use a specially designed capacitor selector unit. This selector has a variable capacitance and increasing amounts of microfarads are put in the circuit (in series) until the correct voltage reading is reached. The capacitance registered on the selector indicates the capacitor size that should be put in the circuit.

Some of the capacitors have mechanical connectors (machine screws) while some have solder-type leads. It is necessary to use a small electric soldering copper or soldering gun to connect the second type.

Some running capacitors that used polychlorinated biphenyl dielectric (PCB) fluid are still in use. This fluid is dangerous. Do not open the shell of this capacitor. If the shell is accidentally pierced or broken, be very cautious not to touch the fluid or breathe its fumes. Dispose of capacitors containing this dielectric fluid according to local environmental rules. The manufacturer can help.

7.18.5 Servicing External-Drive Motors

Troubles found in external-drive motors are few and they can be classified as electrical troubles or mechanical troubles. The electrical troubles found in electric motors may be:

- An open circuit, short circuit, or ground may occur in the field windings. In these cases, replacing the motor is recommended.
- Frequent starting of the motor may result in overheating. Overheating the capacitor may cause the switch and the insulation to fail. If the motor will not start until the pulley is spun, but has the characteristic ac hum, it is a sign that the capacitor or the centrifugal switch points have failed. It is easy to replace the old capacitor with a good one of the same capacity.

  If the motor still will not start, the trouble is probably in the centrifugal switch. If the points are dirty, pitted, or dark from overheating, do not try to repair them. Filing or sanding does little good, as the contact material is worn away. Repaired points usually last just a few hours and a call back may have to be made. It is best to replace them.

Mechanical troubles are almost the same in all external-drive motors:

- There is a possibility that the centrifugal switch used for connecting and disconnecting the capacitor and/or the starting winding may become worn. Replacement of the switch is necessary, in such cases. **Figure 7-63** shows a centrifugal switch for a capacitor-start electric motor.

- Other troubles may include bearing wear, end play, excessive vibration, misalignment of the motor with the compressor, and improper air gap between the rotor and stator.

Information on the repair and testing of motors is covered in Chapter 15.

7.18.6 Servicing Fan Motors

The most common fan motor troubles are:

- Loose connections.
- Dry bearings.
- Worn bearings.
- Burned-out motor.
- Loose fan.
- Out-of-balance fan.
- Fan blades touching the housing.

Loose or dirty connections causes too much voltage drop at the motor. The fan motor loses speed, hums loudly, and overheats. A sensitive voltmeter or an ohmmeter will quickly locate the faulty connection. Do not rely on visual inspection.

A dry bearing causes the same symptoms. However, this condition will last only a short time before the bearings will either seize (bind) or become badly worn.

Occasionally, the end play of the rotor becomes excessive (**Figure 7-64D**). This excessive end play causes the rotor to shift back and forth. As the rotor shifts, it produces a distinct knock. Occasionally, the bearing inserts (bushing) in a reconditioned motor are out of position. This may force the rotor out of the magnetic center along its shaft.

When the motor is running, it should float between the extremes of its end play. You may check this by lightly touching the end of the rotor shaft with a wooden stick as the motor is running. It should move back and forth and then settle in between the extremes of the end play.

If the rotor cannot assume its magnetic center, it will hum excessively and heat. When running, the heaviest magnetic flow from the stator tries to line up with the heaviest magnetic flow from the rotor, B. This aligning must take place with end play clearance at D and D^1. The total clearance is usually about .030″ or 1/32″. **Figure 7-65** shows the bushing and thrust bearing washer on a fan motor.

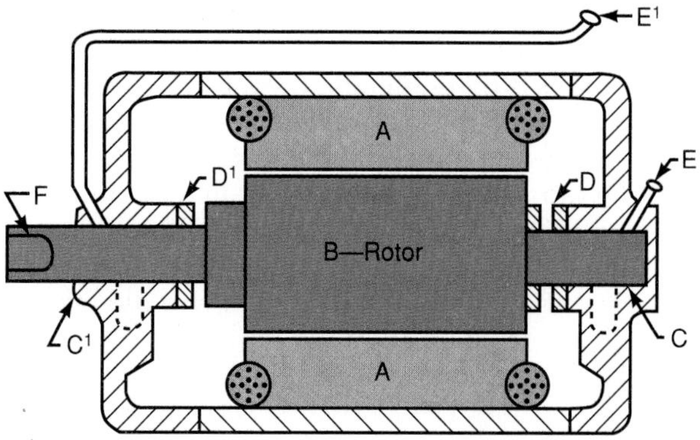

Figure 7-64. *Rotor running in its magnetic center. A—Stator. B—Rotor. C—Bearings. D—End play bearings. E—Oil cups. F—Pulley setscrew contact surface. E and E^1—Oilers for bearings.*

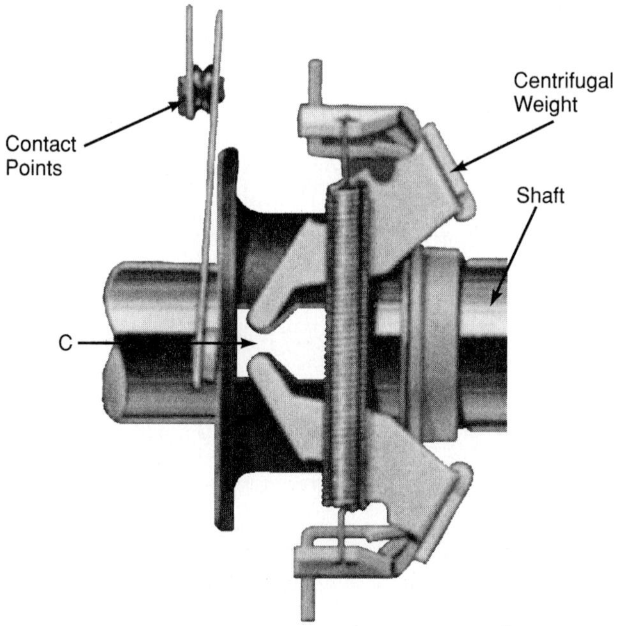

Figure 7-63. *Centrifugal switch mechanism used on external-drive compressor motor. Centrifugal weights and spool, C, revolve with motor shaft (rotor). At about 75% of full operating speed, centrifugal weights cause spool, C, to move to right. This allows contact points to open, breaking the starting winding circuit. (MagneTek)*

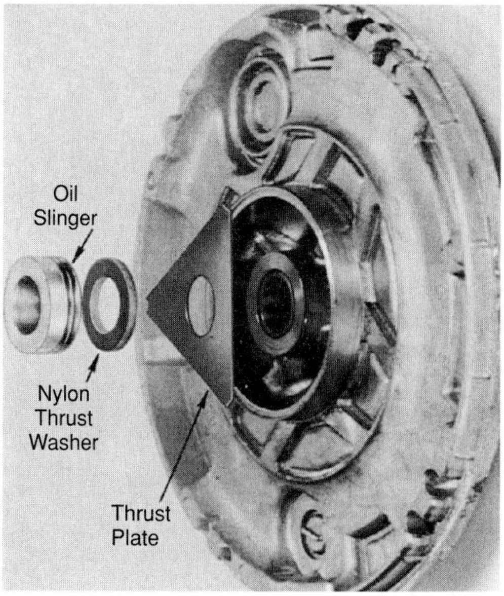

Figure 7-65. *Fan motor bearing and thrust bearing washer. (General Electric Co.)*

A rattle in the fan motor may sometimes be nothing more than a loose fan on the motor shaft. The noise can be remedied by tightening the setscrew that fastens the fan hub to the shaft. Smaller fans have either a round shaft or a flat spot milled on the shaft.

If the fan is abused, the blades may be forced out of position. One or more blades may vibrate. The easiest repair is to replace the fan. Attempts to balance the blades are difficult unless special static and dynamic balancers are available.

If the fan blades touch the fan housing, the motor may be out of line. The shroud or housing may also be bent. The contact spot is usually easily detected. It is remedied by moving the fan on the shaft or moving the shroud or housing. Do not bend the fan blades, as this will cause the unit to vibrate.

7.18.7 Motor Lubrication

External-drive motors may be lubricated in various ways. It depends on the type of bearing used and the position of the motor. External-drive motors using bronze bushings, plain or sleeve, may be lubricated in two different ways:

- Wick system.
- Slip ring system.

The ***wick system*** uses a well or reservoir in the end bell. A wick (cotton or wool yarn) carries oil from the well to the bushing and shaft. This system allows long intervals between servicing bearings and prevents the bearings from getting too much oil. This type of lubrication is shown in **Figure 7-66.**

Motors with this lubrication system have the cotton or wool yarn saturated with oil when shipped from the factory. However, before starting the motor, the oil wells should be filled. Add the amount of oil designated or until oil appears in the lower oil level cup.

If the bearing is to be removed from the shaft or if the bushing is to be removed from the end bell, the yarn should be lifted clear of the bearing. This prevents the yarn being forced between the shaft and the bearing upon replacement.

When replacing the yarn, pack equal amounts on each side of the bearing, and over the slot of the bearing so the spring on the oil well cover will push the yarn down on the shaft. Wick-lubricated bearings should be oiled with one or two drops every six months.

Some larger refrigeration motors use the ***slip ring lubricating method.*** A brass ring rests on the motor shaft through a slot in the top of the bearing. The ring is large enough to dip into the oil pocket below. As the motor shaft turns, the ring turns slowly and the wet portion lubricates the bearing.

Be sure to check the rings when working on these motors. Use a medium viscosity nondetergent oil such as SAE 20 or SAE 30 (200 to 300 viscosity).

Some motors use ball bearings, as shown in **Figure 7-67.** These bearings are grease-lubricated. Most are sealed and do not need any lubrication service. Some, however, come with grease cups and can be lubricated with a grease gun.

These motors, when new, have enough grease in the bearings to lubricate them for several months. A small amount of grease should be added every two or three months. Use a high-grade "medium" grease on fully enclosed motors. Too much grease may cause the bearings to overheat.

The life of bearings depends, to a considerable extent, on cleanliness. Use only clean grease and keep all dirt out of bearings. Clean all fittings before using a grease gun.

Most greases and oils oxidize and will collect dirt while in use. When a motor is reconditioned, the old lubricant must be discarded. The lubricated portions must be thoroughly cleaned, and new lubricant should be used.

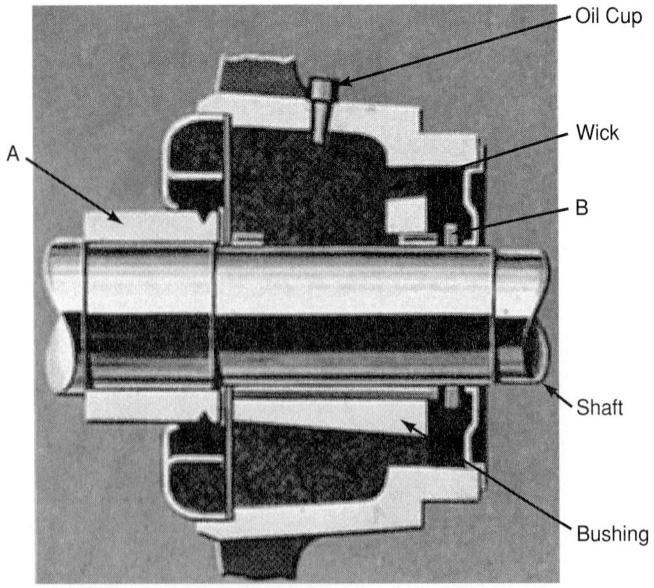

Figure 7-66. *Motor end bearing which uses wick oiler. Oil slingers A and B prevent oil from leaking into motor or out along motor shaft.*

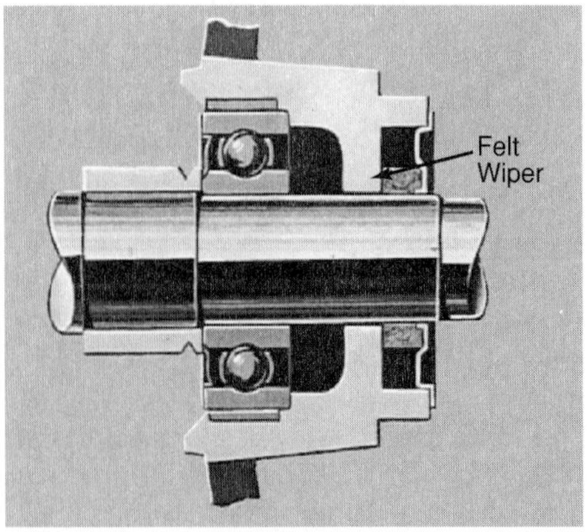

Figure 7-67. *Motor shaft mounted on grease-lubricated ball bearings. Felt wiper keeps out dirt and dust.*

Another method of lubrication presently used is the ***oilless bushing.*** In this arrangement, the shaft passes through a sintered (porous bronze) bushing. The bushing has been impregnated with oil at the factory. The total tolerance between the shaft and bushing is less than .001″. It may have a tolerance of as little as .0003″. As this tolerance increases due to wear, the motor will become noisy.

The oilless bushing is considered to be permanently lubricated. It is often used on fan motors and on other low-horsepower applications.

Fan motors may become very cold when idle. The bearing oil may become very thick. The motor (usually low torque) will start with difficulty. It may even burn out or activate the overload switch. Be sure to use oil that will remain quite fluid at temperatures the bearings may reach during idle time (that is, 0°F [−18°C] to −40°F [−40°C]).

7.18.8 Servicing Motor Bearings

In any service to motors, bearings should be checked to see if they are worn. If the rotor is hitting the stator, bearings are worn out and must be replaced.

Clearance between rotor and stator varies from .015″ to .030″, depending on the size of the motor. This clearance should be the same all the way around the rotor. A heavy rumbling sound at starting usually indicates that the bearing is badly worn, even though the rotor may not be touching the field poles.

Bearings are usually made of phosphor bronze and are pressed into the end brackets or end bells. Sometimes they are locked in place by a pin pressed through the bearing housing and into the bearing. The bearing must always be pressed inward to remove it.

Take care not to put an out-of-line force on the end bell when pressing bearings out. This would probably crack the end bell. A special tool, **Figure 7-68,** can be used to remove or install bushings or sleeve bearings.

After the new bearing is pressed into the bracket, the bearing must be reamed. It is best to ream the two in-line with adjustable reamers.

The surface of the shaft in contact with the bearing must be perfectly smooth. A scored shaft may be repaired in a lathe using a grinder mounted on the toolpost.

If a bearing is overheating, any one of the following may be the cause:

- Oil too heavy.
- Oil too thin. Select a good grade of mineral lubrication oil. It should not be greatly affected by a change in temperature. It also should not foam or bubble too freely.
- Dirt or grit in the oil.
- Belt too tight.
- Pulley hub rubbing against the bearing.
- Motor not properly lined up, causing the armature shaft shoulder to pull on, or be pushed against, one bearing.

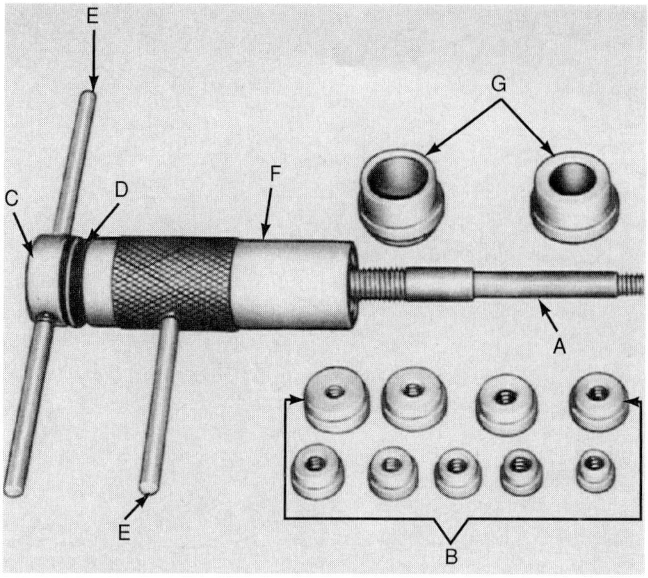

Figure 7-68. *Bushing and bearing tool kit for removing and replacing motor bearings or bushings. A—Tool shaft. B—Engaging taps. C—Tension nut. D—Thrust washer. E—Tension nut handle. F—Bearing tool housing. G—Adaptor sleeves. (Grainger, Div. of W.W. Grainger, Inc.)*

7.18.9 Cleaning Motors

While a motor is in service, it should be cleaned regularly. Dust and lint in the motor will prevent proper air circulation. Compressed air or a hand bellows should be used frequently to blow dirt out of the motor.

Any oil, which may overflow from the bearings, should be wiped off. A little attention will result in efficient operation. The motor will give good service for many years.

If the motor must be dismantled, all parts should be carefully cleaned. This should be done before being worked on or reassembled. Cleaning fluids that will not harm the electrical insulation material or the technician's health must be used. There are many cleaning fluids on the market. Be sure to check the one being used for safety.

All cleaning fluids should be used in well-ventilated and fireproof surroundings. Use only enough cleaning fluid for the job. Too much fluid may be dangerous. Always avoid using carbon tetrachloride, as the fumes are very toxic and can be fatal.

Motor interiors are well designed but rough handling may damage them. **Figure 7-69** illustrates typical internal wiring construction.

7.19 Pulleys

Motor shaft pulleys or sheaves are available in many sizes and types of construction. Some are made of cast iron and some of steel stampings. They come in various shaft sizes and diameters. The most popular shaft

Figure 7-69. *A worker assembling a motor. The exterior is the stator of the motor. The wire that is being insulated will form the running winding. (A-1 Compressors, Inc.)*

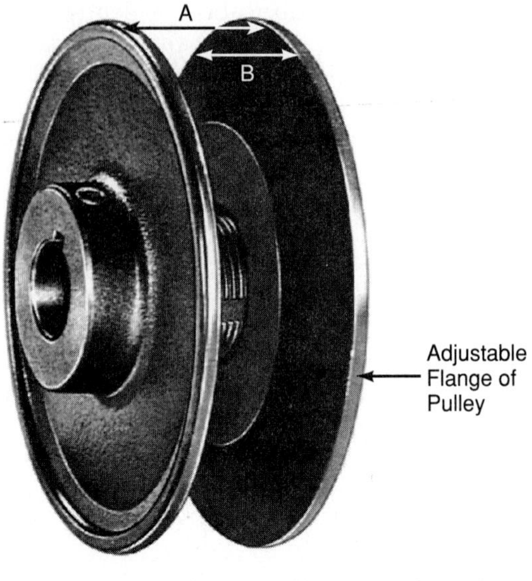

Figure 7-70. *Adjustable V-pulley is used to change speed of belt-driven appliances. A—Moving the adjustable flange away from other flange widens the V and gives the effect of using smaller diameter pulley. B—Narrowing V by moving flange in gives effect of larger diameter pulley. (Maurey Mfg. Corp.)*

sizes for fractional horsepower motors are 1/2″, 5/8″, and 3/4″ diameter. Practically all pulleys have a keyway and a setscrew. Pulley diameters vary from 3″ to 38″.

The V-belt pulleys come in two popular widths. The A width is for belts up to 1/2″ width. The B width fits belts 1/2″ to 21/32″ wide.

Multiple-groove pulleys are available for units with two or more belts to drive the flywheels. Some air conditioning units use step pulleys for driving the air movement fan. By changing the belt from one groove to another, the speed of the fan can be changed.

Special variable-pitch pulleys are also available. These are made with half of the pulley threaded on the hub of the other half, as shown in **Figure 7-70.** A setscrew locks the variable half in place when it is properly adjusted.

By turning the variable half, the V-groove can be widened. This allows the belt to ride closer to the hub, thus reducing the speed of a driven flywheel or fan. The speed of the driven unit can be varied by as much as 30% using these pulleys. **Figure 7-71** illustrates a double-groove, variable-pitch pulley.

Bushings are available to adapt large-bore pulleys to small shafts. For example, a bushing can reduce a 3/4″ bore to a 1/2″ bore.

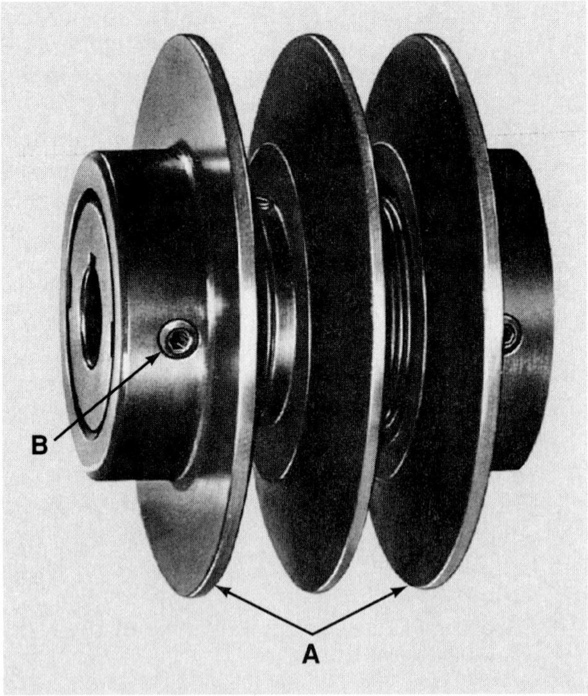

Figure 7-71. *Adjustable pulley with double V-groove. Both grooves are adjustable and must be evenly spaced or one belt will take all the load. A—Adjustable pulley halves. B—Setscrew. (Maurey Mfg. Corp.)*

7.20 Belts

The V-belt is the most popular way to drive the external-drive compressor and large fans. These belts are made in layers of rubber, fabric, and cord. Some belts are a mixture of natural and synthetic rubber. **Figure 7-72** illustrates the most commonly used belts for refrigeration and air conditioning—the standard multiple-cord industrial system and the narrow industrial system.

Belts are made in many lengths from 15″ length to 660″ length. (***Outside length*** is the distance around the

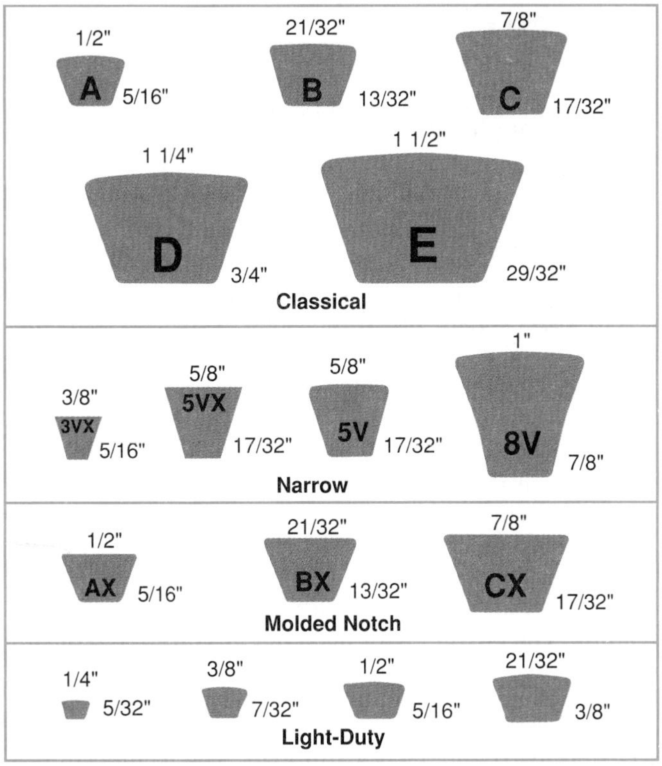

Figure 7-72. *Above are industry-accepted cross sections and dimensions for industrial belts. (The Gates Rubber Company)*

outside of the belt.) This length is quickly determined by using a steel tape, cloth tape, or a special belt-measuring fixture.

Most belts fall into one of four standard widths. The industry designation for these widths are: Classical, Narrow, Notched, and Light-Duty (see **Figure 7-72**). Measurement is made at the belt's greatest width.

When the motor is belted to the driven machine, both shafts must be parallel. This will make the belt ride properly on the pulleys.

New rubber and new cording design and material have been developed. With these, belts with a small cross section may be used. When installing belts, be careful to adjust them for proper tension and alignment. They should be snug, but not tight. The correct tension is the lowest tension at which belts will not slip when the drive is under full load. On modern power transmission equipment, tensions are higher and are more critical. Therefore, the old hand or finger deflection method is not recommended. A commercial tension gauge will provide optimum performance.

The compressor flywheel and the motor pulley must be in-line with each other in two different ways. Such alignment will give long life to the belt and to the electric motor. First, the centerline of the compressor must be parallel with the centerline of the electric motor shaft. Secondly, the pulley grooves must be in line with each other.

A poorly aligned belt will shorten the life of the motor. The motor is not designed to stand an excessive end load. A noisy, poorly operating motor may be the result. **Figure 7-73** shows a tool which may be used in adjusting and aligning belt drives.

Automobile air conditioner belts are designed with exacting standards to transmit power to the compressor. It is very important to follow factory instructions when adjusting these belts. See Chapter 28.

Figure 7-74 illustrates how a ***serpentine belt*** transmits power from a single primary pulley to numerous

Figure 7-73. *Belt adjusting and aligning tool. Tool is inserted between pulleys to tension belt. A—Pulley. B—End with flywheel. C—Tool.*

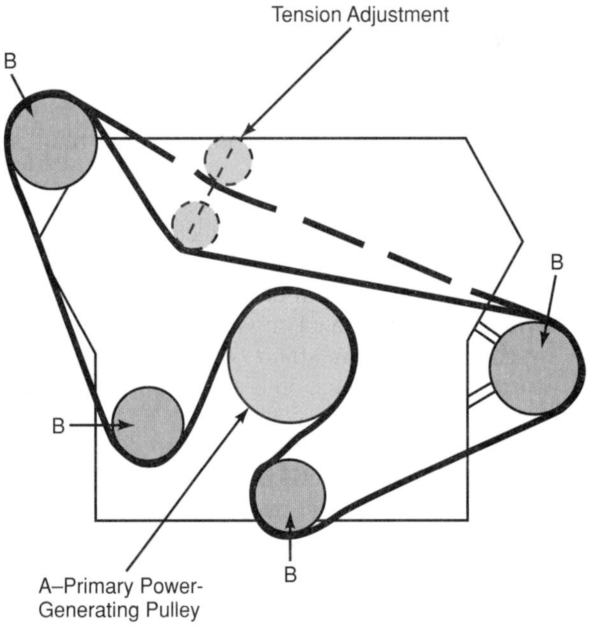

Figure 7-74. *Typical use of a serpentine belt, which transmits power from primary power-generating pulley, A, to secondary pulleys, B. Notice the tension adjustment.*

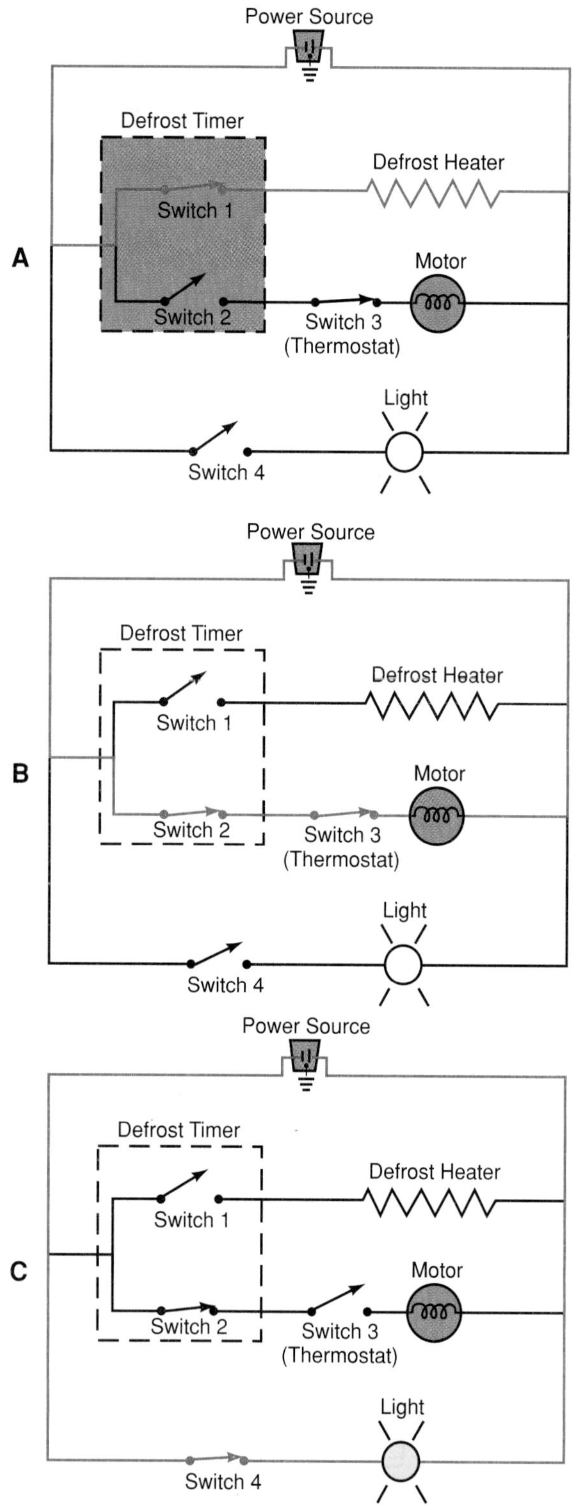

Figure 8-5. *Ladder diagram for a domestic refrigerator. A—Defrost heater is turned on by Switch 1, and compressor motor is turned off by Switch 2. B—Compressor motor is on when Switches 2 and 3 are closed. C—Light is on when Switch 4 is closed.*

system and injury to people. Refrigeration and air conditioning control systems are often divided into the following three categories:

1. ***Conditioned area:*** The area in which the temperature, pressure, and humidity are being controlled.
2. ***Controlling instrument:*** Instrument that is responsive to changes. This is done with sensing devices, thermostats, motor controls, pressurestats, humidistats, and air distribution controls.
3. ***Operating device:*** The mechanism that directly affects actual conditions and is regulated by the controlling instrument. Examples of this are valves, dampers, fans, and compressors.

A control system that constantly corrects the condition is a ***closed-loop*** (feedback) ***control system.*** In the operation of a typical refrigerator, the thermostat is the controlling instrument. The compressor is the operating device. The space inside the refrigerator is the conditioned area. As the temperature in the space becomes too warm, the thermostat turns on the compressor. The compressor then circulates the refrigerant so that it can cool the space. After the space reaches the desired temperature, the controlling instrument turns off the compressor. This cycle is repeated as the controlled space continues to warm up and cool down.

8.3.1 Types of Controlling Instrument Action

The basic types of controlling instrument action include:

- Two-position "on-off."
- Timed "on-off."
- Variable.
- Proportional.
- Proportional with automatic reset.

A description of each type of action is as follows:

Two-position on-off control is the most common type of action. The control turns the operating device either "on" or "off." A typical example of this type of action is the operation of a refrigerator. The thermostat turns the compressor "on" to cool the refrigerated space when it gets too warm. It turns the compressor "off" after the space reaches its desired temperature. The following terms define the operating characteristics of a two-position on-off control device:

Range represents the temperature or pressure at which a control is set to operate. Adjusting the range (Section 8.3.7) changes the setpoint level.

For example:

	Thermostat "On" Temp.	**Thermostat "Off" Temp.**
Original setting	72°F	77°F
Setting after + 3°F range adjust	75°F	80°F

Desired differential is the difference between the high and low temperature or pressure setpoints. Adjusting the differential increases or decreases this difference (Section 8.3.8).

Thermostatic Control Mechanism Construction—Vapor Pressure

Many mechanisms are used in vapor pressure type thermostats. The action of the contact points needs to be quite rapid. If contact points were allowed to open very slowly, there would be electric arcing as the current jumped across the tiny gap. This arcing action would very quickly burn the contact points and ruin their ability to make a good electrical contact.

There are two ways to get this rapid "snap" action. One is through a snap action toggle mechanism. The other is with a permanent magnet. These two mechanisms are shown in **Figure 8-11.**

The toggle mechanism is shown in **Figure 8-11A.** In this mechanism, the fulcrum points are under some pressure. This tends to push them together. As the sensing bulb warms, the bellows expands. The toggle point will move down. The instant it passes the center point, it will snap into the lower position. This closes the contact points.

As the sensing bulb cools and the bellows ***contracts*** (shrinks), the toggle will be moved upward again. As soon as it passes the center point, it snaps open very quickly. An adjusting screw controls the pressure tending to move the fulcrum points together. Increasing this pressure will lengthen the running time. This toggle arrangement gives good snap action.

The system with the magnet snap action is shown in **Figure 8-11B.** The bar carrying the contact points is made of a magnetic metal, such as iron.

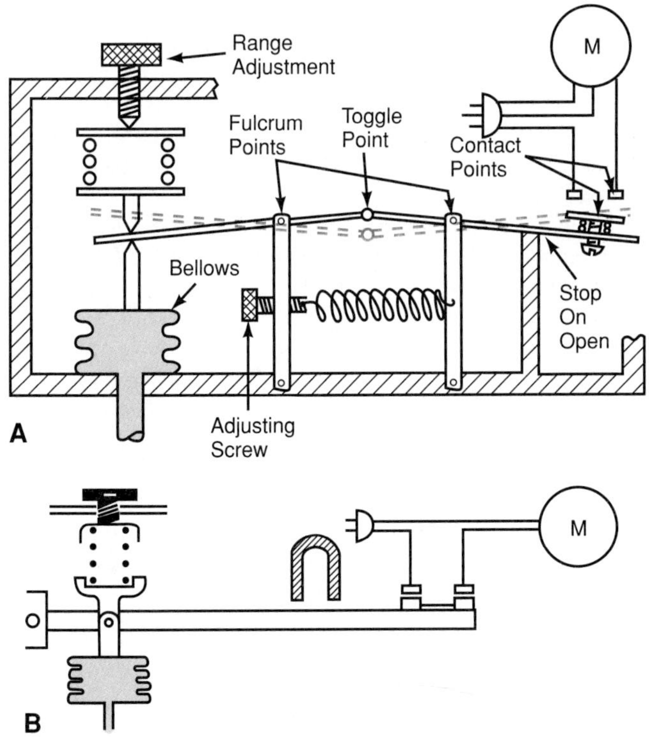

Figure 8-11. *Construction detail drawings show how bellows are connected to snap action motor control switch. A—Toggle snap action. B—Magnet snap action.*

The permanent magnet tries to draw this material toward it. Magnetic lines of force always try to be as short as possible. The pressure in the bellows tends to close the points. Therefore, magnetic effect increases as the iron bar approaches the magnet. This causes a snap action which quickly closes the points.

As the sensing element cools and the bellows contract, pulling the points apart takes force. However, the magnetic pull decreases rapidly as soon as the points separate. This allows a quick opening of the points.

With this type of snap action, the running interval can be shortened by moving the magnet away from the iron bar. It can be lengthened by bringing the magnet closer to the iron bar. The magnet should never touch the bar.

A typical vapor pressure motor control is shown in **Figure 8-12.** This control uses a magnetic snap action. The sensing bulb is at the end of the coiled capillary tube. It has a range control that allows adjusting the cabinet temperature.

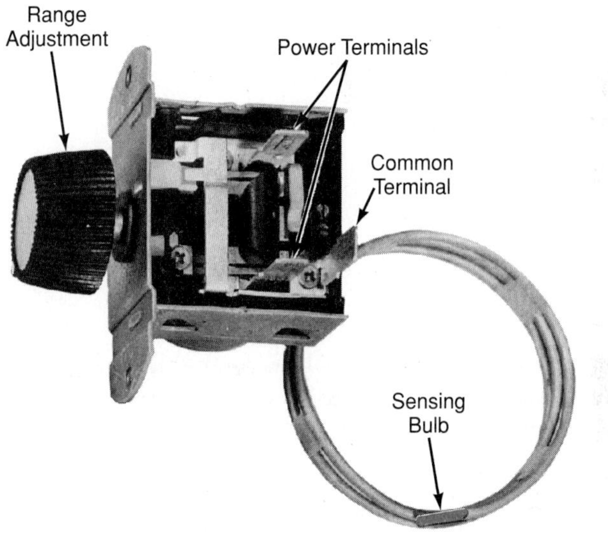

Figure 8-12. *Internal construction of vapor pressure type thermostatic motor control, or "cold control." (Eaton Corp., Controls Div.)*

8.3.3 Temperature Control Principles—Bimetal

Bimetal strip temperature control devices consist of two different metals that are bonded together. The two metals commonly used are copper and steel. Copper has a greater coefficient of expansion than steel. This means that it expands more with temperature increase than steel. This causes the bimetal strip to bend as temperature rises. The bending action of the strip opens and closes the contact points in the electrical circuit. The strip bends, as shown in **Figure 8-13.**

Another common control device is the mercury switch mounted on one end of a coiled bimetal strip. This type of thermostat is most commonly used in air conditioning and heating thermostats. **Figure 8-14**

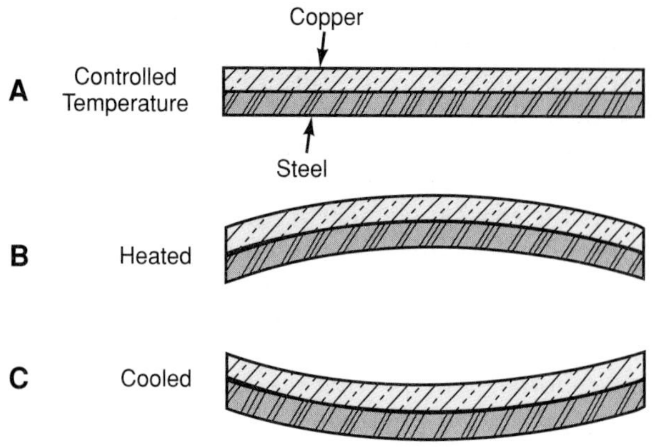

Figure 8-13. *A bimetal strip bends or warps with temperature change. A—Strip at controlled temperature. B—Strip warmed above control temperature. C—Strip cooled below control temperature.*

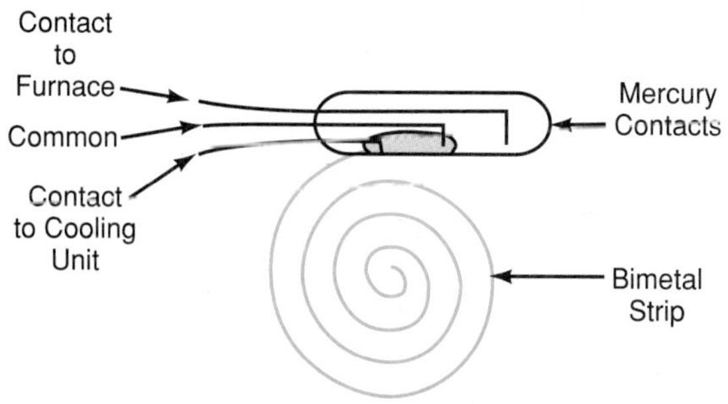

Figure 8-14. *Mercury bimetal type thermostat.*

illustrates a bimetal-coil heating thermostat. As the air surrounding the coil gets warmer, the coil expands. This tilts the bulb upward and causes movement of the mercury. The contacts are no longer closed and the unit is "off." During the cooling cycle, the opposite occurs.

Thermostatic Control Mechanism Construction—Bimetal

The bimetal thermostat is used more often in air conditioning than in refrigeration. The construction of a bimetal thermostat is shown in **Figure 8-15.** These thermostats use bimetal discs for fast snap action. The bimetal control usually consists of a dished disc. Its construction is such that it is dished in one direction when it is cold. As it warms, it suddenly snaps into a dished position in the other direction. Snap action discs are also used on pilot lights and gas burner controls. The bimetal strip serves two purposes. It provides adjustable calibration and a small effective temperature differential.

8.3.4 Temperature Control Principles—Electronic (Solid-State)

Electronic controls have several advantages over other controls. They are compact and reliable, respond rapidly, and lack moving parts. Electronic controls can be identified by their low operating voltages. Most

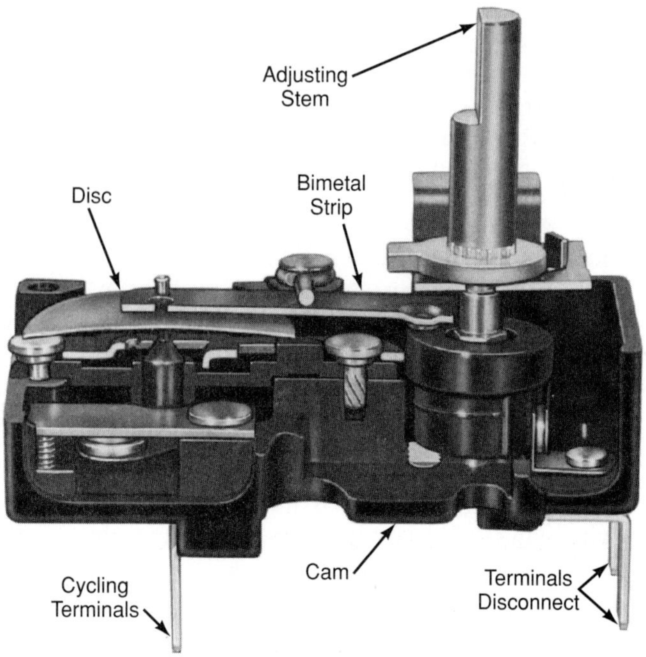

Figure 8-15. *Construction of thermostatic motor control. Note location of bimetal strip and disc. This type of unit is used for control in refrigeration, electric heat, and air conditioning. (Therm-O-Disc, Inc.)*

operate at 5 V to 15 V. A stepdown transformer and rectifying circuit provides this voltage.

The sensing device in an electronic temperature control is usually a ***thermistor.*** This is a device in which the resistance varies as temperature varies. See Section 6.6.7. If thermistor resistance *decreases* as temperature *increases,* it has a negative temperature coefficient (NTC). There are also thermistors with a positive temperature coefficient (PTC). Their resistance increases as temperature increases. PTC thermistors are used in motor safety controls.

Changes in a thermistor's resistance are detected by an electronic circuit called a ***Wheatstone bridge.*** As the resistance changes due to temperature change, the resistances in the bridge become unbalanced. This causes the output voltage of the bridge to change. A simple form of Wheatstone bridge is shown in **Figure 8-16.** The

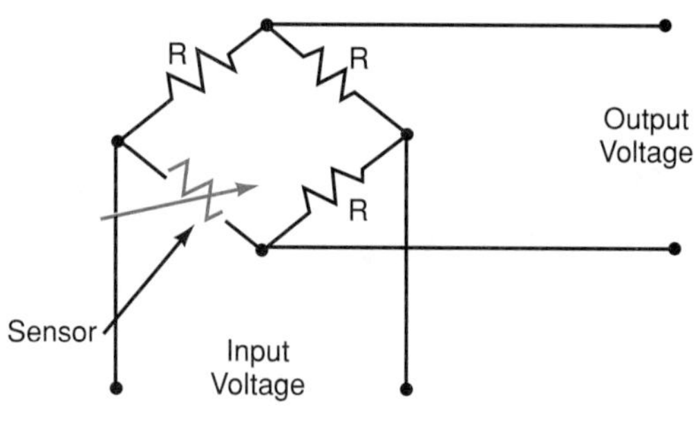

Figure 8-16. *Wheatstone bridge circuit principles. Wheatstone bridge measures changes in sensor resistance.*

output voltage can be amplified to signal the control device. The signal to the control device indicates the action that needs to be taken.

Electronic thermistors have the ability to accurately sense extremely high and low temperatures. These devices are useful in measuring temperatures where conventional methods do not work.

8.3.5 Temperature Alarm System

In some installations, such as food freezers, an electrical alarm system will sound if the temperature in the cabinet rises above an upper safe limit. These systems sometimes operate from the electrical circuit powering the compressor. Others are provided with a dry cell arrangement. If the dry cells are in good condition, the alarm will work even during a power failure.

8.3.6 Installing Thermostats

Thermostats must be correctly installed or the system will not operate accurately or regularly. The electrical connections must be clean and tight. The wire terminal must be large enough to carry the current used in the wire (lead).

Some terminals are connected to posts which either have screws or threaded studs with a nut and washer. See **Figure 8-17.** Wrapping stranded wires around a terminal screw does not make a good or permanent connection. Strands of wire may work loose and ground the wire or short the terminals. Many types of terminals have been developed to make good connections. These quick-disconnect terminals help service technicians remove and replace wire leads quicker. See **Figure 8-18.** The connections should be cleaned with clean steel wool before installing wires.

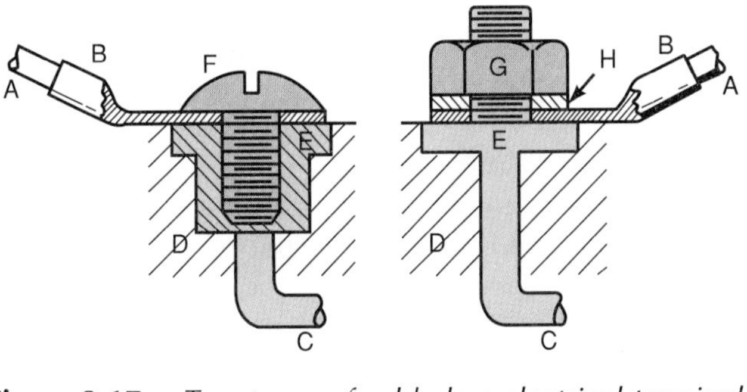

Figure 8-17. *Two types of solderless electrical terminal assemblies. Screw type is at left, nut type at right. A—Wire (lead). B—Wire terminal. C—Wire (lead) to inside of control. D—Insulation. E—Insulated terminal block. F—Machine screw (round head). G—Nut. H—Washer.*

The sensing bulb of the thermostat must be very carefully mounted. It should be attached tightly to the evaporator or the tubing. The sensing bulb and the place to which it is clamped should be cleaned with clean steel wool before assembly. The best place for attaching the thermostat sensing bulb is on the last one-third of the evaporator.

When installing the control, be careful not to bend the capillary tube back and forth. This small copper tube will ***work-harden*** and may break. Also, be sure the capillary tube does not touch any part of the evaporator. If the tubing rubs against any part, it may wear through or work-harden at that spot and crack. If any of the capillary tube is coiled, tape the coil to prevent vibration.

Servicing and Testing a Thermostat

A mixture of crushed ice and water may be used to check and adjust a refrigerator thermostat. With these materials, it is easy to determine the operation of the thermostat at 32°F (0°C), the temperature of melting ice.

To use this method, place the thermostat control bulb in the ice and water mixture. It is a good idea to use a thermometer when making this test. With the control bulb in the ice and water mixture, set the control dial temperature reading to 32°F (0°C). The contact points of the control should be open. After a few minutes, lift the control bulb from the mixture. As the control bulb warms, the points should again close. Use a test cord with a small light bulb to check the opening and closing of the contact points.

To check further, place the control bulb in water and adjust the water temperature to 45°F (7°C). Set the thermostat control to 45°F. At this setting, the points should be open (they begin to open below 45.5°F or 46°F).

Lift the control bulb out of the water and let it warm up for a few minutes. The points should close. If the points do not open and close properly, the thermostat should be replaced.

Freezer cabinet temperatures are usually in the 0°F to −20°F (−18°C to −29°C) range. Ice and water

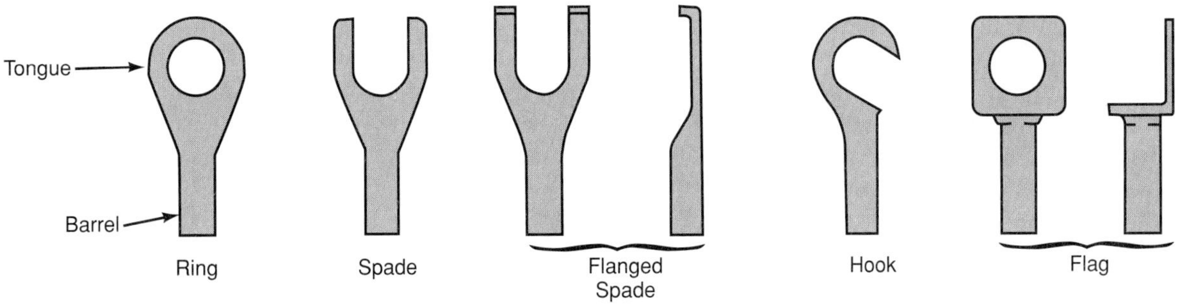

Figure 8-18. *Several types of terminals that are used to connect leads (wires) to terminal posts.*

mixtures cannot be used in setting these thermostats. A thermostatic control tester which uses the refrigerant R-134a or R-22 is recommended. **Figure 8-19** illustrates one of these instruments.

In operation, the freezer thermostat control bulb is placed in the tester. The refrigerant is allowed to flow into the control pressure chamber. By adjusting the refrigerant pressure, it is possible to obtain any desired temperature. Freezer thermostats have a predetermined cut-in and cut-out temperature similar to refrigerators. However, the freezer temperatures are much lower. Also, the differential is usually a little greater, 10°F to 12°F (5.5°C to 6.6°C).

To test a freezer thermostat, place the control bulb in the control tester with the freezer plugged in and operating. Adjust the refrigerant flow and pressure until the thermometer indicates the desired cut-out temperature. If the freezer control cuts out before the desired temperature is reached, the control may be adjusted to cut out at a lower temperature.

Leave the freezer connected and open the cabinet door. Adjust the control tester to the desired cut-in temperature. If the thermostat does not cut in at the desired temperature, an adjustment should be made to the thermostat. If it is not possible to get a satisfactory thermostat adjustment, the thermostat should be replaced. To check the operation of the freezer thermostat, use a recording thermometer. Chart the temperature and time for the freezer over a 24-hour period.

Several kinds of trouble may be encountered in thermostats:

- Corrosion may occur at the contact points, causing a poor electrical circuit. Use a test light to check electrical circuits. Once contact points become worn and cause trouble, it is best to replace them. Any repair would only be temporary. A repeat call (call-back) would soon be necessary.

  Contact points must be clean. Corroded or pitted points (best detected by using an ohmmeter) usually should not be repaired. The control should be replaced. In an emergency, the points can be cleaned with a small, clean file. A single-cut or mill file is best, but even a clean nail file may do a temporary job.

  Thermostats must have good electrical connections. They must also be adjusted to correct temperatures.

  The sensitive element must accurately sense the temperature of the evaporator or the cabinet.

  Electrical connections must be clean and tight. Only metal terminals should be used. Clean the terminals, the terminal jacks, or the screw posts with clean steel wool.
- The overload protection devices may also have dirty contact points. Low current flow may result.
- The power element and bellows may lose their charge. This can be detected by a simple check. If the charge is lost, the bellows are very easily compressed. If the bellows were charged, the pressure inside would probably be around 75 psig (517 kPa) or more. Finger pressure would not affect it. In the event of leakage, replace this part of the control or the complete control.
- Frequently, the control bulb is not attached tightly to the evaporator. A great change of temperature range is needed before the motor will cut in and cut out. The power element must be firmly clamped to the evaporator. Good thermal contact must be obtained between the thermal bulb and the evaporator. Many evaporators have metal sockets into which the thermal elements are inserted. Clean the contact surfaces.

  To adjust to the correct temperature settings, mount a thermometer at the sensing bulb. Then cycle the unit. Obtaining accurate settings is difficult until the unit cycles several times. Time may be saved by using a thermal bath or a thermostat adjusting tool.

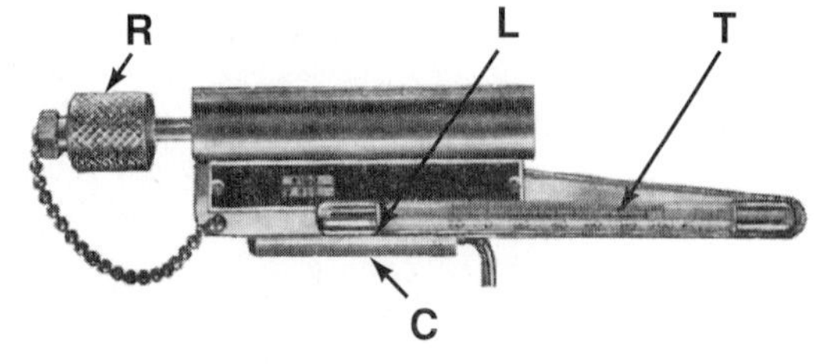

Figure 8-19. *Thermostat testing and adjusting instrument. T—Accurate thermometer. C—Clamp for holding control bulb. R—Fitting for attaching refrigerant cylinder. L—Shallow trough for liquid refrigerant.*

The method of checking and adjusting thermostats is explained in Chapter 12.

8.3.7 Range Adjustment

Range adjustment provides for correct minimum and maximum temperature or pressure in automatically operated systems. For example, the range adjustment will keep a refrigerator between certain temperatures. It also keeps an air-conditioned room between certain temperatures. It will keep an air compressor between certain pressures.

It is very difficult to keep any device at one particular temperature and/or pressure, such as 34.4°F (1.3°C) or 150.2 psig, which is 164.9 psia (1136 kPa). That is why a range adjustment is used.

A range adjustment is shown in **Figure 8-20.** The unit cuts in at 25°F (−3.9°C) and cuts out at 15°F (−9.5°C) on the evaporator temperature. To make the cabinet operate at a warmer temperature, the range adjustment may be changed. The cut-out then becomes 16°F (−8.9°C) and the cut-in 26°F (−3.3°C), as shown in **Figure 8-20A.**

Note that the temperatures are higher, but the temperature distance between the two has not changed. This distance between cut-in and cut-out is still 10°F (5.6°C). The new settings will only slightly affect the running time of the unit. There will be a slight decrease in compressor motor current instead of a decrease in running

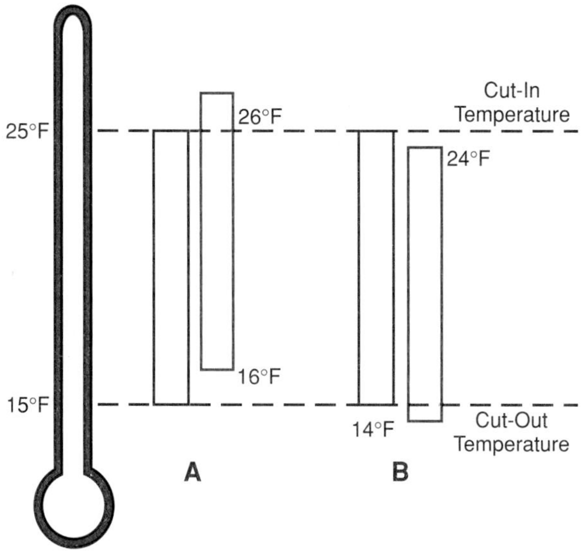

Figure 8-20. *Typical range adjustments. Basic range is shown in black.*

time. This is because the condition desired is not as cold. Therefore, the compressor will not have to do as much work.

The temperature of the refrigerator may be lowered one degree. This is done by making the range adjustment cut in at 24°F (−4.4°C) and cut out at 14°F (−10°C), as shown in **Figure 8-20B.**

Range Adjustment Mechanisms

The ***range adjustment*** is easily recognized. It is an adjustable force pressing directly upon the bellows or diaphragm that operates the switch. This force is exerted on the bellows whether the switch is in either the cut-out or cut-in position.

The adjustable force may be an adjustable weight that always presses against the bellows. More commonly, it is a spiral spring with an adjustable screw. Turning the screw changes the pressure or the tension of the spring. The spring may press or pull directly on the bellows or diaphragm, or on a lever attached to the bellows. See **Figure 8-21.**

Most range adjusting screws have a calibrated dial or a pointer connected to them. This indicates the direction the screw should be turned to provide a warmer or colder setting.

8.3.8 Differential Adjustment

The ***differential adjustment*** controls the temperature difference between the cut-out and the cut-in settings. The differential adjustment is built into the temperature control mechanism. It is not adjusted or changed by the operator when making a temperature selection with the control knob. Any change should be made by a service technician who understands the working of the differential adjustment mechanism.

If the evaporator is set to cut in at 25°F (−3.9°C) and cut out at 15°F (−9.5°C), the difference, or differential, is

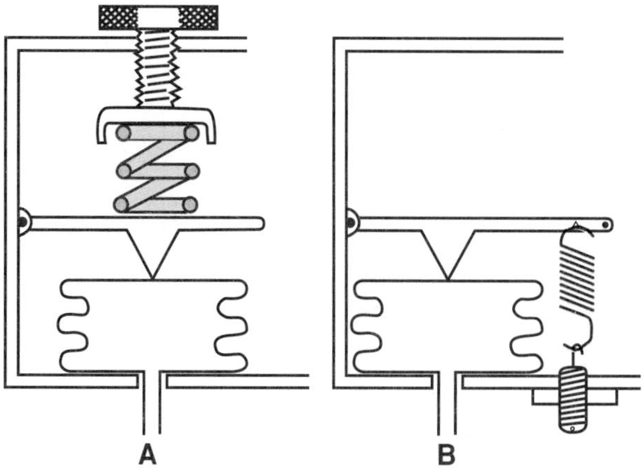

Figure 8-21. *Adjusting range settings of a motor control. A—Compression spring range adjustment. B—Tension spring range adjustment.*

10°F (5.6°C). Whenever the ***differential*** (the distance between the settings) is changed, the range is also changed. If only the range is adjusted, the differential will not be affected.

The thermostat differential for capillary tube systems must be large enough to allow the pressures to equalize. This increases the off cycle, but the change should not be too much. If it is, the fixture temperature will rise too high. A differential of about 14°F (8°C) is common for fresh food refrigerators and about 11°F (6°C) for freezers.

Figure 8-22 shows the types of differential adjustments:

- Cut-in type. The cut-in point may be moved without changing the cut-out as shown at 1 and 2 in **Figure 8-22A.**
- Cut-out type. The cut-out point may be moved without changing the cut-in setting as at 1 and 2 in **Figure 8-22B.**
- Double type (cut-in and cut-out). The cut-in and cut-out settings may be brought closer together or moved farther apart, as shown at 1 and 2 in **Figure 8-22C.**

Cut-in and cut-out differential adjustments may be recognized by the way they affect the operation of the switch mechanism. In the cut-in type, the effect is seen only when the switch is in the cut-out position, ready to snap back into cut-in. In the cut-out type, the effect is seen only when the switch is in the cut-in position, ready to snap to cut-out position.

The third type is usually an adjustable arrangement. This type affects the effort of the toggle to snap off and on. The adjustment affects both the cut-in and the cut-out. They may be adjusted farther apart or closer together.

To get a certain cabinet temperature, it is usually necessary to adjust both the range and the differential. To keep the same average cabinet temperature, the control may be adjusted to 14°F (−10°C) and 25°F (−3.9°C),

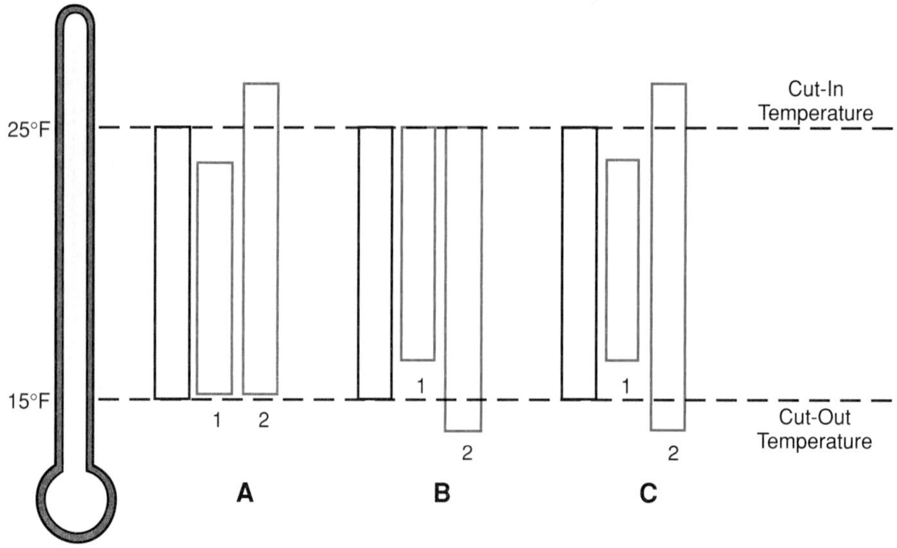

Figure 8-22. *Effect of various types of differential control adjustments. A—Cut-in setting type. B—Cut-out setting type. C—Cut-in and cut-out type. In each illustration, normal setting is shown in black.*

shown in 1 of **Figure 8-23A.** With this setting, the compressor will run longer but will not cycle as often.

If this control range adjustment is set to cut in at 25°F (−3.9°C) and cut out at 13°F (−10.6°C), as shown in 2 in **Figure 8-23A,** the compressor will run longer than normal. The cabinet temperature also will be lower than normal.

By setting the control to cut in at 27°F (−2.8°C) and cut out at 15°F (−9.4°C), as shown in 3 in **Figure 8-23A,** the compressor will run less and the cabinet temperature will be higher.

In each of the above examples, the differential has been 12°F (6.6°C), but the range has changed.

If the control is adjusted to cut in at 24°F (−4.4°C) and cut out at 16°F (−8.9°C), as shown in 1 in **Figure 8-23B,** the cabinet temperature will be normal. However, both the On and Off cycles will be shorter. The cabinet temperature will vary less than normal. The differential is now 8°F (4.4°C). However, the range is the same setting as one which would produce the scale reading in black.

With the control adjusted to cut in at 25°F (−3.9°C) and to cut out at 17°F (−8.3°C), as shown in 2 of **Figure 8-23B,** the cabinet temperature will be warmer than normal. The On and Off cycles will be shorter than normal. Cabinet temperature will vary less than normal. Note that the range has been raised compared to the black scale.

With the control adjusted to cut in at 23°F (−5°C) and cut out at 15°F (−9.4°C), as shown in 3 of **Figure 8-23B,** cabinet temperature will be below normal. However, it will vary less than normal. The On and Off cycles will be shorter than normal. The range is lower than for the black scale.

For a slightly lower cabinet temperature with little or no increase in running time, the cut-out type differential may be adjusted to cut out at 14°F (−10°C), as in

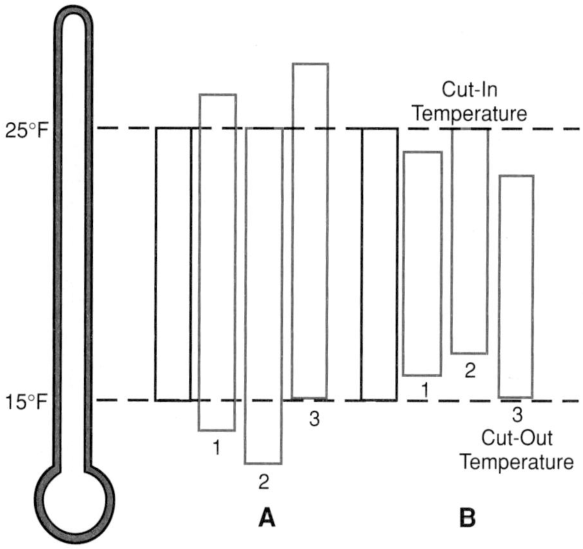

Figure 8-23. *Cut-in, cut-out, and differential adjustment. A1—Differential increased. A2—Range lowered. A3—Range raised. B1—Differential decreased. B2—Range raised. B3—Range lowered.*

1 of **Figure 8-24A.** The range adjustment may be set as shown in 2 of **Figure 8-24A** to decrease running time, compared to 1. The running time for the black scale will also remain the same.

A cut-out differential adjustment, as shown in 1 of **Figure 8-24B,** will give a slightly warmer cabinet temperature and a shorter cycle interval. Also, the cabinet temperature will vary less than for the black scale. A range adjustment, as shown in 2 of **Figure 8-24B,** will provide a normal cabinet temperature with less variance. The cycling interval will be shorter than normal.

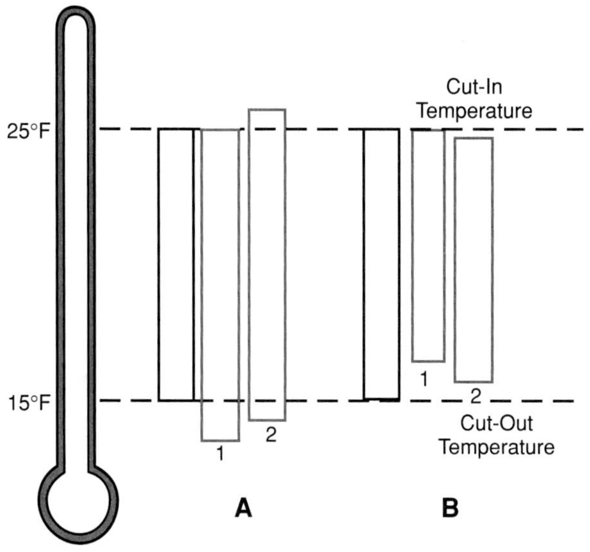

Figure 8-24. *Cut-out differential adjustments. A1—Differential lowered from 15°F to 14°F cut-out. A2—Range is then adjusted to 14.5°F cut-out and 26.5°F cut-in. B1—Differential is changed from 15°F to 16°F cut-out. B2—Range is then adjusted to 15.5°F cut-out and 24.5°F cut-in.*

A typical control adjustment follows:

1. To increase the cycle time when a control with a cut-in differential is used:
   A. Adjust the differential from a 25°F (−3.9°C) cut-in to a 27°F (−2.8°C) cut-in. See 1, **Figure 8-25A.**
   B. Adjust the range to move the settings to 26°F (−3.3°C) cut-in, and 14°F (−10°C) cut-out. See 2, **Figure 8-25A.**
2. To decrease the cycling time (interval) when a control with a cut-in differential is used:
   A. Adjust the cut-in differential to cut in at 23°F (−5°C). See 1, **Figure 8-25B.**
   B. Adjust the range to cut in at 24°F (−4.4°C) and cut out at 16°F (−8.9°C).
3. To decrease the running time with a cut-out differential:
   A. Adjust the cut-out differential to cut out at 17°F (−8.3°C). See 1, **Figure 8-25C.**
   B. Adjust the range to cut out at 16°F (−8.9°C) and cut in at 24°F (−4.4°C). See 2, **Figure 8-25C.**
4. To increase the running time with a cut-out differential:
   A. Adjust the cut-out differential to cut out at 13°F (−10.6°C). See 1, **Figure 8-25D.**
   B. Adjust the range to cut out at 14°F (−10°C) and cut in at 26°F (−3.3°C), as shown in 2 in **Figure 8-25D.**

Differential Adjustment Mechanisms

Two types of differential adjustments use a spring that affects the movement of the bellows. The effect can be just before the contact points cut out or cut in, but not both. This limited action is controlled by a stop on the spring (on the right side in each unit, **Figure 8-26**). **Figure 8-26A** is a cut-in differential. **Figure 8-26B** is a cut-out differential. **Figure 8-26C** is a cut-in differential.

A third type of differential adjustment affects both the cut-out and cut-in. A spring makes it easier for the points to open and close, or more difficult for the control to open or close the electrical contact points. See **Figure 8-26D.**

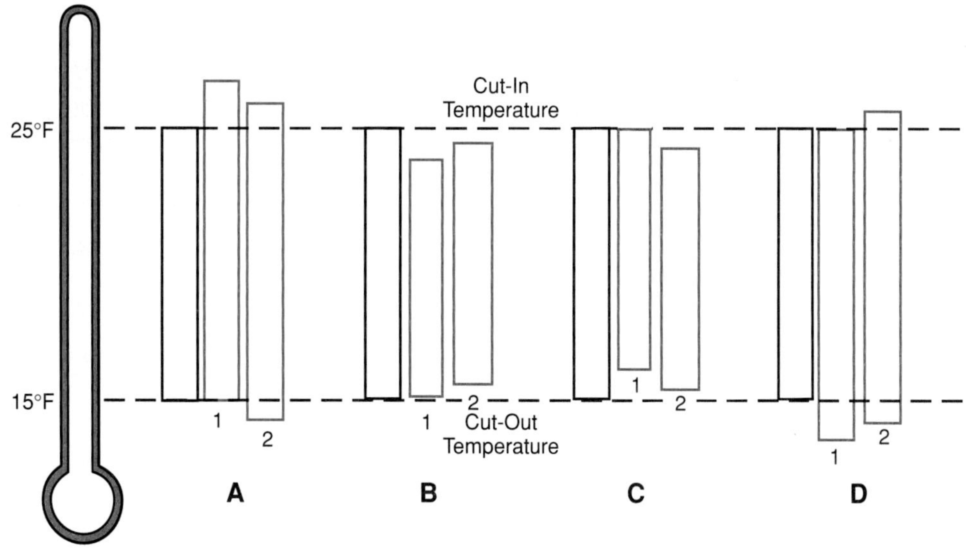

Figure 8-25. *Effect of a cut-in differential adjustment: A1—Cut-in differential raised. A2—Range lowered. B1—Cut-in differential lowered. B2—Range raised. Effect of cut-out differential adjustment: C1—Cut-out differential raised. C2—Range lowered. D1—Cut-out differential lowered. D2—Range raised.*

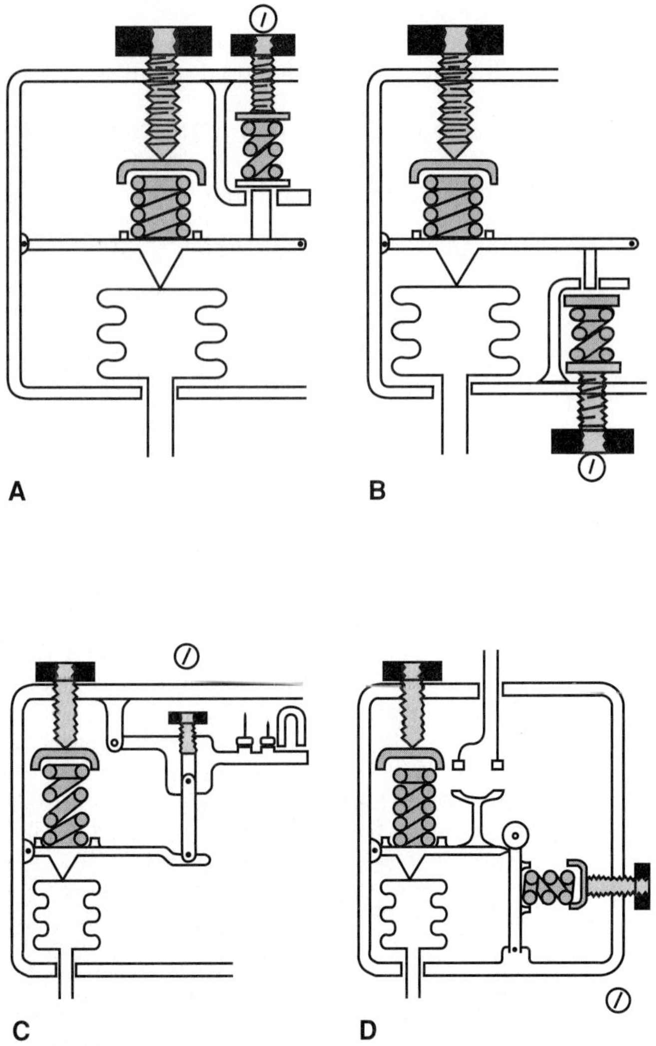

Figure 8-26. *Various types of differential adjustment mechanisms. A—Cut-in type. B—Cut-out type. C—Cut-in type using a slot and magnet. D—Double (cut-in and cut-out) type. 1—Differential adjustment.*

Figure 8-27 shows a mechanism with a cut-in differential. **Figure 8-28** shows a control with a cut-in and cut-out calibration screw and lever, and a range adjustment knob.

Some controls have a small heater unit. The heat from this unit keeps the bellows and diaphragm from becoming too cold. A too-cold thermostat body will not cut in as soon as it should. This may cause erratic cabinet temperatures. A complete control is shown in **Figure 8-29.**

8.3.9 Adjusting Controls

It is advisable to adjust controls as the manufacturer specifies. Some controls are marked "cut-in adjusting screw" and "cut-out adjusting screw." Others mark the adjustment "range."

Some controls have a cut-in differential with the control marked "step one, cut-out adjustment" for one adjusting screw. "Step two, cut-in adjustment" is for a second adjusting screw. In such cases, the cut-in screw is the differential adjustment. The cut-out screw is the range.

If the control has a cut-out differential, the opposite applies. The cut-in will be the range; the cut-out, the differential.

Some controls have a differential adjustment controlling the distance between the cut-out and cut-in. The first setting should be made by using the differential adjustment. This gives the correct distance between the settings. Then the range adjustment is turned to obtain the correct settings.

The adjustment used by the owner is usually a limited range adjustment. However, some models allow the owner to adjust only the cut-out setting. This design ensures a safe cut-in temperature at all times.

Contact points may chatter as they open or close. Operate the control to check for this condition. If there is any visual indication of pitting or burning of points, replace the entire control.

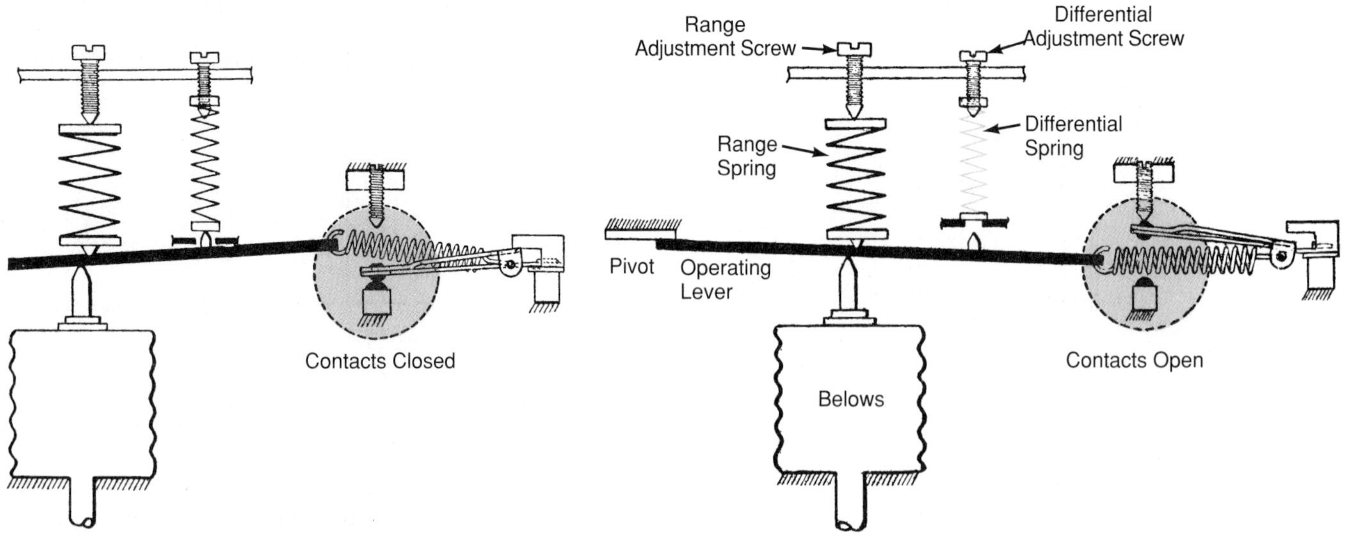

Figure 8-27. *Toggle switch mechanism showing range adjustment and a cut-in type differential adjustment.*

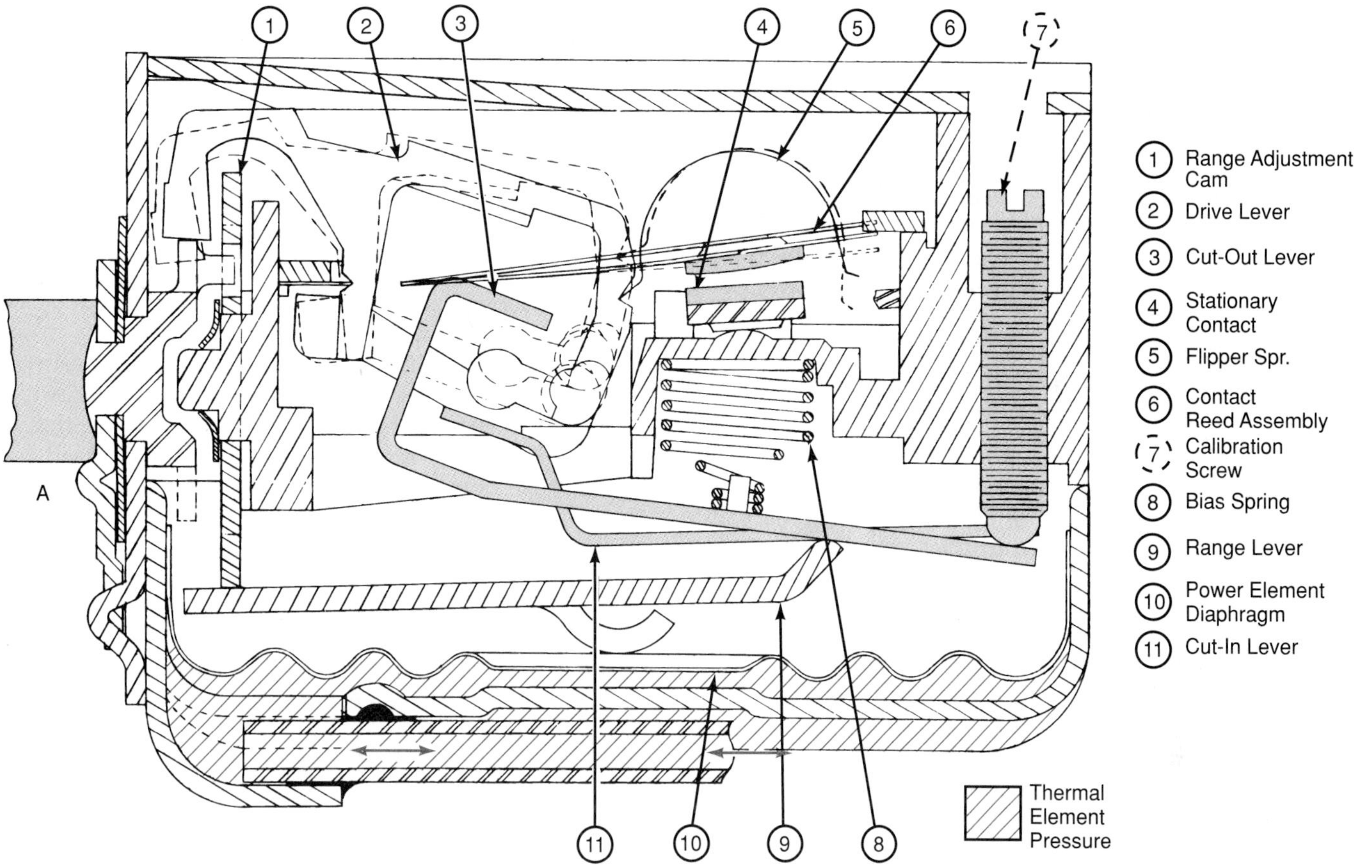

Figure 8-28. *Operating mechanisms and adjustments of a thermostat. 1—Range adjustment (cam). 7—Cut-out differential adjustment. A—Range adjustment knob.*

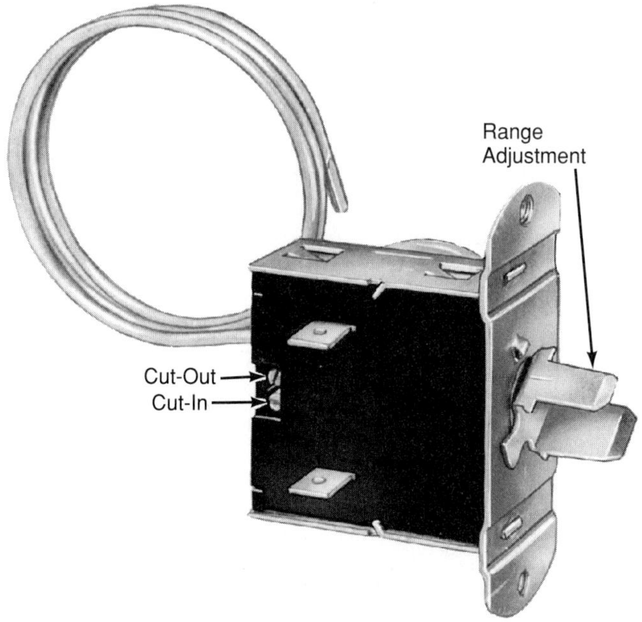

Figure 8-29. *Domestic cabinet thermostat ("cold control"). Range adjustment used by the owner is at right. Note cut-in and cut-out adjustment at left. (Eaton Corp., Controls Div.)*

ELECTRIC CONTROLS MODULE

8.4 Refrigerator and Freezer Controls

Automatic refrigeration and air conditioning is designed to provide correct temperatures with the least attention. To produce these temperatures under all conditions, a refrigerating unit needs more capacity than usual. This unit would over-refrigerate or overheat if it ran all the time.

In the northern temperate latitudes, domestic refrigeration units will run 35% to 40% of the time. In semitropical latitudes, units will run about 50% of the time. Domestic refrigerators usually run 5 to 10 minutes and are idle 10 to 20 minutes.

Most manufacturers design their units to operate only 8 to 14 hours out of each 24. This is about 40% of the time on the average. The 14-hour operating time is based on average use of the cabinet. The ***ambient*** (surrounding) temperature also affects the running time. A refrigerator in a room at 95°F (35°C) will run longer than the same refrigerator operating in a room at 75°F (24°C). Frost-free and some automatic-defrost refrigerators may run longer or more often than conventional older styles. Defrosting energy adds to the heat load.

Domestic refrigerator cabinets usually have a temperature range between 35°F and 45°F (1.7°C and 7.2°C). The adjustment on the motor control allows the owner to select the desired temperature within this range.

Food freezers are usually designed to operate from about 5°F to −30°F (−15°C to −34.4°C). The motor control must be designed for food freezer use. Most of these controls provide for an adjustment in temperature range. The principle of operation of the motor control used on food freezers is exactly the same as that used on domestic refrigerators.

8.4.1 Effect of Altitude on Refrigerator Temperatures

Refrigeration systems with a sensing bulb thermostat calibrated at sea level may be too cold at elevations above 5000′. This is a result of the decreased pressure at higher elevations. Different altitudes do not affect bimetal and electronic thermostats.

At high altitudes, the atmospheric pressure drops. This lowers pressure on the control diaphragm or bellows enough to affect the settings. The altitude adjustment and range control pressure for the bellows or diaphragm should be increased. This will make up for the lower atmospheric pressure.

Figure 8-30 shows such a control and its altitude adjustment table. Note that the dial is divided into 60 equal calibrations. **Figure 8-31** shows two controls and a table of adjustments used to correct each for various altitudes.

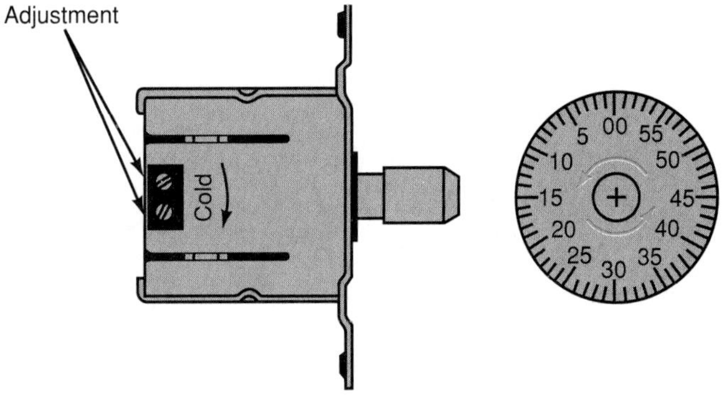

Altitude Correction Both "Cut-In" and "Cut-Out" Screws Must Be Adjusted

Altitude in Feet	Counterclockwise Turns
2,000	7/60
3,000	13/60
4,000	19/60
5,000	25/60
6,000	31/60
7,000	37/60
8,000	43/60
9,000	49/60
10,000	55/60

This scale may be used as a guide for measuring degrees of rotation required for altitude correction. The arrows indicate direction of screw rotation.

Figure 8-30. *A thermostat equipped with altitude adjustment. Table indicates correct setting. (Amana Refrigeration, Inc.)*

8.5 Ice Maker Controls

Many domestic refrigerators are equipped with automatic ice makers. The automatic ice maker is mounted in the freezer compartment. It is designed to produce ice cubes automatically.

A typical ice maker is wired across the line and will harvest in the refrigeration or defrost cycles. The water valve and solenoid valve are in the compressor compartment. A 3/16″ (4.8 mm) polyethylene tube is used for H_2O (water) flow from the water valve to fill the trough. **Figure 8-32** shows the external construction of a typical unit.

The ice mold, **Figure 8-33,** has a thermostat bonded to its front surface. The mold has semicircular compartments. The water enters through the rear and fills each compartment. A mold heater is attached on the lower side of the ice mold. See **Figure 8-34.** The mold heater is wired in series with the ice maker thermostat, which acts as safety device. The heater rests on top of the ice in the storage container. It will stop on top of the ice during either revolution. When a given amount of ice is removed, the ice maker arm lowers and production resumes. When the container is filled, the signal arm raises and the circuit is broken.

The ice maker requires the proper functioning and timing of all components. The following procedures should be identified when servicing an ice maker. Make certain that:

- The ice maker has been properly installed and connected to the water and electric power.
- The freezer compartment is at the proper temperature. Check mold temperature to determine if it is above 15°F (−9.5°C). If the freeze temperature is above 15°F, it is not cold enough to close the ice maker thermostat.
- Several ice making cycles have been completed and the ice maker is in the freezing cycle.
- The ice maker thermostat is a single-throw switch wired in series with the mold heater.
- The ejector blades make two revolutions per cycle, and that ice is not stored on the blades after harvest.
- The water valve solenoid is wired in series with the mold heater.

The operation of the ice maker can best be understood by following its operation step by step. **Figure 8-35** illustrates the cutaway exterior view of the ice maker. **Figure 8-36** shows a freeze cycle. Note that all components are de-energized. The mold is filled with water.

The start of the harvest cycle is shown in **Figure 8-37.** The thermostat is in a closed position because of the ice in the mold. The mold heater and

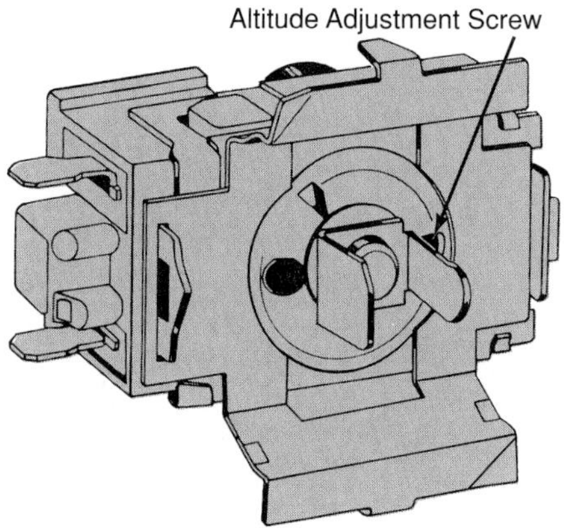

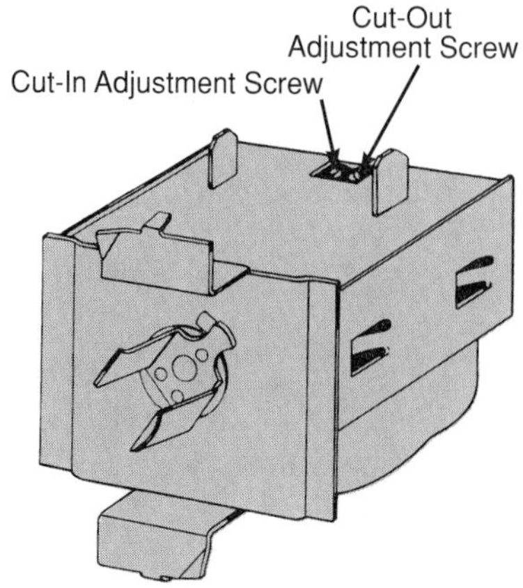

Altitude Above Sea Level—Feet	Constant Cut-In: Altitude Screw Adjustment (Turns Clockwise)
1000	No Change
2000	1/16
3000	1/8
4000	3/16
5000	1/4
6000	5/16
7000	3/8
8000	3/8
9000	—

ALTITUDE ADJUSTMENT — Both Cut-In and Cut-Out Screws Must Be Adjusted		
Altitude Above Sea Level—Feet	Constant Cut-In: Turns Counterclockwise — Cut-In Screw	Cut-Out Screw
2000	1/8 CCW	1/16 CCW
3000	7/32	1/8
4000	5/16	5/32
5000	13/32	7/32
6000	1/2	1/4
7000	19/32	5/16
8000	11/16	3/8
9000	13/16	13/32
10000	15/16	7/16

Altitude Above Sea Level—Feet	Variable Cut-In: Range Screw Adjustment (Turns Clockwise)
1000	3/32
2000	3/16
3000	7/32
4000	1/4
5000	3/8
6000	7/16
7000	15/32
8000	1/2
9000	9/16

Altitude Above Sea Level—Feet	Variable Cut-In: Turns Counterclockwise — Cut-In Screw	Cut-Out Screw
2000	1/16 CCW	1/16 CCW
3000	1/8	1/8
4000	5/32	5/32
5000	3/16	3/16
6000	1/4	1/4
7000	5/16	5/16
8000	3/8	3/8
9000	13/32	13/32
10000	7/16	7/16

Figure 8-31. *Two domestic type thermostats are equipped with altitude adjustments. Left—General Electric. Right—Eaton Corp.*

circulating motor are energized. The ejector blades begin to turn. The motor rotates for a few degrees, **Figure 8-38,** and the timing cam switches the holding switch to its open position. This completes the cycle. The mold heater remains on through the thermostat circuit. During the first half of the cycle, the signal arm is raised and lowered by the timing cam and operates the shut-off switch. In **Figure 8-39,** the ejector blade contacts the ice in the mold and stalls. The mold heater is on. The blade remains in this position until the ice has thawed loose.

In **Figure 8-40,** the first revolution is nearly completed and the timing cam closes the water valve solenoid and its switch. The thermostat is still closed and the mold heater is functioning. Current does not pass through the water valve solenoid and its switch. This is because electrical current follows the path of least resistance.

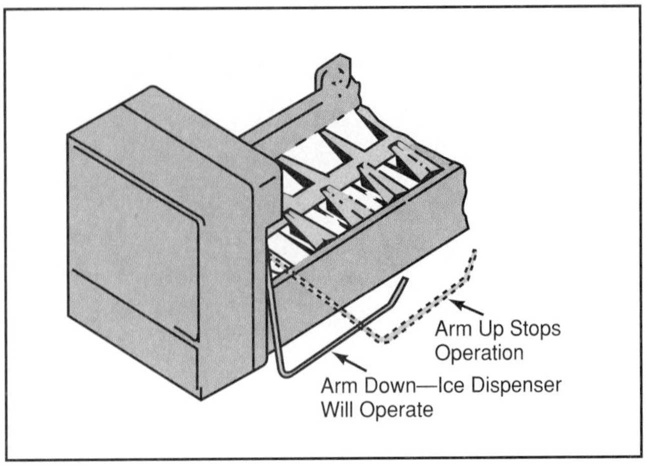

Figure 8-32. *One type of ice cube maker used on domestic refrigerators. Note the signal arm which stops the cycling when the cube bin is full. (Frigidaire Company)*

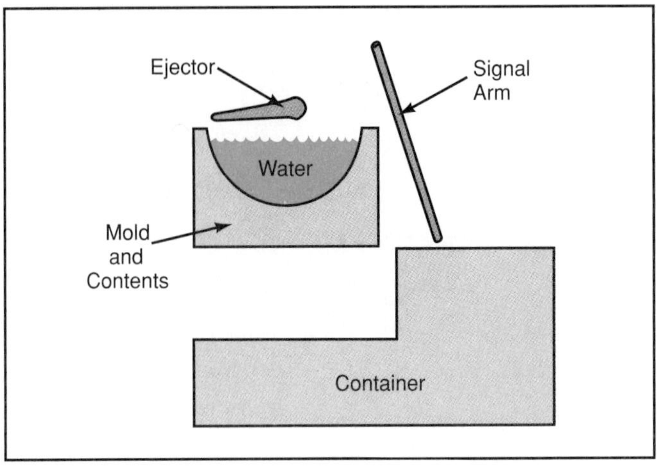

Figure 8-35. *Four basic components of the ice maker. (Frigidaire Company)*

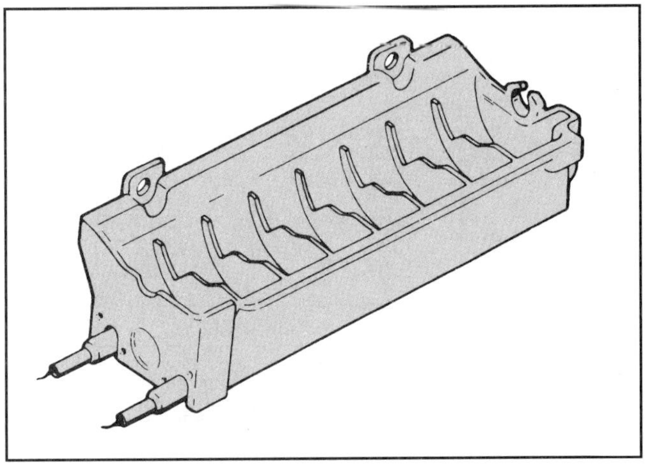

Figure 8-33. *The ice mold, where the ice cubes are formed. (Frigidaire Company)*

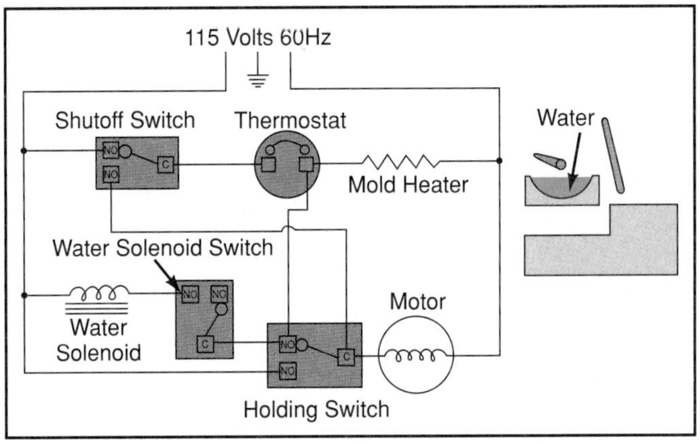

Figure 8-36. *The ice cube freeze cycle. (Frigidaire Company)*

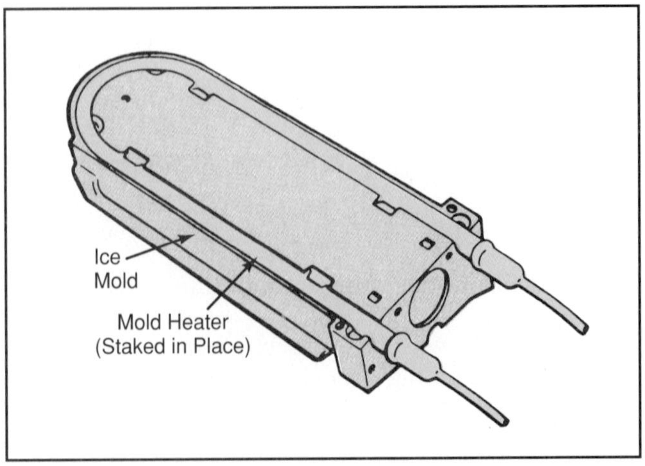

Figure 8-34. *The mold heater is attached to the bottom of the ice mold. (Frigidaire Company)*

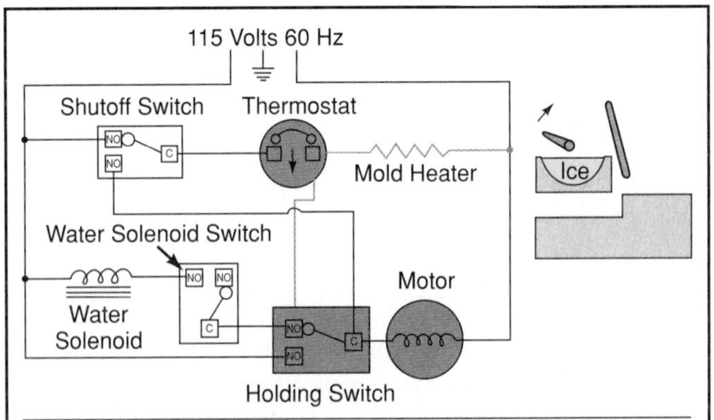

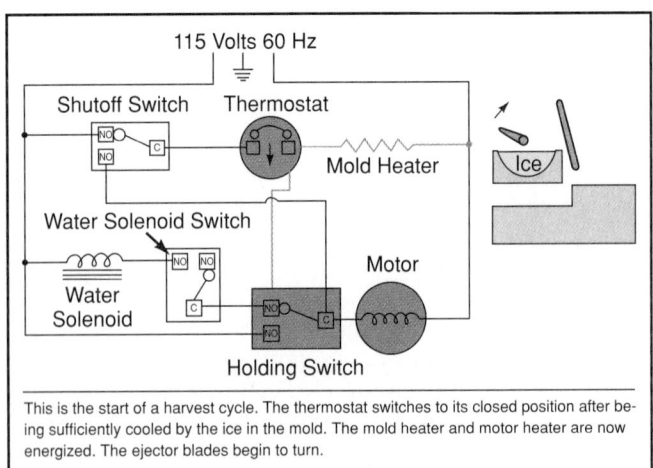

This is the start of a harvest cycle. The thermostat switches to its closed position after being sufficiently cooled by the ice in the mold. The mold heater and motor heater are now energized. The ejector blades begin to turn.

Figure 8-37. *Electrical circuit when ice has formed. (Frigidaire Company)*

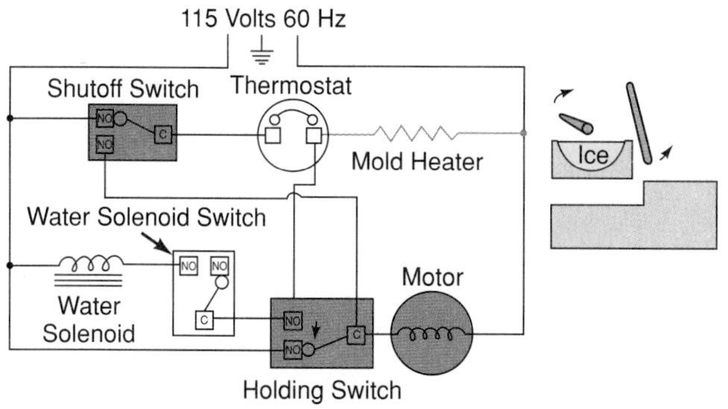

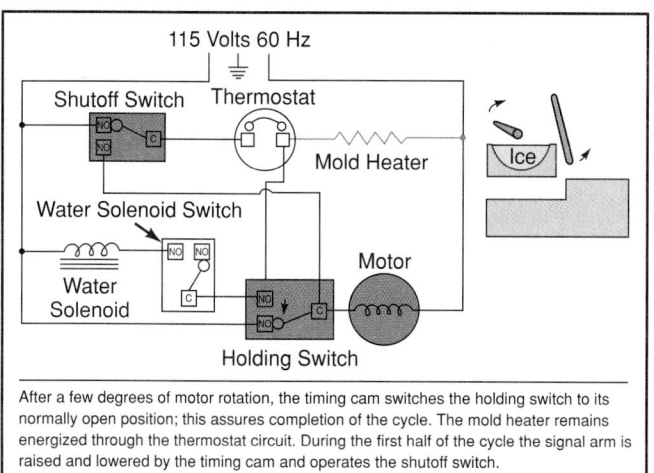

After a few degrees of motor rotation, the timing cam switches the holding switch to its normally open position; this assures completion of the cycle. The mold heater remains energized through the thermostat circuit. During the first half of the cycle the signal arm is raised and lowered by the timing cam and operates the shutoff switch.

Figure 8-38. *Electrical circuit when the holding switch is open. (Frigidaire Company)*

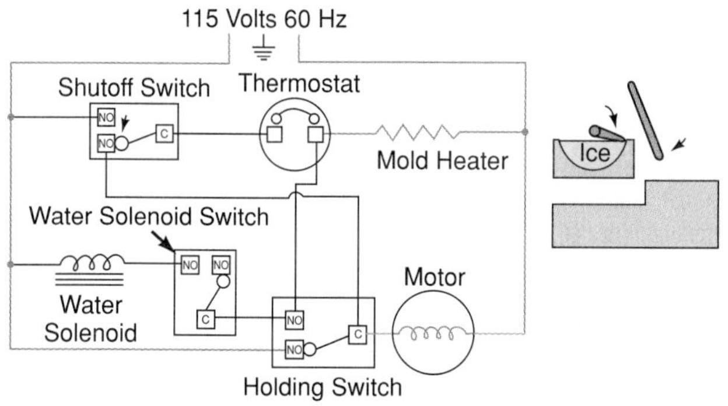

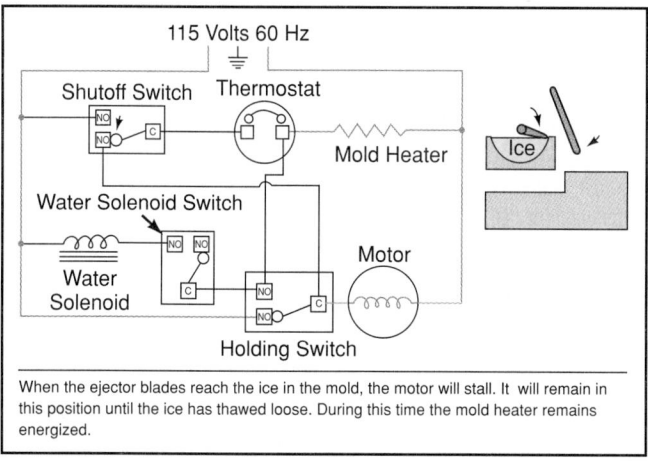

When the ejector blades reach the ice in the mold, the motor will stall. It will remain in this position until the ice has thawed loose. During this time the mold heater remains energized.

Figure 8-39. *The ejector blade touches the ice in the mold. (Frigidaire Company)*

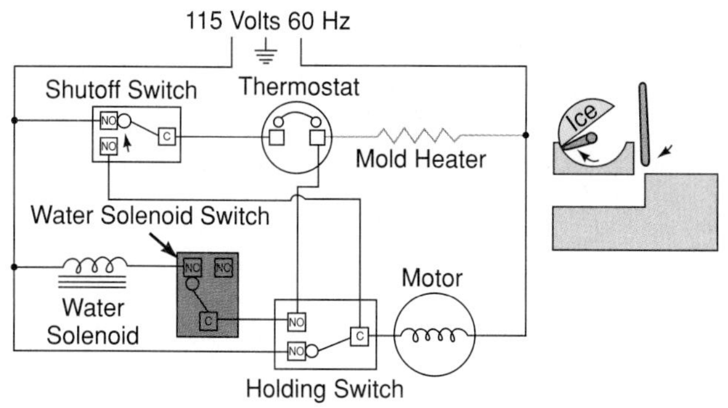

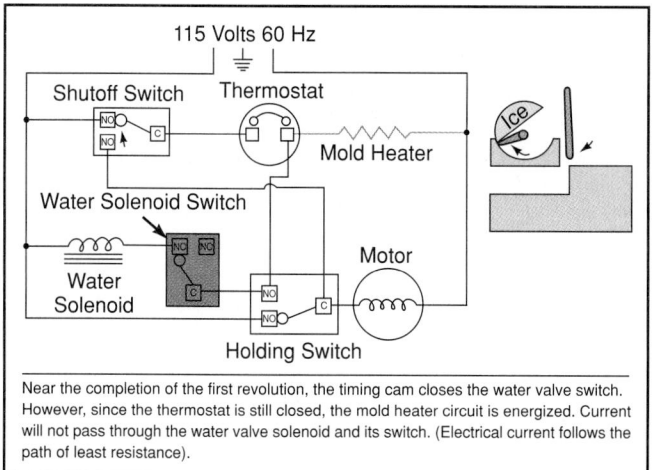

Near the completion of the first revolution, the timing cam closes the water valve switch. However, since the thermostat is still closed, the mold heater circuit is energized. Current will not pass through the water valve solenoid and its switch. (Electrical current follows the path of least resistance).

Figure 8-40. *The first revolution of the ice maker is almost complete. (Frigidaire Company)*

When the first revolution is completed, **Figure 8-41,** the timing cam opens the holding switch. However, the thermostat is still closed, and a second revolution begins.

In **Figure 8-42,** after a few degrees of rotation, the timing cam is closed. The holding switch provides the circuit to the motor, permitting completion of the revolution. The mold heater remains on. The signal arm will raise and lower, again operating the switch. The ice that was harvested during the first revolution is dumped into the container.

During the second revolution, **Figure 8-43,** the mold heater resets the thermostat. With the thermostat on, the mold heater is de-energized. If the container is full, the signal arm will remain in a raised position.

Figure 8-44 shows the near completion of the second revolution. The timing cam closes the water valve switch. The circuit is completed through the water valve solenoid and the mold heater. When the water valve is energized, the valve opens and water fills the mold.

Figure 8-45 illustrates the end of an ejector cycle. The container is full and no other cycles will start until enough ice has been used to lower the signal arm.

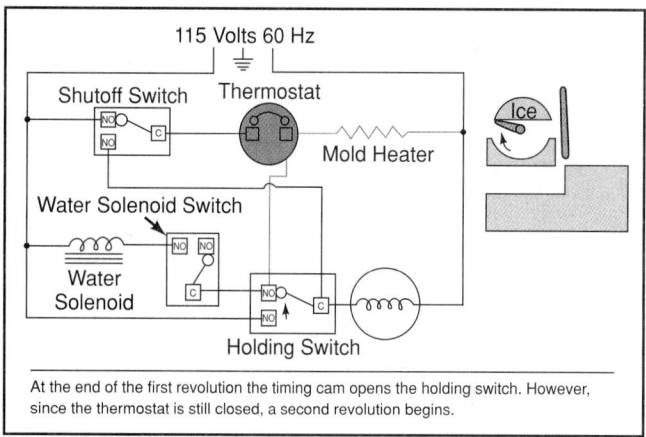

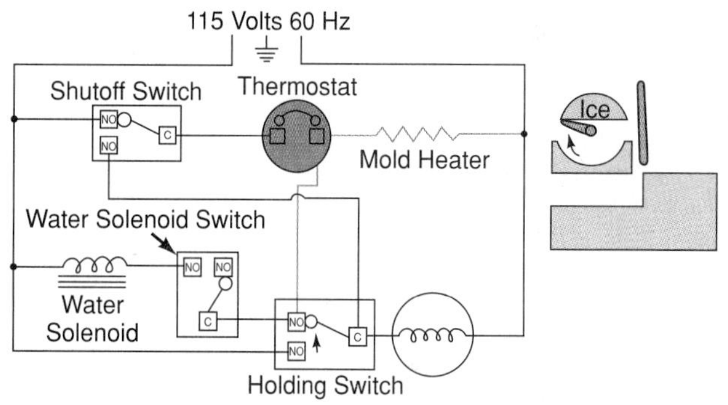

At the end of the first revolution the timing cam opens the holding switch. However, since the thermostat is still closed, a second revolution begins.

Figure 8-41. *At the end of the first revolution, the thermostat remains closed. (Frigidaire Company)*

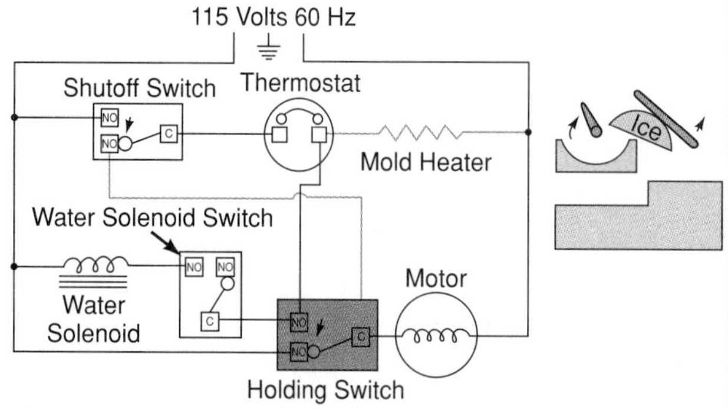

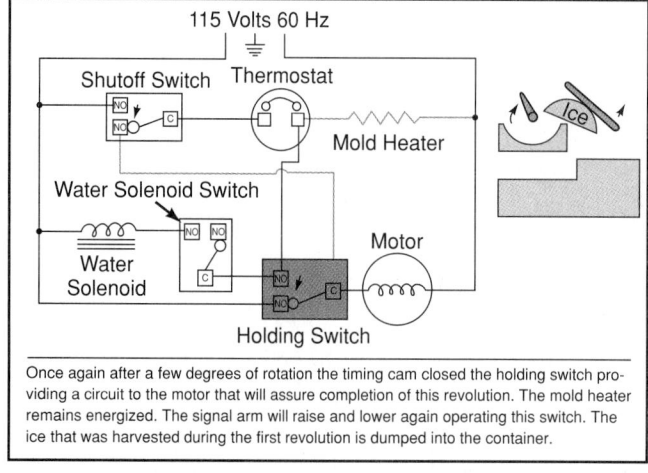

Once again after a few degrees of rotation the timing cam closed the holding switch providing a circuit to the motor that will assure completion of this revolution. The mold heater remains energized. The signal arm will raise and lower again operating this switch. The ice that was harvested during the first revolution is dumped into the container.

Figure 8-42. *The ice harvested during the first revolution is dumped into the container. (Frigidaire Company)*

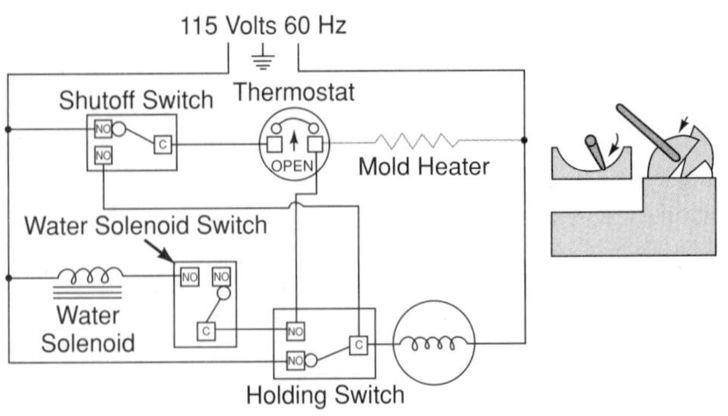

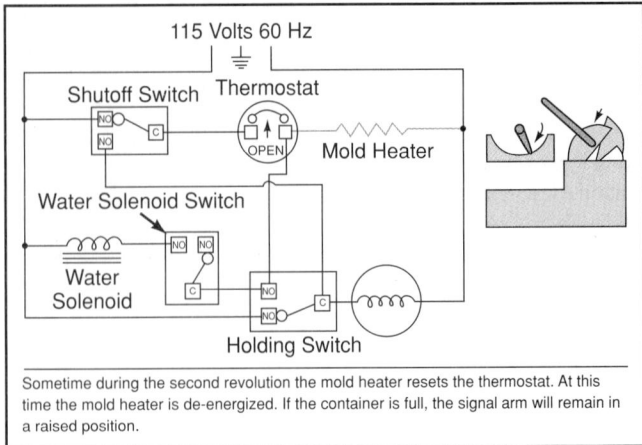

Sometime during the second revolution the mold heater resets the thermostat. At this time the mold heater is de-energized. If the container is full, the signal arm will remain in a raised position.

Figure 8-43. *During the second revolution, the mold heater is de-energized by the thermostat and the ice is dumped. (Frigidaire Company)*

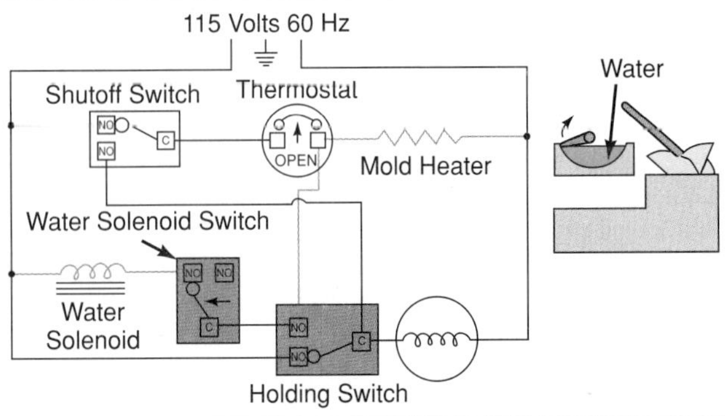

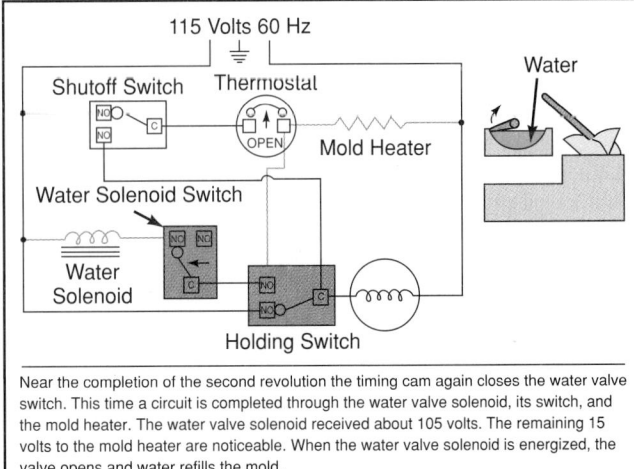

Near the completion of the second revolution the timing cam again closes the water valve switch. This time a circuit is completed through the water valve solenoid, its switch, and the mold heater. The water valve solenoid received about 105 volts. The remaining 15 volts to the mold heater are noticeable. When the water valve solenoid is energized, the valve opens and water refills the mold.

Figure 8-44. *The water valve is energized and opens. Water refills the ice cube mold. (Frigidaire Company)*

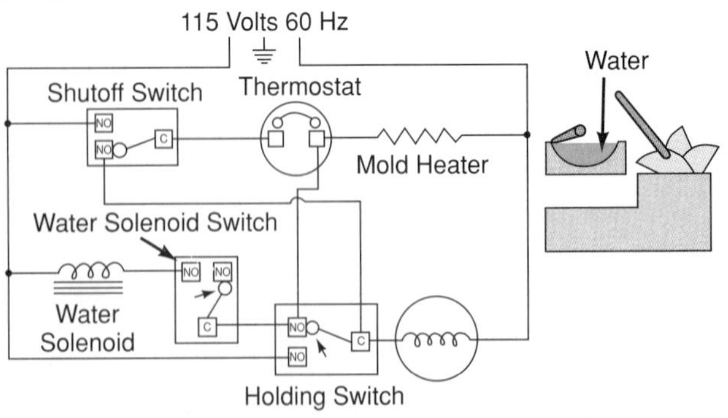

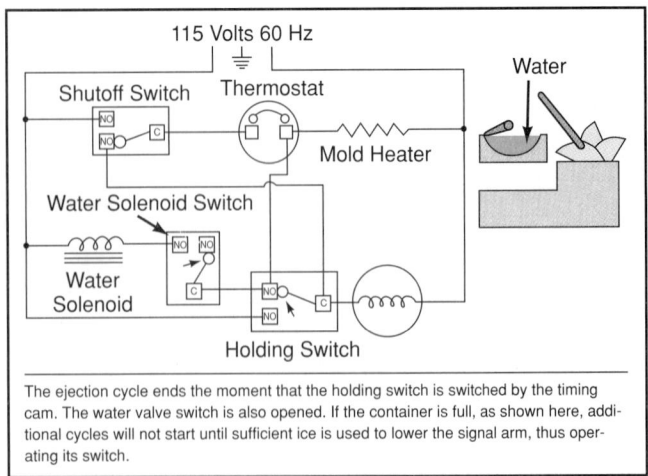

The ejection cycle ends the moment that the holding switch is switched by the timing cam. The water valve switch is also opened. If the container is full, as shown here, additional cycles will not start until sufficient ice is used to lower the signal arm, thus operating its switch.

Figure 8-45. *The container is full of ice cubes. No other cycles will begin until the signal arm is lowered by the removal of ice cubes from the container. (Frigidaire Company)*

8.6 Comfort Cooling Air Conditioning Controls

Comfort cooling air conditioners have two standard controls:

- A thermostat to control the temperature.
- A defrost control to eliminate icing of the evaporator.

The thermostat sensing element is usually located in the return air duct of the air conditioner. When this incoming air cools enough, the thermostat will stop the unit.

A two stage (two-switch) motor control thermostat is shown in **Figure 8-46.** The sensing element reacts to the return air temperature. A rise in temperature will first cause the fan switch to cut in. A further rise in temperature will cause the compressor switch to cut in. As the room temperature drops, this operation will be reversed (the fan is the last to turn off). **Figure 8-47** shows a schematic which helps explain the operation of this control.

A control knob provides an adjustment for setting the range of the air conditioner. If the room is too cold, the air conditioner's running time can be reduced.

The ***defrost control,* Figure 8-48,** prevents ice formation on the air conditioner evaporator. This control opens the circuit if the evaporator temperature is close to freezing. It has a factory adjustment only, and should not require adjusting on the job.

This control, **Figure 8-49,** is a ***single-pole, single-throw (SPST)*** control if used for defrost only. It provides

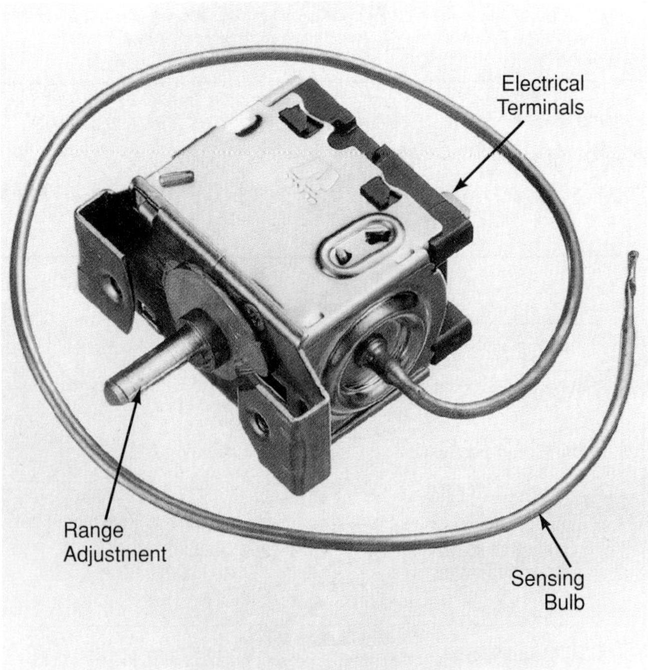

Figure 8-46. *A comfort cooling air conditioner motor control thermostat. It controls operation of fan and compressor. The sensing bulb is located in the return air duct. (Ranco North America)*

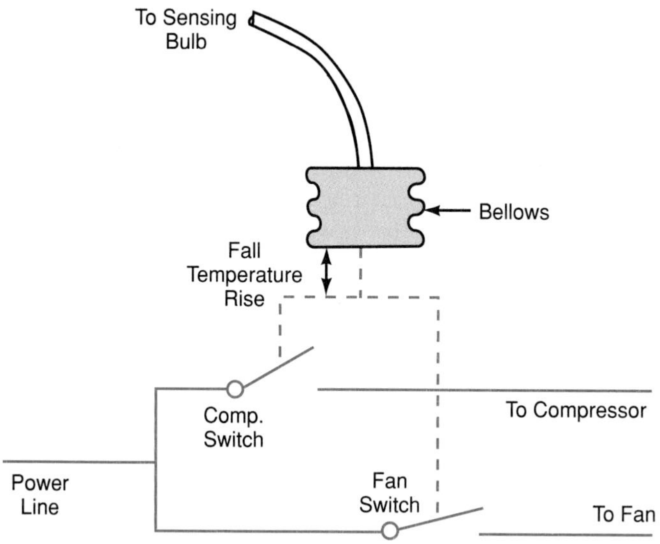

Figure 8-47. *This schematic shows how a room air conditioner thermostat works. (Ranco North America)*

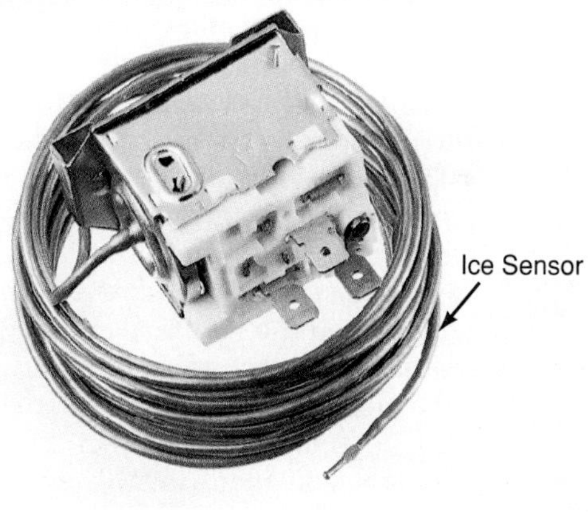

Figure 8-48. *An air conditioner defrost control is used to prevent formation of ice on the evaporator. (Ranco North America)*

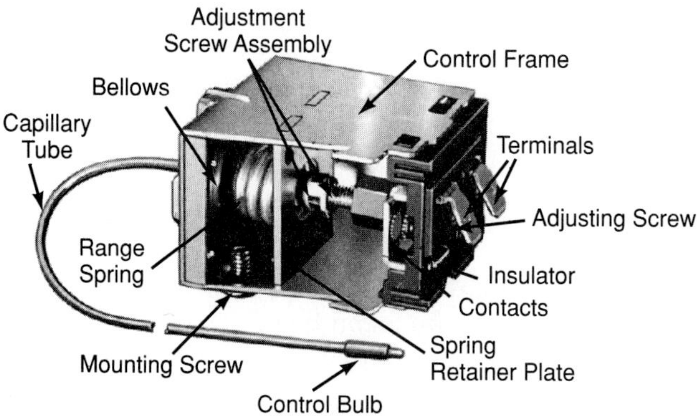

Figure 8-49. *Mechanism used for an air conditioner defrost control. In operation, the control bulb is mounted on the evaporator of the air conditioner.*

defrost by keeping compressor off for sufficient time after sensor shows ice formation temperature. When defrost *heating* units are used on the off cycle, it is a ***single-pole, double-throw (SPDT)*** control.

The same controls are used in refrigerators and heat pumps. **Figure 8-50** is a schematic diagram of the control that shows how it is connected into the air conditioning electrical and refrigerating circuits. In **Figure 8-50A,** it is shown *without* a defrost heater. **Figure 8-50B** shows a circuit *with* a defrost heater.

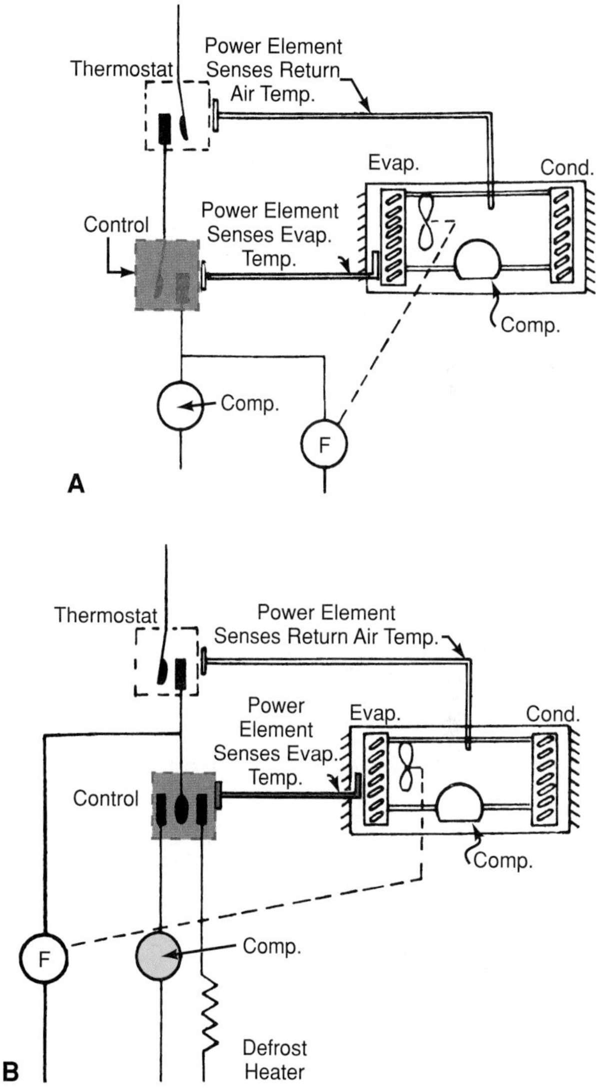

Figure 8-50. *An air conditioning defrost control. A—Control without a defrost heater. B—Control with a defrost heater.*

8.7 Central Air Conditioning Controls

Central air conditioning implies that the system can provide heating, cooling, humidification, and dehumidification. In many systems, electrostatic air cleaning is also provided. Each of these systems has electrical controls.

The heating controls govern the running time of the heating device. They usually operate from a room thermostat. The thermostat, then, must make suitable electrical contact. It will operate the heating mechanism until the desired room temperature is reached. Then, the electrical circuit is opened and the heating device stopped. Details concerning installation and operation are given in Chapters 20, 21, 22, and 23.

Cooling is usually controlled by a room thermostat, as well. This thermostat may be a separate instrument, or may be built in with the heating thermostat. A refrigerating mechanism usually provides cooling. The room thermostat turns on the refrigerating mechanism when the temperature of the room goes above a certain setpoint. It shuts off the refrigerating mechanism when the temperature has been lowered to the desired level.

Automatic humidification controls usually govern the flow of water or steam into a humidifier. If the air is too dry, additional moisture is provided. The ***humidistat*** is usually located in the same room as the temperature controls.

In some central air conditioning systems, the humidistat causes the refrigerating mechanism to operate if the air is too moist. The evaporator is located in the plenum chamber of the central air conditioning system. The cold surfaces of the evaporator will condense out moisture. Thus, the air being circulated is dried.

Electrostatic air cleaners usually operate at the same time as either the heating or cooling mechanisms. Air being drawn through the air conditioner is "scrubbed" while passing through the electrostatic air cleaner.

Requirements of air conditioning systems that use a heat pump are described as follows:

- Air conditioning systems using a heat pump provide a motor control to turn the motor compressor on and off. Usually, the same motor compressor is used for both heating and cooling. Its operation is controlled by the room thermostat.
- During summer cooling, this system works without supplementary controls. During winter heating, however, more controls are required.
- An electric resistance heater is also usually required at the evaporator. This heater will operate if the evaporator surface starts to ice over. (Ice would keep air from flowing through the heater.)
- When it is not possible to extract enough heat from the outside air, some systems use an electric heating unit. The operation of this auxiliary electric heating unit is automatically controlled by the condenser's temperature.

8.8 Water Cooler Controls

Drinking fountains are cooled with small refrigerating mechanisms. The evaporator is placed so that the water flowing to the outlet faucet is cooled to a temperature of about 50°F (10°C).

These water coolers usually provide storage cylinders where water is maintained at a lower temperature. In some cases, ice is formed. The motor control is operated from the accumulator. It operates to keep a "cold reserve" available to supply needed water to the faucet. Details of some of these water coolers are shown in Chapter 14.

8.9 Remote Temperature Sensing Elements

Remote temperature sensing elements are available in several designs. One major difference in type is based on the pressure in the element. There are two common designs:

- *Above-atmospheric-pressure* in the element, used for controlling refrigeration temperatures.
- *Below-atmospheric-pressure* in the element, used for controlling heating units.

The above-atmospheric-pressure element is used for controls which close the electrical circuit on temperature rise. If the element loses its charge, the unit is unable to start. This is known as a "fail-safe" action. It is used where continuous running is harmful, such as in refrigerators and comfort cooling units.

The below-atmospheric-pressure element is used for controls which open the electrical circuit on temperature rise. It is found on electric heating and electric defrost units. If the element loses its charge, the bellows or diaphragm will be unable to contract due to loss of vacuum. The points will then open (they are designed to close on temperature drop and open on temperature rise). As in units above atmospheric pressure, this is a "fail-safe" device. It prevents an electric heating coil from overheating.

The temperature of the mechanism at the point of contact with the sensing element, in most cases, controls the motor operation. This sensing element may be used to stop the motor if the condensing temperature rises too high. Sensing elements may also be used as safety devices to control the motor operation. Commercial installations, in particular, may use them to stop the motor when:

- The flow of water through the condensing unit stops.
- The oil pressure in the compressor is too low.
- The head pressure is too high.
- The desired low temperature in the cabinet is reached.

Many motor controls may satisfy several of these requirements. This means that there may be more than one capillary tube entering the motor control.

8.10 Pressure Motor Controls

A low pressure must be maintained in the evaporator to permit evaporation of the refrigerant at a low temperature. Therefore, automatic control of the motor may be based on pressure differences in the evaporator. This control is used on commercial systems. A bellows-operated low-pressure control is shown in **Figure 8-51.**

This is how it operates: as the evaporator warms, the low-side pressure increases and the bellows expands. The switch is closed, and the motor starts. When the pressure and temperature become low enough, the bellows assembly contracts and contacts open. The motor automatically shuts off.

A pressure motor control with cut-out and cut-in adjustments is shown in **Figure 8-52.** The cut-out and cut-in controls are the only ones needed for full control over the differential and range. The electrical switch may be of either the mercury bulb or the open-contact-point type.

The range adjustment will lower both the cut-in and the cut-out an equal distance if the screw is moved *out* (counterclockwise). Cut-out and cut-in pressure will be raised if turned *in* (clockwise). The spring is under compression and presses on the bellows at all times.

The differential adjustment will raise the cut-out pressure when turned to the right (clockwise). This is the cut-out type differential adjustment. If the spring tension is increased by turning the screw clockwise, it is harder for the bellows to reach its cut-out setting. The spring has no effect on the cut-in setting, however.

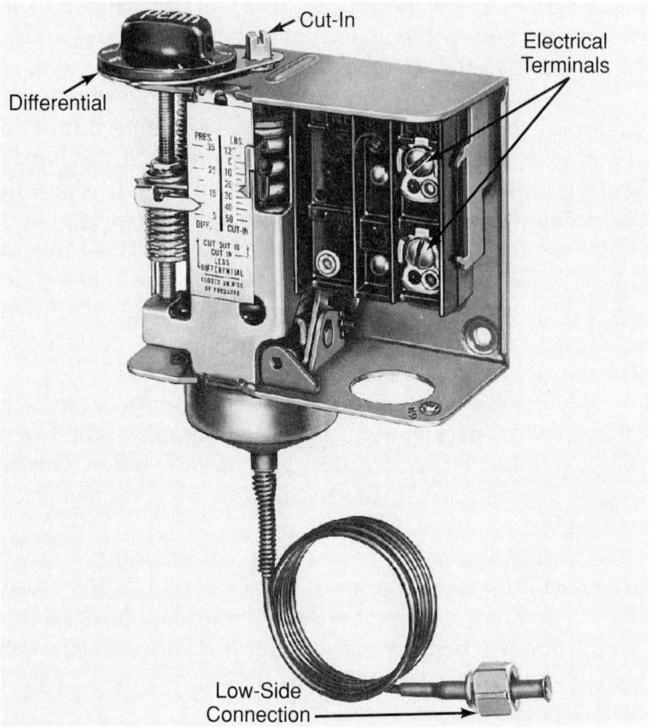

Figure 8-51. *Pressure type motor control with cover removed. The cut-out pressure is the cut-in minus the differential setting. (Johnson Controls, Inc.)*

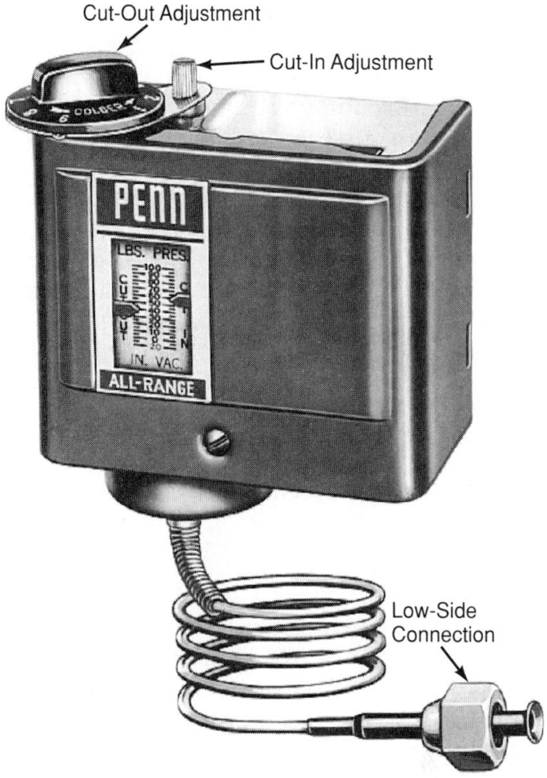

Figure 8-52. *Typical pressure motor control. It is possible to adjust cut-in and cut-out pressure on this motor control. (Johnson Controls, Inc.)*

Some models of pressure control are also equipped to act as a safety device. A bellows construction with a pressure tap to the high-pressure side of the compressor is used. If the compression pressure or head pressure should become too high, the bellows will expand. This movement will open the switch and stop the motor. Such a control is a safety device for the motor. It is especially necessary when a water-cooled unit is used. See Chapter 13 for more information on commercial controls.

The low-pressure type of control is easy to adjust on the job. It can also be easily adjusted with a vacuum pump and compound gauge after removing it from the unit. The adjustment of these controls is important for the satisfactory operation of the unit. The unit operation depends, to a great extent, on proper operation of the controls.

8.11 Motor Safety Controls

Several safety mechanisms are used in the electrical circuit to make sure that the motor compressor is protected. The most common safety controls are:

- Head-pressure safety cut-out.
- Low-pressure safety cut-out.
- Oil-pressure safety cut-out.

One of the most harmful things that can happen to a hermetic system is to have high head (condensing) pressures. These high pressures raise the temperature of the vapor and oil moving past the compressor exhaust valve. This may cause oil and refrigerant breakdown. This condition is worsened if a little moisture and dirt are present. Carbon, acids, and sludges may be formed.

It is important to shut down the system before these dangerous temperatures are reached. A high-pressure safety cut-out is often used for this purpose. If pressure exceeds a certain setpoint, the current to the motor will be shut off. Thus, the motor will be stopped.

Several conditions might cause this control to operate:

- Lack of proper air circulation through the condenser.
- Lack of flow of water through a water-cooled condenser.
- A greatly increased refrigeration load of some kind. See Chapter 13.

A low-pressure switch can be used to cycle a refrigerating system using a TEV ***(thermostatic expansion valve).*** A low-pressure switch can also be used as a safety device. The cooling of a motor compressor depends on the amount and temperature of the suction vapor. If this vapor pressure is too low (system is low on refrigerant or flow is restricted), the motor compressor may overheat and burn out. The low-pressure safety control will stop the motor compressor before it is damaged. The schematic wiring diagram for a low-side or suction pressure control is shown in **Figure 8-53.**

Some larger units use a safety control device connected to the compressor lubrication system. These shut off the unit if the oil pressure decreases. They also shut off if the oil pressure falls below a predetermined safe pressure above the low-side pressure. The construction is similar to a low-pressure control but usually with a fixed or nonadjustable differential. This control has one connection to the oil pump and one connection to the low-pressure side. See Chapter 13.

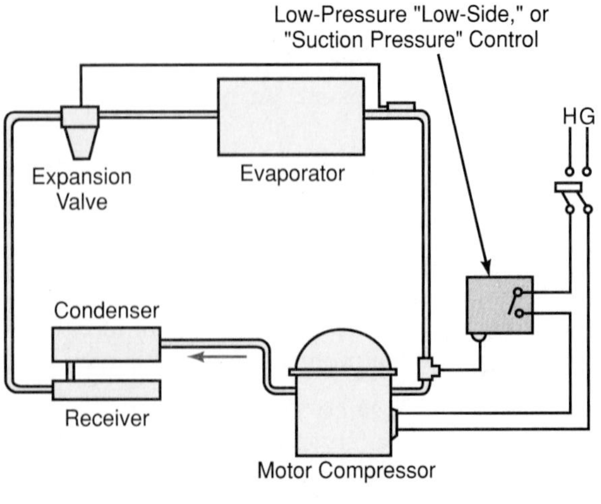

Figure 8-53. *Schematic of a low-pressure motor control installed in a refrigerating system.*

8.11.1 Low-Side Pressure Limiter

A popular refrigerator and air conditioning safety device incorporates a low-side pressure limiter. The condensing unit cannot be overloaded if the low-side pressure is maintained at a low enough level.

Low-side pressure limiters consist of a pressure sensitive element such as a diaphragm or bellows. The element is placed in series with a condensing unit circuit. The electrical circuit will be open if the low-side pressure is higher than a desired limit. Some low-side pressure limiters operate through a relay. The low-side pressure operates the relay. The relay, in turn, controls the electrical circuit.

8.11.2 Overload Protection

Refrigeration and air conditioning units should be connected to separate circuits from the control panel. This applies to both domestic units and commercial units. The fuse or circuit breaker in the individual circuit must have sufficient capacity to permit a continuous flow of current under normal operating conditions. However, it should open the circuit in the event of continuous overload of more than 25%.

At the instant of starting, all motors draw an overload of current. This may amount to 600%. However, this overload lasts for a very short time. The circuit breaker or fuse should not open the circuit during this brief period.

High horsepower motors usually incorporate a starting device in the electrical circuit. This starter does not throw the motor directly on the line. It brings into use a resistance or induction unit. The unit restricts the flow of current at the instant of starting. It allows an increase later as the motor speed increases.

All starting relays have some type of overload protection. The most popular type is a bimetal control in series with the power supply to the motor. A resistance heating unit is alongside the bimetal control. It is also in the series with the power supply. This resistance unit will heat up if the motor is overloaded. The bimetal safety device, reacting to the heat of the resistance unit, will bend. The points at the end of the bimetal strip will open, stopping the motor. The motor will not restart until the safety device cools down.

Three-phase circuits and motor compressors are often used in sizes above 1 hp. The voltages of the three phases should be kept within 10% of each other. This is to prevent damage to the motor and to prevent motor reversing.

Controls are used to open the circuit if the voltage in any of the three circuits changes over 10%. They also open if the motor reverses. Some controls will close the circuit again when normal conditions are reached. The controls usually act from about 0.1 second up to 2 seconds (if desired).

These protective devices are basically voltage-controlled relays. Transformers are used to step down the voltage for each potential relay. The stepped-down voltage is then rectified and fed to the solenoid coil. The same voltage is also fed to the control of a transistor. If the voltage varies too much, the transistor will bypass the solenoid. This causes the circuit to open. **Figure 8-54** shows a three-phase circuit protector.

Thermistors with a positive temperature coefficient (PTC) are also used in electronic motor safety controls. A PTC thermistor is connected in series with the motor windings. The temperature increases when an overload condition exists. This causes the resistance of the thermistor to go up. This increase in resistance limits the current to the motor. After the motor and thermistor cool to a safe temperature, the motor can draw current to start up again. Refer to Section 7.12 for further information concerning motor protection.

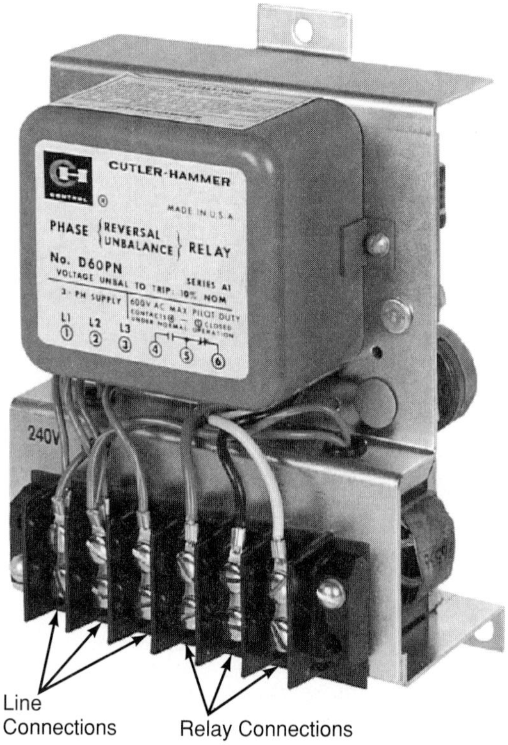

Figure 8-54. *Three-phase protection system with cover removed. (Eaton Corp., Controls Div.)*

8.12 Motor Starting Relays

Some motor controls for hermetic systems are different from those used on external drive systems. Starting relays are found on the outside of hermetic compressor systems. These relays are usually one of the following types:

- Current (magnetic).
- Potential (magnetic).
- Thermal.
- Solid-state electronic.

The relay permits electricity to flow through the starting winding of the motor. This continues until the motor reaches about two-thirds of its rated speed. The relay then disconnects ("opens") the starting winding circuit.

The starting winding should be energized only for three or four seconds at a time. If current flows through it for a longer period, the winding may overheat. Many relays have current and/or thermal protection devices to protect the starting winding from damage.

To operate correctly, the relay must be the right size for the motor. When replacing a relay, the new one must have the same electrical specifications as the original. It is impossible to use open electrical contacts inside a sealed system.

8.12.1 Current (Magnetic) Relay

Current relays are usually found on low-torque, smaller horsepower motors. The current (magnetic) relay uses the electrical characteristics of the motor to operate.

As the rotor picks up speed, magnetic fields build up and collapse in the motor. This produces a counter electromotive force (cemf), or voltage on the running winding. The running winding consumes more current when the rotor is not running, or is turning slowly, than it does at full speed. Current-operated relay switches are used to close and open the starting winding. They operate on the change in current flow of the running winding. This is done as the winding goes from a start condition to run.

The magnetic relay is an electromagnet much like a solenoid. Either a weight or a spring holds the starting winding contact points open when the system is idle. **Figure 8-55** is a schematic of a weight-operated unit. When the motor control (thermostat or pressurestat) contacts close, high current flows in the running winding. The magnetic current relay coil is then heavily magnetized. It lifts the weight or overcomes the spring pressure and closes the contacts.

This action closes the starting winding circuit. The motor will quickly accelerate (speed up) to two-thirds or three-fourths of the rated speed. As it does so, the amperage draw of the running winding of the motor decreases. This decreases the magnetic strength of the magnetic current relay. The decrease is enough to allow the weight or the spring to open the points. **Figure 8-56** shows a magnetic current relay in the closed (starting) position and also in the open (running) position.

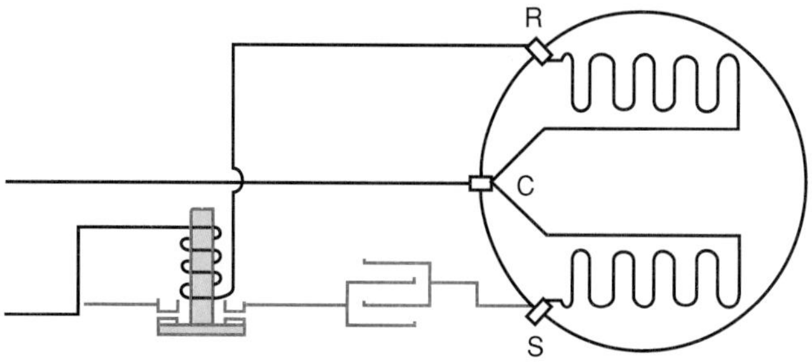

Figure 8-55. *Current-operated relay schematic. Relay is shown in open position. R—Running winding terminal. S—Starting winding terminal. C—Common terminal.*

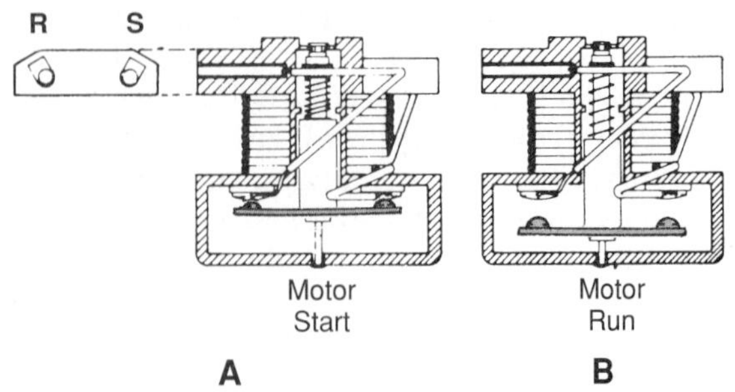

Figure 8-56. *Current (magnetic) relay. A—Relay is in motor starting (closed) position. B—Relay is in motor running (open) position. (Frigidaire Company)*

Current relays are sometimes called *amperage* relays. It is the ampere draw on the circuit that operates the relay. One type of magnetic current control uses a rotary solenoid. This type can be mounted in any position. The weight type must be mounted level.

One type of weight-operated magnetic current relay is shown in **Figure 8-57.** A spring-operated type is shown in **Figure 8-58.** A method of mounting a current-starting relay is shown in **Figure 8-59.**

These relays are available in a number of capacities. The difference between closing amperage and opening amperage settings is small. This small difference in current flow will close the starting circuit. It will then open the circuit when the motor reaches approximately three-fourths speed. **Figure 8-60** is a circuit using a current relay.

The utility companies sometimes reduce line voltage to save current. The current-type starting relay gives some protection to the motor under these conditions. However, there is still some danger of a motor burnout.

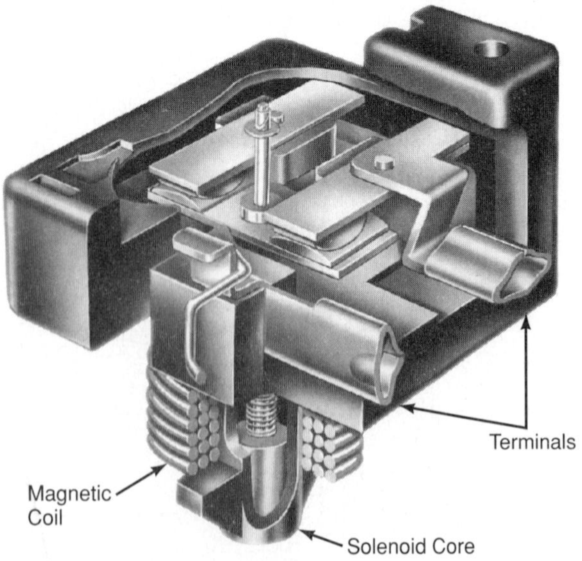

Figure 8-57. *Current type magnetic relay. Plastic housing contains solenoid that has a movable core (plunger). Heavy current draw when starting raises the plunger and closes starting winding circuit. (Tecumseh Products Co.)*

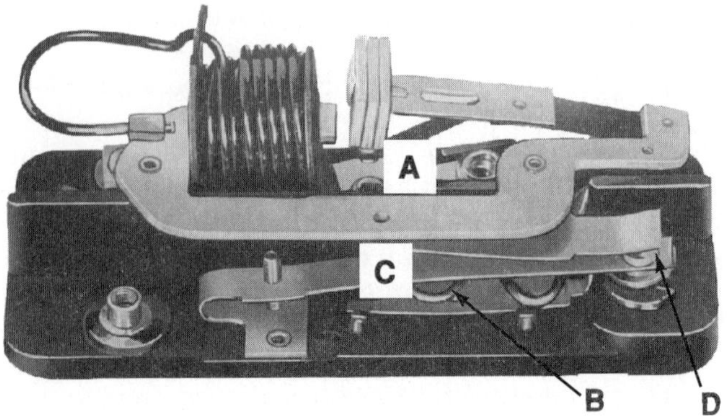

Figure 8-58. *Current-type magnetic starting relay with an overload safety switch. Starting winding points are kept open by means of cantilever spring at A. At instant of starting, magnetism pulls spring down and closes starting points at D. If current draw is too great, the resistance wire at B will heat and cause bimetal strip at C to bend and open the circuit at D.*

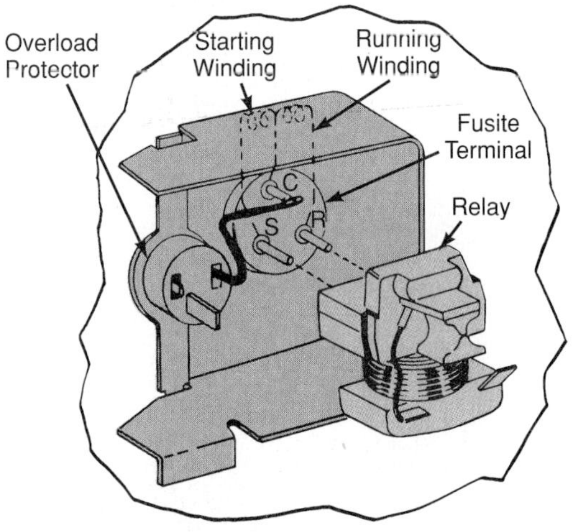

Figure 8-59. *Starting relay mounted on compressor housing. (Frigidaire Company)*

8.12.2 Potential (Magnetic) Relay

Potential relays (voltage relays) are usually used with high-torque, capacitor-start motors. They look somewhat like the amperage relay. However, the operation of these relays is based on the voltage increase. The increase occurs as the unit approaches and reaches its rated speed. **Figure 8-61** is a potential magnetic relay.

The contact points remain closed during the off part of the cycle. This feature is its biggest advantage. If the points are closed as the thermostat closes the power circuit, there will be no arcing of the relay points. Arcing quite often occurs with the current relay.

Figure 8-62 is a circuit diagram of a unit with a potential relay. As the motor speed increases, higher voltage from the starting windings energizes the relay coil, 2 and 5. This opens the normally closed contacts, 1 and 2. The starting windings are therefore dropped out of the

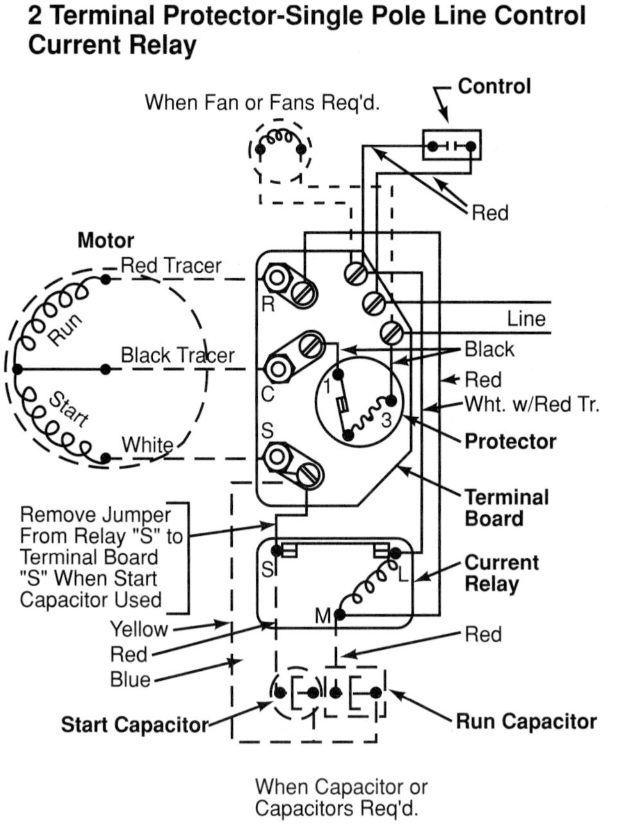

Figure 8-60. *Complete wiring diagram of system using current relay.*

Figure 8-61. *Potential starting relay. A—With casing removed, components are visible. Weight (1) closes the points (3) during the off cycle. On starting, increasing voltage into the coil (2) will pull contact points apart and stop current flow through the starting winding. B—A similar potential relay with casing in place. (White-Rodgers Division, Emerson Electric Co.)*

circuit. The relay coil is connected across the starting winding. The relay coil is made of small wire, so very little current passes through it. This minimizes the heating of the coil and core.

Resistance of the relay coil must be high enough to prevent the contact points from opening before the motor reaches 80% to 90% of full speed. However, such resistance must be low enough to positively open the points and remove the starting winding from the circuit at the right time. If not, the motor will overheat. The relay itself is shown in **Figure 8-63.** The wiring diagram for its installation is shown in **Figure 8-64.**

8.12.3 Thermal Relay

There are two types of thermal relays. One type uses two bimetal strips to control the contact points. The other controls the contact points through a resistance wire under tension.

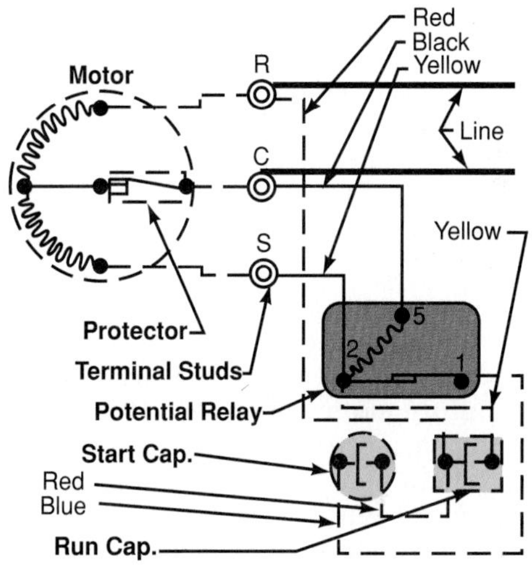

Figure 8-62. *Wiring diagram for a potential (voltage) magnetic starting relay. Note that starting capacitor circuit is opened when relay contacts open, but running capacitor is still connected across starting and running windings in series. (Copeland Corp.)*

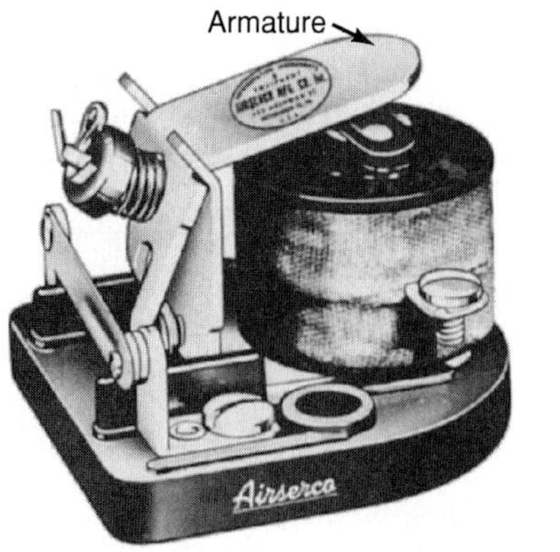

Figure 8-63. *Potential relay. As the armature lever is pulled down by electromagnet, lever at left will open starting circuit points.*

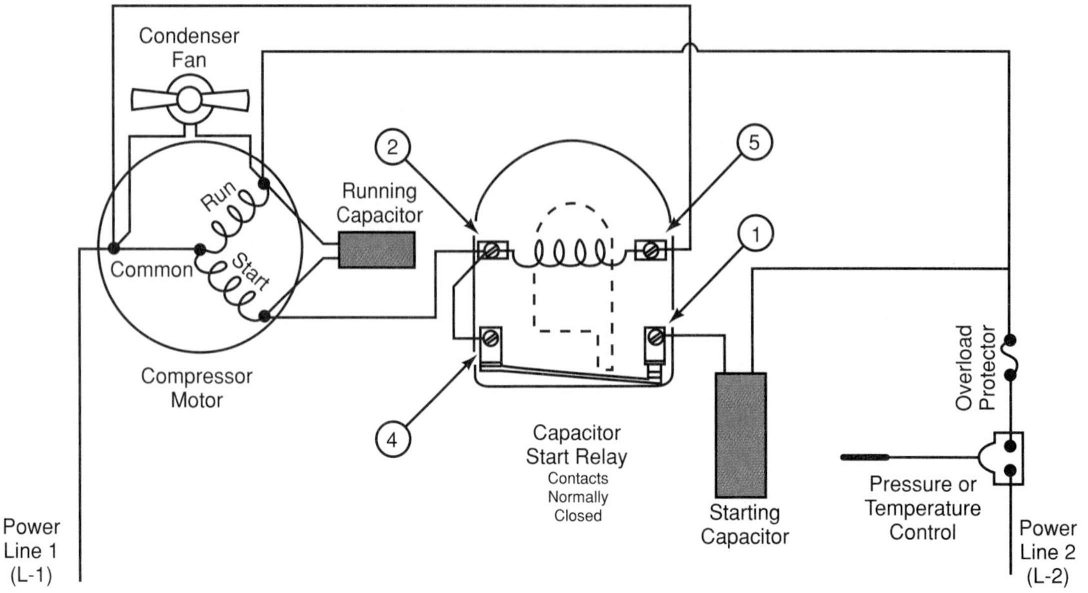

Figure 8-64. *Wiring diagram for potential relay shown in* ***Figure 8-63.*** *Note that this is a capacitor-start, capacitor-run motor.*

In the first type, one strip controls the starting winding and the other, the running windings. See **Figure 8-65.** When cold, both sets of contact points are closed. A resistance wire is mounted near the bimetal strips. It is in series with both the starting and the running winding. It is the right size and distance from the bimetal strip. Therefore, its heat opens the starting winding contact points when the motor reaches its proper operating speed.

This control also serves as a safety cut-out. If the motor should use too much current, the resistance wire will heat the bimetal strip. The heated bimetal strip will open the contact points and stop the motor.

In the second type, the resistance wire is attached in series with both the starting and running windings. The tension of this wire, when cold, keeps both sets of contact points closed. See **Figure 8-66.**

While the current passes through, the resistance wire is heated and expands or stretches. At a predetermined setting, the stretched wire opens the starting winding contacts. This control also serves as a safety cut-out. If the motor should use too much current, the wire will stretch enough to open the running winding

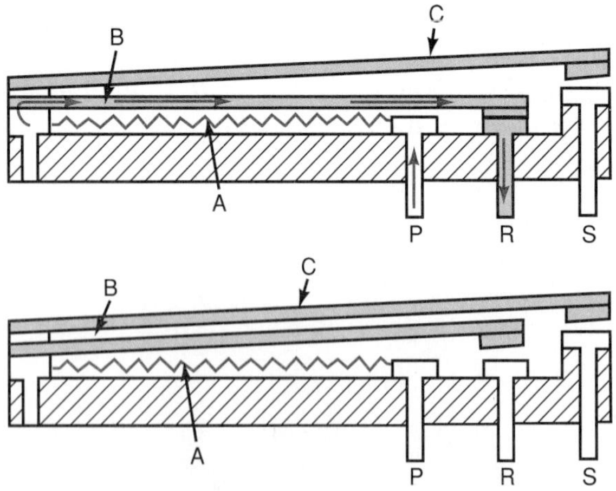

Figure 8-65. *Thermal starting relay using two bimetal strips. A—Heating wire. B—Running winding bimetal. C—Starting winding bimetal. P—Power wire connection. R—Running winding connection. S—Starting winding connection. At top, starting circuit is open, unit is operating on running winding only. At bottom, both circuits are open and have been drawing too much current.*

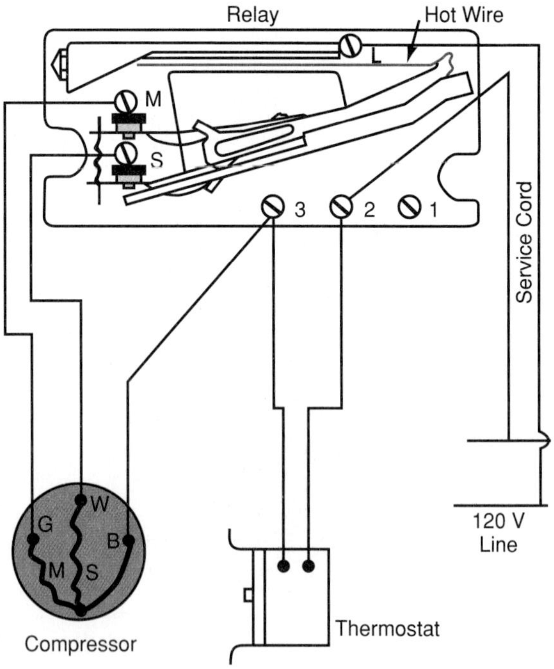

Figure 8-66. *Circuit diagram for thermal (hot wire) relay. Current draw will cause "hot" wire to heat and stretch. Slight heating will cause starting points at S to open. Further heating by too-heavy current will cause points at M to open and stop unit.*

contact points. The complete wiring diagram for a refrigerator using a thermal relay is shown in **Figure 8-67.**

8.12.4 Solid-State Electronic Relays

Relays using solid-state transistors, diodes, silicon-controlled rectifiers, diacs, and triacs are now used to control starting of hermetic motors. Changes in voltage in the motor, as it starts and then gathers speed, are used to open the starting winding circuit at the correct time. These relays are not as sensitive to the size of the motor as other relays. The same solid-state unit can be used for motors varying from 1/12 hp to 1/3 hp.

The solid-state positive temperature coefficient resistor is also called a PTCR. It has a thermally operated overload protector and capacitor. See **Figure 8-68A.** Note that the wiring diagram indicates a compressor start circuit. When this circuit is first energized, the solid-state relay has low resistance. The resistance is 3 to 12 ohms (Ω). Therefore, the running and starting windings are used to start the compressor. The run capacitor is bypassed by the PTC relay. It has no function during the compressor starting. See **Figure 8-68B.** When the solid-state relay reaches a sufficient temperature, it changes to a very high resistance. At a resistance of 10 kΩ to 20 kΩ, it switches off the starting windings. The run capacitor is now in series with the starting windings. The compressor run circuit is shown in **Figure 8-68C.**

Servicing: Checking and Testing Relays

In general, relays should be replaced, not repaired. The service technician's job is mainly to determine if the relay is defective. If so, it is replaced with an exact duplicate. **Figure 8-69** shows the wiring connections to three types:

- Klixon magnetic.
- General Electric magnetic.
- Delco hot wire.

The wire size, the contact point area, and the spring tension or weight plus the air gaps, must be accurately set for each unit. A slight difference in weight or spring tension, for example, might result in a difference of 100 motor revolutions.

The most effective way to determine if the relay is causing the trouble is to first check the other parts of the circuit. Check the motor, the capacitor, the overload cutout, and the thermostat. Only if these parts test all right should the relay be replaced.

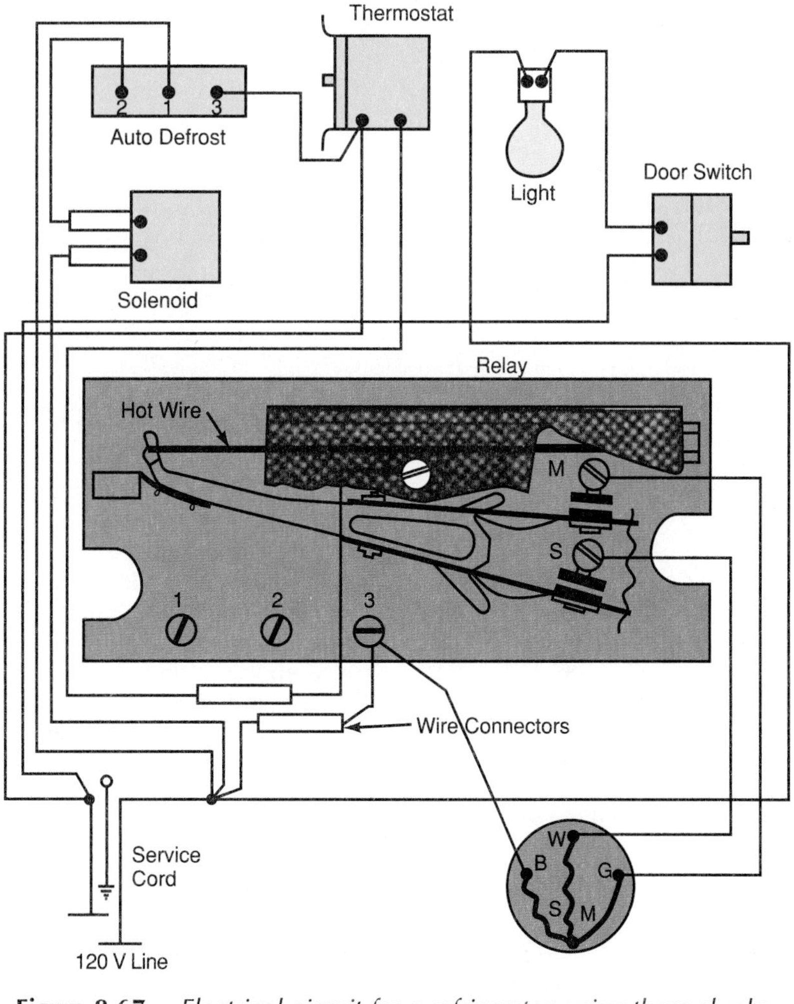

Figure 8-67. *Electrical circuit for a refrigerator, using thermal relay.*

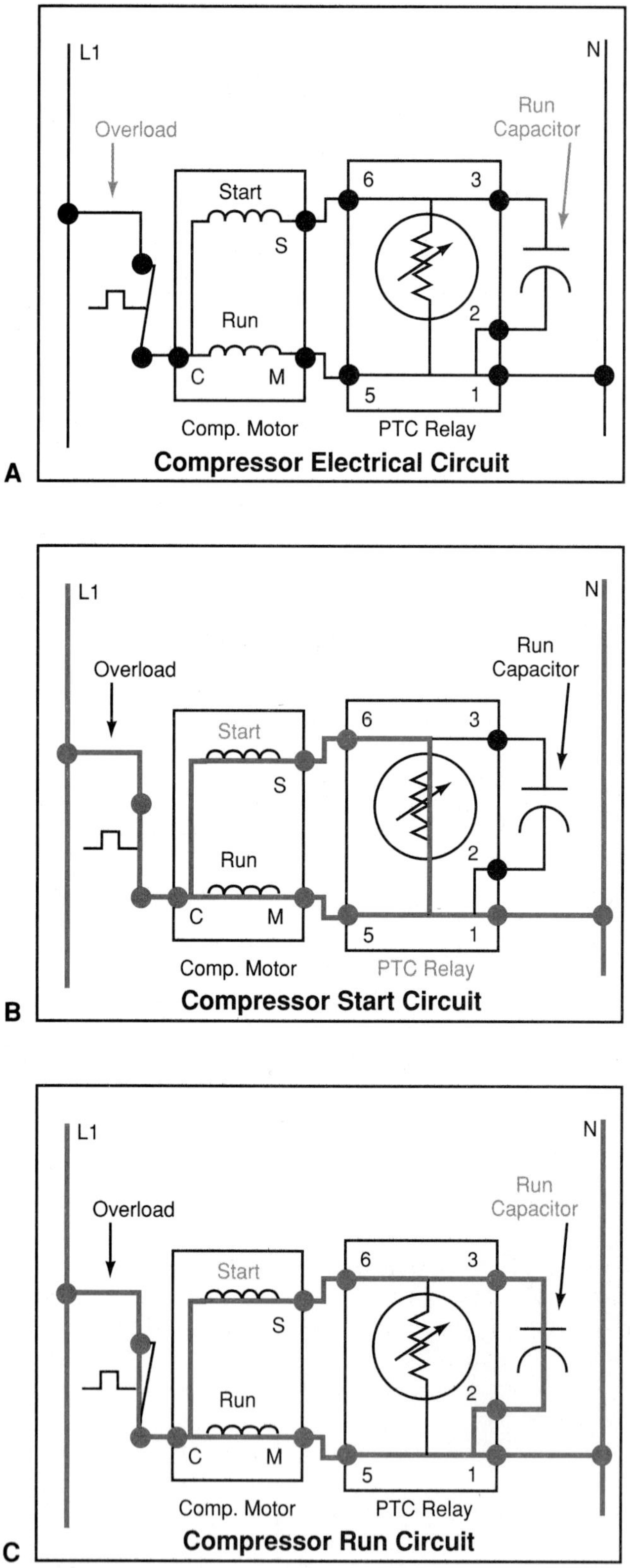

Figure 8-68. *Solid-state positive temperature coefficient resistor (PTCR) with a thermally operated overload protector and a run capacitor. (Frigidaire Company)*

The tester shown in **Figure 8-70** will check out either a current relay or potential relay. It will also check the relay for line voltage. The procedure is simple and quickly done. Connect the numbered leads from the numbered jacks on the tester to the same numbered terminals on the relay. Test on the 120 V switch before using the 208 V to 240 V switch.

Keep the relay cover in place; never discard it. Dust that collects on the contact points will quickly cause them to burn. This results in excessive voltage drop across the points and a control that works poorly.

The weight-type amperage relay must be mounted in a straight up-and-down position. Otherwise, the plunger will rub and bind against the sides of the relay body. The relay should open in about three seconds if it is working correctly. When replacing the relay, it is important to disconnect the power supply. Use the correct size screwdriver. Label each wire as it is disconnected (the terminals are usually numbered). A tag or clip on each wire with the corresponding number makes it easier to connect the new relay. Masking tape and a marking pencil are useful for such labeling.

If an exact potential relay replacement is not available, use one of a lower voltage rating (90% of rating). A relay with a higher voltage rating should not be used. Capacitors can discharge at 300 V or higher. Potential relay points may be burned (fused) by this discharge if the unit short-cycles. To eliminate this trouble, use capacitors equipped with resistors across the capacitor terminals or use a time delay switch to prevent the short-cycling.

Frequently a unit will short-cycle because the thermostat is exposed to vibration (on a shaky wall or a stairs). Be sure to mount the thermostat to something solid. Avoid tapping a relay to check it. Such tapping may cause the points to touch. This brief contact may ruin the points and damage the motor. The relay must function correctly without being tapped or it should be replaced.

8.13 Automatic Defrost Controls

A number of refrigerators have a standard-temperature section and a frozen foods section in the cabinet. These dual-purpose cabinets need a special series of motor controls. First, the controls must give correct temperatures in both sections. Second, they must provide completely automatic defrost. One type of control is shown in **Figure 8-71.** The wiring diagrams for it are shown in **Figure 8-72.**

The basic means of controlling the defrosting interval are:

- An electric timer that defrosts the unit at certain time intervals.
- A device that defrosts the units based on the number of times the refrigerator door is opened.
- A clock that runs only when the unit is running. It keeps the motor off to defrost after a predetermined number of hours of running time.

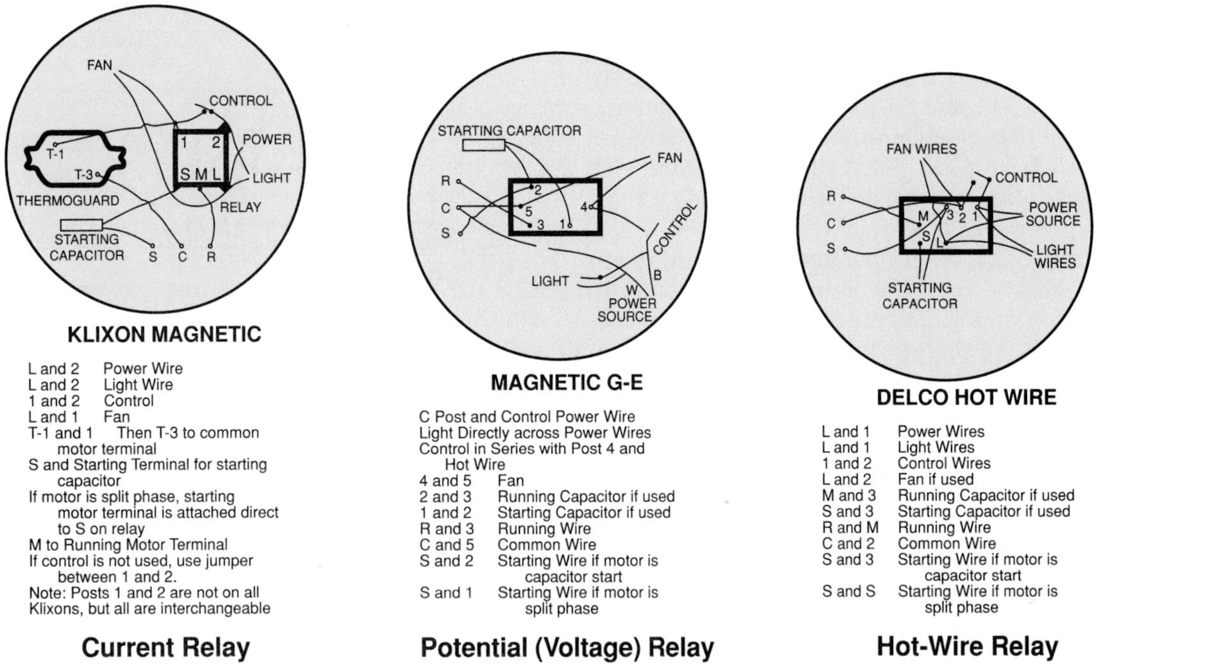

FAN
CONTROL
POWER
LIGHT
T-1
T-3
1 2
S M L
THERMOGUARD
RELAY
STARTING CAPACITOR S C R
KLIXON MAGNETIC
L and 2 Power Wire
L and 2 Light Wire
1 and 2 Control
L and 1 Fan
T-1 and 1 Then T-3 to common motor terminal
S and Starting Terminal for starting capacitor
If motor is split phase, starting motor terminal is attached direct to S on relay
M to Running Motor Terminal
If control is not used, use jumper between 1 and 2.
Note: Posts 1 and 2 are not on all Klixons, but all are interchangeable
Current Relay
STARTING CAPACITOR
FAN
R
C
S
CONTROL
LIGHT
W
B
POWER SOURCE
MAGNETIC G-E
C Post and Control Power Wire
Light Directly across Power Wires
Control in Series with Post 4 and Hot Wire
4 and 5 Fan
2 and 3 Running Capacitor if used
1 and 2 Starting Capacitor if used
R and 3 Running Wire
C and 5 Common Wire
S and 2 Starting Wire if motor is capacitor start
S and 1 Starting Wire if motor is split phase
Potential (Voltage) Relay
FAN WIRES
CONTROL
POWER SOURCE
LIGHT WIRES
STARTING CAPACITOR
DELCO HOT WIRE
L and 1 Power Wires
L and 1 Light Wires
1 and 2 Control Wires
L and 2 Fan if used
M and 3 Running Capacitor if used
S and 3 Starting Capacitor if used
R and M Running Wire
C and 2 Common Wire
S and 3 Starting Wire if motor is capacitor start
S and S Starting Wire if motor is split phase
Hot-Wire Relay

Figure 8-69. *Circuit diagram for three starting relays.*

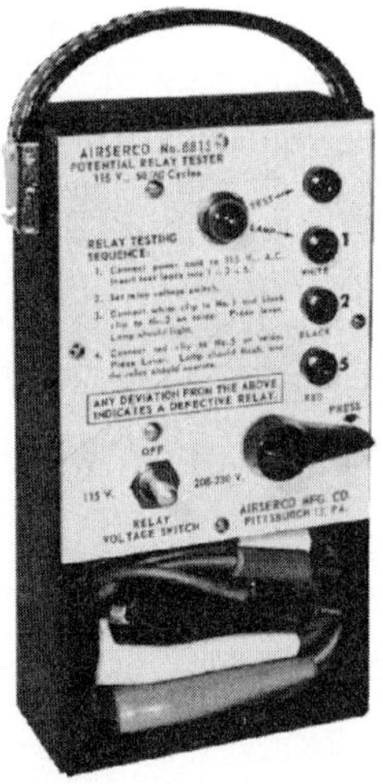

Figure 8-70. *Motor starting relay tester will test both current and potential type relays. It may be used on both 120 V and 240 V circuits.*

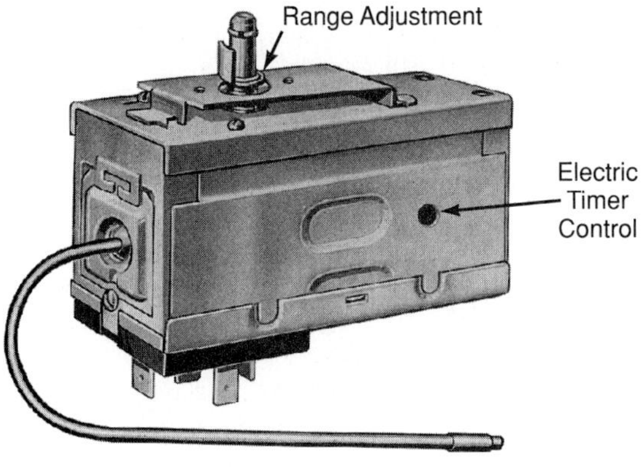

Figure 8-71. *Fully automatic defrost control.*

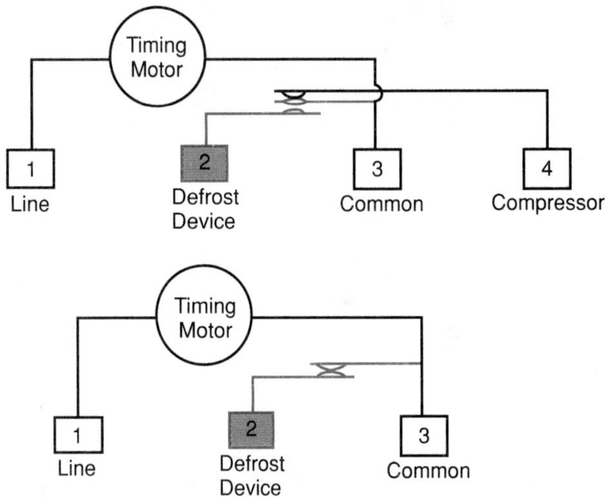

Figure 8-72. *Wiring diagram of automatic defrost control. Control terminals are labeled. Above, electric defrost system breaks motor circuit during defrost time. Lower diagram is for hot-gas defrost system, which has compressor running during defrost cycle.(Ranco North America)*

- A no-frost system that uses forced convection evaporators and defrosts these evaporators during each off-portion of the operating cycle. Either hot-gas or electric heating elements are used.

For example, a defrost system that uses a timer to operate the defrost cycle every six hours works this way:

1. It shuts off the compressor and the evaporator fans and starts the electric heaters. The heaters will be on for about 16 minutes.

2. It then shuts off the electric heaters and starts the compressor.
3. Evaporator fans start after compressor has run about four minutes. The unit returns to normal operation.

Figure 8-73 shows such a timer. A wiring diagram for a three-step defrost method is shown in **Figure 8-74.** A thermostat controlling the on-and-off circuit for the electric heater elements is shown in **Figure 8-75.**

One control has a timer-operated cam, as shown in **Figure 8-76.** This returns the refrigerating circuit to normal operation after a certain time interval. This occurs even though the thermostat may not call for cooling. This device prevents too long a defrost interval and also serves as a safety device.

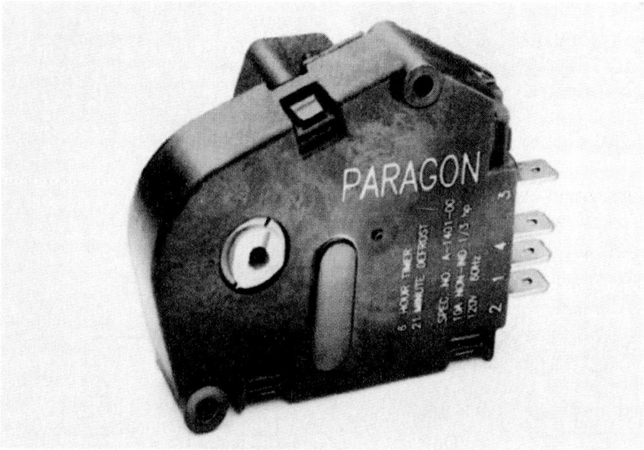

Figure 8-73. *Defrost timer used to operate defrost cycle in three-step no-frost refrigerator. Note connecting terminals. (Sealed Unit Parts Co., Inc.)*

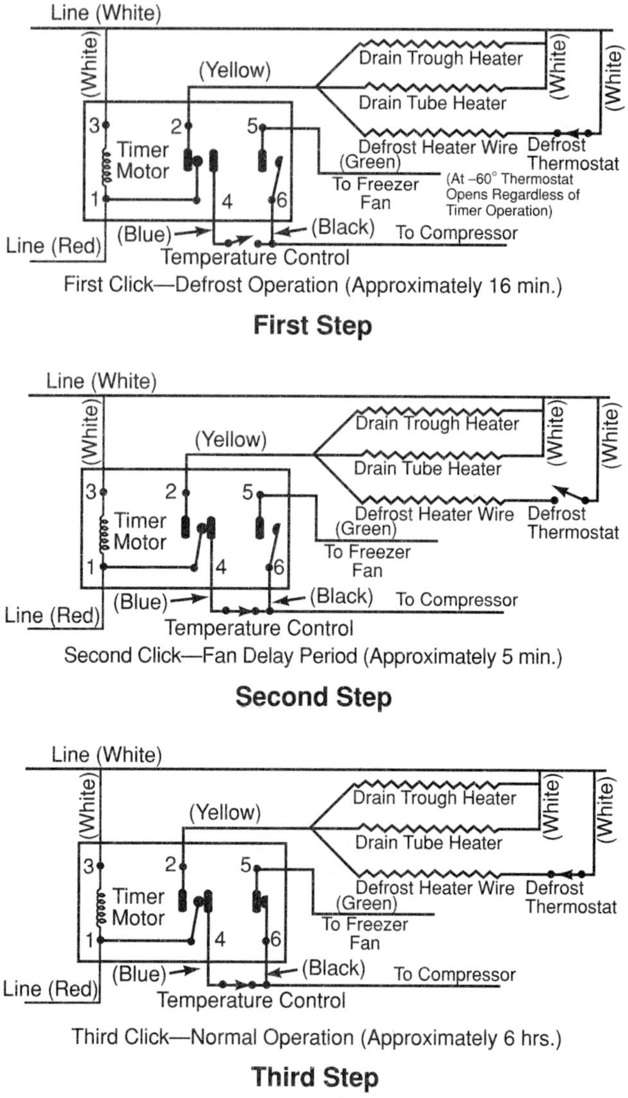

Figure 8-74. *Electric circuits used in the three-step defrost method. First step—Timer stops compressor and freezer fan, then closes circuit to three heaters. Second step—Timer stops heater and starts compressor, but freezer fan does not start. Third step—Freezer fan circuit closes.*

8.14 Semiautomatic Defrost Controls

Some domestic refrigerators use semiautomatic defrosting controls. These devices do two things:

- Defrost the unit when the owner presses the button.
- Return to regular operation automatically after the unit has defrosted.

A second system raises the range a fixed amount when a button is pressed. The evaporators will run at a temperature warm enough to permit defrosting. However, they will still give satisfactory refrigeration. Pulling on the button returns the system to regular operation.

Refrigerators with manual defrost frequently use a double capillary tube control, **Figure 8-77.** This control has a power element for normal cycling. It has another element as a cut-in control for the defrost. Defrosting starts when the control knob is pushed in. This movement opens the motor circuit and closes the circuit to either a defrost solenoid (hot-gas system) or to electric heater elements.

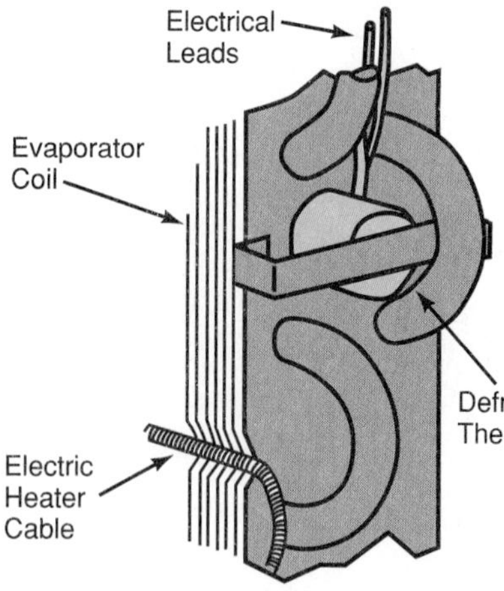

Figure 8-75. *Bimetal defrost thermostat. Control closes at 20°F (–6.7°C) and opens at 50°F (10°C) during defrost.*

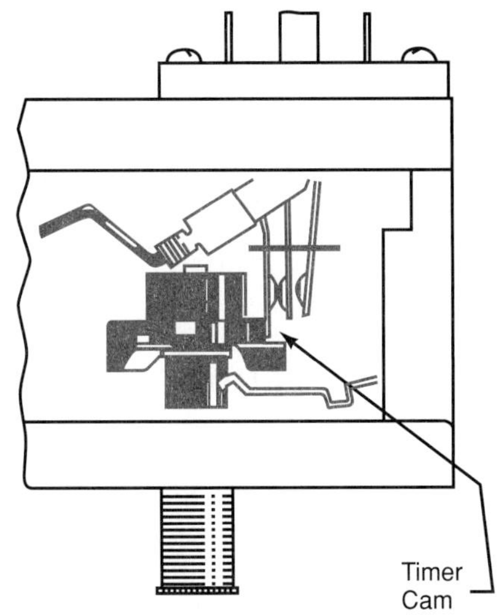

Figure 8-76. *Timer-operated cam which returns unit to normal cycling after about 45 minutes.*

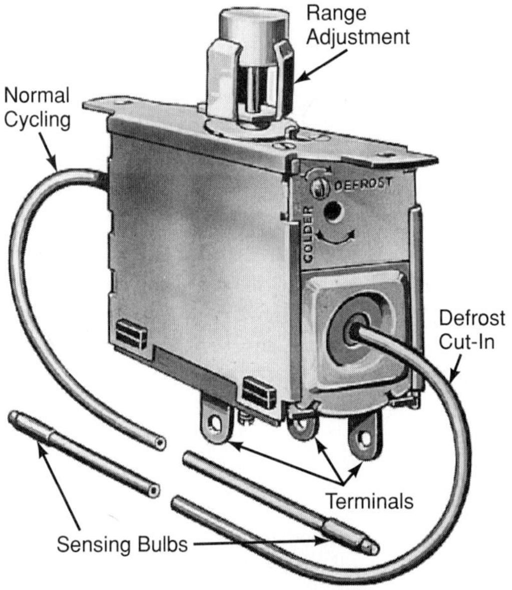

Figure 8-77. *Combination temperature and defrost control.*

When the coils are defrosted, the defrost capillary tube will create enough bellows pressure to open the defrost circuit. Once again, the control connections will return to the motor circuit.

A wiring diagram of how this type of control is connected to a hot-gas defrost system is shown in **Figure 8-78.** The circuits used for an electric defrost system are shown in **Figure 8-79.**

8.15 Hot-Gas Defrost Controls

The hot-gas method of defrosting uses a solenoid valve to open and close the bypass from the compressor discharge to the evaporator. **Figure 8-80** shows the

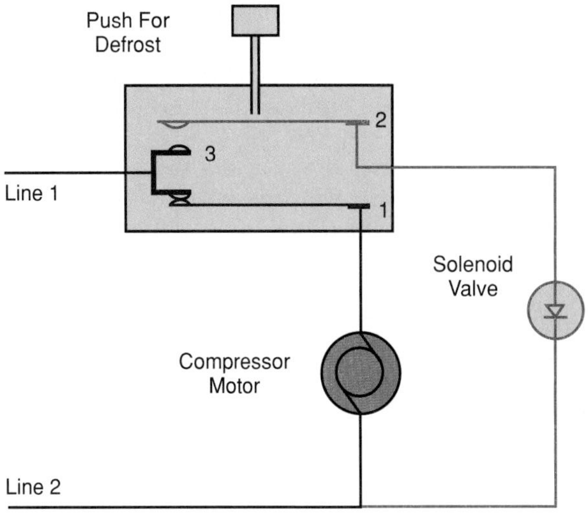

Figure 8-78. *Wiring diagram for "hot-gas" defrosting device which uses combination control. Control is shown in refrigerating position. Defrost system has solenoid valve that controls flow of hot gas.*

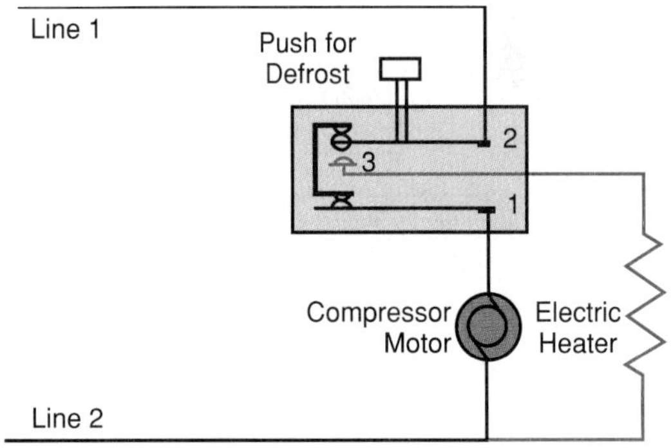

Figure 8-79. *Wiring diagram for electric heater defrost system using combination control. Control is shown in refrigerating position. Heater defrost circuit is shown in red.*

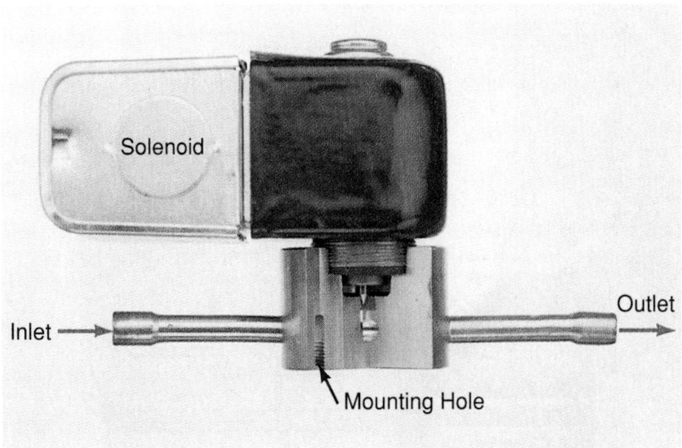

Figure 8-80. *Solenoid valve which may be used with thermostats for either hot-gas defrosting or secondary system control. (Alco Controls Division, Emerson Electric Company)*

valve equipped with connector lines. It is similar to the solenoid valves used to control refrigerant flow in secondary systems of two-temperature refrigerators. The valve must be mounted in a vertical position to work correctly.

Figure 8-81 shows the solenoid valve installed in a hot-gas defrost system. **Figure 8-82** is an installation in a bypass refrigeration system.

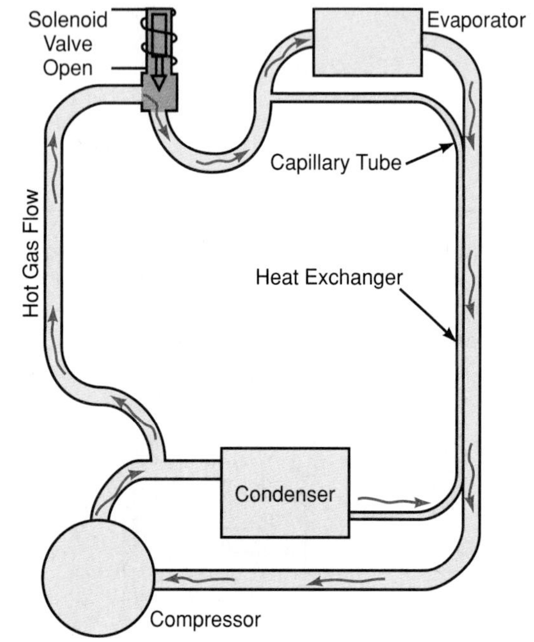

Figure 8-81. *Hot-gas defrost system using solenoid valve. Illustration shows valve open and hot gas passing from compressor to evaporator.*

Some refrigerators use two separate controls. One control is the regular thermostat, while the other control is the defrost control.

A single control is shown in **Figure 8-83.** It can be used with either the hot-gas or electric defrost systems. The control is designed with a vacuum in the sensitive element. If the bellows loses its charge, the pressure will rise and open the circuit (a fail-safe control).

8.16 Ice Bank Controls

Many types of coolers, vending machines, and other medium-temperature refrigeration systems use ice banks to provide reserve cooling capacity. During periods when the system cooling demand is low, a bank of ice is formed around the evaporator, **Figure 8-84.** When demand increases, the cooling offered by the ice assists the refrigeration system.

A typical system of this type is found in some drink dispensers. Ice is built up during evening or low-activity hours. It is then used to cool beverages during high-activity periods. Another use for this type of refrigeration would be in an air conditioning system used on a periodic basis. An example might be a system in a church building. Ice would be built up during the week and used to cool the building on the weekend.

A control is needed to limit the amount of ice that is formed. A conventional temperature control cannot distinguish between ice and water, since both can be present at 32°F (0°C). An ice bank control has been developed. It can determine the presence of ice by sensing a pressure change.

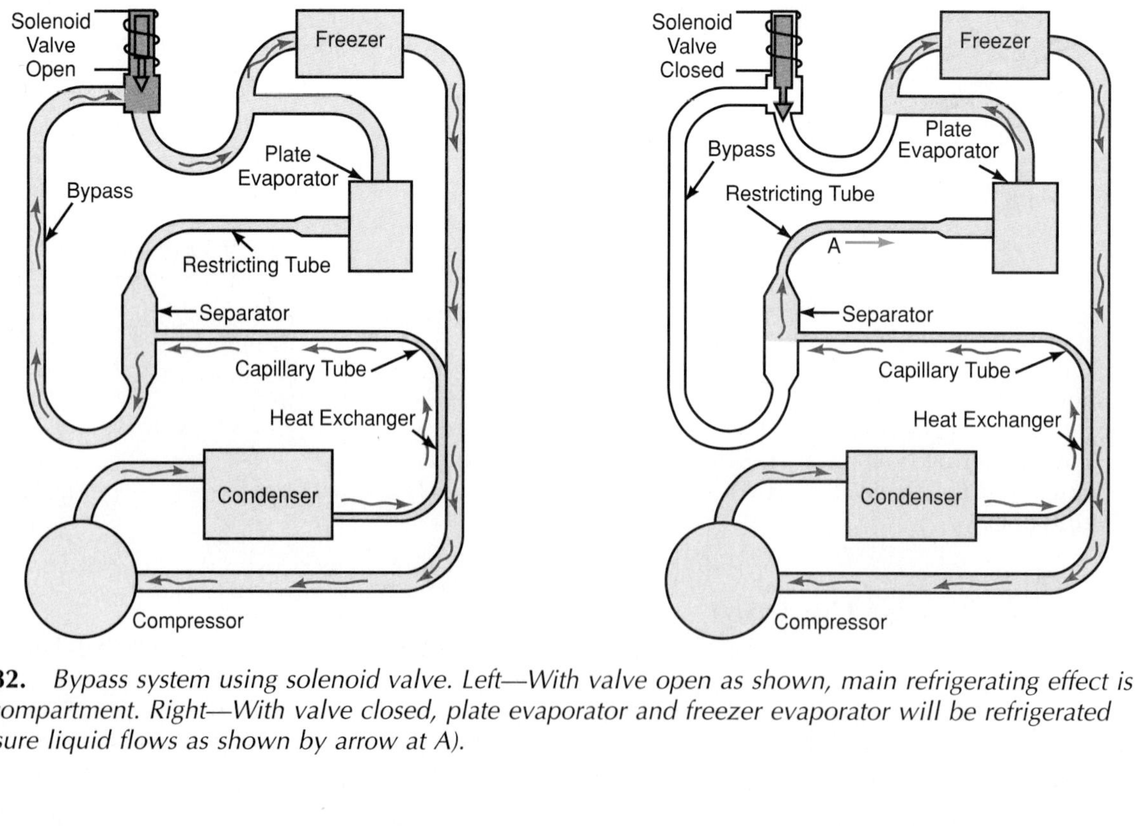

Figure 8-82. *Bypass system using solenoid valve. Left—With valve open as shown, main refrigerating effect is in freezing compartment. Right—With valve closed, plate evaporator and freezer evaporator will be refrigerated (low-pressure liquid flows as shown by arrow at A).*

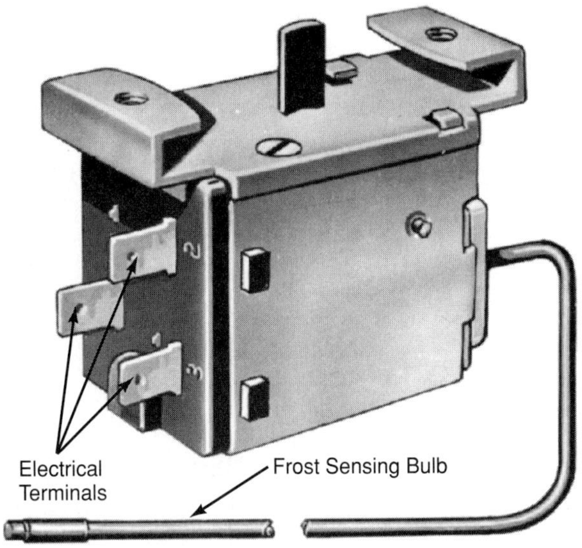

Figure 8-83. *Defrost control used in addition to main temperature control. (Ranco North America)*

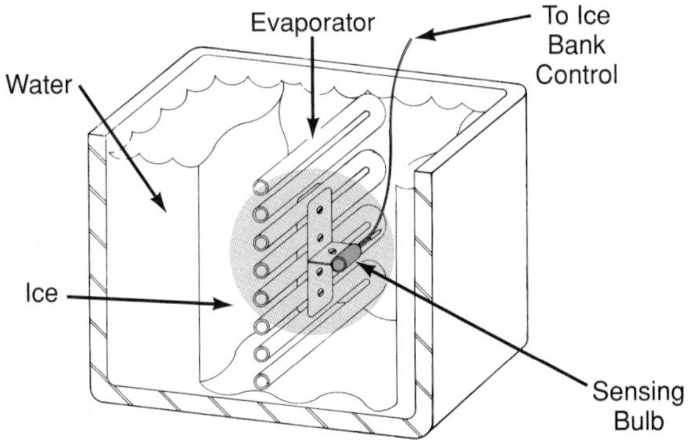

Figure 8-84. *Electric ice bank control. Bulb senses temperature of the ice, which can be below 32°F (0°C). If the temperature is below 32°F, ice bank control turns compressor off. If the temperature is at 32°F, the bulb senses the pressure increase due to ice crystals forming in bulb. Again, the control turns compressor off. (Ranco North America)*

This type of control uses a sensing bulb consisting of two compartments divided by a membrane. See **Figure 8-85.** One side contains water. When frozen, the water expands and flexes the membrane. The other side contains a liquid that transmits the membrane movement up to the head of the sensing bulb. This, in turn, operates a control switch. The switch turns off the compressor, resulting in the melting of some of the ice. The cycle for making and detecting ice can then begin again.

Figure 8-86 shows how an ice bank control is mounted to control the size of the ice bank. Ice forms approximately 1/8″ (3 mm) beyond the outermost mounting of the sensing bulb before the compressor is turned off.

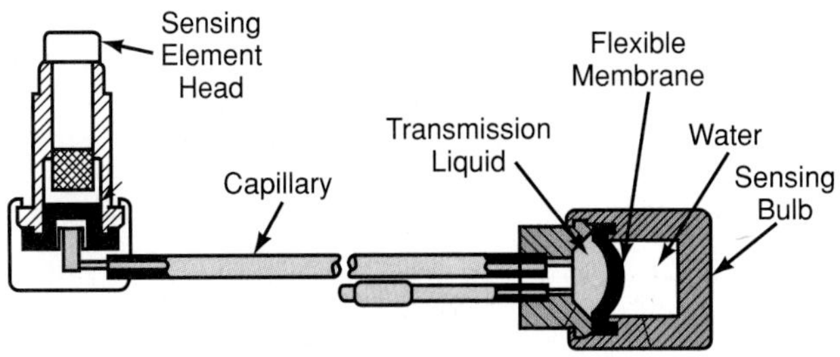

Figure 8-85. *Sensing element bulb. Note transmission liquid in the capillary tube. (Ranco North America)*

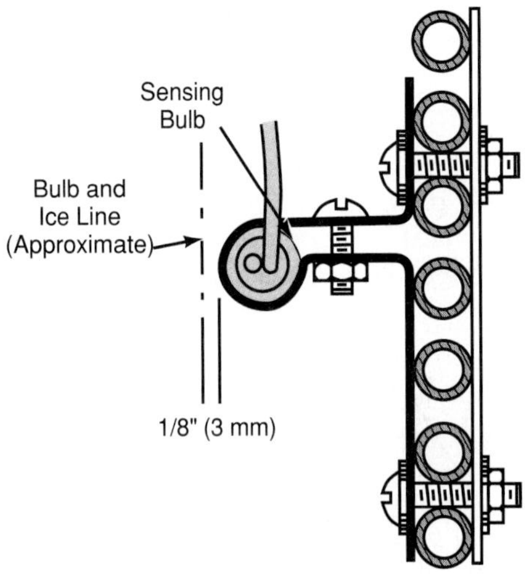

Figure 8-86. *Cross section of an evaporator with sensing bulb mounted. (Ranco North America)*

8.17 De-Ice Controls

Air-to-air heat pumps that have an outdoor coil sometimes develop an icing problem. Ice accumulates on the outdoor coil during the cold season when the heat pump is used as a heating unit. This ice reduces the heat flow into the outdoor coil. It also tends to block the airflow through the coil.

A special ***de-ice control*** is used to prevent ice accumulation (buildup). The heat pump de-ice control shown in **Figure 8-87** combines the actions of a timer and a thermostat. The timer mechanism will start a defrost cycle unless the thermostat has reacted to a certain temperature level at the outside air coil.

The timer is adjustable for coil defrost cycles of 30, 45, or 90 minutes. The sensitive bulb is cross-charged (see Section 5.1.2) to ensure constant bulb control. This control permits the defrost cycle only if the coil is at 26°F (−3.3°C) or below. If the coil is above this temperature, the defrost cycle is skipped until the next defrost interval. The sensitive bulb is usually mounted at the place where the ice last melts from the coil.

Figure 8-88 shows the sensitive bulb details. The temperature cut-in adjustment, the dimensions, and the

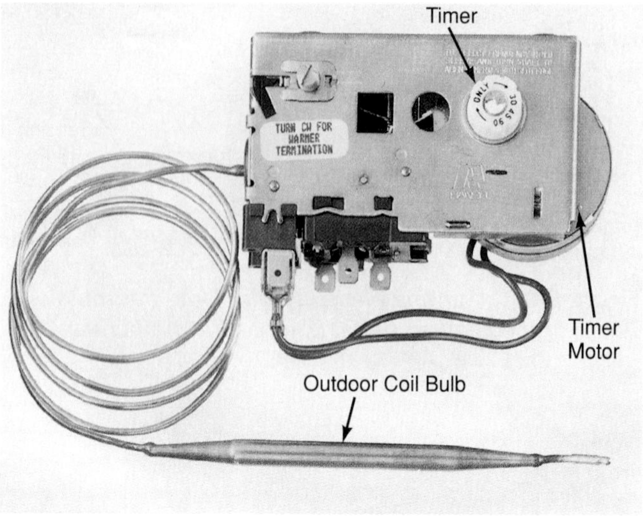

Figure 8-87. *Heat pump de-ice control. Control reverses heat pump cycle and defrosts outdoor coil each 30 to 90 minutes if the outdoor coil is at 26°F (−3.3°C) or below. (Ranco North America)*

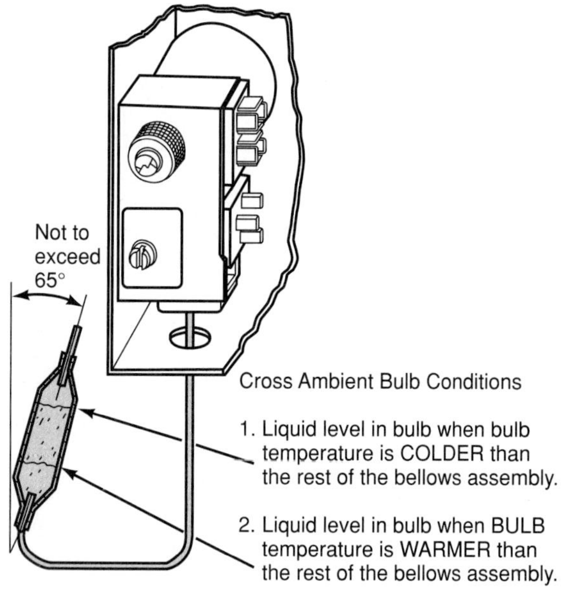

Figure 8-88. *De-icer sensitive bulb operation. Note how temperature-sensitive bulb is positioned.*

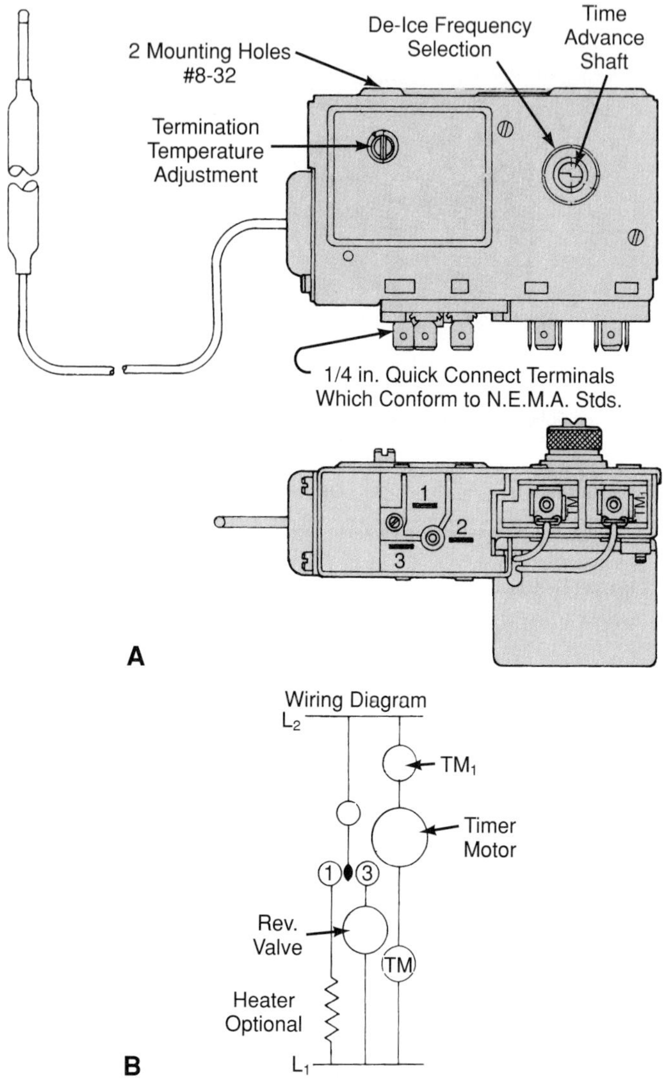

Figure 8-89. *De-icer control. A—Mechanisms of the control and its adjustments. B—Wiring diagram.*

wiring diagram are shown in **Figure 8-89.** Note that the defrost cycle reverses the heat pump. The outdoor coil temporarily acts as a condenser during the defrost cycle.

8.18 Humidity Controls

Humidity controls are used to control humidifiers and dehumidifiers. As humidity changes, the resistance value of the sensor in an electronic humidistat varies accordingly. The change in resistance is detected by a Wheatstone bridge circuit, as shown in **Figure 8-16.**

In humidifiers, the humidistat closes the circuit when humidity *decreases* (drops). See **Figure 8-90.** Chapter 21 explains the design and operation of humidifiers.

On a dehumidifier, the humidistat closes the circuit when the humidity *increases* (rises). See **Figure 8-91.** Chapter 22 explains dehumidifiers in more detail.

8.19 Defrosting Timers

Many companies have made defrosting clocks standard equipment on their domestic refrigerators (Section 8.13). Defrosting clocks ease the burden of the user. They also permit more efficient operation of the refrigerating unit.

Domestic refrigerator timers are built in accordance with the requirements of the manufacturer. Many companies are using electronic controls to meet energy standards. Another type of defrost timer, shown in **Figure 8-92,** is based on the adaptive defrost system rather than

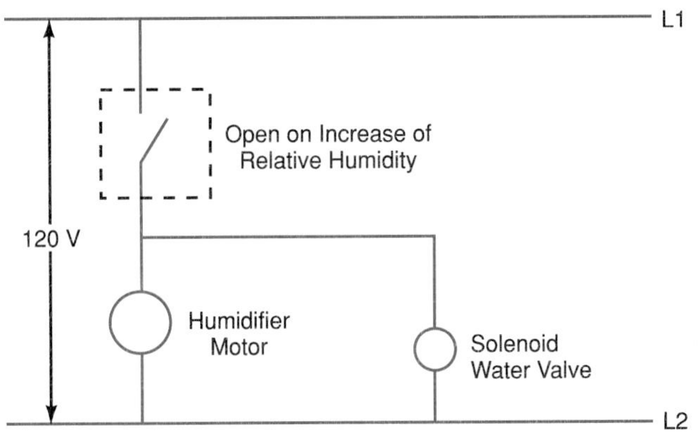

Figure 8-90. *Wiring diagram of humidifier circuit. Control is adjustable to maintain relative humidity between 20% and 80%.*

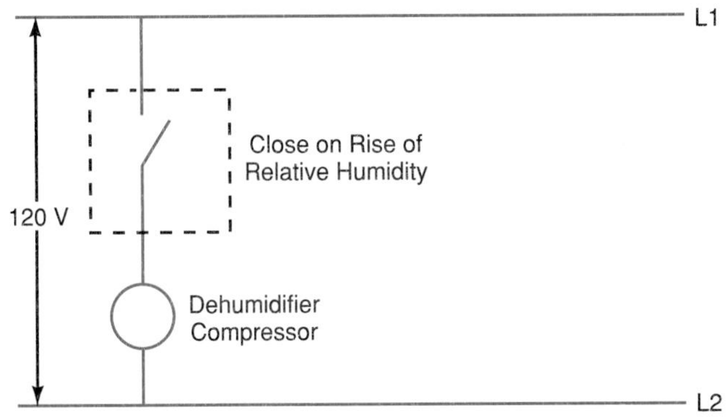

Figure 8-91. *Wiring diagram of dehumidifier circuit. Control is adjustable between 20% and 90% relative humidity.*

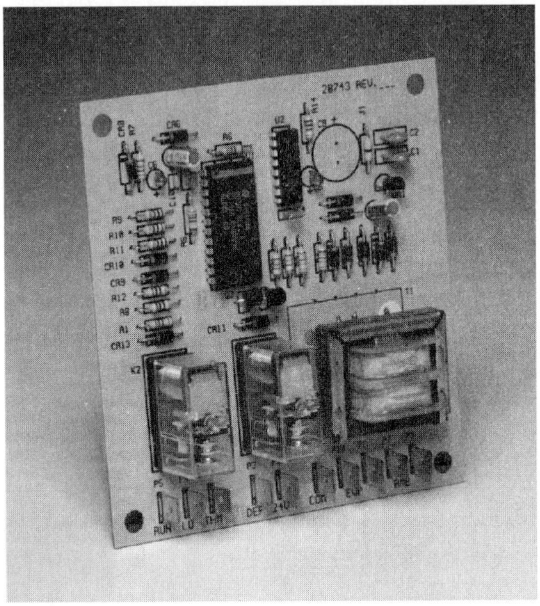

Figure 8-92. *An adaptive defrost timer is used on systems to provide energy savings. Defrost control system monitors variable operating conditions and adjusts necessary defrost interval accordingly. (Paragon Electric Company, Inc.)*

the straight time-based defrost. The adaptive defrost is designed to offer added energy savings to the owner. Such defrost timers are used on both hot-gas defrost systems and electric heating coil systems.

8.20 Review of Safety

Anything in motion, anything that holds back a pressure, anything that can conduct electricity or heat, anything that is rough or sharp, and anything that can be dropped is a potential safety hazard.

Most accidents are the result of carelessness. When a person is concentrating on getting a job done, he or she tends to momentarily neglect safety. Therefore, service technicians must train themselves to do things safely. They must study the job to identify safety problems and their solutions before starting. Think about the safety aspects before each step of a job.

Always disconnect the electrical power. Make sure no one can turn it on while working on the electrical parts of a system. Replace worn electrical wires or wires that have brittle insulation (brittle insulation cracks when the wire is bent into a loop).

Use only screwdrivers with insulated (wood or plastic) handles. Use only wrenches and pliers with insulated handles. This habit is double insurance against shocks.

If you must work in a damp or wet room, stand on a dry, insulated platform. Many technicians carry a 3′ by 3′ (0.9 m by 0.9 m) rubber mat to stand on when doing electrical service work.

The human body is made up of many fluids that have fairly high electrical conductivity. For this reason, the body's electrical resistance is too low to protect against the flow of electrical current. Most of the resistance to electrical flow is on the skin or surface of the skin. Electrical resistance is lowered further if the skin is wet, so a service technician is more apt to receive an electrical shock when perspiring than when dry. Rubber gloves, rubber-soled shoes, or rubber boots provide additional protection.

Before handling wires, terminals, capacitors, or other parts, always check for electrically charged circuits. Use instruments to check the circuit.

Always short-circuit any capacitors in the circuits before working on them. Use a 100 kΩ resistor across the capacitor terminals. This will discharge the capacitor.

Water pipes, radiators, heating plants, and the like are grounded. If the "live" (black) wire contacts a ground, a circuit breaker or a fuse will be blown. Someone touching both a live wire and a ground will receive a very severe shock.

8.21 Test Your Knowledge

Please do not write in this text. Place your answers on a separate sheet of paper.

ELECTRIC CONTROL CIRCUITS MODULE

1. Pressure and thermostatic motor controls are similar because they both operate from ______.
   A. changing pressures
   B. temperature
   C. a capillary tube
   D. All of the above.
2. Thermostatic motor control bulbs are charged with ______.
   A. R-12
   B. volatile liquid
   C. R-134a
   D. nitrogen
3. What controls the temperature difference between the cut-out and cut-in settings in a temperature control mechanism?
   A. Differential adjustment control.
   B. Range adjustment.
   C. Thermostatic adjustment.
   D. None of the above.
4. In a temperature control mechanism, what provides an increase or a decrease in the desired temperature?
   A. Differential adjustment control.
   B. Range adjustment control.
   C. Temperature control.
   D. None of the above.
5. What is a common color used for ground wire?
   A. Red.
   B. Black.
   C. Green.
   D. Any of the above.
6. What is the purpose of a ladder diagram?
   A. To identify actual location of specific diagrams.
   B. To help troubleshoot a system.
   C. To provide a standard electrical circuit diagram.
   D. All of the above.
7. How does a bimetal strip respond to an increase in temperature range?
   A. It expands.
   B. It contracts.
   C. It remains the same.
   D. It bends.
8. Which of the following is a type of mechanism used in motor control thermostats?
   A. Sensing bulb.
   B. Bimetal strip.
   C. Solid-state (with thermometer).
   D. All of the above.
9. What is the common size wire used in a domestic refrigerator?
   A. No. 12.
   B. No. 14.
   C. No. 16.
   D. No. 18.
10. What is the common size wire used for heavy duty air conditioners?
    A. No. 12.
    B. No. 14.
    C. No. 16.
    D. No. 18.

ELECTRIC CONTROLS MODULE

11. A motor may be operated by ______.
    A. low-side pressure
    B. high-side pressure
    C. Both A and B.
    D. None of the above.
12. Hermetic motor controls have starting relays to disconnect the starting winding.This occurs when the motor reaches ______ of its rated speed.
    A. 1/3
    B. 1/2
    C. 2/3
    D. 3/4
13. Identify the starting relay used on hermetic compressor systems.
    A. Current or potential (magnetic).
    B. Thermal.
    C. Solid-state electronic.
    D. All of the above.
14. What devices are built into motor controls to protect the motor from using too much current?
    A. Bimetal strips.
    B. Resistance heating unit.
    C. Thermistors with positive temperature coefficient (PTC).
    D. All of the above.
15. The current flow when a motor is starting is ______ when the motor is running.
    A. greater than
    B. less than
    C. the same as
    D. Either B or C.
16. Which of the following is a common starting relay?
    A. Current.
    B. Potential.
    C. Solid-state electronic.
    D. All of the above.
17. When a system is running, in what position are the potential relay contact points?
    A. Midway.
    B. Open.
    C. Closed.
    D. Any of the above, depending on the unit.

18. High head/condensing pressure may cause ______.
    A. an increase in the temperature of the vapor
    B. an increase in the temperature of the oil
    C. an increase in the formation of carbon, acids, and sludge
    D. Any of the above, depending on the unit.
19. If a relay is not operating, which of the following is *not* the proper procedure?
    A. Replacing the relay.
    B. Checking the motor.
    C. Checking the capacitor and overload cut-out.
    D. Checking the thermostat.
20. The term "central air conditioning" indicates that a system can provide ______.
    A. heating and cooling
    B. humidification and dehumidification
    C. electrostatic air cleaning
    D. All of the above.

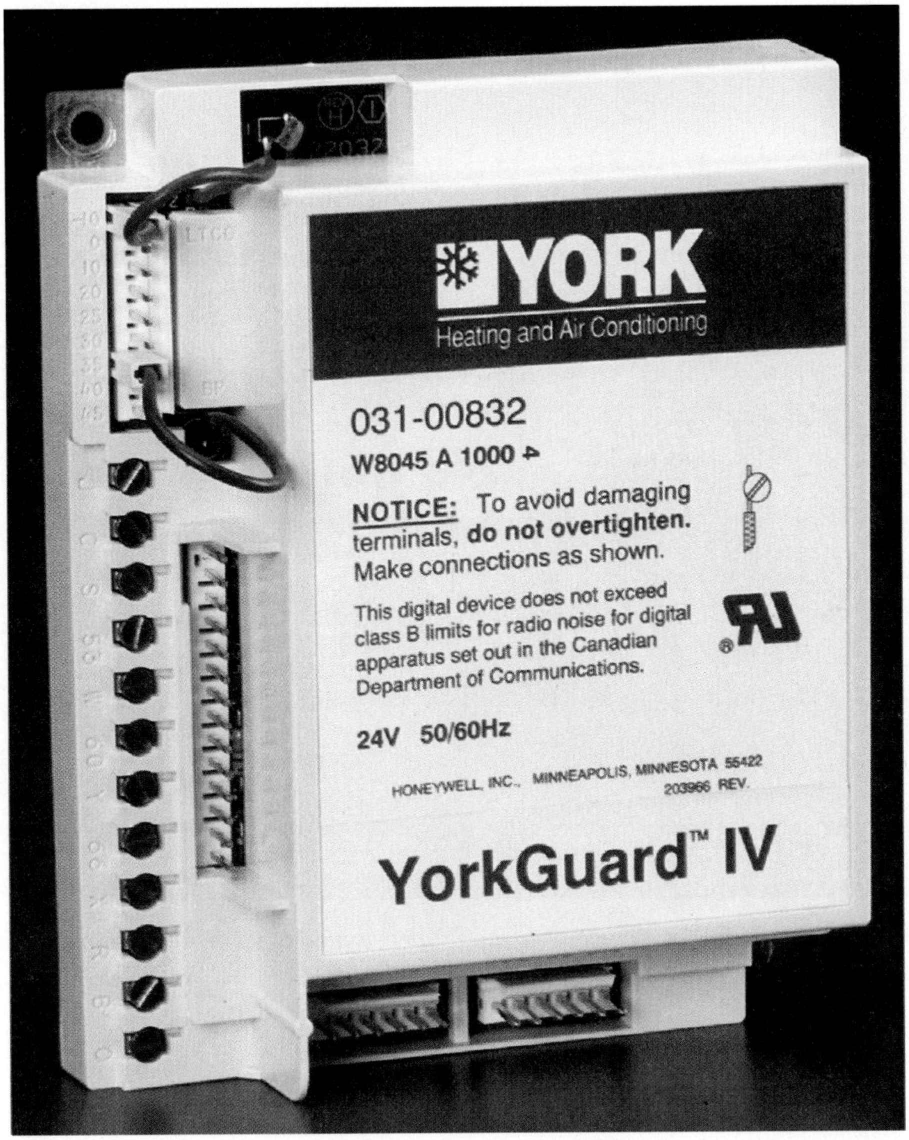

Microprocessor control used on heat pumps. This solid-state logic module controls the quantity and source of back-up heat and reduces the number of defrost cycles and shutdowns on operations in the event of a power loss or when pressures or refrigerant temperatures are abnormal. (York International Corp., Unitary Products Group)

Common refrigerant cylinders. Refrigerants and proper storage of them are covered in Chapter 9. (National Refrigerants, Inc.)

Chapter 9

REFRIGERANTS

Key Words:

azeotropic
chlorofluorocarbons (CFCs)
enthalpy
flammability
hydrochlorofluorocarbons (HCFCs)
hydrofluorocarbons (HFCs)
ozone
retrofitting
toxicity
zeotropic

Learning Objectives:

After studying this chapter, you will be able to:

- Understand the differences between CFCs, HCFCs, and HFCs.
- Correctly identify and classify common refrigerants by their numbers.
- List the necessary properties of refrigerants.
- Read a pressure-temperature curve and identify the proper refrigerant.
- Demonstrate ability to read pressure-enthalpy diagrams.
- Discuss properties of different refrigerants and their applications in a system.
- Demonstrate handling of refrigerant cylinders and identify color codes.
- **Follow approved safety procedures.**
- **Identify the safety procedures for using refrigerant cylinders.**

9.1 Refrigerants and the Ozone Layer

The word "ozone" has become a part of our everyday terminology. A very thin layer of the earth's upper atmosphere contains ozone. The ozone layer acts as a filter for the sun's ultraviolet rays. This protects human, plant, and sea life from the damaging effects of these rays.

Scientists have found that releasing chlorofluorocarbons (CFCs) from some refrigerants can harm the ozone layer. The CFCs destroy this protective layer of the earth's atmosphere. This concern has developed into what we refer to as the EPA (Environmental Protection Agency) regulations. These regulations identify the types of refrigerants that can be produced. They also regulate how the refrigerants will be used. For more information, see Chapter 10.

Most refrigerants commonly used today are classified into four areas:

- Chlorofluorocarbons (CFCs).
- Hydrochlorofluorocarbons (HCFCs).
- Hydrofluorocarbons (HFCs).
- Refrigerant blends (azeotropic and zeotropic).

9.1.1 Identifying Refrigerants by Number and Color Code

Refrigerants are identified by number. The number follows the letter R, which means refrigerant. This identifying system has been standardized by the American Society of Heating, Refrigerating and Air-Conditioning Engineers (ASHRAE). Refer to Chapter 31 for a summary of the ASHRAE numbering system. You should become familiar with refrigerant numbers, as well as with the names.

Refrigerant cylinders are often color coded to permit easy identification of the refrigerants they contain. This practice helps to prevent accidental mixing of refrigerants within a system. Always read the label and identify the refrigerant before using a cylinder. The color code shown is not a requirement for all manufacturers. Popular refrigerants, with their R-numbers and cylinder color codes, are shown in **Figure 9-1.** Cylinders for *recovered* refrigerants are gray with yellow ends.

Cylinder Color	Number	Refrigerant Name	Chemical Composition	General Application
Orange	R-11	Trichloromonofluoromethane	CFC	Used in centrifugal chillers for large applications
White	R-12	Dichlorodifluoromethane	CFC	Versatile, widely used in reciprocating and rotary-type equipment; household and industrial applications.
Light blue	R-13	Monochlorotrifluoromethane	CFC	Low-temperature refrigerant used in low stage of cascade systems.
Coral	R-13B1	Bromotrifluoromethane	CFC	Medium- to low-temperature applications with one or two stages of compression.
Light green	R-22	Monochlorodifluoromethane	HCFC	Residential, commercial, and industrial applications.
Light gray	R-23	Trifluoromethane	HFC	Low-temperature refrigerant to be used as replacement in low stage of cascade system.
Purple	R-113	Trichlorotrifluoroethane	CFC	Low capacity centrifugal chillers.
Dark blue	R-114	Dichlorotetrafluoroethane	CFC	Principally used with chillers for higher capacities.
Light gray	R-123	Dichlorotrifluoroethane	HCFC	Serves as a replacement for R-11 in centrifugal chillers.
Deep green	R-124	Chlorotetrafluoroethane	HFC	Medium-pressure refrigerant for chiller applications. Used in marine applications.
Medium brown or Tan	R-125	Pentafluoroethane	HFC	Substitute for use in low-temperature refrigeration applications.
Light (Sky blue)	R-134a	Tetrafluoroethane	HFC	Used in the automobile industry and refrigeration systems in residential, commercial, and industrial applications.
Coral red	R-401A	R-22 + R-152a + R-124	Zeotropic	Substitute for use in most medium-temperature systems.
Mustard yellow	R-401B	R-22 + R-152a + R-124	Zeotropic	Used in transport refrigeration equipment and domestic and commercial refrigerators.
Blue-green (Aqua)	R-401C	R-22 + R-152a + R-124	Zeotropic	Replacement refrigerant in mobile air conditioning.
Light brown	R-402A	R-22 + R-125 + R-290	Zeotropic	Ice machines, food service, vending, supermarket.
Green-brown	R-402B	R-22 + R-125 + R-290	Zeotropic	Supermarket, transport, food service.
Orange	R-404A	R-125 + R-143a + R-134a	Zeotropic	Medium and low temperature applications.
Light gray-green	R-406A	R-22 + R-142b + R-600a	Zeotropic	Used for R-12 retrofit.
Bright green	R-407A	R-32 + R-125 + R-134a	Zeotropic	Used for R-502 retrofit.
Cream	R-407B	R-32 + R-125 + R-134a	Zeotropic	Used for R-502 retrofit.
Chocolate brown	R-407C	R-32 + R-125 + R-134a	Zeotropic	R-22 replacement
Rose	R-410A	R-32 + R-125	Zeotropic	Replacement refrigerant in residential air conditioning applications.
Yellow	R-500	Refrigerants 152A/12	Azeotropic	Used with reciprocating compressors in industrial and commercial applications.
Light purple	R-502	Refrigerants 22/115	Azeotropic	Supermarket freezers and refrigerated cases.
Aquamarine	R-503	Refrigerants 23/13	Azeotropic	Used in low stage of cascade-type systems.
Teal	R-507	Refrigerants 125/143a	Azeotropic	Replacement refrigerant for low-temperature commercial refrigeration applications.
Silver	R-717	Ammonia	Inorganic Compound	Used in large reciprocating compressors and absorption-type systems.

Chlorofluorocarbons = CFC Hydrofluorocarbons = HFCs Hydrochlorofluorocarbons = HCFCs

Figure 9-1. *The most commonly used refrigerants, with their color codes and typical applications.*

9.1.2 CFC Refrigerants

The first halogen-based refrigerants (fluorinated hydrocarbons) were developed over sixty years ago. These refrigerants are composed of chlorine, fluorine, and carbon, and are called ***chlorofluorocarbons (CFCs).***

These refrigerants are low in toxicity, noncorrosive, and compatible with other materials. They are not flammable or explosive, but sizable quantities must not be released where there is a flame or electric heating element. Heat can cause them to break down into their elements, causing harm to human tissue. They are particularly harmful to the respiratory system. Common CFC refrigerants are:

- R-11
- R-12
- R-113
- R-114
- R-115

CFCs are thought to be one of the major causes of ozone depletion. By international agreement, they have not been manufactured since 1995. However, they are still widely used in existing residential units.

Due to laws forbidding release of CFCs to the atmosphere, new procedures and equipment have been developed. These are used to recover, recycle, and reclaim refrigerants containing CFCs. For further information regarding the EPA Act that governs the use of refrigerants, see Chapter 10.

9.1.3 HCFC Refrigerants

Hydrochlorofluorocarbons (HCFCs) are molecules composed of methane or ethane in combination with a halogen. This makes up a new molecule that is considered to be partially halogenated.

The HCFCs have shorter lives and cause less ozone depletion than the fully halogenated CFCs. Therefore, they have reduced potential for global warming. HCFCs such as R-22 and R-123 are considered to be interim refrigerants. They will be used until suitable replacements are available. The EPA requires the phaseout of HCFCs by the year 2030.

9.1.4 HFC Refrigerants

Hydrofluorocarbons (HFCs) include such refrigerants as R-134a and R-124. They are different from chlorofluorocarbons—they contain one or more hydrogen atoms and no chlorine atoms. HFCs are considered to have zero potential for ozone depletion. They have only a slight effect on global warming.

R-134a is typically used in new systems that are specifically designed for its use. The concept that R-134a is an easy replacement for R-12 is not correct, however. When using R-134a in retrofitting a system, numerous items must be considered. (***Retrofitting*** is the updating of an existing system to new standards.) R-134a refrigerants will not readily mix with mineral oils or alkylbenzene lubricants. Synthetic oils must be used for lubrication of hydrofluorocarbons; existing oils must be replaced.

The use of the proper recovery unit is necessary for the removal of R-12. There are also a number of other factors to be considered. These include system performance, hardware changes, and existing material and lubricant compatibility. Prior to retrofitting a system, the technician should always check with the manufacturer to be certain that it is proper.

9.1.5 Refrigerant Blends (Azeotropic—Zeotropic)

Another more recent category is that of refrigerant ***blends,*** commonly referred to as "azeotropic" and "zeotropic." The use of refrigerant blends is increasing. ***Azeotropic*** blends do not change or separate in composition when used in refrigeration systems. ***Zeotropic*** refrigerants are also blends comprised of various refrigerants. When used in a refrigeration system, their volumetric composition and saturation temperature do change.

See Sections 9.5.5 and 9.5.6 for more information on these refrigerants.

9.2 Requirements for Refrigerants

A fluid used as a refrigerant should have certain properties:

- It must follow the standards set forth by the EPA (Environmental Protection Agency).
- It should be nontoxic (not harmful if inhaled or spilled on the skin) and nonpoisonous.
- It should be nonexplosive.
- It should be noncorrosive.
- It must be nonflammable.
- It should make leaks easy to detect and locate. (See Chapter 12 for leak detection methods and use of instruments.)
- It should operate under low pressure (have a low boiling-point).
- It should be a stable gas.
- It should permit refrigerator or compressor parts moving in the fluid to be easily lubricated.
- It should have a high liquid volume per pound to provide durable refrigerant controls.
- It should have a high latent heat per pound to produce good cooling effect per pound of vapor pumped.
- It should have a low vapor volume per pound. This will reduce compressor displacement needed.
- It should have as little pressure difference as possible between evaporating pressure and condensing pressure. This increases pumping efficiency.
- It should meet the requirements of all current EPA rules and regulations.

Normal pressures in the refrigeration system should be kept as close to atmospheric pressure as possible. Excessive differences may cause leaks, overwork the compressor, and decrease the efficiency of the valves.

The standard comparison of refrigerants, as used in the refrigeration industry, is based on specific evaporating and condensing temperatures. The evaporating temperature is 5°F (−15°C) and the condensing temperature is 86°F (30°C). In this chapter, each refrigerant discussed is compared on this basis.

9.3 Use of Pressure-Temperature Curves

The pressure-temperature curves in **Figures 9-2** and **9-20** show the refrigerants in a state of equilibrium. Refer to **Figure 1-28A.** The vertical scale is temperature in °F; the horizontal scale is pressure in psig.

The pressure of the refrigerant at any particular temperature can be found by using these curves. Read horizontally from the temperature reading until the curve of the particular refrigerant is reached, then move directly down to the pressure reading.

For example, the vapor-pressure of R-12 refrigerant at a temperature of 100°F (38°C) is 132 psia (911 kPa) or 117 psig (807 kPa). The temperature is always the temperature of the refrigerant. The same curve may be used to determine both the condensing and evaporating temperatures and pressures. The condensing values for both temperature and pressure are higher than the equilibrium state.

When using this chart, keep several things in mind:

- The temperature of the refrigerant in the evaporator is about 8°F to 12°F (4°C to 7°C) colder than the evaporator when the compressor is running.

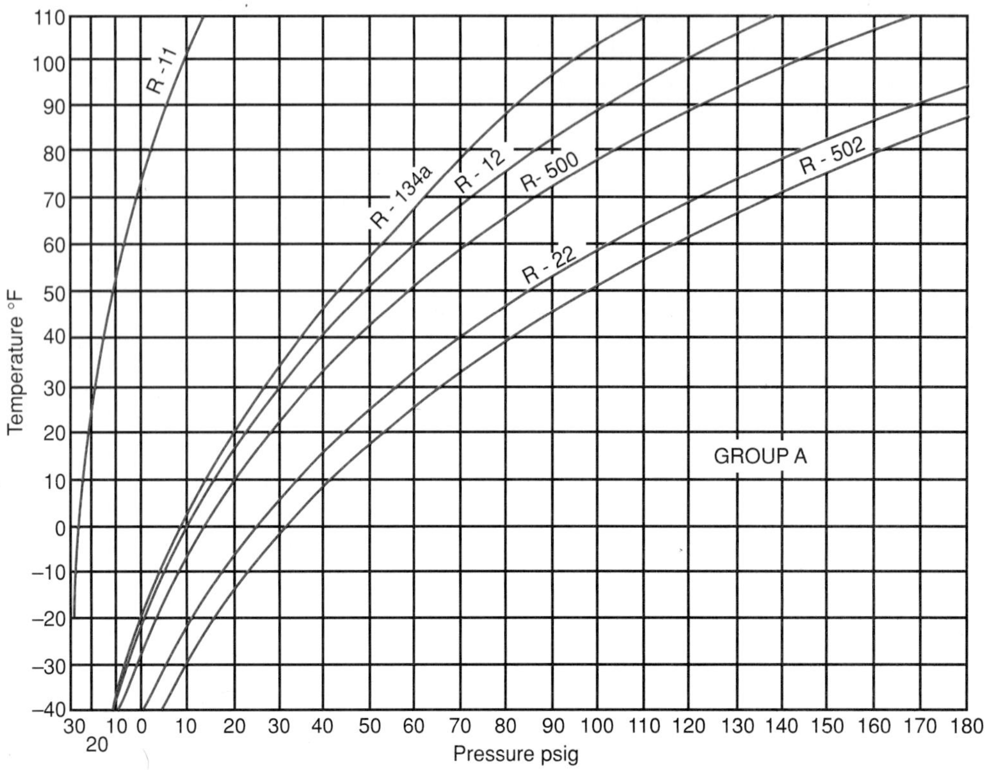

Figure 9-2. *Pressure-temperature curves for popular Group A refrigerants. R-11 is called a low-pressure refrigerant; R-502 is a medium-pressure refrigerant.*

- The temperature of the refrigerant in the evaporator is the same as the evaporator temperature when the compressor is not running.
- The temperature of the refrigerant in an air-cooled condenser is approximately 30°F to 35°F (17° C to 19°C) warmer than the room temperature.
- The temperature of the refrigerant in a water-cooled condenser is approximately 20°F (11°C) warmer than the water temperature at the drain outlet.
- The temperature of the refrigerant in the condenser will be about the same as that of the cooling medium after the unit has been shut off for 15 to 30 minutes.

9.3.1 Standard Evaporator and Condenser Temperatures

In domestic refrigerators, the service technician usually adjusts the controls using a temperature of 5°F (−15°C) in the evaporator. Also, the controls are adjusted to give a standard condensing temperature of 86°F (30°C). A cycle with an 80°F temperature difference is shown in **Figure 9-3.**

9.4 Grouping and Classification of Refrigerants

Refrigerants have been cataloged by various organizations. They have arrived at similar conclusions as to toxicity and flammability levels of a refrigerant. ***Toxicity*** is the ability of a refrigerant to be harmful or lethal with acute or chronic exposure. This exposure may be by contact, inhalation, or ingestion.

Some of these organizations are the ASHRAE (American Society of Heating, Refrigerating and Air-Conditioning Engineers), HMIS (Hazardous Material Identification System), NFPA (National Fire Protection Association), NRSC (The National Refrigeration Safety Code), and NBFU (National Board of Fire Underwriters).

The ASHRAE toxicity classification of refrigerants is indicated by assigning a letter, A or B. Class A refrigerants are those that have not been identified as having a toxicity level. Class B refrigerants have a toxicity level that has been identified. The standard used to determine this was concentration at or below 400 ppm (parts per million).

The ***flammability*** classification is indicated by number: 1 (no flammability identified), 2 (lower flammability), or 3 (high flammability).

Figure 9-4 shows the grouping and classification of some common refrigerants. A more extensive explanation on the subject of classification of refrigerants is available. Refer to *ANSI/ASHRAE Standard 34-1992* and addenda *ANSI/ASHRAE 34a-34j.*

9.5 Group A Refrigerants

Important characteristics of the most-used Group A refrigerants are explained at length in this chapter.

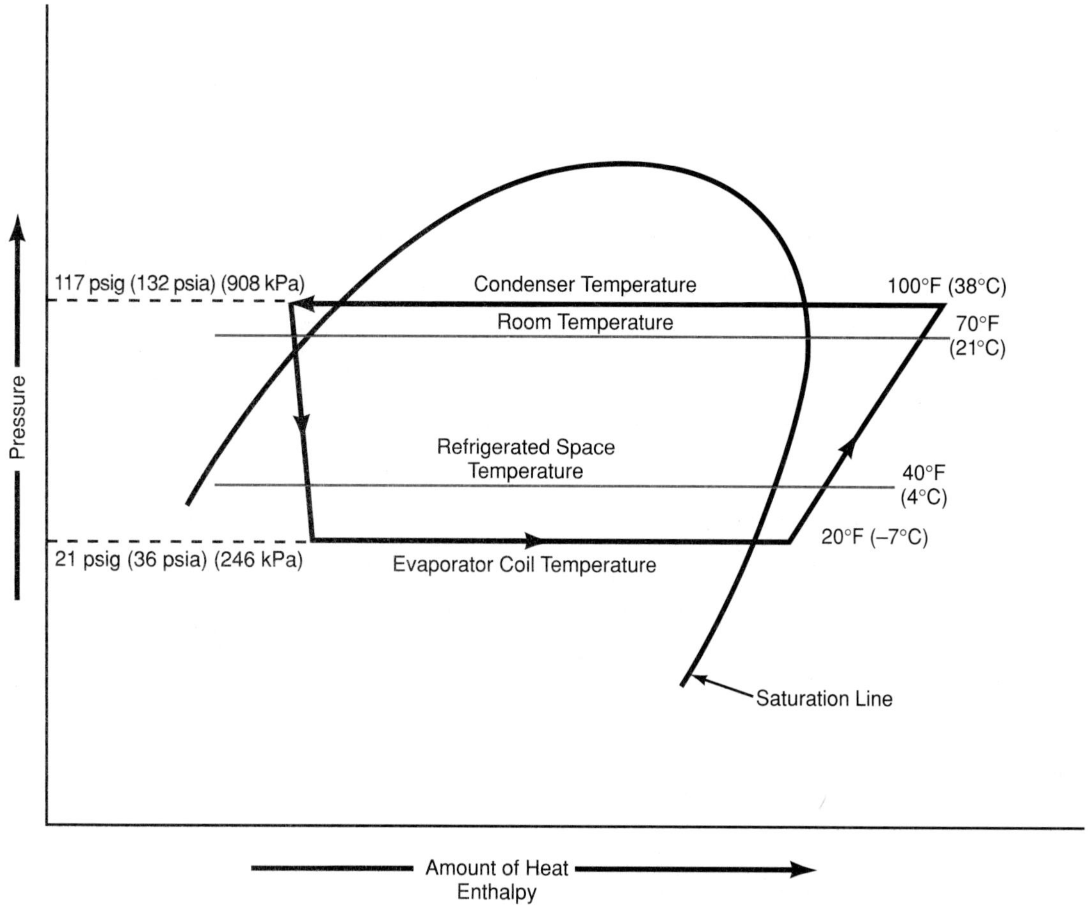

Figure 9-3. *Typical refrigeration cycle with temperatures of one evaporator and condenser shown in comparison to room and refrigerated space temperatures.*

	ASHRAE Safety Classifications	
Refrigerant No.	**Toxicity**	**Flammability**
R-11	A	1
R-12	A	1
R-22	A	1
R-123	A	1
R-124*	A	1
R-125*	A	1
R-134a	A	1
R-401A	A	1
R-406A	A	2
R-500	A	1
R-502	A	1
R-503	—	—
R-507	A	1
R-717	B	2
R-744	A	1

*Denotes toxicity classification is based on recommended exposure limits provided by chemical suppliers. This rating is provisional and will be reviewed when toxicological testing is completed.
—(No rating listed)

Figure 9-4. *Safety classifications for some popular refrigerants, as listed by the American Society of Heating, Refrigerating and Air-Conditioning Engineers (ANSI/ASHRAE Standard: Number Designation and Safety Classification of Refrigerants).*

(Pressure-temperature curves for six common Group A refrigerants are in **Figure 9-2.**)

Refrigerants in this group may be used in the largest quantities in any installation. The allowable quantities are specified by the *American Standard Safety Code for Mechanical Refrigeration.* The amounts are:

- Up to 20 lbs. in hospital kitchens.
- Up to 20 lbs. in residential air conditioning systems.
- Up to 50 lbs. in residential use if precautions are taken.
- Up to 50 lbs. (indirect system) in public assemblies.

Some refrigerants in Group A are:

R-11	Trichloromonofluoromethane (CCl_3F)
R-12	Dichlorodifluoromethane (CCl_2F_2)
R-22	Monochlorodifluoromethane ($CHClF_2$)
R-134a	Tetrafluoroethane (CF_3CH_2F)
R-500	73.8% R-12 and 26.2% R-152a
R-502	48.8% R-22 and 51.2% R-115
R-507	45% R-125 and 55% R-143a
R-744	Carbon dioxide CO_2

Chlorofluorocarbons (R-11, 12, 113, 114, and 115) in Group A refrigerants are being replaced by manufacturers. Alternative refrigerants are being used. This is in accordance with EPA rulings concerning the effects of CFCs on the earth's ozone layer.

9.5.1 R-11 Trichloromonofluoromethane (CCl_3F)

R-11 is a synthetic chemical product which can be used as a refrigerant. It is stable, nonflammable, and nontoxic. This means it will not burn and is not poisonous. It is considered to be a low-pressure refrigerant. It has a low-side vacuum of 24″Hg (610 mmHg) at 5°F (−15°C). The high-side pressure is 18.3 psia (126 kPa) or 3.6 psig (94 kPa) at 86°F (30°C). The latent heat at 5°F (−15°C) is 84.0 Btu/lb. (195 kJ/kg).

This refrigerant is extensively used in large centrifugal compressor systems. As much as 35 lbs. of this refrigerant may be used for each 1000 ft^3 (28m^3) of air conditioned space. (This would be a room about 10′ (3m) × 12.5′ (3.8m) × 8′ (2.4m). Leaks may be detected by using a soap solution, halide torch, or an electronic detector. The cylinder code color of R-11 is orange.

Section 9.6.1 describes the properties of R-123, an HCFC refrigerant used as a replacement for R-11. This is in keeping with EPA rulings concerning CFCs.

9.5.2 R-12 Dichlorodifluoromethane ($CCl_2 F_2$)

R-12 is a colorless, almost odorless liquid. It has a boiling point of −21.7°F (−29°C) at atmospheric pressure. It is nontoxic, noncorrosive, nonirritating, and nonflammable.

Chemically, it is inert at ordinary temperatures, and thermally stable to above 800°F (427°C). This temperature is well above the safe operating temperatures of most refrigerating mechanism materials and lubricants. A table of properties of R-12 is shown in **Figure 9-5.**

R-12 has a relatively low latent heat value. In the smaller refrigerating machines, this is an advantage. The large amount of refrigerant circulated permits using less sensitive and more positive operating and regulating mechanisms. It has been widely used in reciprocating, rotary, and large centrifugal compressors. It operates at a low but positive head and back pressure, with a good volumetric efficiency.

R-12 has a pressure of 26.5 psia (183 kPa) or 11.8 psig (81 kPa) at 5°F (−15°C). It has a pressure of 108 psia (745 kPa) or 93.3 psig (644 kPa) at 86°F (30°C). The latent heat of R-12 at 5°F (−15°C) is 68.2 Btu/lb. (159 kJ/kg). Latent heat is the difference between the last two columns of the table in **Figure 9-5.**

An R-12 leak may be detected by means of a soap solution, a halide lamp, colored oil added to the system, or an electronic leak detector. See Chapter 12 concerning the use of leak detectors.

Water is only slightly soluble in R-12. At 0°F (−18°C), R-12 will hold only six parts per million of water by weight. The solution formed is only very slightly corrosive to metals commonly used in refrigerator construction. The addition of mineral oil to the refrigerant has no effect on the corrosive action. It does lessen the amount of discoloration caused by the free water. R-12 is more critical as to its moisture than R-22 and R-502.

R-12 is soluble in oil down to −90°F (−68°C). This helps the oil flow in very cold evaporators. The oil will begin to separate at this temperature. Because it is lighter than the refrigerant, it will collect on the surface of the liquid refrigerant.

R-12	Pressure		Volume Vapor	Density Liquid	Heat Content Btu/lb.	
Temp °F	Psia	Psig	Cu. Ft./Lb.	Lb./Cu. Ft.	Liquid	Vapor
−150	0.154	29.61*	178.65	104.36	−22.70	60.8
−125	0.516	28.67*	57.28	102.29	−17.59	63.5
−100	1.428	27.01*	22.16	100.15	−12.47	66.2
− 75	3.388	23.02*	9.92	97.93	− 7.31	69.0
− 50	7.117	15.43*	4.97	95.62	− 2.10	71.8
− 25	13.556	2.32*	2.73	93.20	3.17	74.56
− 15	17.141	2.45	2.19	92.20	5.30	75.65
− 10	19.189	4.49	1.97	91.69	6.37	76.2
− 5	21.422	6.73	1.78	91.18	7.44	76.73
0	23.849	9.15	1.61	90.66	8.52	77.27
5	26.483	11.79	1.46	90.14	9.60	77.80
10	29.335	14.64	1.32	89.61	10.68	78.335
25	39.310	24.61	1.00	87.98	13.96	79.9
50	61.394	46.70	0.66	85.14	19.51	82.43
75	91.682	76.99	0.44	82.09	25.20	84.82
86	108.04	93.34	0.38	80.67	27.77	85.82
100	131.86	117.16	0.31	78.79	31.10	87.03
125	183.76	169.06	0.22	75.15	37.28	88.97
150	249.31	234.61	0.16	71.04	43.85	90.53
175	330.64	315.94	0.11	66.20	51.03	91.48
200	430.09	415.39	0.08	60.03	59.20	91.28

*Inches of mercury below one atmosphere.

Figure 9-5. *Properties of liquid and saturated vapor of R-12. Note pressures corresponding to standard evaporating temperature of 5°F (−15°C) and condensing temperature of 86°F (30°C). (DuPont Company)*

The pressure-heat *(enthalpy)* diagram for this refrigerant is shown in **Figure 9-6.** See Section 1.36 for an explanation of enthalpy. A metric unit pressure-enthalpy diagram for R-12 is shown in **Figure 9-7.** It is safe to use 30 lbs. of R-12 for each 1000 ft^3 of air-conditioned space.

A typical R-12 cycle for a frozen foods unit is shown in **Figure 9-8.** The chart shows conditions, with an R-12 system (0°F evaporator, 80°F condenser). Line A-E-B represents a 0°F evaporator at 23.9 psia, 9.2 psig; a condenser at 80°F, 98.9 psia, 84.2 psig.

The refrigerant leaving the metering device at Point C, and entering the evaporator at Point E, has a quality of 26%. This means 26% of the refrigerant was flashed off to keep the refrigerant temperature and pressure constant. The refrigerant has a heat of 26 Btu, as shown at Point D. The refrigerant leaving the evaporator at Point B has a heat of 77 Btu. This is shown at Point F.

At this point, the refrigerant leaves the compressor at Point C with a heat of 86 Btu, as shown at Point I.

To show the efficiency of the system:

Point F: 77 Btu leaving evaporator
− Point D: 26 Btu entering evaporator
= 51 Btu/lb.(119.5 kJ/kg) of refrigerant effect

The line G-C represents the condition in the condenser. From G to H, superheat is removed from the vapor. From H to C, the vapor is condensed back to liquid. The total heat removed from the condensor is equal to:

Point G: 86 Btu leaving compressor
− Point C: 26 Btu leaving condenser
= 60 Btu/lb. (140.6 kJ/kg) total heat removed by condenser

To calculate efficiency, the refrigerant effect is divided by the heat removed, as follows:

51 Btu/lb.(119.5 kJ/kg) refrigeration effect
60 Btu/lb.(140.6 kJ/kg) removed from condenser
51 ÷ 60 = .85 × 100 = 85 (85% efficiency)

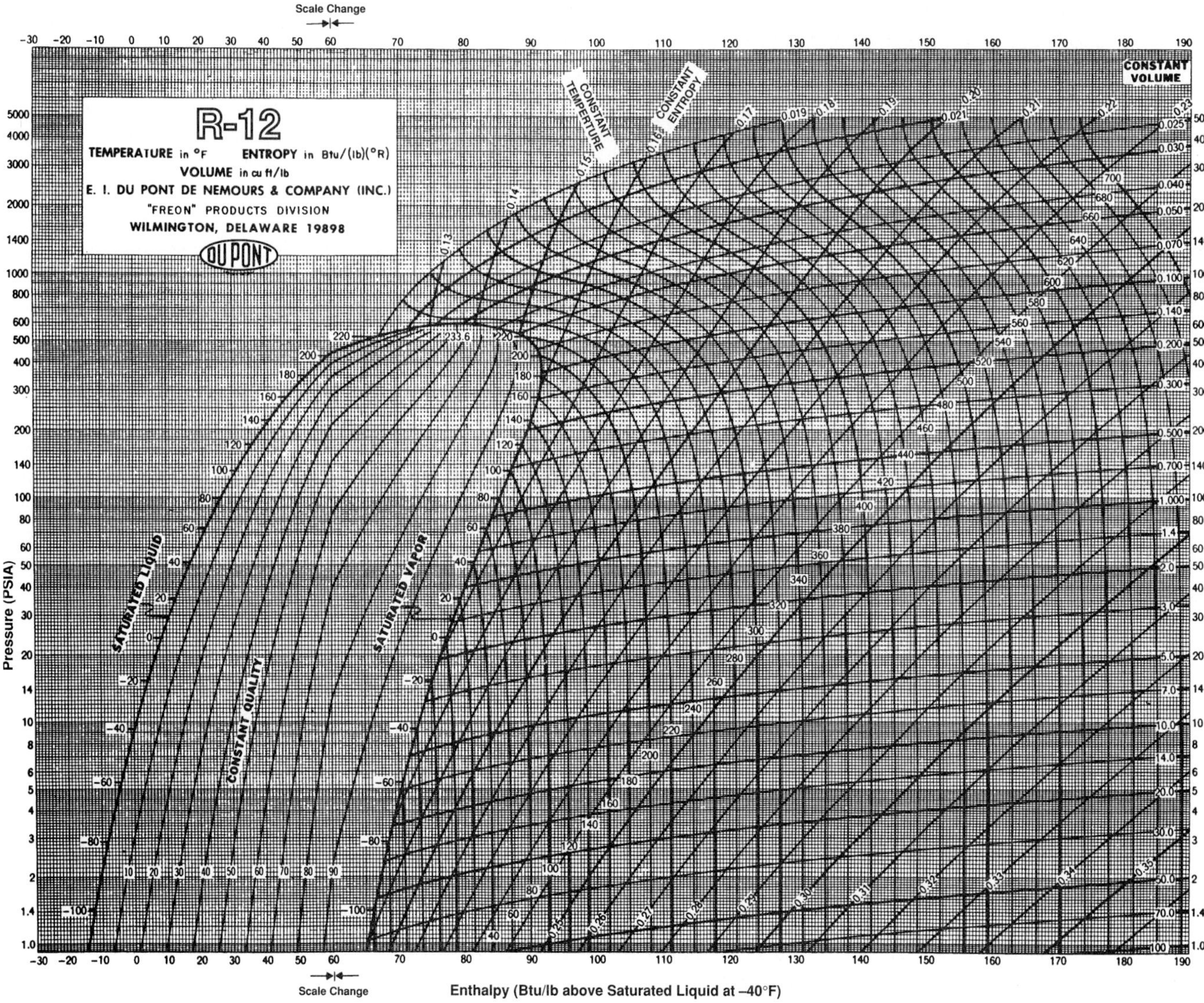

Figure 9-6. *Pressure-enthalpy diagram for R-12. Note that 0 of enthalpy scale is taken at −40°F (−40°C). (DuPont Company)*

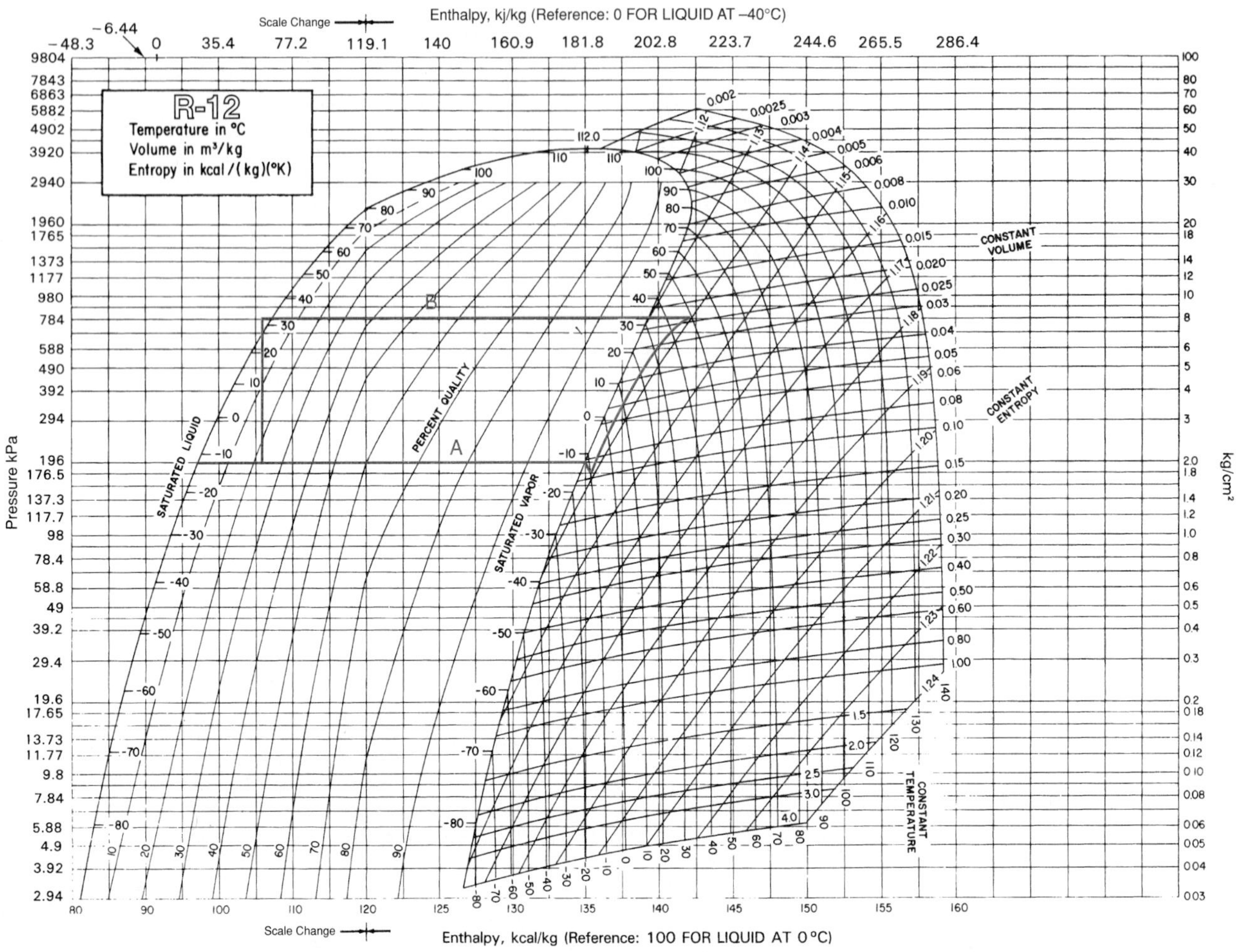

Figure 9-7. *Pressure-heat diagram for R-12 expressed in metric units. The standard refrigerating cycle of evaporating temperature is shown at A and condensing temperature at B. (Adapted from DuPont Company)*

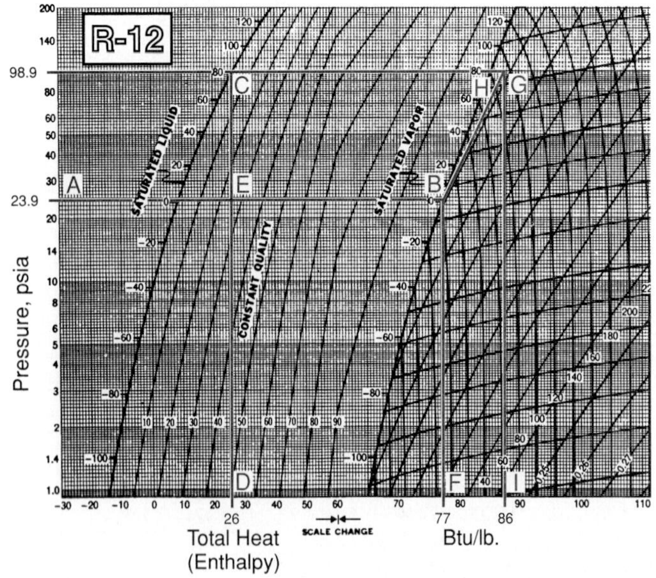

Figure 9-8. *Pressure-enthalpy diagram indicating a typical R-12 cycle for a freezer. Note the pressure range from 23.9 psia to 98.9 psia, and the enthalpy (heat) change of 26 to 86.*

R-12 is available in a variety of cylinder sizes. The cylinder code color is white. This is an important fact to remember when purchasing or using refrigerants. R-134a is an HFC refrigerant for use as a replacement for R-12. This is in keeping with EPA rulings concerning CFCs. Section 9.5.4 describes the properties of R-134a.

9.5.3 R-22 Monochlorodifluoromethane ($CHClF_2$)

R-22 is an HCFC refrigerant. It is a synthetic refrigerant developed for installations that need a low evaporating temperature. (***Synthetic*** means that it is made by humans and not found in nature.) It is referred to as "monochlorodifluoromethane" and also as "chlorodifluoromethane." **Figure 9-9** shows the properties of R-22.

R-22 has been successfully used in air conditioning units and in household refrigerators. It is also used in nonindustrial heat pumps and positive displacement chillers. One application is in fast freezing units which

R-22	Pressure		Volume Vapor	Density Liquid	Heat Content Btu/lb.	
Temp °F	Psia	Psig	Cu. Ft./Lb.	Lb./Cu. Ft.	Liquid	Vapor
−150	0.272	29.37*	141.23	98.24	−25.97	87.52
−125	0.886	28.12*	46.69	96.04	−20.33	90.43
−100	2.398	25.04*	18.43	93.77	−14.56	93.37
− 75	5.610	18.50*	8.36	91.43	− 8.64	96.29
− 50	11.674	6.15*	4.22	89.00	− 2.51	99.14
− 25	22.086	7.39	2.33	86.48	3.83	101.88
− 15	27.865	13.17	1.87	85.43	6.44	102.94
− 10	31.162	16.47	1.68	84.90	7.75	103.46
− 5	34.754	20.06	1.52	84.37	9.08	103.96
0	38.657	23.96	1.37	83.83	10.41	104.47
5	42.888	28.19	1.24	83.28	11.75	104.96
10	47.464	32.77	1.13	82.72	13.10	105.44
25	63.450	48.75	0.86	81.02	17.22	106.84
50	98.727	84.03	0.56	78.03	24.28	108.95
75	146.91	132.22	0.37	74.80	31.61	110.74
86	172.87	158.17	0.32	73.28	34.93	111.40
100	210.60	195.91	0.26	71.24	39.27	112.11
125	292.62	277.92	0.18	67.20	47.37	112.88
150	396.19	381.50	0.12	62.40	56.14	112.73

*Inches of mercury below one atmosphere.

Figure 9-9. *Properties of liquid and saturated vapor of R-22. Note pressures corresponding to standard evaporating temperature of 5°F (−15°C) and condensing temperature of 86°F (30°C).*

maintain a temperature of −20°F to −40°F (−29°C to −40°C). See the temperature conversion table in Chapter 31. To obtain these low temperatures, it is not necessary to use R-22 at below-atmospheric pressures. R-22 is used with both reciprocating and centrifugal compressors.

R-22 is stable and is nontoxic, noncorrosive, nonirritating, and nonflammable. It has a boiling point of −41°F (−41°C) at atmospheric pressure. The normal head pressure at 86°F (30°C) is 173 psia (1194 kPa) or 158 psig (1090 kPa). This is shown in the table in **Figure 9-9.** The evaporator pressure of R-22 is 43 psia (297 kPa) or 28 psig (193 kPa) at 5°F (−15°C). The latent heat of R-22 at 5°F (−15°C) is 93.2 Btu/lb. (217 kJ/kg). Latent heat is the difference between the last two columns in **Figure 9-9.**

Water mixes more readily with R-22 than with R-12, by a ratio of 3:1. This is 19.5 ppm (parts per million) by weight. Water must be kept at a minimum. ***Desiccants*** *(driers)* should be used to remove most of the moisture. Because of the ability of water to mix with R-22, larger amounts of desiccant are needed.

R-22 has good solubility in oil. This solubility remains high down to about 16°F (−9°C). The oil will remain fluid enough to flow down the suction line at temperatures down to −40°F (−40°C). However, at or slightly below this temperature, the oil begins to separate from the refrigerant. Because oil is lighter, it will collect on the surface of the liquid refrigerant.

Leaks may be detected with a soap solution, halide torch, or an electronic leak detector. Some of the properties of R-22 are shown in **Figure 9-10.** A metric unit pressure-enthalpy diagram for R-22 is shown in **Figure 9-11.** The cylinder code color of R-22 is light green.

9.5.4 R-134a Tetrafluoroethane (CF_3CH_2F)

R-134a (Ethane 1,1,1,2 - Tetrafluoro) is an HFC refrigerant. It is used as a replacement for R-12 (a CFC refrigerant). It is used in centrifugal, reciprocating, rotary screw, and scroll compressors.

R-134a is nontoxic, noncorrosive, and nonflammable. **However, exposure to concentrations over 75,000 ppm may cause cardiac irregularities.**

R-134a has a boiling point of −14.9°F (−26.1°C). Its auto-ignition temperature is 1418°F (770°C). Its ozone depletion level is 0. The coefficient of performance for R-134a is slightly lower than that of R-12. The solubility of R-134a in water is 0.11% by weight at 77°F (25°C). The critical temperature of R-134a is 252°F (122°C). The cylinder color code is light blue.

Refrigerant 134a is not compatible with the mineral-based refrigerant oils and lubricants presently used for R-12. It is compatible with polyol ester oil. Check with manufacturer for exact specifications.

Numerous design changes have been developed and are being implemented for use with R-134a. These include a 30% increase in condenser and evaporator sizing, a change in desiccant type (from silicone gel to molecular sieve), the use of smaller hoses, and 30% increase in control pressure regulations.

Leaks of R-134a can be detected by use of:

- Soap solution.
- Fluorescent dyes.
- Ultrasonic leak detectors.
- Halogen-selective detectors.
- Electronic leak detectors.

R-134a is presently being used as a standard refrigerant in vehicular air conditioning. It has been named as a substitute for a wide range of applications. These

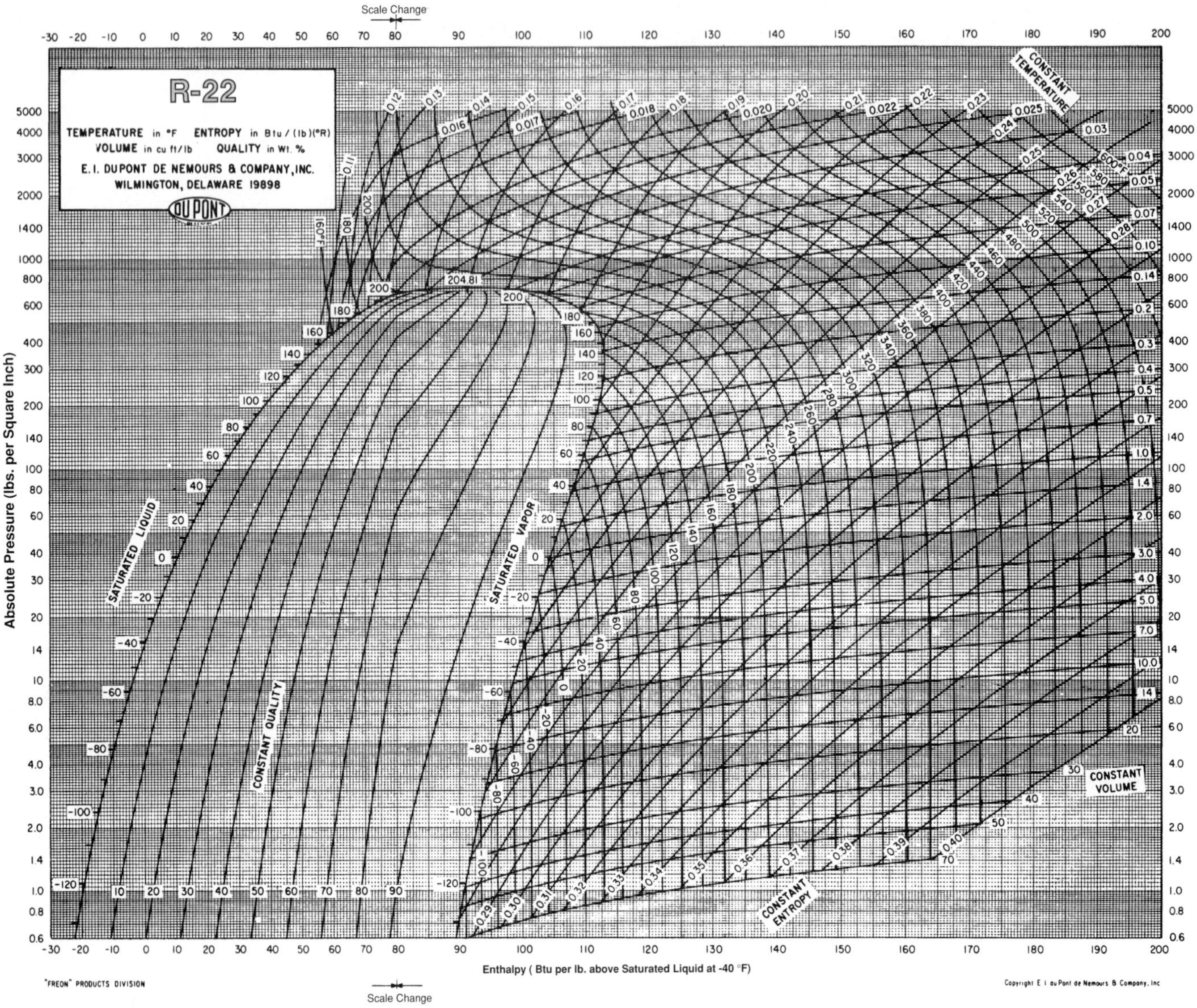

Figure 9-10. *Pressure-enthalpy diagram for R-22. Note that 0 of enthalpy scale is taken at −40°F (−40°C). Also note that heat (enthalpy) scale changes at 80 Btu/lb. to help technician read superheat values more easily. (DuPont Company)*

include air conditioning and refrigeration systems in residential, commercial, and industrial applications.

The properties of the liquid and saturated vapor of R-134a are shown in **Figure 9-12.** A pressure heat-enthalpy diagram for R-134a is shown in **Figure 9-13.** See Section 1.36 for an explanation of enthalpy. (For pressure-heat enthalpy diagram in metric units, see Chapter 31.)

9.5.5 Azeotropic Refrigerants

The following azeotropic definitions apply to dual-component mixtures. However, mixtures with three or more components ("ternary mixtures") have similar characteristics.

An ***azeotropic refrigerant mixture*** (ARM) is a multi-component working fluid of specific composition. At atmospheric pressure, this composition will not change when it evaporates or condenses. R-500 and R-502 are examples of azeotropic refrigerants. R-500 consists of CFC-12 and HFC-152a. R-502 consists of HCFC-22 and CFC-115.

Azeotropic refrigerants are patented refrigerants. The manufacturing process is rather complicated. Service technicians should never attempt to make their own mixtures.

☐ Refrigerant R-500 (R-152a + R-12) (CCl_2F_2 + CH_3CHF_2)

R-500 is an azeotropic mixture of 26.2% R-152a and 73.8% R-12. It is used in both industrial and commercial applications. However, it is used only in systems with reciprocating compressors. It has a fairly constant vapor-pressure temperature curve. This curve is different from the vaporizing curves for either R-152a or R-12.

R-500 offers about 20% greater refrigerating capacity than R-12 (when used for the same purpose and with the same size motor). The evaporator pressure of R-500

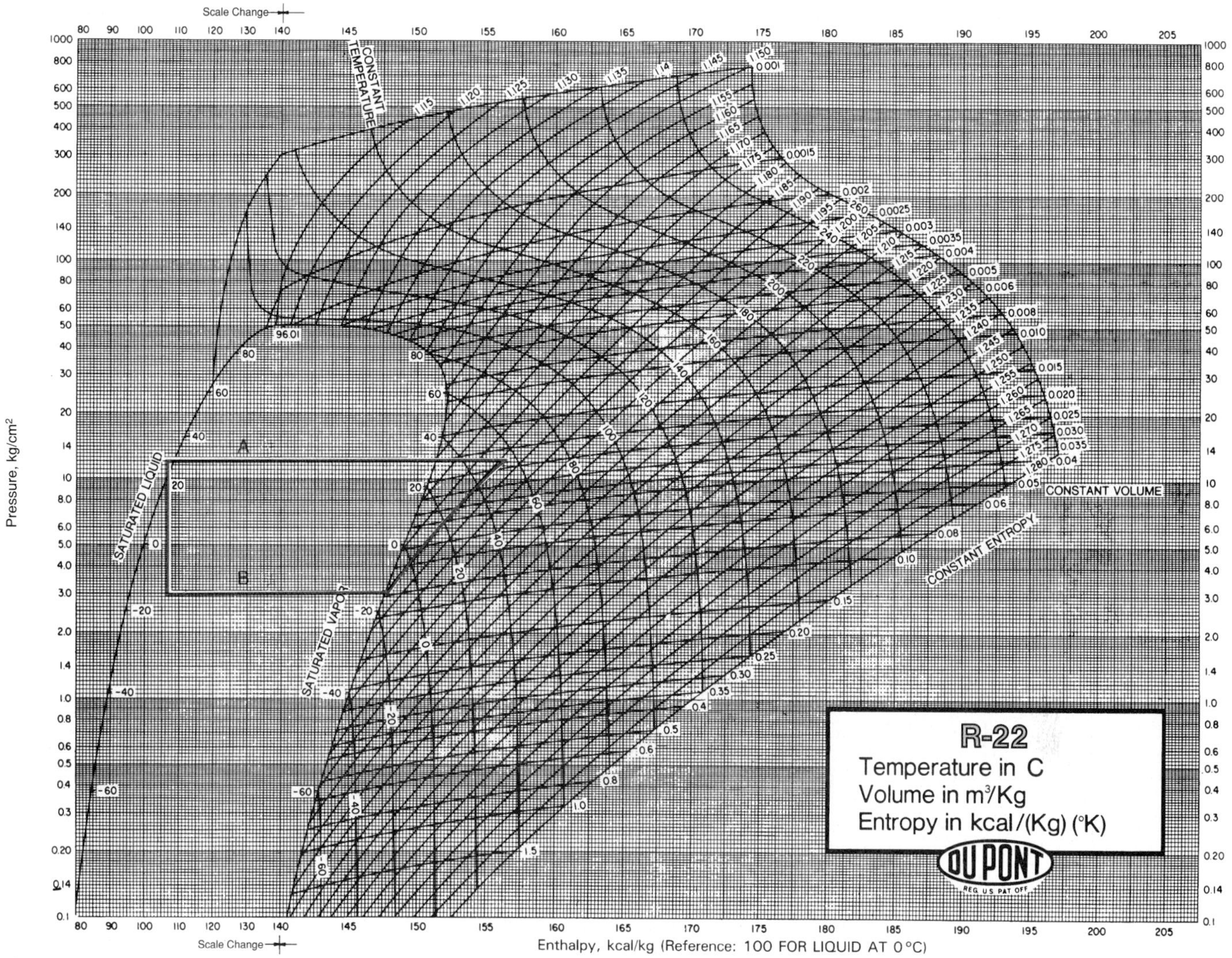

Figure 9-11. *Pressure heat diagram for R-22 expressed in metric units. The standard refrigerating cycle of evaporating temperature is shown at A and condensing temperature is shown at B. (DuPont Company)*

R-134a Temp °F	Pressure		Volume cu.ft./lb.	Density lb./cu.ft.	Heat Content (Enthalpy) Btu/lb.	
	Absolute	Gauge	Vapor	Liquid	Liquid	Vapor
–150	0.07107	29.776 *	457.0719	102.344	–32.781	80.212
–125	0.28333	29.344 *	123.4418	99.641	–25.383	83.716
–100	0.89915	28.090 *	41.5241	96.891	–17.939	87.245
– 75	2.3866	25.062 *	16.5646	94.087	–10.472	90.760
– 50	5.4966	18.730 *	7.5560	91.220	– 2.995	94.248
– 25	11.2964	6.9214*	3.8338	88.278	4.503	97.721
– 15	14.6686	0.0555*	2.9960	87.078	7.518	99.109
– 10	16.6293	1.9334	2.6610	86.472	9.030	99.804
– 5	18.7906	4.0947	2.3705	85.862	10.546	100.499
0	21.1665	6.4706	2.1177	85.248	12.067	101.195
5	23.7710	9.0751	1.8969	84.630	13.593	101.891
10	26.619	11.923	1.7036	84.007	15.125	102.587
25	36.773	22.078	1.25178	82.110	19.763	104.677
50	60.032	45.335	0.78067	78.836	27.666	108.149
75	93.080	78.384	0.50743	75.387	35.851	111.553
86	111.321	96.626	0.42412	73.799	39.560	113.004
100	138.28	123.58	0.33993	71.701	44.393	114.782
125	198.27	183.57	0.23204	67.679	53.385	117.660
150	276.12	261.42	0.15914	63.126	62.989	119.879
175	375.69	360.99	0.10705	57.601	73.581	120.788
200	502.54	487.85	0.06542	49.439	86.528	118.155

*Inches of mercury below one atmosphere.

Figure 9-12. *Properties of liquid and saturated vapor of R-134a. Note pressures corresponding to standard evaporating temperature of 5°F (–15°C) and condensing temperature of 86°F (30°C). (ICI Americas, Inc.)*

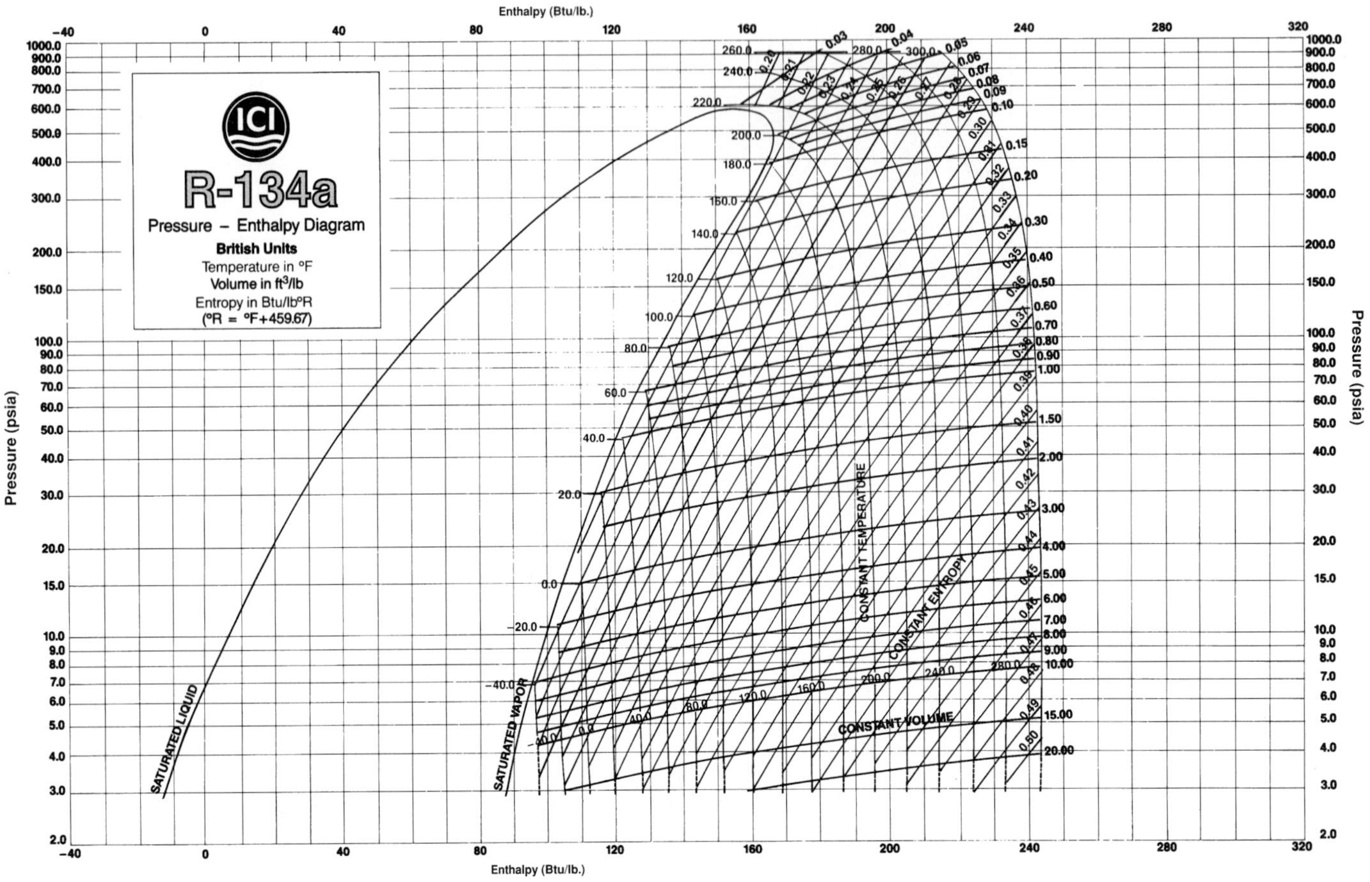

Figure 9-13. *Pressure-enthalpy diagram for R-134a. (ICI Americas, Inc.)*

is 31.1 psia (215 kPa) or 16.4 psig (113 kPa) at 5°F (−15°C). It has a boiling point, at atmospheric pressure, of −28°F (−33°C). Its condensing pressure is 128 psia (883 kPa) or 113 psig (780 kPa) at 86°F (30°C). Its latent heat at 5°F (−15°C) is 82.5 Btu/lb. (192 kJ/kg). This is shown in the table in **Figure 9-14.**

R-500 can be used whenever a higher capacity is needed than can be obtained with R-12. There is little change in condensing temperatures, as shown in **Figure 9-15.** R-500 is also recommended where electrical service varies from 60 cycle to 50 cycle (Hz).

The ***solubility*** (mixing with or going into solution) of water in R-500 is highly critical. R-500 has fairly high solubility with oil. Leaks are detected using halide leak detectors, electronic leak detectors, soap solution, or colored tracing agents.

Servicing refrigerators that use this refrigerant does not present any unusual problem. Since water is quite soluble in this refrigerant, it is necessary to keep moisture out of the system. This is accomplished by careful dehydration and by using driers. The cylinder code color of R-500 is yellow.

Refrigerant R-502 (R-22 + R-115) ($CHClF_2$ + $CClF_2CF_3$)

R-502 is an azeotropic mixture of 48.8% R-22 and 51.2% R-115. It has been used since 1961. It is a nonflammable, noncorrosive, practically nontoxic liquid. R-502 is a good refrigerant for obtaining medium and low temperatures. It is suitable where temperatures from 0 to −60°F (−18 to −51°C) are needed. It is often used in frozen food lockers, frozen food processing plants, frozen food display cases, and in storage units for frozen foods and ice cream. It is only used with reciprocating compressors.

Its boiling point is −50°F (−46°C) at atmospheric pressure. The condensing pressure is 191 psia (1318 kPa) or 177 psig (1221 kPa) at 86°F (30°C). Its evaporating pressure at 5°F (−15°C) is 50.6 psia (349 kPa) or 35.9 psig (248 kPa). Its latent heat at −20°F (−29°C) is 70.8 Btu/lb. (165 kJ/kg). This is shown in the table of properties in **Figure 9-16.** The latent heat of R-502 at 5°F (−15°C) is 67.3 Btu/lb. (157 kJ/kg).

R-502 combines many of the best properties of both R-12 and R-22. It gives a machine the approximate capacity of R-22. The condensing temperature of a system is just about the same as one using R-12. A pressure-enthalpy diagram of the refrigerant is shown in **Figure 9-17.**

A condensing temperature of 30°F (−1°C) is common for the frozen food applications mentioned above. When R-502 is used in these applications, the life of the compressor valves and other parts is increased. Better lubrication is possible because of increased viscosity of the oil at a lower condensing temperature. With R-502, it is possible to eliminate liquid injection for compressor

R-500	Pressure		Volume Vapor	Density Liquid	Heat Content Btu/lb.	
Temp °F	Psia	Psig	Cu. Ft./Lb.	Lb./Cu. Ft.	Liquid	Vapor
−40	10.95	7.62*	4.0	84.28	0.00	87.74
−30	14.10	1.22*	3.15	83.35	2.38	89.04
−20	17.92	3.23	2.52	82.40	4.79	90.31
−10	22.52	7.82	2.03	81.44	7.22	91.57
0	27.98	13.3	1.66	80.46	9.71	92.81
5	31.07	16.4	1.501	79.96	10.96	93.42
10	34.43	19.7	1.36	79.46	12.23	94.03
20	41.96	27.3	1.13	78.45	14.79	95.22
30	50.70	36.0	0.94	77.41	17.40	96.39
40	60.75	46.1	0.79	76.34	20.05	97.53
50	72.26	57.6	0.67	75.26	22.75	98.64
60	85.33	70.6	0.57	74.14	25.48	99.71
70	100.1	85.4	0.48	72.98	28.28	100.75
80	116.7	102.0	0.42	71.80	31.12	101.75
86	127.6	113.0	0.38	71.06	32.85	102.33
90	135.3	121.0	0.36	70.56	34.01	102.70
100	155.9	141.0	0.31	69.28	36.97	103.60
110	178.8	164.0	0.27	67.95	40.00	104.44
120	204.1	189.0	0.23	66.55	43.10	105.22
130	231.9	217.0	0.20	65.08	46.29	105.91
140	262.4	248.0	0.17	63.51	49.58	106.51

*Inches of mercury vacuum.

Figure 9-14. *Properties of liquid and saturated vapor of R-500. Note pressures corresponding to evaporating temperature of 5°F (−15°C) and condensing temperature of 86°F (30°C). (Allied Signal, Inc.)*

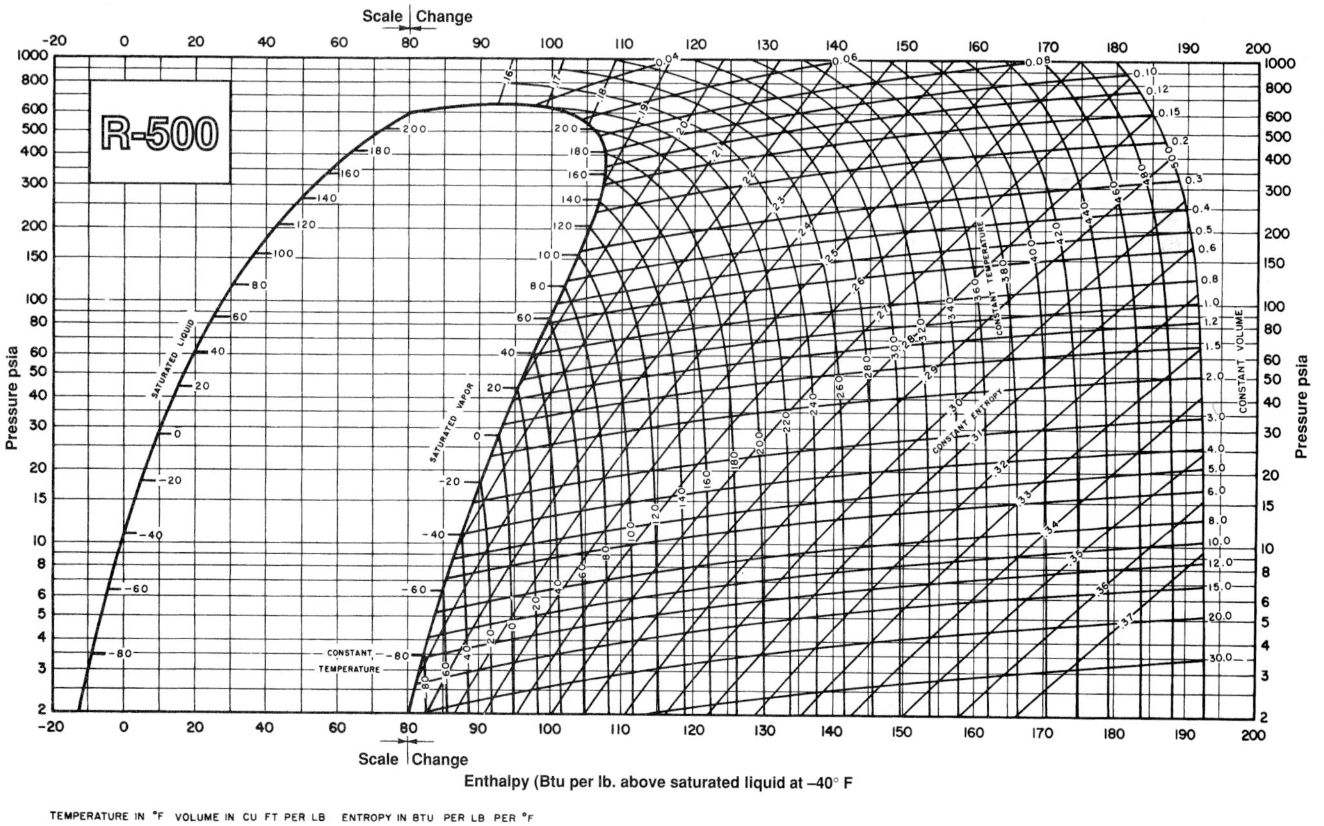

Figure 9-15. *Pressure-enthalpy diagram for R-500. Note that 0 of enthalpy scale is taken at −40°F (−40°C). Also note that enthalpy scale changes at 80 Btu/lb. to help technician read superheat values.*

R-502	Pressure		Volume Vapor	Density Liquid	Heat Content Btu/lb.	
Temp °F	Psia	Psig	Cu. Ft./Lb.	Lb./Cu. Ft.	Liquid	Vapor
−100	3.261	23.281*	10.461	97.857	−12.548	65.885
− 75	7.281	15.097*	4.959	95.234	− 7.597	68.919
− 50	14.602	0.190*	2.596	92.513	− 2.251	71.928
− 25	26.817	12.121	1.465	89.673	3.496	74.866
− 20	30.006	15.310	1.317	89.088	4.693	75.442
− 15	33.480	18.784	1.187	88.496	5.905	76.012
− 10	37.256	22.560	1.073	87.898	7.133	76.577
− 5	41.349	26.653	0.973	87.293	8.376	77.137
0	45.775	31.079	0.881	86.681	9.633	77.690
5	50.553	35.857	0.801	86.062	10.906	78.237
10	55.697	41.001	0.731	85.434	12.193	78.777
15	61.225	46.529	0.666	84.797	13.494	79.310
20	67.155	52.459	0.612	84.152	14.809	79.836
25	73.503	58.807	0.557	83.497	16.138	80.353
50	112.12	97.42	0.367	80.058	22.977	82.800
75	163.81	149.11	0.248	76.269	30.122	84.958
86	191.28	176.59	0.210	74.453	33.359	85.789
100	230.89	216.19	0.171	71.967	37.563	86.711
125	316.04	301.35	0.118	66.838	45.361	87.834
150	423.06	408.35	0.079	60.092	53.850	87.757
160	473.38	458.69	0.066	56.429	57.732	87.013

*Inches of mercury below one atmosphere.

Figure 9-16. *Properties of liquid and saturated vapor of R-502. Note pressures corresponding to standard evaporating temperature of 5°F (−15°C) and condensing temperature of 86°F (30°C). (DuPont Company)*

cooling, because of the lower condensing pressure. Such cooling is often necessary with R-22.

R-502 has all the qualities found in the other halogenated (fluorocarbon) refrigerants. It is nontoxic, nonflammable, nonirritating, stable, and noncorrosive. Leaks are detected with soap solution, halide torch, or electronic leak detector.

R-502 will hold 1.5 times more moisture at 0°F (−18°C) than R-12 (12.0 ppm). R-502 has fair solubility in oil above 180°F (82°C). Below this temperature, the oil tries to separate. It tends to collect on the surface of liquid refrigerant. However, oil is carried back to the compressor at temperatures down to −40°F (−40°C). Special devices are sometimes used to return the oil to the compressor. The cylinder code color of R-502 is orchid.

R-507, an HFC refrigerant, is used to replace R-502. This is in keeping with EPA rulings concerning CFCs. R-125, pentafluoroethane, is an HCFC refrigerant for use in stores and supermarkets to replace R-502.

☐ Refrigerant 503 (R-23 + R-13) (CHF_3 + $CClF_3$)

Refrigerant R-503 is an azeotropic mixture of 40.1% R-23 and 59.9% R-13. It is a nonflammable, noncorrosive, practically nontoxic liquid. Its boiling temperature at atmospheric pressure is −126°F (−88°C). This is lower than either R-23 or R-13. Its evaporating pressure at 5°F (−15°C) is 266 psia (1835 kPa) or 252 psig (1739 kPa). Its critical temperature is 67°F (20°C). The critical pressure is 632 psia (4361 kPa) or 592 psig (4085 kPa).

This is a low-temperature refrigerant. It is good for use in the low section of cascade systems. These require temperatures in the −100°F to −125°F (−73°C to −87°C) range. Properties of R-503 are shown in **Figure 9-18.** A pressure-heat enthalpy diagram for this refrigerant is shown in **Figure 9-19.** The latent heat of vaporization at atmospheric pressure (−127°F or −88°C) is 77.2 Btu/lb. (180 kJ/kg). The latent heat at 5°F (−15°C) is 48.9 Btu/lb. (114 kJ/kg).

R-503 system leaks may be detected using a halide torch, soap solution, or an electronic leak detector. This refrigerant will hold more moisture than some other low-temperature refrigerants. All low-temperature applications must have extreme dryness. Moisture not in solution with the refrigerant may form ice at the refrigerant control devices.

Oil does not circulate well at low temperatures. Cascade systems are usually fitted with oil separators and other devices for returning the oil to the compressor. This is also true for other low-temperature units. The code color for R-503 cylinders is aquamarine.

☐ Refrigerant 507 (R-125 + R-143a) (CHF_2CF_3 + CF_3CH_3)

R-507 is an azeotropic mixture of 45% R-125 and 55% R-143a. It is an HFC refrigerant used for low- and medium-temperature applications as a replacement for R-502.

It is colorless, nonflammable, and has a slight ethereal odor. The boiling point of R-507 is −52.1°F (−46.7°C). The critical temperature is 160°F (70.9°C). R-507 has a slightly higher capacity than R-502. It is not compatible with mineral oil. Polyol ester oil is a suitable lubricant. The color code for R-507 is teal blue.

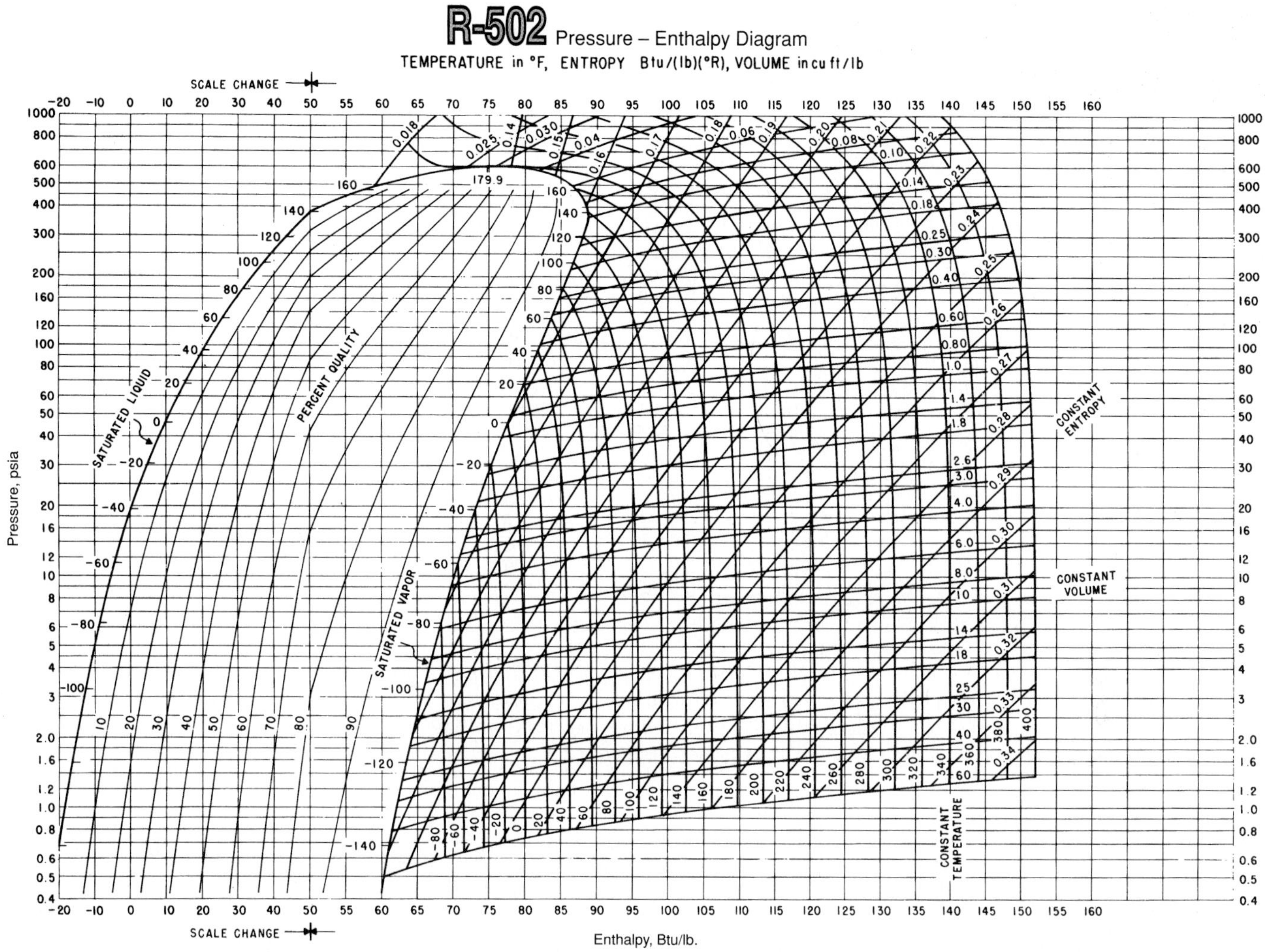

Figure 9-17. *Pressure-enthalpy diagram for R-502. Note that 0 of enthalpy scale is at −40°F (−40°C). Also note that scale changes at 50 Btu/lb. above saturated liquid at −40°F (−40°C) to permit easier reading of superheat values. (DuPont Company)*

Leaks can be detected by:

- Soap solution.
- Electronic leak detectors.
- Halogen-selective detectors.
- Fluorescent leak detectors.

9.5.6 Zeotropic Refrigerants

Zeotropic refrigerants are working fluids with two or more components. For example, R-401A is a zeotropic refrigerant, composed of HCFC-22, HFC-152a, and HCFC-124. The components have different vapor pressures and boiling points. When the fluid evaporates or condenses, the liquid and vapor components will have different compositions.

Zeotropic refrigerants are patented refrigerants. The manufacturing process is rather complicated. Service technicians should never attempt to make their own mixtures.

Refrigerant 401A (R-22 + R-152a + R-124)

R-401A is a zeotropic mixture of 53% HCFC-22, 13% HFC-152a, and 34% HCFC-124. It is an alternative for CFC-12. It is used in most medium-temperature systems. Examples of such systems are: walk-in coolers, food and dairy display cases, vending machines, etc. Cylinder color code for R-401A is coral red.

R-401A can be used with alkylbenzene (AB), polyol ester (POE), or other oil mixtures indicated by the equipment manufacturer. The types of leak detectors used on this refrigerant are:

- Halogen-selective (identifies fluorine, chlorine, bromine, and iodine compounds).
- Compound-selective (identifies specific compound in the refrigerant).
- Fluorescent additive.

Refrigerant 406A (R-22 +R-600a + R-142b)

R-406A is a zeotropic blend of 55% R-22, 4% R-600a (isobutane), and 41% R-142b. It is compatible with R-12 systems, and is soluble in mineral and alkyl benzene oils. It is intended to work well with existing hoses, valves, and seals found in R-12 systems. However, it is always important to check with the manufacturer of the system

R-503	Pressure		Volume Vapor	Density Liquid	Heat Content Btu/lb.	
Temp °F	Psia	Psig	Cu. Ft./Lb.	Lb./Cu. Ft.	Liquid	Vapor
−140	9.234	11.1 *	4.123	93.49	−26.45	52.88
−130	12.98	3.49*	2.998	92.47	−23.93	53.84
−120	17.83	3.13	2.227	91.39	−21.36	54.77
−110	23.98	9.28	1.685	90.25	−18.77	55.66
−100	31.64	16.9	1.296	89.05	−16.16	56.52
− 90	41.05	26.3	1.012	87.78	−13.53	57.35
− 80	52.42	37.7	0.8008	86.44	−10.87	58.13
− 70	66.00	51.3	0.6409	85.02	− 8.19	58.86
− 60	82.05	67.4	0.5182	83.52	− 5.49	59.54
− 50	100.8	86.1	0.4227	81.93	− 2.76	60.16
− 40	122.6	108	0.3474	80.25	0.00	60.72
− 30	147.6	133	0.2872	78.46	2.81	61.20
− 20	176.2	161	0.2387	76.56	5.66	61.60
− 10	208.6	194	0.1991	74.52	8.59	61.89
0	245.3	231.0	0.1664	72.33	11.60	62.05
5	265.9	251.5	0.15275	70.99	13.17	62.04
10	286.4	272.0	0.1391	69.65	14.74	62.04
20	332.6	318	0.1160	67.35	18.05	61.82
30	384.1	369	0.0962	64.45	21.60	61.31
40	440.6	426	0.0793	61.12	26.03	60.45
50	503.3	489	0.0640	57.09	26.69	58.95
60	574.8	560	0.0485	51.40	34.32	55.77

*Inches of mercury vacuum.

Figure 9-18. *Properties of liquid and saturated vapor of R-503. Note that to operate above 0°F (−18°C), condenser pressure in excess of 231 psig (1695 kPa) is required. Note pressure corresponding to condensing temperature of 5°F (−15°C). R-503 is used in cascade system with evaporating temperature as low as −50°F (−46°C). Also note that 0 of liquid heat content is taken at −40°F (−40°C). (Allied Signal, Inc.)*

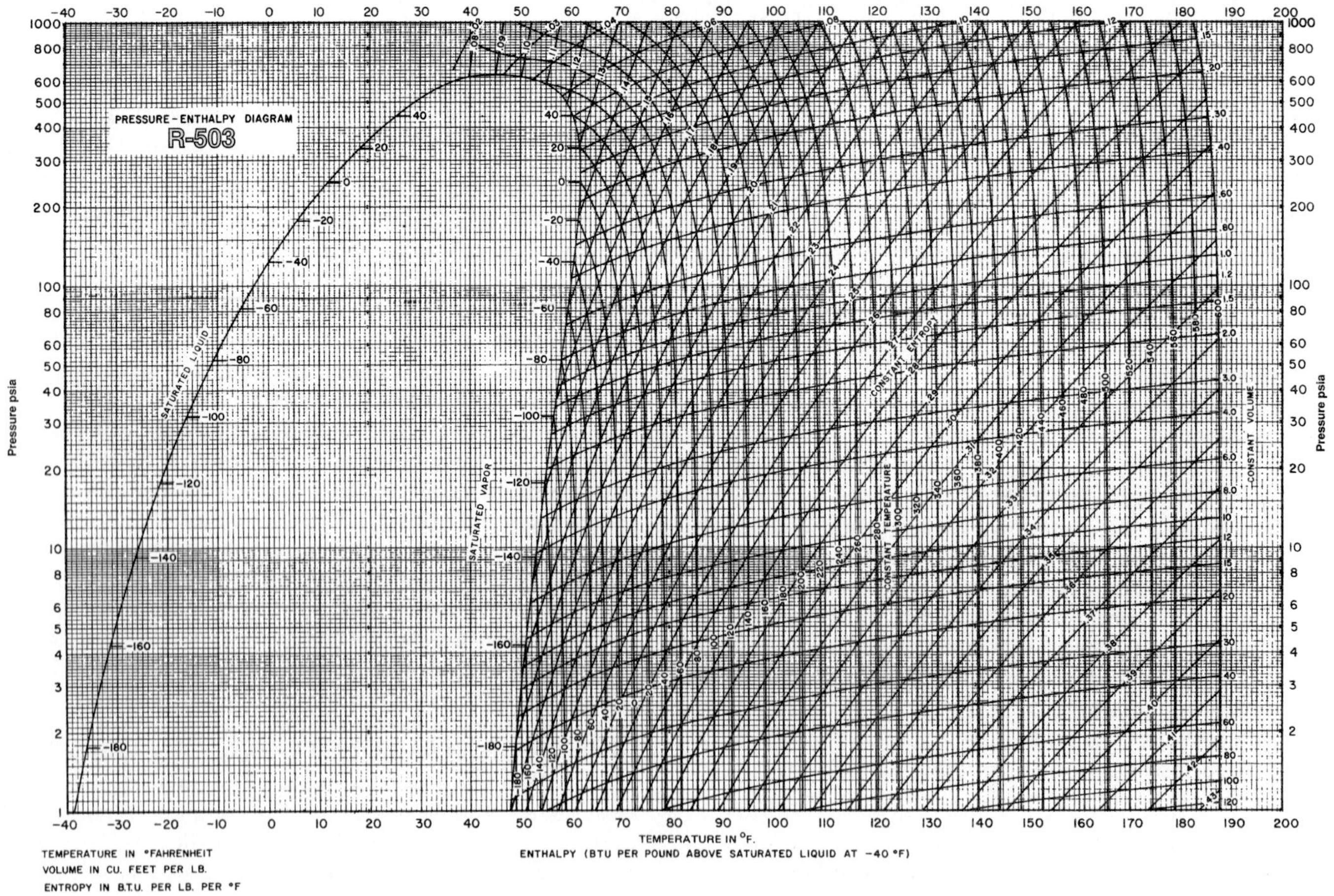

Figure 9-19. *Pressure-enthalpy diagram for R-503. (Allied Signal, Inc.)*

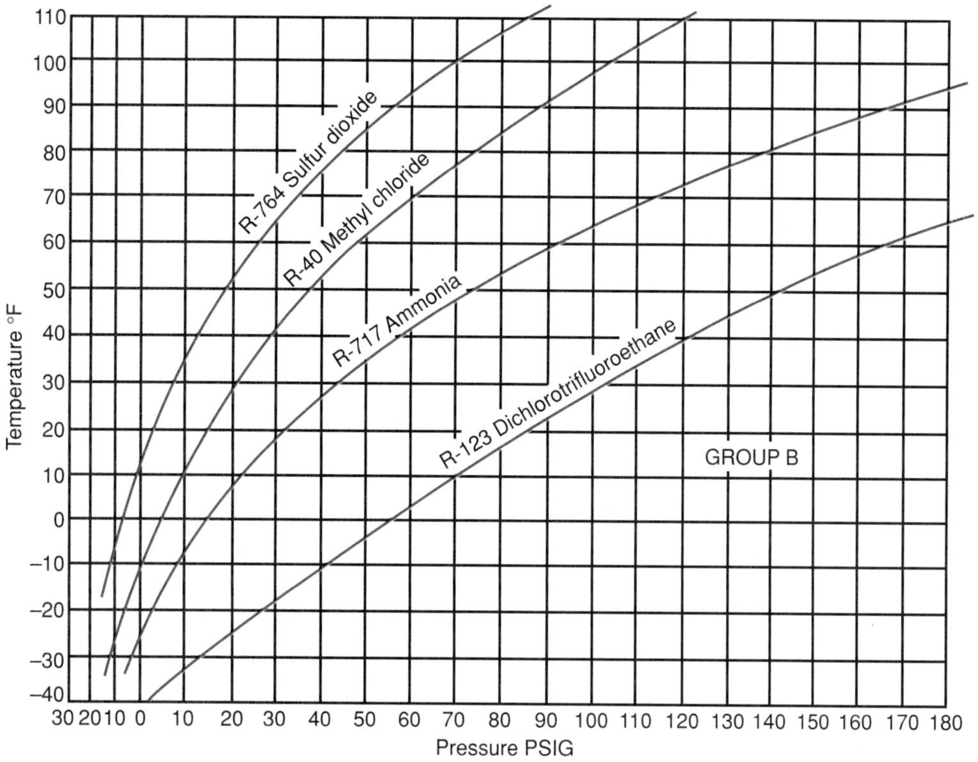

Figure 9-20. *Saturation temperature-pressure curves for some Group B refrigerants.*

under repair. This is to assure that using R-406A (or any alternate refrigerant), will not void the equipment warranty. Note that R-406A has a flammability classification of 2. (See **Figure 9-4** and Section 9.7.) Cylinder color code for R-406A is light gray-green.

9.6 Group B Refrigerants

The Group B refrigerants are toxic. They have the ability to be harmful or lethal. Acute or chronic exposure–by contact, inhalation, or ingestion–must be avoided. Some of the refrigerants in this group include:

R-40 Methyl chloride (CH_3Cl)
R-123 Dichlorotrifluoroethane ($CHCl_2CF_3$)
R-717 Ammonia (NH_3)
R-764 Sulfur dioxide (SO_2)

Pressure-temperature curves for some Group B refrigerants are shown in **Figure 9-20.**

R-717 was one of the first refrigerants used. However, with the exception of absorption refrigerators, it is used today only in large industrial installations.

At one time, R-764 was the refrigerant most used in domestic refrigerators. R-764 and R-40 are seldom used today. However, some sulfur dioxide- (R-764) and methyl chloride- (R-40) charged units still in use.

Properties of R-123 and R-717 are described in the following paragraphs. Additional information about Group B refrigerants can be found in Chapter 31.

9.6.1 R-123 Dichlorotrifluoroethane ($CHCl_2CF_3$)

Refrigerant 123 is an HCFC used as a replacement for R-11. It is used in centrifugal compressors and in foam-blowing applications. R-123 is similar to R-11. It is a low-boiling-point, nonflammable liquid with low chemical reactivity.

R-123 has a boiling point of 82.2°F (28°C). It has a critical temperature of 363°F (184°C). It is colorless, odorless, and has an ozone depletion level of 0.016. Properties of R-123 are shown in **Figure 9-21.** Toxicity levels are shown in **Figure 9-4.** A pressure-heat enthalpy diagram for R-123 is shown in **Figure 9-22.** (See Chapter 31 for pressure-heat enthalpy diagram in metric units.)

R-123 has a greater coefficient of performance than R-11. It is compatible with mineral oil and alkyl benzene oil. The color code for R-123 is light gray.

9.6.2 R-717 Ammonia (NH_3)

R-717 is commonly used in industrial systems. It is a chemical compound of nitrogen and hydrogen (NH_3). Under ordinary conditions, it is a colorless gas. Its boiling temperature at atmospheric pressure is −28°F (−33°C) and its melting point from the solid is −108°F (−78°C).

The low boiling point makes it possible to have refrigeration at temperatures considerably below zero. This is accomplished without using pressures below atmospheric in the evaporator. Its latent heat is 565 Btu/lb. (1310 kJ/kg) at 5°F (−15°C). Thus, large refrigerating effects are possible with relatively small sized

R-123

Molecular Formula	$CHCl_2CF_3$
Molecular Weight	152.91
Normal Boiling Point, (°F)	82.2
Vapor Pressure[a], (psia)	13.24
Liquid Density[a], (lb./cu.ft.)	91.29
Vapor Thermal Conductivity[a] (Btu in/hr ft² °F)	0.0722
Liquid Viscosity[a], (cP)	0.481
Solubility in H_2O[a], (ppm)	2100.
H_2O Solubility[a],(ppm)	662.
Enthalpy of Vaporization (Btu/mol)	24.93
Vapor Flammability[a,b,d]	
Lower limit (vol. %)	none
Upper limit (vol. %)	none
Flash Point [c,d], (°F)	none

[a]Measured at 77 °F.
[b]Flame limits determined using ASTM E 681 with electrically activated match ignition.
[c]No flash point via open cup (ASTM D 1310-67) or closed cup (ASTM D 56-82).
[d]Flammability properties of these materials are not intended to reflect the fire hazards of any resultant cellular or foamed plastic products.

Figure 9-21. *Properties of the liquid and saturated vapor of R-123. (Allied Signal, Inc.)*

machinery. Condensers for R-717 are usually of the water-cooled type, although air-cooled condensers are being developed. The evaporator pressure at 5°F (−15°C) is 34.3 psia (237 kPa) or 19.6 psig (135 kPa). The condenser pressure is 169 psia (1166 kPa) or 155 psig (1170 kPa) at 86°F (30°C). See **Figure 9-23.**

R-717 is somewhat flammable. With the proper proportions of air, it will form an explosive mixture. Accidents from this source, however, are rare.

While not classed as poisonous, the effect of ammonia on the respiratory system is violent. Only very small quantities of it can be breathed safely. The strongest concentration that can be tolerated is about 0.35 volumes per 100 volumes of air. Because of its pronounced and distinguishable odor, R-717 is easily detected in the air.

At 3 to 5 ppm, ammonia is identified by smell. At 15 ppm, the odor is quite irritating. At 30 ppm, the service technician will need a respirator. Exposure of 5 minutes to 50 ppm is the maximum allowed by OSHA. It becomes a hazard to life at 5000 ppm, and is flammable at 150,000 to 270,000 ppm.

Always stand to one side when operating an ammonia valve. A small stem leak may burn and damage the eyes. It may also cause an almost instant loss of consciousness. Wear a tight-fitting mask. R-717 leaks may be quickly and easily detected. In the presence of a sulfur candle or sulfur spray vapor, white smoke-like fumes will form.

A recognized safety code is available for the use and handling of ammonia in refrigerating systems. This code is supplied by The International Institute of Ammonia Refrigeration.

R-717 attacks copper and bronze in the presence of a little moisture. However, it does not corrode iron or steel. It presents no special problems in connection with lubrication unless extreme temperatures are encountered. R-717 is lighter than oil and there is no separation of the two. Excess oil in the evaporator may be removed by opening a valve in the bottom of the evaporator. The solubility of oil in liquid R-717 is only 20 ppm at 5°F (−15°C). At 86°F (30°C), it is only 125 ppm. R-717 vapor is extremely soluble in water. It is used in large compression machines using reciprocating compressors. It is also used in many absorption-type systems. The basic properties of R-717 are shown in Chapter 31.The cylinder color code for R-717 is silver.

9.7 Combustible Refrigerants

Certain refrigerants may form a flammable mixture when mixed with air. The ASHRAE Safety Classification is 2 (low flammability) or 3 (high flammability). (See Section 9.4.) Some of the refrigerants in these groups are:

R-30	Methylene chloride (CH_2Cl_2)—Group 2
R-40	Methyl chloride (CH_3Cl)—Group 2
R-50	Methane (CH_4)—Group 3
R-170	Ethane (C_2H_6)—Group 3
R-290	Propane (C_3H_3)—Group 3
R-406A	Zeotropic mixture (R-22, R-600a, R-142b)—Group 2
R-600	Butane (C_4H_{10})—Group 3
R-717	Ammonia (NH_3)—Group 2

Details concerning R-406A and R-717 were discussed in preceding sections. The remaining refrigerants in this classification are no longer in common use. Their characteristics are not covered in detail in this chapter. See Chapter 31 for more information.

9.8 Expendable Refrigerants

An expendable refrigerant cools a substance or an evaporator, then is released to the atmosphere. The refrigerant is used only once. It is not collected and recondensed, as is the case with the usual compression system. Systems using expendable refrigerants are sometimes referred to as chemical refrigeration or open-cycle refrigeration systems. Refrigerants of this type have a low boiling temperature.

The most common expendable refrigerants are:

- Liquid nitrogen (R-728). Boiling temperature at atmospheric pressure, −320°F (−196°C).
- Liquid helium (R-704). Boiling temperature at atmospheric pressure, −452°F (−269°C).
- Carbon dioxide (R-744). Boiling temperature in either the solid or liquid state, at atmospheric pressure, −109°F (−78°C).

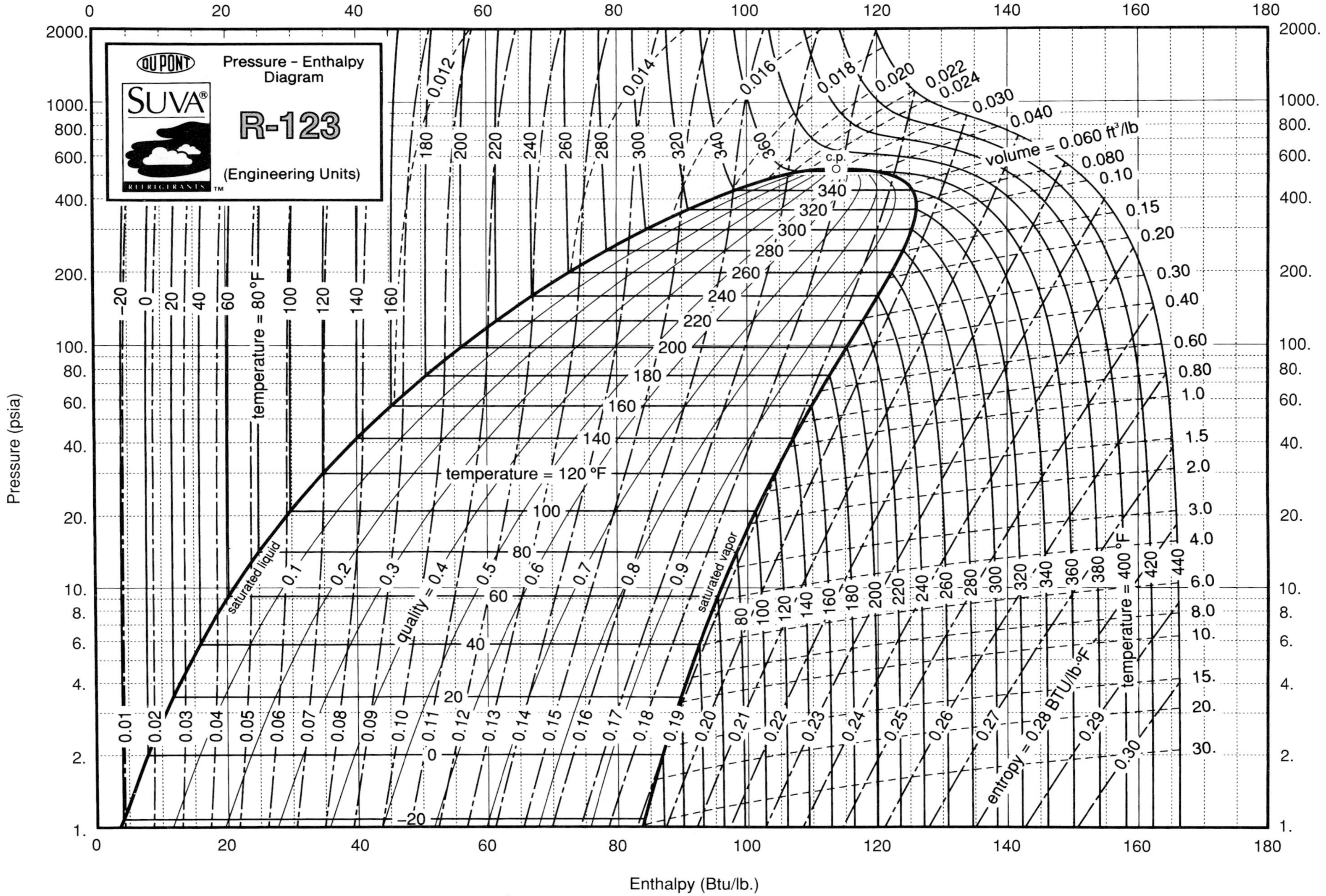

Figure 9-22. *Pressure-heat enthalpy diagram for R-123. (DuPont Company)*

R-717	Pressure		Volume Vapor	Density Liquid	Heat Content Btu/lb.		Latent Heat Btu/lb.
Temp. °F	Psia	Psig	Cu. Ft./Lb.	Lb./Cu. Ft.	Liquid	Vapor	
−100	1.24	27.4*	182.4	45.52	−63.3	572.5	635.8
− 75	3.29	23.2*	72.81	44.52	−37.0	583.3	620.3
− 50	7.67	14.3*	33.08	43.49	−10.6	593.7	604.3
− 35	12.05	5.4*	21.68	42.86	5.3	599.5	594.2
− 25	15.98	1.3	16.66	42.44	16.0	603.2	587.2
− 20	18.30	3.6	14.68	42.22	21.4	605.0	583.6
− 15	20.88	6.2	12.97	42.00	26.7	606.7	580.0
− 10	23.74	9.0	11.50	41.78	32.1	608.5	576.4
− 5	26.92	12.2	10.23	41.56	37.5	610.1	572.6
0	30.42	15.7	9.12	41.34	42.9	611.8	568.9
5	34.27	19.6	8.15	41.11	48.3	613.3	565.0
10	38.51	23.8	7.30	40.89	53.8	614.9	561.1
15	43.14	28.4	6.56	40.66	59.2	616.3	557.1
20	48.21	33.5	5.91	40.43	64.7	617.8	553.1
25	53.73	39.0	5.33	40.20	70.2	619.1	548.9
35	66.26	51.6	4.37	39.72	81.2	621.7	540.5
50	89.19	74.5	3.29	39.00	97.9	625.2	527.3
75	140.5	125.8	2.13	37.74	126.2	629.9	503.7
86	169.2	154.5	1.77	37.16	138.9	631.5	492.6
100	211.9	197.2	1.42	36.40	155.2	633.0	477.8
125	307.8	293.1	0.97	34.96	185.1	634.0	448.9

*Inches of mercury below one atmosphere.

Figure 9-23. *Properties of R-717 (ammonia). Note high latent heat. Latent heat is difference of columns 6 and 7.*

9.9 Water as a Refrigerant

Water is never used in the compression-cycle refrigerating mechanism. However, it is the refrigerant for steam jet refrigeration for air conditioning systems. (See Chapter 20). At atmospheric pressure, water boils at 212°F (100°C). One pound of water absorbs 970 Btu in changing from a liquid at 212°F to a vapor (steam) at 212°F. One kilogram of water absorbs 2260 kJ in changing from a liquid at 100°C to a vapor at 100°C. The usual temperature range in using water as a refrigerant is above 45°F (7°C). Water, in changing from a liquid to a vapor, absorbs a considerable amount of heat.

The heat absorbed can be stated in a more familiar way, using the concept of latent heat. The latent heat of water at 212°F (100°C) is 970 Btu/lb. (2260 kJ/kg). To see how effective water can be, compare this with latent heats for standard refrigerants.

The volume of vapor formed is large. At 45°F (7°C), one pound of water will turn into 2040 ft^3 (57.7 m^3) of vapor. Water vaporizing at 29.6″ Hg. vacuum, or at 0.15 psia (7800 microns), produces a refrigeration temperature of 45°F (7°C).

9.10 Food Freezants

When processing frozen foods, it is best to complete freezing in the shortest possible time. Many commercial food freezing companies submerge the food to be frozen in a liquid refrigerant. The United States Department of Agriculture has approved certain refrigerants for this purpose. They are of a high purity and are designated as food freezants.

This method is very rapid. Heat transfer from the food to the liquid is much faster than if it were surrounded by air at the same temperature. The refrigerant used in this process does not affect the wholesomeness of the food. See Chapter 14 for description of the equipment required for the use of food freezants.

9.11 Cryogenic Fluids

Use of cryogenic fluids is becoming quite common in modern industry. These fluids range in temperature from −250°F (−157°C) to absolute zero (−459.69°F or −273°C). This is called the cryogenic range. Chapter 31 has a listing of these temperature ranges using various temperature scales.

Such low temperatures may be easily reached by evaporating cryogenic fluids. Common cryogenic fluids are:

R-702 Hydrogen
R-704 Helium
R-720 Neon
R-728 Nitrogen
R-729 Air
R-732 Oxygen
R-740 Argon

Figure 1-26 lists the boiling temperature of cryogenic fluids. Containers must be able to withstand extremely low temperatures without losing their strength. Insulation is very heavy, since the temperature of the fluids inside the container is very low.

Small containers are of a vacuum bottle-type construction. Pressure is kept at a relatively low level corresponding to the fluid's vapor pressure.

No attempt is made to seal the fluid in a pressure-tight container. As a result, some of the fluid is continually boiling. This maintains the rest of the fluid at a very low temperature. The vapor from the boiling fluid is allowed to escape.

In general, fluids for low-temperature application are expendable. This means that they are only used once and the vapor vented to the atmosphere.

There are certain cautions which must be observed by anyone handling these fluids:

Do not attempt to use any of these fluids in any container or mechanism that was not designed for its use.

Cryogenic fluids must never be allowed to touch the skin. Such a contact would result in immediate freezing of the flesh. A person handling cryogenic fluids must have his or her entire body protected by suitable clothing, helmets, gloves, and the like.

For further information on cryogenic fluids and temperatures, see Section 1.37 and **Figure 1-26.** Also refer to Chapter 31.

9.12 Refrigerant Cylinders

There are three types of refrigerant cylinders:

- Storage cylinders.
- Returnable service cylinders.
- Disposable (throwaway) cylinders.

The cylinders are made of steel or aluminum. Larger ones usually have a fusible plug safety device threaded into the concave bottom. This protects against overheating or excessive pressure. A valve at the top provides a connection for charging or discharging.

Regulations for cylinders are prescribed by the Department of Transportation (DOT). These regulations ensure the safety of those working with cylinders containing refrigerants. The DOT regulation requires that cylinders which have contained a corrosive refrigerant must be checked every five years. Cylinders containing noncorrosive refrigerants must be checked every ten years. Cylinders over a 4 1/2″ (114 mm) diameter and 12″ (305 mm) long must have a pressure release protective device. This can be a fusible plug or a spring-operated relief valve.

9.12.1 Storage Cylinders

It is cheaper to purchase refrigerants in 100 lb. and 150 lb. cylinders. These become storage cylinders, and often are positioned upside-down with the valve at the bottom. This makes charging service cylinders much easier.

Transferring refrigerant from the large cylinder to the smaller service cylinder should be done carefully. A record should be kept of the quantity of refrigerant removed. To the total figure, add 3% to account for vapor losses.

Storage cylinders should be dated and stamped with a DOT stamp . No cylinder should be used beyond six years from the date shown on it. Refrigerant manufacturers request that all cylinders be returned to them every six months, or sooner. This is necessary so that valve fittings and the complete cylinder may be carefully checked. This service helps to assure safe cylinders. **Use a hoist to lift and move cylinders weighing over 35 pounds (16 kg).**

Storage cylinders are fitted with a valve, and usually protective cap. The cap may be screwed over the valve for shipment. This is shown in **Figure 9-24.** The cylinder valves are usually of the packed one-way type. They should receive the same care as the service valves of a refrigeration system. The packing nut should be kept tight unless the valve is being used. The refrigerant opening should be sealed with a plug or cap when not in use.

Figure 9-24. *Common refrigerant storage cylinders. (National Refrigerants, Inc.)*

9.12.2 Service Cylinders

The service technician carries small (4 lb. to 25 lb.) refillable service cylinders. They are used to charge refrigerating systems. The cylinder valve is usually fitted with 1/4″ (6.4 mm) male flare. The service cylinders are usually filled from the storage cylinders located at the shop.

Caution: Never completely fill a cylinder with liquid refrigerant. Allow space for expansion. Liquid refrigerant expands with an increase in temperature. A cylinder completely filled with cold or cool refrigerant will burst if allowed to warm up. The safe limit is 80% full.

Service cylinders should be weighed before and after filling. In this way, the amount of refrigerant in the cylinder may be readily determined. During charging, the service cylinder should be placed on an accurate scale. Only the specified weight of refrigerant should be charged into it.

Returnable Service Cylinders

Most refrigeration supply houses provide service cylinders on an exchange basis. Empty cylinders are returned and full ones are provided as a replacement. The supply house then arranges to refill the empty service cylinder.

Figure 9-24 illustrates several common refrigerant storage cylinders. Some refrigerant service cylinders are fitted with a carrying handle. The reclaiming of refrigerants has increased the use of returnable cylinders. See Chapter 10.

A new liquid-vapor valve, **Figure 9-25,** is available on standard size cylinders. This valve enables the service technician to charge a system in the usual manner. The system is charged from a standard valve fitting. It may be charged either as vapor or liquid without inverting the cylinder.

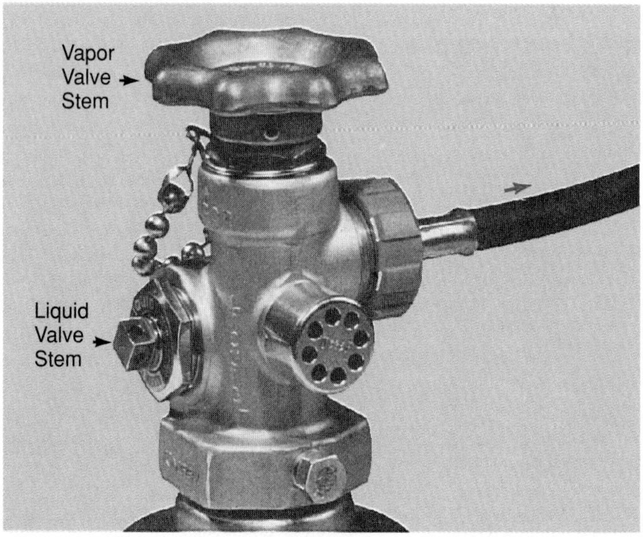

Figure 9-25. *Refrigerant storage cylinder has valve that uses two valve stems. Refrigerant may be drawn from cylinder either as a liquid or vapor by using correct valve stem.*

To transfer refrigerant either as a liquid or vapor, use either of the two valve stems. The vapor valve stem is located at the top of the valve and is marked "vapor." The liquid valve stem, marked "liquid," is located on the side of the valve. It is attached to a tube which extends inside to the bottom of the cylinder. The valve has only one valve stem wheel. This prevents opening both stems at the same time. A standard hood cap is placed over the valve when it is not in use. **Figure 9-26** is a cross section of the liquid-vapor valve.

9.12.3 Disposable Cylinders

Many popular refrigerants are available in disposable ("throwaway") cylinders. These contain small quantities of refrigerant, from a few ounces up to 50 lbs. These containers are easy to handle and they eliminate the problem of refilling. **Figure 9-27** illustrates a popular disposable cylinder.

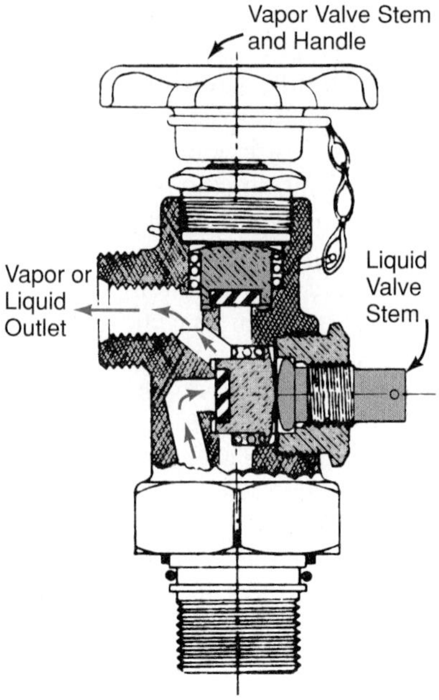

Figure 9-26. *Cross section of a refrigerant cylinder valve which allows either refrigerant liquid or refrigerant vapor to leave the cylinder.*

Most disposable cylinders are fitted with relief valves. Usually these are located in the valve body. Some "throwaway" refrigerant containers are sealed cans. The top is made so that a special service valve may be tightly clamped on. When clamped on the can, this valve can be made to puncture it. This provides a means of drawing refrigerant from the can. **Figure 9-28** illustrates such a valve.

Caution: Disposable cylinders should never be recharged. They should not be used to store refrigerant removed from a system.

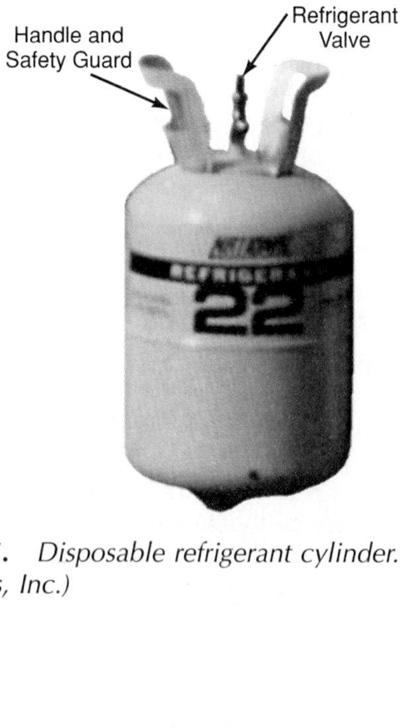

Figure 9-27. *Disposable refrigerant cylinder. (National Refrigerants, Inc.)*

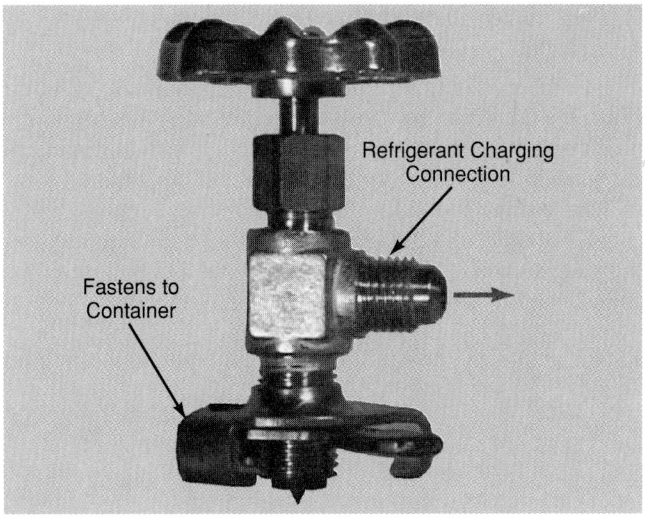

Figure 9-28. *Hand valve which may be attached to throwaway refrigerant containers. Valve clamps to top of refrigerant container.*

9.13 Use of Pressure-Temperature Tables

The pressure-temperature relationship of a refrigerant under saturated conditions can be shown in a table. (***Saturated conditions*** means holding as much vapor as the refrigerant is able).

The table also shows the volume of one pound of the vapor at that temperature. The latent heat, specific heat, and density of the liquid are also shown.

Tables of these values for some popular refrigerants are shown in **Figures 9-5, 9-9, 9-14, 9-16, 9-18,** and **9-23.** This information is of great value to service technicians and engineers.

To use such a table, find the temperature being investigated in the vertical left-hand column. Move across the columns horizontally to find the pressure.

9.14 Head Pressures (High Side)

Pressures will vary with refrigerants. In air-cooled condensers, the head pressure should be between 30°F (17°C) and 35°F (19°C) higher than the ***ambient*** (surrounding) temperature of the air passing over the condenser.

In a water-cooled condenser, the head pressure should correspond to a temperature 15°F (8°C) to 20°F (11°C) above the exhaust temperature of the water. Using this information, the correct head pressure for common refrigerants may be found. Refer to **Figures 9-2** and **9-20.** Read across the chart.

In all cases, the condensing temperature will rise until the heat loss from the condenser equals the heat input into the condenser. If condensing pressure is too high, the compressor has to work too hard. Too much vapor will be left in the compressor clearance pocket. This lowers its volumetric efficiency. The temperature of the exhaust vapor will be too high and may cause oil deterioration. Usual causes of above-normal head pressures are listed below:

- A noncondensable vapor or gas, such as air, trapped in the condenser. Head pressure will be the sum of the refrigerant vapor pressure plus the air pressure. This is due to Dalton's Law.
- An overcharge of refrigerant in systems using a low-side float, an expansion valve, or a thermostatic expansion valve. Some of the heat-radiating space in the condenser will fill with liquid refrigerant and reduce the condenser's heat-radiating ability.
- Either the inside or the outside of the condenser is dirty. This dirt will act as an insulator, lowering the heat-radiating capacity of the condenser. Then, the condenser temperature will rise.
- Air or water movement through the condenser is reduced by blocked passages or poor water flow. When this occurs, there will not be enough heat-removing material to transfer heat from the condenser.
- A restriction in the system, such as a clogged capillary tube or a stuck refrigerant control. This may temporarily cause a high head pressure.
- An above-normal low-side pressure. When this occurs, the head pressure will be higher than normal.

9.15 Refrigerator Temperatures

The low-side pressure in a refrigerating system determines the temperature in the evaporator.

You must first determine the desired temperature of the cabinet or fixture. Then, you must adjust the motor control until this temperature is maintained. However, there are many cases where a certain evaporator temperature and a cabinet temperature relationship should exist.

Cabinet temperatures are fairly standard. **Figure 9-29** shows recommended temperatures for some common fixtures (cabinets). The recommended temperature for various applications is shown in **Figure 9-30.**

It is necessary to have the correct-size evaporator for the temperature desired. If the evaporator is too large, temperature will be above normal. If the evaporator is undersized, temperature will be below normal. The evaporator will have a lower temperature than the fixture temperature (a temperature difference is needed for heat flow).

Normally, the refrigerant will be 10°F (6°C) colder than the evaporator temperature when the unit is running. The refrigerant and the evaporator will become the same temperature during the off cycle. The evaporator surface temperature depends on its size. It also depends on the rate at which heat is being removed from the fixture.

Fixture (Cabinet)	Temp. °F	Temp. °C
Back Bar	37–40	3–4
Beverage Cooler	37–40	3–4
Beverage Precooler	35–40	2–4
Candy Case (Display)	60–65	16–18
Candy Case (Storage)	58–65	15–18
Dairy Display Case	36–39	2–3
Double Display Case	36–39	2–3
Delicatessen Case	36–40	2–4
Dough Retarding Refrigerator	34–38	1–3
Florist Display Refrigerator	40–50	4–10
Florist Storage Case	38–45	3–7
Frozen Food Cabinet (Closed)	–10 to –5	–23 to –21
Frozen Food Cabinet (Open)	–7 to –2	–22 to –19
Grocery Refrigerator	35–40	2–4
Retail Market Cooler	34–39	1–3
Pastry Display Case	45–50	7–10
Restaurant Service Refrigerator	36–40	2–4
Restaurant Storage Cooling	35–39	2–3
Top Display Case (Closed)	35–42	2–6
Vegetable Display Refrigerator (Closed)	38–42	3–6
(Opened)	38–42	3–6

Figure 9-29. *Recommended fixture (cabinet) temperatures.*

Application	Temp. °F	Temp. °C
Service	34–38	1–3
Meats	30–34	–1 to 1
Bananas	60–65	16–18
Fresh Meats	28–32	–2 to 0
Aging Room	30–34	–1 to 1
Chill Room	35–39	2–3
Curing Room	32–36	0–2
Freezer Room	–15	–26
Poultry	30–34	–1 to 1
Vegetables, Fresh	36–42	2–6
Ice Cream Hardening	–25	–32
Ice Cream Storage	–20 to –10	–29 to –23
Plants and Flowers	38–50	3–10
Fur Storage	33–37	0–3
Locker Room	–5 to 0	–21 to –18

Figure 9-30. *Recommended temperatures for various refrigeration applications.*

The temperature of a typical frosting-type domestic evaporator will vary from 0°F to 25°F (−18°C to −4°C), and the refrigerant temperature will be about 10°F (6°C) lower than this. It will be in the range of −10°F to 15°F (−23°C to −9°C) while the unit is running. The table in **Figure 9-31** gives the pressure corresponding to the evaporating temperatures and condensing temperatures for eleven popular refrigerants.

To change psig values to the metric equivalent (kilopascals or kPa), first change psig to psia by adding 14.7 to the psig value. Then, divide the psia value by 14.7 and multiply by 101.3. For simplicity, most technicians just multiply psia by 6.9 to obtain the kPa value.

Example: R-12 at 5°F (−15°C) has a pressure of 11.8 psig. The pressure in psia is: 11.8 + 14.7 = 26.5 psia. The metric pressure is:

$$26.5 \div 14.7 \times 101.3 = 182.6$$
$$= 183 \text{ kPa}$$

Using 6.9 directly: $26.5 \times 6.9 = 182.9 = 183$ kPa.

9.16 Refrigerant Applications

Some popular refrigerant applications are shown in **Figure 9-32.** One type of refrigerant may be used in a number of applications. Some refrigerant applications recommended for different types of compressors are shown in **Figure 9-33.**

The type of refrigerant to be used in a given system is determined by the manufacturer. Several items are considered in the selection of the refrigerant:

- The *system capacity,* governed by the refrigerant boiling point.
- The *volume of the vapor* pumped to provide the necessary refrigeration.
- The *latent heat* of the refrigerant.
- The *operating temperatures* required.
- The *size* of the equipment.

9.17 Changing/Identifying Refrigerants

Due to the phase-out of CFC refrigerants, it may be necessary to change the type of refrigerant in a unit. This is referred to as ***retrofitting.***

The identification of the type of refrigerant used in a system may be difficult. It is normally accomplished by checking the manufacturer's tags on the equipment. Identification by color or smell is difficult. The only exceptions are R-764, sulphur dioxide, and R-717 (ammonia). **Sniffing of refrigerants can be deadly.**

A refrigerant identification instrument is shown in **Figure 9-34.** This instrument indicates whether the system has R-12, R-134a, or another mixture of refrigerants. The instrument can be used on residential, commercial, or automotive systems, or on refrigerant cylinders. It is used extensively in automotive air conditioning, where it is difficult to determine if an R-12 system has been converted to R-134a or another type of refrigerant.

If it is necessary to change refrigerant, certain system components must be modified. This is determined by the original design of the system. The technician should contact the manufacturer prior to changing the type of refrigerant or refrigerant controls.

Often the refrigerant control will have to be changed and the oil replaced. The proper new controls and oil are determined by the new refrigerant to be used. Also, a new filter-drier would be installed in the system. *Always use the proper recovery/recycle equipment when changing refrigerants.* See Chapter 10.

9.18 Amount of Refrigerant Required in a System

The amount of refrigerant that should be used varies with the type of system. Some systems are not sensitive to the amount of refrigerant. These include low-side float, automatic expansion valve, and thermo-

PSIG	Temperature, °F								
	REFRIGERANT - (Sporlan Code)								
	MP39 (X) or 401A (X)		HP62 (S) or 404A (S)		HP80 (L) or 402A (L)		AZ-50(P) or 507(P)	124 (Q)	125
	Dew point	Bubble point	Dew point	Bubble point	Dew point	Bubble point			
5 *	−23		−57		−59		−59	3	−63
4 *	−22		−56		−58		−57	4	−61
3 *	−20		−54		−56		−56	6	−60
2 *	−19		−53		−55		−55	7	−58
1 *	−17		−52		−54		−53	9	−57
0	−16		−51		−53		−52	10	−56
1	−13		−48		−50		−50	13	−53
2	−11		−46		−48		−47	16	−51
3	−9		−43		−45		−45	18	−49
4	−6		−41		−43		−43	21	−46
5	−4		−39		−41		−41	23	−44
6	−2		−37		−39		−39	26	−42
7	0		−35		−37		−37	28	−40
8	2		−33		−36		−35	30	−39
9	4		−32		−34		−34	32	−37
10	6		−30		−32		−32	34	−35
11	8		−28		−30		−30	36	−33
12	9		−27		−29		−29	38	−32
13	11		−25		−27		−27	40	−30
14	13		−23		−26		−25	41	−29
15	14		−22		−24		−24	43	−27
16	16		−20		−23		−23	45	−26
17	17		−19		−21		−21	46	−24
18	19		−18		−20		−20	48	−23
19	20		−16		−19		−18	49	−22
20	21		−15		−17		−17	51	−20
21	23		−14		−16		−16	52	−19
22	24		−12		−15		−15	54	−18
23	25		−11		−14		−13	55	−16
24	27		−10		−12		−12	57	−15
25	28		−9		−11		−11	58	−14
26	29		−8		−10		−10	59	−13
27	30		−6		−9		−9	61	−12
28	32		−5		−8		−8	62	−11
29	33		−4		−7		−6	63	−10
30	34		−3		−6		−5	65	−8
31	35		−2		−5		−4	66	−7
32	36		−1		−4		−3	67	−6
33	37		0		−2		−2	68	−5
34	38		1		−1		−1	69	−4
35	39		2		0		0	71	−3
36	40	30	3		0		1	72	−2
37	42	31	4		1		2	73	−1
38	43	32	5		2		3	74	0
39	44	33	6		3		4	75	0
40	45	34	7		4		5	76	1
42	46	36	8		6		6	78	3
44	48	38	10		8		8	80	5
46	50	40	12		10		10	82	7
48		42	14		11		12	84	8
50		44	16		13		13	86	10
52		45	17		14		15	88	11
54		47	19		16		16	90	13
56		49	20		18		18	91	15
58		50	22		19		19	93	16
60		52	23		20		21	95	17
62		53	25		22		22	97	19
64		55	26		23		24	98	20
66		56	27		25		25	100	22
68		58	29		26		27	101	23
70		59	30	29	27		28	103	24
72		61	32	31	29		29	104	26
74		62	33	32	30		30	106	27
76		64	34	33	31		32	107	28
78		65	35	34	32	30	33	109	29
80		66	37	36	34	31	34	110	31
85		69	40	39	37	34	37	114	33
90		73	42	42	40	37	40	117	36
95		76	45	44	42	40	43	120	39
100		78	48	47	45	43	46	123	42
105		81		50	48	45	48	126	44
110		84		52	50	48	51	129	47
115		87		55		50	53	132	49
120		89		57		53	56	135	51
125		92		59		55	58	138	54
130		94		62		57	60	140	56
135		96		64		60	62	143	58
140		99		66		62	64	145	60
145		101		68		64	67	148	62
150		103		70		66	69	150	64
155		105		72		68	71	152	66
160		108		74		70	73	154	68
165		110		76		72	74	157	70
170		112		78		74	76	159	72
175		114		80		75	78	161	73
180		116		82		77	80	163	75
185		117		83		79	82	165	77
190		119		85		81	83	167	79
195		121		87		82	85	169	80
200		123		88		84	87	171	82
205		125		90		86	88	173	83
210		127		92		87	90	175	85
220		130		95		91	93	178	88
230		133		98		94	96	182	91
240		136		101		97	99	185	94
250		140		104		99	102	188	97
260		143		107		102	105	192	99
275		147		111		106	109	196	103
290		151		115		110	112	201	107
305		155		118		114	116	205	111
320		159		122		118	120	209	114
335		163		126		121	123	213	118
350		167		129		125	126	217	121
365		170		132		128	129	221	124

*Inches mercury below one atmosphere

PSIG	Temperature, °F				
	REFRIGERANT - (Sporlan Code)				
	22 (V)	502 (R)	12 (F)	134a (J)	717 (A)
5 *	−48	−57	−29	−22	−34
4 *	−47	−55	−28	−21	−33
3 *	−45	−54	−26	−19	−32
2 *	−44	−52	−25	−18	−30
1 *	−43	−51	−23	−16	−29
0	−41	−50	−22	−15	−28
1	−39	−47	−19	−12	−26
2	−37	−45	−16	−10	−23
3	−34	−42	−14	−8	−21
4	−32	−40	−11	−5	−19
5	−30	−38	−9	−3	−17
6	−28	−36	−7	−1	−15
7	−26	−34	−4	1	−13
8	−24	−32	−2	3	−12
9	−22	−30	0	5	−10
10	−20	−29	2	7	−8
11	−19	−27	4	8	−7
12	−17	−25	5	10	−5
13	−15	−24	7	12	−4
14	−14	−22	9	13	−2
15	−12	−20	11	15	−1
16	−11	−19	12	16	1
17	−9	−18	14	18	2
18	−8	−16	15	19	3
19	−7	−15	17	21	4
20	−5	−13	18	22	6
21	−4	−12	20	24	7
22	−3	−11	21	25	8
23	−1	−9	23	26	9
24	0	−8	24	27	11
25	1	−7	25	29	12
26	2	−6	27	30	13
27	4	−5	28	31	14
28	5	−3	29	32	15
29	6	−2	31	33	16
30	7	−1	32	35	17
31	8	0	33	36	18
32	9	1	34	37	19
33	10	2	35	38	19
34	11	3	37	39	20
35	12	4	38	40	21
36	13	5	39	41	22
37	14	6	40	42	23
38	15	7	41	43	24
39	16	8	42	44	25
40	17	9	43	45	26
42	19	11	45	47	28
44	21	13	47	49	29
46	23	15	49	51	31
48	24	16	51	52	32
50	26	18	53	54	34
52	28	20	55	56	35
54	29	21	57	57	37
56	31	23	58	59	38
58	32	24	60	60	40
60	34	26	62	62	41
62	35	27	64	64	42
64	37	29	65	65	44
66	38	30	67	66	45
68	40	32	68	68	46
70	41	33	70	69	47
72	42	34	71	71	49
74	44	36	73	72	50
76	45	37	74	73	51
78	46	38	76	75	52
80	48	40	77	76	53
85	51	43	81	79	56
90	54	46	84	82	58
95	56	49	87	85	61
100	59	51	90	88	63
105	62	54	93	90	66
110	64	57	96	93	68
115	67	59	99	96	70
120	69	62	102	98	73
125	72	64	104	100	75
130	74	67	107	103	77
135	76	69	109	105	79
140	78	71	112	107	81
145	81	73	114	109	82
150	83	75	117	112	84
155	85	77	119	114	86
160	87	80	121	116	88
165	89	82	123	118	90
170	91	83	126	120	91
175	92	85	128	122	93
180	94	87	130	123	95
185	96	89	132	125	96
190	98	91	134	127	98
195	100	93	136	129	99
200	101	95	138	131	101
205	103	96	140	132	102
210	105	98	142	134	104
220	108	101	145	137	107
230	111	105	149	140	109
240	114	108	152	143	112
250	117	111	156	146	115
260	120	114	159	149	117
275	124	118	163	153	121
290	128	122	168	157	124
305	132	126	172	161	128
320	136	130	177	165	131
335	139	133	181	169	134
350	143	137	185	172	137
365	146	140	188	176	140

*Inches mercury below one atmosphere

Figure 9-31. *Temperature-pressure chart which may be used to determine operating pressures for various refrigerants. (Sporlan Valve Company)*

Refrigerant Type	Appropriate Lubricant			
R- 11			MO	
R- 12	POE	AB	MO	
R- 13	POE	AB	MO	
R- 22	POE	AB	MO	
R- 23	POE			
R-123	POE	AB	MO	
R-124	POE	AB		
R-125	POE			
R-134a	POE			PAG*
R-176			MO	
R-401A	POE	AB		
R-401B	POE	AB		
R-401C	POE	AB		
R-402A	POE	AB		
R-402B	POE	AB		
R-403B	POE	AB	MO	
R-404A	POE			
R-407A	POE			
R-407B	POE			
R-407C[1]	POE			
R-410A[1]	POE			
R-500	POE	AB	MO	
R-502	POE	AB	MO	
R-503	POE	AB	MO	
R-507	POE			
R-717			MO	

*PAG is used primarily for automotive applications as a lubricant with R-134a.
[1]This has been proposed for addition to ASHRAE Standard 34-1992.

Figure 9-35. *Appropriate lubricants for use with various refrigerants. POE = Polyol ester. AB = Alkylbenzene. MO = Mineral oil. PAG = Polyalkylene glycol.*

It may be impossible to remove *all* moisture from a refrigerant. However, the amount of moisture must be kept very low. The maximum amount of moisture allowed will vary with the kind of refrigerant and the low-side temperature.

Most refrigerant manufacturers supply refrigerants that are dry (virtually free of moisture). The moisture content never exceeds five ppm (parts per million). Liquid refrigerants can hold more moisture in solution as the low-side temperature rises. This enables the refrigerant to circulate without danger of the moisture separating from it. Moisture that separates may freeze or form harmful compounds. For example, R-12 is safe to use at 20°F (−7°C) with 17 ppm moisture content. At 0°F (−18°C), it is safe to use with 8.3 ppm. At −20°F (−29°C) it is safe to use with 3.8 ppm. At −40°F (−40°C), it is only safe with 1.7 ppm.

A table of safe moisture content for certain refrigerants is shown in **Figure 9-36.** Any amount of moisture at or above the value of the "wet color" will be harmful to the system. The service technician must depend on the moisture indicator to determine the amount of moisture in the system.

If the moisture indicator shows a "wet color," a new drier should be installed in the line. The system should then be operated until the moisture indicator indicates a "dry color." It may sometimes be necessary to replace the drier several times to remove sufficient moisture from the system.

Refrigerant	Dry Color	Wet Color
12	Below 5	Above 15
22	Below 30	Above 100
502	Below 15	Above 50

Figure 9-36. *Safe ("dry color") and unsafe ("wet color") moisture content for three types of refrigerants. Dry color column shows allowable water concentration in parts per million at 75°F (24°C). Wet color column shows concentrations that will cause problems. When these quantities are present, water will start freezing in the low side. This may block automatic expansion valves.*

When servicing a system, avoid exposing cold internal parts to air. Moisture from the air will condense on the parts and get inside the system. Warm the parts to room temperature with a heat lamp before opening the system.

9.21 Review of Safety

Wear goggles and gloves at all times, especially when charging or discharging refrigerant. These will protect the eyes, skin, and hands in case of a sudden leak.

Liquid refrigerant on the skin may freeze the skin surface and cause frostbite. If this should happen, quickly wash away the refrigerant with water. Treat the damaged surface for frostbite. Any accident involving refrigerants should be immediately referred to a doctor.

Refrigerants R-717 and R-764 are very irritating to the eyes and lungs. The service technician must always avoid exposure to these refrigerants.

Refrigerant oil contained in a hermetic compressor which has had a burnout may be very acidic. This oil should never be allowed to touch the skin. It may cause an acid burn.

When a leak is suspected, thoroughly ventilate the room before working on the unit. Many refrigerants have no disagreeable odor. It is possible to work in an area without being aware that there is a considerable amount of refrigerant vapor present. Also, many refrigerants are heavier than air and will replace the air in a room. This can be very dangerous. The air you breath must contain at least 9% oxygen. If it does not, you will lose consciousness. Instruments are available to warn when the air's oxygen content is below a safe level. Every worker must know how to read instruments which warn of a poor oxygen level. Practice some of the steps which should be taken if a warning is given.

Sniffing of a refrigerant can cause death.

Always use the proper recovery/recycle equipment when changing refrigerants.

Always check for recommended operating pressures for each refrigerant. Install gauges to find the pressures in the system.

To avoid mixing refrigerants, always check the refrigerant R-number before charging. Make certain no lighted flames are near a system that is suspected of having a bad fluorocarbon refrigerant leak. The refrigerant may break down and produce dangerous gases.

Always charge refrigerant vapor into the low side of the system. Liquid refrigerant entering a compressor may injure the compressor. It may cause the unit to burst.

Moisture should not be allowed to enter a refrigerating system. It is likely to cause considerable damage to the system. All parts of refrigerating mechanisms must be kept dry at all times. Containers of oil must always be kept tightly sealed. This avoids the possibility of the oil absorbing moisture from the air.

Make sure that a refrigerant service cylinder is *never* completely filled with liquid refrigerant. If a service cylinder is completely filled with liquid refrigerant and allowed to become warm, the hydrostatic pressure inside the cylinder will cause it to burst. Always check the DOT cylinder stamp to make sure it is a safe cylinder.

Refrigerant cylinders should always be stored in a cool, dry place. Carefully assemble fittings and tubing to cylinders. Stripped threads are dangerous and costly. Never use refrigerant cylinders as supports or rollers.

Refrigerant cylinders should be used only for storing the refrigerant marked on the label. Using the cylinder for compressed air is very dangerous. The cylinder may explode when exposed to these high pressures.

9.22 Test Your Knowledge

Please do not write in this text. Place your answers on a separate sheet of paper.

1. Why is it necessary to be cautious when handling contaminated refrigerant oil?
   A. It is a controlled substance according to the EPA.
   B. It may be acidic.
   C. It may stain clothes.
   D. All of the above.
2. Which color cylinder designates R-134a?
   A. Yellow.
   B. White.
   C. Sky blue.
   D. Light green.
3. Which of the following refrigerants is not composed of CFCs?
   A. R-11.
   B. R-22.
   C. R-12.
   D. R-113.
4. Which of the following is a means for locating R-502 leaks?
   A. Soap solution.
   B. Halide torch.
   C. Electronic leak detector.
   D. All of the above.
5. What is a common head pressure for air-cooled R-12 refrigerating systems?
   A. 92 psig.
   B. 126 psig.
   C. 11.8 psig.
   D. None of the above.
6. What is the low-side pressure at 5°F (−15°C) for R-22?
   A. 28.3 psig.
   B. 155.7 psig.
   C. 11.8 psig.
   D. None of the above.
7. Group B refrigerants are ______.
   A. toxic
   B. harmful
   C. lethal
   D. All of the above.
8. Which of the following is *not* an oil suitable for use with R-134a?
   A. Mineral oil-based lubricant.
   B. Polyol ester.
   C. Alkyl benzene.
   D. Polyalkylene glycol.
9. Three commonly used azeotropic refrigerants are ______.
   A. R-502, R-500, and R-507
   B. R-717, R-500, and R-502
   C. R-401A, R-401B, and R-401C
   D. R-134a, R-502, and R-500
10. An azeotropic blend does not change in composition when it ______.
    A. evaporates
    B. condenses
    C. Neither A nor B.
    D. Both A and B.
11. Which type of refrigerant has not been produced since 1995?
    A. HCFCs.
    B. CFCs.
    C. Azeotropics.
    D. Zeotropics.
12. Refrigerants containing chlorofluorocarbons are being replaced because of rulings issued by ______.
    A. EPA (Environmental Protection Agency)
    B. American Standard Safety Code for Mechanical Refrigeration
    C. ASHRAE (American Society of Heating and Air-Conditioning Engineers)
    D. Both A & B.
13. Which of the following is an HFC refrigerant?
    A. R-114.
    B. R-134a.
    C. R-22.
    D. R-113.
14. R-123 serves as a replacement for R-11 in which type of application?
    A. Centrifugal chillers.
    B. Automotive.
    C. Residential.
    D. Commercial.

15. Which of the following is a refrigerant blend?
    A. Azeotropic.
    B. Chlorofluorocarbons (CFCs).
    C. Hydrofluorocarbons (HFCs)
    D. Hydrochlorofluorocarbons (HCFCs)
16. What does the word "ozone" describe?
    A. A thin layer in the earth's upper atmosphere.
    B. A protective layer for the earth.
    C. A filter for the sun's ultraviolet rays.
    D. All of the above.
17. Refrigerants are classified into which areas?
    A. CFCs, HCFCs.
    B. CFCs, HCFCs, azeotropic.
    C. CFCs, HCFCs, HFCs.
    D. CFCs, HFCs, HCFCs, blends (azeotropic and zeotropic).
18. Which of the following is an HCFC refrigerant?
    A. R-114.
    B. R-22.
    C. R-113.
    D. All of the above.
19. How can the refrigerant temperature in an air-cooled condenser be determined?
    A. By adding 30°F to 35°F (17°C to 19°C) to the desired condenser temperature.
    B. By subtracting 25°F (13°C) from the present condenser temperature.
    C. Either A or B.
    D. By taking the operational pressure.
20. Why should refrigerant cylinders be filled to only 80% of capacity?
    A. Liquid refrigerant expands with an increase in temperature.
    B. A refrigerator cylinder completely filled with cold refrigerant will burst if it warms up.
    C. Liquid refrigerant will foam in the cylinder.
    D. Both A & B.

Chapter 10

REFRIGERANT RECOVERY/RECYCLING/ RECLAIMING

Key Words:

Air-Conditioning and Refrigeration Institute (ARI)
chlorofluorocarbon (CFC)
Department of Transportation (DOT)
Environmental Protection Agency (EPA)
halogenated chlorofluorocarbon (HCFC)
liquid transfer method
reclaiming
recovery
recycle
refrigerant management system
vapor recovery method

Learning Objectives:

After studying this chapter, the technician will be able to:

- Describe the effect of chlorofluorocarbon (CFC) refrigerants on the ozone layer in the atmosphere.
- Understand the Environmental Protection Agency (EPA) rules governing fully halogenated refrigerants (CFCs).
- Follow the EPA regulations regarding recycling of refrigerants.
- Discuss the proper procedures to recover, recycle, and reclaim chlorofluorocarbon refrigerants (CFCs).
- Identify the various types of refrigerant recovery and recycling equipment and their use.
- Follow the procedures as set forth by the Department of Transportation regarding the transportation of refrigerant cylinders and drums.
- **Follow approved safety procedures.**

10.1 Chlorofluorocarbons (CFCs), Hydrochloroflurocarbons (HCFCs), and the Ozone Layer

The ozone layer is a fairly thin layer of the earth's upper atmosphere. It is approximately thirty-five miles above the ground. It is often called a screen or shield. The ozone layer is credited with protecting the earth from the damaging ultraviolet rays of the sun. The ozone layer functions as a filter for the sun's ultraviolet rays. It protects all life forms on the earth from the damaging effects of the sun, **Figure 10-1.** Destruction of this shield by the release of ***chlorofluorocarbons (CFCs)*** into the atmosphere is of great concern. The use of chlorofluorocarbon refrigerants is, therefore, also of concern.

CFCs are a family of chemicals containing chlorine, fluorine, and carbon. Chlorofluorocarbon R-12, R-11, and others are used as refrigerants. They are also used as blowing agents for the manufacture of insulation, packaging, etc. The stability and chlorine contents of these compounds cause depletion of the ozone layer.

Figure 10-1. *The depletion of the ozone layer can cause problems for plant and sea life. (General Filters, Inc.)*

Countries throughout the world have passed legislation preventing the use of chemicals that affect this layer. The United States ***Environmental Protection Agency (EPA)*** has also enacted regulations. These regulations state that fully halogenated CFC refrigerants must be phased out by the turn of the century. These refrigerants include:

R-11	(trichloromonofluoromethane)
R-12	(dichlorodifluoromethane)
R-113	(trichlorotrifluoroethane)
R-114	(dichlorotetrafluoroethane)
R-115	(chloropentachloroethane)

Production phaseout of these refrigerants is scheduled to be reduced from current production levels as follows:

1996	60% reduction.
1997	85% reduction.
1998	85% reduction.
1999	85% reduction.
2000	Total phaseout.

Recycling of refrigerants used for air conditioning in vehicles (autos, trucks, etc.), is mandatory.

The schedule for hydrochlorofluorocarbons (HCFCs—R-22, R-502, and others; see Section 9.1.3) phaseout is as follows:

2015	Production freeze and use limitations.
2020	Prohibited for new air conditioning and refrigeration use.
2030	Total phaseout.

However, the phaseout of a refrigerant can be changed by the EPA. An example of this was the total phaseout of R-12 in 1996.

Penalties and fines for violating these provisions are rather severe. The EPA is authorized to seek legal action against any person violating these provisions.

Studies have indicated that HCFC-22 and R-502 are considered less of a problem than fully halogenated chlorofluorocarbons. The latter are made of hydrocarbon molecules. In these, all the hydrogen atoms have been replaced by the halogen atoms "chlorine" and "fluorine." R-22 and R-502 are not fully halogenated. R-22 has two fluorine atoms, one chlorine atom, and a hydrogen atom. They are connected to a carbon atom. These molecules will tend to break down in the lower atmosphere before getting to the stratosphere. In the stratosphere, the chlorine can damage the ozone layer. Therefore, nonfully halogenated refrigerants like R-22 and R-502 cause much less ozone damage.

Research continues to determine how existing equipment may be modified to accept the newer alternative refrigerants, such as R-134a and R-123, that have little or no effect on the ozone. R-134a is currently being used in automotive air conditioning systems. R-123 is in use in new commercial applications.

Figure 10-2 shows a comparison of the physical properties for present refrigerants, possible replacements, and their uses.

10.2 Recovery, Recycling, Reclaiming of Refrigerants

Laws preventing release of CFC refrigerants into the atmosphere have resulted in new procedures. Methods of recovering, recycling, and reclaiming these refrigerants have been developed. The industry has adopted specific definitions for these terms:

- ***Recovery:*** To remove refrigerant in any condition from a system and store it in an external container. This may be done without necessarily testing or processing the refrigerant in any way.
- ***Recycling:*** To clean refrigerant for reuse by oil separation and single or multiple passes through devices such as replaceable core filter-driers. These devices reduce moisture, acidity, and matter. This term usually applies to procedures implemented at the field job site or at a local service shop.
- ***Reclaim:*** To reprocess refrigerant to new product specifications by means which may include distillation. This will require chemical analysis of the refrigerant to determine that appropriate product specifications are met. Reprocessing procedures are usually available only at a reprocessing or manufacturing facility. This also includes on-site or local service shops that are equipped with highly technical equipment.

Any technician opening a system for servicing or refrigerant disposal must use certified recovery equipment. ***Refrigerant recovery management equipment,*** **Figure 10-3,** is divided into three categories:

- ***Recovery:*** A unit that recovers or removes the refrigerant.
- ***Recovery/Recycle:*** A unit that will recover and recycle the refrigerant.
- ***Reclaim:*** A unit that will reclaim the refrigerant within the EPA standards.

The technician must always follow local, state, and EPA rules and regulations when working with refrigerants. One of the primary guidelines in its proper use is the utilization of refrigerant recovery/recycle equipment. While this is standard procedure, it is *not* indicated in all of the servicing descriptions throughout the book.

10.3 Refrigerant Recovery Equipment

Recovery machines are available in various designs. The basic small units are intended for use with R-12 and act as recovery stations. See **Figure 10-4.** Cylinders used for recovered refrigerants are gray with yellow ends.

PHYSICIAL DATA

	113	11	114	12	500	22	502	13	503	HCFC 141b	HCFC 123	HFC 134a
Chemical Formula.	$C_2Cl_3F_3$	CCl_3F	$C_2Cl_2F_4$	CCl_2F_2		$CHClF_2$		$CClF_3$		CCl_2FCH_3	$CHCl_2CF_3$	CF_3CH_2F
Molecular Weight.	187.4	137.4	170.9	120.9	99.3	86.5	111.6	104.5	87.5	116.95	152.91	102.03
Boiling Point @ 1 Atmos. (°F).	117.6	74.9	38.8	–21.6	–28.3	–41.4	–49.8	–114.6	–126.1	89.7	82.2	–15.08
Freezing Point @ 1 Atmos. (°F).	–31	–168	–137	–252	–254	–256	—	–294	—	–154.3	–160.6	–141.9
Critical Temperature, (°F).	417	388	294	234	222	205	180	84	67	410.4	363.2	214
Critical Pressure, (psia).	499	640	473	597	642	722	591	561	632	673.0	540.0	589.8
Saturated Liquid Density @ 86 °F*	96.8	91.4	89.8	80.8	71.2	73.0	74.5	82.4Φ	78.5Φ	76.31	90.41	74.17
Specific Heat of Liquid @ 86 °F (Btu/lb. °F).	0.22	0.21	0.24	0.24	0.30	0.31	0.30	0.24Φ	0.28Φ	0.35	0.21	0.36
Specific Heat of Vapor at constant pressure. (Cp), at 86 °F and 1 Atmos. (Btu/lb. °F).	0.15[1]	0.14	0.17	0.15	0.18	0.20	0.17	0.13Φ	0.14Φ	0.17[1]	0.17	0.21
Specific Heat Ratio of Vapor (k = Cp/Cv) at 86 °F.	1.08[1]	1.13	1.08	1.14	1.14	1.18	1.14	1.18Φ	1.21Φ	—	1.10	1.12
Flammability and Explosivity.	None	None	None	None	None	None	None	None	None	7.6 – 17.7†	None	None
Toxicity Rating**	4-5	5	6	6	5	5	5	6	6	Not Avail.	Not Avail.	Not Avail.

genetron 113
Trichlorotrifluoroethane
Used in low capacity centrifugal chiller packaged units. Operates with very low system pressures, high gas volumes.

genetron 11
Trichlorofluoromethane
A centrifugal refrigerant with low operating pressures. Gives higher capacity than Genetron 113. Is also used as a secondary coolant in low temperature systems. A popular choice for use in thermal insulation construction projects.

genetron 114
Dichlorotetrafluoroethane
Intermediate in pressure and displacement. Principally used with centrifugal compressors for higher capacities or for lower evaporator temperature process type applications. Also used in foam applications.

genetron 12
Dichlorodifluoromethane
A very versatile and widely used refrigerant. Common in reciprocating and rotary type equipment. For all types of applications, household to industrial. Also employed in some centrifugal designs and in several special applications such as sterilant gas, blowing agents and aerosols.

genetron 500
Azeotrope
An azeotrope of genetron 12 which has slightly higher vapor pressures and provides higher capacities from the same compressor displacement.

genetron 22
Chlorodifluoromethane
As a refrigerant, operates with higher system pressures but offers low compressor displacement requirement. Popular in residential, commercial, and industrial applications. Used as a blowing agent in aerosols and as an intermediate to produce fluoropolymers.

genetron 502
Azeotrope
An azeotrope of CFC-115 and HCFC-22 which is especially suited to low evaporation temperature applications. Handles high temperature lifts well and simultaneously provides capacity gains.

genetron 13
Chlorotrifluoromethane
A specialty low temperature refrigerant used in the low stage of cascade systems to provide evaporator temperatures in the range of –100 °F.

genetron 503
Azeotrope
An azeotrope of CFC-13 and HFC-23 which is used in the low stage of cascade type systems where it provides gains in compressor capacity and in low temperature capability.

HCFC 141b
Dichlorofluoroethane
HCFC-141b presents significant opportunities as a blowing agent alternative for use in rigid board, foam systems, flexible foam and other end-use applications.

HCFC 123
Dichlorotrifluoroethane
As a leading candidate in the next generation of blowing agents, HCFC-123 may offer effective solutions in such diverse applications as rigid board and foam systems insulation. Also may be used in centrifugal refrigeration equipment and in specialized solvent applications.

HFC 134a
Tetrafluoroethane
A hydrofluorocarbon with an ozone depletion potential of zero. HFC-134a holds great promise as a CFC substitute for a wide range of air conditioning and refrigeration systems in residential, commercial, and industrial applications.

DISCLAIMER
All statements, information, and data given herein are believed to be accurate and reliable but are presented without guaranty, warranty, or responsibility of any kind, express or implied. Statements or suggestions concerning possible use of our products are made without representation or warranty that any such use is free of patent infringement and are not recommendations to infringe any patent. The user should not assume that all safety measures are indicated, or that other measures may not be required.

* (lbs./cu. ft.)
** (based on Underwriters' system)
[1] @ 0.2 Atmos. press.
Φ @ –30°
† Upper and lower vapor flammability (Vol. %)

Figure 10-2. *Physical properties of refrigerant. Note the difference between R-12 and R-134a. (Allied Signal, Inc.)*

Figure 10-3. *Refrigerant recovery management system. Unit A is a recovery/recycling unit which handles R-12, R-22, R-500, and R-502. Unit B is used for R-12 as a recovery/recycling unit. Unit C is a recovery unit for R-12. (Robinair Division, SPX Corporation)*

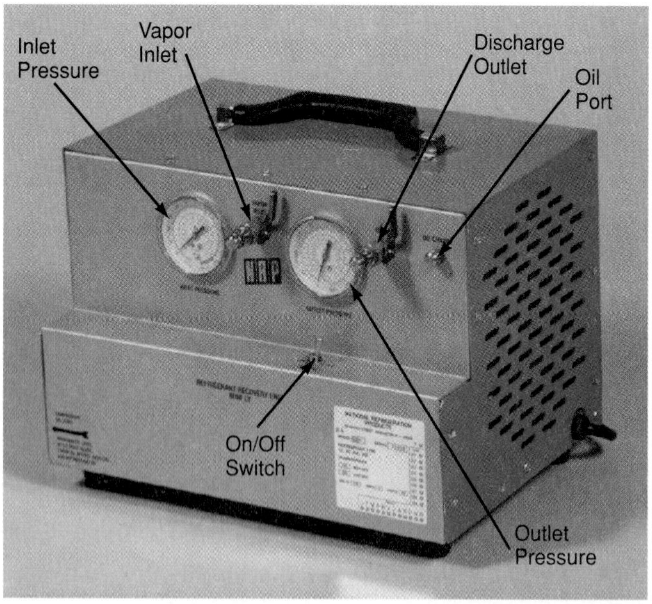

Figure 10-4. *Small recovery unit weighs approximately 44 lb. and operates on 115 volts. (National Refrigeration Products)*

The refrigerant is removed from the system in its present condition and stored in a disposable or transferable cylinder. The refrigerant then can be recycled at the service center. It may also be sent to a reclaiming station and used at a later date.

The ***vapor recovery method*** is used to remove refrigerant with some small recovery equipment. Using this equipment, the technician can remove refrigerant from light commercial, automotive, residential, and appliance applications. Refrigerant is removed from the system by using the vapor within the system and the pumping power of a recovery machine. See **Figure 10-5.**

Recovery is similar to evacuating a system with the vacuum pump. Procedures vary with each manufacturer. Basically, the hose is connected from a low-side access port to the recovery unit suction valve. Once the exhaust hose is attached, the recovery device is turned on and recovery begins. Some units have a signal device to indicate when the recovery is completed. This means that no more vapor is being processed by the recovery equipment. In some instances, the recovery device automatically closes off the vacuum system.

When the recovery is completed, the low-side isolation valve is shut off. The system should sit for at least five minutes. If the pressure rises to 10 psi or more, it may indicate pockets of cold liquid refrigerant throughout the system. It may then be necessary to start recovery again.

It is much faster to recover liquid refrigerant rather than vapor refrigerant. Many machines are designed to remove liquid refrigerant using standard refrigerant cylinders. Some small transfer units use special recovery cylinders. These allow the technician to remove liquid and vapor refrigerant.

Figure 10-6 shows a procedure for removing the refrigerant using a ***liquid transfer method.*** This type of recovery unit requires a cylinder with two ports. The transfer unit pumps the refrigerant vapor from the top of the cylinder and pressurizes the air conditioning unit. Pressure difference between the cylinder and the unit transfers the liquid refrigerant to the cylinder. Once the liquid has been removed, the remaining vapor is removed by changing the hook-up.

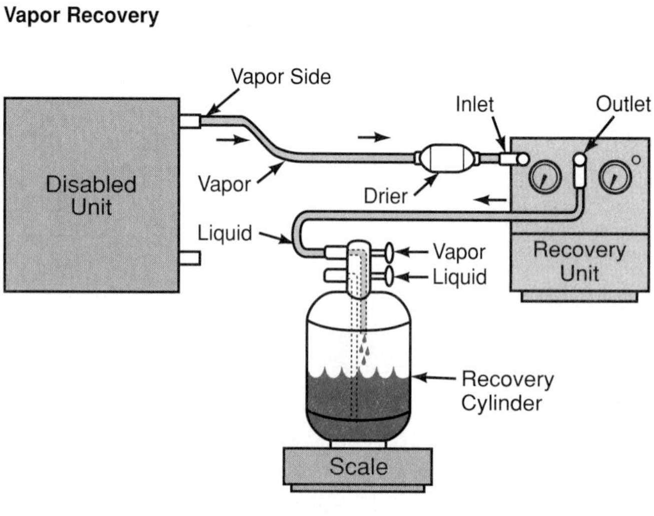

Figure 10-5. *The recovery of vapor refrigerant from a system. Note the direction of refrigerant flow from the disabled unit to the transfer unit. (National Refrigeration Products)*

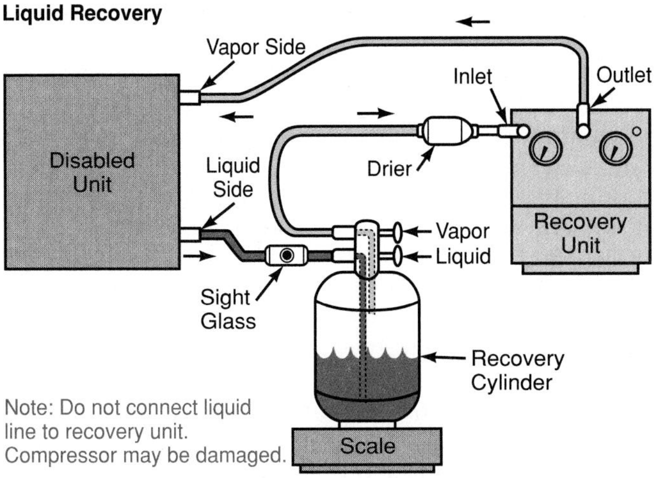

Figure 10-6. *The recovery of refrigerant from a system using a liquid transfer method. Note the direction of refrigerant flow from the disabled unit to the transfer unit. (National Refrigeration Products)*

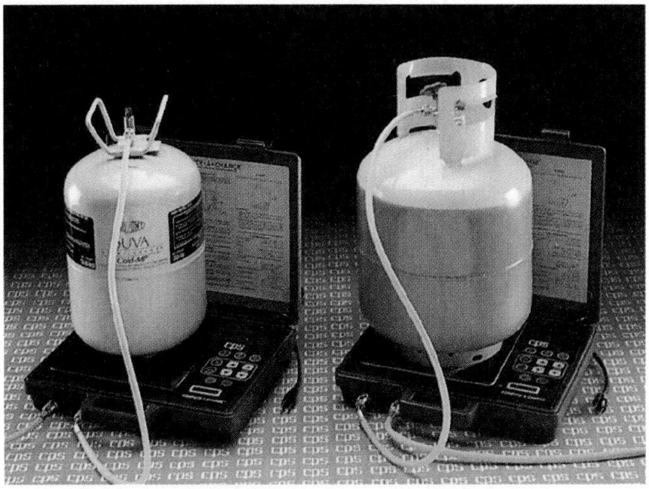

Figure 10-7. *The instrument on the left is set to automatically charge a system to a programmed amount. On the right, the instrument is set up to automatically remove a predetermined amount of refrigerant as set on the keyboard. (CPS Products, Inc.)*

The compressor oil from the recovery unit should be changed after recovery from a burned-out system. Compressor oil should also be changed before recovery of a different refrigerant. The drier must be replaced and the transfer unit and hoses evacuated before transferring a different refrigerant.

The technician should make certain that the vessels being filled are not overfilled—80% capacity is normal. As the cylinder is filling, the pressure should be watched. In a recovery unit with a moisture-indicator sight glass, any changes that occur should be noted.

If the system used only recovers refrigerant, recharging can be accomplished in many ways. A computer charging or discharging system is shown in **Figure 10-7.** This allows you to charge a system through the use of a computerized scale. A predetermined weight is set on the keyboard. The maximum gross capacity is 110 lb. or 50 kg. The system can be switched from U.S. conventional to SI metric. The hold key allows the technician to interrupt the charging or recovery cycle. This can be done without losing the program.

Some complete recovery systems have a recovery cylinder supplied by the manufacturer. Replacements are readily available. See **Figure 10-8.** This type of cylinder has an integrated level switch. The level switch is electrically connected to the recovery unit. It automatically shuts off the unit when the recovery cylinder is 80% full. Shutting off the system eliminates the need for scales or other weight devices.

The type of unit used determines whether you can service a system. The units can be divided into two categories—manually portable and those that are wheeled onto the project. **Figure 10-9** shows a portable unit that can be carried to a rooftop installation for servicing. **Figure 10-10** illustrates a unit that must be wheeled on the site.

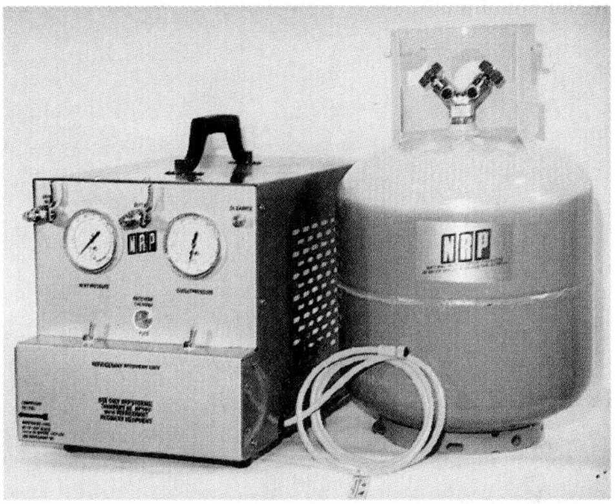

Figure 10-8. *Recovery system using a recovery cylinder that has a level switch. (Note the color-coded cylinder.) It is connected directly to the recovery unit. The recovery unit will shut off electrically when the cylinder is 80% full. (National Refrigeration Products)*

10.4 Refrigerant Recycling Equipment

In the past, refrigerant was typically vented into the atmosphere. Using current technologies, this refrigerant can now be recovered and recycled. However, old or damaged chlorofluorocarbons cannot be reused simply by removing them and compressing them to a vapor. The vapor, to be reused, must be clean. Recovery/recycling machines recover and clean the refrigerant on site or at a local service shop.

Figure 10-10 shows an automatic refrigerant recovery/recycling unit. This unit has an automatic microprocessor-controlled operation. The technician selects the type of refrigerant in the system and presses the corresponding button. This unit can be used on R-12, R-22, R-500, R-502, or R-134a. This system is designed

Figure 10-9. *Service technician carrying recovery/recycling equipment to service a rooftop unit. (Recycling Specialists International)*

Figure 10-10. *A computerized moveable recovery/ recycling unit. The unit is capable of being used for refrigerant recovery in either liquid or vapor state. (Carrier Corporation, Replacement Components Division)*

for a maximum size storage cylinder of 50 lb. (22.7 kg). The push-button control at the top of the unit provides the technician with numerous operations. The system can be used in any recovery, recycling, or recharging operations.

Recycling, as performed by most of the machines on the market today, reduces the contaminants. This is done through oil separation and filtration. The refrigerant is cleaned, but not necessarily to the manufacturer's original specifications of purity. **Figure 10-11** illustrates a system mechanism capable of handling R-12, R-22, R-500, and R-502.

Many of these units, known as ***refrigerant transfer units,*** are designed to pump down the system. This provides an on-site recycling machine that returns the recycled refrigerants to the same system. Some of the units separate the oil and acid and measure the oil in the vapor. The used refrigerant can be processed by the recycling machine to make it usable again. Replaceable-core filter-driers or other devices reduce moisture, particles, acidity, etc. Oil separation of the used refrigerant is achieved by one or more passes through the unit. The ***single-pass recycling machine*** processes refrigerant through a filter-drier and/or uses distillation. It makes only one trip from the recycling process through the machine and into the storage cylinder. The ***multiple-pass recycling machine*** recirculates refrigerant through the filter-drier many times. After a given period of time or number of cycles, the refrigerant is transferred into the storage cylinder.

The following guidelines apply to this type of recovery/recycling equipment:

1. Properly maintain the recovery/recycling equipment per the manufacturer's guidelines. Change filters as recommended. Check the system and the recycling equipment for leaks.
2. Use recovery/recycling equipment and procedures when, in the past, refrigerant would have been exhausted into the atmosphere.
3. Keep the refrigerant contained and keep the air out. The procedure of connect-and-charge-and-disconnect is no longer needed. Most new units have

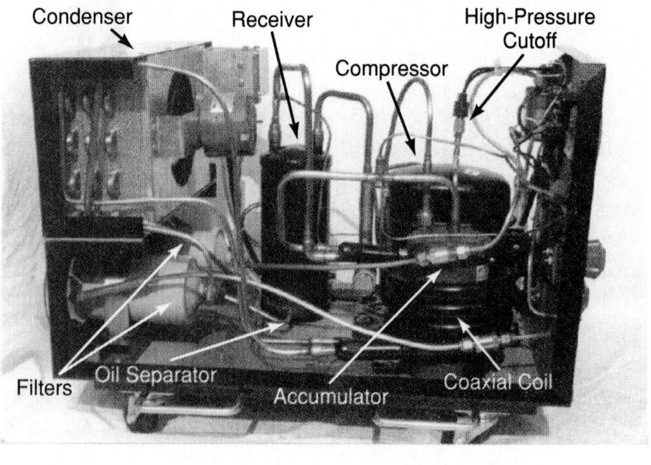

Figure 10-11. *Refrigerant recovery station mechanism. Note that the unit has two filters. (Thermal Engineering Company, Division of Seakay Co., Inc.)*

shutoff valves. These operate automatically as the hose is connected or disconnected.

4. Obtain an approximate appropriate vacuum and run a leak check. See **Figure 10-12.** Repair all leaks.
5. Use basic principles of refrigerant flow and heat transfer to speed the recovery process. When transferring from one container to another, transfer liquid from one tank to the other, if possible. This will allow transferring all of the liquid from one tank to the other without frosting the tank.
6. Always use appropriate refillable containers. Fill to 80% (maximum) of volume with liquid. Do not use disposable or unapproved containers.
7. Do *not* mix refrigerants. Mark containers. Thoroughly clean containers and all fittings upon completion.

Figure 10-13 illustrates a refrigerant recovery/recycling unit called a ***refrigerant management system.*** This unit weighs approximately 59 lb. and has an internal storage capacity of 10 lb. It is operated as a recovery unit and has a 1/2 hp compressor. Its recovery ability is approximately two pounds per minute for R-12, R-22, R-500, and R-502.

The refrigerant management system has a low-pressure gauge, a high-pressure gauge, and a high-pressure condenser gauge. The front side of the unit has a refrigerant scale for various refrigerants. There is also a refrigerant-level indicator with a refrigerant column sight glass.

Figure 10-12. *Technician using an ultrasonic leak detector. (Amprobe Instrument)*

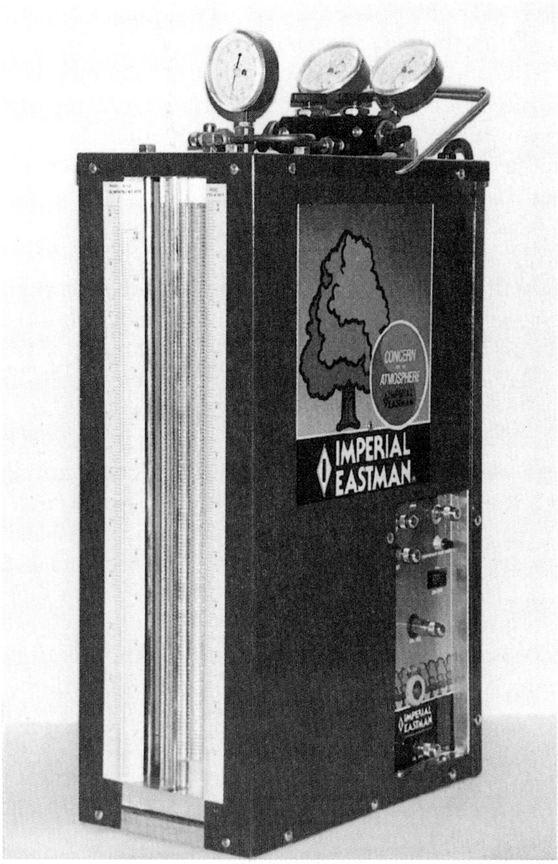

Figure 10-13. *Portable refrigerant recovery/recycling equipment. The unit has a built-in holding tank. (Imperial Eastman, Imperial Division)*

The lower front section has a low-side access port, high-side access port, vacuum discharge port, compressor power switch, and the recovery compressor oil fill. An oil sight glass indicator and a recovery compressor oil drain are at the bottom.

When hooking up the refrigerant management systems, the high-side and low-side reclaiming should be used. This procedure avoids restrictions through the refrigerant control, expansion valve, cap tube, or orifice restrictor. Recovery from one side only may result in excessive recovery time or incomplete refrigerant recovery. Therefore, the hoses are connected to the high- and low-side of the recovery system. Then they are connected through the high- and low-side of the refrigeration system. Under no circumstances should liquid be removed from the system on a continual basis. The system is designed for vapor recovery. The initial recovery of high-side pressure "refrigerant" would be approximately 200 psig.

As the unit operates and vapor recovery takes place, liquid refrigerant will eventually appear in the column. The recovery is complete when the lowest vacuum on the low-side gauge is achieved. The liquid refrigerant in the column then stops rising.

This recovery procedure should not be confused with the procedure used to evacuate a system using a vacuum pump. This recovery unit operates with an acceptable range of 15″ to 20″ of vacuum for refrigerant removal.

A recovery/reclaiming unit being used on a large rooftop unit is shown in **Figure 10-14.** Note the filter-drier in the technician's hands. The filter-drier is changed whenever the unit is connected to a new air conditioning system. This prevents mixing of refrigerants.

Many recovery/recycling units provide flexibility of use and ease of operation. This is necessary for performing difficult service calls in a short period of time. **Figure 10-15** illustrates a recovery/recycling unit in which the unit and the cylinders remain in the truck. An example of such a unit is shown in **Figure 10-16.** The hoses are connected to the unit. The technician then brings the gauge manifold hose to the unit location. The maximum length of the hose is 100′.

The entire recovery/recharge operation is done by opening and closing of the valves. The unit can be operated in both the liquid and vapor cycles. The hose does not have to be removed. This system includes a refrigerant compressor and a vacuum pump working in series. Through this process, a deep vacuum is readily attained.

Figure 10-14. *Service technician using a recovery/reclaiming unit. (National Refrigerants, Inc.)*

10.5 Refrigerant Reclaiming Procedure

Reclaiming is the reprocessing of a refrigerant to original production specifications. This must be verified by chemical analysis. In order to accomplish this, the machine must meet the SAE standards and remove 100% of the moisture and oil particulates. Many recovery/recycling machines cannot guarantee that the refrigerant will be returned to its original specifications and, therefore, cannot be regarded as a true reclaiming unit.

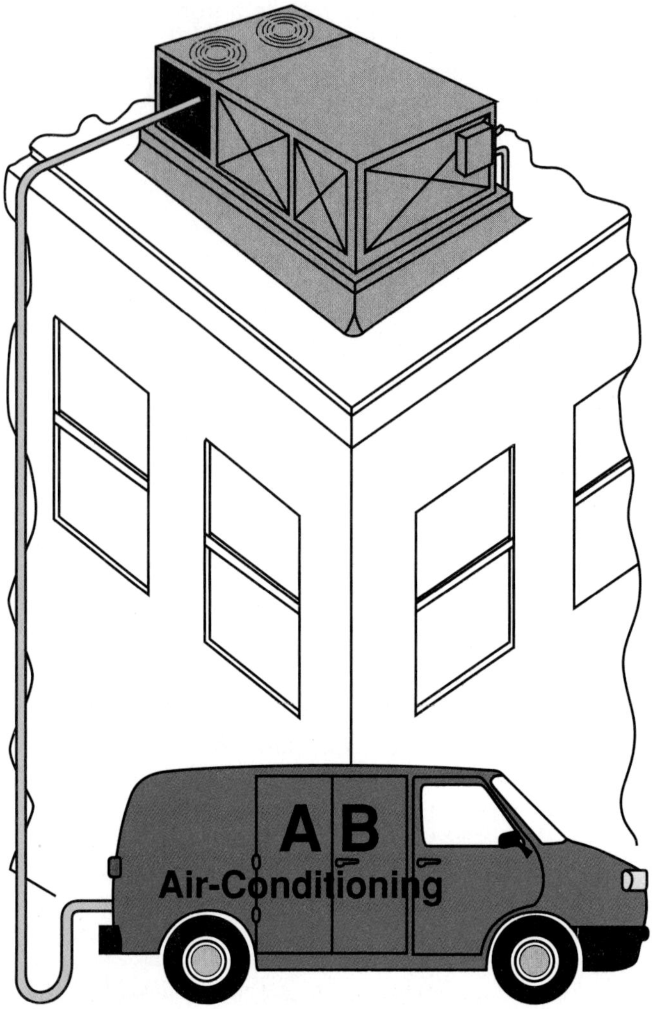

Figure 10-15. *A recovery/recharging unit being used on a rooftop unit. The maximum distance possible is 100′.*

In order to clean the used refrigerant, an on-site recycling station must meet certain requirements. It must be able to provide separation of oil, acid, hard particle contaminants, moisture, and air.

A commercial reclaiming unit is shown in **Figure 10-17.** This type of unit is available for use with R-12, R-22, R-500, and R-502. It is designed for the continuous use required on a long-run recovery/recycling procedure.

The operation of the system can best be described as follows:

1. The refrigerant is introduced into the system as either vapor or liquid.
2. Refrigerant is violently boiled at high temperature under extremely high pressure.
3. Refrigerant then enters a large, unique separator chamber where the velocity is radically reduced. This allows the vapor, at high temperature, to rise. During this phase, contaminants—copper chips, carbon, oil, acid, and all other contaminants—drop to the bottom of the separator. They will be removed during the "oil out" operation.
4. The distilled vapor passes to the air-cooled condenser and is converted to liquid.

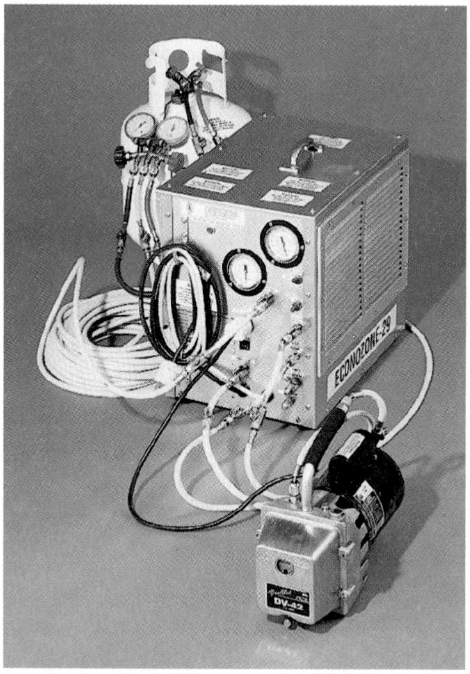

Figure 10-16. *Refrigerant recovery unit that uses a vacuum pump as part of its total system. (Refrigerant Management Systems, Inc.)*

5. The liquid passes into the on-board storage chamber(s). Within the chamber(s), an evaporator assembly lowers the liquid temperature. It is lowered approximately 100°F (56°C) to a subcooled temperature of 38°F (3°C) to 40°F (4°C).
6. A replaceable filter-drier in this circuit removes the moisture as well as the microscopic contaminants.
7. Chilling the refrigerant also facilitates the transfer to any external cylinders which are at room temperature.

Numerous refrigerant manufacturers and others have set up refrigerant recovery/reclaiming services. These provide a way to dispose of used refrigerant and obtain pure replacements as needed. You must use ***Department of Transportation (DOT)***-approved returnable cylinders and tags. See **Figure 10-18.** Standard cylinders will hold approximately 100 lb. of used refrigerant and oil. Other containers can range from 40 lb. to one ton. You then use a refrigerant transfer unit designed to pump down the system. See **Figure 10-19.** This unit may also be used for charging a system.

The positive-displaced compressed air machine removes both liquid and vapor. The refrigerant is reprocessed to designated purity specifications.

On large commercial installations, you are provided with sample cylinders. These are sent back to a reclaiming center. This is to obtain refrigerant analysis of contaminants prior to evacuation.

After being approved for reclaiming, the refrigerant is removed. See **Figure 10-20.** You then take the refrigerant to the service center. From there it is shipped back to the company. The company processes it accordingly and returns it for future sale as a used refrigerant.

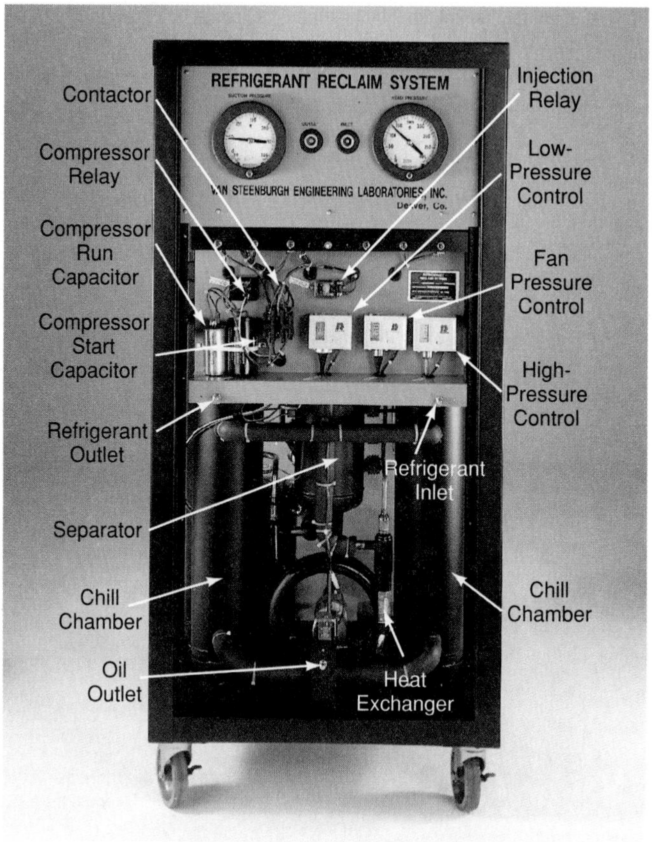

Figure 10-17. *Refrigerant reclaiming system. Note the location of the refrigerant inlet and outlet between suction pressure and head pressure gauges. (Van Steenburgh Engineering Laboratories, Inc.)*

Reprocessing may be used for low-pressure refrigerants R-11 and R-113. It may also be used for high-pressure refrigerants R-12, R-22, R-114, R-500, and R-502.

Company standards vary in regards to the type of vessel used to transport the refrigerant. Some accept 55 gallon, 10-gallon minimum, etc. See **Figure 10-21.** Each manufacturer has a procedure which must be followed, **Figure 10-22.** Numerous documents are required by each company. See **Figure 10-23.**

A reclamation company also provides a solution for the disposal of unwanted refrigerant. Disposal of refrigerants can only be accomplished by incineration at 1200°F (649°C). Currently, only a limited number of plants in the United States are equipped to do so.

10.6 Retrofit

The phasing out of CFC-12 has required use of a replacement refrigerant. The process of preparing a system for use with a replacement refrigerant is known as ***retrofitting*** the system. As mentioned previously, one of the more popular replacement refrigerants is HFC-134a.

The mineral oil commonly used in existing refrigerant systems is not soluble in HFC-134a. The system does not return the proper amount of oil to the compressor. This can lead to a decrease in performance. Therefore,

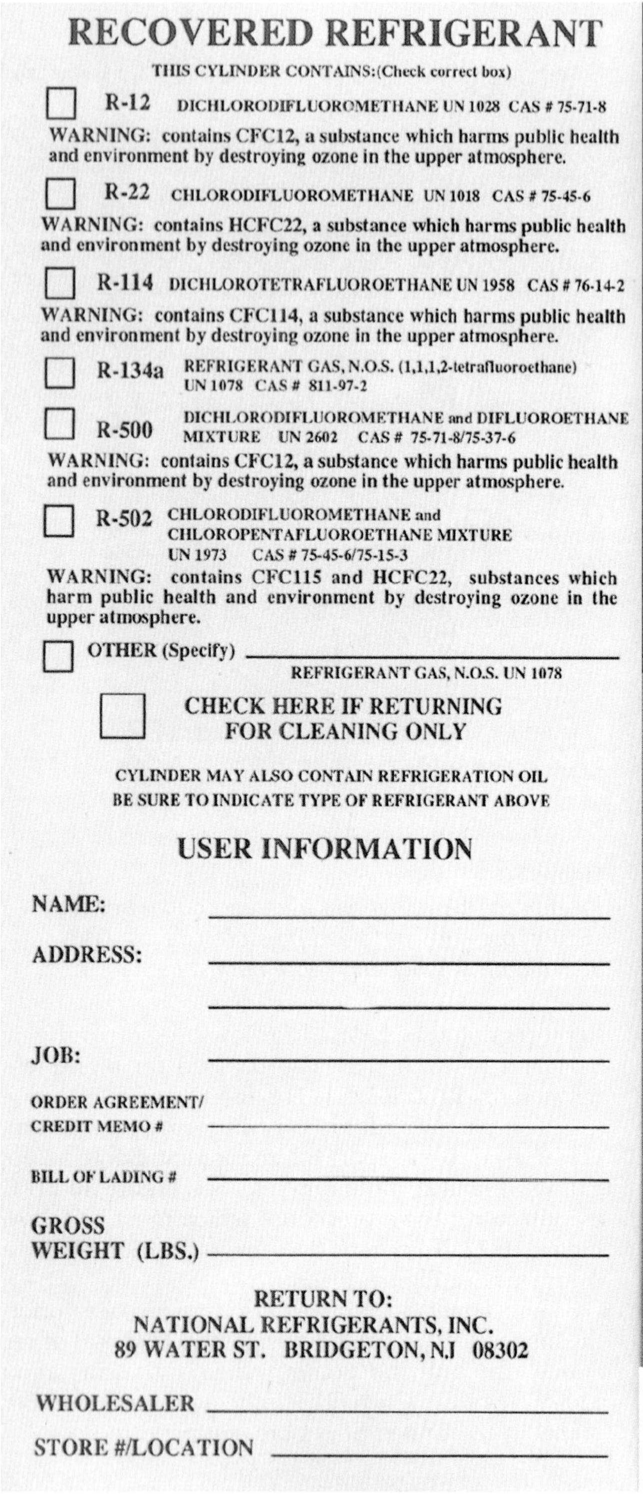
RECOVERED REFRIGERANT

THIS CYLINDER CONTAINS:(Check correct box)

☐ R-12 DICHLORODIFLUOROMETHANE UN 1028 CAS # 75-71-8

WARNING: contains CFC12, a substance which harms public health and environment by destroying ozone in the upper atmosphere.

☐ R-22 CHLORODIFLUOROMETHANE UN 1018 CAS # 75-45-6

WARNING: contains HCFC22, a substance which harms public health and environment by destroying ozone in the upper atmosphere.

☐ R-114 DICHLOROTETRAFLUOROETHANE UN 1958 CAS # 76-14-2

WARNING: contains CFC114, a substance which harms public health and environment by destroying ozone in the upper atmosphere.

☐ R-134a REFRIGERANT GAS, N.O.S. (1,1,1,2-tetrafluoroethane) UN 1078 CAS # 811-97-2

☐ R-500 DICHLORODIFLUOROMETHANE and DIFLUOROETHANE MIXTURE UN 2602 CAS # 75-71-8/75-37-6

WARNING: contains CFC12, a substance which harms public health and environment by destroying ozone in the upper atmosphere.

☐ R-502 CHLORODIFLUOROMETHANE and CHLOROPENTAFLUOROETHANE MIXTURE UN 1973 CAS # 75-45-6/75-15-3

WARNING: contains CFC115 and HCFC22, substances which harm public health and environment by destroying ozone in the upper atmosphere.

☐ OTHER (Specify) ____________
REFRIGERANT GAS, N.O.S. UN 1078

☐ CHECK HERE IF RETURNING FOR CLEANING ONLY

CYLINDER MAY ALSO CONTAIN REFRIGERATION OIL
BE SURE TO INDICATE TYPE OF REFRIGERANT ABOVE

USER INFORMATION

NAME: ____________

ADDRESS: ____________

____

JOB: ____________

ORDER AGREEMENT/
CREDIT MEMO # ____________

BILL OF LADING # ____________

GROSS
WEIGHT (LBS.) ____________

RETURN TO:
NATIONAL REFRIGERANTS, INC.
89 WATER ST. BRIDGETON, NJ 08302

WHOLESALER ____________

STORE #/LOCATION ____________

A

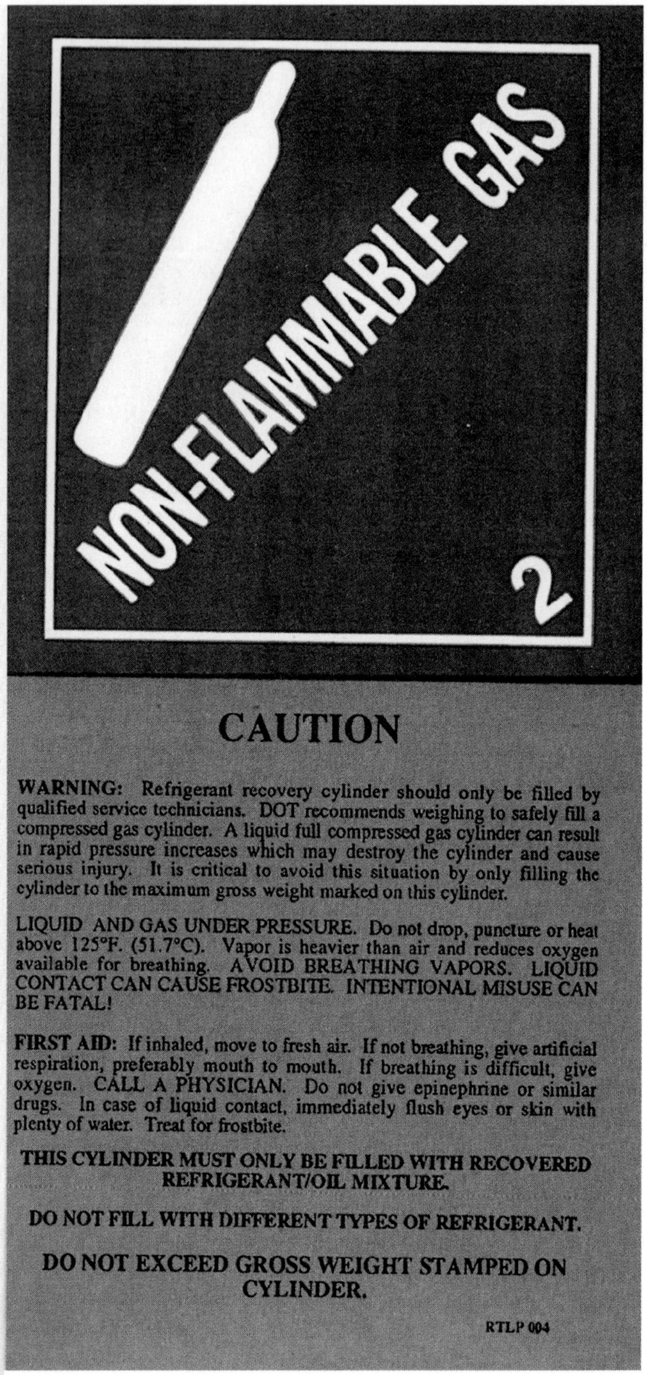

CAUTION

WARNING: Refrigerant recovery cylinder should only be filled by qualified service technicians. DOT recommends weighing to safely fill a compressed gas cylinder. A liquid full compressed gas cylinder can result in rapid pressure increases which may destroy the cylinder and cause serious injury. It is critical to avoid this situation by only filling the cylinder to the maximum gross weight marked on this cylinder.

LIQUID AND GAS UNDER PRESSURE. Do not drop, puncture or heat above 125°F. (51.7°C). Vapor is heavier than air and reduces oxygen available for breathing. AVOID BREATHING VAPORS. LIQUID CONTACT CAN CAUSE FROSTBITE. INTENTIONAL MISUSE CAN BE FATAL!

FIRST AID: If inhaled, move to fresh air. If not breathing, give artificial respiration, preferably mouth to mouth. If breathing is difficult, give oxygen. CALL A PHYSICIAN. Do not give epinephrine or similar drugs. In case of liquid contact, immediately flush eyes or skin with plenty of water. Treat for frostbite.

THIS CYLINDER MUST ONLY BE FILLED WITH RECOVERED REFRIGERANT/OIL MIXTURE.

DO NOT FILL WITH DIFFERENT TYPES OF REFRIGERANT.

DO NOT EXCEED GROSS WEIGHT STAMPED ON CYLINDER.

RTLP 004

B

Figure 10-18. *Tags used on returnable cylinders with recovered refrigerant. A—Note that the technician must indicate the type of refrigerant in the cylinder and whether the cylinder is being returned for cleaning only (as well as other identifying information). Note the warning, which is required by the EPA. B—Nonflammable gas classification tag for each one-half ton tank, as required by Department of Transportation (DOT). (National Refrigerants, Inc.)*

the existing mineral oil must be replaced by an ester-based, HFC-134a-compatible lubricant (see Chapter 9).

In addition to the potential lubricant problem, other modifications may be needed. These include adjustment of the expansion device; replacement of the driers; and replacement of O-rings, gaskets, and other nonmetallic parts in the system. The proper procedure is available by contacting the equipment manufacturer. The following

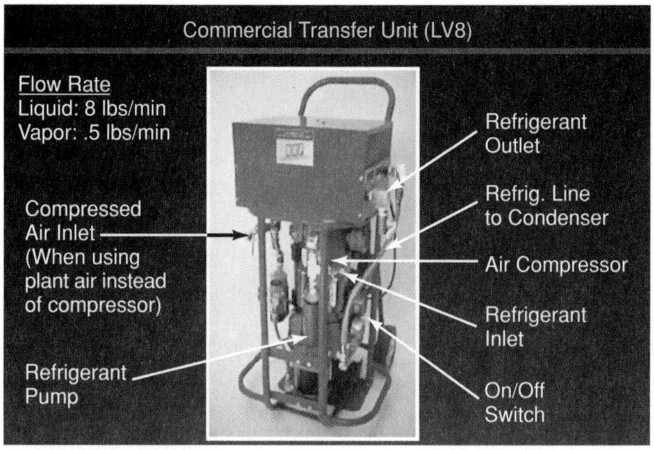

Figure 10-19. *Refrigerant transfer unit used for recovering refrigerants from commercial installations. This air-driven unit will recover from 100 lb. to 1500 lb. (National Refrigeration Products, Inc.)*

Figure 10-20. *Refrigerant being pumped into tanks from a large air conditioning system. Technician is using a large-capacity recovery unit that will recover 1000 lb. or more. (National Refrigerants, Inc.)*

items are needed in order to obtain information about the proper procedure:

- A detailed assessment of the use of HFC-134a in the system.
- Identification of all modifications that must be accomplished before retrofit.
- Identification of the correct ester-based oil to be used.

Retrofit procedures vary, depending on several factors, including the type of refrigerant, the type of the

Figure 10-21. *Department of Transportation (DOT)-approved recovery cylinders and drums—40 lb., 125 lb., 1000 lb., and 2000 lb. cylinder for pressurizing refrigerants and 100 lb., 200 lb., and 650 lb. drums for low-pressure refrigerants. (National Refrigerants, Inc.)*

FILLING PROCEDURE FOR RECOVERING REFRIGERANT INTO CYLINDERS ONLY

1. Visually inspect the cylinder to be filled. Strictly follow all DOT requirements for inspection of refrigerant cylinders.
2. Place the cylinder on a scale. Note empty weight of cylinder to determine Maximum Gross Weight.
3. Connect transfer hoses to the cylinder. Make certain they are leak free. If at all possible, change hoses when recovering different types of refrigerants to avoid contamination by unintentionally mixing refrigerants.
4. Open the cylinder outlets and begin the transfer process following manufacturer's instructions for the recovery unit.
5. DO NOT LEAVE THE CYLINDER UNATTENDED. Watch the scale closely. DO NOT OVERFILL. Do not exceed the gross weight limit. Do not fill more than 80% by volume. It is illegal to transport an overfilled cylinder!
6. When the scale reaches the gross weight limit—stop the transfer process. Tightly close all valves and other outlets.
7. Disconnect the transfer hose. AVOID CONTACT WITH LIQUID REFRIGERANT/OIL MIXTURES. Immediately replace all caps and other cylinder closures.
8. Weigh the cylinder. Write the weight on ALL appropriate forms and on the cylinder HANG TAG.
9. Completely fill out the cylinder HANG TAG attached to the cylinder. BE SURE THE HANG TAG INDICATES THE CORRECT REFRIGERANT IN THE CYLINDER. *IT IS ILLEGAL TO TRANSPORT A CYLINDER WITHOUT CORRECTLY IDENTIFYING THE CONTENTS.* (Including an empty cylinder)
10. There will be a cylinder cleaning charge for cylinders returned less than 50% full. Check off the "For Cleaning Only" box on the hang tag. There will be a container handling fee for overfilled containers as determined by NRI's Cylinder Weight Chart.

Figure 10-22. *Procedures that must be followed to obtain service from a refrigeration reclaiming company. (National Refrigerants, Inc.)*

equipment, and type of refrigerant recovery equipment used. **Figure 10-24** illustrates one company's guidelines for retrofitting existing equipment from CFC-12 to HFC-134a.

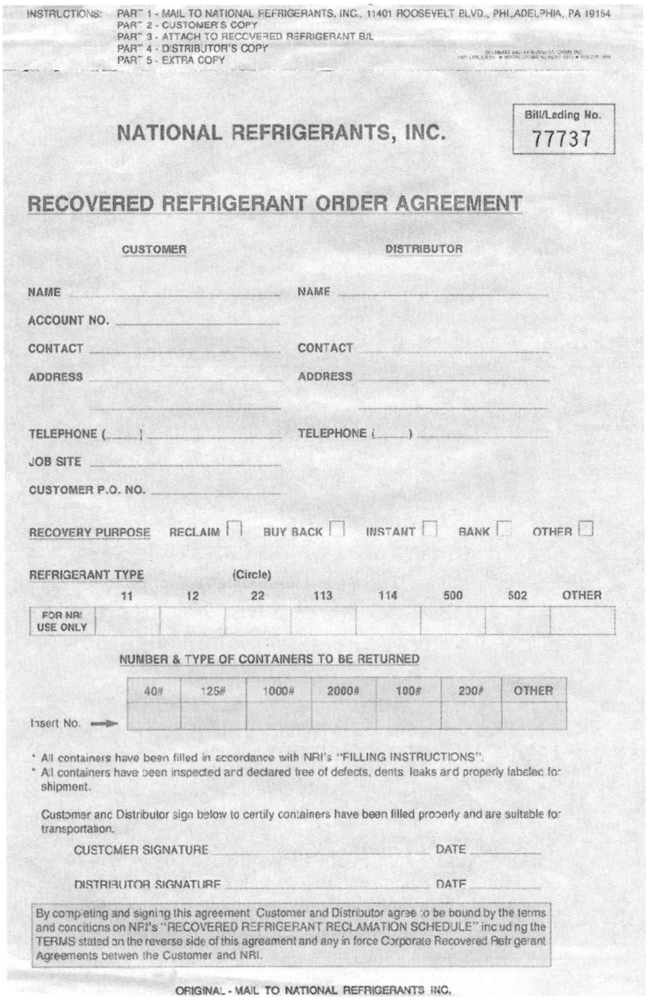

INSTRUCTIONS: PART 1 - MAIL TO NATIONAL REFRIGERANTS, INC., 11401 ROOSEVELT BLVD., PHILADELPHIA, PA 19154
PART 2 - CUSTOMER'S COPY
PART 3 - ATTACH TO RECOVERED REFRIGERANT B/L
PART 4 - DISTRIBUTOR'S COPY
PART 5 - EXTRA COPY

NATIONAL REFRIGERANTS, INC.

Bill/Lading No. 77737

RECOVERED REFRIGERANT ORDER AGREEMENT

CUSTOMER | DISTRIBUTOR

NAME ____ NAME ____
ACCOUNT NO. ____
CONTACT ____ CONTACT ____
ADDRESS ____ ADDRESS ____
TELEPHONE ( ) ____ TELEPHONE ( ) ____
JOB SITE ____
CUSTOMER P.O. NO. ____

RECOVERY PURPOSE RECLAIM ☐ BUY BACK ☐ INSTANT ☐ BANK ☐ OTHER ☐

REFRIGERANT TYPE (Circle)

	11	12	22	113	114	500	502	OTHER
FOR NRI USE ONLY								

NUMBER & TYPE OF CONTAINERS TO BE RETURNED

	40#	125#	1000#	2000#	100#	200#	OTHER
Insert No. ➔							

* All containers have been filled in accordance with NRI's "FILLING INSTRUCTIONS".
* All containers have been inspected and declared free of defects, dents, leaks and properly labeled for shipment.

Customer and Distributor sign below to certify containers have been filled properly and are suitable for transportation.

CUSTOMER SIGNATURE ____ DATE ____

DISTRIBUTOR SIGNATURE ____ DATE ____

By completing and signing this agreement Customer and Distributor agree to be bound by the terms and conditions on NRI's "RECOVERED REFRIGERANT RECLAMATION SCHEDULE" including the TERMS stated on the reverse side of this agreement and any in force Corporate Recovered Refrigerant Agreements betwen the Customer and NRI.

ORIGINAL - MAIL TO NATIONAL REFRIGERANTS INC.

A

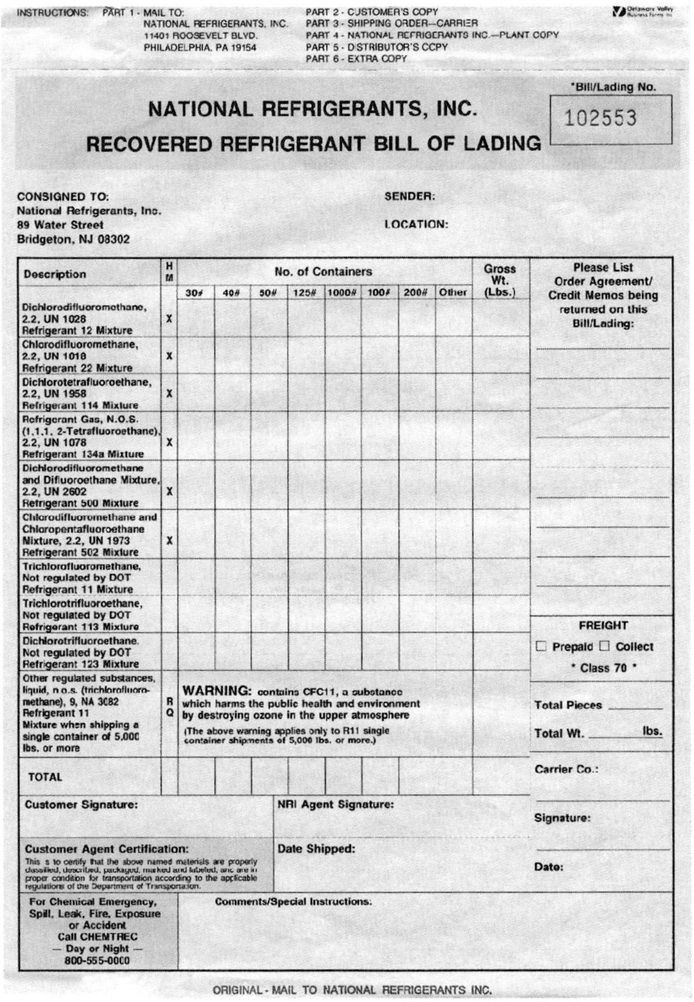

INSTRUCTIONS: PART 1 - MAIL TO: NATIONAL REFRIGERANTS, INC. 11401 ROOSEVELT BLVD. PHILADELPHIA, PA 19154
PART 2 - CUSTOMER'S COPY
PART 3 - SHIPPING ORDER—CARRIER
PART 4 - NATIONAL REFRIGERANTS INC.—PLANT COPY
PART 5 - DISTRIBUTOR'S COPY
PART 6 - EXTRA COPY

NATIONAL REFRIGERANTS, INC.
RECOVERED REFRIGERANT BILL OF LADING

*Bill/Lading No. 102553

CONSIGNED TO:
National Refrigerants, Inc.
89 Water Street
Bridgeton, NJ 08302

SENDER:

LOCATION:

Description	HM	No. of Containers 30#	40#	50#	125#	1000#	100#	200#	Other	Gross Wt. (Lbs.)
Dichlorodifluoromethane, 2.2, UN 1028 Refrigerant 12 Mixture	X									
Chlorodifluoromethane, 2.2, UN 1018 Refrigerant 22 Mixture	X									
Dichlorotetrafluoroethane, 2.2, UN 1958 Refrigerant 114 Mixture	X									
Refrigerant Gas, N.O.S. (1,1,1, 2-Tetrafluoroethane), 2.2, UN 1078 Refrigerant 134a Mixture	X									
Dichlorodifluoromethane and Difluoroethane Mixture, 2.2, UN 2602 Refrigerant 500 Mixture	X									
Chlorodifluoromethane and Chloropentafluoroethane Mixture, 2.2, UN 1973 Refrigerant 502 Mixture	X									
Trichlorofluoromethane, Not regulated by DOT Refrigerant 11 Mixture										
Trichlorotrifluoroethane, Not regulated by DOT Refrigerant 113 Mixture										
Dichlorotrifluoroethane, Not regulated by DOT Refrigerant 123 Mixture										
Other regulated substances, liquid, n.o.s. (trichlorofluoromethane), 9, NA 3082 Refrigerant 11 Mixture when shipping a single container of 5,000 lbs. or more	RQ	WARNING: contains CFC11, a substance which harms the public health and environment by destroying ozone in the upper atmosphere (The above warning applies only to R11 single container shipments of 5,000 lbs. or more.)								
TOTAL										

Please List Order Agreement/ Credit Memos being returned on this Bill/Lading:

FREIGHT
☐ Prepaid ☐ Collect
* Class 70 *

Total Pieces ____
Total Wt. ____ lbs.
Carrier Co.: ____
Signature: ____
Date:

Customer Signature:

NRI Agent Signature:

Customer Agent Certification:
This is to certify that the above named materials are properly classified, described, packaged, marked and labeled, and are in proper condition for transportation according to the applicable regulations of the Department of Transportation.

Date Shipped:

For Chemical Emergency, Spill, Leak, Fire, Exposure or Accident Call CHEMTREC — Day or Night — 800-555-0000

Comments/Special Instructions:

ORIGINAL - MAIL TO NATIONAL REFRIGERANTS INC.

B

Figure 10-23. *A—Order agreement used with reclaiming company. Note line marked "Recovery Purpose." B—Recovered refrigerant bill of lading. (National Refrigerants, Inc.)*

10.7 CFC Recovery/Recycle/Reclaim Safety and Standards

By law, a technician must now be certified to open a system or purchase refrigerants. Major topics that may be covered in certification programs include:

- CFC storage and handling.
- Transportation.
- Recovery equipment and procedures.
- Hazardous waste handling, storage, and disposal regulations.

You must fully understand the safety involved in handling and storage of refrigerant. See **Figure 10-24.** Certification programs approved by the EPA are also being offered, **Figure 10-25.** A number of areas are covered by most EPA-approved seminars and workshops:

- Procedures for removal.
- Basic field testing of refrigerant for purity.
- Isolation of system components to prevent refrigerant venting.
- Leak detection, isolation of leaks, and leak repairs.

It is your responsibility to use safe practices and procedures. This includes the replacement of both suction line and liquid line driers. If the system only has one filter-drier, install another filter-drier in the opposite side. This will aid the refrigeration purification process.

Recycled refrigerants follow a standard that has been set by the ***Air-Conditioning and Refrigeration Institute (ARI),*** *Standard ARI-700-88.*

10.8 Mobile Air Conditioning

Clean Air Act Section 609 establishes the requirements regarding mobile air conditioners. No person may perform service for consideration on any motor vehicle air conditioner without properly using approved refrigerant recycling equipment. **Figure 10-26.** Also, no

RETROFIT	
1.	Consult original equipment manufacturer (OEM) for specific equipment modifications.
2.	Isolate CFC-12 charge from the compressor by either pumping down into the system's receiver or transferring into a recovery cylinder.
3.	Drain the mineral oil from the system, especially the compressor, all low points, and the oil separator, if fitted.
4.	Replace filter/dryers.
5.	Charge the system with appropriate grade EMKARATE RL ester lubricant. (Refer to technical bulletin on EMKARATE RL ester lubricants for more information.)
6.	Pull a vacuum, hold for a minimum of one hour, and charge the system with CFC-12.
7.	Run the system on CFC-12 and ester lubricant to allow sufficient time for the ester lubricant and mineral oil to mix totally with the CFC-12 charge. Running time will vary from several hours to one week, depending on the system.
8.	Drain the ester lubricant and recharge with a fresh charge of EMKARATE RL ester lubricant. Repeat this procedure until the level of residual mineral oil is less than the OEM recommended amount. Most OEMs are now recommending less than 5% residual mineral oil content. Although some systems can tolerate higher levels of oil, a significant quantity of mineral oil may have adverse effects on the heat exchanger performance. In an HFC-134a system, the mineral oil will tend to drop in the evaporator and reduce the overall efficiency of the heat transfer surfaces.
9.	Recover and reclaim the CFC-12.
10.	Perform any equipment modifications necessary (as recommended by the OEM). This may include changing the liquid and suction line filter-driers, adjusting the expansion valves, resetting the controls, and possibly modifying the compressor.
11.	Leak test and evacuate the system.
12.	Charge with KLEA 134a.

Figure 10-24. *Retrofit procedures recommended by a manufacturer of refrigerants and refrigeration oils. (ICI Americas, Inc.)*

person may perform such service unless properly trained and certified. This requirement is now effective for all establishments performing such service regardless of the number of vehicles serviced.

Further regulations have been and are being developed concerning:

- Governing stationary air conditioner and refrigeration equipment.
- Disposal of CFCs during service or repair.
- Certification of recovery/recycling equipment.
- Certification of technicians.
- Record keeping.

10.9 Review of Safety

Refrigerants used in refrigeration and air conditioning present no problems in normal use and handling. However, they should always be used in the proper manner to avoid potential hazards. Most refrigerants have low boiling points, thus protective clothing and eye protection should always be used to avoid frostbite. Liquids with higher boiling points can cause skin irritation. At atmospheric pressure, the refrigerant vaporizes readily. If refrigerant contacts the skin, the latent heat of vaporization removes the heat from the skin. Frostbite can occur. If refrigerant comes in contact with your eyes, it may freeze them, causing blindness. Accidents involving refrigerants should be immediately referred to a doctor.

The technician should make certain refrigerant service cylinders are not filled completely with liquid refrigerant. Refrigerants should be stored only in cylinders with labels indicating the specific refrigerant.

SAFETY RECOMMENDATIONS	
1.	Only fill cylinders which are currently DOT-approved for fluorocarbon refrigerants. Always inspect the cylinder for pressure rating and latest hydrostatic test date. Be sure to thoroughly check each cylinder for dents, gouges, bulges, cuts, or any other imperfections which may render it unsafe to hold refrigerant for storage or transportation.
2.	It is highly recommended to read the Air-Conditioning and Refrigeration Institute "Guideline K—Guideline for Cylinders for Recovered Fluorocarbon Refrigerants."
3.	Be sure all connections are made tight before transferring refrigerant into cylinders. Be sure all closures are made tight on the cylinder immediately after filling.
4.	Always use a scale when filling any cylinder. DO NOT OVERFILL.
5.	CAUTION: Liquid refrigerant can cause frostbite if skin contact occurs. Be aware that the refrigerant/oil being removed from a system may contain contaminants which may be harmful to breathe or contact with the skin. Always provide fresh air when working in enclosed areas. Avoid breathing vapors. Always wear safety glasses and gloves (cold resistant for pressurized refrigerants and rubber-type for R-11, R-113, or R-123). Avoid contact with clothing.

Figure 10-25. *Standard safety recommendations to be followed when removing refrigerant from a system. (National Refrigerants, Inc.)*

Figure 10-26. *Service technicians taking a motor vehicle refrigerant recycling and recovery examination. (National Institute for Automotive Service Excellence [ASE])*

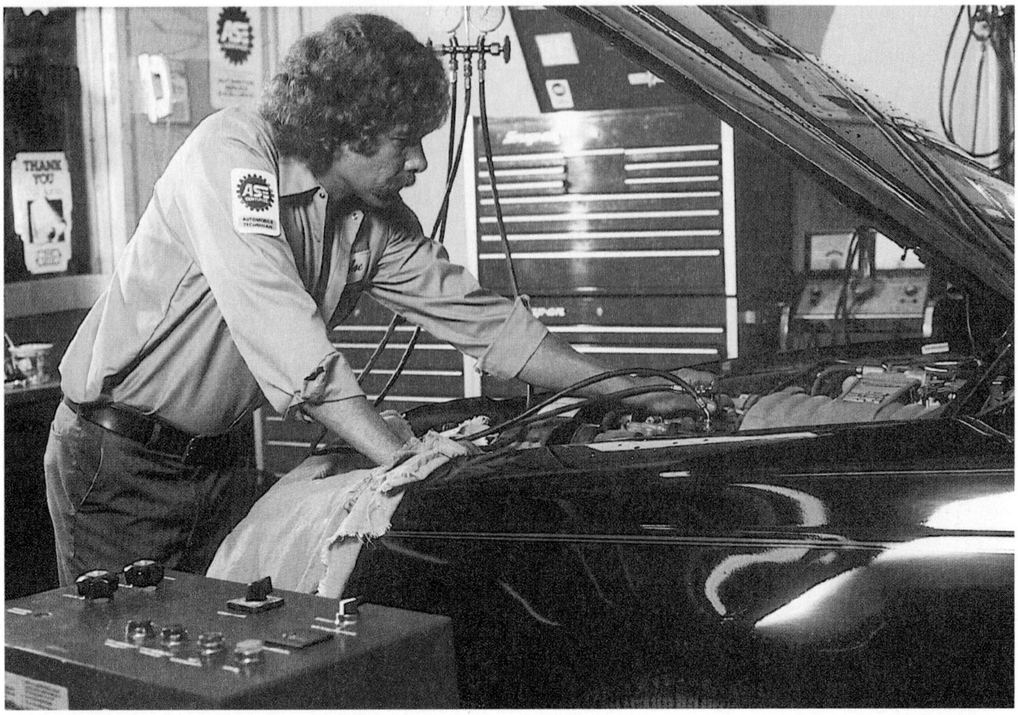

Figure 10-27. *Automotive service technician servicing an automotive air conditioning system in keeping with the EPA regulations. (National Institute for Automotive Service Excellence [ASE])*

Fluorocarbon vapors are heavier than air and tend to accumulate in low areas. They replace the air in the room. The person breathing this will lose consciousness. Instruments that measure the percent of oxygen in the air are available and should be used. Air should contain a minimum of 19.1% oxygen. Exposure to refrigerants should be avoided. The inhalation of excessive amounts can lead to possible cardiac arrest and death.

10.10 Test Your Knowledge

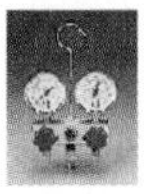

Please do not write in this text. Place your answers on a separate sheet of paper.

1. What is the ozone layer?
   A. A thin layer of the earth's upper atmosphere.
   B. The area located approximately 35 miles above the ground.
   C. The area which protects the earth from ultraviolet rays of the sun.
   D. All of the above.
2. _______ refrigerants have the greatest negative effect on the ozone.
   A. CFC
   B. HCFC
   C. Azeotropic
   D. All of the above.
3. What is the EPA?
   A. A thin layer of the earth's upper atmosphere.
   B. A type of CFC.
   C. A governing agency which restricts use of CFCs.
   D. None of the above.
4. What is meant by the term "recovery"?
   A. To clean refrigerant for reuse.
   B. To remove refrigerant and store it in an external container without testing or processing.
   C. To reprocess refrigerant.
   D. None of the above.
5. What is meant by the term "recycling"?
   A. To clean refrigerant for reuse.
   B. To remove refrigerant and store it in an external container without testing or processing.
   C. To reprocess refrigerant.
   D. None of the above.
6. What is meant by the term "reclaim"?
   A. To clean refrigerant for reuse.
   B. To remove refrigerant and store it in an external container without testing or processing.
   C. To reprocess refrigerant.
   D. None of the above.
7. The primary method of removing refrigerant using recovery equipment is the _______ method.
   A. liquid recovery
   B. vapor recovery
   C. Either method is acceptable.
   D. Method of recovery varies based upon type of refrigerant used.

8. When recovery of the refrigerant has been completed, how can the system be checked to see if there are pockets of cold liquid refrigerant?
   A. If the pressure rises to 5 psi or more with the system shut off for five minutes or more.
   B. If the pressure drops after the system is shut off for five minutes or more.
   C. If the pressure rises to 10 psi or more with the system shut off for five minutes or more.
   D. None of the above.
9. A refrigerant cylinder or vessel should contain up to ______% of the maximum capacity.
   A. 90
   B. 100
   C. 70
   D. 80
10. Refrigerant contaminants are removed by ______.
   A. oil separation
   B. filtration
   C. Both A and B.
   D. None of the above.
11. Refrigerant in a system should be disposed of by ______.
   A. venting it into the air
   B. bleeding off one-half of the charge into the air and recovering the remainder
   C. using recovery/recycling equipment
   D. None of the above.
12. What percent of the moisture and oil particulates must be removed from CFC refrigerants to meet the SAE standards?
   A. 50%.
   B. 75%.
   C. 100%.
   D. None of the above.
13. How are unwanted refrigerants disposed of?
   A. They cannot be disposed of according to the EPA.
   B. They may be vented into the air.
   C. Incineration at 1200°F.
   D. None of the above.
14. The major function of an on-site recycling station is to ______.
   A. recover and recycle used refrigerants
   B. provide separation of oil and particles
   C. provide separation of moisture and air
   D. All of the above.
15. In a system that contains CFCs, ______ drier(s) should be replaced or added to aid in the purification process.
   A. suction
   B. liquid line
   C. Both A and B.
   D. None of the above.
16. What type of identification must be on recovery cylinders and drums used for transporting fluorocarbon refrigerants?
   A. Tags.
   B. Nonflammable classification tags.
   C. Used refrigerant identification tags.
   D. All of the above.
17. Why is a small sample of refrigerant from a large commercial installation sent to a reclaiming center prior to removal?
   A. To obtain an analysis of contaminants.
   B. To determine whether it can be reclaimed.
   C. Both A and B.
   D. To determine the refrigerant and oil contained.
18. ______ may occur if an excessive amount of refrigerant is inhaled while being removed from a system?
   A. Cardiac arrest
   B. Asthma-like symptoms
   C. Allergic reactions
   D. All of the above.
19. What happens if the refrigerant/oil being removed from a system comes in contact with the skin?
   A. It can cause skin cancer.
   B. It can cause an allergic reaction.
   C. It can cause frostbite.
   D. All of the above.
20. Who may repair or service an automotive air conditioner?
   A. Anyone who has passed the RSES CM exam.
   B. Anyone who has completed a two-year vocational course.
   C. Anyone who has passed the EPA certification test and is trained.
   D. All of the above.

Interior view of an upright refrigerator-freezer.(Amana Refrigeration, Inc.)

Chapter 11

DOMESTIC REFRIGERATORS AND FREEZERS

Key Words:

ambient compensator
cabinet
cold ban
colloids
enzymes
evaporation
featheredging
freezer burn
insulation
mechanism
perimeter drier

Learning Objectives:

After studying this chapter, you will be able to:

- Discuss the construction of domestic refrigerators and freezers.
- Describe the mechanisms and cabinets for different types of refrigerators and freezers.
- Compare standard circuit diagrams and ladder diagrams of a system.
- Discuss differences in circuits for manual defrost and automatic defrost systems.
- Demonstrate how to repair damaged cabinet finishes.
- Identify the various types of condensing units and evaporators used in domestic systems.
- **Follow approved safety procedures.**

A modern domestic refrigerator or freezer consists primarily of three parts:

- The cabinet.
- The mechanism (condensing unit and evaporator).
- The electrical circuit.

The ***cabinet*** contains and supports the evaporator and condensing unit. It also supplies shelving and storage space for the foods or beverages.

The ***mechanism*** consists of the condensing unit and the evaporator. In the evaporator, the liquid refrigerant expands and becomes a vapor. This vapor absorbs heat from the foods or beverages in the cabinet. The condensing unit removes the heat absorbed in the evaporator. The liquid refrigerant then returns to the evaporator to repeat the refrigerating cycle.

The ***electrical circuit*** includes all circuits, relays, overcurrent protection, and other devices which direct the flow of current throughout the refrigerator and freezer. The circuit extends from the grounded extension cord and plug to the final load either inside or outside of the cabinet.

Carefully review **Chapter 4** of this text before continuing study of this chapter.

11.1 Preserving Foods by Refrigeration and Freezing

Foods (vegetables and fruits) last longer when kept at temperatures just above freezing. These temperatures slow down oxidation of the food. This reduces the multiplication of the bacteria in the cells and fibers. It also reduces the ***evaporation*** (loss of fluid) from the food.

11.1.1 How Cold Preserves Food

Food has cells, enzymes, colloids, water, and a few microorganisms. If it is not kept cold, food will spoil. ***Enzymes*** are tiny particles of matter that exist in food substances. Enzymes, which cause food spoilage, are controlled by low temperatures. To preserve some foods for long periods (a year or more) temperature must be well below 0°F (−18°C). For best results, it should be −20°F (−29°C).

Enzymes are not destroyed by fast freezing. Their growth rate is slowed down by the low temperatures, however. They seem to stimulate organic change. However, enzymes are destroyed by pasteurization.

Colloids are found in flesh foods. They are tiny cells in meats, fish, and poultry. If they are abused in any way, such as cell disruption (breaking), the food quickly becomes rancid (spoiled). Colloids are considered to be cell "containers" or "capsules." If the container is broken, the food rapidly deteriorates. Meat, poultry, and fish have important colloidal (miniature cell) changes. The changes can be slowed by low temperatures.

Water in food forms ice crystals when frozen. Fast freezing produces small ice crystals and is less damaging to food. Slow freezing allows time for larger crystal growth. The larger the ice crystals, the more the food cell walls are damaged.

11.1.2 Storage of Fresh Foods in the Refrigerator

The air in a fresh food refrigerator is always quite dry. Any moisture in the refrigerator collects and condenses on the evaporator surfaces. Therefore, food containers should be covered and as airtight as possible to keep food moist.

The refrigerator cabinet temperature should be kept at 35°F to 45°F (2°C to 7°C). Most fresh foods may be kept from three days to a week at the above temperatures. Unfrozen meat and fish should be stored at as close to 32°F (0°C) as possible. The recommended storage temperatures for various foods are listed in Chapter 16.

11.1.3 Storage of Frozen Food in the Freezer

The air in a food freezer, as in a refrigerator, is very dry. Any moisture in the air of the freezer quickly condenses on the evaporator surfaces. It is very important, therefore, that all frozen foods be packaged in moisture-resistant containers.

When packaging food for the freezer, as much air as possible should be removed from the packaging. Frozen food packages must be tightly sealed. Ordinary paper is too porous for freezer use. If not properly packaged, frozen food will develop freezer burn. ***Freezer burn*** is indicated by a change in color of the food. Food value is not affected, but there is a change in color and outside appearance.

Most frozen foods may be kept for several weeks at 0°F to −10°F (−18°C to −23°C). Food to be kept for a year or more should be frozen at −20°F (−29°C) or lower. Some frozen foods keep better than others; beef keeps better than pork. Commercial systems for food freezing are explained in Chapter 14.

11.2 Refrigerator and Freezer Insulation

Insulation lines the walls of the refrigerator and the freezer cabinet. Insulation prevents heat from leaking through the walls and into the cabinet. The most common insulation materials used in household refrigerators and freezers are urethane foam or fiberglass. Other insulating materials are used in some commercial and industrial systems. Tables giving the insulating properties of various materials are shown in Chapter 31.

11.3 Refrigerator—Single-Door, Manual Defrost

A simple fresh-food refrigerator consists essentially of an evaporator. The evaporator may be placed across the top of the cabinet. It can also be placed in one of the upper corners of the cabinet. **Figure 11-1** is a typical installation. The evaporator is across the top of the cabinet. There is a small area to store frozen food for a short period of time. The condensing unit is in the bottom of the cabinet.

Some additional food storage is provided in the door. A vegetable crisper drawer is located below the bottom shelf. A butter conditioner is located in the door. This conditioner has a door, closing it off from the refrigerated compartment. The butter is kept here at a slightly higher temperature than other food in the cabinet.

Figure 11-1. *Compact refrigerator. Note the thermostat control knob and the vegetable crisper. The evaporator is located at the top of the refrigerator. (Whirlpool Corporation)*

11.3.1 Cabinets

Refrigerator cabinets are made of pressed steel. The seams are welded. The outside shell must be smooth and vaporproof. The inner shell provides a surface for the interior finish of the cabinet. It also provides brackets for mounting shelves, lights, thermostats, temperature controls, etc.

Insulation is installed between the outer and the inner shell. Urethane foam, when used, is expanded in this space. This allows it to fit with no crevices or open places. The hinge arrangement is usually a part of the outside shell and door assembly.

In the simple refrigerator, the cabinet provides a space for the evaporator along the top or in the upper corner. The cold air from the evaporator flows by natural circulation through the refrigerated space. The shelves are constructed so that air can circulate freely past the ends and sides. In such an installation, there is no need to use a fan. The crisper for fresh vegetables is usually located in the bottom shelf of the refrigerator. It generally has a cover in order to maintain fairly high humidity around the vegetables.

In most cases, the light switch is located at the hinge side of the door. The light is turned on and off as the door is opened and closed.

Heat flow from the outer to the inner shell at the door opening can be reduced. A connecting trim of a special piece of plastic is usually used for this purpose. This plastic is sometimes called a ***cold ban,*** being a poor conductor of heat. This trim is usually attached to the refrigerator liner and shell with trim clips. The finish on refrigerators is usually a good grade of baked-on enamel. It is on both the outside and inside of the cabinet. Porcelain enamel is found on steel cabinet liners.

11.3.2 Mechanisms

A simple refrigerator mechanism consists of a hermetic compressor. It is placed in the cabinet base. The condenser is either at the bottom or at the back of the cabinet. An evaporator is placed inside the cabinet at the top. A typical mechanism is illustrated in **Figure 11-2.** In this illustration, the parts are shown in diagram fashion.

The cycle operation is as follows:

1. The liquid refrigerant (usually R-12 or R-134a) enters the evaporator.
2. The refrigerant boils and absorbs heat in the evaporator. The vapor is drawn through the suction line back to the compressor.
3. In the compressor, the vapor is compressed to a high pressure, and by doing so, its temperature is increased. The compressed vapor flows through the high-pressure vapor line and into the condenser. In this case, the condenser is a vertical, natural draft, wire-and-tube type.
4. In the condenser, the high-pressure, high-temperature vapor gives up its heat to the surrounding air. The vapor is condensed back to a liquid. The liquid is shown in the bottom of the condenser.
5. Liquid refrigerant then flows through the filter-drier. It enters the capillary tube at point H. The capillary tube refrigerant control is attached to the suction line at the heat exchanger.
6. The warm refrigerant passes through the capillary tube. Some of its heat is given up to the cold suction line vapor. This increases the heat-absorbing ability of the liquid refrigerant slightly. It increases the superheat of the vapor entering the compressor.
7. The low-pressure liquid now enters the evaporator and the cycle is repeated.

This is the simplest type of automatic domestic refrigerator. These refrigerators are manually defrosted. It will be necessary to remove frost as it builds up on the evaporator. The ice accumulation on the evaporator greatly reduces the refrigeration effect.

There are two common methods for manually defrosting these refrigerators:

- The refrigerator is turned off and allowed to remain off overnight. A drip pan is used to catch the condensation that comes from defrosting the refrigerator.
- The refrigerator is turned off and a pan of hot water is placed in or near the evaporator. This will quickly remove the frost. The refrigerator can be returned to normal service in a few minutes. *Never use a metal scraper on an evaporator. There is danger of puncturing it.*

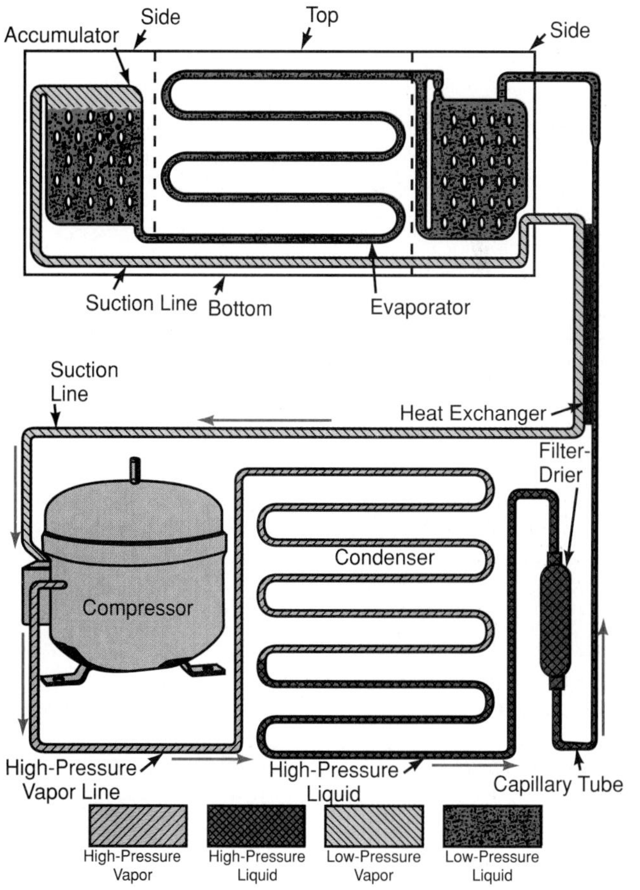

Figure 11-2. *Cycle diagram for a refrigerator.*

Evaporator surfaces and inner refrigerator surfaces should be cleaned each time the mechanism is defrosted. A solution of baking soda and water is best.

11.3.3 Electrical

The electrical supply comes through the grounded extension cord and plug, **Figure 11-3.** The electrical circuit then supplies current to the panel disconnect. Two separate circuits lead away from this panel. One circuit supplies current to the cabinet light. The light is controlled by the cabinet switch. The cabinet switch is in series with the light. The light comes on when the refrigerator door is opened and is turned off when the door is closed.

The second circuit brings current to the motor compressor. The thermostat in **Figure 11-3** is in series with this circuit. It controls the compressor operation.

The temperature in the cabinet is controlled by a thermostat. When the temperature reaches a predetermined point, the thermostat completes the circuit through the motor. The compressor runs and the refrigeration cycle goes into operation. The temperature inside the cabinet is lowered to the minimum desired temperature. The thermostat then turns off (opens) the current and the motor compressor stops.

A motor control thermostat is shown in **Figure 11-4.** The thermostat control knob is attached to the control adjustment. The altitude adjustment is turned clockwise as the altitude increases. Refer to Chapter 8 for instructions on how to adjust these thermostats.

Most hermetic refrigerators use a starting relay. It is usually mounted on the body of the motor compressor. These starting relays also provide overload protection for the motor. The overload protection contains a resistor, which is wired in series with the running current. Should the current draw be too great (overload), the resistor will heat up. This causes a bimetal contact to break the circuit.

Figure 11-5 illustrates the electrical circuits for a refrigerator. The starting relay connects both the starting winding and the running winding to the power circuit. When the compressor motor reaches about 75% running speed, the relay disconnects the starting winding from the power circuit.

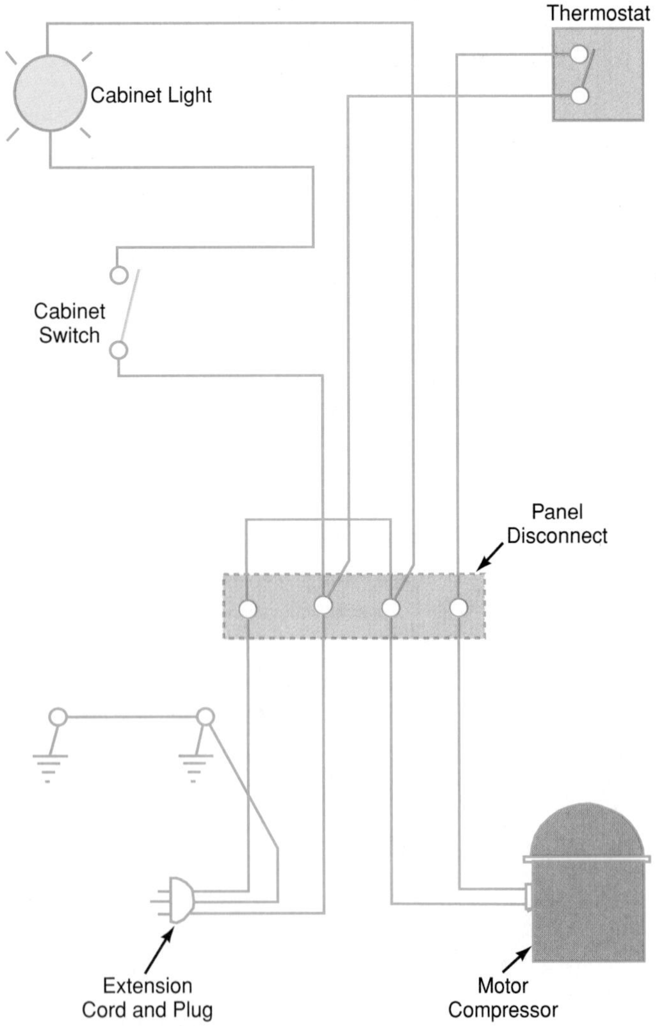

Figure 11-3. *Wiring diagram for a refrigerator.*

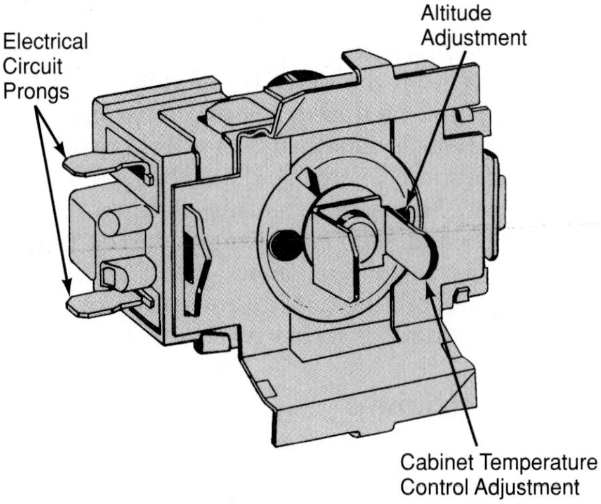

Figure 11-4. *Thermostat for a refrigerator.*

11.4 Refrigerator-Freezer—Manual Defrost

A refrigerator-freezer, **Figure 11-6,** consists essentially of two refrigerated spaces. It has a freezer compartment across the top of the cabinet for frozen foods. The temperature in this compartment is kept at approximately 0°F (−18°C). A refrigerator compartment is located below the freezer compartment for fresh foods. The refrigerator compartment maintains a temperature of about 35°F to 45°F (2°C to 7°C).

Each of these compartments has a separate door. The condensing unit is usually in the bottom of the cabinet. The condenser is either in the bottom or at the back.

Refrigerator-freezers provide shelves in both compartments. A butter conditioner is usually located in the door of the refrigerator compartment. Doors of both

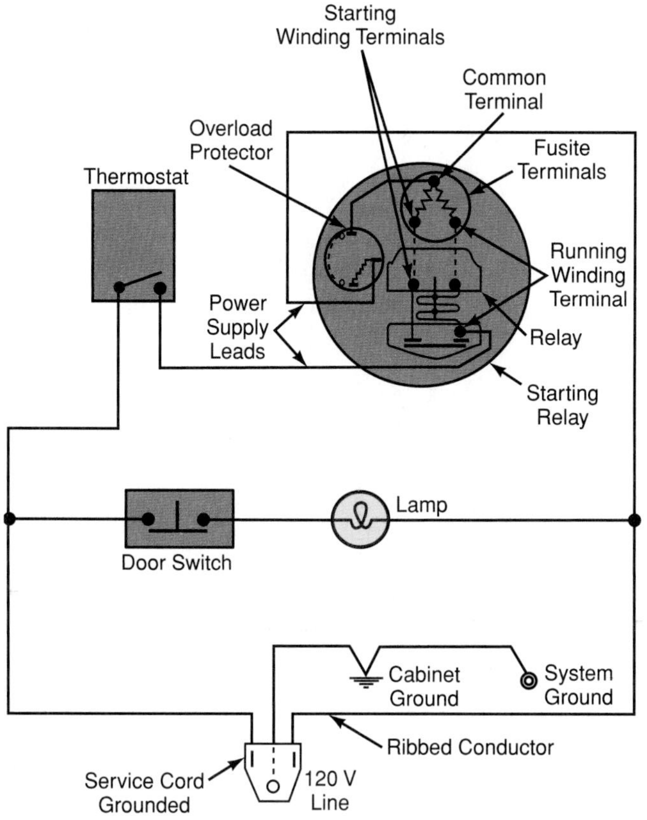

Figure 11-5. *Motor control relay for a refrigerator. A current-type relay is shown here using the running winding current. The diagram shows electrical connections to the thermostat and motor control relay. This motor compressor relay has an overload protector.*

Figure 11-6. *Double-door refrigerator-freezer with freezer at the top. Note the evaporator in the upper part of the refrigerator compartment. (Whirlpool Corporation)*

compartments are often fitted with narrow shelves for storage of small containers.

11.4.1 Cabinets

Cabinet construction for refrigerator-freezers is similar to that of simple refrigerators. However, thicker insulation of the cabinet is required. This added insulation maintains the lower temperatures necessary in the freezer compartment. A separate freezer door is provided to maintain low temperatures when the refrigerator door is opened.

The motor control, temperature control, light switch, shelf support, and crisper are the same as those on the fresh food refrigerator. The finish is usually a good grade of lacquer. Some cabinets have porcelain-finished interiors.

11.4.2 Mechanisms

Refrigerators with a freezer compartment have a hermetic compressor in the base of the cabinet. The condenser is either at the bottom or the back of the cabinet. The liquid refrigerant flows from the capillary tube into the evaporator in the freezing compartment.

A typical mechanism is illustrated in **Figure 11-7.** The refrigerant charge is usually sufficient to keep the freezer evaporator, A, filled. Enough is needed for spillover from the freezer evaporator, B. This evaporator is usually located at the side of the refrigerator compartment. It has a rather large accumulator. Any possible spillover from this enters a third evaporator, C. The third evaporator is usually located at the side of the refrigerator compartment. The third evaporator is also fitted with an accumulator. This assures that all refrigerant is evaporated before vapor is allowed to enter the suction line.

From this accumulator, vaporized refrigerant is drawn back to the suction line into the compressor. The vapor is compressed and pumped first into a small condensing coil, D. From here the high-pressure vapor is pumped through a loop in the base of the compressor. This serves as an oil cooler. From here the compressed vapor flows into condenser, E, at the bottom or at the back of the refrigerator.

At this point, the heat of the vapor is radiated to the surrounding air. The refrigerant is then condensed back to a liquid. The liquid flows from the bottom of the condenser through the filter-drier. It then flows into the capillary tube attached to the suction line. The capillary tube controls the refrigerant flow into the freezer evaporator, A. The cycle is then repeated.

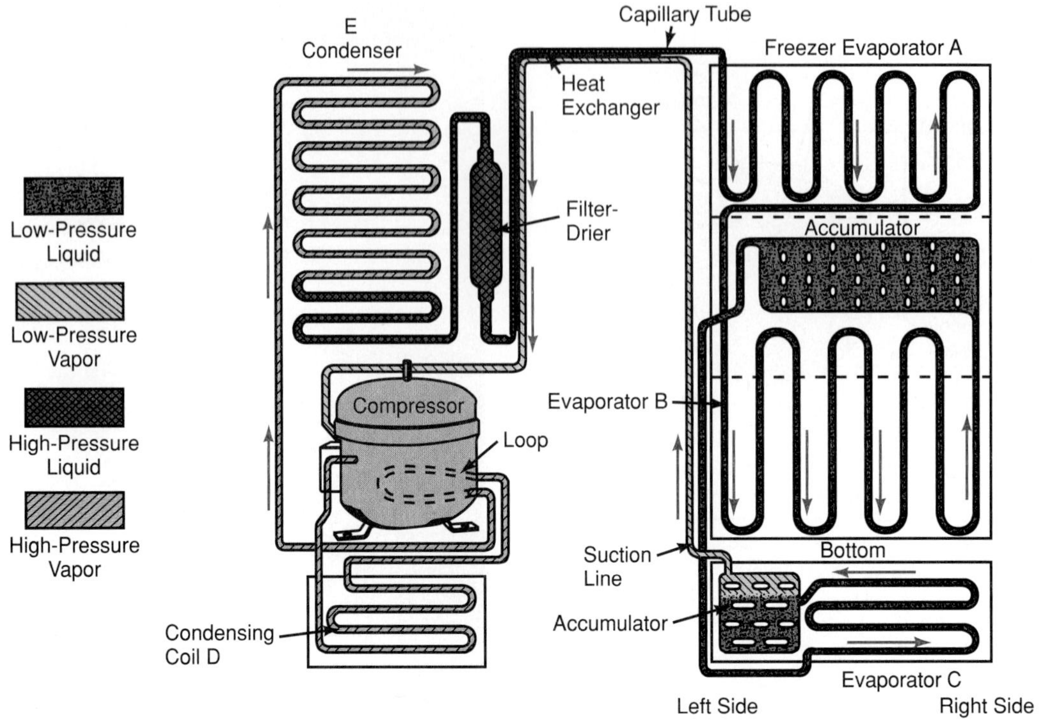

Figure 11-7. *Cycle diagram for a manual defrost refrigerator-freezer.*

11.4.3 Electrical Circuits

The electrical supply is fed through the grounded extension cord and plug at 1, **Figure 11-8.** The electrical circuit supplies current to the panel disconnect at 2.

From the panel-mounted disconnect, two separate circuits are provided. One circuit supplies current to the cabinet light at 3. It is controlled by cabinet switch, 4. The cabinet switch is in series with the light. The light comes on or goes off as the refrigerator door is opened and closed.

The second circuit goes to the motor compressor. The thermostat, 5, is in series with the circuit. It controls the compressor operation, 6.

When the cabinet temperature reaches a predetermined temperature, the thermostat completes the circuit through the motor. The compressor runs and the refrigeration cycle begins. The temperature inside the cabinet is brought down to the desired temperature. The thermostat then turns off the current and the motor compressor stops.

Two additional electrical devices are usually found on this type of refrigerator. One is an electrical resistance heat wire, also called a ***perimeter drier.*** It is shown at 7 and is located in the trim of the freezer door. This heater operates all the time. It provides enough "warming effect" to stop condensation on the exterior of the cabinet. It also prevents condensation around the freezer compartment door.

The second electrical device is an ***ambient compensator*** (shown at 8). This is also an electrical resistance heat wire. The ambient compensator provides a continuous small heat flow into the refrigerator compartment. This causes the refrigerator to cycle if the ambient room

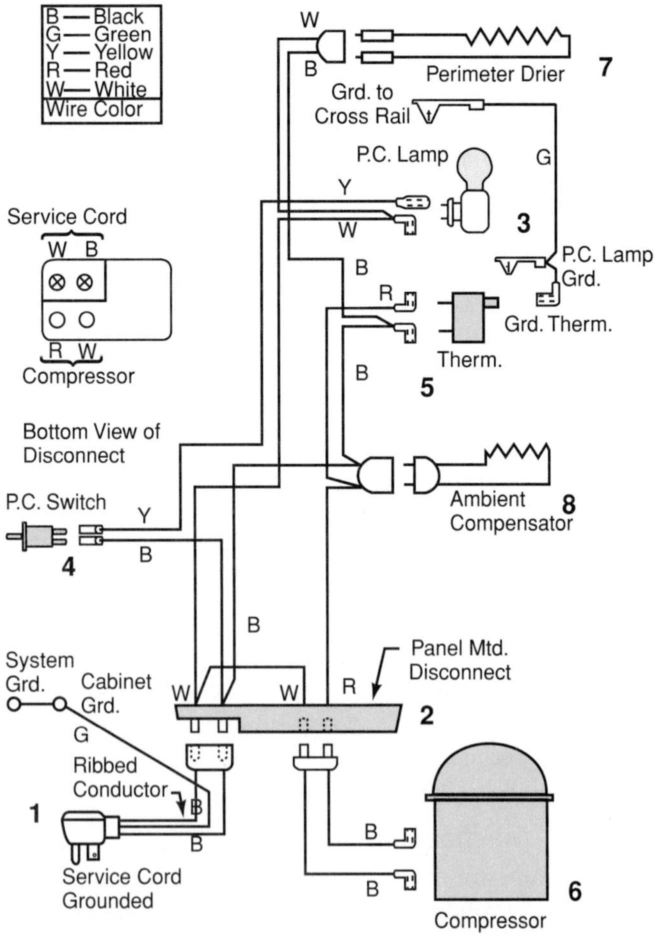

Figure 11-8. *Wiring diagram for a manual defrost refrigerator-freezer.*

temperature drops below the normal thermostat setting. The ambient compensator draws 15 VA to 20 VA. It is attached to the insulation side of the refrigerator compartment thermostat. It is energized only on the Off cycle (thermostat contacts open).

Figure 11-9 illustrates the electrical circuits for this refrigerator. The area near the letter A shows the starting relay. The starting relay connects both the starting winding and the running winding to the power circuit. It disconnects the starting winding when the compressor motor reaches about 75% running speed.

This starting relay also provides overload protection for the motor. The overload protector contains a resistor in series with the running current. In the event the current draw is too great (overload), the resistor will heat up. This will cause a bimetal contactor to break the circuit. The ambient compensator is shown at B. The perimeter drier is shown at C.

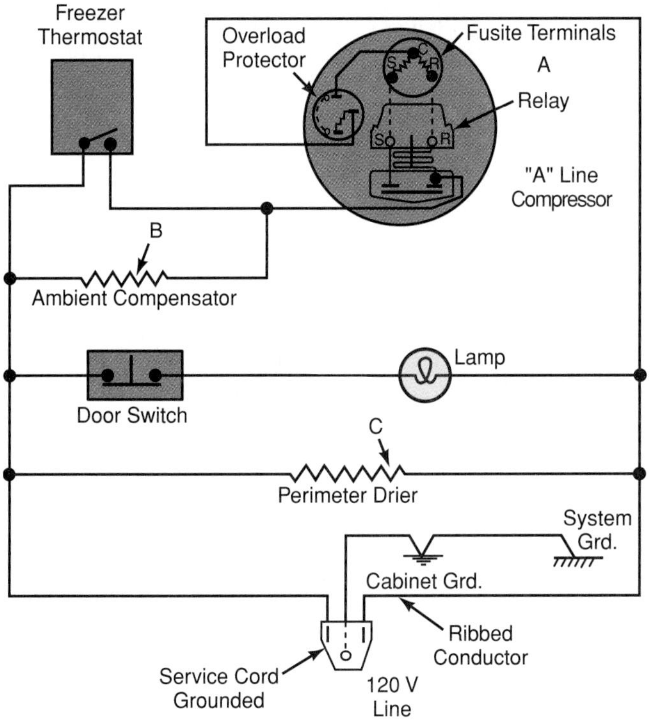

Figure 11-9. *Motor control relay for a manual defrost refrigerator-freezer. This ladder diagram illustrates the electrical connections to thermostat and motor control relay. The motor compressor relay shown here is provided with an overload protector*

11.5 Refrigerator-Freezer—Automatic Defrost

All air contains some moisture. Air can come in contact with an evaporator surface, which is below the freezing temperature. Moisture will condense and form ice on the evaporator.

Frequently, it is necessary to defrost the evaporator in order to maintain good refrigerating efficiency. This applies to the evaporator for both the refrigerator and the freezer compartments.

The owner usually considers it a chore to manually defrost the refrigerator. As a result, most refrigerators provide a system for automatic defrosting. There are two basic systems used in automatic defrosting. The hot gas system accomplishes this through the use of solenoid valves. Heat from vapor in the compressor discharge line and condenser is used for evaporator defrosting. The other system uses electric heaters to melt the ice on the evaporator surface. A section in this chapter will be devoted to each of these systems.

11.5.1 Refrigerator-Freezer—Electric Heater Automatic Defrost

Figure 11-10 illustrates a typical refrigerator with a freezer compartment. This refrigerator uses an automatic defrost. As in most domestic refrigerators, the condensing unit is mounted in the cabinet base. Some refrigerators have the condensing unit located at the back of the cabinet. Shelving is provided in both the refrigerator compartment and the freezer compartment.

Some refrigerators operate on what is called "frost-free" or "no-frost" cycle. In these refrigerators, the evaporator is located outside the refrigerated compartment. During the running part of the cycle, air is drawn over this evaporator. It is forced into the freezer and refrigerator compartment by a motor-driven fan. During

Figure 11-10. *Automatic defrost refrigerator-freezer. Evaporator in freezer compartment serves as fast-freezing shelf. Automatic ice cube maker is located in freezer compartment. Refrigerator compartment provides butter conditioner, fresh meat storage, and vegetable crisper. (Whirlpool Corporation)*

the Off part of the cycle, these evaporators automatically defrost.

Such refrigerators may use a single evaporator for both the freezer and the refrigerator compartments. However, a separate evaporator may be used in each compartment for some models. Evaporator condensation, which melts during the Off cycle, is carried to an evaporating pan. It may be carried to a collecting surface directly over the compressor and condenser. The heat from the compressor evaporates this moisture, allowing it to return to the room's atmosphere. There is never any visible frost accumulation in this type of frost control.

Cabinets

Cabinet construction for automatic defrosting refrigerator-freezers is different than manual defrost models. The condensation, which collects on the evaporator, must be melted from time to time. Different methods are used for disposing of this condensation. The cabinet must have tubing to conduct this moisture to the top of the motor compressor. Usually, it is collected on a plate or surface just over the motor compressor. This surface, or plate, is heated by the motor compressor and by heat from the condenser. The moisture then evaporates and goes back into the room.

Some refrigerators use electric defrost. Provisions must be made in the cabinet for housing electric heaters and their controls.

Figure 11-11 shows a refrigerator-freezer. It provides other temperatures besides that of the freezer and the refrigerator compartments. The air circulation provides lower temperatures to areas reached first. It provides higher temperatures to those areas reached last.

The evaporator is beneath the fast-freezing shelf at the bottom of the freezing compartment. All of the refrigerating effect comes from this evaporator.

The air circulation system, consists of a fan, ducts, and a damper. These are located at the back of the cabinet. It provides the necessary airflow to give the temperature desired in each compartment. The condensing unit is located in the bottom of the cabinet.

Mechanisms

Some mechanisms provide more than two temperatures in the refrigerator with a freezer compartment. A typical mechanism of this type is shown in **Figure 11-12.**

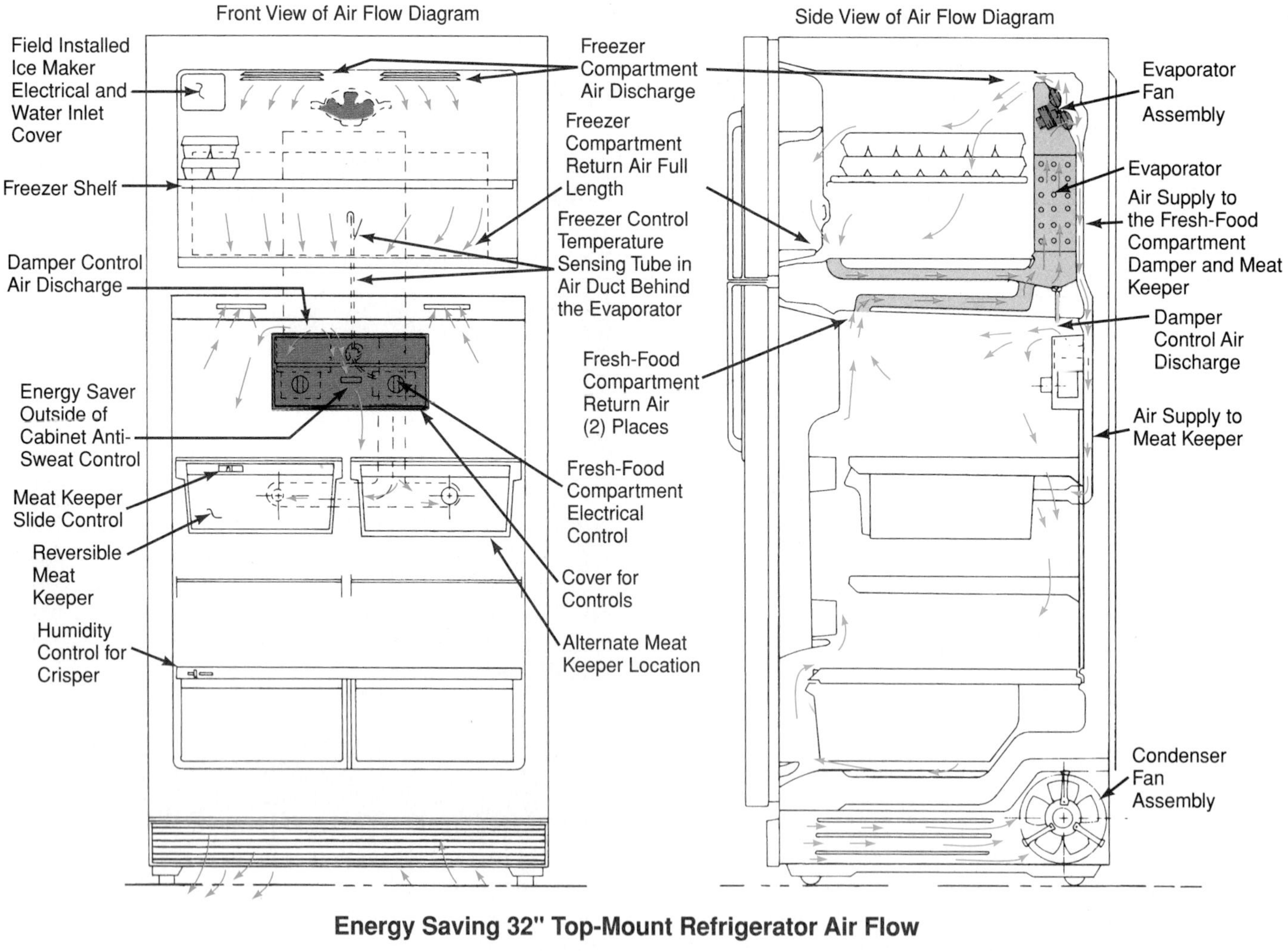

Figure 11-11. *Air circulation in refrigerator-freezer. Evaporator fan forces circulation of cooled air through various compartments. Dampers control air duct openings and thus, control cabinet temperatures. (Amana Refrigeration, Inc.)*

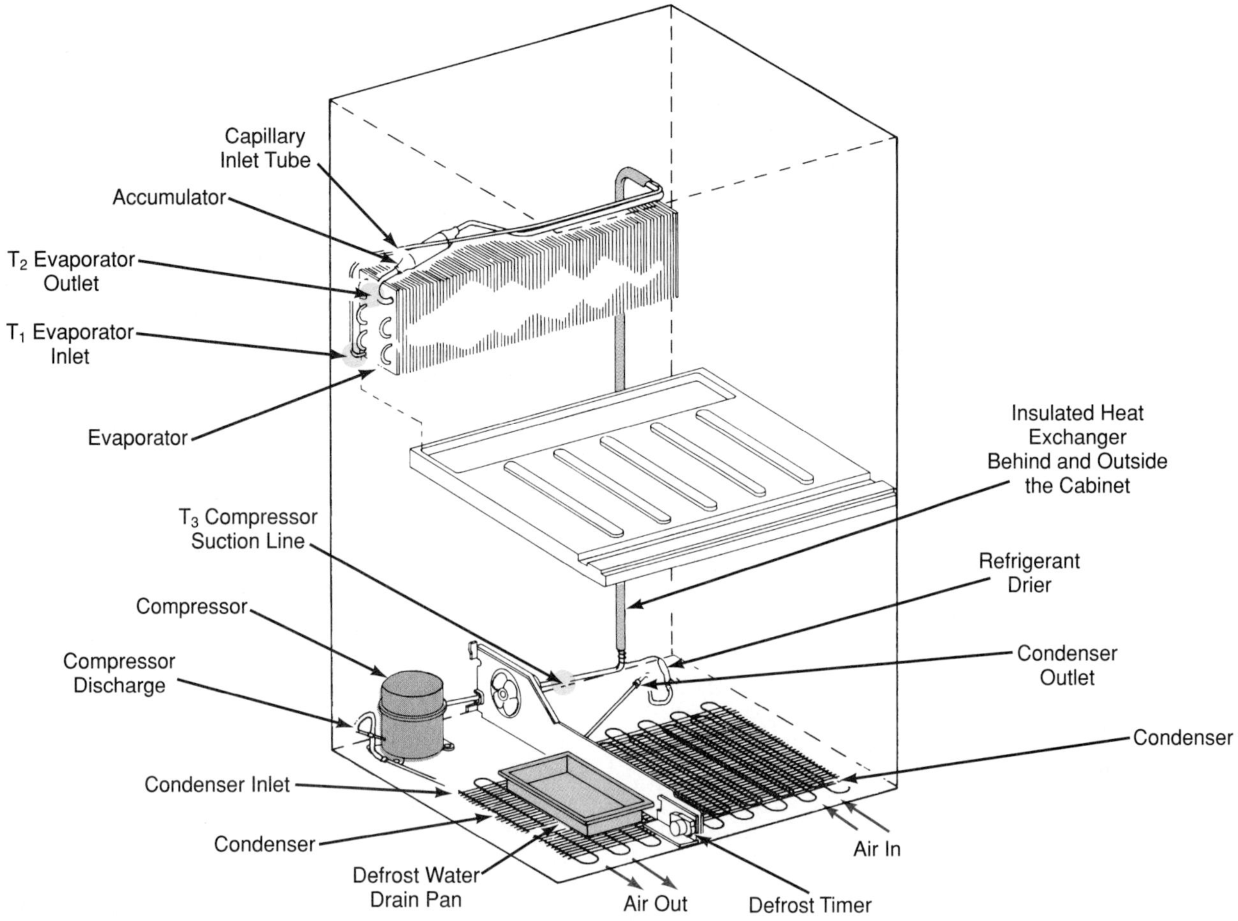

Figure 11-12. *Cycle diagram for automatic electric defrost refrigerator-freezer. Correct operation may be checked by determining temperatures at test points, T_1, T_2, and T_3. Recommended operating temperatures are: T_1 = −15°F to −13°F (−26°C to −25°C), T_2 = −15°F to −14°F (−26°C), T_3 = 80°F to 103°F (27°C to 39°C). Temperature T_3 is high because the capillary tube is in contact with suction line to transfer excess condenser heat to the suction line. This increases the superheat at the compressor inlet. (Amana Refrigeration, Inc.)*

Evaporator

The evaporator is located at the back of the shelf. It separates the freezer compartment from the refrigerator compartment. The refrigerant used is R-12 or R-134a. Refrigerant evaporation in the evaporator provides the heat absorption (cooling) required in the cabinet. Usually, a motor-driven fan forces air over the evaporator surface. Air is forced through the various ducts. This provides all the necessary refrigerator temperatures for the compartments.

Motor Compressor

The suction line from the evaporator extends down the wall of the cabinet. It extends to the inlet side of the hermetic motor compressor in the cabinet base.

Condenser

The condenser is a wire-and-tube type. Forced air circulation is provided by a motor and fan. They are located at the back of the compartment containing the compressor and the condenser.

Capillary Tube

The refrigerant is condensed in the condenser. It flows through a high-side filter-drier into a capillary tube. The capillary tube is attached to a section of the suction line. This provides a heat exchange between the capillary tube and the suction line. The refrigerant from the capillary tube then flows into the evaporator. The cooling cycle is completed.

Features

Only the refrigerator compartment has a light. It is operated by a switch, which is activated by the door movement. The butter conditioner temperature is slightly above the cabinet temperature. There are control dampers for the refrigerator and the freezer compartments. These dampers regulate flow of cold air from the evaporator.

Heaters

Several heating devices are used as driers. An electrical resistance heater is located at the top of the cabinet, inside the outer case. This keeps the outside of the cabinet warm so that it will not collect condensation during damp days.

A second drier wire is placed inside the center mullion to keep its surface dry. A third heating device is around the freezer flange (freezer door opening). A fourth heater is placed in the drip pan. It evaporates the condensation which flows into the drip pan after automatic defrost.

A "power-saver" switch is located inside the cabinet. It provides a means of disconnecting these heaters when temperature and humidity conditions allow it. Normally, the heaters are in continuous operation.

Automatic Defrost

The evaporator is automatically defrosted by an electric resistance heater. This occurs every six hours of compressor running time. The heater is located in the fin area on the underside of the evaporator. A timer activates the switch, which turns on the defroster. A thermostat attached to the evaporator opens the defrost heater circuit. This ends defrosting when the evaporator temperature reaches 50°F (10°C). (Temperature may be ±6°F [3°C].) After 28 minutes from the start of the defrost cycle, the timer restores the compressor operation. It also restores the air-circulating fan operation. The defrost terminator contacts close at 20°F (−7°C), ±8°F (4°C). The temperatures of the cabinet are regulated by the temperature control. This control is mounted in the rear wall of the freezer compartment. The temperature of the evaporator tubing near the end of a running cycle may vary. It may vary from −13°F to −25°F (−24°C to −32°C). The difference between the evaporator inlet and outlet temperature will not vary more than about 3°F (2°C).

Electrical Circuits

Electricity is supplied through the grounded extension cord and plug. These are shown at 1 in **Figure 11-13.** The compressor and defrost timer are energized directly from the grounded extension cord.

A machine compartment connector block is located in the bottom of the cabinet. (This block is shown at 2.) It provides electrical connections to the following: evaporator fan, cabinet light, defrost heater, defrost terminator, mullion heater, freezer flange heater, and power saver switch. If there is an automatic ice maker, it is wired into the same connector block. A ladder diagram of the wiring circuit is shown at the top of **Figure 11-13.**

11.5.2 Refrigerator-Freezer—Hot Gas Automatic Defrost

Hot gas automatic defrost is accomplished through the use of solenoid valves. Heat from the vapor in the discharge line and condenser is used to defrost the evaporator. **Figure 11-14** illustrates a refrigerator-freezer equipped with hot gas defrost.

Cabinets

These cabinets provide the usual facilities for both fresh- and frozen-food storage. The thermostat is located in the refrigerator compartment.

The defrost timer and the solenoid valve are housed in the cabinet base. Many refrigerators pipe compressed vapor through tubing on the inner surface of the outside shelf. This is done particularly around the door openings to prevent condensation from forming around the front of the cabinet. This uses a perimeter hot-tube system. The perimeter hot tube is part of the high-side of the system. It provides heat needed to prevent condensation from forming around the front flanges of the cabinet. The use of the perimeter hot-tube system eliminates the need for an electrical heater. See **Figure 11-15.**

The function of these coils is to keep the surface of the refrigerator cabinet dry. They are sometimes referred to as ***driers.*** The same term may be applied to electric heating units installed in the cabinet for the same purpose. This use of the word "drier" should not be confused with the term drier as applied to liquid and suction lines.

Mechanisms

The mechanism in **Figure 11-16** is typical of those used in refrigerators using hot gas defrost.

Evaporator

Two evaporators are used with R-12 or R-134a as their refrigerant. Evaporator A is located in the freezer compartment. It extends along the back, ends, and bottom of the compartment. Refrigerant (both vapor and liquid) leaving this evaporator flows into evaporator, B. This evaporator is located at the back of the refrigerator compartment. It has a rather large accumulator since all refrigerant should leave the refrigerator compartment evaporator as vapor.

Motor Compressor

Vapor is drawn back to the compressor. It is compressed and discharged directly into a water evaporating plate and coil assembly. This is located over the compressor. It serves to evaporate the moisture drained from the evaporators during the defrost cycle. This compressed vapor next flows through the oil cooler line in the bottom of the compressor.

Condenser

From the oil cooler line, vapor flows to the vertical wire-and-tube condenser. Heat is given off to the surrounding air. The compressed vapor returns to a liquid state. From the bottom of the condenser, the liquid flows through the filter-drier.

Capillary Tube

From the filter-drier, the liquid then flows into the capillary tube refrigerant control, which is soldered to the suction line. This serves as a heat exchanger. It reduces the temperature of the liquid refrigerant in the capillary tube and increases the superheat of the vapor in the suction line. From the capillary tube, the refrigerant enters the freezer evaporator at a reduced pressure. It evaporates and absorbs heat from the inside of the cabinet, thus completing the cycle.

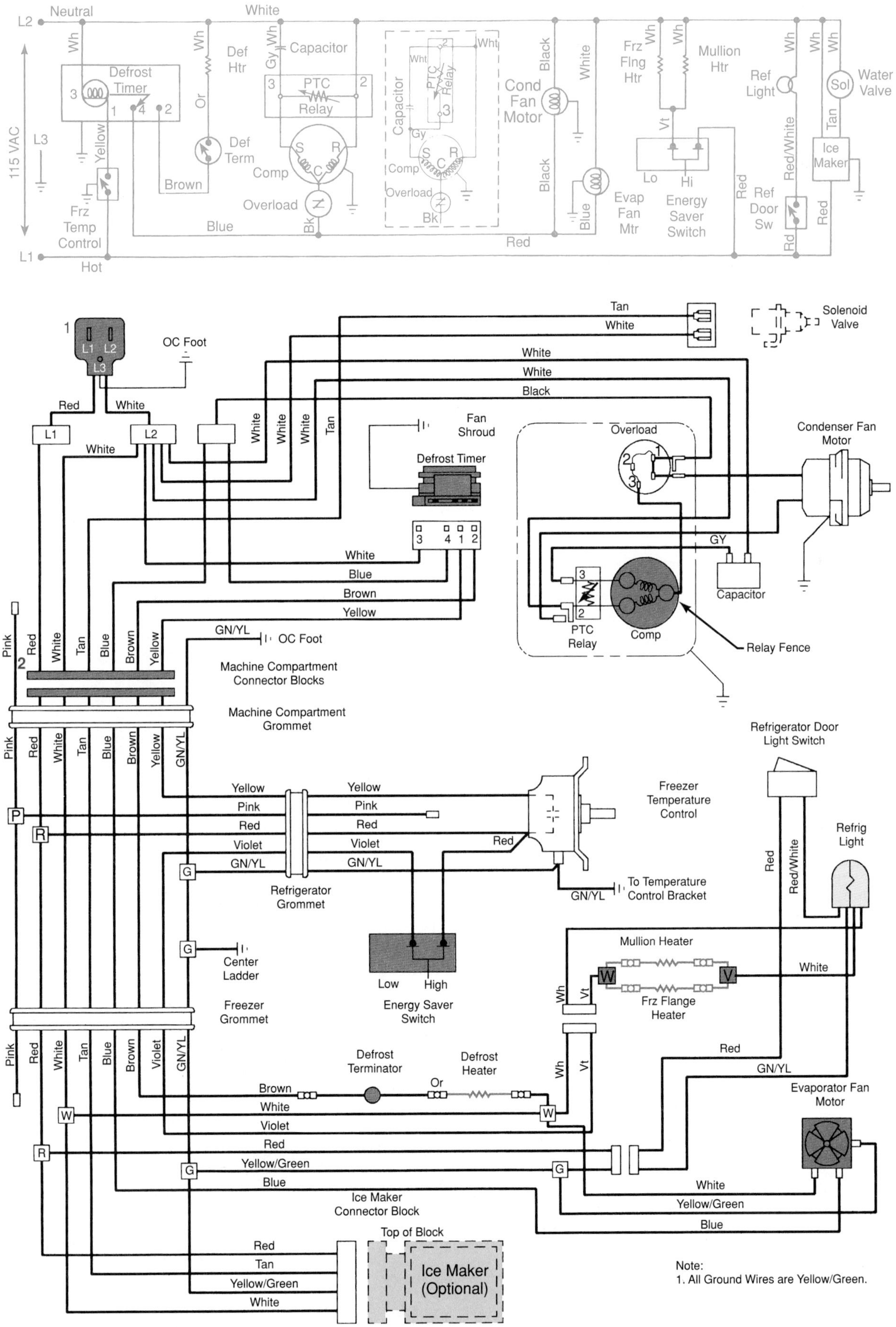

Figure 11-13. *Wiring diagram for automatic electric defrost refrigerator-freezer. Note the ladder diagram at the top. Often, ladder diagrams proceed from the top of the page to the bottom. (Amana Refrigeration, Inc.)*

Figure 11-14. *Hot gas defrost refrigerator-freezer. Evaporator extends along back and ends of freezer compartment and lower shelf. Thermostat, butter conditioner, meat keeper, and vegetable crispers are in refrigerator compartment. (Whirlpool Corporation)*

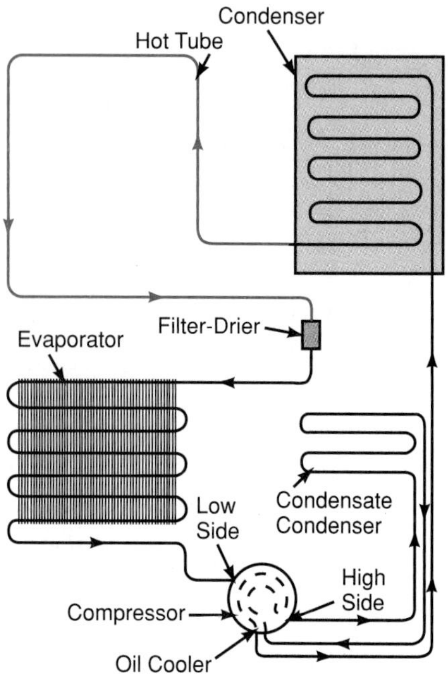

Figure 11-15. *Perimeter hot-tube system. (Frigidaire Company)*

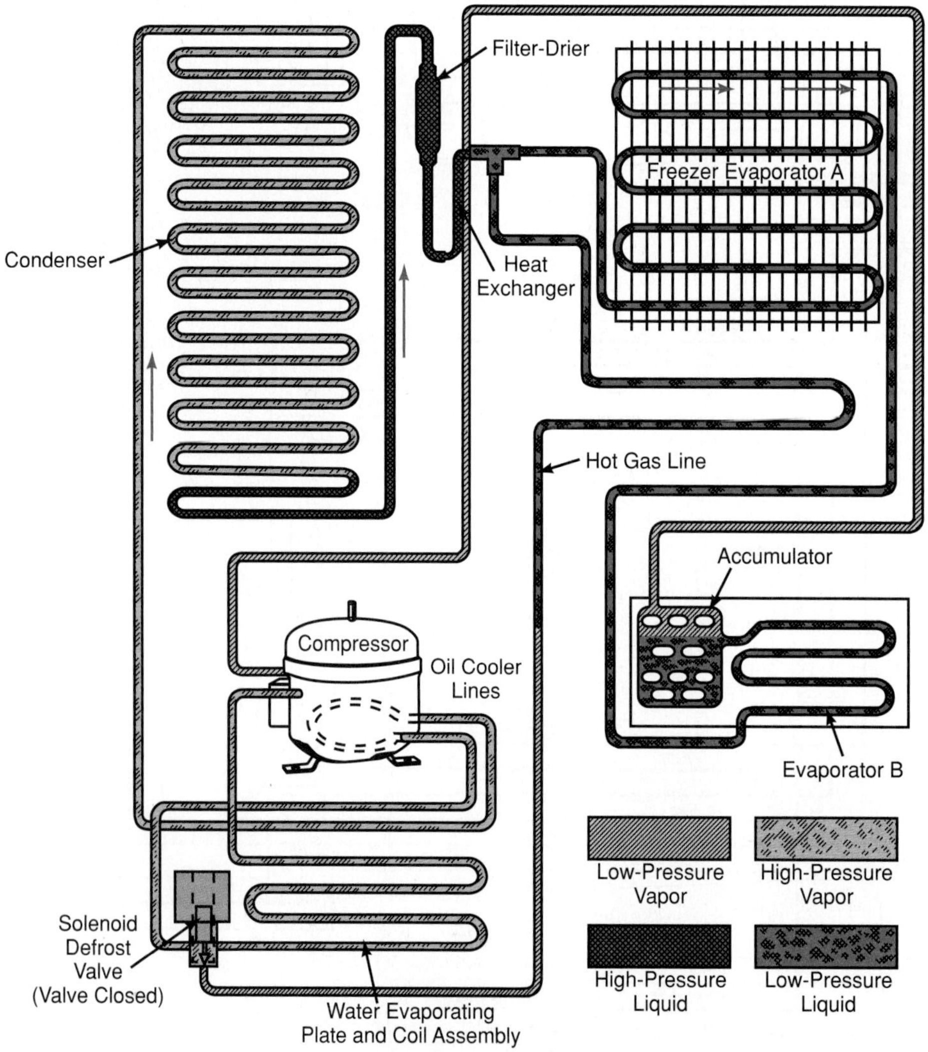

Figure 11-16. *Refrigerating cycle for hot gas defrost refrigerator-freezer. The solenoid valve is closed and closes off the bypass line.*

Hot Gas Defrost

Figure 11-17 shows the solenoid valve open and the hot gas defrost cycle in operation. The vapor from the evaporators is drawn into the compressor. It discharges into the water evaporating plate, helping to heat this surface. Then, hot, compressed vapor flows back through the drain sump bypass line and flows into the freezer evaporator. From there, the vapor flows into the refrigerator compartment evaporator through the accumulator. It then flows back into the suction side of the compressor. This is because the heat is supplied directly to the inner surfaces of the evaporators.

A defrost timer, **Figure 11-18,** controls the solenoid defrost valve. The timer is located at the back lower-left corner of the refrigerator cabinet. It is driven by a self-starting electric motor geared to turn the shaft slowly. The shaft completes one revolution every eight hours of compressor operation. When the timer opens the solenoid valve, the defrost cycle continues for approximately 17 minutes. Then, the solenoid valve closes and the refrigerating cycle operates. The freezer compartment fan is not operating during the defrost cycle.

Electrical

Electricity is supplied through the grounded extension cord and plug. These are shown at 1 in **Figure 11-19.** The electrical supply goes to the panel-mounted disconnect at 2. Various circuits are fed from this disconnect. The defrost solenoid is shown at 3. The defrost solenoid, defrost timer, and the compressor are energized from the four right-hand connections. With an automatic ice maker, the water valve, 4, is energized through the two terminals as indicated.

In addition to the hot gas defrost, three electrical resistance heaters are in the system. The ambient compensator at 5 is attached to the insulation side of the refrigerator compartment. The perimeter drier, 6, is installed in the trim of the freezer door. The drain sump heater prevents freezing of condensation from the

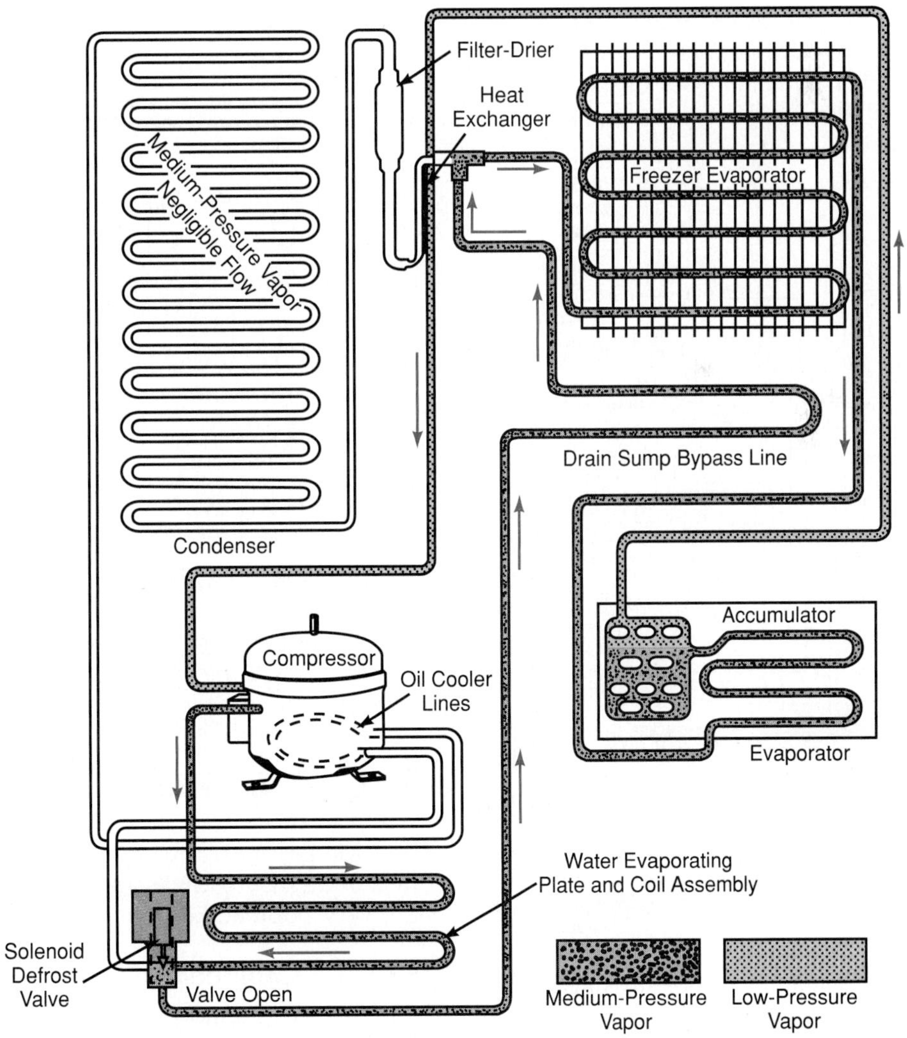

Figure 11-17. *Defrost cycle of a hot gas defrost refrigerator-freezer. The solenoid valve is open and hot compressed vapor is traveling through the bypass line. Heat from the compressed vapor keeps drain sump free of ice. The heat melts ice from both the freezer evaporator and refrigerator evaporator.*

Figure 11-18. *Automatic defrost timer. A—Timer is connected into motor compressor circuit and provides both a clock mechanism and switching mechanism. The clock operates the switch mechanism in such a way that, after eight hours of compressor operation, fan motor is turned off and defrost solenoid is energized. B—Timer motor.*
(Paragon Electric Co., Inc.)

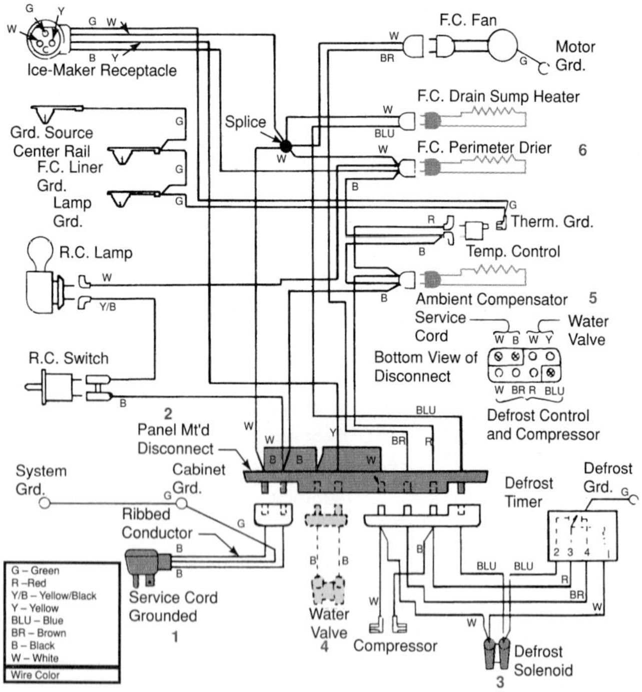

Figure 11-19. *Wiring diagram for a hot gas automatic defrost refrigerator-freezer. In the illustration, F.C. means freezer compartment. R.C. means refrigerator compartment.*

evaporator as it moves down the drain tube to the evaporating pan.

This wiring diagram provides a complete grounding system. Each metal part of the mechanism and cabinet has its own ground. (See Section 6.7.2.) **Figure 11-20** is a schematic of this wiring diagram.

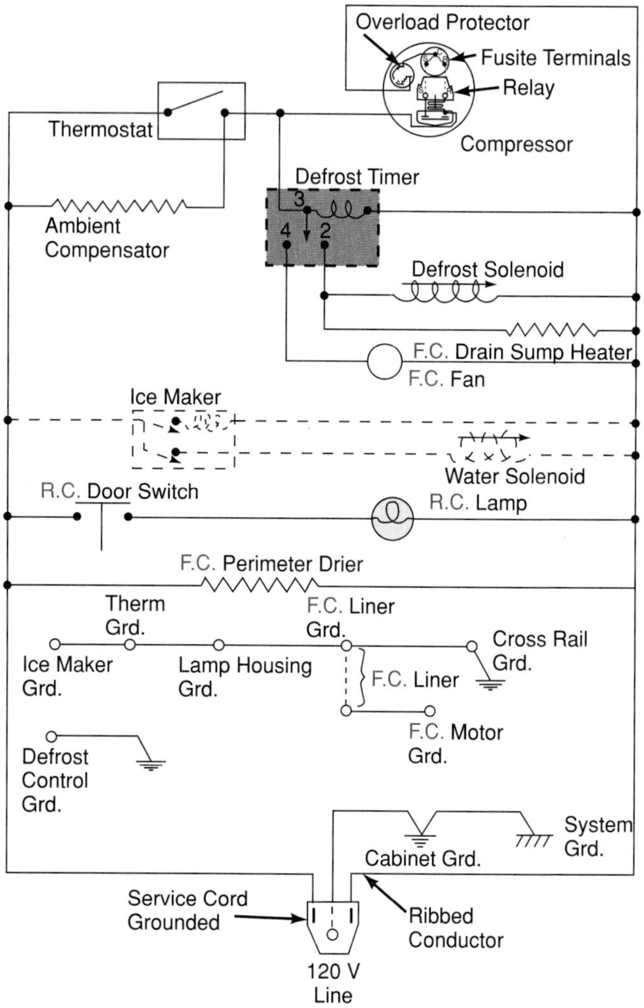

Figure 11-20. *Ladder wiring diagram for an automatic defrost, hot gas refrigerator-freezer. Note the provisions for required grounding of the various components. In the illustration, F.C. means freezer compartment. R.C. means refrigerator compartment.*

11.6 Refrigerator-Freezer—Frost-Free

Modern refrigerators are designed to eliminate the task of defrosting the evaporator. Frost-free refrigerators are very popular. Section 11.7 explains how they work.

11.6.1 Cabinets

The refrigerator-freezer, **Figure 11-21,** stores frozen food at the top and fresh food at the bottom. The evaporator is in the upper back part of the cabinet. The

Figure 11-21. *Frost-free refrigerator-freezer. Freezer compartment is cooled by very cold air from evaporator, which enters through freezing compartment grille. Condenser is cooled by air which circulates in and out through bottom grille. Evaporator fan is turned off automatically when freezer compartment door is opened. (Whirlpool Corporation)*

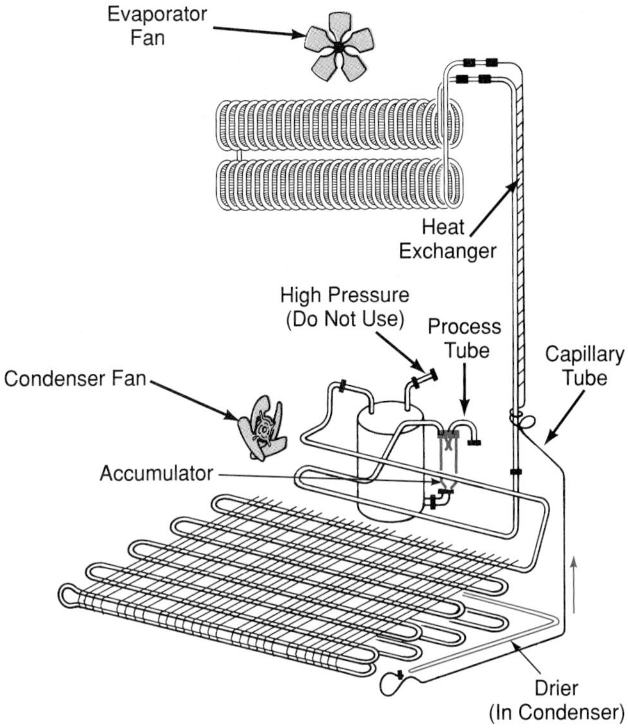

Figure 11-22. *Cycle diagram showing construction of frost-free refrigerator-freezer. Note the accumulator and the drier. (General Electric Co.)*

condenser is along the lower back part. A fan moves cold air from the freezer compartment evaporator into the refrigerator compartment. Another fan circulates room air through the grille at the cabinet bottom and over the condenser. With this type of condenser, it is not necessary to provide any clearance space at the sides or top of the cabinet for air circulation.

Both doors are held shut by magnets. This maintains a tight seal without the need for a mechanical latch. The cabinet is commonly mounted on wheels, making it easy to roll the refrigerator out for cleaning.

11.6.2 Mechanisms

Figure 11-22 is a diagram of the mechanism for this refrigerator-freezer. The compressor, condenser, capillary tube, heat exchanger, and evaporator are identified. Note that fans are used on both the condenser and evaporator.

11.6.3 Electrical Circuits

Electricity is supplied through a grounded extension cord and plug shown at 9 in **Figure 11-23.** There are two electrical resistance heating elements used in this cabinet.

The evaporator defrost is electrically heated. The freezer door contains a recess and duct heater. The heater helps eliminate moisture. In this illustration, the resistance of the various electrical components is indicated in ohms (Ω). This electrical circuit allows you to easily connect an automatic ice maker to the cabinet.

11.7 Refrigerator-Freezer—Frost-Free, Side-by-Side

The side-by-side refrigerator-freezer arrangement is very popular. The freezer compartment stores frozen foods at a temperature of 0°F (−18°C) or below.

With advancements in electronics technology, more features are being incorporated into the refrigerator-freezer unit. Besides ice maker hardware and freezer controls, there may be cold water, crushed ice, and liquid dispensers. These dispensers are built into the door. See **Figure 11-24.** In some models, the refrigerator-freezer is being used as a kitchen entertainment center. They may have radio and cassette systems, digital clocks with timers, alarms, etc. Some units also have electronic monitor consoles that control all accessories and electrical current.

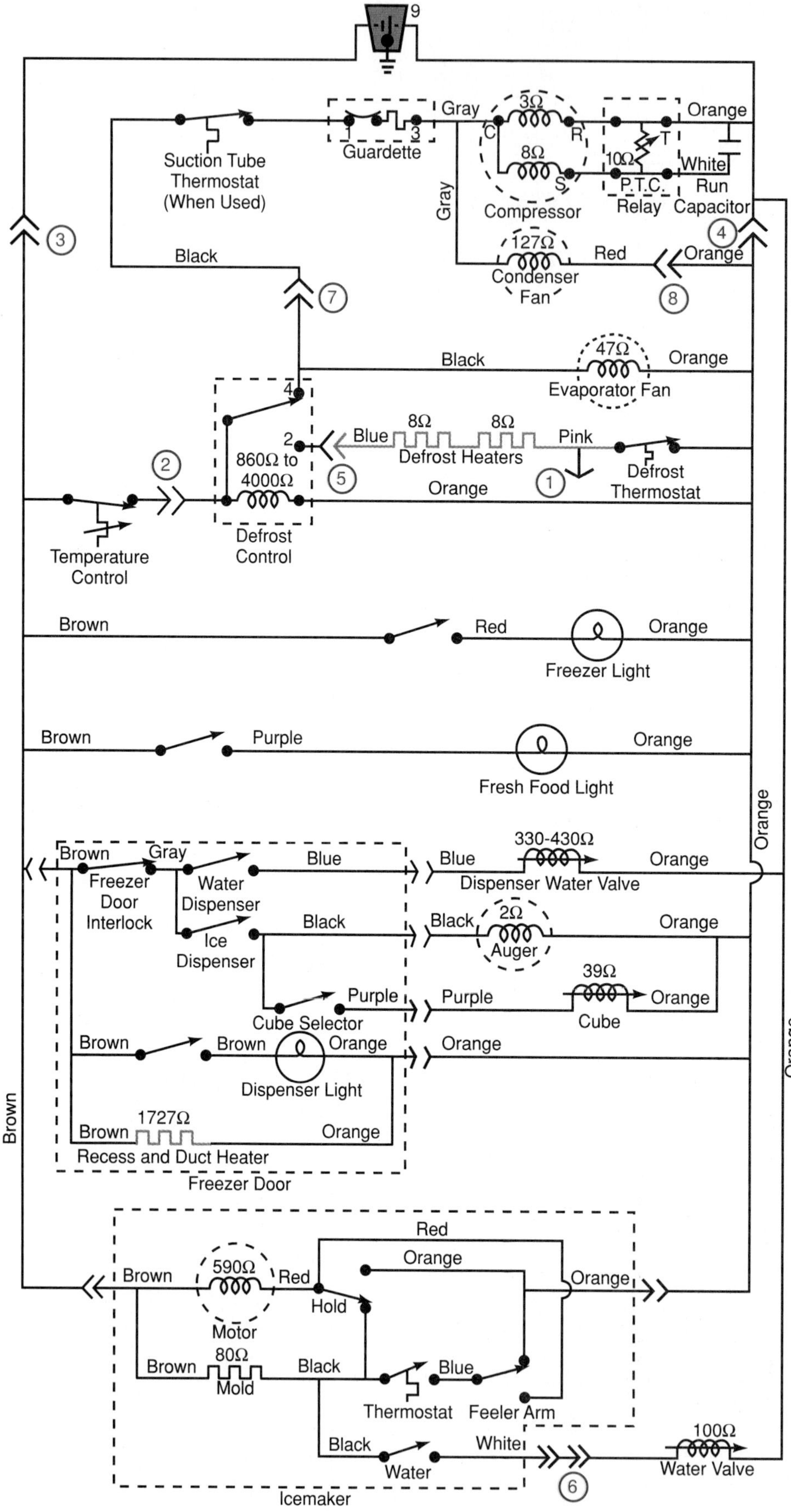

Figure 11-23. *Ladder wiring diagram for frost-free refrigerator-freezer. The freezer compartment is located above and the refrigerator compartment below. (General Electric Co.)*

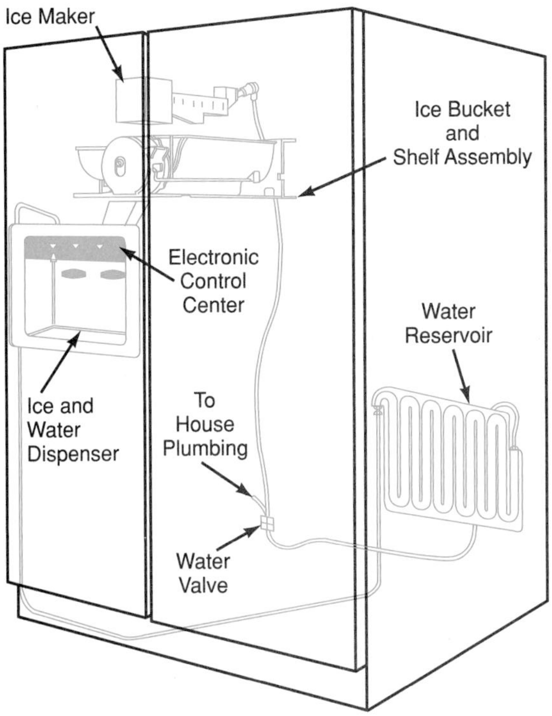

Figure 11-24. *Chilled water and ice maker system construction for side-by-side refrigerator-freezer. (General Electric Co.)*

11.7.1 Cabinets—General Electric

The General Electric side-by-side refrigerator-freezer is shown in **Figure 11-25.** This unit provides an automatic ice and water dispenser. They are controlled by an electronic monitor. The freezer compartment is on the left side. The refrigerator compartment is located on the right side. The freezer compartment stores foods at a satisfactory temperature of 0°F (−18°C) or below. Temperature controls for both compartments are located at the top of the refrigerator compartment.

The refrigerator compartment defrosts during every Off cycle. The freezer evaporator defrosts for about 25 minutes. This occurs after a six-hour accumulated compressor running time. The freezer defrost timer is located on the cabinet front near the bottom. Moisture from the evaporator surface flows down to pan resting on top of the condenser. Heat from the condenser evaporates the moisture.

The cabinet is commonly fitted with rollers, which makes it easy to move.

Figure 11-25. *Side-by-side frost-free refrigerator-freezer with automatic ice maker. (General Electric Co.)*

11.7.2 Mechanisms—General Electric

The evaporator, compressor, and condenser are at the back of a frost-free, side-by-side refrigerator-freezer. A fan circulates air over the condenser. This air enters and leaves through the bottom grille.

A fan circulates very cold air from the evaporator into the freezer compartment. Some cold air flows from the freezer compartment into the refrigerator compartment. This airflow is controlled by damper arrangements. Warmer air in the refrigerator compartment returns to the evaporator compartment. The airflow is shown in **Figure 11-26.**

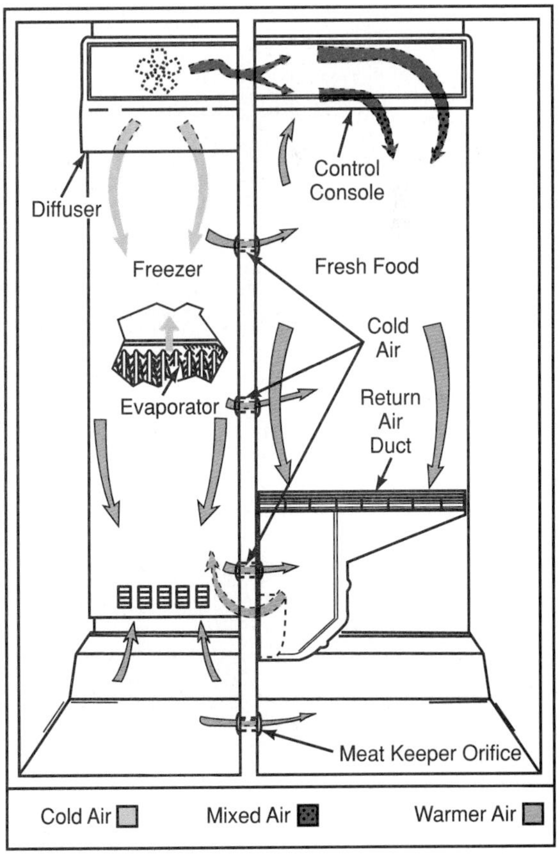

Figure 11-26. *Airflow diagram for frost-free, side-by-side refrigerator-freezer. (General Electric Co.)*

Refrigerant control is accomplished by capillary tube. The capillary tube is attached to the suction line as shown in **Figure 11-27.**

An electronic control console is located at the front of the freezer door, **Figure 11-28.** It is above the ice and water dispenser. To operate the control console, touch the desired selection pad. The identified pads in **Figure 11-28** perform the following functions:

1. Allows you to obtain ice cubes or crushed ice.
2. Dispenser light for nighttime lighting of controls.
3. The door alarm. This signals when either the refrigerator or freezer door is open longer than 30 seconds.
4. Door open monitor. Indicates when refrigerator or freezer door is open or ajar.
5. Warm temperature monitor. Indicates when the freezer temperature has been above normal for one to four hours.
6. Normal indicator, that indicates there are no failures detected by the diagnostic system.
7. Diagnostic code will be flashed on the control console when an abnormal condition exists. If more than one code function requires service, the highest priority is displayed until corrected. Then, the next highest will occur. The following codes are listed in priority:
   FF—Frozen foods should be checked for thawing.
   PF—Power has been interrupted.
   CI—Ice maker is not operating properly.
   dE—Defrost has not been detected for an extended amount of time.
   CC—Freezer temperature is warm.
8. System check/reset reviews all five diagnostic codes. It will identify any abnormal conditions that may exist. If none exist, the normal light will appear briefly.

The electronic control console is connected to the wiring in the system as shown in **Figure 11-29.**

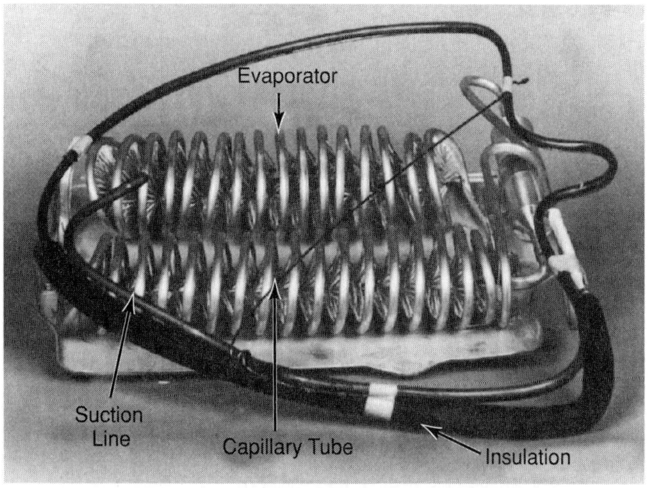

Figure 11-27. *Capillary tube and suction line connected to an evaporator. (General Electric Co.)*

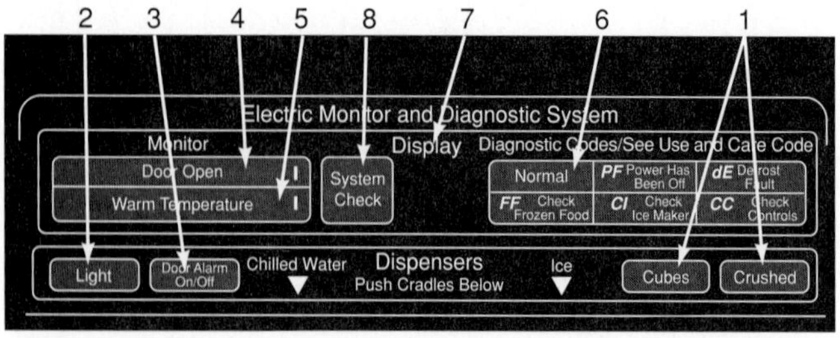

Figure 11-28. *Electronic control monitor diagnoses various operating difficulties. The monitor is used by owners and service technicians to assess problems. (General Electric Co.)*

11.7.3 Electrical Circuits—General Electric

A grounded extension cord and plug supply electricity to the refrigerator-freezer. **Figure 11-30** is a wire harness diagram for a side-by-side model. Individual circuits can be worked on any time since wiring is not "foamed" in place. Note that there are heat elements for defrost heaters and a recess and duct heater. Color coding locates the wire going to each of the electrical circuits.

The automatic ice maker controls are plugged into an ice maker receptacle. This eliminates the need to disturb any circuits when installing or removing the ice maker.

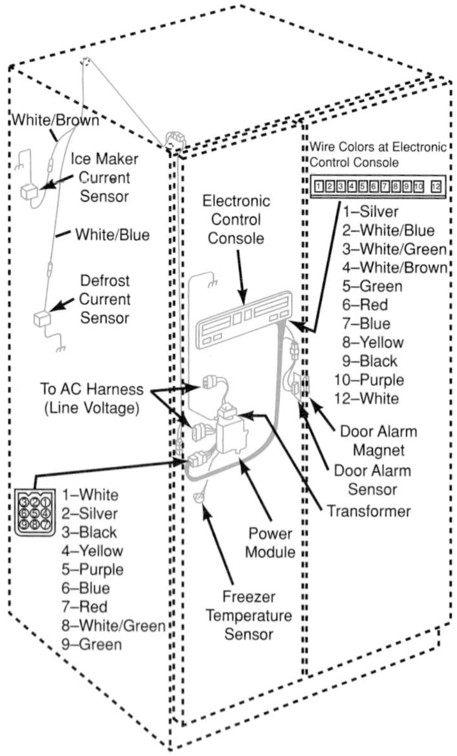

Figure 11-29. *Refrigerator with electronic control console. Note various control sensors. (General Electric Co.)*

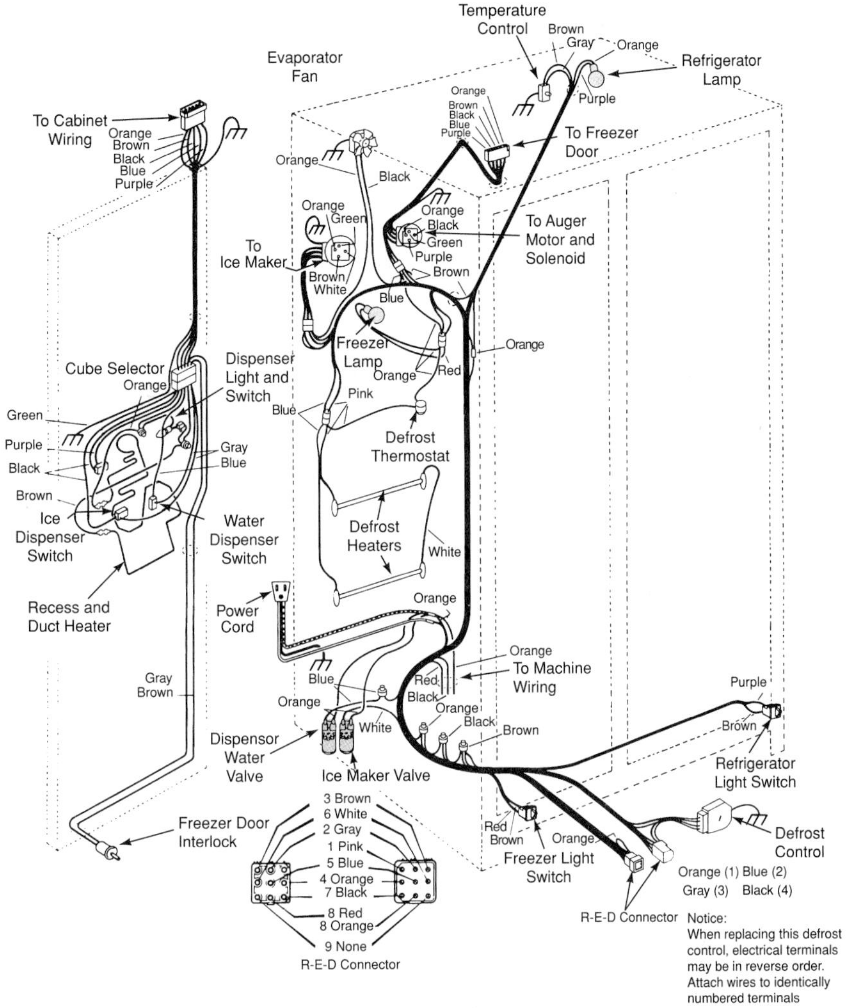

Figure 11-30. *Wiring harness diagram of side-by-side refrigerator-freezer with automatic ice maker. (General Electric Co.)*

11.7.4 Cabinet—Amana

The refrigerator-freezer in **Figure 11-31** has a refrigerator compartment on the right and a freezer compartment on the left. Freezer compartment door shelves provide convenient storage space for small items. The center door on the left side contains an external, automatic dispensing unit for ice cubes and water. See **Figure 11-32.**

Airflow through the cabinet is shown in **Figure 11-33.** A fan draws air through the single evaporator and distributes it throughout the cabinet. Dampers control temperatures in the various parts of the cabinet.

Door switches control two lights. One is in the refrigerator compartment and one is in the freezer compartment. A resistance wire or wires are placed inside the center mullion. They are also placed around the door openings to control condensation.

Other features of the cabinet include magnetic door gaskets on four sides and base rollers for easy moving of the unit.

Figure 11-31. *Side-by-side refrigerator-freezer with ice and water dispenser. Freezer temperatures are controlled by thermostat in left side in upper-left compartment. Refrigerator temperature thermostat is in right side. (Amana Refrigeration, Inc.)*

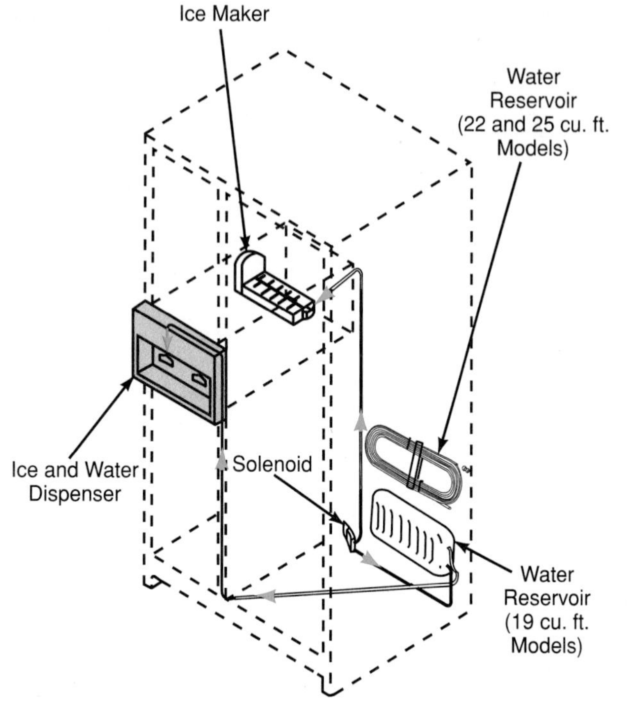

Figure 11-32. *Side-by-side refrigerator-freezer with ice and water dispenser. Note the location of the water reservoir in the refrigerator side and also the water flow circuit. (Amana Refrigeration, Inc.)*

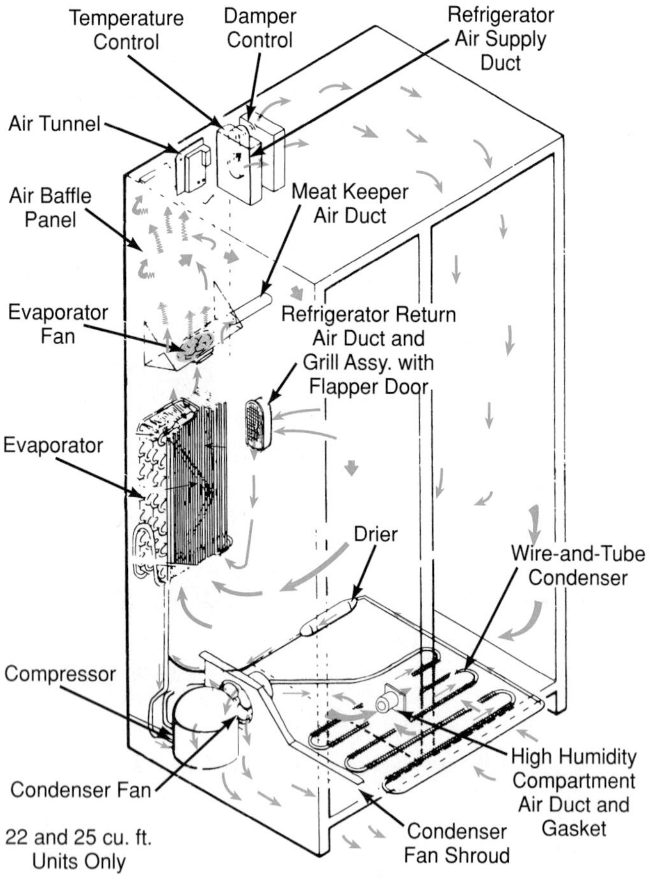

Figure 11-33. *Air movement pattern in side-by-side refrigerator-freezer. (Amana Refrigeration, Inc.)*

11.7.5 Mechanisms—Amana

The evaporator is behind the freezer compartment. The compressor and condenser are in the bottom. See **Figure 11-34.** Air circulated over the condenser by a fan enters and leaves through the bottom grille. A fan on the evaporator circulates very cold air in the freezer compartment.

Damper arrangements allow some of this cold air to flow into the refrigerator compartment. The refrigerator compartment acts as a return air duct. Air is returned from the freezer compartment back into the evaporator compartment.

Refrigerant control is by capillary tube. This tube is attached to the suction line as in **Figure 11-34.** It is called the heat exchanger.

The refrigerator automatically defrosts every six hours of compressor running time. The defroster is an electric heater attached to the evaporator. It is energized by a switch which is turned on and off by a timer. A defrost terminator (thermostat) is attached to the left end

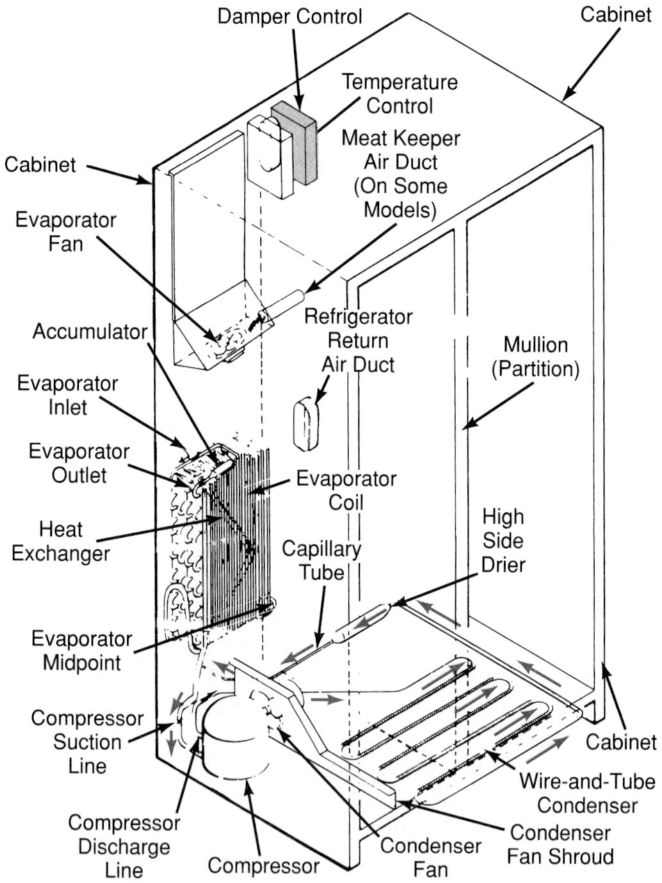

Figure 11-34. *Construction of side-by-side refrigerator-freezer. Note that condenser lies flat underneath bottom of refrigerator. Compressor is under freezer compartment at back. Evaporator is behind freezer compartment. The various cabinet temperatures are obtained by use of dampers, which control flow of cold air into the compartments. (Amana Refrigeration, Inc.)*

plate of the evaporator. It opens the heater circuit at approximately 50°F (10°C). After 28 minutes from the start of the defrost cycle, the timer restores unit operation. The compressor and air circulating fan operation is restored. The defrost terminator contacts close (reset) at about 20°F (−7°C).

11.7.6 Electrical Circuits—Amana

Electricity is supplied through a grounded extension cord and plug. Note the electrical circuits for this refrigerator-freezer in **Figure 11-35.** Wiring is located in the foamed-in-place insulation.

There are a large number of heaters in the electrical circuit. These include: defrost heaters, mullion heater, butter cavity heater, freezer flange heater, water tank heater, freezer control duct heater, and ice chute heater.

An auxiliary heater is foamed-in-place next to the connected heater. If the original heater fails, the auxiliary may be connected into the circuit. This is done by disconnecting the existing plastic-covered male terminal. It is then connected to the extra plastic-covered male terminal.

The electrical connections are located behind the evaporator covers. These heaters are in series with a switch on top of the freezer control. They may be turned off if the refrigerator is operated in extremely low-humidity areas. This is called a *power-saver switch.*

A color code identifies the wires going to each of the electrical cabinet's circuits. The automatic ice maker controls are plugged into an ice maker receptacle. This eliminates the need to disturb any circuits to remove or restore the ice maker.

11.8 Solid-State Ice Maker

In some models of refrigerator-freezers, solid-state controls are used for the automatic ice cube maker. They use transistors, diodes, relays, and other semiconductor components reviewed in Chapter 6. These controls are then assembled into a single printed circuit board. See **Figure 11-36.** Inputs are fed into the circuit board through an edge connector. Inputs include existing conditions such as ice level, temperature, power, etc. Outputs are also fed through the edge connector. Outputs include motor power, switch signals, etc.

The circuitry on the board requires two types of voltages:

- Low voltage dc signals that operate the semiconductor devices. The temperature sensor and the ice ejection cycle controls are functions directly controlled by such signals.
- A 120 V ac line voltage to operate the drive motor and water valve.

A thermistor is located in the front of the ice cube tray. It monitors the temperature at that location. The thermistor senses temperatures below 13°F (−11°C). It will send a signal back to the solid-state control circuit. The control system then triggers a relay, which completes the circuit to drive the motor. This initiates the harvest, or ice cube ejection, cycle.

The mechanical ice ejection process is accomplished using a cam gear mechanism. The unit in **Figure 11-37** produces ice cubes. The ice is deposited in an ice bucket or ejected through a freezer door duct. General operation of the ice maker is controlled through a series of switches. The switches (located in the line voltage circuit) control the motor and water valve. These switches are used to control and monitor the ice ejection process. They usually operate through a sliding pin setup. The pin follows a cam molded into the front surface of the cam gear. The solid-state ice cube maker also has an ice level switch. It stops the ice maker from ejecting new ice if the container is full.

The specific operation of a residential ice cube maker is explained in Section 8.5.

11.9 Chest-Type Freezers

Chest-type freezers have certain advantages. Cold air is heavier than warm air. The cold air in this type of freezer does not "spill out" when the lid is opened. This stops a considerable amount of moisture from entering the cabinet. There is little air change when the cabinet is opened.

Certain features may be added to chest-type freezers to make them more convenient to use than upright freezers. They are usually fitted with baskets that may be lifted out. Thus, access is provided to frozen-food packages near the bottom. Also, the lids usually have a counterbalancing mechanism, which makes them easy to open. A light in the lid gives good illumination. The chest-type freezer is an economical type of food-freezing appliance.

Most chest-type freezers require a manual defrost. Little moisture enters the freezer, however. Therefore, defrosting is usually not needed more than once or twice a year.

Defrosting may be accomplished best by unplugging the condensing unit. Then, remove the stored food. Place an electric space heater or a bucket or two of hot water inside. With the cabinet closed, the ice will soon drop away from the evaporator surface. It will be easy to remove.

Most chest-type freezers have a drain. This makes removing the moisture from the cabinet quite easy. Remaining moisture must be wiped out of the cabinet.

Cabinets are available in various capacities. Height and width are quite uniform. However, the length will vary with the capacity of the freezer.

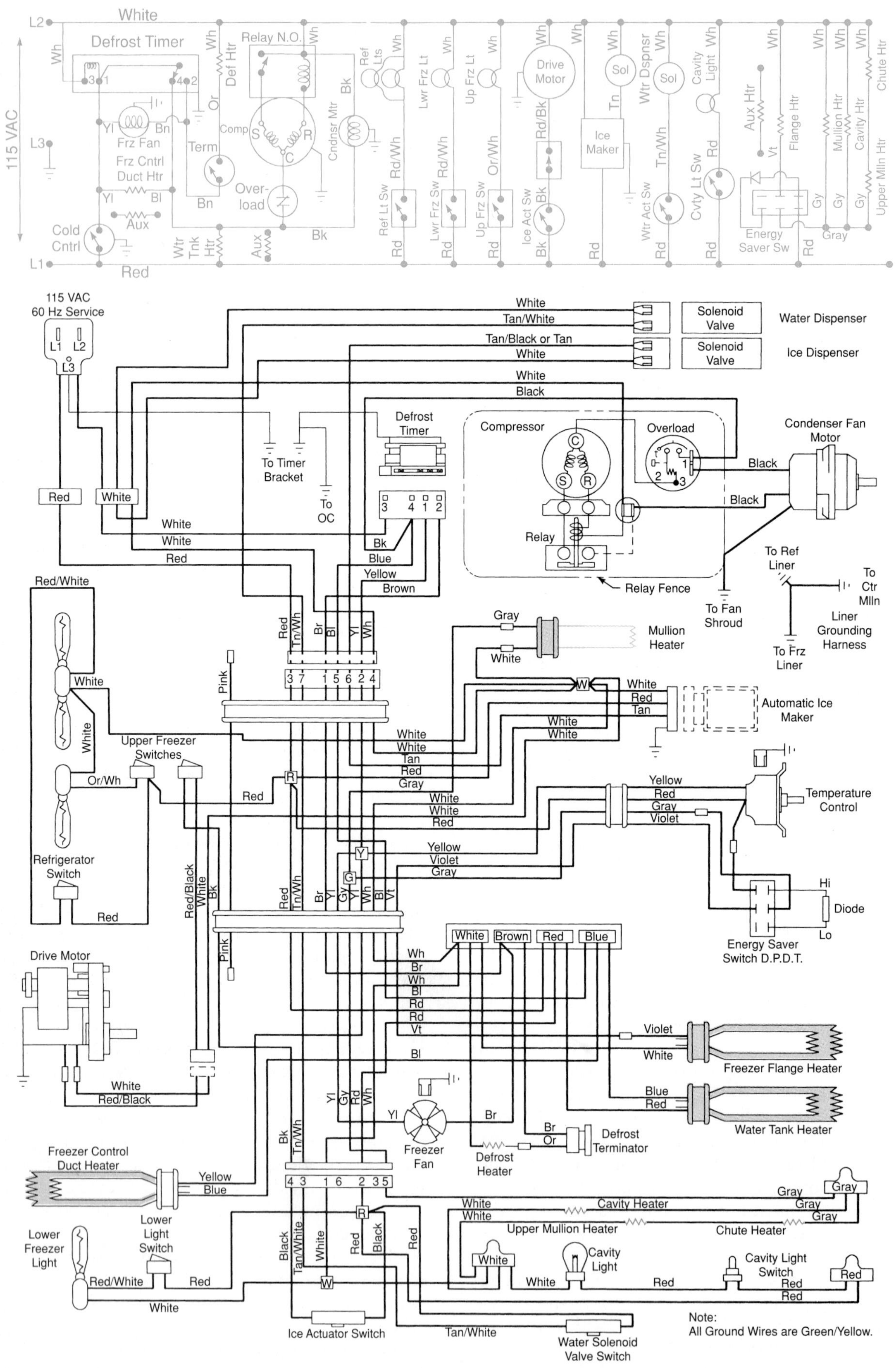

Figure 11-35. *Top portion is ladder wiring diagram for side-by-side refrigerator with automatic defrost and automatic ice maker. (Amana Refrigeration, Inc.)*

Figure 11-36. *Printed circuit board for solid-state ice maker. (Frigidaire Company)*

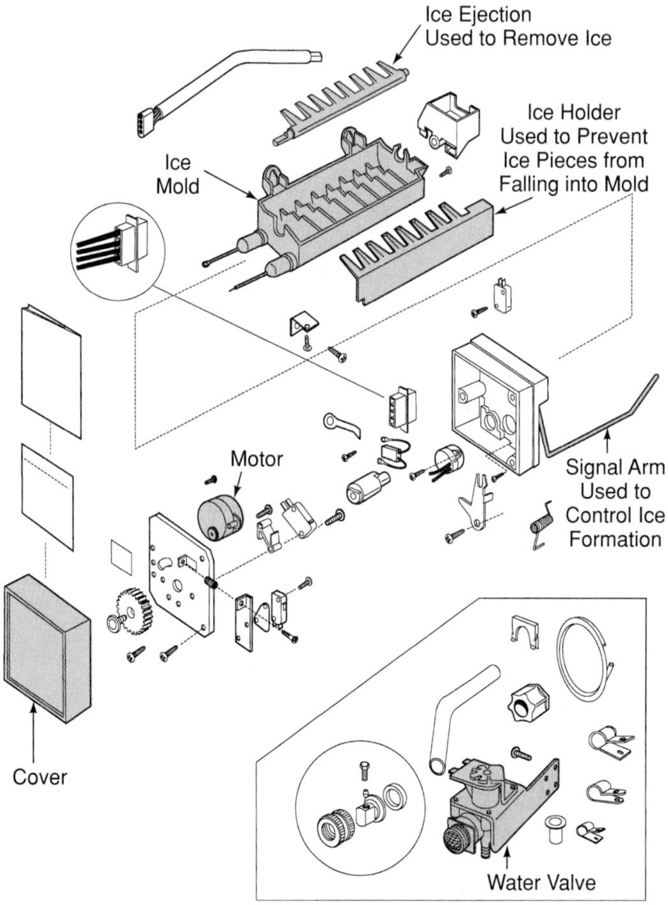

Figure 11-37. *Construction of ice dispenser. (Frigidaire Company)*

11.9.1 Cabinet, Mechanisms, and Electrical Circuits—Kelvinator

The outer and inner shells of the chest-type freezer, **Figure 11-38,** are metal. The evaporator surrounds the

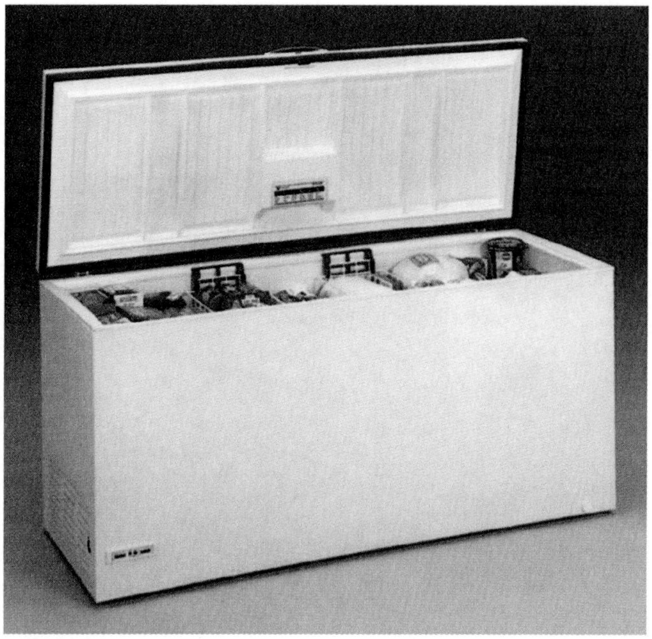

Figure 11-38. *Chest-type food freezer. The use of wire baskets makes it easy to reach items stored in the bottom of the freezer. (Frigidaire Company)*

inner liner and is attached to it. The condenser is attached to the inside of the outer shell. It completely surrounds the cabinet.

The hermetic compressor of a chest-type freezer is shown in **Figure 11-39.** The liquid refrigerant flows through the capillary tube and into the evaporator. There, the refrigerant evaporates and cools. The compressor draws the vaporized refrigerant through the compressor. It pumps the vaporized refrigerant into the precooler condenser on the back freezer wall. Here, it releases part of its latent heat of vaporization and sensible heat of compression.

Chest-type freezers are manually defrosted. Therefore, condensate (water) is usually drained out through the bottom or side of the cabinet. See **Figure 11-40.**

From the precooler condenser, the refrigerant passes back to the machine compartment. It passes through the oil cooling coil in the compressor dome. See **Figure 11-41.** Here, additional heat is picked up from the oil. The compressed vapor then flows back to the main condenser. There, additional heat is released to the atmosphere. The refrigerant condenses from a high-pressure vapor to a high-pressure liquid.

The condenser tubes are in contact with the outer shell of the cabinet. Thus, heat from the condenser passes into the outer shell and warms it slightly. This causes a natural flow of warm air upward over the cabinet shell, preventing sweating. The liquefied refrigerant collects in the bottom of the condenser tubing. It flows into the filter-drier, moves into the capillary tube, and into the evaporator. The cycle then repeats. The cycle is shown in **Figure 11-42.**

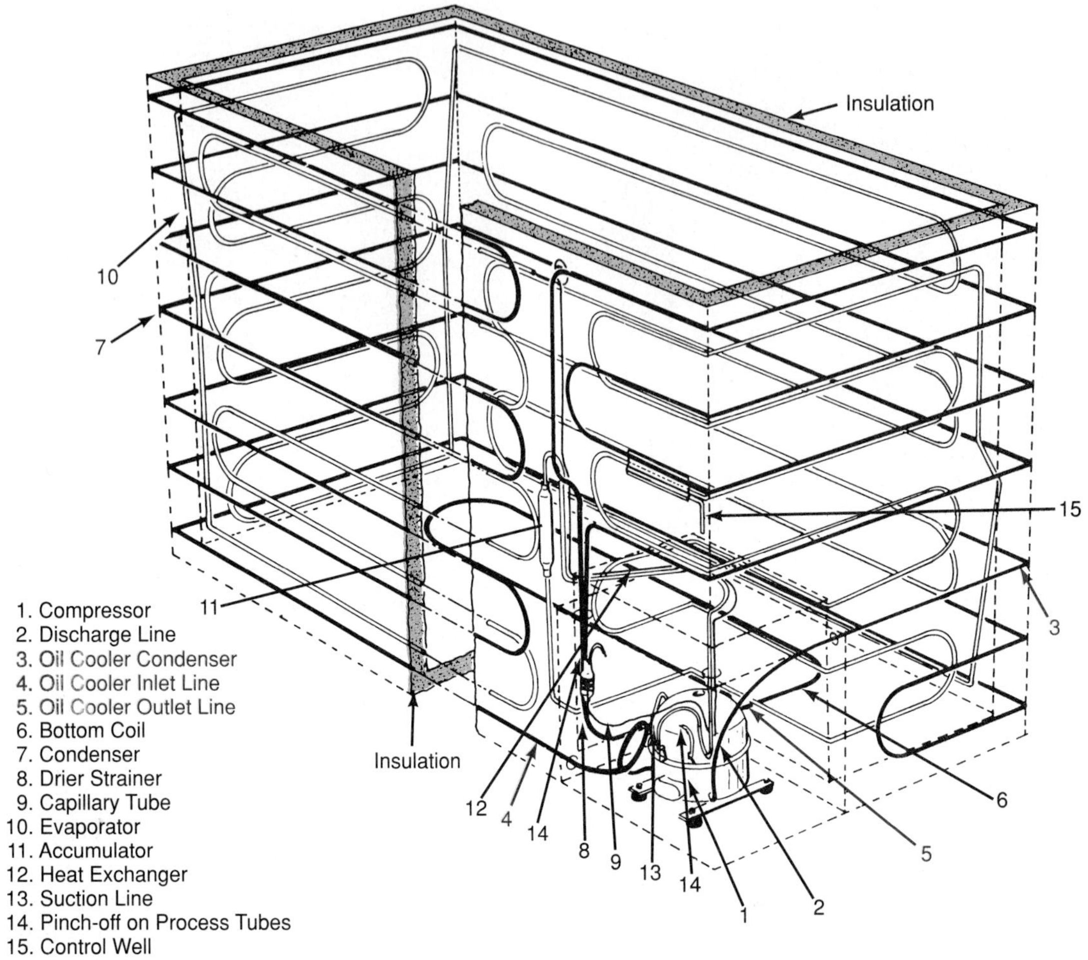

Figure 11-39. *Chest-type freezer evaporator and condensing unit. Note special oil cooler condenser, 3, at right of cabinet, oil cooler inlet line, 4, and oil cooler outlet line, 5.*

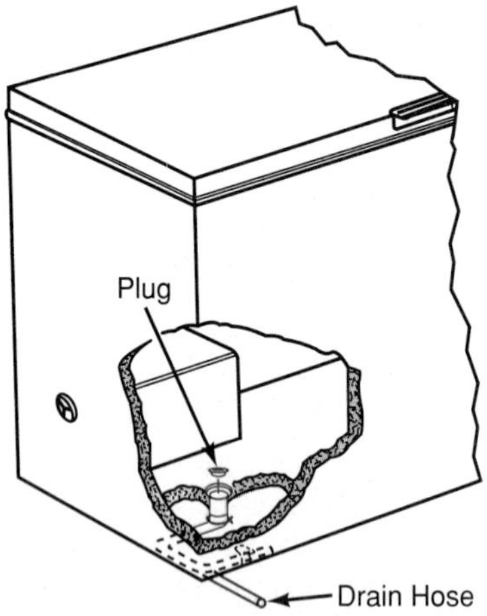

Figure 11-40. *Drain system for a chest-type freezer. (Frigidaire Company)*

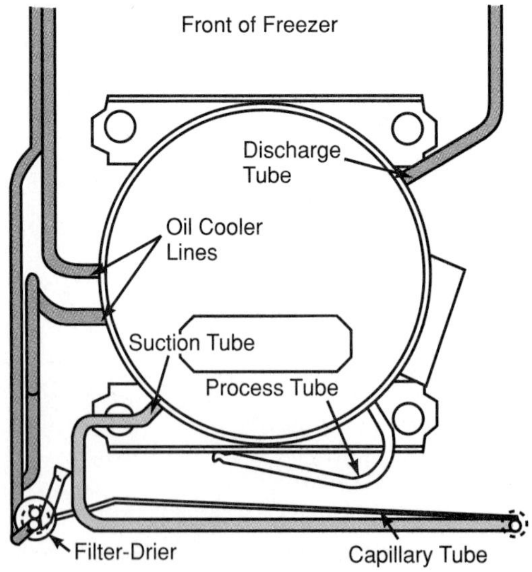

Figure 11-41. *Compressor dome showing oil cooler connections.*

Figure 11-43 shows a schematic wiring diagram for this freezer model. Electrical power is supplied through a grounded three-prong extension plug and cord. The starting relay and overload protector are attached to the motor by a "push-on" mount. See **Figure 11-44.** Note that all parts of the mechanism and cabinet are grounded. The thermostat is at the end of the cabinet near the top of the compressor compartment. The dial is marked for Off, Normal, and Cold positions.

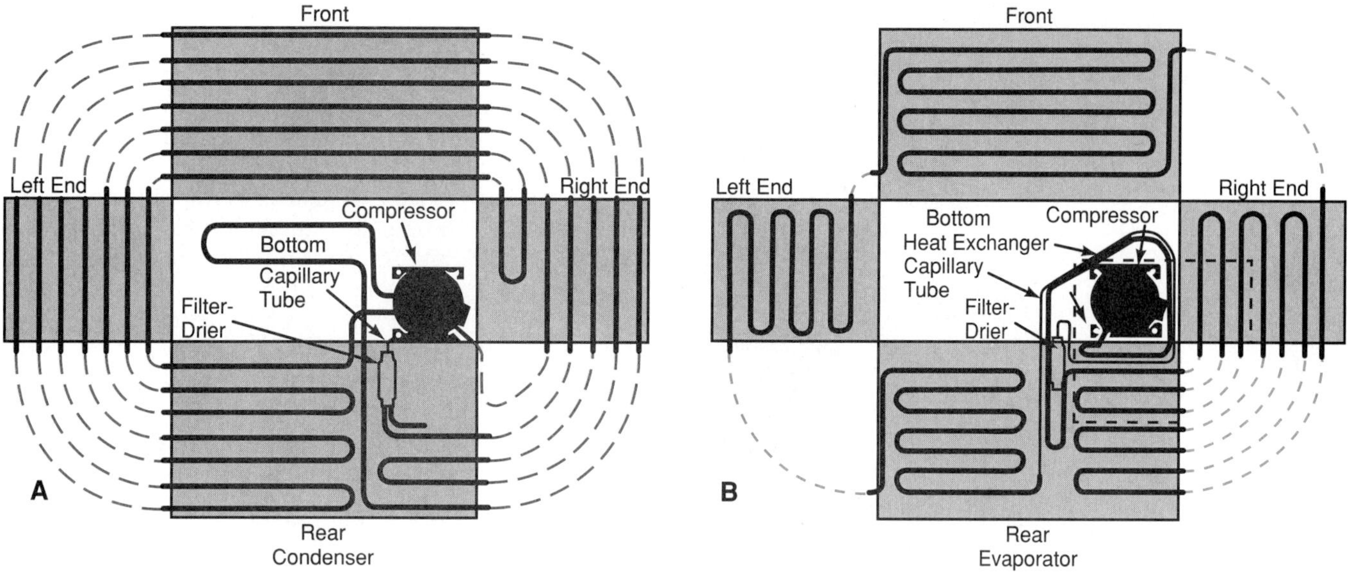

Figure 11-42. *Refrigeration cycle diagram for a chest-type freezer. A—High side of cycle. The heat absorbed in the evaporator is now released by the condenser into the surrounding atmosphere. B—Low side of cycle. Heat is absorbed by the evaporator in the cabinet. (Frigidaire Company)*

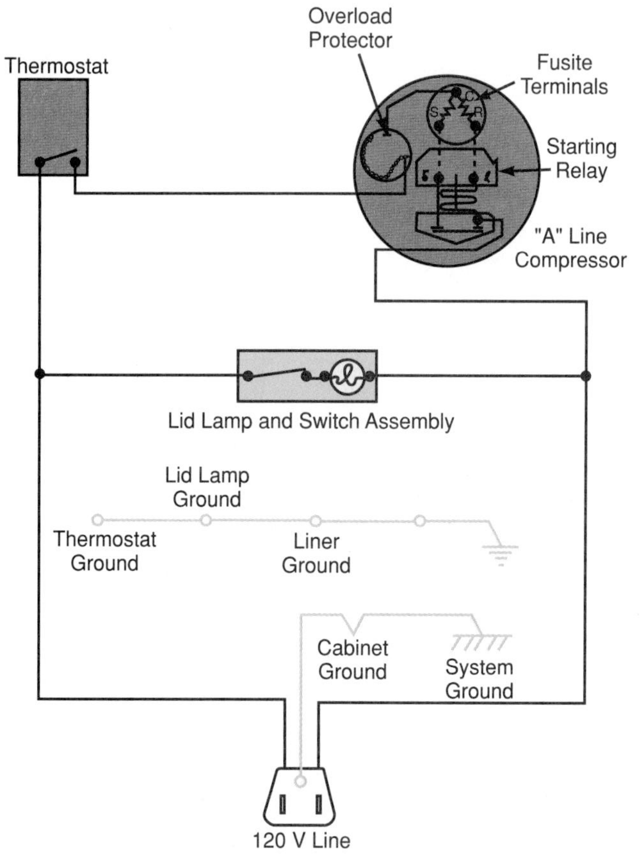

Figure 11-43. *Schematic wiring diagram for chest-type freezer.*

11.9.2 Cabinet, Mechanisms, and Electrical Circuits—Amana

Chest-type freezers are available in various capacities from 7 ft^3 to 28 ft^3. **Figure 11-45** shows a 10 ft^3 model.

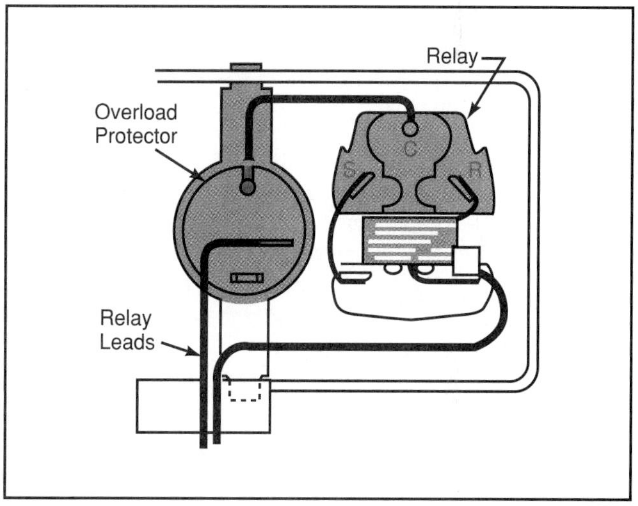

Figure 11-44. *Motor starting relay and overload protector. Starting relay is "push-on" type that is mounted directly to the compressor start (S) and run (R) terminals; C is the common terminal. The overload protector is inserted in the same way. (Frigidaire Company)*

The evaporator surrounds the inner metal lining and is attached to it. The schematic, **Figure 11-46,** shows the evaporator and the low-side refrigerant circuit.

When the unit operates, refrigerant flows through the capillary tube into the evaporator tubes. The evaporator tubing is attached to the cabinet's inner lining. This allows the entire inner surface of the cabinet to be cooled. The refrigerant will be nearly all evaporated by the time it has passed through the evaporator tubes. Any remaining liquid flows into an accumulator placed at the end of the coil. The entrance to the accumulator is from

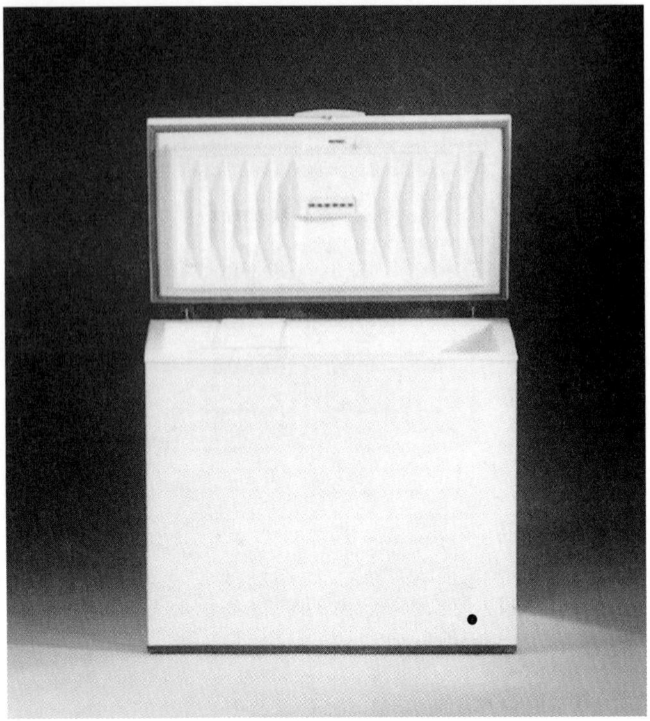

Figure 11-45. *Chest-type freezer. Power-on indicator light is at lower-right corner. (Frigidaire Company)*

the bottom. Thus, the accumulator holds the liquid refrigerant until it is entirely evaporated.

The accumulator outlet is at the top. It leads into the suction line, which connects to the inlet side of the compressor. The vapor goes to the compressor and is compressed to the high-side pressure.

The condenser is attached to the inside of the outer shell. It completely surrounds the cabinet. See **Figure 11-47.** High-temperature vapor from the discharge side of the compressor flows through a precooler. The precooler starts at the top of the back of the cabinet and zigzags across the back of the cabinet down to the compressor. The loop in the bottom of the compressor is immersed (completely covered) in oil. Here, the partially cooled refrigerant picks up some heat from the oil. This helps lower the oil temperature. From here, the high-pressure vapor is carried to the top of the cabinet. It zigzags across the ends and front of the cabinet and returns to the filter-drier. This completes the high-side circuit.

Electrical power comes through a grounded three-prong extension plug and cord. See the wiring schematic in **Figure 11-48.** Note the warning light connected into the electrical circuit. This warning light only indicates whether or not the electrical circuit is "hot." It does not indicate whether cabinet temperature is satisfactory.

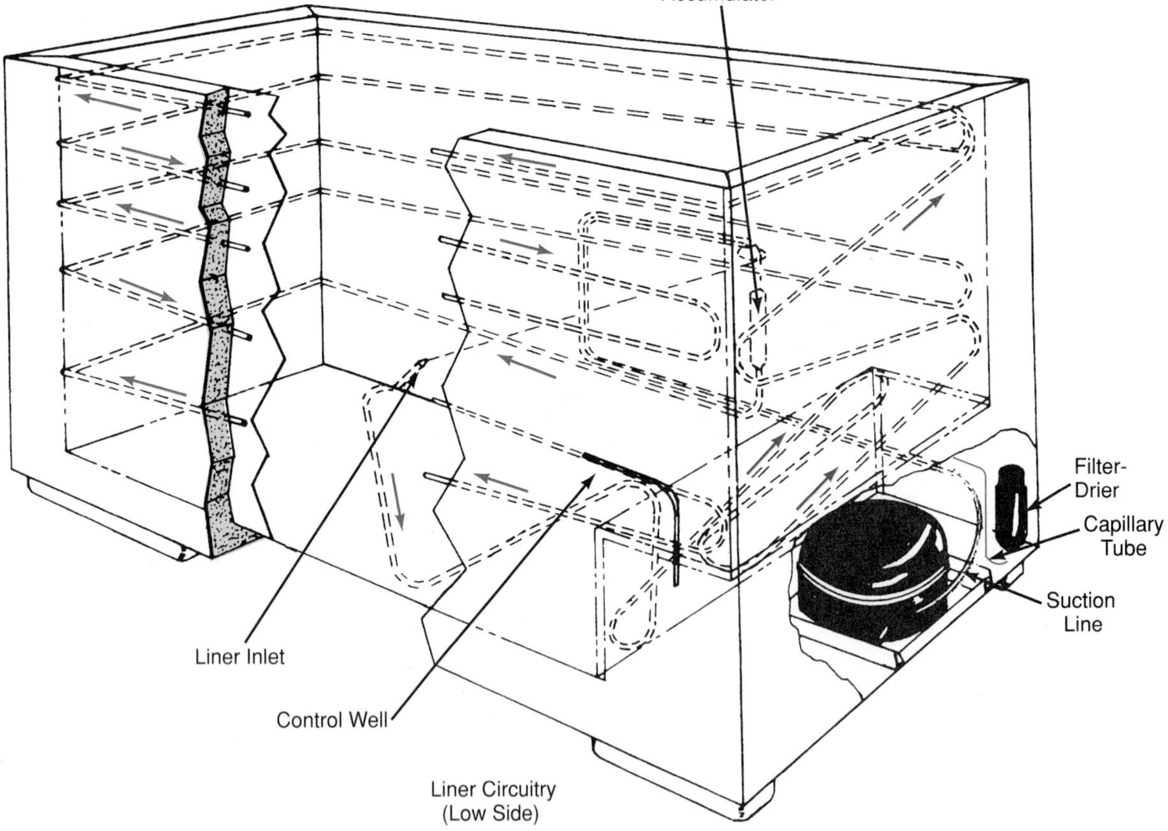

Figure 11-46. *Chest-type freezer evaporator coil. Evaporator is attached to cabinet inner lining. Capillary tube refrigerant control extends from bottom filter-drier to inlet of evaporator coil.*

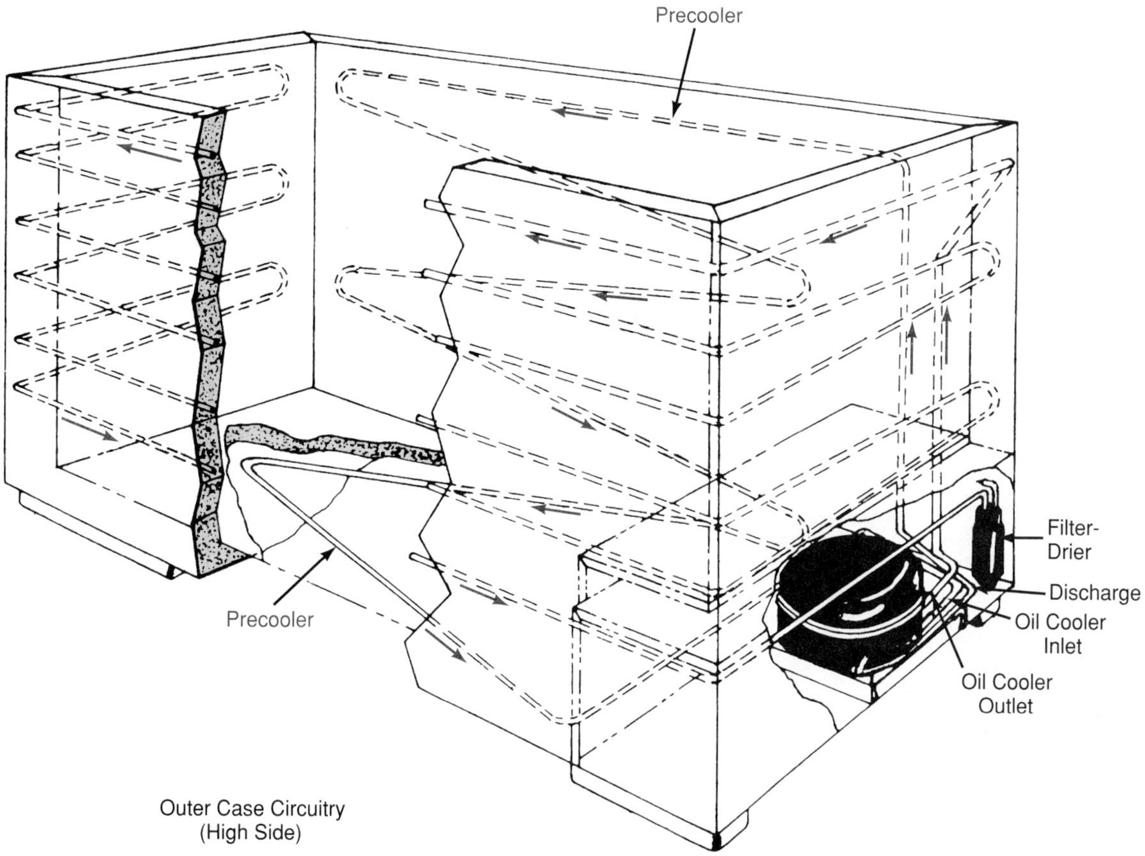

Figure 11-47. *Chest-type freezer condenser. The coil is attached to the inner surface of the outer shell. Note the precooler coil.*

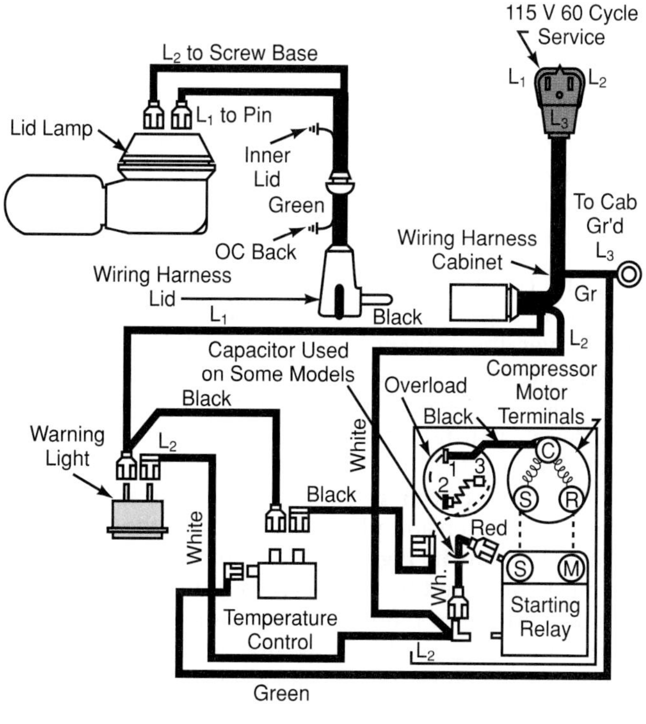

Figure 11-48. *Schematic wiring diagram for a chest-type freezer. (Amana Refrigeration, Inc.)*

The temperature control is wired into the motor circuit. It controls the running time of the compressor to maintain the desired cabinet temperature. Normal operating temperature range for this freezer should be between −10°F and 6.5°F (−23°C and −14°C). All assemblies are grounded through the green wire. This is connected to the grounding terminal of the attachment plug.

11.10 Upright Freezers

The upright freezer makes storage and removal of frozen food convenient. Frost-free and automatic defrost mechanisms have made these freezers very practical. General construction is very similar to the upright refrigerator-freezer. However, insulation may be a little heavier and the motor control will be different.

Cabinets are available in varying capacities. However, the range is not as extensive as for chest-type freezers. Capacity is lower because of height limitations.

11.10.1 Cabinet, Mechanisms, and Electrical Circuits—Frigidaire

A popular upright freezer, **Figure 11-49,** has outer and inner shells of enameled steel. The evaporator is lo-

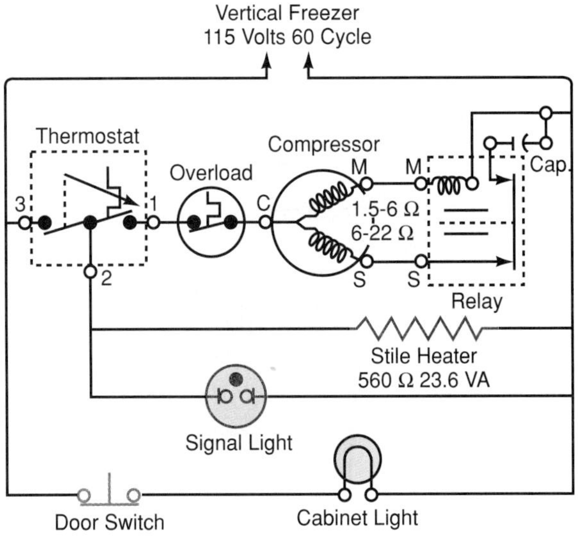

Figure 11-54. *Ladder wiring diagram for a manual defrost upright freezer. Note the stile heater and the signal light, which indicates power is on.*

carried into the cabinet quickly. Articles should be removed from or placed in the cabinet quickly.

The cabinet must be kept clean on the outside as well as on the inside. The condenser and motor compressor should be wiped clean at least every six months. A vacuum cleaner may be used for cleaning the lint from the condenser. The door gasket should be checked for tightness periodically.

11.12 Ice Accumulation in Cabinet Insulation

Ice accumulation is one of the main problems encountered with improperly or carelessly assembled units. A refrigerator or freezer may have an air leak in the outer casing (shell). This allows moisture from the atmosphere to enter and condense in the insulation. This condition will cause considerable problems in a freezer. Ice buildup also reduces the insulating ability of the cabinet, thus causing the unit to run more. If the condition is severe, the condensing unit may run continuously.

Ice accumulation in a freezer may be indicated by a cold spot or condensation on the outside surface. To eliminate the unwanted ice, shut off the freezer and allow it to warm up for a few days. The unwanted ice will melt and drain.

If cold spots or condensation appear, remove the breaker strips. Insert lightly packed, fine fiberglass insulation to fill air pockets.

Some freezers provide an opening in the inner lining. Any moisture in the insulation is allowed to escape into the freezer compartment. There it condenses on the cold surface of the evaporator and keeps the insulation dry.

11.13 Butter Conditioner

Various devices are used to soften butter stored in the refrigerator cabinet. The most common is a

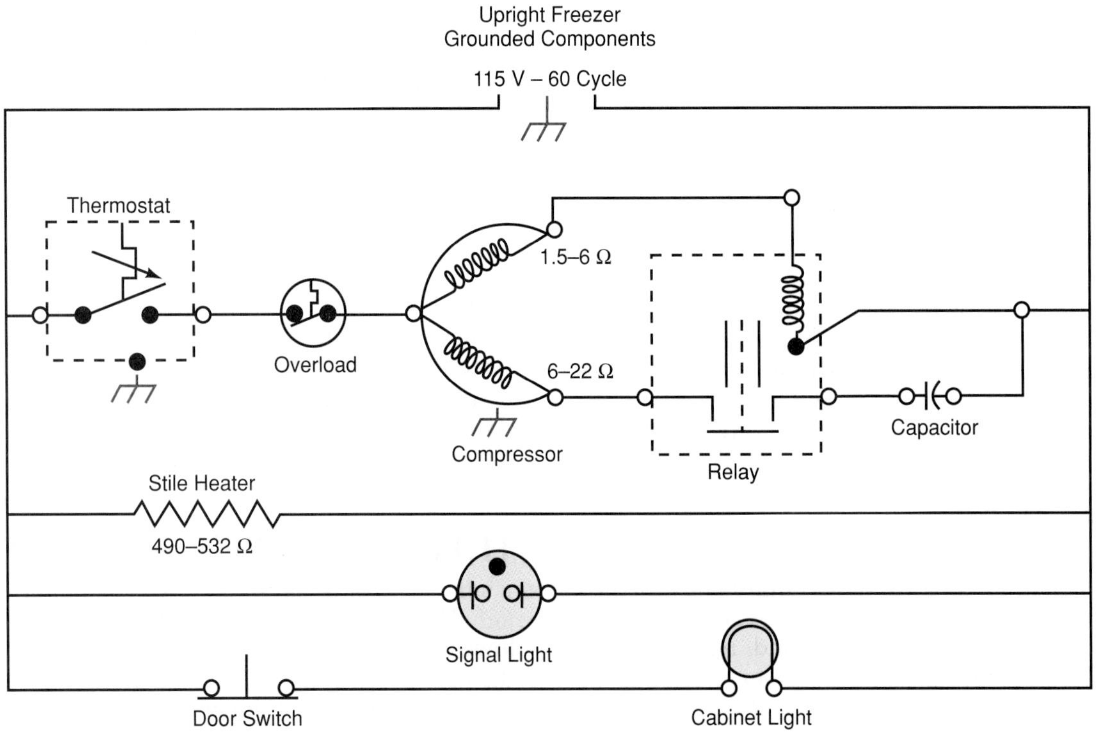

Figure 11-55. *Ladder wiring diagram for an upright freezer. Middle conductor of power cord is grounded by a wire network to all metal parts of the cabinet, hardware, and refrigerating system.*

recess in the refrigerator door, which will hold a quarter pound (or more) of butter. The recess is separated from the refrigerator compartment by a small door. The recess is close to the outer shell of the door. Therefore, there is little insulating effect between butter and room temperature. Some refrigerators provide an electrical resistance unit to aid in warming the butter compartment.

11.14 Cabinet Hardware

Cabinet hinges usually use ball or nylon bearings. These require little or no lubrication. The hinges usually are adjustable to fit the doors to the cabinet properly. **Figure 11-56** illustrates a hinge commonly used in upright refrigerators.

Many old refrigerators used complicated door latch mechanisms. They were designed to draw the door firmly into place and securely latch it. In 1958, a federal law was enacted that required a refrigerator cabinet be designed to be opened from the inside. Opening the door from the inside must require no more than a 15-1b. force. Most modern refrigerators, therefore, use magnets to close and hold the door shut. No positive latching mechanism is used.

Refrigerator cabinets should be carefully leveled so that the shelves will not be tilted. Most have adjustable feet for this purpose. The feet should be adjusted so the door's weight swings it closed from any open position. **Figure 11-57** gives a detail drawing of a leveling adjustment.

Many modern refrigerator-freezers are mounted on rollers. They help in moving the unit for cleaning and servicing. The front rollers are usually adjustable. After the refrigerator is in place, these rollers are used to level it. See **Figure 11-58.** It is adjusted to touch the floor using an adjusting screw. This keeps the refrigerator from moving once it is in place.

Breaker strips (usually made of plastic) connect the metal outer shell (case) to the metal liner where the cabinet contacts the door. Some refrigerators have a plastic inner liner instead of a metal liner. These breaker strips are molded and shaped in such a way that they snap in place. See **Figure 11-59.**

A wide-blade putty knife may be inserted to remove the outer edge of the breaker strip. (Wrap the putty knife blade with tape to prevent scratching the parts.) Pressing with the heel of your hand will sometimes separate the joint.

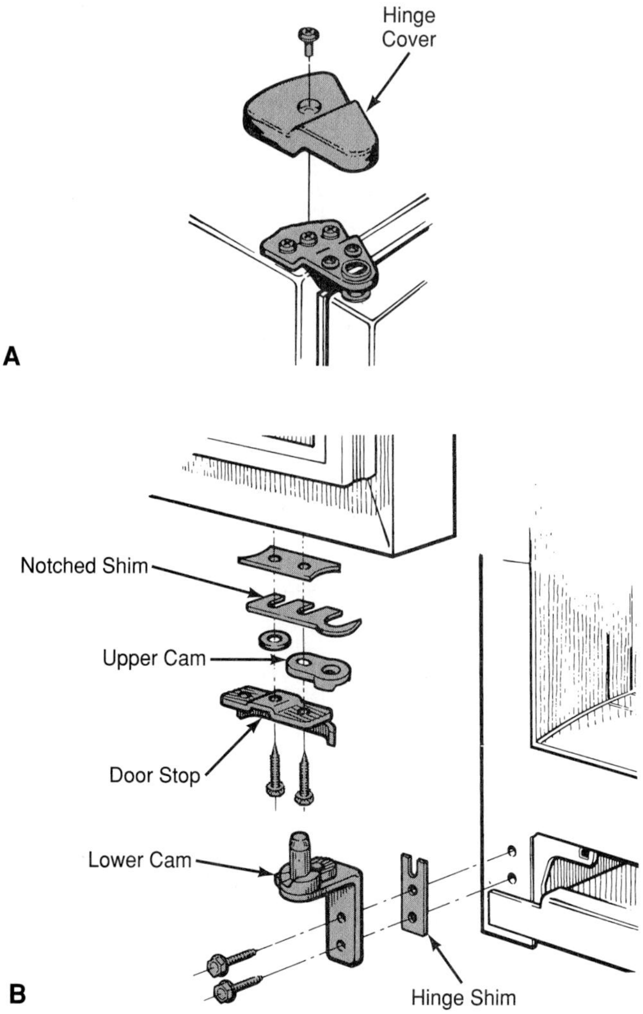

Figure 11-56. *Hinges for refrigerator or upright freezer. A—Top hinge. B—Bottom hinge. (General Electric Co.)*

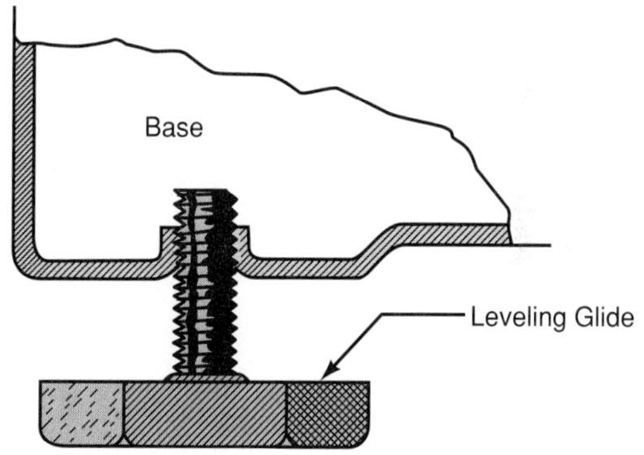

Figure 11-57. *Detail of leveling glide.*

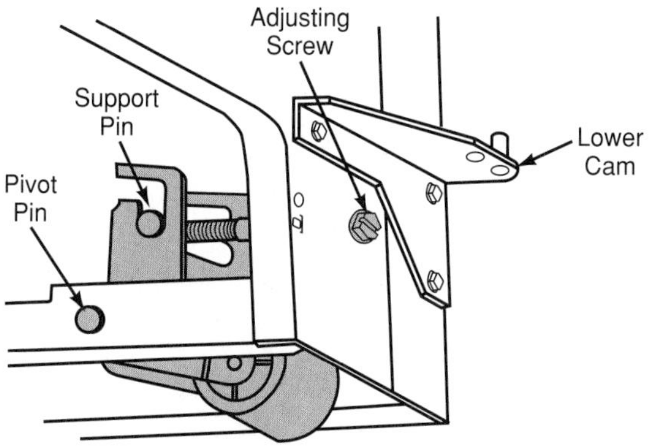

Figure 11-58. *Adjustable front roller assembly for refrigerator cabinet.*

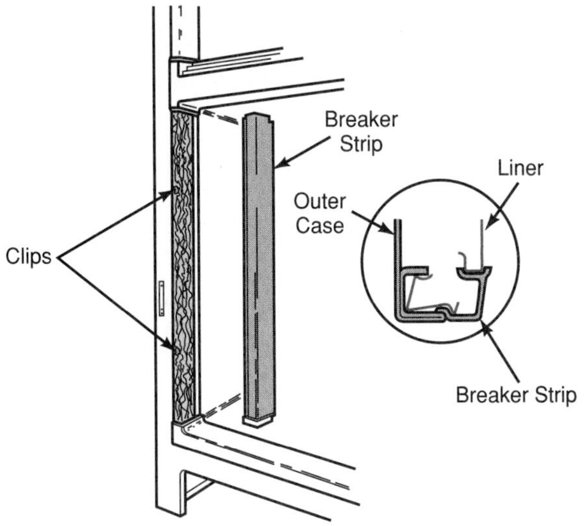

Figure 11-59. *Typical breaker strip. Note clips on outer case. (General Electric Co.)*

The liner should be room temperature to 90°F (32°C) (never warmer) to make it flexible. Use a heated, damp, washcloth to warm it.

Some breaker strips are installed before the foamed-in-place insulation is put in. They are fused to the foam insulation. These strips must be broken or cut to be removed.

Chest-type freezers usually have the lid counterbalanced. The lid need not be held when storing or removing items. These counterbalances are adjusted so that the final closing is by the lid's own weight.

Very little cabinet hardware repair is done on modern refrigerators and freezers. If a latch or hinge fails, replace it rather than repair it.

11.15 Cabinet Gaskets

Door gaskets are usually made of flexible vinyl. They generally have an air cushion design. See **Figure 11-60.** Most gaskets have magnets built into the vinyl to hold the door closed.

If the gasket does not provide an airtight seal, the unit must work harder. It must counteract the warm air leakage through the gasket. Therefore, there is more wear on the unit and thus, increased operating costs.

Door seals may be checked using a .003″ thick plastic feeler gauge. A thin piece of paper may also be used. To check the fit of the gasket, insert the gauge or paper at several places around the door opening. It should require a little pull on the gauge if the gasket is properly fitted. If the gauge falls out of its own weight, the gasket is not fitted properly.

Door gaskets may be quickly and easily checked as follows:

1. Open the refrigerator door about half way. Note carefully the pull required to open it. Allow it to remain open for approximately 10 seconds.

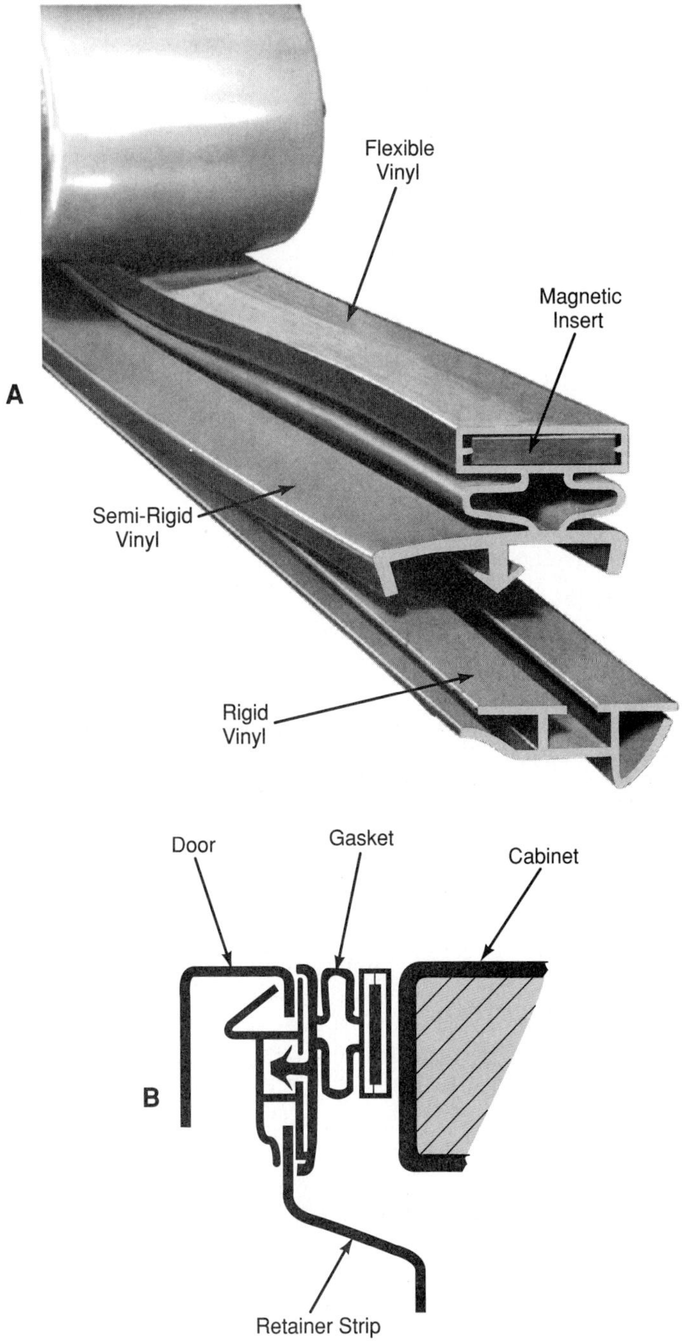

Figure 11-60. *Replacement door gaskets. A—Magnetic door gasket being rolled onto retainer strip. B—Cross section of magnetic gasket and retainer strip assembly installed in door. (Jarrow Products, Inc.)*

2. Close the refrigerator door and allow it to remain closed for about 15 seconds.
3. Open the refrigerator door and note carefully the pull required. It should require more effort to open the door now than at the first opening.

Cold air comes out when the door is opened. Warm air replaces it. When the door is closed, this warmer air is cooled and contracts. The result is that the pressure inside the cabinet is slightly less than the atmospheric pressure in the room. Timing is an important element in this test, as the pressures tend to quickly balance.

On some refrigerators, the hinges may be adjusted to correct a poorly fitted gasket.

Figure 11-61 illustrates a door that is either warped or has improperly adjusted hinges. A warped door can usually be straightened by twisting. To check for hinge adjustment, use the test described for proper gasket seal.

Door gaskets deteriorate with age. If the material has become hard, cracked, or broken, it should be replaced.

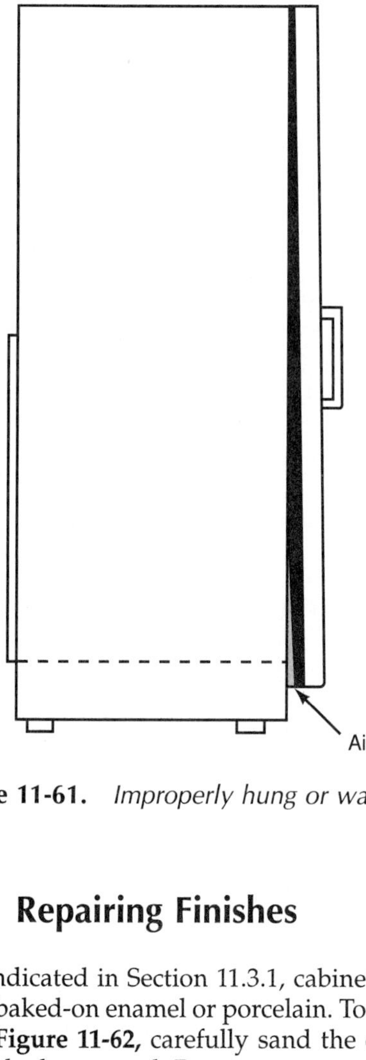

Figure 11-61. *Improperly hung or warped door.*

11.16 Repairing Finishes

As indicated in Section 11.3.1, cabinet finishes may be either baked-on enamel or porcelain. To repair enamel finishes, **Figure 11-62,** carefully sand the damaged area down to the bare metal. Be sure to remove all wax and rust.

Sand the edges of the damaged area with fine waterproof paper (6/0). The old finish should slope evenly from the surface to the center of the damaged area. This result is called ***featheredging.***

Carefully clean the sanded surface. Apply metal primer to the exposed bare metal with a brush or a spray gun. After the primer has dried thoroughly, sand the area again using 6/0 waterproof paper. Use soapy water as a sanding lubricant. Dry thoroughly.

Spray or brush on the enamel. Blend the new coat to the old finish as smoothly as possible. Allow the enamel to dry thoroughly. Then sand with 6/0 paper until the edges are invisible.

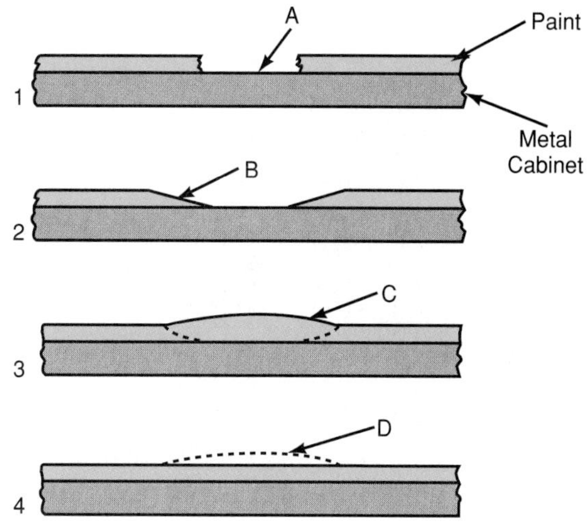

Figure 11-62. *Steps in finishing a chipped surface: A—Paint chipped from cabinet. B—Surface sanded in preparation for repairing. C—New finish applied. D—Excess finish must be sanded level.*

For a gloss finish, rub the entire surface, including the patched area. Use rottenstone, paraffin oil, or rubbing oil. When the desired gloss has been reached, wipe off the rottenstone and oil. Use a moistened cloth or chamois.

Paint spraying must be done in a fireproof, well-ventilated room or booth. The air must be free from dust. The compressed air used in the spraying gun must be free of moisture and oil. Be sure to follow the paint manufacturer's recommendations when refinishing a cabinet surface.

To repair porcelain finish, a special porcelain patching material should be used. Due to the inherent nature of porcelain, its color shade will vary somewhat. A patching kit may be obtained with several colors.

The surface to be patched must be cleaned and warmed. Apply the patching material with a small, fine brush or air brush it. When dry, the finish should be smoothed with waterproof abrasive paper. It should be polished with a soft cloth or with rottenstone and rubbing oil.

11.17 Cabinet Thermometers

The recommended temperature range for a refrigerator compartment is between 35°F (2°C) and 45°F (7°C). The recommended temperature range for the freezer compartment is 0°F to −10°F (−18°C to −23°C). Many types of thermometers are used to monitor the cabinet temperature. **Figure 11-63** displays a thermometer which indicates temperatures within these ranges.

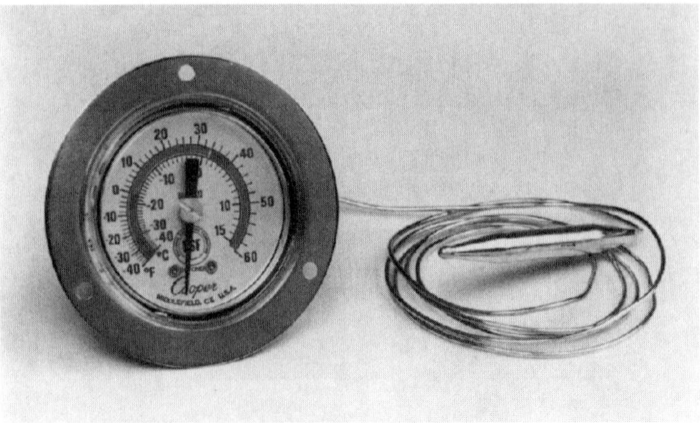

Figure 11-63. *Refrigerator-freezer cabinet thermometer. Scale reads from −40°F to 60°F (−40°C to 15°C). (Cooper Instrument Corporation)*

11.18 Review of Safety

Refrigerators and freezers must be carefully handled. This prevents damage to the cabinets or mechanisms or injury to those handling them.

Do not place your hands near revolving fans.

Always disconnect electrical power before working on system electrical parts. Open the switch or pull the plug. This prevents unpleasant and perhaps fatal shocks.

The electrical system must be grounded properly to the receptacle if an approved three-wire grounded plug is not used. Figure 6-86 and Figure 12-3 show how to check for proper grounding and connect a safe ground wire.

Frost and ice should be removed by heating (hot water or electric heat). To avoid puncturing the evaporator, never use a pointed or sharp metal tool to remove ice from it. One may puncture the refrigerating unit.

Any spray painting must be done in an approved spray room or booth.

If a refrigerator is taken out of service, the door must be removed immediately. Do not let an out-of-service refrigerator stand where children may play in it. The children could suffocate within the cabinet in only a few minutes.

11.19 Test Your Knowledge

Please do not write in this text. Place your answers on a separate sheet of paper.

1. Which of the following is *not* a safety device in the electrical circuit of a hermetic domestic refrigerator?
   A. Starting relay.
   B. Overload protector.
   C. Grounded plug.
   D. Motor control thermostat.
2. Which of the following is *not* a technique or system which may be used to defrost refrigerators?
   A. Cold gas system.
   B. Manually defrost.
   C. Hot gas system.
   D. Electric heaters.
3. Defrost water is removed during the defrost cycle in an automatic defrost refrigerator by ______.
   A. the water settling in the accumulator
   B. an electric heater under the drip pan
   C. the heat from the motor compressor evaporating this moisture
   D. the defrost drier
4. The suction line and the capillary tube are sometimes soldered together to ______.
   A. avoid vibration
   B. serve as a heat exchanger
   C. evaporate moisture drained from the evaporator
   D. increase the temperature of the liquid refrigerant in the capillary tube
5. The best temperature range for a residential chest-type freezer is ______.
   A. 0°F to 20°F (−17.8°C to −6.7°C)
   B. 32°F to 40°F (0°C to 4.4°C)
   C. −10°F to 6.5°F (−23°C to −14°C)
   D. −30°F to −32°F (−34.4°C to −36°C)
6. Electrical resistance wire is used to help evaporate condensation in all but which one of the following areas?
   A. Door openings.
   B. Crisper.
   C. Outside the cabinet.
   D. Inside the mullion tube.
7. Which of the following would *not* be a location for the condensing unit?
   A. Bottom.
   B. Back.
   C. Cabinet base.
   D. Cabinet top.
8. Which of the following is *not* true about the cabinet door?
   A. They are made of flexible vinyl.
   B. They have an air cushion design.
   C. They can be checked with a 1/3″ plastic feeler gauge.
   D. Many have magnets built into the vinyl.
9. Which of the following is *not* a possible evaporator location in the cabinet?
   A. Across the bottom.
   B. Across the top.
   C. In the upper-left corner behind the freezer compartment.
   D. In the upper-right corner behind the freezer compartment.
10. ______ is used as insulation in modern freezers.
    A. Glass wool or mineral wool
    B. Granulated cork or rock wool
    C. Urethane foam or fiberglass
    D. Celotex or rubber

11. What is meant by "featheredging" when repairing an enamel or lacquer finish?
    A. Using a feather to lightly paint the surface.
    B. Sanding the edge of the old finish to a slight slope.
    C. Feathery light strokes with a paint brush.
    D. Using a spray gun in a light, feathery pattern.
12. The ______ controls the flow of the hot gas through the evaporator for defrosting.
    A. electric defrost timer, which trips a solenoid valve
    B. defrost valve
    C. heat exchanger
    D. capillary tube
13. Ice formation on an evaporator is controlled by ______.
    A. manually shutting off the unit
    B. the use of a hot gas system
    C. the use of electric heaters
    D. All of the above.
14. How is the water in a frost-free side-by-side refrigerator-freezer removed?
    A. Accumulated in pan under condenser and evaporator.
    B. Electrical heating wires cause it to evaporate.
    C. Fan is used to cause air movement.
    D. All of the above.
15. What kind of motor control is used on most freezer units?
    A. Motor starting relay.
    B. Overload protector.
    C. Thermostat.
    D. All of the above.
16. How does a no-frost refrigerator prevent sweating and frosting?
    A. Electrical heating wires.
    B. Air circulation.
    C. Plate on top of compressor for water.
    D. All of the above.
17. When a chest-type freezer has an ice buildup, it lowers its efficiency. How often should it be defrosted?
    A. Six times per year.
    B. Eight times per year.
    C. Twice per year.
    D. Only when there is an ice buildup.
18. ______ voltage is required on a printed circuit board of a solid-state controlled ice maker to operate the semiconductor devices.
    A. High-voltage ac
    B. Low-voltage dc
    C. High-voltage dc
    D. Low-voltage ac
19. What is the most common type of refrigerant control used on domestic units?
    A. Thermostatic expansion valve (TEV).
    B. Automatic expansion valve (AEV).
    C. Capillary tube.
    D. High-side float.
20. What controls the temperature of the butter compartment?
    A. Thermostatic temperature control.
    B. Electrical resistance wire.
    C. Recessed open compartment.
    D. Normal refrigerator airflow.

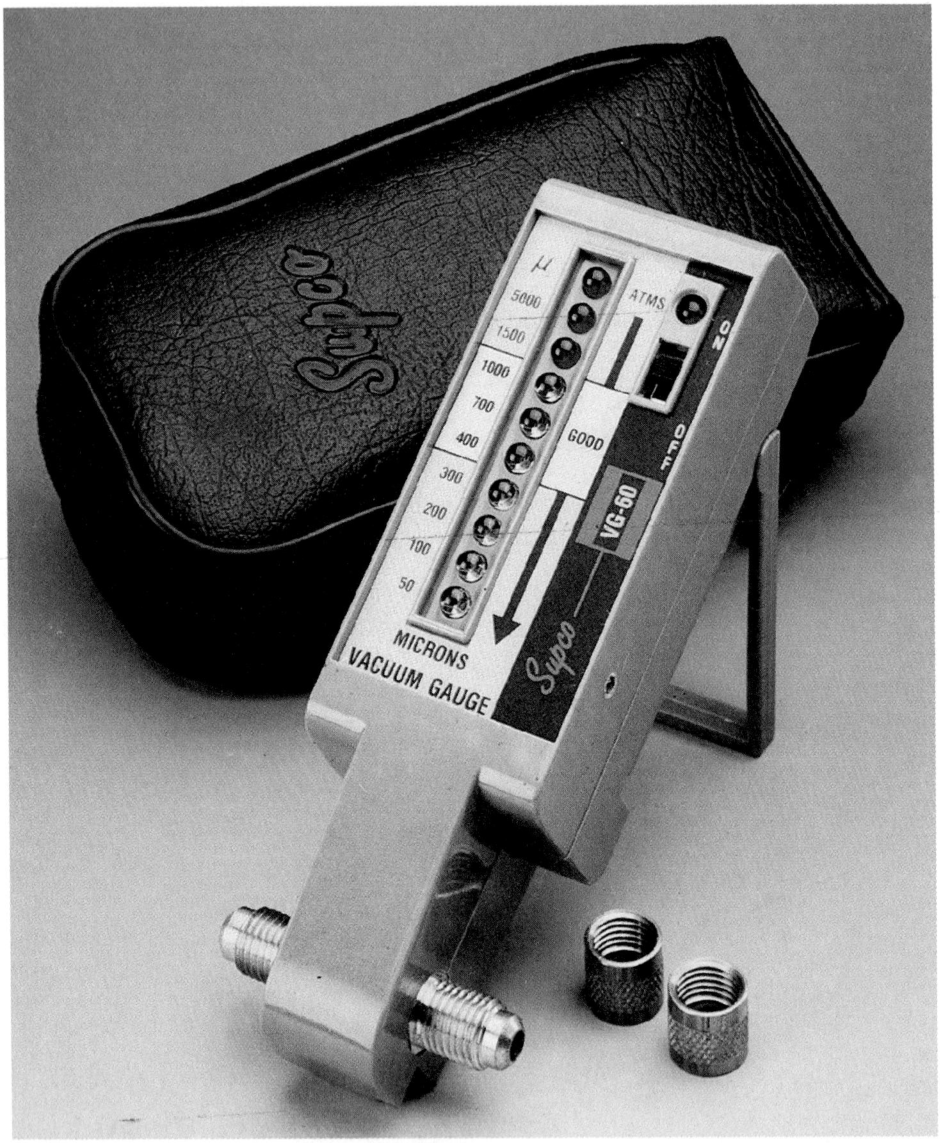

Solid-state thermistor vacuum gauge used to measure vacuum level of a system. (Sealed Unit Parts Co., Inc.)

Chapter 12

SERVICING AND INSTALLING SMALL HERMETIC SYSTEMS

Key Words:

capacitor
core valves
electronic leak detector
fluorescent leak detection
gauge manifold
halide torch
hermetic system
high-pressure gauge
piercing valve
voltmeter

Learning Objectives:

After studying this chapter, you will be able to:

- Select proper tools and instruments needed for installing or servicing domestic and small commercial systems.
- List supplies needed on a typical installation or service call.
- Service internal and external mechanisms using the proper tools and materials.
- Locate areas causing noise and make necessary adjustments to the system.
- Recognize trouble signals.
- List common external service operations.
- Start a stuck compressor.
- Demonstrate proper use of piercing valves.
- Check for restrictions in the system.
- Troubleshoot common refrigeration problems.
- Use the proper methods and equipment for checking electrical systems of refrigerators and freezers.
- **Follow approved safety procedures.**

R-12 has been the primary refrigerant used in domestic refrigerators and freezers for many decades. As of 1996, R-12 was no longer manufactured. R-134a is now the major refrigerant used in domestic equipment. However, units with R-12 will be operational for many years to come and will require repair. Therefore, you must learn how R-12 units operate and how to service them.

12.1 Instruments, Tools, and Supplies

Tools and supplies used by a technician are listed in Chapter 2. In addition, some special instruments, tools, and supplies are needed for domestic refrigerator service work. These are listed in the following paragraphs.

12.1.1 Instruments

Many special testing instruments are needed for refrigeration service work. Their use will be explained as the various service operations are described. **Figure 12-1** shows a small hermetic system. A gauge manifold, vacuum pump, and charging cylinder are connected to it. Some of the basic instruments required are listed below:

Basic Instruments

- Pressure recorder.
- Temperature recorder.
- Off-On recorder.
- Watt recorder.
- Electronic sound tracer.
- Electronic leak detector.
- Compound gauge.
- Pressure gauge.
- Thermometer with a range of −20°F to 212°F (−29°C to 100°C).
- Voltmeter.
- Ammeter.
- Ohmmeter.
- Test light (incandescent, 120 V, 100 W).

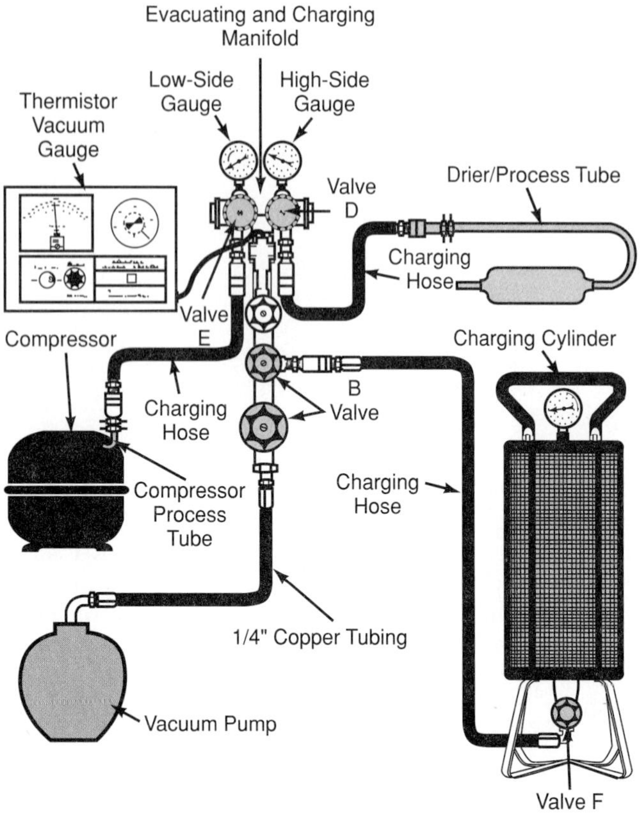

Figure 12-1. *Typical installation of service equipment manifold on a hermetic system. Note refrigerant cylinder, vacuum pump, and valve arrangement. Also note adaptor connection to compressor process tube and to drier/process tube. (Amana Refrigeration, Inc.)*

12.1.2 Tools

Refrigerant Tools

- High-vacuum pump.
- Recovery/recycling unit.
- Service cylinders for R-12, R-22, R-502, R-134a.
- Purging line (1/4″ diameter by 15′, equipped with hand shut-off needle valve and check valve).
- Capillary tube cleaner.
- Capillary tube sizing kit.
- Soldering-brazing torch, (LP fuel-air, acetylene-air, or oxyacetylene).
- Hand vacuum cleaner.
- Gauge manifold.
- Process tube adaptors.
- Bending springs.

Wrenches

- Set of 3/8″ drive sockets (12 point, 7/16″ to 1″), with 3/8″ drive torque handle, speed handle, swivel handle, and T-handle.
- Adjustable open-end wrench (8″).
- Set of Allen setscrew wrenches.
- Refrigeration ratchet wrench, 3/16″, 7/32″, and 1/4″ with square openings.
- Set of 15° open-end wrenches (1/2″, 3/4″, 7/8″, and 1″)
- Box-end wrench (1/2″).
- T-socket wrench (1/2″).

Pliers

- Combination (6″).
- Wire cutter.
- Slim nose.

Hammers

- Claw hammer.
- Rubber mallet.

Tubing Tools

- Bending springs (for 1/4″, 3/8″, and 1/2″ OD tubing).
- Flaring tool (3/16″ to 1/2″ capacity).
- Tubing cutter.
- Pinch-off tool.
- Swaging tool.

Screwdrivers

- Standard screwdrivers (3″, 6″, and 8″) with insulated handles.
- Phillips screwdrivers (3″, 6″, and 8″) with insulated handles.

12.1.3 Supplies

Some special supplies will be needed by a service technician who makes house calls to answer customer complaints. In general, these supplies will include:

- Wire solders (60/40 and 95/5).
- Soldering flux.
- Silver brazing wire (30%-60% silver, no cadmium).
- Phosphorous-copper alloy wire.
- Brazing flux.
- Steel wool.
- Medium-grade sandpaper.
- Plastic tape.
- Refrigerant recovery unit.
- Disposable cylinders of refrigerant, (R-12, R-22, R-134a, and R-502).
- Coils of soft copper tubing (1/4″, 5/16″, 3/8″, and 1/2″).
- Copper pipe (as needed).
- Capillary tubing.
- Filter-drier cartridges.
- Refrigerant oil (150 and 300 viscosity).
- Refrigerant oil can (spout type).
- Cleaning cloths.
- Relays.
- Capacitors.
- Motor controls.
- Refrigerant controls.
- Overload protectors.
- Light switches.
- Sealing compounds.
- Driers (flared and soldered fittings).

- Filter-driers (flared and soldered fittings).
- Sight glasses (flared and soldered fittings).
- Flared fittings (SAE)—all sizes and shapes.
- Soldered fittings—all sizes and shapes.
- Piercing valves and valve adaptors.
- Valve cores.

12.2 Installing Refrigerators and Freezers

Correct installation is very important to the proper operation of a refrigerator or freezer. This includes leveling of the cabinet, providing correct electrical power, and assuring good ventilation. The manufacturer ships the units carefully crated. The unit is also shipped with full written instructions. These instructions include information on how to move, uncrate, and install the unit.

12.2.1 Uncrating a Refrigerator or Freezer

As noted, the carton usually has printed instructions on safe handling and uncrating. These instructions should be carefully followed. Many dealers uncrate the cabinet at the store. Others do it just outside the home. (Most crates are too large to fit through household doors.)

Certain areas of the cabinet may be easily damaged in moving or uncrating. They are:

- Bottom. The condensing unit may be damaged.
- Back. The refrigeration condenser may be damaged. (Not all cabinets have condensers in the back.)
- Door. The door may be forced out of line or buckled.

There are two shipping methods used for motor compressor domes. One involves removable shipping bolts. They are used when the dome is mounted on or suspended from springs. The other is loosening the dome shipping bolts two or three turns. This is done when the dome is mounted on synthetic flexible grommets.

Figure 12-2 shows a hand truck with a hold-on strap. The side rails may be used as skids to aid in moving the appliance in and out of the truck and in and out of the building.

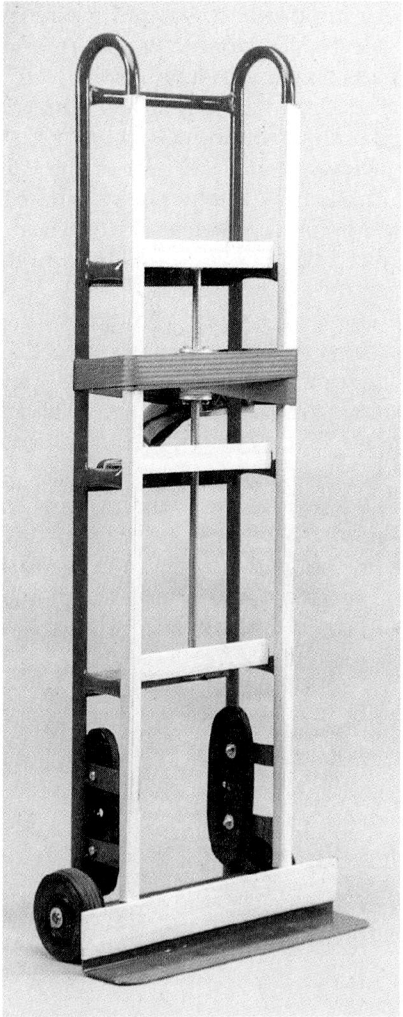

Figure 12-2. *Appliance hand truck. Strap is wound around appliance and is tightened by a wrench powering a ratchet gear. (W.W. Grainger, Inc.)*

12.2.2 Properly Positioning a Refrigerator-Freezer

Locate the refrigerator-freezer where it will not be in direct sunlight, if possible. It should not be near an oven, heat radiator, or warm air register. Precautions should be taken if the unit will be located near an oven or radiator. An aluminum foil strip should be hung covering the side next to the heat source.

The room should be large enough to provide enough air to cool the condenser. The preferred size is 100 ft^2. See Section 12.2.4.

12.2.3 Electrical Supply

The electrical outlet for the refrigerator-freezer must provide the correct electrical supply. Be sure to read the electrical ratings on the appliance. Check these against the electrical supply provided at the wall outlet. The modern household refrigerator-freezer may need more current than older, simpler refrigerators and freezers.

There should be a separate circuit from the fuse or circuit breaker box to the refrigerator-freezer outlet. Avoid using an extension cord between the refrigerator power cord and the wall outlet. The resulting voltage may be too low.

Voltage at the refrigerator outlet can be quite easily checked with a ***voltmeter.*** The circuit capacity (wire size, etc.) is checked as follows: If, at the instant of starting, the voltage at the refrigerator outlet drops more than 10 V, the wiring in the circuit is not heavy enough. A flicker in the lights at the instant the refrigerator starts is a sure sign of a poor electrical supply.

It is very important to ground the refrigerator. All removable electrical parts—fans, thermostats, timers, etc., are already safety grounded. If the wall outlet has a three-prong socket and the unit has a matching power cord plug, there is grounding. Otherwise, a wire must be attached between a metal part of the cabinet and a good ground, such as a water pipe. The type of plug used on the appliance's power cord indicates the voltage and grounding. See Figure **6-104** for the different plug designs.

Always check for proper grounding in the outlet box supplying current to the unit being serviced. Take a voltmeter reading from the "live" wall receptacle connection to the receptacle ground connection. A full voltage reading will indicate that the outlet is properly grounded. See Figure 12-3A.

A second way to check for proper grounding is shown in **Figure 12-3B.** Insert the leads of a 120 V incandescent light into the live terminal and the ground terminal of the receptacle. The lamp should light and should have normal brilliance.

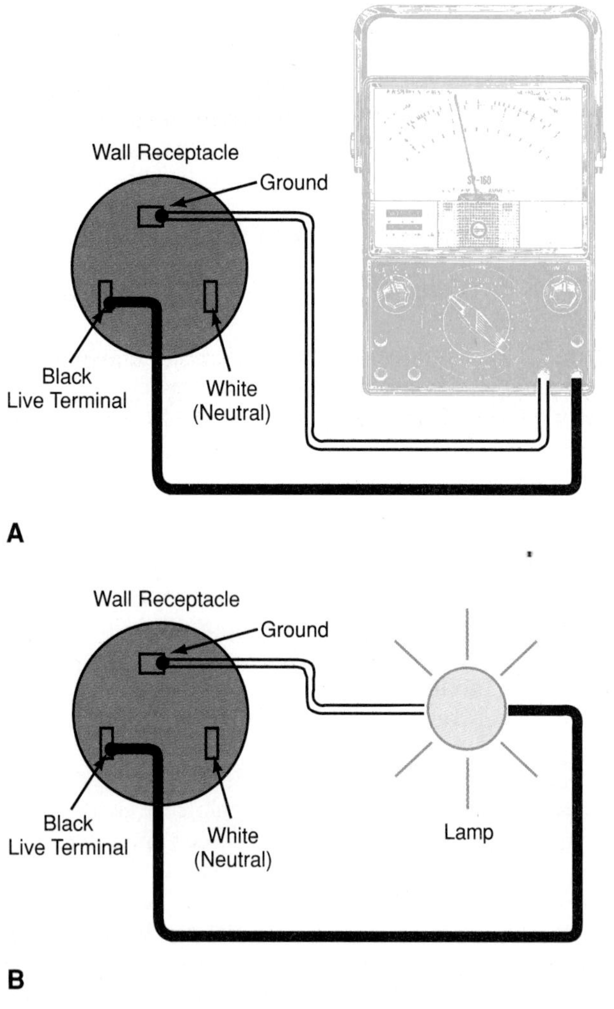

Figure 12-3. *How to make sure that ground terminal is properly connected to a ground. A—Using a voltmeter. B—Using a 120 V, 100 W incandescent lamp. (A.W. Sperry Instruments, Inc.)*

12.2.4 Providing Proper Ventilation

Since domestic refrigerators are air-cooled, proper ventilation is very important. Yet, many kitchens are designed without adequate space for air movement around the appliance. In these installations, the refrigerator requires cooling fans. The fans draw cool air in at the floor level, circulating it over the condenser. The warm air is then exhausted back into the kitchen at, or near, floor level. Nothing should be placed in front of these openings to block airflow.

On many domestic refrigerators, condensers are mounted on the back. Some are protected by a shroud. The shroud provides a chimney effect, increasing the rate of airflow over the condenser. With this type of condenser, air circulation must be provided at the bottom, back, and top.

Many freezers and some refrigerators use the outside shell as the condenser surface. In these, at least 2" (51 mm) of space must be allowed between the refrigerator cabinet and surrounding surfaces.

Leveling the Unit

The refrigerator must be carefully leveled during installation. A spirit level should be used. First check the floor where the rear supports or legs of the refrigerator are to rest. If it is not level, use wood spacers. Usually, the front supports are adjustable. They may be used to properly level the cabinet. **Figure 12-4** illustrates the manner of adjusting these levelers.

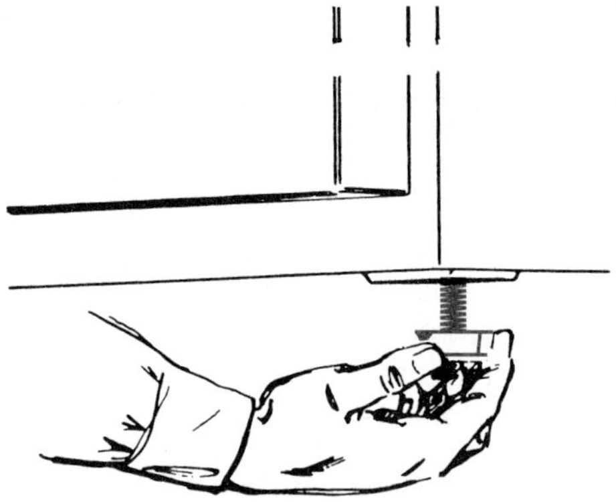

Figure 12-4. *Adjusting a leveler glide.*

12.2.5 Starting a Refrigerator-Freezer

Test the wall outlet with a voltmeter to be certain there is electrical power. Put the temperature control in the off position. Connect the electrical cord to wall outlet. Then test the unit for running before moving the refrigerator into position.

When starting the refrigerator for the first time, set the temperature control at its middle range. After a few hours of operation, check the thermometer in the fresh foods (refrigerator) compartment. Adjust the temperature control setting to the customer's requirements.

If the refrigerator does not start, make sure that the electrical circuit is in good condition. Then check for mechanical trouble. Open and close the doors to make sure that the interior lights are functioning properly.

Note: If a refrigerator-freezer using a capillary tube refrigerant control is stopped and then started immediately, it may fail to operate. This is not necessarily a malfunction. The motor in this type of refrigerator provides insufficient starting torque to overcome high head pressure. Disconnect the refrigerator for a few minutes. This allows the refrigerant pressure time to balance between the high and the low sides. Then start the system.

12.2.6 Installing an Ice Cube Maker

Many domestic refrigerators have automatic ice cube makers. See Chapter 11. These units are connected to a cold water line by a coil of 1/4″ copper tubing. Refer to **Figure 12-5.**

Before putting the refrigerator into place, run the copper tubing to the nearest cold water line. (Cabinet partitions or the floor may have to be drilled.) Mount a tap valve on the water line. Connect the tubing to the valve (usually with a compression fitting). Connect the other end of the tubing to the refrigerator water line fitting. Leave several large loops of the tubing in back of the refrigerator. This will allow the refrigerator to be moved out of its wall recess for cleaning and servicing.

Turn the tap valve stem in slowly and pierce the cold water pipe. Check for water leaks. Gently move the refrigerator back into its recess. Be careful to avoid kinking or buckling the tubing.

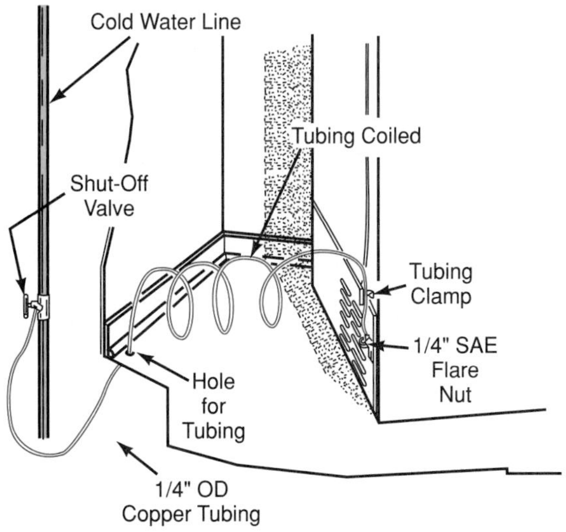

Figure 12-5. *Typical water line installation for automatic ice cube maker. Note that tubing is coiled to permit moving refrigerator without disconnecting tubing. (General Electric Co.)*

12.2.7 Shutting Down a Refrigerator-Freezer

When shutting down a refrigerator-freezer for a period of time, certain precautions should be taken to prevent rusting and remove odor. After the electrical plug has been removed or current switched off, defrost time is needed. Allow several hours for the unit to completely defrost.

When the defrosting is complete, remove water. Wash the inside of the cabinet with a solution of baking soda and water and thoroughly dry it. A portable heater set inside the cabinet will speed up this step. Leave the doors or lids ajar slightly to allow circulation of air during the shutdown period.

Caution: Federal law states that the door must be removed from an out-of-service refrigerator or freezer. This law resulted from children being suffocated when hiding or playing in an unused or carelessly discarded unit. When taking a refrigerator or freezer out of service, always remove the cabinet door.

12.3 Troubleshooting the Hermetic Refrigerator-Freezer

An unsatisfactory refrigerator temperature or operating condition usually indicates a fault in the mechanism. Such conditions may include:

- Refrigerator does not run.
- Refrigerator runs all the time; temperatures are too cold.
- Refrigerator runs all the time; temperatures are too warm.
- Refrigerator runs all the time; temperatures are satisfactory.
- Refrigerator cycles, but food compartment is too warm; freezing compartment is satisfactory.
- Refrigerator cycles, but freezing compartment is too cold.
- Motor control cuts out.
- Refrigerator cycles satisfactorily; refrigeration is poor.
- Refrigerator cycles, but does not freeze ice cubes.
- Refrigerator cycles, but too much ice accumulates on the evaporator.
- Refrigerator mechanism is very noisy.

Figure 12-6 is a chart listing common troubles, their causes, symptoms, and remedies. This chart should be considered a general guide only. It does not apply to all units.

Experience and judgment are needed to find the cause of poor operation. Always remember, cooling occurs only when the pressure is low enough and if liquid refrigerant is present. For example: An evaporator has the correct low pressure but is warm. This indicates there is no liquid in the evaporator. Another example: If a drier is frosting, there is liquid and also a low pressure in it. The drier is partially clogged.

TROUBLESHOOTING CHART

TROUBLE	COMMON CAUSE	REMEDY
1. Unit will not run.	Blown fuse.	Replace fuse
	Low voltage.	Check outlet with voltmeter, should check 115V plus or minus 10%.
		If circuit overloaded, either reduce load or have electrician install separate circuit.
		If unable to remedy any other way, install auto-transformer.
	Broken Motor or Temperature control.	Jumper across terminals of control. If unit runs and connections are all tight, replace control.
	Broken relay.	Check relay, replace if necessary.
	Broken overload.	Check overload, replace if necessary.
	Broken compressor.	Check compressor, replace if necessary.
	Defective service cord.	Check with test light at unit; if no circuit and current is indicated at outlet, replace or repair.
	Broken lead to compressors, timer or cold control.	Repair or replace broken leads.
	Broken timer.	Check with test light and replace if necessary.
2. Refrigerator section too warm.	Repeated door openings.	Instruct user.
	Overloading of shelves, blocking normal air circulation in cabinet.	Instruct user.
	Warm or hot foods placed in cabinet.	Instruct user to allow foods to cool to room temperature before placing in cabinet.
	Poor door seal.	Level cabinet, adjust door seal.
	Interior light stays on.	Check light switch; if faulty, replace.
	Refrigerator section airflow control.	Turn control knob to colder position. Check airflow heater.
		Check if damper is opening by removing grille. With door open, damper should open. If control inoperative, replace control.
	Cold control knob set at too warm a position, not allowing unit to operate often enough.	Turn knob to colder position.
	Freezer section grille not properly positioned.	Reposition grille.
	Freezer fan not running properly.	Replace fan, fan switch, or defective wiring.
	Defective intake valve.	Replace motor compressor.
	Air duct seal not properly sealed or positioned.	Check and reseal or put in correct position.
3. Refrigerator section too cold.	Refrigerator section airflow control knob turned to coldest position.	Turn control knob to warmer position.
	Airflow control remains open.	Remove obstruction.
	Broken airflow control.	Replace control.
	Broken airflow heater.	Replace heater.
4. Freezer section and refrigerator section too warm.	Fan motor not running.	Check and replace fan motor if necessary.
	Cold control set too warm or broken.	Check and replace if necessary.
	Finned evaporator blocked with ice.	Check defrost heater thermostat or timer. Either one of these could cause this condition.
	Shortage of refrigerant.	Check for leak, repair, evacuate and recharge system. Recover/recycle refrigerant.
	Not enough air circulation around cabinet.	Relocate cabinet or provide clearances to allow sufficient circulation.
	Dirty condenser or obstructed condenser ducts.	Clean the condenser and the ducts.
	Poor door seal.	Level cabinet, adjust door seal.
	Too many door openings.	Instruct customer.
5. Freezer section too cold.	Cold control knob improperly set.	Turn knob to warmer position.
	Cold control capillary not properly clamped to evaporator.	Tighten clamp or reposition.
	Broken cold control.	Check control. Replace if necessary.

Figure 12-6. *Chart lists some common hermetic system troubles, their causes, and suggested remedies.*

TROUBLESHOOTING CHART

TROUBLE	COMMON CAUSE	REMEDY
6. Unit runs all the time.	Not enough air circulation around cabinet or air circulation is restricted.	Relocate cabinet or provide proper clearances around cabinet – remove restriction.
	Poor door seal.	Check and make necessary adjustments.
	Freezing large quantities of ice cubes, or heavy loading after shopping.	Explain to customer that, heavy loading causes long running time.
	Refrigerant charge.	Undercharge or overcharge – check, evacuate and recharge with proper charge.
	Room temperature too warm.	Ventilate room as much as possible.
	Cold control.	Check control; if it allows unit to operate all the time, replace control.
	Defective light switch.	Check if light goes out. Replace switch if necessary.
	Excessive door openings.	Instruct customer.
7. Noisy operation.	Loose flooring or floor not firm.	Tighten flooring or brace floor.
	Tubing contacting cabinet or other tubing.	Move tubing.
	Cabinet not level.	Level cabinet.
	Drip tray vibrating.	Move tray – place on styrofoam pad if necessary.
	Fan hitting liner or mechanically grounding.	Move fan.
	Compressor mechanically grounded.	Replace compressor mounts.
8. Unit cycles on overload.	Broken relay.	Replace relay.
	Weak overload protector.	Replace overload protector.
	Low voltage.	Check outlet with voltmeter. Underload voltage should be 115 V plus or minus 10%. Check for several appliances on same circuit or extremely long or undersized extension cord being used.
	Poor compressor.	Check with test cord and also for ground before replacing.
9. Stuck motor compressor.	Broken valve.	Replace motor compressor.
	Insufficient oil.	Add oil; if unit still will not operate, replace motor compressor.
	Overheated compressor.	If compressor faulty for any reason, replace motor compressor.
10. Frost or ice on finned evaporator.	Broken timer.	Check with test light and replace if necessary.
	Defective defrost heater.	Replace heater.
	Defective thermostat.	Replace thermostat.
11. Ice in drip catcher.	Defective drip catcher heater.	Replace heater.
12. Unit runs all the time, temperature normal.	Ice builds up on the evaporator.	Check door gaskets – replace if necessary.
	Control bulb on themostat not in contact with evaporator surface.	Place control bulb in contact with the evaporator surface.
13. Freezer runs all the time. Temperature too cold.	Faulty thermostat.	Check thermostat – test and replace if necessary.
14. Freezer runs all the time. Temperature too warm.	Ice buildup in insulation.	Remove breaker strips, stop unit, melt ice and dry insulation, seal outer shell leaks and joints and then assemble.
15. Rapid ice buildup on the evaporator.	Leaky door gasket.	Adjust door hinges. Replace door gasket if cracked, brittle or worn.
16. Door on freezer compartment freezes shut.	Faulty electric gasket heater.	Use alternate gasket heater or install new one.
	Faulty gasket seal.	Inspect and check gasket. If worn, cracked or hardened, replace it.
17. Freezer works then warms up.	Moisture in refrigerator.	Install drier in liquid line.
18. Gradual reduction in freezing capacity.	Wax buildup in capillary tube.	Use capillary tube cleaning tool or replace capillary tube.

Figure 12-6. *Continued.*

12.3.1 Ice on the Evaporator

A large ice buildup in the food storage space acts as insulation. This may result in poor cooling. The cause is usually a leaky door seal (gasket). In a frost-free or automatic defrost refrigerator, ice buildup indicates that the defrost is not operating.

Many refrigerators with a separate freezer compartment door have an electric heater around the door opening. This keeps ice or moisture from forming there. If this heater is not working, ice buildup may keep the door from closing properly. Moisture may then enter and collect on the evaporator. The refrigerator wiring circuit diagram indicates whether or not a ***door heater*** is used on the unit.

A flat slip of paper may be inserted between the door and the cabinet. It should be held tightly when the door is closed. If this paper can be pulled out easily, the gasket is not tight enough. It may be that the door seal has lost its flexibility (life) or is broken. In either case, the gasket should be replaced. In some cases, the hardware (latch and hinges) can be adjusted to obtain a better seal.

12.3.2 Moisture and Ice in the Cabinet Insulation

The presence of moisture and ice in the cabinet insulation is serious. It means there is an air leak in the outside cabinet seal or shell. The leak allows warm, moist air to enter this space. When the warm air strikes the cold inner liner, it gives up its moisture.

When this occurs in the refrigerator cabinet, the insulation becomes wet. It will lose its heat-insulating qualities. There will be two indications of this trouble:

- The condensing unit will run more often than normal.
- The outside surface of the refrigerator will feel colder than normal wherever the insulation is wet.

In a freezer compartment, the condensed moisture will form ice in the insulation. The symptoms will be much the same as in a refrigerator. If this condition continues, enough ice will soon build up to cause the sides of the cabinet to buckle. The leak in the outside cabinet surface must be located and completely sealed.

Most freezers provide a small opening through the inner lining. This connects the insulated area with the inside of the freezer cabinet. The temperature inside of the freezer is much lower than the insulation temperature. Therefore, any moisture will tend to escape through this small opening. It will then condense on the evaporator surface.

12.3.3 Temperature-Pressure Conditions

Before servicing a refrigerator, you should know:

- Normal temperature in the evaporator during the operating cycle.
- Normal pressure on the low-pressure side during the operating cycle.
- Normal temperature of the condenser during the operating cycle.
- Normal pressure on the high-pressure side during the operating cycle.

The above conditions depend upon the refrigerant being used. Temperature-pressure properties of refrigerants vary.

Figure 12-7 lists the average temperature-pressure conditions for the evaporator and condenser of a domestic refrigerator-freezer. When using the table, be sure that the correct ambient temperature is used.

The most common hermetic refrigerant is R-12. Due to the EPA ruling, R-12 is being replaced by R-134a. See Section 9.5.4 and Chapter 10. In low-temperature (frozen food) units and air conditioners, R-22 is quite popular.

To determine the operating temperatures, a temperature chart recorder with vapor-filled bulbs may be used. **Figure 12-8** illustrates such a thermometer in use. A compound gauge and a ***high-pressure gauge*** may be used to determine the operating pressures. **Figure 12-9** shows the position of gauges for a refrigeration system.

Temperature-Pressure Conditions in the Refrigerator Mechanism

	R-12 Ambient Temp.	
	70°F	90°F
Refrigerator Evaporator Temperature		
Start of Cycle	15	15
Middle of Cycle	5	5
End of Cycle	0	0
Refrigerator Evaporator Pressure, psig		
Start of Cycle	12	12
Middle of Cycle	8	8
End of Cycle	5	5
Condenser Temperature		
Start of Cycle	70	90
Middle of Cycle	100	120
End of Cycle	100	120
Condenser Pressure, psig		
Start of Cycle	70	85
Middle of Cycle	120	158
End of Cycle	120	158

Figure 12-7. *Average temperature and pressure conditions in a domestic refrigerator using R-12 refrigerant. The unit has a freezer compartment.*

Figure 12-8. *Temperature chart recorder using vapor-filled bulbs and systems. Recorder is being used to test operating temperatures. It has two sensing bulbs. One is placed in freezer compartment and one in refrigerator compartment.*

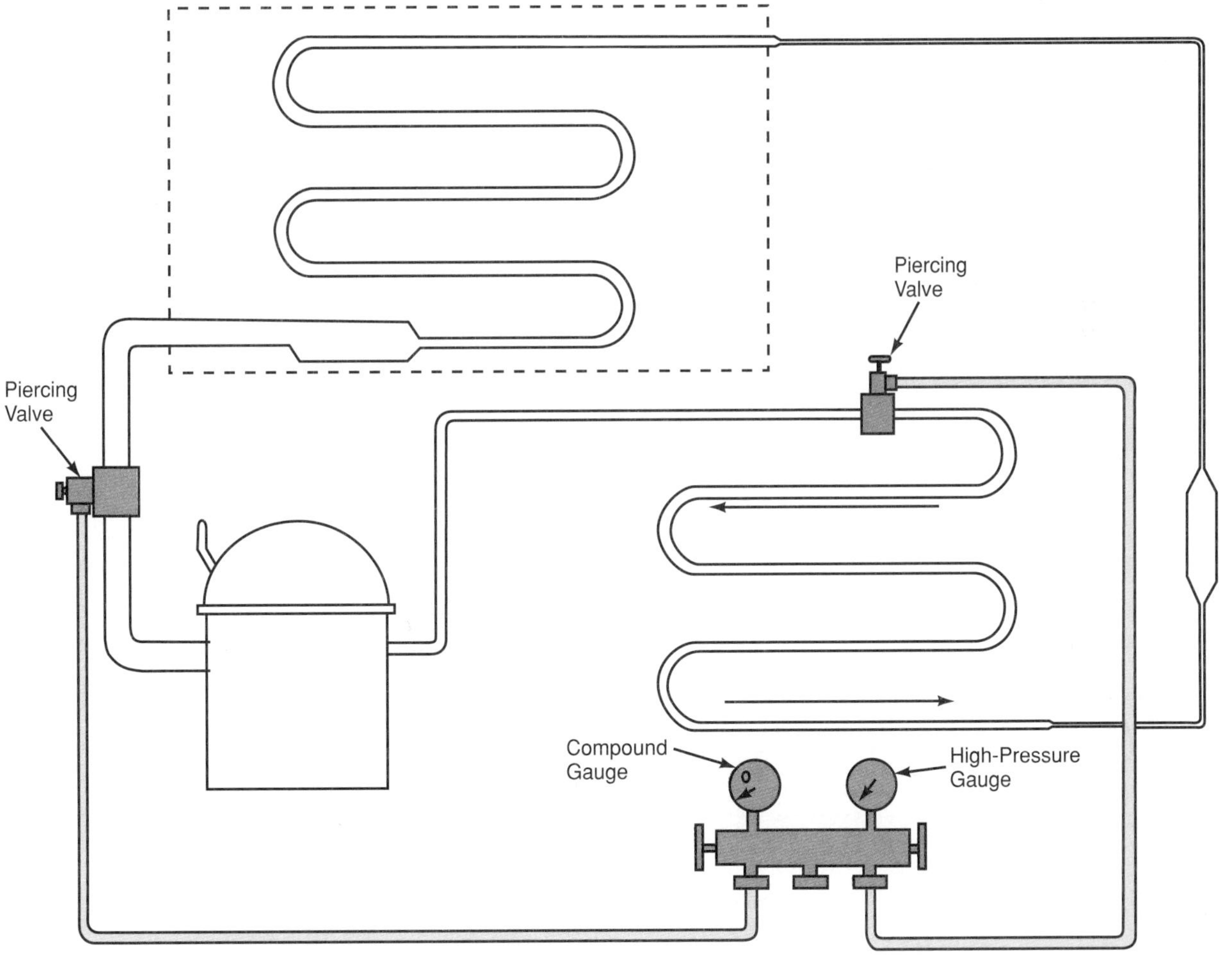

Figure 12-9. *High-pressure gauge and compound gauge are connected to the hermetic system with service lines. Piercing valves are used to open service lines to hermetic system.*

12.3.4 Locating and Eliminating Noises

Most audible noise in the refrigerator will come from rattles. Noise sources may be:

- Loose baffles or ducts.
- Tubing touching something while vibrating.
- A ***listing*** (leaning to one side) of the condensing unit, caused by an uneven floor.
- Fan and motor vibration.
- A loose evaporator unit door.
- Loose articles on shelves.
- Shelves not seated properly on supports.

An ultrasonic leak detector, like the one shown in **Figure 12-10** can be used to isolate many noises. (***Ultrasonic*** refers to sound above the human hearing range.)

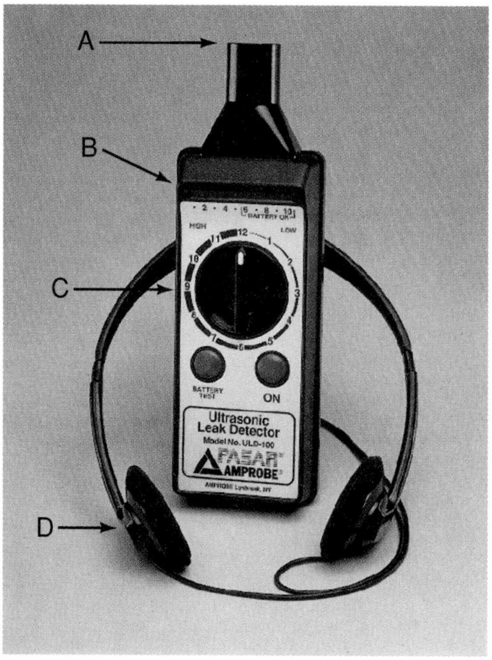

Figure 12-10. *Electronic sound detector. A—Sensor horn. B—Display panel (lights up dependent upon frequency of sound). C—Sensitivity dial is set to identify sound. D—Earphone. This unit can also be used to identify refrigerant leaks. (Amprobe Instrument)*

Sounds generated by leaks or other defects can be located and the cause of the problem determined. **Figure 12-11** shows a simple, homemade stethoscope that can be used to pinpoint the source of a rattle or other noise.

Noise originating within the unit may indicate that it is laboring too hard. To determine this, test the electrical load with an ammeter or wattmeter. An overloaded unit can sometime be identified by its starting behavior. Three seconds to operate the relay is an average time. A slower start indicates an overload.

Tubing that rattles against refrigerator parts should be carefully bent away from contact. The tubing may be rigid, but may have a vibration or hum (harmonic vibration). This noise can be reduced by clamping rubber blocks on the tubing. See **Figure 12-12.** Loose baffles and ducts can be secured with self-tapping sheet metal screws.

The sound-tracing device may indicate that the noise is coming from inside the motor compressor. In this event, no service operation should be attempted until you study Sections 12.12, 12.12.1, and 12.15.1.

Another very helpful device for inspecting a refrigerator is an adjustable mirror on a long flexible extension. A mirror on a rod is shown in **Figure 12-13.**

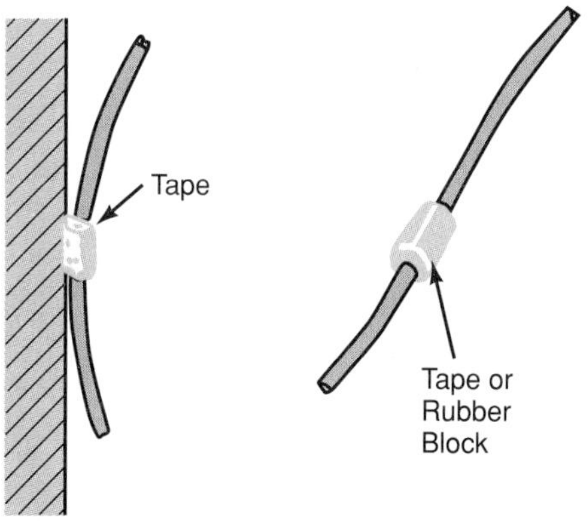

Figure 12-12. *Two ways are shown to reduce noise caused by vibrating tubing. A—Wind tape around tubing where tubing touches cabinet. B—Put tape or rubber block on tubing in center of vibrating section.*

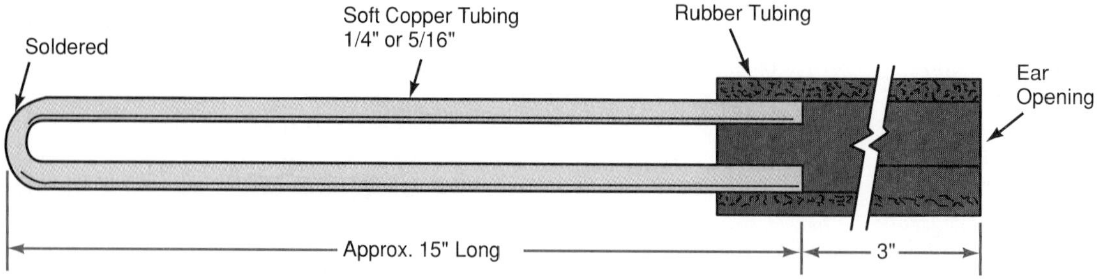

Figure 12-11. *Simple stethoscope made from section of 1/4" copper tubing and rubber tubing. It is excellent for locating source of many noises. Place closed end against the mechanism and then put rubber tube opening to ear. (Use care around moving parts.)*

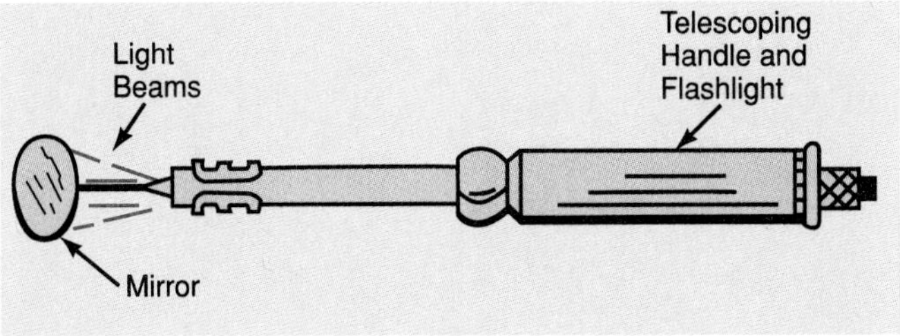

Figure 12-13. *An inspection mirror is often needed to see into hard-to-reach places. It is used to check fan alignment, motor condition, tubing location, cleanliness of condenser, condition of evaporator, etc. Avoid touching moving parts with mirror.*

12.3.5 Cycling Time for Refrigerators and Freezers

Cycling time on home refrigerators and freezers cannot be given in definite terms. This will vary depending on various factors. These include the amount of storage space being used, outside box temperature, and compressor condition.

Placing warm food in the cabinet to be frozen will also affect the cycling time. The condensing unit may run about one-third of the time. In other words, it may run 5 minutes and be off 10 minutes. It may run an hour and be off two hours. The important point is that any unusual change in cycling time should be investigated immediately. It may indicate that trouble is developing in the system.

12.4 Hermetic Servicing Guide

To service refrigerating units, you must know how they should perform when in good condition. **Figure 12-14** shows the performance data of a typical domestic unit.

Always check the system data before trying to locate the cause of the trouble. System data is usually located on the identification plate mounted on the motor compressor. The data for a 1/3 hp two-pole motor unit used in a combination refrigerator-freezer is shown in **Figure 12-15.**

Section 12.3 and **Figure 12-6** provide information which you should understand before beginning any service operation. Servicing of hermetic refrigerators may be divided into three major areas:

- External servicing.
- Internal servicing.
- Overhaul of hermetic system.

12.5 External Servicing Operations

Some of the more common external service operations are as follows:

- Cabinet hardware.
- Cleaning.
- Noise (rattles).
- Ice cube maker.
- Electrical.
  - Power-in circuit.
  - Thermostat.
  - Defrost thermostat.
  - Interior light and circuit.
  - Fan motor and circuit.
  - Damper controls.
  - Motor compressor (relay and overload protector, capacitor, or motor terminals).
  - Defroster.
  - Defroster control and circuit.
  - Defroster heater coil.
  - Cabinet heaters.
  - Evaporator fan and circuit.
  - Condenser fan and circuit.
  - Light circuit.
  - Butter conditioner circuit.

These external mechanisms, electrical wiring, and electrical parts can all be checked for operation quickly. This is done by inspection and by using a volt-amp-ohmmeter.

	70°F Ambient Temperature	90°F Ambient Temperature	100°F Ambient Temperature
Cabinet Temperature	38°F	40°F	47°F
% Operating Time	38	62	100
Cycles Per Hour	3	2	None
kWh/24 hr.	3.8	6.0	9.9
Control Position	4	4	4
Evaporator Air Temperature	1.5°	−1°	0°
Suction Pressure (Min-Max)	2" Hg–13 psig	0–13 psig	13–20 psig
Watts (Complete System)	390 ± 20	395 ± 20	395 ± 20

Figure 12-14. *Chart shows operating characteristics of 18 ft^3 combination refrigerator-freezer which has 1/3 hp two-pole motor compressor. Note kWh (averaged over 24 hrs.) changes as ambient temperature changes.*

	DATA
REFRIGERANT	R-12
CHARGE (IN OUNCES)	10 1/2
COMPRESSOR hp	1/3
COMPRESSOR SPEED rpm	3450
RUNNING AMPERES	5.6
VOLTAGE	120
PHASE	SINGLE

Figure 12-15. *Refrigerant and electrical data which is typical of information found on identification plate mounted on motor compressor.*

Other external service troubles can be located by checking for:

- Ice on evaporator.
- Frost or sweat on suction line.
- Warm or hot discharge line.
- Ice or sweat on driers.
- Dirty condensers.

12.5.1 Diagnosing External Troubles

It is important to locate the trouble and determine the cause accurately. Some hermetic units are needlessly replaced because the internal mechanism is believed to be faulty. The real trouble, however, was in the external devices.

The trouble could be in some part of the external electrical circuit. For example:

- Power-in connections.
- Thermostat.
- Wire terminals.
- Relay.
- Capacitor (if the unit has one).

Each of these devices should be checked carefully before the unit itself is considered faulty. These parts can be checked best by removing them from the wiring system. Either of the following can then be done:

- Parts can be checked independently.
- A part can be temporarily replaced by a test part of the proper size. The unit may then be checked to see if it will run.

Figure 12-16 shows a plug-on thermal motor protector. This protects the motor from overheating, resulting from a locked rotor or running overload conditions. In addition, a start relay PTCR (Positive Temperature Coefficient Resistor) is used. **Figure 12-17** shows various electrical circuits typical of a domestic refrigerator.

Electrical connections must be clean and tight. If loose or dirty, they often overheat. This high temperature will discolor the connection. The connection may be darkened by oxidation. A blue or greenish tint indicates overheating and corrosion. If the surrounding insulation is charred, overheating has occurred.

Troubles such as open circuits and grounded electrical wires are easily checked with a test light. A test

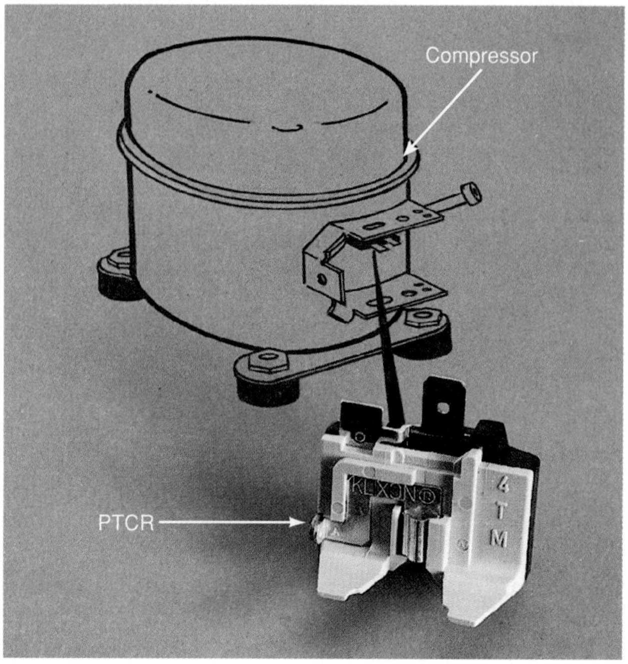

Figure 12-16. *This PTCR (Positive Temperature Coefficient Resistor) start relay is mounted directly on the motor compressor. (Texas Instruments, Inc.)*

cord can be used to check four-pole motors. However, two-pole motors should be tested only by using a proper size relay in the circuit. These motors overheat if the starting circuit is connected for more than two or three seconds.

Carefully review Chapters 6, 7, and 8 before attempting to locate trouble in electrical units. When checking and servicing hermetic electrical circuits, first check the outlet electrical supply. Check the appliance voltage specifications.

Using a voltmeter, test the open circuit voltage. Next, plug in the appliance and check the voltage again while it is running. The open circuit voltage is likely to be slightly higher than with the motor running. However, this difference should not be more than 5 V. A difference of 10 V or more indicates serious trouble:

- An overload.
- Something wrong in the motor windings.
- Poor wiring to wall outlet.

Most refrigerators and freezers have a wiring diagram attached to the back. Locate this diagram and check each circuit independently. If the compressor fails to start, follow these steps:

1. Find out if electricity is reaching the motor compressor.
2. If it is, check the starting relay and circuit protectors. See Chapter 8.
3. Disconnect all wiring from the motor compressor.
4. Check the motor compressor with a manual start test cord. Such a test cord is shown in **Figure 12-18B.** The ground clip, labeled 4, must be fastened to the

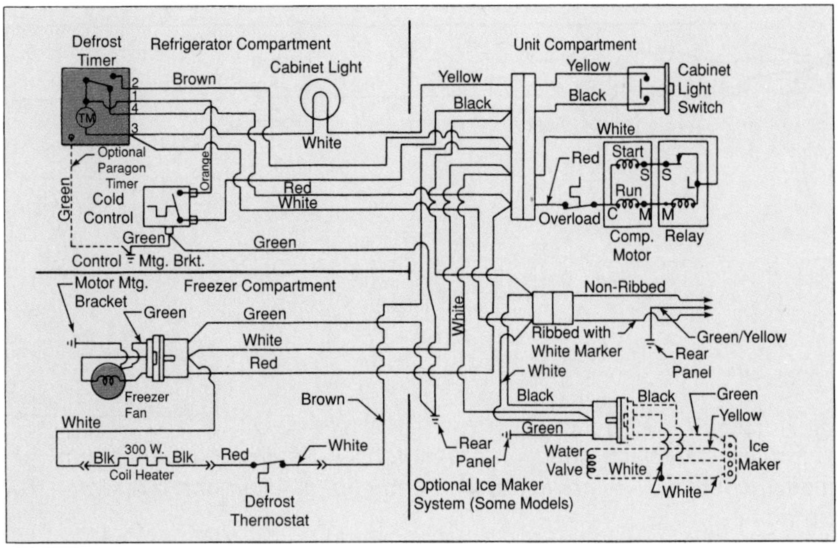

Figure 12-17. *Wiring diagram of the domestic single-door automatic defrost refrigerator with a defrost timer and freezer fan. Note color coding of wiring to aid you in tracing the circuits. (Frigidaire Company)*

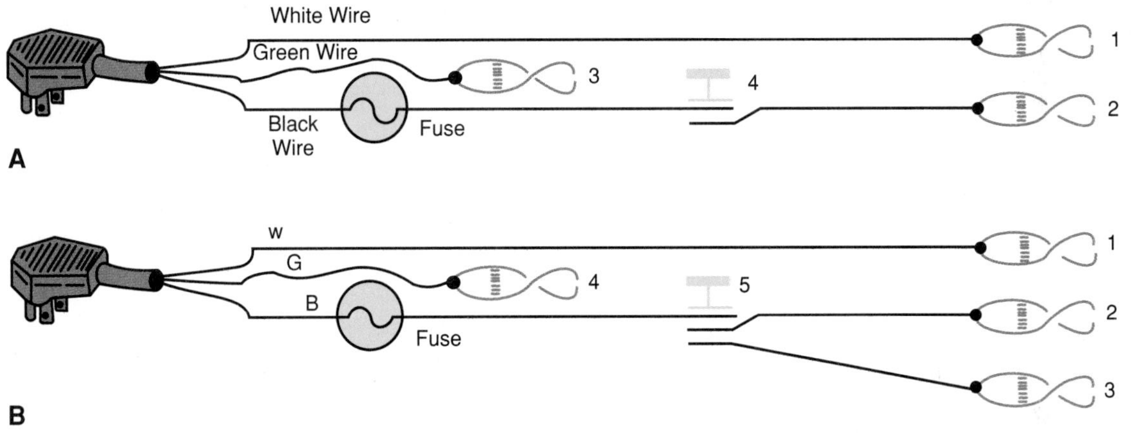

Figure 12-18. *Test cords. A—Cord for testing fan motors. Clip 1 goes to winding common terminal. Clip 2 attaches to winding power terminal. Clip 3 goes to ground. No. 4 is start and run switch. B—Test cord for hermetic motors. Clip 1 goes to hermetic compressor common terminal. Clip 2 attaches to run terminal. Clip 3 goes to start terminal. Clip 4 goes to ground. No. 5 is start and run switch. (A.W. Sperry Instruments, Inc.)*

dome. **All clips or connectors should be plastic-coated, or should have a rubber boot over them to protect you from shocks.**

When the manual switch is pressed, the running circuit closes first. Then the starting circuit closes. After one to two seconds, lift the switch button just enough to open the starting winding. If the motor operates correctly, the problem is in the external circuit. The fuse is located in the black wire of the three-prong plug.

A capacitor-start, induction-run motor is tested as shown in **Figure 12-19.** The ***capacitor*** should be replaced by a new one of the same voltage and microfarad rating. An extra clip wire or lead is needed. Operate the switch as explained in the preceding paragraph. **After testing is completed, remember to short the testing capacitor using a 20,000 Ω (20 kΩ), 2 W resistor. This will eliminate the possibility of electrical shock.**

The test for a capacitor-start, capacitor-run motor is shown in **Figure 12-20.** Use new capacitors of the same rating as the ones on the system being tested. Operate the switch as previously explained. If the motor compressor works, check the electrical system up to the compressor. If the motor does not operate when tested, further motor checks are needed. This is explained in Chapter 7.

The test cords shown in **Figures 12-18, 12-19,** and **12-20** can be used for checking continuity and grounding by replacing the fuse with a lightbulb.

The trouble may be the evaporator motor or the condenser fan motor. These motors are usually replaced if found to be faulty. Before removing a fan from a motor shaft, mark the position of the fan hub on the shaft. This ensures that the fan is located correctly on the new shaft.

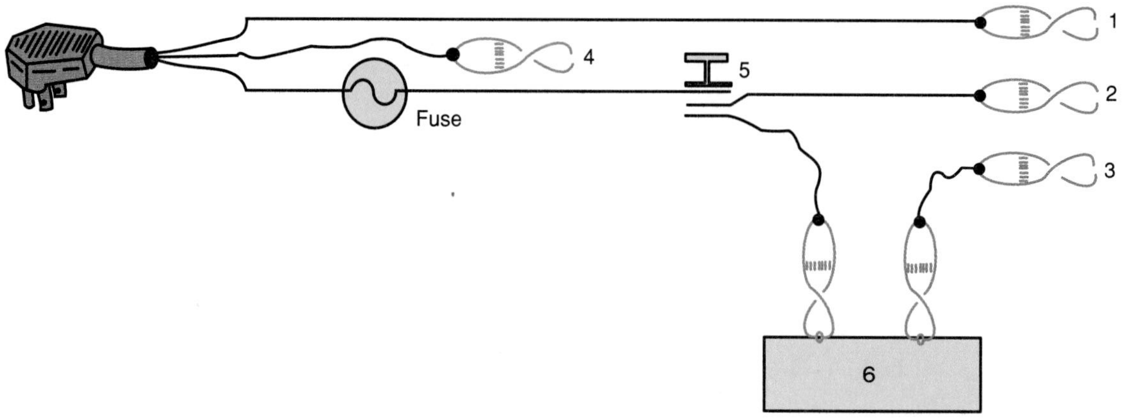

Figure 12-19. *Test cord for capacitor-start, induction-run motor. 1—Common winding terminal. 2—Running winding terminal. 3—Starting winding terminal. 4—Grounding wire and clip. 5—Start and run switch. 6—Starting capacitor. (A.W. Sperry Instruments, Inc.)*

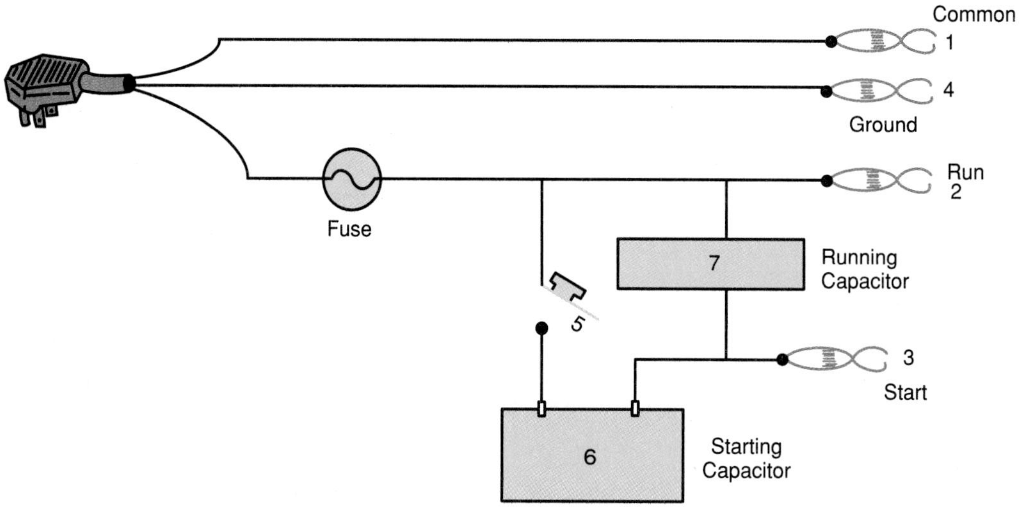

Figure 12-20. *Test cord used to check capacitor-start, capacitor-run hermetic motor. 1—Common winding terminal. 2—Running winding terminal. 3—Starting winding terminal. 4—Ground wire and clip. 5—Manual switch. 6—Starting capacitor. 7—Running capacitor. (A.W. Sperry Instruments, Inc.)*

Electrical failure in the mullion heater may cause a door gasket to freeze to the cabinet. The heater must be checked with a continuity light or an ohmmeter. Locate the circuit in the wiring diagram. Disconnect both ends of the mullion heater leads. Then, test the heater for continuity.

If the mullion heater is defective, look for a second (extra) heater in the insulation. Most cabinets have one. Test it also, and install it if it is operating properly. If there is no extra heating unit, install one of the same wattage (volt-ampere) rating.

If the problem is a faulty wire, use a stiff steel wire to pull new wiring through foamed-in-place insulation. If necessary, drill a hole (up to 1/2″) in the back of the refrigerator to help feed wires. Seal the hole after the wire or wires are pulled through.

If the cabinet temperature is not responding properly to the temperature control, the control mechanism may be faulty. It is advisable to use a recording thermometer. Record temperatures over a 24-hour period. The temperature-time chart will show if the appliance is operating properly.

Electrical Troubleshooting

The General Electric Co. has developed a rapid method for checking electrical circuits on its refrigerators. This method (which is called the "RED system") checks circuits for the defrost system, thermostats, heaters, fans, and compressor. Simpler circuits, such as lights, are not included in this system.

A multiple-circuit connector, mounted at the front bottom of the cabinet, is used. See **Figure 12-21.** When this connector is separated, the tester is connected to it. See **Figure 12-22.** A diagram of the circuits to be tested is shown in **Figure 12-23.** Note the number code for test connections.

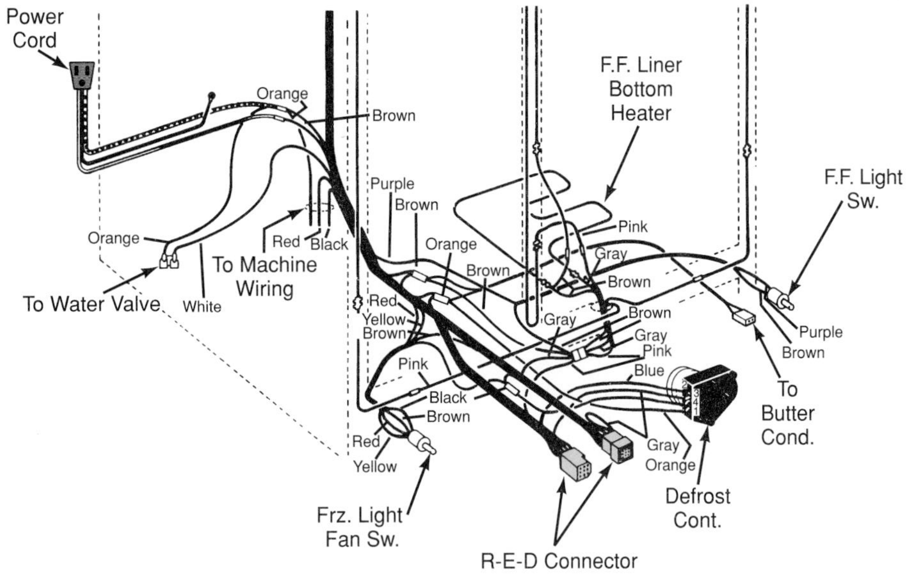

Figure 12-21. *Multi-circuit connector used for RED system simplifies testing of electrical circuits on some General Electric refrigerators.*

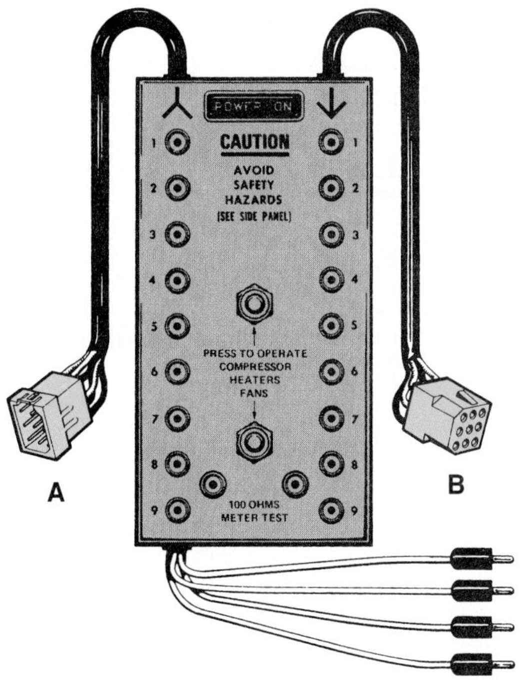

Figure 12-22. *The RED test instrument and adaptor. A—Male connector to electrical system. B—Female connector to electrical system. (General Electric Co.)*

Always turn the thermostat to the off position before separating the connector. The symbol for the female connectors is >—, and for the male connectors, →. After connecting the tester and turning the power on, the "power on" light should be lit. If not, the power circuit needs repair. Refer to the General Electric data and diagrams when performing this test.

All Tecumseh motor compressor terminals are set up to read "Common-Start-Run" (reading from left to right, going down a line at a time). See **Figure 12-24.**

Starting a Stuck Compressor

A unit may not start when connected to electric power. There may be a small piece of dirt in the compressor, or perhaps the unit has not been run for a long time. To start such a compressor, three methods may be tried:

- Disconnect the wiring to the motor compressor. Connect a test cord into the electrical circuit of the main winding. Use an extra capacitor, connecting it into the circuit as shown in **Figure 12-25.** Turn on the power from one to three seconds. The extra capacitor will try to reverse the compression rotation. *Caution: This capacitor cannot be left in the circuit more than a second or two. It will cause the motor to overheat.* If the compressor is successfully reversed, remove the reversing capacitor. Try to operate the compressor, using the normal electrical circuit. A compressor that will not start after three or four such reversal attempts usually requires rebuilding or replacement.
- Another method is to connect the 120 V motor compressor into a 240 V power circuit, using a starter cord. *Be careful—press the pushbutton switch for only a second at a time (count "one thousand one") to avoid motor damage.* The extra voltage may break the stuck compressor loose. If not, the motor compressor will have to be replaced.
- An extra-torque method is to connect a 240 V, 100 microfarad (μF) start capacitor across the terminals of the run capacitor. The connection must be for no more than one second. This may free the compressor.

Short-Cycling

Short-cycling means that the unit starts and stops too frequently. Causes may be:

- Thermostat not mounted securely.
- Loose connections in the starting relay.

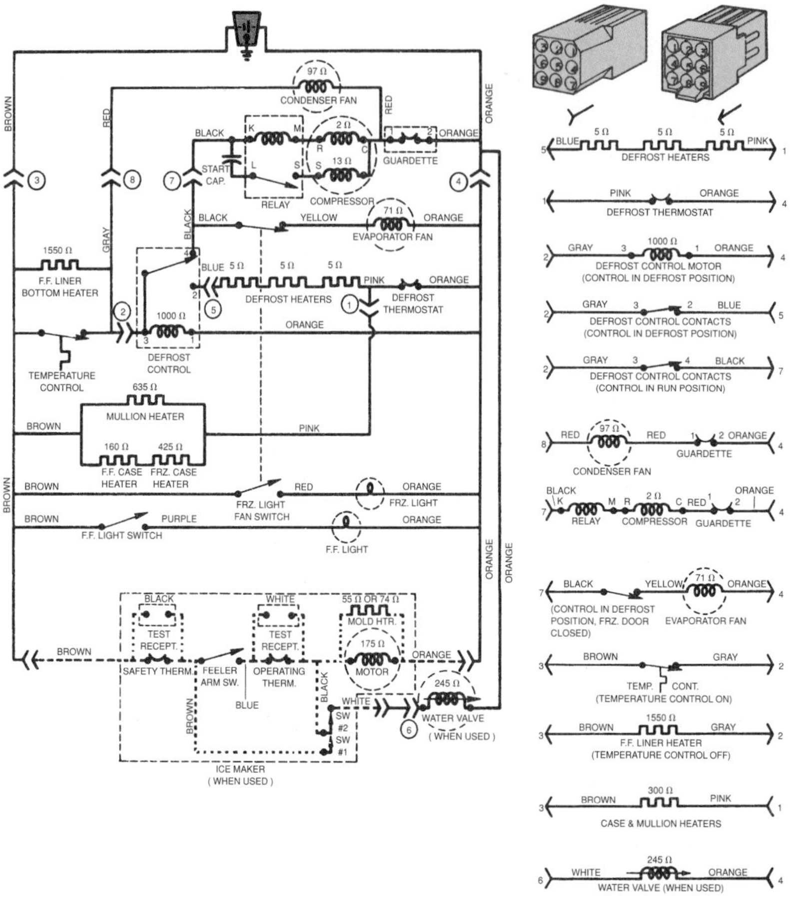

Figure 12-23. *Diagram of refrigerator circuits with number code at points where RED instrument and adaptors connect into circuits. Note: Numbers at either end of circuits shown at right correspond to terminals in male and female plugs above. (General Electric Co.)*

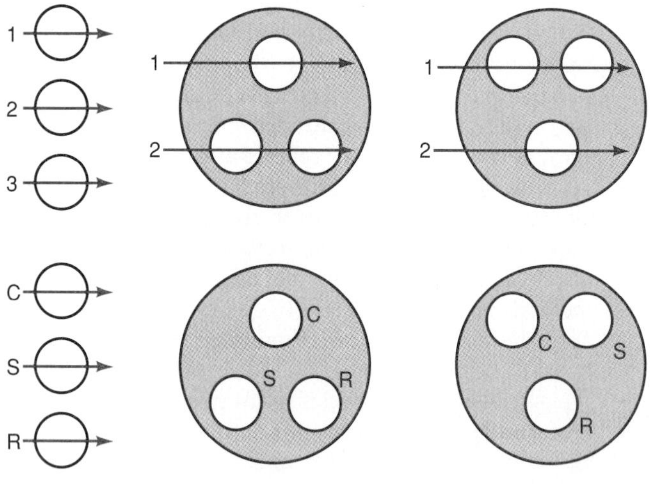

Figure 12-24. *Tecumseh motor compressor terminals are easy to identify. Referring to illustration above, read from left to right and then from top to bottom. Arrows show direction to read. Numbers show order of reading arrows.*

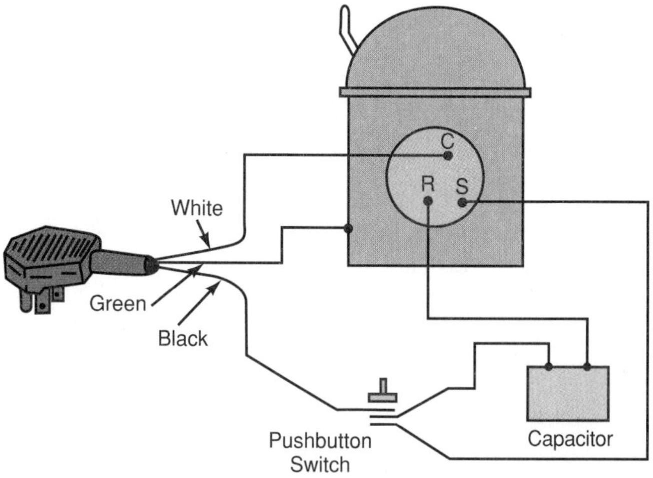

Figure 12-25. *Capacitor is being used in the running winding circuit to reverse rotation of "stuck" motor compressor. C—Common terminal. R—Running winding. S—Starting winding.*

Do not tap a relay to check it. The jar may cause points to touch. This short contact may ruin the points and injure the motor. If the relay will not function correctly without being tapped, it should be replaced.

Cleaning the External Mechanism

The hermetic refrigerating system is basically a heat-transfer mechanism. Air must circulate around and through the unit and condenser to carry away the heat. Dirt and lint act as insulation, decreasing heat transfer. Therefore, the condenser and compressor unit must be kept as clean as possible. For economical operation and long life, completely clean them about every three months.

The hermetic mechanism can be cleaned where it is. This is done by using a small vacuum cleaner. A special vacuum cleaner nozzle with a brush attachment can also be used. The vacuum cleaner keeps lint from circulating and settling on the floor. It is also quicker and more thorough than hand brushes or cleaning cloths. If a brush or cleaning cloth is used, place a paper or cloth underneath the unit to catch loosened dirt and lint.

Units that use condenser fans should be disconnected from electrical power before being cleaned. Sometimes it is necessary to partially remove the unit to do a good cleaning job.

In the shop, high-pressure air, nitrogen, or carbon dioxide is often used for cleaning purposes. Lint or other dirt is blown from between the fins or coils. Areas that otherwise would be difficult to reach are cleaned with this method. **Goggles should always be used and there should be good ventilation.**

12.6 Internal Service Operations

Internal service operations are those that do any of the following:

- Remove any part of the hermetic system.
- Determine whether there is air in the system.
- Discover a lack of refrigerant.
- Check if there is a clogged filter-drier or capillary tube.

For these jobs, attach gauges and servicing devices. These include vacuum pumps, refrigerant cylinders, and so forth.

Before attempting any field service operations that require opening the system:

- Thoroughly clean all connections and valve fittings.
- Install a valve adaptor or piercing valve.
- Install a gauge manifold.
- Install refrigerant recovery equipment.

Probably the most frequent service operations will be the following:

- Locating and repairing refrigerant leaks.
- Purging, charging, and recovery of refrigerant.
- Cleaning or replacing the capillary tube.
- Replacing a compressor.
- Replacing a filter-drier on the high side.
- Installing a filter-drier on the low side.
- Evacuating the system
- Adding oil.
- Using a high-vacuum pump.
- Replacing an evaporator and/or condenser.

These same service operations can also be performed in the shop. Shops have better facilities to do the job. Before performing any service operation, you should study this chapter carefully.

12.6.1 Diagnosing Internal Troubles

There are many ways to find the cause of trouble inside a small hermetic system. This is done by using gauges, thermometers, and electrical instruments, combined with careful observation. A service technician should be able to locate the cause of almost every problem in a system.

The evaporator may be partially frosted, while another part is heavily frosted. This indicates a lack of refrigerant.

A sweating or frosted suction line indicates that liquid refrigerant is in the suction line. There may be a broken thermostat or too much refrigerant (if a capillary tube is used).

Internal electrical troubles, involving the motor and connections, are very rare. (They occur in about 3 out of 1000 cases.) Most internal problems come from air and moisture in the motor compressor. This causes corrosion and, eventually, a burnout.

If liquid refrigerant reaches the compressor, it may remove the oil. The liquid evaporates in the crankcase and carries the oil with it into the condenser. Valves may be broken as the compressor tries to pump oil or liquid refrigerant.

A restriction may occur in the capillary tube, filter-drier, or screen on the high side. This will be indicated by continuous running, no refrigeration, and a condenser cooler than normal.

The following paragraphs discuss most of the reasons refrigeration systems will not operate correctly. The description of the repair and testing of refrigeration systems will follow.

Moisture in the Refrigerant Circuit

Moisture in the refrigerant system will cause the unit to malfunction. The moisture forms ice in the refrigerant control at the point where liquid refrigerant is expanding into the evaporator. Ice closes the opening, blocking flow into the evaporator.

This condition can be recognized by several observations:

- The system will completely defrost. Then the ice which caused the blockage will disappear. The unit will work properly again. However, the unit will only work until ice forms again at the refrigerant control.

- Pressure decreases in the suction line. The compound gauge shows a steady decrease over several hours (even to a vacuum). Then, pressure suddenly becomes normal again. This odd cycle will keep repeating.
- Warming the refrigerant control by using a safe resistance heater (hot pad) or radiant heat bulb during system shutdown will cause the ice to melt. Should the system then begin to work properly, there is definitely moisture in the refrigerant.

Moisture in the refrigerant circuit also creates corrosion problems within the system. This occurs when refrigerants react with water molecules to form acids. The acids increase the amount of corrosion in the system.

Moisture in the refrigerant circuit may be removed by putting a drier in the liquid line. The procedure is as follows:

1. Install the gauge manifold.
2. Recover the refrigerant.
3. Dry and clean filter-drier connections.
4. Apply flux.
5. Heat the connections.
6. Remove the old drier.
7. Install the new drier.
8. Braze the connections.
9. Test for leaks.
10. Evacuate the system.
11. Charge the system.
12. Warm the refrigerant control enough to melt the ice.

The drier will absorb this moisture as it circulates.

There are certain substances that can be placed in the refrigerant circuit to keep moisture from forming ice. A filter-drier is the best solution, however. It prevents circulation of the moisture through the system. It also reduces the chance of oil breakdown (sludge and acid).

Wax

Manufacturers have removed as much wax as possible from refrigeration oil. Refrigeration oil is discussed in Section 9.19. Although manufacturers have removed most of the wax, small amounts still remain.

Some oil circulates with the refrigerant. Sudden expansion at the refrigerant control, accompanied by low temperature and pressure, causes some wax to separate from the oil. The wax collects in the refrigerant control. In time, it will build up sufficiently to restrict flow or completely clog the control.

Presence of a restriction can be checked with the aid of a piercing valve. Look for the following:

- A pressure test shows low-side pressure to be very low.
- Liquid refrigerant shows up in the condenser.
- The unit does not produce any refrigeration at all.

Always clean or replace a clogged valve or capillary tube. The ice or wax should be kept locked in the refrigerant control being removed. This can be done most easily by packing the control in dry ice before removal. Another method is to very quickly open the joints after the unit is discharged. Use the same steps described in Section 12.13, but repair and/or replace the refrigerant control. When servicing frozen foods equipment, use only the best low-wax oil.

Shortage of Refrigerant

Shortage of refrigerant is a common source of poor refrigeration. If a shortage of refrigerant is found, there is often a leak. Small systems have only one or two pounds of refrigerant. Therefore, even the smallest leak will soon cause poor refrigeration. A leak with a loss rate as low as one ounce per year can be located and must be repaired.

A lack of refrigerant will be shown by:

- A low-side pressure that is below normal.
- An evaporator (or the outlet end of the evaporator) that is warm.
- A high-side pressure that is below normal.
- A piercing valve mounted on the outlet of condenser that, when opened, allows only gas to escape. (It should be liquid.)

12.7 Gauge (Service) Manifold Types and Construction

Make all gauge and service connections to a hermetic system through a ***gauge (service) manifold.*** See **Figure 12-26.** There are two basic types of manifolds: the standard manifold and the block manifold. Both are available with front or side wheels. Most domestic refrigerators and freezers with hermetic systems do not provide for gauge connections. Therefore, special attaching devices must be installed in order to use the gauge manifold. A cutaway view of the type of manifold in **Figure 12-26C** is shown in **Figure 12-27. Figure 12-28** shows the various uses and valve positions of a gauge manifold. **Figure 12-29** shows the setup for charging a system.

Separate gauges and hand valves can be used with the special service valves mounted on hermetic systems. However, these devices must be removed and others used to perform other operations. These operations include purging a system, checking pressures, or evacuating and charging the system.

A gauge manifold with two gauges, two hand valves, and three separate lengths of flexible refrigerant tubing will enable you to perform these operations more easily. The three flexible hoses are equipped with 1/4″ (6 mm) flare fittings with synthetic rubber gaskets. Thus, connections can be made pressure-tight with finger pressure alone. **Figure 12-30** shows design and construction of a flexible service and refrigerant line.

You must learn how to use the gauge manifold. The hand valves on the manifold can be used for most

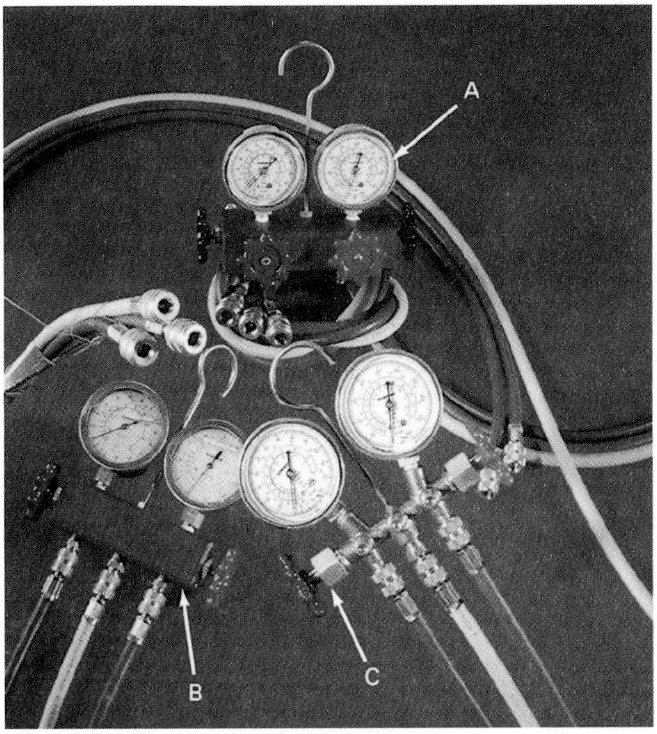
A
B
C

Figure 12-26. *Three types of gauge manifolds: A—Four-way manifold, front and side, used to do testing, evacuation, and recharging of a system without having to switch hoses. B—Aluminum bar manifold. C—Side wheel manifold. (Robinair Division, SPX Corporation)*

Figure 12-27. *This cutaway view of a gauge manifold shows hand valves, gauges, and refrigerant openings. Gauges will always show a pressure reading. When low-pressure hand valve is turned all the way in, the low (evaporator) pressure can be checked. When high-pressure hand valve is turned all the way in, high (condensing) pressure can be checked. When both valves are open (turned out by twisting to the left), high-pressure vapor will flow into low side. When only low-pressure valve is open, you can charge the system or evacuate it, put oil in system, or clean it. (Uniweld Products, Inc.)*

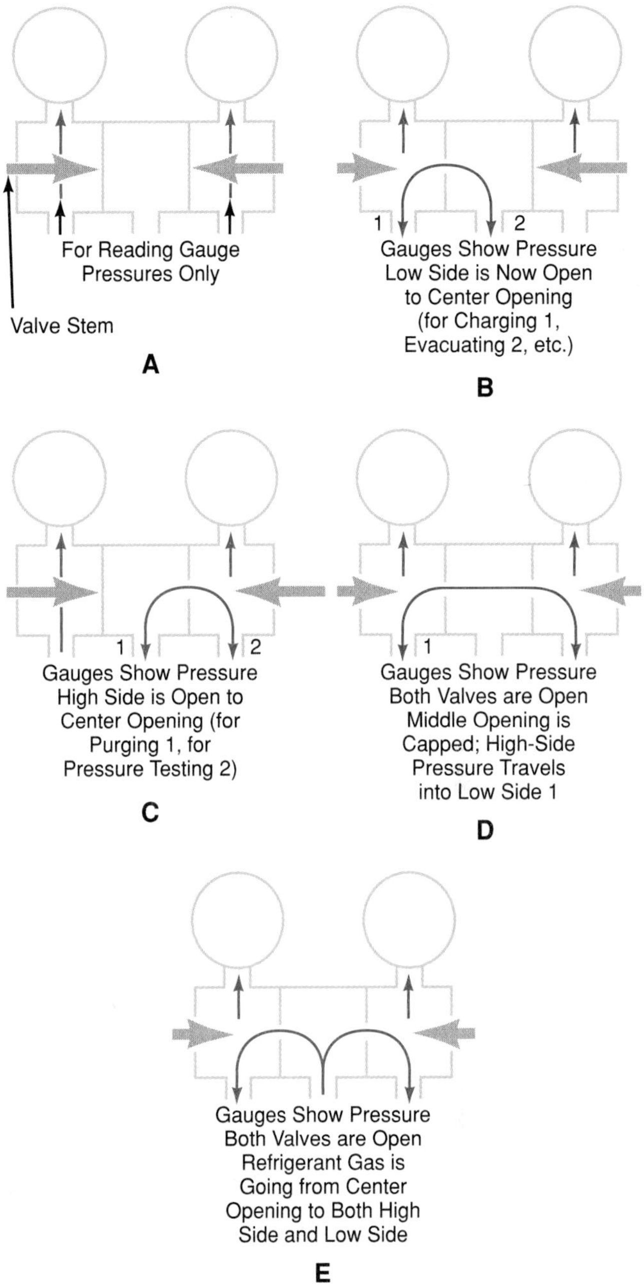
For Reading Gauge Pressures Only
Valve Stem
A
1
2
Gauges Show Pressure Low Side is Now Open to Center Opening (for Charging 1, Evacuating 2, etc.)
B
1
2
Gauges Show Pressure High Side is Open to Center Opening (for Purging 1, for Pressure Testing 2)
C
1
Gauges Show Pressure Both Valves are Open Middle Opening is Capped; High-Side Pressure Travels into Low Side 1
D
Gauges Show Pressure Both Valves are Open Refrigerant Gas is Going from Center Opening to Both High Side and Low Side
E

Figure 12-28. *How to use hand valves of gauge manifold for various service operations.*

operations. **Figure 12-31** shows a gauge manifold used with a vacuum pump. It is used to produce a vacuum, install a part, and charge a system.

12.7.1 Connecting a Gauge Manifold

To check the pressure in a system, gauges must be connected to the system. The connection must be made without allowing air, moisture, or dirt to enter. The procedure for connecting gauges to a system depends on the system design. It is different for each system, as shown in **Figure 12-32.**

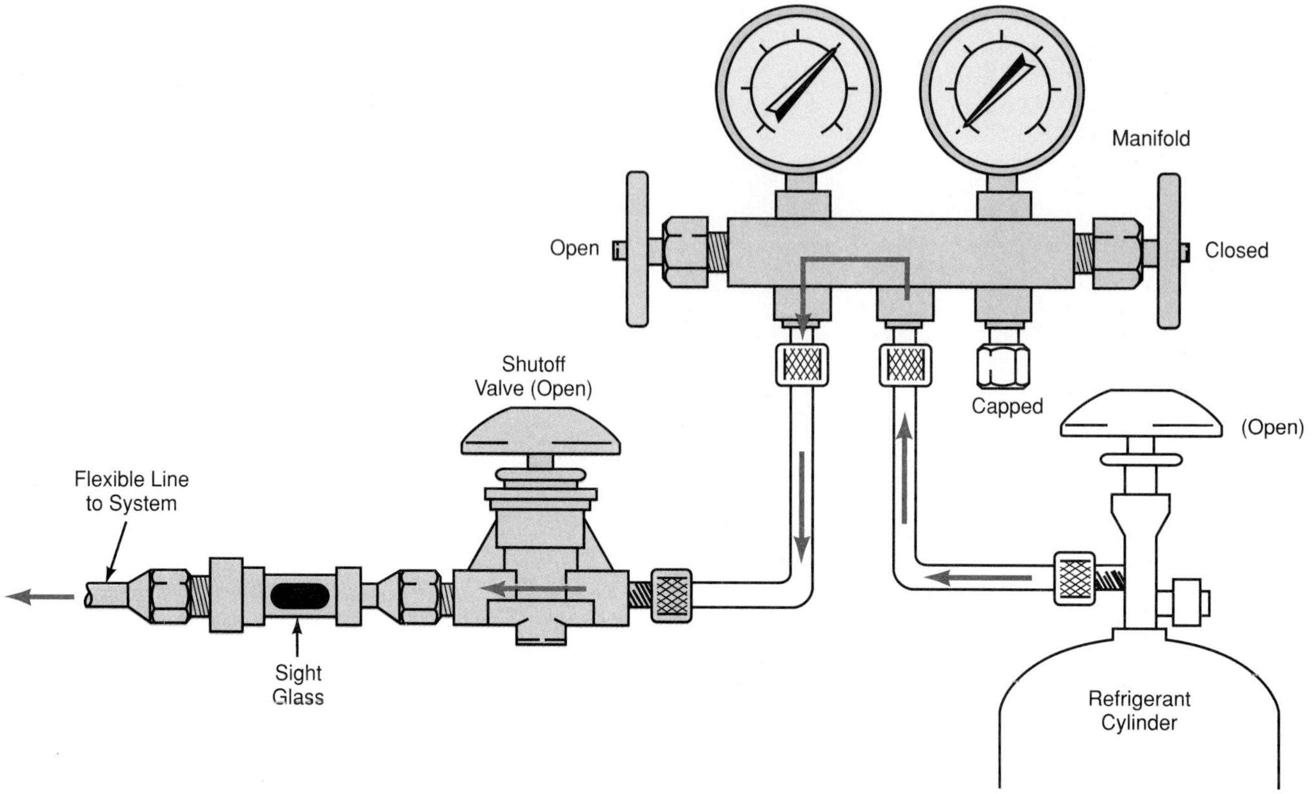

Figure 12-29. *Gauge manifold fitted with flexible refrigerant tubing. This installation is used when charging a system.*

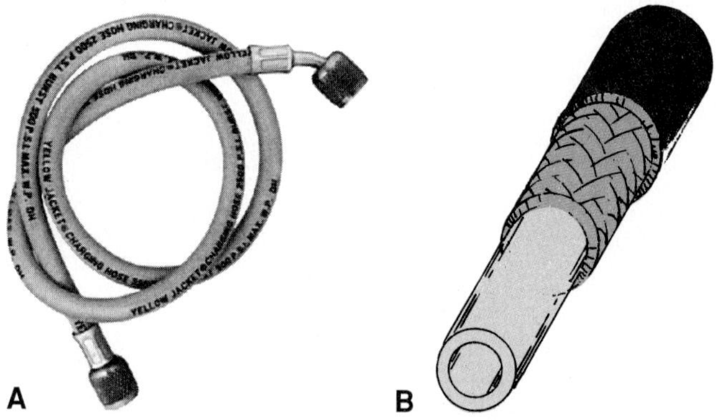

Figure 12-30. *Flexible service and refrigerant line. A—Flexible charging hose with external flare connection. B—Cutaway showing wall construction of flexible hose. (Ritchie Engineering Company, Inc.)*

- Some systems have both a suction service valve and a discharge service valve.
- Some have a suction service valve adaptor mounted on the compressor.
- Some do not have any service valves, but *do* have a process tube.
- Some have a process tube that is too short or inaccessible. In such systems, a piercing valve is used. It is attached to either the liquid line, the suction line, or both.

The system in **Figure 12-32A** (two service valves) is the easiest for attaching gauges. It also permits checking both the low-side pressure and the high-side pressure. This system is most common on commercial systems. Its installation and use is described in Chapter 15.

The system in **Figure 12-32B** (valve adaptor) is described in Section 12.8.1. The system in **Figure 12-32C** (process tube) is described in Section 12.8.2. The piercing valve **(Figure 12-32D)** is described in Section 12.8.3.

When connecting refrigerant lines or gauge manifolds to any refrigerating system, keep the system clean. The lines, gauges, and manifold must be free of dirt, moisture, and air. The manifold should be purged with the same refrigerant used in the system. The manifold and connecting lines must be purged before the system service valve is opened. They must also be purged before using a piercing valve stem to open the tubing.

Figure 12-33 illustrates the most popular way to purge the service lines. Loosen the line fitting on the system service valve at C, then open valve B. Open cylinder line valve E just a little. Repeat the same procedure for valves D, A, and E. The cylinder refrigerant will purge all the lines and the manifold of air and moisture.

Usually only one connection is made to the system. This connection is to the low or *suction* side, at valve C. The flexible line between B and C is connected to the system valve, C. However, the use of the gauge manifold allows checking both low-side pressure and the high-side pressure. (For the low-side pressures, valve C is open and valve B is closed. For high-side pressures, valve D is open and valve A is closed.)

The manifold also allows you to charge a system. Valves C and B are open. The cylinder valve, E, is opened

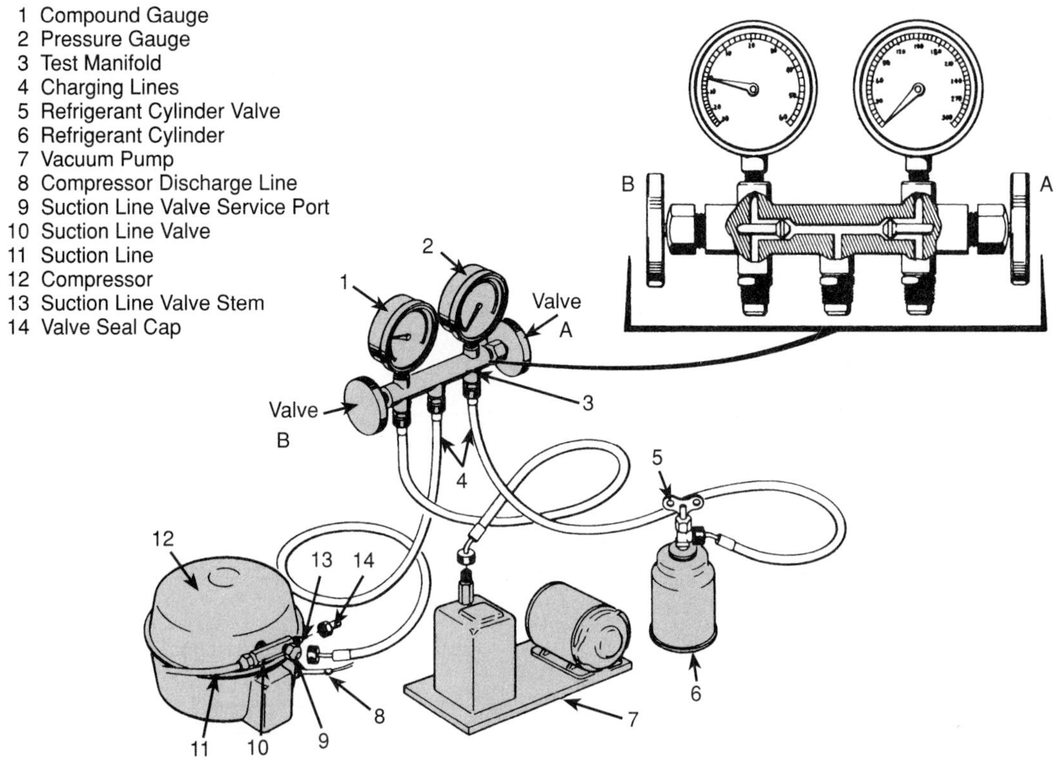

Figure 12-31. *A drawing of a system being evacuated. The illustration also indicates another means of utilizing a manifold gauge. The vacuum pump is attached to the suction service valve line (B). A vacuum will be created through the center service line, which is connected to the suction service valve. Once the vacuum is obtained, the small service cylinder charges the system up to 0 psig or slightly above. The compressor can then be removed from system and the rest of system sealed off to avoid air entering into it.*

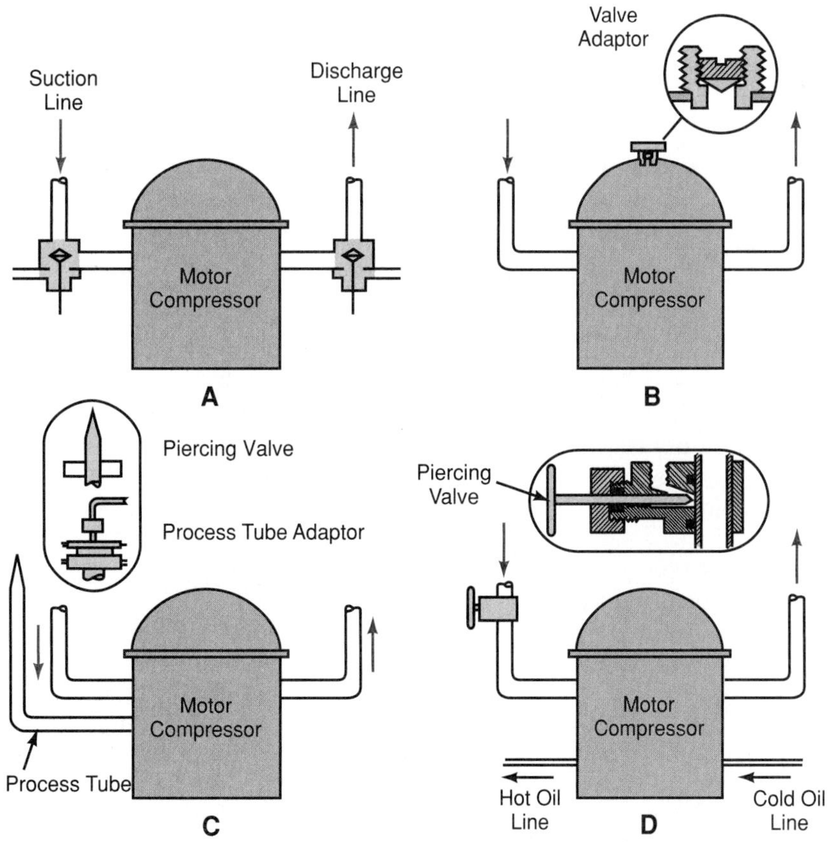

Figure 12-32. *Four different methods for connecting a gauge manifold to a hermetic system.*

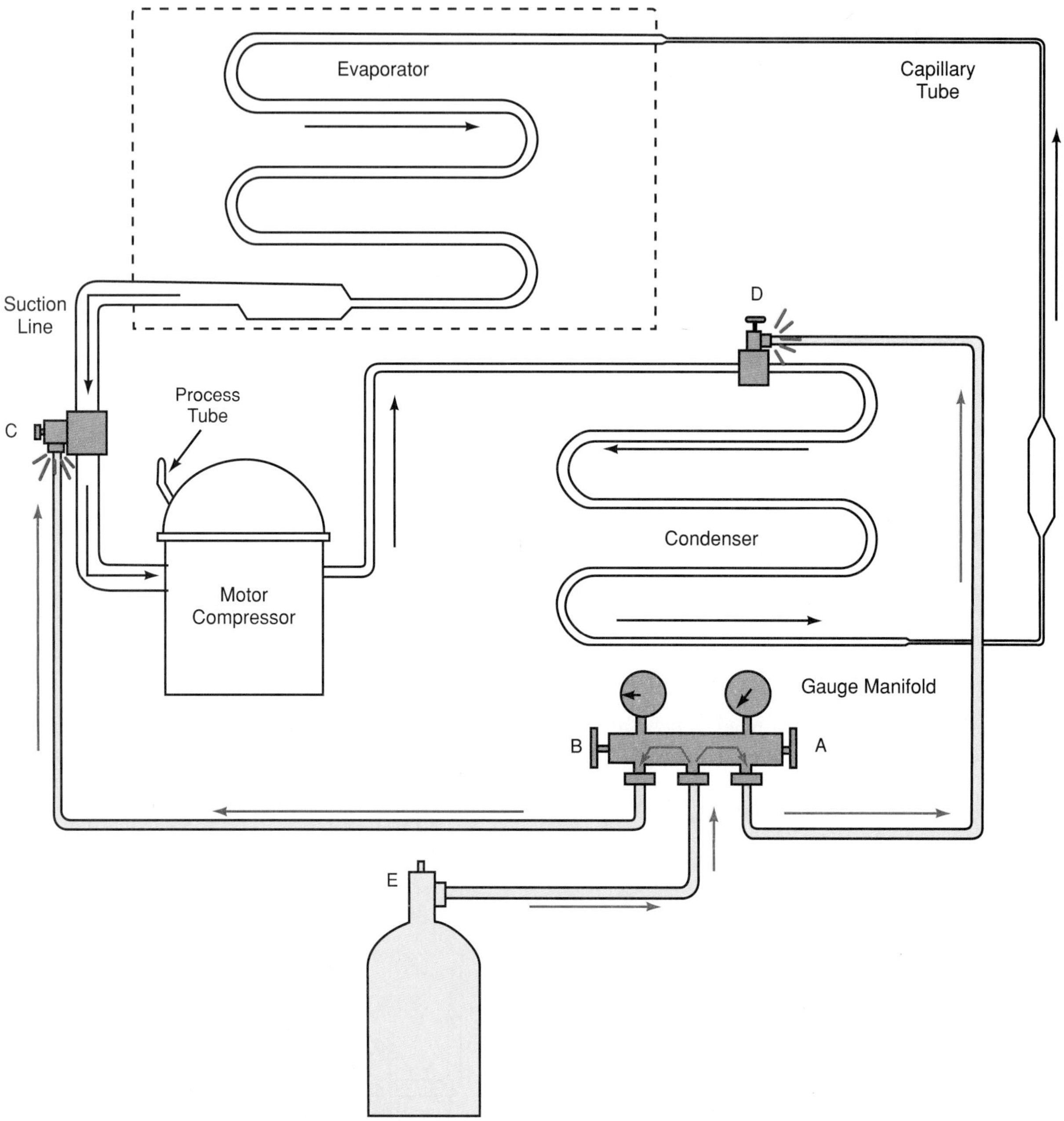

Figure 12-33. *Valves on gauge manifold are opened to purge service lines. Fittings at C and D are loosened to allow air in service lines to leak out.*

slowly. The manifold can also be used to evacuate the system. A vacuum pump line is connected to the middle connection of the gauge manifold. Valve C is open and valve B is opened.

When checking high-side pressure, use a piercing valve if the system is already charged. If system has just been assembled and not charged, braze a process tube into the condenser line.

If the unit will run, operate the system after installing the gauge manifold. The system should be operated through at least three cycles. Carefully record the suction pressures, condensing pressures, evaporator temperature, and the condenser temperature. It is helpful to record a table similar to **Figure 12-14** and/or **Figure 12-15.** The data may be used for future reference.

12.8 Hermetic Service Valves and Adaptors

Most hermetic refrigerators do not have service valves. Some have fittings to which valves may be attached for service operations. The valves are removed when the service work has been completed.

Others have neither service valves nor provision for fitting valves to them. For such units, it is necessary to fit and attach valves to the mechanism. Attachments of various types are available from refrigeration supply wholesalers.

Some hermetic mechanisms have a process tube. **Figure 12-34** shows one being used. Note the hand valves and process tube adaptor fitting. Note also that a charging cylinder is used.

Service valves mounted on a hermetic system may be used for many purposes:

- To check the internal pressures.
- To discharge the system or add refrigerant.
- To add oil.
- To evacuate the system.
- To make it easier to replace driers, motor compressors, evaporators, and refrigerant controls.
- To recharge the system.

Usually, a flexible charging line is connected to the service valve adaptor. It is also attached to either a hand valve or a service manifold mounted on the other end of this tubing. This makes service easier. **Figure 12-35** shows this type of service connection set up for charging a hermetic system.

The valve should be loose at the attachment point (the suction line in this illustration). Use vapor from the cylinder to blow out (purge) the lines. The gauge may be located on the compressor dome, suction line, or process tube.

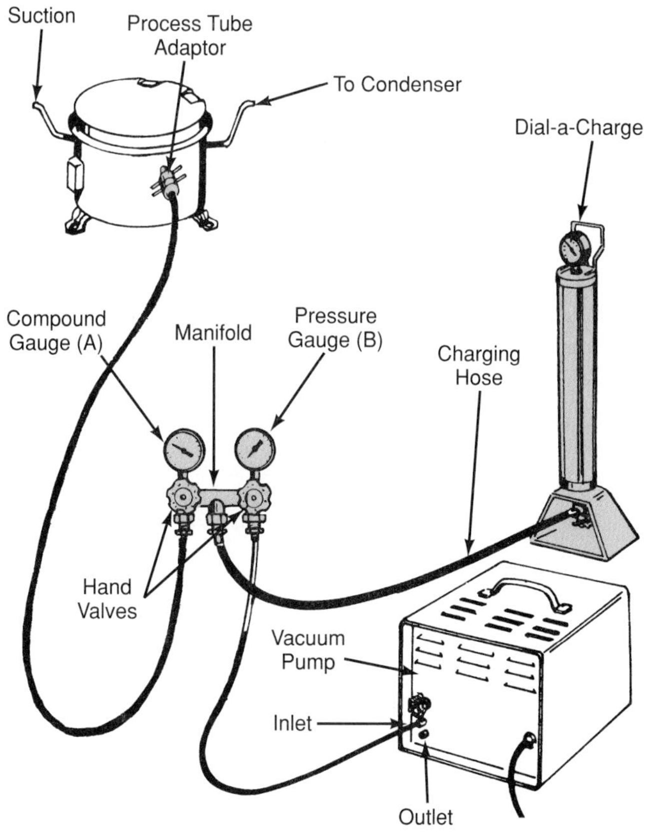

Figure 12-34. *System being charged with process tube adaptor. Note use of a charging cylinder. (Frigidaire Company)*

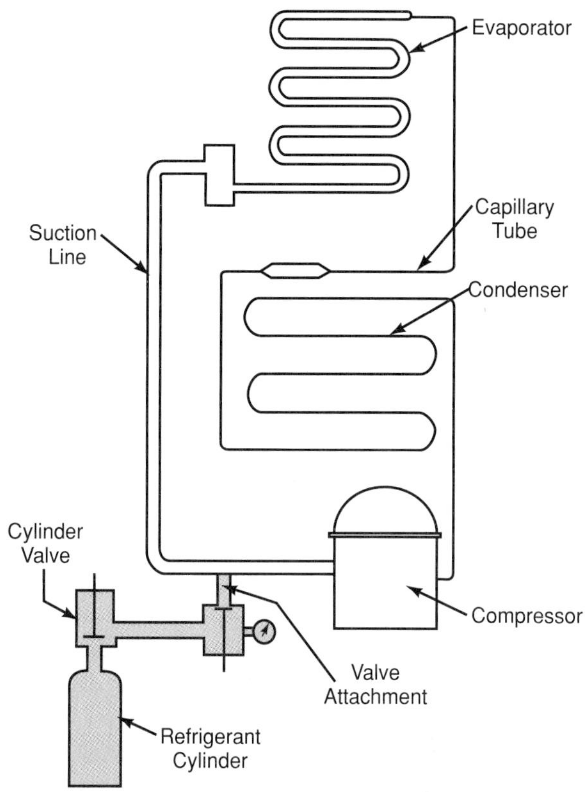

Figure 12-35. *Charging a hermetic system with a flexible charging line attached to a service valve adaptor. Service valve adaptors may be attached to the suction line, process tube, or compressor dome.*

12.8.1 Systems with Valve Adaptors

Valve adaptors are one way of connecting gauges and charging cylinders to a hermetic system. **Figure 12-36** shows the part of the adaptor which is fastened to the compressor dome. The adaptor has a removable service valve, as shown in **Figure 12-37.** The valve and the adaptor connect together as shown in **Figure 12-38.** A service valve with two openings is shown in **Figure 12-39.**

The adaptor provides a means of operating the small needle valve mounted on the motor compressor. It also provides an opening for a service gauge or a gauge manifold connection. Synthetic or copper gaskets are used to seal the valve joints. An assortment of valve adaptors is shown in **Figure 12-40.**

The following procedure should be followed when using valve adaptors:

1. Clean the outside.
2. Remove the dust cap from the adaptor mounted on the motor compressor.
3. Choose the correct valve stem drive.
4. Push the service valve stem forward in the body of the valve attachment.
5. Engage the valve stem in the valve adaptor needle.
6. Thread the valve adaptor unit into the attachment body.
7. Use good gaskets.
8. Before opening the valve adaptor needle, tighten the packing unit around the valve stem.

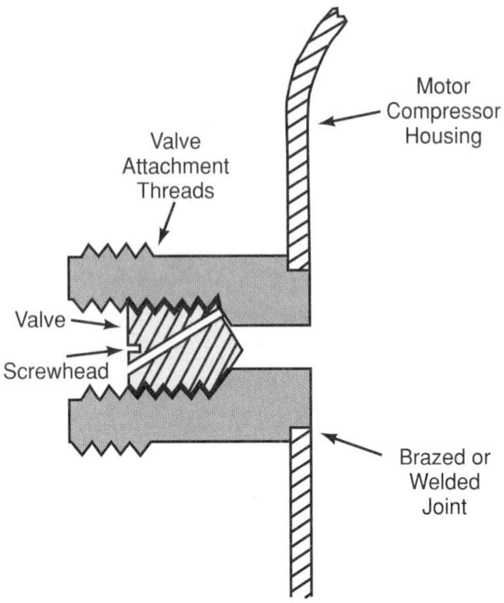

Figure 12-36. *Service valve assembly for hermetic unit. Valve attachment must be fastened in place before valve can be opened. If it is not, refrigerant will escape.*

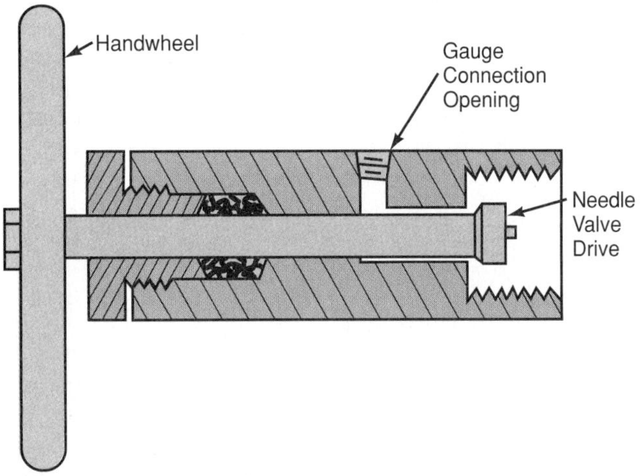

Figure 12-37. *Service valve attachment is installed on valve adaptor, which is fastened to the motor compressor dome.*

Bleed the passages (valves, gauges, and flexible lines). Purge the assembly using the same refrigerant as in the system. Leave the flexible line fitting loose at the valve attachment. After purging, tighten the loose connection. Always test the assembly for leaks using a refrigerant pressure of 15 psig to 20 psig (30 psia to 35 psia or 207 kPa to 242 kPa).

12.8.2 Process Tube and Adaptors

The process tube can be adapted for service of systems by installing a piercing valve on the process tube. Piercing valves are discussed in Section 12.8.3.

The manufacturer uses a process tube to evacuate, test, and charge the new unit. The tube left in the system may be used by a service technician. An extension

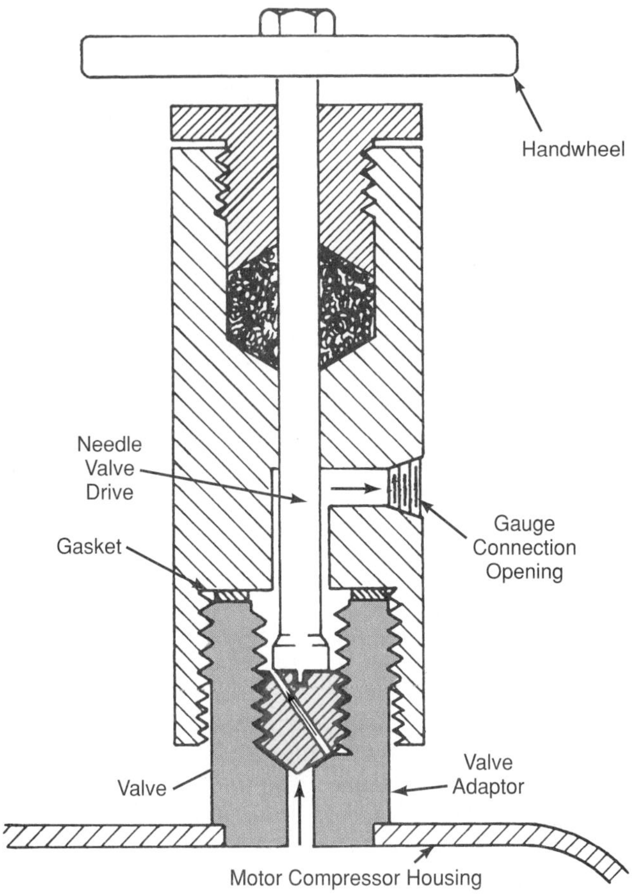

Figure 12-38. *Valve attachment as it would appear in cutaway when connected to a valve adaptor on the motor compressor housing.*

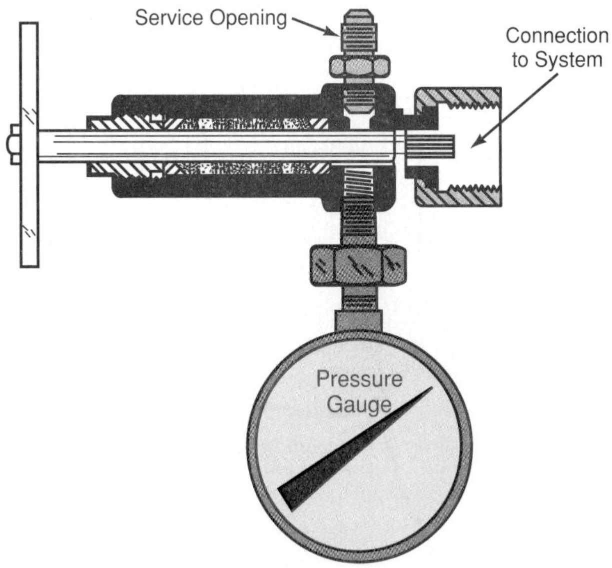

Figure 12-39. *Service valve attachment. Note that there are two openings. One may be used for the pressure gauge and the other for performing service operations such as discharging, charging, and adding oil. (Fedders North America)*

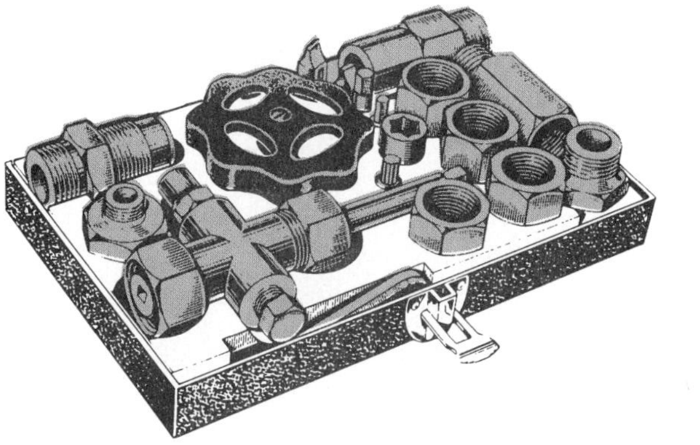

Figure 12-40. *Valve kit and adaptors that can be used on various makes of semihermetic or hermetic refrigeration units.*

may be brazed to it or a process tube adaptor mounted on it. This becomes a means of attaching a manifold. An adaptor kit is shown in **Figure 12-41.** The adaptor enables use of the process tube without soldering an extension or flaring the tubing. It provides a positive seal. See **Figure 12-42.** The adaptors are of various sizes. The tool may be used on 3/16″, 1/4″, 5/16″, or 3/8″ copper tubing.

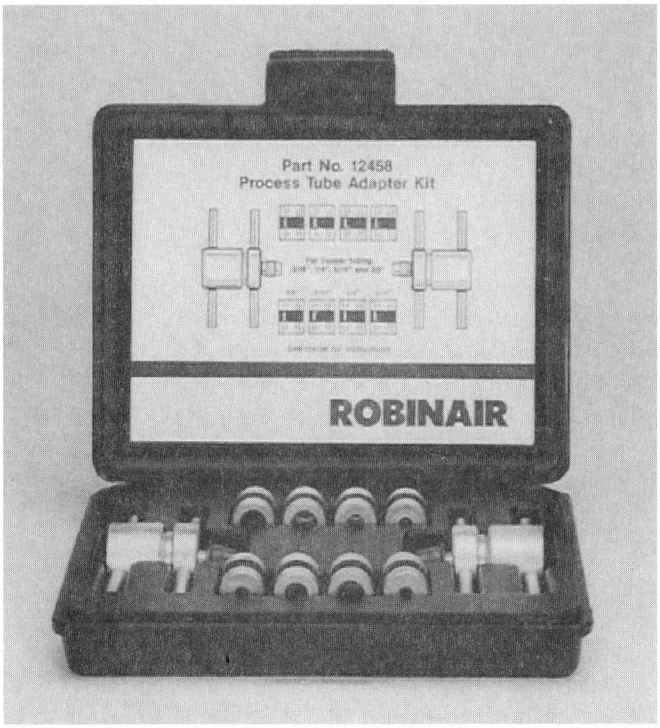

Figure 12-41. *Process tube adaptor and kit. (Robinair Division, SPX Corporation)*

Pinch-Off Tool

The pinch-off tool is used wherever it is necessary to seal off soft copper tubing (up to 3/8″ OD). **Figure 12-43** shows one type of pinch-off tool. It has a screw-type action shaft with a ball bearing on the end that presses against the tube. The tool is placed over the copper tubing in the same manner as a tubing cutter. The tubing should be slowly compressed by turning the pinch-off tool handle clockwise.

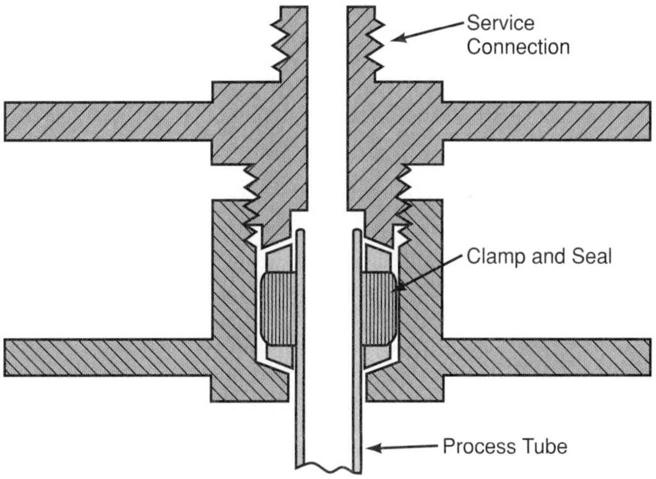

Figure 12-42. *Process tube adaptor.*

Figure 12-43. *Pinch-off tool is usable on tubing up to 3/8″ OD. This makes good seal and also keeps tubing strong at the pinch-off point. Tubing at A is shown before pinching; tubing at B has been pinched with the tool.*

As the handle is turned, the ball bearing presses into the tubing. It compresses the tubing against the die on the bottom of the tool. A permanently pinched line is produced. See **Figure 12-44.** Take care that the pinch-off tool is not rotated too far or excessive pressure applied. Leave the tool in place until the adaptor is removed and the tubing end is sealed by brazing. The pinch-off tool may be used when an emergency arises that requires isolation of parts. This may occur in situations such as a bad leak. Pinching lines is a practice to be used in cases of emergency only. Some technicians follow this practice needlessly and it leads to future trouble.

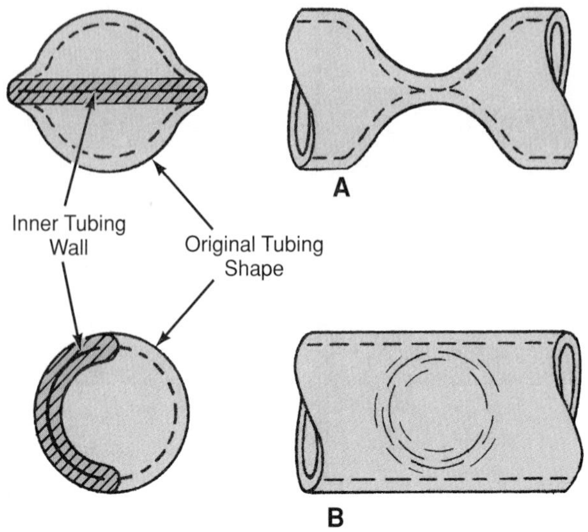

Figure 12-44. *Action of pinch-off tools. A—Pinch-off made with pliers-type tool. B—Pinch-off made by tool shown in* ***Figure 12-43.***

12.8.3 Piercing Valves

A popular way to gain access to a hermetic system is to mount service piercing valves. These may be mounted on the suction tubing, discharge tubing (tubing to condenser), or both. The piercing valve may also be mounted on the process tube. A piercing valve is shown in **Figure 12-45.** Many designs of tubing-mounted piercing valves have been developed. However, there are two general types: bolted-on and brazed-on.

Figure 12-46 shows cross sections of two types of bolted-on piercing valves. These valves are available in several sizes for various sizes of tubing.

Tubing should be straight and round. It should be carefully cleaned. (Do not scratch the tubing while cleaning.) Make sure there are no dents in it. Check to see if there is enough space to operate the attachment valve. Check also to determine if connecting tubing can be easily mounted on the attachment valve.

Figure 12-45. *Bolted-on piercing valve installed on the high side of a system as a permanent valve. (Sealed Unit Parts Co., Inc.)*

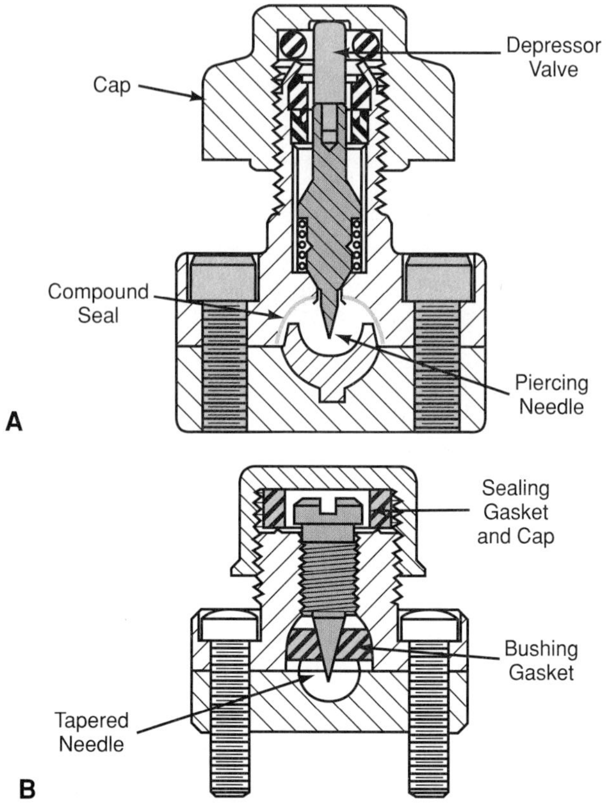

Figure 12-46. *Cross sections of two bolted-on piercing valves. A—Valve is bolted to line by two socket head screws. Note the use of the special compound seal. B—Bolted-on tubing-mounted service valve. A gasket seals the hole made by the piercing valve. (Watsco Components, Inc.)*

Put a little clean refrigerant oil on the tubing. Be sure that the synthetic sealing washer is in place and that the needle-point piercing valve stem is all the way out. Mount the valve on the tubing. Tighten the unit clamping screws evenly. These valves are usually left in place on the system. The attachment valve design and construction is similar to those shown in **Figures 12-38** and **12-39.** Two types of service valve attachments are shown in **Figure 12-47.**

The second type of piercing valve is brazed-on. See **Figure 12-48.** The braze-mounted type (saddle design) is safe to use. Neither the suction tubing nor the condenser tubing have liquid in them. Therefore, they may be heated to a brazing temperature. However, make sure there are no flammables or soft-soldered joints close to the brazing.

Be sure the tubing is straight and round at the brazing point. Clean both the saddle and tubing mating surface with clean sandpaper or clean steel wool. Remove the piercing valve stem and the gasket from the saddle. Put clean, fresh brazing flux on the saddle (outer edges). Or, use a phosphorous-copper brazing filler rod.

If flux is used, follow manufacturer's specifications. If phosphorous-copper brazing filler rod is used, how-

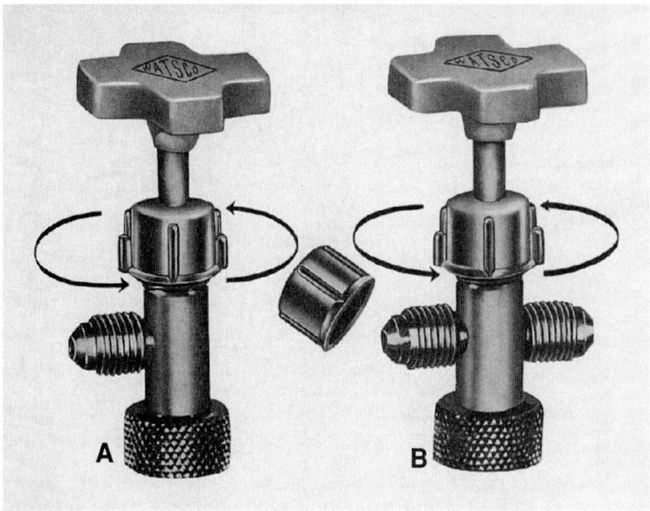

Figure 12-47. *Two types of service valve attachments used with tubing-mounted piercing valve adaptors. A—Valve with one 1/4" male flare service opening. B—Valve with two 1/4" male flare service openings. (Watsco Components, Inc.)*

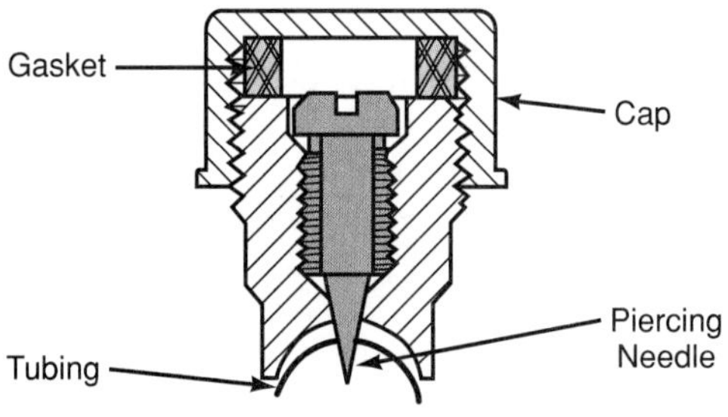

Figure 12-48. *Braze-mounted tubing piercing valve. Note use of sealing gasket. A preformed brazing ring is usually used for brazing alignment and proper metal flow. (Watsco Components, Inc.)*

ever, flux is unnecessary. This is because the phosphorous in the brazing material deoxidizes the copper surface.

Mount the saddle on the tubing. Determine if there is room (clearance) for mounting the service valve on the tubing mounting valve. Heat both the tubing and the saddle until the filler rod material flows around the saddle.

The saddle must not move or shift during the brazing or while the brazed joint is cooling. Some technicians hold the saddle in place with a small C-clamp during the brazing operation.

Do not overheat the tubing. It may be weakened to the point of failure and burst. Wear goggles during the brazing operation.

Inspect the brazed joint carefully. Use a mirror to check hard-to-see edges. After the brazed joint has cooled, install the piercing needle and gasket. The unit is then ready for the installation of the service valve

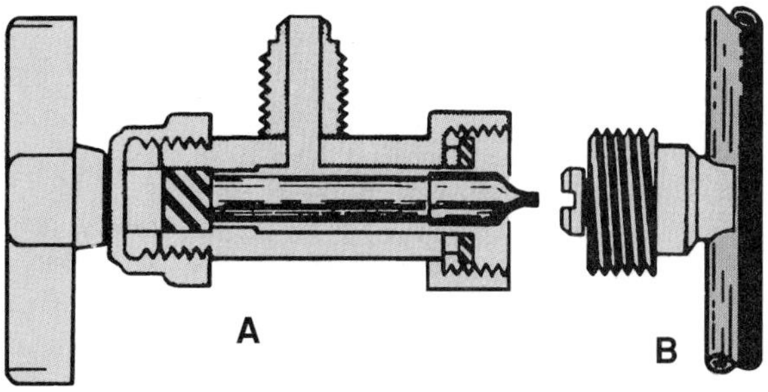

Figure 12-49. *This piercing valve, brazed onto line, may be used on hermetic refrigerator systems. Part A can be removed after servicing to discourage tampering with system. Part B remains on system. Threaded cap on the piercing valve is used to protect threads and prevent tampering.*

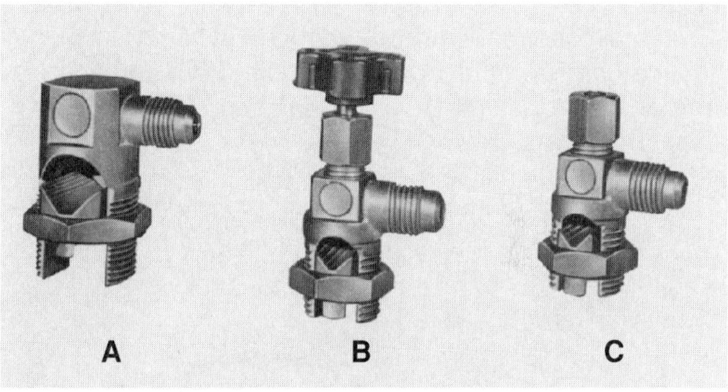

Figure 12-50. *Three types of piercing valves. A—Charge-and-tap valve. B—Hand-valve type. C—Line-tap type with hexagonal wrench. (Robinair Division, SPX Corporation)*

attachment. See **Figure 12-49.** Three gasket-type piercing valves are shown in **Figure 12-50.**

12.8.4 Core Valves

Many systems use a Schrader core valve to gain access to a hermetic system. See **Figure 12-51.** This type is similar to the valve cores used in automobile tires.

A clamp-on core valve adaptor is shown in **Figure 12-52.** The flexible service tubing fitting or the service valve mounted on this fitting has a pin. The pin depresses the core valve stem as the fitting or service device is mounted. Some valve adaptors are threaded to the system. Others are brazed or clamped to the tubing.

Some technicians use a service valve attachment that mounts on the Schrader valve adaptor. This device has a long stem to remove the valve core while evacuating the system. The core is loosened to allow more flow of vapor when drawing a vacuum. Vacuum lines and fittings should be as large as possible. **Figure 12-53** shows the advantage of removing the valve core while evacuating.

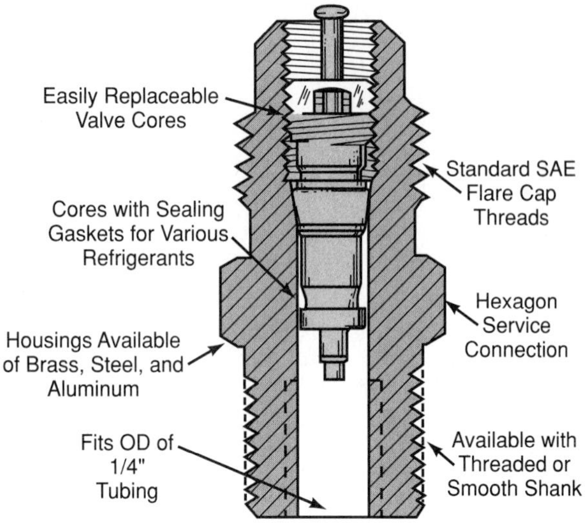

Figure 12-51. *This Schrader valve fitting may be used to connect service lines to a hermetic system. When service line is mounted on this fitting, a pin depresses (forces inward) the stem of the valve core. This opens the system for service.*

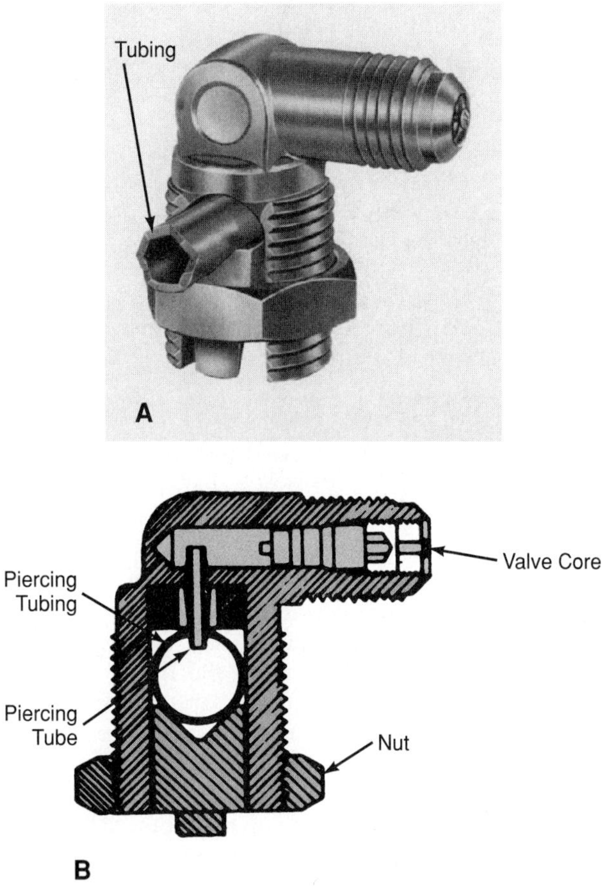

Figure 12-52. *This valve-core access valve adaptor clamps onto tubing. A—Piercing valve mounted on tubing. B—Cross section of same adaptor. Passage is opened when valve core stem is depressed by service line fitting. (Robinair Division, SPX Corporation)*

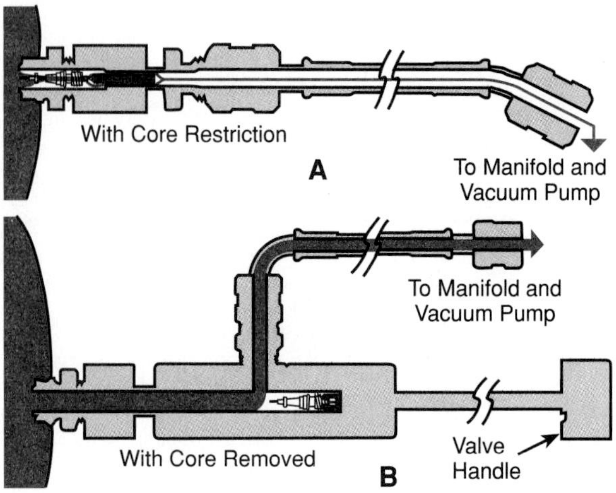

Figure 12-53. *Evacuating passages are larger when the valve core is removed from its fitting. A—Small flow with core in place. B—Large flow with core removed.*

12.9 Locating Refrigerant Leaks

Methods of testing for leaks vary with the refrigerant used. However, all methods have one procedure in common: applying pressure to the system with an inert gas, such as nitrogen or carbon dioxide. At the start of testing, a ***positive pressure*** (greater than atmospheric pressure) of 5 psig to 30 psig (20 psia to 45 psia or 138 kPa to 310 kPa) is necessary throughout the circuit. If no leaks are found, test again at or above the normal condensing pressure for the refrigerant used. For example, using R-12, this pressure would be 90 psig to 135 psig (105 psia to 150 psia or 725 kPa to 1035 kPa).

Check for leaks before the unit is evacuated. Moisture could enter the system through a leak during evacuation or pump-down. Always use the proper recycle/recover equipment when locating and repairing leaks.

Many companies recommend using the refrigerant in the system to test for leaks. A sensitive leak detector should be employed. An electronic model that can be used for R-134a and R-12 is shown in **Figure 12-54.** If a leak is found and repaired, the complete unit must be rechecked. This provides a check for the repair and will reveal any additional leaks.

12.9.1 Pressure-Testing for Leaks

With proper care, nitrogen or carbon dioxide may be used safely when pressure-testing for leaks. The pressure in the nitrogen cylinder is about 2000 psig (14 MPa). In a carbon dioxide cylinder, it is about 800 psig (6 MPa). **A pressure reducing device that has both a pressure regulator and a pressure relief valve must always be used when testing with either of these two gases.** A recommended pressure regulating device is shown in **Figure 12-55.**

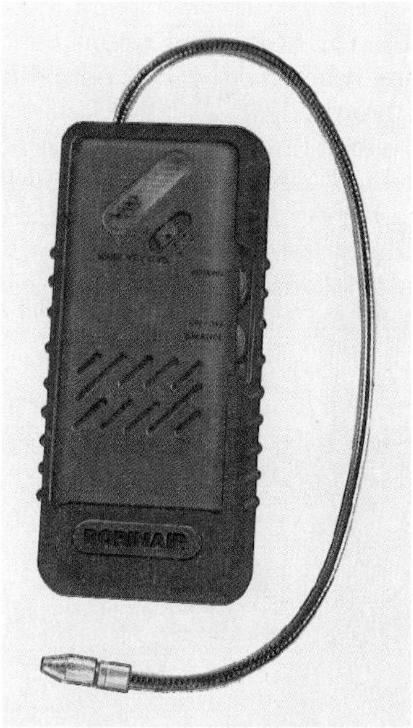

Figure 12-54. *An electronic leak detector that can be used for R-12 or R-134a. The unit is capable of detecting leaks smaller than 1/2 oz. per year. (Robinair Division, SPX Corporation)*

A refrigerating system can explode if pressure is allowed to build up in the system. Many accidents have been caused by using too much testing pressure.

Before using nitrogen or carbon dioxide to test a system, look at the system nameplate. In most cases, it will give recommended testing pressures. **If these pressures are not known, never test all or part of a hermetic system at a pressure over 170 psig (185 psia or 1280 kPa).** See Chapter 15 for information on pressure-testing commercial systems.

Caution: Never use oxygen or acetylene to develop pressure when checking for leaks. Oxygen will cause an explosion in the presence of oil. Acetylene will decompose and explode if it is pressurized over 15 psig to 30 psig (30 psia to 45 psia or 210 kPa to 310 kPa).

12.9.2 Leak Detecting Devices

Refrigerant system leaks are usually very tiny and require sensitive detecting devices. Commonly used devices include bubble solutions, fluorescent dyes, refrigerant dyes, halide torch, and electronic detection. New detectors may be used on R-134a, R-123, and the other new alternative refrigerants. Each method has its advantages. The various methods are reviewed in the following paragraphs.

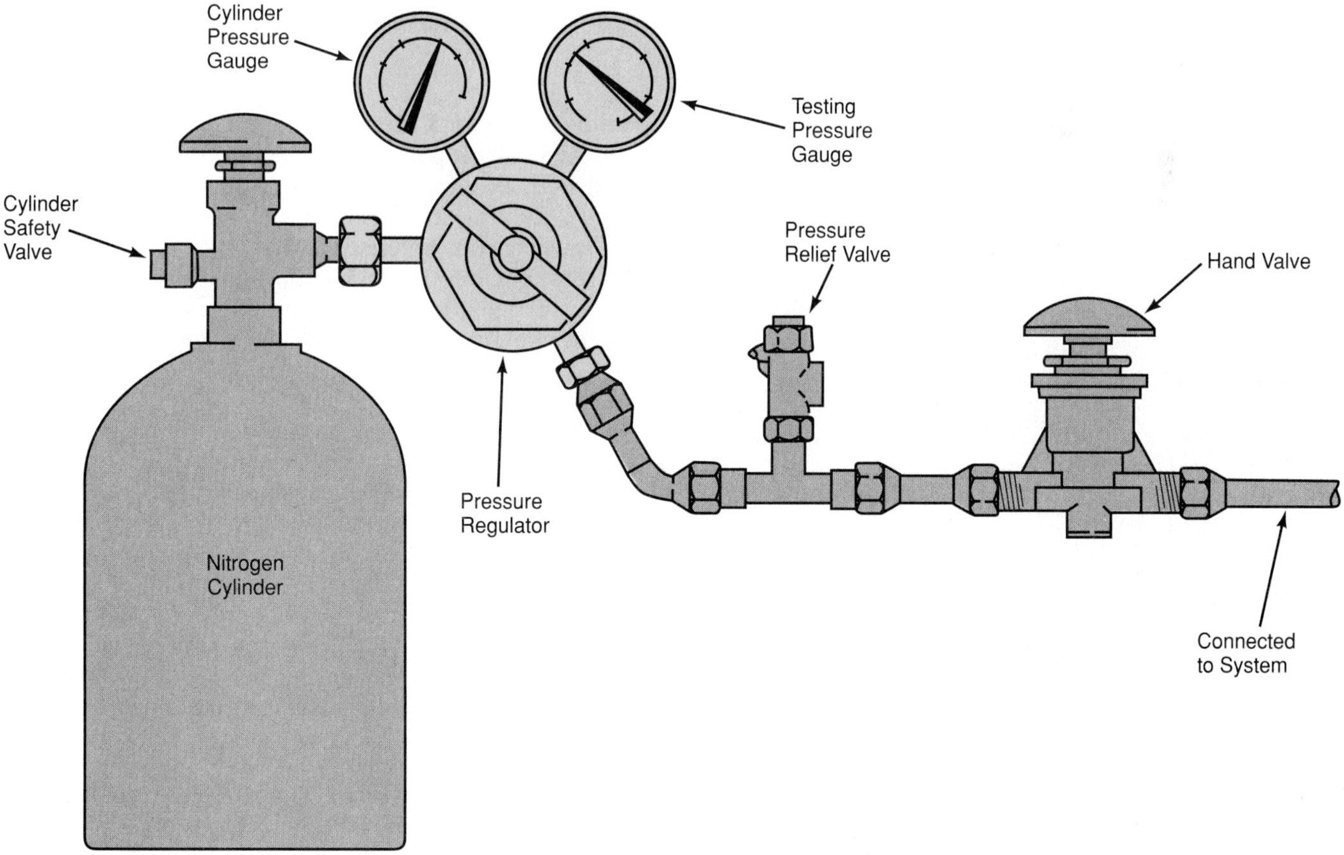

Figure 12-55. *Pressure regulator system. Note that both a regulator and pressure relief valve are used.*

Bubble Solutions

The bubble method of leak detection, employing a water-soap solution, is commonly used. This solution is brushed over an area suspected of leaking. Gas coming through the solution will cause bubbles.

Patented solutions provide a stronger, longer-lasting bubble film than the soap solution. For this reason, they are more popular than soap. **Figure 12-56** illustrates the action of one of these solutions in the presence of a leak. The bubble solution should be wiped off the tubing or fitting after checking for leaks.

The bubble method has certain advantages, compared to the use of instruments. These include its ease of use, low cost, and ease of application. A disadvantage is that larger leaks will blow through the solution and no bubbles will appear.

The halide torch and electronic leak detector are difficult to use around urethane insulation. Urethane uses refrigerant as the expander. Therefore, such detection devices show a leak trace all the time. In such cases, the bubble test is best.

Refrigerant Dye and Fluorescent Leak Detecting

Refrigerant dye is another tool used for locating leakage problems. Refrigerant dye in a system produces a bright red color at the point of leakage. Most leaks show up in a short time. However, a period of up to 24 hours may be necessary in some cases.

In most systems, the entire refrigerant charge must be replaced with refrigerant containing the dye. This is necessary in order to achieve maximum leak detection. The dye method is dependent upon the oil circulation rate. Therefore, it may take a long period of time (up to 24 hours) to indicate leaks.

The ultraviolet fluorescent leak detection procedure is another method used. A fluorescent additive is circulated through the system. The refrigerant leak is found by scanning the system with an ultraviolet light, **Figure 12-57.** This method may be used with a variety of refrigerants, including R-134a.

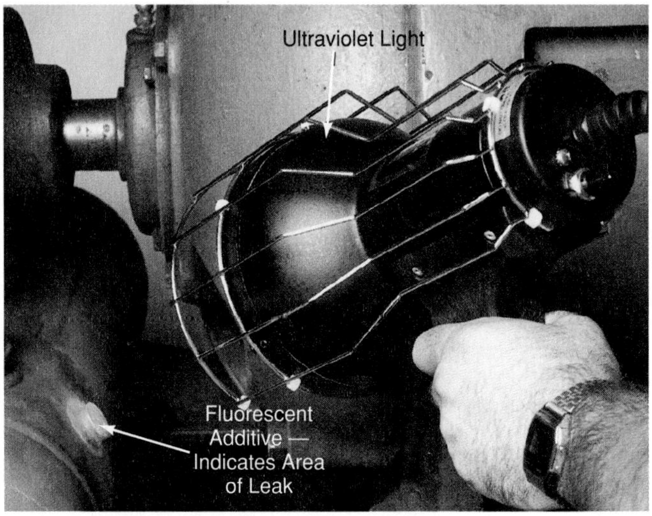

Figure 12-57. *Fluorescent leak detection system in use. (Spectronics Corporation)*

Figure 12-56. *Bubble leak test being made on an evaporative coil. Test solution is placed on the connection. Bubbles will indicate a refrigerant leakage. (Refrigeration Technologies)*

Halide Torch Leak Detector

Alcohol, propane, acetylene, and most other torches burn with an almost colorless flame. A flame will continue to be almost colorless if a copper strip is placed in it. However, the tiniest quantity of a halogen refrigerant, brought into contact with the heated copper, will cause the flame to change to a light green color. (Halogen refrigerants include R-12, R-22, R-11, R-500, R-502, etc.) This principle is used in halide torches to detect leaks in refrigeration systems.

A halide torch leak detector is shown in **Figure 12-58.** The torch burner is shown at the top. One end of a rubber tube is connected to the base of the burner. The other end is free to be moved about to various parts of the system. The rubber tube will draw air from its open end into the burner.

As the open end of this tube nears a leaking connection, it draws up some of the leaking refrigerant vapor. As the vapor contacts the burner, the flame color immediately becomes green, indicating a leak.

Electronic Leak Detector

Three commonly used types of electronic leak detectors are electrochemical sensor, ultrasonic, and dielectric. The electrochemical sensor consists of a ceramic layer covered by a reactive element maintained at high

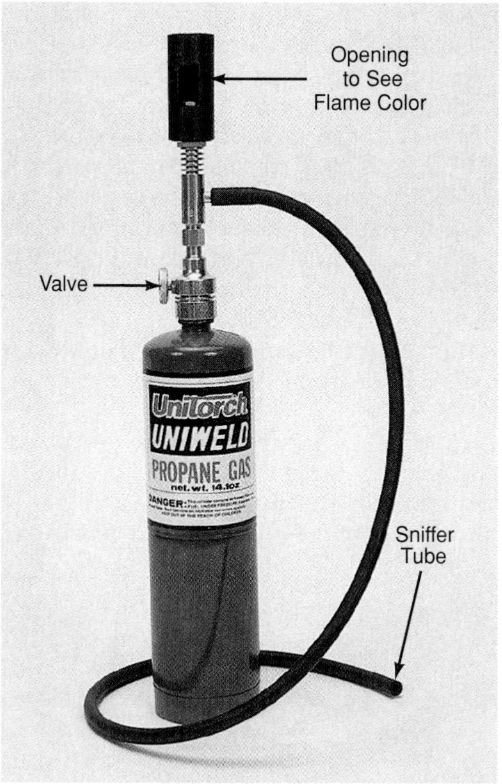

Figure 12-58. *Halide torch used to test for leaks. Green flame showing in burner opening will indicate leak at the sniffer tube opening. (Uniweld Products, Inc.)*

temperature by a built-in heating element. See **Figure 12-59.** Contact with a halogen-bearing gas causes an electrical current to flow to a collection electrode. It is not necessary to reset the detector for different refrigerants. These sensors provide similar responses to CFCs, HCFCs, HFCs, and refrigerant replacement blends. Therefore, the operator need not determine the refrigerant in use. The dielectric leak detector operating principle is based on the different heat conductivity of different gases. Some detectors are based on the dielectric difference of gases. The gases are run between the plates of a capacitor. The gases act as the dielectric (insulator) for each capacitor. Different frequencies of an oscillator indicate a leak.

In operation, the detector is turned on and adjusted in a normal atmosphere. The leak-detecting probe is then passed over surfaces suspected of leaking. If there is even a tiny leak, the refrigerant is drawn into the probe. The new vapor changes the resistance in the circuit. The detector will emit a piercing sound, or a light will flash, or both.

The electronic detector is probably the most sensitive of any of the leak-detecting devices. It detects all halogenated refrigerants except R-14. **The electronic leak detector should not be used in areas containing explosive or flammable vapors.** It uses transistorized circuitry powered by batteries. The plastic tip guard should be used only in situations that might contaminate the sensing tip.

Figure 12-59. *A service technician using an electronic leak detector to check for refrigerant leaks in a rooftop central air conditioning unit. The leak detector has a heated diode sensor tip that automatically adjusts to any refrigerant, and is thus equally sensitive to R-12, R-134a, R-11, or any other refrigerant material. (Leybold Inficon, Inc.)*

When using an electronic leak detector, minimize drafts. Shut off fans or other devices that cause air movement. Always position the sniffer below the suspected leak. Since it is heavier than air, refrigerant drifts downward.

Move the tip slowly, at a rate of about one inch per second. (This can be measured by moving tip an inch after each "one thousand" verbal count.) A tip adjusted in ambient air will only buzz. The instrument will squeal when tip sniffs refrigerant. Remove the plastic tip and clean it before each use. Avoid clogging with dirt and lint.

Units are available that are sensitive to many types of refrigerants including: R-134a, R-12, R-22, and R-500.

Ultrasonic leak detectors have also become widely used in the industry. These units use headphones and a portable, hand-held detector. Ultrasonic frequencies are sound waves that are beyond the range of human hearing. Ultrasonic leak detectors detect the sound that a vapor makes as it is escaping from a pressurized system.

12.10 Repairing Leaks

To repair a leak, remove and recover the refrigerant from that part of the system. (In some cases, you will have to empty the complete system.) Check the pressure to be sure it is 0 psi (neither pressure nor vacuum in the system).

If possible, avoid soldering or brazing a system with refrigerant in it. Heat may cause a breakdown of the refrigerant. A method of soldering or brazing parts to a system involves the use of nitrogen. See **Figure 12-60.** The nitrogen cylinder is connected to the process tube. The system is pressurized and checked for leaks. When the leak is found, you can braze the area. The system should be rechecked for leaks before removing the nitrogen cylinder.

Refrigerant systems are made of copper, steel, and/or aluminum materials. Leaks may start in any part of the system. The repair depends on the material that has failed or on the combination of materials at the leak.

To find out what metal is used, scrape the surface. Steel is gray-white. It is hard and magnetic. (Use a small magnet to test.) Copper is reddish in color when scraped and is nonmagnetic. Aluminum is white, soft, and nonmagnetic. Steel and copper may be brazed; aluminum may be aluminum-soldered or brazed. Aluminum may be resistance-welded to steel or copper. It may also be repaired with epoxy cement.

Leaks most often are found at tubing connections. If they occur at a flared connection, the following are possible causes:

- The tube flare is not correct.
- The flare nut has not been tightened securely.
- The threads are stripped.

It is best to replace a leaking fitting by making a new flare. Use new flared fittings. Leakage at a brazed or silver-soldered connection can be repaired by cleaning, coating with flux, and reheating. Steel tubing usually has a lengthwise seam. This seam must be clean for brazing. Clean by wire-brushing lengthwise or file off enough metal to remove the seam depression.

If the fitting has been taken apart, reflux and assemble. Heat the connection and solder or braze it in place. *Avoid overheating other parts of the system. Never heat a drier. Moisture will be driven out into the system.* It is better to cut tubing with pliers or tube cutter.

Use a fire-resistant sheet material as a protective barrier between flammable surfaces and open flame. See **Figure 12-61.** This type material is also used when the tubing is next to a metal side.

Aluminum evaporators can be repaired with epoxy cement. Follow instructions supplied by the manufacturer.

Check for leaks before the unit is evacuated. Moisture could enter the system through a leak during evacuation or pump-down. Always use the proper recycle/recover equipment when locating and repairing leaks.

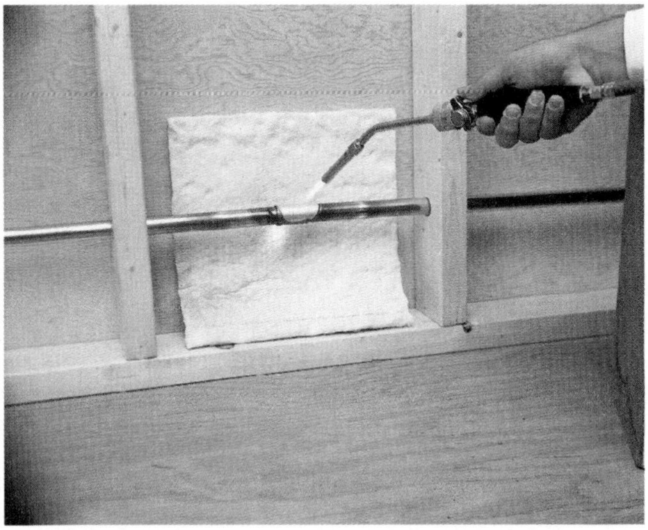

Figure 12-61. *Tubing joint prepared for brazing. Note use of fire-resistant material to protect nearby materials from heat of flame. (Uniweld Products, Inc.)*

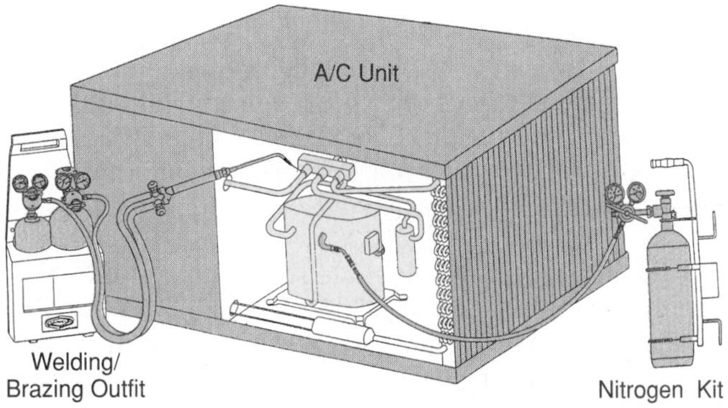

Figure 12-60. *Nitrogen is used in the system to check for leaks and when brazing. A pressure of 2-3 psi is maintained when brazing with nitrogen. (Uniweld Products, Inc.)*

12.11 Charging a Hermetic System

If testing indicates a lack of refrigerant, there is a system leak. Be sure the leak is repaired before adding refrigerant.

A hermetic unit needs refrigerant if there is:

- A partially frosted evaporator.
- A low head pressure.
- A low pressure on the low-side.
- A leak.
- The unit is running too frequently.

One or more of the listed conditions can indicate the system needs charging with refrigerant. Remember that a pressure difference is needed to move the refrigerant from the cylinder into the system. The cylinder is at higher pressure; the system is at low pressure.

The following procedure is used to add refrigerant:

1. Evacuate the system.
2. Connect a refrigerant cylinder to the charging manifold. Charge only with correct refrigerant vapor.
3. The refrigerant cylinder may be heated with warm water or an electric heater insert. **Temperature must not exceed 120°F (49°C). Never use an open flame for heating.**
4. Install pressure gauges and valves as described and explained in Section 12.7.1.

The service connection lines must be clean and free of air (which carries moisture). Clean the lines by flushing refrigerant through them before recharging.

When charging a unit that is already partially charged, add a small quantity of vapor refrigerant. The unit should then be allowed to cycle. A proper charge is indicated best by the frost on the evaporator. When frost starts to come down the suction line, purge out a little of the refrigerant. The system then will have the correct charge.

If a system has been evacuated first, it can be charged by replacing the evacuating pump with a refrigerant cylinder. You may use valves to close off the pump and then open the connection to the cylinder.

If the system has not been evacuated, purge both service lines and manifold. This is done by loosening the service line at the piercing valve. Then, open the left-side manifold valve. When the cylinder valve is opened, the refrigerant vapor will purge air, moisture, and dirt out of the two lines and the manifold. Tighten the service line at the piercing valve.

Compressor—Running Test 1

Start the unit. Open the line service valve, gauge manifold valve, and refrigerant cylinder valve. Watch the low-side pressure gauge. A pressure of not more than 5 psig to 25 psig (20 psia to 40 psia or 140 kPa to 280 kPa) should be created. This pressure is controlled by adjusting the refrigerant cylinder valve. Allow the refrigerant charge to enter the system for about 3 to 5 minutes. Connection should be as shown in **Figure 12-62.**

After the time lapse mentioned, close the gauge manifold valve. Allow the unit to operate and check the frost line on the evaporator. If the frost line is inadequate, repeat the charging for short intervals, checking after each. The frost line must not go beyond the accumulator in the suction line.

When the proper amount of frost has been observed, close the refrigerant cylinder valve, adaptor valve, and gauge manifold valve. After closing all valves, follow these steps:

1. Close the adaptor valve, if one was installed on the suction line. Leave it mounted for future service operations.
2. Check for leaks using a leak detector.

For a system that uses a separate process tube, follow these steps:

1. Pinch the process tube between the compressor and the adaptor valve with a pinch-off tool. See **Figure 12-63.**
2. Remove the adaptor valve. Flatten the tube end by crimping, and braze the end of the tubing.

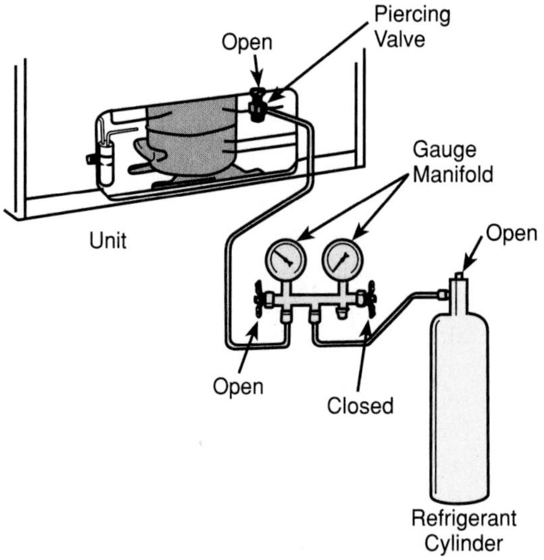

Figure 12-62. *Recharging setup after adjustment of refrigerant cylinder valve. Note open piercing valve, left-side manifold valve, and refrigerant cylinder valve.*

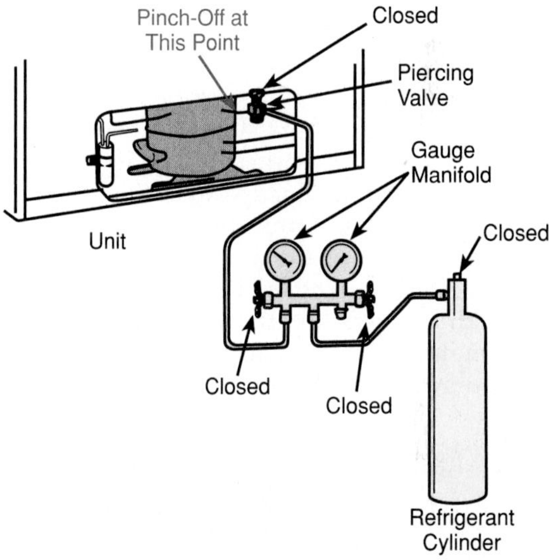

Figure 12-63. *System after piercing valve is closed along with refrigerant cylinder valve and gauge manifold valves. Note pinch-off point.*

Compressor—Running Test 2

The procedures that follow rely primarily on temperature as an indicator of correct charging. Refrigerants charged into hermetic refrigeration units must be of top quality. Always transfer refrigerants in chemically clean cylinders and lines. Always keep the charging cylinder at room temperature or warm it only with warm water.

There is a charging device that is mounted between the refrigerant cylinder and the low side. The device allows liquid refrigerant to flow into it from the cylinder. The refrigerant vaporizes inside this device. Always charge a unit (except large commercial units) with refrigerant vapor. *Never charge liquid refrigerant into the low side of a domestic or small commercial unit.*

When the charging device is first used, the process tube will sweat. It may even frost a little. As the system becomes fully charged, this sweating and frost will go away. This is because the suction pressure is higher. The system is now correctly charged.

A fast-reading dial thermometer provides a second check on the correct charge. Suction line temperature should be about 20°F higher at a distance of 6″ to 10″ from the compressor than at the evaporator outlet.

If the temperature is lower, liquid refrigerant may enter the compressor, causing damage. If the temperature is higher, the motor compressor may overheat and burn out.

Evacuation Data

The following pertains to the charging device mounted between refrigerant cylinder and the low side. A check valve in the charging unit allows you to evacuate the system easily with this unit attached to the line. The charging unit is available in three capacities: less than 1 hp; 1 hp to 4.75 hp; and 5 hp to 10 hp. The correct size must be used.

Compressor—Running Test 3

Systems that use a capillary tube must have an exact refrigerant charge. If the system is overcharged, the evaporator will be overcharged or flooded.

The use of accumulator spaces at the outlet of these evaporators relieves the problem somewhat. However, you must be careful of the amount charged into these systems. A common method is to slowly charge these systems with refrigerant in the vapor state. This is done until suction line starts to sweat and/or frost back. Then they are purged a little at a time until the "frost back" disappears.

Charging with Exact Amount

Another method is to completely discharge the system. Then, recharge with a cylinder containing exactly the right amount of refrigerant. This amount is determined by the manufacturer's recommendations.

Conclusions

The following points are important and bear repeating:

- Charge a system into the low side, if possible.
- **Refrigerant should be put into the system in vapor form. Forcing liquid refrigerant into the system may damage the compressor and injure you.**
- Remember that if a system is short of refrigerant, there is a leak. Locate and correct the leak before the system is charged.
- Use the proper recovery/recycling equipment when locating and repairing leaks.

When adding refrigerant to a system, remember that some of the oil will dissolve in the refrigerant. If the unit becomes noisy soon after adding refrigerant, oil should be added.

12.11.1 Charging with Portable Charging Cylinder or Digital Scale

A charging cylinder with a glass-tube liquid level indicator is shown in **Figure 12-64A.** This allows you to transfer refrigerant into a system and measure the amount on a scale. Some cylinders are electrically heated. This speeds up the evaporation and maintains pressure in the cylinder.

This process of heating a cylinder is usually done with an electrical insert. In some cases, the compressor itself is heated with a heat gun. The refrigerant and oil will circulate and be purged more easily. **In both cases, the required temperature and pressure safety controls must be provided. It is therefore extremely important to use a pressure control relief valve and a thermostat.**

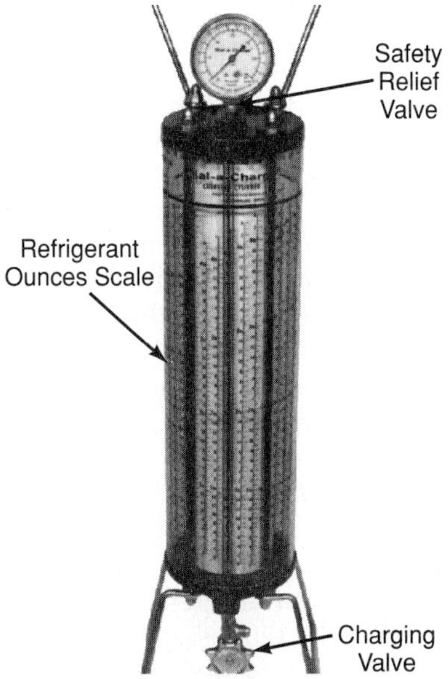

A

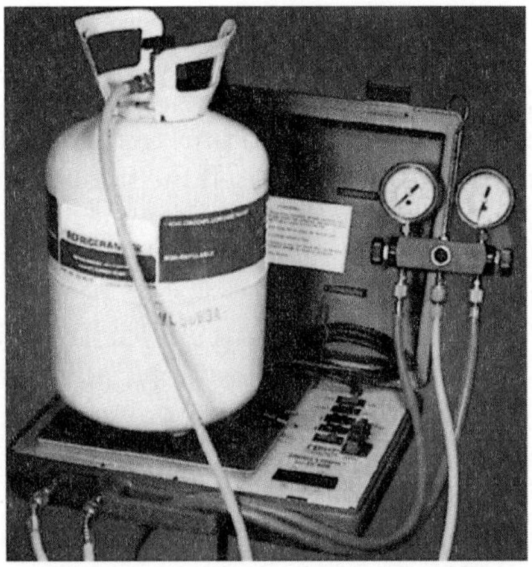

B

Figure 12-64. *Instruments for charging a system. A—Portable charging cylinder may be used to accurately charge hermetic systems. Cylinder is precisely marked in ounces of refrigerant. (Robinair Division, SPX Corporation) B—Digital scale weighs refrigerant. (CPS Products, Inc.)*

The system has a pressure gauge and a hand valve on the bottom. They are used for filling the charging cylinder or for charging liquid refrigerant into a system. There is also a valve at the top of the cylinder. This valve is used for charging refrigerant vapor into the system. (This is the best and safest method.) **Figure 12-64B** shows a digital weighing scale. See Chapter 10 for additional information on the use of a digital scale for system charging.

The following steps are recommended for use of a portable charging cylinder after evacuation. **Wear goggles** and follow these steps:

1. Attach a line from the charging cylinder to the center of the gauge manifold. Then purge with the fitting loose at the center part of the gauge manifold. See **Figure 12-65.** Tighten this connection.
2. Open the piercing valve or valve adaptor and gauge manifold valve.
3. Crack the charging cylinder valve and allow the refrigerant to enter the system. See **Figure 12-66.** Know what the new scale reading on the tube must be to put in the correct charge.
4. When the correct amount of refrigerant has entered the system, close the cylinder valve. The amount can be checked by reading the scale on the charging cylinder.
5. Close the piercing valve or the valve adaptor and the gauge manifold valves. See **Figure 12-67.**
6. Use a pinch-off tool to close off the process tube between the compressor and the valve adaptor. Leave the pinch-off tool in place until the tube end has been brazed.
7. Remove the piercing valve or the valve adaptor.
8. If a piercing valve was used, cut off the part of tubing with hole in it. Use a pipe cutter. **Wear goggles!**
9. Crimp the end of the process tube.
10. Braze the end of the process tube. **Wear goggles!** Check the system for leaks.

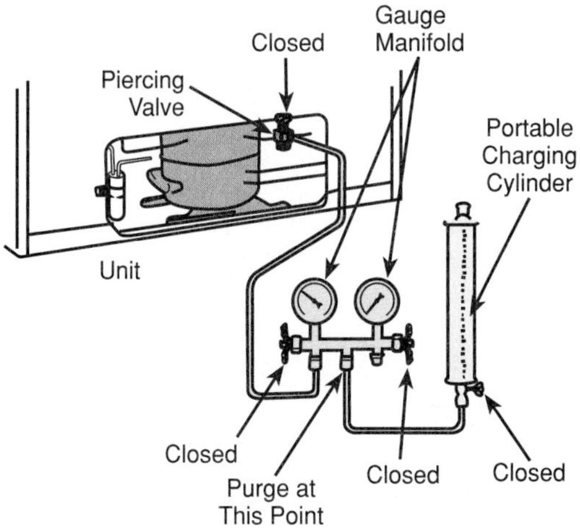

Figure 12-65. *When portable charging cylinder is used, purge charge line (after evacuating the system) by leaving fitting loose at center part of gauge manifold. If system is not evacuated, purge all lines up to piercing valve.*

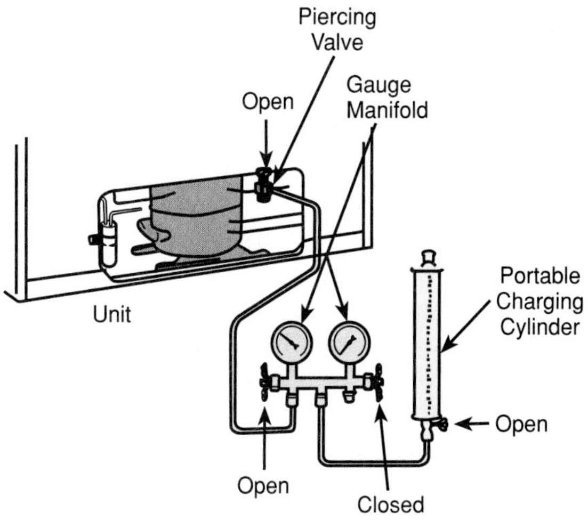

Figure 12-66. *Proper hookup for using portable charging cylinder to charge refrigerant into system.*

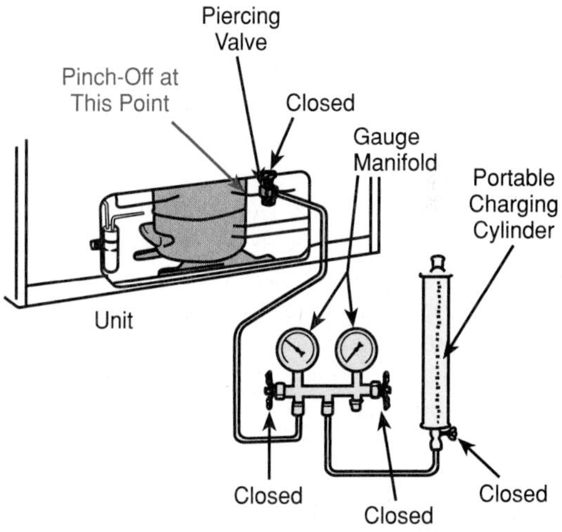

Figure 12-67. *Proper steps and valve settings for removing servicing equipment after system has been charged.*

12.11.2 Adding Oil to the System

The correct amount of oil in a system is very important. Lack of oil will shorten the life of the mechanism, increase friction, and cause noise. An overcharge will cause the compressor to pump excessive amounts of oil. This will reduce its refrigerant-pumping capacity. It will also subject the compressor valves to severe strain.

Refrigerant oils are available in several ***viscosities*** (ratings of ease of flow at different temperatures). Be sure to follow the manufacturer's viscosity recommendations. On a service call, add oil only if there is a sign of oil leakage.

It is rarely necessary to add oil to a hermetic system. However, leaking refrigerant always carries some oil with it. This lost oil should be replaced. If the hermetic unit is completely equipped with service valves, oil may be added using the conventional method. That is, oil can be siphoned or poured in.

If the system has had a low-side leak, moisture and air may have entered. In this case, it is best to replace the compressor oil. Measure the amount removed and replace it with the same amount of clean, dry oil. See Chapter 15. The unit should be charged in much the same way as when adding refrigerant to the system.

Figure 12-68 illustrates a practical transparent charging cylinder with a chart for accurate measurement. It measures the amount of refrigerant or oil charged into a system by volume. This method provides greater accuracy. Use clean lines. Purge lines of air with clean refrigerant.

There are several ways to connect a transparent charging cylinder to a system. **Figure 12-68** shows the tube connected directly to the low side. Here, a compound gauge and process tube adaptor are used. The process tube is brazed to the suction line. **Figure 12-69** shows a method of using a vacuum pump and gauge manifold to evacuate for repairing procedures.

A pump may be used to put oil into a system. See **Figure 12-70.** The charging lines must be purged to remove air, moisture, and dirt. This hand pump can build up pressures as high as 300 psig (315 psia or 2200 kPa). Oil can be forced into the system even when the system is under pressure.

Using the single-service-line technique, add oil as shown in **Figure 12-71.** The correct amount of oil is put into the service cylinder. A small amount of refrigerant (the same type used in the system) is also put into the cylinder. This creates a pressure. The cylinder is inverted. It is connected from the cylinder valve to the valve attachment by clean lines. Be sure the cylinder pressure is higher than the system pressure. Then open the cylinder valve and valve attachment. Oil will be forced into the system.

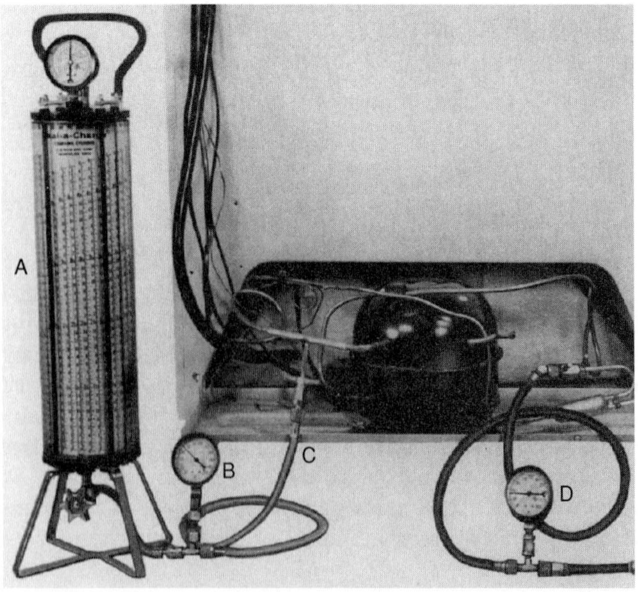

Figure 12-68. *Charging cylinder connected to low side. A—Charging cylinder. B—Compound gauge. C—Line adaptor. High-pressure gauge is connected to the outlet of the filter-drier by a tube adaptor. D—High-pressure gauge.*

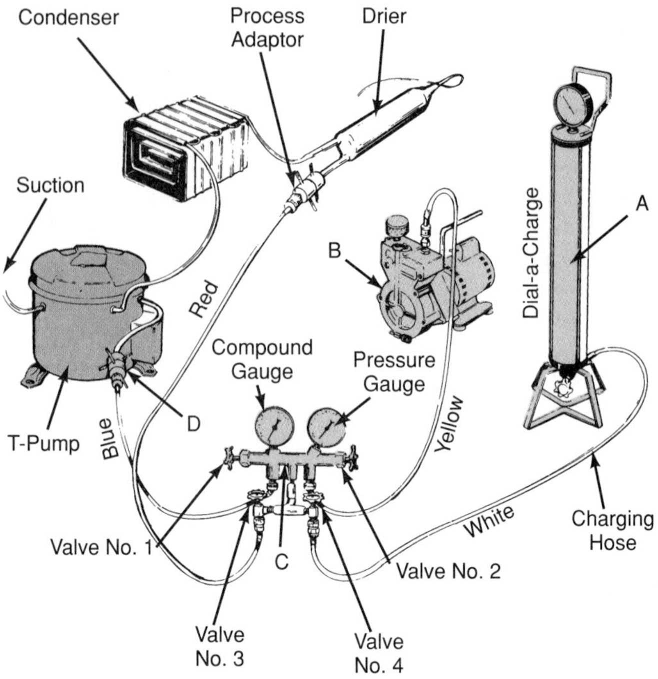

Figure 12-69. *Component parts for a general unit used for evacuating and charging a system. A—Charging cylinder. B—Vacuum pump. C—Gauge manifold. D—Process tube adaptor. (Frigidaire Company)*

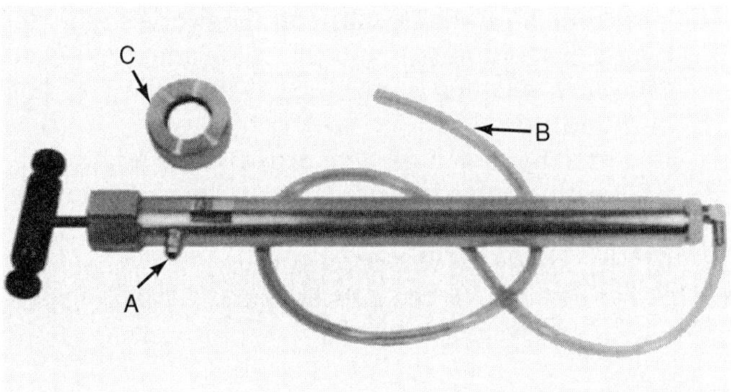

Figure 12-70. *Hand pump used to force oil into system. A—Opening to oil charging line. B—Plastic tubing for drawing oil from refrigerant oil storage can. C—Rubber adaptor for oil can. (Robinair Division, SPX Corporation)*

12.12 Diagnosing Component Problems

Before removing any hermetic system component, be certain that it is the cause of the problem. The parts that may cause the trouble are:

- Motor compressor.
- Filter-drier.
- Refrigerant control (capillary tube or AEV).
- Hot gas defrosting valve.

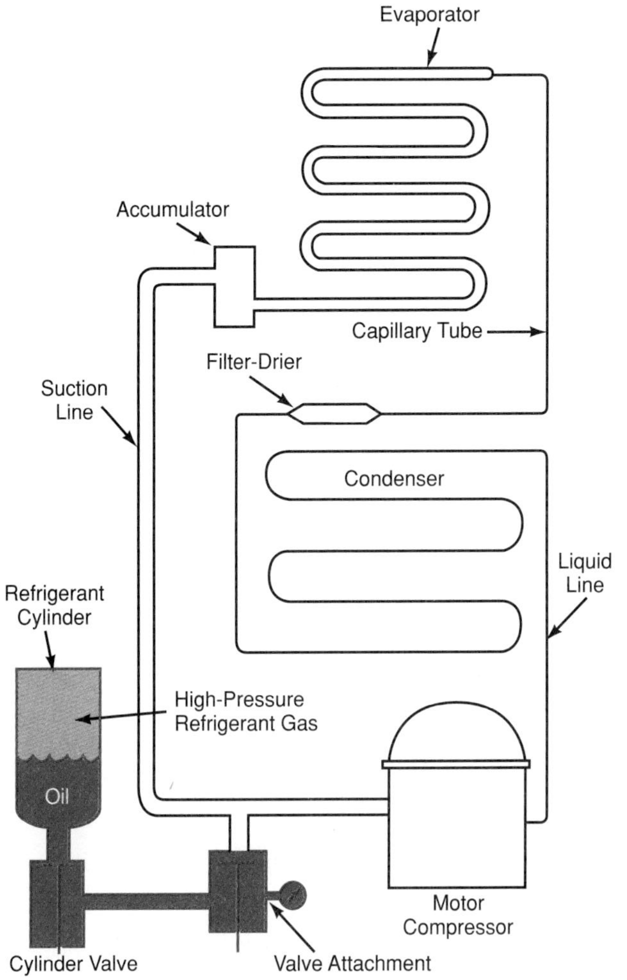

Figure 12-71. *Single-service-line technique can be used for adding oil to small hermetic system. First, place correct amount of oil in service cylinder, along with small amount of refrigerant to create pressure. Then turn service cylinder upside down. Its pressure must be higher than system pressure, so that oil will enter system.*

12.12.1 Locating Motor Compressor Faults

Replacing the motor compressor is the most costly service repair. Be sure that it is a bad motor compressor before replacing it. The fault may be either electrical or mechanical. When a motor compressor is in good condition, the most common reason for replacement is an electrical fault.

Other electrical problems often mislead a service technician into thinking the motor compressor is at fault. To check it, first clean the outside of the compressor dome. Then, remove the cover over the electrical connections. Disconnect the system wiring from the compressor: relay, capacitors, overload, and all. Use an ohmmeter to check motor windings for continuity, shorts, and grounds. See Chapter 7. If unit checks out correctly, connect a starting circuit to the motor compressor. Use the correct size of capacitors and overload cutout. Connect as shown in **Figure 12-72.**

If system starts and operates correctly with these manual-start electrical connections, the problem is in the external system. (It may be in the wiring, thermostat, relay, or overload.) If the internal electrical motor is faulty, the motor compressor must be replaced. If the electrical system operates correctly, the compressor may not be pumping.

The best check of the motor compressor is its volt-ampere (watt) reading at normal low-side and high-side pressures. If the volt-ampere reading is below the motor rating, the pump may be worn out. To check the compressor's pumping ability, install a piercing valve and a gauge manifold. Then, pinch the suction line as shown in **Figure 12-73.** Next, run the unit to determine how much of a vacuum it will pull. (It should pull between 25″ Hg and 28″ Hg [7 kPa and 17 kPa] of vacuum.) Stop the compressor. A weakening of the vacuum is indicated when it is moving toward 20″ Hg (34 kPa), then toward 10″ Hg (68 kPa). The weakening of the vacuum indicates the exhaust valves of the compressor are leaking. The motor compressor must be replaced or overhauled.

Another method is to install a piercing valve on the process tube. Then, connect one end of flexible service line to the valve adaptor. Connect the other end of that line to the compound gauge end of manifold. Purge these lines to clean them. Then, pinch the suction line and start the unit. The motor compressor should pull 16″ Hg of vacuum (47 kPa) in about two minutes. (Do not run any longer without cool suction vapor flowing. The motor will overheat.) Stop the compressor and observe. If low-side pressure increases, the compressor valves are leaking.

12.12.2 Capillary Tube Service

Capillary tubes must be correctly sized. They must have the correct inside diameter (ID) and length for the system capacity and the desired evaporator temperatures. Cut capillary tubing by filing a notch around it. Then break the tubing by small back-and-forth bending motions. A tube cutter will change the ID too much.

Figure 12-74A shows a capillary system of correct design. The system in **Figure 12-74B** has too much resistance in the tube. It is either too long or it has an undersized inside diameter. The drawing also illustrates what happens if the system has a starved evaporator. This may be due to a partially clogged filter-drier or capillary tube.

The amount of refrigerant in a capillary tube system is critical. Refer to **Figure 12-75.** Notice the change in head pressure as the charge of refrigerant changes. If the system is undercharged as at C, the evaporator will not receive enough refrigerant. The system may run all the time. If the system is overcharged, the liquid refrigerant may flow down the suction line. This may cause oil-pumping in the motor compressor. The suction line will sweat and even frost up all the way to the motor compressor.

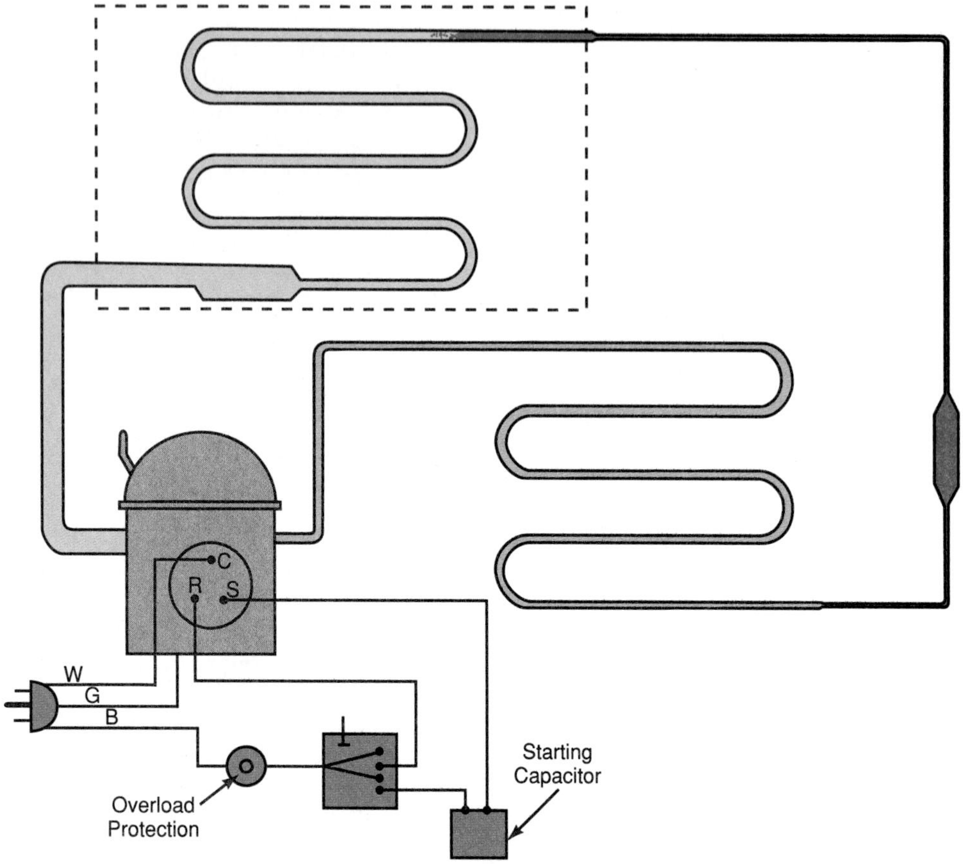

Figure 12-72. *Testing capacitor-start motor compressor after removing all electrical leads and connecting test cord.*

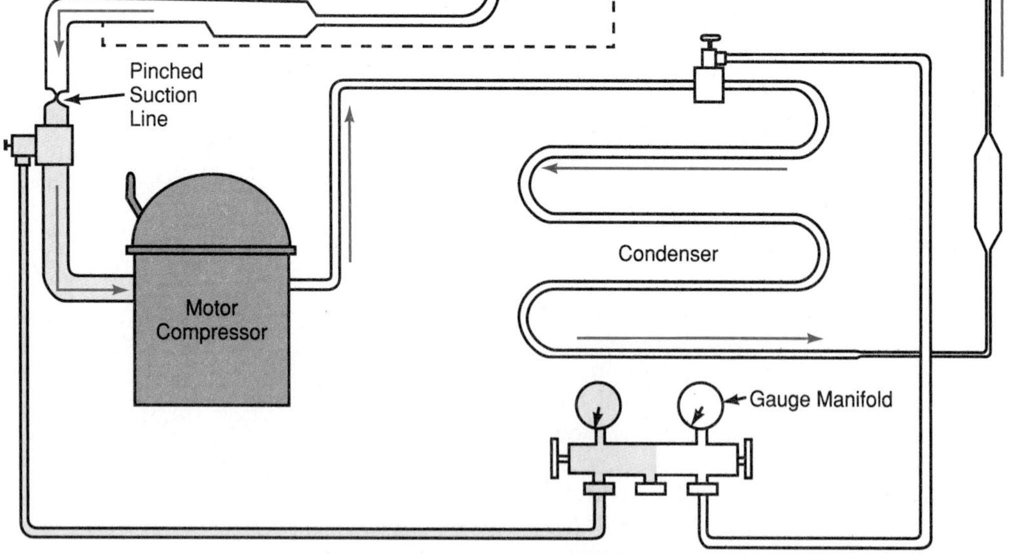

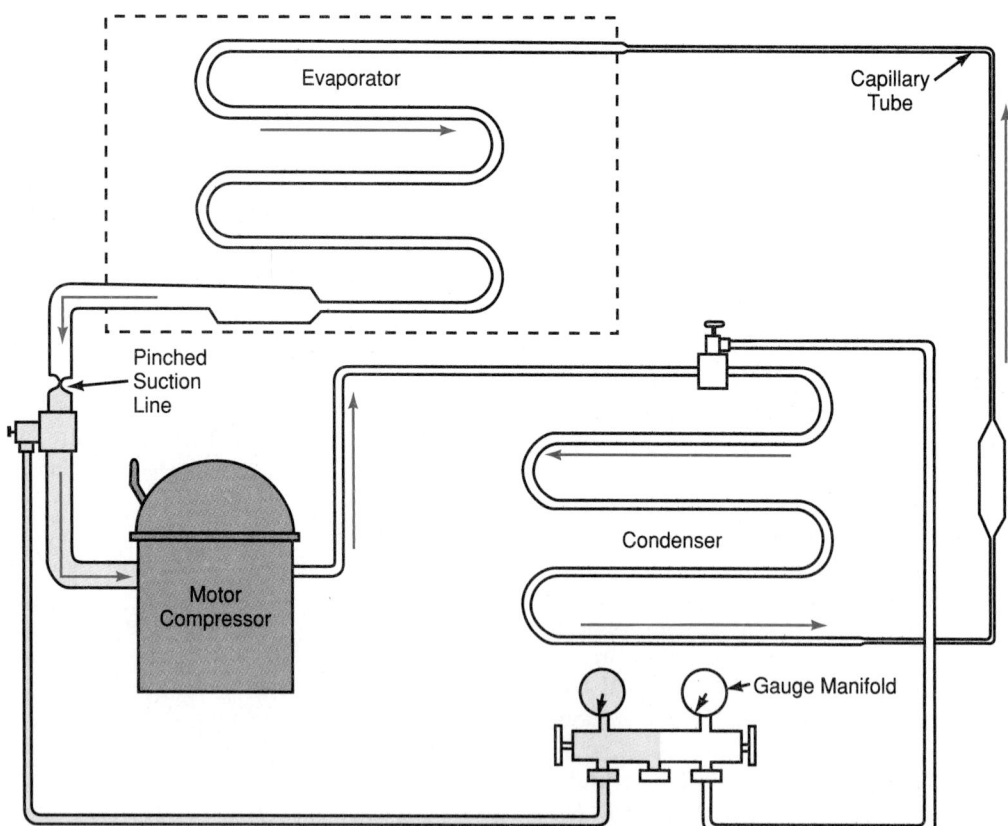

Figure 12-73. *A compressor's pumping capacity can be tested by pinching the suction line. This is a very short run test or motor will overheat. Compressor should pull 25" to 28" of vacuum in a few seconds. Yellow area indicates vacuum in line.*

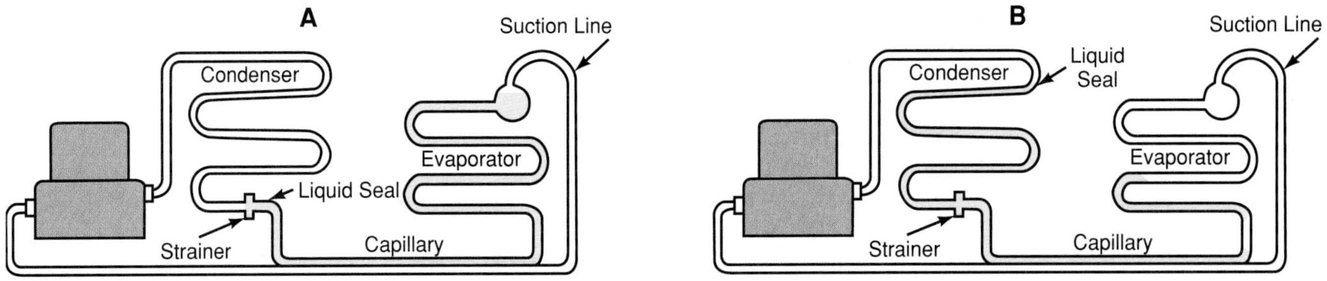

Capillary selected for capacity balance conditions. Liquids seal at capillary inlet but no excess liquid in condenser. Compressor discharge and suction pressures normal. Evaporator properly charged.

Too much capillary resistance—liquid refrigerant backs up in condenser and causes evaporator to be undercharged. Compressor discharge pressure may be abnormally high. Suction pressure below normal. Bottom of condenser subcooled.

Figure 12-74. *Effects of correct and incorrect capillary tube installation. A—Correct installation, operation normal. B—Incorrect installation. There is too much resistance in tube, so the evaporator is "starved."*

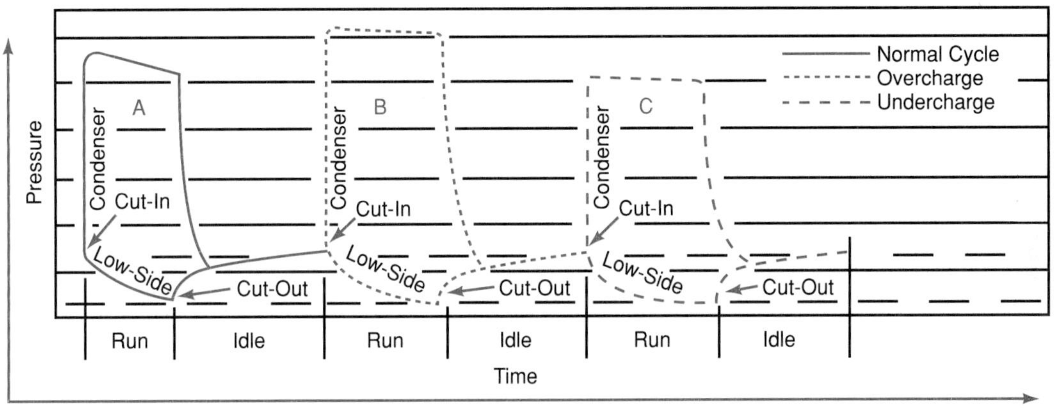

Figure 12-75. *Pressure-time cycle diagrams for three conditions in capillary tube system. A—Normal charge. B—Overcharge. C—Undercharge. Overcharge will usually cause frosted or sweating suction line.*

12.12.3 Checking for Restricted System

To check the capillary tube, run the system for a few minutes. Stop the unit and listen where the capillary tube enters the evaporator. If there is no hissing sound, the capillary tube is clogged.

Heat the evaporator end of the capillary tube with a rag and warm water. *(Do not use a flame.)* If the clogging is from ice, there will be a hissing sound as it melts. A clogged strainer or capillary tube will fill the condenser with all the refrigerant. The motor compressor may stop or it may overload during a start-up.

There is a quick test to determine if the system is short of refrigerant or if the filter-drier or capillary tube is clogged. First, install a piercing valve on the suction line. You may also install it on the end of the process tube of the compressor. Purge the service lines using vapor from a refrigerant cylinder. Open the piercing valve. If the low side is in a vacuum, the system has a restriction or is low in refrigerant.

If the defrost system is not working (ice buildup), the evaporator is not working. Check the evaporator fan. If it is working, check the defrost system. (Inspect and electrically test defrost resistance wire.) Then check as follows:

1. Install another piercing valve at the condenser outlet. See **Figure 12-76.** Open the valve just a little. Allow some refrigerant to escape.
2. If no vapor escapes or if only vapor escapes, the system is short of refrigerant.
3. If liquid refrigerant escapes, the filter-drier or the capillary tube is clogged.
4. To determine which component is clogged, discharge the refrigerant. Clean the connection between the drier and the capillary tube. Flux it, heat it, and separate the capillary tube from the filter-drier. The system can now be checked to find out if either the drier or capillary tube is clogged.
5. To find out which is at fault, see **Figure 12-77.** Put some vapor refrigerant in the system. Open valves C, A, and B. If filter-drier is open, vapor will come out opening D. If capillary tube is open, a small flow (because tube is small) will come out its opening, E. If either one is clogged, there will be no flow through it.

Clogged units must be replaced. Sometimes clogged capillaries can be opened with a high-pressure hydraulic pump. (See Section 12.13.6.)

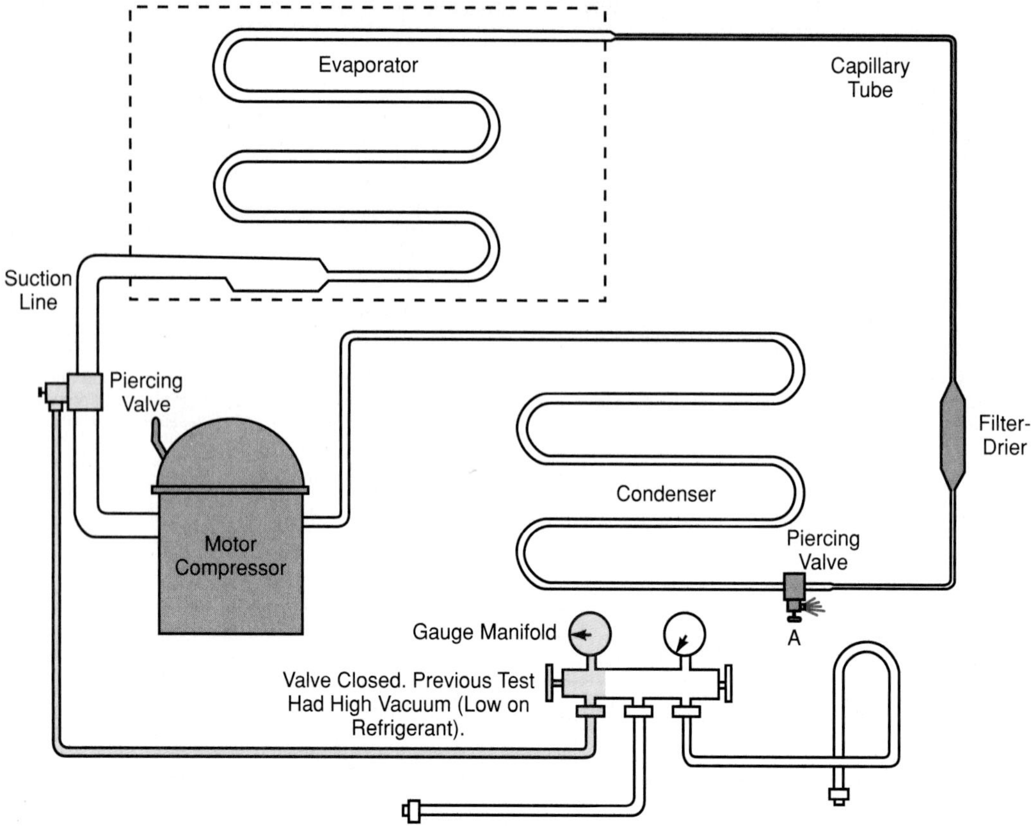

Figure 12-76. *Testing for a clogged drier or capillary tube or for lack of refrigerant. With the compressor running and valve A open, escaping vapor means there is lack of refrigerant. If liquid refrigerant escapes, the filter-drier or capillary tube is clogged.*

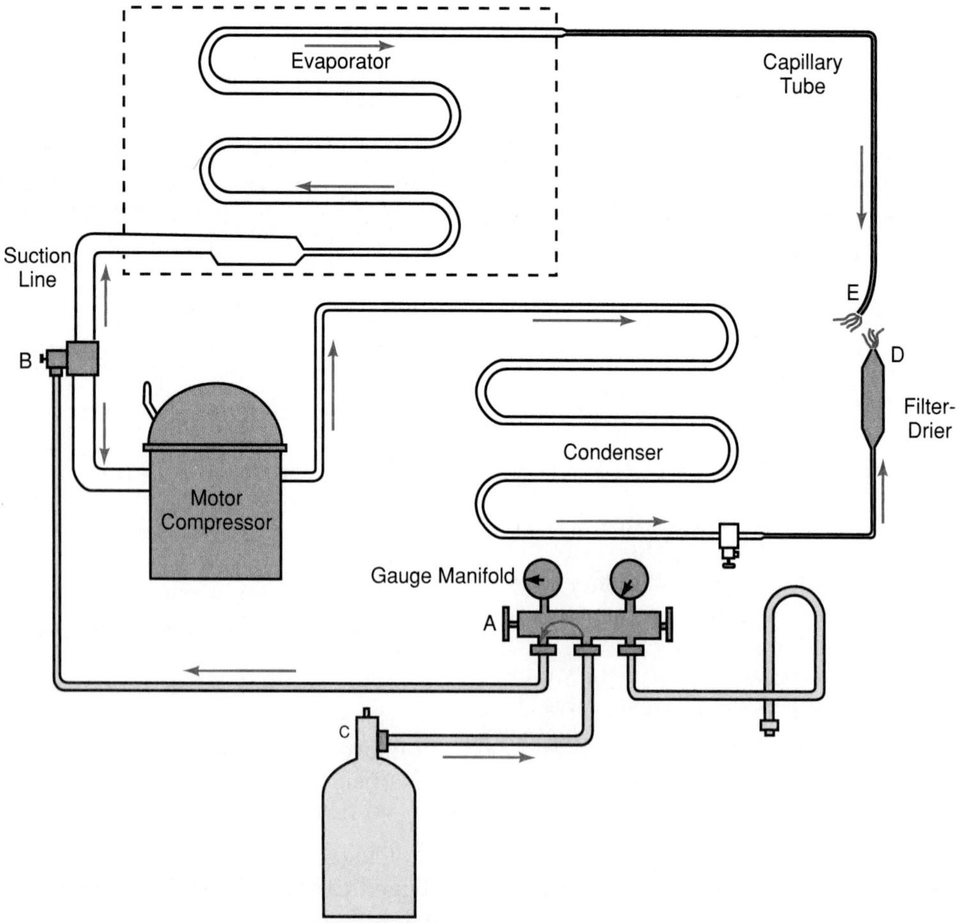

Figure 12-77. *To check whether the capillary tube or the filter-drier is clogged, separate and check flow at D. Charge some vapor refrigerant into system by way of valves C, A, and B.*

12.12.4 Filter-Drier Service

A filter-drier should be replaced whenever a new motor compressor is installed, or if the filter is clogged. Install driers and filters in the refrigerant circuit to keep the system clean and dry inside. A combination filter-drier is shown in **Figure 12-78.**

A solid moisture absorbent will usually do a satisfactory job. Silica gel, alumina gel, and synthetic silicates are excellent moisture absorbers. A bead silica gel gives good results. A filter-drier with a purge valve built in is shown in **Figure 12-79. Figure 12-80** shows the same unit installed.

Never use a liquid drying agent in a unit equipped with a solid ***desiccant*** (drying chemical). The liquid dryer chemical will release the moisture already trapped in the drier.

Likewise, a solid drier should not be put in a system that is already using a liquid drier. To avoid this danger, all systems should be labeled indicating which drying agent is used. The moisture absorbent properties of these desiccants are shown in **Figure 12-81.**

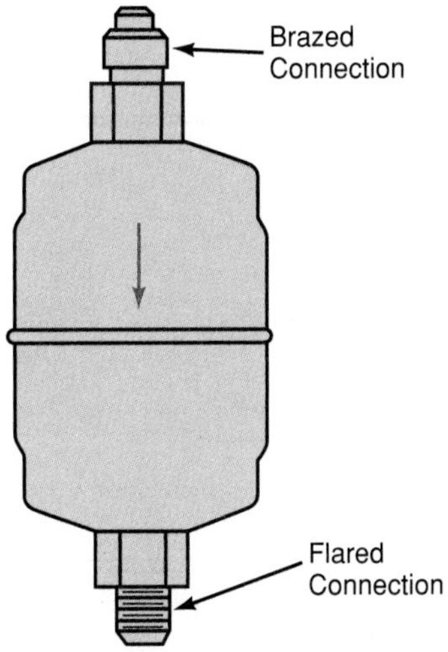

Figure 12-78. *Filter-drier of the type that is used on small hermetic systems. Arrow shows direction of refrigerant flow.*

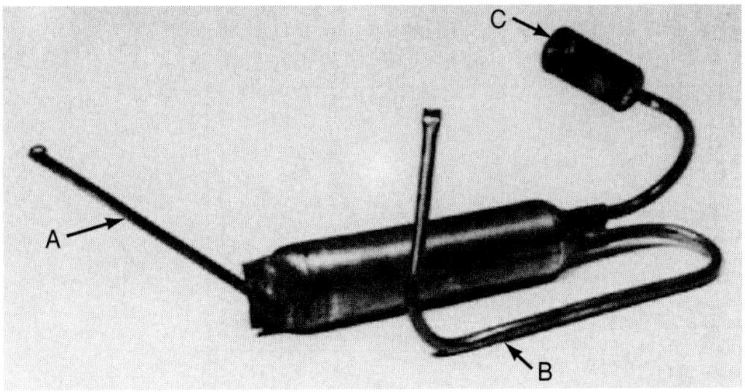

Figure 12-79. *Replacement filter-drier. A—Outlet tube. B—Inlet tube. C—Purge valve. (General Electric Co.)*

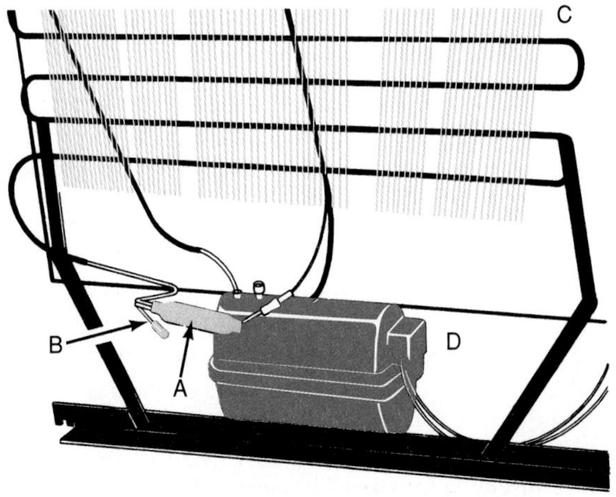

Figure 12-80. *Replacement filter-drier installed. A—Filter-drier. B—Purge valve. C—Condenser. D—Compressor.*

Desiccant (Drying Chemical)	Mesh	Absorption from Liquid Percent of Weight of Desiccant
Silica Gel	8–20	16
Activated Alumina	8–10	12
Synthetic Silicates	8–20	16

Figure 12-81. *Moisture-absorbing ability of some desiccants in driers. Some driers have mixtures of these chemicals. These driers may be used with R-12, R-22, R-500, and R-502.*

The Refrigeration Electrical Manufacturer's Association has recommended that the capacity of dehydrators be rated. **Figure 12-82** shows the recommended volume of drying agents according to horsepower of the unit. All driers are sealed by the manufacturer. Do not remove the sealing caps until just before installation.

Driers absorb water faster at lower temperatures. If at all possible, the drier should be installed just ahead of the refrigerant control. If the filter-drier accidentally becomes heated, the moisture it has absorbed may be driven out. The moisture will recirculate with the

Drier Sizes — Domestic	
Cu. In.	**Capacity, hp**
2	1/8
3	1/6 to 1/4
6	1/4 to 1/2
9	1/2 to 3/4

Figure 12-82. *Drier capacities in cubic inches recommended for systems with various horsepower ratings.*

refrigerant. A position just before the refrigerant control can have advantages:

- It is likely that both the filter-drier and refrigerant control will become heated at the same time. This reduces the chance that ice will form in the refrigerant control.
- The filter-drier is kept far away from the heated tubing of the condenser.

A filter-drier has an arrow stamped or cast on the body. This arrow indicates the direction in which the refrigerant should flow. Be sure it is installed properly. Filter-driers may be installed with either flared or brazed connections.

Whenever a system is opened, always install a new filter-drier in the liquid line. Many service technicians install two filter driers on a system after repairing it. One is placed on the high side just before the refrigerant control. Another is placed on the low side, between the evaporator and compressor. This improves the chance of removing all moisture or contaminants which may have entered the system during servicing.

12.12.5 Hot Gas Defrost Problems

The hot gas valve is usually operated by a solenoid. Problems can occur with a solenoid in two ways:

- The solenoid valve is stuck closed. If the evaporator is overloaded with ice, the solenoid electric coil may have failed (open circuit) or the timer may not be operating correctly. Both of these problems can be checked electrically.
  If the electrical system is operating correctly, the problem is probably a stuck valve stem in the solenoid. Rap the valve body while the defrost timer switch is closed. If the valve breaks loose, you can hear the surge of hot gas. The line between the solenoid and the evaporator will also become warm to the touch.
- The solenoid valve is stuck open, as in **Figure 12-83.** If the evaporator and the line between the hot gas solenoid and evaporator are warm, the valve is stuck open. Refer again to **Figure 12-83.** (Hot gas is shown in color.) Again, rap the valve sharply while the timer is on open circuit. If the valve closes, the low-side pressure will start to decrease

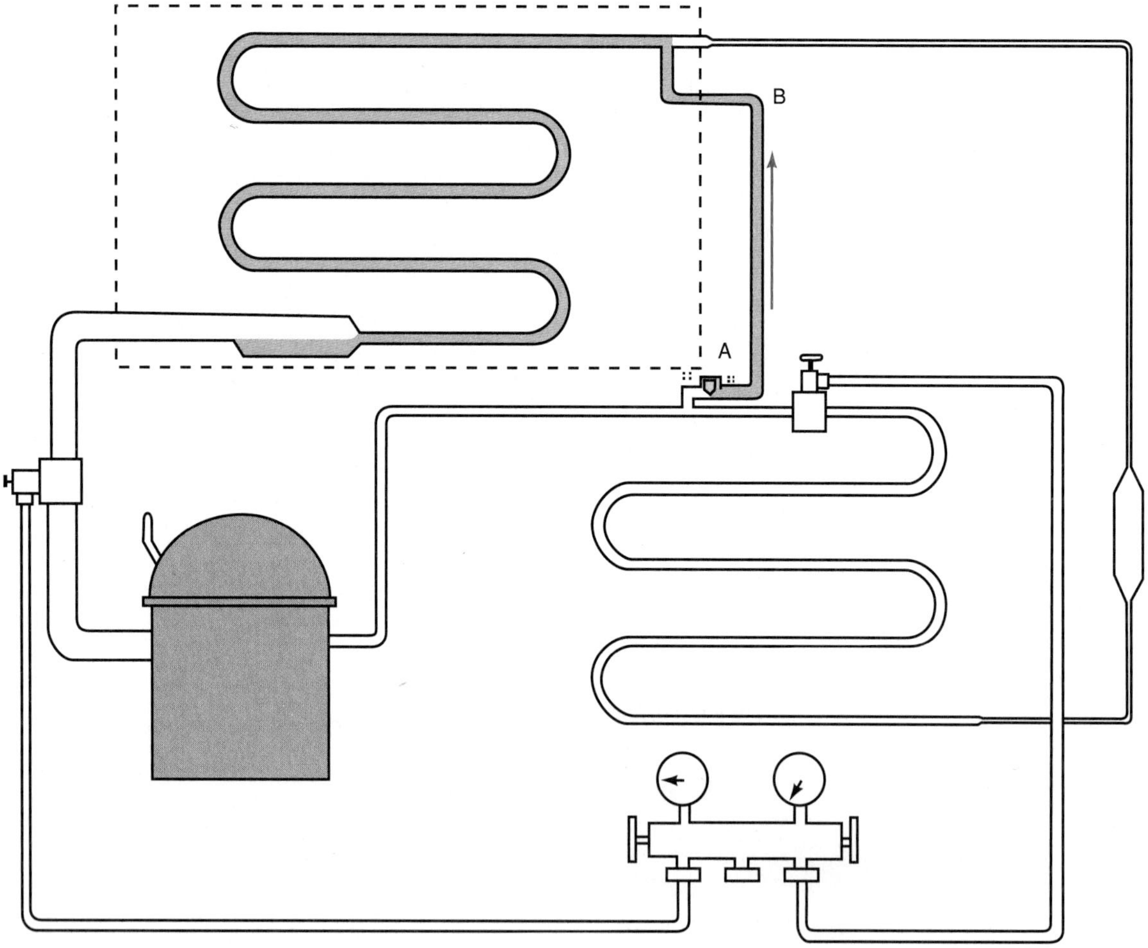

Figure 12-83. *Hot gas defrost system with solenoid valve stuck open. A—Solenoid valve. B—Hot gas line.*

immediately. The evaporator will start to cool and frost.

If the valve still does not operate after trying these solutions, it must be removed and replaced.

12.13 Dismantling a System

It is sometimes necessary to replace parts of a system. Some parts that may need replacing include: the motor compressor, condenser, capillary tube, evaporator, accumulator, and filter-drier. The system should be prepared as follows:

1. Disconnect the electrical circuit.
2. Carefully clean all surfaces. It is good to clean the entire mechanism. This will reduce the chance of dirt and other contamination entering the system.
3. Install service valve and gauge manifold.
4. Remove and recover the refrigerant. The system must be purged in accordance with Environmental Protection Agency (EPA) regulations. Use a recovery/recycling unit. See Section 12.13.1 and Chapter 10.
5. Cut the tubing and remove the part to be replaced.

12.13.1 Removing Refrigerant

Purging or venting into the atmosphere is illegal. This is due to damage to the ozone layer caused by the release of CFC refrigerants. When installing or servicing refrigerant units, the proper use of a recovery/recycling unit is required. See Chapter 10.

Always work in a well-ventilated area. Be sure to wear goggles!

Figure 12-84 illustrates a technician using a refrigerant recovery system. The operation of refrigerant recovery equipment is different for each type of unit. The procedures are sequential steps you must follow. **Figure 12-85** diagrams the basic steps for the use of this unit. After steps 1, 2, and 3, the pressure reading determines your next action. If pressure is 0 psi, proceed with steps 4, 5, 6, 7, 8, 9, and 10. If the pressure is *not* 0 psi, follow steps 4A, 5A, 6A, 7A, and 8A. For additional information on the operation of recovery/recycling units, see Chapter 10.

12.13.2 Removing a Motor Compressor

Follow this procedure to remove a motor compressor:

1. Disconnect the electrical circuit.
2. Install the gauge manifold. Use a piercing valve if necessary.
3. Recover the refrigerant.
4. Disconnect the lines. **Wear goggles!**
   A. Clean both the suction and discharge tubing on straight sections near the compressor. Use a tube cutter. *Plug the lines at once.*

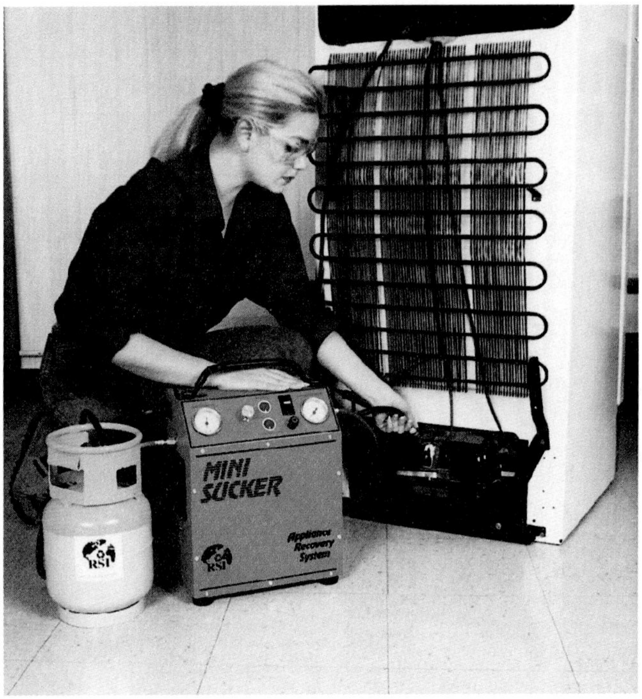

Figure 12-84. *A refrigerant recovery system being used on a domestic unit. (Recycling Specialists International)*

   B. Clean the tubing or fittings at the compressor. Put brazing flux on the connection. Heat the joint and pull the tubing out of the fittings. *Plug openings immediately.*
5. Remove the motor compressor.
6. If the motor compressor has oil cooler lines, they must be pinched, then cut with a tubing cutter. The compressor tubing openings should be sealed.

The unit is now ready to have a replacement motor compressor installed.

Cause of Motor Compressor Burnouts

A strong pungent refrigerant odor may be present when a piercing valve is opened slightly. This is a certain indication of a burnout. Much study has been done on why motor compressors burn out. Moisture, dirt, and air in the system are possible causes. Another cause may be too much current flow from inaccurate safety devices in the electrical circuit. Additional reasons may be a stiff compressor, low voltage, or a lack of refrigerant (poor motor cooling).

High head pressure is one of the most frequent reasons for motor burnout. This pressure creates very high temperatures as the vapor passes the compressor discharge valves. The high temperature increases chemical action. This adds to or creates new corroding elements in the system. Oil breaks down and forms carbon and sludge. If the temperature at the discharge line to the condenser reaches 350°F (177°C), oil breakdown is taking place.

It is very important that the condenser be large enough. It must be clean and the air should flow over it

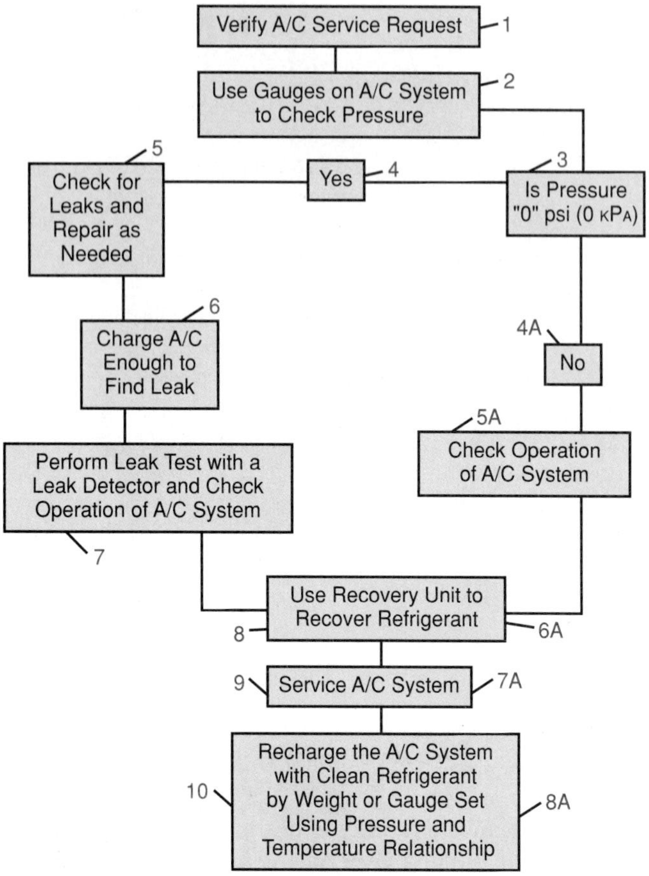

Figure 12-85. *Procedures for using a refrigerant recovery system. Pressure reading at Step 3 determines course of action.*

efficiently. (Fans, fan motors, ducts, air-in, and air-out must all be in good condition.)

You should check the head pressure of each unit. All the necessary things must be done to bring this pressure down. Purge the system through the high-side line to the manifold gauge. Completely clean the condenser with high-pressure gas (air, carbon dioxide, nitrogen). See Section 12.9.1. **Wear goggles!** Brush the condenser to remove dust and dirt. Use long bristle brushes and a vacuum cleaner. Check the air-in and air-out passages. They must be in good condition.

Cleanup after Motor Burnout

When a motor begins to burn out, it overheats. This overheating will cause the refrigerant to break down. If moisture is present, overheating can cause the formation of hydrochloric and hydrofluoric acids. Oil in this condition is said to be "acidic." The acid will cause insulation on motor windings to deteriorate and increase the motor temperature. Eventually, the motor windings will short-circuit and burn out.

If a system has a motor compressor burnout, refrigerant controls should be repaired or replaced. (This includes AEV, solenoid valves, reversing valves, etc.) Flush the system with R-11 or the same refrigerant used in the system.

Do not touch the oil from a burned-out motor compressor. It will cause a severe acid burn! Caution: Wear goggles and rubber gloves.

If oil cooler lines must be cut, be even more careful. Do not allow oil to run on the floor. Trap it in glass containers.

The burnout can be mild or severe. If severe, the oil will be black and acidic with a pungent, very unpleasant odor. If mild, oil will be clear but there will be a pungent odor and a mild acidic condition. Many acid test kits are available for determining the amount of contamination. See Chapter 15 for further information. If oil is clean and odor-free, there is no burnout. The trouble is mechanical.

After replacing a motor compressor, install two filter driers. See **Figure 4-9.** One should be in the suction line between the evaporator and the compressor. The other should be between the condenser and the liquid refrigerant line. (If a capillary tube is used, the filter-drier should be just before the capillary tube.)

The system may be flushed using a recovery unit that incorporates the use of filters. Normally the process is performed more than once to ensure good flushing. R-11 or the refrigerant in the system may be used.

You may also purge the system using a recovery unit. Use the same type of refrigerant that is in the system. This will ensure that all R-11 is removed from the system. **Wear goggles!**

12.13.3 Installing a Motor Compressor

The motor compressor being installed should be an exact replacement. It must have the same capacity as the one being replaced. The motor compressor must be designed for the low-side pressure desired (low, medium, or high).

This is the recommended procedure for installing a replacement motor compressor:

1. Carefully clean about 2″ (50 mm) at the ends of the suction, discharge, and oil lines.
2. Carefully clean the suction, discharge, and oil line connections on the motor compressor.
3. If a piercing valve and valve adaptor are not already installed, do so. Connect the gauge manifold. (See Section 12.8.3.)
4. Connect the lines. Use adaptor fittings, lengths of tubing, or an expander, if necessary. The expander will allow you to telescope the tubes from the motor compressor into the suction and discharge lines. See **Figure 12-86.** A punch-type swaging tool may be used if an expander is not available. Telescope the tubes together, using as little flux as possible. Braze the connections.

   When brazing, keep the heat away from other brazed joints. Use wet cloths or special heat-absorbing compounds to protect the other joints. Metal sheeting will protect the cabinet, wires, and plastic parts from the flame. CAUTION: The replacement motor compressor or stub lines may be smaller in diameter than the cabinet unit suction

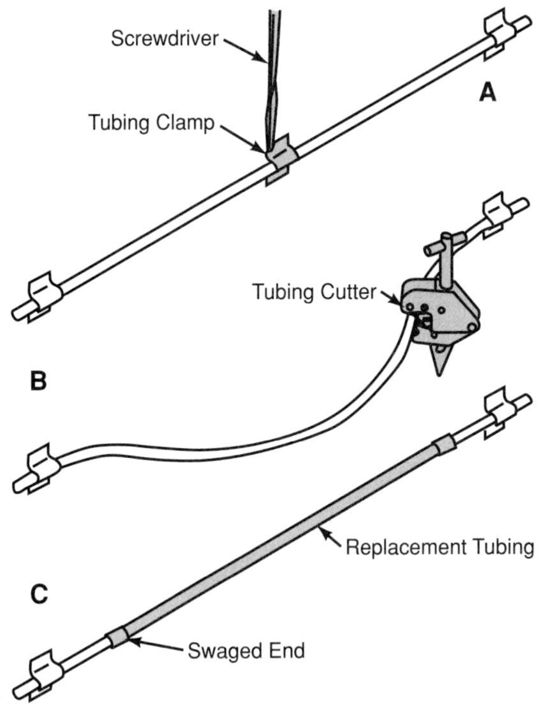

Figure 12-86. *Replacing tubing using swaged ends. A—Removing damaged tubing. B—Cutting old tubing with tubing cutter. C—Installing swaged-end replacement tubing. Joints are brazed. (Frigidaire Company)*

and/or discharge line. This indicates the replacement motor compressor may be too small.

5. Cut the liquid line between the condenser and the refrigerant control. Install a filter-drier of the proper capacity. (See **Figure 12-82.**)
6. Put some vapor refrigerant into the system (about 25 psig [40 psia or 280 kPa]). Check for leaks. When using a recovery unit, recover the refrigerant and prepare to vacuum.
7. Connect a vacuum pump to the system through the gauge manifold. Draw as high a vacuum as possible. Now, with the vacuum pump operating, hold this high vacuum for at least an hour. You may also put refrigerant in again and then evacuate it—the three-step method. (Another way to check for leaks is to close off the connection to the vacuum pump. Observe whether the vacuum on the system remains constant or not. If it doesn't, it indicates that there is a leak in the system.)
8. Charge a small amount of R-11 into the system and purge. This will help clean out the refrigerant lines.
9. Charge the lines with a small amount of the refrigerant used in the system. Adjust the pressure to atmospheric or very slightly above.
10. Reconnect the electrical circuit.
11. Charge the system with the correct amount of refrigerant. It is best to overcharge slightly and purge to the desired quantity. The use of the service manifold makes this operation quite easy.
12. Close all manifold valves, plug in the electrical circuit, and operate the system through several cycles.
13. Adjust the refrigerant charge as described in Section 12.11.
14. Remove the gauge manifold and recovery unit. Carefully seal all openings into the system, depending upon the gauge manifold connections used.
15. It is good practice to place a recording thermometer in the refrigerator cabinet. This enables the operation to be checked continuously for at least 24 hours.

Figure 12-87 shows a replacement motor compressor. It has a suction line, discharge line, and a discharge process tube. In **Figure 12-88,** the process tube has a tubing extension brazed to it.

In many instances, the replacement motor compressors burn out soon after they are installed. Most repeated burnouts are due to the system containing moisture or not being clean enough. Installing both a suction line filter and a high-side filter will reduce repeat burnouts.

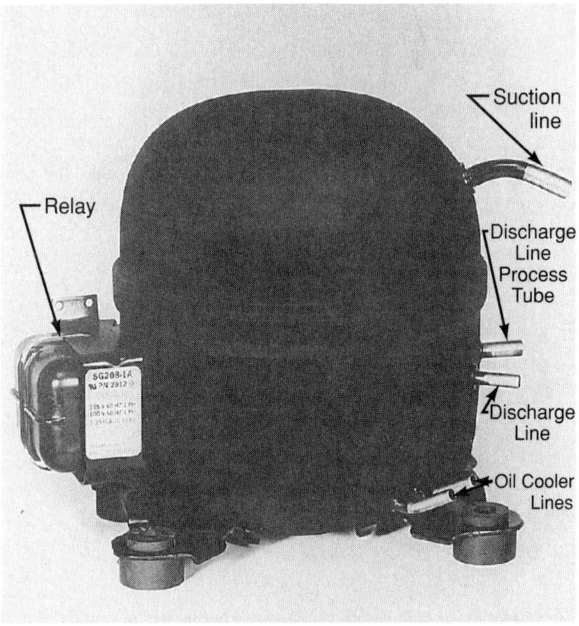

Figure 12-87. *A typical replacement motor compressor. (Americold Compressor Division of White Consolidated Industries, Inc.)*

12.13.4 Condenser Leak Repairs

Condenser troubles are usually caused by leaks or by lint and dirt accumulation on the outside. You may be able to repair leaks without bringing the unit into the shop. He or she must use certified, self-contained recovery equipment. A technician must install piercing valves, recover refrigerant, repair the leak, test, evacuate, and charge. If the leak is not repairable, the condenser must be replaced.

Figure 12-88. *Motor compressor replacement installed, showing suction and discharge lines, process tube with extension, and relay. (Americold Compressor Division of White Consolidated Industries, Inc.)*

12.13.5 Repairing Evaporators

Evaporators may be made of either stainless steel or aluminum. For stainless steel, repairs are made by either brazing or welding with a tungsten arc-inert gas system. For aluminum evaporators, repairs may be made by soldering, brazing, or welding, but the use of epoxy is common. Follow the procedure recommended below for repairing leaks in each type:

Stainless Steel

1. Locate the leak.
2. Remove the refrigerant using recovery/recycling unit.
3. Clean the metal around the leak.
4. Purge nitrogen (very low pressure) while brazing or welding.
5. Polish the weld or clean the brazed joint.
6. Test for leaks.

Aluminum

1. Locate the leak.
2. Remove the refrigerant using recovery/recycling unit.
3. Clean around the leak. The surface oxide is hard and must be removed. Repair right after cleaning as the oxide surface forms quickly. (Sand, file, and then clean with epoxy cleaner.) If leak is a large hole, use a clean small screw or metal plug to fill most of the opening.
4. Mix the epoxy and catalyst.
5. Apply with mixing spatula. (Be sure there is no positive pressure in the system. System must be open to atmosphere at some other opening.)
6. Allow at least an hour for hardening.
7. Sand the patch to a smooth finish.
8. Test for leaks. If system still leaks at repaired joint, remove epoxy by filing and/or grinding. Then install a new patch. Aluminum foil can be used with the epoxy cement to strengthen the joint. It can also improve the appearance of the repair. Another method of repair is to heat the tubing and apply a paste mix (or stick) of special epoxies and resins.

Avoid brazing aluminum tubing, as it overheats too easily. Too much of the tubing will be annealed (softened), weakening the tubing walls. Aluminum tubing is usually 3003 alloy. However, 5005 and 1100 alloy are also used.

Aluminum solder, containing 92% to 100% zinc, is used. The melting temperatures are 700°F to 800°F (370°C to 430°C). Do not use a flux. Aluminum may also be welded. Use an inert gas tungsten arc welding system (sometimes called TIG or GTAW). A tube coupling used to join aluminum and copper is described in Chapter 2.

12.13.6 Servicing Capillary Tubes

The capillary tube is replaced using the same steps as in Section 12.13. One step is different—the capillary tube and filter-drier are removed by loosening the brazed joints.

It is sometimes possible to repair a capillary tube by cleaning it. Disconnect the capillary tube at both ends if possible, then proceed as follows:

1. Attach the capillary tube cleaner to the outlet end of the tube.
2. Build up pressure on the tube to force the wax and dirt out. **Figure 12-89** illustrates a capillary tube cleaner with an adaptor fitting. These cleaners are capable of building up pressure to a possible 20,000 psia (140 MPa).
3. After the capillary tube has been cleaned, continue to flush it out thoroughly. If there is a solid obstruction, a capillary tube kit should be used to remove it, or the entire capillary tube should be replaced.
4. Install a new filter-drier and reconnect the capillary tube into the system. If needed, new tubes must have the same inside diameter (ID) and the same length as the one removed. **Figure 12-90** shows a tool for measuring tubing ID.

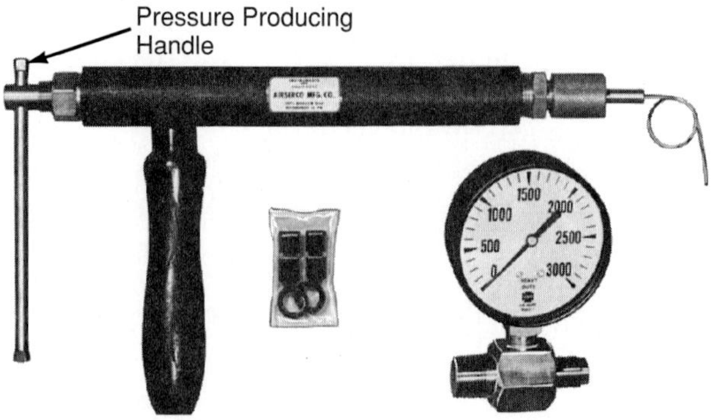

Figure 12-89. *Cleaner used to clear obstructions from capillary tubes. It has a capillary tube adaptor fitting. (Airserco Mfg. Co.)*

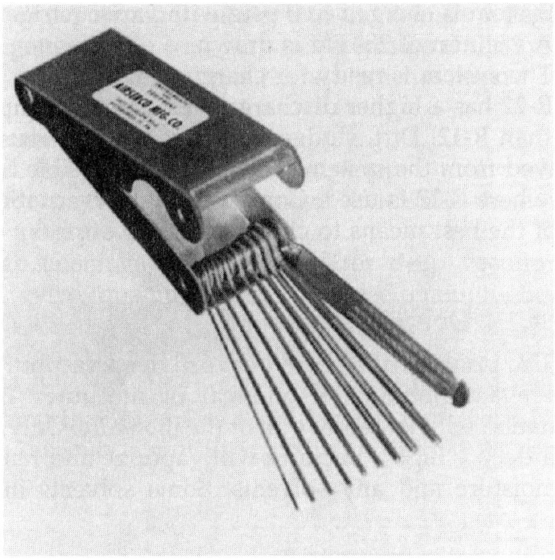

Figure 12-90. *A capillary tube sizing kit. The gauge measures 18 ID sizes, and includes a file for nicking and cutting capillary tube. (Airserco Mfg. Co.)*

If wax caused the tube to become clogged, remove the oil from the system. Replace it with fresh, clean, wax-free oil.

Never use a liquid drier (mainly methanol) to stop moisture from freezing at the refrigerant control. These "antifreeze" substances do not remove the moisture. They merely keep it in circulation. In many cases, they may damage the motor insulation.

There are patented replacement capillary tubes on the market that may be used when necessary. These tubes may be fitted with a calibrated wire inside to provide the proper refrigerant control, or the capillary tube itself is adjustable. See **Figure 12-91. Figure 12-92** shows a replacement capillary tube with adaptor fittings.

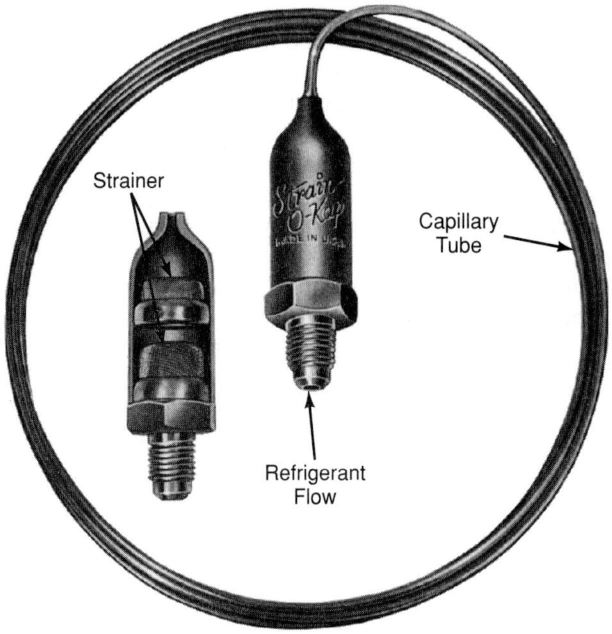

Figure 12-91. *Capillary tube, complete with strainer, meters refrigerant flow. Different strainer size is used for 1/20 hp to 1/4 hp and 1/3 hp to 1 hp units. Capillary tube can be installed by either soldering or by using standard capillary tube fitting. (Watsco Components, Inc.)*

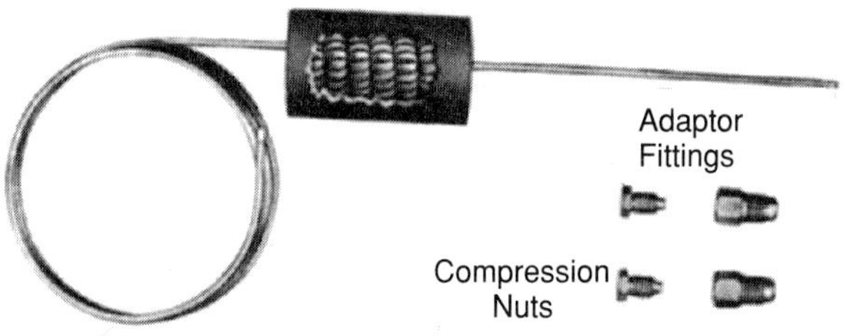

Figure 12-92. *Capillary tube refrigerant control replacement kit.*

12.14 Evacuating a System with a Vacuum Pump

You should use a vacuum pump to remove vapor and moisture from the system .The most common type is called the ***two-stage rotary vacuum pump.*** See **Figure 12-93.** It has two chambers and obtains a deep vacuum. This type of pump is capable of reducing pressure down to the extremely low level of 0.01 micron. However, it is seldom used at these levels in the field. The manufacturer's recommended vacuum is 250 microns to 50 microns, depending on the type of system.

The removal of moisture from the system is accomplished by reducing the pressure. Any moisture will boil (vaporize) and be removed by the gas. A system opened for any type of repair must be completely evacuated. This is necessary to remove air and moisture.

Two different evacuation methods are used:

- Deep vacuum.
- Triple evacuation.

Figure 12-93. *A two-stage rotary vacuum pump. (Robinair Division, SPX Corporation)*

will take out more of the air. Repeat the charging and evacuating. Only about 0.01% of the air will then remain in the system.

If a good vacuum pump is used and draws 1 mm to 2 mm of vacuum (1000 to 2000 microns), the system will need to be partially charged and then evacuated again. However, if a deep vacuum pump draws 50 microns to 100 microns and holds this pressure, the system is clear of moisture and air.

12.14.2 Triple Evacuation

The system should be pressure-tested before triple evacuation. Use either dry nitrogen or dry carbon dioxide. **Use these gases only with a pressure regulator and a large capacity pressure relief valve. This should be set to release at 175 psig (1300 kPa).** Test the system at 150 psig (165 psia or 1140 kPa). The system should hold this pressure after the gas valve is closed for several hours. There should be no decrease in pressure.

Never heat nitrogen or CO_2 cylinders. The maximum cylinder temperature should be 110°F (43°C). A test setup for a triple evacuation is shown in **Figure 12-96.**

Unit evacuating and drying is a very important part of system assembly work. The system should be as clear of air, moisture, solvents, and other foreign matter as possible. It should be as close to 100% clear as possible. Remember that the most careful evacuating and purging will not clean a unit that was carelessly assembled with dirt in the system. See Chapter 10 for recovery and recycling procedures.

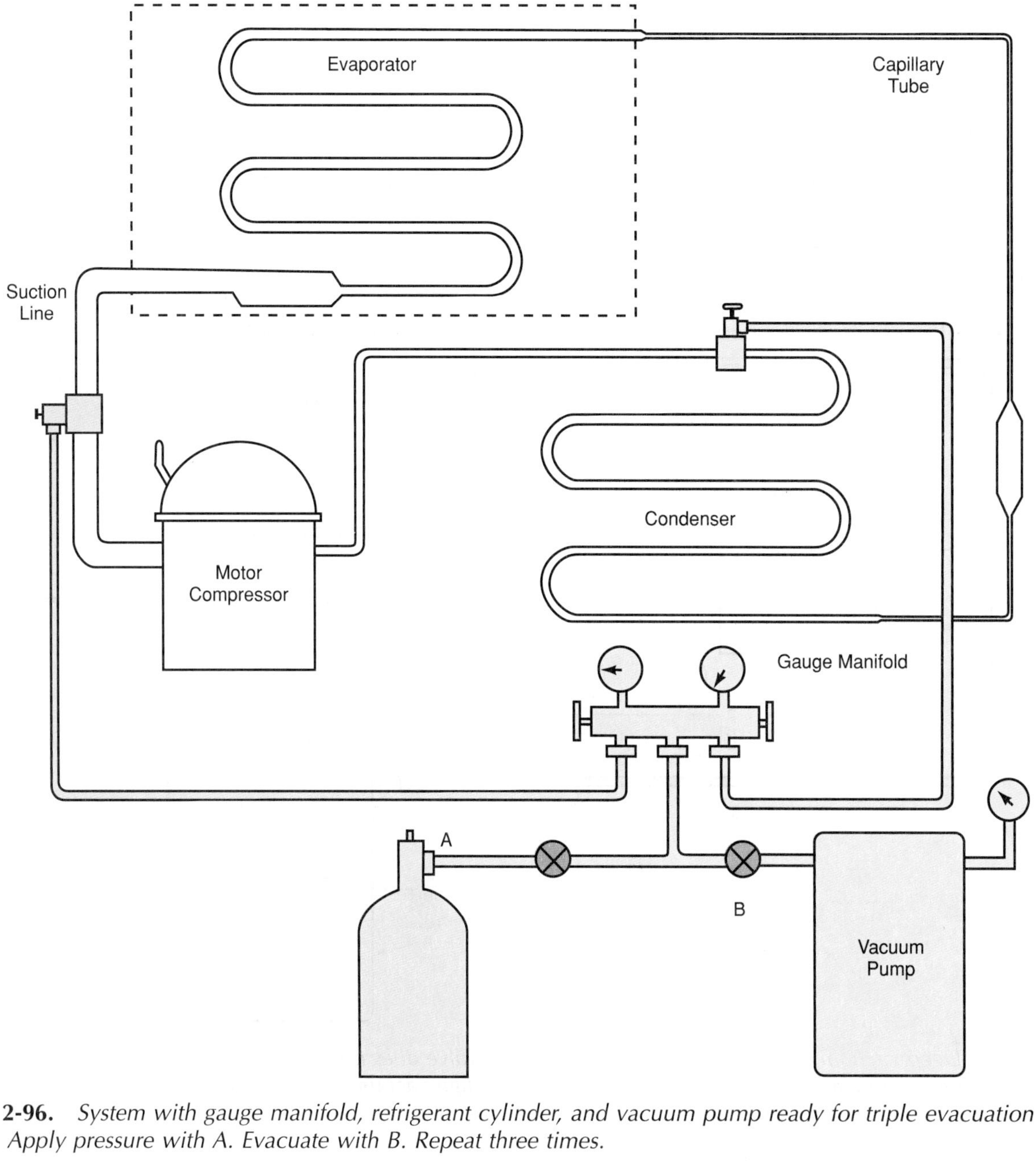

Figure 12-96. *System with gauge manifold, refrigerant cylinder, and vacuum pump ready for triple evacuation process. Apply pressure with A. Evacuate with B. Repeat three times.*

12.14.3 High-Vacuum Pumps

Standard reciprocating-type air compressors do not create a vacuum that is high ("deep") enough to evacuate systems being serviced. Nor are ordinary refrigeration compressors designed to produce the necessary deep vacuum. There are two main types of vacuum pumps, single-stage and two-stage.

Single-stage vacuum pumps are used when the triple evacuation method is employed. ***Two-stage vacuum pumps*** are used when the deep-vacuum (high vacuum) method is used. There are two designs in high-vacuum pumps:

- Rotary pump (oil-sealed).
- Vapor pump (diffusion type).

The rotary pump should be able to pull a 50-micron vacuum pressure. (A 50-micron vacuum is equal to 0.05 torr. A ***torr*** equals 1 millimeter.)

The rotary pump uses two rotors in series (compound pump). See **Figure 12-97.** Most refrigeration compressors pull about 50 torr to 80 torr (50 000 microns to 80 000 microns or 50 mm to 80 mm).

To evacuate, proceed as follows:

1. Pull vacuum with a low-vacuum pump (50 torr to 80 torr).
2. Switch to a high-vacuum pump.
3. Repeat pull-down to 50 microns. A rise (in three minutes or more) to 300 microns indicates that the refrigeration system is dry and evacuated.

Use copper tubing or special metal hose for vacuum pump connections. If standard synthetic charging or servicing hose is used, the sections may collapse at this high vacuum. Also, the synthetic tubing material is ***permeable*** (will allow the passage of gases).

Portable two-stage, high-vacuum pumps are available. These will draw down to less than 1 micron of mercury column. A micron is 1/1000 millimeter (0.001 mm). One inch is equal to 25.4 mm. Therefore, 25 400 microns equal 1″. One micron is close to a perfect vacuum.

A high-vacuum pump will produce a vacuum lower than 29″ Hg (23 mm or 23 000 microns, refer to **Figure 12-98**). Vacuums lower than 29″ Hg are necessary to completely ***dehydrate*** (remove moisture from) the system. Portable high-vacuum pumps are shown in **Figures 12-98** and **12-99.**

Oil in Vacuum Pumps

The oil sight port ("sight glass") permits checking both the oil level and the oil color. This special oil should be replaced frequently, since oil in single-stage pumps rapidly becomes dirty when water and solvent vapor is in the system. Water in the oil will:

- Raise the oil level.
- Turn the oil white and foamy. If this dirty oil is left in the pump, sludge will form in the system.

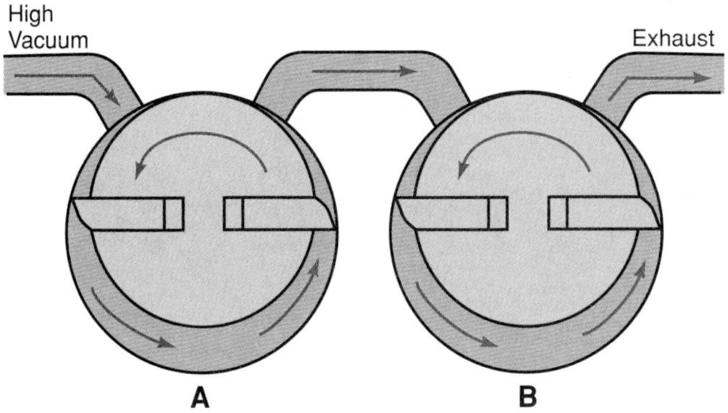

Figure 12-97. *Two-stage rotary high-vacuum pump. A—First stage. B—Second stage.*

Figure 12-98. *Portable high-vacuum pump mounted on stand for transport. Note flexible metal evacuating line. This is a large capacity line and will not collapse. Two hand valves control evacuating operation. (Airserco Mfg. Co.)*

Figure 12-99. *Portable high-vacuum pump. It may be used to remove air or refrigerant from system. (Airserco Mfg. Co.)*

For good results, change oil before each system pump-down, or test pump with valves closed and vacuum gauge connected. If a pump will not pull down to high vacuum, change the oil.

Compound (two-stage rotary) ***pumps*** need oil changes after about 10 pump-downs. Most of these pumps will pull a vacuum of 20 microns or better.

High-Vacuum Gauges

To measure deep ("high") vacuum, an electronic or a solid-state thermistor vacuum gauge is used. A regular compound gauge cannot read accurately to micron levels. A high vacuum gauge is shown in **Figure 12-100.** A vacuum from 29.25″ Hg (about 17 000 microns) to 29.9″ Hg (540 microns) is necessary to allow the moisture inside the system to evaporate at room temperature.

The inside design of a vacuum gauge tube used on an electronic mechanism is seen in **Figure 12-101.** The sensing element is a thermocouple. It is important to use the tube in an upright position to keep out foreign matter.

The gauge's electrical circuitry is shown in **Figure 12-102.** The filament gets warmer as air pressure around it decreases. (There are fewer air molecules to remove heat.) The thermocouple gets warmer and the increase in emf is registered on the meter. The meter is calibrated in microns. The vacuum dial scale is shown in **Figure 12-103.**

If vacuum gauge reading levels off at 5000 microns, ice or free water is in the system. Ice may be located by a cold spot, frost, or sweat on outside of system. Stop the pump and allow the ice to melt, or use radiant heat. *Never allow the system pressure to enter a high-vacuum gauge.* Ice or free water will damage the gauge.

The thermocouple vacuum gauge has two parts, a meter and a tube that threads into the refrigeration system. The tube should be right-side-up with threads down. The tube sometimes collects vapors and/or oil. It

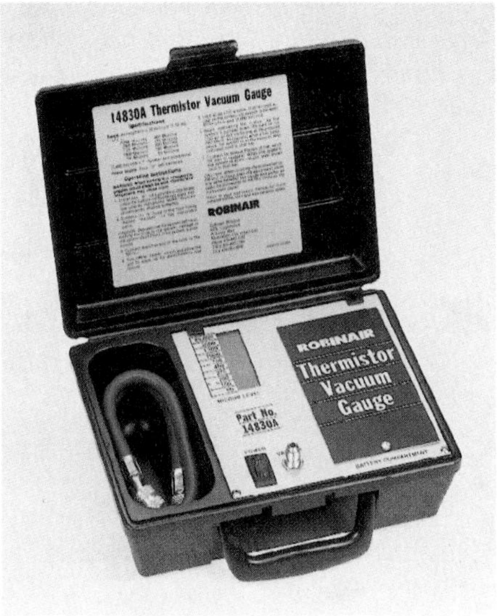

Figure 12-100. *Solid-state thermistor vacuum gauge used to measure the vacuum level of a system. (Robinair Division, SPX Corporation)*

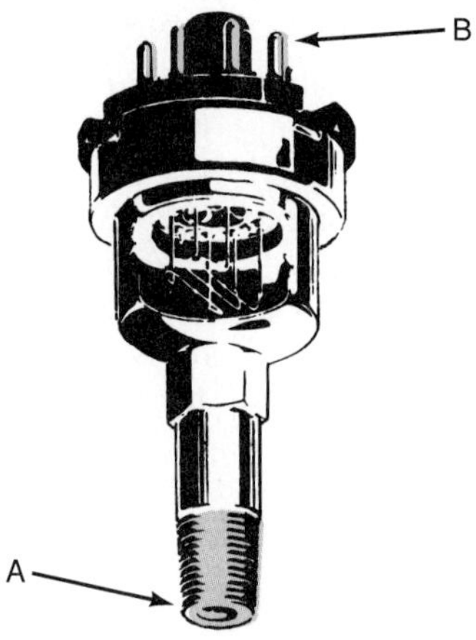

Figure 12-101. *Internal construction of high-vacuum gauge tube. A—Pipe-thread connection to system manifold. B—Electrical connections to vacuum gauge. (Airserco Mfg. Co.)*

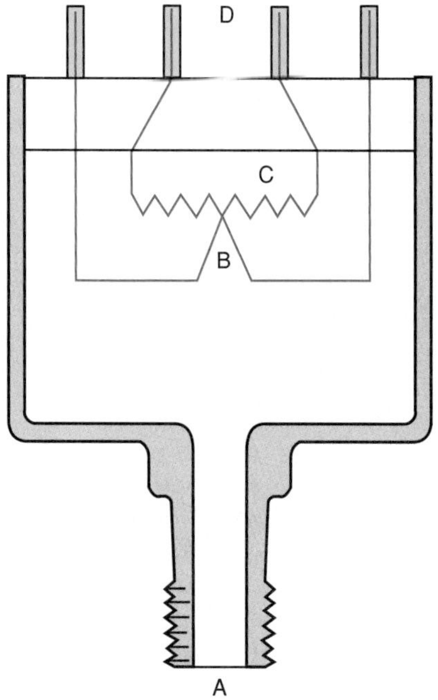

Figure 12-102. *Schematic internal design of high-vacuum gauge. A—Connection to system being checked. B—Thermocouple. C—Resistance unit. D—Electrical leads.*

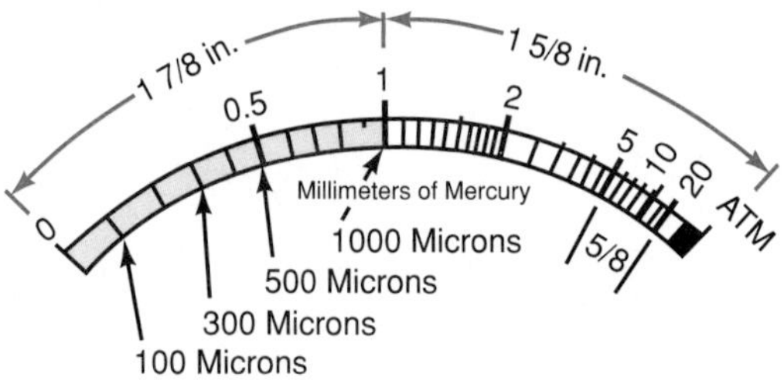

Figure 12-103. *Dial scale of high-vacuum gauge. Note that pressure between 0 microns and 1000 microns has been greatly expanded for easy reading. (Airserco Mfg. Co.)*

can be cleaned by putting a solvent in the opening with an eyedropper. R-23 is a good cleaner. Clean as follows:

1. Fill.
2. Rock tube gently.
3. Empty.
4. Repeat Steps 1–3 two or three times.
5. Clean with an alcohol rinse.

12.14.4 Using a Vacuum Pump

Clean the instrument dial cover with soap, water, and facial tissues. The crystal over the dial is plastic and solvents will fog it.

If a system has a leak, the vacuum produced will not be a high vacuum. The dial needle will rise steadily when the valves are closed. Moisture in the system will also cause the vacuum produced to be less than desired. When the valve is closed, the dial needle will rise and level off. It levels off at a pressure corresponding to the water vapor pressure at that temperature. **Figure 12-104** shows the two conditions of a pressure-time graph. These are made with the valve to the vacuum pump closed.

The following procedure ensures complete dehydration (drying out) of a wet domestic system.

1. Connect a 250 VA (250 W) lamp (in place of a fuse) in series with a test cord.
2. Attach the cord to the compressor run and common terminals.
3. Connect to a 120 V outlet.

Since the lamp functions as a resistor, this procedure allows only about 30 V to go through the running windings. The compressor windings are warmed. Any moisture that may be trapped in the system is vaporized. *The compressor should not run during this operation.*

If a manifold is used, it must be degassed. This is accomplished by pulling a vacuum on it for several days. The procedure will remove all vapors absorbed in the walls and cracks of the manifold. (These residual gases may make you believe there is a leak in the system.) Evacuate the system from both the high side and low side.

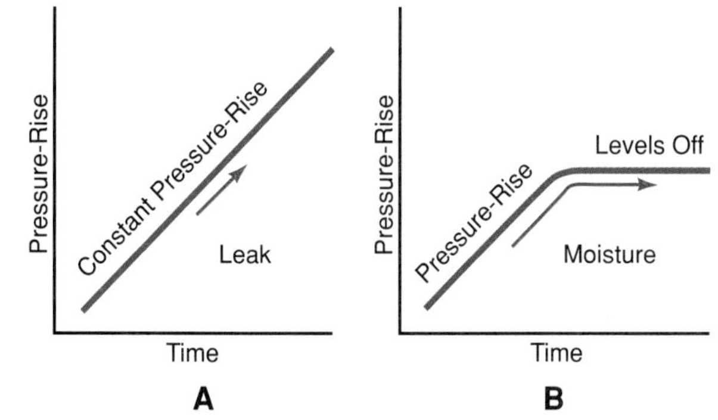

Figure 12-104. *A leak or moisture will affect high-vacuum gauge readings. A—Shows effect of a leak. B—Shows effect of moisture in system. (Airserco Mfg. Co.)*

Use as large a diameter vacuum line as possible. Keep the vacuum line as short as possible. Vacuum pump sizes are given in cubic feet per minute (cfm):

- 1.5 cfm (good for 3-to 5-ton domestic systems).
- 3–5 cfm (5- to 100-ton medium systems).
- 10–15 cfm (large systems over 100 tons).

Always ***break the vacuum*** (allow air into) of a vacuum pump before storing it. Otherwise the cylinder will fill with oil and become oil-locked (hard to turn).

Figure 12-105 is a chart of the various pressures that are possible with different types of pumps. Note the

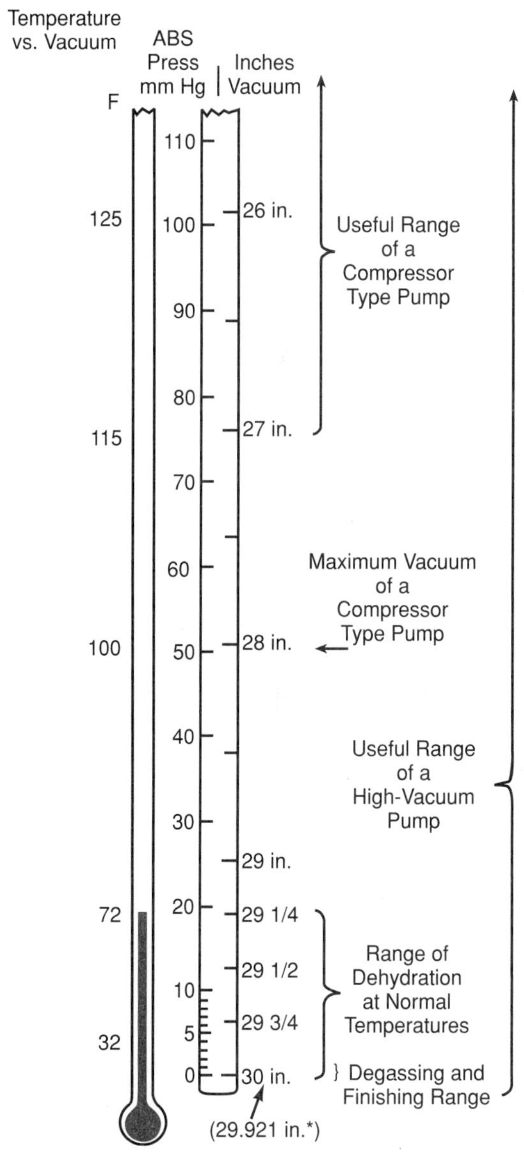

*Note: 29.921 is from 76.000 cm divided by 2.540 cm per in.

Figure 12-105. *Pressure scales used during high-vacuum evacuation of system. Temperature scale shows evaporating temperature of water at these pressures. For example, water will evaporate (boil) at 72°F (22°C) at 20 mm Hg pressure (20 000 microns or 3 kPa). (Airserco Mfg. Co.)*

evaporating temperature of water at the various pressures.

During manufacture, systems are evacuated to 50 microns to 100 microns before oil is added. In the field, oil is already in the unit. Evacuating to 50 microns may cause some of the oil to vaporize.

The motor compressor depends on gas flow to cool the motor windings and compressor. Therefore, you should avoid using it as a vacuum pump. The motor compressor may heat up and be damaged.

12.15 Overhauling a Hermetic System

A complete overhaul of a hermetic system is usually done in a specialty repair shop. The shop must be well equipped. Some of the equipment needed includes:

- A compressor opener.
- A welding system (a tungsten arc welder).
- A cleaning booth.
- A paint booth.
- Benches and vises.
- Storage cabinets.
- A lathe and surface grinder.
- A recovery/recycling unit.
- A compressor test bench.
- A deep vacuum system.
- Compressed air (dry and clean).

12.15.1 Removing the System

Sometimes, mechanical or electrical trouble cannot be fixed easily. You may have to remove the unit from the cabinet and send it to a shop for repair.

It is important to protect the refrigerating unit while it is being moved. Cradles or crates should be used to hold mechanisms. Wooden frames and C-clamps will hold the parts down. This will keep them from being damaged while in transit.

Moving the complete cabinet in a truck requires shrouding it in a special padded blanket. The procedure for removing the unit from the cabinet varies. Some evaporators are removed from the rear of the cabinet. Others are removed from the front (by way of the cabinet door). Cabinets that have the evaporator removed from the rear are not difficult to dismantle. Skill and patience are required to remove the unit from the front of the cabinet The compressor and condenser are sometimes fastened to the rear of the cabinet by four to six mounting screws and bolts. Be careful not to damage the mechanism while removing these devices. Be careful not to kink or buckle the refrigerant lines when removing the mechanism.

The refrigerant lines are sometimes run just under the door frame breaker strips. These breaker strips must be removed in order to remove the lines. Do this carefully, since the strips are brittle. If allowed to come up to room temperature, they will be more flexible. These strips are usually made of a plastic material. Several other designs are shown in Chapter 11. There are several methods used to fasten the breaker strips to the cabinet shell and the liner. On double-door cabinets, the strips between the two compartments must be removed before the unit can be removed.

12.16 Review of Safety

Working safely implies three things:

- **Safety to you, the service technician.**
- **Safe handling of tools, instruments, and equipment.**
- **Proper preservation of food so that it is kept in a safe condition.**

There are a few great hazards to a technician in refrigeration service work. The following items are some of the more common of these hazards to remember.

Good housekeeping is very important. Keep the work area clean. Keep oil and water off the floor.

The refrigerator is usually motor-driven. Therefore, electrical supply to the motor and the controls presents some hazards. If the system is not properly insulated and handled, you may receive a dangerous electrical shock.

Always disconnect the electrical circuit or make sure all electrical devices are safe before starting on a job. It is best to remove rings and wristwatches when working on electrical equipment. An electrical short across a ring or wristwatch can cause a severe burn.

Electrical shocks occur when a technician comes in contact with an electrical current and a ground. Avoid working on any electrical circuit if standing on a damp floor. Also, do not work on electrical circuits if one hand is touching a water pipe.

In some refrigerating systems, the condensing pressures run quite high. Head pressure becomes high if the condenser is not losing enough radiating heat. It also becomes high if there is a restriction because of dust or lint. This can become dangerous if it reaches the bursting point. Watch the gauges carefully. The system should shut off if the pressure is too high.

A hot compressor, exhaust line, or condenser can cause burns.

When heating a charging cylinder or compressor with an electrical insert or heat gun (to speed up vaporization for a better purging result), always use a pressure control relief valve and thermostat. These will provide the proper temperature and pressure safety controls. (Refer to Section 12.11.)

Freezing of skin is a hazard when handling some refrigerants. If the liquid refrigerant spills on skin, rapid evaporation will lower the skin temperature. It may be lowered to considerably below the freezing temperature. Liquid refrigerant on the face or eyes is very dangerous. Always wear goggles or a face shield when handling liquid refrigerants.

Dropping heavy objects on your feet or toes is another potential hazard. This can be avoided by using

proper trucking and hoisting equipment. You should wear safety shoes with metal tips to protect your toes.

Arm and leg muscles must be used correctly when lifting heavy objects. This will prevent back injury.

Fire is always a potential hazard. Never use gasoline or any other flammable material when cleaning.

Use face shield and rubber gloves when handling oil which may be acidic. (This often occurs in hermetic systems where motor compressor has burned out.) If burned with acidic oil, wash the affected area with water, apply ice, and see a physician immediately.

There is some danger in using common tools such as screwdrivers and wrenches. If they should slip, knuckles may be skinned. Be careful to handle tools correctly.

Vacuum pumps can be a hazard. Clothing may catch on the rotating shaft, fans, belts, or other parts. If a hand is drawn into a pulley, a severe injury may result.

General safety precautions include the following:

1. Wear goggles when working on a charged system. When using a flame, be sure there are no flammable materials nearby. Be sure the pressure in the system is close to atmospheric pressure before opening it.
2. For every operation, a technician must be trained to first ask, "Is there a possible safety hazard in this operation?" Proper precautions should be taken to reduce any hazard.
3. Never, under any circumstances, use carbon tetrachloride for cleaning. Its use is illegal. Its effects are cumulative in the human body; continued use may be fatal.
4. Avoid testing with excessive pressures. Test systems at a maximum of 170 psig (165 psia or 1140 kPa). Always wear goggles when testing.
5. Recover/recycle all refrigerants.

12.17 Test Your Knowledge

Please do not write in this text. Place your answers on a separate sheet of paper.

1. What indicates an electrical failure in a door mullion heater?
   A. Frozen to cabinet.
   B. Sweats.
   C. Electrical shock.
   D. All of the above.
2. Which of the following is *not* a way to prepare a hermetic mechanism for removal?
   A. Use crates or cradles to protect the system.
   B. Use C-clamps to hold down all parts during transit.
   C. Disconnect power and remove.
   D. Vent all refrigerant.
3. What is the most popular method of sealing tubing joints when servicing hermetic systems?
   A. Soldering.
   B. Brazing.
   C. Bonding.
   D. All of the above.
4. What trouble may be indicated by a warm evaporator line?
   A. A shortage of vapor.
   B. An excess of refrigerant.
   C. A shortage of refrigerant.
   D. Excess vapor.
5. A frosted suction line with an excessive low-side pressure is the indication of what trouble?
   A. Presence of liquid refrigerant in the suction line.
   B. A lack of refrigerant.
   C. Air in the compressor.
   D. A restriction on the high side.
6. To what part of the system is the high-pressure gauge usually connected?
   A. The evaporator.
   B. The low side.
   C. The high side.
   D. None of the above.
7. Which of the following is *not* a problem indicating that a hermetic unit needs refrigerant?
   A. A partially frosted evaporator.
   B. Low head pressure.
   C. Low low-side pressure.
   D. High head pressure.
8. What is a micron?
   A. 1/1000 of a millimeter.
   B. 23 mm.
   C. 0.5 millimeters of mercury.
   D. None of the above.
9. At what temperature does water evaporate with a 28″ Hg vacuum?
   A. 100°F (38°C).
   B. 250°F (121°C).
   C. 150°F (66°C).
   D. None of the above.
10. Which of the following would *not* be the possible result of installing a capillary tube of a longer length?
    A. Too much capillary resistance.
    B. Not enough capillary resistance.
    C. Liquid refrigerant backing up into the condenser.
    D. Starving of the evaporator.
11. Which type of pump is needed to produce a high vacuum?
    A. A two-stage rotary pump.
    B. A single-stage pump.
    C. A single-stage vapor pump.
    D. A two-stage pump, either rotary or vapor.
12. Where are piercing valves installed?
    A. Suction tubing.
    B. Discharge tubing.
    C. Process tubes.
    D. All of the above.

13. When should you install a new liquid line filter-drier?
    A. When a new motor compressor is installed.
    B. When the filter is clogged.
    C. Once a year during system maintenance.
    D. Both A and B.
14. If the low-side pressure reads 28″ Hg vacuum, what is wrong?
    A. A shortage of refrigerant or a restriction.
    B. A leak in the tubing.
    C. Liquid and refrigerant in the suction line.
    D. Both B and C.
15. If the suction line is frosted, the trouble could be ______.
    A. a shortage of refrigerant
    B. too much refrigerant
    C. a broken thermostat
    D. Both B and C.
16. What is an indication of a hot gas valve that is stuck in the closed position?
    A. The evaporator is warm.
    B. The evaporator is overloaded with ice.
    C. A hissing sound.
    D. The high pressure is decreasing quickly.
17. Is it necessary to check the service lines and gauge manifold for leaks?
    A. Yes.
    B. No.
    C. Only if they are over five years old.
    D. Only if there is a hissing sound.
18. What type of gauge is used to read a 25-micron pressure?
    A. An electronic high-vacuum gauge.
    B. A single stage pump.
    C. A two-stage pump.
    D. A rotary pump.
19. Is the refrigerant charged into the system in the liquid form or vapor form?
    A. Liquid.
    B. Vapor.
    C. This depends upon the location where you install the piercing valve.
    D. Either liquid or gas is acceptable.
20. How can a compressor which is stuck be started?
    A. Rewire circuit, temporarily adding a reverse-start capacitor.
    B. Connect to 240 V.
    C. Connect to 240 V with a 100-mfd start capacitor in circuit.
    D. Any of the above.

Chapter 13

COMMERCIAL SYSTEMS

Modules:

Key Words:

air defrosting
auger
evaporative condenser
fill
sight glass
surge tank
sweet water bath
thermostatic motor control
tube-within-a-tube condenser
water-cooling tower

Learning Objectives:

After studying this chapter, you will be able to:

- Explain the differences between the mechanism of commercial refrigeration systems and domestic systems.
- Compare the differences between various commercial mechanisms.
- Describe how each mechanism (condenser, evaporator, and compressor) operates.
- Discuss the theory and operation of control devices.
- **Follow approved safety procedures.**

COMMERCIAL SYSTEMS MODULE

13.1 Construction of Refrigerating Mechanisms

Operating fundamentals of domestic refrigeration systems also apply to commercial systems, **Figure 13-1.** However, many commercial systems using mechanical cycle mechanisms differ from the domestic mechanism. These differences are chiefly in the following:

- The number of evaporators connected to a single condenser.
- Electrical voltage.
- Compressor design and size.
- Condenser unit design and size.
- Motor controls, both temperature and pressure.
- Refrigerant controls, both liquid and vapor.
- Piping.
- Variety of evaporator designs.
- Defrosting systems.
- Variety of refrigerants used.

13.2 Mechanical Cycle

Some large commercial compressor systems are semi-hermetic. These may also be serviced. Many single unit applications use a completely hermetic system. Such single unit applications include bottle coolers, beverage dispensers, and ice cream cabinets. Parts of the system are similar to the designs shown in Chapter 11. The design of the cabinet varies with its application.

In some cases, the cycles are more complicated. They may include unloading and automatic defrost devices, multiple evaporators, additional controls, and more complicated piping.

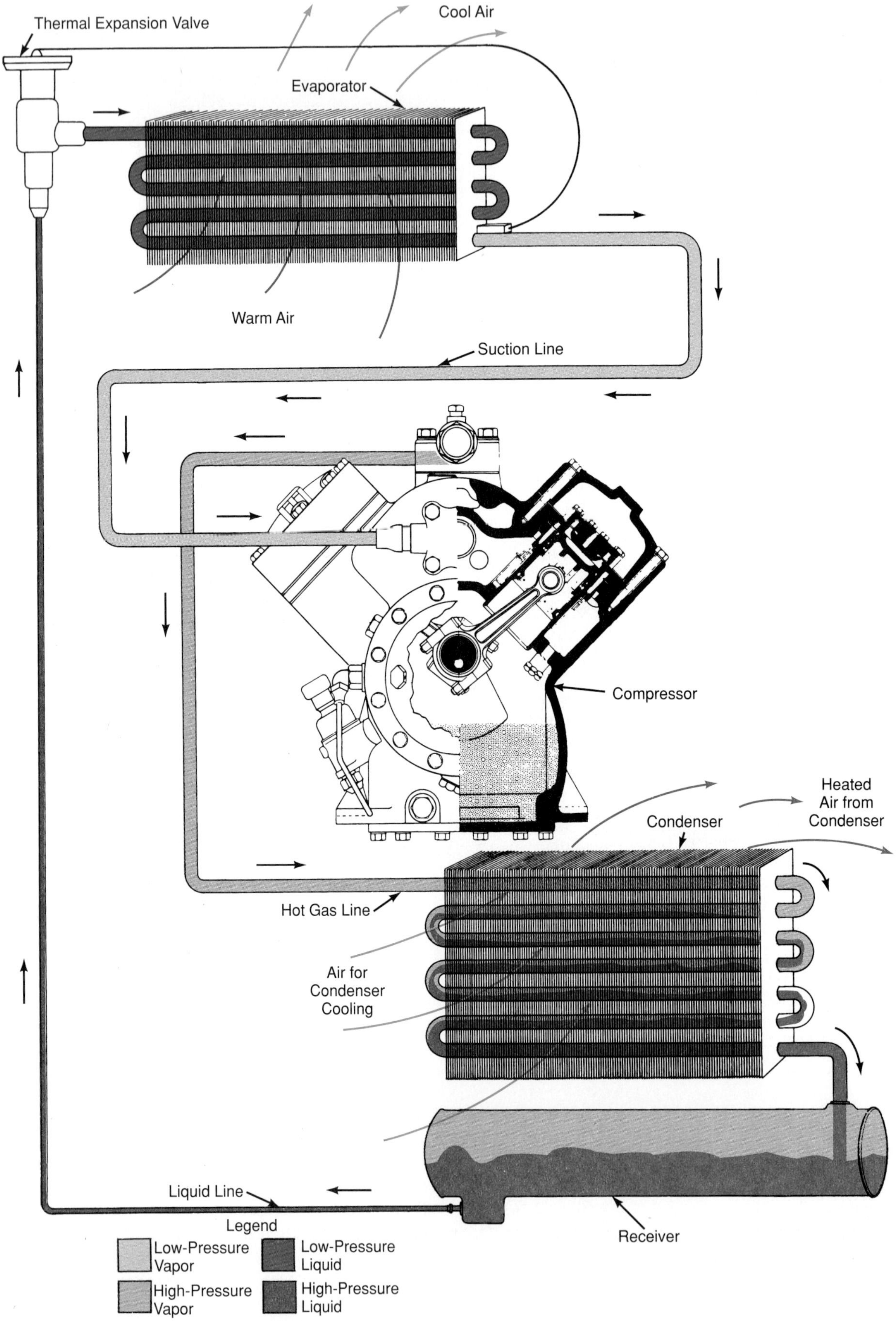

Figure 13-1. *A serviceable commercial system with air-cooled condenser, thermostatic expansion valve, and V-type compressor. Note size of compressor. (Carrier Corp., Subsidiary of United Technologies Corp.)*

13.3 Complete Mechanical Mechanism

Figure 13-2 shows a single-unit mechanism. It includes:

- High-pressure side:
  A. Compressor, usually hermetic.
  B. Condenser, usually air-cooled.
  C. Liquid receiver, when a thermostatic expansion valve or automatic expansion valve is used.
  D. High-pressure safety motor control.
  E. Liquid line with drier and sight glass.

The refrigerant control is at the division point between the low side and the high side of the system. It consists of an automatic thermostatic expansion valve or capillary tube.

- Low-pressure side:
  A. Evaporator.
  B. Low-pressure or temperature motor control.
  C. Suction line—some with filter-driers and surge tanks.

The multiple mechanism is shown in **Figure 13-3.** It includes:

- High-pressure side:
  A. Compressor, often with an oil separator.
  B. Condenser, water- or air-cooled.
  C. Liquid receiver.
  D. High-pressure motor control.
  E. Liquid lines with a drier and a sight glass.
  F. Water valve, used with a water-cooled unit.

The refrigerant control is the division point between the high-pressure side and the low-pressure side.

- Low-pressure side:
  A. Refrigerant controls, two or more. These are usually thermostatic expansion valves.
  B. Evaporators, two or more. These may be natural convection, forced convection, or submerged.
  C. Motor control, usually operated by pressure.
  D. Suction lines with drier and suction pressure regulator.
  E. Two-temperature valves for multiple temperature installation.
  F. Surge tanks for reducing rapid pressure changes.
  G. Check valves for multiple temperature installations.

There are many varieties of commercial systems. A water chiller system is shown in **Figure 13-4.**

Subcooling is used on low-temperature units such as display cases, freezers, etc. The process of subcooling reduces the refrigerant temperature in the liquid line below the saturated temperature. The lower the temperature in the liquid line, the greater the system's heat removal capacity. This will result in a more efficient system.

Subcooling is accomplished by refrigerating the liquid line on a low-temperature system. A high-temperature system, such as an air conditioning unit, is used. High-temperature systems remove Btus three times more efficiently than low-temperature refrigeration systems. The high-temperature air conditioning and the low-temperature freezer cases work together. The two systems increase the overall efficiency of the refrigeration process.

Subcooling systems can also be added to existing refrigeration systems in supermarkets. **Figure 13-5**

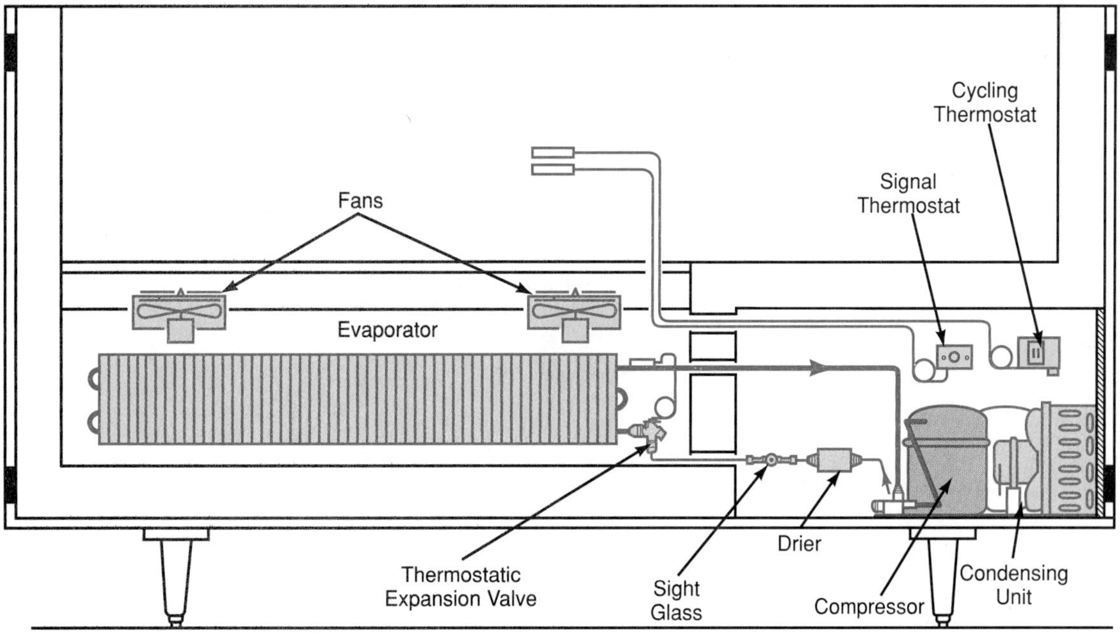

Figure 13-2. *This open-top display cabinet has a self-contained refrigeration unit. (Danfoss Automatic Controls, Division of Danfoss, Inc.)*

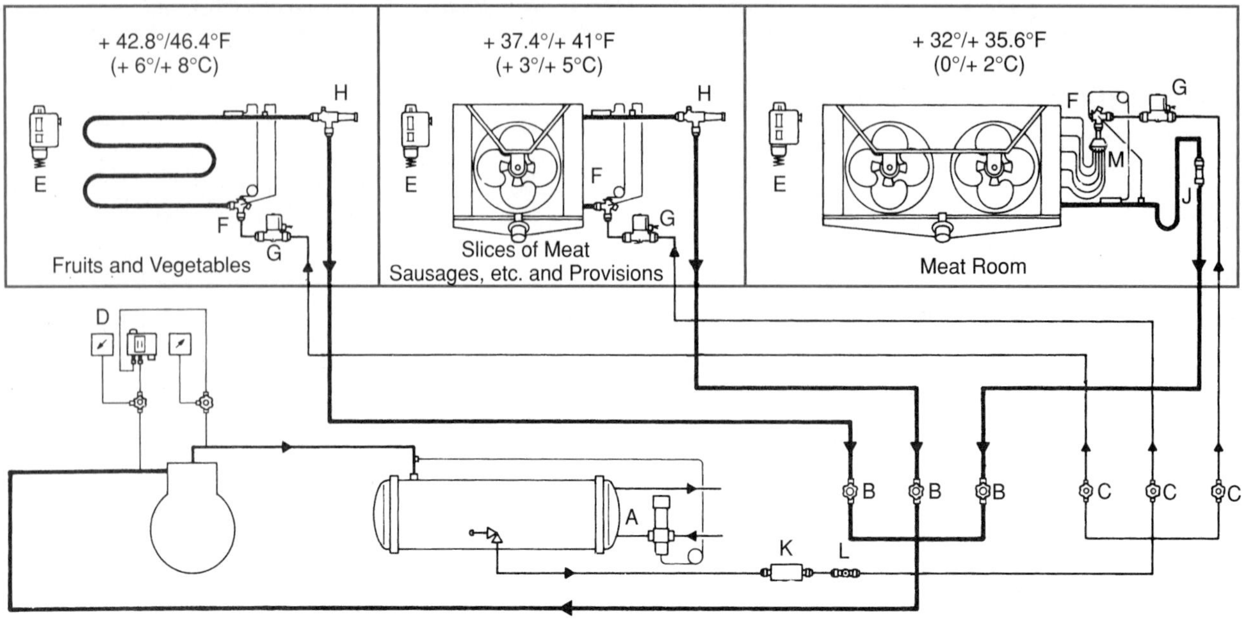

Figure 13-3. *A multiple evaporator system. A—Water valve. B—Suction line shutoff valves. C—Liquid line shutoff valve. D—Low- and high-pressure motor control. E—Thermostats. F—Thermostatic expansion valves. G—Liquid line solenoid valves. H—Two-temperature valves. J—Check valve. K—Drier. L—Sight glass. M—Distributor.*

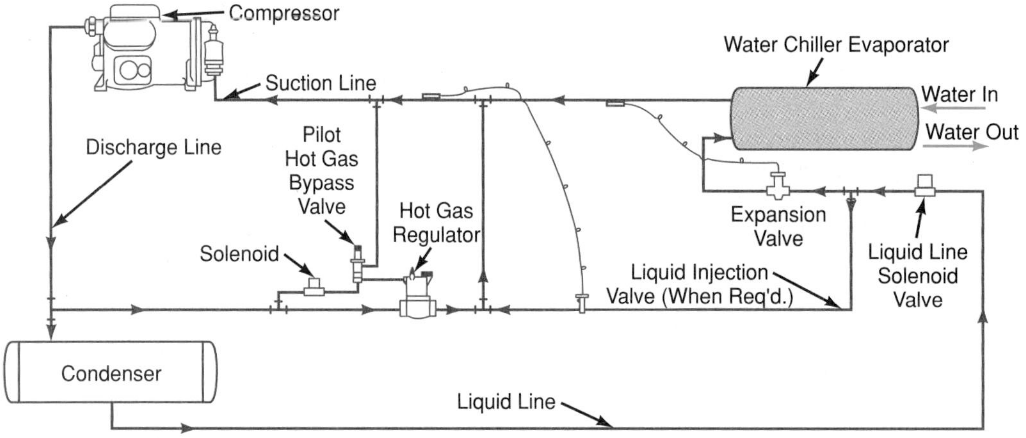

Figure 13-4. *Schematic of a water chiller. Hot gas bypass keeps low-side pressure high enough to prevent freezing of chilled water. Liquid injection keeps suction vapor cool enough to prevent overheating of compressor motor.*

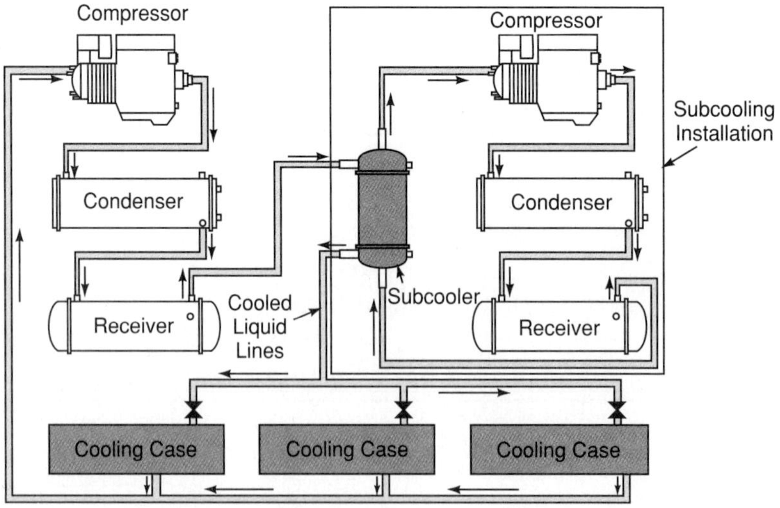

Figure 13-5. *This subcooling installation uses existing air conditioning system. Subcooler was installed to cool liquid lines leading to the two display cases. (Standard Refrigeration Co.)*

shows a subcooler installation. The subcooler is installed into the store's present high-temperature air conditioning system.

Condensers are normally mounted on a steel base. In the external drive unit, the motor is mounted outside the compressor. It drives the compressor either directly or with one or more belts as in **Figure 13-6.**

In the hermetic unit, the motor is connected directly to the compressor. **Figure 13-7** shows a cutaway and the internal parts of a unit. The crankcase and system pressures are equalized upon start-up. This is done by means of the crankcase pressure equalizing tube and spinner tube assembly. Equalizing the pressure prevents the oil from leaving the compressor during start-up.

For large commercial installations, compressors are made with three, four, five, six, seven, or more cylinders.

Compressors are also named after their cylinder arrangement. Some of these are vertical single, horizontal single, 45° single (inclined), vertical two cylinder, V-type two cylinder, W-type three cylinder, radial three cylinder, vertical four cylinder, and V-type four cylinder.

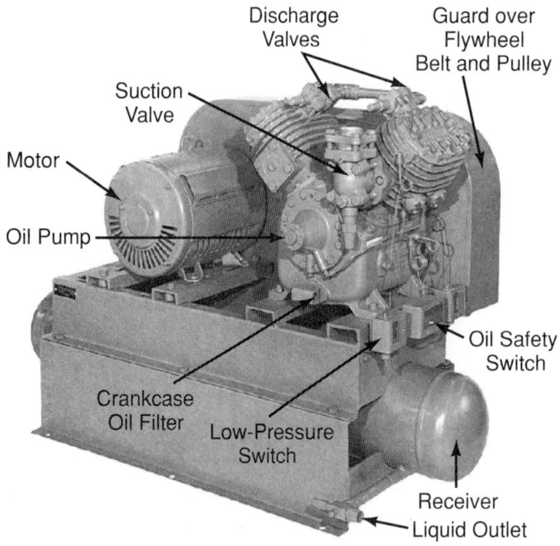

Figure 13-6. *Condenser with belt-driven, four-cylinder, air-cooled compressor. Unit is mounted on steel base. (Grasso, Inc.)*

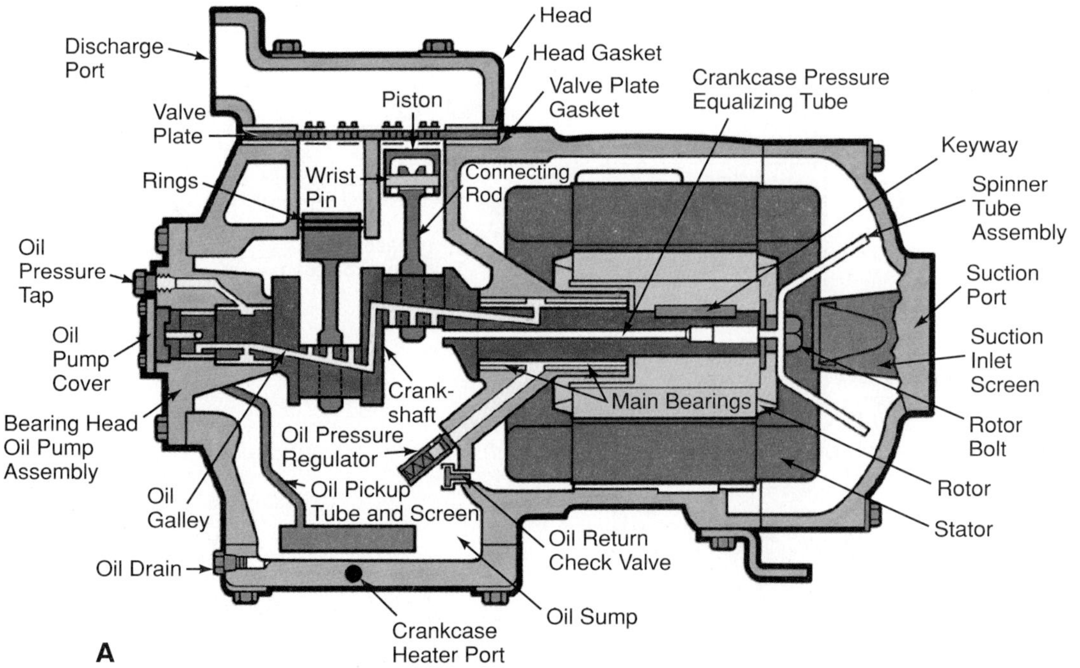

Figure 13-7. *Six-cylinder hermetic unit. A—Cutaway view. Note the oil passage system used to lubricate the unit. B—External view of the system. (Carlyle Compressor Company, Division of Carrier Corporation)*

There are also many crankshaft arrangements. The inside construction of a multiple-cylinder, serviceable hermetic motor compressor is shown in **Figure 13-8.**

A six-cylinder compressor with an external drive motor is shown in **Figure 13-9.** It has a crank throw type crankshaft. A steel frame is used to hold the shell and tube condenser on the hermetic unit, **Figure 13-10.**

13.3.1 Commercial Hermetic Units

Several companies now produce hermetic units of over 20 horsepower. Some have a bolted assembly. These are often called "field serviceable" or "accessible." Some units are sealed in a welded casing. Both types are equipped with service valves. They may be connected

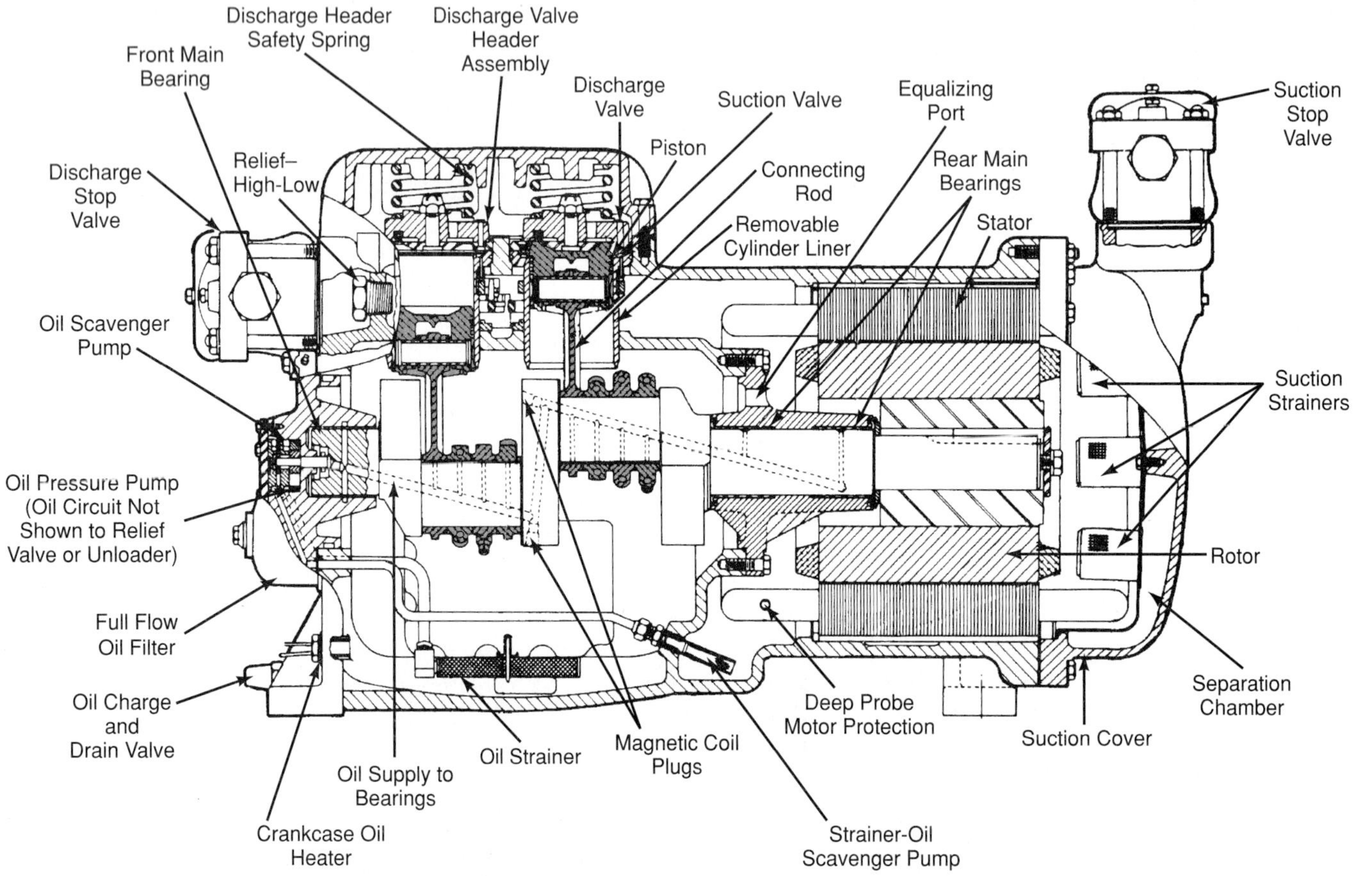

Figure 13-8. *Bolted type field serviceable eight-cylinder hermetic motor compressor.*

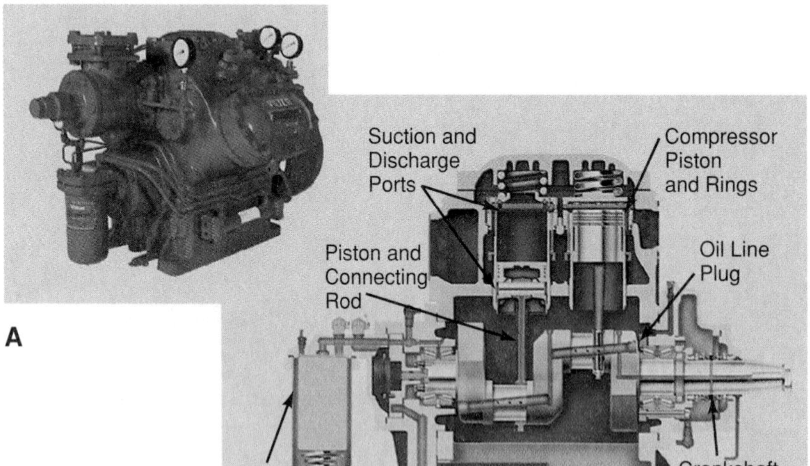

Figure 13-9. *Serviceable six-cylinder W-type compressor designed for external belt-driven or direct-driven engine. A—External. B—Cutaway view. (Vilter Mfg. Corp.)*

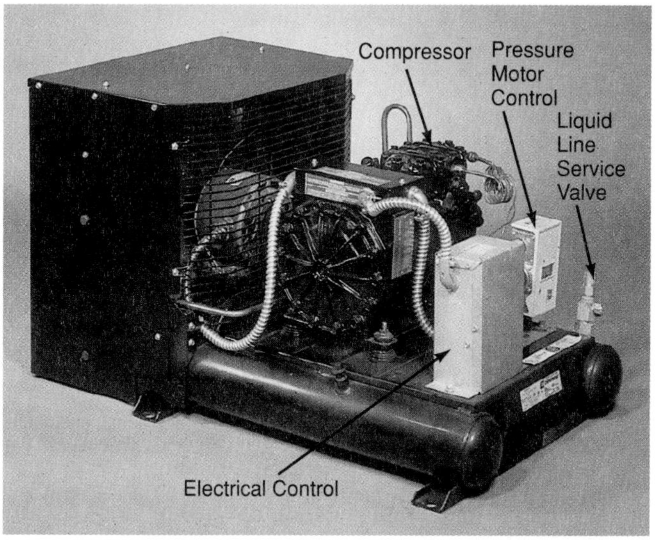

Figure 13-10. *Serviceable commercial hermetic condenser. Note the electrical panel. Wiring is enclosed in the flexible metal conduit because the unit is spring-mounted to allow movement. Note, also, the pressure motor control mounted next to the electrical control. (Copeland Corp.)*

to any type of evaporator and used for many different applications.

An advantage of hermetics in the commercial field is the elimination of the crankshaft seals and belts. Any trouble in the compressor mechanism involves both the compressor and the motor. Therefore, the technician working on these hermetic units must keep moisture and dirt out of the system.

An outdoor hermetically sealed air-cooled condenser is shown in **Figure 13-11.** It has a fan condenser, shroud, and service valve. The inside of a welded hermetic motor compressor is shown in **Figure 13-12.** Smaller units have single-phase motors. Units over 1/2 hp generally have three-phase motors, **Figure 13-13.**

Figure 13-11. *Commercial hermetic compressor condenser with air-cooled condenser. (Dairy Equipment Co.)*

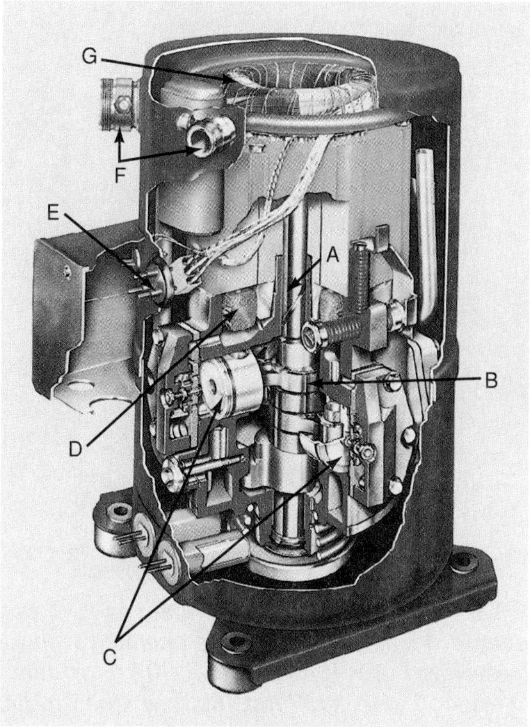

Figure 13-12. *Two-cylinder hermetic motor compressor. This type is used on air conditioning, heat pump, and commercial condensers. A—Crankshaft. B—Connecting rod. C—Pistons. D—Motor windings. E—Electrical terminals. F—Suction and discharge openings. (Tecumseh Products Company)*

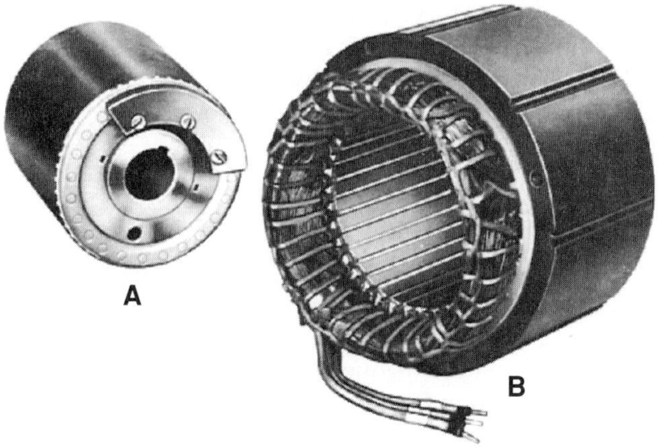

Figure 13-13. *Three-phase motor is used in larger hermetic refrigeration units. A—Rotor. B—Stator. (Emerson Electric Co.)*

The condensers may be installed in many different ways. Some are mounted on the roof. Others are mounted on the same floor level with the evaporator. These units are located in different rooms or outside the building. See **Figure 13-14.** Note the suction service valve on the compressor, the liquid receiver service valve on the receiver, and the forced-air convection condenser.

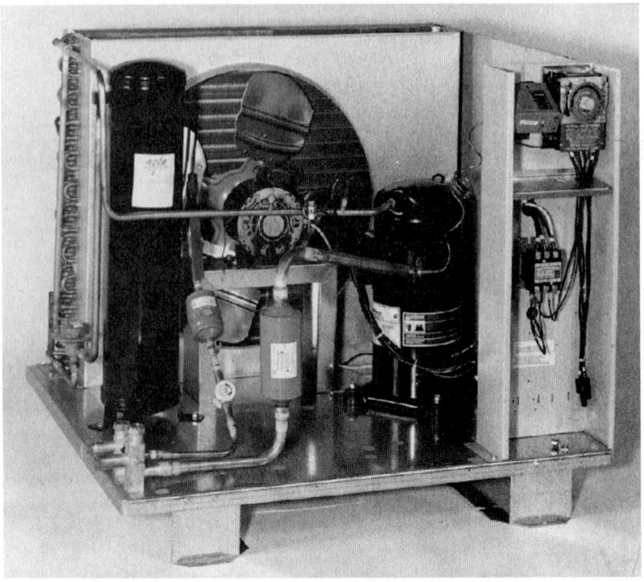

Figure 13-14. *A condensing unit that can be installed in numerous locations. This unit is operated with a scroll compressor and uses R-404A and R-507 refrigerant and polyol ester oil. (Heatcraft Inc., Refrigeration Products Division)*

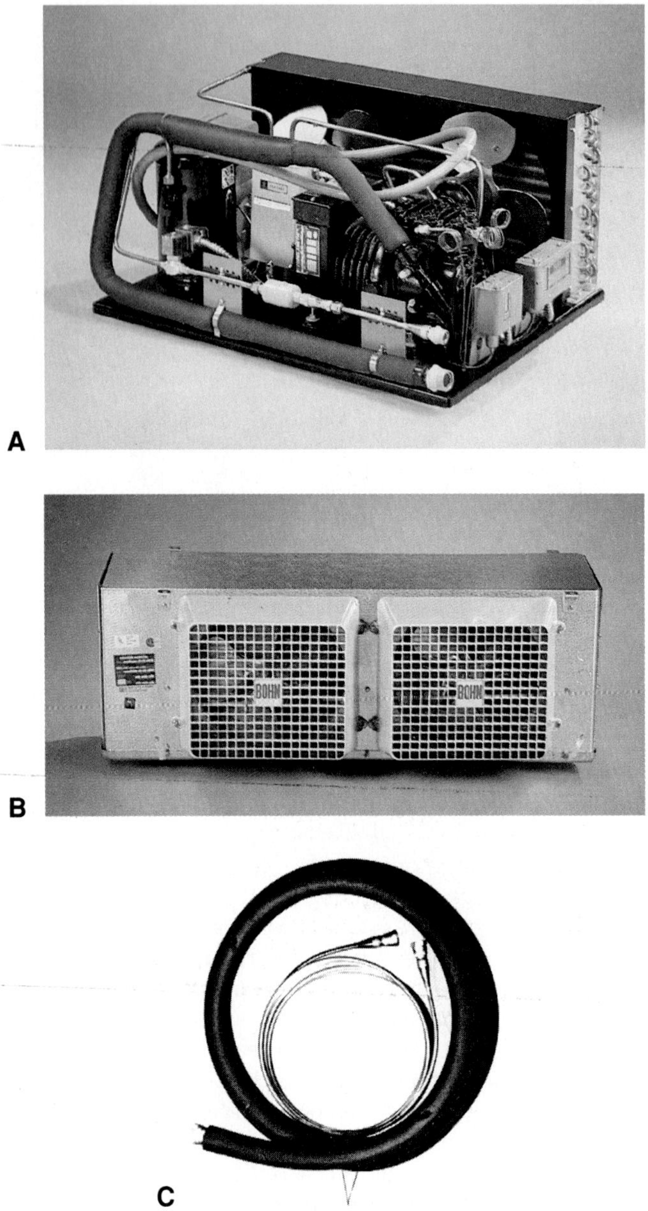

Figure 13-15. *A preassembled commercial refrigeration system including condensing unit (A), evaporator coil (B), and precharged refrigeration lines (C). (American Panel Corp.)*

A factory-assembled condenser and evaporator combination is shown in **Figure 13-15A.** Matched with an evaporator assembly, **Figure 13-15B,** it can meet a wide range of cooling needs. This built-up split refrigeration system is designed to operate under various product loads. Factory-installed accessories include prewired controls and fully charged interconnecting refrigeration lines, **Figure 13-15C.**

For large installations, two-motor compressor designs have been developed to provide greater refrigeration capacity. They are:

- Tandem assembly motor compressors.
- Parallel assembly motor compressors.

The tandem design connects two motor compressors together at the motor end. See **Figure 13-16.** These units can be run separately for low load or together for full load. However, if one motor compressor fails, the complete system must be shut down during the service time.

A remote air-cooled air conditioning compressor is shown in **Figure 13-17.**

Parallel design connects two or more units in parallel by piping, **Figure 13-18.** The units also require a compressor oil piping system. This ensures that all the compressors have the correct oil amount in each crankcase while operating.

Figure 13-16. *Tandem motor compressor assembly doubles refrigerating capacity of unit. (Copeland Corporation)*

13.3.2 Outdoor Air-Cooled Condensers

Outdoor air-cooled condensers save space when air conditioning commercial buildings and homes. Air-cooled units save the cost of plumbing for water circuits. They are also used where chemicals in the water make

Figure 13-17. *Twin parallel units with three condensing fans. System frequently used in convenience stores, specialty areas, frozen food applications, and delicatessens. (Tyler Refrigeration Corporation)*

water cooling impractical. These units may be mounted on the roof, the outside wall, or at ground level.

In such cases there are four major provisions:

- There must be a head pressure control if the unit is exposed to outdoor weather that may go below the operating cabinet temperature.
- A method of preventing short cycling must be designed into the system.
- A means must also be provided to prevent dilution of the compressor oil by liquid refrigerant.
- The completed condenser must be constructed and installed so it is virtually weatherproof.

Low ambient temperatures will cause low head pressures. This pressure may drop so low it may even stop the flow of refrigerant. Four different methods will maintain pressure:

- Partially fill the condenser with liquid refrigerant.
- Stop or slow the condenser fans.
- Partially or completely close the ambient air louvers.
- Heat the condenser.

Outdoor units require about 1000 cfm (cubic feet per minute) of condenser air circulation per horsepower. They are less costly to operate than indoor air-cooled units.

A dual-compressor model, **Figure 13-19,** has two completely separate refrigeration systems. They range from 6 tons to 35 tons and provide what is known as a partial standby system. The basic operation is that of a two-stage compressor, with each compressor activated individually from a two-stage space thermostat. It provides a two-stage cooling and a two-stage heating system, with automatic changeover.

The basic operation is as follows: System No. 1 turns on the first stage of the thermostat. If the load is light, the system carries the load. The compressor cycle is on and off at the call of the thermostat. An example of this would be in cool mornings, late evenings, cloudy days, etc. This would cause the thermostat to signal for a compression cycle.

System No. 1 may not be adequate to handle the load. In this event, the room thermostat, **Figure 13-20,** automatically turns on the second compressor. It will

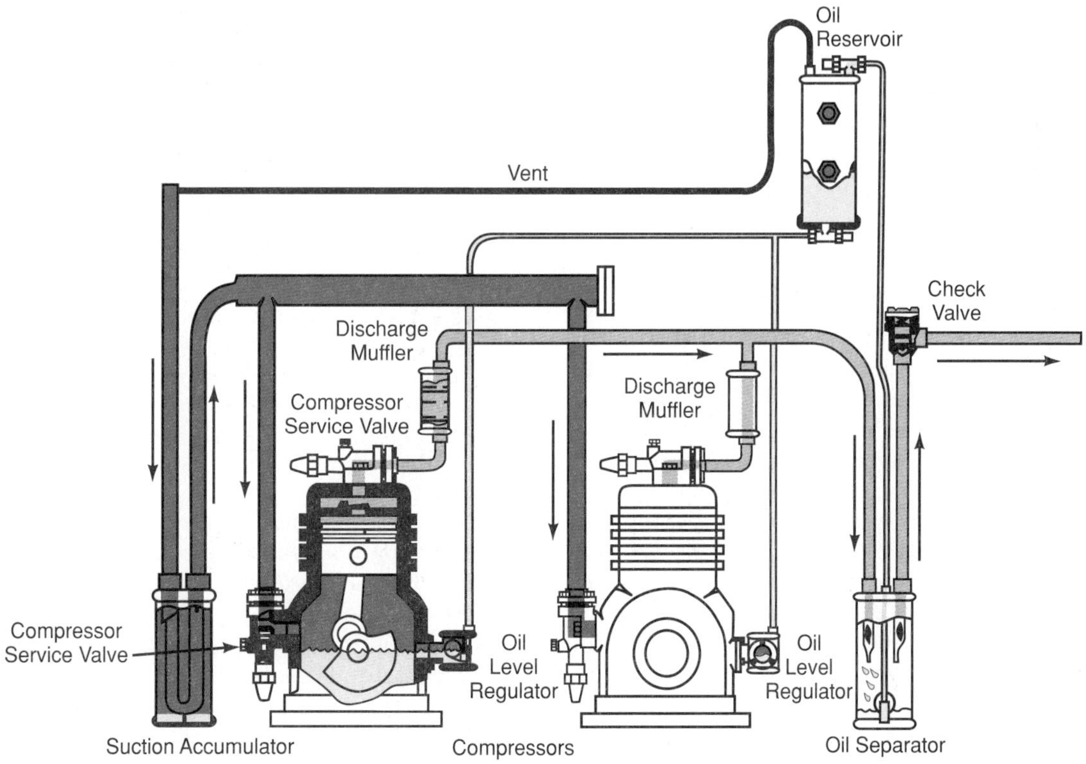

Figure 13-18. *Simple diagram of parallel motor compressor assembly. Two motor compressors are used. Note oil separator, oil reservoir, venting, suction accumulator, and oil level regulator. (AC&R Components, Inc.)*

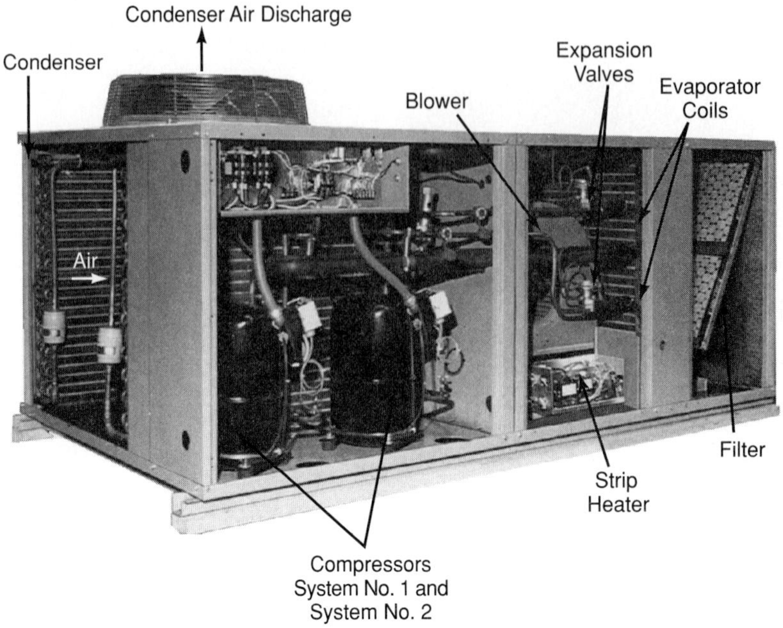

Figure 13-19. *Outdoor packaged unit. Note twin compressors for Systems 1 and 2 and dual expansion valves. (Addison Products Company)*

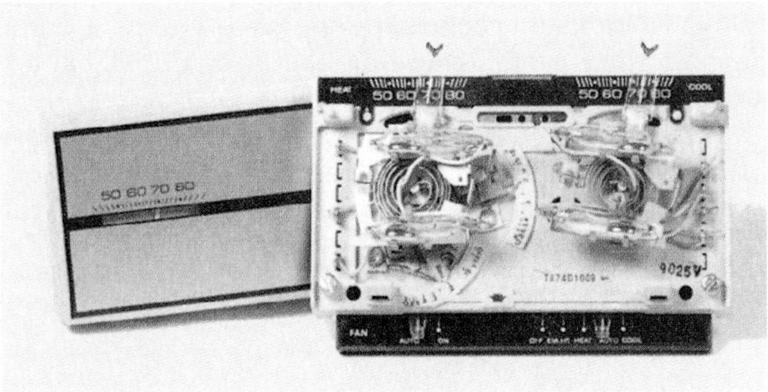

Figure 13-20. *Thermostat used with single- or dual-compressor air conditioner has two-stage cool, two-stage heat, automatic change-over, and dual setpoint. (Addison Products Company)*

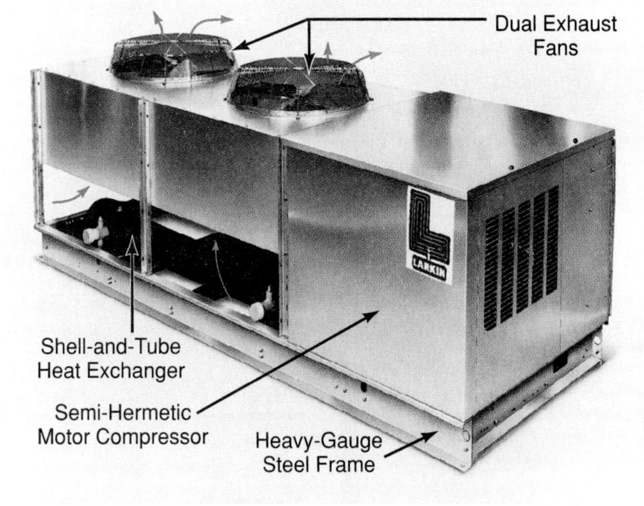

Figure 13-21. *Air-cooled packaged water chiller. (Heatcraft, Inc.)*

then signal on and off to carry the rest of the load while Compressor No. 1 runs constantly, removing the moisture from the air. When the load drops and the facility temperature is lowered, Compressor No. 2 shuts down. Compressor No. 1 then cycles to carry the reduced load.

Figure 13-21 illustrates an air-cooled, packaged water chiller with a shell and tube heat exchanger.

The main concern with outdoor units is keeping the thermostatic expansion valve operating at full capacity during cold weather. Capacity depends on the pressure difference across the valve. If condensing pressure for R-12 reduces from 102 psi, 90°F (32°C) to 56 psi, 30°F (−1°C), the valve capacity will drop. (Not enough liquid refrigerant will flow.) The refrigerated fixture temperatures may then rise too high. Also, a small pressure difference may cause short cycling of the condenser.

Condensing temperatures and pressures may be kept at proper operating levels by a design change. The unit is made to nearly fill the condenser tubes with liquid. Just enough condensing surface is left to maintain the pressure.

Figure 13-22 shows how a check valve and limiter valve maintain a head pressure in the condenser in low outdoor temperatures. The pressure on the limiter valve on the outlet of the condenser will not open the valve until the condensing pressure reaches the proper level. When pressure rises, the limiter valve will open and allow liquid to leave the condenser.

The installation specifications must be carefully checked. The receiver must hold enough liquid refrigerant to flood most of the condenser in the winter. It must also safely hold the refrigerant during the warm season.

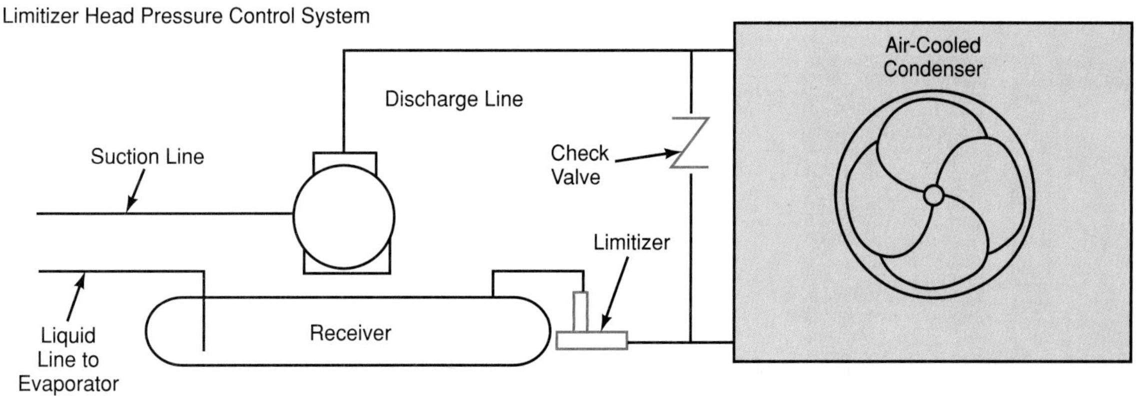

Figure 13-22. *Schematic of pressure control system on outdoor condensing system. The check valve and limiter valve ensure good condensing pressure during cold weather.*

Another way to maintain pressures is to close the condenser housing airflow louvers as the pressure drops. A pressure-sensitive device does this. It is connected into the condenser tubing as shown in **Figure 13-23.** This head pressure device will move the rod out as pressures increase. The linkage will open the louvers, as **Figure 13-24** illustrates. As head pressures decrease, the rod will move back and start to close the louvers.

The condenser fans may either operate when louvers are closed or they may be shut off when the louvers near the closing point. A condenser with an ambient-temperature-controlled, adjustable damper is shown in **Figure 13-25.**

Other systems shut off the condenser fan when the condenser pressure falls to a minimum level. See **Figure 13-26.** Controls lower the fan speed when the head

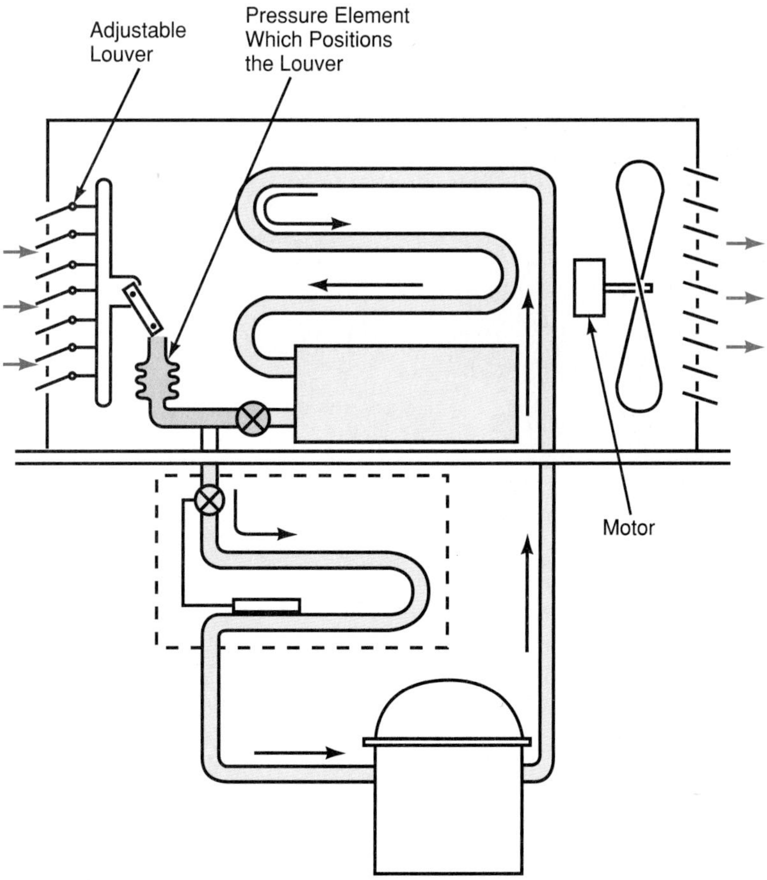

Figure 13-23. *An air-cooled condenser with pressure-operated louver. As condensing pressure decreases, louver will start to close, reducing condenser airflow.*

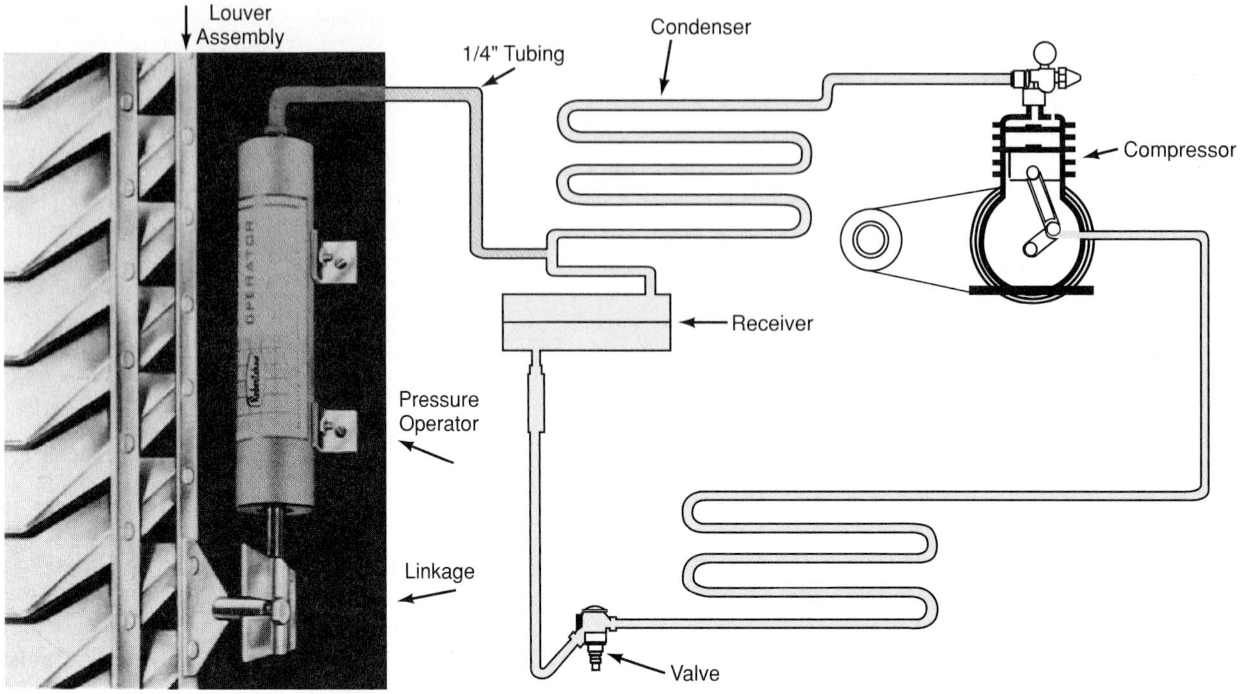

Figure 13-24. *Positioning cylinder (pressure operator) at left will react to pressure from the condenser. An increase in pressure moves cylinder and rod (connected to linkage) and causes louvers to open. (Robertshaw Controls Co.)*

Figure 13-25. *This 3 hp air-cooled condenser is equipped with an adjustable louver to vary airflow as ambient temperature changes. (Tyler Refrigeration Corp.)*

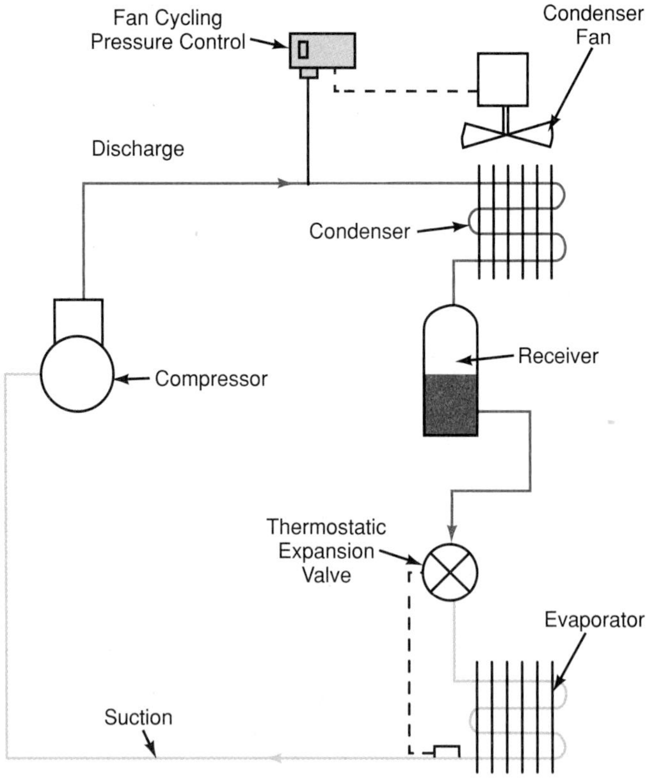

Figure 13-26. *This schematic shows refrigeration cycle with a fan-cycling pressure control that senses condenser pressure. (Ranco North America)*

pressure drops. Electrically controlled modulated fan speeds are used for this purpose. This system operates with a thermistor on the condenser and a special fan motor. See **Figure 13-27.**

Electric heating elements are sometimes placed in or around the receiver. This keeps the receiver temperature warmer than the cabinet temperature. If it were allowed to become too cold, the receiver would act like a condenser.

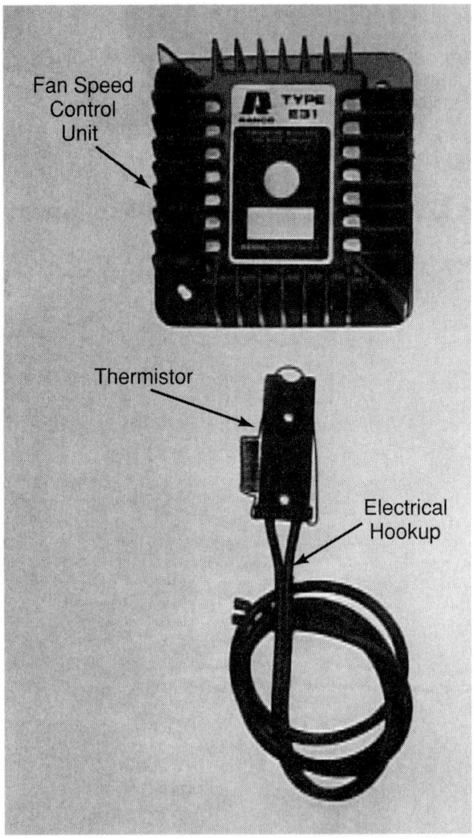

Figure 13-27. *A fan speed control unit may be equipped with a thermistor. (Ranco North America)*

Some systems use a bypass from the compressor to the receiver. This bypass feeds hot refrigerant vapor to the receiver to keep it warm. The bypass has a check valve mounted in it to ensure one-way refrigerant flow.

The compressor must be kept warm enough during cold weather. This prevents dilution of the oil by the liquid refrigerant. Heat is provided by electric heating elements in or around the motor compressor. They are thermostatically operated to energize the heating element at about 50°F (10°C). This heater usually has a 100 W to 200 W capacity.

The outdoor unit may not function adequately in strong winds. Under these conditions, the built-in capacity to overcome low ambient temperatures may not be sufficient. Windy conditions can prevent damper and fan operation. The cooling effect may be more than the electric heating element can overcome. The unit must be installed in a position to avoid the harmful effects of any high-velocity cold winds. It should also be weatherproofed as well as possible—particularly the electrical installation. See **Figure 13-28.**

Head pressure control valves are often used. These are usually thermostat operated. A low-pressure switch may not cut in because condenser pressures are below cut-in pressures. All refrigerant may then transfer to the condenser. This occurs because it will then be the coldest part of the system. A check valve is often used in the condenser outlet. It prevents flow of refrigerant to the cold receiver.

Figure 13-28. *A—Internal view of complete heating, air conditioning, and ventilating system. B—System shows the compressor and precharged refrigerant lines. (Tyler Refrigeration Corp.)*

As noted before, some systems flood the condenser with liquid refrigerant during low ambient temperatures. This maintains high condensing pressures. When the condenser pressure decreases, one valve closes the outlet of the condenser. The condenser then begins to fill with liquid refrigerant. This action reduces the condensing surface, and condensing pressure will rise.

Figure 13-29 shows a system with a valve that opens as the receiver pressure falls. Hot gas is allowed to bypass into the receiver (at about 20 psi pressure difference). This raises the receiver pressure and increases the flow of liquid refrigerant to the evaporators.

The valve has two openings, B and C. As one closes, the other will open. If located in a cold place, the receiver may need an electric heating element. This helps the system operate efficiently. These valves must be sized to the capacity of the system. Avoid using excessive pressures when testing for leaks. The valve bellows may suffer damage. Keep the pressure at or below 200 psi.

The system usually is charged with twice as much refrigerant as would be needed without the condenser flooding feature. Year-round operation requires a receiver with capacity to store all this extra refrigerant during the summer. Also, for service purposes, the receiver should be twice the normal size. This will enable it to hold all the refrigerant.

The compressor may also collect small amounts of liquid refrigerant during the Off cycle. A trap may be needed in the compressor discharge line. An inverted trap may be necessary at the condenser outlet.

13.3.3 The Compressor

Commercial compressors are of two general types:

- External drive.
- Hermetic.

Commercial compressor designs are explained in Chapter 4. A typical small, belt-driven, external drive compressor is pictured in **Figure 13-30.**

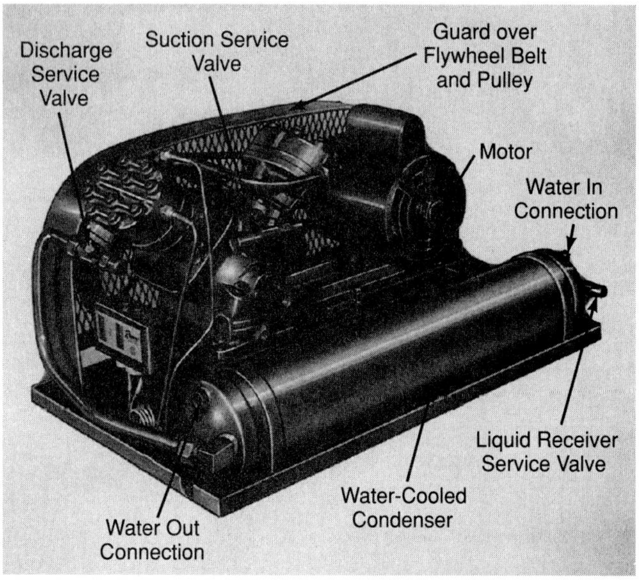

Figure 13-30. *Condenser with belt-driven, two-cylinder, air-cooled compressor. It uses a water-cooled condenser. (Frick Co.)*

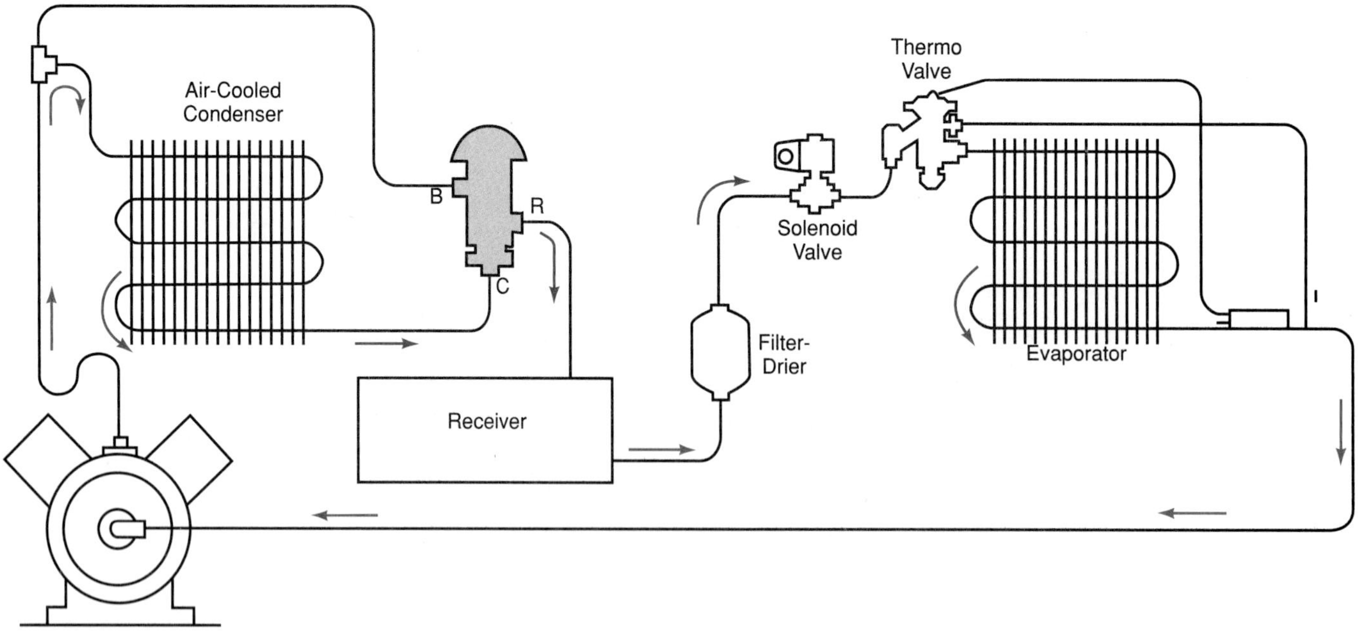

Figure 13-29. *Schematic shows condenser pressure-control valve. Compressor discharge pressure is above valve setting. Opening at B is closed and flow is through condenser and openings C and R. During cold weather, opening at C is closed and openings at B and R are open. Hot gas then flows directly from the compressor into the receiver. (Alco Controls Div., Emerson Electric Co.)*

There are several types of commercial hermetic compressors. **Figure 13-31** illustrates some parts of a bolted (serviceable) hermetic two-cylinder compressor. It has a force-feed lubrication system.

Some large units have either hydraulic or electric unloading devices to control the number of cylinders pumping. The higher the load, the more cylinders used to pump the vapor. **Figure 13-32** is a schematic of the oil circuit. The illustration describes how it is controlled to operate the compressor unloader.

Another type of hermetic compressor is the welded motor compressor design, nonfield serviceable. These units are built in sizes from 1/6 hp up to 20 hp. Internal design varies with size and manufacturer. Some are spring mounted internally, while others use outside (external) mounting springs. The smaller units usually have one cylinder. Larger units (1/2 hp and up) have two or more cylinders. Small-unit motors may be either two- or four-pole (single-phase). Three-phase motors are generally used in larger units.

Many combinations, types, and sizes of compressors can be used to provide the pumping for a great variety of evaporator types and sizes. Each compressor has a minimum and maximum:

- Revolutions per minute (rpm) for efficiency.
- Compression ratio (a maximum pressure difference between low side and high side).
- Discharge temperature.
- Volume of gas it can pump.

Before using a compressor, the manufacturer's operating specifications must be known. See Chapter 16.

Many low-temperature systems use a cascade system. The first stage compressor may be a reciprocating unit, but rotary units are also used. The rotary compressor pressure limit is about 45 psi across the compressor. It works very well with a compression ratio of about 4:1. It also works well with a discharge temperature of about 200°F (93°C).

The rotary has a high volumetric efficiency. A check valve is usually placed in the discharge. It prevents back-up of refrigerant during off cycle. A check valve should be placed in the oil lines for the same reason.

Compressors can have from one to twelve cylinders. There are many different cylinder arrangements: vertical, V, W, Y, X, or radial. See **Figure 13-33.**

Internal unloaders are usually operated by oil pressure. A spring holds the intake valve open until the oil pressure builds up. This causes all intake valves to operate. It is also used to reduce pumping capacity during low-load periods. Solenoid valves are mounted in the oil lines to unloaders. When the solenoid closes, the oil pressure drops in the unloader. The intake valves are kept open. See **Figure 13-34.**

Low-side pressure switches operate the solenoids. A timer bypass pressure switch operates the system at full capacity for about a minute each hour or two. External unloaders use a bypass to the evaporator inlet, ensuring suction vapor is cool (de-superheating).

13.3.4 Air-Cooled Condenser

Air-cooled condensers are quite common in commercial systems. Cooling water may be too expensive or corrosive. Smaller units use static condensers with thermal airflow.

Larger condensers may be cooled by a big fan built onto the motor or into the compressor flywheel on

Figure 13-31. *Bolted hermetic motor compressor. Motor is in back of compressor. Note the use of temperature limit controls and oil pressure sensing safety control. (Copeland Corp.)*

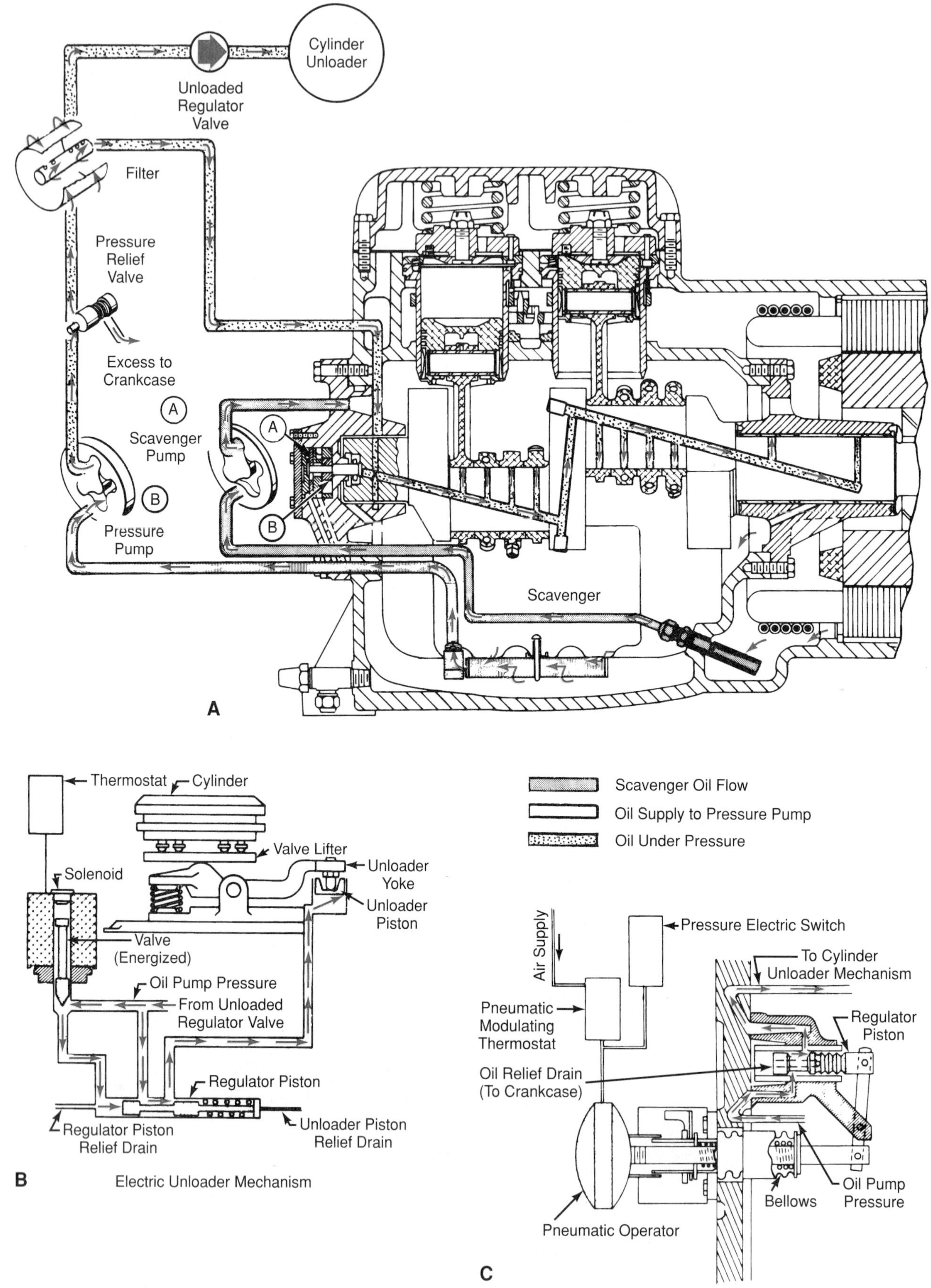

Figure 13-32. *Oil circuit of an eight-cylinder compressor with oil take-off to operate compressor unloader. A—Oil circuit. B—Electrical control. C—Pneumatic control.*

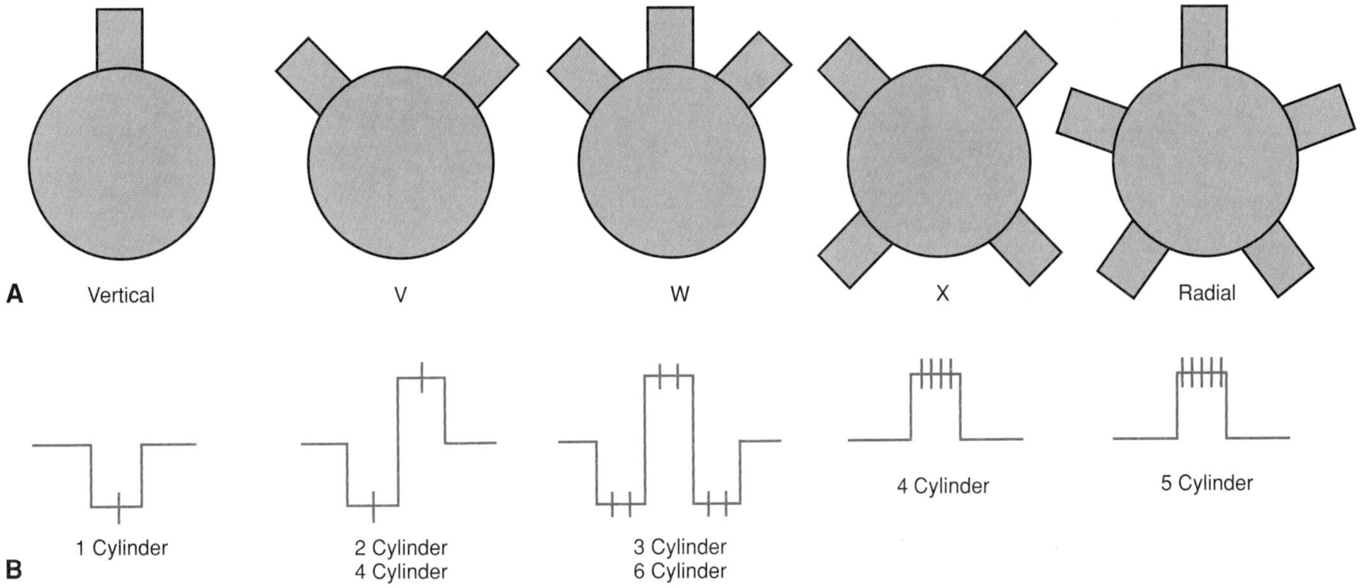

Figure 13-33. *Cylinder arrangements used for reciprocating compressors. A—Cylinder arrangement. B—Crankshaft configuration. Vertical markings on crankshaft indicate number of rods attached to each throw (offset).*

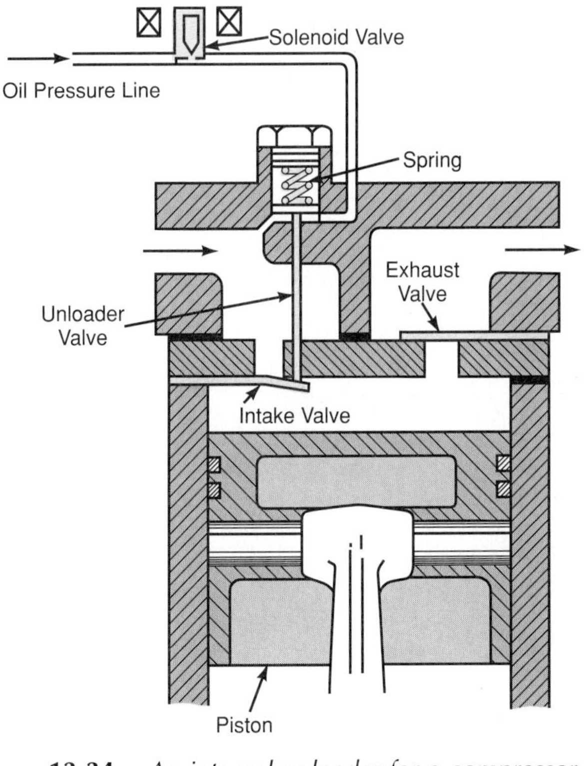

Figure 13-34. *An internal unloader for a compressor. During starting, the capacity needs to be reduced.*

external drive units. Larger hermetic units use separate motors to drive the fans. See Chapter 9 for determining condensing temperatures and pressures for an air-cooled condenser.

The fan efficiency may be increased by placing a metal shroud around an air-cooled condenser. In **Figure 13-35,** an air-cooled condenser uses two fans. Air can be drawn, induced (led into), or forced through the condensers. These condensers have fins and frequently use a double or triple row of tubes. Many fin arrangements and constructions have been used. See **Figure 13-36.**

To cool the compressor head and valves, a double air-cooled condenser is sometimes used. Refrigerant leaves the compressor and passes through one condenser. Then it is led back through the motor compressor to help cool it. From there it goes into the second condenser, where it is condensed (cooled) into liquid.

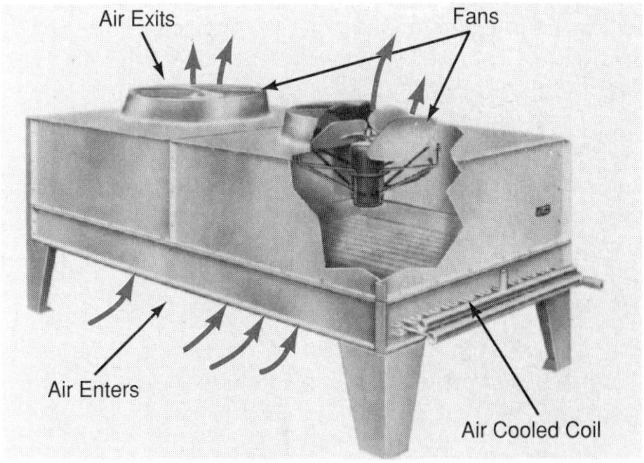

Figure 13-35. *Air-cooled condenser cutaway shows fan, motor, mount, and condenser coil arrangement. Air enters from beneath and leaves at top.*
(BOHN Refrigeration Products, a Unit of Heatcraft, Inc.)

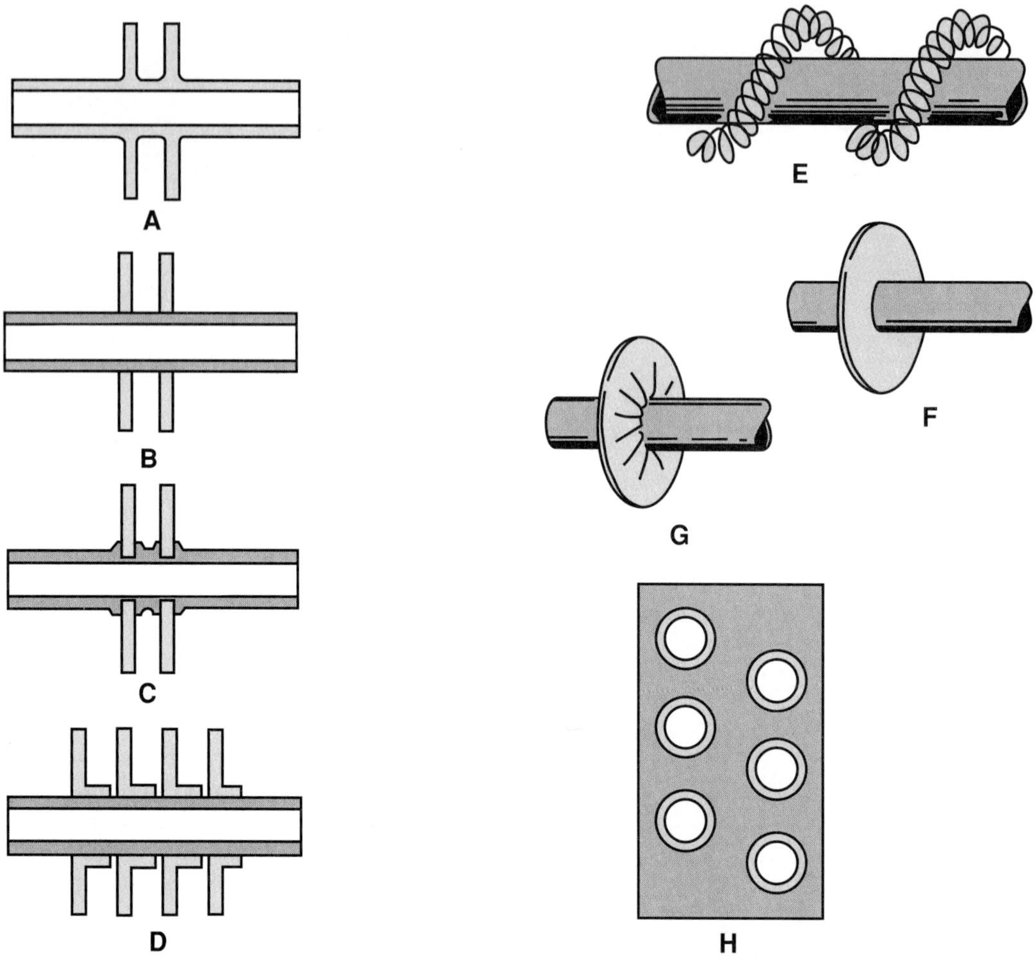

Figure 13-36. *Extended surface fin arrangements used in air-cooled condensers. A—Fin is part of tubing. B—Fin pressed on tubing. C—Fin fastened by crimped tubing. D—Fin flange pressed on tubing. E—Coiled wire used as fin. F—Circular fin. G—Crimped circular fin. H—Large multiple tubing fin.*

13.3.5 Outdoor Air-Cooled Condensers

Systems with the motor compressor and the liquid receiver indoors may use outdoor air-cooled condensers. Only the condenser is placed outdoors. The compressor discharge line carries the hot high-pressure vapor to the outdoor air-cooled condenser. Condensed liquid is piped back into the building.

These units use the same devices described in Section 13.3.2 to protect the condenser and to maintain good head pressures in low air temperatures. They provide protection in temperatures as low as 50°F (10°C) or lower.

13.3.6 Water-Cooled Condenser

Many large commercial refrigerating units use a water-cooled condenser. This condenser is built in three styles:

- Shell and tube.
- Shell and coil.
- Tube-within-a-tube.

In the first type, the refrigerant vapor goes directly from the compressor into a tank or shell. At the same time, water travels through the tank or shell in straight tubes. The second type also uses a shell. However, the water travels through the shell in coils of tubing.

The third type uses two pipes or tubes—one inside the other. The refrigerant passes one way through the outer pipe. The condenser water flows in the opposite direction through the inner tube.

Water velocity should be 7-10 fps (feet per second). If flow is too fast, water may remove the oxide coating, causing pitting. If the water velocity drops to 3 fps, scaling will occur.

Water-cooled compressors are sometimes used with water-cooled condensers. The water flow, with few exceptions, is through the condenser first. It then flows through the cylinder head, and finally into the drain. Water flow may be regulated by an automatic water valve. See Section 13.11.3.

Shell and Tube Condenser

Shell and tube condensers are cylinders usually made of steel with copper tubes inside. Water circulates through the tubes, condensing hot vapors in the cylinder into a liquid. The bottom part of the shell serves as the liquid receiver. See **Figure 13-37.**

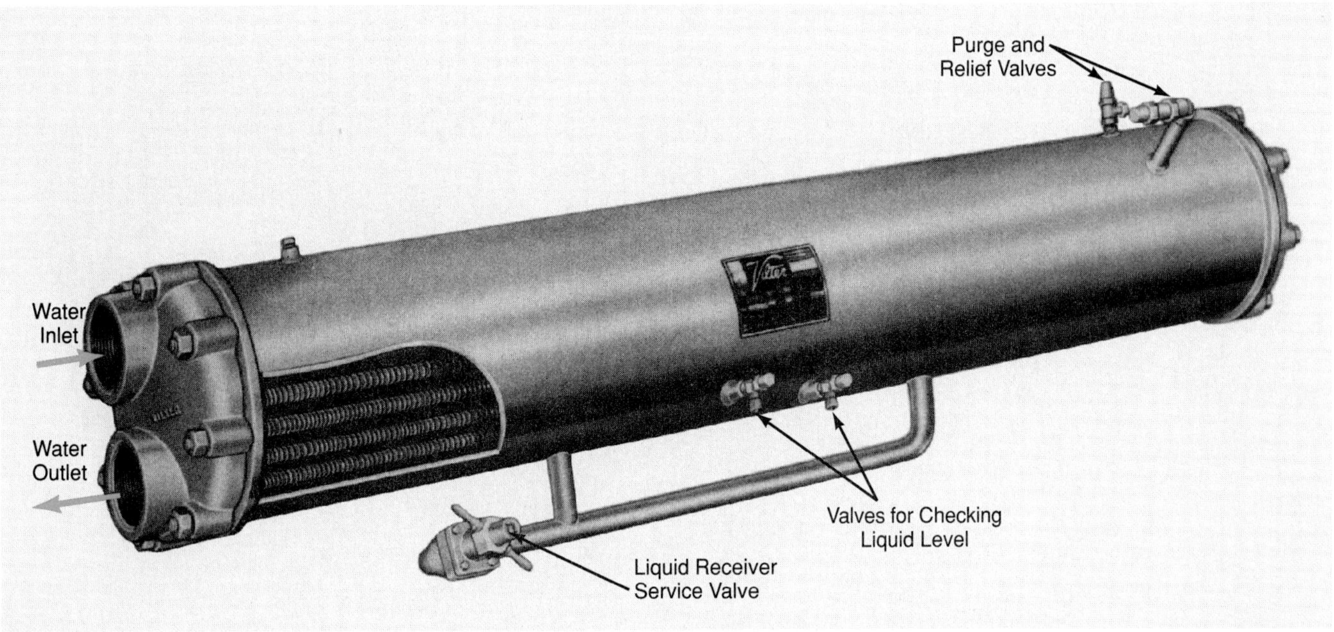

Figure 13-37. *A typical shell and tube condenser-liquid receiver. Note that water pipes run straight through and are finned for better heat transfer. Receiver ends are removable to allow cleaning access. (Vilter Mfg. Corp.)*

The shell and tube condenser has some advantages. It is compact, needs no fans, and combines the condenser and receiver in one. It uses numerous straight tubes inside the receiver with a water manifold on both ends. When these manifold ends are removed, the water tubes can easily be cleaned of deposits. See **Figure 13-38.** This is sometimes called a shell and pipe condenser.

Figure 13-38. *Cleaning a shell and tube condenser. Flexible shaft on power cleaner rotates at high speed inside a nylon casing. High-pressure water moves from the machine to the cleaning brush or tool. (Goodway Technologies Corp.)*

Shell and Coil Condenser

The ***shell and coil condenser*** is very much like the shell and tube water-cooled condenser. It has a coil of water tubing inside the shell rather than a straight tube. It is often used in smaller commercial units. See **Figure 13-39.**

The shell and coil condenser is less costly to manufacture. However, it cannot be cleaned mechanically. The water tube must be cleaned with chemicals.

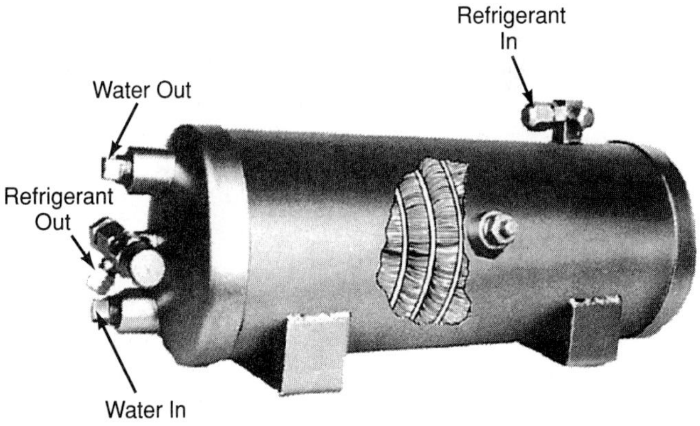

Figure 13-39. *Shell and coil water-cooled condenser. Condenser also serves as liquid receiver. (Refrigeration Research, Inc.)*

Tube-within-a-Tube Condenser

The ***tube-within-a-tube water-cooled condenser*** is popular because it is easy to make. Water passing through the inside tube cools the refrigerant in the outer tube, **Figure 13-40.** The outside tubing is also cooled by air in the room. Double cooling improves efficiency.

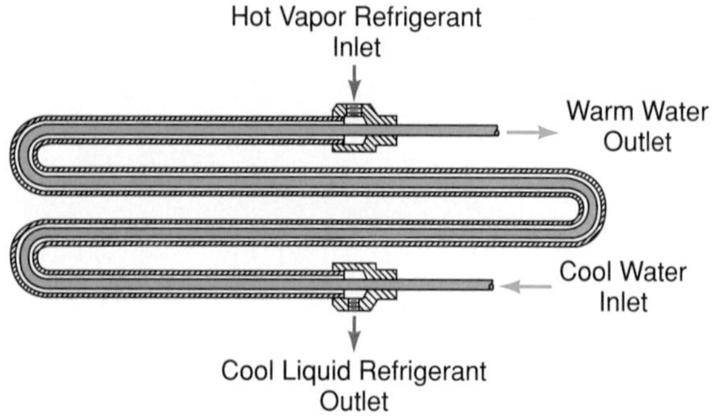

Figure 13-40. *Tube-within-a-tube condenser. Water flows through inner tube. Water flow is opposite vapor flow.*

This type of condenser may be constructed in a cylindrical, spiral, or rectangular style. See **Figure 13-41.**

The inner tube, **Figure 13-42A,** shows a 6-lead and an 8-lead grooved inner tube. This design increases heat transfer. **Figure 13-42B** shows an available option of a double-walled inner tube with a grooved design. This design achieves venting of refrigerant vapor in the event of a leak.

Water enters the condenser at the refrigerant outlet. It leaves the condenser at the point where the hot vapor from the compressor enters. This is called counterflow design. The warmest water is adjacent to the warmest refrigerant. The coolest refrigerant is next to the coolest water. This condenser can also be made with hard copper pipe. See **Figure 13-43.**

The rectangular tube-within-a-tube condenser uses a straight, hard, copper pipe with manifolds on the ends. When the manifolds are removed, the water pipes may be cleaned mechanically.

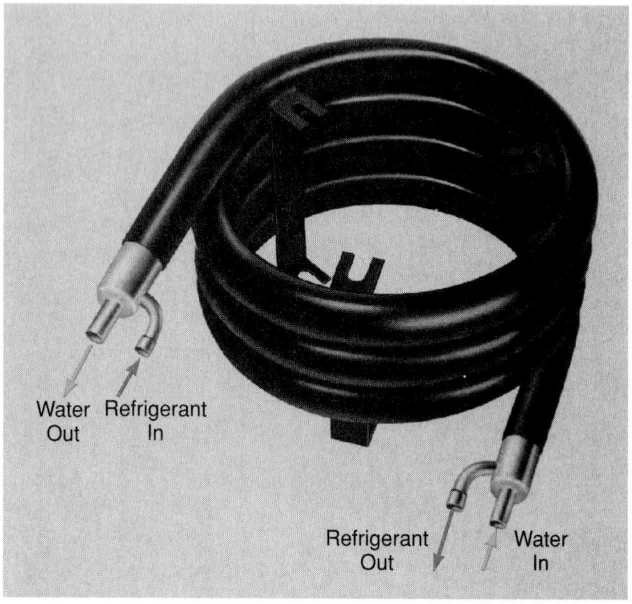

Figure 13-41. *This tube-within-a-tube condenser is shaped into a spiral. (Packless Industries)*

13.3.7 Cooling Towers

In some areas, water contains chemicals making it unsuitable as a coolant. In other localities, water may be very scarce, expensive, or its use may be limited by law.

Water-cooling towers save on water consumption. These towers serve the same purpose as the spray towers used in large industrial refrigeration systems.

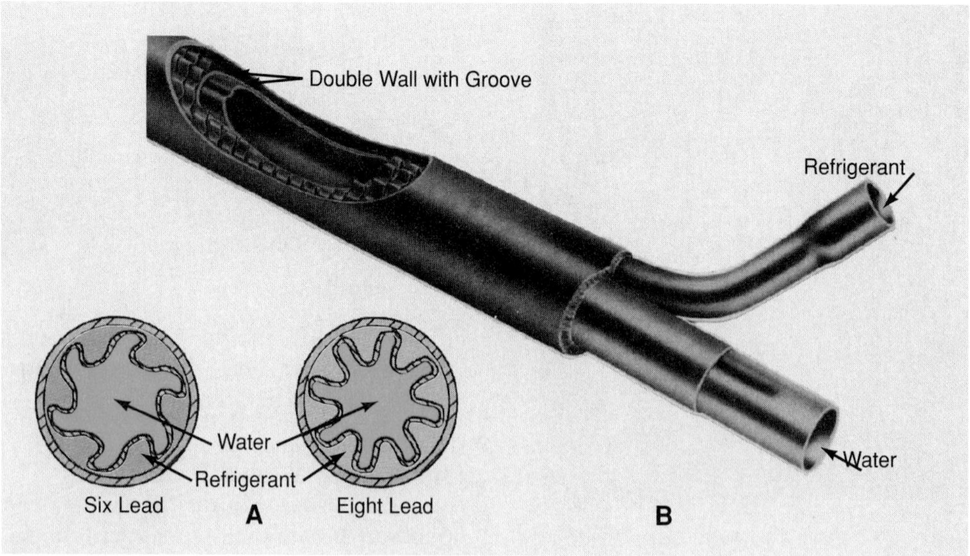

Figure 13-42. *Two more views of tube-within-a-tube condensers. A—End view shows six- and eight-lead (groove) designs to increase heat transfer from the refrigerant to the water. B—Cutaway of a double-walled tube-within-a-tube design. (Packless Industries and Edwards Engineering Corp.)*

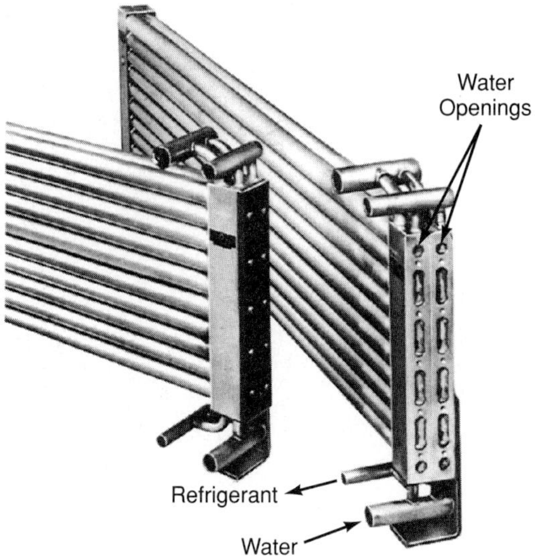

Figure 13-43. *Tube-within-a-tube condenser designed to permit cleaning of water tubes (inner tube). Clean-out plate is removed in right-hand view.*

A variety of cooling tower designs are shown in **Figure 13-44.** Cooling towers can be rather noisy. They should be located away from noise-sensitive areas such as offices, restaurants, and living quarters.

The following factors affect the performance of a cooling tower:

- Design conditions.
- Humidity requirements.
- Tower heat load.
- Design wet bulb temperature.
- Water quality.

One system connects the condenser water lines to a water coil in an enclosure. A pump forces the water through the condenser and then through the coil in the tower. The tower coil is pierced with holes, and the water is sprayed into the enclosure.

Air, rushing through the sprayed water, evaporates some of it. Evaporation cools the remaining water to the outdoor temperature or even lower (wet bulb temperature).

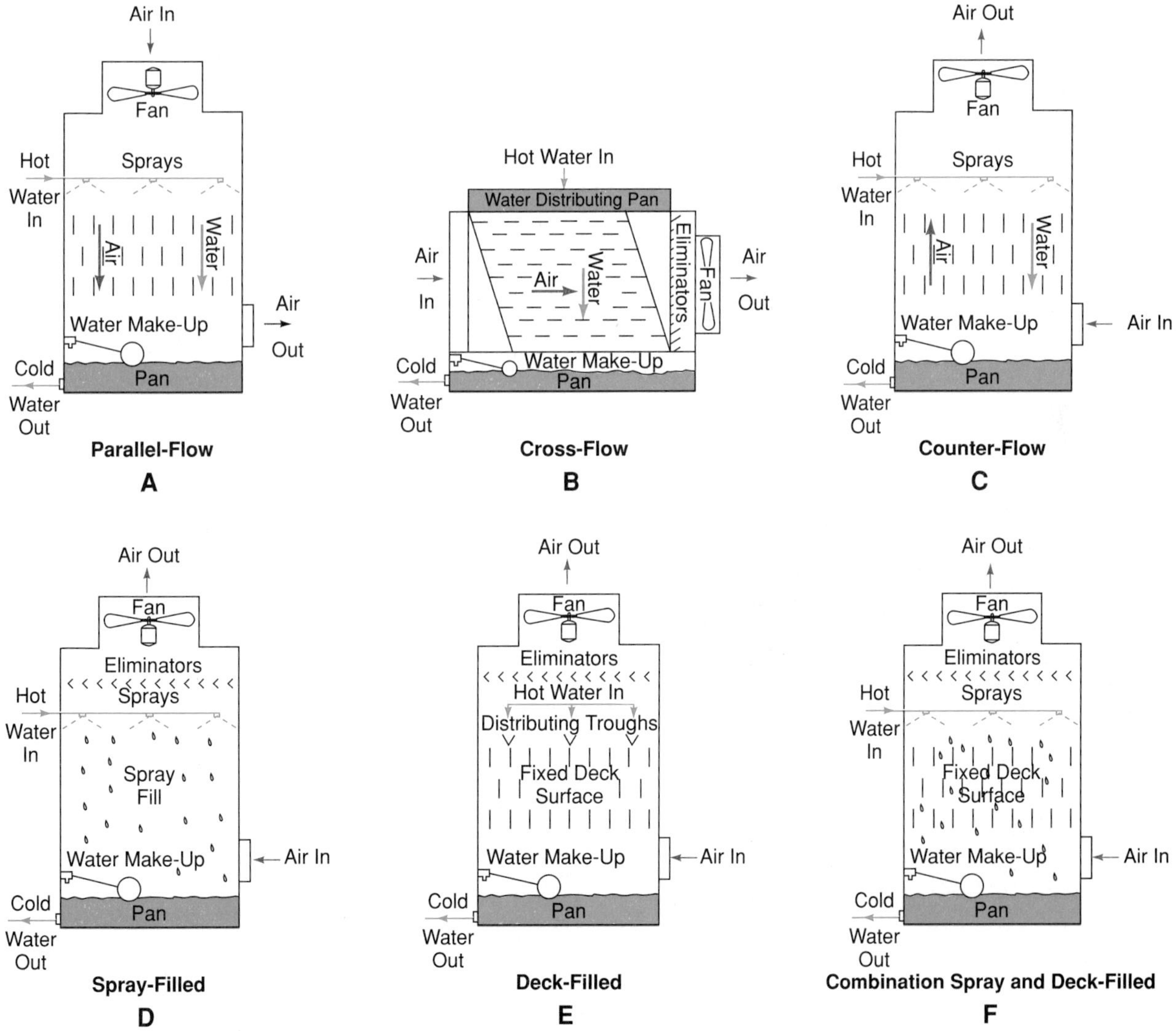

Figure 13-44. *Different types of cooling towers. A, B, and C show three different airflow patterns. D, E, and F show different methods to vaporize some of the water for cooling.*

In some systems, motor-driven fans control airflow through the cooling tower. **Figure 13-45** shows a cooling tower fan that is constructed primarily of fiberglass.

A cross-flow cooling tower that takes the water from the heat source through an inlet on the side of the unit to the hot water distribution basins on each side is shown in **Figure 13-46.** Gravity flow nozzles distribute the water evenly over the wet deck surface. Air is drawn through the air inlet louvers and across the wet deck surface, causing a small portion of the water to evaporate, removing the heat from the remaining water. Cooled water then flows into the lower sump and returns to the heat source.

Cooled water collects in the bottom of the enclosure. It passes through a screen that removes leaves or other foreign material. Then it is recirculated through the condenser.

A float-controlled valve in the lower water pan adds more water as needed. This float operates like a refrigerant low-side float mechanism.

A drain continually bleeds some water out of the water pan. This keeps water hardness to a minimum. Chemicals may be added to retard rust formation, algae, fungus growth, and the like.

Recently, cooling towers have been linked to the spread of Legionnaire's disease. This disease is caused by the Legionella bacteria. Several precautionary measures are being recommended to help eliminate this problem. Cooling towers may be placed downwind. Chloride compounds may be used as disinfectants on a monthly maintenance schedule. **Many facilities are proactively testing for Legionella. When counts are high, actions may be taken to reduce levels before disease occurs.** See Chapters 19 and 23.

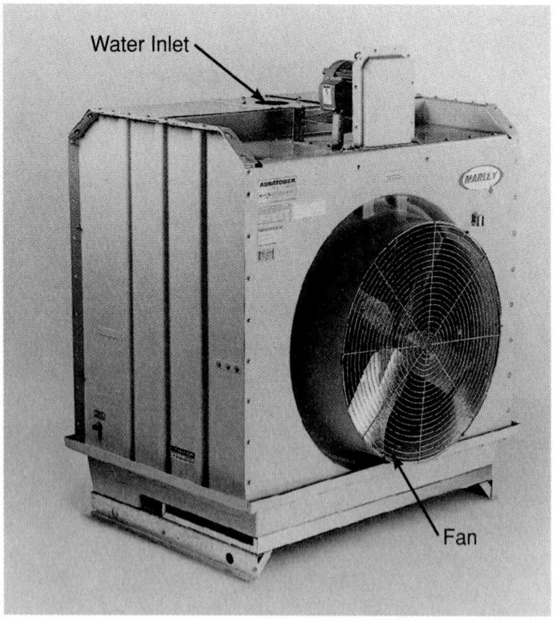

Figure 13-45. *Water flows into chamber at top of cooling tower and then by gravity through nozzles. Propeller fans provide air flow. (The Marley Cooling Tower Company)*

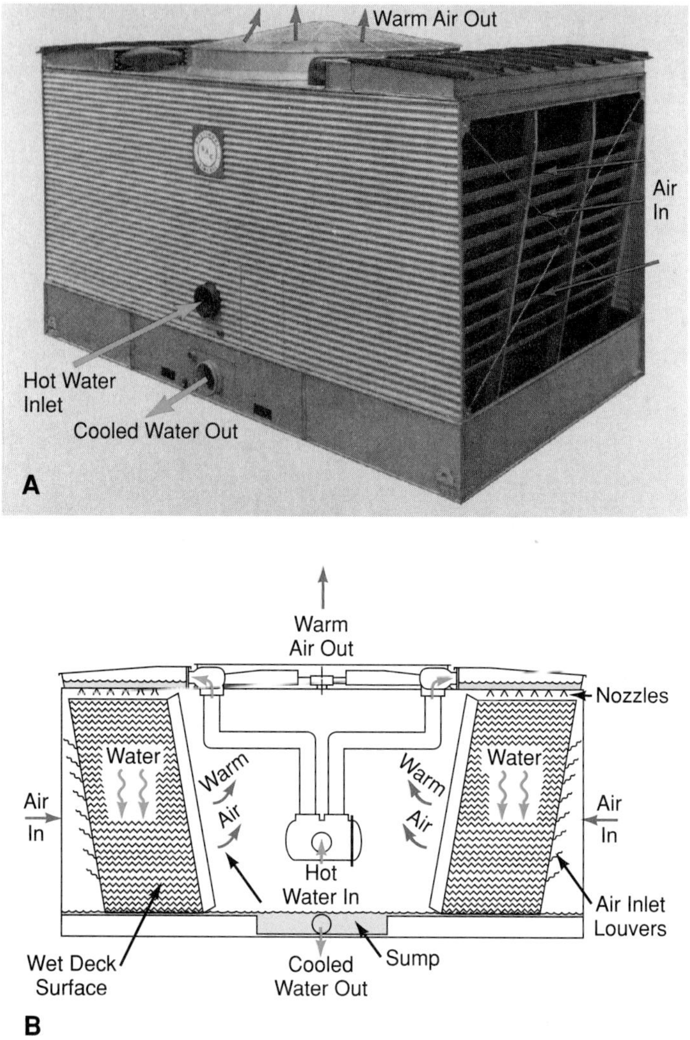

Figure 13-46. *A—An industrial cross-flow cooling tower. B—Diagram of the cooling tower, indicating the flow of air and its operation. (Baltimore Aircoil Company)*

Cooling towers are made of corrosion resistant materials. Among these are steel (zinc-dipped after assembly), copper, stainless steel, plastic, or treated wood.

The more water surface in contact with the air flowing through a cooling tower, the more efficient the cooling action. Most towers have some arrangement so that water flows over materials in thin films. This material is usually called ***fill.*** Fills are made of many materials: metal fins, wood slats, plastic, asbestos-plastic, and asbestos-cement. The shapes of the surfaces vary from Z-shaped, honeycomb, embossed, flat sheet to corrugated sheet. The cellular (honeycomb) fill is becoming very popular. The distribution system (nozzles, troughs, V-notches) must be kept clean and must distribute the water evenly to prevent scale buildup.

Ordinarily, cooling towers are in no danger of freezing while in operation. However, electric heat will keep the water temperature up during shutdowns. Immersion and convection heaters have been used for this purpose. An electric heater may also be installed in the pump circuit.

Hot water or steam can be used to prevent reservoir freeze-up. Pipes may need insulation or electric heater tape.

Overflow pipes in the cooling tower carry excess water to the building drain system. Some have airflow control to prevent freezing if wet or dry bulb temperature goes below 32°F (0°C). An air outlet thermostat operates the fan dampers.

Use only coarse screens on pump inlets. Be sure all suction lines are below water level in the cooling tower. Otherwise, air may enter the suction line. The resulting drop in pump volume can cause pump damage.

Pump outlets, on the other hand, need fine screens. The water pump should push water through the system. This will prevent low water pressures in the condenser tubes or pipes.

Cooling towers evaporate about two gallons of water every hour for each ton of refrigeration capacity. A gallon of water weighs about 8.3 lb. About 1000 Btu are needed to evaporate 1 lb. of water. Thus, to evaporate a gallon of water, 8.3 × 1000 or 8300 Btu are required. 2 × 8300 or 16,600 Btu are required to evaporate two gallons of water. For details of tower sizes and capacities, see manufacturers' catalogs.

13.3.8 Evaporative Condensers

The ***evaporative condenser*** system carries the refrigerant into a condenser. The condenser is in an enclosure much like a cooling tower. In this system, as "evaporative" indicates, water is sprayed or drips over the condenser. This cools it. The water cycle is in the condenser cabinet only. See **Figure 13-47.**

Usually, the evaporative condenser is mounted outdoors. However, it may be used indoors if air ducts are provided to the outside.

Some systems pump water to a trough above the condenser. The water then drips over coils as air is forced through them. A thermostat can be used to control the water flow. A fan blows air over the condenser whenever the condenser is operating.

The condenser is cooled by air alone until the condenser temperature reaches 80°F (27°C) or more. Water cooling is then turned on by a thermostat.

Another method is to subcool the refrigerant as it leaves the receiver. The liquid line goes through (into and out of) the evaporative condenser. Temperature of the refrigerant can be dropped 10°F (6°C) by subcooling. When the temperature reaches 45°F (7°C) or lower, the water is shut off. However, the condenser can still carry the load as an air-cooled condenser.

Some have water reservoirs inside the building. The reservoir must be large enough to hold all the water in the system. This allows protection from freezing weather. Fan dampers are sometimes used to decrease airflow as outside temperature drops.

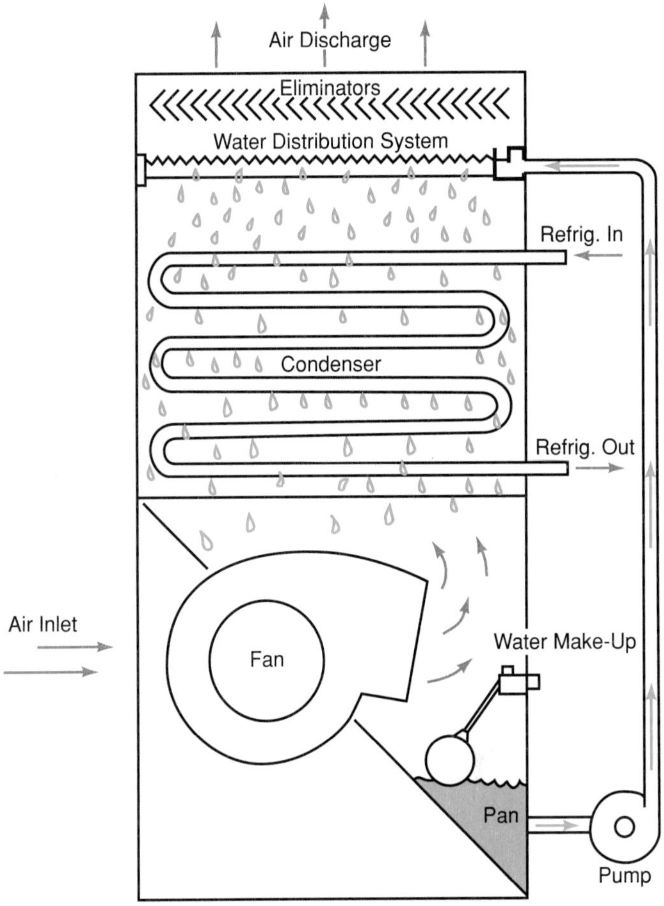

Figure 13-47. *Another evaporative condenser design. Note air and water flow. Water make-up is controlled by float valve. (Baltimore Aircoil Company)*

13.3.9 Liquid Receiver

The ***liquid receiver*** is a welded steel tank usually equipped with two service valves. One is a liquid receiver service valve mounted between the liquid receiver and the condenser. The other is located between the receiver and the liquid line (king valve). These two valves enable the technician to disconnect the liquid receiver from the system separately.

Receivers should have safety devices. A thermal release plug provides minimal safety. Some receivers have both thermal and pressure releases. See Section 13.12.4. A special line should be installed on relief valves to the refrigerant recovery system.

Receivers may be mounted either vertically or horizontally. **Figure 13-48A** illustrates a vertical receiver. **Figure 13-48B** illustrates a horizontal receiver with inlet valve, outlet valve, and safety valve. The horizontal style usually hangs underneath the compressor and motor frame. Some are provided with a device—sight glass, magnet floats or valves used for determining the level of the liquid refrigerant.

Liquid receivers with a water coil inside have a shell just like those with no water coil. However, it is usually larger for the same size compressor. The receiver should be large enough to hold all the refrigerant in the system.

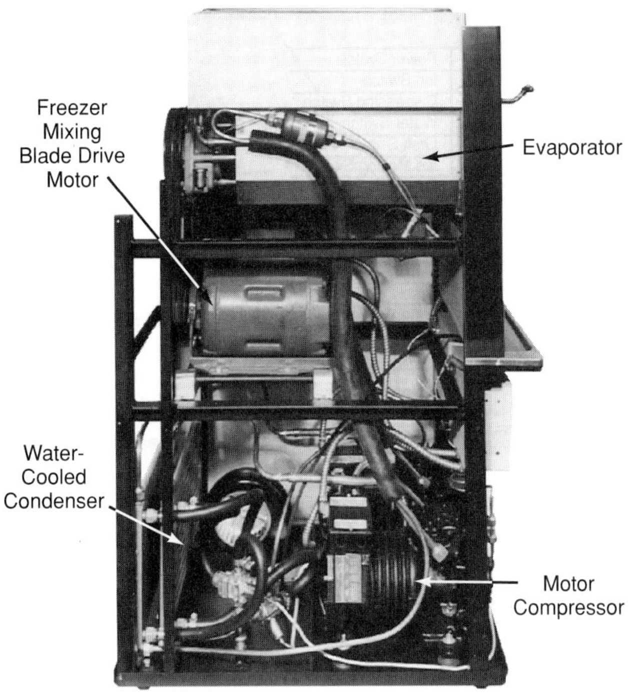

Figure 13-71. *View of freezer-dispenser with part of case removed. (Taylor Company)*

13.4.9 Evaporator Defrosting

Many evaporators operate at temperatures below freezing. The demand for open display cases and frozen food requires these low-temperature systems. The evaporators operate at refrigerant temperatures of 0°F (−18°C), −10°F (−23°C), and even −20°F (−29°C). Blower evaporators are often used.

Low temperatures and small fin spacings make frequent defrosting necessary. Frost accumulation would otherwise soon clog the evaporator. Other types of evaporators also need defrosting even though not so frequently. It is desirable that this be done with very little rise in fixture temperature.

Defrosting is usually automatic. Some evaporators defrost during each Off part of the cycle. On others, a timer control is used. This control may turn on the defrosting mechanism once a day, or it may be turned on after a given number of hours of compressor operation.

There are six defrosting methods:

- Hot refrigerant vapor system.
- Nonfreezing solution system.
- Water system.
- Electric heater system.
- Reverse cycle defrost system.
- Warm air system.

These defrosting methods either heat the evaporator from the inside or outside to melt the frost.

It is important to clean the evaporator, drain pans, and drain lines frequently.

Hot Gas Defrost System

In a hot gas system, hot refrigerant vapor is pumped directly through the evaporator tubing. The system has a refrigerant line running directly from the compressor discharge line to the evaporator. This line is sometimes connected between the thermostatic expansion valve or the capillary tube and the evaporator. The line is opened and closed by a solenoid shutoff valve.

At the predetermined time (usually 12 midnight or 1 A.M.), the timer closes a circuit. This starts the compressor, opens the solenoid valve, and stops the evaporator fan motors. Hot compressed vapor rushes through the evaporator and warms it. It then returns to the compressor along the suction line. See **Figure 13-73.**

Such a system will usually defrost the evaporator in 5 to 10 minutes. Defrost water must be kept from freezing in the drain pan and tube. Therefore, part of the hot gas defrost line may be installed under the drain pan and the drain pipe. Otherwise, a small electric heater may be installed in this location.

Refrigerant that condenses during the defrost cycle should be evaporated. The following steps explain how this is done:

1. A defrost bypass puts some hot gas into the suction line to vaporize any liquid refrigerant. Some hot gas bypass systems also use a liquid injection system. A TEV is mounted on a line between the

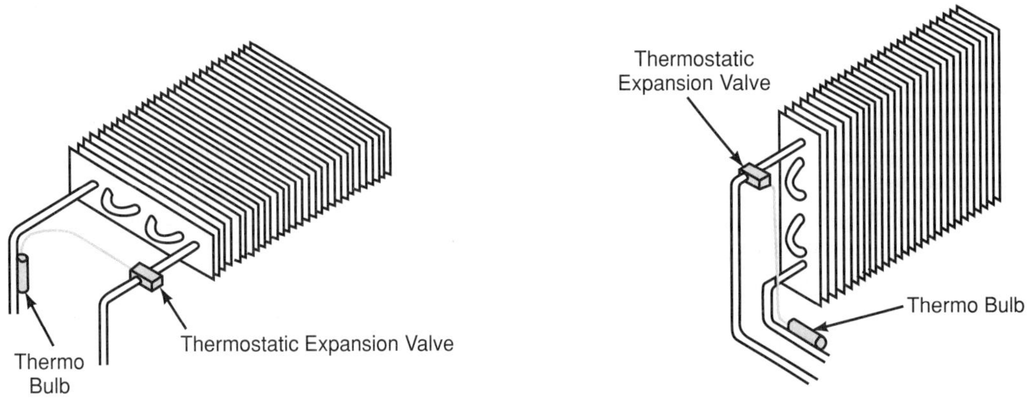

Figure 13-72. *Recommended expansion valve mounting for two different evaporator installations.*

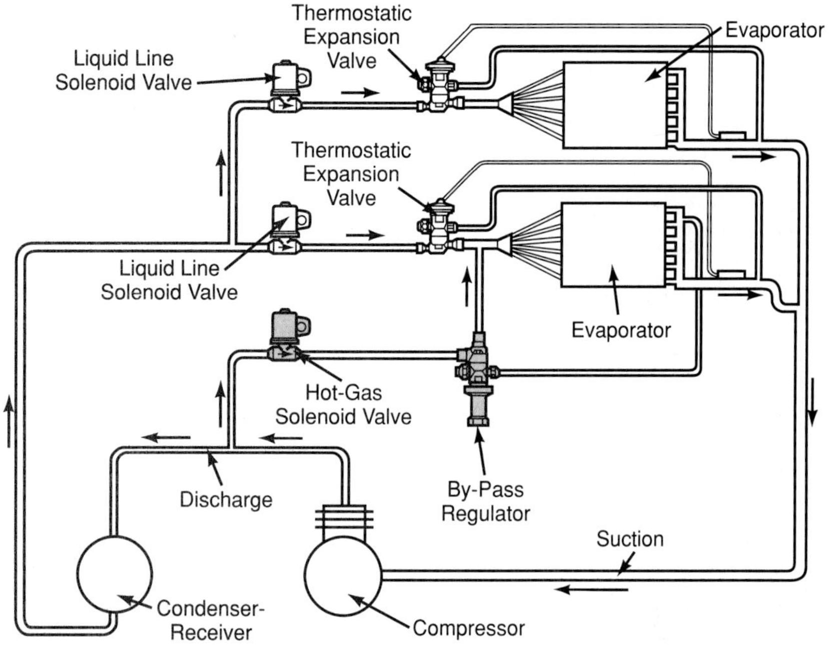

Figure 13-73. *Schematic of typical "hot gas" bypass from compressor discharge to evaporator inlet. Upper evaporator is nonfrost type. (Alco Controls Div., Emerson Electric Co.)*

liquid line and the suction line. Its sensing bulb is mounted on the suction line. If gas returning to the compressor becomes warm, the TEV will open. The liquid refrigerant will mix with the hot bypass gas. The mix is kept at 45°F (7°C) to 65°F (18°C). This ensures that the compressor will be cooled. A solenoid valve in the bypass TEV inlet shuts off the TEV when the system is operating normally or on a full load.

2. Heat is applied to vaporize the returning refrigerant. Electric heat is used in some cases.
3. A special blower-evaporator may be installed in connection with the suction line. This, plus air forced over the re-evaporator, allows only vapor to return to the compressor. The blower works only while the unit is on defrost. It is best to always use an accumulator mounted in the suction line. It will trap liquid refrigerant and vaporize it before it reaches the compressor.
4. Sometimes the hot gas is fed backwards into the evaporator. Then the condensed refrigerant bypasses the TEV by means of a check valve. This forces the liquid into the receiver by way of the liquid line.

A system for forcing the hot gas backwards through the evaporator is shown in **Figure 13-74A.** The defrost cycle is shown in **Figure 13-74B.**

A timer starts the defrost action. A termination thermostat returns the system to normal refrigerating. When defrosting starts:

1. A solenoid valve opens a line from the top of the receiver to the suction line.
2. A holdback valve reduces the high-pressure gas as it goes into the compressor.
3. A three-way solenoid closes the suction line to the compressor. It also opens a valve allowing hot gas up the suction line to the evaporator. The hot gas warms the evaporator and then condenses.
4. The condensed refrigerant bypasses the TEV through a check valve. It travels to the receiver by way of the liquid line.
5. A check valve in the condenser drain tube keeps refrigerant from backing up into the condenser.
6. A pressure regulating valve maintains proper hot gas pressures and temperatures. This valve is located at the top (entrance to the condenser). **Figure 13-75** shows the wiring diagram.

One type of hot gas bypass valve has a connection to the suction line. This valve is adjusted by the low-side pressure of the vapor going to the compressor. The valve shown in **Figure 13-76** is pressure operated. It opens wider as suction pressure drops and starts to close as low-side pressure rises. The pilot-controlled valve is installed in a system as shown in **Figure 13-77.**

Another type has an adjustable bellows in the sensing element to change the opening pressure:

Opening Pressure	PSI	Adjustment Range
R-12	30	25-35
R-22	58	50-65
R-500	38	32-44

The valve may send hot condenser gas into the evaporator. This occurs if the evaporator refrigerant temperature reaches freeze-up temperatures such as 26°F (−3°C). It is used where there are intervals of low-heat load conditions. This system is used on medium tonnage units (5 to 30 tons).

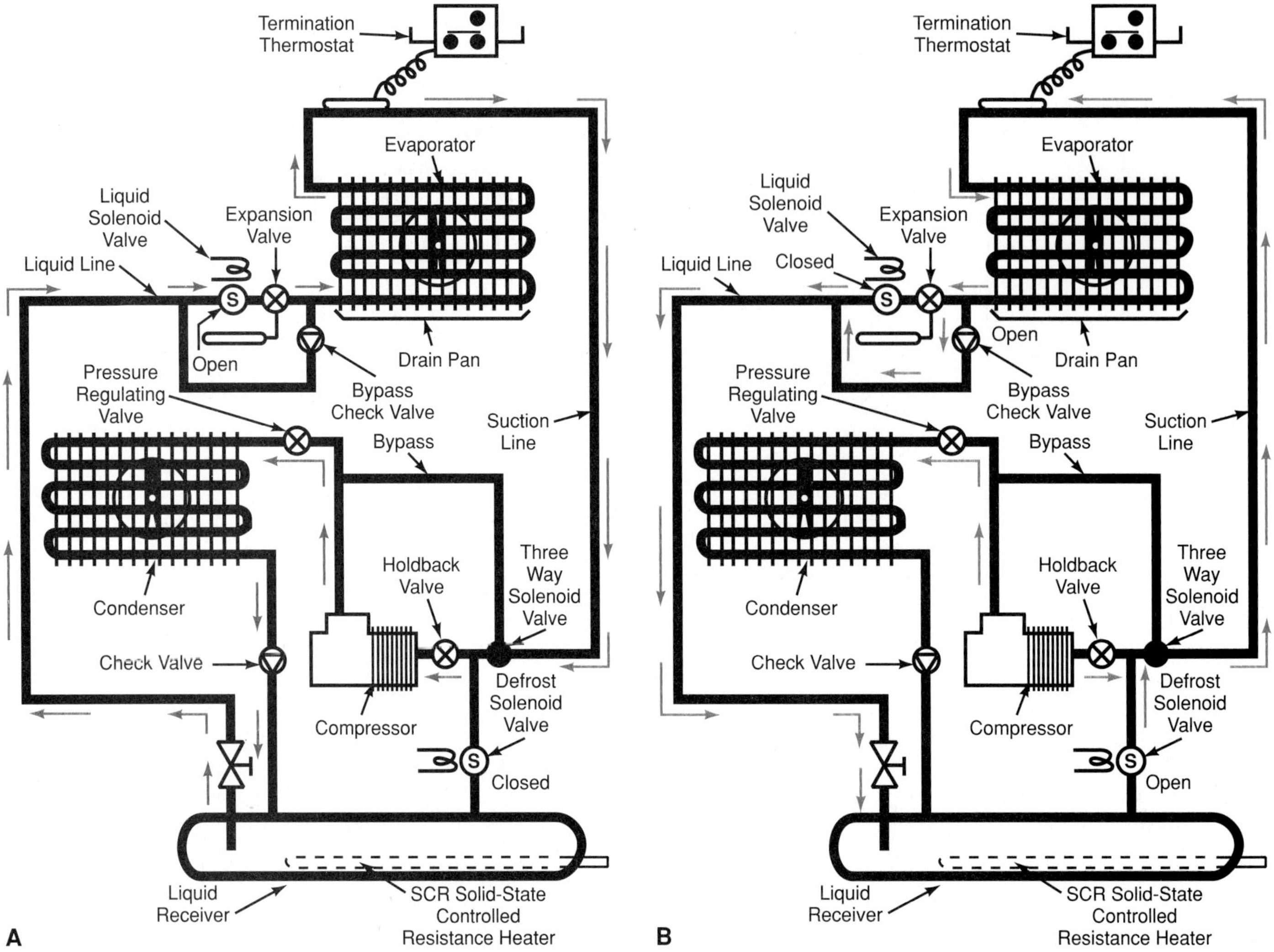

Figure 13-74. *Cycles for normal two-pipe hot gas defrosting system. A—Cooling cycle. Note direction of refrigerant flow through compressor and to the condenser. B—Defrost cycle. Note that hot gas from compressor and receiver is traveling back to the evaporator by way of the suction line. Condensed liquid refrigerant is traveling around TEV by way of the check valve and is returning to the receiver through the liquid line. Heater in receiver provides more hot gas for defrosting.*

Slugs of liquid refrigerant must not enter the compressor, where they would cause damage. Reevaporation should be almost complete before the refrigerant reaches the compressor.

In multiple systems having several evaporators, evaporators should be defrosted one at a time. The others are used to evaporate the liquid coming from the defrosting evaporator. This method ensures that no liquid refrigerant will reach the low side of the motor compressor. See **Figure 13-78.**

A valve may take the place of the two suction line valves on each evaporator. See **Figure 13-79.** The evaporator gas at point A travels into the valve during normal operation. The hot gas travels out of the valve at A during defrosting. The internal construction of the valve can be seen in **Figure 13-80.** When the pilot solenoid valve at A is energized, it opens. This allows high-pressure gas from the discharge connection to push down on the double valve. This action closes the top valve, B. It stops flow from evaporator into suction line, and opens bottom valve, C. High-pressure hot gas flows from the discharge connection up into the evaporator.

Nonfreezing Solution Defrost System

The "hot fluid" defrost system has been used for years. It has a container in which a brine (a nonfreezing solution) is stored. The refrigerant vapor from the compressor is pumped through this heat storage container. It then goes to the condenser.

The brine in the container may also be electrically heated. Such heating is provided during the normal running (freezing) part of the refrigerating cycle.

When the refrigerating system shuts off, the defrost timer closes a solenoid valve. This valve is located in a line running from the liquid line to the evaporator. This is the beginning of the defrost cycle. The evaporator fan is usually shut off. The brine solution is pumped through its own piping along the drain line, drain pan,

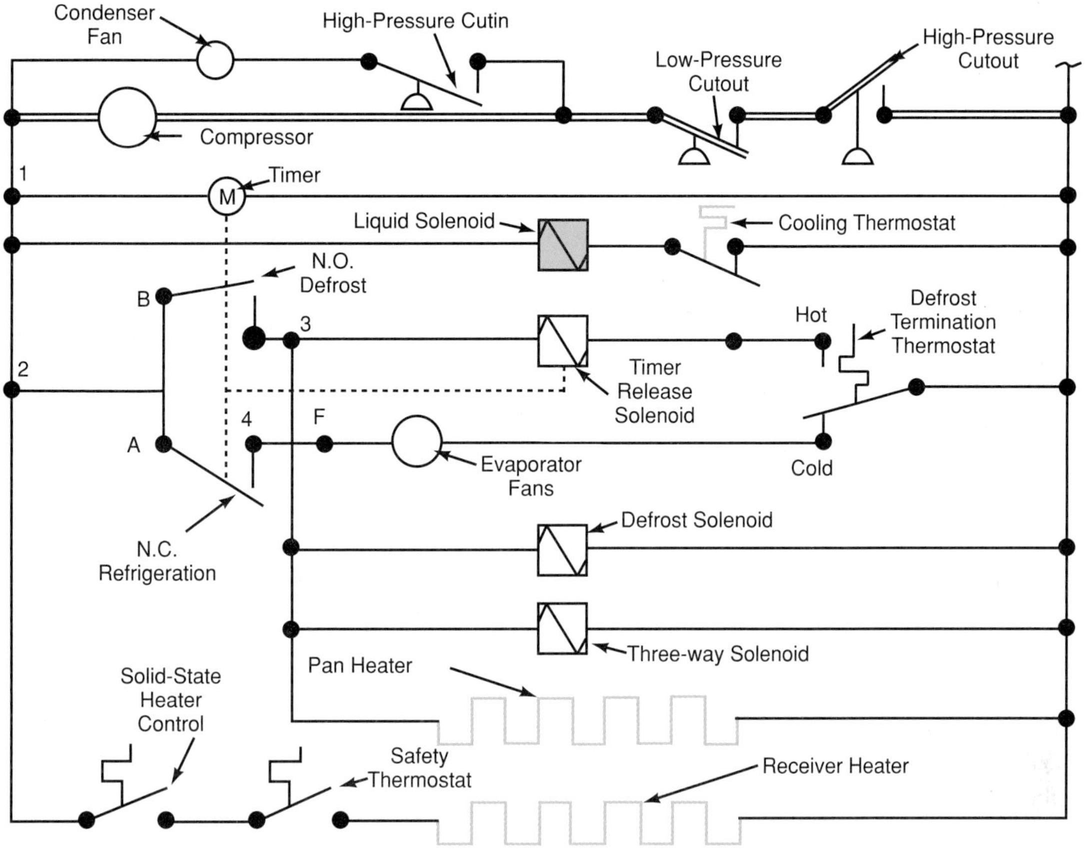

Figure 13-75. *Wiring diagram of a two-pipe hot gas defrosting system. Note the electric pan heater and electric receiver heater. Note, also, that cooling thermostat only controls liquid line solenoid.*

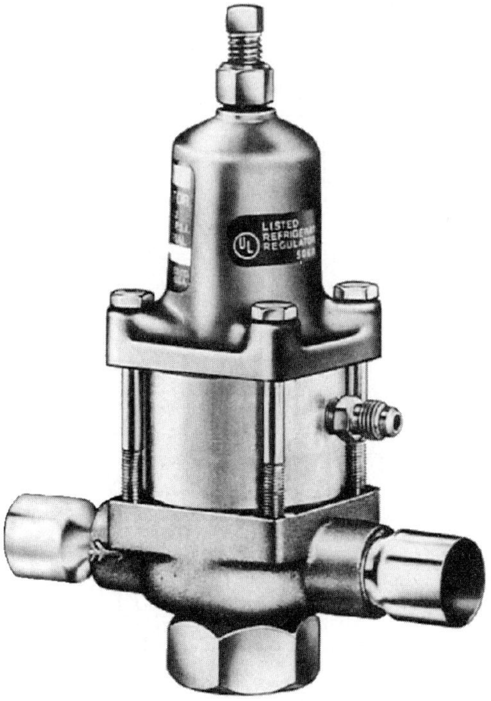

Figure 13-76. *Hot gas bypass valve. Its hot gas capacity using R-12 is 4 to 13 tons. Using R-22, it is 7.4 to 24 tons. (Refrigerating Specialties Div., Parker-Hannifin Corp.)*

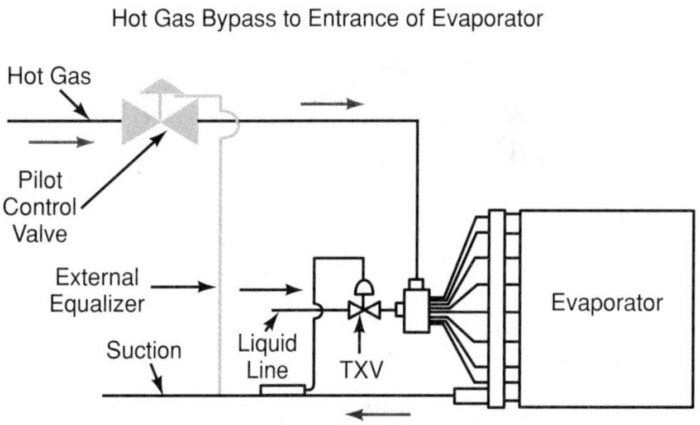

Figure 13-77. *Schematic diagram of hot gas bypass regulator. Note external equalizer line. It operates valve dependent on suction line pressure. (Refrigerating Specialties Div., Parker-Hannifin Corp.)*

and evaporator. Then it returns to its container. **Figure 13-81** shows such a defrost cycle.

Water Defrost Systems

The water defrost system runs tap water over the evaporator when the system is off. This is done either manually or automatically. During this operation, the evaporator louvers are closed. The water is warm

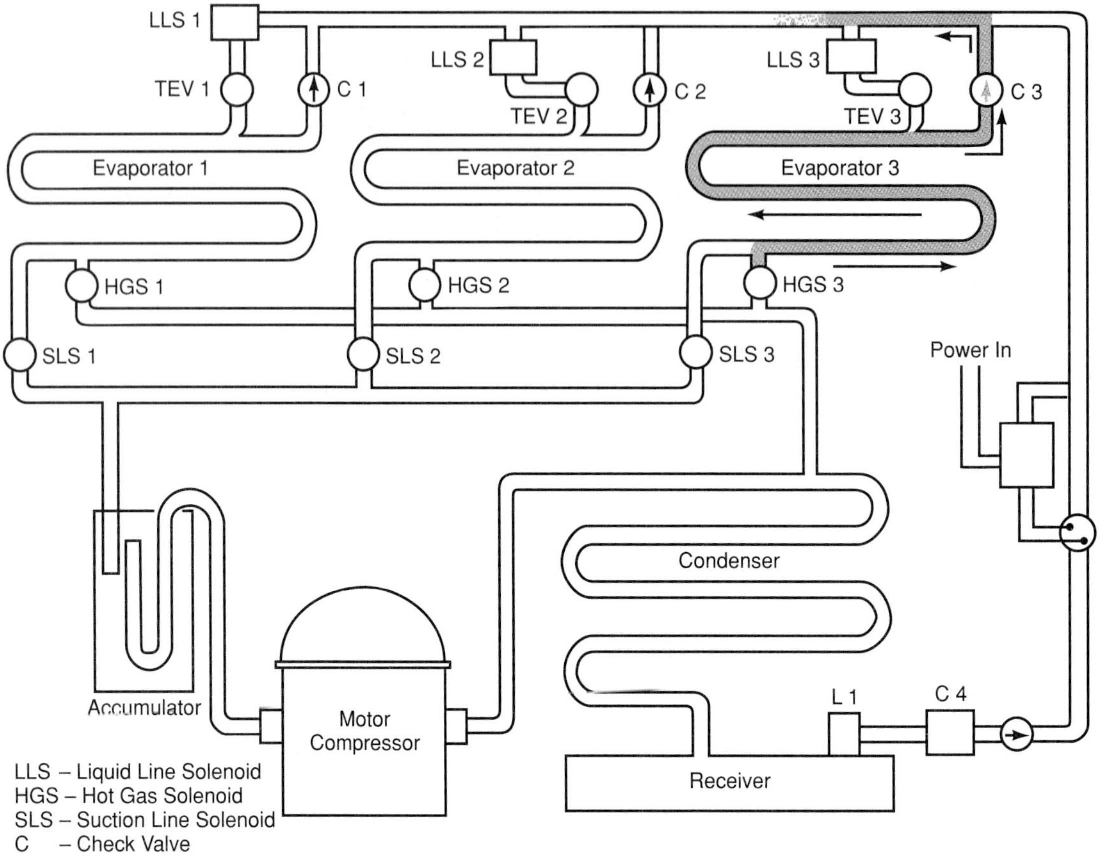

Figure 13-78. *"Hot gas" defrost system for multiple system. Timer controls each evaporator defrost at a different time. Liquid solenoid L1 closes, partially pumps down system. Hot gas solenoid on one evaporator opens suction line, solenoid valve closes and hot gas rushes into evaporator backwards. It enters liquid line through bypass check valve and feeds liquid refrigerant to other two evaporators. If Evaporator 3 is to defrost, Solenoid SLS3 closes, HGS3 opens, and LLS3 closes. Hot gas flows into Evaporator 3, condenses as it melts frost; then condensed liquid goes into liquid line by way of Check Valve C3. This liquid then moves into Evaporators 1 and 2 through TEVs 1 and 2.*

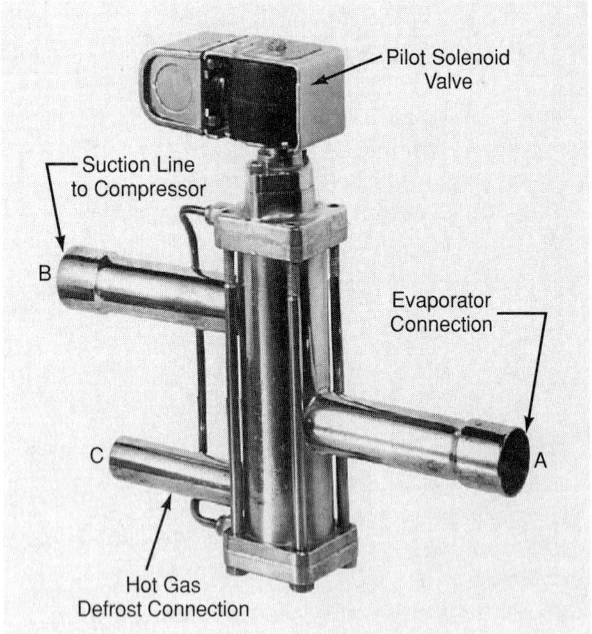

Figure 13-79. *Hot gas defrost solenoid valve. (Fluidex Division, Parker-Hannifin Corp.)*

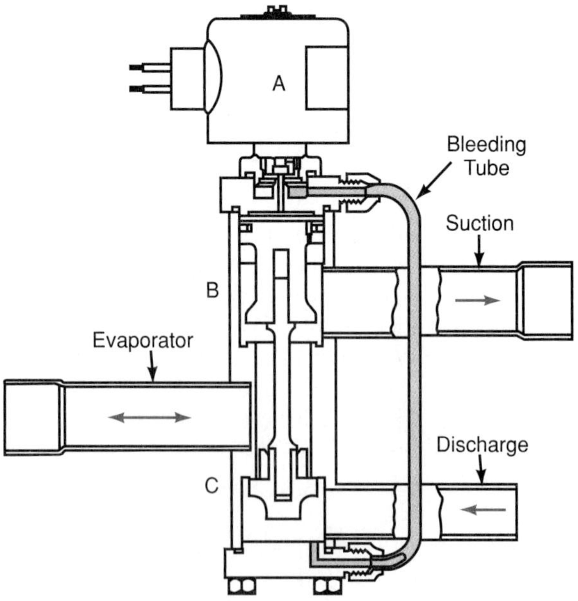

Figure 13-80. *Internal construction of a hot gas defrost solenoid valve. Note small tubing connection from discharge connection to pilot valve bleed. (Fluidex Division, Parker-Hannifin Corp.)*

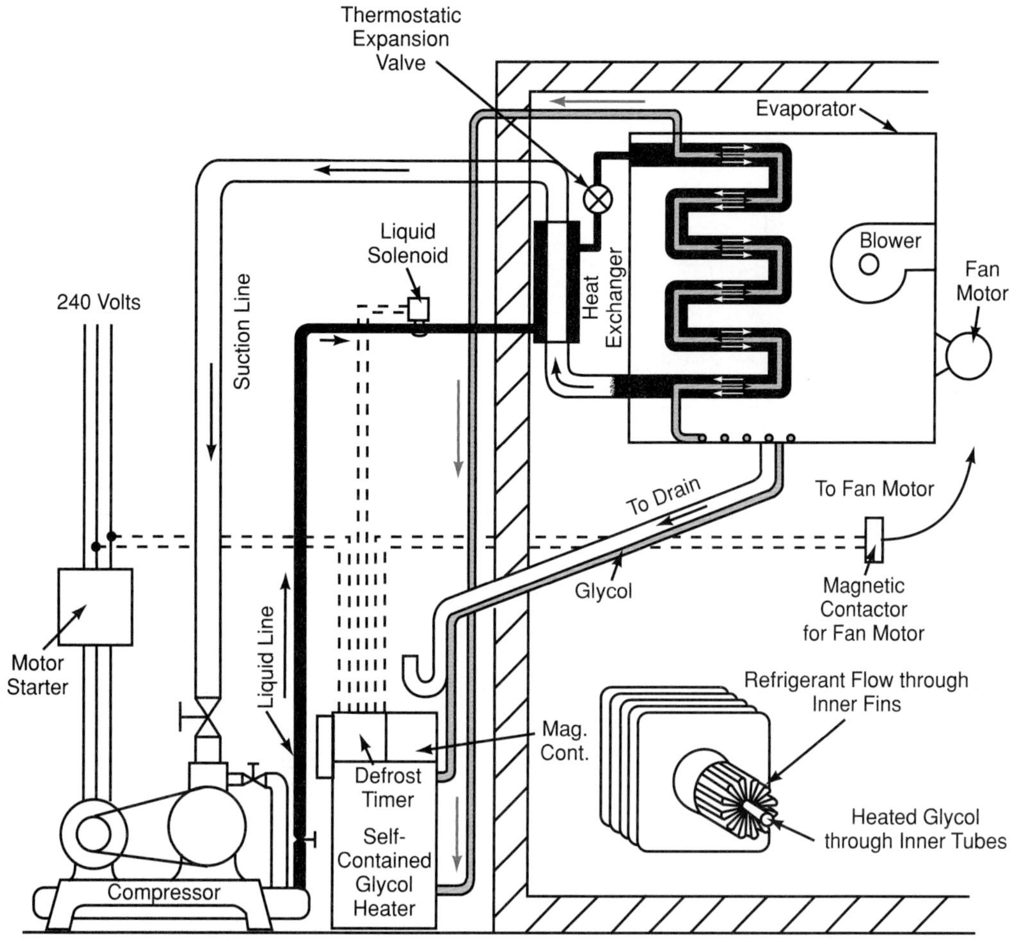

Figure 13-81. *Nonfreeze solution defrosting system. During defrost, glycol solution is pumped through inner tubing of evaporator and along the drain piping.*

enough to melt the ice. It then drains away into the evaporator drain pan. Drainage from the water lines must be complete before the unit is turned on or the water will freeze.

The water may be sprayed over the evaporator. It may be fed to a pan located over the evaporator instead. Holes in the pan feed the water evenly over the evaporator.

An electric timer provides automatic operation. **Figure 13-82** demonstrates the principle of water defrost. The two types of manual water defrost are shown. It also shows one automatic defrost system.

Special systems have been designed to defrost by spraying a brine over the evaporator. A pump may be employed to recirculate a lithium chloride brine. Eliminator plates are needed to prevent brine spray from passing into the refrigerated space.

Electric Heater Defrost System

Electric heat is popular for defrosting low-temperature evaporators. Heating coils are installed in the evaporator, around it, or within the refrigerant passages.

One method uses resistance wire heating elements mounted underneath the evaporator, under the drain pan, and along the drain pipe. A timer stops the refrigeration unit and closes the liquid line. It then pumps the refrigerant out of the evaporator. Then the blowers and the electric heaters are turned on.

The heaters melt the frost from the evaporator and the water drains away. The evaporators become warm enough to ensure that all frost is gone. A thermostat on the evaporator returns the system to normal operation.

"Pump down" is a control system. The thermostat operates a solenoid in the liquid line. At the same time, a low-pressure switch operates the compressor. Its purpose is to prevent flow of liquid refrigerant from the evaporator to the compressor. It is especially important in cases of electric defrost.

1. When the thermostat is satisfied, it opens and the liquid line solenoid closes.
2. The compressor continues to run and removes the refrigerant vapor from the evaporator and suction line.
3. When the proper low-side pressure is reached, the low pressure switch opens and the compressor stops.

There should be very little refrigerant in the compressor oil. When a pump-down system is used

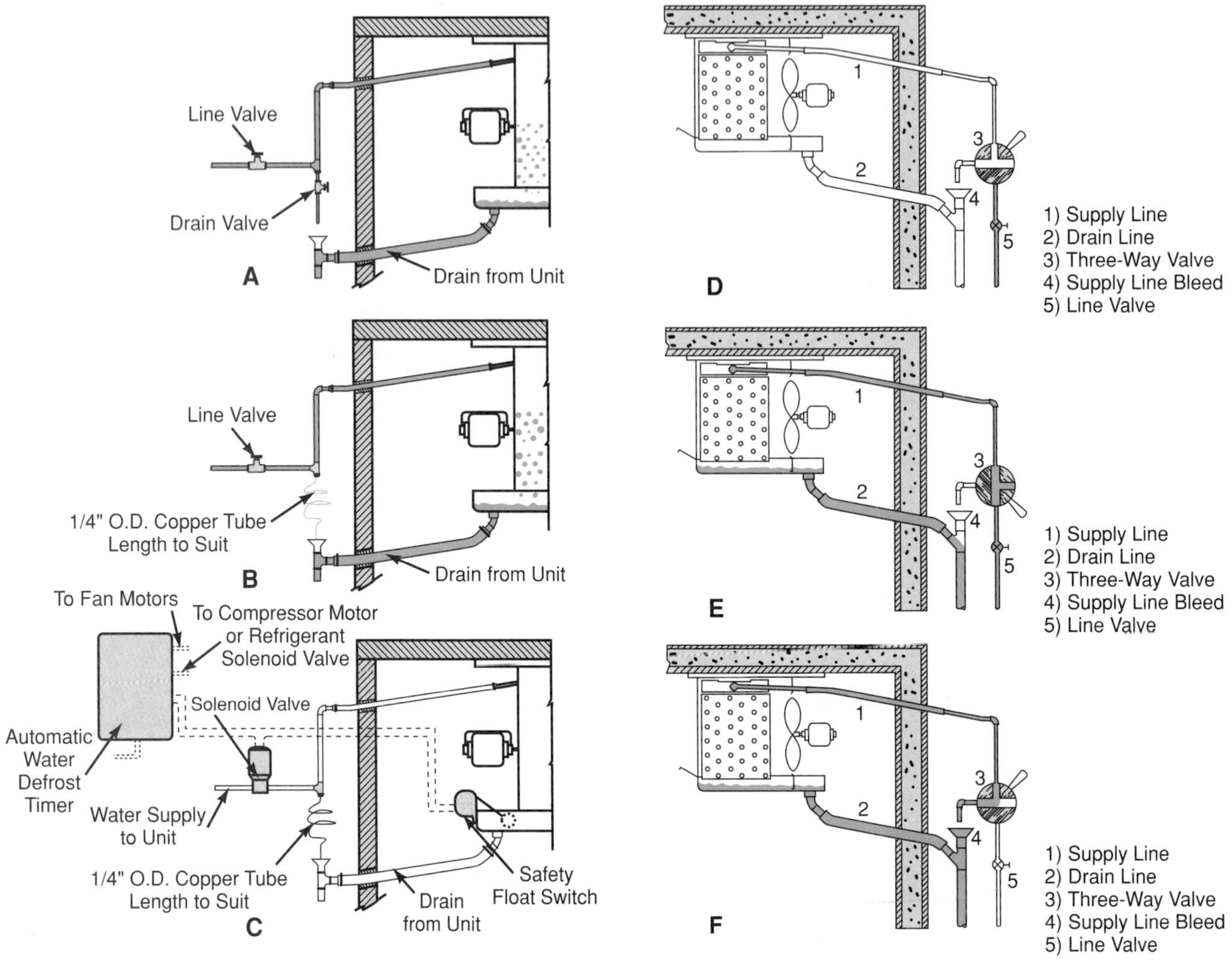

Figure 13-82. *Water spray defrost system schematics. Three methods of operation are shown. A—Manual defrost and manual drain. B—Manual defrost and automatic drain. C—Automatic defrost. There are three steps in the defrost cycle for a manual defrost and manual drain defrost system (View A): D—During refrigeration cycle. E—Defrost operation. F—Water lines and drain being cleared of water at end of defrost operation.*

for each cycle, crankcase heaters may be needed. These heaters will drive the refrigerant during the Off cycle.

"Pump out" has a similar purpose. However, an extra relay is wired into the compressor circuit in parallel to the normal relay. It is connected to the thermostat circuit. The extra relay operates the start button on the normal starting relay. The compressor, therefore, cannot restart until the thermostat points close.

Another electric defrost system uses an immersion electric heater to heat a separate charge of refrigerant. The warm refrigerant circulates around the evaporator in its own passageways. The evaporator is warmed and the system defrosted. This happens while the unit is turned off.

Still another way of using the electric heater defrost system is with a double-tube evaporator. The evaporator refrigerant passes through the passageway between the tubes during normal refrigeration. Electric heating elements are inserted in the center tube. In the defrost operation, the system is stopped and the electric heating elements are turned on. See **Figure 13-83.** Thereby, the evaporator tubes cause defrosting from the inside.

Warm Air Defrosting

Where there is enough of it, warm air can be used to defrost low-temperature evaporators. Cabinet air can be used for defrosting if it is at the correct temperature.

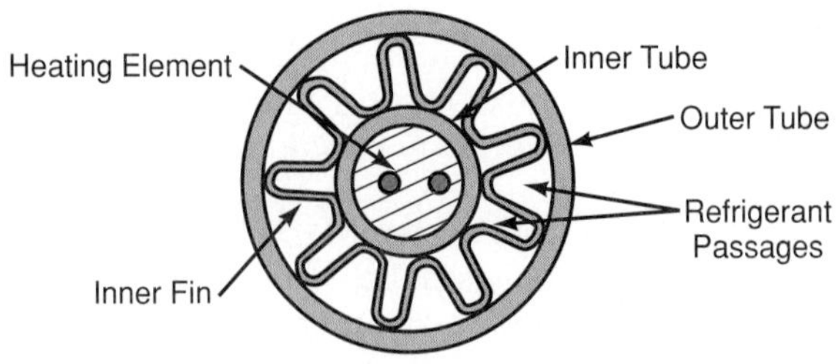

Figure 13-83. *Cross-sectional view of electric defrost system having electric heating elements installed in evaporator tubing.*

The cycles must be frequent enough and long enough to defrost the evaporator completely. Some installations bring in outside air for defrosting. A controlled duct system with blowers and fan is used.

13.4.10 Heat Exchangers

A heat exchanger mounted in the suction and liquid line has three advantages:

- It subcools the liquid refrigerant and increases operating efficiency.
- It reduces flash gas in the liquid line.
- It reduces liquid refrigerant in the suction line.

A heat exchanger is shown in **Figure 13-84.** Heat is transferred from the warmer liquid in the liquid line to the cool vapor coming from the evaporator. **Figure 13-85** shows the outside appearance of a heat exchanger.

Liquid cooled 10°F to 20°F (5°C to 11°C) at the prevailing head pressure absorbs more latent heat. This occurs as it changes to a vapor in the evaporator.

The reduction of flash vapor (sometimes called "flash gas") is important. Flash gas (vaporized refrigerant) comes from the sudden change of some of the liquid to a vapor. This happens as the refrigerant passes through the refrigerant control. The valve capacity is reduced, increasing low-side pressure drop. The amount of heat each pound of refrigerant absorbs as it evaporates is also reduced. The "flash gas" cools the remainder of the liquid to the evaporating temperature.

The heat exchanger also helps prevent sweat backs or frost backs on the suction line. Low temperature liquid refrigerant in the returning suction vapor will evaporate in the heat exchanger. This occurs as the refrigerant absorbs heat from the liquid line.

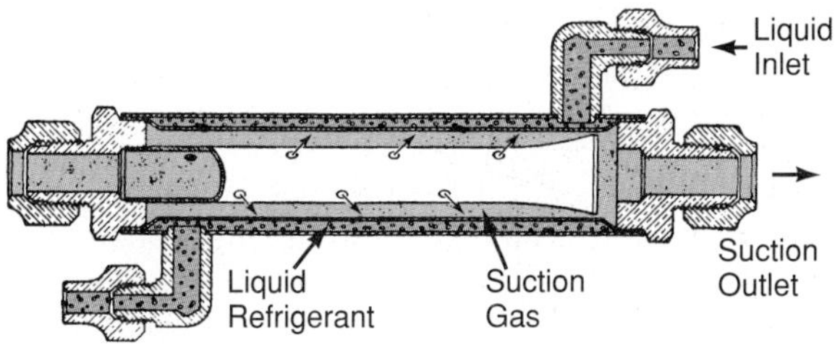

Figure 13-84. *Cross section of heat exchanger used on commercial systems. Note flared connections.*

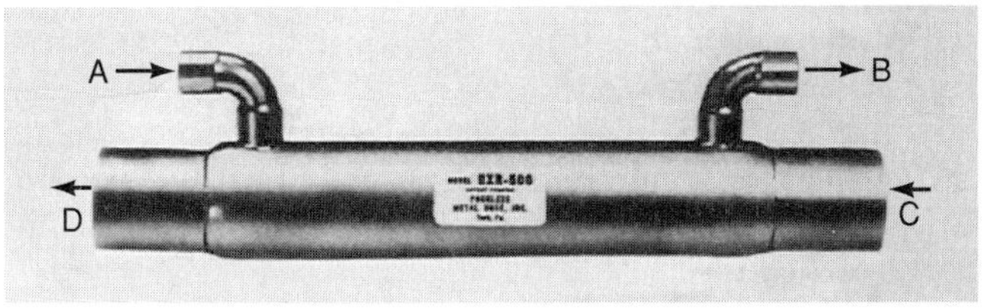

Figure 13-85. *Heat exchanger, external view. Brazed connections are used. A—Liquid in. B—Liquid out. C—Suction vapor in. D—Suction vapor out. (Packless Industries)*

The subcooled liquid in the liquid line reduces the chance of flash gas forming in the liquid line. This is especially true on warm days or if the liquid line has a long vertical run. Pressure drop in the suction line portion of the heat exchanger should not be over 2 psi (14 kPa).

COMMERCIAL SYSTEMS—CONTROLS MODULE

13.5 Refrigerant Controls

Five refrigerant controls can be used for installations involving one evaporator and one condenser. These include: thermostatic expansion valves, automatic expansion valves, high-side floats, low-side floats, and capillary tubes. These are explained in Chapter 5.

In multiple installations, two types of refrigerant controls may be used. They are the low-side float and the thermostatic expansion valve. Thermostatic expansion valves are used extensively but there are also some low-side float systems.

Some of the thermostatic expansion valves have a large capacity. Usually they are constructed with a pilot valve operating a larger valve.

The thermostatic expansion valve is explained in Chapter 5. Technicians should study its design, operation, installation, care, and repair before proceeding with this chapter.

13.6 Motor Controls

Two basic types of motor controls are used in commercial refrigeration:

- Thermostatic.
- Pressure.

These are the same ones used in domestic refrigeration. Large systems use magnetic starters operated by motor controls.

In multiple evaporator commercial work, pressure motor controls are used quite often because:

- The low-side pressure is an indication of the temperature in the evaporators.
- One control works well regardless of the number of evaporators connected to it.

The controls provide both range and differential adjustments. Explanation of various range and differential adjustments are found in Chapter 8. **Figure 13-86** shows the internal construction of such a control.

When larger motors start, the current draw is more than control contacts can handle. A motor starter is necessary for single-phase ac motors over 1 hp. Three-phase ac motors also require a starter. The motor control operates the relay in the starter.

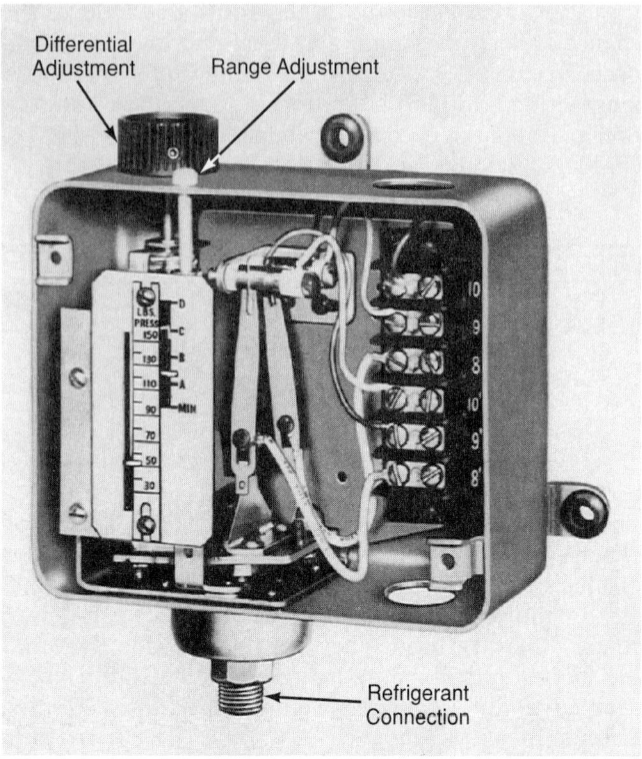

Figure 13-86. *Cover has been removed from the pressure motor control. Note adjustments, electrical connections, and pressure scale. (Johnson Controls, Inc.)*

Electrical work should be done by a licensed electrician. The work should comply with local electrical codes. **Figure 13-87** lists the recommended average pressure motor control settings. Both temperature control and defrost control are listed.

13.6.1 Pressure Motor Control

The pressure motor control is usually mounted on the condenser. It is operated by low-side pressure. Some companies suggest connecting the control into the low-side suction line. The control should be placed about 10′ to 15′ from the compressor. The vibration effect on the control will be reduced.

The range settings vary with the application. Cut-out pressure should be set about 10°F (6°C) below the desired evaporator outside surface temperature. Cut-in pressure should be about the same as the highest allowable evaporator temperature. See **Figure 13-88.**

The differential setting will vary, depending on the temperature accuracy wanted. A wide pressure difference will allow some variation in cabinet temperature. It will also lengthen the operating cycle interval of the condenser. (This means the compressor would not run as often.) A differential set to close limits will maintain a more uniform cabinet temperature. However, it will shorten the cycling interval of the condenser. The unit would run more often. Pressure difference between cut-in and cut-out point varies with the refrigerant used. Common pressure difference is about 20 psi for R-12, 22 psi for R-22, 16 psi for R-500, and 25 psi for R-502.

13.6.2 Thermostatic Motor Control

The ***thermostatic motor control*** is like the pressure motor control in design. However, the sensing bulb and capillary tube are different. **Figure 13-89** shows a control with the cover removed.

This control is generally used in large single installations. However, satisfactory setups have been made in multiple installations. Hopefully, when the controlled cabinet is at the desired temperature, the others are also. These controls are also used together with a solenoid valve. Thus, each separate cabinet in a multiple installation may be controlled.

Some are made with a very close differential such as 1°F (0.5°C). These are used for certain display cases, bulk milk coolers, frost alarms, liquid chillers, and refrigerated trucks.

Thermostatic motor controls are popular in brine cooling installations. The sensing bulb is submerged in the brine. Ice cream cabinets are a typical example. In single cabinet installations, the sensing bulb is usually mounted in the cabinet 4′ up from the floor. It is located between the cold and warm air flues and at least 2″ from the wall.

Some are wall mounted in walk-in coolers, meat storage rooms, warehouses, and florist cabinets. Some have double-throw contacts (SPDT). With these, the control may also operate other devices (fans and defrost systems).

13.6.3 Safety Motor Controls

Commercial and domestic controls differ in that many commercial electrical systems also use the following safety devices:

- A high-pressure safety cutout.
- An oil pressure safety cutout.

The high-pressure safety device is a bellows built into the control. It is connected to the high-pressure side of the system, **Figure 13-90.** It is often connected to the cylinder head. This permits easy disconnecting of the control from the system.

The head pressure may become too high from air in the system. Or, if the condenser water is shut off, head pressure will increase. The bellows will expand if head pressure becomes too high. The bellows is attached to the plunger, which is pushed against the switch, shutting off the motor.

Action of the high-pressure safety device prevents the buildup of dangerous pressures within the system. It also prevents ruining the motor through overloading and overheating.

The control is usually set to cut out at 20% above normal head pressure. In R-12 systems, the control is set at about 150 psi to 160 psi. With R-22, it is set at 260 psi to 270 psi. With R-502, it is set at 280 psi to 290 psi and with R-500, it is set at 190 psi to 200 psi.

Recommened Case Temperature & Defrost Control Settings

Tyler Defrost Controls: TC = Straight Time Clock. TG = Temperature Guard with current sensing relays resetting control. TS = Time Solenoid reset, similar to TG and used on walk-in cooler coils. Multi-circuit timers are used with Parallel Compressor units. Termination is by temperature or straight time.

Case Models	Disch Air Temp.	Defrost Htr. Amps. 8'	12'	Press. Cont. R-22 Cutin	Cutout	EPR Set'g R-22	Defrost Control PerDay, FailSafe, Term Temp Elect	HotGas
DMF, XDF(S) XDFC(2)	−15°	6.9	10.3	15-22	4-8	7.0	1@60/50°	2-3@16-20/70-75°
XDWFC	−15°	13.8	20.6	15-22	4-8	7.0	1@46/50°	2-3@16-20/70-75°
XDWFCE	−17°	6.9	—	15-22	4-8	7.0	1@46/50°	2@16-20/70-75°
XD7FE (F/F)	−15°	13.0	—	15-22	4-8	7.0	1@36/50°	2-3@16-20/70-75°
FFJ	−15°	13.8	20.6	15-22	4-8	7.0	1@60/50°	2-3@16-20/70-75°
FFJE	−15°	8.6	—	15-22	4-8	7.0	1@60/50°	2-3@16-20/70-75°
FFJG[1]	−15°	13.8	20.6	15-22	4-8	7.0	1@60/50°	2-3@20-25/50-54°
FFJEG[1]	−15°	8.6	—	15-22	4-8	7.0	1@60/50°	2-3@20-25/50-54°
XDFI(S) XDFIC2	−25°	13.8	20.6	9-15	1-5	2.0	1@36/50°	2-3@16-20/70-75°
XDWFIC	−25°	13.8	20.6	9-15	1-5	2.0	1@46/50°	2-3@16-20/70-75°
XDWFICE	−27°	6.9	—		1-5	2.0	1@46/50°	2-3@16-20/70-75°
XD7FIE (I/C)	−25°	13.0	—	9-15	1-5	2.0	1@36/50°	2-3@16-20/70-75°
FZJ	−25°	27.6	41.2	9-15	1-5	2.0	1@36/50°	2-3@16-26/70-75°
FZJE	−25°	8.6	—	9-15	1-5	2.0	1@36/50°	2-3@16-26/70-75°
FZJG[1]	−25°	27.6	41.2	9-15	1-5	2.0	1@36/50°	2-3@25-30/50-54°
FZJEG[1]	−25°	8.6	—	9-15	1-5	2.0	1@36/50°	2-3@25-30/50-54°
D6F (3 Phase)	−10°	15/Lg	23/Lg	13	7	10.0	2@40/50°	3@16-20 60/30°
T5FG (F/F)	−4°	9.6	17.3	16	8	12.0	1@60/50°	2@18-20/70°
T5FG (I/C)	−12°	9.6	17.3	12	4	8.0	1@60/50°	2@20-26/70°
DFRG5 (F/F)	−10°	4.5	Rem	10	6	9.0	2@36/50°	2@15

Defrost Check List

1. Check to see that defrost contactor is wired to cases on that condensing unit, then check defrost time and number.
2. Check heater amps during and after defrost.Is coil clear at termination? Does limit switch open too soon?
3. Check waste outlet for proper hook-up (max. of 12' of 1" pipe 1/4" per foot slope) and that it isn't frozen by refrigeration line contact.

Good Housekeeping—Essential for Good Refrigeration!

Good housekeeping is not only necessary for sanitation, but it contributes measureably to reliability and to the quality of refrigeration. A good maintenance program is essential. Meat cases should be cleaned thoroughly once a week, other cases at least twice a year. Clean the interior with germicidal detergent. RINSE: Flush the waste outlet with hot water.

OTHER POINTS: Eliminate drafts over open cases. (Maximum allowable draft—50 FPM) DO NOT BLOCK AIR DUCTS. OBSERVE LOAD LINES!! Cases must be level to operate properly.

Figure 13-87. *Recommended motor control pressure settings. These are recommended for various case applications. Pressures are in psi. It may be necessary to change these settings somewhat for a particular installation. (Tyler Refrigeration Corp.)*

* Inches Vacuum **Application**	R-12 Out	R-12 In	R-22 Out	R-22 In	R-502 Out	R-502 In
Freezer–Open Type	7*	5	4	17	9	23
Freezer–Closed Type	1	8	11	22	17	29
Ice Cube Maker–Flooded or Dry Type Coil	4	17	16	37	22	47
Sweet Water Bath–Soda Fountain	21	29	43	56	52	67
Showcase–Frost Cycle	10	25	25	50	32	60
Showcase–Defrost Cycle	18	34	39	64	49	76
Beer, Water, Milk Cooler	19	29	40	56	47	67
Walk-In Cooler–Defrost Cycle	12	35	29	66	37	77
Ice Cream Trucks, Hardening Rooms	2	15	12	33	17	42
Vegetable Display–Defrost Cycle	11	35	27	66	35	77
Eutectic Brine Tank, Ice Cream Truck	1	4	11	16	17	22
Reach-In Cooler–Defrost Cycle	18	36	39	68	47	79
Beer Coolers–Blower Dry Type	15	34	33	64	42	76
Beer Coolers–Bare Pipe Dry Type–Frost Cycle	12	27	29	53	37	64
Instantaneous Beer Coolers	12	29	29	56	37	67
Retail Florist Box–Blower Coil	26	42	51	77	61	88

Figure 13-88. *Typical refrigeration applications and low-side pressure motor control settings.*

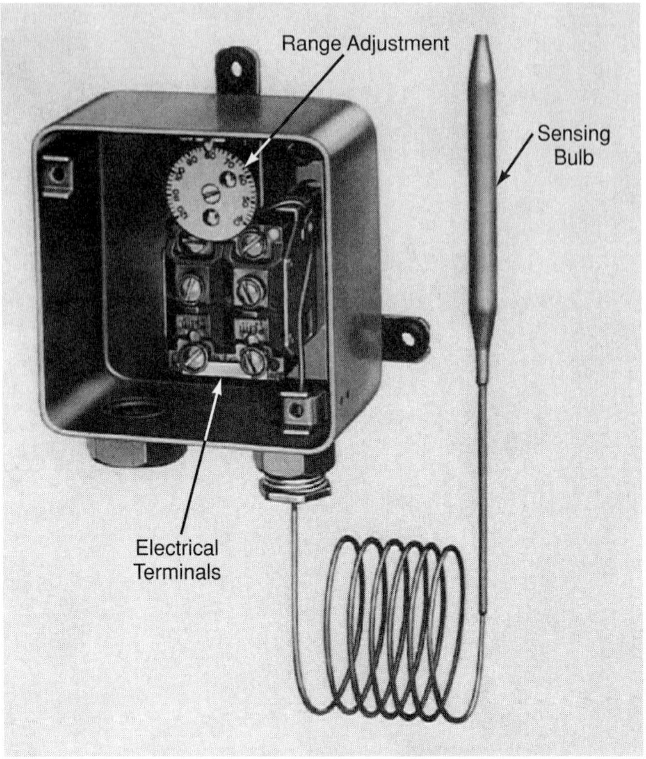

Figure 13-89. *Thermostatic motor control with cover removed. Note temperature range dial (Fahrenheit scale) and electrical terminals. (Johnson Controls, Inc.)*

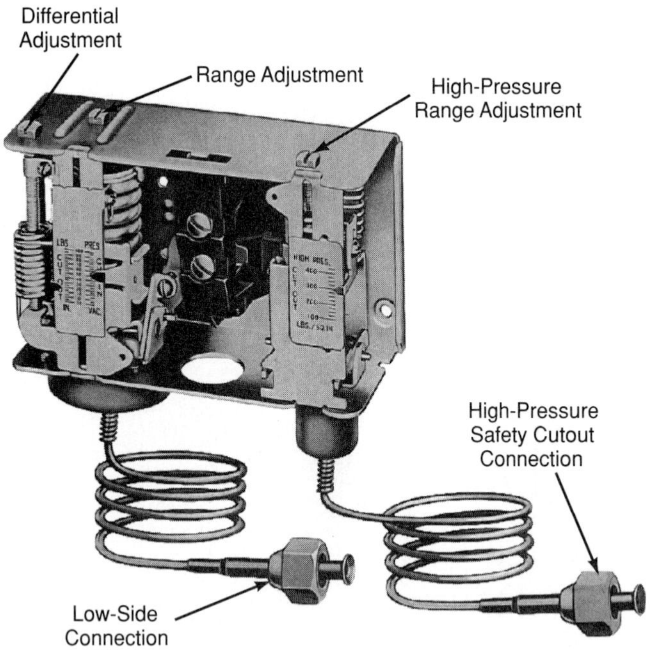

Figure 13-90. *This pressure operated motor control also has a high-pressure safety cutout. Note that it has three adjustments. (Johnson Controls, Inc.)*

The oil pressure safety cutout will shut off the electrical power if the oil pressure fails or drops below normal. It is a differential control, using two bellows. One bellows responds to the low-side pressure and the other responds to the oil pressure. The oil pressure must always be above the low-side pressure for oil to flow. See **Figure 13-91.**

The wiring diagram for an oil pressure safety control is shown in **Figure 13-92.** The control opens the circuit if the pressure difference between the two

Figure 13-91. *Commercial system pressure control with oil pressure safety cutout. It operates on the difference between refrigerant and oil pressures. (Johnson Controls, Inc.)*

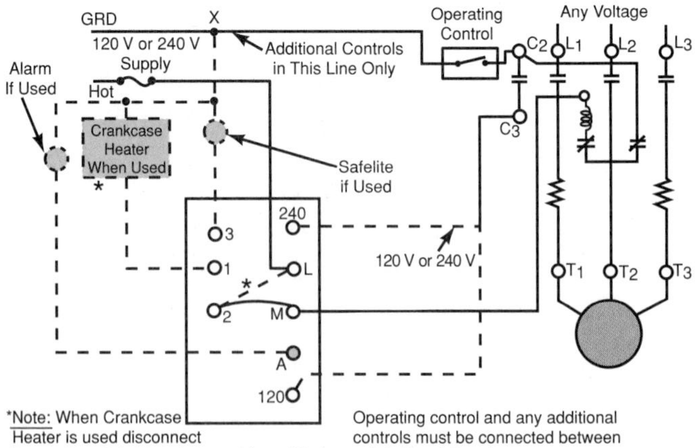

Figure 13-92. *Wiring diagram shows oil pressure safety cutout at A. Motor is three-phase. Note provisions for adding alarms, safety light and crankcase heater. (Johnson Controls, Inc.)*

bellows drops below the required oil pressure. Large commercial systems use this type. In some systems, the control points are in the compressor motor circuit. In other cases, the points will close and current is sent through a bimetal strip. (Or, it will be sent through a resistance heater near the bimetal strip.) If this strip heats up before the pressure returns to normal, the power will be disconnected. An oil pressure safety control is shown installed on a motor compressor in **Figure 13-93.**

Refrigerant level may be kept within safe limits by a float switch. The float may be used for signaling or it may actually control the refrigerant level.

The switch may be used to control the liquid level in:

- Flooded surge drums.
- Flooded shell and tube chillers.
- High- and low-pressure receivers.
- Intercoolers.
- Transfer vessels.
- Various kinds of accumulators including liquid recirculating types.

If the refrigerant level is too high, the float switch closes an electrical circuit. The circuit acts to allow a refrigerant flow out of the control device. If the refrigerant level is too low, the float switch actuates a circuit that allows refrigerant to flow into the system. A float control switch is illustrated in **Figure 13-94.**

13.6.4 Motor Starters

The pressure or temperature control contacts can safely carry a limited amount of current. This holds true for both open and sealed types. The National Electric Code and local electric codes usually dictate the limitations of these controls. However, these same commercial controls can handle larger motors (larger loads) with the help of a device called a magnetic starter (contactor).

The magnetic starter is an electromagnetic device. The magnetism is controlled by the electricity that flows through the motor control. The magnetism attracts a piece of steel (or armature). When this armature moves, it closes large contact points. These contact points safely carry the large current flow needed for the large motors.

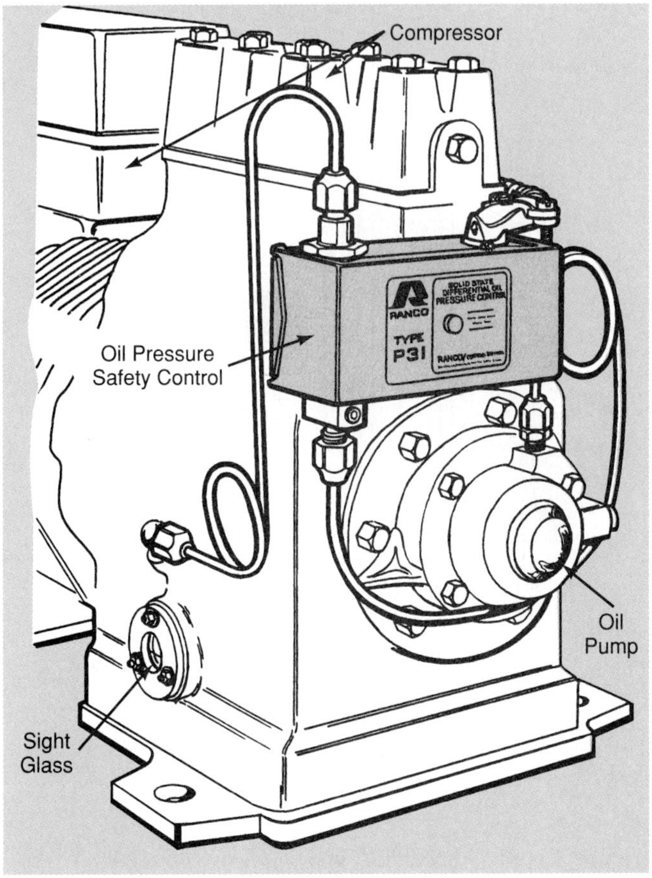

Figure 13-93. *Oil pressure safety control assembly must be installed level in a motor compressor.*

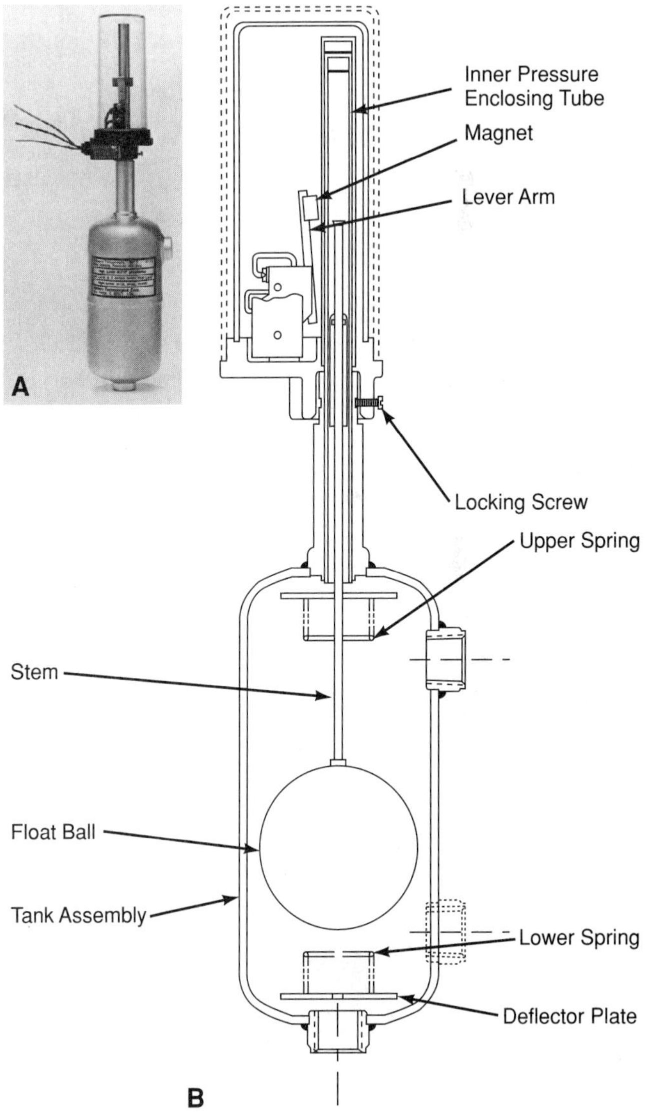

Figure 13-94. *Float-operated switch designed to control level of liquid refrigerant in system. A—Exterior view. B—Cross-section. Note float ball and electrical switch with magnet control. (Hanson Technologies Corp.)*

Figure 13-95 is a schematic wiring diagram of a magnetic starter.

These starters are mounted in an approved metal box with a safety access door. Some units incorporate a manual shutoff switch, fuses, and an overload thermal safety breaker switch.

The safety switch is operated by a heating element. This is located in the motor circuit black lead inside the contactor box or starter. Should the motor demand too much current (shorts, grounds, or overloads), the heater will bend a thermal bimetal strip in the control circuit. This will open the electromagnet circuit. This action opens the main switch.

Figure 13-96 shows a wiring diagram of a 120-240 V single-phase system using a magnetic starter. A wiring diagram for a three-phase system is shown in **Figures 13-97** and **13-98.**

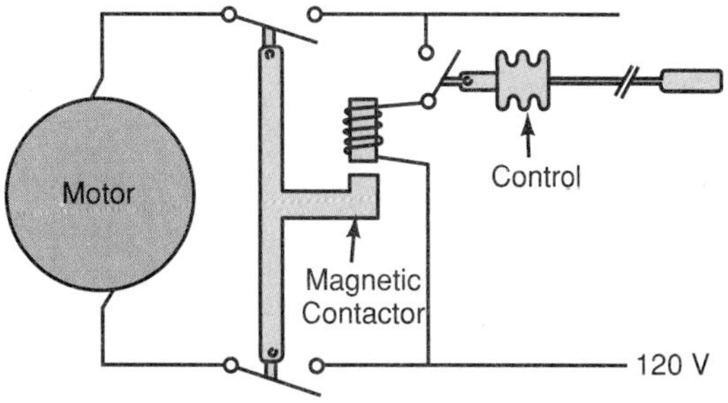

Figure 13-95. *Electrical schematic diagram for an automatic control on a magnetic starter. Circuit allows high current flow to motor without overloading control contact points.*

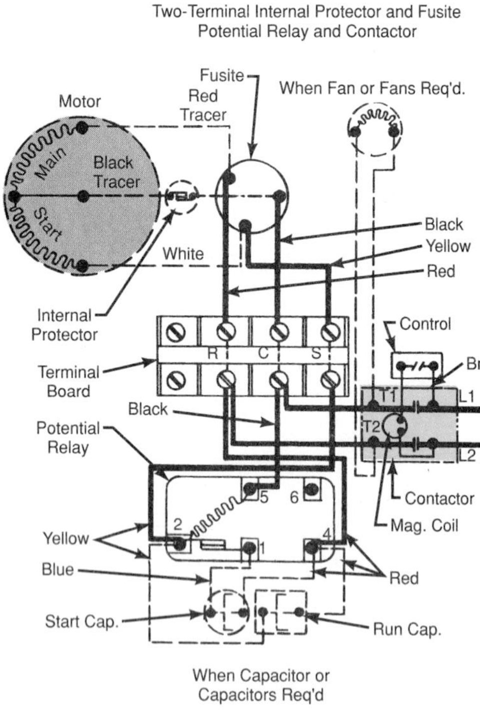

Figure 13-96. *Wiring diagram for a 120-240 V single-phase system. Note use of contactor (motor starter).*

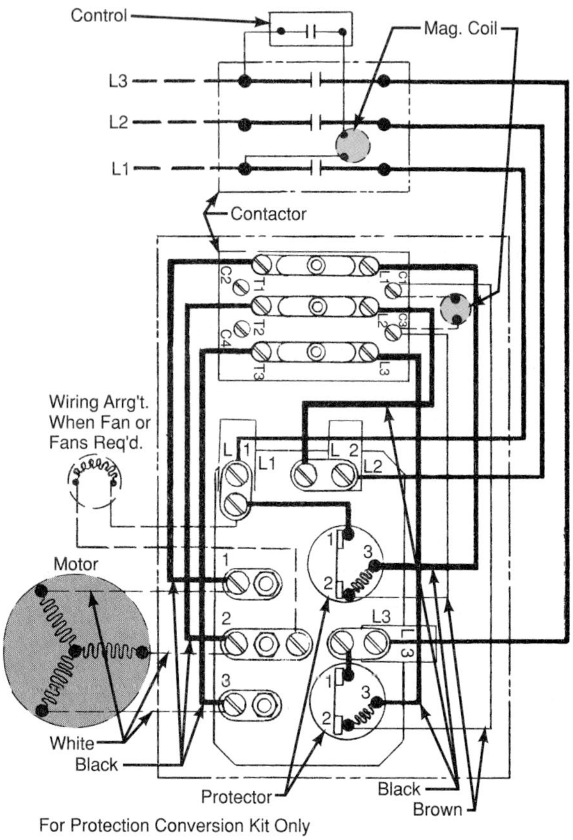

Figure 13-97. *Wiring diagram for a three-phase system using a magnetic starter. Control is wired in series with magnetic coil of contactor (starter).*

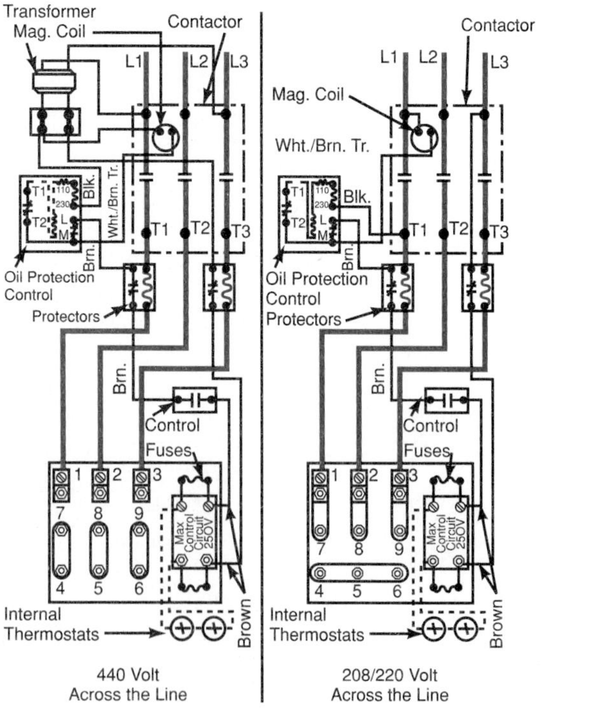

Figure 13-98. *These wiring diagrams show 440 V and 208-220 V circuits designed for three-phase power. Lines L_1 through L_3 each carry one leg of the three-phase voltage. (Copeland Corp.)*

13.7 Ice Maker Controls

Automatic ice cube makers or ice flake makers have additional refrigerant and motor controls. They have controls located in the bin to stop the system when the storage bin is full. It shuts off the machine until some of the ice is removed or melts. Devices used include:

- Mechanical levers.
- Temperature controls.

The mechanical type has a lever or a diaphragm. When this is contacted (pressed) by accumulated ice, it opens a switch and stops the unit.

The temperature control shuts off the unit when the control bulb directly contacts the ice. Both controls are located at the top of the ice bin.

The wiring diagram, **Figure 13-99,** is from a system that freezes cubes and then removes (harvests) them from the freezing grid. The refrigerating system has a hot gas defrosting system, as shown in **Figure 13-100. Figure 13-101** is an actual wiring diagram. It shows the interlocking of the controls, devices, and wiring.

13.8 Vending Machine Controls

Most vending machines that use refrigeration operate automatically. These machines can perform several operations:

- Heat, cool, and select foods.
- Accept coins to activate the dispenser system.

Some of the units automatically heat (if necessary). They may also move the items being dispensed such as bottles, bulk fluids, and packages of ice cream. Thermostats, relays, microswitches, positioning motors, and solenoids are used.

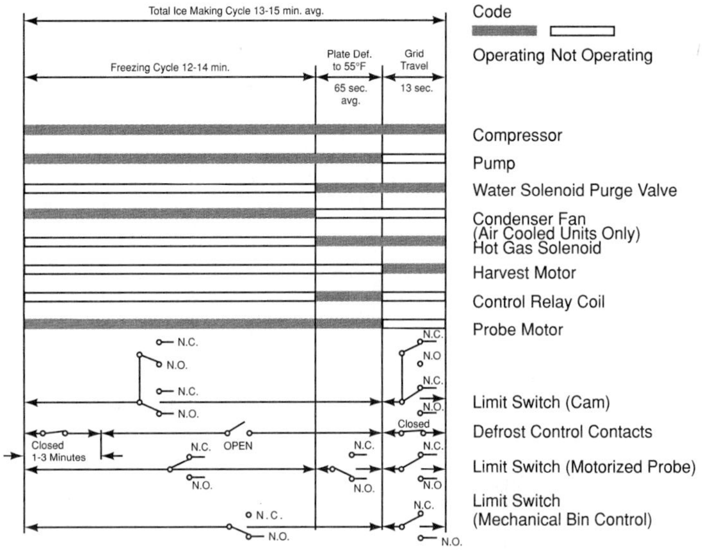

Figure 13-99. *This wiring diagram is for an ice cube maker that uses harvest motor to remove cubes from the freezing grid. (Ice-O-Matic)*

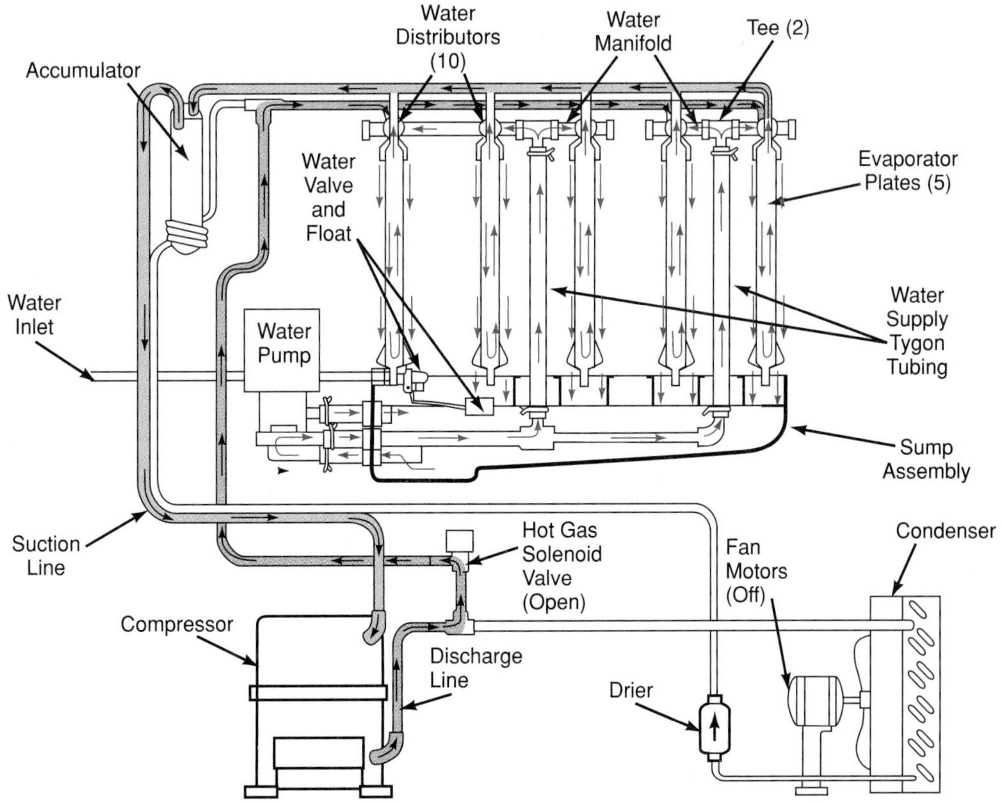

Figure 13-100. *This ice cube maker refrigerating system uses a hot gas defrosting system. (Scotsman Ice Systems)*

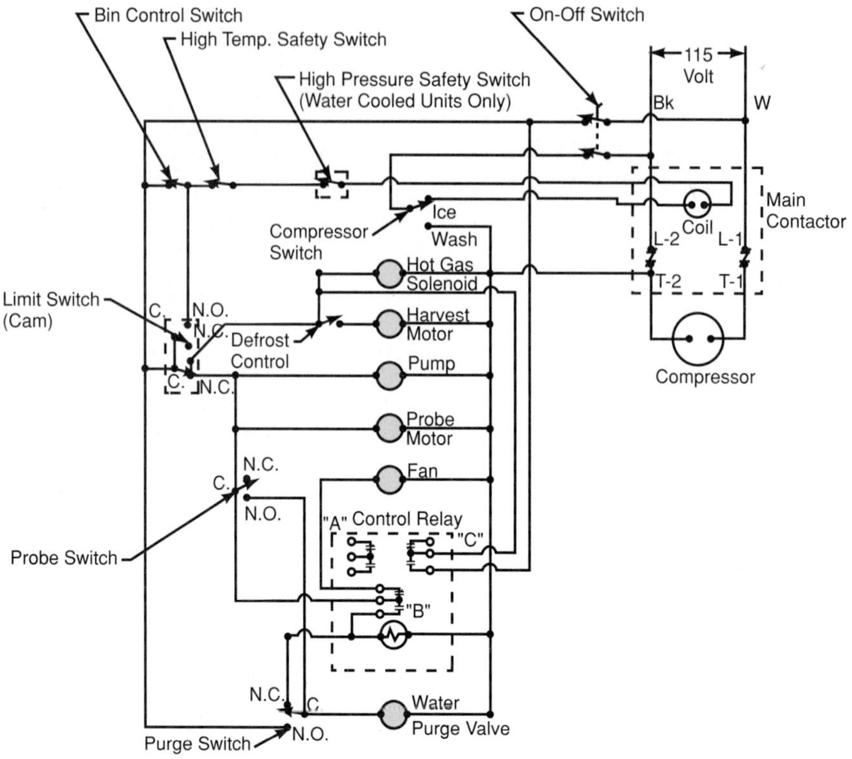

Figure 13-101. *Note the number of circuits shown in this ladder diagram: hot gas defrost, probe motor, harvest motor, fan, water purge valve, and a water pump. (Ice-O-Matic)*

The automatic operation of these units involves a vending motor, magnets, signal lights, and relays. Thus, an elaborate wiring system is necessary. **Figure 13-102** shows eight parallel circuits being used in one dispenser. The evaporator fan operates continuously.

13.9 Defrost Timers

Most automatic defrosters need an automatic device to start the defrost cycle. The unit shown in **Figure 13-103** uses a system of spring-loaded levers. These are activated by trippers positioned in the 24-hour dial when defrosting is desired. These levers operate the switches.

Some time clocks are connected directly to electric power. They will cause defrost at intervals necessary to keep the system working well. Each evaporator design has its own requirements for good operation. Some need to be defrosted during each cycle; some every few hours. Others need defrosting no more than once a day. Length of defrost can be adjusted.

Some timers are connected to electric power in parallel with the motor. The clock mechanism registers only the running time of the condenser. These mechanisms then start the defrost cycle after a set running time.

The timer wiring differs with the type of defrost system. In one hot gas system, the timer energizes the solenoid bypass valve. It stops the fan motors, energizes auxiliary electric heater elements, and runs the compressor. It also may be used to prevent the start of the

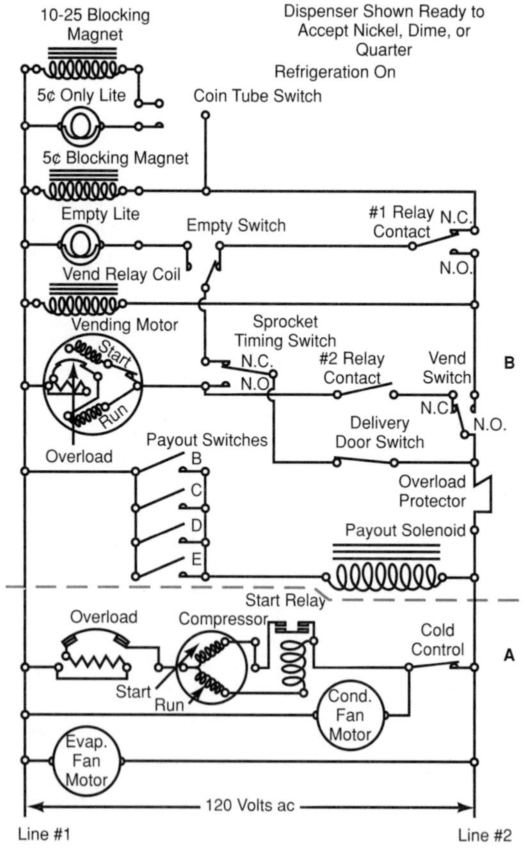

Figure 13-102. *Ladder diagram for refrigerated bottled beverage vending machine. A—Refrigerating unit wiring. B—Vending wiring.*

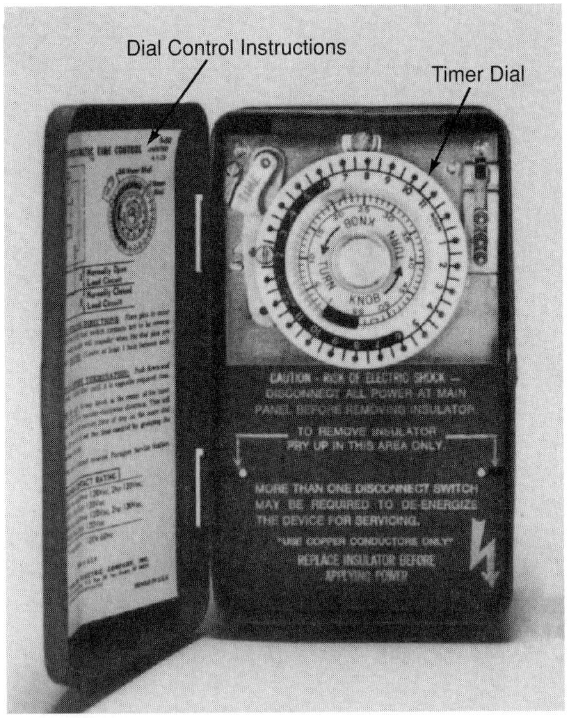

Figure 13-103. *Time switch used for controlling defrost cycles in commercial systems. Wiring diagram and instructions are located inside the cover. (Paragon Electric Co., Inc.)*

normal cycle until the low-side pressure is normal. Some basic electrical circuits using timer controls are shown in **Figure 13-104.**

Another type of automatic timer for defrosting is shown in **Figure 13-105.** A timer starts the defrost cycle. The temperature bulb returns the unit to normal operation when the evaporator temperature is above 32°F (0°C).

The timer in **Figure 13-106** can be used with either air defrost or electric heat. It uses the timer motor to start the defrost action. It also uses a pressure control connected to the low-pressure side. This returns the system to normal operation. The electrical diagrams are shown in **Figure 13-107.**

Some commercial installations use a modular multiple circuit timer, **Figure 13-108,** for defrost control. It is adjustable for handling from 1 to 12 operations during the defrost initiation. The defrost termination control is adjustable from 6 to 106 minutes in one-minute increments. **Figure 13-109** shows a time-terminated hot gas defrost system during defrost cycle. The system has a compressed thermostat bypass. **Figure 13-110** shows an electrical panel with four timers and six motor starters (contactors).

Units with transistorized solid-state circuitry are also used. A thermistor may control the defrost cycle in a display case. The thermistor measures the temperature difference of air moving through the evaporator. This control replaces a timer as defrost is controlled only by demand. Sufficient temperature differences between the

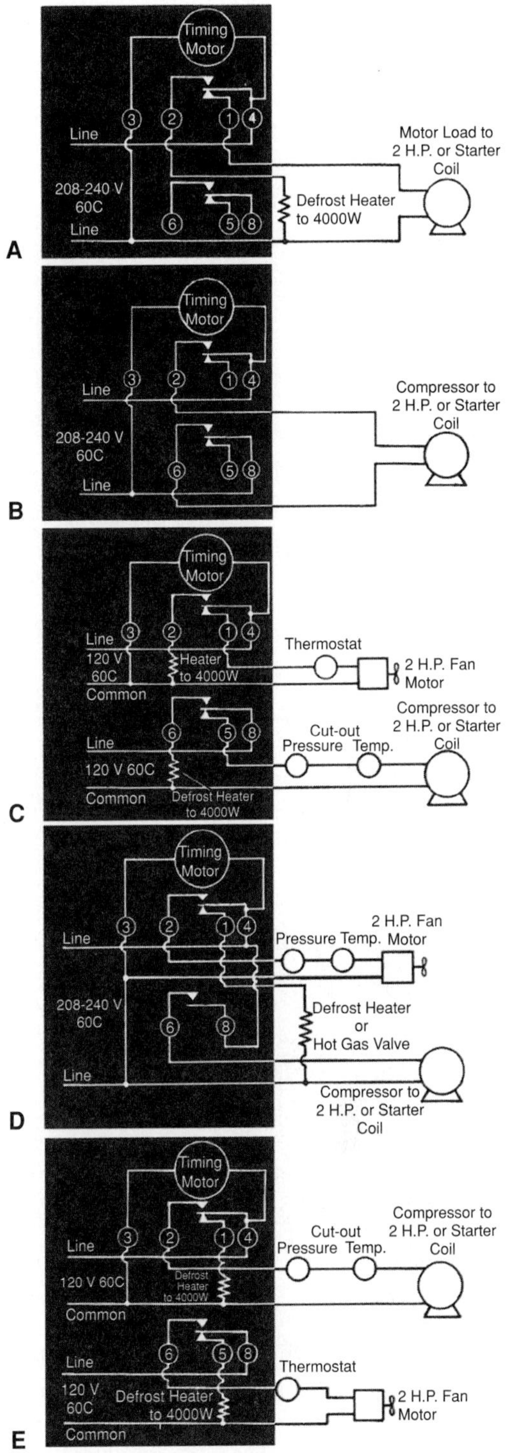

Figure 13-104. *Wiring diagrams for several types of defrost control arrangements. A—Circuit controlled by an SPDT (single-pole, double-throw) switch that activates defrost heaters as it shuts off refrigerating unit. B—Circuit controlled by a DPST (double-pole, single-throw) switch that only shuts off refrigerating unit. C—Circuit controlled by a DPDT (double-pole, double-throw) switch. It shuts off compressor and fan and turns on two defrost circuits. D—Circuit for delayed fan shutoff during defrost and for turning on one defrost circuit. E—Circuit controlled by a DPDT (double-pole, double-throw) switch for delayed fan shutoff and two defrost circuits.*

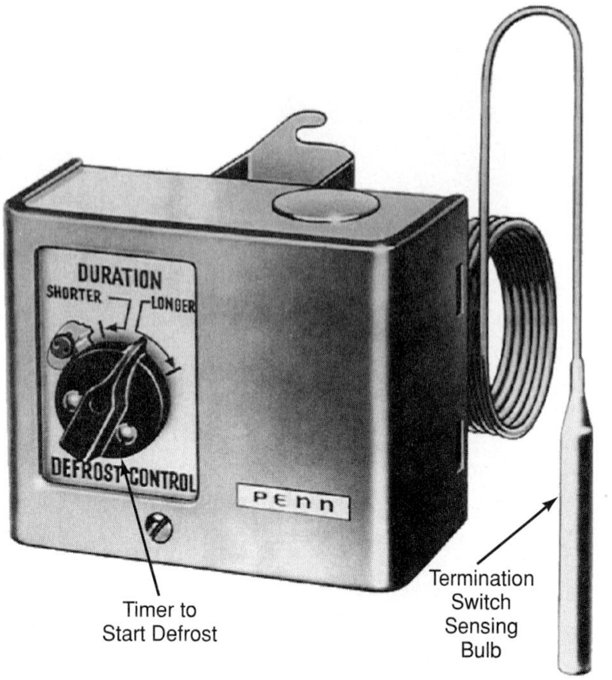

Figure 13-105. *A timer and thermal bulb combination for controlling defrost cycle. Timer starts defrost action. Thermal sensing bulb, located on the evaporator, returns the system to normal operation after frost has melted. (Johnson Controls, Inc.)*

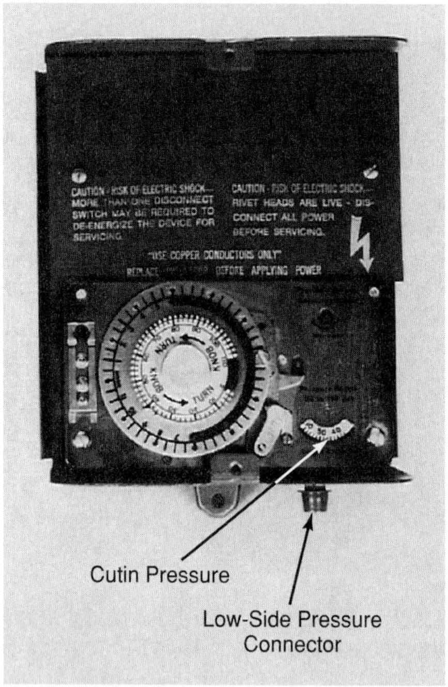

Figure 13-106. *Defrost timer with a low-side pressure-operated switch. Timer starts the defrost action and low-pressure switch returns system to normal operation. (Paragon Electric Co., Inc.)*

air entering and leaving the evaporator trigger a thermistor to signal. If this temperature difference becomes more than 20°F (11°C) to 30°F (16°C), this signal (electrical pulse) will start the defrost cycle. A standard thermostat returns the system to a normal cycle. This occurs when the evaporator temperature measures about 40°F (4°C).

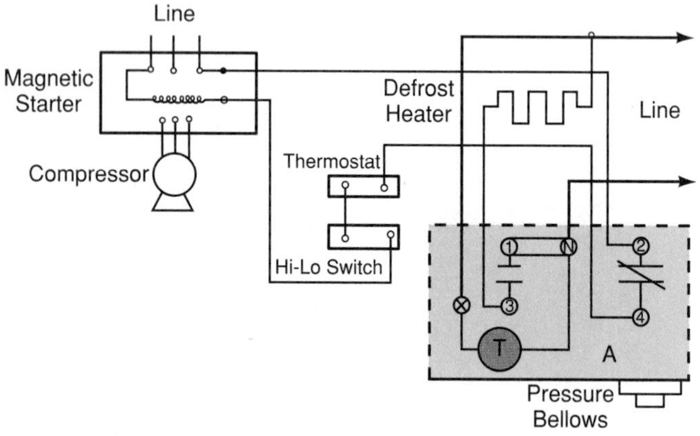

Figure 13-107. *Wiring diagram for a defrost control where timer motor starts the defrost and a low-pressure switch returns the system to normal operation. A—Timer/pressure-operated control. Timer motor at left, marked "T," is connected to line power. Pressure switch is connected to compressor circuit. (Paragon Electric Co., Inc.)*

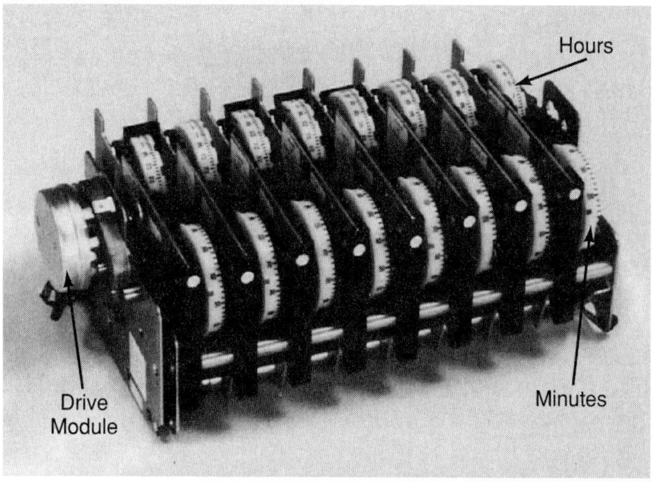

Figure 13-108. *This single-drive timer module has eight different circuits. Time setting controls are arranged in hours and minutes. (Paragon Electric Co., Inc.)*

Figure 13-111 shows an electronic modular electric circuit defrost control. It is used on parallel refrigeration systems in supermarkets. The unit can be used for the specific defrost control needs of the systems. Outputs may be selected as defrost, fan delay, or master hot gas defrost system.

An electronic refrigeration defrost control is shown in **Figure 13-112.** This unit allows a selection from 12- to 24-hour clock format. It allows up to eight defrost starts per day. A temperature sensor controls the compressor during the refrigeration cycle. Cut-in and cut-out temperatures from −40°F to 87°F (−40°C to 30°C) are

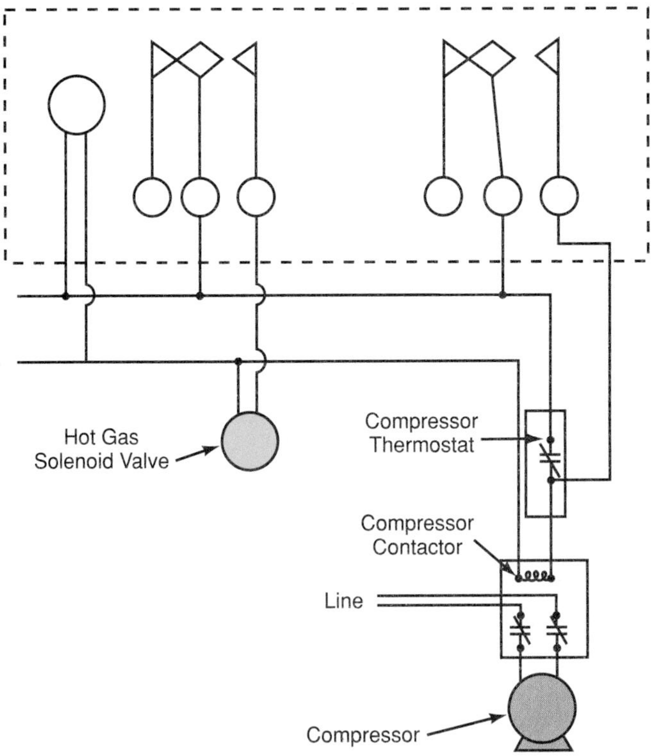

Figure 13-109. *Typical wiring diagram for modular multiple-circuit defrost control. Note location of hot gas solenoid valve module. (Paragon Electric Co., Inc.)*

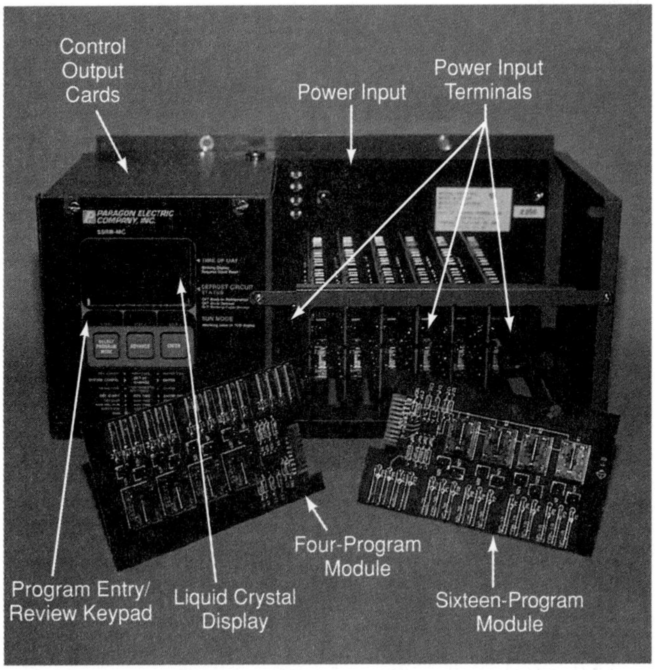

Figure 13-111. *Programmable defrost control uses a three-button keypad and provides 24-hour time-of-day format. Individual programs can be inserted into control output through the use of output cards. (Paragon Electric Co., Inc.)*

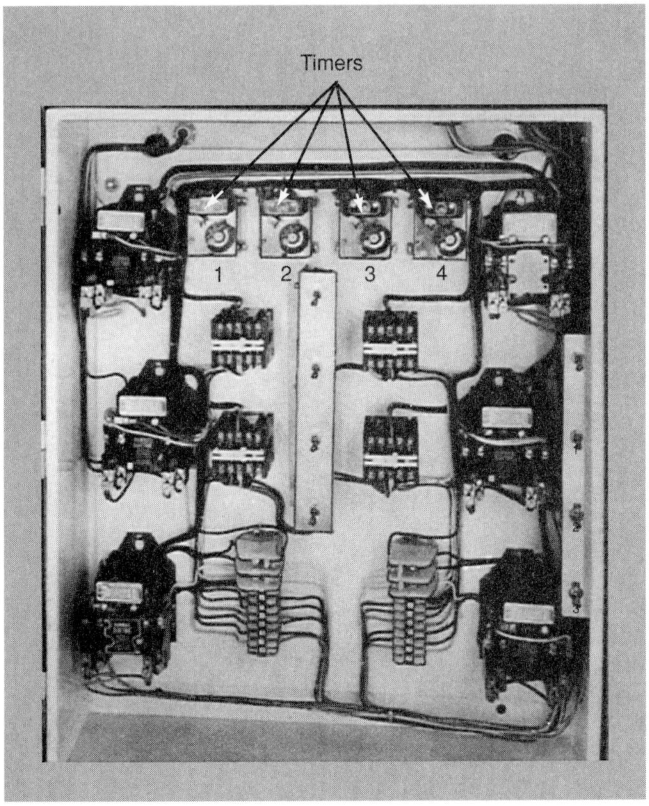

Figure 13-110. *Electric control panel with four timers. Each controls production of 10 tons of ice on each cycle.*

Figure 13-112. *Electronic refrigeration defrost control. Unit is mounted indoors in a clean environment, free of contaminants such as dirt and moisture. (Paragon Electric Co., Inc.)*

allowed. The digital display will indicate the refrigerated space temperature. An alarm relay is energized after the door is ajar.

13.10 Valves, Pressure Regulating

Commercial systems use many types of pressure regulating valves. Some of these valves control:

- Evaporator pressure (two-temperature valves).
- Crankcase pressure.
- Discharge bypass pressure with solenoid valve control for pull down (service), to prevent freezing.
  A. Some into suction.
  B. Some into evaporator.
- Head pressure control valve.

Some of these valves have Schrader service connections for gauge mounting.

13.10.1 Valves, Two-Temperature

In many multiple installations, different temperatures are maintained in evaporators connected in the same system. Thermostatic expansion valves may be used if the temperature differences are not over 5°F (3°C). However, in some instances, the temperature differences are too great. An example of this would be a storage cabinet and an ice cream cabinet combination. A two-temperature valve is then put into the warmest evaporator suction line. This prevents pressure of the warmest evaporator from going below a safe setting.

The controlled evaporator(s) should not have more than 40% of the system's total load. If a controlled evaporator is too large, erratic cycling will result. (See surge tanks, Section 13.10.3.) If the controlled load is more than 40%, separate condensers should be used.

Types of Two-Temperature Valves

Two-temperature valves are sometimes called constant pressure valves or pressure-reducer valves. They are also used to ensure a constant low-side pressure. The valves have a bellows or diaphragm, a needle, and a seat. These are arranged so the bellows are operated by the pressure in the warmest evaporator.

As the compressor pumps the low side down to the desired pressure, the bellows shuts off the valve. This action stops the pressure in the warmest evaporators from going below the pressure desired. Pressure in the evaporator builds up from vaporizing the refrigerant. The bellows then opens the valve, passing vapor on to the compressor.

The pressure on the surface of liquid refrigerant determines the temperature at which it will evaporate. The suction line valve controls the temperature of the evaporator to which the line is attached. This is true even when the compressor suction pressure is considerably below the evaporator pressure.

Two general types of two-temperature valves are:

- Pressure operated.
  A. Metering.
  B. Snap-action.
- Temperature operated.
  A. Sensing bulb and bellows.
  B. Thermostat and solenoid.

Metering Type Two-Temperature Valve

The metering, two-temperature valve acts more as a throttling device than as a shutoff valve. See **Figure 13-113. Figure 13-114** shows a cross section of such a valve.

Some of these valves have a gauge opening. This allows the technician to check and adjust the warmer evaporator's pressure. Having no differential, it opens and closes when the pressure varies only a fraction of a pound. (A differential means one pressure to open it and a different pressure to close it.) The bellows pressure area and the valve area are equal. Thus, only the adjustment spring and the warm evaporator pressure changes can operate the two-temperature valve.

The valve openings must be large enough to offer efficient vapor flow. Many of these metering controls have a small adjustment range. They are especially designed to maintain pressures just above crankcase pressures. See **Figure 13-115. Figure 13-116** shows a

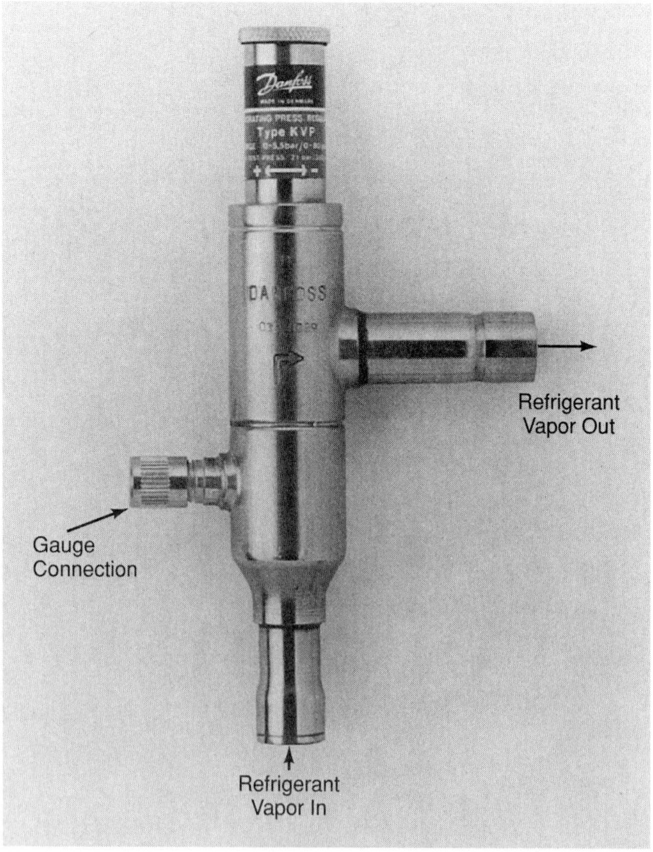

Figure 13-113. *Metering two-temperature valve. Note the different connections. (Danfoss Automatic Controls, Division of Danfoss, Inc.)*

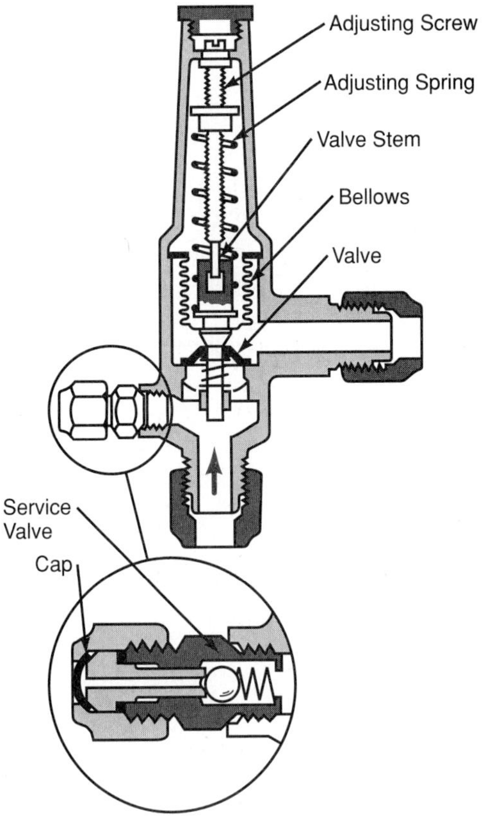

Figure 13-114. *Cross section through metering two-temperature valve.*

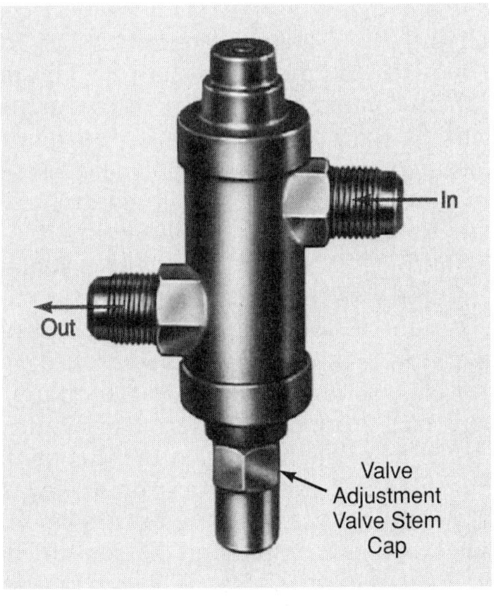

Figure 13-115. *Metering two-temperature valve with flare fitting connections. This pressure-operated throttling valve is an example of the metering two-temperature valve.*

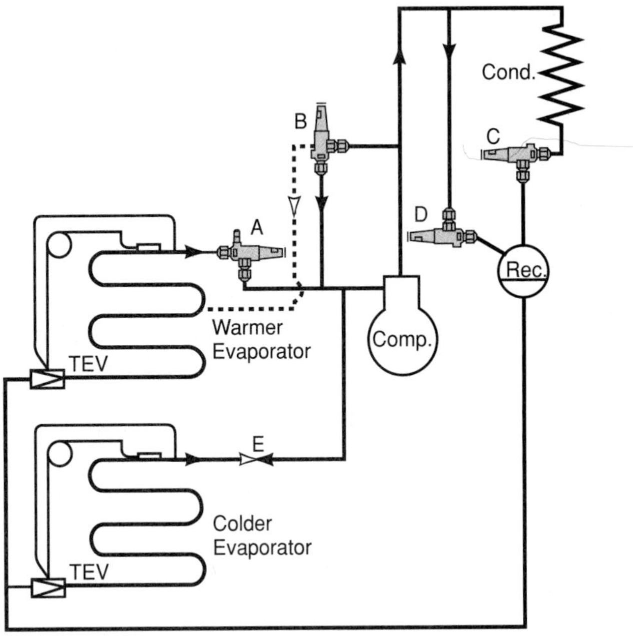

Figure 13-116. *A two-evaporator system equipped with several pressure control valves. A—Two-temperature valve (evaporator pressure regulator). B—Condenser bypass valve. C—Condenser pressure regulating valve. D—Capacity regulating valve. E—Check valve. (Danfoss Automatic Controls, Division of Danfoss, Inc.)*

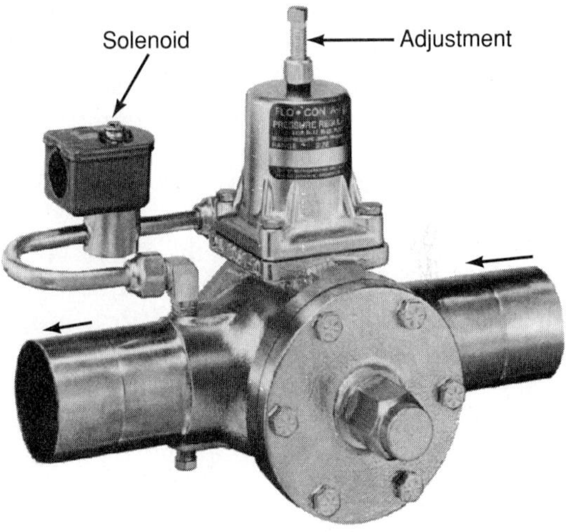

Figure 13-117. *A large capacity evaporator pressure regulator. Note use of metallic fittings for brazed connections to suction line pipe. (Refrigerating Specialties Div., Parker-Hannifin Corp.)*

two-evaporator system equipped with several pressure-controlled valves.

Large systems must rely on forces other than springs to control pressure for efficient operation. The large capacity two-temperature valve in **Figure 13-117** has a plastic seat. It has connections for brazing to the suction line.

Snap-Action Two-Temperature Valve

When a snap-action valve closes, a decided pressure rise occurs in the warm evaporator before the valve opens again. This pressure-operated valve has a definite cut-in pressure and temperature. It is often used when defrosting is wanted on each cycle.

A snap-action two-temperature valve is normally used with multiple evaporator systems. It is used on systems that do not operate at a wide temperature difference. Such systems include walk-in coolers and display cases. The valve should be located on the suction line to the display case.

Thermostatic Two-Temperature Valve

Another type of two-temperature valve has a temperature control. It is built much like a thermostatic expansion valve and works much the same. It operates from the temperature of the evaporator or temperature of the air entering or leaving. It has a capillary tube and a sensing bulb much like the thermostatic expansion valve. A bellows moves a rod as different pressures are created in the sensing bulb.

When the evaporator becomes cool enough, the cooling sensing bulb lowers pressure in the bellows. The bellows then contracts. This pulls on the valve plunger and shuts off the valve. With the valve closed, pressure cannot drop any lower in the evaporator. Then, the valve controls the minimum temperatures of the evaporator.

As the evaporator warms, so does the sensing bulb. The increase in pressure is transmitted to the bellows. It expands, pushing the plunger so that the valve opens. With the valve open, the compressor, once more, draws vaporized refrigerant from the evaporator. This type of valve is always located in the suction line of the warmest evaporator.

Figure 13-118 shows a regulator that responds to the temperature of the air. It does so as the air leaves the evaporator and enters the fixture or case. It may also respond to the temperature of the air entering the evaporator.

It has a sensitive element located in this airstream. The valve body is mounted in the suction line. A liquid line connection is also made to the valve body. This provides the pressure needed to open the main valve in the TPR (temperature pressure regulator). See **Figure 13-119.**

As the temperature sensing bulb warms, its increase in pressure closes a small valve. Then the high-side pressure in the valve decreases. This allows the main piston

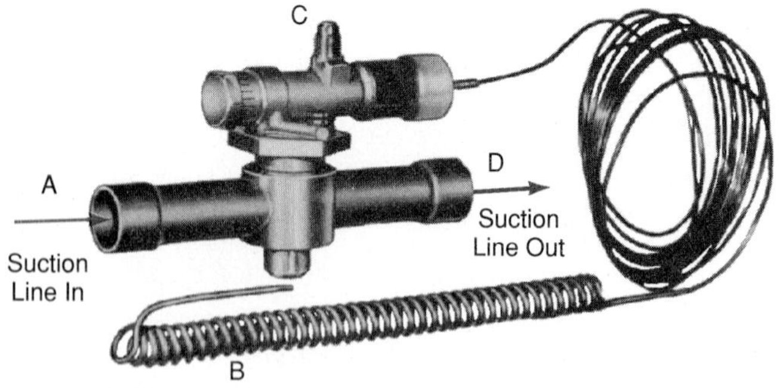

Figure 13-118. *Temperature controlled suction line metering valve. A—Suction line from evaporator. B—Temperature sensing bulb. C—High-side pressure connection. D—Suction line to compressor.*

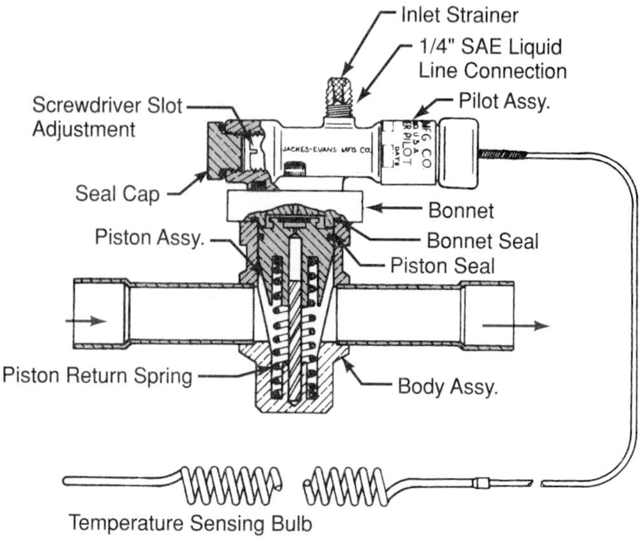

Figure 13-119. *Temperature controlled suction line evaporator pressure control.*

to open the suction line more. As the sensing bulb cools, the high-pressure pilot valve is opened. The main piston is forced down, closing the suction line passage partially or completely (modulates).

Solenoid Two-Temperature Valve

A thermostat connected in series with a solenoid valve also provides various fixture temperatures in multiple installations.

The solenoid shutoff valve is usually placed in the liquid line of the evaporator it controls. It has an electrical connection to a thermostat, as shown in **Figure 13-120.**

The thermostat is operated by the fixture temperature. When the fixture reaches the correct temperature, the thermostat opens. The electric solenoid loses its magnetism and the valve closes. No more refrigerant is fed to the evaporator. The cabinet will gradually warm up until the thermostat points close. The solenoid valve then opens, and refrigeration starts again.

This system of refrigeration control is based on fixture temperature. The condenser is controlled by a pressure motor control. The motor will not stop until all the fixtures are cooled to their correct temperature. Some systems have the solenoid valve in the suction line. This prevents removing the refrigerant from the evaporator.

This system uses a normally open solenoid and a thermostat. It breaks the circuit on temperature drop and closes it on temperature rise. When magnetized, the valve closes. See **Figure 13-121.**

A solenoid valve is also used to stop low-pressure side flooding during the Off cycle. This solenoid is also located in the liquid line. It is electrically connected in parallel with the pressure motor control, **Figure 13-122.**

13.10.2 Check Valves

Check valves are an important function in refrigerating systems. They prevent flow of liquid and/or vapor refrigerant in the wrong direction. They are used in two-

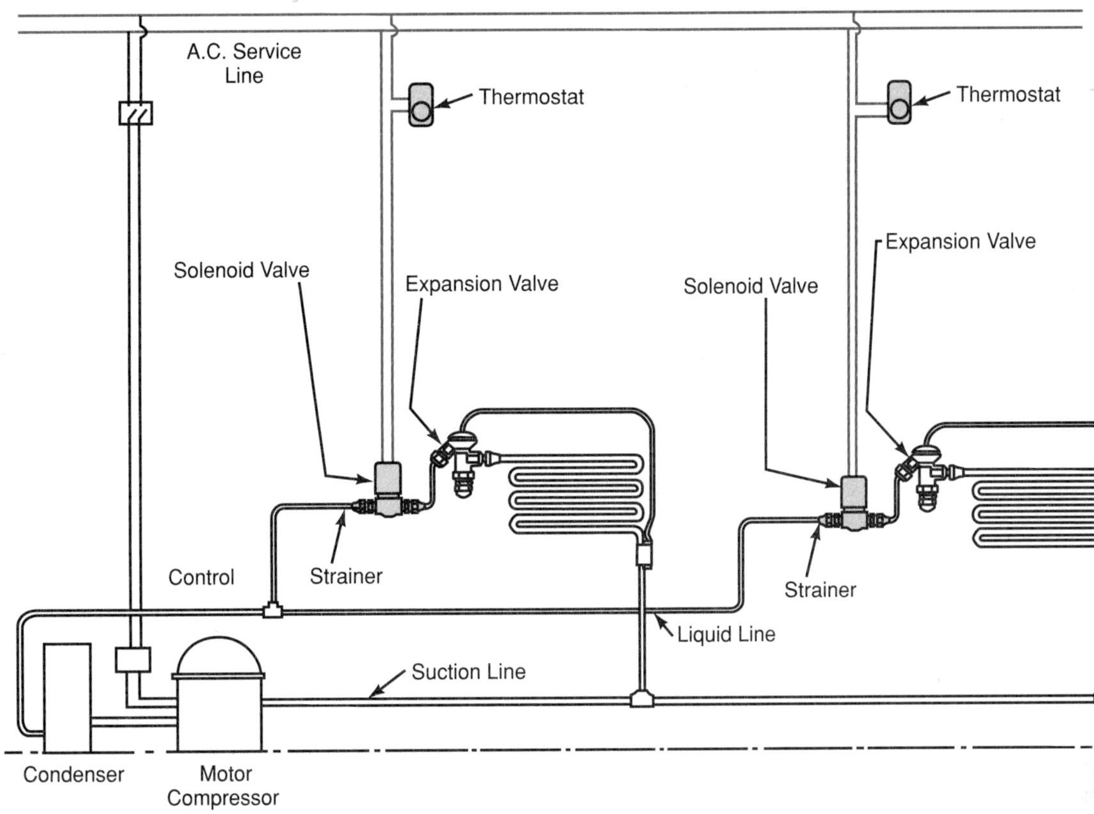

Figure 13-120. *Installation of solenoid valves. Each solenoid controls the cabinet temperature of a different cabinet. Valves and thermostats are using line voltages.*

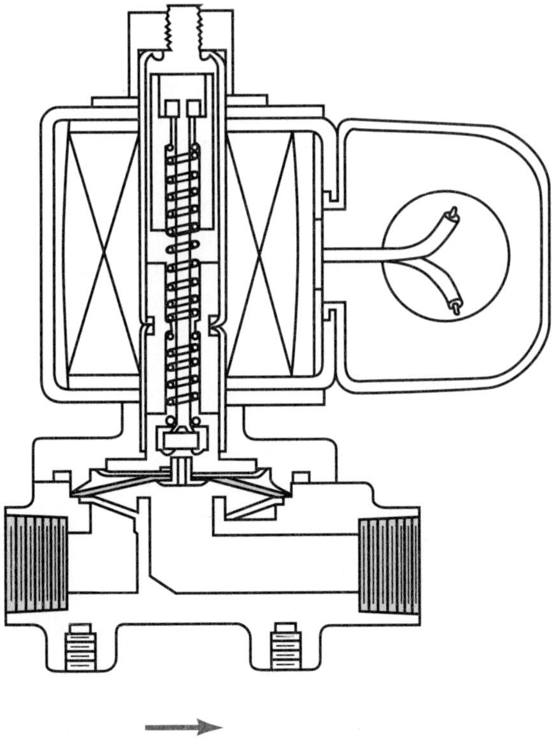

Figure 13-121. *A solenoid valve designed to open the valve when magnet is not energized and close the valve when magnet is energized. Note the threaded pipe connections at either end. (Fluidex Div., Parker-Hannifin Corp.)*

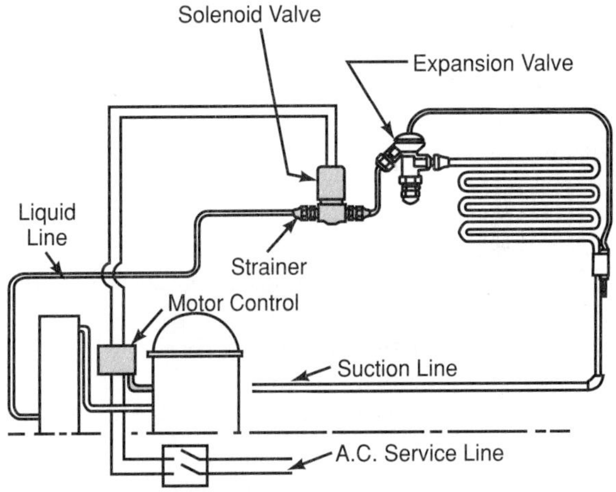

Figure 13-122. *Solenoid valve located in liquid line and controlled by pressure motor control.*

temperature installations and in defrost systems. They are used to prevent vapor passage during Off cycles.

In multiple installations, one condenser is connected to several evaporators. Each evaporator is at a different temperature. Two-temperature valves are used to obtain the desired temperatures. Check valves, **Figure 13-123,** are sometimes put in the suction line of the coldest evaporators. They prevent excess warming of the cold evaporator during the Off cycle.

After the condenser has stopped, one of the two-temperature valves may open before the condenser turns

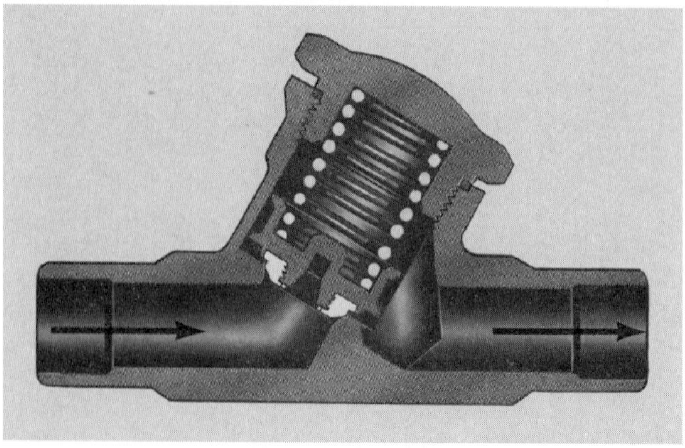

Figure 13-123. *Suction line check valve is used on colder evaporators in multiple systems. (Superior Valve Co., Division of AMCAST Industrial Corporation)*

on. The low side may then flood with warm refrigerant vapor. This vapor will also travel along the suction line to the coldest evaporator.

If it enters the evaporator, it will start condensing and releasing its latent heat. This will make the cold evaporator defrost or, at least, warm up somewhat. The check valve installed in the suction line of the coldest evaporator only allows vapor to be drawn from this evaporator.

This check valve must have a tight seat and it must open easily. If it is small or opens with difficulty, it will act as a throttling device and cause too large a pressure drop. The result will be poor refrigeration in the coldest evaporator.

Figure 13-124 shows a large-capacity check valve. Reverse-cycle systems and some hot gas defrost systems use check valves.

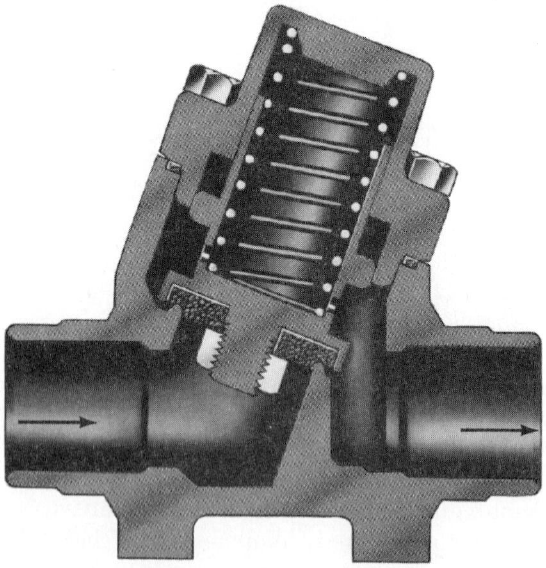

Figure 13-124. *Large-capacity check valve. Note double cylinder design which provides smoother operation. (Superior Valve Co., Division of AMCAST Industrial Corporation)*

Check valves can be a source of noise in a refrigerator. They open and close with a metallic click or bang. There is also noise associated with the inefficient operation of a valve. A "hammering" noise occurs when the valve does not completely close off the reverse flow. In a hydronic system, this is referred to as "water hammer." Valves have been designed to minimize this noise problem.

13.10.3 Surge Tanks

Multiple temperature installations may short cycle a pressure-controlled condenser. This may occur when the opening and closing of the two-temperature valves cause pressure fluctuations. Pressure is said to fluctuate when it rises and falls repeatedly and in an uncertain pattern.

The following conditions can cause a short cycle:

- If the two-temperature valve is closed and the condenser cools the lowest temperature evaporator enough to open the pressure motor control, the condenser will stop.
- If, just after it stops, a two-temperature valve that controls one of the warmer evaporators opens, the low-side pressure will rise rapidly. The pressure turns on the condenser and, thus, causes a short cycle.

To eliminate this trouble, a ***surge tank*** or a larger cylinder may be installed in the main suction line just ahead of the compressor. The surge tank shown in **Figure 13-125** is large enough to absorb a pressure buildup. Thus, if the unit stops and a two-temperature valve opens, the low-side pressure cannot increase quickly. Therefore, the unit will not short cycle. The capacity of the surge tank is great enough to absorb a large volume of vapor. It thereby slows down the rapid pressure changes. Such changes would make the motor control turn off and on. The line connected to the bottom of the tank leads to the compressor and helps return the oil to the compressor.

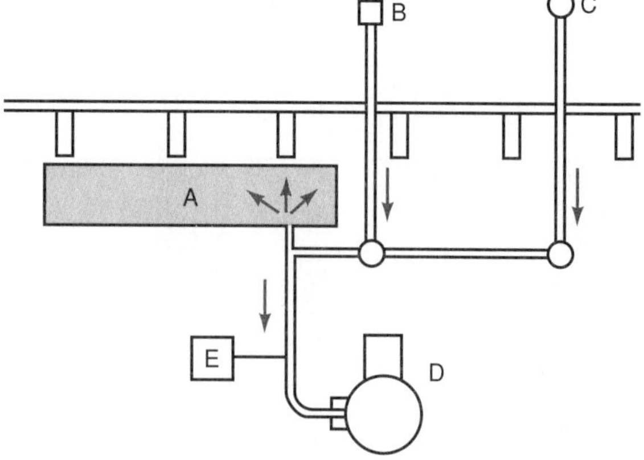

Figure 13-125. *A surge tank installation. A—Surge tank. B—Check valve. C—Two-temperature valve. D—Compressor. E—Motor control.*

13.11 Compressor Protection Devices

Reciprocating compressors are damaged when liquid refrigerant accidentally flows into the compressor from the suction line.

The refrigerant must be in a vaporous state. The vapor temperature must be higher than that of the evaporating liquid in the evaporator. This increase in temperature means the vapor is superheated.

There are many devices for preventing or minimizing suction line liquid refrigerant from entering the compressor:

- An accumulator in the suction line.
- Hot-gas bypass valves to move hot gas into the suction line. There, the gas can evaporate any liquid.
- Temperature-sensing devices and solenoid valves.
- Heat exchangers to warm the suction line vapor-liquid.
- Electrical heaters to warm the suction line vapor-liquid.
- An evaporator (blower coil) in the suction line.

Drops and slugs of liquid refrigerant may travel in the suction line. This can occur when a system uses a hot gas defrost or has a sudden load change. Many systems have an accumulator in the suction line. This reduces the danger of liquid refrigerant flow into the compressor, **Figure 13-126.**

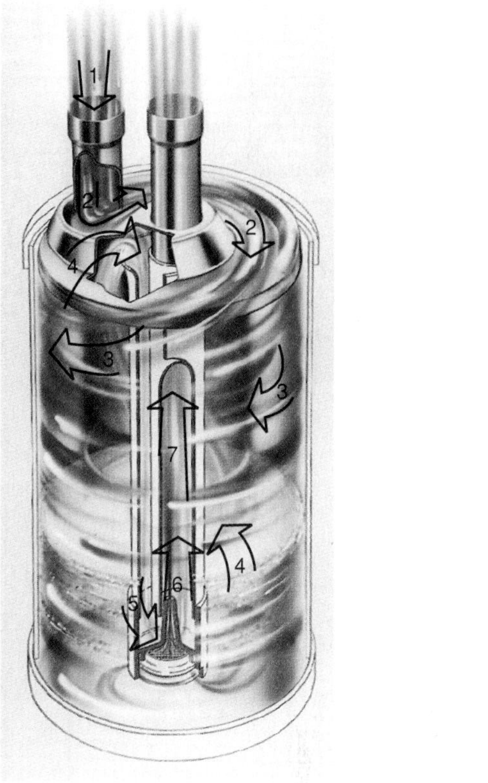

Figure 13-126. *Suction line accumulator: (1) Mixture of refrigerant vapor liquid and oil enters. (2) Swirling motion created on entering mixture. (3) Liquid strikes inside wall. (4) Refrigerant vapor and mist drawn upward, vertical motion, and then downward into tubing. (5) Flow turn 180° and upward through orifice, drawing measured amount of liquid refrigerant and oil from bottom. (6) Combination of refrigerant vapor, compressor lubricating oil, and refrigerant flow vertically, forming a mist before entering (7), the compressor suction. (Tecumseh Products Company)*

The accumulator will lengthen the cycling interval (gas storage). It usually has aspirating (suction) devices to return oil. **Figure 13-127** shows the internal design of an accumulator.

The hot gas bypass device depends on the use of a temperature sensor attached to the suction line. This sensor controls a solenoid valve. This valve opens when there is danger of liquid refrigerant flowing into the compressor. Hot gas is allowed to flow into the suction line.

Liquid line-suction line heat exchangers are discussed in Sections 4.5 and 13.4.10.

Electrical heaters may be attached to the suction line. A temperature-sensing element turns on the current when heating of the suction line is needed.

The blower coil is usually operated by a temperature-sensing element attached to the suction line. The suction line temperature indicates when there is a danger of liquid refrigerant entering the compressor. The fan in the blower coil will then be turned on.

A temperature-operated device or a pressure device must detect the presence of liquid quickly. It must turn on mechanisms to stop the liquid from reaching the compressor. The best sensor is a thermistor. It reacts very quickly. It can be connected to an alarm circuit or operating circuit. The compressor will be stopped before it can be damaged.

13.11.1 Oil Control Systems

Since they must be kept lubricated, compressors run in oil. A certain amount of this oil leaves the compressor with the refrigerant vapor. It is important that this oil be prevented from moving through the system.

Various devices are designed to collect the oil and return it to the compressor. Most industrial HVAC systems use one or more of three basic oil control components. They may use an oil level regulator, an oil reservoir, and an oil separator.

An ***oil level regulator*** controls the level of oil within the compressor. It usually uses a float-type mechanism. This allows oil flow to the compressor only when the float indicates that the oil level is low. See **Figure 13-128.**

An ***oil reservoir*** in the system has a dual purpose. It holds the compressor's oil supply and the regulator draws from it to replenish the oil in the compressor. Oil trapped by the separator is returned to the reservoir until it is needed. The oil reservoir may contain two sight glasses for observation of oil level. See **Figure 13-129.** The oil reservoir also contains a flare fitting for adding oil to the system.

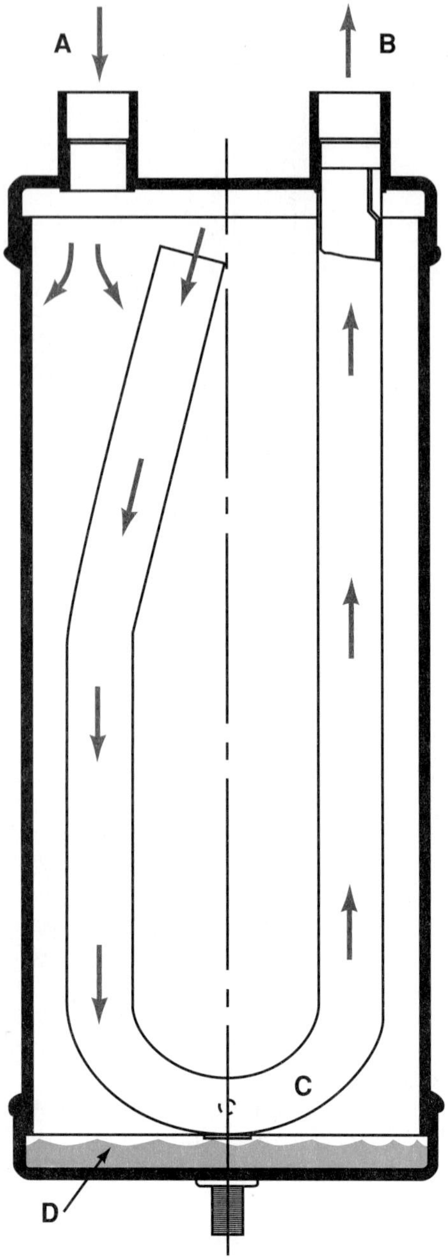

Figure 13-127. *Inside of an accumulator. A—Suction gas in. B—Suction gas out. C—Oil return aspirator hole. D—Liquid refrigerant trapped until it can evaporate. (Virginia KMP Corp.)*

Refrigeration systems work best when the oil is kept at a proper level in the compressor. Oil in the condenser and evaporator will reduce efficiency of the unit.

It is important to keep the oil from circulating in low temperature installations. It thickens at low temperatures and becomes difficult to move out of the evaporator.

Oil separators remove oil from the hot compressed vapor as the vapor leaves the compressor. The oil separates because the vapor flow slows down as it arrives in the separator. The oil will collect in the separator until a certain level is reached. Then, a float opens a needle valve and the oil returns to the compressor crankcase.

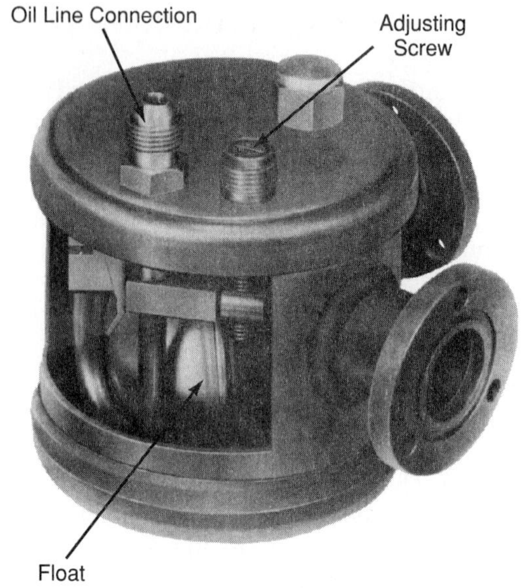

Figure 13-128. *The oil level regulator controls the oil level in a compressor crankcase. Float-operated valve holds excess oil until the oil level in the compressor crankcase drops. (AC & R Components, Inc.)*

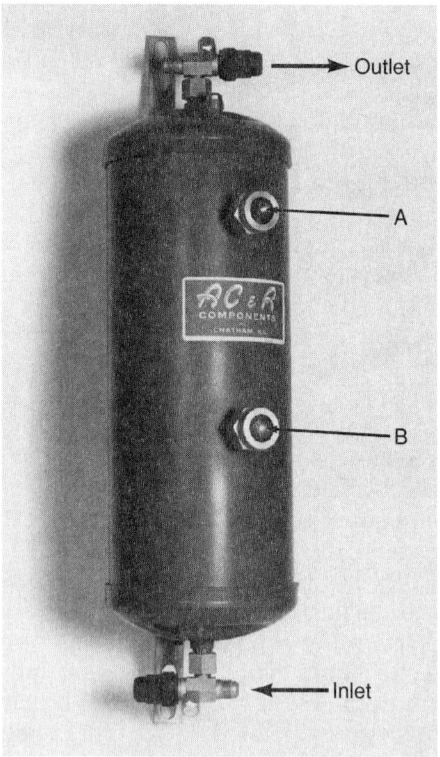

Figure 13-129. *Oil reservoir holds standby oil as part of the oil control system. Note sight glass ports, A and B, to observe oil level. (AC & R Components, Inc.)*

Oil separators are also placed between the compressor and the condenser, **Figure 13-130.** The separator is insulated to prevent it from acting as a refrigerant condenser. It would otherwise pass heat to the surrounding air.

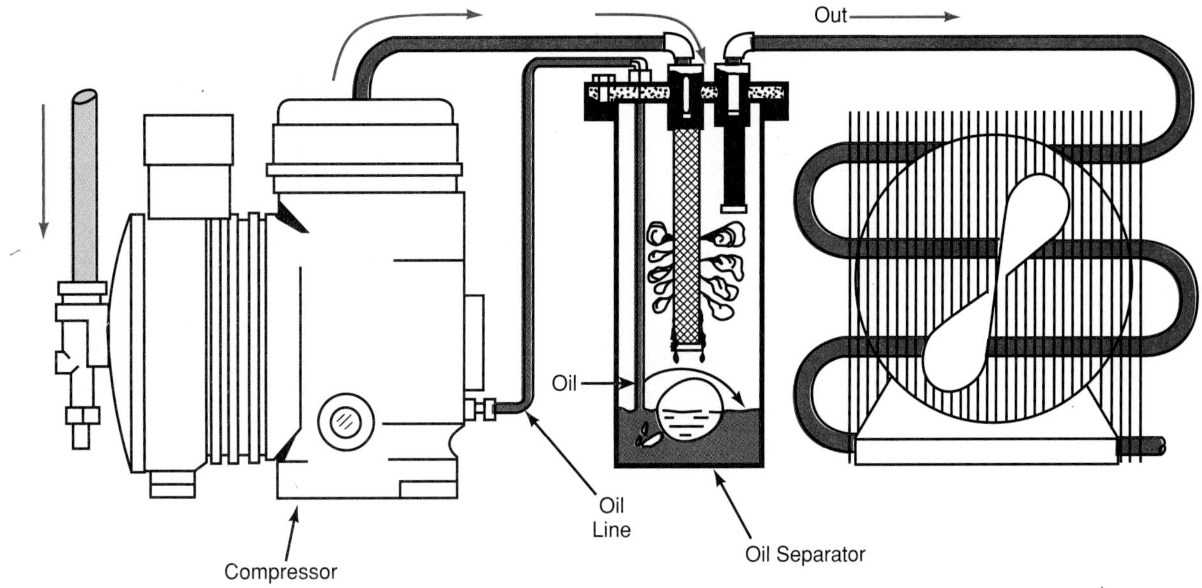

Figure 13-130. *Oil separator installation. Oil is removed from high-temperature, high-pressure refrigerant and returned to compressor.*

Many oil separators are serviceable (bolted construction). See **Figure 13-131.** On hermetic systems, the oil-return line is usually connected to the suction line near the motor compressor.

Liquid refrigerant may collect in the oil separator during long Off cycles or during long manual shutdown. This liquid refrigerant returns by way of the oil return line to the compressor. This may cause oil pumping resulting in damage to the compressor. A check valve in the vapor outlet of the oil separator will reduce this danger. A filter in the oil return line will help keep the oil clean.

A solenoid valve in the oil return line is sometimes used. It allows oil or refrigerant to return to the crankcase during the Off cycles. A thermostat controls the solenoid. The thermostat will close only when the oil separator is warm (100°F to 130°F [38°C to 54°C]).

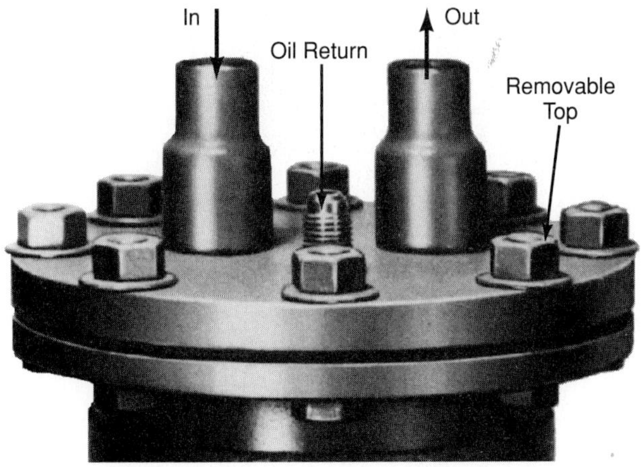

Figure 13-131. *Oil separator with bolted assembly top, making it cleanable and serviceable.*
(AC & R Components, Inc.)

Figure 13-132 shows an oil separator used in large systems. A helical oil separator is located in the upper section. It provides a centrifugal flow path for oil separation with low pressure drop. The vapor enters the system on the left-hand side. It proceeds through the oil separator and through the center tube. From there it goes to the outlet at the top of the housing. The oil is accumulated at the bottom. It returns through the lower outlet to the compressor.

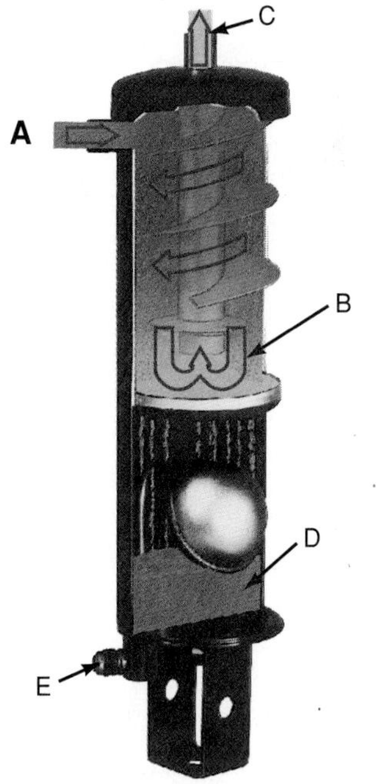

Figure 13-132. *Helical oil separator. Refrigerant and oil enter at A and swirl in circular motion. Refrigerant follows Path B to C. The oil drops down and accumulates at Point D. It exits at Point E.*
(AC & R Components, Inc.)

Refrigerant entering the oil separator from the compressor encounters the edge of the helical oil separator. The vapor oil mixture is forced along the spiral path of the helix. The heavier oil particles spin to the perimeters, where a screen layer is located. The screen layer serves a dual function: It is both an oil stripping and draining medium. Oil flows down the outer shell through a baffle and into the oil collector area. The collector is located at the bottom of the separator. The oil-free refrigerant vapor exits through a fitting at the bottom. It returns back to a crankcase.

Some compressors use an oil reservoir as a holding vessel for standby oil. This is necessary for the operation of air conditioning and refrigeration oil control systems. The unit has two valves. The one on top is an oil reservoir, receiving oil from the separator. The bottom valve distributes oil to the oil level regulators.

13.11.2 Compressor Low-Side Pressure Control Valves

Starting a compressor places a heavy load on the motor. It has to overcome inertia of the moving parts. (Objects at rest tend to stay at rest.) It must also overcome high crankcase pressure. (In fact, crankcase pressure may be at its highest at starting.) It is important, then, to use higher horsepower. Even so, the motors are usually taxed to the limit at the moment of starting. This is especially true when starting against normal or above-normal head pressures.

A compressor low-side pressure control is used on some installations. This keeps crankcase low-side pressures at a reasonable level, even though the rest of the low-side pressure may be high. It never permits the crankcase pressure to exceed a certain safe value, and is known as a reverse metering, two-temperature valve.

Crankcase pressure-regulating valves are needed where the compressor runs too long before the low side drops to a pressure that does not overload the compressor. The suction line pressure may exceed the safe pressure. In this event, the valve shuts the suction line off from the compressor.

The valve body is usually made of brass and the diaphragm or bellows of phosphor bronze. The needle and seat are usually made of wear-resisting steel alloy. See **Figure 13-133. Figure 13-134** shows the valve in a typical installation, located in the suction line between the evaporator and the compressor. On some applications, other system components must be located after the valve.

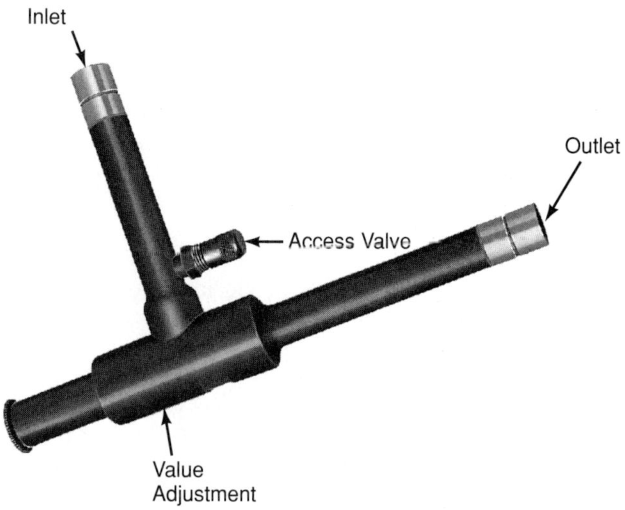

Figure 13-133. *Adjustable crankcase pressure-regulating valve with adjustable pressure. Arrow indicates flow of refrigerant. (Alco Controls Div., Emerson Electric Co.)*

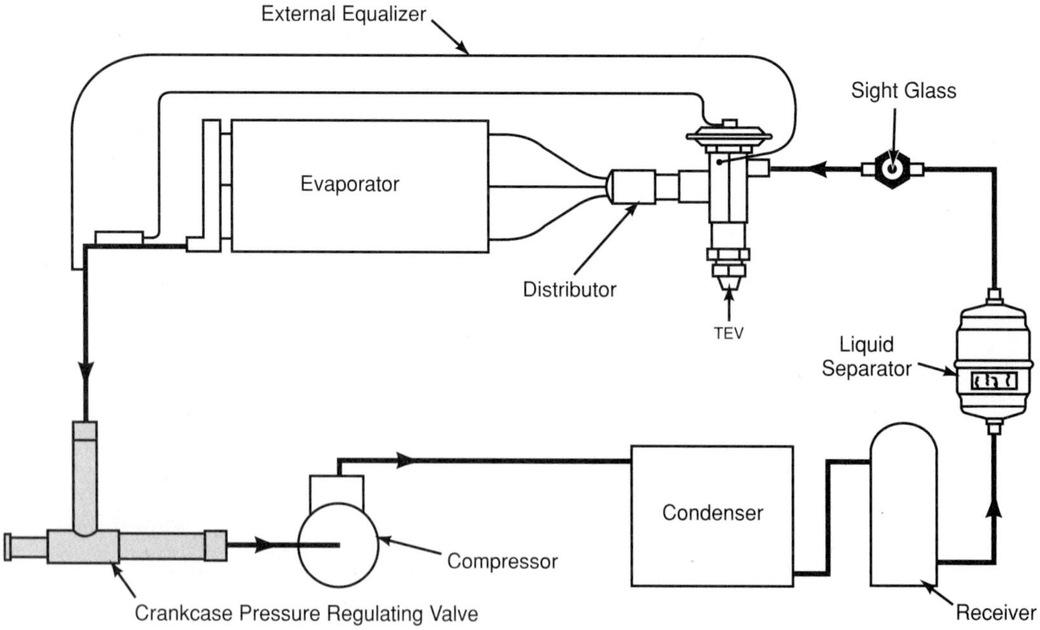

Figure 13-134. *The crankcase pressure-regulating valve is located in the suction line between the evaporator and compressor. (Sporlan Valve Co.)*

The pressure-regulating valve prevents overloading of the compressor motor. It eliminates the crankcase pressure during and after the defrost cycle or after the normal shutdown period.

The crankcase pressure-regulating valves close as the outlet pressure increases. They also respond to the compressor, crankcase, or suction pressure. They close on the rise of the outlet pressure. **Figure 13-135** illustrates the operation of this valve. Inlet pressure is exerted on the underside of the bellows and the top side of the seat disc.

A system suction line with a pilot-operated suction regulator valve is shown in **Figure 13-136.**

13.11.3 Water Valves

Some larger commercial units use water-cooled condensers. With such units, good, inexpensive water must be available. Driving the condenser requires less power than for the same size air-cooled installation. This is due to better heat transfer and lower condenser temperatures and pressures. The saving in electrical power helps compensate for the cost of water used for cooling.

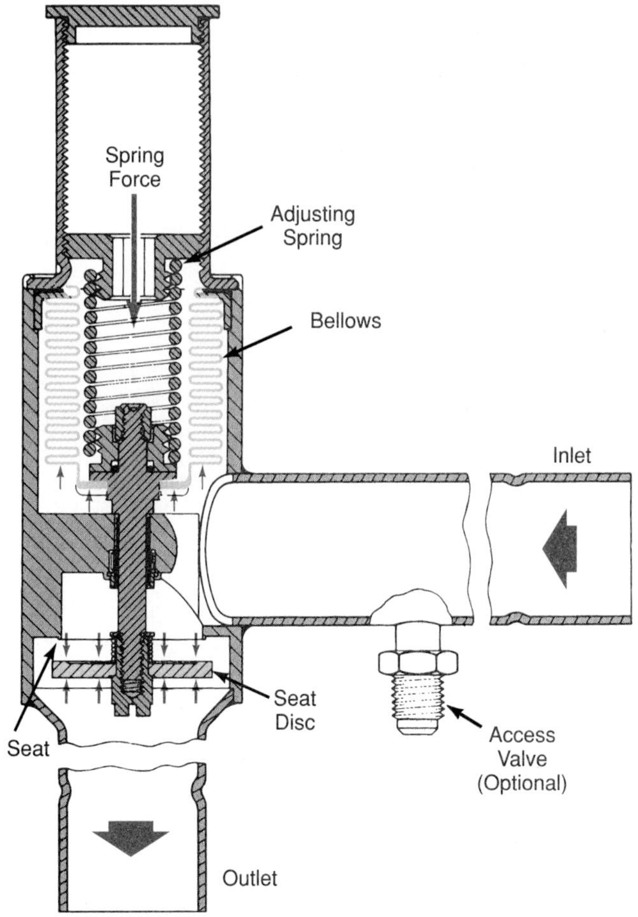

Figure 13-135. *Cutaway of crankcase pressure-regulating valve. Note inlet pressure on bellows and on seat disk. Also note outlet pressure on seat disc. Arrows indicate which direction vapor is flowing. (Sporlan Valve Co.)*

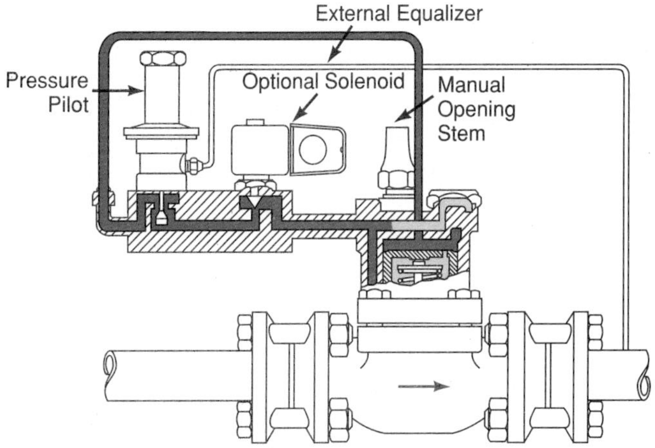

Figure 13-136. *Pilot-operated compressor low-side pressure control. Pilot valve releases pressure above main piston when compressor pressure reaches safe level. This opens main valve, allowing evaporator vapor to move to compressor. (Alco Controls Div., Emerson Electric Co.)*

The water valve turns the water on and off as needed. However, it also varies the amount of water as required. Three types of water valves are used:

- Electric.
- Pressure.
- Thermostatic.

It is good practice to install a strainer in the water inlet to the valve. See **Figure 13-137.**

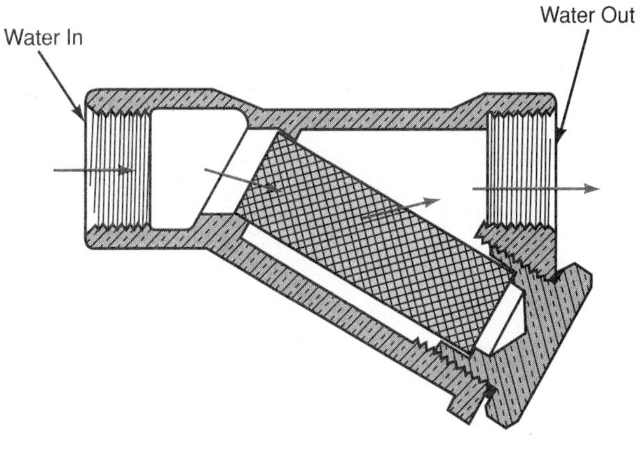

Figure 13-137. *Water line strainer. Screen is removable for cleaning. (Superior Valve Company, Division of AMCAST Industrial Corporation)*

Electric Water Valve

Electrically operated water valves are of two principal types: solenoid-activated and motor-operated.

A water valve is located between the water supply and the condenser. Usually, it is mounted on the condenser base. The moment the motor starts, this valve opens. When the motor circuit is opened, the solenoid is de-energized and the valve closes. See **Figure 13-138.**

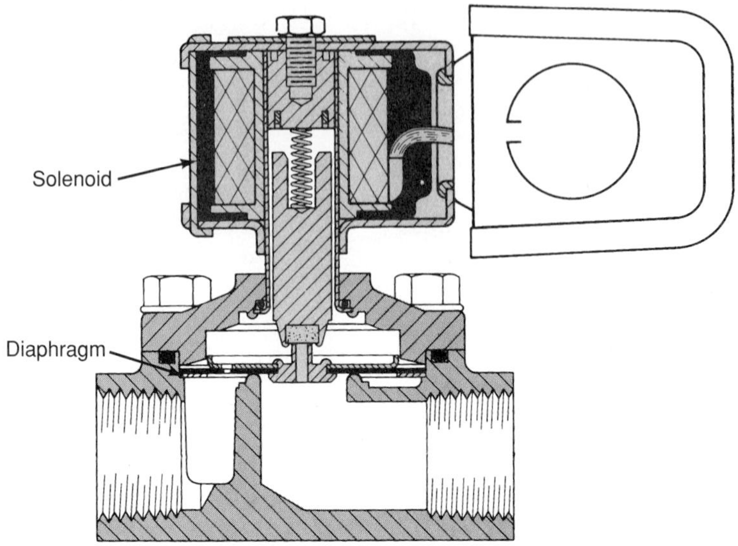

Figure 13-138. *Solenoid-operated water valve with diaphragm. (Sporlan Valve Co.)*

Electric water valves consume a small amount (6 W to 10 W) of current while in operation. **Figure 13-139** illustrates two typical electric water valve circuits. One uses a low-voltage solenoid valve. The other uses a 120 V solenoid valve. Most valves require 120 V.

A solenoid water valve of larger capacity is shown in **Figure 13-140.** The body of the valve is brass. It is made with either threaded or soldered connections. The plunger is made of noncorrosive steel. The valve seats are usually made of brass or bronze. A special rubber composition is used for the valve face.

Water flow is constant in this type of control. The valve stem is loosely connected to the plunger. This permits a shock action to open the valve. Gravity and water pressure close the valve when the power is shut off. Large-volume water flow may also be controlled by motor-operated valves, **Figure 13-141.** The electrically controlled water valve may be removed or replaced without disturbing the refrigeration system.

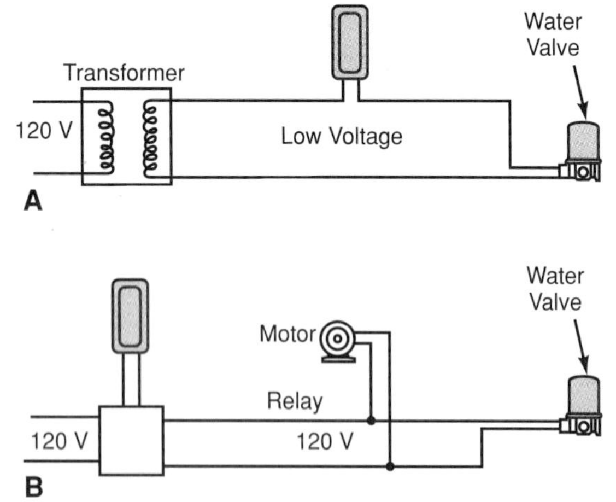

Figure 13-139. *Solenoid-operated water valve wiring diagrams. A—Solenoid uses low voltage. B—Solenoid uses 120 V.*

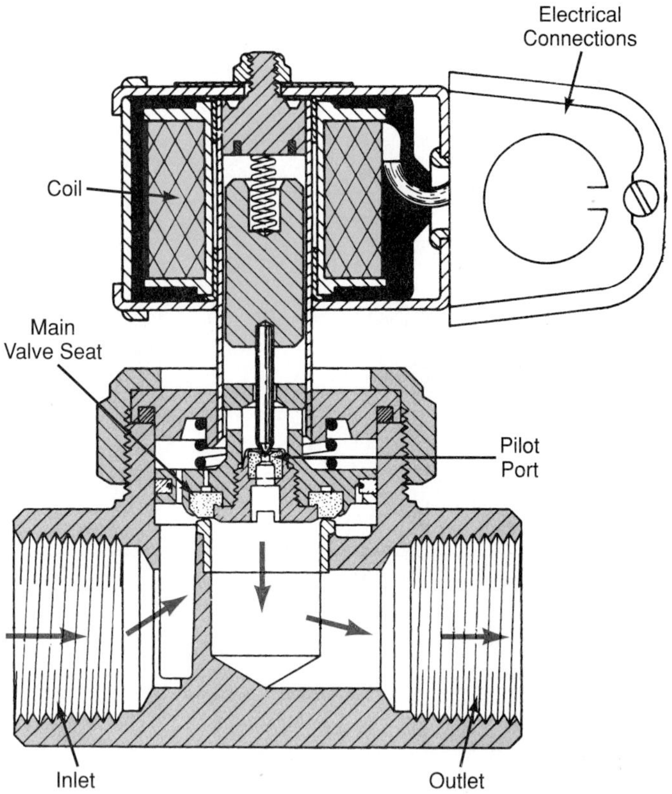

Figure 13-140. *Pilot-operated electric water valve used on large installations. Note that solenoid valve, when open, only decreases water pressure above large piston. (Sporlan Valve Co.)*

The inside of the motor-actuated water valve is shown in **Figure 13-142.** Pipe joints are unions to enable easy removal of the valve. The screen may be serviced

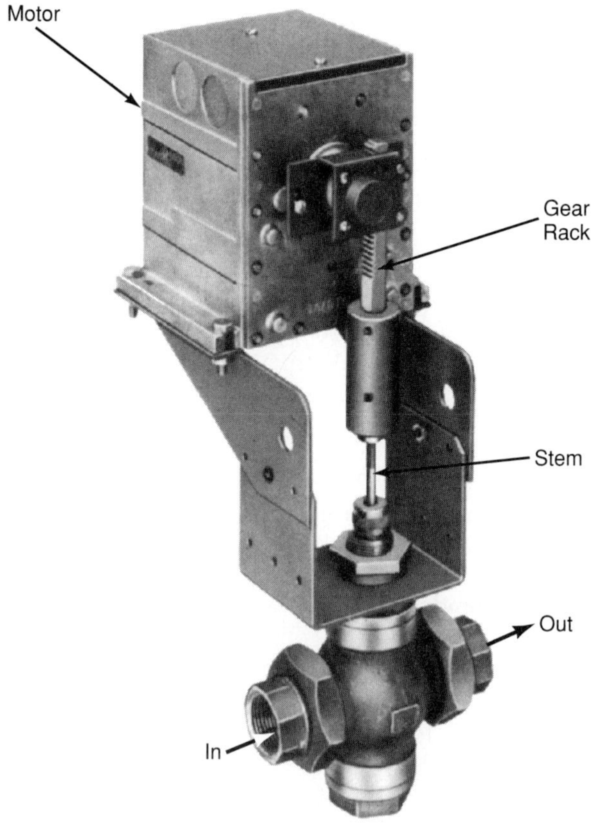

Figure 13-141. *Motorized water valve. Motor raises or lowers valve stem. Note gear rack connected to stem. (Johnson Controls, Inc.)*

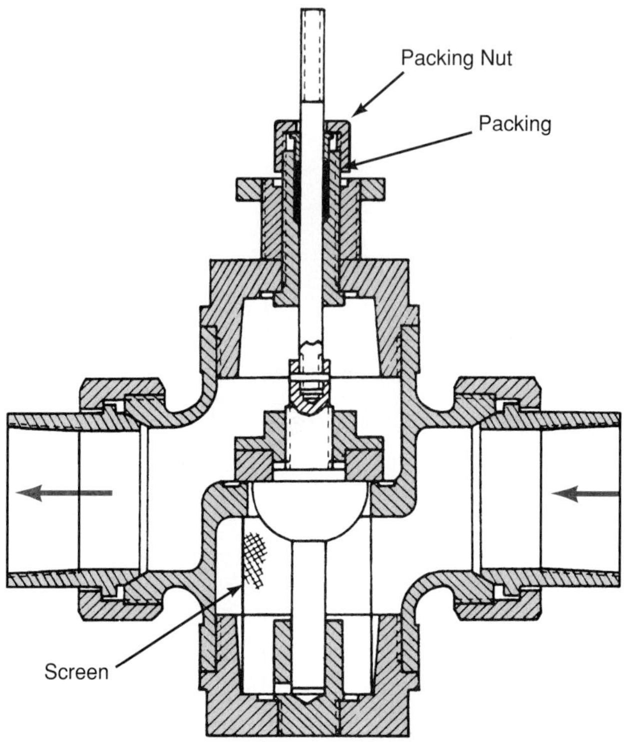

Figure 13-142. *Water valve body of motor-actuated valve. Note direction of flow, screen, valve stem packing, and packing nut.*

by removing the cap on the bottom of the valve. These valves have capacities varying from 1/2″ to 4″ pipe size.

Pressure-Operated Water Valve

Pressure-operated water valves are the most popular type of water valve. It is a bellows attached to the high-pressure side, preferably to the cylinder head. This bellows operates the water valve, as shown in **Figure 13-143.**

As condenser pressure rises, the bellows in the water valve contracts. The valve is opened by any of various mechanisms, depending on the specific water valve. Water flows into the condenser to cool the compressed vapor. The valve opens the water circuit only when the water is needed—as the pressure rises. It will keep increasing the water flow just as long as there is an increase of high-side pressure.

These valves may be adjusted using a heavy spring that presses against the bellows. The valves are set to open at definite head pressures. The pressure depends on the temperature of the water and the refrigerant used. See Chapter 15.

Some pressure-controlled water designs require opening the system to remove the valve. Others may be removed without disturbing the refrigeration system.

Water flow can be modulated (adjusted) with this valve. As the condensing pressures and temperatures

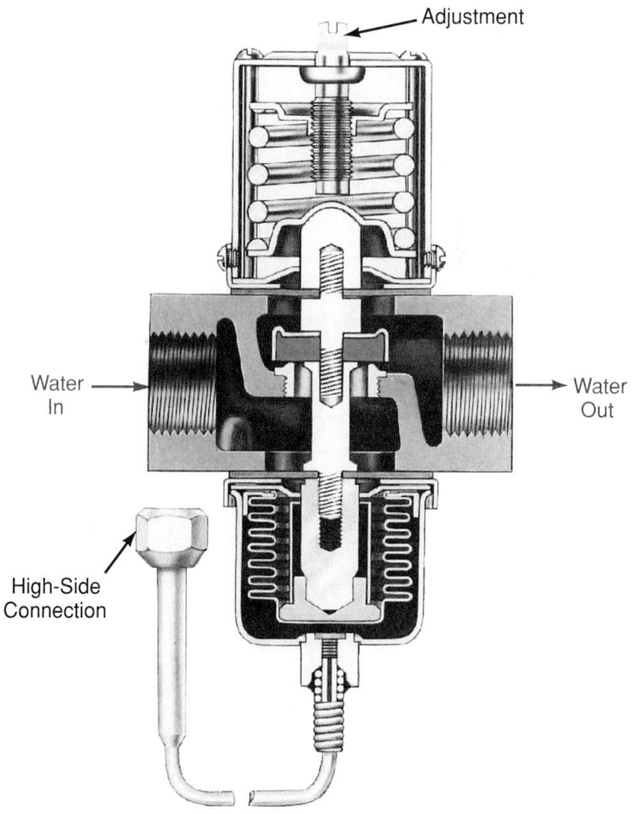

Figure 13-143. *Pressure-controlled water valve is connected to high-pressure side at compressor head. Rate of water flow is adjusted by spring pressure (top) on valve. (Johnson Controls, Inc.)*

increase, the valve opens more. When the pressures and temperatures drop, the water flow decreases.

The valve faces are a hard rubber composition, Bakelite, or fiber. The seat is usually made of copper or brass. The valves are equipped with either a packing gland or a bellows. This is located where the water stem goes into the water valve body. The packing must be adjusted occasionally to keep it from leaking.

These valves usually do not depend on the pipe for support. They do, however, have a mounting arrangement or flange. The inlet and the outlet are clearly labeled. Valves are usually threaded for standard pipe connections. Most are constructed so that water pressure tends to keep the valve closed. **Figure 13-144** shows a large-capacity valve used on 1″ lines. This valve has a gear mechanism for adjusting the pressures. The pressure-operated double water valve in **Figure 13-145** controls flow in two separate circuits.

Thermostatic Water Valve

The thermostatic water valve is controlled by the temperature of the exhaust water. The valve is identical to the pressure water valve except that it has a thermostatic element connected to the bellows operating the valve. See **Figure 13-146.** The element is charged with a volatile liquid. The power bulb is mounted in the condenser water line. Pressure is created by the volatile liquid in the bulb. This pressure opens the valve when the condenser water becomes warm. It closes the valve as the water cools.

13.12 Manual Valves

Manual servicing valves used on commercial refrigerating systems are used to:

- Determine the operating pressures.
- Charge or discharge a system.
- Remove any part of the system without disturbing the other parts.

These hand valves and service valves must resist corrosion. They must also withstand frequent opening and closing without leaking. Valve stems and packing must be handled with care.

13.12.1 Condenser Service Valves

Many condensers are equipped with both two-way and one-way service valves. See Chapter 2. Some of these valves are quite large. The valve stems may be 3/8″ across flats, and larger. This is because liquid lines are as large as 3/8″ OD.

Some of the larger systems may be equipped with additional service valves. Separate valves may be used for installation purpose and for servicing. Many systems have a valve between the condenser and the liquid receiver. The gauge connections may be 1/4″ or 1/8″ pipe. Some systems use Schrader valves to connect gauges and to perform service operations. See Chapter 12.

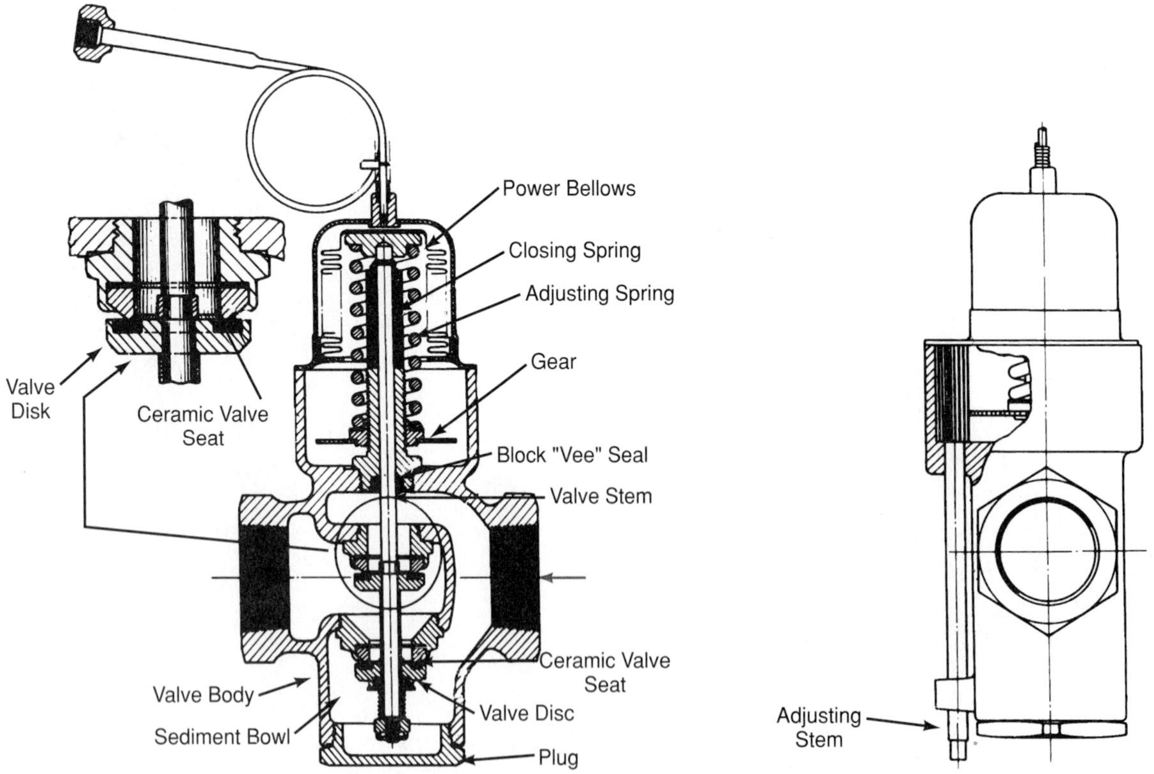

Figure 13-144. *Large capacity pressure-operated water valve. Double valve and seat arrangement balances force from water pressure. One valve is opened and the other is closed by water pressure.*

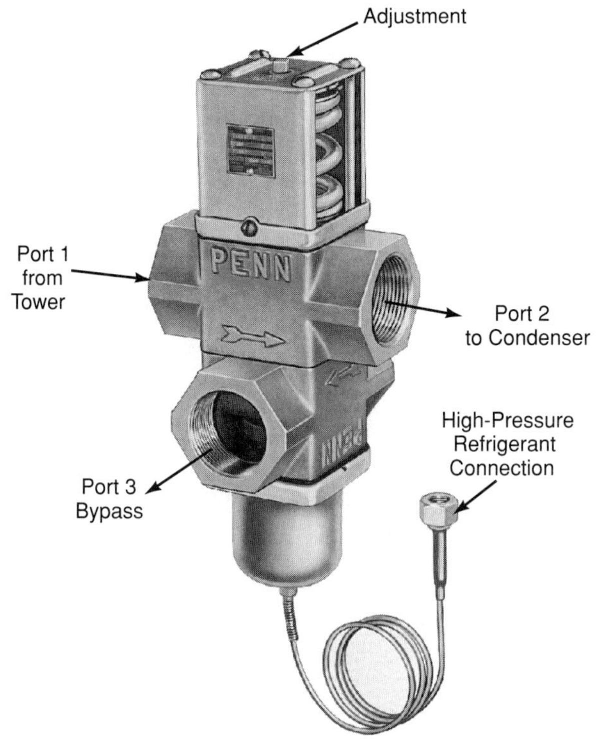

Figure 13-145. *This double water valve is pressure-operated. Note direction of water flow on each valve body. (Johnson Controls, Inc.)*

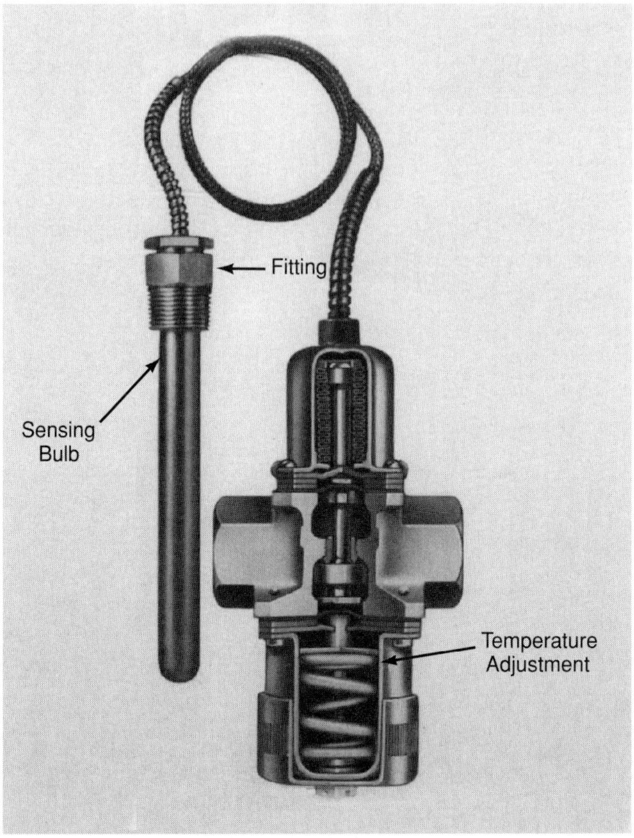

Figure 13-146. *Thermostatic water valve with sensing bulb and temperature adjustment. It is fitted inside water outlet piping using fitting.*

13.12.2 Manual Installation Valves

In addition to the usual service valves, multiple installations are usually equipped with shutoff valves. See **Figure 13-147.** These valves operate by hand. They must be located so that they may be easily turned. They may be classified as riser or manifold valves.

In multiple installations, the suction line should run from the compressor to a manifold. The individual suction lines for each evaporator should go from this manifold to the evaporators.

Between each of these suction lines and the manifold is a hand-operated shutoff valve mounted into the manifold. This valve permits any one of the suction lines to be closed without interfering with the operation of the others. A similar manifold device is also provided for the liquid line. These valve groupings are usually mounted in a cabinet or on a special valve board near the condenser.

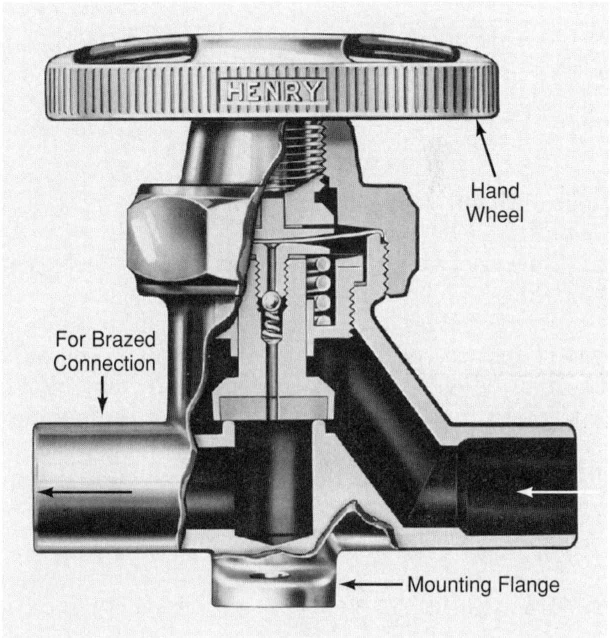

Figure 13-147. *Manual shutoff valve used on multiple installations. Valve uses diaphragm in place of packing. Piping openings are in line. (Henry Valve Co.)*

13.12.3 Riser Valves

A ***riser valve*** is another type of shutoff valve. It is hand-operated with three openings to which refrigerant lines may be connected. Two openings are in line with each other on opposite sides of the valve. The third is a little closer to the valve wheel, at right angles to the other two openings.

Turning the hand valve in closes the opening at right angles to the other two. This construction permits mounting of the valve in either a liquid or suction line. The technician can then connect another evaporator to it. This may be shut off from the remainder of the system. To do this, the valve is turned in all the way. **Figure 13-148** shows a multiple installation using two

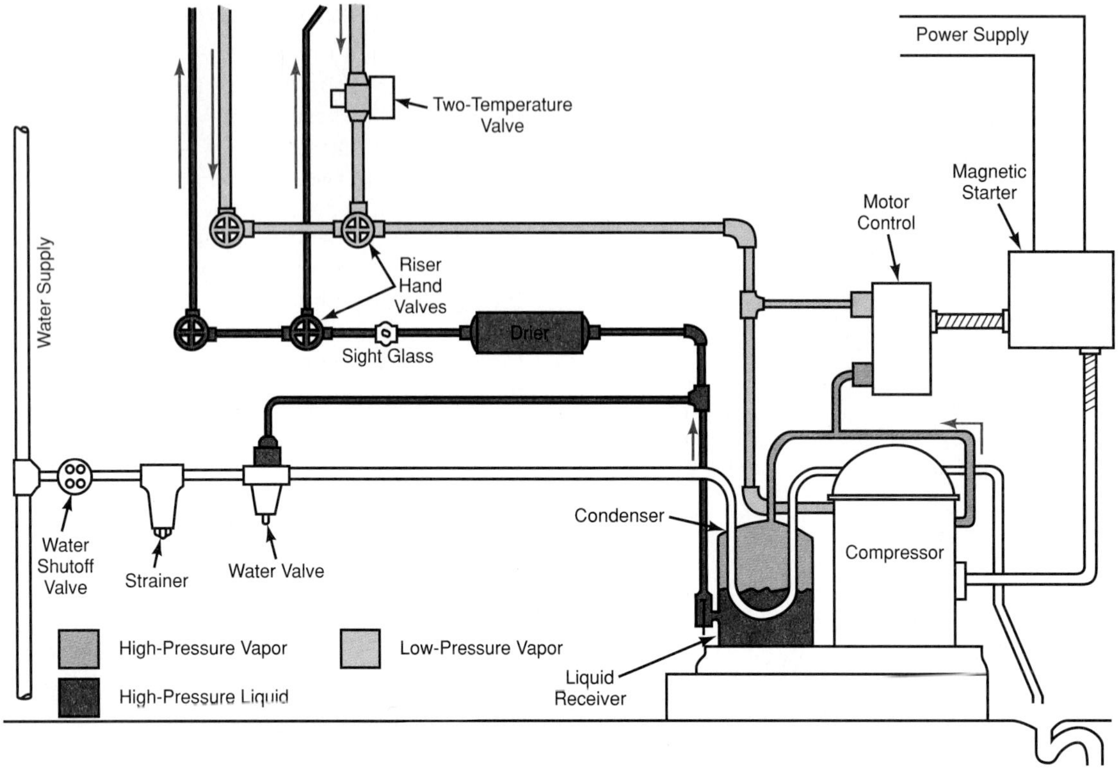

Figure 13-148. *Typical multiple installation showing location of important parts. This installation uses four riser valves.*

liquid line riser valves. Two suction line riser valves are also used.

Service valves are usually made of drop-forged brass to reduce seepage through the valve. The valve stem may be either brass or steel. Valves may also be the packless type. These use a bellows or a diaphragm as a sealing device rather than packing. Some valves are self-seating. This means the valve is easily seated again by tapping the valve stem into the seat. The valve seat is made of a soft lead alloy or Monel metal.

13.12.4 Relief Valves

A refrigerating system, regardless of size, is a sealed system. It is a pressurized container. The pressures vary, but high pressures could cause some part of the system to explode. This might occur during shutdowns, fires, extreme temperature conditions, or with faulty electrical controls.

To prevent dangerous pressures, relief valves are mounted on the units. They are usually on the liquid receiver. Hand valves must not be placed between the system and the relief valve. The National Refrigeration Code and most local codes require valves under the following conditions:

- If the unit is greater than a certain tonnage.
- If the amount of refrigerant exceeds specified minimums.
- If the internal volume is large enough.

Hand valves must not be placed between the system and the relief valve.

The relief devices are of three principal types:

- Fusible plug.
- Rupture disk.
- Spring-loaded valve.

The fusible plug is shown in **Figure 13-149.** It is threaded for connecting to the liquid receiver. A flared fitting is used to connect the purge line used to carry the released refrigerant outdoors. Low-temperature alloy in the plug will melt if receiver temperature rises

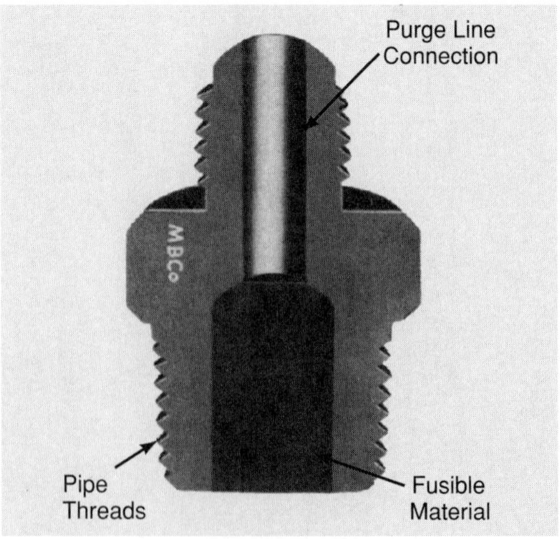

Figure 13-149. *Fusible plug for liquid receivers. Note flared fitting at outlet for connecting purge line that carries refrigerant outdoors. (Mueller Refrigeration Products Co., Division of Mueller Industries, Inc.)*

above a certain temperature. When the alloy melts, all the refrigerant will be released.

The rupture disc is shown in **Figure 13-150.** Similar in appearance to the fusible plug, it has a thin metal disc. This disc will burst before the pressure in the system reaches dangerous levels. It, too, is threaded into the liquid receiver. It is connected to a purge line that carries the released refrigerant outdoors.

The spring-loaded safety valve reseals itself or closes when the pressure drops to a safe limit. **Figure 13-151** shows such a valve. Note the spring loading. The relief pressure is adjustable but, once set, the valve is sealed to prevent tampering. This is shown in **Figure 13-152.** It is important that relief settings not be adjusted in the field. The seal must not be broken. If it is, the valve should be replaced with a correctly adjusted and sealed valve. Pressure relief valves usually close at 10 to 20% below their opening pressure.

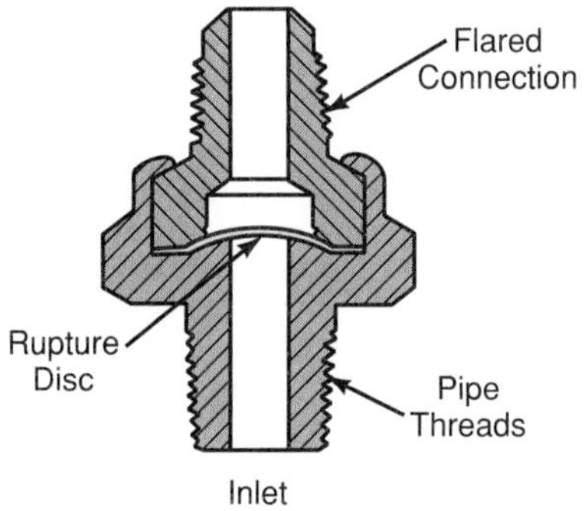

Figure 13-150. *A safety head (rupture disc) for R-12. Disc is made of silver. Safety head is available with various rupture pressures. (Reprinted by permission of the American Society of Heating, Refrigerating, and Air-Conditioning Engineers, Atlanta, GA, from the 1994 ASHRAE Handbook—Refrigeration)*

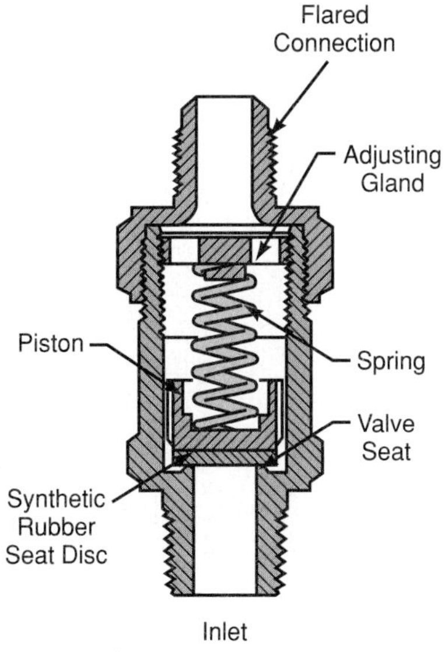

Figure 13-151. *Spring-loaded pressure relief valve. It uses a synthetic rubber seat and is made in several pressure ranges.*

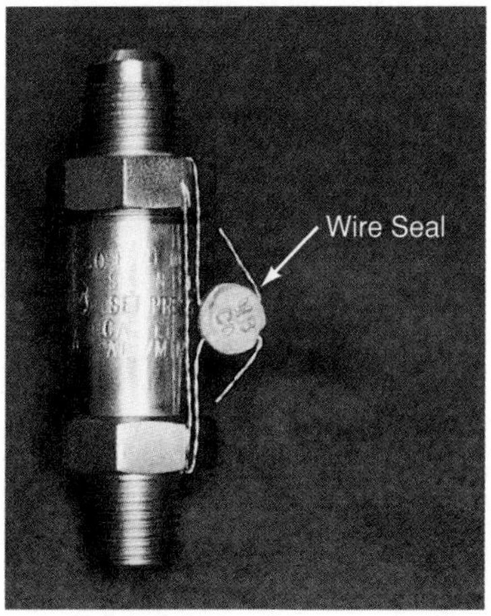

Figure 13-152. *Adjustable pressure relief valve. Wire seal is used to prevent tampering. (Mueller Refrigeration Products Co., Division of Mueller Industries, Inc.)*

13.13 Refrigerant Lines

Hard-drawn copper pipe is usually used to carry the refrigerant around the system. This pipe is furnished in iron pipe sizes. The fittings are not interchangeable with tubing sizes. See Chapter 2 for copper tubing sizes. Streamline brazed connections are used to connect the fittings to the pipe.

The National Refrigeration Code and local codes require the use of hard copper pipe. Soft copper tubing is permissible at the condenser end of the lines. It is also permissible in the fixtures. However, even these short lengths should be eliminated wherever possible.

The appearance of an installation is important. Therefore, the piping should be put in as neatly as possible.

13.13.1 Vibration Absorbers

Vibration absorbers may be installed in the compressor suction and discharge lines near the condenser. These will reduce any condenser vibrations that might travel into the lines. **Figure 13-153** shows construction of a vibration absorber. **Figure 13-154** shows two designs of vibration absorbers.

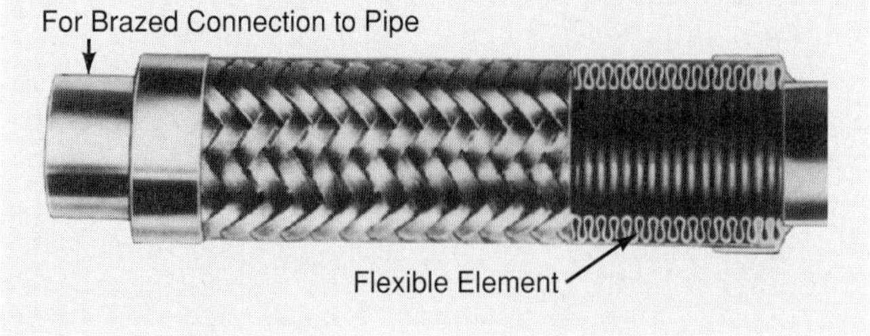

Figure 13-153. *Vibration absorbers, such as above, are installed in discharge and suction lines to prevent condenser vibration from traveling through lines. (Y/P Products, Inc.)*

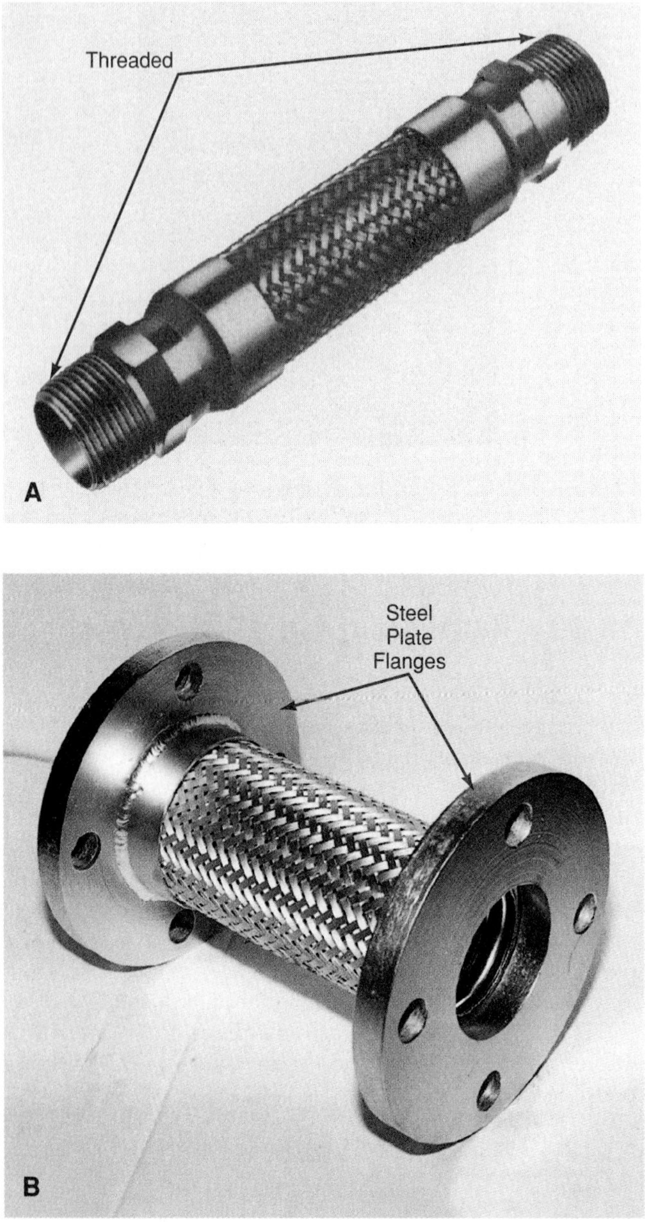

Figure 13-154. *Two designs of vibration absorbers. A—Threaded connection type. B—Steel plate flange pump connector. (Y/P Products, Inc.)*

For good sound absorption, two vibration absorbers are placed in each line. One is placed vertically and one horizontally.

The flexible absorber should be fastened to the unit or to a wall at the end pointing away from the vibration source. This fastening will prevent the vibration traveling along the pipe. It is important not to stretch, compress, or twist the vibration absorber when installing it.

13.13.2 Mufflers

Compressor pressure pulses are noisy and tend to follow the refrigerant lines. Most domestic refrigerating systems have small ***mufflers,*** **Figure 13-155.** They are

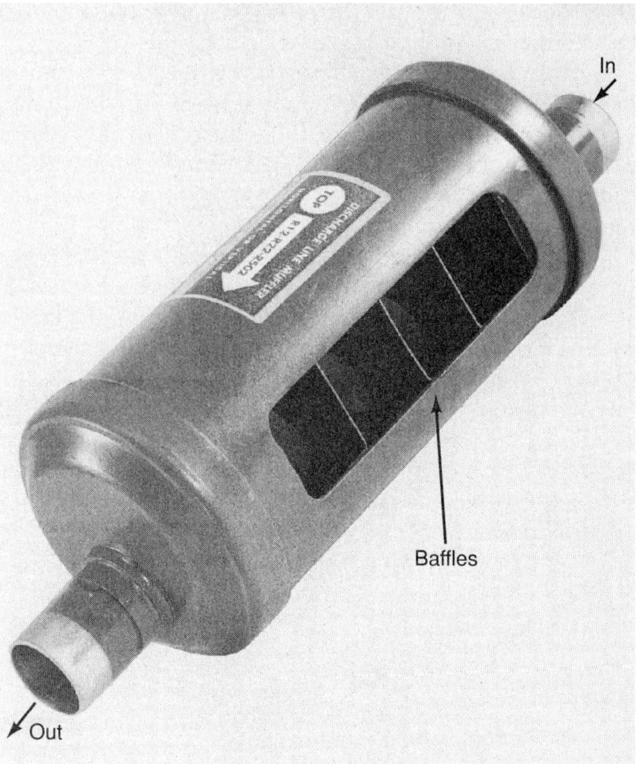

Figure 13-155. *Refrigerant gas muffler designed for vertical or horizontal mounting on discharge line. (AC & R Components, Inc.)*

built into the refrigerant circuits to break up the pressure pulses. They are usually in the compressor suction and discharge lines within the hermetic unit dome.

Most commercial refrigerating systems also use mufflers, especially systems for comfort cooling in air conditioning. Mufflers are installed near the condenser, usually vertically, to provide efficient oil movement.

13.13.3 Sight Glasses

Sight glasses are usually installed in liquid lines of commercial installations. The sight glass will show bubbles if the system is low on refrigerant.

The sight glass may show a few bubbles when the system first starts. Or, they may appear just as the system stops. These are normal equalizing actions and do not indicate a shortage of refrigerant.

Bubbles may also show a restriction in the circuit ahead of the sight glass. This may be due to a partially clogged drier, screen, or filter. The sight glass has long extensions that permit soldering or brazing the joints without injury to the sight glass. Some sight glasses are transparent, as in **Figure 13-156.**

A sight glass may be used on a large liquid line. This involves installing a smaller parallel flow pipe, as shown in **Figure 13-157.** If bubbles are in the liquid, some will travel through the sight glass tube and indicate a refrigerant shortage.

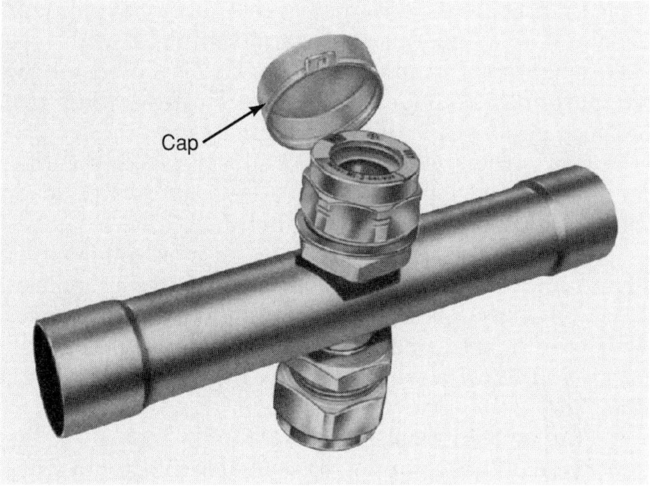

Figure 13-156. *Transparent sight glass. Cap is removed from top only in this view. When both caps are removed, technician can easily see any bubbles present by looking through liquid refrigerant. (Henry Valve Co.)*

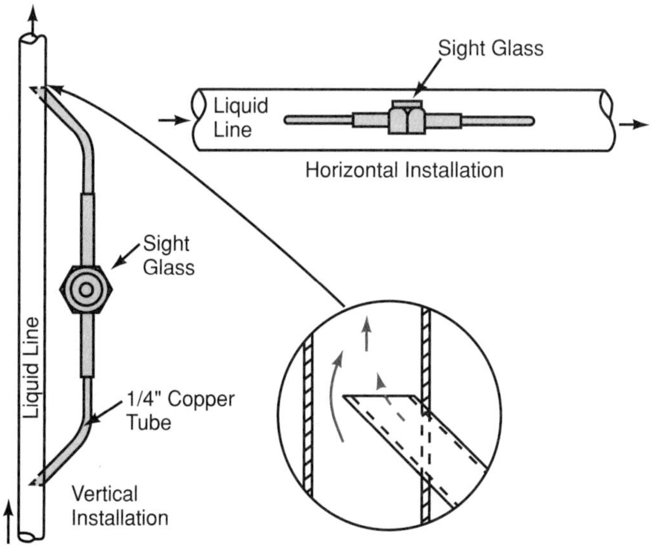

Figure 13-157. *Method of installing parallel sight glass in large liquid line. Joints are usually brazed.*

Electronic Sight Glass

Electronic sight glass can be used to determine levels of refrigerant. See **Figure 13-158.** This type of sight glass is clamped on a refrigeration line. It uses ultrasonic sound waves to detect the bubbles in the liquid flow. The unit emits an audible signal when bubbles are seen, indicating the system is low on refrigerant.

These units can be permanently installed to provide continuous monitoring. They may also be used by the technician when charging a system.

Moisture Indicators

Many sight glasses have a moisture-indicating chemical built into the sight chamber. The chemical will change color if there is moisture in the system. **Figure 13-159** shows a two-purpose sight glass.

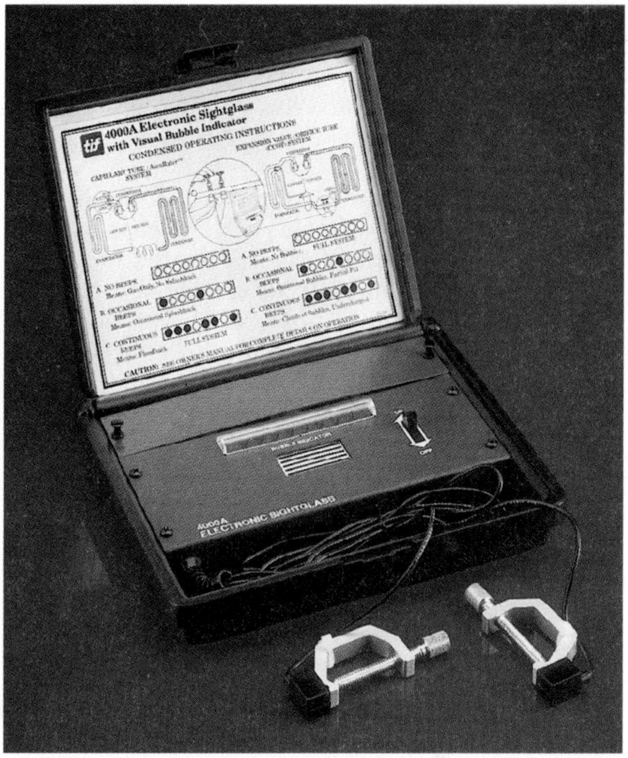

Figure 13-158. *Electronic sight glass generates a tone when bubbles are present in a liquid line. Also note the light bar that illuminates a visual duplication of actual bubbles. (TIF Instruments, Inc.)*

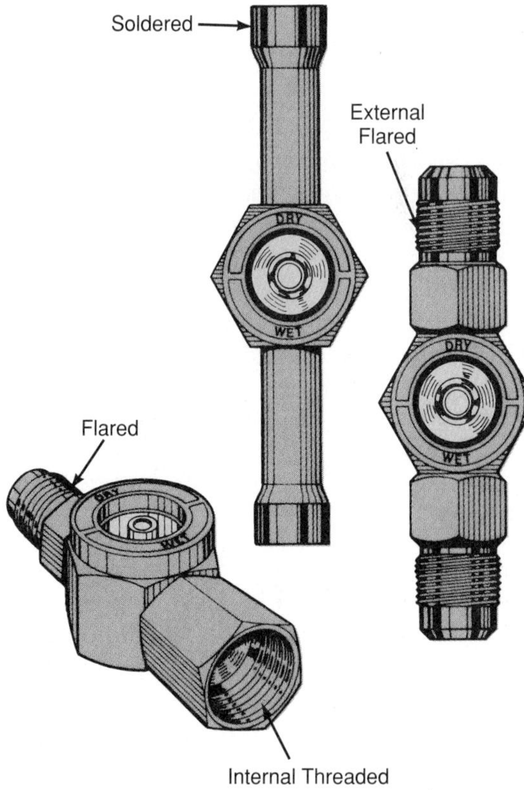

Figure 13-159. *A sight glass may also have a moisture indicator. Note methods of connection. (Virginia KMP Corp.)*

The chemical substance will turn pink if moisture is present in a system charged with either R-11, R-12, R-13, R-113, or R-114. It stays blue if there is a safe minimum amount of moisture. With a system charged with R-22, R-500, or R-502, the chemical is green when dry, and pink when wet.

With some moisture indicators, the words "wet" and "dry" appear when the chemical changes color. Temperature is important. The higher the liquid temperature, the higher the moisture content needed to produce a color change. An indicator, if hot, can show "dry" even though the system has too much water. See **Figure 13-160.** For accurate indicators, the liquid line should be as near 75°F (24°C) as possible.

R-502	PPM (Parts per Million)	
Degrees °F	Turn Blue (Dry) Below	Turn Pink (Wet) Above
75 100 125	5 10 15	15 30 45
R-12	**PPM (Parts per Million)**	
Degrees °F	Dry	Wet
75 100 125	5 10 15	15 30 45
R-11–R-500	**PPM (Parts per Million)**	
Degrees °F	Dry	Wet
75 100 125	30 45 60	120 180 240

Figure 13-160. *Effect of temperature on moisture indicators. Note that the amount of water can increase as the temperature increases and indicator will still show a "dry" condition. Technician should install indicator where liquid line will remain cool.*

It takes an hour to get a good reading. However, about eight hours are needed for the indicator to give an accurate color signal.

Oil may turn the moisture indicator tan. Flushing the indicator with clear refrigerant will remove the color. However, if it continues to turn tan, the system has too much oil.

Alcohol placed in the system to absorb the moisture will affect the operation of the moisture indicator. Too much water or alcohol in the system will wash the chemicals off the indicator surface. The indicator will need replacing after the system is dried and the alcohol removed.

13.13.4 Filter-Driers (Liquid Line)

The efficient operation of a commercial system depends greatly on the internal cleanliness of the unit. Only clean, dry refrigerant and clean, dry oil should circulate in the system.

Practically all dirt and water must be removed. Contaminants must be trapped in some part of the system where they cannot do harm. Devices used for this purpose may be in separate units. However, they may be built into a single unit which filters and adsorbs. (Adsorption is the ability to collect substances on a surface in a condensed layer.) Screens, filters, and water adsorbents are used as part of the filter-drier.

A common method of removing moisture is with a liquid line drier, **Figure 13-161.** Enough drying material must be used for both the high and low moisture ranges. If it is fully activated, it can keep the refrigerant both clean and dry. Driers are usually installed in the liquid line.

The conventional straight-through drier is a cylinder made of brass, copper, or steel. It is filled with a desiccant chemical such as activated alumina or silica gel. These chemicals can adsorb 12% to 16% of their weight in water. Both ends of the cylinder usually contain filter elements. The end caps are fitted with either flare or soldered connections.

One design of liquid line drier allows the casing to stay in the line. Only the drier cartridge needs to be changed. **Figure 13-162** shows a drier that uses one replacement cartridge. **Figure 13-163** shows a split replacement cartridge.

Figure 13-161. *Filter-drier designed for use in liquid lines. It has a solid core of desiccant. Connections are flared. (Sporlan Valve Co.)*

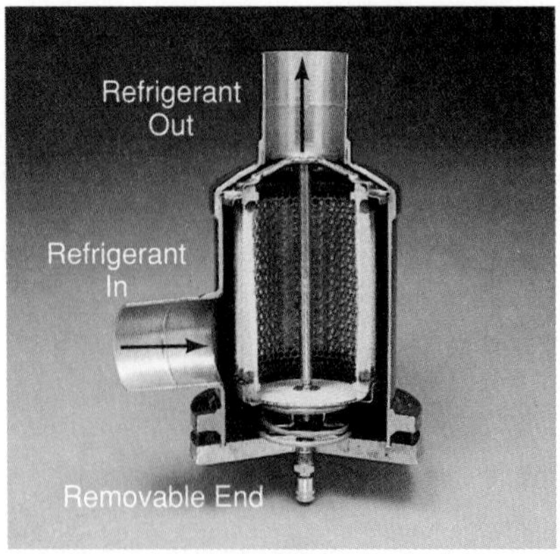

Figure 13-162. *Commercial filter-drier with removable end. (Alco Controls Division, Emerson Electric Company)*

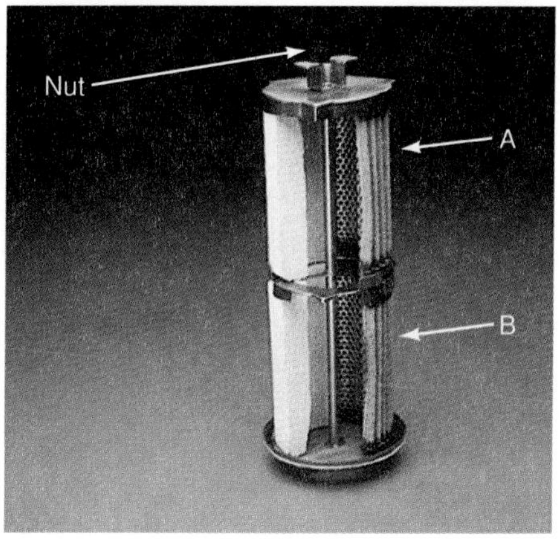

Figure 13-163. *This replacement cartridge is made in two sections (A and B). Note the nut at end of shaft that is removed when replacement cartridges are inserted. (Alco Controls Division, Emerson Electric Company)*

The refrigerant should be dried as follows:

- Less than 15 parts of water per million if R-12 is used.
- Less than 25 ppm if R-22 or R-500 is used.
- 5 ppm for R-502.

Corrosion begins at:

- 15 ppm of water in R-12.
- 120 ppm for R-22 or R-500.
- 15 ppm for R-502.

Refrigerant with the indicated safe amount of moisture avoids many problems in the system. Experience shows that corrosion, oil breakdown, and motor burnouts are almost eliminated if the guidelines are followed.

Cleaning a refrigeration system involves four basic tasks:

- Removing water.
- Removing acid.
- Filtering out circulating solids.
- Measuring when the drying job is completed.

Driers will handle the first three tasks. A moisture indicator is required for the fourth.

Driers should be left in the system permanently since oil loses its moisture slowly. Also, insulation in hermetic compressors and in small crevices may release moisture for long periods. A drier is like a sponge. However, it can become saturated if the drier is too small. This will leave the refrigerant wet. A moisture indicator is the only sure means of recognizing a wet condition.

R-22 driers must be three to five times as large as those needed for an equal quantity of R-12. The greater the ability of a refrigerant to hold water, the larger the drier required. R-500 driers need to be as large as R-22 driers. R-502 driers need to be as large as R-12 driers.

13.13.5 Filter-Driers (Suction Line)

Filter-driers are often mounted in the suction line, preventing foreign particles from entering the compressor. Particles over 5 microns in size, as well as acids, sludge, and moisture are also prevented from entering the compressor. Strainers (screens) are usually made of Monel metal. See **Figure 13-164.**

Only two things should be allowed inside a refrigeration system: clean, dry refrigerant and good, dry oil. A system which is clean, dry, and acid-free will run almost indefinitely. Corrosion, freeze-ups, oil breakdown, or hermetic motor burnouts should not occur. In such a system, there is nothing to filter and plugging is impossible. A clean, dry, acid-free system remains factory bright and trouble-free in operation.

A suction line filter-drier should be replaced if pressure drop is excessive. The table below indicates the maximum pressure drop allowable before replacing filter-driers. If the pressure drop exceeds this amount, the filter-driers should be replaced.

	Low Temperature Units (psi)	Medium Temperature Units (psi)	High Temperature Units (psi)
R-12 and R-500	2	6	8
R-22 and R-502	3	9	14

Replace for R-22 and R-502 refrigerants if pressure drop exceeds 3 psi for low-temperature units, or up to 14 psi for high-temperature units.

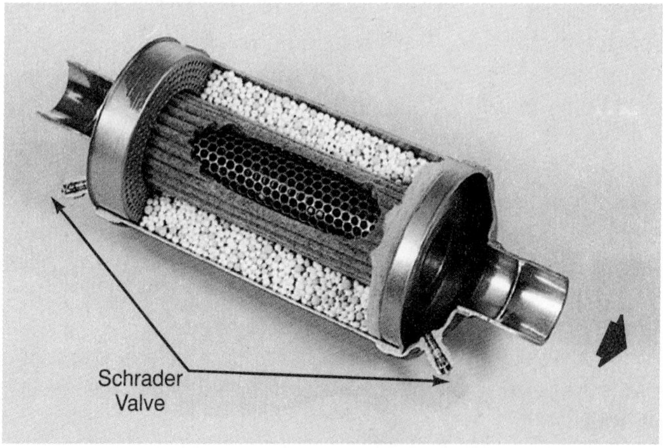

Figure 13-164. *Suction line filter-drier. Note Schrader valves at both ends. These are used to check the pressure drop through the filter-drier. (Alco Controls Division, Emerson Electric Company)*

13.14 Engine-Driven Systems

Natural gas, gasoline, and propane engines may be used to drive refrigerating compressors. A variable compressor speed produces flexible capacity. There is a comparatively low operating cost. Such units are available in four to 75 ton capacities. Engine-compressor units of one to five tons capacity are available for use on truck units. They are also used for air conditioning.

Pressure controls are usually used. The pressure control is connected to the engine's throttle. It is placed in the low-side suction line. The linkage is such that as the suction pressure increases, the engine's throttle is opened. This increases the compressor speed to increase the rate of refrigeration. As the temperature in the evaporator drops, the engine will slow down. This should result in a balance between the engine's speed and low-side pressure. It will give the desired temperature in the refrigerated space.

13.15 Review of Safety

Commercial systems vary considerably in size.

Small, self-contained units must be handled with care. These handling methods have been described in previous chapters.

As the units become larger, safety precautions become increasingly important. The investment in the machines is greater and repairs are more costly. The large machines are also more dangerous. The energy output of larger moving parts and larger refrigerant containers is potentially dangerous.

Do not close the compressor discharge valve on a 10-ton capacity unit while it is operating. It would almost instantly ruin the compressor or rupture a gasket. Carelessly opening a receiver valve may cause the loss of hundreds of pounds of refrigerant. It could also injure the service technician. Trapping liquid refrigerant in any part of the system with no gas space may cause sufficient hydraulic pressure to burst the container.

Inside pressures must be atmospheric before any part of the system is opened. The technician must also be sure no liquid is present before opening any system part. Goggles should *ALWAYS* be worn when working on any unit.

Open electrical circuits and lock switches before working on a system if no power is needed.

Always follow local and national refrigeration and electrical codes when servicing all systems. Follow OSHA standards.

It is not safe to work on any part of the system unless the following are known:

- **Pressure and temperature.**
- **The condition of the refrigerant (liquid or vapor) inside that part.**
- **The fundamentals of working on that system. See Chapters 12 and 15.**

Pressure and temperature relief devices on the units protect the equipment, user, and service technician. They should be frequently checked for accuracy and kept in good operating condition.

Never use cylinder oxygen to test any device for leaks. Use either refrigerant, carbon dioxide, or nitrogen. See Section 12.9.1

13.16 Test Your Knowledge

Please do not write in this text. Place your answers on a separate sheet of paper.

COMMERCIAL SYSTEMS MODULE

1. What is commonly used to produce flake ice?
   A. Electrical grids.
   B. Compression of ice.
   C. Scraper on evaporator.
   D. Auger.
2. In a hot gas defrost system, which part(s) of the system is (are) heated?
   A. Condenser.
   B. Drain pan.
   C. Evaporator.
   D. Both A and C.
3. What type of metal is used in an all-metal pressure evaporator?
   A. Copper and cast aluminum.
   B. Copper and cast iron.
   C. Copper and bronze.
   D. Copper.
4. What is the counterflow principle?
   A. Refrigerant passes in the opposite direction of the air.
   B. Refrigerant flows in the opposite direction of the water.
   C. The condenser uses a counterflow fan.
   D. Refrigerant flows in the same direction of the water.
5. Why do liquid receivers have safety devices?
   A. To provide a pressure release.
   B. To provide a thermal release.
   C. To provide a safety release.
   D. All of the above.
6. What does the inner tube of a *tube-within-a-tube* contain?
   A. Refrigerant.
   B. Water.
   C. Nitrogen.
   D. Any of the above.
7. What is the advantage of forced circulation evaporators?
   A. They cool the cabinet quickly.
   B. They do not require baffles.
   C. They take up little space.
   D. All of the above.
8. An evaporator condenser operates as follows:
   A. Water is sprayed onto the condenser.
   B. Air is blown across the condenser.
   C. Heat from the evaporator is absorbed.
   D. All of the above.

9. What is the advantage of a water-cooled condenser?
   A. It has a lower operating head pressure.
   B. It provides greater system capacity.
   C. It reduces wear on the system's moving parts.
   D. All of the above.
10. What is the advantage of an air-cooled condenser?
   A. It has a lower operational cost.
   B. It is easily installed.
   C. It may be used where sufficient air is available.
   D. All of the above.

COMMERCIAL SYSTEMS—CONTROLS MODULE

11. Vibration dampers are not installed near the ______.
   A. compressor discharge line
   B. compressor suction line
   C. condenser
   D. evaporator
12. Which type of water valve will not vary the water flow as the refrigeration load changes?
   A. Electric.
   B. Pressure.
   C. Thermostatic.
   D. All of the above.
13. What percentage of refrigeration load may be placed on an evaporator controlled by a two-temperature valve?
   A. 30%.
   B. 40%.
   C. 50%.
   D. 60%.
14. If an oil separator is not insulated, it will ______.
   A. act as a refrigerant condenser and expel heat
   B. not return oil to the condenser
   C. not remove oil from the hot vapor
   D. All of the above.
15. In a multiple evaporator installation, what type of motor control is most frequently used?
   A. Thermostatic.
   B. Pressure.
   C. Defrost motor control.
   D. Either A or B.
16. In a multiple installation that has two-temperature valves, the check valve should be placed in the ______.
   A. suction line of the coldest evaporator
   B. high-side line of the warmest evaporator
   C. suction line of the warmest evaporator
   D. high-side line of the coldest evaporator
17. What type of water valve is commonly used for commercial installations?
   A. Electric.
   B. Pressure.
   C. Thermostatic.
   D. All of the above.
18. What is the basic type of two-temperature valve?
   A. Pressure-operated.
   B. Temperature-operated.
   C. Electrically operated.
   D. Both A and B.
19. Why are high-pressure motor cutouts used with water-cooled condensing units?
   A. Danger of inadequate water supply.
   B. Possible water restrictions.
   C. Possible excessive head pressure.
   D. All of the above.
20. To provide easy removal of the high-pressure motor control, where should it be connected?
   A. Cylinder head of the compressor.
   B. Suction side of the compressor.
   C. After the condenser.
   D. Before the condenser.

produce. Those contents are then transferred to a walk-in storage cabinet overnight. Therefore, temperatures may be kept at 40°F (4°C) to 45°F (7°C) in both compartments.

Evaporators used in these installations must be narrow. They are made with fins as small as 1 1/4″ (32 mm) wide. Some of the shelf evaporators are the plain tubing type. The evaporators are usually connected in series.

Many of these display cases are now using blower evaporators for cooling, since they take little space. Because of the circulating air, they provide even refrigeration temperatures throughout the display case.

14.5.3 Open Display Case

For easier customer self-service, open display cases are commonly used in supermarkets. These cases may have storage space in the base of the unit. Space at the top is open. The walls, or the upper part of the walls, may be enclosed in three to four layers of glass.

The higher temperature case is used for fresh meats and dairy products. These cases do not present any special evaporator problems. Blower evaporators are used. Ducts carry the cold air through grilles. The grilles are at the rear of the case at the level of the refrigerated foods. See **Figure 14-16.** The warm air returns down the front of the case. An open display case for meat and delicatessen products is shown in **Figure 14-17.** Note location of the food display in relation to the evaporator.

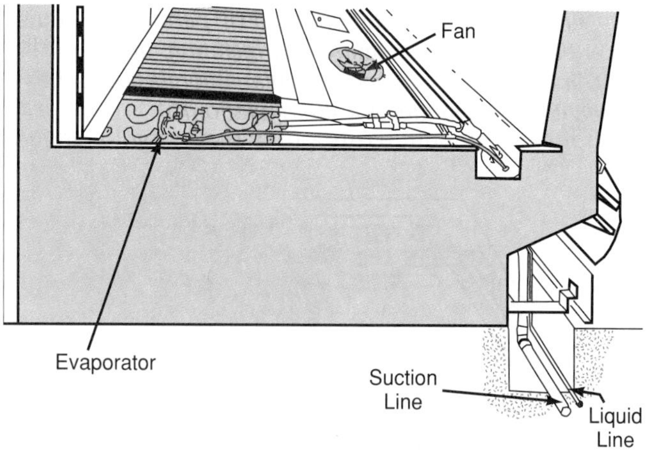

Figure 14-16. *Display case installation. Note trough in floor. It provides space to run refrigeration piping and electrical conduit.*

Many supermarkets have open display cases for produce. These cases are kept at about 40°F (4°C) and at a high humidity. See **Figure 14-18.** If dry air circulates over the contents, it will remove some of the moisture. This would spoil the appearance and decrease the weight of the produce. A cross section of an air curtain open display case is shown in **Figure 14-19.** An installed, operating unit is shown in **Figure 14-20.**

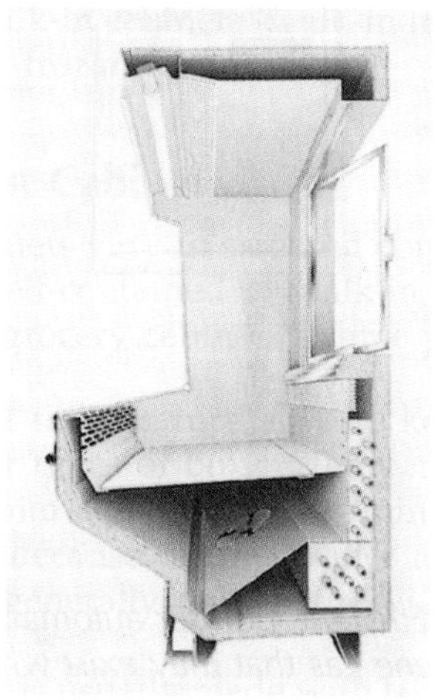

Figure 14-17. *Cross section of open meat display case. Note evaporator located to right of meat tray. (Tyler Refrigeration Corp.)*

Figure 14-18. *Open display case with canopy. Canopy mirror (at back) helps display produce. (Tyler Refrigeration Corp.)*

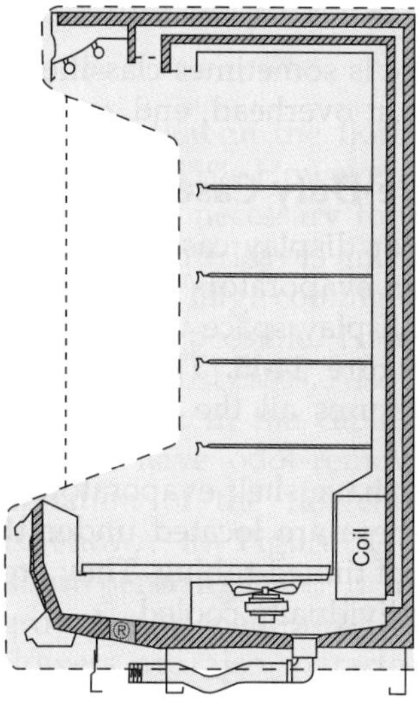

Figure 14-19. *This case is designed for display of dairy products and delicatessen items. (Kysor/Warren Division of Kysor Industrial Corp.)*

Figure 14-20. *Air curtain display case. Units like this are designed for delicatessen and dairy foods. (Kysor/Warren Division of Kysor Industrial Corp.)*

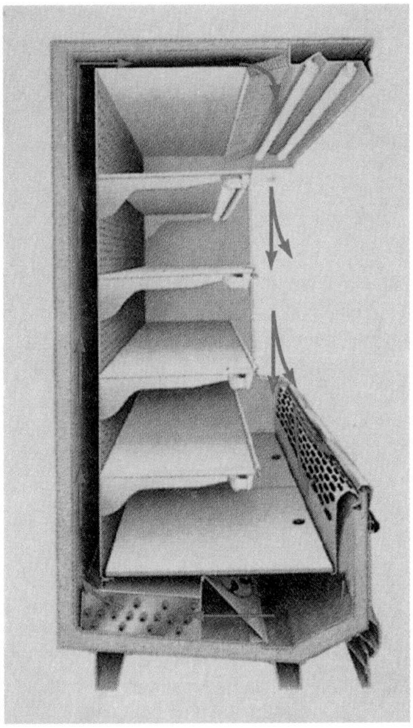

Figure 14-21. *Cross-sectional view is of open display case using blower evaporator. Note airflow pattern which carries cooled air over displayed foods. (Tyler Refrigeration Corp.)*

The airflow patterns in these cases can be monitored. An airflow meter or chemical smoke may be used for this purpose. The airflow curtain should not touch the shelves or the products. A cross section through a dairy/deli open display case is shown in **Figure 14-21.** Some of the electrical circuits of an open display case are shown in **Figure 14-22.**

14.5.4 Open Frozen Food Display Case

An open frozen food display case is shown in **Figure 14-23.** Storing and displaying frozen foods in either open or closed cabinets presents some problems. Temperatures near 0°F (−18°C) must be maintained. The evaporators must operate at −10°F to −15°F (−23°C to −26°C). Heater wires are installed along those parts of display cases where condensation from the air might collect.

Frozen food storage and display cases are constructed in both chest and upright cabinet styles. **Figure 14-24** shows an upright case.

Since temperatures are very low in these cases, openings are fitted with gaskets or seals. For the same reason, insulation is thick and carefully hermetically sealed. Chest types are popular, because the top openings prevent cold air spillage when the case is opened.

Open cases must be protected from drafts produced by grilles, unit heaters, and fans. Drafts will interfere with the air curtain of the case. This will cause higher operating costs and defrosting problems. Several cases are usually connected end-to-end in supermarkets. The total electrical load must be carefully checked to provide enough service.

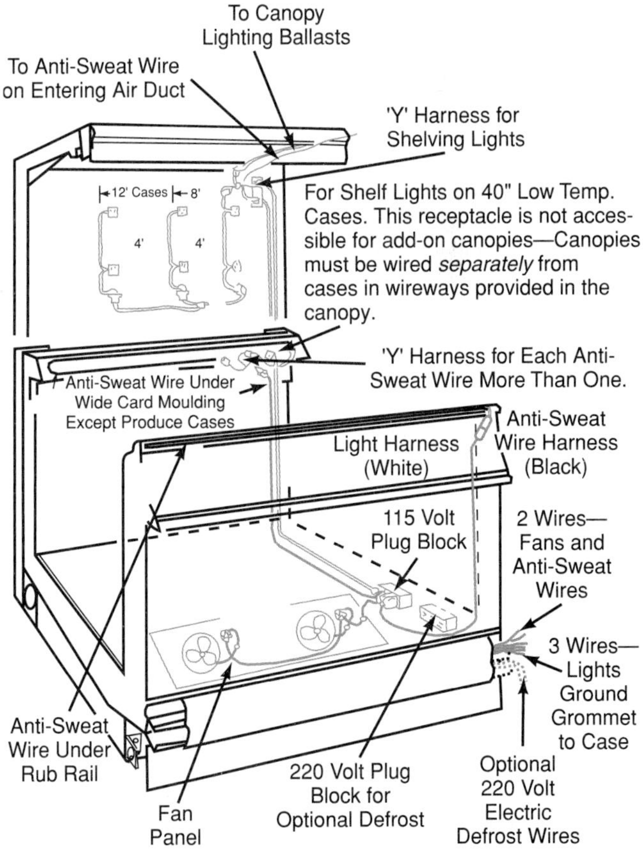

Figure 14-22. *Electrical wiring system of a display case is designed to handle lighting, heating elements, and fans. This system is typical of most open display cases.*

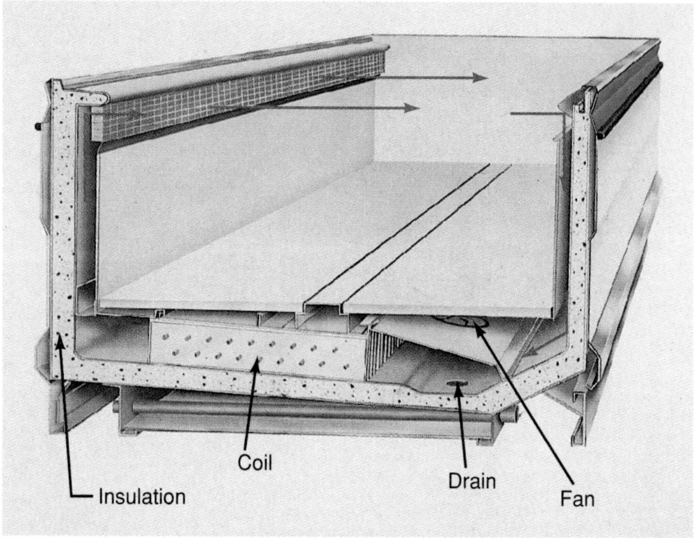

Figure 14-23. *Construction details of open frozen foods case. Notice airflow curtain across top of case as shown by arrows. (Tyler Refrigeration Corp.)*

Figure 14-24. *An upright frozen foods display case. Note, especially, the fans and airflow pattern.*

The need to maintain a low temperature presents a difficult evaporator defrosting problem. The evaporator must be defrosted at least once a day. This must be done quickly to prevent too much warming of the case. The defrosting is done automatically. A timer is used to operate a hot gas defrosting system or an electric heater defroster device. See Chapter 13 for details concerning these systems.

Some cabinets use two or three air curtains. The principle of operation is shown in **Figure 14-25.** The construction of such a cabinet is shown in **Figure 14-26.**

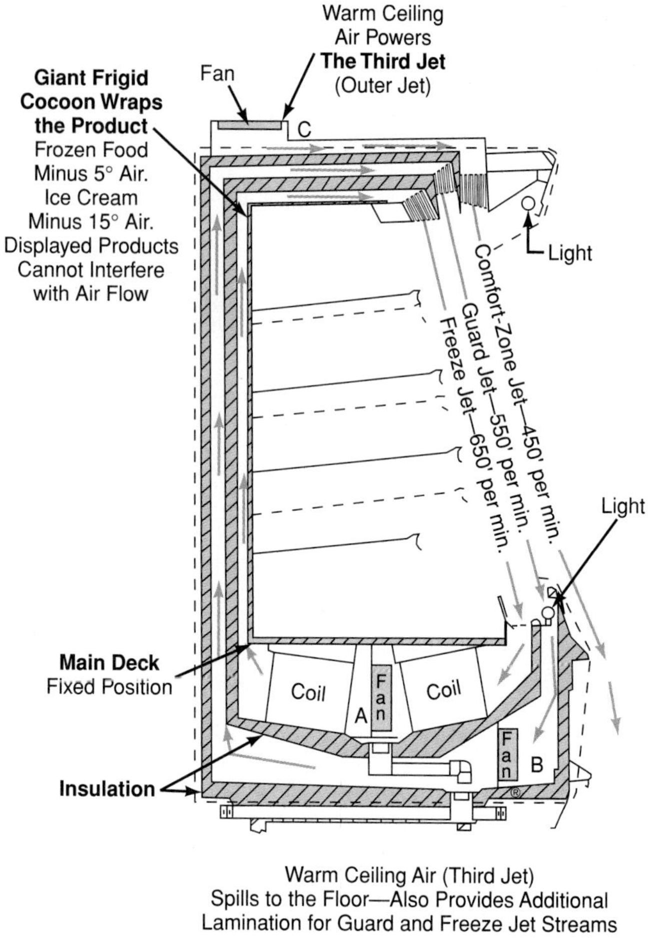

Figure 14-25. *A frozen foods case with three curtains of air. Note the three separate fan systems. Inner curtain is powered by fan at A, middle curtain by fan at B, outer curtain by fan at C. (Kysor/Warren Division of Kysor Industrial Corp.)*

14.6 Frozen Food Storage Cabinet

The frozen food storage cabinet may be either a chest or upright type insulated with 4″ to 6″ (10 cm to 15 cm) of polyurethane. Stainless steel is often used for the inner liner. The outer liner is usually aluminum or stainless steel. The doors or access openings also are heavily insulated. Double gaskets are usually provided for better sealing. See **Figure 14-27.**

Display cabinets that use temperatures below the dew point of the room require heaters. These are needed to prevent moisture on the glass. (***Dew point*** is the temperature at which moisture will form on a surface.) Frequently these are electrical resistance strip heaters under the surface of the cabinet. Electrical resistance strip heaters are also used around door frames and other parts which sweat.

These cases are for storage purposes only. The food is moved to display cases as needed. Storage cabinets operate at 0°F (−18°C) or lower. The refrigerating mechanism is normally installed on top of the cabinet.

1 Refrigerated Duct Honeycomb.
2 Cold Air Honeycomb Anti-Sweat Heater.
3 Nozzle Heater.
4 Guard Duct Honeycomb.
5 Canopy Honeycomb.
6 Fluorescent Bulb.
7 Canopy Fan Blade.
8 Canopy Fan Motor.
9 Ballast (2-Lamp).
10 Top Shelf Assembly (Accessory Shelf—High or Low Front).
12 Center Shelf Assembly (High or Low Front).
14 Bottom Shelf Assembly (High or Low Front).
16 Refrigerated Air Duct Anti-Sweat Heater.
17 Return Air Grille Anti-Sweat Heater.
18 Thermopane Cap Anti-Sweat Heater.
19 Return Air Grille.
20 Expansion Valve.
21 Raceway Bumper Rail.
22 Kick Plate.

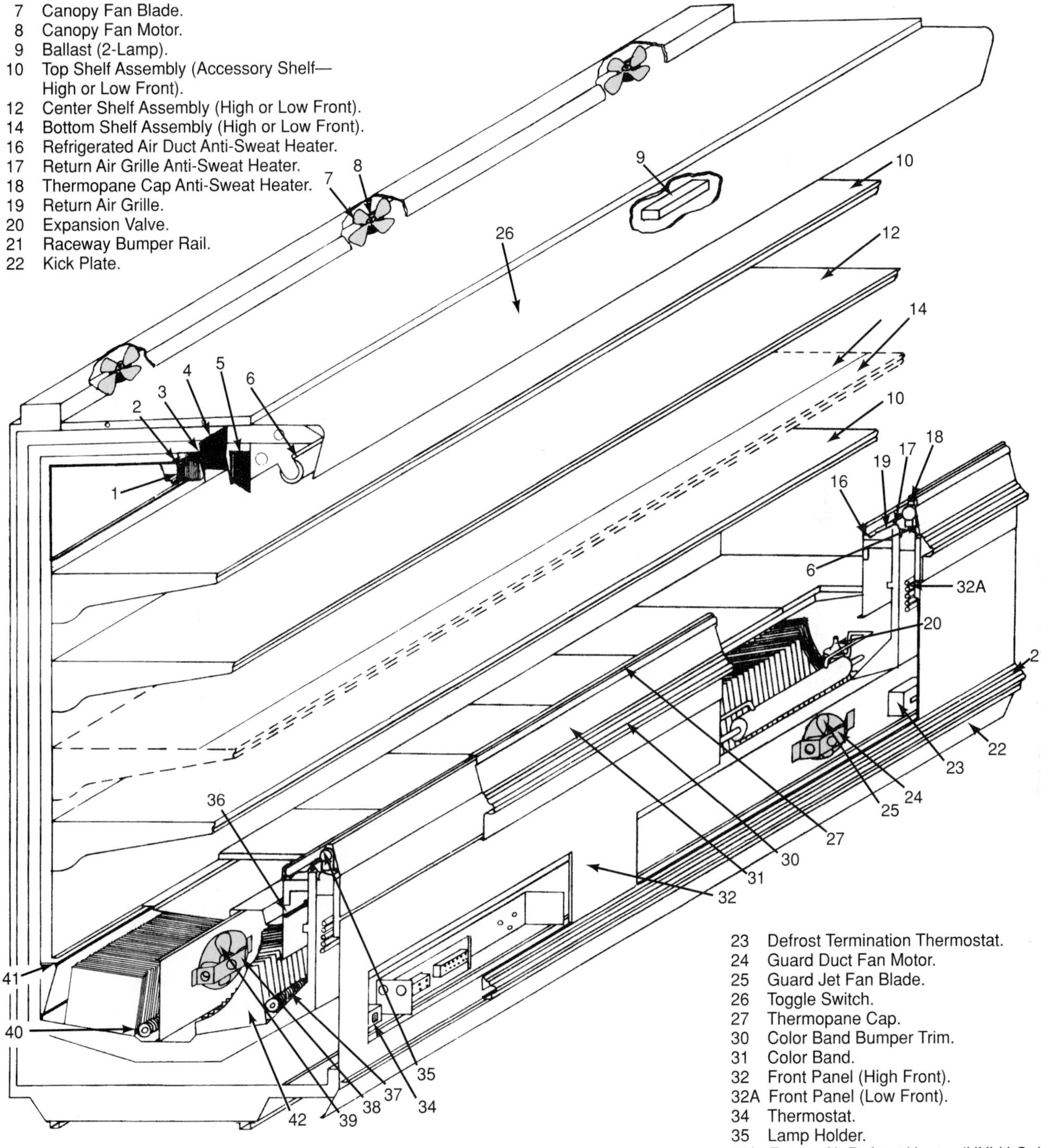

23 Defrost Termination Thermostat.
24 Guard Duct Fan Motor.
25 Guard Jet Fan Blade.
26 Toggle Switch.
27 Thermopane Cap.
30 Color Band Bumper Trim.
31 Color Band.
32 Front Panel (High Front).
32A Front Panel (Low Front).
34 Thermostat.
35 Lamp Holder.
36* Return Air Defrost Heater (HXLH Only).
37 Front Coil Defrost Heater.
38 Cold Air Duct Fan Motor.
39 Fan Blade.
40 Drain Area Defrost Heater.
41 Rear Wall Defrost Heater.
42 Evaporator.
*For HXLI
36 Dual Voltage Return Air Heater.

Figure 14-26. *Construction details of open frozen foods case. Three air curtains protect food.*

Figure 14-27. *A frozen food closed-door display case. Note the flow of the air. (Tyler Refrigeration Corporation)*

14.7 Fast-Freezing Case

Cases used for freezing foods rapidly are similar to storage cases. However, temperatures are maintained at about −20°F (−29°C). Also, food is placed as close to the freezing plates as possible. Some cases use refrigerated shelves to provide more heat-transfer surface.

14.8 Ice Cream Cabinet

Ice cream cabinets have a steel framework with a sheet metal exterior. Modern cabinets use polyurethane insulation about 3″ (8 cm) thick.

The size of the sleeve or tank holder is standard. Therefore, construction of the various makes of bulk ice cream cabinets is similar. The size ranges from one to twelve sleeves.

The bulk ice cream cabinet should be kept at approximately 0°F (−18°C). If the temperature is too cold, it is difficult to scoop out the ice cream. Also, too-low temperatures tend to crystallize the ice cream.

Dry-type evaporators are used with a capillary tube, a thermostatic expansion valve, or an automatic expansion valve refrigerant control. Some of the evaporators are tinned tubing wrapped around and soldered directly to the sleeves. Others are sheet metal with refrigerant passages formed in them. The sleeve covers are also standardized by manufacturers. (***Sleeve covers*** are tops that must be raised to get to the ice cream.) Since sleeve openings are at the top, there is no spilling of cold air when the cabinet is opened.

Some ice cream cabinets are self-contained. The refrigerating machine or condensing unit is built into one end of the cabinet. Other cabinets are made with the condensing unit separate (remote type).

Besides the chest type ice cream cabinets, upright (**Figure 14-28**) and open display types are used. Packaged ice cream should be kept at about −10°F (−23°C) in order to retain its firmness.

Figure 14-28. *Two-door upright ice cream display case. (Nor-Lake, Incorporated)*

14.9 Soda Fountain

The soda fountain provides a way to store and dispense ice cream, cold water, beverages, and syrup. It also makes and stores ice. It provides a compact, attractive unit for these purposes. The design also makes serving easy. A built-in ice cream cabinet usually occupies one part of the fountain. Another part contains a water-cooling mechanism.

Evaporator outlet tubing from the ice cream cabinet and the drinking water cooler passes around the beverage syrup containers. This keeps them relatively cool. Beverages are cooled to the same temperature as the water.

Syrups should be kept at about 45°F (7°C). Water temperature should range between 32°F and 50°F (0°C to 10°C). Ice cream, as mentioned, should be kept between 0°F and 10°F (−18°C and −12°C).

Soda fountains are often difficult for the technician to service. Since they are very compact, they leave little space in which to work.

Figure 14-29 shows the complete cycle of a soda fountain refrigerated by one condensing unit. The soda fountain has an ice cream compartment, bottle compartment, syrup rail, and beverage coolers. It uses a thermostatic expansion valve. The ice cream evaporator has a check valve in the suction line. The syrup and bottom compartment evaporator temperature is controlled by the two-temperature valve. The beverage cooler has a solenoid liquid line shutoff valve. Note the sight glass, heat exchanger, and drier mounted in refrigerant lines near the condensing unit.

Many restaurants, soda fountains, and other public areas use drink dispensers. A dispenser for ice and various beverages is pictured in **Figure 14-30. Figures 14-31** and **14-32** show the fluid circuits and system schematic diagrams for a beverage dispenser.

14.10 Dispensing Freezers

Special applications of refrigerating systems are required in soft ice cream-making machines, also known as ***dispensing freezers.*** Temperature requirements will vary depending on the product:

- Frozen carbonated beverages, 25°F (−4°C).
- Slushes, 28°F (−2°C).
- Fruit or water ices, 10°F to 20°F (−12°C to −7°C).
- Soft-serve ice creams, 21°F (−6°C).
- Milkshakes, 27°F (−3°C).
- Sherbets, 20°F (−7°C).

This unit uses a large refrigerating machine to cool or fast-freeze the mix. It is then fed to the freezing cylinder. The same or a separate motor drives the stirring mechanism or dasher. A 1/2 hp (373 W) refrigerating unit can fast-freeze one gallon of custard in about six minutes.

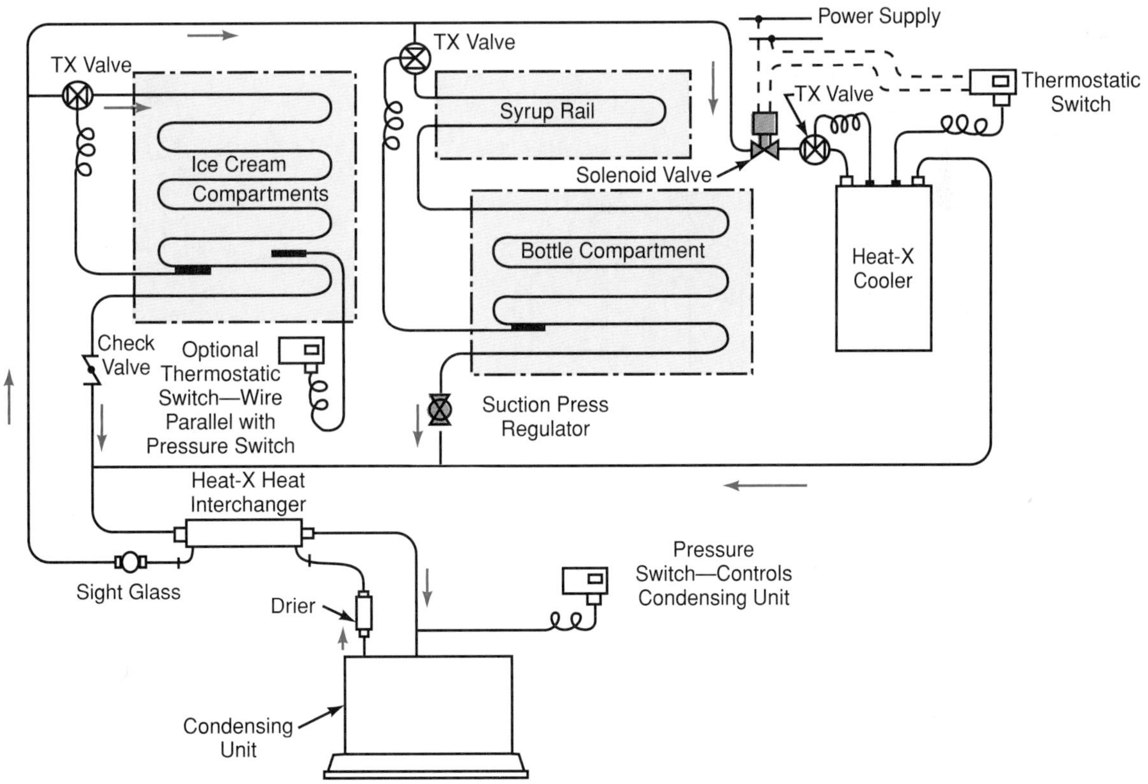

Figure 14-29. *Complete soda fountain cycle diagram. Note solenoid valve and suction pressure regulator, and how these valves control temperature in various parts of the installation. (Dunham-Bush, Inc.)*

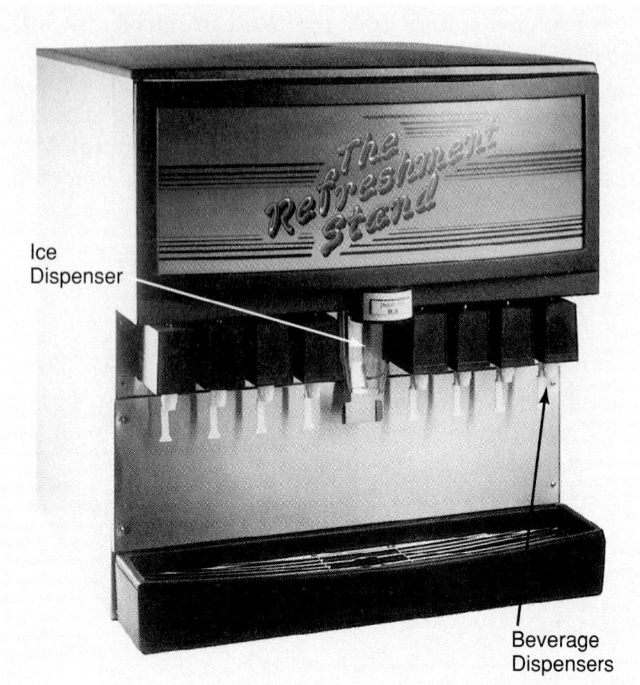

Figure 14-30. *Ice and beverage dispenser. (Scotsman Ice Systems)*

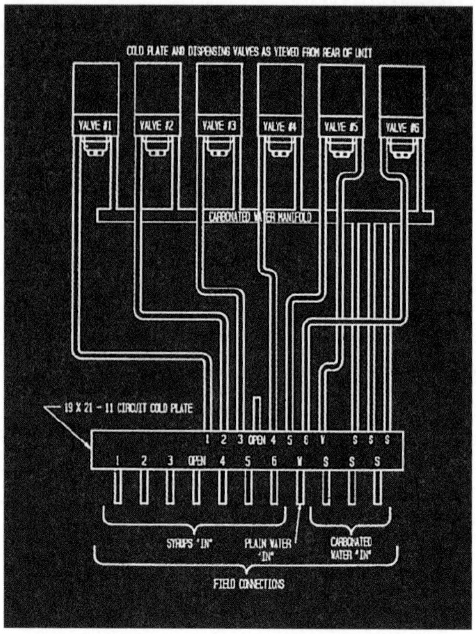

Figure 14-31. *Fluid circuits of drink dispenser. Carbonated water and syrups are mixed in electric heads. Note post-mix valve, which combines chilled syrup and carbonated water; also, separate ice dispensing system. (Scotsman Ice Systems)*

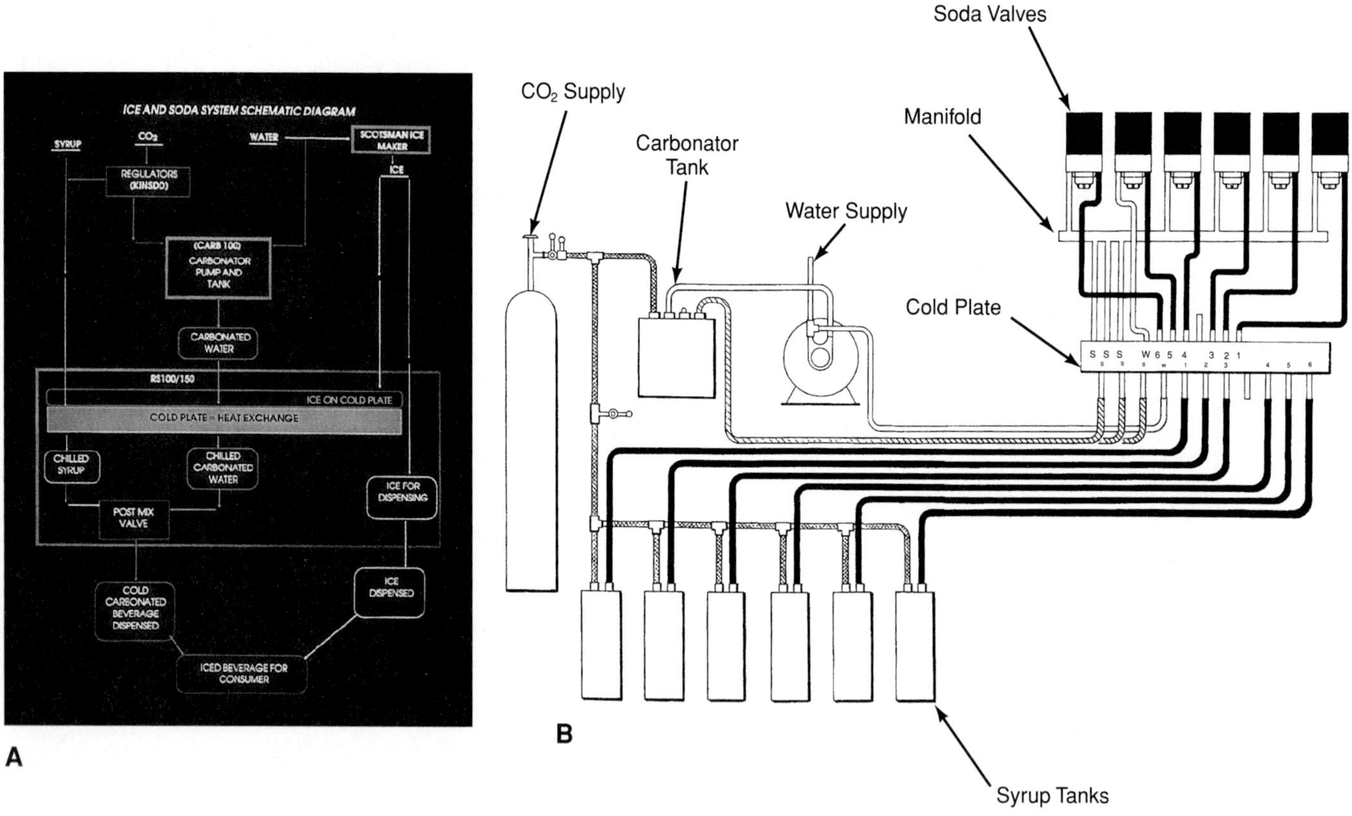

Figure 14-32. *Beverage dispenser diagrams: A—Fluid circuit diagram. Beverage syrup, plain water, and carbonated water inlets pass through the cold plate, then to the dispensing valves, where carbonated manifold is located. Ice dispensing system is separate from soda system. B—Soda system schematic. (Scotsman Ice Systems)*

A combination unit is illustrated in **Figure 14-33.** **Figure 14-34** shows the construction of a freezing dispenser cabinet.

Some units operate continuously. This is because the mix in the freezer often has to be kept within a narrow temperature range. It must be kept within 1°F or 2°F (0.5°C to 1°C) of its correct temperature. Thermostatic expansion valves are usually used as the refrigerant control.

The machine is usually adjusted to deliver the custard or sherbets at 20°F (−7°C). A total of 3 hp (2238 W) can operate the refrigerating unit and drive the dasher. A unit with a capacity of 12.5 gal. (48 L) per hour usually has a 2 hp (1492 W) dasher motor. It has a 3 hp (2238 W) refrigerating unit. The large units are water-cooled.

The quality of the mix is very important. Many problems thought to be in the refrigeration system have turned out to be poor mixes. A frozen carbonated beverage is about one part syrup and four parts carbonated water. It is served at 22°F to 26°F (−6°C to −3°C). Health codes require keeping both the mix containers and the freezing cylinder sanitary. A mix storage or supply cabinet usually has three units to store the mix until it is used.

A commercial application of the dispensing freezer is the "shake maker." This machine is filled with the desired shake mix. The temperature of the "shake maker" is controlled with the use of a thermostat. The temperature control is based on the consistency of the mix. As the mix is frozen, the torque required to drive the agitator increases. An idler on the belt drive side is connected to a microswitch. When the mix is frozen to the desired consistency, the belt tightens, moving the idler. This opens the microswitch which shuts off the machine. See diagram in **Figure 14-35.**

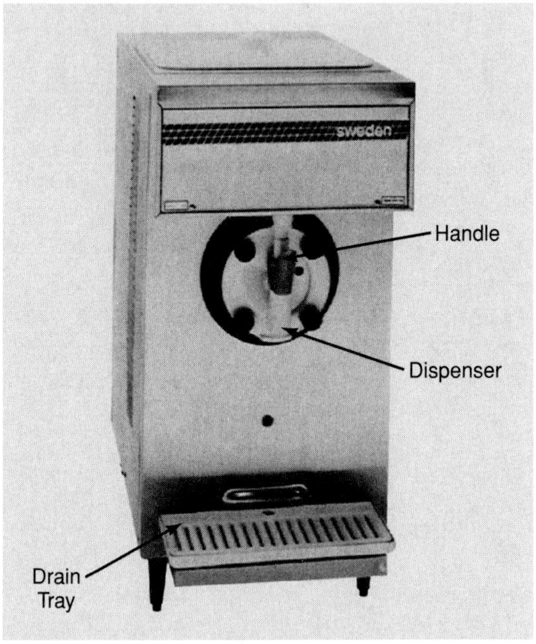

Figure 14-33. *Air-cooled soft-serve dispensing unit. Note names of its various parts. (Sweden-Alco Dispensing Systems, A Div. of Alco Foodservice Equipment Co.)*

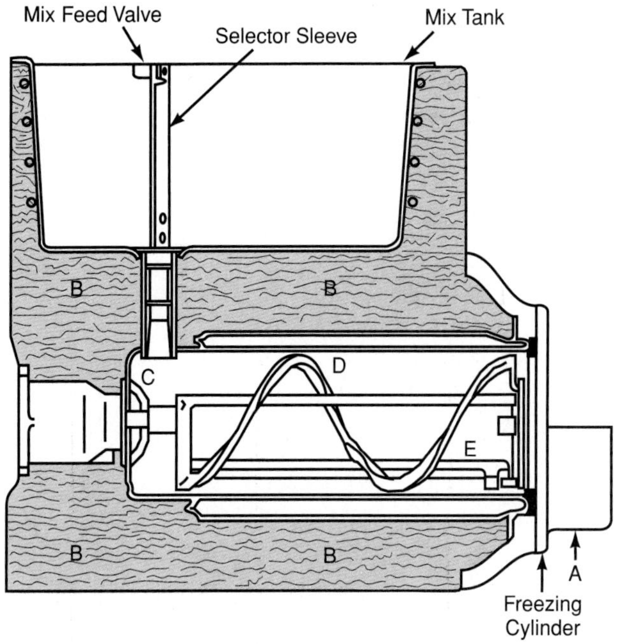

Figure 14-34. *A dispensing freezer. A—Dispensing opening. B—Heavy insulation around freezing cylinder. C—Rear product seal. D—Dasher. E—Scraper blade. (Sweden-Alco Dispensing Systems, A Div. of Alco Foodservice Equipment Co.)*

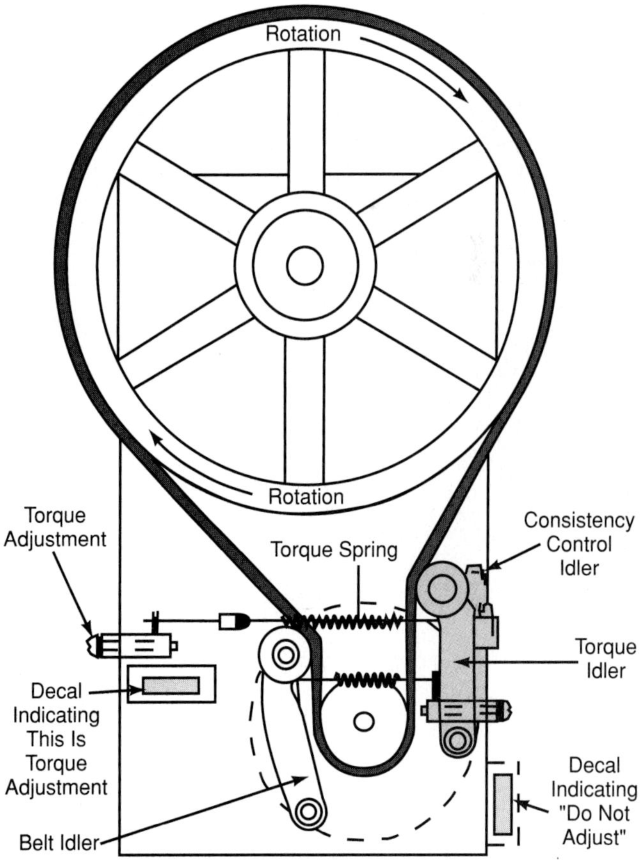

Figure 14-35. *Shake-maker consistency control. As mix hardens, tight side of belt moves consistency control idler and finally opens motor circuit. (Sani-Serv)*

14.11 Modular Refrigeration Systems

Many party/specialty stores use flexible refrigeration systems, **Figure 14-36.** The glass door storage components can be used in numerous combinations with the refrigeration units. The refrigeration unit shown cools up to four storage units. The system has forced air circulation, automatic defrost, adjustable temperature control, and automatic condensate evaporator. **Figure 14-37** illustrates an under-counter refrigeration unit used in delicatessens and small restaurants.

Figure 14-36. *Modular refrigeration system with cooler unit between cabinets. (Stevens Lee Company, Silver King Division)*

Figure 14-37. *Under-counter refrigeration unit. (Stevens Lee Company, Silver King Division)*

14.12 Water Cooler

The water cooler cabinet usually has a sheet metal housing attached to a steel framework. Inside this sheet metal housing there is usually a condensing unit located near the floor. Above it is the water-cooling mechanism. The latter is the only part insulated with foamed plastic. The insulation usually is specially formed. It is between 1″ and 2″ (4 cm to 5 cm) thick. This cabinet provides easy removal of one or more sides to access the interior. The water cooler basin is generally porcelain-coated cast iron, porcelain-coated steel, or stainless steel.

Some water coolers also have a heater providing hot water. Some also have a refrigerated compartment for storing milk, soft drinks, or food.

Heat exchangers are frequently used on water coolers. Some make use of the low temperature of the waste water to precool the fresh water line to the evaporator. The temperature of the cooled water is usually maintained at 50°F (10°C). The thermostat controls the temperature of the cooled water.

A water cooler that has a supply and drain connection must be installed according to the National Plumbing Code and local codes. The plumbing should be concealed. A hand shutoff valve should be installed in the fresh water line. A drain pipe, at least 1 1/4″ (3 cm) in diameter, should be provided. The bubbler opening must be above the drain. This eliminates accidental siphoning of the drain water back into the fresh water system.

The tap water model uses a variety of evaporator designs. **Figure 14-38** shows a combination evaporator water-cooling tank. **Figure 14-39** is the wiring diagram for a water cooler having a water heating service.

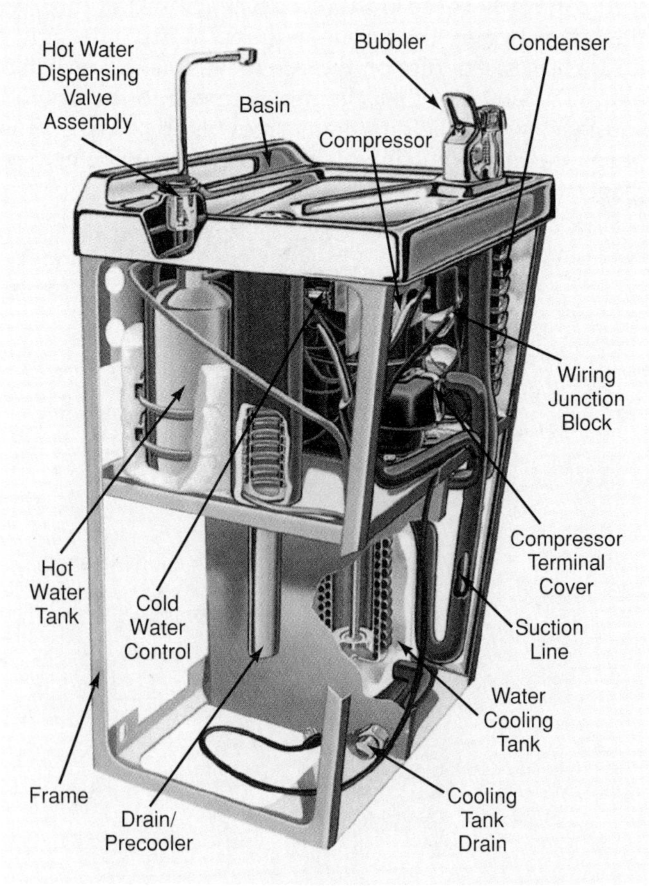

Figure 14-38. *A self-contained water cooler. Note combination hot water and cooler. (Ebco Manufacturing Company)*

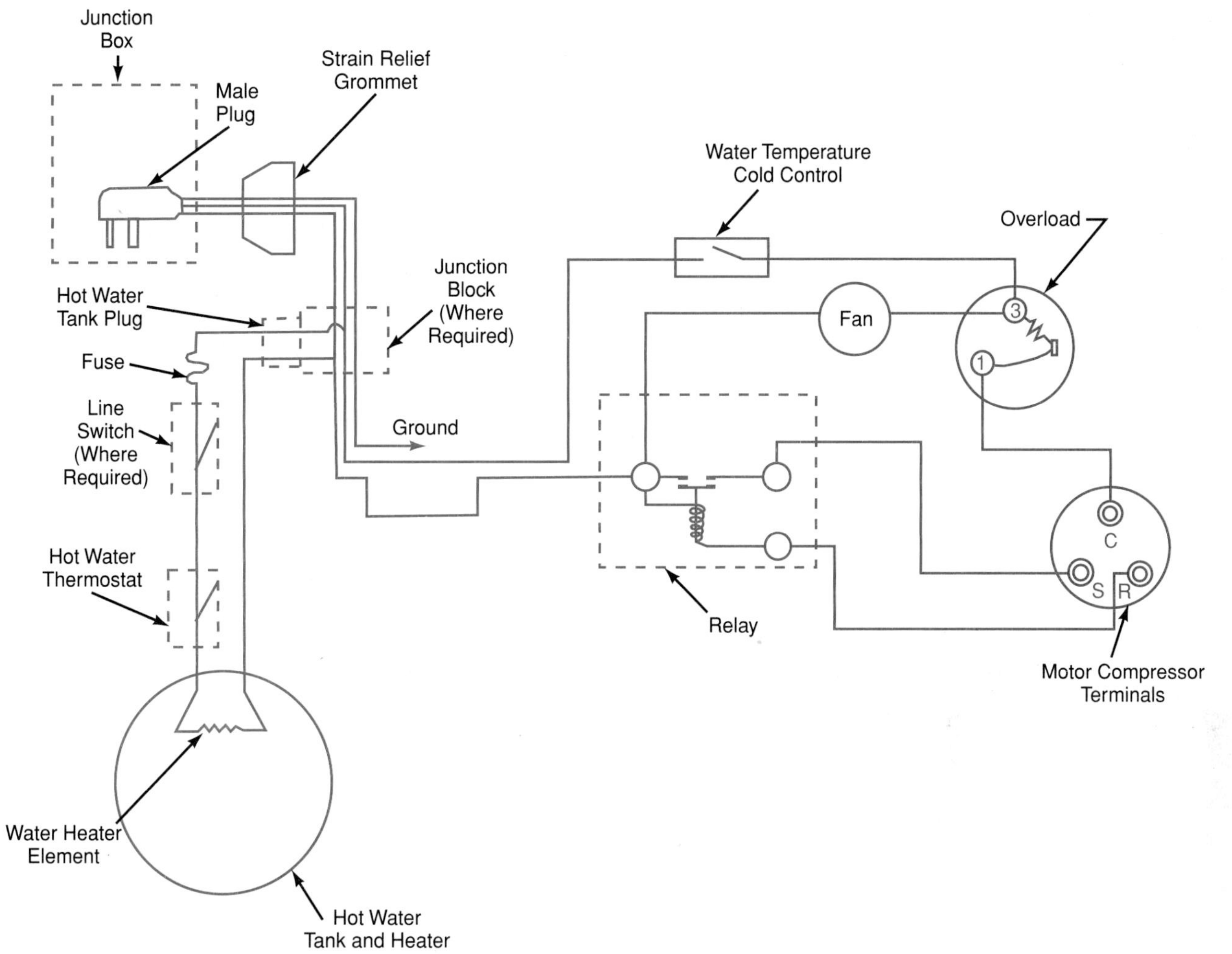

Figure 14-39. *Wiring diagram for water cooler which also provides hot water. The hot-water system, is located on the left side, the cool-water system is on the right. (Elkay Mfg. Co.)*

Water temperatures are set by a person checking a cooler. In a heat-treating area of a factory, cooled water should be 50°F to 55°F (10°C to 13°C). For offices, the temperature should be 50°F (10°C). **Figure 16-47** lists temperatures.

Multiple water coolers, instead of individual ones, are popular for certain applications. Bubbler construction may vary. These coolers have one large condensing unit supplying refrigeration to many "bubblers" (water fountains). They are often used for large business establishments, office buildings, or factories.

Some water coolers use a self-contained water supply, **Figure 14-40.** Large glass containers of water, delivered on a regular basis, are used for the supply. The cooling system is similar to that used in other water cooler models.

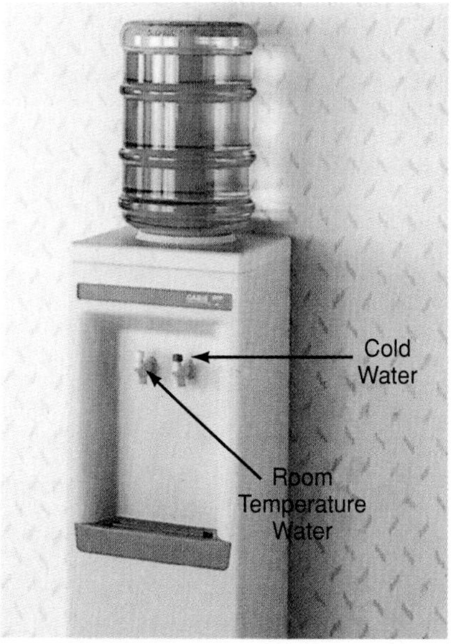

Figure 14-40. *Water cooler with self-contained water supply. One outlet is for chilled water, the second is for room-temperature water. On some units, the second outlet provides heated water. (Ebco Mfg. Co.)*

14.13 Automatic Ice Maker

Automatic ice makers are now widely used commercially. Several different types are available. They automatically control water feed and freeze the water. They empty the ice into storage bins and shut down when the storage space is full. The ice formed is clear and sanitary, since only flowing water is used. Cloudy ice cubes are caused by entrapped air.

Floats and solenoids control water flow. Switches operate the storing action when ice is made. The ice is removed from the freezing surfaces in various ways. These include use of electrical heating elements, hot water, hot gas defrosting, or mechanical devices.

Urethane foam, polystyrene, or fiberglass may be used for cabinet insulation. Freezing surface and bin storage basin are made of stainless steel. Some are self-contained, while others use remote condensing units. Both supply and drain plumbing are needed.

Capacity can vary among units from a few pounds up to many tons per day. Capacities, of course, decrease as water temperature and/or ambient air temperature increase. Ice cube maker water circuits and ice-freezing parts should be cleaned yearly.

A typical automatic ice cube maker is shown in **Figure 14-41.** Another unit, **Figure 14-42,** automatically makes chipped or flaked ice. Each of these units produces approximately 900 lb. (408 kg) of ice a day. **Figure 14-43** shows a combination nugget ice and tap water dispensing unit used in restaurants.

Figure 14-42. *Flake or chipped ice unit. It has a 1/2 hp (373 W) air-cooled condensing unit. It will store 255 lb. (115 kg) of ice and produce 442 lb. (199 kg) of ice in 24 hours. (Scotsman Ice Systems)*

14.14 Vending Machines

Refrigerated vending machines for foods or beverages are becoming increasingly popular. They automati-

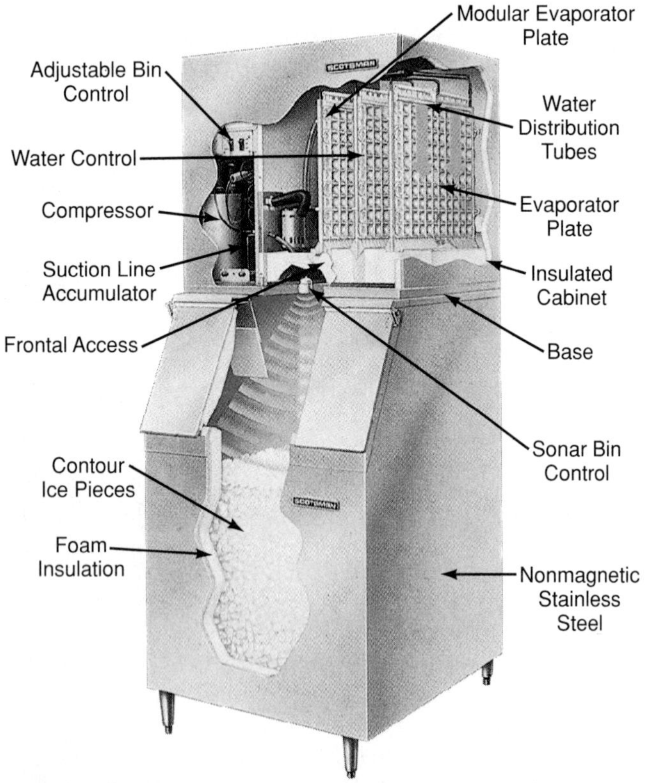

Figure 14-41. *Automatic ice maker. (Scotsman Ice Systems)*

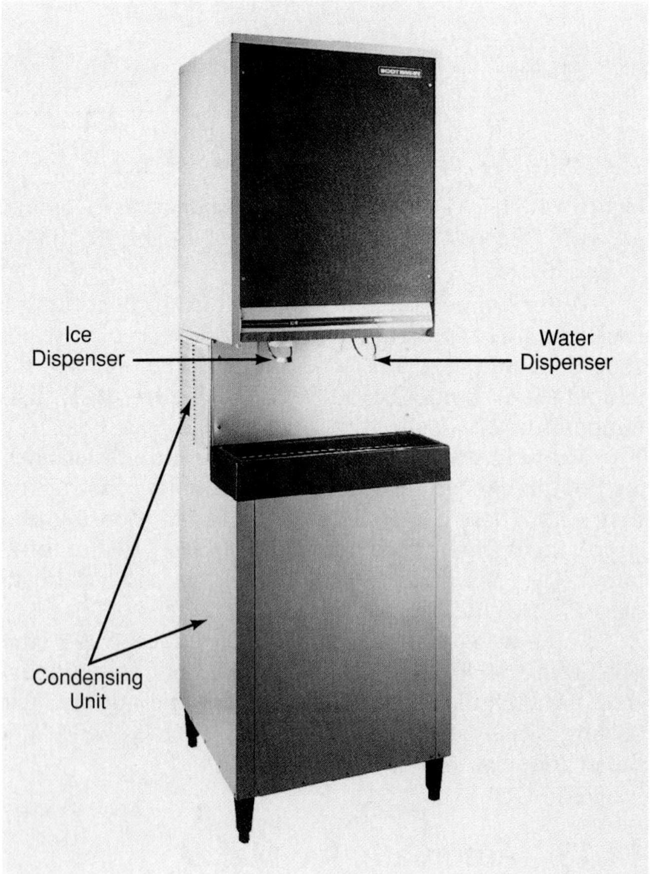

Figure 14-43. *A unit that dispenses nugget ice and tap water. The technician will need to be familiar with the condensing unit as well as the ice and water dispensers. (Scotsman Ice Systems)*

cally dispense cold drinks with ice and canned or bottled soft drinks. They are also used to dispense ice cream, cold food, cold milk, and frozen desserts. See **Figure 14-44.**

Refrigerating systems employed in these units are typical. Most use capillary refrigerant controls, hermetic motor compressors, and defrosting devices. The electrical system generally is part of the overall system. It electrically transfers the material and operates the coin and currency devices.

Vending machines have many components necessary for dispensing:

- Coin and currency devices (acceptors, rejectors, changers, and steppers and accumulators).
- Carbon dioxide systems.
- Cup dispensers.
- Heating systems.
- Refrigerating systems that are usually air-cooled using fan evaporators. Some have defrost systems.
- Transfer systems.

14.15 Milk Cooler

Milk must be cooled within an hour after being taken from the cow. By law, it must be cooled to 50°F (10°C), then stored at 40°F (4°C) or lower. Bulk-type coolers will handle this specific cooling requirement.

Bacterial growth in milk is dramatically affected by temperature. During a 24-hour period, bacteria count will increase as follows:

- To 2400 at 32°F (0°C).
- To 2500 at 39°F (4°C).
- To 3100 at 46°F (8°C).
- To 11,600 at 50°F (10°C).
- To 180,000 at 60°F (16°C).
- To 1,400,000,000 at 86°F (30°C).

The stainless steel bulk cooler shown in **Figure 14-45** has an evaporator in the base. Coolers of this type have a 450 gal. to 6000 gal. (1703 L to 22 712 L) capacity. R-22 is the refrigerant used in this type of system.

The condensing unit is mounted outside the milk room, and is usually air-cooled. It should not be put in the vacuum pump room. Air flowing over the condenser can be drawn from the milk room, however. In winter, where permitted, this warm air can be ducted to heat the milk room.

Coolers should always be properly grounded. Use a separate safety ground wire the same size as the power wires or larger. These units also typically have a crankcase heater.

14.15.1 Milk Dispensers

Many food service places dispense milk from bulk containers. Cans or plastic bags holding 3 gal. to 5 gal. (11 L to 19 L) of milk are installed in dispensers. These units meet all health and sanitation codes. Milk is kept at about 36°F (2°C) by a hermetic refrigerating system. Reserve milk containers are kept in a walk-in cooler or in a milk storage refrigerator.

14.16 Bakeries

Many raw products used by bakeries must be refrigerated to preserve or improve quality. Bakeries must refrigerate:

- Perishable raw materials.
- Perishables during interrupted processing.
- Finished products.

Frozen ingredients must, of course, be stored. Even water and flour used for breadmaking require cooling during certain periods of the year. Health codes require these practices. Creams and custards can be kept longer at a cool temperature. **Figure 14-46** shows a reach-in cabinet designed for baking use.

Both normal and low-temperature refrigeration is used. Normal refrigeration is suitable for butter, eggs, coconut, cream, fat, meat, margarine, nuts, and yeast, and for dough retardation. Low-temperature systems are needed to freeze baked goods that are sold frozen. For example, bread that is fast-frozen to −1°F (−18°C) will remain fresh for almost a month.

Figure 14-44. *Coin-operated vending machine for refrigerated can and bottle beverages. Cabinet doors are open to show storage compartment, controls, and refrigeration mechanisms.*

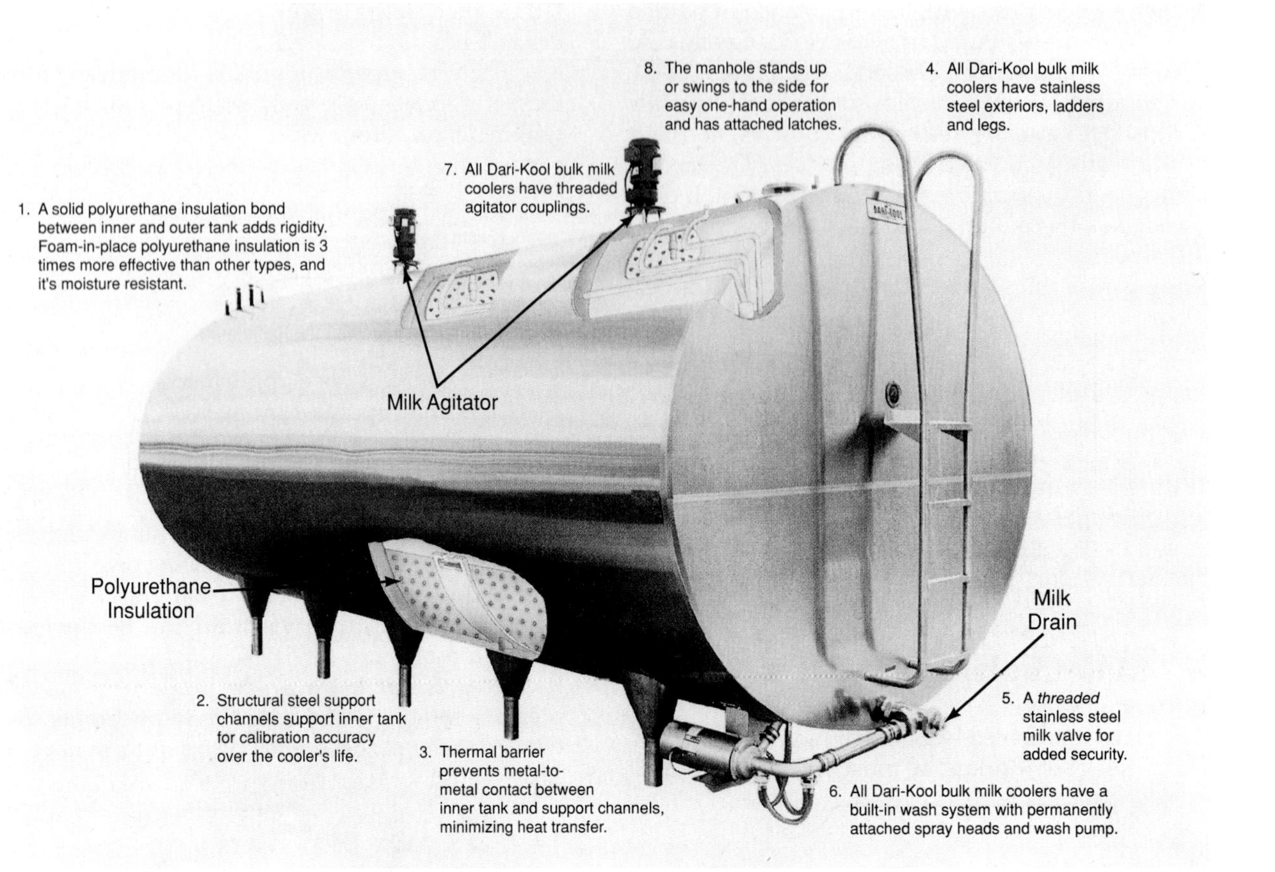

8. The manhole stands up or swings to the side for easy one-hand operation and has attached latches.
4. All Dari-Kool bulk milk coolers have stainless steel exteriors, ladders and legs.
7. All Dari-Kool bulk milk coolers have threaded agitator couplings.
1. A solid polyurethane insulation bond between inner and outer tank adds rigidity. Foam-in-place polyurethane insulation is 3 times more effective than other types, and it's moisture resistant.
Milk Agitator
Polyurethane Insulation
Milk Drain
2. Structural steel support channels support inner tank for calibration accuracy over the cooler's life.
3. Thermal barrier prevents metal-to-metal contact between inner tank and support channels, minimizing heat transfer.
5. A threaded stainless steel milk valve for added security.
6. All Dari-Kool bulk milk coolers have a built-in wash system with permanently attached spray heads and wash pump.

Figure 14-45. *Bulk milk cooler. Direct expansion evaporator is located in the base of the unit. (Dairy Equipment Co.)*

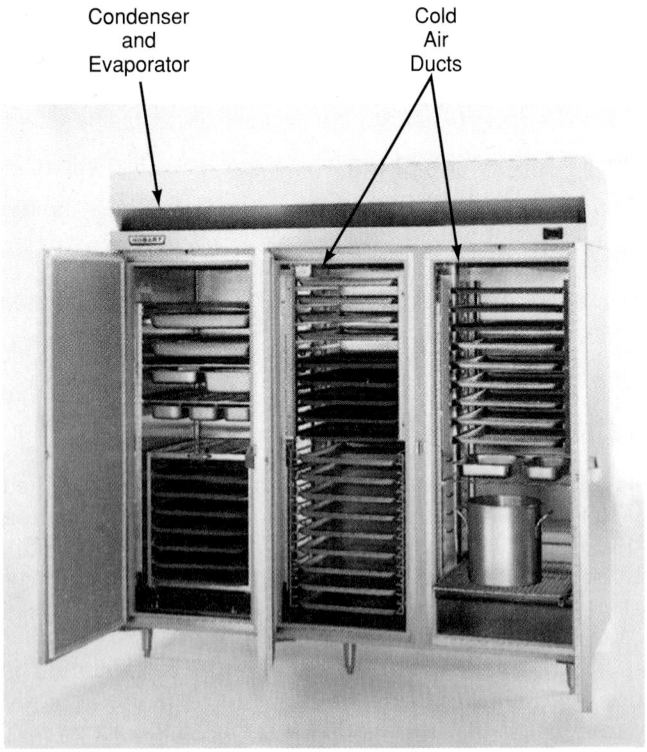

Figure 14-46. *This reach-in cabinet is designed for bakery use. It has a top-mounted condensing unit and blower evaporator. (Hobart Corp.)*

Controlled temperature and humidity is important in many baking processes. Therefore, air conditioning using refrigerating equipment is also found in bakeries.

14.17 Refrigerant-to-Water Heat Recovery System

A heat recovery system produces and stores hot water. It does this by transferring the heat from the condenser to cold water. See **Figure 14-47.** This type of unit is adaptable to most refrigeration systems. The tanks are constructed of vertical plates. These plates allow the refrigerant to flow through them. A rapid transmission of heat to the water is thereby provided. This system is often used as a supplement to existing hot water systems. Examples of such use would be in supermarkets, restaurants, and other commercial applications.

14.18 Laboratory Refrigerated Incubators

An interesting application of refrigeration helps maintain correct temperatures in incubators. **Figure 14-48** shows an incubator that can maintain constant

Figure 14-47. *A two-unit refrigeration system, shown at left, is connected to a water heat recovery system at right. (Dairy Equipment Co.)*

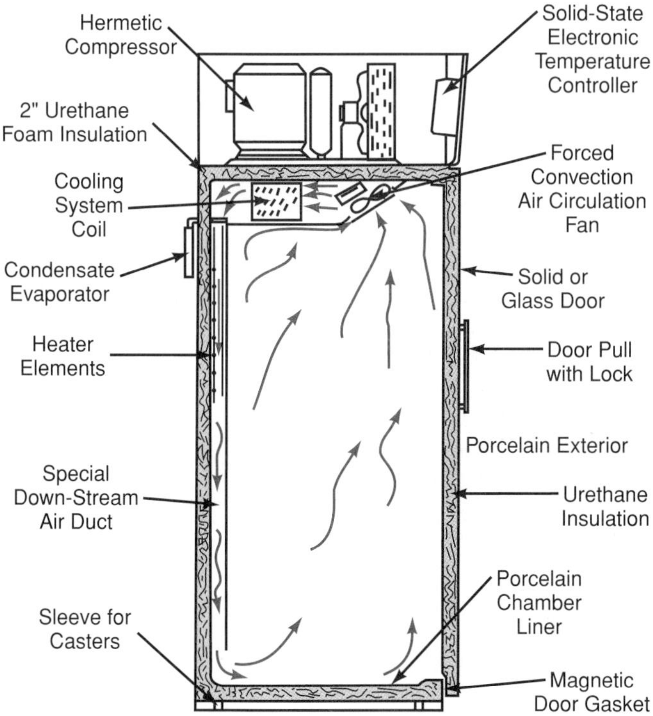

Figure 14-48. *Incubator has hermetic refrigerating system and electric heat. Temperatures are accurate within 1°F (0.5°C). (Rheem Mfg. Co., Scientific Products Div.)*

temperatures. A refrigerating system is used for cooling, and electric elements for heating. Such a unit can maintain any constant temperature between 36°F (2°C) and 158°F (70°C).

14.19 Industrial Applications

Refrigeration has a variety of applications in manufacturing processes. Smaller units are automatic. Three common uses are:

- Cooling of water which, in turn, cools electrodes on resistance welders.
- Cooling of quenching liquids used to cool metals in heat-treating applications.
- Cooling compressed air.

Since moisture may rust and corrode air tools or spoil paint spraying, compressed air must be *dry*. Air, therefore, is cooled to keep its dew point above the coldest spot in the air system. (The dew point is the water condensing temperature.) This prevents moisture in the lines.

A refrigerating system cools air to bring it below its ***critical pressure dew point.*** Then it is reheated. Pressure dew points are about 50°F (28°C) above atmospheric dew points at 100 psi (700 kPa) air pressure. Generally, the compressed air is cooled to about 35°F to 50°F (2°C to 10°C). A 3000 cfm (1.42 m^3/s) air compressor will need about a 20 ton (55.2 kW) capacity refrigerating machine.

A refrigerating system located near flammable or explosive materials must be explosion-proof. Sealed lights and sealed contact points are required.

Refrigerators and freezers of all types are used in research. They are used to maintain constant temperature, constant humidity, and low-temperature control. Low-temperature units capable of maintaining −140°F (−96°C) are available. These systems usually use 5″ (13 cm) of insulation and a cascade of systems. (A ***cascade*** is two systems connected in a series.)

14.19.1 Industrial Freezing of Foods

Industrial freezing of food is performed in two principal types of establishments:

- Processing plants.
- Locker plants.

Processors of frozen foods have freezing centers in many large food-producing areas. For example, fish is packed and frozen along a seacoast and then shipped to all parts of the country.

A ***locker plant*** is a smaller unit designed to prepare, freeze and store various products. Refrigerating equipment in processing and locker plants varies considerably. However, the plan for freezing the food is similar.

Figure 14-49 shows the flow of produce through a typical plant. The food is weighed and checked for purity and suitability for freezing. Then, it moves to the processing room. There, meats are cut, fowl cleaned and dressed, vegetables blanched, and the various items packaged. The packed foods are next sent to the freezing section where they are completely frozen and readied for storage.

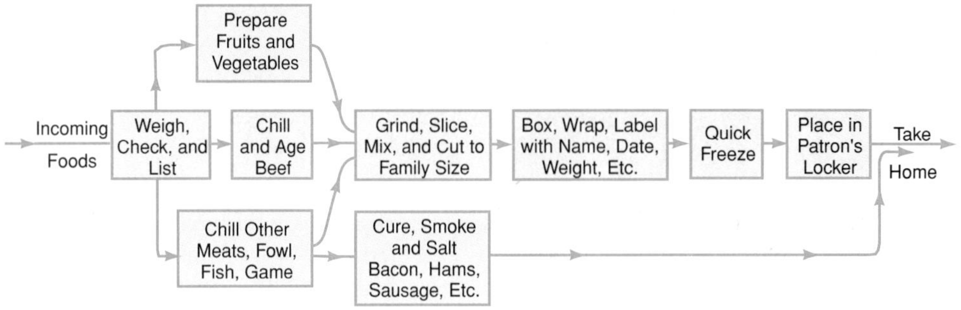

Figure 14-49. *Flow chart of food moving through preparation process in a freezing plant.*

High humidity is very important in rooms where food is cured and stored. Meat tastes better and keeps its weight if relative humidity is kept close to 100%. Temperature should be near 39°F (4°C). This gives the best humidity results with a nonfrosting evaporator.

Processing plants freeze food rapidly—they expose as much of the food as possible to the lowest possible temperature. A fast-freezing system usually uses a track to hold the produce as it moves through an ultra-low-temperature chamber for fast freezing. See **Figure 14-50.**

Some fast food-freezing systems use liquid nitrogen or carbon dioxide. This turns perishable fresh food into long-lasting frozen food. This process is commonly referred to as ***cryogenic food freezing.*** Temperatures of −320°F (−196°C) are obtained; freezing is instantaneous. This method of quick freezing causes little or no damage to the food. See Chapter 18.

14.19.2 Freeze Drying

Freeze drying consists of the following steps:

1. Food is frozen.
2. The frozen food is put into a vacuum chamber.
3. The food is then heated to change its frozen water (ice) directly into a vapor. It does not go through the liquid state. (Below 4.7 mm Hg, water changes state from a solid to a gas without passing through a liquid state.) Freeze drying is usually done at a pressure of between 50 microns and 500 microns.
4. The vapor is condensed as frost on a cold evaporator. This occurs at about −40°F (−40°C). The above step saves pumping about 100,000 ft^3 (2832 m^3) of water vapor. (This would be the amount needed to form 1 lb. or 0.5 kg of vapor at 50 microns.)

After the product has dried, the evaporator is exposed to heat at atmospheric pressure. The frost melts and is removed. Dry nitrogen is used to raise the

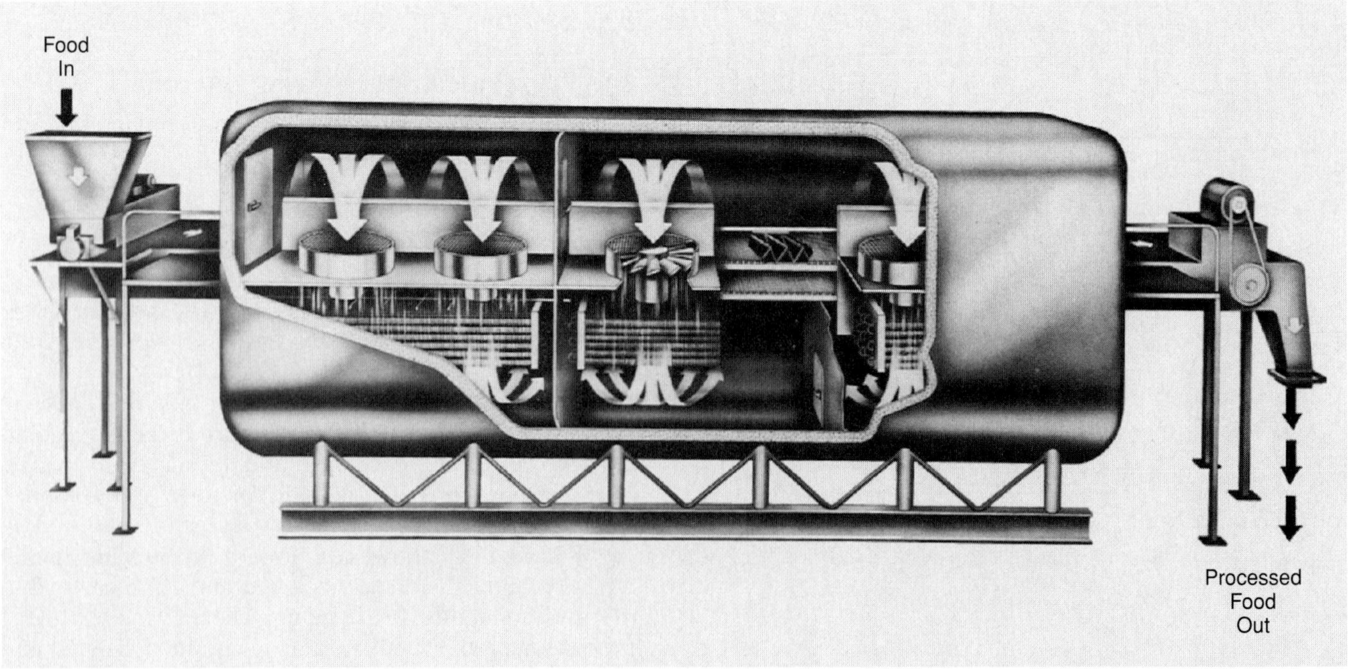

Figure 14-50. *Automatic track food freezer. Air in the chamber is about −30°F (−34°C). Food moves from left to right on the conveyor belt.*

chamber pressure from 50 microns to atmospheric pressure.

14.19.3 Industrial Storage of Frozen Foods

Storage requirements for most frozen foods are about the same. A temperature of 0°F to −20°F (−18°C to −29°C) is desirable, with a variation of 2°F to 3°F (1°C to 2°C) normally allowed. A high temperature differential causes harmful "breathing" and volume changes. Humidity in storage rooms should be as high as possible. Storage areas may be cooled by blower evaporators, direct contact plates, or brine coils.

In smaller plants, frozen food is stored in lockable drawers in the refrigerated area. The customer enters the area and removes the frozen food from the drawer. Other locker plants are constructed so that frozen foods are delivered from a storage area. The customer does not need to enter the cold area.

14.20 Review of Safety

Always check a unit before working on it. This is the best means of protecting yourself and the public. Be sure the mechanism and cabinet are installed according to local and national codes. These include fire, electrical, plumbing, building, and safety codes.

Remember, any code violations must be reported to the appropriate authorities. Such violations might result in serious damage to either the physical structure or to personnel.

Move cabinets on rollers. Avoid pinching hands or allowing the cabinets to fall on your feet. Wear steel-toed safety shoes.

Use extreme caution when moving large commercial cabinets. Use handling equipment appropriate for the job and always work carefully. Floor load limits should be checked before heavy commercial equipment is installed. Wiring and plumbing must be adequate, securely mounted, and made of the correct materials.

Health Department regulations apply to all equipment which is used around food and beverages. These regulations must be carefully followed.

14.21 Test Your Knowledge

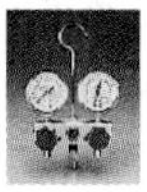

Please do not write in this text. Place your answers on a separate sheet of paper.

1. What method is frequently used in frozen food cases to prevent sweating?
   A. Small electrical heaters.
   B. Maintaining temperature difference of 10°F (5.5°C) from desired temperature.
   C. An air curtain.
   D. Any of the above.
2. What are the recommended temperatures for a florist cabinet?
   A. 38°F and 40°F (3°C and 4°C).
   B. 35°F and 40°F (2°C and 4°C).
   C. 38°F and 35°F (3°C and 2°C).
   D. None of the above.
3. Why are display case lights sometimes located outside the case?
   A. For decorative display.
   B. So that heat generated by the lights will not increase the operational load.
   C. To prevent an increase in humidity.
   D. None of the above.
4. How many different air streams are common in a frozen foods case?
   A. One comfort zone curtain.
   B. Two comfort zone curtains and a guard curtain.
   C. Three: a comfort zone curtain, a guard curtain, and a freeze curtain.
   D. Four comfort zone curtains, a guard curtain, a freeze curtain, and a lower curtain.
5. What causes cloudy ice cubes?
   A. Entrapped air.
   B. Moisture.
   C. Rapid freezing.
   D. Slow freezing.
6. What is the proper temperature for a water cooler in a heat-treating room of a factory?
   A. 50°F to 55°F (10°C to 13°C).
   B. 55°F to 60°F (13°C to 17°C).
   C. 50°F to 60°F (10°C to 17°C).
   D. 60°F to 65°F (17°C to 18°C).
7. What method is frequently used to prevent bacteria and mold growth in walk-in cabinets?
   A. Ultraviolet lamps.
   B. Lower temperatures.
   C. Variable temperatures.
   D. Change in humidity.
8. In a produce storage cabinet, how should the humidity be kept?
   A. High.
   B. Moderate.
   C. Low.
   D. Same humidity at the store.
9. What is the proper temperature for a water cooler in an office building?
   A. 50°F (10°C).
   B. 55°F (13°C).
   C. 60°F (17°C).
   D. 65°F (18°C).
10. At what temperature should packaged ice cream be kept?
    A. −10°F (−23°C).
    B. 0°F (−18°C).
    C. 20°F (−7°C).
    D. 25°F (−4°C).

11. What is the relative humidity necessary for fresh foods?
    A. 65°F (18°C).
    B. 70°F (21°C).
    C. 75°F (24°C).
    D. 80°F (32°C).
12. What is the purpose of an activated carbon air filter?
    A. To remove odor.
    B. To reduce mold growth.
    C. To prevent contamination.
    D. All of the above.
13. Which temperature requirement for producing the specified item is *incorrect*?
    A. Slushes: 30°F (−1°C).
    B. Soft serve: 21°F (−6°C).
    C. Milkshakes: 27°F (−3°C).
    D. Fruit or water ices: 10°F to 20°F (−12°C to −7°C).
14. What are the two basic types of ice makers?
    A. Flake and cube.
    B. Chip and flake.
    C. Square cube and round cube.
    D. Crushed and block.
15. What are the recommended temperatures for a walk-in cooler containing meat or fresh produce?
    A. 35°F to 40°F (2°C to 4°C).
    B. 38°F to 40°F (3°C to 4°C).
    C. −10°F to 0°F (−23°C to −18°C).
    D. 32°F to 40°F (0°C to 4°C).
16. Why do refrigerated cabinets have dual-pane windows?
    A. To provide insulation.
    B. To prevent condensation.
    C. To improve the visibility of contents.
    D. All of the above.
17. What types of codes must be followed when installing a water cooler that has supply and drain connections?
    A. National Plumbing Code.
    B. Local plumbing code.
    C. Neither A nor B.
    D. Both A and B.
18. Why do display cabinets using temperatures below the dew point of the room require heaters?
    A. To prevent moisture between glass panes.
    B. To prevent overheating.
    C. To prevent overcooling.
    D. All of the above.
19. Where is the condensing unit usually located in a total unit?
    A. Outside the cabinet.
    B. In the base of the cabinet.
    C. On top of the cabinet.
    D. Both B and C.
20. Where are multiple condensing units usually located?
    A. In the back of the store.
    B. In the servicing area.
    C. In a separate area.
    D. All of the above.

Chapter 15

SERVICING AND INSTALLING COMMERCIAL SYSTEMS

Modules:

Key Words:

dry evaporator
gauge manifold
motor burnout
scale deposits
service valve
torque stand
water hammer

Learning Objectives:

After studying this chapter, you will be able to:

- List and describe the types of commercial installations.
- Explain the difference between noncode and code installations.
- Demonstrate troubleshooting techniques.
- Explain the proper procedures used in replacing or repairing defective commercial components.
- Demonstrate procedures for evacuating and recharging a refrigerating system.
- Write service estimates.
- Remove, test, repair, or replace various compressors.
- **Practice approved safety procedures.**

Installing and servicing commercial units is a very important part of the refrigeration industry. If equipment fails, companies may suffer severe losses.

INSTALLING COMMERCIAL SYSTEMS MODULE

15.1 Types of Commercial Installations

Commercial refrigeration installations vary considerably. Following are the classifications of installations commonly used:

- Self-contained unit:
  A. Hermetic.
  B. Conventional.
- Remote condensing unit:
  C. Single cabinet.
  D. Multiple cabinet.

An installation may be as small as a 1/20 hp self-contained unit. It may be as large as a 140 hp unit. Larger units must be assembled on the premises.

Multiple units must be installed to handle the refrigeration load efficiently. At the same time, it must eliminate hazardous conditions that might lead to accidents. Two major concerns should be durability and neatness. Many cities have laws and codes covering certain refrigeration systems. Furthermore, most refrigeration manufacturing companies have rules for installing their equipment.

Some cities and rural communities are not restricted by code. In such areas, installations are usually made at minimal cost. One must, therefore, consider two types of installations:

- Code.
- Noncode.

Where there is no local code, the code of the nearest city should be followed.

15.1.1 Code Installation

Most localities have definite codes, or rules, governing the installation of refrigerating equipment. Domestic systems and some other small-capacity, self-contained systems are usually not included. This is because these units use only small quantities of refrigerant.

In certain commercial installations, there are requirements to ensure uniform performance and safe installations. These include units assembled on the premises and units with horsepower or refrigerant needs that exceed certain limits. Codes also protect the purchaser from careless installations.

The following are some points found in most codes:

- Only licensed refrigeration contractors may install commercial equipment.
- A permit must be obtained for each installation.
- Each installation must be inspected by civic authorities.
- Lines must be labeled to identify the refrigerant used.
- Certain safety devices must be installed in the system.
- The condensing unit must be installed in a safe place.
- Electrical and plumbing work must conform to code and be done by licensed electricians and plumbers.
- Systems must be tested under pressure on both high side and low side. All systems must be free of leaks.

The codes are based on experiences with many installations. Codes provide safety for installer, owner, user, and public. They should be carefully followed.

15.1.2 Noncode Installations

Definite procedures should be followed when assembling a system. This safeguards against mistakes and reduces careless errors. Refrigeration service departments claim that carelessness causes more than 90% of the servicing difficulties.

Assume the units are of a size that will efficiently handle the heat load. The installation must then take full advantage of the design. Tubing, safety valves, and protective devices should be placed into the system. They will produce operating efficiency, permanency, and safety.

All refrigeration installation and service work should be performed with correct tools. **Figure 15-1** shows some that are commonly used. Chapters 2 and 12 describe in more detail such tools and supplies.

15.2 Installing Condensing Units

You must first determine where to place the condensing unit. This location should be as close to the cabinets as possible. A central location is best.

Installation should progress in the following order:

1. Put cabinets in place.
2. Locate place for condensing unit and install it.
3. Install evaporators.
4. Install valves and controls.
5. Install tubing.
6. Check for leaks.
7. Evacuate system per EPA regulations.
8. Charge system.
9. Start system.
10. Check operation of unit and get 24-hour temperature and pressure records of unit in operation.

Cabinets are bulky and sometimes difficult to handle. A dolly, as shown in **Figure 15-2,** is handy for cabinet moving.

It is recommended that the condensing unit be put in the basement. It may also be placed in a room next to the room containing the cabinet. Avoid putting the condensing unit where it will be exposed to the sun or to low or freezing temperatures. Locating condensing units in the same room with the counters and cabinets is not recommended because of the heat and noise they produce.

There will be some vibration produced by the running condensing unit. **Figure 15-3** illustrates a vibration-absorbing compressor mounting. Protect the condenser by putting a heavy wire cage around it. Install a valve and accessory board on the wall just above the condensing unit. This will support valves, drier motor controls, two-temperature valves, electrical boxes, and service instruction cards.

In some cases, the condenser is installed away from the compressor. In these instances, traps are used in the compressor discharge line. The traps keep the oil in the condenser and away from the compressor discharge valve. Discharge lines should also be slanted down.

Most codes require the condensing unit to be placed where it cannot be damaged. Some 1 hp to 5 hp condensing units can be converted easily from indoor to outdoor installations. These units can be used on most coolers and freezers regardless of their location. These new units are designed for use with R-404A, which is a replacement for R-502. These units are approved by the Underwriters Laboratories test for outdoor installation without an enclosure. This mean the units can be installed with no covering and still meet UL standard code requirements. The unit also can be installed inside a building meeting all UL approved standards. See **Figure 15-4.**

If required, a protective cage should be placed around small units. Or, they should be placed in a separate, well-ventilated room. Windows provide this ventilation for smaller units, while a forced exhaust is needed for larger units. This will permit escape of refrigerants should the unit develop a leak. Screen the openings to prevent insects and other objects from entering. Larger units must also be protected from fire damage by fire-resistant self-closing doors. The condensing unit must be electrically grounded.

Shipping blocks are used to protect units using motors and fans to cool the condenser. The shipping blocks must be removed before the unit can be installed. Spin the fan by hand to be sure it runs freely.

To prevent violent rupturing or an explosion of the condensing unit due to excessive pressure, the code specifies:

- High-pressure cut-outs to stop the motor.

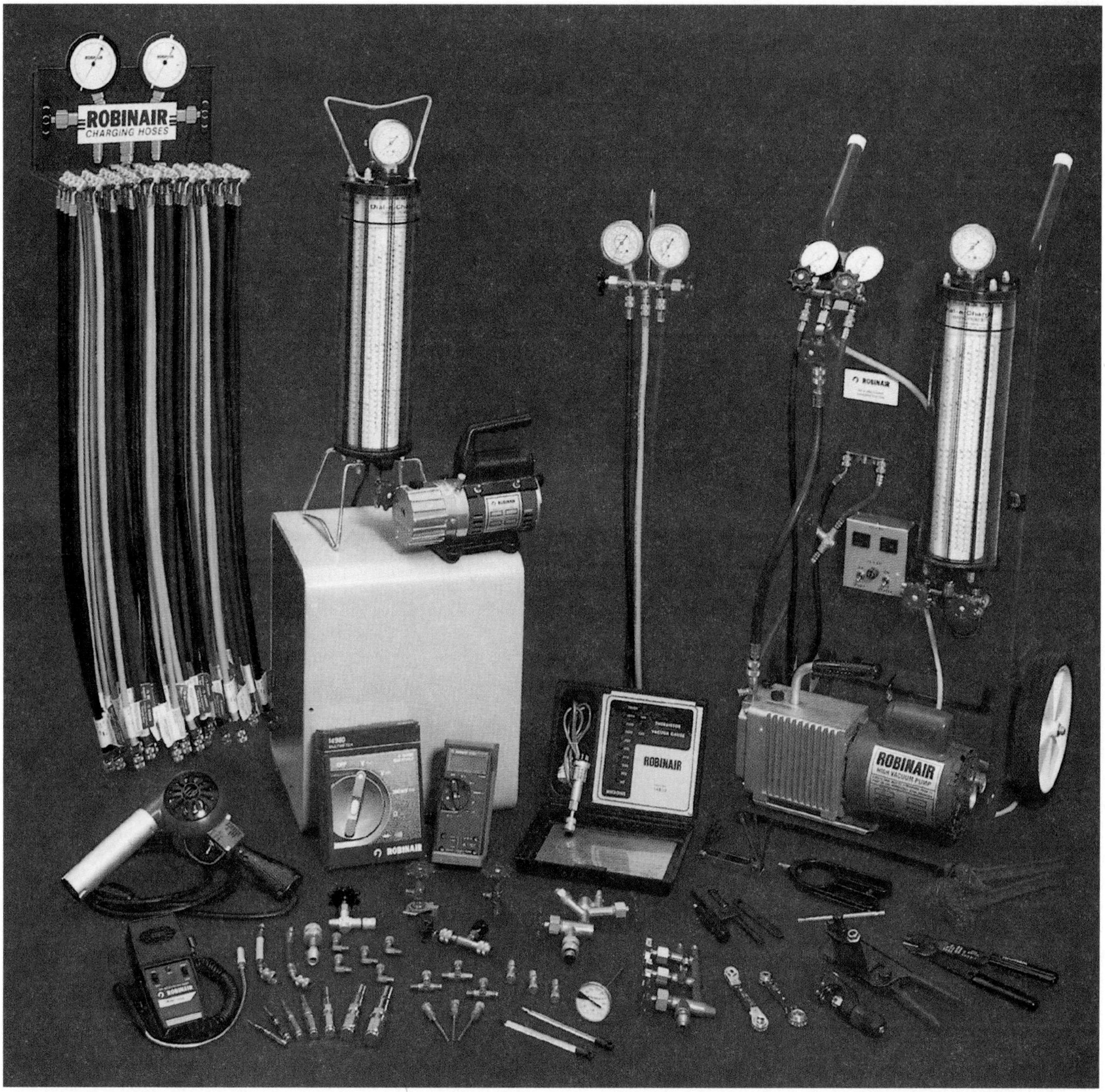

Figure 15-1. *The installation and servicing of commercial refrigeration units is similar to that of domestic units discussed earlier. The tools and instruments listed in Chapters 2 and 12 can also be used for commercial systems. Pictured here are some of the commonly used tools you should have on hand at any installation or service call. From actually handling the tools and studying Chapters 2 and 12, many of the tools will be recognizable. (Robinair Division, SPX Corporation)*

- Pressure relief valves or rupture disks to dissipate discharge slowly. These safety openings are piped outside by way of copper pipe connected by brazed joints. Spring-loaded safety valves and fuse plugs are used where the unit may become overheated due to fire.

All refrigerant lines should be permanently labeled with signs identifying refrigerant.

Install the condensing unit so that all parts are accessible for maintenance and service. Keep them away from the sun and away from other heat sources, such as steam pipes, hot air grilles, and ovens.

Water lines connected to the unit are either soft copper or flexible plastic pipe. Allow enough piping to permit some movement of the condensing unit.

Be careful! If a system has too much oil, it will pump liquid oil and be damaged.

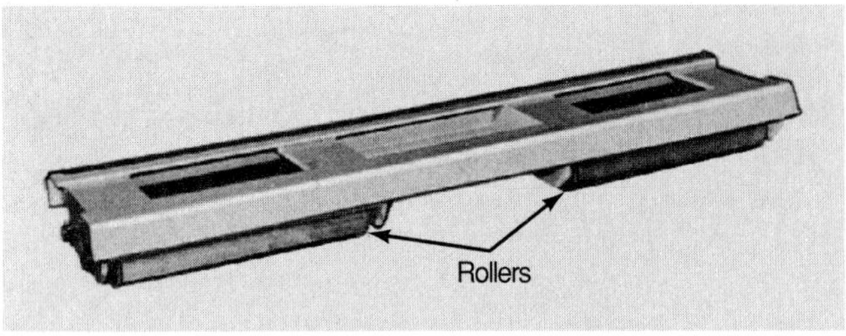

Figure 15-2. *One or more of these dollies under a cabinet enables you to easily move and locate heavy, bulky cabinets.*

15.3 Installing Evaporators

Evaporators for commercial refrigeration installations should be carefully mounted, delicately leveled, and firmly fastened. Different evaporators use different mountings. **Figure 15-5** shows a shelf evaporator used in a display case.

Blower evaporators are usually equipped with hanging brackets, as in **Figure 15-6.** Evaporators are usually fastened to the ceiling in florist cabinets and walk-in coolers. A plumb line is used to locate the holder positions. A cardboard template is useful for locating the mounting brackets.

The hanger is attached to the ceiling of the cabinet. Hydraulic or pneumatic adjustable-height platforms lift the evaporators. They hold the evaporators in place until they can be fastened. The unit should be carefully checked after mounting to make sure it is level.

Display counter evaporators are usually supported by stands. These stands or brackets should be provided with leveling adjustments in all directions.

All natural convection evaporators should be properly baffled. **Figure 15-7** shows one type of baffle and the hangers that fasten it to the evaporator.

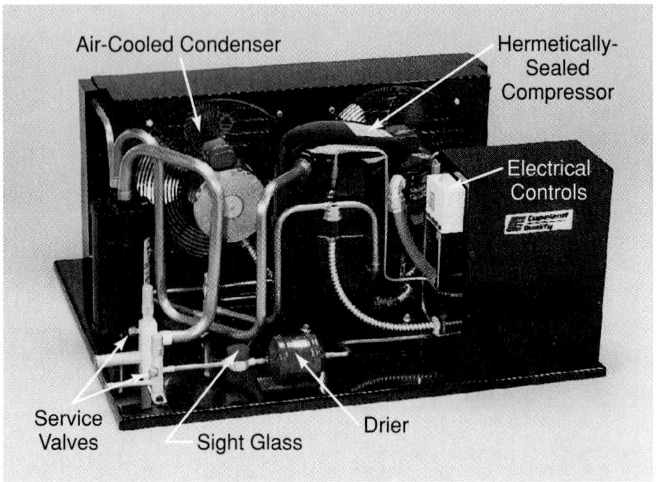

Figure 15-4. *Note pre-assembled unit in the installation above. (Copeland Corporation)*

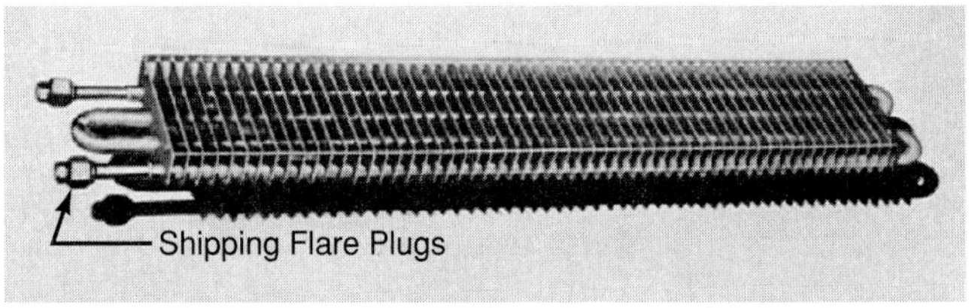

Figure 15-5. *Shelf evaporator. Aluminum fins and copper tubing are mechanically bonded. Fins are offset. (Peerless of America, Inc.)*

The code recommends refrigerant limits for evaporators. These are based on the exposure of people to the refrigerant in case of leaks. Some large evaporators installed in air ducts must be cooled with a brine rather than refrigerant.

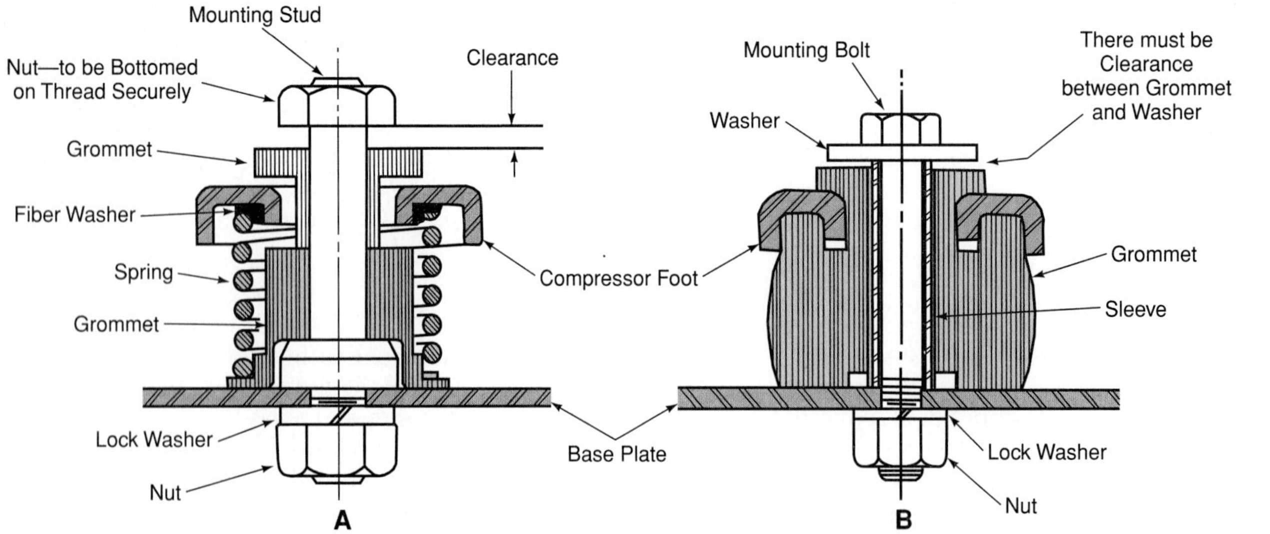

Figure 15-3. *Hermetic motor compressor mountings are designed to absorb vibration. A—Synthetic rubber grommet and spring. B—Synthetic rubber grommet only. (Tecumseh Products Co.)*

Figure 15-6. *Small blower evaporator. Evaporator must be mounted level for efficient operation and good drainage. (Peerless of America, Inc.)*

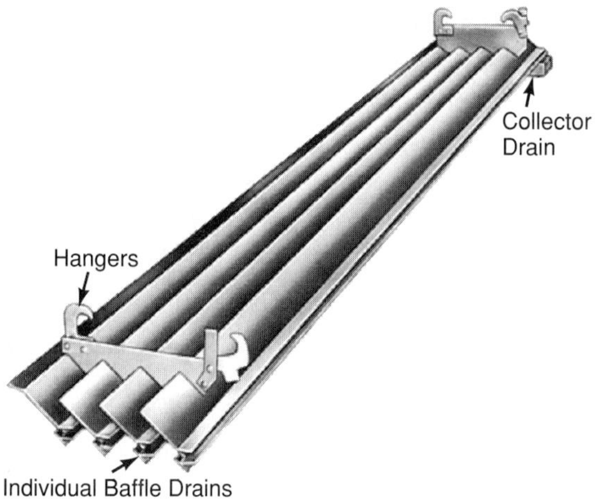

Figure 15-7. *Evaporator baffle and drain pan. It is compact, permits good circulation and condensate drainage.*

The evaporator must be electrically grounded if it has a motor and fan. A commercial evaporator using a motor-driven fan is shown in **Figure 15-8.**

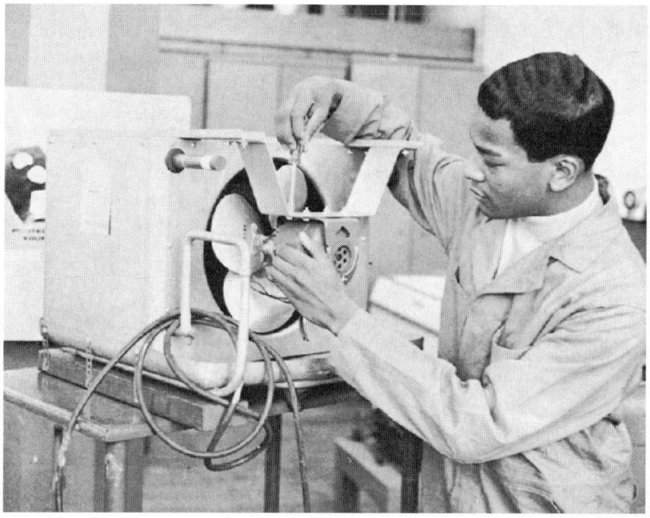

Figure 15-8. *Mounting motor fan on commercial evaporator. (Detroit Public Schools)*

15.4 Installing Refrigerant Piping

Code specifications require strong piping for refrigerant lines. These should be type K (the strongest) or type L. Piping should be protected by adequate guards. Some codes recommend at least .065″ wall thickness where hard copper pipe is exposed. Joints in the refrigerant piping must be placed so they can be easily inspected. The joints must be made with strong fittings. The brazing material used must be of excellent quality.

The code recommends that piping always be supported by the building structure. This will prevent pinching or crimping the piping. Pipe should not be run across joists or studs where unsupported sections can be damaged. The piping must be at least 7 1/2′ above floor level when crossing a room. **Figure 15-9** shows a piping system for a roof-mounted refrigerating system.

The suction line should be mounted with a slight drop in horizontal runs toward the compressor. This provides for proper oil return.

It is important to install flexible sections in the pipe for noise control. This is also necessary for piping that experiences rapid temperature changes (defrosting hot gas lines). See **Figure 15-10.**

15.4.1 Fittings

Commercial system capacity has increased steadily during the past few years. This is especially true with comfort cooling installations in air-conditioning units and in supermarkets.

Liquid and suction line sizes in these installations may be as large as 6″ OD. Brazed joints with sweat fittings are used. These fittings, described in Chapter 2, are made of drop-forged or extended copper. They have a recess large enough to receive the hard copper pipe.

It is important that the brazing of hard copper pipe joints be expertly done. Otherwise, considerable trouble may result (bad joints and leaks). Hard drawn copper pipe is seamless. It usually has greater wall thickness than annealed copper tubing. It comes in 10′ or 20′ lengths rather than in rolls. Ends are either capped or plugged. When making an installation of this kind, use fluxes and solders recommended by the manufacturer. Joining surfaces should be clean and ends of the tubing should be square. This will prevent flux or joining metal from running into the tubing.

Welding Equipment

Gas welding equipment needed for refrigeration installation consists of the following: oxygen cylinder, acetylene cylinder, regulators and gauges, hose, and a

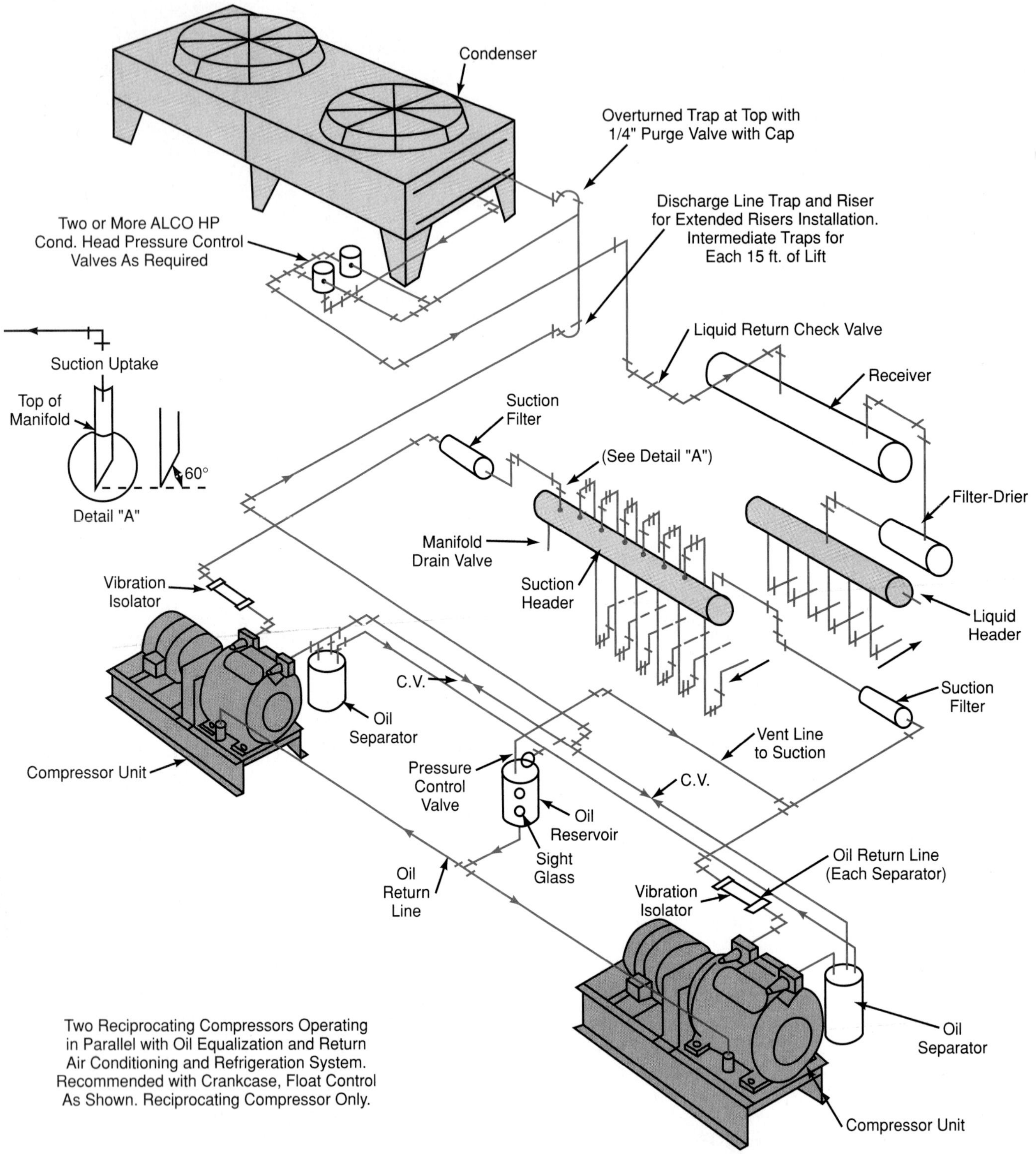

Figure 15-9. *Schematic piping diagram for commercial refrigeration system. It uses a roof-mounted, air-cooled condenser, two motor compressors, and suction and liquid header, each connected to six refrigerant lines. (Dunham-Bush, Inc.)*

torch. **Figure 15-11** illustrates a welding and cutting unit.

Before doing any welding, the local welding code should be thoroughly studied and understood. Never operate a welding outfit near inflammable material. Never use excessive pressures with any gas. A severe explosion may result.

Caution: Never use oxygen, acetylene, or any other welding fuels to develop pressure in refrigeration tubing, piping, or equipment. Carbon dioxide, nitrogen, and

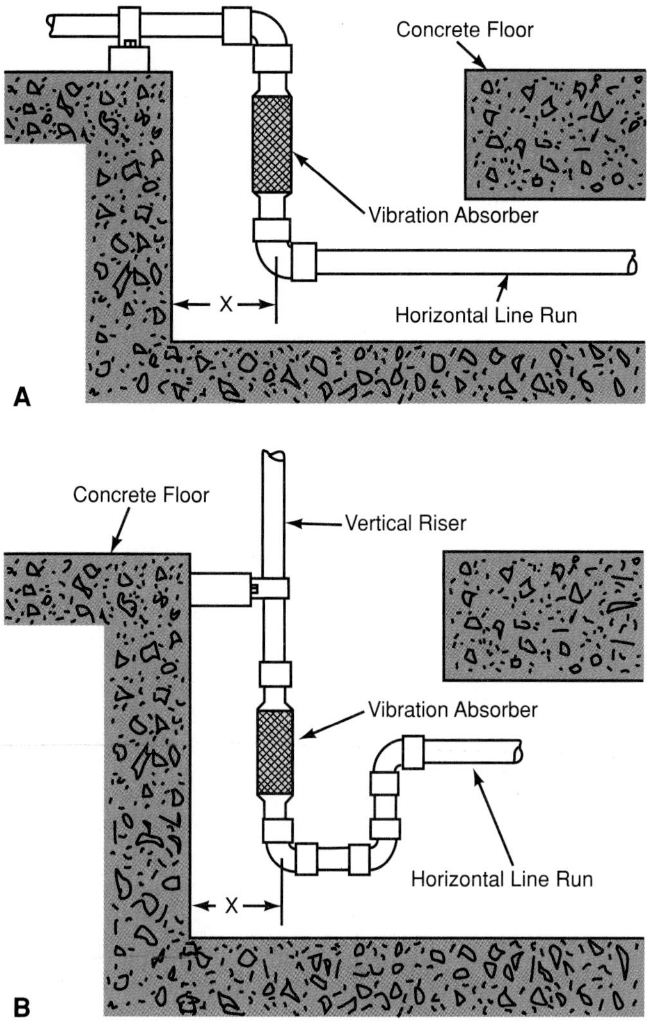

Figure 15-10. *Two instances showing how flexible vibration absorbers are used. Allow a space of 1 1/4" at X for each 100' per 100°F (55°C) temperature change. A—Horizontal piping. B—Vertical piping.*

argon are safe if used with a pressure regulator and a pressure relief valve. They may be used for developing pressures in refrigeration lines.

Brazing Equipment

A refrigeration service technician uses brazing for many jobs. Air-acetylene torches furnish a clean flame at a temperature of 2500°F (1400°C). With compressed air, the torch flame temperature is about 2500°F to 2800°F (1400°C to 1500°C). Acetylene is supplied in cylinders of 10 ft^3 or 40 ft^3 capacity for portable welding.

Detailed instructions on the construction and use of acetylene-air brazing equipment are in Chapter 2. **Figure 15-12** shows a basic refrigeration and air conditioning kit. It provides the proper tools for soldering, cutting, and leak detection.

It is important to follow these safety precautions:

Acetylene pressure should never be over 15 psi (207 kPa). Higher pressure may cause an explosion because acetylene is not stable at higher pressures.

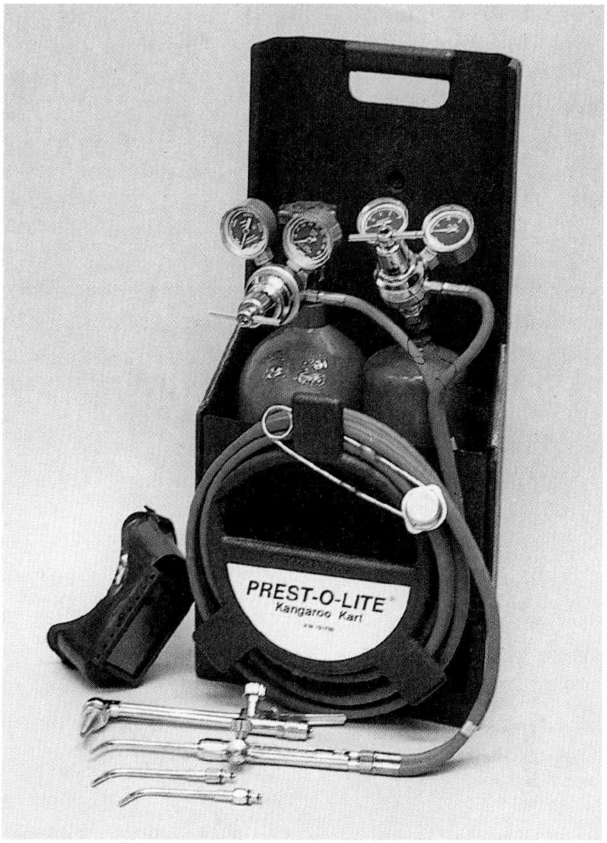

Figure 15-11. *Welding and cutting unit on portable cart. Note various types of interchangeable tips for proper flame size and heat. (PREST-O-LITE®, Product of the ESAB Group, Inc.)*

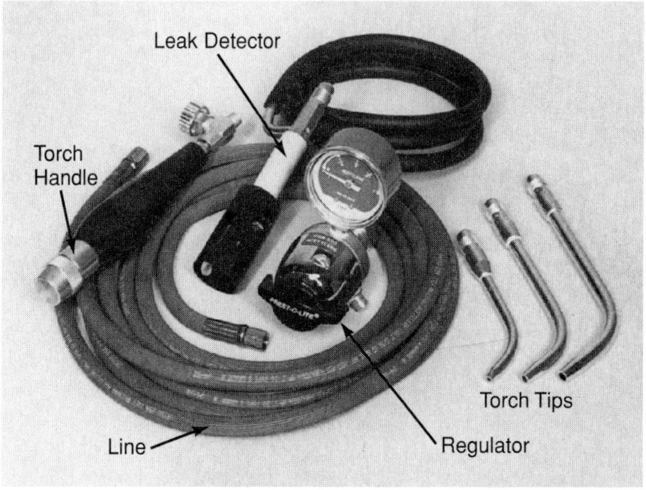

Figure 15-12. *Combination soldering, cutting, and leak detection tools. (PREST-O-LITE®, Product of the ESAB Group, Inc.)*

Always use the cylinder in a vertical position. It has a porous filler in which the acetylene is dissolved. If the cylinder is lying down while in use, some acetone may flow out. Acetone causes a dirty flame and may grease up the regulator and valves.

Fuel air torches that use propane or other high-temperature fuel are also used. They are light and work in any position. See **Figure 15-13.** Propane fuel is usable down to −10°F (−23°C).

For safety, keep the flame away from any combustible substance. Such substances include oil, wood, paper, paint, cleansing fluids, and methyl chloride. It also includes barrels or cylinders that may have contained flammable material at one time. Use sheet metal or board to protect surfaces, such as when assembling piping along a wall. This will prevent discoloring or scorching when using a torch.

Always light the torch with a flint lighter. Using matches or a cigarette lighter places your hand too close to the flame.

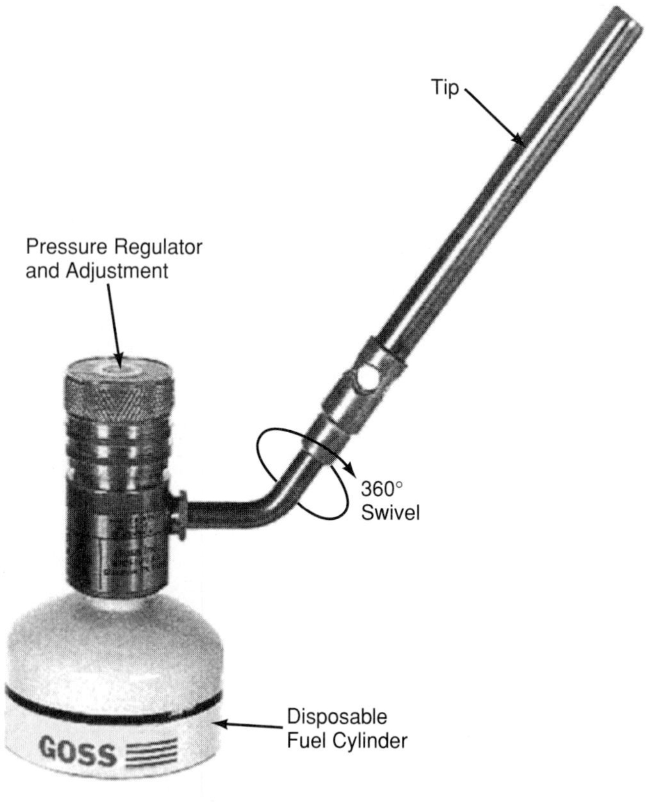

Figure 15-13. *Soldering and brazing torch. (Goss Inc.)*

15.4.2 Installing Tubing

Tubing in noncode installations is usually run along the walls and ceiling. Supports are used at intervals frequent enough to keep tubing straight and firmly fastened. Special clamps are available as tubing fasteners. A galvanized 1/2″ conduit clamp, **Figure 15-14,** is sufficient for most situations. The tubing should be insulated or protected from these clamps. Short wrappings of plastic tape will prevent chafing and galvanic action. (***Galvanic action*** is erosion of material caused by two different metals touching in moist air.)

Tubing running through a floor or wall should be protected. Short runs of conduit or flexible metal tubing (Greenfield) provide sufficient protection. The ends should be sealed with a sealing compound to avoid chafing and other troubles. See **Figure 15-15.** In all cases, tubing should be run horizontally and vertically with neat bends.

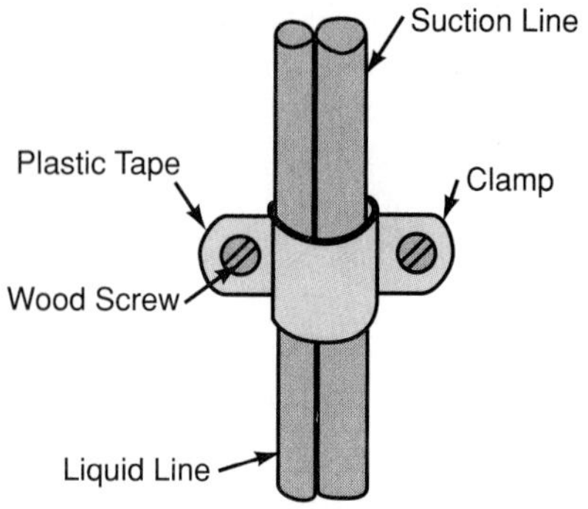

Figure 15-14. *A method of fastening tubing to the wall.*

The liquid line presents no difficulties regarding slope and position. Suction lines must drain toward the compressor. Low spots in the suction lines will accumulate return oil. The oil may eventually form a liquid slug in the tubing. Such a slug carried to the compressor will produce disturbance in the crankcase and may cause temporary oil pumping.

Sometimes tubing must slant upward from the evaporators to the condensing unit. If so, the tubing run should have a steady downward slant to a certain point. Then a short U-bend (two street ells) should be made. The U-bend will act as an oil trap. This will ensure a positive return of oil to the compressor.

Never run tubing near sources of heat, such as hot water lines, steam lines, or furnaces. Heat will reduce efficiency.

Copper tubing usually comes in 50′ coils. It is dehydrated and sealed at the ends by the manufacturer. In the average small commercial installation, 1/4″ tubing is used for the liquid line and 1/2″ for the suction line. See Chapter 16 for correct line sizes.

During installation, tubing should be kept clean. Never put it aside with ends open unless it will be used immediately. A practical method of installing tubing is

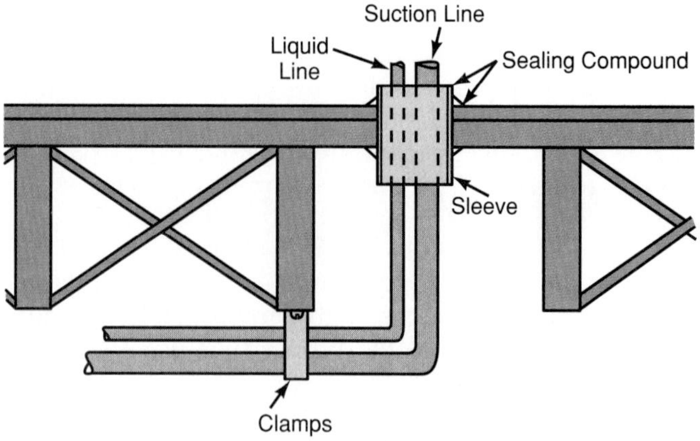

Figure 15-15. *Use a sleeve to protect tubing if it goes through a wall or floor.*

to uncoil 10′ at a time. Unroll the coil along the floor. Then run the tubing up through floor openings from underneath, gradually working it into place.

Quarter-inch tubing is not difficult to install. However, 1/2″ diameter must be carefully handled to prevent buckling when it is bent. A tube bender should be used. See Chapter 2.

In noncode installations where individual suction and liquid lines run to main lines, T-fittings may be used. The valves for shutting off the individual evaporators may be located near the evaporator.

Valves, driers, or other heavy objects should not be supported by tubing. These items should be mounted on the wall or some other support. These connections may be of the SAE flare type or streamline soldered (sweat) fittings.

Newly installed tubing should be sealed immediately after flare or streamline connections are made. This will keep the tubing clean. The tubing should be attached permanently to the supports along which it runs.

Always try to run tubing so that supports will protect the tubing from accidents. Many technicians use a sponge rubber covering over the tubing. The covering serves both as a protection and as insulation, **Figure 15-16.** This covering must be placed on the tubing before assembly unless the insulation is split.

Tubing connections or valve installations are commonly covered with insulating tape, as in **Figure 15-17.** Tubing should be placed where it will not be damaged by handling of nearby articles. Do not put loops or unsupported bends in the tubing except at the condensing unit.

Install a drier and sight glass in the liquid line at the condensing unit. A vibration eliminator, as shown in **Figure 15-18,** should be included in both lines. If a suitable vibration eliminator is not available, an alternate method may be used. This method is to make one horizontal loop of soft copper tube in the suction and liquid lines. Carefully study the manufacturer's installation and service procedures. This will ensure that every assembly and adjustment is made correctly.

Copper tubing normally comes with no special

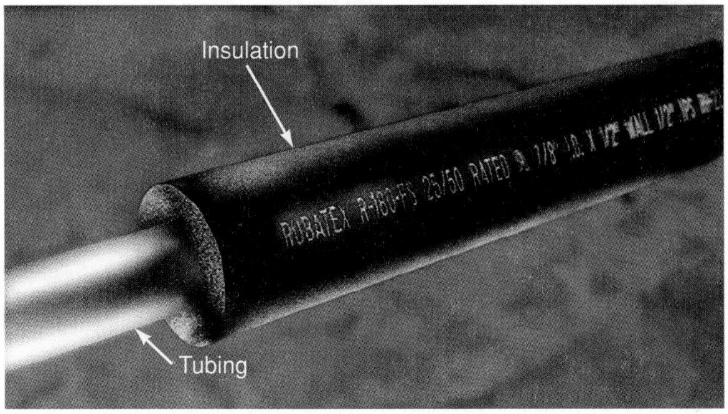

Figure 15-16. *Rubber insulation mounted on suction line. Insulation is usually installed before tubing or pipe connections are made. (Rubatex Corp.)*

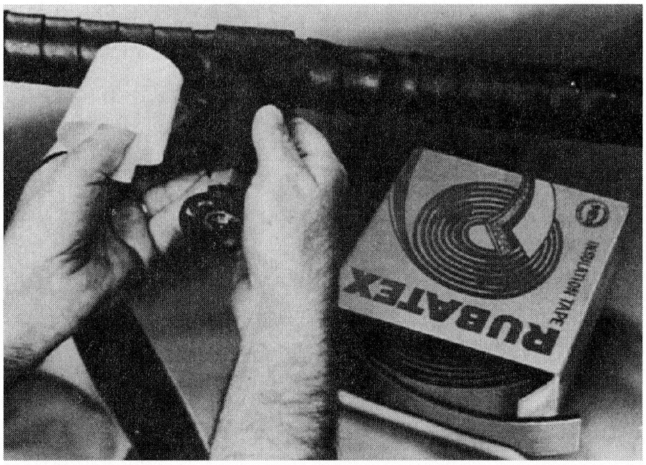

Figure 15-17. *Foamed plastic insulating tape is wrapped around valve to prevent sweating or frosting. (Rubatex Corp.)*

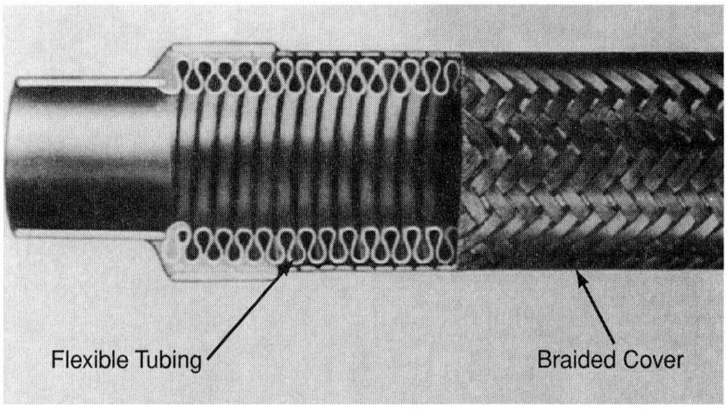

Figure 15-18. *Flexible tubing is used for vibration damping. One of these is usually mounted in suction line near condensing unit. (ANAMET Industrial, Inc., An ANAMET Company)*

finish provided for the inside or the outside surface. Such tubing will corrode if it is run through liquid, food, or beverages. It will also corrode if it is run through air saturated with acid fumes or corrosive elements. Where sanitation is of primary importance, tubing with a tinned surface may be used.

Many local codes require the use of special tubing for conveying beverages between kegs and dispensers. Stainless steel is usually specified.

15.4.3 Multiple Evaporator Piping

There are two common methods of installing piping in a multiple installation. In one method there is a common liquid line and common suction line. The various evaporator liquid and suction lines tap into it at the most convenient points.

Another method is to use a clustering system. Here, various lines are brought to a common point. At this point, they are connected through a hand valve to a manifold. A large suction and liquid line is run from the manifold to the compressor. This method is not always practical. An example would be where one evaporator is at some distance from the box. In this situation, a duplication of long runs would be required. The manifold of a code installation is located on a wall near the condensing unit.

Keep in mind that every fitting and bend used in the lines cuts the efficiency of the installation. Limit their number as much as possible.

Inner parts of valves, driers, filters, and sight glasses must be removed while brazing tubing to them. Or, the part may be wrapped with a wet cloth or some heat absorber. Do not allow moisture to enter the valve.

15.4.4 Service Valves

Service valves must be leakproof where the valve stem goes into the valve. The packing is made of lead, graphite, and other materials. Packings differ with different valve designs. When replacing them, the proper packing must be used.

A replacement service valve is shown in **Figure 15-19.** It is a four-bolt mounting style service valve. Surfaces are slightly recessed for ease of bolt tightening.

Practically all service valves have a drop-forged brass body and a steel stem. Stems have a tendency to rust and score the valve gland or packing. Always clean and oil a valve stem before turning it. See **Figure 15-20.**

This corrosion, especially in damp locations, can be reduced. Fill the valve body with clean and dry refrigerant oil before replacing the plug. This should be done each time the service valve is used. This oil used should be the specified refrigerant oil for that machine.

Service valves on commercial installations must be kept in good condition. You can do three things to assure good service and valve life:

- Fit the wrench to the valve stem.
- Maintain the packing so that the service valve will not leak.
- Oil the threads of the gauge connections each time gauges are used.

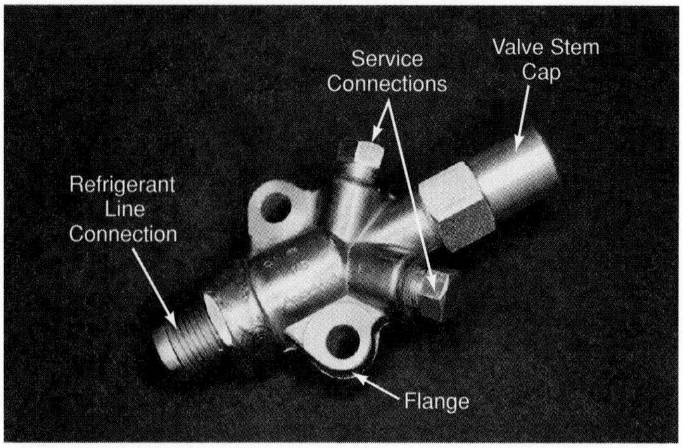

Figure 15-20. *Refrigeration unit service valve with two service openings, two-bolt flange, and valve stem cap. (Mueller Refrigeration Products Co., Division of Mueller Industries, Inc.)*

Occasionally, after a period of use, these service valves must be replaced. Pipe threads in the valve gauge openings may become worn and leak. This results from frequent mounting of flexible line fittings. The fittings inserted in these gauge openings may be given a thin coat of solder. In this manner, trouble can often be eliminated.

When cracking the valve, always use a fixed wrench (not a ratchet wrench). This is done so the valve may be quickly closed again if necessary.

Occasionally a service valve will be found in such condition as to be useless. In such cases, remove the refrigerant or isolate it in another part of the system and replace the valve.

Many external drive systems have a liquid receiver service valve (LRSV). See **Figure 15-21.** Some of the LRS valves are three-way. This enables you to charge liquid refrigerant into the system.

When using a system service valve, remove the valve cap—if the valve has one. Loosen the service valve packing nut one turn. Next, clean the valve stem before

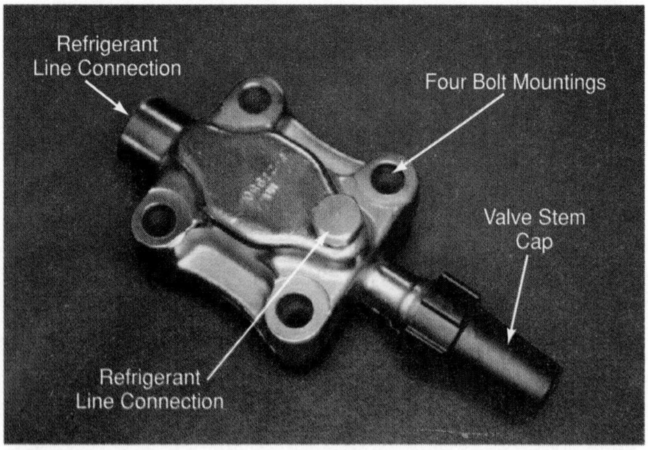

Figure 15-19. *Compressor service valve, four-bolt style. Note refrigerant line connection. (Mueller Refrigeration Products Co., Division of Mueller Industries, Inc.)*

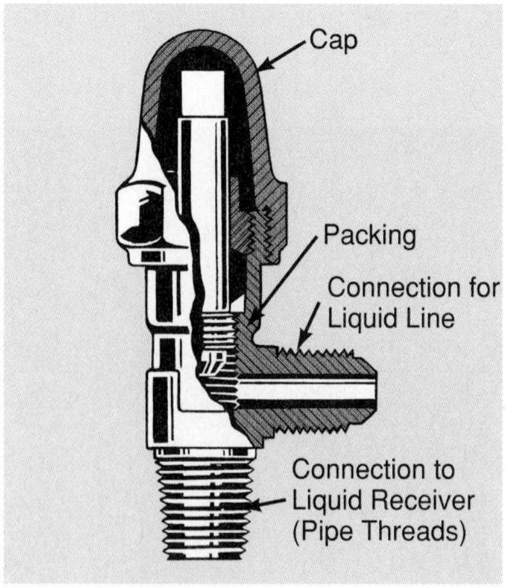

Figure 15-21. *A liquid receiver service valve. (Superior Valve Company, Division of AMCAST Industrial Corporation)*

turning it. Turn the valve stem back in about 1/16 of a turn. Tighten the packing nut and replace the valve stem cap.

Turning the valve stem back in slightly prevents the valve from "freezing" against its seat. Such a condition sometimes leads to broken valve stems.

When installing the gauge opening plug, tighten the plug firmly. Never tighten a cold gauge plug into a hot service valve. This may result in freezing of the plug to its seat. When using a service valve wrench on these valves, apply the turning force gradually. Adjustable end or fixed open end wrenches are not recommended for service valve stems. Only special socket wrenches called keys are to be used.

If the gauge plug is frozen in the service valve, it can be loosened. First, heat the outside of the service valve body with flame from a torch. Be careful not to overheat. This heating will cause the valve body to expand. As a result, it will weaken the body thread grip on the plug. The wrench can then be used to loosen the valve stem.

Access valves are often used at the evaporator outlet or the liquid line inlets. They are placed just ahead (downstream) of the refrigerant control and on both sides of the automatic valves in the system. (These valves include the solenoid, bypass valves, hot gas defrosting valves, and driers.) The need for more convenient service outweighs the policy of having a minimum of connections.

15.4.5 Refrigerant Line Valves

All code line valves are to be hand operated. These valves must be constructed so that anyone may shut them off manually. There must be no need for special tools to operate the valve. There is usually a handwheel mounted permanently on the valve. These valves are provided with brackets that may be attached firmly to a panel. Two styles are available:

- Three-way valves.
- Two-way valves.

One type of three-way valve shuts off just one of the three connections to the valve. The other two, remaining uncontrolled, permit the passage of refrigerant to the rest of the system.

The other type valve is two-way. It stops the flow of refrigerant when turned in clockwise. **Figure 15-22** shows a two-way valve for brazed joints.

15.5 Electrical Connections

Units must have enough of the correct electrical power to operate motor, controls, and solenoids. Chapters 6, 7, and 8 explain electrical fundamentals and electrical power variables.

Before installing a condensing unit, check the electrical voltage, cycle, and phase of the compressor motor. Be sure they are the same as the electrical power source.

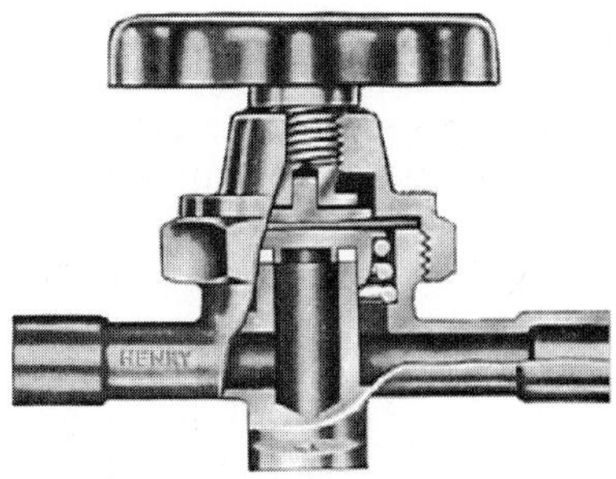

Figure 15-22. *Two-way hand valve designed for brazed connections. (Henry Valve Co.)*

Smaller units—1/8 hp to 1/4 hp—usually use 120 V single-phase ac. Medium size units—1/4 hp to 1 hp—may use 208 V or 240 V single-phase or three-phase ac. Larger units may use 240 V or 440 V three-phase ac.

Both evaporator and condenser fans must be checked to be sure the identification plate data matches power available. This is true as well for solenoid valves and controls.

Wire size is important. Wire should have a capacity 50% over the load it will carry.

Condensing units have wiring diagrams either fastened to them or supplied in the shipping crate. **Figure 15-23** shows a typical wiring diagram for a single-phase

Two Terminal Protector-Single Pole Line Control Current Relay

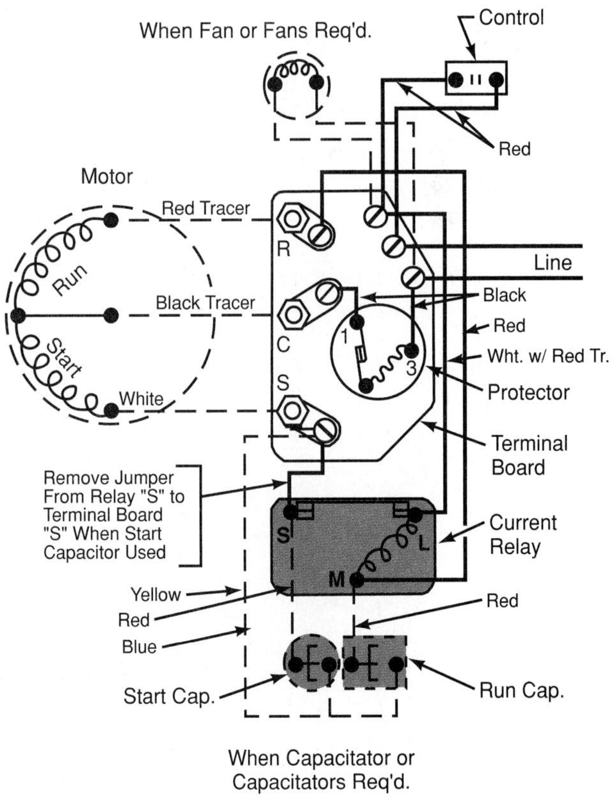

Figure 15-23. *Wiring diagram of single-phase unit using current relay, run capacitor, and start capacitor. (Copeland Corp.)*

circuit. **Figure 15-24** shows a three-phase refrigeration system wiring diagram. Correct connections are extremely important. **Do not turn on the electrical power until the circuits are correct and all connectors are clean and tight.**

Overloaded electrical circuits are dangerous. They may cause unit burnouts or electrical wiring fires.

A complete condensing unit installation is shown in **Figure 15-25.** Use an ohmmeter to check all circuits for continuity before turning on the power.

15.6 Testing Code Installations

Code authorities require that permits be obtained before an installation can be made. A permit is also needed before performing a major service operation on commercial units. Specifications of the proposed job must be presented. Permits are not issued unless the specifications presented meet code requirements.

On completion of the work, an inspector is called. Approval must be given by the inspector before the unit may be run. Some codes require that refrigeration installation personnel be licensed.

Figure 15-25. *Commercial unit with dual compressors. A—Vibration eliminators. B—Sight glass. C—Filter-drier. (ANAMET Industrial, Inc., An ANAMET Company)*

The inspector checks the installation to see if all the work has been done according to specifications and code. Then the system is tested primarily for leaks and safety. This requires building up the normal pressures in the system's high and low sides. Nitrogen is usually

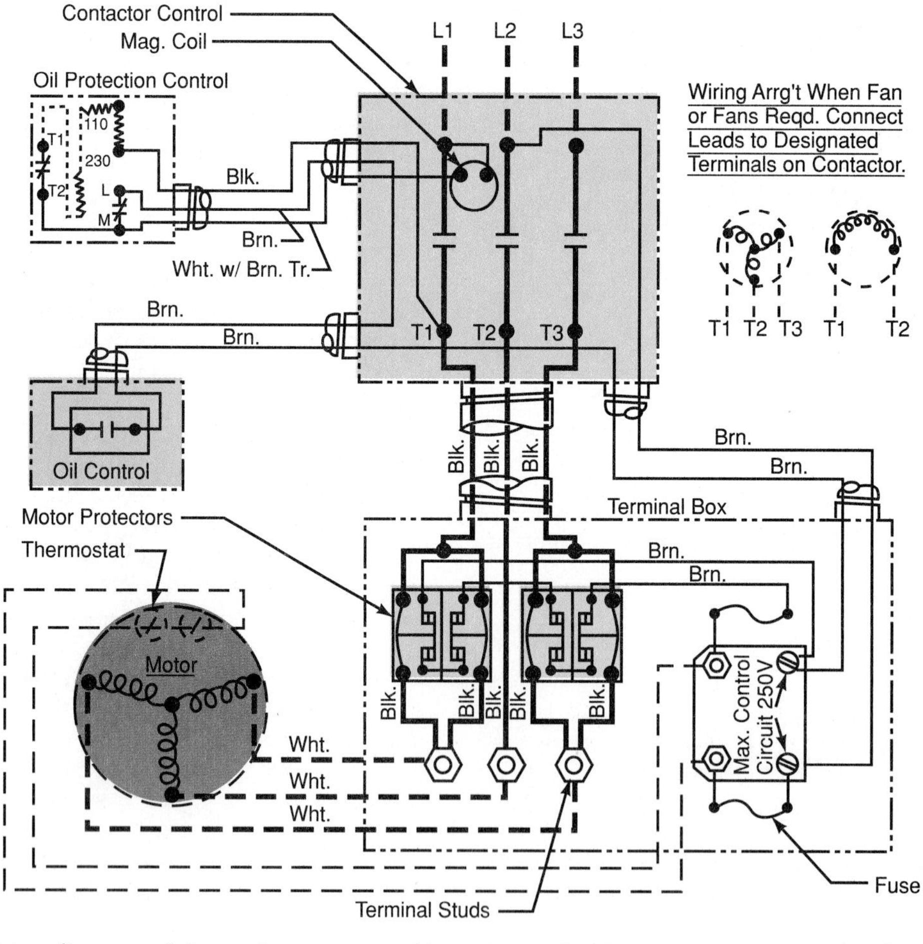

Figure 15-24. *Wiring diagram of three-phase motor refrigeration unit. Note contactor control, oil protection control, and motor circuit protectors.*

used for this. These pressures vary with the kind of refrigerant in the system. **Figure 15-26** gives the recommended minimum test pressures to be used for each refrigerant.

Minimum Design—Testing Pressure			
	Low Side	High Side	
Refrigerant		Evap. or Water-Cooled	Air-Cooled
R-12	85	127	169
R-22	144	211	278
R-500	102	153	203
R-502	162	232	302
R-717	139	215	293

Figure 15-26. *Recommended minimum design testing pressures are based on Safety Code for Mechanical Refrigeration.*

Low-side and high-side test pressure are usually specified by the code. One must avoid using higher pressures because the system may rupture.

The same refrigerant that will be used in operation is first charged into the system in a vapor form. This creates a low pressure (20 psi to 50 psi or 240 kPa to 450 kPa). The system is then tested for leaks. Large leaks are easily detected at these low pressures. If no leaks are found, nitrogen is used to build up to code pressure.

A technician or inspector must be very careful. First, the refrigerant cylinder should be disconnected to prevent nitrogen backing up into it. Second, there must be a hand shutoff valve, a pressure regulator, a pressure gauge, and a pressure relief valve in the nitrogen charging line. The relief valve should be adjusted to open 1 psi or 2 psi above the test pressure. A safe method for using nitrogen to pressurize a system is shown in **Figure 12-55.**

After pressures are built up in the piping, each joint should be rapped. A rubber mallet is used on each brazed or mechanical joint. This will make sure the joint will be leakproof under working conditions. (Paint or flux may otherwise temporarily stop a leak.) If no leaks are found, leave the nitrogen pressures in the system for 24 hours.

With nitrogen, the soap bubble test is used to check for leaks. To use a halide torch or electronic leak detector, mix some R-12 or R-22 with the nitrogen. (Mix about 1/4 lb. per ton of refrigeration.)

If a brazed joint leaks, take it apart to repair it. Put flux on the joint before heating it to keep the brazing material clean. Take the joint completely apart and then assemble and braze again.

If no leaks are indicated at the pressures established, the inspector sometimes checks the system subjected to a vacuum pressure. If this vacuum is maintained over a certain period of time, the installation is approved.

After approval of the system, you should record the unit's running behavior. It should be recorded for at least 24 hours. **Figure 15-27** shows a recording thermometer. Any variations in temperature reveals need for adjustments.

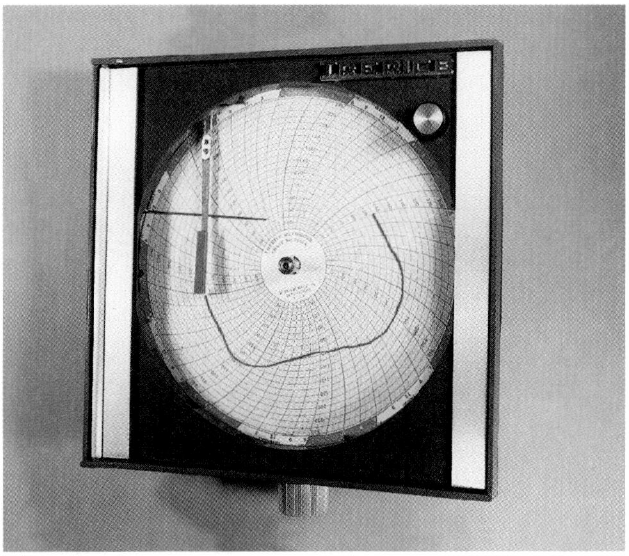

Figure 15-27. *A 24-hour, 7-day recording thermometer. Chart should be dated and kept for future reference. (H.O. Trerice Co.)*

15.6.1 Gauge Manifold

The service ***gauge manifold*** is very useful for service technicians. With this piece of equipment, they are able to check low- and high-side pressures. They can charge and discharge a system, add oil, and bypass the compressor. They are also able to unload gauge lines of high-pressure liquid and vapor, as well as perform many other operations without replacing regular gauges. See Chapter 12 for gauge manifold information.

A typical manifold is shown in **Figure 15-28.** It has two gauge openings, three line connections, and two shutoff valves. The shutoff valves separate the outside connections from the center line connection. **Figure 15-29** shows a compound gauge having a pressure scale and three different refrigerant temperature scales.

In **Figure 15-28,** a 1/4″ copper tubing or a flexible line connects the manifold to the SSV. (The SSV, suction service valve, is shown at D.) It also connects the manifold to the DSV (discharge service valve) at C. Most service valves on the compressor have 1/8″ FP (female pipe) gauge openings. Therefore, two 1/8″ MP (male pipe) by 1/4″ MF (male flare) half unions are installed in the service valves. Before removing the pipe line plugs and installing the fittings, check that the compressor service valve stems are turned all the way out. Be sure that the outside of the valve is clean.

Lines from the manifold are attached to these fittings. The line attached to the SSV at D should be left one to two turns loose. The line to the DSV should be tightened. Then open both the manifold valves at A and B 1/4 to 1/2 turn. Cap the middle opening, E.

Now turn the DSV C stem in 1/8 to 1/4 turn for just a moment. (This is called "cracking the valve.") A surge of high-pressure refrigerant will then rush through

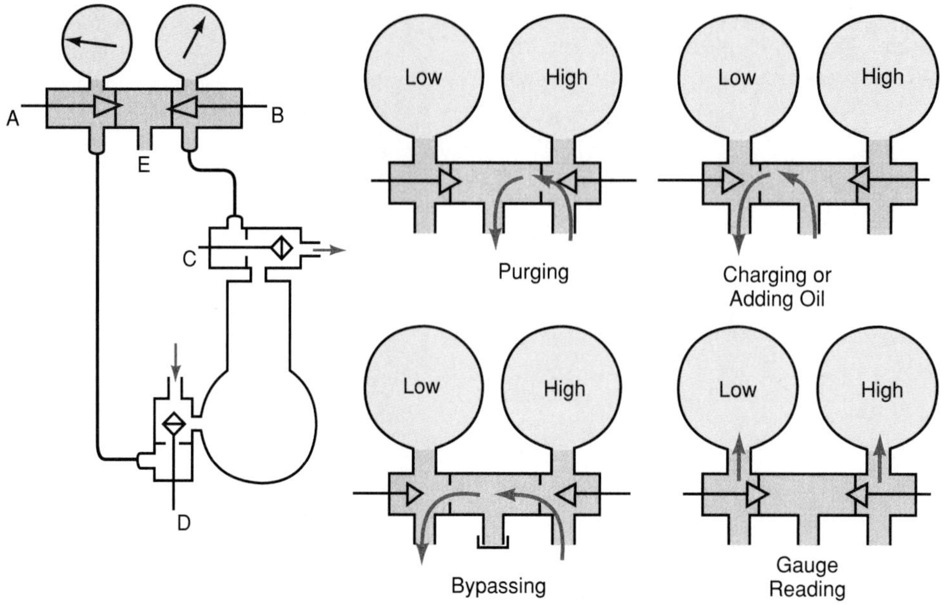

Figure 15-28. *Schematic of gauge manifold installation on external drive compressor with service valves. At left, both manifold valves are turned all the way in. The system is pumping vapor, and both the low- and high-side pressures are being read. A—Manifold suction valve. B—Manifold discharge valve. C—Compressor discharge service valve. D—Compressor suction service valve. E—Service opening.*

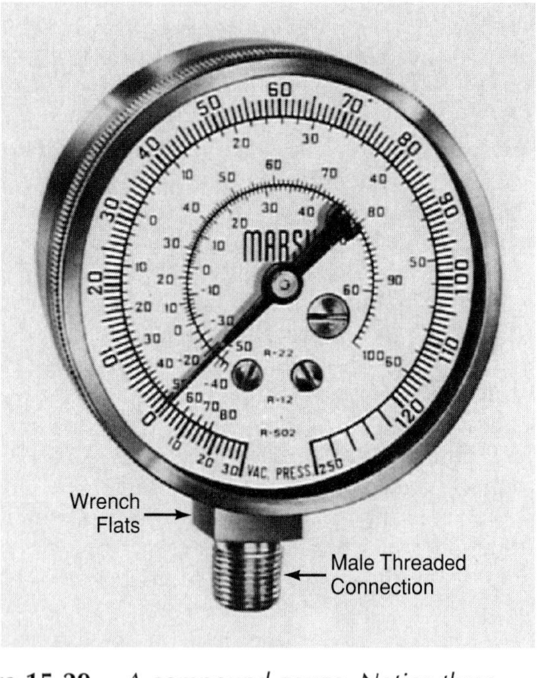

Figure 15-29. *A compound gauge. Notice three refrigerant temperature scales. These are for R-12, R-22, and R-502.*

the lines and the manifold. A small amount will purge to the atmosphere at the loose connection at D, the SSV. This connection may then be tightened. Purging is necessary to remove air and moisture from the manifold and lines.

Carefully test for leaks while the manifold and its line are under high pressure. Correct any leak immediately.

Various service and testing may be performed after the testing manifold has been installed.

- Observe operating pressures by:
  1. Closing valve A.
  2. Closing valve B.
  3. Cracking open back seat of valve C.
  4. Cracking open back seat of valve D.
- Charge refrigerant into system by:
  1. Connecting refrigerant cylinder to E (vapor only).
  2. Opening valve A.
  3. Closing valve B.
  4. Closing front seat of valve D slowly.
- Purge condenser by:
  1. Closing valve A.
  2. Opening valve B.
  3. Cracking open valve C.
- Charge liquid refrigerant into high side by:
  1. Connecting refrigerant drum to E.
  2. Closing valve A.
  3. Opening valve B.
  4. Mid-positioning valve C.
- Build up pressure in low side for control setting or to test for leaks by:
  1. Sealing E with seal cap.
  2. Opening valve A.
  3. Opening valve B.
  4. Back seating then cracking open valve C.
  5. Mid-positioning valve D.
- Charge oil into compressor by:
  1. Connecting oil supply to E.
  2. Opening valve A.
  3. Closing valve B.
  4. Turning valve D all the way in.

After completing service operations, the manifold is removed from the system. This must be done without losing refrigerant or admitting air. Turn the DSV at C all the way out. Then open both manifold valves A and B 1/4 to 1/2 turn. This arrangement will move all the high-pressure refrigerant from the line and the high-pressure gauge. It will put it into the low side. Now turn the SSV stem at D all the way out. Turn both manifold valve stems all the way in. Remove the lines from the service valve. Use soft synthetic fittings for finger tight connections. See **Figure 15-30.** Remove the fittings from the service valves. Install the service valve gauge opening plugs and tighten them. Immediately plug the lines and all other openings on the manifold. This is necessary to keep out dirt, moisture, and air. See **Figure 15-31.**

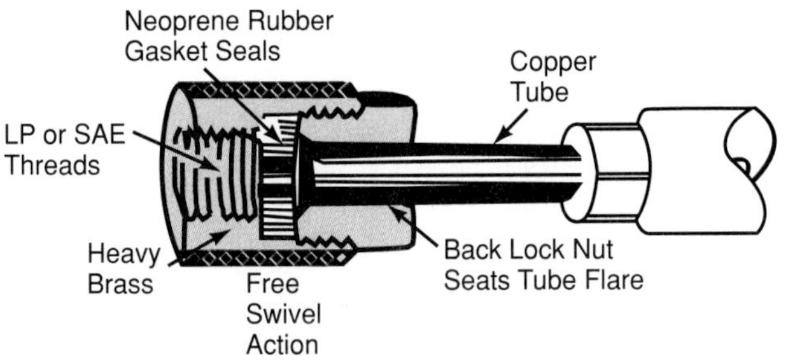

Figure 15-30. *Speed coupling used on charging and purging lines. Synthetic rubber gasket produces leakproof joint when connection is finger tightened.*

Figure 15-31. *Gauge manifold set, equipped with three flare plug hose holders. These allow the flexible lines to be attached when not in use. This method is used to keep service lines clean. (Robinair Division, SPX Corporation)*

15.6.2 Testing for Leaks

Before trying to locate leaks, build up a pressure in all parts of the system. Use carbon dioxide or dry nitrogen through the system and a small amount of the refrigerant (R-12, R-22, etc.). Two methods may be used:

- Using an inert gas.
- Using refrigerant under pressure.

In case a low-pressure refrigerant is used, some other gas may be used for testing. This is also true if the local code specifies a pressure test above the refrigerant's vapor pressure. Carbon dioxide, nitrogen, or argon are satisfactory. However, the pressure may be dangerous. See Section 12.9.1.

Caution: Never use oxygen, air, or any flammable gas for this purpose. An explosion may occur.

This testing should include the liquid line, suction line, and all other parts installed. The only exception is a new condensing unit. It has been pressure tested at the factory. Install a high-pressure gauge only. (A compound gauge may be ruined by the pressure.) Build up a pressure of 30 psig to 100 psig (45 psia to 115 psia or 311 kPa to 794 kPa). Then close the cylinder valve.

If the gauge shows no drop in pressure after an hour, raise the test pressure. It should be raised to 170 psig (185 psia or 1277 kPa). Then test the system again. Do not exceed the pressures prescribed by the code. Too high pressure may rupture some part of the system. If pressure shows no decrease after 24 hours, the system is safe to operate.

Purge the test gas from the system. Evacuate by the deep vacuum, two- or three-step vacuum method, and charge the system. The unit should be ready to operate.

Testing for leaks using the system's own refrigerant is the most common noncode practice. It is convenient; there is no need for an inert gas cylinder. Also, leak testers should be a standard part of your tool kit.

To do this, proceed as follows. Install a pressure gauge in the system. The liquid line valve should be opened just enough to build up a 15 psig to 30 psig (30 psia to 45 psia 207 kPa to 311 kPa) pressure throughout the system.

Test for leaks using one or more of the following:

- Soapsuds.
- Halide torch.
- Electronic leak detector.
- Liquid tracer.
- Ultraviolet fluorescent leak detectors. (These are used on large systems for the detection of refrigerant leaks.)

If no leaks are detected at low pressure, increase system to full pressure of refrigerant (vapor only) and test again.

If a leak is found, recover the system to atmospheric pressure. Open the system at the leak point and inspect all parts. Replace any defective parts, clean, and assemble. If a soldered or brazed joint is leaking, flux, heat, and take it completely apart. Clean and assemble, then repeat the leak detecting procedure. If no leaks are found, this part of the unit is ready to operate.

When blowing out lines and pressure testing with nitrogen or carbon dioxide, use an accurate pressure regulator. Use a relief valve designed to open at 180 psi (1346 kPa). The pressure should not exceed 170 psig (185 psia or 1277 kPa) when testing with CO_2 or nitrogen. See Chapter 12 for safe use of high-pressure gases.

Some motor compressor domes are designed to operate under low-side pressure. Be sure to read the product information to find the maximum safe pressure.

15.7 Evacuating System

When the system is tested, remove all air and moisture from the system. Air is pumped out of the lines and the evaporator with a vacuum pump.

Avoid pumping refrigerant vapor into the room where the condensing unit is located. Refrigerant vapors may harm people in the room and will interfere with leak detecting. Follow EPA regulations.

Connect a gauge manifold. Open both the discharge service valve and the suction service valve. Then pump a vacuum on the complete system. Air being removed will be discharged through the vacuum pump.

Figure 15-32 shows a gauge manifold and vacuum pump connected to a small commercial hermetic system. See Chapter 12 for evacuating methods.

A 3 ft^3/min. vacuum pump is large enough for systems up to 10 hp. Pressure drop is very important! The service lines must be as large and as short as possible.

Evacuation takes eight times longer with a 1/4″ line than it does with a 1/2″ line. It takes twice as long through a 6′ line as through a 3′ line.

Use heat lamps, electric heaters, and blowers to provide heat. Avoid using a torch flame. It may cause local high temperatures which may decompose oil, insulation, and refrigerant.

When the pump is shut off (valve closed), pressure will rise a little due to pressure drop equalizing. Take a reading one minute after closing valve and again 30 minutes later. If there is no pressure rise, the system is sealed and free of moisture.

15.8 Checking System before Starting

The motor control should be adjusted before the system is put in operation. The settings of the motor controls will vary with the demands of the cabinets. They will also vary with the different kinds of refrigerant used. See Chapter 8.

If it is a water-cooler condensing unit, be sure the water is turned on and check fuses in the electric circuit for proper size.

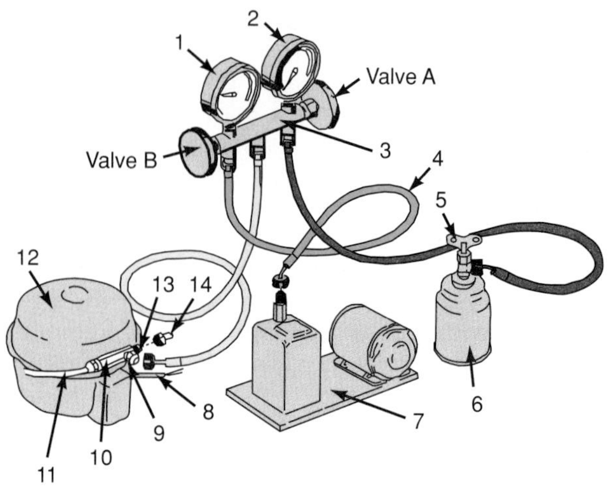

1. Compound Gauge
2. Pressure Gauge
3. Test Manifold
4. Charging Lines
5. Refrigerant Cylinder Valve
6. Refrigerant Cylinder
7. Vacuum Pump
8. Compressor Discharge Lines
9. Suction Line Valve Service Port
10. Suction Line Valve
11. Suction Line
12. Freezer Compressor
13. Suction Line Valve Stem
14. Valve Seal Cap

Figure 15-32. *Gauge manifold, vacuum pump, and refrigerant cylinder connected to small commercial hermetic motor compressor.*

For the first 24 to 48 hours of unit operation, it is good to use recording instruments. Recording thermometers, voltmeters, and ammeters should be installed on a unit. Records will make adjustments easier.

15.9 Charging Commercial Systems

When charging a system, refer to the manufacturer's directions if they are available. The manufacturer has designed and tested the products under various operation conditions. Specific charging procedures have been developed. In general, there are two basic methods used to charge a system:

- Low-side method.
- High-side method.

In the low-side method, charging small quantities of refrigerant into commercial and domestic systems is similar. It is usually done by charging into the low side (vapor method).

To charge a commercial external drive system equipped with service valves, the storage cylinder should be attached to the gauge manifold. See **Figure 15-33.** Evacuating and charging apparatus combinations are popular with service technicians. **Figure 15-34** illustrates a combination vacuum pump and charging unit.

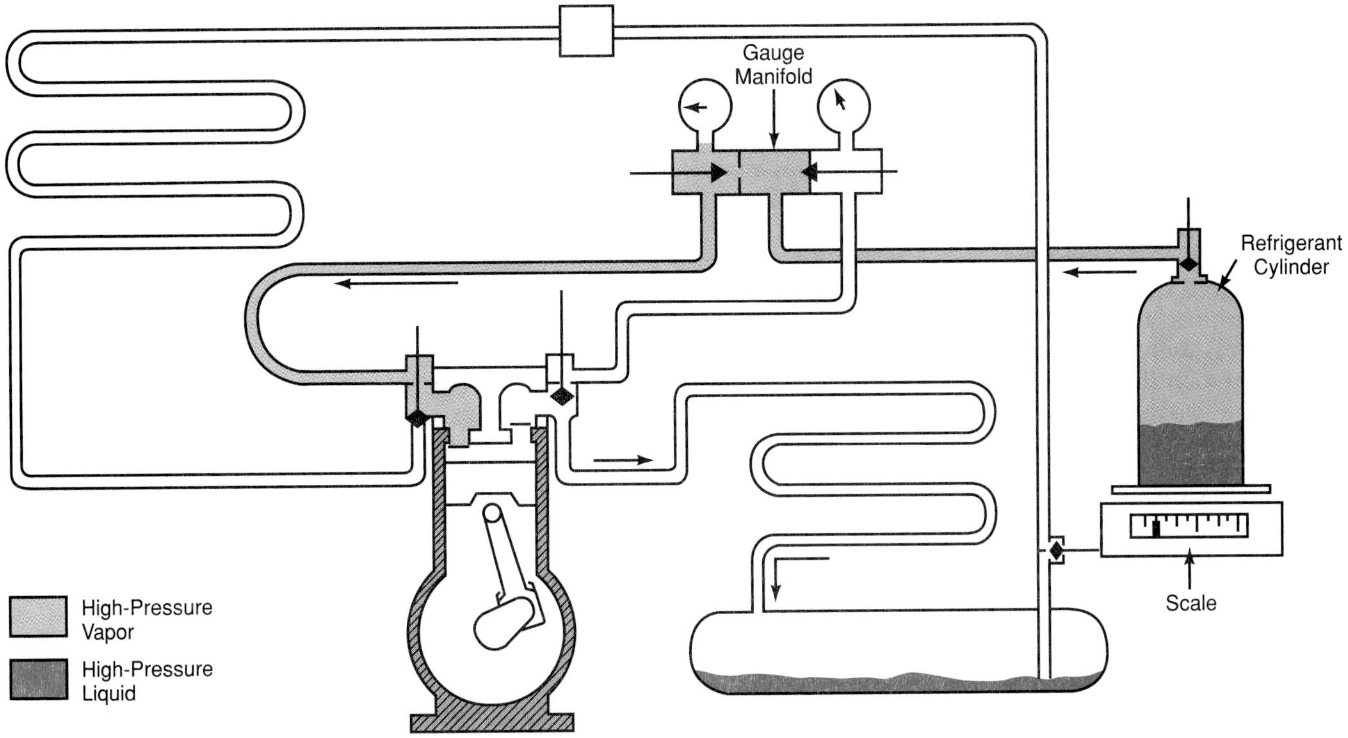

Figure 15-33. *Method of charging small external drive system with refrigerant vapor. Refrigerant cylinder is connected to manifold center opening. After purging charging lines, the SSV is turned almost all the way in. Unit is started and cylinder valve is opened just enough to keep low-side pressure within normal operation safe limits. Scale indicates amount of refrigerant being put into system.*

It is equipped with charging cylinder, gauge manifold, vacuum pump, and vacuum gauge.

Charging lines must be clean and purged to rid them of air and moisture. Connections must be tested for leaks prior to the actual charging operation. **Remember to wear goggles when transferring refrigerants.**

In the low-side method, the service cylinder serves as a temporary evaporator in the system. As the compressor runs, it will remove refrigerant vapor from the cylinder and the evaporator.

Charging may be speeded up by partly closing the suction service valve. This reduces flow from the regular evaporators and speeds evaporation from the service cylinder. Hot water may be applied to the service cylinder to help speed the evaporation. **Never use a torch to warm a cylinder.** The low-side pressure should be kept at normal levels. Too high a pressure may overwork the compressor. Pressures which are too low may cause oil pumping.

The low-side method ensures clean refrigerant due to the distilling action during evaporation of the refrigerant. You must be present at all times during the charging. A service cylinder must not be left connected into a system.

It is very important that liquid refrigerant not be allowed to reach the compressor. The liquid is not compressible. Compressor valves, and even the bearings and rods, may be ruined if the compressor pumps liquid.

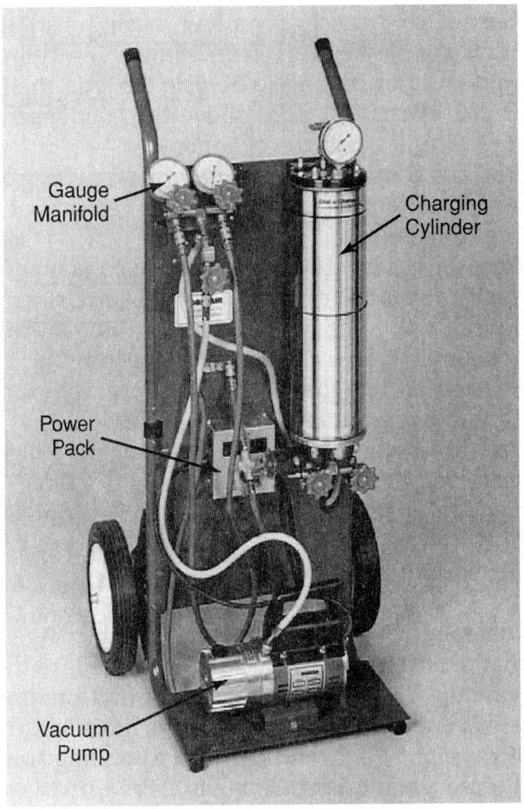

Figure 15-34. *Combination vacuum pump and charging unit. (Robinair Division, SPX Corporation)*

	R-12		R-22		R-502	
Unit Size	Flooded	Dry	Flooded	Dry	Flooded	Dry
1/2 hp	3	1 1/2	3	1 1/2	3	1 1/2
1 hp	6	3	6	3	6	3
1 1/2 hp	9	4 1/2	9	4 1/2	9	4 1/2
2 hp	12	6	12	6	12	6

Figure 15-36. *Approximate pounds of refrigerant that may be safely added to a system that is low on refrigerant.*

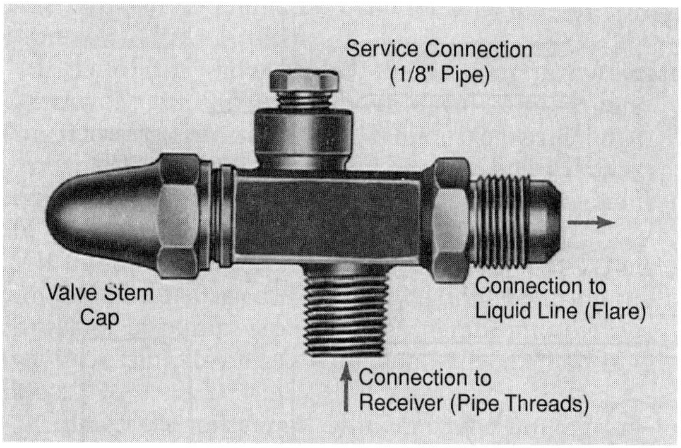

Figure 15-37. *Liquid receiver service valve with service connection.*

of each individual expansion valve should also be checked. You must also determine if the TEV adjustment is correct for each evaporator. Suction line frosting or sweating indicates whether the TEV is opened too far or not far enough. Another important procedure at this time is determining if the system has enough refrigerant.

Test for leaks after the unit has operated for 24 hours. Maintain records for use during future maintenance or service operations.

SERVICING COMMERCIAL SYSTEMS MODULE

15.11 Servicing Commercial Units

Modern commercial refrigerating units are available in a great variety of forms. Chapters 13 and 14 describe the design, construction, and operation of various mechanisms.

Small units are serviced in the same way as the domestic systems. An example of a small commercial unit is a self-contained beverage cooler of hermetic design. Details are given in Chapter 12.

Some commercial installations use an external drive system with motors and belts. Many use hermetic condensing units.

In most communities, the local refrigeration code controls the servicing of large commercial systems. Major repairs or changes to a commercial system can only

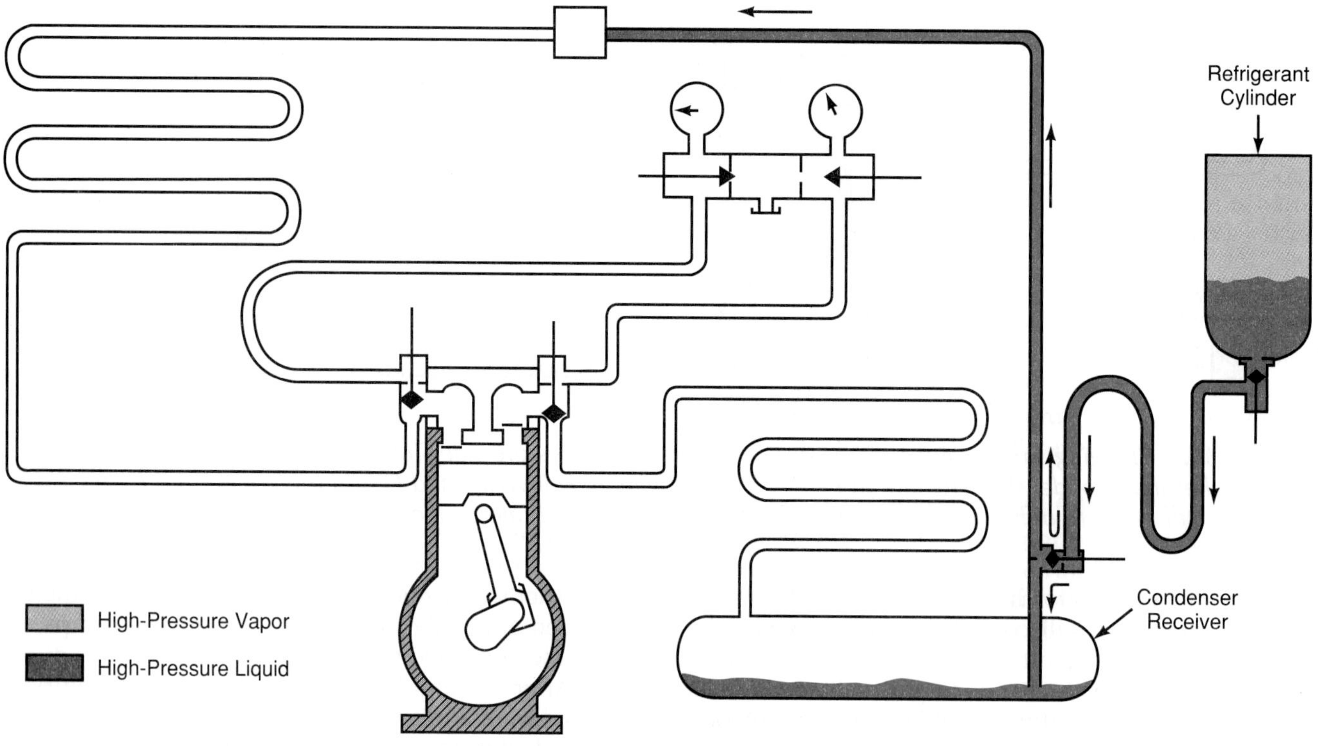

Figure 15-38. *Charging system through two-way liquid receiver service valve.*

be done by licensed contractors. When completed, their work must be checked by the local community refrigeration inspector. Plumbing and electrical service work should be performed by licensed plumbers and electrical contractors.

The servicing of commercial installations is much like working on domestic units. However, the use of multiple evaporators on a single compressor is common. Unloading and defrosting systems add to service complications.

The troubles encountered come under various headings. Examples include no refrigeration, continuous running, high cost of operation, poor refrigerating temperatures, and frosted suction lines.

Figure 15-39 illustrates a service technician measuring the pressure at a suction line service valve. He also is checking the evaporator boiling temperatures from the suction line. (A digital thermometer is mounted to the line.)

15.12 Service Equipment

When servicing refrigeration units, a set of quality, well-maintained tools is required. Also, complete records of each job should be kept in an orderly manner. Most companies provide a panel truck or pickup truck equipped with major items such as:

- Vacuum pump, recovery/recycling unit.
- Tubing and piping.
- Combination soldering, brazing, and welding outfit.
- Supply of replacement parts and materials.
  A. Controls.
  B. Fittings.
  C. Oil.
  D. Refrigerant.

Figure 15-39. *A service technician checking system pressures and temperatures, utilizing the proper tools to service the system. (Reproduced with permission of Fluke Corporation)*

- Leak detectors—especially electronic tester.
- Electrical testing instruments.

A service technician is usually expected to furnish his or her own hand tool kit. Chapter 2 describes many of the tools needed. You should:

- Keep tools clean. This will result in better and faster work and extend tool life.
- Keep tools together on the job—either in a tool kit or in the truck. Tools should be organized and arranged neatly.
- Use good lighting. Keep an extension cord and light that can be safely mounted.

15.13 General Service Instructions

Common sense is needed to service, troubleshoot, and diagnose a refrigerating system. A thorough knowledge of refrigeration fundamentals is also essential. To operate correctly, a system must have the following capabilities:

- Cooling (low side).
  A. Enough liquid refrigerant must be in the evaporator.
  B. Evaporator pressure must be low enough so that the liquid will boil at the correct temperature.
  C. Heat from the items being cooled must transfer to the liquid refrigerant in the evaporator.
- Condensing (high side).
  A. Vapor must be pumped into condenser at the correct pressure and temperature.
  B. Heat must be removed from condenser (clean condenser, airflow, or water flow).
  C. There must be enough vapor space (heat transfer surface) in the condenser.
- Refrigerant flow in liquid line. Line must be large enough. There must be minimum restrictions (pinched pipe, partially clogged screens, filters, or drier). Only liquid refrigerant should be in the liquid line.
- Vapor and oil flow in the suction line. Only a small pressure drop is allowable. The screens and drier must not be restricted in any way.

The diagnosing starts with the owner's report. Then you should check the low-side and high-side pressures and the evaporator temperature. Check the sight glass for bubbles. Feel the suction line. It should be cool. Feel the liquid line. It should be the temperature of the surrounding air (ambient).

Refrigeration equipment that has been exposed to flooding must be carefully reconditioned before attempting start-up.

Clean and dry all of the outside of the equipment. Use a detergent and bacteria cleanser. Replace all open motors or have them completely reworked.

Replace all external electrical parts. If attempting to clean and reuse them, an electric insulation leak inhibitor must be used.

Replace capacitors, relays, overload devices, and limit switches. Clean compressor terminals and spray with electrical insulation leak inhibitor. Check the electrical system completely with an ohmmeter. Check especially for grounds.

15.13.1 Removing System Parts

When part of a system needs service, empty the refrigerator cabinet. Place the contents to one side and cover them. Spread a drop cloth around and under the mechanism.

Be careful of all surfaces. Chipping or cracking may necessitate replacing a complete panel. Do not soil enamel finishes with oil or grease.

Tools and materials should be in a safe place to prevent injury from tripping. Always arrange for good lighting.

When removing any part of a system, follow these general steps:

1. Recover all refrigerant from the part to be opened.
2. Balance pressures in parts just evacuated to 0 psi (101.3 kPa).
3. Isolate parts to be opened from the rest of the system.
4. Clean and dry joints to be opened.
5. All refrigerant openings should be immediately plugged as soon as they are opened.

Refrigerant is removed by first installing a gauge manifold in the system. The service valves are properly adjusted and the compressor operated. **Figure 15-40** shows an elementary unit with the location of the three main service valves marked.

Removal of any part of the refrigerant may be accomplished as follows: A low pressure (less than atmospheric) is drawn on the part to be dismantled. This is done in order to evaporate the refrigerant from it. Then pressure is equalized to 0 psig (14.7 psia or 101.3 kPa). This is called balancing with atmospheric pressure.

The low pressure removes the refrigerant. The equalizing or balancing prevents air rushing into the mechanism when the system is opened. This last step is very important.

To begin refrigerant removal, close the inlet service valve to the part to be removed. Run the compressor until the gauge shows a 0 psig (14.7 psia or 103.5 kPa) or a slight vacuum. Stop the compressor. Then open the inlet service valve until the gauge reads zero. Close the inlet service valve to the part. Close the outlet service valve to the part. Clean and dry the joints. Remove the part. Always plug all refrigerant openings immediately after removing the part. This is necessary in order to keep out dirt and moisture.

For example, suppose you wished to remove the compressor evaporator or TEV. The refrigerant, then, is stored in the liquid receiver. The liquid receiver service valve is closed. The compressor is run until no liquid refrigerant is in the liquid line, evaporator, or suction lines. See **Figure 15-41.**

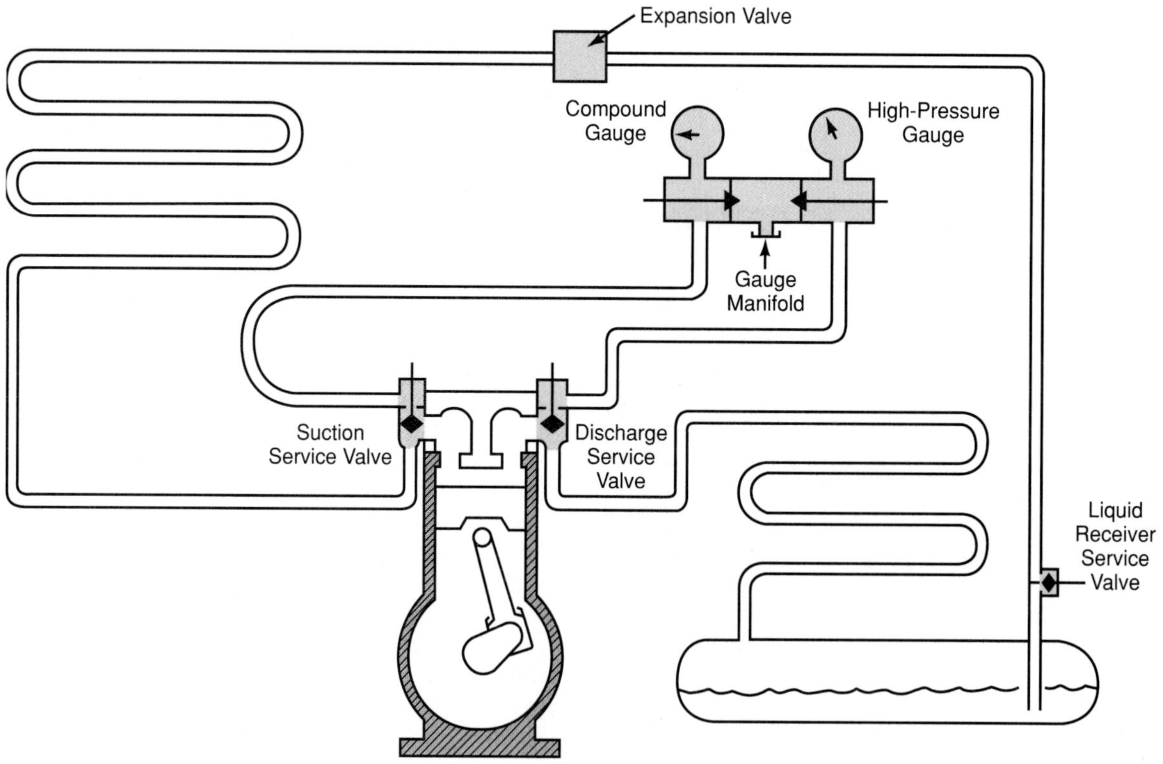

Figure 15-40. *Elementary system shows location of gauges and service valves.*

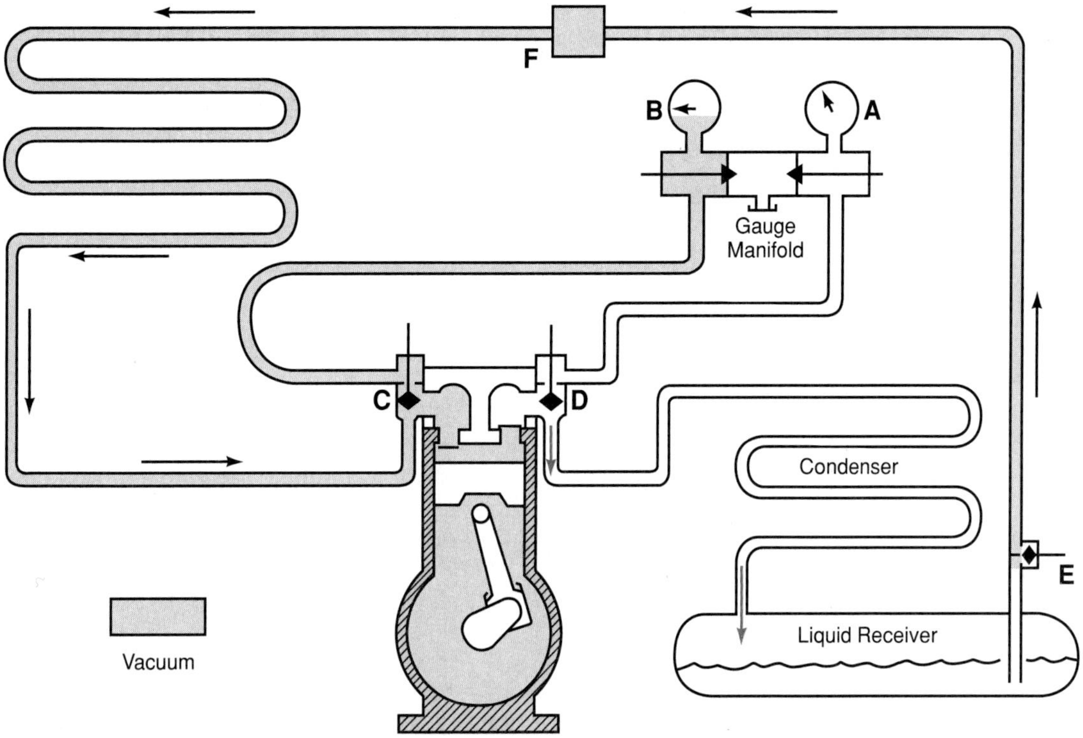

Figure 15-41. *Compressor is evacuating liquid line, evaporator, suction line, and compressor crankcase. Refrigerant is being stored in condenser and liquid receiver. Valves A and B are closed, valves C and D are in mid-position, and valve E is closed.*

Internal parts of the machine must be kept as chemically clean as possible. Moisture can freeze in low temperature passages and may cause acids and sludge to spread. Dirt (solids) clog screens and causes wearing of control valves, compressor valves, and seats.

15.13.2 Removing Refrigerant from a System

If there is no place to store the refrigerant in the system, the refrigerant must be:

- Stored in an outside cylinder.
- Purged; removed from a system by recovery/recycling equipment. Refer to **Figure 15-42.**

To remove refrigerant from a system:

1. Attach a line from a storage cylinder (if one is to be used) to the middle opening of the manifold.
2. Purge the line leading from the cylinder. (Seal the line at the cylinder and leave it open at the manifold.)
3. Crack the cylinder valve. Escaping gas will force the air out of the line.
4. Seal the line at the manifold. Close the compressor discharge line by turning the discharge service valve all the way in. Test for leaks and start the compressor. The pressure should not exceed the normal condensing pressure for the particular refrigerant. Excessive head pressures may be avoided by cooling the refrigerant cylinder with ice or water. They may also be avoided by running the compressor intermittently. (This means intervals of running interrupted by short stop periods.)

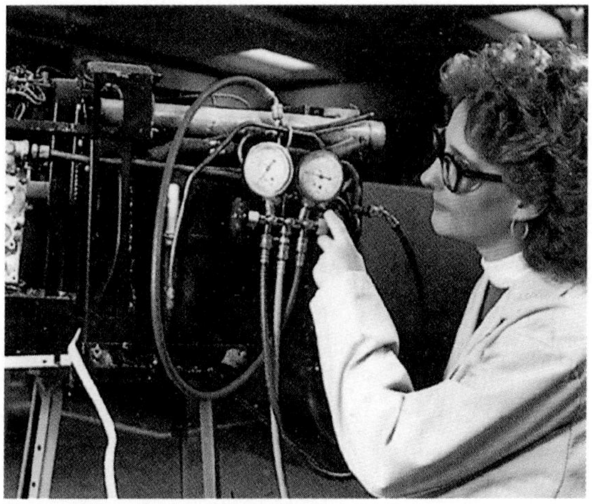

Figure 15-42. *Service technician removing refrigerant from a residential air conditioning system. Note the proper use of safety glasses. (Southeast Oakland Vocational Education Center)*

5. Allow the compressor to run with all but the discharge service valve open.
6. Shut the compressor off after a constant low pressure has been maintained for several minutes. Never allow the system to pump oil. The hydraulic pressures may cause serious damage to the compressor and lines.
7. The operation may be quickened by cautiously applying heat to the liquid receiver and to the evaporator. This should be done cautiously. Use a heat

lamp or warm water. *Never use a torch, as it may melt the fuse plugs and brazed joints.* Never allow any part or spot to become too warm to touch with the hand.

8. When all refrigerant is pumped from the system and placed in a storage cylinder, stop the compressor.
9. Open the low-side manifold valve until the compound gauge indicates 0 psi (14.7 psia or 103.5 kPa). This returns enough vapor refrigerant into the system to balance the pressure in the entire system. Any part of the system may now be removed. **Wear goggles!** Always leave the gauge manifold connected to the system until the system is opened. The pressure in the system must be known. As mentioned before, clean and dry all the connections to be opened.
10. Upon removal of any parts, the refrigerant openings should immediately be plugged.

Exhausting of refrigerant to the air depends on refrigerant type and laws pertaining to it. Unnecessary release of chlorofluorocarbons (CFCs) into the atmosphere is of concern to all service technicians. Local laws and regulations must be followed when releasing any type refrigerant into the atmosphere.

When purging refrigerant into a clean refrigerant cylinder, a purging line should be attached to the manifold center opening. The line is made of 1/4″ copper tubing. It should have a hand needle valve and check valve mounted in it. These should be mounted in the manifold. The hand needle valve should be located between the check valve and the manifold. During purging, the manifold high-pressure valve is opened.

The purpose of the hand valve is to control the amount of gas purged. The check valve prevents air or moisture backup into the unit after completely purging.

All refrigerants being purged have an oil content. Therefore purging should be done into an oil trap. Always purge in a well ventilated space.

This method cannot be used for ammonia (R-717), because of the odor. See Section 9.6.2. Most communities forbid purging refrigerant into a sewer system. Check local codes.

Refrigerant to be returned to the unit may be stored temporarily. It may also be stored temporarily if facilities are available for distilling it. A clean refrigerant cylinder should be used for storing. Remember that the refrigerant will always have an oil content. Some large companies save all refrigerant, redistill it, and process it for further use. This is good practice from the standpoint of both economy and ecology.

Federal laws govern chemical substance disposal. Refrigerant disposal is strictly classified by the following regulations:

- The United States Resources Conservation and Recovery Act (RCRA) of 1976.
- 1984 amendments to the above act.
- EPA regulations, 1990.

Large quantities of refrigerants must be stored in steel drums and moved to registered waste disposal sites. See Chapter 10 for further details concerning refrigerants containing CFCs.

15.13.3 Moisture in System

Many troubles in refrigerating systems may be traced to the presence of moisture. Moisture circulates in the presence of oil and refrigerant at high temperatures (compressor and condenser). During circulation through the system, moisture has many complex effects.

Moisture may freeze at the refrigerant control orifice, eventually clogging it. In hermetic systems it may also cause a chemical breakdown. This occurs between the oil, the refrigerant, and the motor winding insulation. Acids that ruin the motor windings are created. Immediate action must be taken to remove all moisture or to make the moisture harmless. This will eliminate many troubles.

If moisture content is high enough and temperature low enough, ice separates from refrigerants. The occurrence and extent of this separation depends on refrigerant type and temperature, **Figure 15-43.**

Many service technicians install large driers in a system on a temporary basis. These units quickly clean the system and remove moisture. **Figure 15-44** shows a drier installed in a liquid line.

To remove the unit, close valve No. 1. Wait until most of the refrigerant is removed from the drier. Then close valve No. 3, and open valve No. 2. The system is now back in operation. Then close valves No. 4 and

Temp.,	Solubility, ppm by Weight								
°F	R-11	R-12	R-13	R-22	R-113	R-114	R-123	R-134a	R-502
80	113	98	—	1350	113	95	900	1300	580
70	90	76	—	1140	90	74	770	1100	490
60	70	58	44	970	70	57	660	880	400
50	55	44	—	830	55	44	560	730	335
40	44	32	26	690	44	33	470	600	278
30	34	23.3	—	573	34	25	400	490	225
20	26	16.6	14	472	26	18	330	390	180
10	20	11.8	—	384	20	13	270	320	146
0	15	8.3	7	308	15	10	220	250	115
−10	11	5.7	—	244	11	7	180	200	90
−20	8	3.8	3	195	8	5	140	150	69
−30	6	2.5	—	152	6	3	110	120	53
−40	4	1.7	1	120	—	2	90	89	40

Data on R-134a adapted from Thrasher *et al.* (1993) and Allied-Signal Corporation. Data on R-123 adapted from Thrasher *et al.* (1993) and DuPont Company. Remaining data adapted from DuPont Company and Allied-Signal Corporation. Used by permission.

Figure 15-43. *Solubility of water in the liquid phase of different refrigerants, in parts per million (ppm), at various temperatures. The greater the solubility, the less water separation from the refrigerant in the system, and the less ice formed. (Reprinted by permission of the American Society of Heating, Refrigerating, and Air-Conditioning Engineers, Atlanta, Georgia, from the 1994 ASHRAE Handbook—Refrigeration)*

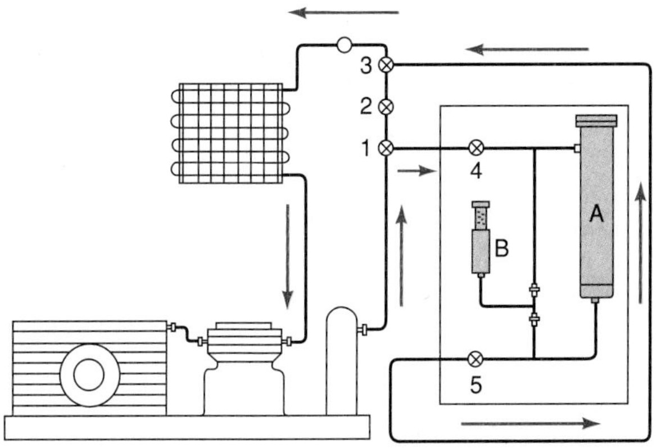

Figure 15-44. *Master drier and filter installed in liquid line. A—Drier. B—Moisture indicator. Valves at 1 and 3 are two way. Valves at 2, 4, and 5 are one way. To operate, close valve at 2. Open valves at 1, 4, 5, and 3. Red arrows show direction of refrigerant flow.*

No. 5. Remove the connecting lines and cap the openings. Be careful; there may be high pressure in the lines and some liquid.

Use of Driers (Dehydrators) and Filters

Many service calls are in response to problems caused by moisture. Sufficient moisture will form ice in capillary tubes and expansion valves and plug them.

There are a number of ways to remove moisture, including:

- Vacuum.
- Slow steady flow of hot, dry air.
- Several alternate moderate vacuums broken with dry refrigerant.
- Use of filter-driers. (Chapters 12 and 13 describe driers in detail.)

In all commercial systems, liquid line driers are a standard part of the installation. Drier size is usually based on the following:

- The horsepower of the system.
- Type of refrigerant.
- Use of system (air conditioning, commercial, or low temperature).

A 30 in^3 drier will serve for a 1-hp R-12 or R-500 air conditioning system. A 100 in^3 unit (3″ dia. by 10″ long) will serve the following:

- A 5-hp air conditioning system, a 3-hp commercial, or a 3-hp low temperature unit having R-12 or R-500 refrigerant.
- A 10-hp air conditioning system, a 5-hp commercial or a 5-hp low temperature unit using either R-22 or R-502.

Driers will hold more moisture as their temperature lowers. Install the drier as close to the refrigerant control as possible.

Remember, a drier can heat up due to shortage of refrigerant or high temperature. It may then release some of its absorbed moisture.

Keep the drier sealed until it is installed. If left open, it will absorb moisture from the air.

A drier becomes warmer as refrigerant flows and as it absorbs water from the refrigerant. This warming may be used as an indication that the drier is absorbing water. A sight glass with a moisture indicator will determine whether the system is dry.

Install the sight glass in the liquid line between the filter-drier and the refrigerant control. As the unit starts, bubbles will usually appear in the sight glass. This is normal. If bubbles continue, the system is low on refrigerant or the filter-drier is partially clogged. If not working, the filter-drier will feel cooler than normal to the touch because evaporation is taking place.

When installing the sight glass moisture indicator, avoid overheating. A temperature of 275°F (135°C) or higher will affect the chemical. Put a cold pack on the sight glass indicator when brazing connections near it. Avoid getting flux inside the sight glass indicator. Flux chemicals may affect the moisture indicator chemicals.

All driers use screens to trap solids in the refrigerant. These screens or strainers are of several types. The screens are usually made of bronze, brass, stainless steel, or monel wire. They should be 100 to 120 mesh. That is, there should be 100 openings along a 1″ rule length (10,000 holes per square inch). Popular screens are 100 by 90, 100 by 100, 120 by 108, and 120 by 120. The wire is usually .004″ to .005″ diameter. In this size wire, the openings are about .005″ square.

Felt filters may be used. These are about 1/8″ thick and are made of special material. Wool batt is also used as a filter in some driers. One company has a specially processed coarse cotton yarn wound in a diamond pattern over a metal frame. One of the latest filters makes use of powdered metal pressure castings.

Driers and filters installed in the suction line protect the motor compressor from burnout. They should always be used after a hermetic motor burnout. Dirt in a system may cause serious damage. The small solid particles act as an abrasive. They will wear the needles and seats of refrigerant controls, valves, and valve seats of compressors, bearings, and pistons of compressors. These particles may also wear through the motor insulation and cause a burnout.

Figure 15-45 shows a filter for a large system. It has a bolted assembly to permit replacement of the filter element. The service connection allows you to check the pressure drop. The gauge must not read more than 2 psig (17 psia or 117 kPa) higher than the compressor compound gauge while the unit is running. If the gauge has a higher reading, the filter element should be replaced.

A suction filter-drier is a must if oil is added to a system. The oil additives may react and make a sludge. All of the common refrigerants can be successfully dried with a drier in either the liquid or suction line.

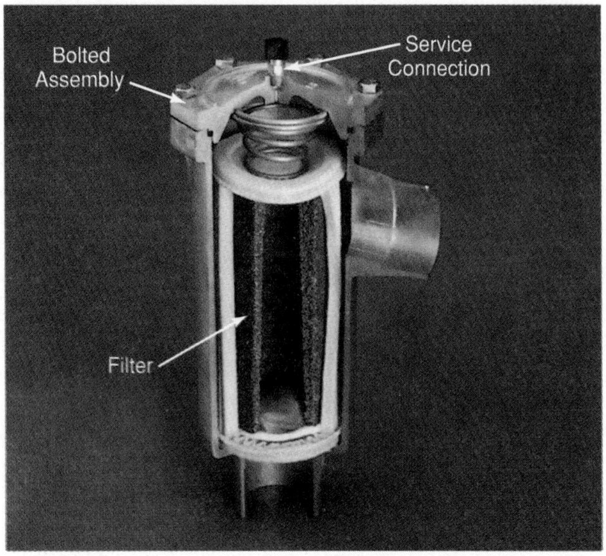

Figure 15-45. *Suction line filter with bolted flange construction permits easy replacement of filter element. Service connection is equipped with Schrader valve. (Superior Valve Company, Division of AMCAST Industrial Corporation)*

15.14 Servicing Condensing Units

Condensing units come under several divisions:

- Open (external drive) compressor.
- Serviceable hermetic motor compressor (field serviceable compressors).
- Welded hermetic motor compressor (nonfield serviceable compressors).

Compressor types may be:

- Reciprocating.
- Rotary.
- Centrifugal.
- Screw.
- Scroll.

Condensers may be:

- Air-cooled.
- Water-cooled.

The variety of mechanisms and applications is a great challenge to the service technician. Fortunately, there are certain basic problems all these condensing units have in common:

- Compressor efficiency.
- Condenser efficiency.
- Refrigerant charge.
- Refrigerant cleanliness.
- Electric circuit problems.

The compressor may be tested for efficiency as described in Chapter 12.

Symptoms of an air-cooled condenser having a lack of refrigerant are explained in Chapter 12. Water-cooled condensers present a different problem. Temperature rise of water as it travels through the condenser should be limited. Water flow should be adjusted so temperature does not exceed a 15° rise. The water passages must be clean.

If the unit is belt driven, belts should be checked for alignment and tautness.

A decidedly metallic pounding sound occurring regularly in the compressor should be looked into carefully. Check for low oil level or worn parts.

The amount of refrigerant in the system should be checked carefully. The motor control should be inspected to determine whether it trips freely. The points—if any—must be clean. Dirty or pitted contact points should be replaced.

15.14.1 Checking Refrigerant Charge

The correct refrigerant charge is very important. Several methods may be used to determine if a refrigerator has enough refrigerant.

In undercharged systems, the motor operates continuously. The motor compressor is overloaded, and there is poor refrigeration. A lack of refrigerant is indicated by an increase in liquid line and drier temperatures. A heated drier will release some of its moisture and cause a wet system.

Overcharge will cause excessive head pressure in TEV systems. Liquid refrigerant will be forced into the compressor in capillary tube systems.

A dry or expansion valve system is more difficult to check for refrigerant amounts. The appearance of the valve body may be the first sign of low refrigerant. Under normal conditions the valve body frosts over evenly. It frosts evenly as far back as the liquid line nut. When there is too little refrigerant, the expansion valve body next to the liquid line will not frost. This frost method cannot be used for above-freezing evaporator operation conditions.

Checking Refrigerant Charge: Liquid Receiver, Condenser

To determine the refrigerant amount, check the actual amount of refrigerant in the liquid receiver and condenser. One way to find this out is to determine the high-side head pressure. In a water-cooled unit, head pressure should correspond to refrigerant temperatures about 10°F (6°C) higher than the temperature of the water leaving the condenser. The water temperature, in this case, should be checked as it leaves the condenser. It should not be checked at the end of a long drain pipe. Head pressure as much as 10 psi (69 kPa) below normal indicates lack of refrigerant.

Checking Refrigerant Charge: Sight Glass

A popular way to check for sufficient refrigerant charge is by using a sight glass in the liquid line. The sight glass can be installed into the line (pipe fitting type) or clamped onto the line (electronic type). See Chapter 13 for details of both types. A sight glass allows checking for the presence of bubbles in the liquid line. Bubbles indicate insufficient refrigerant.

Figure 15-46 shows a sight glass that also indicates moisture. **Figure 15-47** pictures a see-through sight glass used in larger liquid lines.

At low head pressure, bubbles may appear regardless of the amount of refrigerant in the system. If no bubbles appear in the sight gauge, the machine probably has enough refrigerant.

However, bubbles may appear if the line ahead of the sight gauge has a restriction. They will appear even though there is sufficient refrigerant in the system. If possible, the sight glass should be mounted between the drier and the liquid line. It should be located upstream from the drier.

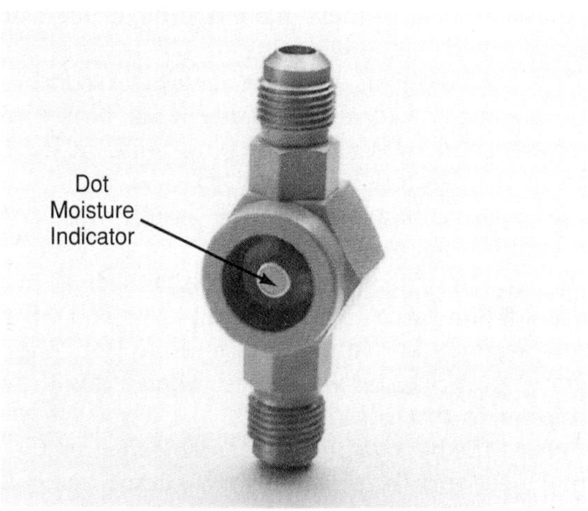

Figure 15-46. *Sight glass that also indicates if refrigerant in system is wet or dry. Green indicates a dry system. Yellow indicates wet system. Bubbles indicate lack of refrigerant. (Sporlan Valve Company)*

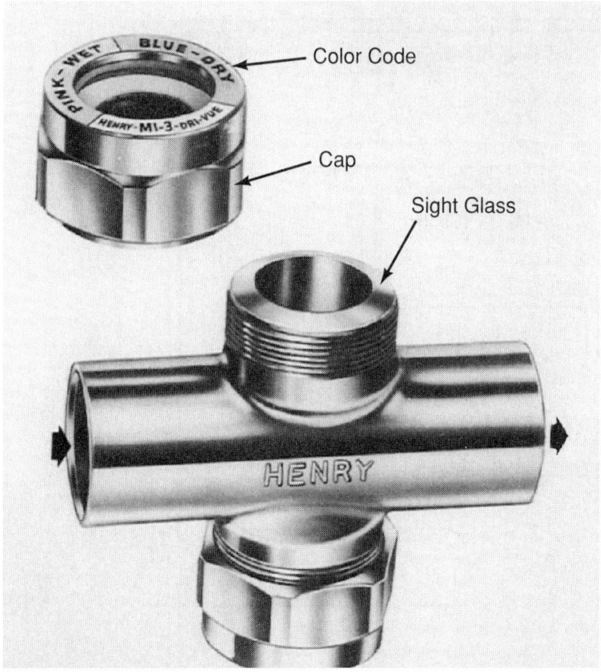

Figure 15-47. *Double port or "see-through" sight glass for larger liquid lines. Unit also indicates dryness of refrigerant. Device has brazed connections. Note dryness code on the seal cap (pink for wet, blue for dry). (Henry Valve Co.)*

Liquid Level Indicators

Some machines are equipped with refrigerant liquid level indicators. Petcocks are mounted in the side of the liquid receiver at definite heights. If liquid refrigerant comes out when the petcock is opened, refrigerant level is at least up to this height. Two petcocks are usually provided. When opened, vapor should come from the top petcock. Liquid refrigerant should come from the lower one.

Temperature Survey

Another method is used to determine the amount of refrigerant in the liquid receiver water-cooled unit (where the water coils are located within the receiver). The temperature difference is determined at two different points of the receiver shell. Part of the receiver will contain hot vapor and part will have cold liquid refrigerant. The resulting temperature difference may be easily checked by feeling the receiver with the hand.

A check of the system's components can help diagnose troubles. A quick check can be achieved by using a surface temperature probe and meter. See **Figure 15-48.**

Figure 15-48. *Temperature survey being accomplished with the use of a temperature survey instrument. (Reproduced with permission of Fluke Corporation)*

Using this instrument, you can accomplish a quick survey of:

- Compressor oil pump and head temperature.
- Evaporator coils.
- Suction line temperatures.
- Discharge line temperature.
- Condenser coils and liquid line temperatures.
- Evaporator fan motor temperature.

In this way, you can do a complete system scanning. The temperatures of various components can be determined without having to touch individual components.

Determining Quantity of Liquid in Receiver

One more method for finding the quantity of liquid refrigerant in the liquid receiver is as follows. The cooling water to the condenser is turned off. The compressor is allowed to operate. If the liquid line warms up quickly, it indicates insufficient refrigerant. Another indication is a change in head pressure after the water is shut off. It rises quickly. A rapid head pressure drop indicates too little liquid refrigerant in the liquid receiver.

Lack of refrigerant is most likely due to a leak in the equipment. A careful check should be made of all joints and parts that could possibly leak. This should be done before the unit is recharged and put into service. See Chapter 12.

15.14.2 Checking External Drive Compressors

In most cases of refrigeration failure, the compressor should be checked first. You should also look for other indications of problems while checking the compressor. However, the satisfactory operating condition of the compressor should be determined first.

The amount and the condition of oil in the compressor is important. Some compressors have an oil level sight glass or port. Others have a plug located at the proper oil level.

Two of the most common causes of compressor trouble are faulty valves and faulty seals. Noisy valves may be detected by a sharp clicking noise in the compressor as it operates.

Leaky valves may be detected as follows:

1. Install the gauge manifold and test for leaks. Turn the suction service valve stem all the way in to close the suction line. Turn the power on and off for a few seconds at a time. This is done until the danger of pumping oil is stopped. Then allow the compressor to run, **Figure 15-49.**
2. Record the best vacuum obtainable against the normal head pressure for the refrigerant being used. Also record the time. The compressor should produce a vacuum greater than 20″ Hg (34 kPa) against normal head pressure. If it does not, in most cases it should be overhauled.

A worn piston or cylinder is indicated by a clicking noise. This noise is somewhat duller than the noisy valve

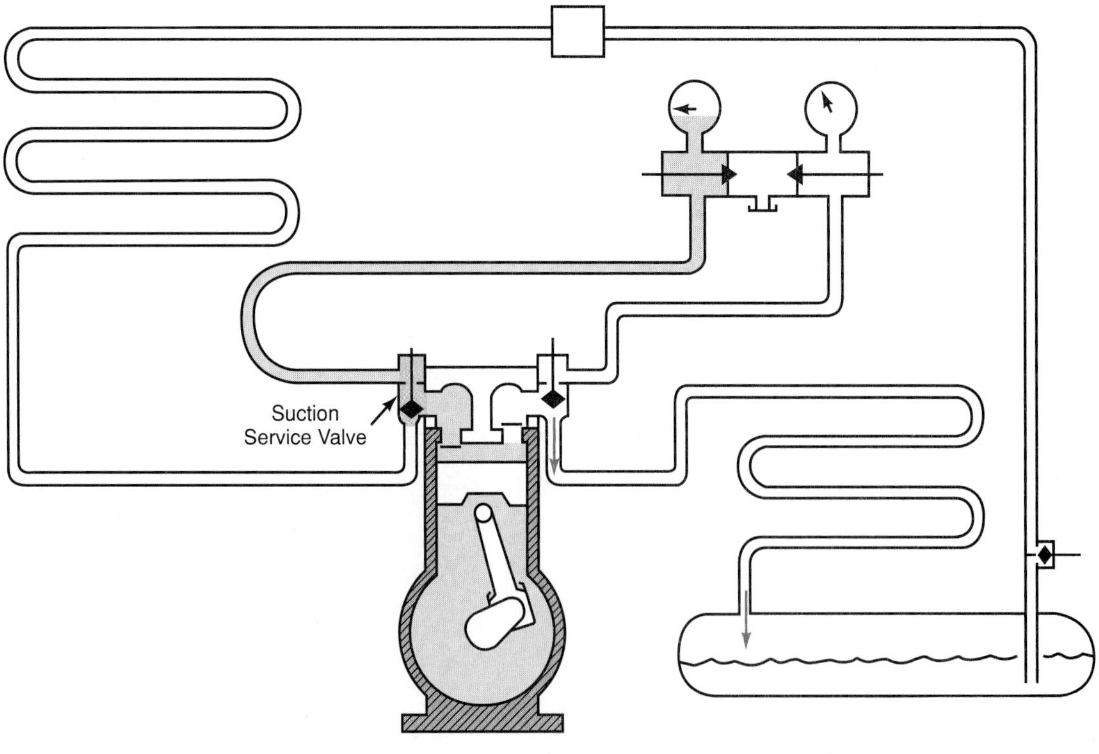

Figure 15-49. *Efficiency test for an external drive compressor. Close suction service valve and run compressor. Pumping efficiency will be indicated by maximum vacuum obtainable and time it takes to develop this vacuum.*

indication mentioned before. Worn connecting rods and main bearings are quite noisy when the compressor runs with a low-suction pressure.

The compressor must pump a specified quantity of gas at a certain pressure difference. This is necessary to do the required work. It is difficult to check this, so the methods just described are used as secondary checks. Some repair shops use a shop-mounted tank. The compressor pumps air into these tanks while being tested. The time required to pump to 150 psi (1140 kPa) is recorded for each size compressor. In this way, a relative volumetric efficiency check is possible.

Compressor testing methods:

- Ability to produce a vacuum.
- Ability to hold high pressure.
- Ability to hold both vacuum and head pressure.

If the compressor exhaust valve leaks, there are two ways to check it:

- The high head pressure will leak back through the exhaust valve. A pressure above atmospheric on the compound gauge will be produced. Therefore, the compound gauge creeping above 0 psi (101.3 kPa) when the compressor is idle indicates the exhaust valve needs repair. A leak on the low side (seal) of the compressor will only cause pressure to rise to approximately 0 psi.
- Turn the discharge service valve stem all the way in. If the discharge valve leaks, the pressure—as indicated on the high-pressure gauge—will decrease. This is especially true if the compressor is turned by hand. (The pressure will drop as the piston goes down.) Thus, if the gauge pressure fluctuates considerably, it indicates a leaky exhaust valve. If pressure merely increases and does not drop back much, the exhaust valve is not leaking.

The inability of the compressor to produce a high vacuum indicates an intake valve leak. However, the vacuum produced is maintained after the compressor is shut off provided the exhaust valve is holding.

Inability of the compressor to produce a high vacuum may also be due to other factors. These include too thick a gasket or worn piston and rings. Lack of oil will also result in poor pumping ability.

The following method is used to determine whether the crankshaft seal is leaking. Close the suction service valve. Pump as high a vacuum as possible on the crankcase of the compressor. Then turn the discharge service valve all the way in, as shown in **Figure 15-50.** Keep the compressor running. The head pressure is indicated on gauge at A. It will gradually increase with the running of the compressor if there is a low-side leak. This indicates that gas or air is being drawn in on the compressor low side.

Traces of oil at the seal or on the floor underneath indicate a seal leak. Leak detectors may also be used to check for a seal leak. (Crankcase pressure must be over 0 psig.)

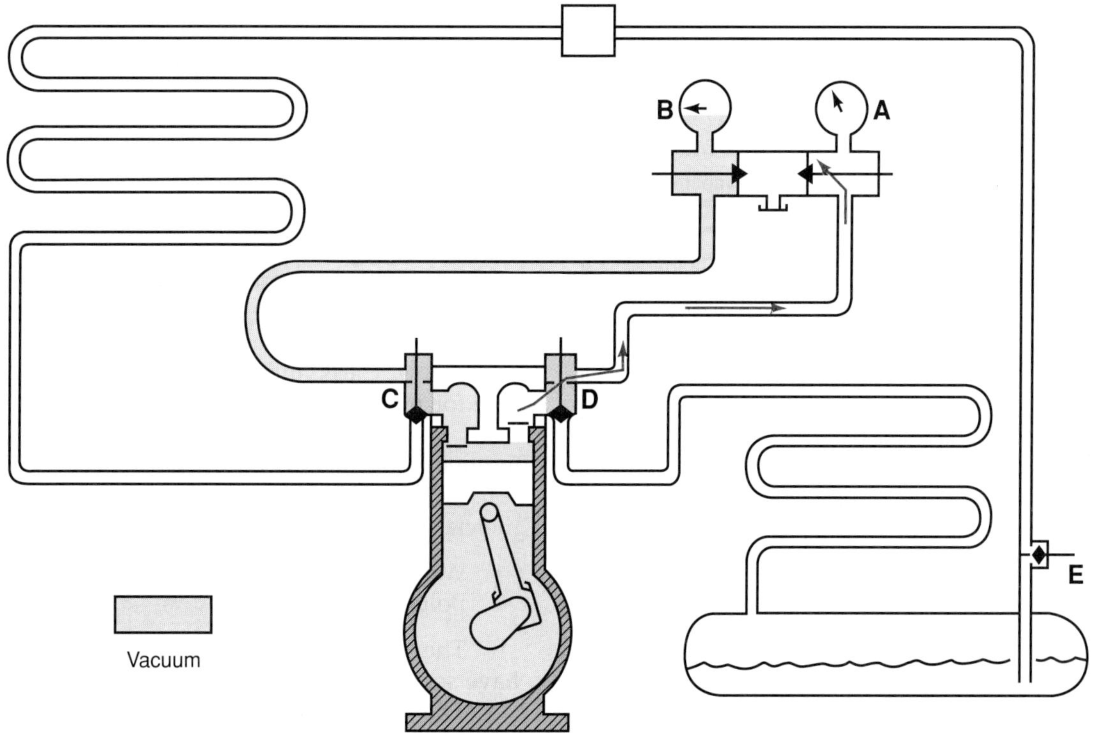

Figure 15-50. *Testing compressor for low-side leaks. This is done by determining best vacuum it can create with suction service valve at C turned all the way in. Then discharge service valve at D must be turned all the way in. If air or vapor are entering compressor, high pressure will increase.*

against freezing when working on water coolers and water chillers.

For an overcharged condition, purge the system while compressor is stopped.

Remember that the purpose of the condenser is to remove heat. The condenser will fail to do its job if the heat transfer surfaces are inefficient. The heat removing medium (air or water) must also be at correct temperature and volume.

Servicing Air-Cooled Condensers

The correct pressure in an air-cooled condenser may be determined as follows. First, add 30°F or 35°F (−1°C to 1.6°C) to the air temperature. This gives the refrigerant temperature on the inside of the condenser. Then, using this corrected temperature, refer to the refrigerant charts for the correct head pressure. See Chapter 9.

Air may be in the system if pressure is above normal and enough air movement exists. (Both air inlets and air outlets must be free.) There is also a chance that the unit is overcharged. If the pressure is below normal, it is possible that the unit is undercharged.

Most commercial air-cooled condensers use forced convection. They have one or more fans for moving air through the condenser. Larger fans are either belt driven or direct driven. Fans, motors, and belts need regular maintenance and service. Belt service operations are described in Chapters 7 and 23.

Multiple fan condensers sometimes have sequenced fans. When more condensing is needed, all fans operate. As the condensing load decreases, first one fan shuts off, then the second, and so on. Sequence controls should be checked if the fans fail to operate.

Other systems use variable speed fans. Some outdoor air-cooled condensers have thermostat controlled louvers. They are powered to partly close or completely close as the outdoor temperature decreases. If the head pressure is too low, the refrigerant control capacity is decreased.

Removing Air-Cooled Condensers

A condenser may need replacing due to leaks or other problems. If so, it must be removed from the system. Before removal, the liquid refrigerant must be taken out of the condenser. Also, the pressure must be adjusted to atmospheric pressure.

Close the valve between the condenser and the liquid receiver. Purge the condenser by removing the gauge plug from the discharge service valve. Then open the valve until the condenser pressure is down to atmospheric pressure. **Be sure to wear goggles!**

Be careful. There may be oil in the condenser. It is best to connect a purge line to the gauge opening of the valve. Run the purge line into a container to trap the oil.

If there is no shutoff valve between the condenser and the liquid receiver, refrigerant can be saved by pumping it into a cylinder. See **Figure 15-55.** The refrigerant cylinder usually has to be cooled during this operation. Otherwise, its pressure will quickly rise to dangerous levels. Run the compressor intermittently. Place the cylinder in a tub or bucket of running cold water or ice water.

Some service technicians put a spare condenser between the compressor and the service cylinder. This speeds the condensing operation.

The liquid receiver may need to be heated to vaporize the liquid in it. Use warm water. **Never use an open flame.**

To remove a condenser, first clean the condenser as well as possible. Brushes, vacuum cleaner, air or nontoxic refrigerant jets, carbon dioxide, and nitrogen jets can be used.

Most air-cooled condensers are housed in a protective shroud. This shroud also serves as an air duct. On some of the larger units these sheet metal parts are heavy. Handle with care. Gloves and safety shoes are recommended. Be sure to save sheet metal screws and/or assembly bolts in a container.

Fans, fan brackets, belts, and motors may need to be removed on some units. These parts should be labeled and stored for reuse. Be sure the fan blades are not nicked or bent. This may put them out of balance and decrease their efficiency.

If electrical connections are removed, label them. Use masking tape and a marking pencil.

Always clean the connections before disconnecting the condenser from the unit. Immediately plug the refrigerant openings. This keeps the internal refrigerant passages clean and prevents oil spills when moving the condenser.

Avoid damage to condenser fins. Wood or cardboard protectors taped over the corners of the fins will provide protection. Because fins are sharp, always use gloves when lifting or carrying a condenser.

Repairing Air-Cooled Condensers

A leaking condenser can be repaired. First clean the condenser and flush the inside of the refrigerant tubes. If a brazed joint is leaking, clean the outside of the joint. Then put flux on it, heat it, and take the joint apart.

Clean the brazed surfaces, flux the male part of the joint. Then assemble, support the joint, and braze. Remove the flux.

If a tube is cracked, remove the damaged part and replace with new tubing section. Braze the new part in place. Fins may be straightened using a fin comb, **Figure 15-56.** To test a condenser, plug one end of the condenser. Connect a refrigerant cylinder to the other end. Build up a refrigerant vapor pressure in the condenser and test for leaks. Use one of the following methods:

- Bubble test.
- Halide torch.
- Electronic leak detector.
- Immerse condenser in water.

Inspect fittings (and flares, if used). These connections must be in good condition.

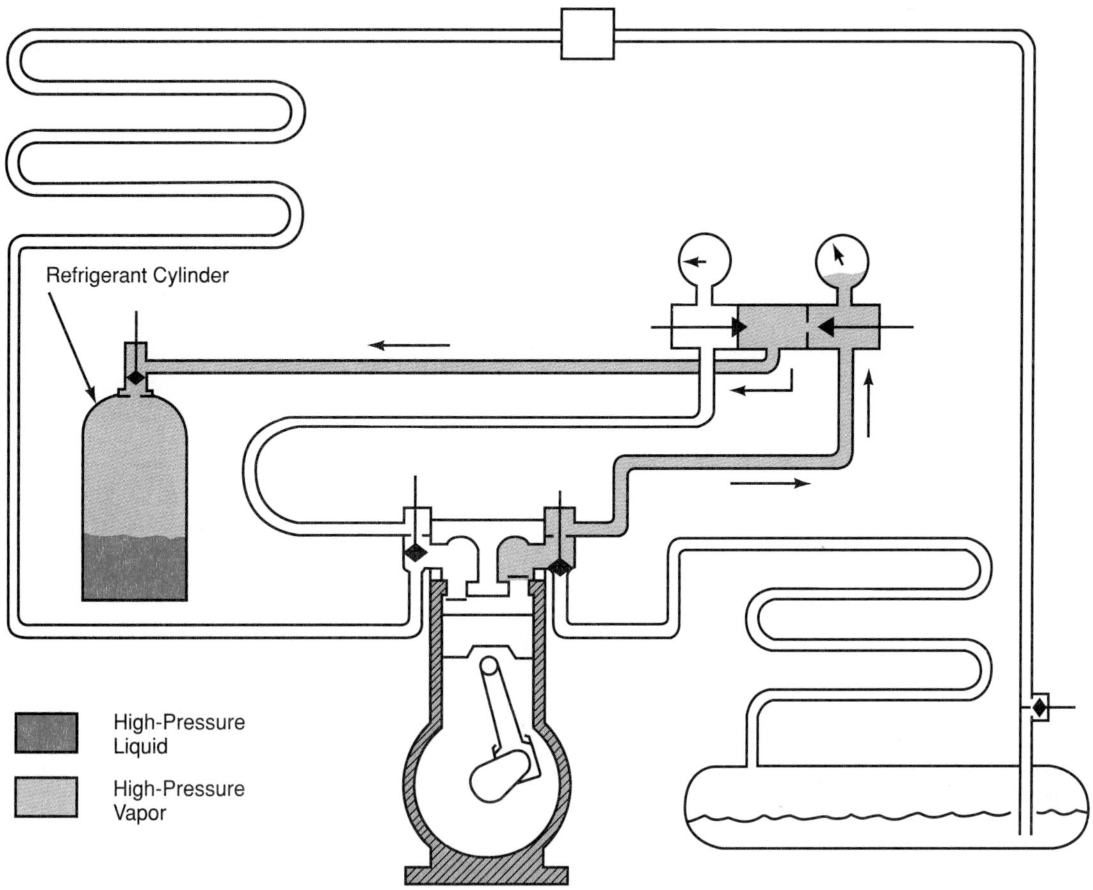

Figure 15-55. *One way to discharge a system. Refrigerant is being pumped into refrigerant cylinder.*

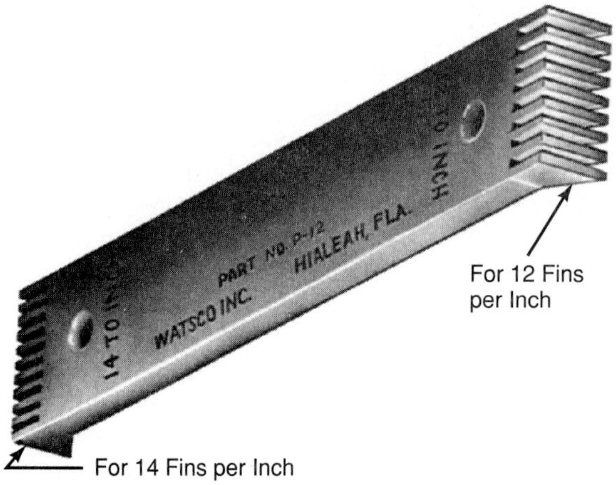

Figure 15-56. *Plastic comb used to straighten condenser and evaporator fins. (Watsco Components, Inc.)*

Installing Air-Cooled Condensers

Always protect the condenser fins and condenser tubing, including return bends. Mount the condenser securely in its frame. Install it as level as possible. Connect the condenser to the compressor and liquid receiver, if used.

Flare connectors should be carefully aligned. Fittings should be aligned so they are not under tension or forced in any way. The fittings may be out of line or under strain. If so, threads on the fittings or the flare may be damaged. Brazed connections must also be carefully aligned before brazing.

The surface to be brazed should be cleaned. Just before assembling the joint, use clean steel wool, wire brush, or dry sandpaper. Flux should be put on the outside of the male part of the fitting only. The joint should be supported during the brazing operation. See Chapter 2 for instructions on brazing. Use metal sheets to protect other parts of the unit from the brazing flame.

The air that is in the condenser can be purged. Use a small amount of refrigerant presently in the system and carbon dioxide or dry nitrogen. Build up a pressure of 15 psig (30 psia or 207 kPa) in the condenser and test for leaks. If a leak is found, do not attempt to patch the leak. Instead, take the joint completely apart and test again. If no leak is found, increase the pressure to approximately 100 psig to 170 psig (115 psia to 185 psia or 794 kPa to 1277 kPa) and test for leaks again.

If no leaks are found, install fans, belts, and motors needed to operate the system. All these parts should be cleaned prior to assembling.

Run the system. **Keep clear of the fan, belts, and pulleys as they may cause serious injuries.** Check refrigerant charge and operation of the unit. Test for leaks again. (Shut the unit off to stop airflow.) If the unit has no leaks, install the shroud or casing. These parts must be securely fastened, or annoying rattles and inefficient airflow may result.

Servicing Water-Cooled Condensers

Water removes heat from metal surfaces about 15 times more rapidly than air can. Therefore, water-cooled condensers are much smaller than air-cooled condensers. Since water is usually colder than air, the condenser temperature and pressure can be lower.

Usually, a water temperature rise is permitted as it goes through the condenser. If the water inlet temperature is high, the condenser temperature will be high. Take care to avoid freezing the water circuit of water-cooled units.

If the unit is water-cooled, add 10°F to 15°F (6°C to 8°C) to the water temperature as it leaves the condenser. This will determine what the refrigerant temperature should be. The correct head pressure may then be taken from a refrigerant chart. The unit must be running for these conditions to hold true.

If the head pressure exceeds this value by more than 5 lb., stop the compressor. Purge the system through the discharge service valve gauge opening for 10 seconds to 15 seconds. Then run the condensing unit again.

The trouble may be due to excess refrigerant or air in the system. To determine this, stop the unit and purge as before for 15 seconds to 20 seconds. If the pressure drops somewhat, the trouble was air in the system.

If the pressure does not drop, continue to purge the unit. Purge until that part of the condenser and liquid receiver full of liquid refrigerant cools. Do not allow the temperature to go lower than 32°F (0°C). The water tubes may freeze and burst.

Some condensing units have small valves that can be used to check liquid levels. A liquid level sight glass is installed in some large units. It is connected to the top and bottom of the receiver. When the pressures equalize, the liquid level in the sight glass will equal that in the receiver. This will quickly reveal the level of the refrigerant in the condenser and liquid receiver. One may easily judge whether this is the correct amount.

Always leak test the water leaving the condenser. There may be a leak between the water tubing and the refrigerant in the system.

Erratic refrigeration and constant oil slugging or pumping in the compressor indicates excess oil in the system. This frequently occurs at start-up.

A common water-cooled condenser problem is the formation of deposits from water on the tubing walls. Minerals normally found in solution in water are drawn to the walls of the condenser tubing. These materials include carbonate, sulfate, lime, iron, etc. This electrical process occurs because of opposite charges of the minerals and the tubing. These deposits then act as an insulating layer. See **Figure 15-57.** If this layer cannot be removed, the condenser must be replaced.

Two different operations are needed to clean a condenser:

- Preventing scale.
- Cleaning the system.

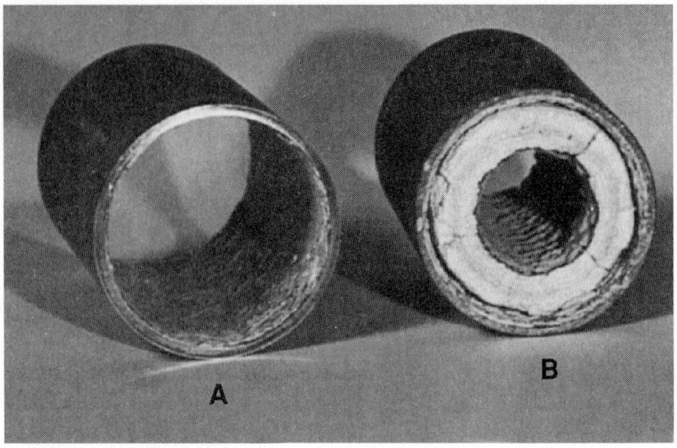

Figure 15-57. *Scale deposits. A—Tubing with treated water. B—Tubing using hard, untreated water. (Scale Free Systems, Inc.)*

A sulfuric acid-chromate solution is good scale prevention in open systems such as cooling towers and evaporative condensers. However, chromates are poisonous, and no significant amounts should be allowed in waterways. **Wear goggles, rubber gloves, and a rubber bib apron!**

Sulfuric acid in weak solution reacts with steel. Sulfuric acid in strong solution reacts with copper. It is best not to use it for scale prevention.

Careless mixes of sulfuric acid and water can ruin copper piping system and steel structures. Some technicians do use acids to descale, but it must be done rapidly. Then it must be cleaned as quickly as possible. **Wear goggles and rubber gloves!**

For cleaning condensers, use only prepared chemicals. Carefully balanced cleaners, such as inhibited muriatic acid or sulfuric acid, work well. They will not damage the equipment, if properly used.

A corroded condenser can be recognized by checking the liquid line temperatures of the refrigerant. A corroded condenser will produce a hot liquid line in a water-cooled condenser. (This holds true provided the refrigerant amount is correct and other troubles just mentioned are not found.) Temperature and pressure in the condenser will be considerably higher than expected. Eliminate all other possible causes for excessive head pressure. When these possibilities have been eliminated, a badly corroded or dirty condenser is probably the cause.

Soft ***scale deposits*** can be removed from some water-cooled condensers with a power-driven wire brush. See **Figure 15-58.** This can usually be done without removing the condenser from the system. However, the water circuit must be closed and the unit shut down. Always use new gaskets and tighten assembly cap screws evenly.

There are "scale-free" systems which effectively eliminate the electrical process that causes deposits. **Figure 15-59** illustrates a scale-free system. It exposes an electrolyte to the water in the system being treated. This

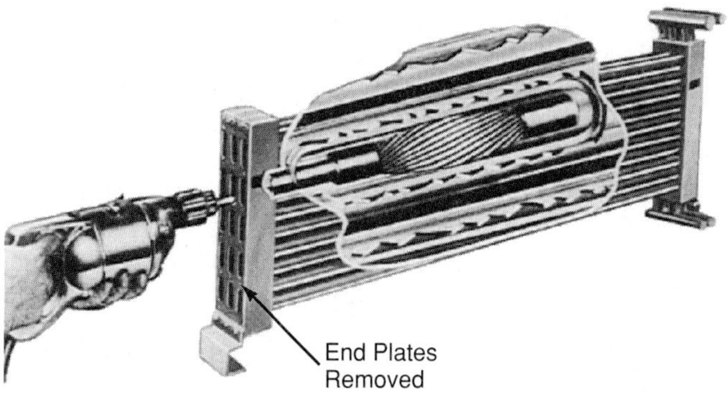

Figure 15-58. *Tube-within-a-tube condenser being cleaned with power-driven wire brush. (Standard Refrigeration Co.)*

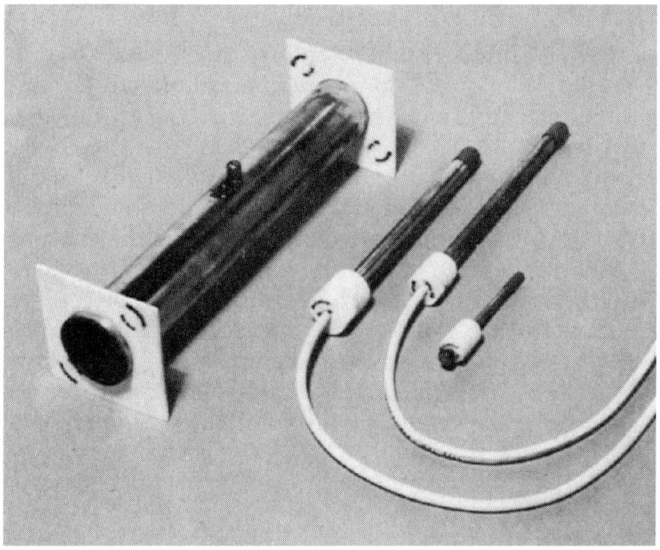

Figure 15-59. *Four different size units used in equipment that has water as a heat exchanger. These units will remove and prevent the formation of scale. (Scale Free Systems, Inc.)*

is done through the use of a positive ground rod. This process, in turn, picks up the electrical energy from the water. It is grounded to the outside of the boiler condenser. Thus, the inside tubing is kept free from scale. This electrical energy also causes existing scale or deposits to go back into solution. They are then carried through the system and discarded.

Removing Water-Cooled Condensers

A leaking condenser should be removed and repaired or replaced. Removing a water-cooled condenser is similar to removing an air-cooled condenser. However, the water line must be disconnected. To do this, first close off the water circuits. Then disconnect the water lines and drain all the water from the condenser. (It may freeze.) Remove the refrigerant from the condenser and isolate the condenser. Do not allow freezing temperatures.

If a pressure-operated water valve is used, leave it installed in the system, if possible. This is necessary because of its refrigerant tubing connection.

Clean the outside of the condenser, wipe away the water, and clean the condenser connections. Dry the connections thoroughly. If the connections are mechanical, use wrenches of proper size. **Wear goggles!** Plug the refrigerant openings at once using good plugs. Either synthetic rubber expanding plugs or flared plugs are recommended.

If the condenser is heavy, use a lifting device. When a lifting device is not available, have two or more individuals move it.

Repairing Condensers and Receivers

When a condenser or receiver malfunctions, replacing the unit is usually cheaper than repairing it. Welding or brazing is sometimes used to repair leaks, but it requires careful work. If possible, weld or braze all joints to ensure a lasting, leakproof joint.

A welded-shell condenser can be cut open. A new water coil can then be installed and the shell rewelded. However, this is done only in an extreme emergency because pressure vessels should be made under well-controlled conditions. Repairs should be made by a pressure vessel certified welder. After being repaired, the vessel must be tested at twice the operating pressure.

Liquid receivers in most commercial systems serve as refrigerant storage cylinders. They usually have a welded steel shell.

The shell and coil condenser also develops leaks. (This type of condenser has a water coil built into the liquid receiver.) There is a corrosive action of the water and refrigerant under certain conditions. Eventually, the copper tubing used to carry the water may be corroded. The leaking tube lets refrigerant from the system into the cooling water. This type of leak may be found by checking for refrigerant at the water drain. Leaks sometimes occur at the joints where the water-cooling coil attaches to the liquid receiver. Such damage may be due to abuse or to corrosive action. In such a case, the condenser should be replaced with a new one.

If new parts cannot be found, a fairly satisfactory repair may be made as follows:

1. The water tubing—if corroded within the liquid receiver—must be removed.
2. The liquid receiver should be mounted on a lathe. The end of the receiver should be cut open. This permits removing the old water coil and putting in a new one. The replacement unit must have the same length of tubing as the one removed. Otherwise, the capacity of the condenser will be changed. The new coil is usually made up by winding it on a drum mounted on a lathe. The tubing is then put in the liquid receiver and the joints are brazed.
3. After the interior has been thoroughly cleaned, the end may be replaced and welded.

Large liquid receivers are equipped with safety release valves. Receiver repair should be attempted only with permission of local inspectors.

Installing Water-Cooled Condensers

Installing a water-cooled condenser is similar to the installation of an air-cooled unit. The condenser mounting, joint brazing, and leak testing should be done with the same care. On a water-cooled unit, the leak test should include the exhaust water.

Connect water lines according to the local plumbing code. Test the water circuit for leaks also. All parts should be cleaned before assembly. Be sure to test for leaks after the unit has run for a few hours.

Servicing Water Valves

Water-cooled condensers often require attention because of incorrect water flow. This trouble is sometimes due to the water valve or to the screens in the water circuit. See Chapter 13 for details on how these valves are built.

The water valve provides water while the unit is running. It stops the water flow when the unit is idle. Some trouble caused by water valves are:

- Inadequate water flow.
- Excess water flow.
- Water flow does not stop when the unit is idle.

A water control valve will only operate correctly if installation is correctly made. Water must also be clean.

Restricted Water Flow

If water flow is inadequate, the cause might be:

- Leaking valve.
- Clogged screen.
- Chattering valve.
- Valve incorrectly adjusted.
- Sediment-bound valve.
- Leaking bellows.

In addition to these possible problems, the water-cooling system may also be the cause. Some water-cooled systems use a length of rubber or plastic hose between sections of the water pipe. This hose is run along the wall and the condensing unit water lines. It eliminates the transmission of the condensing unit vibration into the building's plumbing system. It also prevents damage to tubes caused by this vibration.

The cold water inlet hose connection presents no difficulties. This hose will ordinarily give many years of service. However, the water outlet hose may decompose and clog.

Occasionally, someone may partially or completely shut off the water supply. This happens when the hand-operated valve installed in the system is closed. You should always put signs near the shutoff valves, warning of the effect on the unit if these valves are closed.

There are two common signs that indicate troubles with water circulation. One is the lack of cooling in the condensing unit. The other is too great a consumption of water.

If the water circulation is stopped, the refrigerating system will start to short cycle. The high-pressure switch causes the short cycling. These systems are always provided with a high-pressure safety motor control for this purpose. As the head pressure of the machine builds up (due to a lack of cooling), the control opens a switch, stopping the motor. Once the motor is stopped, the head pressure drops rapidly. This permits the high-pressure control to turn the motor on again.

Short cycling will continue unless the trouble is remedied. Such a condition is a severe strain on the motor. Furthermore, it does not provide satisfactory refrigeration.

Excess Water Flow

Excess water flow will give satisfactory refrigeration, but more water will be used than needed. This will increase the cost of operation. Three things may cause this condition:

- Water pressure too high.
- Water valve leaking.
- Water valve incorrectly adjusted.

Water pressure that is too high is seldom found, unless the water supply pressure is uncontrolled. If found in one machine, it may be true for all the machines in that locality.

The condition may also be due to either a leaking valve or one that is open too far. To determine which is true, first, find out if anyone has worked on the machine. If not, and if trouble has just started, there is usually a leaking water valve. A further indication of this is a continuous flow of water on the Off cycle. A leaking water valve is often caused by foreign matter between the valve and seat. This material can usually be dislodged by flushing the valve. The flushing may be done by prying the valve open.

Tracing Water Circuit Troubles

If there are water circuit troubles, the water valve may be at fault. Or, the problem may be at some other part of the water circulating system. The exact source of the trouble must be located. To do so, disconnect the joints where the hose fastens to the system.

Disconnecting the water outlet pipe will indicate if the water is flowing as far as this point. If so, the connection should be resealed. The other end of the pipe must be disconnected from the wall pipe. If water does not flow up to this point, the trouble is probably in the drain pipe.

To determine whether water is coming as far as the water valve, disconnect the pipe used as the inlet connection. If water flows through the pipe, the water valve and the condenser proper are the problem. To check these sources, reseal the inlet pipe to the system and disconnect the water valve from the condenser.

The water may flow through the water valve, but not the condenser. If so, the trouble is a major one. Replacement or cleaning of the condenser water tubes is needed. The water may not flow through the water valve. In this case, the water valve must be disconnected and repaired or replaced.

Water hammer is a very noisy condition that is easily recognized. A single, distinct thump (rap) is heard in the pipes just as a valve closes. Generally, this condition can be corrected by placing a short, vertical pipe in the water line just ahead of the valve. It provides an air column, absorbing the shock of the sudden stopping of water flow.

Removing Water Valves

To remove an electrical solenoid water valve from the system, open the hand switch that controls the circuit to the motor. Remove the water valve wires from the motor circuit. If these wires are soldered and taped, you must unsolder or cut them. Most are slip-on connections.

Before disconnecting the water valve from the water system, shut off the water supply. This is done using the hand valve. If no replacement valve is available, temporarily connect the water system without it. The water flow can be regulated with the hand shutoff valve.

Some pressure-operated water valves are difficult to remove from the system because the valve is connected to the high-pressure side of the condensing unit. The pressure tube for these valves is usually connected into the compressor cylinder head. However, some manufacturers connect this tube into the liquid line. Sometimes this tube has a hand shutoff valve. If so, removal of the valve is simple.

If the tube is connected to the cylinder head of the compressor, the following procedure is suggested:

1. Install the gauge manifold.
2. Turn suction service valve all the way in.
3. Run compressor until pressure in crankcase reaches 0 psi. Note: Be very careful of oil pumping, which sometimes occur before 0 psi is reached.
4. Heat the water valve line and the water valve bellows carefully with a heat lamp. Do this for three or four minutes until both are quite warm to the touch. This operation will move the liquid refrigerant that has condensed in this tube and valve back into the condensing unit. Then, only a small quantity of high-pressure vapor will be left in this tube.
5. Turn the discharge service valve all the way in.
6. This bypasses high pressure in the manifold and water valve refrigerant line into the low side. Open both manifold valves.
7. Clean joints to be opened.
8. Disconnect pressure tube from water valve. **Wear goggles!**
9. Plug the refrigerant pressure tubing openings immediately.
10. Gently heat the water valve again. Often, a quantity of liquid refrigerant becomes oil-bound within the bellows chamber. It releases with explosive force a few minutes after the valve is opened to the atmosphere. Heating will drive it out.
    Note: Be very careful not to point the refrigerant openings toward anyone. There is danger of being hit by the refrigerant. It is best to wrap the refrigerant openings with several layers of heavy toweling. The toweling will absorb any refrigerant being thrown from the mechanism.
11. Shut off water supply.
12. Disconnect water valve from water line. Replace it with a new valve or connect water lines directly. Now the water valve is ready to be dismantled and repaired.

Some water valves permit removing valve body without disturbing the refrigerant connections. To disconnect one of these valves, shut off the water. Disconnect the valve body from water lines and bellows body.

Thermostatic water valves or motorized water valves are easily removed. Only electrical connections need to be broken and the water circuit closed.

Repairing Water Valves

It is better to replace a worn water valve than to repair it. Sometimes a water valve only needs cleaning. A muriatic acid solution and wire brushes work best. Use the same precautions as when cleaning the water tubes of a condenser.

If no replacement valve is available, the valve and valve seat must be repaired. The valve seat is brass in most cases. It may be lapped in a manner similar to lapping a compressor valve seat.

The valve is usually made of fiber, rubber, or Bakelite material. It should be replaced. In an emergency, however, this valve may be trued-up. This is done by using a fine grade of sandpaper backed up by a level surface.

Occasionally, a packing gland is used to seal the joint. This is done where the valve stem passes into the valve body proper. The packing is usually composed of graphite, lead, and other materials. If the packing nut is turned all the way down and this joint still leaks, replace the packing.

Electric water valves may have faulty electrical coils (either shorted or with an open circuit). Replace the coil using one with the same electrical properties (voltage and wattage). Thermostatic water valves may lose the element charge. If so, replace them.

Installing and Adjusting Water Valves

After cleaning and repairing a pressure-operated water valve, test and adjust it. Only after testing and adjusting should it be placed in service. If maximum water supply temperature is 75°F (24°C), adjust the valve to open at these pressures:

87 psig (102 psia or 704 kPa) for R-12
92 psig (107 psia or 735 kPa) for R-134a
144 psig (159 psia or 1097 kPa) for R-22
152 psig (167 psia or 1152 kPa) for R-717
112 psig (127 psia or 876 kPa) for R-500
173 psig (188 psia or 1297 kPa) for R-502
182 psig (197 psia or 1355 kPa) for R-404A

If the water inlet temperature is not 75°F, adjust valve to its correct opening pressure. Refer to **Figure 15-60.**

A valve should also be tested for leaks while it is being adjusted. To do this, connect an air pressure line to the valve's water inlet. The pressure operating bellows controls the water flow. To test it, connect another air line and a pressure gauge to this fitting. No air should flow

	Water Temperature, °F										
	50	55	60	65	70	75	80	85	90	95	100
Refrigerant	Head Pressure, (psig)										
R- 12	56	62	68	74	80	87	93	101	108	117	125
R-134a*	55	61	68	76	84	92	101	110	120	131	142
R-22	95	104	113	123	133	144	155	158	180	194	208
R-717	98	108	119	130	140.5	152	164	177	191	205	220
R-500	71	78	86	94	103	112	121	131	142	153	165
R-502	116	126	137	148	160	173	186	200	214	230	246
R-404A*	120	131	143	155	168	182	196	212	228	245	263

*(*Information provided by DuPont Company)*

Figure 15-60. *Table of head pressures for systems with various refrigerants at various inlet water temperatures.*

through the water valve until correct control bellows pressure is reached. Adjustment may be made to obtain this condition.

After installing the water valve, check it for leaks (both water and refrigerant), outlet water temperature, water flow, and condensing pressure.

Servicing Cooling Towers

Evaporative condensers and cooling towers collect deposits from the cooling water. These deposits must be removed periodically or they will act as insulation. Deposits may be reduced by using water softening chemicals. Such chemicals can be bought from wholesale supply companies.

Chemicals in water are measured by a pH factor. The scale of pH is from 1 to 14. An acid solution is indicated at 1 through 7. A basic condition is indicated at 8 through 14. Chemicals may be added to the water to create a pH of 7 or 8 in the water.

The water temperature is important when testing for pH (acidic or basic). The warmer the water, the more active the reaction to probes or color testers. It is best to test the water between 70°F and 80°F (21°C to 27°C). Water near boiling is about 15% more active. Very cold water (near freezing) will give readings about 5% below the true value.

Chemicals may be used to lessen algae, mold, and slime growths. Deposits can be removed by scraping or by using a weak acid solution. This should be followed by a soda solution rinse and wash. A water-cooled condenser must be protected from freezing temperatures. An electric heater or automatic drain controls provide such protection.

When shutting down a water-cooled condensing unit, condenser coils must be completely emptied of water. This is important in case the unit is exposed to freezing temperatures. This may be done by blowing out the coils with air, nitrogen, or carbon dioxide. Do not exceed a 60 psig (75 psia, 449 kPa to 518 kPa) pressure or the system may be damaged.

The water drain valves should be left open. This will allow drainage of residual water in the piping. Be sure the drain plug of the circulating pump is removed and left loose.

Cooling towers need regular maintenance. Once a year, repair corrosion spots. It is good practice to do the following monthly:

- Inspect fan and motor bearings and oil sleeve. Grease (with water inhibitor) ball bearing.
- Inspect belt tightness and alignment and adjust, if necessary.
- Clean strainer.
- Clean and flush pump.
- Inspect water level; adjust float, if necessary.
- Inspect spray nozzles and clean if necessary.
- Inspect water level bleed; it must be working.
- Inspect air inlet screens and clean, if necessary.
- Inspect water for algae, leaves, or other dust particles.

During cool weather, cooling towers and evaporative condensers create a fog exhaust. The humid air leaving the unit condenses enough moisture to create this fog. (Moisture is determined by the dew point in the air.) The fog can be reduced by cutting down cooling tower operation in cold weather. It can also be reduced by increasing airflow and reducing water flow.

Plan and install cooling tower drain lines as carefully as refrigerant and water lines. They must be large enough to be easily cleaned and must have clean-out connections. The joints must be leakproof. Piping must have a downward slope (1/4″ per foot for horizontal runs). If a rise is unavoidable, a pump must be used.

All lines exposed to freezing temperatures should be insulated and heated. Heating tape can be used. Most outdoor air-cooled condensers, cooling towers, and evaporative condensers use motor-driven fans. Motor bearings, belts, and fan bearings are exposed to wide temperature changes, moisture, and dirt. Plain bearings require a special lubricant that will not wash out under moist conditions. Silicone oils and greases are adequate. The same lubricants are also good for ice makers, water pumps, and hydronic heating pumps.

Good city water systems deliver water with about 120 parts per million (ppm) of dissolved solids. This water can be cycled through a cooling tower only about six times. After that, scale will start to form. For this reason, most water must be treated. Moreover, there must be a bleed-off device to keep the evaporating water system clean.

Water that evaporates should be in vapor form. Small droplets of water exhausted from the cooling tower contain solids, adding to air pollution.

One of the newer ways to treat water is electrostatically. A device exposes untreated water to an electrostatic force. The force loosens the bond in the scale forming chemical. The loosened bond prevents scale from forming and prevents impurities from combining. It even loosens scale already formed on metal surfaces.

The basic theory behind water treatment is to reduce the ion content in the water. It must be reduced to a condition where salts and foreign matter are electrically neutral. These materials then remain in solution.

The electrostatic treatment must be adjusted to the raw water condition. Electrical instrumentation and chemical analysis are used to determine the adjustment.

15.14.4 Servicing Ice Makers

Ice makers have refrigerating systems similar to other cooling applications. However, they have special water circuits, defrosting devices, and ice cube or flake moving devices. **Figure 15-61** shows a typical ice cube maker and dispenser.

Those parts of the ice cube maker, which are in contact with water, should be cleaned about once a month. Special commercial cleaners for ice makers are available.

Figure 15-61. *An ice cube maker and dispenser frequently used in hotels. (Hoshizaki America, Inc.)*

You should be familiar with some of the basic terms used in servicing ice makers:

Fill cycle: When a unit fills to a given level with water.
Harvest cycle: The removal of ice cubes from the evaporator by warming to 48°F (8.9°C). This occurs when the hot gas valve, located in the refrigeration system, opens.
Freeze cycle: The formation of ice on a surface.
Spray tubes: Water distributor, or spray tubes, are two separate water circuits. The center circuit is a series of small holes. These direct the inlet water down between the evaporator plates during the harvest cycle. The two outer rows are large holes. They direct the water over the evaporator freezing plate during the freeze cycle.

This system can be broken down into four simple steps:

1. Fill cycle **(Figure 15-62).** When power is on, the water valve is opened, and the unit begins to fill. A float valve switch indicates when the proper amount of water is in the system.
2. Harvest period **(Figure 15-63).** When the compressor starts, the hot gas valve opens. The water valve remains open. The evaporator warms. A thermistor is located on the suction line. When it measures a 48°F (8.9°C) temperature, the defrost cycle is stopped.
3. Freeze cycle **(Figure 15-64).** The water allowed to go through the spray tubes determines the amount of ice produced. The water level is determined by the float switch. Heat is removed from the water by the refrigerant passing through the evaporator. As this occurs, ice forms, creating the ice cubes. See **Figure 15-65.**
4. Harvest-pump-out-cycle. Upon completion of the freeze cycle, the float switch and hot gas valve open. The Harvest begins. Then the unit goes through a defrost cycle.

The basic operational sequence chart is shown in **Figure 15-66.**

The electrical circuit controls needed with automatic ice makers vary with each machine. The electrical circuit for the unit just described is shown in **Figure 15-67.**

Ice machines use a number of different refrigerants. R-22 and R-502 are most common in older machines. Now R-404A is also being used in newer machines.

15.14.5 Servicing Direct Expansion Evaporators

Dry evaporators are coils using an automatic expansion valve (AEV) or a thermostatic expansion valve (TEV) refrigerant control. See Chapter 5.

These evaporators must be clean both inside and out for good heat transfer. They must contain the proper amount of liquid refrigerant at the proper vapor

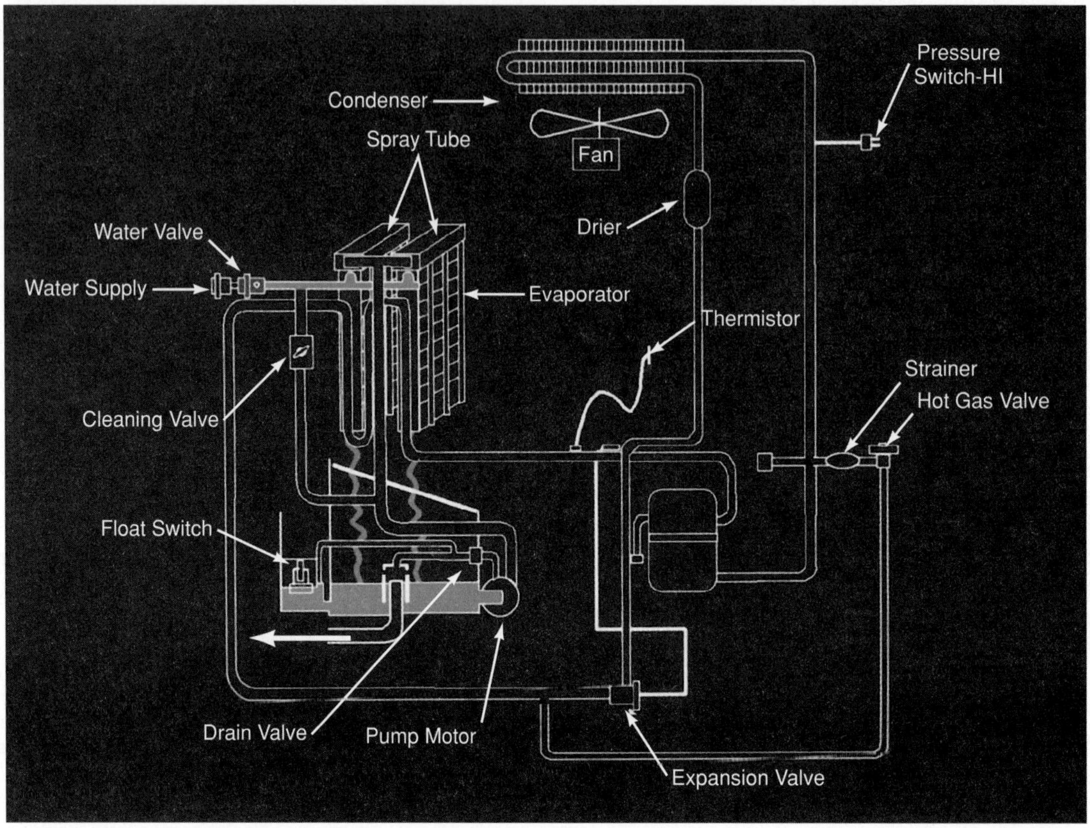

Figure 15-62. *A one-minute fill cycle. Note the water valve is opened, and the flow of water. (Hoshizaki America, Inc.)*

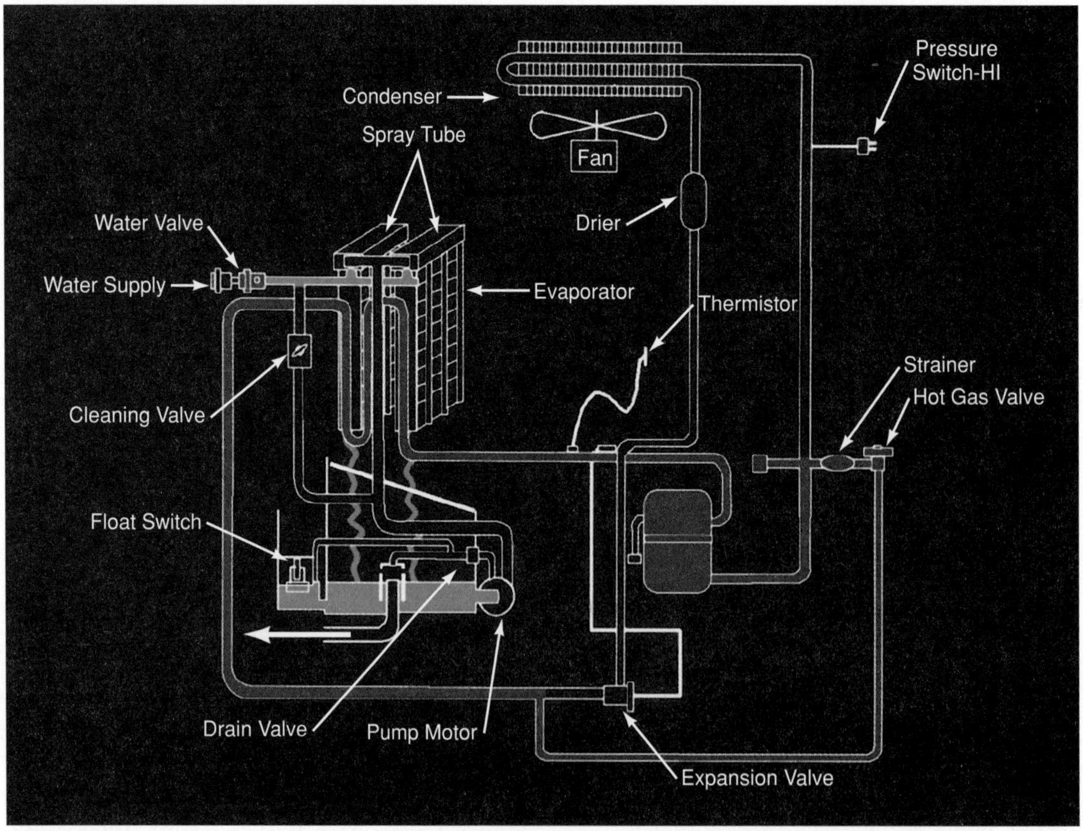

Figure 15-63. *The initial harvest cycle. The hot gas valve and water valve are open. (Hoshizaki America, Inc.)*

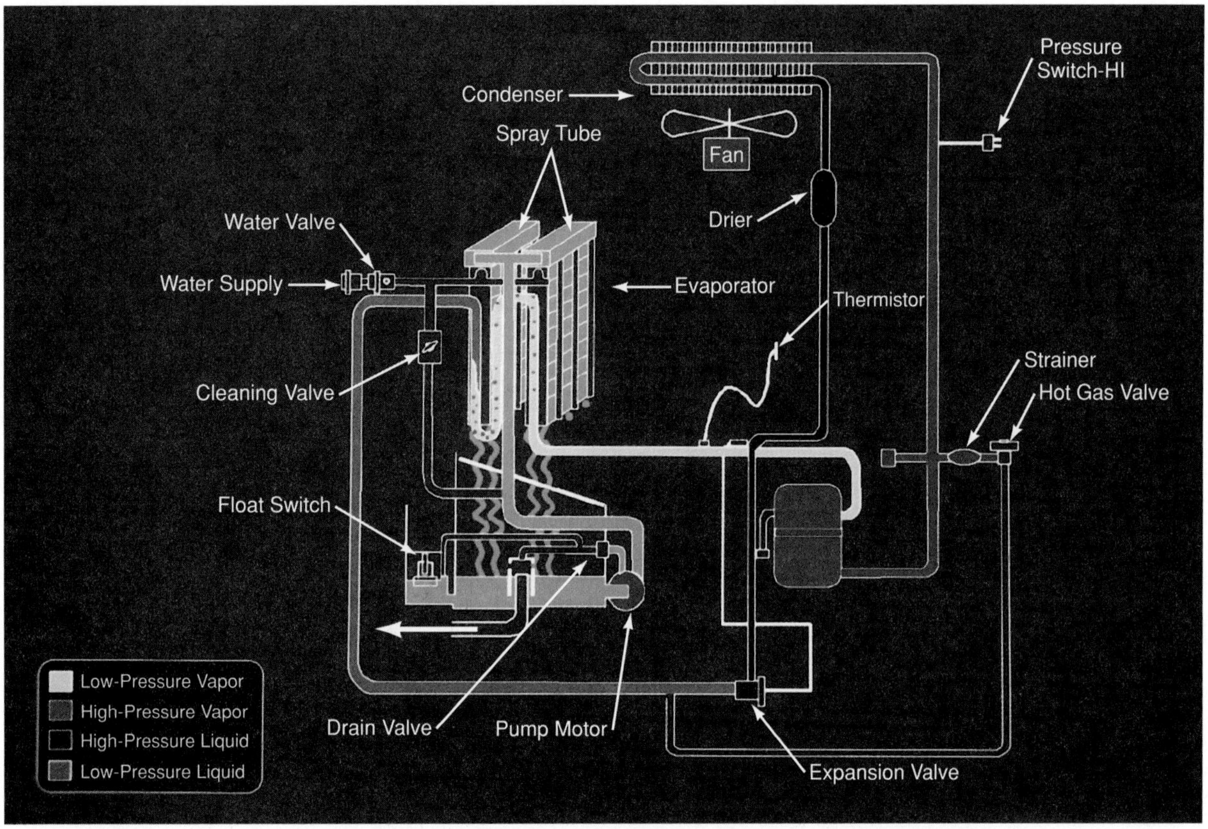

Figure 15-64. *Freeze cycle. The water is in the spray tubes and the system is cooling. (Hoshizaki America, Inc.)*

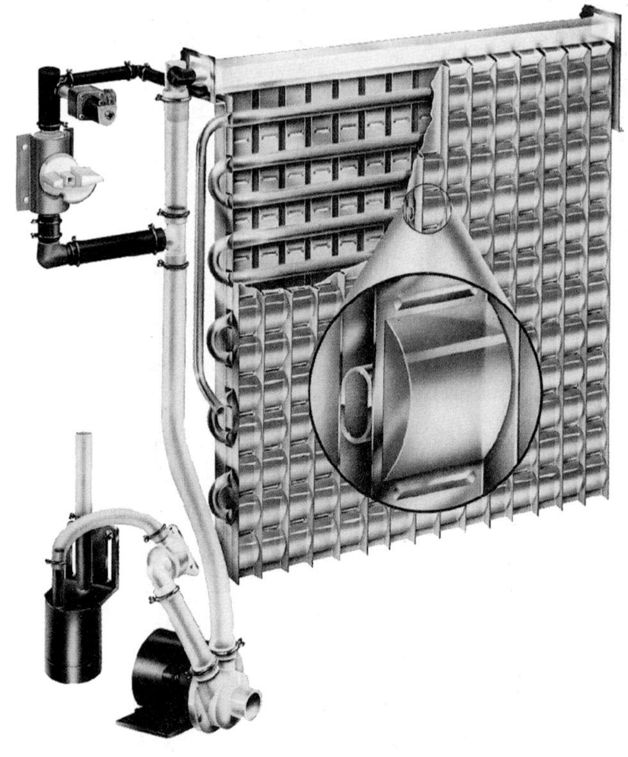

Figure 15-65. *Ice being released from spray tubes. (Hoshizaki America, Inc.)*

pressure to provide required cooling. Air or water being cooled must flow in and out of the evaporators efficiently. The evaporator must be leakproof and of the proper size.

You should check these conditions. Ideally, pressure at the evaporator inlet (just after the AEV or TEV) should be measured. Pressure should also be measured at the outlet.

However, most service technicians check only the low-side pressure at the compressor suction service valve. They think that the evaporator pressure is close to this pressure. The pressure drop due to friction in the tubing and bends can be checked. This is done by reading the low-side pressure when the unit is running. Then it is read again just as the compressor stops. The rise in pressure is the pressure drop. Normally this pressure drop will be 2 psi to 3 psi.

Evaporator temperature should also be checked. Thermometers can be mounted on tubing using spring-loaded, clip-on thermometer holders. Superheat setting of the thermostatic expansion valve can be checked using thermometers. The best setting is when the bulb temperature varies the least while the unit is running. Location of the liquid in the evaporator can also be determined this way.

Frost accumulation acts as an insulation and also tends to reduce the airflow. Accumulation near the TEV usually means too great a superheat adjustment along

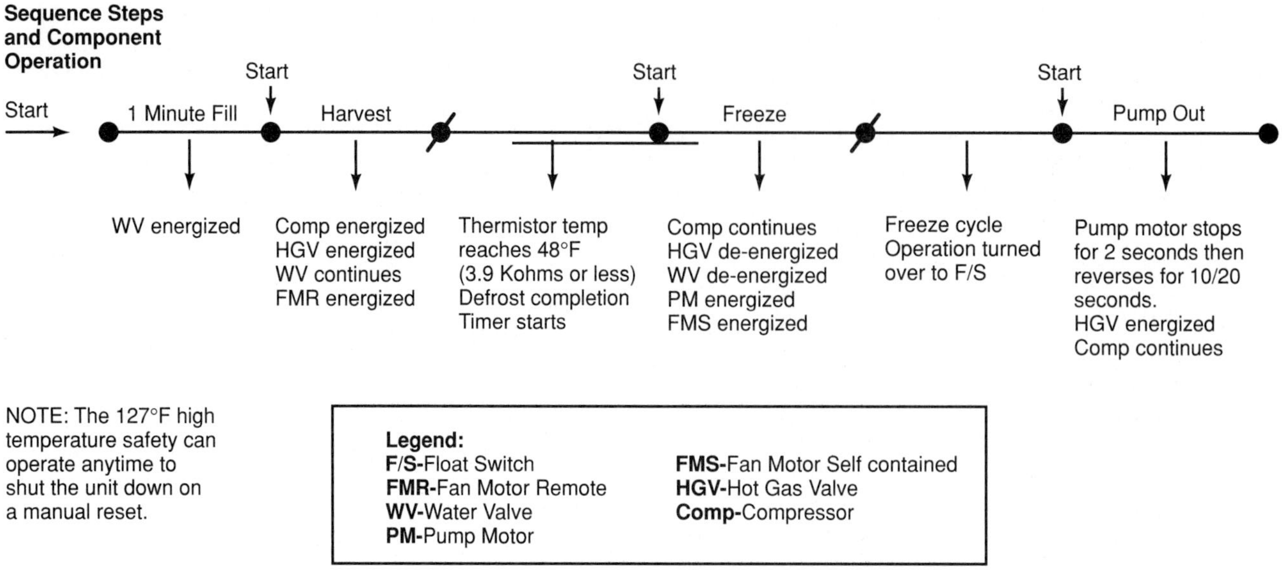

Figure 15-66. *The basic operational sequence of the ice machine. Note that the legend explains the various letters. (Hoshizaki America, Inc.)*

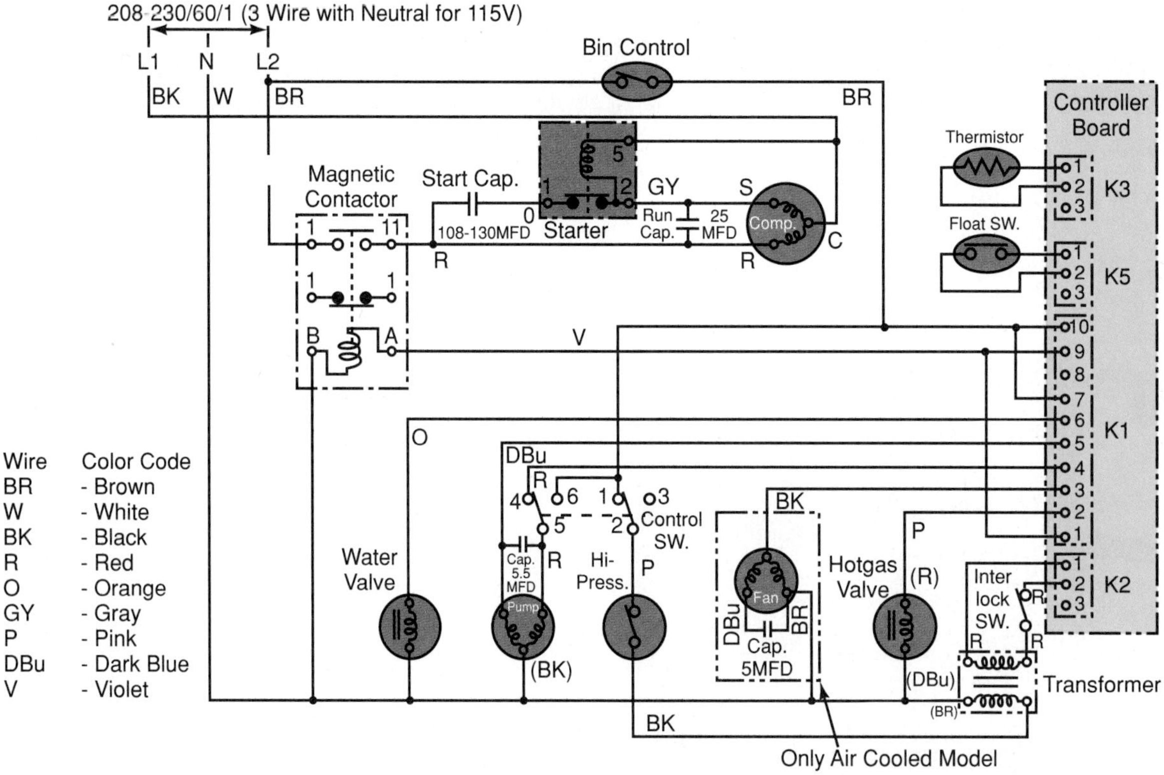

Figure 15-67. *Electrical wiring diagram for ice maker. Note the controller board. This is a solid state control board that is the connector to all electrical components. (Hoshizaki America, Inc.)*

with low-suction pressures. Spotty frost usually means uneven airflow over the evaporator. Otherwise, it may indicate that some defrosting elements are not working.

Airflow through the evaporator can be checked with an ***anemometer.*** This is also called an air velocity meter. See Chapter 19. If the air outlet or inlet is too small, the evaporator will be starved. Air temperatures at the inlet and outlet can also be checked. The air temperature will usually drop about 15°F (8°C) as it passes through the evaporator.

When checking for leaks, the fan and unit should be shut off. The low-side pressure should be at least 15 psig (30 psia or 207 kPa) when testing for leaks.

Removing Evaporator Units (Dry System)

When removing a dry evaporator, first install a gauge manifold and test for leaks. Start the compressor and close the LRSV (liquid receiver service valve). Run the compressor until atmospheric pressure or a constant vacuum has been produced. Continue running the compressor until the evaporator and liquid line are warm. At this point, all the liquid refrigerant is removed. To speed the operation, heat the evaporator carefully with a heat lamp or hot water. Never allow it to get more than warm to the touch.

A balanced (atmospheric) pressure in the evaporator may be obtained by warming the unit. Or, it is reached by bypassing high pressure back through the gauge manifold. **Figure 15-68** shows an evaporator being pumped down. Turn the suction line service valve (A) all the way in, closing the suction line after balancing.

Check the suction line. If there is a suction pressure regulator or a solenoid valve, it should be open.

A hot gas injection unit or liquid injection unit may be connected to the suction line. If so, be sure the solenoid control valves for these units are shut off.

The system may have hand shutoff valves for each evaporator of a multiple installation. If so, use these valves instead of the compressor service valves. Close the liquid line valve first. Then pump refrigerant out of the evaporator. Be sure there is 0 psi or slightly more in the evaporator. Close the suction line hand valve and the evaporator is ready for removal.

Shut off the electric power to the fan and liquid line solenoid valve (if there is one). Remove the casing or shroud of the evaporator carefully. If electric defrost elements are mounted in or on the evaporator, disconnect them.

Clean and dry the suction line where it is connected to the evaporator. Also clean the inlet connection. Then unfasten the suction line and liquid line from the evaporator. Plug the openings with appropriate fittings.

Repairing Direct Expansion Evaporator

Evaporator repairs are usually limited to:

- Repairing leaks.
- Repairing or replacing fittings.
- Straightening fins.
- Replacing defrosting elements.
- Repairing or replacing hangers.
- Repairing or replacing fins and/or motors.

Where leaks occur, completely dismantle that part and clean the surfaces. If it is a brazing repair, follow the procedure explained in Chapter 2. Always anneal an old tube before flaring it. Fins can be straightened using a fin comb or wide-jaw pliers.

Electrical defrosting elements should be checked for continuity. Terminals and insulation should be inspected also. Rusty or bent hangers and abused hanger assembly bolts should be replaced.

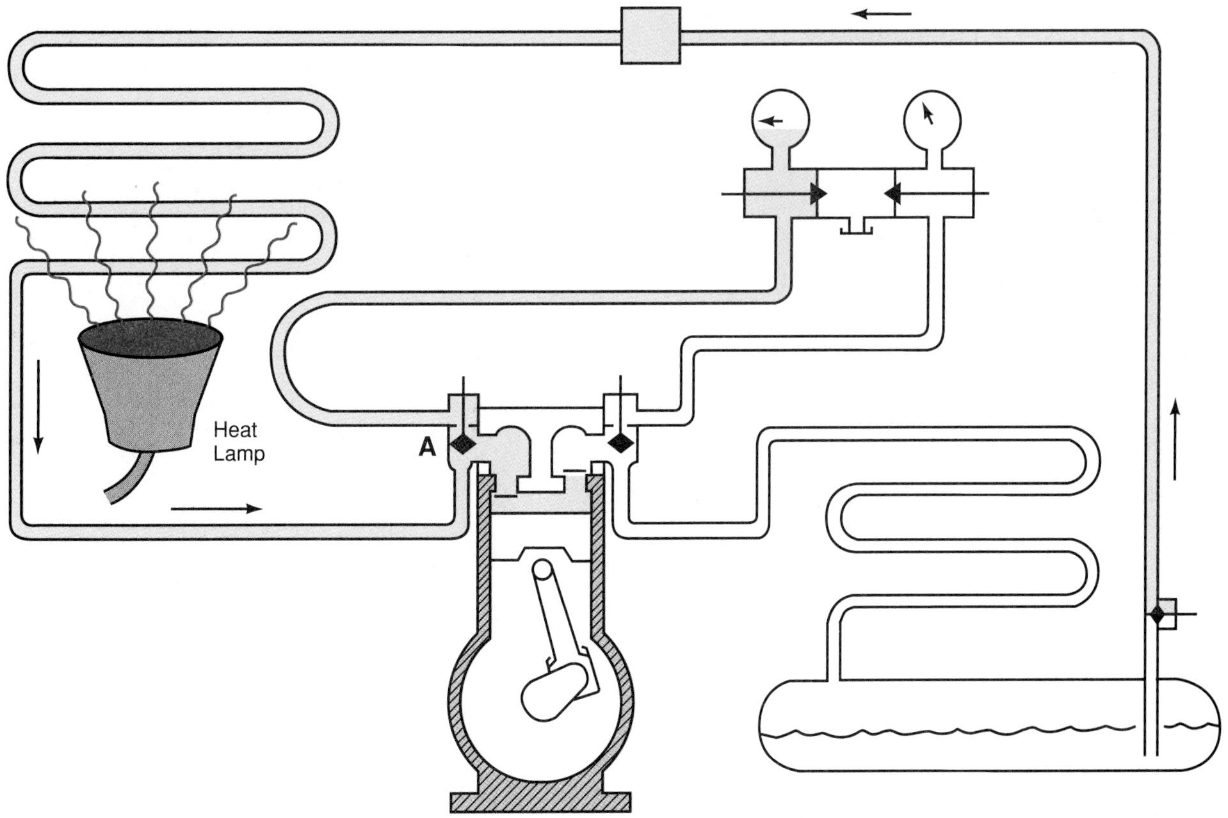

Figure 15-68. *Liquid line, evaporator, and compressor crankcase being pumped down. Refrigerant is being pumped into condenser and receiver.*

Check for the following:

- Fan and motor for vibration.
- Tightness of the fan on the motor shaft.
- Motor end play.
- Motor bearing wear.
- Condition of lubricant.

Small faulty motors should be replaced. Larger motors can be rebuilt. (See Chapter 7.)

All parts should be cleaned before assembly, **Figure 15-69.** The evaporator is usually assembled on the job. If leaks have been repaired, the evaporator should be leak tested before it is installed.

Installing Direct Expansion Evaporators

If the evaporator has been removed, the following reassembly procedure should be followed to ensure proper operation. Bolt the evaporator back into the refrigerator and level it. Then, remove the plugs on the refrigerant openings. Attach the liquid and suction lines to the unit. Be careful that no moisture enters the lines during these operations. It is good practice to dry the surfaces of lines and evaporator before removing seals.

The bolted thermostatic expansion valve (TEV) should be installed using a new gasket. Connect a vacuum pump and evacuate the evaporator, suction line, and compressor. Evacuate the evaporator a second time. Test for leaks with 5 psig to 25 psig (20 psia to 40 psia or 138 kPa to 276 kPa) pressure before operating the evaporator. Test for leaks again at ambient refrigerant pressure (high-side pressure).

The evaporator may be dehydrated more completely. This is accomplished by heating it to a fairly high temperature as it is evacuated. It should be heated to 175°F to 200°F (79°C to 93°C). This drives out any moisture that may be present. Heat lamps may be used for this purpose.

After installing the evaporator and testing for leaks, install the defrost unit electrical connection. Install the fan and motor. The electrical connections should be tight and moisture-proof. Operate the defrost unit and the fan.

Assemble the casing or shroud. Start the unit and check for normal operation.

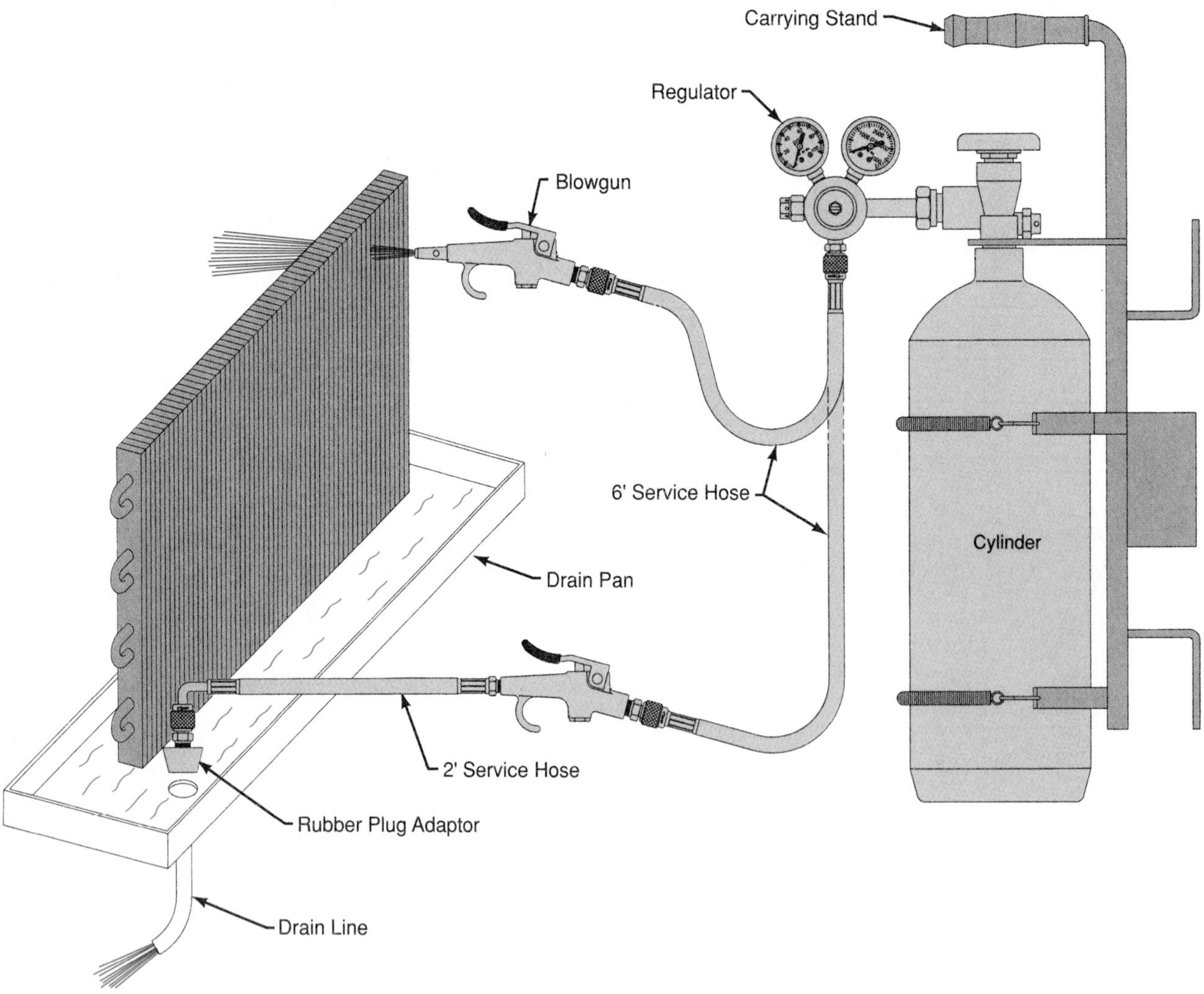

Figure 15-69. *A portable pressure cylinder being used as a blowgun to clean an evaporator and drain line. (Uniweld Products, Inc.)*

15.14.6 Dry Evaporator Refrigerant Controls

The operation of refrigerant controls in commercial systems is described in Chapter 5.

All types of refrigerant controls may be found in commercial systems. Systems having a self-contained hermetic unit may be serviced as described in Chapter 12.

Multiple evaporator installations, however, and larger commercial units, have other features that you must know.

Servicing Thermostatic Expansion Valves

Dry systems which use the thermostatic expansion valve refrigerant control may have a single evaporator or multiple evaporators. The design and operation of this valve is explained in Chapter 5.

You should check the complete system, install the gauge manifold, and check for pressure. Check the amount of refrigerant through the sight glass. The following should also be checked:

- Liquid line solenoid valves.
- Two-temperature valves.
- Hot gas or liquid injection systems.
- Driers (both suction line and liquid line).
- Temperature and pressure motor controls.
- Electrical supply.

You can check the evaporator by appearance, sound, and temperature of the expansion valve. If the evaporator frosts back so the suction line is frosted:

- The needle may be leaking.
- The control may be adjusted for too little superheat.
- The valve may have the incorrect thermal bulb charge.
- The valve orifice may be too large.
- The power element may be attached loosely to suction line.
- The power element may be located in too warm a position.
- Dirt in the system may be holding the valve open.
- An external equalizer valve may be needed.
- The screen may be clogged.

It is difficult to determine which of these troubles is responsible for the problem. The best method is to systematically check everything it could be. Rule out the possible causes one by one.

The power element location is easily checked and its attachment to the suction line noted. Recommended procedure is to place the thermostatic bulb on top of the suction line. This is a better location than beside or below it. In this position, liquid in the bulb makes good thermal contact with the suction line. **Figure 15-70** shows two recommended thermal bulb locations.

One rarely finds a thermostatic control in which someone has tampered with the adjustment. Therefore, do not attempt to readjust the control at first. Check for other troubles instead. A leaking needle or valve cannot be repaired. It should be replaced.

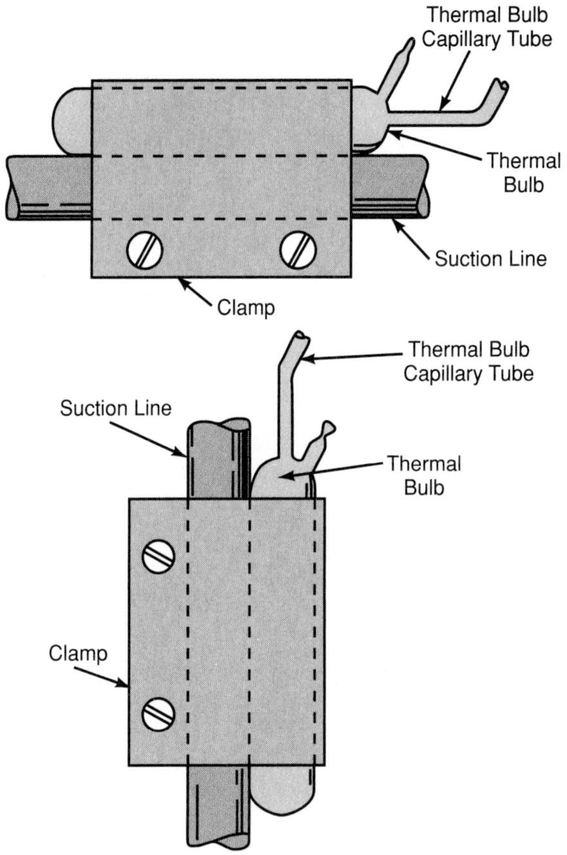

Figure 15-70. *Recommended locations for thermostatic bulbs on suction lines to obtain best operation. Bulb should be on top of horizontal suction line. It should have closed end on bottom when mounted on vertical suction line.*

A starved evaporator gives poor cabinet temperatures while frosting unevenly or not sweating properly. This problem may be due to the following:

- Clogged screen in expansion valve, which may give no refrigeration.
- Loss of refrigerant from power element in expansion valve, which will give erratic (undependable) refrigeration.
- Moisture in the system, which may sometimes give good refrigeration and then no refrigeration. The moisture will freeze in the orifice to expansion valve and close it, then defrost.
- Wax from the oil in the valve. Its presence means that the oil used was for a different temperature range. Otherwise, it may have been improperly prepared for refrigeration service.
- Needle stuck shut. (This is a rare occurrence.)
- Under-capacity valve orifice. It is usually recommended that a new valve be installed when an expansion valve gives trouble.

Sweating or frosted suction line beyond the thermal bulb position indicates too much refrigerant flow. This condition may be caused by:

- Thermal bulb loose from suction line.

- Thermal bulb in warm airflow.
- TEV orifice is too large.
- TEV needle stuck open.
- Undersize evaporator.
- Thermal bulb has wrong charge.
- Pressure drop in evaporator is too great.

If the bulb is loose, incorrectly mounted, or in a warm air stream, remove it. Then clean it and the tubing. Remount it firmly and insulate, if necessary.

An oversize TEV should be replaced with one of the correct size. When a needle is stuck open, the best remedy is to replace the valve.

Only rarely one finds the following:

- An undersize evaporator.
- A TEV with the wrong charge (replace).
- Too great a pressure drop (replace evaporator).

Removing Expansion Valves

A faulty expansion valve should be removed and replaced with one in good condition. The troublesome valve may then be checked in a shop equipped for this purpose. In this way the trouble can be accurately determined.

Remove the valve using the procedure in Section 15.14.5.

Repairing Clogged TEV Screens

A clogged screen may be easily detected. Indications are poor refrigeration, sweating, or frosting near the TEV only and no refrigerant sound.

Removing an expansion valve having a clogged screen means removing both liquid and vaporized refrigerant from lines to be opened. The liquid line may be carefully heated with a heat lamp. This drives the liquid refrigerant back to nearest shutoff valve. This valve then is closed. The evaporator is already evacuated (indicated by a warm evaporator). Suction line valve may be closed if low-side pressure is at atmospheric pressure or higher. The screen may be removed after cleaning and drying the connections.

Clean the screen or replace with a new one. This is important. Fine-mesh copper or stainless steel screens are cleaned well with air pressure and safe solvent. However, the best way is by heating it. This must be very carefully performed or the screen will be burned. Never allow an expansion valve into service without a screen in the liquid line entrance.

Install the screen, assemble the TEV, and install it in the system. Then evacuate the air and test for leaks. Return the evaporator to normal operation by opening the suction and liquid line valves.

Installing Expansion Valves

Mount the expansion valve and evacuate the liquid line, evaporator, and suction line. Carefully test for leaks by first purging, then building up a refrigerant vapor pressure. Install the fan and motor, if used. Open liquid receiver valve or liquid line hand valve. Start the compressor and observe its operation.

In multiple systems, install all dry evaporators using expansion valves with individual shutoff valves. These should be installed for both the liquid and suction lines to each evaporator.

Adjusting thermostatic expansion valves for too great a difference between various cabinet temperatures causes erratic operation. This is particularly true in the evaporators that are closed off the most. To fix this, install one or more two-temperature valves in the suction line.

Use gauges and thermometers to check for superheat setting. When the unit is operating correctly, connect the defrost. If used, install casing and shroud.

In multiple commercial installations, using finned evaporators, the expansion valve is sometimes attached to the evaporator with an SAE flared connection. The flare nut in such an installation must be shellacked or sealed from moisture. This is done after the installation is made and before the unit starts to operate. Otherwise, ice may form between the nut and the tubing. In a short time the tube will collapse or break. This condition can also occur where the suction line fastens to the evaporator. **Figure 15-71** shows the pinching operation of ice formation between the flare nut and tubing.

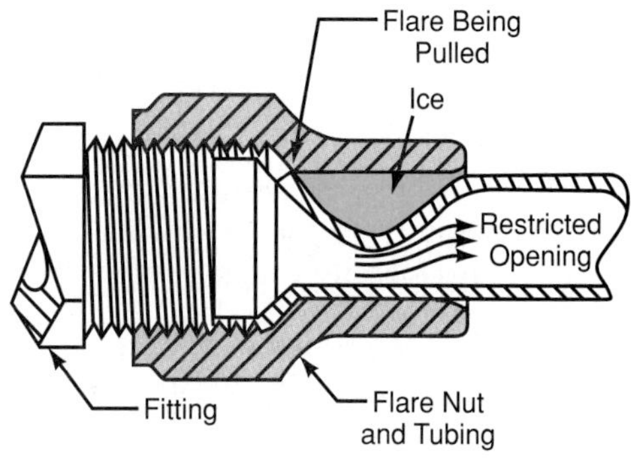

Figure 15-71. *Illustration shows what happens when moisture freezes between nut and tubing.*

Other methods have been devised to stop moisture from collecting behind flare nuts. One has a rubber seal at the end of the flare nut. Another method consists of drilling holes through the flare nut. The idea is that moisture will drain out. If ice does form, it will release its pressure through the holes in the nut. Otherwise, pressure would be against the tubing.

Short shank flare nuts should be used in places where frosting occurs. **Figure 15-72** shows a short flare nut with openings across the threads for moisture escape. As mentioned, the best type of connection for cabinet interiors is the brazed flanged connection.

Adjusting Thermostatic Expansion Valves

The sensing element of the thermostatic expansion valve should be clamped to the suction line. This is done at the point where it attaches to the evaporator. It is possible to make bench adjustments of this valve.

In service, the thermostatic expansion valve may be adjusted as follows:

1. With the refrigerating condensing unit operating, note the temperature of the evaporator.

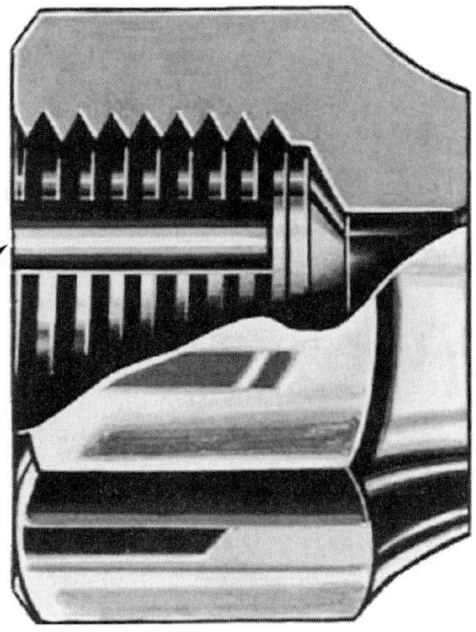

Figure 15-72. *Special flare nut used to prevent ice accumulation between flare nut and tubing. (Superior Valve Company, Division of AMCAST Industrial Corporation)*

2. If the evaporator is too warm, adjust the control to allow more refrigerant into evaporator.
3. If the suction line frosts up, adjust the control to reduce the refrigerant flow. Most technicians adjust the thermostatic expansion valves according to the MSS (minimum stable signal) point. See Chapter 5. The amount of superheat is properly taken care of by this adjustment.

Servicing Two-Temperature Valves

As explained in Chapter 13, the two-temperature valves are automatic. They maintain a higher refrigerant pressure in one or more evaporators of a multiple evaporator system. This pressure is higher than pressures in the remainder of the system.

Four types of two-temperature valves are:

- Metering.
- Snap-action.
- Thermostatic.
- Solenoid valve, which is thermostat controlled.

The four troubles commonly encountered with pressure valves are:

- Leaky needle.
- Valve stuck shut.
- Valve out of adjustment.
- Frost accumulation on bellows.

If the valve is leaking, the warmer evaporator will be too cold. There will be danger of freezing. Cold temperatures may also be due to other causes, including adjusting the two-temperature valve too close and at too low a pressure.

To determine which of the two troubles is present, check to see if the valve has been adjusted recently. If the valve has not been tampered with, the trouble is probably a leaking needle. If the valve has been tampered with, it must be readjusted. A thermometer is used to get the correct evaporator temperature. The adjusting nut should not be turned more than a half-turn at a time. A 15-minute interval should be allowed between each adjustment. This permits the evaporator to completely respond to the new pressure.

A low-pressure gauge should be installed in the evaporator side of the valve. This allows for accuracy in making adjustments on two-temperature valves. Sometimes such a gauge opening is available, but in many cases it is not. When installing two-temperature valves, you should install a shutoff valve. It should have a gauge opening in the suction line. This will permit the use of a gauge to check the evaporator low-side pressure. This process will make service work on such valves easier.

It is rare to have a valve become stuck shut. It is easily recognized by a lack of cooling in the warmer evaporator. There will be a lack of adequate refrigerant supply to the two-temperature valve (but not through it). To check refrigerant supply, crack the flare nuts on the high-pressure side of the two-temperature valve. Some of these two-temperature valves are provided with screens. A clogged screen will be indicated by a warmer cooling condition of the evaporator. Symptoms will be similar to the stuck needle condition.

Frost accumulates on the bellows if the valve is located in or near a freezing compartment. The valve should be removed from the freezing compartment. The bellows should be covered with light grease.

The same troubles are encountered with the snap-action, two-temperature valve as with the metering valve. However, most snap-action valves have a gauge connection. This gauge connection makes adjustment of the valve simple. Therefore, it can be determined if the valve is leaking or out of adjustment.

Thermostatic two-temperature valves have the same troubles, causes, and remedies as the other two. In addition, they have problems resulting from the thermostatic element. These troubles are the same as mentioned for thermostatic expansion valves. They are:

- Loss of charge from thermostatic element.
- Frost on bellows.
- Poor contact between power element and evaporator.
- Pinched capillary tube.
- Improper adjustment.

Problems are checked similar to the way the thermostatic element in thermostatic expansion valves is checked.

The warmer evaporator has an electric solenoid valve and thermostat. The following troubles may occur:

- Needle stuck open.
- Needle stuck shut.
- Thermostat troubles:
  A. Points stuck together.
  B. Open circuit.
  C. Out of adjustment.
- Poor wiring.
- Open solenoid winding.
- Burned-out solenoid winding.

The solenoid valve should be replaced if the needle is not working properly. When a solenoid coil is replaced, be sure to use the correct coil voltage (24 V, 120 V, or 240 V).

Installing Two-Temperature Valves

The pressure-operated two-temperature valves should be installed in the suction line of the warmer evaporators. Their operation is usually not affected by the distance from the evaporator. Therefore, they may be connected at any place on the suction line. Frequently, valves of this kind are found in soda fountains. Solenoid valves for two-temperature operations are mounted in the liquid line of the warmer evaporators.

This type of installation may be improved by mounting a check valve in the suction line of the coldest evaporator. This prevents higher pressure gases from backing up into it. Also, a surge tank should be mounted near the condensing unit. It should be connected between the compressor and the main suction line. This will cut down on fluctuations (rapid up and down changes) of the low-side pressure.

In most code installations, the two-temperature valve and check valve must be mounted near the condensing unit. The surge tank must be mounted on the condensing unit base.

Servicing Hot Gas Bypass Valves

There are two types of hot gas bypass valves:

- Solenoid, timer on—thermostat off.
- Solenoid pilot-operated pressure valve, timer on—thermostat off.

If the evaporator is not defrosting, check the timer. If it is not running, determine the problem and repair or replace. If working properly, turn the timer to the On position. Then check the hot gas line. If it does not become warm, check the electrical circuit with a voltmeter. This will determine whether there is power to the solenoid.

If the timer is faulty, check the points. If they are corroded, replace them. If the solenoid coil is faulty, replace it. To determine condition, disconnect wires after shutting off the power. Then check for open circuit or grounds.

If the valve is sticking, rap the body of the valve. Note whether or not this action allows the hot gas to flow. If the valve core remains stuck, the valve must be replaced.

To replace the valve, refrigerant must be removed from the system. (Valve is connected to high side and low side.) Balance the pressures. Remove the valve and install a new one. Test the system for leaks, evacuate it, and recharge.

Removing Service Valves

Occasionally a service valve stem will break or the threads will strip. The valve must be replaced.

If it is the suction service valve, remove all refrigerant from the evaporator unit (unless the evaporator has hand shut-off valves), then balance the pressure.

To remove a discharge service valve or a liquid receiver service valve, first remove refrigerant from the entire system. Do not pinch the lines to replace valves, as weakened tubing will cause trouble in the future. These valves have been successfully replaced by supercooling the system refrigerant with dry ice. Dry ice is packed around the refrigerant containing parts of the system. When the gauges show atmospheric pressure, the system can be opened.

Servicing Solenoid Valves

Solenoid valves develop both electrical and refrigerant troubles. The electrical connections may be dirty or loose. If so, the coil may not create enough magnetism to raise the valve. Usually, the valve should be mounted with the coil on top and the valve level. If not, they may stick or chatter. Solenoid valves will sometimes develop a leaky needle and seat. In this case, the valve must be replaced. It is important that the solenoid be of the proper voltage and amperage rating. Otherwise, the coil may burn out. A 120 V valve cannot be connected to a 240 V circuit.

To remove a valve from the system, use the same pump-down procedure as for a TEV.

15.14.7 Servicing Motor Burnouts

A ***motor burnout*** is possible with any system using a bolted assembly or welded motor compressor. An acid condition may develop due to moisture, dirt in the system, and excessive temperatures. (The warmest spot is usually the motor windings.) In time, the condition could lead to a motor winding short (burnout). See Chapter 12.

To prevent motor burnout, the system must be kept free of moisture and dirt. To detect acid formation, the system should be checked regularly:

- Use a sight glass with moisture indicator.
- Take an oil sample often, test it with an oil test kit. The oil sample must be kept sealed until tested. See Section 15.14.2, **Figure 15-54.**

If a burnout occurs, the motor compressor must be replaced. Recover the refrigerant for possible recycling.

In large systems equipped with shutoff valves, it may be possible to save the refrigerant. Purge the motor compressor. **Use goggles, rubber gloves, and work in ventilated space. The oil may be acidic and cause serious burns—do not get on skin. The fumes may also be irritating and toxic.**

The system can be reconditioned by flushing the complete system with nitrogen. Use regulated CO_2 pressure to circulate the refrigerant or use a separate pump.

After cleaning, install the new motor compressor and then test for leaks. Evacuate the system to 50 microns to 500 microns. Install a drier in the suction line. Test for leaks and evacuate again. Charge the system.

Connect the electrical wires. Never solder leads to the compressor terminal. The glass may crack or the terminal may come loose, causing a leak.

Start the unit and check operation. Make frequent oil acid tests. Replace the drier if the oil sample is discolored or shows an acid trace.

Adding Oil to System

A compressor that runs too warm or is noisy may lack oil. There are several ways to add refrigerant oil. Remember, the oil and all the oil transfer equipment must be clean and dry. The oil must be the proper viscosity for the compressor, the refrigerant, and the low-side temperature.

- The most rapid method is as follows: Attach tubing equipped with a hand valve to the middle opening of the gauge manifold. Purge the tubing. Then immerse it in a clean, dry container of refrigerant oil. Run the compressor. Draw a vacuum on the low side by turning the suction service valve all the way in. Oil will be drawn into the crankcase. It is important that some of the oil in the glass container be left there. This is so the filling tube is always immersed in the oil. Otherwise, air will be drawn into the system. Glass containers are used so you can observe how much oil has been added. Not over 1/4 pint of oil should be added at a time to smaller units.
- Another method of adding oil to the system is to evacuate the crankcase, then equalize pressures, remove the oil plug of the crankcase housing, and add the oil. Replace the oil plug and then evacuate the compressor.
- Oil can also be forced into the system by putting the oil in a service cylinder first. (Draw in by using an evacuated cylinder.) Then build up a pressure in the cylinder with refrigerant vapor through the gauge manifold. Invert the cylinder. By using the low-side manifold valve, the oil can be forced into the compressor.
- Special pumps are available to hand pump oil into a compressor. This can even be done against a high low-side pressure. Some compressor oil circulates around the system with the refrigerant. Therefore, oil must be added to the compressor if refrigerant lines are over 30′ long. (This includes both suction and liquid lines.) Add about 3 fl. oz. of oil for each 10′ of tubing installed.

Servicing External Drive Motors

Motors used on external drive commercial systems usually vary in size. They may be anywhere from 1/12 hp for fans to 15 hp for compressors. Air conditioning systems require motors of 1/3 hp to 25 hp. These motors are connected to either 120 V-240 V single-phase or 240 V three-phase lines.

In addition to the condensing unit motors, commercial systems use motors for other purposes. Motors are used for fans, water pumps, and mixers in ice cream machines.

Many localities require a licensed electrical contractor to remove, repair, and install motors. You must be able to diagnose motor troubles.

Motor problems can be traced to:

- Mechanical troubles.
- Electrical troubles.

Mechanical troubles include faults in the bearings and pulleys, misalignment and excessive end play. Electrical troubles may be further classified as either internal or external.

The sound of the motor can indicate trouble. Under normal conditions a motor will make a steady low hum. A problem exists if there are erratic beats in the humming or rotors chatter. In either case, the trouble could be any of the following:

- Worn bearings.
- Rubber armatures.
- Dry bearings.
- Lack of voltage.

With the motor running, rotor position should be between the two extremes of the rotor end play. If so, it means that the rotor is trying to assume magnetic center. Because it cannot, it is running inefficiently. The end play should never exceed 1/16″. This may be adjusted by using end play washers, which may be obtained at electrical supply houses.

Adequate lubrication of the motor bearings is necessary. Normally motors should be oiled twice each year. Use electric motor oil (SAE 30). Too much oil is as bad as too little oil. Most motors are equipped with overflow openings. These openings eliminate most of the effects of excess oil. Many lubricants are now available in aerosol cans.

Bearing temperatures should be checked after operation. Use a thermometer.

Thoroughly clean the motor occasionally. Dust, dirt, and grease accumulations should be removed from inside and outside the motor. Clean commutators and brushes, if used. They must make good contact.

The brush throw-out mechanism should move freely. Brush releasing mechanisms of small motors can be checked by mounting a V-belt on the motor pulley. A load is put on the motor by pulling on the belt. A ***torque stand***, or ***dynamometer***, is the best means of determining the real capacity of a motor.

There are various causes of a noisy motor, including a loose pulley, a loose fan on the pulley, or a loose flywheel. These items should be checked when a noise complaint is received.

Fan motors are usually of the shaded-pole type, and their most common trouble is worn bearings. Many of these motors are designed to not need lubrication. However, practice has proven that many do need oiling.

Always be sure the motor is wired correctly. Voltage must be sufficient. Always test a motor for grounds and always ground the frame of the motor.

Removing Electric Motors

To remove an electric motor, check for a fuse or circuit breaker in the system. Then disconnect the circuit. Disconnect the power line and remove the wires from the motor terminals.

Label the terminals for easier assembly later. Next, loosen hold-down bolts that attach the motor to the base.

If it is a belt-driven unit, remove the belt from the flywheel first. Then remove it from the pulley. The motor can then be lifted out. Do not allow the fan to hit the condenser or catch on the belt. Use a puller to remove the pulley from the motor shaft after loosening the lockscrew. (An example of a belt-driven unit is a compressor drive or large fan motor.)

Fan motors are sometimes difficult to remove. It is best to loosen and remove the fan. Generally, the fan hub is locked in place on the motor shaft with an Allen setscrew. **Figure 15-73** shows an Allen setscrew wrench set designed to work in hard-to-reach places.

Installing External Drive Electric Motors

External drive compressor motors may either drive a compressor directly or use a belt. In either case, the motor must be carefully aligned when installed. Always lock out the power before starting work.

After loosely installing the hold-down bolts, install the belts. Use a lever to move the motor until the belts are tight and the motor is in line. While holding the motor in this position, tighten at least two of the hold down bolts. Then tighten the others. Motors are usually fastened to their mounting base with nut-bolt-washer combinations.

Usually, either the motor base or its base float have slots for the bolts. One must reach under the mounting base to hold the bolt in place. It is often very difficult to put a wrench on it. Some technicians use caulking compound to hold the bolts in place until motor is installed and the washers and nuts are started on bolts.

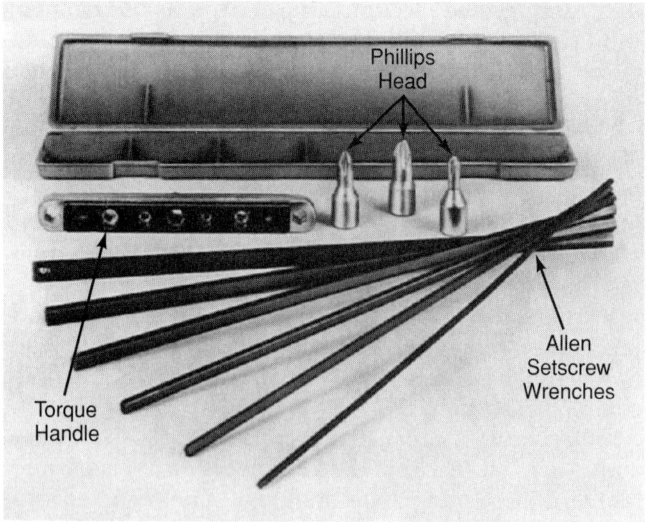

Figure 15-73. *Set of Allen setscrew wrenches and Phillips screwdrivers with turning handle. These tools can reach screws in recess up to 9" in depth. (Watsco Components, Inc.)*

In direct drive units, set the motor on its part of the stand. Then assemble and install the coupling. Check the alignment carefully. The motor shaft center must be the same height as the compressor shaft center. The two shafts must be in alignment when looking down on them. A dial gauge should be used to check the alignment. Install and tighten the bolts. Water pump motors require the same care.

If possible, always turn the motor by hand. This will determine if the assembly will rotate freely. Do this before unlocking the power.

Servicing Hermetic Motors

Servicing hermetic motors is basically the same as servicing conventional motors. Chapter 7 describes the principles, design, construction, operation, and servicing of such motors.

As with all energy devices, first check that the power has all the correct characteristics. Voltmeters, ammeters, demand meters, wattmeters, and power factor meters can be used for this purpose. Check the electrical properties of the motor. Then check the power to make sure the motor is receiving the correct voltage and current.

Voltage must be within 10% of the motor rating. (A 120 V motor should have a minimum voltage of 108 V.) This voltage should be read with the motor running. Voltage higher than required is not critical unless it is 20% over the rating.

The correct amount of current is vital to good operation. A terminal ammeter (shunt unit) should never be connected across the line (in parallel). It must always be put in series (interrupt one wire only) with the electrical device being checked.

Carefully check external wiring and electrical controls for correct operation before assuming motor fault. An ohmmeter is recommended to check continuity in these circuits. Check each circuit separately. Disconnect if there is a chance of parallel circuits. Be sure the power is locked off.

Capacitors should be checked with a capacitor tester, as shown in **Figure 15-74.** Study Sections 6.5.7, 7.18.3, and 7.18.4. Do not test capacitors by shorting after charging, as this method is not accurate enough.

The motor should be checked for:

- Open circuits. (Motor should be cool or internal overload in open position will give false readings.)
- Shorted windings. (Ohmmeter readings should be compared to manufacturer's specifications.)
- Grounded windings. First check with ohmmeter, then with a 500 V circuit tester. Insulation breaks can only be checked accurately with this high voltage tester. Handle carefully to avoid shocks.

Companies report that many motor compressors returned labeled "faulty motor" actually have good motors. This false diagnosis indicates the need for careful checking.

Figure 15-75 is a list of typical motor and circuit troubles, their causes, and their remedies.

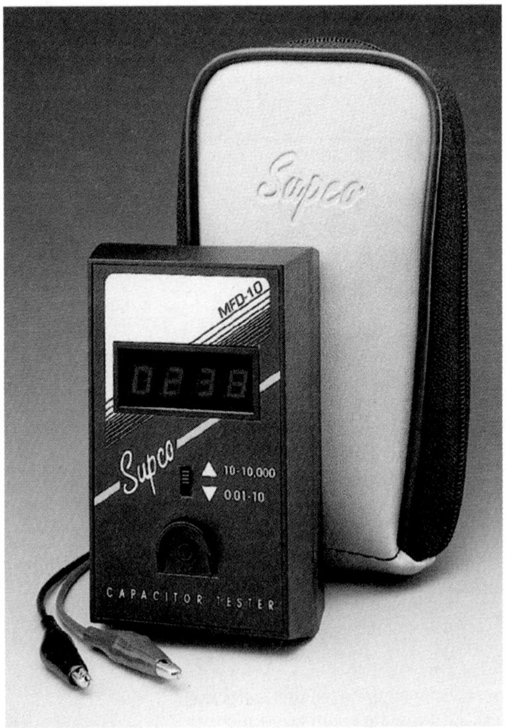

Figure 15-74. *Digital capacitor tester used to detect open or shorted diodes and capacitors. (Sealed Unit Parts Co., Inc.)*

Removing Hermetic Motor Compressors

If the motor is definitely faulty in a hermetic motor compressor, it must be removed. Removal procedure follows:

1. Remove the refrigerant as described in Chapter 12 and in this chapter for serviceable hermetics.
2. Open the main circuit switch and lock the switch in the open position. Tag the switch to inform others why the switch is locked open.
3. Disconnect the wires from the motor compressor (label them or use color code).
4. Clean the outside of the motor compressor.
5. Disconnect lines. The type of disconnect depends on assembly. Unbolt service valves, open brazed joints by heating or cutting the lines. Only use tube cutter to avoid getting chips into the system. **Wear goggles!**
6. Remove motor compressor. Do not tilt, or oil may be spilled. Avoid lifting if it is heavy—use a lifting machine (tripod or fork lift).
7. Plug refrigerant openings.

Installing Hermetic Motor Compressors

A replacement hermetic motor compressor is usually furnished with the following:

- A starting relay.
- Capacitors.
- Overload protectors.
- Other accessories.

Use an exact replacement. These motor compressors are designed for either low, medium, or high low-side pressures.

Use the same type as the one removed. The refrigerant openings are closed with service valves. Otherwise, the unit may be provided with short tubing ends. These are brazed in place, with the ends of the tubing crimped and brazed.

Carefully mount the motor compressor in place. Use all safety precautions (lifting, safety shoes, protecting floors, and equipment). Install the mounting bolts. The springs, grommets, and hold-down bolts must be in the correct position.

Install the electrical devices (overload and starting relay) and the electrical wires. All connections must be clean and tight. All wires, including insulation and wire terminals, must be in good condition.

Avoid connecting aluminum wires to copper wires or copper terminals. Rapid corrosion takes place.

Install suction and condenser lines. Units using service valves are installed the same way as described for a conventional compressor.

If brazed connections are used, identify the tubing stubs. Distinguish between the suction connections, discharge connections, process tube, and oil cooler connections. Cut the tubing stubs with a tube cutter. Select connector fittings or swage the tubing. Flux the outside of the dome. Connect the system lines to the compressor lines. Clean the inside of the suction line and condenser line. Braze the joints. (The area must be well ventilated during brazing.) Install a liquid line drier and a suction line drier. Clean the joints with warm water to remove flux.

Install a gauge manifold. (Use a charging stub or a valve mounted on the suction line and liquid line.)

Evacuate the system. Charge with vaporized refrigerant sufficient to build a 15 psi pressure. Test for leaks and repair any, if found. Evacuate to a 50 micron to 500 micron range for several hours. Then charge the system.

Turn on the power, run motor, check the temperatures, pressures, and electrical power. If any operation is not normal, be sure to diagnose it. Then remedy it before leaving the job. Remove the service connections and braze these joints. Clean up the area.

15.14.8 Servicing Outdoor Air-Cooled Condenser Controls

Certain devices maintain good head pressures when outdoor air-cooled condensers are exposed to colder temperatures (50°F to −10°F (10°C to −23°C)). These are explained in Chapter 13.

Service of these units depends on the type. In all cases, condenser operation can be checked as follows: The condenser air inlet or outlet is blocked with cardboard. This raises the pressure. Then the cardboard is removed to lower the pressure. Use the following procedures according to type:

- Modular fans—These units have electrical troubles. Either the fan motor or the solid-state control is

Hermetic Compressor Service Chart

Problems and Causes	Remedies
Compressor will not start – no hum.	
1. Open line circuit. 2. Protector open. 3. Control contacts open. 4. Open circuit in stator.	1. Check wiring, fuses, receptacle. 2. Wait for reset – check current. 3. Check control, check pressures. 4. Replace stator or compressor.
Compressor will not start – hums intermittently (cycling on protector).	
1. Improperly wired. 2. Low line voltage. 3. Open starting capacitor. 4. Relay contacts not closing. 5. Open circuit in starting winding. 6. Stator winding grounded (normally will blow fuse). 7. High discharge pressure. 8. Tight compressor. 9. Weak starting capacitor or one weak capacitor of a set.	1. Check wiring against diagram. 2. Check main line voltage, determine location of voltage drop. 3. Replace starting capacitor. 4. Check by operating manually. Replace relay if defective. 5. Check stator leads. If leads are all right, replace compressor. 6. Check stator leads. If leads are all right, replace compressor. 7. Eliminate cause of excessive pressure. Make sure discharge shut-off and receiver valves are open. 8. Check oil level – correct binding condition, if possible. If not, replace compressor. 9. Replace.
Compressor starts, motor will not get off starting winding.	
1. Low line voltage. 2. Improperly wired. 3. Defective relay. 4. Running capacitor shorted. 5. Starting and running windings shorted. 6. Starting capacitor weak or one of a set open. 7. High discharge pressure. 8. Tight compressor.	1. Bring up voltage. 2. Check wiring against diagram. 3. Check operation – replace relay if defective. 4. Check by disconnecting running capacitor. 5. Check resistances. Replace compressor if defective. 6. Check capacitance – replace if defective. 7. Check discharge shutoff valves. Check pressure. 8. Check oil level. Check binding. Replace compressor if necessary.
Compressor starts and runs but cycles on protector.	
1. Low line voltage. 2. Additional current passing through protector. 3. Suction pressure too high. 4. Discharge pressure too high. 5. Protector weak. 6. Running capacitor defective. 7. Stator partially shorted or grounded. 8. Inadequate motor cooling. 9. Compressor tight. 10. Unbalanced line (three-phase). 11. Discharge valve leaking or broken.	1. Bring up voltage. 2. Check for added fan motors and pumps connected to wrong side of protector. 3. Check compressor for proper application. 4. Check ventilation, restrictions, and overcharge. 5. Check current – replace protector if defective. 6. Check capacitance – replace if defective. 7. Check resistances; check for ground – replace if defective. 8. Correct cooling system. 9. Check oil level. Check for binding condition. 10. Check voltage of each phase. If not equal, correct condition of unbalance. 11. Replace valve plate.
Starting capacitors burnout.	
1. Short cycling. 2. Prolonged operation on starting winding. 3. Relay contacts sticking. 4. Improper relay or incorrect relay setting. 5. Improper capacitor. 6. Capacitor voltage rating too low. 7. Capacitor terminals shorted by water.	1. Reduce number of starts to 20 or less per hour. 2. Reduce starting load (install crankcase pressure limit valve), increase voltage if low – replace relay if defective. 3. Clean contacts or replace relay. 4. Replace relay. 5. Check parts list for proper capacitor (mfd.) rating and voltage. 6. Install capacitors with recommended voltage rating. 7. Install capacitors so terminals will not be wet.
Running capacitors burnout.	
1. Excessive line voltage. 2. High line voltage and light load. 3. Capacitor voltage rating too low. 4. Capacitor terminals shorted by water.	1. Reduce line voltage to not over 10% above rating of motor. 2. Reduce voltage if over 10% excessive. 3. Install capacitors with recommened voltage rating. 4. Install capacitors so terminals will not be wet.
Relays burnout.	
1. Low line voltage. 2. Excessive line voltage. 3. Incorrect running capacitor. 4. Short cycling. 5. Relay vibrating. 6. Incorrect relay.	1. Increase voltage to not less than 10% under compressor motor rating. 2. Reduce voltage to maximum of 10% above motor rating. 3. Replace running capacitor with correct mfd. capacitance. 4. Reduce number of starts per hour. 5. Mount relay rigidly. 6. Use relay recommended for specific motor compressor.

Figure 15-75. *Troubleshooting guide for hermetic compressor service.*

faulty. First check the motor for mechanical faults, such as a loose fan or stuck rotor. Then check for electrical power to the motor with a voltmeter. If it has power, shut off power and check the motor windings or ground. Check the sensing device of the solid-state control for correct position and fastening. If the control is faulty, replace it.
- Controlled louvers—Check operation by blocking air over condenser to find out if louvers will operate automatically. The power element is condenser-pressure-operated (operated by head pressure) and may be faulty. If so, refrigerant must be removed from the system (unless there is a shutoff valve). Care must be taken if the line is pinched to remove this control. The pinched part of the line must be well-supported to prevent a break.

15.14.9 Servicing Liquid Lines

The liquid line contains several important items needing inspection:

- Size of the liquid line.
- Hand shutoff valves.
- Sight glass.
- Moisture indicator.
- Screen filter.
- Drier or dehydrator.
- Vibration absorber.
- Connections.
- Solenoid valves.
- Joints.
- Pinched or buckled pipe or tubing.

When diagnosing system troubles, first determine if each part is the proper capacity. The liquid line should be as large as the condensing unit liquid receiver valve connections. Check to be sure that reducer fittings have not been used. See Chapter 16 for recommended liquid line sizes. Many large units use various sizes of liquid lines.

At least one end of the liquid line valves should be open when servicing the unit. Otherwise, a temperature rise may create a high hydrostatic pressure, which may burst the line. Be especially careful if a solenoid valve, clogged screen, or clogged dehydrator is in the line.

Check the condition of the full length of line. The line must be protected from abrasion and abuse as objects are moved. The line should be well supported along its full length.

Take care when admitting liquid refrigerant into a liquid line through the liquid receiver service valve. Always open the valve slowly. A sudden rush of liquid may injure the screen or pack the desiccant in the drier so firmly that it will clog.

Test all joints for leaks. Check the temperature of the liquid line. It should be close to room temperature along its full length.

A lower-than-normal temperature at the outlet indicates a partially clogged screen or drier. Sweating and even frosting may be seen. There will also be an excessive pressure drop. This may cause bubbles in the sight glass.

15.14.10 Servicing Suction Lines

Servicing the suction line is much like servicing the liquid line. Parts to be inspected are:

- Size of the suction line.
- Hand shutoff valves.
- Vibration absorber.
- Check valves.
- Two-temperature valves.
- Constant-pressure valves.
- Filter-drier.
- Muffler.
- Accumulator.
- Joints.
- Pinched or buckled suction line.

Suction line size is important. If it is too small, it will cause too much pressure drop. High gas velocities will cause noise. The line should be the size of the suction service valve connection, or it should be the size of the suction line connection on the hermetic dome.

On multiple installations, the suction line is smaller for each evaporator than the main suction line. For example, the most remote evaporator may have a 1/2″ OD size. The next would be 5/8″ OD, then 1″ OD, then 1 1/2″ OD. The line at the compressor may be 2″ OD. See Chapter 16.

Pressure drop should be checked by installing one gauge at the most remote evaporator and one at the compressor. Record the low-side pressure when the unit is running. Do so again just as the unit stops. The difference is the pressure drop. The pressure drop should be approximately 2 psi (14 kPa). If more, line sizes should be increased.

The pressure change across two-temperature valves and constant pressure valves should be checked. Their pressure drops are separate from the line pressure drops.

Also determine the pressure drop across the suction line filter-drier. A large pressure drop here indicates a partially clogged filter-drier. In this case, the drier should be replaced. Test for leaks with at least 15 psi pressure in the lines.

15.14.11 Servicing Electrical Circuits

More and more electrical devices are being used on refrigerating systems. Some of these are electrical defrost systems, solenoid valves, multiple units with electrical interlocks, crankcase heaters, internal motor winding protectors, and various other accessories. You must be knowledgeable about electrical devices and electrical circuits. Chapters 6, 7, and 8 explain the fundamentals of electricity, electric motors, and electric controls.

It is important that you have a wiring diagram of the system being serviced. Certain items should always be checked:

- Are the wires large enough?
- Is there electrical power up to the machine?
- Is the voltage correct? (It must not be too low.) See **Figure 15-76.**
- Is the current draw correct? See **Figure 15-77.**

With the power on, the current draw and the voltage can be checked. If the unit will not run, turn off the power and then check the circuits for continuity. Use an ohmmeter. See **Figure 15-78.**

Locate the break in the circuit continuity by measuring sections of the circuit with the ohmmeter. If the motor hums but will not start, check the starting capacitor. Use a capacitor tester. Do not start a unit without having overload cut-out in the circuit.

15.14.12 Assembling Refrigeration Systems

When the repairs have been made and the parts tested, the system must be properly installed.

Four fundamentals must be followed when installing parts of a refrigerating mechanism:

- Clean and dry each part to be put into the system.
- Purge and evacuate that part of the system which has been opened, using the proper equipment.
- Test for leaks.
- Start and adjust the unit.

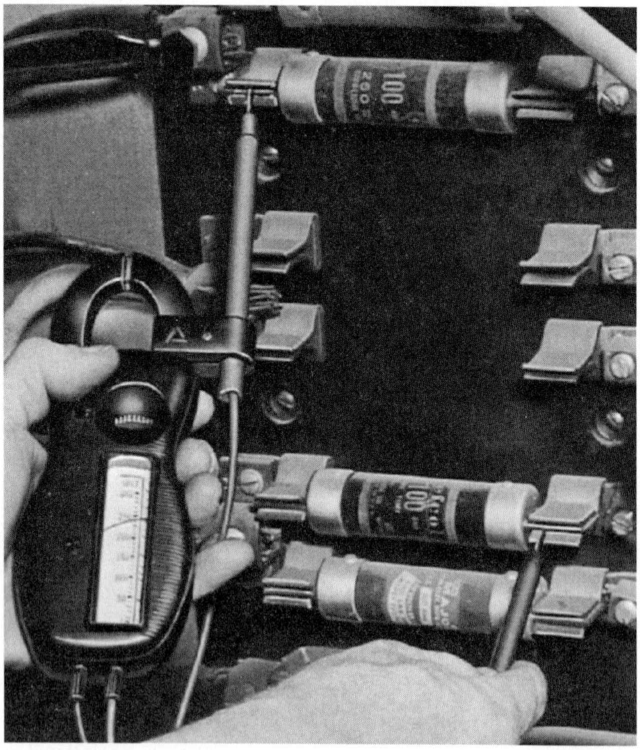

Figure 15-76. *Checking voltage at main load center. (Amprobe Instrument)*

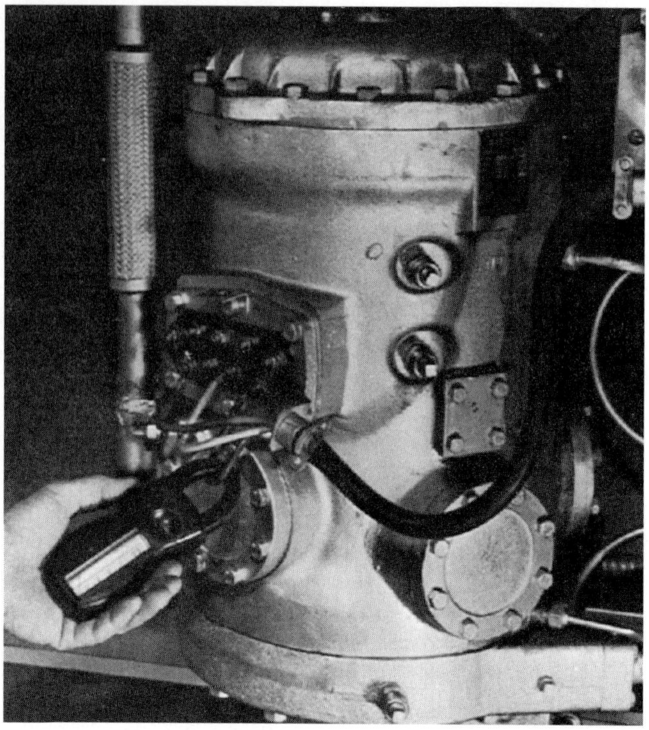

Figure 15-77. *Measuring current flow to motor. Current flow should not be more than motor rating. (Amprobe Instrument)*

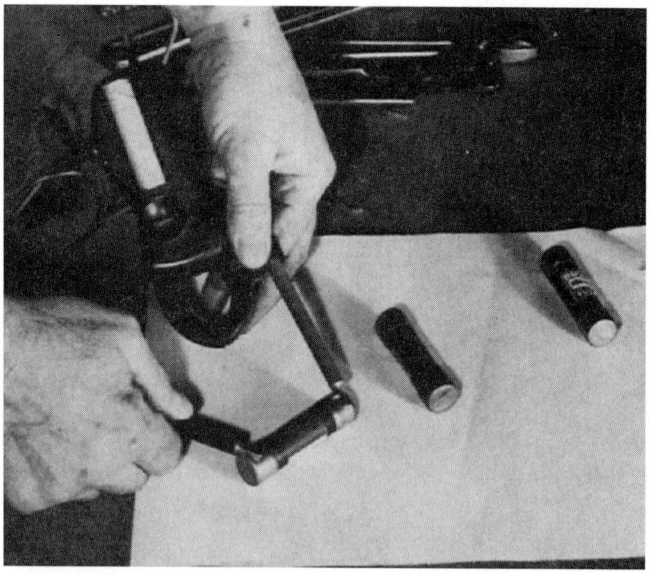

Figure 15-78. *Using an ohmmeter to check continuity of fuses. (Amprobe Instrument)*

Assembling a refrigerating mechanism in the field without foreign matter entering the system is difficult. It is almost impossible to keep dirt and moisture out of the system. Filter-driers are needed to remove this matter. They are installed in both the liquid line and suction line of the system.

A service call should be made within the next day or two after repair. You should check on the operating pressures and the general condition of the refrigerator. If at all possible, connect temperature and pressure recorders into the system for 24 hours.

15.15 Service Notes

Pinching lines is a practice to be used in cases of emergency only. Some service technicians follow this practice needlessly. This only leads to future trouble. Almost all systems are provided with enough valves to service them. However, pinching lines used to service hermetic units, prior to brazing stub, is common practice.

Remember when adding refrigerant to a system that some of the oil will be dissolved in the refrigerant. If a unit becomes noisy soon after the refrigerant is added, refrigerant oil should be added.

If heated while in the system, driers will release water to the system. (This may occur because of refrigerant shortage, for example.) Install a new drier if the moisture indicator signals moisture.

Crankshaft seals may leak if the compressor has been idle for a long time. Turn the compressor over by hand a few times. This allows oil to seep between the rubbing metal surfaces. Also, put an ounce of special refrigerant detergent oils into the crankcase. This will help eliminate this problem.

15.16 Summary of Refrigerator Mechanism Servicing

If this chapter has been studied carefully, one should recognize the following important things:

- Liquid refrigerant must be removed from that part of the mechanism to be overhauled.
- It is necessary to equalize pressures in the unit before dismantling.
- All refrigerant openings should be plugged immediately after dismantling.
- Put in new gaskets wherever gaskets are used.
- When reassembling, remove all of the air and moisture from the lines. Also, remove air moisture from whatever part has been exposed to the air. This may be done by purging, or by deep evacuating.
- Keep the inside of the system clean.
- To remove any part, close the nearest valve between that part and the liquid receiver. Evacuate the unit, by means of the compressor, into the condenser and liquid receiver.

It is usually advisable to take a large overhaul to the shop. In the meantime, replace the unit with a temporary one. By doing so, the owners may have the use of the system.

Before servicing a system, you should review the system wiring diagram and service manual. The complex nature of some of the systems requires that a service manual be used.

15.17 Periodic Inspections

The capital investment in a commercial refrigeration installation amounts to hundreds of dollars. Because of the way most mechanisms are built, one system problem may cause others. It is important, therefore, that all commercial machines be completely inspected periodically. You should use a systematic method of doing this. In this way, no detail will be overlooked. All inspections should cover such things as:

- Electrical connections.
- Motor and safety devices.
- Compressor noises.
- Amount of refrigerant.
- Dryness of refrigerant.
- Oil level.
- Water flow.
- Gas leaks.
- Coil conditions.
- Supports for tubing.
- Coil supports.
- Cleanliness.

For a conventional condensing unit, you should check:

- Belt condition.
- Belt alignment.
- Belt tightness.

For a hermetic condensing unit, you should check:

- Overload cutout.
- Relay.
- Capacitors.

Preparing a check sheet is recommended to prevent forgetting items. One copy should be given to the owner.

15.18 Locating Troubles

Methods of testing to locate sources of trouble are based on the mechanism's operating principles. Checking pressures, temperatures, running time, and current or voltage helps to pinpoint the faulty part.

You must have a thorough knowledge of refrigeration fundamentals and cycles. Only then can he or she become reliable and competent at trouble tracing and repair. Troubles in a refrigerator mechanism must be located before dismantling. This keeps the cost of servicing at a minimum and ensures proper operation of the unit after repair and assembly.

Methods of locating troubles vary with the type of system. Locating troubles differs in systems using direct expansion from those using a capillary tube. The call for service should indicate what the trouble is. The owner will probably say that it costs too much to operate or that the unit is not freezing but runs continuously; or, the complaint may be that it is freezing but runs continuously. From these complaints, you should usually get an idea what the trouble is. Always verify these statements by checking over the refrigerator before attempting any troubleshooting or service work.

In trouble tracing, first classify the type of service call. Then determine what caused the trouble described in the service call. The following troubleshooting pointers have been prepared to help you. Naturally, it is impossible to give every detail. However, once a technician learns the method of tracing trouble, there should be no difficulty.

Check all of the following in a refrigerating mechanism before deciding what the trouble is:

- Low-side pressure.
- High-side pressure.
- Temperature of evaporator.
- Temperatures of liquid and suction lines.
- Amount and dryness of refrigerant.
- Running time of mechanism.
- Probability of leaks.
- Noise.

Several basic fundamentals make locating trouble easier. When there is poor, or no refrigeration, either or both of two things can be wrong:

- There is little or no refrigerant.
- The pump is not moving the refrigerant. Pressures are not correct.

If there is no refrigerant, there will be no liquid refrigerant in the evaporator. Refrigerant has leaked out or is being held in a certain part of the system. It could be held due to clogged needles, clogged screens, and pinched lines. Clogging causes a high vacuum reading on the low side.

A lack of refrigerant throughout the system causes a hissing sound at the refrigerant control. This indicates the refrigerant passages are not closed. The sight glass will show bubbles.

The hissing sound at the refrigerant control always indicates a lack of refrigerant. The dry gas going through the restriction causes the gas noise.

If the pump is not functioning, the low-side pressure will be above normal. The condenser and discharge line from the compressor will be below normal temperature.

To determine what is responsible for a poor condensing condition, install the gauge manifold. Then determine the head pressure. Compare this pressure with what the pressure should be for the refrigerant being used.

15.18.1 Low or No Refrigeration—Unit Runs Continuously

Direct Expansion Unit (TEV)

If the unit has lost all refrigerant, there is, naturally, no refrigeration. To test for this, install gauges and determine evaporating or low-side pressure. If this pressure is normal, the unit probably has little or no refrigerant. If the compound gauge indicates a high vacuum (20″ Hg. (34 kPa) or more) it means:

- The expansion valve is adjusted so that it draws this vacuum.
- The expansion valve is frozen closed.
- The system has a clogged screen.

Clogging can be caused by moisture freezing at the refrigerant control and stopping refrigerant flow. The results are the same as a needle stuck closed or a clogged screen. There is one difference. After the system warms above 32°F (0°C) at the valve, frozen moisture will melt. Normal refrigeration will return.

A new drier must be installed if moisture is the problem. If it is suspected that moisture in the valve has caused clogging, heat the valve. This moisture problem occurs with all refrigerants that do not chemically combine with the water, including R-12, R-22, R-500, and R-502. A high vacuum may also be caused by a clogged or restricted suction line filter. In either case, it is not allowing refrigerant to flow through. There will be a high pressure in the evaporator. This causes continuous running with little or no refrigeration.

If the compound pressure gauge shows a high pressure on the low side (this is a pressure that does not allow the refrigerant to evaporate at a low temperature), an expansion valve may be stuck open or out of adjustment. If this is the trouble, there may also be a frosted or sweating suction line. This simply shows that the refrigerant is going into the low side too fast. The liquid will flood both the evaporator and the suction line.

High pressure may also be due to an inefficient compressor. If the expansion valve is stuck open, there may be dirt on the needle. To remedy this, the valve may be flushed. This is done by alternately opening and closing the liquid receiver service valve. Otherwise, use the liquid line hand shutoff valve in the same manner. Surges of liquid will rush past the expansion valve needle, cleaning it.

Capillary Tube

No refrigerant can pass into the evaporator under the following conditions:

- If the capillary tube is restricted or completely clogged.
- If there is moisture frozen in the tube.
- If the screen is clogged.

This stoppage will cause a high vacuum reading. The evaporator will be warm, and the condenser will be cool. There will be a normal or low head pressure. If refrigerant is low, the capillary tube will hiss as the compressor shuts off. In addition, low-side and high-side pressure will be below normal.

An above-normal low-side pressure and a normal or below-normal head pressure indicates an inefficient pump.

15.18.2 No Refrigeration—Unit Does Not Run

With no cooling and the unit not running, the trouble may be in the electrical circuit. A test light or a voltmeter will reveal if power is being supplied to the motor.

If there is power, the motor may be burned out or the circuit open. An ineffective temperature control may prevent the motor from starting. (An ineffective temperature control might be a leaking power element, for instance.) It should be checked as described in Chapter 8.

Further circuit checking may show a manual switch or overload in the Off position. Otherwise, an internal overload switch may be open.

If the motor compressor is hot, this may be the trouble. It should be checked with a test light or an ohmmeter. Basically, the problem is in the motor or the electrical power circuit.

15.18.3 Normal or Excess Refrigeration—Motor Running Continuously

The motor compressor may run continuously and the unit may produce normal or too much refrigeration. The probable cause of the trouble is a faulty motor control. It will not cut out at the correct temperature. Remember, there is a relationship between the control cut-out point and the low-side evaporating pressure. The control cannot stop the electric motors if its setting is lower than the evaporating pressure.

Direct Expansion System (TEV)

An undercharged system may only provide enough refrigerant to fill the evaporator partially. It may be insufficient to operate the temperature control cut-out point. As a result, there may be normal refrigeration, but the unit never shuts down. This trouble is indicated by the way frost or sweat collects on the evaporator outlet tubing. If it reaches as far as the TEV sensitive bulb, there is a refrigerant shortage. Another indication of refrigerant shortage is a warm liquid receiver and liquid line. The sight glass should show bubbles.

The trouble may also be caused by an overcharge of refrigerant or air in the condenser. Excessive head pressures lower the efficiency of the compressor so much that continuous operation results.

A leaking expansion valve will sometimes give normal refrigeration. However, the pressure cannot drop to the point where the motor control will cut out the motor. An improperly adjusted expansion valve may cause the same trouble, creating a frosted or sweating suction line. It also will cause an above normal, fluctuating low-side pressure.

An inefficient compressor is another possible cause and can be checked with gauges. Expansion valve troubles and the compressor troubles may be checked as mentioned earlier in this chapter.

Capillary Tube

If there is too much refrigerant, the excess will collect on the low side. It may then enter the suction line. This liquid may prevent the compressor from producing a low enough pressure to operate the thermostat. Therefore, the unit will run continuously. It will produce either a normal refrigeration effect or, more likely, excessive refrigeration.

A slight lack of refrigerant will cause a partially refrigerated evaporator. The refrigerated part may not be close enough to the thermostat. Then it will not cause the thermostat to shut off the motor.

15.18.4 Short Cycling

Short cycling means that the system runs and then stops every few minutes. It is necessary to locate the control that is turning the system on and off:

- Temperature control.
- Overload controls.
- High-pressure safety control.
- Oil pressure safety control.

Rapid pressure rise may occur on the low side where an expansion valve is leaking. Short cycling may result. This leak will cause a frosting or sweating of the suction line.

Most units are equipped with an overload safety device in the electrical system. These devices stop the motor if it becomes too hot or uses too much current. They will then restart the motor after it cools. A temperature control with a small differential also causes a short cycling.

In refrigerators using pressure motor controls, short cycling may be caused by a leak in the refrigerant control or by poorly seated compressor valves. In either case, low-side pressure will rise rapidly during the Off cycle. This will cause the motor to start.

Machines equipped with a high-side pressure safety control will sometimes short cycle. This occurs if the condensing pressure becomes too high because of a high condensing temperature.

15.18.5 Noisy Unit

Noise can come from three principal sources: the compressor, the electric motor, and the mounting of the condensing unit. A compressor is noisy when it pumps oil. It is also noisy when valves, piston pins, connecting rods, and pistons have become worn. When the compressor gets very warm, it will sometimes develop knocks. These are usually hard to remedy, but may be caused by a lack of oil.

The metal shaft seal used on open compressors occasionally becomes noisy, giving out a shrill squeal. This is usually caused by a lack of oil at the seal. If not remedied, it will soon score the seal and cause it to leak.

Conventional electric motors may have noises such as fan roar, bearing squeak, or motor rumble. Occasionally, if the motor is loaded too heavily, the starting winding does not cut out. It will cause continuous noisy operation. If allowed to run this way, the motor will burn out. End play in an electric motor is necessary. However, too much will cause a dull knock.

In a conventional unit, belt noise may result from a dry belt. It may be stopped by using a dressing recommended for belts. Pulleys that are out of line may also cause noise. This may be remedied by realigning the pulleys. Do not use oil.

Sometimes the whole machine unit will vibrate excessively. This produces a rumbling sound as the unit runs, shaking the cabinet disagreeably. This is probably due to poor mounting or spring suspension. Another cause may be too little movement in the suction and liquid lines. Some obstruction may have been put in the compartment. The obstruction destroys the action of the shock and the noise-absorbing mounting of the condensing unit.

Excessive head pressure will make a unit vibrate more than normal. A badly worn needle or seat sometimes makes a chattering noise while the unit is operating.

15.19 Refrigeration Service Contracting

It is good business to offer contracts for maintenance and service. Many large companies have developed such contracts. Even larger independents are now offering their customers this type of servicing. See Chapter 29.

The usual contracting plan offers a definite monthly or weekly rate. For this amount, the service company agrees to keep the refrigerating mechanism in good condition. This charge may or may not cover parts. Contracts may be on a time and materials basis.

The success of such a plan depends on large volume. Large volume will offset cost of maintaining extremely bad installations. Two features of a service contract may appeal to the purchaser. They are the 24-hours availability service clause and an absolute guarantee of work done.

If one has a service contract, a procedure sheet or record sheet should be used. It will prove service and ensure thorough inspection. This check sheet should indicate the date, the name of the technician making the call, and the following checklist:

- Test for leaks.
- Check refrigerant charge:
  A. Head pressure.
  B. Low-side pressure.
- Check oil charge.
- Check water valve.
- Check water drain.
- Check and lubricate motor.
- Check belt condition and tension.
- Clean evaporator.
- Clean condenser.
- Straighten fins.
- Voltage reading.
- Wattage reading.
- Check circulating fans.

Service records are absolutely essential if one wishes to establish a permanent business. These records should contain details of ownership, machine, type of work done, and materials used. This record enables "check backs" if the system does not operate correctly. Furthermore, it establishes sales prospects as systems become older.

15.19.1 Service Estimates

Many organizations operating refrigerating equipment ask for bids when repair, replacement, or service is required. See Chapter 29. A service organization bidding on this work needs someone who specializes in estimating such work. This specialist should be thoroughly acquainted with cost of materials, service problems, and labor costs. The individual must be able to judge time necessary to do the repair.

A pleasing personality combined with rapid and accurate estimating ability is essential. Records kept of service and maintenance work are used as a guide in making estimates. Needless to say, estimates must include overhead expenses. Such expenses include rent, equipment obsolescence, office and shop services, and advertising.

15.20 Refrigerant Recovering and Recycling

Federal law has resulted in a decline in production of fully halogenated chlorofluorocarbon refrigerants. See Chapter 9 for details. This legislation has created the need for refrigerant recovery and reclamation procedures. CFCs R-12, R-22, R-500, and R-502 cannot be reused from old or damaged refrigeration systems. The vapor must be cleaned.

In the past, the refrigerants had been vented to the atmosphere. Now the governmental emission standards concerning chlorofluorocarbons prohibit this. The refrigerant is now recovered and recycled by the use of recovery management systems. See **Figure 15-79.** The refrigerant is processed through two filters. These filters are accessible for replacement at the top of the unit. An oil trap removes contaminated oil from the system's refrigerant and can be drained. A sight glass indicates refrigerant condition after all filtering and oil separation has been completed. When the system is completely evacuated, the reclaimer shuts down automatically on low suction pressure. The unit does not have internal storage capacity. It must be connected to refillable storage cylinders.

Today's recycling reduces contaminants through oil separation and filtration. Many of these units are designed to pump down the system. The recycled refrigerants are then returned to the same system.

See Chapter 10 regarding standard procedures for using refrigerant recovery, recycling, and reclamation equipment.

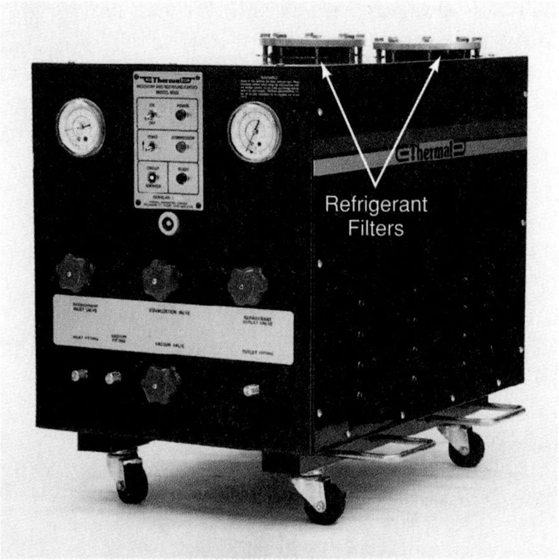

Figure 15-79. *Refrigerant recovery system designed to be used with systems containing R-12, R-22, R-500, and R-502, in either liquid or vapor form. (Thermal Engineering Company, Division of Seakay Co., Inc.)*

15.21 Review of Safety

A refrigeration service engineer must always be alert to safety concerns. Systems have hazards arising from pressure, electricity, power devices, heat, flames, heavy objects, and climbing.

The safety program must include:

- Safety for the mechanism.
- Safety for the items being refrigerated.
- Safety for the operator, the installation, the technician, and people near the mechanism.

All parts in a refrigerating system must be absolutely clean before they are installed in the system.

Always know what is inside a pressure vessel and know the pressures. Always wear goggles when working on a pressure vessel (refrigerating unit). Also wear goggles when there is danger of flying particles.

Never breathe fumes of any kind. Do not neglect the use of the gas mask when working in a refrigerant-laden atmosphere. This protective measure also applies to fumes from cleansing bath, soldering, brazing, and welding. The human body can and will get rid of certain amounts of strange chemicals and fumes. However, some chemicals and fumes accumulate in the body. There may be no ill effects felt for years. Good ventilation is of vital importance.

Avoid exposure to electrical shocks. Keep open electrical terminals covered. Do not work on electrical circuits in damp or wet surroundings.

Avoid spilling liquid refrigerant on any fixture, finished surface, or floor. It may ruin the finish. Avoid contact with the liquid refrigerant, especially on the body and eyes. A freeze burn will result.

Put guards on powered moving objects such as flywheels, belts, pulleys, and fans. Use the leg muscles, not the back, when lifting.

Have a fire extinguisher handy before using flame for leak testing, soldering, brazing, or welding. Remove all combustibles from the area and provide good ventilation. A dry chemical fire extinguisher can be used on all types of fires. When using a flame to perform tests or make repairs, protect surrounding objects and surfaces. This can be done with sheet metal or some other flame-resistant shield.

Always test used compressor oil for acid content before allowing any to touch the skin. A severe acid burn may result.

Never use air, oxygen, or any fuel gases for developing pressure in a system. If gas, other than refrigerant, is desired, use carbon dioxide, nitrogen, helium, or argon. Use them at controlled pressures, with a pressure relief valve. See Section 12.9.1.

Always wear goggles when handling refrigerants, or when opening a refrigerating mechanism. Also use goggles at any time when there is danger from flying liquids. Wear rubber gloves when handling substances which may have an acid content. Never use carbon tetrachloride for any cleaning operation.

Use care when handling capacitors. A charged capacitor can deliver a severe shock.

The Occupational Safety and Health Act (OSHA) is now in effect with regulations. It is also known as the William Steiger Act of 1970. Most of the regulations became mandatory in March, 1973.

Many regulations are based on standards developed by:

- The American National Standards Institute (ANSI).
- National Fire Protection Association (NFPA).
- Walsh Healy Act (noise).
- The Service Contract Act.
- American Society for Testing and Materials (ASTM).
- The American Conference of Governmental and Industrial Hygienics (ACGIH).

The Act covers almost every safety standard including:

- Walking and working surfaces.
- Means of exit (egress).
- Powered platforms.
- Occupational health and environment control.
- Hazardous materials.
- Personal protective equipment.
- General environment controls.
- Medical and first aid.
- Fire protection.
- Compressed gases.
- Material handling.
- Machinery and machine guarding.

Any employee can ask for an inspection. If violations are found, the inspector can impose fines. If safeguards are present but the employee does not use them, the employer will still be fined. However, an employer has excellent grounds for releasing an employee who is not using the safety precautions.

Some of the OSHA priorities during inspection are:

- Asbestos.
- Carbon monoxide.
- Cotton dust.
- Lead.
- Silica.

Asbestos and carbon monoxide are of special importance to refrigeration and air conditioning.

Some other OSHA concerns are:

- Beryllium.
- Heat.
- Mercury.
- Ultraviolet radiation.
- Fibrous glass.
- Trichloroethylene.
- Chromic acid.
- Parathion.

Heat, ultraviolet radiation, fibrous glass, and trichloroethylene are also important safety considerations.

The Threshold Limit Values (TLV) are important. Usually this is the upper safety limit for the hazard over an eight-hour exposure period. Most have been established by ANSI (American National Standards Institute).

Noise is important. The OSHA lists 90 dB (decibels) as maximum for eight hours of exposure, 92 dB for six hours, 95 dB for four hours.

Above all else, report any injury—no matter how slight—in writing. Send the report to your employer and keep a copy.

15.22 Test Your Knowledge

Please do not write in this text. Place your answers on a separate sheet of paper.

INSTALLING COMMERCIAL SYSTEMS MODULE

1. What safety devices are used to prevent explosions in the condensing unit?
   A. High-pressure cutouts.
   B. Pressure relief valves.
   C. Rupture disks.
   D. Any of the above.
2. What is the purpose of the short U-bend when tubing is slanting upwards from an evaporator to the condenser?
   A. It provides strength to the connection.
   B. It acts as a moisture trap.
   C. It acts as an oil trap.
   D. All of the above.
3. What is the purpose of safety relief valves on condensing units?
   A. To provide rerouting of the refrigerant from the condensing unit to the evaporator, when necessary.
   B. To provide for pressure gauge readings.
   C. To provide release of refrigerant due to overheating by fire.
   D. All of the above.
4. The most common method of installing piping in a multiple installation is ______.
   A. common liquid line
   B. common liquid line and common suction line
   C. clustering system
   D. common suction line
5. Suction lines from the evaporator to the compressor should be ______.
   A. straight
   B. towards the compressor
   C. away from the compressor
   D. Any of the above.
6. Water lines on a condensing unit are composed of ______ plastic.
   A. soft copper or flexible
   B. hard copper or hard
   C. brass or flexible
   D. flexible
7. What is the maximum number of openings a three-way valve can close at one time?
   A. One.
   B. Two.
   C. Three.
   D. Any of the above.
8. What type of control is used on a condenser to prevent violent rupturing or an explosion due to excessive pressure?
   A. High-pressure relief valve.
   B. Rupture disk.
   C. Safety cut-off valve.
   D. Both A and B.
9. The suction line has a slight slope to ______.
   A. prevent liquid condensation
   B. aid liquid return
   C. aid oil return to the compressor
   D. prevent moisture from accumulating on the low side
10. What leaves the system when it is purged?
    A. Refrigerant.
    B. Moisture.
    C. Air.
    D. All of the above.

SERVICING COMMERCIAL SYSTEMS MODULE

11. What is the purpose of a sight glass on a compressor?
    A. To check the moisture content.
    B. To check the oil level.
    C. To check the low-side pressure.
    D. All of the above.

12. What does the sight glass frequently indicate?
    A. Low refrigerant.
    B. Moisture in the system.
    C. Refrigerant flow.
    D. Both A and B.
13. ______ may cause hermetic motor compressor burnout.
    A. Moisture in the system
    B. Dirt in the system
    C. High temperatures
    D. All of the above.
14. How is oil put into a compressor?
    A. Draw a vacuum and add it to the low side.
    B. Produce a low pressure to draw it in.
    C. Balance the pressure to draw it in.
    D. Any of the above.
15. On multiple installations, the suction line is ______.
    A. larger than the main suction line
    B. smaller than the main suction line
    C. the same size as the main suction line
    D. graduated in size, with each smaller than the preceding one
16. Evaporative pressure must be ______ so that the liquid boils at the correct temperature.
    A. low enough
    B. high enough
    C. twice the low-side pressure
    D. None of the above.
17. In a multiple evaporator system, the pressure-operated two-temperature valves are located in the ______.
    A. liquid line of the warmer evaporator
    B. suction line of the warmer evaporator
    C. high-pressure line between the compressor and condenser
    D. liquid line of the coolest evaporator
18. A ______ pressure drop is allowed in the suction line.
    A. large
    B. small
    C. moderate
    D. Any of the above.
19. A hissing sound at the refrigerant control indicates ______.
    A. excessive refrigerant
    B. moisture in the system
    C. lack of refrigerant
    D. Both B and C.
20. What indicates a lack of refrigerant in a TEV installation?
    A. Unit shuts down.
    B. The sight glass is clear.
    C. There is a knocking sound at the refrigerant control.
    D. There is frost or sweat at the outlet of the evaporator to the sensing bulb of the TEV.

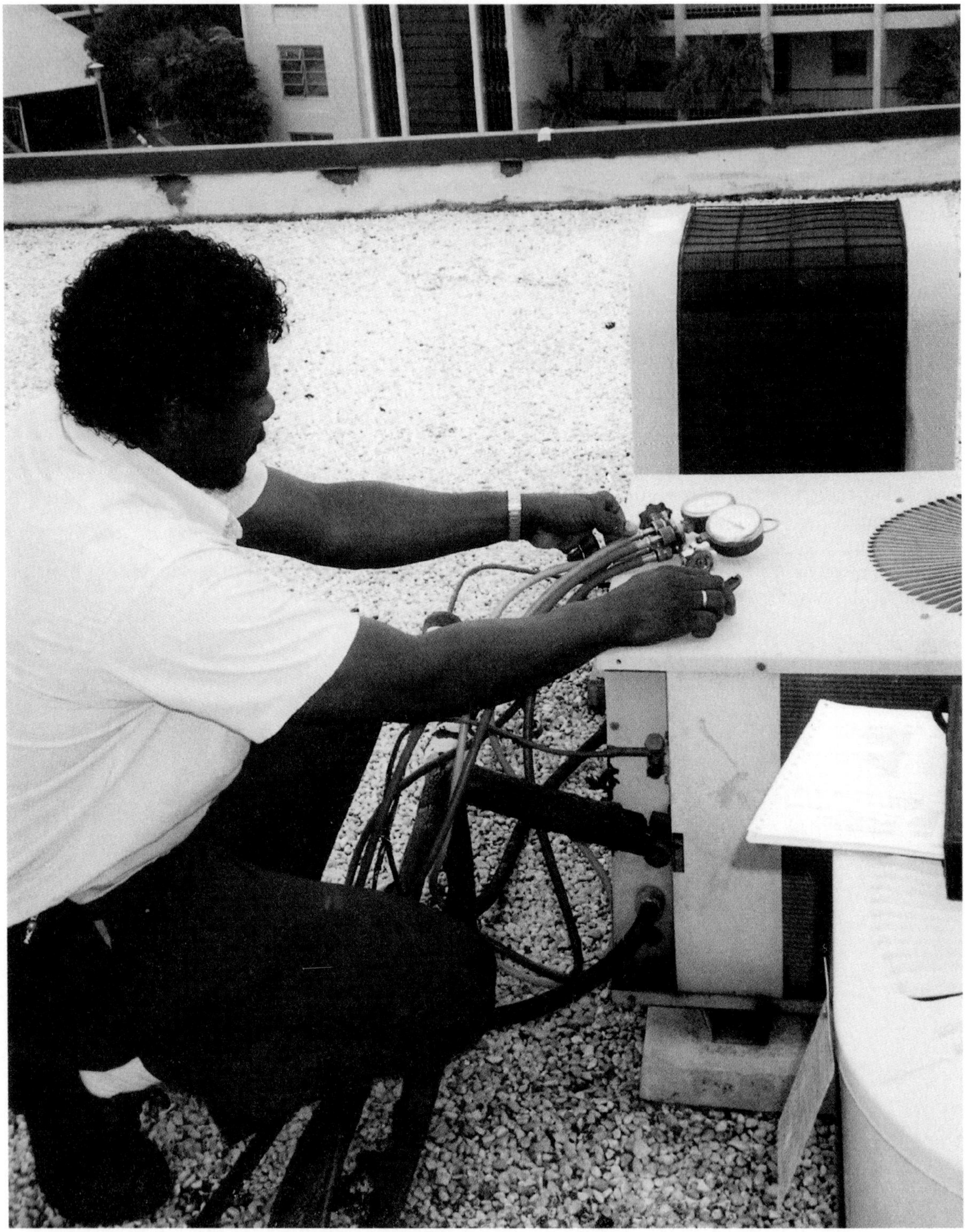

Technician servicing a small commercial rooftop unit. (Superior Contract Services, Inc.)

Chapter 16

COMMERCIAL SYSTEMS—HEAT LOADS AND PIPING

Modules:

Key Words:

bypass cycle
cascade system
coefficient of performance (COP)
heat exchanger
heat leakage load
heat load
pressure-heat diagram
seasonal energy efficiency ratio (SEER)
service load
specific heat
thermodynamics
vapor velocity
viscosity
volumetric efficiency

Learning Objectives:

After studying this chapter, you will be able to:

- Discuss system balance and explain four important factors in balancing commercial systems.
- Explain and calculate heat loads.
- List the individual loads that make up the total heat load.
- Demonstrate proper use of tables in computing heat loads.
- Correctly size system components using manufacturers' tables.
- Discuss and calculate seasonal energy efficiency ratio (SEER).
- **Follow approved safety procedures.**

HEAT LOADS MODULE

To have good refrigeration system performance, four main items must be ***matched*** (balanced or made equal to each other):

- Heat load. Determine the total amount of heat that must be removed for each 24 hours. Amount of heat is measured in British Thermal Units (Btu). One Btu is the amount of heat needed to raise the temperature of 1 lb. of water one degree Fahrenheit (°F).
- Condensing unit. Determine what size condensing unit is needed to handle the heat load. To do this, determine whether the unit is to run 16, 18, or 20 hours out of each 24.
- Evaporator. Determine evaporator capacity needed to handle the heat load. The evaporator can remove heat only while the condensing unit is running. Therefore, its capacity must be based on the same hours of operation as the condensing unit.
- Total system. Consider water supply, temperature control devices, refrigeration line sizes, air circulation, and humidity control. Correctly install the parts according to code.

When determining the heat load, two main factors must be considered:

- Heat leakage into the cabinet. Heat leakage is affected by the amount of exposed surface, the thickness and kind of insulation, and the temperature difference between inside and outside of cabinet.
- Usage or service heat load of the cabinet. This load is determined by the temperature of articles put into the refrigerator, their specific heat, generated heat, and latent heat, as the requirements demand. Another consideration is the nature of the service needed. This includes air changes (determined by the number of times per day that doors of the refrigerator are opened) and heat generated inside by fans, lights, and other electrical devices.

Total heat load is the sum of the wall heat transmission load, the air change load, the product load, and miscellaneous loads.

The selection of a condensing unit is made from manufacturers' tables of condensing unit capacities. Evaporators are selected from manufacturers' specifications for capacities to balance the capacity of the condensing unit. Also affecting the selection of the evaporator are the capacity of the refrigerant control, type of temperature control, arrangement for air circulation, and specific duty. The installation of all commercial refrigeration equipment involves a technical understanding of the variables. This is a determining factor in the operation of the system.

16.1 Heat Load

The total heat load consists of the amount of heat to be removed from a cabinet during a certain period. It is dependent on two main factors:

- Heat leakage load.
- Heat usage (service) load.

The ***heat leakage load*** or heat transfer load is the total amount of heat that leaks through the walls, windows, ceiling, and floor of the cabinet per unit of time (usually 24 hours).

The ***heat usage (service) load*** is the sum of the following heat loads per unit of time (usually 24 hours):

- Cooling the contents to cabinet temperature.
- Cooling of air changes.
- Removing respiration heat from fresh or "live" vegetables and from meat.
- Removing heat released by electric lights and motors.
- Removing heat given off by people entering and/or working in the cabinet.

In this chapter, heat load calculations are in the U.S. conventional system (pounds, Btu, °F, feet, etc.). Conversion factors used to make these calculations in the SI metric system (kilograms, kilocalories, °C, centimeters, etc.) are explained in Chapter 31.

16.1.1 Heat Leakage Variables

Research organizations, manufacturers, and refrigeration associations have determined the amount of heat leakage through walls and other heat loads. Charts and tables based on these calculations are used by engineers and technicians.

Five factors (variables) that affect heat leakage are:

- Time. The longer the period of time, the more heat will leak through a certain wall. The standard time unit is the 24-hour period in refrigeration situations. A one-hour period is used in air conditioning situations.
- Temperature difference. The difference in temperature is important in the heat leakage into a container. The greater the temperature difference, the more heat will leak or transfer through the wall. Compare this idea to pressure: the more pressure, the more water will flow through an opening. The room temperature usually chosen is the average summer temperature. In the United States, this varies between 90°F and 105°F (32°C and 40°C). See **Figure 16-1.** This value can be reduced to 75°F or 80°F (25°C or 27°C) if the room is air-conditioned.
- Thickness of insulation. The thicker the insulation, the less heat will flow through it. Twice as much heat will leak through a wall with 1″ (2.5 cm) insulation than through a wall having 2″ (5 cm) insulation.
- Kind of insulation. The kind of insulation or the material used is important. Expanded polystyrene (foam), for instance, will insulate approximately six times better than wood. Some insulations, however, are more costly than others.
- External area of cabinet. The more area through which heat may leak, the greater the heat flow. This is similar to water flow, in which the size of a pipe determines how much water will flow through it. The bigger the pipe, the more water will flow.

The common unit used for determining the heat flow is the total square foot area. This area is always measured on the outside of the cabinet.

Summer Design Temperatures					
State	Design Dry Bulb °F	°C	State	Design Dry Bulb °F	°C
Alabama	95	29	Massachusetts	90	32
Alaska	74	23	Michigan	88	31
Arizona	105	41	Minnesota	90	32
Arkansas	98	37	Mississippi	97	36
California:			Missouri	98	37
lower	86	30	Montana	88	31
middle	94	34	Nebraska	97	36
upper	83	28	Nevada	95	35
Colorado	92	33	New Hampshire	90	32
Connecticut	88	31	New Jersey	92	33
Delaware	93	34	New Mexico	95	35
Dist. of Col.	94	34	New York	90	32
Florida:			North Carolina	95	35
upper	96	36	North Dakota	93	34
lower	93	34	Ohio	90	32
Georgia	95	35	Oklahoma	102	39
Hawaii	87	31	Oregon	90	32
Idaho	94	34	Pennsylvania	92	33
Illinois:			Rhode Island	87	31
upper	95	35	South Carolina	95	35
lower	97	36	South Dakota	95	35
Indiana	95	35	Tennessee	96	36
Iowa	95	35	Texas	101	38
Kansas:			Utah	95	35
upper	97	36	Vermont	87	31
lower	100	38	Virginia	95	35
Kentucky	95	35	Washington	90	32
Louisiana	98	37	West Virginia	94	34
Maine	88	31	Wisconsin	90	32
Maryland	94	34	Wyoming	90	32

Figure 16-1. *Table of summer design temperatures. These may be used as ambient temperatures when calculating heat leakage loads.*

16.1.2 K Factor

To bring together the values just discussed, standards have been developed for use by refrigerating companies. In preparing these standards, the variables have been reduced to unit values. The unit values, in turn, are used to indicate heat leakage of the wall.

The unit or basic values are the ***thermal conductance (K)*** obtained for an area of insulation one square foot in size, one inch thick, with a temperature difference of 1°F over a period of time of either one hour or 24 hours. Values obtained represent the amount of heat flow through the insulation under these conditions.

Unit values vary with the kind of insulation. This material has no air film or liquid film on either side. The symbol is K.

If the insulation is less than or more than 1″ thick, the heat leakage will be different. In this case, the symbol is K_T.

Example:

$$K_T = \frac{K_1}{\text{thickness}}$$

By definition:

K_T = conductance, total
K_1 = conductance for 1″ thickness

If thickness is 2″:

$$K_T = \frac{K_1}{2} \qquad K_T = \frac{1}{2}K_1$$

If thickness is 1/2″:

$$K_T = \frac{K_1}{\frac{1}{2}} \qquad K_T = K_1 \div \frac{1}{2}$$

$$K_T = K_1 \times 2$$
$$K_T = 2K_1$$

A special formula is needed to find heat leakage or thermal conductance through a composite wall, **Figure 16-2**. The formula for computing the KT or total heat conductance factor follows:

Resistance to heat flow is known by the symbol R. If the same heat is flowing through two substances, the total resistance is equal to the resistance of each substance.

$R_T = R_1 + R_2$
R_T = resistance total
R_1 = resistance of substance 1
R_2 = resistance of substance 2

R is the ***reciprocal*** (inverse) of K, or R = 1/K. In the formula, it would be:

$$\frac{1}{K_T} = \frac{1}{K_1} + \frac{1}{K_2}$$

K_T = conductance, total
K_1 = conductance, substance 1
K_2 = conductance, substance 2

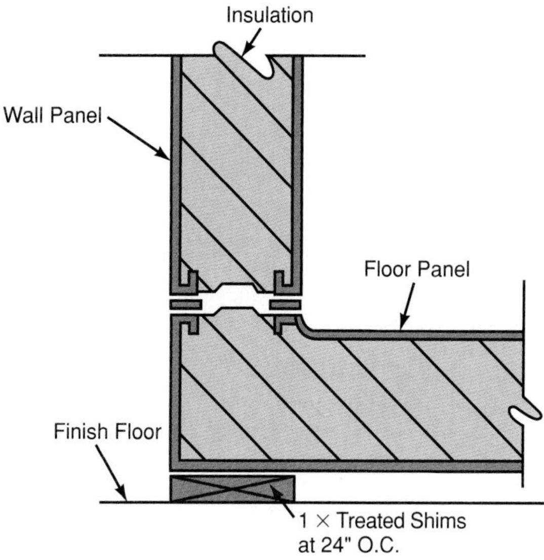

Figure 16-2. *Cross section of a walk-in cooler wall. Note system for sealing joint. (Tyler Refrigeration Corp.)*

Example:

$K_1 = .6$ and $K_2 = .2$

$$\frac{1}{K_T} = \frac{1}{.6} + \frac{1}{.2}$$

$$\frac{1}{K_T} = \frac{1}{\frac{6}{10}} + \frac{1}{\frac{2}{10}}$$

$$\frac{1}{K_T} = \frac{10}{6} + \frac{10}{2} \quad \text{or} \quad \frac{1}{K_T} = \frac{5}{3} + \frac{5}{1}$$

$$\frac{1}{K_T} = \frac{5}{3} + \frac{15}{3} \quad \text{or} \quad \frac{1}{K_T} = \frac{20}{3}$$

$$K_T = \frac{3}{20} \qquad K_T = .15$$

$$R = \frac{1}{K_T} \qquad R = \frac{1}{.15}$$

$$R = \frac{1 \times 100}{.15 \times 100} = \frac{100}{15} = 6.7$$

If a wall is made up of three different materials, **Figure 16-3,** the overall heat leakage (K) is found as follows:

$$K_T = \frac{1}{\frac{\text{thickness of material 1}}{\text{conductivity factor material 1}}} + \frac{1}{\frac{\text{thickness of material 2}}{\text{conductivity factor material 2}}} + \frac{1}{\frac{\text{thickness of material 3}}{\text{conductivity factor material 3}}}$$

where

Th_1 = thickness of material 1
Th_2 = thickness of material 2
Th_3 = thickness of material 3

and

K_1 = conductivity factor for material 1
K_2 = conductivity factor for material 2
K_3 = conductivity factor for material 3

then the formula becomes:

$$K_T = \frac{1}{\frac{Th_1}{K_1} + \frac{Th_2}{K_2} + \frac{Th_3}{K_3}}$$

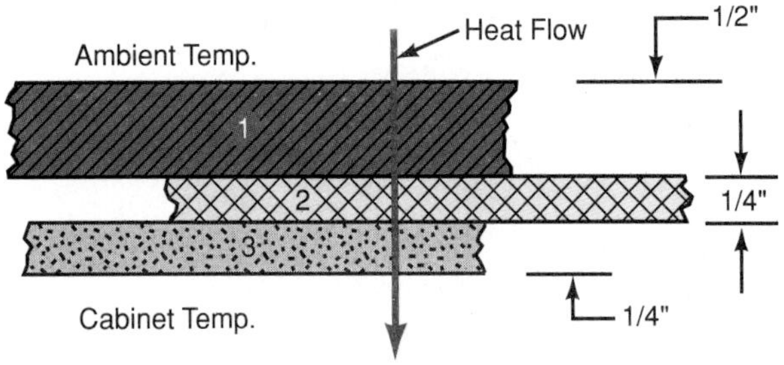

Figure 16-3. *A composite insulated panel or wall. Insulating materials, labeled 1, 2, and 3, are three different types and thicknesses of materials.*

To solve for the conductivity for the panel shown in **Figure 16-4,** proceed as follows:

$$K_T = \frac{1}{\frac{\text{thickness A}}{\text{K for wood}} + \frac{\text{thickness B}}{\text{K for Celotex®}} + \frac{\text{thickness C}}{\text{K for polyurethane}}}$$

From Chapter 31, K values are:

Material	K value
Wood	= .80 for 1″ thickness
Celotex®	= .31 for 1″ thickness
Polyurethane	= .160 for 1″ thickness

However, in the example, the wood is 1/2″ thick, the Celotex® is 1/4″ thick, and the polyurethane is 1/4″ thick.

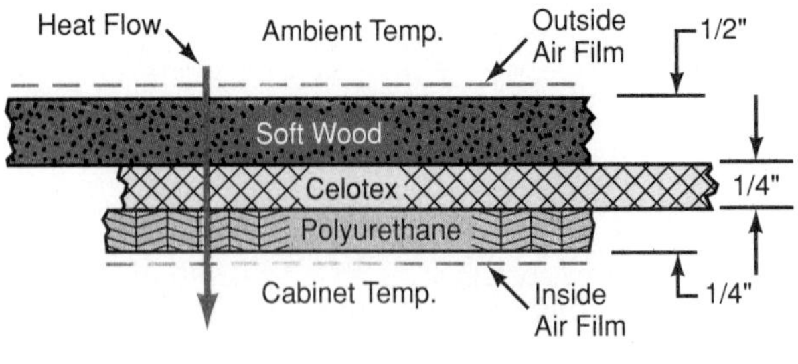

Figure 16-4. *Cross section shows a composite insulating panel made up of various materials at specified thicknesses with an air film on both sides.*

Substituting these values in the previous formula:

$$K_T = \frac{1}{\frac{.5}{.80} + \frac{.25}{.31} + \frac{.25}{.160}}$$

$$= \frac{1}{.625 + .807 + 1.563} = \frac{1}{3.0}$$

K_T = .333, which is the conductivity for the panel in Btu/ft^2/hr./°F

An air film that clings to the outer and inner surfaces of the cabinet adds to the insulating value of the cabinet walls. This added resistance to heat transfer is calculated in the following formula. In this formula, the outside air film (F_o) is considered to have a heat transfer value of 6.0. The inside-wall air film (F_i) has a value of 1.65:

K_T = unit of conductivity for materials of a composite nature
U = unit of conductivity for materials of a composite nature plus the effect of the air clinging to both the outside (F_o) and the inside (F_i) walls.

If the insulating value of the air clinging to the walls is considered, the formula becomes:

$$U = \frac{1}{\frac{1}{F_o} + \frac{Th_1}{K_1} + \frac{Th_2}{K_2} + \frac{Th_3}{K_3} + \frac{1}{F_i}}$$

If the value of F_o = 6.0 and the value of F_i = 1.65, then the problem in **Figure 16-4** may be solved as follows:

$$U = \frac{1}{\frac{1}{6.0} + \frac{.5}{.80} + \frac{.25}{.31} + \frac{.25}{.160} + \frac{1}{1.65}}$$

$$= \frac{1}{.166 + .625 + .807 + 1.563 + .606}$$

$$= \frac{1}{3.767} = .27 \text{ Btu/ft}^2\text{/hr./°F}$$

The value of K, as computed in the previous problem, is .33. The value of U, as computed in this problem, is .27. This shows the additional insulating effect of the air films.

This type of computation is complicated and slow. For this reason, standard tables for computing heat leakage are generally used.

16.1.3 Air Change Heat Load

Air that enters a refrigerated space must be cooled. Air has weight and it also contains moisture. When air enters a refrigerated space, heat must be removed from it.

By ***Charles' Law,*** air which enters and is cooled reduces in pressure. If the cabinet is not airtight, air will continue to leak in. The actions of material moving in or out of the cabinet, and a person going into or leaving a cabinet, result in warm air moving into the space. Each time a service door or a walk-in door is opened, the cold air inside, being heavier, will spill out the bottom of the opening. This allows the warmer room air to move into the cabinet. This air movement is sometimes called ***infiltration.***

Figure 16-5 shows accepted air change volume values for refrigerated cabinets of various internal volumes. **Figure 16-6** shows the total heat (sensible + latent) to be removed from this air. It depends on various outside conditions and refrigerator temperatures.

16.1.4 Product Heat Load

Any substance which is warmer than the refrigerator it is placed in will lose heat. This will continue until the substance cools to the refrigerator temperature.

Three kinds of heat removal may be involved:

- Specific heat.
- Latent heat.
- Respiration heat.

The total ***product heat load*** would be the sum of these three heat loads. An example of all three heat loads

Volume (ft^3)	Air Changes per 24 hr.	Volume (ft^3)	Air Changes per 24 hr.
200	44.0	6,000	6.5
300	34.5	8,000	5.5
400	29.5	10,000	4.9
500	26.0	15,000	3.9
600	23.0	20,000	3.5
800	20.0	25,000	3.0
1,000	17.5	30,000	2.7
1,500	14.0	40,000	2.3
2,000	12.0	50,000	2.0
3,000	9.5	75,000	1.6
4,000	8.2	100,000	1.4
5,000	7.2		

NOTE: For heavy usage, multiply the above values by 2. For long storage, multiply the above values by 0.6.

Figure 16-5. *Average air changes per 24 hours for storage rooms. Values take into account door openings and air filtration. (Reprinted by permission of the American Society of Heating, Refrigerating, and Air-Conditioning Engineers, Atlanta, Georgia)*

Heat Removed in Cooling Air to Storage Room Conditions, Conventional Units (Btu/ft³)

Storage Room Temp, °F @ 80% RH	Temperature of Outside Air, °F: 85		90		95		100	
	Relative Humidity, %: 50	60	50	60	50	60	50	60
65	0.45	0.64	0.68	0.91	0.93	1.20	1.21	1.51
60	0.66	0.85	0.89	1.12	1.14	1.41	1.42	1.71
55	0.85	1.04	1.08	1.31	1.33	1.60	1.61	1.91
50	1.03	1.22	1.26	1.49	1.51	1.78	1.79	2.09
45	1.19	1.39	1.43	1.66	1.68	1.94	1.95	2.25
40	1.35	1.55	1.59	1.81	1.83	2.10	2.11	2.41
35	1.50	1.70	1.74	1.96	1.99	2.25	2.26	2.56
30	1.64	1.84	1.88	2.10	2.13	2.39	2.40	2.70

Storage Room Temp, °F @ 80% RH	Temperature of Outside Air, °F: 40		50		90		100	
	Relative Humidity, %: 70	80	70	80	50	60	50	60
25	0.39	0.43	0.69	0.75	2.02	2.24	2.54	2.84
20	0.52	0.56	0.82	0.89	2.15	2.38	2.68	2.97
15	0.65	0.69	0.95	1.01	2.28	2.50	2.80	3.10
10	0.77	0.82	1.08	1.14	2.40	2.63	2.93	3.22
5	0.89	0.94	1.20	1.26	2.52	2.75	3.05	3.34
0	1.01	1.05	1.31	1.38	2.64	2.86	3.16	3.46
–5	1.13	1.17	1.43	1.49	2.76	2.98	3.28	3.58
–10	1.24	1.29	1.55	1.61	2.88	3.10	3.40	3.70
–15	1.36	1.41	1.67	1.73	2.99	3.22	3.52	3.81
–20	1.48	1.52	1.78	1.85	3.11	3.34	3.64	3.93
–25	1.60	1.64	1.90	1.97	3.23	3.45	3.75	4.05
–30	1.72	1.76	2.03	2.09	3.35	3.58	3.88	4.17

Figure 16-6. *Chart gives total heat removed to cool storage room air under varying conditions of humidity and temperature. Values of heat removed are in Btu/ft³. (Reprinted by permission of the American Society of Heating, Refrigerating, and Air-Conditioning Engineers, Atlanta, Georgia)*

would be moist head lettuce at 55°F (13°C) being put into a 35°F (2°C) refrigerator. The lettuce must be cooled (specific heat) to 35°F (2°C). Some of the moisture on the lettuce will evaporate and collect on the evaporators (latent heat). The lettuce, being a live vegetable, would absorb carbon dioxide and release oxygen. This change would release heat energy (respiration heat). Meats go through a slow bacteriological change. During this action, heat is released (another instance of respiration heat). Some examples of heat loads:

- **Specific heat.** Bottled beverages at 50°F (10°C) are placed in a 35°F (2°C) refrigerator. This action is a specific heat problem. If the bottles or bottle cartons were moist, there would be a moisture-evaporating (latent heat) problem.
- **Latent heat.** If meat at 50°F (10°C) is placed in a refrigerator, and cooled (frozen) to 0°F (−18°C), the meat initially cools to about 27°F (−3°C), where it freezes, then cools to 0°F (−18°C). The latent heat of freezing of the meat is considerable. For fresh lean beef, it is 100 Btu/lb. **Figure 16-7** shows the specific heat and latent heat of various refrigerated products. In addition, it recommends storage temperatures and relative humidity.
- **Respiration heat.** If lettuce is stored at 40°F (4°C), each pound will release 7.99 Btu/24 hr. or 15,980 Btu/ton. **Figure 16-7** shows the respiration heat for some of the more common vegetables and fruits.

16.1.5 Miscellaneous Heat Load

All sources of heat not covered by heat leakage, product cooling, and respiration load are usually listed as ***miscellaneous heat loads.*** Some of the more common miscellaneous heat loads are: lights, electric motors, people, defrosting heat sources, and the sun (solar heat).

- Lights located in the refrigerated space will release heat. For example, a 100 W lamp will give off 342 Btu in one hour or

  342 × 24 = 8208 Btu/24 hr.

  If the workday is eight hours (only time light is on), the heat load would be

  342 × 8 = 2736 Btu/24 hr.

- On the average, electric motors release 2550 Btu/hp/hr. The amount of heat released also depends on motor efficiency. The larger the motor, the more efficient it is. **Figure 16-8** shows the heat given off by motors and the devices they drive. Forced convection evaporators usually have motors and fans. Therefore, the total heat release of such a motor is about 4600 Btu/hp/hr. for sizes from 1/8 hp to 1/3 hp. For example, a 1/8 hp motor-fan would release:

  4600 Btu/hp/hr.
  4600 Btu/h × 1/8 × 1
  = 4600 + 8 = 575 Btu/h

  575 × 24 = 13,800 Btu/24 hr. if motor runs continuously.

- People inside a refrigerated space release heat at varying rates. This depends on what they are wearing (insulation), the temperature of the cabinet, and on how hard they are working. **Figure 16-9** shows a range from 720 Btu/h/person at 50°F (10°C) to 1400 Btu/h/person at −10°F (−23°C). For example, if one person worked in a 30°F (−1°C) refrigerator for eight hours, the heat load would be:

  950 Btu/h × 8 hr. = 7600 Btu

- Many refrigerating units have defrosting heat sources, especially if the fixture temperature is 32°F (0°C) or lower. Whether the defrost heat source is electric, hot gas, or water, the defrosting operation adds heat to the interior of the refrigerator. The amount of heat is difficult to determine because most of the defrosting heat is removed in the defrost drain water. Add approximately 10% of the defrosting heat input as part of the heat lead.
- If part of the refrigerator is exposed to the sun, the heat from that source must be considered. Add the following to the room or ambient temperature: for a dark surface, add about 10°F (6°C). If it is a medium-colored surface, add 5°F (3°C). If a light surface, add 3°F (2°C) to the ambient temperature.

16.1.6 Cabinet Areas

The area of a cabinet is measured from the outside. There are six surfaces: four walls, the ceiling, and the floor. Usually the floor and ceiling have the same area. Opposite walls are the same area, also. To determine the total outside area:

1. Multiply width by length, then multiply by two. These areas are the areas of the floor and ceiling of the cabinet.
2. Multiply width by height, then multiply by two. These areas are the areas of ends of the cabinet.
3. Multiply length by height, then multiply by two. These areas are the areas of sides of cabinet.
4. Add these three values to determine the total external area of the cabinet.
   By formula:

   L = Length W = Width H = Height
   W × L × 2 = area of ceiling and floor
   W × H × 2 = area of ends
   L × H × 2 = area of sides
   Total external area = sum of the three areas

Most companies compute total area based on the outside of the cabinet. The exterior is easier to measure and the results are on the safe side.

After computing the external area of the cabinet, subtract the window area to obtain the area of the insulated surface. Window areas are calculated from the measurements of the outside edges of the window

Heat Load of Various Refrigerated Products

Product	Quick Freeze Temp.	Storage Temp.		Humidity % RH	Specific Heat		Latent Heat	Freezing Point	Respiration BTU/lb. per Day
		Long	Short		Above Freezing	Below Freezing			
Apples	–15	30-32	38-42	85-88	0.92	0.39	91.5	28.4	0.75
Asparagus	–30	32	40	85-90	0.95	0.44	134.0	29.8	....
Bacon, Fresh		0-5	36-40	80	0.55	0.31	30.0	25.0	....
Bananas		56-72	56-72	85-95	0.81	....	108.0	30.2	4.18
Beans, Green		32-34	40-45	85-90	0.92	0.47	128	29.7	3.3
Beans, Dried		36-40	50-60	70	0.30	0.237	18	....	....
Beef, Fresh, Fat	–15	30-32	38-42	84	0.60	0.35	79	....	....
Beef, Fresh, Lean	–15	30-32	38-42	85	0.77	0.40	100	....	....
Beets, Topped		32-35	45-50	95-98	0.90	....	....	26.9	2.0
Blackberries	–15	31-32	42-45	80-85	0.89	0.46	125	28.9	....
Broccoli		32-35	40-45	90-95	0.93	....	....	29.2	....
Butter	+15	....	40-45	....	0.64	0.34	15	15.0	....
Cabbage	–30	32	45	90-95	0.93	0.47	130	31.2	....
Carrots, Topped	–30	32	40-45	95-98	0.87	0.45	120	29.6	1.73
Cauliflower		32	40-45	85-90	0.90	....	....	30.1	....
Celery	–30	31-32	45-50	90-95	0.95	0.48	135	29.7	2.27
Cheese	+15	32-38	39-45	....	0.70	....	....	....	....
Cherries		31-32	40	80-85	0.85	....	118	28.0	6.6
Chocolate Coatings		45-50	....	....	0.3	....	....	....	....
Corn, Green		31-32	45	85-90	0.86	....	....	29.0	4.1
Cranberries		36-40	40-45	85-90	0.91	....	....	27.3	....
Cream		34	40-45	....	0.88	0.37	84	....	....
Cucumbers		45-50	45-50	80-85	0.93	....	....	30.5	....
Dates, Cured		28	55-60	50-60	0.83	0.44	104	....	....
Eggs, Fresh	–10	30-31	38-45	....	0.76	0.40	98	31.0	....
Eggplants		45-50	46-50	85-90	0.88	....	....	30.4	....
Flowers		35-40	....	85-90	....	....	....	....	....
Fish, Fresh, Iced	–15	25	25-30	....	0.82	0.41	105	30.0	....
Fish, Dried		30-40	....	60-70	0.56	0.34	65	....	....
Furs		32-34	40-42	40-60	....	....	....	....	....
Furs, To Shock		15	15						
Grapefruit		32	32	85-90	0.92	....	111	28.4	0.5
Grapes		30-32	35-40	80-85	0.92	....	111	27.0	0.5
Ham, Fresh		28	36-40	80	0.68	0.38	87	....	....
Honey		31-32	45-50	....	0.35	0.26	26	....	....
Ice Cream	–20	....	0-10	....	.5-.8	0.45	96	....	....
Lard		32-34	40-45	80	0.52	0.31	90	....	....
Lemons		55-58	....	80-85	0.91	0.39	190	28.1	0.4
Lettuce		32	45	90-95	0.90	....	....	31.2	8.0
Liver, Fresh		32-34	36-38	83	0.72	0.42	94	....	....
Lobster, Boiled		25	36-40	....	0.81	0.42	105	....	....
Maple Syrup		31-32	45	....	0.24	0.215	7.0	....	....
Meat, Brined		31-32	40-45	....	0.75	0.36	75.0	....	....
Melons		34-40	40-45	75-85	0.92	0.35	115	28.5	1.0
Milk		34-36	40-45	....	0.92	0.46	124	31.0	....
Mushrooms		32-35	55-60	80-85	0.90	....	....	30.2	....
Mutton		32-34	34-42	82	0.81	0.39	96	29.0	....
Nut Meats		32-50	35-40	65-75	0.30	0.24	14	20.0	....
Oleomargarine		34-36	....	....	0.65	0.34	35	15.0	....
Onions		32	50-60	70-75	0.91	0.46	120	30.1	1.0
Oranges		32-34	50	85-90	0.89	0.40	91.0	27.9	0.7
Oysters		....	32-35	....	0.85	0.45	120.0	....	....
Parsnips	–30	32-34	34-40	90-95	0.82	0.45	120.0	28.9	....
Peaches, Fresh		31-32	50	85-90	0.92	0.42	110	29.4	1.0
Pears, Fresh		29-31	40	85-90	0.90	0.43	106	28.0	6.6
Peas, Green		32	40-45	85-90	0.80	0.42	108	30.0	....
Peas, Dried		35-40	50-60	....	0.28	0.22	14	....	....
Peppers		32	40-45	85-90	0.90	....	....	30.1	2.35
Pineapples, Ripe		40-45	50	85-90	0.90	....	127	29.9	....
Plums		31-32	40-45	80-85	0.83	....	115	28.0	....
Pork, Fresh		30	36-40	85	0.60	0.38	66	28.0	....
Potatoes, White	–30	36-50	45-60	85-90	0.77	0.44	105	28.9	0.85
Poultry, Dressed	–10	28-30	29-32	....	0.80	0.41	99	27	....
Pumpkins		50-55	55-60	70-75	0.90	....	....	30.2	....
Quinces		31-32	40-45	80-85	0.90	....	....	28.1	....
Raspberries		31-32	40-45	80-85	0.89	0.46	125	30.0	3.3
Sardines,Canned		....	35-40	....	0.76	0.410	101	....	....
Sausage, Fresh		31-36	36-40	80	0.89	....	....	....	....
Sauerkraut		33-36	36-38	85	0.91	0.47	128	....	....
Squash		50-55	55-60	70-75	0.90	....	....	29.3	....
Spinach		32	45-50	85	0.92	....	....	30.8	....
Strawberries	–15	31-32	42-45	80-85	0.92	0.48	129	30.0	3.3
Tomatoes, Ripe		40-50	55-70	85-90	0.95	....	135	30.4	0.5
Turnips		32	40-45	95-98	0.90	....	....	30.5	1.0
Veal	–15	28-30	36-40	....	0.71	0.39	91	29	....

Figure 16-7. *Temperature, specific heat, and latent heat data for some common foods. These can be used in determining heat loads. (Dunham-Bush, Inc.)*

Heat Equivalent of Electric Motors			
	Connected Load in Refrigerated Space[a]	Motor Losses Outside Refrigerated Space[b]	Connected Load Outside Refrigerated Space[c]
Motor hp	Btu/hp • h	Btu/hp • h	Btu/hp • h
1/8 to 1/3	4600	2550	2100
1/2 to 3	3800	2550	1300
5 to 20	3300	2550	800

[a]For use when both useful output and motor losses are dissipated within refrigerated space; motors driving fans for forced circulation unit coolers.
[b]For use when motor losses are dissipated outside refrigerated space and useful motor work is expended within refrigerated space; pump on a circulating brine or chilled water system; fan motor outside refrigerated space driving fan circulating air within refrigerated space.
[c]For use when motor heat losses are dissipated within refrigerated space and useful work expended outside of refrigerated space; motor in refrigerated space driving pump or fan located outside of space.

Figure 16-8. *Heat released by operating electric motors. Note different operating conditions and how they affect heat released. (Reprinted by permission of the American Society of Heating, Refrigerating, and Air-Conditioning Engineers, Atlanta, Georgia, from the 1994 ASHRAE Handbook—Refrigeration)*

Heat Equivalent of Occupancy	
Refrigerated Space Temperature, °F	**Heat Equivalent/ Person, Btu/h**
50	720
40	840
30	950
20	1050
10	1200
0	1300
–10	1400

Note: Heat equivalent may be estimated by $q_p = 1295 - 11.5t$ (°F)

Figure 16-9. *Heat released by a person in the cooled space. (Reprinted by permission of the American Society of Heating, Refrigerating, and Air-Conditioning Engineers, Atlanta, Georgia, from the 1993 ASHRAE Handbook—Fundamentals)*

frame. They must be considered separately. Total external area minus the window area equals the insulated area.

Use the accompanying table to find the amount of heat that will leak through the insulation per square foot of area per 24 hours for that particular type of wall construction for the temperature difference. Consider a wall made of steel paneling on both sides with a 4″ slab of cork insulation (or its equivalent) in between. The table in **Figure 16-10** will reveal that, at a temperature difference of 60°F (95°F – 35°F), 108 Btu will leak through every square foot during a 24-hour period. Expanded polystyrene insulation would either be 33% less or about 2 1/2″ thick for the same heat loss.

Some synthetic material insulation values are:

	K
Expanded rubber, rigid	0.22
Glass fiber, organic-bonded	0.25
Expanded polystyrene (extruded), plain	0.25
Expanded polystyrene (extruded), R-12 expanded, 1″ thick or greater	0.19
Expanded polystyrene, molded beads	0.28
Expanded polyurethane, R-11 expanded, 1″ thick or greater	0.16
Silica aerogel, loose fill	0.17

The glass leakage table will give values for the heat leakage through one square foot of glass. If the cabinet has double glass, 660 Btu will leak through at a temperature difference of 60°F. Adding the two heat leaks will give the total heat leakage into the cabinet.

Example:

A walk-in cooler is 10′ × 9′ × 8′ (3 m × 2.7 m × 2.4 m) high. It has two double-pane glass windows 1 1/2′ × 2′ (0.46 m × 0.30 m). The box is kept at 35°F (2°C) in a room with a summer design temperature of 95°F (35°C). The wall construction consists of 4″ (10 cm) cork (or its equivalent) with metal on each side (or it could be 2 1/2″ [6 cm] of expanded polystyrene). The windows are of double-pane construction. The temperature difference is 95°F – 35°F = 60°F (16°C).

Solution:

Walls:

$10 \times 9 \times 2 = 180$ ft^2 (ceiling and floor)
$9 \times 8 \times 2 = 144$ ft^2 (ends)
$10 \times 8 \times 2 = 160$ ft^2 (sides)
484 ft^2 of total area

Windows:

$1\ 1/2 \times 2 \times 2 = 6$ ft^2 of window
$484 - 6 = 478$ ft^2 of insulated wall

From table, **Figure 16-10:**

1 ft^2 of the wall allows transfer of 108 Btu/24 hr.
108×478 ft^2 = 51,624 Btu/24 hr. through the walls

From table, **Figure 16-10:**

1 ft^2 of window allows transfer of 660 Btu/24 hr.
660×6 ft^2 = 3960 Btu/24 hr. through the windows,

. . . or a total heat leakage of:

51,624 + 3960 = 55,584 Btu/24 hr.

16.1.7 Cabinet Volume

Cabinet volume is the volume based on the inside dimensions of the cabinet. This volume is used to help find out the air changes and product load.

Heat Gain Factors (Walls, Floor, and Ceiling)

Btu per ft² per Day

Insulation	Temp. difference (ambient temp. minus storage temp.), °F																	
Cork or equivalent in.	1	40	45	50	55	60	65	70	75	80	85	90	95	100	105	110	115	120
3	2.4	96	108	120	132	144	156	168	180	192	204	216	228	240	252	264	276	288
4	1.8	72	81	90	99	108	117	126	135	144	153	162	171	180	189	198	207	216
5	1.44	58	65	72	79	87	94	101	108	115	122	130	137	144	151	159	166	173
6	1.2	48	54	60	66	72	78	84	90	96	102	108	114	120	126	132	138	144
7	1.03	41	46	52	57	62	67	72	77	82	88	93	98	103	108	113	118	124
8	0.90	36	41	45	50	54	59	63	68	72	77	81	86	90	95	99	104	108
9	0.80	32	36	40	44	48	52	56	60	64	68	72	76	80	84	88	92	96
10	0.72	29	32	36	40	43	47	50	54	58	61	65	68	72	76	79	83	86
11	0.66	26	30	33	36	40	43	46	50	53	56	60	63	66	69	73	76	79
12	0.60	24	27	30	33	36	39	42	45	48	51	54	57	60	63	66	69	72
13	0.55	22	25	28	30	33	36	39	41	44	47	50	52	55	58	61	63	66
14	0.51	20	23	26	28	31	33	36	38	41	43	46	49	51	54	56	59	61
Single glass	27.0	1080	1220	1350	1490	1620	1760	1890	2030	2160	2290	2440	2560	2700	2840	2970	3100	3240
Double glass	11.0	440	500	550	610	660	715	770	825	880	936	990	1050	1100	1160	1210	1270	1320
Triple glass	7.0	280	320	350	390	420	454	490	525	560	595	630	665	700	740	770	810	840

Note: Where wood studs are used multiply the above values by 1.1

Figure 16-10. *Heat gain factors for walls, floor, and ceiling. (Reprinted by permission of the American Society of Heating, Refrigerating, and Air-Conditioning Engineers, Atlanta, Georgia)*

In the sample cabinet, which is 10′ (3 m) long × 9′ (2.7 m) wide × 8′ (2.4 m) high, the walls are 4″ (10 cm) thick. Therefore, the internal or inside length is 10′ minus 8″. (There is a wall at each end.)

10′ − (4″ + 4″) = inside length
or 10″ − 8″ = inside length
or 9′-4″ = inside length
or 9 1/3′ = inside length

The same method is used for the internal width and height. Calculated, the internal dimensions become: length, 9 1/3′; width, 8 1/3′; and height, 7 1/3′.

The inside volume = 9 1/3 × 8 1/3 × 7 1/3

$$= \frac{(9 \times 3) + 1}{3} \times \frac{(8 \times 3) + 1}{3} \times \frac{(7 \times 3) + 1}{3}$$

$$= \frac{27 + 1}{3} \times \frac{24 + 1}{3} \times \frac{21 + 1}{3}$$

$$= \frac{28}{3} \times \frac{25}{3} \times \frac{22}{3} = \frac{15{,}400}{27}$$

$$\frac{15{,}400}{27} = 570.4 = 570 \text{ ft}^3$$

Usable inside volume is the total inside volume minus shelves, racks, and evaporator space. To be on the safe side, the total internal volume is used when figuring heat loads. **Figure 16-11** uses a net (internal) volume.

16.1.8 Total Heat Load

Information given in the previous paragraphs can be illustrated by the following problem:

Problem:

The metal sheathed walk-in cabinet is 10′ long × 9′ wide × 8′ high with 4″ thick walls. It is in an 85°F (29°C), 80% ***relative humidity (RH)*** room. It cools 2000 lb. of fresh beef from 60°F (15.6°C) to 35°F (1.7°C) each day. The evaporator has two 1/8-hp motors and the cabinet has two 40-watt lamps (operating 8 hours each day). One person works in the cabinet 8 hours each day.

What is the total heat load?

1. Heat leakage load—55,584 Btu/day (refer to the end of Section 16.1.6).
2. Air change load = volume × air changes/24 hr. × Btu/ft³ (heat to be removed, cooling air from 85°F [29°C] 80% RH to 35°F [2°C] 60% RH).
   Air change load = 570 (see Section 16.1.7) × 23 (see **Figure 16-5**) × 1.70 (see **Figure 16-6**) = 22,287 Btu/day.
3. Product load = weight × spec. heat (see **Figure 16-7**) × temp. difference.
   Product load = 2000 lb. × .77 spec. heat × 25°F (14°C) temp. difference = 38,500 Btu/day.
4. Miscellaneous load =
   A. The two motors' loads (continuous operation) = No. of motors × Btu/hp/hr. (see **Figure 16-8**) × hp × hr.
   2 × 4600 × 1/8 × 24 = 27,600 Btu
   B. The two lamps' loads = No. of lamps × watts × hr. of operation × 3.42 Btu/W
   2 × 40 × 8 × 3.42 = 2189 Btu
5. Occupancy load = No. of persons × hours of work × heat equivalent per hour (see **Figure 16-9**.)

$$1 \times 8 \times \frac{950 + 840}{2}$$

$$1 \times 8 \times \frac{1790}{2}$$

$$1 \times 8 \times 895 = 7160$$

Therefore:

Total heat load = 55,584 + 22,287 + 38,500 + 27,600 + 2189 + 7160 = 153,320 Btu/24 hr.

$$\text{Per 16 hr.} = \frac{153{,}320}{16\text{ hr.}} = 9582.5$$ (3/4-ton load) Btu/16 hr. based on 16 hours of system operation.

The 16 hours of running time will provide a system with 50% reserve capacity. If 30% reserve capacity is desired, select equipment which will operate 18 hours per day to handle the load. If the fixture is in an air-conditioned room, and the ambient temperature is 75°F (24°C) year-round, select equipment to run 20 hours per day.

16.1.9 Determining Heat Leakage Using Tables (Short Method)

A method used by some manufacturers to determine heat leakage into a cabinet is shown in **Figure 16-10.** The table in **Figure 16-11** gives the external area and internal volume of a cabinet. This is based on actual experiments and investigations.

To use the tables, proceed as follows (using a walk-in refrigerator box as a sample problem): Visit the establishment and obtain all the data possible about the cabinet and the service. Determine the exterior dimensions of the box, the window dimensions, and also the number of window panes. Then, determine the type and thickness of the insulation. It is also necessary to determine how much business the user does, and the temperatures desired in the cabinet. The average summer temperatures for the locality and the highest possible water temperature (if a water-cooled installation is to be made) must also be known.

Specification sheets are available for tabulating data needed for the selection of proper equipment. A sample sheet is shown in **Figure 16-12.**

Using cabinet size 9′ × 10′ × 8′ without windows, **Figure 16-11** shows that the area is 484 ft^2. In **Figure 16-10,** note that the Btu leakage/ft^2/24 hr. (4″ thickness, 60° temperature difference) is 108 Btu.

- Heat leakage = total ft^2 × leakage/ft^2
- Heat leakage = 484 × 108 = 52,272 Btu/24 hr.

16.1.10 Determining Usage Load Using Tables (Short Method)

The total heat load of the refrigerator cabinet depends upon the heat leaking through the walls and windows. It is also affected by the heat to be removed from articles in the cabinet, the air changes, and other sources of heat. This heat is called ***heat usage,*** or ***service load.*** Heat usage is caused by changes of air in the cabinet,

Cabinet Areas, Volumes, and Thickness of Insulation

Cabinet		Internal Volume 8 ft. High Capacity (ft^3)								Internal Volume 10 ft. High Capacity (ft^3)						
	Outside	Wall Thickness							Outside	Wall Thickness						
Lg. & Wd.	ft^2	4"	4 1/2"	5"	6"	7"	8"	10"	ft^2	4"	4 1/2"	5"	6"	7"	8"	10"
5× 5	210	137	131	124	112	101	90	71	250	174	167	159	144	131	117	93
5× 6	236	169	161	154	140	127	114	91	280	215	206	197	180	164	148	120
5× 7	262	201	194	184	168	153	138	111	310	256	248	236	216	198	179	146
5× 8	288	233	224	214	196	179	162	131	340	296	286	274	252	232	210	172
6× 6	264	209	201	193	175	160	145	119	312	266	256	247	225	207	188	156
6× 7	292	248	238	228	210	193	176	146	344	316	304	292	270	249	228	192
6× 8	320	286	277	267	245	226	207	173	376	364	353	342	315	292	269	228
6× 9	348	325	315	305	280	259	238	200	408	414	402	390	360	335	309	263
6×10	376	364	353	343	315	292	269	227	440	463	451	439	405	378	350	299
6×12	432	444	432	419	385	358	331	281	504	555	546	536	495	463	430	370
7× 7	322	294	283	272	254	234	214	180	378	374	361	348	326	302	278	237
7× 8	352	341	329	317	294	273	252	214	412	434	420	406	378	353	328	281
7× 9	382	386	374	362	334	312	290	248	446	492	477	463	430	403	377	326
7×10	412	433	420	407	374	346	318	282	480	551	536	521	481	448	413	371
7×12	472	527	512	497	454	414	374	348	548	670	653	635	583	535	486	458
8× 8	384	394	382	369	343	320	296	253	448	501	487	473	441	413	385	333
8× 9	416	448	434	420	392	367	341	294	484	570	553	538	504	474	443	386
8×10	448	503	587	471	441	413	385	335	520	641	748	603	567	534	500	441
8×12	512	610	591	573	539	506	473	417	592	776	755	734	692	653	615	548
8×14	576	718	697	675	637	594	561	499	664	914	889	864	818	768	730	656
9× 9	450	510	489	469	448	420	392	341	522	649	623	600	576	543	510	449
9×10	484	570	554	537	504	473	443	386	560	725	706	686	647	612	575	508
9×12	552	694	674	654	616	581	545	476	636	883	859	836	792	752	708	626
9×14	620	814	793	771	728	687	647	566	712	1035	1011	987	935	888	840	745
10×10	520	638	620	602	567	534	500	440	600	870	790	770	729	680	650	579
10×12	592	776	755	734	693	655	617	547	680	988	962	939	890	847	802	720
10×14	664	912	889	866	818	775	733	654	760	1158	1132	1110	1050	1005	954	860
12×12	672	946	919	893	848	804	760	680	768	1203	1172	1144	1090	1038	988	895
12×14	752	1110	1086	1052	1001	951	900	809	856	1411	1382	1348	1289	1230	1170	1060
14×14	840	1304	1269	1235	1180	1126	1072	968	952	1660	1619	1568	1518	1458	1394	1272

Figure 16-11. *Table of cabinet external areas and internal volumes (capacity).*

Refrigeration Sales Engineer's Data Sheet

Name ______________ Type of Business ______________ Date ________

Address ______________ City ______________ Zone ______ State ________

Person Contacted ______________ Title ______________ Phone ________

Fixture No. 1-Make ________ Fixture No. 2-Make ________ Fixture No. 3-Make ________

Use ______ Model ______ Use ______ Model ______ Use ______ Model ______

Temperature ________ ________ ________

Humidity ________ ________ ________

Width ________ ________ ________

Length ________ ________ ________

Height ________ ________ ________

Construction ________ ________ ________

Insulation:

Kind ________ ________ ________

Thickness ________ ________ ________

Glass:

Area ________ ________ ________

No. of Panes ________ ________ ________

Produce ________ ________ ________

Lights ________ ________ ________

Motors ________ ________ ________

Sun Load ________ ________ ________

No. of People in Refrigerator ________ ________ ________

Unusual Temperatures ________ ________ ________

Unusual Service ________ ________ ________

Remarks ________ ________ ________

Use Reverse Side for Sketch of Installation

Salesman ______________

Figure 16-12. *This data sheet is typical of those used by sales engineers in recording information prior to a refrigeration installation.*

produce to be cooled, lights and motors which may be used inside the box, and the occupancy of the box.

Refrigeration equipment manufacturers have developed a standard that gives a fairly accurate estimate of the usage heat load. With this method, the cabinet is classified according to the type of service to be performed: florist's cabinets, grocery boxes, normal market coolers, fresh meat cabinets, and restaurant short-order cabinets. From experience, these companies have found that cabinets used for the same general line of business hold rather closely to the same usage heat load.

This load depends on four basic factors:

- Temperature difference between exterior and interior of cabinet.
- Volume of cabinet (internal).
- Type of service.
- Time.

It is possible to determine the usage heat load of an installation. You must know the amount of food put into the refrigerator, how many times the door is opened, and how long the employees are inside the cabinet. This is a difficult process. If not carefully done, errors are bound to appear in the results.

Data in the tables are based on 1 ft^3 content at various temperature differences. To determine the usage heat load, proceed as follows:

1. Use temperature difference of the same value used for heat leakage into the cabinet.
2. Figure the volume of the cabinet from inside dimensions.
3. Determine the type of service for which the cabinet is being used, such as average, heavy, or long storage. The service load is also dependent on cabinet size. The smaller the cabinet, the more heat load is caused by service. A case for meat storage, for instance, may be in a small neighborhood store or in a supermarket.
4. Time (24 hours).
   A. After the total volume of the box has been found (from **Figure 16-11**), the load for each cubic foot is determined by using the table in **Figure 16-13.**
   B. If the 9′ × 10′ × 8′ cabinet appears to have average service with a temperature difference of 60°F (33°C), and has a volume of 570 ft^3, the amount of heat to be removed from each cubic foot will be 71 Btu/24 hours. Multiply this value by the total volume in cubic feet, and a fairly accurate estimate of the service or usage load may be obtained. The table in **Figure 16-13** gives the heat usage over a period of 24 hours, since this time is the established standard.

Heat usage = usage Btu/ft^3 × volume in ft^3.

Heat usage = 71 × 570 = 40,470 Btu/24 hr.

Another table, based on type of usage of the cabinet, is shown in **Figure 16-14.** Its values can be substituted for the values in **Figure 16-13** when exact use of cabinet is known.

16.1.11 Total Heat Load using Tables

The ***total heat load*** is the sum of the heat leakage load and the heat usage load.

Total heat load = heat leakage + heat usage

From the previous example:

Total heat load = heat leakage + heat usage
= 52,272 Btu/24 hr. (Section 16.1.9)
+ 40,470 Btu/24 hr. (Section 16.1.10)
Total heat load = 92,742 Btu/24 hr.

The addition of the heat leakage and usage will give the total heat load upon the cabinet for a certain set period of time. This value may be listed either as Btu/24 hr. or Btu/h.

Btu/h = Btu/24 hr. ÷ 24
= 92,742 ÷ 24
= 3864 Btu/h

However, in refrigeration applications, the unit to be installed should be big enough to remove this heat in less than 24 hours. This gives extra capacity for heavy loads and wear in the unit. It allows time for defrost operations.

For fixtures above 32°F (0°C), it is generally understood that the system should operate 16 hours out of 24 (two-thirds of the time). For fixtures below 32°F (0°C), unit should operate 18 hours out of 24 (three-fourths of the time).

For example, if the system were to operate 16 hr./day, a service technician would divide the total Btu heat load for a 24-hr. period by 16.

For 16-hr. running:

$$\frac{92{,}742}{16} = 5796 \text{ Btu/h for each hour of running}$$

For 18-hr. running:

$$\frac{92{,}742}{18} = 5152 \text{ Btu/h for each hour of running}$$

16.2 Thermodynamics of the Refrigeration Cycle

Thermodynamics is the science which deals with the relationships between heat and mechanical action. The refrigeration compression cycle is based on thermodynamics.

Every technician should know how to determine what size compressor is needed to produce a certain amount of refrigeration. You should also know how large a motor is needed to drive this compressor. To understand how these values are determined, the heat behavior of the refrigerant must be understood.

The refrigerant cycle is simple. The refrigerant is let into the evaporator in the liquid state and near room temperature. Some of the refrigerant vaporizes under the low pressure in the evaporator. It cools the remainder to the desired refrigerating temperature. Then, as the remainder of the refrigerant evaporates, it removes heat from the evaporator and, therefore, from the cabinet. The total amount of heat absorbed is the total ***latent heat of vaporization.*** The amount of heat absorbed from the cabinet and evaporator is the ***effective latent heat.***

The refrigerant vapor formed during evaporation passes down the suction line. As this happens, the vapor decreases a little in pressure (usually 2 psi [14 kPa]). It increases in temperature about 10°F (6°C). The warming up or increasing in temperature of the refrigerant after it has vaporized is called "superheating of the vapor." The degree of ***superheat*** is the difference between the temperature of the vapor at the compressor and its evaporating temperature.

Usage Heat Gain, Btu/24 hr. for 1 ft³ Interior Capacity

Volume (ft^3)	Service*	Temperature Difference °F (Ambient Temp. Minus Storage Room Temp.)										
		1	40	50	55	60	65	70	75	80	90	100
20	Average	4.68	187.	234.	258.	281.	305.	328.	351.	374.	421.	468.
	Heavy	5.51	220.	276.	303.	331.	358.	386.	413.	441.	496.	551.
30	Average	3.30	132.	165.	182.	198.	215.	231.	248.	264.	297.	330.
	Heavy	4.56	182.	228.	251.	274.	297.	319.	342.	365.	410.	456.
50	Average	2.28	91.	114.	126.	137.	148.	160.	171.	182.	205.	228.
	Heavy	3.55	142.	177.	196.	213.	231.	249.	267.	284.	320.	355.
75	Average	1.85	74.	93.	102.	111.	120.	130.	139.	148.	167.	185.
	Heavy	2.88	115.	144.	158.	173.	188.	202.	216.	230.	259.	288.
100	Average	1.61	64.	81.	84.	97.	105.	113.	121.	129.	145.	161.
	Heavy	2.52	101.	126.	139.	151.	164.	176.	189.	202.	227.	252.
200	Average	1.38	55.	69.	76.	83.	90.	97.	103.	110.	124.	138.
	Heavy	2.22	90.	111.	122.	133.	144.	155.	166.	178.	200.	222.
300	Average	1.30	52.0	65.	71.5	78.	84.5	91.	97.5	104.	117.	130.
	Heavy	2.08	83.2	104.	114.	125.	135.	146.	156.	166.	187.	208.
400	Average	1.24	49.6	62.	68.2	74.4	80.6	86.8	93.	99.2	112.	124.
	Heavy	1.96	78.4	98.	108.	118.	128.	137.	147.	157.	176.	196.
500	Average	1.21	48.4	60.5	66.6	72.6	78.7	84.7	90.7	96.8	109.	121.
	Heavy	1.87	74.8	93.5	103.	112.	122.	131.	140.	150.	168.	187.
600	Average	1.17	46.8	58.5	64.	70.	76.	82.	88.	94.	105.	117.
	Heavy	1.85	74.0	92.5	102.	111.	120.	130.	139.	148.	167.	185.
800	Average	1.11	44.4	55.5	61.1	66.6	72.2	77.7	83.3	88.8	100.	111.
	Heavy	1.76	70.4	88.0	96.8	106.	115.	123.	132.	141.	158.	176.
1,000	Average	1.10	44.0	55.0	60.5	66.	71.5	77.	82.5	88.	99.	110.
	Heavy	1.67	66.8	83.5	91.9	100.	108.	117.	125.	134.	150.	167.
1,200	Average	.995	39.8	49.8	54.7	59.7	64.7	69.7	74.7	79.6	89.6	99.5
	Heavy	1.58	63.2	79.0	86.9	94.8	103.	111.	119.	126.	142.	158.
1,500	Average	.920	36.8	46.0	50.6	55.2	59.8	64.4	69.	73.6	82.8	92.
	Heavy	1.50	60.0	75.0	82.5	90.0	97.5	105.	113.	120.	135.	150.
2,000	Average	.835	33.4	41.8	45.9	50.1	54.3	58.5	62.7	66.8	75.2	83.5
	Long storage	.775	31.0	38.8	42.6	46.5	50.4	54.3	58.1	62.	69.8	77.5
3,000	Average	.750	30.0	37.5	41.3	45.0	48.8	52.5	56.2	60.0	67.5	75.0
	Long storage	.576	23.0	28.8	31.7	34.6	37.3	40.3	43.2	46.1	51.8	57.6
5,000	Long storage	.403	16.1	20.2	22.2	24.2	26.2	28.2	30.2	32.2	36.3	40.3
7,500	Long storage	.305	12.2	15.3	16.8	18.3	19.8	21.4	22.9	24.4	27.5	30.5
10,000	Long storage	.240	9.6	12.0	13.2	14.4	15.6	16.8	18.0	19.2	21.6	24.0
20,000	Long storage	.187	7.48	9.35	10.3	11.2	12.2	13.1	14.0	15.0	16.8	18.7
50,000	Long storage	.178	7.12	8.90	9.79	10.7	11.6	12.5	13.4	14.2	16.0	17.8
75,000	Long storage	.176	7.04	8.80	9.68	10.6	11.5	12.3	13.2	14.1	15.8	17.6
100,000	Long storage	.173	6.92	8.65	9.52	10.4	11.2	12.1	13.0	13.8	15.6	17.3

*For average and heavy service, product load is based on product entering at 10° above the refrigerator temperature; for long storage the entering temperature is approximately equal to the refrigerator temperature.
Where the product load is unusual, do not use this table.

Figure 16-13. *Table for determining usage heat gain. It uses average storage, heavy storage, and long storage variables.*

The compressor then takes the slightly superheated vapor and converts (compresses) it. It is compressed to a high-temperature, high-pressure vapor. This condensing temperature sometimes becomes as high as 250°F (121°C), depending upon the refrigerant and the conditions.

The superheated vapor passes to the condenser. If its temperature is higher than ***ambient*** (water or air) temperature, it transfers enough of its heat to the air or water to cool to its vapor pressure-temperature. If the vapor pressure-temperature is above that of the water or air temperature, the refrigerant vapor starts losing some latent heat of evaporation. The quantity of heat it loses determines the amount of the vapor that will condense into a liquid. After it has become liquid, the refrigerant cools down close to the ambient temperature.

The refrigerant then goes to the refrigerant control, where the pressure is reduced. The refrigerant cools as it vaporizes (becomes "flash gas"). The rest vaporizes to remove heat from the cabinet. Thus the refrigerant cycle is repeated. **Figure 16-15** shows this cycle taking place. It also indicates the temperatures in various parts of the refrigerating system.

Temperature Difference in °F	Use of Refrigerator			
	Florist	Grocery or Normal Market	Market with Heavier Service or Freshly Killed Meats	Restaurant Short Order
40°	40.0	65.0	95.0	120.0
50°	50.0	80.0	120.0	150.0
60°	60.0	95.0	145.0	180.0
70°	70.0	114.0	167.0	210.0
80°	80.0	130.0	190.0	240.0
90°	90.0	146.0	214.0	270.0

Figure 16-14. *This chart is based on usage heat gain in Btu/24 hr. for one ft^3 interior capacity. (Reprinted by permission of the American Society of Heating, Refrigerating, and Air-Conditioning Engineers, Atlanta, Georgia)*

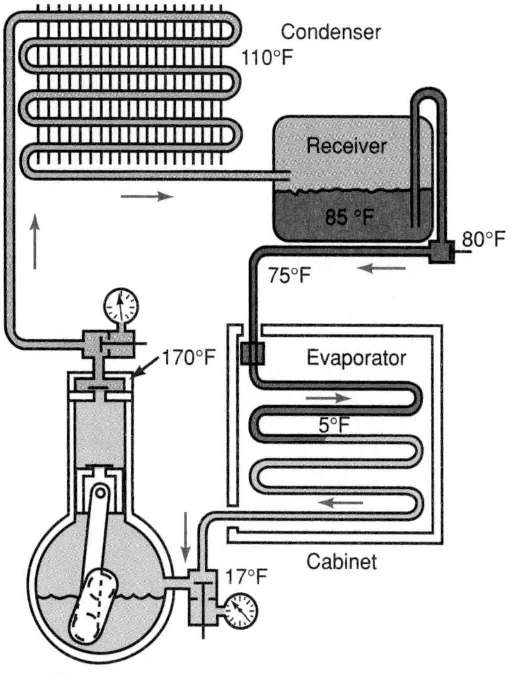

Figure 16-15. *A refrigerating system schematic shows the approximate temperatures of refrigerant in various parts of the system.*

16.2.1 Pressure-Heat Diagram

The following discussion of refrigerant behavior is based on one pound of refrigerant, regardless of its state (liquid or vapor). The discussion deals only with the pure refrigerant. It does not include the effect of lubricating oils and other influences.

Figure 16-16 charts the behavior of one pound of refrigerant in a refrigerating machine. The horizontal scale shows the amount of heat present in one pound of refrigerant at all times and under all conditions. The vertical scale shows the pressure imposed upon it.

The graph shown in **Figure 16-16** is commonly called a ***pressure-heat chart.*** It is also called a pressure-

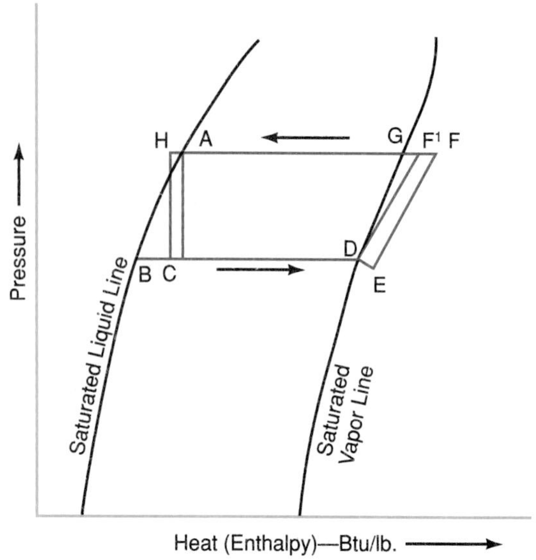

Figure 16-16. *Pressure-heat diagram for a refrigerant. Saturated liquid curve represents heat in liquid refrigerant at various pressures before it will start vaporizing. Saturated vapor curve represents division between superheated gas and point where gas starts condensing into liquid.*

enthalpy chart. The heat (Btu) in the pound of refrigerant is usually measured from saturated liquid refrigerant at −40°F (−40°C). The pressures will be different for each kind of refrigerant.

Note in the pressure-heat chart that, as the refrigerant vaporizes at the lower constant pressure, it passes horizontally from B to D. This line indicates the vaporization of the refrigerant from a liquid into a vapor in the evaporator. The distance D to E represents the heating of this vapor into a superheated condition as it passes down the suction line. Note that only a few Btu of heat have been added and that the pressure has decreased a little.

Point E represents the condition of the vapor when it moves into the compressor and is compressed. It is then compressed to F. Note how the pressure increases rapidly and how a few Btu of heat are added to the vapor as the vapor is compressed from E to F. The vapor leaving the compressor is considerably superheated. See Section 16.2.6.

Point F represents the condition of the vapor as it leaves the exhaust valve of the compressor. The distance between F and G is the cooling down of this superheated vapor to the point where it starts to condense.

At G, the vapor has no superheat and is 100% saturated vapor. The line G to A represents the condensation of the refrigerant in the condenser from a vapor into a liquid.

Point A represents the amount of heat in the liquid and the pressure imposed on the liquid as it forms in the condenser. From A to H is the loss of heat from the liquid as it passes along the liquid line to the refrigerant control. This action (sometimes called ***subcooling***) occurs because the liquid refrigerant cools to room temperature.

Line H to C represents the throttling of the liquid upon passing through the refrigerant control orifice. The cycle is now ready to be repeated for the pound of refrigerant.

Note that the distance C to D does not represent the *total* latent heat of the liquid at the low-side pressure condition. This means that R-12, which has a latent heat of 70 Btu/lb. at 5°F (−15°C), will not remove all of this heat from the evaporator. Some of it (approximately 19 Btu) is used to cool down the remaining liquid refrigerant to the 5°F (−15°C) temperature. The 51 Btu remaining is called the ***effective latent heat*** or the ***effective refrigerating capacity.*** This means that 19/51 of the pound, or 37% of the R-12, flashes into gas at the refrigerant control.

The pressure-heat chart in **Figure 16-16** shows physical property changes in a pound of a refrigerant as it passes around the refrigerating cycle. A thorough knowledge of this chart is helpful to the engineer and service technician.

16.2.2 Pressure-Heat Areas

The pressure-heat chart in **Figure 16-17** is divided into three main areas. To the left of the saturated liquid line, all of the refrigerant is liquid. Between the saturated liquid line and the saturated vapor line, the refrigerant is a mixture of liquid and vapor, as shown in the boxes in the drawing.

Close to the saturated liquid line in **Figure 16-17,** the refrigerant is almost all liquid. Close to the saturated vapor line, the refrigerant is almost all vapor. Note that the vapor in the area to the right of the saturated gas line is in a superheated condition.

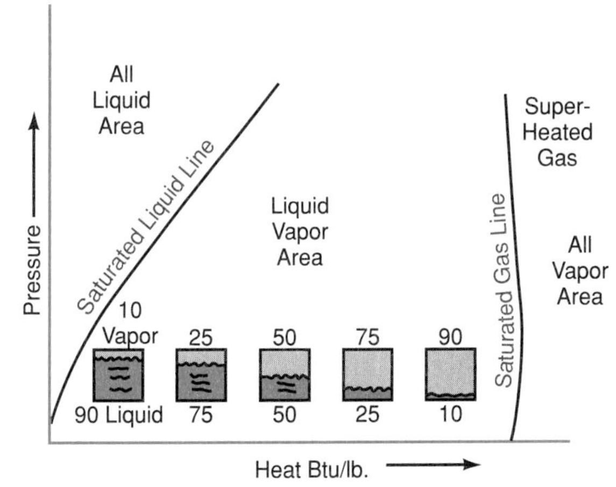

Figure 16-17. *Pressure-heat diagram. Note that as heat is subtracted, refrigerant becomes a liquid. As heat is added, refrigerant becomes a vapor.*

16.2.3 Constant Value Lines of Pressure-Heat Chart

Many facts can be read from the chart in **Figure 16-18.** Along any vertical line, such as A, the heat in one pound of refrigerant is ***constant*** (the same). Along any

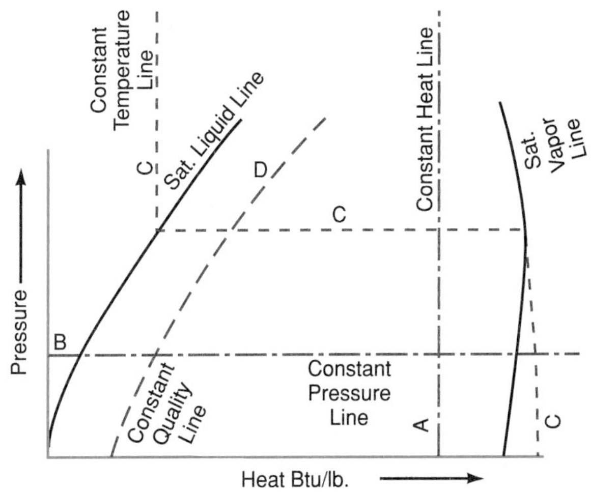

Figure 16-18. *Pressure-heat diagram. Line A indicates constant heat condition with pressure change. Line B shows constant pressure and condition of refrigerant with changing heat content. Line C indicates constant temperature with changing pressure and heat.*

horizontal line, such as B, the refrigerant has constant pressure. Along line C, the temperature reading is constant. It is almost vertical in the liquid area. It is horizontal in the liquid and vapor area. It slopes down and to the right in the superheated vapor area.

Refrigerant quality means how much of the one pound of refrigerant is liquid and how much is vapor. "Ten percent quality" means that the pound of refrigerant is 10% vapor and 90% liquid. The line showing the same or constant quality is at D.

16.2.4 Effect of Pressure on Latent Heat

The value of the latent heat of the refrigerant when vaporizing or condensing differs at different pressures. At the lower pressure (vaporizing), **Figure 16-19,** the ***total latent heat*** to be added to (absorbed by) the liquid refrigerant to vaporize is the greatest. It is more than that needed to be subtracted or removed from it to condense it into a liquid at the higher pressure. This is because the liquid, when formed, is at a high temperature. It is higher than the liquid at the vaporizing pressure. This difference is represented by the specific heat of the refrigerant multiplied by the temperature difference for the two conditions.

The ***effective latent heat*** of the refrigerant, when vaporizing, consists of:

- The total latent heat at the low pressure.
- Minus the difference in the heat content of the liquid at the high (condensing) pressure.
- Minus the heat content of the liquid at the low (evaporating) pressure.

The high-pressure liquid, when throttled in the refrigerant control, must be cooled down. It must be cooled down to a low-pressure-temperature liquid before it can vaporize. It can then remove heat from the surrounding substances. Part of the liquid vaporizes so

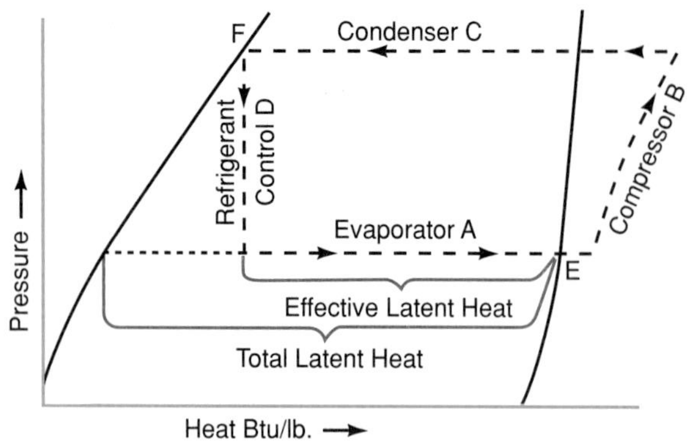

Figure 16-19. *Pressure-heat diagram. At A, liquid is boiling in evaporator and pressure is constant as heat is being added to refrigerant. At B, the compressor raises pressure to condensing pressure; heat of compression is added, and temperature rises. At C, the condenser cools hot vapor to saturated vapor line, then condenses vapor into a liquid. Pressure is constant and heat is removed. At D, the refrigerant control quickly reduces pressure to evaporating or low-side pressure. Temperature of remaining liquid drops as some liquid refrigerant "flashes" into vapor.*

it can cool the remaining liquid to the lower temperature. The vapor formed during this operation is called ***flash gas.***

The effective latent heat is the total heat at E in **Figure 16-19** minus the heat of the liquid at F. The effective latent heat is an average value because low-side pressure and high-side pressure vary somewhat during the operation of the system. Other conditions, such as oil in the system, and system efficiencies affect the ideal cycle.

For R-12, the total latent heat of vaporization at 5°F (−15°C) is 79 Btu/lb. However, its actual heat-absorbing ability or effective latent heat is only about 51 Btu/lb. These values are based on the standard evaporating temperature of 5°F (−15°C) and standard condensing temperature of 86°F (30°C). This reduction in heat absorption is due to the fact that some heat is absorbed by the flash gas when refrigerant pressure is reduced at the refrigerant control.

16.2.5 Saturated Vapor

A ***saturated vapor*** is a refrigerant vapor under conditions which permit some of the vapor to condense when a little heat is removed from it. Another definition: Saturated vapor is a substance in vapor form in the presence of some of its own liquid. For example, a saturated vapor is the vapor in a refrigerant cylinder that is half-full of liquid refrigerant.

When the refrigerant vaporizes in the evaporator, it is (at first) saturated vapor. However, as this vapor passes down the suction line to the compressor, it usually becomes warmer by 5°F to 15°F (3°C to 8°C). This increase in temperature is called "superheating the low-side vapor." That is, it is raising the vapor above its saturated condition for this pressure. The vapor will now obey Charles' or Boyle's Laws. See Chapter 31. In **Figure 16-17,** any vapor in the space marked "liquid vapor area" is saturated vapor.

16.2.6 Superheated Vapor

Under certain conditions, a vapor becomes a ***superheated vapor.*** The volume and/or pressure of the vapor decreases when some heat is removed from it. This occurs with no condensation. For example, the vapor that the compressor handles is always superheated. This can change when a condition arises in which the last drop of liquid refrigerant evaporates as the refrigerant travels past the exhaust valve.

After the low-pressure superheated vapor enters the compressor, it is compressed. The energy put into it greatly increases the temperature and pressure on the vapor. The amount of superheat increases.

Superheating of the vapor lowers the efficiency of a machine. The less superheating that takes place, the more efficient the machine will be. The heat added to the vapor is the mechanical energy of compression being converted into heat energy. By knowing how much heat has been added, a technician may calculate the size of the motor necessary to drive the compressor.

There are three places in the refrigeration cycle where superheating usually takes place:

- In the suction line, a low-temperature superheat.
- In the compressor.
- In the top part of condenser, a high-temperature superheat.

These three conditions are shown in **Figure 16-20.**

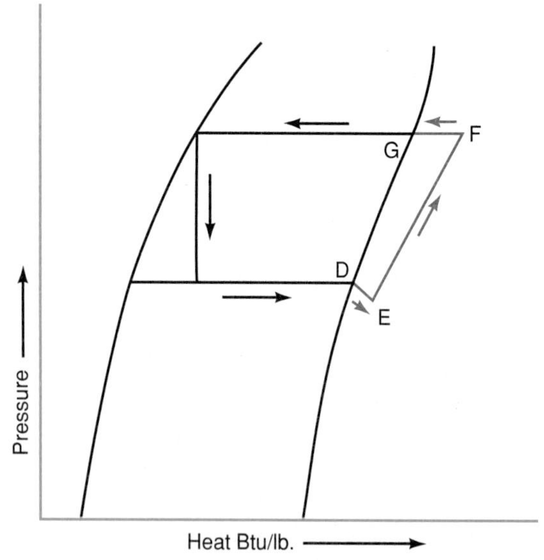

Figure 16-20. *Superheated vapor sections of refrigeration cycle. From D to E, superheat is added to one pound of refrigerant vapor as it travels through suction line from evaporator to intake valve of compressor. From E to F (the condition at exhaust valve), superheat is added as vapor is compressed. From F to G, superheat is removed in top portion of condenser.*

16.2.7 Specific Heat

The *specific heat* of a substance is the amount of heat necessary to raise the temperature of one pound of that substance 1°F. Substances may exist in three different states (solid, liquid, and vapor). Every substance has three different values for its specific heat. This depends on whether it is a solid, a liquid, or a vapor.

There are two kinds of specific heat of vapor:

- Specific heat when under a *constant pressure.*
- Specific heat when confined to a *constant volume.*

A vapor under constant pressure has a greater specific heat value than the same vapor under constant volume. The vapor heated with a constant pressure upon it will expand and do external work such as increasing the size of a balloon. This external work naturally requires an additional quantity of heat.

When a compressor compresses one pound of the refrigerant vapor, it does not add heat to it under a constant pressure or constant temperature condition. This state in the compressor is called adiabatic compression. ***Adiabatic compression*** means that no heat has been removed from or added to the vapor as it was compressed. Actually, a refrigeration compressor operates almost adiabatically because the compression takes place so rapidly.

The specific heat of liquid refrigerants varies considerably, depending on pressure imposed upon them. As shown in **Figure 16-21,** the pressure to which the liquid refrigerant is subjected in the condenser, after it has condensed at D, must be determined. This must be done before calculating how much heat must be removed from one pound of the liquid at this temperature to further cool it to room temperature C.

Example:

One pound of R-12 condensing at 125°F (52°C), and then cooled to 75°F (24°C) in the line, must lose:

Heat of liquid at 125°F = 37.28 (See **Figure 9-5**)

Heat of liquid at 75°F = 25.20

37.28 − 25.20 = 12.08 Btu/lb. must be removed.

After the refrigerant passes through the throttling valve, it is subjected to a lower evaporating pressure. The specific heat of liquid under the evaporating pressure, shown at B, must be determined. This will indicate how much heat must be removed to cool it down to vaporizing temperature. If refrigerant goes through the refrigerant control while at its condensing temperature, it will have A to D liquid specific heat/lb.

16.2.8 Practical Pressure-Heat Cycle

The refrigeration cycles described in previous paragraphs are based on 5°F (−15°C) evaporating temperature and 86°F (30°C) condensing temperature. Practical cycles for various other refrigerating systems are somewhat different. For example:

- With an air-cooled condensing unit used for long term frozen food storage, the evaporator refrigerant temperature will be −10°F (−23°C).

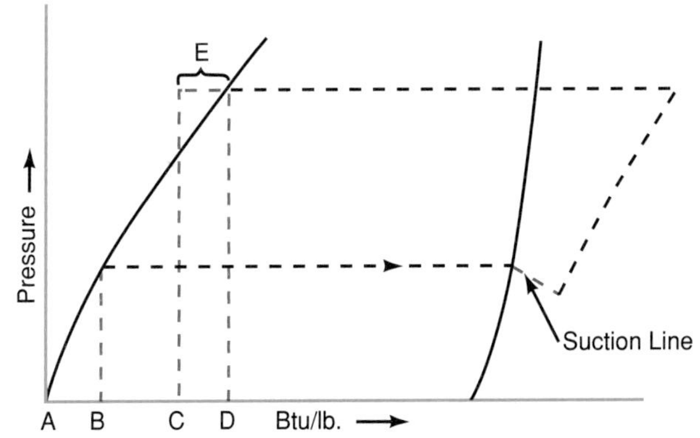

Figure 16-21. *Pressure-heat diagram. By using heat exchanger or installing liquid and suction lines together, heat increase in suction line vapor comes from heat decrease in liquid line. E indicates cooling of liquid in receiver, liquid line, and by heat exchangers. Note the gain in effective latent heat from C to D.*

- With summer-design ambient temperatures of 95°F (35°C), the condensing temperature will be 95°F (35°C) + 30°F (17°C), or 125°F (52°C).

In **Figure 16-22,** note the difference in compressor performance and refrigerant temperatures between suction line refrigerant temperatures entering the compressor at 0°F (−18°C) and at 60°F (16°C). NOTE: Surface of suction line must be above dewpoint temperature of air or the suction line will sweat, collect moisture, and/or collect frost.

An air conditioning (comfort cooling) cycle is shown in **Figure 16-23.**

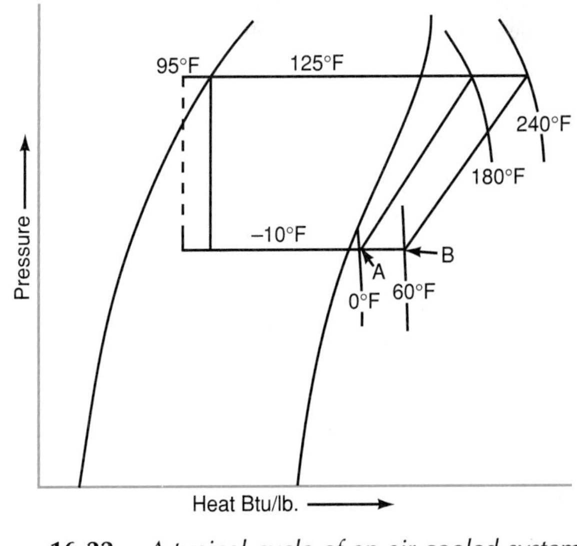

Figure 16-22. *A typical cycle of an air-cooled system for refrigerating frozen foods. Cabinet is kept at 0°F (−17.8°C) (refrigerant, −10°F [23.3°C]). Air temperature is 95°F (35°C). Note effect of the temperature change of suction vapor entering the compressor at point A (0°F [−17.8°C]) and point B (60°F [15.6°C]).*

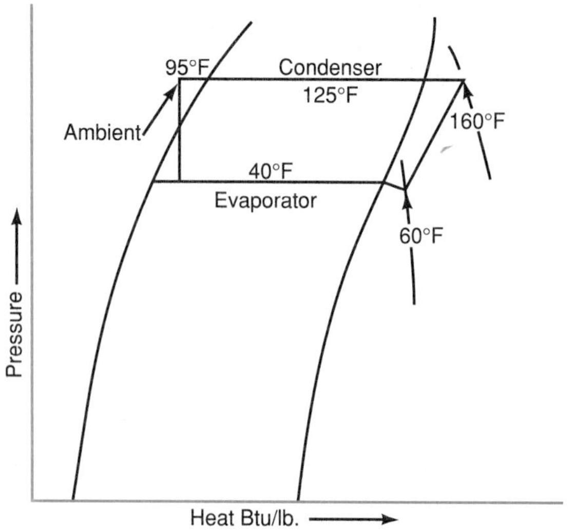

Figure 16-23. *Typical air conditioning comfort cooling cycle with an evaporator temperature of 40°F (4.4°C) and an ambient temperature of 95°F (35°C).*

16.3 Evaporator and Condensing Unit Capacities

After calculating heat load, it is necessary to decide what size evaporator will give enough refrigeration. Some important things to remember are:

- The evaporator removes heat from the cabinet only when the condensing unit is running.
- The refrigerating unit usually runs from 16 to 20 hours out of each 24 hours. This means that the unit must have a refrigerating capacity in 16 hours of operation equal to the total heat load in 24 hours. If the refrigerator is installed in an air-conditioned room, the room temperature can be lowered. Otherwise, the running time of the unit can be increased (for example, from 16 hours to 18 or 20 hours).

The evaporator capacity is determined by three conditions:

- Cabinet temperature.
- Refrigerant temperature.
- Space allowed for the evaporator.

Condensing unit capacity depends on three conditions, as well:

- Low-side pressure.
- Condensing cooling medium (air or water).
- Size of compressor and condenser.

It is more important to balance the capacity of the evaporator to the capacity of the condensing unit than to the heat load of the cabinet. When balancing the capacity of the condensing unit and the evaporator, calculations for each must be based on the same low-side pressure.

The same low-side pressure is used because:

- The capacity of the evaporator increases as the evaporator temperature decreases.
- The capacity of the condensing unit decreases as the low-side pressure decreases. See **Figure 16-24.**

Figure 16-25 shows that this particular evaporator matches the condensing unit at a low-side pressure of 31 psi. The evaporator/condenser combination will remove 12,500 Btu/h when the evaporator refrigerant temperature is 32°F (0°C) and the pressure is 31 psi. The temperature difference for a 42°F (6°C) cabinet in this case is 10°F (6°C).

The manufacturers of evaporators and condensing units list the capacities of their products in Btu for either 1 hour, 16 hours, or 18 hours of operation. **Figure 16-26** lists typical condensing unit capacities at condensing temperatures of 110°F (43°C) and 120°F (49°C). **Figure 16-26** also gives four different evaporating temperatures of −30°F, −15°F, 20°F, and 40°F (−34°C, −26°C, −7°C, and 4°C).

Note that only at a 40°F (4°C) evaporator refrigerant temperature does the ***horsepower*** (hp) equal the tonnage capacity. Referring to **Figure 16-26,** for example, a 3-hp condensing unit, operating at 40°F evaporator refrigerant temperature and 110°F (43°C) condensing temperature, has the capacity of 36,300 Btu/h. The

Condensing Unit		Evaporator	
Low-Side Temp.	Btu/h	Temp. Diff.	Btu/h
40	6650	1	400
35	6100	10	4000
30	5600	12	4800
25	5100	15	6000
20	4650	300 ft² surface natural convection evaporator.	
15	4200		
10	3800		
5	3400		
0	3000		
− 5	2600		
−10	2250		
−15	1900		
−20	1550		
−25	1250		
−30	950		
90°F ambient air. Add 6% for 10 °F drop in air temperature, subtract 6% for each 10 °F rise in ambient temperature. Liquid line 1/4". Suction line 5/8". Approximately one hp.			

Figure 16-24. *Condensing unit and evaporator capacity variations with temperature and pressure. Evaporator temperature difference is the difference between refrigerant temperature and cabinet air temperature.*

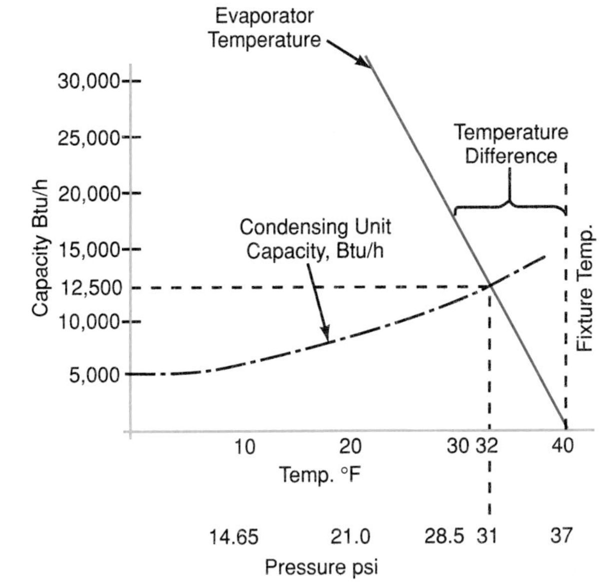

Figure 16-25. *This graph shows relative effect on evaporator and condensing unit capacity of different temperatures inside the evaporator.*

capacity of a 1-ton machine is 12,000 Btu/h. Therefore, 36,300 ÷ 12,000 = 3 ton (plus) capacity.

16.3.1 Condensing Unit Capacities

When choosing a condensing unit, first decide if it is to be water-cooled or air-cooled. Next, determine whether it is to be a hermetic unit or an external drive unit. Then, find out what electric power is available (120 V, 208 V, or 240 V, single-phase, or 220 V or 440 V, three-phase).

Figure 16-27 is a table of data on a typical external drive condensing unit. The air-cooled unit is used with a 1 hp motor and a 2 cylinder compressor. It has a 2″ (5 cm) bore and a 2″ (5 cm) stroke.

If the evaporator was selected because of a 15°F (8°C) temperature difference and a 16-hour running time, the cabinet will operate at 38°F (3°C). Therefore, the refrigerant temperature will be 38°F minus 15°F, or 23°F (−5°C). If it is an R-12 refrigerant unit, the condensing unit capacity must be matched to the evaporator capacity at a low-side pressure of 23 psig (38 psia or 262 kPa).

The table, **Figure 16-26,** lists the average hourly capacity of various horsepower capacity condensing units. This clearly shows the increase in capacity of a condensing unit as the low-side pressure (and temperature) increases, providing the head pressure remains fairly constant. It also shows the effect of condensing temperature (and pressure) on the capacity of a condensing unit. The capacity of hermetic condensing units can also be found in tables provided by manufacturers.

16.4 Evaporator Installations

Practically all refrigerator cabinets are built to be used with mechanical refrigeration. These cabinets are

	Average Compressor Capacities Btu/h							
	Evaporating Temperatures (°F)							
	–30°		–15°		+20°		+40°	
	Condensing Temperature (°F)							
hp	110°	120°	110°	120°	110°	120°	110°	120°
2	5,200	4,500	9,100	8,200	18,000	16,800	22,800	21,200
3	9,000	8,300	14,300	13,200	22,100	20,800	36,300	34,200
5	14,200	12,500	24,800	22,400	41,700	39,300	62,400	58,500
7 1/2	25,000	20,000	31,000	28,000	53,000	48,000	87,000	81,700
10	31,000	26,000	43,600	44,800	81,000	75,000	120,000	112,000
15	42,600	37,500	74,400	67,200	111,000	102,000	171,600	160,000
20	56,000	44,700	82,000	71,000	154,000	142,000	235,000	218,000
25	70,000	56,000	96,000	85,000	188,000	174,000	283,000	263,000
30	80,000	67,000	116,500	102,500	225,000	210,000	349,000	324,000
40	94,000	75,000	155,000	135,000	325,000	306,000	439,000	406,000
50	122,000	100,000	188,500	159,500	375,000	350,000	585,000	550,000
60	168,000	134,000	240,000	220,000	450,000	420,000	710,000	670,000
70	196,000	156,000	272,000	239,000	571,000	534,000	800,000	742,000
75	210,000	167,000	291,000	256,000	582,000	542,000	855,000	795,000
80	224,000	178,000	310,000	273,000	622,000	578,000	900,000	842,000
90	252,000	201,000	349,000	307,000	750,000	700,000	1,027,000	955,000
100	280,000	223,000	388,000	341,000	777,000	723,000	1,170,000	1,100,000

NOTE: The above figures are only approximate and based on catolog ratings of leading compressor manufacturers. For precise figures, refer to catolog of your compressor manufacturer.

Figure 16-26. *Condensing unit capacities in Btu/h at various evaporator temperatures and at two different condensing temperatures.*

Condensing Unit Capacities		
Compressor rpm	Refrigerant (Evaporator) Temperature	Btu/h
475	45	11,000
	40	10,200
	35	9,370
	30	8,530
	25	7,740
540	25	8,700
	20	7,960
	15	7,200
	10	6,430
	5	5,740
	0	5,100
665	0	5,780
	– 5	5,700
	–10	4,430
	–15	3,820
	–20	3,280
	–25	2,800

Figure 16-27. *Btu/h capacity table for a 1 hp condensing unit which is air-cooled and designed to operate at 90°F ambient temperature at various low-side temperatures.*

specially designed to improve the cooling efficiency of the cabinet with evaporators. You should know the theory of air circulation in the cabinet. This will allow you to understand what is supposed to happen during the operation of the unit. The following deals with the study of efficient baffling and correct air circulation in cabinets.

Baffles are surfaces, or air ducts, which increase the efficiency of the airflow through the evaporator and throughout the cabinet. They direct the airflow so that it is carried all around the interior of the box. There are no dead air spots or warm air spots.

A typical evaporator and baffle arrangement is shown in **Figure 16-28.** The colder air coming from the evaporator is made to flow down the center of the cabinet. The warmer air is directed up the walls and back to the evaporator. The design must be scientifically proportioned to ensure that air circulation is unrestricted. No objects should block the airflow.

Any horizontal baffle or drain pan should be insulated. The top surface may be in contact with cold air. The under part of the baffle may be in contact with relatively warm air. If it were not insulated, this temperature difference could cause condensation. ***Eddy currents*** (small circular flows of air) would also disturb airflow in the cabinet.

Multiple-baffled evaporators of the natural convection type are often used. The air flows around the cabinet because of the relative weights (density difference) of the cold air and the warm air. Warm air is lighter per cubic foot of volume and, therefore, rises in the box. This natural circulation must not be blocked or the box temperature will not be constant. Baffling the evaporators

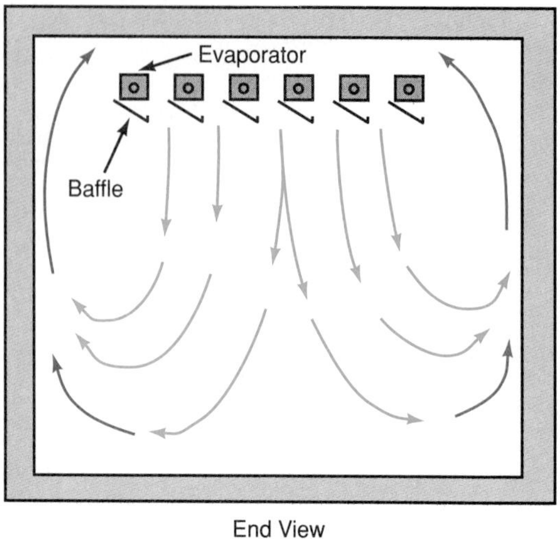

Figure 16-28. *Airflow pattern for a typical refrigerant cabinet. Baffles direct cold air downward while warm air rises and circulates through the evaporator.*

promotes this natural circulation of the air and speeds it up. Baffles may also serve as drain pans.

16.4.1 Evaporator Locations

Many cabinets do not have room for overhead evaporators. A walk-in cooler with an exterior height of approximately 7′-6″ (2.3 m) is too low. This height requires the use of a blower evaporator. There are two common types of mountings:

- Evaporators may be placed in an upper corner of the cabinet, as far away as possible from the entrance door. See **Figure 16-29.**
- Wall evaporators may be mounted against the wall opposite the windows or the reach-in doors of the cabinet, **Figure 16-30.**

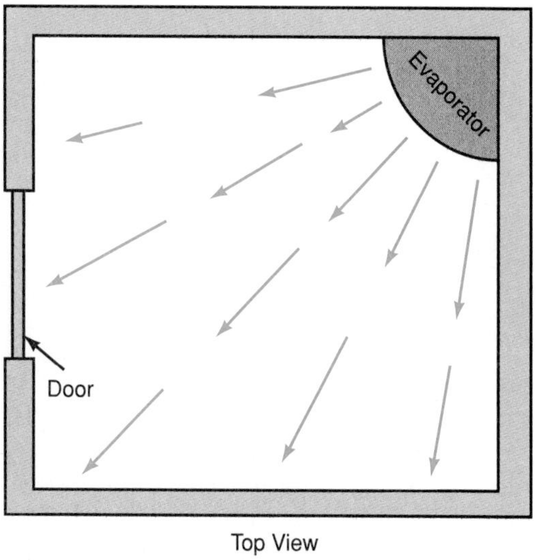

Figure 16-29. *Blower evaporator mounted in upper corner of walk-in cooler. Note air distribution relative to door.*

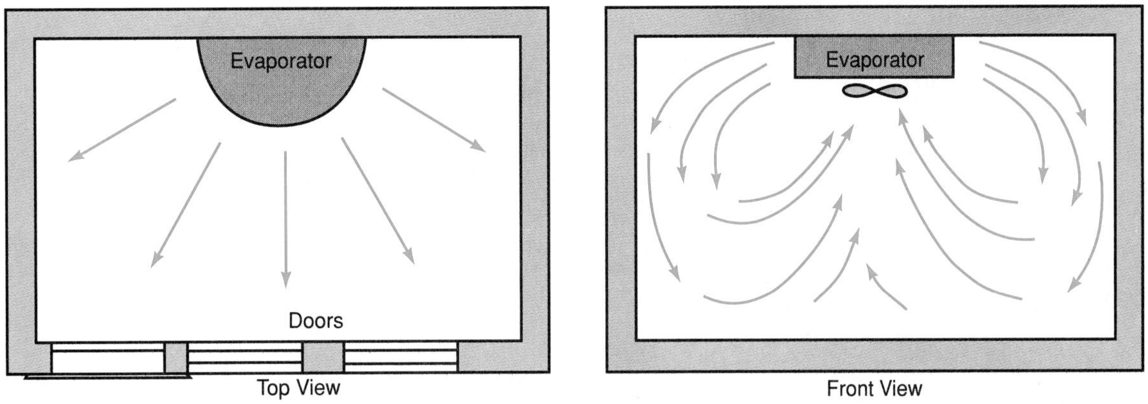

Figure 16-30. *Note airflow patterns with blower evaporator mounted on center back wall of walk-in cooler.*

The evaporators are shrouded and have a motor-driven fan to circulate air. They also have built-in drain pans. Also see **Figures 16-31, 16-32,** and **16-33.**

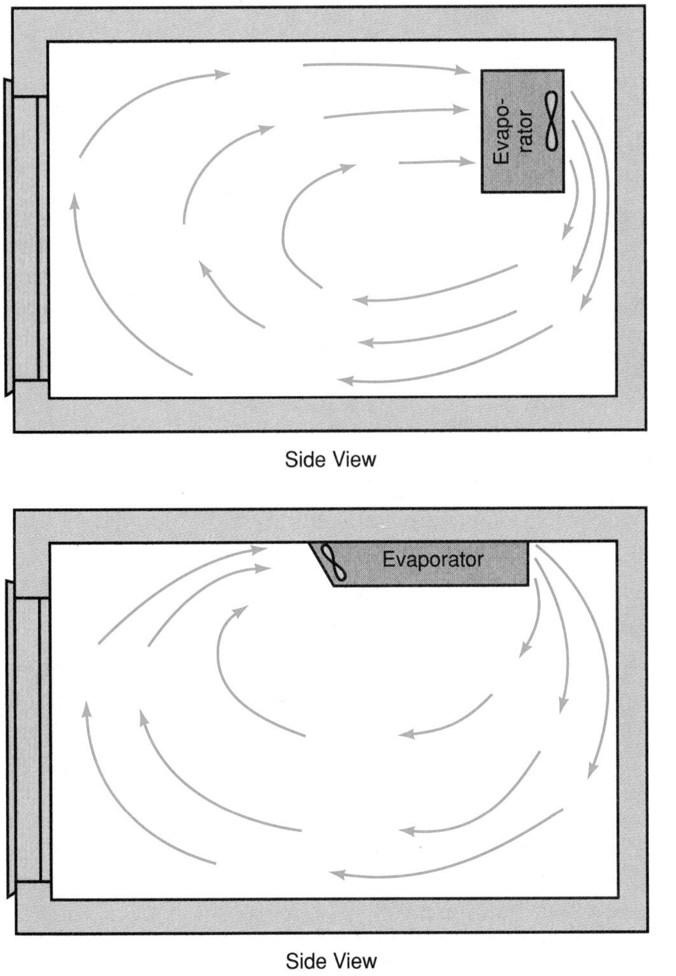

Figure 16-31. *Where there is too little clearance for overhead installation, blower evaporators may be installed on walls of walk-in refrigerators. Installation should always be opposite reach-in doors or windows.*

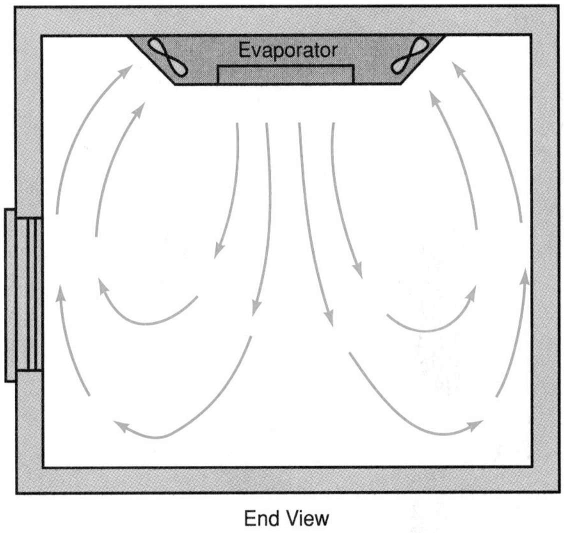

Figure 16-32. *Two motor-driven fans reduce the distance air must circulate in cabinet.*

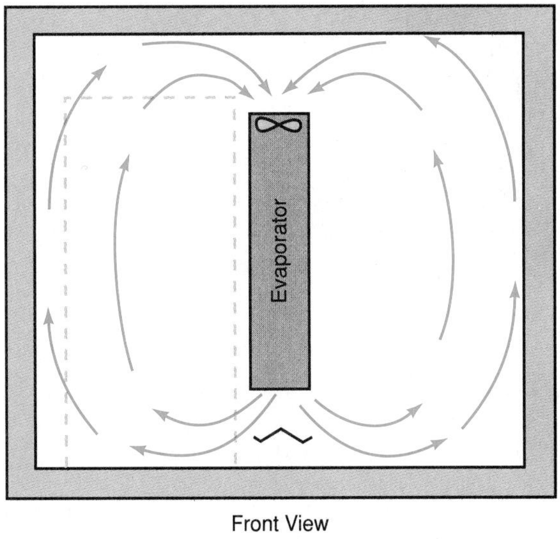

Figure 16-33. *Air circulation pattern for a reach-in cabinet with blower evaporator mounted behind mullion or door frame.*

16.4.2 Evaporators for Display Cases

Evaporator designs for closed- and open-type display cases have to overcome a difficult air circulation problem. The cases are narrow and there is less room for the evaporator. Many of the cases are open. **Figure 16-34** shows a single evaporator installation for a double-duty case.

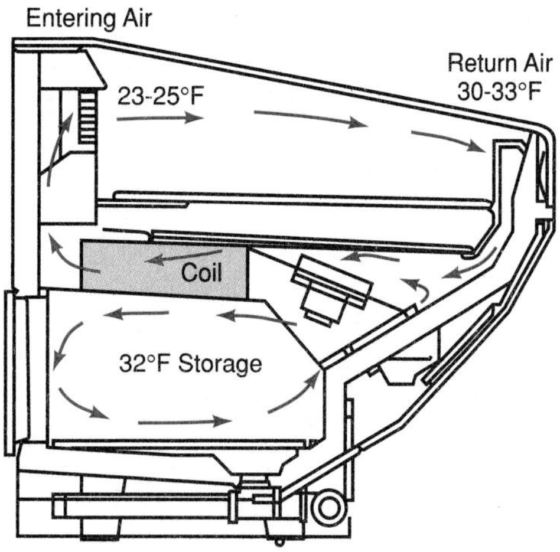

Figure 16-34. *A cross section shows the airflow pattern of a double-duty case. Note the temperature differences. (Tyler Refrigeration Corp.)*

A display case using a blower evaporator is shown in **Figure 16-35.** Note the location of the motor-driven fan. A design for an open display is shown in **Figure 16-36.** Notice how ducting is used to control the flow of the refrigerated air.

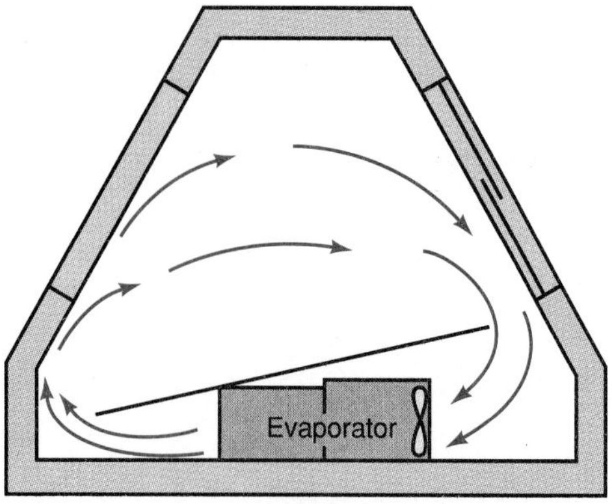

Figure 16-35. *This diagram of a closed display case shows a blower evaporator located below the display shelf.*

16.4.3 Evaporator Types

Many kinds of evaporators have been used in mechanical refrigeration. However, two basic types are in use:

- The air-cooling evaporator is used to directly cool the air within the cabinet.

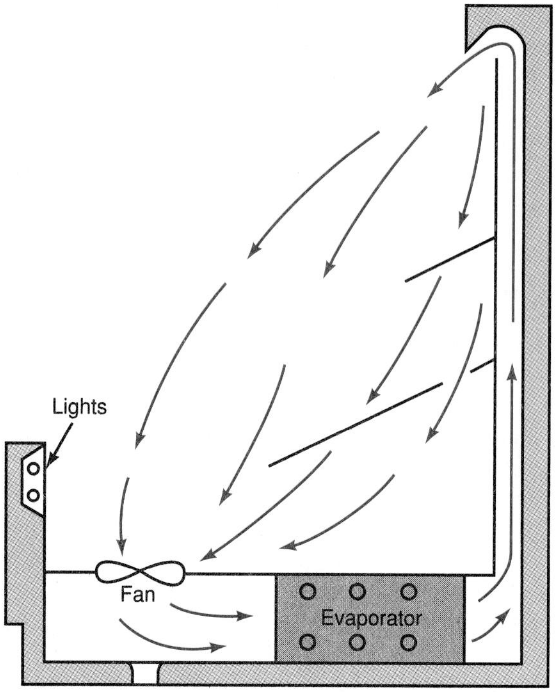

Figure 16-36. *Open display case with a blower evaporator located in the base of the cabinet.*

- The liquid-cooling evaporator is used to cool a liquid, which may be consumed or used to cool other substances.

The air-cooling evaporator is classified as the dry type. It may be the frosting, defrosting, or nonfrosting type. Some air-cooled units are forced-circulation evaporators. Liquid-cooling evaporators are of two types: the submerged evaporator and the tube-within-a-tube evaporator.

In commercial refrigeration, certain kinds of evaporators are more popular than others. Among these are the dry nonfrosting air-cooling evaporator, forced circulation dry evaporator, and submerged flooded or dry evaporator. See Chapter 13.

16.4.4 Air-Cooling Evaporator Theory

The theory of the air-cooling evaporator involves heat transfer from the air circulating over the evaporator to the refrigerant. As the warmer air comes in contact with the evaporator, air molecules strike the fins and release some of their energy (transfer some heat) to the fin. This heat, in turn, travels through the fins. Then it travels through the tubing. It contacts the liquid refrigerant on the inside of the tubing, transferring the heat energy to the liquid.

Because air density is low, heat transfer in air is slow. After reaching the metal, the heat travels efficiently and quickly. Upon reaching the interior surface of the tubing, it again has difficulty in reaching the refrigerant in the system. Gas bubbles clinging to the internal surface, along with an oil film, reduce the heat flow.

16.4.5 Evaporator Capacities

One of the laws of thermodynamics (heat in action) is that heat always flows from an object at a higher temperature to an object at a lower temperature. As in the case of heat leakage, the amount of heat transfer depends on five variables:

- Area.
- Temperature difference.
- Heat conductivity of the material.
- Thickness of material.
- Time.

The kinds of materials used in evaporators are of utmost importance. They must be good heat conductors.

Heat may be transmitted through various materials. For air evaporators, **Figure 16-37,** the heat must pass through air film, A, on the metal surface. Then it travels through the metal tubing and the oil or liquid refrigerant film on the inside of the evaporator tubing to the refrigerant, B.

If the air is moved rapidly, heat flow to the metal is greater. More air contacts the metal per unit of time and the air film is thinner. If the oil or refrigerant film is moved faster, the film will be thinner. This is due to greater movement. Generally, the denser the fluid, the greater the heat flow. The faster the fluid motion, the greater the heat flow.

The ***U-value*** (heat transferability) for several types of evaporators follows:

- Natural convection evaporators—about 1 Btu/ft^2/hr./°F.
- Blower evaporators—about 3 Btu/ft^2/hr./°F.
- Liquid-cooling evaporators—about 15 Btu/ft^2/hr./°F.

These values are fairly accurate if the temperature difference is taken between the average air or liquid temperature and the refrigerant temperature. See points T_1 and T_2 in **Figure 16-37.** For natural convection evaporators, the temperature difference usually selected is 10°F (6°C). For example, a 45°F (7°C) cabinet would have a refrigerant temperature in the evaporator of 35°F (2°C).

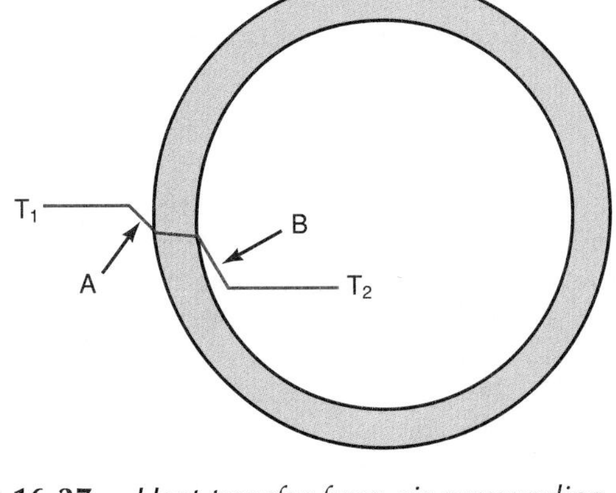

Figure 16-37. *Heat transfer from air surrounding an evaporator, A, to the evaporator, then through the metal to the refrigerant inside (B). T_1 is cabinet air temperature; T_2 is refrigerant temperature.*

The smaller the temperature difference, the higher the relative humidity can be. For example, a 10°F to 12°F (6°C to 7°C) temperature difference will keep a 75% to 90% RH, while a 20°F to 30°F (11°C to 17°C) temperature difference will keep a 50% to 70% RH.

If the evaporator refrigerant temperature goes below 28°F (−2°C), frost will form on the evaporator. The cycle off time must be long enough to permit defrosting, with the air at 35°F (2°C). If refrigerant temperature goes below 28°F, a defrost system must be used.

16.4.6 Evaporator Area

When calculating the capacity of evaporators, it is best to rely on the manufacturer's specifications. They obtain their heat capacity values from actual experiments. Such things as poor circulation, frosted fin condition, air turbulence around the evaporator, and even the amount of moisture in the air will affect the capacity of the evaporator.

To calculate the external surface area of an evaporator, consider the following:

- Both surfaces of the fin.
- Outside surface of the tubing (disregard the area where it comes in contact with the fins).
- External surface of the tubing bends.

Example:

Find the area of a 10′ (3 m) long evaporator with 6″ × 8″ (15 cm × 20 cm) fins .025″ (.06 mm) thick, having 1/2″ (13 mm) fin spacings and using two 5/8″ (15 mm) tubes 4″ (10 cm) apart. See **Figure 16-38** and proceed as follows:

1. Area of one fin:
   Each fin is 8″ × 6″
   8″ × 6″ = 48 in^2
   There are two sides to each fin:
   48″ × 2 = 96 in^2
2. However, there are two 5/8″ diameter holes in each fin. Therefore:
   Fin area = area of each side of fin minus area replaced by tubing

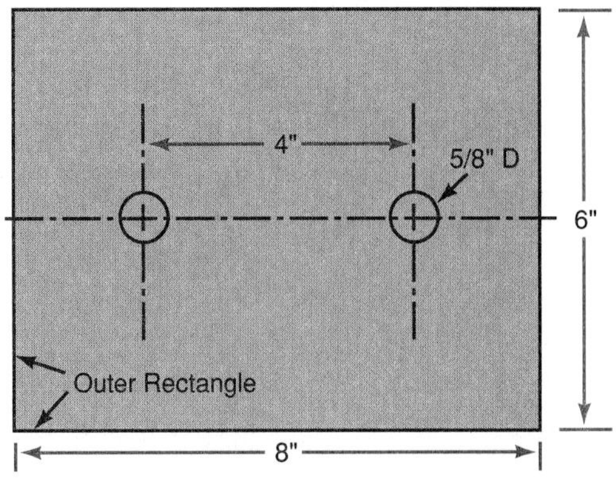

Figure 16-38. *Evaporator fin specification for sample problem. Fin is .025″ thick.*

Area of a circle = πR^2
(R = radius and π = 3.1416)
2R = diameter = D

$$\pi\left(\frac{D}{2}\right)^2 = \pi\frac{D^2}{2^2} = \pi\frac{D^2}{4}$$

$$\text{Area} = \frac{\pi \times D^2}{4} = \frac{\pi \times \left(\frac{5}{8}\right)^2}{4}$$

$$= \frac{\pi \times \left(\frac{25}{64}\right)}{4} = \frac{\pi \times 25}{4 \times 64} = \frac{3.1416 \times 25}{4 \times 64}$$

$$= \frac{78.54}{256} = .307 \text{ in}^2$$

However, each hole takes out *two* fin surfaces this size:

.307 × 2 = .614 in^2

There are two holes in each fin:

.614 in^2 × 2 = 1.228 in^2

3. Outer rectangle fin edge area, **Figure 16-38:**

28″ × 0.025 = 0.70 in^2
The 0.70 in^2 adds to the fin area.

Find the area for one fin as in the following process:

Heat removing area for one fin
= 96 in^2 − 1.228 + 0.70 = 95.472 in^2

4. The total number of fins:

10′ × 12″/ft. = 120″ long
2 fins/in. = 120 × 2 = 240 fins
240 fins + 1 extra fin at end = 241 fins

5. Total fin area = 95.472 × 241 = 23,008.7 in^2
6. Area of two tubes 5/8″ D; 10′ long:

10′ × 12″/ft. = 120″
There are two tubes: 120″ × 2 = 240″

The circumference of the tube is:

Cir. = π × diameter
= π × 5/8″
= 3.1416 × 5/8″
= 1.9635″

The area = length × circumference
= 240″ × 1.9635″
= 471.24 in^2

7. Actual tube area is decreased by thickness of the fins:

Fin contact area = π5/8 × .025 × 241
= 1.9635 × .025 × 241 = 11.83 in^2
There are two holes: 2 × 11.83 = 23.66
Actual tube area: 471.24 − 23.66 = 447.58 in^2

8. Tube bend area = length of bend × circumference length × number of bends (1)
(2″ radius)

Length = π × D = 3.1416 × 4″ = 12.5664
but it is only 1/2 of a circle, so 12.5664 ÷ 2 = 6.2832

Circumference of tube =

5/8″ × 3.1416 = .625 × 3.1416 = 1.96
Tube bend area = 6.2832 × 1.96 = 12.3 in^2

9. Total area:

23,008.7 = total fin area
447.6 = actual tube area
+ 12.3 = tube bend area
23,468.6 in^2

or, in square feet 23,468.6 ÷ 144 = 162.98 ft^2

The evaporator is unable to remove heat from the cabinet when the compressor is not running. Therefore, the heat-removing capacity of the evaporator is calculated on the same running time as that of the compressor.

Heat transfer problem:

What is the capacity of a natural convection evaporator having an external area of 15 ft^2 and a refrigerant temperature of 22°F (−16°C), if the average cabinet temperature is 42°F (6°C)?

Solution:

First, it is known that 1 ft^2 of evaporator surface will handle 1 Btu/hr./°F. The refrigerant temperature is 22°F (−6°C). With the air temperature passing over the evaporator at 42°F (6°C), the temperature difference is 20°F (11°C). If 1°F temperature difference will handle 1 Btu, 20°F (11°C) temperature difference will handle 20 Btu. Multiply this value by the number of square feet to find the total capacity of the evaporator per hour. In this case: Btu = 15 × 20 = 300 Btu/h.

The maximum effective distance that the fin should extend from a natural convection evaporator is 3″ (8 cm). **Figure 16-39** shows an evaporator constructed of rippled fin surfaces and with a flange contact between the prime surfaces (tubing) and the fin. A typical evaporator capacity table for nonfrosting evaporators is shown in **Figure 16-40.**

16.4.7 Evaporator Design

Many types of evaporators are available. Some common combinations of material used are: copper tubing and aluminum fins, copper tubing and copper fins, and steel tubing and aluminum fins (for ammonia R-717).

Usually, the evaporator fins are securely bonded to the tubing. However, construction varies. Some manufacturers make the fin to fit the tubing with a drive fit. Others use some mechanical device to attach the fins firmly to the tubing. Some expand the tubing with a mandrel or by hydraulic pressure. This will force it against the fin. See **Figure 16-41.** A method of bonding

Figure 16-39. *Section through tubes and fins of an evaporator. A—Tubing (primary surface). B—Fins (secondary surface).*

tubing to off-center fins is shown in **Figure 16-42.** Another method of mounting fins on tubing while spacing them is shown in **Figure 16-43.**

Fin spacing varies between 1/2″ and 1 1/2″ (1 cm and 4 cm) for natural convection evaporators, and 1/16″ to 1/4″ (2 mm to 6 mm) for forced convection. This spacing is a means of varying the capacity of the evaporator. It is also used to compensate for the depth of the evaporator. The deeper the evaporator, the greater the fin spacing (to minimize air restriction). Evaporators which have 6″ to 8″ (15 cm to 20 cm) depth usually have 1″ (2.5 cm) spacing. Those with 18″ to 20″ (38 cm to 51 cm) depth have 1 1/2″ (4 cm) spacing. Fin spacing of 1″ or less is said to decrease air turbulence. The tubing used usually is 5/8″ (16 mm) OD. However, 3/4″ (19 mm) OD tubing is used in large evaporators.

Some companies use one continuous piece of tubing for the complete evaporator. Others make the bends separately and then braze straight lengths to them. Some manufacturers use devices inside the tubing to swirl the refrigerant. This improves heat transfer to the boiling refrigerant.

Fittings which connect tubing to the evaporator are usually 1/2″ (13 mm) OD tubing brazed to the 5/8″ tubing. It is then flared with an external nut mounted on it. Some manufacturers use a 1/2″ male flare fitting brazed to the end fin. **Figure 16-44** shows two types.

16.4.8 Forced-Circulation Air-Cooling Evaporator Capacities

A forced-circulation evaporator is one having an electric fan mounted near it to increase airflow. Velocities of 44 ft./min. (feet per minute) to 2000 ft./min. are permissible, with 1000 ft./min. being the average. Sometimes, variable speed fan motors are used. Draining facilities for condensation removal must be built into the unit.

In a forced-circulation unit, a large amount of air strikes the evaporator per unit of time. The capacity of the evaporator in Btu per square foot per hour per degree Fahrenheit (Btu/ft^2/hr./°F) is increased remarkably. The values vary with the air speed. The table in **Figure 16-45** gives these values.

Evaporators have fins, with the spacing of the fins varying considerably. See Section 16.4.7. The fins must be kept straight and equally spaced or airflow will be reduced. See Chapter 15 for service instructions.

16.4.9 Liquid Evaporator Capacities

The capacity of a liquid evaporator, regardless of type, is calculated by using a base factor of about 10 Btu/ft^2/hr./°F to 120 Btu/ft^2/hr./°F. The U-value (heat transferability), however, varies with the fluid velocity, evaporator construction, and total temperature difference. **Figure 16-46** shows the average values for a certain evaporator.

Some evaporators are used to cool a brine solution. Since this solution never freezes, there are no frost accumulation problems. Many submerged evaporators use a fresh water bath to provide an ice holdover around the evaporator. This maintains good capacity during peak loads.

The capacity of an evaporator will vary, depending upon whether or not it is the frosting type. Ice formation around the evaporator reduces its capacity, because the heat must go through this extra material. Furthermore, it makes the heat travel through one extra contact surface. This, too, reduces efficiency. Liquid cooling evaporators are used for beverage cooling, for the water chillers used on some air conditioners, and for cooling brine solutions.

16.4.10 Special Evaporator Capacities

Many different types of evaporators are in use. Among them are brine spray units, intermediate refrigerant evaporators, and cold plates. They may be constructed from cast metal or iron pipe. Whatever the type, they have the same problems as those previously discussed.

The metal used as the refrigerant-carrying device does not have much effect upon the heat transfer capacity of the evaporator. Its conductivity is greater than the contact between the evaporator and the air. The capacity of a cast metal or iron pipe evaporator may be calculated exactly the same as the method described in Section 16.4.6.

In brine spray systems, the brine is forced through a pipe extending into the cooling chamber. The pipe is perforated with a number of fine holes. The brine sprays out of these holes. It mixes with the air flowing over the baffle, removing the heat.

Model ELC

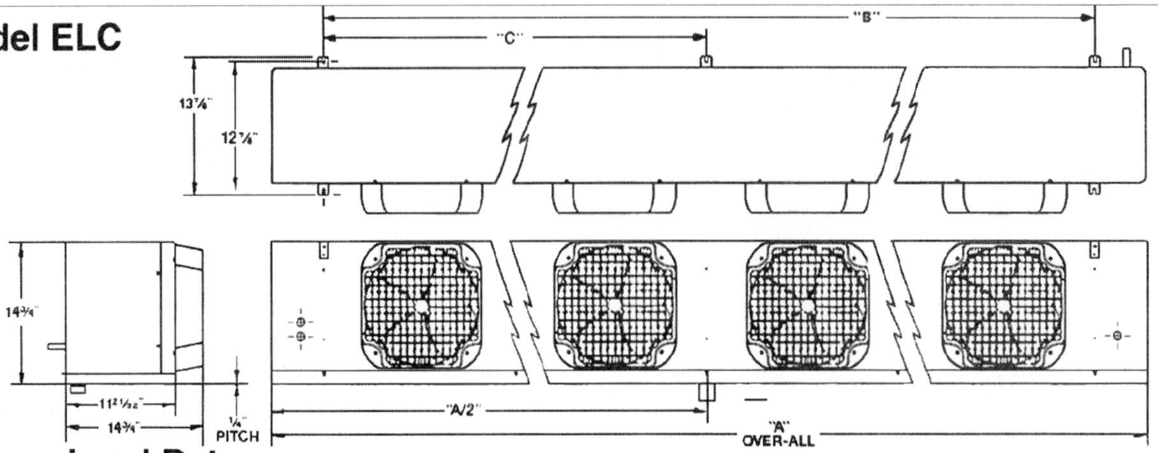

Dimensional Data

ELC MODEL NO.	CONNECTIONS				DIMENSIONS			APPROX. SHIP WT. LBS.
	INLET	SUCTION	EXT. EQUALIZER	DRAIN	A	B	C	
041F	1/2" FN	5/8" ODS	1/4" FN	3/4" FPT	32-1/4"	21"	–	52
056F	1/2" FN	5/8" ODS	1/4" FN	3/4" FPT	44-1/4"	33"	–	65
071F	1/2" FN	7/8" ODS	1/4" FN	3/4" FPT	44-1/4"	33"	–	74
096F	1/2" FN	7/8" ODS	1/4" FN	3/4" FPT	52-1/4"	41"	–	118
122F	1/2" FN	1-1/8" ODS	1/4" FN	3/4" FPT	72-1/4"	61"	–	138
143F	1/2" FN	1-1/8" ODS	1/4" FN	3/4" FPT	72-1/4"	61"	–	149
163F	1/2" FN	1-1/8" ODS	1/4" FN	3/4" FPT	92-1/4"	81"	40-1/2"	163
184F	1/2" FN	1-1/8" ODS	1/4" FN	3/4" FPT	92-1/4"	81"	40-1/2"	171
245F	7/8" ODS	1-1/8" ODS	1/4" FN	3/4" FPT	132-1/4"	121"	60-1/2"	210
286F	7/8" ODS	1-1/8" ODS	1/4" FN	3/4" FPT	132-1/4"	121"	60-1/2"	262

Performance Data

ELC MODEL NO.	CAPACITY BTUH				AIR FLOW CFM	FAN		AIR THROW FEET	MOTORS					HEATERS		
	-20F	-10F	+20F									SHADED POLE	PSC	AMPS		WATTS
	@ 10° TD	@ 10° TD	@ 10° TD	@ 15° TD		QTY	DIA		HP	RPM	BTUH HEAT	MOTOR AMPS†	MOTOR AMPS†	208-230/60/1	208-230/60/3	
041F	4100	4305	4715	7073	825	1	12"	50	1/20	1550	220	0.9	0.4	4.4	3.9	1006
056F	5600	5880	6440	9660	1565	2	12"	50	1/20	1550	440	1.8	0.8	7.0	6.2	1605
071F	7100	7455	8165	12247	1485	2	12"	50	1/20	1550	440	1.8	0.8	7.0	6.2	1605
096F	9600	10080	11040	16560	1565	2	12"	50	1/20	1550	440	1.8	0.8	9.0	8.1	2060
122F	12200	12810	14030	21045	2340	3	12"	50	1/20	1550	660	2.7	1.2	13.3	11.9	3067
143F	14300	15015	16445	24668	2320	3	12"	50	1/20	1550	660	2.7	1.2	13.3	11.9	3067
163F	16300	17115	18745	28118	3120	4	12"	50	1/20	1550	880	3.6	1.6	16.3	14.5	3738
184F	18400	19320	21160	31740	3120	4	12"	50	1/20	1550	880	3.6	1.6	16.3	14.5	3738
245F	24500	25725	28175	42262	4575	6	12"	50	1/20	1550	1320	5.4	2.4	28.3	25.4	6517
286F	28600	30030	32890	49335	4575	6	12"	50	1/20	1550	1320	5.4	2.4	28.3	25.4	6517

Figure 16-40. *Manufacturers develop performance and dimensional data for evaporators. The tables shown are for use with a system using R-22 or R-502 refrigerants. (Heatcraft, Inc.)*

The capacity of these systems is calculated on the temperature difference between the air in the box and the brine. It is safe to assume a high efficiency of heat transfer for brine spray installations.

The Baudelot Cooler (milk cooler) runs water, or the liquid to be cooled, over refrigerant-cooled pipes or plates. The liquid being cooled is in the open. It can be easily controlled. Icing is not critical. Therefore, the liquid can be cooled close to its freezing temperature. See Section 16.13.2.

The cooling capacity of cold plates may also be determined by the area, temperature difference, and the materials. The heat transfer for air to metals is about 1 Btu/ft^2/hr./°F. Cold plates usually contain a eutectic

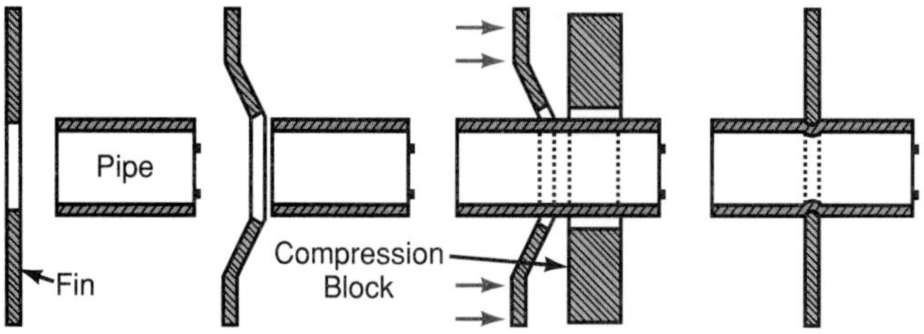

Figure 16-41. *Evaporator fins can be mechanically bonded to tubing. Fin is dished to expand tube hole before tubing is inserted. Mandrel flattens fin, firmly pressing fin into tubing surface.*

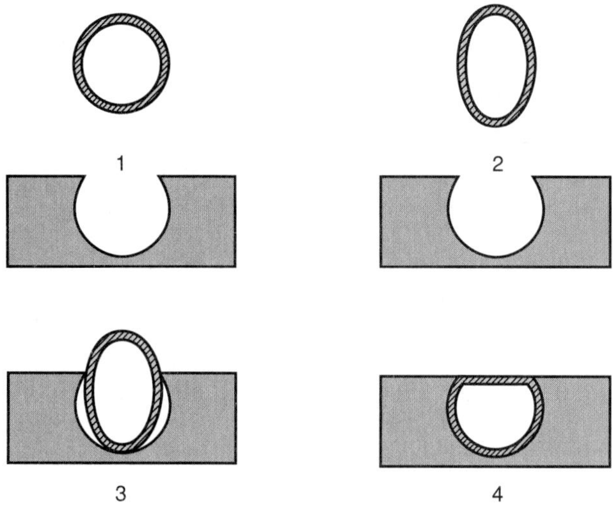

Figure 16-42. *Note method of mechanically bonding tubing to off-center fins. 1—Original tubing and fin. 2—Tubing is formed into elliptical shape. 3—Tubing is inserted into fin opening. 4—Fixture holds fins while tube is pressed into fin-opening shape. (Peerless of America, Inc.)*

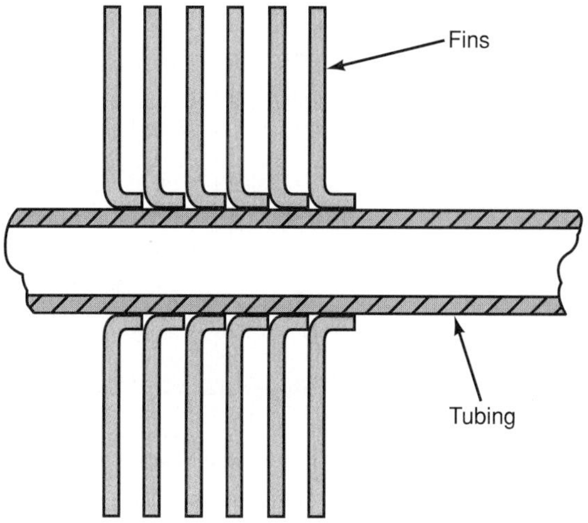

Figure 16-43. *How flanged fins can be mounted on a section of tubing. Flanges automatically space fins.*

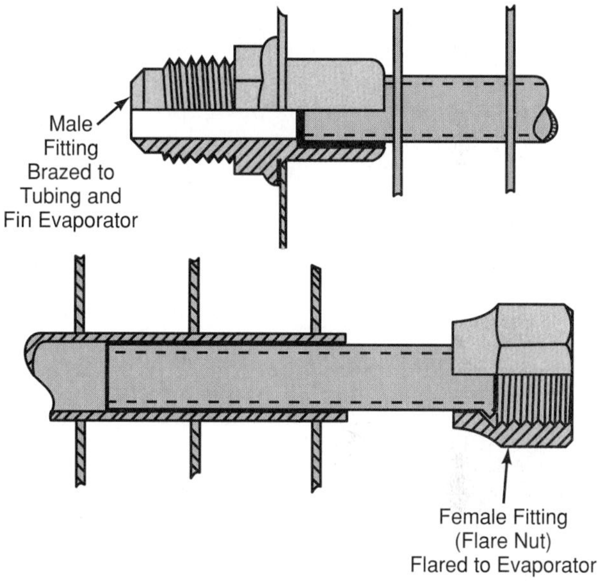

Figure 16-44. *Construction of coupling devices (fittings) used in attaching refrigeration lines to evaporator.*

solution. This solution freezes at a certain temperature. The plate remains at this constant temperature until all the eutectic solution has melted. The refrigerating system may or may not be operating at the time the eutectic solution is absorbing heat (melting).

16.5 Water-Cooling Loads

Finding the refrigeration load of a water-cooling installation is a combination of a specific heat and a heat leakage problem:

1. Water is cooled to temperatures which vary upward from 35°F (2°C). The amount of heat removed from the water to cool it to a certain temperature is a specific heat problem.
2. Water, being maintained at these low temperatures, results in a heat leakage from the room into the water. This part involves the heat leakage portion of the installation.

The two major items to be solved in a water-cooling installation are:

1. How much water is to be consumed at the temperature difference desired. **Figure 16-47** gives the values of these two variables.
2. Drinking water temperatures should be regulated based on the type of work the consumers are doing. The heavier the work, or the warmer the room temperature, the warmer the drinking water must be.

The amount of water consumed varies quite a bit in different applications. By using the table in **Figure 16-47,** the exact heat load is easily determined.

Example:

Item 4 on the table points out that for heavy manufacturing, drinking water should be kept within 50°F to 55°F (10°C to 13°C). Also, 1/4 gal. (1 L) of water per hour per person will be consumed.

Forced Circulation Evaporator Capacities

Surface in ft²			Parallel Paths	Cooling Capacities (Btu/h)		Motor		Fan	
Tube	Fin	Total		15 °F td*	25 °F td*	hp	Speed rpm	Diameter	cfm
2.94	32.39	35	3	2200	4500	1/80	1800	12"	620
5.88	63.0	68	5	4100	7200	1/80	1800	12"	465
7.94	70.06	78	3	5200	9000	1/10	1140	15 1/2"	1200
13.2	116.8	130	5	8300	12300	1/10	1140	15 1/2"	1000
14.7	83.3	98	3	6500	10500	1/8	1140	17"	2020
17.2	145.8	163	5	10000	14000	1/8	1140	17"	1715

*Temperature Difference

Figure 16-45. *Table lists forced-circulation evaporator capacities.*

Evaporator Heat Transfer Data

Water Velocity (ft./min.)	Total Temperature Difference				
	6	8	10	12	15
	Transfer (Btu/ft²/hr./°F)				
150	67	76	83	90	97
200	83	95	103	110	118
250	97	109	115	122	129
300	103	115	123	130	138

Figure 16-46. *Heat transfer in Btu/ft²/hr./°F for typical flooded liquid evaporator using 5/8" OD tubes. Total temperature difference is between refrigerant and water temperatures.*

Usage	Final Temp. Required (°F)	Total Amount of Water Used and Wasted
1. Office Building—Employees.	50	1/8 gallon per hour per person
2. Office Building—Transients.	50	1/2 gallon per hour for each 250 persons per day
3. Light Manufacturing. .	50 to 55	1/5 gallon per hour per person
4. Heavy Manufacturing.	50 to 55	1/4 gallon per hour per person
5. Restaurant	45 to 50	1/10 gallon per hour per person
6. Cafeteria.	45 to 50	1/12 gallon per hour per person
7. Hotels	50	1/2 gallon per room (14 hr. day)
8. Theaters	50	1 gallon per hour per 75 seats
9. Stores.	50	1 gallon per hour per 100 customers per hour
10. Schools	50 to 55	1/8 gallon per hour per student
11. Hospitals.	45 to 50	1/12 gallon per day per bed

NOTE—Total amount of water used and wasted varies with type of installation and kind of service. This table will serve as a basis for determining cooler capacity required.

Figure 16-47. *Recommended values for providing cooled drinking water for various public places and places of work. (Temprite Div. of Elkay Mfg. Co.)*

A production foundry may be classed as heavy manufacturing. If the foundry employs 50 workers for a period of eight hours, the water load per day would be $50 \times 8 \times 1/4$, or 100 gal. (379 L) of water to be cooled each eight hours.

If the incoming water is at a temperature of 75°F (24°C) in the pipes, it must be cooled 20°F (11°C) to reach 55°F (13°C). Since there are 8.34 lb. of water in 1 gallon, the specific heat load can be computed as follows:

Formula:

Btu = specific heat × weight × temp. difference

Solution:

$$\text{Btu} = 1 \times 100 \times 8.34 \times 20$$
$$= 16{,}680 \text{ Btu/100 gal. water}$$

The heat leakage is determined by the external area of the insulated parts of the system. Insulation 1" to 3" (2.5 cm to 7.5 cm) thick is common for water-cooling insulations. Ice water insulation is standard at 1 1/2" (4 cm). Heat leakage for water-cooling installations is calculated the same as heat leakage for cabinets. The example just described deals only with a unit installation.

Many water-cooling installations involve the circulation of refrigerated water to various fountains. To maintain satisfactory temperatures, the heat leakage load is figured on the basis of gallons of water per hour circulated through the system. The table in **Figure 16-48** illustrates another method of computing this load.

16.6 Ice Cream Freezing and Storage Load

Ice cream is manufactured from milk solids, fat, sugar, gelatin, and water. After mixing, the product iscooled to about 27°F (−3°C), then frozen. Next, it is

Gallons per Hour to Be Circulated per 100 Feet of Pipe to Hold Temperature Rise Within 5°F*

Pipe Size	Temperature Difference Between Room and Circulating Water							
	20°	25°	30°	35°	40°	45°	50°	55°
1/4"	4.8	6.3	7.3	3.2	9.7	10.9	12.3	13.6
3/8"	5.5	6.8	8.2	9.6	11.0	12.3	13.7	15.0
1/2"	6.3	8.0	9.5	11.2	12.7	14.3	16.0	17.6
3/4"	6.7	8.4	10.1	11.8	13.5	15.2	16.9	18.6
1"	7.3	9.1	10.9	12.8	14.6	16.5	18.5	20.6
1 1/4"	8.6	10.4	12.5	14.6	16.6	18.7	20.4	22.1
1 1/2"	9.0	11.2	13.5	15.7	18.0	20.2	22.5	24.7

*Add This Amount to Usage.

Figure 16-48. *This table shows heat gain through insulated cold-water pipes. Add the amounts to the usage in* ***Figure 16-47.*** *(Temprite Div. of Elkay Mfg. Co.)*

cooled rapidly to approximately −20°F (−30°C). Brick ice cream is maintained between 0°F to 5°F (−18°C to −15°C). Bulk ice cream is held at 5°F to 12°F (−15°C to −11°C).

The heat values of various ice creams vary. See **Figure 16-7.** Specific heat of the mix before freezing is 0.80; latent heat, at 27°F (−3°C), is about 96 Btu/lb. Specific heat of frozen ice cream is 0.45 Btu/lb. Weight of the original mix is about 9 lbs./gal. On freezing, it expands and comes to a density of 5 lbs./gal., if flavored. If the ice cream contains fruits or nuts, it comes to 6 lbs./gal.

Normally, an ice cream cabinet is designed to hold brick ice cream and flavored ice creams next to the evaporator. However, many cabinets need separate evaporators for the two types. **Figure 16-49** shows an ice cream display cabinet. Metal-finished cabinets with 2″ to 3″ (5 cm to 7.5 cm) of urethane are used.

The ice cream is delivered at the correct temperature. The heat load, therefore, is composed only of leakage and air entering when the covers are removed. The table in **Figure 16-11** may be used to calculate heat leakage. Use the outside area and add 20% to take care of the cover openings.

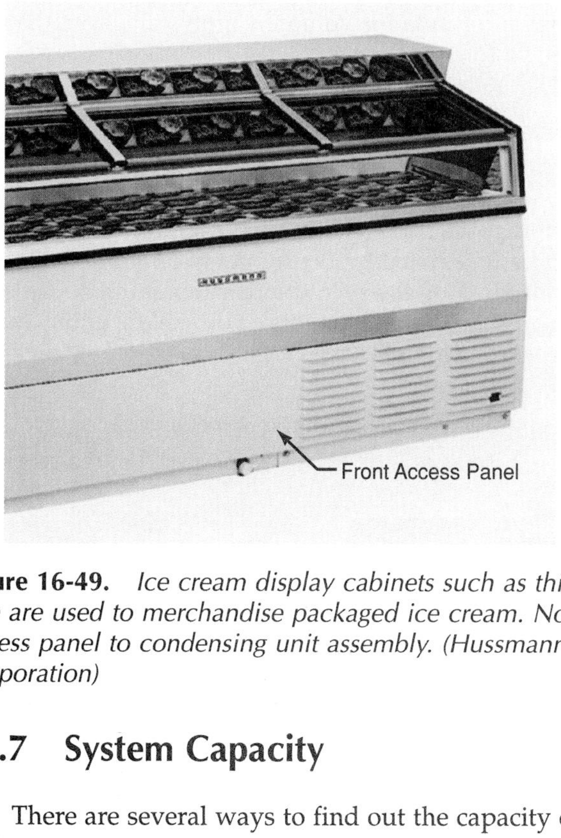

Figure 16-49. *Ice cream display cabinets such as this one are used to merchandise packaged ice cream. Note access panel to condensing unit assembly. (Hussmann Corporation)*

16.7 System Capacity

There are several ways to find out the capacity of a system:

- Measure the refrigerant flow:
  - Amount (weight).
  - Flowmeters.
- Measure the heat gain of the condenser cooling water.
- Measure the heat gain of the air flowing over the condenser.

Next, record the condensing temperature and pressure, and the evaporating temperature and pressure of the refrigerant in the cycle. Use the refrigerant's pressure-heat diagram to find the Btu pickup in the evaporator.

If the pounds per hour of refrigerant flow is known:

1. Draw the system cycle on the pressure-heat diagram.
2. Multiply the Btu/lb. heat pickup by the pounds circulated to find the Btu removal capacity of the system.
   Note: Take the temperature of the refrigerant in the liquid line just before it reaches the refrigerant control. Use a good thermometer in a thermometer well on the line or a thermocouple thermometer fastened to the line.

Determine the heat gain in water:

lb./hr. × temp. change × spec. heat = Btu lb./hr.
= gal./hr. × 8.34 lb./gal. (1 gal. water = 8.34 lb.
Temp. change = temp. out minus temp. in)
× specific heat = 1

Example:

A system uses 300 gal./hr. of 70°F (21°C) water that warms to 80°F (27°C) at the condenser outlet. How much heat is being removed?

(300 gal. × 8.34) × (80°F to 70°F) × 1 = Btu
2502.00 × 10 × 1 = Btu
25,020 = Btu

A one-ton machine would be about 12,000 Btu/h. Then, 25,020 divided by 12,000 equals approximately two. So this will be approximately a two-ton machine.

Measure air temperatures into and out of the condenser. Measure the average air velocity in feet per hour in or out (smoothest airflow). Take as many as 16 readings at different places across the airflow; then average them. (See Chapter 19.)

Measure the condenser's length and width in inches.
Air temp. rise = air temp. out − air temp. in
Condenser area = width × length =

$$\frac{\text{Area in in}^2}{144} = \text{area in ft}^2$$

Volume of air = area in ft^2 × average velocity (ft./hr.)
About 14 ft^3 of air = 1 lb.
Spec. heat of air = .24 Btu/lb. °F
Btu = weight of air/hr. × spec. heat × temp. change

$$= \frac{\text{volume of air/hr.}}{14\ \text{ft}^3/\text{lb.}} \times .24\ \text{Btu/lb. °F} \times (\text{temp. out minus temp. in})$$

Example:

An air-cooled condenser has an air velocity of 300 ft./min., an air-in temperature of 80°F (26.7°C) and an air-out temperature of 90°F (32.2°C). The condenser measurements are 25″ × 40″ (64 cm × 102 cm). Calculate the Btu/hr. released by the condenser.

Btu = weight of air/hr. × spec. heat × temp. change

Weight of air:

1. Condenser area = 25 × 40 = 1000 in^2

$$\frac{1000}{144} = 6.94\text{ft}^2$$

2. Volume = 6.94 ft^2 × (300 ft./in. × 60 min./hr.)
 = 6.94 × 18,000

Volume = 124,920 ft^3/hr.

Weight = 124,920 ÷ 14 ft^3/lb.
 = 8922.86 lb. of air/hr.

Btu = 8922.86 × .24 × (90 − 80)
 = 8922.86 × .24 × 10
 = 8922.86 × 2.4
 = 21,414.86 Btu (about 1 3/4-ton system)

16.8 Compressor Capacities

The ***compressor*** is the heart of the refrigerating machine. It uses mechanical energy, produced by an electric motor, to pump refrigerant through the cycle. The refrigerant picks up heat at one place and releases it at another place.

The most efficient construction possible is to have a compressor just large enough to handle the necessary amount of refrigeration. If the compressor is too large, energy is lost in excess friction, starting, etc. If the compressor is too small, it will not produce enough refrigeration.

Basically, the compressor must remove the vapor from the evaporator fast enough to enable the refrigerant to vaporize at the correct low pressure. To do this, it must remove the refrigerant vapor as fast as heat goes into the evaporator to vaporize the refrigerant.

The method of determining compressor size may be simply stated: an evaporator is usually designed to remove the 24-hour heat load in a 16- or 18-hour running period. The amount of time depends on the factors described in previous paragraphs.

Assume that the effective heat removing ability of the refrigerant is 60 Btu/lb. As each pound of refrigerant vaporizes in the evaporator, it picks up 60 Btu of heat from the evaporator. To remove this much heat, the compressor must handle all the vapor formed. Refrigerant tables give these values, called "specific volumes." The specific volume values mean that at certain pressures one pound of refrigerant, as it is vaporizing, will form a certain number of cubic feet of vapor.

Assume, for example, that 1 lb. of R-12, vaporizing at 9.17 psig (23.87 psia or 164.7 kPa) and 0°F (−17.8°C), forms 1.637 ft^3 of vapor in 10 minutes. The compressor, then, must remove 1.637 ft^3 of vapor from the evaporator in the same period.

The compressor size needed to do this depends on the volume of vapor pumped per revolution of the compressor. This is determined by the ***bore*** (cylinder diameter), ***stroke*** (distance traveled by a piston), number of cylinders, speed of the compressor (rpm), and its volumetric efficiency. As the crankshaft of the compressor completes one revolution, the piston moves from the lowest point of its travel (bottom dead center) to the highest point (top dead center) and back to the lowest point again.

On the downstroke, low-pressure vapor is drawn into the cylinder. It fills up the space between the top of the piston and the head of the cylinder. The piston compresses this vapor on the upstroke. It pushes the vapor through the exhaust valve into the high-pressure side of the system. The volume handled is the volume displaced by the piston on each stroke (movement from bottom dead center to top dead center). This volume may be calculated by the following formula:

$$V = \pi\frac{D^2}{4} \times S \times N \times R$$

V = Volume in cubic inches
D = Diameter of cylinder in inches
S = Length of stroke in inches
N = Number of cylinders
R = revolutions per minute (rpm)

This formula for volume simply calculates the area of the piston head,

$$\frac{\pi D^2}{4},$$

and multiplies it by the length of the stroke. The answer is the displacement volume in cubic inches. Next, this figure is multiplied by the number of cylinders. Then it is multiplied by the revolutions per minute of the compressor. This gives the total volume in cubic inches pumped per minute.

Example:

How much vapor will a two-cylinder compressor pump if it has a 2″ bore, a 2″ stroke, and operates at 400 rpm?

Formula:

$$V = \pi\frac{D^2}{4} \times 2 \times 2 \times R$$

Solution:

$$V = \frac{3.1416 \times 4}{4} \times 4 \times 400 = 3.1416 \times 4 \times 400$$

$$= 3.1416 \times 1600 = 5026.56\ \text{in}^3/\text{min.}$$

1728 in^3 = 1 ft^3

The volume in ft^3 will be:

$$V = \frac{5026.56}{1728} = 2.9\ (\text{plus})\ \text{ft}^3/\text{min.}$$

Compressor problem:

Calculate the bore and stroke of a two-cylinder compressor operating at 1750 rpm, which will compress in 10 minutes the refrigerant vapor formed by vaporizing

10 lb. of R-22 at 5°F (−15°C). Note from the table that vaporized R-22 at 5°F (−15°C) occupies 1.2434 ft^3/lb.

Solution:

$$V = 12.434\ ft^3 = 12.434 \times 1728\ in^3$$
$$= 21{,}485.95\ in^3$$

$$\text{Volume pumped per min.} = \frac{21{,}485.95}{10}$$
$$= 2148.6\ in^3/min.$$

Using the formula:

$$V = \pi\frac{D^2}{4} \times S \times N \times R = 2148.6\ in^3$$

Compressors are usually designed with a bore equal to the stroke (S = D). So the formula becomes:

$$V = \pi\frac{D^3}{4} \times N \times R$$
$$D^3 = \frac{V \times 4}{\pi \times N \times R}$$
$$D^3 = \frac{2148.6 \times 4}{\pi \times 2 \times 1750} = \frac{2148.6 \times 2}{\pi \times 1750}$$
$$D^3 = \frac{2148.6}{\pi \times 875}$$
$$D^3 = .8\ in^3$$
$$D = .93'' \text{ (approx.)}$$

Therefore, this compressor has a bore of .93″ and a stroke of .93″. However, this value is the theoretical amount of vapor pumped by the compressor. The actual amount will be less. It will depend on the volumetric efficiency of the compressor. See Section 16.8.1.

16.8.1 Volumetric Efficiency

Volumetric efficiency is the ratio between the volume actually pumped per revolution, divided by the volume calculated from the bore and stroke.

If the refrigerating unit is maintaining a 150 psig (165 psia or 1139 kPa) head pressure and a 0 psig (14.7 psia or 101 kPa) low-side pressure, the following things happen:

1. When the piston is on its upward stroke, it compresses this 0 psig vapor. It is compressed until the pressure of the vapor in the cylinder reaches 150 psig. When this pressure is reached, the vapor should start passing through the exhaust valve into the condenser. However, in addition to reaching high-side pressure, it must overcome exhaust valve spring force or weight. This means a slight additional increase in pressure is required.
2. After the piston reaches top dead center, there is still a little vapor remaining at a high pressure between the piston head and the exhaust valve. The space it occupies is necessary for clearance between the piston and cylinder head. Otherwise, the piston would pound against the cylinder head at top dead center.
   The residue of vapor is under a high pressure (150 psig or more). The piston moves down to receive a new charge of vapor. The remaining high-pressure vapor expands and partially fills the cylinder chamber. This ***residual*** (remaining) vapor decreases the amount of vapor that may move into the chamber from the low-pressure side of the system. The necessary space, or "clearance volume," varies between 4% and 9% of the piston displacement.
3. At speeds of 300 rpm to 3400 rpm or more, the piston is traveling so fast that the inertia or weight of the vapor prevents it from filling the cylinder chamber completely. Therefore, there are losses. The pressure in the cylinder may be 2″ Hg. (95 kPa), instead of 0 psig. The resistance to vapor flow through the valve openings is called "wire drawing." The pressure in the cylinder never gets as high as the pressure in the suction line during the suction stroke. The higher the speed of the compressor, the less vapor will be pumped per stroke.
4. The exhaust valve offers a restriction to the vapor flow. The intake valve with its force, and the weight of the valve parts, also offers resistance.
5. The compressor runs at a warm temperature. Some of this heat warms the vapor as it enters the cylinder, causing it to expand. This also keeps a complete load of vapor from entering the cylinder.
6. Other losses, such as the leaking of the vapor past the piston and rings into the crankcase, further explain why compressors cannot pump the amount of vapor calculated by the bore and stroke formula.

Small compressors used in domestic refrigeration have a bore and stroke of about 1 1/2″ (4 cm). Their volumetric efficiency varies between 40% and 75%, with 60% being an average value. The larger commercial compressors, depending on their size and speed, have volumetric efficiencies of between 50% and 80%. The average value is 70%. A volumetric efficiency value of 60% to 65% should be used if the unit is air-cooled.

Example:

If the compressor described in Section 16.8 has a volumetric efficiency of 60%, the size of the compressor would be increased as follows:

$$D^3 = .8 \text{ (as calculated in Section 16.8)}$$
$$D^3 = \text{corrected} = \frac{.8}{.60} = 1.33\ in^3$$
$$D = \text{cube root of } 1.33$$
$$= 1.1 \times 1.1 \times 1.1$$
$$= 1.33''$$
$$D \text{ corrected} = 1.1''$$

Therefore, a bore and stroke of 1.1″ × 1.1″ would be required. Note that the correction for volumetric efficiency was made on the displacement volume of the cylinder, not on the calculated bore and stroke.

The volumetric efficiency of a compressor depends on the difference between its low-side pressure and its high-side pressure. For instance, the compressor in an R-12 system used for domestic purposes will be more efficient than if it were converted into an ice cream system. The decrease in low-side pressure from 10 psig (25 psia or 173 kPa) down to 10″ Hg. (68 kPa) with the same head pressure reduces the actual pumping capacity of the compressor. The low-pressure vapor expands when the cylinder is filled at low-side pressure. Therefore, only a small amount (by weight) is pumped. All of the various items affecting efficiency, such as increasing head pressure, increasing speed, using thicker gaskets, and overheating the compressor, will reduce the compressor's pumping efficiency.

16.8.2 Coefficient of Performance

The cooling effect in Btu values in a refrigeration cycle compared to the Btu equivalent of the energy put into the system is called the coefficient of performance. ***Coefficient of performance (COP)*** is the ratio of output divided by input. In refrigeration work, the output is the amount of heat absorbed by the system. The input is the amount of energy required to produce this output.

For example, if one pound of refrigerant has an effective latent heat of 50 Btu, and the compressor pumping energy is equivalent to 10 Btu/lb., the coefficient of performance is 50 to 10 or 5:1.

The heat input by the compressor is less than the electrical energy put into the motor. The motor is not 100% efficient, and there are also compressor friction losses. Usually, the overall coefficient of performance will be approximately 60% of the theoretical COP. The actual coefficient, then, is approximately 3:1, rather than 5:1.

This means, for example, that three times more heat would be obtained from a heat pump by using the compressor than by using electricity to produce the heat. This explains the advantage of a heat pump. It also explains why hot gas defrost is used in some large systems. The cost of the extra piping and valves is soon recovered in the savings in cost of defrosting.

16.9 Cascade System

To produce extremely low temperatures efficiently, two refrigerating systems may be used instead of one. The two systems are connected in series. The resulting arrangement is called a ***cascade system.*** That is, the evaporator of the higher pressure-temperature cycle (first or low stage) removes the heat from the condenser of the lower pressure-temperature cycle (second or high stage). **Figure 16-50** shows the principle of this type system on a pressure-heat diagram.

Many cascade systems use a different refrigerant for the low-temperature system than for the high-temperature system. Actually, the evaporator of the high-temperature system must remove all the condensing heat of the low-temperature system. A^1 to B^1, as shown in **Figure 16-50,** should equal E to D.

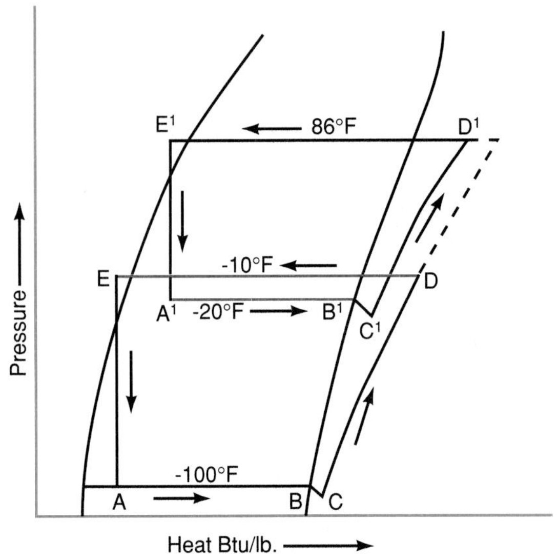

Figure 16-50. *Pressure-heat diagram for cascade system used to obtain ultralow temperatures. Evaporator A^1 to B^1 removes heat from condenser D to E.*

16.10 Two-Stage Compressor

Some refrigerating systems, especially ultralow-temperature systems, use two compressors connected in series. They pump the very low pressure suction line vapor up to the condensing pressure and temperature condition. In the first stage, typically, a large cylinder pumps the vapor up to a midpoint on the compression curve. Then, the compressed vapor is cooled but stays vaporized. The second cylinder compresses the cooled intermediate vapor to the final pressure-temperature condition. **Figure 16-51** shows an approximate pressure-heat cycle.

Two-stage compressors are used when the compression ratio is more than 10:1. That is, if the low-side pressure is 0 psig and the head pressure is 210 psig, the ratio is:

$$\text{Head pressure abs/Low-side pressure abs} = \frac{210 + 15}{0 + 15} = \frac{225}{15} = 15:1$$

In this case, a two-stage compressor would be used.

$$\text{Stage 1: } \frac{45 + 15}{0 + 15} = \frac{60}{15} = 4:1$$

$$\text{Stage 2: } \frac{210 + 15}{45 + 15} = \frac{225}{60} = 3.75:1$$

16.11 Bypass Cycle

The hot gas bypass may be used for any of these purposes:

- Defrosting evaporators.

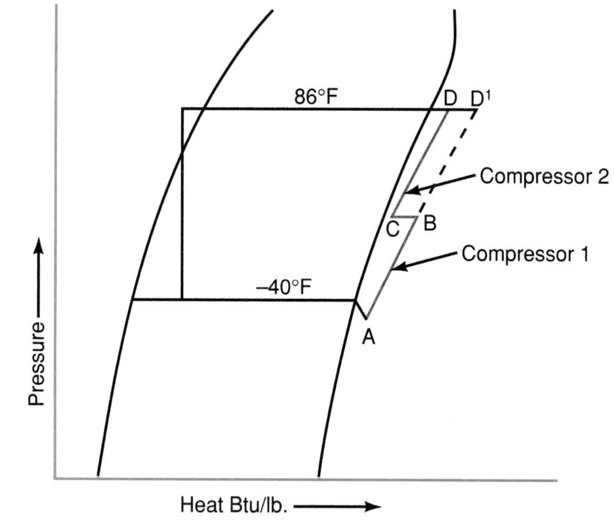

Figure 16-51. *Cycle of a two-stage compressor system. Compressor 1 (first stage) compresses vapor from A to B. Vapor is cooled in heat exchanger (air to water) from B to C. Compressor 2 (second stage) then compresses vapor to condensing pressure from C to D. This action reduces amount of heat of compression at final stage D to D^1 and also reduces superheat temperature at compressor 2 exhaust valve.*

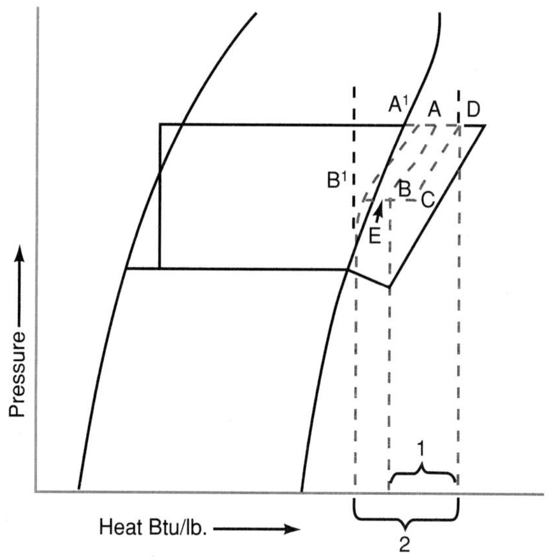

Figure 16-52. *The hot gas defrosting cycle. Heat lost from D to B, as shown at 1, is heat used to defrost the evaporator. If defrost hot gas is cooled too much (as at B^1), it will become partly liquid. This could cause liquid to enter the compressor, so an accumulator is put at E to ensure that only vapor can reach compressor. Defrost cycle is A to B to C to D. Maximum heat for defrosting is shown at 2, unless bypass gas is allowed to condense and is then vaporized (in another evaporator of a multiple system or in a special defrost evaporator).*

- Preventing suction pressure from going too low (when cooling load decreases).
- Keeping liquid refrigerant from entering compressor.

Many refrigerating systems use an automatic bypass system. Two automatic bypass types are:

- Hot gas bypass.
- Liquid bypass.

The hot gas bypass cycle used to defrost an evaporator is shown in **Figure 16-52.** The hot gas is traveling through the evaporator from A to B to C.

The cycle for hot gas bypass for low-pressure control is shown in **Figure 16-53.** The bypass line (controlled by a solenoid valve and a pressure valve) is piped from the hot gas part of the condenser into the suction line near the compressor. The bypass circuit is controlled by a pressure control connected to the suction line. The bypass action will return the compression line to approximately C^1 to D^1. The four horizontal evaporator lines represent how the low-pressure side changes from cut-in pressure to cut-out pressure.

A liquid bypass cycle is shown in **Figure 16-54.** If the low-side pressure drops to B^1 because the evaporator is not picking up enough heat, the suction pressure control will open a solenoid valve. This permits liquid refrigerant to bypass the refrigerant control and evaporator. It feeds directly into the suction line at B.

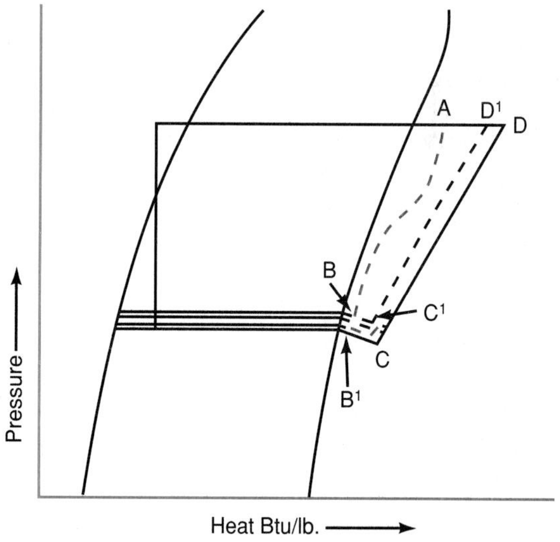

Figure 16-53. *Hot gas bypass system allows low side to maintain normal low-side pressure. If pressure tends to drop to point B^1, bypass circuit A to B opens and brings low-side pressure up to normal at C^1. Without bypass, low-side pressure would tend to operate at C.*

The liquid evaporates quickly and maintains a definite low-side pressure.

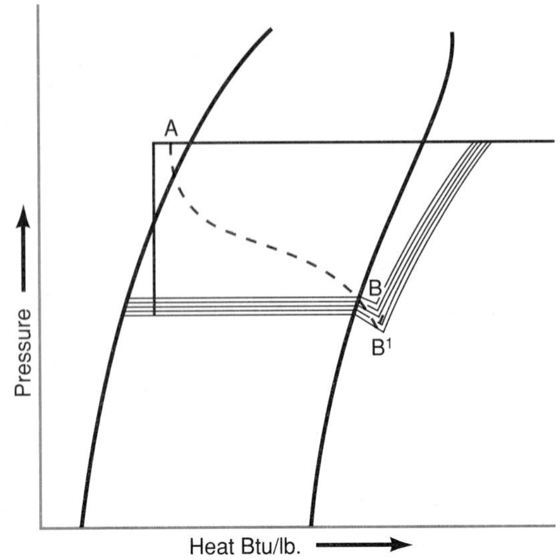

Figure 16-54. *Liquid refrigerant bypass A to B is used to maintain normal low-side pressures. The five parallel evaporator pressure lines and compression lines show changes in the cycle as the low-side pressure changes from cut-in to cut-out pressure.*

16.12 Motor Sizes

The size of an electric motor necessary to drive a compressor in a refrigerating machine may be calculated in two different ways:

- ***Mean effective pressure method.*** The horsepower (hp) of the motor may be calculated by compressor and the mean effective pressure (mep) of the vapor in the compressor.
- ***Heat input method.*** Motor size may be determined by using the amount of heat added to the vapor in the compressor as being the energy taken out of the motor.

16.12.1 Mean Effective Pressure Method

The ***mean effective pressure*** (MEP) of the vapor is the median (average) pressure bearing down on the piston head. It is the pressure to be overcome by the electric motor when driving the compressor.

The MEP of the vapor is determined by a formula that uses:

- Low-side pressure.
- High-side pressure.
- Ratio of specific heat at constant pressure to specific heat at constant volume C_P/C_V for the refrigerant used.

The formula for determining MEP will be found in Chapter 31.

16.12.2 Heat Input Method

The amount of heat (in Btu) added to a vapor when it is compressed by the compressor can be determined from the pressure heat charts. Refer to the R-12 pressure heat chart in Chapter 9, and note that approximately 10 Btu are added to the one pound of gas if the low-side pressure is 10.81 psig (25.5 psia or 176 kPa) and the high-side pressure is 93.2 psig (747 kPa).

It is known that 2545.7 Btu/h is equal to 1 hp and is also equal to 746 W.

Example:

Calculate the hp required to drive the compressor just discussed. Add 10 Btu/lb. Suppose 10 lb. of refrigerant is compressed in two minutes.

Solution:

The Btu rate per hour = 100/2 × 60 = 3000 Btu/h
The hp required (Btu method) 3000/2456 = 1.18
or, in round numbers, 1.2 hp
hp = 1.2 hp

This is the mechanical equivalent of the heat energy put into the vapor. If the compressor friction were zero, this would be the size of the motor necessary to drive the compressor. However, about 50% must be added to this value to allow for compressor friction and motor losses. Therefore, this system would require about:

$1.2 \times .50 = .6$
$1.2 + .6 = 1.8$ or 2 hp

16.12.3 Motor Efficiency

Theoretically, an electric motor should produce 1 hp of mechanical energy for every 746 watts of electrical energy put into it. That is, a 1-hp electric motor on a 120 V circuit should only consume 6.8 A. Such efficiency, however, is not possible because of bearing friction, magnetic eddies, magnetic air gaps, and the power factor of the motor.

The ***efficiency*** of the motor is the mechanical energy delivered at the motor shaft divided by the power input to the motor.

As the size of the motor increases, the efficiency increases. For small domestic motors of approximately 1/6 hp, the efficiency of the motor is only 40% to 60%. Friction losses and air gap losses are constant, even though the size of the motor increases. Large motors have an efficiency of 90% to 95%.

Example:

The motor example used in Section 16.12.2.

Solution:

With 1.8 hp needed, the electrical input to the motor would need to be more. If 1.8 hp is 75% of motor input, the motor input would be:

Total input × .75 = 1.8 hp

$$\text{Total input} = \frac{1.8}{.75} = 2.4 \text{ hp}$$

2.4 × 746 W/hp = 1790 W input

16.13 Condenser Capacities

The calculation of the heat transfer capacity of a condenser is similar to the problem of figuring the capacity of an evaporator. The condenser must remove the heat from the vapor rapidly. In a given unit of time, as much vapor should condense in the condenser as is being pumped into it by the compressor. When this condition is reached, the head pressure will have built up. The temperature rises to the point where the heat removed will equal the heat put into the condenser.

The problem of figuring the capacity of the condenser varies according to the type of condenser being used. Condensers may be divided under the following headings:

Air-cooled:

- Plain tubing.
- Finned tubing:
  - Natural convection.
  - Forced convection.

Water-cooled:

- Tube and shell type.
- Pipe and shell type.
- Tube-within-a-tube type.

The methods of calculating condenser capacities are explained in Sections 16.13.1 and 16.13.2.

16.13.1 Air-Cooled Condenser Capacities

The capacity of an air-cooled condenser may be calculated by:

- Using the total external area of the condenser to compute its heat-dissipating ability.
- Basing computations upon the frontal area of the condenser.

Using the total external area of the condenser for dissipating heat depends upon the following variables:

- External area.
- Temperature difference.
- Time.
- Air velocity.

Using these values, the capacity of an air-cooled condenser varies between 1 and 4 Btu/ft^2/hr./°F. The effect of air velocity is to increase the condenser's capacity as the air speed is increased. The fans drive air through the condenser at speeds of between 400 ft./min. and 1000 ft./min. When an air speed of 400 ft./min. is used, a 2.5 Btu/ft^2/hr./°F value will be found satisfactory. This value will increase up to approximately 4 Btu with a 1000 ft./min. air velocity. The calculation of the area of the condenser is the same as that for a finned evaporator. (See Section 16.4.6.)

Example:

A condenser has 60 ft^2 of surface with a heat removal rate of 2.5 Btu/ft^2/hr./°F. What refrigerant temperature is necessary to dissipate 5000 Btu/h if the room temperature is 75°F?

Formula:

$$\text{Area} \times \text{Btu/ft}^2\text{/hr./°F} \times \text{temp. diff.} = \text{Btu/h}$$

Solution:

$$60 \times 2.5 \times \text{temp. diff.} = 5000$$
$$150 \times \text{temp. diff.} = 5000$$
$$\text{temp. diff.} = \frac{5000}{150}$$
$$\text{temp. diff.} = 33.3\text{°F.}$$

Assuming an ambient temperature of 75°F, the refrigerant temperature = 33.3 + 75 = 108.3°F.

The same problem using a 75 ft^2 condenser:

$$75 \times 2.5 \times \text{temp. diff.} = 5000$$
$$\text{temp. diff.} = \frac{5000}{187.5}$$
$$\text{temp. diff.} = 26.7\text{°F.}$$

Therefore:

The refrigerant temperature = 26.7 + 75 = 101.7°F.

The heat to be removed by the condenser for each pound of vapor is the heat content of the vapor as it leaves the compressor, minus the heat of the liquid at the condensing pressure.

If a condenser is under-capacity, the compressor head pressure will rise proportionally in order to dissipate the required amount of heat. Therefore, condensers of various sizes can be used with the same compressor. If too small a condenser is used, it will result in a decrease in compressor efficiency. There will be an increase in motor load and a decrease in the life of the unit. The examples given illustrate this principle.

When the capacity of the condenser is based upon frontal area, the air being blown through the condenser is removing heat only from the surface which it strikes directly. Also, the turbulent flow against the rear surfaces makes the heat removal from these surfaces negligible.

A single-row tube condenser has a total area 20 times its frontal area. The capacity per square foot of frontal area naturally is greater than the value stated. It is between 6 Btu/ft^2/hr./°F and 10 Btu/ft^2/hr./°F, depending on the air speed. For air cooling, the dry bulb temperature of the room should be used in the calculation.

16.13.2 Water-Cooled Condenser Capacities

The capacity of a water-cooled condenser is high because of the good thermal contact between the cooling medium and the refrigerant. Different types of water-cooled condensers are in common use. See **Figure 16-55.** Capacity will vary with the type used.

The heat transfer varies directly with the amount of water passed through the condenser. If the water flow is fast, more heat will be removed; if the flow is slow, heat removal will be less. At 50 ft./min., water will remove about 185 Btu/ft^2/hr./°F. At 200 ft./min., the water will remove about 330 Btu/ft^2/hr./°F. The heat-removing capacity of these units varies between 30 Btu/ft^2/hr./°F and 50 Btu/ft^2/hr./°F in the smaller machines. For machines of one-ton capacity or more, this value

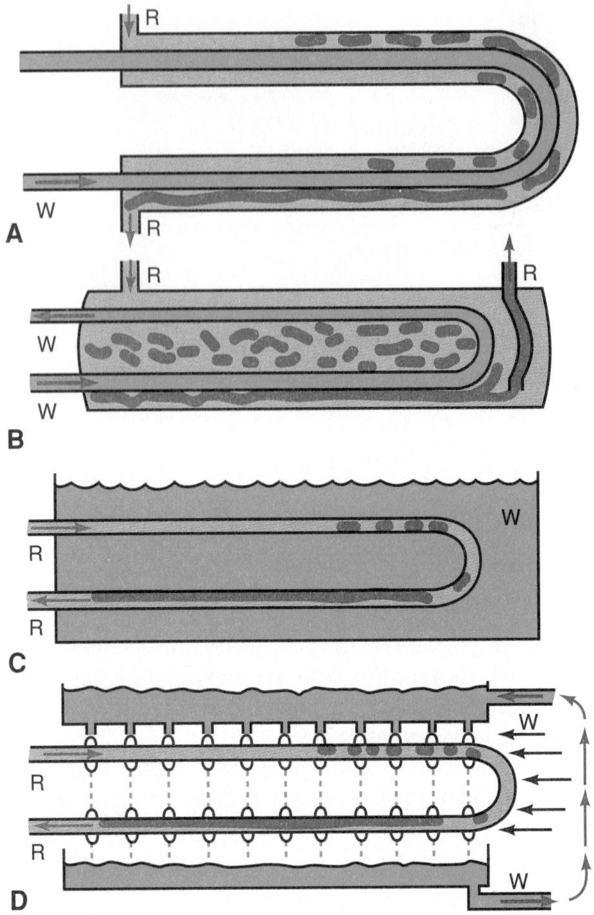

Figure 16-55. *Water-cooled heat exchangers. A—Tube-within-a-tube. B—Shell and tube. C—Tank. D—Baudelot. R—Refrigerant. W—Water.*

may be increased up to 90 Btu/ft²/hr./°F or 330 + 90 = 420 Btu/ft²/hr./°F.

In addition to heat removal by water, the air-cooling surface of the condenser also must be calculated to reach the correct capacity. Include such things as outside area of the shell, or the outside area of the refrigerant tubing in a tube-within-a-tube type.

To determine the temperature difference between the cooling medium (water) and the refrigerant, use the following factors: For refrigerant temperature, use the saturation temperature of the refrigerant at the existing head pressure. For water temperature, take the average between the water-in and water-out temperatures.

Example:

A shell and tube type water-cooled condenser is required to remove 5000 Btu/hour. How much tubing 3/8″ OD must be put into the receiver to remove this heat if the water supply is 70°F and the outlet water is 80°F? How many gallons of water per hour must be circulated?

Consider the refrigerant temperature at 100°F. Assume the heat-removing capacity of the condenser to be 40 Btu/ft²/hr./°F.

Formula:

Tube area = condenser capacity = area (ft²) × temp. diff. °F × Btu rate × time

Area = circumference × length

Area = π D × length

$$\text{Area in square feet} = \frac{\pi \text{D} \times \text{length (in.)}}{144}$$

$$\text{Temp. difference} = 100°\text{F} - \left(\frac{80 - 70}{2}\right)$$
$$= 100 - 75 = 25$$

Btu = area × temperature difference × Btu/ft²/hr./°F × hr.

$$5000 = \frac{\pi \frac{3}{8} \times \text{length (in.)}}{144} \times 25 \times 40 \times 1$$

$$\frac{5000}{25 \times 40} = \frac{\pi \frac{3}{8} \times \text{length (in.)}}{144}$$

$$\frac{5000 \times 144}{25 \times 40} = \pi \frac{3}{8} \times \text{length}$$

$$\text{or length} = \frac{5000 \times 144}{25 \times 40 \times \pi \frac{3}{8}} = \frac{720}{\pi \frac{3}{8}}$$

$$= \frac{720}{3.1416 \times .375} = \frac{1920}{3.1416}$$

$$= 611''$$
$$= 51' \text{ approximately}$$

The amount of water circulated = sp. heat × wt. × temp. diff. = Btu

1 × wt. × 10 = 5000

wt. = 500 lb./hr

1 gal. of water weighs 8 1/3 lb.

$$\frac{500}{8\ 1/3} = 60 \text{ gal. of water/hr.}$$

Cooling towers are a very efficient type of water-cooled condenser. One pound of evaporating water removes about 1050 Btu and cools the remaining water. Ideally, the evaporating water will cool it to the wet bulb temperature. See Chapter 19. However, practically, it cools the remaining water to some temperature above the wet bulb temperature. This can be measured with a thermometer.

About 3% of the water is evaporated and must be made up by using a float valve water feed as a control. The float valve also makes up for run-off water which is also about 3% (used to keep water chemicals to a minimum). The water pump used to circulate this water should be about 1% of the condensing unit horsepower, or about 2% if the water pipes are long (high total head).

16.13.3 Liquid Receiver Sizes

Liquid receivers for a commercial system should be 15% larger than the total liquid volume in the system. This practice is recommended for service operations. It is a safety measure if the refrigerant circuit should become restricted. (The restriction could be a clogged filter or a clogged screen.)

Figure 16-56 shows recommended minimum sizes of receivers based on horsepower capacity of the system. The receivers may have to be larger than this. It depends on the refrigerant, piping lengths, and other factors.

Liquid receivers are a service addition to a system. The systems would operate efficiently without them. However, practical problems of refrigerant reserve and convenient service storage makes them a common part of most systems.

Recommended Liquid Receiver Volumes					
hp	Volume (in^3)	Weight (lbs.) Refrigerant R-12	R-22	R-500	R-502
1/2	150	6.8	6.2	5.9	6.3
3/4	225	10.3	9.3	8.9	9.4
1	300	13.7	12.4	11.9	12.9
1 1/2	450	20.5	18.6	17.9	19.3
2	600	27.4	24.8	23.8	25.8
3	750	35	32	31.8	33.0
5	900	41	37	35.5	38.5
7 1/2	1500	70	64	61.6	66.0

Figure 16-56. *Minimum net recommended liquid receiver volume for four common refrigerants: R-12, R-22, R-500, and R-502. (Reprinted by permission of the American Society of Heating, Refrigerating, and Air-Conditioning Engineers, Atlanta, Georgia)*

16.14 Servicing: Refrigeration Troubleshooting

To locate trouble, you must determine what is going on inside a refrigerating system. The system is sealed, so you use gauges to check the pressure. Thermometers are used to measure evaporator, line, and condenser temperatures. He or she also uses the system sight glass to check the amount of refrigerant and its dryness.

Much of the investigation has to be by logic. Technicians need to know what is supposed to be going on inside a system. Then they must be able to visualize the behavior of the refrigerant and what each part of the system is supposed to do. The pressure-heat diagram provides considerable aid in this area.

The following paragraphs show the effect of some of the more common troubles on the pressure-heat cycle.

16.14.1 Effect of Lack of Refrigerant

If the system is undercharged, each pound of refrigerant does not completely liquefy before it passes through the refrigerant control. This is shown at A in **Figure 16-57.** The result is threefold:

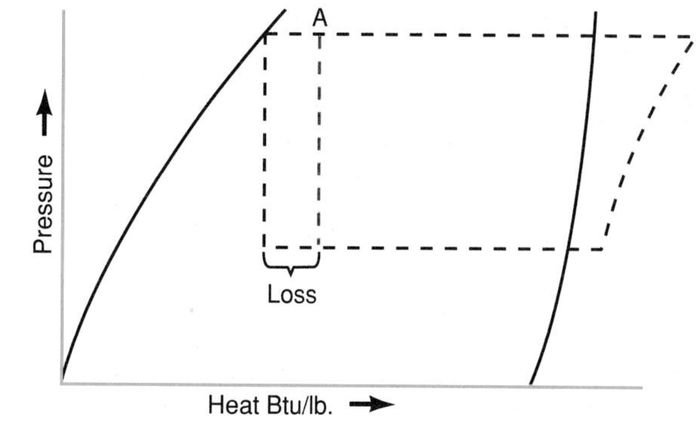

Figure 16-57. *Pressure-heat diagram shows effect of insufficient refrigerant in system. Note loss of effective latent heat. This loss means unit will have to run longer to remove same amount of heat.*

1. Effective latent heat is reduced by the amount indicated by the "loss." Therefore, refrigeration is poor.
2. Some vapor now passes through the refrigerant control, reducing the refrigerant control capacity.
3. This vapor, passing between the needle and seat at a high velocity, increases wear on the refrigerant control needle and seat.

16.14.2 Effect of Air in System

Air in the refrigerating system increases the total head pressure. Total head pressure will equal the refrigerant condensing pressure plus the pressure of the air in the condenser. The refrigerant will have to condense at the higher temperature and pressure.

Because total head pressure is higher, the compressor has to pump the vapor to a higher temperature and pressure. The extra work performed by the compressor is illustrated in **Figure 16-58.** The heat added to do this is a "loss." The cylinder head (especially the exhaust valve) and the top tube of the condenser will be at above-normal condensing temperatures. This may also harm the oil.

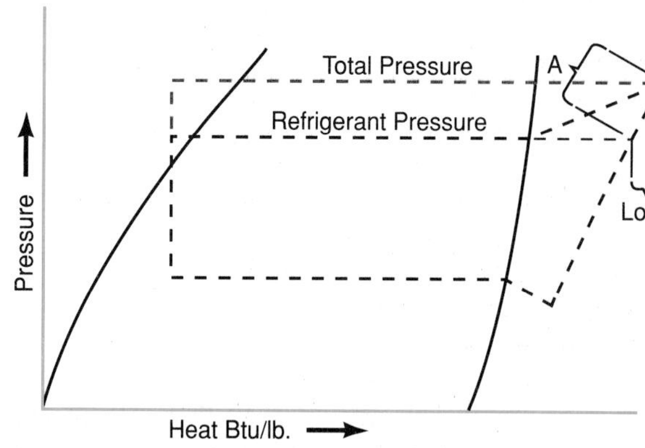

Figure 16-58. *Pressure-heat diagram shows effect of air in system. Point A indicates increase in cylinder head and exhaust valve temperature. The "loss" is heat energy put into the vapor by compressor. The waste is in the extra electrical energy used by the electric motor.*

16.14.3 Effect of Heat Exchanger

After suction line vapor leaves the evaporator, it travels down the suction line into the compressor. During this part of the cycle, the vapor usually warms up somewhat. This is shown at A in **Figure 16-59.**

The low-pressure vapor picks up heat in most cycles. System efficiency can be improved if excess heat is removed during some part of the cycle. The heat exchange is done by putting the suction line in contact with the liquid refrigerant line just before the liquid goes into the refrigerant control.

Figure 16-59 illustrates the removal of heat at B. The result is a gain in effective latent heat. There is a reduction in "flash gas." This will serve to increase the life of the refrigerant control.

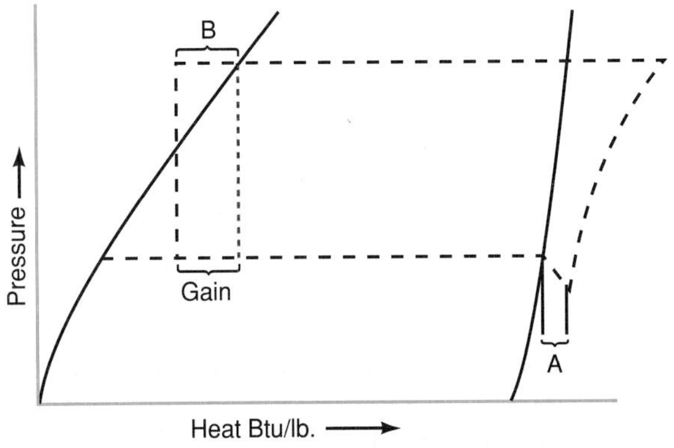

Figure 16-59. *Pressure-heat diagram shows effect of use of heat exchanger. At A, the diagram shows a slight amount of decrease in intake pressure and an increase in temperature. At B, the amount of effective heat gain is shown. In addition, there is a reduction of flash gas, which improves operation of refrigerant control. Heat exchanger also minimizes chance of liquid refrigerant in the suction line reaching the compressor.*

16.14.4 Excessive Condensing Pressure

If the condenser is undersized or dirty (internally or externally), the head pressure and condensing temperature will rise. **Figure 16-60** shows a cycle diagram in which condensing pressure is above normal. The higher temperature will cause the compressor to pump to this higher pressure and temperature. The added heat of compression is shown as a "loss" at A. If the liquid does not subcool to room temperature, there is an added loss in effective latent heat. There is an increase in flash gas. See B in **Figure 16-60.**

16.14.5 Excessive Suction Line Pressure Drop

If the pressure of the vapor going into the compressor decreases, the compressor will pump less weight of vapor per stroke. Therefore, less will be pumped per minute. The less vapor pumped, the lower the capacity of the system.

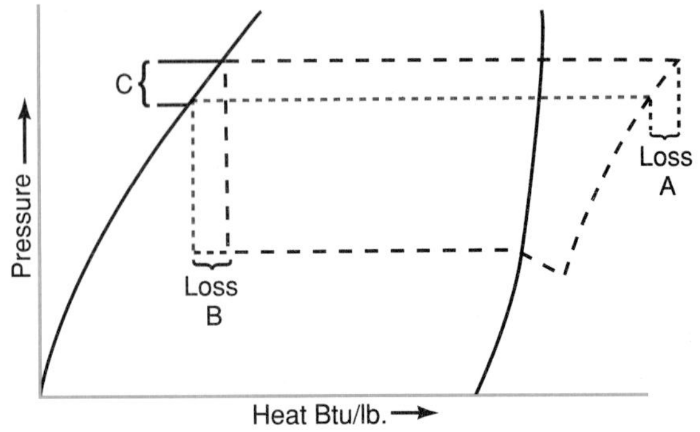

Figure 16-60. *Pressure-heat diagram shows effect of a dirty or undersize condenser or an above-average room temperature. A shows loss due to unnecessary added heat of compression. At B, diagram shows loss in effective latent heat of liquid. Loss due to work done to compress vapor at higher pressure is shown at C.*

Figure 16-61 shows the effect of excessive suction line pressure drop on the cycle. As the volume of the vapor increases, there is more volume per pound of vapor. The vapor picks up heat, increasing its temperature. Therefore, the exhaust valve temperature increases. The condenser must remove more heat from each pound of vapor. When the exhaust valve temperature becomes too high, there is danger that the oil will deteriorate. The effect of a partially clogged filter-drier in the suction line is shown in **Figure 16-62.**

LINES AND PIPING MODULE

16.15 Refrigerant Lines and Piping

The lines on refrigerating machines must be large enough to handle the amount of the liquid or vapor required. These include the liquid lines, suction lines, compressor discharge lines, and "hot gas." To calculate the capacities of these lines, first determine the maximum velocity allowed in the line. Once the amount of liquid or vapor to be handled is known, the internal cross section of the line may be calculated.

Generally speaking, for R-12, R-22, R-500, and R-502, liquid velocities are about 100 ft./min. Suction lines have about 1500 ft./min. Discharge lines and hot gas lines have about 3000 ft./min.

Figure 16-63 is a graph of refrigerant line capacities for R-12. For a 6-ton unit with a −40°F (−40°C) evaporating temperature and a velocity of 2000 ft./min., suction line diameter is 3/8″ (9.5 mm). Liquid line diameter is 3/4″ (19 mm). Discharge line diameter is 7/8″ (22 mm).

Figure 16-64 is a graph for R-22, and **Figure 16-65** is for R-134a.

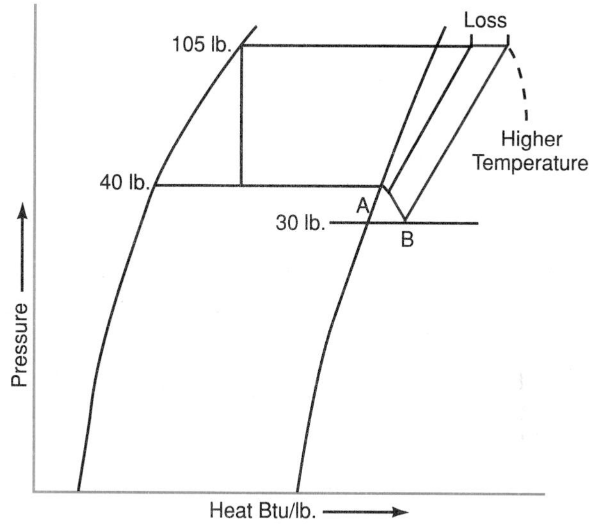

Figure 16-61. *Typical refrigeration cycle is shown when there is an excessive pressure drop and temperature rise in suction line A to B. This causes excessive temperature at the exhaust valve and cylinder head.*

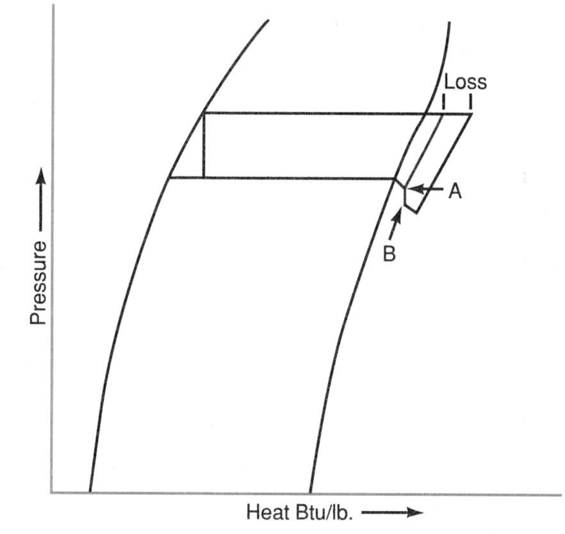

Figure 16-62. *Effect of partially clogged suction filter and/or drier. Drastic pressure drop occurs from A to B.*

16.15.1 Refrigerant (Liquid) Line Capacities

Velocities in a refrigerating unit liquid line vary with the density and viscosity of the liquid. These velocities may vary between 50 ft./min. and 200 ft./min., depending on the refrigerant used. (R-12 should have velocities no greater than 100 ft./min.)

Example:

If 75 in^3 of liquid is used per minute, the internal cross-sectional area of the liquid line required to keep line velocity at 100 ft./min. or below would be:

$$\text{Cross-sectional area} = \frac{\text{volume in in}^3\text{/min.}}{\text{velocity in in./min.}} = \frac{75}{(100 \times 12)}$$

$$\text{Cross-sectional area} = .063\ \text{in}^2$$

$$\text{Area} = \pi\frac{D^2}{4} \quad D^2 = \frac{\text{area} \times 4}{\pi} \quad D = \sqrt{\frac{\text{area} \times 4}{\pi}}$$

$$\text{The inside diameter} = \sqrt{\frac{.063 \times 4}{\pi}} = .28''$$

Use 3/8″ OD tubing (ID = .307).

It is important that refrigerant-carrying lines have sufficient capacity. The cost of increasing the tubing size is small compared to the total cost of the machine. There is no real necessity for calculating tubing size to close limits. A liquid line at least a 10% to 20% oversize is recommended.

If the liquid line is too small, or has too many restrictions, pressure drop may reduce refrigerant flow capacity of the refrigerant control below the capacity of the evaporator. Extremes are revealed by sweating or frosting liquid lines when excessive pressure drops occur. These drops are caused by partially clogged driers and strainers or pinched lines.

The pressure drop in a liquid line carrying R-12 is shown in **Figure 16-66.** Note that if a 1/4″ OD liquid line were used for a 12,000 Btu/h or one-ton load, the pressure drop would be .42 psi/ft. (9.5 kPa/m). A 100′ equivalent-length liquid line would then have a total pressure drop of 42 psi (290 kPa). A 10 m line has a drop of 95 kPa. (***Equivalent length*** is the actual length of the piping, plus the pressure drop in the bends and fittings, as expressed in feet.) Looking ahead, **Figure 16-73** lists values that should be added to the tube length for each fitting or valve.

If normal head pressure were 125 psig (140 psia or 966 kPa), the pressure in the liquid line near the thermostatic expansion valve would be 125 minus 42, or 83 psig (98 psia or 676 kPa). At 83 psi, the boiling temperature is 79°F (26°C) and sweating might occur in a humid 90°F (32°C) room.

In large systems, keep in mind how much refrigerant is stored in the liquid line. This amount also affects pressure based on the weight of the liquid (static head), as shown in **Figure 16-67.**

Bends and fittings increase the resistance to the fluid flow ***(friction).*** There may be as much resistance to flow in a 90° elbow as in five feet of straight tubing of the same size. The friction resulting from bends, fittings, and normal fluid flow through the tubing must be calculated when figuring fluid velocities.

In multiple installations with series-connected evaporators, the liquid line usually varies in diameter as the number of evaporators it feeds changes. See **Figure 16-68.** Note in the diagram that two 1/2″ OD liquid lines

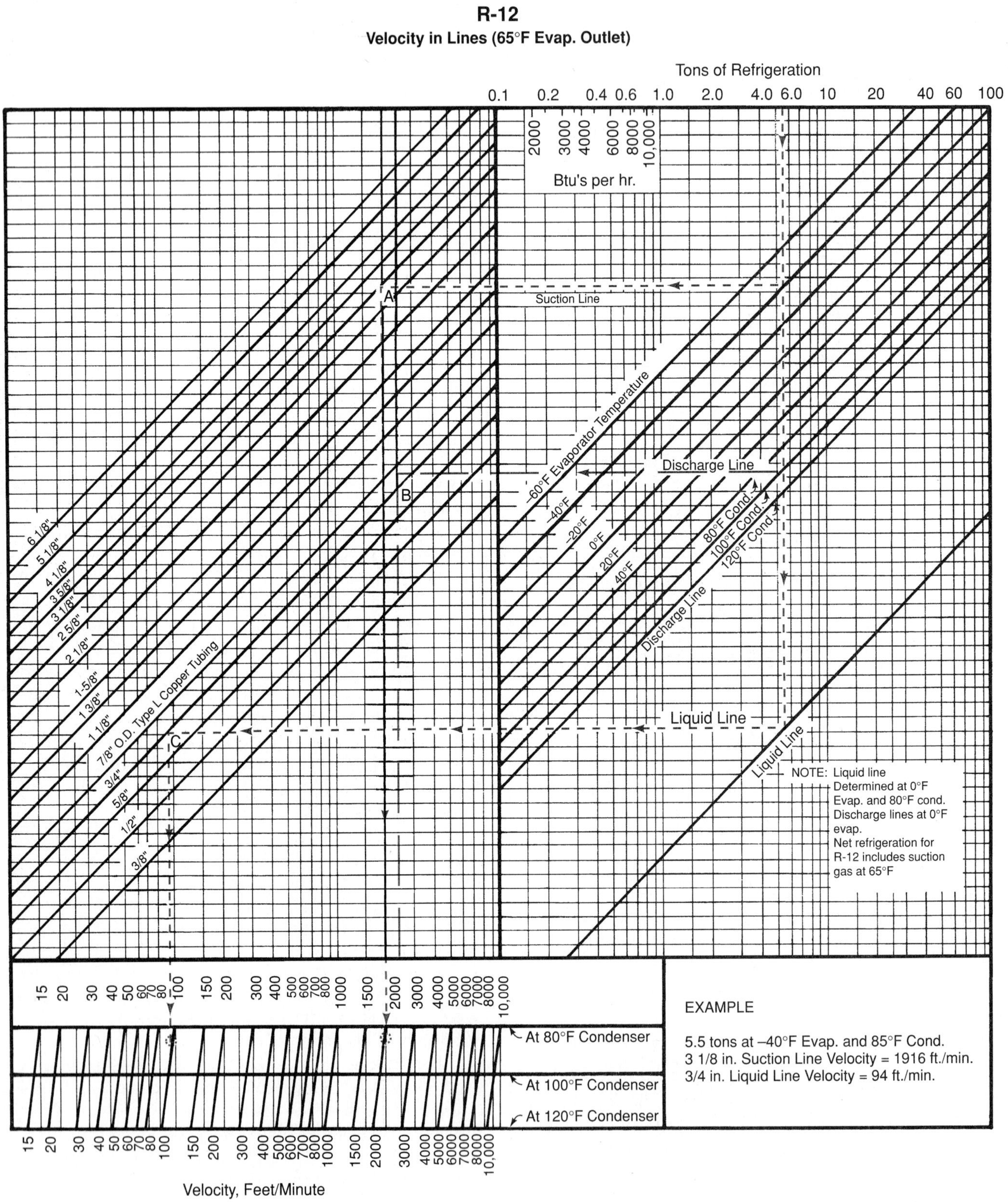

Figure 16-63. *Chart shows liquid and suction line sizes for R-12 systems from 0.1-ton to 100-ton capacity. (DuPont Company)*

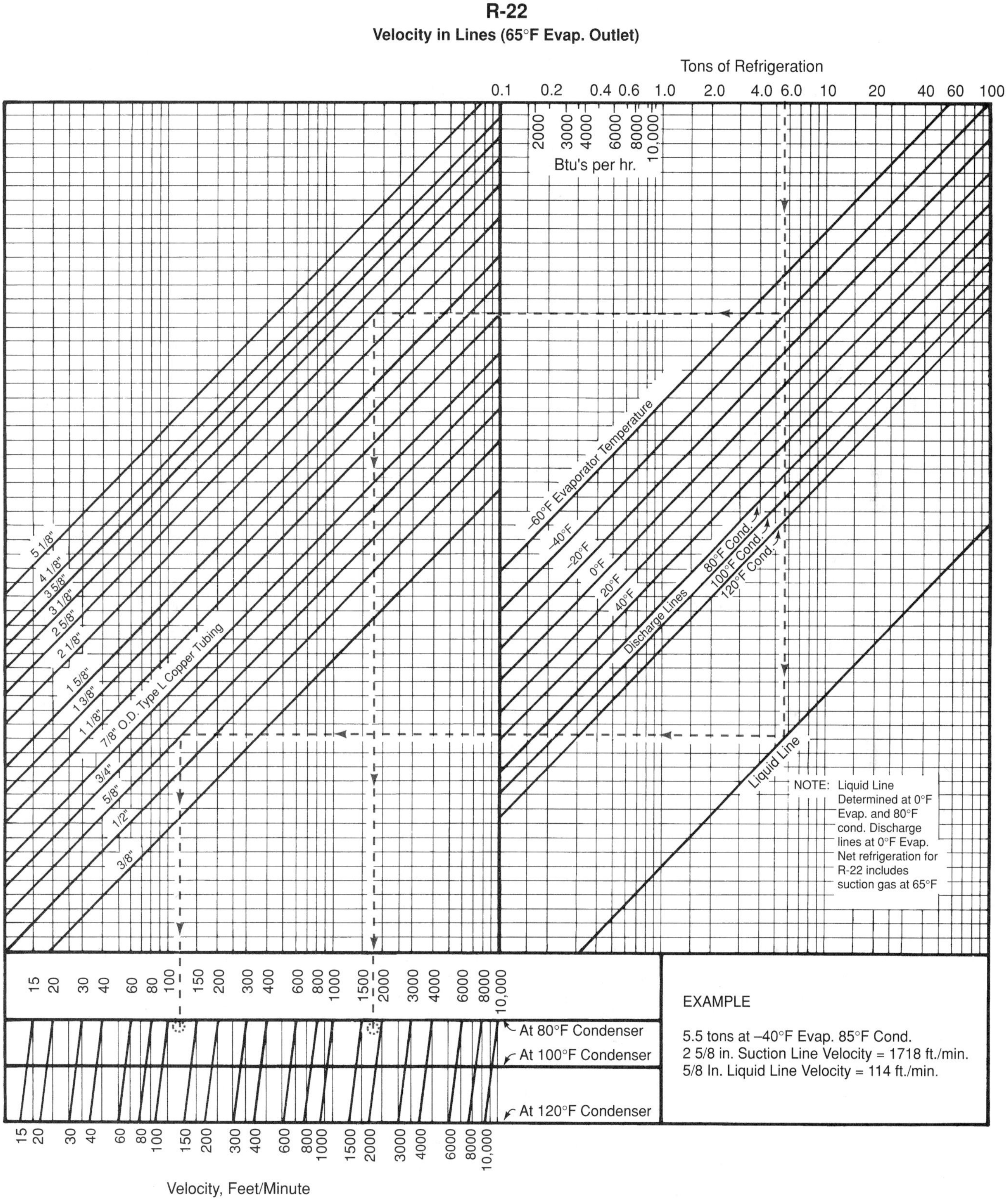

Figure 16-64. *Chart used to find liquid and suction line sizes for systems using R-22. (DuPont Company)*

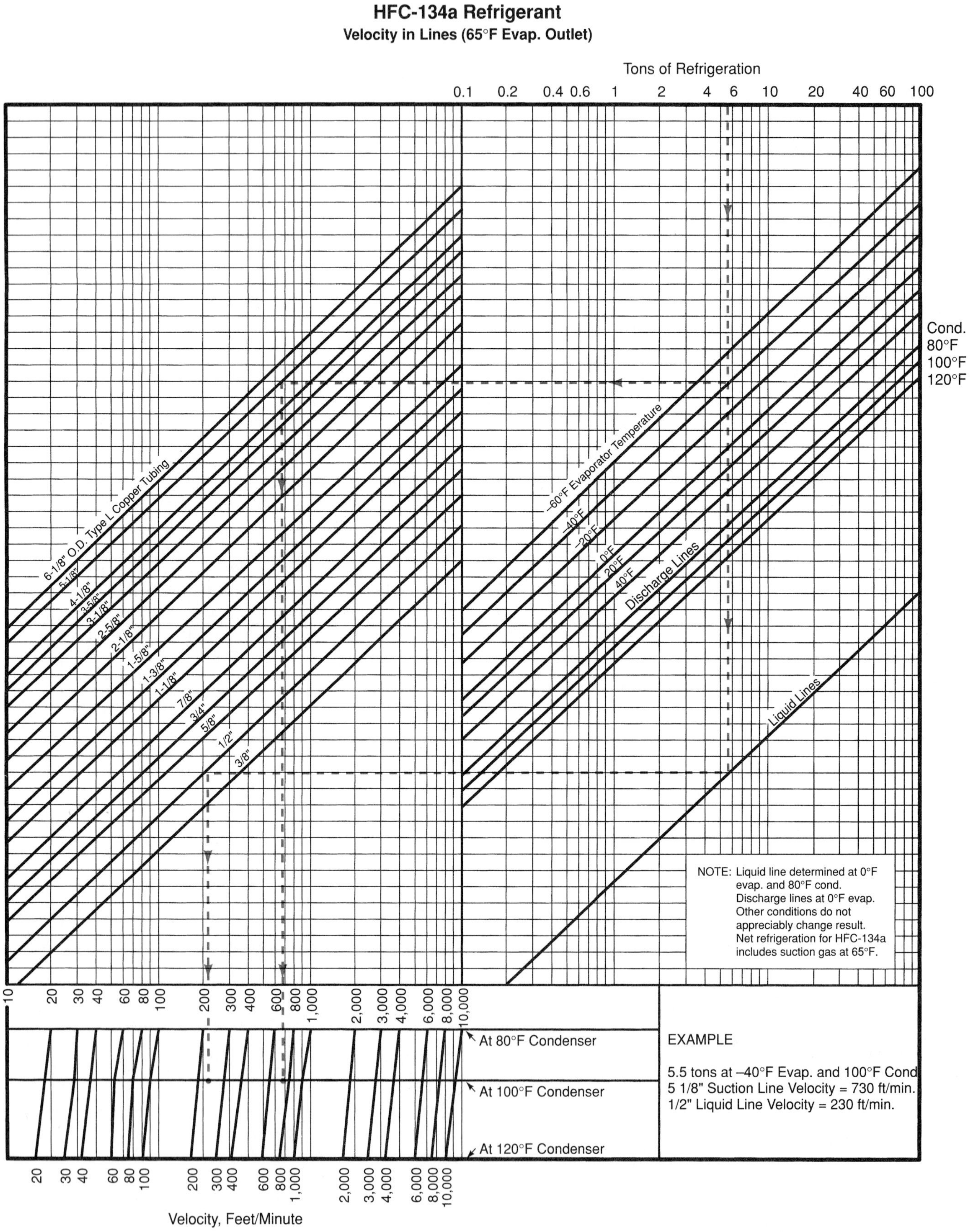

Figure 16-65. *Chart for finding liquid and suction line sizes for systems using R-134a. (DuPont Company)*

	Tube Size OD						
Load Btu/h	1/4"	3/8"	1/2"	5/8"	3/4"	7/8"	1 1/8"
3,000	.035						
6,000	.120	.011					
9,000	.250	.021					
12,000	.420	.036					
18,000		.075	.010				
24,000		.127	.016				
36,000		.260	.033	.012			
48,000		.450	.054	.020	.010		
60,000			.080	.030	.014	.009	
84,000			.150	.054	.025	.015	
120,000			.280	.100	.049	.028	.009
240,000				.350	.160	.095	.029
360,000					.340	.200	.058
480,000						.340	.100

Figure 16-66. *Pressure drop in an R-12 liquid line in psi per foot of tubing. Table is based on size of tube and load in Btu/h. Total pressure drop will be indicated pressure drop per foot multiplied by length in feet.*

¼"	⅜"	½"	⅝"	¾"	⅞"	1 ⅛"
.015	.043	.086	.134	.202	.269	.458

Figure 16-67. *Refrigerant charge in pounds per foot of liquid line. (Dunham-Bush, Inc.)*

do not feed from a 1" OD line. Instead, the cross-sectional areas are added: *Line 1 area + line 2 area = line 3 area*. The wall thickness is ignored.

$$\frac{\pi D_1^2}{4} + \frac{\pi D_2^2}{4} = \frac{\pi D_3^2}{4}$$

$$D_1^2 + D_2^2 = D_3^2$$

$$\sqrt{D_1^2 + D_2^2} = D_3$$

$$\sqrt{\left(\frac{1}{2}\right)^2 + \left(\frac{1}{2}\right)^2} = D_3$$

$$\sqrt{\frac{1}{4} + \frac{1}{4}} = D_3$$

$$\sqrt{\frac{2}{4}} = D_3$$

$$\sqrt{\frac{1}{2}} = D_3$$

$$\sqrt{.5} = D_3$$

$$.70 = D_3$$

Unless it has a considerable ***static head*** (vertical run), the liquid line presents no problem except size. In case of a large static head, pressure at the refrigerant control end of the liquid line must be high enough to maintain pressures above the flashpoint of the refrigerant at the temperature at the refrigerant control. Due to the weight of the liquid refrigerant, the pressure in a vertical liquid line will drop. The amount of the pressure drop per foot of vertical rise is shown in **Figure 16-69.**

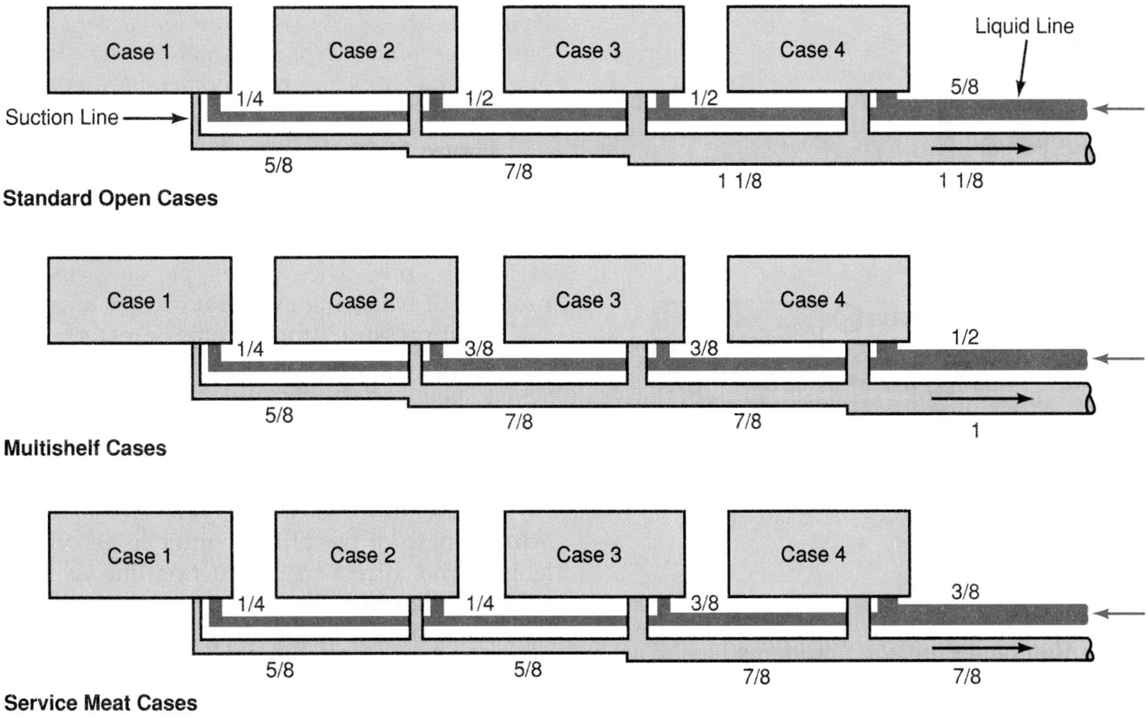

Figure 16-68. *Liquid line sizes. The size of liquid line must increase as the number of evaporators increases. Line sizes shown are for normal temperature refrigeration installations. Note change in diameter of liquid lines.*

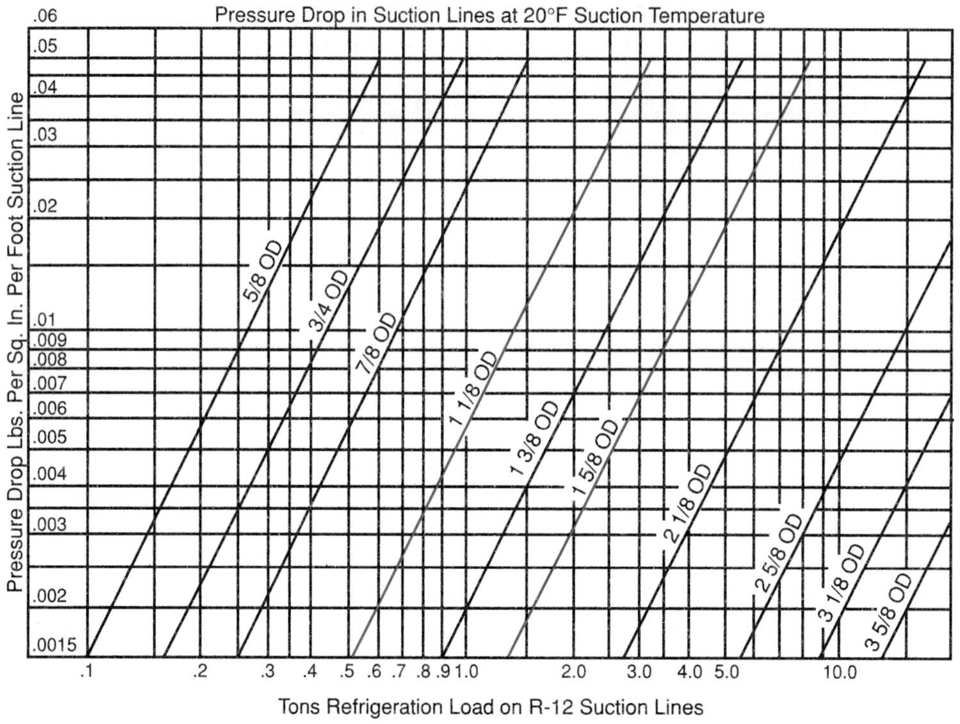

Figure 16-75. *Graph of suction line capacities for R-12 refrigerant. Choose suction line size by using a pressure drop of .02 to .03 psi per foot. Chart is based on 20°F. If lower temperatures are used, larger suction lines are needed, and vice versa.*

3 tons of capacity, depending on the pressure drop. However, the best choice would probably be between .02 psi and .03 psi pressure drop per foot. This pipe should be used for 2 ton to 2 1/2 ton units. If the suction temperatures are lower or higher than 20°F (−7°C), the pressure drops must be corrected. See **Figure 16-76.** Denser vapor should have lower velocities, and vice versa. **Figure 16-77** shows a method of finding suction line, discharge line, and liquid line sizes.

The capacity of the installation is usually known in Btu/h or in tons of refrigeration. Correct suction line size can be estimated by first getting an approximate size from **Figure 16-75.** Next, calculate the equivalent length of pipe from **Figure 16-73.** Then correct for temperature using the factors given in **Figure 16-76.**

Example:

To determine suction line size for a 5-ton system at 0°F, allow for a total 2 psi pressure drop. The suction line has 30′ of straight run, six 90° elbows, one tee, and one valve, at 0°F. Assume that 1 1/8″ OD suction line is to be used:

Suction line	30′
6 elbows × 2	12′
1 tee × 4	4′
1 valve × 4	4′
The equivalent length	50′

If a total of 2 psi pressure drop is desired, then
$2 \div 50 = .04$ psi per foot.

However, this line operates at 0°F instead of 20°F. Therefore, $.04 \div 1.38 = .029$ psi per foot of length. Now, referring to **Figure 16-75,** a 5-ton load with a .029 psi per foot pressure drop needs a 1 5/8″ OD pipe for the suction line.

Correction Factors for Other Suction Temperatures										
Suction Temp.	−40	−30	−20	−10	0	10	20	30	40	50
Correction Factor	2.70	2.28	1.90	1.62	1.38	1.18	1.00	0.88	0.75	0.64

Figure 16-76. *Table of correction values for pressure drops in a suction line. If suction temperatures exceed 20°F, pressure drop is decreased. If temperatures are 0°F, pressure drop increases by 1.38. Equivalent length is to be multiplied by correction factor.*

It is important to always use piping as large as the fittings on the evaporator and compressor. If a compressor suction line connection is 1″ OD, it is advisable to use this size piping. If the liquid receiver liquid line connection is 1/2″ OD, use this size.

16.15.3 Suction Line Problems

The oil return to the compressor by way of the suction line is a critical problem in a refrigeration system.

- Returning oil is a mixture of oil and refrigerant.
- The mixture thickens as the temperature drops.
- The mixture travels mainly along the inside wall of the tubing or piping.
- The mixture travels by gravity and by the action (velocity) of the refrigerant vapor.

	Recommended Refrigerant Line Sizes					
	Compressor Capacity (Btu/h)	Length of Run				
		15 Ft.	25 Ft.	35 Ft.	50 Ft.*	100 Ft.*
		Tube Dia.	Tube Dia.	Tube Dia.	Tube Dia.	Tube Dia.
Suction Line	18,500–20,000	5/8	5/8	5/8		
	20,000–22,000	5/8	5/8	5/8		
	22,000–24,000	5/8	5/8	5/8	3/4	3/4
	24,000–34,000	5/8	5/8	3/4	3/4	3/4
	38,000–40,000	3/4	3/4	3/4	7/8	7/8
	40,000–44,000	3/4	7/8	7/8	7/8	7/8
	44,000–51,000	7/8	7/8	7/8	7/8	7/8
	53,000–66,000	7/8	7/8	7/8	1 1/8	1 1/8
Liquid Line	18,500–20,000	5/16	5/16	5/16		
	20,000–22,000	5/16	5/16	5/16		
	22,000–24,000	5/16	3/8	3/8	3/8	3/8
	24,000–34,000	5/16	3/8	3/8	3/8	3/8
	38,000–40,000	5/16	3/8	3/8	3/8	3/8
	40,000–44,000	3/8	3/8	3/8	3/8	3/8
	44,000–51,000	3/8	3/8	3/8	3/8	3/8
	53,000–66,000	1/2	1/2	1/2	1/2	1/2
Discharge Line	18,500–20,000	5/16	3/8	3/8		
	20,000–22,000	3/8	3/8	3/8		
	22,000–24,000	3/8	3/8	3/8	1/2	1/2
	24,000–34,000	3/8	3/8	1/2	1/2	1/2
	38,000–40,000	3/8	1/2	1/2	1/2	1/2
	40,000–44,000	3/8	1/2	1/2	1/2	1/2
	44,000–51,000	3/8	1/2	1/2	1/2	5/8
	53,000–66,000	1/2	1/2	5/8	5/8	3/4

(1) These recommendations are based on the use of standard refrigeration tubing with .028 or .032 wall thickness.
(2) Line sizes listed are outside tube dimensions.
(3) These suggestions do not include consideration for additional pressure drop due to elbows, valves or reduced joint sizes.
* = Add 3 fluid ounces for each 10 ft. of pipe over 35 ft.

Figure 16-77. *Suction line, discharge line, and liquid line sizes. Selection is based on system capacity, length of pipe, and use of R-12.*

Observe these cautions:

- Keep the mixture as fluid as possible.
- Slope the horizontal suction line downward toward the compressor.
- It is important to leave enough refrigerant ***vapor velocity*** to push the mixture along the pipe.

The velocity in a horizontal line should be at least 500 ft./min. to 750 ft./min. If the refrigerant vapor must flow up a suction line, the velocity in the vertical tubing or pipe must be at least 1000 ft./min. to 1500 ft./min. (It must overcome both gravity and viscosity.)

Be sure to install small U traps at the base of the vertical up-flow suction lines. These traps prevent a large slug of oil returning to the compressor during start-up of system.

The ***viscosity*** of refrigerant oil determines how easily it flows. Viscosity is measured with an instrument called a Saybolt Universal viscosimeter, and is expressed in SSU (Saybolt Seconds Universal) units. See Chapter 31. Oil containing dissolved refrigerant has a lower viscosity (flows easier). As the oil travels in the suction line, it becomes warmer (suction superheat). It also loses some of its dissolved refrigerant. Tests show that the viscosity of the oil actually increases as it travels through the suction line toward the compressor. The longer the suction line, the more careful you must be to provide proper suction line velocities and oil traps. In low-temperature systems, the refrigerant dissolved in the oil is the one main factor that keeps viscosity low enough to allow the return of oil.

Example:

150 SSU oil at −20°F has a viscosity of 100,000 SSU. With R-12 dissolved in it, however, 150 SSU oil has a viscosity of about 50 SSU and it flows with relative ease. As temperatures in the suction line rise, viscosity of the oil increases (as refrigerant content decreases). Then, it finally starts to decrease (as the oil becomes warmer).

Refrigerant vapor velocity will vary in the suction line as the heat load changes. At maximum heat load, the amount of vapor produced will be maximum. Vapor velocities will be high and the oil return will be good. However, as the vapor volume decreases, the compressor ***unloads*** (one or more compressors stop if it is a compound system, or one or more cylinders of a modulated compressor stop pumping). Vapor velocity will drop and oil return will be more difficult.

One solution to suction line problems under varying heat loads is to use a double suction line. One line would have an oil trap and one would not. See **Figure 16-78.** When the system is at full capacity, suction lines A and B will carry the refrigerant vapor at about 1500 ft./min. As load decreases and compressor pumping is reduced, the vapor velocity will slow. The oil trap, C, will fill with oil, closing line B. Now the vapor velocity in suction line A will be high enough to carry the oil back to the compressor. When the system returns to full

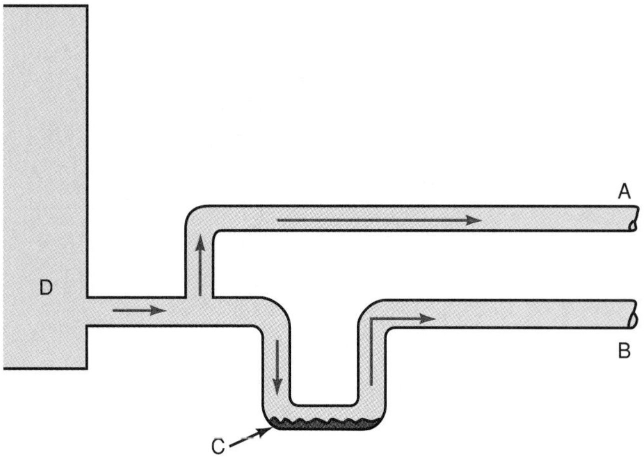

Figure 16-78. *Double suction line. A—Suction line direct to compressor. B—Suction line with oil trap. C—Oil trap. D—Evaporator.*

capacity, the oil at C will be moved back to the compressor by way of line B.

Remember, all horizontal suction lines must slope *toward* the compressor at about 1/4″ per 10′ (6 mm per 3 m) of run. If the suction line *rises* (leaves the compressor vertically for a distance), a trap must be installed at the bottom of the rise. See **Figure 16-79.**

Discharge lines rising from compressors to remote condensers must also have an oil trap if the vertical rise is 8′(2.5 m) or more. The trap keeps oil from returning by gravity flow to fill the space above the exhaust valves of the compressor during the off cycle. This condition could damage the valves when the compressor first starts up. See **Figure 16-80.** Like a suction line, a horizontal discharge line must slope toward the condenser about 1/4″ per 10′ length of pipe.

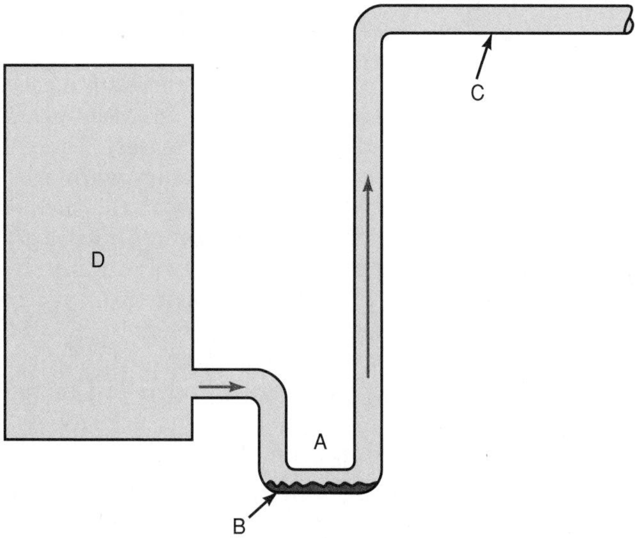

Figure 16-79. *On a suction line with a vertical rise, an oil trap must be installed at the low point of the rise so that oil cannot enter the compressor. A—Oil trap. Note that it is below the compressor suction line outlet. B—Collected oil. C—Horizontal pipe must slope toward the compressor about 1/4″ for every 10′ of pipe. D—Evaporator.*

Systems using compound compressors have an oil balance to maintain. Otherwise, one compressor may collect too much and will then pump oil. Another compressor, too low on oil, may be damaged. **Figure 16-81** shows oil piping that provides equal distribution of oil.

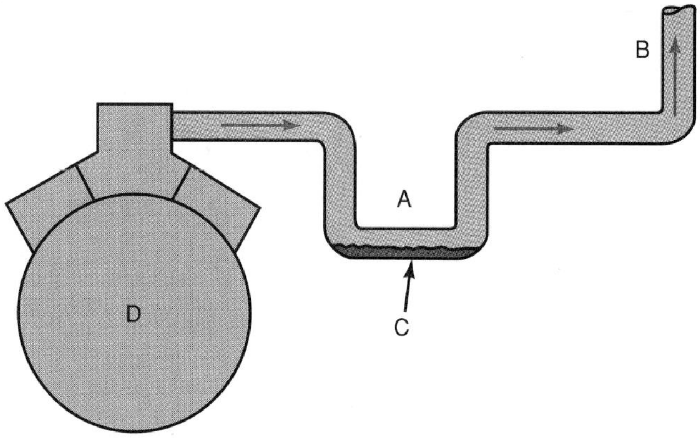

Figure 16-80. *An oil trap installed in the discharge line of a compressor with a remote condenser keeps oil from draining back to exhaust valves of the compressor. A—Oil trap. B—Discharge line. C—Oil collects here during off cycle. D—Compressor.*

16.15.4 Discharge Line Piping

When compressor discharge vapor is piped to a remote condenser, the condenser may become warmer than the compressor during the off-part of the cycle. When this happens, refrigerant vapor may move back from the condenser. It will condense in the head of the reciprocating compressor.

If the compressor discharge valves leak, liquid refrigerant will collect in the cylinders. This could result in the compressor pumping liquid refrigerant on start-up. This condition would reduce lubrication of the pistons and valves. It could even break them. If the compressor valves do not leak, the collection of liquid refrigerant in the cylinder head may cause damage when the compressor starts up. This is due to the dynamic hydraulic pressure on the compressor head and piping. A check valve installed in the discharge line near the condenser will eliminate this potential danger.

16.15.5 Refrigerant Control Capacity

Two popular refrigerant controls are the thermal expansion valve (TEV) and the capillary tube. Both control refrigerant flow to the evaporator from the liquid line.

The size of the TEV's ***orifice*** (opening) is controlled by a needle. It must be carefully calculated. See **Figure 16-82.** TEV orifice size depends on the shape of the opening and the viscosity of the liquid passing through it. The pressure difference as the fluid passes through the orifice is also a factor.

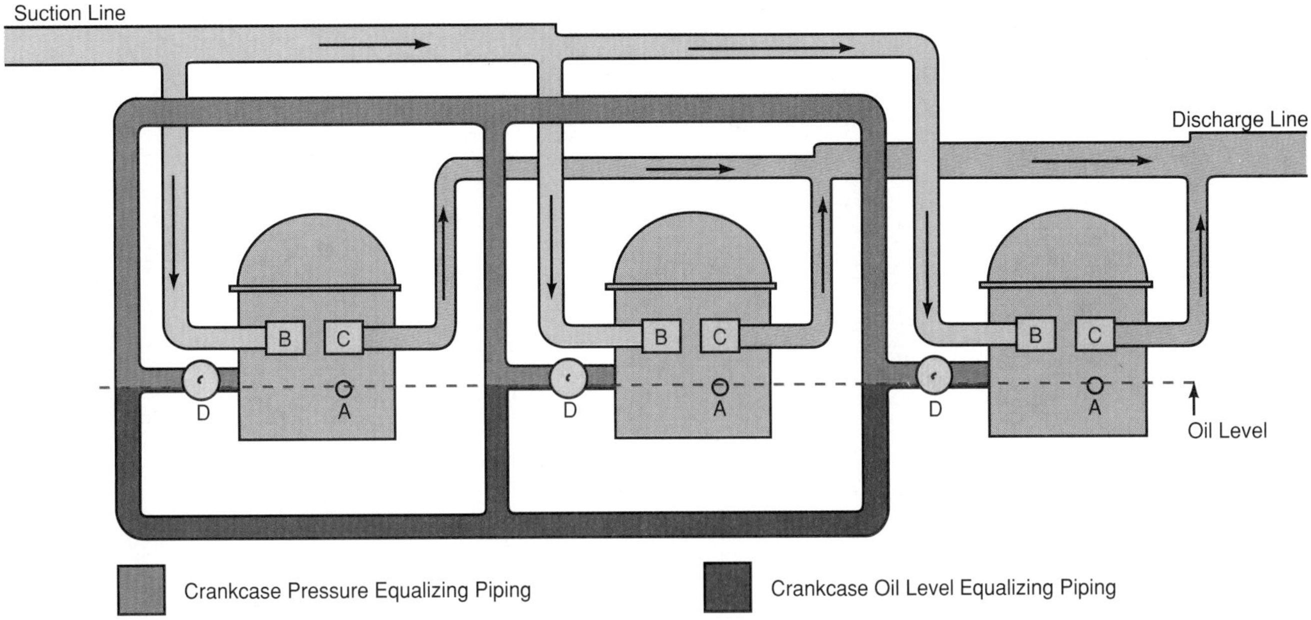

Figure 16-81. *Diagram shows a piping system used to keep an equal amount of oil in each motor compressor crankcase. A—Oil level sight glass (one on each compressor). B—Suction service valve. C—Discharge service valve. D—Shutoff valve (closed only when removing motor compressor).*

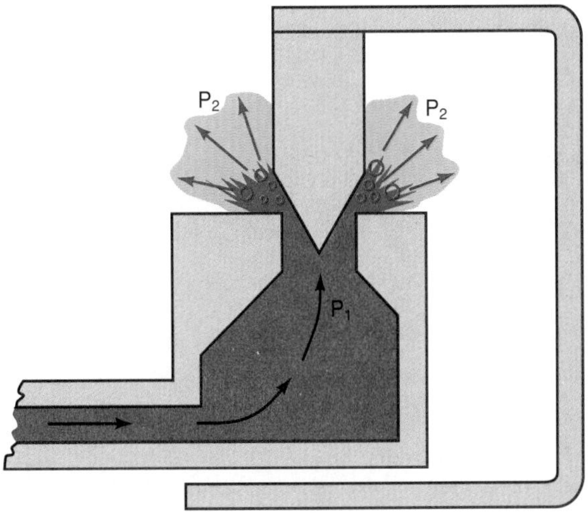

Figure 16-82. *Action of refrigerant as it passes through orifice of an automatic or thermostatic expansion valve. Liquid refrigerant at P_1 (high-side pressure) is forced through orifice and almost at once reaches P_2 (low-side pressure). As pressure changes, some liquid (about 30%) instantly changes into vapor (flashes). This vaporizing action cools remaining liquid to evaporator refrigerant temperatures.*

TEV orifice sizes have become fairly standard, however, for domestic and commercial machinery. Orifice openings of 0.093″ and .156″ have become most popular. If large orifices are needed, multiple installations of expansion valves are used.

If the orifice is *undersize* (too small), not enough refrigerant can pass through the valve. The evaporator will be starved. Also, the evaporator pressure will drop too rapidly.

If the orifice size is *oversize* (too large), the valve will feed too much refrigerant too fast. This will cause a "sweat back" or "frost back" down the suction line. The resulting "hunting" or "searching" action will cause alternate flooding and starving of the evaporator.

The pressure difference is important. As the difference increases, TEV capacity increases. Therefore, if head pressure is high, the valve may feed refrigerant too fast. It may cause a sweat back or frost back. If the pressure is too low, the valve will feed too little refrigerant and the evaporator will starve.

Causes of low pressure may be:

- Head pressure is low.
- Liquid line is too long or has too many bends or fittings.
- Liquid line is too small.

Capillary tube capacity is determined by the pressure difference, length of the tube, and inside diameter of the tube. **Figure 16-83** lists suggested capillary tube sizes for two different temperature applications. It covers low-temperature and medium-to-high temperature evaporators, based on capacity of the system in Btu/h.

16.16 Seasonal Energy Efficiency Ratio (SEER)

The ***SEER*** (seasonal energy efficiency ratio) is a combination of the ***EER*** (energy efficiency ratio) and the COP (coefficient of performance). Both ratings, SEER

Recommended Capillary Tube Length and Diameter					
Compressor Capacity (Btu/h)	Condenser Type	Normal Evaporating Temperature			
		–10 to +5	+5 to +20	+20 to +35	+35 to +50
R-12 Low Temperature					
200–300	Static (Fan)	16' – .026"	10' – .026"		
300–400	Static (Fan)	12' – .026"	12' – .031"		
400–700	Static	12' – .031"	12' – .036"		
	Fan	10' – .031"	10' – .036"		
700–1100	Static	12' – .036"			
	Fan	10' – .036"			
1100–1300	Static	10' – .036"			
	Fan	8' – .036"			
1300–1700	Static	12' – .042"			
	Fan	10' – .042"			
1700–2000	Static	12' – .049"			
	Fan	10' – .042"			
2000–3000	Fan	10' – .054"	15' – .059"		
3000–4000	Fan	10' – .059"	12' – .064"		
4000–4500	Fan	12' – .064"	12' – .070"		
4500–5000	Fan	10' – .070"	12' – .080"		
5000–7000	Fan	10' – .059" (2 pcs.)	12' – .064" (2 pcs.)		
7000–9000	Fan	10' – .064" (2 pcs.)	10' – .070" (2 pcs.)		
9000–12,000	Fan	10' – .070" (2 pcs.)	12' – .080" (2 pcs.)		
12,000–15,000	Fan	10' – .070" (3 pcs.)	12' – .080" (3 pcs.)		
R-22 Low Temperature					
1000–2000	Fan	10'–.036"	12'–.042"		
2000–3000	Fan	12'–.042"	15'–.049"		
3000–4000	Fan	10'–.054"	15'–.059"		
4000–5000	Fan	10'–.064"	15'–.070"		
R-12 Medium and High Temperature					
1400–1600	Fan		12'–.036"	8'–.036"	8'–.042"
1600–1800	Fan		10'–.036"	12'–.042"	
1800–2500	Fan		12'–.042"	12'–.049"	8'–.049"
2500–3500	Fan		10'–.042"	10'–.049"	
3500–4000	Fan		12'–.049"	10'–.054"	
4000–5000	Fan		10'–.054"	10'–.059"	
5000–6000	Fan		12'–.059"	12'–.064"	
6000–7000	Fan		10'–.059"	10'–.064"	12'–.070"
7000–10,000	Fan		12'–.070"	12'–.080"	
			12' – .054" (2 pcs.)	10' – .059" (2 pcs.)	
10,000–13,000	Fan		12' – .059" (2 pcs.)	10' – .064" (2 pcs.)	
13,000–16,000	Fan		12' – .070" (2 pcs.)	10' – .080" (2 pcs.)	
16,000–25,000	Fan		12' – .080" (2 pcs.)	10' – .085" (2 pcs.)	
25,000–40,000	Fan		10' – .070" (4 pcs.)	12' – .080" (4 pcs.)	
40,000–60,000	Fan		10' – .070" (5 pcs.)	12' – .080" (5 pcs.)	

Figure 16-83. *Capillary tube sizing. Length and diameter of the capillary tube is based on the kind of refrigerant, system capacity, type of condenser, and evaporator temperature. These capacities are based on the assumption that not less than 3' of the capillary tube is attached to the suction line (heat exchanger). There is no subcooling of the liquid below ambient temperature. (Tecumseh Products Co.)*

and EER, are used for refrigeration and air conditioning units sold for household use in the United States. This rating was set by the U.S. Department of Commerce. It is indicated on the machine (see **Figure 16-84**). An energy efficiency ratio is used to evaluate an HVAC unit much the same as an mpg (miles per gallon) rating is used to evaluate automobiles. The higher the value, the more efficient the machine.

The EER or energy efficiency ratio is the rated cooling capacity of a unit in Btu/h divided by the electrical power in watts. The EER is calculated as follows:

Divide the cooling in Btu/h by the power input in watts. These numbers are for any given set of rating conditions. This is expressed in Btu/h/W.

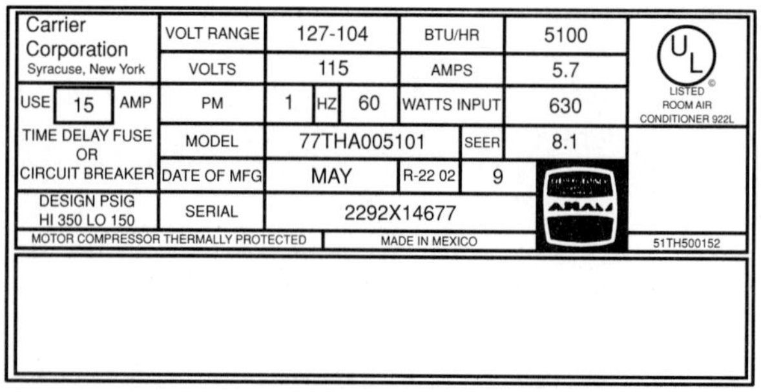

Figure 16-84. *A label like this one, indicating cooling capacity and efficiency, must be carried by all air conditioning and refrigeration units. (Carrier Corp., Subsidiary of United Technologies Corp.)*

The formula is:

$$EER = \frac{\text{Btu/h (cooling) output}}{\text{power input (watts)}}$$

The SEER is similar to the EER. It is computed in the same way, but the wattage is adjusted to be more realistic.

The SEER system uses the *actual* wattage. When a unit starts, it draws locked rotor amps. Therefore, more power is consumed when the unit starts than when it is fully operational. The more frequent the cycles, the more power used. The SEER wattage is adjusted to include the average number of starting and running cycles. The wattage on the name plate, **Figure 16-84,** is the average starting and running watts.

The SEER is the total cooling accomplished by a unit during its normal annual usage, divided by the total electric energy input in watt-hours during this time. The formula is:

$$SEER = \frac{\text{Sum of Btu/h (cooling) outputs at all test conditions}}{\text{Sum of all watt-hour inputs at all test conditions}}$$

For the label shown in **Figure 16-84,** the SEER was calculated as follows:

$SEER = 5100 \text{ Btu/h} \div 630 \text{ watts} = 8.09$

The Coefficient of Performance (COP) of a machine can be found by multiplying the SEER by a factor of 0.293:

$COP = SEER \times 0.293.$

The COP for the machine indicated is:

$COP = 8.09 \times 0.293 = 1.37$

Example:

Find the COP and SEER for a refrigerator that has a cooling capacity of 10,000 Btu/h and requires 800 W of electrical energy:

Solution:

$SEER = 10{,}000 \div 800 = 12.5$
$COP = SEER \times 0.293 = (12.5) \times (0.293) = 3.66$

Example:

What is the SEER of a 2-ton air conditioner that requires 1.5 kW of electrical power?

Solution:

A ton of refrigeration is 12,000 Btu/h. The cooling rate is then 2 × 12,000 = 24,000 Btu/h. The SEER is then:

$SEER = 24{,}000 \div 1500 = 16.$

16.16.1 Converting SEER of Heat Pump Used in Heating Cycle to COP of Same Heat Pump Used in Cooling Cycle

Sometimes it is necessary to convert the SEER of a heating machine to the COP of a cooling machine. Suppose the heat output of the heat pump is 50,000 Btu/h and the electrical input is 4 kW.

$SEER = 50{,}000 \div 4000 = 12.5$
$COP = (SEER \times 0.293) - 1$
$COP = (12.5 \times 0.293) - 1 = 3.66 - 1 = 2.66$

The formulas are more useful with all numbers in Btu/h:

Q_{HOT} = Heat given off at the high temperature
Q_{COLD} = Heat removed from the cool space
$Q_H - Q_C$ = W (Work done by compressor plus heat given off by compressor to refrigerant)
$SEER_Q = Q_H/W = 50{,}000/4000 \times 0.293 = 3.66$
$COP = Q_C/W = 36{,}340/4000 \times 0.293 = 2.66$

The number Q_C was obtained as follows:

$$Q_C = Q_H - W = 50{,}000 - 4000 \text{ (3.415 Btu/h/W)} = 50{,}000 - 13{,}660 = 36{,}340 \text{ Btu/h}$$

Note that W = 13,660 Btu/h

$COP = EER_Q - 1$

For additional information, see Chapter 31.

16.17 Review of Safety

Excessive temperatures and pressures are dangerous. You cannot guess the sizes of piping, condenser, evaporator, and motor to be used on a system. Use manufacturers' specification sheets and recommendations in all cases.

Carefully compute the size of each item according to the methods described in this chapter. Improper sizing of the unit, or any part thereof, may create damaging or dangerous conditions.

Always carefully review any calculations related to pressures, velocities, and capacities. An error might cause too high a pressure or a system failure.

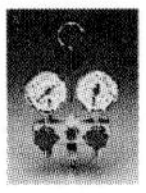

16.18 Test Your Knowledge

Please do not write in this text. Place your answers on a separate sheet of paper.

HEAT LOAD MODULE

1. The evaporator capacity is determined by ______.
   A. cabinet temperature
   B. refrigerant temperature
   C. space allowed for evaporator
   D. All of the above.
2. What does the term "superheat of the vapor in the suction line" mean?
   A. Additional heating of the vapor after it has changed state.
   B. Additional heating of the vapor before it has changed state.
   C. Less heating of the vapor after it has changed state.
   D. None of the above.

3. The specific heat of frozen ice cream is ______ Btu/lb.
   A. 0.45
   B. 0.40
   C. 0.32
   D. 0.50
4. What is a Btu?
   A. The amount of heat required to raise one pound of water 1°F.
   B. The total amount of heat absorbed in a change of state.
   C. The amount of heat necessary to raise one pound of a substance 1°F.
   D. The total amount of heat to be removed from a given area in a specific time.
5. The coefficient of performance is the ______.
   A. ratio of output divided by input
   B. amount of heat absorbed by the system and the energy required to produce it
   C. cooling effect in Btu values compared to the Btu equivalent of the energy to produce it
   D. All of the above.
6. Which of the following descriptions of heat flow is *incorrect?*
   A. Heat travels from food to air fairly slowly.
   B. Heat travels from air to evaporator slowly.
   C. Heat travels into the refrigerant slowly.
   D. Heat travels through the evaporator metal slowly.
7. What is the coefficient of performance as applied to refrigerators?
   A. Ratio of output divided by input.
   B. The amount of heat absorbed by the system.
   C. The amount of energy required.
   D. Cooling effect versus the energy required to produce it.
8. As the refrigerant temperature decreases, the evaporator capacity ______.
   A. decreases
   B. increases
   C. remains the same
   D. fluctuates
9. What is a total heat load?
   A. The total amount of heat to be removed from a given area in a specific time.
   B. The total amount of heat absorbed in a change of state.
   C. The amount of heat required to raise a substance 1°F.
   D. The amount of heat required to raise one pound of water 1°F.
10. What is latent heat?
   A. The total amount of heat to be removed from a given area in a specific time.
   B. The total amount of heat absorbed in a change of state.
   C. The amount of heat required to raise a substance 1°F.
   D. The amount of heat required to raise one pound of water 1°F.

LINES AND PIPING MODULE

11. In an SEER rating, what wattage is used?
   A. Wattage used in starting.
   B. Total running wattage.
   C. Average yearly running wattage.
   D. Average of yearly startings and runnings.
12. In SEER readings, the efficiency of a unit is determined by ______.
   A. dividing the EER by Btu/h
   B. dividing Btu/h output (cooling) by power input (watts)
   C. dividing the power input (watts) by the Btu/h output (cooling)
   D. multiplying the power input (watts) by the Btu/h output (cooling)
13. The velocity of the refrigerant in the liquid line varies with ______ of the liquid.
   A. density
   B. viscosity
   C. Both A and B.
   D. None of the above.
14. What is the difference between SEER and EER?
   A. There is no difference.
   B. EER is more accurate.
   C. SEER is more accurate.
   D. .05% per watt.
15. Where is the oil trap installed on the vertical rise of a suction line?
   A. The high side of the rise.
   B. The low point of the rise.
   C. At mid-point.
   D. Any of the above.
16. The higher the SEER rating, ______.
   A. the more efficient the unit
   B. the less efficient the unit
   C. the larger the unit
   D. None of the above.
17. SEERs are used to designate the ______.
   A. size of the compressor
   B. size of the unit
   C. lowest temperature that the unit will produce
   D. None of the above.
18. In a multiple four-evaporator liquid line installation, flash gas goes to ______.
   A. the nearest evaporator
   B. the most remote evaporator
   C. all evaporators equally
   D. all evaporators in proportion to the distance of each
19. Liquid line and suction line sizes may best be determined by the use of ______.
   A. charts
   B. compound gauge readings
   C. ambient temperature
   D. temperature that will be maintained in a cooled area
20. Horizontal suction lines should slope toward the compressor at approximately ______.
   A. 1/2″ for every 10′
   B. 1/4″ for every 10′
   C. 1/4″ for every 8′
   D. 3/4″ for every 8′

Chapter 17

ABSORPTION SYSTEMS—PRINCIPLES AND APPLICATIONS

Key Words:

absorption
adsorption
ammonia
carbon monoxide
continuous absorption system
effectiveness
energy efficiency
fuse plug
generator
gravity flow
hot gas method
intermittent absorption system
lithium bromide

Learning Objectives:

After studying this chapter, you will be able to:

- Explain the difference between absorption and compression refrigeration systems.
- Explain and describe the operation of each type of absorption system.
- Demonstrate procedures for servicing absorption systems.
- **Follow approved safety procedures.**

Previous chapters have focused on vapor-*compression* systems consisting of a cooling and refrigeration system. The *absorption* system is different from the compression system. It uses heat energy instead of mechanical energy. This heat energy is used to create the conditions necessary to complete a refrigeration cycle. Vapor compression systems use a compressor and a refrigerant to create a cooling effect. Absorption cooling uses a chemical process to change low-temperature, low-pressure vapor into a high-pressure vapor. The refrigerants most commonly used in absorption systems are ammonia and water. The absorption system may use any number of heat sources. These include natural gas, LP (liquefied petroleum) gas, kerosene, steam, or electricity (resistance heating).

An absorption system has few moving parts. Smaller units have moving parts only in the heat source valves and controls. Some larger units also use circulating pumps and fans.

17.1 The Absorption System

The condenser, receiver, and evaporator (cooling coil) are similar to those in a compression system. The compressor, however, is replaced by a heater and generator. Systems shown have been simplified by leaving out various controls. These will be covered later. **Figure 17-1** illustrates a basic absorption system of the liquid absorbent type. It uses a water-cooled condenser.

Figure 17-2 illustrates the fundamentals of a basic absorption system. This diagram shows the solid absorbent type.

17.2 Types of Absorption Systems

Absorption systems are based on combinations of substances which have an unusual property: One substance will absorb the other without any chemical action taking place. It will absorb the other substance when cool and release it when heated. If the substance is a solid, the process is sometimes called ***adsorption.*** If the substance is a liquid, the process is called ***absorption.***

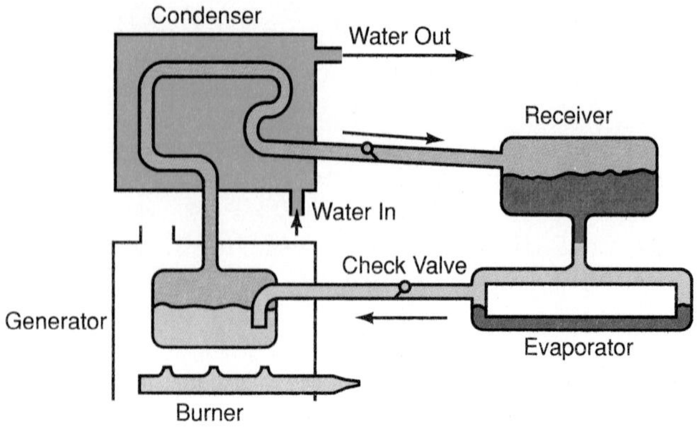

Figure 17-1. *Simple liquid absorbent refrigerating unit.*

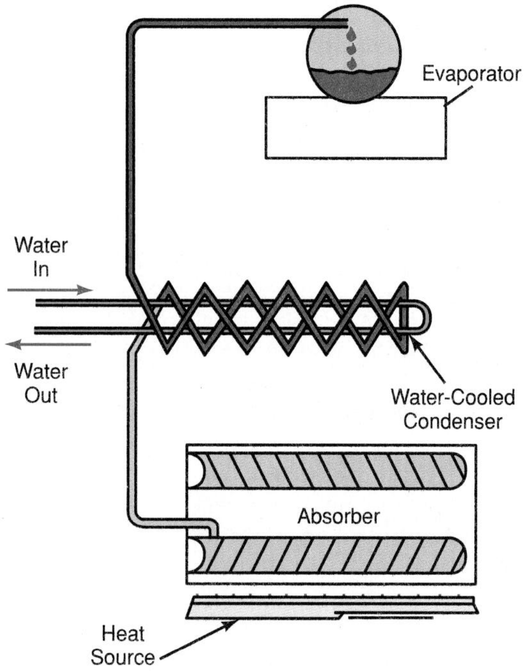

Figure 17-2. *Elementary solid absorbent refrigeration unit. Note water-cooled condenser.*

There are two types of absorption refrigerators. One uses a solid adsorbent material. The other uses a liquid absorbent.

Absorption systems are further classified as:

- Intermittent systems.
- Continuous systems.

Absorption systems have had several applications:

- Domestic.
- Recreational vehicles.
- Hotel rooms.
- Industrial.
- Air conditioning.

Absorption systems are also identified by heat source:

- Kerosene.
- Natural gas.
- Steam.
- Electrical heat.
- Solar energy.

Some absorption units used in family trailers and mobile homes may be heated electrically. Others may use LP gas as fuel.

17.3 Principle of the Solid Adsorption System

Solid adsorption systems operate on principles discovered in 1824 by the British scientist Michael Faraday. Through experiments, he succeeded in liquefying ***ammonia.*** Up to that time, scientists had believed ammonia to be a "fixed" gas. It was considered impossible to change it to either a solid or a liquid.

He exposed the ammonia vapor to a powder, silver chloride. When the silver chloride had taken all the vapor it could adsorb, he applied heat. This resulted in the formation of a liquid. However, when the heat was removed, he discovered that the liquid soon began to "boil." It vaporized, drawing heat from its surroundings. The present-day adsorption system uses this same phenomenon. Faraday's experiment is described in Chapter 3 and **Figure 3-21.**

17.4 Efficiency of Absorption Systems

In absorption cooling systems, performance is evaluated in two ways:

- ***Energy efficiency.*** This is the cooling effect produced divided by the heat energy supplied to the absorber.
- ***Effectiveness.*** This is the cooling effect produced divided by the work equivalent to the heat supplied to the absorber.

The second evaluation helps in comparing absorption systems with vapor compression systems. (In compression systems, the input energy is work, not heat.)

17.5 Principle of the Intermittent Absorption System

For localities having neither gas nor electricity as a power source, the refrigerator cycle known as the Superfex and Trukold cycle is convenient. The cycle used in this ***intermittent absorption system*** is similar to the Faraday principle, but has some different features.

In **Figure 17-3,** ammonia is mixed with water in a sealed tank or generator. Underneath, a kerosene burner heats it. Heat from the burner drives the ammonia, in vapor form, out of the mixture. This vapor is forced up the pipe and through a condenser. This is immersed in water from a tank on top of the refrigerator.

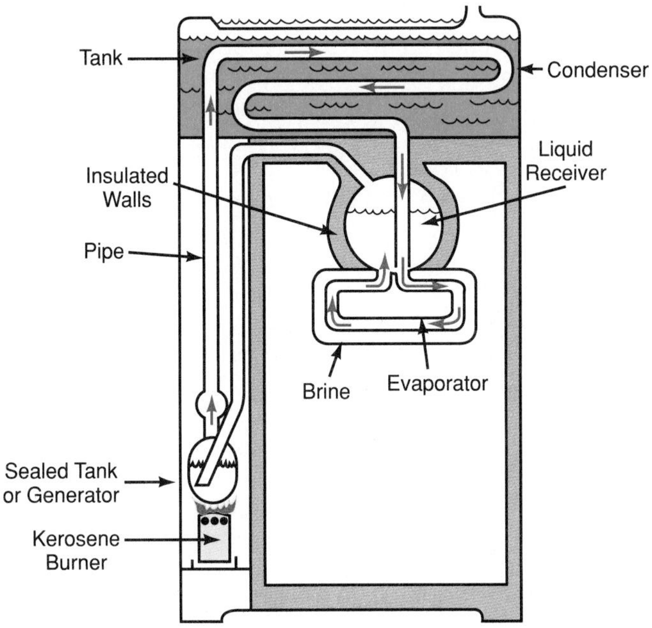

Figure 17-3. *Generation, or heating interval, in typical intermittent-type absorption refrigerator.*

The water has a cooling effect. This causes the ammonia vapor to return to a liquid (condense) at the high generating pressure. This liquid ammonia drops through a pipe into the liquid receiver. From here it passes to the evaporator. The evaporator is surrounded by a brine. The liquid receiver is insulated to prevent this container from overcooling the food compartment by acting as the evaporator.

The process continues for a short time until the kerosene is used up and the burner goes out. The absorber cools to room temperature. Meanwhile, the ammonia evaporates at a low temperature in the evaporator. This occurs because, as the generator cools, it tends to reabsorb the ammonia vapor. In turn, this reduces the pressure and permits the liquid ammonia in the evaporator to boil at low temperature. This evaporation causes the cooling effect or *refrigeration* required to preserve the food compartment contents.

In other words: Heat from the burner drives the ammonia from the generator into the evaporator. This requires only a short time. The ammonia in the evaporator vaporizes and passes back to the generator slowly over a period of 24 to 36 hours. The vaporization of the ammonia in the evaporator produces a refrigerating effect.

A depression in the top of the condenser tank may be filled with water. This is done for additional efficiency in hot climates, or for handling extra large loads. The water will evaporate rapidly and aid the cooling of the condenser.

Absorption mechanisms are provided with a fuse plug. The ***fuse plug*** *releases the charge from the mechanism if the temperature of the unit becomes excessive.* (Excessive temperatures would be 175°F to 200°F [79°C to 93°C], as would be encountered in a fire.) *This would prevent the mechanism from exploding.*

17.6 Principle of the Continuous Absorption System

The absorption system uses ammonia, water, and hydrogen. When it provides refrigeration constantly, it is called a ***continuous absorption system.*** A continuous refrigerating cycle operates automatically through the use of automatic controls.

Many companies have variations of the basic system. However, the principle of operation remains the same. The burner is lighted and its heat applied to the ***generator*** (shown at 1 in **Figure 17-4**). Ammonia vapor is then released from the solution. This hot vapor passes upward through the percolator tube at 2. This solution is carried off to the upper level of the separator at 3.

Most of the liquid solution settles in the bottom of the separator and flows into the absorber. The hot ammonia vapor is light. It rises to the top of the tube, marked 4, into the condenser. The hot ammonia vapor

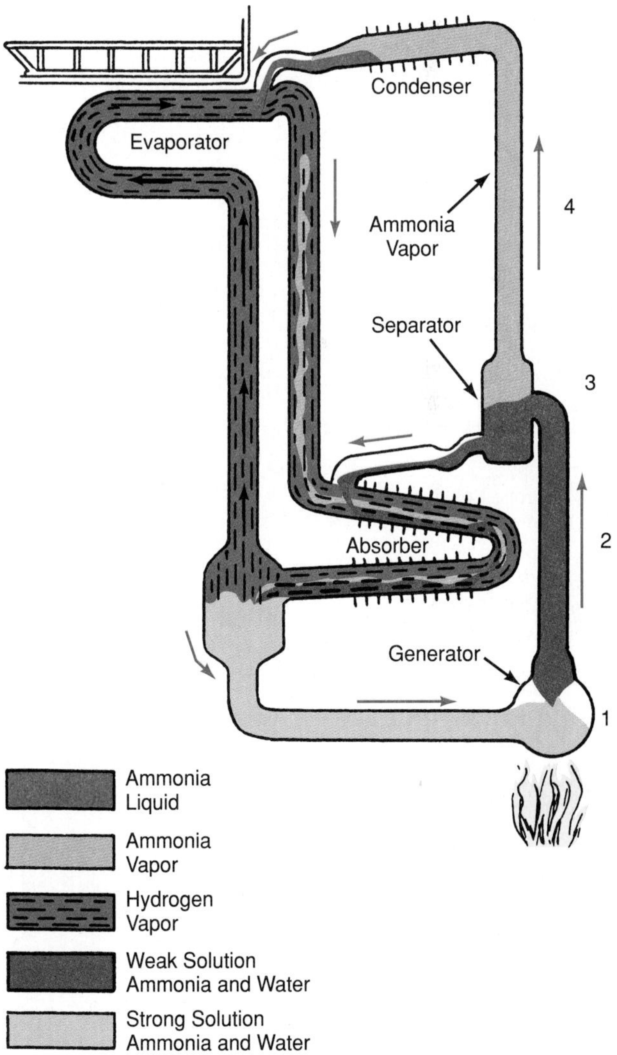

Figure 17-4. *Air-cooled continuous refrigeration cycle. Note water circuit, ammonia flow, and hydrogen circuit.*

then condenses into a liquid. The ammonia is now in a pure state and it flows by gravity into the evaporator.

The ammonia flows through the liquid ammonia tube and spills into the evaporator. (This is because a liquid will always seek its own level.) In the evaporator, it forms large shallow pools on a series of horizontal baffle plates.

The large amounts of hydrogen gas fed to the evaporator permit the liquid ammonia to evaporate. This evaporation occurs at a low pressure and at a low temperature (Dalton's principle). During this evaporation process, the ammonia absorbs heat from the food compartment of the refrigerator. It causes the water in the ice cube containers to freeze. The more hydrogen and less ammonia, the lower the temperature. The evaporated ammonia mixes with the hydrogen gas.

Meanwhile, a weak solution of ammonia and water flows by gravity from the separator, at 3. It flows down to the top of the absorber. *Note: A "weak" solution has little absorbed ammonia vapor. A "strong" solution is one in which a great deal of ammonia vapor has been absorbed.* At the top of the absorber, the solution meets the mixture of hydrogen gas and ammonia vapor coming from the evaporator. The weak and fairly cool solution absorbs the ammonia vapor. The hydrogen gas is left free since it will not mix with water. Since the hydrogen is also very light, it now rises to the top of the absorber. From there, it returns to the evaporator.

The absorber has fins for air cooling. Cooling the weak solution helps it to reabsorb ammonia gas from the ammonia vapor-hydrogen gas mixture. As the water reabsorbs the ammonia vapor, considerable heat is liberated. The air-cooled fins remove this heat to permit refrigeration to continue. The liquid ammonia and water mixture flows back to the generator. There, its cycle begins again.

The apparatus is a welded assembly. There are no moving parts to wear out or get out of adjustment. The total pressure throughout the cycle is about 400 psig (415 psia or 2864 kPa). This is at a room (ambient) temperature of 100°F (38°C). Therefore, construction must be rugged to ensure a long life.

To produce a temperature of 0°F (−18°C) in the evaporator, the ammonia must boil at 15.7 psig (30.4 psia or 209.8 kPa). This means the hydrogen must make up the remainder of the pressure. (This would be 384.3 psig [2651.7 kPa], if the total pressure is 400 psig.) This refrigerator is considered to be unique among domestic refrigerators.

17.6.1 Continuous Absorption System with Pump

The continuous absorption refrigerating system with a pump, **Figure 17-5,** uses ammonia as the refrigerant. It uses an aqueous ammonia solution as the

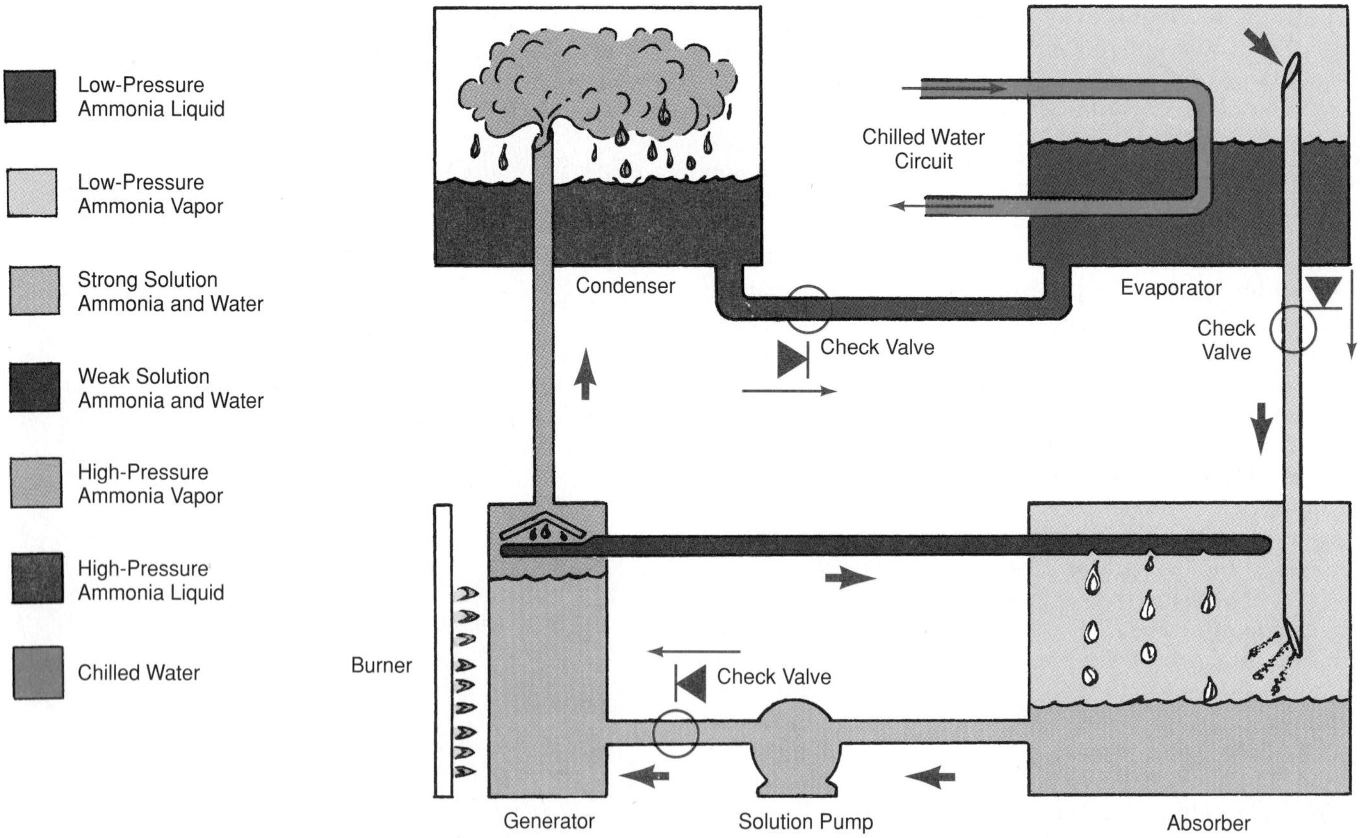

Figure 17-5. *Continuous absorption system uses pump to maintain pressure difference between low-pressure side and high-pressure side of system. Same pump transfers strong-in-ammonia, weak-in-water solution. (Robur Corporation)*

absorbent. Any means of heating can be used. However, natural gas, steam, or LP gas are the most popular.

The system operates under two pressures. The high-side pressure is from 200 psig to 300 psig (215 psia to 315 psia or 1484 kPa to 2174 kPa). The low-side pressure is 40 psig to 60 psig (55 psia to 75 psia or 380 kPa to 518 kPa). High and low sides are separated by check valves, liquid traps, a pump, or other controlling devices. The operational system can be divided into four sections including generator, condenser, evaporator, and absorber.

The generator in **Figure 17-5** is heated by a vertical burner. Heat causes the liquid to boil and the ammonia in the liquid turns into vapor. The vapor will rise up through the tube to the air-cooled condenser. In the condenser, heat from the vapor is removed by cooler air passing across it. The vapor will condense into a liquid, which then acts as the refrigerant.

The liquid refrigerant now passes, at a high pressure, to the evaporator. In the evaporator, water carrying heat from the cooled area passes through tubes. The heat from the water tubes is transferred to the refrigerant liquid. The water in the tubes returns to the area that needs to be cooled. The water is at a low temperature so it can absorb heat from the area that is to be cooled.

Heat that the refrigerant has absorbed from the chilled water circuit causes it to boil and turn into a vapor. This vapor refrigerant is drawn back to the solution-cooled absorber. From there, the heat is sent to the outside air.

The liquid refrigerant is then pumped back (preheated) by the solution pump to the generator. There, the procedure is repeated.

17.6.2 Cooling Process

Absorption refrigeration is widely used in recreational vehicles and residential applications where electricity usage is important. The absorption system is unique in refrigeration, since it involves no moving parts and is virtually noiseless. A typical unit is shown in **Figure 17-6.** Continuous absorption types of refrigerators have four main sections. They are the generator (boiler), shown as A; condenser, D; evaporator, E; and absorber, J. The four sections are connected by steel tubes. The entire system is welded together. The necessary heat for generation is applied at the generator (boiler). The heat can be provided by either a gas burner or an electric heating element.

The system is charged with ammonia, water, and hydrogen. This combined solution is at a pressure that will allow the ammonia to condense at room temperature.

When the unit begins operation, some of the ammonia is in a rich solution with water in the vessel. Heat is applied to the boiler, A, raising the solution temperature to 350°F (177°C). Some of the ammonia rises through the vapor pump. The liquid falls back through the boiler and liquid heat exchanger. It then flows back to the vessel, as a weak solution, through the absorber, J.

When leaving the vapor pump, B, the ammonia vapor is about 300°F (149°C). The ammonia gas is mixed with steam and the water is condensed out of the solution in the rectifier, C.

The pure ammonia vapor then flows through the condenser, D, at room temperature. After the ammonia condenses, the liquid falls in the precooler, I. As the liquid ammonia reaches the evaporator, E, the solution begins to evaporate into the hydrogen. This cools the freezer section to between −24°F and 0°F (−31°C and −18°C).

The hydrogen-ammonia mixture then drops back down through the return pipe. It enters the reservoir to begin another cycle.

Note that the entire cycle is carried out entirely by ***gravity flow*** of the refrigerant. *It is important that the unit remain in a level, upright position. The heat generated in the absorber must be removed. The heat removed by the condenser must be carried away by the surrounding atmosphere.*

17.6.3 Automatic Defrosting

Automatic defrosting on domestic continuous absorption units can be done by the ***hot gas method.*** Hot gas is brought from the generator directly to the fresh food evaporator. (It is not brought to the freezer evaporator.) This hot gas melts the ice on the evaporator fins. Defrost water runs into a drip tray. The operation is controlled by a siphon tube in the boiler (generator). See **Figure 17-7.**

The bypass pipe outlet from the siphon chamber is closed during normal operation. This is caused by a strong ammonia-water liquid solution, as shown in A. During a normal cycle, the solution collects slowly in the chamber. This continues until it reaches the siphon tube outlet, about every 15 to 25 hours. See B. The siphon action then empties the siphon system of its liquid, C. Hot gas is allowed to go directly from the generator to the evaporator, as shown in D. This circulation will continue for about 30 minutes. The solution then fills the siphon system again and covers the bypass pipe outlet, E. Hot gas circulation repeats when siphon chamber liquid level rises enough to repeat the siphon action.

17.6.4 Continuous Absorption Systems Construction

There are several types of continuous absorption systems. **Figure 17-8** illustrates a typical two-door domestic refrigerator with a freezer compartment. This unit uses LP gas or electricity as its source of heat. The internal system with ice formation in the freezer is shown in **Figure 17-9.**

Some units are heated with fossil fuels. Some are heated with electricity. Either fuel gas or electricity can be used to heat the unit in **Figure 17-10.** Electricity may be either 12 V (battery) or 120 V (house current).

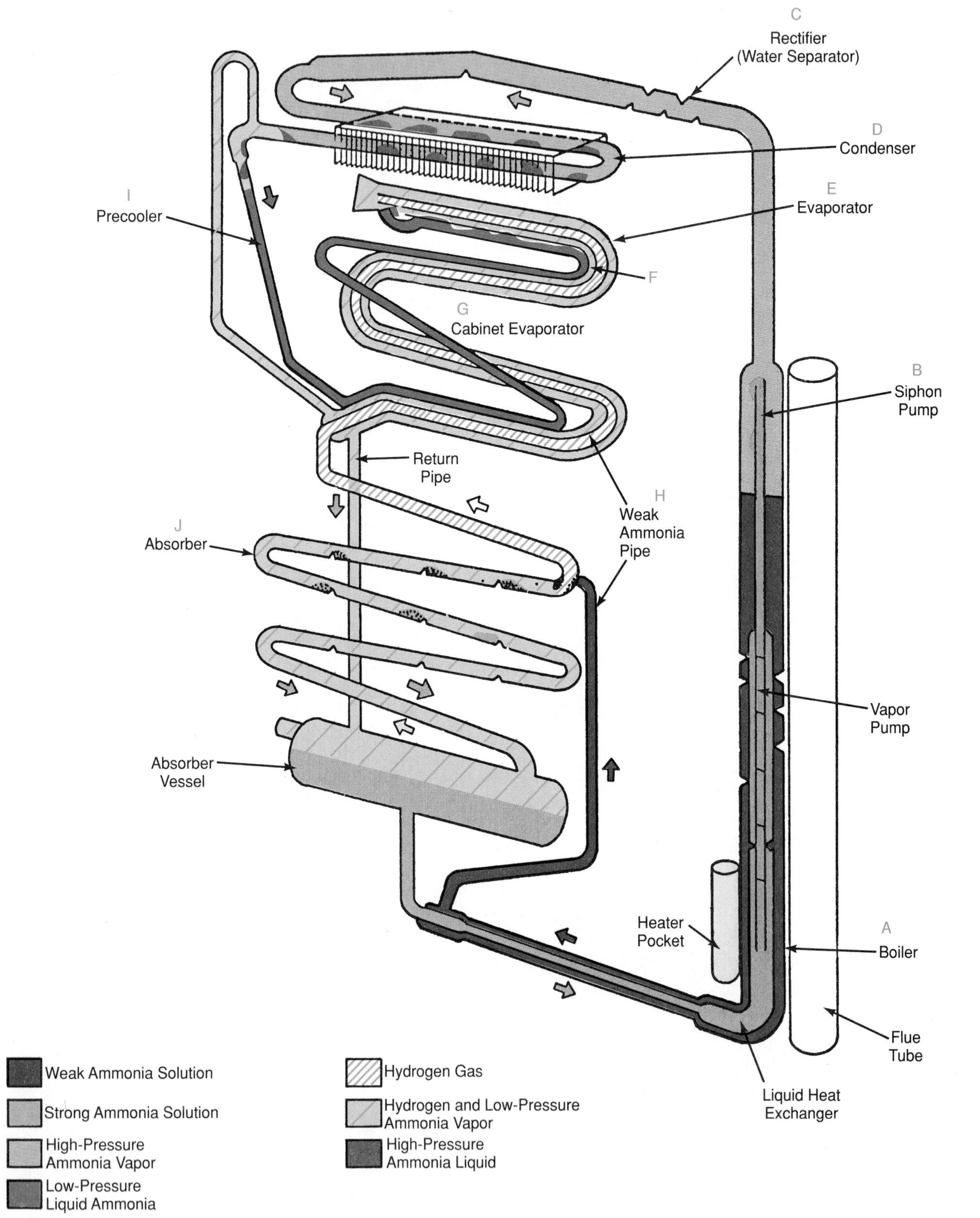

Figure 17-6. *This absorption system is used in recreational vehicles. Leveling of system is not as critical as for older machines. Internal siphon pump helps reduce liquid blockage in cooling coil. System can operate on any heat source, such as 120 V ac, 12 V dc, propane, butane, or kerosene. (Robur Corporation)*

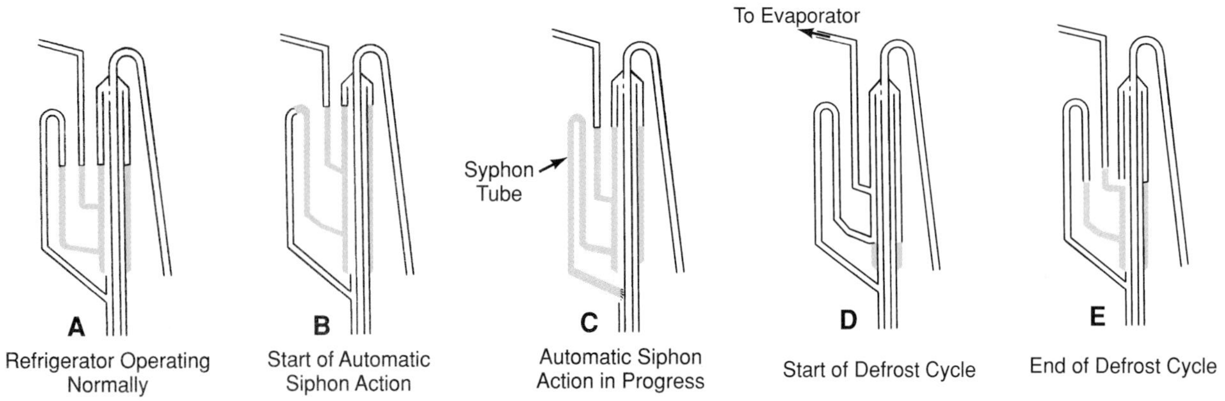

Figure 17-7. *One type of automatic defrost system for domestic continuous cycle absorption system. Liquid gradually builds up in right side of siphon tube. Every 15 to 24 hours, it spills over and allows hot vapor to move into higher temperature evaporator. Defrosting stops when enough liquid collects in outer tube to close tube to evaporator. (Electrolux AB)*

Figure 17-8. *A domestic absorption refrigeration system, working on LP gas or electricity. Note series of controls at base: electricity or gas selector dial, intermittent spark ignition for reignition of gas flame in emergencies, and thermostat. (SIBIR)*

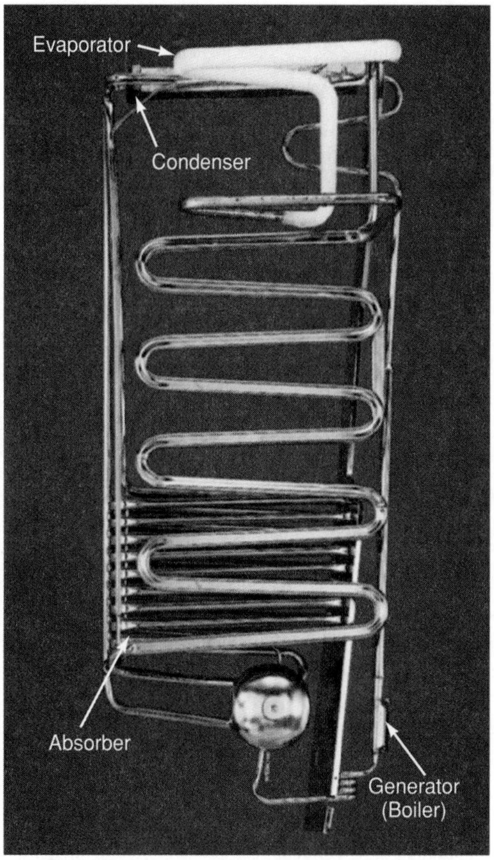

Figure 17-9. *Domestic absorption system. It contains four main sections. Note ice formation at evaporator. (SIBIR)*

A switch, as shown in **Figure 17-11,** is used to change over from 12 V to 120 V.

Combination gas/electrical systems use two thermostats. One operates when gas is used for fuel, the other when electricity is used. **Figure 17-12** shows wiring diagrams for 120 V and combination 12 V and 120 V systems.

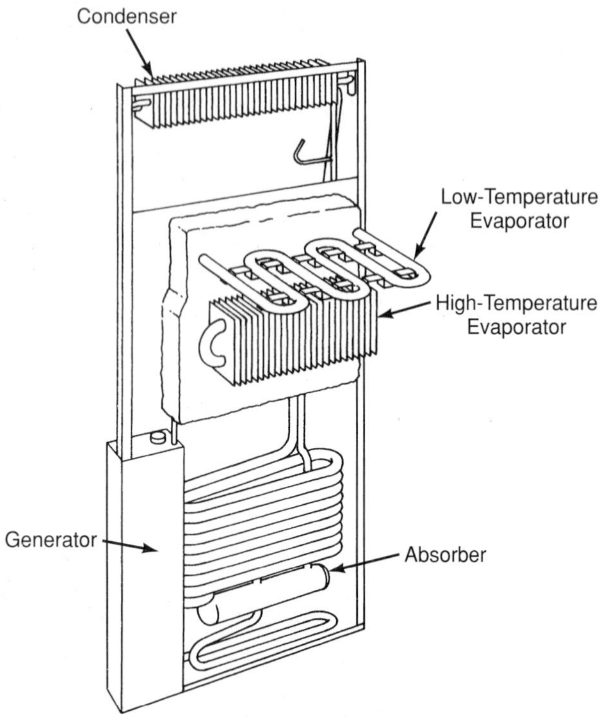

Figure 17-10. *Continuous operation absorption unit can be heated by fuel gas or electricity.*

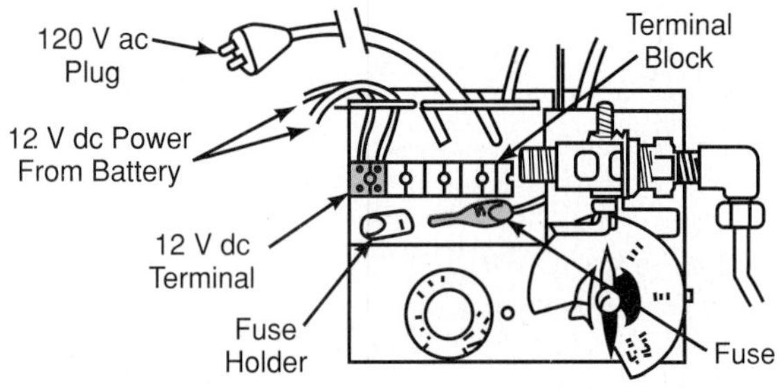

Figure 17-11. *Electrical switch and fuse used to shut off electrical power or change from 120 V to 12 V. If switch is on 12 V when system is connected to wall outlet, fuse will open circuit (blow) to protect the heater from damage. (Robur Corporation)*

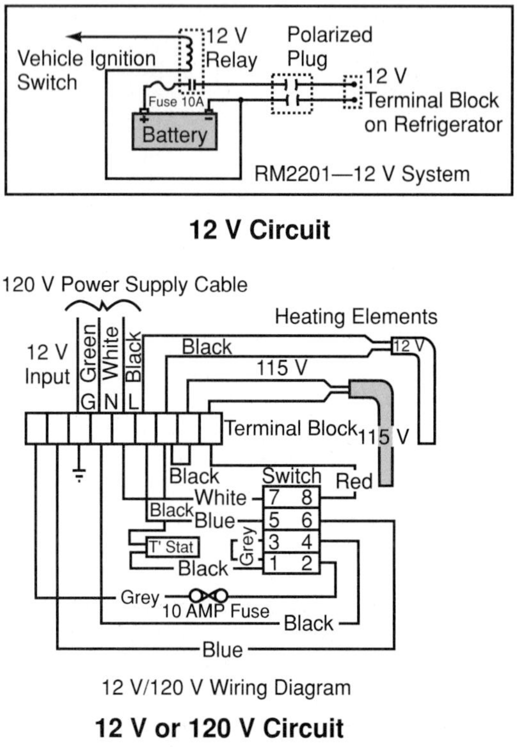

Figure 17-12. *Wiring diagrams for an electrically heated continuous operation absorption system. (Electrolux AB)*

17.7 Installing an Absorption Refrigerator

Installation of the absorption refrigerator depends on where it is to be placed. It will depend on whether it is in a home, recreational vehicle, or mobile vehicle.

The absorption refrigerator, like all refrigerators, moves heat. Heat is moved from the inside of the cabinet to the outside of the cabinet. This warmed air must be removed from near the cabinet. This will allow cooler air to continue to receive heat from the condensers.

Kerosene, natural gas, or LP gas, when burned, forms carbon dioxide gas and steam vapor. Both these gases are harmless. **However, if the burner is not burning all the fuel, carbon monoxide may be formed. Carbon monoxide is dangerous!**

The gas supply line from the house gas piping to the refrigerator must be tested for leaks. Use only soap suds! Gas pressure adjustments for minimum flame and maximum flame must be carefully made. A water column manometer is usually used. See Chapter 21 for more information on fuel gas servicing. The electrical service must be carefully checked, as well.

Some city codes require that the system's fuse plug opening be vented to the outside. This prevents any chance of discharging the refrigerant into the house.

Inside the mechanism, liquids flow by gravity. The absorption unit must be carefully leveled, or movement of liquids and gases will be uncertain. Some units have a level-indicating device. This allows for easier positioning during installation and for quick level checks during maintenance.

Due to the fact that warm gases rise, the absorption cabinets must have proper airflow space. This space is necessary beneath, in back of, and over the top of the cabinet. For good airflow past the burner, combustion products must be moved away from the cabinet. Always provide enough air inlet and exhaust for proper combustion, condenser, and absorber cooling.

The condenser, absorber, and flues of the unit must be kept clean to allow proper airflow and flue gas flow. The condenser, burner, and absorber should be cleaned at least twice a year. It may be cleaned more often if necessary.

17.7.1 Continuous System Gas Supply

Three different gases are used as fuel for gas refrigerator units. These fuels are discussed in more detail in Chapters 21 and 31. Heating quality of each is given as follows:

- Manufactured gas (500 to 600 Btu/ft^3).
- Natural gas (1000 to 1100 Btu/ft^3).
- Liquid petroleum LP gas (2500 to 3200 Btu/ft^3).

The fuel most often used for absorption system refrigerators is natural or LP gas. A clean fuel, such as gas, prevents carbon monoxide formation or carbon deposits in the burner. The cleaner the fuel, the less frequently the burners will need cleaning.

Gas refrigerators should be supplied with fuel under steady pressure. The burner should be designed for the type of gas being used. In case of burner difficulty, check to make sure the correct burner has been installed. The gas must be filtered before entering the burners. It must have a pressure regulator to provide a constant, unchanging pressure on the burner. **Figure 17-13** is a diagram showing suitable gas connections. Before installing these refrigerators, it is important to know the local codes governing such installations.

17.7.2 Continuous System Controls

Service on the absorption refrigerator is generally limited to the heating controls and air circulation equipment. Adjustments must be made carefully, since they determine the efficiency of the unit.

Heating gas valves in the absorption system automatically control the amount of heating gas burned. The systems use a continuous flow, and the flame sizes vary depending on the demand.

An electronic ignition, similar to that in a conventional gas furnace, lights the flame. Some units also have an automatic flame re-light system. All continuous systems need gas volume and safety controls. A bulb pressure-temperature control, located at the evaporator, regulates the amount of gas burned. It senses the refrigerator needs. **Figure 17-14** shows a cross section. Evaporator temperature affects flame size.

Successful operation of domestic refrigerators depends largely on the way the automatic control valves work. You must be thoroughly familiar with their operation and servicing.

Heat energy for the continuous operation absorption unit is usually supplied by a gas burner. See **Figure 17-15.** Several different methods have been used for regulation and control of this gas. A manual shutoff valve, a strainer, and a pressure-regulating valve are used. They are placed between the gas main and the operating controls.

As noted, the unit operates on the continuous cycle and heat is continuous. However, the size of the gas flame must be automatically controlled. This is needed to take care of changes in demand on the refrigerator itself. The variation in flame size is made possible by the use of a control valve. The control valve is operated by a power element located at the evaporator. As the refrigerator warms up, gases in the power element expand. They press on a diaphragm in the control valve to open the gas control. This allows more gas to flow to the burner. The large flame speeds up the refrigerating cycle. It continues to speed up the cycle until the evaporator has cooled.

As the evaporator cools, the power (control) element on the evaporator will cool. The pressure on the gas valve will be reduced. This closes the heating gas opening and reduces the size of the flame. Turning the adjustment clockwise (inward) increases the gas supply and produces more refrigeration.

A safety valve will shut off gas in case the flame goes out. This valve is a thermoelectric type, **Figure 17-16.** A thermocouple is placed close to the flame. It

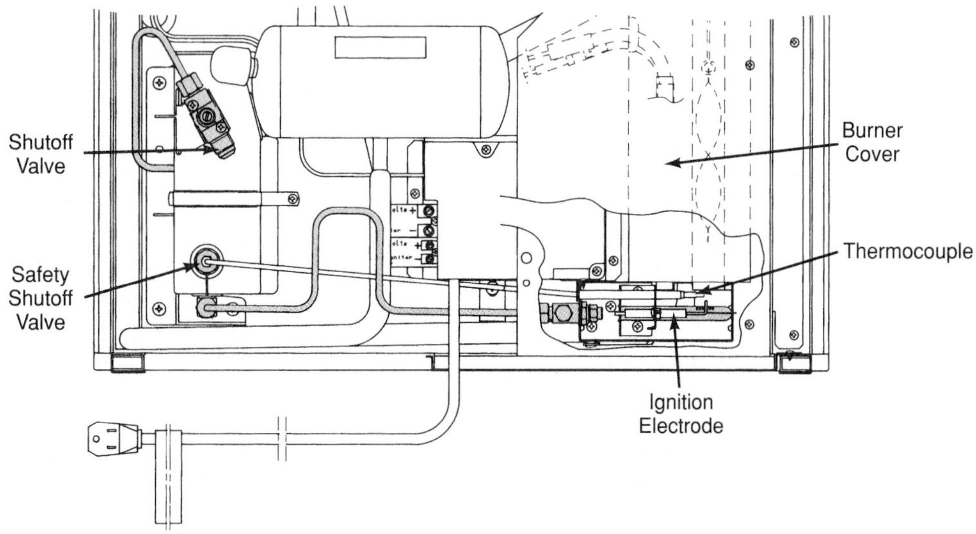

Figure 17-13. *Gas connections as used on absorption-type refrigerators. (Robur Corporation)*

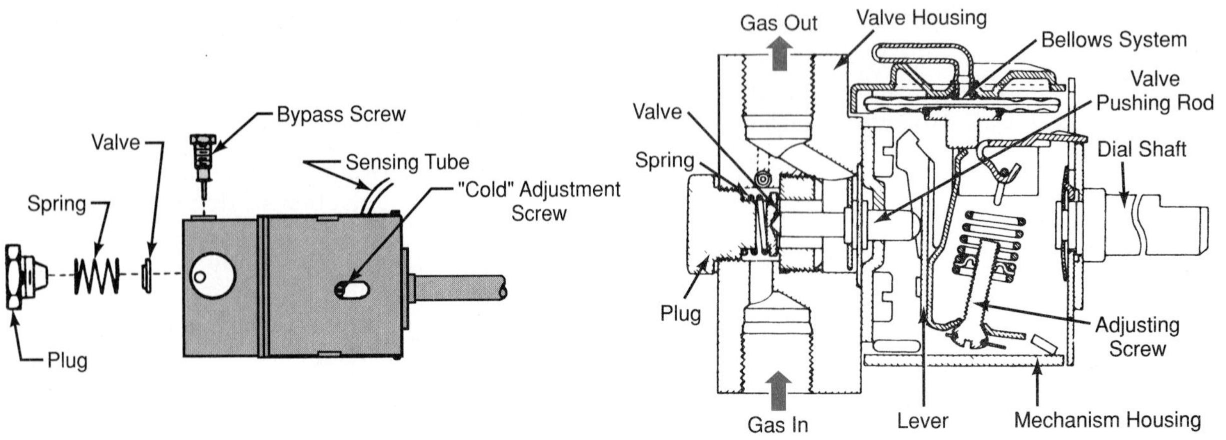

Figure 17-14. *Thermostat used in gas-fired small continuous operation absorption systems. Valve has bypass screw opening that allows a small flame to burn even when the main valve is closed (thermostat on zero when sensitive bulb is 40°F [4°C] or colder).*

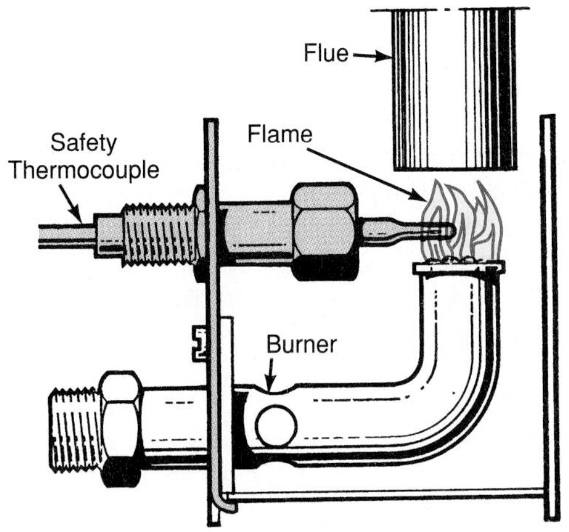

Figure 17-15. *Burner, flue, and thermocouple safety device.*

remains hot as long as the gas is ignited. If the flame goes out, the thermocouple will cool. On cooling, it stops creating electricity. The magnetic coil will lose its strength. This will allow the spring to close the valve. This action completely shuts off the supply of the gas. If this happens, the gas must be reignited by pushing in the button on the manual valve opener. This will hold the valve open. Then, to ignite the fuel gas, operate the spark lighter mounted on the burner housing. See **Figure 17-17.**

Each burner unit has an automatic pressure control for constant gas pressure. Through this valve, changes in gas supply pressures are kept minimal.

Newer units may also have a tilt control system. This system senses when the unit is not level and diverts heat away from the generator. When the level is corrected, proper heating begins again. This is helpful in situations where level changes occur constantly. Constant level changes may result in blockage or a shutdown of the absorption system.

17.7.3 Pressure Regulating Valves

The pressure-regulating valve supplies a steady flow of gas to the burner. Without a regulating valve, changing gas pressure would change the flame and it might go out. The pressure regulator both reduces the pressure and provides a constant gas pressure, **Figure 17-18.** LP gases may not need a pressure regulator at the refrigerator. A pressure regulator mounted on the LP cylinder will perform the same duty.

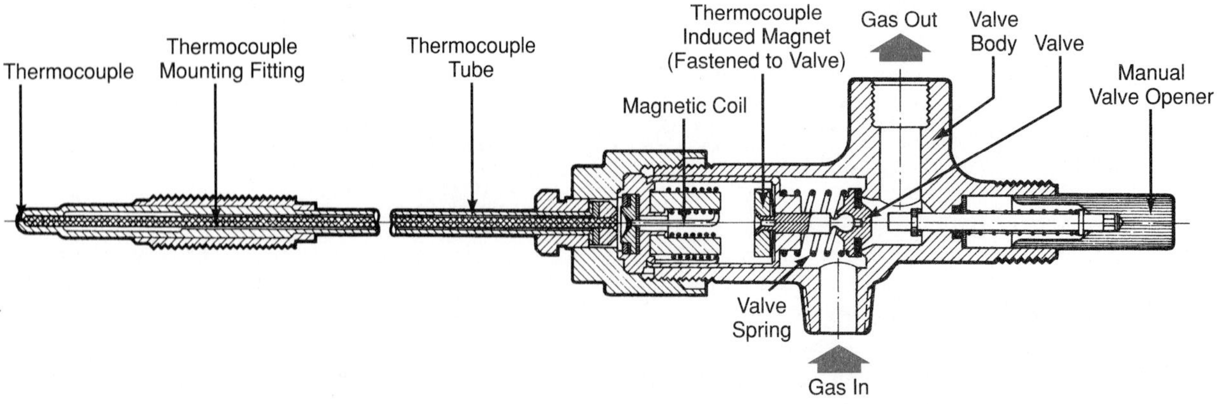

Figure 17-16. *Safety shutoff valve for fuel gas-heated absorption system.*

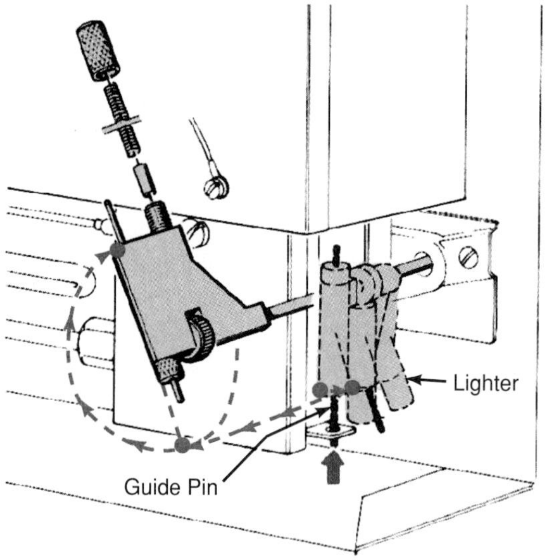

Figure 17-17. *Flint gas ignitor in recreational vehicle refrigerating unit.*

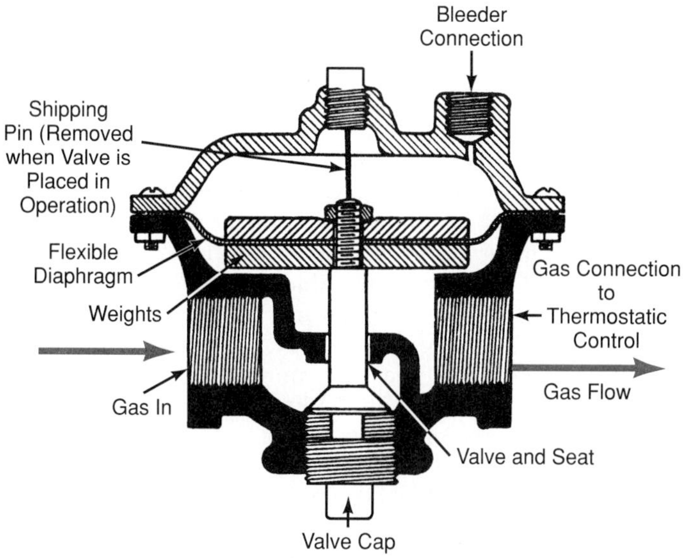

Figure 17-18. *Pressure regulating valve.*

The pressure regulator operates much the same as an expansion valve. Pressure at the outlet presses against a synthetic rubber or fabric-reinforced diaphragm. If the pressure begins to drop, the diaphragm will move. This will open the gas valve to allow more gas to flow. The increased gas flow will press the diaphragm up, closing the valve. The pressure regulator should be accurate to about 0.01″ (.25 mm) of water pressure.

Pressures the regulator must maintain vary from 1.6″ to 3.9″ (40 mm to 99 mm) of water. Pressure needs vary with gas flow in cubic feet per hour. This gas flow is controlled by the orifice size in the burner.

Pressure also varies with the density or specific gravity of the gas. The greater the gas flow, the greater the pressure needed. Pressures must be adjusted to within 0.1″ (2.5 mm) water pressure for good results.

17.8 Portable Absorption Refrigerators

The portable refrigerator shown in **Figure 17-19** is a self-contained continuous absorption type. It uses propane gas, or either 120 V ac or 12 V dc as a heat source to the generator.

The refrigerator's pull-down time is from two to six hours, depending on the ambient temperature. The small propane cylinder will provide approximately 70 hours of continuous operation. The gas cylinder, electrical connections, and controls for this refrigerator are shown in **Figure 17-20.**

When operating from a propane cylinder:

1. Attach gas connection to gas inlet fitting.
2. Open valve on gas bottle.
3. Press the red button on the safety valve. Hold it for 10 to 15 seconds. (This clears air from the gas line.)

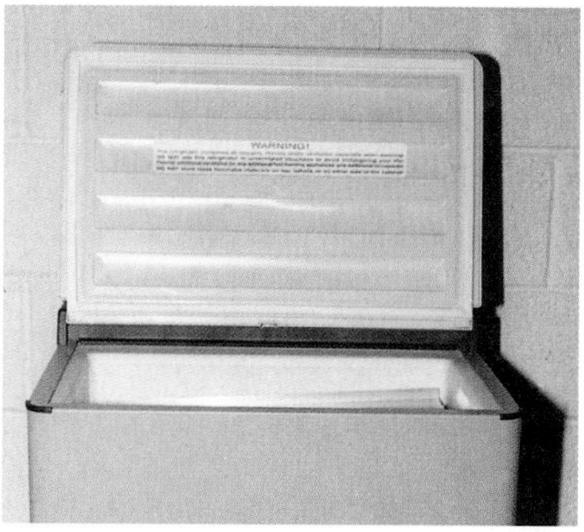

Figure 17-19. *Portable absorption refrigerator which can use propane gas cylinder or electric heating element or either 120 V ac or 12 V dc as its energy source. (Robur Corporation)*

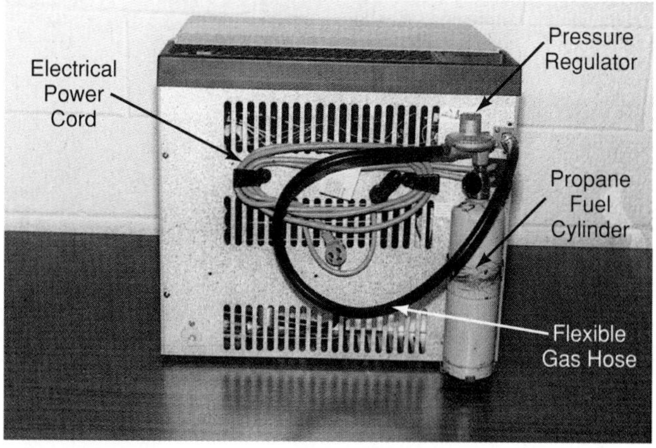

Figure 17-20. *Portable absorption refrigerator. Refrigerator can be operated with either gas flame or electric heating unit of 12 V or 120 V. (Robur Corporation)*

4. Light match. Press red button again and apply flame to burner. Keep depressed for 20 seconds after the burner is lit.

If the flame goes out, the safety valve will automatically shut off the gas supply. A pressure regulator will maintain an 11″ (279 mm) water column pressure on the burner. When using the gas burner, be sure no combustible material or vapors are near refrigerator.

For electrical operation, separate leads are supplied for the 120 V ac and the 12 V dc connections.

The cabinet must be completely sealed on the top, sides, and bottom. This prevents dangerous flue gases from entering the vehicle. Combustible surfaces should be covered with fire-resistant material. Keep them at least 1″ (25 mm) away from the refrigerator mechanism on the sides. They should be at least 7″ (180 mm) away from the top.

An absorption system requires a long cooldown time after being started. The cooldown time usually required is 8 to 10 hours. Manufacturers recommend starting the unit the night before it is to be loaded with food.

17.9 Absorption Refrigerators for Mobile Homes

Mobile homes and travel trailers often use an absorption-type refrigerator. The units are usually designed with both an electric heating element and a gas burner. These are used to heat the generator of the continuous unit. Gas heat is used when electricity is not available. The refrigerator must be mounted level. If not, gravity-controlled flow of fluids will not function properly.

The installations must be carefully designed with sufficient air ventilation to cool the condenser. They must also provide air for the flame and outside exhaust for combustion gases.

The units must also remain as level as possible when being moved. The refrigerator must be securely fastened. An access or service door must be provided on the outside of the vehicle. This door usually serves as the inlet vent as well. There should also be an exhaust vent. Refer to **Figure 17-21.** Park the vehicle so that winds do not blow directly against the vents on the outside.

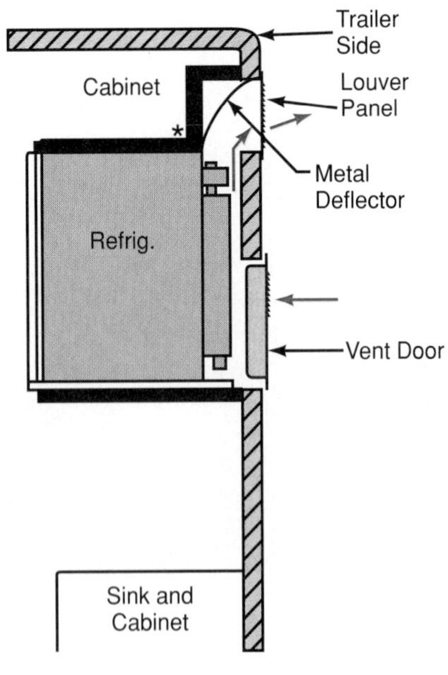

Figure 17-21. *Gas-fired absorption system with side vent for use in mobile homes.*

17.10 Residential Absorption Air Conditioners

The number of absorption-type air conditioning systems used in residential and commercial buildings has increased. Their basic operation is similar to that described in Section 17.6.1. One of the major changes for such use is the addition of a pump. The pump transfers the weak solution from the absorber to the high side of the cycle. Either a hydraulic diaphragm pulse pump or an electric motor-driven magnet pump is sealed in the system.

A typical chilled water residential air conditioning system is shown in **Figure 17-22.** The generator has a gas burner that heats a mixture of ammonia and water. The boiling point of ammonia is lower than that of water. Therefore, it becomes a vapor and flows through the line marked 1. It then flows through the rectifier to the condenser as a high-temperature, high-pressure gas.

As outside air passes over the condenser, it removes heat from the ammonia. The ammonia condenses to a liquid and passes through the line marked 2, to the precooler. The precooler acts as a heat exchanger. It reduces the temperature of the liquid ammonia before the ammonia reaches the evaporator. It also heats the cold ammonia vapor leaving the evaporator through line 3.

Pressure drops as liquid ammonia leaving the precooler passes through a restrictor into the evaporator. Here, it picks up heat from the chilled water circuit and boils to ammonia vapor. Ammonia vapor in line 3 passes through the precooler to the absorber heat exchanger. In the generator, most of the ammonia boils out, leaving a weak solution. This solution leaves the generator at a high pressure. It passes through restrictors which meter flow and separate high- and low-pressure sides of system.

There are restrictors in line 4 leading to the absorber heat exchanger. Here, solution temperature is lowered more by heat transfer. At the absorber end, heat is removed by outside air. Ammonia absorption is completed.

The solution travels from the absorber, through the condenser, to the solution pump through line 5. The pump forces it at a high pressure to the rectifier through line 6. It picks up heat from the surrounding hot ammonia vapor along the way. Continuing to the absorber heat

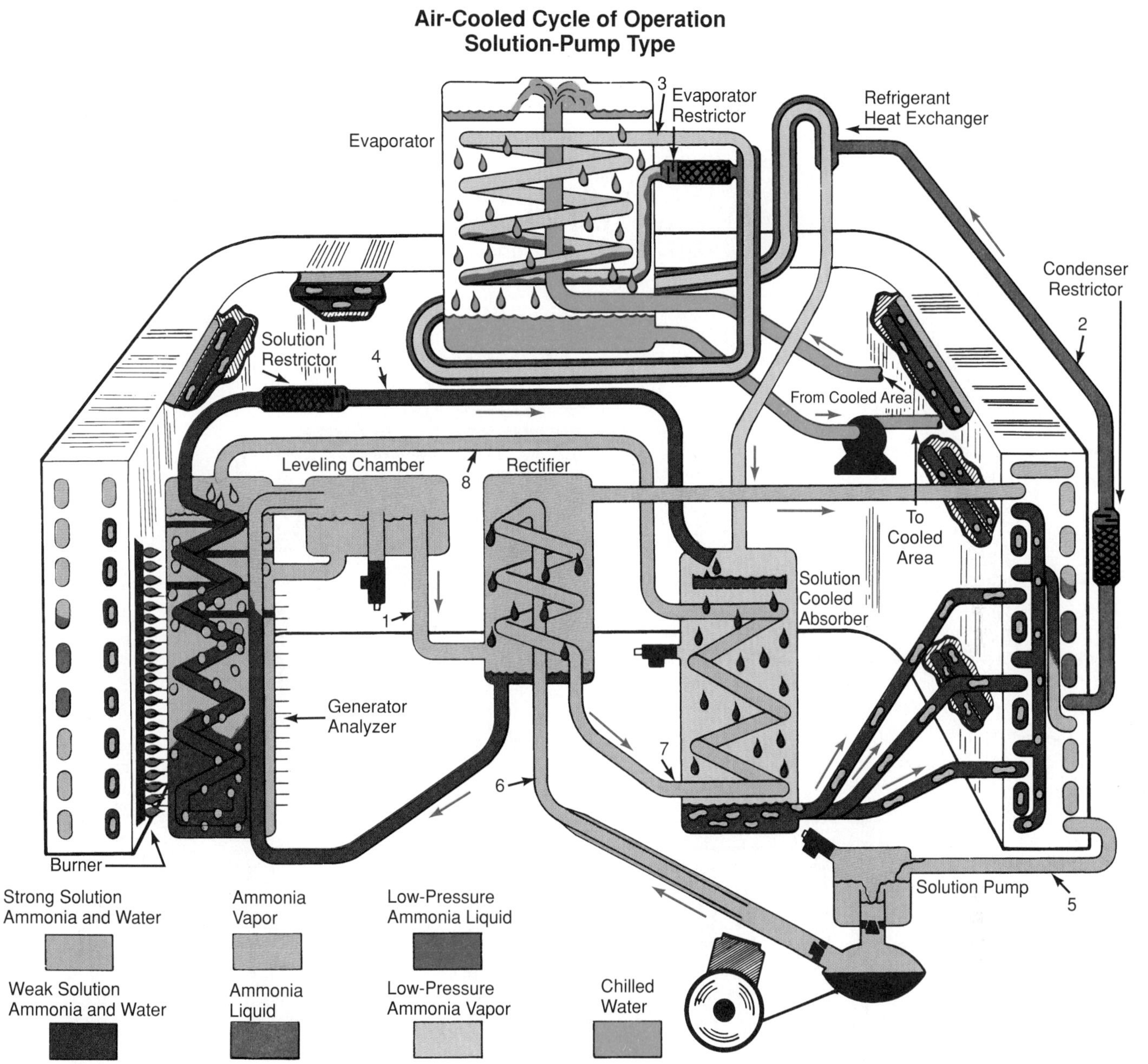

Figure 17-22. *Absorption refrigeration cycle. This refrigeration cycle uses pump to circulate ammonia liquid through cycle. (Robur Corporation)*

exchanger through line 7, more heat is picked up. This preheated solution returns through line 8 to the generator. The cycle begins all over again. The same action occurs in larger units except that two burners may be used.

Another system cycle of slightly different design is shown in **Figure 17-23.** Weak solution (strong in ammonia content) is returned to the generator by the solution pump. The burner heats it and drives off the vapor. The vapor, a mixture of ammonia and water, flows through the generator assembly. It comes into direct contact with the weak solution coming into the generator. During the process, the vapor is partially purified. This occurs as it comes into contact with the cooler weak solution. This solution is flowing through a coil in the opposite direction.

The rectifier contains rings. As the ammonia vapor passes through, it comes into direct contact with the rings. The rings have a large surface contact area. Thus, they help remove any water vapor left in the ammonia vapor.

The purified ammonia vapor then flows into the condenser tube. It is cooled by air moving across the condenser. The hot ammonia vapor in the condenser gives up heat to the flow of air. By this process, it is liquefied.

Liquid refrigerant leaves the condenser and passes through the first restrictor. There, pressure and temperature drop somewhat. It then flows through the outer tube of the liquid suction heat exchanger. Heat is given up to cooler ammonia vapor in the inner tube.

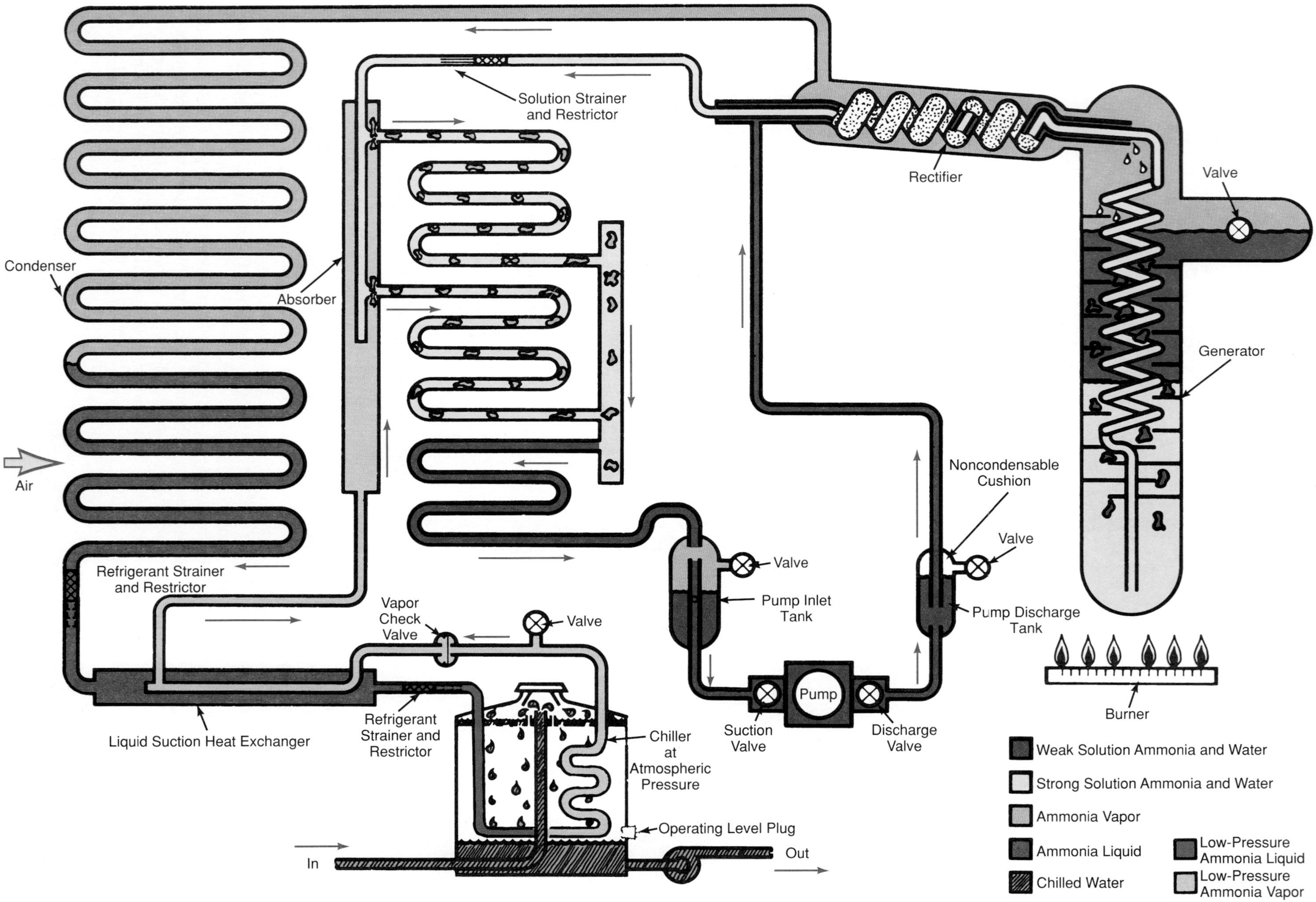

Figure 17-23. *Absorption cycle used for small air conditioning systems. Notice service valves on system.*

(Cooler ammonia vapor is flowing through the inner tube in the opposite direction.)

From the liquid suction heat exchanger, the liquid ammonia enters the second restrictor. There, pressure and temperature are further reduced. The low-pressure, low-temperature liquid refrigerant then flows through the chiller coil (evaporator). Here, a glycol and water solution cascades down across the evaporator. It gives up heat and vaporizes liquid refrigerant.

The refrigerant vapor leaves the chiller. It then passes through the inner tube of the liquid suction heat exchanger. In passing, it picks up heat from the liquid refrigerant flowing through the outer tube in the opposite direction. Finally, the vapor enters the absorber header. This completes the refrigeration circuit.

Hot, strong solution is left behind as ammonia vapor is driven out of the weak solution in the generator. This solution then passes up through a coil in the generator. The solution then leaves the generator and enters the inner coil of the rectifier. Here, the strong solution comes into thermal contact with the weak solution. (The weak solution is flowing in the opposite direction through the outer coil.) As the strong solution leaves the rectifier, it passes through the strong-solution restrictor. This restrictor reduces the solution from high-side to low-side pressure.

As the strong solution leaves the restrictor, it enters the absorber header. This is a tube-within-a-tube. The inner tube contains the strong solution. The outer tube contains ammonia vapor which is being returned from the chiller (evaporator). The strong-solution tube contained within the absorber header has two small holes. These holes allow strong solution to flow out into direct contact with the surrounding ammonia vapor. At this point, the strong solution and ammonia vapor begin to form a weak solution.

Strong solution and ammonia vapor then leave the absorber header. They leave by way of the two tubes and begin to flow into the absorber. Throughout the absorber, the strong solution completely absorbs the vapor, forming a weak solution.

As the weak solution leaves the absorber, it is picked up by the solution pump. This will move the solution back to the high-pressure side of the unit. As the weak solution leaves the solution pump, it passes through the rectifier. Heat is picked up from strong solution vapor flowing in the opposite direction.

Preheating the weak solution as it flows through the rectifier reduces the heat input required. The overall efficiency of the cycle is increased. The weak solution leaves the rectifier and drips back into the generator analyzer assembly to start another cycle.

17.10.1 Residential Absorption Air Conditioner Installation and Construction

A typical gas and water single unit zone application is shown in **Figure 17-24.** The unit is installed on a concrete slab. The installation is similar to that of a standard air conditioning unit. As in the standard unit, the supply and return connections are covered with flexible hose. The hose is secured with stainless steel clamps to prevent sound transmission. The coil for this application is located on the typical air conditioning-type gas furnace. The coil may need to be located in the hot air

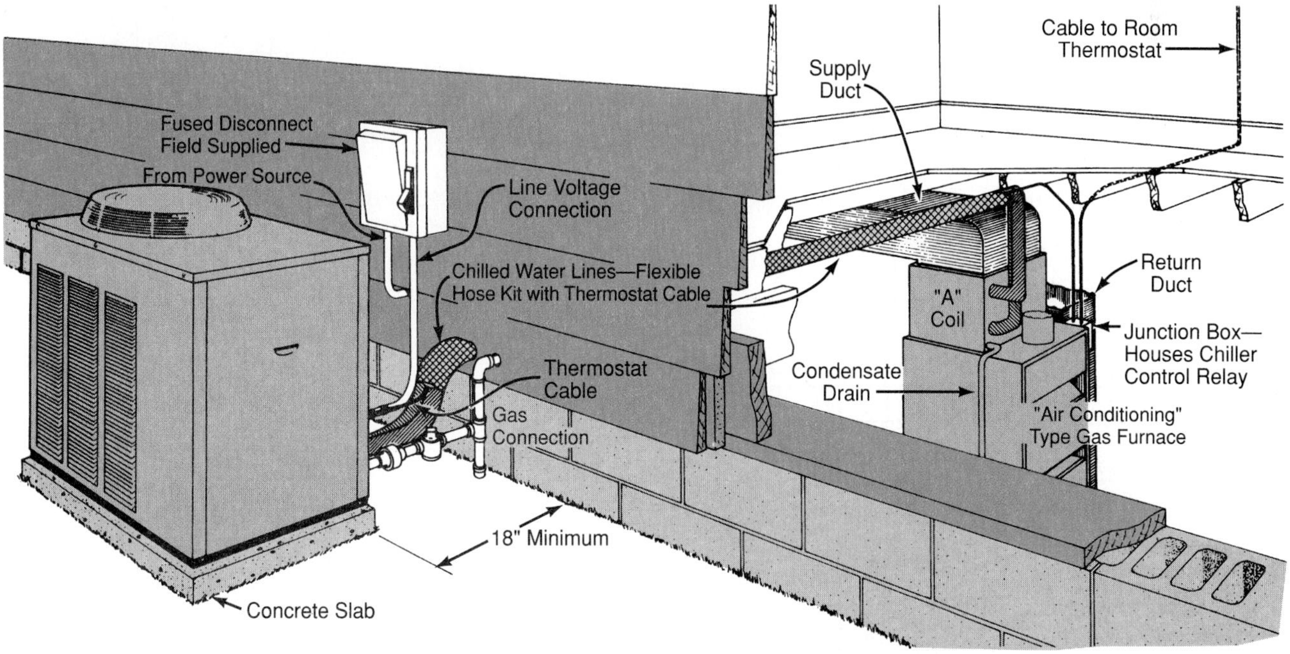

Figure 17-24. *Single-unit, single-zone application of absorption cooling system combined with a basic air-conditioning type gas furnace. Note the various insulated lines from the cooling system to the furnace. (Robur Corporation)*

stream of the furnace. In this event, plastic piping should not be attached directly to the coil. Copper piping extensions should be used in this area. The condensate drain lines must also be insulated. All piping should comply to local codes.

A control transformer applies the power for the control circuit. The circuit includes gas valve, direct spark ignitor and time delay switch for both heating and cooling. An electromagnetic ignition system provides a direct spark ignition of the main burner. It has a ten-second flame detection system. The system has a solid-state time delay. This prevents continuous circulation of both refrigeration solution and chilled water. Circulation is delayed for three minutes and fifteen seconds after the thermostat has been satisfied. The basic system has numerous safety controls. These include flame detectors, chilled water switches, and heating-side operating controls. These are in addition to basic safety controls.

The capacity of air-cooled chillers varies with ambient air temperature and leaving chilled water temperature. Capacity characteristics of the units are shown in **Figure 17-25.**

In the self-contained unit, **Figure 17-26,** the insulated evaporator cools a glycol and water solution. The solution then circulates through a heat exchange coil in the furnace bonnet. It may also be used in a separate air circulation system within the building. The absorption-system is shown in **Figure 17-27.** Controls are shown in **Figure 17-28.**

The absorption system can also be utilized as part of a multiple unit load system. Typical applications are in large office buildings, strip malls, and efficiency apartment complexes. An internal temperature-sensing device monitors the return water temperature. Temperatures higher than normal indicate increased need for cooling. Another unit is ignited to bring the chilled water temperature down to proper level. The chilled water is directed by means of valves to the proper fan coil. The proper zone control is thereby accomplished.

The low pressure and temperature of the chilled water system allows the use of PVC. Gas units are single-phase power, 115 V or 230 V. Individual units range from 3 tons to 25 tons.

These systems are made of steel and aluminum. Use of copper or copper alloys is very dangerous. An explosion may result.

Figure 17-26. *Exterior view of gas-fueled absorption system used for residential air conditioning. (Robur Corporation)*

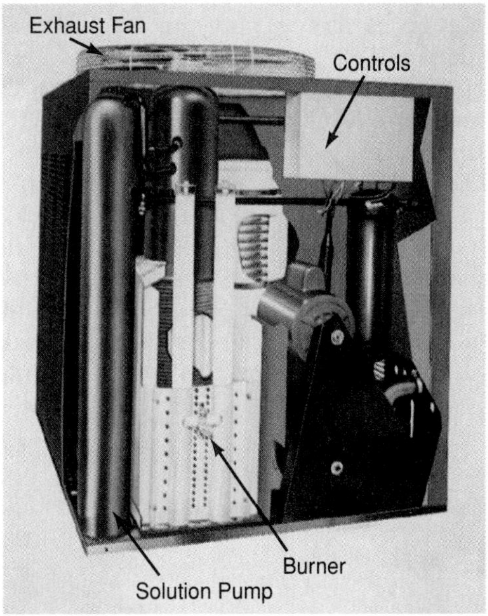

Figure 17-27. *Absorption system air conditioner with housing removed. (Robur Corporation)*

Capacities in Thousands of Btu/h

Leaving Chilled Water Temperature	Air Temperature Entering Condenser (°F)											
	3-Ton Unit				4-Ton Unit				5-Ton Unit			
	90	95	100	105	90	95	100	105	90	95	100	105
50°F	36.76	36.28	35.10	33.64	49.01	48.37	47.37	44.85	61.26	60.46	58.50	56.07
48°F	36.72	36.18	34.92	32.65	48.96	48.24	47.13	43.53	61.20	60.30	58.20	54.42
46°F	36.65	36.04	34.50	31.32	48.86	48.04	46.57	41.76	61.08	60.06	57.50	52.20
44°F	36.54	35.82	33.72	29.60	48.72	47.76	45.51	39.45	60.90	59.70	56.20	49.32
42°F	36.36	35.53	32.30	27.43	48.48	47.37	43.60	36.57	60.60	59.22	53.84	45.72
40°F	36.00	35.03	30.60	23.40	48.00	46.70	41.30	31.20	60.00	58.38	51.00	39.00

Figure 17-25. *Capacities of air-cooled chillers in thousands of Btu/h. Note that capacities vary with ambient air temperature and the temperature of chilled water exiting the unit. (Robur Corporation)*

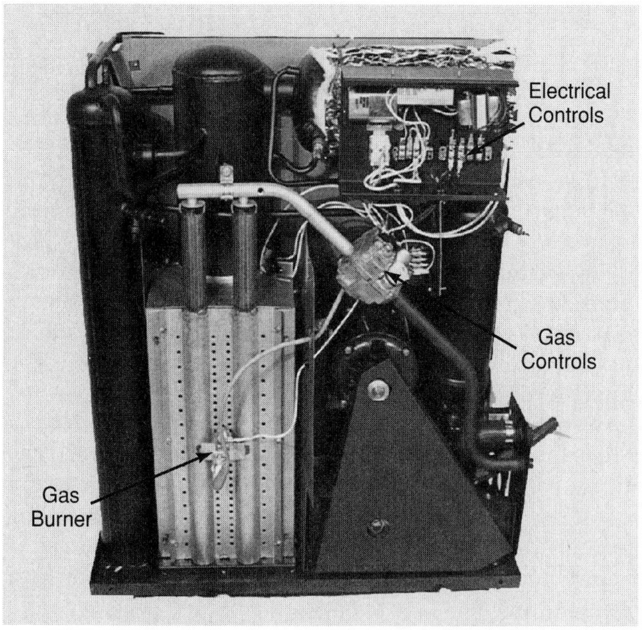

Figure 17-28. *Controls of absorption air conditioner are easy to reach for service when housing is removed. (Robur Corporation)*

17.10.2 Residential Absorption Air Conditioner Service

Most residential absorption systems are serviceable. They are equipped with service valves. However, you should be trained by the manufacturer before attempting to service these systems.

Ammonia is toxic and flammable when mixed at certain ratios with air. Wear a face shield or safety goggles. Ammonia reacts with some metals. Use only steel or aluminum tubing, gauges, fittings, and manifolds.

Figure 17-29 shows a system equipped with four service valves. Valves A and D are on the low-pressure side. Valves C and E are on the high-pressure side of the system. Valve C is not shown. Valves D and E are gauge mounts for checking pressures on the system. Refer back to **Figure 17-23.** It also shows the location of the service valves.

The system is charged with a solution of ammonia, distilled water (pH 6.0 +), and a corrosion inhibitor. A solution cylinder is used to charge the system with the solution. This cylinder usually has about a 45 1b. capacity. The solution charge is about 35 lb. of distilled water. It contains inhibitor and 15 lb. of anhydrous ammonia.

The solution cylinder is filled with distilled water and inhibitor (yellow in color). It is put in through the fill plug. If a white precipitate forms in the solution, discard it and make a new batch.

The solution cylinder is then charged with anhydrous ammonia. Anhydrous ammonia is available in 25 1b. cylinders. The cylinder has both a vapor and a liquid valve. **Figure 17-30** shows the charging arrangement. Note that the charging line is connected to the liquid

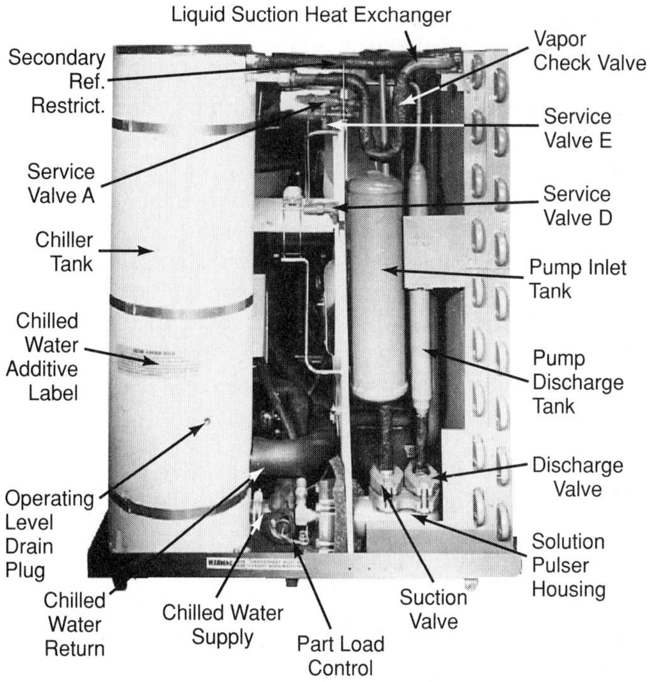

Figure 17-29. *Absorption system for air conditioning. Service valve A is mounted on solution-cooled absorber. Service valve D is mounted on pump inlet tank. Service valve E is mounted on pump discharge tank. Service valve C is not shown.*

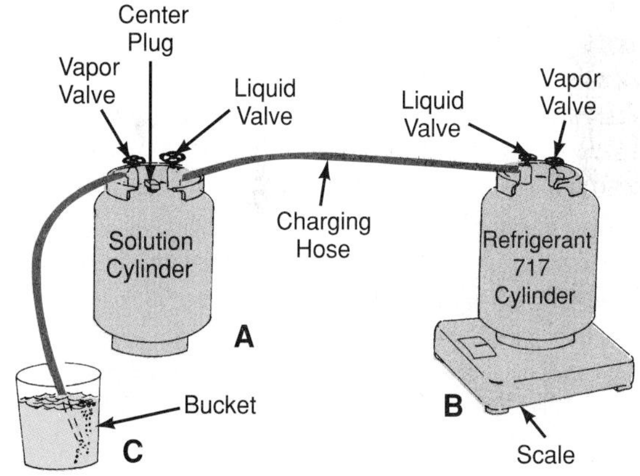

Figure 17-30. *Cylinders set up for recharging an absorption system. A—Solution cylinder has liquid valve, vapor valve, and center plug. It is being charged with liquid ammonia from cylinder at right (1 lb. ammonia for each 2 lb. distilled water). B—Ammonia cylinder. C—Pail holds water and a purge line connected to gas valve of cylinder A. Purging decreases pressure in A to allow flow from B. (Robur Corporation)*

valve of the anhydrous ammonia cylinder. The pail is partly filled with water. It is connected to the vapor valve of the solution cylinder. Vapor is purged from the solution cylinder if necessary to lower the pressure. (This will enable anhydrous ammonia to flow into the solution cylinder.) The water in the pail will absorb the small amount of ammonia purged.

The system must have the correct pressures and the correct amount of anhydrous ammonia. It must also have the correct amount of distilled water and inhibitor. To check pressures, install a 100% steel constructed gauge manifold, steel lines, and steel fittings. **Do not use copper or brass.** A steel gauge manifold is shown in **Figure 17-31**. Note the steel manifold. It is constructed of standard steel fittings and steel valves. The manifold operates exactly like the manifolds described in Chapters 12 and 15.

The four service valves on the system are used for a number of service operations.

Valve A:

- Checks absorber pressure (low-side pressure).
- Purges ammonia vapor.
- Adds ammonia liquid or vapor.
- Adds solution.
- Reduces system pressure to atmospheric pressure.

Valve C:

- Checks high-side pressure.
- Checks solution level.
- Removes excess solution.
- Adds solution after repairs.
- Reduces system pressure to atmospheric.

Valve D:

- Purges air.
- Adds air.
- Adds solution.
- Removes solution.

Valve E:

- Removes large amounts of solution.
- Determines if discharge chamber has proper amount of noncondensables.

Figure 17-31. *All-steel gauge manifold is connected to valve C (high-pressure side) and valve A (low-pressure side). Valve E is for removal and checking of solution.*

17.11 Commercial Absorption System

Absorption systems are used successfully for air conditioning comfort cooling installations. Such systems may also be used for heating. Some units use the ammonia-water-hydrogen continuous cycle. Others use water as the refrigerant, and various chemicals as the absorber.

A system using water as the refrigerant and lithium bromide as the absorber is shown in **Figure 17-32.** Steam heat applied to the generator percolates water vapor (red dots) and weak solution up to the separator. The liquid lithium bromide (shown in black) then flows by gravity through the heat exchanger. It flows to the absorber where it absorbs the evaporated water. The strong solution (black dots) settles to the bottom of the absorber. It returns to the generator after passing through the heat exchanger. The pressure difference is maintained by the pressure head of the lithium bromide liquid.

The water vapor (red dots) in the separator rises to the condenser. There, it is condensed and becomes water. The water flows by gravity through an orifice into the evaporator. The water evaporates at a low temperature due to a near-perfect vacuum in the system. The water vapor is absorbed by the lithium bromide (black). Note that the absorber and condenser are both cooled by water coils. The condenser water is then taken to a cooling tower. It is cooled there and used over again. The condensing pressure is about 50 mm to 60 mm Hg (about 1 psia or 6.9 kPa). The evaporating pressure is 8 mm to 10 mm Hg (about 0.17 psia or 1.2 kPa). Lithium chromate is often used as a corrosion inhibitor. A typical cooling tower is shown in **Figure 17-33.** More detail is shown in Chapter 13.

17.12 Absorption Unit for Air Conditioning and Heating

The application of absorption refrigerating systems in comfort cooling air conditioning and heating is increasing. Absorption units have advantages in solar energy systems. Solar energy, as a source of heat, can cool buildings when used in absorption systems. Installations using steam heat can use it as a heat source for absorption cooling in the summer.

These systems are also used to produce chilled water. The chilled water, in turn, may be used for various purposes. It may be used as quenching baths and drinking water. It may also serve as a special coolant to lower the working temperature of welding tips. An absorption system for chilling water and heating is shown in **Figure 17-34.** The cooling cycle is shown in **Figure 17-35A.**

The refrigerant which is dispersed in the evaporator extracts the heat from the chilled water. It is then vaporized. The chilled water then passes through the system.

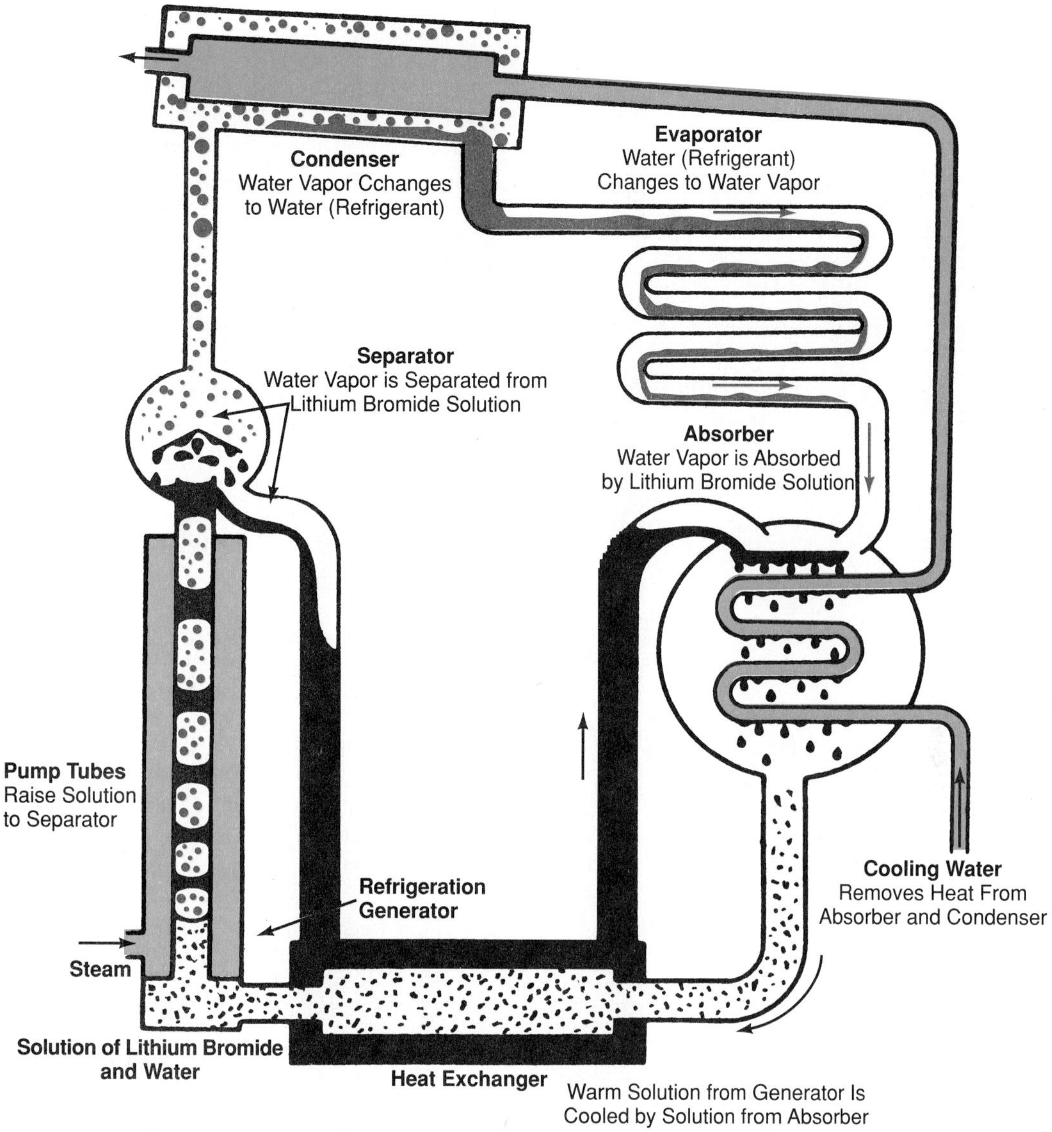

Figure 17-32. *Absorption refrigeration cycle which uses water as refrigerant and lithium bromide as absorbent.*

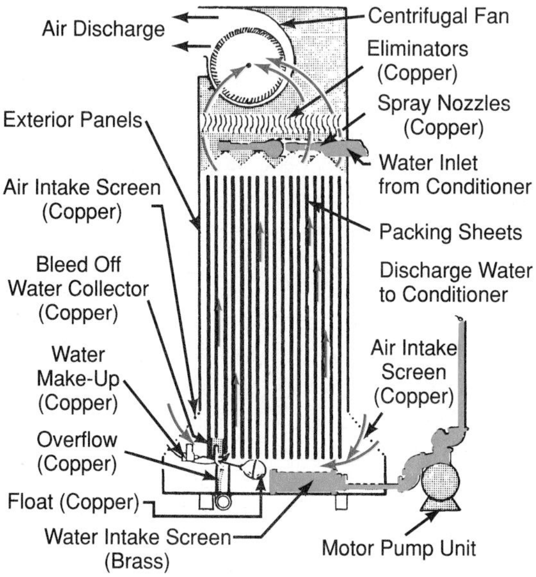

Figure 17-33. *Cooling tower used to cool condenser and absorber cooling water. Tower evaporates about 15% of condenser water. In doing so, it cools rest of water down to wet bulb temperature of air. The tower consists of water sprays, packing sheets, overflow tubes, make-up water float valve, and water pump. Eliminator plates keep water from being drawn into fan. Air enters at bottom and leaves at top.*

The heating cycle is shown in **Figure 17-35B.** During the heating cycle, the evaporator functions as a condenser. Hot water in the evaporator tubes absorbs the heat given off during condensation of the refrigerant. The heated water is circulated throughout the system. Absorption system heat can be supplied by exhaust gas from another industrial or residential system. This results in large efficiency gains.

Figure 17-34. *A large capacity, double-effect absorption chiller and heater. System produces both cooling and heating.*

17.13 Servicing Absorption Refrigerators

When servicing gas-fired absorption refrigerators, be sure the gas supply is at the correct pressure. Check the gas pressure using a water-filled manometer, as shown in **Figure 17-36.** The safety valve body has a manometer connection. The amount of gas fed to the refrigerator may be checked by the flame size.

The flue must be kept clean to allow good transfer of heat. Brushes should be used to clean the flue. Fins on the ammonia condenser must be cleaned periodically. This will ensure good heat removal from these surfaces.

If a service call indicates that the refrigerator is too cold, check the temperature control dial. It may be set too cold. The evaporator unit temperature may be lower than that indicated by the temperature control dial setting. A time-temperature graph of the evaporator temperature should be taken.

Perhaps the most common service call will be "little or no refrigeration." Following are some possible causes:

- Overloaded cabinet.
- Improper condensing temperatures.
- Little or no heating of the generating unit.
- The gas supply has been turned off or restricted. If the line becomes clogged, there will be low consumption of gas. There is, of course, little or no refrigeration. This trouble may be traced by checking the pressures at the burner.
- Restricted or dirty gas flue.

Gas-fired and kerosene-fired refrigerators are equipped with flues. They direct the hot gases around and away from the generating units. Occasionally, a flue may be restricted because the refrigerator is too close to the wall. A flue also may be obstructed by objects blocking the opening or falling into it. Flues must be kept clean to ensure proper functioning of the refrigerator.

After a period of operation, the generator flue will normally become coated with sooty deposits. When this occurs, rapid transfer of heat from the gas flame to the generator is impossible. This soot deposit must be removed periodically. (Usually once every one to two months is sufficient.) Frequent cleaning also reduces gas consumption of the unit.

When scraping the generator flue or removing soot from any surface of the generator, take care to prevent damage to the surface. Always put papers or cloth under the refrigerator when cleaning flues.

If either the condenser or absorber is dirty or lint-covered, poor refrigeration will result. This is due to poor airflow around these components. Wipe, brush, or vacuum away these accumulations.

A temporarily unused absorption unit may not freeze. Make sure the burner is lit. Air may have filled the gas line; it may take several tries to light the burner. If the burner is lit, the problem is likely due to blockage within the unit. Manufacturers recommend tilting to remove the blockage. After 10 minutes of operation, tilt the refrigerator to the right for about 30 seconds. Then tilt it to the left for 30 seconds. Do this three to four times, then put the unit back in the upright, level position. If it still does not cool, replace the cooling unit.

If the system is overheated, the pipe going to the condenser will overheat. The percolation pump will stop working. If the paint on the pipe to the condenser is blistered, overheating has taken place. To remedy this problem, shut off the heat and allow the system to cool. Then turn the unit upside down several times to put the

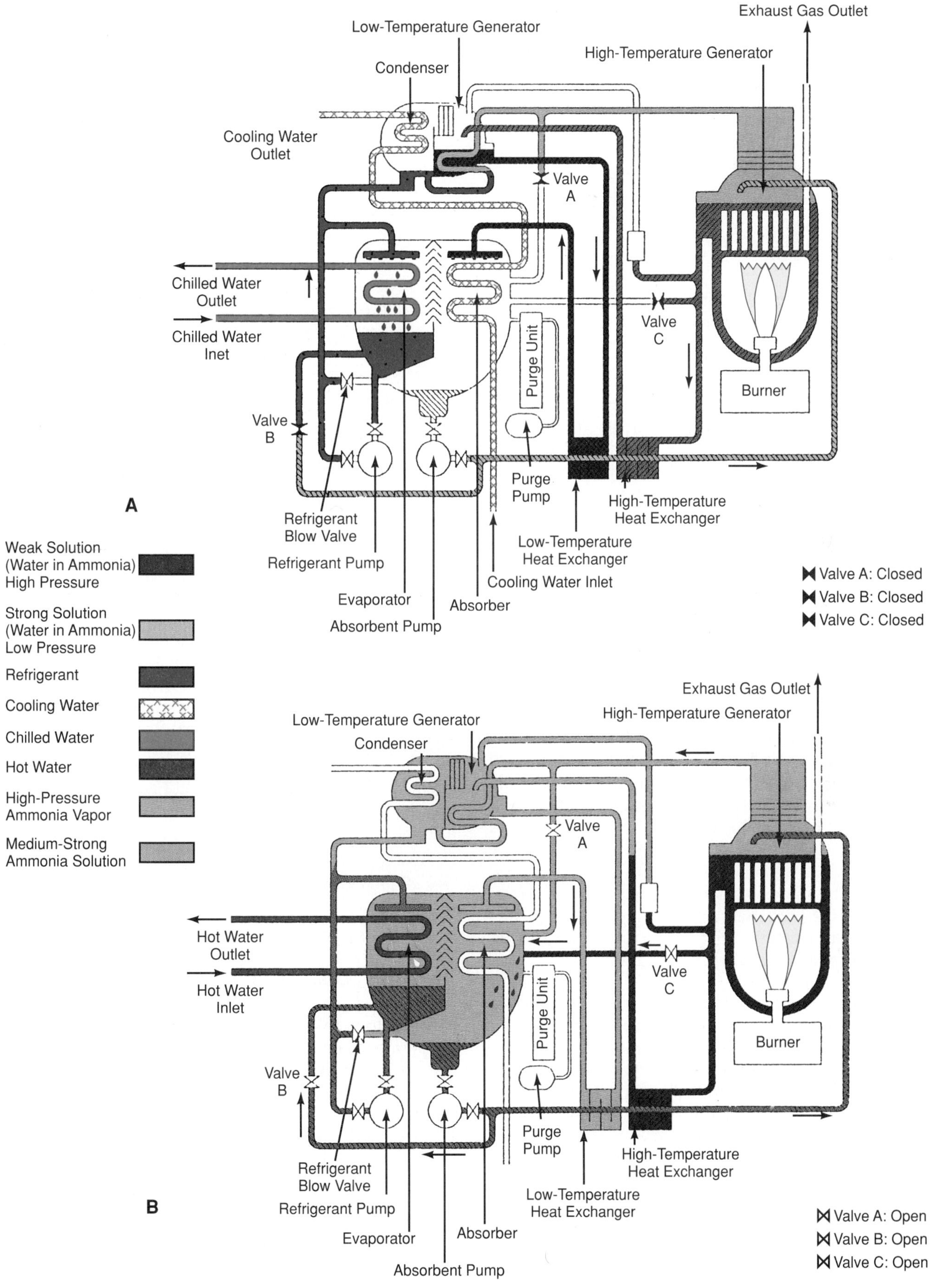

Figure 17-35. *Schematic diagrams of an absorption system. A—Cooling cycle of chiller/heater. B—Heating cycle of chiller/heater.*

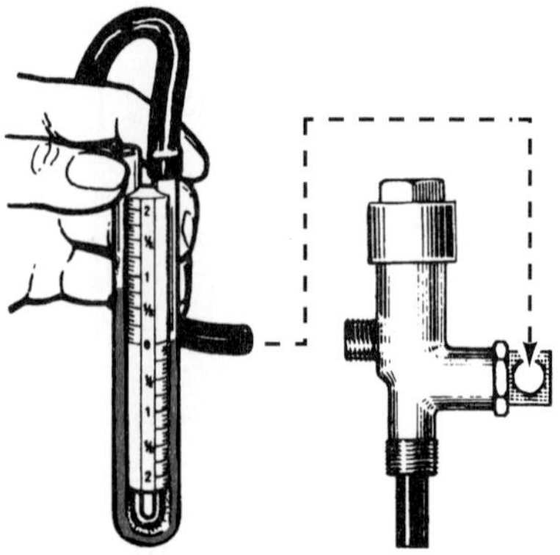

Figure 17-36. *Water-filled manometer measures gas pressure to burner. (General Electric Co.)*

fluids in their proper places. Restart the unit with a lower heat input to the boiler (generator).

If there is a leak, a yellow deposit will collect at the point of leak. If the leak occurs at the evaporator, an ammonia odor will be noticeable. A burning sulfur candle will produce white smoke at an ammonia leak.

17.13.1 Servicing Lithium Bromide Systems

Lithium bromide is a nontoxic, nonflammable, nonexplosive, and chemically stable substance. It is used as a liquid. It can be handled in open containers but becomes corrosive when exposed to air. It may irritate skin, eyes, and mucous membranes. Octyl alcohol is sometimes added to reduce surface tension of lithium bromide. (It acts as a wetting agent.)

Sixty-five percent lithium bromide by weight will start to crystallize at 110°F (43°C). Solution must not be allowed to reach high concentrations or low temperatures which allow crystallization.

The typical charge is a:

- Lithium bromide solution 120 gal.
- Inhibitor 1 pt.
- Refrigerant (water) 35 gal.
- Octyl alcohol (2-ethylhexanol) 1 gal.

The solution becomes thicker as the amount of lithium bromide increases. This will cause a greater temperature difference between refrigerator temperature and chilled water temperature. If solution concentration gets too high, the refrigerant will turn solid and must be dissolved. If the absorber becomes too cold (below 85°F or 29°C), solidification can also occur.

Note: In lithium bromide systems, "strong" (concentrated) solution means strong in ability to absorb. "Weak" (dilute) means weak in its ability to absorb.

The system is charged with R-13 vapor (not soluble in water) to test for leaks. An electronic leak detector is used. Then the system is evacuated completely. Helium may also be used for leak testing. However, it requires the use of a special detector or soap bubbles.

Evacuation of the system is necessary, after the system is opened, for two reasons:

- To be able to reach 40°F (4°C).
- To remove noncondensables.

Evacuation is needed if the system pressure is 1" Hg. (25,400 microns) or more. The pressure is determined by a manometer connected by means of a service valve.

17.14 Review of Safety

The refrigerant most commonly used in the small absorption refrigerating units is ammonia. Its odor is pungent (sharp or irritating) and tends to restrict breathing. It is toxic and injurious to the skin and eyes. Avoid puncturing the system or creating too high a pressure in the system. A leak may result.

Caution: Never cut or drill into an absorption refrigerating mechanism. The high-pressure ammonia solutions are dangerous. They may cause blindness if the fluid gets into the eyes.

Many of the absorption units are heated with LP gas or natural gas. The gas piping system must be leakproof. Always use soapsuds to check for leaks. Never use an open flame, such as a match. An explosion may occur. The burner flues should be cleaned periodically or a poor flame may result.

The flame safety device should be checked. To do this, smother the flame and check to determine if the safety valve closes.

The condenser duct system and the condenser should be cleaned at least every six months. Otherwise excessive condenser pressures may result.

Some absorption systems use electrical current as well as fuel gas. The usual precautions should be used in handling these circuits.

A circuit may become grounded to the cabinet frame or part of the mechanism. This could result in a shock. To eliminate this danger, it is a good idea to ground these refrigerators.

17.15 Test Your Knowledge

Please do not write in this text. Place your answers on a separate sheet of paper.

1. _______ provides heat energy in a continuous absorption system.
   A. Kerosene
   B. Propane gas
   C. Electricity
   D. Any of the above.
2. What type of heat source may be used on an absorption system?
   A. Electric heat or kerosene.
   B. Natural gas or LP gas.
   C. Steam or solar energy.
   D. All of the above.

3. ______ discovered the adsorption principle.
   A. John Dalton
   B. Michael Faraday
   C. Thomas Edison
   D. Jacob Perkins
4. Where are kerosene-fired intermittent absorption refrigerators used?
   A. Where there is no natural gas.
   B. Where there is no electricity.
   C. Where there is no propane.
   D. All of the above.
5. ______ has been used in absorption refrigerators to absorb refrigerant gas.
   A. Water
   B. Lithium bromide
   C. Silver chloride
   D. All of the above.
6. In a continuous absorption cycle refrigerator, what happens when the temperature is lowered in the refrigerator?
   A. The flame increases.
   B. The flame decreases.
   C. The ammonia vapor decreases.
   D. None of the above.
7. Why must the absorption unit be level?
   A. Unit depends on gravity flow for efficiency.
   B. Heat generated in the absorber must be removed.
   C. Heat removed by the condenser must be carried away to the surrounding atmosphere.
   D. All of the above.
8. What is provided on an absorption system to release the refrigerant in case of high temperatures?
   A. Three-way service valve.
   B. Fuse plug.
   C. Pressure safety valve.
   D. None of the above.
9. In a lithium bromide-water absorption system, the refrigerant is ______.
   A. lithium bromide
   B. water
   C. lithium bromide and water
   D. ammonia
10. In a continuous absorption cycle, what occurs when the flame is increased?
    A. The gas consumption increases.
    B. The system produces more cooling.
    C. The system efficiency is reduced.
    D. The system temperature decreases.
11. A ______ solution is used inside a continuous absorption cycle refrigerator.
    A. lithium bromide and water
    B. ammonia and water
    C. lithium bromide and ammonia
    D. None of the above.
12. Upon what does the liquefaction of the refrigerant depend?
    A. Pressure in the heating cycle.
    B. Pressure in the cooling cycle.
    C. Pressure in the cooling and heating cycle.
    D. None of the above.
13. What type of energy does an absorption system use?
    A. Mechanical.
    B. Heat.
    C. Water.
    D. Air.
14. What purpose does hydrogen serve in the continuous absorption system?
    A. It prevents overcooling.
    B. It permits the liquid ammonia to evaporate at a low temperature and pressure.
    C. It prevents overheating.
    D. It causes the liquid ammonia to evaporate at a high temperature.
15. What is the purpose of the insulation on the storage cylinder or receiver in the intermittent absorption refrigerator?
    A. To prevent overcooling.
    B. To prevent condensation from forming.
    C. To provide proper internal temperatures needed for a change of state.
    D. All of the above.
16. For what purpose is lithium bromide used in an absorption system?
    A. It is used as an absorbent where water is the refrigerant.
    B. It is used as an absorbent where ammonia and water are the refrigerant.
    C. It is used where ammonia is the refrigerant.
    D. It can be used in any of the above systems.
17. In a five-ton, ammonia-water air conditioning unit, the solution pump is used to move the liquid from the ______.
    A. absorber to the generator
    B. condenser to the evaporator
    C. generator to the condenser
    D. All of the above.
18. How is defrosting accomplished in an absorption system?
    A. Hot gas.
    B. Reverse cycle.
    C. Intermittent operation.
    D. None of the above.
19. ______ causes the liquid refrigerant to flow in a domestic absorption system.
    A. Gravity
    B. Pressure
    C. Evaporation
    D. Condensation
20. What may cause too little refrigeration in a continuous system?
    A. Little heating of the generating unit.
    B. Improper condensing temperature.
    C. Restricted or dirty gas flue.
    D. All of the above.

A diffuser air suit used for extremely hot conditions. The suit is cooled by a vortex tube cooling device. (ITW Vortec Corporation)

Chapter 18

SPECIAL REFRIGERATION SYSTEMS AND APPLICATIONS

Key Words:

cryogenics
eutectic plates
expendable refrigerants
freeze drying
heat pipe
immersion freezing
jet pump
modules
multistage systems
snow making
Sterling cycle
thermoelectric refrigeration
vortex tube
wick

Learning Objectives:

After studying this chapter, you will be able to:

- Discuss the operation and application of expendable refrigerant systems.
- Define and discuss principles and operation of systems of liquefying gas, thermoelectric refrigeration, vortex tube cooling and heating, jet cooling, multistage cooling systems, heat pipe, immersion freezing, cryogenic refrigeration, and the Sterling cycle.
- Discuss the numerous types of truck refrigeration.
- Explain the process of snow making.
- **Follow approved safety precautions.**

18.1 Transportation Refrigeration

Transportation refrigeration can be divided into four categories: conventional, expendable, ice, and dry ice. The expendable, ice, and dry ice systems are individually designed to maintain the proper temperature of the cargo.

There is little difference between truck systems and commercial refrigeration systems. The truck systems are designed for various ambient temperatures and operating temperatures. The units must also be designed to withstand the stress and vibrations that will occur. Depending upon the system type and size, the truck unit may also utilize back-up units.

18.1.1 Truck Refrigeration

Truck refrigeration requires special trailer bodies and refrigeration units. Such bodies are 9′ to 18′ (3 m to 6 m) long. They use a 1 1/2 hp to 2 hp (1120 W to 1490 W) refrigerating system.

Bodies should be light and well-insulated. Constant vibration and rough handling might destroy the insulating value of the walls if the body is not constructed soundly.

Figure 18-1 illustrates a refrigeration system used on a trailer system. The main components are the compressor, air-cooled condenser, expansion valve, and direct expansion evaporator. These systems commonly use R-134a as a refrigerant. **Figures 18-2** and **18-3** illustrate the cooling and heat-defrost cycles of a diesel-powered unit.

Various insulating materials are used. Most trailers have all-metal bodies with various insulation thickness, depending on the application. Foamed-in-place insulation is most often used. **Figure 18-4** shows this insulation being installed in the side of a trailer body.

There are numerous applications for trailer refrigeration. Each application must be studied before a temperature may be recommended. A truck using dry ice for refrigeration must be insulated for −109°F (−78°C). An ice cream truck must be insulated for −15°F (−26°C). Fresh foods require insulation for only 32°F to 35°F (0°C to 2°C) temperatures. Fresh produce, flowers, and fruits need accurate control of temperature, humidity, and

Figure 18-1. *Refrigerated trailer. Refrigeration system may be either electric, gasoline, or diesel-powered. (Carrier Transicold Division, Carrier Corp.)*

ventilation. These materials have a tendency to lose fluids to the surrounding air.

There are four main truck refrigeration systems:

- Mechanical.
  A. Blower system.
  B. Hold-over eutectic plate.
- Expendable refrigerant (liquid nitrogen and liquid carbon dioxide).
- Ice.
- Dry ice.

Mechanical refrigeration is similar, in most cases, to typical refrigerating units. The major difference is the compressor drive. Two common drives are:

- Engine-driven electric generator and motor.
- Separate gasoline or diesel engines.

Electric generators and motors use standard voltages, cycles, and phases. These permit the unit to be plugged into a wall outlet in the garage, and are useful if the trailer must be kept cold while off the road. Units driven by gasoline engines are automatically controlled to start and stop as the system requires.

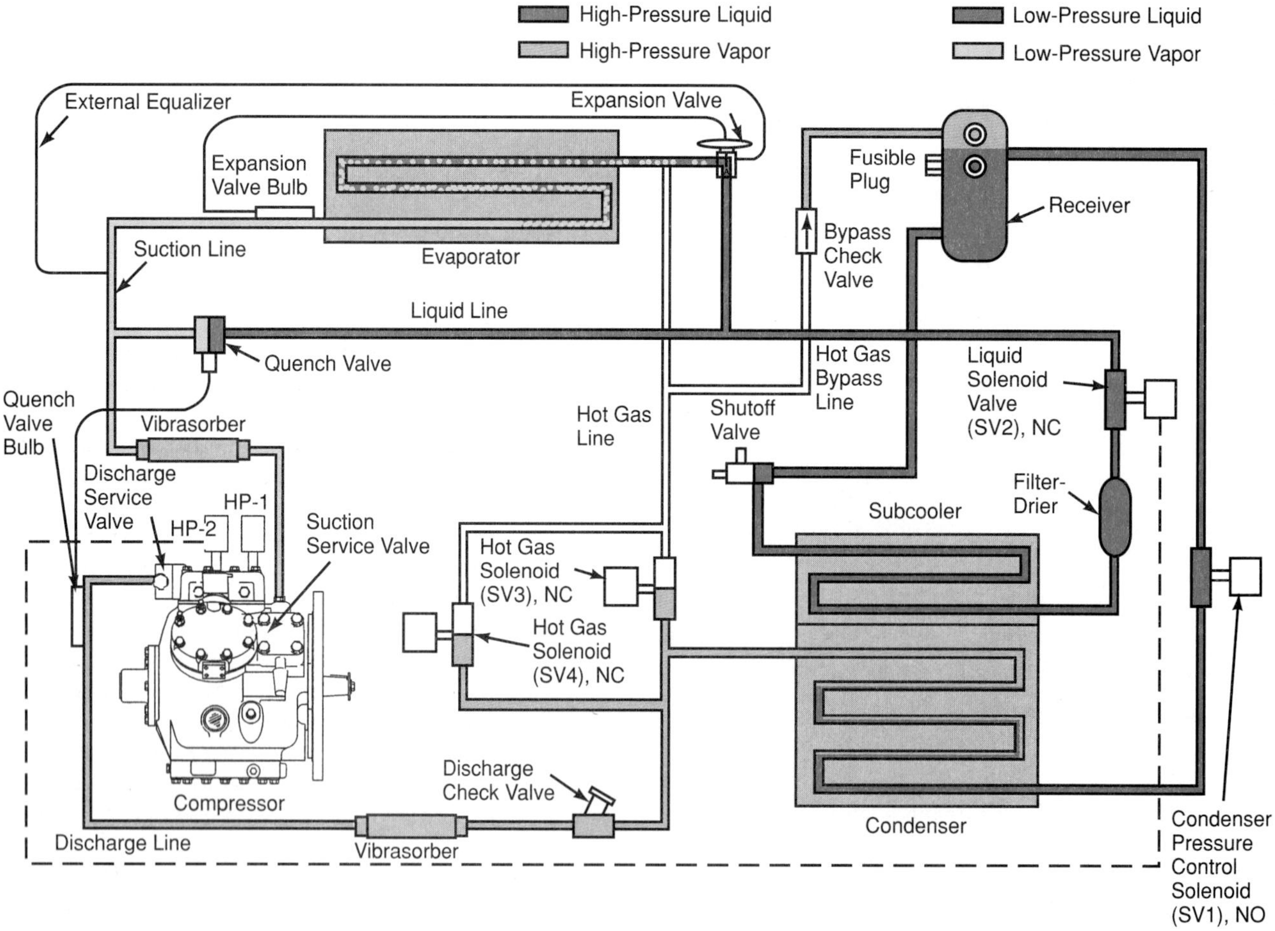

Figure 18-2. *Diagram of a cooling system using a diesel-powered gas engine, showing the cooling cycle. (Carrier Transicold Division, Carrier Corp.)*

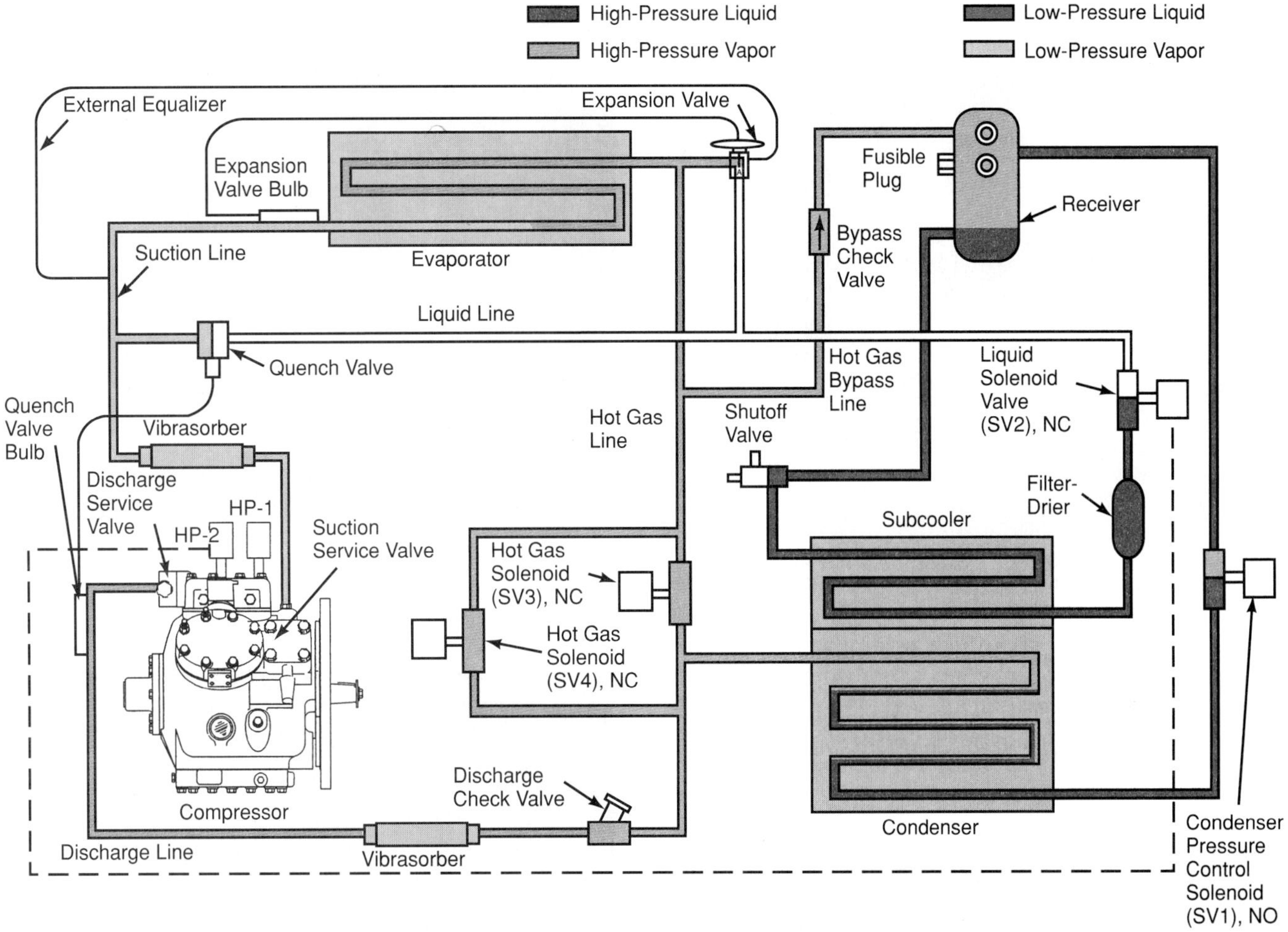

Figure 18-3. *Diagram of the heat/defrost cycle of a cooling system that uses a diesel-powered gas engine. (Carrier Transicold Division, Carrier Corp.)*

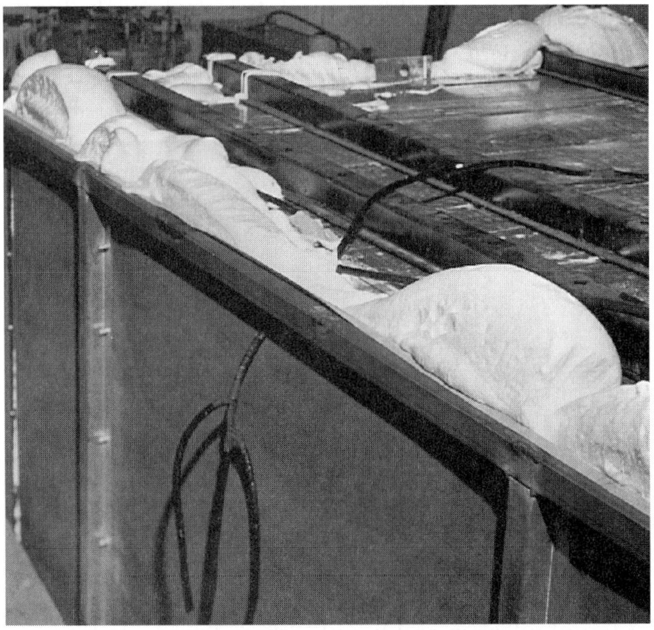

Figure 18-4. *Truck body is being insulated with plastic foam insulation. Excess will be trimmed away.*

Some vans and short trucks use a cube-style, or thinline, wall-mounted evaporator. A roof-mounted or body-mounted condenser and engine-driven compressor are used. There is also a remote Cab Command control module. See **Figure 18-5.** The remote in-cab module includes solid state temperature controller, temperature selector, and manual defrost operations.

Figure 18-6 shows the inside of the standby motor-driven compressor compartment. In addition, a compressor mounted above (and driven by) the engine may be used. See **Figure 18-7.** Most of these systems use a hot gas defrost. Some larger commercial units utilize diesel fuel. In some systems a generator provides power for the evaporator fans. **Figure 18-8** illustrates a typical finished installation mounted over the truck cab.

Eutectic plates operate without the use of mechanical assistance during normal operating. See **Figure 18-9.** A small electrical condensing unit is normally connected at night. This freezes the eutectic solution storing enough energy for positive cooling throughout the day. The plates act basically like a battery with stored energy. The eutectic plates are constructed of small diameter tubing for circulating refrigerant. Fins are attached to the

Spray System

Transport vehicles are also cooled by spraying liquid nitrogen or carbon dioxide directly into the refrigerated space. The nitrogen turns into vapor inside the cargo area. **Figure 18-12** shows a complete spray system with the cylinder inside the truck body.

The liquid spray method has many of the same parts as the cold plate method. (These include, for example, liquid containers, control box, and fill box.) It also requires additional devices not necessary in the plate method, such as spray headers, emergency switches, and safety vents. Another type of spray cooling system is the horizontal cylinder system shown in **Figure 18-13.**

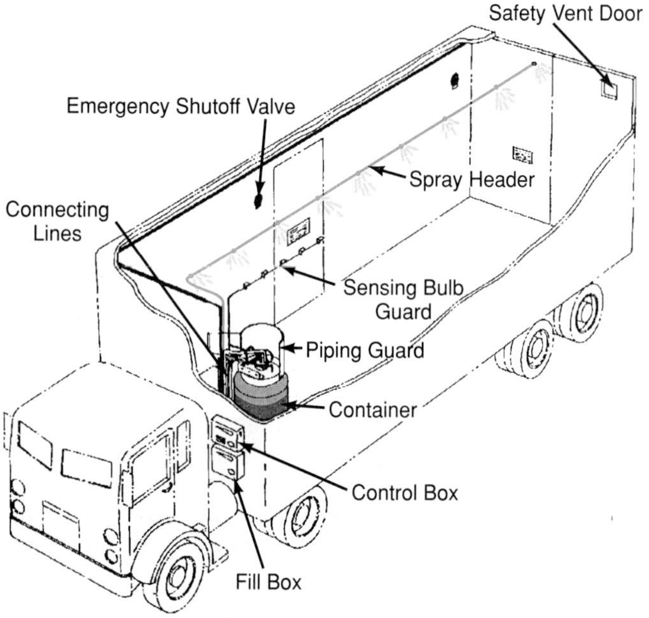

Figure 18-12. *An expendable refrigerant system for a refrigerated truck. Liquid nitrogen is in the insulated container, installed vertically inside the truck body.*

This is how the system operates:

1. Liquid nitrogen is pumped into the storage cylinder by way of the fill box.
2. When the containers are filled and cargo space is loaded, the temperature is selected at the main control. A temperature sensing device anticipates temperature changes in the cargo space.
3. When cargo temperature rises above the setting, the temperature controller opens the liquid line solenoid valve. This allows liquid refrigerant to enter into the spray header. There, it becomes a vapor and maintains the desired temperature.

Some units have two or more containers—a primary container and a secondary container. These are filled in series. As the first or primary container is filled, liquid refrigerant will overflow. It will flow into the second container.

The spray header system is a perforated pipe. It is usually mounted along the roof at the center of the cargo compartment. The nitrogen tanks are equipped with safety valves, which are needed if the container pressure should rise above 22 psi (152 kPa). Nitrogen would then be released to the outside of the vehicle.

Some truck bodies have electrical heaters. These are used when the truck is exposed to temperatures below required cargo space temperature.

Always read the warning signs on refrigerated vehicles before entering them. Liquid nitrogen is released from the spray nozzles at subzero temperatures. If it were to hit part of the human body, flesh would be frozen instantly. Be sure no living animal or human is in the refrigerated space when the doors are closed.

Spray cooling systems using nitrogen or carbon dioxide have additional advantages. Beyond ease in providing necessary refrigerating temperatures, these gases replace oxygen in the storage space. The inert atmosphere preserves fruits, vegetables, meats, and fish, both in transit and in storage.

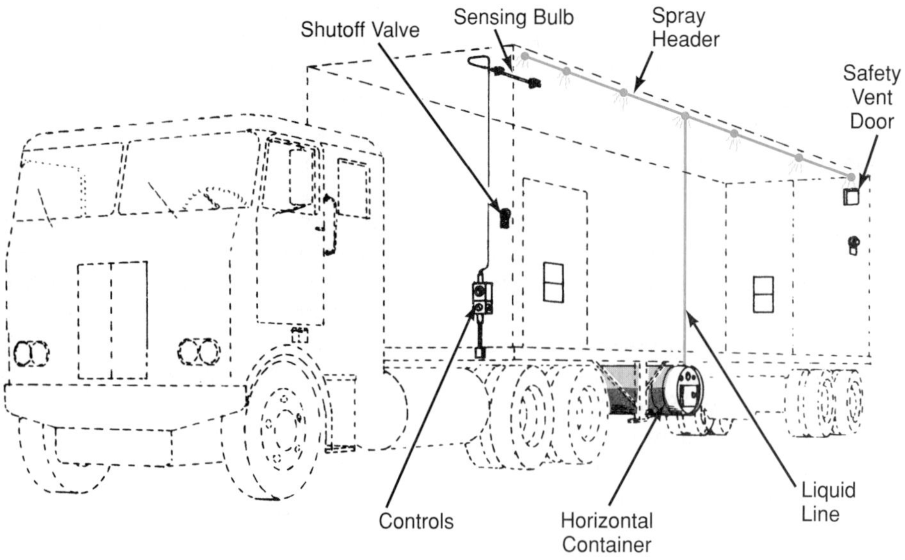

Figure 18-13. *An expendable refrigerant system using liquid nitrogen.*

The inside of the expendable refrigerant systems must be kept free of dirt and moisture. To pressure test the piping, use only nitrogen or helium. See Section 12.9.1.

18.1.5 Refrigerated Containers

Refrigerated containers are used aboard ships, trucks, railroad cars, and airplanes. Perishable commodities gathered in fields or orchards are stored immediately in refrigerated containers. Some containers may be as large as 8′ × 8′ × 20′. For long-distance shipment, the containers may require a refrigeration mechanism. An evaporator and condensing unit may be used.

Condensing units may be driven either with an electric motor or an internal combustion engine. For short distances, a cold plate with a eutectic solution is used. See **Figure 18-14.** (See Chapter 31.)

The eutectic solution is frozen by the evaporator, which is part of the cold plate. It is not necessary to operate the condensing unit during the short trip. The eutectic solution provides considerable cooling for the short time the shipment is in transit.

Containers aboard ships may be refrigerated by being connected to the ship's central refrigerating system. A refrigerating mechanism using an electric motor and condensing unit may be plugged into an outlet. It will then be driven by power aboard the ship. In such cases, the condensing unit is usually water-cooled. For warehouse or dock storage, the local power supply may be used for container condensing units.

Figure 18-14. *Cross sections of a KOLD-HOLD plate eutectic evaporator. (Tranter, Inc., Edgefield Division)*

18.2 Mobile Air Conditioners

Mobile air conditioners are used in industry as emergency or standby air conditioning. They are used to cool and dehumidify manufacturing processing and storage installations and control centers. They are also frequently used for cooling during construction, temporary loss of existing cooling systems, comfort conditioning during athletic or sports events in large astrodomes, and back-up cooling for computer room air conditioners. The cooling capacities vary from 2-ton to 60-ton units. Units are air- or water-cooled and fully self-contained. They operate in high temperatures and in dirty, dusty, or corrosive environments. They are transported on wheels. Thus, they are easily moved from one point to another. A mobile air conditioner is shown in **Figure 18-15.**

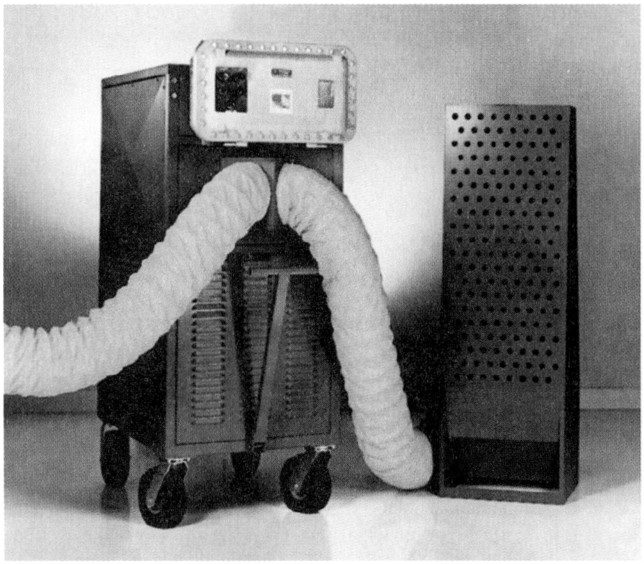

Figure 18-15. *Mobile air conditioner with air diffuser at right. (Scientific Systems Corporation)*

18.3 Thermoelectric Refrigeration

The thermoelectric process removes heat from one area and puts it in another area. Electrical energy is used as a "carrier," rather than refrigerant. It has been used mainly in portable refrigerators, luxury stationary domestic refrigerators, and water coolers. It is also used for cooling scientific apparatus used in space explorations and in aircraft.

Another application of the thermoelectric principle is in computer systems for electronic component cooling. In general, these components are cooled using fans and finned components. In larger systems, chilled water or refrigerant circulation may be necessary. However, as computer systems become smaller, thermoelectric cooling systems are becoming more common.

Thermoelectric cooling requires none of the conventional equipment necessary in a vapor system. There is no compressor, evaporator, condenser, or refrigerant. In fact, there are no moving parts. The unit is silent, compact, and requires little service. **Figure 18-16** shows an electrical circuit diagram for a typical thermoelectric cooler power supply.

The input of 120 V ac is stepped down in a transformer to 20 V ac. This current passes through a rectifier and is changed to 20 V dc. The direct current is then passed through the thermoelectric module. The junction inside the refrigerator becomes cold and the junction outside the refrigerator is warmed. Principles covering operation of the thermoelectric refrigerator are explained in Chapter 6.

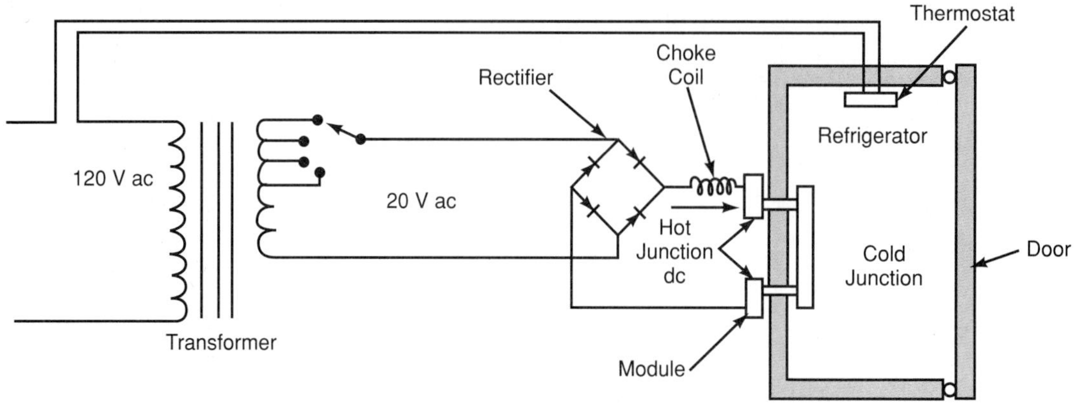

Figure 18-16. *Electrical circuit for thermoelectric refrigerator.*

Thermoelectric cooling units, when used in refrigeration, are called ***modules.*** A module consists of several cold and hot junctions in series. The diagram of such a module is shown in **Figure 18-17.** The letters *P* and *N* do not refer to current polarity positive (+) and negative (−). In thermoelectric units, *P* and *N* refer to the properties of the semiconductor materials. The materials are also designated as positive or negative, depending on how the semiconductor electrons behave under the influence of current flow. Construction of a module attached to a refrigerator is shown schematically in **Figure 18-18.** Direction of the direct current flow into the module determines whether the junction is warmed or cooled.

The reversing switch is usually made a part of the electrical circuit. In this way, the cabinet may be caused to either cool or warm the food.

Thermoelectric refrigerators have a low coefficient of performance (COP), but they are also very versatile. They can operate on 12 V dc power or from a 110 V ac power adaptor. This has made the units popular for picnics and camping use. One of these refrigerators is illustrated in **Figure 18-19.**

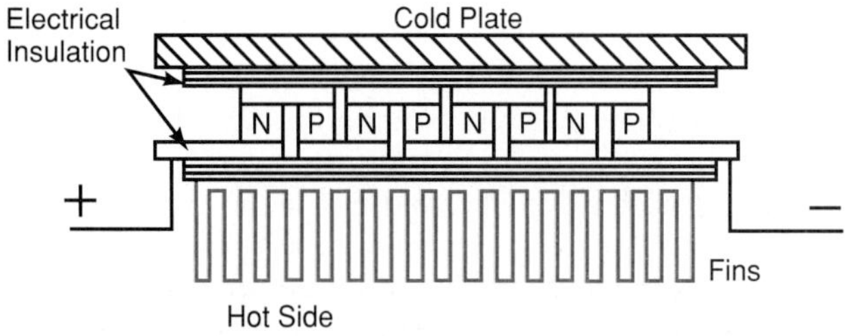

Figure 18-17. *Assembled thermoelectric module. Note use of fins (hot side) to speed up removal of heat that has been absorbed from surface of the cold plate. (Koolatron Industries)*

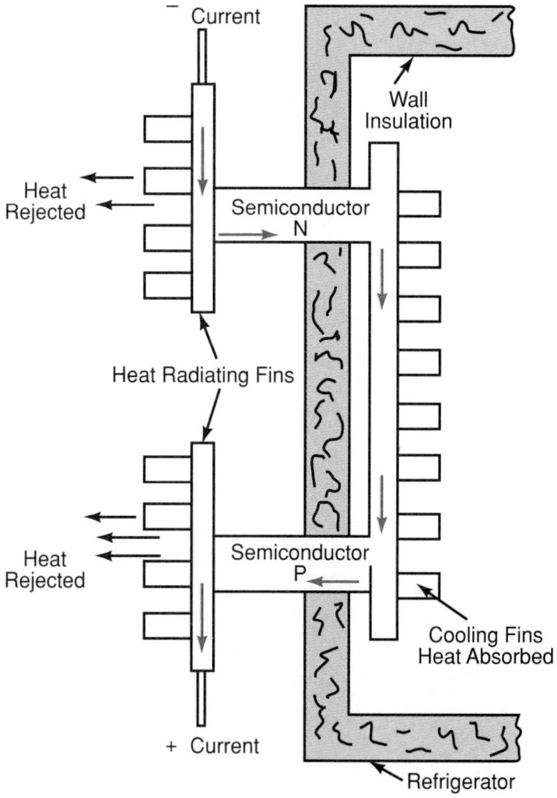

Figure 18-18. *Diagram of simplified thermoelectric system used in cooling small areas. Note flow of current to produce cooling within box.*

18.4 Vortex Tube

Fluid (or air) that rotates around an axis, like a tornado, is known as a ***vortex.*** A ***vortex tube*** creates a vortex from compressed air and separates it into two air streams—one hot and one cold. See Chapter 20.

Vortex tubes can be easily adjusted to deliver cold air down to −50°F (−46°C) or hot air up to 250°F (121°C) when supplied with 100 psig (794 kPa) compressed air at 70°F (21.1°C). (These temperatures, however, cannot be arrived at simultaneously.)

Figure 18-20 is a schematic drawing of a vortex tube. Compressed air enters the vortex generation chamber, which is proportionately larger than the hot (long) tube, where it causes air to rotate. The rotating air is discharged tangentially (moving in a straight line) along the

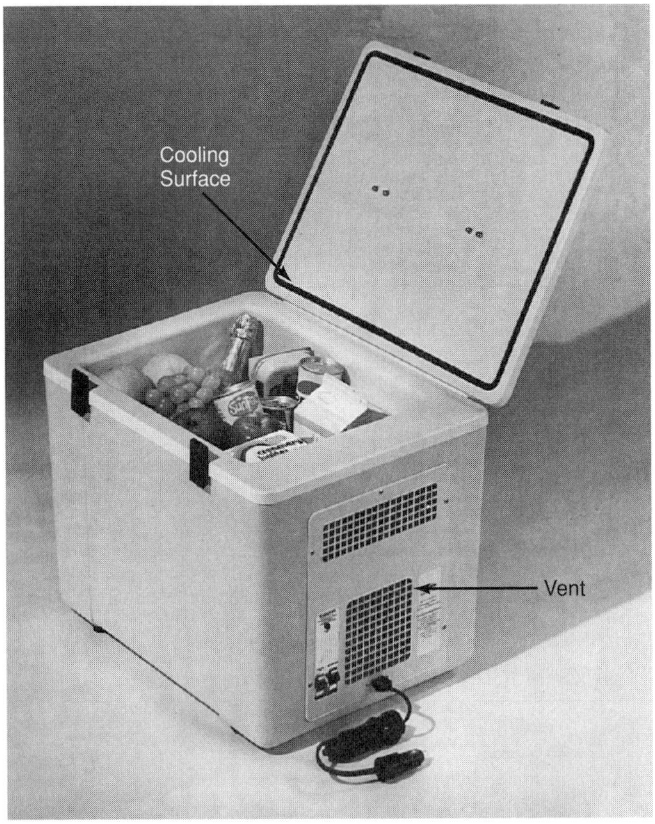

Figure 18-19. *Small portable refrigerator operates from normal 120 V household current or from 12 V battery. Airflow through vent keeps junctions cool.*

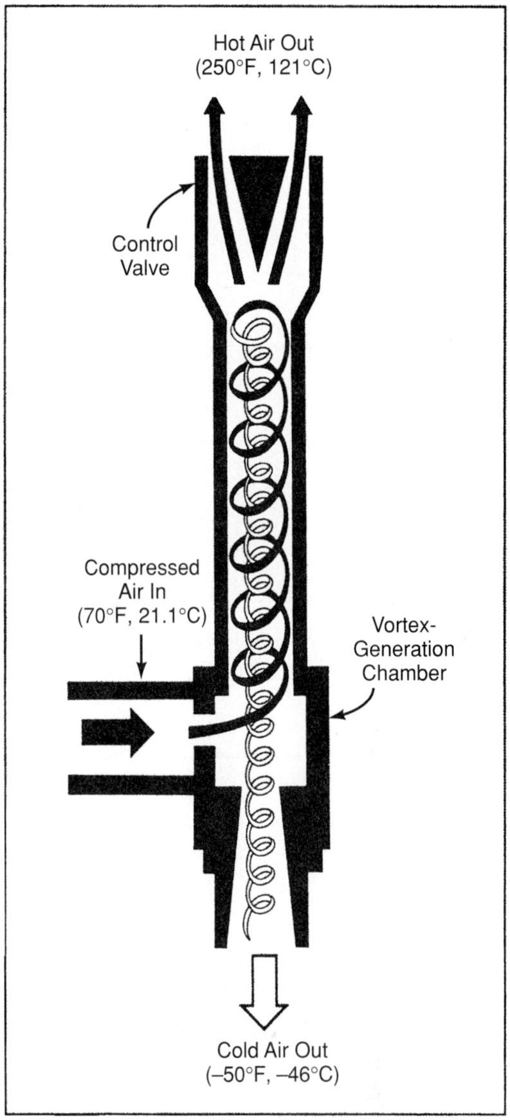

Figure 18-20. *Schematic drawing of a vortex tube. (ITW Vortec Corporation)*

inside surface of the tube. As the compressed air expands into the tube, its pressure will decrease while the velocity of the air stream assumes near sonic speed (1000 rpm). The air follows a centrifugal (away from the center) path down the inside surface of the tube. There a portion of the air stream is released through the control valve as hot air. The remainder is forced to form a cold counterflowing air stream up the center of the tube.

Vortex tube cooling has many different applications. **Figure 18-21** illustrates four different types of vortex tubes. The vortex tube is commonly used in various manufacturing processes. **Figure 18-22** illustrates its use in machining operations to cool metals and plastics. It is also used as a means for setting solders and hot metals, and curing adhesives. In scientific work, it is used to dehumidify gas samples, **Figure 18-23.** It is also used in cooling environmental chambers and electronic cabinets. See **Figure 18-24.**

A common application of vortex tube cooling is the cooling of clothing. A diffuse-air vest, worn under a protective cover, is frequently used. The tempered air is distributed over the upper body, where it is needed most. The vest has tiny holes, which allow cool air from the vortex tube to escape over the upper body. **Figure 18-25A** shows a worker wearing the vest while spraying in a confined area. **Figure 18-25B** shows a worker wearing a personal cooling system under his welding leathers. If the entire body is to be cooled or heated, a total air respiratory system is used.

18.5 Jet Cooling Systems

A ***jet pump*** consists of a centrifugal pump and ejector. It can replace the compressor in some air conditioning and refrigeration systems. Two applications involve ***steam jets*** and ***refrigerant jets.*** The refrigerant jet uses R-11 or R-12 (or recent replacement refrigerants) as the working fluid.

Steam jet operation is based on the fact that water under high vacuum boils at a relatively low temperature. This causes evaporation to occur and reduces the temperature. Chapter 20 shows a table of water boiling temperatures under various vacuum pressures.

Water is the refrigerant in this type of system. Therefore, only temperatures down to about 40°F (4°C)

Figure 18-21. *Four vortex tubes—600 BtuH, 2500 BtuH, 1500 BtuH, and 400 BtuH. (ITW Vortec Corporation)*

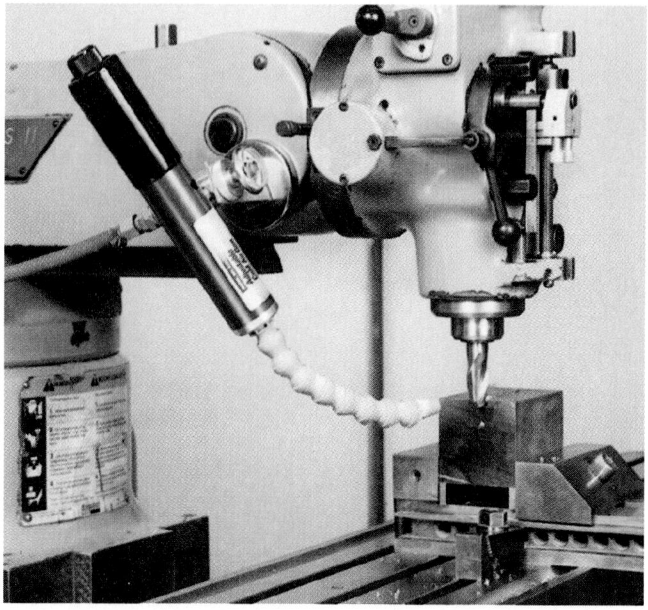

Figure 18-22. *Cold air gun replaces mist coolants. (ITW Vortec Corporation)*

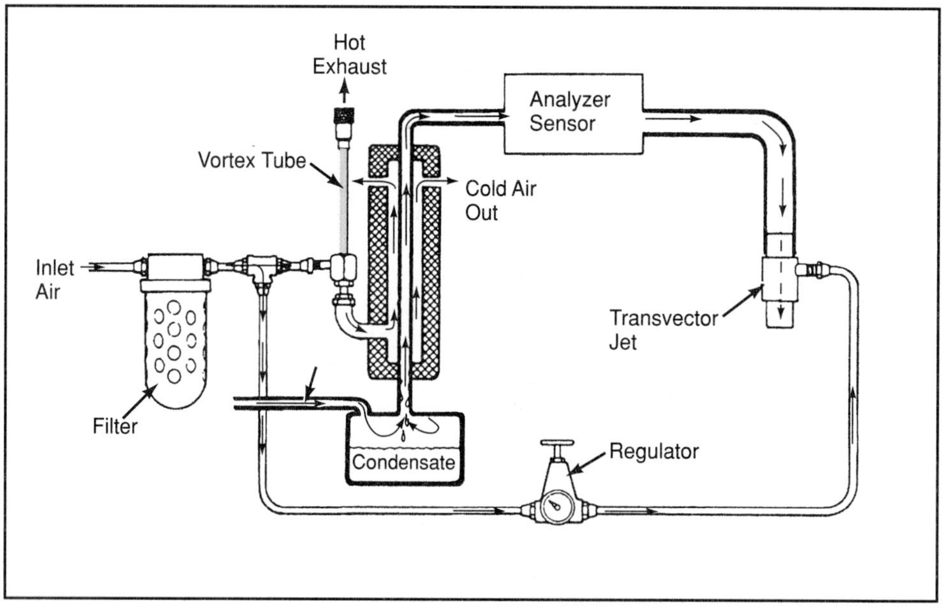

Figure 18-23. *Vortex tube used to cool flue gas samples for off-stack analysis. (ITW Vortec Corporation)*

are possible. Exhaust steam from a high-pressure steam operating machine is often used for a steam jet.

In a refrigerant jet, refrigerant vapor at high pressure and temperature flows through the ejector nozzle to the condenser. **Figure 18-26** shows a diagram of an air conditioning system using a refrigerant jet pump. These units are inefficient and require a large condenser to remove heat. This type of cooling system is found in commercial installations where waste heat is available.

Both of the above systems show how energy conservation can be incorporated into high waste applications.

18.6 Multistage Systems—Cascade and Compound

Ultracold temperatures are not economically obtainable using a single-stage system. The compression ratios may be too high to obtain necessary evaporating and condensing vapor temperatures. In such cases, a ***multistage system*** is used.

The name, multistage, applies to any refrigeration system with more than one stage of compression. There are two general types: cascade and compound.

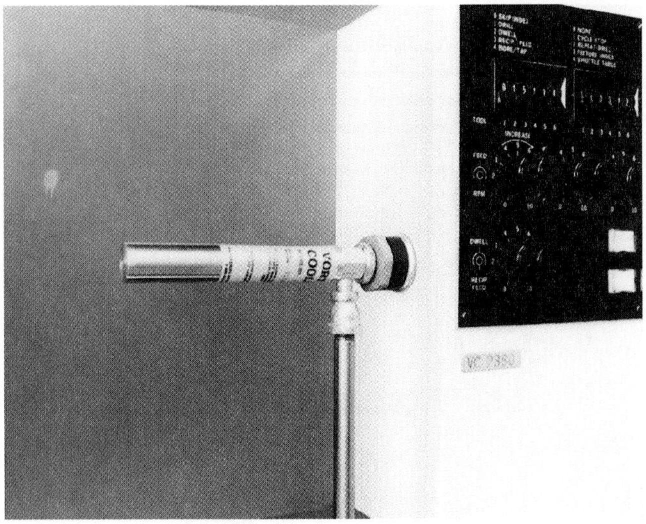

Figure 18-24. *Vortex cooler installed in a NEMA 4 / 12 cabinet used to house electronics. (ITW Vortec Corporation)*

In the ***cascade system,*** **Figure 18-27,** two separate refrigerant systems are interconnected. The evaporator from one unit is used to cool the condenser of the other unit. This allows one unit to operate at a lower temperature and pressure than possible with the same size single-stage system.

Cascade systems may be used to produce temperatures below −250°F (−157°C). See ***Cryogenics*** in Section 1.32. This system is actually two independent units. It allows, if desired, the use of two different refrigerants.

Compound systems obtain low temperatures by using several compressors connected in series to the same refrigeration system. **Figure 18-28** is a diagram of such a unit. It can increase the performance and efficiency of low-temperature refrigeration systems.

Vapor in a low-temperature system has a high specific volume. In a single-stage system, this would require longer-than-normal compressor piston strokes operating at high speeds. Because of the temperature it would also reduce the volumetric efficiency. The cost of operation would be too great. In the compound system, the first stage compressor is larger than the secondary stage compressor. In each stage, the compressor gets smaller because the higher the stage, the more dense the vapor.

Compound refrigeration systems using two-stage compression equipment can produce temperatures from −20°F to −80°F (−29°C to −62°C). Three-stage systems (three compressors in series) can maintain temperatures of −135°F (−93°C).

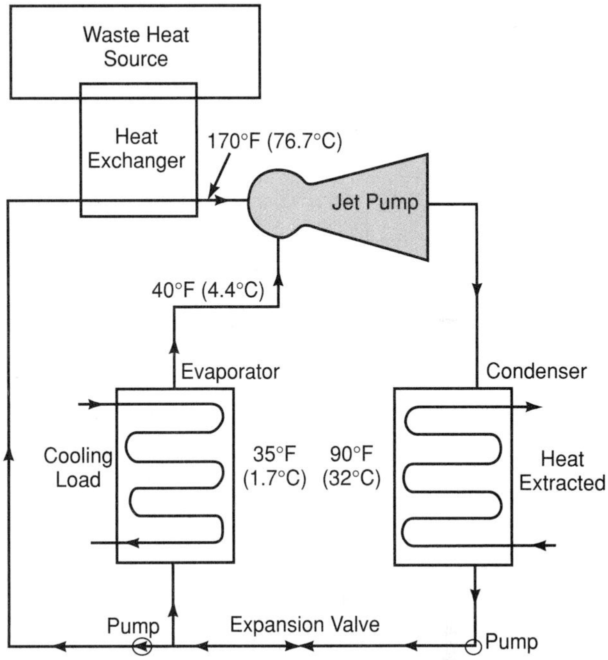

Figure 18-26. *Refrigerant jet pump air conditioning system.*

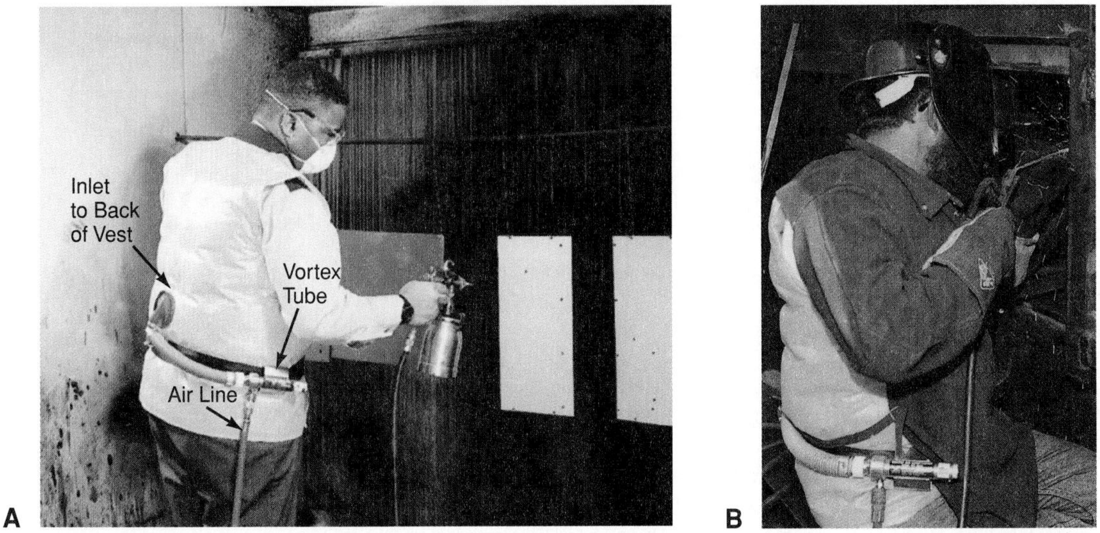

Figure 18-25. *Vortex personal air conditioners provide relief from hot environments. A—Worker using a diffuser air vest in a hot, confined area. B—A vortex cooling system used under welding leathers. (ITW Vortec Corporation)*

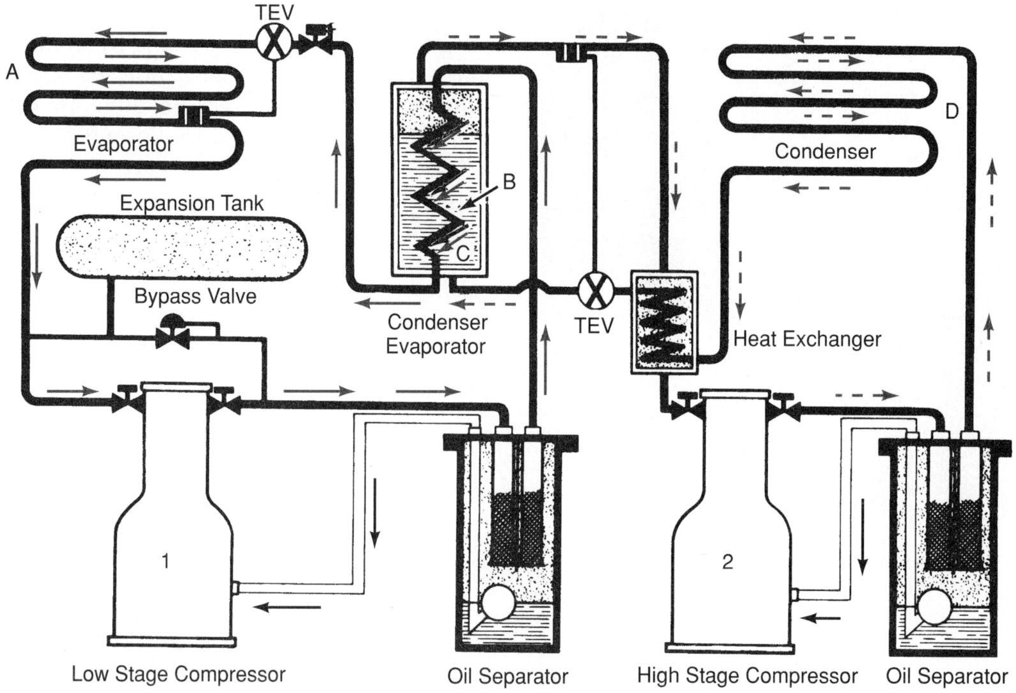

Figure 18-27. *Cascade refrigerating system. Condenser B of system No. 1 is being cooled by evaporator C of system No. 2. This arrangement enables ultracold temperatures in evaporator A of system No. 1. Condenser of system No. 2 is shown at D. Note use of oil separators to minimize circulation of oil.*

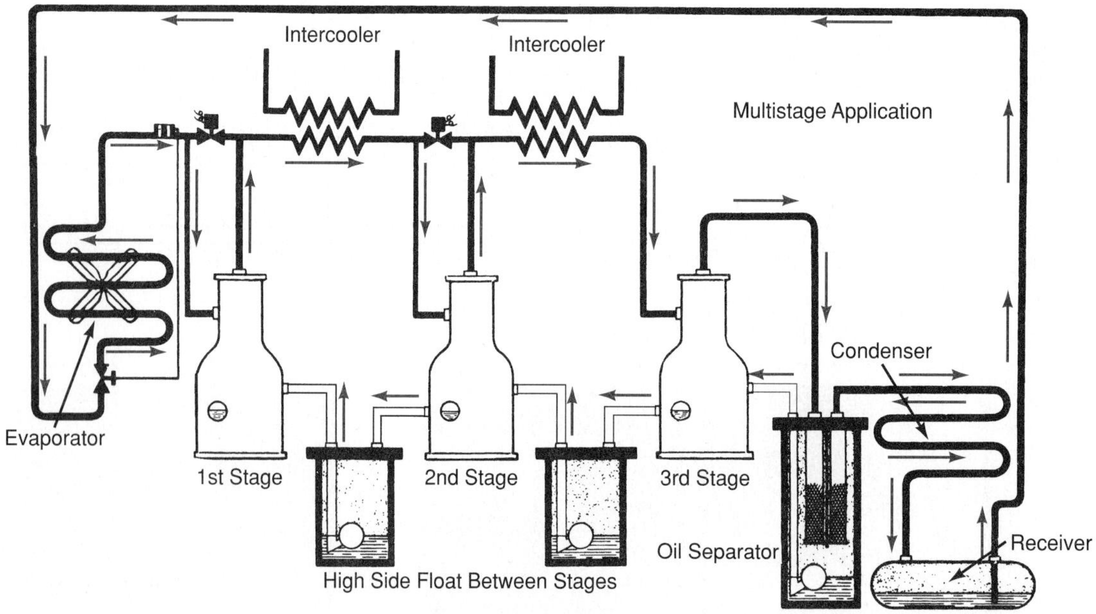

Figure 18-28. *Multistage refrigerating system using three compressors (stages). Compressor No. 1 pumps vapor into intercooler and then into intake of compressor No. 2. This operation is repeated between second and third stages. In third stage, refrigerant vapor is further cooled and travels to evaporator for specific cooling use.*

18.7 Snow Making

Many ski slopes add to the natural snow on their runs with artificial snow. Artificial snow is a water spray into which compressed air is added. In this way, the water is broken up into a fine mist and freezes very rapidly. See **Figure 18-29.**

The low temperature required for snow making is not created by refrigeration mechanism. It comes from the surrounding atmosphere. The temperature should be 30°F (−1°C) or lower. With low relative humidity, snow may be made at temperatures as high as 35°F (2°C). This is possible because, with low humidity, some of the water evaporates. In doing so, it absorbs heat. This reduces

Figure 18-29. *Snow-making equipment being used on a ski slope. The system breaks up the water into a fine mist, which freezes rapidly. (Larchmont Engineering)*

the temperature of the water droplets below the freezing temperature. Ice crystals (snow) are formed.

In general, the ground must be frozen before snow making is practical. Normally, manufactured snow only supplements natural snow.

18.8 Heat Pipe

R.S. Gaugler developed the basic principle of the heat pipe in 1942. It has been used in aerospace work as well as in industrial and domestic applications. Its purpose is to transfer heat from one location to another.

The ***heat pipe*** is an evacuated, hermetically sealed chamber containing volatile working fluid. Its inside walls are lined with a porous substance called a ***wick.*** **Figure 18-30** shows a cross-section of such a pipe.

Heat is applied to one end (heat in). The liquid vaporizes and vapor flows to the opposite end. When the heat is removed, the vapor condenses into a liquid again. The liquid returns to the evaporator, or hot end, of the pipe through the wick, completing the cycle.

Return of fluid as a liquid to the evaporator is accomplished by gravity flow. As the liquid is vaporized in the evaporator section, voids are left in the wick's porous structure. Liquid travels from the condenser section to fill these voids by capillary action. This motion is a result of the attractive forces between the liquid and the wick material and the surface tension of the liquid.

The heat pipe has been used in aerospace projects where it cools instruments or devices. Another successful application has been found in cooking. A heat pipe (pin) may be inserted into a piece of meat. When placed in a conventional oven, it cooks the meat from the inside.

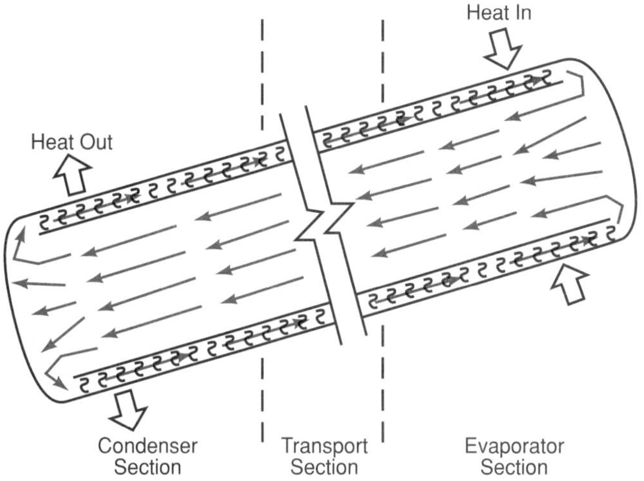

A—Side View

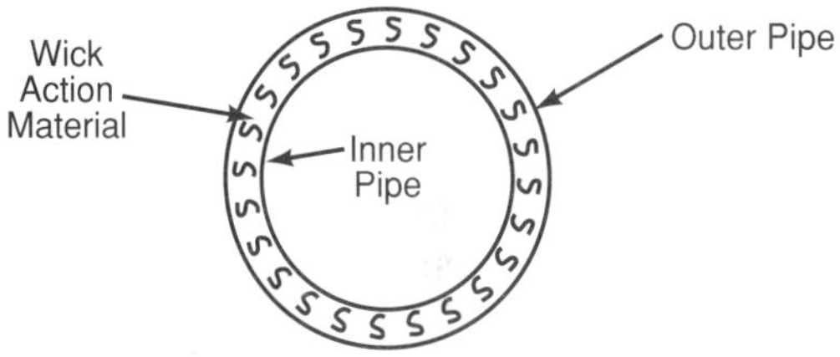

B—Pipe Cross Section

Figure 18-30. *A—Basic design of heat tube. Fluid evaporates at the evaporator section as heat travels into tube. Vapor moves the length of the inner tube to the condenser, where fluid releases enough heat to become liquid. The liquid then travels through the wick (by capillary action) back to evaporator. Transport section does not gain or lose heat. B—Cross section shows end view of heat tube.*

Another heat pipe application is for heat recovery from a furnace or boiler flue. With the heat pipe, the high stack temperature in a flue can provide additional heat, useful in heating basements and garages. **Figure 18-31** shows how a heat pipe can be used to recover heat. The heat is recovered from warm exhaust air during the heating season and cooled exhaust air in the cooling season.

The Alaskan pipeline provides another good application for heat pipes. The pipeline transports oil above ground across Alaska, Canada, and the United States. The pipeline itself is supported on pilings that have been driven into the ground. The designers faced the problem of freezing and thawing. This freezing and thawing would shift the pilings, subsequently causing cracks in the pipeline. Their solution was to incorporate liquid ammonia heat pipes into the pilings. In cold weather, the ammonia vapor at the top of the heat pipe is cooled. It then condenses and flows down below ground level. The

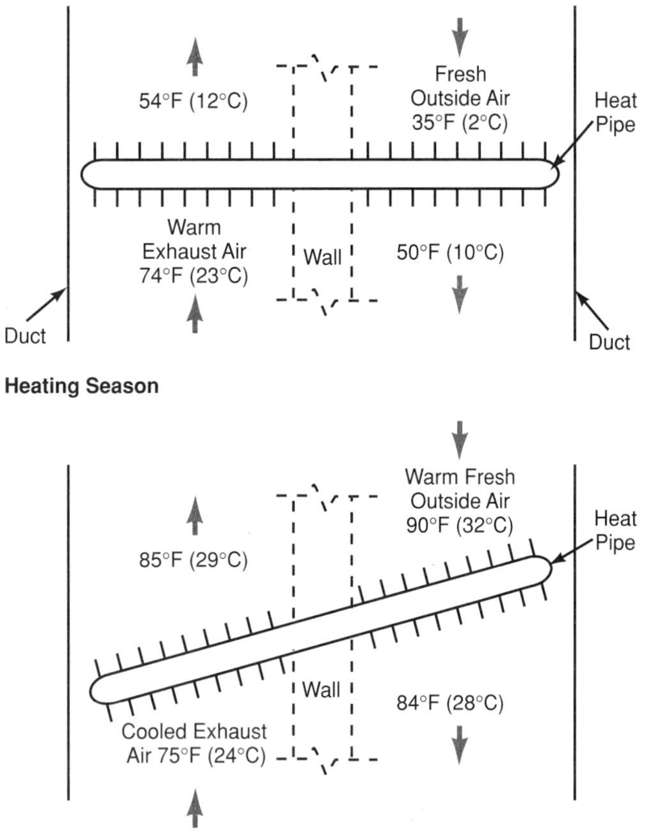

Figure 18-31. *Heat pipe application. Pipe is used to transfer heat between building exhaust air and fresh incoming air.*

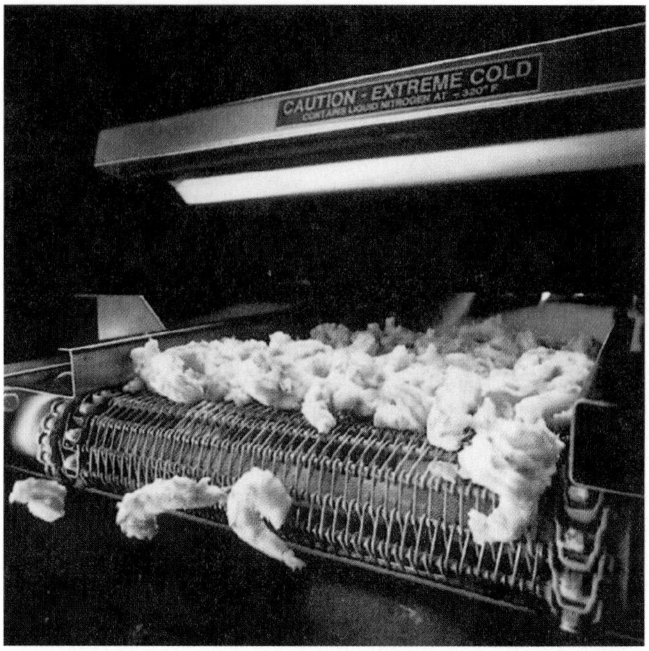

Figure 18-32. *Liquid nitrogen immersion freezing unit used to freeze cooked shrimp. (Compressed Air Magazine)*

liquid ammonia, below ground level, evaporates and removes heat from the surrounding ground. The area surrounding the piling remains in a deep freeze even throughout the warmer months.

18.9 Immersion (Fast Freeze)

Immersion freezing consists of dipping articles to be frozen into liquid refrigerant. For fast freezing of food, the usual refrigerant is specially prepared liquid R-12 or R-22. For some very low-temperature applications, liquid carbon dioxide or liquid nitrogen may be used. See **Figure 18-32.**

The temperature is so low that refrigerant should never be allowed to touch the worker. It would result in immediate freezing of the skin.

In this freezing system, some liquid refrigerant boils and is vaporized in absorbing heat from the food. Some larger installations recover the vaporized R-12 or R-22. It is recovered by using a refrigerated, finned coil placed over the liquid refrigerant. This will condense the vapor. It may be collected and returned to the refrigerant storage tank. With liquid carbon dioxide or nitrogen, however, the refrigerant is expendable.

18.10 Cryogenic Refrigeration

Cryogenic temperatures are between −459.7°F (absolute zero) (−273°C) and −250°F (−157°C). Many substances are completely different in character at those temperatures. For example, a steel plate at −350°F (−212°C), when dropped on the floor, will shatter like glass.

Gases such as oxygen, nitrogen, argon, hydrogen, and helium can be liquefied and separated at such temperatures.

Three methods are used to create these low temperatures:

- Expansion process with heat exchangers (Joule-Thompson process).
- Expansion process with heat exchangers and with gases performing work.
- Multiple cascade system.

The science of cryogenics has produced many useful instant freezing and refrigeration techniques. Nitrogen at −350°F (−212°C) is used to instantly freeze plant and animal tissue or other fast decaying items. In this way they are preserved for future study. Liquid oxygen is a convenient way to store millions of tons of rocket fuel. It is also used as a subzero chiller for steel, making it less prone to warpage.

Liquid nitrogen, poured out into more common ambient temperatures, immediately becomes a vapor. The temperature of this vapor is −320°F (−196°C). This process is used in some cargo trucks requiring deep-freeze conditions. It is used in immersion freezing, **Figure 18-33.** It is also used to "freeze dry" foods. ***Freeze drying*** is

Figure 18-33. *Cryogenic food freezing in batch production. (Compressed Air Magazine)*

a cryogenic process where foods are instantly frozen. A strong vacuum is used to remove all ice crystals. The resultant food, if kept completely dry, no longer requires any refrigeration.

Special care must be taken when operating and servicing cryogenic equipment. Liquids are ultracold and will severely injure anyone coming into contact with them. Certain parts of the system are also ultracold and could cause injury.

Caution: You should always wear goggles and gloves when servicing any part of cryogenic equipment. Cryogenic temperatures are very low. Liquid refrigerant will instantly freeze any part of the body it is allowed to touch.

In the presence of such low temperatures, special materials must be used in construction. These include nickel (stainless) steels, copper alloys, and aluminum metals. Electrical devices must be of special design and construction.

Lubrication, too, is a special problem because of the extreme cold. If any is used, it must be a dry lubricant (molybdenum sulfide, for example).

18.11 Sterling Cycle

This refrigerating cycle was originally developed in 1816 by Robert Sterling. The Sterling cycle was adapted for refrigeration by John Herschel in 1834. The first practical machine was built in 1845. It is now being used in some refrigeration installations which operate at −110°F (−79°C) down to −300°F (−184°C). This cycle, when used as a three-stage system, can produce temperatures down to −450°F (−268°C). These compact units use helium and hydrogen gases.

The ideal system will pick up heat only at the lowest temperature. It will be discarded only at the highest temperature. There can be no heat gain or heat loss between these two temperatures. The Sterling cycle is almost as good. It conserves the energy and uses it in another part of the cycle.

The system uses one cylinder and two pistons with a stationary regenerator between the pistons. See **Figure 18-34.** In view A, the right piston is stationary and next to the regenerator. The left piston is at the beginning of compression. Then in view B, the left piston compresses the gas. The gas is cooled (no temperature rise).

In view C, the left piston now completes its stroke. At the same time, the right piston moves to the right. Thus, the volume of trapped gas remains the same. The regenerator collects heat during this operation. In view D, piston No. 2 moves to the right. The gas cools by expansion. This gas is heated by warming the right side of the cylinder. Finally both pistons move together and return to their position in view A.

During this time, all the gas passes through the regenerator. Heat is added to the gas from the regenerator during this action.

In practice, the left end of the cylinder is water-cooled. The right end of the cylinder is the cooling unit. The regenerator must allow gas to pass through and it must have a good heat-absorbing ability.

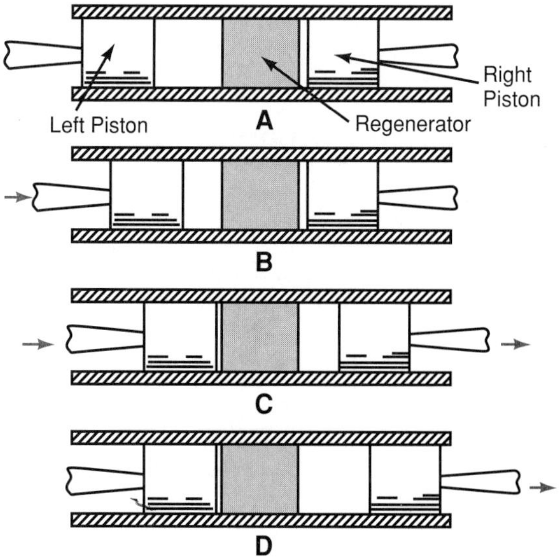

Figure 18-34. *Basic actions of Sterling Cycle system. A—Start position. B—Compressing gas and cooling. C—Moving gas through regenerator. D—Expanding gas (it absorbs heat). Pistons now return to positions in A.*

18.12 Review of Safety

When working with electrical equipment, the equipment being tested must always be fully grounded. Grounding may be through either the use of a polarity plug or a ground wire. See Chapter 8. Local and national refrigeration and electrical codes should be followed when servicing and installing units.

Before checking a thermoelectric system, get a wiring diagram of the system. Use the diagram to be certain that the polarity is not reversed.

Certain precautions must be taken when checking expendable refrigerant systems. Make certain safety doors are opened and the truck body vented before entering the conditioned space. Before entering a unit being cooled, make certain that all refrigerant flow control valves are closed.

As repeated throughout the text, always wear goggles when checking a unit that uses refrigerant.

Before working on any part of an expendable refrigerant system, you must know the required pressures. The nature of the safety valves and controls used in the system must also be known.

As indicated in Section 18.6 and Section 18.10, temperatures in the cryogenic range are below −250°F (−157°C). These temperatures are dangerous. The rate that heat is removed from the body surface at these temperatures is great. Flesh may be severely frozen before you feel the cold.

Remember, most expendable refrigeration systems use carbon dioxide or nitrogen. Humans and animals cannot live in atmospheres of either of these substances.

In handling any type of refrigerants, the operator should wear gloves, and face shield or goggles. Make sure that no liquid refrigerant is ever allowed to touch the skin. Handling such refrigerants as liquid nitrogen, liquid air, and liquid carbon dioxide is particularly dangerous.

18.13 Test Your Knowledge

Please do not write in this text. Place your answers on a separate sheet of paper.

1. Which of the following is an advantage of a thermoelectric system as compared to a compression system?
   A. It is silent.
   B. It has no moving parts.
   C. It requires little service.
   D. All of the above.
2. Which type of current does a thermoelectric system use?
   A. Alternating current.
   B. Direct current.
   C. It depends upon the use of the system.
   D. None of the above.
3. How can a thermoelectric module used for cooling be converted into a heating unit?
   A. It cannot be converted to a heating unit.
   B. Add an ac adaptor.
   C. Reverse the current flow through the junction.
   D. None of the above.
4. What is the main refrigerant used in an expendable system?
   A. Nitrogen.
   B. R-12.
   C. R-134a.
   D. None of the above.
5. ______ is a type of expendable refrigerant system available today.
   A. Cold plate cooling
   B. Spray cooling
   C. Both A and B.
   D. None of the above.
6. One application of a cascade system is to ______.
   A. provide a compound system
   B. operate a single-stage low-temperature system
   C. obtain a temperature of −300°F (−184°C)
   D. obtain a temperature of −250°F (−157°C)
7. What is the least number of compressors a multistage system may use?
   A. Four.
   B. Two.
   C. Three.
   D. None of the above.
8. In a cascade system, the ______ of the ______ pressure system cools the ______ of the ______ pressure system.
   A. condenser, lower, evaporator, higher
   B. condenser, higher, evaporator, lower
   C. evaporator, higher, condenser, lower
   D. None of the above.
9. What type of chemical is used in expendable refrigerant spray systems?
   A. Ammonia.
   B. Carbon dioxide.
   C. Nitrogen.
   D. Both B and C.
10. How is the fluid in a heat pipe returned to the evaporator, or heat source?
    A. Through internal system pressures.
    B. Through gravity return.
    C. By wick action.
    D. None of the above.
11. What determines the amount of hot air released from a vortex tube system?
    A. The generation chamber.
    B. The control valve at the end of the cold outlet.
    C. The jet pump.
    D. The control valve at the end of the hot air tube.
12. ______ is used as a coolant in a vortex tube system.
    A. Moving air
    B. R-12
    C. Ammonia
    D. A solution of water and brine

13. What supplies the energy to a steam jet cooling system?
    A. A compressor.
    B. Steam under pressure.
    C. Nitrogen.
    D. Electricity.
14. Which of the following power sources drive truck refrigeration compressors?
    A. Separate gasoline engines.
    B. Engine-driven electric generator and motor.
    C. Separate diesel engines.
    D. All of the above.
15. A steel block that has been cryogenically frozen to −350°F (−212°C), when dropped on the floor, will ______.
    A. break in two
    B. develop cracks in it
    C. shatter like glass
    D. Any of the above.
16. Nitrogen tanks are equipped with safety valves in case container pressure rises above ______.
    A. 30 psi (207 kPa)
    B. 25 psi (172 kPa)
    C. 22 psi (152 kPa)
    D. 100 psi (689 kPa)
17. You should use ______ to pressure test piping in an expendable refrigerant system.
    A. nitrogen
    B. helium
    C. carbon monoxide
    D. Either A or B.
18. Thermoelectric refrigeration uses ______ as a "carrier."
    A. refrigerant
    B. electrical energy
    C. system pressures
    D. None of the above.
19. Thermoelectric refrigerators have a ______ COP.
    A. low
    B. high
    C. average
    D. fluctuating
20. With low relative humidity, snow may be made at temperatures as high as ______.
    A. 40°F (4°C)
    B. 38°F (3°C)
    C. 32°F (0°C)
    D. 35°F (2°C)

Installation and servicing of home air conditioners is one of the many careers for a person with knowledge and hands-on training in refrigeration and air conditioning. (Lennox International, Inc.)

Chapter 19

FUNDAMENTALS OF AIR CONDITIONING

Modules:

Key Words:

anemometer
contaminants
desiccants
dew point
heat sink
humidity
Indoor Air Quality (IAQ)
ozone
pitot tube
pollen count
psychrometry
Sick Building Syndrome (SBS)
velocimeter
wind chill index

Learning Objectives:

After studying this chapter, you will be able to:

- Explain the principles of air conditioning.
- Discuss the physical principles of air movement and humidity.
- List the important factors involved in the operation of an air conditioning system.
- List and explain the factors of air conditioning that affect comfort and health, and the methods of conditioning air for these purposes.
- Understand what various instruments, such as psychrometers, dry bulb thermometers, hygrometers, pitot tubes, recorders, manometers, and barometers, are used for.
- Read and interpret psychrometric charts and scales.
- **Follow approved safety procedures.**

AIR MOVEMENT AND MEASUREMENT MODULE

19.1 Definition of Air Conditioning

The American Society of Heating, Refrigerating and Air-Conditioning Engineers (ASHRAE) defines air conditioning as: "The process of treating air so as to control simultaneously its temperature, humidity, cleanliness, and distribution to meet the requirements of the conditioned space."

As the definition indicates, the important actions involved in the operation of an air conditioning system are:

- Temperature control.
- Humidity control.
- Air filtering, cleaning, and purification.
- Air movement and circulation.

Winter heating conditions require automatic control of the heating source to maintain desired room temperatures. Humidity control for winter conditions usually requires the addition of moisture by a humidifier.

Summer cooling conditions require automatic control of the refrigerating system to maintain the desired room temperatures. Humidity control for summer conditions requires dehumidifiers, which pass air to be cooled over cold evaporator surfaces.

In general, air filtering is the same for both summer and winter. Air filtering equipment usually consists of very fine porous substances. Air is drawn through them to remove contaminating particles. Filters using activated carbon and electrostatic precipitators may be added to the usual filtering mechanisms to improve air cleaning. Air pollutants, and methods for removing them from the air, will be covered in later paragraphs.

Many industries air condition their plants for two reasons: for the comfort provided and for more complete control of manufacturing processes and material. Better control of manufacturing temperatures and relative humidity improves the quality of the finished product.

19.2 Air—Atmosphere

Air is an invisible, odorless, and tasteless mixture of gases that surround the earth. Air surrounding the earth is called the ***atmosphere.*** It extends above the earth about 400 miles and is divided into several layers. The layer closest to the earth, which extends from sea level to 30,000′, is called the ***lower atmosphere.*** The ***troposphere*** extends from 30,000′ to 50,000′. The layer extending from 50,000′ up to 200 miles is called the ***stratosphere.*** The layer beyond 200 miles is called the ***ionosphere.***

Air is a mixture of oxygen, nitrogen, carbon dioxide, hydrogen, sulfur dioxide, and water vapor (moisture). It also contains a very small percentage of rare gases. **Figure 19-1** gives the percentages of these gases, both by volume and by weight.

- Oxygen—The atmosphere is approximately 23% oxygen by weight. Oxygen readily combines with many substances. When fuels such as wood, coal, or oil are burned, their carbon and hydrogen combine with oxygen in the atmosphere forming carbon dioxide and water, respectively. The oxygen in the atmosphere is replenished by plants, which absorb carbon dioxide and release oxygen.
- Nitrogen—About three-fourths of the earth's atmosphere consists of nitrogen. Nitrogen is a gaseous element that does not readily combine with other substances. If combined with other elements, nitrogen is usually unstable. Nitrogen is combined commercially with hydrogen to form ammonia, the basis of most fertilizers and an important refrigerant (NH_3 [R-717]). Liquid nitrogen obtained by cooling of air is also a special purpose expendable refrigerant.
- Carbon dioxide—Carbon dioxide makes up 0.03% to 0.04% of the atmosphere. Carbon dioxide is a combination of carbon and oxygen. Absorbed by growing plants, it becomes one of the "building blocks" in the development of plant cells.
- Hydrogen—Hydrogen (H_2), a very light gas, does not show in weight percentage. Hydrogen is present in most fuels. When burned, it combines with oxygen to form water (H_2O) in steam and vapor form.

		Dry Air	
Name	Chemical Symbol	Amount by Weight %	Amount by Volume %
Nitrogen	N_2	75.47	78.03
Oxygen	O_2	23.19	20.99
Carbon Dioxide	CO_2	.04	.03
Hydrogen	H_2	.00	.01
Water	H_2O	.00	.00
Dust	—	.00	.00
Rare Gases	—	1.30	.94

Figure 19-1. *Gases and substances that make up air in the atmosphere.*

- Sulfur dioxide—Sulfur dioxide is the most common gaseous contaminant. It is formed by combustion of fuels that contain sulfur. Many large power plants now have facilities for removing sulfur from these fuel sources and sulfur dioxide from the stack gases.
- Water vapor (moisture)—The amount of water vapor in the atmosphere varies with the temperature. It is not indicated in percentage, but rather by the term "relative humidity."
- Rare gases—Rare gases make up from 0.9% to 1.3% of the atmosphere by weight. These gases include neon, argon, helium, krypton, and xenon.

In addition to these substances, air contains a variety of contaminants. These vary considerably from region to region and with time. However, contaminants in the air are of great importance in air conditioning.

19.3 Physical Properties of Air

Air has weight, density, temperature, specific heat, and heat conductivity. In motion, it has momentum and inertia. It holds substances in suspension and in solution.

Air pressure at the earth's surface is due to the weight of air above the earth. Air pressure decreases as altitude increases. This is due to the reduction of the weight of the air above. Air presses against the earth at sea level with a pressure of 14.7 psi (101 kPa).

Since air has weight, energy is required to move it. Once in motion, air has energy of its own (kinetic energy). The weight of moving air turns windmills. The mills convert the kinetic energy to mechanical energy.

According to Bernoulli's Equation, increasing the velocity decreases the pressure. In a tornado, the velocity is very high, reducing the pressure.

Tiny dust particles may be picked up and held in suspension in moving air. They may remain in suspension for long periods of time.

The density of air varies with the atmospheric pressure and humidity. One pound of air at standard conditions (14.7 psi, 69.8°F [21°C]) occupies 13.341 ft^3. One kilogram of air occupies 0.83285 m^3. Air has a density of 0.07496 $lb./ft^3$ (1.2007 kg/m^3).

Air temperatures may be measured with either the Fahrenheit scale or the Celsius scale. Under ordinary conditions, the familiar glass-stemmed thermometers are satisfactory. Expanding metals (solids) such as bimetal strips or rods are also used. When measuring very low temperatures, thermocouple thermometers or resistance temperature detectors are used. Thermocouple thermometers may be used for measuring high temperatures. Thermistor thermometers and pyrometers are also popular.

The specific heat of air is the amount of heat required to raise the temperature of one pound of air one degree Fahrenheit or one kilogram of air one degree Celsius. The specific heat of air at sea level is 0.24 Btu per pound.

Air is a poor conductor of heat. For this reason, air spaces are often used for insulating purposes.

For computation purposes, certain pressure, temperature, and density values are required. The requirements are defined under Standard Air in Chapter 31.

19.3.1 Humidity

Humidity is the presence of moisture or water vapor in the air. The amount of moisture that the air will hold depends on the air temperature. Warm air will hold more moisture than cold air.

The amount of humidity in the air affects the rate of evaporation of perspiration from the body. Dry air causes rapid evaporation. This makes the surface feel cool. Moist (humid) air prevents rapid evaporation of perspiration. This makes it feel warmer than the temperature indicated by a thermometer. Remember that this moisture (humidity) is in vapor form, and is invisible.

Relative Humidity

Relative humidity (rh) is a term used to express the amount of moisture in a given sample of air. It is compared with the amount of moisture the air would hold if totally saturated at the temperature of the sample. Relative humidity is stated in a percentage (such as 30%, 75%, and 85%).

A water vapor saturation curve, **Figure 19-2,** is a graph showing the amount of water that air can hold at different temperatures. Point B contains 111 grains of moisture per pound of dry air at 85°F (29.4°C). The saturated condition at C for the same temperature is 183 grains of moisture per pound of air.

Therefore, the relative humidity at Point B is:

$$\frac{111}{183} \times 100 = \text{rh}$$

$$0.606 \times 100 = \text{rh}$$

$$61\% = \text{rh}$$

The line from A to B represents what happens when saturated air is warmed. Point D represents what happens when saturated air is cooled. The distance D to E represents the moisture condensed out of the air. (Saturated air at the same temperature will hold only 66 grains of moisture.) The amount condensed is:

$$111 - 66 = 45 \text{ grains.}$$

A typical outdoor condition in winter is represented at Point F. Air is taken indoors at 30°F (−1°C) and 100% relative humidity. It holds 24 grains of moisture. If this air is heated to 75°F (24°C) and no moisture is added, its new condition will be as shown at G. The saturated condition at Point G would be 131 grains. The original air had only 24 grains of moisture. Therefore, the relative humidity is (24 ÷ 131) × 100 = 18%.

Indicators of Low Humidity

Low atmospheric humidity will be indicated by an increase in the amount of noticeable electrostatic energy. As one moves about and touches grounded metal objects, a spark jumps from the hand or fingers to the object. Also, human hair tends to become unmanageable. Furniture joints shrink and become loose. Woodwork, such as doors and floors, crack open. The surface of the skin and membranes in the nose become dry. To feel more comfortable, the temperature may need to be raised.

Humidity Measurement

A hygrometer is an instrument used to measure moisture in the air. See **Figure 19-3.** The operation of these instruments depends upon the use of some moisture-absorbing substance. These substances change their shape or size, depending upon the relative humidity of the atmosphere. Human hair, wood, and fibers may be used.

It is also possible to measure relative humidity electronically, using a substance in which the electrical conductivity changes with the moisture content. Such an instrument is shown in **Figure 19-4.** In operation, the

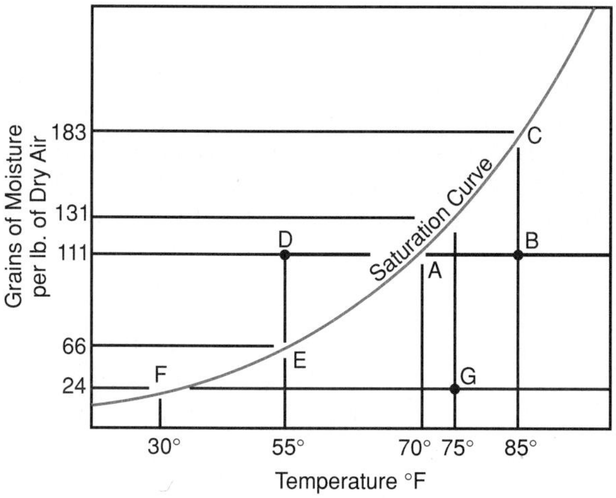

Figure 19-2. *Typical water vapor saturation curve for air. As temperature increases, amount of moisture that air will hold also increases.*

Figure 19-3. *Wall-type hygrometer and temperature indicator is calibrated in percent of relative humidity. (Abbeon Cal, Inc.)*

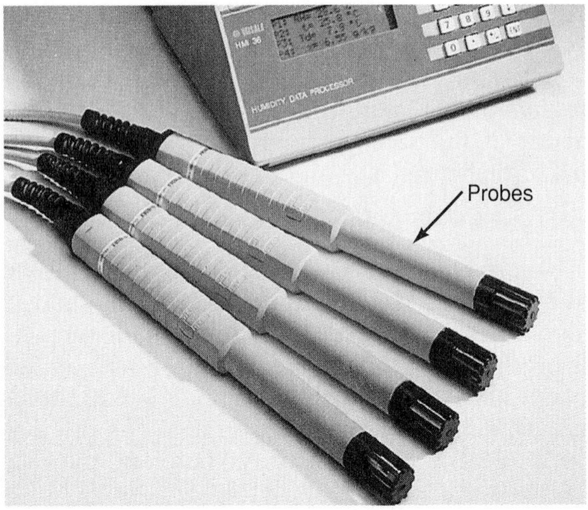

Figure 19-4. *Relative humidity meter contains a microprocessor. Unit measures relative humidity and related quantities and temperatures in specially air conditioned areas such as computer rooms, and laboratories. Unit can have up to four probes to measure the relative humidity and temperature in ambient air. From these measurements, the humidity data processor calculates the dew point, fixing ratio, and absolute humidity. (Vaisala, Inc.)*

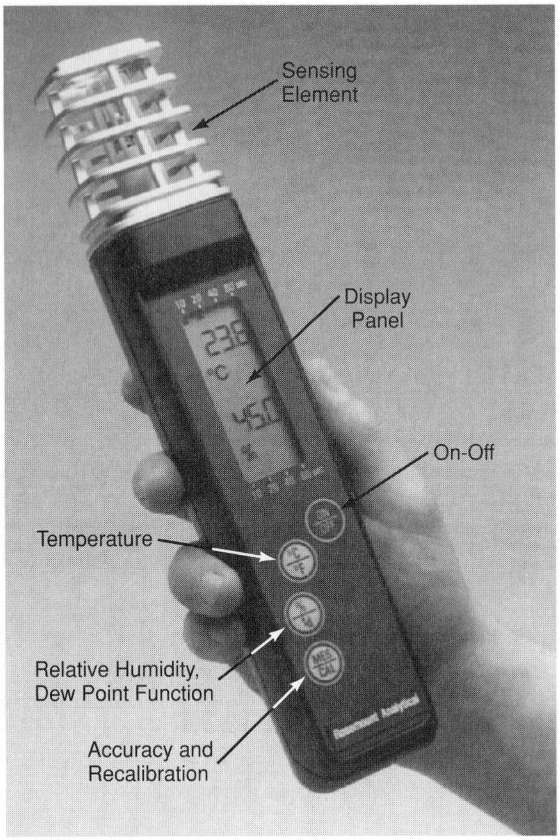

Figure 19-5. *Electronic hygrometer with buttons. (Rosemount Analytical, Inc.)*

sensing element is placed in the space in which the relative humidity is to be measured.

Figure 19-5 illustrates an easy-to-use electronic relative humidity measuring instrument. The meter measures temperature, relative humidity, and dew point. The reading will then appear. Another button is used to measure the accuracy of the readings. It is also used to recalibrate when necessary.

When measuring pressure and temperature, it is sometimes helpful to have an extended reading in a controlled space. A seven-day recorder that indicates the moisture and temperature is shown in **Figure 19-6.** When using this instrument, refer to a psychrometric chart to find the relative humidity.

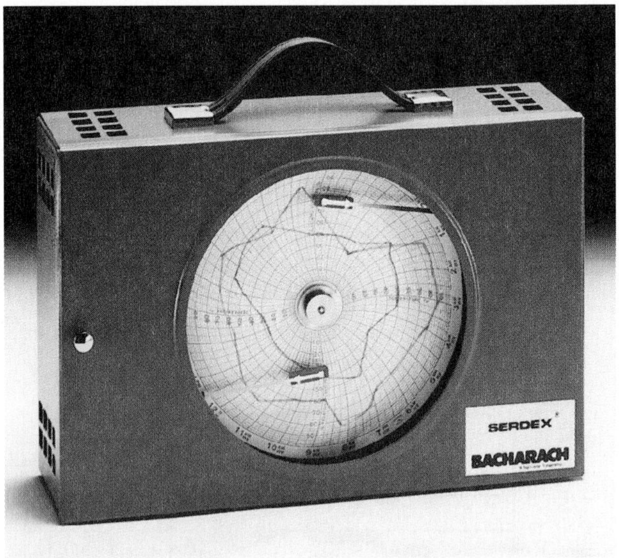

Figure 19-6. *A seven-day humidity/temperature recorder. The unit has a temperature range from −35°F to 130°F (−37°C to 54.5°C). Note marker pens in red and blue; one indicates the humidity and the other indicates the temperature. (Bacharach, Inc.)*

Hygroscopic Substances—Desiccants

Substances that have the ability to absorb moisture from the air are called ***desiccants.*** Some common desiccants are: activated alumina, silica gel, calcium sulfate, and zeolites. Many desiccants can be reactivated (dried out) by heating.

Many instruments are packaged in containers with a package of desiccant. The desiccant tends to absorb the moisture in the container. It keeps the instrument dry to reduce corrosion.

Humidity Controls

Health studies indicate that humidity control is an important factor in air conditioning. Humidity controls operate during the winter heating season to add moisture to the air. They keep the humidity at a satisfactory level.

Humidity controls operate in the summer to remove moisture from the air. To remove moisture, the humidity control usually operates an air bypass. This varies the airflow over the evaporators. These controls usually operate electrically to regulate solenoid valves or dampers. The control element may be a synthetic (made by humans) fiber or human hair. These elements are sensitive to the amount of moisture in the air. **Figure 19-7** shows construction principles of a humidity control device.

Thermo-humidigraphs (temperature and humidity recorders) may be fitted with alarms. These will alert attendants if the temperature or humidity fails to remain at the proper level. This is helpful in computer rooms and other installations that require close humidity control. **Figure 19-8** shows a thermo-humidigraph fitted with alarms.

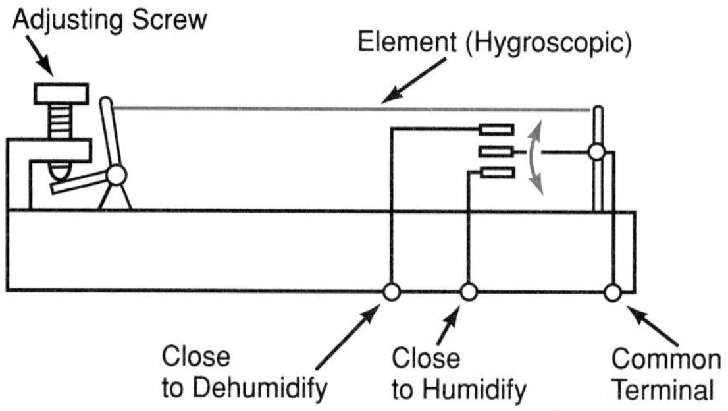

Figure 19-7. *Schematic diagram of relative humidity control, showing operating mechanism.*

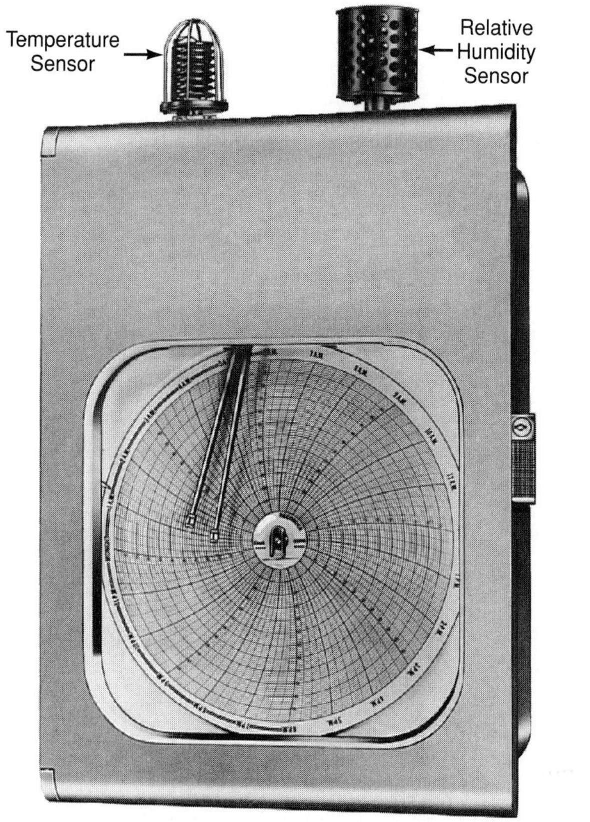

Figure 19-8. *Temperature-relative humidity recorder can be fitted with contact points. Points may be connected to electric alarm to provide signal if temperature or relative humidity is not kept within required limits. (Bristol Babcock, Inc.)*

19.3.2 Air Temperature

The behavior of air varies with its temperature. As previously stated, the higher the temperature, the greater its ability to hold moisture.

Air temperature is measured with a thermometer. Several different scales are used, including Fahrenheit, Celsius, and Kelvin. (See Section 1.6.)

Dry Bulb Temperature

Human comfort and health depend a great deal on the air temperature. In air conditioning, the air temperature indicated usually is the ***dry bulb temperature*** (db). It is taken with the sensitive element of the thermometer in a dry condition. This is the temperature normally reported.

Wet Bulb Temperature

If a moist wick is placed over a thermometer bulb, the evaporation of moisture from the wick will lower the thermometer reading. This temperature is known as the ***wet bulb temperature.*** If the air surrounding a wet bulb thermometer is dry, evaporation from the moist wick will be rapid. If the air is quite moist, evaporation will be slower. **Figure 19-9** compares dry bulb temperature and wet bulb temperature.

When the air is saturated with moisture, no water will evaporate from the cloth wick, and the wet and dry bulb temperatures are identical. However, if the air is not saturated, water will evaporate from the wick. In doing so, it will lower the wick temperature. Then heat will flow from the mercury to the wet wick and the reading will be lower.

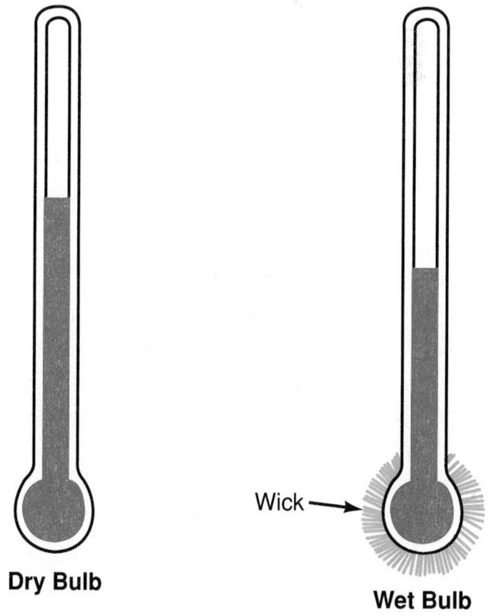

Figure 19-9. *Dry bulb and wet bulb thermometers. Note that temperature shown on wet bulb thermometer is considerably lower than dry bulb thermometer.*

The wet bulb reading depends on how fast the air passes over the bulb. Speeds up to 5000 ft./min. (60 mi./hr.) are best. Also, the wet bulb should be protected from heat radiating surfaces. This includes radiators and electric heaters. Errors as high as 15% may be made if the air movement is too slow, or if too much radiant heat is present.

19.3.3 Psychrometric Properties of Air

Psychrometry is the science and practice of dealing with air mixtures and their control. The science deals mainly with dry air and water vapor mixtures. **Figure 19-10** shows a computer terminal used to determine and control the condition of the air in a large building complex.

Psychrometry deals with the specific heat of dry air and its volume. It also deals with the heat of water, heat of vaporization or condensation, and the specific heat of steam in reference to moisture mixed with dry air.

Tables and graphs have been developed to show the pressure, temperature, heat content (enthalpy), volume of air, and steam content of air. A pressure of 29.92″ Hg (76 cm Hg) is used as the standard atmospheric pressure.

Psychrometer

Airflow over a wet bulb thermometer should be quite rapid to ensure accuracy. A sling psychrometer is often used to whirl a pair of thermometers, one dry bulb and one wet bulb. See **Figure 19-11.** To operate, saturate the wick on the wet bulb and whirl. When the mercury stops dropping, read the two thermometers. Place the wet bulb temperature over the dry bulb temperature scale on a slide rule. Arrow will indicate the relative humidity.

A battery-operated digital sling psychrometer is shown in **Figure 19-12.** It has illuminated thermometer scales and a fan that draws air over the thermometer sensitive bulbs.

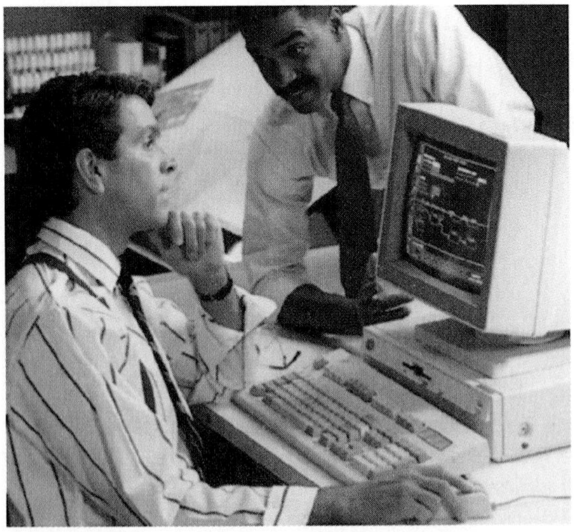

Figure 19-10. *Computerized building automation system controls energy, comfort, fire, and security. (Johnson Controls, Inc.)*

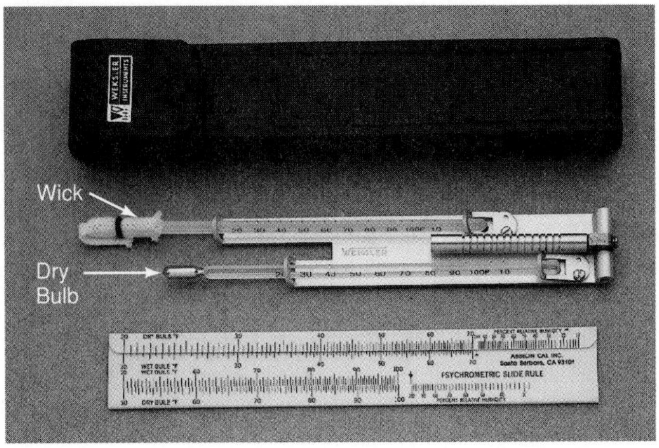

Figure 19-11. *Sling psychrometer. Dry bulb and wet bulb temperatures are used to determine relative humidity. (Abbeon Cal, Inc.)*

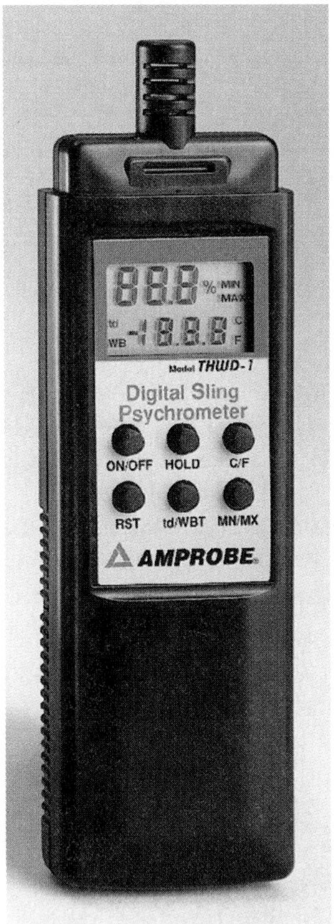

Figure 19-12. *Digital sling psychrometer. Unit is battery operated and provides readings in digital form in either °F or °C. (Amprobe Instrument)*

Psychrometric Charts

The psychrometric chart is a graph of the properties (temperature, relative humidity, etc.) of air. It is used to determine how these properties vary as the amount of moisture (water vapor) in the air changes. A basic psychrometric chart is shown in **Figure 19-13.** The horizontal scale (abscissa) is the dry bulb temperature. The vertical scale (ordinate) represents water vapor pressure.

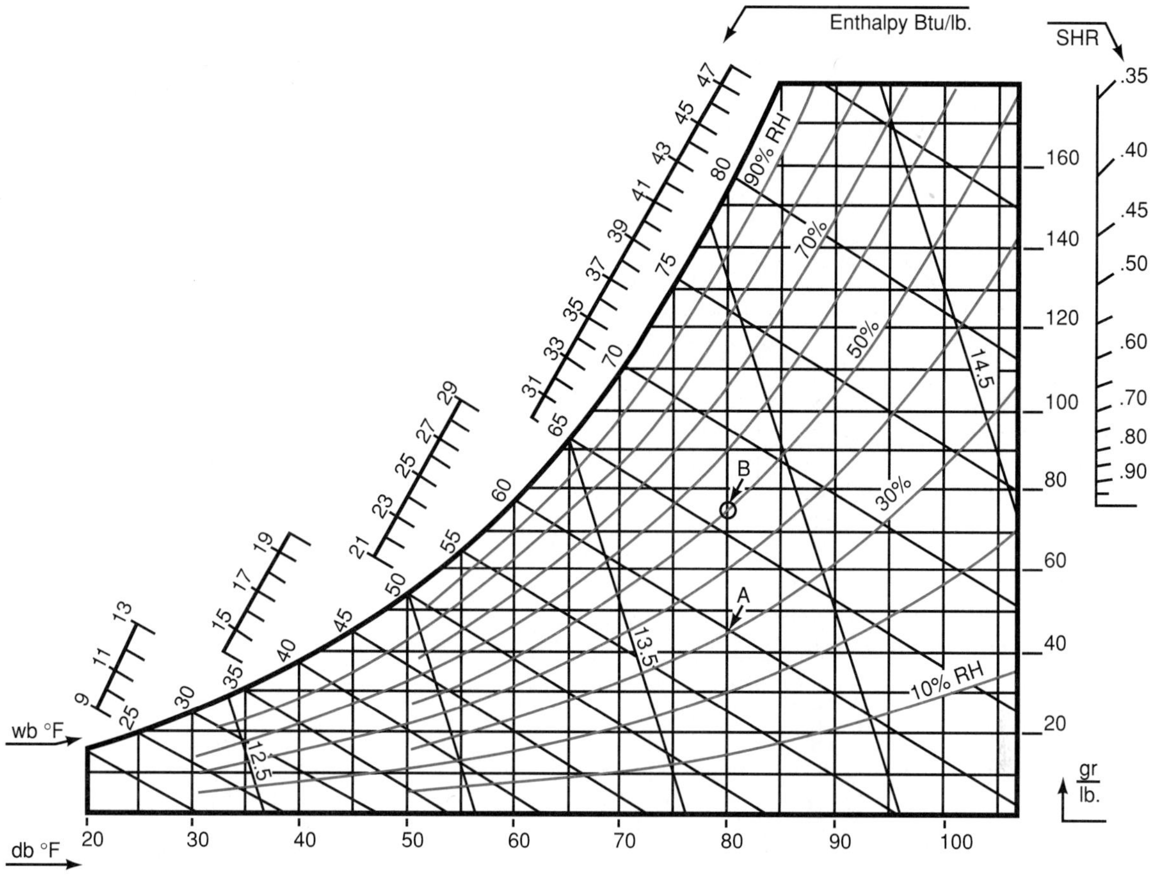

Figure 19-13. *Psychrometric chart. Red lines indicate relative humidity, in percent. Dry bulb (db) temperature is shown at bottom. Wet bulb (wb) temperature is on uppermost curve. Right side gives grains/lb.. Point A indicates relative humidity of 30%. Note reading of 13.5 ft^3 near center. This is volume of 1 lb. of air at a given temperature and humidity. Also note Sensible Heat Ratio (SHR) scale, each line of which angles away from central comfort zone point B. (Reprinted from Air Conditioning Contractors of America's [ACCA] Basic Installation Manual by permission of ACCA)*

Figure 19-14 shows lines on the psychrometric chart that represent constant conditions. The chart shows a line of constant dry bulb temperature. This is always a vertical line. **Figure 19-14B** shows a line of constant wet bulb temperature. This is also the line of constant enthalpy. **Figure 19-14C** shows a line of constant water vapor pressure (lb. water/lb. dry air or water grains/lb. dry air). **Figure 19-14D** shows a line of constant relative humidity (%). The 100% relative humidity line is also known as the dew point or saturation temperature line.

Each point on the psychrometric chart represents air at a specific set of conditions. The following examples refer to the points A through D on **Figure 19-15.**

Example:

Dry bulb temperature is 75°F (24°C): If wet bulb temperature is 60°F (16°C), what is the relative humidity?

Follow the vertical line corresponding to the 75°F (24°C) dry bulb temperature. Then follow the 60°F (16°C) wet bulb temperature line. These lines cross each other at Point A. This point is just above the 40% relative humidity line. Therefore, the correct answer would be about 41% relative humidity.

Example:

A sample of air has a dry bulb temperature of 80°F (27°C) and a relative humidity of 60%. Determine the dew point.

Find where the 80°F (27°C) dry bulb line crosses the 60% relative humidity line. This point is labeled B. If the air represented by this point were cooled without a change in moisture content (represented on the psychrometric chart as a horizontal line), the dew point line would be intersected at about 65°F (18°C). This is labeled as point C.

Therefore, 65°F (18°C) is the dew point for the sample of air. The 80°F (27°C) temperature and the 60% relative humidity could represent a typical summer evening. Dew would appear on surfaces when the 65°F (18°C) temperature was reached.

Example:

Find the relative humidity when the dry bulb temperature is 75°F (24°C) and the humidity (or water vapor pressure) is 100 grains per pound of dry air.

First, find the vertical line representing a constant dry bulb temperature of 75°F (24°C). Follow that line

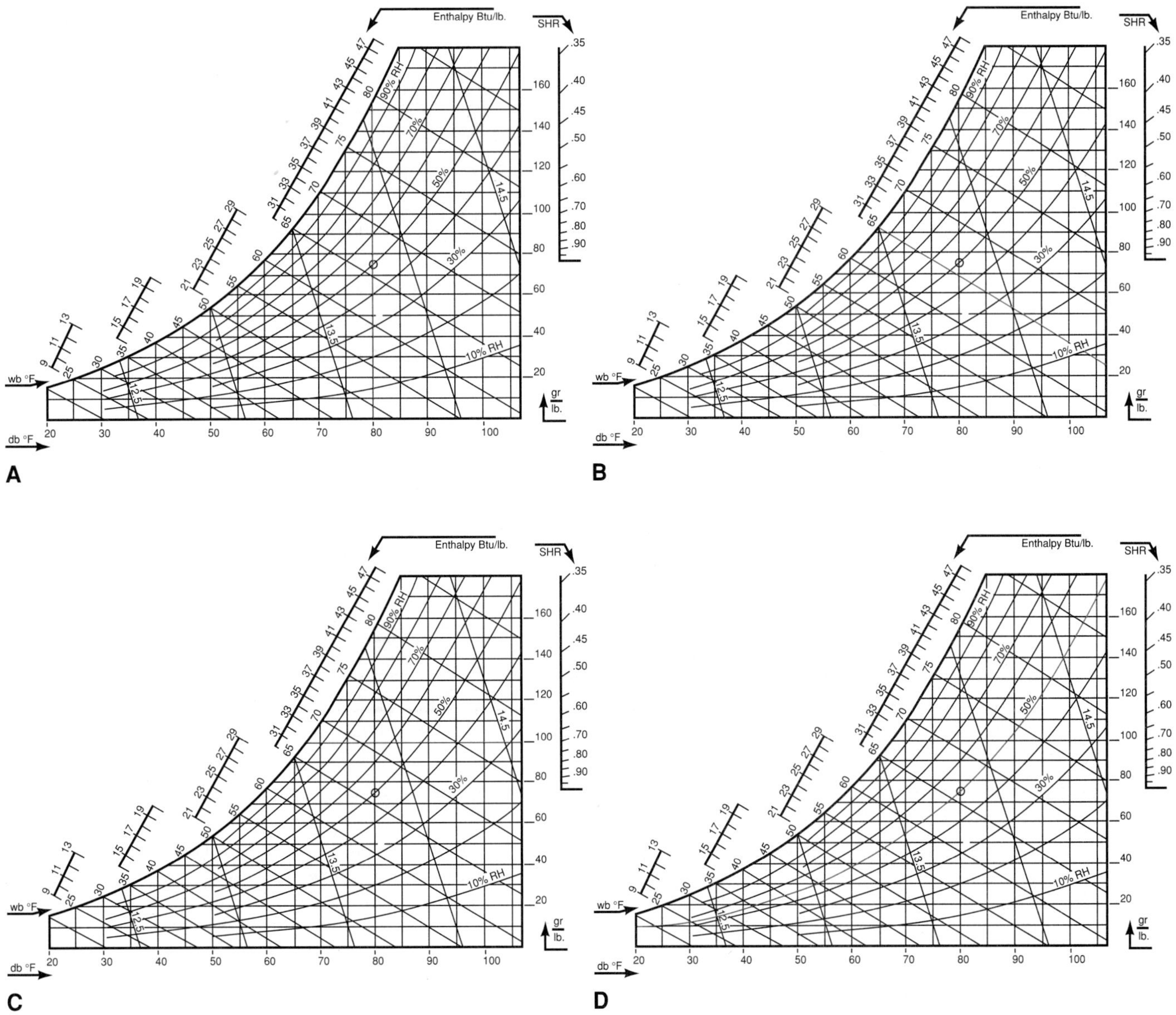

Figure 19-14. *Constant psychrometric conditions. A—80°F constant dry bulb temperature. B—65°F constant wet bulb temperature. C—100 grains of water/lb. of dry air constant water pressure. D—50% constant relative humidity.*

until it crosses the horizontal line representing 100 grains of moisture per pound of dry air. The intersection point is labeled Point D. This point falls between the 70% and 80% relative humidity lines. The answer would be a relative humidity of about 77%.

This chart should be studied carefully. It provides a simple way for determining the various conditions of air. Remember, the warmer the air, the more moisture it will hold. Also, as pressure is reduced, air absorbs more moisture.

Using the Psychrometric Chart

Many air conditioning problems in this text will involve the use of a psychrometric chart. A chart can show what is happening during a specific heating, ventilating, and air conditioning (HVAC) process.

Psychrometric charts give a considerable range of temperature and humidity conditions. The human body will be comfortable under a variety of temperature and humidity combinations. This is shown in **Figure 19-16.** Most people are comfortable in an atmosphere with the relative humidity between 30% and 70% and the temperature between 70°F and 85°F (21°C to 29°C).

The HVAC industry exists because nature does not always provide these ideal conditions. The HVAC system must modify existing conditions. To accomplish this, heating, cooling, humidification, and dehumidification processes are used.

These processes can be modeled on the psychrometric chart. See **Figure 19-17.** For example, point A in the figure shows a dry bulb temperature of 40°F and a relative humidity of 30%. The desired condition is point B, 75°F and 50% relative humidity. The HVAC system must provide the processes represented by the colored lines connecting points A and B. This is a good way to visually show what the system capability must be.

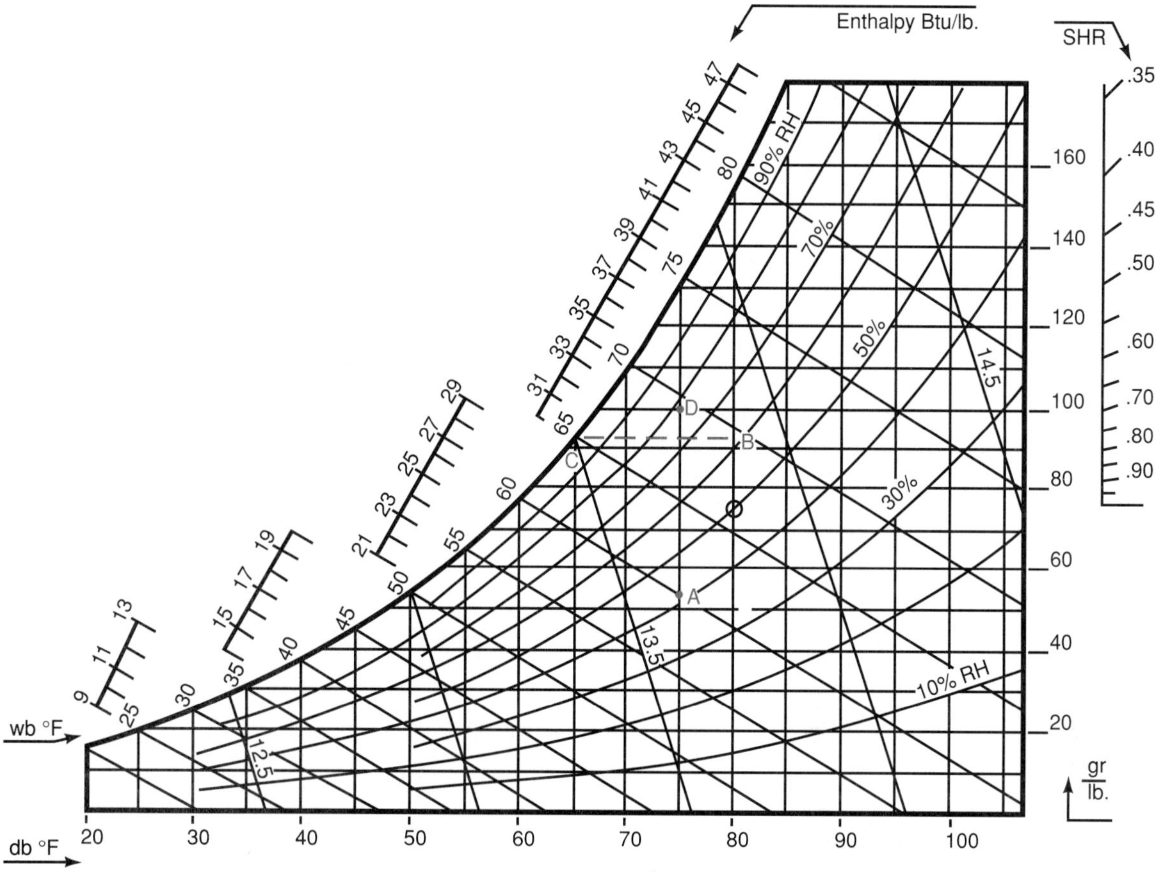

Figure 19-15. *Psychrometric chart showing specific conditions.*

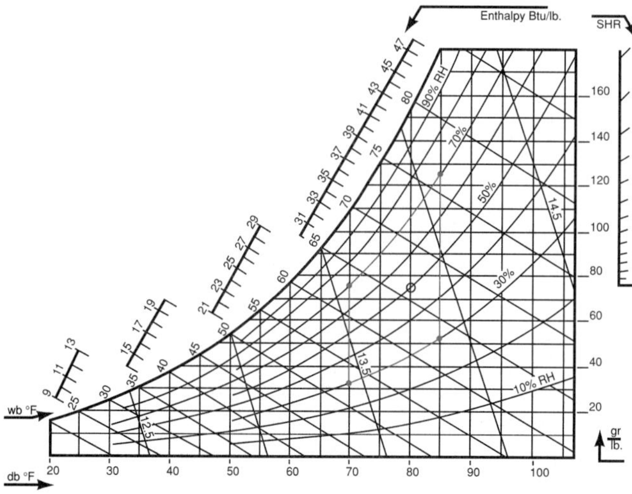

Figure 19-16. *Psychrometric chart showing comfort zone.*

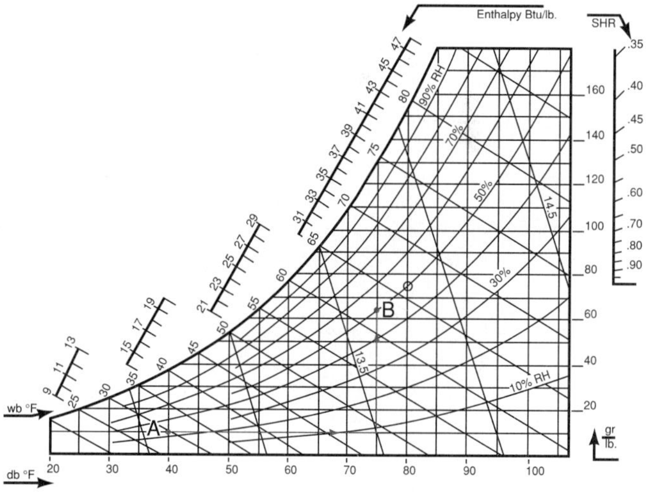

Figure 19-17. *HVAC process modeled on psychrometric chart.*

The psychrometric chart can then be used to plot the actions of the evaporators, heaters, and chillers in an HVAC system. Further study of psychrometrics can result in equations representing all the processes used in the conditioning of air. This gives scientists and engineers the basics for design and evaluation of new HVAC systems.

19.3.4 Dew Point

Dew point is the temperature below which water vapor in the air will start to condense. It is also the 100% humidity point. The relative humidity of a sample of air may be determined by its dew point. Several methods may be used to find the dew point.

The dew point can be determined with fair accuracy by the following method. A volatile fluid is placed in a bright metal container. The fluid is then stirred with an air aspirator. A thermometer is placed in the fluid to indicate the temperature of both the fluid and the container. While stirring, a mist or fog appears on the outside of the metal container. The temperature at which this appears is the dew point. **Flammable or toxic volatile fluids must not be used for this experiment.**

An instrument used to determine dew point is illustrated in **Figure 19-18.** This unit can measure dew points from room temperatures down to −80°F (−62°C).

A sample of air is pumped into the observation chamber of the instrument. The pressure is above atmospheric. A pressure ratio gauge adjusts for this pressure. Then the valve is manipulated to exhaust the air. The observation window will indicate a fog when the sample is cooled to its dew point. This window is lighted and a "sunbeam" effect is noted if any fog exists. The pressure ratio determines the dew point temperature.

A window during the winter heating season offers a good example of dew point. **Figure 19-19** shows the surface temperature that will cause condensation (dew point) for various humidity.

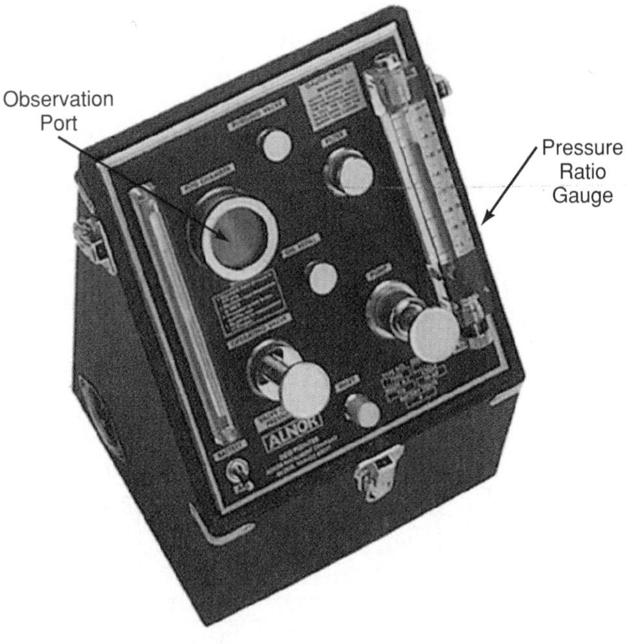

Figure 19-18. *Instrument for determining dew point temperature. Note observation port. (Alnor Instrument Co.)*

19.4 Vapor Barriers

Water vapor flows easily through all porous substances. Water in vapor form remains a vapor as long as its temperature is above the dew point. However, when its temperature drops to the dew point, the water vapor will condense into droplets.

Relative Humidity of Air (Percent)	Dry Bulb Temperature of Surface When Condensation Starts	
	70°F (21°C) Air Temp.	80°F (27°C) Air Temp.
100	70	80
90	67	77
80	64	73
70	60	69
60	56	65
50	51	60
40	45	54
30	37	46
20	28	35

Figure 19-19. *Table gives temperature to which surface must be cooled to have condensation start. Table is based on ambient temperature of air at either 70°F or 80°F (21°C or 27°C).*

In modern housing, water vapor is kept from passing through walls and toward surfaces where it might condense. This is done by using moisture-proof materials, such as aluminum foil and plastic sheeting to form a vapor barrier. The barrier keeps water vapor from passing from warm surfaces to cold surfaces. Vapor barriers should always be installed on the warm side of a heated space.

Lack of proper vapor barriers is often indicated by peeling paint near kitchen and bathroom areas. The moisture from these areas travels through the walls. When it contacts the cold undersurface of the paint, droplets of water are formed, causing the paint to peel.

19.5 Air Movement

Air movement affects comfort. Cool, dry air circulated past a warm body will speed heat flow from the body. Evaporation will increase. This tends to cool the body.

During cold days, a person exposed to outside atmospheric conditions often feels much colder than the thermometer shows. This chilling effect is due to wind velocity and relative humidity. The term "wind chill" applies to this uncomfortable feeling.

Air movement in a conditioned space is also very important. Air movement is necessary to supply fresh air to a controlled space. If the air moves too fast (a draft), a person feels uncomfortable. If the air movement is too slow, the air becomes stale (contaminated) and lacks oxygen.

19.5.1 Air Velocity Measurement

Outside air velocity (wind) is measured in miles per hour (mph) or knots.

Air velocity is usually expressed in feet per minute (fpm). It is possible to calculate the volume of air flowing through the duct in cubic feet per minute (cfm). To do this, multiply the air velocity by the cross-sectional area of a duct.

If air flows through the occupied space at more than 15′ to 20′ per minute, occupants of the space will feel a draft. It is difficult to develop accurate instruments to measure drafts. The usual method is to use a smoke generator and a stopwatch. The flow of the smoke through the space being analyzed is timed.

Several methods are used to measure air velocity:

- Anemometer (rotating).
- Anemometer (hot wire).
- Velocimeter (swinging vane).
- Velocity pressure (pitot tube).

The rotating anemometer, the direct-reading velocimeter, and the pitot tube are not accurate at very low air velocities.

Anemometer—Rotating and Hot Wire

A small propeller placed in an airstream will revolve as air flows past the blades. An instrument connected to the propeller can measure the flow. See **Figure 19-20.** Devices of this type are called ***anemometers.***

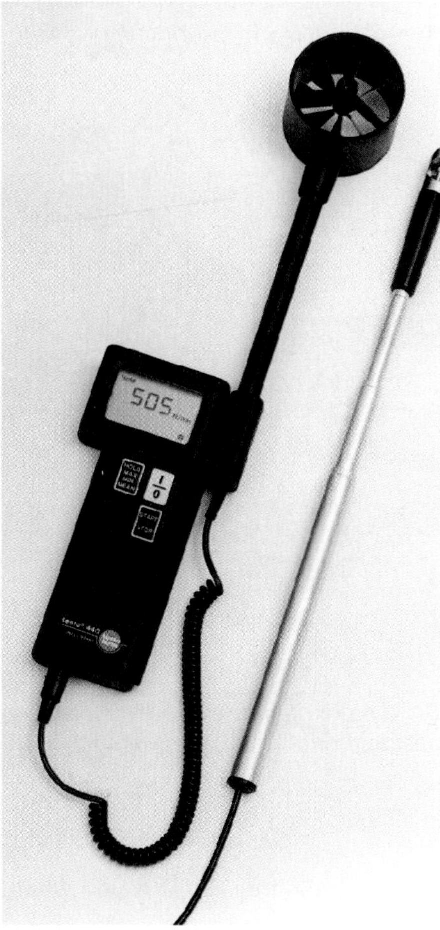

Figure 19-20. *Handheld digital anemometer used to measure airflow. Large diameter vane for airflow range of 40–4000 fpm. The small diameter vane (range 80–8000 fpm) is designed for duct insertion. (Testoterm, Inc.)*

Anemometers generally have a start lever and a return-to-zero lever. To use the instrument, carefully place it in the airstream at right angles to the airflow. Allow it to reach a constant speed (about one minute), then trip the registering mechanism. At the same time, start a stopwatch. Record the reading and the time. From this data, compute the velocity of the air in feet per minute. Divide the number of feet by the elapsed (passed) time. For example: If the reading is 236 for 1/2 min., the velocity will be 472 ft./min. It is advisable to take several readings and compute the average to ensure greater accuracy.

Another type of anemometer is shown in **Figure 19-21.** The dial will indicate airflow from HVAC grilles in cubic feet per minute (cfm). The operator now can calculate the number of Btu going into the space through each grille. The airflow is multiplied by the appropriate temperature factor. The device takes account of the grille area which is entered as data into the instrument.

The operation of the hot wire anemometer depends upon the cooling effect of air flowing over an electrically heated wire. A hot wire instrument is shown in **Figure 19-22.**

Velocimeters (Swinging Vane)

In using a swinging vane ***velocimeter,*** incoming air pushes on a small vane. It tilts at different angles as the

Figure 19-21. *Anemometer reads air velocity in cfm. (TIF Instruments, Inc.)*

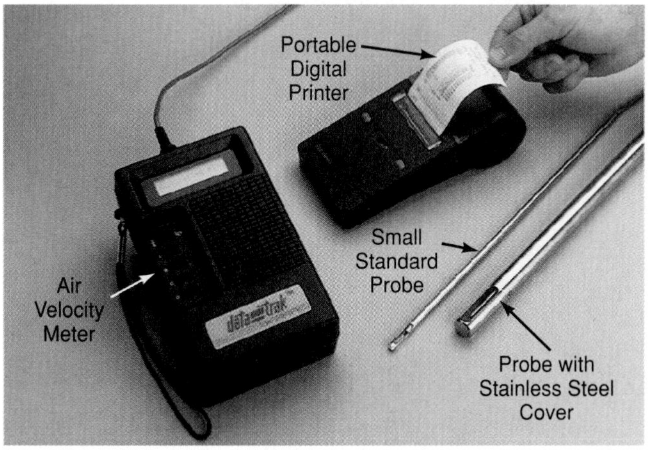

Figure 19-22. *Portable air velocity meter, which provides velocity averaging, with reading in conventional or metric units. (Sierra Instruments, Inc.)*

air velocity increases. The instrument is put directly in the airstream, **Figure 19-23.**

Special jets are used for velocity readings when it is difficult to place the instrument in the airstream. They adapt the instrument to these conditions.

The direct-reading instrument can be used to measure duct air velocities, **Figure 19-24.** Note that a special jet is attached to the air inlet of the instrument with a flexible tube.

The velocimeter is also used to measure air velocities in main ducts and branch ducts. See **Figure 19-25.** An instrument of this type is necessary to balance air distribution systems.

The instrument is calibrated for use at a temperature of 68°F. Corrections must be made if the duct temperature is not at 68°F. Formula for correction:

$$\text{fpm} = \left[\frac{(460 + T)}{(460 + 68)}\right] \times \text{instrument reading}$$

(T = temperature Fahrenheit of air in duct.)

Velocity-Pressure (Pitot Tube)

The velocity-pressure method of measuring air velocity uses an instrument called a ***pitot tube.*** See **Figure 19-26.** Air contacting the nose of the pitot tube creates a total pressure. The outer tube, with the holes on the side, measures the static pressure. When these two pressures are connected to the end of a manometer, the difference is the velocity-pressure. This pressure difference is measured in inches of water. An inclined manometer, **Figure 19-27,** is used with the pitot tube.

Formula:

Velocity = 4010 × square root of velocity-pressure (in inches of water)

Figure 19-23. *Airflow velocimeter provides fast measurements. Instrument is held in the air stream. (Alnor Instrument Co.)*

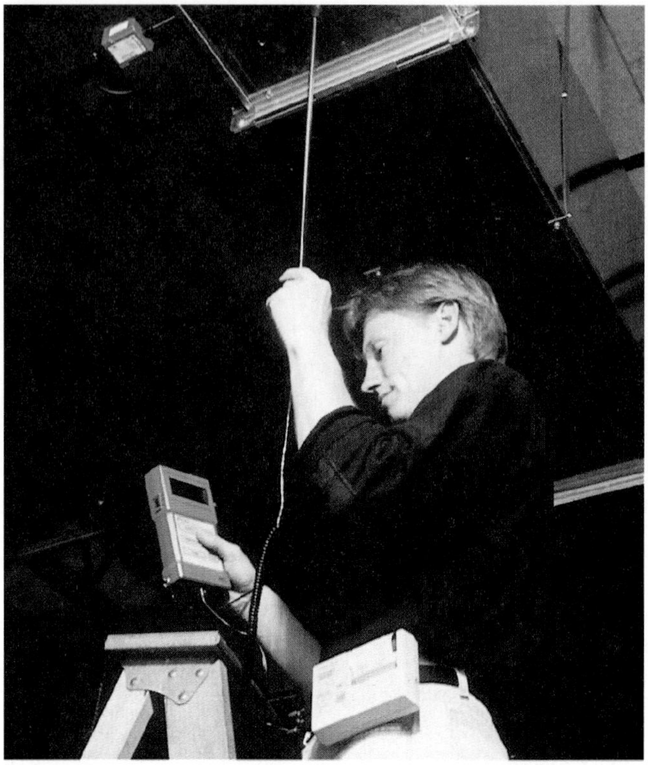

Figure 19-24. *Technician measuring air velocity at duct. (TSI Incorporated)*

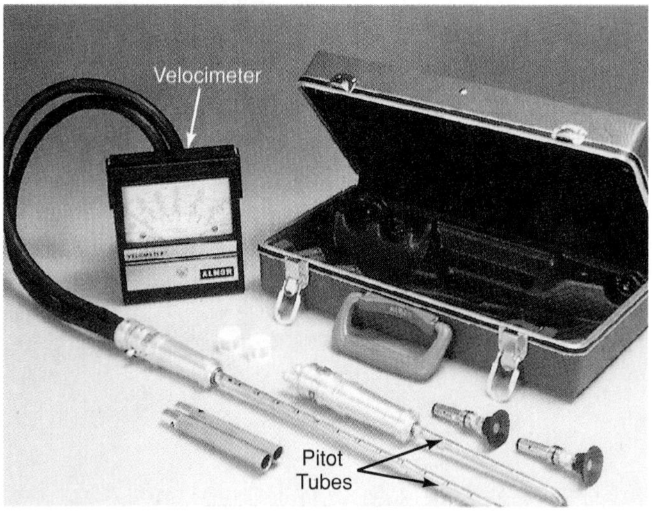

Figure 19-25. *Direct-reading airflow meter determines air velocity inside a duct. (Alnor Instrument Co.)*

Example:

If the velocity-pressure is 1″ of water, what is the velocity?

Solution:

Velocity = 4010 ($\sqrt{1''}$)
Velocity = 4010 × 1
Velocity = 4010 ft./min.

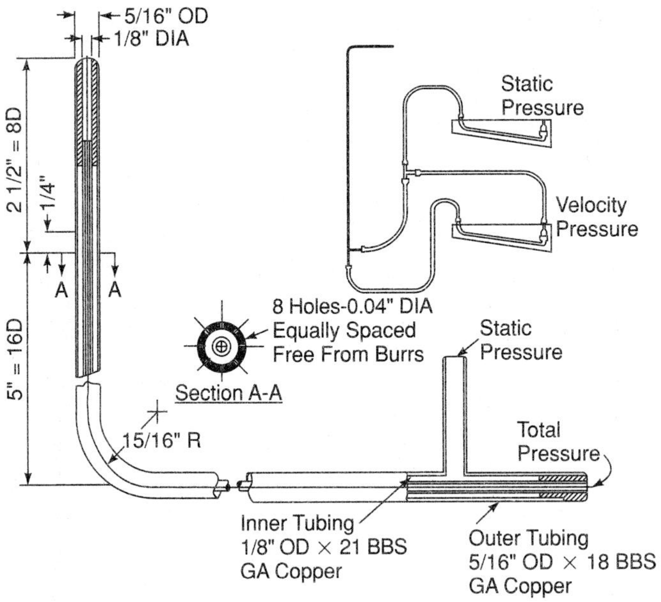

Figure 19-26. *Pitot tube connected to two inclined manometers. The unit measures air velocity. (Reprinted by permission of the American Society of Heating, Refrigerating, and Air-Conditioning Engineers, Atlanta, Georgia, from the 1993 ASHRAE Handbook—Fundamentals)*

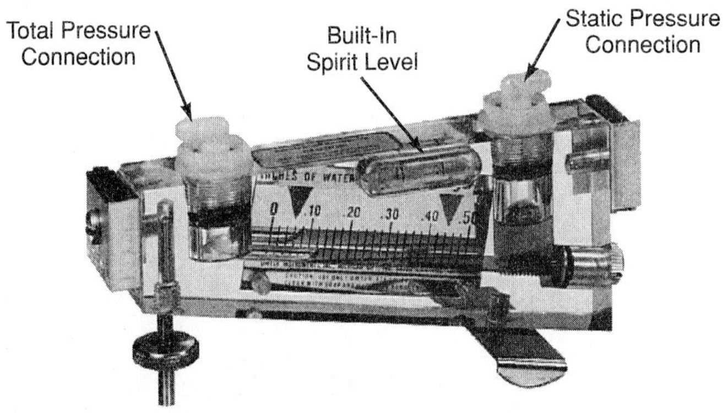

Figure 19-27. *Inclined gauge for use with pitot tube. This gauge may also be used for measuring filter pressure drops. Liquid level must be adjusted to a zero reading to level the unit. (Dwyer Instruments, Inc.)*

Example:

If the velocity-pressure is 0.25″ of water, what is the velocity?

Solution:

Velocity = 4010 ($\sqrt{0.25}$)
Velocity = 4010 × 0.5
Velocity = 2005 ft./min.

The constant 4010 is for standard conditions. Other values are shown in **Figure 19-28.** The constant changes are based on density of the air. The manometer must be mounted level to obtain accurate readings.

To obtain correct velocity readings, take several readings in various parts of the duct. Average the readings. The recommended method to use is shown in

Volume ft³/lb.*	Velocity Constant
11.5	3720
12.1	3818
13.2	3980
13.4**	4010
14.1	4118
15.1	4260
16.2	4410
17.1	4530

*Values for any conditions may be read from the psychrometric chart.
**Standard.

Figure 19-28. *Velocity correction factor changes with change in air density (effect of temperature and altitude).*

Figure 19-29. Sensor locations for measuring velocity are shown for both rectangular and circular ducts. The location of each of the points is as recommended by the American Society of Heating, Refrigerating and Air-Conditioning Engineers (ASHRAE).

Rotating, turbulent airflow can affect the ability of the pitot tube to measure the true pressures. The pitot tube should be used only where the duct is very long. The length of the duct downstream of the measuring location should be a minimum of 10 times the duct diameter. If precise measurements are required, air straightening vanes should be located upstream from the pitot tube.

19.5.2 Ventilation

Ventilation is a term applied to changing the air in a building. In any space occupied by people, breathing reduces the oxygen content. Activities in the space may add some pollutants to the environment. The most economical way to maintain health and comfort conditions is by replacing the air. This is done by bringing in outside air to ventilate the space.

Sometimes it is desirable to quickly replace all the air in the confined space. This is done by opening windows and doors, flushing the space completely with 100% outside air.

Most heating systems provide continual replacement of air. A small percent of the air in the conditioned space is continually replaced. This is done by slowly exhausting some of the air and bringing in fresh air from outside. In homes, the gradual change may not be noticeable. There is always a small amount of air entering and leaving the home. This movement of air occurs through the cracks around windows and doors. It also takes place through doorways each time doors are opened.

Some building materials are porous (contain many tiny holes). Thus, a large amount of air may filter in and out of a building. The amount of infiltration depends upon the wind velocity and temperature difference inside and outside the building. Air tends to enter the

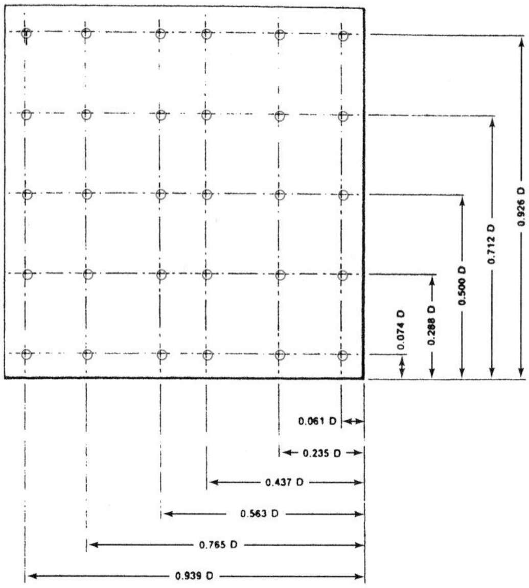

NO. OF POINTS OR TRAVERSE LINES	POSITION RELATIVE TO INNER WALL
5	0.074, 0.238, 0.500, 0.712, 0.926
6	0.061, 0.235, 0.437, 0.563 0.765, 0.939
7	0.053, 0.203, 0.366, 0.500, 0.634, 0.797, 0.947

LOG TCHEBYCHEFF RULE FOR RECTANGULAR DUCTS

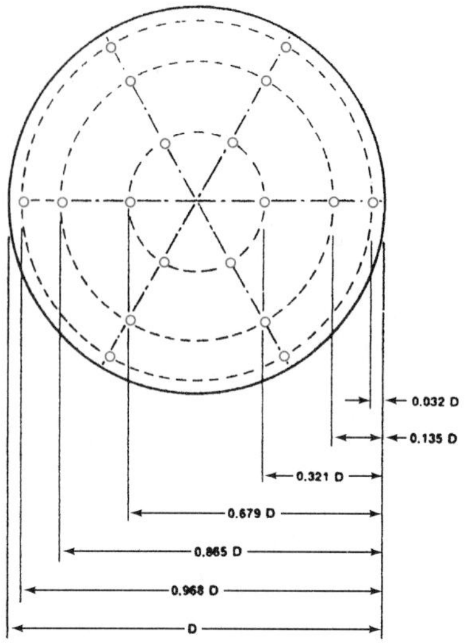

NO. OF MEASURING POINTS PER DIAMETER	POSITION RELATIVE TO INNER WALL
6	0.032, 0.135, 0.321, 0.679, 0.865, 0.968
8	0.021, 0.117, 0.184, 0.345, 0.655, 0.816, 0.883, 0.981
10	0.019, 0.077, 0.153, 0.217, 0.361, 0.639, 0.783, 0.847, 0.923, 0.981

Figure 19-29. *Locations for measuring points in duct. Average of readings will produce average duct velocity. (Reprinted by permission of the American Society of Heating, Refrigerating, and Air-Conditioning Engineers, Atlanta, Georgia, from the 1993 ASHRAE Handbook—Fundamentals)*

building on the upwind side and leave the building on the downwind side.

Since warm air is lighter than cold air, it tends to rise in a room. In buildings with more than one story, the warm air rises from the lower to upper floors. The rising air will create a slight pressure that causes some warm air to escape through the upper surfaces of the building. This lost air is replaced by cold air entering at the lower levels.

With summer air conditioning, the opposite situation occurs. Cold air tends to flow downward and may leave the building at the lower levels. The cold air is replaced by warmer air entering at the upper levels.

Whenever air is exhausted from a space, it must be replaced. The air brought in from outdoors must be cleaned and adjusted to have the same temperature as the indoor air.

Replaced air is conditioned to provide a comfortable room environment. This conditioned air brought into the room is called "make-up air."

A structure that keeps inside air pressure slightly above atmospheric pressure has a positive pressure. A structure that maintains an air pressure slightly below atmospheric pressure has a negative pressure.

Fuel-burning furnaces, stoves, and fireplaces operating in the winter tend to cause a negative pressure. There will be a considerable amount of air leakage through the walls and cracks. Positive pressure can be maintained only if a fan or blower of some kind is used to bring in fresh air.

With heated, tall structures, upper rooms may be slightly above atmospheric pressure. The lower rooms may be slightly below atmospheric pressure because warm air rises.

19.6 Climate

Climate is defined as the weather conditions of a region. These conditions include temperature, humidity, sunshine, pressure, and air movement.

Outdoor climate, of course, cannot be affected much by air conditioning (heating, cooling, and humidifying). In an enclosed space, however, these factors may be controlled. An "indoor" climate can be provided to meet any desired condition.

Indoors, the factors that determine comfort can be completely controlled. There is a definite relationship between comfort and the temperature, humidity, and air movement conditions. **Figure 19-30** illustrates the constant comfort condition with varying temperatures and humidity. Most homes and workplaces are completely air conditioned.

Increasing the air movement tends to give a cooling effect on the human body. If the heating system provides too much air movement (over 15 to 20 fpm), a temperature increase may be necessary to help maintain a comfortable indoor climate.

Weather is the conditions in the atmosphere. These include temperature, wind velocity and direction,

102	99	97	95	92	89	86	84	82	80
96	95	92	89	87	85	83	81	79	77
92	89	87	85	83	82	80	78	76	75
87	85	83	81	80	78	77	75	74	72
82	80	79	78	76	75	73	72	71	69
76	76	74	73	72	71	70	69	68	67
72	71	70	69	68	68	67	67	66	64
68	67	66	66	65	65	64	64	63	63
62	62	62	62	61	61	61	61	60	60
58	58	58	58	58	58	58	58	58	58
53	53	53	53	53	53	53	53	53	53
47	47	47	47	47	47	47	47	47	47

Temperature: 100, 90, 80, 70, 60, 50

Relative Humidity: 90, 70, 50, 30, 10

Comfort Zone

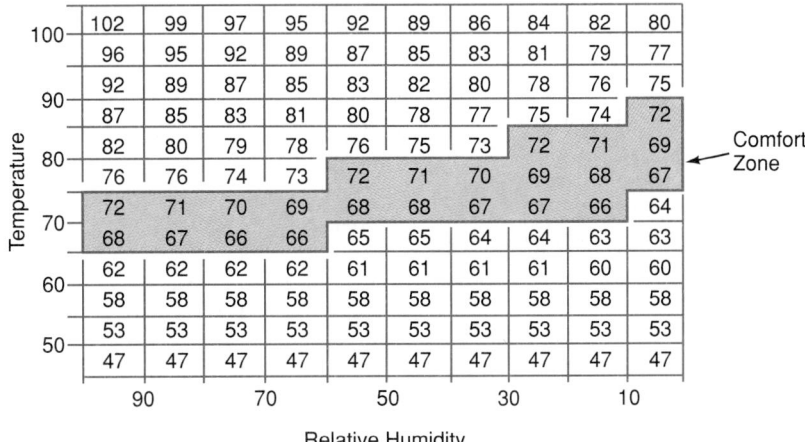

Figure 19-30. *Equivalent temperatures (similar to effective temperature). Note comfort zone. Area inside red lines indicates usual temperature and relative humidity range in which most people are comfortable. Note that with high relative humidity, the comfort zone has lower temperatures.*

clouds, moisture, and atmospheric pressure. Weather affects the need for and requirements of air conditioning.

19.6.1 Air Temperature

Air temperatures in the United States vary from a low of about −55°F (−48°C) to a high of around 120°F (49°C). The normal, desirable temperature is 72°F (22°C).

Normally, the temperature of the human body is 98.6°F (37.0°C). Skin temperature is lower, about 91°F (33°C). In temperate zones, the average atmospheric temperature in winter is below the body temperature. Clothing is required to help conserve body heat. Also, heat needs to be added to the occupied space for the comfort of its occupants.

The human body loses heat easily when the air temperature falls below 98.6°F (37°C). The body may also lose heat at air temperatures above 98.6°F (37.0°C) through the evaporation of perspiration from the body.

In order to maintain comfortable temperatures, air must be heated or cooled. The specific heat of dry air is 0.24 Btu per lb. Energy is required to bring about the desired temperatures for heating or cooling.

Degree Days

Degree days is a measure used to help indicate the heating or cooling needed for a given region. Calculations are based on a temperature of 65°F (18°C). The degree day is computed as follows. The mean (average) of the highest temperature and the lowest temperature is taken for a day. Then this average is subtracted from 65°F (18°C).

Formula:

$$65°F - \left[\frac{\text{(high temp. + low temp.)}}{2}\right]$$

Example:

The lowest recorded temperature for a certain day was 28°F (−2°C). The highest recorded temperature for the same day was 36°F (2°C).

Solution:

$$65°F - \left[\frac{(28°F + 36°F)}{2}\right]$$

$$65 - 32 = 33 \text{ degree days (F)}$$

If the temperature conditions went on for two days, the result would be 66 degree days. If Monday has 30 degree days and Tuesday has 20, the result is 50 degree days for the two days.

Solution:

In Celsius degrees:

$$-2 + 2/2 = 0°C \text{ or } 18°C - 0°C = 18 \text{ degree days (C)}$$

Degree days may be added by weeks, months, or for a season. This will give a comparison of the heating needs for different years.

Sun Heat Load Fundamentals

Radiant heat (light) from the sun furnishes a tremendous amount of heat energy. If a glassed-in surface is exposed to this light energy, the energy will enter a space and become heat. Glass is a rather poor conductor of heat. Therefore, heat that enters as a light ray is trapped in the room as heat energy.

The surfaces of buildings exposed to sunlight are also heated by the sun's rays. This heat source must be considered when designing heating and cooling requirements of air conditioning systems. Many building materials are poor conductors of heat. There is a lag or delay between the time the radiant heat energy strikes the building surface and the time the heat enters the conditioned space.

Color has a considerable effect on the amount of heat absorbed from the sun's rays. Black and red absorb much more heat than white and yellow. Likewise, surfaces that radiate heat are much more efficient if painted dark rather than light colors. Light-reflecting surfaces, such as polished metal, chrome, and nickel plate, do not absorb heat easily. They do not radiate heat efficiently from their surfaces.

19.6.2 Wind

The Beaufort Scale is frequently used by the United States Weather Bureau in indicating wind velocity. **Figure 19-31** gives wind velocity values and effects according to the Beaufort Scale.

Increasing the wind velocity increases the heat loss of a heated structure. The calculated heat load for a structure should include maximum wind velocity expected for the area.

During the winter months, the ***wind chill index*** ("chill factor") combines temperature and wind speed, **Figure 19-32.** The chill factor is calculated and released by the United States Weather Bureau.

The chill factor is based on both temperature and wind speed. For example: At a temperature of 0°F (−18°C) and a wind speed of 10 mph, the chill index temperature is −22°F (−30°C). Human flesh exposed to the atmosphere freezes at −25°F (−32°C).

19.7 Heat Insulation

In extremely hot or cold climates, it is desirable to use materials that do not transfer heat readily. This will maintain desired air conditioning temperatures more economically. One method is to reduce heat conductivity. Usually, spaces in the structure can be filled with insulating material to help prevent the flow of heat through the structure. Modern buildings are usually insulated with one of the following: mineral wool, expanded mica, balsam wool, urethane, and sometimes cork in either sheet or granular form.

19.7.1 Heat Sink

A warm body radiates heat rays. There are two common effects:

- Heat rays that strike another surface of the same temperature will reflect back. There is no increase or decrease in heat in the body being struck by the radiation.

  Imagine that the body giving off the heat rays is totally surrounded by a surface at the same temperature as that body. Then all the surfaces, including the central body, receive back as much radiation as they give off. All the surfaces and the central object remain at the original temperature.

Beaufort Number	Wind Velocity		Observed Wind Effects	Terms Used in USWB Forecasts
	Mph	Knots		
0	Less than 1	Less than 1	Calm; smoke rises vertically	Light
1	1–3	1–3	Direction of wind shown by smoke drift; but not by wind vanes	
2	4–7	4–6	Wind felt on face; leaves rustle; ordinary vane moved by wind	
3	8–12	7–10	Leaves and small twigs in constant motion; wind extends light flag	Gentle
4	13–18	11–16	Raises dust, loose paper; small branches are moved	Moderate
5	19–24	17–21	Small trees in leaf begin to sway; crested wavelets form on inland waters	Fresh
6	25–31	22–27	Large branches in motion; whistling heard in telephone wires; umbrellas used with difficulty	Strong
7	32–38	28–33	Whole trees in motion; inconvenience felt walking against wind	
8	39–46	34–40	Breaks twigs off trees; generally impedes progress	Gale
9	47–54	41–47	Slight structural damage occurs; leaves, branches blown from trees	
10	55–63	48–55	Seldom experienced inland; trees uprooted; considerable structural damage occurs	Whole Gale
11	64–72	56–63	Very rarely experienced; accompanied by widespread damage	
12 or Higher	73 or Higher	64 or Higher	Very rarely experienced; accompanied by widespread damage	Hurricane–Typhoon

Figure 19-31. *Beaufort Scale of wind velocity.*

Wind Chill Index

Wind Velocity in mph		Ambient Temperature														
	°F	40	35	30	25	20	15	10	5	0	−5	−10	−15	−20	−25	−30
	°C	4	2	−1	−4	−7	−9	−12	−15	−18	−21	−23	−26	−29	−32	−34
		Equivalent Temperature in Still Air														
Calm (0)	°F	40	35	30	25	20	15	10	5	0	−5	−10	−15	−20	−25	−30
	°C	4	2	−1	−4	−7	−9	−12	−15	−18	−21	−23	−26	−29	−32	−34
5	°F	37	33	27	21	16	12	7	1	−6	−11	−15	−20	−26	−31	−35
	°C	3	1	−3	−6	−9	−18	−14	−17	−21	−24	−26	−29	−32	−35	−37
10	°F	28	21	16	9	2	−2	−9	−15	−22	−27	−31	−38	−45	−52	−58
	°C	−2	−6	−9	−13	−17	−19	−23	−26	−30	−33	−35	−39	−43	−47	−50
15	°F	22	16	11	1	−6	−11	−18	−25	−33	−40	−45	−51	−60	−65	−70
	°C	−6	−9	−11	−17	−21	−24	−28	−32	−36	−40	−43	−46	−51	−54	−57
20	°F	18	12	3	−4	−9	−17	−24	−32	−40	−46	−52	−60	−68	−76	−81
	°C	−8	−11	−16	−20	−23	−27	−31	−36	−40	−43	−47	−51	−56	−60	−63
25	°F	16	7	0	−7	−15	−22	−29	−37	−45	−52	−58	−67	−75	−83	−89
	°C	−9	14	−18	−22	−26	−30	−34	−38	−43	−47	−50	−55	−59	−64	−67
30	°F	13	5	−2	−11	−18	−26	−33	−41	−49	−56	−63	−70	−78	−87	−94
	°C	−16	−15	−19	−24	−22	−32	−36	−41	−45	−49	−53	−57	−61	−66	−70
35	°F	11	3	−4	−13	−20	−27	−35	−43	−52	−60	−67	−72	−83	−90	−98
	°C	−11	−16	−20	−25	−29	−33	−37	−42	−47	−51	−55	−58	−64	−68	−72
40	°F	10	1	−6	−16	−22	−29	−36	−45	−54	−62	−69	−76	−87	−94	−101
	°C	−12	−17	−21	−26	−30	−34	−38	−43	−48	−52	−56	−60	−66	−70	−74

COOL · COLD · VERY COLD · BITTER COLD · EXTREME COLD

Figure 19-32. *Wind chill index reveals that increasing wind velocity greatly increases chill effect. With thermometer reading of 0°F (−18°C) and wind velocity of 20 mph, effect on human body is same as it would be if person were in temperature of −40°F (−40°C). (Michigan Farmer)*

- If the radiant heat strikes a surface colder than the radiating body, the heat rays do not all bounce back. Some of the radiant heat is absorbed into the colder surface. This surface becomes what is called a ***heat sink.***

On cold days, heat in a heated room flows from the room through cold window surfaces. A person sitting close to a window under such conditions will feel cold. This is due to the loss of heat into the heat sink. The same will also be true if the walls of the room are not well insulated. If the walls are cold, they become a heat sink.

19.7.2 Stratification

If there is no air movement within a room, the air may tend to stratify. That is, the cold air will sink to the floor and the warmer air will rise to the ceiling. A certain amount of air movement is needed in the room to prevent this.

If the thermostat is located in the upper part of a room with no air movement, the temperature difference (because of stratification) will be more noticeable.

19.7.3 Heat Exchange

Methods of heat transfer are covered in Chapter 1. However, additional information on heat exchange may be necessary in the study of air conditioning. Four types of heat exchange are possible:

- ***Radiation***—Radiant heat exchange means that a body is radiating heat. If heat being radiated strikes a body or substance at a lower temperature, this heat is lost to the lower temperature substance. If a body is surrounded by surfaces at a higher temperature, its temperature will increase.
- ***Convection***—Convection is the transfer of heat from one body to another through a medium, normally air or water. Conventional ovens heat by convection—the oven transfers heat to the food through the air in the oven.
- ***Evaporation***—Evaporative heat exchange takes place from the human body. Moisture is fed to the skin from the sweat glands. Evaporation of this moisture tends to lower the skin temperature. The evaporative moisture constitutes a considerable heat exchange from the human body. Evaporative heat exchange can be considered a form of convection. (The evaporated moisture is carried away along with its heat content.)
- ***Conduction***—Conduction is the transfer of heat between molecules or bodies in direct contact with one another. If one end of a steel rod is heated, the other end will also warm because the molecules in the rod conduct the heat.

intensity of odor-causing molecules in trace concentrations. The portable odor monitor uses a highly sensitive metal oxide thermal conductivity sensor. Numerous readings are initially made to set the quality standards or acceptable odor levels. Once this is established, the monitor will then identify the odor level at any time.

Ozone

Ozone is a form of oxygen photochemically produced in nature. The chemical formulas for oxygen and ozone are O_2, and O_3, respectively.

In the upper atmosphere, ozone is made by ultraviolet light reacting with oxygen. It may also be produced by an electric discharge in air (lightning).

Ozone is a disinfectant. It is sometimes used to purify water or to maintain a sterile atmosphere. It also removes odors in such places as cold storage rooms and hospitals.

No universal agreement exists concerning the benefits or possible hazards in using ozone in conditioned spaces. A very small amount of ozone in the air is considered to be beneficial. However, in larger concentrations, it may be detrimental to health. Ozone may be one of the reasons for irritation caused by smog. There is no evidence that ozone is accumulatively harmful.

An ozone concentration of 0.1 parts per million (ppm) is generally considered the maximum permissible for eight-hour exposure. For continuous occupancy, ozone should not exceed 0.01 ppm. The effect doubles for each 15°F (8°C) increase in temperature. The use of some electronic air cleaners may slightly increase ozone content.

Reliable instruments have been developed for measuring ozone concentration. A typical ozone monitor is illustrated in **Figure 19-35.**

Pollen

During certain seasons, some plants create a concentration of pollen in the atmosphere, irritating many people. The most troublesome plants are ragweed, timothy, goldenrod, and roses.

The ***pollen count*** may be determined by exposing an adhesive-coated surface to the atmosphere for 24 hours. The number of pollen grains in a square centimeter determines the pollen count for the past 24 hours.

19.8.2 Indoor Air Quality

Indoor Air Quality (IAQ) has recently become a concern because of the improved building standards. This has resulted in increased insulation and reduced energy-consuming ventilation systems. The combination of these factors has led to an increase in indoor air quality complaints. See **Figure 19-36.**

Studies have shown that productivity can be increased by 15% by improving the working environment.

IAQ problems are commonly classified as one of the following: Sick Building Syndrome (SBS), Building Related Illness (BRI), or Multiple Chemical Sensitivity (MCS). The majority of IAQ concerns stem from poor ventilation, poor filtration, and contaminated HVAC systems.

Sick Building Syndrome (SBS)

One of the most common causes of IAQ complaints is ***Sick Building Syndrome (SBS).*** It occurs when approximately 20% of the building's occupants complain of drowsiness, fatigue, eye and skin irritations, or respiratory problems. These symptoms frequently disappear when the individual is removed from the environment.

SBS is the presence of any combination of the following:

- Poor temperature/humidity control.
- Poor ventilation.
- Improper maintenance.
- Airborne chemicals or pollutants.
- Excess noise.
- Improper system design.
- Poor lighting.

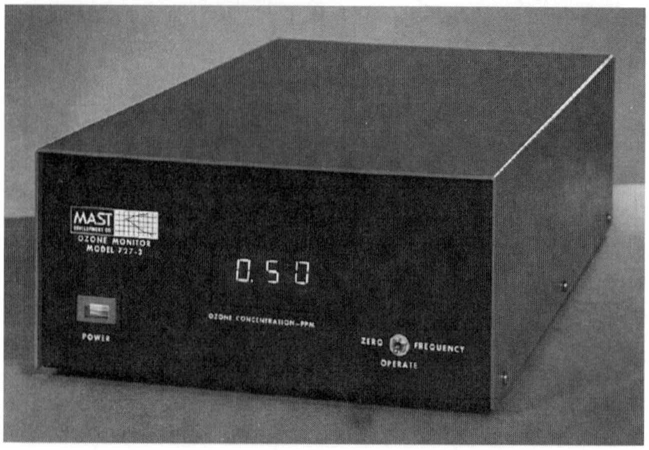

Figure 19-35. *Ozone monitor indicates ozone content of area in 0 to 9.99 parts per million (ppm). (Mast Development Co.)*

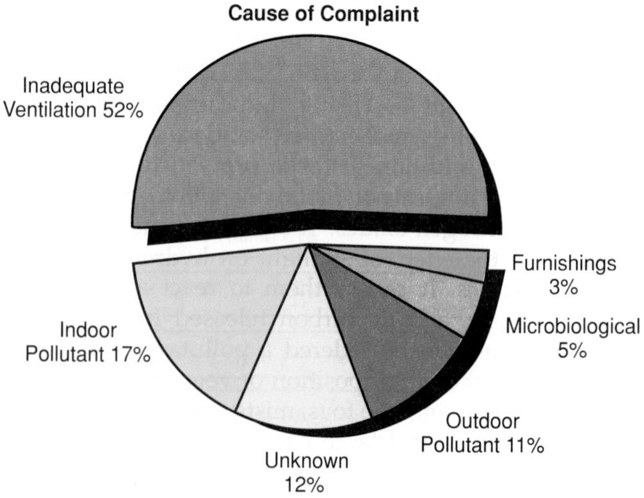

Figure 19-36. *General causes of IAQ problems. Note that 52% of IAQ difficulties stem from poor ventilation.*

Building Related Illness (BRI)

Exposure to an airborne agent leading to a diagnosable illness is referred to as ***Building Related Illness (BRI).*** BRI health problems do not disappear when the occupant moves to a more favorable environment. Example of BRI include Legionnaire's disease, colds, flu viruses, tuberculosis, measles, and smallpox. Illnesses caused by BRI can permanently damage health and, in extreme cases, may be fatal.

There are numerous causes of BRI:

- Viruses spread by the airflow system.
- Stagnant water.
- Toxins and allergens.
- Radon.
- Airborne biological agents (spores, fungus).

Multiple Chemical Sensitivity (MCS)

A very small portion of the population has experienced ***Multiple Chemical Sensitivity (MCS).*** These individuals appear to have abnormal sensitivity to chemicals in the environment.

19.8.3 Indoor Air Contaminants

There are three major indoor air contaminants: asbestos, bioaerosols, and radon.

Asbestos is a silicate that occurs in fiber bundles. It is known for its strength and fire resistance. It was often used in old commercial buildings. Exposure to asbestos normally occurs in four settings: production (mining), materials production (insulation, brake linings), construction, and removal. Asbestos is a cancer-causing agent.

Most agencies agree that asbestos that has not deteriorated should be left alone. Removal of asbestos should only be done by professional asbestos abatement companies. Sampling for asbestos is accomplished through environmental monitoring, including visual assessment and, if needed, bulk sampling.

Bioaerosols are airborne microorganisms derived from viruses, bacteria, fungi, protozoa, mites, and pollen. They are found both indoors and outdoors. Excessive moisture indoors increases the growth of some microorganisms. Certain types of humidifiers, water spray systems, and wet porous surfaces act as breeding grounds. Microorganisms in the indoor environment may cause allergic building-related illness (BRI).

Legionnaire's disease is caused by bioaerosols. The Legionella bacteria may be found in cooling towers, evaporative condensers, and domestic water systems.

Inadequate preventative system maintenance may provide nutrients for growth of bacteria such as Legionella. The design of the equipment and proper maintenance of the systems may reduce risk.

The HVAC system should be checked whenever medical evidence indicates the presence of diseases (such as humidifier fever, allergic asthma, and allergic rhinitis, etc.). The system must be checked to determine if there are any microorganisms. An initial walk-through is recommended, with inspection for possible reservoirs and sites of contamination. If a site is located, a sample should be obtained and analyzed.

Radon is an odorless, tasteless, radioactive gas that occurs naturally in soil and rocks. Radon is formed from the natural decay of uranium and is found in some industrial wastes.

Radon has been shown to cause lung cancer. When inhaled, it settles in a person's lungs. The radioactive particles then begin to damage the lung tissue. Radon gas comes from the soil and rock upon which the building is built. It may enter a building through small cracks in concrete floors, floor drains, sump pumps, and pores in hollow-block walls. See **Figure 19-37.**

Radon may be detected by using a charcoal canister or a small container called an alpha track detector. See **Figure 19-38.** The detectors are usually sent to a laboratory for analysis. If results show the presence of radon, entry points of the gas must be located and repaired.

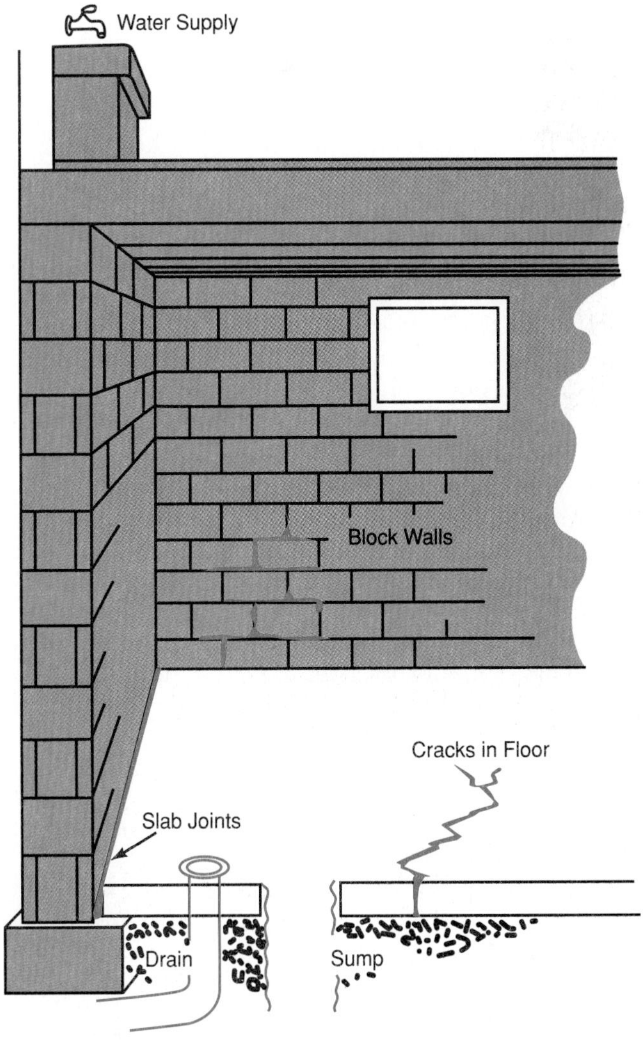

Figure 19-37. *Common entry points for radon into a home. (United States Environmental Protection Agency, U. S. Dept. of Health and Human Services)*

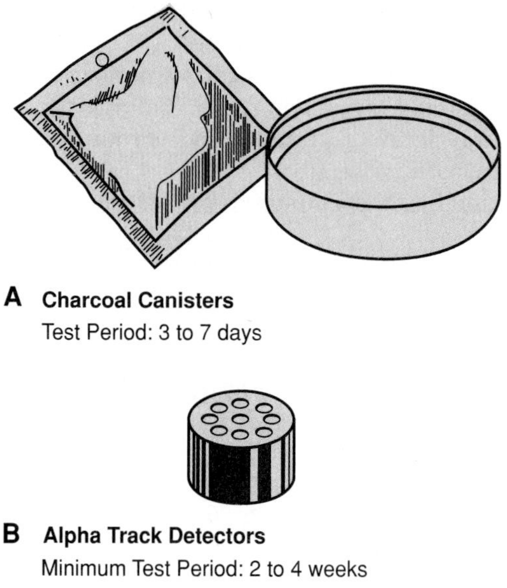

Figure 19-38. *Two common radon leak detectors. A—Charcoal canister. B—Alpha track detector. (United States Environmental Protection Agency, U. S. Dept. of Health and Human Services)*

Carbon Dioxide (CO_2)

Carbon dioxide (CO_2) is inhaled and exhaled by humans. The concentration of CO_2 in exhaled breath is fairly constant, approximately 3.8%. Once the CO_2 leaves the mouth, it mixes with the surrounding air.

When people exhale CO_2, they also exhale other gases, odors, bacteria, and viruses. When these build up in a space due to poor ventilation, poor air quality results. When this occurs, occupants often complain of fatigue, headaches, and general discomfort. Carbon dioxide does not create these symptoms. However, high CO_2 concentrations indicate that these other contaminants may also be present.

Diagnosing Indoor Air Contamination

A procedure for diagnosis and lessening of indoor air contamination is shown in **Figure 19-39.** When evaluating the ventilation, ASHRAE Standard 62–1989 should be used.

1. Obtain a description of symptoms from the occupants. Some of the areas of concern are physical symptoms, odors, and frequency and time of their occurrence.
2. Determine the possible sources. Examine the ventilation system and carefully review potential sources of contaminants. If the source is not yet evident, continue to Step Three.
3. Take an air pollutant sampling and perform a chemical analysis. Sampling should be done at the indoor location and at an outdoor location near the system air inlet. Building-related contaminants usually peak in the morning after the system has been inactive throughout the night. Occupant-related contaminants usually peak in late afternoon. Following the interpretation of data, proceed to Step Four.

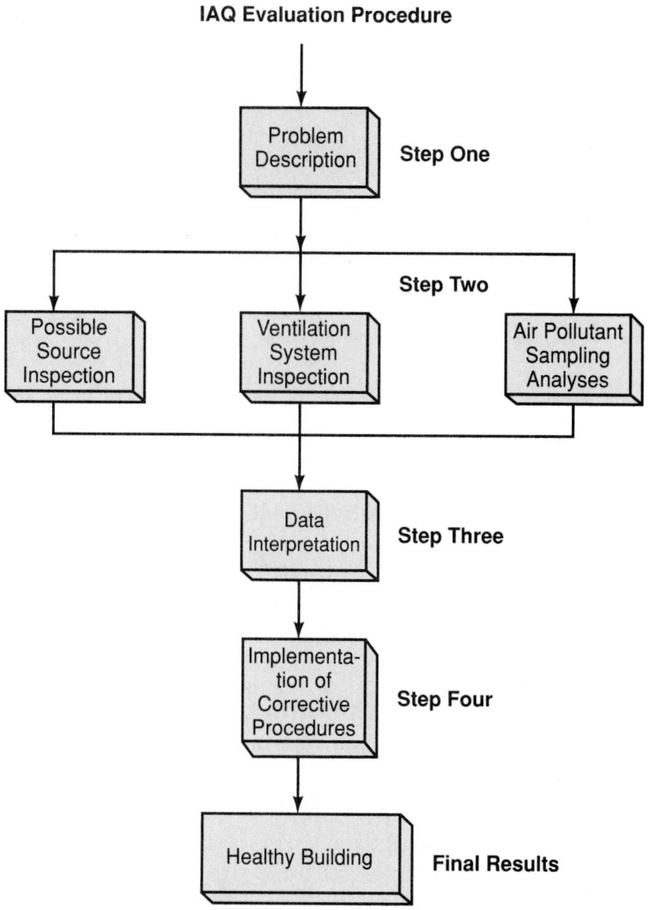

Figure 19-39. *The four-step method for evaluating the IAQ of a building problem.*

4. The proper procedures are implemented to correct the problem. Possible solution for IAQ problems may include increased ventilation, air cleaning, and control of the problem areas. These corrections should lead to a healthy building.

Servicing Ventilation Systems

Numerous methods have been developed to measure the amount of required indoor air circulation. One type, a portable unit, measures the carbon dioxide level and determines proper area ventilation requirements. See **Figure 19-40.**

Complete IAQ evaluators (demand control ventilation) are permanently installed in large buildings. See **Figure 19-41.** This type unit is installed in the duct. **Figure 19-42** is a wall-mounted carbon dioxide controller. This also is used for demand control ventilation. When the monitor reads a set amount, the system then transfers these CO_2 readings to the central system. The central system makes the necessary adjustments.

A similar system has an electrochemical cell signal that monitors and tracks air contaminants that cause Sick Building Syndrome. See **Figure 19-43.**

A variety of monitors are available for detection and measurement of air quality contaminants. These microprocessor-based instruments detect and measure noise, heat stress, and indoor air quality.

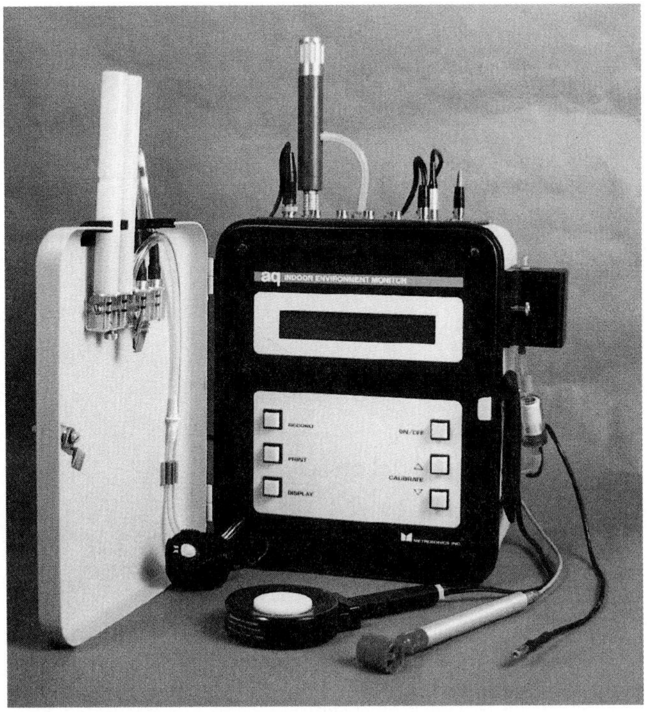

Figure 19-40. *A portable indoor environment monitor. Used to identify ventilation and sick building syndrome problems. (Metrosonics, Inc.)*

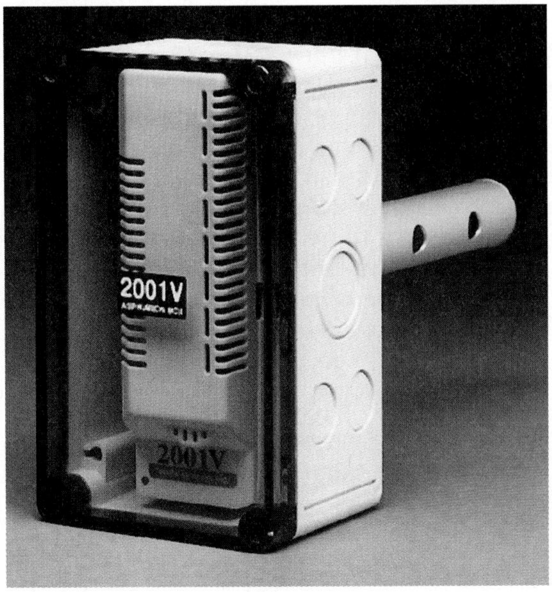

Figure 19-41. *An in-duct carbon dioxide monitor. (Telaire Systems, Inc.)*

Figure 19-42. *Carbon dioxide controller with digital CO_2 indicator. (Telaire Systems, Inc.)*

Measuring Filter Efficiencies

Two common methods of measuring filter performance are described in ASHRAE Standard 52–76:

- Atmospheric dust spot efficiency.
- Synthetic dust weight arrestance.

Atmospheric dust spot efficiency measures the ability of the filter to remove atmospheric dust. It is the measurement of flow rates on both sides of the filter being tested.

Special filter paper targets are located on both sides of the tested filter. The efficiency is calculated based on the quantity of air drawn through the target filter, the amount of light transmitted through the target filter, and the difference in light transmission for the two paper filter targets. Amount of air and light transmission through the target filter papers decreases during the test due to a buildup of dust.

Synthetic dust weight arrestance is a measure of a filter's ability to remove synthetic dust from test air. It is calculated based upon the weight of the synthetic dust that passes through the filter being tested. This weight is compared to the weight of the amount fed into the filter.

Another filter efficiency test is called the ***DOP Smoke Penetration Method.*** The name of the test comes from the name of the testing particles DiOctylPhthalate. This test is used mainly with high efficiency filters. Particles of 0.3 microns are sprayed into the inlet duct of the filter being tested. Small sample white filters collect some of this dust from the airstream ahead of the filter. Other sample filters collect dust from the air leaving the filter. The difference in the sampling filters, by color or weight, determines the filter efficiency. For example, the

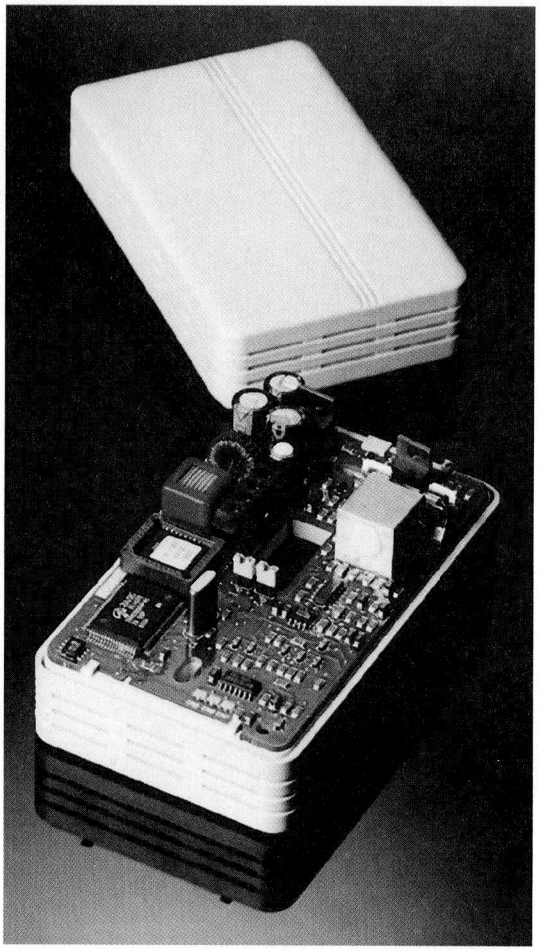

Figure 19-43. *Indoor air evaluator. This instrument assesses the ventilation quality by detecting, measuring, and recording carbon dioxide, temperature, and relative humidity. It has an optional electrochemical cell for signaling the presence of additional air contaminants. (HyCal Unit of General Signal)*

outlet sample filter may collect only 1% of the particles as the sample filter in the inlet. This filter would then have 99% efficiency.

Duct Cleaning

Awareness of indoor air quality is increasing. More people are considering duct cleaning as a possible solution to solving IAQ problems.

Numerous areas where the duct system can be entered should be selected. A three-step process is used for treatment of the ductwork.

1. A ***duct sweeper*** is rotated through the sides of the ducts. This releases dust, mold, and mildew.
2. A high-velocity commercial vacuum removes loose particles from the duct work, **Figure 19-44.**
3. A microbial biocide is sprayed into the cleaned duct system. This helps prevent mold and mildew build-up.

Filters are checked, cleaned, or replaced if necessary. This will reduce future contamination.

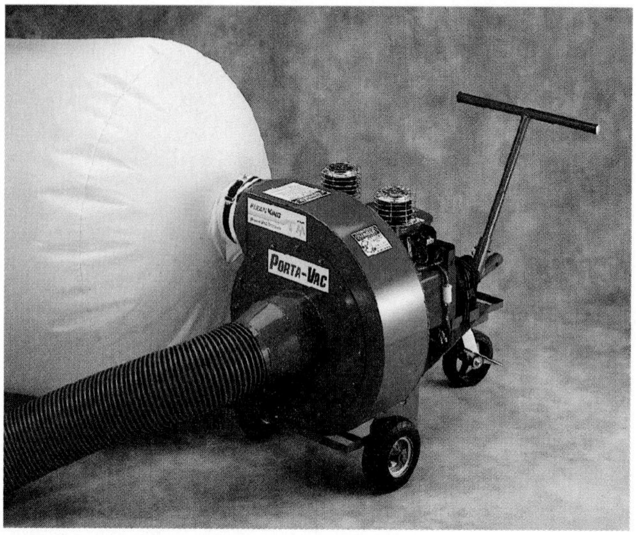

Figure 19-44. *A small residential dust cleaning vacuum. (Wm. W. Meyer & Sons, Inc.)*

19.9 Commercial and Residential Air Quality Systems

Development of complete indoor air quality systems for residential homes has occurred in recent years. An example is shown in **Figure 19-45.** This type of complete system includes an air conditioner, furnace, humidifier, electronic air-filter, and an energy recovery vent heater. The unit is a controlled ventilation system. It reduces pollen, dust, odors, and other pollutants. The unit exhausts the stale humid air. Approximately 70% of the existing heated air is used for recirculation. Additional components, alarm systems, and lighting may be added. The components in this type system are coordinated to ensure adequate ventilation, heating, and cooling.

Indoor air quality systems are also available for commercial uses. See **Figure 19-46.** Such systems ensure the delivery of the correct amount of outdoor air. They

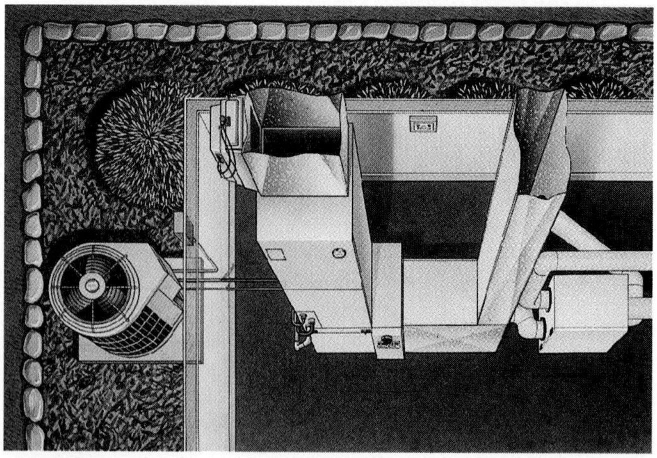

Figure 19-45. *Overhead view of a complete indoor air quality system. (Bryant Air Conditioning/Heating)*

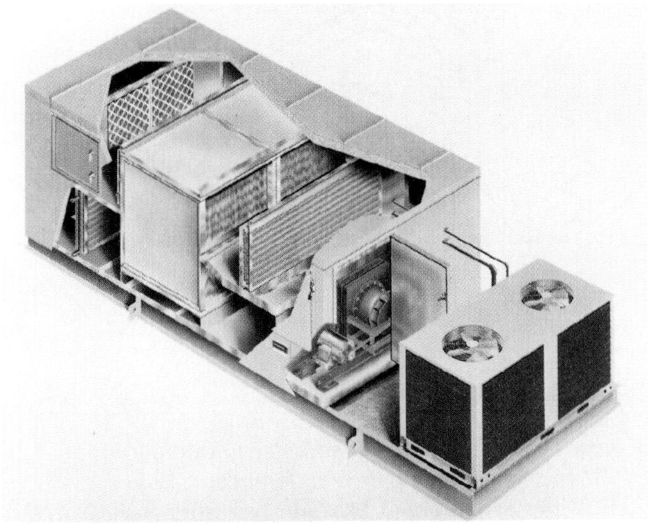

Figure 19-46. *A commercial indoor quality system. Note extensive infiltration system. (Des Champs Laboratories, Inc.)*

control space humidity and building pressure. These units increase building efficiency, improve indoor air quality, and increase comfort.

19.9.1 Thermometers—Air Conditioning

Common thermometers used in refrigeration and air conditioning service are described in Section 2.10.1. Special thermometers are needed to accurately determine the operating temperatures.

Two types of electric thermometers are shown in **Figure 19-47.** These units may be either battery or 120 V ac powered. The probe reacts quickly and accurately. The scale is calibrated in both Fahrenheit and Celsius degrees.

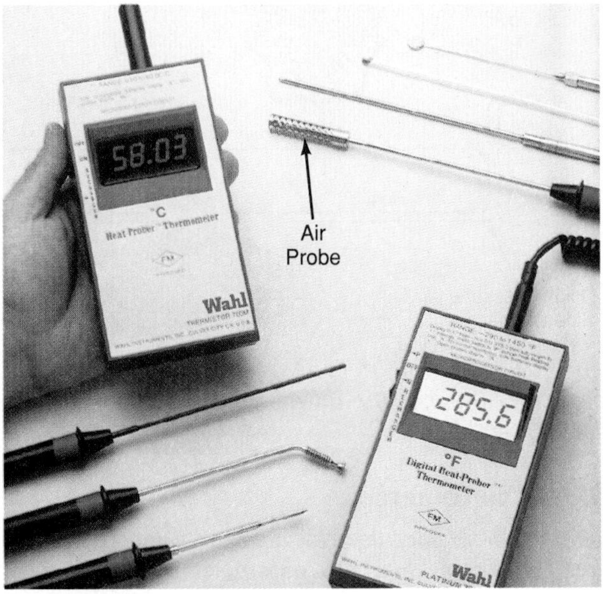

Figure 19-47. *Digital sensor probes for measuring temperature. Left—Thermistor-type thermometer. Right—Platinum-RTD (Resistance Temperature Detector) thermometer. Note the air probe with ventilated shield around tip. (Wahl Instruments, Inc.)*

A recording thermometer helps locate malfunctions by making 24-hour or 7-day temperature records. **Figure 19-48** illustrates a recording-type thermometer.

A wet globe thermometer is shown in **Figure 19-49.** This has been developed to measure overall comfort conditions in hot workplaces. The instrument consists of a 2 3/8″ hollow copper sphere. The sphere is painted black

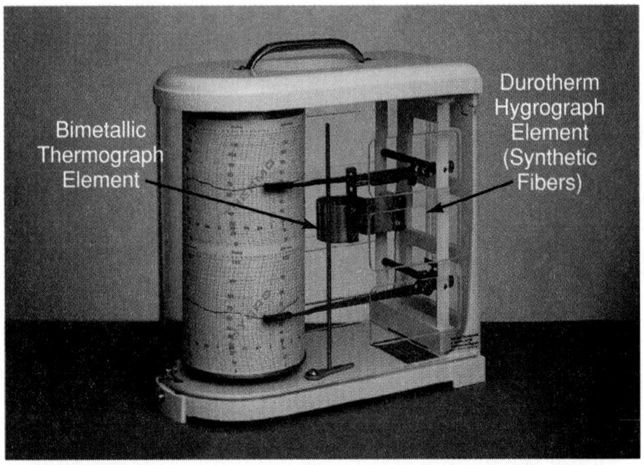

Figure 19-48. *Thermo-hygrograph showing location of the temperature and humidity elements. Arm at top records the temperature; arm at bottom records the humidity. (Abbeon Cal, Inc.)*

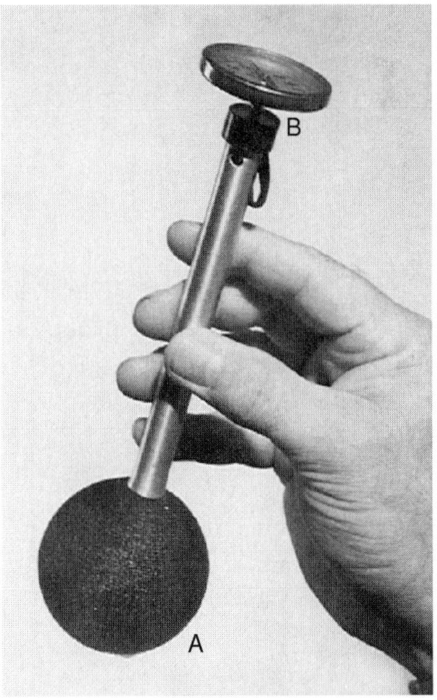

Figure 19-49. *Wet globe thermometer. Black-cloth-covered sphere will reach a temperature that is the balance of wet bulb cooling ability, outgoing radiation, and incoming radiation. A—Cloth-covered sphere. B—Dial thermometer. (BOTSBALL, Howard Mfg. & Consulting, Inc.)*

and covered with a double layer of black cloth. A 5″ aluminum tube connects the copper sphere. This tube is filled with water and is capped at the other end. It keeps the globe wet. A dial thermometer stem passes through the centerline of the water reservoir tube and into the globe. When placed in a hot area, the globe is warmed by heat radiating from the hot surfaces. It is cooled by evaporation from the globe surface. The evaporation depends on air velocity and relative humidity.

The wet globe reaches an equilibrium temperature after a few minutes. At this time, the heating and cooling effects are in balance. The dial thermometer reading will indicate the wet globe temperature. The wet globe thermometer provides an excellent index of human responses to heat.

19.9.2 Manometers

The principle of operation of the manometer is explained in Chapter 1. A manometer used in air conditioning work is shown in **Figure 19-50.** Flexible tubing enables it to be rolled or folded into a small space for carrying.

Figure 19-51 illustrates a method of connecting a manometer to an air duct to determine its pressure. To measure duct pressures, a water manometer is usually needed. The scale is usually movable, making it easier to adjust for the neutral point. Sudden pressure changes must be avoided or the liquid may be forced out of the manometer.

Some manometers measure the pressure difference between two different places in a duct. An example of this is a manometer used to measure the pressure drop across a filter in an airflow system.

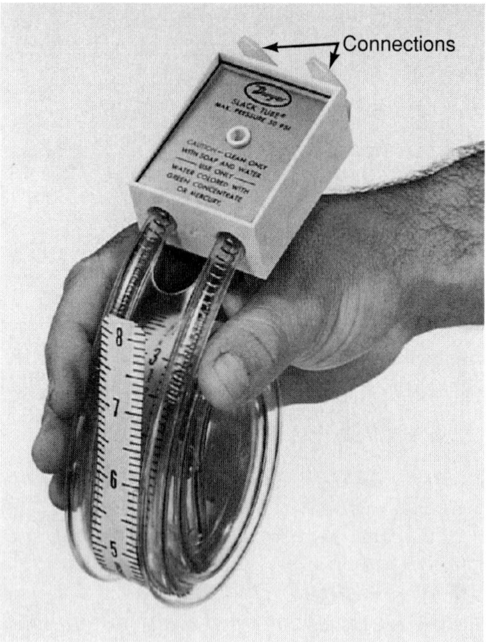

Figure 19-50. *A manometer is used to measure air pressure in ducts. The flexible tube permits easy storage. (Dwyer Instruments, Inc.)*

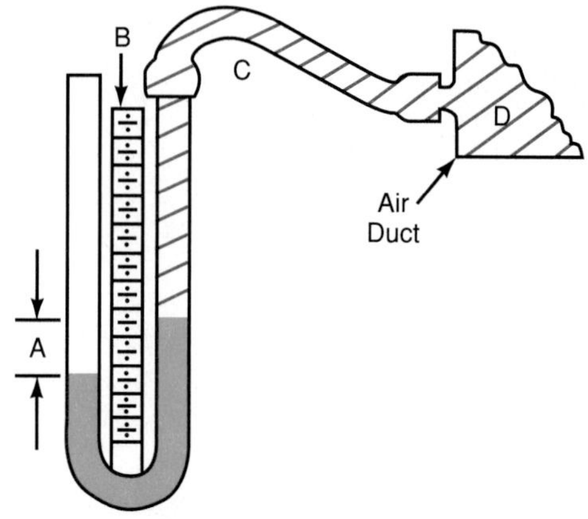

Figure 19-51. *Simple manometer in operation. A—Pressure is indicated by difference in liquid level in two sides of manometer. Usually, pressure is measured in inches. B—Scale in inches. C—Rubber connecting tube. D—Pressure being measured.*

Manometer scales are based on the following data:

14.7 psi = 29.92″ Hg = 34′ water
1″ Hg = .491 psi
1 psi = 2.035″ Hg
1 psi = 2.31′ water
1′ water = .432 psi
1″ water = .036 psi

Figure 19-52 shows a dial-type manometer in use. The two probes allow a comparison of readings.

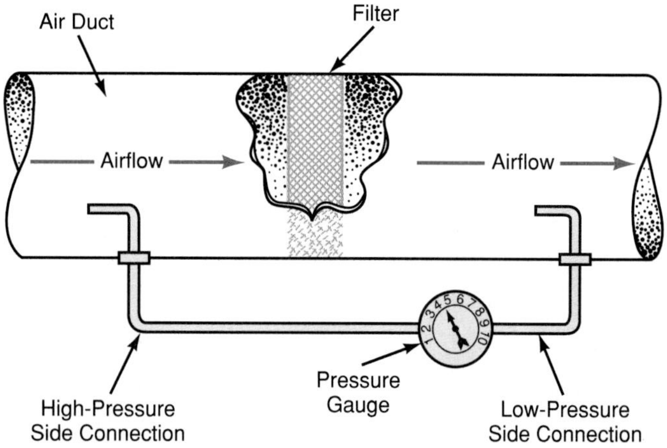

Figure 19-52. *One of many uses of a dial manometer. The greater the pressure difference, the more resistance (clogging) at filter.*

19.9.3 Barometers

Barometers are used to measure atmospheric pressure. The simple mercury barometer is illustrated and explained in Section 1.11.2. Barometers used in air conditioning measure pressure by the deflection of a bellows or diaphragm. A recording barometer commonly used in air conditioning work is shown in **Figure 19-53.** This instrument has a selectable rotation period of 1 day, 7 days, or 31 days.

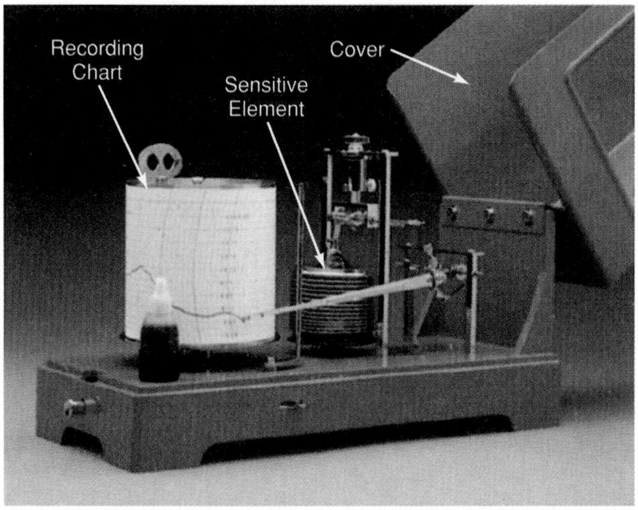

Figure 19-53. *A 24-hr. recording barometer (barograph). (Qualimetrics, Inc.)*

19.10 Comfort Conditions

Comfortable conditions result from a desirable combination of temperature, humidity, air movement, and air cleanliness. However, one may have comfort under varying values of these factors. For instance, high relative humidity tends to be uncomfortable. However, it may be counteracted by a relatively low temperature and rapid air movement. In many homes in wintertime, compensation is made for a low relative humidity. This is done by an increase in room temperature and slight air movement. **Figure 19-54** illustrates what is commonly accepted as the comfort zone for various conditions.

A more technical graph showing the comfort zones for both winter and summer is illustrated in **Figure 19-55.** This comfort zone represents a considerable area.

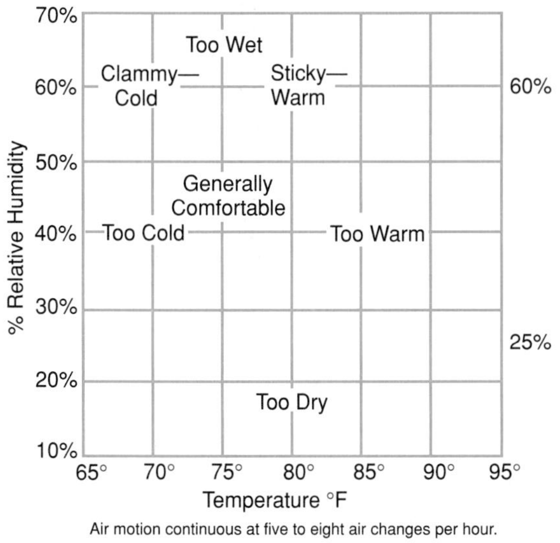

Figure 19-54. *Indoor comfort chart. Most people will feel comfortable at temperature and relative humidity indicated in center. (Lennox International, Inc.)*

However, any point in this area gives approximately equal comfort under equal conditions of clothing and work. These areas are sometimes defined as effective temperature (ET). ***Effective temperature*** is the combined effect of dry bulb temperature, wet bulb temperature, and air movement, that provides an equal sensation of warmth or cold.

In the summertime, air conditioned buildings are usually kept at temperatures approximately 10 to 15 degrees F below the outside temperature. Some people are quite sensitive to thermal shock when entering or leaving an air conditioned space. This danger is lessened if the difference between inside and outside temperatures is reduced. It can also be reduced by using a sweater or coat when indoors.

Figure 19-55 indicates the comfort range for most people. In summer, this is between 72°F (22°C) db (dry-bulb temperature) and 90% rh (relative humidity) up to 87°F (31°C) db and 23% rh. In winter, it is between 66°F (19°C) db and 70% rh to 80°F (27°C) db and 20% rh.

The average person is most comfortable if the skin surface temperature is 91°F (33°C). This skin temperature is usually maintained in cold weather by wearing adequate clothing. In hot weather, it is maintained by the evaporation of moisture (sweat) and by radiation from the skin surface.

The skin temperature may drop below this figure in hot, humid weather. This occurs because of rapid evaporation of moisture from the skin surface. However, the person is not uncomfortable because body heat is being released by moisture evaporation from the skin surface.

Temperature-related illnesses are called "thermal disorders." In cold climates, it is possible for the body temperature to drop a few degrees below normal due to lower metabolism.

High temperatures may cause human illness, especially if there is also high humidity. Heat stress is being investigated by the Occupational Safety and Health Administration. The measure of heat stress is done using a 6″ black copper sphere. A dry bulb thermometer is inserted until the sensitive bulb is at the center of the sphere. At 79°F (26°C) WBGT (wet bulb globe temperature), a person should only work half time for the first five days. Salt tablets should be taken moderately if the WBGT is 79°F (26°C) or higher.

19.10.1 Comfort-Health Index (CHI)

The American Society of Heating, Refrigerating and Air-Conditioning Engineers recognizes a Comfort-Health Index. **Figure 19-56** indicates the temperature, sensation, and effect on physiology and health of the body. This chart indicates that the human body attempts to adjust exposure to very hot conditions. The body attempts to adjust by increasing sweating and the flow of blood. These physiological conditions may result in an increased danger of heat strokes and cardiovascular difficulty.

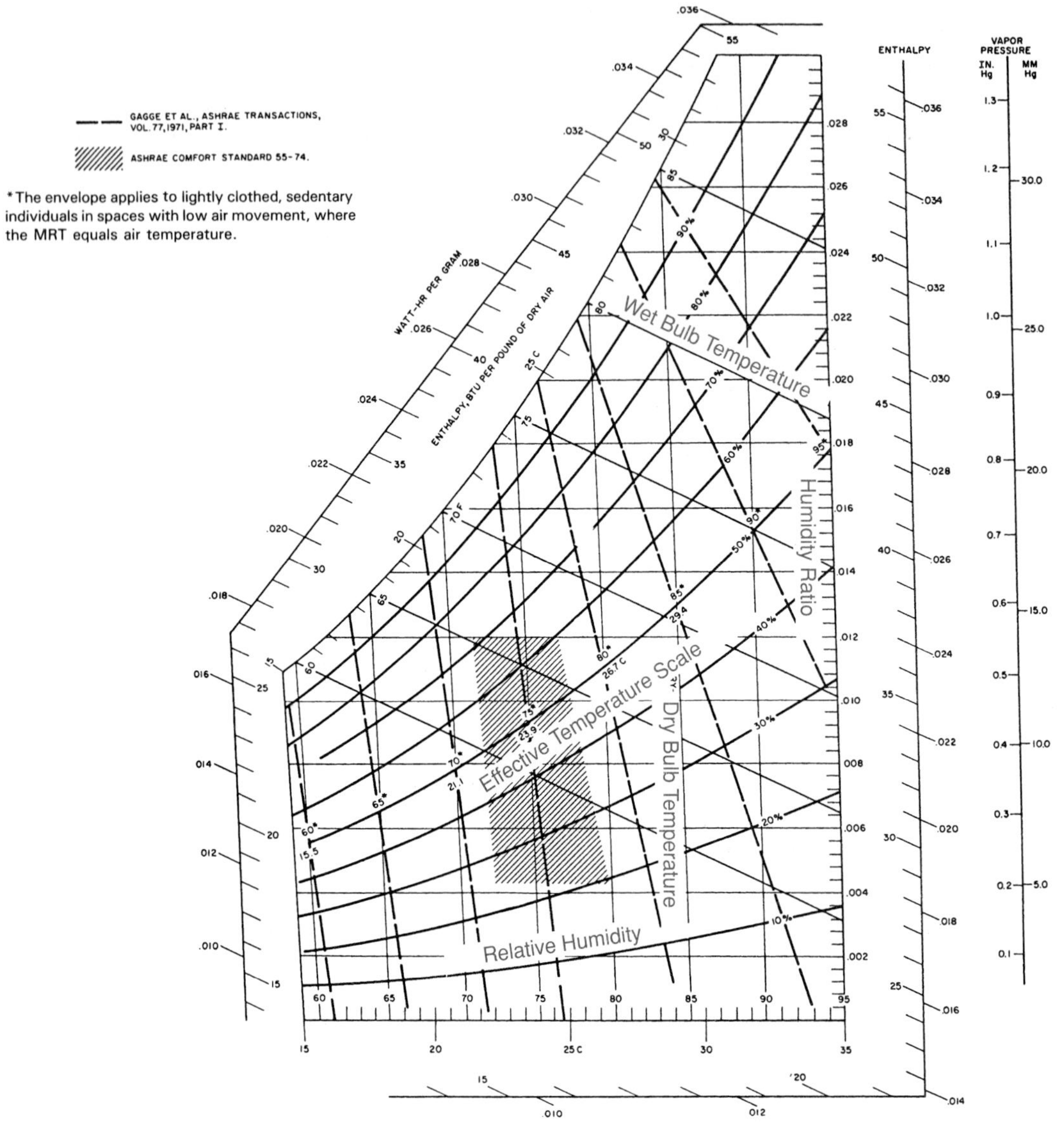

Figure 19-55. *Graph of comfort zone. Note dry-bulb and wet-bulb temperature lines, and relative humidity line. (Reprinted by permission of the American Society of Heating, Refrigerating, and Air-Conditioning Engineers, Atlanta, Georgia)*

The chart shows that, at comfortable temperatures, there is no sensation of either warmth or cold. Also, there are no apparent physiological effects. Moving down in temperature to very cold conditions, the body is uncomfortable. Also, physiologically, the body attempts to correct this condition by shivering. From the standpoint of health, this may cause an increase in deaths, particularly in older people.

19.11 Noise

The air conditioning system must handle both the heat load and fresh air required. It must do these functions in a manner that will not be annoying to the occupants. In addition to drafts, a further source of annoyance is objectionable noise.

Noise is unwanted sound. Complaints of unpleasant noise due to air conditioning systems are common. The noise problem can be divided into three types:

- Noise source.
- Noise carrier.
- Noise amplification or reflection.

The noise source is an audible vibration. This vibration may start in the heating unit, cooling unit, fan mechanism, air turbulence, duct panels, duct hangers, or grilles.

Sound or noise is produced by movement of an object. This movement may be caused by vibration of the object or air movement against the object, such as in air conditioning ducts. A vibrating duct panel will create alternate waves of low-pressure and high-pressure air, producing a sound similar to the hum of a mosquito.

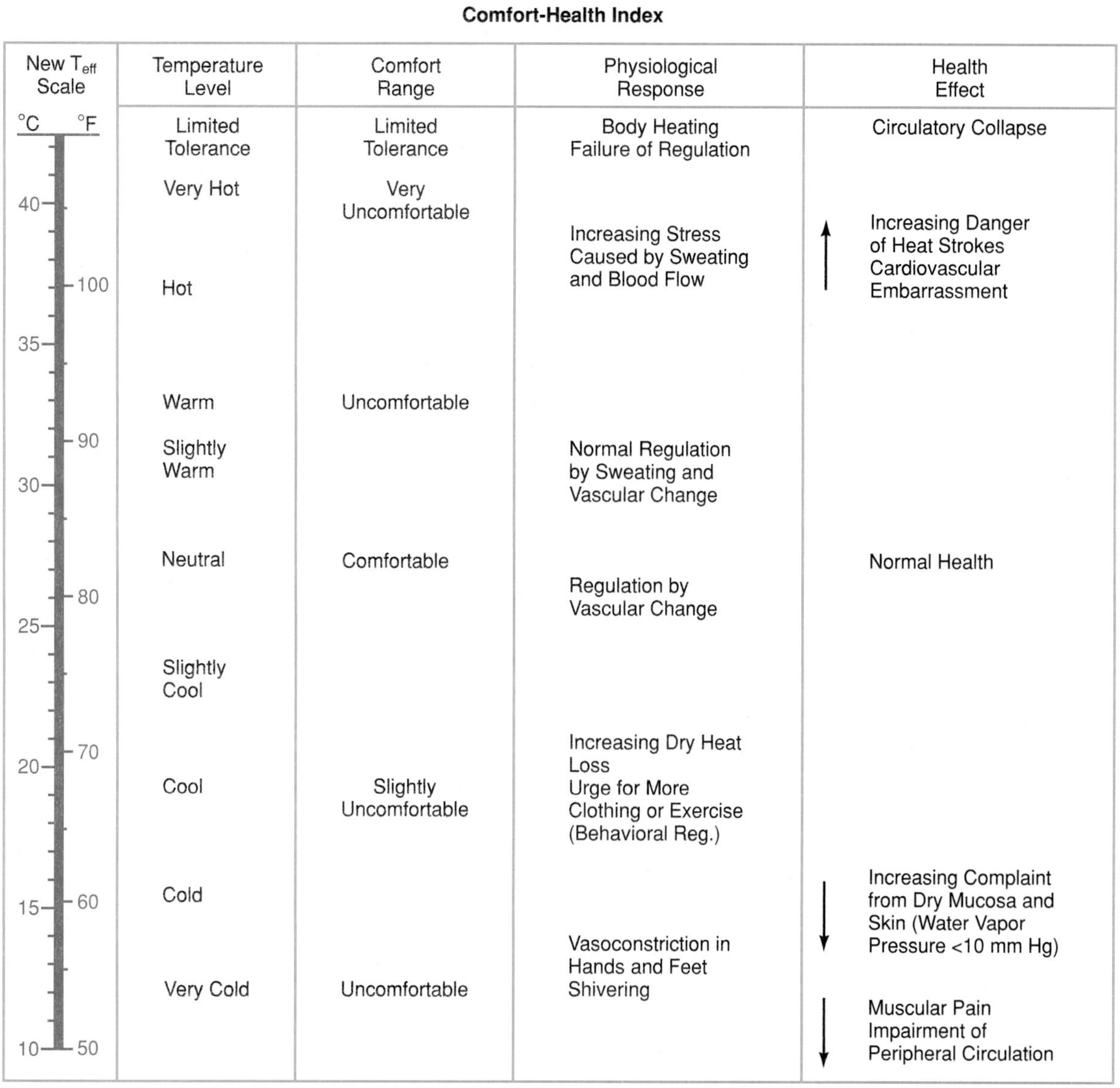
Comfort-Health Index
New T_{eff} Scale
°C °F
40
100
35
90
30
80
25
70
20
60
15
10 50
Temperature Level
Limited Tolerance
Very Hot
Hot
Warm
Slightly Warm
Neutral
Slightly Cool
Cool
Cold
Very Cold
Comfort Range
Limited Tolerance
Very Uncomfortable
Uncomfortable
Comfortable
Slightly Uncomfortable
Uncomfortable
Physiological Response
Body Heating Failure of Regulation
Increasing Stress Caused by Sweating and Blood Flow
Normal Regulation by Sweating and Vascular Change
Regulation by Vascular Change
Increasing Dry Heat Loss Urge for More Clothing or Exercise (Behavioral Reg.)
Vasoconstriction in Hands and Feet Shivering
Health Effect
Circulatory Collapse
Increasing Danger of Heat Strokes Cardiovascular Embarrassment
Normal Health
Increasing Complaint from Dry Mucosa and Skin (Water Vapor Pressure <10 mm Hg)
Muscular Pain Impairment of Peripheral Circulation

Figure 19-56. *Comfort-Health Index indicates sensory, physiological, and health responses to prolonged exposures. (Reprinted by permission of the American Society of Heating, Refrigerating, and Air-Conditioning Engineers, Atlanta, Georgia, from the 1993 ASHRAE Handbook—Fundamentals)*

Another common complaint is noise caused by high speed air traveling through the ducts, causing air turbulence. Often, this is the result of an undersize unit or duct, in which the blower has been speeded up in an attempt to make up for the unit's small size.

Noise or vibration carriers are rigid structures. They carry vibrations to places where they may be annoying. Floors, ceilings, ducts, doors, and pipes may carry these vibrations.

Noise amplifiers or reflectors are usually hard, smooth surfaces. Walls, ceilings, floors, and furnishings may pick up a small vibration. They will reflect it at a frequency and direction so that all or parts of the space become uncomfortable.

Soft fabrics such as drapes and curtains and fabric-covered furniture are noise absorbers. Felt-lined or soft-insulation-lined ducts also absorb noise.

Some communities have codes regulating how noisy a mechanism may be. A city may limit the decibel of a window or outdoor condensing unit. One city, for example, limits level to 50 decibels at a 10′ distance. More sound-deadening devices may be needed to meet this standard.

The air velocity may depend on the type of building being air conditioned. Where noise is a factor, the velocity should be kept to a minimum. If the velocity cannot be decreased, noise may be reduced by other means. An acoustical discharge chamber may be used. The ducts may be lined or wrapped with sound-absorbing material, such as felt.

19.11.1 Noise Measurement

Sound waves are rapid changes of air pressure. The amplitude or strength of the sound pressure waves may be measured. Sound strength and sound pressure level (SPL) are rated in decibels (dB). Two words are used in connection with the definition of sound:

- *Sound strength.* This is the total amount of sound, in decibels, coming from a unit.
- *Sound pressure.* This is the strength, in decibels, of sound after it travels a specified distance from a source.

For instance, a truck engine gives off a measurable sound strength at the engine. Sound, after it has traveled 50′ to a storefront, is measured in sound pressure.

The number of vibrations per second of sound waves is measured with an amplifying microphone. The international unit for sound frequencies is the Hertz (Hz), cycles per second (cps).

Sound travels in waves through the air in all directions. Its strength diminishes with the distance from the source. The first measurement of sound strength is usually taken about three feet from its source. The amplitude of the sound waves will be reduced by the cube of the distance that the receiving or recording instrument is away from the sound source.

Sound pressure is shown in **Figure 19-57.** Increasing the sound frequency tends to increase the apparent loudness. This is because the human ear does not respond equally to all frequencies. This effect is illustrated in **Figure 19-58.** For most people, sounds in the 1000 Hz to 4000 Hz range are easiest to hear.

In measuring the loudness of sound, the meters generally read in dB(A) or dBA. The dBA scale loudness means that a standard "A" filter has been placed in the microphone circuit. The filter reduces the intensity of the low frequencies. **Figure 19-59** illustrates the effect of using an "A" filter in the microphone circuit.

A comparison of loudness measurements using the dB and the dBA scales shows the filter's effect. See **Figure 19-60.** Note that equal loudness requires a higher dB rating in the lower frequencies than the dBA rating in the lower frequencies.

Laws regulating permissible sound or noise levels are usually written around the "A" scale. The Walsh-Healy Act limits the time workers may be exposed to various sound levels. A table of these limits is shown in **Figure 19-61.**

Source	Sound Pressure Pa	Sound Pressure Level dB re 20μPa	Subjective Reaction
Military jet takeoff at 100 ft	200	140	Extreme danger
Artillery fire at 10 ft	63.2	130	
Passenger's ramp at jet airliner (peak)	20	120	Threshold of pain
Loud rock band	6.3	110	Threshold of discomfort
Platform of subway station (steel wheels)	2	100	
Unmuffled large diesel engine at 130 ft	0.6	90	Very loud
Computer printout room	0.2	80	
Freight train at 100 ft	0.06	70	
Conversational speech at 3 ft	0.02	60	
Window air conditioner	0.006	50	Moderate
Quiet residential area	0.002	40	
Whispered conversation at 6 ft	0.0006	30	
Buzzing insect at 3 ft	0.0002	20	
Threshold of good hearing	0.00006	10	Faint
Threshold of excellent youthful hearing	0.00002	0	Threshold of hearing

Figure 19-57. *Sound pressure measured in pascals (Pa). (Reprinted by permission of the American Society of Heating, Refrigerating, and Air-Conditioning Engineers, Atlanta, Georgia, from the 1993 ASHRAE Handbook—Fundamentals)*

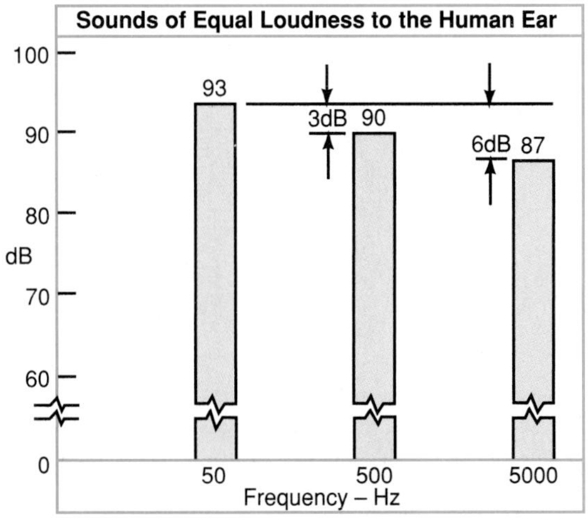

Figure 19-58. *Effect of frequency on the ability to hear sounds. These three sounds appear of same strength to human ear, yet decibel rating for each of the frequencies is different.*

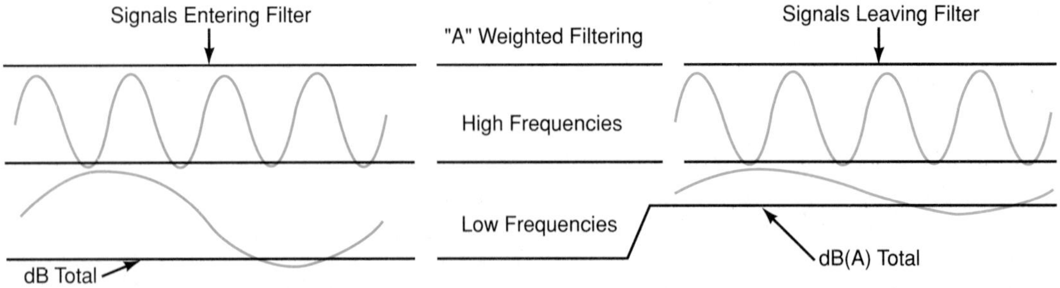

Figure 19-59. *Effect of using an "A" filter in a microphone circuit. Left—Unfiltered sound waves. Right—Sound waves after passing through the "A" filter. Note that only low frequency waves have been reduced in amplitude. (Vickers, Inc.)*

Loudness Scales

Frequency (Hz)	dB	dB(A)
5000	87	87
500	90	87
50	93	87

Recommended Maximum Sound

Frequency (Hz)	Decibels	
	Night	Day
63	66	76
125	59	69
250	52	62
500	46	56
1000	42	52
2000	40	50
4000	38	48
8000	37	47

Figure 19-60. *Top—Effect of using "A" filter on sound loudness for the three frequencies shown in* ***Figure 19-58.*** *Bottom—City code maximum noise level. (Milwaukee, Wisconsin)*

Tolerance to Sound

Noise Level (dBA)	Maximum Exposure (hrs.)
90	Unlimited
90 to 92	6
92 to 95	4
95 to 97	3
97 to 100	2
100 to 102	1.5
102 to 105	1
105 to 110	0.5
110 to 115	0.25
Above 115	None

Figure 19-61. *Limits of human tolerance to sound. Note that there is no limit if dBA is 90 or below. Also, the unprotected ear should not be exposed to sound over 115 dBA. (Walsh-Healy Act)*

An instrument for measuring sound level is illustrated in **Figure 19-62.** The printer provides a hard copy (printout) of the noise data. Based on OSHA (Occupational Safety and Health Act) standards, the instrument measures continuous, intermittent, and impulse noises in the range from 80 dBA to 130 dBA.

The meter, **Figure 19-63,** makes sound level measurements over a range of 40 dB to 140 dB. It provides an output jack which permits the use of a meter with a recorder. The unit has a jack for use with an "A" filter.

Usually, a noise is made up of a number of different vibrations per second. (It is not a pure tone.) It is possible, with the use of filters, to separate noise into octave bands. The sound level in each octave band is recorded in decibels. By adding the decibels for all the bands, the sones value is obtained:

- ***Octave:*** A series of eight tones extending from a given tone to a tone on the eighth degree above it.
- ***Sone:*** A calculated sound loudness rating.

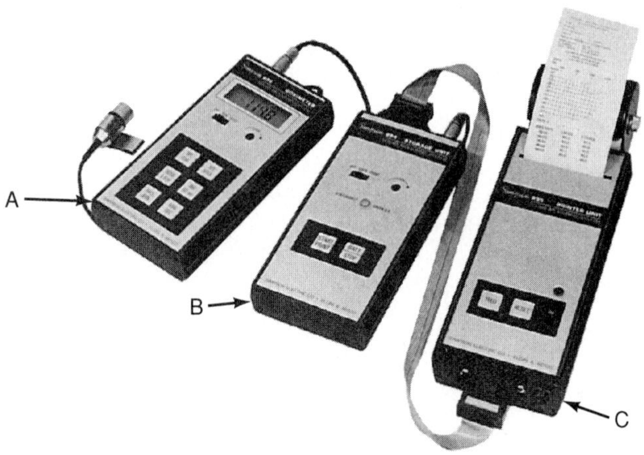

Figure 19-62. *Instruments used for measuring sound level. A—Noise dosimeter. Unit measures sound level in dBA. B—Memory storage unit. C—Printer that gives printout of noise data. (Simpson Electric Co.)*

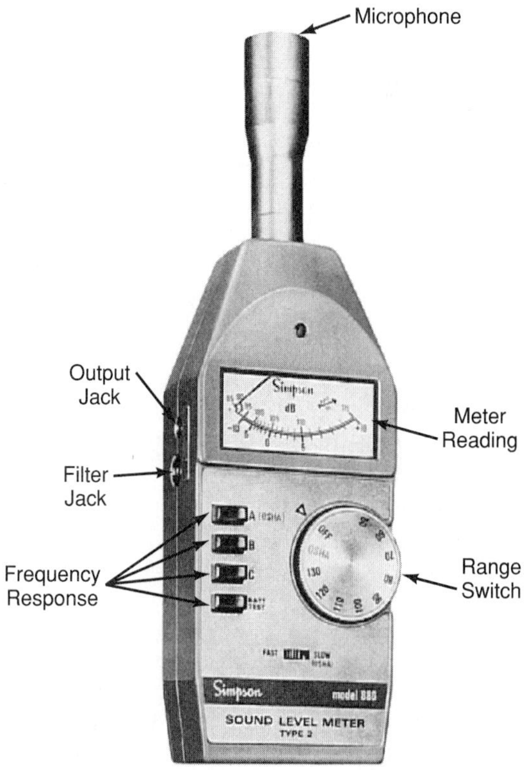

Figure 19-63. *Sound level meter measures sound pressure level over range of 40 to 140 dB. (Simpson Electric Co.)*

The sone rating is particularly useful in comparing machine noise levels.

Noise sources include fans, compressors, high velocity air, and motors. High-velocity refrigerant flows (especially at sharp pipe turns) may also produce noise. **Figure 19-64** illustrates some typical ratings and their corresponding sound pressure ratings.

Sound pressures in the 0 dB to 90 dB range generally are not objectionable. Sound levels in the 100 dB to

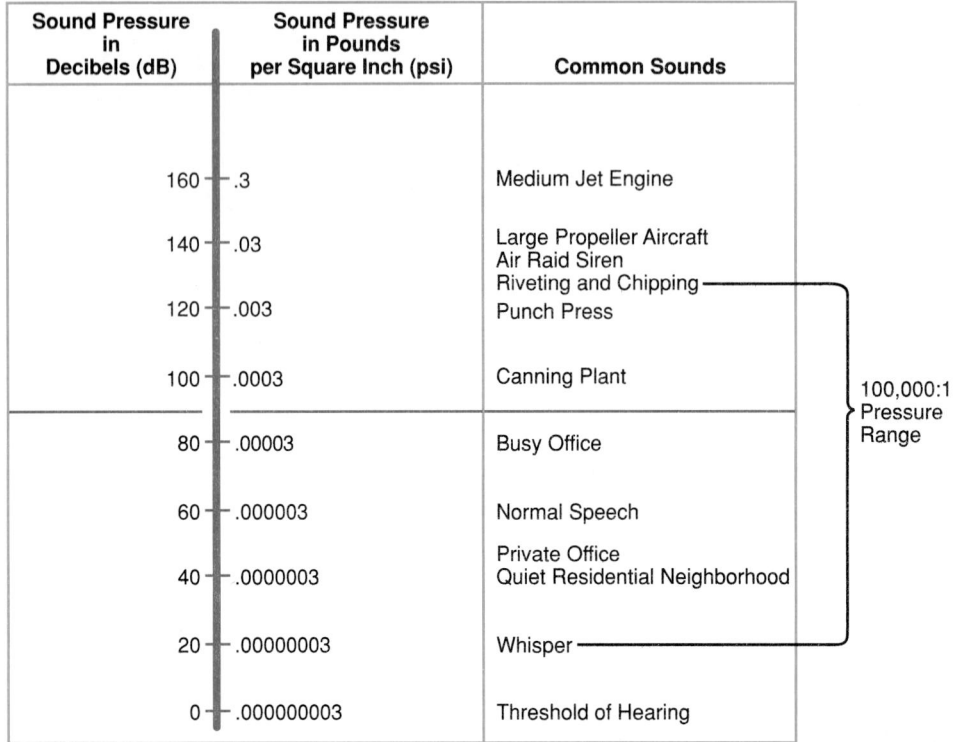

Sound Pressure in Decibels (dB)	Sound Pressure in Pounds per Square Inch (psi)	Common Sounds
160	.3	Medium Jet Engine
140	.03	Large Propeller Aircraft Air Raid Siren Riveting and Chipping
120	.003	Punch Press
100	.0003	Canning Plant
80	.00003	Busy Office
60	.000003	Normal Speech
40	.0000003	Private Office Quiet Residential Neighborhood
20	.00000003	Whisper
0	.000000003	Threshold of Hearing

100,000:1 Pressure Range (Riveting and Chipping to Whisper)

Figure 19-64. *Decibel (dB) rating of some common sounds. Line indicates usual upper limit to which human ear may be continuously exposed. (Vickers, Inc.)*

160 dB range are very objectionable. Sounds in the upper end of this range are particularly objectionable.

To protect the hearing of people in noisy situations, hearing protectors are used. See **Figure 19-65.** These protectors will reduce the level of the noise. The federal safety law known as the Occupational Safety and Health Act (OSHA) requires the use of hearing protectors in certain places. Always wear them when working on or near noisy machinery.

Figure 19-65. *Hearing protector should be used when working on or near source of painful noise. (David Clark Co., Inc.)*

19.12 Review of Safety

Air to breathe should be as clean as possible and have the correct oxygen content. It is absolutely essential that no fumes (products of combustion) mix with the air being sent to rooms. Air conditioners must provide enough fresh air to keep oxygen content within allowable limits.

Kitchen ventilating fans tend to produce a slightly lower pressure in a house. Leakage into the house must make up for this exhausted air. If the air pressure is too low, products of combustion may be drawn back into the house. It is wise to provide some air leakage into the house. Air leakage can be designed into the heating, ventilating, and air conditioning (HVAC) system.

Work or experiments performed in connection with this chapter require the use of instruments. If the instruments are connected to an electrical or compressed air supply line, carefully check the installations to prevent possible danger to the operator handling these supplies.

Instruments are very delicate. They must be handled carefully. They must never be dropped and many must be kept in an upright position. Some instruments, such as hygrometers and psychrometers, are made by using glass tubing. Use care in handling these instruments. Do not break the tubing.

Many of these instruments are very expensive. If connected incorrectly or not handled carefully, the instrument may not read accurately, or it may be severely damaged. The person handling the instrument must use great care.

19.13 Test Your Knowledge

Please do not write in this text. Place your answers on a separate sheet of paper.

AIR MOVEMENT AND MEASUREMENT

1. Which of the following is *not* a function of a complete air conditioning system?
   A. Temperature and humidity control.
   B. Air filtering, cleaning, and purification.
   C. Air movement and circulation.
   D. All of the above are functions of a complete air conditioning system.
2. _______ make up the living portion of the atmosphere.
   A. Oxygen, nitrogen, carbon monoxide, and hydrogen
   B. Oxygen, nitrogen, carbon dioxide, hydrogen, and sulfur dioxide
   C. Oxygen and hydrogen
   D. Oxygen, nitrogen, and carbon dioxide
3. In what form does water exist in the air?
   A. Solid.
   B. Liquid.
   C. Vapor.
   D. None of the above.
4. Dry bulb and wet bulb temperature indicate _______.
   A. air temperature
   B. relative humidity
   C. amount of moisture and heat in the air
   D. All of the above.
5. How may moisture be removed from the air?
   A. Condensation.
   B. Absorption using desiccants.
   C. Both A and B.
   D. None of the above.
6. What happens to the moisture absorption properties of air as the temperature decreases?
   A. It is able to absorb increasingly less moisture.
   B. It is able to absorb increasingly more moisture.
   C. Its moisture absorption properties remain the same.
   D. None of the above.
7. Does vegetation provide oxygen in the atmosphere?
   A. Yes.
   B. No.
   C. Only indoors.
   D. Only outdoors.
8. When air is heated, the relative humidity _______.
   A. increases
   B. decreases
   C. remains constant
   D. may increase or decrease, depending on the air temperature
9. The percentage of oxygen in the atmosphere is _______% by volume.
   A. 20.99
   B. 15.98
   C. 25.98
   D. None of the above.
10. A pitot tube involves what three pressures?
    A. Static, total, and velocity.
    B. Static, absolute, and atmospheric.
    C. Critical, head, and maximum operating.
    D. None of the above.

AIR QUALITY MODULE

11. Under what conditions are most people comfortable in the summer?
    A. 72° db and 90% rh to 87° db and 23% rh.
    B. 65° db and 80% rh to 85° db and 20% rh.
    C. 68° db and 75% rh to 80° db and 30% rh.
    D. None of the above.
12. Dust particles are over _______ microns in diameter.
    A. 60
    B. 300
    C. 600
    D. None of the above.
13. Noise resulting from high-speed air traveling through ducts may be due to _______.
    A. undersize units
    B. undersize ducts
    C. blow speed-up to compensate
    D. All of the above.
14. Air quality monitors are available to detect and measure _______.
    A. heat stress
    B. noise
    C. indoor air quality
    D. All of the above.
15. _______ is the cause of more than half of IAQ (Indoor Air Quality) complaints.
    A. Radon
    B. Asbestos
    C. Bioaerosols
    D. Inadequate ventilation
16. Building-related contaminants usually peak in the _______.
    A. morning
    B. night
    C. afternoon
    D. evening
17. Sound may be described by _______.
    A. sound vibration
    B. sound strength
    C. sound pressure
    D. Both B and C.
18. The chemical formula for ozone is _______.
    A. O_1
    B. O_2
    C. O_3
    D. None of the above.

19. The Comfort-Health Index indicates which of the following?
    A. Temperature.
    B. Sensation.
    C. Effect on physiology and health of the body.
    D. All of the above.

20. Productivity can be increased by ______% when employees work in an "ideal environment."
    A. 15
    B. 25
    C. 30
    D. 50

Chapter 20

BASIC HEATING AND AIR CONDITIONING SYSTEMS

Key Words:

cooling tower
dehumidifier
electrical resistance heating
forced-air heating
gravity heating system
heat exchanger
heat pump
humidifier
hydronic heating
oil burner
products of combustion
radiant heating
steam jet

Learning Objectives:

After studying this chapter, you will be able to:

- Discuss the design of different air conditioning systems.
- Name and describe various systems and explain their differences by type of fuel or energy used, by transfer medium, and by type of heating/cooling components used.
- Explain in specific detail how each system operates.
- List various types of systems and explain their applications.

This chapter introduces the basic types of heating, ventilating, and air conditioning (HVAC) systems. Detailed application, installation, maintenance, and servicing data of each type of system will be explained in later chapters.

20.1 Gas-Fired Gravity Heating System

A gas-fired gravity furnace is shown in **Figure 20-1.** Fuel gas, controlled by a pressure regulator, is fed to the burner in the ***gravity heating system.*** Fuel is under constant low pressure. An atmospheric-type burner is used. Fuel may be natural gas, propane, LP gas, or artificial gas.

The room thermostat controls the operation of the burner through a solenoid-controlled gas valve. A pilot light burns continuously. It ignites the gas in the burner whenever the solenoid valve opens the gas line. A thermocouple is connected in series with a safety gas control solenoid. It shuts off the gas if the pilot light goes out and the thermocouple cools. New systems use electronic ignition, rather than a pilot light.

The ***products of combustion*** (burning) are carbon dioxide and water vapor. They flow through the stack into the chimney. An air break helps keep a constant pressure in the combustion chamber.

Heat made in the ***heat exchanger*** (combustion chamber) moves by conduction through the chamber wall. It is carried or radiated into the air surrounding the combustion chamber. The heated air around the heat exchanger naturally rises. It flows through the warm air ducts into the rooms through openings or grates called ***warm-air registers.*** As air cools in the rooms, it becomes heavier. It flows down through the cold-air duct and back into the bottom of the furnace.

A **high-limit control** or "safety stat" is located in the bonnet of the furnace. The ***bonnet*** is the sheet-metal chamber where heat collects before being distributed. See **Figure 20-2.** The high-limit control will automatically shut off the gas if the bonnet temperature rises above the control temperature setting.

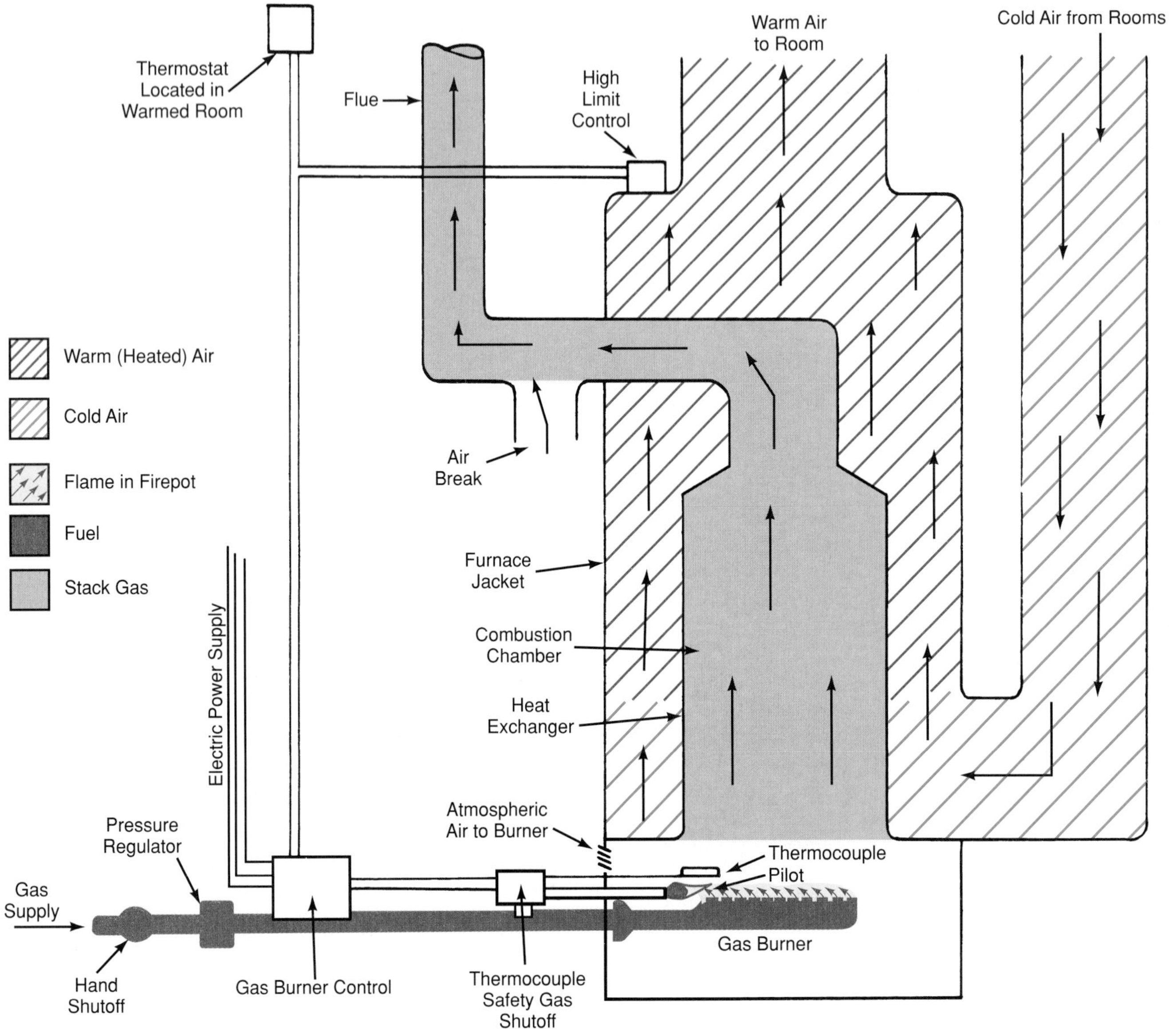

Figure 20-1. *A gas-fired gravity furnace. Air heated in furnace rises. Colder air from rooms sinks to take its place. This natural convection circulates air through rooms. In newer furnaces, the constantly burning pilot light has been replaced by an electronic ignition system as an energy conservation measure.*

A room ***thermostat*** checks the room temperature, and responds as needed. The thermostat operation keeps the room temperature within about 2°F (1.1°C) of the desired temperature.

20.2 Gas-Fired Forced-Air Heating

In this heating system, **Figure 20-2,** fuel gas is fed to the burner under constant low pressure controlled by a pressure regulator. The fuel gas is burned in a power-type burner.

The room thermostat controls the operation of the burner through a solenoid-controlled gas valve. The thermostat also turns on a combustion air blower. The blower forces air into the combustion chamber. A pilot light, electric spark, or hot surface ignites the gas in the burner. This occurs at the instant the solenoid valve opens the gas line. The power burner starts.

In pilot light systems, the pilot flame heats a thermocouple which, in turn, controls a safety shutoff valve. A thermocouple solenoid is located above the pilot light. It will shut off the gas control if the pilot light goes out.

Heat generated in the combustion chamber is conducted through the wall and radiated into the air surrounding the combustion chamber. As the air heats, it rises and warms the bonnet fan control. A high-limit control will automatically shut off the burner if the bonnet temperature exceeds the control setting. The high-limit control is sometimes a part of the bonnet fan control.

When bonnet temperature reaches the control setting, the fan in the cold-air duct return starts. It moves

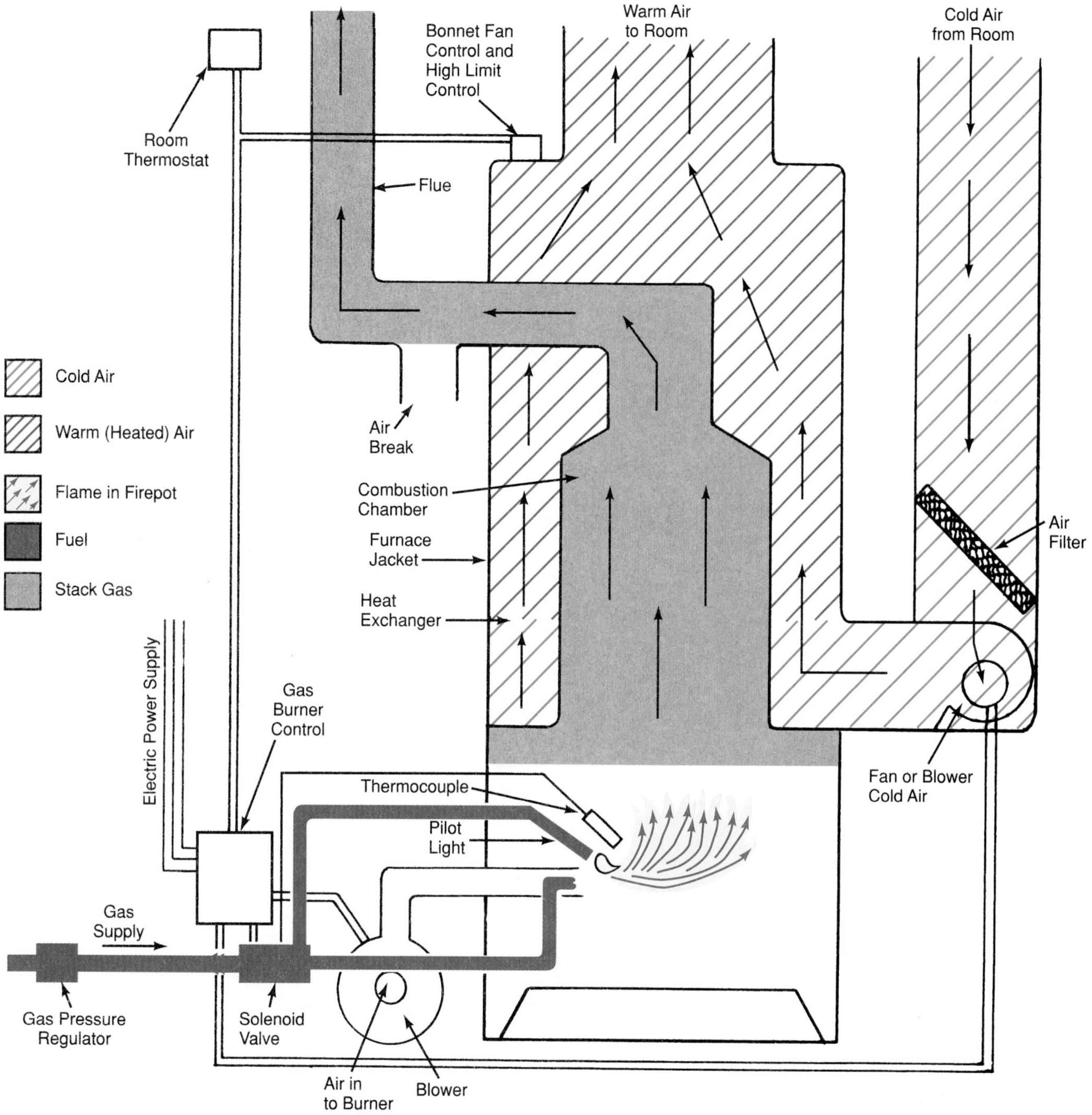

Figure 20-2. *Forced air circulation heating system using fuel gas with a power burner.*

air through the heating system. This air is drawn from the cold-air register in the floor above. It is drawn through the air filter, and then through the furnace. Warm air is distributed through the ducts and warm-air registers or diffusers. It enters the space to be heated.

20.3 Gas-Fired Hydronic Heating

This heating system burns fuel gas under low and constant pressure. The system is shown in **Figure 20-3.** An automatic pressure regulator maintains constant gas pressure on the atmospheric type burner.

The heat generated in the combustion chamber is carried through the ***boiler*** wall. It is carried into the water. Water temperature in the boiler is controlled by a temperature- and pressure-sensing element. This is located in the top of the boiler. This sensing element is connected into the electrical system. It turns the burner on when the temperature drops below the required level. It also turns off when the temperature rises to this level. A pilot light or electronic ignition system ignites the burner. The products of combus-

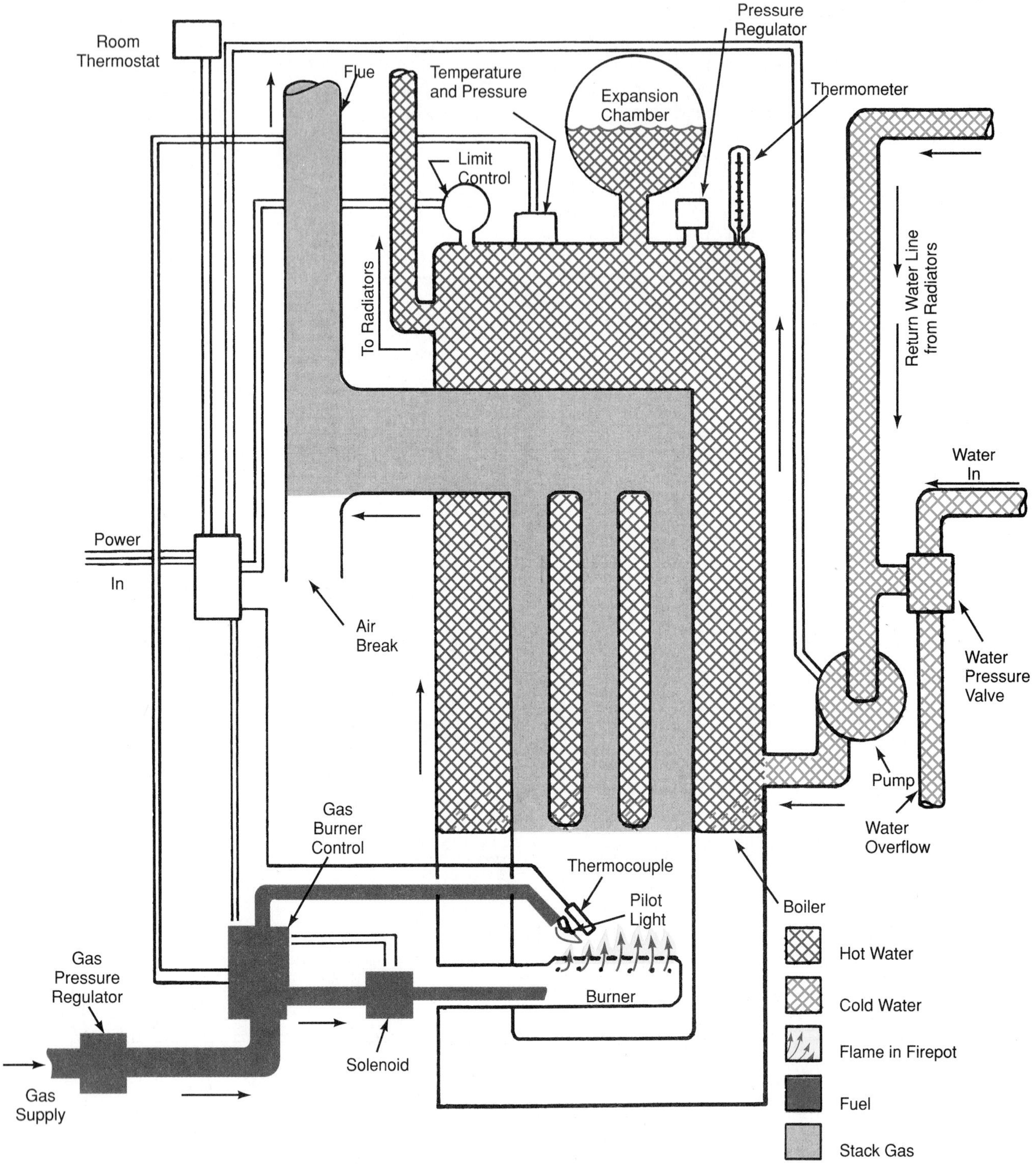

Figure 20-3. *Hydronic heating system using atmospheric fuel gas burner as source of heat.*

tion flow through the stack into the chimney. An air break or draft diverter is installed in the stack. It helps maintain a constant pressure in the combustion chamber.

The room thermostat controls the operation of the water pump. The pump circulates the warm water through the room radiators. Then, it returns the water to the boiler.

A high-limit control is attached to the warm water outlet of the boiler. The high-limit control automatically shuts off the gas if the water temperature or pressure get too high. The system also has a pressure relief valve. This valve prevents the buildup of dangerous pressures in the boiler. An ***expansion tank*** permits water volume to expand and contract as it heats and cools. The expansion tank should be located at the highest place in the heating system. It is possible to use a pressure regulator in place of an expansion tank.

20.4 Oil-Fired Forced-Air Heating

Where fuel oil is the heat source, a gun-type oil burner throws a flame into a ***firepot*** (combustion chamber). The firepot is lined with ***refractory*** (fire-resistant) material. **Figure 20-4** shows this type of furnace.

Fuel oil is stored in a tank, either inside or outside the building. It is pumped into the burner nozzle under a pressure of about 100 psig (115 psia or 790 kPa). Before reaching the burner, the oil is drawn through a filter.

A room thermostat controls operation of the burner. When heat is required, the thermostat operates a relay. It closes the electrical circuit to the burner motor. When the burner starts, a high-voltage transformer is connected into the electrical circuit. Sparks jump the gap of two electrodes located at the edge of the burner nozzle. The fuel spray from the burner nozzle is ***atomized*** (broken into small drops). The spark ignites this atomized fuel, causing a continuous flame in the firepot. The flame will burn as long as the burner is operating.

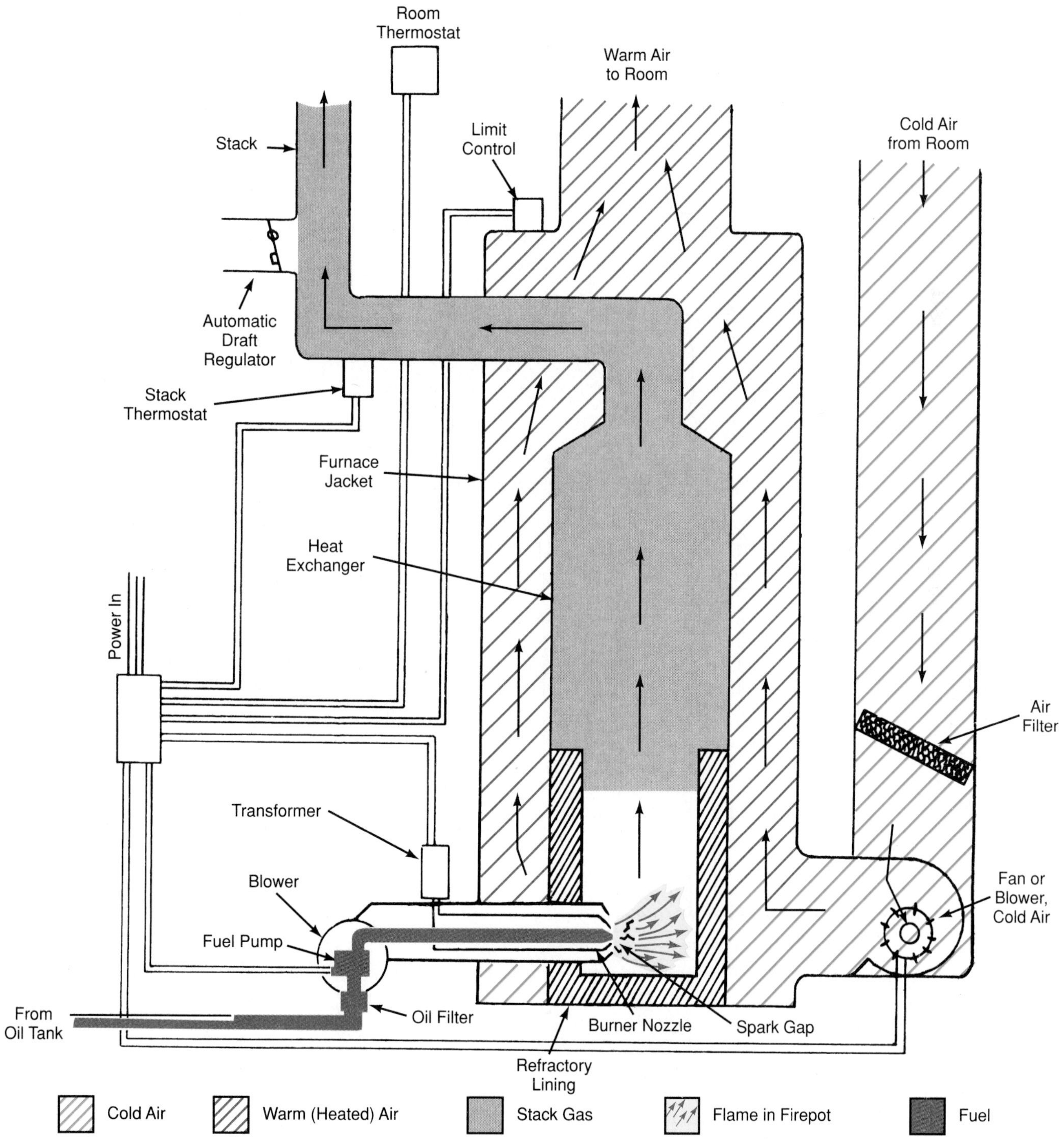

Figure 20-4. *Forced-air furnace with a gun-type oil burner as source of heat.*

A stack thermostat senses the temperature of the gases leaving the furnace. If, for any reason, the atomized fuel is not ignited after a few seconds of pump operation, the thermostat will detect this and stop the pump. Normally, a manual reset will have to be operated before it will cycle again.

A temperature-sensing device in the furnace bonnet will start the blower or fan in the cold-air duct. This happens as soon as the bonnet temperature reaches its desired setting. A temperature-controlled limit switch also is placed in the bonnet. It will open the circuit and stop the burner if the bonnet temperature goes too high.

20.5 Oil-Fired Hydronic Heating

In oil-fired hydronic heating systems, fuel oil is burned in a gun-type burner. See **Figure 20-5.** Burner operation is the same as that in an oil-fired forced-air furnace. The firepot is lined with refractory material. Fuel oil is stored in a tank which may be located outside the building. It is pumped into the burner nozzle under a pressure of about 100 psig (115 psia or 790 kPa).

As the burner starts, a high-voltage transformer is connected into the electrical circuit. Sparks jump across

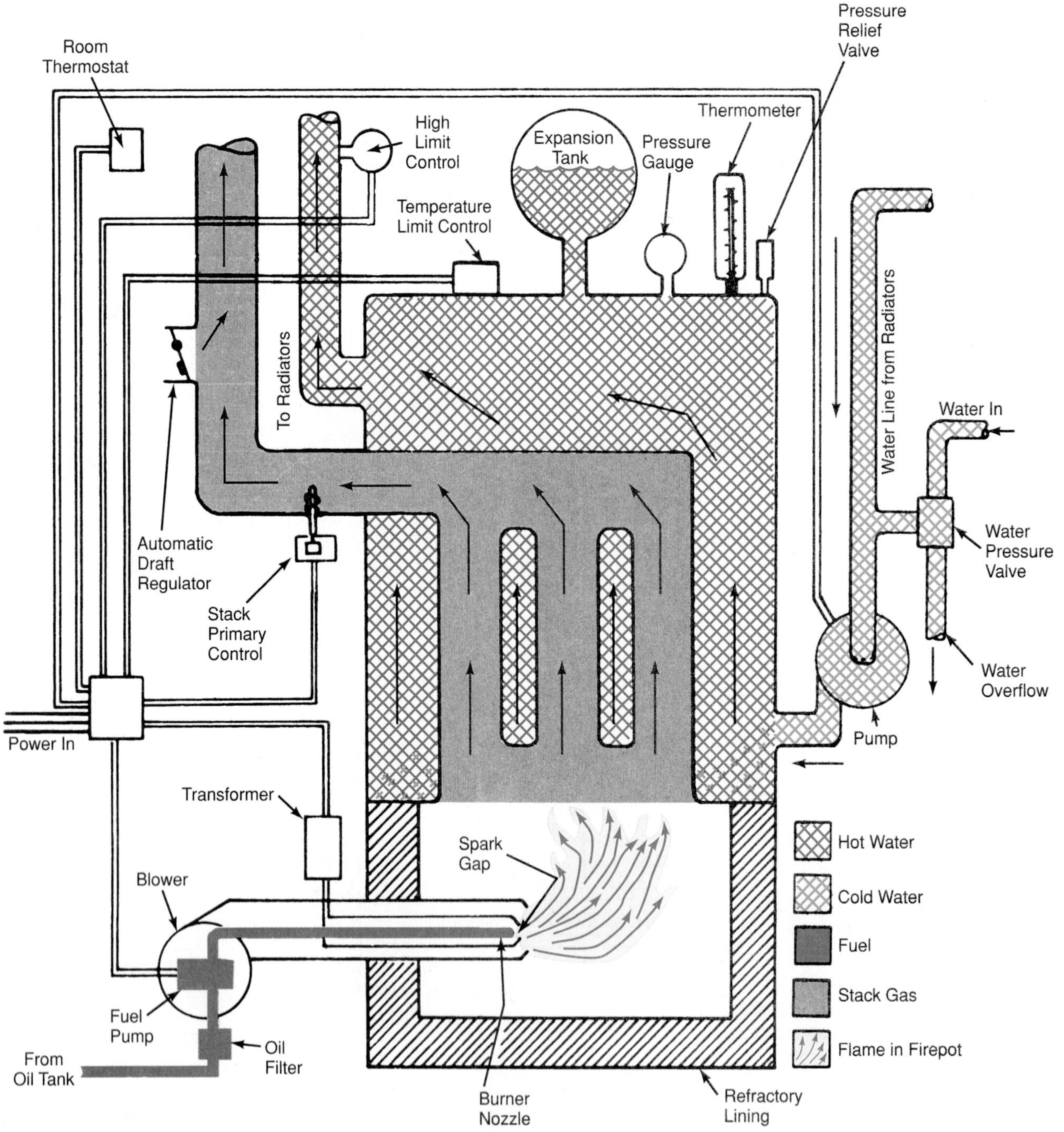

Figure 20-5. *Hydronic heating system with a gun-type oil burner as source of heat.*

the spark gap just at the edge of the burner nozzle fuel spray. This spark ignites the atomized fuel. It causes a flame in the firepot as long as the burner is operating.

A stack "stat" senses the temperature of the gases leaving the furnace. If the atomized fuel is not lighted after a few seconds of pump operation, the pump will stop. A manual reset will have to be operated before it will cycle again.

Heat from the combustion chamber is conducted through the boiler wall into the water. The gases from burning fuel flow through the stack into the chimney. An automatic draft regulator helps maintain a constant pressure in the firepot.

A room thermostat controls the water pump. The pump circulates the warm water through room radiators and returns it to the boiler. Boiler water temperature is controlled by a temperature- and pressure-sensing element in the boiler top. This sensing element is connected into the electrical system. It turns the burner on when the temperature drops below the required level. It shuts the burner off when the temperature reaches the desired level.

A high-limit control is attached to the boiler chamber. Sometimes, this control is attached to the warm water outlet of the boiler. The high-limit control automatically shuts off the fuel if the water temperature or pressure get too high. A pressure relief valve is also mounted on the boiler. This valve keeps pressures down to a safe level.

An expansion tank is used to take care of expanding (warm) or contracting (cool) water. Air in the tank acts as a cushion.

20.6 Electrical Resistance Heating

When an electrical resistance heating system is installed, heating units are located in each room. Electrical power is brought to the units from a power panel. The power supply is usually 240 V. One advantage of this system is that each room is regulated by its own thermostat. There are four different ways that the control may be accomplished. These different ways are illustrated in **Figure 20-6** at A, B, C, and D.

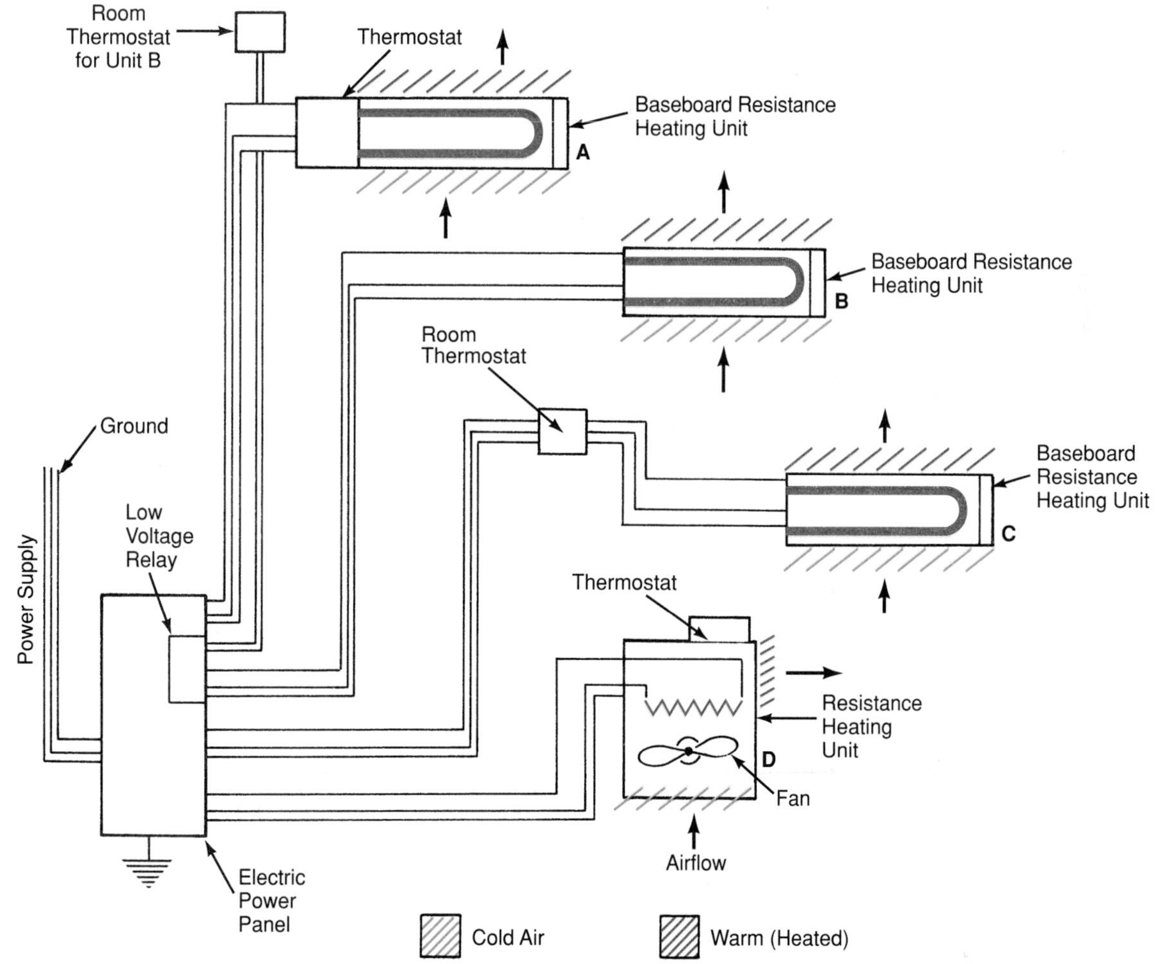

Figure 20-6. *Four different types of electric heating systems. These units provide separate temperature control for each room.*

At A, electrical power is connected directly from the panel to the baseboard heating unit. The heating unit has an individual thermostat attached to it. Electrical power is supplied to the thermostat. If heating is required, the thermostat completes the connection from the power supply to the resistance heating unit.

At B, a room thermostat controls the electrical supply at the power panel. When heat is required, a relay in the power panel connects the baseboard resistance unit to the electrical supply.

At C, a room thermostat is mounted on the wall. The power supply from the power panel is connected through the thermostat to the baseboard heating unit.

In A and C, all current used by the heater flows through the thermostat points. In B, the room thermostat is handling only a small amount of low-voltage current. A relay connected to the thermostat is located in the power panel. This relay switches the current to the room resistance heater.

At D, the unit is a resistance heater and a fan. The thermostat controlling this unit is usually mounted on top of the heater. It controls the current to both the heating unit and the fan. All of the fan and heater current goes through the thermostat points. The fan may be set to run whenever the heating unit is on or to run independently.

The electric heating units are always grounded. The green wire usually indicates the cabinet ground. **Circuit breakers should be installed in the power line to each electrical resistance heating unit. No other appliances should be connected into these circuits.**

20.7 Radiant Heating

Radiant heating provides a very comfortable living environment. It has little or no equipment in sight. The most common type consists of electric heating wires embedded in the floor, ceiling, or walls. They may also be in some combination of these three locations. **Figure 20-7** illustrates a typical radiant heating installation.

With this installation, the surface is slightly warmed by the electric wires. The amount of heat radiated by this type of system can heat a room. A thermostat, mounted on the wall, controls the current flow through the heating wires. The room temperature is thereby controlled, as well.

This type of heating works the opposite of a heat sink. (See Section 19.7.1.) In the heat sink, heat radiated from the human body is lost to surrounding surfaces. With radiant heating, heat radiated from the surrounding surfaces is absorbed by the human body. The body is comfortable in ambient temperatures somewhat lower than 66°F (19°C).

Walls, floors, or ceilings used in radiant heating systems must be heavily insulated to maintain the surface temperatures. Often, radiant heat installations are supplemented by other heat sources. They provide a

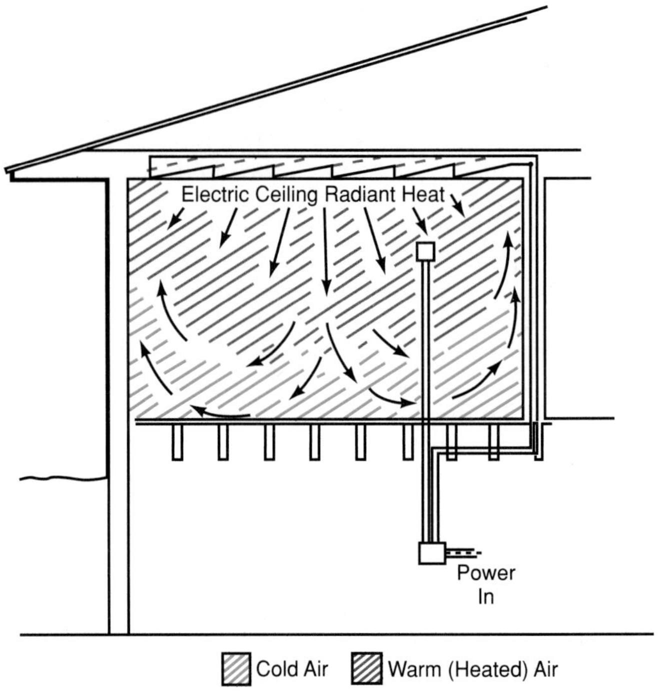

Figure 20-7. *Radiant heat is supplied by electrical resistance heating wires embedded in ceiling plaster. Heavy insulation is needed with a radiant heat installation.*

complete heating and air conditioning system that has enough heating capacity.

20.8 Air-to-Air Heat Pumps

The ***heat pump*** is a heat-moving mechanism used in homes and in some industries. Heat is *absorbed* in an evaporator in one location. It is *released* through a condenser in another location. The system can reverse its operation. The evaporator becomes the condenser and the condenser becomes the evaporator. Heat flow is reversed. Thus, using a special reversing valve, the mechanism either heats or cools the conditioned space. *The flow through the compressor is always in the same direction.*

Figure 20-8A shows the flow through the valve causing the conditioned space to be heated. **Figure 20-8B** shows the valve in position to cool the conditioned space.

Heat pumps use compression-type refrigerating mechanisms. These are similar to a food-refrigerating or air conditioning mechanism. Heat pumps have two heat transfer surfaces. One is located inside the conditioned space and the other out-of-doors.

On the heating cycle, **Figure 20-8A,** the outdoor coil becomes an evaporator. The indoor coil becomes the condenser. In operation, liquid refrigerant enters the outdoor coil. It picks up heat from the air and is vaporized. The vapor is drawn into the compressor. It is compressed

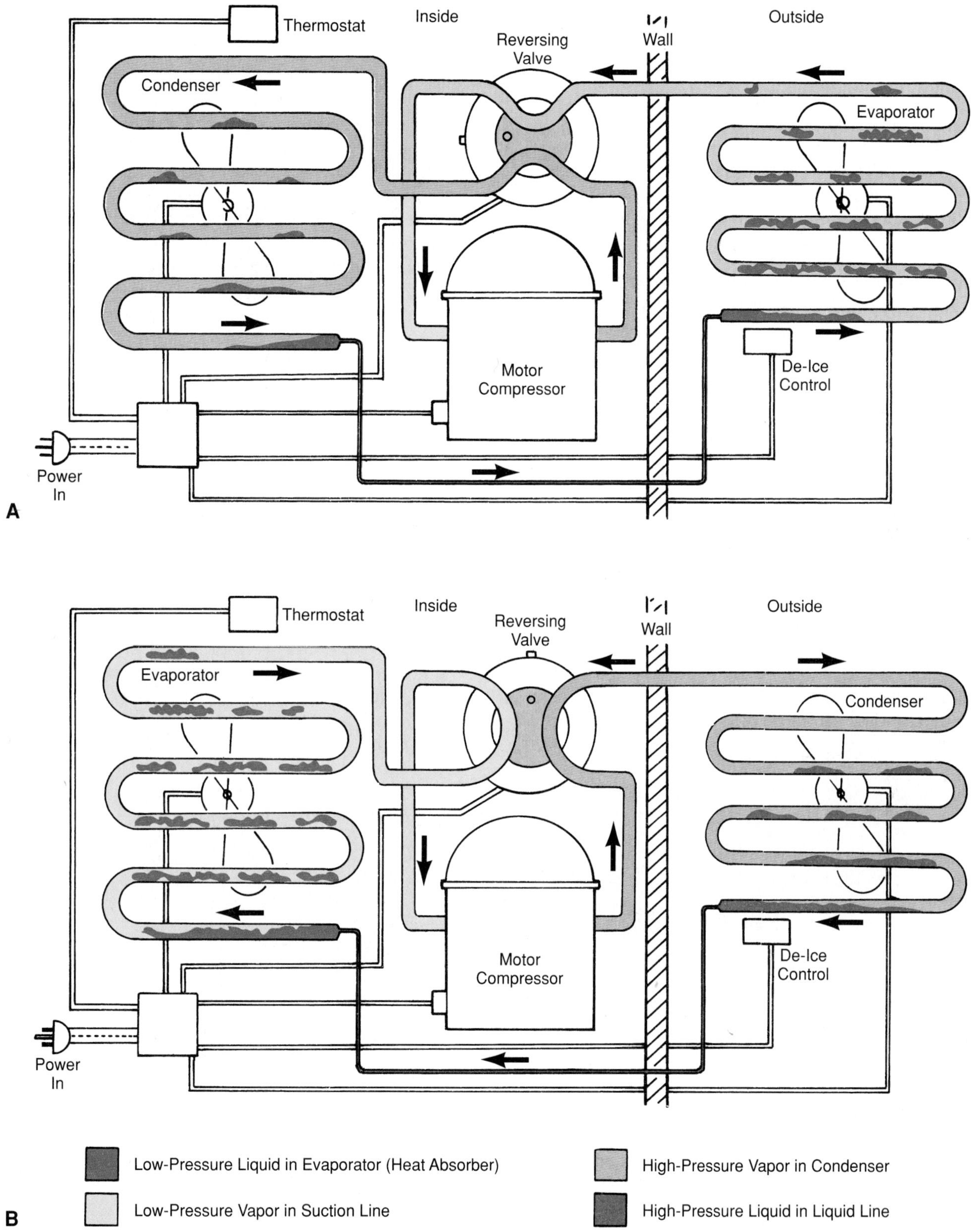

Figure 20-8. *Air-to-air heat pump. A—Heating cycle. Reversing valve is set so that coil on outside acts as an evaporator. Heat absorbed in evaporator is released by condenser inside house. B—Cooling cycle. Valve is set so that coil on inside acts as an evaporator. Heat absorbed in evaporator is released by condenser outside house.*

to a high temperature and is pumped into the indoor coil. Since its temperature is higher than the indoor temperature, heat is released into the room.

Compressed refrigerant vapors will condense upon giving up their heat of vaporization. They will return to a liquid state. The liquid then flows back through the capillary tube into the evaporator. The cycle is repeated. Since the outdoor coil is colder than the outdoor surrounding air, ice may form on it. This occurs if the outdoor temperature is rather low. Therefore, outdoor coils are fitted with de-icing controls. These controls operate in either of two ways when ice forms:

- They automatically turn on electric heating units.
- They turn off the compressor, allowing the evaporator surface to warm up and melt the ice.

On the cooling cycle, **Figure 20-8B,** the coil in the conditioned space becomes an evaporator. Refrigerant flows through the capillary tube into the evaporator. The liquid refrigerant boils, absorbing heat. Vapor from the boiling refrigerant is drawn into the compressor. There, it is compressed. The heated vapor is pumped into the outdoor coil, which has become a condenser.

The air surrounding the outdoor coil is cooler than the compressed vapor in the coil. Therefore, the compressed refrigerant vapor gives up its heat to the outside air. It condenses and flows to the bottom of the condenser as liquid refrigerant. From here, it flows through the capillary tube into the bottom of the evaporator. From this point, the cycle is repeated. Motor-driven fans on both coils aid heat flow from coil surfaces.

Air heat pumps are also used as part of a ductless split air conditioning system. See **Figure 20-9.** This type of system uses a single outdoor condensing unit. The outdoor heat pump unit is installed on a pad. Tubing and wiring are run to the exact position where the indoor units will be placed. This arrangement allows up to three indoor units, using only one condensing unit.

The location of the indoor units can vary, depending upon needs. Each indoor unit has its own independent controls. Some systems use remote controls, similar to those used with television sets. The remote ductless multiple systems are used in new and retrofitted offices, homes, and motels. This is frequently done when conventional window or wall units are not desirable or applicable.

The main advantage is that it serves multiple independent rooms with only one condensing unit. Heat pump installations are ideal where winter heat loads are almost the same as summer cooling loads. Air-to-air installations are most satisfactory when the ambient air temperature in the winter remains above (or only occasionally drops below) the freezing temperature.

20.8.1 Auxiliary Electric Heaters

Air-to-air heat pump installations operate efficiently when the outside air temperature is above freezing. However, when the outside temperature drops down to or below freezing, efficiency drops off rapidly. To make up for this efficiency loss, the indoor section often has auxiliary electric resistance heating units. When the thermostat calls for more heat than the heat pump can deliver, these elements turn on.

Heat pump operation for the heating and cooling cycle is the same as explained in Section 20.8. **Figure 20-10** is a heat pump cycle diagram. The unit shown has an auxiliary electric heating system. Note the resistance heating units in the indoor section.

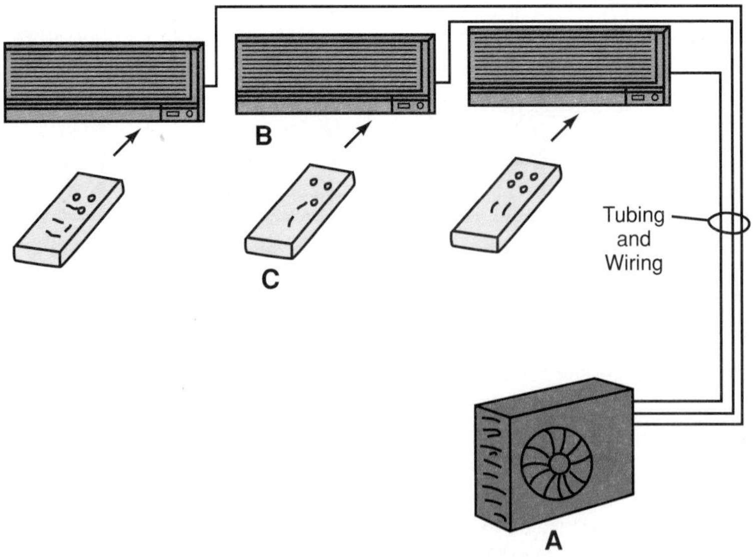

Figure 20-9. *A multizone ductless heat pump system. A—Single outdoor condensing unit. B—Three individual indoor units. C—Individual remote controls for the indoor units.*

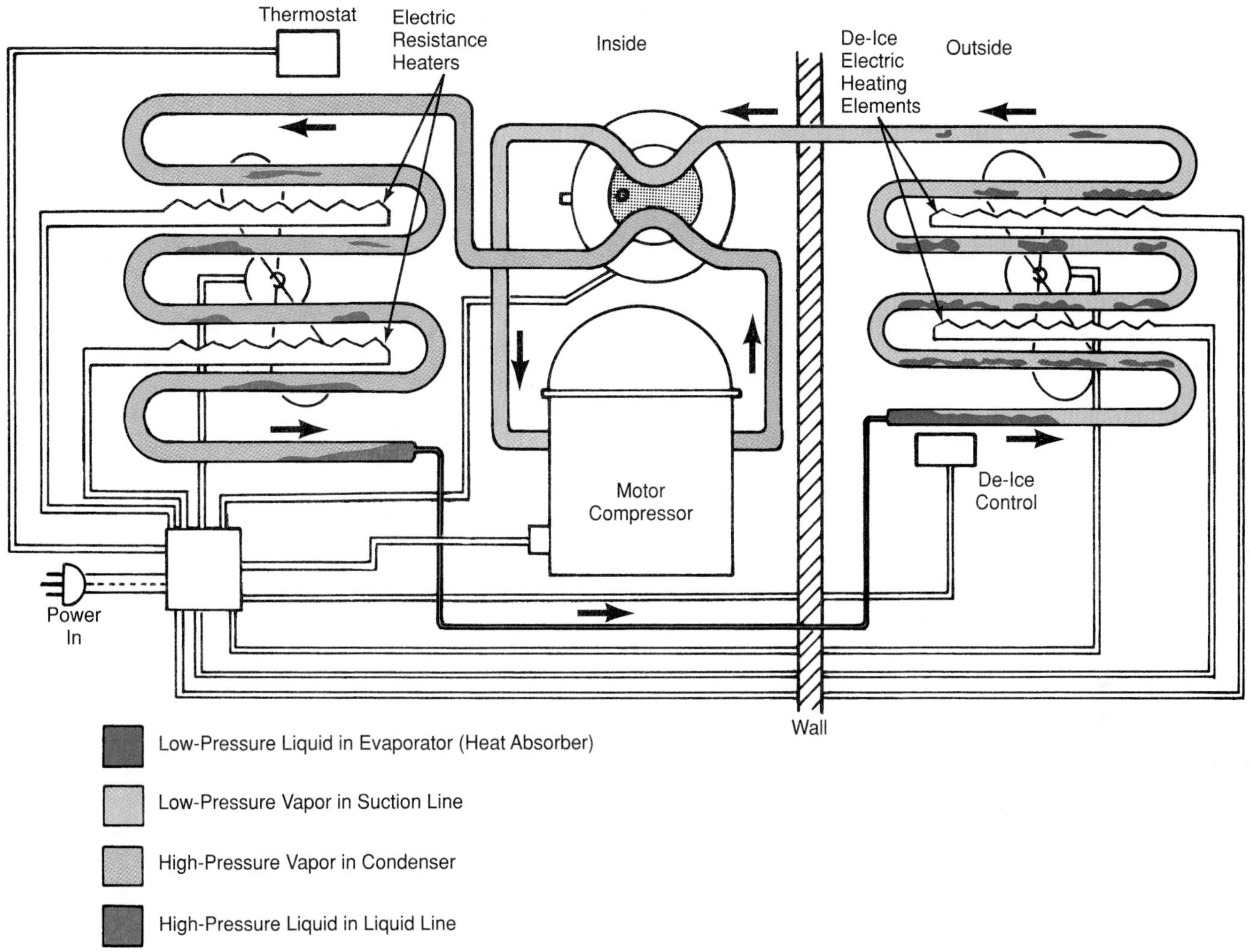

Figure 20-10. *Air-to-air heat pump with electric resistance heating elements. Heating cycle is on in this diagram. Electric resistance heaters provide additional heat if needed.*

20.9 Geothermal Heat Pump Systems

As explained in Section 20.8.1, the air-to-air heat pump efficiency depends greatly on outdoor temperature. To improve this efficiency, some installations use a coil buried in the ground. The coil is buried below the frost line. This ground coil is used rather than a coil in the atmosphere. Systems that use ground or water for their operation are referred to as ***geothermal systems.*** If the coil is long enough and is buried at some depth, heat pump efficiency may be very good. A schematic diagram of the cycle is shown in **Figure 20-11A.**

On the heating cycle, liquid refrigerant flows through a refrigerant control and into the ground coil. Since the refrigerant in the ground coil is under low pressure, it boils. It absorbs heat from the ground surrounding the coil.

The vaporized refrigerant is then drawn into the compressor. It is compressed and discharged into the condenser. In this case, the condenser is the heating coil for the system. The condenser changes the vaporized refrigerant to a liquid. The refrigerant gives up its heat to the room air. The liquid refrigerant returns to the refrigerant control to repeat the cycle.

The same mechanism may be used to cool the building in summer. The cycle is reversed to move heat from the building to the outdoors. In this case, the inside coil serves as the evaporator. The ground coil becomes the condenser. The ground absorbs the heat from the vaporized refrigerant. **Figure 20-11B** illustrates the heat pump with the valves set for cooling the conditioned space.

The four-way valve is electrically controlled by the thermostat. If heat is called for, the valve will allow fluid flow, as indicated in **Figure 20-11A.** If cooling is needed, the flow will be as shown in **Figure 20-11B.** In each case, the refrigerant flow through the compressor is in the same direction. The suction side and discharge side of the compressor are always the same. The cycle change is accomplished entirely by operation of the four-way valve.

The ground coil may be placed in water, such as a spring or flowing well with water at about 50°F (10°C). Some installations have successfully used a coil placed in the bottom of a lake.

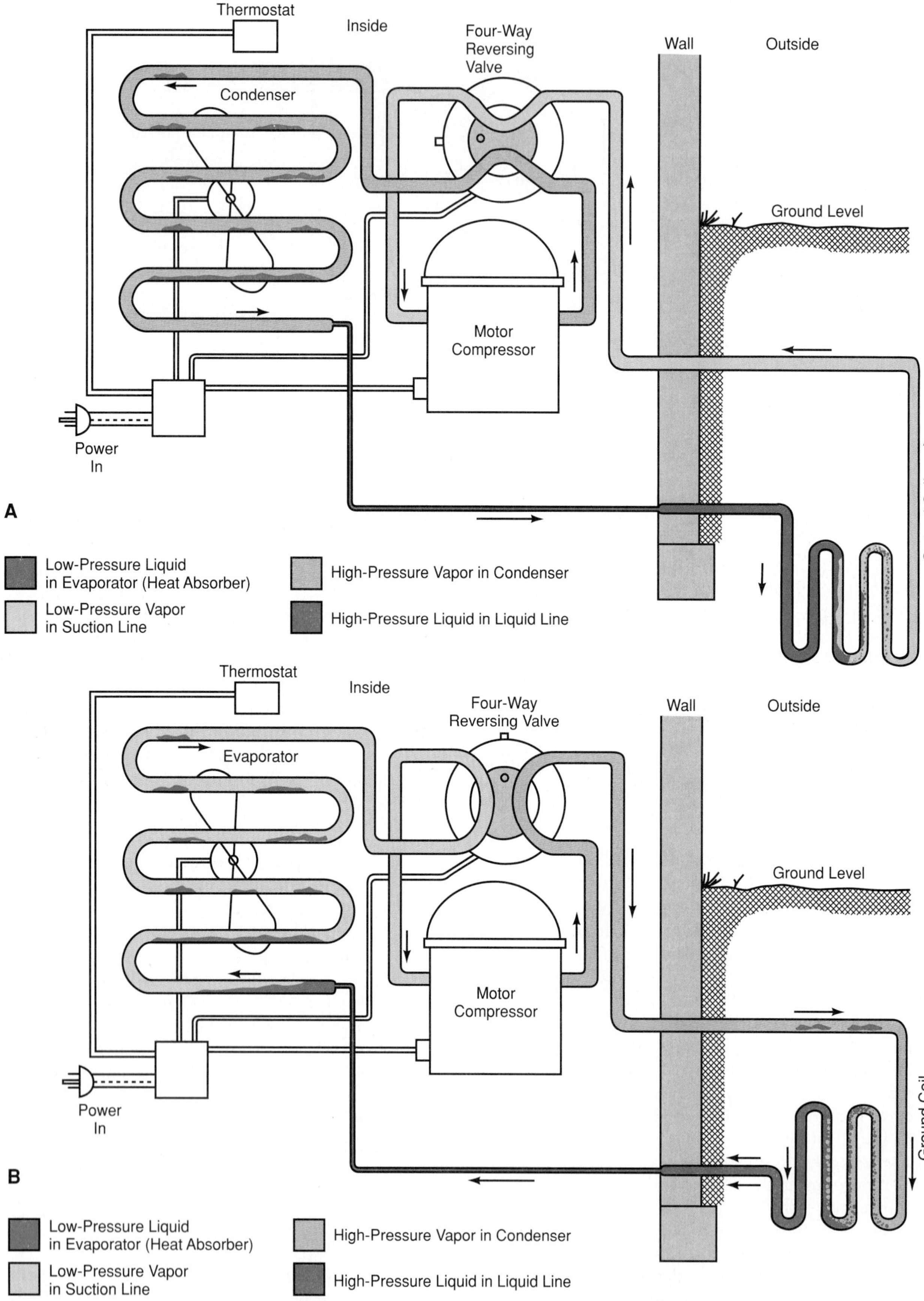

Figure 20-11. *Heat pump using a ground coil (or coil in a well or lake). A—Heating cycle. B—Cooling cycle.*

20.10 Room Humidifiers

A room ***humidifier*** is used to maintain ***relative humidity*** (percentage of moisture in the air). The humidifier is housed in a cabinet located in the room or space in which humidity is to be increased. The cabinet is supplied with air-in and air-out louvers. A fan circulates air through the cabinet. A rotating screen or filter (wetted surfaces) dips into a pan or trough filled with water. It then exposes the wetted surfaces to the airstream. A typical humidifier is shown in **Figure 20-12.**

An electric heating element is sometimes used to warm the water for greater evaporation. The controls consist of an On-Off switch and a relative humidity control. An indicator light signals the need for refilling the water supply trough. In some installations, the humidifier may be connected to building water supply and refilled by means of a valve controlled by an automatic float mechanism.

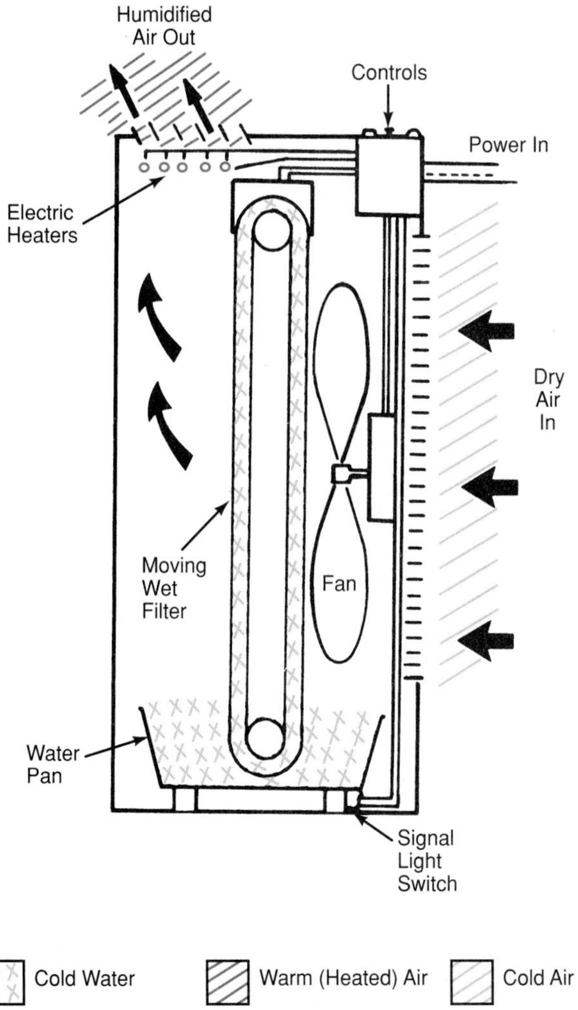

Figure 20-12. *Room humidifier. The wet filter (porous belt) slowly moves through water in water pan. Fan forces air through the wet belt and relative humidity increases. Some units have electric heaters to reheat humidified air. The humidistat controls operation of the unit. A signal light goes on when water pan is empty.*

20.11 Room Dehumidifiers

Typically, a ***dehumidifier*** consists of a hermetic compressor, condenser, and evaporator using a capillary tube refrigerant control. See **Figure 20-13.** In the schematic diagram, dark red indicates high-pressure liquid, and dark blue, low-pressure liquid. Light blue indicates low-pressure vapor; and light red indicates high-pressure vapor.

Liquid refrigerant collects in the lower coils of the condenser. It flows through the filter into the capillary tube. Then it moves into the evaporator, which is under low pressure. In the evaporator, the liquid refrigerant boils rapidly. It picks up heat from the evaporator surface. A motor-driven fan forces large amounts of air through the evaporator.

Due to the low evaporator temperature, moisture carried in the air ***condenses*** (is changed to a liquid) on the evaporator surfaces. The moisture drips to the bottom of the evaporator and into the condensate trough. Air flowing through the evaporator is both cooled and dehumidified. Cooled air is then forced through the condenser. There, it cools the condenser and again picks up heat. Therefore, the air leaving the dehumidifier is about the same temperature as it was when it entered. However, air discharged by the dehumidifier has a lower relative humidity than the intake air.

Low-pressure vapor is drawn from the evaporator through the suction line to the compressor. It is again compressed to high-side pressure and is forced into the condenser. Here, it is cooled and becomes a liquid. The cycle is repeated.

In addition to an On-Off switch, dehumidifiers usually have two other controls. One is a ***humidistat,*** used to regulate humidity. It permits the dehumidifier to operate until the desired relative humidity is reached. The control then shuts the machine off. The other is a frost control element. This is placed in the suction line between the evaporator and the compressor. It stops the motor compressor when a sufficiently high temperature is reached. Therefore, the evaporator will not freeze over and stop the flow of air through it.

In the drawing, arrows in black show the direction of airflow through the dehumidifier. A fan is commonly used to move the air.

20.12 Room Air Conditioners

Window or through-the-wall air conditioners used for room cooling consist of three basic parts:

- A hermetic compressor.
- A condenser.
- An evaporator using a capillary tube refrigerant control.

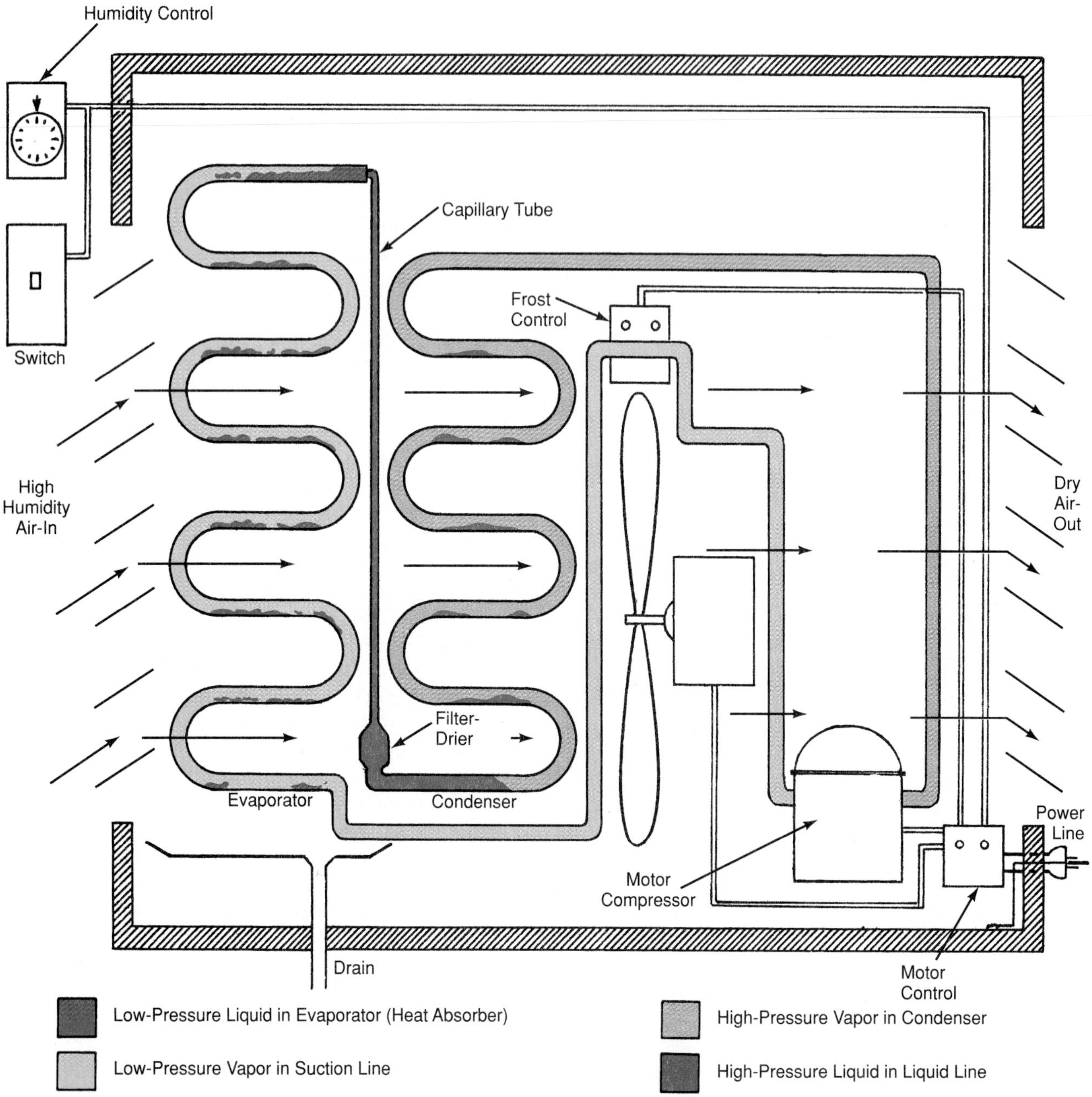

Figure 20-13. *Room dehumidifier. Room air is cooled as it flows through evaporator. Water vapor is condensed on evaporator surface and drains away. Air is reheated as it flows through and cools the condenser.*

A schematic diagram is shown in **Figure 20-14.** Dark red indicates high-pressure liquid refrigerant and dark blue, low-pressure liquid refrigerant. Light blue indicates low-pressure vapor and light red, high-pressure vapor.

Liquid refrigerant collects in the lower coils of the condenser. It flows through the capillary tube refrigerant control into the evaporator. When the unit is in operation, the evaporator is under low pressure. The liquid refrigerant rapidly boils and picks up heat from the evaporator surface. A motor-driven fan draws air from inside the room, and pulls it through a filter. The fan forces the air over the evaporator. Here, the air is cooled and goes back into the room. Arrows in **Figure 20-14** show the airflow pattern.

Low-pressure vapor is drawn from the evaporator through the suction line back to the compressor. Compressed to the high-side pressure, the vapor is then forced into the condenser. There, it is cooled and condensed to a liquid. The cycle begins again. An adjustable thermostat, mounted on the control panel, provides the necessary control. The thermostat has an On-Off switch.

The compressor and condenser are in the part of the unit that projects outside the building. The compressor compartment fan draws in outdoor air. The fan circulates air over the condenser, and discharges it outside.

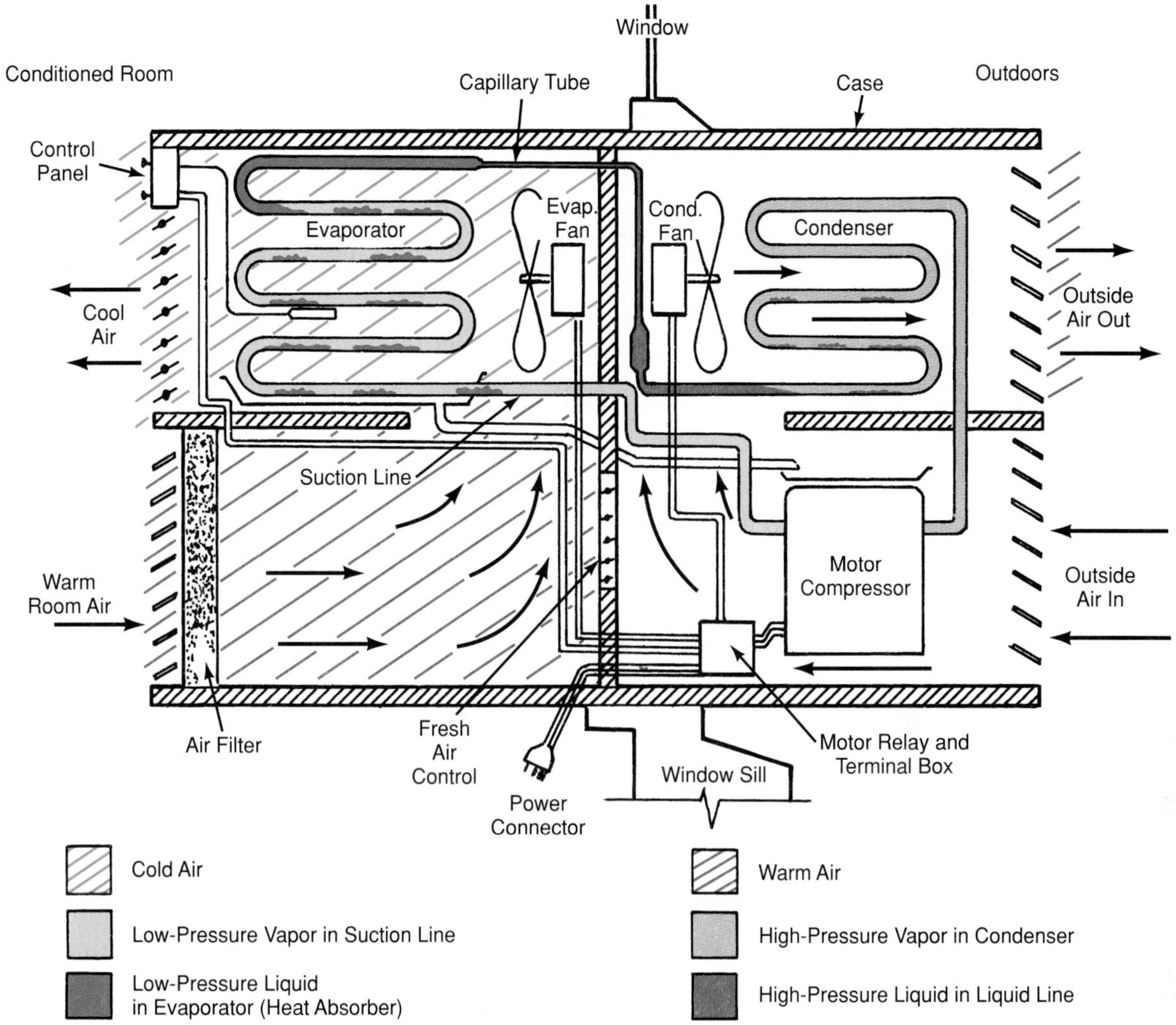

Figure 20-14. *Room air conditioning comfort cooling system. This unit mounts in a window.*

Air flowing through the evaporator (inside the cooled space) is cooled and, to some extent, dehumidified. Moisture that collects on the evaporator drains to a drip pan. In some machines, it flows into a pan in the compressor compartment. As the moisture evaporates, it helps to cool the compressor and condenser.

20.12.1 Room Air Conditioners with Electric Heat

As noted in the preceding section, a room air conditioner usually consists of a hermetic motor, compressor, condenser, evaporator, and capillary tube refrigerant control. In some models, electric resistance heating units are included for cold weather use, as shown in **Figure 20-15.** During cold weather:

- The refrigerating mechanism is turned off.
- The electric resistance heating units are turned on.
- The room air fan is turned on. The same fan circulates warm air in cold weather and cooled air in warm weather.

These air conditioners are usually connected to 240 V circuits. A control provides a choice of temperatures.

20.13 Central Air Conditioners

Fuel gas is used for heating in this air conditioning system. The gas is burned in an atmospheric burner. A compression system using an A-frame evaporator in the furnace plenum chamber provides cooling. **Figures 20-16** and **20-17** show the system in heating and cooling operations.

The condensing unit is located outside the building. A single combination heating and cooling thermostat is often used. A humidistat controls the relative humidity in the conditioned space. In winter, a humidifier in the plenum chamber adds moisture to the heated spaces. Details of the humidifier are shown in **Figure 20-18.** Summer humidity is controlled by condensation of moisture on the evaporator. A drain removes this moisture.

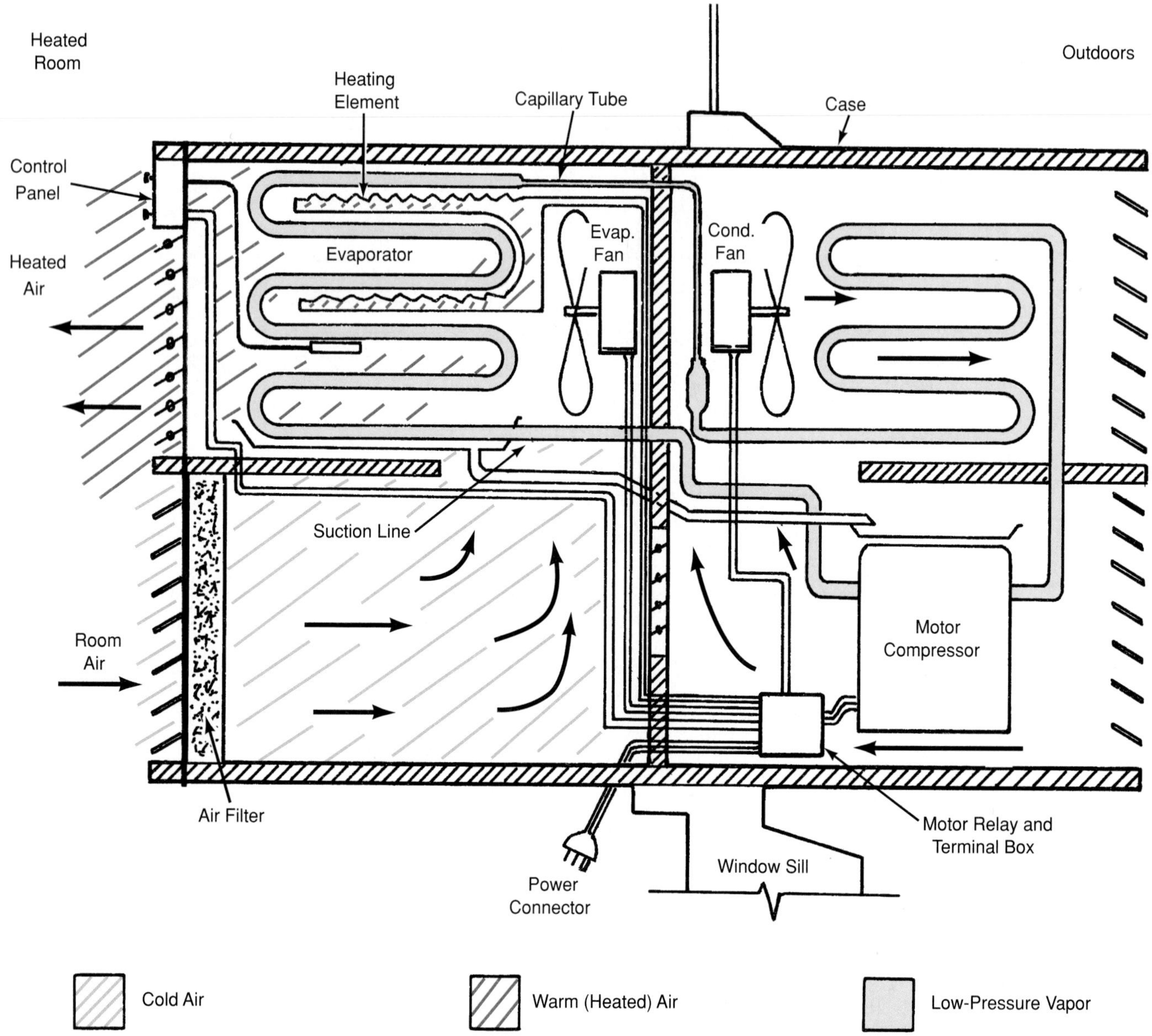

Figure 20-15. *Room air conditioner with electric heating elements. These provide heat during cold weather. Unit is shown heating air in winter.*

Warm air from the furnace is forced into the rooms by a blower. This is located beneath the filter in the cold air return. A control in the top of the furnace turns on the blower when the desired bonnet temperature is reached. It also turns off the furnace if bonnet temperature goes higher than a predetermined setting. This is a safety device to keep the furnace from overheating. Electrical power to the furnace is turned on and off by a control panel. The panel is located on the outside wall of the furnace.

An arrangement is sometimes provided to bring in outside fresh air as needed. This may be controlled either thermostatically or manually.

A pilot light, when used, is controlled by a thermocouple. The thermocouple is connected in series with a solenoid valve in the gas supply line. If the pilot light goes out, the gas to the burner will be shut off.

In summer, a centrally-located thermostat may call for cooling. The same distribution system airflow used for heating is used for cooling. However, forced air passes across the cooled evaporator, instead of through a heated chamber. This lowers the temperature of the air. At the same time, it removes some moisture to reduce the humidity. The filter in the incoming air duct cleans the air before it reaches the blower.

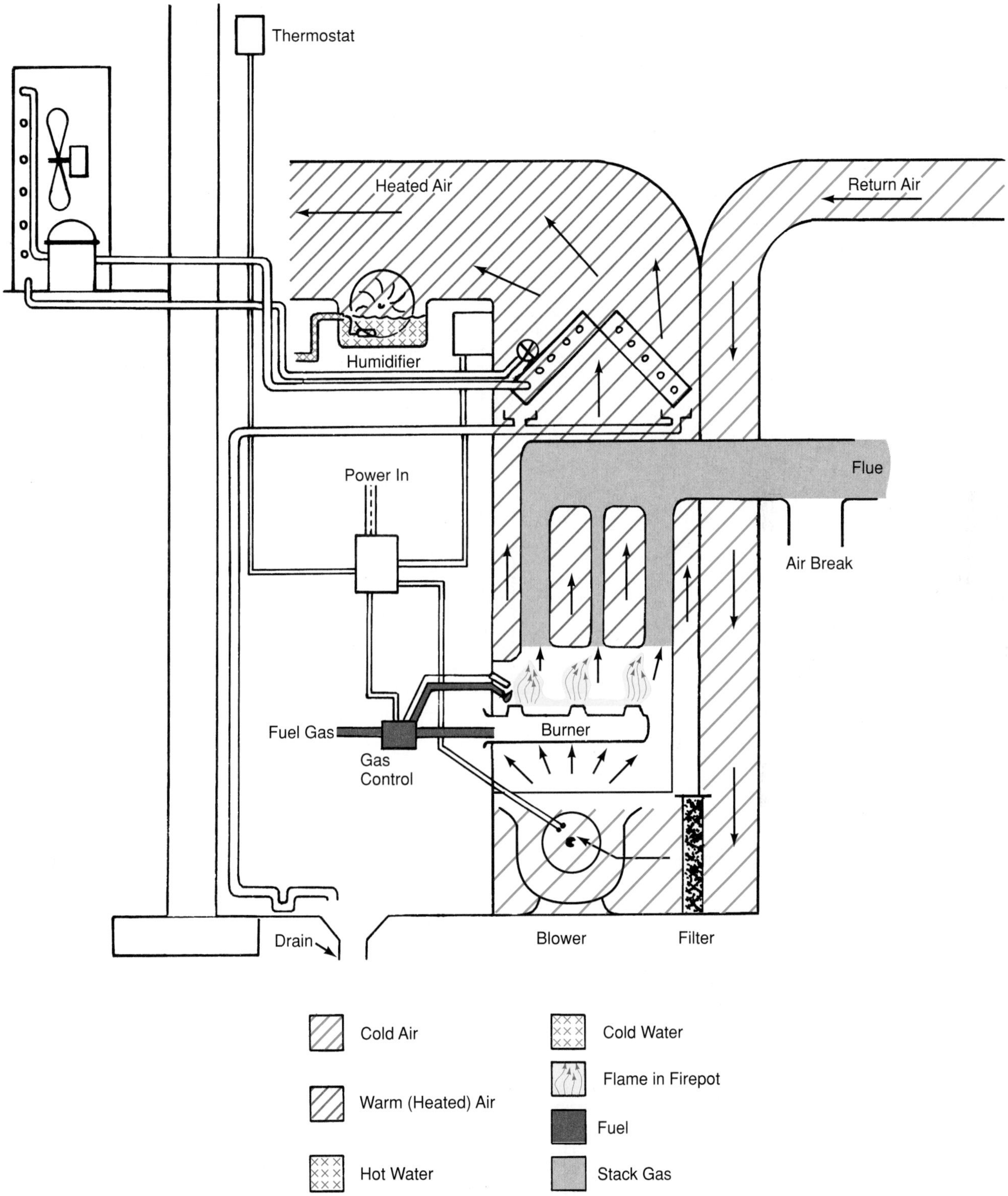

Figure 20-16. *Complete air conditioning unit provides both heating and cooling. Winter operation is shown. Winter heating is supplied by gas burner. Humidity is supplied by a humidifier in the plenum chamber. The same blower and filter are used for both summer and winter operation.*

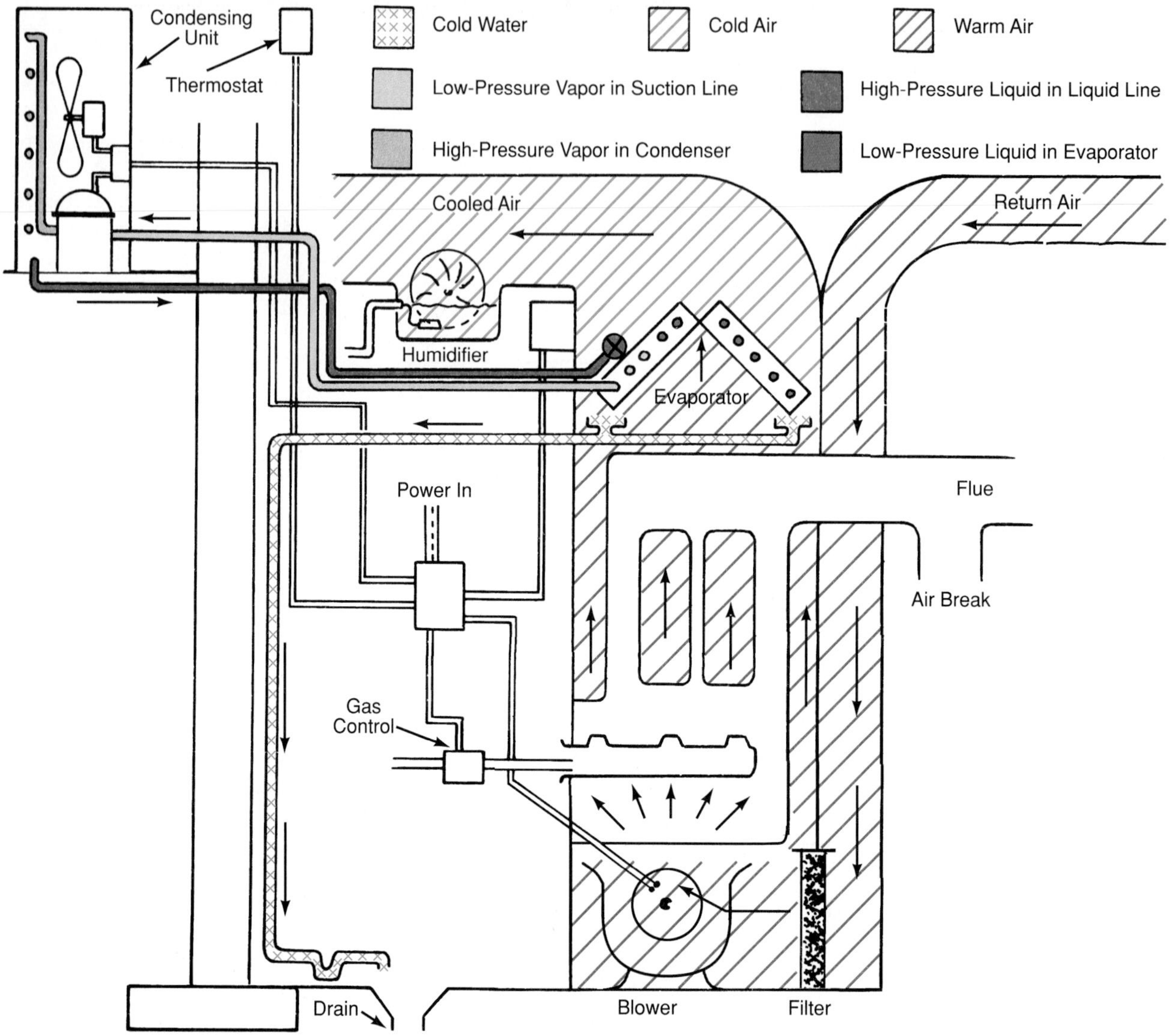

Figure 20-17. *Complete air conditioning system during summer operation. An A-frame evaporator in the plenum cools air forced through it by blower. Outside condensing unit disposes of heat absorbed in the evaporator. Summer humidity is removed by condensing of moisture on the evaporator surface. A drain tube carries away condensed moisture.*

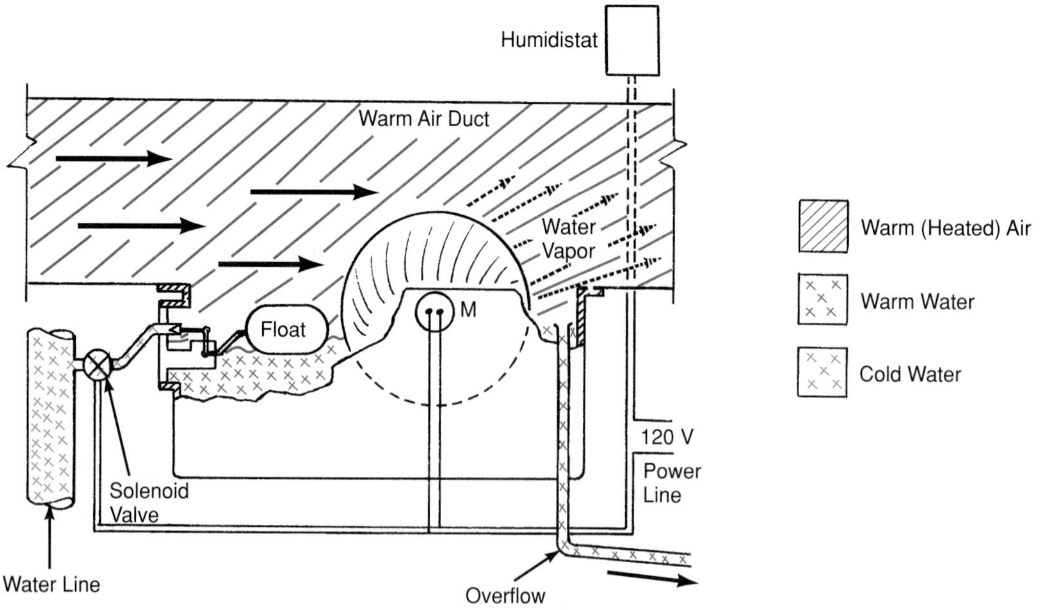

Figure 20-18. *Warm-air-duct humidifier. The water level is controlled with a float. A humidistat operates the motor and solenoid water valve. Broken arrows indicate air with moisture.*

20.14 Absorption Cycle Systems

Most large absorption cycle air conditioning systems use water as the refrigerant. A lithium bromide (LiBr) water solution is used as the absorber. **Figure 20-19** is a schematic diagram of such a system.

Steam or hot water heats the water and lithium bromide solution. The water turns to water vapor, and is then condensed by a water-cooled condenser. The water flows into the evaporator. There, it evaporates and is absorbed by the lithium bromide at the absorber.

Three pumps maintain the pressure difference. Pump 1 moves the solution which is strong in LiBr. (Sometimes this solution is described as "weak in water.") The solution is moved back to the absorber. There, more strong LiBr solution is removed from the concentrator. Pump 2 recycles the water not evaporated in the evaporator back to the spray heads. Pump 3 moves the solution which is weak in LiBr up to the concentrator.

The temperature changes in the system are marked on the drawing. Cooling water leaves the evaporator at 44°F (7°C). It travels through the cooling coils located in the rooms to be air conditioned (comfort-cooled). The water then returns to the evaporator at 54°F (12 °C). The lithium bromide solution always stays in liquid form. The condenser cooling water also cools the absorber.

20.15 Evaporative Condensers

Many air conditioning systems use water-cooled condensers. An ***evaporative condenser,*** **Figure 20-20,** may be used to cool the condenser vapor.

This system uses a conventional motor compressor, condenser, liquid receiver, drier, thermostatic expansion

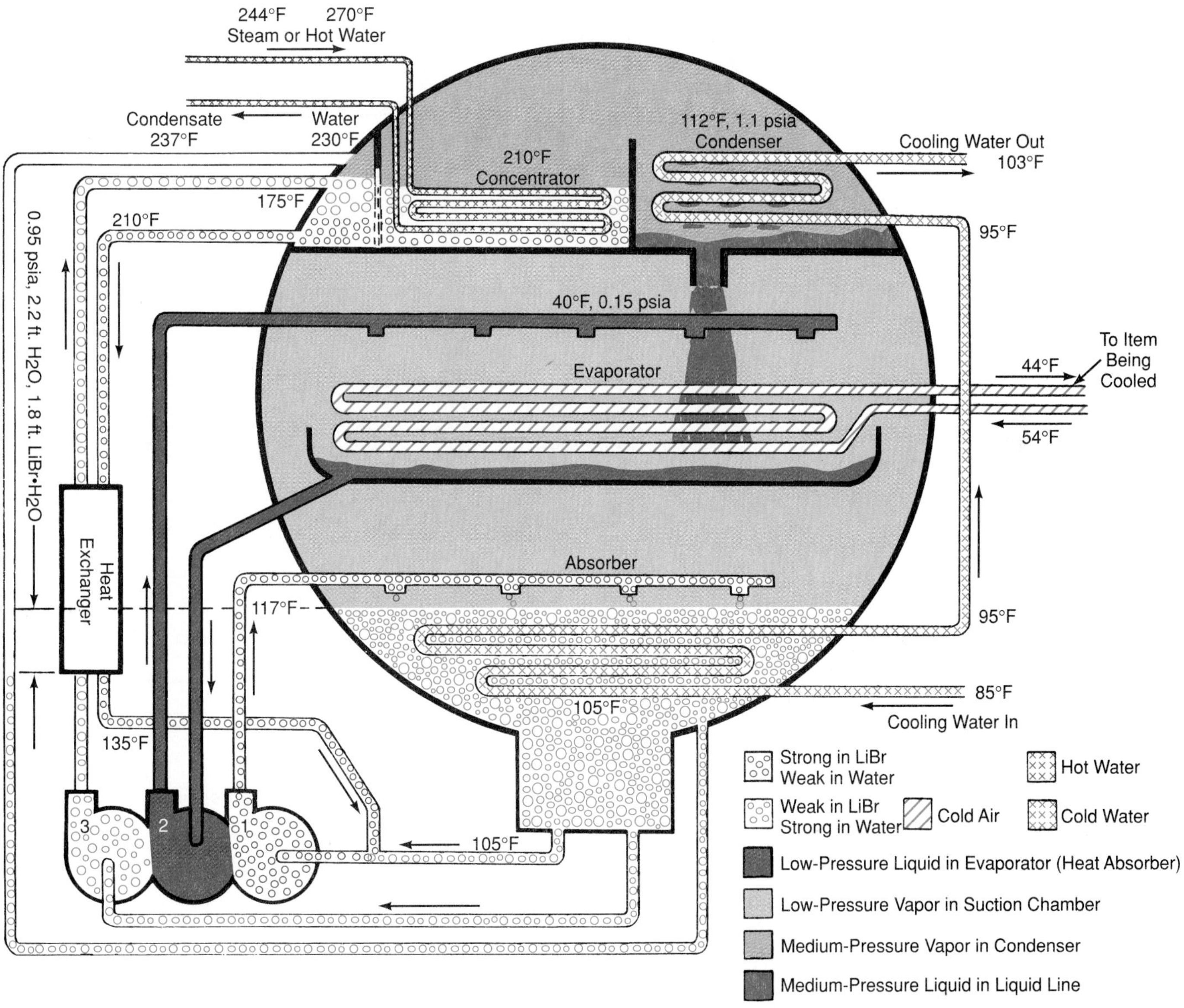

Figure 20-19. *Absorption-type air conditioning system. It uses water as the refrigerant and lithium bromide solution as the absorber. (The Trane Co.)*

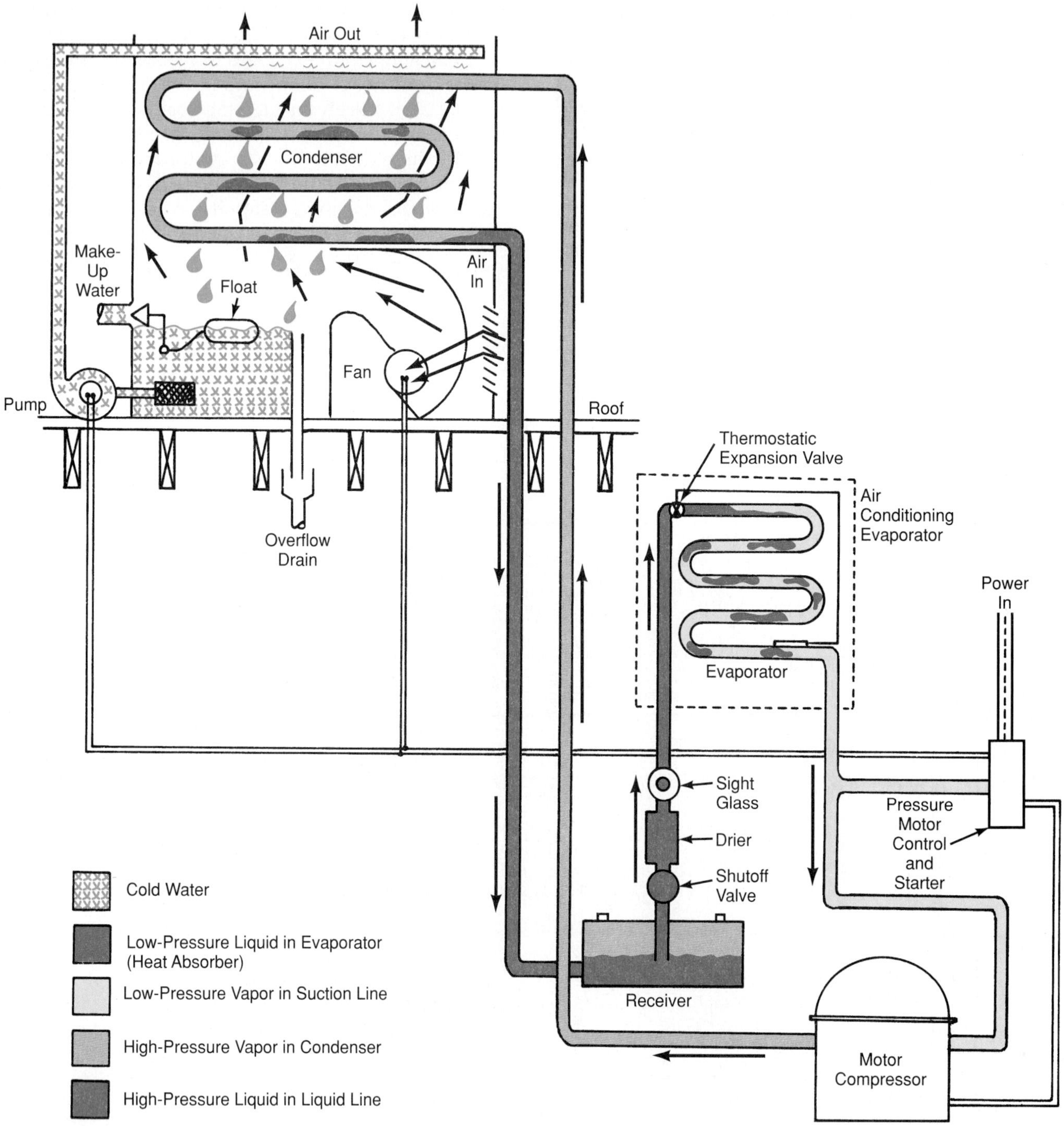

Figure 20-20. *Air cooling unit which uses an evaporative condenser. Note that condenser is located outside of conditioned space.*

valve, and evaporator. The hot compressed refrigerant vapor is piped to the evaporative condenser. The condenser is usually located on the roof or outside the building, as shown.

In this mechanism, the water supply is piped to a holding tank. A float mechanism maintains a constant level of water in the tank. A water pump circulates and sprays water over the refrigeration condenser.

A fan draws in air through the side of the evaporative condenser housing. It forces the air upward through the top. The water droplets are cooled by evaporation and then flow over the condenser. Some water is used up by the evaporative process. This is automatically replaced by flow into the holding tank. A pressure motor control is used on the refrigeration motor compressor in this instance.

20.16 Cooling Towers

Many refrigeration and air conditioning systems have water-cooled condensers. These are very efficient and do not take very much space. Often, water-cooled condensers have tap water circulated through them. This water is then discharged into the sewer. Such an arrangement uses large amounts of water and may be expensive. Moreover, many places do not allow use of tap water for cooling air conditioner condensers.

In such cases, cooling towers can be employed to cool the water. The cooled water is recirculated through the compressor's outer shell. Some makeup water will be required to replace the water lost by evaporation. A schematic of a water cooling tower is shown in **Figure 20-21.**

The cooling tower is a housing or shed into which air is drawn. It has a water spray arrangement and baffles. The sprayed water is exposed to the stream of air and becomes cool. A float mechanism is connected to the water spray. It maintains a constant level in the water reserve tank. The pump circulates the cooled water through the refrigerant condenser. Water is sometimes sprayed over the baffles.

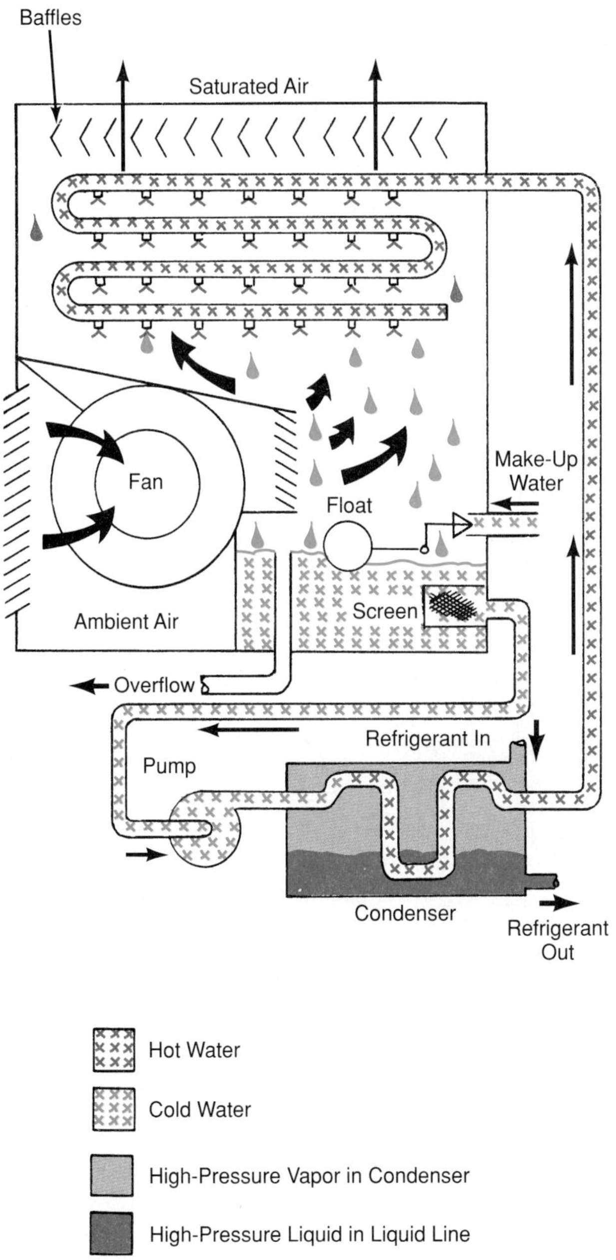

Figure 20-21. *Typical cooling tower application. Recirculated water is cooled several degrees in tower. Water is then circulated through refrigerant condenser to cool it.*

Cooling towers are available in a great range of sizes. Small ones may be used to cool the water-cooled condensers for home air conditioners. Very large ones are required for cooling the condensers in large steam power plants. **Figure 20-22** shows a complete system.

20.17 Steam Jet Cooling

Steam jet cooling uses water as the refrigerant. Pressure on the surface of water is reduced to lower its boiling temperature. This is shown in the table in **Figure 20-23.** At a pressure of 0.2 psia (10 mm Hg), the boiling temperature of water is 53°F (12°C).

A steam jet is shown in **Figure 20-24.** An ejector draws water vapor from the surface of the water in the evaporator. This causes the pressure in the evaporator to drop. The ejector reduces the pressure in the evaporator. The pressure is reduced until the water will vaporize at the desired temperature. While vaporizing, it absorbs heat and cools the rest of the water in the evaporator.

Steam pressure at the ejector nozzle should be about 150 psia (1030 kPa). Sometimes the steam is condensed at another location for the following reasons:

- To recover some heat from the steam.
- To recover the water in the steam.
- To reduce the pressure so that it will not "back up" into the steam jet cooling chamber.

The pressure in the condenser is not shown in the illustration. It will be about 3 psia (21 kPa, 160 mm Hg). The pressure, 3 psia, corresponds to the steam condensing temperature, 141°F (61°C). See **Figure 20-23.**

Evaporation of some of the water in the evaporator reduces the temperature of the remaining water. Pumps circulate this cold water, at 40°F to 70°F (4°C to 21°C), to the area to be cooled.

These systems usually have a capacity of 100 tons or more. This capacity provides for a large supply of steam under a fairly high pressure. It also provides for a large supply of water for cooling the condenser.

Steam jet systems are often used in air conditioning. They cool water used in certain chemical plants for gas absorption. The cooling temperatures provided by the steam jet mechanism are usually between 40°F and 70°F (4°C and 21°C). Temperatures below 40°F (4°C) are impractical due to the danger of freezing.

One application of steam jet cooling is removing water from diluted solutions that contain juices. Orange juice can be concentrated in this way. Steam jet cooling does this by replacing the water with orange juice and not providing makeup water. The process does not boil

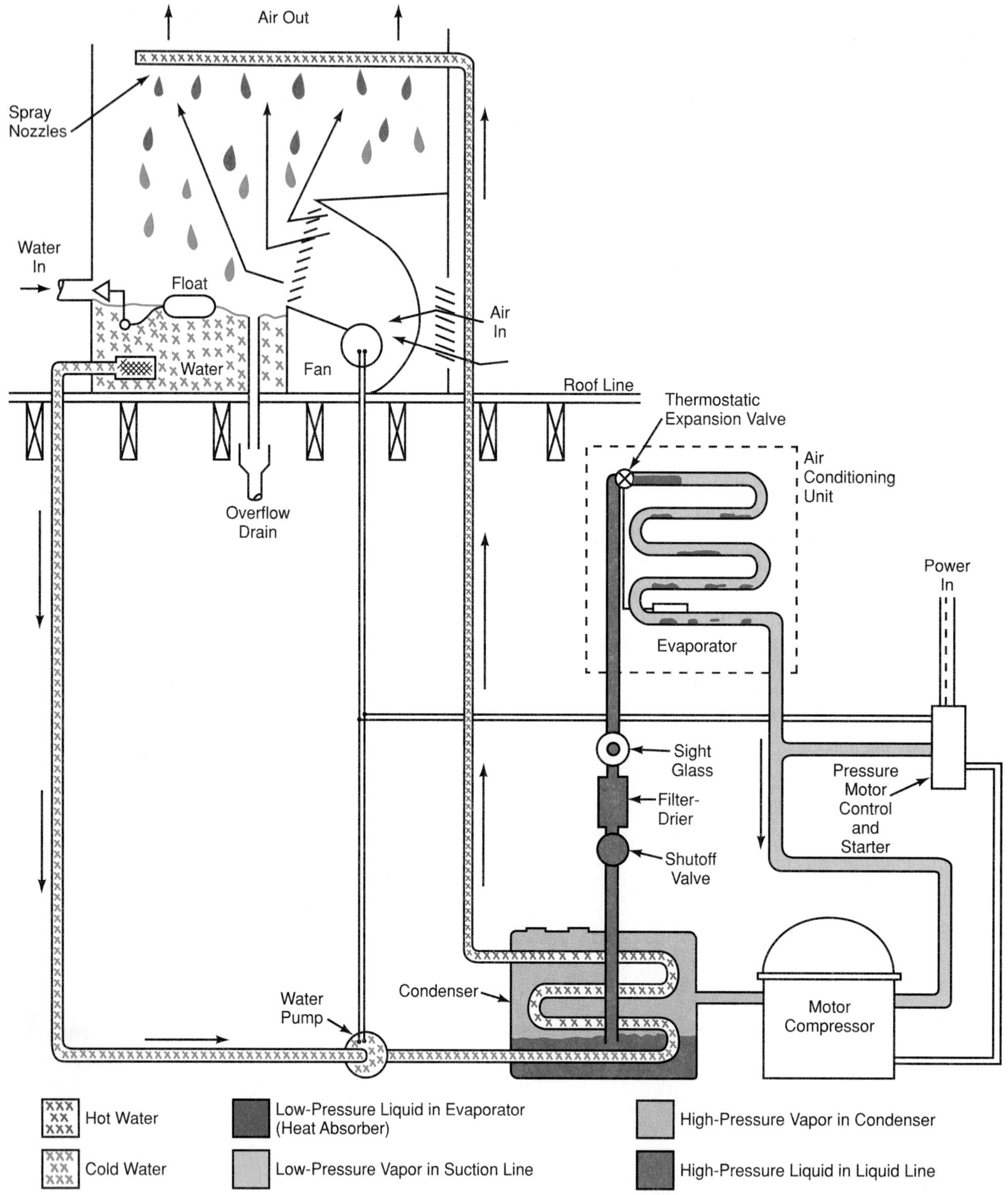

Figure 20-22. *An air conditioning comfort-cooling installation using cooling tower.*

juice at temperatures near 212°F (100°C). Therefore, vitamins are kept at full strength.

20.18 Vortex Tube Cooling

An interesting device, the ***vortex tube,*** uses compressed air to produce low temperature. Pressurized air is directed smoothly along the inner surface of a tube or cylinder. One end of the tube is completely open. The other end is closed off except for a small-diameter tube. During operation, warm air leaves through the unrestricted (wide open) end. Cold air leaves through the small tube at the other end.

Figure 20-25 shows the operation of the vortex tube at D, C, and E. There are three openings. One is an inlet

Water Boiling Temperatures

psia	Boiling Temperature		psia	Boiling Temperature	
	°F	°C		°F	°C
.1	35	2	5.	162	72
.2	53	12	6.	170	77
.3	64	18	7.	177	81
.4	73	23	8.	183	84
.5	80	27	9.	188	87
.6	85	29	10.	193	89
.7	90	32	11.	198	92
.8	94	34	12.	202	94
.9	98	37	13.	206	97
1.	102	39	14.	209	98
2.	126	52	14.7	212	100
3.	141	61	15.	213	101
4.	153	67	20.	228	109

Figure 20-23. *Table shows boiling temperature of water at various pressures. Note that pressures are in pounds per square inch absolute (psia). Atmospheric pressure is 14.7 psia (101.3 kPa).*

while the other two, as explained before, are outlets. The inlet opening at C is a jet nozzle. It is connected to the compressed air source B. This jet injects the air into the tube at an angle.

The air swirls rapidly in a corkscrew pattern inside the large tube. This is due to the jet design and the high pressure of the air. Both openings, D and E, are at or near atmospheric pressure. The temperature of the air, as it leaves through tube E, will be greatly reduced. Temperatures below zero are obtained with this device.

The principle of operation of the vortex tube is as follows. High-velocity air stays in the outer circles of tube D, as shown in **Figure 20-25.** Low-velocity air stays near the center of the vortex. The air molecules are moving slower in the low-velocity air than in the high-velocity region. Thus, the center of the vortex is cooler. (Temperature is a measure of the average speed of molecules—lower *speed* means lower *temperature.*)

There is a reason that the low-velocity air stays near the center of the vortex. Circles near the center are very small. Air cannot achieve much speed due to the center area becoming more and more like a point. At the exact center, all velocities are zero. Another way to say this is that near the center, each circle of the airstream consists of opposite flows on opposite sides of the circle. These opposite flows cancel each other.

The vortex system is useful where both fresh air and cooling are desired at one time. A large quantity of compressed air must be used.

A typical application of the vortex tube principle is in the cooling of protective suits. These are used by industrial workers who must work in toxic atmospheres or in very hot places. Suits cooled by vortex tubes are used when a worker must wear heavy protective

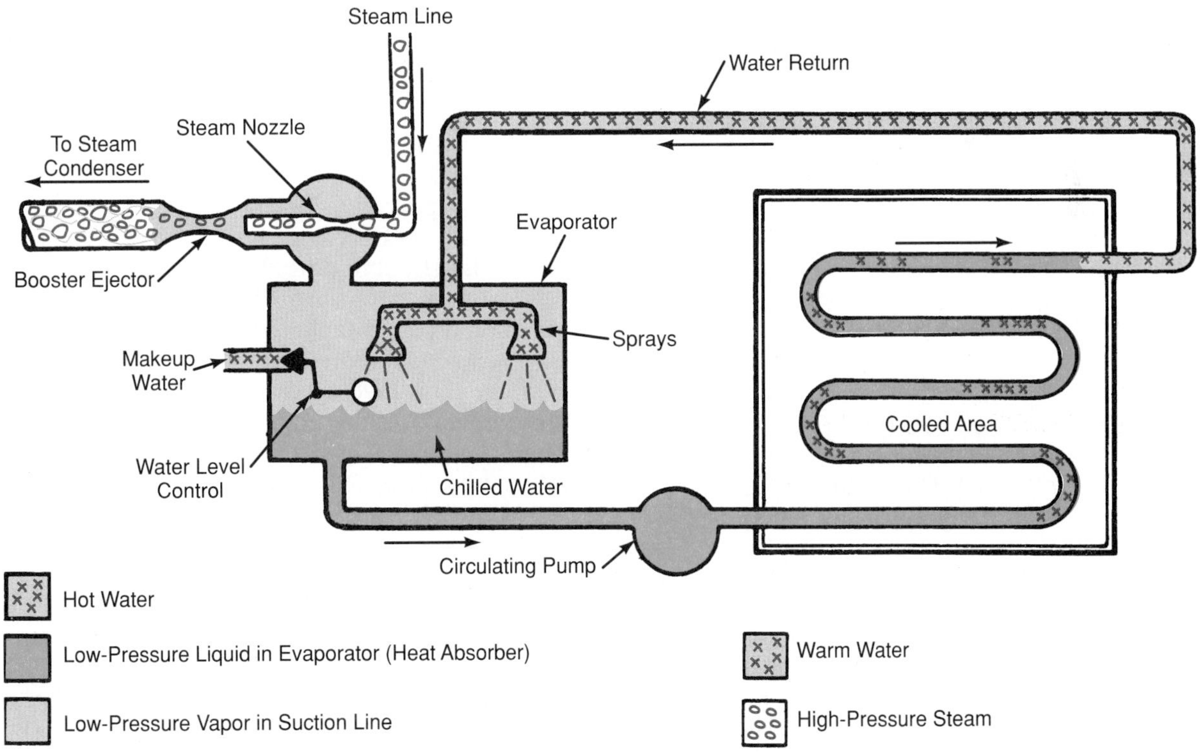

Figure 20-24. *Steam jet refrigeration. Steam escaping through nozzle in the ejector causes low-pressure condition over surface of water in the evaporator. This low pressure causes the water to evaporate rapidly, absorbing heat and reducing temperature of water in the evaporator. Chilled water is circulated where needed for cooling. If items in cooled area are above 57°F (14°C), some liquid water will turn into water vapor inside the cooled area coil. If sprayheads are restricters, no boiling occurs in cooled area.*

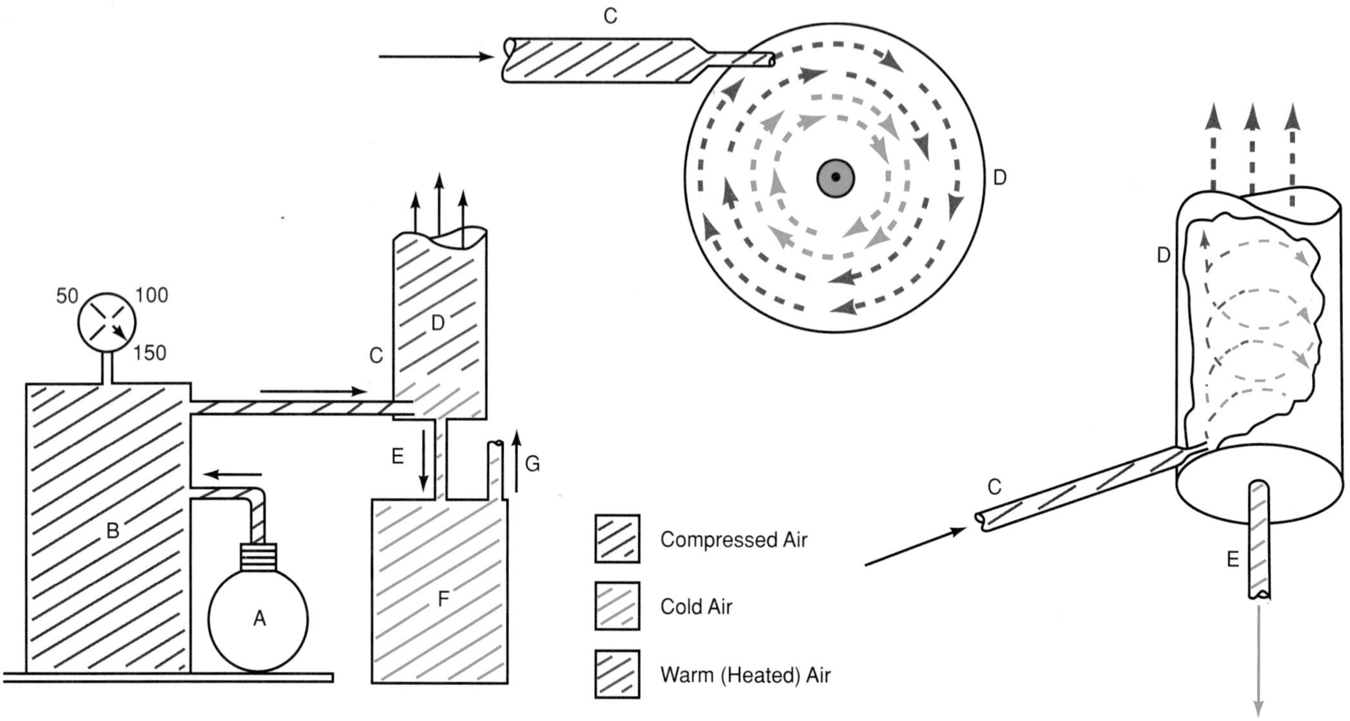

Figure 20-25. *Simplified illustration of vortex tube refrigeration. A—Air compressor. B—Compressed air storage tank. C—Compressed air nozzle. Cold air is produced at E, and flows into space F, which is to be cooled. Extra air from space F is pushed out through passageway G. Warm air from outer circles of vortex region is expelled through large tube D.*

clothing. (Examples would be sandblasting, cleaning a brick building, etc.) Usually these devices are needed only when the worker is performing a particular job.

Vortex tube devices are designed to operate on a continuous basis. No thermostatic control of any type is used. However, there is a manual control. This control allows the vortex tube operator to adjust outlet air temperature as needed.

20.19 Evaporative Cooling

Evaporative cooling is often used in climates with bright sunshine and low relative humidity. For example, evaporative cooling can maintain a safe temperature for plants growing in greenhouses. See **Figure 20-26A.**

One end of the greenhouse has a lattice of fibers such as ***excelsior*** (fine curled wood shavings). Water pipes with small holes are along the top of the lattice. The water from the small holes flows downward, wetting the fibers. A fan at the other end of the greenhouse draws air through the lattice. The air is discharged outdoors.

The air entering the greenhouse will be cooler than the outside air. This is due to the evaporation of the water on the lattice surfaces. Plants do best under conditions of high relative humidity. Therefore, this type of cooling is ideal. It provides a high relative humidity as well as a lower temperature inside the greenhouse.

Foundries sometimes use this system in situations where a few degrees of cooling are desirable. In such installations, the evaporation takes place in a structure on the roof of the building. The cooled air is brought down into the work area.

A shallow pool of water on a flat roof may be used for cooling purposes. This pool would be 2″ or 3″ (50 mm or 75 mm) deep. Where there is bright sunshine much of the day, the building heat load may be high. If the relative humidity is usually low, the ponded roof will maintain a comfortable indoor temperature.

Wet roof cooling is at times referred to as "swamp cooling." This method is used primarily where ambient air temperatures are high and relative humidity is low. There are numerous methods of wet roof cooling. Some systems are designed so that the roof is a trough or pond. The entire surface remains wet and the water is replaced as it changes phases.

Another method is to spray the roof with fine droplets of water. See **Figure 20-26B.** Roof spraying is based on the fact that water will change its state when sprayed on a hot roof. In doing so, it absorbs a relatively large amount of heat from the roof's surface. Evaporation continues as long as the roof's surface temperature is greater than ambient air wet-bulb temperature This will reduce the heat gain through the roof. The surface temperature varies, depending on the latitude, month, etc.

Various controls in the system monitor the roof surface temperature. They control the amount of water that is sprayed. It can be evaporated from the roof according

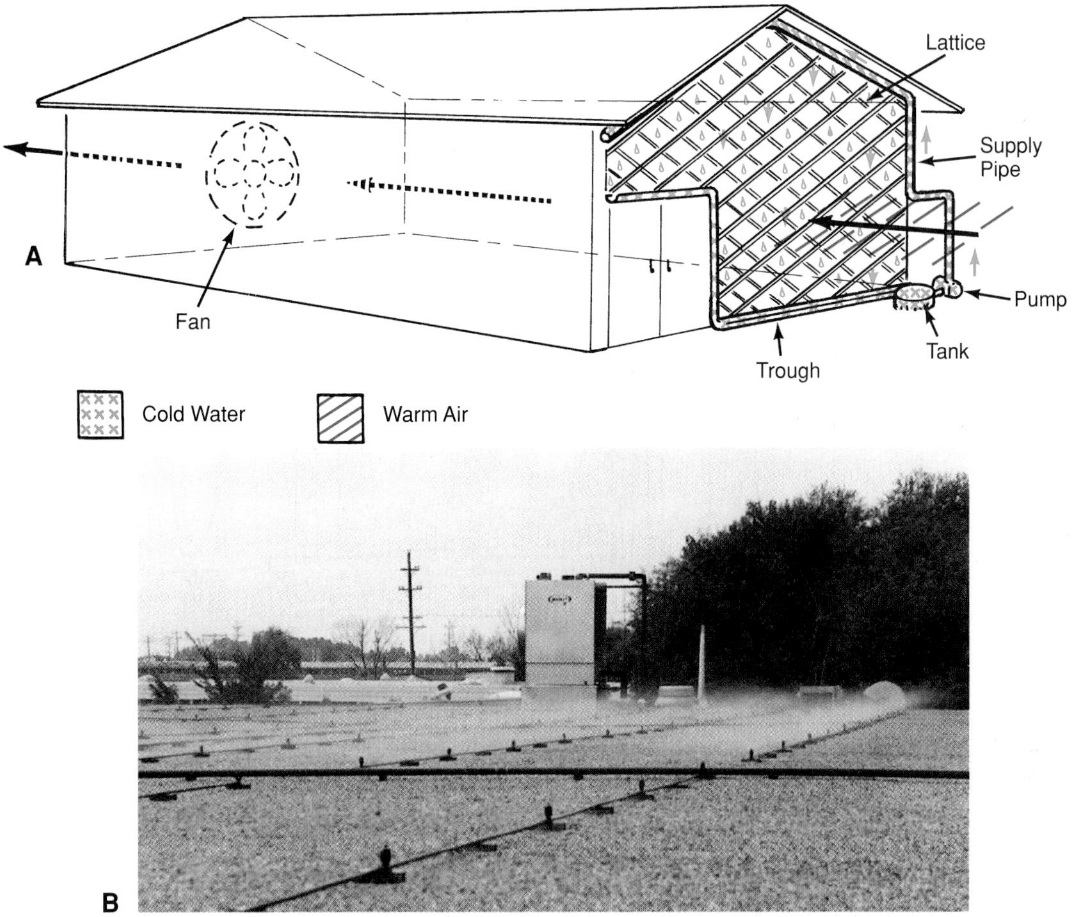

Figure 20-26. *Evaporative cooling. A—Evaporative cooling system used in a greenhouse. B—Lowering the roof surface temperature through the use of a fine mist of water on the roof surface. (Sprinkool Systems, Inc.)*

to its temperature at that time. The amount of water used is coordinated with the roof temperature and ambient air temperature. This allows the water to evaporate without any overflow or stagnation.

The higher the temperature, the greater the amount of water vapor that can exist with air in the atmosphere at any relative humidity. The evaporation will continue even when the relative humidity is 100%. The heat source (roof) that is in contact with water must be higher than the wet bulb of the air/vapor combination adjacent to the surface. The roof cooling system should maintain a roof surface temperature close to the wet-bulb temperature. This, therefore, minimizes the heat flow through the roof.

The evaporative roof cooling can be used as a single system or as part of a mechanical air conditioning system or energy management program.

20.20 Automobile Air Conditioning

Automobile air conditioners use a refrigerating mechanism to cool the air inside the vehicle. The components of an automobile air conditioning system include:

- Refrigerating compressor driven by the engine.
- Condenser located in front of the radiator.
- Liquid line to the refrigerant control.
- Evaporator.
- Fan.
- Duct system to circulate the air inside the vehicle.

Temperature control is based chiefly on the temperature of the air flowing through the evaporator. **Figure 20-27** illustrates a typical automobile air conditioner.

Air conditioning a moving vehicle presents some problems not found in the usual residential or commercial installation. The compressor is driven by the engine. Therefore, its speed will change as the engine speed changes. The system cooling capacity can take care of the cooling load under the most unfavorable temperature and speed conditions. As a result, under normal driving conditions, the system has much more capacity than needed.

This problem is solved with a magnetic clutch in the compressor drive pulley hub. This clutch is controlled by a thermostat. When passenger compartment temperature is down to the desired level, the thermostat releases this clutch. The compressor stops turning. When

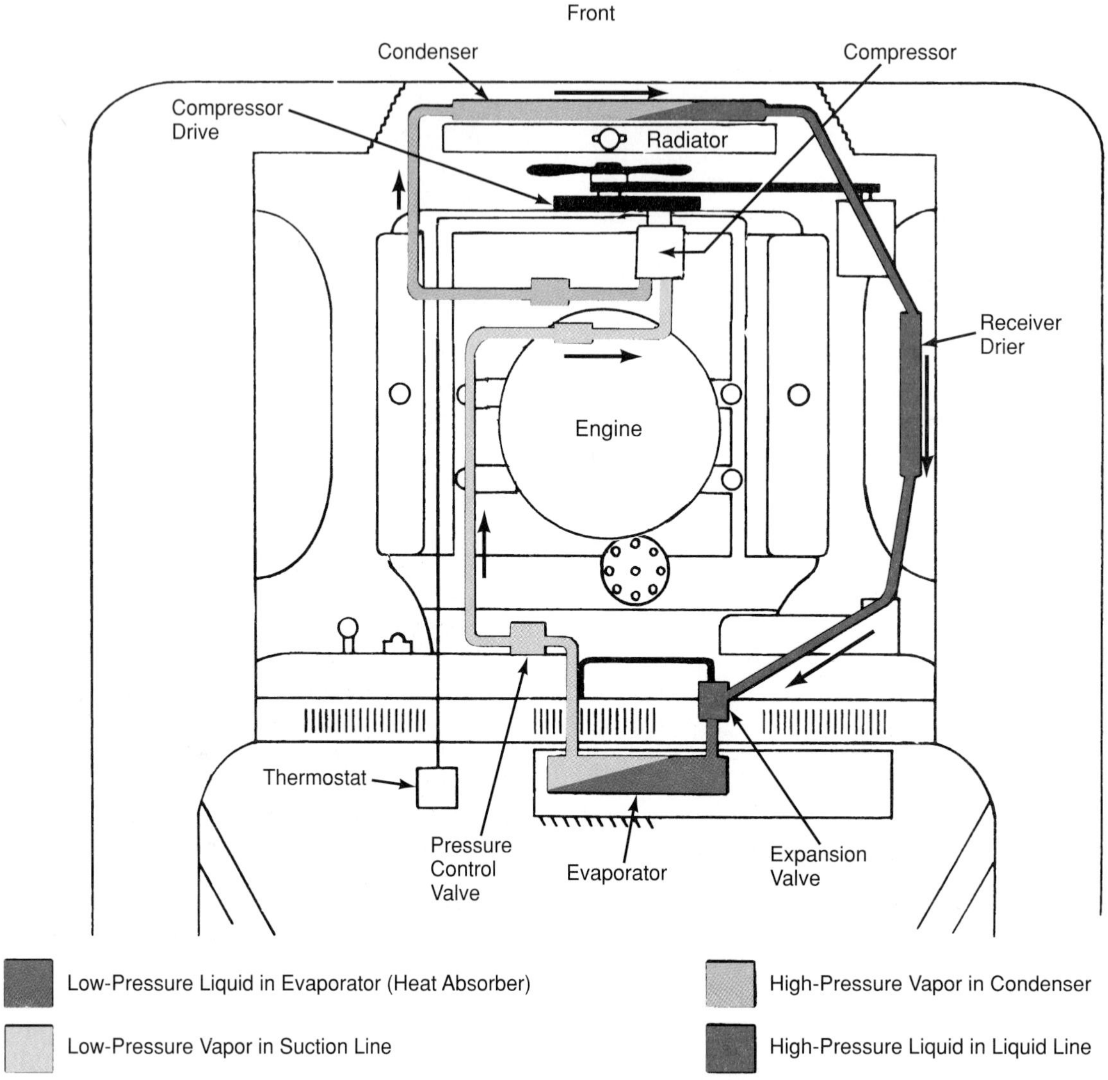

Figure 20-27. *Automobile air conditioner. Compressor is driven by the engine. Condenser coil is located ahead of the automobile radiator. Evaporator is located in duct system in the passenger compartment.*

cooling is again needed, the magnetic clutch engages the compressor. The compressor draws power from the engine when the magnetic clutch is engaged. A cutout switch disengages the compressor when more power is needed for acceleration.

There is still another problem with air conditioning an automobile. The evaporator condenses moisture from the air. If the evaporator temperature is maintained at or below freezing temperature, this moisture will freeze and adhere to the evaporator. Soon, the evaporator will be completely frozen over. No air will be able to circulate through it. This problem is solved by placing a suction throttling valve in the suction line. This valve keeps the evaporator pressure slightly above the pressure at which the boiling refrigerant temperature in the evaporator would cause moisture to freeze to the evaporator surface.

The preceding description of automobile air conditioning has been kept brief. See Chapter 28 for full details concerning automobile air conditioning.

20.21 Review of Safety

Before installing, maintaining, or servicing air conditioning systems, it is important to be experienced. There should be some on-the-job training as an assistant to a service technician. Regional and local codes and regulations must be known before working on air conditioning systems.

Safety is the first consideration when working with equipment which has electrical circuits, fuels, vapor, or liquids under pressure.

You should always use instruments to check equipment. To be reasonably certain of your safety, the voltage and pressure must be known.

Avoid using an ignition source when there is any chance of a fuel being present. A flame, a friction spark, or an electric spark (from opening or closing a switch) can ignite a fuel vapor-air mixture, causing an explosion. This could be damaging to property and cause injury or death.

Chapters 21, 22, 23, 24, 26, 27, and 28 describe specific safety precautions relating to air conditioning systems.

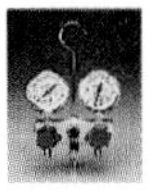

20.22 Test Your Knowledge

Please do not write in this text. Place your answers on a separate sheet of paper.

1. Can an air filter be used on a forced-air heating plant?
   A. Yes, in the cold air duct.
   B. No.
   C. Yes, in the warm air duct.
   D. Both A and C.
2. A ______ senses whether a pilot light is on or off.
   A. pilot light
   B. solenoid valves
   C. thermocouple solenoid
   D. combustion air blower
3. A hydronic heating system ______.
   A. contains a water pump
   B. uses circulating water to carry heat from the furnace
   C. has no moving parts
   D. Both A and B.
4. Which of the following is the usual method of heating water in a hydronic system?
   A. Fuel gas.
   B. Oil.
   C. Solar.
   D. Both A and B.
5. Why is an automatic draft control needed in an oil burner installation?
   A. To draw in the fuel oil and filter it.
   B. To maintain a constant pressure in the firepot.
   C. To reduce drafts in the boiler chamber.
   D. All of the above.
6. The heating elements in radiant heating installations are usually placed in the ______.
   A. ceilings
   B. walls
   C. floors
   D. All of the above.
7. What are the basic components of a through-the-wall room air conditioner?
   A. Hermetic compressor, condenser, evaporator.
   B. Semi-hermetic compressor, condenser, evaporator.
   C. Compressor, fan, evaporator.
   D. Compressor, fan, condenser.
8. In a window or through-the-wall air conditioner, heat absorbed by the evaporator is ______.
   A. reabsorbed by the refrigerant
   B. returned to the compressor
   C. released by the condenser located outside
   D. None of the above.
9. In a central air conditioning system, where is the condensing unit usually located?
   A. Inside the building.
   B. Outside the building.
   C. Next to the compressor.
   D. None of the above.
10. In an air-to-air heat pump installation, where is the heat for warming the room obtained?
    A. From the indoor air.
    B. From the outdoor air.
    C. From both indoor and outdoor air.
    D. None of the above.
11. In automobile air conditioning, the compressor is usually driven by ______.
    A. electricity
    B. the fuel
    C. the engine
    D. None of the above.
12. In automobile air conditioning, what is the role of the magnetic clutch on the compressor drive pulley?
    A. To connect or disconnect the compressor and maintain proper temperature.
    B. To provide ventilation.
    C. To remove excess moisture.
    D. All of the above.
13. Will a dehumidifier operating in a room change the room's air temperature?
    A. Yes, by approximately 2°F (1.1°C).
    B. Yes, by approximately 4°C (2.2°C).
    C. No, the air temperature will remain about the same.
    D. None of the above.
14. In a heat pump installation, the direction of refrigerant flow is reversed through the ______ and the ______ when the cycle is reversed from heating to cooling.
    A. compressor, evaporator
    B. compressor, condenser
    C. evaporator, condenser
    D. None of the above.
15. Steam jet refrigeration is most used in ______.
    A. automotive plants
    B. production of metals
    C. chemical plants
    D. All of the above.
16. The cooling fluid (water) in steam jet refrigeration operates at about ______°F to ______°F (______°C to ______°C).
    A. 0, 45 (−18, 7)
    B. 35, 45 (2, 7)
    C. 32, 62 (0, 17)
    D. 40, 70 (4, 21)
17. ______ is a common application of the vortex tube cooling system.
    A. Food processing
    B. Production of metals
    C. Chemical production
    D. Ventilation, heating, or cooling of special work uniforms

18. What is an advantage of electrical resistance heating?
    A. It is very cost-effective.
    B. Each room is regulated by its own thermostat.
    C. It is not necessary to place an electrical resistance unit in each room.
    D. Air conditioning may be easily added.
19. Identify the benefit(s) of radiant heating.
    A. There is little or no equipment in sight.
    B. It provides a very comfortable environment.
    C. There is very little insulation required.
    D. Both A and B.
20. A ______ solution is commonly used in a large absorption system.
    A. R-134a
    B. R-12
    C. lithium bromide/water
    D. ammonia

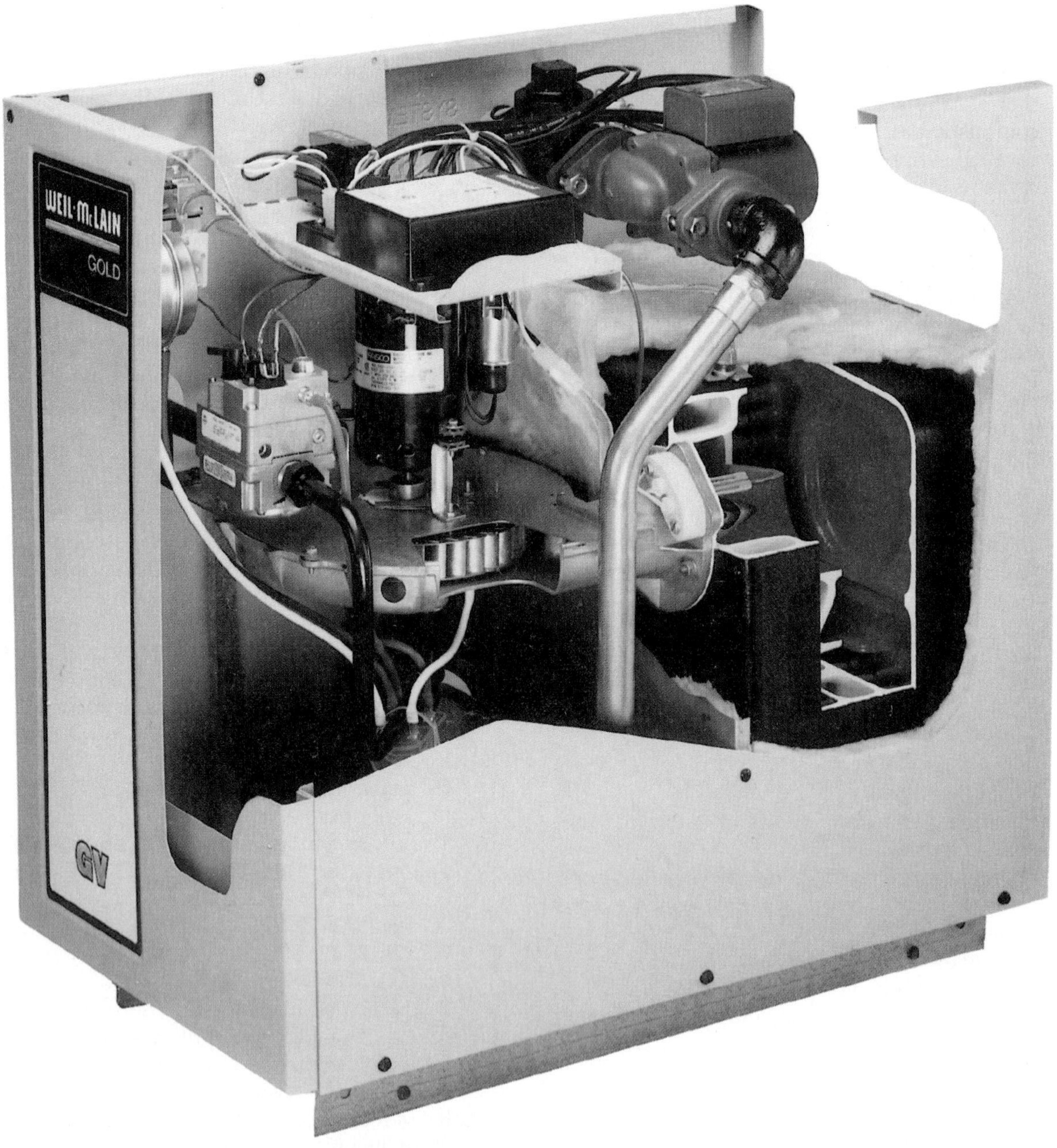

Gas-fired hot water boiler. (Weil-McLain, A United Dominion Co.)

Chapter 21

HEATING AND HUMIDIFICATION SYSTEMS

Modules:

Key Words:

Annual Fuel Utilization Efficiency (AFUE) rating
Combined Annual Efficiency (CAE) ratio
combustion
downflow furnace
embrittlement
hydrocarbons
ignition systems
primary air
pulse combustion process
radiant heat
refractory cement
secondary air
upflow furnace

Learning Objectives:

After studying this chapter, you will be able to:

- Describe the combustion process.
- Identify the basic components of a forced-air system.
- Define the difference between a mid-efficiency and a high-efficiency furnace.
- Explain furnace operation.
- Identify the three types of ignition systems.
- Explain service procedures for various types of furnace systems.
- Compare gas, oil, hydronic, and electric heating systems.
- Identify and service various types of humidifiers.
- **Follow approved safety procedures.**

GAS HEATING SYSTEMS MODULE

21.1 Types of Systems

There are many types of heating systems: forced-air, hot water, radiant, infrared, and unit heaters. Sources of heat may be classified by types of fuels—oil, gas, or electric. See **Figure 21-1.** Gas and oil systems include forced-air, hot water, and unit heaters. Electric systems consist of infrared and radiant systems. Forced-air, hot water/hydronic, and unit heaters are the most commonly used residential systems. Radiant and infrared systems are often used for commercial purposes.

21.2 Combustion

Combustion is the process by which the energy contained in a fuel is converted to heat and light energy. Fuels most commonly used are natural gas, LP gas, or oil. Recently, electricity (not a fuel, but an *energy source*) has become a popular heat source. The physical properties of the fuel are considered when determining requirements for combustion.

The three basic elements necessary for combustion to produce a flame are fuel, heat, and oxygen. See **Figure 21-2.** Combustion occurs when the chemicals in the fuel combine with oxygen.

Heat: Combustion can occur only when the air-to-fuel mix is within an acceptable range. During this process, the energy contained within the fuel is changed into heat and light. A given amount of oxygen per part of fuel is necessary. For combustion to take place with natural gas, the mixture must be between 4% to 14% gas (96% to 86% air). An amount of gas less than 4% would be too lean. An amount of gas above 14% creates a mixture that is too rich. Each fuel source has unique air-fuel mix requirements for combustion. See **Figure 21-3.**

Oxygen: The amount of oxygen required varies with each fuel. ***Primary air*** is the air which is mixed with the fuel prior to ignition. See **Figure 21-4.** Sufficient oxygen

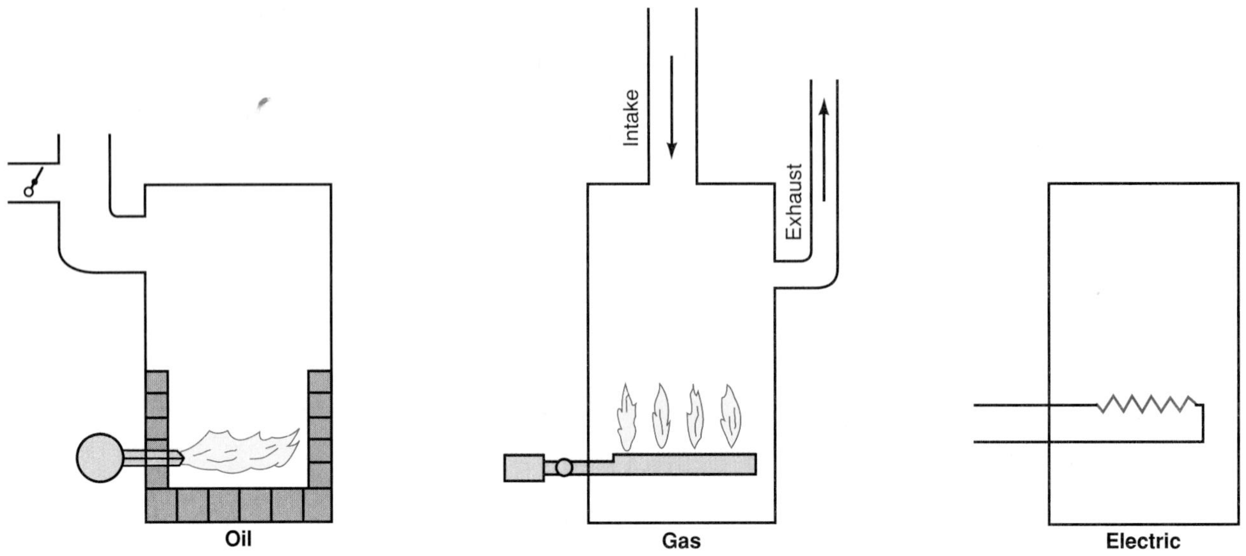

Figure 21-1. *Three popular types of heat sources used in furnaces.*

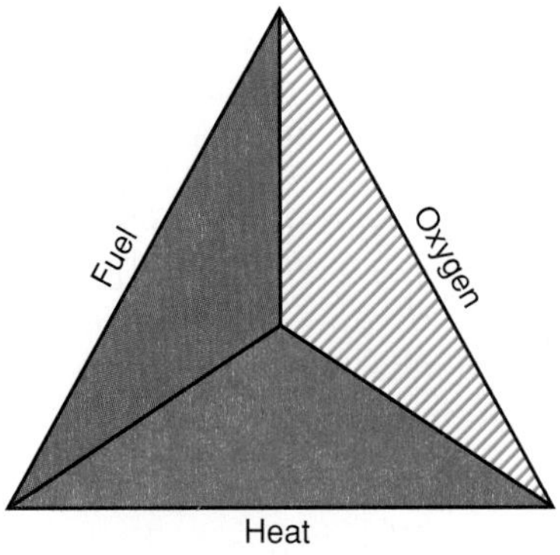

Figure 21-2. *The combustion triangle. All three components are necessary for combustion to take place.*

must be available to support combustion. The oxygen is consumed during the combustion process and must consistently be replaced. ***Secondary air*** is the air added to the flame after ignition to maintain combustion. See **Figure 21-5.** The secondary air surrounds the outermost area of the flame and maintains combustion.

Fuel: All fuels contain hydrogen and carbon atoms in varying amounts. These are referred to as ***hydrocarbons.*** The hydrogen atoms burn more quickly and at a lower temperature than the carbon atoms. See **Figure 21-6.** The hydrogen burns first, using the air (oxygen) it needs for complete combustion. Its flame is of a bluish color. Carbon particles must reach the outside edge of the flame for complete combustion to occur. They are slower-burning but, at a high temperature, will produce a bright yellow light.

Flame color will indicate when the proper amount of primary air is present in a gas furnace. The flame has a bright blue inner core, surrounded by light blue. This indicates there is no unburned carbon. ***Incomplete combustion*** occurs when the flame does not receive enough oxygen to finish the combustion process. If there is

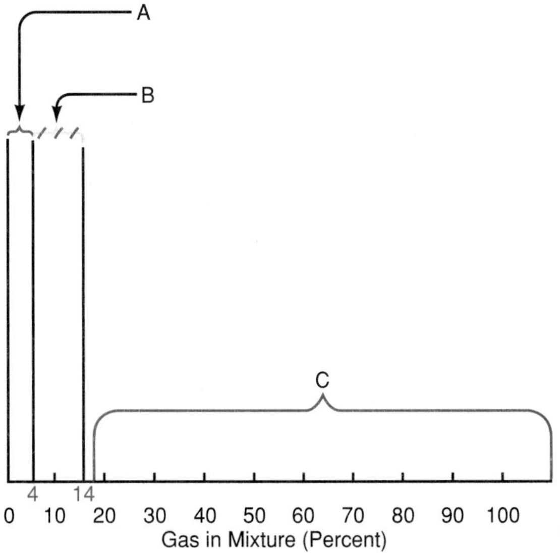

Figure 21-3. *Percentages of air and fuel required for combustion to occur. A—Mixture is too lean. It will not burn or explode. B—Flammable mixture. It will burn or explode, depending upon accumulation and ignition time. C—Mixture is too rich. It will burn only if secondary air is added.*

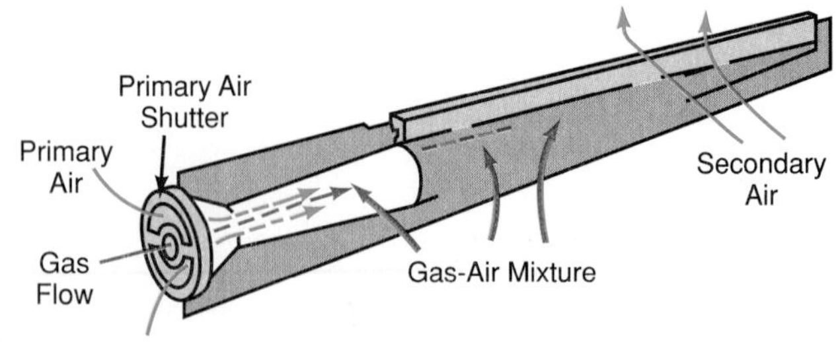

Figure 21-4. *Primary air is mixed with fuel prior to ignition. Secondary air is required after ignition to maintain combustion.*

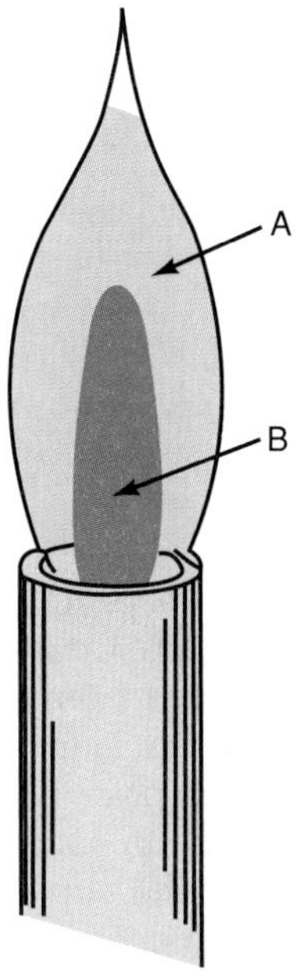

Figure 21-5. *Secondary air surrounds outermost areas of flame and maintains complete combustion. A—Combustion, secondary air and gases. B—Initial combustion, primary air and gas.*

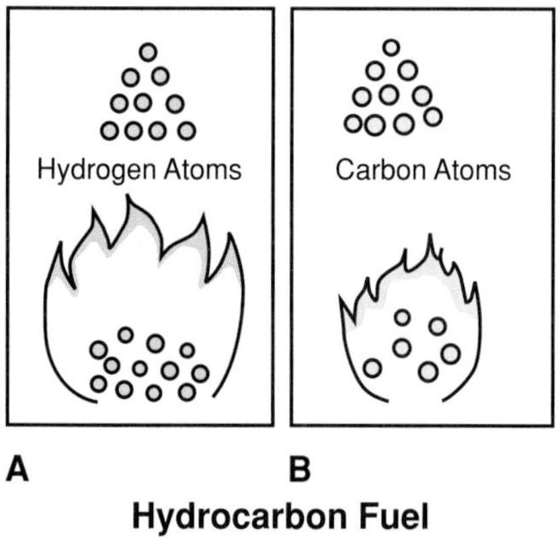

Figure 21-6. *Burning of hydrogen and carbon atoms. A—Hydrogen atoms burn quickly. B—Carbon atoms burn slowly.*

insufficient primary air, the flame tip will be yellow. See **Figure 21-7.** This is due to carbon particles that are not burned. **A byproduct of incomplete combustion is carbon monoxide. This can be dangerous to a building's occupants as well as to the furnace. Therefore, all service calls must include inspection of the flame during operation.**

21.3 Fuel Gases

There are three types of fuel gases:

- Natural.
- Manufactured.
- Liquefied petroleum (LP gas).

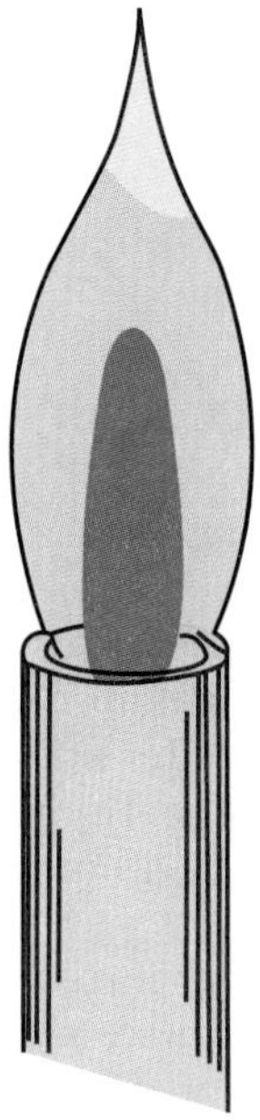

Figure 21-7. *Incomplete combustion. A lack of primary air causes yellow-tipping of the flame.*

Natural gas is obtained from gas deposits in the ground. Manufactured gas is made by distilling or "cracking" coal or oil, and by other processes. Natural gas consists of about 84% CH_4 (methane) and 16% C_2H_6 (ethane). It has a heating value of 1000 Btu/ft^3 to 1100 Btu/ft^3. To burn 1 ft^3 of natural gas, 8 ft^3 of air is required. However, some excess air is needed. Therefore, about 11 ft^3 of air is used for each cubic foot of natural gas. (This provides 30% excess air.)

As the gas burns, it yields about 1 ft^3 carbon dioxide and 1 ft^3 nitrogen. It also yields 2 ft^3 of water vapor and about 25% to 50% excess air. If there is insufficient primary air, the burning will produce a yellow flame (indicating incomplete combustion). (Refer to Section 21.5.) If there is too much primary air, the flame will be noisy. It will jump around above the burner. If the stack has CO (carbon monoxide), more primary air or secondary air is needed. (Secondary air quantity is usually fixed.)

Manufactured (coal) gas varies in content. It contains about 50% H_2 (hydrogen), 8% CH_4 (methane), and other gases. It has a heating value of approximately 500 Btu/ft^3 to 600 Btu/ft^3.

Liquefied petroleum gas (LP gas) usually is propane with a little butane added. It can be liquefied, stored, and transported in cylinders or tanks. LP gas vaporizes easily and is changed into its gaseous form before it is burned. It has a heating value varying between 2500 Btu/ft^3 and 3200 Btu/ft^3.

Propane boils at −40°F (−40°C) at atmospheric pressure. Butane boils at 32°F (0°C) at atmospheric pressure. It is fed to the burner at about 11″ water column (WC) pressure.

Propane and butane are dangerous if carelessly used. Both are heavier than air and will collect in the firebox

or in the basement. All these gaseous fuels have an odor added. If an odor is detected, the gas is present. Immediately shut off the main fuel valve and thoroughly vent the area.

21.4 Basic Forced Air Components

Furnace construction includes a combustion chamber (except in electric furnaces), a fuel feeding and burning device (burner), a flue or outlet, and a heat exchanger. The combustion chamber must be leakproof and must provide efficient heat transfer. It must be able to change from room temperature to almost 2000°F (1093°C) without expansion or contraction stresses.

Basic components of forced air systems are consistent regardless of the furnace's efficiency. The following are found in a typical forced air furnace:

Indoor Blower Motor: The blower motor is responsible for creating air flow through the heat exchanger and the duct work. See **Figure 21-8.** Two basic types of electric motors (up to 3/4 hp) are commonly used for direct-drive or belt-driven residential systems. In direct-drive applications, the blower and motor turn at identical speeds (rpm). In belt-driven operations, the use of larger or smaller pulleys will increase or decrease motor speed. More recent technology allows the blower motor to vary speed as needed. The speed may be varied to meet the space heating requirements. Solid-state controls constantly monitor the heat output, firing rate, and blower speed.

Combustion Blower Motor: The combustion blower motor **(Figure 21-8)** provides two functions. It exhausts flue gases and pre-purges the heat exchanger. There are two types of combustion blower motors used: induced-draft and forced-draft. ***Induced-draft*** blower motors pull and exhaust the products of combustion. ***Forced-draft*** combustion blowers force air into the combustion chamber.

Heat Exchangers: The primary exchanger **(Figure 21-9)** transfers heat on its surface to the conditioned space. It can be constructed from a variety of materials that provide different heat transfer rates. A heat exchanger should be tested prior to each heating season for possible cracks. Combustion gases may escape through any cracks into the supply air. This condition could be fatal to occupants. Leak testing may be accomplished by the use of sodium spray along with a halide torch or electronic leak detector. High-efficiency furnaces may make use of a secondary heat exchanger **(Figure 21-9).** It allows additional heat transfer to the conditioned space.

Combination Gas Valve: Some residential and light commercial furnaces use the automatic combination gas

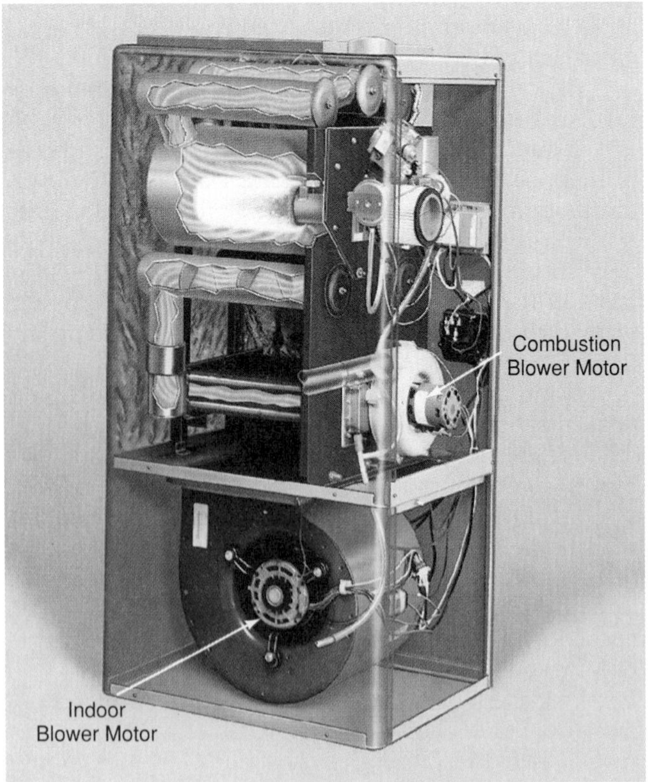

Figure 21-8. *Condensing gas furnace. Note the combustion chamber location. (Thermo Products, Inc.)*

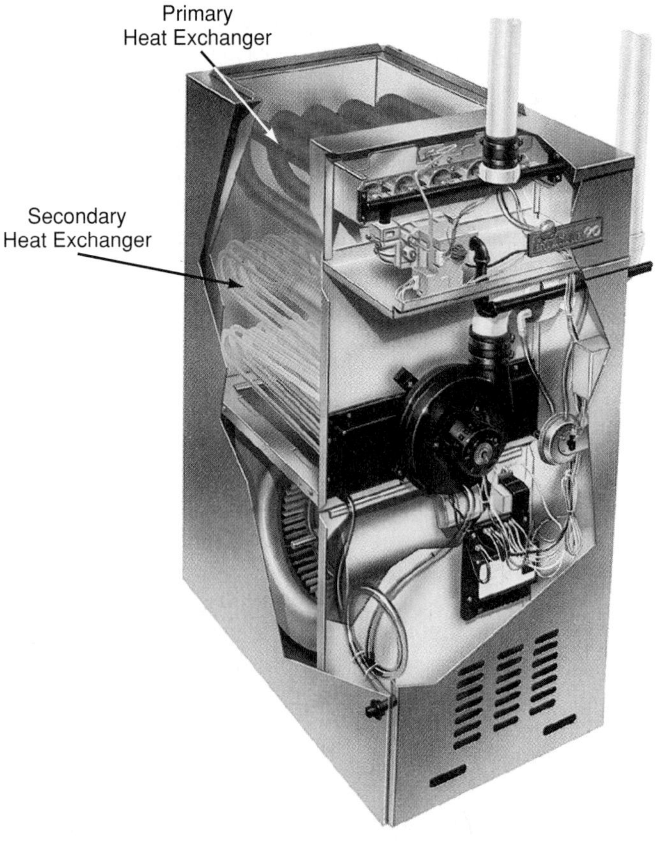

Figure 21-9. *High-efficiency furnace with primary and secondary heat exchangers. (Consolidated Industries Corporation)*

valve. See **Figure 21-10.** However, these units are being replaced by more energy efficient systems. These include the direct or hot-surface ignition systems. Combination gas valves combine the functions of the pressure regulator and gas valve into one unit. They consolidate the following features: controls and gas supply for the pilot, adjustment and safety shutoff for the pilot, and the pressure regulator. These valves use a manual on-off control. The gas flow is controlled electrically.

Burner: Gas combustion occurs at the burners. There are three types of burners: slotted, ported, and ribbon. See **Figure 21-11.** They are usually constructed of steel and may be of varying lengths and sizes. Burners supply the premixed fuel along with primary air to the furnace. Some burners have primary air shutters. These aid in regulation of the air mixture. See **Figure 21-4.**

21.4.1 Warm Air Duct System

The forced warm air system or mechanical warm air system uses a motor-driven blower to increase the flow of heated air to the needed areas. This system usually includes an air filter.

A system of ducts is used with this heating arrangement. The duct system delivers warm air to the various spaces to be heated. The ducts are carefully sized to provide the correct amount of heat to each room. See Chapter 23 for duct sizing. Another system of ducts returns the cool air to the furnace for reheating. A humidifier is usually installed in the airflow to provide the correct relative humidity conditions. See Section 21.41.

All ducts collect dust and dirt during use. They should be vacuum cleaned every five years. Special equipment designed for this purpose should be used.

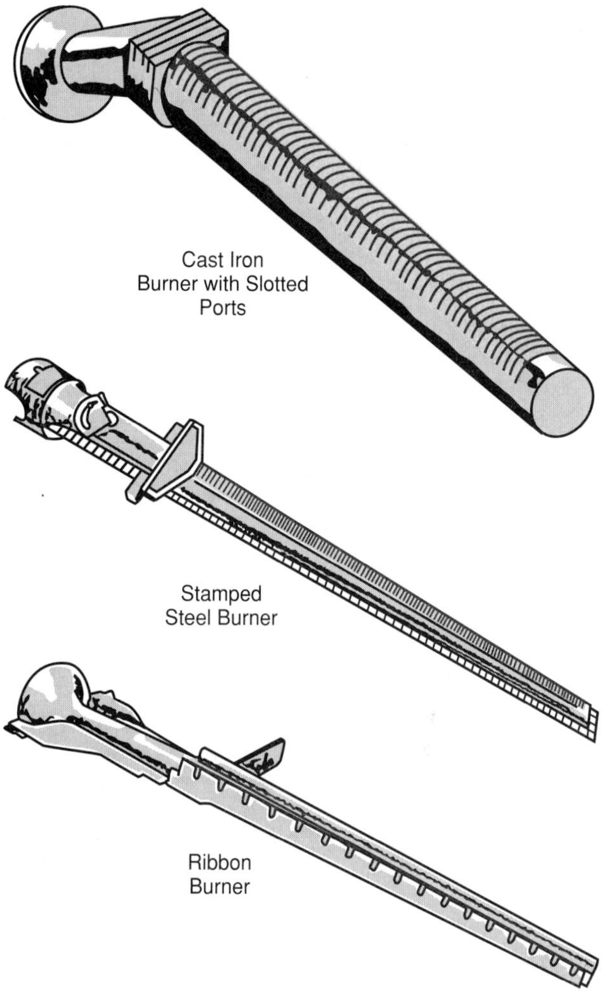

Figure 21-11. *Three types of burners. (Carrier Corporation)*

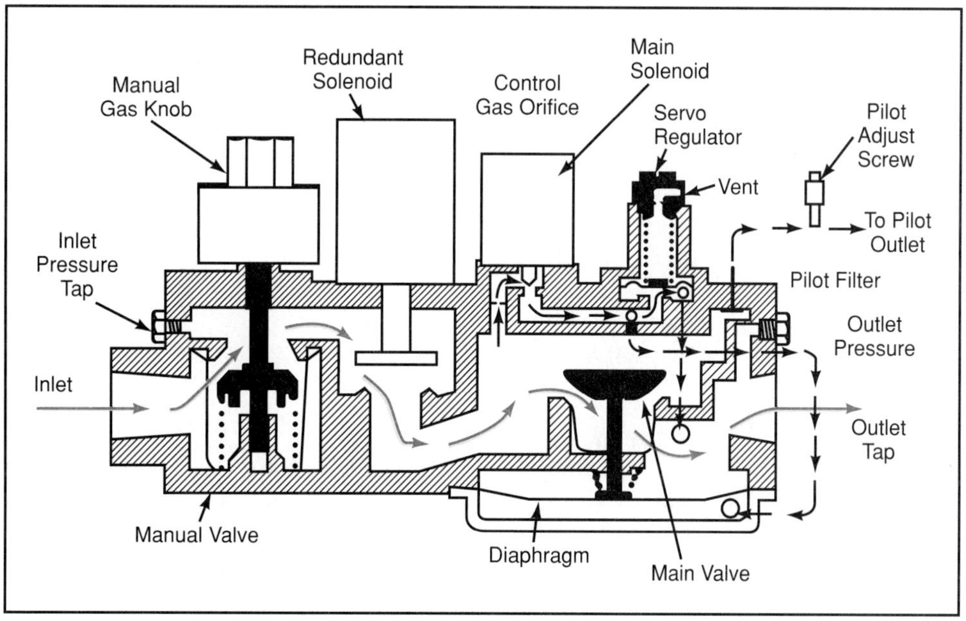

Figure 21-10. *Gas flow through a combined gas valve used in a standing pilot system. (Inter-City Products Corporation)*

21.5 Gas Burners

Gas burners usually have a simple design. Gas is fed through an orifice and mixed with a certain amount of air (primary air). This mixture passes to the burner head. There, combustion takes place and the gas mixes with the secondary air. As much as 35% excess air is fed to the burner to ensure thorough combustion.

There are several types of gas burners:

- Atmospheric injection.
- Luminous flame.
- Power burner.

These burners are used in both furnaces and unit heaters. See Section 21.9.

Furnace-type gas burners may be used with warm air, hot water, or steam systems. The warm air furnace may be one of the following types:

- Airflow up through heat exchanger.
- Airflow down through heat exchanger.
- Airflow across the heat exchanger.

As altitude increases, the size of the burner orifice must be smaller. There is less air and, as a result, less oxygen. Therefore, less fuel must be fed through the orifice at one time. The capacity of the furnace will decrease as the altitude increases. A burner orifice correction must be made when the altitude is over 2000′ (610 m).

The burner system consists of a manual shutoff valve, pressure regulator, automatic shutoff valve, control valve, manifold, burner spuds (nozzles) and adaptors, orifices, primary air inlet, burner head, and pilot valve.

Some gas burners feed the fuel gas and primary air mixture through a manifold to a series of holes, to a series of narrow slots, or to produce a large flame hitting a target (inshot burner). See **Figure 21-12.**

The ***power burner*** uses a blower to force both primary air and secondary air into the burner. See **Figure 21-13.** The burner tube usually has angular vanes to spin or twirl the flame for more efficient burning. A power gas burner is shown in **Figure 21-14.**

Commercial buildings often have combination gas furnaces and cooling systems installed on the roof. Many use fans to force or draw air through the combustion chamber. These fans must be kept clean. No stack is needed. Fans reduce the chance of downdrafts. Filters are used to clean the combustion air, and they require service. The heat exchanger surface should be cleaned each year.

To check gas pressure to the manifold, the main burner must be operating. Install a water manometer by attaching it to the manifold gauge opening located on the outlet of the pressure regulator (1/8″ pipe). See Section 21.12.

It is possible to use LP gas as a standby fuel source in place of natural gas. The LP gas must be mixed with air before it is burned in the heating device. Mixing of the proper amount of air with 1 ft^3 of LP gas (in vapor form)

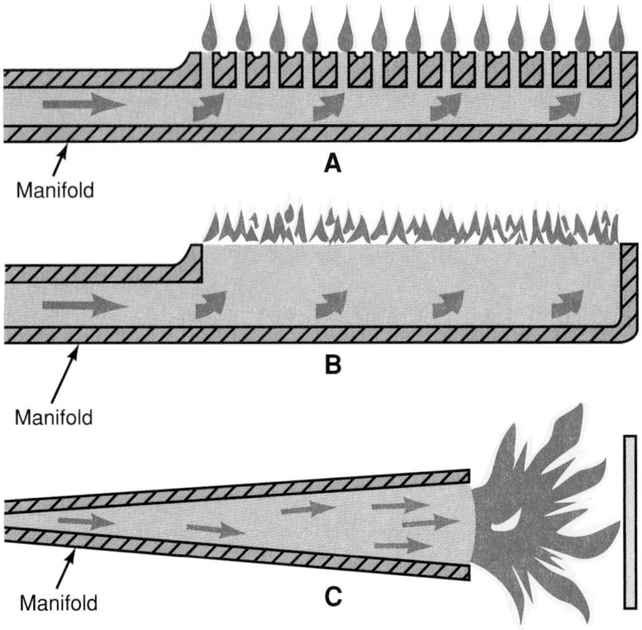

Figure 21-12. *Three gas burners of atmospheric type. A—Burner has holes. B—Burner has slots. C—Burner flame hits target plate.*

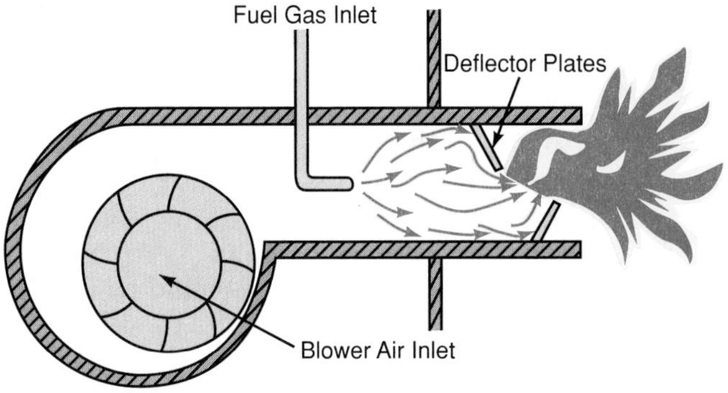

Figure 21-13. *Operation of a power burner.*

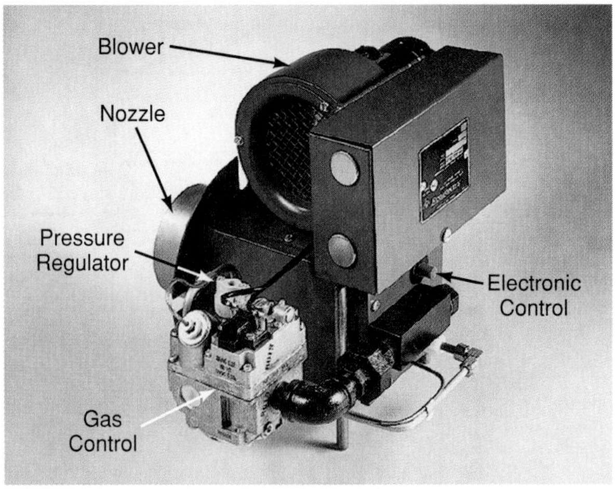

Figure 21-14. *Power burner gas conversion replacement for oil type burner. (Midco International, Inc.)*

allows use of the same primary and secondary air openings. The same burner holes or slots may also be used. The furnace needs no adjustment for the air and LP gas mixture. The final mixture should have about 1000 Btu/ft^3 for use in a natural gas furnace. **Figure 21-15** illustrates the difference between a natural gas flame and LP gas flame.

21.6 Furnace Types and Construction

Furnace design is based upon several factors. These include the fuel used, the space available, and the heat transfer medium. The heat transfer medium may be air, water, or steam. There are three types of furnace construction: upflow, downflow, and horizontal. Application and space demands determine the type used.

The ***upflow furnace*** is the most common type, due to building construction. The upflow takes air and forces it up through the bottom of the furnace. In **Figure 21-16,** note the small blue-colored motor between the blower motor and gas burners. This is referred to as an *induced blower motor.* It increases the system's efficiency. The air is next forced through the heat exchanger and into the ductwork.

The ***downflow furnace*** is also referred to as a ***counterflow*** type. It forces return air from the top of the furnace downward through the heat exchanger. See **Figure 21-17.**

The ***horizontal furnace*** requires air to flow horizontally through the heat exchanger. See **Figure 21-18.**

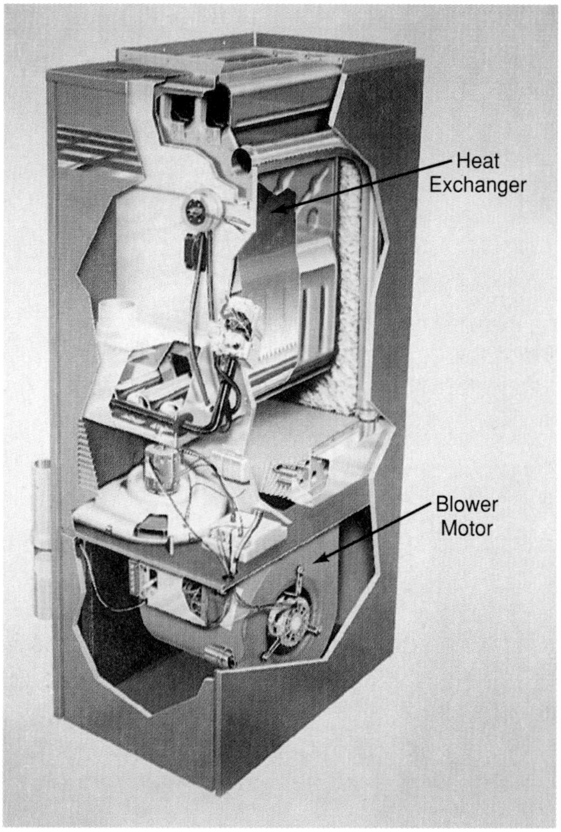

Figure 21-16. *Upflow gas furnace. Return air is induced by the blower motor. Conditioned supply air leaves heat exchanger. (Heil Heating and Cooling Products, Inter-City Products Corporation)*

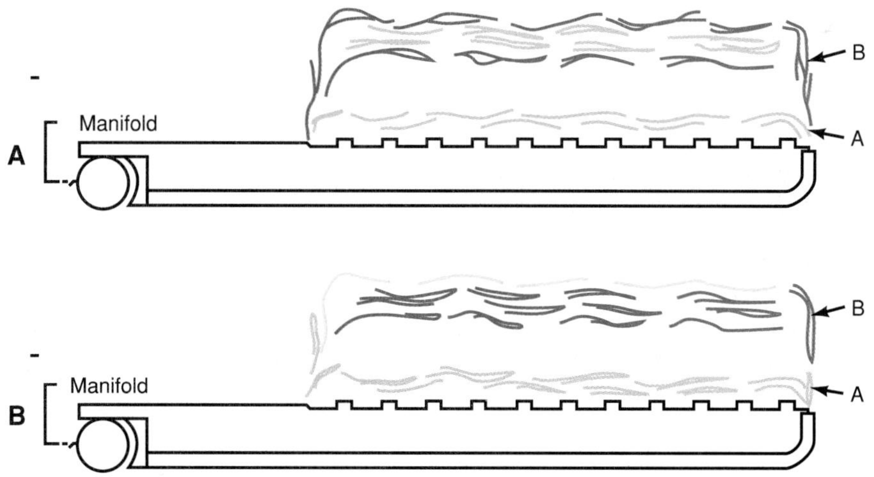

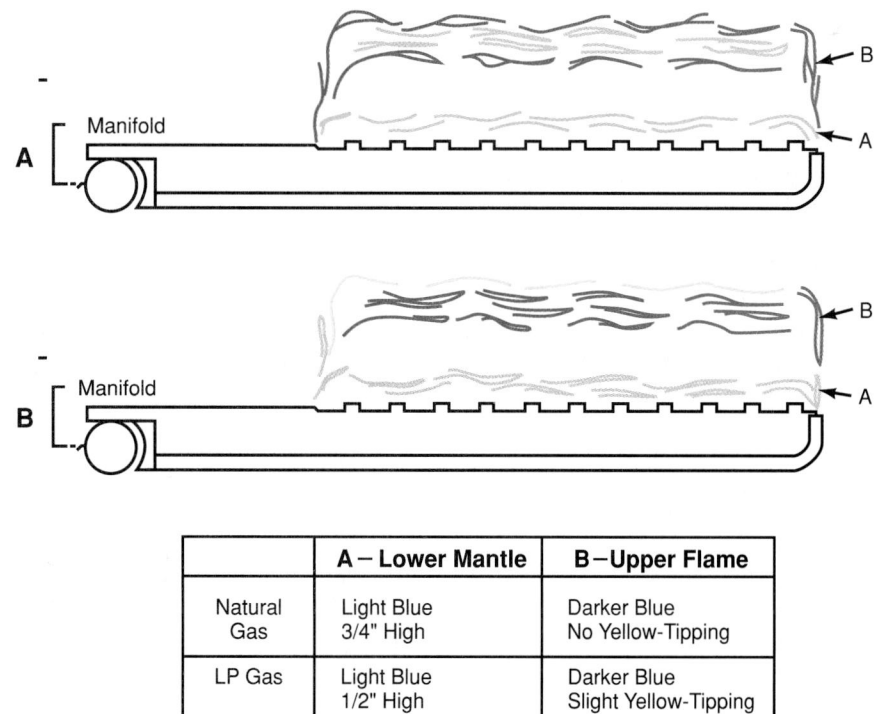

	A – Lower Mantle	B – Upper Flame
Natural Gas	Light Blue 3/4" High	Darker Blue No Yellow-Tipping
LP Gas	Light Blue 1/2" High	Darker Blue Slight Yellow-Tipping

Figure 21-15. *Main burner flames. A—Flame produced when burning natural gas. B—LP gas flame. (Armstrong Air Conditioning, Inc.)*

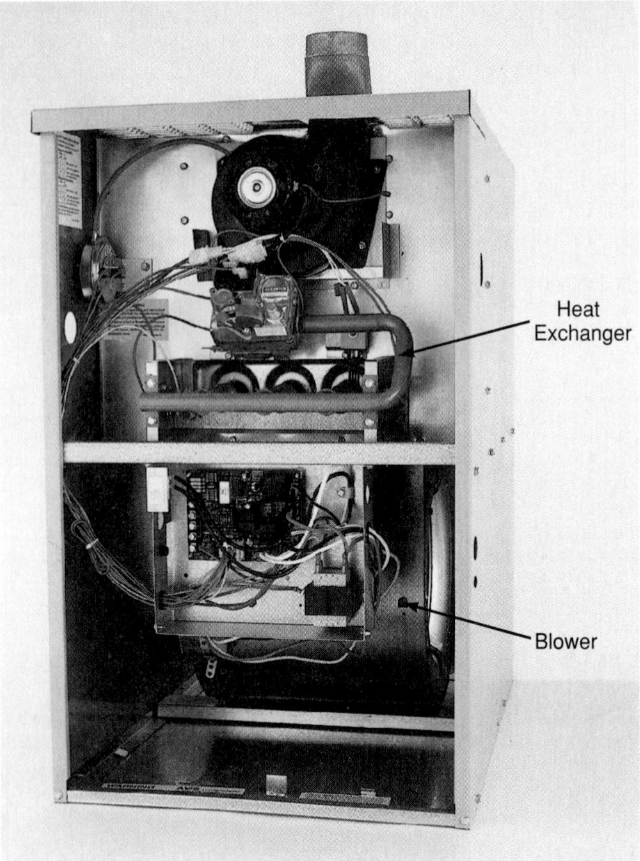

Figure 21-17. *A downflow 80% efficiency gas furnace. Note the position of the blower motor. Return air is forced down from the top of the furnace. (Evcon Industries, Inc.)*

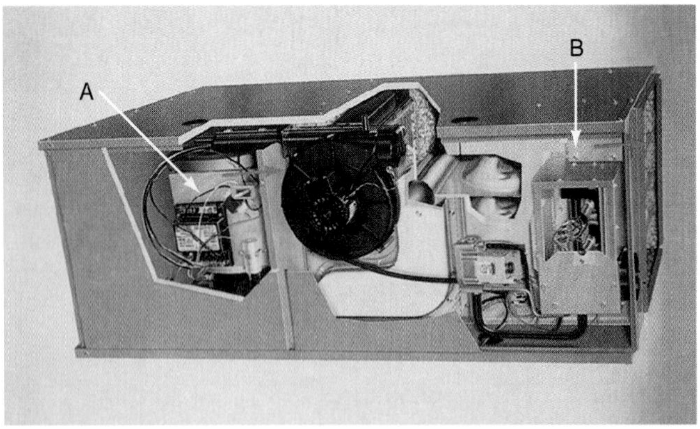

Figure 21-18. *Horizontal furnace. A—Return air is induced by blower motor. B—Conditioned supply air leaving heat exchanger. (Inter-City Products Corporation, USA)*

Return air is pulled through the end opposite the heat exchanger.

All three types require installation of a cold air return. This return supplies the furnace with air to be reheated or cooled.

Most furnaces are of steel construction. The blower compartment is lined with insulation to reduce Btu loss. The insulation also serves as a noise barrier. Furnace manufacturers provide a section where an air filter may be installed. A ***safety interlock switch*** is often provided inside the blower access door (panel). **When the door or panel is opened or removed, this switch de-energizes electrical power to all components. The possibility of personal injury due to moving parts is thereby eliminated.**

21.7 Gas Furnace Efficiency

Many new furnaces being introduced to the market have very high efficiency ratings. High-efficiency furnaces use less fuel and produce more heat. The measurement of efficiency for furnaces is the ***Annual Fuel Utilization Efficiency (AFUE) rating.*** The AFUE compares the yearly or annual energy output to the annual energy input. The higher the AFUE rating, the more efficient and cost-effective the furnace. The federal government's minimum efficiency standard is 78% AFUE. High-efficiency furnaces have ratings above 84% AFUE. **Figure 21-19** shows an approximate yearly operating cost for a variety of furnace efficiencies. This design provides the maximum efficiency available today. Increased efficiency is accomplished through the use of new technologies. High-heat-transfer exchangers may be used. There may be design improvements in flue pipe routing. Secondary heat exchangers may be used. These changes extract as much heat as possible from the system. The temperature at which flue gas is emitted to the atmosphere is thereby reduced. The reduction in flue gas temperature results in condensation of water vapor. This water must be allowed to flow freely to an approved drain. See Section 21.12. High-efficiency furnaces commonly use PVC (polyvinyl chloride) piping for flue exhaust and for air intake. See **Figure 21-20.**

Mid-efficiency furnaces have an AFUE range of approximately 79% to 83%. This type of furnace uses high-heat-transfer heat exchangers with induced draft motors. These features help increase the AFUE rating. Most mid-efficiency furnaces have a flue gas outlet of only three

Furnace Efficiency	Approximate Yearly Operating Cost						
60%*	400	500	600	700	800	900	1000
65%*	365	460	550	640	735	825	915
70%*	340	425	510	595	675	760	845
75%*	315	395	470	550	630	710	785
80%	295	365	440	515	585	660	735
90%	255	320	385	450	515	580	640
95%	240	305	365	425	485	545	605

*Furnaces no longer manufactured after Jan. 1, 1992.

Figure 21-19. *Estimate of yearly operating costs for a number of furnaces of various efficiencies. Units less than 80% efficient are no longer manufactured.*

Figure 21-20. *A high-efficiency upflow furnace. Note that the exhaust and intake connections are PVC. (Heat Controller, Inc.)*

inches. This is due to the heat exchanger design and the decrease in the combustion products. (Products of combustion will include carbon dioxide and water vapor.) These furnaces may use a Class B chimney or an approved flex chimney liner as the main chimney. See Section 21.10. You should check with heating inspectors of local municipalities when installing chimneys. Some manufacturers have designed their mid-efficiency furnaces to be low-profile. They are designed with the option of being installed either vertically or horizontally. See **Figure 21-21.** The installation of all furnaces should conform to the recommendations of the manufacturer. They must also conform to local restrictions.

21.7.1 Mid-Efficiency Furnaces

Mid-efficiency furnace components **(Figure 21-22)** vary from those of the high-efficiency furnace in two ways. There is no secondary heat exchanger, and there must be a chimney for exhaust gases. Since hot flue gases are sent directly through the chimney, much heat energy is lost.

The mid-efficiency furnace heating section consists of burners, a steel heat exchanger, and a venting system. See **Figure 21-17.** When the thermostat calls for heat,

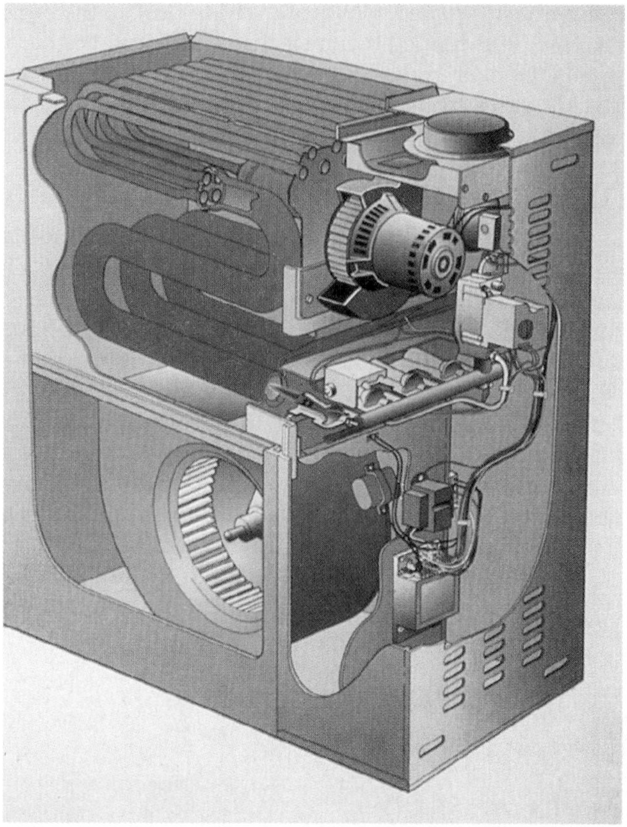

Figure 21-21. *This high-efficiency furnace may be installed in an upflow, upflow/horizontal, or downflow/horizontal configuration. (Consolidated Industries Corporation)*

Figure 21-22. *An 80% mid-efficiency upflow furnace. (Evcon Industries, Inc.)*

a gas valve is opened. Gas is allowed to flow to the burners. The gas is ignited by an electric spark, and the flame warms the heat exchanger, which in turn warms the air. The hot gases from combustion collect at the top of the heat exchanger. They are then routed to a vent.

21.7.2 High-Efficiency Furnaces

The ***high-efficiency furnace*** is the most popular among the new furnaces on the market. These furnaces have Annual Fuel Utilization Efficiency (AFUE) ratings from 84% to beyond 95%. The increasing price of fuel makes the high-efficiency models attractive to the homeowner.

Central to high-efficiency furnace design are the flue piping, heat exchanger, and secondary heat exchanger. A traditional chimney is not needed for combustion. Outside air enters through a polyvinyl chloride (PVC) pipe run through the wall or roof.

The high-efficiency furnace, **Figure 21-23,** uses a gas flame to heat the primary heat exchanger. The combustion blower motor draws the hot combustion gases through the primary heat exchanger. Hot gases produced by combustion are not vented out the chimney. Instead, the hot gases are drawn down by the combustion blower motor into a secondary heat exchanger. There, the latent heat is exchanged, lowering flue stack temperatures and thereby condensing water vapor. For this reason, high-efficiency furnaces are often referred to as "condensing furnaces."

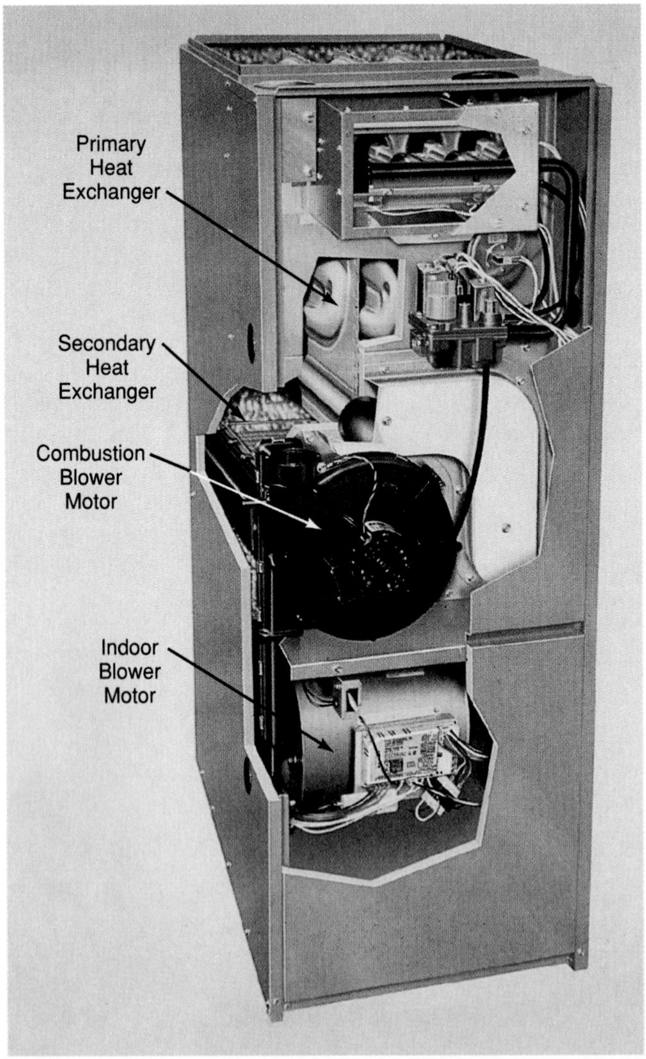

Figure 21-23. *Components of a high-efficiency upflow furnace. (Inter-City Products Corporation, USA)*

The indoor blower forces circulating room air across the primary and secondary heat exchanger. The air picks up the latent heat of vaporization in the flue gases. This latent heat would have been lost in a conventional gas furnace. Hot air created by the secondary heat exchanger is circulated through the furnace duct system. The exhaust gases are low in temperature. They are vented outdoors through a PVC pipe attached to the exhaust vent. Primary air for combustion is introduced from the outdoors through PVC piping.

21.7.3 Combined Heating and Hot Water System

A high-efficiency home heating system has been designed that provides both space and water heating. An example of this system is the "Complete-Heat" system. Combined heating system units are rated according to a ***Combined Annual Efficiency (CAE) ratio.*** The Complete-Heat system has a CAE of 90%.

The system produces up to three to four times more hot water than conventional water heaters. The Complete-Heat unit can be vented vertically or horizontally with a 2″ PVC pipe. Therefore, a separate water heater vent system and a vertical chimney are not necessary. Only outdoor air is used for combustion.

The system contains two modules. The heat module acts as storage. An air-handling module provides comfort conditioning for the home. See **Figure 21-24.** Upon a call for heat, a pump in the air-handling module draws hot water from the heat module and circulates it. Heat from the water is transferred through use of the heat exchanger and the blower. The water then returns to the heat module, where it is reheated. Hot water necessary for domestic use is drawn directly from the heat module.

Figure 21-25 is a wiring diagram of the "Complete-Heat" system. The diagram includes a legend with component identification.

21.8 Pulse Combustion Furnace

The ***pulse combustion process*** was first discovered in the 1900s. However, it has only recently been put to use in the heating industry. The concept is used in forced air condensing furnaces, **Figure 21-26.** The concept is different from that used in the conventional atmospheric burner furnace. The pulse furnace does not have an open flame, pilot burner, main burner, or conventional flue or chimney.

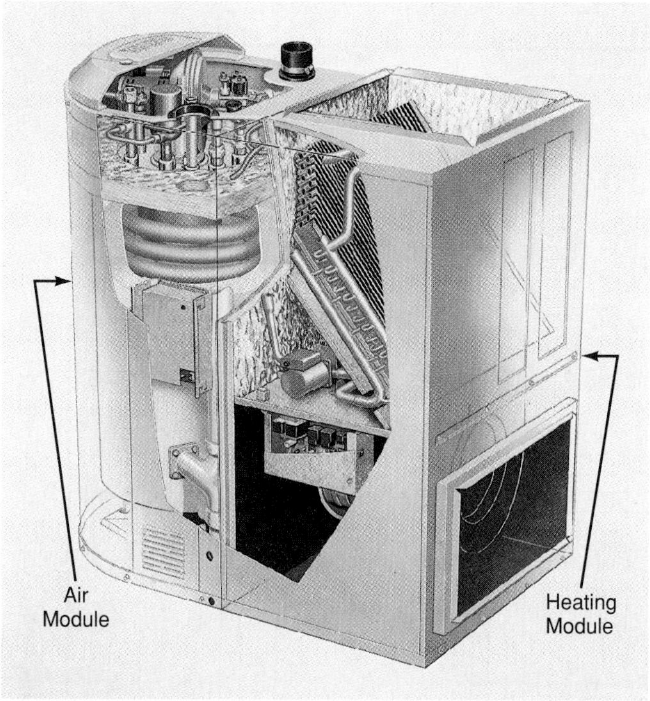

Figure 21-24. *The "Complete-Heat" system. Note the air module at left and heating module at right. (Lennox International, Inc.)*

In the pulse combustion process, the combustion air is drawn 100% from the outdoors. It is brought into the unit through a 2″ PVC pipe. The combustion is started in an enclosed chamber from a direct spark ignition device. See **Figure 21-27.** A sensor checks if ignition has begun. Five ignition trials are allowed before the sensor closes the gas valve and control circuit. The sensor also checks for loss of combustion and will shut down the system. Many other safety sensors are part of the intake and exhaust outlets.

After ignition, the heat from combustion passes through the combustion chamber, tail pipe, exhaust decoupler, and heat exchanger coil. All of the above are located in the system's airstream. In the process, exhaust temperatures drop from 1200°F to 350°F (650°C to 180°C). As exhaust gases are forced through the fin-and-tube heat exchanger, water vapor is condensed out. The latent heat of combustion is recovered. Exhaust gases and condensate at temperatures from 100°F to 120°F (40°C to 50°C) are vented into a plastic "T" connection. The condensate exits from one side of the "T" into a 1/2″ plastic condensate drainpipe. The exhaust exits from the top of the "T" into a 2″ PVC pipe that is vented outside. A conventional chimney is not required.

Figure 21-28 illustrates the basic pulse combustion process:

1. The gas and air supply enter the combustion chamber and mix.
2. To start the cycle, the spark ignitor is turned on. It ignites the gas and air mixture. This creates the initial combustion and is referred to as one "pulse."
3. The positive pressure from the resulting combustion closes the flapper valves. Exhaust gases are forced through a tail pipe. These combustion products are vented outdoors through a 2″ PVC pipe installed vertically or horizontally.
4. The venting of the exhaust gases creates a negative pressure in the chamber. This opens the flapper valves, drawing in more gas and air for the next ignition.
5. At the same time, some of the pressure pulse is deflected back from the tail pipe. This causes the new gas and air mixture to ignite. This is referred to as another "pulse." After the first few seconds, the spark ignitor and air blower are turned off, because the combustion process is self-sustaining. No spark is needed.

Steps 4 and 5 are repeated 60 to 70 times per second. This forms consecutive "pulses" of 1/4 to 1/2 Btu each.

Before and after each heating cycle, a small blower purges (flushes out with air) the combustion chamber. This provides fresh air for the next mixture of air and fuel gas.

21.9 Unit Heaters

Many stores, commercial buildings, and factories use unit heaters to heat certain rooms or spaces. These heaters can be gas-fired, oil-fired, or use hot water coils or steam coils. They are suspended from the ceiling. The heaters use a motor-fan to force the heated air in a controlled direction. Many units have adjustable louvers to help direct airflow. The different sizes handle from 300 cfm to about 6000 cfm. Their capacities range from 20,000 Btu/hr. to 360,000 Btu/hr. Gas-fired and oil-fired unit heaters require a flue to carry the products of combustion outdoors.

Special unit heaters are available for vertical downward warm airflow. They are also used for high velocity airflow (about 2500 fpm) against large door openings. This helps to keep out cold outside air. Usually, they are mounted about three feet above and four feet away from the opening. They are often used on car and truck doors or shipping and receiving doors. The unit is operated by either a door switch or a thermostat connected in parallel. Doors less than 8′ high and 10′ wide usually are not protected by these devices. Some users operate the fan in summer to help keep out dust and insects.

21.10 Venting of Furnaces and Chimney or Exhaust Gases

There are many factors to consider when venting a furnace. The following areas are critical to effective venting: furnace capacity, type of piping, air flow, length of run, heat load, and number of elbows. The manufacturer's recommendations should be checked when

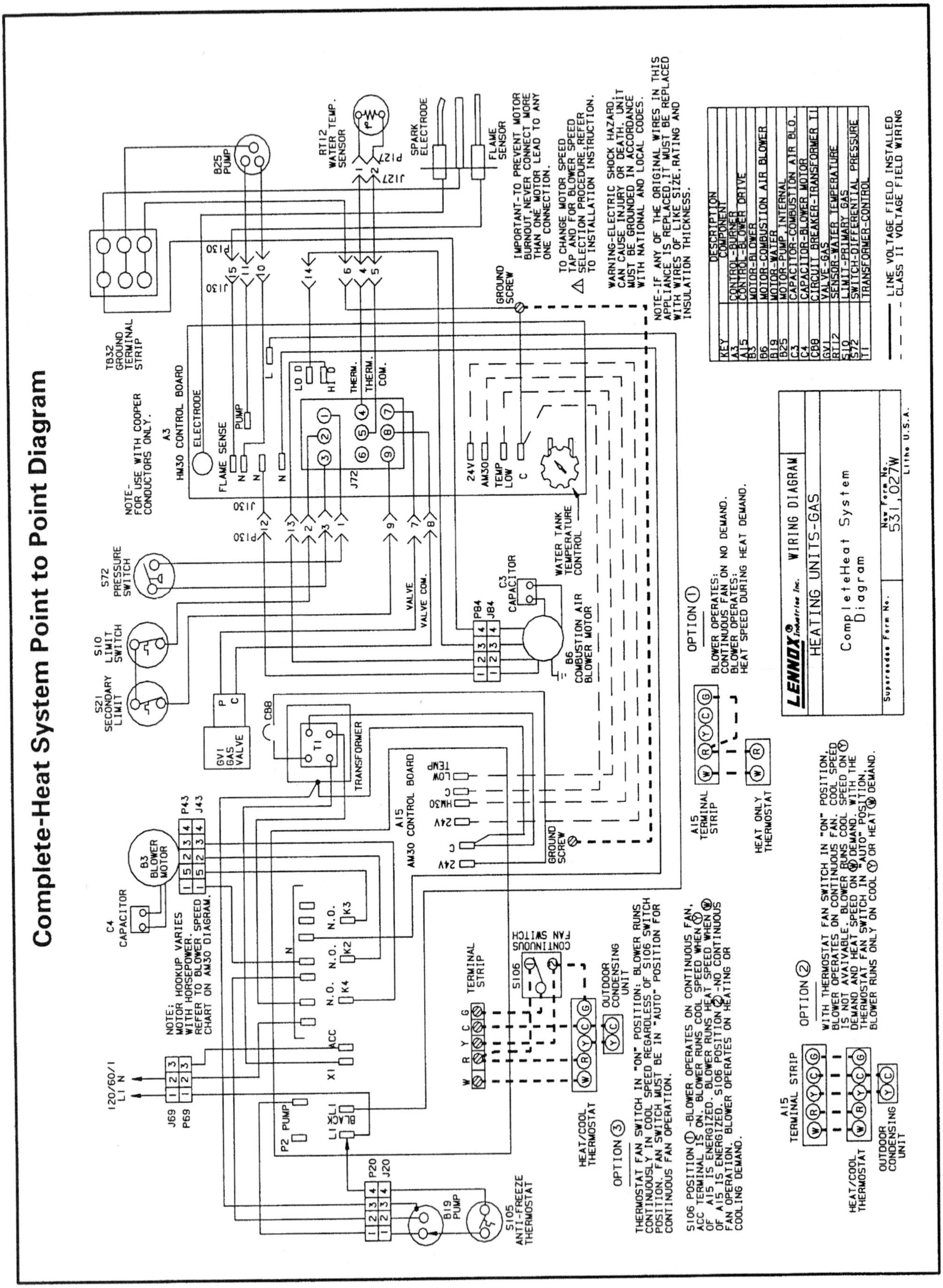

Figure 21-25. *Wiring diagram for the heating system shown in* **Figure 21-24.** *Key for components is at lower right. (Lennox International, Inc.)*

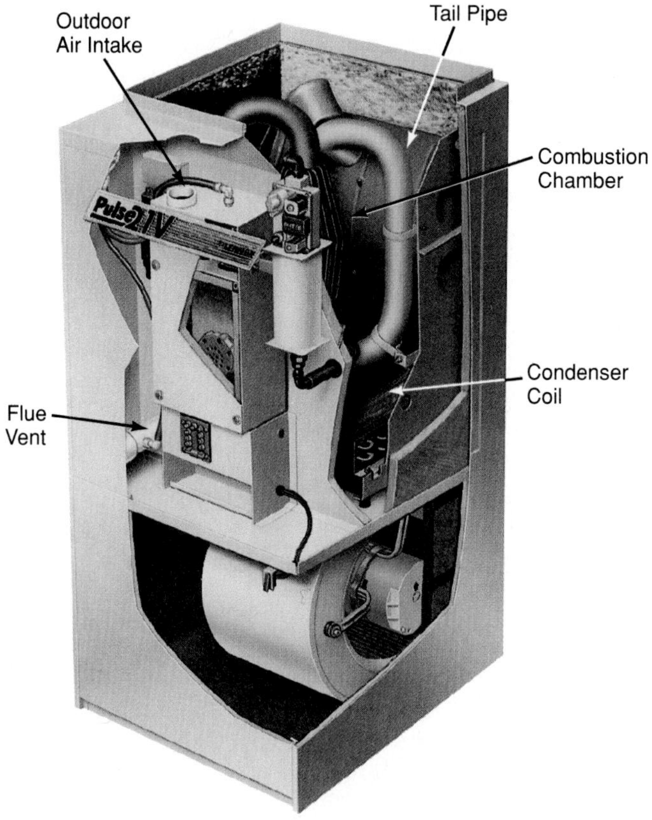

Figure 21-26. *Pulse gas furnace upflow model. (Lennox International, Inc.)*

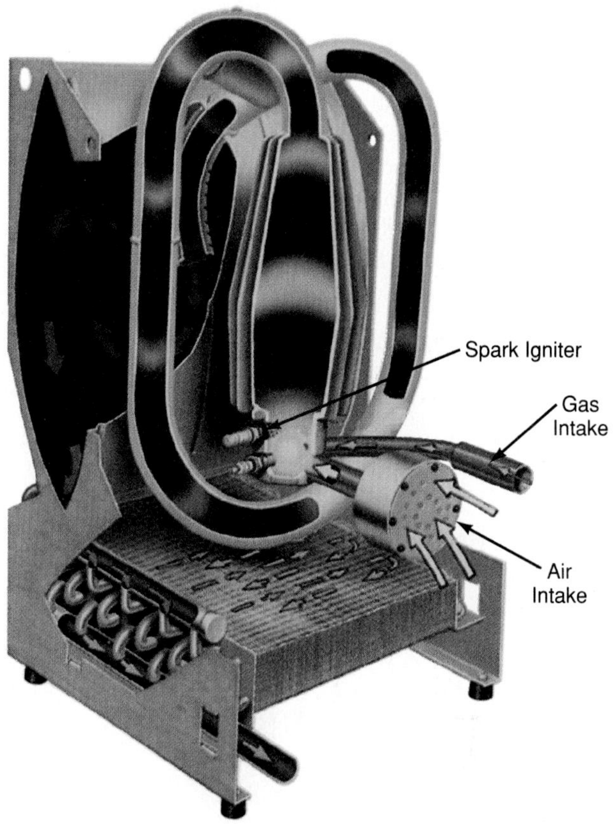

Figure 21-27. *Pulse combustion process heat exchanger assembly. (Lennox International, Inc.)*

determining venting practices. Consult the local municipality to confirm code requirements.

The flow of combustion gases out of a flue and chimney affects heating system efficiency. Fuel losses of 4% to 15% are possible. Combustion gases leaving the chimney can also contribute to air pollution.

Combustion gas flow affects the amount of air entering the furnace for combustion purposes. This flow is affected by the pressure difference between the combustion air entering the furnace and combustion air leaving the flue or chimney. Both the pressure in the building and the atmospheric pressure affect the flow of combustion gases. The temperature of the combustion gases also has an effect. (If it is too cold, the flow will be slow. If it is too hot, the flow will be fast.)

To remove the exhaust gases (the products of combustion) from a conventional furnace, the furnace must be vented. This vent system consists of a flue pipe that joins a chimney vented to the outside. The flue pipe reaches a temperature of between 500°F to 600°F (260°C to 320°C). Therefore, 30% to 40% efficiency of the furnace is exhausted out the chimney.

It is important to distinguish types of chimneys that are still commonly used. They are Class "A" (masonry type); Class "B" (double-wall metal type); PVC (polyvinyl chloride type); and an approved "metal flex liner" that is inserted into the entire length of Class "A" masonry chimneys. See **Figure 21-29.**

When installing a liner system **(Figure 21-29D)**, a clear and unobstructed chimney is necessary. Care must be taken to clear the chimney, removing any tar or creosote. There should be no obstructing mortar. Check the chimney for any cracked, loose, or missing bricks. Make any chimney repairs necessary to provide safe internal

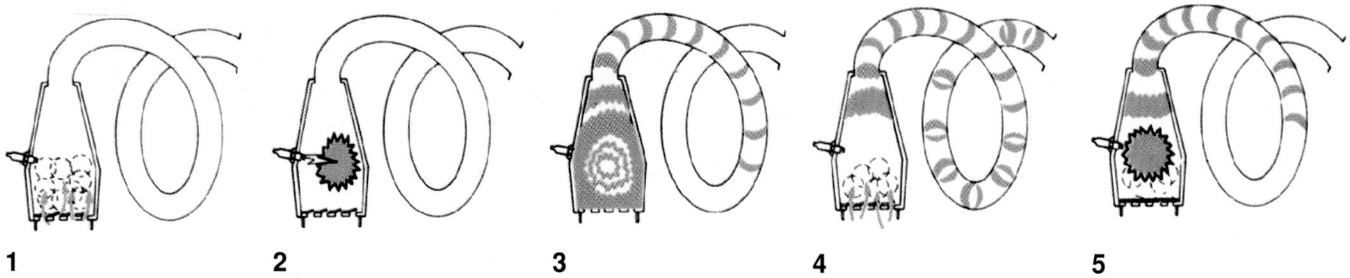

Figure 21-28. *Pulse combustion cycle. (Lennox International, Inc.)*

and external conditions. The existing chimney must provide at least 1/2″ (1 mm) space between liner and masonry inner wall. The diameter of the liner must be proper to ensure good exhaust. Installation instructions will provide specifics regarding sizing and additional installation information. Always check for proper venting following installation and start-up.

In the new high-efficiency furnaces, the conventional chimney is not needed. Flue gases leave the furnace at 115°F to 118°F (46°C to 48°C). This is a result of the high efficiency heat exchanging processes. Exhaust gases are vented to the outside through a plastic (PVC) pipe. Outdoor air is brought into the combustion chamber through a PVC pipe.

When installing PVC piping for high-efficiency condensing furnaces, the furnace may be vertically or horizontally vented. See **Figure 21-29A** and **B.** When venting a horizontal furnace, the manufacturer's procedures

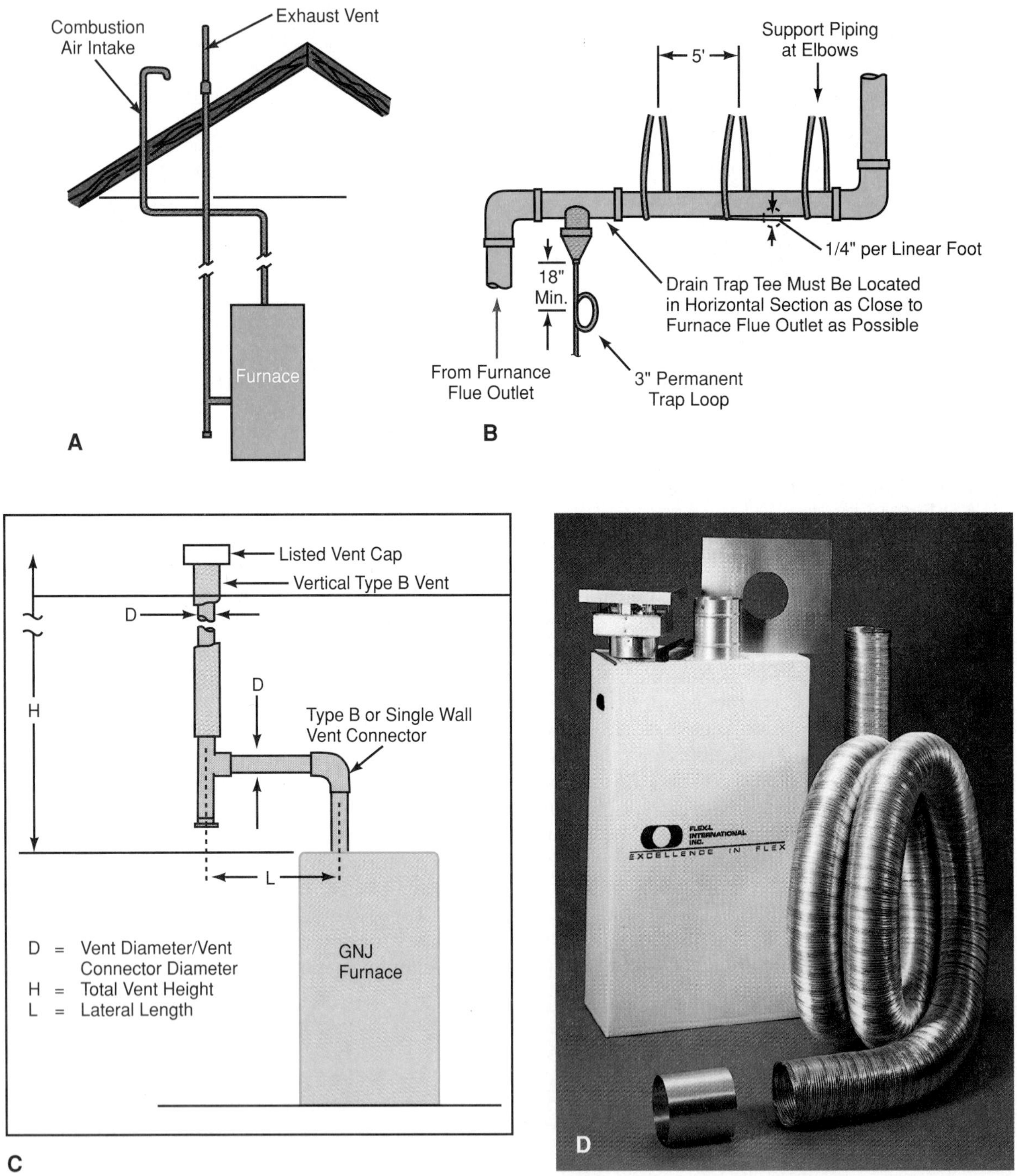

Figure 21-29. *Venting for high technology furnace. A—Combustion air intake and exhaust vent installed vertically. B—Horizontal venting, displaying proper strap and drain tee location. C—Class "B" double-wall metal liner. (Comfortmaker GNJ, Inter-City Products Corporation) D—Flexible liner that can serve a masonry chimney. (Flex-L International, Inc.)*

must be followed. The venting pipe should use the shortest route possible, with the fewest elbows. When they protrude through an exterior side wall, dimensions between the intake and exhaust vents must be as accurate as possible. They should be made according to the manufacturer's specifications. Height termination above grade level is also of great concern.

Outside air is used for intake combustion air. Due to lowered flue gas temperature, a drain line is needed to dispose of condensate. The furnaces must be individually vented and not combined with any other appliance. All PVC connections must be leak-tight. See **Figure 21-30.**

The following optional devices are found on some old chimney installations using Class "A" and Class "B" venting:

- A bimetal damper, shown in **Figure 21-31.** This is installed in the flue outlet opening or as a section of flue piping. It is thermally operated and requires no motors or switches for operation.
- A vent damper, shown in **Figure 21-32.** This device is installed like the bimetal damper but is different in operation. It is electrically operated and is wired to open on call for heat. It closes when the thermostat is satisfied. With this device, there is a faster response to opening and closing the damper.

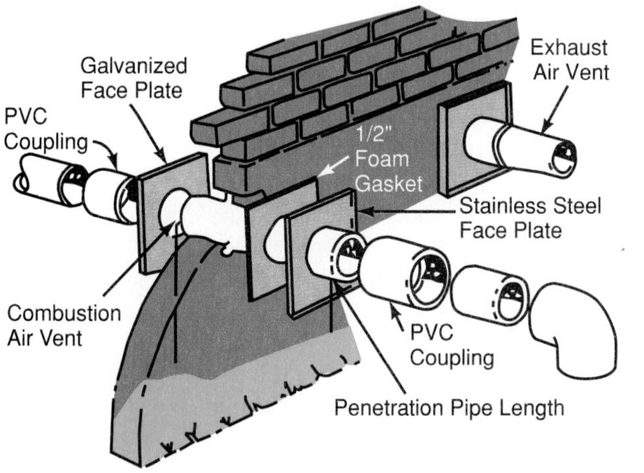

Figure 21-30. *Side wall vent. (Comfortmaker, Inter-City Products Corporation)*

Figure 21-31. *Bimetal quadrant draft control. (American Metal Products Co., a Masco Company)*

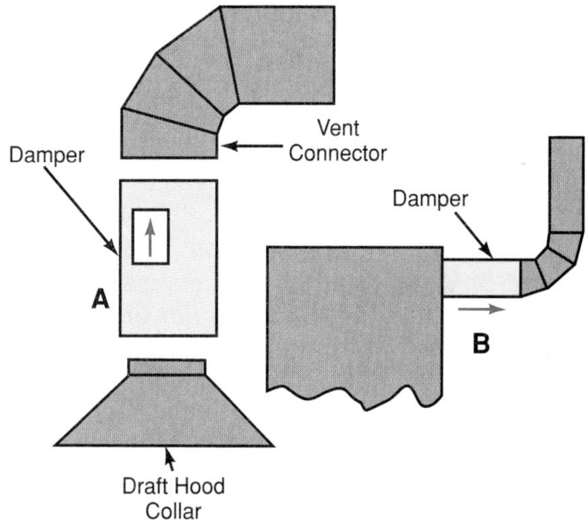

Figure 21-32. *Vent damper installation. A—Vertically between draft hood collar and vent collar. B—Horizontally between furnace and vent connection. (American Metal Products Co., a Masco Company)*

21.10.1 Make-Up Air Units

Current codes and regulations in many states require the use of make-up air control units. See **Figure 21-33.** ***Make-up air units*** help to solve negative pressure problems created by tightly constructed homes. They build up a slight pressure in the home. This prevents infiltration around windows and doors and assists in eliminating drafts. Furnace efficiency is improved through providing proper air for combustion.

Make-up air units deliver controlled, fresh air to the furnace. This air is then cleaned by filters and heated or cooled as necessary. The air is then circulated throughout the duct system.

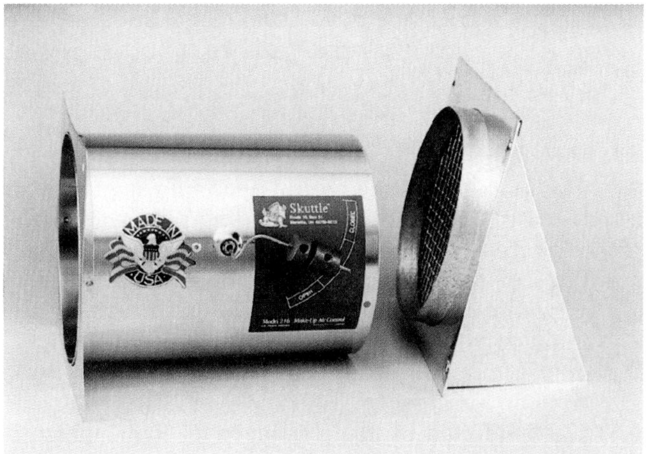

Figure 21-33. *A make-up air control provides controlled fresh air automatically. These systems assist in solving negative pressure problems. (Skuttle Mfg. Co.)*

21.11 Ignition Systems

The three basic types of ignition systems in use today are:

- Pilot-light.
- Direct-spark.
- Hot-surface.

The most common ignition system used on furnaces for many years was the ***standing-pilot*** or "pilot light." In this system, a pilot assembly and thermocouple was needed. The pilot light remained on constantly, even when the thermostat was not calling for heat. See **Figure 21-34.** There is now an awareness that fuel resources are being diminished. A higher technology design in furnaces is also available. These factors have led to the replacement of the standing-pilot ignition system. Today, gas furnaces commonly use either a "direct-spark" type or "hot-surface" type ignition system.

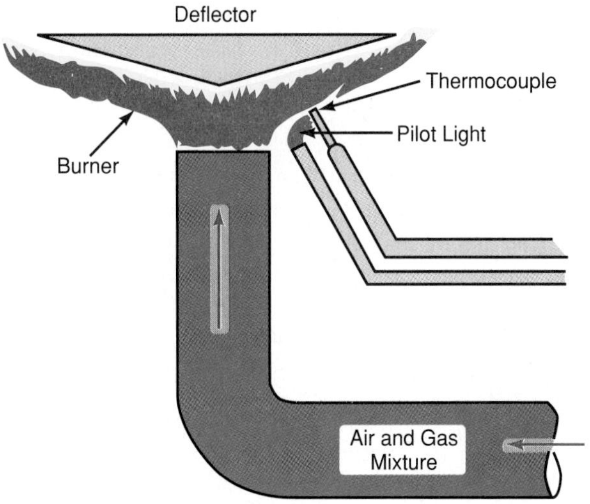

Figure 21-34. *Schematic drawing of safety thermocouple and pilot light for a gas furnace. Thermocouple generates a small electric current which actuates a control. Control will shut off gas supply to the furnace if the pilot light is extinguished.*

21.11.1 Standing-Pilot System

The following is the typical operating sequence for a gas furnace equipped with a standing-pilot system. Refer to **Figure 21-35** as the sequence is followed.

1. On call for heat, thermostat receives 24 V from transformer. The thermostat closes, energizing the main valve and combination gas valve.
2. At that time, the main valve opens and allows gas to main burners.
3. Gas is ignited by the existing pilot light, which is maintained by a thermocouple.
4. Heat is produced within the heat exchanger. As heat transfer occurs, and at a predetermined fan setting, the fan switch contacts close.

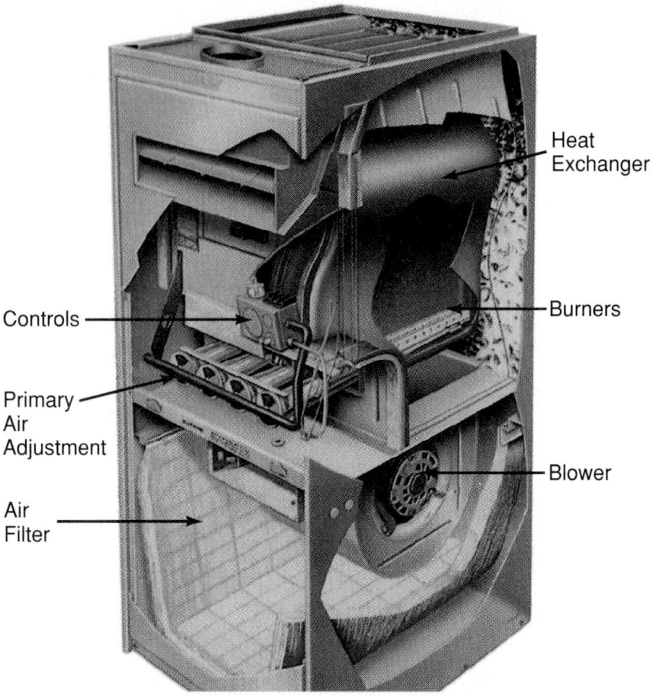

Figure 21-35. *Four-burner gas-fired warm air furnace. (Lennox International, Inc.)*

5. This starts the indoor blower motor at the heating speed. If unsafe levels of heat exist, the limit control will open its contacts. The gas valve will be de-energized.
6. When thermostat is satisfied, its contacts open. The main gas valve is de-energized, and gas flow to the burners is shut off.
7. Pilot light still remains powered by the thermocouple.
8. As heat decreases within the heat exchanger, the fan switch opens its contacts. The blower motor is shut off.

21.11.2 Direct-Spark System

The ***direct-spark ignition system*** is most widely used today. In this system, a spark is created to ignite the gas-air mixture. These systems are produced by several manufacturers, and their operations have some similarities. Direct-spark ignition system components include a flame sensor, gas valve, electrode assembly, and relight control. See **Figure 21-36.** The pilot light for a direct-spark is created by a relight control. It generates a spark to ignite pilot gas. There is no remaining pilot once the thermostat has been satisfied.

21.11.3 Hot-Surface System

Many furnace manufacturers use a "hot-surface" system. The ***hot-surface ignition system*** includes a combination gas valve, a sensor ignitor, and control module. See **Figure 21-37.** The heat produced by the ignitor lights the main gas flow through the burner.

Hot-surface ignition makes use of a silicon carbide ignitor that is heated to a red-hot condition. This silicon carbide element then lights the main burner. As in direct spark ignition, there is no remaining pilot once the thermostat has been satisfied. See Section 21.14.1.

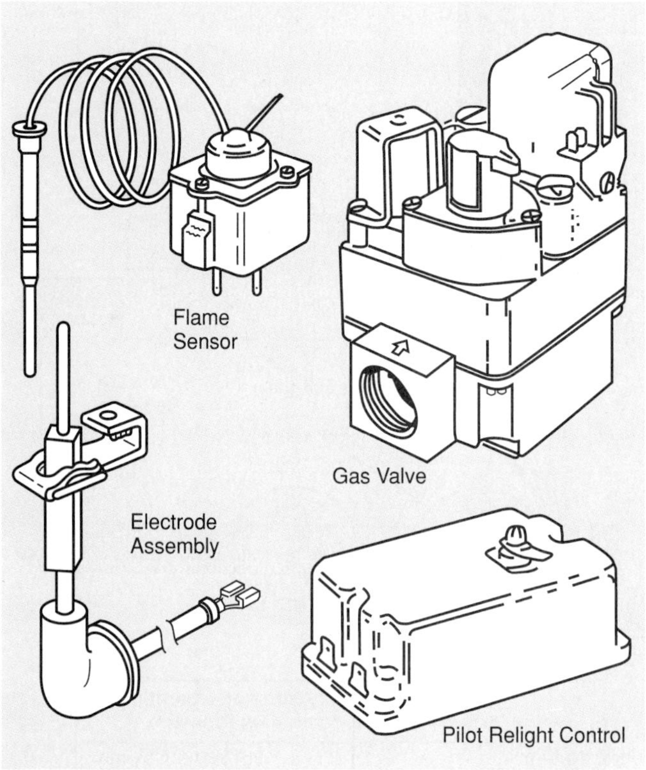

Figure 21-36. *Components of a direct-spark ignition system. (White-Rodgers Division, Emerson Electric Co.)*

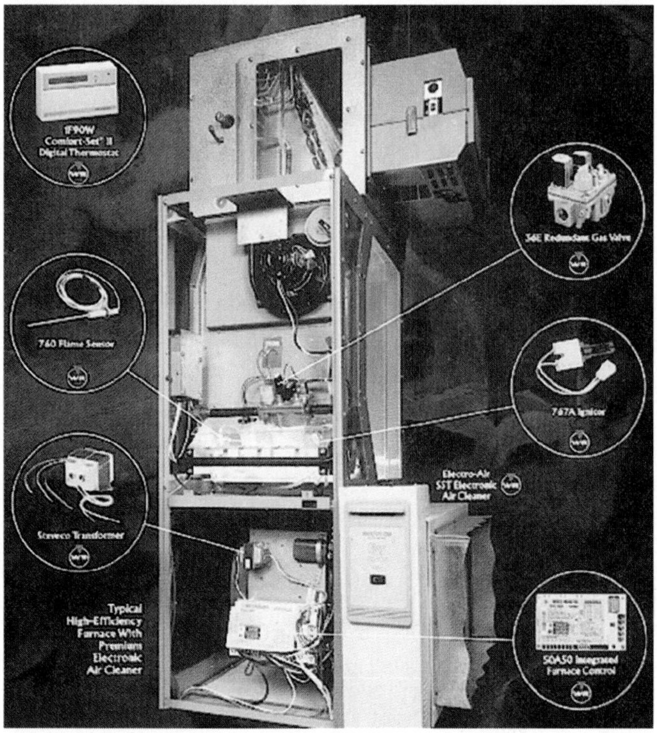

Figure 21-37. *Components of a hot-surface ignition system. (White-Rodgers Division, Emerson Electric Co.)*

21.12 Piping and Gas Pressure

Most cities have code requirements covering the installation of heating equipment. It is important to know this code and carefully follow it. **Gas is dangerous.** See Section 21.3.

Black iron pipe is the type used for natural gas applications. This pipe is treated so that it will not flake or allow contaminants to flow with the gas. The installation contractor must correctly abide by all local and state laws regarding piping size. The piping must meet the demands of the applications within the building. When large gas supplies are needed, contacting the gas utility company may be necessary. The utility company may have to install a specific meter and/or regulator. This may be necessary in order to meet full load demands of the gas supply.

Gas piping for most residential homes is sized for 1″ nominal pipe. It is first reduced to 3/4″, then to 1/2″ for connection to the appliance. The drawing in **Figure 21-38** illustrates the use of pipe compound on the threads to assure a leakproof connection. **Figure 21-39** illustrates a drip leg, which is a code requirement. The drip leg functions as a trap. It collects possible contaminants that may flow with the gas.

There are many methods and aids to help in sizing gas pipe. **Figure 21-40** is a chart that may be used to size residential gas piping.

The utility gas company will install their meter on the template bracket once the entire gas piping has been completed and connected to the appliance. The utility company will usually test the gas piping for a piping system leak. If one is found, the installing contractor is informed of its location. **When you check for a gas leak, always use soap bubbles. Never use an open flame.**

Once the gas piping system has been completed and is found to be leakproof, the gas pressure is checked through the use of a manometer. It is checked in the main line and at the appliance for proper operating pressure. See **Figure 21-41A.** Common operating gas pressure for

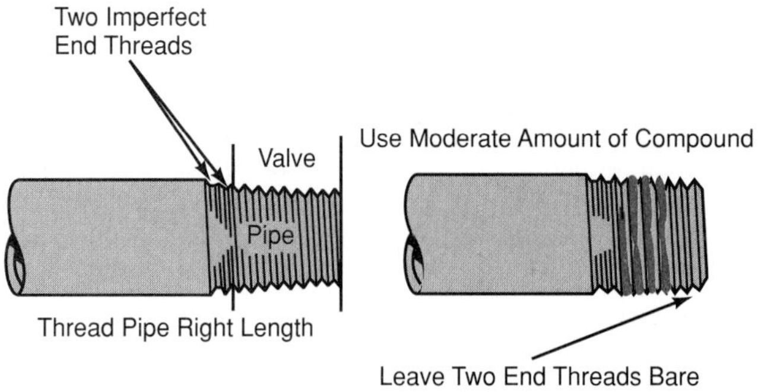

Figure 21-38. *The proper way to put pipe compound on threads. (Honeywell, Inc.)*

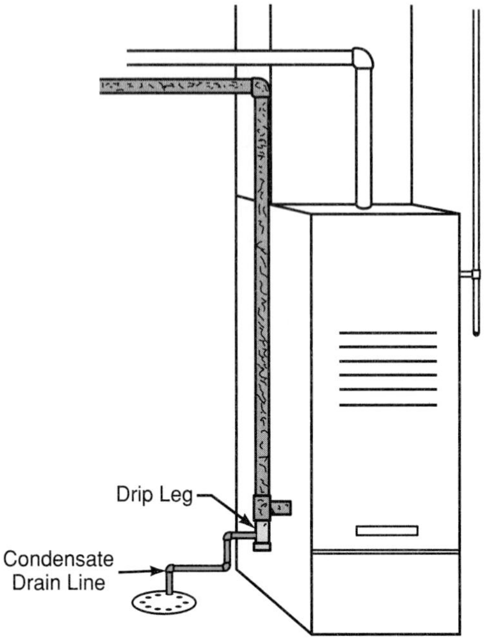

Figure 21-39. *Drip leg installed in gas pipe to furnace. Drip leg will trap dirt and moisture. Also note the condensate drain line. (Bacharach, Inc.)*

natural gas furnaces is 3.5″ water column (WC). For LP gas equipment, it is 11″ WC. See **Figure 21-41B.** Always check the manufacturer's information for the correct operating procedures.

21.13 Start-Up Check Sheet

When installing a new heating system, it is important that certain information be recorded. A sample copy of a form developed for this purpose is shown in **Figure 21-42.** Upon completion of a job, a copy should be given to the homeowner. One copy should be retained by you. This information will be valuable during future service calls. It can serve as baseline data during troubleshooting.

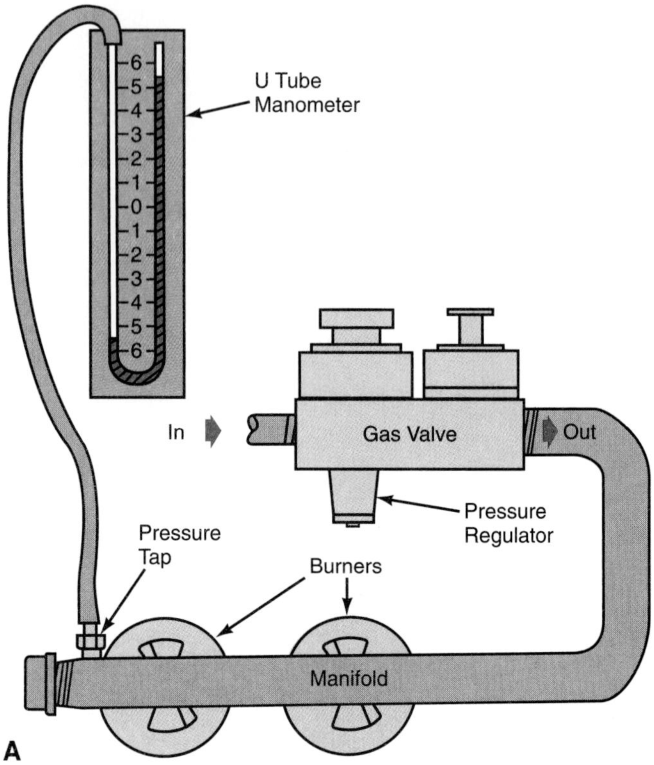

A

B

Gas	Common Operating Gas Pressure
Natural	3.5" Water Column
LP	11.0" Water Column
Mixed	3.5" Water Column
Manufactured	2.5" Water Column

Figure 21-41. *Gas operating pressures. A—Correct use of a manometer for checking operating pressure. B—Operating pressures for various types of fuel gases.*

Nominal Iron Pipe Size, in.	Internal Diameter, in.	Length of Pipe, ft.													
		10	20	30	40	50	60	70	80	90	100	125	150	175	200
1/4	0.364	32	22	18	15	14	12	11	11	10	9	8	8	7	6
3/8	0.493	72	49	40	34	30	27	25	23	22	21	18	17	15	14
1/2	0.622	132	92	73	63	56	50	46	43	40	38	34	31	28	26
3/4	0.824	278	190	152	130	115	105	96	90	84	79	72	64	59	55
1	1.049	520	350	285	245	215	195	180	170	160	150	130	120	110	100
1 1/4	1.380	1050	730	590	500	440	400	370	350	320	305	275	250	225	210
1 1/2	1.610	1600	1100	890	760	670	610	560	530	490	460	410	380	350	320
2	2.067	3050	2100	1650	1450	1270	1150	1050	990	930	870	780	710	650	610
2 1/2	2.469	4800	3300	2700	2300	2000	1850	1700	1600	1500	1400	1250	1130	1050	980
3	3.068	8500	5900	4700	4100	3600	3250	3000	2800	2600	2500	2200	2000	1850	1700
4	4.026	17,500	12,000	9700	8300	7400	6800	6200	5800	5400	5100	4500	4100	3800	3500

Notes: 1. Capacity is in cubic feet per hour at gas pressures of 0.5 psig or less and a pressure drop of 0.5 in. of water; Specific gravity = 0.60.
2. Copyright by the American Gas Association and the National Fire Protection Association. Used by permission of copyright holder.

Figure 21-40. *Pipe sizing chart for residential gas piping. (Reprinted by permission of the American Society of Heating, Refrigerating, and Air-Conditioning Engineers, Atlanta, Georgia, from the 1993 ASHRAE Handbook—Fundamentals)*

START-UP CHECK SHEET

Owner Name: ____________________ Dealer Name: ____________________

Address: ____________________ Address: ____________________

City, State, Zip: ____________________ City, State, Zip: ____________________

Model Number: ____________________ Serial Number: ____________________

Type of Gas: Nat: ____________ LPG:

Gas Valve ____________ Line Pressure ____________ Manifold Pressure ____________ Input ____________

Thermostat: ____________ Subbase: Heat Anticipator Setting: ____________

Blower Motor H.P. ____________________ Voltage: ____________________

Cooling Coil Size: ____________________ Model Number: ____________________

Temperature Rise: Supply Air: ____________ °F — Return Air: ____________ °F = °F

External Static Pressure (in. W.C.) Supply ____________ Return ____________ Total ____________

Calculated or Measured CFM: Heating ____________ Cooling ____________

Temperature Rise: Return Air: ____________ °F Supply Air: ____________ °F

Filter Size and Type: ____________________

Burner Flame Properly Adjusted? ____________________

Drip-Leg Installed Prior to Gas Valve? ____________________

Blower Speed Checked? ____________ Blower Properly Lubricated? ____________

All Electrical Connections, Mounting Screws, etc. Tight? ____________________

Adjustment Required: ____________________

Date of Installation? ____________________

Date of Start-Up: ____________________

____

____

____

____

Figure 21-42. *Suggested start-up check sheet. (Comfortmaker GNJ, Inter-City Products Corporation)*

21.14 Service Check List

The following are basic checks you should implement during a service call:

- Visually check all electrical components for loose wiring or defective (cracked) wires.
- Make sure electrical power is available at the furnace by checking the circuit breaker position.
- Check the disconnect switch at the furnace to be sure it is on correct position.
- Make sure the blower door is secure. It depresses the interlock switch that permits the furnace to operate.
- If standing-pilot ignition is used, check for a burning pilot light.
- Check condition of the air filter.
- Check condition and adjustment of the fan belt (if a belt-driven motor is used).
- Check for any accumulation of dust/dirt on the blower motor and blower cage.
- Start furnace operation by placing the thermostat in a heat-demand setting.
- Observe burner start-up and operation.
- Observe fan motor start-up.
- Allow the furnace operating (burning) cycle to run ten minutes. Then, check for airflow through room registers.
- Turn the thermostat to a lower setting to shut off burner operation.
- Check that burner flames are off and pilot light remains on (if a standing-pilot system is used).
- Listen carefully to determine if blower shuts off shortly after burner flames are extinguished.

When servicing any type or make of heating equipment, always refer to manufacturer's manuals for that equipment.

21.14.1 Ignition Systems Troubleshooting

Both direct-spark and hot-surface ignition systems are dependent upon a solid-state ignition control module. This module sends and receives messages to/from other components within the ignition system. There is one main difference in operation between the two systems. In the direct-spark system, a pilot light must be established prior to main burner operation; in the hot-surface system, there is no open flame. The operational sequences of both hot-surface and direct-spark systems are listed below.

The operational sequence of the hot-surface ignition system is as follows:

1. When there is a need for heat, the thermostat informs the control module. The control module is a self-diagnostic device. See **Figure 21-43.** It immediately starts to perform a self-check. If the module senses a problem externally or internally, it responds by flashing an indicator light. See chart, **Figure 21-44.** The control module will check for closed limit contacts on the pressure switch. It will also check for normally open contacts.

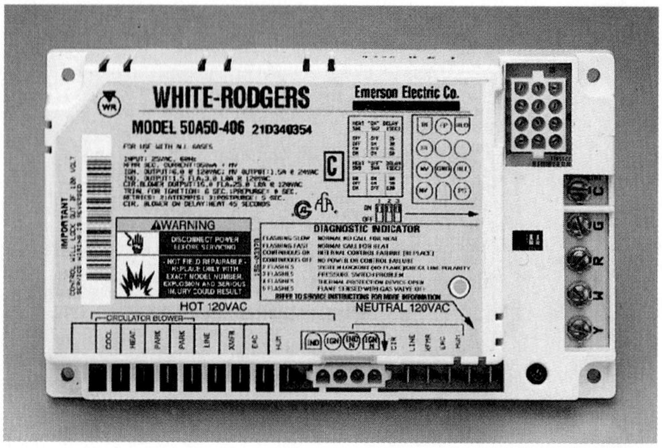

Figure 21-43. *Control module for hot-surface ignition system. Note the diagnostic indicator with color codes to assist in troubleshooting. (White-Rodgers Division, Emerson Electric Co.)*

1 flash—system lock-out due to retry
2 flashes—pressure switch stuck closed
3 flashes—pressure switch stuck open
4 flashes—open high limit switch
5 flashes—open rollout switch
6 flashes—grounded sensor
continuous flash—flame sensed with no call for heat must interrupt 120 volt power supply for 1 second.

Figure 21-44. *Sequence of operation for a hot-surface ignition system. (Comfortmaker GNJ, Inter-City Products Corporation)*

2. The induced draft blower will then start and purge the system for 30 seconds.
3. After the pre-purge, the silicon carbide ignitor becomes energized. This occurs for about 17 seconds before the gas valve opens.
4. When the gas valve is energized, gas flows to the burner and the heated ignitor. The gas is ignited.
5. During normal operation, the ignitor becomes de-energized approximately 4 seconds after the gas valve is energized.
6. Through the flame sensor, the control module must detect main burner operation within 7 seconds. If it does not, the gas valve will become de-energized. After burner flames have been sensed, the fan blower motor is energized. This is accomplished through a time delay within the control module.
7. When the thermostat becomes satisfied, the gas valve is de-energized, along with the inducer blower. The adjustable delay-to-fan-off sequence begins to de-energize the fan blower.

A troubleshooting flow chart for hot-surface ignition is shown in **Figure 21-45.** In this example, the ignitor is not glowing red. Follow through the chart, checking each item. When arriving at "Disconnect electric power to system," you would proceed to do this.

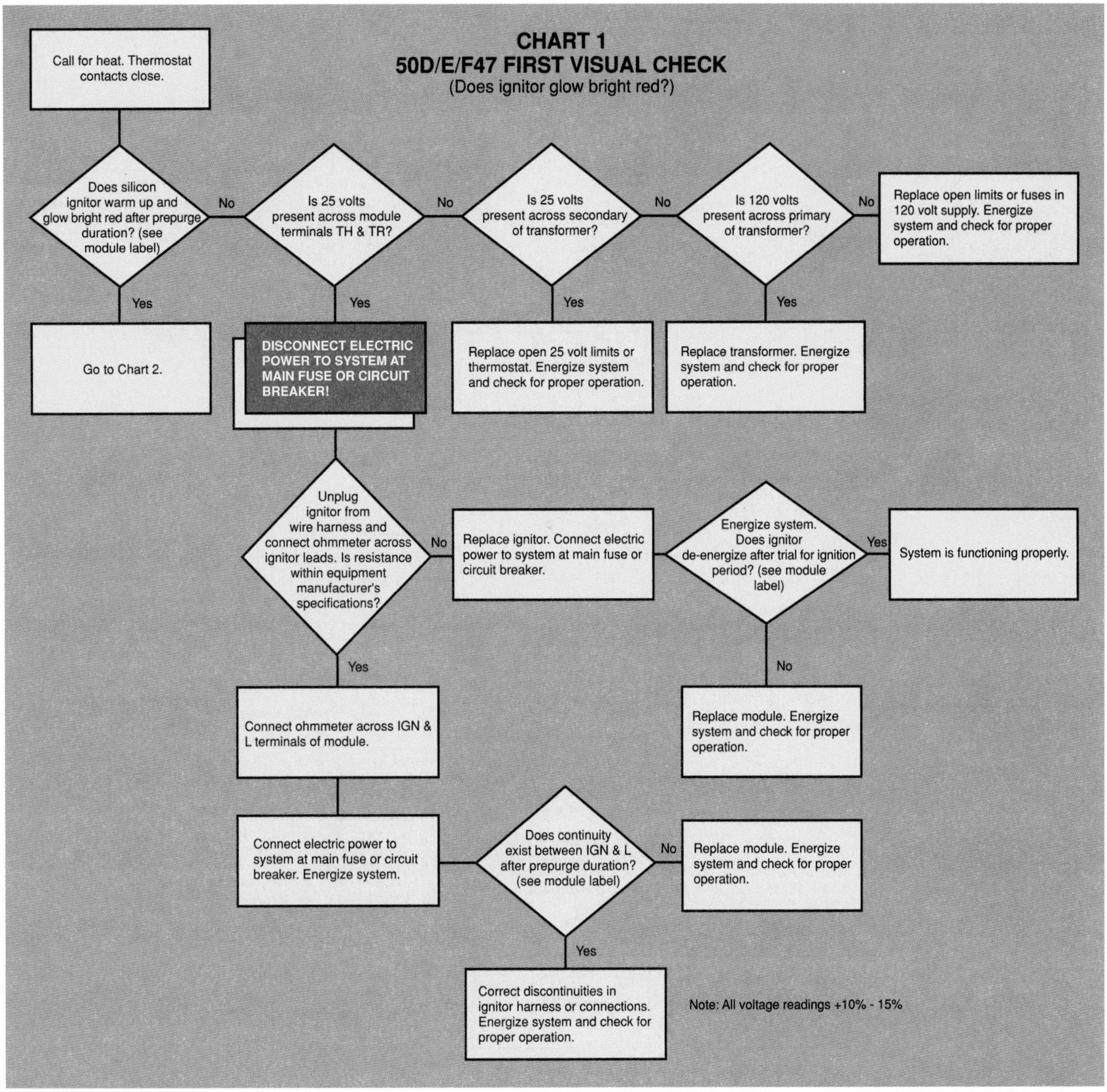
CHART 1
50D/E/F47 FIRST VISUAL CHECK
(Does ignitor glow bright red?)
Call for heat. Thermostat contacts close.
Does silicon ignitor warm up and glow bright red after prepurge duration? (see module label)
No
Is 25 volts present across module terminals TH & TR?
No
Is 25 volts present across secondary of transformer?
No
Is 120 volts present across primary of transformer?
No
Replace open limits or fuses in 120 volt supply. Energize system and check for proper operation.
Yes
Go to Chart 2.
Yes
DISCONNECT ELECTRIC POWER TO SYSTEM AT MAIN FUSE OR CIRCUIT BREAKER!
Yes
Replace open 25 volt limits or thermostat. Energize system and check for proper operation.
Yes
Replace transformer. Energize system and check for proper operation.
Unplug ignitor from wire harness and connect ohmmeter across ignitor leads. Is resistance within equipment manufacturer's specifications?
No
Replace ignitor. Connect electric power to system at main fuse or circuit breaker.
Energize system. Does ignitor de-energize after trial for ignition period? (see module label)
Yes
System is functioning properly.
Yes
Connect ohmmeter across IGN & L terminals of module.
No
Replace module. Energize system and check for proper operation.
Connect electric power to system at main fuse or circuit breaker. Energize system.
Does continuity exist between IGN & L after prepurge duration? (see module label)
No
Replace module. Energize system and check for proper operation.
Yes
Correct discontinuities in ignitor harness or connections. Energize system and check for proper operation.
Note: All voltage readings +10% - 15%

Figure 21-45. *Part I of a three-part visual check procedure for a hot-surface ignition system. (White-Rodgers Division, Emerson Electric Co.)*

There are other charts provided by the manufacturer for possible problems.

Direct-spark ignition units produce an electric spark. This spark ignites pilot gas to create a pilot light. The operational sequence of direct-spark ignition is as follows:

1. The thermostat calls for heat.
2. The relight control is energized and starts to produce a voltage spark.
3. The pilot-redundant gas valve is also energized within the combination gas valve. This will allow gas to flow to the pressure switch and to the pilot orifice.
4. The spark created by the relight control is carried by the electrode. There, it ignites the gas to create a pilot flame. The pilot flame now will heat up the mercury flame sensor tip.
5. After approximately 45 seconds, the flame sensor energizes the main valve within the gas combination valve. In turn, all burners would begin operating. When the thermostat is satisfied, it will de-energize the ignition system. The pilot light will go out until the thermostat once again calls for heat.

A troubleshooting flowchart for direct-spark ignition is shown in **Figure 21-46.** The flowchart details the first visual checks to be completed.

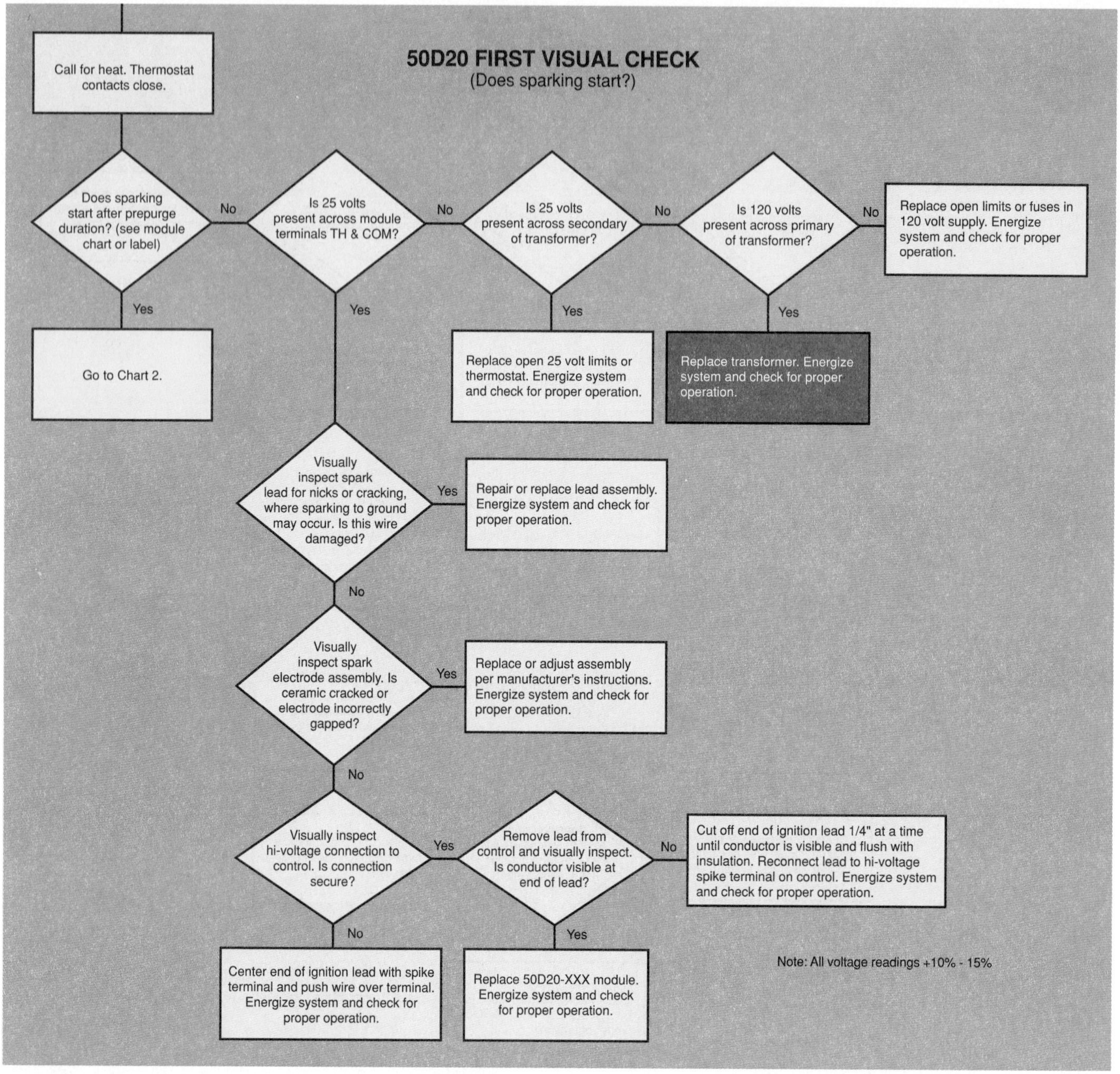

Figure 21-46. *Visual check flowchart for a direct spark ignition system. As illustrated, the problem is in the transformer. (White-Rodgers Division, Emerson Electric Co.)*

21.14.2 Flame Troubleshooting

Proper start-up and operation of a furnace requires the correct fuel-air mixture, gas pressure, and flame.

Primary air promotes complete combustion. The amount of primary air entering the burner is controlled by adjusting a shutter. If a burner does not have a shutter adjustment, check the manufacturer's literature.

A proper flame has a soft blue color without evidence of yellow tipping or lifting. See **Figure 21-15A.** A flame should not flash back, pop, float, or roll out. The following is a list of improper flame conditions, possible causes, and corrective actions.

Flame Condition	Possible Cause	Corrective Action
Yellow Flame	Lack of primary air.	Reset primary air and check for blockage.
Lifting Flame	Gas velocity faster than speed gas can burn.	Reduce input gas or primary air.
Popping	Flashback during shutoff; burning continued.	Increase gas pressure, reduce primary air, reduce orifice, check gas valve and burner.

Flame Condition	Possible Cause	Corrective Action
Floating Flame	**Very dangerous!** Incomplete combustion causing release of carbon monoxide that can create serious health and safety problems.	Check gas flow, primary air components dealing with secondary air, and check flues and burners.
Roll-Out	**Very dangerous!** Blocked flue, poor draft, incomplete air supply, burner over-firing.	Check gas flow, flues, burner, secondary air sources, and primary air.

21.14.3 Electrical Safety

Furnaces operate through use of electricity. Since electricity can be dangerous, electrical safety is a concern. It is important to keep in mind the following when servicing a system:

- **When a fuse or circuit box is turned off, lock it.**
- **An improperly used jumper wire can damage components and injure you. Use them only with knowledge and caution.**
- **Never work on a live unit, unless the problem requires checking a live circuit.**
- **Use only tools with insulated handles.**
- **Cautiously survey your work environment for dangers.**
- **Never assume a unit is without power simply because it is not running.**
- **Always be aware of local codes and follow them.**
- **The combination of a spark and gas is dangerous!**

HYDRONIC HEATING MODULE

21.15 Hydronic Heating System

Hydronic systems use hot water to carry heat to occupied spaces. Such systems have been used for many years. In some systems, the hot water circulates by thermal convection. The circulating water is under atmospheric pressure, and an expansion tank allows changes in volume.

Most hot water systems, however, use a circulating pump to increase water flow. It carries more heat per unit of time to the room heat transfer units. Having a pump in the system permits the use of a smaller boiler. **Figure 21-47** shows a cross section of a gas-fired domestic water boiler. This particular unit can be vented in two different ways: a natural venting process, and a power-vented package. The natural venting may be vented into fireclay tile-lined masonry chimneys. The power vent is designed to vent directly through the wall.

A commercial hydronic boiler is shown in **Figure 21-48.** Switches on the burner control permit the choice

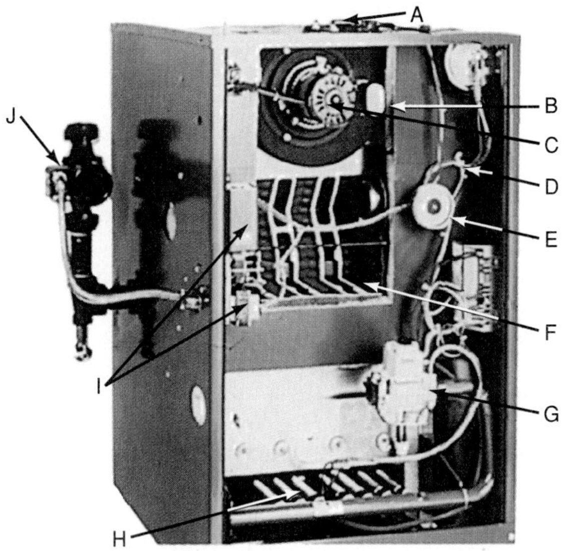

Figure 21-47. *Cross section view of a gas-fired hot water boiler. A—Nonmetallic venting. B—Insulated jacket. C—Draft inducing fan. D—Labeled wiring. E—Pressure-temperature gauge. F—Cast-iron sections. G—Gas valve. H—Stainless steel burners. I—Controls. J—Self-lubricating circulator. (Burnham Corporation)*

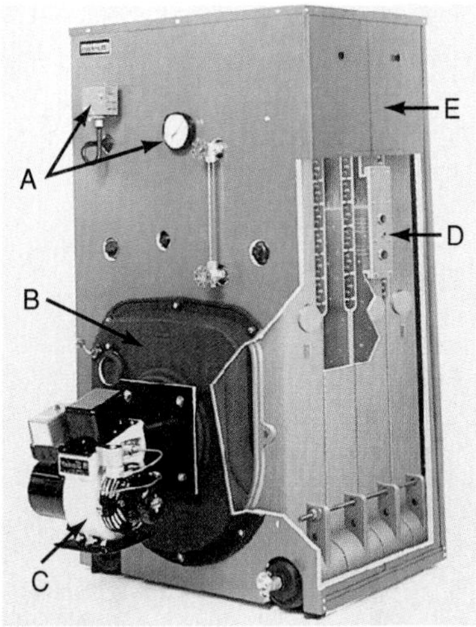

Figure 21-48. *Commercial steam boiler. A—Controls for adjustment and maintenance. B—Burner mounting plate with flame observation port. C—Burner. D—Tankless heater. E—Rear flue outlet. (Burnham Corporation)*

of fuel being burned, heating oil, gas, or a gas/oil combination.

Floor radiant heat systems are basically hydronic heating systems. They can be divided into three specific areas:

- Heat source. This is any device that is used with any kind of fuel to make warm water.
- Heated water is directed from the source to the radiant zone. This is done with pumps and controls. They are supported by various valves and gauges.

- Water is pumped through the radiant zone by means of a hose supply and return manifold.

Hydronic under-floor heating systems are designed in numerous ways. See **Figure 21-49.** Three basic concepts are:

- In a concrete floor, tubing is tied to the reinforcing mesh before the slab is poured.
- On second floor installations, the tubing and heat emission plates are attached to the ceiling joists of the first floor, or to sheet rock.
- Tubing is installed over suspended wood floors and covered with a lightweight concrete.

Figure 21-50 shows a circulating pump of the centrifugal type. A shaft seal is located where the pump shaft leaves the casing.

The hydronic pump is shown in **Figure 21-51.** The unit is a three-piece assembly. The motor is a separate

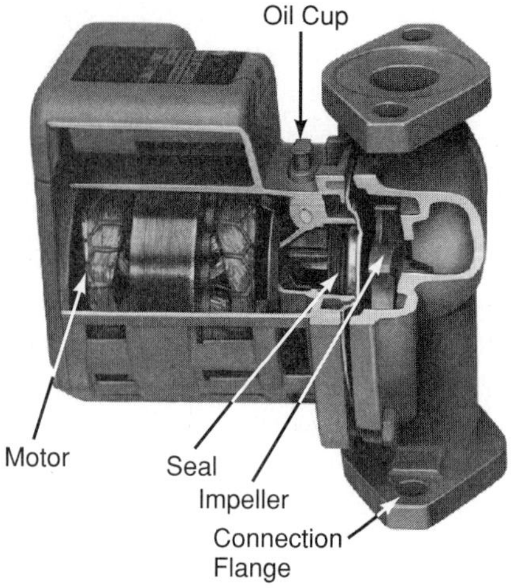

Figure 21-50. *Cross section of a centrifugal pump used in a hydronic system. (Courtesy of ITT Fluid Handling Sales)*

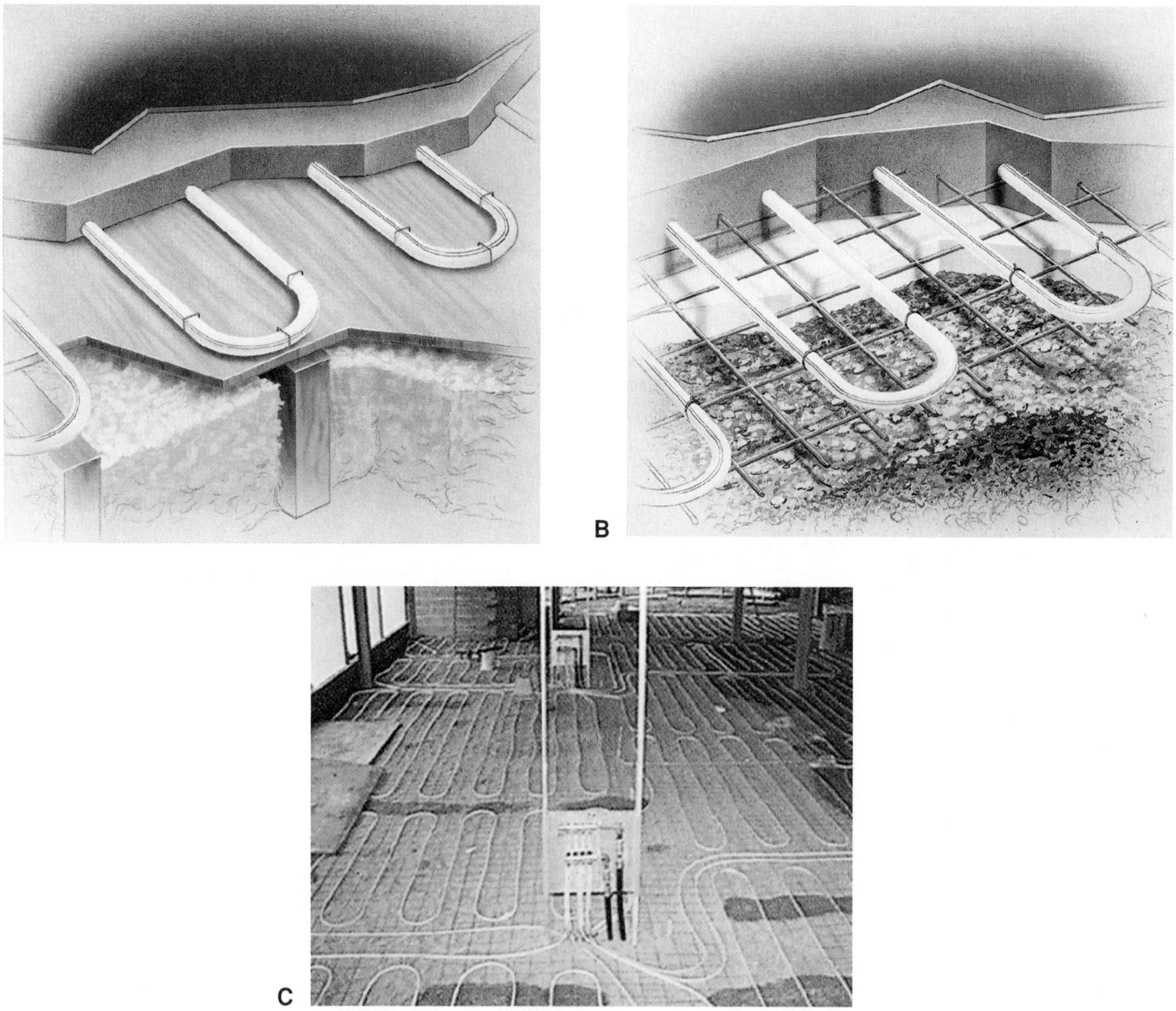

A

B

C

Figure 21-49. *Hydronic heating system. A—Application over wood floor. B—Application in a concrete floor. C—Large commercial application on concrete slab floor. (Wirsbo Company)*

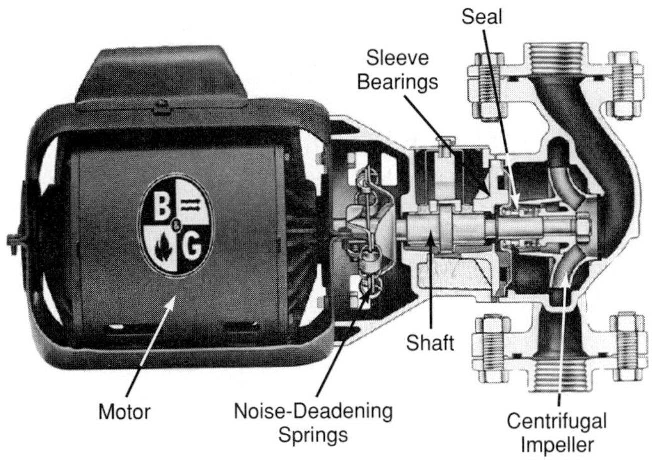

Figure 21-51. *A three-piece oil-lubricated booster pump used on hydronic systems. (Courtesy of ITT Fluid Handling Sales)*

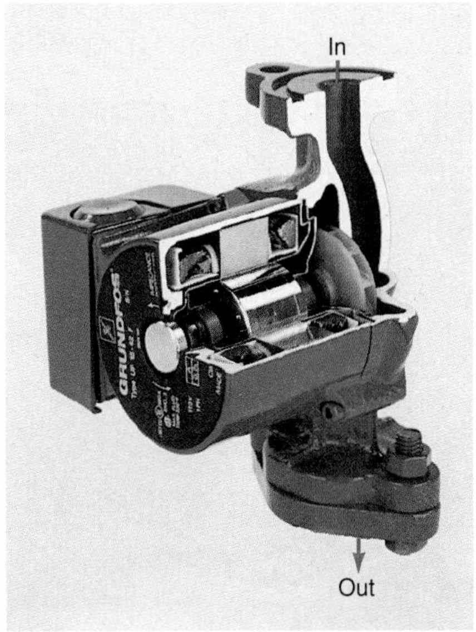

Figure 21-52. *Internal view of hydronic pump. Note the water inlet and outlet. (Grundfos Pumps Corporation)*

section. The shaft and seal are housed in a second unit, which can be removed and replaced or repaired. The front (impeller) section is the housing for the flow of water. Specific components are:

- The motor.
- A noise-dampening coupler to give the unit a quiet operation.
- The bearings and shaft alignment, which provides constant circulation of oil over the bearing surfaces.
- The pump shaft.
- The seal, which can withstand a wide range of water temperatures and pressures. It can also withstand additives and dissolved solvents that are often found in hydronic systems.
- Centrifugal impeller, which prevents accumulation of air at seal bases.

Figure 21-52 shows the inner structure of a hydronic pump. To remove the pump, turn the shutoff valves all the way in. A piping arrangement when one pump is used is shown in **Figure 21-53.**

Water heating systems may use one of several different temperature control devices:

- Single control, which starts and stops the pump.
- Zone control, using two or more controls. Each operates one pump.
- Individual radiator controls for specific rooms.

A system with three zones is illustrated in **Figure 21-54.** It includes a pressure relief valve, which is required in all pressurized heating systems.

Figure 21-55 illustrates a hydraulic type radiator control. Temperature control is adjustable. Note that the thermostatic radiator valve operates the flow valve. The sensitive bulb is located in the cold air inlet to the radiator. It is a modulating (variable volume flow) control that can be used for either steam systems or hot water (hydronic) systems.

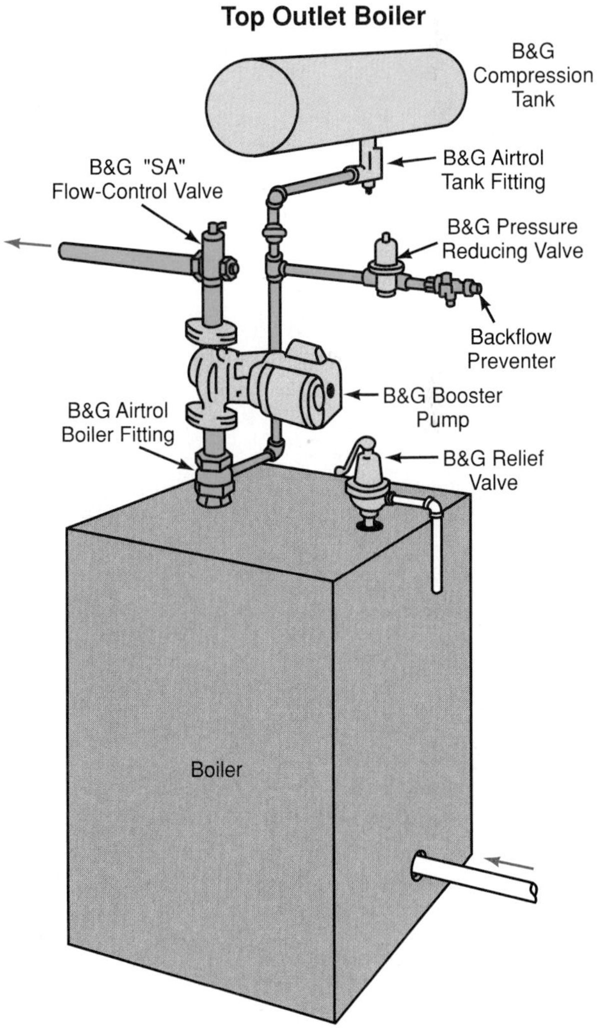

Figure 21-53. *Diagram of a hydronic system, showing piping and location of pump, compression tank, and relief valve. (Courtesy of ITT Fluid Handling Sales)*

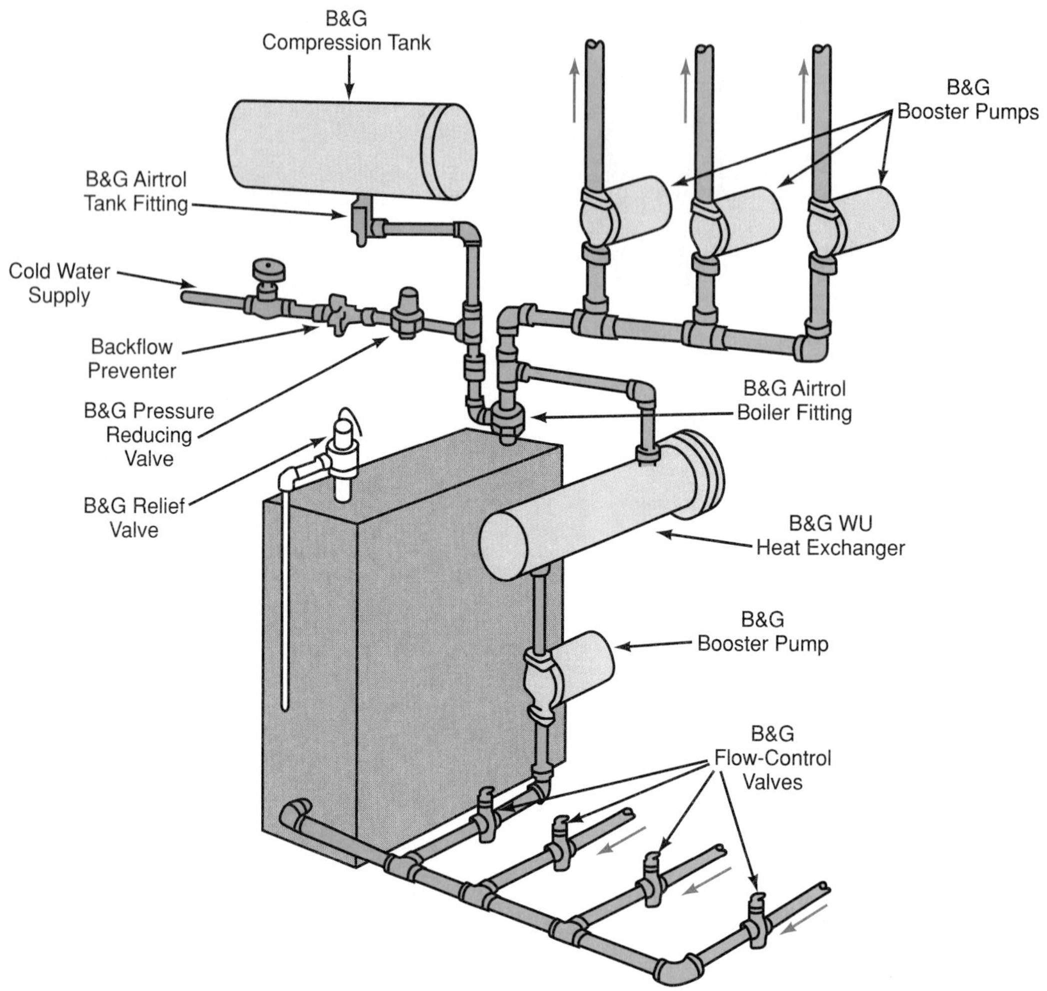

Figure 21-54. *Three-zone hydronic system. Pumps in the system are discharging out of the boiler. A fourth booster pump circulates water through the heater to provide hot water. (Courtesy of ITT Fluid Handling Sales)*

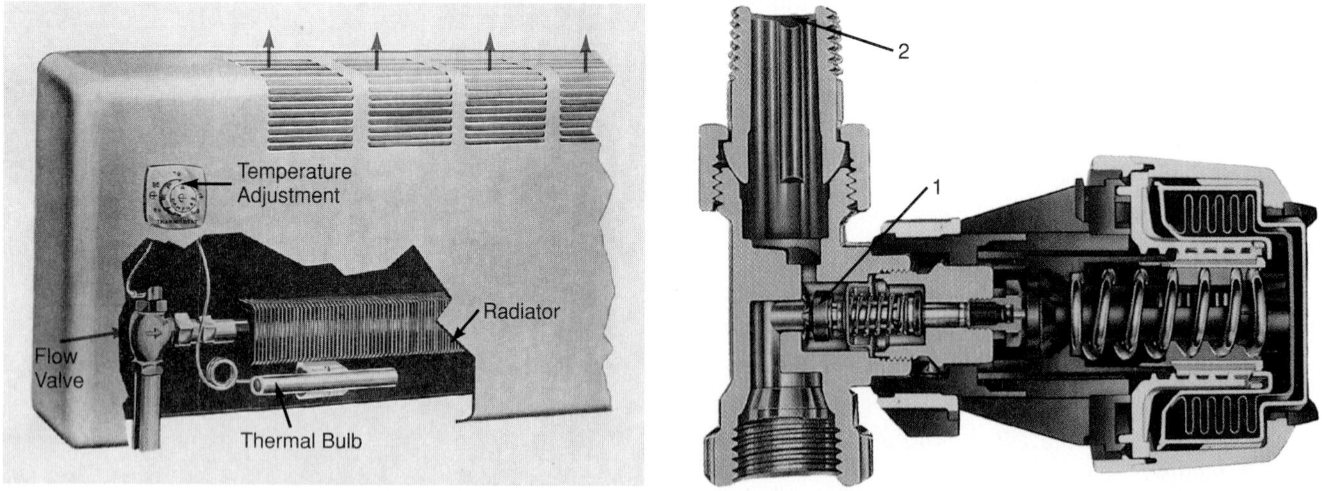

A B

Figure 21-55. *Individual thermostatic radiator valve commonly used for one-pipe or two-pipe water heating systems in residential, commercial, or institutional buildings. A—Thermal control valve mounted on convector heating system. This control provides individual room temperature control with either circulating hot water (hydronic) or steam as heat source. B—Thermostatic radiator valve. 1) Drop in ambient temperature causes the valve to open. 2) This increases supply of hot water or low-pressure steam entering the radiator. Rise in ambient temperature causes valve to close, decreasing supply of hot water or low-pressure steam entering radiator. (Danfoss Automatic Controls, Division of Danfoss, Inc.)*

A hot water system using a flow switch control system is shown in **Figure 21-56.** The main circuit is shown in solid lines. If the water flow ceases, the flow switch will open the electrical circuit and shut off the burner. Some systems use a recirculating system within the boiler. This is shown in the broken line piping diagram. This recirculation of water maintains a more constant water temperature within the boiler. If water flow stops in this circuit, the flow switch shown will shut down the system.

Figure 21-57 shows a hot water system used to heat incoming outside air in a ventilating system. If the water flow stops, a flow switch shuts the damper and turns on the alarm. Sometimes it will shut off the burner, depending on the number of other radiators.

A system that mixes outside air within return air is shown in **Figure 21-58.** The recirculated air and fresh air intake are controlled together. This ensures proper air conditions in the occupied space.

In many commercial and industrial buildings, it is desirable to maintain different temperatures in different rooms or areas. This is also desirable in some homes. For instance, the bedrooms may be kept at a different temperature than the living room. The laundry

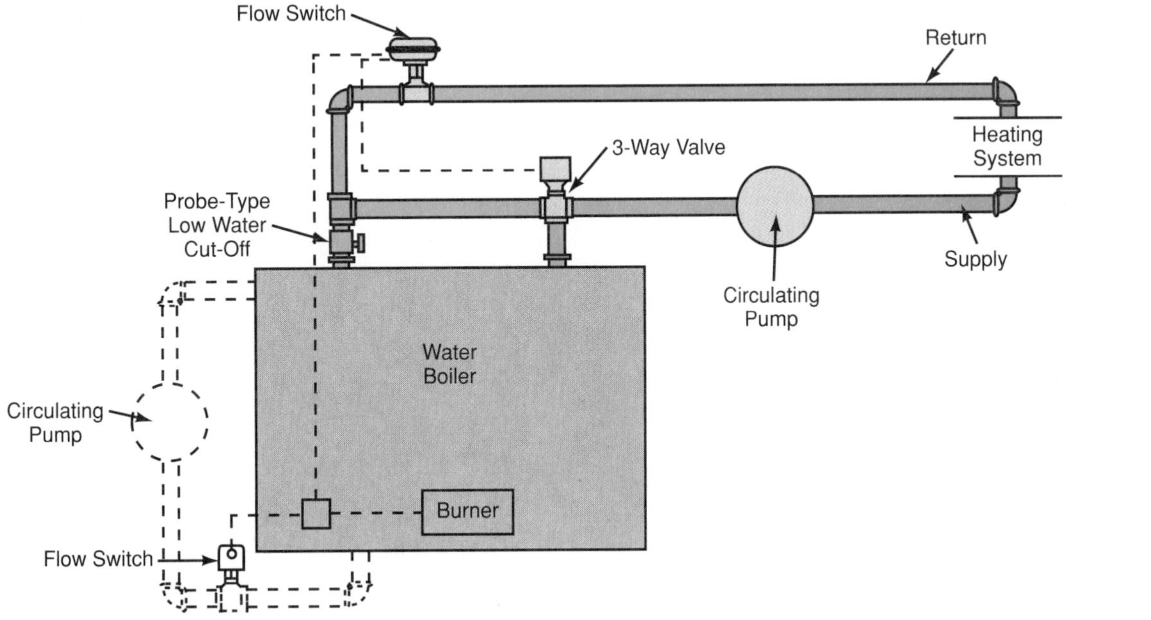

Figure 21-56. *Schematic diagram of hot water system using circulating pump, three-way valve, and flow switch. (ITT McDonnell & Miller)*

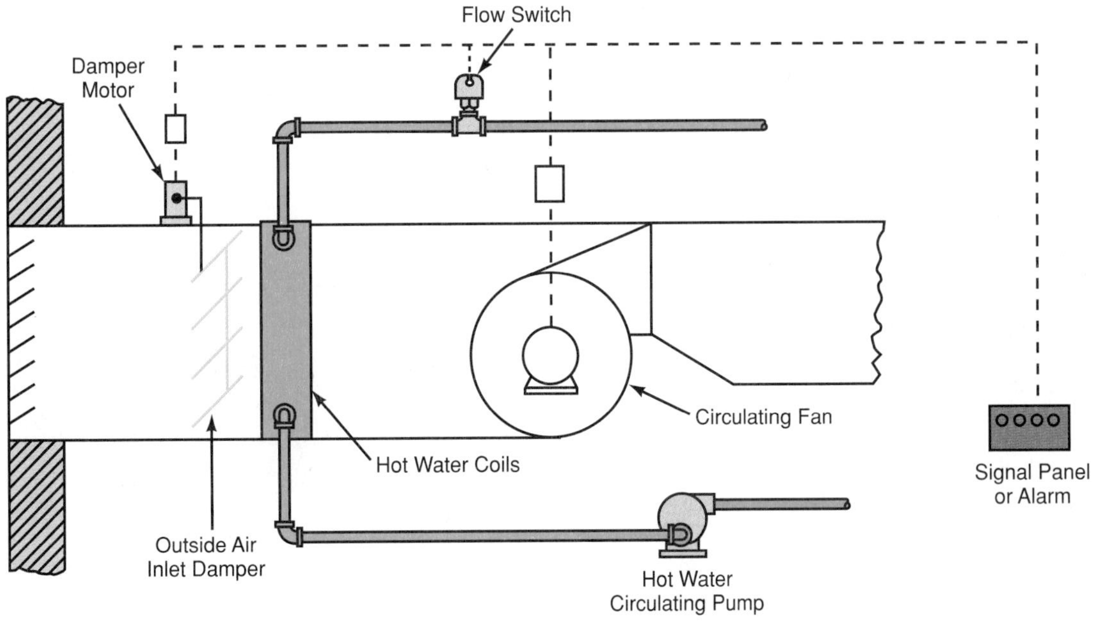

Figure 21-57. *Water heating system used to heat incoming air of ventilating system. Note motorized damper control, flow switch, and signal panel. (ITT McDonnell & Miller)*

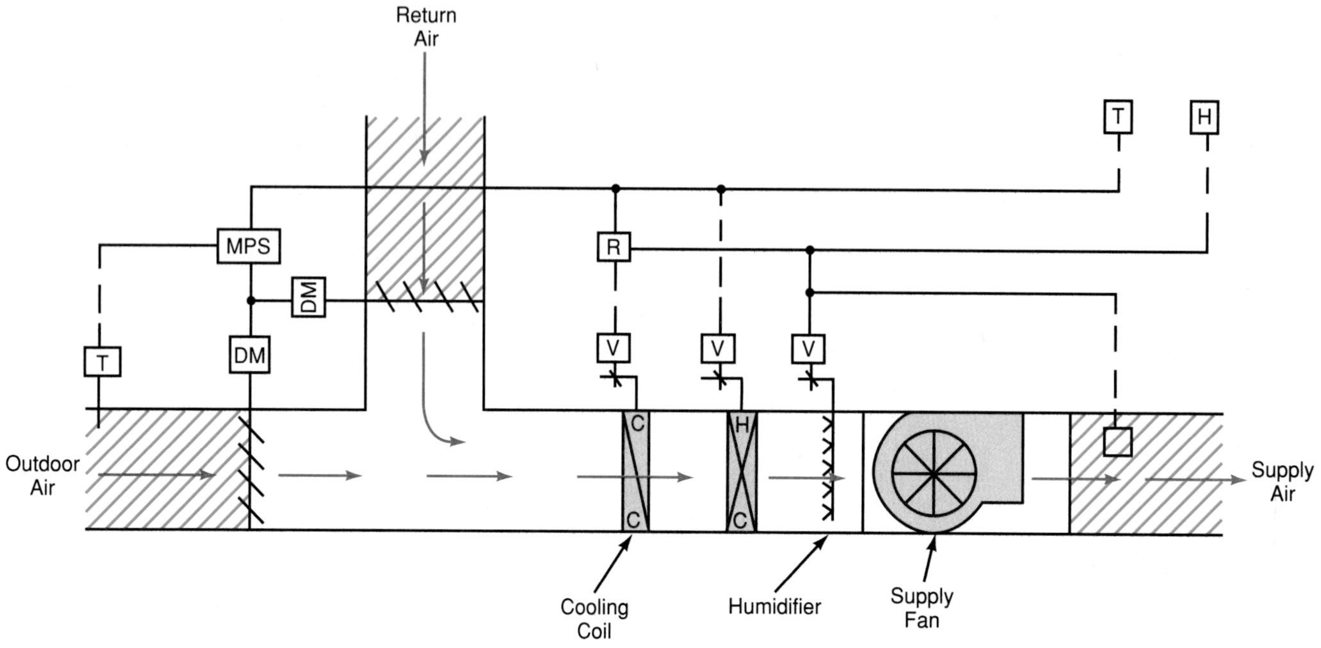

Figure 21-58. *Diagram of controlled air system. Outdoor air and return air are proportionately mixed to obtain correct atmosphere. (Reprinted by permission of the American Society of Heating, Refrigerating, and Air-Conditioning Engineers, Atlanta, Georgia)*

room may be kept at a different temperature than the kitchen. Thermostatic control of heating or cooling media maintains different temperatures in these areas.

21.15.1 Hydronic System Fluids

Water for hydronic systems usually has substances added. These substances lower the freezing temperature and raise the boiling temperature. These substances may also keep the water from forming deposits in the pipe. Tap water has impurities that may cause scale, corrosion, or embrittlement.

Scale is formed from salts in the water. The salts settle on metal surfaces as the water passes through temperature changes. These salts should be removed before the water enters the heater. As an alternative, chemicals can be added to form a sludge with these salts. The salts can then be purged from the system.

The salts are usually calcium carbonate, calcium sulfate, calcium chloride; magnesium carbonate, magnesium sulfate, magnesium chloride; sodium carbonate, sodium hydroxide, or silica oxide. Iron and manganese may also form deposits in the boiler.

Corrosion takes place when the water is acidic or when certain gases are dissolved in the water. Corrosion is reduced by neutralizing the acid condition with an alkali. It is also reduced by removing the gases by deaeration. (***Deaeration*** is the release of gases dissolved in a liquid.) Chemical scavengers and corrosion inhibitors are also used.

Water can be contaminated by such dissolved gases as hydrogen sulfide, carbon dioxide, or oxygen, or by foaming caused by organic matter and oil. Some companies specialize in the treatment of boiler water. Substances can be added to water to reduce the effects of impurities. These are available at contracting supply houses.

Embrittlement causes metal failure along drum seams, under rivets, and at tube ends. Water may flash to steam through small leaks in these highly stressed areas. This allows any sodium hydroxide in water to concentrate. See **Figure 21-59.**

Embrittlement may be slowed by:

- Maintaining low alkalinity (low hydroxide).
- Avoiding leaks at stressed metal.
- Using special inhibiting agents.

Safe levels of impurities in boiler water for various chemicals are shown in **Figure 21-60.** The pH level should be about 10. See Section 15.14.3.

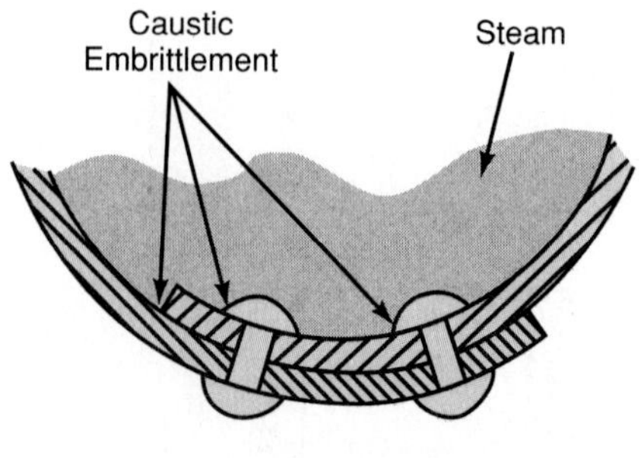

Figure 21-59. *Cross section view of steam container (pipe or boiler) shows where caustic embrittlement can take place. Activity is greatest where metal is stressed and where caustic material (sodium hydroxide) can collect.*

Maximum Allowable Impurities in Boiler Water

Chemical Name	Chemical Symbol	ppm
Sodium Sulphite	Na_2SO_3	1.0
Sodium Chloride	NaCl	10.0
Sodium Phosphate	Na_3PO_4	25.0
Sodium Sulphate	Na_2SO_4	25.0
Silica Oxide	SiO_2	0.20
Total Dissolved Solids		50.0

Figure 21-60. *Maximum allowable amount of certain impurities in good quality boiler water, listed in parts per million (ppm).*

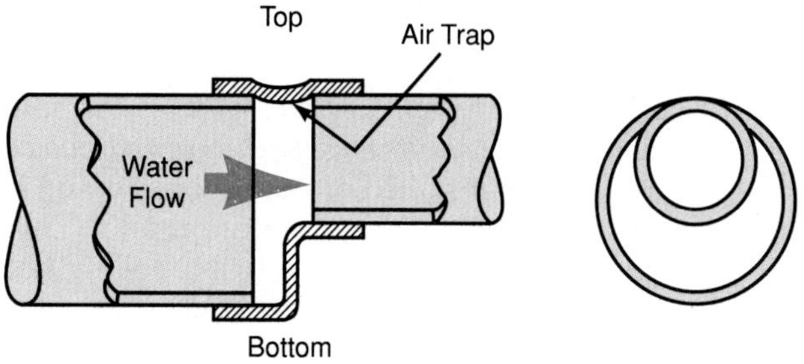

Figure 21-61. *Eccentric fitting in water circulation systems used to reduce danger of air pockets when pipe size is reduced.*

21.15.2 Hydronic System Operating Sequences

Hydronic systems are controlled in four basic ways:

- The heat is turned on at the same time the pump is turned on.
- The heat is on continuously and the pump is cycled to provide heat.
- The pump is operated continuously and the zone valves are cycled.
- The zone valves are cycled and turn on the pump or heater.

In forced circulation closed systems, the high-temperature water is above atmospheric pressure and smaller pipes can be used. The heat load is based on a 20° temperature differential between water-in and water-out. The top of the tubing should be even. It is best to use eccentric reducer fittings. **Figure 21-61** shows the reason for using eccentric reducer fittings.

Two-pipe systems are the most common:

- Direct return (less piping). Each circuit is a different design.
- Reverse return. Easier to balance; each pipe circuit is the same length. See **Figure 21-62.**

When installing water heating systems, you must allow for pipe expansion. Expansion for steel is 3/4″/100′/100°F. For copper, it is 1 1/16″/100′/100°F. Where riser pipes connect to a horizontal run, allow for a flexible joint. Also, install an expansion joint at the boiler.

A hydronic system may deliver hot water to more than one heating unit ***(radiator).*** If there are two or more radiators, the water flow must be balanced. Each radiator must receive its design quantity of water per unit of time. This balancing is usually done by installing gate valves in the piping. Opening and/or closing these valves reaches the balance. Flow meters are used to accurately measure quantity of flow.

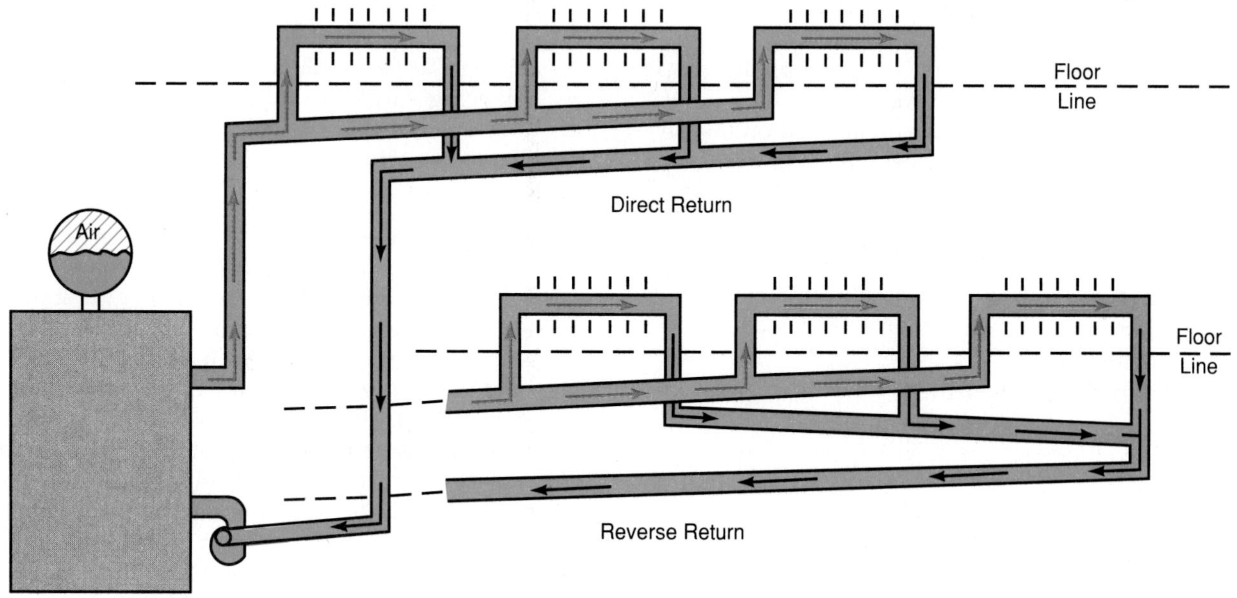

Figure 21-62. *Horizontal pipes in hydronic steam system should always slope toward boiler. Amount of slope is exaggerated in drawing. Total length of pipe run for each radiator (both hot water and return water) should be of equal length. Note use of equal total lengths in reversed return system.*

21.16 Installing Hydronic Systems

The boilers for a hydronic systems must be mounted level. All local code requirements must be checked and followed.

After a system is installed, it should be checked before it is put into service. The procedure used to check the new boiler is commonly referred to as "boiling out." The water to be used should be analyzed. A preventative maintenance solution should be added.

The fluxes, pipe joint compounds, and cutting oils sometimes form gases in a system. Dirt, sand, steel thread chips, solder bits, or sawing and filing chips may erode the system. They may also clog screens. Dirt and chips also ruin valves and pump seals.

Fill the system with water. Add about one pound of trisodium phosphate for each 50 gallons of water. Circulate for about four hours, then drain. Clean the screens and fill the system with water. It is then ready to operate. A card with a record of past maintenance should be included. A list of future treatment procedures should also be included.

Unless the pump seals leak and the vents are used quite often, avoid treating the water with chemicals. The chemicals may injure seals and valves. Avoid using phosphates and polyphosphates. Do not use over 300 ppm of chromates or over 500 ppm of nitrites.

Organic growth can be controlled by using sodium pentachlorophenate. It is best to consult a water treatment expert before attempting boiler water treatment.

21.17 Servicing Hydronic Systems

Air in a hydronic system is a common cause of trouble. Air will cause noise in the system. It will also reduce the system water supply, and interfere with water circulation. The air acts as a brake on the circulation of water. Series systems can be purged of air by putting an outlet hose in a bucket. When bubbles stop appearing, the air has been removed.

Other systems must have a manual or automatic air vent at the high points in the system. See **Figure 21-63.** A stand-pipe (drain) should be provided for each air vent.

Thermostats are another common source of heating problems, for these reasons:

- Vibration.
- Poor contacts.
- Broken wires.
- Improper temperature settings.

Other common problems include:

- Air in hydronic systems using natural gas. Pump motor failure and water leaks around the pump packing, due to air in the system.
- Noise from water pump motor and pump bearings made of carbon or Teflon® (water-lubricated).
- Cavitation (localized gaseous condition in a system).
- Turbulence.
- Poor supports (noise).

Figure 21-63. *Automatic air vent for water heating systems (also on chilled water systems) is installed at high point of convector, baseboard units, or radiator. (Maid-O'-Mist Div.)*

Hydronic heating problems include:

- Uneven heating. (Room-by-room heat loss calculation is needed. Balancing valves are needed.)
- Velocity.
- Vibration of parts of system.
- Pipe expansion noises. (Do not clamp tightly; pipes should not touch edges of openings through floors.) Expansion joints are of considerable help.

Some controls needed are:

- Air control devices (very much needed).
- Air vents (piping pitched up to the vent). Automatic air vents are a problem source.
- Balancing valves (to control flow rates).
- Check valves that prevent reverse flow during low heat-load periods.
- Gate valves.

Venting a system requires more water to be added to it. This water contains air in solution, plus some corrosive chemicals. It is best to trap this new air in the compression tank. Avoid venting a compression tank, if at all possible. The air compressed in this tank permits the water volume in the system to expand and contract. It also permits higher water temperatures.

When the system is working properly, you should be able to place your hand on the compression tank. Higher temperatures indicate problems, and the relief valve may open. The compression tank water is cooler. Therefore, it absorbs air before the water flows into the boiler. When in the boiler, the water is heated and the air is released. Some air may go into the pipes and end up in the vents. The tank may gradually lose its air, then the relief valve will open. If the relief valve spills water on each heating cycle, the compression tank lacks air. Most air is released from the water when the water velocity is lowest, and where the temperature is highest (in the boiler).

Another problem occurs when the heat is off. The water will cool and the pressure will drop. The reducing valve will allow more fresh water (containing more air) into the system. If this continues, the inside of the system will become corroded from the chemicals that were in the make-up water.

A safety caution: cut the exhaust end of the pipe from the relief valve at an angle to prevent someone plugging it or capping it.

21.17.1 Preparing a Boiler System for the Heating Season

Use the following procedure to check a boiler system before the heating season begins:

1. Clean burner (gas or oil).
2. Clean nozzle of oil burner. Use a cloth and solvent. Do not use a wire brush (bristles may scratch the orifice).
3. Clean and adjust electrodes. Inspect the insulation. Replace if cracked.
4. Clean flame detector lens. Operate by closing the fuel valve. Detector controls should lock out (lock in the off position).
5. Clean pilot light if a gas burner is used.
6. Tighten all connections.
7. Oil motor.
8. Check motor temperatures. If warmer than normal, cleaning or new bearings may be necessary.
9. Inspect tubes for soot or fly ash. Clean.
10. Inspect breeching (top of boiler flue). Clean.
11. Cycle controls. Shut off water feed to check for low water cutoff.
12. Operate all valves and cocks to check operating condition.
13. Operate safety and relief valves.

21.18 Steam Heating Systems

Steam heating is a means of distributing heat to occupied areas. Steam is generated in a boiler. The steam (vapor), being lighter, travels to the upper parts of the piping circuit. This steam is at 212°F (100°C) or higher, except in vacuum systems (which are rare). As it releases its heat to the occupied area, the steam condenses. The condensed water, being heavier, returns to the boiler. The steam releases about 1000 Btu for each pound that condenses. The heat exchange devices located in the room are called ***radiators.***

Two basic systems are in use. The ***single-pipe system*** uses one pipe to carry steam to the radiators. The same pipe is used to return the condensate. The ***two-pipe system*** uses one pipe to carry the steam to the radiator and another pipe to return the condensate.

For domestic applications, these systems operate at low pressures or at a partial vacuum. The units are tested at 50 psig (65 psia or 450 kPa) for safety purposes. Commercial and industrial systems use progressively higher pressures that approach 1000 psi (7 MPa). A steam boiler requires pressure safety valves, a water level gauge, a pressure gauge, and a temperature gauge. **Figure 21-64** shows a steam boiler.

A water-heating unit (tankless heater) in a boiler is shown in **Figure 21-65.** The unit heats only as much water as is needed; water is heated as it flows, instead of being stored.

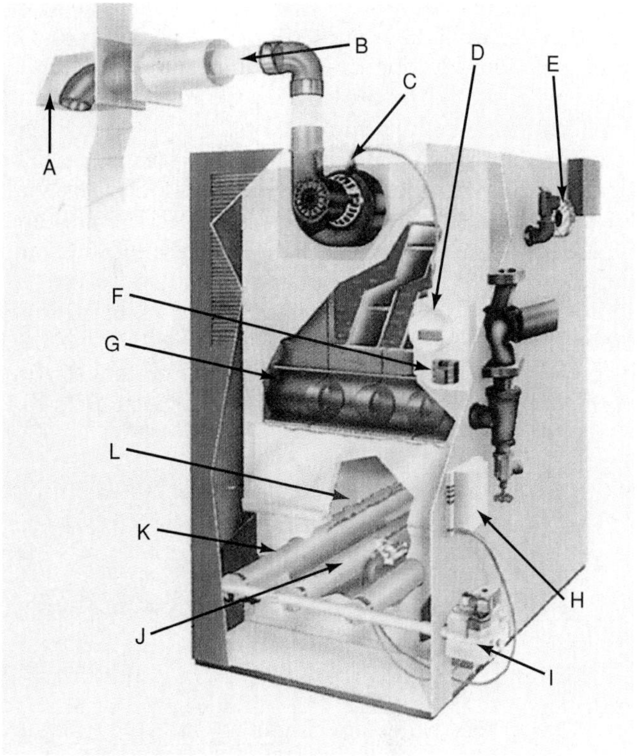

Figure 21-64. *Gas-fired direct-vent hot water boiler. A—Outdoor cap. B—Vent piping. C—Exhaust fan. D—Positive vent safety shut-off. E—Pressure relief valve. F—Relay operated safety circuit. G—Heat exchanger. H—Primary ignition module. I—Safety gas valve. J—Electronic pilot ignition sensor. K—Stainless steel burners. J—Flame. (Utica Boilers, Inc.)*

Figure 21-65. *Water heater being mounted in steam boiler. (Weil-McLain, a United Dominion Company)*

21.19 Steam Heating Installation

Steam heating systems vary with each installation. When room radiators are used, the system does not produce humidity. The air is not humidified or cleaned. Air circulation is by thermal movement (hot air rises). Separate humidity and cleaning systems must be used.

Many installations have the radiator installed in a forced convection duct. This duct system may also have a filtering system and humidifying system.

All steam heating systems must be installed according to code regulations. As with hot water boilers, the steam boiler must be mounted level. The pipes must be mounted with a slope down to the boiler. The piping must be designed to provide for expansion. Air vents must be located at the high points of the system. Each radiator should have an air vent and a steam trap. The steam trap keeps the steam in the radiators. It only allows condensate to return to the boiler. There are three types of steam traps: mechanical, thermostatic, and impulse.

After installation, the system must be flushed to remove all dirt. The system must be leak-tested as specified by code before it may be operated. After system startup, perform the safety inspection detailed in Section 21.20.1.

21.20 Servicing a Steam Heating System

Servicing a steam heating system should be done with great care. Escaping steam or condensate can cause severe burns. A boiler will explode if steam pressure is permitted to exceed the boiler safe pressure.

Check the water level gauge and the pressure gauge. Shut down the system at once if the water level does not show in the level gauge, or if the pressure gauge is above normal.

Servicing of the gas burner has been explained in previous paragraphs. Section 21.28.1 suggests components to check in oil burner servicing.

If one radiator is cool while the others are hot, it is not receiving steam. This problem may be caused by:

- Thermal valve to radiator is closed (thermostat for valve or valve not working).
- Radiator may be air-bound (air vent not working).
- Radiator may be filled with condensate (steam trap not working).

Lightly tapping the control with a rubber mallet may loosen the valve. This may allow the system to work momentarily. Then:

1. Shut down the system.
2. Reduce pressures to atmospheric by purging. **BE CAREFUL! Stand to one side to prevent burns from steam or condensate water.**
3. Replace thermal valve, air vent, or steam trap.
4. Start up system.

If the system is low on water and the boiler is hot, cool the boiler. This must be done before opening or repairing the boiler feed valve. Cold water may crack a hot boiler section. Some boilers are made with tubing heat exchangers. These boilers may use either a water tube or a finned tube design.

Short cycling times cause rapid temperature changes in the boiler. After several thousand cycles, the boiler walls or tubes may crack.

21.20.1 Steam Heating Safety Inspection

A steam heating system should be checked once each month during the heating season. Check the water level. The water level gauge must show the boiler water level at one-third to one-half full. If the water level is higher, drain the system to the correct lower level. If the water level is low, but still shows in the level gauge, add water. Water should be added until the level is correct. Close the fill valve completely. If no water shows in the water level gauge, shut the system off at once. Add water only after the boiler has cooled.

The water level sight glass may show water if its openings are clogged. Trust the reading only if the system is clean and the sight level glass and its connections have been cleaned recently. Water level petcocks (when used) are a good way to check the sight level glass tube reading. **Be sure the petcocks do not spill steam or hot water on anyone. Wear goggles!**

Operate the safety (relief) valve to be sure it is not stuck closed. **Again, be careful not to allow the escaping steam to hit anyone. Wear goggles!** If no steam or water comes out, shut off the system at once. Have the relief valve serviced or replaced.

Check the low level shutoff by opening the drain valve. The shutoff should operate right away by cutting off all fuel and electrical power. If the shutoff does not operate, shut down the unit and service the low water level shutoff.

OIL FURNACES MODULE

21.21 Fuel Oils

Fuel oils vary considerably. Generally, they contain about 85% carbon (C) and 12% hydrogen (H). Various other elements are present in the remaining 3%. During combustion, carbon and hydrogen combine with the oxygen (O) in the air. This produces carbon dioxide (CO_2) and water (H_2O). Fuel oil grades are established by the U.S. Department of Commerce. They conform to ASTM (American Society for Testing and Materials) specifications. A very low sulfur (S) content is very important. This is because the sulfur turns into corrosive gases and liquids.

Fuel oil grades 1 and 2 are used in domestic and small commercial furnaces. Grade 1 is used in pot-type oil burners. Grade 2 is the most popular domestic fuel oil (about 140,000 Btu/gal.). It has a flash point of 100°F (38°C). Its Saybolt viscosity is 37.9 (compared to Grade 4 with a viscosity of 125). Grades 4, 5, and 6 are used in industrial applications. They provide slightly more heat per gallon.

Products of combustion should be carbon dioxide (CO_2) and water in vapor form. Actually, there is also sulfur dioxide (SO_2) and **(Very dangerous!)** carbon monoxide (CO) present. About 106 lb. of air is required for each gallon of Grade 2 fuel oil consumed. Multiplying 106×14 (ft^3 per lb.) equals about 1500 ft^3. This is the quantity of air that must be fed to a furnace for each gallon of fuel oil consumed. This means that 1500 ft^3 of air must enter the building for each gallon burned. (This would be about each two hours of oil burner running time for the average home.)

The 106 lb. of air equal about 1500 ft^3 of air. Of the 106 lb. of air, about 84 lb. is nitrogen. Nitrogen does not produce burning. It acts as a gas to lower the temperature. It wastes some heat as it is warmed and goes up the chimney. Of the 106 lb. of air, about 22 lb. is oxygen (O) that combines with the oil to form about 20 lb. of CO_2 and 9 lb. of water (steam).

When perfect combustion takes place, about 15% of the flue gas volume is CO_2. This level is not reached with oil burners. Heavy carbon molecules (soot- and smoke-formers) are in the oil. Excess air is used to burn this carbon more completely. Excess air is air which is passing through the flue in excess of that needed for complete combustion. It is frequently expressed as a percentage of the air required for complete combustion. The excess air fed to the firepot should be sufficient to produce CO_2 at about 10%. If 100% excess air is used, the CO_2 content reduces to 7.5%. CO_2 is measured first because it is easier to measure.

In the flame, the hydrogen always burns first. It burns completely, then the carbon starts to burn. This action causes pulsations (pressure waves) in the flame. Also, the carbon first turns into CO, then into CO_2. This, too, may cause pulsation.

Combustion gases vary. For good combustion, excess air must be used. Therefore, considerable nitrogen (from the air) goes up the stack. Some oxygen, carbon dioxide, steam, and impurities also go up the stack. About 2000 ft^3 of air (providing 400 ft^3 of oxygen) is used per gallon of oil. These gases may be moved up the stack in the following ways:

- By natural convection (common in domestic and small commercial units).
- By forcing with a fan or blower (forced draft).
- By drawing the gases up the chimney (induced draft).

It is important to keep flue gases warm. Otherwise, condensation will take place in the stack and flue. This causes severe corrosion. One corrosion agent will be sulfurous acid (H_2SO_3). It corrodes steel rapidly and discolors brick and stone. Most good fuel oil additives will:

- Reduce sulfur dioxide by about 50%.
- Keep the flue and chimney cleaner (65% cleaner).
- Cause less opaque (visible) smoke.
- Reduce soot blowing of tubes.

Proper flame appearance is luminous, mainly yellow. If there is incomplete combustion, the flame is dull orange or red. Sulfur trioxide is more odorous than sulfur dioxide. It is minimized by using additives and a higher temperature.

An oil furnace in good condition should not release visible smoke from the flue, chimney, or stack. However, there may be soot deposits and fly ash. These should be removed annually. The soot may be removed by using air pressure, mechanical cleaning, vacuum cleaning, or chemicals.

Oil sludge, which clogs filters and nozzles, may be caused by bacteria. The bacteria will multiply if water is present in the oil. An additive can be used to kill the bacteria.

The combustion chamber must be kept in good condition. Deposits on the refractory (asbestos cement) lining must be kept to a minimum. Fuel oil additives reduce deposits in the combustion chamber, heat exchangers, and flue.

Number 2 fuel oil viscosity changes from between 50 and 100 at 0°F (−18°C) to between 35 and 45 at 70°F (21°C). This means that gun-type oil burners may have pumping and combustion problems when the oil is cold.

The use of Number 2 fuel oil distillate has increased in commercial buildings and in industry. This is the result of efforts to reduce air pollution. This grade burns more completely (and thus, is cleaner) than Number 3, 4, 5, and 6 fuel oils. The Number 1 and Number 2 oils are called distillates. They are products of a distillation or cracking process at the oil refinery. This means that they were vaporized, then condensed in the refining system.

21.22 Oil Furnaces

The most common type of oil burner is the gun type. See **Figure 21-66.** Rotary-type and pot-type oil burners are rarely used.

The ***gun-type burner*** forces oil under pressure through an orifice of a controlled size. The oil is broken into finely divided particles ***(atomized)***. It is mixed with air, and forced into the combustion chamber by a blower. **Figure 21-67** shows a gun-type oil burner. The unit has an air-injection combustion head with radial air-injection holes in the nozzle.

Remember that oil will not burn while it is in the liquid form. To burn, it first must be ***vaporized*** (turned into a gas). To vaporize oil, heat must be added (latent heat of vaporization). The oil turns into a gas more

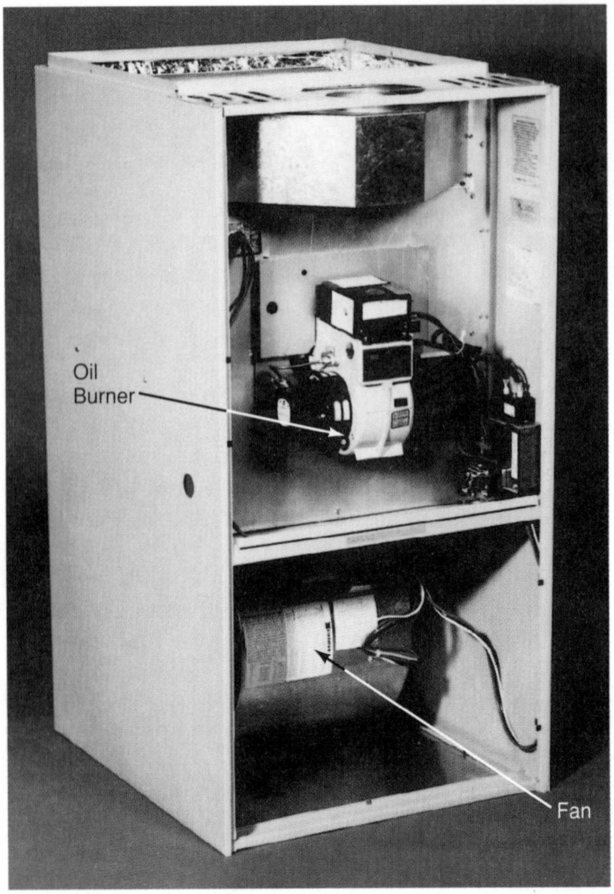

Figure 21-66. *Conventional forced-air oil furnace. (Heat Controller, Inc.)*

Figure 21-67. *Base-mounted gun-type oil burner. A—Combustion head nozzle with air-injection holes. B—Electrical control, pulse ignition transformer, recycle primary control, and cad cell flame detector. C—Oil pump. (Carlin Combustion Technology, Inc.)*

quickly and easily if it is finely divided (sprayed). This spraying process is called atomizing. The minimum atomization pressure is 75 psig (90 psia or 620 kPa). Gun-type oil burners atomize oil by forcing it into a twisting, spiraling, turbulent air stream. A small heat source (an electric spark) turns a few of the finely divided particles into gas so the burning will start.

Some large industrial furnaces use combination oil and gas burners. Switches on the burner control permit the choice of fuel.

Pulsation in an oil furnace is usually caused by positive pressure in the combustion chamber. (There is not enough draft.) Draft should be 0.02″ (0.5 mm) water pressure. Too much positive pressure may be caused by:

- Too much air (air shutter open too far).
- A chimney that is too small, partially blocked, or not tall enough. The chimney must be 2′ (61 cm) higher than the highest point of the building.
- A faulty nozzle (distorted flame pattern).

Oil on the floor of the furnace room is dangerous. It may be caused by an air leak in the oil suction line. (The air causes drip at the nozzle.) It may be due to loose compression fittings or unions, or by a pump seal leak. **Compression fittings should never be used on oil systems.**

Check for air in the system by connecting a pressure gauge. If the gauge needle fluctuates, it signifies that there is air in the system. For small leaks, put oil outlet tube in a bottle of oil. Bubbles will indicate an air leak in the suction line.

If there is soot in the boiler flue passages, clean the passages and the blower tube. Also, clean blower blades with a brush.

An oil furnace blowback is usually caused by delayed ignition. The most common reasons for blowback are:

- Electrodes improperly spaced.
- Distorted pattern away from electrodes.

If the oil nozzle is in poor condition, replace it with an exact replacement unit. (Use the same orifice, spray angle, and solid or hollow cone as originally used.)

Large oil burner furnaces may have a metal combustion chamber. Smaller furnaces use ***refractory cement*** liners for the combustion chamber. This cement consists of 80% dry asbestos and 20% Portland cement. Enough water is added to make the mixture workable. Avoid putting this cement on the edge of or inside of the air cone. (Cement will change the air pattern and cause inefficient burning.) It is best to fill the tube with a rag while applying the cement. Many refractory liners are preformed, then installed. **Figure 21-68** shows a burner installed in a boiler.

21.23 Gun-Type Oil Burners

Gun-type oil burners are available in two types:

- High-pressure.
- Low-pressure.

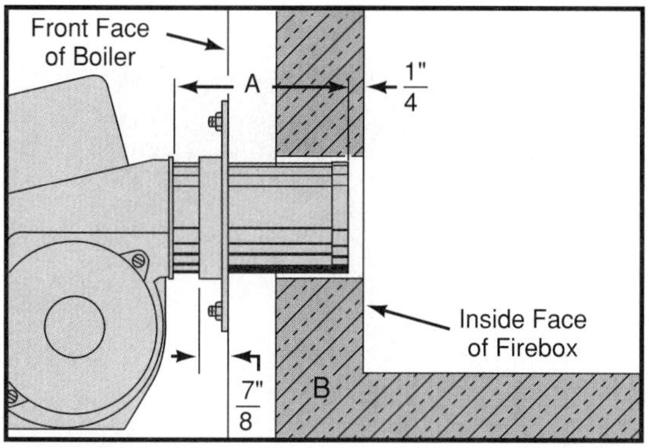

Figure 21-68. *Typical gun-type oil burner installation in boiler. A—Length of air tube. B—Refractory insulation. (R.W. Beckett Corp.)*

In the high-pressure type, oil is fed to a nozzle. It is under a pressure of from 100 psig to 300 psig (115 psia to 315 psia or 795 kPa to 2175 kPa). Air is forced into the furnace through a tube that surrounds this nozzle. Usually, the air is twisted in one direction. The oil spray is given a twist in the opposite direction. **Figure 21-69** shows a gun-type oil burner. The nozzle should be carefully centered in the housing. Two electrodes are located near the front of the nozzle. The ignition transformer provides a high voltage spark between these electrodes. **Figure 21-70** shows the nozzle and electrode assembly.

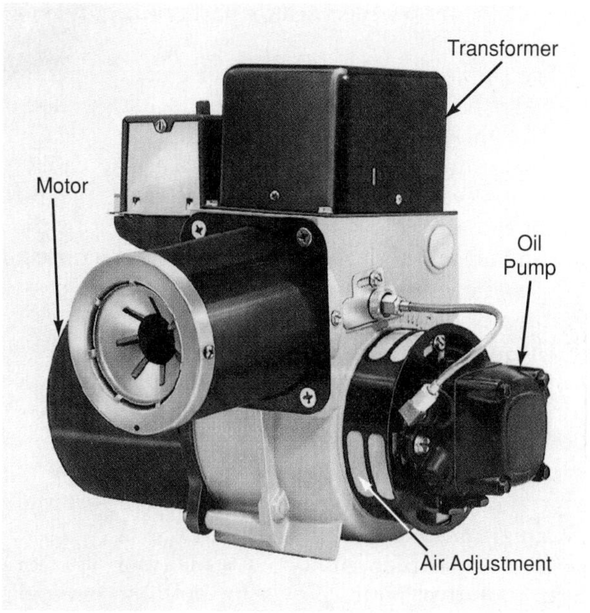

Figure 21-69. *Flange-mounted gun-type oil burner. (R.W. Beckett Corp.)*

The low-pressure type burner uses oil at 1 psig to 4 psig (16 psia to 19 psia or 110 kPa to 130 kPa). Oil is mixed with air before it reaches the nozzle.

The main parts of a gun-type burner are: motor, oil pump, fans, nozzle, choke, air tube, and ignition system.

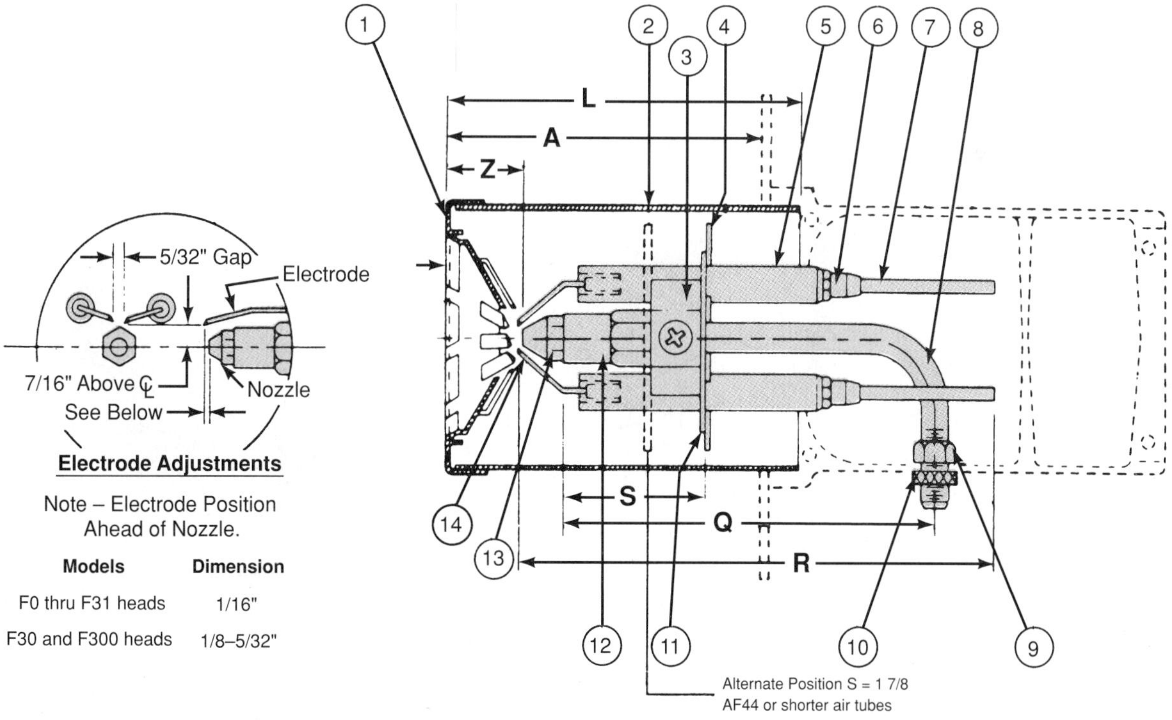

Figure 21-70. *Design of air tube, oil nozzle, and electrode assembly of gun-type oil burner. 1—Burner head. 2—Air tube. 3—Electrode clamp. 4—Centering spider. 5—Porcelain. 6—Electrode rod extension adaptor, as required. 7—Electrode rod extension, as required. 8—Nozzle line and vent plug. 9—Bulkhead fitting kit. 10—Locknut bulkhead fitting. 11—Static plate, static plate holding screws. 12—Nozzle adaptor, single. 13—Nozzle. 14—Electrode rod and tip. (R.W. Beckett Corporation)*

Figure 21-71 shows the various parts of a gun-type oil burner.

The ***choke*** is a tapered-down or smaller opening at the end of the air tube. (The air tube is a large tube that is part of the pair marked No. 1 in **Figure 21-71.**) The choke is located just past the oil nozzle. The choke has swirl strips (vanes). They increase the twist and turbulence of the air. This results in better mixing of the oil spray and air for more efficient burning. The flame shape can be changed by moving the choke closer to or away from the nozzle.

The air tube has a disk mounted inside it. This disk disturbs the airflow and creates air turbulence for better mixing. This disk is usually called the static pressure disk. The oil moving through the nozzle travels through very small holes. These holes are drilled at an angle to the nozzle. As the oil travels through the holes, it is given a twisting movement.

The oil burner motor is usually a split-phase 1/6 hp unit. It provides power for both the fan and the fuel pump. The motor is electrically connected to the master oil burner stack control. It uses 120 V, 60 Hz electricity. **Figure 21-72** shows an adjustable air band that adjusts the firing range of the burner.

Motor speeds may be 1750 rpm (at 60 Hz) fan speed and pump speed, or 3450 rpm (at 60 Hz) fan speed and pump speed. This depends on whether the motor is a 4-pole or 2-pole type. On a 50 Hz input, a 2-pole motor will run at 2850 rpm.

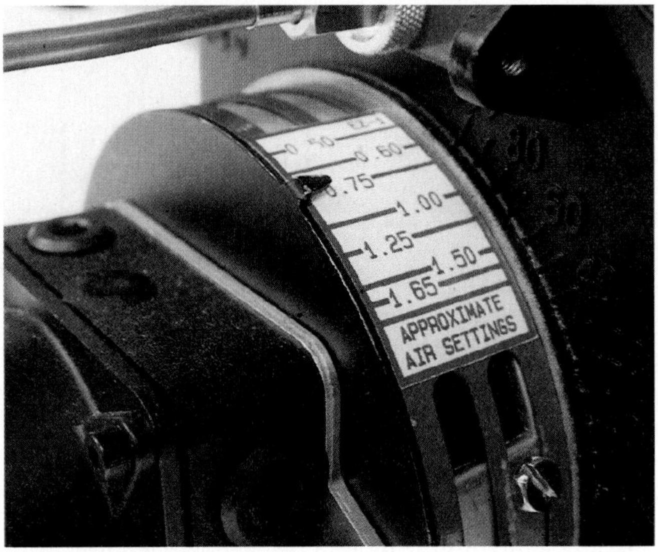

Figure 21-72. *Adjustable air band. The desired firing rate is set on air scale with adjustable arrow. This provides instant air setting for firing rate of burner. (Carlin Combustion Technology, Inc.)*

The fan is usually a radial flow type with adjustable air inlet openings. The openings are adjusted until the flame burns a yellow color.

Some excess air is needed to ensure enough oxygen for proper combustion of the oil. (The flame action

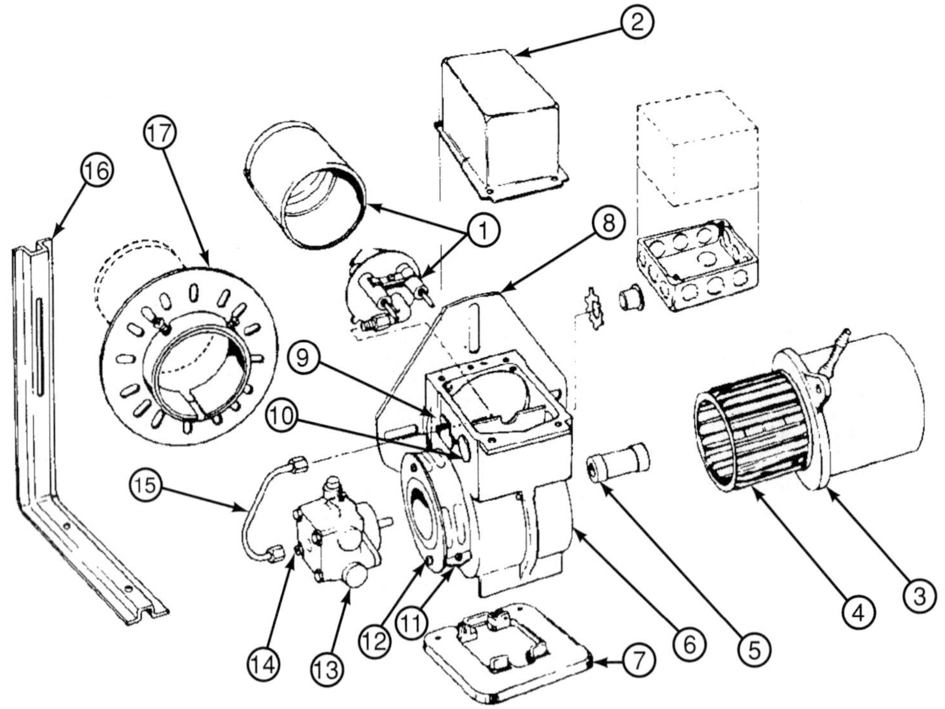

Figure 21-71. *Exploded view of gun-type oil burner. 1—Air tube combination. 2—Ignition transformer. 3—Drive motor. 4—Blower wheel. 5—Flexible coupling. 6—Burner housing assembly, with inlet bell. 7—Pedestal support. 8—Unit flange, or square plate. 9—Nozzle line escutcheon plate. 10—Hole plug-wiring box. 11—Bulk air band. 12—End air shutter. 13—Fuel unit. 14—Pump outlet fitting. 15—Connector tube assembly. 16—Extended pedestal. 17—Adjustable mounting flange. (R.W. Beckett Corporation)*

is fast!) The excess air also allows for airflow decrease in the period between furnace inspection and cleaning. (Fan blades pick up lint, causing airflow to decrease.) Excess air will slow down evaporation of oil droplets.

Adjusting an oil burner to obtain a proper flame can be misleading. A flame may look like it needs more air when it does not. This may be due to a dirty nozzle or impingement (striking the choke vanes). It may also be due to oil leakage during the "off" part of the cycle.

The only good way to check an oil burner is with instruments. Use a CO_2 analyzer, oxygen analyzer, smoke test, or other method.

Safety devices are installed to avoid spraying unburned oil into a furnace (a dangerous situation). Safety devices also prevent continuous oil pump operation in case of ignition failure or if the oil flame goes out. The stack control is one method. See Chapter 26 for oil burner controls and wiring circuits.

Figure 21-74. *Single-stage rotary fuel pump for gun-type oil burners. A—Shaft. B—Shaft seal. C—Pump rotor. D—Pump housing. (Suntec Industries, Inc.)*

21.24 Gun-Type Oil Burner Pumps

Several types of fuel oil pumps are used in gun oil burners. These include the gear type and the rotary type. The rotary oil pump is more common, but the gear type is more easily serviced.

21.24.1 Rotary-Type Fuel Oil Pump

Rotary pumps come in either single-stage, **Figure 21-73,** or two-stage models. The internal construction of a single-stage rotary fuel oil pump is shown in **Figure 21-74.**

The oil supply system carries fuel oil from the tank through a filter in the line. It is carried through the inlet

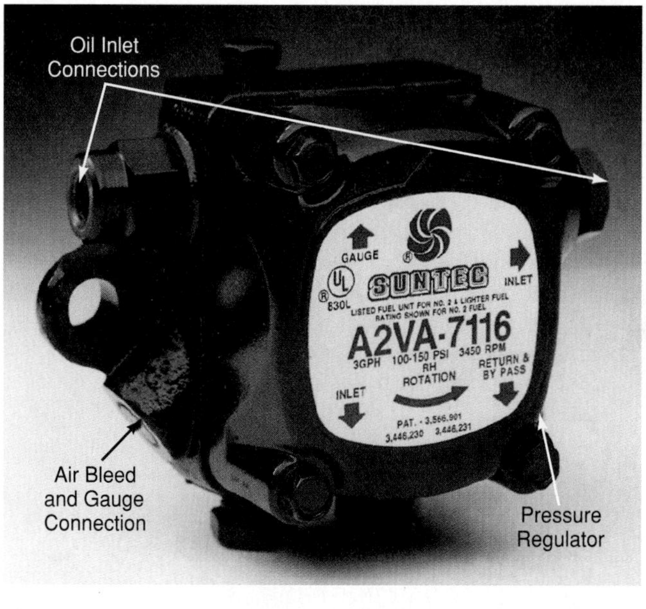

Figure 21-73. *Single-stage gun-type oil burner fuel pump. Note arrow on nameplate showing shaft rotation direction. (Suntec Industries, Inc.)*

screen to the pump. From there it goes into the pressure regulator and relief valve. The pump rotates counterclockwise and oil flow is from top to bottom in **Figure 21-73.** Note the shaft direction arrow on the pump nameplate. The oil leaves the lower central axis of the pressure regulator and passes to the gun nozzle. This is shown in **Figure 21-75.**

Another design for a single-stage rotary fuel oil pump is shown in **Figure 21-76.** The inlet is at the bottom left. The return/bypass is at the bottom right. The outlet is at the top left.

Many systems are equipped with a two-stage fuel oil pump when the two-pipe system is used. Part of the oil is returned to the fuel tank. The principle of operation is shown in **Figure 21-77.**

Selection of a single-stage or two-stage pump is determined by the limitations of the oil itself. Most single-stage pumps are sold for inlet vacuums of 7″ Hg (78 kPa) single-pipe. (For two-pipe, inlet vacuums would be 10″ Hg [68 kPa].) When fuel oil is subjected to a vacuum in excess of 10″ Hg, it starts to come apart. At 15″ Hg (51 kPa), it is a mixture of foam and clean oil. A two-stage pump is shown in **Figure 21-78.** It is ported so that, at 15″ Hg, the second (pressure stage) inlet is submerged in clean oil. The intake from the tank is at the top. Oil passes through the first stage of the pump into the regulator. From there, it passes back to the tank. The second stage removes only oil from the strainer chamber. The oil is pumped into the nozzle. Excess oil is returned to the strainer chamber in some pumps. It is returned back to the tank in others.

Details of the relief valve are shown in **Figure 21-79.** Oil pressure creates a force against the piston. When this force equals the compression spring force, the piston moves down. This permits oil to flow back into the pump inlet.

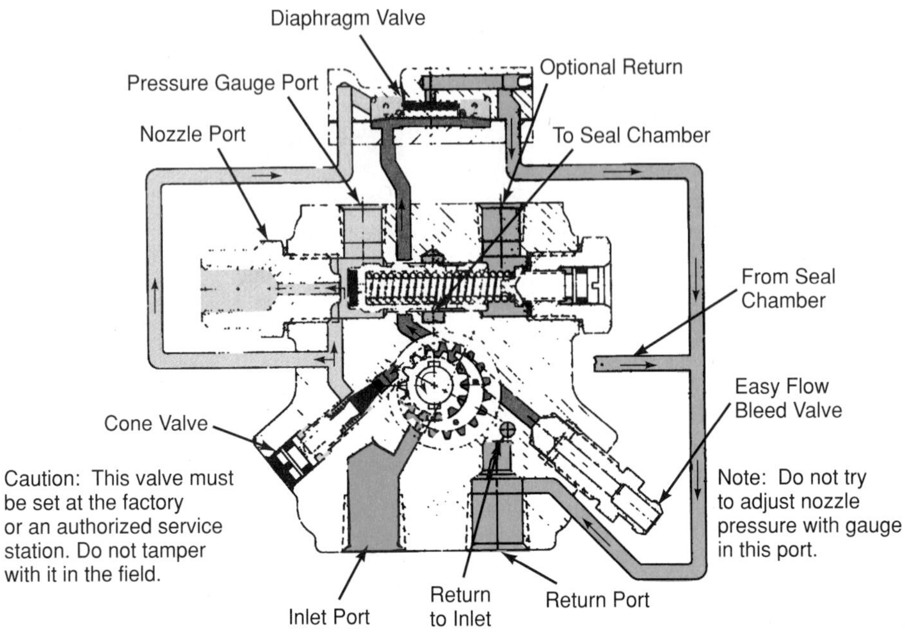

Figure 21-75. *Diagram of oil flow through a single-stage oil pump. The four colors represent the different pressure levels. Green—Suction or inlet pressure. Red—Gear pressure. Yellow—Nozzle pressure. Blue—Return pressure. (Suntec Industries, Inc.)*

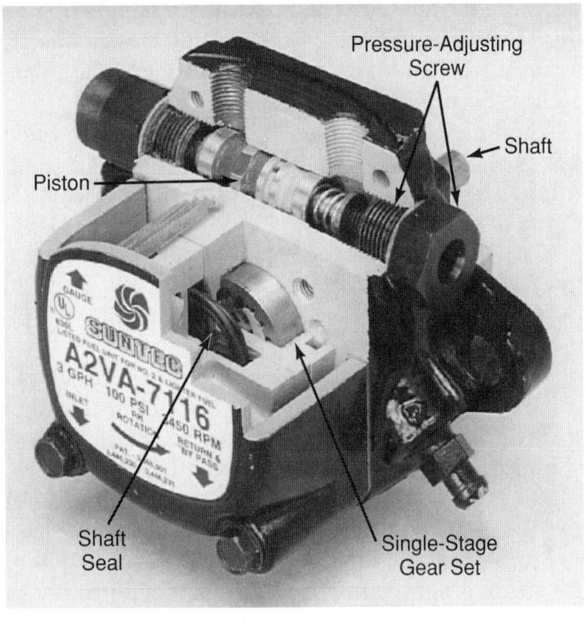

Figure 21-76. *Cutaway of single-stage rotary fuel oil pump for oil burners. (Suntec Industries, Inc.)*

21.24.2 Gear-Type Fuel Oil Pump

The gear-type oil pump is available in both single-stage and two-stage models. **Figure 21-80** shows the external appearance of a single-stage gear pump. Single-stage pump operation is shown in **Figure 21-81.** In A, the fuel oil enters the single-stage unit and fills the front chamber. The rotating blades filter the oil. This occurs as it passes from the front chamber to the suction sides of the gears. In B, the oil then goes from the lower suction side to the upper pressure side of the gears. It flows into the valve. At a predetermined pressure, the valve piston moves. The oil flows out the nozzle port. In C, the surplus oil returns to the front chamber through the surplus return passage. Oil lubricating the internal shaft and seal returns to the front chamber through the seal drain.

Nozzles are generally of two types. The 80° type gives a hollow-cone spray pattern. These have a capacity of 0.75 gph to 1.75 gph (gallons per hour). The 60° hollow-cone type has a capacity of 1.75 gph to 12 gph. A nozzle line heater **(Figure 21-82)** may be used on some oil burners. It maintains oil temperature and viscosity during off cycle in ambient situations. The oil is heated to 120°F to 130°F (48°C to 54°C) at the oil nozzle. This improves the ignition and overall combustion.

The gun-type oil burner is an efficient heating unit. However, it must be properly maintained to give peak performance. An experienced technician should check, clean, and adjust the system each year. Some of the important items to check are shown in **Figure 21-83.**

21.25 Electrical (Transformer-Electrode) Ignition

Gun-type oil burners generally use electrical ignition. Ignition controls are described in Chapter 26. The system includes a transformer and two electrodes. The transformer is mounted on the oil burner. It transforms 120 V ac to about 10,000 V ac. The ignition system must raise the oil temperature to 700°F (370°C) for

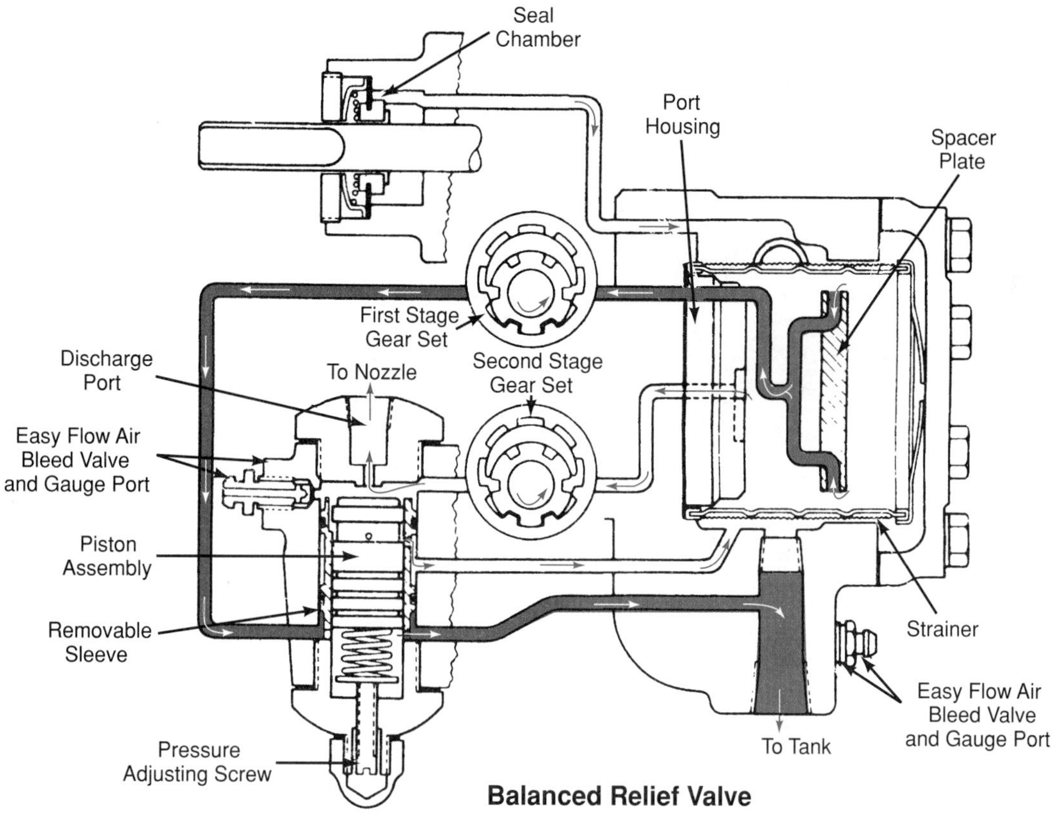

Figure 21-77. *Two-stage pump for oil burners, used with two-pipe system from storage tank. Note the flow of some oil back to the storage tank.*

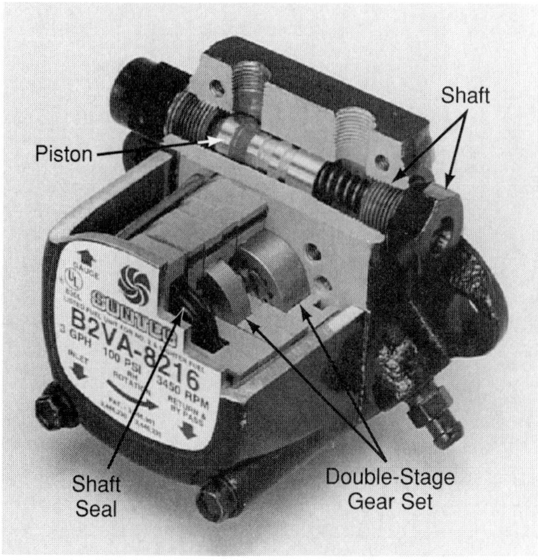

Figure 21-78. *Two-stage fuel oil pump used with two-pipe system from storage tank. (Suntec Industries, Inc.)*

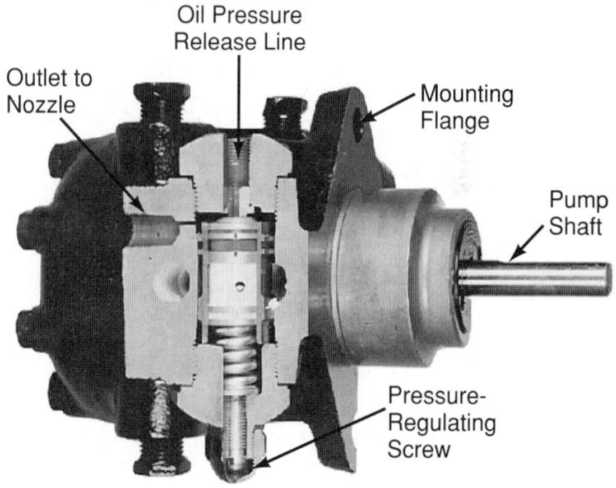

Figure 21-79. *Cutaway view of relief valve for a fuel oil pump. (Suntec Industries, Inc.)*

burning to occur. The electrodes, made of stainless steel, are mounted in ceramic insulators. No part of an electrode should be closer than 1/4″ (6 mm) to any metal part.

The electrode ends are positioned in front and above the nozzle. The atomized oil swirls out of the nozzle and mixes with the turbulent air. At the same time, a spark jumps between the electrode ends and ignites the mixture. The ignition may be continuous while the oil burner is in operation, or the ignition may operate only until the fuel ignites.

The electrode gap should be between 1/8″ and 3/16″ (3 mm and 5 mm). The electrode ends should be approximately 1/2″ to 5/8″ (13 mm to 16 mm) above the nozzle, and 5/16″ to 1/2″ (8 mm to 13 mm) in front of the nozzle. For over-45° nozzles, this last dimension should be approximately 1/2″ (13 mm). For 30° nozzles,

Figure 21-80. *Single-stage gear pump for gun-type oil burners. (Suntec Industries, Inc.)*

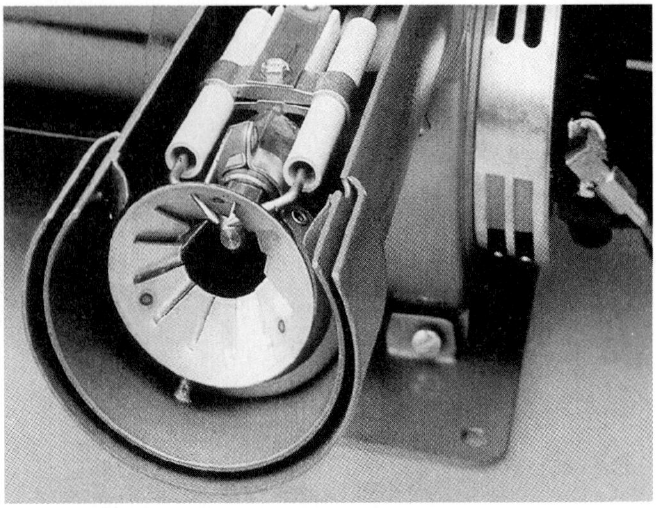

Figure 21-82. *Nozzle line heater attached to combustion head. (Carlin Combustion Technology, Inc.)*

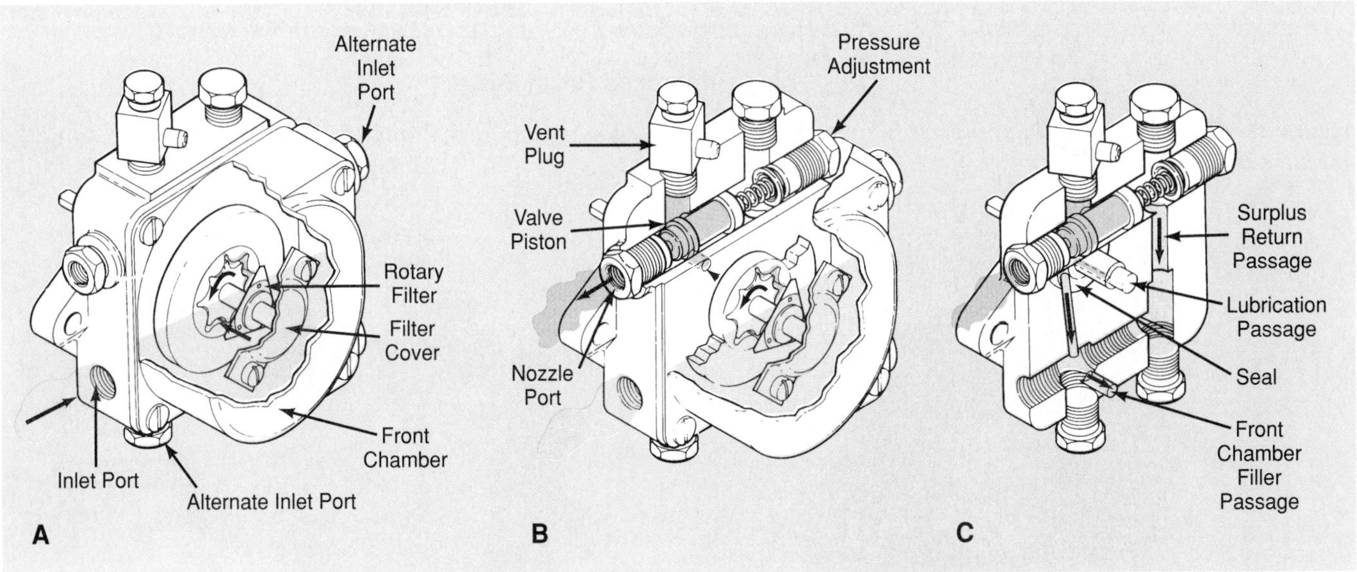

Figure 21-81. *Single-stage gear pump unit operation. (Webster Heating)*

the electrode ends should be 5/16″ (8 mm) in front of the nozzle. Refer to the manufacturer's service manual for exact setting specifications. The porcelain insulators must be kept clean or the high voltage will short. The spark should jump a 1″ (25 mm) gap with the blower off.

A flame inspection mirror, **Figure 21-84,** can be used to observe ignition and spray action. This is done to check if operation is normal.

Delayed ignition and a puffback may be due to any of the following:

- Weak ignition.
- Wrong position of the electrodes.
- Poor insulation.

Figure 21-85 shows the interior of an ignition transformer case. A transformer and line voltage testing instrument is shown in **Figure 21-86.**

A ***puffback*** is the ignition of a large amount of vaporized oil in the firepot. Sometimes, it will blow soot into the furnace room and into the living quarters. This makes a major cleaning job necessary. In no case should the electrode ends be touching the oil spray. If they do, the electrodes will become carbon-coated.

Moisture in the oil will retard combustion and may even cause a "flameout." When a flameout occurs, combustion stops but fuel continues to feed into the combustion chamber. If moisture is causing poor combustion, continuous ignition is usually recommended.

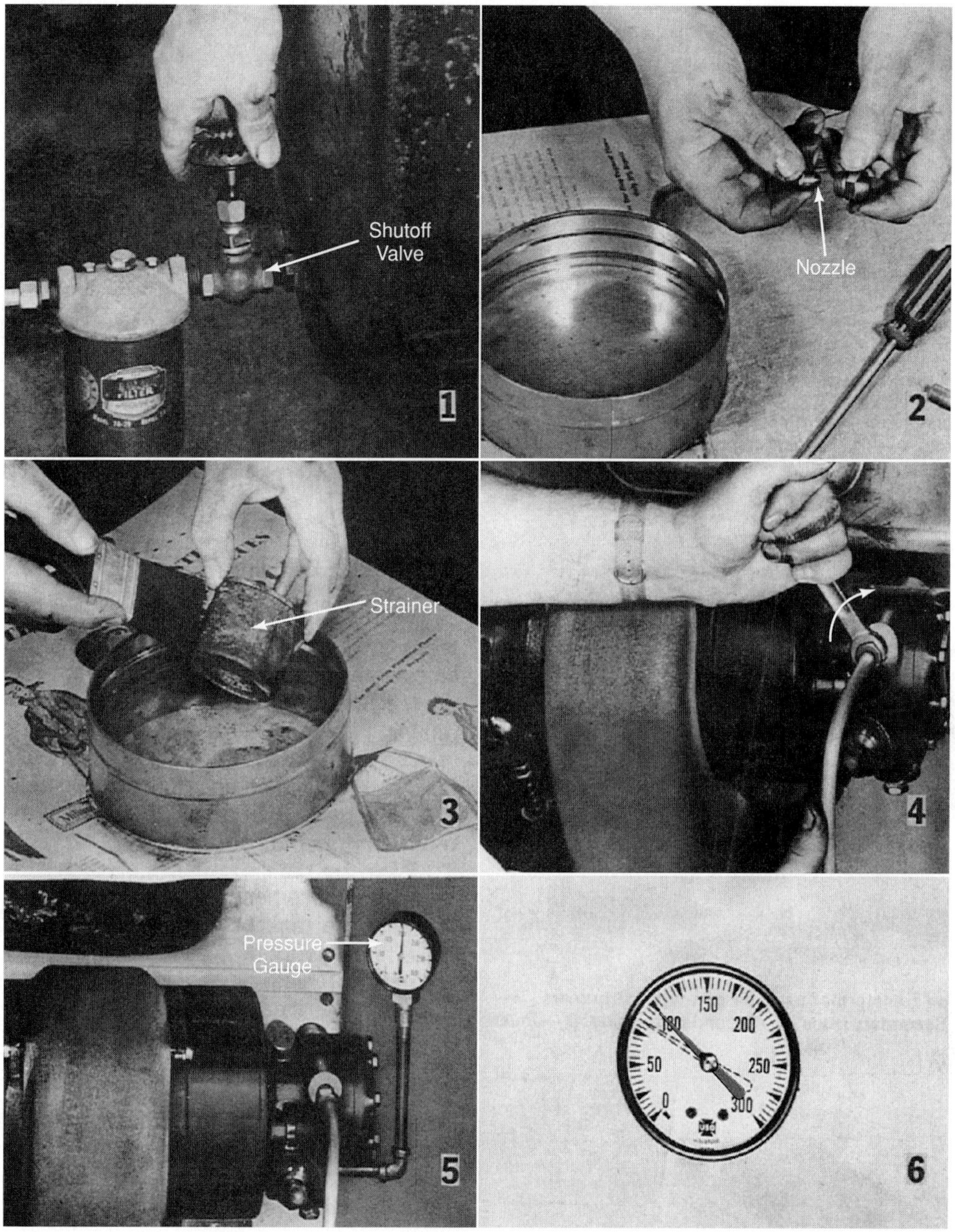

Figure 21-83. *Six main items to be checked at start of each heating season. 1—Shutoff valve and line filter. Replace filter cartridge. 2—Check and clean nozzle assembly. Follow manufacturer's recommendations. 3—Clean strainer using clean fuel oil or kerosene. 4—Check connections for tightness. 5—Insert pressure gauge into pressure port. Start burner and adjust pressure setting to manufacturer's specifications, usually about 100 psig (115 psia or 790 kPa). 6—Pressure gauge reading for correct pressure setting. (Suntec Industries, Inc.)*

21.26 Primary Control

The primary control unit has two main functions. First, it provides the means for operating the unit. It receives the low-voltage signal from the room thermostat. This causes the burner motor and ignition transformer to operate. Its second function is that of safety. If the burner does not ignite, or the flame is extinguished, it will trip the primary relay safety switch. The primary relay must be reset manually.

21.27 Oil Tank Installation

Oil burners must be installed with great care. A complete installation includes a 200 gal. to 1000 gal. storage tank, hand shutoff valve, filter and trap combination, and copper tubing oil line. **Figure 21-87** shows an installation for a one-pipe system. The storage tank is located in the room with the furnace. **Remember that the storage tank should be at least 7′ (2 m) away from the furnace.** In this installation, oil feeds by gravity to the

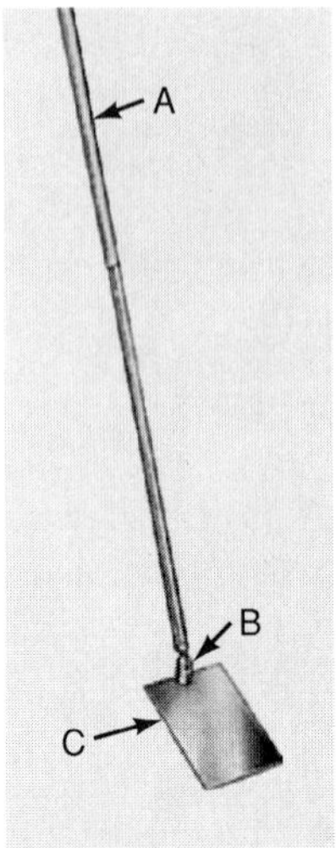

Figure 21-84. *Inspection mirror used to check nozzle condition, electrode condition, and flame. A—Telescoping handle. B—Hinge. C—Metal mirror.*

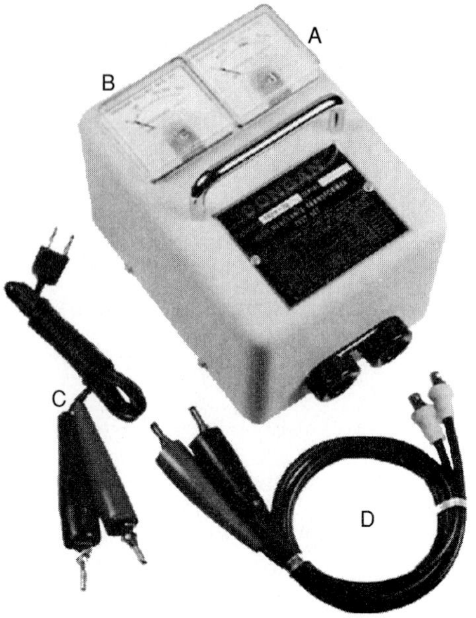

Figure 21-86. *Ignition transformer testing instrument. A—High-voltage meter. B—Line-voltage meter. C—Line testing leads. D—Transformer high-voltage testing leads. (Dongan Electric Mfg. Co.)*

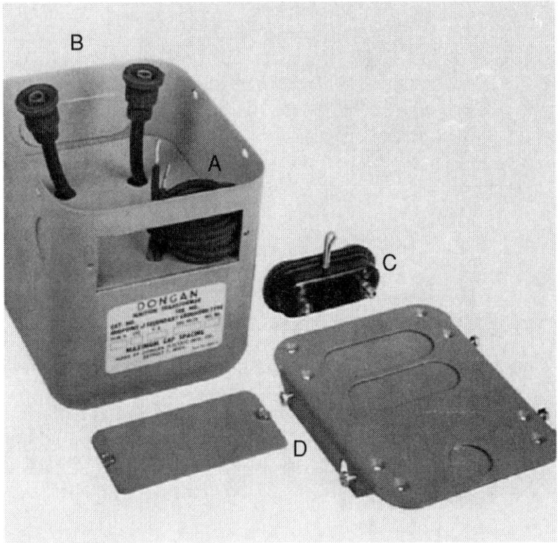

Figure 21-85. *Ignition transformer used on gun-type oil burners. A—Primary leads. B—Secondary leads. C—Secondary bushing. D—Cover plates.*

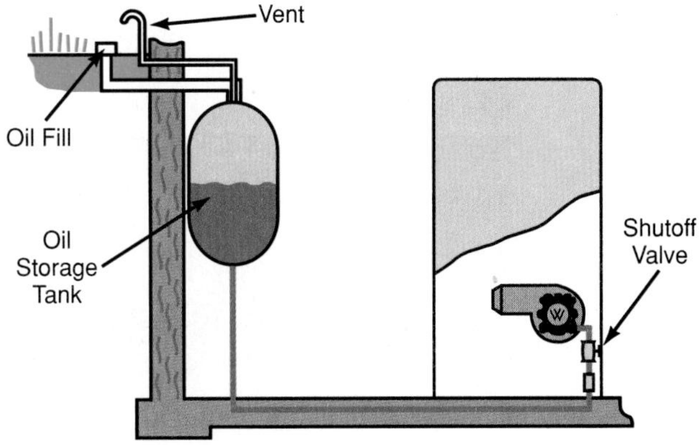

Figure 21-87. *Typical gun-type oil burner installation. Note fill pipe, vent pipe, and oil line installation. (Webster Heating)*

oil burner. The storage tank should be elevated less than 25′ (7 m) above the burner. This will keep the feed line pressure below 10 psig (25 psia or 170 kPa). Otherwise, a pressure-reducing valve must be used.

In some installations, the fuel tank is placed outside the building, either above ground or underground. Some local codes allow only 3 psig to 5 psig (18 psia to 20 psia or 120 kPa to 140 kPa) on the pump inlet. The two most common installations are:

- Placing the tank above the oil burner, as in a residence with a basement. See **Figure 21-88.**
- Placing the tank below the oil burner level, as in a home without a basement. See **Figure 21-89.**

These installations should have the tank located within a reasonable distance of the oil burner. On runs of 50′ to 100′ (15 m to 30 m), 3/8″ (10 mm) tubing should be used. For runs of 200′ to 300′ (60 m to 90 m), 1/2″ (13 mm) tubing should be used. The manufacturer's specifications should be checked if the oil must be raised above the tank.

The oil tank should be installed with a slight slope away from the oil line connection. This provides a low spot in the tank for dirt and water to collect. The vent pipe is very important:

- It provides atmospheric pressure inside the tank and permits volatiles to escape.
- It must be designed with a 180° bend (to keep out dirt and rain).
- It must have an opening that is above the highest possible snowfall or other blockage.

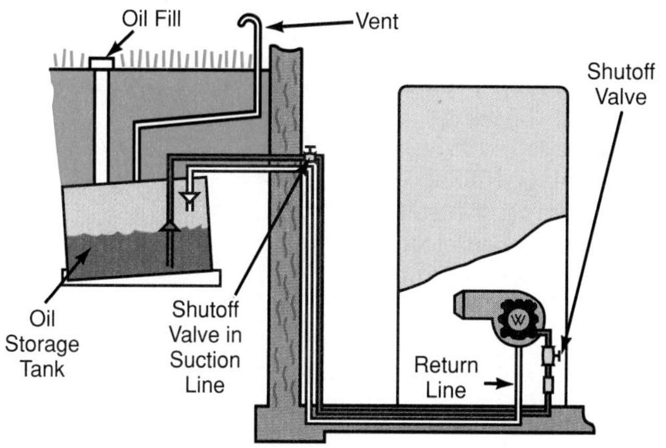

Figure 21-88. *Gun-type oil burner installation with storage tank installed underground, but above oil burner. Note two oil lines. (Webster Heating)*

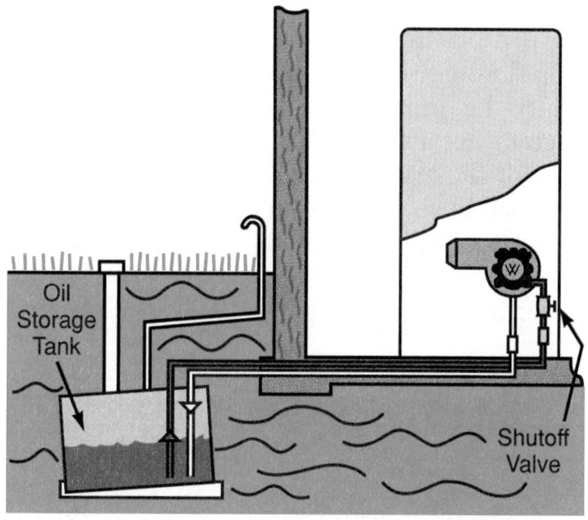

Figure 21-89. *Gun-type oil burner installation with storage tank underground, below the level of oil burner. (Webster Heating)*

The oil fill cap should always be in place, except when filling the tank. The cap helps keep the fuel oil clean and reduces the chances of fire or explosion.

Always use a thread compound on the pipe threads of the fittings. This compound should be of the oil-resistant, non-hardening type. During storage and while installing tubing, keep the tubing ends sealed with tape. This will keep out dirt and moisture. Remove the tape at the time the tubing is connected. The 3/8″ or 1/2″ (10 mm or 13 mm) OD copper tubing is attached to the fittings with standard SAE 45° flares. Flaring techniques are described in Chapter 2. Tight, leakproof connections are essential. Never use Teflon® tape.

The oil lines in the tanks should be mounted so the tubing opening is 3″ to 4″ (7.5 cm to 10 cm) above the bottom of the tank. The return oil line, if the system has one, does not need to go near the bottom of the tank for light oils (Number 1 or Number 2). It should go to within 4″ to 5″ (10 cm to 13 cm) of the tank bottom for heavy fuel oil. This will keep the oil more fluid. **Never use compression fittings with fuel oil systems.**

21.28 Oil Burner Installation

The oil burner installation must be made in accordance with local codes and the manufacturer's instructions. The burner must be the correct height above the bottom of the combustion chamber. Burners are mounted either on adjustable legs or on a flange bolted to the furnace. The burner air tube and nozzle must be inserted into the combustion chamber an exact distance. The distance must be exactly as the manufacturer recommends. The opening into the furnace must be carefully sealed to prevent air leaks. (Some have flange adaptors.)

The nozzle must be the correct size, and it must be in good condition. The nozzle orifice (hole) size and the amount of oil pressure determines the rate at which fuel oil burns. This also determines therefore, the rate at which heat is produced. The size of the nozzle selected must match the heating requirements of the heated space. If the nozzle is too small, the burner may not heat the space adequately. If the nozzle is too large, the burner may turn on and off quite frequently.

Nozzles are usually supplied with a fine filter at their inlet. The filter is designed to eliminate the possibility of dirt entering and plugging the nozzle.

Be careful not to twist the tube or move the nozzle tube out of line. A tool for safely removing the nozzle is shown in **Figure 21-90.**

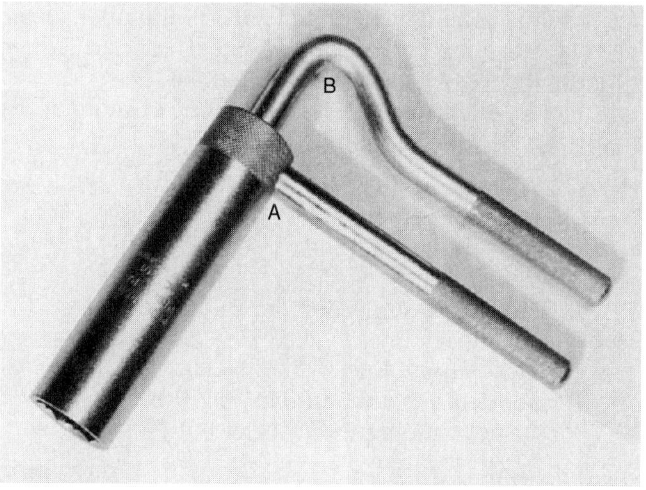

Figure 21-90. *Tool for removing and installing nozzles. A—Nozzle tube socket wrench and handle. B—Nozzle socket wrench and handle. (Monarch Nozzle Company)*

21.28.1 Starting an Oil Burner after Installation

Before starting the oil burner, air should be removed from the lines and the pump. A vent plug (air bleeder fitting) is mounted in the pump housing. Usually, this vent plug seals the port used for pressure gauge installations.

If enough oil and air collects in the combustion chamber and ignites, anything may happen. There may

be a puff of flame or an explosion. The explosion may be forceful enough to wreck the building, and maim or kill. Inspect the firepot. If oil is present, shut off oil valves and vent the combustion chamber. Remove the oil by means of a suction pump, rags, etc. Continue removal until all danger of oil fumes is gone. Use an explosion-proof flashlight.

Air in an oil line will form bubbles, which could result in:

- Oil not being pumped.
- Blowbacks.
- Flame failures.

The line must be completely purged of air. A two-pipe system reduces the chance of air remaining in the system. However, air can still be trapped in high spots in the line. A leak in the oil line will almost always cause air-in-line troubles.

Always check the fuel oil nozzle to be sure it is the correct size. Also be sure that it is in the center of the gun air duct. The electrodes must be kept clean and in correct relation to the nozzle. **Figure 21-91** shows a typical oil burner nozzle. These nozzles come in various capacities. All are based on gallons per hour (gph) at 100 psig (115 psia or 790 kPa). Capacities range from 0.40 gph to 28 gph. Some nozzles are large enough to feed 100 gph. A 1 gph nozzle delivers 140,000 Btu/hr. Thus, if the overall efficiency is 60%, the useful heat would be 84,000 Btu/hr. Poor oil delivery may be the result of clogging. The main filter, the pump filter, or the nozzle filter may be partially clogged. Check all three filtering devices when servicing the unit.

Flame failure may be caused by one or more of the following:

- Oil tank out of oil.
- Oil tank not vented.
- Clogged filter in oil line.
- Ice in fuel line.
- Loose oil line connection (air in line).
- Dirt in supply line.
- Water in supply line.
- Loose wiring or connections.
- Motor not running (check reset button).
- Defective pump.
- Pump losing prime.
- Changing pressure or low pressure at pump (slipping coupling).
- Clogged nozzle.
- Damaged nozzle.
- Improperly installed bypass plug.
- No spark at electrodes:
    - Loose wiring.
    - Bad transformer.
    - Low voltage.
    - Crack in electrode porcelain.
    - Electrodes carboned.
    - Electrode spacing too far or too close.
    - High-voltage wiring loose.

Proper flame appearance is luminous (mainly yellow). If there is insufficient air, the flame turns dull orange or red. There may also be smoky tips to the flame. **Figure 21-92** illustrates a properly adjusted flame.

The draft in the firepot is measured by the air pressure drop (in the firepot). It should be about 0.02″ to 0.05″ (0.5 mm to 1.3 mm) WC. (Use an inclined water tube manometer.) See **Figure 1-16.** This check will also help determine if the automatic draft is working properly.

Some oil burner motors are reversible. **Figure 21-93** shows the method of reversing one type of oil burner motor. Controls for oil burners are described in Chapter 26. Testing instruments are described in Chapter 23.

Figure 21-92. *Gun-type oil burner flame. (Carlin Combustion Technology, Inc.)*

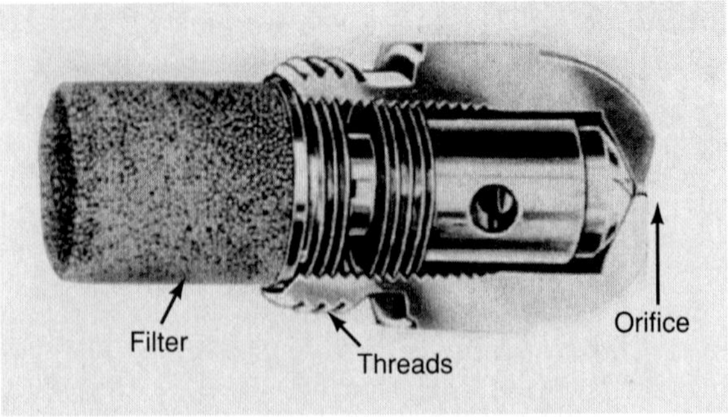

Figure 21-91. *Stainless steel nozzle used with gun-type oil burners. Note fine filter at nozzle inlet. (Monarch Nozzle Company)*

Figure 21-93. *Method of reversing direction of rotation of an oil burner motor. This is accomplished by reconnecting two wires, as shown by dashed lines.*

Inspect the electrode wires. If they are cracked or brittle, replace the wires. Inspect the electrode tubular ceramic insulators. If the ceramic tubes are cracked, replace them.

Soot deposits in a combustion chamber typically collect when the unit first starts. The more often the unit starts, the greater the soot deposit. A correctly sized oil burner that operates less frequently will deposit less soot in the furnace and stack.

If an oil line is dirty or clogged, blow it out with nitrogen gas. Always use a pressure regulator and relief valve. ***Never* use compressed air or oxygen! A violent explosion may result.**

21.29 Servicing a Fuel Oil Burner

Oil burner problems, symptoms, and possible causes are given in the following troubleshooting outline.

I. Burner motor does not start, starts and locks out (CAD cell shuts off the control switch), or cycles.
   A. Does not start.
      1. Relay does not close (will not close or contacts dirty).
      2. Safety lockout stays open.
      3. Bad relay coil.
      4. Low voltage.
      5. Open high limit control.
      6. Broken wires or loose connections.
      7. Relay transformer open.
      8. Thermostat open (dirt on contacts, loose or dirty connections).
      9. Stack switch open.
      10. Heat sensing contacts out of place or open.
      11. Motor overload open (burned out, or has dirty contacts).
   B. Starts, but locks out.
      1. No fuel oil out of nozzle.
         - Clogged.
         - Pressure too low.
         - Pump not working.
         - Loose motor coupling.
         - Air leaks in fuel line.
         - Fuel oil line hand valve closed.
         - Strainers or screens clogged (filter, pump screen, or nozzle strainer).
         - The pressure regulator in the pump body is stuck open.
         - Vent on fuel oil tank closed.
         - Empty fuel oil tank.
      2. Fuel oil coming out of nozzle but no ignition.
         - Electrodes not positioned correctly.
         - Insulators cracked.
         - Ignition wires worn, loose, or with dirty connections.
         - Transformer not operating.
         - Primary wires worn, loose, or with dirty connections.
         - Low line voltage.
      3. Fuel oil to nozzle, ignition OK, but no flame.
         - Clogged nozzle.
         - Clogged nozzle strainer.
         - Nozzle loose.
         - Pressure too low.
         - Fuel oil too heavy (wrong oil or too cold).
         - Excessive air or too much draft.
         - Electrodes in wrong position.
      4. Flame burns only a few seconds.
         - Flame sensor not in correct position.
         - Stack switch not operating correctly.
         - Excessive air or air too cold.
         - Flame is too lean.
   C. Cycles, but not on lockout.
      1. Thermostat differential too close.
      2. Anticipator set too close.
      3. Limit switch set too low.
      4. Overfired (reaches high limit temperature too quickly).

II. Burner does not operate correctly.
   A. Smoke, soot, odors, and/or pulsating sound.
      1. Wrong oil pressure.
      2. Flame touches combustion chamber.
      3. Not enough draft.
         - Dirty chimney.
         - Draft control is out of adjustment or it is stuck open.
         - Dirty flue.
         - Either the combustion chamber or the heat exchanger leaks.
      4. Poor mixing of air and oil.
         - Nozzle is worn, loose, dirty, or drips.
         - Oil pressure too low or high.
         - Poor air velocity and turbulence.
         - Not enough air (shutter closed too much, fan binding, or tight bearings).
   B. Puffs back.
      1. Water in oil.
      2. Delayed ignition.
         - Electrodes not positioned correctly or loose.
         - Insulators carbonized.
         - Nozzle worn, loose, dirty, or drips.
         - Voltage drop when burner starts.
         - Oil pressure too low or too high.
         - Transformer leads loose or dirty.
         - Transformer not operating correctly.
         - Excessive air or high draft.
   C. Noise.
      1. Loose fan.
      2. Loose shutter.
      3. Worn pump.
      4. Dirty strainer.
      5. Air in oil line.
      6. Transformer hum.
      7. Draft control vibrates.
      8. Motor coupling worn.
      9. Motor and pump not lined up correctly.
      10. Relay contacts not seating tightly.
      11. Oil suction line restricted.
      12. Motor mounting loose.

13. Tight motor bearings.
14. Tank hum.

D. Fuel oil consumption is too high.
1. Nozzle is worn, loose.
2. Combustion chamber is dirty.
3. Too much combustion air (heat escapes up flue due to high flow of flue gases).
4. Poor mixing of air and oil.
5. Not enough draft over fire.
6. Air leaks into combustion chamber.
7. Oil pressure too high or too low.
8. Overfired, as noted by a high stack temperature.

There are two ways to check combustion efficiency. The first is carbon dioxide (CO_2) analysis of flue gas. The second is by the temperature of flue gases.

1. Use a carbon dioxide analyzer to check a sample of flue gas. It should be 10% to 12% CO_2 without visible smoke. If the reading is too low, it means too much air:
   - With 6% CO_2, 155% excess air is used.
   - With 8% CO_2, 85% excess air is used.
   - With 10% CO_2, 50% excess air is used.
   - With 12% CO_2, 26% excess air is used.
2. If the temperature of the flue gas is too high, a considerable amount of heat is being wasted. If the flue gas temperature is too low, water vapor condenses in the flue or chimney. The small amount of sulfur will form sulfurous acid (H_2SO_3), which is very corrosive.

   With a 10% to 12% CO_2 stack analysis, combustion efficiency is as follows:

Temp. °F	Temp. °C	Percentage of efficiency
1000	538	65 to 69
800	427	70 to 73
600	316	76 to 79
400	204	82 to 84

These are flue gas temperatures minus furnace room air temperature.

ELECTRIC HEATING MODULE

21.30 Electric Heat

The use of electricity to heat homes, stores, commercial buildings, and factories is steadily growing in popularity.

Some advantages of electric heat are:

- Low first cost.
- Electric heating devices need no oxygen and, therefore, no air supply.
- Highest temperatures needed are below the ignition temperature of most materials. Therefore, the system is considered safer than other heating systems.
- Due to the absence of combustion and combustion gases, there is less danger of toxic conditions arising. No chimney is needed.
- Equipment normally requires less space than other systems.
- Individual room temperature control is easily obtained.
- Very clean operation.

The electric heating process also has some disadvantages, however:

- The cost per unit of heat is higher than for some other fuels.
- Humidity control problems may occur.
- Added electrical circuits are required.

21.31 Principles of Electric Heating

Electricity is a form of energy. Since one form of energy can be changed to another form of energy, electrical energy can be changed to heat energy.

Heating with electricity can be done either directly or indirectly:

- Direct heat.
  - Resistance heating, accomplished by passing a fluid over an electrically heated element. (Air is the fluid, although some units use water.)
  - Radiant heating, accomplished by heating an element to a temperature high enough to give off heat.
  - Thermoelectric heating. See Chapter 18.
- Indirect heat, by using a heat pump. See Chapter 24.

21.32 Applications of Electric Heating

Electric heating has a wide range of applications. It may be used for heating operations in industrial processes. Examples would be fast drying of paints and melting of low-temperature metals or metal alloys. It has been used for domestic and commercial cooking and baking. It is often used for providing hot water.

For residential use, electric heating has a growing use as:

- The only source of heat.
- Supplementary heat, even though some other system provides most of the heat in the residence. It is also used where the heat load is low. Heat pump systems may use electric resistance heating. It is used where the heating energy required during cold weather is more than the system can supply.
- Resistance heating provides heat in parts of a building that are not well-heated. It may be used in

areas of a building where heating by the standard system would be unsafe. Resistance heating also may be used to heat additions to buildings. The present system might not have enough capacity to heat the addition. Perhaps the extension of the present system would be too costly.

21.32.1 Electric Furnaces

Electricity may be used to provide heat for warm air or hydronic central heating systems. These furnaces have capacities ranging from 34,000 Btu/hr. (10 kW) up to 120,000 Btu/hr. (35 kW).

A forced-air furnace that uses tubular heating elements is shown in **Figure 21-94.** This furnace can be installed for upflow, downflow, or horizontal airflow. The heating elements are used in stages of 5 kW or 10 kW. The elements are sequenced (turned on one at a time). The electrical circuits are shown in **Figure 21-95.** Note that the incoming power is 240 V, while controls are 24 V.

A hydronic boiler is shown in **Figure 21-96.** This boiler has a capacity of from 34,130 Btu/hr. (10 kW) to 81,900 Btu/hr. (24 kW). It is a very efficient unit. It measures about 10 1/2″ wide by 21 3/4″ high, and 23 3/4″ long (about 25 cm × 50 cm × 60 cm).

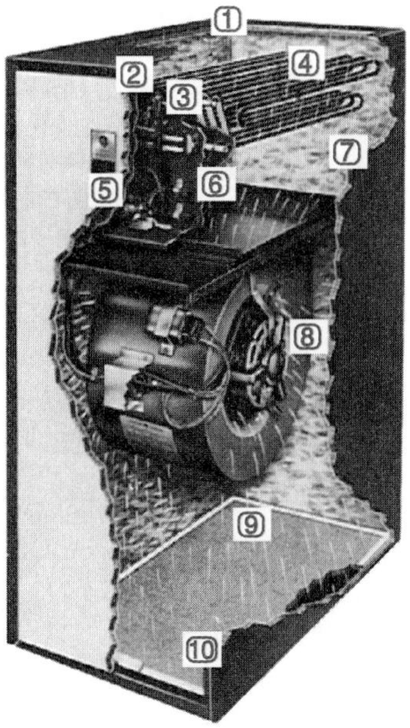

Figure 21-94. *A forced-air furnace with tubular electric resistance heating elements. 1—Warm air outlets. 2—Insulated jacket. 3—Fan and limit controls. 4—Electric resistance heating elements. 5—Operating controls. 6—Fan outlet. 7—Heating chamber. 8—Fan and motor. 9—Filter. 10—Furnace base.*

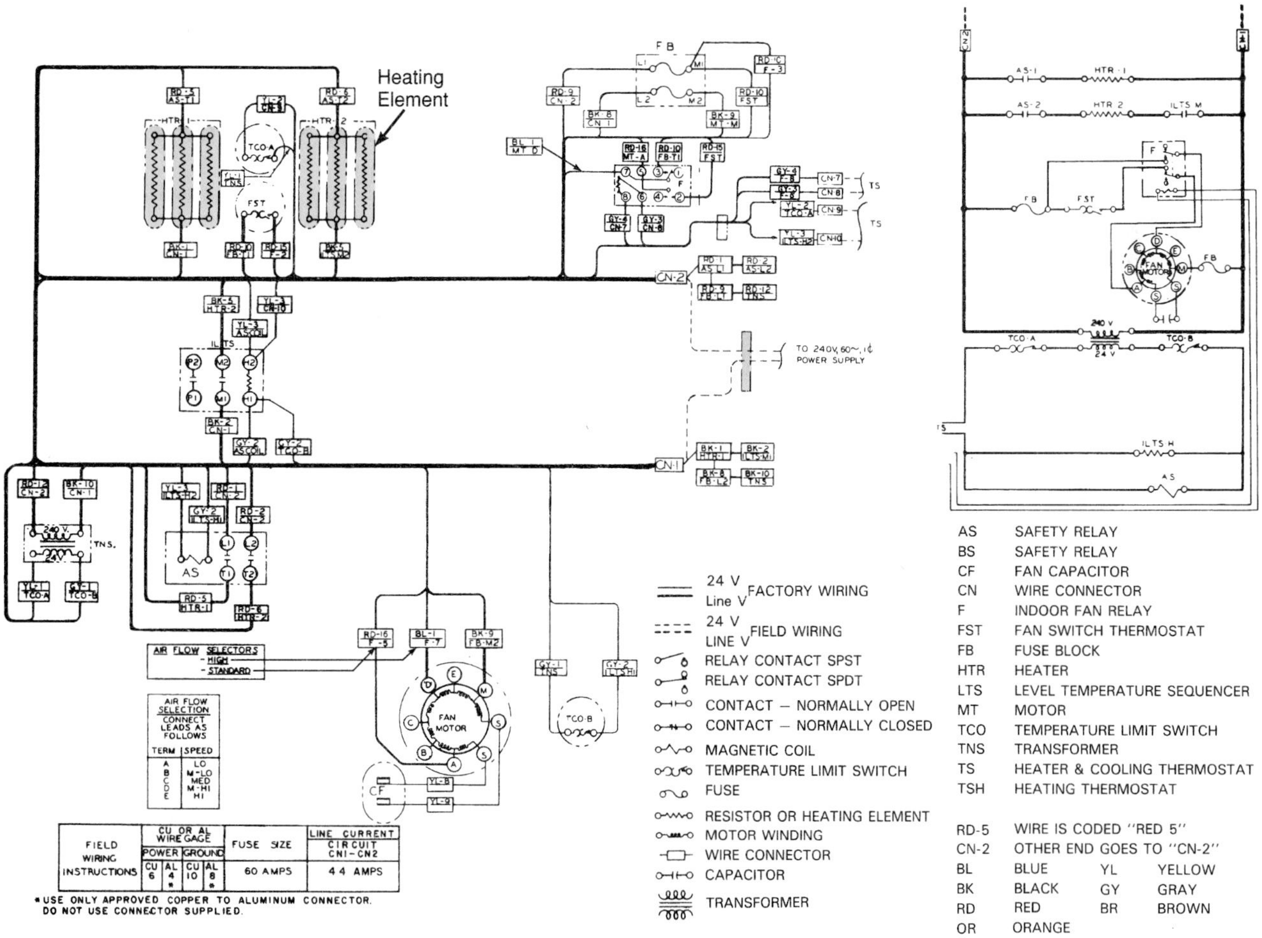

Figure 21-95. *Wiring diagram of electric furnace, with schematic. (General Electric Co.)*

ers are using solid fuel. It may provide some or all of their heating energy needs. Commercial systems are available which replace natural gas or oil heating sources with wood- or coal-burning sources. This can be found in hydronic systems as well as forced-air systems.

The use of wood or coal may be more economical in some situations. However, it usually requires a considerable amount of attention by the homeowner. It takes time to load the fuel, stoke the fire, and clean the ashes. Also, the flue and chimney must be frequently and carefully maintained. Solid fuels burn dirtier than conventional fuels. This means there is more soot in the flue and chimney.

A solid fuel heating unit is commonly used as part of a conventional furnace system. An example of this is a system that uses a heat exchanger in the fireplace. This heat exchanger is connected in series with conventional heating system ductwork in the home. When a fire is in the fireplace, the heat exchanger draws latent heat. (This latent heat is from water vapor in the fireplace flue gas.) The warm air is distributed through the forced air ductwork.

21.39 Solar Heating

All forms of energy originate from the sun. The amount of energy produced by the sun and the growing need to conserve fossil fuels makes solar heating desirable in some areas. Solar energy is used for heating, cooling, and domestic water heating. It is an alternative to using other energy resources.

The use of solar energy in heating systems has developed slowly. Solar heating development began in southern California during World War I. It was used there in military camps. It reappeared as a means of water heating in the 1930s. The use of solar energy expanded in the late 1970s and 1980s. At that time, federal programs became available for solar heating and cooling.

In its simplest form, a solar heating/cooling system is one that reduces conventional fuel consumption. Solar energy heating systems use equipment to collect, store, and distribute solar heat. The two basic types of solar systems are passive and active. An example of a ***passive solar heating system*** is the sun's rays entering a south window in cold weather. The sun's rays increase the room temperature.

Active solar heating systems have a number of different sections. These are necessary to utilize the heat in a given area. The solar heat must be collected, stored, and distributed. The active system is the concept that is most used for heating and cooling.

Solar heating and cooling has increased in recent years in areas where sun rays are available for extended periods of time. More detailed information on solar energy can be found in Chapter 25.

21.40 Cogeneration

Many large industrial plants, commercial buildings, and shopping malls have installed cogeneration systems. These systems both generate electricity and supply heat. This type of system has been installed where there is a need for heating year-round.

Most installations use gas turbines, **Figure 21-107,** to generate electricity. The waste exhaust heat is used to produce steam for industrial processes or building heat. The exhaust gases leave the turbine at temperatures of about 700°F (370°C). These temperatures are high enough to run a boiler at 75% of its usual efficiency.

In some installations, the waste heat is used to drive absorption cooling systems. These are called on-site generating systems, combined-cycle plants, or total energy systems. See Chapter 24 for further information.

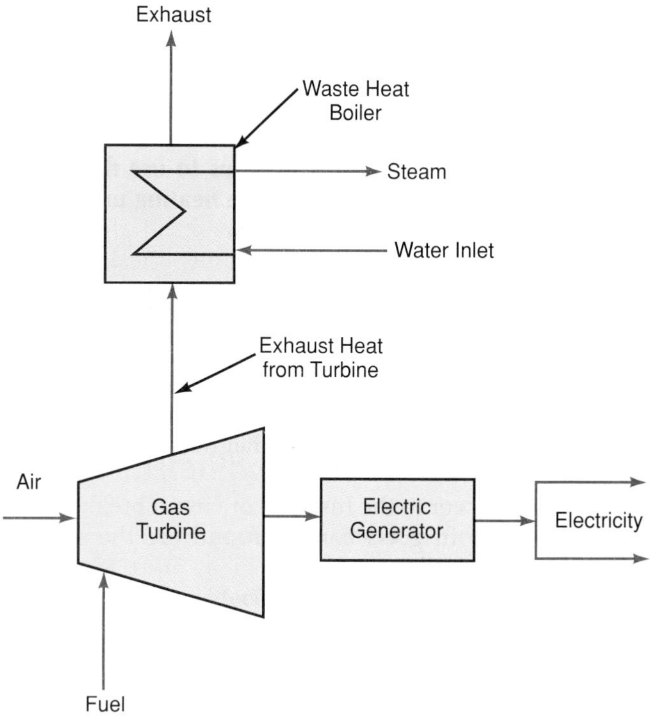

Figure 21-107. *Simple cogeneration system. Waste heat from gas turbine is used to produce steam, which may be used for heating or for absorption cooling systems.*

HUMIDIFICATION MODULE

21.41 Humidifiers

When air is heated, it can absorb more water vapor. Human comfort requires a relative humidity (RH) of about 35%. When outside air at 30°F (−1°C) and 90% RH is heated to 72°F (22°C), its RH drops to about 18%. See **Figure 21-108** for information from a psychrometric chart.

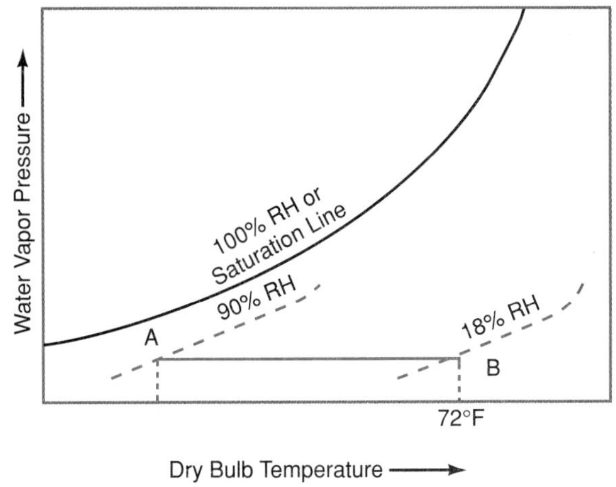

Figure 21-108. *Graph showing decrease in relative humidity. A—Starting point. Air sample is heated from 30°F (–1°C) at 90% RH, to 72°F (22°C) at 18% RH. B—Result for process carried out without adding any water vapor to air sample.*

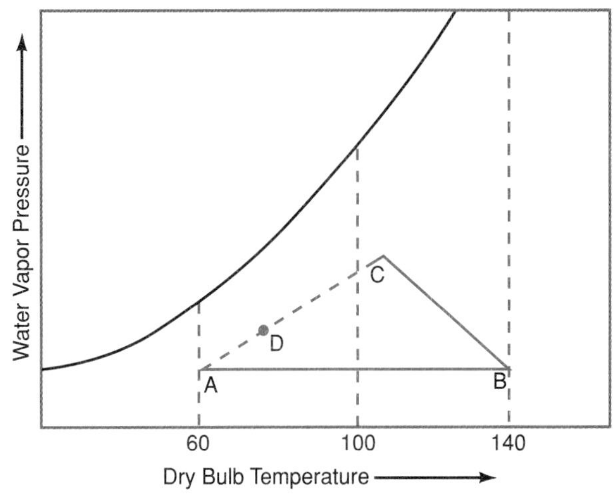

Figure 21-109. *Psychrometric chart depicts warm air recirculating heating cycle. A—Cold air return. A to B—Heating in furnace. B to C—Humidifying air. C to A—Mixing of air with room air. D—Final condition after mixing.*

A dry atmosphere causes dry skin and breathing dryness. There is a loss of moisture from hygroscopic materials. Examples of this are natural wood fibers (wood furniture and woodwork) and most foods. Dry wood cracks and furniture joints get loose. Dry air also creates static electricity conditions. Moisture must be added to the air. Typical indoor moisture sources are plumbing devices, cooking, and perspiration. In addition, moisture in the air can be increased and controlled by using humidifiers.

Humidifiers add water vapor (low-temperature steam) to the air. If the return to a warm air furnace is about 60°F (16°C) and 25% RH, and the furnace heats the air to 140°F (60°C), a humidifier may be used to add moisture to the warmed air. This heated air is then mixed with the air in the room. In **Figure 21-109,** A to B indicates the air being heated. From B to C, this warm air is passing over the humidifier (total heat is constant). Between C and A, the heated and the humidified air are mixed. D indicates the final condition of the air as it is delivered to the conditioned space. Remember, it requires about 1000 Btu to vaporize each pound of water.

Most humidifiers in warm air systems are part of the furnace or of the ductwork. **Figure 21-110** shows the operating principle of a humidifier. Evaporation takes place as the heated air passes through the evaporator pad. The pad is wetted by water metered through a solenoid valve. The water is then distributed over the pad. Water that has not evaporated flows to the bottom of the humidifier and is drained out. Humidified air is then returned to the heating system and enters the living area. This type of unit has a small fan motor and solenoid water valve. **Figure 21-111** shows the exterior of this type of humidifier.

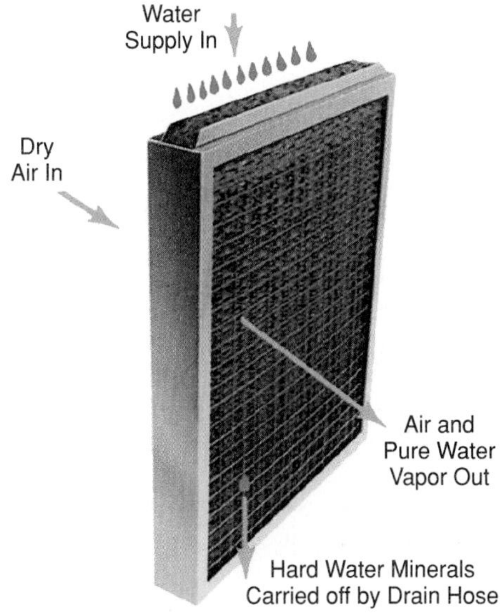

Figure 21-110. *Water flow through a basic filter humidifier. The flow of air is governed by the furnace blower and the operation of a small fan motor. (General Filters, Inc.)*

Some humidifiers use a vibrating object to atomize water (break it up into tiny droplets). The vibrating object can be a ***piezoelectric crystal.*** This operates at a high frequency of about 100 kHz (100,000 cycles per second). See **Figure 21-112.** "Piezo" is pronounced "pea-ay-zo." "Piezoelectric" means vibration due to electric waves in a flat crystal plate. The unit produces moist air that feels cool on contacting the human body. This is because partially evaporated droplets are vaporized on contact with a warmer object. If air cools the body too much, the air

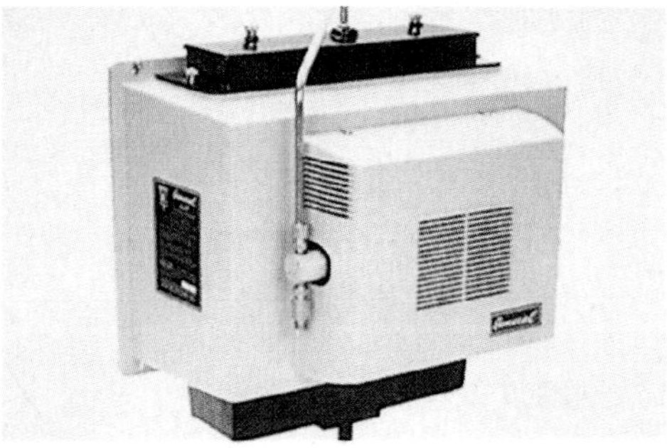

Figure 21-111. *Humidifier with evaporator pad. Unit has a solenoid valve to evenly distribute water over pad and has a motor and fan for air movement. (General Filters, Inc.)*

must be heated. This occurs in the building heating system before distribution.

Humidifiers can be easily added to warm air heating systems. A separate cabinet-type humidifier is needed with some heating systems, however. They are used with hydronic heating, steam heating systems, and most electric heating systems. These separate humidifiers are necessary if the needed relative humidity is to be maintained.

21.42 Types of Humidifiers

Humidifiers of various types are used. They include:

- Plate type (low-capacity).
- Rotating drum type (for restricted spaces).
- Rotating disk type.
- Fixed filter type.
- Fan type.
- Plenum / warm air duct type (slings the water).
- Plenum / duct electric type.
- Ultrasonic (piezoelectric) type.

A ***humidistat*** is used in a humidifier to control the level of relative humidity. Too much humidity may cause swelling of hygroscopic materials (wood products). It may cause condensation on cold surfaces such as windows, window frames, and doors. Water vapor may also condense on the inner surface of outside walls.

Excess humidity should be avoided because mold and rot can occur at 70% RH. Wet objects take a long time to dry. Wood frames rot from water dripping off cold glass window panes.

If outside air at 70% RH filters into a building and is heated to 72°F (22°C), this heated air will have the relative humidity shown in **Figure 21-113.** If outside air at 90% RH and 30°F (−1°C) is heated to 72°F (22°C), it will have a relative humidity of 19%. If it starts at 100% and 30°F (−1°C), it will be 20.7% at 72°F (22°C). All water services in the home increase the relative humidity. Perspiration and respiration of humans and animals also increases the relative humidity. Even if these moisture resources double the relative humidity, it would still be too low. This is the case except for 30°F (−1°C) outside air. The chart in **Figure 21-113** can be used at outside relative humidities other than 70%. Use the following

	Relative Humidity (Percent)		
Outside Temperature	Outside RH	Inside RH	Safe Inside RH
−10	70	2	20
0	70	5	25
10	70	7	30
20	70	11	35
30	70	15	40

Figure 21-113. *Chart shows relative humidity change in air as it is brought into building from outside; also recommended inside relative humidity based on outside temperature. Building inside temperature is 72°F (22°C). Example: 0°F (−18°C) outside air at 70% RH when brought into building and heated to 72°F (22°C) will have RH of 5%. It should have RH of 25%.*

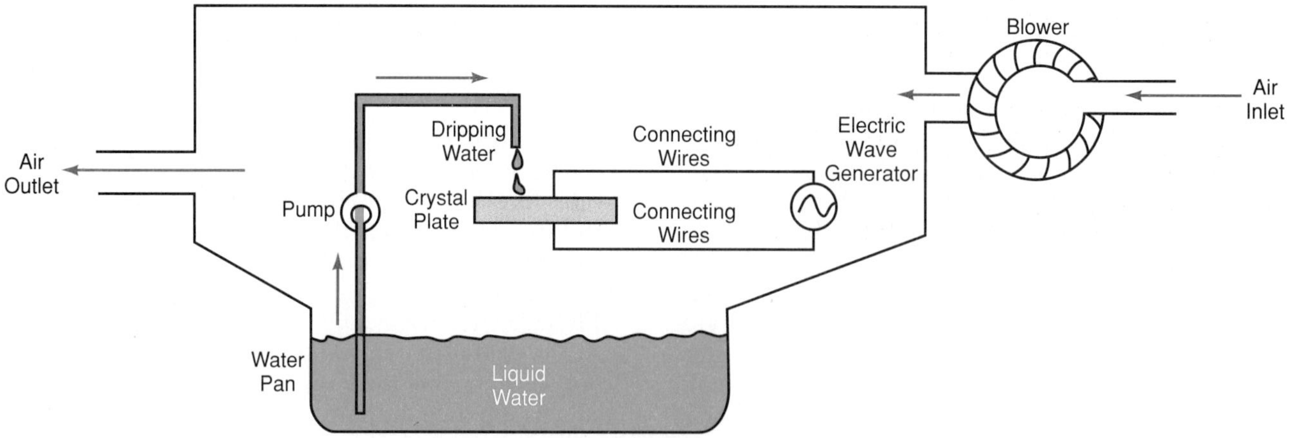

Figure 21-112. *Piezoelectric crystal vibrates to break water into mist. Note that the crystal plate size is exaggerated for clarity.*

example with 100% outside RH (converting the first line of the chart):

100%/70% × Inside RH = 1.43 × 2 = 2.86%.

Humidifiers are used to add the needed moisture (water vapor) to the room air. A ranch home 25′ × 60′, with 8′ ceilings, has a volume of 12,000 ft^3. It will require about four (tight building) to 16 (loose building) gallons of added moisture per day. This depends on the number of changes of air in the building:

- A tight building has 0.5 air changes/hour.
- An average building has 1.0 air changes/hour.
- An average loose building has 1.5 air changes/hour.
- A loose building has 2.0 air changes/hour.

Figure 21-114 shows a humidifier with float mechanism. It has built-in adjustable thermostat which senses duct temperature. It will electrically turn the unit off and on to control the humidifier output.

Regular water (city mains or well water) contains various types of foreign matter. If water is to be used in a humidifier, this foreign matter must be treated or removed. The perfect water for humidifiers would be distilled water. Distilled water does not put any foreign vapors into the air. In addition, it leaves no deposits in the humidifier or in the duct system.

The quality of the water varies with its source:

- ***Soft water.*** Natural, untreated water with low mineral content has about 5 grains of hardness per gallon and no chlorides. The natural source is rain water.
- ***Softened water*** is treated by the ion exchange process (water softener). This removes hardness and minerals. The unwanted mineral ions (charged atoms) are replaced with water-soluble sodium salts.
- ***Demineralized water*** has been treated to remove minerals.
- ***Medium hard water*** (well water) is untreated water with 5 to 15 grains/gal. mineral content.
- ***Very hard water*** (well water) is untreated water with over 15 grains/gal. of mineral content. Hard water is tested with a substance called "green soap." Green soap will not form suds in water that has 10 grains/gal. or more of minerals.

21.42.1 Electric Humidifiers

Most heated buildings need humidifying. A water-vapor-producing device is usually needed during the winter season in the temperate zones. An electric humidifier may be used for this purpose.

Buildings using steam, hydronic, or warm-air-heated equipment installations can use an electric humidifier. Warm air heating units sometimes are equipped with a humidifier. It uses the warm air as a source of heat. However, if the humidifier water is separately heated, the amount of humidification can be more accurately controlled.

The electric humidifier has the advantage of ease of installation and flexibility of location. It has accurate relative humidity control. **Figure 21-115** shows an electric humidifier designed for use in a duct or plenum chamber. This unit has a 1400 W electric heating coil. This coil is controlled by an adjustable humidistat. Water level is controlled by a pan-type float. The float operates a switch in the solenoid water valve circuit. **Figure 21-116** shows the construction of the electric humidifier.

If the relative humidity is too *high,* a ***dehumidifier*** may be needed. (See Chapter 22.) An electrically-heated house or business facility, for example, tends to have too much relative humidity. The building is very

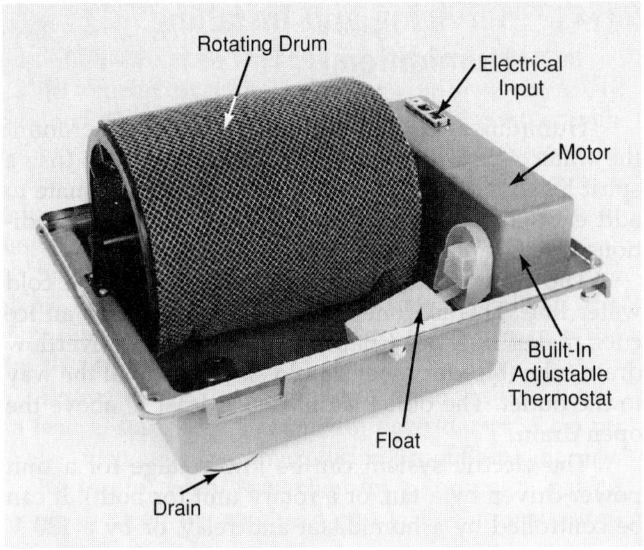

Figure 21-114. *Humidifier designed to fasten to bottom of supply duct. (General Filters, Inc.)*

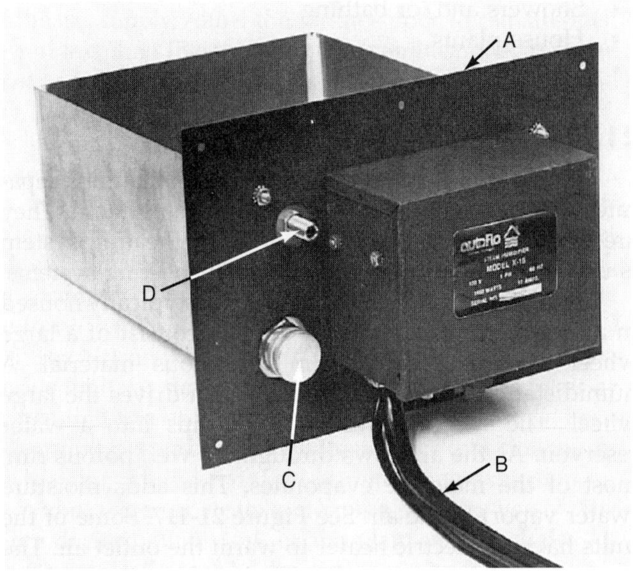

Figure 21-115. *Electric humidifier designed for installation in warm air duct. A—Plate for mounting on duct. B—Electrical line and control cord. C—Drain cap. D—Water inlet. (Auto Flo)*

heat source off, of course.) The switch should shut off the fuel and electrical power. Be careful when removing water which might be under pressure. (Removal is called a "blowdown.") Let the unit cool for one hour before opening valves.

Never use compression fittings on an oil-burning system.

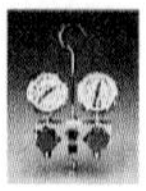

21.45 Test Your Knowledge

Please do not write in this text. Place your answers on a separate sheet of paper.

GAS HEATING SYSTEMS MODULE

1. Combustion is a process by which the energy in a fuel is converted to _______ energy.
   A. heat
   B. light
   C. electrical
   D. Both A and B.
2. All fuels contain _______ amounts.
   A. hydrogen and carbon atoms in equal
   B. hydrogen and carbon atoms in varying
   C. oxygen and carbon dioxide atoms in equal
   D. oxygen and carbon dioxide atoms in varying
3. The possible cause of a yellow flame is _______.
   A. overbalance of primary air
   B. flashback during shutoff
   C. burner overfiring
   D. lack of primary air
4. A possible remedy for a lifting flame would include _______.
   A. increasing gas pressure
   B. increasing primary air
   C. reducing input gas or primary air
   D. All of the above.
5. Which of the following is *not* a basic component of a forced-air system?
   A. Combustion blower motor.
   B. Gun-type burner.
   C. Heat exchanger.
   D. Combination gas valve.
6. High-efficiency furnaces _______.
   A. have an AFUE rating of 84% and above
   B. have secondary heat exchangers
   C. use less fuel and produce more heat
   D. All of the above.
7. The "Complete-Heat" unit _______.
   A. does not require a separate water heater vent or vertical chimney
   B. has a CAE of 90%
   C. contains three modules: the heat module, the water module, and the air handling module
   D. Both A and B.
8. A pulse furnace has a(n) _______.
   A. open flame
   B. direct-spark ignition device
   C. main burner
   D. pilot burner
9. Which of the following is *not* a concern critical to venting?
   A. Capacity of furnace.
   B. Airflow.
   C. Number of elbows.
   D. Whether the piping must be vertically or horizontally vented.
10. Which of the following is true regarding standing-pilot furnaces?
    A. There is no combination gas valve.
    B. The limit control energizes the gas valve.
    C. The pilot light is maintained by a thermocouple.
    D. As heat decreases within the exchanger, the fan switch closes its contacts.

HYDRONIC HEATING MODULE

11. When using a concrete slab floor with hydronics, the tubing is placed _______ the concrete.
    A. under
    B. in
    C. above
    D. Any of the above.
12. Which of the following temperature control devices is used with water heating systems?
    A. Single control.
    B. Zone control.
    C. Individual radiator controls.
    D. Any of the above.
13. Which of the following is *not* a means of controlling a hydronic system?
    A. Zone valves are cycled and turn on the pump or heater.
    B. Pump is operated continuously and the zone valves are cycled.
    C. Heat is on continuously and the pump is cycled.
    D. All of the above are means of controlling a hydronic system.
14. Scale is formed by _______ in the water.
    A. salt
    B. corrosion
    C. sludge
    D. embrittlement
15. Prior to putting a new boiler into service, you must _______.
    A. use alkali to neutralize acid
    B. boil out the boiler
    C. add calcium carbonate
    D. None of the above.
16. Why must you flush the system before putting it into service?
    A. To remove salts.
    B. To remove fluxes, dirt, sand, and chips.
    C. To ensure there are no leaks.
    D. To treat the system.

17. To flush out a boiler, 1 pound of trisodium phosphate should be added for each ______ gallons of water for ______ hours.
    A. 5, 10
    B. 100, 8
    C. 50, 4
    D. 10, 1
18. Organic growth may be controlled through the use of ______.
    A. chemicals
    B. sodium pentachlorophenate
    C. phosphates
    D. All of the above.
19. At what temperature is the steam in a steam-heat system?
    A. 212°F (100°C).
    B. 102°F (38°C).
    C. 120°F (49°C).
    D. None of the above.
20. How often should a steam heating system be checked?
    A. Once per year.
    B. Once per month.
    C. Once per week.
    D. As needed.

OIL FURNACES MODULE

21. What grade fuel oil is most commonly used in domestic gun-type burners?
    A. Grade 1.
    B. Grade 2.
    C. Grade 4.
    D. None of the above.
22. About ______ pounds of air is required for each gallon of Grade 2 fuel oil consumed.
    A. 100
    B. 1500
    C. 106
    D. None of the above.
23. Which of the following is *not* true regarding oil furnaces?
    A. Oil burns best in a liquid form.
    B. Oil burns best when atomized.
    C. The most common type of oil burner is the gun type.
    D. In the blower, the oil mixes with air.
24. Too much positive pressure is caused by ______.
    A. not enough air
    B. a chimney that is too tall
    C. a faulty nozzle
    D. None of the above.
25. Which of the following is a true statement regarding gun-type oil burners?
    A. High-pressure type feeds oil at 200 psi to 300 psi.
    B. Low-pressure type uses oil at 1 psi to 5 psi.
    C. The nozzle on a high-pressure type must be carefully centered.
    D. All of the above.
26. Which of the following statements is true regarding electrical ignition?
    A. It is commonly used on oil burners.
    B. It includes a transformer and three electrodes.
    C. It must raise the oil temperature to 1000°F.
    D. All of the above.
27. Which of the following is *not* a cause of delayed ignition and puffback?
    A. Wrong position of electrodes.
    B. Moisture in the oil.
    C. Poor insulation.
    D. Weak ignition.
28. An oil storage tank should be at least ______ feet from the furnace.
    A. 7
    B. 10
    C. 5.5
    D. 12
29. Which items determine the rate at which fuel oil is burned?
    A. Size of the hole in the nozzle.
    B. Amounts of oil pressure.
    C. Length of the nozzle.
    D. Both A and B.
30. Which of the following is *not* a result of air in the oil line?
    A. Flashback.
    B. Oil not being pumped.
    C. Blowbacks.
    D. Flame failure.

ELECTRIC HEATING MODULE

31. The higher the temperature of the heating element, the ______ the space it needs to occupy.
    A. smaller
    B. larger
    C. wider
    D. cooler
32. One watt equals ______ Btu/hr.
    A. 3415
    B. 3.415
    C. 34.15
    D. None of the above.
33. In order to modify an oil or gas system to electric resistance heating, ______.
    A. have metal window frames and insulate walls and ceilings
    B. have wood or plastic window frames and dehumidify
    C. have metal window frames and dehumidify
    D. have wood or plastic window frames and insulate wall, ceilings, floors, and basement
34. Which of the following is *not* a common type of electric resistance heating wire?
    A. Closed wire.
    B. Open wire.
    C. Open ribbon.
    D. Tube-encased wire.

35. Which type of control is recommended for use with baseboard electric heat?
    A. Relief valve.
    B. Safety limit switch.
    C. Flow switch.
    D. Pressure-reducing valve.
36. The total furnace output of an electric furnace should be _______ percent _______ the design load.
    A. 10, over
    B. 10, under
    C. 20, over
    D. 5, over
37. Heating elements in an electric furnace vary from _______ each.
    A. 10 kW to 20 kW
    B. 15 kW to 20 kW
    C. .5 kW to 1.5 kW
    D. 5 kW to 10 kW
38. Electric duct heaters are used primarily as _______.
    A. space heaters
    B. supplementary heat sources
    C. auxiliary heaters
    D. All of the above.
39. Radiant heat decreases as the square of the distance. That is, an object twice as far away from the radiant heat source will receive _______ as much heat.
    A. one-fourth
    B. one-half
    C. one-eighth
    D. twice
40. Which of the following is true of electric heat?
    A. Operates below ignition temperature of most materials.
    B. Humidity control is easily maintained.
    C. Operational costs are lower than other fuels.
    D. Increase in toxic indoor air conditions.

ALTERNATIVE HEATING METHODS MODULE

41. Which of the following is used for solid fuel heating?
    A. Coal and gas.
    B. Wood and oil.
    C. Oil and coal.
    D. Coal and wood.
42. Which statement is true about solid fuel heating?
    A. It is cleaner than conventional fuel.
    B. It is dirtier than conventional fuel.
    C. It produces less soot in the flue.
    D. Both B and C.
43. What is the function of solar energy system equipment?
    A. To collect heat.
    B. To store heat.
    C. To distribute heat.
    D. All of the above.
44. When a heat pump system is used for cooling, what is the outdoor coil?
    A. Condenser.
    B. Evaporator.
    C. Neither of the above.
    D. Both A and B.
45. Heat pumps alternate the flow of refrigerant by the use of a _______ valve.
    A. three-way
    B. thermostatic expansion
    C. four-way
    D. bypass
46. The term "geothermal heat pump" describes a system in which the evaporator/condenser is in _______.
    A. the ground
    B. a well
    C. a lake
    D. Any of the above.
47. The outdoor coil of a heat pump can be used for the _______.
    A. condenser
    B. evaporator
    C. Both of the above.
    D. None of the above.
48. When the heat pump system is used for heating, what is the outdoor coil?
    A. Evaporator.
    B. Condenser.
    C. None of the above.
    D. Both A and B.
49. How many basic types of solar energy systems are there?
    A. One.
    B. Two.
    C. Three.
    D. Four.
50. Cogeneration systems are mainly used when there is a need for _______.
    A. year-round heating
    B. year-round cooling
    C. short heating seasons
    D. short cooling seasons

HUMIDIFICATION MODULE

51. Human comfort requires a relative humidity (RH) of approximately _______.
    A. 25%
    B. 30%
    C. 35%
    D. 40%
52. A dry atmosphere causes _______.
    A. static electricity
    B. dry wood
    C. skin dehydration
    D. All of the above.

53. What do humidifiers add to the air?
    A. Water vapor.
    B. Low-temperature steam.
    C. Cool vapor.
    D. Both A and B.
54. What is used to control the level of relative humidity?
    A. Furnace.
    B. Air conditioner.
    C. Humidistats.
    D. All of the above.
55. Excessive humidity may cause ________.
    A. swelling of wood
    B. condensation on cold surfaces
    C. condensation on exterior walls
    D. All of the above.
56. If the outside relative humidity is 70%, the temperature is 0°F (−18°C), and the building is heated to 72°F (22°C), what is the indoor relative humidity?
    A. 2%.
    B. 5%.
    C. 7%.
    D. 11%.
57. Water softeners remove ________.
    A. hardness
    B. minerals
    C. soluble sodium salts
    D. Both A and B.
58. What type of water is best used in a humidifier?
    A. Very hard water.
    B. Medium hard water.
    C. Distilled water.
    D. Demineralized water.
59. A humidifier is controlled by a ________.
    A. humidistat and relay
    B. 120 V humidistat
    C. thermostat
    D. Both A and B.
60. Dehumidifiers are used when ________.
    A. relative humidity is too high
    B. relative humidity is too low
    C. indoor and outdoor temperatures vary
    D. the indoor temperature is too low

Top—Window air conditioning units provide cooling for a relatively small area. (Whirlpool Corporation) Bottom—This total environmental center provides heating, air conditioning, and relative humidity control for large commercial structures. (Tyler Refrigeration Corp.)

Chapter 22

COOLING AND DEHUMIDIFYING SYSTEMS

Key Words:

console air conditioner
dehumidifier
evaporative comfort cooling
humidistat
packaged terminal air conditioner
window unit

Learning Objectives:

After studying this chapter, you will be able to:

- Explain the principles of air conditioning.
- Classify types of comfort cooling systems and recognize variations between them.
- Properly install and service window air conditioners.
- Properly install and service console air conditioners.
- **Follow approved safety procedures.**

Refrigeration is the basis of comfort cooling air conditioning. All comfort cooling systems use one of the standard refrigerating cycles and standard refrigerants. Standard types of compressors, condensers, piping, refrigerant controls, motor controls, and evaporators are used. The refrigerant evaporating temperature of most comfort cooling systems is 40°F to 50°F (4°C to 10°C). This chapter describes most designs that use mechanical refrigeration for cooling and dehumidifying.

22.1 Principles of Atmosphere Cooling

As explained in Chapter 19, comfort depends on temperature and relative humidity. Certain industrial operations also depend on temperature and relative humidity. For example, large industrial computer installations operate in special temperature and relative humidity conditions. Manufacturers of medicines and biological products require special conditions. Industrial operations dealing with hygroscopic (water-absorbing) materials and processes require controlled relative humidity.

The first mechanical atmosphere cooling and humidity control used water to both reduce the temperature of the air and dehumidify it. Air was passed over water-cooled coils. When the relative humidity was not a concern, air was passed through water sprays.

Figure 22-1 shows the operation of a typical cooling unit. Return air is mixed with some fresh air. Then this air mixture is filtered and cooled. Moisture is removed before it is redistributed into the building.

Cool air leaving the evaporator is at 100% relative humidity. This saturated air, as it mixes with air in the conditioned space, warms up somewhat. Thus, relative humidity is brought down to a comfortable level.

There is a more positive way to control relative humidity. Bypassing some of the return air into the air conditioner outlet warms up the cooled air before it leaves the duct system. This method is shown in **Figure 22-2.**

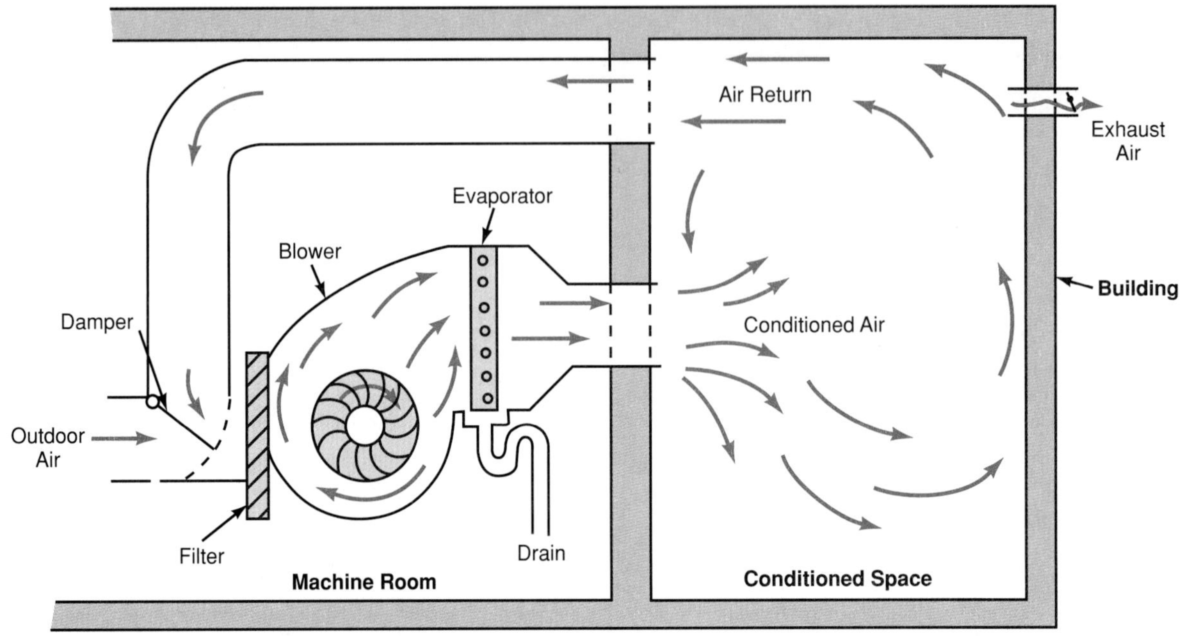

Figure 22-1. *Basic operation of conditioner that cools, recirculates, and removes moisture from outdoor air. A centrifugal blower is used.*

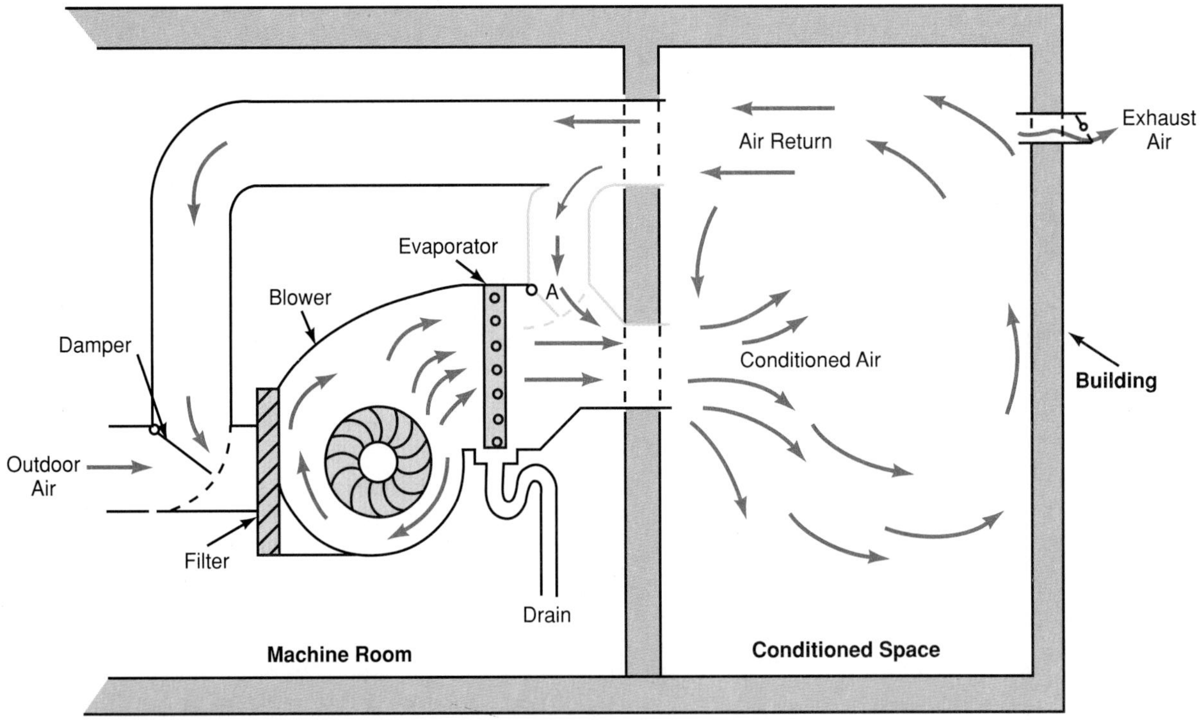

Figure 22-2. *Comfort cooling installation with return air bypass, A, for relative humidity control.*

22.1.1 Cooling Cycle

In a cooling cycle, the dry bulb (db) temperature of the air is lowered. When this happens, as in **Figure 22-3,** A to B, the relative humidity increases. Some moisture should be removed to make this air comfortable. Moisture can be removed by either of two methods:

- Dehydrate the air with chemicals.
- Cool the air down to the saturation curve at C. Then remove moisture by condensing it on a cool surface. See curve C to D, **Figure 22-3.** The distance from C to D is the drop in vapor pressure or grains of moisture removed.

Reheating along a horizontal line, D to E, will decrease the humidity. However, usually the air leaving at D is mixed with the room air. The room air is at some

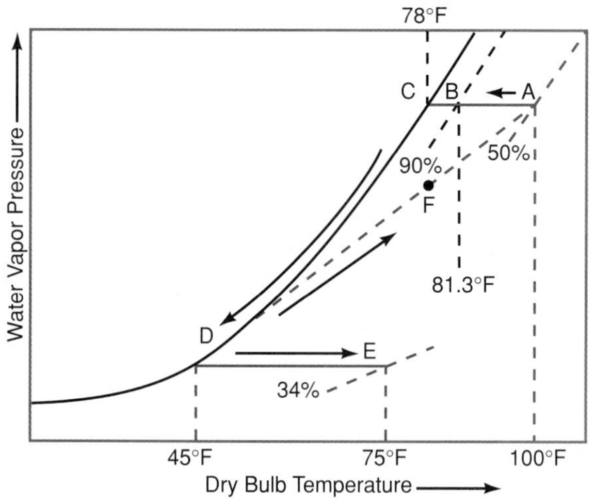

Figure 22-3. *Cooling cycle on psychrometric chart. A—Condition of outside air. B—Partly cooled air. C—Air cooled to saturation. D—Air cooled to remove some moisture. E—Dehydrated air reheated. F—Result of mixing treated and untreated air.*

intermediate condition between 81°F and 100°F (27°C and 38°C). The mixture meets on the line between point D and point A. If a third of the air, by weight, is passed through the evaporator, the mixed air temperatures will be a third of the way from A to D—that is to F.

22.1.2 Evaporative Cooling

In dry climates, ***evaporative comfort cooling*** is desirable and practical. If air at 105°F (41°C) and 20% relative humidity is moved rapidly over water at the same temperature, some of the water will evaporate. The remaining water can cool to as low as 55°F (13°C). Normally, the water will cool down to a range of about 65°F (18°C) to 70°F (21°C). Other air forced around the water container can then be distributed into the space to be cooled. It will cool down the air leaving the unit to approximately 75°F (24°C) or 80°F (27°C) at about 40% relative humidity. The conditioned room would be quite comfortable. **Figure 22-4** illustrates such a system.

22.2 Comfort Cooling Systems

Several types of comfort cooling systems are in common use. They can be classified by arrangement of the mechanism:

- Self-contained coolers.
  - Window units.
  - Package terminal units.
  - Console units.
  - Multizone ductless split systems.
- Remote (controlled from a distance). Remote units are of two types:
  - The condensing unit is remote. The evaporator is installed in the room to be conditioned or in the main duct.
  - The central air conditioning plant. The condensing unit and the evaporator are installed away from the place being conditioned. Cooled brine or water is circulated to heat exchangers in the spaces to be conditioned.

22.2.1 Self-Contained Comfort Coolers

All self-contained air conditioners provide cooling during the hot season. Some models also can provide heating during the cold season. See Chapter 24.

Both types have a complete refrigeration plant, including condensing unit, refrigerant valves, evaporators, and filters. Individual room thermostats provide control.

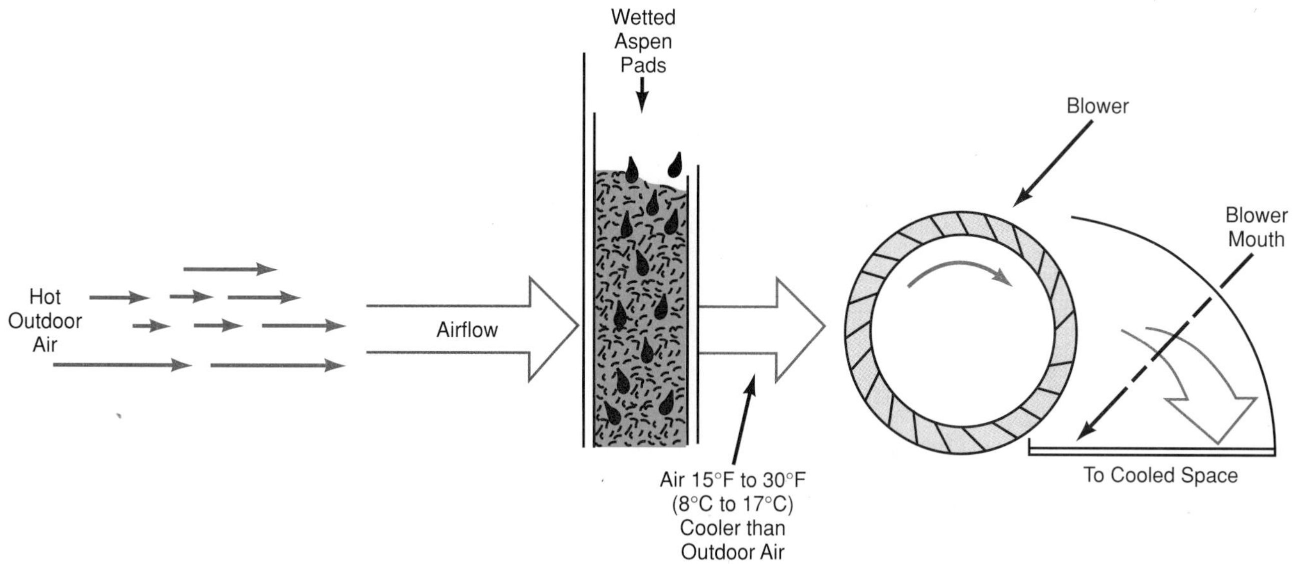

Figure 22-4. *Evaporative cooling process. Outside air filters through a water-saturated evaporative medium, is cooled by evaporation, and is circulated by a blower.*

Window units are air-cooled. They are easily mounted, and operate from 120 V or 240 V single-phase circuits. Their capacities vary from 4000 Btu/hr. to 40,000 Btu/hr.

Console units may be either air-cooled or water-cooled. They may be installed in the room to be conditioned or in an adjacent room. If in an adjacent room, short ducts deliver air and provide for return air.

Window Units

The window or wall-mounted comfort cooler is very popular, **Figure 22-5.** The ***window unit*** mounts on a windowsill, and installation is relatively easy. The condenser is located in the section of the cabinet that is outside the building. Outside air is forced over the condenser by a fan. Inside the room, another fan draws air in through a filter and forces it over the evaporator. The two airflow fans may be driven by the same motor or each may have its own motor. The internal construction of a unit is shown in **Figure 22-6.**

Window units are available in several types. One type cools and filters the air and has a fresh air intake. Another also has an electrical resistance heating unit to furnish heat. A third type uses a reverse cycle system (heat pump). This permits the use of the refrigerating units both for comfort cooling and heating. See Chapter 24.

Window units may be obtained to fit double-hung windows or casement windows. Units can be installed in special wall openings.

A detailed schematic is provided in **Figure 22-7.** Evaporator condensate is often drained to the base of the motor compressor and the condenser. There, it helps to cool these parts. A capillary tube or a bypass AEV refrigerant control is usually used.

Some units change the cooled airflow from side to side as the unit runs. This is done by rotating angle deflector plates. The plates are mounted on a shaft. The shaft is turned by an air-operated rotor in the exhaust air, **Figure 22-8.**

Figure 22-5. *Window air conditioner. (Whirlpool Corporation)*

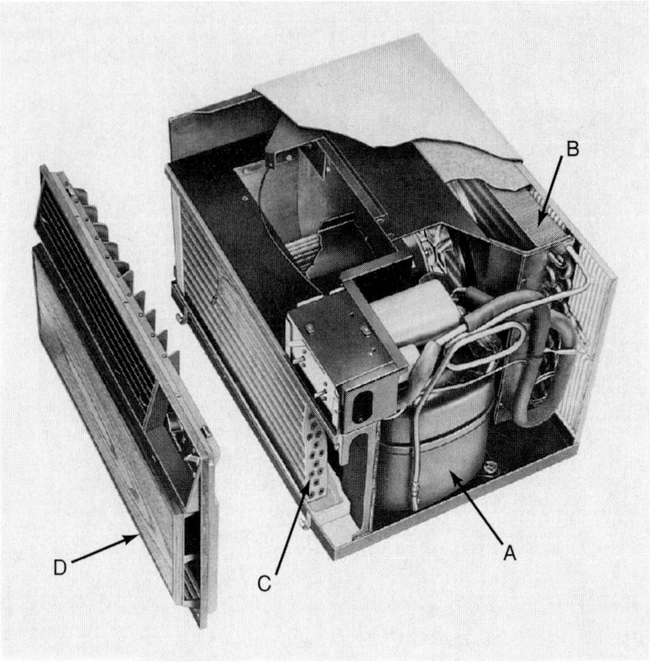

Figure 22-6. *Internal construction of a window unit. A—Hermetic motor compressor. B—Condenser. C—Evaporator. D—Removable front panel. Return grille, filter, and outlet grille are built into front panel.*

The systems are controlled by thermostats. The sensitive bulb is usually mounted at the inlet of the evaporator. A differential of about 5°F (3°C) is normal.

If the part of the bulb farthest from the evaporator is insulated, the bulb will respond better to evaporator temperatures. It will cool sooner and stop the unit before it overcools. It will also stop the unit if the evaporator ices up. The unit will be prevented from starting again until the ice melts.

Room air conditioners have been designed with handheld infrared controllers, **Figure 22-9.** Indicator lights identify the operating mode and temperature setting. A handheld remote control can be used to operate the system within a room. The user can program a clock, operating mode, fan speed, temperature control, and start/stop times.

Caution: If a unit is shut off and then immediately turned on again, it may stall. The motor compressor may be damaged. Let high pressures dissipate.

Installing Window Units

Window units must be installed with the outside tilted down for condensate drain. They must be securely fastened in place to prevent the unit from falling out of the window. All edges should be sealed to minimize air infiltration. The window must be secured in the proper position.

Window units are installed into the opening using metal plates, rubber gaskets, and sealing compounds.

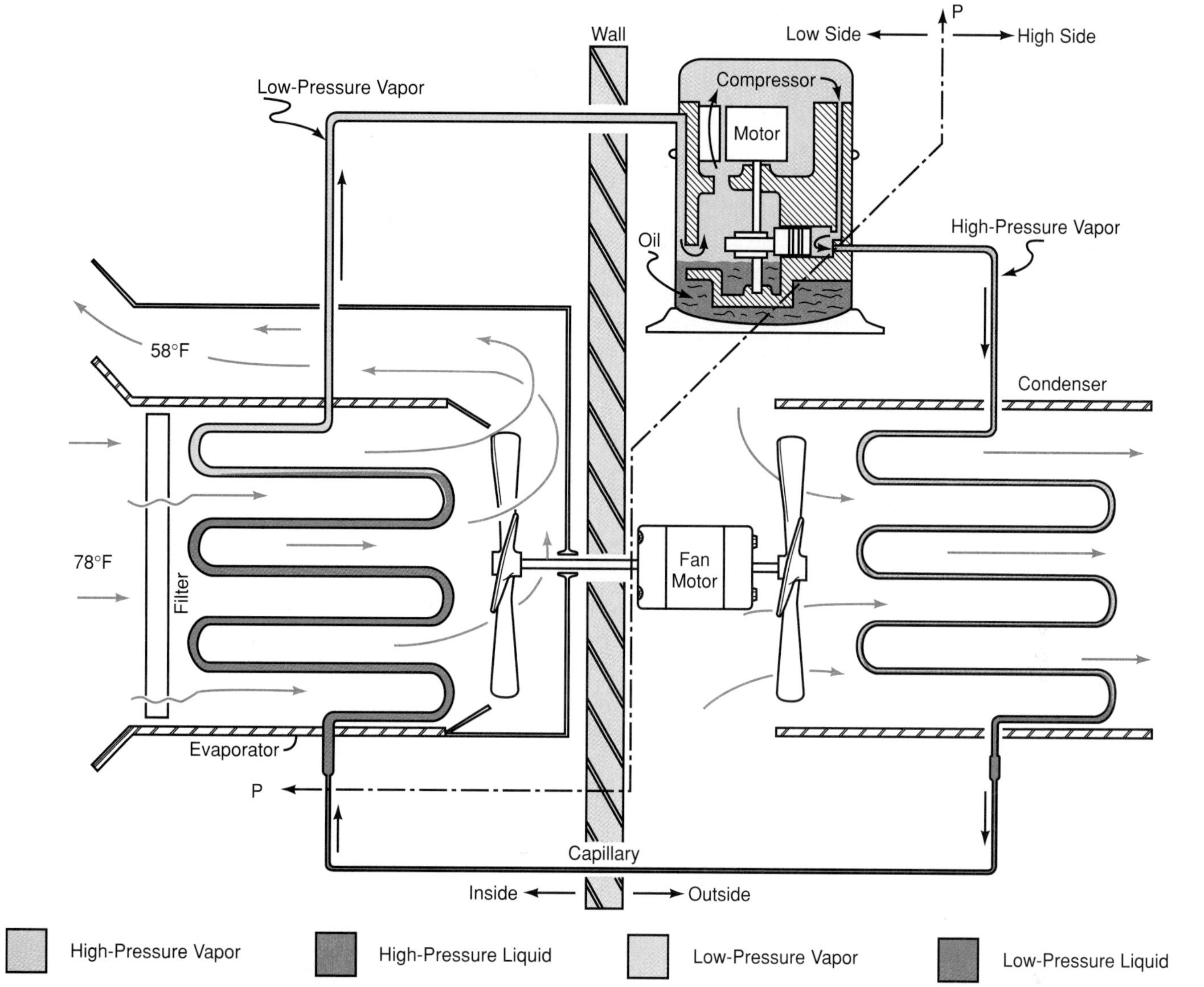

Figure 22-7. *Drawing of window or in-the-wall comfort cooling unit. Note conditioned airflow and separate condenser airflow.*

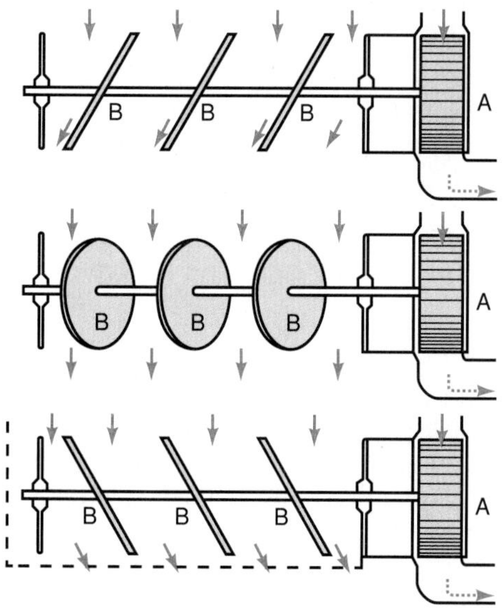

Figure 22-8. *Oscillating grille deflector. Grille is behind conditioned air outlet for an air conditioner that continually sweeps air from side to side. Air turbine, A, revolves. As it does, it turns deflector plate at B. Top—Air is deflected to left. Middle—Air flows directly ahead. Bottom—Air flows to right.*

The unit and the parts needed to install it are shown in **Figure 22-10. Figure 22-11** shows a windowsill with a leveling bracket and security bracket mounted. Another method of bracing and leveling a comfort cooling unit is shown in **Figure 22-12.**

The unit housing should be adjusted to tilt downward about 1/4″ on the outside. This is enough to provide condensate drainage. A sponge rubber or plastic strip is usually placed between the housing and the windowsill to help make a leakproof joint. The sill brackets

Figure 22-9. *Window air conditioner with electronic control panel and remote control. (Whirlpool Corporation)*

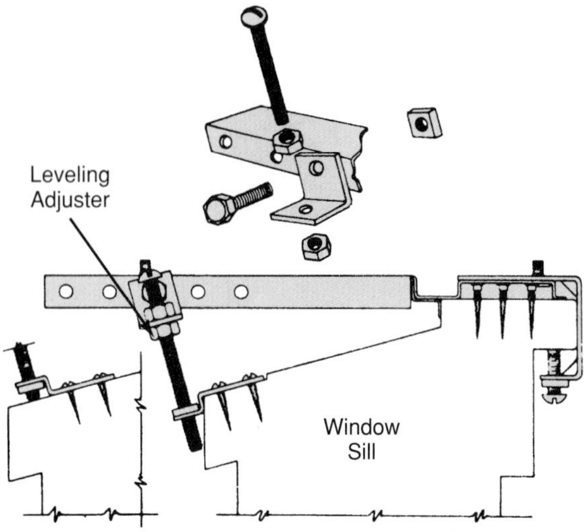

Figure 22-11. *Bracing is used to hold window air conditioner on sill.*

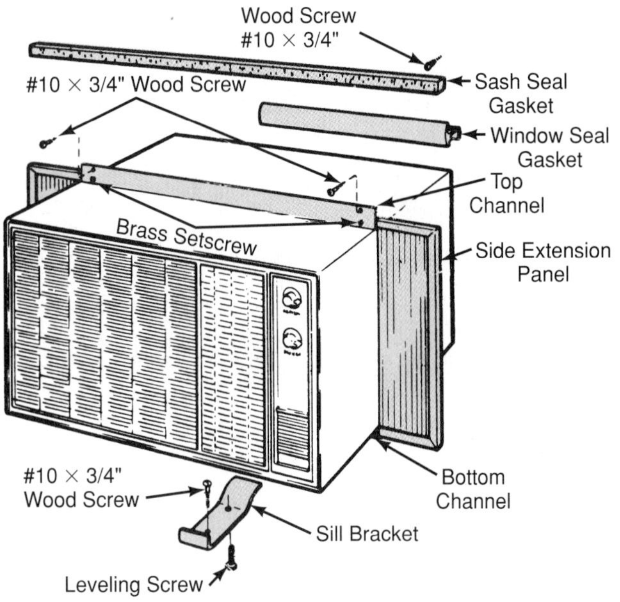

Figure 22-10. *Window unit showing necessary parts to mount unit safely and seal openings.*

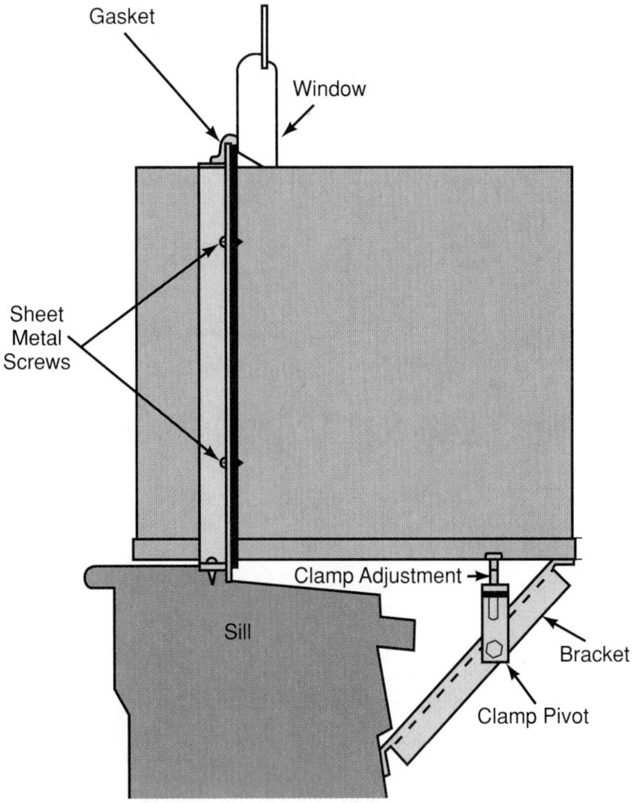

Figure 22-12. *Window unit with indoor flush mounting. The sheet metal screws hold the side closure panels.*

and unit housing are installed. Then the rubber seal strips and the filler boards are put in place, **Figure 22-13.** Manufacturers each have slightly different methods for mounting window units.

Where the lower sash is raised to make room for the air conditioner, an air gap will exist between the two sashes. This opening may be sealed with a sponge rubber or Styrofoam strip, **Figure 22-14.** Notches help make the strip fit.

The housing must be securely fastened before the unit is put in place. Filler boards are placed between the unit housing and the side of the window. They are usually sealed with sponge rubber strips or Styrofoam. This is held in place with sheet metal screws and with spring clips. **Figure 22-15** shows one method of installing them. A typical window unit installation in a casement window is shown in **Figure 22-16.**

The inside mechanism is heavy. It should be moved using a dolly or special carrier. Avoid moving or lifting

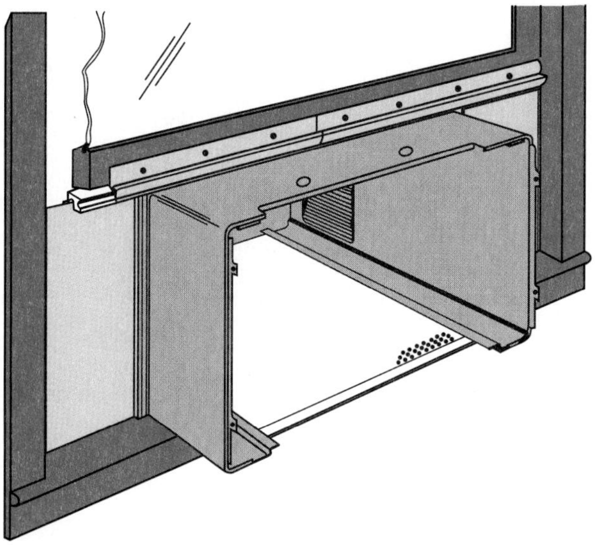

Figure 22-13. *Window air conditioning casing installed, showing rubber seal strips and filler boards.*

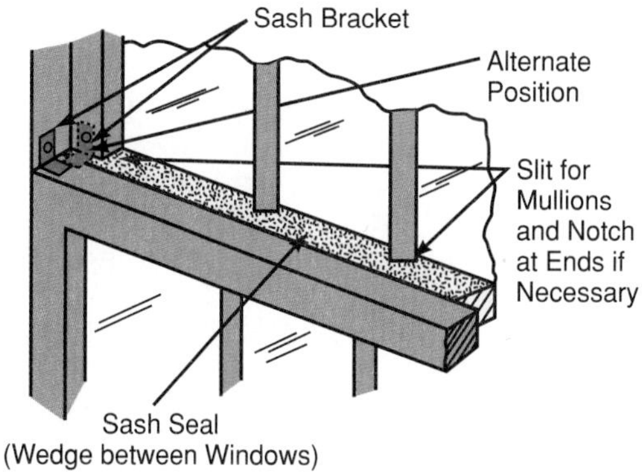

Figure 22-14. *Sponge rubber seal placed between upper edge of lower sash and upper sash of double-hung window. Sash bracket keeps lower sash locked.*

the unit by using the tubing or coils as hand grips. Carry the unit by holding onto the bottom pan.

Avoid forcing the unit into the casing. As the unit moves into the casing, check that refrigerant lines and wiring are free and clear. The front grille, filter, and control knobs are easily installed.

As a final step, check all joints for tightness. Caulk seams which show light leak or they may not be airtight.

When making the electrical hookup, use a separate circuit. A polarized plug (one with a ground wire) is required.

Thermostats are used with most window units. They are adjustable to cut out between 56°F (13°C) and 60°F (16°C). The cut-in adjustment is between 77°F (25°C) and 80°F (27°C). Their differentials vary between 3°F (2°C) to 8°F (4°C). If a thermostat fails, the unit will not start. To test the operation of a thermostat, cover the air outlet and air inlet with a cloth. The air will now recirculate into the unit. The temperature will quickly drop to the cutout temperature. Use a thermometer.

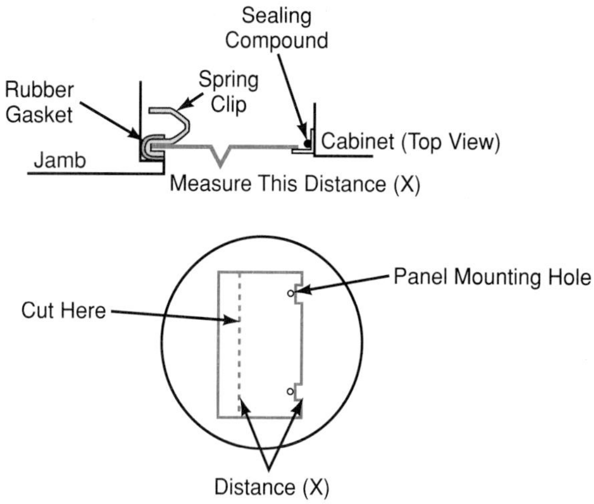

Figure 22-15. *Filler panel between unit housing and window casing.*

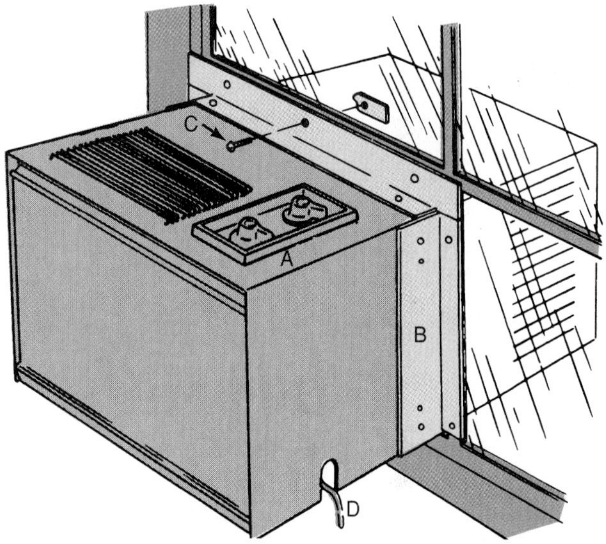

Figure 22-16. *Method of mounting window unit in casement window. A—Controls. B—Angle plate, usually enameled steel, fastens to both casing and window frame. C—Machine screw holds angle plate to window frame. D—Electrical cord. (Frigidaire Company)*

Units that mount through the wall are popular in new apartment units. There is no interference with windows, and comfort cooling can be provided as desired. **Figure 22-17** shows a typical installation.

Servicing Window Units

Servicing window units is similar to the servicing of hermetic refrigerating units. Chapters 12 and 15 describe most of the servicing operations.

Some of the external service operations are as follows:

- Semiannual cleaning or replacement of the filter (usually done by the owner). **Figure 22-18** shows a filter design.

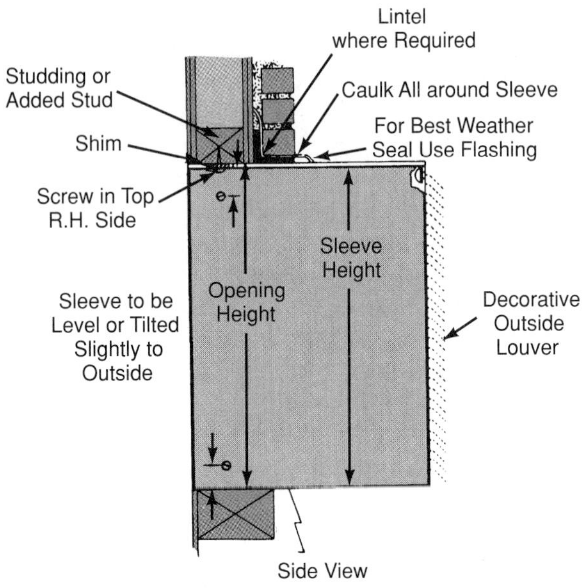

Figure 22-17. *Typical through-the-wall unit installation.*

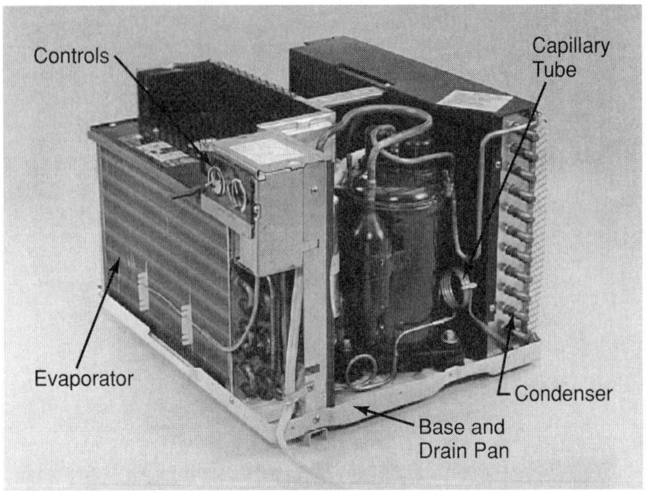

Figure 22-19. *Window air conditioner unit removed from its cabinet and ready for cleaning. (Carrier Corporation, Subsidiary of United Technologies Corporation)*

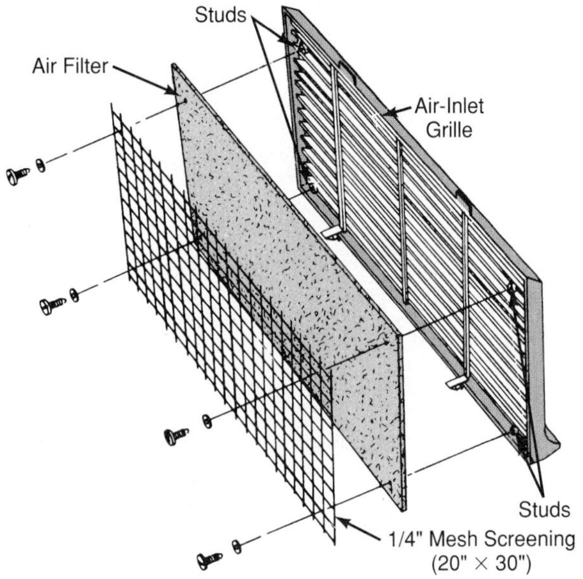

Figure 22-18. *Typical filter installation for window air conditioner.*

- Annual cleaning of the evaporator, condenser, fan blades, fan motor, motor compressor, and casing. The unit is removed from its casing for these operations, as shown in **Figure 22-19.**
- Inspect fan motor or motors and lubricate them unless they have hermetic bearings. Always wipe away excess oil. Oil mist on the fan blades collects lint and reduces air movement efficiency.

Place a tarpaulin or newspapers on the floor. Remove or tie back curtains or drapes before cleaning the unit. Use a commercial model vacuum cleaner with a brush-equipped nozzle to clean the cabinet inside.

Finned evaporators and condensers are difficult to clean. The fin spacing prevents the vacuum brush from reaching the lint and dirt. In such cases, plastic blades and a powerful vacuum will remove most of the dirt. Dirt must be removed if the unit is to continue working well. Never use metal blades for cleaning; they may cause leaks.

Servicing the unit outdoors is more desirable. A powerful water and detergent spray can then be used for cleaning. Coils must be cleaned thoroughly. Fins, if bent, should be straightened.

When servicing fan motors, make certain that fans are tight on the shaft. They should be carefully positioned in the shroud for efficient air movement. Avoid bending the fan blades or twisting them. An off-balance fan will soon wear out the motor bearings. It will be noisy because of vibration. Replace an abused fan.

Inspect the drain. It must be clean. Remove lint from the drain hole and tube using a soft wire. Check all bolts, nuts, and screws for tightness.

Before replacing the unit in the cabinet, run it to check for noise. Find its source and stop the noise.

Always put a cloth over the air conditioner outlet when it is first started after cleaning. Loosened dirt not removed by the vacuum cleaner will be blown out of the adjustable grille.

The wiring of a window unit is very similar to other refrigerating units. See **Figure 22-20.** External electrical servicing procedures are usually the same as for domestic and commercial units except the following:

- Fan motors usually have two or three speeds.
- Some systems have three capacitors: starting capacitor, running capacitor, and fan motor capacitor.

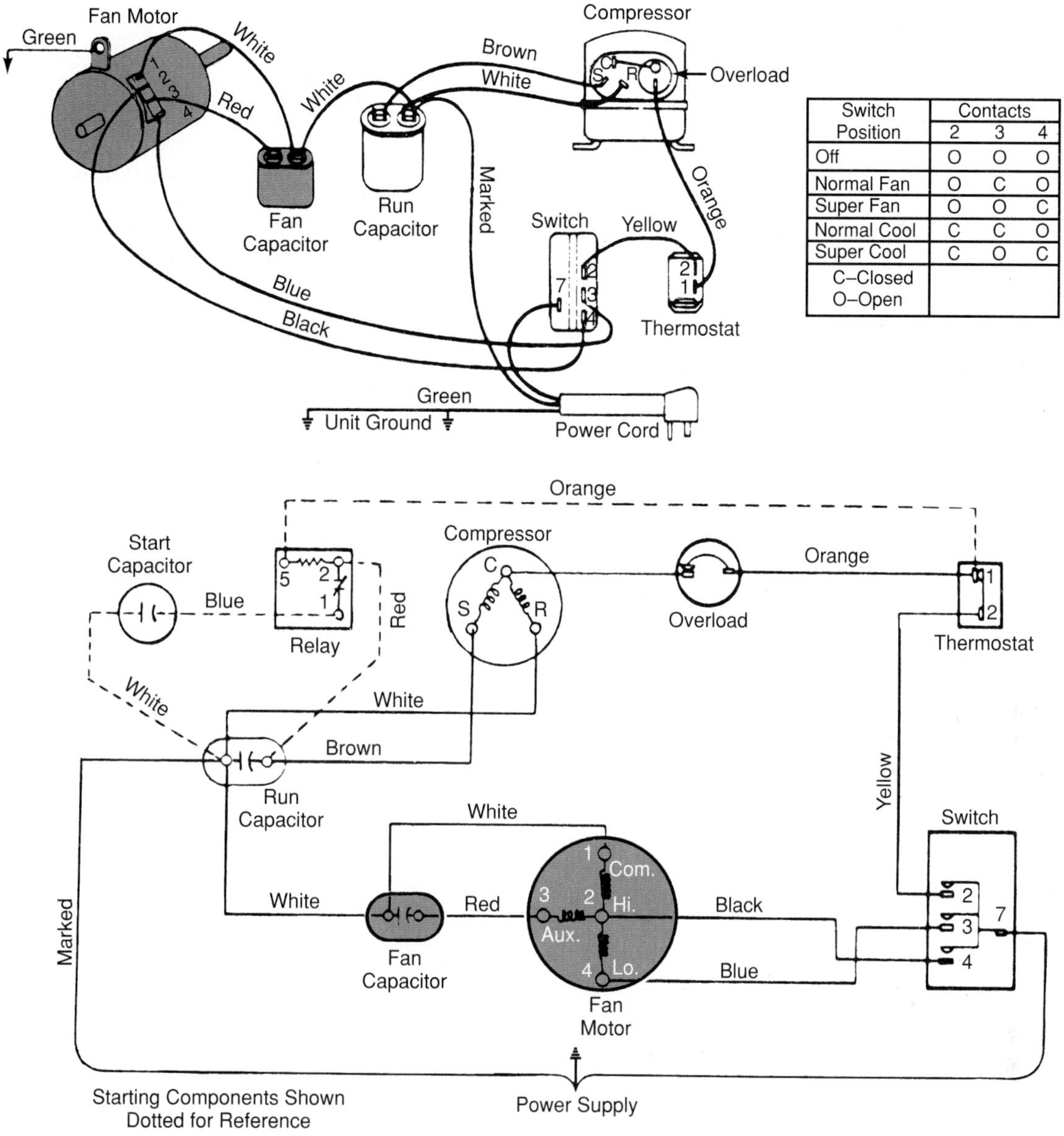

Switch Position	Contacts 2	3	4
Off	O	O	O
Normal Fan	O	C	O
Super Fan	O	O	C
Normal Cool	C	C	O
Super Cool	C	O	C
C–Closed O–Open			

Figure 22-20. *Wiring diagram of 120 V window air conditioner with 7000 Btu/hr. capacity. Note two-speed fan motor and fan capacitor. (Fedders North America)*

Shown in **Figure 22-21** is a unit with a starting capacitor and a running capacitor. An arrangement with three capacitors is shown in **Figure 22-22.** Position of the fan control switch, thermostat, capacitors, and wiring are seen in **Figure 22-23.**

Figure 22-24 shows a wiring diagram for a three-speed fan motor system. The system has a 21,000 Btu/hr. (6.15 kW) capacity using a 240 V circuit.

Testing of outside electrical parts is described in Chapters 6, 7, and 8. Fan and compressor motor testing is described in Chapter 7. All but the motor compressor can be repaired on-the-job. Before doing internal service work, be sure the malfunction is not in the external circuit. Test for power. Check the thermostat, the relay, the capacitors, and the overload protectors (both electrical and temperature).

Troubles inside the unit may include:

- Lack of refrigerant.
- Stuck compressor.
- Inefficient compressor
- Clogged refrigerant circuit.
- Short circuit, open circuit, or grounded motor windings.

The motor condition can be checked with a continuity light or with an ohmmeter. To check for lack of refrigerant or clogged refrigerant lines, installing service valves may be necessary. This can be achieved by

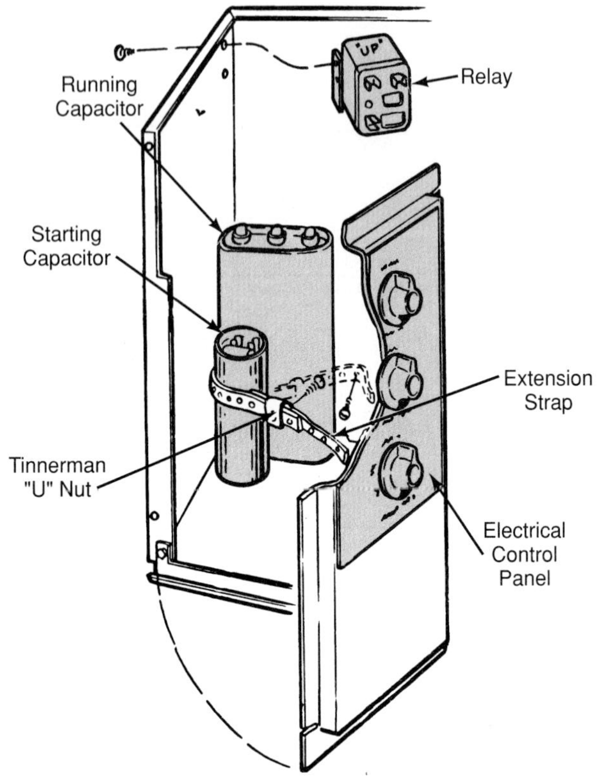

Figure 22-21. *Window unit showing location of parts such as capacitors, control panel, and relay.*

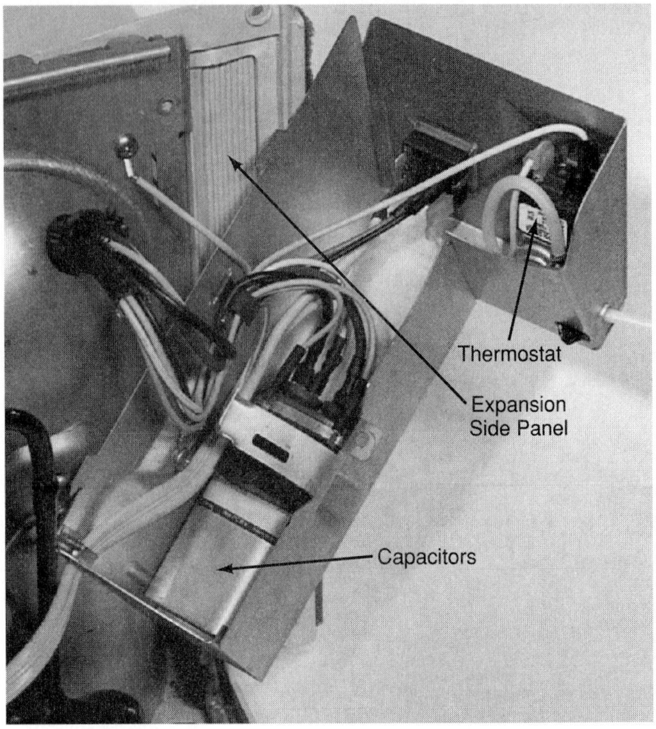

Figure 22-23. *Window unit showing location of capacitors and thermostat. Note the expansion side panel. (Carrier Corporation, Subsidiary of United Technologies Corporation)*

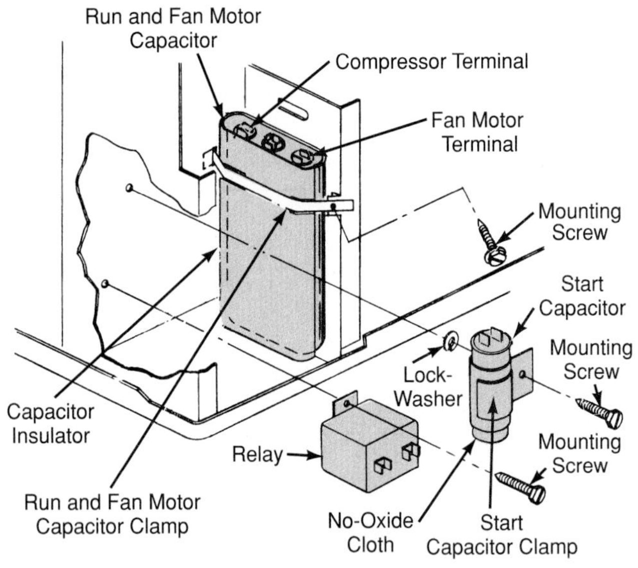

Figure 22-22. *Arrangement of start, run, and fan motor capacitors. Note that fan motor capacitor and compressor capacitor are in the same container.*

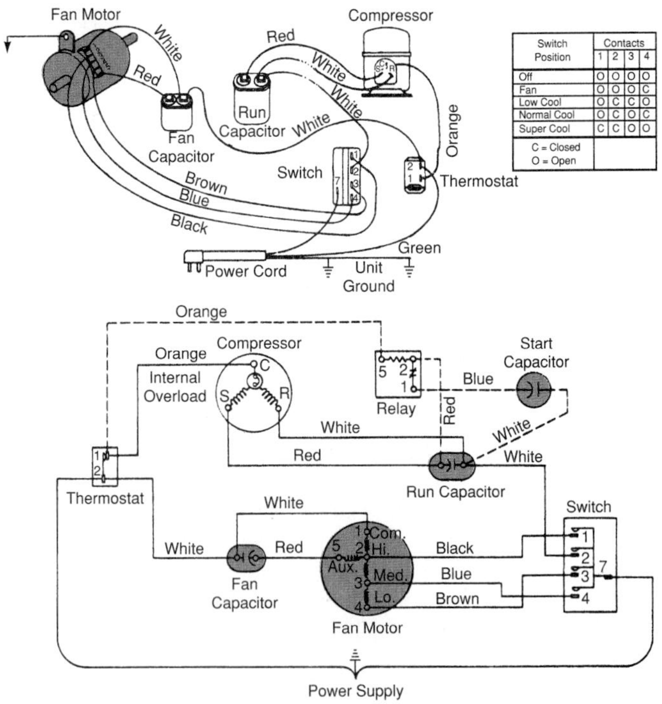

Figure 22-24. *Wiring diagram of three-speed fan system. (Fedders North America)*

piercing or by using a sweat-on valve. Then install a gauge manifold, as shown in **Figure 22-25.**

Determining (diagnosing) trouble is explained in Chapters 12 and 15. The unit should be moved to the shop if the motor compressor needs repairs. The motor compressor can be replaced on the owner's premises. If the unit lacks refrigerant, locate the leak and repair it before recharging.

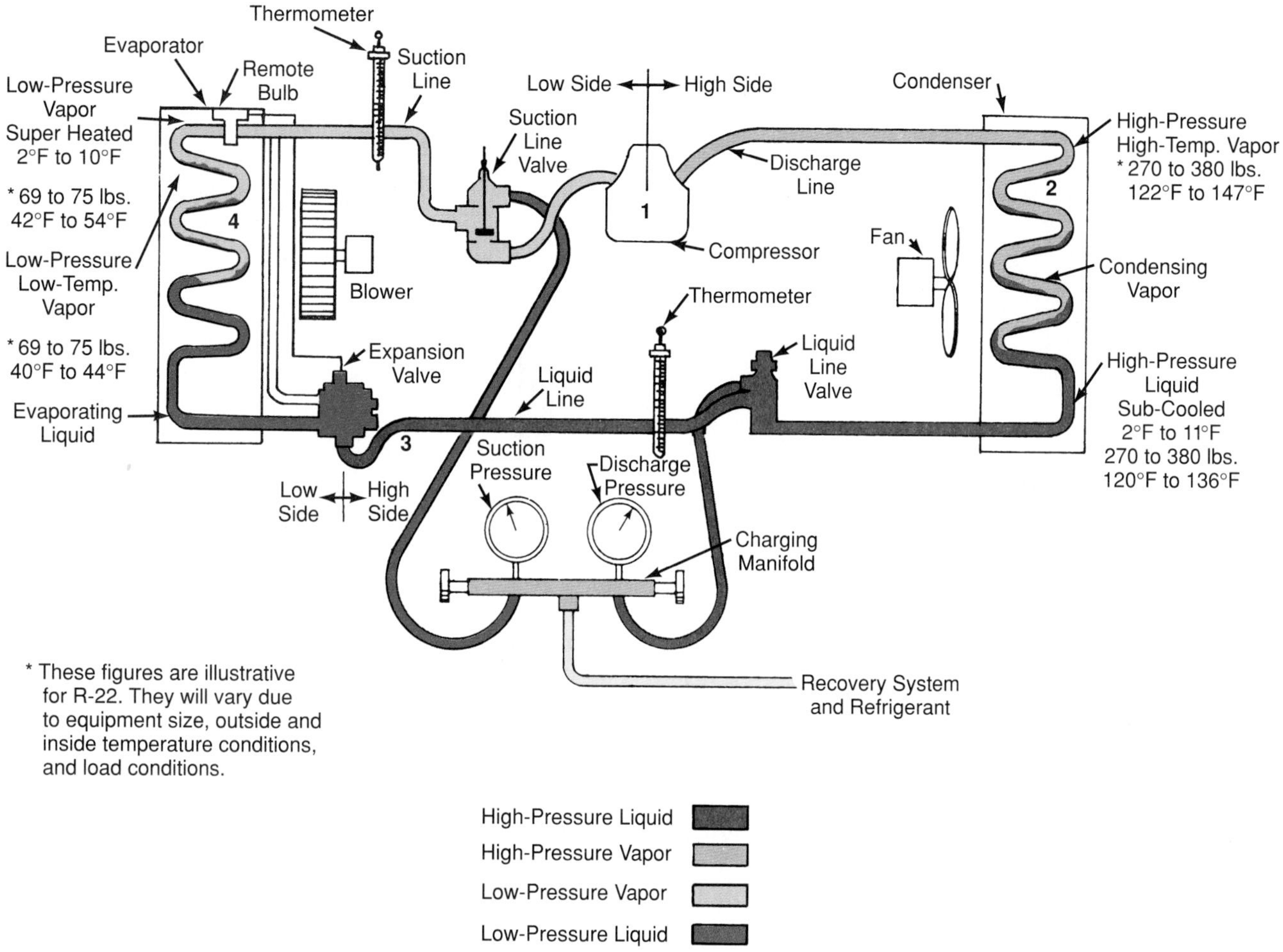

Figure 22-25. *Air conditioning unit cycle diagram showing gauge manifold installed. (Goodman Manufacturing Corporation)*

Many window air conditioners use PSC (permanent split capacitor) compressor motors. These motors do not use a relay for starting. If supplied voltage is low (10% or more), they will start with great difficulty. Many service technicians install a starting capacitor and a relay to overcome this problem. The capacitor and the relay must be exactly the right size for the motor.

The best way to determine the correct size is to follow the manufacturer's recommendation. If this is not available, the table in **Figure 22-26** will help in selecting the correct electrical starting system.

When the motor compressor reaches a satisfactory speed, the starting capacitor needs a relay. This relay will open the starting capacitor circuit. Special starting kits, **Figure 22-27,** are available for PSC motor compressors.

Window units are often removed during the winter season. However, if this is inconvenient, the unit can be winterized. The air inlet and the outlet grilles can be blocked with cardboard or flexible plastic sheeting. A storm sash can be custom built to fit around the air conditioner. Plywood held in place with caulking or rubber grommets can also be used.

Recommended Ratings for Add-On Starting Capacitors

Running Capacitor Microfarads	Special Starting Capacitor Microfarads (Add-On)*
20	18
25	18 or 25
30	25
35	25
40	25 or 45
45	45
50	45

*Most manufacturers specify Microfarad rating of special start capacitor for each of their units. Follow manufacturer's specifications.

Figure 22-26. *Table of special starting capacitor sizes to be used on PSC (permanent split capacitor) motor compressor when starting difficulties are found. PSC motors are introduced in Section 7.8.3.*

Replacement capillary tubes for window units must be very accurately selected by size. **Figure 22-28** gives correct sizes when R-22 refrigerant is used.

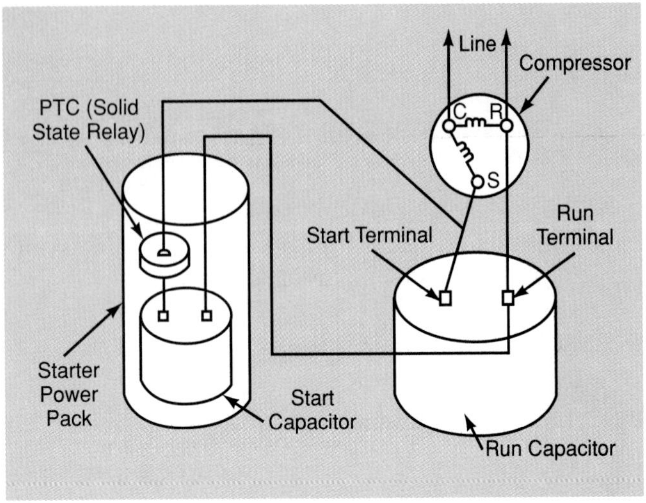

A

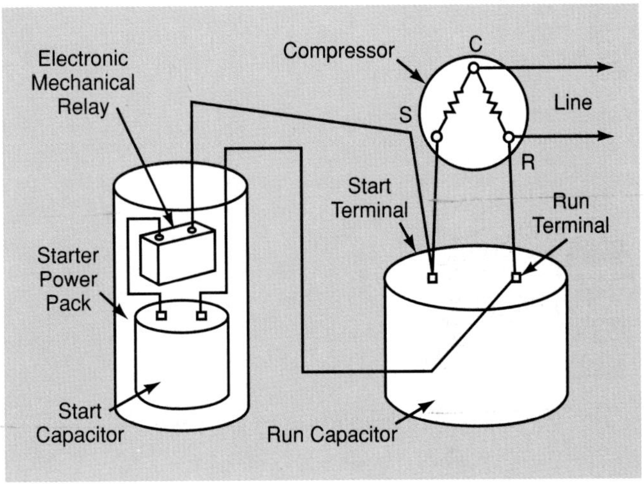

B

Figure 22-27. *Starting kits for PSC motors. A—This starting kit uses a PTC (positive temperature coefficient) solid state relay. As current enters the PTC, the resistance greatly increases and literally chokes off current to the starting winding. This sharp increase in resistance occurs within 1/2 second. B—This starting kit is a combination start capacitor and electronic-mechanical relay with normally open contacts. When the internal coil is energized, a solid state timer deactivates the coil in about 1/2 second. This type of relay does not depend upon "pick-up" or "drop-out" voltages as the common potential relay operates. (Sealed Unit Parts Co., Inc.)*

If the window unit drips water into the room, it is not correctly installed. Check the slope of the unit from inside to outside with a spirit level. It must slope to the outside (condenser edge) about 1/4". Condensate water will then run to the depression of the unit base under the condenser fan and condenser. Make sure the drain hole is open, then level the unit along its other dimension. Finally, recheck the unit installation for airtight sealing in the window opening.

Compressor Capacity Btu/Hr.	Capillary Size		Coil Circuits	
	Short	Long	3/8" Tube	1/2" Tube
4500	36 in. × .042	80 in. × .049	1	
5000	25 in. × .042	64 in. × .049	1	
5500	20 in. × .042	52 in. × .049	1	
6000	40 in. × .049	75 in. × .054	1	
6500	35 in. × .049	65 in. × .054	1	
7000	28 in. × .049	52 in. × .054	1	
8000	36 in. × .054	65 in. × .059		1
9000	28 in. × .054	48 in. × .059	2	1
10,000	36 in. × .059	64 in. × .064	2	1
11,000	28 in. × .059	50 in. × .064	2	1
12,000	40 in. × .064	68 in. × .070	2	1
13,000	32 in. × .064	56 in. × .070	2	1
14,000	44 in. × .070	70 in. × .075	2	1
15,000	36 in. × .070	56 in. × .075	3	2
16,000	30 in. × .070	48 in. × .075	3	2
17,000	38 in. × .075	65 in. × .080	3	2
18,000	35 in. × .075	55 in. × .080	3	2
19,000	28 in. × .075	48 in. × .080	3	2
20,000	40 in. × .080	58 in. × .085	3	2

Figure 22-28. *Capillary tube sizes for window air conditioners that use R-22 refrigerant. Coil circuits are the size of tubing in evaporator. Each unit has one capillary tube. (Tecumseh Products Co.)*

Packaged Terminal Air Conditioners

The ***packaged terminal air conditioner*** is a combined heating and cooling system. It is designed to service an individual room or zone. This type of unit is frequently used for year-around comfort in installations requiring zone heating and cooling. Examples include hotels, motels, apartments, dormitories, shops, and offices. See **Figure 22-29.**

Many of the units are designed to fit through a standard 42" × 16" wall space. This frequently represents the same size that would be used for installing an electrical resistance or heat pump unit. **Figure 22-30** illustrates a packaged terminal air conditioner that provides gas

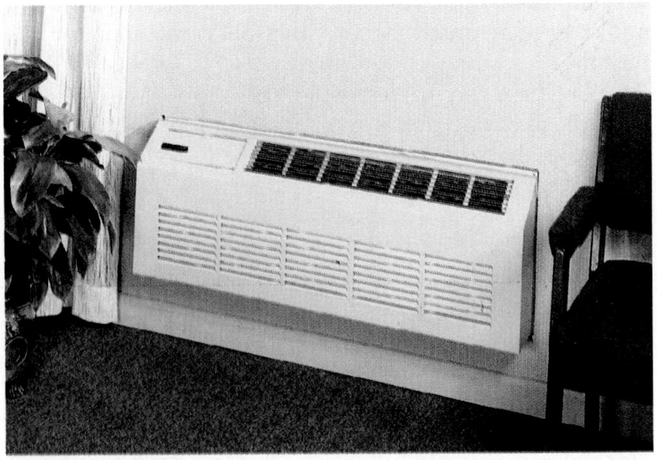

Figure 22-29. *Combination heating and cooling system frequently used for commercial installation in offices and apartments. (Suburban Manufacturing Company)*

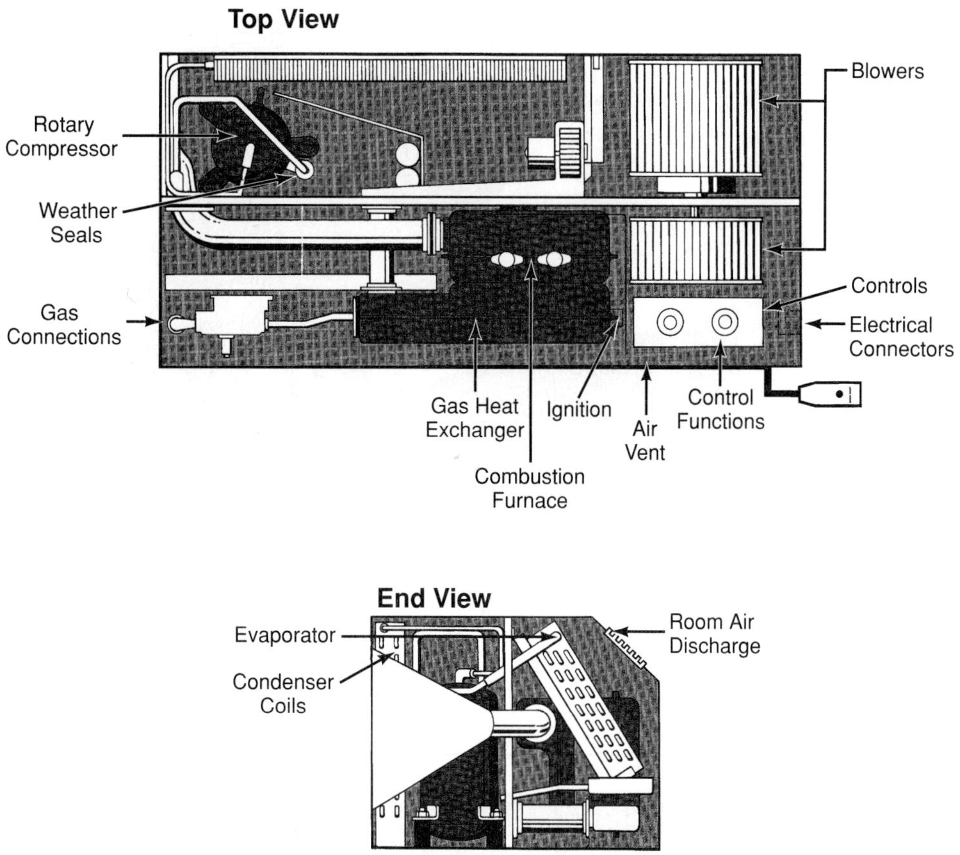

Figure 22-30. *Packaged terminal heating and cooling system. (Suburban Manufacturing Company)*

heat. The unit has a sealed combustion furnace that burns only the outside air. A solid state electronic hot surface ignition is used. Heating is provided by a gas-fired heat exchanger with the electronically controlled pilotless ignition system.

The refrigeration system utilizes a hermetically sealed rotary compressor. Airflow system includes an airflow fan and a combustion fan for the evaporator and condensing units.

Basic controls include a heating and cooling thermostat. A rotary control is used for heating, cooling, and fan operation. Units are normally operated with natural gas or LP gas. The system uses a capillary tube metering device. This type of unit is normally operated with R-22 or R-134a. For servicing, the entire unit will slide out of the wall cabinet.

Multizone Ductless Split System

Multizone ductless systems are popular for new and retrofit office use. They are frequently used in legal and medical offices, motels and homes without ducts.

The basic components of the system include a single outdoor condenser, three independent evaporators, and individual evaporator temperature control. The condensing unit is located outside on a slab. The lines to the evaporators enter the building at the desired locations, **Figure 22-31.**

The new system now gives a ductless system owner four options for obtaining air conditioning. These are the use of the window unit, the wall unit, console unit, or the ductless unit. Some of the primary advantages of the ductless system are as follows:

- A single condensing unit is used with three independent evaporators.
- Each evaporator temperature is maintained individually, providing different temperatures for three different offices.
- Most units are equipped with a remote wireless control for temperature.
- The units can be installed in walls or ceilings.

Console Air Conditioners

In ***console air conditioners,*** entire systems are mounted in a cabinet. They vary in capacity from 2 hp to 10 hp. Such units are often used in small commercial establishments such as restaurants, stores, and banks. Console molds may have either water-cooled or air-cooled condensing units. Air-cooled models, needed in some localities because of water restrictions, must have air ducts to the outdoors for condenser cooling.

Figure 22-32 shows a water-cooled console unit. Return air enters the lower grille. Cooled air is discharged at the upper grilles. Ducts can be connected to portions or all of the upper sections. These are needed when partitions interfere with cooled air distribution. The condensing unit is mounted in the bottom of the console. Air blowers are in the middle. The evaporator is in the top of the cabinet.

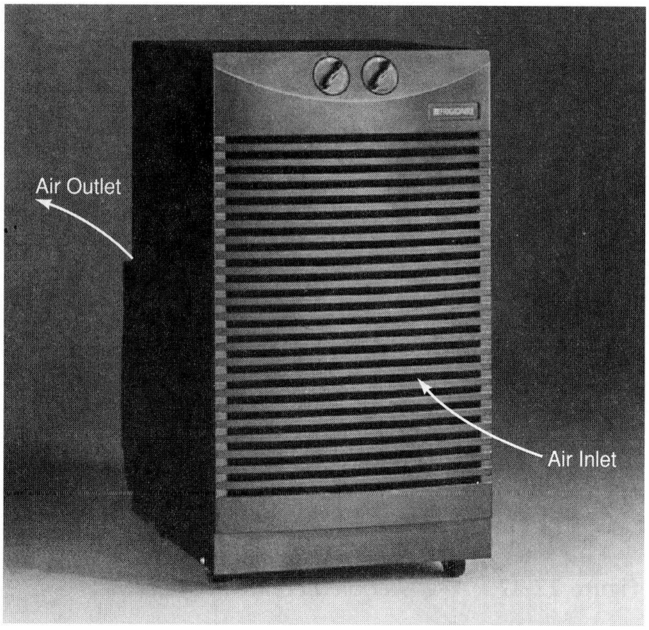

Figure 22-39. *Outside view of dehumidifier. Air enters in front of cabinet and is forced out at back. Note arrows. (Frigidaire Company)*

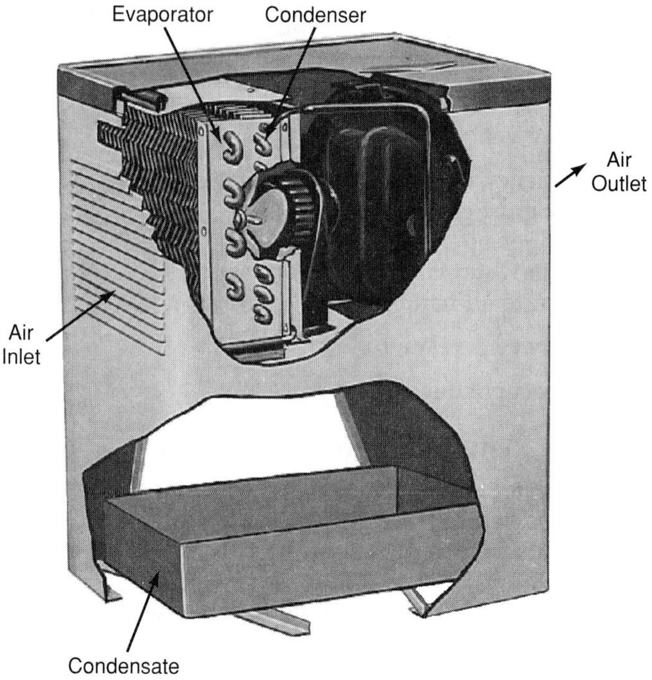

Figure 22-40. *Section view of dehumidifier. Air enters the dehumidifier and passes through the evaporator, in front of the fan. This is the cooling or dehumidifying coil. It then passes through the next coil, which is the condenser, and reheats. Condensate pan or direct drain can be used to remove condensate.*

evaporator. The cooled air is then moved over the condenser, **Figure 22-40.** It reheats the air to a reasonable relative humidity.

The device is used to "dry" the air. It is useful in basements and other damp places.

These units usually have a ***humidistat.*** This is a device that senses moisture in air. A container is used to collect the condensate or it is removed through a drain tube. **Figure 22-41** shows a humidistat and other controls.

In some installations, certain chemicals are used to absorb moisture from the air. The chemicals are usually cycled so moisture from the air is first absorbed into the chemical. Then the chemicals are heated. Moisture driven from the chemical is exhausted. The chemicals are ready to absorb moisture once more.

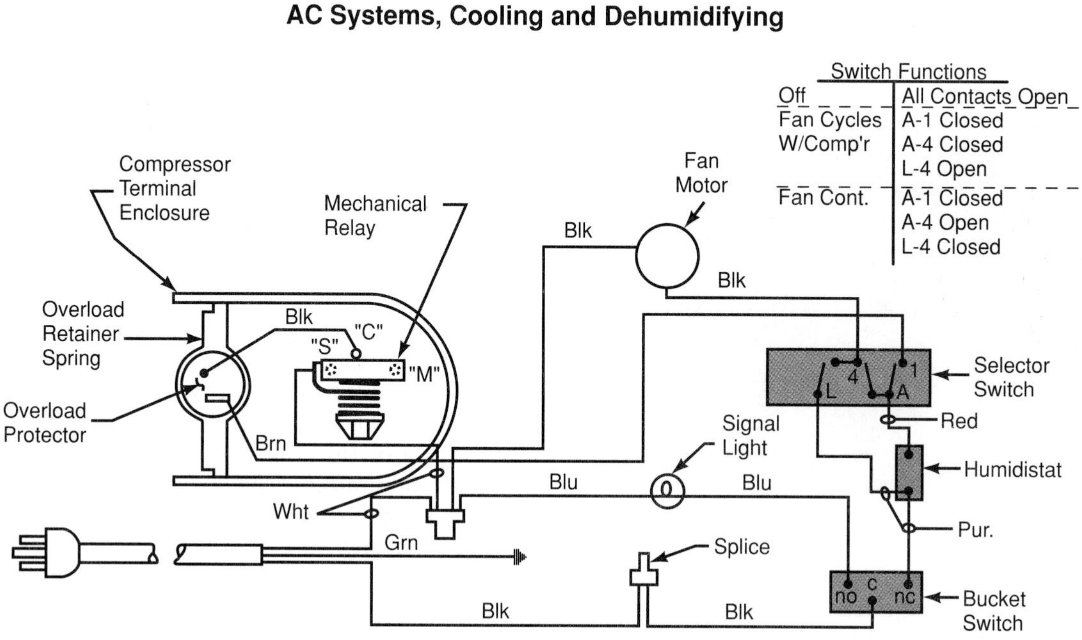

Figure 22-41. *Wiring diagram for dehumidifier. Note selector switch, humidistat, and bucket switch.*

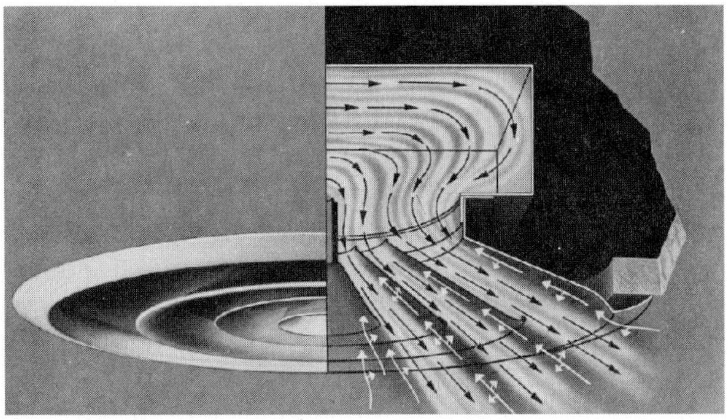

Figure 23-4. *Airflow and air mix of ceiling grille. Black arrows indicate airflow from duct. White arrows indicate room air moving into grille to mix with duct air. (Anemostat Products Div.)*

23.2 Air Circulation

In warm air heating, three basic systems are used to circulate the air:

- Gravity.
- Intermittent forced air.
- Continuous forced air.

The gravity system is no longer popular. Too much energy is lost before the air gets to the room being heated.

Most installations use the intermittent forced air system. A thermostat in the furnace plenum chamber is used to control the fan.

Becoming more popular is the continuous blower system. It provides a more constant temperature in rooms.

Systems designed to provide cooling as well as heating need additional capacity to move air. A cubic foot of cooled air will not change room temperature as much as a cubic foot of warmed air. The reason for this is the temperature difference. Warmed air comes out of the duct many degrees warmer than the room air it is replacing. Cooled air is not all that much cooler than the room air it is replacing. Therefore, greater quantities need to be moved into the room to get the desired effect. Also, the cooled air has a high relative humidity. People have difficulty getting cool when the relative humidity is high. More airflow is needed to cool than to heat.

The high volume of cooled air needed is somewhat offset by the lower cooling load. In average conditions, a 30% to 50% airflow increase is required. Either of two methods will increase the airflow when needed:

- Use a two-speed blower motor (for directly-driven blowers).
- Install a two-speed pulley on the motor if a belt-driven unit is used.

23.2.1 Room Air Movement

Air entering a conditioned space through ducts must circulate without causing annoying drafts. This depends on the number and size of the air inlet grilles. It also depends on the velocity of the air moving through them.

Air delivered to the room from the supply duct, moving at a velocity of 150 ft. per minute or more, is called ***primary air.*** The primary air pushes against and mixes with air already in the room. The distance the air from the grille travels before it slows down to 50 ft. per minute (terminal velocity) is called the ***throw.*** The outlet velocity is the speed of the duct air as it leaves the grille.

The overall size of the grille is not important. The total area of the air openings in the grille determines the grille capacity. The spread of the air that leaves the grille is very important. Return air grilles should be located where room air has the slowest movement.

23.2.2 Return Air Ducts

Return air ducts are important. Airflow through these ducts is almost always from the "pulling" action of a fan or blower. If the return airflow does not match the airflow into a room, the flow of air will not be properly balanced.

If there is more return air than entering air, the room may have a negative pressure. Thus, more entering air will be needed by this room. In turn, other rooms may starve for air. During the heating season, rooms starved for air will be too cold.

Duct return grilles should be placed in the stratified (stagnant) air zone of a room. During the heating season, this area is along the floor. During the cooling season, this place is near the ceiling. Ideally there should be two places for return air grilles. In all cases, the place is the maximum distance from the inlet grilles.

23.3 Basic Ventilation Requirements

As noted before, air is a mixture of gases. Normally air contains about 21% oxygen. A human system requires that a certain oxygen content be contained in the air:

- To maintain life.
- To be comfortable.

If a room is tightly sealed, any human in that room would slowly consume the oxygen. The amounts of carbon dioxide, water vapor, and various impurities would also increase. This could cause drowsiness or even death.

Human living space must have air with a good oxygen content. This air must be kept at a reasonable temperature. It is very important that fresh air be admitted to provide the oxygen.

In the past, this fresh air entered the space by infiltration (leakage). Infiltration normally occurs through door and window openings and cracks in the structure. However, modern construction is reducing this air leakage. The air conditioning apparatus, then, must furnish fresh air. Modern units have a controlled fresh-air intake.

This fresh air is conditioned and mixed with the recirculated air before it reaches the room.

Some conditioned air leaves a building through doors, windows, and other construction joints. Some also leaves through the same openings (exfiltration). Any kind of exhaust fan removes conditioned air.

It is best to bring in replacement fresh air through an air system. When this is done:

- The air can be cleaned.
- The air can be cooled or heated.
- A positive pressure can be maintained in the building to help keep out airborne dirt, dust, and pollen. (A negative pressure reduces the efficiency of exhaust fans and of a fuel-fired furnace.)
- A definite amount of fresh air (makeup air) is brought in for health purposes (oxygen content).

Certain building areas should have slightly less positive pressure than the rest of the building. A lower positive pressure (10% to 15% less than the rest of the building) reduces the spread of odors. Such areas would include the kitchen, lavatories, and where certain industrial operations produce fumes.

The amount of fresh air required depends on the use of the space and the amount of fresh air admitted by infiltration. One basic rule is to provide at least 15 cfm of fresh air per person. This will provide enough oxygen and will remove carbon dioxide. Six people occupying a 10,000 ft^3 space would need 90 cfm of fresh air (6×15 cfm = 90 cfm). It would take 10,000/90 = 111 minutes (1.85 hours) to completely replace the air in the space. This is rather slow.

Remember, however, that the purpose is not to replace the air quickly. It is costly to use heat as fast as needed to replace the air immediately.

The air can be handled either to produce positive or negative pressure in a building. (Positive pressure is higher than atmospheric pressure. Negative pressure is below atmospheric pressure.) A positive pressure will eliminate infiltration of air from outside or from other spaces. It is done by using special air intakes to the blowers. A positive pressure assures that all air entering a building can be filtered. It is cleaned before reaching the occupied space. Negative pressure increases the infiltration at windows and doors. This air is untreated and may be dirty.

Residential homes that use fuel-burning furnaces need air for combustion. Combustion air, leaving through the chimney, might create a slightly negative pressure inside the house. See **Figure 23-5.**

The amount of impurities in the air may be great enough to require air cleaning. (Odor, smoke, and bacteria may be a part of such impurities.) The remedy may be either ventilation, using fresh air, or improved air cleaning.

Ventilation is usually based on air changes per hour for the conditioned space. In a 1000 ft^3 space, for example, three changes per hour would mean 3000 ft^3/hour or 50 cfm. Three changes every hour is the minimum for a residence during the heating season. As high as 12

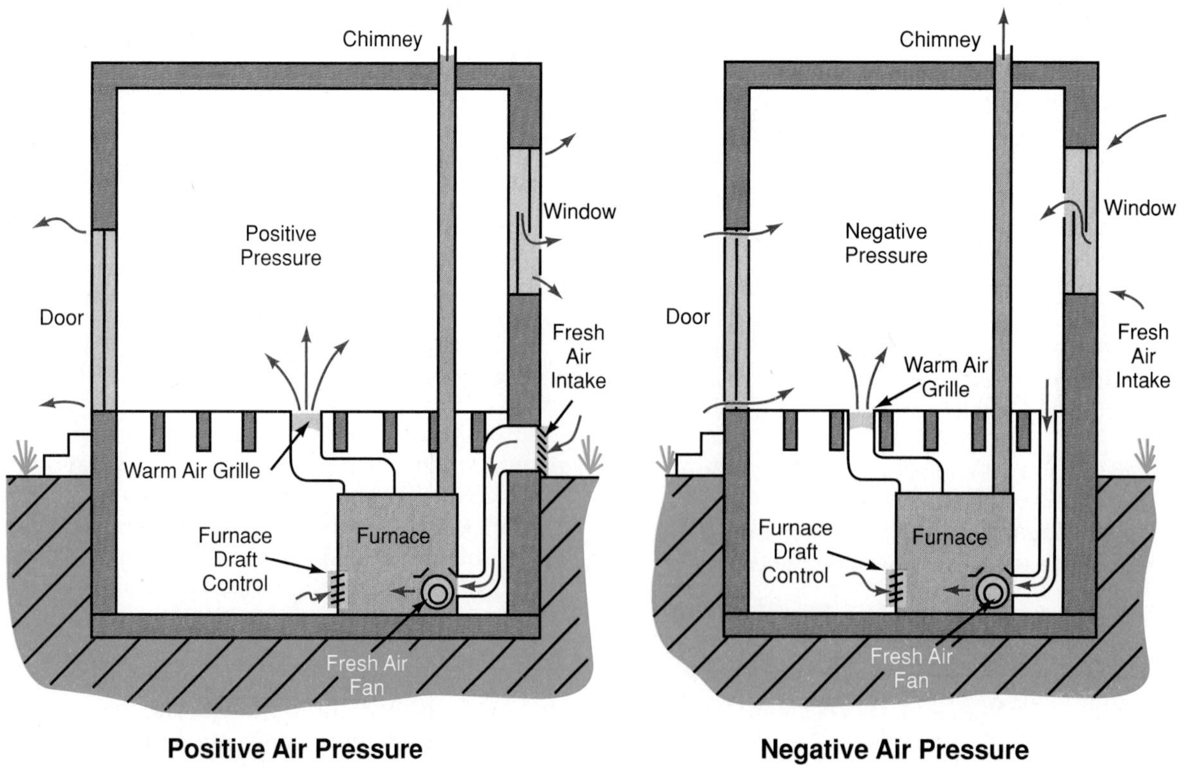

Figure 23-5. *Simplified diagram of airflow into and out of a building during the heating season. Red arrows indicate airflow.*

changes per hour (in the above case, 200 cfm) are recommended for cooling in public assembly. **Figure 23-6** shows typical air changes for both the heating and cooling seasons.

Use	Air Changes/Hour	
	Heating	Cooling
Homes	3–6	6–9
Offices Stores	5–8	6–12
Public Assembly	5–10	6–12

Figure 23-6. *Recommended air changes for various types of occupancy with 1000 ft^3 of space per room.*

It is good practice to keep the air blowers running all the time. They provide good ventilation to all parts of the building. Variable speed blowers are sometimes used. They provide more air movement when the heating or cooling system is running. Less movement is provided when the systems are off.

An adequate air supply is the best way to control comfort. Body comfort is controlled by evaporation, convection, radiation, and respiration. Therefore, the temperatures of the walls, floors, and ceilings must be controlled. Enough air must also be supplied to promote good respiration, evaporation, and convection.

Where specific conditions are unknown, it is best to design for 2 cfm/ft^2 or 12 changes/hr. It is important to remember that people occupying a closed space give off considerable heat. A sleeping person gives off about 200 Btu/hr., while a person doing heavy work gives off up to 2400 Btu/hr. One Btu = 252 calories.

Another way to determine ventilation requirements is to design for 4 cfm to 6 cfm of fresh air per person. About 25 cfm to 40 cfm recirculated air should be provided per person. This means the system should handle a total of 29 cfm to 46 cfm per person (1 cfm = 0.0283 m^3/min.).

23.3.1 Attic Ventilation

It is important to properly ventilate the attic space. The air within the attic affects the conditions of the structure. The attic may develop a mildew odor or become extremely hot. To prevent this, some type of ventilation is required.

An unventilated attic could reach 150°F (66°C) on a hot summer day. This would make it difficult to maintain comfortable temperatures in the structure.

A common method of ventilating an attic is to use louvers or vents. This allows the fresh air to enter and the internal air to escape. At the same time, water, insects, and other objects are prevented from entering.

Many buildings use exhaust fans. These fans remove extremely hot air that collects in attic spaces. Some attic fans may also be used to bring cool evening air into a building.

These fans have large capacities ranging from 1000 cfm to 4000 cfm. It is desirable to have a fan large enough to make the needed air changes. A complete change of air should be made in the building every 8 to 10 minutes.

Example:

A building measures 30′ × 60′ with an 8′ ceiling.

Volume = length × width × height
= 60′ × 30′ × 8′
= 14,400 ft^3

An exhaust fan of 1440 cfm capacity will change air every 10 minutes.

ft^3 of space/cfm
= 14,400 ft^3/1440 cfm = 10 minutes/change

23.3.2 Basement Ventilation

Basements tend to be cool and damp in the summer. Therefore, mold and odors are problems. An exhaust fan will reduce the dampness and mold growth. This fan should remove basement air from the floor level and exhaust it outdoors.

One method of installing such a fan is to remove a basement window pane and install an exhaust fan with an inlet duct leading down to floor level.

23.4 Air Ducts

To deliver air to the conditioned space, air carriers are needed. These carriers are called ***ducts.*** Ducts are made of sheet metal or some structural material that will not burn (noncombustible).

Ducts work on the principle of air pressure difference. If a pressure difference exists, air moves from higher pressure areas to lower pressure areas. The greater this pressure difference, the faster the air will flow.

Ducts are made of many materials. Pressure in the ducts is small, so materials with a great deal of strength are unnecessary. Originally, hot air ducts were thin, tinned sheet steel. Later, galvanized sheet steel, aluminum sheet, and insulated ducts were developed. They are made from materials such as fiber-board. Passageways formed by studs or joists are sometimes used for return air. This can be done where a fire hazard does not exist.

There are three common classifications of ducts:

- Conditioned-air ducts.
- Recirculating-air ducts.
- Fresh-air ducts.

Ducts are round, square, or rectangular. **See Figure 23-7.** Round ducts are more efficient. That is, less material is needed for the same capacity as a square or rectangular duct. Resistance to airflow is also less. Since

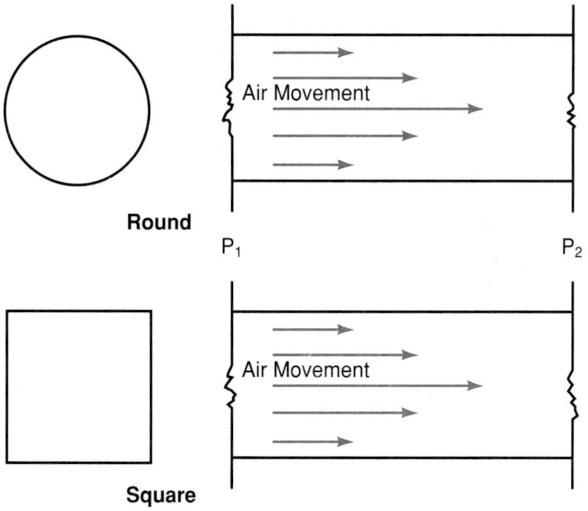

Figure 23-7. *Duct shapes vary. Arrows indicate that the closer air is to the duct wall, the slower the flow. P_1 and P_2 indicate pressure along a duct. To create movement, P_1 must be greater than P_2.*

round ducts have less surface area, they have a much lower heat loss or gain.

The square or rectangular duct conforms better to building construction. It fits into walls and ceilings better than round ducts. It is easier to install rectangular ducts between joists and studs.

Tables have been developed to compare carrying capacities of rectangular and round ducts. See **Figure 23-8.** There are several round duct equivalent sizes from which to choose. The one selected depends on the one side dimension desired. For example, ducts 11″ high may be desired to improve appearance. They may be needed to fit in between joists (a 14″ distance).

The quantity of sheet metal used for a rectangular duct may be a deciding factor. The amounts of material for round and rectangular ducts can be compared. The perimeter of a circle is called the circumference, and is equal to the diameter multiplied by π (pi, which is 3.1416). The perimeter of a rectangle is simply the lengths of all four sides added together (or twice the width multiplied by twice the length).

Formula:

Circumference = Diameter × π

Example:

Determine the perimeter of an 18″ round duct.

Solution:

18″ × 3.1416 = 56.55″

Formula:

Perimeter (Rectangle) = 2W + 2L

Example:

A 17″ × 16″ rectangular duct has a capacity equal to an 18″ round duct. Determine how much more material is needed for the rectangular duct.

Solution:

Perimeter = 2 × 17″ + 2 × 16″
= 34″ + 32″
= 66″

Perimeter − Circumference = 66″ − 56.55″
= 9.45″

In other words: For every foot of length, the rectangular duct requires 12″ × 9.45″ or 113.4 in^2 more material than the round duct.

23.4.1 Types of Duct Systems

Air ducts deliver air to a room or rooms. They then return the air to the heating (furnace) or cooling (evaporator) system. **Figure 23-9** shows a typical residential air duct system.

There are several types of supply duct systems. Some installations are a combination of the systems:

- Individual round pipe system.
- Extended plenum system.
- Reducing trunk system.

Return air systems are of two types:

- Single return system.
- Multiple return system.

The return systems can also be combinations of the two systems. **Figure 23-10** shows the basic designs. Some types of duct connections are shown in **Figure 23-11.**

Duct systems may be installed in basements, crawl spaces, and attics. They may be installed in concrete floors (slabs) of homes without basements. **Figure 23-12** shows an overhead duct system.

In basements, the main duct is run across and just under the floor joists. Branch ducts are then run between the joists to the grille or diffuser openings. Return ducts usually use the joists and the floorboards for three sides of the branch ducts (panned joist space). Then the main return duct is run alongside the conditioned air duct.

Ducts are usually insulated if the duct system is in a crawl space or attic. Otherwise heat loss would be too great.

Ducts installed in a concrete slab are usually made of metal, plastic, or ceramic. The branch ducts usually connect to a perimeter duct. Grilles or diffusers are connected to the perimeter duct at intervals along the floor. See **Figure 23-13.** A downflow furnace is used, as shown in **Figure 23-14.**

Some buildings are designed with unit ventilators. These ventilators automatically draw air from the conditioned space. At the same time, they bring in makeup air from the outside.

False ceilings are often used to conceal piping, wiring, heat exchangers, and ducts. The false ceiling may have holes to allow movement of conditioned air into the room below. Diffusers may also be used. Space between

Circular Equivalents of Rectangular Duct for Equal Friction and Capacity

Lgth Adj	Length of One Side of Rectangular Duct (*a*), in. 4.0	4.5	5.0	5.5	6.0	6.5	7.0	7.5	8.0	9.0	10.0	11.0	12.0	13.0	14.0	15.0	16.0
3.0	3.8	4.0	4.2	4.4	4.6	4.7	4.9	5.1	5.2	5.5	5.7	6.0	6.2	6.4	6.6	6.8	7.0
3.5	4.1	4.2	4.6	4.8	5.0	5.2	5.3	5.5	5.7	6.0	6.3	6.5	6.8	7.0	7.2	7.5	7.7
4.0	4.4	4.6	4.9	5.1	5.3	5.5	5.7	5.9	6.1	6.4	6.7	7.0	7.3	7.6	7.8	8.0	8.3
4.5	4.6	4.9	5.2	5.4	5.7	5.9	6.1	6.3	6.5	6.9	7.2	7.5	7.8	8.1	8.4	8.6	8.8
5.0	4.9	5.2	5.5	5.7	6.0	6.2	6.4	6.7	6.9	7.3	7.6	8.0	8.3	8.6	8.9	9.1	9.4
5.5	5.1	5.4	5.7	6.0	6.3	6.5	6.8	7.0	7.2	7.6	8.0	8.4	8.7	9.0	9.3	9.6	9.9

Lgth Adj	Length of One Side of Rectangular Duct (*a*), in. 6	7	8	9	10	11	12	13	14	15	16	17	18	19	20	22	24	26	28	30	Lgth Adj
6	6.6																				6
7	7.1	7.7																			7
8	7.6	8.2	8.7																		8
9	8.0	8.7	9.3	9.8																	9
10	8.4	9.1	9.8	10.4	10.9																10
11	8.8	9.5	10.2	10.9	11.5	12.0															11
12	9.1	9.9	10.7	11.3	12.0	12.6	13.1														12
13	9.5	10.3	11.1	11.8	12.4	13.1	13.7	14.2													13
14	9.8	10.8	11.4	12.2	12.9	13.5	14.2	14.7	15.3												14
15	10.1	11.0	11.8	12.6	13.3	14.0	14.6	15.3	15.8	16.4											15
16	10.4	11.3	12.2	13.0	13.7	14.4	15.1	15.7	16.4	16.9	17.5										16
17	10.7	11.6	12.5	13.4	14.1	14.9	15.6	16.2	16.8	17.4	18.0	18.6									17
18	11.0	11.9	12.9	13.7	14.5	15.3	16.0	16.7	17.3	17.9	18.5	19.1	19.7								18
19	11.2	12.2	13.2	14.1	14.9	15.7	16.4	17.1	17.8	18.4	19.0	19.6	20.2	20.8							19
20	11.5	12.6	13.5	14.4	15.2	16.0	16.8	17.5	18.2	18.9	19.5	20.1	20.7	21.3	21.9						20
22	12.0	13.0	14.1	15.0	15.9	16.8	17.6	18.3	19.1	19.8	20.4	21.1	21.7	22.3	22.9	24.0					22
24	12.4	13.5	14.6	15.6	16.5	17.4	18.3	19.1	19.9	20.6	21.3	22.0	22.7	23.3	23.9	25.1	26.2				24
26	12.8	14.0	15.1	16.2	17.1	18.1	19.0	19.8	20.6	21.4	22.1	22.9	23.5	24.2	24.9	26.1	27.3	28.4			26
28	13.2	14.5	15.6	16.7	17.7	18.7	19.6	20.5	21.3	22.1	22.9	23.7	24.4	25.1	25.8	27.1	28.3	29.5	30.6		28
30	13.6	14.9	16.1	17.2	18.3	19.3	20.2	21.1	22.0	22.9	23.7	24.4	25.2	25.9	26.6	28.0	29.3	30.5	31.7	32.8	30
32	14.0	15.3	16.5	17.7	18.8	19.8	20.8	21.8	22.7	23.5	24.4	25.2	26.0	26.7	27.5	28.9	30.2	31.5	32.7	33.9	32
34	14.4	15.7	17.0	18.2	19.3	20.4	21.4	22.4	23.3	24.2	25.1	25.9	26.7	27.5	28.3	29.7	31.0	32.4	33.7	34.9	34
36	14.7	16.1	17.4	18.6	19.8	20.9	21.9	22.9	23.9	24.8	25.7	26.6	27.4	28.2	29.0	30.5	32.0	33.3	34.6	35.9	36
38	15.0	16.5	17.8	19.0	20.2	21.4	22.4	23.5	24.5	25.4	26.4	27.2	28.1	28.9	29.8	31.3	32.8	34.2	35.6	36.8	38
40	15.3	16.8	18.2	19.5	20.7	21.8	22.9	24.0	25.0	26.0	27.0	27.9	28.8	29.6	30.5	32.1	33.6	35.1	36.4	37.8	40
42	15.6	17.1	18.5	19.9	21.1	22.3	23.4	24.5	25.6	26.6	27.6	28.5	29.4	30.3	31.2	32.8	34.4	35.9	37.3	38.7	42
44	15.9	17.5	18.9	20.3	21.5	22.7	23.9	25.0	26.1	27.1	28.1	29.1	30.0	30.9	31.8	33.5	35.1	36.7	38.1	39.5	44
46	16.2	17.8	19.3	20.6	21.9	23.2	24.4	25.5	26.6	27.7	28.7	29.7	30.6	31.6	32.5	34.2	35.9	37.4	38.9	40.4	46
48	16.5	18.1	19.6	21.0	22.3	23.6	24.8	26.0	27.1	28.2	29.2	30.2	31.2	32.2	33.1	34.9	36.6	38.2	39.7	41.2	48
50	16.8	18.4	19.9	21.4	22.7	24.0	25.2	26.4	27.6	28.7	29.8	30.8	31.8	32.8	33.7	35.5	37.2	38.9	40.5	42.0	50
52	17.1	18.7	20.2	21.7	23.1	24.4	25.7	26.9	28.0	29.2	30.3	31.3	32.3	33.3	34.3	36.2	37.9	39.6	41.2	42.8	52
54	17.3	19.0	20.6	22.0	23.5	24.8	26.1	27.3	28.5	29.7	30.8	31.8	32.9	33.9	34.9	36.8	38.6	40.3	41.9	43.5	54
56	17.6	19.3	20.9	22.4	23.8	25.2	26.5	27.7	28.9	30.1	31.2	32.3	33.4	34.4	35.4	37.4	39.2	41.0	42.7	44.3	56
58	17.8	19.5	21.2	22.7	24.2	25.5	26.9	28.2	29.4	30.6	31.7	32.8	33.9	35.0	36.0	38.0	39.8	41.6	43.3	45.0	58
60	18.1	19.8	21.5	23.0	24.5	25.9	27.3	28.6	29.8	31.0	32.2	33.3	34.4	35.5	36.5	38.5	40.4	42.3	44.0	45.7	60
62		20.1	21.7	23.3	24.8	26.3	27.6	28.9	30.2	31.5	32.6	33.8	34.9	36.0	37.1	39.1	41.0	42.9	44.7	46.4	62
64		20.3	22.0	23.6	25.1	26.6	28.0	29.3	30.6	31.9	33.1	34.3	35.4	36.5	37.6	39.6	41.6	43.5	45.3	47.1	64
66		20.6	22.3	23.9	25.5	26.9	28.4	29.7	31.0	32.3	33.5	34.7	35.9	37.0	38.1	40.2	42.2	44.1	46.0	47.7	66
68		20.8	22.6	24.2	25.8	27.3	28.7	30.1	31.4	32.7	33.9	35.2	36.3	37.5	38.6	40.7	42.8	44.7	46.6	48.4	68
70		21.1	22.8	24.5	26.1	27.6	29.1	30.4	31.8	33.1	34.4	35.6	36.8	37.9	39.1	41.2	43.3	45.3	47.2	49.0	70
72			23.1	24.8	26.4	27.9	29.4	30.8	32.2	33.5	34.8	36.0	37.2	38.4	39.5	41.7	43.8	45.8	47.8	49.6	72
74			23.3	25.1	26.7	28.2	29.7	31.2	32.5	33.9	35.2	36.4	37.7	38.8	40.0	42.2	44.4	46.4	48.4	50.3	74
76			23.6	25.3	27.0	28.5	30.0	31.5	32.9	34.3	35.6	36.8	38.1	39.3	40.5	42.7	44.9	47.0	48.9	50.9	76
78			23.8	25.6	27.3	28.8	30.4	31.8	33.3	34.6	36.0	37.2	38.5	39.7	40.9	43.2	45.4	47.5	49.5	51.4	78
80			24.1	25.8	27.5	29.1	30.7	32.2	33.6	35.0	36.3	37.6	38.9	40.2	41.4	43.7	45.9	48.0	50.1	52.0	80
82				26.1	27.8	29.4	31.0	32.5	34.0	35.4	36.7	38.0	39.3	40.6	41.8	44.1	46.4	48.5	50.6	52.6	82
84				26.4	28.1	29.7	31.3	32.8	34.3	35.7	37.1	38.4	39.7	41.0	42.2	44.6	46.9	49.0	51.1	53.2	84
86				26.6	28.3	30.0	31.6	33.1	34.6	36.1	37.4	38.8	40.1	41.4	42.6	45.0	47.3	49.6	51.7	53.7	86
88				26.9	28.6	30.3	31.9	33.4	34.9	36.4	37.8	39.2	40.5	41.8	43.1	45.5	47.8	50.0	52.2	54.3	88
90				27.1	28.9	30.6	32.2	33.8	35.3	36.7	38.2	39.5	40.9	42.2	43.5	45.9	48.3	50.5	52.7	54.8	90
92					29.1	30.8	32.5	34.1	35.6	37.1	38.5	39.9	41.3	42.6	43.9	46.4	48.7	51.0	53.2	55.3	92
96					29.6	31.4	33.0	34.7	36.2	37.7	39.2	40.6	42.0	43.3	44.7	47.2	49.6	52.0	54.2	56.4	96

Figure 23-8. *Chart determines sizes of rectangular ducts needed to equal carrying capacity of round ducts. To use, find diameter of round pipe in chart. Then find one side of rectangular duct by reading upward toward the top of the chart. Find other side by reading left to "Lgth Adj" column. (Reprinted with permission of the American Society of Heating, Refrigerating, and Air-Conditioning Engineers, Atlanta, Georgia)*

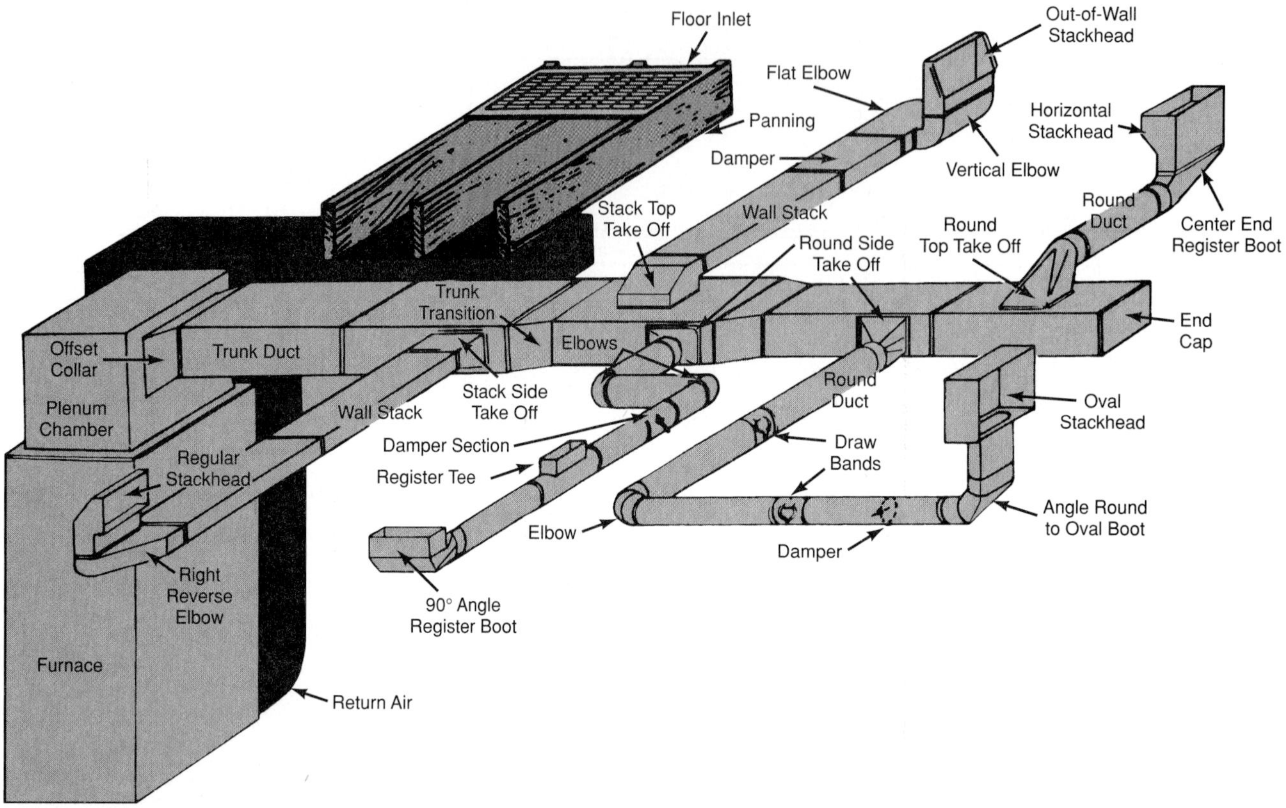

Figure 23-9. *Residential layout showing most sheet metal parts used in a duct system. (Reprinted from Air Conditioning Contractors of America's [ACCA] Basic Installation Manual,* by permission of ACCA)

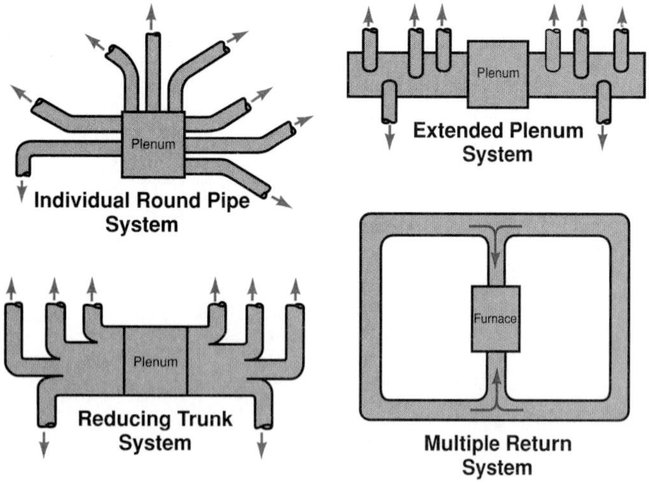

Figure 23-10. *Top views of duct systems.*

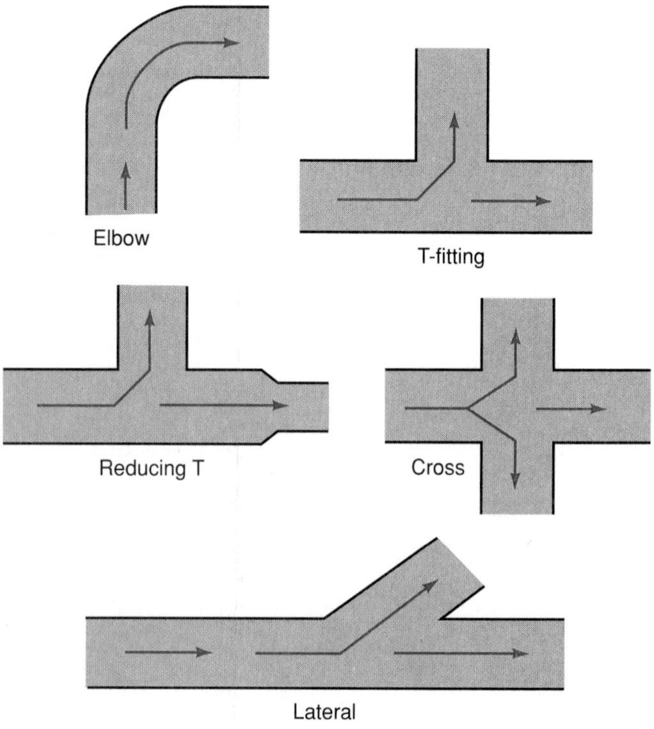

Figure 23-11. *Typical branch duct designs.*

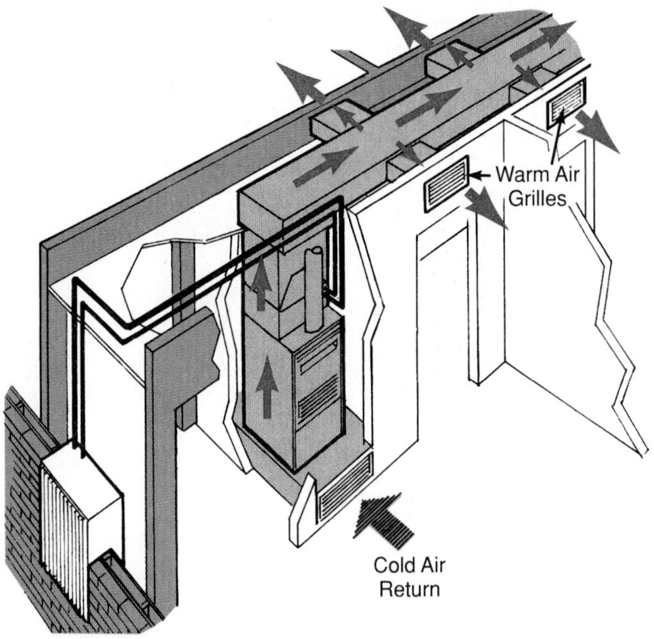

Figure 23-12. *Overhead extended plenum duct system. (Lennox International, Inc.)*

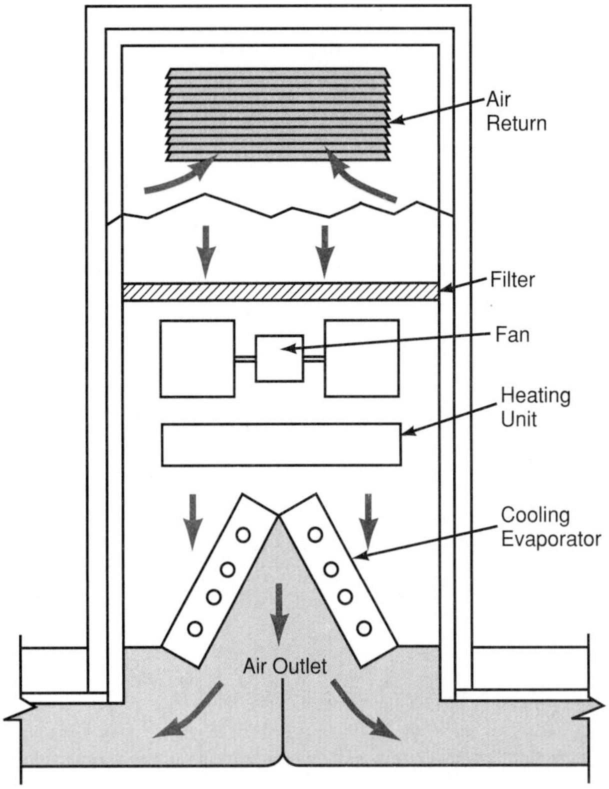

Figure 23-14. *Downflow furnace.*

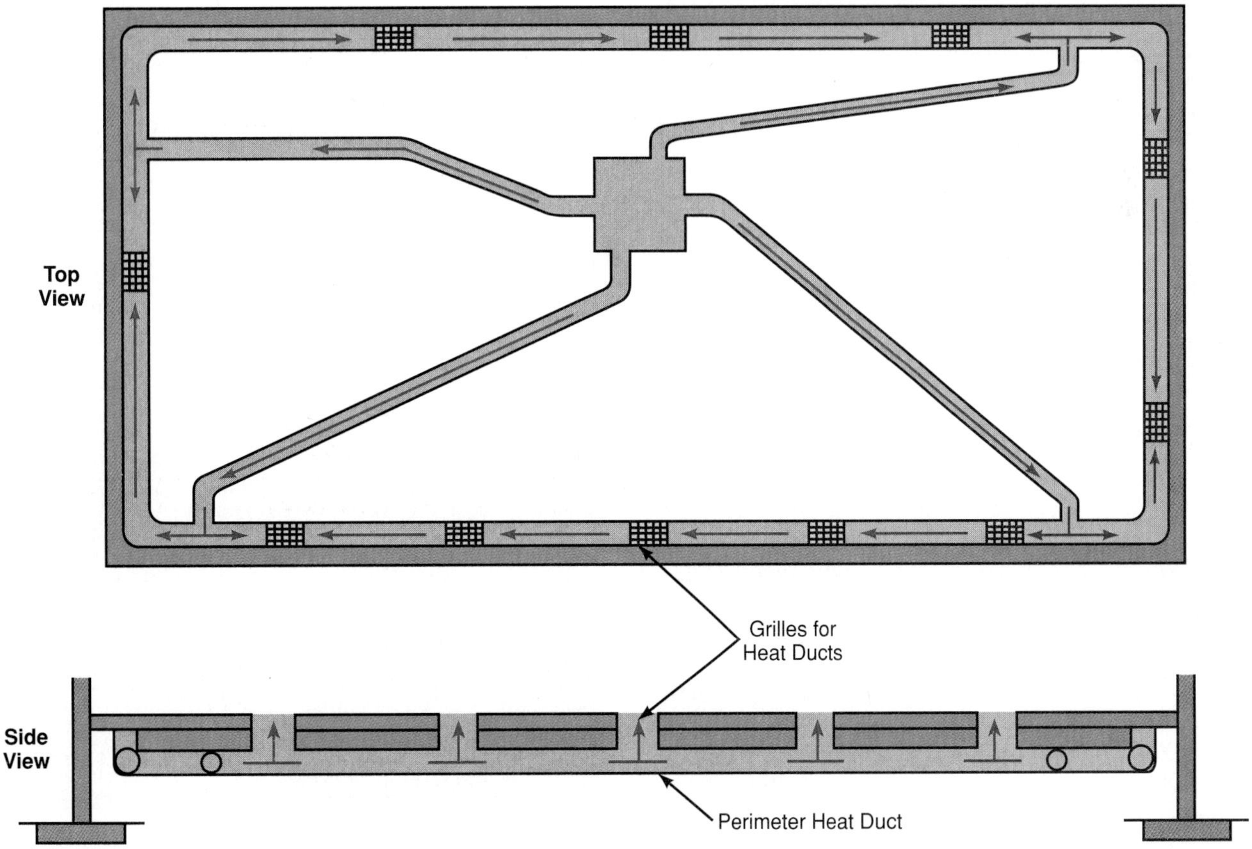

Figure 23-13. *Top view of perimeter loop system for conditioned air ducts. Ducts are cast into concrete floor slab of building.*

the false ceiling and the real ceiling may be used as a fresh air plenum chamber. Some systems use heated panels to provide either radiant heat or radiant cooling. See **Figure 23-15.**

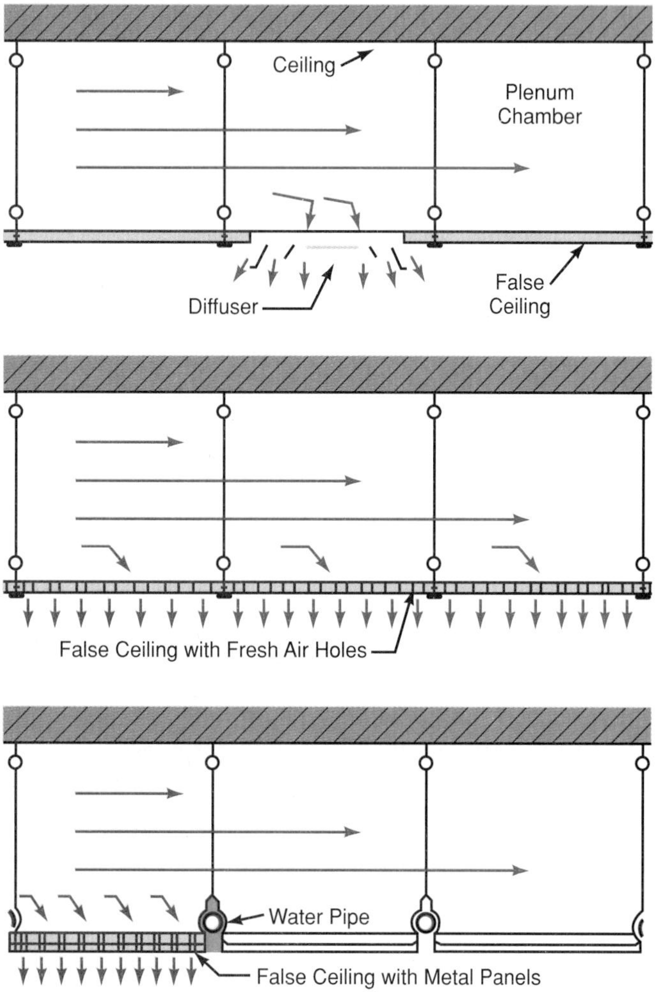

Figure 23-15. *False ceiling can be used to form plenum chamber for either cool or warm air. In the bottom drawing, water pipes heat or cool the metal panels.*

23.4.2 Duct Construction

Ducts may be made of metal, wood, ceramic, or plastic materials. Metal ducts are used for warm air distribution and for exhaust air ducts. The metal is usually sheet steel coated with zinc (galvanized steel). Some ducts are made of aluminum to reduce weight. Sheet lead is used when the duct must carry corrosive gases.

Ducts made of aluminum, glass fiber, or plastic are not approved by some codes due to fire hazard. Many flexible ducts made of wire and fabric are also not permitted if fire could spread through them to other areas.

Figure 23-16 is an illustration of a translucent air duct. The translucent duct is mainly used in areas where the duct will be visible.

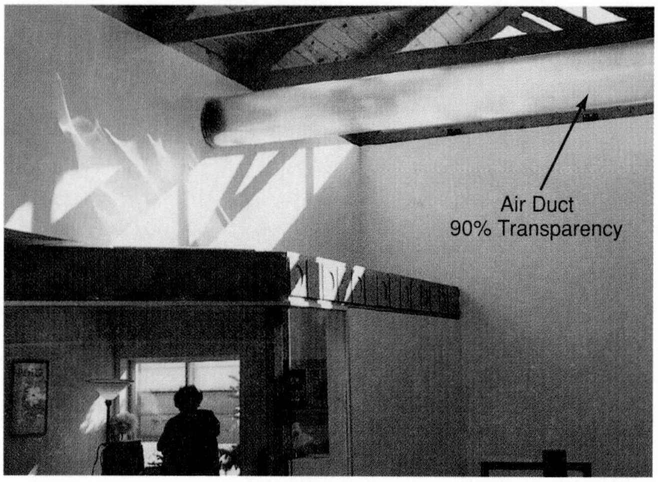

Figure 23-16. *Translucent air duct, approximately 0.040″ thick and weighing 1/2 lb. per lineal foot. (Solar Components Corporation)*

Sheet metal brakes and formers are used for making ducts. Elbows and other connections such as branches are designed using geometric principles. Sheet metal work is a specialty trade and is done by skilled sheet metal workers.

Sheet metal ducts expand and contract as they heat and cool. Fabric joints are often used to absorb this movement. Fabric joints can also be used where ducts fasten to a furnace or air conditioner. This would prevent most fan and furnace noise from traveling along the duct metal. However, most duct joints are made of sheet metal.

The length of the duct can be altered in many ways. **Figure 23-17** illustrates a combination air electrical cutting tool in use. It is being used by a technician to alter the length of a circular duct. This type of tool is used for cutting ducts up to 3/16″ thick.

Several types of sheet metal joints have been developed. See **Figures 23-18** and **23-19.** The joint should be airtight and strong. Many joints are riveted for added

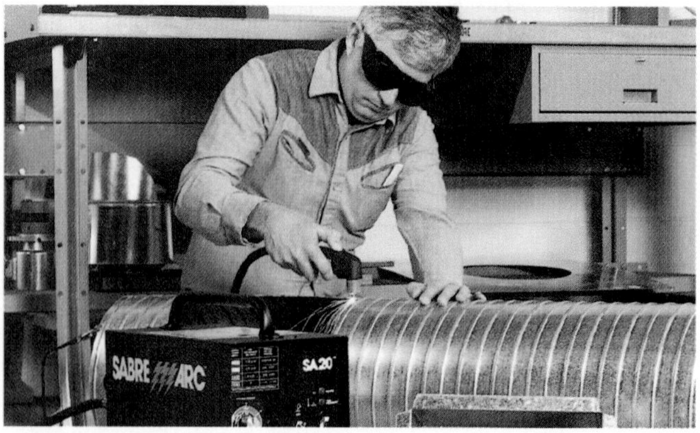

Figure 23-17. *Service technician using electrical air combination cutting tool to alter length of circular duct. (Hypertherm, Incorporated)*

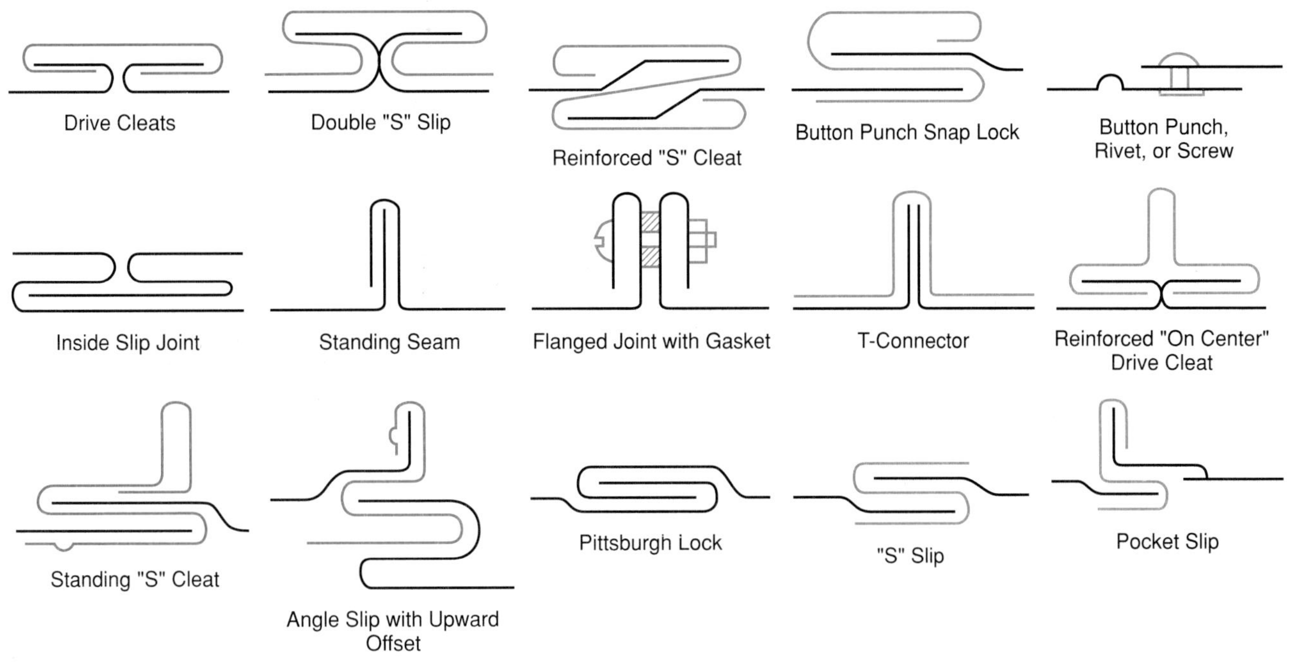

Figure 23-18. *Some sheet metal duct joint designs.*

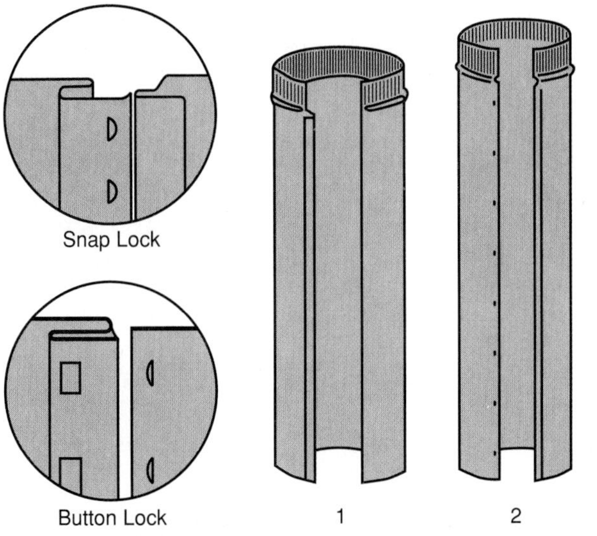

Figure 23-19. *Two seamlock methods. 1—Snap lock. 2—Button lock. (Reprinted from Air Conditioning Contractors of America's [ACCA] Basic Installation Manual, by permission of ACCA)*

strength and tightness. **Figure 23-20** shows a tool used to rivet from one side only.

To use this tool, use the following procedure:

1. Drill a hole.
2. Place the rivet blank in the hole.
3. Attach the tool.
4. Work the handle, which withdraws the expander.
5. When the rivet head is completely formed, the expander stem will break off flush with the inside surface of the rivet.

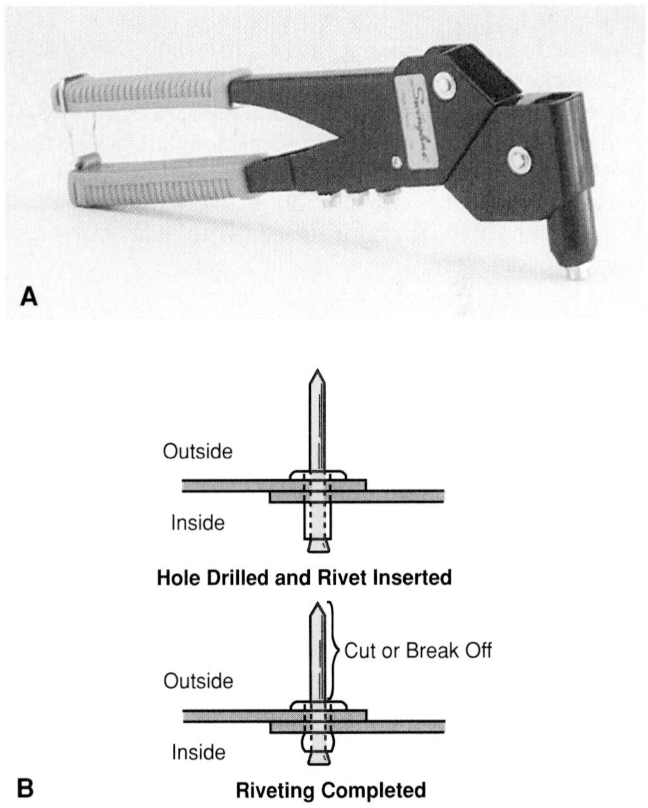

Figure 23-20. *Tool used for riveting sheet metal and duct work in places where it would be difficult to install a solid rivet, such as areas with access only to the outside. Rivet is often called a "blind" rivet. A—Riveting pliers. B—Riveting procedure. Portion of rivet remaining outside of the duct is either cut or broken off. (DESA International)*

Many of these joints are also sealed with special duct tape to make them leakproof. Sealants are put in the duct seam for the same purpose.

Duct joints in forced warm air systems should be as leakproof as possible. Leaking joints cut down on air volume delivered at the end of long runs. Connect all duct sections with good sheet metal joints.

Leaks may also make cold air ducts inefficient. Joints and seams should also be sealed the same as warm air ducts.

If duct systems are noisy, the air velocity may be too high. There may be turbulence, and metal edges may produce wind noise. Some ducts are lined with an acoustic material to provide a more quiet duct system. Duct sections equipped with silencers are also available. Such units act as mufflers.

Many ducts are insulated—either on the inside or outside—to reduce noise as well as heat transfer. See **Figure 23-21.** The insulation is fastened to the duct with adhesives. In some cases, metal clips hold the insulation in place. **Figure 23-22** shows a sheet metal connector insulated on the inside.

Round ducts made of plastic and spring steel wire are often used. They need no elbows and are easily installed around other piping, beams, and joints. This type of duct generally relies on self-inflation. This should be considered when determining system requirements. The plastic used will not burn.

Some ducts are made of nonmetals such as fiberglass, urethane, plastic, or sheetrock. See **Figures 23-23** and **23-24. Figure 23-23B** shows a flexible duct. A flexible bend allows it to be used for gradual bends. This is needed when connecting air ducts to diffusers or when routing air ducts through spaces with many obstructions. A duct with a branch and a vane air diverter is shown in **Figure 23-25.**

Duct material must not release any loose material into the airstream. Ducts made of asbestos or fiberglass should be plastic-coated. **Asbestos fibers and fiberglass particles released into the air are harmful to health.** Some ducts are made with a sheet metal inner wall and a sheet metal outer wall. They have insulation and sound absorbent material in between.

Any transition in the size of a duct should be tapered. When reducing the size, the taper should be approximately 1″ for every 4″ in length. When enlarging, use a change rate of 1″ in every 4″ to 7″ of length.

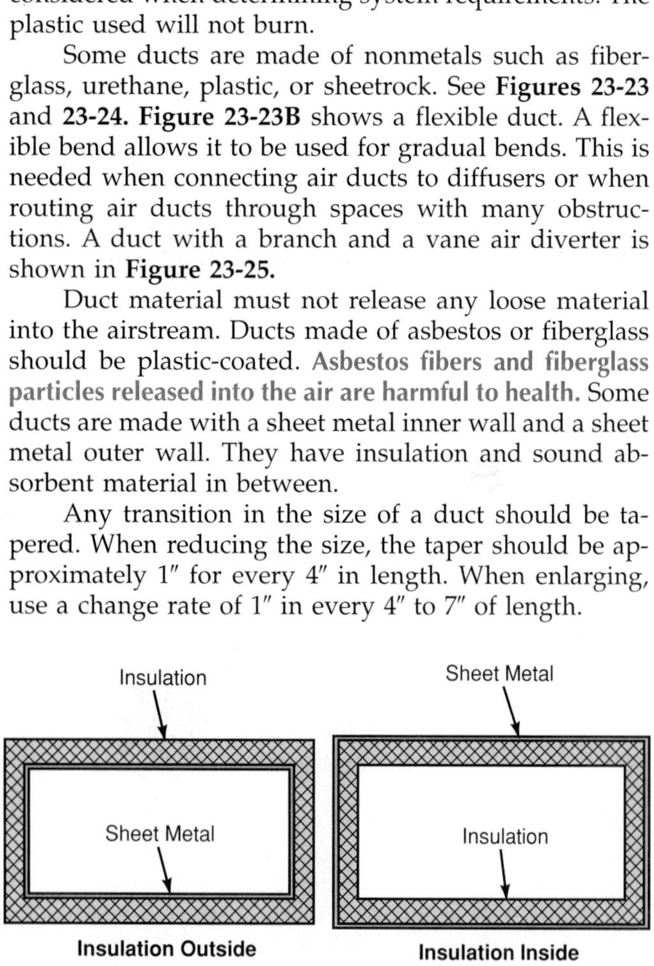

Figure 23-21. *Insulated sheet metal duct.*

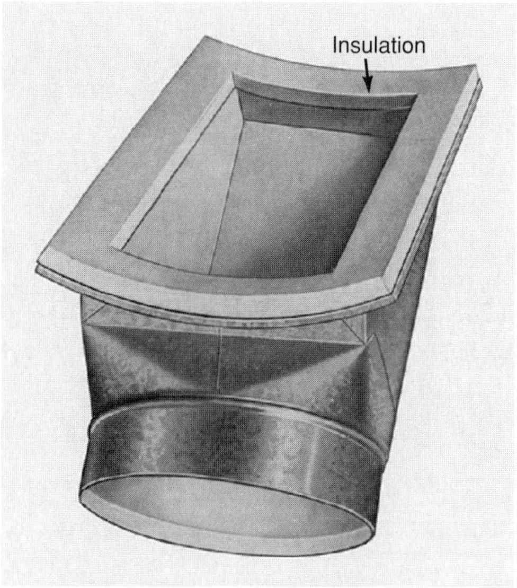

Figure 23-22. *Round duct branch elbow made of sheet metal with inside insulation. (Seal-Tite)*

A

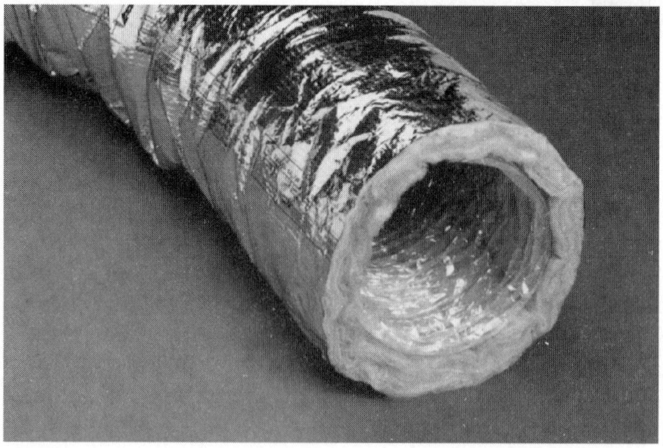

B

Figure 23-23. *Nonmetal duct materials. A—Resin bonded fiberglass air duct system used in residential and commercial cooling, heating, or dual-temperature air circulation systems. (Knauf Fiber Glass GmbH) B—Flexible duct insulated with fiberglass insulation. (Owens-Corning Fiberglas Corp.)*

Figure 23-24. *Flexible fiberglass duct being installed into rectangular duct air circulation system. (Owens-Corning Fiberglas Corp.)*

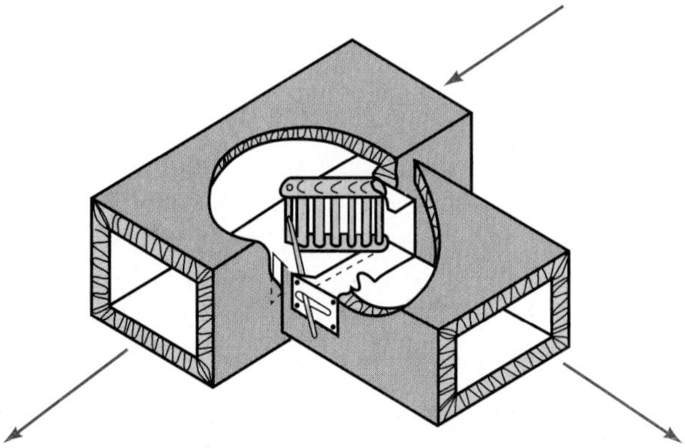

Figure 23-25. *Fiberglass duct connection with an adjustable vane air diverter.*

Rectangular ducts must be carefully chosen. The width should not be much more than the height. If it is, the amount of metal needed for the cross-sectional area becomes excessive. The ratio of the wide side to the narrow side is called the aspect ratio. For example, if a duct is 18″ wide and 6″ deep, the aspect ratio is 18/6 or 3 to 1.

To show how much more metal is used as the aspect ratio increases: Assume a duct is 10″ × 10″ (aspect ratio 1:1). It has a cross-sectional area of 100 in^2 while the distance around (perimeter) is 10″ + 10″ + 10″ + 10″ = 40″. Another duct of 100 in^2 is made 20″ wide and 5″ deep. It has a perimeter of 20″ + 5″ + 20″ + 5″ = 50″ or a 25% increase in metal for each unit of length. Sheet metal contractors classify ducts by their aspect ratios and estimate costs by class. **Figure 23-26** shows some common aspect ratios.

A duct should not be placed in a stud or joist space when its size is equal to the inside dimensions of the space. There must be room for fittings and allowance for alignment. The recommended clearance varies with the type of system. Ducts from solar collectors (or heat exchangers) have more clearance. **Figure 23-27** illustrates a satisfactory duct clearance between floor joists.

Duct Class (Aspect Ratio)	Wide Side (In.)	Perimeter (In.)
1	6–18	24–72
2	12–24	36–72
3	26–40	70–106
4	24–88	60–220
5	48–90	116–216
6	90–144	210–336

Figure 23-26. *Duct aspect ratio. This ratio is determined by dividing width of duct by its height.*

Figure 23-28 gives the recommended gauge thickness for various sizes of round ducts. **Figure 23-29** lists recommended gauge thickness for various size rectangular ducts. Ducts going through floors should be protected on the corners by angle iron. Large sections of sheet metal or ducts must be cross-broken to reduce panel vibration. See **Figure 23-30.**

The use of controlled fresh air is increasing. It has been found that natural infiltration is unreliable and too variable. This fresh air must be filtered. The outdoor intake end of the fresh air duct should be fitted with a strong screen and bar combination. This will keep out debris, birds, animals, and insects.

Air ducts present a potential Indoor Air Quality (IAQ) problem. They can become a breeding surface for

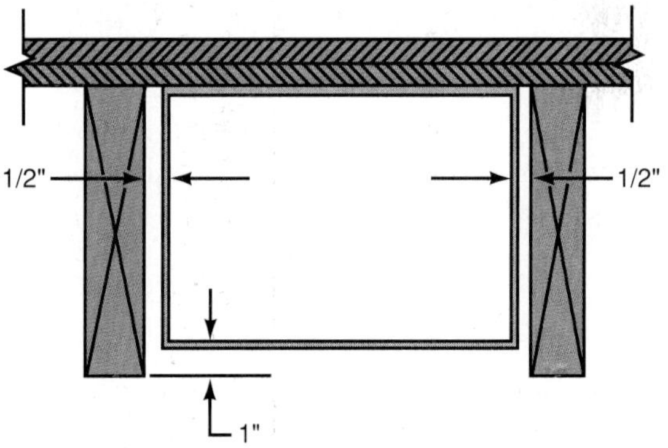

Figure 23-27. *Duct installation between floor joists. Note clearance space allowed between duct and joists.*

Round Duct Diameter	Commercial Sheet Steel, Galvanized (Gauge)	Residential Sheet Steel, Galvanized (Gauge)
Up to 12"	26	30
13" to 18"	24	28
19" to 28"	22	
27" to 36"	20	
35" to 52"	18	

Figure 23-28. *Recommended gauge thickness for round metal ducts.*

Rectangular Ducts Wide Side	Commercial		Residential
	Sheet Steel, Galvanized (Gauge)	Aluminum	Sheet Steel, Galvanized (Gauge)
Up to 12"	26 Gauge	.020"	28 Gauge
13"–23"	24	.025"	26
24"–30"	24	.025"	24
31"–42"	22	.032"	
43"–54"	22	.032"	
55"–60"	20	.040"	
61"–84"	20	.040"	
85"–96"	18	.050"	
Over 96"	18	.050"	

Figure 23-29. *Recommended gauge thickness for rectangular metal ducts.*

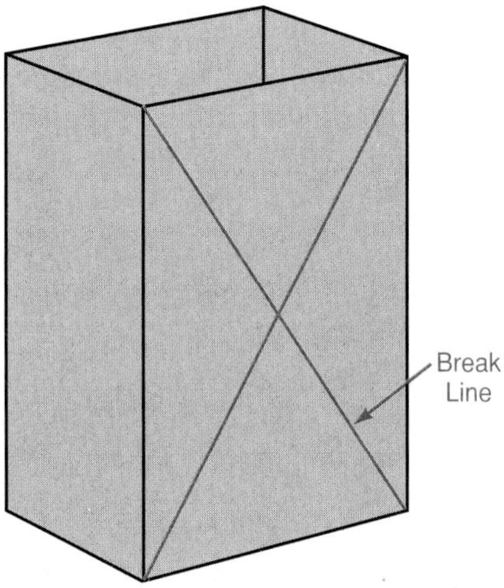

Figure 23-30. *Break lines (ridges) are formed on large sheet metal panels or large ducts to make them more rigid.*

bacteria, mold, mildew, and fungi. Portable systems have been developed to meet the demand for duct cleaning. See Chapter 19.

23.4.3 Diffusers, Grilles, and Registers

Room openings to ducts serve two purposes:

- To control the airflow.
- To keep large objects out of the duct.

There are several different openings:

Registers deliver concentrated airstreams into a room. Many have one-way or two-way adjustable airstream deflectors.

Grilles control the air-throw distance, height, and spread, as well as the amount of air. Grilles offer some resistance to airflow, blocking about 30% of the air. For this reason, the duct cross section is usually enlarged at the grilles. This enlargement also slows air movement and reduces noise. **Figure 23-31** shows a typical warm air grille.

Grilles have many different designs. Some are fixed and can direct air in only one direction. Others are adjustable and can be set to send air in different directions.

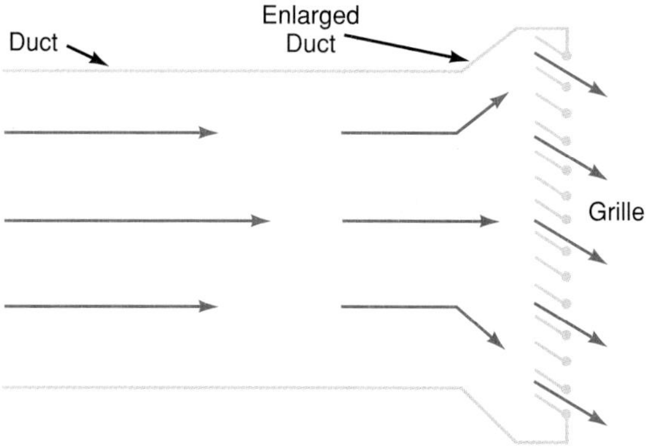

Figure 23-31. *Typical warm air grille. Many are supplied with controls for adjusting size of opening and redirecting airflow.*

A rectangular grille with adjustable direction airflow vanes is shown in **Figure 23-32.** Some grille designs have air vents to direct the airflow three different directions at one time. See **Figure 23-33.** Others are made with adjustable damper horizontal flow vanes, and vertical flow vanes. A ceiling mounted rectangular air diffuser, **Figure 23-34,** directs air in four directions along the ceiling.

Diffusers deliver widespread, fan-shaped flows of air into a room. Some diffusers cause the duct air to mix with some room air in the diffuser. Grilles are usually used as covers for the return air duct and have no adjustments.

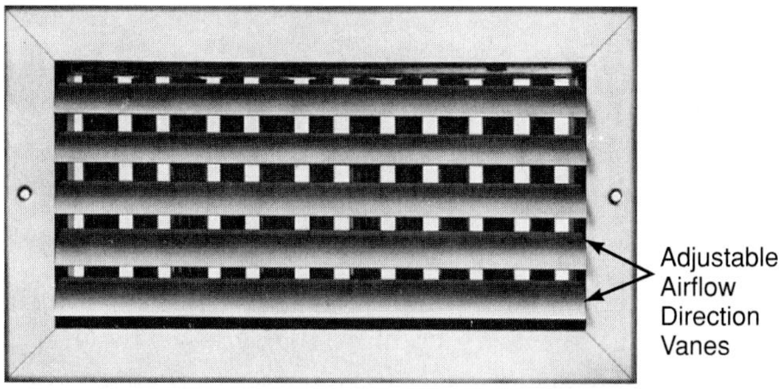

Figure 23-32. *Rectangular grille. (Hart & Cooley, Inc.)*

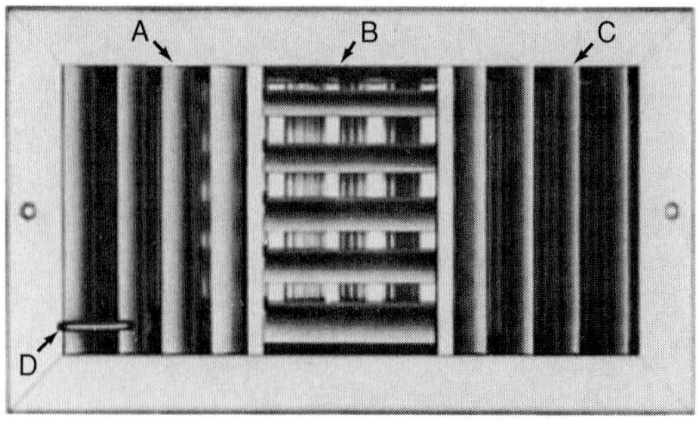

Figure 23-33. *Rectangular grille with adjustable airflow vanes. A—Vanes sweep flow in horizontal direction. B—Vertical flow. C—Horizontal flow. D—Adjustment.*

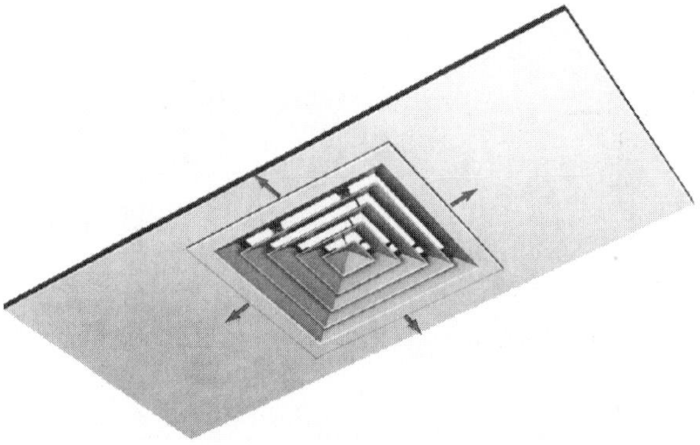

Figure 23-34. *Ceiling-mounted air diffuser. Air flows four ways.*

23.4.4 Dampers

Controlling airflow is necessary in forced air systems. Without this control, some spaces would receive too much air while others would receive too little. One method of getting even air distribution is through the use of duct ***dampers.***

These dampers balance airflow. They can also close or open certain ducts for zone control. Such a damper is shown in **Figure 23-35.** This damper is called a multiple-blade damper. There are two types of multiple-blade dampers. These are the parallel-blade (as shown in **Figure 23-35**) and the opposed-blade (as shown in **Figure 23-36**). Some dampers are located in the diffuser or grille while some are in the duct itself. There are three types, all shown in **Figure 23-36**:

- Butterfly.
- Multiple-blade.
- Split damper.

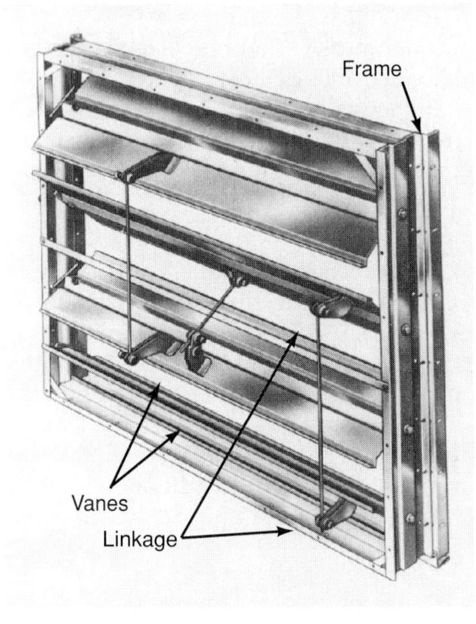

Figure 23-35. *Four-vane adjustable damper controls airflow in duct. These dampers may be operated either manually or automatically. (Honeywell, Inc.)*

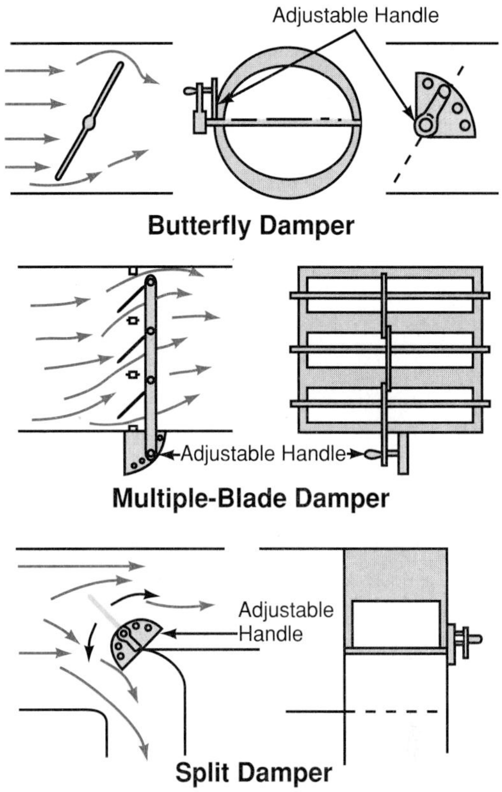

Figure 23-36. *Three types of dampers. These dampers adjust airflow volumes to help achieve a balanced system.*

When installing a damper, always draw a line on the end of the damper shaft that extends out of the duct, showing damper position. See **Figure 23-37.** Labels may help show the position of the damper blades.

For accurate air control, these dampers should be tight-fitting with minimum leakage. Many are automatically controlled for either zone heating or cooling. Automatic controls are used for humidity control, or for

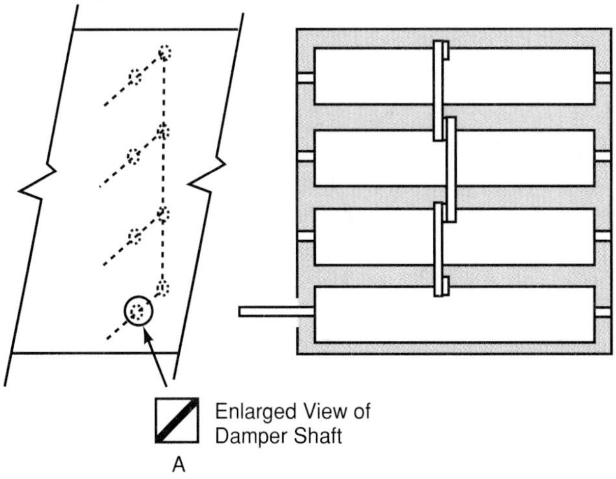

Figure 23-37. *Mark the shaft extending out of duct to indicate position of hidden damper blades inside duct. Attach labels for CLOSED and OPEN positions. A—File cut or hacksaw cut shows position of damper blades.*

temperature control. Automatic controls are also used to mix two airflows for either fresh air or recirculated air mixes. **Figure 23-38** is a diagram of typical damper installations. These multiple dampers are interlocked by controls to provide different mixes and to maintain correct total airflow.

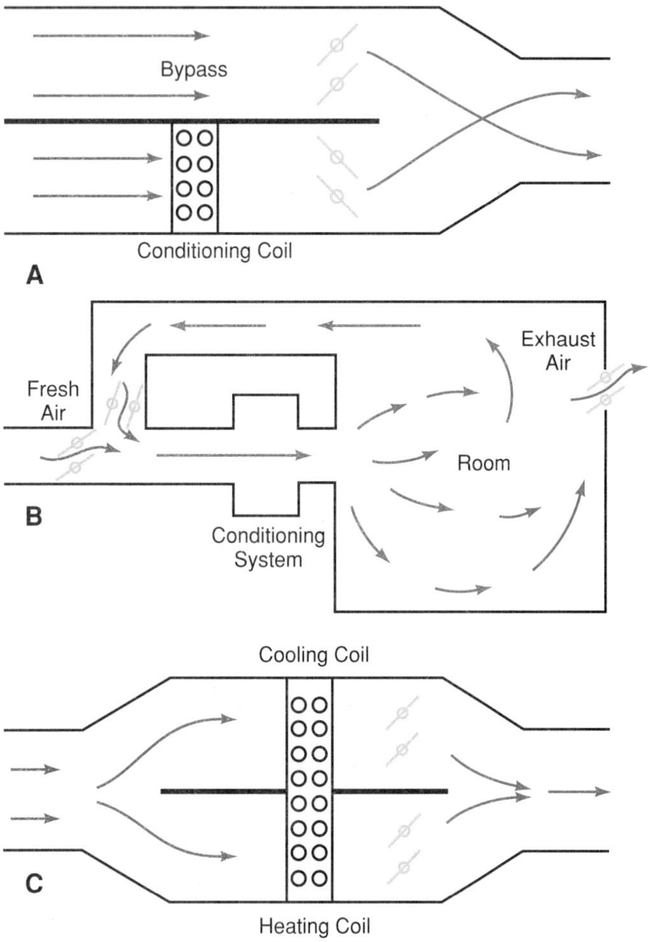

Figure 23-38. *Dampers control airflow. A—Bypass to control temperature and relative humidity. B—Dampers to control fresh air, exhaust air, and bypass. C—Controlling air over either cooling or heating coil.*

23.4.5 Fire Dampers

Automatic fire dampers, **Figure 23-39,** should be installed in all vertical ducts in commercial and industrial buildings. Ducts, especially vertical ones, will carry fumes and flames from fires. These dampers should be inspected and tested at least once a year to ensure that they are in good operating condition.

Ducts going into or through a fire wall must have fire dampers. These openings are rated as follows:

- Class A—Opening will hold back a fire indefinitely.
- Class B—Fire dampers may be used when a two- to four-hour hold is required. This class is approved for most general installations.
- Class C—Fire dampers are used where a one-hour hold is required.

Figure 23-39. *Fire damper in closed position. (Air Balance Inc.)*

The following dampers are fail-safe units:

- Spring-loaded to close.
- Weight-loaded to close.

Vertical shafts serving two or more floors must be closed in a fire partition. Fire dampers must be used. Ducts of less than 20 in^2 (129 cm^2) area do not require a fire damper.

Fire dampers are usually held open by a fusible link. Heat will melt the link. The damper will close either by gravity, weights, or springs. See **Figure 23-40.** Some fire dampers have electronic sensors that operate a closing mechanism. The damper blade latches may use power to close the damper. Electric power devices may also be used to close the damper.

Smoke dampers use a photoelectric device to detect smoke. An electronic device will trip a holding device and the damper will close.

Before making a duct installation, local smoke and fire damper regulations must be checked.

23.4.6 Elbows

Elbows are bends (90° unless otherwise specified) in the ducts. Airflow direction is diverted from a straight line using elbows. Air has inertia. That is to say, air has weight and it obeys Newton's laws of motion. Once set in motion, air tends to continue on in the direction it is moving. In addition, air is compressible.

It takes energy to make airflow change its direction. The air wants to flow in a straight line. On turns, therefore, it crowds against the outside, as shown in **Figure 23-41.**

23.4.7 Fans

Air movement is generally produced by some type of fan. Usually, fans are located in the inlet of the air conditioner. Air can be moved by creating positive pressure.

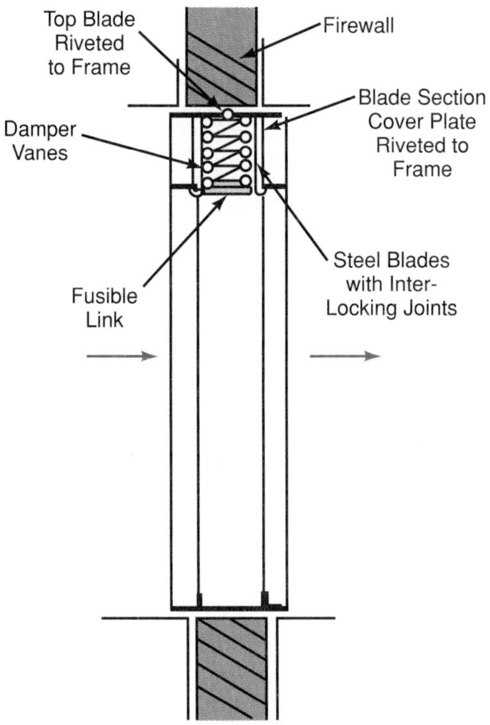

Figure 23-40. *Fire damper in open position. If link is heated, it will melt and vanes will fall, closing duct. (Air Balance Inc.)*

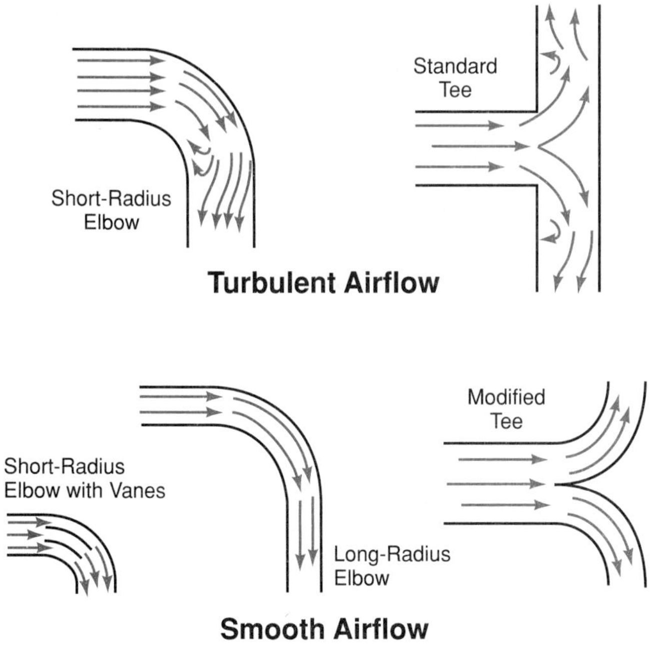

Figure 23-41. *Airflow in some duct bends and elbows.*

It can also be moved by creating negative pressure. All fans produce both conditions. The air inlet to a fan is negative pressure. The exhaust of the fan is positive pressure. See **Figure 23-42.** The air feed into a fan is called induced draft and the exhaust from a fan is called forced draft.

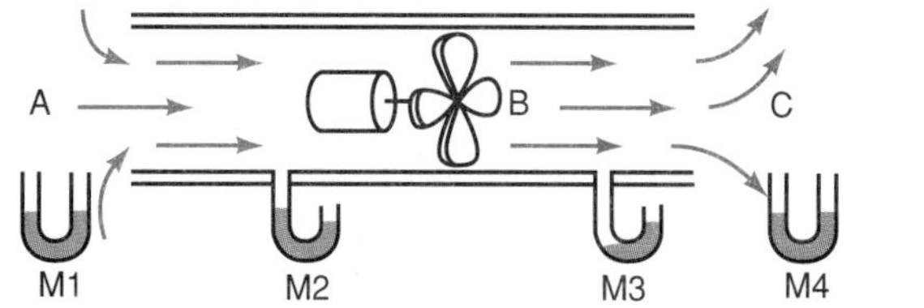

Figure 23-42. *Pressure conditions in simple duct and fan installation. A—Intake. B—Fan and motor. C—Exhaust. M1—Atmospheric pressure. M2—Negative pressure. M3—Positive pressure. M4—Atmospheric pressure.*

Fans are constructed of metal and plastic. There are several types of fans, but the two most popular types are:

- Axial flow (propeller).
- Radial flow (centrifugal flow).

If air flows along the direction the axle is pointing, it is called axial flow. If air flows at right angles to the axle (radius), it is called radial flow. This is shown in **Figure 23-43.**

A three-blade axial flow fan is shown in **Figure 23-44.** The rotation is clockwise. The axial flow fan is

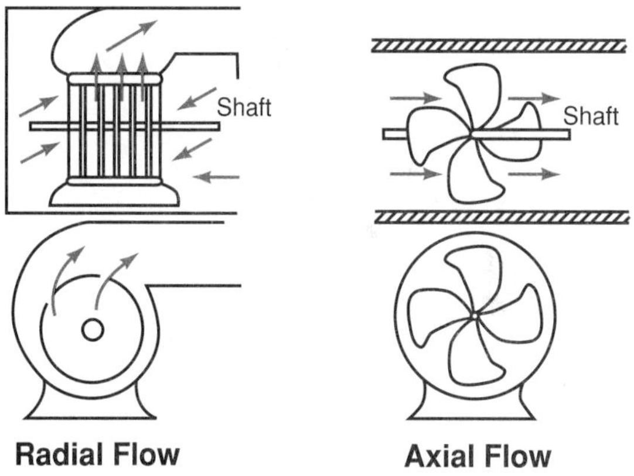

Figure 23-43. *Principal types of fans.*

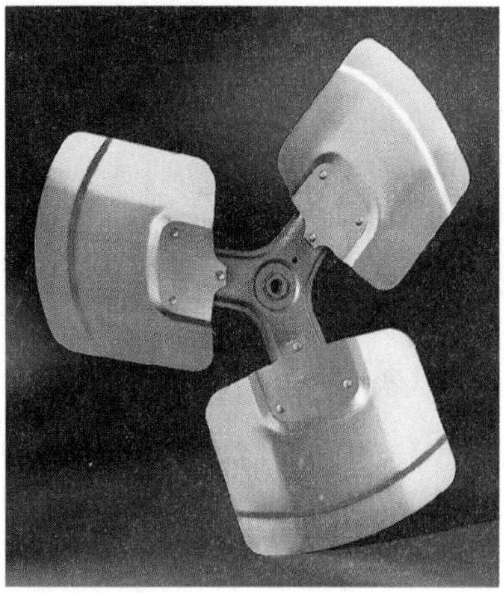

Figure 23-44. *Blade for axial flow fan. (Revcor, Inc.)*

usually direct-driven by mounting fan blades on the motor shaft. These fan blades should be handled carefully. If they are bent or twisted, the fan should be replaced.

Part of a radial flow fan is shown in **Figure 23-45.** A belt-driven radial flow fan assembly is shown in **Figure 23-46.** The radial flow fan is most often used on large installations. It is either directly driven or belt-driven. A belt-driven unit is shown in **Figure 23-47.**

These units are made in four classes of air systems. These are listed by the Air Movement and Control Association, Inc. (AMCA):

Low pressure, Class I	Up to 3 3/4″ (9.5 cm) water column total pressure.
Medium pressure, Class II	3 3/4″ to 6 3/4″ (17 cm).
High pressure, Class III	6 3/4″ to 12 1/4″ (31 cm).
High pressure, Class IV	Over 12 1/4″ (31 cm).

Figure 23-45. *Rotor for radial flow fan. (Torrington Research)*

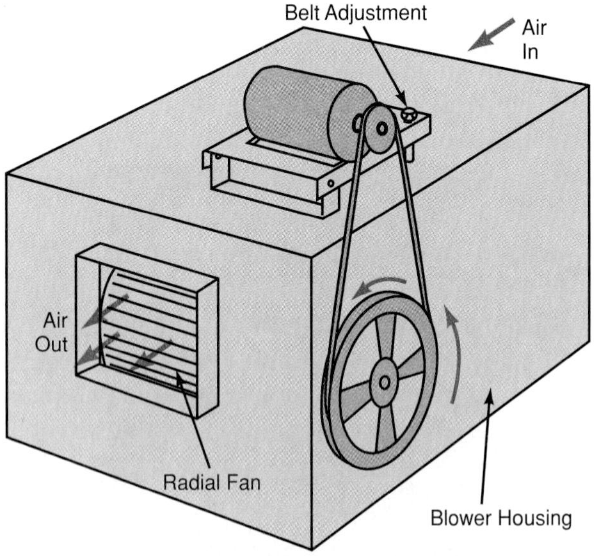

Figure 23-46. *Typical blower housing using belt-driven radial flow fan. Note direction of rotation of fan pulley and belt adjustment.*

Figure 23-47. *Radial flow belt-driven fan shown in complete assembly including housing and filters.*

Radial flow fans are made in different designs:

- Backward-inclined blades (small units).
- Forward-inclined blades (large units). See **Figure 23-48.** Static pressure increases as the square of the ratio of increase of the volume of airflow.

1000 cfm to 2000 cfm	= two times more flow
$2^2 = 2 \times 2$	= four times more static pressure.

This means that the static pressure is the square of the ratio change of rpm. **Figure 23-49** shows how to measure the pressure of a fan.

To determine the fan capacity for a furnace, use the following formula:

$$\text{cfm} = \frac{\text{Btu/hr. output of furnace}}{1.0505} \times \text{air temperature rise in °F}$$

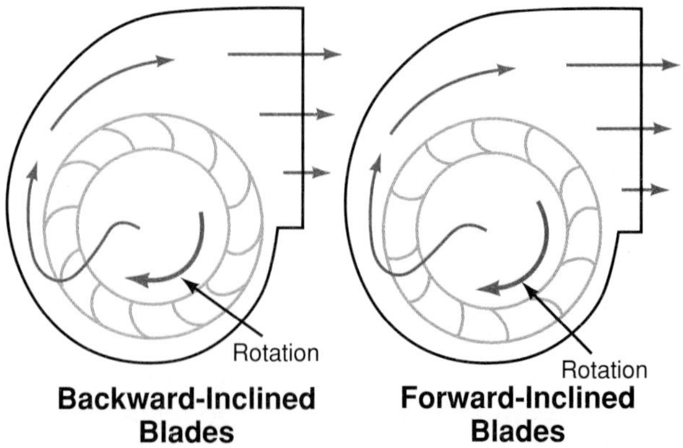

Figure 23-48. *Radial flow fan designs.*

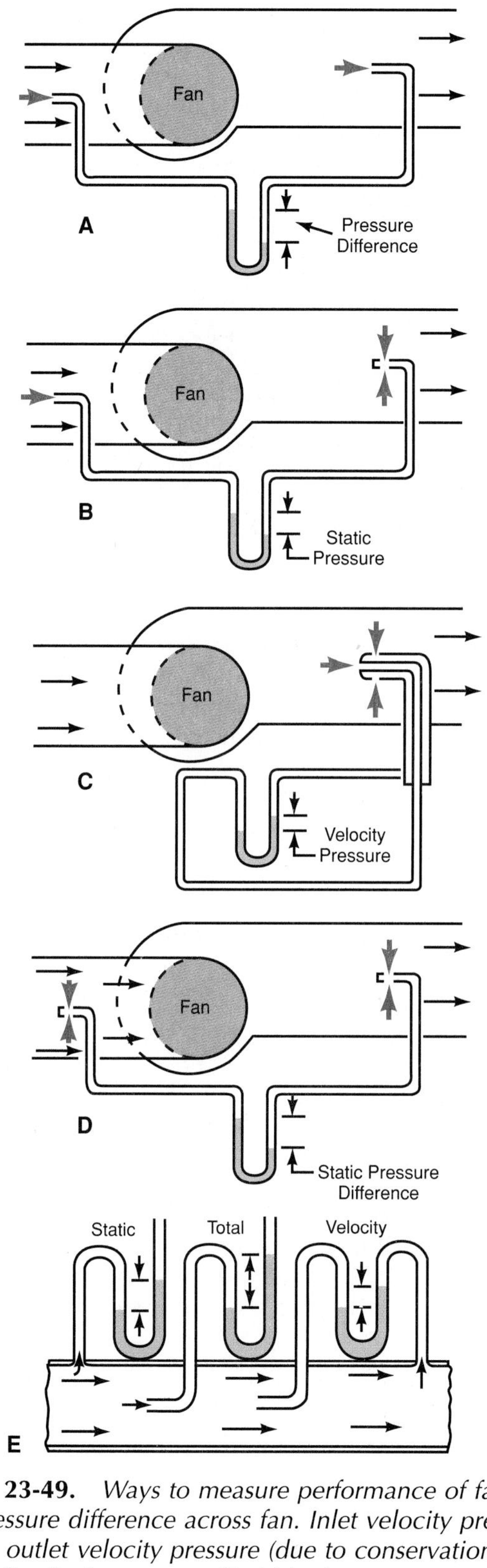

Figure 23-49. *Ways to measure performance of fan. A—Pressure difference across fan. Inlet velocity pressure equals outlet velocity pressure (due to conservation of mass). B—Static pressure. C—Velocity pressure. Pressures are related as follows: $C = B - A$ (A, B, C are the pressures from views A, B, C). Note also that $B = C + A$, which means that B acts like a total pressure. D—Static pressure difference. E—Some technicians measure pressure relative to room air pressure. Note that $V = T - S$.*

The air temperature rise in °F is determined by taking the warm air supply plenum temperature. The cold air return temperature is then subtracted. The 1.0505 factor is the total external static pressure.

For cooling, use 400 cfm/ton of capacity. The total pressure drop in the ducts should be about 0.2″ of water column. (This equals the pressure rise across the furnace.) For furnaces with a cooling unit, the fan should provide 0.4″ to 0.5″ of water column. This leaves 0.2″ for all but the cooling unit.

If possible, belt tension should be on the lower belt section. This will provide more efficient belt drive, as shown in **Figure 23-50.** Belt tension is about right when it can be pushed out of line a distance equal to its width. Fan speed can be varied by using adjustable (variable pitch) pulleys. The one shown in **Figure 23-51** is for a two-belt drive.

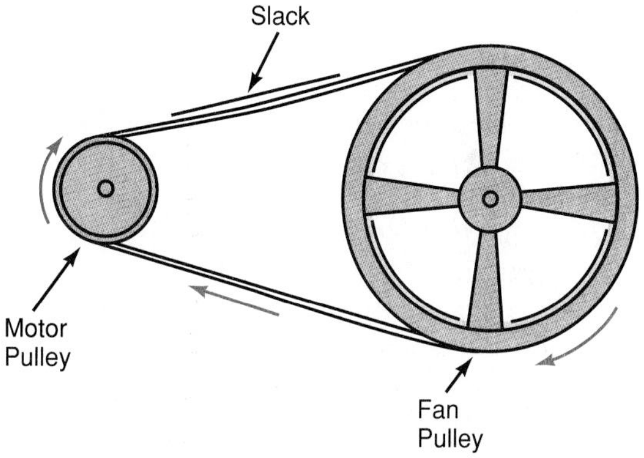

Figure 23-50. *Motor pulley and fan pulley. Note tension on lower part of belt and slack at upper. This is how properly tensioned belt should look in operation. One should be able to push slack area down about 1/2″. Some technicians allow slack area to sag an amount equal to width of belt.*

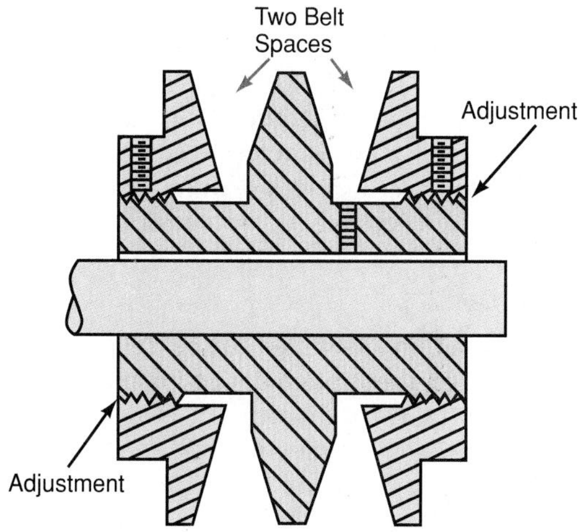

Figure 23-51. *Variable pitch pulley. Note location of setscrews. Pulley is used on dual belt system and can provide fan variation of about 100 rpm through adjustment.*

All fans collect lint and dirt. This reduces the efficiency of the fan. Dirt should be removed every six months. Remove the fan and scrape, rub, or vacuum the dirt off the blades. Blades will collect dirt even more quickly. If fan bearings are oiled too much, this extra oil coats the fan blades. Each bearing requires but one or two drops of oil each year. See Chapter 7 for more fan motor information.

23.5 Special Duct Problems and Duct Maintenance

Ducts that are not concealed may need to be painted for good appearance. If the ducts are made of galvanized (zinc-coated) steel, paint will not stick very well. A treatment with vinegar or other weak acid will etch the surface. The paint will then stick. If any ducts are outdoors, the weather will etch the surface. The duct may then be painted. Some contractors store ducts outside until they are needed, making use of the natural weathering.

Duct systems should have no "dead ends" where dust and stagnant air may collect. If a duct must have dead space, a duct door for cleaning may be needed. Any section of duct, straight or not, will collect some dust. Reversing the direction of air flow can loosen some of the dust. The dust can then be caught in a filter.

Aluminum-to-steel joints encourage corrosion. This is due to electro-chemical action between the two different metals. Joints between different metals should be avoided. However, if one must be made, slip a thin strip of zinc or magnesium into the gap between the two different metals. The zinc or magnesium will corrode before the aluminum or steel duct can be damaged.

City and state codes require smoke detectors in residential or commercial structures. These detectors must be properly located to be effective. High-velocity air from ducts can keep the detectors from working. When a long horizontal run of duct ending in an outlet grille is not concealed, the smoke detector should be hung on the underside of the duct. It should be located just behind the outlet grille. With other duct designs, such as residential wall, floor, or ceiling grilles, the duct is concealed. In these systems, smoke detectors should be as far away from the grilles as possible. Detectors should never be put in corners where there is almost no airflow.

23.5.1 Noise

Air distribution systems must be designed to circulate clean air. Yet, the movement of air must not annoy occupants or cause them discomfort. Two common problems that must be overcome are objectionable noise and drafts. Noise is produced by movement or vibration of an object.

Three factors affect noise. These are:

- Noise source.
- Noise carriers.
- Noise amplifiers or reflectors.

The noise source is a vibration that is loud enough to be heard. This vibration may originate in the heating system, cooling system, or fan mechanism. A vibrating duct panel will create alternate waves of low-pressure and high-pressure air, producing sound.

Duct noise can be very disturbing. Noises may be:

- A high pitch sound. This is usually caused by an air velocity that is too high. It may be due to air hitting sharp metal edges.
- A low pitch rumble. This is usually caused by fan and motor sounds traveling along the duct system.
- A popping sound when the unit starts or stops. This may be caused by expansion or contraction of the duct as it warms up or cools.

To locate the source of the high-pitched sound, remove the grille or diffuser. If the noise stops, it is caused by sharp edges in the grille. If it continues, the air velocity is too high (use an anemometer to measure), or there is a sharp edge in the duct system. Locate the problem and correct it.

Decibel meters measure noise level. See Chapter 19. Decibels (dB) refer to the frequency of the pressure fluctuations in the air and the amplitude or size of these vibrations. Airborne sound is usually expressed in cycles per second (cps).

Both mechanisms and the motion of matter cause noise in systems. Sources include:

- Fans.
- High-velocity air traveling through ducts and causing turbulence.
- Motors.
- Sucking and throbbing noises produced by compressors.
- High-velocity refrigerant flow, especially at sharp bends in piping.

Noise caused by high-speed air is often the result of an undersize unit or duct. Noise problems develop when the blower has been sped up to compensate for this lack of capacity.

Noise or vibration carriers are rigid structures. They carry vibrations to places where they may be annoying. Floors, ceilings, ducts, doors, and pipes may carry vibrations.

Hard, smooth surfaces in the conditioned space often reflect or amplify sound. Walls, ceilings, floors, and furnishings may pick up a small vibration. They may reflect it at a frequency and direction so that all parts of the space are made uncomfortable. Problems involving acoustics (absorption and reflection of sound) are constantly being studied and improved.

Soft fabrics, such as drapes and curtains, and fabric-covered furniture are noise absorbers. Felt-lined, soft-insulation-lined, and covered ducts also absorb noise.

23.5.2 Drafts

It is relatively simple to provide properly-sized ducts and fans. They must be large enough to give the correct amount of air for conditioning. The air must enter and circulate to all parts of the room. It must not interfere with flow of air to the air return. There should not be objectionable drafts or noise.

Air moving past people faster than 25 ft./min. (about 1/4 mph), creates an annoying draft. This means that air should not flow faster than 1/4 mph through the length of a 25′ living room. If it does, an uncomfortable feeling will result. For a grille outlet designed to throw the air into the room a distance of 8′ to 13′, a velocity of 500 ft./min. (about 5.5 mph) is needed. Therefore, to keep that part of the space occupied by humans at a 25 ft./min. velocity, grille or outlet locations must be carefully selected.

The location of the air returns is important when moving air across a long space at a reasonable velocity. Air returns for a long room should be located on the side opposite where air enters the space.

Air returns should be located high on the wall for warm air return (cooling season). They should be low on the wall (or in the floor) for cold air return (heating season). **Figure 23-52** shows some typical airflow patterns. The return air openings often used are shown in **Figure 23-53.**

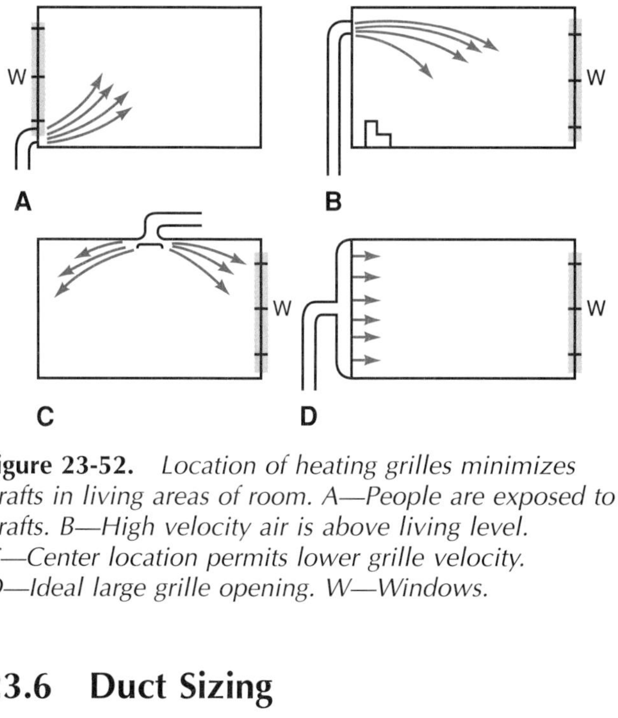

Figure 23-52. *Location of heating grilles minimizes drafts in living areas of room. A—People are exposed to drafts. B—High velocity air is above living level. C—Center location permits lower grille velocity. D—Ideal large grille opening. W—Windows.*

23.6 Duct Sizing

To determine duct sizes, the air volume to be delivered must be known. This volume depends on the amount of heat the air must deliver and remove.

Normally, these calculations are not difficult. However, conditions may complicate the problem. An example is split heating systems where the same duct is delivering both heated and cooled air. (A split system can use radiators or convectors, or use hydronic baseboards. It can have a separate mechanical ventilation system.) The amount of air delivered must always equal or exceed the minimum fresh air ventilation requirements.

To reduce duct size and to save space, smaller size ducts are now being used. Such ducts operate with about twice the normal air velocity. This increase in pressure and velocity requires more powerful fans. In turn, there is more noise.

In the colder climates, the size of the ducts must be based on the heating needs. In warmer parts of the country, they must be based on cooling needs. Where both heating and cooling are required, the same duct system serves both. To compensate for the higher air volumes needed for air conditioning, the system capacity is increased by increasing the fan speed about 20%.

23.6.1 Air Volumes for Heating

If the furnace is the only source of heat for a room, only three factors must be known to calculate the air volume:

- Heat load.
- Room temperature.
- Duct temperature.

The ***heat load*** can be determined by methods described in Chapter 27. The room temperature is decided by the designer. Normally, the temperature is 72°F (22°C).

The duct temperature is more difficult to decide. If low duct temperatures are used, large air volumes will be necessary to carry enough heat. If high duct temperatures are used, the furnace must operate with higher chimney (stack) temperatures. The ducts may have to be insulated.

Engineers recommend that the grille temperatures be at least 125°F (52°C). The duct air temperature should be near 140°F (60°C). The lowest temperature at which these results are possible depends on the duct lengths. The specific heat of air is 0.24 Btu/lb. °F. The weight of air needed is easily found by using the specific heat equation:

Formula:

Heat load =
specific heat × wt. of air × temperature difference

Example:

A room has a heat load of 20,000 Btu/hr. What is the weight of air needed for one hour with a room temperature of 72°F (22°C), and a duct temperature of 140°F (60°C)?

Solution:

20,000 Btu/hr = 0.24 Btu/lb. °F × wt. of air × (140 − 72°F)
20,000/0.24 × 68 = wt. of air
20,000/16.32 = wt. of air
1225.5 lb. = wt. of air per hour
Divide by 60 to obtain lb./min.
20.4 lb. = wt. of air per minute

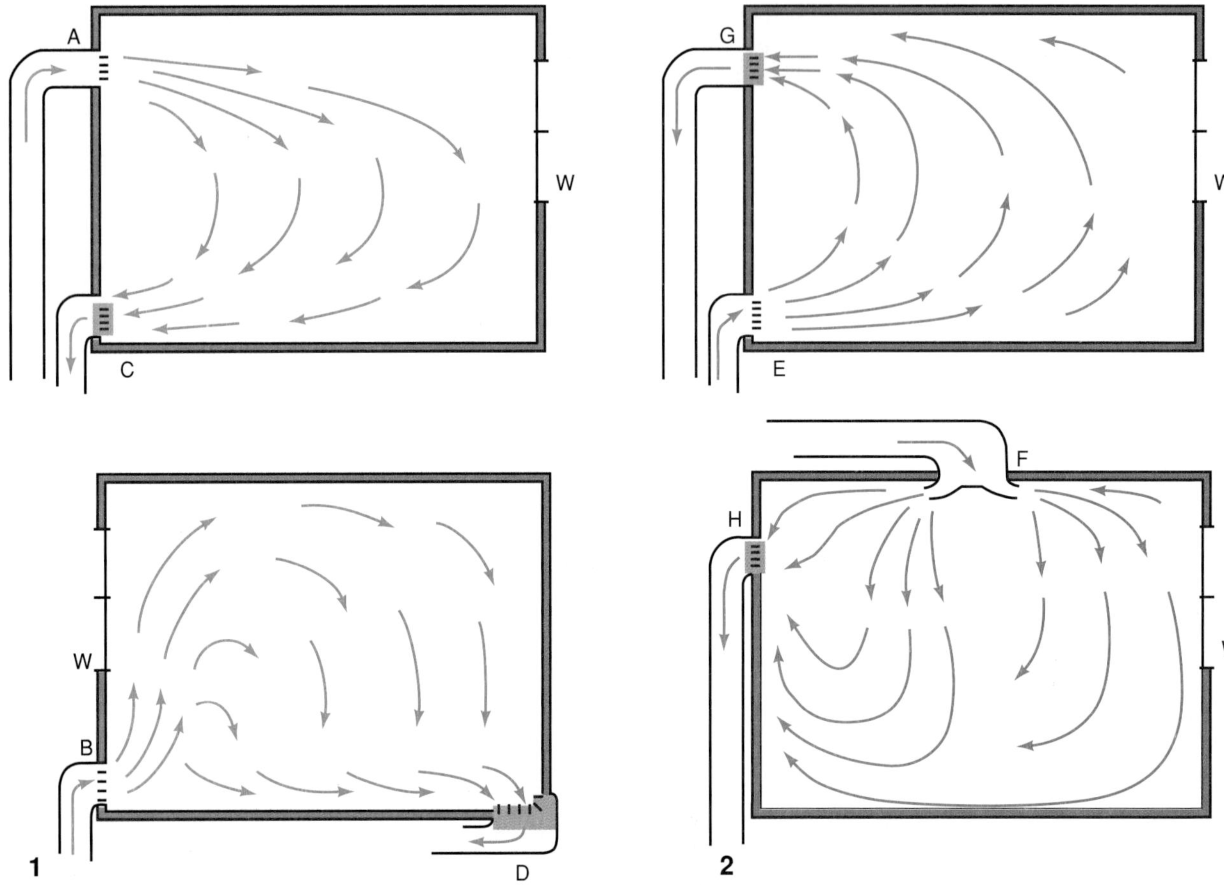

Figure 23-53. *Location of return air grilles in residential installation. 1—For heating: A and B warm air in, C and D cold air return. 2—For cooling: E and F cold air in, G and H warm air return. W—Windows. Some backflow into grille in bottom right view is useful for air mixing. Note that these short rooms can have both air outlet and air inlet on same side of room.*

The air volume must be determined before proceeding. To find the volume, first find the volume of one pound of air at the room and duct temperature.

A chart may not read as high as 140°F (60°C). Then the volume is calculated using Charles's Law (Chapter 31), knowing the volume at 72°F (22°C) dbt and 50% relative humidity (13.55 ft^3). Obtain this value from a psychrometric chart. One pound of air has volume of 15.28 ft^3. One kilogram of air has volume of 0.9539 m^3.

Formula:

$$\frac{V_1}{V_2} = \frac{Ta_1}{Ta_2}$$

Where: V_1 = Volume of air at temperature Ta_1
V_2 = Volume of air at temperature Ta_2
Ta_1 = Absolute temperature of air at volume V_1
Ta_2 = Absolute temperature of air at volume V_2

Example:

Determine the volume of 1 lb. of air at 140°F.

Solution:

$$\frac{13.55 \text{ ft}^3}{V_2} = \frac{(460 + 72)\text{ R}}{(460 + 140)\text{ R}}$$

$$\frac{13.55 \text{ ft}^3}{V_2} = \frac{532}{600}$$

$$13.55 \times \frac{600}{532} = V_2$$

$$\frac{8130}{532} = V_2$$

15.28 ft^3/lb. = V_2 (15.294 ft^3 if convert from Section 23.1.1)

The volume of air per min. is 20.4 lb./min. × 15.28 ft^3/lb.

20.4 lb./min. × 15.28 ft^3/lb. = 311.7 cfm

Now, to determine the duct size, two separate items must be considered. If the space is limited, the area of the duct may be fixed. For example, if the duct is to run between studs in a partition, the space available is no more than 14″ × 2 1/3″. (This fits between 2 × 4 studding on 16″ centers.) Such a duct has an area of:

14″ × 3 1/4″ = 45.5 in^2 or 293.5 cm^2

Convert this to square feet, dividing by 144:

45.5 in^2 ÷ 144 ft^2/in^2 = 0.316 ft^2 (0.029 m^2)

The velocity multiplied by area (0.316 ft^2), must provide a volume of 311.7 cfm.

Since the velocity is not known, it must be found by dividing the volume by the area:

Velocity = 311.7/0.316 = 986 ft./min. or 300 m/min.

This velocity would produce air turbulence noise. Therefore, two ducts measuring 14″ × 3 1/4″ must be used.

The resulting velocity will now be 493 ft./min. (150 m/min.). This should be satisfactory.

Air duct calculations are available from manufacturers and other suppliers. These provide you with the correct calculations to determine the correct velocity for a given size duct.

23.6.2 Air Volume for Cooling

To get the air volume needed, large ducts and lower air pressure (velocity) should be used. Therefore, less power will be needed. Duct cost is a one-time cost. A larger fan and motor, however, means continuous higher power costs. The higher velocity needed with smaller ducts also creates more noise.

A short method may be used to determine air volume. For each square foot of floor space, excluding basement, use 1 cfm. If a home has 1500 ft^2 (139 m^2), based on outside dimensions, the fan capacity should be 1500 cfm (43 m^3/min.).

The typical uninsulated home may need 12,000 Btu of cooling per hour (1 ton) for each 400 ft^2 (37 m^2) of floor space. To determine the cooling loads:

The 1500 ft^2 (139 m^2) should then use
1500/400 × 12,000 Btu = 45,000 Btu/hr. (4 ton).

The number of air changes are important. There should be 6 to 10 air changes per hour. The volume of the 1500 ft^2 room with an 8′ ceiling is 1500 × 8 = 12,000 ft^3 (340 m^3). The 1500 cfm fan will move 1500 × 60 min./hr. = 90,000 ft^3/hr.

90,000 ft^3/hr./12,000 ft^3 = 7.5 air changes per hour

Return air ducts must be as efficient as the delivery air ducts. The return air is warmer and is at a lower pressure. Therefore it occupies more volume. Return ductwork should be about 20% larger in cross-section area than the delivery duct.

If the delivery duct to a room has a cross section of 30 in^2 (194 cm^2), the return air duct should be:

30 × 120% = 30 × 1.2 = 36 in^2 (232 cm^2).

The air volume for cooling is calculated in much the same way as for heating. Use the same specific heat equation. Knowing heat load, specific heat of air, and temperature difference, the weight of air needed can be determined. If the weight is known, the air volume can be determined. The duct sizes can then be selected.

Formula:

$$\text{Btu} = \frac{\text{specific heat}}{\text{of air}} \times \text{wt. of air} \times \frac{\text{temperature}}{\text{difference}}$$

Example:

Suppose a room has only 400 ft^2 (37 m^2) of floor area. Recall that a room with 400 ft^2 has a heat load of one ton of cooling per hour (12,000 Btu/hr.). With a specific heat value for air equal to 0.240 Btu/lb. °F, the air temperature in the duct and the air temperature in the room must be determined. If the designer wants 75°F (24°C) in the room and the comfort cooling unit is designed for 65°F (18°C) duct air temperature:

$$12{,}000 \text{ Btu/hr.} = 0.24 \text{ Btu/lb.°F} \times \text{wt. of air} \times (75°F - 65°F)$$

$$12{,}000 \text{ Btu/hr.} = 0.24 \times \text{wt. of air} \times 10$$

$$\frac{12{,}000}{0.24 \times 10} = \text{wt. of air}$$

$$\frac{12{,}000}{2.4} = \text{wt. of air}$$

$$5000 \text{ lb./hr.} = \text{wt. of air}$$

$$\frac{5000}{60} \text{ min./hr.} = 83.3 \text{ lb./min. } (37.8 \text{ kg/min.})$$

At 60°F (16°C) db and 60% relative humidity, 13.4 ft^3 (0.38 m^3) of air weighs 1 lb.

To change 83.3 lb./min. to cfm:

cfm = 83.3 × 13.4 = 1116 cfm or (32 m^3/min.)

To determine the duct size for this volume of air moving into the room at 500 ft./min., look ahead to **Figure 23-55.** Locate the point where the 1120 cfm line crosses the 500 ft./min. velocity line. This point will show the several alternative sizes:

- A 20″ (51 cm) round duct (at 0.018″ of friction loss). Note: 20.3″ must be rounded down to the next duct size that is available.
- A 22″ × 16″ (56 cm × 41 cm) rectangular duct (using conversion table in **Figure 23-8**).
- A 28″ × 13″ (71 cm × 33 cm) duct.

23.6.3 Duct Calculations

In some heating or cooling systems, a duct serves more than one room. The duct must be designed so that each room served receives the correct amount of air. If the distribution is not balanced, one room will be too warm while another will be too cold.

There are two methods for calculating the proper size plenum chambers, main ducts, branch ducts, and grilles:

- Unit pressure drop system.
- Total pressure drop system.

Unit Pressure Drop System

Air forced through a duct follows the path of least resistance. Many duct systems have several openings (grilles) for the air to escape from the duct. A duct with low resistance will allow most of the air to flow through it. Ducts with higher resistance will not carry the correct amount of air.

In the past, many duct installations were made that fed too much air to some rooms and did not heat or cool other rooms sufficiently.

The unit pressure drop calculating system uses the same pressure drop for each length of duct throughout the system.

For example: Suppose that the total heat load during the heating season is 80,000 Btu/hr. There are six rooms with heat loads as follows:

Living Room	= 25,000 Btu/hr.
Dining Room	= 15,000 Btu/hr.
Kitchen	= 5000 Btu/hr.
Bathroom	= 8000 Btu/hr.
Bedroom No. 1	= 15,000 Btu/hr.
Bedroom No. 2	= 12,000 Btu/hr.

To determine the air volume needed to heat these six rooms, refer back to Section 23.6.2. Recall that the specific heat of air is 0.24. The volume of one pound of air is 15.28 ft^3. Therefore:

$$\text{ft}^3/\text{Btu for a } 68°\text{F (38°C) change} = \frac{15.28}{0.24 \times 68} = 0.936\ \text{ft}^3/\text{Btu}$$

$$= 0.936\ ([\text{ft}^3/\text{hr.}]/[\text{Btu/hr.}])$$

Divided by 60 min./hr. = 0.0156 ([cfm]/[Btu/hr.])

Knowing the amount of heat that must be carried to each room per minute, the air volumes required per minute for each room can be calculated:

Living Room	25,000 × 0.0156 = 390 cfm
Dining Room	15,000 × 0.0156 = 234 cfm
Kitchen	5000 × 0.0156 = 78 cfm
Bathroom	8000 × 0.0156 = 124.8 cfm
Bedroom No. 1	15,000 × 0.0156 = 234 cfm
Bedroom No. 2	12,000 × 0.0156 = 187.2 cfm

The total air volume is 1248 cfm.

To determine duct sizes that will handle the air volumes listed above, airflow data is needed. **Figure 23-54** describes how to read the values shown in a typical friction air chart. **Figure 23-55** is a friction air chart for straight ducts. Values were obtained by research. These charts have four variables:

- Friction loss in inches of water on the vertical scale (equal value lines are horizontal).
- Cubic feet of air/min. on the horizontal scale (equal value lines are vertical).

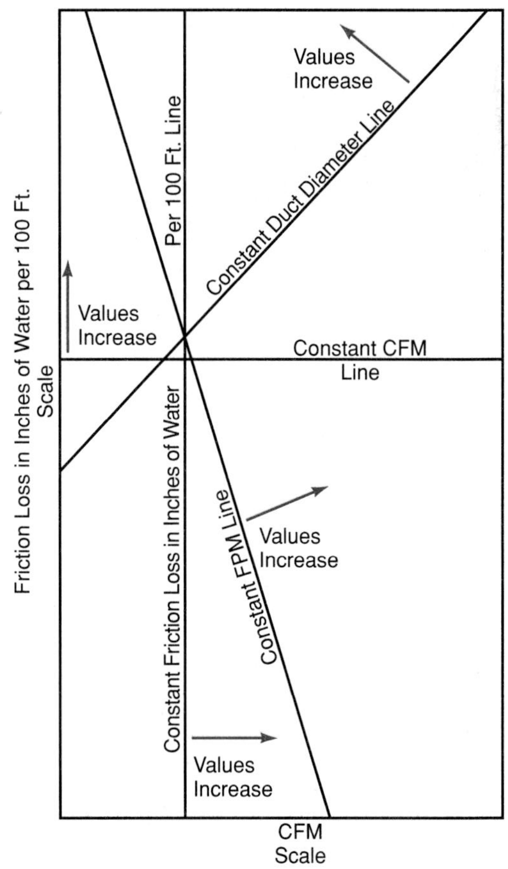

Figure 23-54. *Diagram shows how to read value lines in **Figure 23-55.** Air volume per minute is read off scale. By locating air volume per minute lines, one can find duct diameter, velocity, and friction by following along proper lines to scales on edges.*

- Velocity on scale lines that slant down to right.
- Round duct diameter, which is on scale lines that slant down to left.

To continue with the problem, the main duct must handle 1248 cfm. A friction loss of 0.04″ water column per 100′ (1 mm water column per 30 m) should be used. This will keep the velocity to a low noise level.

On the chart, these two values meet. They show that the velocity will be 700 ft./min. (213 m/min.). The main round duct will be 18″ (46 cm) in diameter.

Using the same friction loss for the branch ducts, the round duct sizes are:

Living Room	= 530 ft./min. and 12″ dia. (30.5 cm)
Dining Room	= 480 ft./min. and 10″ dia. (25.4 cm)
Kitchen	= 350 ft./min. and 7″ dia. (16.3 cm)
Bathroom	= 400 ft./min. and 8″ dia. (20.3 cm)
Bedroom No. 1	= 480 ft./min. and 10″ dia. (25.4 cm)
Bedroom No. 2	= 440 ft./min. and 9″ dia. (23.0 cm)

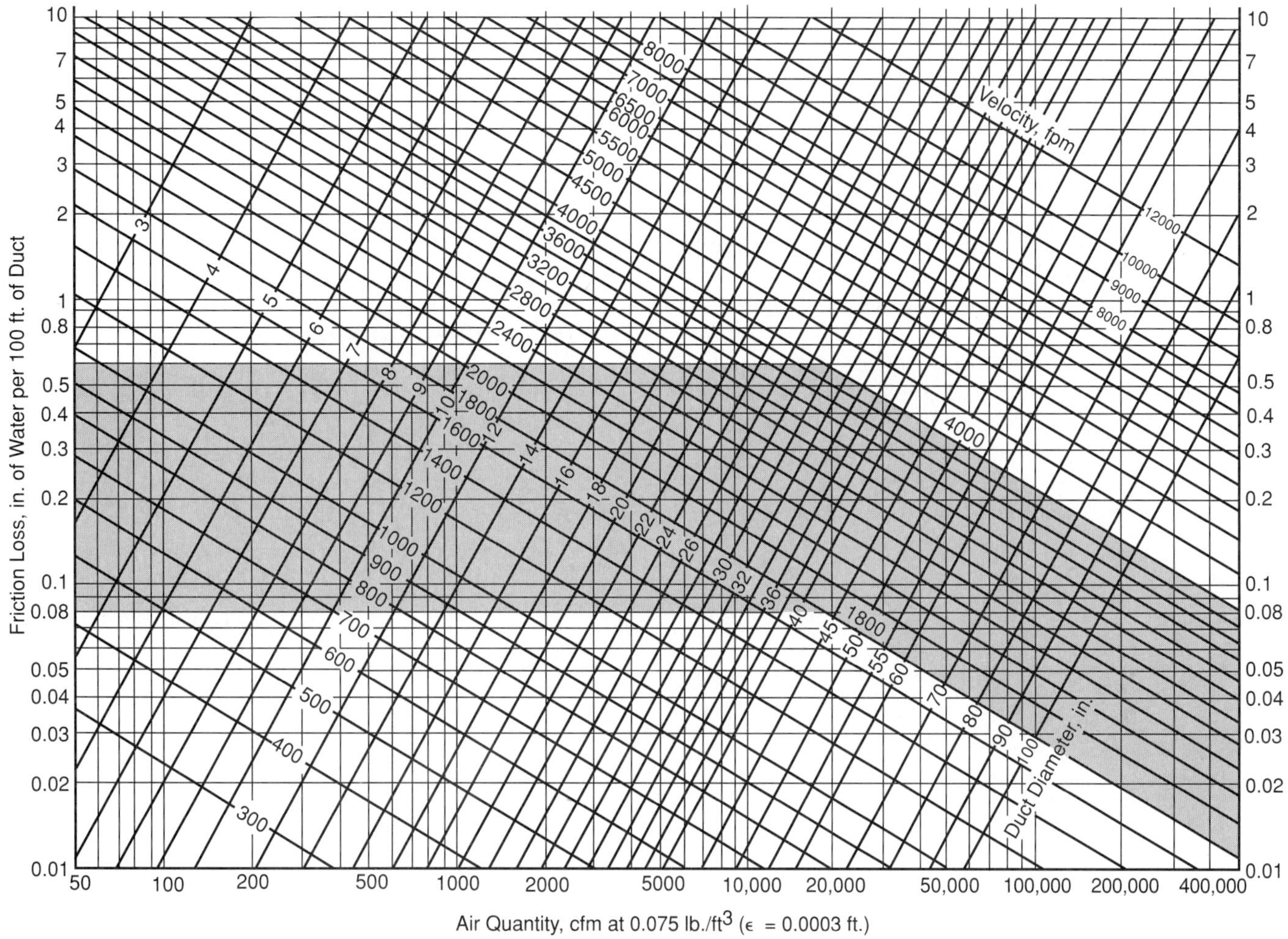

Figure 23-55. *Friction chart for airflow in ducts. Low-volume airflow is indicated by the 50 to 2000 cfm range. High-volume airflow is in the range above 2000 cfm. (Reprinted with permission of the American Society of Heating, Refrigerating, and Air-Conditioning Engineers, Atlanta, Georgia, from 1993 ASHRAE Handbook—Fundamentals.)*

These velocities are reasonably low and the system would work. However, the total pressure drop system is a more accurate method.

Total Pressure Drop System

A more accurate method of calculating proper sizes of ducts is the pressure drop system. It is based on having the same total pressure drop from the fan to each outlet. **Figure 23-56** shows the duct system used with the rooms as calculated in the unit pressure drop example. It is good practice to letter each duct size.

Suppose that the following air volumes must be carried:

Duct	*Air volume*
A	1404 cfm
B	930
C	790
D	526
E	439
F	220
G	220
H	474
I	211

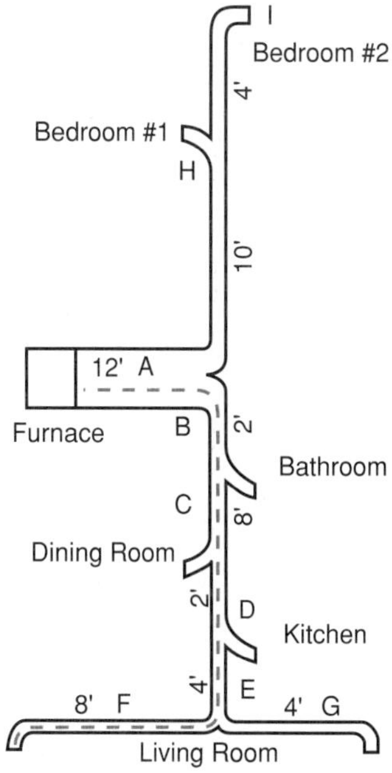

Figure 23-56. *Typical duct installation. Longest air path is shown in red.*

Each duct preceding an outlet must have a correct, equal amount of total air pressure drop. This ensures that the correct air volume leaves each outlet.

The method followed is to determine the longest and most complicated duct. This combination includes ducts A, B, C, D, E, and F.

Assume a total pressure drop of 0.04″ (0.1 cm) of water. This total means that the pressure drop to each room outlet must be 0.04″. For example, the opening to the bathroom is the shortest overall distance. It must have the same total pressure drop as the longest run through F.

It is important to consider bends and elbows when determining pressure drop. The pressure drop of one elbow is equal to 10 diameters of the duct. Assume the following: There is one large bend above the furnace. The grilles are located at the 7′ level in the room.

The total length of duct A, B, C, D, E, and F is approximately:

Elbow (19″ × 10″)	15′ (4.6 m)
A	12′ (3.8 m)
Elbow (16″ × 10″)	13′ (4.0 m)
B	2′ (0.6 m)
C	8′ (2.4 m)
D	2′ (0.6 m)
E	4′ (1.2 m)
Elbow (9 1/2″ × 10″)	8′ (2.4 m)
F	8′ (2.4 m)
Elbow	8′ (2.4 m)
Vertical rise	7′ (2.1 m)
Elbow	8′ (2.4 m)
Total	95′ (28.9 m)

The elbow that is 16″ in diameter is important in learning how calculations are done. Section B has a flow of 930 cfm. This is about 2/3 of 1404. (1404 is the total flow. The total flow passes through section A, the duct that is 19″ in diameter. The other 1/3 goes to section H, which carries 474 cfm. Note that 930 + 474 = 1404.) The flow through section B is 2/3 of the total. Therefore, the cross-sectional area of B should be about 2/3 of that for A. The diameter for B is found as follows:

$$B = \sqrt{(2/3 \times 19'' \times 19'')} = 15.51''$$

This value of the diameter for duct B is then rounded to 16″. (This will account for different velocities of air in each duct.)

$\left(B = \sqrt{\dfrac{730 \text{ ft./min.}}{670 \text{ ft./min.}} \times \dfrac{2}{3} \times 19'' \times 19''} = 16.2''\right)$ By using similar steps for 220 cfm, the method gives 9 1/2″ as a duct diameter $\left(\sqrt{\dfrac{730}{460} \times \dfrac{220}{1404} \times 19'' \times 19''} = 9.5''\right)$. The numbers are also obtained from the graph in **Figure 23-55** with 930 cfm, 220 cfm, and 0.04″ of water pressure drop. The 0.04″ is used, since each duct acts like it extends back to the furnace, as if it had its own isolated duct wall inside the main duct. (Such ducts would all be parallel inside the main duct. The "isolated" ducts would add up to exactly the cross-sectional area of the main duct.) One can imagine each duct extending 100′ back, so that it will provide 0.04″ of loss per 100′.

The preceding column of numbers is added. The total length (equivalent) is 95′. Recall that the total pressure drop was 0.04″. The 0.04″ drop occurred over 95′. This 0.04″ must be converted to the pressure drop for 100′. If a 95′ section gives 0.04″, then a 100′ section gives

$$\frac{100}{95} \times 0.04 = 1.05 \times 0.04 = 0.042'' \text{ (0.107 cm) water}$$

However, more important than the factor related to a 100′ length factor is the pressure drop provided by each section of the longest duct. Pressure drop in each part equals total pressure drop multiplied by the ratio of length of part to longest equivalent length.

$$\text{Pressure drop for each part} = 0.04 \times \left(\frac{\text{Length of part}}{\text{Total equivalent length}}\right)$$

Elbow (19 × 10)	$= 0.04 \times \frac{15}{95}$	= .0063″
A	$= 0.04 \times \frac{12}{95}$	= .0050″
Elbow (16″ × 10)	$= 0.04 \times \frac{13}{95}$	= .0055″
B	$= 0.04 \times \frac{2}{95}$	= .0008″
C	$= 0.04 \times \frac{8}{95}$	= .0034″
D	$= 0.04 \times \frac{2}{95}$	= .0008″
E	$= 0.04 \times \frac{4}{95}$	= .0016″
Elbow (9 1/2 × 10)	$= 0.04 \times \frac{8}{95}$	= .0034″
F	$= 0.04 \times \frac{8}{95}$	= .0034″
Elbow	$= 0.04 \times \frac{8}{95}$	= .0034″
Vertical rise	$= 0.04 \times \frac{7}{95}$	= .0029″
Elbow	$= 0.04 \times \frac{8}{95}$	= .0034″
Total		= .0399″ = 0.04″ (0.1 cm)

The pressure drop in each part of the longest duct is thus determined. Now determine the pressure loss up to each branch duct. Then, from this value and the length of the branch duct, determine the pressure loss per 100′ for the branch duct. For example, consider the kitchen duct:

The pressure loss up to the kitchen branch duct is the sum of all pressure losses along the way. Thus,

0.0063 + 0.0050 + 0.0055 + 0.0008 + 0.0034 + 0.0008 =0.0218, which is the total pressure drop.

If the total pressure drop to the outlet at the kitchen must equal 0.04″ water, then 0.0400 − 0.0218 = 0.0182 as the pressure drop in the kitchen branch. Assuming 87 cfm volume (526 − 439 = 87) and 0.04″ pressure drop, the kitchen branch has length equal to the following:

Elbow (6.9″ × 10)	= 6′ (1.8 m)
Riser	= 7′ (2.1 m)
Elbow (6.9″ × 10)	= 6′ (1.8 m)
Total	= 19′ (5.7 m)

The number 0.04″ is used because the kitchen branch acts like an isolated duct extending back to the furnace. One allows a kitchen-to-furnace length anywhere from 52′ to 100′. The error in making the allowance is only about 11% (17′ replaces 19′). The graph has been used only to find 6.9″ for the elbow diameter. The 7′ rise is given.

If the pressure drop in 19′ is 0.0182, the pressure drop per 100′ =

0.0182 × 100/19 = 1.82/19 = 0.096″ water/100′

From the graph, using a volume of 87.73 cfm and the resistance of 0.096, the following data is obtained: Size = 5.8″ (14.7 cm) dia. Velocity = 530 ft./min. (161.5 m/min.).

Compare this method with the unit pressure drop values. Notice how these values differ. One calculates comfort cooling air in about the same way. If the cold air duct is exposed to warm, moist air, condensation may form on the outside surface. This condensation may cause corrosion. Moisture may also drip on structural parts and cause damage. In such cases, ducts should be insulated.

Balancing the System

Balancing means sizing the ducts and adjusting the dampers to be sure that each room receives the correct amount of air. Conditioned air must be fed in the proper amounts to each room in multiple room systems. Also, the correct amount of air must be returned. If the system is not balanced, rooms will maintain different temperatures. Some ducts will be noisy. Some will have incorrect relative humidity, and some will have stale air.

In order for a system to have ***total air balance*** (TAB), the air velocity leaving each grille must be measured. Then, the "free" area the grille or diffuser has must be determined. The "free" area is the actual size of the air openings.

To balance a system do the following:

- Inspect the complete system; locate all ducts, openings, and dampers.
- Open all dampers in the ducts and at the grilles.
- Check the velocities at each outlet.
- Measure the "free" grille area.
- Calculate the volume at each outlet.

Velocity × area = volume
ft./min. × area in $in^2/144$ =
cfm (volume per minute)

- Total the cfm
- Determine the floor areas of each room. Add to determine total area.
- Find out the proportion each room should have.

Area of room/Total floor area ×
total cfm = cfm for room

- Adjust duct dampers and grille dampers to obtain these values.
- Recheck all outlet grilles.

In some cases, it may be necessary to overcome excess duct resistance. To do this, an air duct booster must be installed. These are fans that increase airflow when a duct is too small or too long. They may also be used where a duct has too many elbows. A booster fan is shown in **Figure 23-57.**

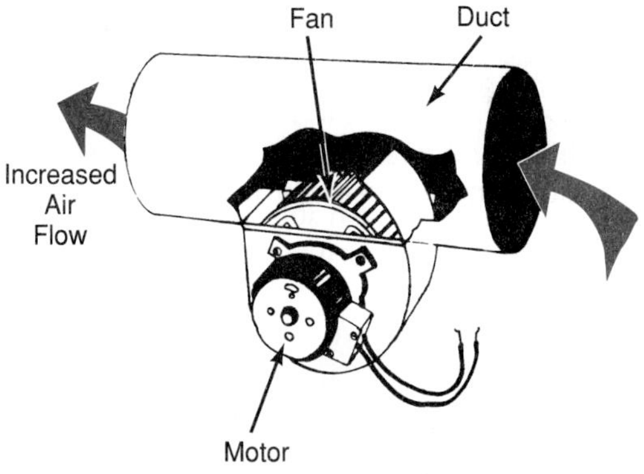

Figure 23-57. *Duct fan increases airflow in air ducts. (Tjernlund Products, Inc.)*

An effective but simpler technique can be used to balance airflow to the different rooms:

1. Mount accurate thermometers in folded cardboard as shown in **Figure 23-58.** Place one of these thermometers in each room. Locate them on a table away from sunlight, lamps, or any extra heat source.
2. Adjust dampers until each room has the temperature desired. Allow several hours for the system to adjust to any damper change.

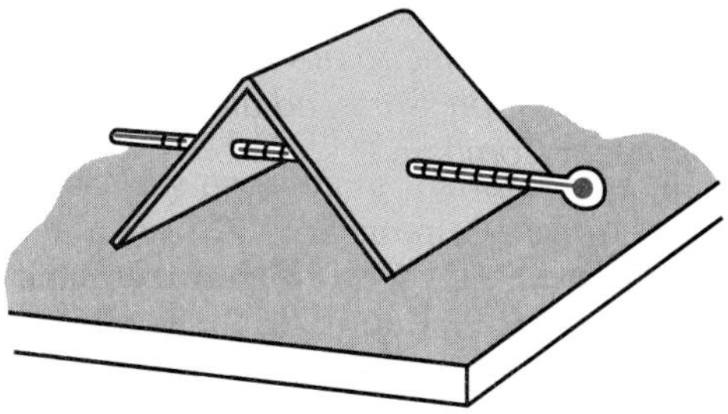

Figure 23-58. *How to use glass stem thermometer to measure room temperature. One is placed in each room. A folded piece of heavy paper is used to keep the thermometer from touching the tabletop.*

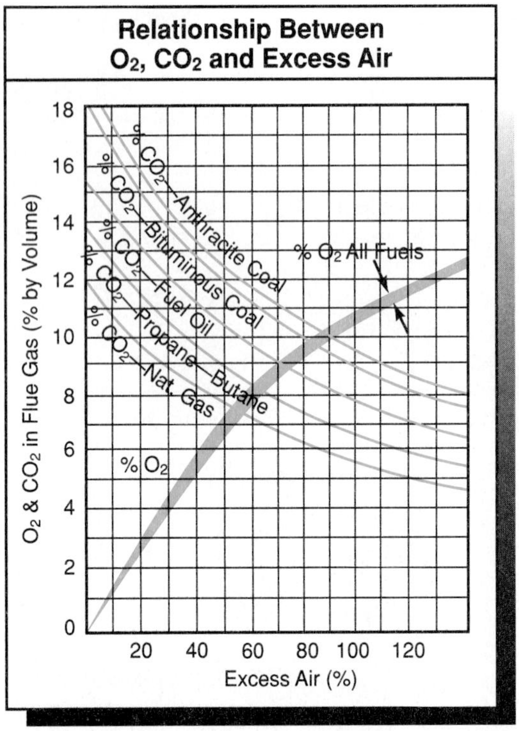

Figure 23-64. *Graph shows changes in stack carbon dioxide for various fuels and the amount of oxygen as the amount of excess air changes from 0 to 140%. (Bacharach, Inc.)*

The graph shows these percentages for five different types of fuel. When the CO_2 amount has been determined, the technician can find the stack temperature next. The combustion efficiency of the furnace is determined through the use of an efficiency calculator. See **Figure 23-65.**

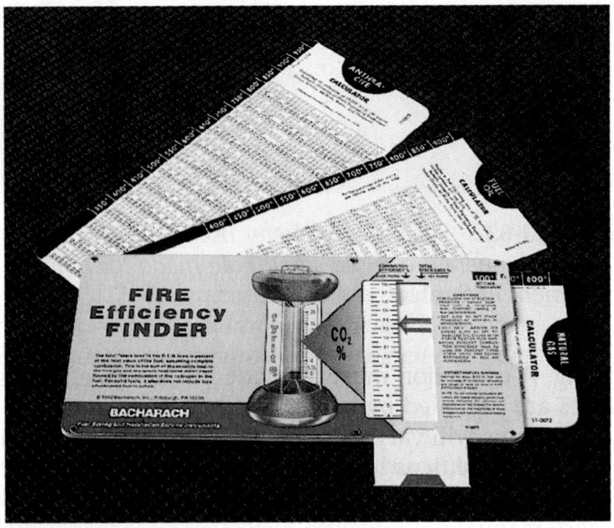

Figure 23-65. *Combustion efficiency slide calculators are used to determine the efficiency of different types of fuel. The three charts shown are to be used with anthracite coal, fuel oil, and natural gas. (Bacharach, Inc.)*

The probe of this portable digital analyzer is inserted into the air being sampled. Analyzers are also available, which display carbon monoxide levels. See **Figure 23-66.** These analyzers digitally display readings of the carbon monoxide levels.

Figure 23-66. *Service technician measuring carbon monoxide level of residential plants. (Bacharach, Inc.)*

Some analyzing instruments use electronic sensors, circuits, and indicators. **Figure 23-67** shows an oil burner combustion testing kit. These instruments are used to test for air pollution, boiler efficiency, and stack gas analysis. They use thermal conductivity and a combustion chamber with catalysts to measure the gas sample.

Figure 23-68 shows an orsat apparatus. An orsat apparatus consists of pipettes and a burette. (Pipettes are glass tubes with narrow ends. A burette is a graduated glass tube that shows liquid or gas quantity.) Some instruments use air to carry the sample; some use helium, argon, or nitrogen. These instruments can accurately analyze for hydrogen, oxygen, carbon monoxide, methane, ethane, and carbon dioxide.

Separate silica gel and molecular sieves are used to partition and adsorb certain gases. A pump is used to move the gases. A recorder records the results. In about three minutes, an accurate flue gas analysis is given for oxygen, carbon monoxide, and carbon dioxide.

Smoke Test

A smoke test is an excellent way to check combustion efficiency. Air-fuel ratio, primary air, secondary air, and draft are checked. Several different tests may be used.

Figure 23-76. *Large commercial electronic air cleaner. (Trion, Inc.)*

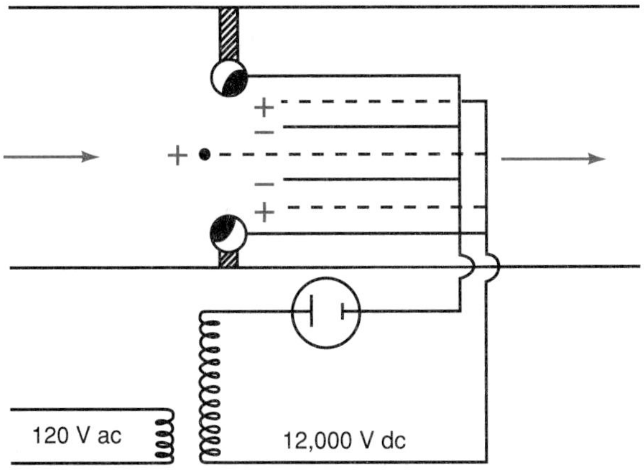

Figure 23-77. *Electrical circuit and airflow in simple electrostatic-type air filter for forced-air furnace.*

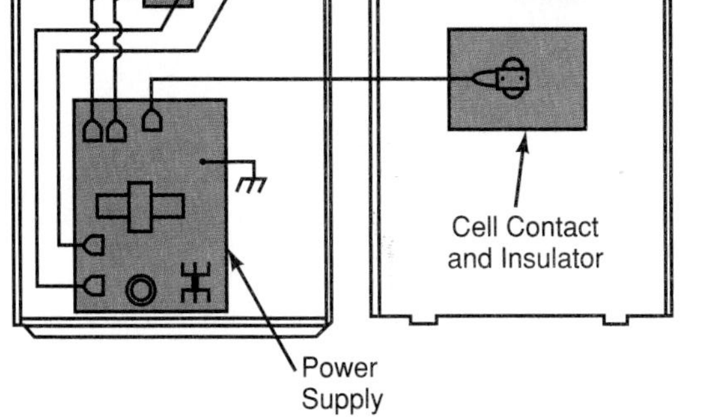

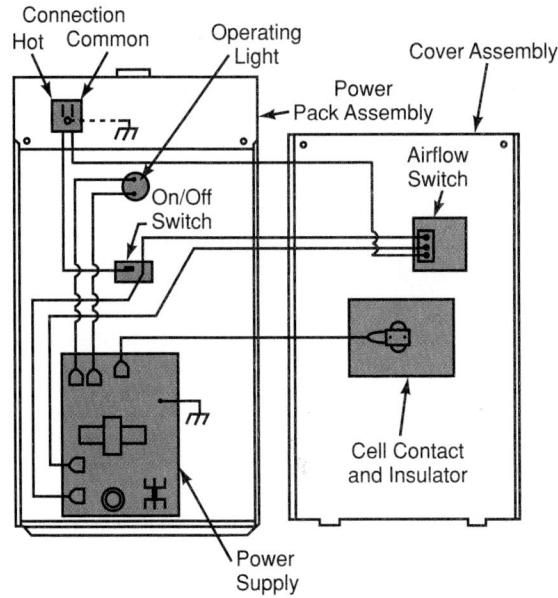

Figure 23-78. *Wiring diagram of electronic air filter. (White-Rodgers Div., Emerson Electric Co.)*

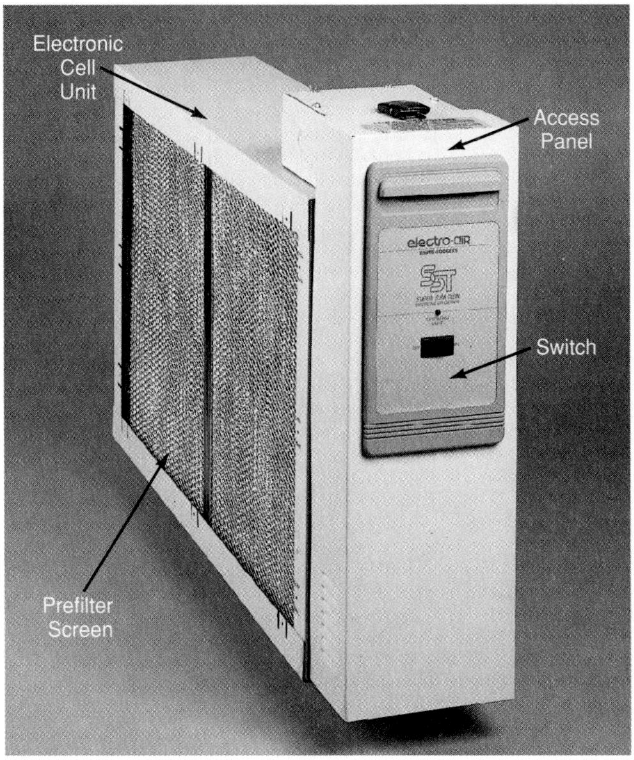

Figure 23-79. *Electronic filter designed for installation in air duct. (White-Rodgers Division, Emerson Electric Co.)*

Figure 23-80 shows different filter positions at a furnace or in room air ducts.

The electronic cells and protective screens must be cleaned every two to three months. Collected material is black in color. Continue cleaning with hot water until cleaning water is clear. Use recommended detergents and water-detergent solutions.

Electrostatic air cleaners are used in ceilings and in portable units. Airborne particles circulated by heating and air conditioning systems are removed by electronic air cleaners. This includes smoke, dust, mold, spores, pollen, and bacteria. See **Figure 23-81.**

1. Internal fan draws the polluted air into the electrostatic air cleaner.
2. Prefilter screen filters and collects large particles.
3. Ionizing section has fine wires that give the incoming airborne particles a positive charge. These particles are attached to the collecting area.
4. The collecting cell's negatively-charged plates hold the particles. (This is how magnets attract and collect iron filings.)
5. The clean, purified air is recirculated throughout the room.

A ventilation system may be used as another means of controlling indoor air from excessive humidity. (This may be due to bathrooms, laundry rooms, and shower rooms, causing condensation on windows and walls.) The ventilating system may also control odors (from cooking, smoking, or other household activities). A central ventilation with heat recovery system is shown in

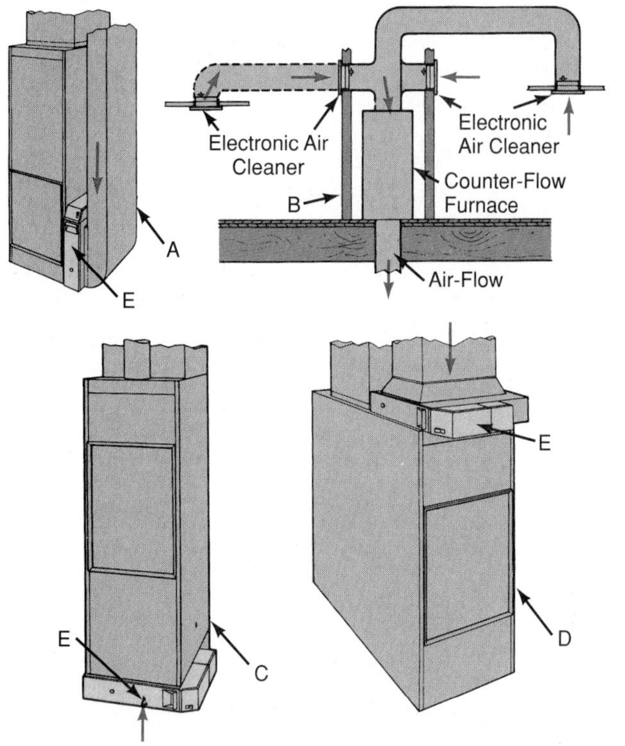

Figure 23-80. *Various installation positions for electronic filter. A—Cold air return from the side. B—Independent room units. C—Upflow. D—Downflow. E—Electronic filter. (Honeywell, Inc.)*

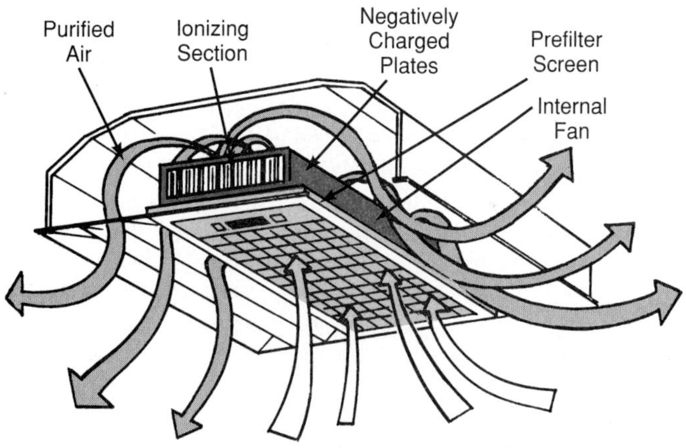

Figure 23-81. *Electrostatic design ceiling filter. (Tectronic Products Company, Inc.)*

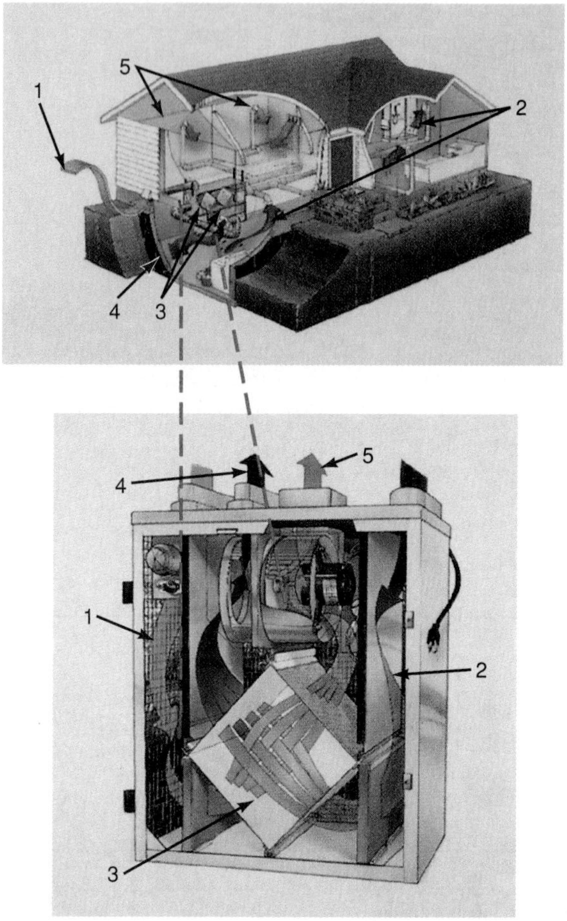

Figure 23-82. *Recovery ventilators in a normal installation. 1—Outdoor air enters through vent, filter, and heat exchanger core. 2—Fan draws exhaust air from bathrooms, laundry, etc., and through alternate passages of the core. 3—Core allows two air streams to pass close to each other, and heat is transferred. Energy transfer warms or cools incoming air. 4—Exhaust air continues through ducting to the outside. 5—Fresh air is distributed throughout the home. (Conservation Energy Systems)*

Figure 23-82. The heat recovery ventilation system is designed to bring air in from the outside. The system preheats it in the winter or cools it in summertime. It is designed to blend outdoor air into the home by the use of a fan.

Electronic filters have four main parts:

- Frame.
- Power supply.
- Prefilter and airflow distributor.
- Electronic cell.

The unit must be installed level and plumb for proper drainage and more efficient airflow. The hot water line to the washer should have a strainer.

Airflow should be evenly distributed across the face of the air cleaner for maximum efficiency. If the unit is near an airflow elbow, use movable air vanes or baffles.

Excessive lint interferes with electronic air cleaner (EAC) operation. A fine-mesh screen or filter should be installed ahead (upstream) of the filter.

Servicing Electronic Filters

Electronic air cleaners (EAC) need service in the following situations:

- Unit does not arc.
- Meter (if used) reads low.
- Trouble lights remain on.
- Strong ozone odor is detected.
- Rooms are dusty and dirty.
- The unit arcs constantly.

Check with the owner on the last cleaning of the filter. Be sure the power doors or panels are closed. Be sure the power switch is on. Check the fuses. Check the meter readings. (Refer to the service manual.)

Meters will show if:

- Conditions are normal.
- Filter is dirty.
- Filter is wet. (Operate "dry" switch.)
- There is an electrical failure.

Some meters show if electrical trouble is in the power source circuit or high-voltage circuit.

If the trouble is in the high-voltage circuit: Inspect and electrically test the "power pack" capacitors and collecting cell ionizing wires. **Figure 23-83** shows an electronic filter with the collector cell and power factor being checked. A kilovoltmeter probe is being used to check the high voltage circuit. This is done to determine the location of the faulty cell.

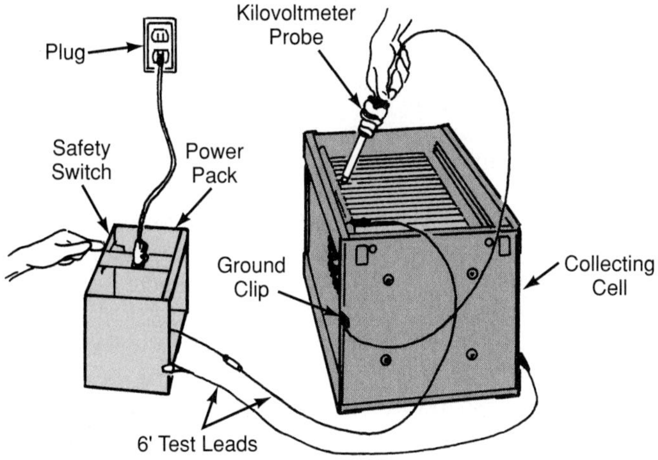

Figure 23-83. *Electrostatic filter being checked. Note test leads from power pack to collecting cell. Make sure that power pack is functioning prior to testing collecting cell. (White-Rodgers Division, Emerson Electric Co.)*

In the power pack, inspect the low side first using a test light or voltmeter. Test each part, starting with the wall outlet or power source. Then check the transformer, rectifiers (ac to dc), and the capacitor resistors. (The resistors should discharge the capacitors in about 10 seconds.) The capacitor can be checked by replacement. **Figure 23-84** shows tools and instruments used during servicing.

In the collector section, inspect for the following: bent plates, plates out of position, dirt bridging the gap between ionizing wires and the plates, broken insulators, and broken wires. Plates must be straight. Remove and replace broken insulators and ionizing wires.

The building and the complete air-handling system should be inspected. New carpeting, for example, may temporarily cause an overload on the filter. Leaking duct systems and untreated concrete floors are all unusual

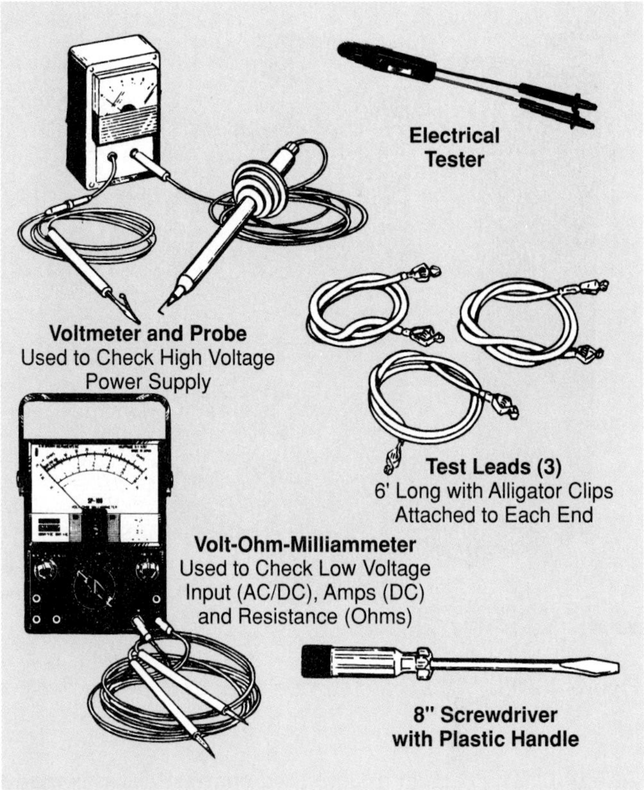

Figure 23-84. *Instruments, leads, and tools used to test electrical circuits of electronic air filter. (A.W. Sperry Instruments, Inc.)*

high-load conditions. Dusty construction work in the vicinity may also overload the unit.

Electronic air cleaners should be cleaned on a planned schedule. Frequent washing of a unit is not harmful. A neglected unit, however, will not clean air effectively.

The filter washing procedure for units without built-in wash systems is as follows:

1. Turn the electronic air cleaner, furnace, and blower power off.
2. Remove lint screens and ionizing collecting cells. Do not run the system without replacing the screens and cells.
3. Clean, using hot soapy water, and rinse thoroughly.
4. Replace lint screens and ionizing collecting cells after they have dried thoroughly. If they are not properly dried, electrical arcing may occur when system is turned on.

A properly operating unit will be indicated by black water when the cell is cleaned. A properly operating unit allows only fine white dust to leave the ducts. If a cheesecloth placed over a grille becomes discolored, the EAC is not working properly.

Carbon Filters

A filter made of activated carbon will remove solid particles, odor-causing gases, and bacteria. This type of filter is being used in air conditioners and in refrigerators. **Figure 23-85** shows an activated carbon filter

Figure 23-85. *Activated carbon filter and air purifier.*

assembly. The carbon in activated charcoal form is made from various substances, including such materials as carbon from refining petroleum and coconut shells. This charcoal will adsorb as much as 50% of its weight in foreign gases.

It is possible to recycle used carbon filters. However, this is usually done by the manufacturers. They remove the carbon and process it for reuse.

23.8.2 Dirt on Walls and Drapes

Dirt collecting on walls, ceilings, and drapes in a conditioned space is always a problem. In most cases, this dirt does not come from the ducts. It is already in the room. Room air movement (convection air current) is responsible for carrying it to the room surfaces.

The collection of dirt around warm-air grilles is called ***thermal precipitation.*** Warm air coming out of the grille picks up dirt from the room air. As this air hits cooler surfaces, the dirt settles (precipitates) on the surfaces. This precipitation takes place on windows also. The cooler the surface, the more dirt it collects. Therefore, insulation and storm sashes reduce the amount of dirt settling out of the air.

Clean rooms are now in use for surgery and research. They are also used for manufacturing, repairing, and servicing critical items such as instruments. Computer disk drives benefit from clean rooms.

A clean room maintains the "clean" requirements in four ways:

- Extremely high efficiency filters.
- Laminar airflow.
- Anti-contamination devices.
- Higher pressure in the clean room than outside the room.

Ideally, the future of residential living will include air cleaning similar to "clean room" standards. Home computers require clean conditions.

23.8.3 Water Sprays

Large air conditioners use ***water sprays.*** These remove wettable solid contaminants, liquid contaminants, and water-soluble gas contaminants from the air. Some of these gases are sulfur dioxide, nitrogen oxides, and carbon monoxide. Water does not remove soot.

Usually the water is sprayed in a pattern that produces 100% duct cross-section coverage. A drain pan catches the water. Eliminator plates in the duct collect any water droplets which travel down the duct. The water drain pan is usually equipped with a float-controlled makeup water connection.

A continual overflow drainoff is used. It removes dust and dirt as it collects on the surface of the water. Water in the drain pan is recirculated by a centrifugal pump. A screen is located at the pump inlet to prevent dirt particles from clogging the spray nozzles. These air washers are popular during the heating season.

Care must be taken to avoid freezing temperatures in the spray chamber. A preheat coil is usually used to keep the temperatures above freezing. The water spray, in addition to cleaning the air, also serves as a humidifier.

Comfort cooling systems that condense moisture out of the air use wet evaporator surfaces for filtering. A rinse device is used to remove the collected dirt.

23.8.4 Odor

Vapors and odors frequently form a large part of atmospheric air contaminants. Most odors are gases, and filters (even electrostatic ones) will not remove them.

Some odors can be removed by cooling the gases to their condensation or freezing temperature. Some are removed by oxidation (and by ultraviolet ray treatment).

Others can be removed by combining them with other chemicals. They may be diluted with air, or absorbed into a liquid. They may also be adsorbed into a solid.

Both vapors and odors can be removed with activated charcoal. Activated alumina with potassium permanganate is also used.

23.8.5 Ultraviolet Light

Ultraviolet light of 14,000 microwatt/cm^2 (duct cross section) will kill most bacteria. One should avoid looking at or being exposed to these rays. The effects are harmful when one is exposed to them for any length of time.

The lamps are installed in the return air duct. Access doors must be equipped with safety shut-off switches in case someone opens these doors. The rays must cover the full cross-section of the duct to be effective.

Ultraviolet lights are used to reduce bacteria, mold spores, viruses, and other microorganisms. Types of microorganisms and quantity killed depends on the length of exposure to the ultraviolet rays.

Ultraviolet light duct fixtures are installed in the duct with an interlocking access door. See **Figure 23-86.** The access door is required for scheduled cleaning and changing of the ultraviolet lamps. A view port may be installed in the duct for visual checking. It has plastic and/or glass lenses that absorb ultraviolet rays.

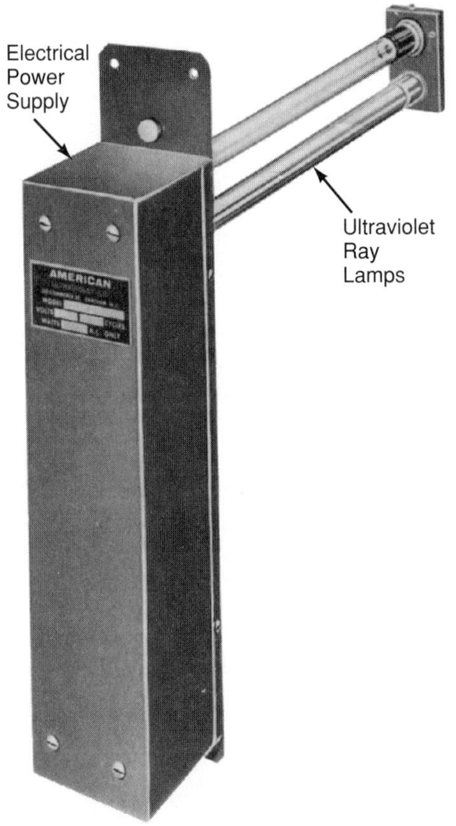

Figure 23-86. *Ultraviolet ray unit designed for duct installation. (American Ultraviolet Co.)*

The fixtures are available in various sizes. For maximum protection, install the lamp tubes across the full cross section of the duct. See **Figure 23-87.**

Units are normally equipped with lamps that do not produce ozone. Applications that require both odor control and germicidal protection are installed using ozone-producing lamps. The ozone level should not exceed 0.05 ppm in a 24 hr. period.

Lamps should be installed as close as possible to the duct where air enters the room. Fixtures may also be installed in the return ducts of recirculating systems.

Lamps should remove 90% of the bacteria in domestic and commercial systems. They should remove 99% in hospital operating and recovery rooms. Systems that remove more than 99% of the bacteria are costly. They do not allow any human exposure.

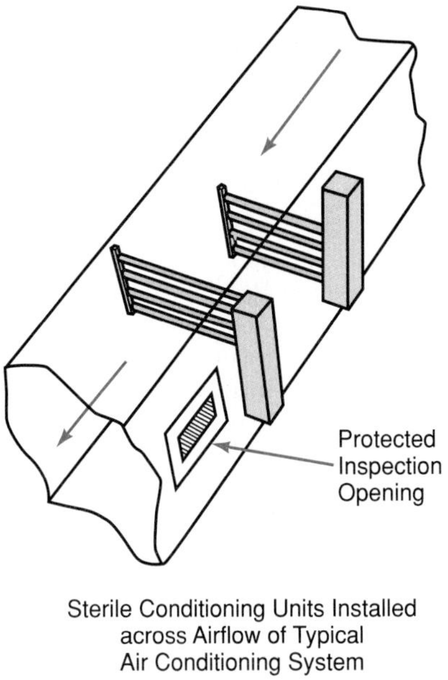

Figure 23-87. *Two banks of ultraviolet lamps installed in a cold air return duct. Rays are directed the length of the duct.*

The useful life of a lamp is approximately 7500 hours. The lamps should be cleaned at least once a month.

23.8.6 Air Curtains

Air curtains are often used at garage doors, delivery docks, little used doorways, and similar boundaries. Air curtains may be used on doors opened frequently.

A powerful blower is connected to a source of warm or cold air. As a door is opened, this air is directed through openings. These openings provide narrow streams of warm or cooled air across the entire door area. This air flow stops any natural flow of air from the building to the outside. It also stops any natural air flow from the outside air into the building.

23.9 Review of Safety

It is especially important to remember that conditioned air must contain enough oxygen to support life. The carbon dioxide content must be kept to a minimum.

Always be careful when working with or handling metal duct material. Use gloves with metal inserts when handling the material. Use stepladders with nonskid bases.

Be sure that the pressure in a duct is low before opening a duct door. If the door bursts open, it may injure someone.

Fans, motors, and belts are potential safety hazards. When these units are operating, protective shields or guards should be provided for protection. When adjusting, be sure the main power switch is off. It must be locked in the off position before handling.

Be careful that objects do not fall into a revolving fan. The object (nut, bolt, tool) may become a dangerous projectile. Before turning on the power, always spin a fan by hand. This will help check if it is free to move.

Electrostatic air filters require high voltage to charge the dust particles. Before servicing electrostatic air filters, be sure that the current is turned off. Be sure the resistors have dissipated the capacitor charge.

All air conditioning equipment is provided with safety controls. These cut out the burner if the bonnet (plenum chamber) temperatures get too high. If filters become clogged, the airflow may be reduced. This may cause the temperature limit control to cut off the heat source.

Fan belts and fan motors sometimes fail. A failure in the fan drive will result in an overheated furnace. Check to make sure that the temperature limit controls are in satisfactory condition.

Avoid exposure to ultraviolet lights. Eyes may be damaged by ultraviolet light.

Always determine the pressure in the system part on which work is to be done. Always measure the temperature of the system parts which will be repaired, adjusted, or touched. Do not guess pressure or temperature.

Use instruments to check electrical circuits. Never assume that the power is off.

Use a master instrument to check any test instrument that has been dropped. If necessary, have it repaired.

Carbon monoxide (CO) is dangerous! It is always present in the combustion gases, especially if combustion is not complete.

Palladium chloride may be used to measure the presence of CO. To use it, place a small amount (the size of a dime) on 2″ × 2″ plastic tabs. The substance will darken when exposed to CO. Amounts of CO can be determined by color:

30 ppm to 70 ppm will cause slight darkening.
80 ppm to 120 ppm cause a grey color.
Over 130 ppm result in a black color.

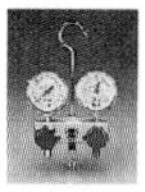

23.10 Test Your Knowledge

Please do not write in this text. Place your answers on a separate sheet of paper.

AIR DISTRIBUTION MODULE

1. When a comfort cooling air duct has condensation on its outer surface, the inner duct air _______.
   A. is cooler
   B. is warmer
   C. has moisture
   D. Any of the above.
2. Negative pressure is _______.
   A. excess humidity which causes constant dehumidification
   B. a room or building with pressures slightly above atmospheric pressure
   C. a room or building with pressure slightly below atmospheric pressure
   D. None of the above.
3. What is the volume of 1 lb. of 72°F (22°C) air at atmospheric pressure if it has 50% relative humidity?
   A. 13.05 ft^3.
   B. 13.55 ft^3.
   C. 7.54 ft^3.
   D. None of the above.
4. What is the specific heat of dry air?
   A. 11.58 Btu/lb.°F.
   B. 0.08645 Btu/lb.°F.
   C. 14.10 Btu/lb.°F.
   D. None of the above.
5. Should blowers run constantly?
   A. Only in humid conditions.
   B. Yes.
   C. No.
   D. Only in weather above 70°F (21°C).
6. In a six-duct outlet system, what is meant by a total pressure drop system?
   A. The fan pressure drop to each outlet will be the same.
   B. The fan pressure drop to each outlet will vary, with the last one being less.
   C. There will be no air movement, pressure drop, at last outlet.
   D. None of the above.
7. Smaller ducts may be used to distribute heated air and maintain good heating season temperatures by _______.
   A. increasing the rate of flow through the ducts
   B. using larger registers
   C. providing additional clearance for air in the ducts
   D. Both A and B.
8. _______ draft is air movement caused by creating a reduced pressure within an air blower. _______ draft is air movement caused by an increased pressure produced by an air blower.
   A. Induced, Reduced
   B. Induced, Forced
   C. Forced, Induced
   D. None of the above.
9. Air ducts carry _______.
   A. warm air or cooled air
   B. fresh makeup air and exhaust air
   C. air from the conditioned spaces back to the conditioned system
   D. All of the above.
10. How can galvanized steel duct be prepared for painting?
    A. Treat with vinegar or weak acid.
    B. Sand with a fine grade sand paper.
    C. Use a primer.
    D. All of the above.

AIR MEASUREMENT AND CLEANING MODULE

11. What is the voltage of the ionizing wires of an electric air filter?
    A. 120 V.
    B. 12,000 V.
    C. 1,200 V.
    D. Any of the above may be used.
12. When clean, 1″ adhesive filters are ______% efficient.
    A. 70
    B. 84
    C. 90
    D. 100
13. The principal impurity removed by an activated carbon filter is ______.
    A. carbon monoxide
    B. moisture
    C. solid particles and odor-causing gases and bacteria
    D. All of the above.
14. How are most electronic filters cleaned?
    A. Simply vacuum.
    B. Clean with R-12.
    C. Both A and B.
    D. Wash with a water and detergent solution.
15. What is the CO_2 content of oil furnace flue gas at 100% excess air?
    A. 10%.
    B. 8%.
    C. 100%.
    D. 25%.
16. ______ may cause the temperature limit control to shut off the heat source in an electronic filter.
    A. Lack of heat
    B. A faulty circuit
    C. Poor air flow in the system
    D. All of the above.
17. Ultraviolet lights are used to ______.
    A. kill bacteria and mold
    B. kill viruses
    C. remove odors from ducts and duct air
    D. All of the above.
18. When using a CO_2 indicator to measure combustion efficiency by a reduction in volume, how will the contraction of the flue gas impact the CO_2 reading?
    A. The reading will be higher than a true reading.
    B. The reading will be lower than a true reading.
    C. The reading will be accurate.
    D. The reading will fluctuate due to expansion and contraction.
19. A draft can be detected ______.
    A. through the use of an electronic draft detector
    B. with a smoke candle or smoke source
    C. by positioning oneself in various places around the room
    D. Any of the above.
20. Water sprays in a large air conditioning unit are used to ______.
    A. remove contaminants from the air
    B. increase the relative humidity
    C. remove soot
    D. Both A and B.

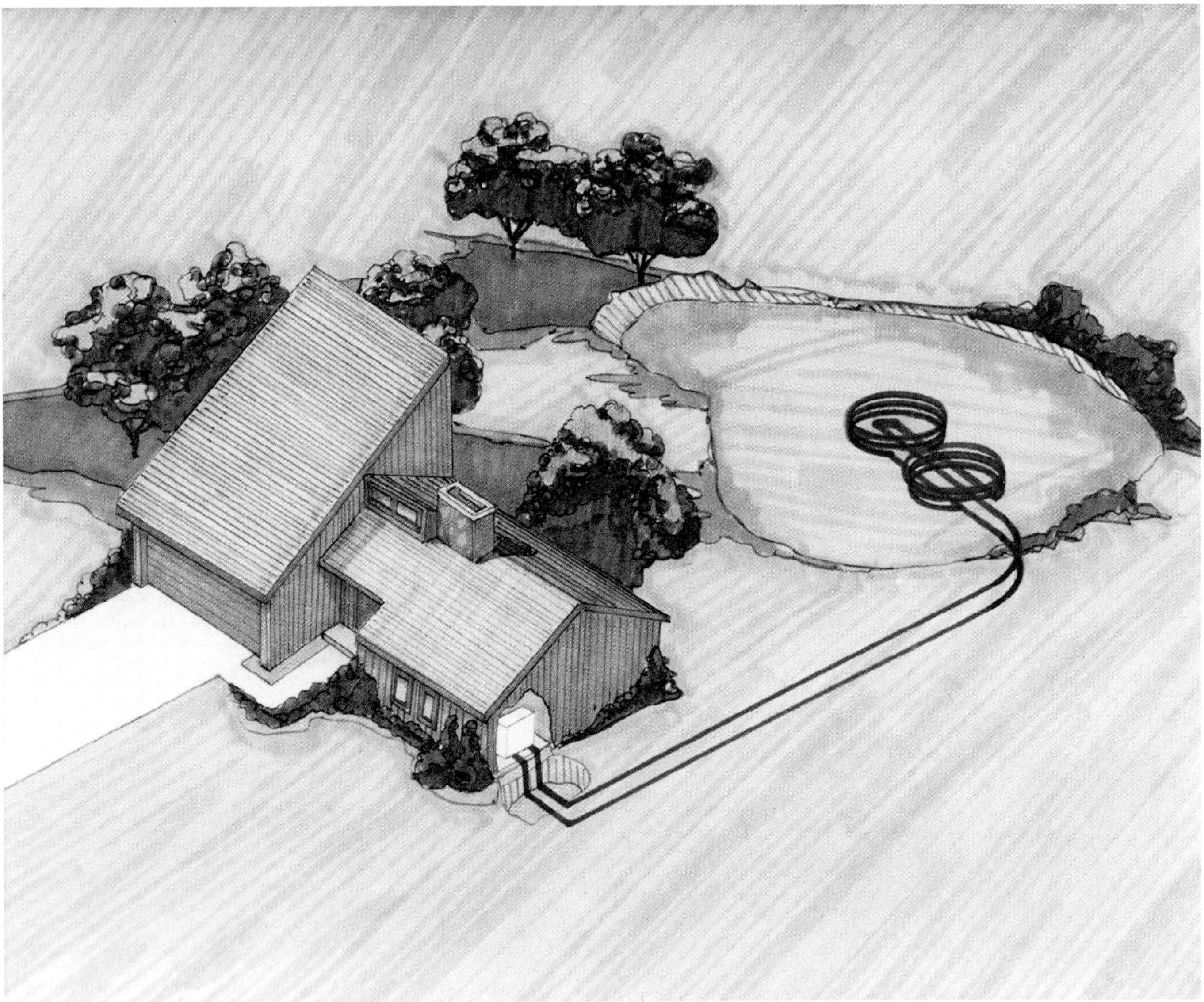

Geothermal systems deliver savings and comfort to a home. This geothermal system—a closed loop system—utilizes a pond for heat transfer. The coiled pipe is placed on the bottom of a pond (or lake) where it transfers heat to or from the water. A 1/4- to 1/2-acre pond is adequate for a typical 2500 ft^2 home. (WaterFurnace International, Inc.)

Chapter 24

HEAT PUMPS AND COMPLETE AIR CONDITIONING SYSTEMS

Modules:

Key Words:

absorption-type chiller
air coil
chillers
compression-type chiller
district heating and cooling systems
geothermal (ground) coil
geothermal heat pump
high-pressure chiller
liquid floodback
low-pressure chiller
natural gas heat pump
reversing valve
scroll compressor
unitary systems
water coil

Learning Objectives:

After studying this chapter, you will be able to:

- Describe the basic operation of a heat pump and list its principal parts.
- Explain the operation of the heat pump reversing valve.
- Identify, on a heat pump diagram, the changes in refrigerant flow when the reversing valve is switched from heating mode to cooling mode, or vice versa.
- Discuss types of outdoor coils and their applications.
- Explain the difference between geothermal heat pumps and air source heat pumps.
- Perform routine maintenance and service on heat pumps.
- Explain total energy systems.
- Perform routine maintenance on residential central air conditioning systems.
- Identify the two basic types of chillers.
- **Follow approved safety procedures.**

HEAT PUMPS MODULE

24.1 Heat Pump Theory

Heat pump theory rests on the principle that heat will move from a *higher* temperature to a *lower* temperature. Thus, a heat transfer coil kept at a lower temperature than its surroundings will pick up heat.

If the evaporator of a refrigerating system were mounted outdoors and operated at a refrigerant temperature of 0°F (−18°C), it would remove heat from the air even if the outside temperature were only 10°F to 15°F (−12°C to −9°C). If the evaporated refrigerant is then compressed to a temperature of 120°F to 140°F (49°C to 60°C), the hot refrigerant will release heat to the indoor space where the condenser is located.

The evaporator and condenser functions can be switched using a system of valves. When the condenser becomes the evaporator, heat can be removed from the living zone during hot weather and discharged outdoors. **Figure 24-1** shows the pressure-heat diagram for a heat pump.

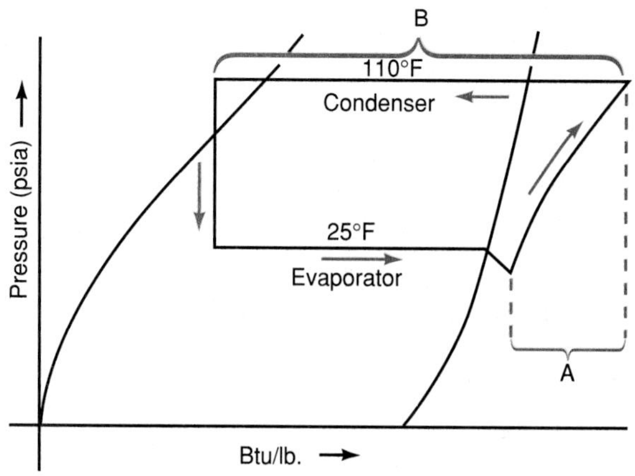

Figure 24-1. *Typical heat pump cycle used for heating. Refrigerant evaporates outside at 25°F (−4°C) in 35°F (2°C) air. Same refrigerant condenses at 110 °F (43°C) in condenser in air duct. A—Heat of compression. B—Heat released to living space.*

The evaporator and the condenser are both heat-transfer devices. They can be used for cooling (picking up heat) or heating (releasing heat). The idea of using a refrigeration unit as a heating mechanism was first proposed by the Scottish scientist William Thomson (Lord Kelvin) more than 100 years ago. A period of more than 30 years elapsed before such a device was actually built.

The heat pump is sometimes called a "reverse-cycle" mechanism. However, the cycle is not actually reversed. Only the evaporator and condenser functions are interchanged. Therefore, the name "reverse-cycle" is not technically correct.

24.2 Types of Heat Pumps

Two terms are used in reference to heat pump source of operation. They are air-to-air heat pumps, and geothermal (ground source) heat pumps.

The ***air-to-air heat pump*** is the most popular type. Its source of heating and cooling is the outdoor air. Its exterior appearance is similar to that of the conventional complete air conditioning system.

The ***geothermal heat pump*** uses the underground (earth or water) temperature to produce the desired temperature. Although outdoor temperatures may vary, the underground temperatures remain relatively constant all year round. This allows a geothermal system to produce the desired heating and cooling temperature year round.

The term "environmental comfort system" is sometimes used to describe a heat pump system. This is based on the fact that a heat pump operates from the environment. It has no negative effects on its surroundings.

The heat pump may be used for many purposes. Water heating and heat recovery from industrial processes are two examples. Defrosting evaporators by means of the hot gas method is a form of heat pump application. Both compression systems and absorption systems can be adapted as heat pumps.

Figure 24-2A illustrates a typical geothermal heat pump installation operating on a heating cycle. **Figure 24-2B** shows the same heat pump operating on a cooling cycle. The basic principles are described in Chapter 20.

Self-contained systems of 2 tons to 25 tons are common, and large systems of 100-ton to 1000-ton capacity are in use. Window units of 1/2-ton to 2-ton capacity are also available.

24.3 Heat Pump Operation

Operation of the heat pump is like any other compression cycle. The principal parts of the systems are:

- Compressor.
- Condenser.
- Liquid line.
- Two refrigerant controls.
- Evaporator.
- Suction line.
- Motor control.
- Reversing valve.
- Two check valves.

Notice that two check valves, a reversing valve, and two refrigerant controls are needed in the heat pump. This allows the unit to change from summer cooling to winter heating. **Figure 24-3** shows a heat pump in the cooling cycle. The four-way reversing valve allows the unit to operate in a conventional cooling mode. High-pressure vapor is emitted from the compressor and then travels to the condenser. It then passes to the evaporator, where it cools the indoor area. It finally returns to the compressor.

In the heating cycle, the four-way valve reverses the path. The compressed vapor is emitted to the indoor coil. This provides warm air to the heating system.

24.3.1 Heating and Cooling Cycles

The heat pump operates in two different cycles:

- Heating cycle.
- Cooling cycle.

The same mechanism is used for both cycles. However, the travel of refrigerant is reversed to change from cooling to heating. **Figure 24-4** shows a basic heat pump system. Hand valves or thermostatically controlled valves may be used to reverse the cycle.

During the heating cycle, heat is removed from the ambient (surrounding) air. It is released inside the building. Heating is not usually needed until the outdoor temperature is less than 65°F (18°C). For example, in the conditions shown below, the outdoor coil will act as an evaporator. It will pick up heat from outdoors. This heat is released in the building, as shown in **Figure 24-5.**

Outside (Ambient) Conditions		**Inside Conditions**	
Temp.	Humidity	Temp.	Humidity
50°F (10°C)	80%	72°F (22°C)	50%

The heat pump heating cycle becomes less efficient as the outdoor temperature drops below freezing. As the outside temperature decreases, heat load increases. This creates problems in colder climates where temperatures drop to 20°F (−7°C) or lower. **Figure 24-6** shows such a condition. The outside temperature at 20°F requires a refrigerant temperature of 0°F (−18°C).

Note that Point A has increased with very little increase in B. This means that the energy efficiency ratio in heat units (EER_Q) is less than in **Figure 24-5.** With the refrigerant boiling at 0°F (−18°C), the evaporator will frost rapidly. Some means of frequent defrosting would have to be found.

When ambient conditions are within 10°F (6°C) and 10% relative humidity (RH) of the inside conditions, very little treatment is needed. This is due to the heat lag and time lag in controlling the variables. For example, the sun's heat on a south wall in the morning may correct a problem of low temperature by afternoon.

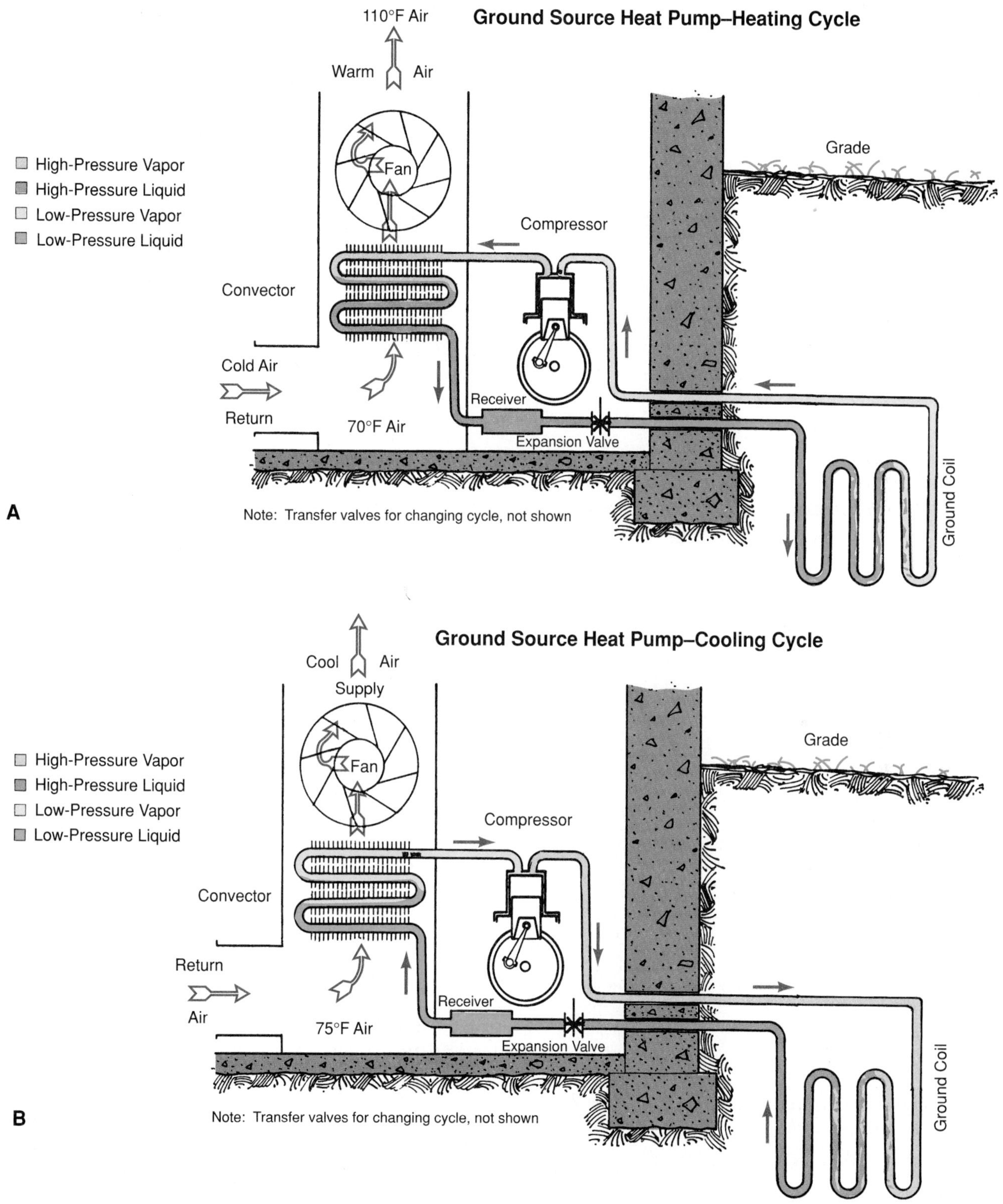

Figure 24-2. *Geothermal heat pump operation. A—Ground coil heat pump operating on heating cycle. B—Ground coil heat pump operating on cooling cycle.*

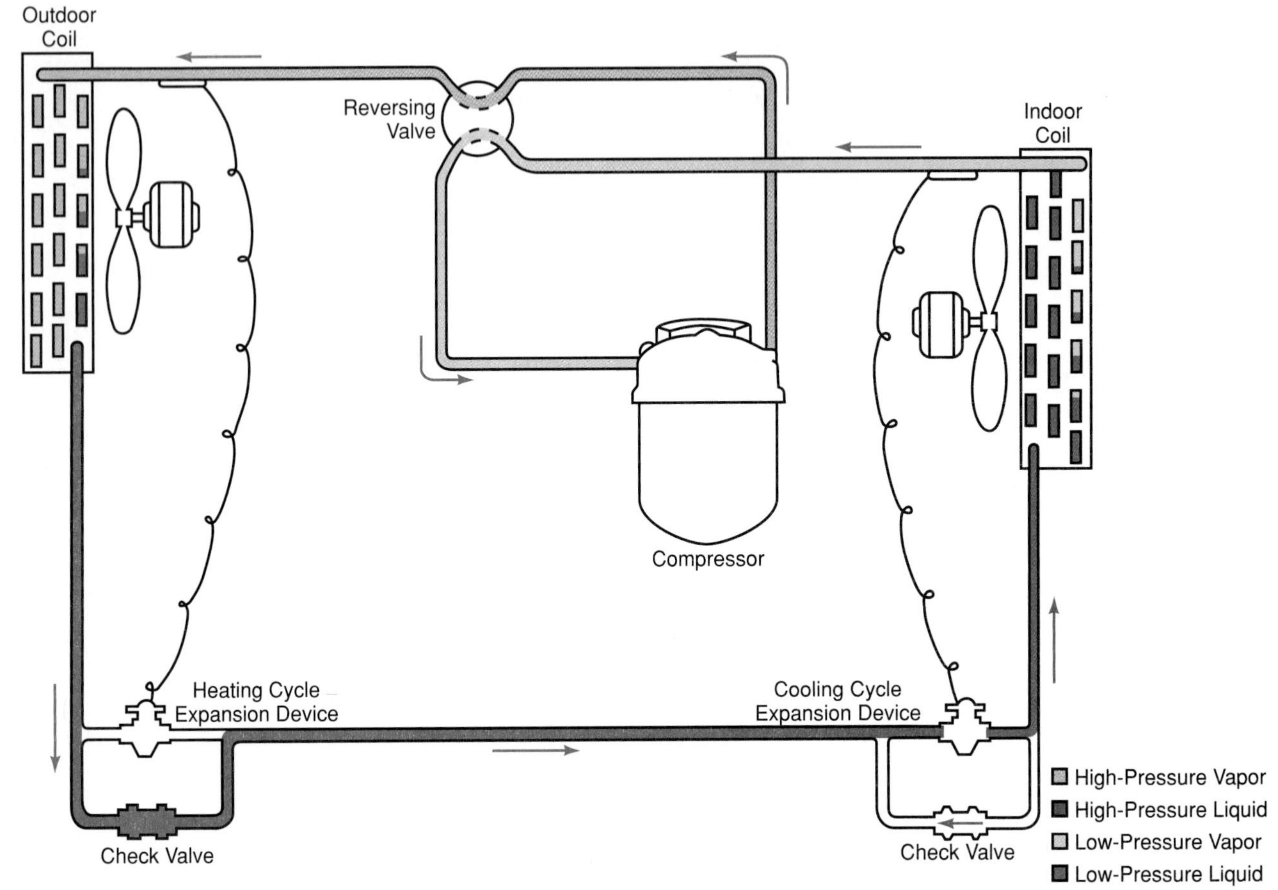

Figure 24-3. *Heat pump. Both heat transfer coils are blower coils. TEV refrigerant controls are used. Note four-way reversing valve. System is operating as comfort-cooling unit with outdoor coil as condenser and indoor coil as evaporator.*

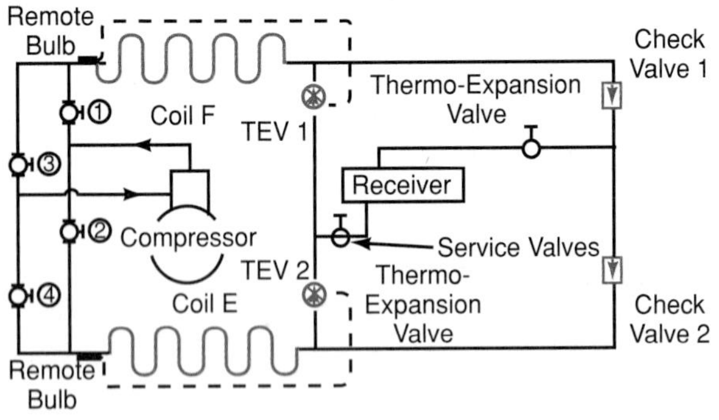

Figure 24-4. *Heat pump has hand valves to permit manual changing of system from heating to cooling. Coil F is the outside coil and coil E is the inside heat transfer surface. During cooling season, valve 1 is open, valve 2 is closed, valve 3 is closed, valve 4 is open. Check valve 1 is open, check valve 2 is closed. TEV 1 is not working, TEV 2 is working. During heating season, valve 1 is closed, valve 2 is open, valve 3 is open, valve 4 is closed. Check valve 1 is closed, check valve 2 is open, TEV 1 working, TEV 2 is not working. (Alco Controls Div., Emerson Electric Co.)*

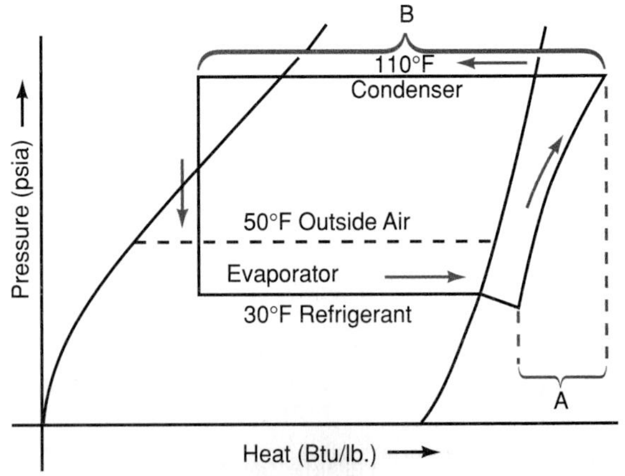

Figure 24-5. *Pressure-heat chart for heat pump serving as heating system with outside temperature at 50°F (10°C). Refrigerant is evaporating at 30°F (−1°C). Refrigerant is condensing at 110°F (43°C). The Btu ratio of B to A is Energy Efficiency Ratio in Heat Units (EER_Q). EER_Q is B/A, which is about 4.*

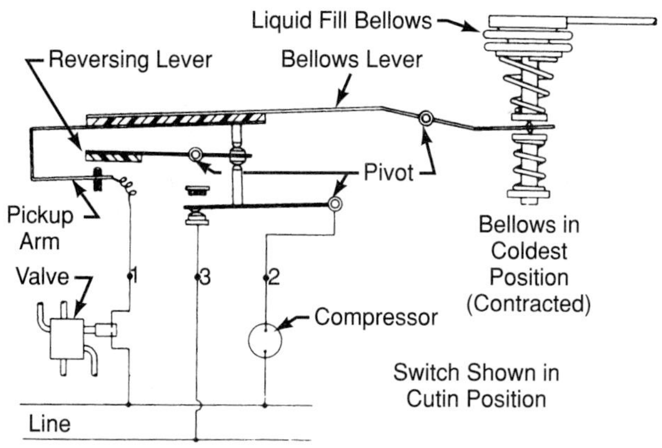

Figure 24-14. *Electrical connections during heating cycle of heat pump.*

Other units use a four-way valve to reverse the flow of refrigerant. See **Figure 24-16.** The valve is operated by the movement of one valve stem. This stem closes and opens several ports in one valve body. Its operation may be manual or electrical. The whole system is easily reversed with one of these valves. They are popular in small tonnage conditioners, such as window units and other air-to-air systems.

One kind of four-way valve, operated by pressure, is shown in **Figure 24-17.** A solenoid valve controls the pressure at the top portion of the reversing valve. When the solenoid pilot valve is de-energized, **Figure 24-17A,** the compressor high-side pressure pushes on the top of the four-way valve. The reversing valve slider is lowered, producing the heat cycle. The indoor coil is the condenser; the outdoor coil is the evaporator.

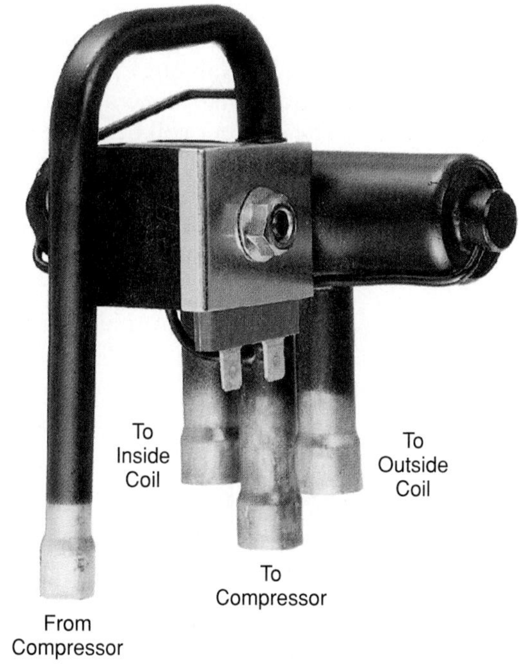

Figure 24-16. *Heat pump four-way reversing valve. (Ranco North America)*

The solenoid pilot valve circuit may be closed either manually or by a thermostat. The solenoid valve is then lifted, and the slide inside the four-way valve comes up. Refrigerant flow to the coils is reversed.

The heat pump has a thermostatic expansion valve and a check valve on each coil. When the system is reversed, the refrigerant flow bypasses the thermostatic expansion valve of the coil serving as the condenser.

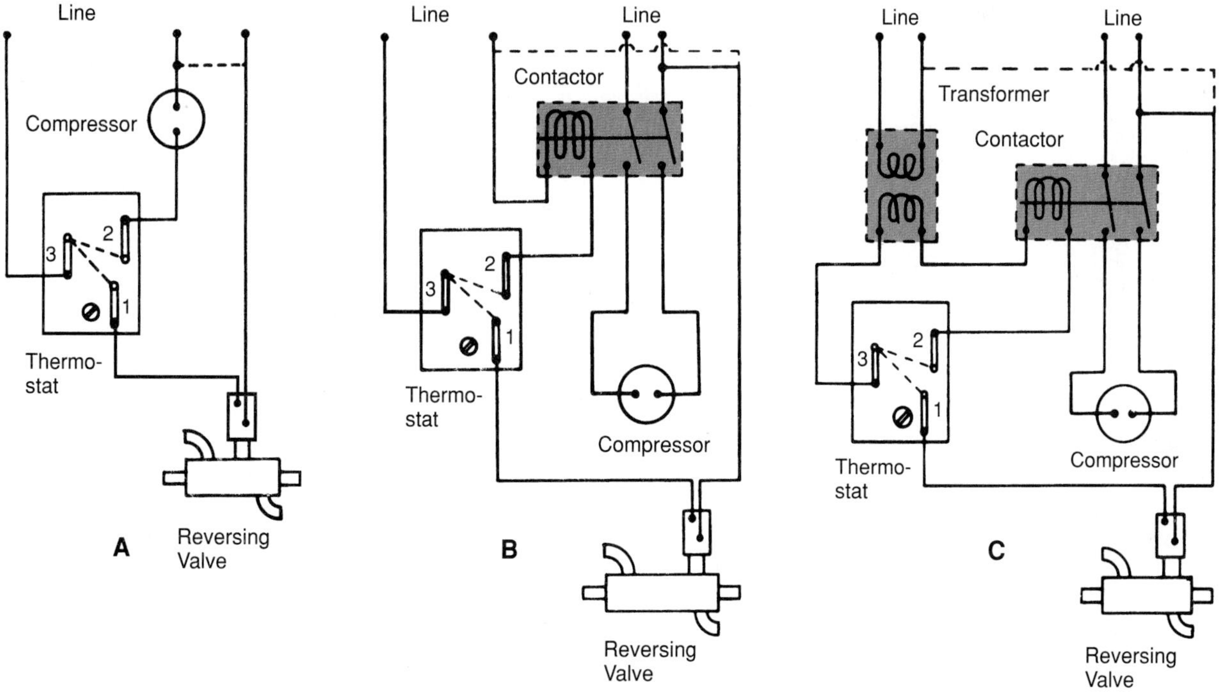

Figure 24-15. *Three wiring diagrams for heat pumps. A—Thermostat controls motor circuit. B—Circuit uses line-voltage coil magnetic contactor. C—Circuit uses low-voltage coil magnetic contactor.*

During the cooling cycle, the upper thermostatic expansion valve is used. The refrigerant cannot travel through the companion check valve.

Figure 24-17B illustrates this same system operating as a cooling unit. The outdoor coil becomes the condenser and is releasing heat to the outdoors. The indoor coil becomes the evaporator.

Another style of reversing control uses a lever inside a solenoid pilot valve. This is shown in **Figures 24-18B** and **24-18C.** The pilot valve controls a four-way slide. The solenoid coil is energized on the heating cycle. The pilot slide pivots and changes the flow of pressure. Pressure moves the main slide, as seen in **Figure 24-18A.** In the cooling cycle, the solenoid is not energized. **Figure 24-19A** shows the reversing valve installed in a heat pump system. The system is in a cooling cycle. When the valve is in position for a heating cycle, the system is set up as in **Figure 24-19B.**

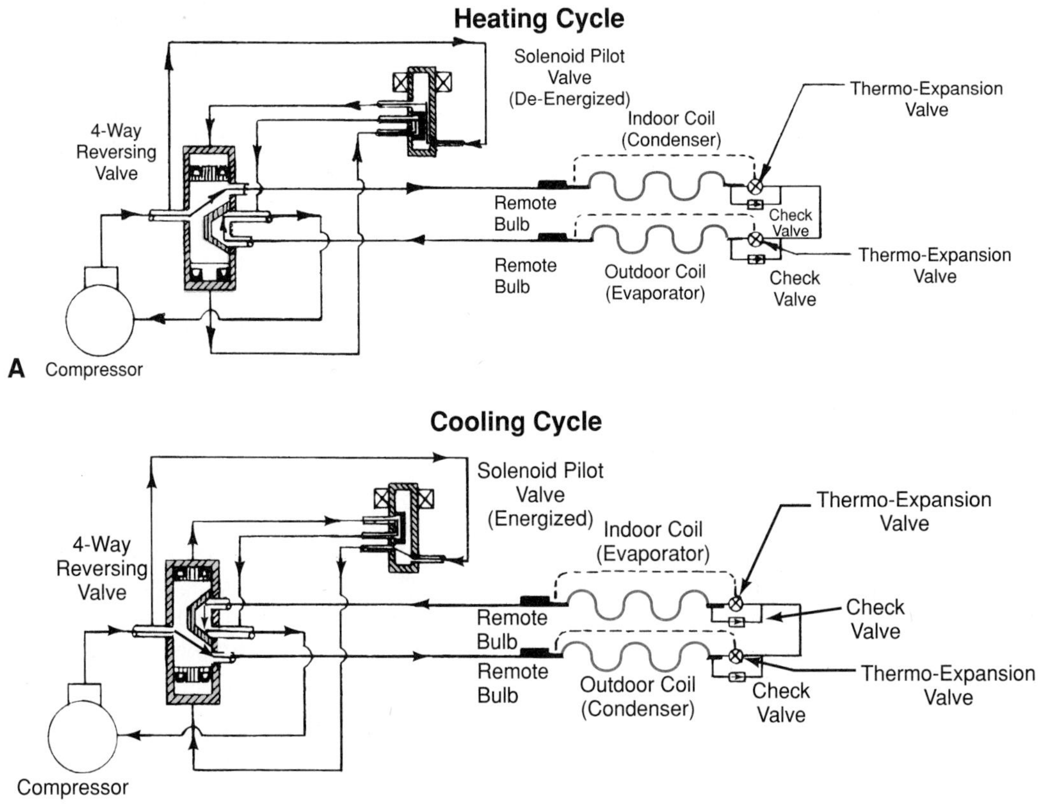

Figure 24-17. *Heat pump diagram. System uses four-way reversing valve. A—Solenoid pilot valve is de-energized during heating cycle. B—Four-way reversing valve in use operating on cooling cycle. Note how functions of indoor and outdoor coils reverse from heating cycle shown in A. (Alco Controls Div., Emerson Electric Co.)*

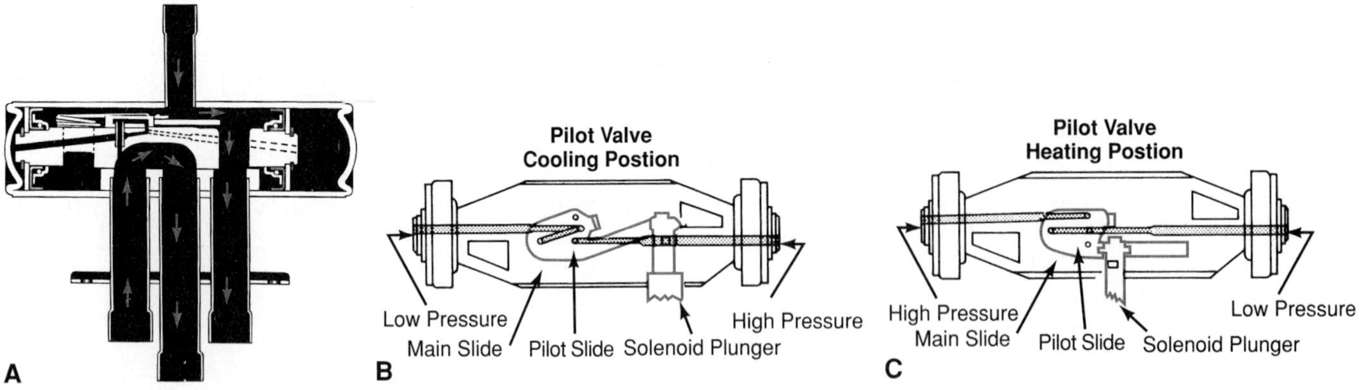

Figure 24-18. *Cutaway of pilot valve mechanism used to operate four-way reversing valve. Note position of solenoid plunger in heating position when solenoid is energized. A—Side view of four-way valve. B—Pilot mechanism and piston in cooling position. C—Pilot mechanism and piston in heating position. Note how high pressure changes place with low pressure. Holes in pilot slide lead to compressor inlet and outlet.*

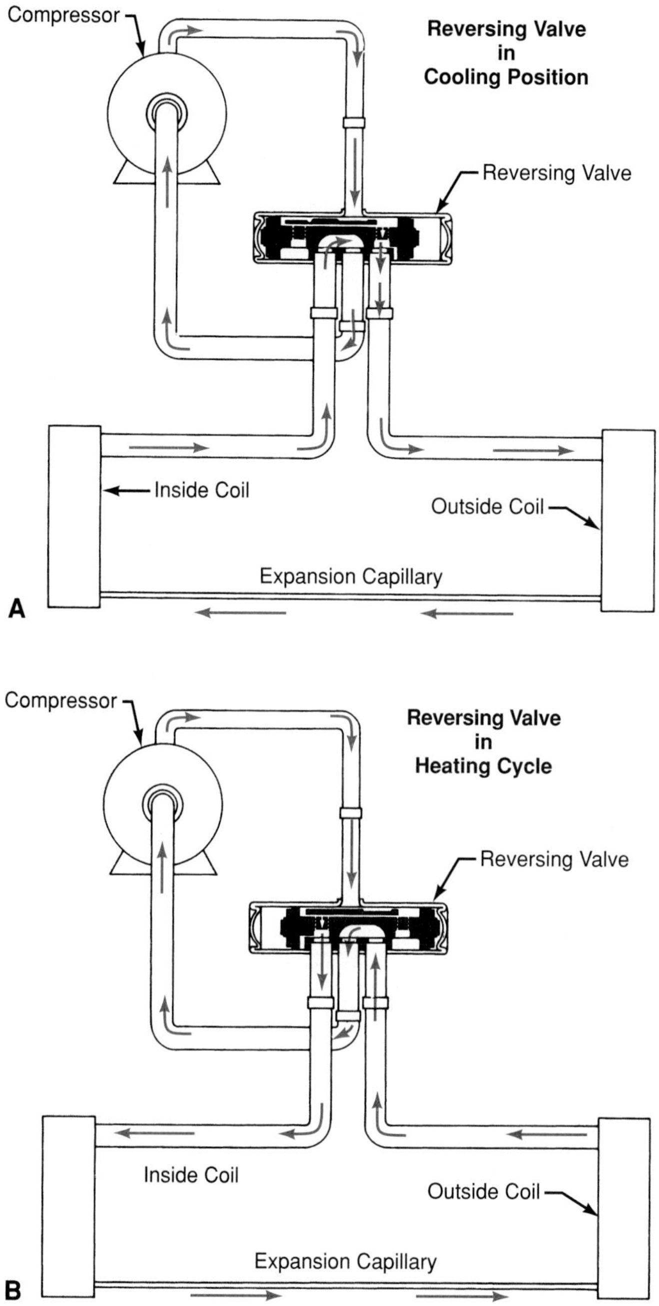

Figure 24-19. *Heat pump cycle uses reversing valve. A—Note flow of refrigerant when valve is set for cooling cycle. B—Heat pump circuit with reversing valve set for heating cycle.*

Figure 24-20 shows a main sliding valve with bleed holes. With the bleed hole system, one less pipe is needed to the pilot valve.

Some systems use a flow check piston as an expansion valve instead of a capillary tube. A sketch of a flow check piston assembly is shown in **Figure 24-21.** Liquid flow from the condenser is metered into the distributor of the evaporator through a hole in the center of the piston. With flow in this direction, the flow check piston provides an even distribution of refrigerant into the evaporator. If the flow is reversed, the piston moves back. This permits a free flow of refrigerant without throttling.

This device is useful where two flow check pistons are used in the line between the evaporator and condenser. In heat pumps used for heating in winter, **Figure 24-22,** and cooling in the summer, the flow is reversed. One flow check piston acts as the expansion device while operating in the heating mode. The other operates as the expansion valve when flow is reversed (in the cooling mode).

Figures 24-23A and **24-23B,** respectively, are diagrammatic line drawings of the heating cycle and the cooling cycle. These drawings show a number of components that are not standard for all systems. They are identified on the drawings. Note that the difference between the cooling and heating cycles is the direction of refrigerant flow. The four-way reversing valve sets the path for the refrigerant. When the refrigerant flow changes, the function of the indoor and outdoor coils is reversed. When the system is used for heating, a high-pressure vapor enters the indoor coil. This produces a warm coil. When the system is used for cooling, a low-pressure vapor enters the indoor coil. This produces a cool coil.

24.4.5 Condenser and Evaporator Coils

The coil mounted inside the structure is normally a standard finned coil with a blower.

The outside coil is available in several designs. As noted earlier, the type used is determined by the substance into which the coil releases heat or from which it picks up its heat. Coils are classified by the type of external heat source used to obtain heat of evaporation. They include:

- Air coil.
- Water coil.
- Ground coil.

Air Coil

The least expensive heat pump outdoor coil is one that picks up heat from (or releases heat to) the outside air. It is also the easiest to install. In climates where outdoor temperatures range from 20°F (−7°C) to 110°F (43°C), the air coil has a number of advantages.

The air coil is a standard heat transfer coil. It has tubing for a primary surface with extended fins bonded to the tubing. Along with a blower, it is mounted in a housing that protects it from the weather. The housing may be mounted on the outside wall of the building, **Figure 24-24.** It may also be mounted on the roof, or in a shelter near the building.

A system with the coil inside the furnace and the heat pump outside the building is called a ***split system.*** **Figure 24-25A** shows a typical wiring and piping diagram for a split system heat pump. The indoor component is a vertical upflow-type unit. The flow is up from the basement. **Figure 24-25B** shows the exterior system.

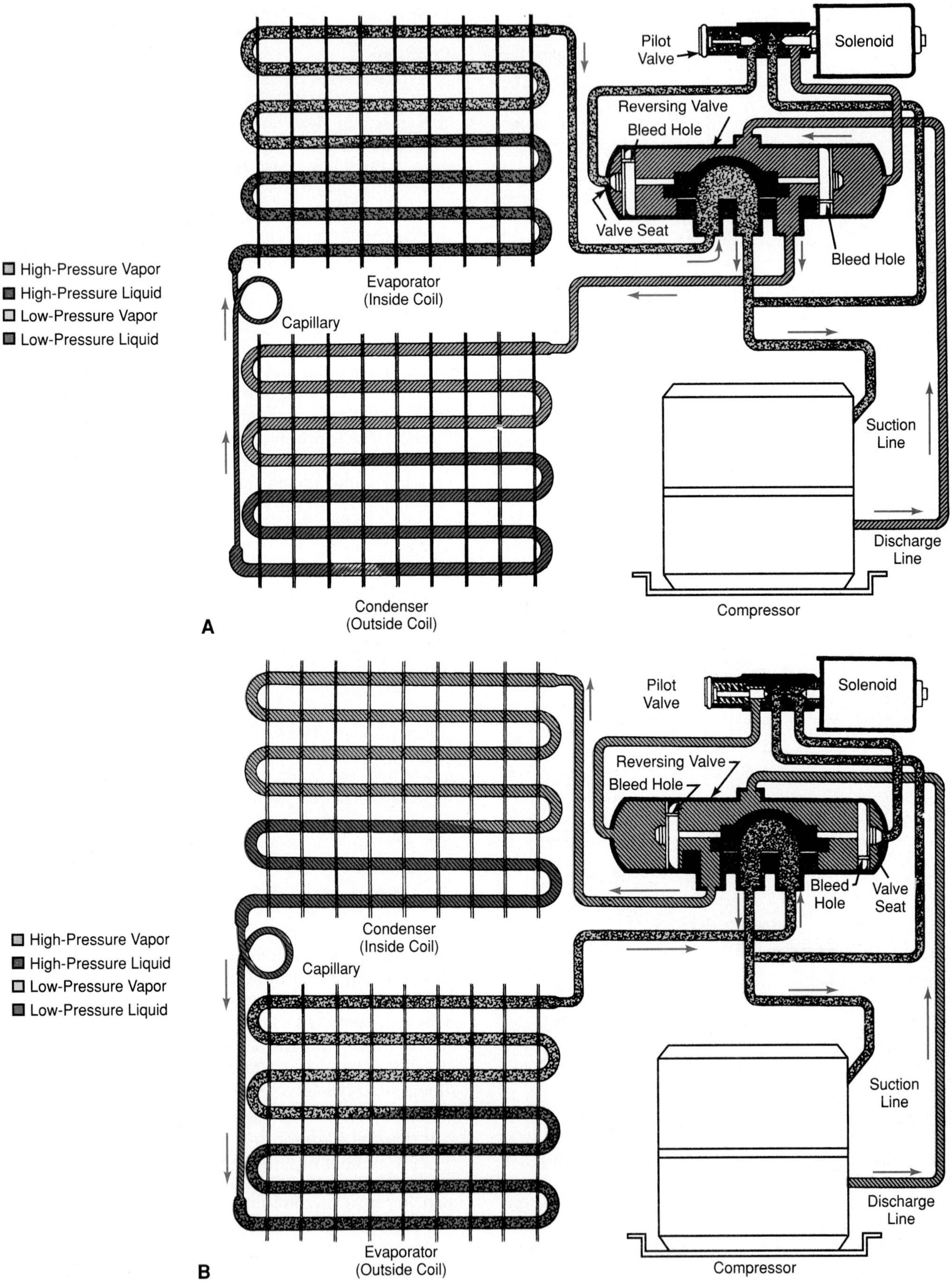

Figure 24-20. *Heat pump uses capillary tube refrigerant control and solenoid-valve-controlled four-way valve. A—Indoor coil is serving as evaporator. Note bleed holes in main slide. With bleed hole system, only three tubes to pilot slide are needed. B—Capillary tube-type heat pump using indoor coil as condenser.*

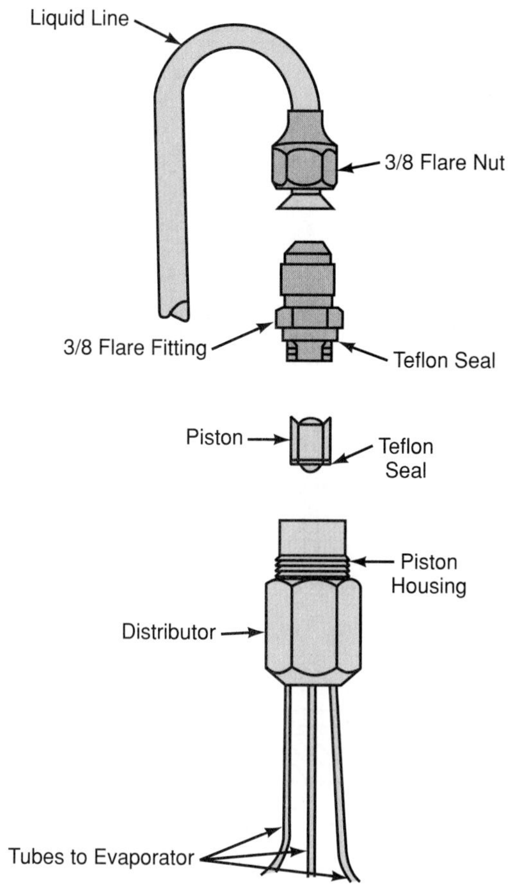

Figure 24-21. *Flow check piston. Refrigerant flows from the liquid line through the piston, into the distributor, and through tubes leading to the evaporator. (Rheem Mfg. Co., Air Conditioning Div.)*

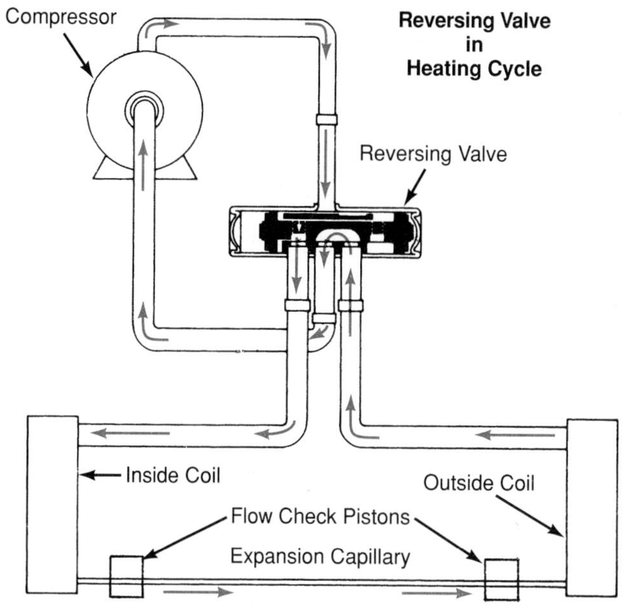

Figure 24-22. *Heat pump circuit with flow check pistons set for heating cycle.*

The outdoor unit has a vertical air discharge. **Figure 24-25C** displays five different types of split system installations and the airflow for each.

During the heating cycle, the outside coil may frost and ice up in certain weather. To prevent this, most heat pumps have a special thermostat mounted on the coil. If frost or ice starts to form, this control will shut off the system. Electric heating coils will start a defrost action. In one method, the cycle will reverse long enough to defrost the coil. Some systems use a timer. It operates the system on defrost for a time. During defrost, the drain also must be heated. This is done to make sure drain water can flow away from the unit.

Water Coil

A heat pump becomes more efficient as the condenser and evaporator temperature approach each other. During the heating cycle, the pump works better using a source warmer than the ambient air. In some installations, the outside coil is lowered to the bottom of the lake, **Figure 24-26.** Such a location for the coil provides a more consistent temperature.

The temperature of water at its maximum density is 39°F (4°C). This means that the temperature of the water at the bottom of a lake will be 39°F (4°C) or above. If the water cools below 39°F (4°C), it will expand and rise to the surface. A temperature of 32°F (0°C) will cause it to freeze. This is the reason that ice forms on the surface of a lake.

The lake would have to freeze solid before the temperature of the water at the bottom could be lower than 39°F (4°C). Thus, a lake—even one covered with ice—can serve as a reservoir of heat. Well water also may provide an efficient heat pickup unit for heating. It is also a good heat dissipater during the cooling cycle. The cost of the well is a disadvantage. However, this would probably soon be offset by the lower cost of operation.

In one popular method, water is drawn from the well, used, and quickly returned. A tube-within-a-tube or a shell-and-tube heat exchanger is used.

With well water at 60°F (16°C), a condensing temperature of 80°F (27°C) can be used during a cooling cycle. An evaporator temperature of 40°F (4°C) can be used during the heat cycle. Often, flowing wells or springs are used to supply water for heat pumps.

A concern with lake or ground water coil installation has been the water's chemical nature and mineral content. A coil made of cupronickel (a copper and nickel alloy) is corrosion-resistant and desirable for water installations.

Geothermal (Ground) Coil

Regardless of latitude or air temperature changes, the ground temperature at a depth of four to six feet changes very little. Ground temperatures at that depth average 40°F (4°C) to 60°F (16°C).

A coil with sufficient heat transfer surface, buried four to six feet deep, can be used as an outdoor coil for both heating and cooling cycles. **Figure 24-2A** illustrates a ground coil system as it operates during the winter. Ground coils are installed horizontal to the grade,

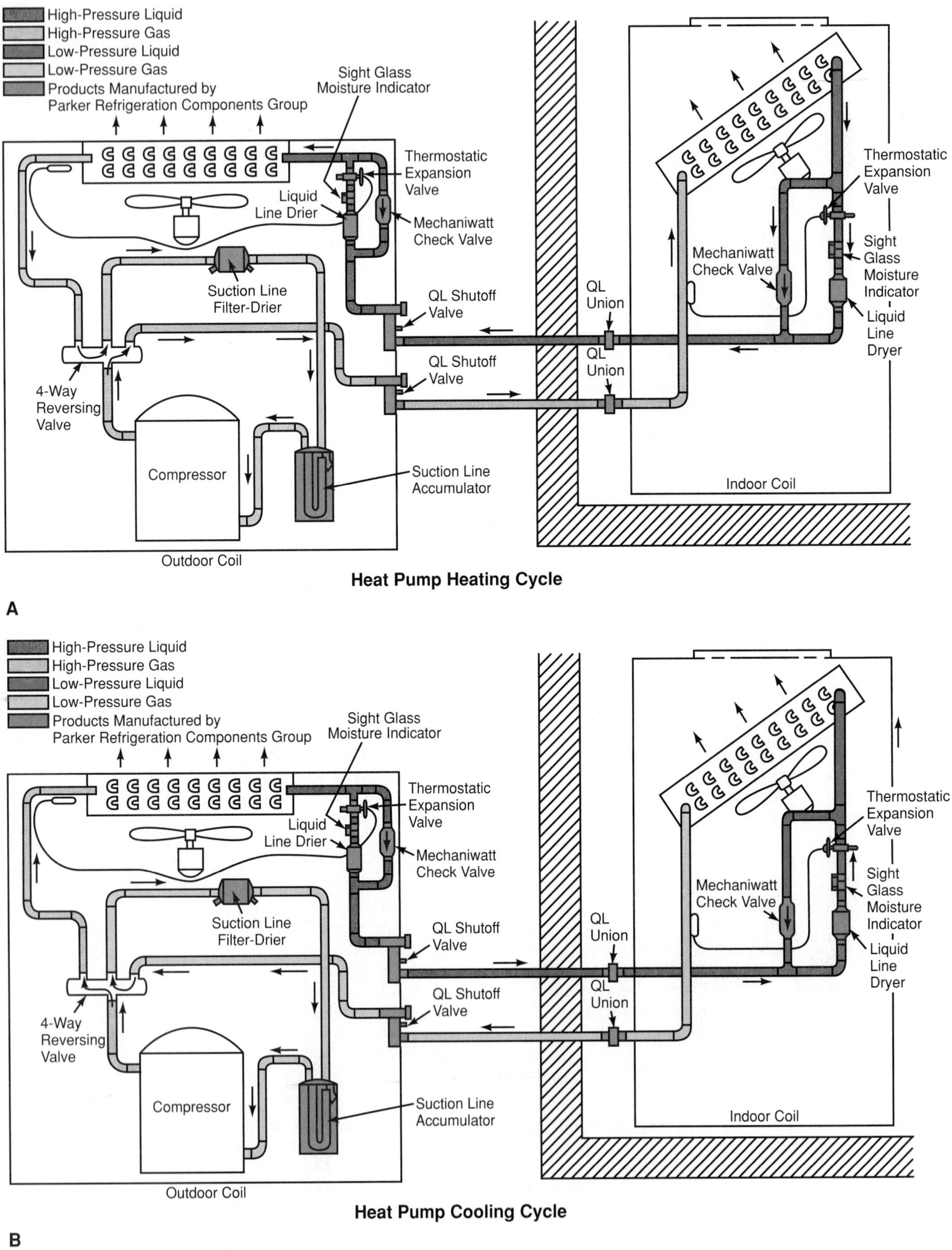

Figure 24-23. *Reversing valve sets the path of the refrigerant through the heat pump to produce a cooling or heating cycle. A—Heating cycle. B—Cooling cycle. (Parker-Hannifin Refrigerants Components Group)*

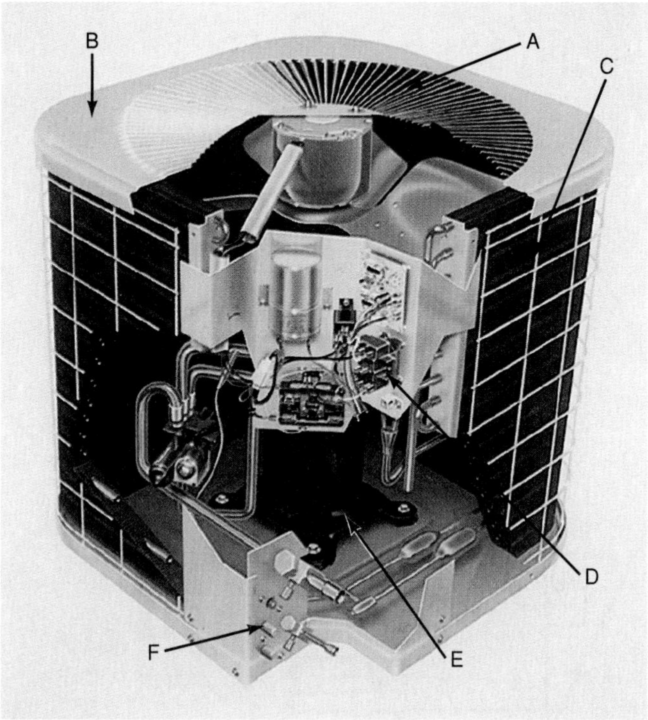

Figure 24-24. *Five-ton heat pump air coil. A—The exhaust air discharges upwards. B—Cabinet. C—Condenser coil. D—Solid-state defrost system to prevent ice build-up. E—Scroll compressor. F—High-pressure switch shuts unit off if pressure gets abnormally high. (Tempstar Heating and Cooling Products, Inter-City Products Corporation)*

although drawings show a simpler view. Also, in a commercial unit, the air return is a split air system. It has a fresh air makeup duct.

Figure 24-2B illustrates the same basic system used as a comfort-cooling mechanism. In practice, valves would be used to send the condensed refrigerant through the liquid receiver. From there the refrigerant would go through the expansion valve from left to right. The reduced-pressure refrigerant would then enter the evaporator in the duct. For example, the receiver would be between the two check valves in **Figure 24-3.**

Installations have been made using a combination of sources. An air coil may be used with a lake coil, well coil, or ground coil. The air coil is used alone when the outdoor temperature permits efficient operation. The outside air may become too cold or too warm. The auxiliary (ground or water) coil may then be connected into the heat pump air coil system.

24.4.6 Heat Pump Defrosting

Defrosting cycles on many heat pumps begin with a 5-minute period of supplementary electric heat. Also in common use today are defrost controls. These are based upon temperature differentials, pressure differentials, or a combined time/temperature differential.

Another unit depends upon the electrical loading of the heat pump's condenser fan motor due to ice blockage. The unit electrically measures the mechanical load of an induction motor. The cut-out signal is obtained from a remote thermistor.

The thermistor is clipped to the liquid line. A pressure or temperature control may also send the signal. The thermistor method is used to maintain a comfortable indoor temperature during the defrosting process. (This is possible because it is not defrosting too long.)

The heat pump cycle reverses for defrosting while the compressor is running. This causes a strain on the compressor and motor. A heat pump on low ambient (outdoor) temperatures cools the compressor motor with liquid refrigerant. (The suction pressure is high enough that liquid remains.) At times, the defrosting system may fail to return the liquid rapidly from the outdoor coil at the start of the defrost cycle. It then fails to provide adequate and rapid defrosting and motor cooling. Hot and cold spots can form on the motor windings and laminated steel cores. This leads to high thermal stresses. Also, when the unit shifts to defrost, a large surge of liquid returns to the compressor. This often causes ***liquid floodback*** or slugging.

Suction line accumulators are used by many manufacturers. These compensate for the liquid line floodback and make the refrigerant return more even. See **Figure 24-27.** The accumulator acts as a receiver for the excess liquid. The refrigerant is returned to the compressor in a vapor form. The vapor is just above its condensing temperature at the low suction pressure.

24.4.7 Supplemental Resistance Heaters

It is standard practice to size heat pumps to handle the cooling load. This size may or may not be enough to handle heating needs.

In southern climates, a unit big enough for cooling can probably supply enough heat with little assistance. However, in very cold weather or in colder areas, additional heat must be supplied from another source. The supplementary electric resistance heating unit is one such source. It requires little space. Blowers distribute the heated air evenly throughout a room. Such units may be installed in the blower-coil section of the heat pump. They may also be installed in the supply air duct of an indoor system.

24.5 Heat Pumps and Solar Heating Systems

Solar heating systems (Chapter 25) and heat pumps are being linked for residential use. Several designs are manufactured.

One system works with air solar collectors. These collect heat, which can be:

- Routed directly into the residential area.
- Stored in a bin of rocks or in a thermal storage tank.

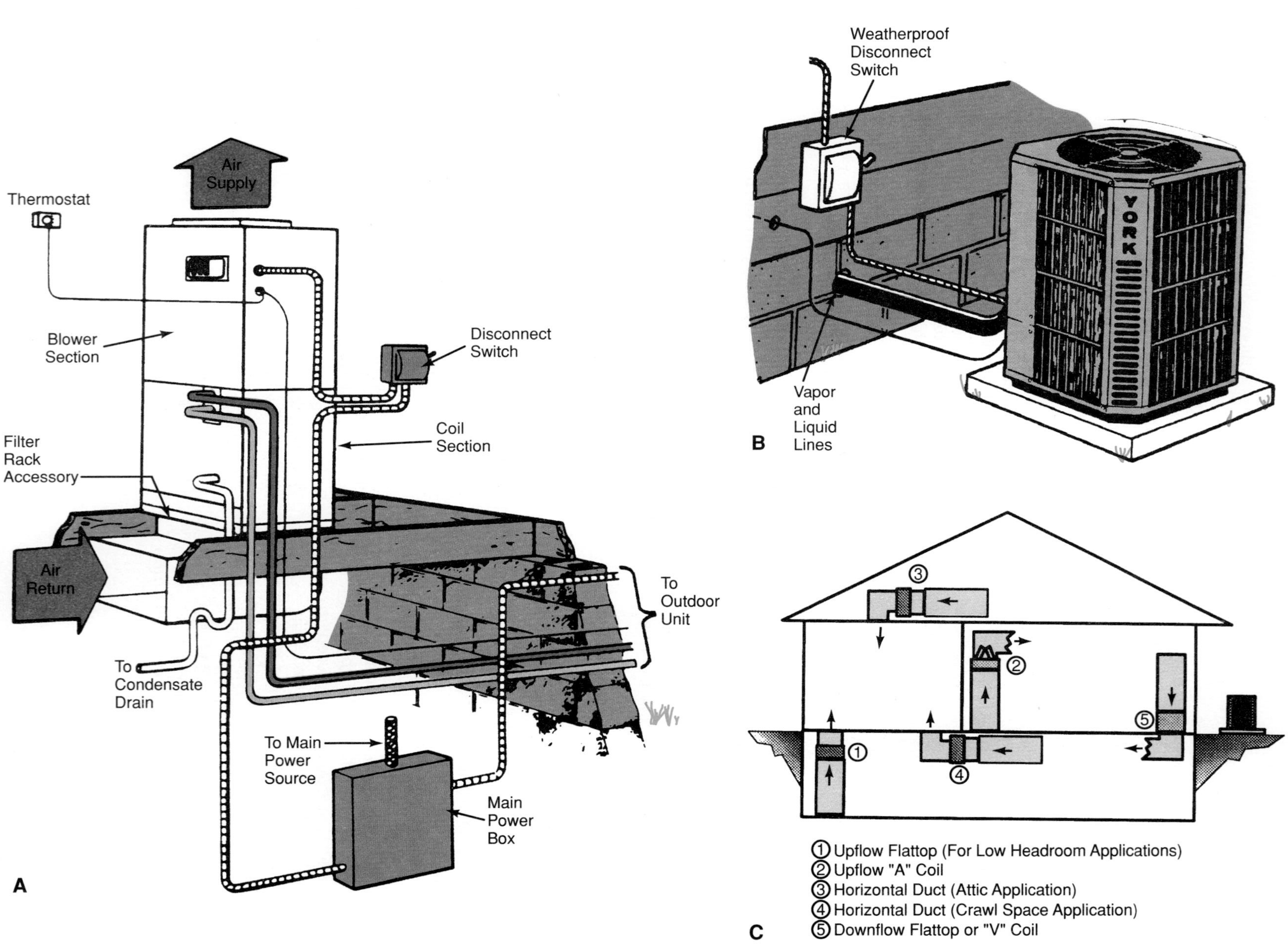

Figure 24-25. *Split system heat pump diagrams. A—Wiring and piping diagram. B—Exterior system. C—Five types of split systems. 1. Upflow flattop (for low-headroom applications). 2. Upflow "A" coil. 3. Horizontal duct (attic application). 4. Horizontal duct (crawl space application). 5. Downflow flattop or "V" coil. (York International Corp., Unitary Products Group)*

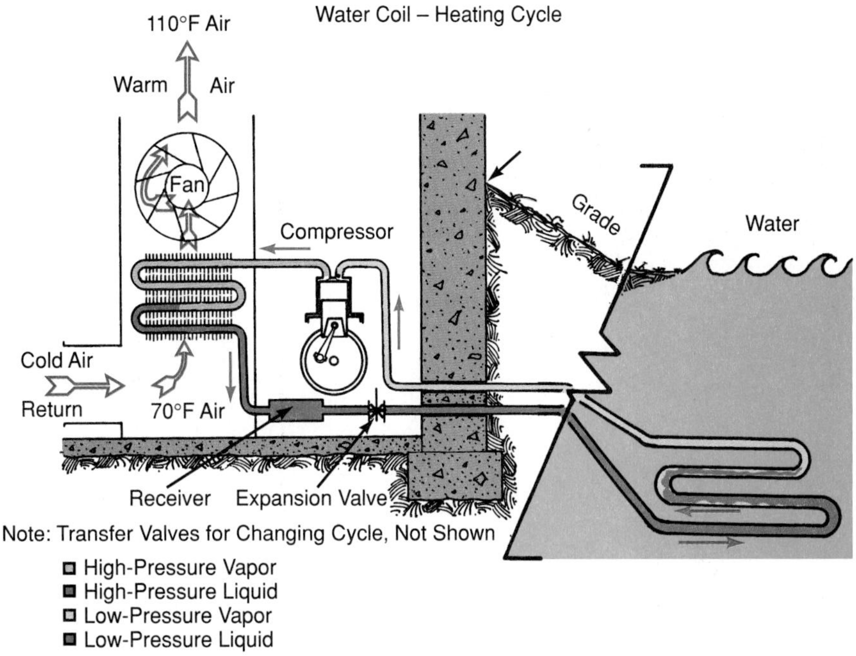

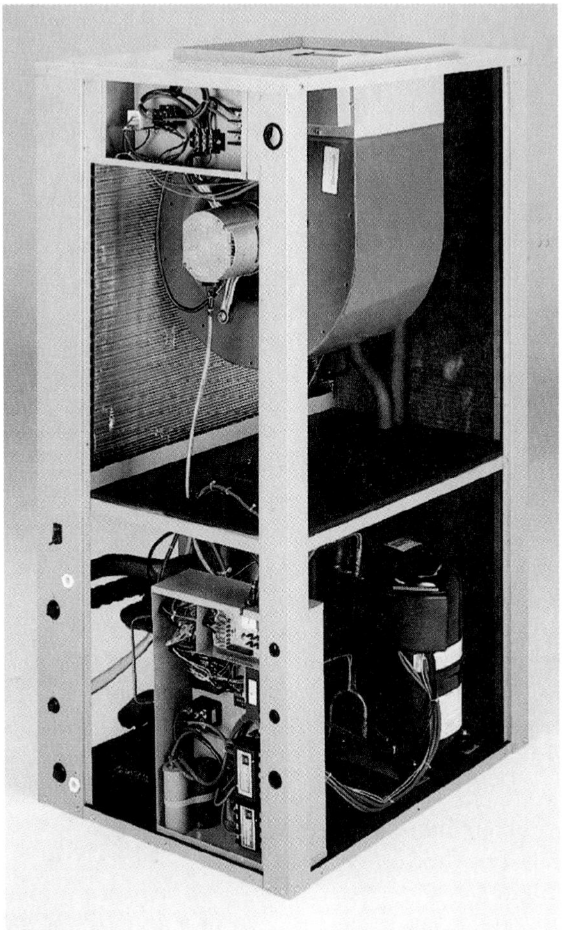

Figure 24-26. *Geothermal heat pump. A—Principle of heat pump using a water coil during heating cycle. B—Cutaway view of a geothermal heat pump that can be used with either a water coil or an earth coil. (Florida Heat Pump Division, a Harrow Company)*

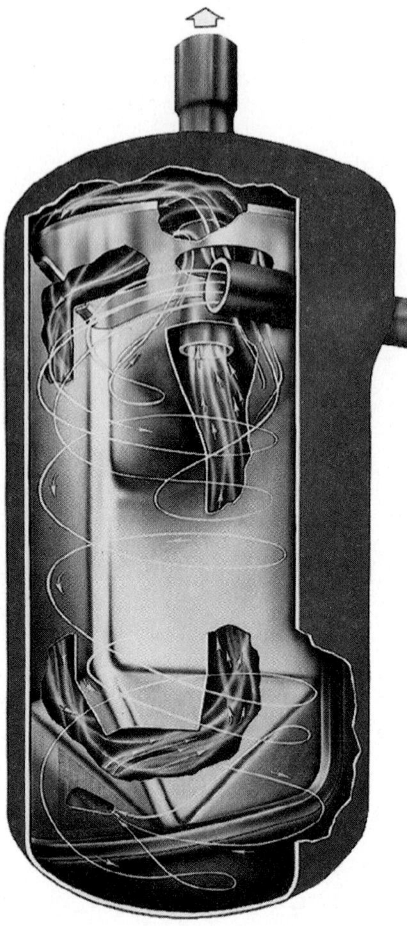

Figure 24-27. *Suction line accumulator. Note that inlet is near top of accumulator and suction line vapor is made to swirl as it leaves accumulator at top. This prevents liquid refrigerant from entering compressor.*

Air in some units gives up heat to a liquid storage area. When there is no direct solar heat, the storage area will supply heat for the living area. See **Figure 24-28.** When the storage temperature drops below 100°F (38°C), the heat pump will operate. It will produce the required temperature.

An air-to-water heat exchanger located at the output of the air collectors can provide domestic hot water heat. See **Figure 24-29.** In summer, the heat pump will operate to provide the desired cooling with a separate outside coil.

24.5.1 Combining Liquid Solar Collectors with Heat Pumps

Some solar heat and heat pump combinations use a fluid to transfer the collected heat. The fluid is typically water with ethylene glycol added.

Typically, a heat pump will operate when the energy efficiency ratio in heat units (EER_Q) is 2.0 or better and the ambient (outdoor) temperature is 20°F (−7°C) or higher. Some heat pumps use supplemental electric resistance heaters when outdoor temperature drops below 25°F (−4°C). This increases operating costs.

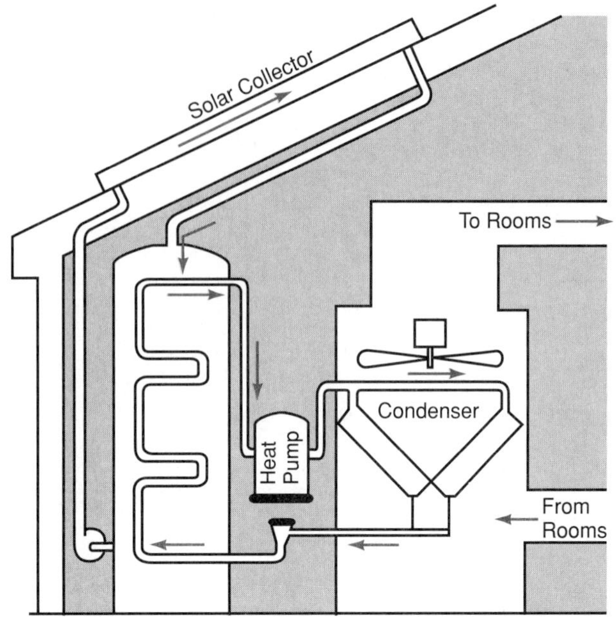

Figure 24-28. *Solar collector is used with heat pump. Solar collector system pump circulates air from solar collector on roof to space surrounding heat exchanger tubes in storage tank. Heat pump evaporator is in the storage tank. Condenser is in plenum chamber of furnace. Heat pump evaporator picks up collected heat in storage tank. Heat pump condenser gives up heat in furnace plenum chamber.*

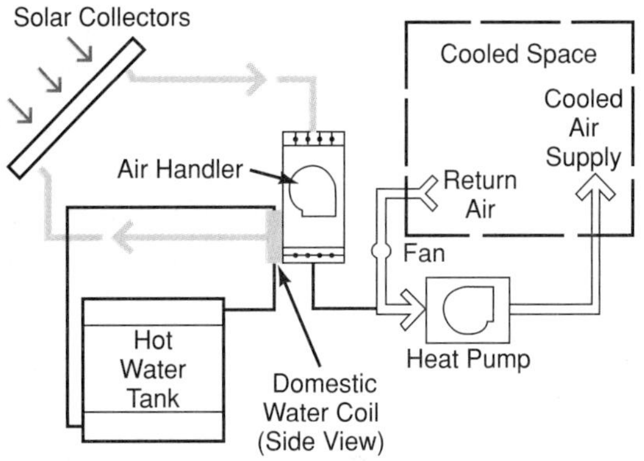

Figure 24-29. *This air solar collector and heat pump system can provide domestic hot water supply. Lower air handler air vanes to heat pump are closed. Hot air goes left to domestic water coil. In summer, cooling from heat pump is separate from solar circulation system. In winter, when heat pump heats room on right, heat pump gets some heat from air which has gone through solar collector.*

The heat pump operates when the temperatures are at or above the ***balance point.*** This is the point at which heat pump heating capacity equals the heat loss of the home.

A liquid solar collector/heat pump combination is shown in **Figure 24-30.** It is designed to provide heating and cooling at minimum operating costs. When the

EER_Q is lower than 2.0 and the ambient temperature is less than 20°F (−7°C), the solar panels will provide the needed heat. The EER_Q is 2.0 at 20°F (−7°C). When the ambient temperature is 35°F (2°C), the $EER_Q = 2.5$. The cost of heating is then 40% of the cost of electric resistance heating.

Solar panels are sized to meet only the supplemental heating needs of the heat pump. The heat from the solar panels is stored in a water storage tank. The stored heat is used when needed. An electric immersion heater provides supplemental heat if needed. The heat pump will draw heat from the solar panels until the outdoor temperature drops to the point at which supplementary heat is required. Then the supplemental heat will be provided by the solar energy that has been collected and stored.

The system is a standard heat pump unit installed in a residential home. The solar panel collector is installed outside, and a hydronic coil is placed in the supply air duct. The hydronic coil receives its heat from the thermal energy storage tank which holds the heated water.

The solar-supplemented heat pump system combines five basic systems:

- A standard heat pump system that will provide cooling when required.
- A standard heat pump system that will produce heat and operate whenever the EER_Q is above 1.0. (The EER_Q is never less than 1.0. For the unit shown in **Figure 24-30,** one assumes that the $EER_Q = 1.0$ for any temperature at 0°F [−18°C] or lower.)
- A solar circuit. This unit consists of solar panels containing a solution of ethylene glycol and water. The solution passes through the shell side of a tube-and-shell heat exchanger. The heat collected by the solar panels is not stored in the heat exchanger.
- A storage circuit. A thermal energy storage tank uses a storage pump to circulate water. The water circulates through the tube side of the heat exchanger. Here the water is heated and returned to the storage tank.
- A hydronic circuit. A hydronic pump circulates the heated water from the storage tank. It is circulated to the hydronic coil in the supply duct.

The EER_Q of a heat pump can be large, although not as large as the theoretical value. The theoretical limit is given in the following formula:

$$EER_Q = \frac{T_H}{T_H - T_C}$$

T_H is the temperature of the condenser and T_C is the evaporator temperature. All temperatures are in degrees Kelvin. As an example, use a hot-side temperature of 72°F (295.2 K). Use a cold-side temperature of 20°F (266.3 K):

$$EER_Q = \frac{295.2\ K}{295.2\ K - 266.3\ K} = 10.2$$

With an EER_Q of 10, the cost of heating with this system is 1/10 of that for electric resistance heating.

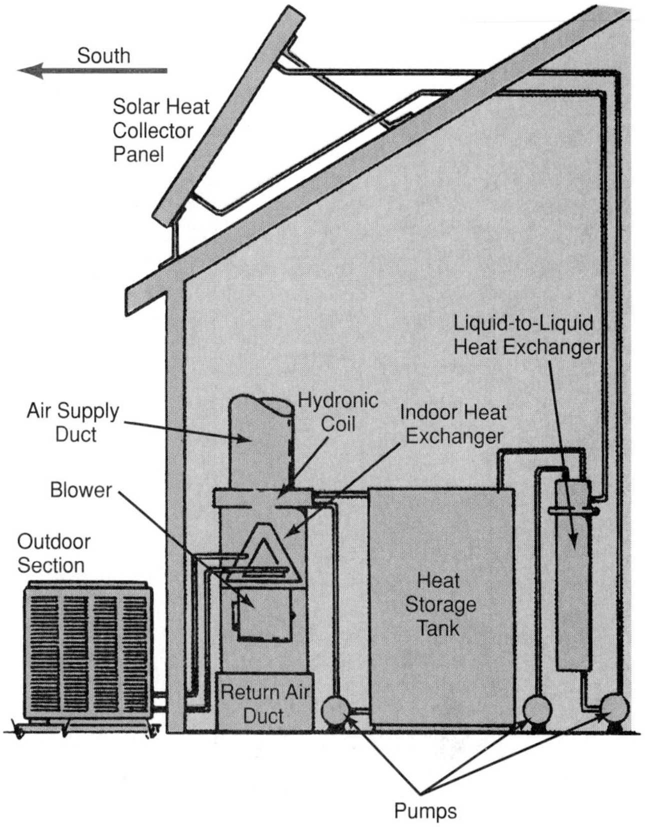

Figure 24-30. *Liquid solar collector/heat pump combination provides both summer cooling and winter heating.*

The seasonal energy efficiency ratio for the heat pump system can also be determined. As previously mentioned, this is the average efficiency of the system on a yearly basis. For more information on how to determine the SEER, see Chapter 31.

24.6 Heat Pump Water Heaters

Heat pumps are replacing water heaters in both domestic and industrial installations. The pump removes heat from the interior of a building. The heat is delivered to a hot water tank. It uses about half the energy of an electric water heater.

The condenser is placed around the outside of the water tank. The evaporator is placed on top of the tank. See **Figure 24-31.** The heat removed from the building must, of course, be replaced by the furnace during the heating season.

Some heat pump water heaters include a heat recovery unit for conditioning ventilation air to a house. In winter, cooled room air leaving the evaporator preheats the colder makeup air required for ventilation.

In summer, if outside air is warmer than room air, the outside air supplies heat to the evaporator. The air cooled by the evaporator may be used to help cool the house.

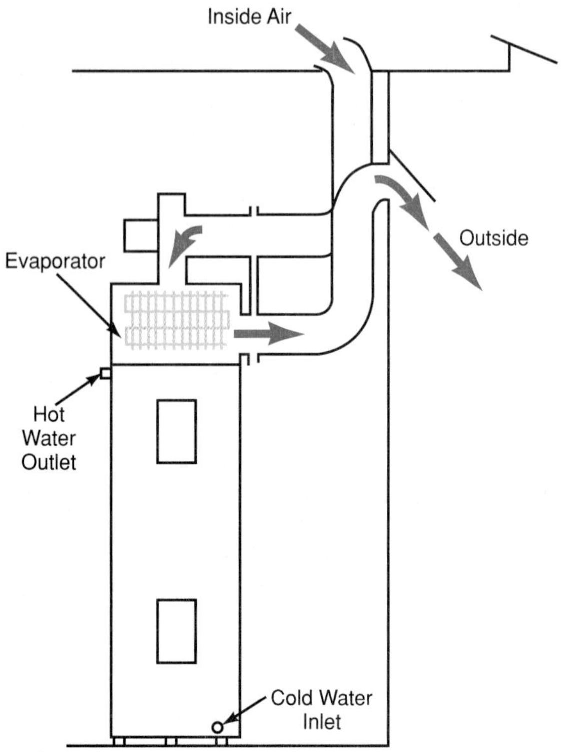

Figure 24-31. *Heat pump water heater has evaporator on top of water tank. Airflow shown is for winter operation. Heat is provided by heat pump condenser inside water tank. (DEC International, Inc.)*

In summer, inside air must leave the house at the same rate as outside air is brought in. The air inside the house can be exhausted directly to the outside. A vent with a natural outward flow can be used. This will save more indoor air than a fan would save. A fan might create too much negative pressure.

Sometimes dehumidification of basement air is required. The evaporator can be used to condense water out of the basement air.

24.7 Natural Gas Heat Pump

Recent technology has introduced heat pump systems powered by engines that run on natural gas, rather than electric motors. Such units provide a lower cost means of operation. See **Figure 24-32A.**

The system shown is powered by a single-cylinder, four-cycle engine that operates on natural gas. The energy generated by the natural gas engine runs the compressor. An innovative concept is the recovery of engine exhaust heat to obtain additional heat in the cool season. The service analyzer, **Figure 24-32B,** can be attached directly to the system by you. It will help to determine what specific services are necessary. The analyzer is connected to the service cable on the system. When it is plugged in, it begins requesting data from the unit. The data appears on the analyzer screen. **Figure 24-33** illustrates the operational data available to a technician.

A

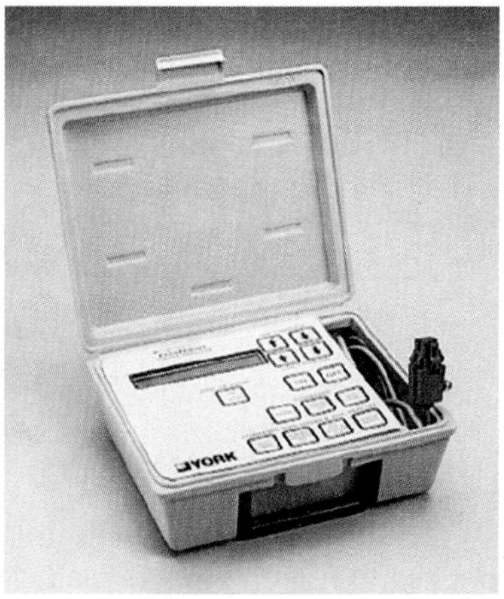

B

Figure 24-32. *A natural gas-powered heat pump system. A—This "Triathlon" system uses a single-cylinder, four-cycle engine. 1. Unit uses 110 V input power. 2. Microprocessor control module. 3. Auxiliary heater. 4. Heavy duty pump. 5. Engine control. 6. Oil sump. 7. Spring mounts used to reduce noise and vibration. 8. Recuperator for reclaiming heat from engine exhaust. 9. Engine. 10. Compressor. B—Service analyzer used with system. (York International Corporation, Unitary Products Group)*

Figure 24-34 illustrates what appears on the analyzer when the fault code data is called up. This indicates one of the possible problems that can be occurring with this equipment. This type of system is used to help obtain the correct solution to a problem. However, it does not give a total diagnosis of the problem.

OPERATIONAL DATA

Outdoor Temperature
Liquid Line Temperature
Engine Coolant Temperature
ID Supply Air Temperature
Commanded Engine Speed
Actual Engine Speed
Control Voltage
System Operating State
Defrost Interval
Time to Next Defrost
Failure Code
SYSTEM Operation Data
Fault Code Data

Figure 24-33. *Operational data that appears on the screen of the service analyzer when the technician presses the proper keys on the unit's keypad. (York International Corporation, Unitary Products Group)*

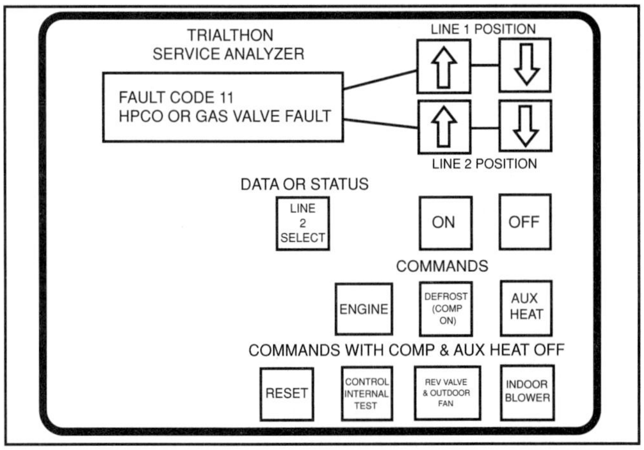

Figure 24-34. *An example of fault code data displayed on the analyzer screen. In this case, the analyzer indicates that the problem might be in the HPCO (high-pressure cutoff) or in the gas valve. (York International Corporation, Unitary Products Group)*

24.8 Heat Pump Installation

A heat pump is nearly the same as a refrigerating system. Most of the installation instructions in Chapters 12 and 15 can be applied to heat pumps as well. The units must be carefully leveled. Electrical service must be of the correct voltage and phase. Wiring must be large enough to carry the load without a critical voltage drop.

In cold climates, the heat pump should be installed above the predictable total snow height for that area. This would be approximately 8″ to 18″ (20 cm to 46 cm) above ground in most localities. In extremely cold climates, a defrost timer should be installed instead of a standard outdoor thermostat. Always advise the owner to shovel the snow away from the condensing unit.

The temperature at the room heating register is usually set at 105°F (41°C). This is much lower than the register temperature (125°F [52°C]) used with oil or gas heating systems. Heat pumps will, therefore, take longer to heat up a cold room than a furnace system would. This causes many first-call customer complaints about new heat pump installations.

Window and wall units require the same casing mounting techniques described in Chapter 22. It is important to closely follow the manufacturer's instructions.

Figure 24-35 pictures a heat pump unit which can be installed in an outer wall of an office or apartment. This type extracts "natural heat" from the outside in the winter for heating the space. The unit cools the space in the summer by absorbing heat from the room. The heat is then discharged to the outside atmosphere.

Through-the-wall heat pumps are easy to maintain. The heat pump chassis may be removed for servicing. The individual heat pump unit is well-adapted to apartment units. The owner of a multiple unit does not have a central air conditioning system. The thermostat for each unit can be set to give the heating or cooling desired. This setting will not affect others living in the building.

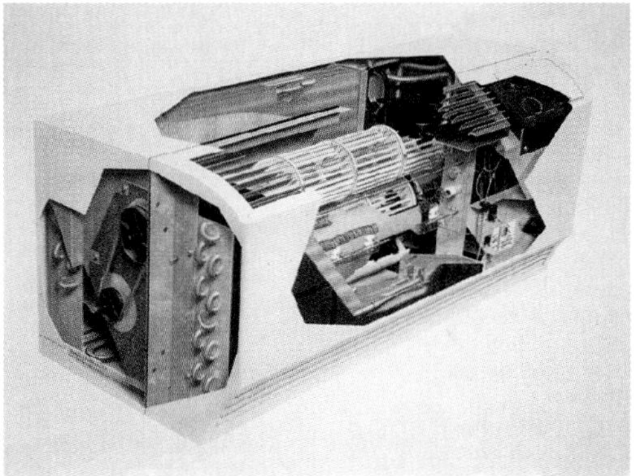

Figure 24-35. *Package terminal air conditioning and heat pump unit that can be installed through wall. Supplementary electrical resistance heating elements maintain desired amount of heating during unusually cold weather. System has a rotary compressor and blower wheels.*

24.9 Troubleshooting Heat Pumps

Figure 24-36 is a troubleshooting chart that is used by Amana Refrigeration, Inc., on its heat pumps. You must be careful that a troubleshooting chart from the manufacturer is specific for the unit. Any additional components must also be taken into consideration. The chart shown is a service problem analysis guide for the specific unit.

SERVICE PROBLEM ANALYSIS GUIDE FOR HEAT PUMP

POSSIBLE CAUSE / DOTS AND STARS IN ANALYSIS GUIDE INDICATE "POSSIBLE CAUSE"	COMPLAINT: NO HEATING OR COOLING												NO COOLING			NO HEATING					TEST METHOD/ REMEDY
Symptom	SYSTEM WILL NOT START	COMPRESSOR WILL NOT START - FANS RUN	COMPRESSOR AND OUTDOOR FAN WILL NOT START	INDOOR FAN WILL NOT START	OUTDOOR FAN WILL NOT START	COMPRESSOR HUMS - WILL NOT START	COMPRESSOR CYCLES ON OVERLOAD	COMPRESSOR IS NOISY	HIGH HEAD PRESSURE	LOW HEAD PRESSURE	HIGH SUCTION PRESSURE	LOW SUCTION PRESSURE	SYSTEM RUNS CONTINUOUSLY - LITTLE COOLING	TOO COOL AND THEN TOO WARM	NOT COOL ENOUGH ON WARM DAYS	COMPRESSOR RUNS CONTINUOUSLY - LITTLE HEAT	SYSTEM RUNS - BLOWS COLD AIR	UNIT DEFROSTS - NO FROST ON COIL	UNIT WILL NOT TERMINATE DEFROST	UNIT WILL NOT DEFROST	
POWER FAILURE	●																				TEST VOLTAGE
BLOWN FUSE	●																				INSPECT FUSE TYPE & SIZE
LOOSE CONNECTIONS	●					●															INSPECT CONNECTIONS - TIGHTEN
SHORTED OR BROKEN WIRES	●	●	●	●	●	●															TEST CIRCUITS WITH OHMMETER
OPEN OVERLOAD		●																			TEST CONTINUITY OF OVERLOAD
FAULTY THERMOSTAT	●		●	●										●							TEST CONTINUITY OF THERMOSTAT AND WIRING
FAULTY TRANSFORMER	●																				CHECK CONTROL CIRCUIT WITH VOLTMETER
SHORTED OR OPEN CAPACITOR		●		●	●	●															TEST CAPACITOR
SHORTED OR GROUNDED COMPRESSOR		●				●															TEST MOTOR WINDINGS
FAULTY COMPRESSOR CONTACTOR		●	●			●	●														TEST CONTINUITY OF COIL AND CONTACTS
FAULTY FAN RELAY				●																	TEST CONTINUITY OF COIL AND CONTACTS
LOW VOLTAGE						●	●														TEST VOLTAGE
SHORTED OR GROUNDED FAN MOTOR				●	●																TEST MOTOR WINDINGS
UNBALANCED POWER – 30																					TEST VOLTAGE
HIGH-PRESSURE CONTROL OPEN																					RESET AND TEST CONTROL
FAULTY REVERSING RELAY			●														●				TEST CONTINUITY OF COIL AND CONTACTS
FAULTY DEFROST RELAY					●														●	●	TEST CONTINUITY OT COIL AND CONTACTS
FAULTY REVERSING VALVE											●		●			●	●		●	●	TEST OPERATION OF VALVE
COMPRESSOR STUCK		●				●															USE TEST CORD
IMPROPER COOLING ANTICIPATOR						●	●							●							CHECK RESISTANCE OF ANTICIPATOR
WRONG TYPE I.D. COIL EXPANSION VALVE																					REPLACE VALVE
EXPANSION VALVE RESTRICTED																					REPLACE VALVE
UNDERSIZED EXPANSION VALVE																					REPLACE VALVE
OVERSIZED EXPANSION VALVE																					REPLACE VALVE
INOPERATIVE EXPANSION VALVE																					CHECK VALVE OPERATION
EXPANSION VALVE BULB LOOSE																					TIGHTEN BULB
SHORTAGE OF REFRIGERANT							●			●		●	●			●			●		TEST FOR LEAKS - ADD REFRIGERANT
RESTRICTED LIQUID LINE							●					●	●		●	●					REPLACE RESTRICTED PART
DIRTY AIR FILTER									●			●	●		●	●					CLEAN OR REPLACE
DIRTY I.D. COIL									●			●	●		●	●					CLEAN COIL
NOT ENOUGH AIR ACROSS I.D. COIL									●			●	●		●	●					SPEED BLOWER, CHECK DUCT STATIC PRESSURE
OVERCHARGE OF REFRIGERANT								●	●		●										RELEASE PART OF CHARGE
DIRTY OUTDOOR COIL									●			●			●			●			CLEAN COIL
NONCONDENSIBLES									●						●						REMOVE CHARGE, EVACUATE, RECHARGE
RECIRCULATION OF CONDENSING AIR									●						●						REMOVE OBSTRUCTION TO AIRFLOW
LOOSE HOLD-DOWN BOLTS								●													TIGHTEN BOLTS
BROKEN INTERNAL PARTS								●													REPLACE COMPRESSOR
INEFFICIENT COMPRESSOR										●	●		●			●					TEST COMPRESSOR EFFICIENCY
LEAKING CHECK VALVE								●			●										REPLACE VALVE
IMPROPERLY LOCATED THERMOSTAT														●							RELOCATE THERMOSTAT
AIRFLOW UNBALANCED														●							REPOSITION AIR VOLUME DAMPERS
UNDERSIZED LIQUID LINE																					REPLACE LINE
UNDERSIZED SUCTION LINE																					REPLACE LINE
FROSTED O.D. COIL																●					CHECK DEFROST CONTROLS
FAULTY DEFROST TIMER																		●		●	TEST TIMER OPERATION
FAULTY 30°/60° CONTROL																		●	●	●	TEST CONTROL
O.D. TEMPERATURE BELOW 15 °F.																					
ADDITIONAL AUX. ELEC. HEAT REQ.																●					REFIGURE HEATING LOAD - ADD HEATERS
BLOWN FUSES IN AUX. HEATER CIRCUIT																●					INSPECT FUSES - CHANGE
O.D. THERMOSTAT SETTING TOO LOW																●					RAISE SETTING

Figure 24-36. *Service analysis guide for heat pumps. An example of how this can be used: Step 1. The complaint is "too cool and then too warm." Step 2. Possible causes are: faulty thermostat; improper cooling anticipator; improperly located thermostat; airflow unbalanced. Step 3. The possible remedies are: test continuity of thermostat and wiring; check resistance of capacitor; relocate thermostat; reposition air volume dampers. (Amana Refrigeration, Inc.)*

The center columns show complaints or problems that might be indicated by the owner. (These include no heating or cooling, no cooling, and no heating.) The specific symptoms are given under each of these headings. The left-hand column indicates the possible cause. The right-hand column indicates the test method and remedy to be used when this occurs.

Using the troubleshooting chart is a relatively simple process. Prior to using the chart, the customer complaint should be considered. This would include complaints such as "System will not start," etc. The chart should be carefully studied for all aspects and the appropriate starred item located. This will help determine the cause and the test method or remedy to be used. For example, following down the "System will not start" column: If the first item "Power failure" is the cause, the test method would be to test the voltage.

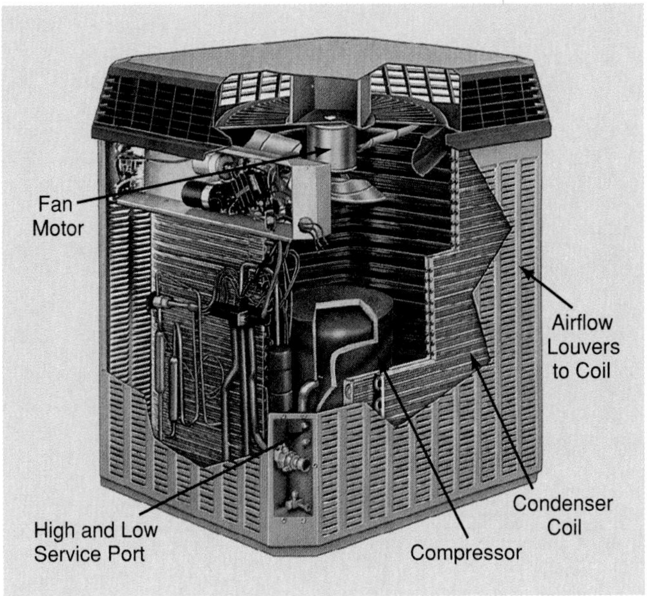

Figure 24-37. *Internal construction of a heat pump condensing unit. (The Trane Company, Unitary Products Group)*

24.10 Servicing Heat Pumps

The heat pump system is a refrigerating unit. Therefore, the service techniques of troubleshooting and repair are much the same as the service techniques described in Chapters 12 and 15.

Routine maintenance requires that voltage, current, and refrigerant pressures be checked. The heat pump should be cleaned. Blow out the coils and fins. Clean the duct passages. Oil the fan bearings and fan motor bearings. Clean the fan blades. The unit should be run on both cycles to check the operation of the reversing valve.

Service calls often start with a customer complaint. Some common complaints are:

- Unit will not operate on either cooling or heating.
- Unit runs too much.
- System is noisy.

You must find the cause of the trouble and repair it. A number of major areas, **Figure 24-37,** may be checked as part of servicing a heat pump. A conventional type of thermostat used on heat pump systems is shown in **Figure 24-38.** Always check the operation of the thermostat before any major changes are made. A leak must be located and repaired. A four-way valve that will not operate must be replaced.

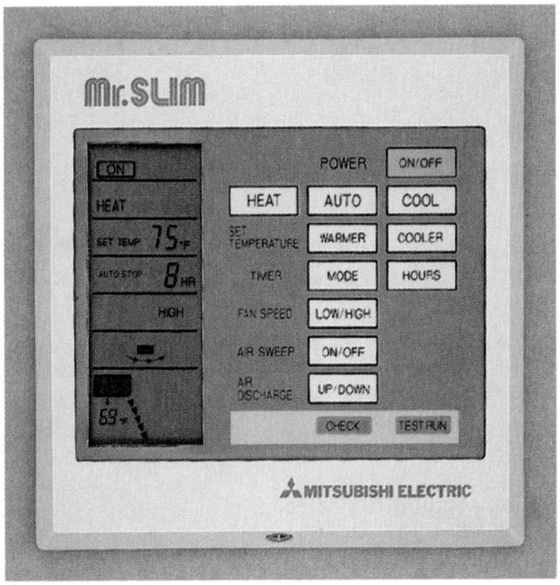

Figure 24-38. *Heat pump thermostat. Note the numerous settings. (Mitsubishi Electronics America, Inc.)*

In servicing or repairing heat pump systems, the system should be simplified, whenever appropriate. This can be done by replacing unidirectional components and check valves with bidirectional components, such as two-way filter-driers, **Figure 24-39.** By installing a bidirectional thermostatic expansion valve and filter-drier, the one-way expansion valves, check valves, and filter-driers can all be replaced with copper tubing. This updates the system. It will then be more trouble-free than in the previous arrangements.

When updating a unidirectional system, whenever possible, replace items as shown in **Figure 24-40.** Note that two existing filter-driers have been replaced with a single bidirectional thermal expansion valve and one two-way filter-drier.

In all refrigeration, air conditioning, and heating service, you must always follow proper safety procedures. **Figure 24-41** illustrates a technician wearing gloves and safety glasses while charging a heat pump system. Certain checks should be routinely made:

- Check defrost control.
- Check and lubricate condenser fan motors.
- Check refrigerant charge.
- Check all electrical connections for tightness.
- Check room thermostat.

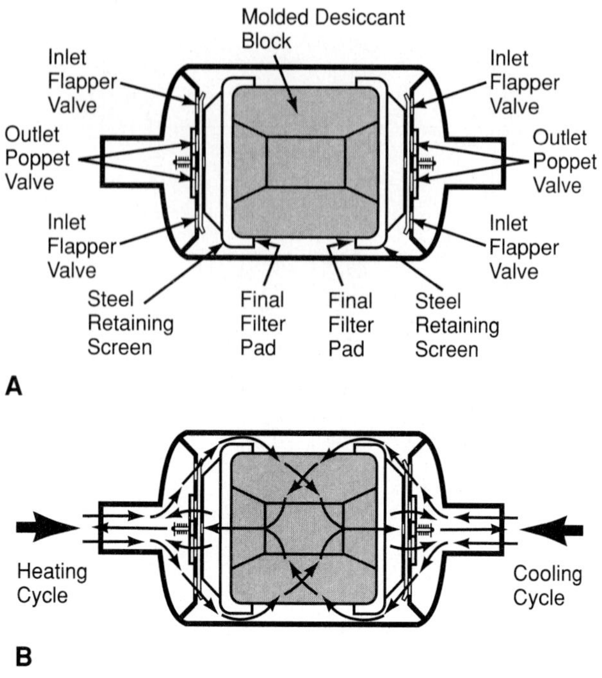

Figure 24-39. *Two-way filter-drier. A—Internal construction. B—Refrigerant flow for heating and cooling cycles. (Alco Controls Division, Emerson Electric Co.)*

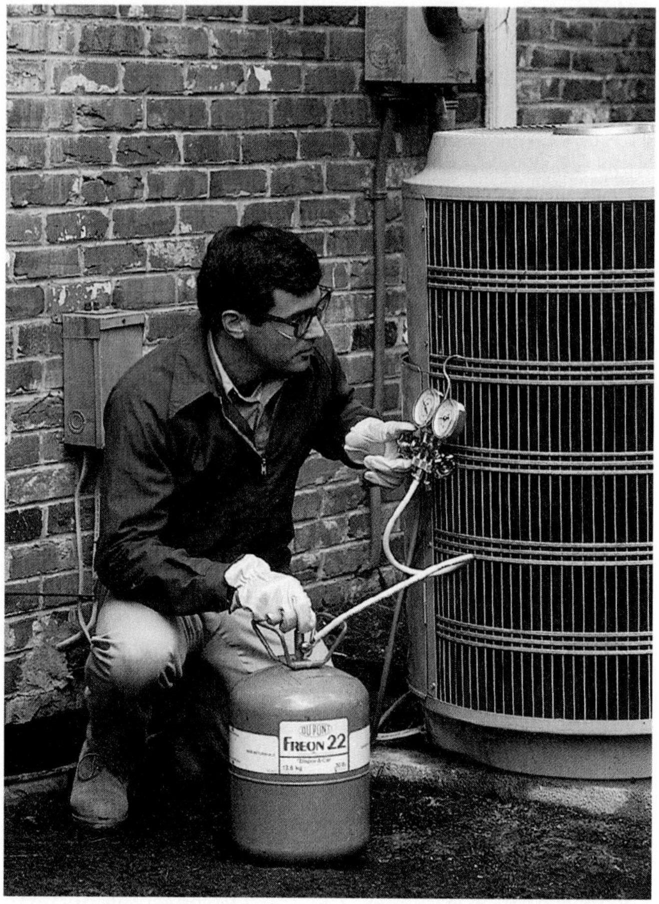

Figure 24-41. *Service technician charging a heat pump. (DuPont Company)*

24.11 Central Air Conditioning Systems

Central air conditioning systems are of two types—unitary and field-erected. Central ***unitary systems*** are ideal for residential air conditioning. They are a complete, manufactured package ready for assembly. All internal wiring and piping has been done. **Figure 24-42** is a unit being assembled at the manufacturer's plant. The condensing unit is located away from the evaporator. There are three evaporator designs in use:

- The A-type evaporator.
- The slant-type evaporator.
- The flat-type evaporator for horizontal airflow.

There are three methods used to install central air conditioning. They are described in the following paragraphs.

In the first method, the condensing unit, evaporator, controls, and tubing are purchased. The conditioner is then assembled on the customer's premises. Most of these systems are installed in forced-air heating systems. **Figure 24-43** is a drawing of such an installation.

These units vary in capacity from 1 1/2 hp (approximately 12,000 Btu/hr. or 12,650 kJ/hr.) to 7 1/2 hp

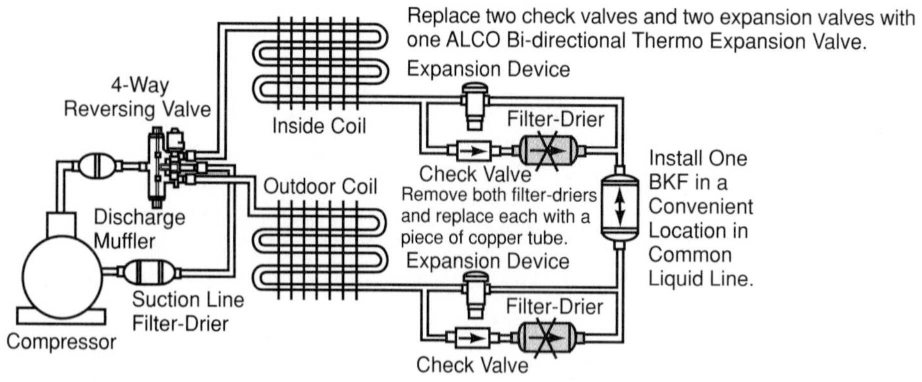

Figure 24-40. *Replacing components with bidirectional units can simplify a system. (Alco Controls Division, Emerson Electric Co.)*

Figure 24-42. *Heat pump being assembled. This worker is checking a drawing as she installs a capacitor. (Bryant Air Conditioning/Heating)*

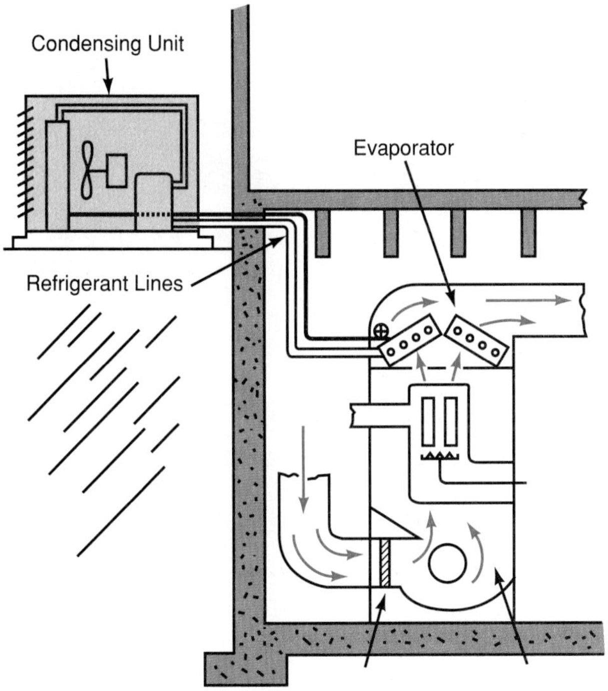

Figure 24-43. *Drawing shows residential central comfort-cooling system installed in forced-air furnace. Condensing unit is mounted outdoors.*

(60,000 Btu/hr. or 63,260 kJ/hr.). Assume the coefficient of performance (COP) is 3.14.

$$COP = \frac{T_C}{T_H - T_C}$$

An oil-fired furnace is shown in **Figure 24-44.** It is complete with a comfort-cooling A-type evaporator, electronic air filter, humidifier, and blower system.

In the second method, the unit is ordered completely assembled and charged. It will include condensing unit, evaporator, controls, and tubing. The complete assembly

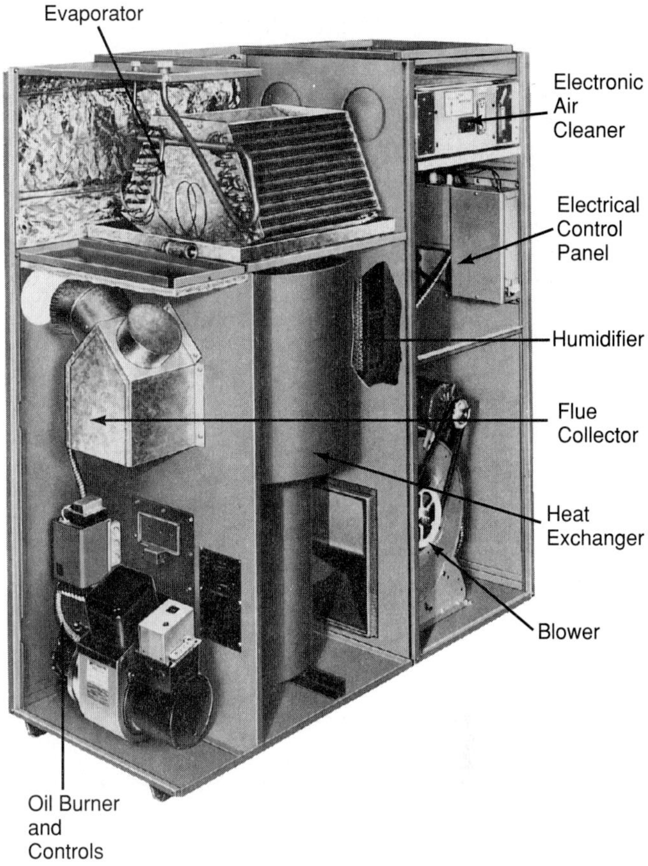

Figure 24-44. *All-season air conditioner has oil burner, evaporator, electronic air cleaner, and humidifier. (Williamson Corporation)*

is shipped as a single package. The tubing needed is usually carefully wrapped around the evaporator.

Work required at the point of installation includes:

1. Uncrating and carefully unwinding the tubing from the evaporator.
2. Installing the condensing unit and evaporator in their proper places.
3. Making the electrical and control connections needed.

This type of system requires very careful handling, since the condensing unit is charged with refrigerant. You must be very careful while uncrating the tubing to avoid kinks. The kinked areas might later crack and leak.

Figure 24-45 shows an A-type evaporator. A condensing unit is shown in **Figure 24-46. Figure 24-47** shows a completely assembled system.

The third method involves getting a completely charged evaporator, condensing unit, and lines. However, the condensing unit, evaporator, and connecting tubing are shipped as separate items. See **Figure 24-48.** The line set uses flare fittings at the evaporator coil. Compression or sweat fittings are used on the condensing unit. Other manufacturers connect the parts with "quick-couplers." These couplers may be connected without losing refrigerant or getting air into the system. The complete system is shown in **Figure 24-49.**

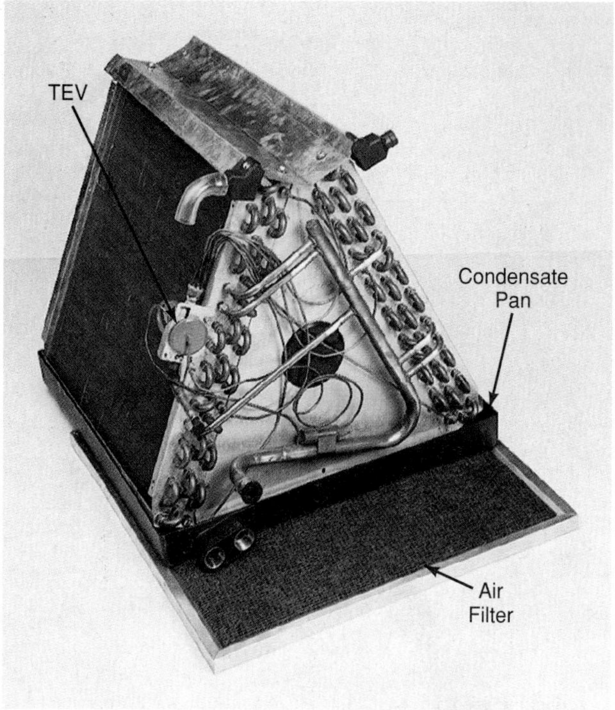

Figure 24-45. *An A-shaped evaporator designed for installation in warm air furnace plenum chamber. (Bryant Air Conditioning/Heating)*

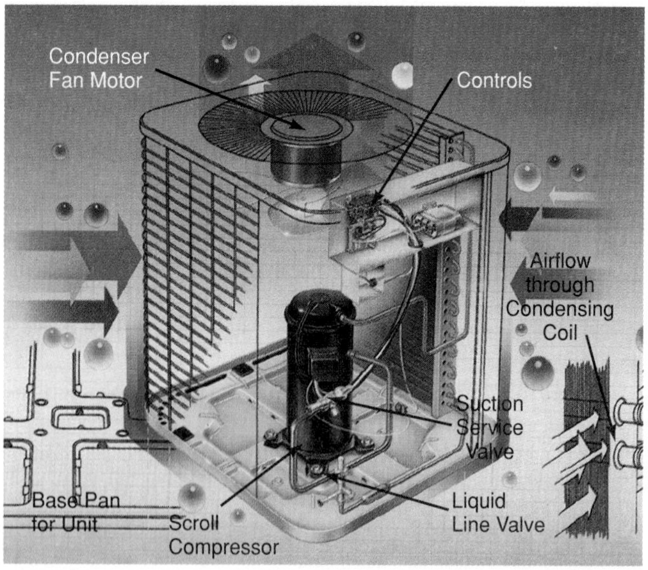

Figure 24-46. *Central air conditioning unit with horizontal inlet air and vertical air discharge. (Bryant Air Conditioning/Heating)*

Some systems use the liquid line as the capillary tube. This larger bore (ID) tubing reduces the chance of clogging from dirt or moisture. *To ensure proper operation, it is very important not to shorten or lengthen this liquid line capillary tube combination during installation.*

Many systems use evaporators with aluminum fins mechanically bonded to copper tubing. Other condensers and/or evaporators are built with aluminum spines

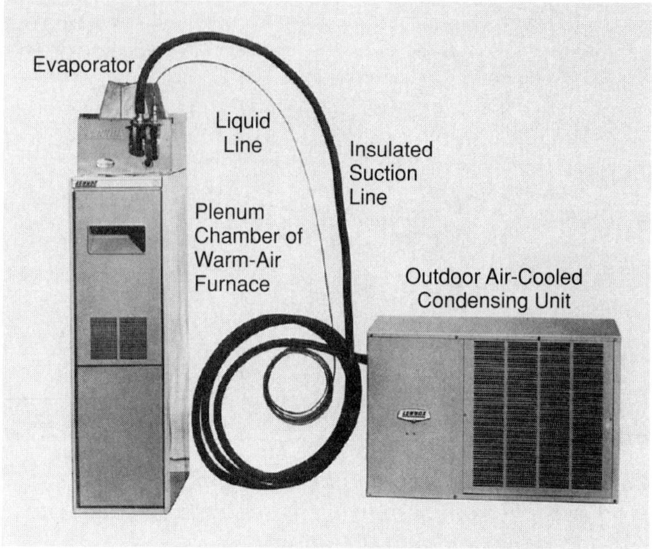

Figure 24-47. *Central comfort-cooling system for a residential installation. (Lennox International, Inc.)*

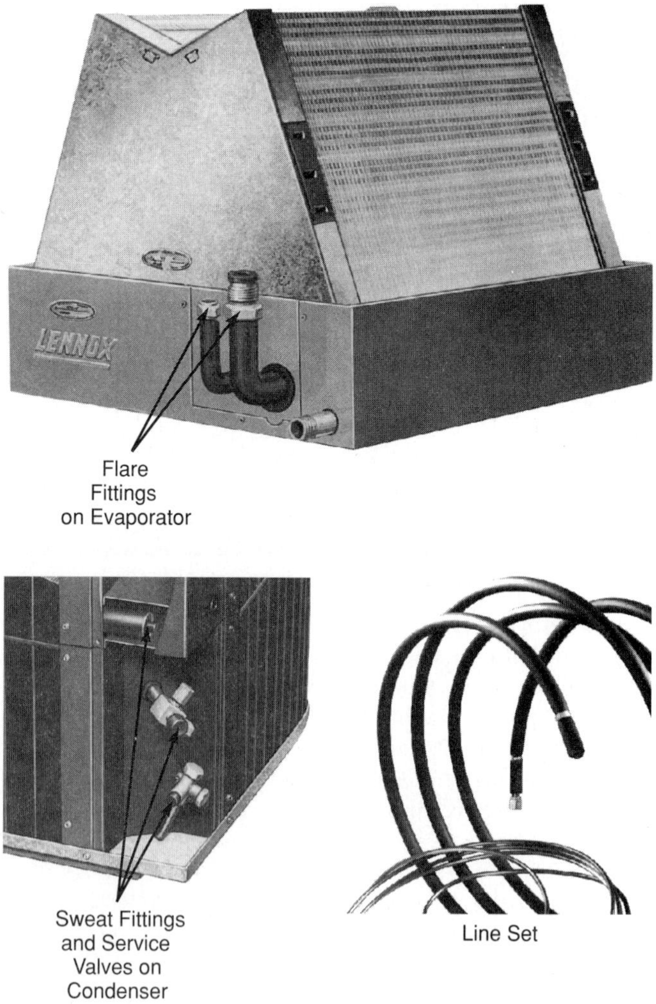

Figure 24-48. *Domestic central comfort-cooling system uses flare fittings and sweat fittings. (Lennox International, Inc.)*

Figure 24-49. *Typical installation of complete air conditioning system can have units at different levels. Evaporator is below condenser for best flow in liquid line. However, since oil may not go back to the motor compressor easily, some installers put U-bend in suction line to assist oil return. U-bend allows oil to collect, so it can be pushed through as a single "plug." (Aeroquip Corp.)*

A
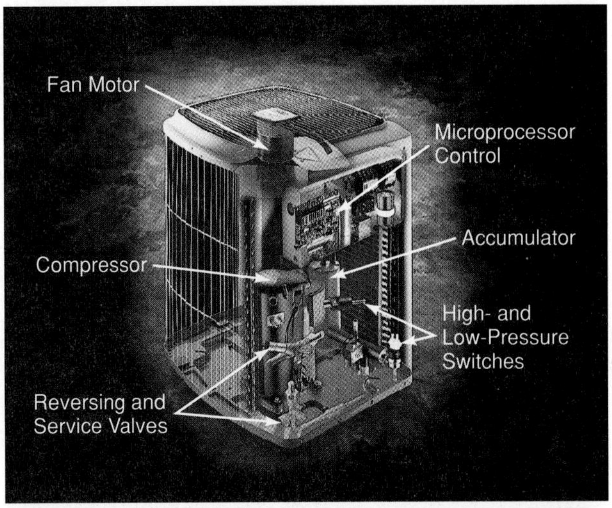

B
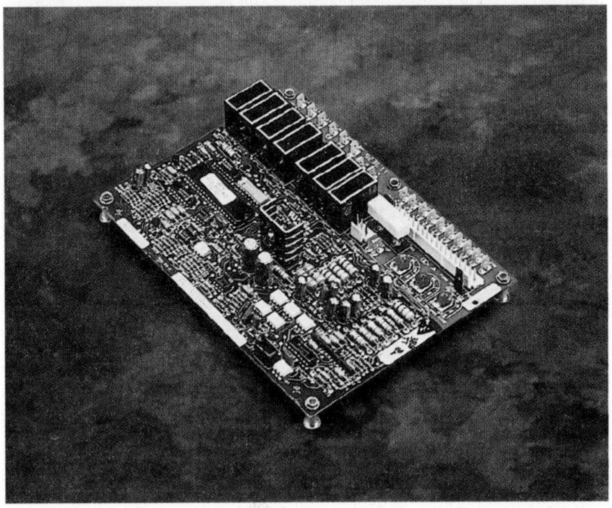

Figure 24-50. *Microprocessor-controlled heat pump. A—Cutaway view. Note the location of the reversing and the service valves. B—Microprocessor board. A technician can use a manufacturer-supplied control reader to obtain information in system performance. (Carrier Corporation, Subsidiary of United Technologies Corporation)*

fastened to aluminum tubing to reduce corrosion. Plastic grilles are often used on the condensing unit to avoid corrosion problems.

Figure 24-50A shows the internal construction of a heat pump condensing unit. The microprocessor control board **(Figure 24-50B)** controls the compressor, outdoor fan speed, and other electronic parts. The control board also provides information on the performance of the system. The unit also has an electronic demand defrost. This prevents ice from building up on the outdoor coil.

An electronic thermostat, **Figure 24-51,** monitors the indoor and outdoor units, at the same time correcting operating speed. The LCD on the thermostat face shows the approximate operating speed, indoor temperature, and outdoor temperature. It also shows whether the unit is in a heating or cooling mode.

24.11.1 Field-Erected Air Conditioning Systems

In ***field-erected air conditioning systems,*** all components are assembled and erected at the spot where the system is to be used. This includes the motor, compressor, receiver, evaporator, piping, and controls. Chapter 15 covers most of the installation procedures.

There is a variety of central field-erected systems. Some are large systems. They heat and cool a number of buildings or various parts of a large building. Other field-erected systems service one domestic building or a small commercial building.

These systems may:

- Cool and/or heat air, which is then pumped by a dual system.
- Cool and/or heat water, which is then pumped to heat exchangers in the conditioned spaces. The cabinets in the spaces have fins, filters, and controls in them. Automatic controls make it possible for a system to change from heating to cooling. Outside air may be used if it is a degree or so above the heating thermostat setting to a degree or so below the cooling thermostat setting. The system then becomes an air distribution and air-cleaning system only. This permits the greatest economy of operation.

Other systems use more fresh air as the outside temperature approaches the temperature desired inside the system. Solid state controls, operated by thermistors, make this possible.

Quick-Connect Couplings

Self-sealing couplings enable manufacturers to produce precharged refrigeration and air conditioning

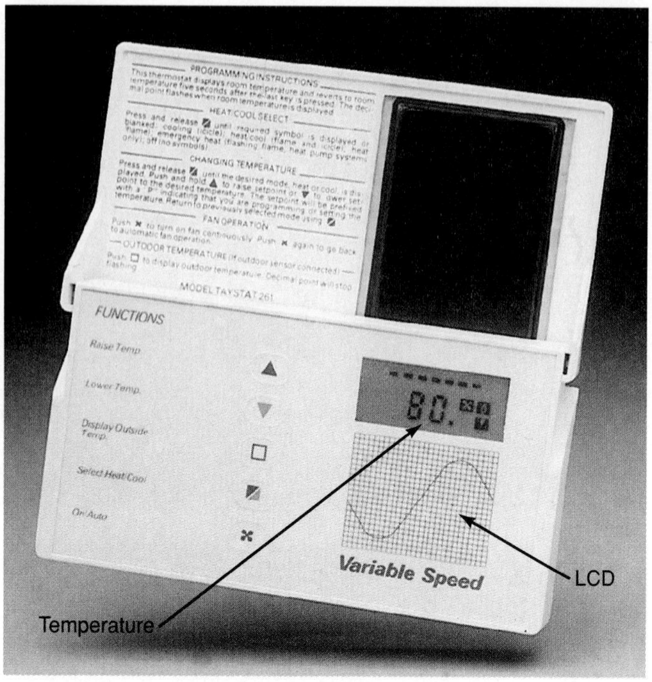

Figure 24-51. *An electronic wall thermostat. Note the location of temperature indicator and LCD, which indicates operating speed, indoor temperature, outdoor temperature, and heating or cooling mode identification. The system can be altered by adjusting the thermostat, raising the temperature, lowering the temperature, taking outside temperature readings on the thermostat, or selecting heating, cooling, continuous fan movement, or automatic fan movement.*

units. Necessary tubing is also provided in separate packages. These separate units may be assembled at the installation site. They are ready to operate without evacuating, charging, or cleaning.

The self-sealing coupling fittings are brazed directly to the tubing. Flared joints are not needed. There are two types of quick-connect fittings:

- Those which can be connected and disconnected many times with very little loss of refrigerant. This type is very seldom used.
- Those which can only be quick-connected once. This type uses diaphragms. When the fittings are attached, the diaphragms are punctured. This allows refrigerant to pass. These couplings cannot be disconnected unless the refrigerant is first removed from the system.

In the first type, when the couplings separate, independent springs force valves in both halves to close. This prevents the escape of refrigerant.

To assemble either type of quick-connect fitting, align the couplings and tighten the coupling nut. This draws the coupling together. The sealing diaphragm of the coupling is pierced internally so the refrigerant can flow. **Figure 24-52** shows three views of a quick-connect

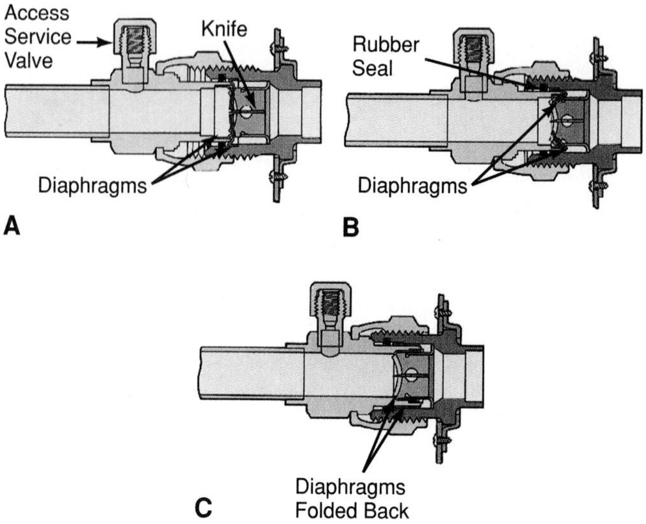

Figure 24-52. *Quick-connect and disconnect coupling with access service valve. Three views are shown: A—Knife edge aligns and begins piercing diaphragm. B—Partially assembled. C—Connected, with refrigerant passage open. (Aeroquip Corp.)*

coupling. **Figure 24-53** shows the exterior view of an assembled quick-connect as it would appear on an installation.

Quick-connect fittings are used mostly on precharged residential air conditioning systems. They are also used on precharged transportation units. Units are usually charged at the factory. The condensing unit, refrigerant lines, and evaporator are charged separately. **Figure 24-54** shows a liquid line that has an access (service) port.

The gasket that joins the quick-connect-disconnect fittings should be covered with clean, dry refrigerant oil just before assembly. Avoid excessive wrench pressure, because a distorted fitting may leak. **Figure 24-55** shows two wrenches being used to tighten the fittings. Most quick-connects and disconnects will reseal themselves several times.

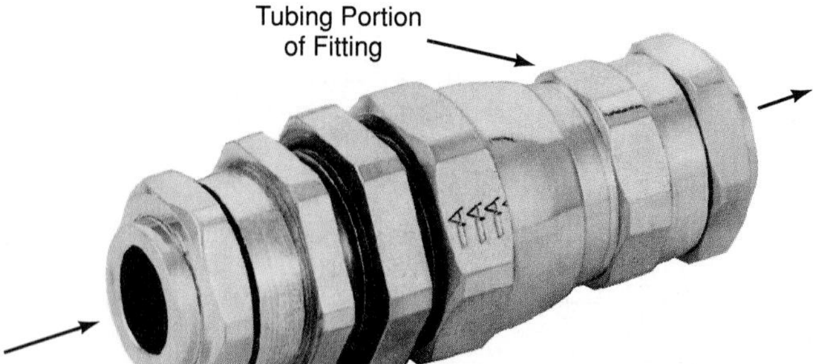

Figure 24-53. *Assembled quick-connect-and-disconnect fitting. (Aeroquip Corp.)*

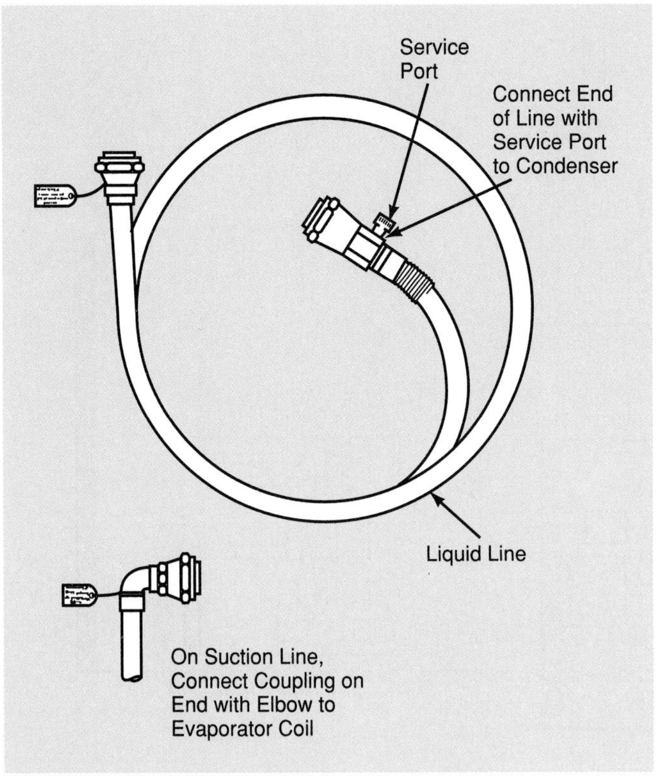

Figure 24-54. *Quick-coupler liquid line. Service port is used for making service manifold high-side pressure gauge connection. (The Coleman Co., Inc.)*

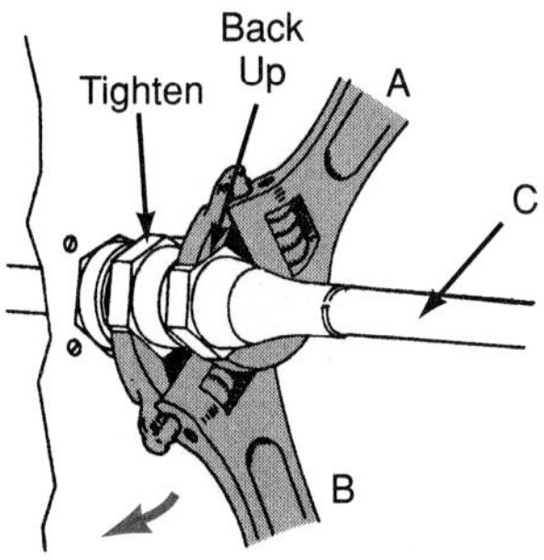

Figure 24-55. *Correct way to tighten a quick coupling. Wrench A is held firmly while wrench B is turned. Note that the sleeve closest to wall mounting turns, but sleeve on tubing C does not turn. Use of two wrenches prevents twisting of tubing.*

24.12 Installing Residential Central Air Conditioning Systems

Cooling systems should be installed only in furnaces less than 15 years old. Older furnaces will need to be replaced. Units are assembled on-site in four steps:

1. Install condensing unit.
2. Install evaporator.
3. Install suction and liquid lines.
4. Install electrical wiring.

The condensing unit uses outdoor air to cool the condenser. Some homes have water-cooled condensers.

Many installation methods are used. Some units are mounted inside the building. Ducts bring outdoor air to the condenser and discharge warm air outside. Some units are mounted on an outside wall. A more popular practice is to mount the unit on a concrete slab or prefabricated slab. This slab is located 12″ to 24″ (31 cm to 61 cm) from the building. A concrete slab at least 4″ (10 cm) thick and reinforced with steel mesh is recommended. **Figure 24-56** shows various installation methods. Outlet air from the condensing unit should move in the same direction as the prevailing summer winds.

Location of the condensing unit is very important. The suction and liquid lines, as well as power lines, should be as short as possible. The condensing unit should be carefully located:

- Away from bedrooms, patios, and neighbors (noise).
- At least 24″ (61 cm) away from wall (air circulation).
- Away from inside corners (air circulation).
- Not under eaves (airflow).
- Beyond the roof overhang (air circulation).

Figure 24-57 shows alternative condensing unit locations. The condensing unit should be mounted level. **Figure 24-58** shows a typical installation with clearance spaces needed.

The evaporator is mounted level and solidly in the furnace bonnet or plenum chamber. Design of the evaporator and its condensate drain depend on the type of furnace (upflow, downflow, or horizontal flow). Removable panels are needed for periodic cleaning and servicing as required. **Figure 24-59** shows a slant evaporator installation with blankoff plates. The plenum chamber is blanked off to make sure all the air goes through the evaporator.

The condensate drain should be piped to an open drain. There should be an air break at the drain. A plastic drain pipe should be kept away from the warmer parts of the furnace. Some technicians install a 4″ U-trap in the drain line to stop airflow through the line. The drain pan is built into the evaporator, as is the drain connection. Check the local building code for proper installation of all units. In some installations, a drain pump must be installed to remove the condensate to the outdoors. **Figure 24-60** shows an A-type evaporator being installed in a furnace plenum chamber.

Suction and liquid lines may have:

- Flared connections.
- Brazed connections.
- Quick-connect-and-disconnect couplings.

The condensing unit in **Figure 24-61** has flared or brazed tubing connections equipped with service valves. The refrigerant control is a thermostatic expansion valve. This unit would be installed as described in Chapter 15. Many condensing units are installed above the evaporator. Therefore, a U-bend should be put in the suction line to assist oil return. The suction line should slope downward slightly toward the condensing unit.

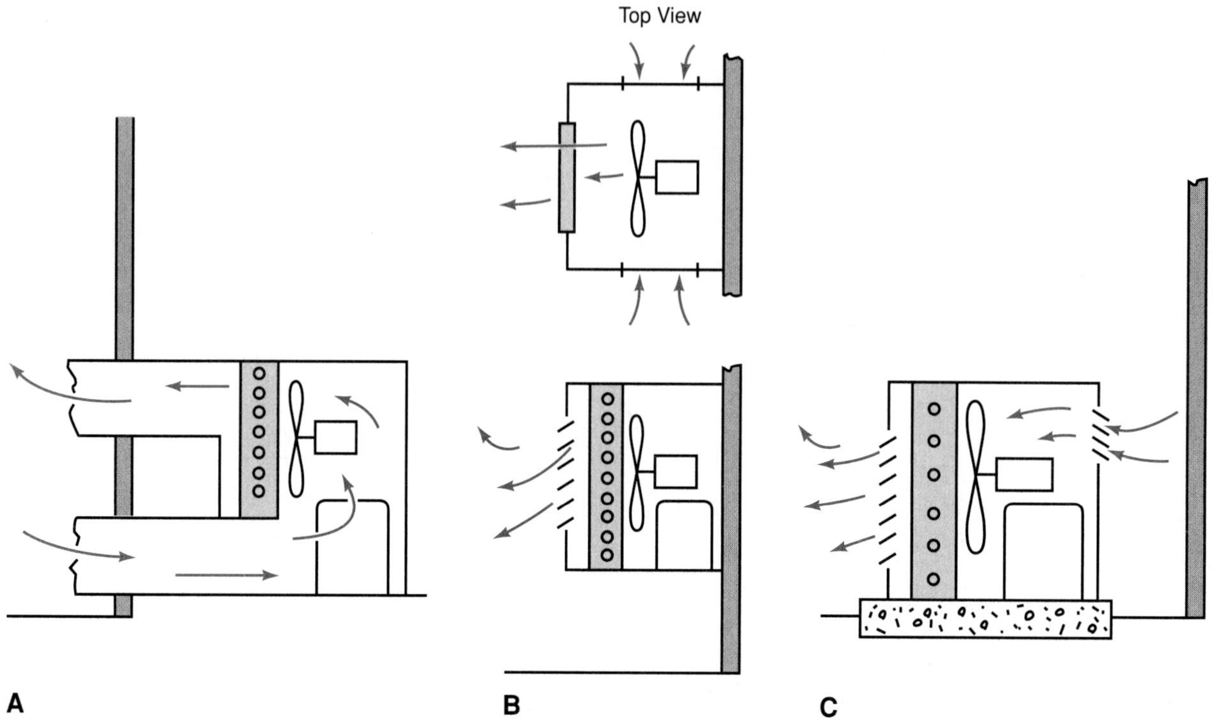

Figure 24-56. *Three types of air-cooled condensing units for residential air conditioning installations. A—Condensing unit inside building. B—Unit hung on outer wall (usually through window). C—Unit mounted on concrete slab outside building.*

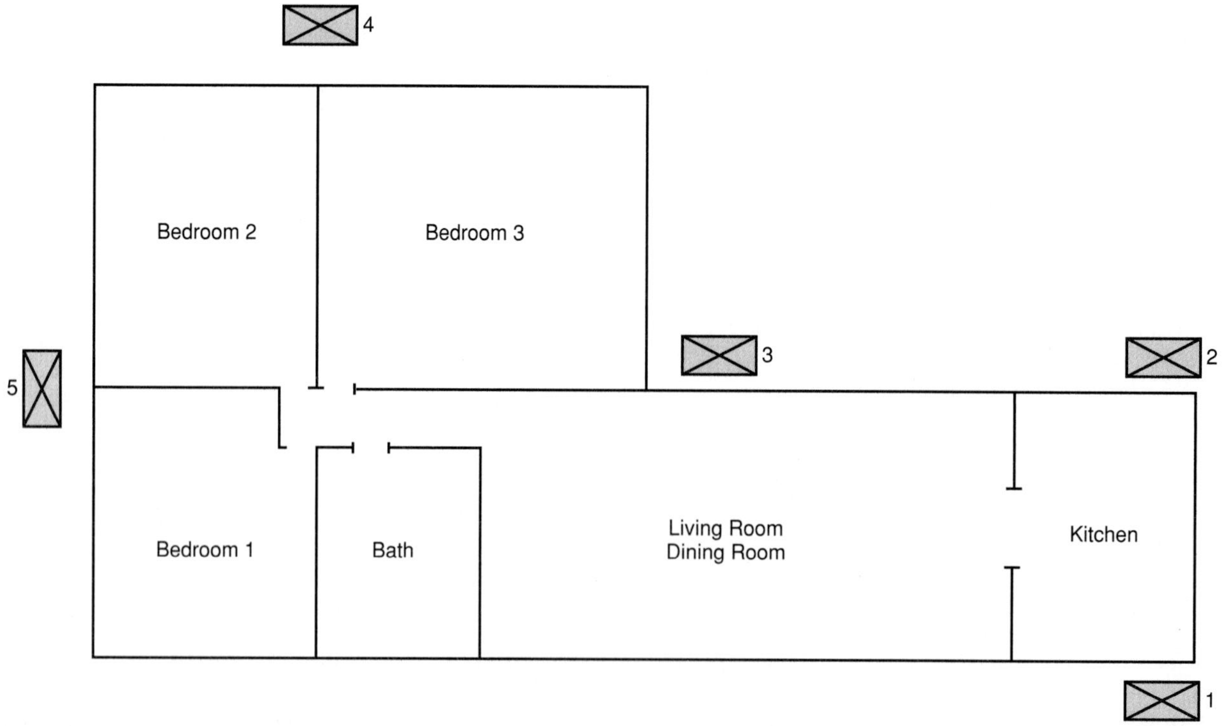

Figure 24-57. *Choose location of outdoor air-cooled condensing unit for best airflow and least noise. 1 and 2 are good positions, 3 is not recommended (it is in air pocket and near a bedroom), 4 and 5 are also too near bedrooms.*

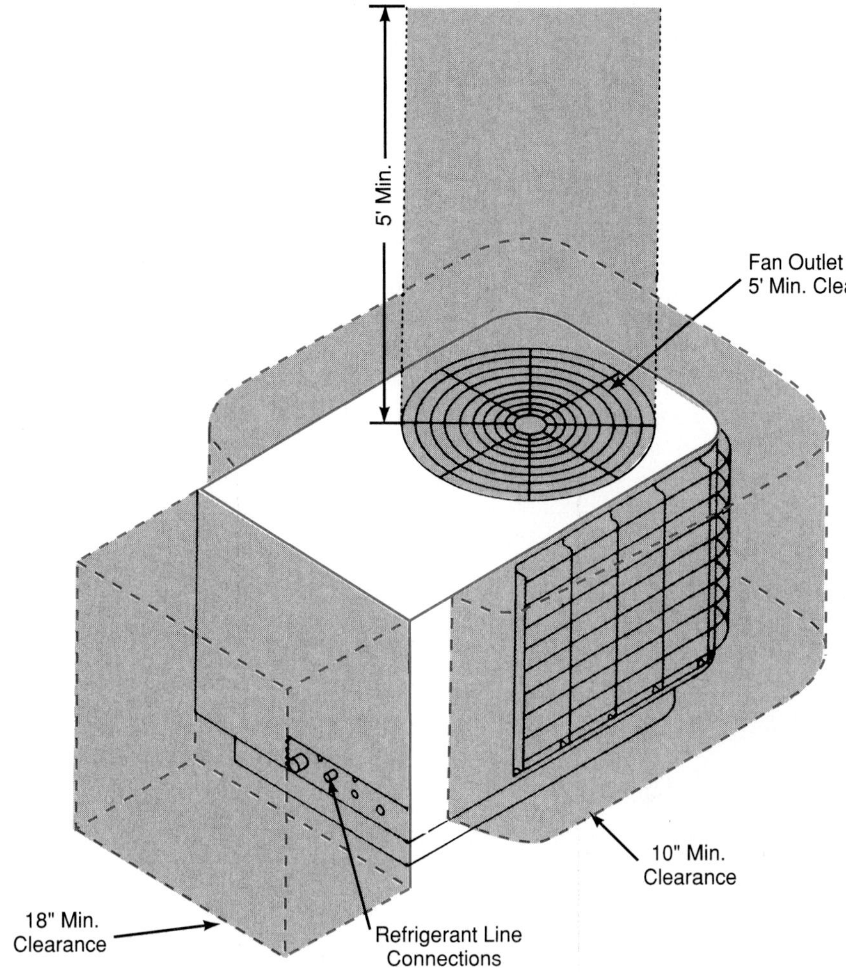

Figure 24-58. *Side view of typical residential heat pump installation shows minimum clearances to maintain around units. (The Coleman Co., Inc.)*

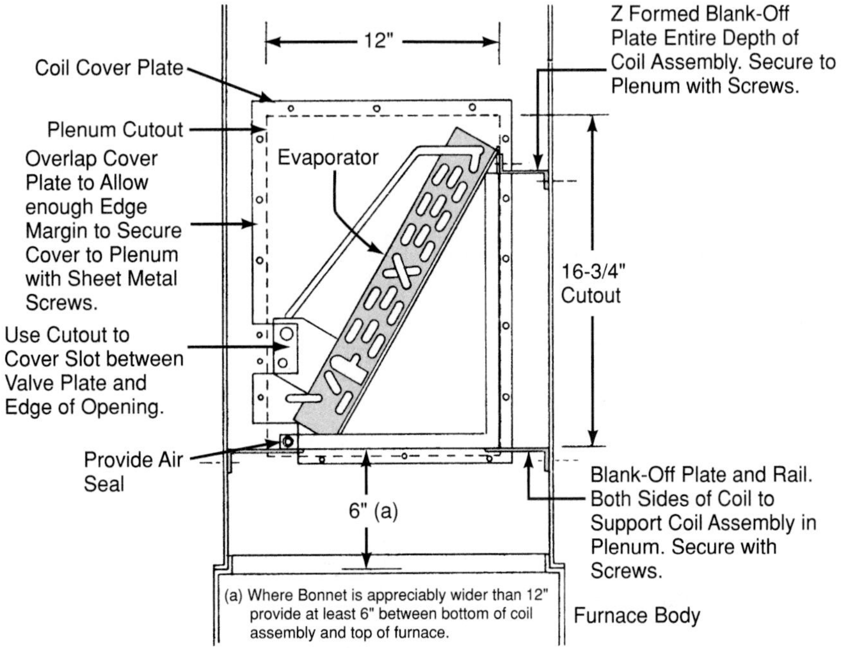

Figure 24-59. *Slant evaporator installed in furnace plenum. Note blankoff plates that make all air go through evaporator.*

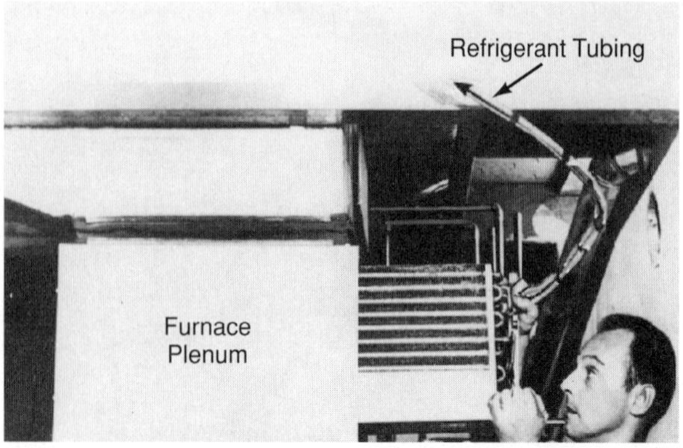

Figure 24-60. *An A-type evaporator being installed in plenum chamber of upflow warm-air furnace. (Fedders North America)*

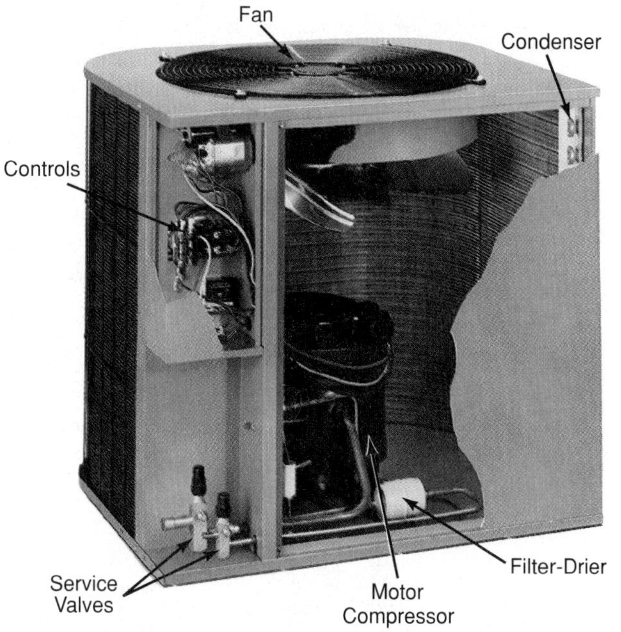

Figure 24-61. *Condensing unit designed for residential central comfort cooling system. (Williamson Corporation)*

A filter-drier and a sight glass should also be put in the liquid line. Many service technicians also place a filter-drier in the suction line. This prevents motor compressor burnouts.

That part of the suction line installed inside the building should be insulated. Use 1/2" line for hot, humid conditions and 1/4" to 3/8" for normal conditions. Without insulation, moisture from the air will condense on the suction line and drip. Some installers insulate all of the suction line. This avoids absorbing heat which will just have to be ejected at the condenser. Some efficiency is lost because of the extra heat. The motor compressor is not cooled as well. Lines should be supported and free of kinks. Openings in the furnace duct and the wall should be sealed. (Use weatherproof, nonhardening sealing compound and tape.)

When the quick-connect system is used, the lines are first installed. The correct length of prefabricated line should be used. When the quick-connects are made, the unit is ready to operate.

In all cases, the system should be thoroughly tested for leaks. This should be done while the pressures inside the system are near ambient temperature-pressure conditions. Use soap bubbles, a halide leak detector, or an electronic leak detector (depending upon the type of refrigerant used).

The electrical installation must follow the wiring diagram furnished with the unit. **Figure 24-62** shows a wiring diagram for a single-phase 240 V ac system.

The electrical circuit must follow the National Electrical Code and local codes. Consult the local electrical utility concerning the primary service capacity. An air conditioning system operating with 240 V uses a #6 wire size. A residential refrigerator operating with 120 V uses a #12 wire.

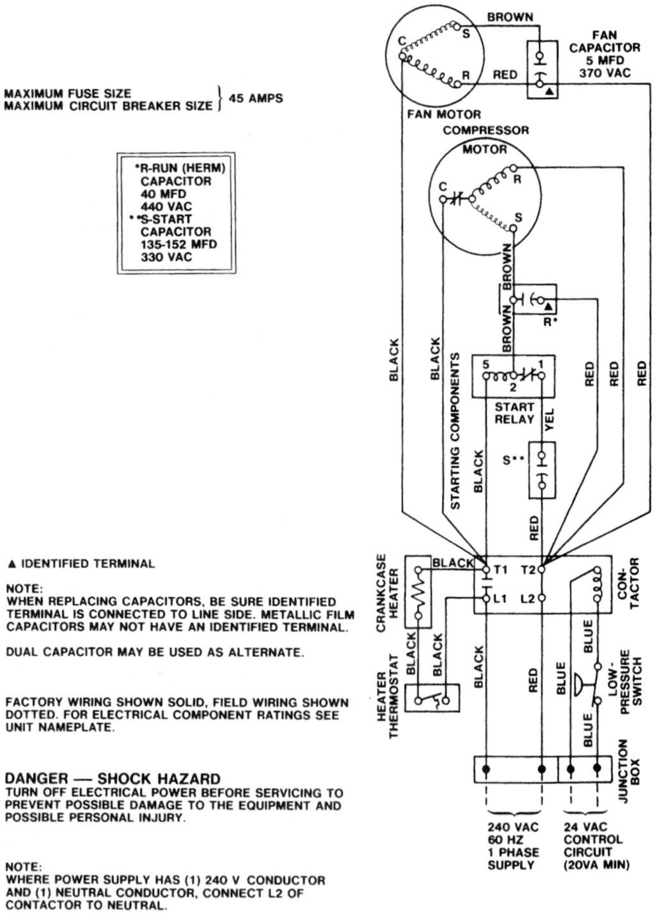

Figure 24-62. *Wiring diagram for central comfort cooling system condensing unit. (The Coleman Co., Inc.)*

24.13 Inspecting Residential Central Air Conditioning Systems

To maintain a heat pump that provides central heating / cooling / humidifying / dehumidifying / ventilating, and keep it economical, certain procedures are required

The complete system should be inspected and serviced each season before it is used.

Services that the owner can perform include:

- Energizing the crankcase heater 24 hours before starting system.
- Cleaning condenser and fans.
- Checking dampers in ducts.
- Replacing filters.
- Lubricating motor and fan bearings.
- Checking fan belt for cracks and glaze (replace if necessary).
- Checking and cleaning drain pans and drain pipe.

Services done by a service technician include:

- Cleaning thermostat points.
- Checking system pressures.
- Checking voltage and full-load amperage.
- Checking refrigerant charge.
- Checking suction line sweating or frosting.

24.14 Servicing Residential Central Air Conditioning Systems

The economical operation and dependability of a system depend on proper servicing. Servicing procedures and troubleshooting diagnosis for residential central systems are similar to those described in Chapters 12 and 15.

Check for leaks, proper refrigerant charge, and malfunctioning refrigerant controls and motor controls. Check also for moisture in the system. Some systems use service valves. **Figure 24-63** shows one type of valve.

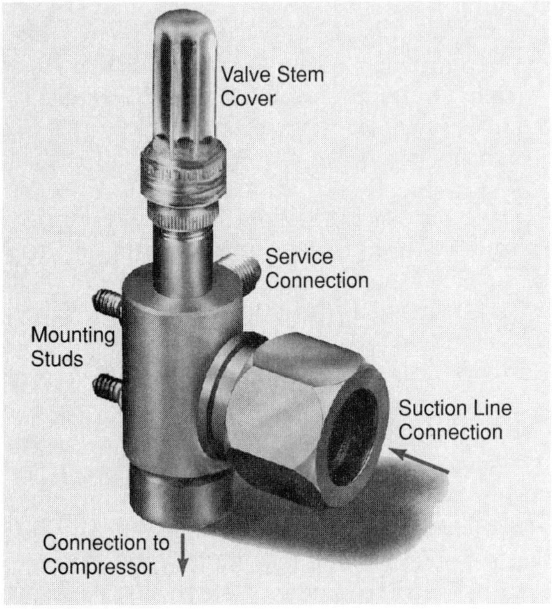

Figure 24-63. *Suction line service valve for residential central air conditioner. (Chatleff Controls, Inc.)*

Figure 24-64 is a cross section showing the internal construction. An access type of service valve with a quick-connect tubing connection is shown in **Figure 24-65.**

Often the same blower, motor, filter, and duct system are used during both the cooling and heating seasons. It is essential that the blower be cleaned once a year. The motor should be lubricated (a few drops of SAE 30 oil) once or twice a year. The filter should be replaced or cleaned at least twice a year. (This should be done at beginning of heating season and beginning of cooling season.)

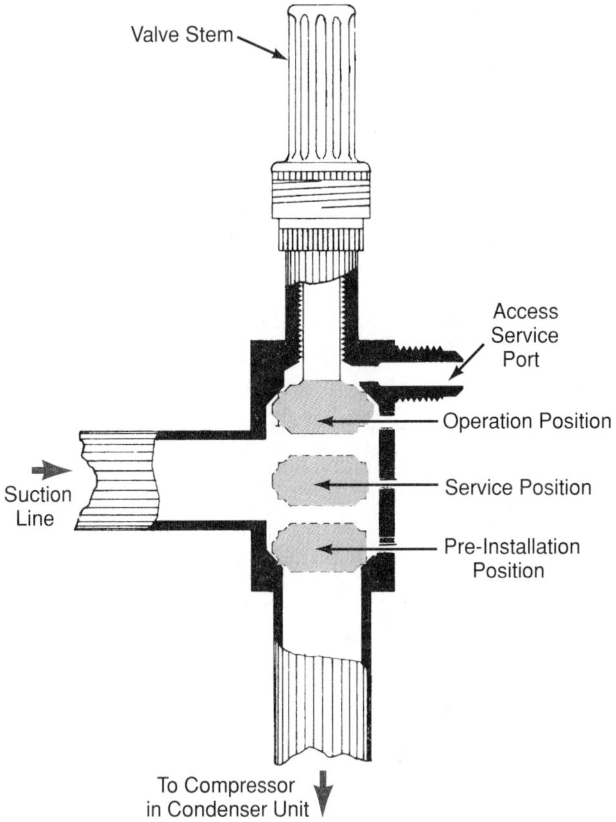

Figure 24-64. *Internal construction of central system air conditioner suction service valve. (Chatleff Controls, Inc.)*

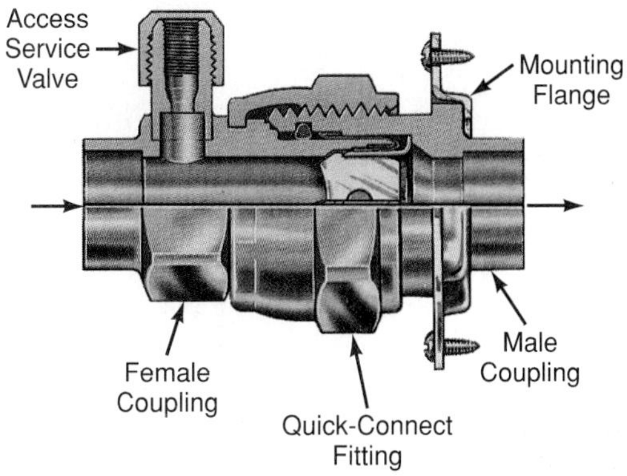

Figure 24-65. *Access-type service valve with quick-connect fitting. (Aeroquip Corp.)*

The motor-blower speed should be increased for summer comfort cooling. Belts on belt-driven blowers should be inspected and replaced if glazed or cracked.

The condensing unit should be cleaned once a year. The condenser, especially, should be blown clean of lint and any bent fins straightened. A carbon dioxide blower and/or vacuum cleaner may be used for cleaning the unit. The A-type evaporator should be cleaned of any lint and its fins straightened, if bent.

Check condensate drainage. Condensate that escapes the drain pan may drip on the furnace heat exchanger, corroding (rusting) it. The best way to clean these coils is as follows: Remove the coil. Plug all connections at once. Then either steam-clean or use hot water and detergent to scrub the fins and tubes. Steam is the best cleaner, although high-pressure warm water and detergent does a fair job.

Do not adjust the thermostat too low for summer cooling. The evaporator may freeze the condensate. This will stop airflow through the evaporator. The evaporator may continue to collect a lot of ice.

A split air conditioning system is shown in **Figure 24-66.** The compressor and condenser are located outside the home. The compressor/condensing unit is connected to the evaporator. The evaporator is installed in the plenum chamber of the warm air furnace.

There may be several reasons for service calls:

- No heat or insufficient heating. See Chapter 21.
- No cooling or insufficient cooling. Refer to Chapters 12, 15, and 22.
- Relative humidity too high. See Chapter 21.
- Air in house is stuffy (stale). See Chapter 23.
- Excessive (indoor or outdoor) noise. See Chapters 21 and 22.
- High cost of operation. Refer to this chapter and Chapters 21 and 22.
- Unit will not start. See Chapters 21 and 22.

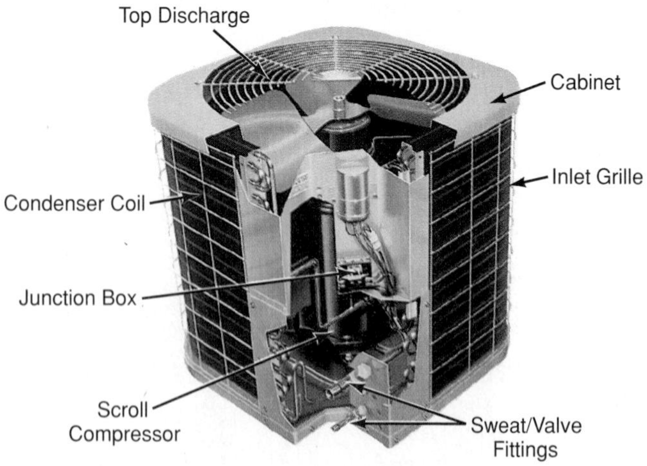

Figure 24-66. *A four-ton central air conditioner using a scroll compressor. Condenser coil has copper tubes. Note junction box where all electrical components and switches are located. (Tempstar Heating and Cooling Products, Inter-City Products Corporation)*

24.15 Air Circulation Systems and Relative Humidity Control

Good, complete air conditioning systems must provide heat, remove heat, and clean and circulate the air. Most systems accomplish all of these tasks. However, many systems do not completely control the relative humidity of the air. To control relative humidity, there must be two devices in the system ready to be used at any time, winter or summer:

- A device to add water vapor to the air (humidifier). See Chapter 21.
- A device to remove water vapor from the air (dehumidifier). See Chapter 22.

Another method is to always have a supply of cool air and warm air. Both cool and warm air should have a normal (50%) relative humidity. By mixing these air volumes, the needed temperature and relative humidity conditions can be produced. On the psychrometric chart, the process moves up or down along a relative humidity line.

Duct systems are used when the heating and cooling system components are ***remote*** (far away from the space to be conditioned). The cost of extending refrigerant lines or having separate relative humidity treatment units may be more than the cost of extending ducts.

Some systems confine the air distribution to the space being conditioned. A cabinet with fans, filters, grilles, and registers is located in the room.

24.15.1 Two-Duct Systems

The ***two-duct system*** uses two supply ducts. One supply duct carries cool dehumidified air and the other warm humid air. These separate airstreams are mixed just before they reach the space to be conditioned. Dampers control and balance air. Therefore, each different space in the building can be conditioned as needed. A single-duct air return is used. See **Figure 24-67.** The mixture of the two airflows takes place in the duct section at B.

The air control is excellent in these systems. However, the ducts are large in cross section and take up valuable space. Some space savings may be obtained by using high-velocity ducts. However, a noise problem may then develop.

24.15.2 Four-Pipe Systems

Four-pipe complete air conditioning systems have a hot water supply pipe, a chilled water supply pipe, and two return pipes. Four-pipe systems are of two types. One has separate heating and cooling coils in the space to be conditioned. The other system uses the same coil for both the warm water and the chilled water.

A system using separate heating and cooling coils is shown in **Figure 24-68.** The heating circuit is completely

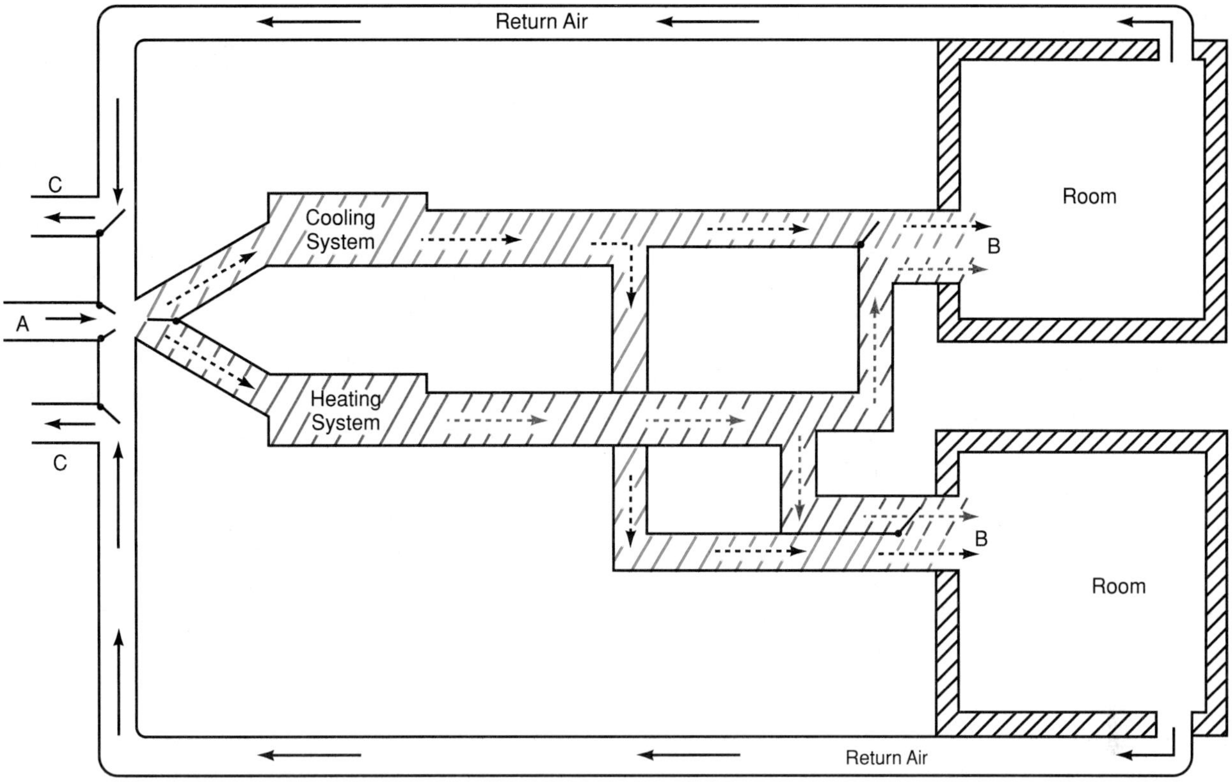

Figure 24-67. *This system has two supply ducts for complete air conditioning. A—Fresh air intake. B—Mixing plenums and diffusers. C—Exhaust air.*

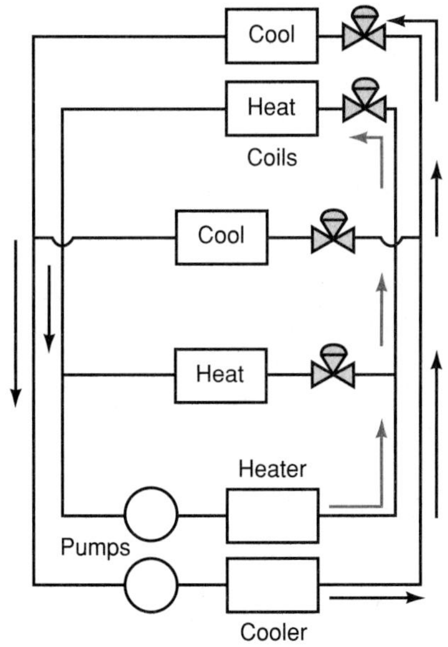

Figure 24-68. *Complete four-pipe system uses separate heating and cooling coils. Only one thermostatically-controlled valve is used for each heat transfer coil. Heating circuit and cooling circuit are separate.*

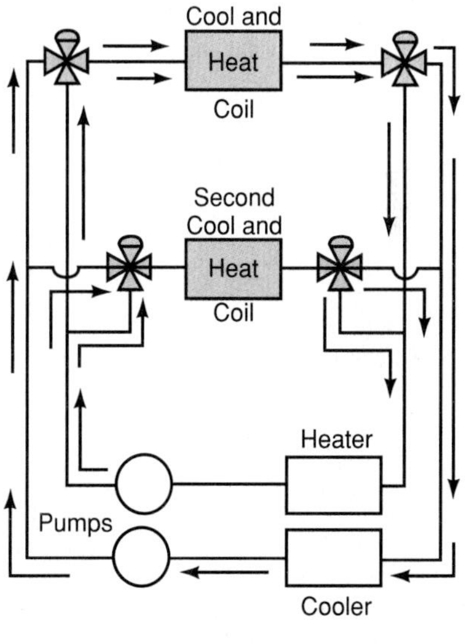

Figure 24-69. *Complete four-pipe system uses water or water-and-glycol mixture to move heat. Same coil is used for heating and cooling. Note four valves (two at each heat transfer unit). These valves, controlled by room thermostats, may be on-off valves or modulating-type valves.*

separate from the cooling circuit. In this case, the heating fluid (heat carrier) may be either water or steam. One two-way valve is used for each coil.

Some systems use the same space heat transfer coil for both heating and cooling. See **Figure 24-69.** Two separate three-way control valves are needed for each heat transfer coil in the conditioned space.

Hot water and chilled water pipes are insulated. The pumps operate only when thermostats in the conditioned spaces call for heating or cooling.

LARGE SYSTEMS MODULE

24.16 Chilled Water Systems

The popularity of ***chillers*** (chilled water systems) is due in large part to their cost-effectiveness. Rather than using only a refrigerant, these systems make use of water. Chillers remove heat from the water, which is then circulated through other components. Since water is fluid and has specific heat value, it makes an excellent ***secondary refrigerant.*** Water is also inexpensive, nontoxic, and largely noncorrosive. Other commonly used secondary refrigerants include: calcium chloride and sodium chloride brines, ethylene and propylene glycols, methanol, and glycerin. A number of refrigerants are used in chillers. Since the EPA rulings, ammonia and the HCFCs R-22 and R-123, and HFC R-134a are the main types used. Those units which employ brine are often used for low-temperature refrigeration. Units using brine may be factory assembled and wired or may be shipped in sections.

Chillers are used primarily for large industrial process cooling and commercial air conditioning. **Figure 24-70** shows a common chilled water system used for commercial air conditioning. This system uses a single compressor and one refrigerant circuit with a water-cooled condenser. Note the microprocessor control center.

The chiller is located on the low side of the system. It is responsible for chilling incoming water. The chilled water is then circulated throughout the building at approximately 45°F (7°C). See **Figure 24-71.**

Figure 24-70. *A water chiller unit with screw-type compressor typical of those used for industrial processes. Some units are part of a complete industrial process heating-and-cooling system. Other units are part of complete building heating-and-cooling system. (Frick Co.)*

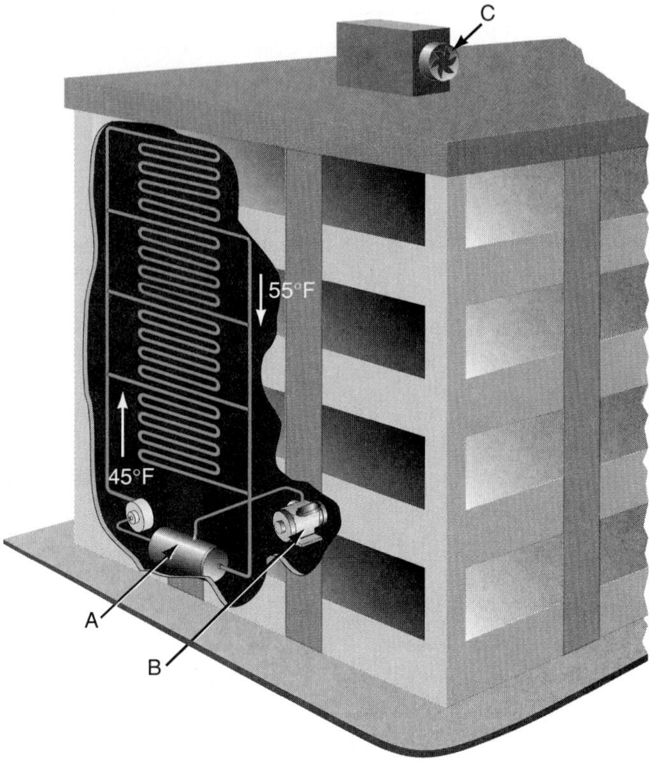

Figure 24-71. *A chiller system operating in a building. The water leaves the chilled water evaporator, A, at 45°F (7°C). It circulates throughout the building, picking up heat, and returns to the compressor, B, at 55°F (13°C). Note cooling tower at C.*

The removal of the interior heat from the building raises the H_2O temperature by approximately 10°F (6°C). The water returns to the chiller at approximately 55°F (13°C). At the chiller, it is once again cooled to 45°F, then recirculated.

In many chiller applications, water-cooled condensers also use a cooling tower. **Figure 24-72** shows the flow of the water in a chiller system with a cooling tower. The water enters the condenser, A, at 85°F (29°C). It picks up heat and leaves approximately 10°F (6°C) higher (95°F [35°C]). The water is then sent to a cooling tower, B, in which the water is sprayed from the top. As the water falls to the bottom, it is cooled by evaporation. Occasionally, forced air fans, C, are used. The water is cooled to 85°F (29°C). It returns through water pumps, D, back to the condenser. **Figure 24-73** is a more descriptive illustration of water flow from the cooling tower to the condenser of the chiller system being used to cool a building.

24.17 Types of Chillers

There are two basic chiller types, absorption and compression. The ***compression-type chiller*** uses the compressor to create a pressure difference between the high side and the low side. This pressure difference

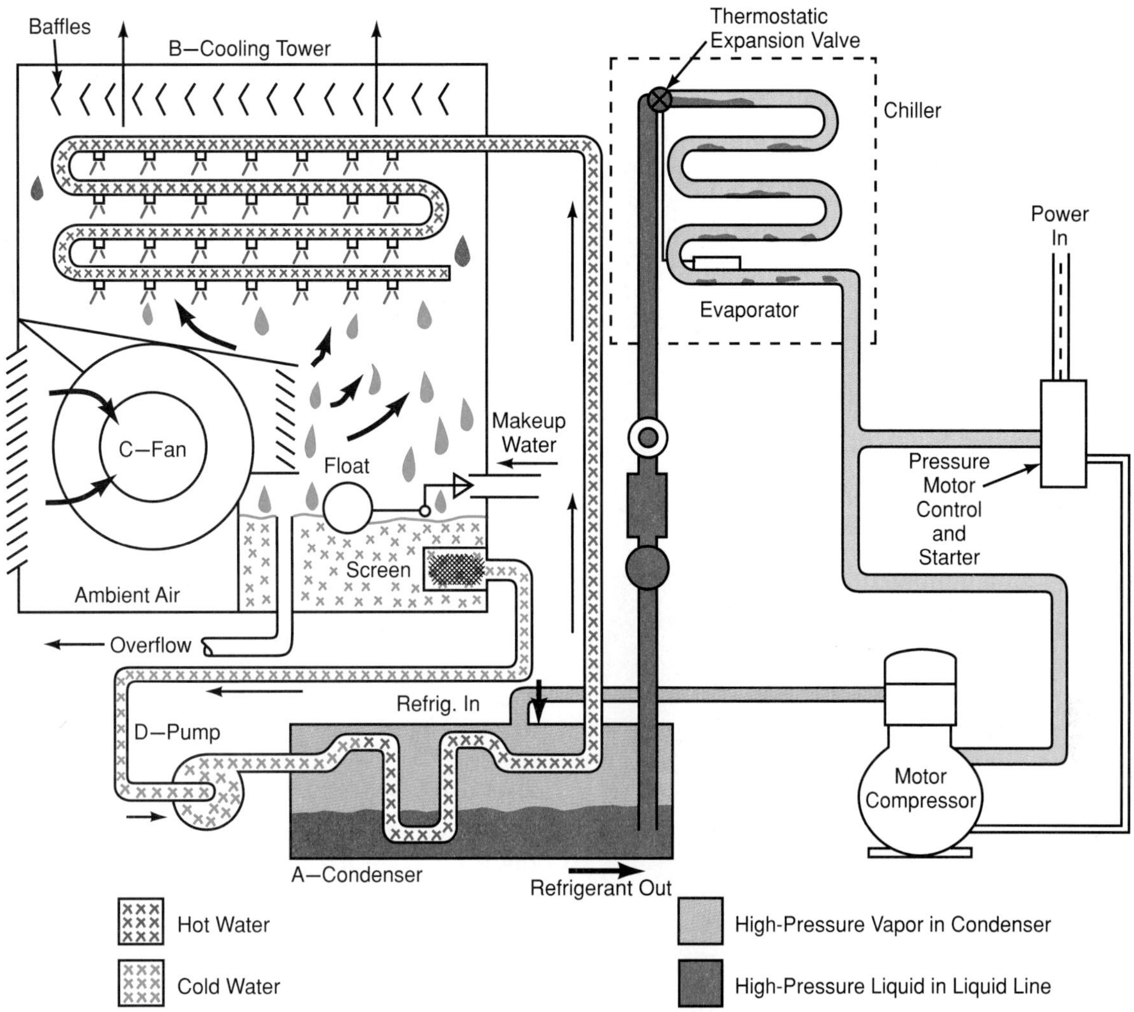

Figure 24-72. *Chilled water application with water-cooled condenser and cooling tower.*

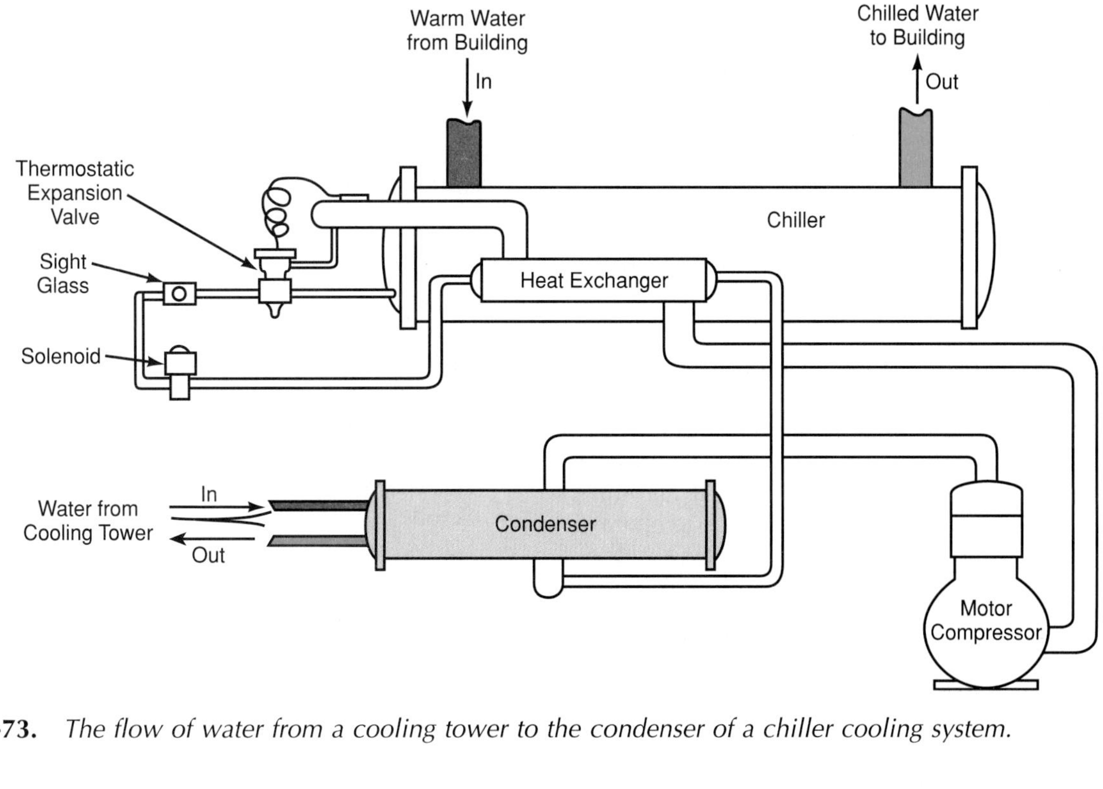

Figure 24-73. *The flow of water from a cooling tower to the condenser of a chiller cooling system.*

causes the refrigerant to boil and condense by lowering the evaporator pressure. Common types of compression chillers include those with centrifugal, scroll, screw, and reciprocating compressors.

The ***absorption-type chiller*** uses a brine solution and water to provide refrigeration or air conditioning. Absorption systems are unique because they do not use a compressor. **Figure 24-74** shows a large-capacity double-effect absorption chiller and heater. Note the control center. It displays performance information: inlet and outlet temperatures of chiller, condenser, and hot water circuits. It also indicates generator pressure and temperature, refrigerant and solution temperatures, and heat input command. The panel also displays the unit's total operating hours, number of starts, and number of purge cycles completed. Information on the operation of the absorption-type chiller can be found in Chapter 17.

Figure 24-74. *Large capacity, double-effect absorption chiller and heater. Note the control center. (York International Corporation, Applied Systems-Parts Division)*

24.17.1 High-Pressure and Low-Pressure Chillers

Chillers are also identified as high-pressure or low-pressure, according to the operational pressure of the system.

High-Pressure Chillers

High-pressure chillers are those units used primarily for producing comfort conditions. The high operational pressure will determine the size of the unit necessary to accomplish the cooling task. An example of this is a 20-story commercial office building. It may require a 400-ton unit to maintain the needed temperatures. This type of installation is classified as a high-pressure chiller system.

Low-Pressure Chillers

The term ***low-pressure chillers*** is used to identify chillers that are used where very low evaporator temperatures are required. Most of the components in a low-pressure chiller are similar to those of the high-pressure system. Therefore, the operation is very similar. High- and low-pressure centrifugal systems are almost identical.

24.18 Basic Components

Chiller systems have many components. These include a refrigerant control device (TEV) and a control center. The common types of compressors used in chilled water systems are: welded hermetic, semi-hermetic, and external direct-drive open. Condenser types include: evaporative, air-cooled, and water-cooled. Evaporators used are usually direct expansion. In a ***direct expansion evaporator,*** the refrigerant will evaporate while flowing through the tubes. Chilled water is cooled as it flows across the outside of the tubes.

A receiver, economizer, or sub-cooler also may be included. Additional optional accessories include an oil cooler, oil separator, purge unit, oil pumps and refrigerant transfer unit. Many chiller systems operate under a vacuum and use purge units. The purge unit removes unwanted air moisture.

24.18.1 Motors

Motors that are used with centrifugal chillers include external-drive and closed-drive types. External-drive motors use the air in a room to cool the motor. The motor which drives the compressor is open or exposed. External-drive units are used in systems from 80 tons to 10,000 tons. Heat from the motor is excessive and must be exhausted from the room. Closed drive motors are hermetically sealed. They are refrigerant-cooled motors. Liquid refrigerant flows around the motor housing. Hermetic units are frequently used in 80-ton to 2000-ton systems.

Another method of circulating refrigerant in a chiller system is by using a gas engine. The use of natural gas engine-driven chillers will provide a fast payback in some areas. The system uses a natural gas engine that drives the compressor. The large units use scroll compressors. In smaller units, reciprocating compressors are used. The system operation, **Figure 24-75,** is the same as for any other chiller unit. The only exception is the drive system.

24.18.2 Compressors

Various types of compressors are used in chillers. Some of the more common types are:

- Reciprocating.
- Scroll.
- Screw.
- Centrifugal.

Additional information on compressors can be found in Chapter 4. The major difference between compressors used in chillers and those used in other applications is size. The chiller compressor is much larger. The

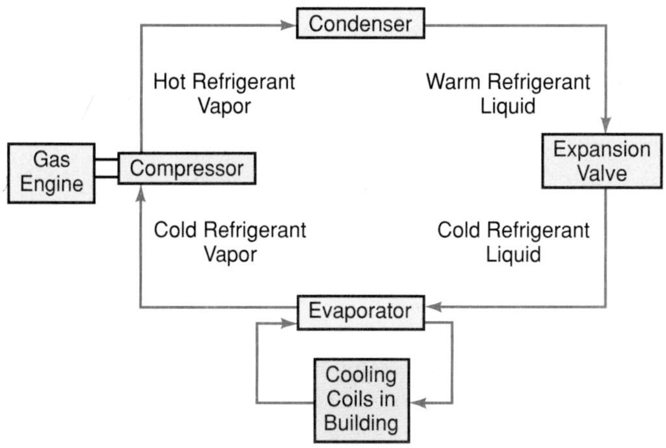

Figure 24-75. *Flowchart for a chiller system that uses a natural gas engine to rotate the compressor. (Tecogen, a Division of Thermo Power Corporation)*

basic operation of a compressor is the same, however, regardless of the unit size.

The type of compressor used in an installation depends on the number of tons of air conditioning needed:

- Up to 25 tons — Reciprocating
- 25 to 80 tons — Screw or Reciprocating
- 80 to 200 tons — Screw, Reciprocating, or Centrifugal
- 200 to 800 tons — Screw or Centrifugal
- Above 800 tons — Centrifugal

Reciprocating

Reciprocating compressors used in chillers operate the same as those in conventional units. ***Reciprocating compressors*** are piston-type or positive displacement compressors. These high-pressure systems are capable of handling up to 200 hp. Reciprocating compressors are the most widely used in all fields of refrigeration. They are very cost-competitive, easy to work on, and inexpensive to replace. However, they are very noisy, and vibration is a concern.

Reciprocating systems use a multiple compressor arrangement. This provides the pumping capacity needed to move large amounts of refrigerant. See **Figure 24-76.** The multiple arrangement provides efficient unloading. For example: A system with two compressors may operate on one compressor while the other compressor is "down." This allows limited maintenance to occur without loss of cooling. Such multiple units may be arranged in a variety of configurations. Additional benefits of multiple compressors include: greater temperature control, less power consumption, less surging at start-up, and extra standby capacity.

Reciprocating compressors are unusual in that pressure rise has little effect on volume flow rate. Therefore, a reciprocating liquid chiller retains nearly full cooling capacity at all times.

Scroll

Scroll compressor chillers are larger than those used in smaller applications. It has become a very widely used

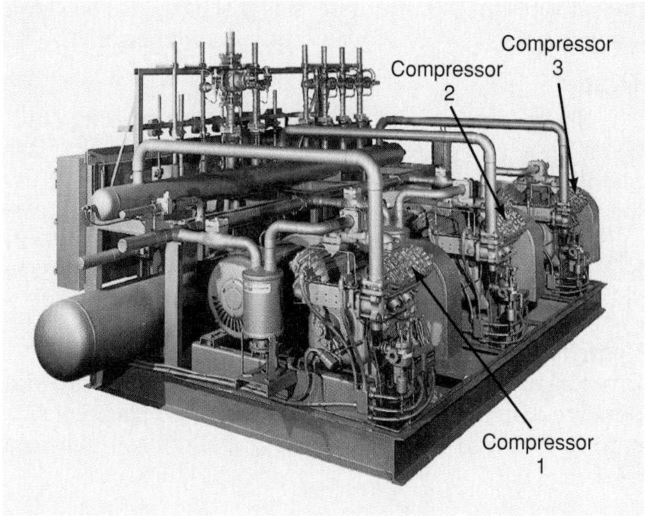

Figure 24-76. *A water-chiller system using three reciprocating compressors. (Grasso, Inc.)*

unit for automotive air conditioning. **Figure 24-77** shows a detailed illustration of a scroll compressor.

The scroll compressor is a positive displacement compressor. It is made up of spiral scrolls which are attached to a flat base. There are two matching scrolls which hold gas pockets when placed together. One scroll orbits around the stationary scroll. As gas enters the scroll, it is progressively compressed into a smaller

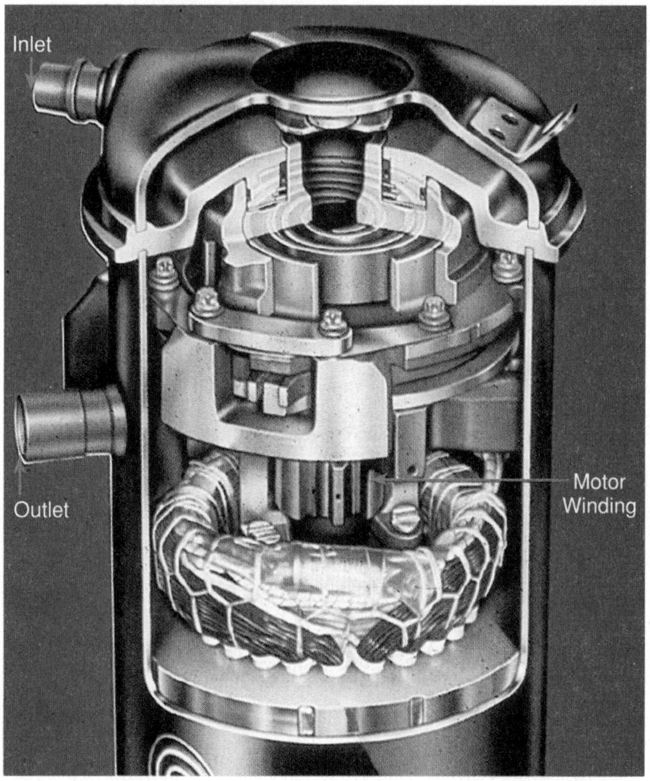

Figure 24-77. *A cutaway view showing details of a scroll compressor. (Copeland Corporation)*

pocket (volume). From there, it is discharged. The scroll compressor is a very efficient, quiet compressor.

Screw

Screw compressors are used in larger capacity chillers. These compressors, with two helical rotors, are mainly used in chilled water systems. A water chiller using a screw compressor with a microprocessor is shown in **Figure 24-78A.** The screw compressor **(Figure 24-78B)** has few moving parts. It is capable of handling large amounts of refrigerant.

Centrifugal

Centrifugal chiller systems **(Figure 24-79)** are not positive displacement systems. Instead, they are considered to be turbo machinery. The impeller is similar to a large fan, which creates pressure. See **Figure 24-80.** It applies centrifugal force to the refrigerant. This moves the refrigerant from the low side to the high side of the system. The centrifugal chiller is flexible under varying load conditions. It produces good efficiencies even when the load is less than 40% of design capacity. Due to their design, centrifugal systems are more efficient than reciprocating compressors.

An *economizer* is often used to allow removal of flash gas in the evaporator. The economizer increases the refrigerant effect of each pound of refrigerant in the evaporator. This allows reduction of the power requirements at the compressor.

The compressor is turned quickly by a gearbox in multiple stages of compression. This provides a pressure difference from the evaporator to the condenser. Increased compressor speeds are accomplished through the gearbox.

Figure 24-79. *A centrifugal compressor used in low-pressure applications. (McQuay International)*

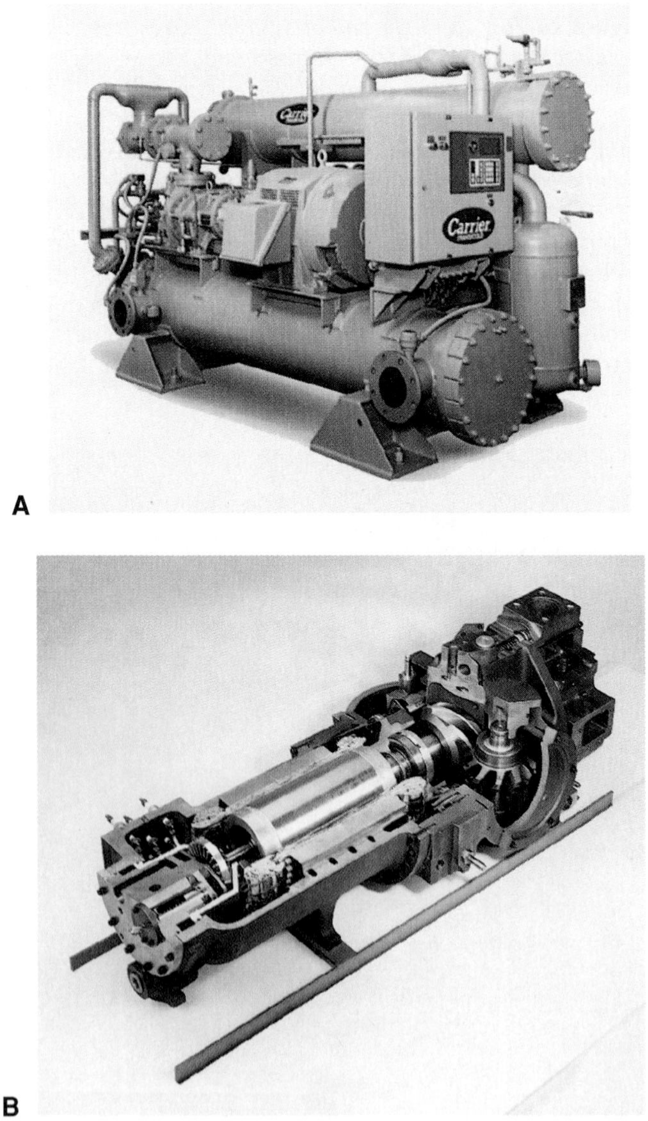

Figure 24-78. *Screw compressors. A—Microprocessor-controlled chiller package using a screw-type compressor. (Carrier Transicold Division, Carrier Corporation) B—Cutaway view of a screw compressor used in a chiller. (McQuay International)*

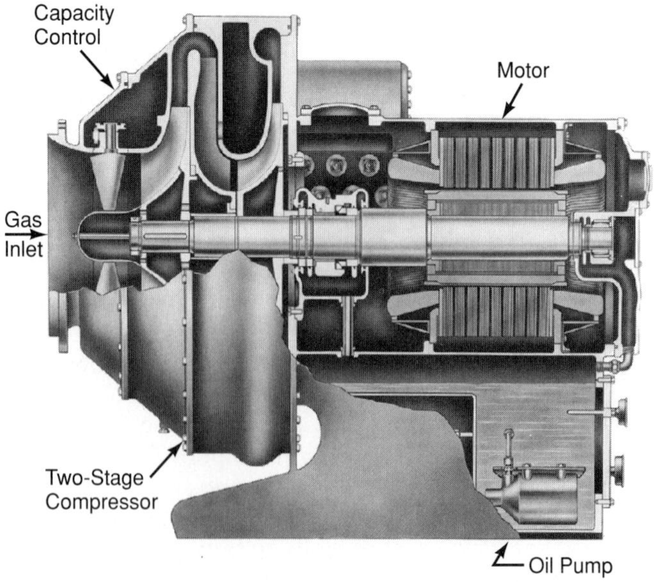

Figure 24-80. *Two-stage centrifugal compressor. Oil pump is driven by separate power source. (Carrier Corporation, Subsidiary of United Technologies Corporation)*

The centrifugal chiller uses large volumes of refrigerant at a low compression ratio. The refrigerants most often used in centrifugal systems manufactured today are R-123 and R-22. R-11 had previously been used.

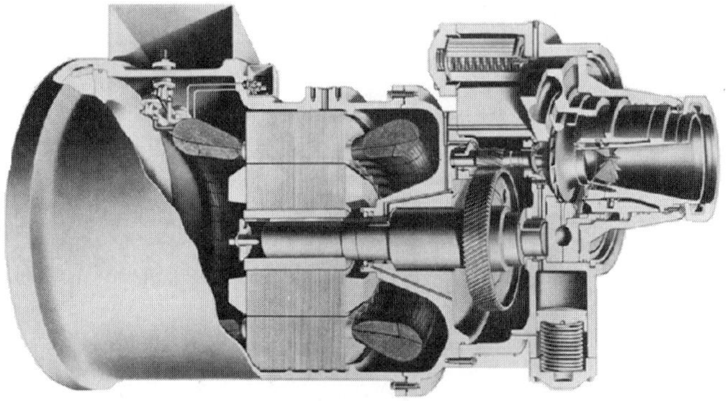

Figure 24-82. *A centrifugal compressor. Note the impeller that is used to create pressure and force. (McQuay International)*

24.19 Comfort Cooling Systems

Many large comfort cooling installations use centrifugal-type compressors. Large centrifugal units are frequently designed with capacities of 100 to 2000 tons. See **Figure 24-81.** These systems use low-pressure refrigerants, and the evaporator operates at below-atmospheric pressure. Both condenser and evaporator are the shell-and-tube type. The compressor is a two-stage centrifugal unit, driven by a hermetically sealed motor. The capacity is controlled by inlet vanes to the two-stage centrifugal unit, driven by a hermetically sealed motor. The capacity is controlled by inlet vanes to the two-stage centrifugal compressor. The vanes are closed to reduce the starting load. **Figure 24-82** shows the compressor construction. This compressor has a forced lubrication system. A separate motor drives the oil pump. The compressor motor is a three-phase unit of 208 V, 240 V, 440 V, 480 V, 550 V, 2300 V, or 4160 V. Note the bolted construction for service purposes.

These systems have an automatic purging device for removing noncondensible gases. Design and construction details of the condenser and the evaporator are shown in **Figure 24-83.** Due to their capacity, these units must have thorough, accurate control.

Persons responsible for the operation of these units should receive thorough training in their correct operation. **Figure 24-84** illustrates the complete wiring and piping system of one of these units. The evaporator, compressor suction line, and chilled liquid lines are always insulated.

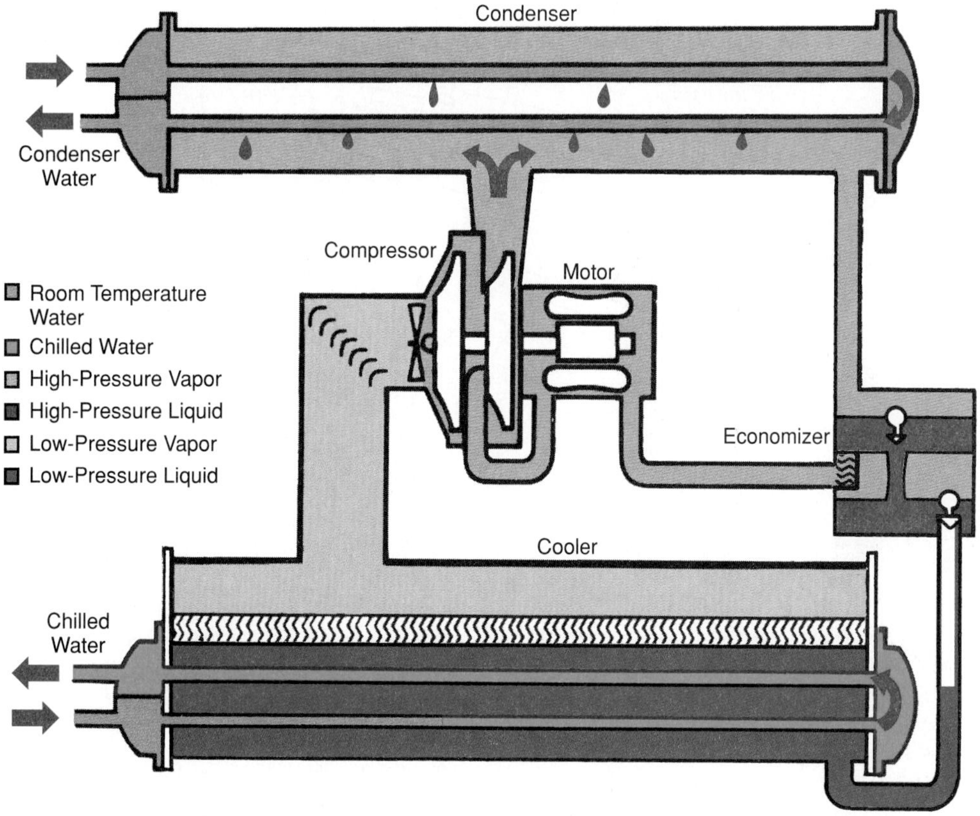

Figure 24-81. *Centrifugal compressor for hermetic chilled water system uses rotors instead of pistons. Chilled water is under great pressure to move it through chemical reaction tank spraybars. (Carrier Corporation, Subsidiary of United Technologies Corporation)*

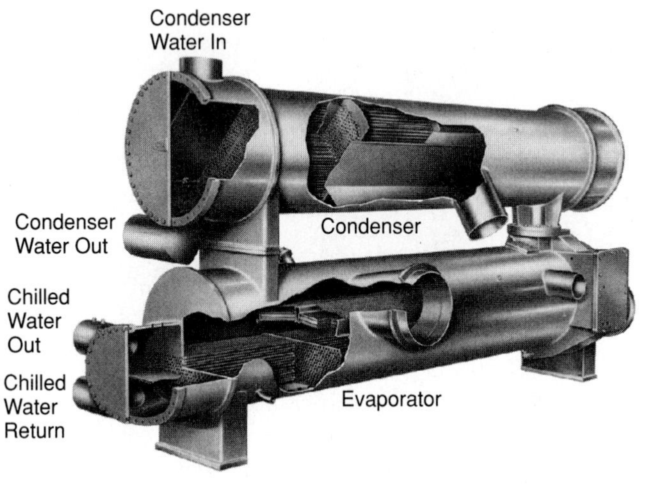

Figure 24-83. *Condenser and evaporator of large chilled water system which uses centrifugal compressor. (Carrier Corporation, Subsidiary of United Technologies Corporation)*

24.20 Rooftop Units

Heat pumps are best used for both heating and cooling. Less equipment is needed for a combined unit. Rooftop systems for heating and cooling were first developed for flat-roofed commercial structures. Such structures include supermarkets and office buildings.

There are three types of rooftop heat pumps:

- Heating only.
- Cooling only.
- Both heating and cooling.

The systems are economical because they are factory-assembled and tested. They are also easily installed once hoisted to the site.

There are some disadvantages to rooftop systems. The roof must be strong enough to carry the extra weight. Openings may develop leaks. Units must be designed for ease of service even in very cold or bad

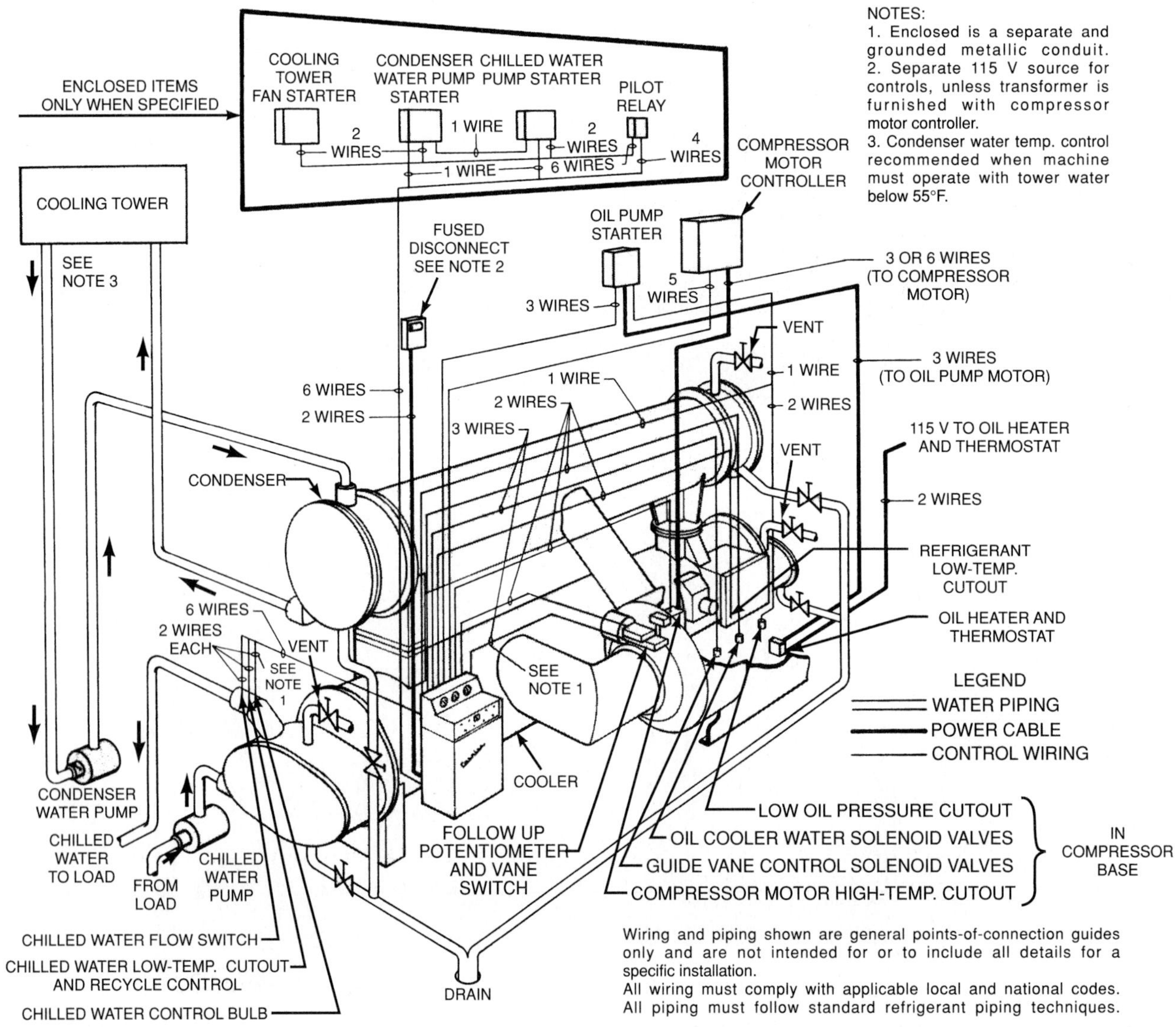

Figure 24-84. *Drawing shows wiring and piping of large-capacity water chiller. (Carrier Corporation, Subsidiary of United Technologies Corporation)*

weather. They are also exposed to winds, which can affect their operation. You must carry tools and supplies to the roof. Equipment must be adequate to service the unit in all kinds of weather.

Figure 24-85 shows a typical rooftop unit used for both heating and cooling. Another assembly is shown in **Figure 24-86.**

Many commercial units come completely equipped with factory-installed heating units. **Figure 24-87** shows a tubular heat exchanger made of aluminized steel. Rooftop gas furnaces usually use an electric ignition system. This is because gas pilot flames tend to blow out during high winds or gusts. These units are completely weatherproofed since they are exposed to all kinds of weather. Access doors for inspection are designed to operate safely under high wind conditions. **Figure 24-88** is a pictorial drawing of the refrigerating mechanism of a rooftop unit. A complete heating and cooling system is shown in **Figure 24-89.**

Figure 24-85. *Rooftop unit provides both heating and cooling. This unit has a gas-fired furnace. Fans move outdoor air through a condenser. (Carrier Corporation, Subsidiary of United Technologies Corporation)*

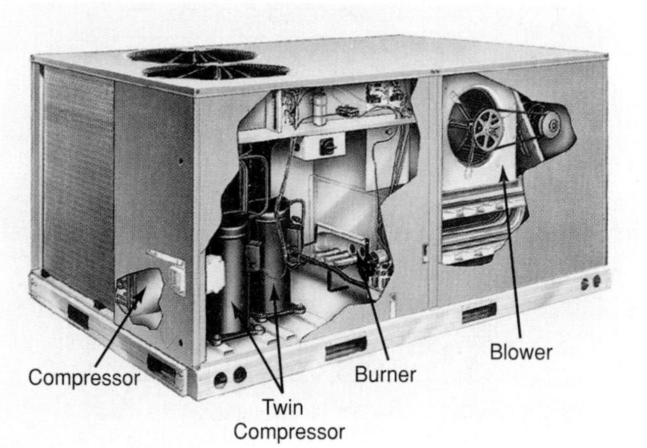

Figure 24-86. *Cutaway view of a rooftop unit. (Carrier Corporation, Subsidiary of United Technologies Corporation)*

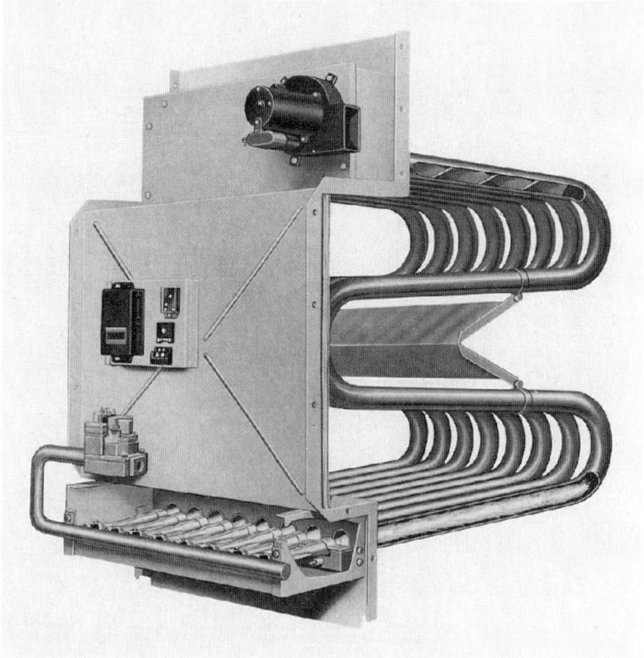

Figure 24-87. *Gas heat section of single-package cooling and gas heating system. (Lennox International, Inc.)*

Rooftop units are sometimes oil-fired. Fuel is stored in an underground tank. A lift pump draws oil to a second, smaller tank. This tank may be at the same level as or slightly lower than the oil burner. Under no circumstances should the tank be more than 10 ft. (3 m) below the burner.

Electric heat may also be used in rooftop units. In **Figure 24-90,** the electric heating elements are partially removed from the housing. These heating elements are available in 20 kW, 40 kW, and 60 kW capacities. Heating elements are carefully protected by ceramic insulators. The circuit has fuses, fusible links, and thermal cutouts.

Air is controlled by dampers. Sensing elements in the mixed-air duct control a motor that operates these dampers:

- The return damper.
- The conditioned air damper.
- The exhaust air damper.
- The fresh air damper.

Figure 24-91 shows the air control system. The air is filtered as in other air circulation systems. See Chapter 23.

24.20.1 Installing Rooftop Units

Rooftop units must be installed using methods that will provide a leakproof roof connection. **Figure 24-92** shows how the rooftop system is connected to a duct

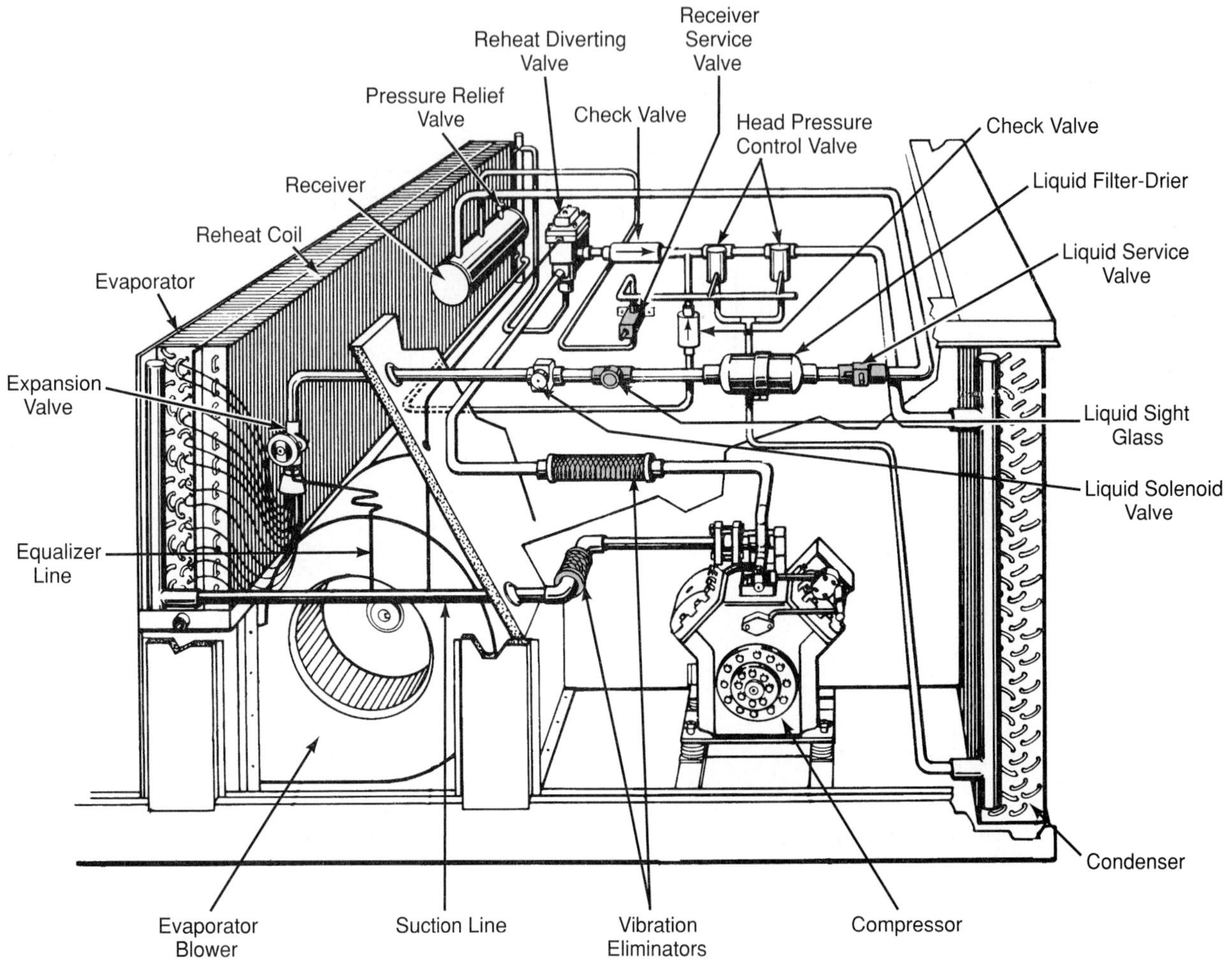

Figure 24-88. *Refrigerating circuit of rooftop system. Note service valves, sight glass, and vibration eliminators.*

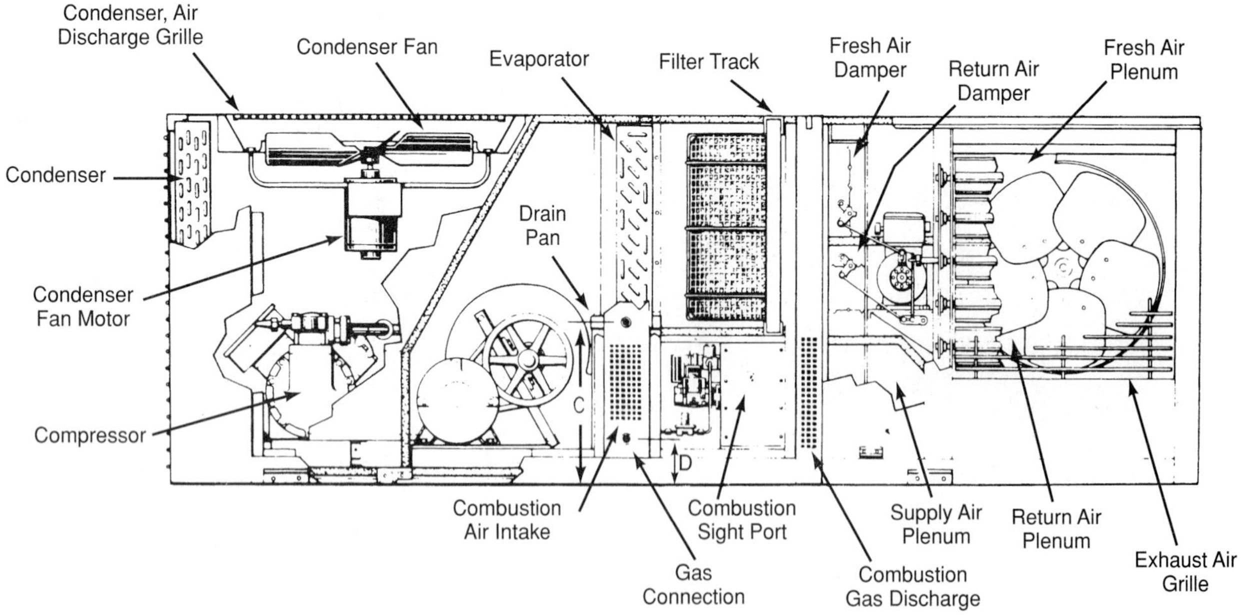

Figure 24-89. *Combination heating and cooling rooftop system.*

Figure 24-90. *Rooftop system is equipped with electric resistance heating elements.*

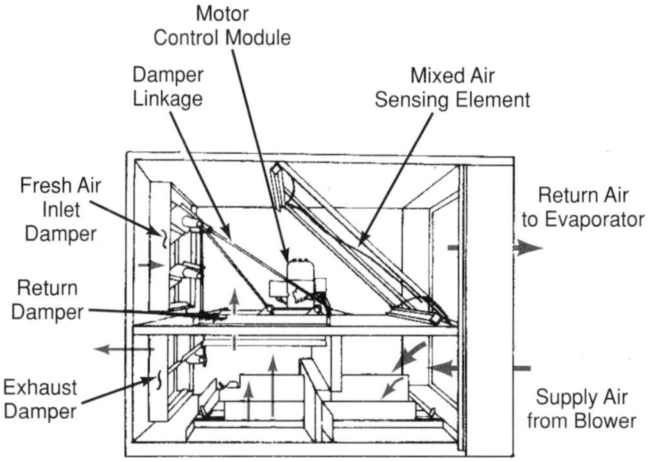

Figure 24-91. *Air-control system for rooftop unit. Dampers are automatically controlled so that they always provide desired air mixture.*

distribution system. The installation is usually made as shown in **Figure 24-93.**

One method used to make the roof leakproof is shown in **Figure 24-94.** Notice the use of acoustical material for deadening sound. The resilient (flexible) gasket is installed between the rooftop base rail and the roof edge. This gasket is very important. There may be leaks if it is not tight. A duct installation connected to a rooftop unit is shown in **Figure 24-95.**

Rooftop units are heavy. Riggers usually must handle their installation, removal, and replacement.

The electrical service must comply with local codes. The control system is usually the responsibility of a refrigeration installation technician. A system using electronic controls is shown in **Figure 24-96.**

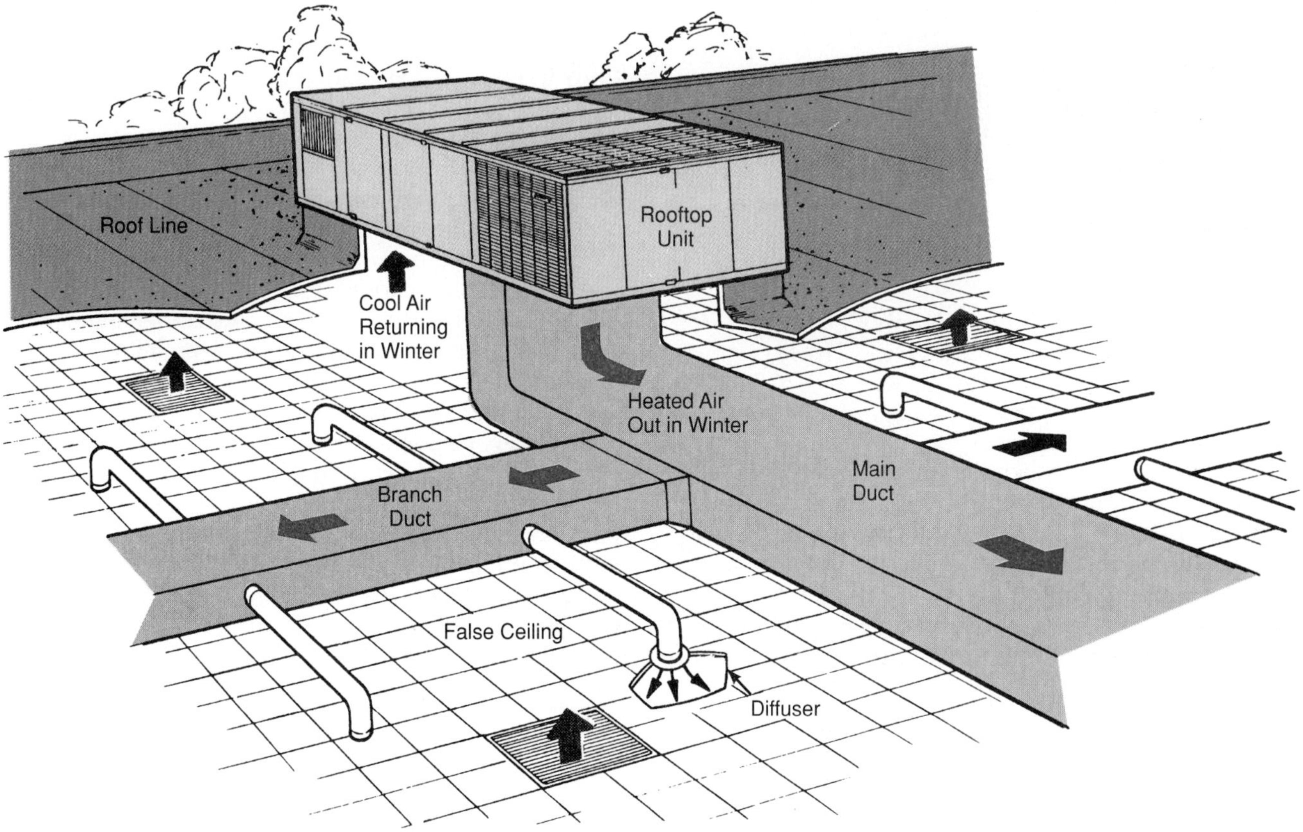

Figure 24-92. *Rooftop unit connected to duct distribution system. (Lennox International, Inc.)*

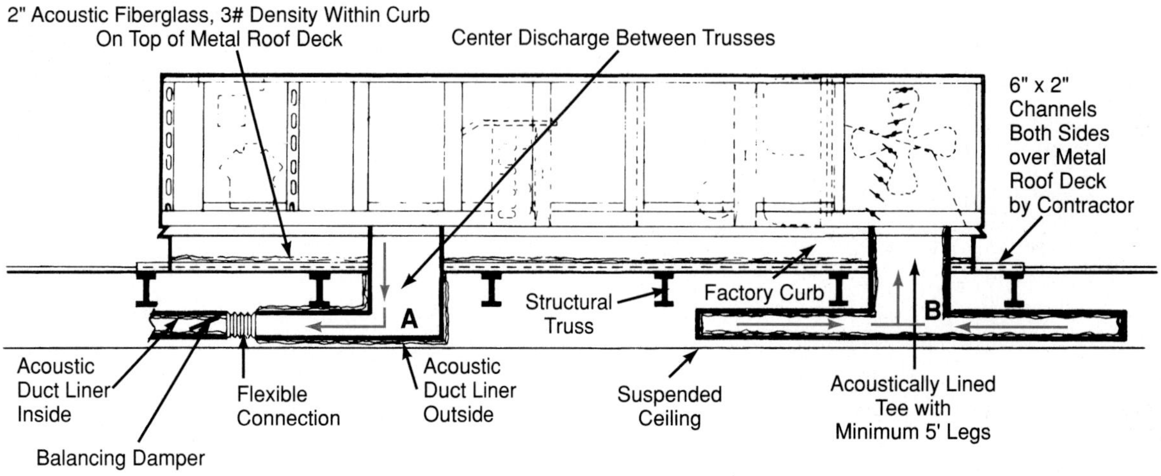

Figure 24-93. *Installation details of rooftop complete air conditioning system. A—Conditioned-air duct. B—Return-air duct.*

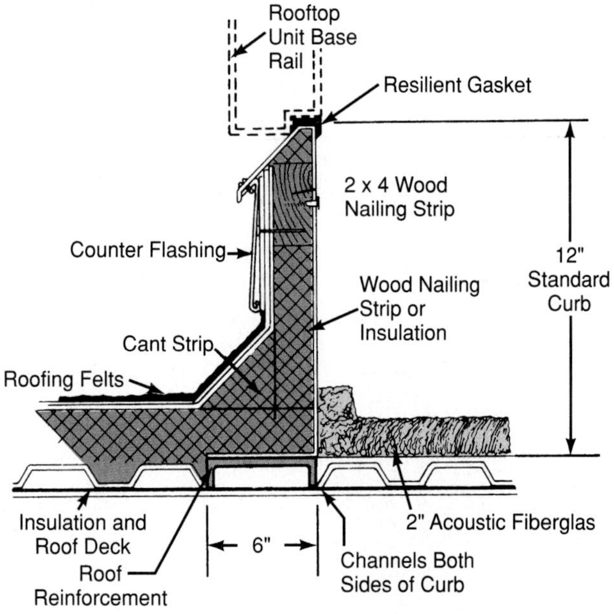

Figure 24-94. *Method used to make roof leakproof when rooftop unit is installed.*

When the rooftop system has a gas furnace, the gas piping must be installed according to the appropriate code. Pipes must be supported and protected. **Figure 24-97** shows a typical installation for a 1 1/4″ gas line. Note that two hand shutoff valves are used. One is located outside the rooftop system casing. It will permit emergency shutdown of the furnace in case of accident or fire.

Some systems use piped-in hot water or steam to provide heating. Piping for these systems must be carefully installed to avoid air traps and freezing. Hot water systems usually use a nonfreezing solution of water and ethylene glycol. **Figure 24-98** shows the piping for a steam system.

24.20.2 Servicing Rooftop Units

Certain chapters have been devoted to servicing different systems. For more information on:

- Refrigerating systems, refer to Chapter 15.
- Heating systems, refer to Chapter 21.
- Blowers and filters, refer to Chapter 23.

However, there are some very important service operations unique to rooftop systems. **It is very important to use all possible safety precautions when climbing to the roof.**

If a portable ladder is used, it must be securely and firmly based on the ground. It must lean against the building at an angle. This will keep it from falling away from the building. It must extend two or three rungs above the roof edge.

You must use both hands on the ladder when climbing or descending. A portable ladder should not be used in a high or gusty wind.

Be especially careful if it is raining or snowing. Lift tool boxes, refrigerant cylinders, and other objects with a rope. The rope should be guided from below.

Extreme caution must be used while working on or near electrical circuits. Hinged panels must be secured in the open position. Otherwise, the wind may swing them violently and injure someone. Loose panels must be held down securely to prevent wind blowing them off the roof. In systems using gas or oil burners, avoid breathing fumes coming out of the flue.

When working on rooftop units in cold and/or windy weather, use gloves and hand warmers. Some prefer portable heaters. Set up a windbreak to provide wind shelter. If the wind is very high, use a safety belt. The belt should be attached to a solid part of the unit.

When working on rooftop units in hot, sunny weather, shield the tools from the sun. Metal may become so hot it will burn your hands. Use gloves when handling the heated panels. Wear a hard hat and goggles. Wear rubber gloves when working on electrical circuits.

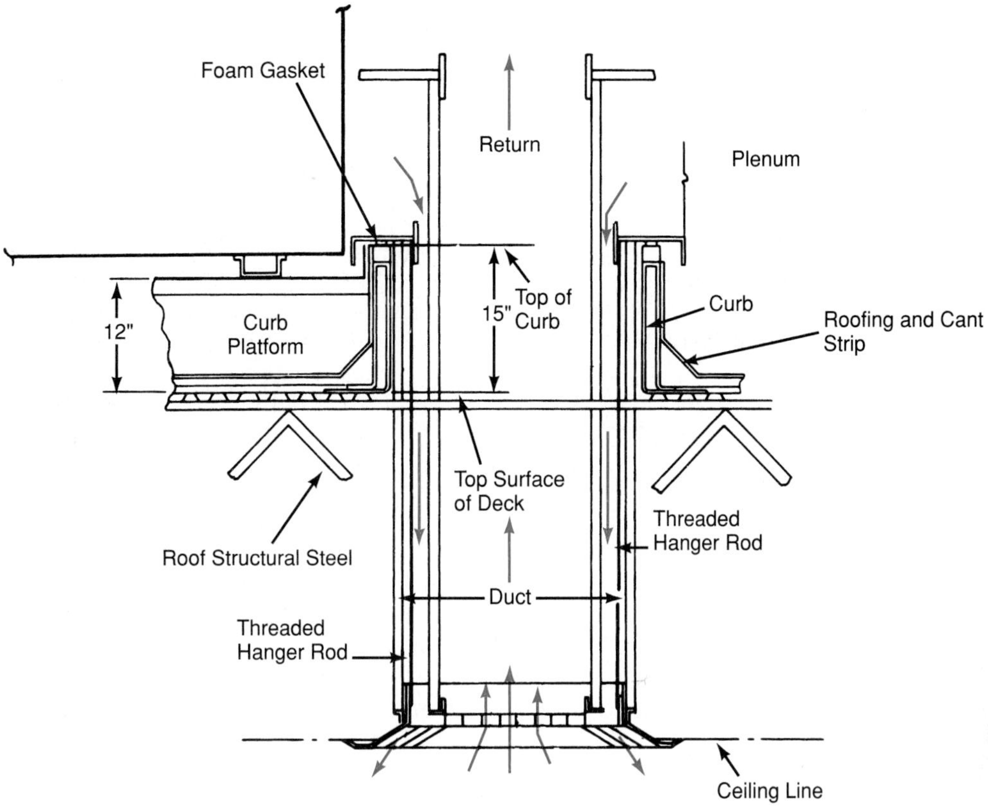

Figure 24-95. *Method used to connect ducts to a rooftop unit.*

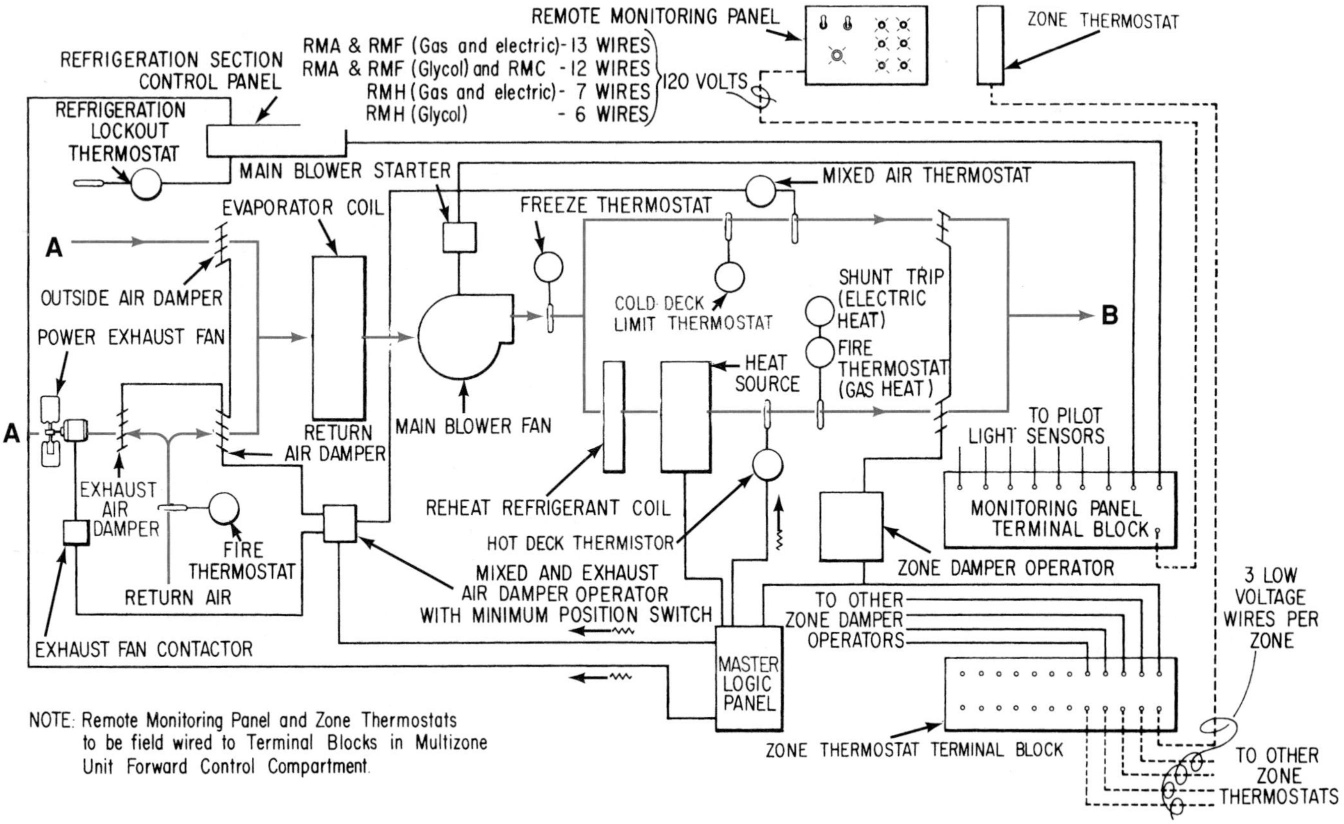

Figure 24-96. *Electronic control system for rooftop unit. A—Return air. B—Conditioned air. Red arrows indicate direction of airflow.*

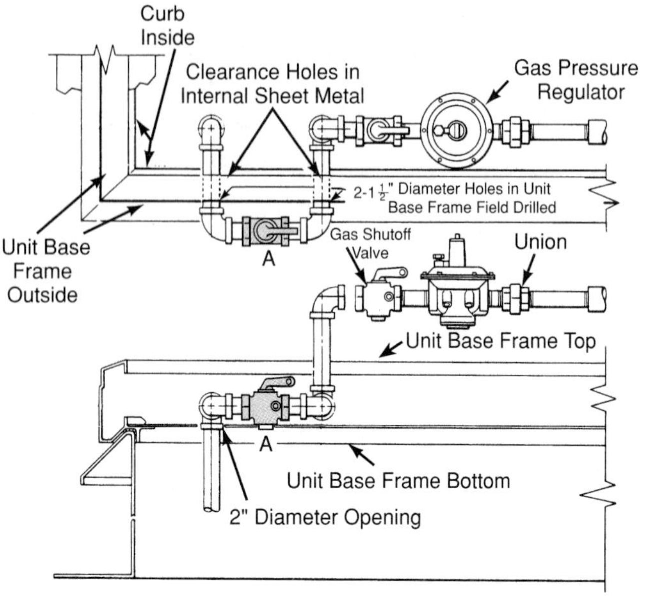

Figure 24-97. *Fuel gas line installation used on rooftop system. A—Hand shutoff valve is mounted outside casing for easy access.*

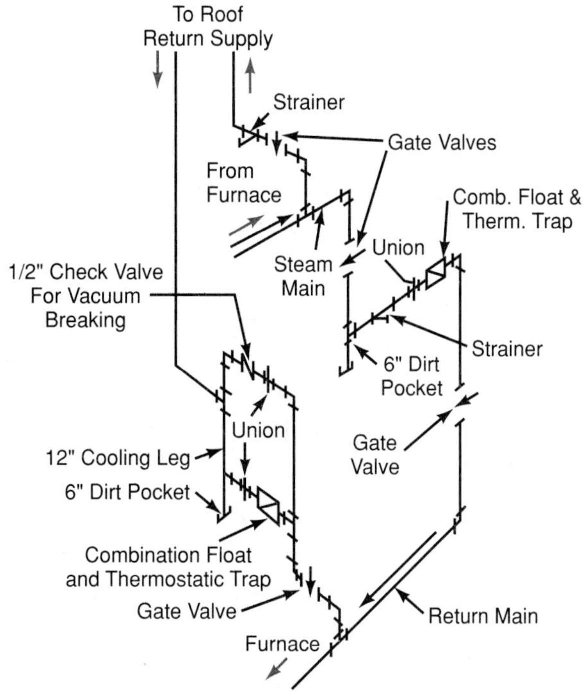

Figure 24-98. *Steam pipe installation for steam heating system used on rooftop complete air conditioning system. More than one steam path is provided. This helps prevent restricted flow if one section is partially closed. Balance is improved.*

24.21 Complete Air Conditioning Systems

The ***complete air conditioning system*** controls the temperature (heating and cooling), relative humidity, air movement, and air cleanliness. The construction and location of the components of the complete air conditioning system will vary, depending upon design. Some of these are described in the following paragraphs.

24.21.1 Through-the-Wall Systems

The complete through-the-wall system (heating, cooling, filtering, and moving air) is mounted in an outer wall. Also called a ***unitary system,*** it is popular in moderate climates. Some models are mounted flush to the outside of the building. Some models extend outside a few inches. They must be mounted firmly to the studding of the building and leveled in all directions. Joints must be weather-stripped and leakproof. Use caulking or gaskets.

Through-the-wall units that also have electric heat usually require 240 V power. Gas-fired or oil-fired systems use 120 V power. Sometimes, a separate 240 V power circuit is used for the refrigerating system.

When cooling the indoor area, these units trap the comfort-cooling condensate. They evaporate the condensate to help cool the air-cooled condenser. The plenum chambers of some units are designed to allow direct airflow into the room. Others use a separate duct for air distribution.

Through-the-wall units must be installed away from combustible material. They provide fresh air intake, combustion air intake, and a chimney flue. They are factory assembled.

24.21.2 Outside Complete Systems

One type of all-season air conditioning for homes has all the systems factory assembled into one unit. This includes the heating and cooling system, fan system, and filtering system. The unit is installed outdoors. Then, it is connected to the duct system of the house or building. A conditioned air duct connection and a return duct connection are all that is needed. These systems are an adaptation of rooftop units made for commercial and industrial use. See **Figure 24-99.**

The advantages are:

- The heating system, cooling system, and fan system are factory-assembled and tested.
- Installation consists of mounting the unit, connecting electric power, gas line, and two duct connections. **Figure 24-100** shows two views of this system.

24.22 Ice-Based Systems

An ice-based system that builds and stores ice is shown in **Figure 24-101.** The system makes use of low-cost electrical energy available during off-peak periods. The stored ice is built on Temp-Plate® evaporator plates located above the water/ice storage tank. When ice reaches approximately 5/16" (8 mm) thickness, it is

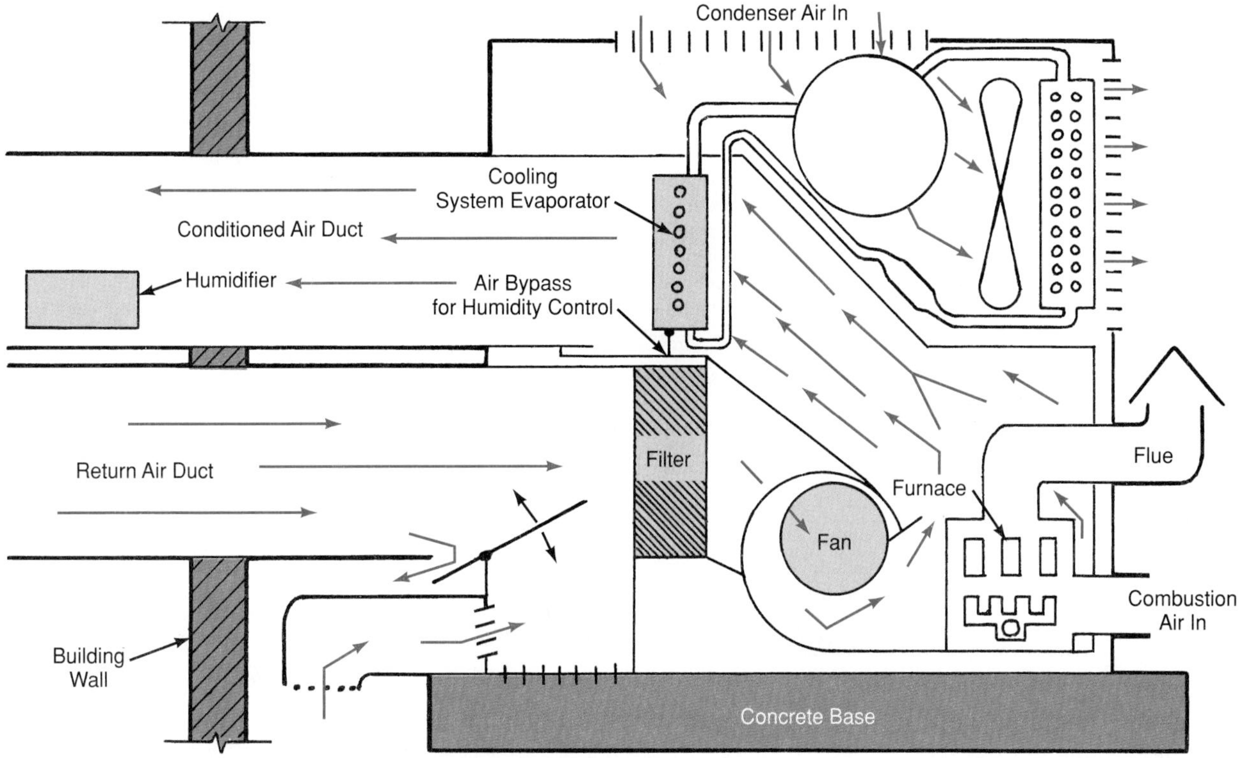

Figure 24-99. *All-season air conditioning system installed outside house or office.*

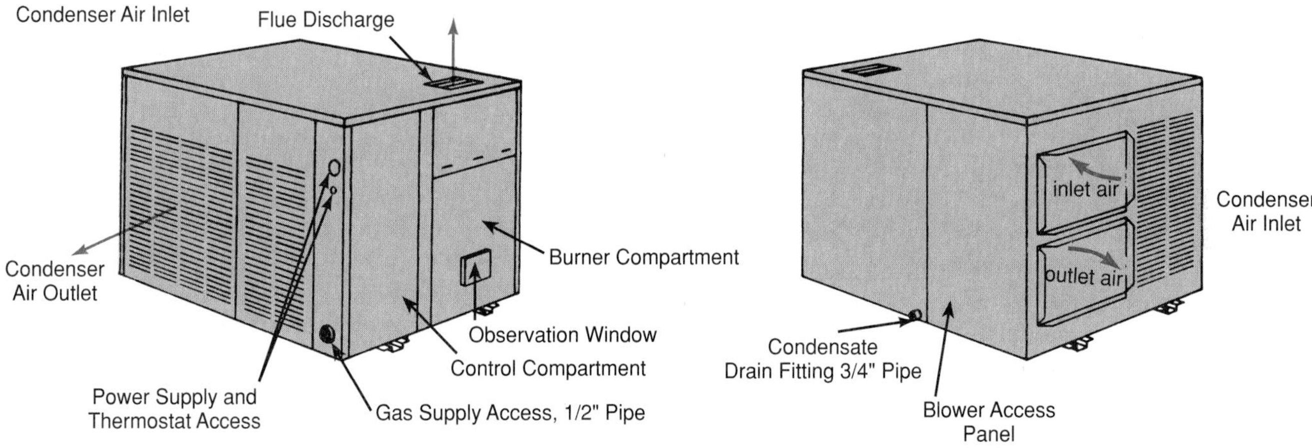

Figure 24-100. *Complete air conditioning system designed for connection to building air conditioning system. (Fedders North America)*

released for storage in the tank. The ice is then also used as a water chiller system.

During the ice-building/chilling mode, **Figure 24-102,** refrigerant is compressed and condensed. It is then passed through a high-side float to a liquid receiver. The refrigerant then passes through a plate-type evaporator and back to the compressor. Water passes over the evaporator. This causes ice to form on the outside surfaces of the evaporator. During the ice harvesting mode, **Figure 24-103,** the refrigerant system uses a hot-gas defrost cycle. Hot refrigerant gas travels through the evaporator for a few seconds. This causes the ice that has built up on the evaporator to fall into the storage tank. The chilled water in the storage tank is then circulated. It may be used for air conditioning or for refrigeration.

24.23 Total Energy Systems

Recently, many large buildings have been constructed using "total energy" or "single energy" systems. In a ***total energy system,*** all the energy-using devices are designed to capture and make use of all the energy of combustion before discarding the by-products. The system generates whatever electricity is needed in the

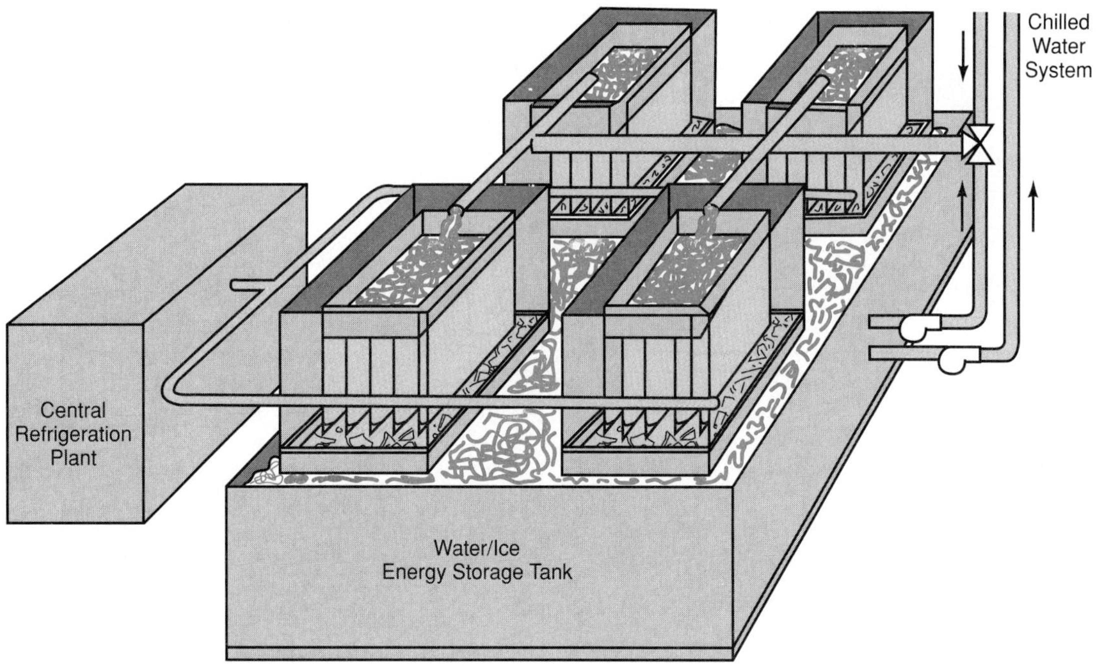

Figure 24-101. *An ice harvester/chiller module with a central refrigeration plant. (Paul Mueller Company)*

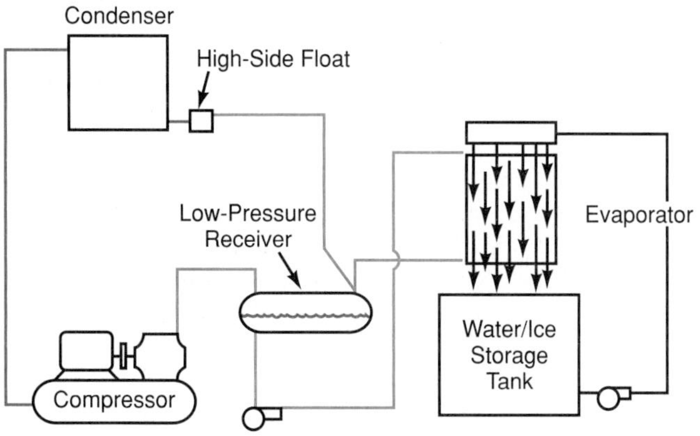

Figure 24-102. *An ice-building/chilling mode. Note flow of refrigerant to produce ice. (Paul Mueller Company)*

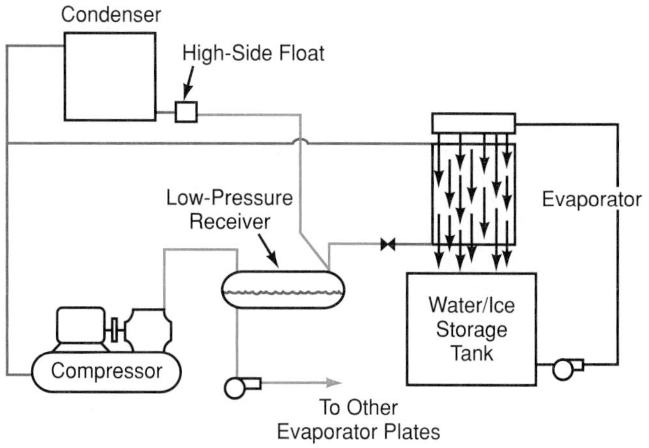

Figure 24-103. *Ice harvesting mode. Note flow of hot refrigerant vapor to evaporator plates, releasing sheets of ice from evaporator. (Paul Mueller Company)*

building. The controls used in this type of system are reviewed in detail in Chapter 26.

There are two advantages of a total energy system. First, a lower fuel rate can be obtained due to the large volume of fuel supplied. Secondly, more complete use is made of the energy released by burning this fuel.

In total energy systems, electricity is generated with a reciprocating gas engine, gas turbine, or steam turbine. The engines drive electric generators in the building. Hot water from the engine's cooling jacket is used as a prime source of heat. The exhaust gas is another major heat source. A gas engine or turbine uses about 33% of its fuel energy to generate electricity. Another 30% ends up in the water-cooling jacket. About 30% is released in the exhaust gases. The remaining 7% is used to heat the lubricating oil or is lost by radiation.

Heat from the cooling jacket water, exhaust gases, and hot lubricating oil can be converted to useful purposes:

- Absorption systems often use the exhaust heat for air conditioning (comfort cooling). At present, exhaust gases cannot be cooled below 250°F (121°C); thus, only about 50% or 60% of the exhaust gas heat can be used.
- Exhaust heat from turbines may also be used to raise the temperature of water in a hydronic heating system during the heating season.
- Heat from the water jacket can be used to furnish hot water.

In total energy systems, heat produced by lighting, solar radiation, people, and outside air are all considered. Each is a heat source in a total energy system. During the heating season, warm exhaust air is used to warm the cold replacement fresh air entering the building. Cooled exhaust air in the summer cools the incoming warm replacement fresh air.

There are many aspects to a total energy system. For example, heat from the return air ducts of the heating system is released through channels in lighting fixtures. The cool air passes across the lighted fixtures, where it is heated. The needed heat for that room is allowed to enter through a duct. The surplus heat is passed back to the primary duct system. It can be used wherever needed in the building.

An example of a total energy system is one used in a multi-apartment complex. A separate hot and chilled water plant is used for each independent apartment complex. A single electric power generator is used for the entire apartment complex.

The hot and chilled water plant provides the heating, air conditioning, and domestic hot water. A four-pipe heating and cooling system is used in each complex. The domestic water is circulated to each use area. It then returns back for reheating. The cooling may use high-efficiency gas-fired/air-cooled absorption chillers.

Three types of systems are in use:

- Fuel is burned to create steam. The steam drives a turbine to power an electric generator.
- Fuel is burned in a gas engine or turbine to drive an electric generator.
- Nuclear fuel is used to create steam.

In most cases, exhaust gases are used to heat one or more of the following:

- Water for heating or for consumption, or both.
- Absorption cooling systems.
- The fresh air intake.

In most cases, cooled or heated exhaust air is used to cool or heat incoming fresh air. Likewise, waste cool or warm water is most often used to cool or heat incoming fresh water. Heat from lighting is sometimes used to heat water or air.

24.24 District Heating and Cooling Systems

Central utilities that provide heating and cooling to full or partial sections of cities and towns have recently been developed. These ***district heating and cooling systems*** are common in Scandinavian countries, Russia, and Eastern Europe. They are becoming more economical and popular in the United States. The largest U.S. system is in the city of St. Paul, Minnesota.

Waste heat from an electric power plant or heat from an incinerator or large boiler facility is recycled. It is used to heat water that is then piped to buildings. In the summer, the waste heat is used in absorption cooling system chillers. The cold water produced is then used for air conditioning.

Another concept of district cooling is in downtown Chicago, Illinois. This system is different in that it produces ice at night. The ice can be used in the daytime for cooling the circulating water. See **Figure 24-104.**

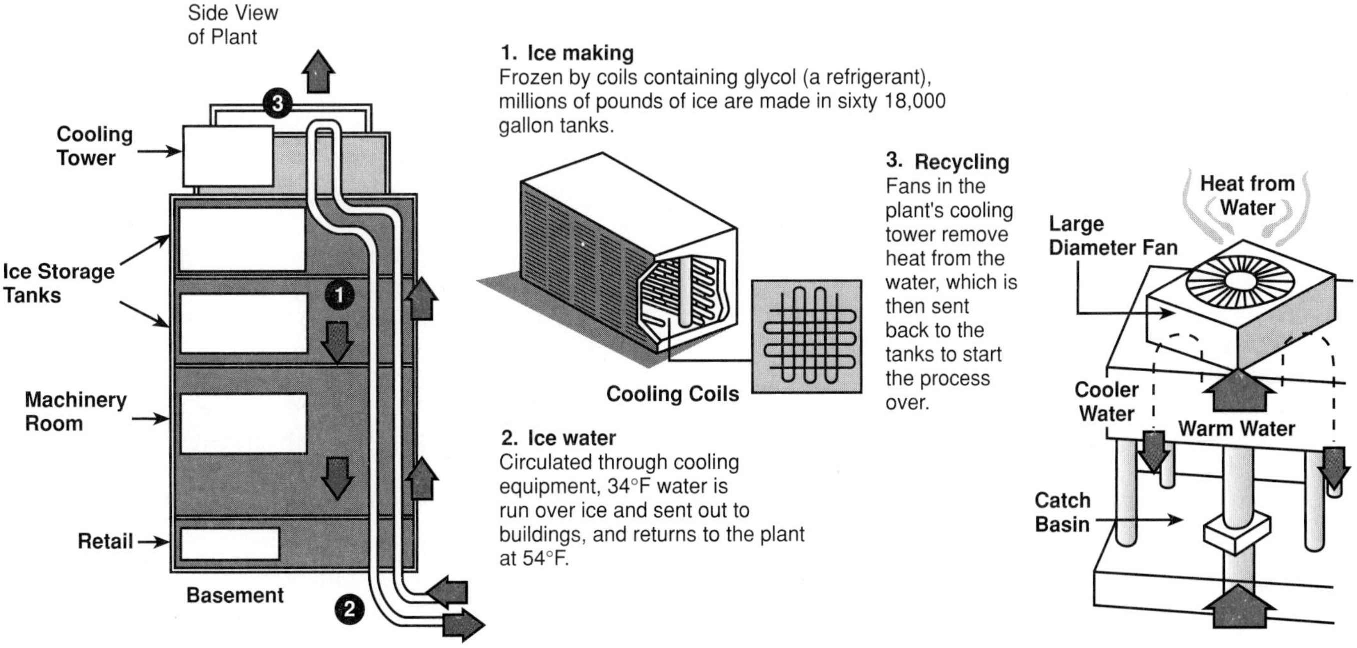

Figure 24-104. *The three basic steps in the operation of a district cooling system.*

The basic procedures areas follows. In the evening (9 p.m.), a large chiller turns on. It begins lowering the temperature in 60 large steel tanks. Each tank holds 18,000 gallons of H_2O containing various chemicals. The refrigerant runs through the coils in each tank. Ice forms around the coils and gradually expands until a tank is almost frozen. At 6 a.m., the chillers shut down. Water begins to flow over the ice. The temperature of the water drops to between 33°F and 34°F (0.6°C and 1.1°C). The water circulates through the building and returns at 54°F (12°C). The heated water is circulated to the roof of the building. The heat is then removed by large fans on the roof, and the water is circulated back to the tanks for the process to begin again.

24.25 Review of Safety

Working on a complete air conditioning system involves most of the same dangers involved in servicing all types of refrigerating systems and heating systems. Always wear goggles when working on these systems. Be sure the electrical service is off (and locked off) before working on the electrical circuits or on electrically powered parts such as the controls and fans. Always install pressure gauges when checking a system, charging it, or adding oil.

When working on rooftop units, use only a ladder in sound condition, and use it safely. Never carry equipment, tools, or supplies up or down a ladder. Lift and lower these items using a rope with a safety tie-down handled by someone on the roof and another person on the ground. Use a guide rope to keep items from swinging or turning while being raised or lowered. Avoid live electrical circuits when the roof is wet or snow-covered.

Avoid having loose panels or loose light materials on the roof during a high or gusty wind. Secure the hinged panels of a rooftop unit to prevent uncontrolled swinging of these panels during a wind.

24.26 Test Your Knowledge

Please do not write in this text. Place your answers on a separate sheet of paper.

HEAT PUMPS MODULE

1. When the air temperature is 10°F (−12°C), heat can be removed by operating an evaporator at a temperature of ______°F (______°C).
   A. −10 (−23)
   B. 0 (−18)
   C. 2 (−17)
   D. 10 (−12)
2. What is/are the operating cycle(s) of a heat pump?
   A. Heating.
   B. Cooling.
   C. Humidification.
   D. Both A and B.
3. The condensing unit in a residential central air conditioning system is installed ______.
   A. indoors
   B. outdoors
   C. on a wall
   D. Any of the above.
4. When does a 2-speed heat pump, operating in a cooling mode, cut into the high speed?
   A. When outside temperature is extremely high.
   B. When outside temperature is extremely low.
   C. When the indoor temperature is lower than the outdoor temperature.
   D. None of the above.
5. What two electrical devices does a heat pump thermostat control?
   A. The TEV and the solenoid.
   B. The motor compressor and the four-way valve.
   C. The capillary tube and the check valve.
   D. None of the above.
6. The solenoid is used to operate the ______.
   A. reversing valve
   B. TEV
   C. check valve
   D. solar circuit
7. What type of heat pump coils are used outdoors?
   A. Air coil.
   B. Water coil.
   C. Ground coil.
   D. All of the above.
8. When tightening a quick-coupler connection, what should be used?
   A. Two wrenches.
   B. A torque wrench.
   C. Pliers.
   D. All of the above.
9. What is the maximum number of check valves that can be used in a heat pump system equipped with thermostatic expansion valves?
   A. One.
   B. Two.
   C. Three.
   D. Four.
10. What is done to prevent the formation of ice on an outdoor coil?
   A. A positive air flow is provided.
   B. The system is recycled every two hours.
   C. Electric resistance heaters are used.
   D. Any of the above.

LARGE SYSTEMS MODULE

11. Reciprocating compressors are ______.
   A. very smooth and quiet operating
   B. positive displacement
   C. negative displacement
   D. quiet units

12. Which type of liquid chiller system is considered to be very efficient?
    A. Scroll.
    B. Screw.
    C. Reciprocating.
    D. Centrifugal.
13. A ______ compressor is used by an absorption chiller.
    A. reciprocating
    B. scroll
    C. screw
    D. None of the above.
14. How are external-drive motors cooled when used on chillers?
    A. Refrigerant cooled.
    B. Exhaust system cooled.
    C. Centered air fan.
    D. Air in the room.
15. At what temperature is the chilled H_2O circulated throughout a building using a chiller system?
    A. 35°F (2°C).
    B. 40°F (4°C).
    C. 45°F (7°C).
    D. 48°F (9°C).
16. The water circulating throughout the interior of a building is warmed by approximately ______°F (______°C).
    A. 5 (2.8)
    B. 10 (5.5)
    C. 12 (6.7)
    D. 15 (8.3)
17. What type of rooftop unit is used in large buildings?
    A. Heating.
    B. Cooling.
    C. Heating and cooling.
    D. Any of the above.
18. Which of the following compressors is commonly used in chilled water systems?
    A. Welded-hermetic.
    B. Semi-hermetic.
    C. Direct-drive external.
    D. All of the above.
19. Screw compressors use ______ helical rotors.
    A. 2
    B. 3
    C. 4
    D. Any of the above.
20. Water is an excellent secondary refrigerant because it ______.
    A. is largely noncorrosive
    B. has specific heat value
    C. is inexpensive
    D. All of the above.

An array of solar cells. Here, the solar arrays are being tested. Note the protective base and coverings used for the solar panels. (PHOTOCOM, Inc.)

Chapter 25

SOLAR ENERGY

Key Words:

active solar energy system
heat pump
N-type semiconductor
P-type semiconductor
passive solar energy system
photovoltaic
selective absorber surface
solar cell
solar collector

Learning Objectives:

After studying this chapter, you will be able to:

- Explain the nature of solar energy and the physics of energy conversion.
- Discuss ways of collecting and converting solar energy for space and water heating and for producing electricity.
- **Follow approved safety procedures.**

The sun radiates energy to space as a result of nuclear fusion. The solar energy (sun) input heats the earth. Solar energy is also one of the causes of weather changes.

The solar heating of the atmosphere creates winds. The sun is therefore one indirect source of wind energy. This energy can be converted to mechanical or electrical energy by wind machines. Ocean waves are caused by solar energy. Hydroelectric energy generated by water turbines at Niagara Falls, Grand Coulee Dam, and Hoover Dam is directly related to the sun. Evaporated water is condensed in the atmosphere, and the resulting rain fills reservoirs. The reservoir water is then used to drive the water turbines in hydroelectric power plants.

Wood burned in stoves and fireplaces gives back solar energy in the form of heat. This solar energy was absorbed by the tree as it grew. Photochemical processes change solar energy to fuels. These are the same processes by which fossil fuels (coal, oil, and natural gas) are formed.

Most energy sources are the indirect result of solar energy conversion processes. This chapter describes methods of converting solar energy directly to useful heat and electrical energy. This energy can then be used as the power source for cooling systems.

25.1 The Nature of Solar Energy

Visible light is only a small fraction of the radiant energy present. Our eyes are sensitive to about 25% of the energy in solar radiation.

Solar energy (solar radiation) is electromagnetic energy. The energy in a beam of radiation can be described by thinking of the beam as a stream of particles called photons. The photons in a beam of solar radiation have a wide range of energies. Some photons have low energy (infrared). Others have high energy (ultraviolet). Every beam of solar radiation has a mixture of many different photon energy levels.

Solar radiation has both particle and wave characteristics. In describing the conversion of solar energy to electricity, it is helpful to refer to the energy of photons.

An electromagnetic wave, as shown in **Figure 25-1,** has both wavelength and amplitude. The wavelength is the distance between two peaks. The amplitude is the height of the wave as shown. Think of the radiation wave as an ocean wave approaching a beach. The energy of a wave for any given amplitude depends upon the wavelength. The shorter the wavelength, the greater the energy of the wave of radiation.

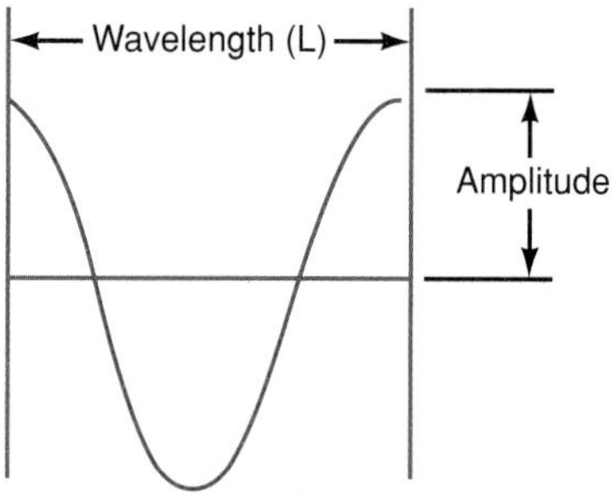

Figure 25-1. *Electromagnetic radiation is described as a wave with wavelength, (L), and amplitude as shown.*

The energy of a single photon can be expressed by the formula:

$$\text{Energy} = E = \frac{hc}{L}$$

h = Planck's constant, which is 6.626×10^{-34} watt seconds squared (W·s²).
c = Speed of light, which is 3.00×10^{8} meters per second (m/s).
L = Wavelength, in meters (m).

The energy is given in watt seconds (W·s) or joules (J) per wave.

In solar energy work, the wavelength of a light wave is usually expressed in microns (one-millionth of a meter). Light radiation with wavelengths greater than 0.73 micron is called infrared radiation. Ultraviolet radiation is radiation with wavelengths less than 0.40 micron. Light radiation with wavelengths between 0.4 micron and 0.73 microns is termed visible light.

Solar energy flux (flow) at different wavelengths as received by the earth is shown in **Figure 25-2.** The area under the curve represents the total energy flow. Outside the atmosphere this is approximately 1.35 kW/m². This is known as the solar constant. The atmosphere absorbs and reflects much of the solar energy due to the presence of water vapor and other gases, particularly carbon dioxide and oxygen.

At the earth's surface, about 45% of the incident radiation is visible light. From outside the atmosphere to the earth's surface, most ultraviolet radiation is absorbed by ozone. The infrared radiation is partly absorbed by both water vapor and carbon dioxide.

The radiation given off by a hot object changes with its temperature. At a low temperature, an object like a stove appears black. This indicates that most of the radiation given off is infrared radiation, outside the visible range. As the object is heated, it gives off more and more radiation in the visible range. It may first appear as a red glow. With further heating, it will change from red to yellow or white. When the object gets hotter, the wavelength at which the most radiation is given off becomes shorter and shorter.

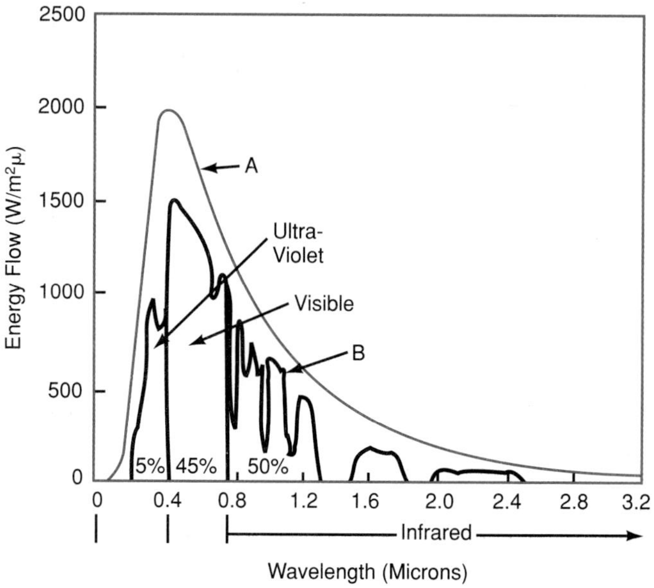

Figure 25-2. *Solar radiation at different wavelengths of light. Curve A illustrates the solar radiation outside the atmosphere. Curve B illustrates an approximate radiation flow at the earth surface, indicating the atmospheric absorption of light. Also indicated are the approximate percentages of energy in the ultraviolet, visible, and infrared radiation regions.*

Figure 25-3 shows how radiation given off by an object varies with temperature. At 10,000°F (5500°C), much of the radiation is in the visible wavelength range. This is the temperature at the sun. At 260°F (130°C), only a small part of the radiation is visible. Most of the solar

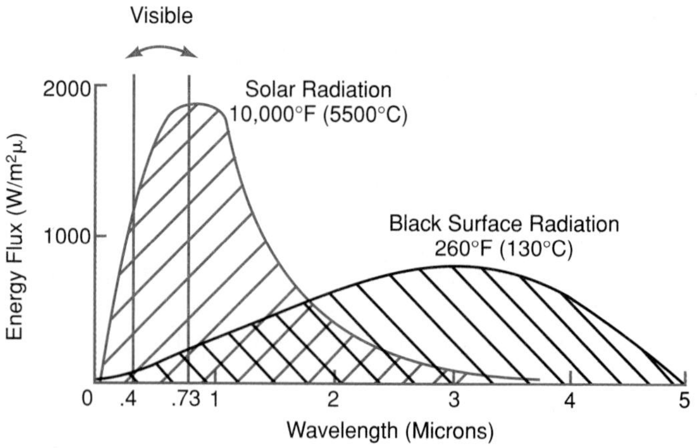

Figure 25-3. *Illustration of the comparison of the energy of radiant rays from the sun with the heat energy of a heated object.*

radiation is at wavelengths less than 3 microns. An object at 260°F gives off much less radiation. However, as is shown, most of this radiation is at wavelengths greater than the visible wavelengths (infrared).

The peak of the solar radiation curve occurs at the wavelengths of yellow light. Thus the sun appears to be yellow. This shift in the distribution of radiation with temperature is critical to the solar collector design.

25.2 Solar Energy Systems

There are two types of solar energy systems, passive and active. ***Passive systems*** depend on solar radiation striking directly on the area to be heated. The best example of passive solar heating is the conventional greenhouse. The energy flows through the glass into the area where the plants are growing. Passive solar energy systems usually require no auxiliary pumps or blowers to distribute the heat collected.

Large window areas on the south wall allow solar energy to be used during winter. This can also be considered a passive solar heating system.

In ***active solar energy systems,*** the solar energy is absorbed into a collector. The energy is then transferred from the collector and stored. It is distributed by an auxiliary circulation system.

Many new residential solar applications use a combination of both passive and active systems. This is referred to as a hybrid solar energy system.

25.2.1 Components of a Solar Energy Heating System

A solar energy heating system must have parts that will provide the following functions:

- Collect solar radiation.
- Circulate heat from the collector and move it to the space being heated.
- Store heat for later circulation when solar radiation is insufficient to heat the space.

A passive solar system has parts that collect the radiation. However, these systems depend on natural radiation. They also depend on convection to circulate the heat through the space being heated.

Active solar heating systems have mechanisms that can store the collected heat. Then, at night or on cloudy days, a circulating system draws heat from storage and moves it to the space.

Existing homes usually require an active solar heating system. However, in the design of new homes, passive systems can be easily incorporated. They require less energy input because pumps and blowers are not needed to circulate heat.

Solar Collectors

The rays from the sun are converted to heat upon striking a dark surface. This applies to visible rays, infrared rays, and ultraviolet rays.

Heat from this solar radiation can be trapped. The simplest trap is an insulated black surface. A surface looks black because it absorbs visible light. Black surfaces will absorb infrared and ultraviolet radiation as well. Coatings of lampblack or fine carbon powder produce surfaces that are nearly black. In the sun they will absorb more radiation. Therefore, they will get much hotter than white or shiny surfaces. A dark surface of about one square meter, exposed to bright sunlight, will absorb about one horsepower (746 watts) of energy.

An object insulated on the back and placed in the sun will absorb radiant energy and get very hot. The temperature will increase until the radiation from the object equals radiation received from the sun. If the object is black, the emitted radiation will be distributed over the spectrum.

The maximum surface temperature a black object can reach in full sunlight is about 253°F (123°C). The temperature may be increased by trapping the radiation emitted by the surface. To accomplish this, a glass cover is placed over the black surface. Low-iron tempered glass is used. It has a higher transmission of heat than float glass. A sketch of a solar collector using this trapping principle is shown in **Figure 25-4.** The glass allows radiation with wavelengths less than about 2 microns to pass through. It is absorbed by the black surface. At temperatures below 253°F (123°C), the black surface emits most of its energy at wavelengths greater than 2 microns. This radiation is not transmitted out by the glass. It is trapped inside the collector. Temperatures of 620°F (327°C) are obtainable in collectors of this type. **Figure 25-5** is a drawing of a typical flat plate solar collector showing construction details.

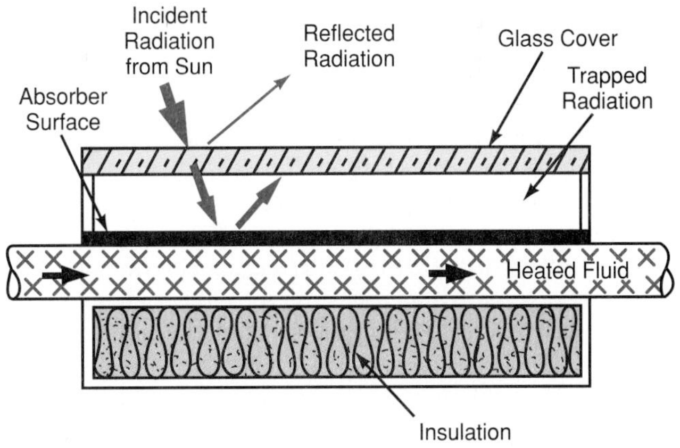

Figure 25-4. *Solar collector elements. Trapping of radiation by glass cover and insulation on the back of the collector is illustrated. Solar radiation passes through the glass, but the radiation from the absorber is trapped.*

Selective Absorber Surfaces

Special absorber surfaces are sometimes used to increase the temperature of a collector. Such a design is often called a ***selective absorber surface.*** This design acts much like the combined glass plate and absorber surface. It absorbs most of the radiation from the sun that

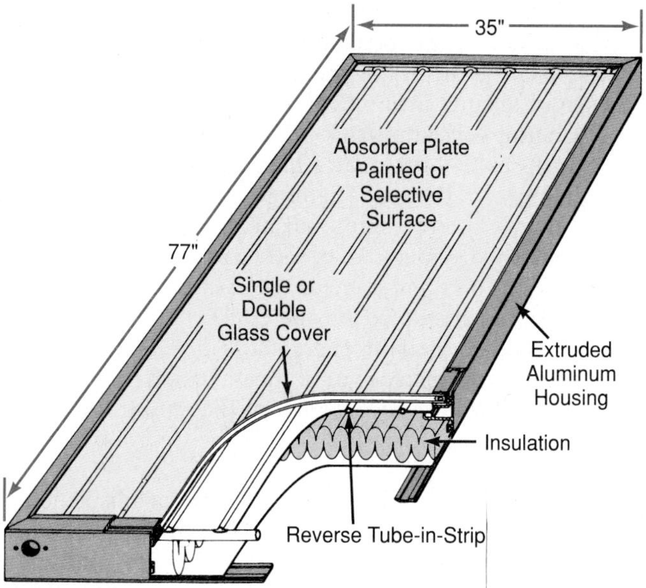

Figure 25-5. *Drawing shows dimensions and design of typical solar collector.*

has wavelengths less than a "cutoff wavelength" (around 2 microns). It does not, however, absorb longer wavelength radiation.

A surface that absorbs poorly at a given wavelength also emits (gives off) poorly at that wavelength. The surface then emits infrared radiation very poorly, and the temperature increases.

Many special paints and surfaces are being developed with this characteristic. They selectively absorb most of the sunlight but not infrared rays. Their performance is specified (rated) by an absorption to emission ratio. This ratio is the absorptivity of sunlight with wavelength shorter than the cutoff wavelength, divided by the absorptivity for longer wavelength radiation. Commercial surface materials are available with an absorption to emission ratio of about 20 to 1.

The way these surfaces work is similar to the absorption of sound by a surface containing small holes. Sound waves with wavelengths smaller than the holes are absorbed by passing into the holes. Sound waves with wavelengths greater than the hole size are reflected.

Selective absorber surfaces do not absorb as much energy as black surfaces at wavelengths shorter than the cutoff wavelength. They may appear gray. However, they do not allow energy at long wavelengths to escape. Most newer solar collectors use selective absorber surfaces rather than black surfaces.

A combined glass cover and selective absorber surface gives the best solar collector performance. Multiple glass covers with special anti-reflecting coatings also increase the radiation trapping efficiency.

Collectors with a single glass cover and a selective absorber surface have been shown to be as efficient in trapping radiation as collectors with two glass covers and a black absorber surface.

The solar collector absorber surface should be black or very dark gray. It should be slightly roughened—somewhat rougher than an eggshell condition. A rough gray surface is a better absorber than a black shiny surface. A smooth and shiny surface would reflect the radiant energy away. It would not absorb it.

The collector surface is usually best applied as an electrodeposition process. In electrodeposition, metallic particles are applied to another metal surface, using an electric current. This process bonds the surface material to the metal conduction plate. Paints are generally not satisfactory, because they will peel and crack at the high temperature of the collector.

Solar Collector Covers

A transparent cover is used on most solar collectors. This cover has three basic functions:

- To protect the absorber surface from the weather.
- To transmit sunlight to the absorber surface.
- To prevent the escape of heat collected by the absorber surface.

Glass is the most widely used cover material. It provides good light transmission and remains clear indefinitely. Typical glass covers are 1/8″ to 1/4″ thick. Special glass materials transmit more sunlight. They allow less collector heat to escape. Iron-free glass is generally used for this purpose.

Glass poses problems in sealing the cover to the other parts of the collector. Glass and metal do not expand at the same rate with temperature change. A flexible sealing material or rubber gasket is necessary at the glass-to-metal joint. This will prevent the glass from breaking as the materials expand and contract.

Plastic cover materials usually soften at high temperatures. Plastic also becomes brittle and opaque from absorption of ultraviolet light.

Some covers are formed in a bubble shape rather than a flat shape. This provides more structural support. It also collects more radiation when the sun is near the horizon. Some special collectors are formed as a tube and are sealed with a vacuum inside. These collectors are very efficient. They do not lose much of the heat collected. **Figure 25-6** shows a typical vacuum tube collector. They are much more efficient at higher temperatures.

Periodic washing of the glass will improve the light transmission. Most installations, however, require washing only once or twice a year.

Angle of the Collector

Ideally, a collector surface would be exactly perpendicular to the rays of the sun. However, a device to provide this aiming—except for extreme conditions—is too expensive. The position of the collector surface, therefore, usually remains fixed. The angle of the sun's rays to the horizon varies during the year. It is necessary to set the collector angle to give the best average collecting effect. The sun angle in midwinter is 66.55° minus the site latitude angle. The angle in midsummer = 113.45° − latitude.

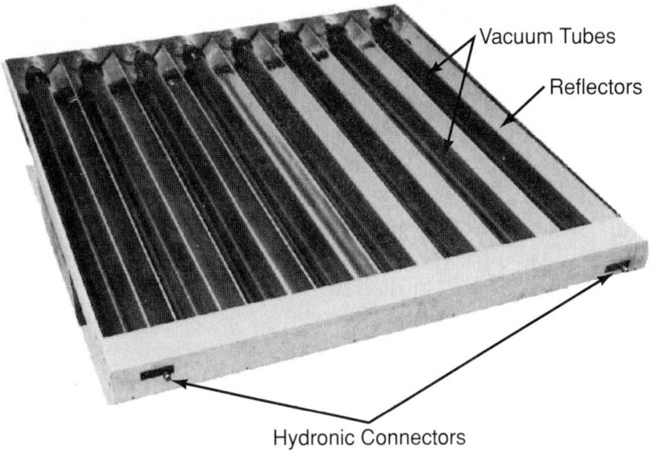

Figure 25-6. *Vacuum tube solar collectors are used for high-temperature application. (General Electric Co.)*

For solar home heating, the angle needs to be positioned to absorb the greatest amount of energy during the heating season. This setting is about 45° to 55° from the horizontal across the mid-section of the United States.

For collectors used during all seasons to provide hot water, the collector angle should be 35° to 45° to the horizontal. For year-round heating, the angle from the horizontal should be the site latitude angle or up to 10° less. The collector should face south. However, an angle from 30° east or west of south only reduces the energy collected by about 10%.

Naturally, it is important that there be no obstructions, such as buildings or trees. These might create a shadow across the collector surface. Simple shadow-indicating devices are available to aid in determining the effect of obstructions at a specific site. Meters are also available to read the solar energy intensity level at a particular location. See **Figure 25-7.**

Heat Insulation of the Collector Surface

A collector surface will both absorb solar energy and release heat energy unless properly insulated. The solar energy coming into the collector is changed to heat. The surface will then radiate heat the same as any other warmed surface. The collector surface needs to be insulated. This will cut down on the radiation of heat to the atmosphere. One or more layers of glass over the front surface may be used to hold in energy as it is converted from solar radiation to sensible heat. The back surface must also be insulated to hold the heat from escaping through the back. About 4″ of conventional insulating material will provide enough insulation on the back surface.

25.3 Solar Energy Storage Systems

Large bodies of water are a natural solar energy storage system. The surface of a lake may be frozen over in the winter. However, in spring, as the sun shines on it, the ice melts and the water warms. This heat is given to the atmosphere in the fall or winter as the water on the surface again freezes.

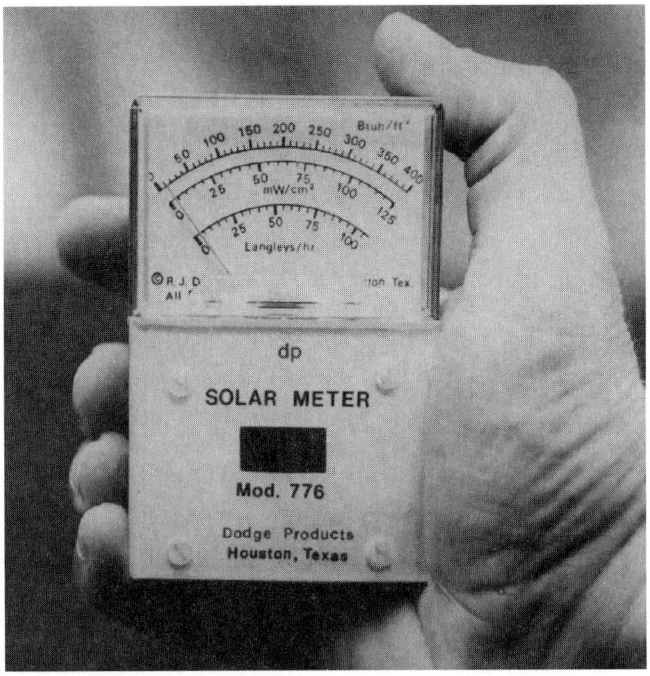

Figure 25-7. *Hand-held solar intensity meter is used to estimate the total heat rate of solar radiation into the heat collector. (Dodge Engineering)*

This system is often used for storing heated water in an insulated container. An example is the heating of water in a tank on the roof of a house. The tank is connected to the water system in the house. Many homes use this system entirely for domestic hot water, particularly in the southern part of the United States.

Rocks may also be used for heat storage. If the sun shines on a stone, the surface will be warmed. It may stay somewhat warm throughout the night.

The specific heat of water is about 1 Btu/lb.°F. The specific heat of the rock is about 0.25 Btu/lb.°F (see Section 1.22). This means that 1 lb. of water will store as much heat as 4 lb. of rocks if they are both heated to the same temperature.

Another type of storage medium used is a series of polypropylene rods in an enclosed honeycomb structure. This allows for efficient movement of air.

Heat can be stored in a chemical. Sodium sulfate crystals are sometimes used. They give up heat when they form crystals from a water solution.

25.3.1 Heat Energy Storage in a Closed Water System

To use solar energy more efficiently, closed system water heating is employed. The system gathers, circulates, and stores absorbed energy in water. The amount of solar energy absorbed depends upon the area and color of the absorbing surface. A complex system uses collector lenses to heat the water to a very high temperature.

As a practical application, a heat-absorbing solar energy system usually consists of:

- A large solar panel facing the sun.
- Water circulated through the panel to storage.
- A large insulated holding tank.

Heat from the water may be used by either of two methods:

- The water may give up its heat directly to the space.
- The water becomes a source of heat for a heat pump. The pump will then be used to transfer the heat to the space as desired.

Ethylene glycol or propylene glycol is added as an antifreeze. In mild climates, a drain-down system is used. This system drains the collector on cold nights.

Heated water may be stored in a large insulated tank. Typical solar heating systems have storage tank capacities from 1.3 to 2.5 gallons for every square foot of collector surface area.

25.3.2 Heat Energy Storage in a Warm Air System

Solar energy may also be absorbed at the collector and transferred to warm air as heat. Air is circulated through the solar panel. The heat is stored in crushed stone. The crushed stone is held in place by an insulated container. The air heated by the sun is circulated through the container. Whenever heat is needed, air in the house is circulated through the heated, crushed stone. The stone then gives up its heat to the house. The crushed stones should be from 1″ to 2″ in diameter. Too much energy is required to circulate the air around smaller stones. On the other hand, stones that are too large will not be heated completely. Only part of the stone will be effective in storing heat.

Figure 25-8 illustrates a rock storage container. The container size is about 0.3 ft^3 to 1.2 ft^3 in volume per square foot of collector area. The height of the container is usually limited to less than 6′. This allows the air to be circulated from the bottom to the top with a small blower.

There must be an empty space at the top and bottom of the container to provide uniform airflow through the rock storage.

The rock storage should be heated by flowing the air from the top to the bottom of the container. When heat is taken from the rocks, airflow should be from the bottom to the top and then out. The temperature of the rocks becomes hotter at the top when being heated. The rocks at the bottom may still be relatively cool. When heat is to be removed, the air should be as warm as possible.

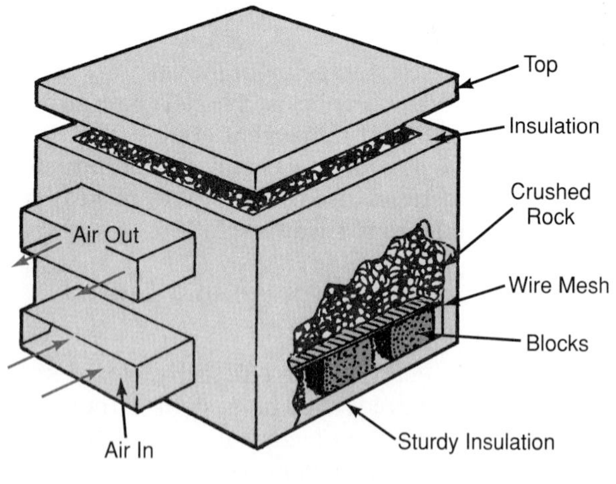

Figure 25-8. *Typical crushed rock heat storage container. Room air, blown in bottom, comes out top heated.*

25.3.3 Solar Space Heating Installations

There are three types of effective solar heating installations in common use:

- ***Natural convection closed water system.*** Heat is absorbed in the collector. Since warm water rises, the storage tank must be located above the collector. The storage tank should be quite heavily insulated to eliminate heat loss. Warm water circulates to heat the desired space through the radiator. It must be located with its inlet about even with the warm water supply. As the heat from the warm water radiates into the room, the water is cooled. It then flows down and is carried back to the bottom of the heat storage tank. There is no temperature control with this system. With bright sunshine the temperature will continue to rise. The heated space could become uncomfortable.
- ***Forced convection in a closed water system.*** In this system, a pump circulates the water. The warm water storage tank may be located without regarding the location of the collector or the radiator. Water is heated in the collector and is circulated to the storage tank and radiator. The circulation through the radiator may be controlled with a thermostat to maintain a certain room temperature. **Figure 25-9** illustrates the components of a forced-convection solar-heated closed water system. The differential thermostat turns off the pump when the solar collector temperature is lower than the temperature in the storage tank. It is usually set to turn on the circulating pump when the collector temperature is 8°F to 10°F (4°C to 6°C) warmer than the water in the storage tank. This prevents the pump from turning on and off frequently, ensuring that cold water from the collector is not mixed in the tank.
- ***Warm air heating system.*** A warm air heating system using forced air circulation is illustrated in **Figure 25-10.** The solar energy is converted to heat in the solar collector. It must be insulated so that only a small portion of the heat is lost. The heated air is circulated through the rock bed. This bed consists of crushed rock approximately 1″ to 2″ in size. Air from the collector is circulated through the rock bed. Heat is stored in the rocks. When heat is

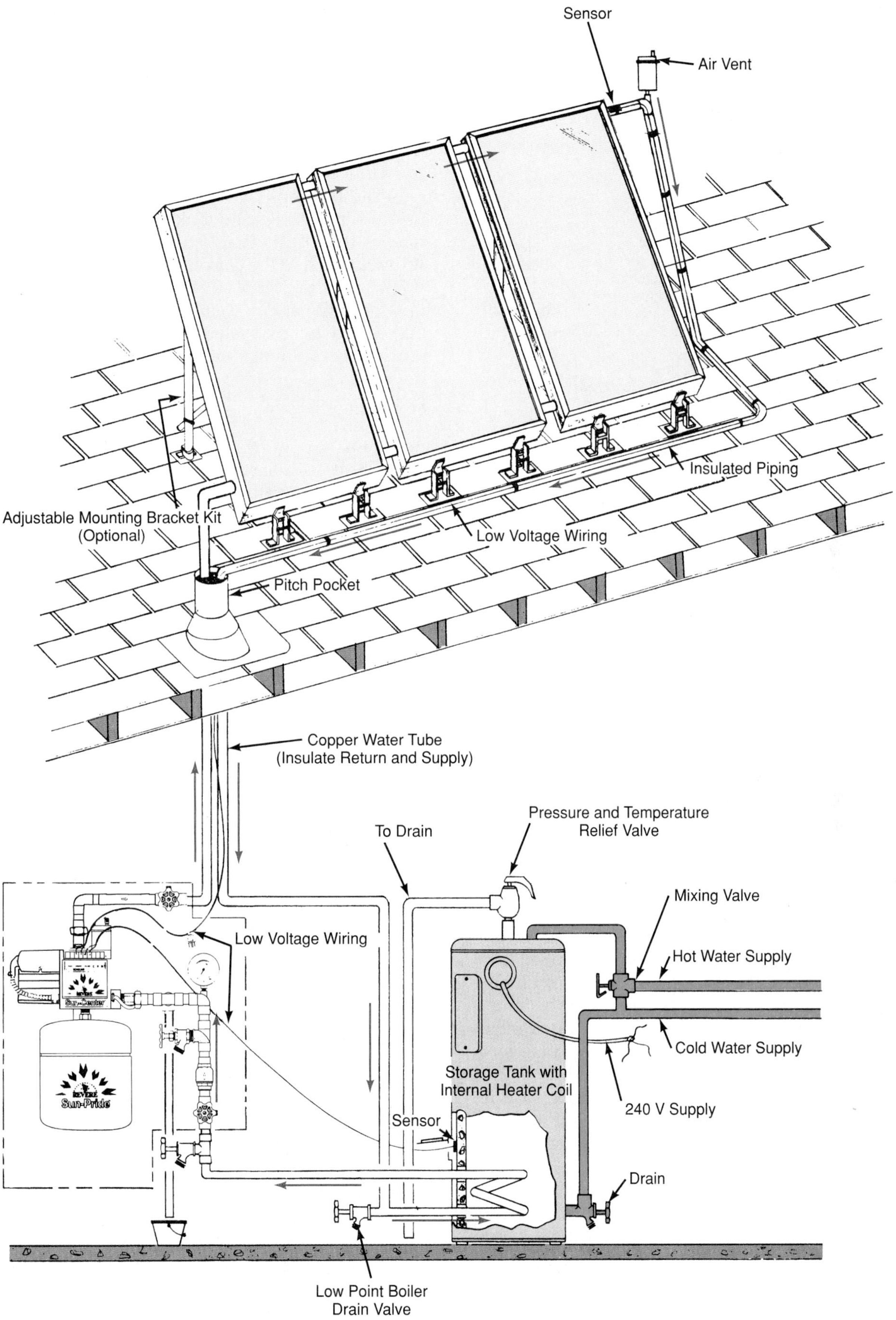

Figure 25-9. *Components in forced convection solar heated closed water system. One tank stores heat and supplies hot water.*

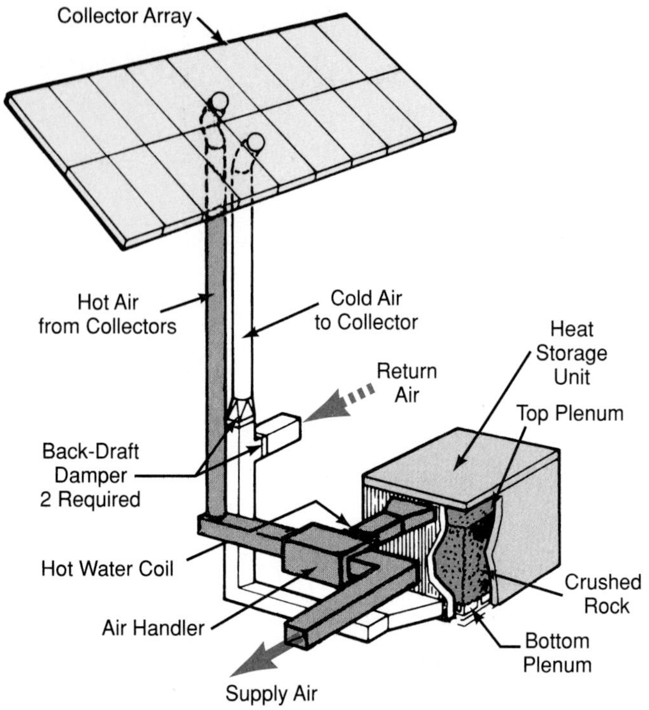

Figure 25-10. *Warm air solar heating system with forced air circulation. Heat is stored in crushed rock.*

needed in the room, air is circulated through the warmed rock bed. The warmed rock bed gives up its heat to the room air. Rocks for this purpose are very carefully washed and cleaned. This makes certain that no dust or sediment will circulate through the room. A thermostat controls the airflow. Supplementary heat, as needed, may be provided as described in Section 25.4.

25.3.4 Solar Domestic Water Heating

Perhaps the most used solar heating system is one which supplies hot water for domestic use. Heated water for dish washing, showering, and washing can be supplied easily. A solar heating system is used to preheat the water. The temperature of water supplied from a well is fairly constant throughout the year, ranging from 40°F to 50°F (4°C to 10°C). Water supply from a large municipal water supply system is drawn either from a large reservoir or storage tanks. In some locations, the water temperature may range from 35°F (2°C) in the winter to 80°F (27°C) in the summer. The heat required to supply domestic hot water at 120°F to 140°F (49°C to 60°C) can vary considerably throughout the year, depending upon geographic location.

Solar domestic water heating systems are more cost effective than space heating. There is a demand for hot water in a home year-round. Space heating is only used in the winter season.

Two types of solar water heating are in common use:

- One-tank system.
- Two-tank system.

The one-tank system consists of a single hot water storage tank. The water is preheated in the solar collector. It is circulated through a heat exchanger in the bottom of the hot water heater tank. An auxiliary heater coil in the top of the tank provides the final heating. It will bring the water to the desired temperature. This system requires replacement of the present hot water tank in retrofit situations.

The two-tank system is shown in **Figure 25-11.** It uses a separate storage tank. Water is preheated in the tank by the solar heated liquid through a heat exchanger. It is circulated to the conventional hot water heater from this storage tank. This system, with a separate storage tank, is easily added to an existing hot water system.

The two-tank system is usually 30% to 50% more efficient than a single-tank system. The cost is usually higher than a one-tank system if a new separate water heater is necessary.

Domestic hot water systems are designed to economically supply from 40% to 75% of the required hot water heat. Auxiliary heating supplies the final heat to reach the temperature required. This is 120°F to 140°F (49°C to 60°C).

Domestic hot water use is approximately 25 gallons per day for each person in a family. Solar collector sizes range from 1/2 ft^2 to 2 ft^2 per gallon of hot water required per day, depending on location. In a two-tank system, the low temperature storage tank is usually half as large as the main storage tank.

25.3.5 Solar Heating of Water for Swimming Pools

Heating of water for swimming pools can be used to extend the swimming season in many areas. The solar heating system for a swimming pool can be as simple and inexpensive as a set of black or transparent plastic bags. The reasons for this simplicity include:

- No storage is required.
- The collector must withstand much lower water pressure than space or domestic water heating systems.
- No freezing protection is necessary.

These swimming pool heaters require only a low pressure pump, plastic solar collectors, and piping.

The size of a solar collector for a swimming pool is usually approximately half of the pool surface area. This will provide good spring and fall heating. The efficiency of swimming pool solar heaters can be 70% to 80%. This is due to the low temperature water that the collectors must supply. The water temperature required is only 80°F (27°C).

25.4 Supplementary Heat

There may be times when the sun does not supply enough heat energy to the collector. In such cases, supplementary heat must be supplied. Electric resis-

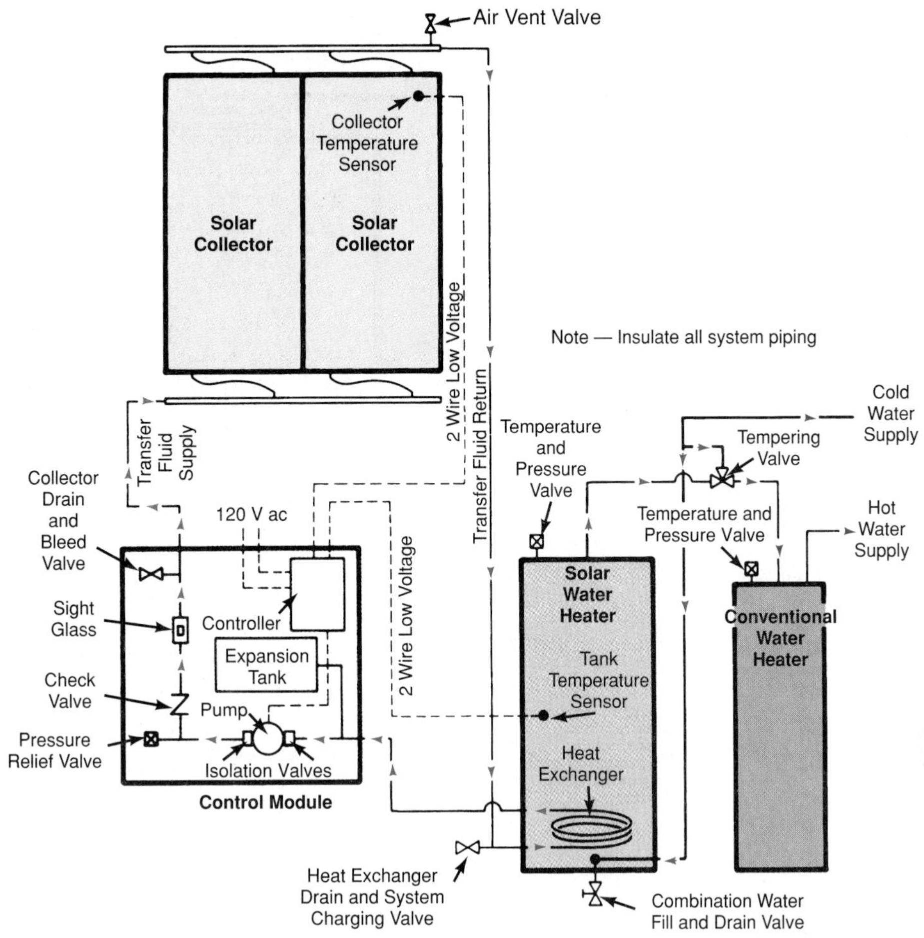

Figure 25-11. *Two-tank domestic hot water heating system. Conventional water heating tank supplies auxiliary heat.*

tance heating, a gas or oil burner, or a heat pump may be used.

In some domestic solar hot water heating systems, the water is preheated by solar energy. It is then heated to the temperature required by a supplementary heater.

25.4.1 Supplementary Electric Heating

Electric resistance heating may be used to supplement solar heat. With a warm air heating system, a direct electric resistance radiator may be installed in the air duct. A thermostat may be used to control the amount of electric heat needed.

In a hot water space heating system, an electrical heating element may be installed in an auxiliary tank. It will provide supplementary heating when the heat stored in the main tank is exhausted.

In some heating situations, individual room resistance heating radiators may be installed.

25.4.2 Supplementary Oil and Gas Heating

If oil or gas is used for supplementary heat, a separate furnace is installed. This is shown in **Figure 25-12.** The burner will be controlled by a thermostat. The thermostat will turn on the burner when heat from the solar source is below the amount of heat required. In systems using warm water radiators, the burners may be used to heat the water in an auxiliary tank.

25.4.3 Heat Pumps

A ***heat pump*** is probably the most efficient method of supplementing solar system heat. The heat pump may also be used with an air conditioning system to remove heat. The heat pump uses a refrigerant fluid the same as an air conditioner. (See Chapter 24.) It can either add heat to a room or absorb heat from a room. **Figure 25-13** illustrates a heat pump installation.

A heat pump with a solar energy collector is the most efficient way to maintain desired temperatures. Heat supplied by the solar collector fluctuates. Theoretically, a heat pump can be efficient on a 4:1 basis. The amount of heat delivered can be four times the heating value of the electrical current required to drive the mechanism. In application, this theoretical value is never reached. A ratio of 3:1 is considered to be very good. In many practical applications, this ratio may not be over 1.5:1.

The heat pump is a natural device to be used in connection with solar heat. It is possible to use solar energy to warm water in a forced convection system. The solar energy may not bring the water temperature up to a level for maintaining a comfortable room temperature.

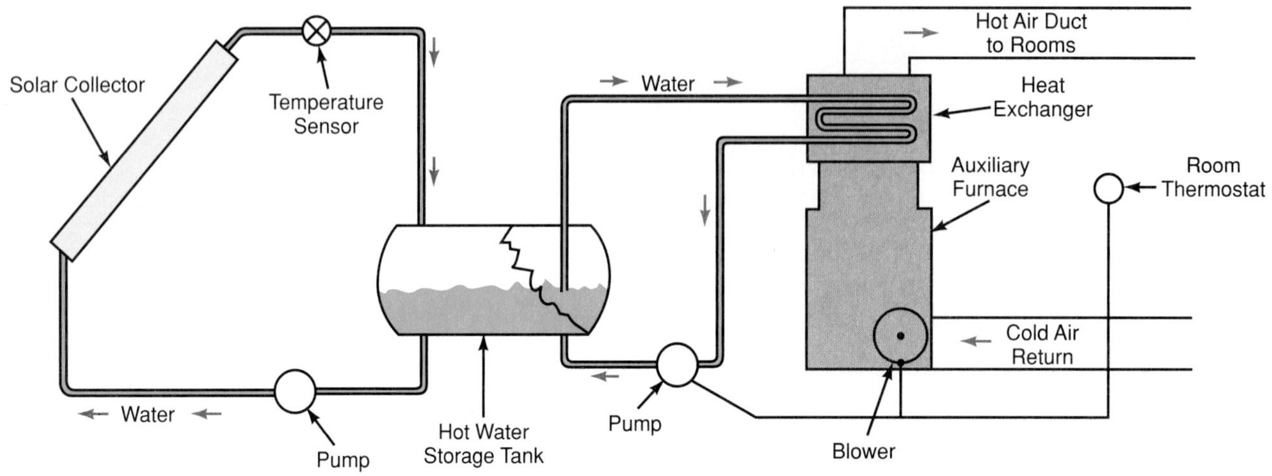

Figure 25-12. *A schematic of a liquid solar collector system used in forced air heating application. The heat exchanger is located in the hot air duct of a conventional furnace, which supplies auxiliary heat.*

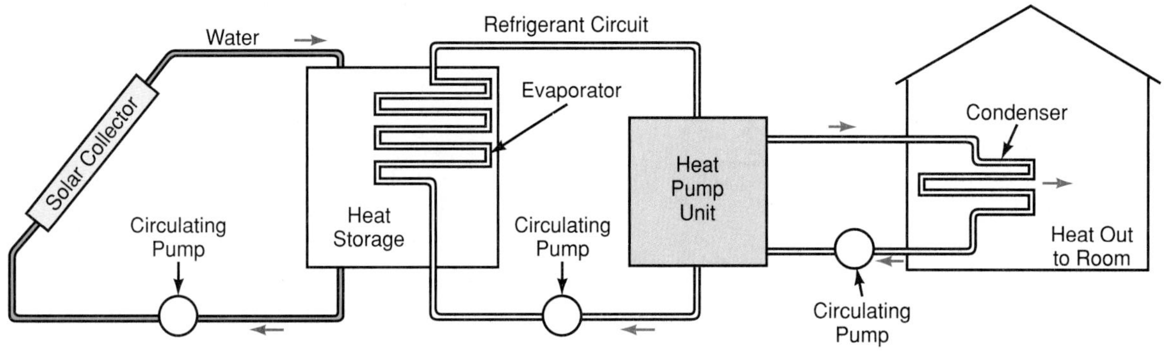

Figure 25-13. *Solar heating with supplementary heat supplied by a heat pump. Heating cycle is shown.*

Through the use of a heat pump, it is possible to extract heat from the heat storage system. The temperature will then be increased to a comfortable level. In this case, electrical energy is being used to supplement the solar energy. However, the heat obtained is two to three times as much as would be obtained by direct electric resistance heating.

25.5 Solar Energy Cooling Systems

Solar energy may also be used for cooling. This is usually done by using absorption system refrigeration. See Chapter 17. Absorption systems require a heat source. The heat is used to drive the refrigerant out of a solution. The sun can supply the heat required to operate absorption cycles.

Usually the solar energy must be concentrated by lenses or mirrors. This will increase the solar radiation temperature to meet the absorption system's requirement.

At present, solar energy is chiefly used for heating. The required mechanisms for cooling are quite expensive.

Experimental systems have been developed using solar energy as the heat source. The system elements are shown in **Figure 25-14.**

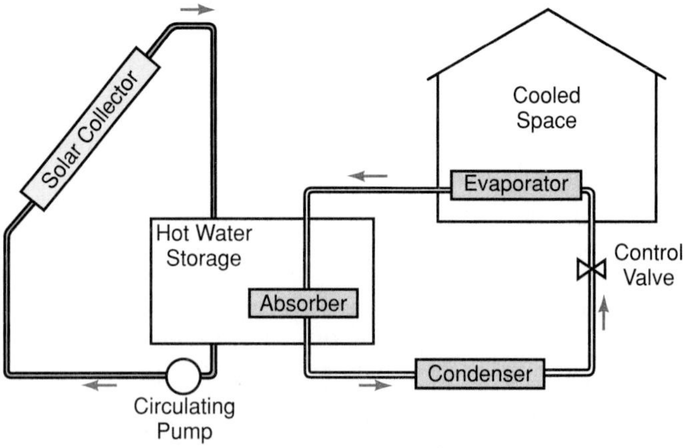

Figure 25-14. *Solar energy air conditioning system utilizing an absorption refrigeration unit. Solar energy supplies heat for the absorption unit. A minimum external power supply is necessary in this cooling unit. Cooling cycle is shown.*

Experiments are also being made with solar dynamic cooling systems. In these systems, the solar energy is converted to electrical or mechanical energy. This energy is then used to drive a compressor. Concentrator collector systems, using lenses or mirrors, are needed. They are required to provide the high temperatures needed for efficient operation.

Solar cooling systems have the highest input when the cooling requirements are greatest. Energy storage requirements are minimized and the energy is available when needed most.

25.6 Converting Solar Energy to Electricity

Solar energy can be converted directly from radiant energy to electricity. The most common device used for this conversion is the ***solar cell.*** (This is also called a ***photovoltaic*** device.) Solar cells, although expensive, are a preferred electrical energy source at remote locations. They are also useful where the power requirements are relatively small. Power for remote weather stations and communications stations is often supplied by solar cells. They are maintenance-free and do not require fuel to supply power. Other applications include situations where a small amount of power is required. These include bridge and pipeline electrochemical corrosion protection systems.

25.6.1 Solar Cell Construction

The components of a typical solar cell are shown in **Figure 25-15.** The principal elements required are two semiconductor materials. These are called N-type semiconductors and P-type semiconductors. An ***N-type semiconductor*** carries current by means of electrons, or negative charges. A ***P-type semiconductor*** carries charges by positive charges, sometimes referred to as "holes." "Holes" refers to the space remaining after an electron has been removed from an atom.

Photons in the light striking a solar cell give up their energy to electrons. This occurs at the junction between the P-type and N-type materials. The electrons with excess energy are then stored in the N material. Electrodes are attached to each material to conduct the charge to an external load circuit.

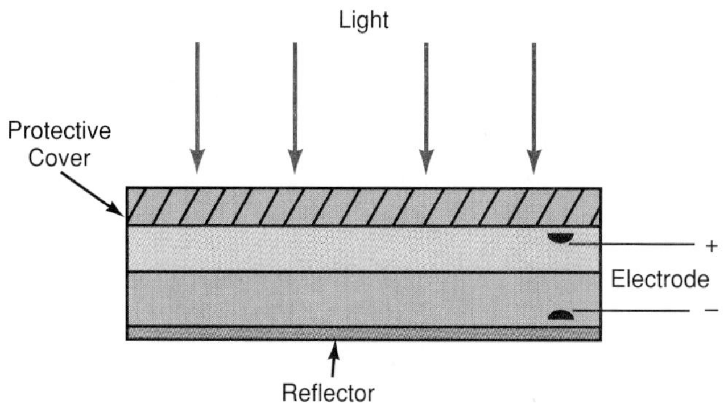

Figure 25-15. *Solar cell schematic indicating the physical components. The sketch is not to scale. The cells are very thin compared to their length and width.*

P-type and N-type materials in most solar cells are mainly silicon. The silicon of the N-type material contains a very small quantity of a material which, compared to silicon, has an excess of electrons. Nitrogen, phosphorus, or arsenic are materials having an excess of electrons.

The silicon of the P-type material has a small quantity of material which, compared to silicon, has a deficit of electrons. Aluminum, boron, gallium, or indium are such materials.

The electrodes on the back are also reflectors. Therefore, any light that goes through the cell is passed back through again. In this way, more light can be absorbed.

The P-type material must also be transparent. This allows the light to reach the junction between the P- and N-type materials.

Power cells are presently made with a thin slice of silicon. A small amount of boron dopant is added to the molten silicon during the crystallization process. This makes the material a P-type semiconductor. A small amount of phosphorous is diffused into the front surface of the wafer. It forms a shallow N-type layer in the silicon. This creates the N-P junction which makes the silicon wafer a solar cell. Metallic contacts are then deposited on the front and back of the cell to allow for connections and improve the flow of current out of the cell. An anti-reflective coating is also applied to improve the absorption of light by the cell.

Finished solar cells are then connected together in series and parallel to provide the desired voltage and current output. The assembly of cells is then encapsulated with weatherproof plastics to a sheet of tempered glass, framed with anodized aluminum. A weather-proof junction box is then added to allow for external connections to the module. The resulting photovoltaic module will withstand even the most severe environments. The design life is in excess of twenty years.

25.6.2 Photovoltaic Solar Cell Applications

Electrically, solar cells operate similarly to batteries. A single cell produces a voltage of approximately one volt. The voltage depends on the solar cell material, ranging from 0.5 V to 1.5 V. For a specific solar cell material, the voltage will remain constant under normal operating conditions. It does not change very much with light intensity. However, under operating conditions, the current produced increases as the sunlight intensity increases. Therefore, the power output increases will also be greater.

In some situations, the sunlight is focused to increase the intensity. The effects of increasing the cell temperature can harm the power output and efficiency. Under these conditions, cooling of the solar cells is beneficial. High intensity solar cells, like those shown in **Figure 25-16,** operate at solar intensities up to 50 times normal sunlight.

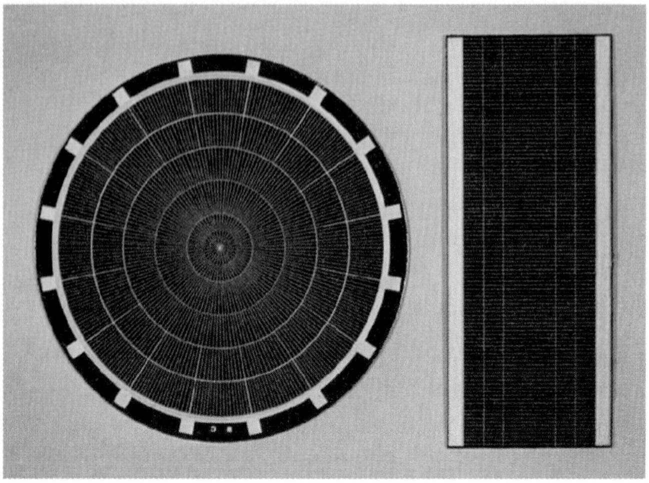

Figure 25-16. *Photograph of typical solar cells which convert solar energy into electrical energy. (Solarex Corporation)*

Solar cells are produced in a wide variety of shapes and sizes for various uses. A 12-volt module is shown in **Figure 25-17.** This particular cell is designed to power a water pump for circulating fluid through hot water solar collectors. They are ideal power sources in situations where power demand increases as sunlight intensity increases. They are useful as a power source in warm climate cooling applications.

Photovoltaic solar cells are used as the primary energy source for most space satellites. They are a key factor in the development of communication satellites used for telephone, radio, and television. Another application is powering hand calculators that operate from interior room lighting as well as sunlight. They are also widely used to provide electricity in remote areas. This provides energy for pumping water, lighting, and telephone communications in such areas.

Figure 25-17. *Photovoltaic module for powering a 12-volt circulating pump. Unit has film of polycrystalline silicon as the absorber. (Solarex Corporation)*

Solar cells are connected in series similar to batteries, in order to obtain higher voltages. The total voltage produced is then the sum of the voltages produced by each cell. The electrical circuit may be as simple as shown in **Figure 25-18.** This indicates a cell connected to a dc motor. The motor would deliver power depending on the sunlight intensity. (At night the motor would not run at all.)

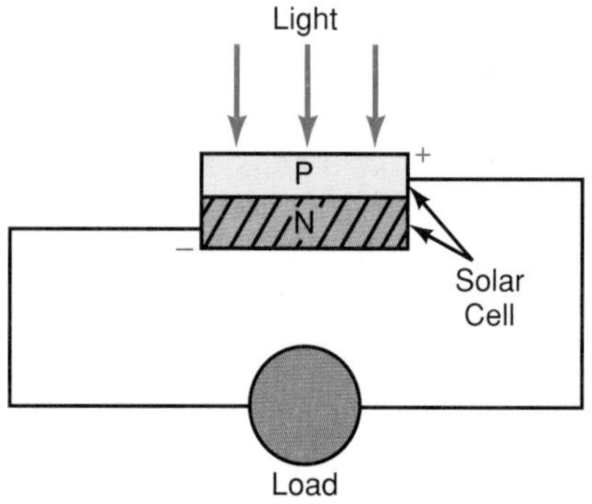

Figure 25-18. *Simple solar cell circuit. Power produced will depend on sunlight intensity.*

Solar cells usually need to be connected to electrical storage devices such as storage batteries. A motor can draw power from the solar cell directly in bright sunlight. When there is little or no sunlight, it draws power from the storage battery.

An electrical circuit to provide controlled power would include a storage battery. It would be connected as shown in **Figure 25-19.**

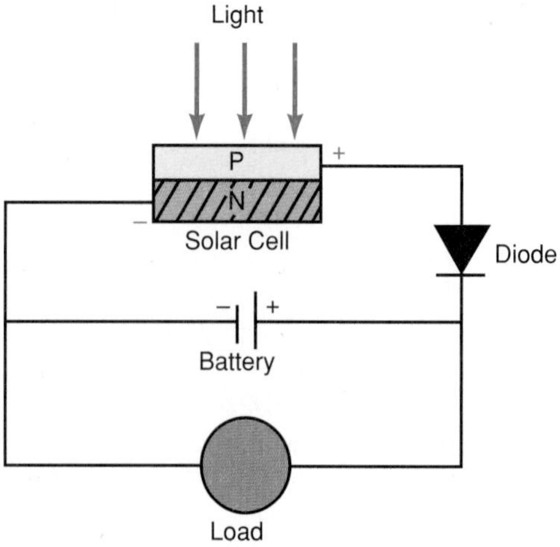

Figure 25-19. *Solar cell circuit provides for electrical energy storage in a battery. However, it still provides a constant flow of energy to a load.*

In electrical energy storage systems using batteries like the one shown, the battery cannot be charged while it is delivering power. Thus, the total power to the load will not be greater than the solar cell or battery alone can provide.

A solar cell produces dc power like a battery. If ac power is required to operate an electrical appliance or ac motor, an inverter must be added to the circuit. Modern inverters use solid state semiconductors to produce ac power from dc power. They have an efficiency of about 90%. Inverters are expensive, however, and dc motors are often used directly to avoid their cost. See Section 6.6.4.

25.6.3 Solar Cell Performance

Solar cell performance depends heavily on sunlight intensity and the solar cell temperature. **Figure 25-20** shows a typical performance curve for a working cell. The electrical power delivered by the cell increases with the sunlight intensity. If intensity continues to increase, the output power eventually levels off as shown. The leveling off is the result of excessive heat buildup in the cell. This heat may be removed by circulating cooling fluid behind the cell or cells.

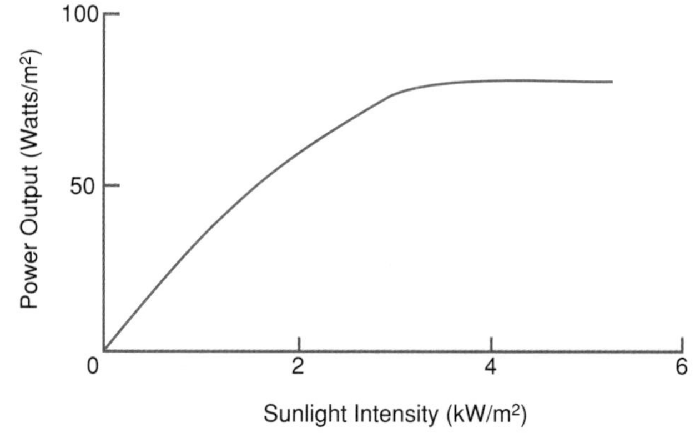

Figure 25-20. *Typical solar cell performance curve. This illustrates how the power output increases as the sunlight intensity increases. At high light intensity, a limit is reached where increased light intensity does not increase output due to heat effects.*

Current output can be increased by using a focusing collector to increase the light intensity on the cell. Presently, most practical applications are designed without a focusing system. Experimental focusing systems with intensity up to 40 times full sunlight have been successful when the cells are cooled. However, most research is directed at producing cells which do not require focusing. A focusing system is very expensive because it must follow or "track" the sun across the sky.

The performance curve shows that power output increases with solar intensity. The accompanying increase in power becomes less with an increase in intensity. Often, manufacturers of solar cells will indicate solar cell efficiency in their specifications. **Figure 25-21** shows how this efficiency varies with solar intensity.

The efficiency is the output power divided by sunlight intensity. This indicates that the efficiency can be very high at very low light intensity. It decreases rapidly as the light increases. Typical solar cells at low light intensity can have an efficiency of 20%. At full sunlight, the efficiency may be only 15%.

Solar energy systems having a concentrating collector with cooled cells can be used to provide both electrical energy and heating. These systems are now in the experimental stage.

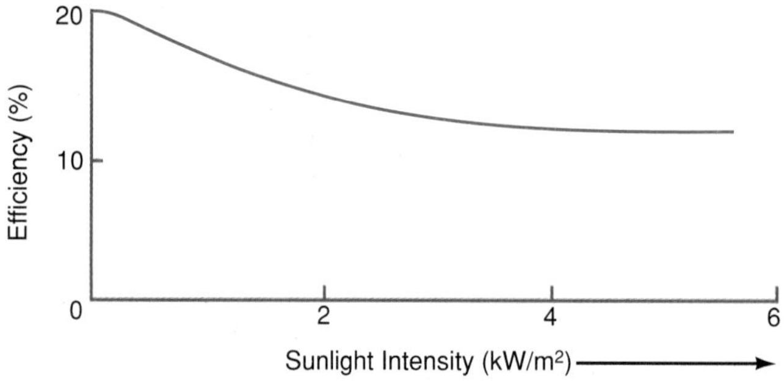

Figure 25-21. *Efficiency of a typical solar cell varies with light intensity.*

25.7 Review of Safety

There may be some hazards connected with the handling of solar energy equipment. Some plumbing connections may be required. If pipes are not well made, or fittings are not tight, some leakage may result.

Some attention needs to be given to the use of dissimilar (unlike) metals in a system. Electrolysis may corrode the two materials at the point where they are in contact.

Some insulating materials used with solar systems are flammable. Use care when soldering, brazing, or welding near these materials.

Caution must be used in domestic solar water heating systems that use ethylene glycol or other additives. Be sure that there are no leaks into the water in the hot water tank. Some protection is provided by a pressure regulation system. It maintains the hot water at higher pressure than the fluid in the solar collector circulation.

When installing solar collectors, protect your eyes from bright sunlight reflecting off surfaces. Some vacuum tube solar collectors can also implode (burst inward) if broken. Wear eye protection when installing collectors.

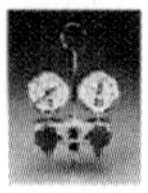

25.8 Test Your Knowledge

Please do not write in this text. Place your answers on a separate sheet of paper.

1. A micron is a(n) _______.
   A. form of solar energy
   B. photon
   C. electromagnetic wave
   D. one millionth (1/1,000,000) of a meter

2. What percent of the sun's rays are visible?
   A. 15%.
   B. 25%.
   C. 50%.
   D. None of the above.
3. Which type of the sun's rays are not visible?
   A. Ultraviolet.
   B. Infrared.
   C. Solar.
   D. Both A and B.
4. A solar collector is a(n) ______.
   A. arrangement of a dark-colored plate and a glass covering
   B. arrangement of a light-colored plate and a glass covering
   C. selective absorber
   D. Both B and C.
5. The ______ the energy of radiation, the ______ the wavelength.
   A. shorter, greater
   B. greater, shorter
   C. stronger, stronger
   D. None of the above.
6. Why is it necessary to provide heat insulation to solar collectors?
   A. To avoid deadly radiation.
   B. To ensure good light transmission.
   C. To keep the collector from releasing its heat.
   D. All of the above.
7. What is the approximate cutoff wavelength of a selective absorber?
   A. 1 milliliter.
   B. 5 microns.
   C. 0.5 microns.
   D. 2 microns.
8. What is the best angle for the sun's rays to strike the absorber?
   A. Exactly perpendicular, 90°.
   B. Exactly parallel, 180°.
   C. 45°.
   D. None of the above.
9. Why are some solar collectors sealed in a vacuum tube?
   A. To avoid loss of radiation.
   B. For increased efficiency.
   C. For structural support.
   D. All of the above.
10. Which of the following is a common way of storing solar energy?
    A. Water.
    B. Rocks.
    C. Polypropylene rods.
    D. All of the above.
11. Solar energy is stored as ______.
    A. heat
    B. light
    C. solar energy
    D. All of the above.
12. Which size stones are most efficient for use as a heat storage system?
    A. Large stones, 6″ to 8″ diameter.
    B. Large boulders, 10″ to 12″ diameter.
    C. 1″ to 2″ diameter crushed stone.
    D. None of the above.
13. ______ pound(s) of water will store as much heat as ______ pound(s) of rock.
    A. Four, one
    B. One, four
    C. One, two
    D. Three, four
14. ______ and water may be used in cold climates to eliminate freezing in the liquid filled container.
    A. Ethylene glycol
    B. Propylene glycol
    C. Sodium chloride
    D. Both A and B.
15. When using water in a heat storage system, how much water volume should be supplied for each square foot of collector surface?
    A. 1.3 to 2.5 gallons.
    B. 2 to 5 gallons.
    C. 4 to 6 gallons.
    D. None of the above.
16. The most efficient domestic warm water system uses ______ tank(s).
    A. one
    B. two
    C. three
    D. None of the above.
17. What is the voltage provided to a solar cell?
    A. 2 to 4 volts.
    B. 0.005 to 0.1 volts.
    C. 0.5 to 1.5 volts.
    D. None of the above.
18. To produce a higher voltage, solar cells should be connected in ______.
    A. series
    B. parallel
    C. combination of series and parallel
    D. None of the above.
19. At what temperature should solar cells be kept?
    A. Cold.
    B. Heated.
    C. In a mid-temperature range.
    D. Any of the above is acceptable.
20. ______ sunlight produces the greatest solar cell efficiency.
    A. High-intensity
    B. Low-intensity
    C. Mid-intensity
    D. The type of sunlight does not affect solar efficiency.

Chapter 26

AIR CONDITIONING AND HEATING CONTROL SYSTEMS

Modules:

Key Words:

controller
humidistat
hydronic
hygroscopic element
infrared heat controls
limit controls
modulate
pneumatic systems
split-system controls
thermostat droop
total energy management system
zone controls

Learning Objectives:

After studying this chapter, you will be able to:

- Define the terms: controls, controller, thermostat, heat anticipator, and humidistat.
- List the various applications of controls in an air conditioning system and tell what conditions they regulate.
- Name the different types of control systems.
- Describe the operation of various devices in the control systems for heating and comfort cooling.
- Explain the application of a computer to air conditioning control.
- Read control circuit diagrams.
- Troubleshoot, service, and repair control systems.
- **Follow approved safety procedures.**

Modern homes, offices, and workplaces are kept comfortable and healthy with air conditioning systems. These systems automatically control temperature, humidity, air movement, air cleaning, and air sterilization. Two important developments have improved heating and cooling systems:

- Automatic controls now operate systems, **Figure 26-1.**
- Electronic circuits control and operate the automatic systems. See **Figure 26-2.**

CONTROL MECHANISMS MODULE

26.1 Controls

Controls are devices that operate or regulate electrical and mechanical systems. They are used on heating, cooling, humidifying, and dehumidifying systems. Controls are also used for combustion and flue systems. Usually, each control device is designed to respond to a certain condition. Examples include devices that regulate:

- Temperature.
- Pressure.
- Liquid or gas flow.
- Liquid level.
- Timed operations.

These controls have made it possible to develop safe, automatic systems.

26.1.1 Controllers and Control Systems

Controllers are groups of controls and circuits that accurately and automatically operate a device. A controller may include primary controls, operating controls, and limit controls. These components will be studied in more detail as parts of typical controllers are discussed later in this chapter.

A ***control system*** includes the controller, the operating device or devices, and the conditioned area. See

Figure 26-1. *Computer systems are now used for automatic HVAC control. (Johnson Controls Inc., Systems and Services Div.)*

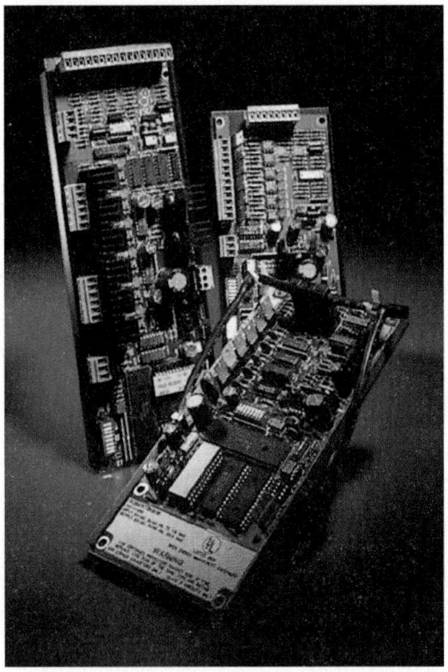

Figure 26-2. *Direct digital modules for building control. Package unit controller will handle both start and stop and staging outputs. Variable air volume (VAV) controller operates terminal units, electric and hot water reheat, as well as fan-powered systems. (Robertshaw Controls Co.)*

Section 8.3. Several types of control systems are in use. They include:

- Electric.
- Pneumatic.
- Electronic.
- Fluidic.
- Combinations of the above.

These control mechanisms automatically turn systems on and off. They modulate (adjust) certain operations and signal conditions. Such mechanisms are ***interlocked*** with safety devices to limit unwanted conditions. Basic principles of many of these devices were described in Chapters 6 and 8.

Control systems in the heating and cooling industry are those which are used to operate heating systems, air conditioning systems, and total energy management systems. There are a number of operating controls necessary in a typical air conditioning system. These controls include:

- Temperature controls.
- Humidifying controls.
- Airflow controls.
- Filter controls.
- Defrost controls.
- Limit or safety controls.

26.2 Thermostats

Probably the most common controller in the heating and cooling system is the thermostat. A thermostat contains an operating control. This control starts and stops the system when preset temperature conditions are reached.

Some thermostats ***modulate*** (increase or decrease an effect), instead of starting and stopping a system. An introduction to the various types of thermostats follows. The basic components of these devices are described, and specific thermostats are reviewed in depth.

26.2.1 Thermostat Types

There are many types of thermostats that control heating and cooling systems. The two basic ones are:

- Heating thermostat.
- Cooling thermostat.

These may be combined into one unit which is called a ***heating-cooling thermostat.*** Some combination thermostats change over automatically; others must be switched manually.

Time-operated thermostats are coupled with a clock mechanism. They will change the on-off settings for night and reset them for day. Some units will also change the on-off settings for different days of the week. (For example, the setting may be changed to "off" during Saturdays and Sundays.)

Automatic furnaces and other heating equipment are controlled by safety devices. They turn the system off or stop an operation if abnormal conditions arise. These devices are mainly electrical, and may be ***activated*** (tripped) by temperature, pressure, or time.

Relays are often used to control full voltages through the use of low-voltage signals. They may also be used to interlock signal devices for safe operation. See **Figure 26-3.**

All systems use room thermostats. Warm air systems also use a bonnet safety thermostat. It will shut off the system if the plenum chamber overheats. If automatic humidity control is desired, humidistats are used.

Oil burners sometimes use a stack thermostat. It will shut off the burner if the stack temperature does not rise within a few seconds after the oil burner is turned on.

Pressurestats shut off steam systems if pressure becomes dangerously high. Each type of heating system has its own special automatic devices.

There are three types of heating thermostats:

- Those that control electrical circuits.
- Those that control air circuits (pneumatic).
- Those that control combination electric and pneumatic circuits.

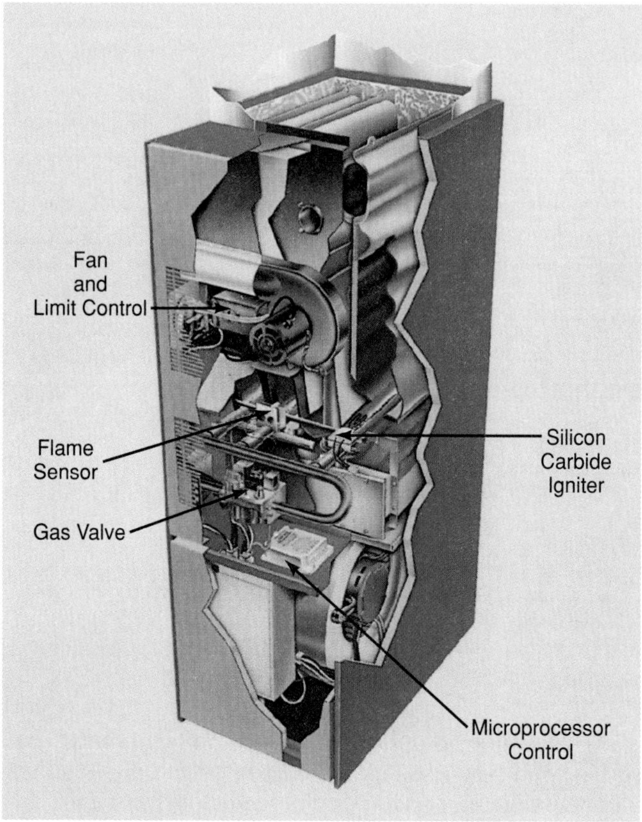

Figure 26-3. *Forced-air furnace. (White-Rodgers Division, Emerson Electric Co.)*

Some of the newer thermostats use solid-state electronic devices. These include triacs, transistors, and amplifiers, and are used to control the functions of systems such as:

- Power circuits.
- Airflow.
- Water flow.
- Steam flow.
- Damper operation.

The solid-state control will provide the best operation for highest efficiency of the system. These same systems will show the conditions of all the variables in the system. The conditions are indicated by instruments, lights, and recorders. Components of a solid-state thermostat usually include a thermistor (temperature sensor), a potentiometer (temperature adjustment device), and an SCR (silicon-controlled rectifier).

There are many types and models of thermostats designed for heating and cooling systems:

- Voltage:
  - Low-voltage types: 24 V.
  - Line-voltage types: 120 V, 120/240 V, or 240 V (also used for 208 V).
- Points:
  - SPST (single-pole, single-throw) two-wire.
  - SPDT (single-pole, double-throw) three-wire.
- Selector switches:
  - None.
  - Winter/summer.
  - Heat/fan.
  - Heat on/heat off.
- Thermometer:
  - Yes.
  - No.
- Solid-state thermostats.

Thermostats are supplied with or without heat anticipators. Their use varies. Some applications are:

- Air (outside, inside, combination).
- Cooling coil.
- Heating coil.
- Fan (exhaust or fresh air).
- Fan (coil).

26.3 Thermostat Operation

The primary device in a thermostat is the component that reacts to temperature change. There are several different types:

- Bimetal strip.
- Rod and tube.
- Bellows or diaphragm.
- Electrical resistance.
- Hydraulic.

Figure 26-4 shows the elements for these five types. Another type of element for heating systems (boilers and hot water) also is shown in **Figure 26-4.**

Thermostats are further classified as:

- Line-voltage thermostats.
- Low-voltage thermostats.

Usually, line-voltage thermostats are used directly in the electrical operating circuit. This is shown in **Figure 26-5.** They are mounted on the baseboard or wall four to five feet above the floor. Most have a manual "off" switch.

Low-voltage thermostats are found in a low-voltage circuit. The circuit is connected to a solenoid coil or relay switch. *These thermostats will be damaged if connected into a line-voltage (120 V) circuit.* They draw electrical power from a stepdown transformer.

If thermostats have a wide range of operation, they should have ambient temperature correction. For example, a thermostat may be designed to operate from 50°F to 250°F (10°C to 121°C). However, it will not be accurate throughout this range unless it has a built-in correction. The correction is needed because all parts of a thermostat expand as they become warmer. The correction can be made by using a temperature-compensating bimetal strip. This strip balances the expansion by moving slightly in the opposite direction.

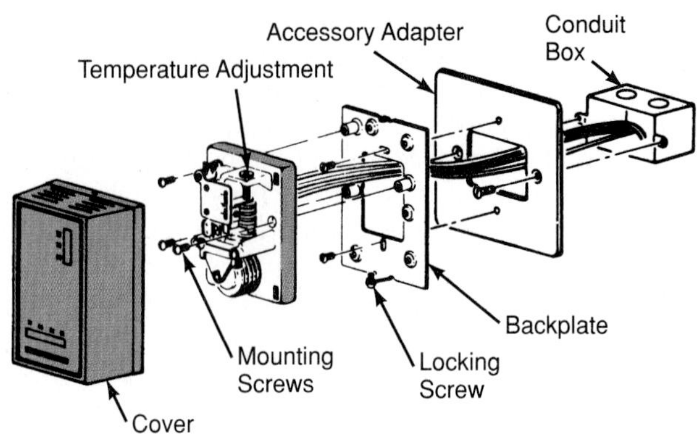

Figure 26-5. *Installation details for line-voltage, heavy-duty thermostat. It controls electric heating circuit. (Honeywell Inc.)*

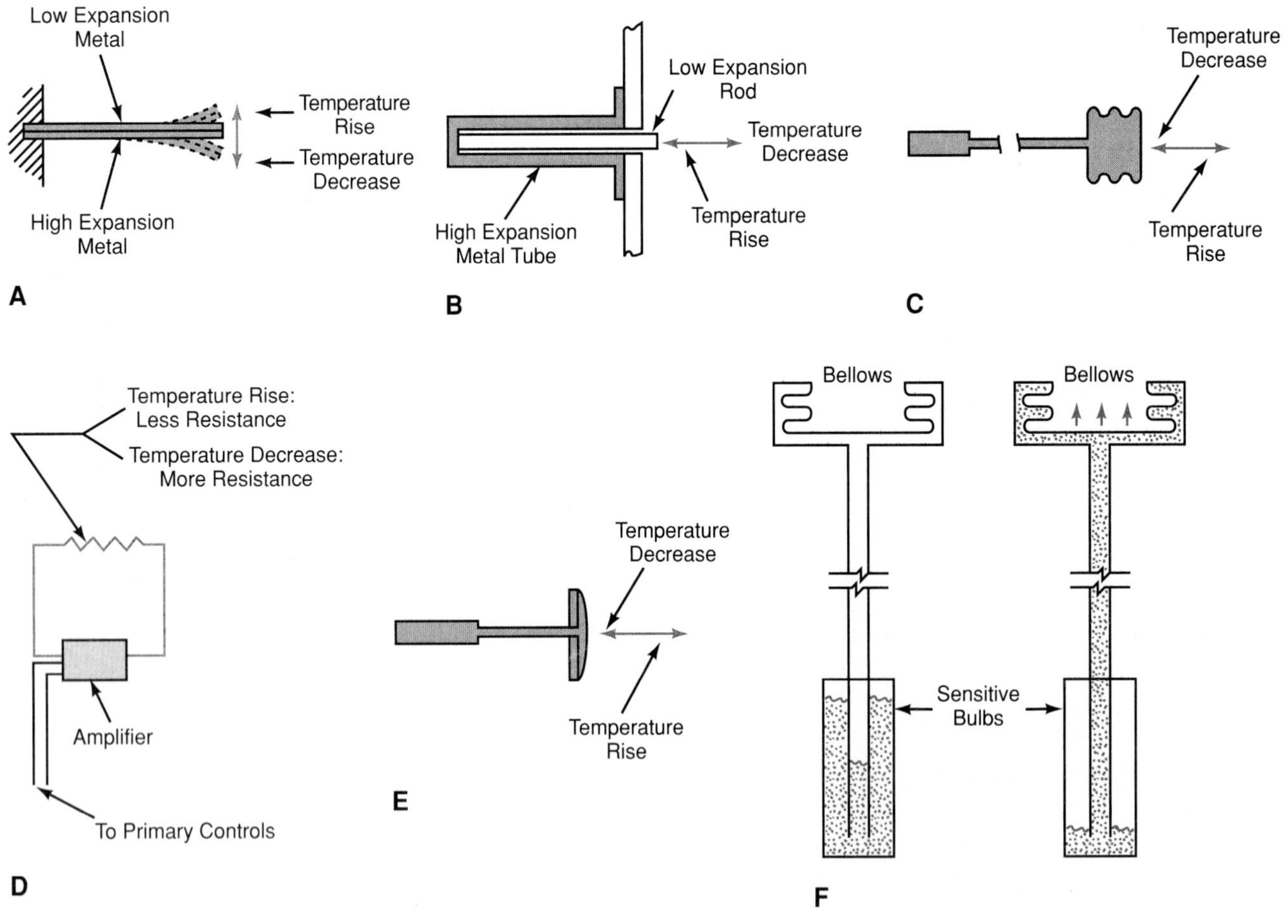

Figure 26-4. *Five basic types of thermostat operating elements: A—Operates on principle that two different metals will expand at different rates. B—Operates on principle that metal will expand when heated. C—Operates on principle that gas will expand with heat. D—Operates on principle of change of resistance in conductor or semiconductor with change in temperature. E—Hydraulically operated diaphragm (100% liquid). F—Temperature-sensing element for heating units such as boilers. Sensitive bulb is placed where temperature is to be controlled. As bulb warms, it forces liquid into bellows area and moves bellows.*

Some three-wire thermostats have two sets of contacts. One set is for "pull-in" and the other is for "hold-in." The design has more stable contact connections.

The bimetal strip is usually wound in a spiral to get more length. This is needed to give more contact movement with each degree of temperature change. These controls are designed to reduce ***contact point bounce,*** which causes arcing. They also reduce contact "walking" (sliding over each other). The latter action also causes arcing.

Mercury contacts are used with the bimetal strip. Sealed glass tubes prevent dust, lint, and oxygen from wearing the contacts. Inside the tube is a ball or globule of liquid mercury. The tube also contains two or more electrical probes that are fused into the glass. **Figure 26-6** shows an SPST mercury switch; **Figure 26-7** shows an SPDT switch.

The mercury tube thermostat wiring diagram shown in **Figure 26-8** is for a zone valve. When connections No. 5 and No. 4 are closed, the motorized zone valve opens. When the thermostat bimetal tilts, the mercury tube and connections No. 5 and No. 6 are closed. The motorized valve closes.

A bimetal-operated thermostat with four mercury switches is shown in **Figures 26-9** and **26-10.** Two are for stage No. 1 cooling and stage No. 2 cooling. The other two are stage No. 1 heating and stage No. 2 heating. The wiring connections for this control are shown in **Figure 26-11.** The schematic wiring diagram is as shown in **Figure 26-12.**

Some heavy-duty thermostats are designed to carry high current. They can carry as much as 16 A (amperes) at 120 V. They will not be damaged by as much as 96 A locked-rotor amperes for a short time.

In the thermometer-type thermostat, two leads are inserted in a mercury-filled glass column. There, they contact the inner tubing in which the mercury column moves. The mercury column rises as the temperature increases. It will close the connection between the two contacts. This closed circuit then activates a warning system.

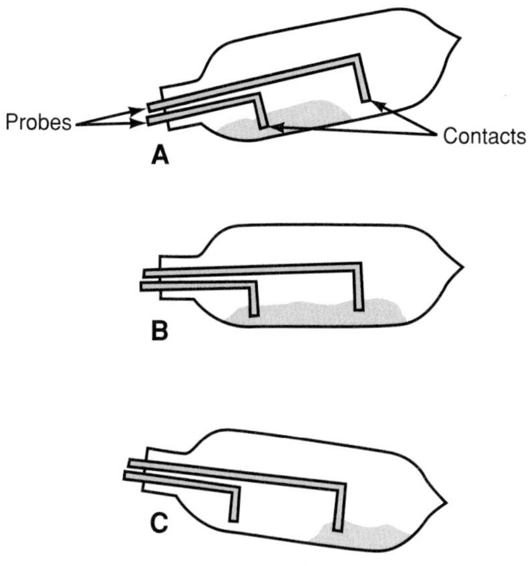

Figure 26-6. *How mercury switch works. Mercury completes circuit between contacts and must be level for proper operation. A—Off for heating. B—On for heating or cooling. C—Off for cooling system.*

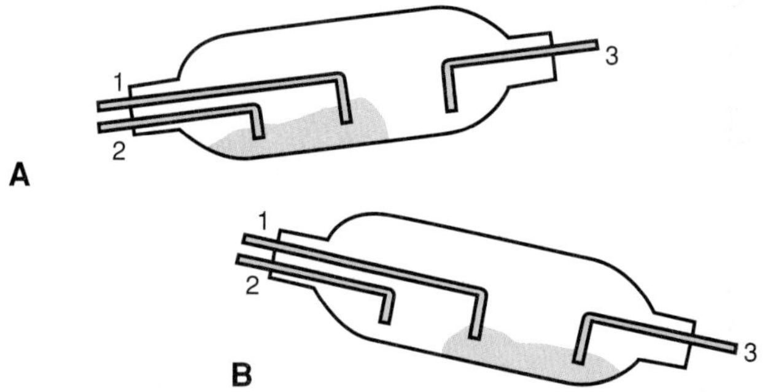

Figure 26-7. *A single-pole, double-throw (SPDT) mercury switch. Mercury tube is usually tilted or moved by bimetal coil. A—Leads 1 and 2 are connected by mercury to complete circuit. B—Leads 1 and 3 are connected by mercury globule.*

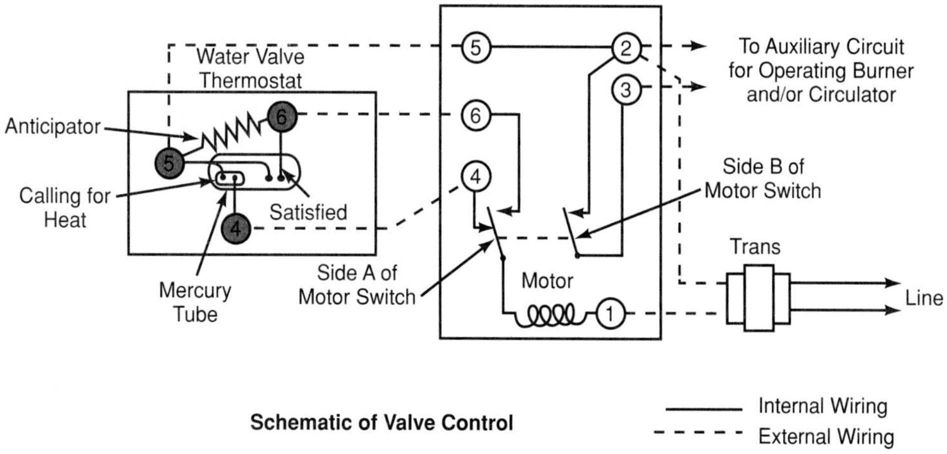

Figure 26-8. *Single-pole, double-throw thermostat mercury switch. This thermostat opens and closes motorized zone valve in hydronic or steam heating system. It can also control zone valve in chilled-water cooling system.*

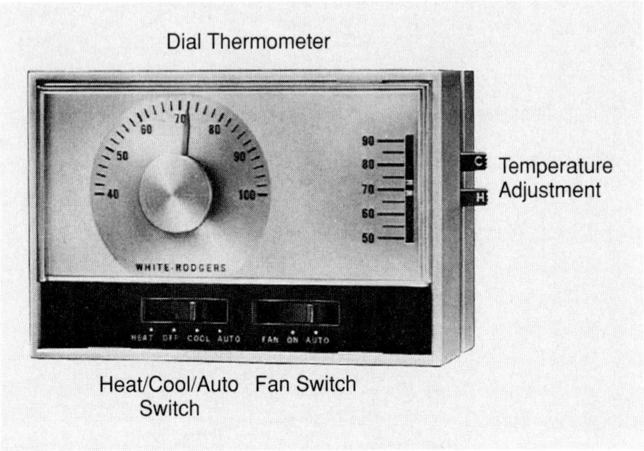

Figure 26-9. *Staging thermostat controls two stages of heating and two stages of cooling. Thermostat temperature adjustment is marked "C" for cooling and "H" for heating. (White-Rodgers Division, Emerson Electric Co.)*

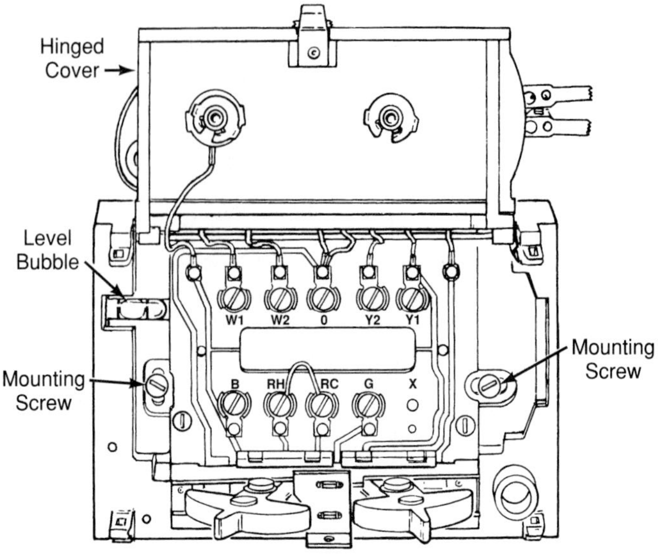

Figure 26-11. *Wiring terminals of a staging thermostat. Note identification on terminals. (White-Rodgers Division, Emerson Electric Co.)*

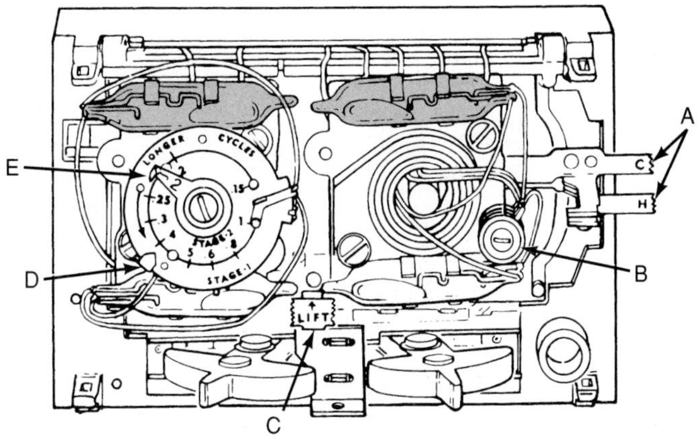

Figure 26-10. *Internal construction of a staging thermostat. Two heating control mercury tubes are on the left. Two mercury tubes controlling cooling are on the right. A—Levers for adjusting temperature. B—Cooling anticipators. C—Tab that must be lifted to raise cover. D—First-stage heat anticipator. E—Second stage heat anticipator. (White-Rodgers Division, Emerson Electric Co.)*

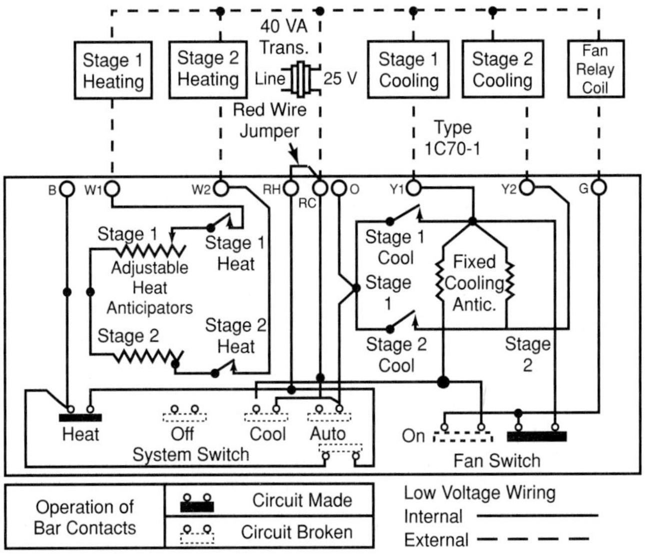

Figure 26-12. *This schematic is for the staging thermostat of a heating and cooling system.*

26.3.1 Line-Voltage Thermostats

Line-voltage thermostats are either 120 V or 240 V. They are designed to control currents as high as 22 A. However, many electric heating experts recommend low-voltage thermostats for most electric heating systems.

Wiring to line-voltage thermostats must be installed according to the local code. Thermostats may have a fan switch mounted on the side. The fan has an off-on-automatic setting. The Off position means the blower is not running. The On position means the fan is in continuous operation. Auto (automatic) means the fan thermostat in the furnace plenum chamber will control the fan.

The thermostat operates the electrical load directly or through a line-voltage operating control. The control is designed for 31 V to 300 V and 24 A maximum for noninductive circuits. (This includes resistance only. No coils, such as solenoids or transformers or motor windings, are in the circuit.)

Some are designed for 31 V to 300 V with six different amperage capacities for inductive circuits (those including solenoids, transformers, and motors). These capacities are: 3, 6, 8, 10, 14, and 16 A at 120 V, full load. The capacities are 18, 36, 48, 60, 84, and 96 A at 120 V locked rotor (starting load) on motor. At 240 V, these ampere ratings are reduced by half. Note the following thermostat classifications and see **Figure 26-13:**

- Single-line break (single-pole) thermostat (heating only or cooling only).
- Heating and cooling.
- Double-line break disconnect thermostat.
- Double-line break cycling thermostat.
- Two-stage thermostat.
- Two-circuit thermostat.

26.3.2 Low-Voltage Thermostats

Low-voltage thermostats carry only a small amount of electrical power. This is an advantage. Line-voltage thermostats have problems with electrical power heating the thermostat itself, opening the contacts too soon. This extra heating temperature rise is called ***thermostat droop.***

Example:

Thermostat is set for 72°F (22°C) and room is at 68°F (20°C). As the room heats up, the thermostat becomes warmer than room temperature because of electric resistance heating of the electrical parts in the thermostat. The thermostat then reaches 72°F (22°C) and cuts off too soon (about the time the room is 69°F or 70°F).

Low-voltage thermostats should use No. 18 AWG (American Wire Gauge) for distances up to 50′ (15 m). For distances over 50′, No. 16 AWG wire is specified.

Many thermostats have three or four wires. Each is a different color. Red, white, green, yellow, and black are popular. *Be sure to connect wires as indicated in the wiring diagrams provided by the manufacturer. This is required.*

Thermostats control the applied energy to low-voltage controls of Class 2 (power) circuits. They are rated as follows:

- 0 V to 15 V, maximum of 5 A
- 15 V to 30 V, maximum of 3.2 A

The thermostat mounting usually requires other parts. The subbase is the part to which the thermostat is attached. It contains the lead terminals. Sometimes it holds other parts, such as switches and indicator lamps. Beneath the subbase is the wall plate. This is the part attached to an outlet box or wall.

A millivolt-rated thermostat may be used with a self-generating electrical source. For example, a multiple thermocouple located near a gas pilot light will generate sufficient current for thermostat operation.

There are two groups and twelve classifications of millivolt thermostats. Group 1 thermostats are those with fixed or adjustable anticipators. Group 2 consists of thermostats without anticipators.

The twelve classifications are:

- Heating only (one-stage).
- Cooling only (one-stage).
- Heating and cooling, manual changeover (one-stage).
- Heating and cooling, auto changeover (one-stage).
- Heating only (two-stage).
- Cooling only (two-stage).
- Heating (one-stage), cooling (two-stage), manual changeover.
- Heating (one-stage), cooling (two-stage), auto changeover.
- Heating (two-stage), cooling (one-stage), manual changeover.
- Heating (two-stage), cooling (one-stage), auto changeover.
- Heating (two-stage), cooling (two-stage), manual changeover.
- Heating (two-stage), cooling (two-stage), auto changeover.

26.4 Heating Thermostats

This paragraph introduces the basic operation of a heating thermostat. Section 26.7 reviews the more recent electronic versions of these thermostats. A residential heating thermostat typically uses open contact points to control either a 24 V or 120 V primary circuit.

Figure 26-14 shows the inside of a typical room thermostat designed for wall mounting. A coiled bimetal strip reacts to temperature change. It has a range of 35°F to 50°F (2°C to 10°C) low setting to 90°F or 95°F (32°C or 35°C) high setting.

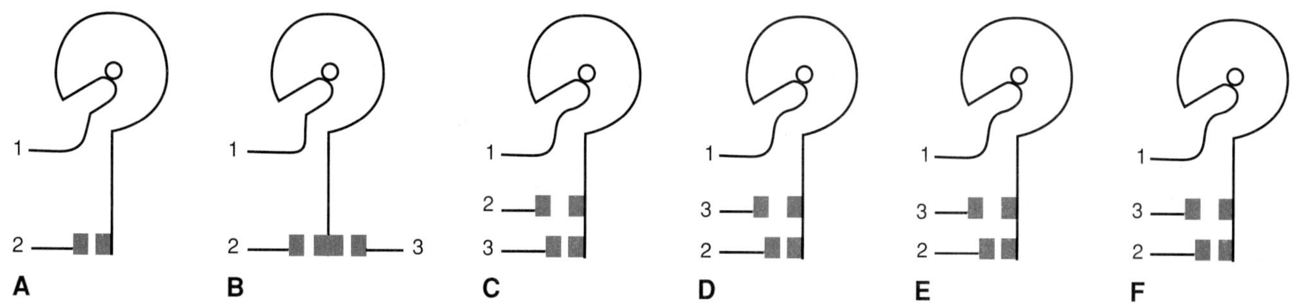

Figure 26-13. *Schematic for thermostat classification. The numeral 1 identifies the common lead, 2 is the first controlled lead, 3 is the second controlled lead. A—Single line-break thermostat. B—Heating-cooling thermostat. C—Double line-break disconnect. D—Double line-break cycling. E—Two-stage thermostat. F—Two-circuit thermostat.*

Figure 26-14. *Inside view of a wall-mounted thermostat shows coiled bimetal element. (Johnson Controls, Inc.)*

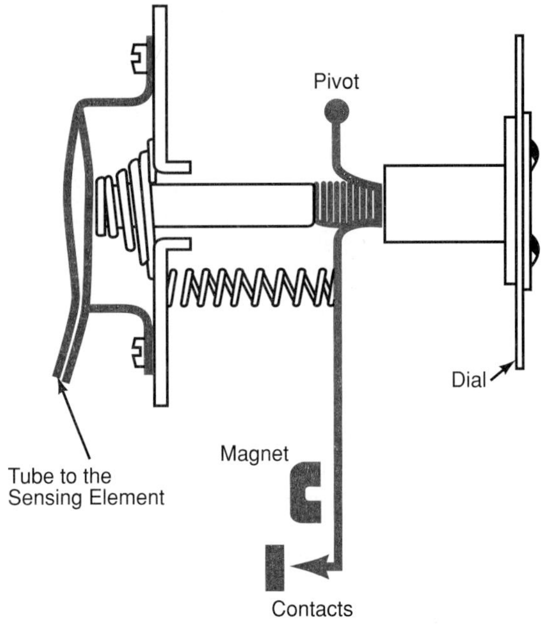

Figure 26-15. *Diagram shows a side view of a hydraulically operated thermostat. (White-Rodgers Division, Emerson Electric Co.)*

These thermostats usually operate on a 0.5°F to 2°F (0.3°C to 1°C) differential or smaller. This allows close temperature control.

Generally, the sensitive element is a bimetal strip. A small magnet is sometimes used to produce snap action in the points. A range adjustment is usually a direct force on the bimetal. The differential usually consists of moving the small magnet closer to or farther from the bimetal strip.

Some thermostats are hydraulically operated. **Figure 26-15** shows a diaphragm mechanism. The diaphragm moves as a liquid expands and contracts. The dial adjusts the setting and the small magnet produces snap action in the thermostat.

Electronic measurement of an electric resistance element may be used for temperature control. It is possible to obtain differentials of .01° using this method.

Thermostats are rated by the voltage they carry. They are also rated by the controls to which they are electrically connected. These controls open the circuit when the temperature rises. They complete the circuit when the temperature falls.

One 24 V thermostat has three wires connected to it. The wires make a contact on temperature drop and close a holding circuit. They will open the circuit on a rise of temperature. This thermostat must be used with a relay. The relay opens and closes the 120 V circuit for an oil burner, gas burner, or electric heater.

Another 24 V thermostat has separate controls. One of these opens the contacts on rise in temperature. The other closes contacts on a drop in temperature. It is used on gas burners, oil burners, or electric heaters.

A third type is a two-wire 24 V thermostat. It controls the 24 V furnace devices directly.

Figure 26-16 shows a wiring circuit for a control using a mercury bulb contact device. The circuit is designed for forced-air heating.

If lint should lodge between the points, an open circuit will result. To clean contact points, use a piece of clean paper. If a mercury tube is used instead of contact points, the thermostat must be carefully leveled. Use a plumb line or spirit level.

Another style of thermostat has been developed for air conditioning systems. These units have easily operated settings. They are designed with all the various types of electrical bases to enable the thermostat to control heating, cooling, or both.

Some heating-cooling systems use a clock-operated thermostat. It changes the temperature range during the

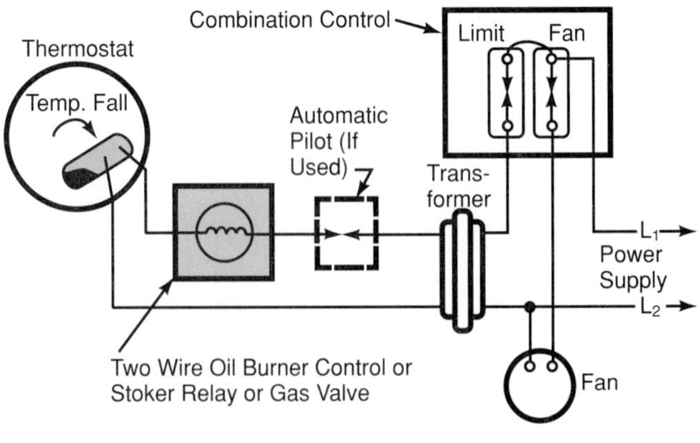

Figure 26-16. *Wiring diagram of a circuit for a heating thermostat. Note that automatic pilot and fuel control are in the low-voltage (24 V) circuit (to left of transformer).*

night. It returns the range to normal during the day. **Figure 26-17** shows a clock-operated thermostat powered on 24 V to 30 V. When programming the thermostat, the display shows the programmed time/temperature settings selected. When programming is completed, the liquid crystal display (LCD) shows time and current room temperature. **Figure 26-18** shows wiring diagrams for three different installations using two- and three-wire clock thermostats.

Frequently, a unit will short-cycle because the thermostat is exposed to vibration. (Examples would be those installed on a shaky wall or a stairway.) Always mount a thermostat to a firm structure.

The thermostats just described are used primarily in warm forced-air systems. The operating features of various thermostats are very similar. In general, a line-voltage type thermostat is used when controlling electric heat. The inside of such a control is operated by a heat sensitive bulb, **Figure 26-19.** Thermostats have also been developed to control gas, oil, or electric furnaces and air conditioners. See **Figure 26-20.**

Units are equipped with safety controls known as ***temperature limit switches.*** Large-capacity units use a relay along with low-voltage thermostats to control high-wattage loads. Some prefer low-voltage thermostats with a relay on all electric heating systems. This reduces "heat lag." The sequence control is sometimes used also. It is made up of a block of relays. Each relay controls a separate heating circuit. The relays are activated one at a time. This sequence timing prevents placing the full electrical load on the main building circuit at one time.

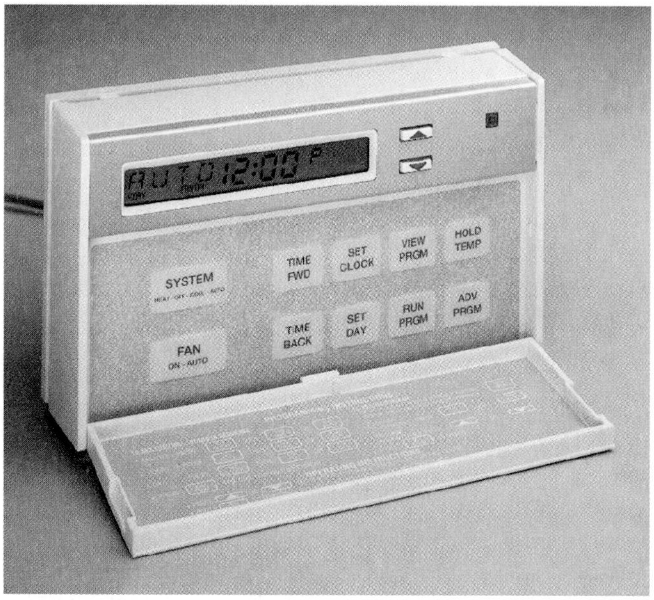

Figure 26-17. *Electronic digital thermostat. This type may be used for either heating or cooling, or for a combination system. Unit has digital display for time and temperature and for fan operation. (White-Rodgers Division, Emerson Electric Co.)*

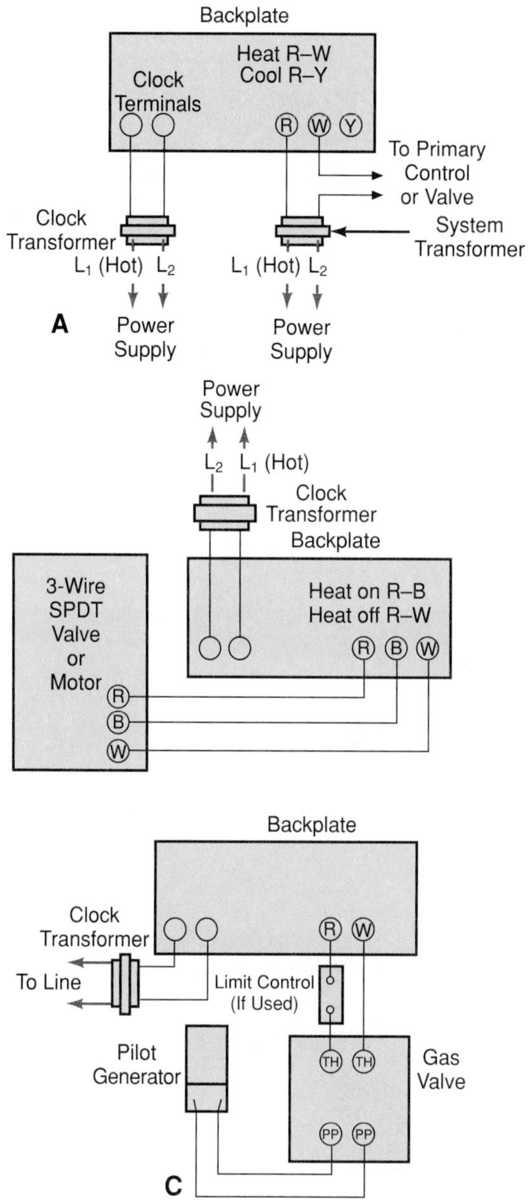

Figure 26-18. *Three different clock thermostat wiring diagrams. A—Two-wire low-voltage system. B—Three-wire system. C—Multivolt pilot generator system. (Honeywell Inc.)*

Some codes require that both leads to the heating element be opened by thermostatic control. **Figure 26-21** shows such a thermostat.

Thermostats mounted on baseboard heaters are usually encased in attractive covers. They have a calibrated dial mounted on the adjustment knob. **Figure 26-22** shows such an installation.

Some line-voltage thermostats produce a wide swing of temperatures. The room temperatures will vary too much. Others may have a temperature droop. The room temperature will go too low before heating effect is felt.

For example, a thermostat is set for 75°F with a 1°F differential. However, the room cools to 70°F and then

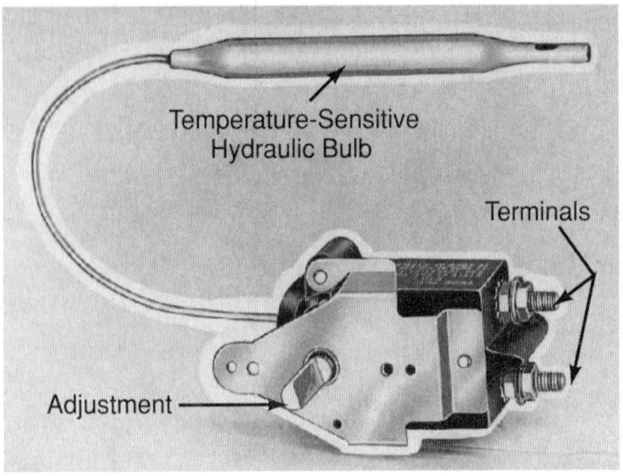

Figure 26-19. *Hydraulic type thermostat used on electric resistance heating units. Rated at 240 V to 277 V, it can control up to 5000 W. Its 1 1/2° differential provides uniform temperature control. (White-Rodgers Division, Emerson Electric Co.)*

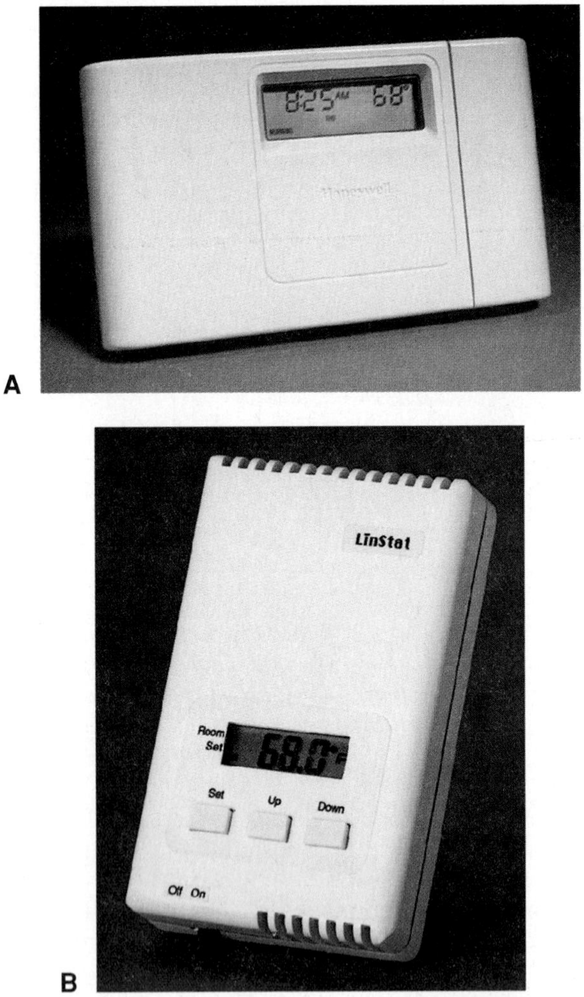

Figure 26-20. *Digital thermostats. A—Thermostat is equipped with a switch on back that can be set to control either 24 V gas/oil systems, or electric furnaces and air conditioners. (Honeywell Inc.) B—Thermostat used in electric heating. It first displays the desired temperature setting, then the actual room temperature. (PSG Industries, Inc.)*

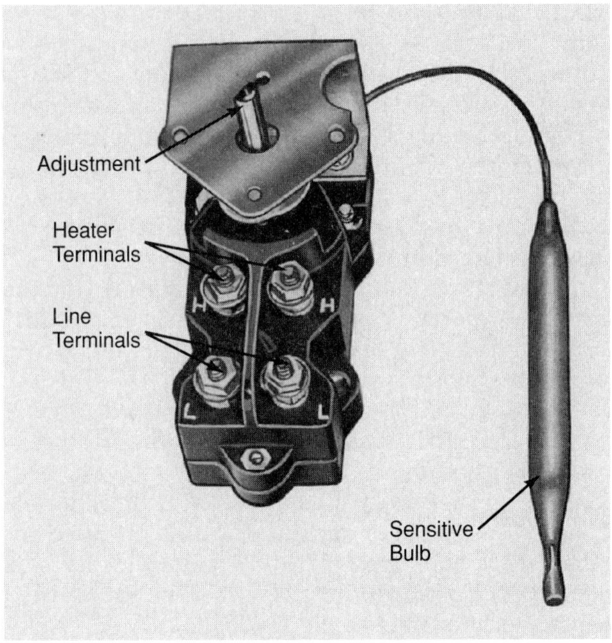

Figure 26-21. *Hydraulic-type thermostat used on electric resistance heating. It controls both electrical leads to heating unit.*

Figure 26-22. *A thermostat designed to be mounted on electric resistance baseboard heater casings. (White-Rodgers Division, Emerson Electric Co.)*

warms to 80°F. The 10°F difference is too much ***swing.*** In another case, the thermostat also is set for 75°F with a 1°F differential. However, the room only heats to 65°F when the thermostat shuts off. The 10°F below normal is a ***droop.*** This condition may be caused by too much heat released in the thermostat from wiring. It may be the result of too much heat from the anticipator. See Section 26.4.1.

When adjusting a thermostat, avoid heating it with your hands or breathing on the element. If the thermostat is warmed above the room temperature, wrong thermostat settings will result.

26.4.1 Heat Anticipators

An interesting problem arises in the operation of room heating thermostats. The room temperature tends to rise above the thermostat setting after the thermostat points open and the burner stops. This action is called ***overshoot*** of the thermostat. This action is the result of the residual heat in the furnace.

Though some heat is generated by the thermostat's electric circuit, a small heating coil may also be used. This heating coil (resistor) is placed in the thermostat. These small resistance coils are usually called ***heat anticipators.*** During the heating cycle, the thermostat is always about 1°F warmer than the room. If the thermostat is set for 74°F (23°C), the thermostat will open when the room temperature is actually 73°F. Then, after the thermostat has opened, room temperature will still rise to 74°F. As noted, this is due to heat in the furnace.

Figure 26-23 shows an electric heating thermostat equipped with an anticipator. The heat anticipator is connected in series with the operating contact points.

A

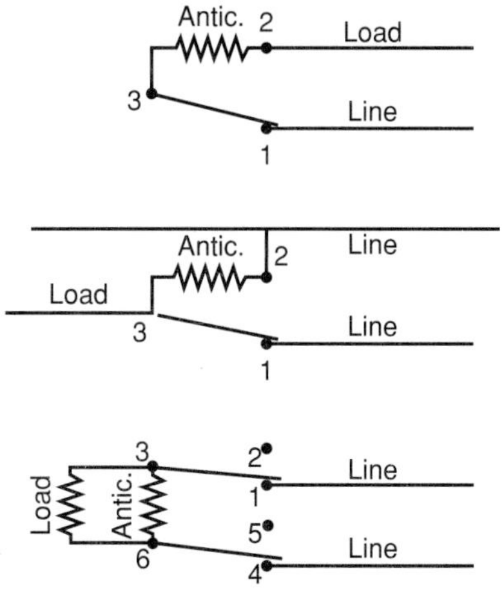

B

Figure 26-23. *Thermostat used in electrical heating system. A—External view. (White-Rodgers Division, Emerson Electric Co.) B—Wiring diagrams for thermostat. Note that it is equipped with an anticipator. The SPST indicates that control has single-pole single-throw switch.*

Heating and cooling anticipators for thermostats are of two types, fixed or adjustable. They are most often used on low-voltage thermostats and on heating-cooling combination thermostats. The anticipator size (current capacity) is related to the current (amperes) flowing through the thermostat contacts. The current varies between 0.15 A and 1.0 A. This depends on the controls connected to the thermostat.

With a fixed anticipator, current flow should equal the control ampere flow. The adjustable anticipator should be regulated to the same ampere flow as the control circuit. The current flow of the control circuit is shown on the label of the control.

An anticipator with a higher rating than the control circuit will warm the thermostat more slowly. (There will be less anticipator heat.) The reverse is also true. Anticipators are used in 24 V systems and also in 750 millivolt systems.

Cooling thermostats may have a heat anticipator connected in parallel with the thermostat points. The anticipator becomes warm during the off cycle and will close the thermostat points about 1°F (0.6°C) warmer than the room temperature. If control is set to cut in at 78°F (26°C), the control will actually start the system at 77°F (25°C). This will allow the system to start cooling the room before it becomes too warm and humid.

Different anticipator heaters are used, depending on the system. **Figure 26-24** shows the outside of a combination thermostat with a heat anticipator. The inside of this thermostat is shown in **Figure 26-25.** The more residual heat in the heating system, the larger the capacity of the anticipator. **Figure 26-26** shows a heating thermostat wiring schematic diagram.

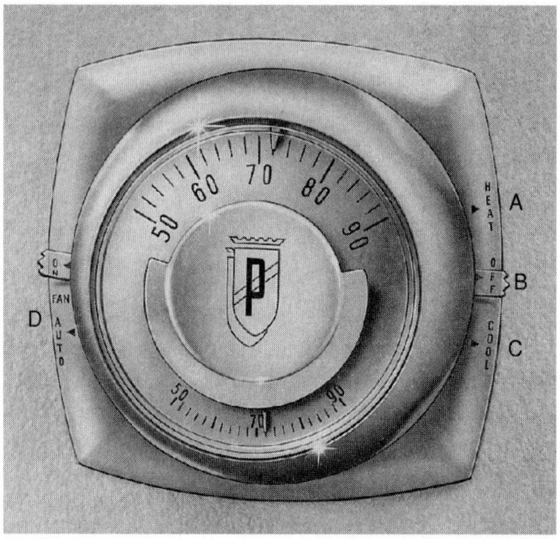

Figure 26-24. *Combination heating and cooling thermostat which has a heat anticipator. A—Switch setting for heating. B—Off position for heating-cooling switch. C—Switch position during cooling season. D—Control switch for fan operation. Top scale is for setting temperature. Scale adjusts for temperatures between 50°F to 90°F. Bottom scale indicates actual room temperature (72°F).*

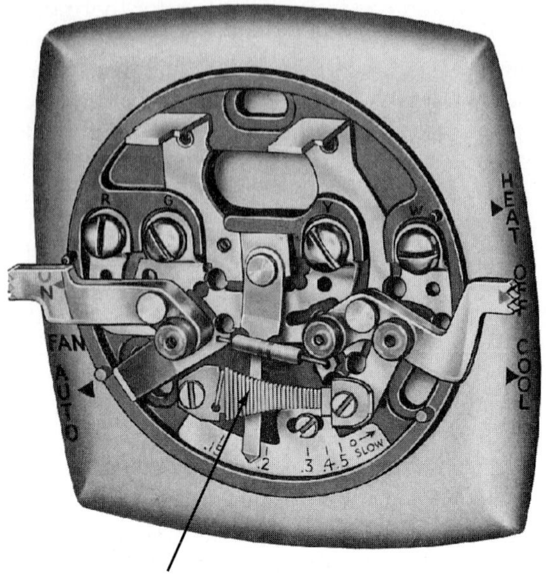

Figure 26-25. *Inside view of thermostat. Note variable heat anticipator. (Johnson Controls, Inc.)*

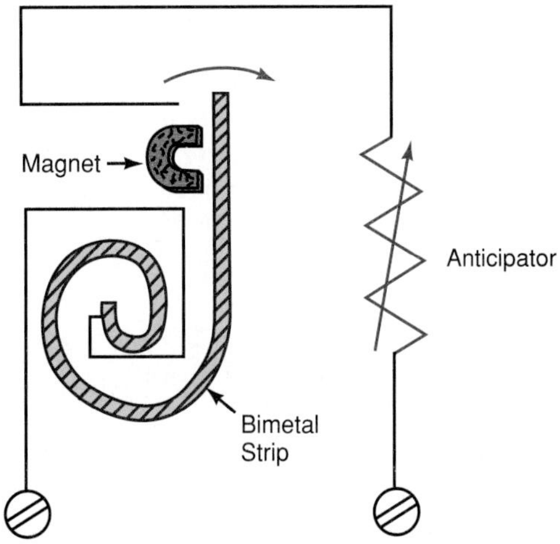

Figure 26-26. *Schematic wiring diagram is of a heating thermostat that is equipped with an anticipator.*

At the end of the off portion of the heating cycle, the thermostat points close. It takes time for the furnace to heat and to move this heat to the room. The room temperature may fall about 1°F during this start-up period. This action is called ***system lag.*** The cut-in temperature is usually set about 1°F above the lowest temperature desired.

26.5 Cooling Thermostats

Comfort cooling thermostats are similar in design to heating thermostats, but work in the opposite manner: the contacts open as the room cools and close as the room warms up.

The popular bimetal thermostats usually have a 1°F (0.6°C) differential. Both the 24 V and 120 V controls are available. Some units have cold anticipators. These small electric resistance elements are in parallel with the bimetal strip. They close the bimetal-controlled points just before the unit reaches the room cut-in temperature.

Figure 26-27 shows a simplified wiring diagram of a cooling thermostat. The cold anticipator warms the bimetal strip during the off cycle. It is bypassed on the on cycle. This heat during the off cycle turns the system on just a little ahead of normal.

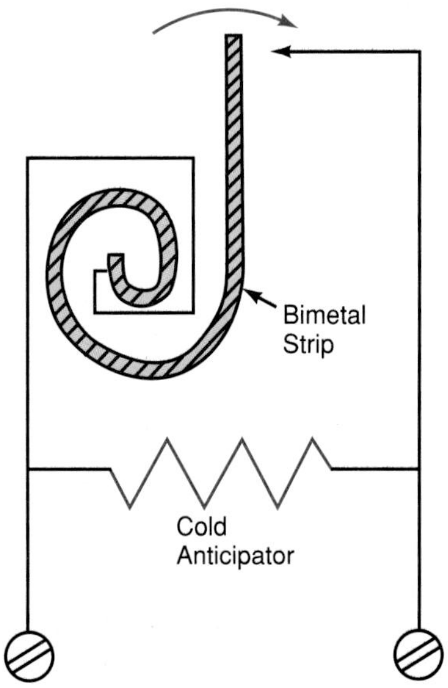

Figure 26-27. *Schematic wiring diagram of cooling thermostat equipped with cold anticipator.*

26.6 Combination Thermostats

Some thermostats have control mechanisms for both heating and cooling. These thermostats are used with heat pumps or with other installations that have both a heating and a cooling system. **Figure 26-28** is a simplified wiring diagram for a combination thermostat. It uses a bimetal strip, heat anticipator, and cold anticipator.

The appearance of a combination heating-cooling thermostat is shown in **Figure 26-29.** This control is for 24 V to 30 V electrical circuits. Inside view is shown in **Figure 26-30.**

Controls are made that go beyond control of cooling and heating. Some regulate humidity, indicate the condition of the filter, or control an odor system. These controls usually operate on a 24 V circuit.

Other combination heating and cooling thermostats use individual system and fan control switches. **Figure 26-31** shows this type thermostat with a calibrated temperature setting dial and a thermometer. The thermostat base, showing the mounting holes, terminals, and switches, is shown in **Figure 26-32.**

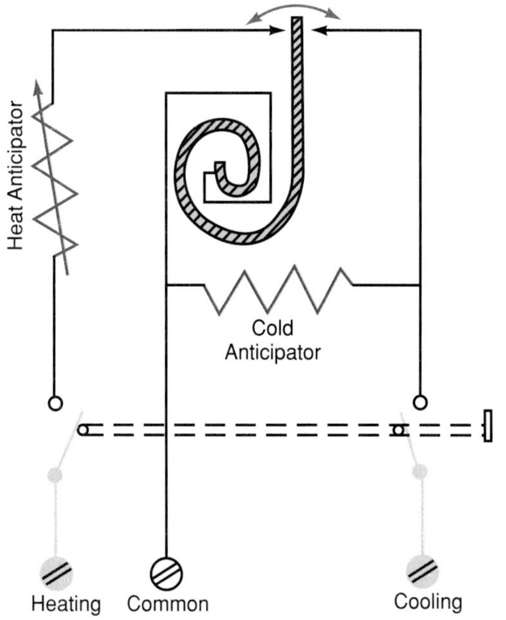

Figure 26-28. *Schematic of bimetal heating-cooling thermostat. Unit is equipped with both a heat anticipator and a cold anticipator.*

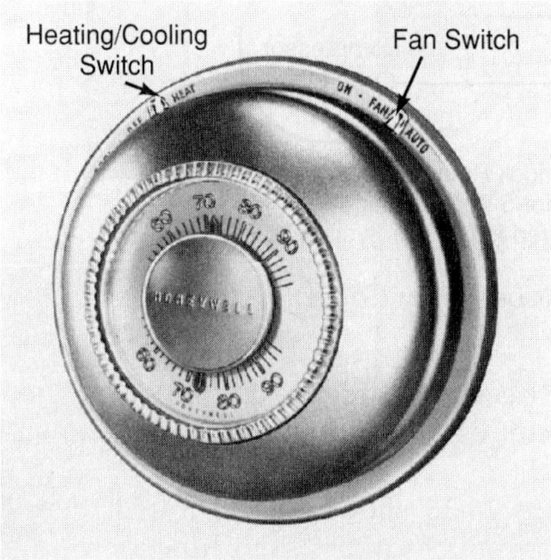

Figure 26-29. *A 24 V thermostat that may be used both for heating and cooling. Range adjustment is on upper temperature scale. Bottom scale indicates room temperature. (Honeywell Inc.)*

An electronic circuit using a combination thermostat is shown in **Figure 26-33.** Five leads go to the thermostat. Used with a modulating gas control, it keeps the heating temperature range within 1°F. Cooling is held within a 3°F (1.6°C) range. The thermostat is shown in **Figure 26-34.** An electronic system is mounted on the thermostat base.

26.7 Electronic Thermostats

Advances in electronic circuitry and controls have revolutionized the thermostat and its capabilities. Elec-

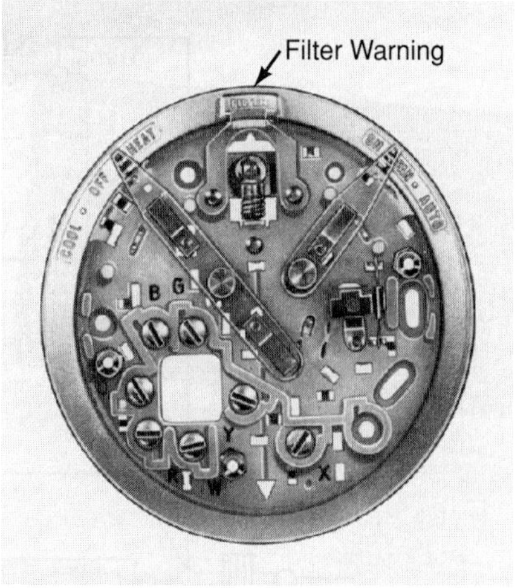

Figure 26-30. *Construction of combination heating-cooling thermostat, shown in* **Figure 26-29.** *Filter warning bulb lights when filter becomes dirty.*

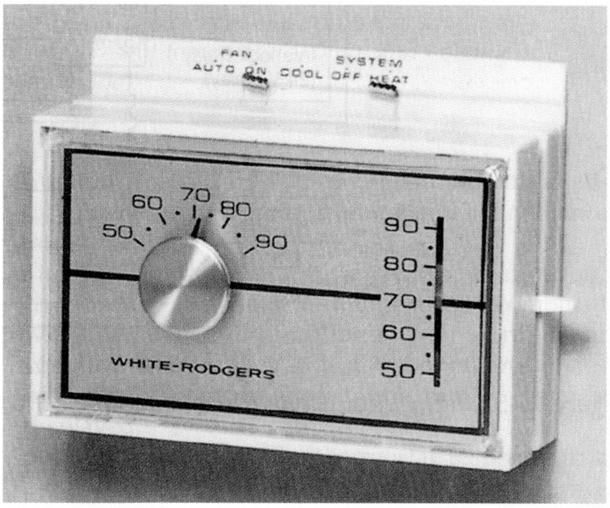

Figure 26-31. *Combination heating-cooling thermostat. Note separate controls for heating, cooling, and fan. (White-Rodgers Division, Emerson Electric Co.)*

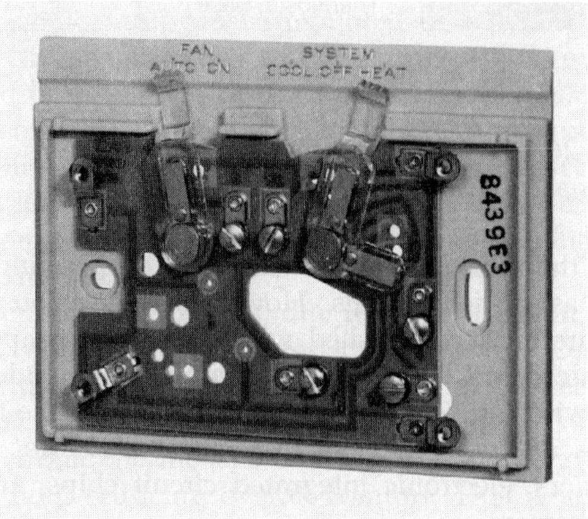

Figure 26-32. *Inside view of combination heating-cooling thermostat shown in* **Figure 26-31.** *(White-Rodgers Division, Emerson Electric Co.)*

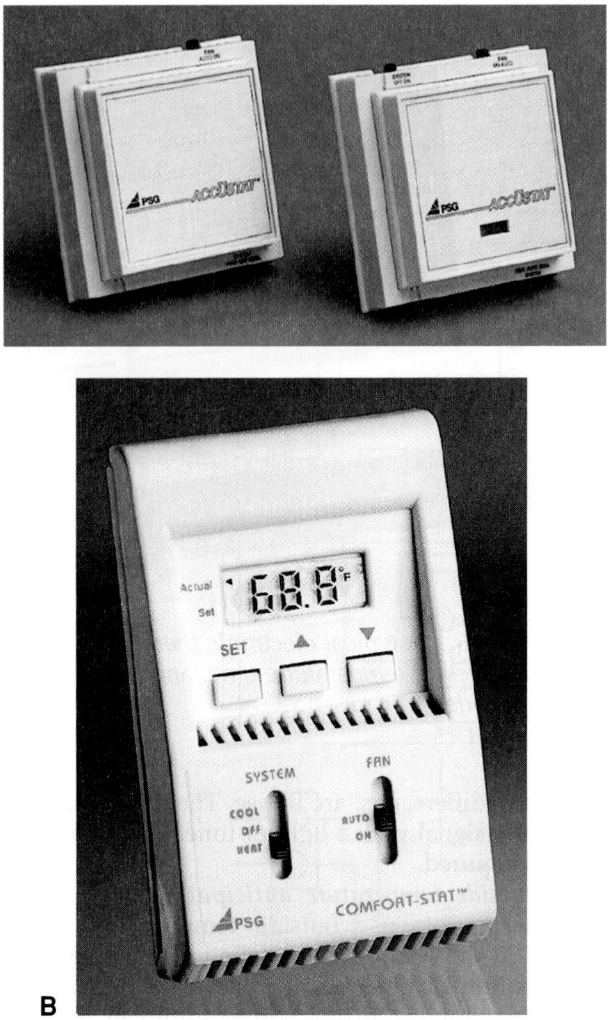

Figure 26-37. *Digital thermostats. A—Remote and control units that can be mounted in separate locations. B—Thermostat with microchip that checks room temperature four times each minute. (PSG Industries, Inc.)*

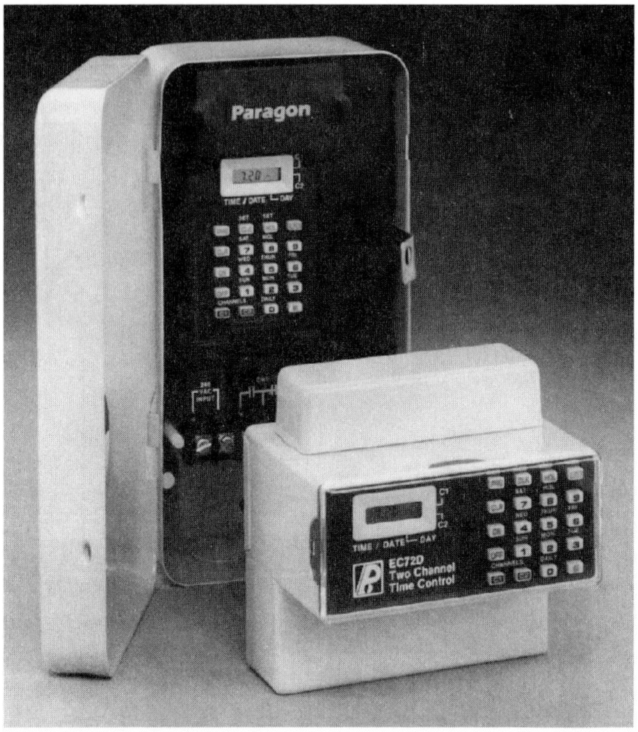

Figure 26-38. *Series timer control used for operating heating or cooling systems for 24-hour or 7-day programming. The keyboard programming and digital display guide the programmer though each operation. (Paragon Electric Co., Inc.)*

26.8 Timer-Thermostats

Many air conditioning installations can be automatically clock-controlled. For example, office cooling systems need not run at normal temperatures during weekends or other non-working hours. Many users want units to start functioning at a certain time before the area is occupied. Heating systems, too, can provide reduced temperatures when the building is not occupied.

Figure 26-38 illustrates an automatic timer that operates on a seven-day schedule. The push buttons are used to set the time and day for system operation.

26.9 Multistage Thermostats

Multistage thermostats are usually designed for low-voltage operation. They will operate two or three circuits in sequence. They control either heating (close on temperature fall) or cooling (close on temperature rise).

Figure 26-39 is a schematic of a multistage thermostat. It is designed to control two stages of heating and one stage of cooling. Three mercury tube (SPST) switches

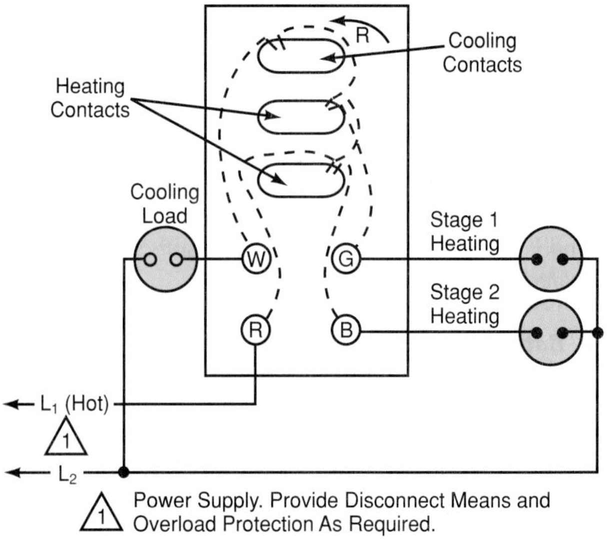

Figure 26-39. *Schematic for a multistage thermostat. It controls two heating stages and one cooling stage. (Honeywell, Inc.)*

are used. Notice the four electrical leads to the thermostat. The thermostat operating element is a bellows. **Figure 26-40** shows a two-stage heating thermostat.

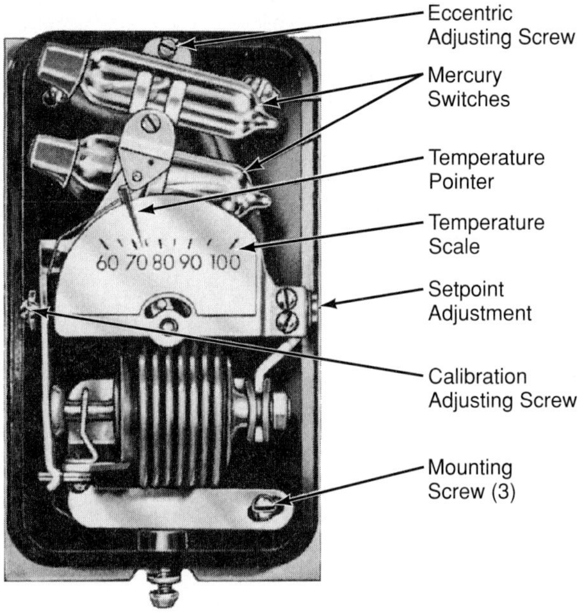

Figure 26-40. *Two-stage thermostat operated by sealed bellows element and using mercury switches. The eccentric adjusting screw adjusts the differential, which is usually 1°F.*

26.10 Hydronic Thermostats

Hot water or hydronic heating systems also require thermostats. While these thermostats provide functions similar to the thermostats discussed, they are constructed differently. Hot water thermostats are more easily adapted to individual room (or radiator) use, **Figure 26-41,** and result in considerable energy savings when compared to a single-control-unit system.

26.11 Portable Thermostats

Portable thermostats are available for control of a furnace or air conditioner. The thermostat electronically activates a responder on the furnace or air conditioner. This can be done up to a distance of about 120′ (36.5 m).

The thermostat may be moved from room to room. It responds to the temperature of that room. Since it is an electronic battery-powered unit, the thermostat has no moving parts. It has an accurate adjustable differential from 0°F to about 5°F (0°C to about 3°C). It has a range of 45°F to 90°F (7°C to 32°C).

Once every minute, the thermostat senses the temperature. If heating (or cooling) is necessary, the thermostat sends a radio signal to the responder.

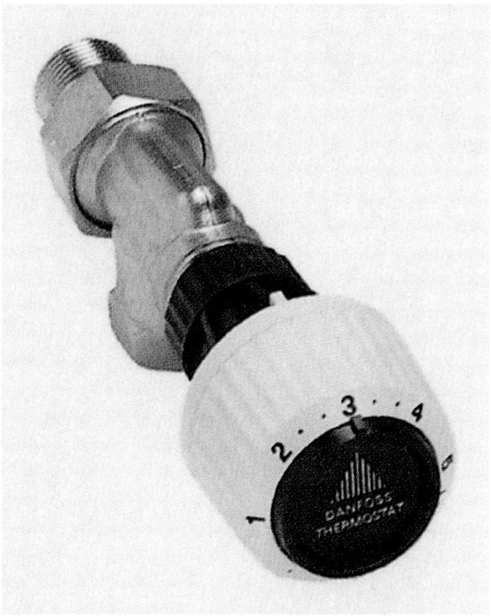

Figure 26-41. *Self-contained thermostatic valve is used for temperature control of individual rooms in a hot water or hydronic heating system. (Danfoss Automatic Controls, Division of Danfoss, Inc.)*

26.12 Thermostat Location and Servicing

The location of a thermostat is important. It should be placed in an average-temperature location. An example would be an inner wall about 5′ from the floor. It should be out of the way of furniture and beyond the reach of small children.

The location of a thermostat should be changed if it is affected by:

- Drafts or dead air spots behind doors and in corners.
- Hot or cold air from ducts.
- Radiant heat from the sun or an appliance.
- Concealed pipes and chimneys.
- Unheated areas behind it, such as an outside wall.

If the heating or cooling system is not working, the thermostat may be at fault. In general, a thermostat provides warning signals. It also comes with a diagnostic chart to follow when troubleshooting, **Figure 26-42.**

Lacking this, a good first step is to check all electrical connections. Make sure that all switches are operating. Then, verify continuity between the thermostat and the furnace or cooling system.

If no problems are found, and the heating or cooling system itself has no apparent problems, the temperature-sensing device within the thermostat is probably at fault. Checking the device externally can be difficult. This condition can best be verified by using another thermostat temporarily set in place. If the alternate thermostat corrects the problem, the

Troubleshooting Guide and Voltage References		
Problem Observed	Possible Cause	Corrective Action
1. No display.	No ac voltage to sub-base.	Measure power at TB-1 pins 1 and 2. Should be 22 V ac to 30 V ac.
	If auxiliary transformer is used.	Check for power output. Check P-3 hook-up for 22 V ac to 30 V ac.
	No power to unit.	Check power hook-up. Check for defective wiring.
	Blown fuse (F-8).	Replace fuse with 1.0 A.
	HVAC transformer defective.	Replace transformer with correct type.
	J-1 not connected to sub-base.	Connect J-1.
	Damaged pins on J-1 connector.	Straighten pins if possible, if broken replace controller.
	Sub-base defect.	Replace sub-base.
	Controller defect.	Replace controller.
2. No temperature display or wrong temperature displayed.	Unit not calibrated.	Calibrate per Section XIII.
	Temperature sensor open.	Replace temperature sensor board.
	Temperature sensor cable damaged or not connected.	Check temperature sensor cable for damage or check for correct connection. If damaged, replace controller.
	Temp cal was used.	Press RESET for 10 seconds.
3. Display on, but none of the function keys work.	Controller defect.	Remove and replace controller.

Figure 26-42. *Troubleshooting guides are normally provided by manufacturers of the thermostat. This excerpt is typical. (DuPont Energy Management Co., Inc.)*

original one is faulty. Usually, the easiest and most economical means of correcting this is to replace the thermostat.

Electronic thermostats are somewhat easier to service . They may provide some simple self-diagnostics (i.e., battery condition, no programmed settings, etc.). In addition, electronic components are usually built onto circuit boards. These can be removed and replaced. See Chapter 6.

26.12.1 Thermostat Guards

One of the biggest problems with thermostats is their ease of adjustment. Whenever there are several persons in the building, there is a chance that two or more of them will change the thermostat setting. These actions may produce uncomfortable results, may be irritating, and may even be dangerous.

One solution is using a thermostat that can only be adjusted with a special key. This key is controlled by one person. Another method is to cover the thermostat with a clear plastic lockable cover.

CONTROL SYSTEM COMPONENTS MODULE

26.13 Controllers

A ***controller*** is a group of controls and circuits used to accurately and automatically operate a device. The various components in a controller are divided into three categories, **Figure 26-43.** These include:

- Primary controls.
- Operating controls.
- Limit controls.

In the heating and cooling industry, these components are combined to form ***control systems.*** These are used to operate heating systems, air conditioning systems, and total energy management systems. The following paragraphs will explain, in detail, some of the basic controller components. Then they will review the more common systems used today.

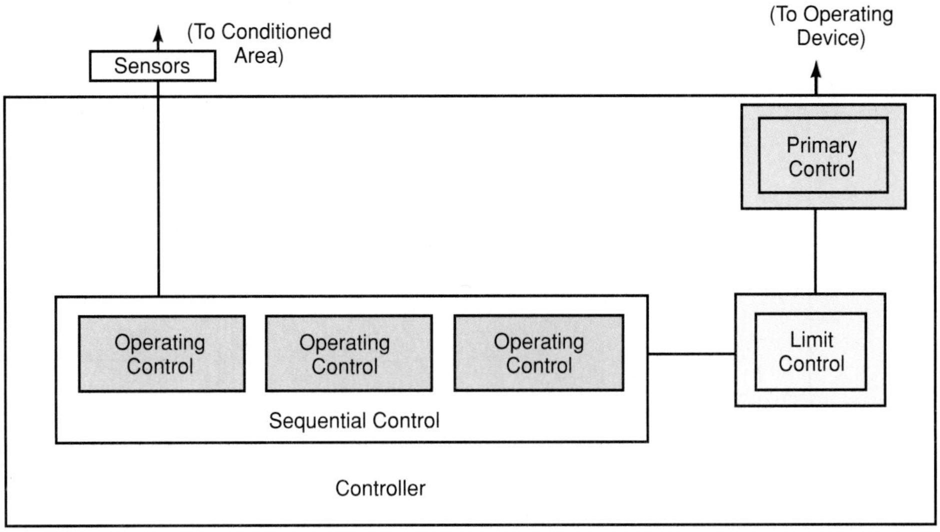

Figure 26-43. *Controller block diagram. The three major control categories (primary, operating, and limit) are indicated. A series of operating controls is shown as a sequential control.*

26.13.1 Feedback

Feedback is the information a controller uses to indicate that a particular operation has occurred. Information concerning the conditioned area is always being monitored by the controller's sensors. This information is then compared to the desired conditions that are set in the controller. The information about the conditioned space is fed back to the controlled device. An example is the signal sent to a damper from a thermostat. The signal might indicate that an adjustment must be made to correct the actual temperature.

Feedback information usually results in one of three actions within the control system. These can be categorized by the type of feedback involved:

- *Positive*—The feedback confirms that the control system is maintaining the proper condition. The system will continue to operate without change.
- *Negative*—The feedback indicates to the control system that its operation must change to maintain a programmed condition.
- *Abort*—This is a type of negative feedback. The control system will revert to a known safe condition or shut down.

In most systems, conditioned space is heated by turning on the heat source. The heat source warms the area. When a preset upper temperature is reached, the heat is turned off. As the area cools to a preset low temperature, the heat source will again turn on. This on-off type of control action alternately heats the area and allows it to cool. A constant temperature is *not* maintained. For a constant temperature, modulating controls are needed.

Modulating controls use feedback. They constantly make smooth continuous changes in the system's operation to maintain a programmed condition. Modulating controls provide a heat input that is proportional to the heat loss. This is a proportional type of control action. The modulating control will maintain a constant temperature in the conditioned area. This type of control may be applied to either heating or cooling. To "modulate" means to vary as follows: 0, 1, 2, 3, 4, 5, etc., in contrast to the on-off ("0, 1, 0, 1") cycle. Solid-state electronics make possible this accurate modulating of temperature.

The modulator is usually controlled by a thermostat, a remote bulb and bellows or diaphragm, pressurestat, or humidistat. These controllers, in most cases, operate a variable potentiometer. The variable resistance of the potentiometer is operated by the controller.

The modulator motor usually combines a reversible capacitor motor with a balancing relay. It also includes a feedback potentiometer and a gear train. See the schematic wiring diagram of the motor control in **Figure 26-44.** A three-wire thermostat controls the current fed to the balancing relay.

Motors are also specially designed to move to only two distinct positions. These two-position motors use cams to open and close switches when the driven mechanism is operated. For such use, motors are usually equipped with electric brakes. They are operated by the motor position switches and lock the rotor in position. **Figure 26-45** shows a solid-state controlled-positioning motor.

26.13.2 Relays

The basic construction and operation of a relay is covered in Chapters 6 and 8. The relay is commonly used as a primary control in a control system. By using relays, current required for system operation does not have to flow through the system controls. Thus, the main circuit can be made as short and as direct as possible. Relays are sometimes called ***contactors.***

The solenoid creates magnetism to close the points in the heating circuit or circuits. It may be energized

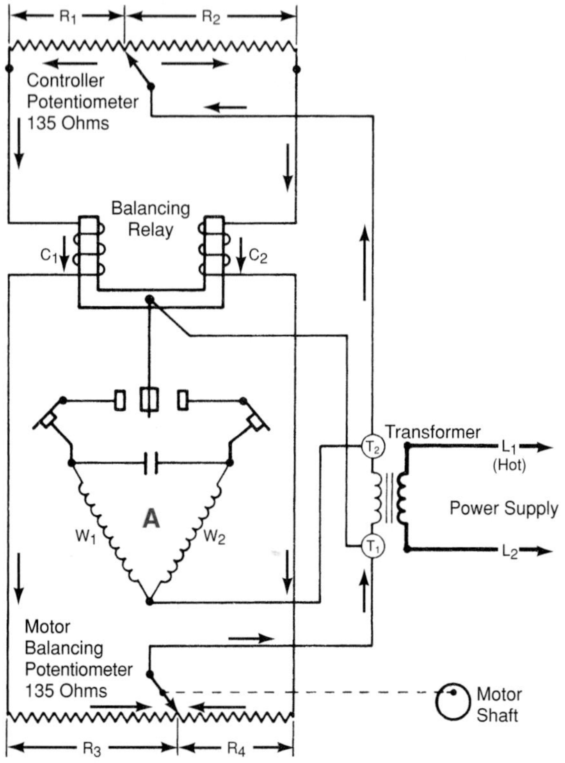

Figure 26-44. *Wiring diagram for modulating electric motor used for controlling valves and dampers. A—Motor windings. (Honeywell Inc.)*

Figure 26-45. *A solid-state controlled-positioning motor. Note motor shaft at left. (Johnson Controls, Inc.)*

(turned on) by 24, 120, or 240 V. Voltage of the thermostat and the relay solenoid coil must be the same. **Figure 26-46** shows a schematic wiring diagram of a relay. This relay controls five separate electric heating circuits and one fan motor circuit.

When the solenoid is energized, it moves (activates) a lever. This lever operates all of the snap switches. The snap switches are calibrated (set) to ***make and break*** (close and open) in sequence. The snap switch controlling the fan motor makes (closes) first and breaks (opens) last.

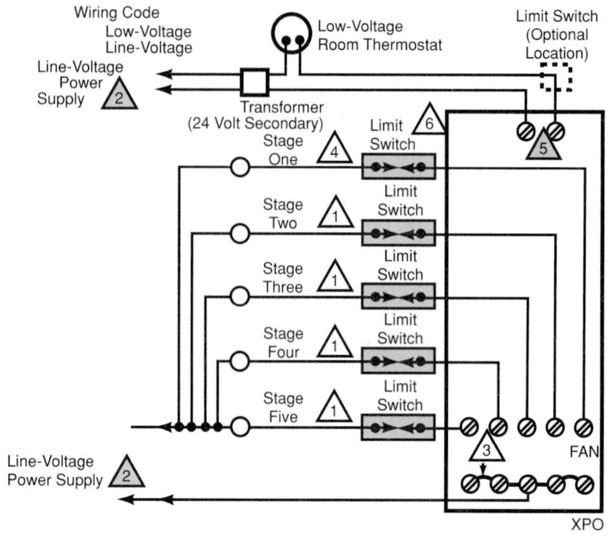

Figure 26-46. *Wiring diagram for relay circuit controlling five electric resistance heaters. At upper left, the "2" indicates the power supply to a stepdown transformer. At lower left, "2" indicates line-voltage power supply. Solenoid coil connections are shown at "5."*

A type of relay commonly referred to in control systems is the ***lockout relay.*** The lockout relay shuts down a circuit. It keeps the circuit from restarting when any of the safety control devices have opened. (Safety control devices would include motor overloads, high- or low-pressure switches, etc.) Repairing the problem and closing the safety control device will not allow the circuit to restart. The power control circuit to the lockout relay must be interrupted. **Figure 26-47** shows a schematic diagram of a lockout relay circuit.

26.14 Primary Controls

Primary controls are devices in a control system that safely turn on and operate the system on command from the operating controls. These primary controls differ depending on the type of heating and/or cooling system used.

The kind of controls used depends on the type of heat energy used: oil, gas, or electricity. Another factor is the kind of heat distribution system used: steam, water, or air. These controls will be explained further.

Figure 26-48 is the wiring diagram for a gun-type oil burner system. Note thermostat, oil burner control, and limit control. **Figure 26-49** shows an oil burner circuit used with a circulation pump (hydronic system). The heavy wires are high-voltage (120 V) lines. The light wires are low-voltage 24 V lines.

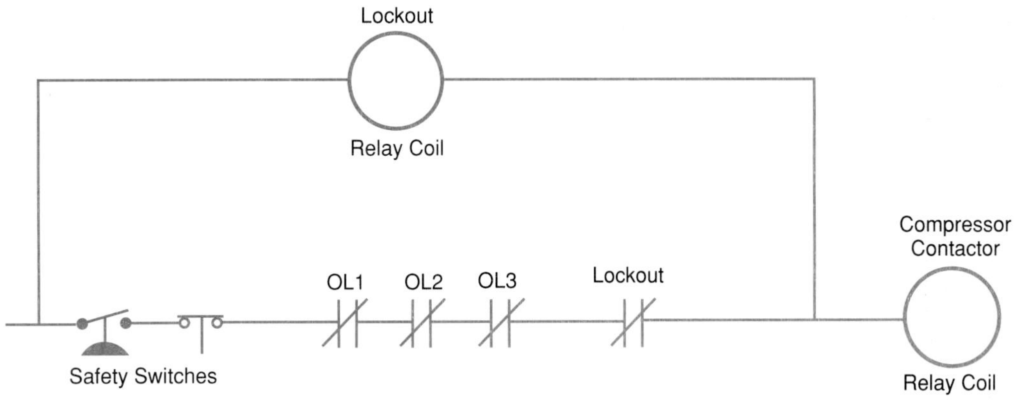

Figure 26-47. *Schematic diagram is for a lockout relay circuit. It closes whenever a safety switch opens.*

Figure 26-50 shows a wiring diagram for a somewhat different gun-type oil burner installation. The circulator motor is turned on by the room thermostat when it calls for heat.

Gas furnaces use many varieties of electrical devices. Some operate on 120 V, some on 24 V. Some operate on current generated by a thermocouple (25 to 700 millivolts [mV]).

The wiring diagram in **Figure 26-51** is designed for a solenoid-operated gas line valve. It is installed on a hydronic heating system.

A wiring diagram as shown in **Figure 26-52** is used on a system with a diaphragm gas valve. The circulator motor operates all the time that the burner is on. **Figure 26-53** shows another circulator motor wiring arrangement.

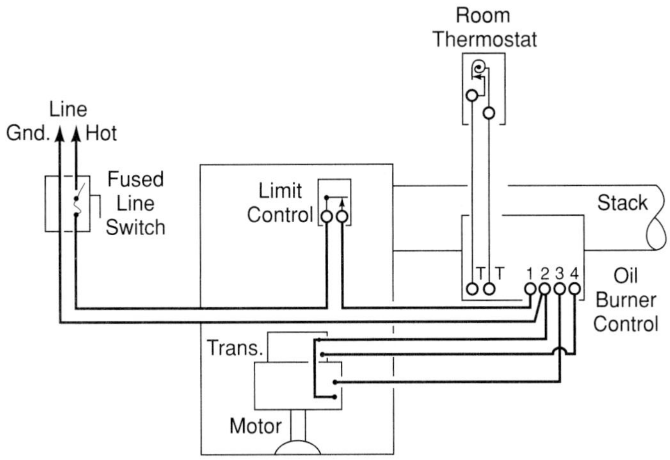

Figure 26-48. *This wiring diagram represents a control system for a gun-type oil burner. (White-Rodgers Division, Emerson Electric Co.)*

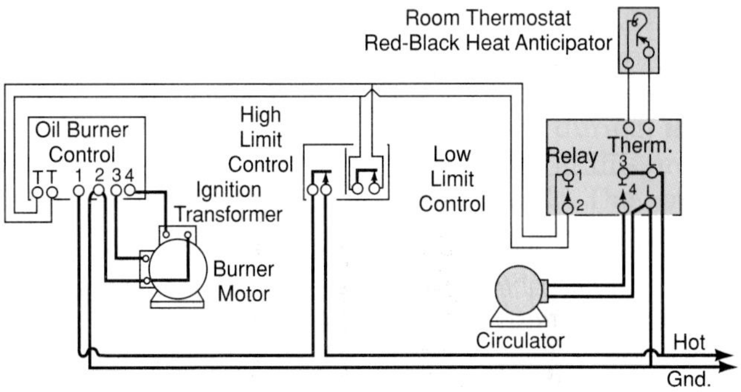

Figure 26-50. *Wiring diagram for gun-type oil burner used in a hydronic installation that has a motor-driven circulator pump. Room temperature controls hydronic pump operation.*

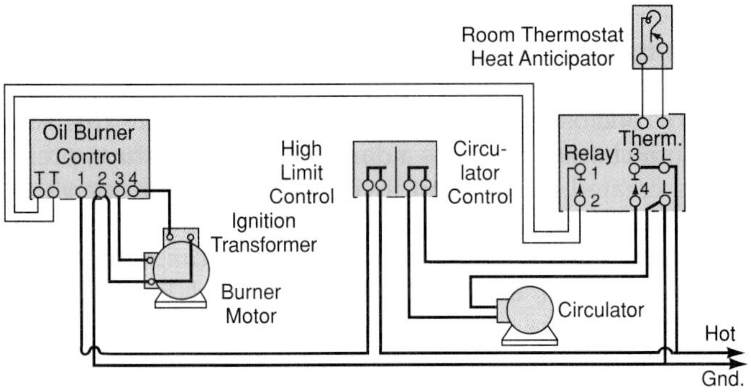

Figure 26-49. *Wiring diagram for gun-type oil burner used in a hydronic installation. Separate thermostat controls water circulating pump.*

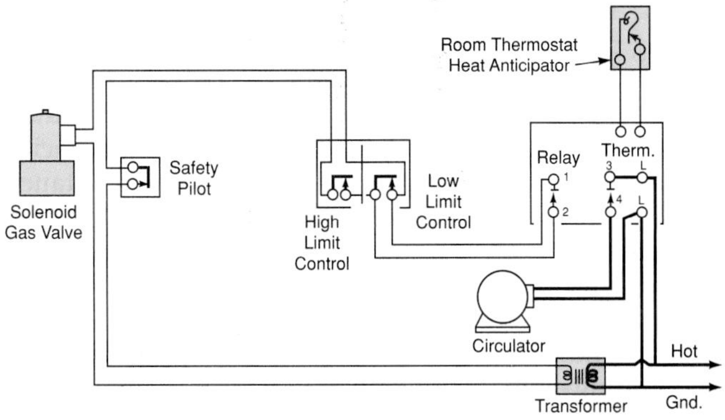

Figure 26-51. *Solenoid-operated gas line valve. This is a low-voltage type. Note stepdown transformer at lower right-hand corner.*

26.18 Control Circuits

A typical control circuit will use many of the components of electrical and electronic circuits covered in Chapters 6 and 8. These components are combined to form primary, sequential, limit, and feedback parts of the total circuit. **Figure 26-58** shows a block schematic of a typical home heating/air conditioning control system. Each block is made up of electronic components arranged to provide the indicated control function.

Heating and cooling control systems use four basic types of electrical circuits. They are classified according to voltage requirements:

- The 120 V circuit. (Occasionally, large commercial and industrial systems use 240 V as well as multiphase circuits.)
- The 24 V circuit. (These circuits often operate relays which, in turn, control the main electrical circuits.)
- The thermocouple circuit. (Operating on a few millivolts to about 700 mV). These are used in some pilot light safety devices and in some gas heating systems.
- The electronic circuit. These use approximately 5 V to operate components such as: thermistors, diodes, and transistors. These circuits are designed with overload safety devices. These may be fuses and/or circuit breakers.

Smaller motors usually have built-in overload protection. Larger motors use external overload devices, usually located in the magnetic starter. *The wiring, components, and protection systems control circuits must conform to local electrical codes.* Section 26.18.1 through 26.18.9 will present some typical control systems and review their basic circuitry.

26.18.1 Gas Furnace Controls

Gas furnaces have primary controls. These ensure safe starting and safe operation of gas burners. Controls used to operate a gas fuel heating system include:

- Room thermostat or thermostat with outdoor air-sensing adjustment.
- Pilot light temperature sensor, such as a thermocouple.
- Limit controls.

Figure 26-59 shows three wiring diagrams for a gas furnace. Note location of the limit controls in the line.

These controls allow gas to flow only if the pilot light is burning. A thermal element is located near the pilot light. It must either produce thermocouple electrical energy or develop a sensitive bulb pressure to open valves that allow main gas flow. See **Figure 26-60.** This thermal element will also shut off the system if the pilot light stops working during the operating cycle of the unit.

A relay will turn on a blower or water pump motor used in the distributing system. Such a relay may have instant action or there may be a short delay.

Primary controls operate on a series of ***electric interlocks.*** All conditions must be safe before the interlocks are in the correct position to allow the system to operate. Mechanical or thermal sensors are usually considered adequate for the smaller capacity domestic burners.

Commercial and industrial systems use more elaborate controls due to the larger flow of fuel. Flame sensors are generally used. These shut off the system by reacting very rapidly should a flame fail. (They will react in a fraction of a second.) The sensors are electronic. They use a flame rod and a photocell. The photocell is either sensitive to the radiant energy or to the ultraviolet

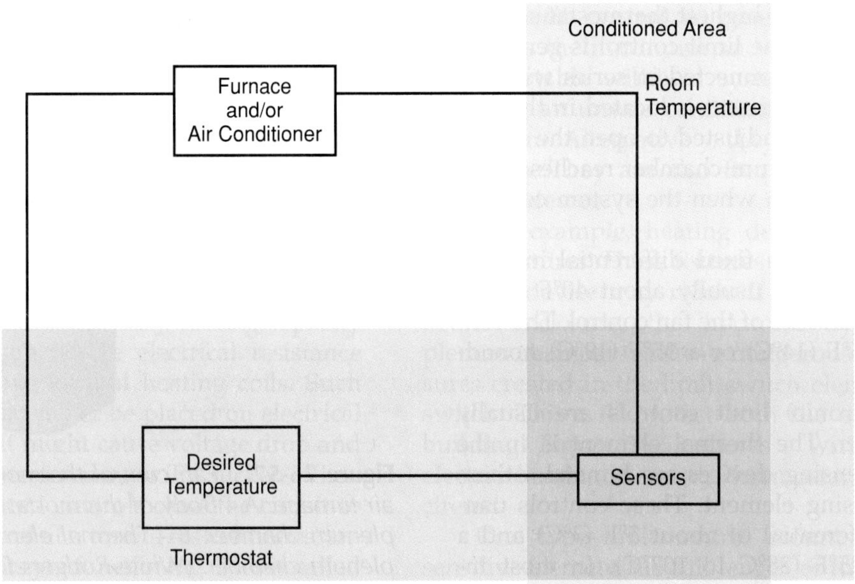

Figure 26-58. *Control system block diagram for a typical home HVAC system. Each block represents specific electronic circuitry.*

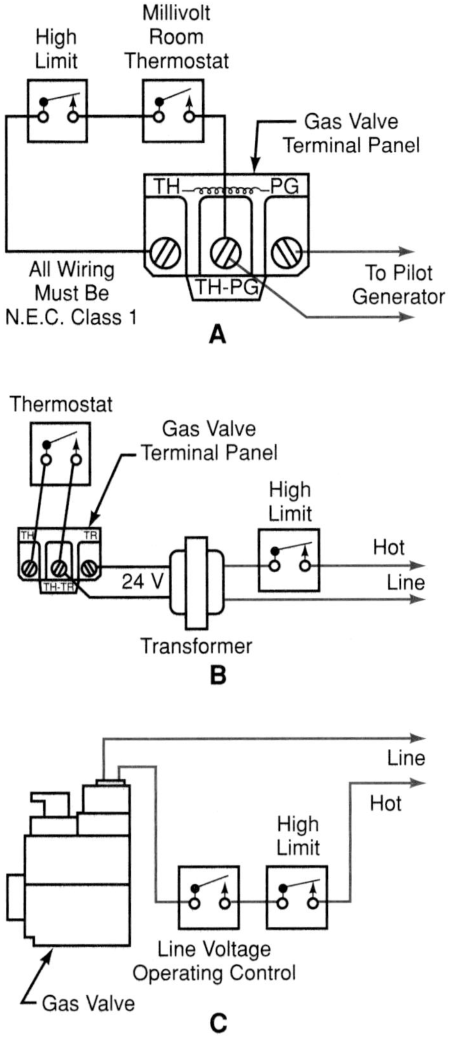

Figure 26-59. *Electrical circuits used with gas controls. A—Millivolt circuit. B—A 24 V circuit. C—A 120 V circuit. A high-limit control is used on each. (White-Rodgers Division, Emerson Electric Co.)*

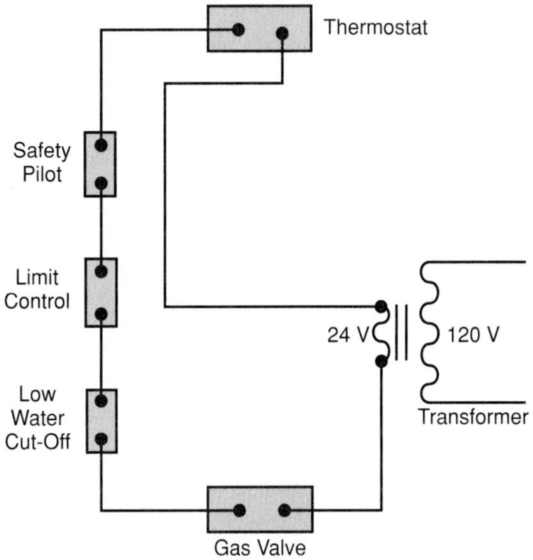

Figure 26-60. *Schematic is for low-voltage electrical circuit used on gas furnace without blower.*

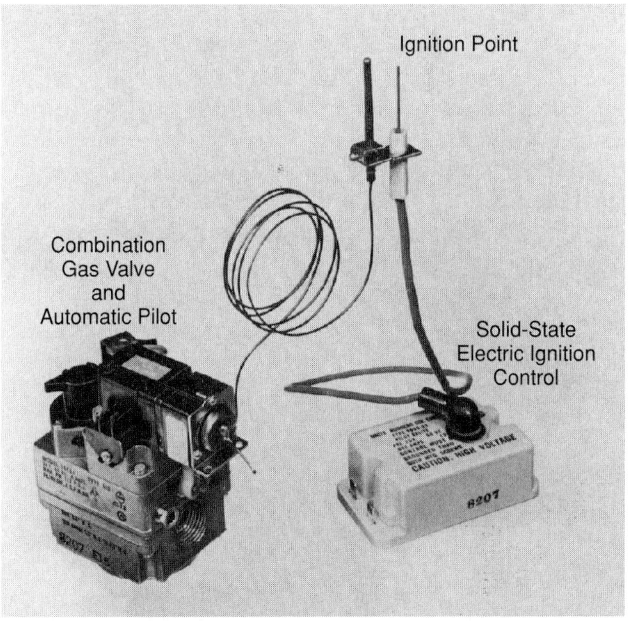

Figure 26-61. *Solid-state electric ignition system for gas furnace burner. (White-Rodgers Division, Emerson Electric Co.)*

rays of the flame. Some installations use a lead sulfide cell. It responds to the infrared rays from the gas flame.

Electric ignition for gas systems is used when the gas furnace is located outside the building. It may also be used in hard-to-reach locations. In this system, an electrical spark ignites the gas at the main burner. It also automatically shuts off the main burner gas supply if the burner does not ignite.

The main gas line to a gas furnace has four valves:

- Hand shutoff valve.
- Pressure regulator.
- Gas flame sensor safety.
- Automatic gas valve operated by thermostat.

The flame sensor probe detects within a fixed period of time whether the flame is operational. If not, it closes off the flow of gas. The ignitor is also turned off, and the controls lock out the system. **Figure 26-61** illustrates the complete gas burner control.

Originally, the four valves listed above were separate units. Presently, the automatic gas valve is made into a single unit for ease of installation. **Figure 26-62** shows a combination valve. The combination gas control is often called a "CGC." **Figure 26-63** is a cross section of the combination gas valve.

The automatic gas valve portion may be operated by a solenoid. It may also be operated by an electrical resistance-heated bimetal blade. A sensing bulb, capillary tube, and bellows combination (hydraulic thermal element) may also be used.

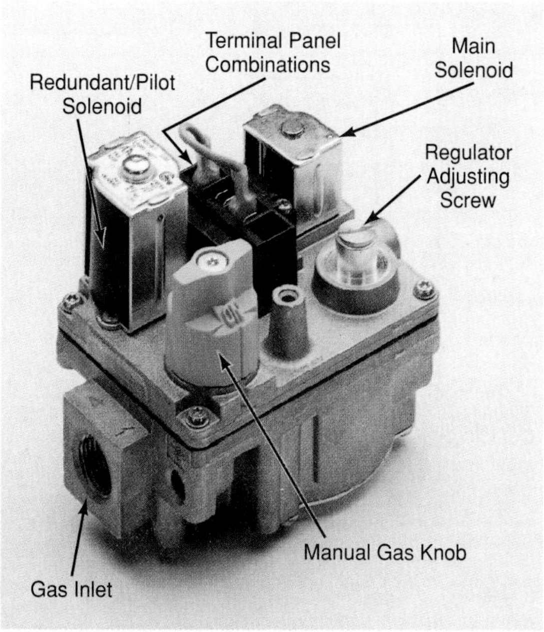

Figure 26-62. *Combination gas valve and automatic pilot. (White-Rodgers Division, Emerson Electric Co.)*

Electronics technology, along with safety regulations, has resulted in many advances in gas furnace control systems. Electronic circuits, sometimes used to control heating systems, can vary (modulate) the gas flame size. The modulating system uses a solid-state thermostat. It also uses a thermistor and several transistors.

The gas flame size depends on a temperature difference between thermostat setting and room temperature. The flame is larger with a greater temperature difference. It gets smaller as room temperature approaches the thermostat setting. The unit starts up the flame at about 20% to 50% of capacity. Then it adjusts the flame to temperature differences. The system's three main control parts are the thermostat, an amplifier, and a modulating gas valve.

Figure 26-64 illustrates a wiring circuit used with a makeup air system. The 120 V ac power is reduced to 24 V ac, using a transformer. A rectifier in the amplifier changes this current to dc. Three control devices are used:

- A remote temperature selector.
- A discharge air sensor.
- A duct stat (for safety).

These controls signal the amplifier. The amplifier then operates the solenoid and modulating regulator on the main gas burner line. **Figure 26-65** shows the amplifier unit. It holds the rectifiers and solid-state components for the control circuits. It also contains an adjustable potentiometer for calibration.

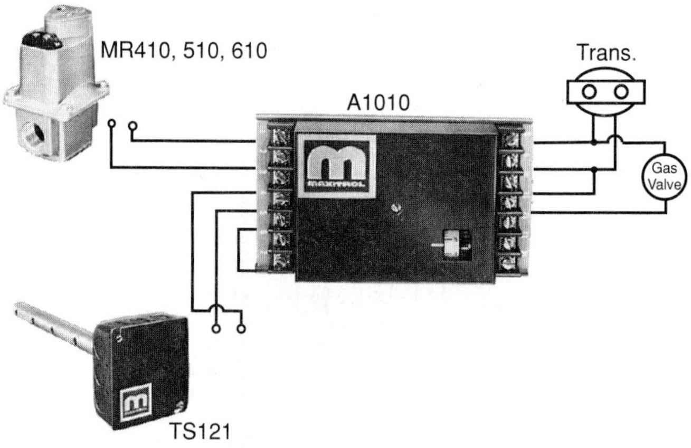

Figure 26-64. *Wiring diagram shows indirect-fired makeup air application. (Maxitrol Company)*

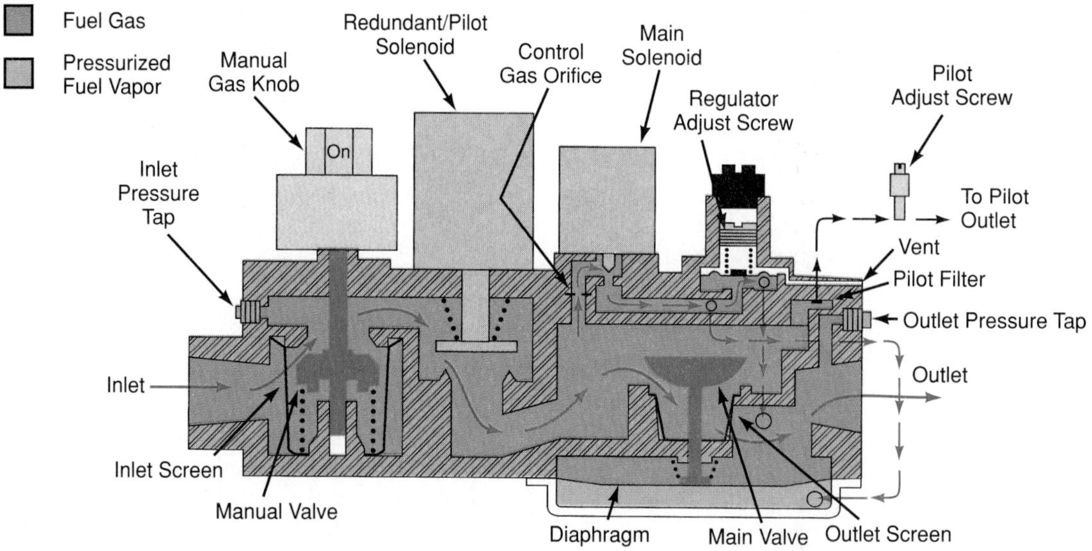

Figure 26-63. *Combination gas control in cross section. It has hand shutoff valve, pilot light control, bypass-operated main valve, and pressure regulator. (White-Rodgers Division, Emerson Electric Co.)*

The remote temperature selector is shown in **Figure 26-66.** This unit contains an adjustable potentiometer. It sets the temperature level of the discharge air. The discharge air temperature is being sensed by a thermistor located in the discharge air sensor, **Figure 26-67.**

The modulator/regulator valve is shown in **Figure 26-68.** This is the valve which varies the gas flow. In addition, an automatic solenoid valve is needed to completely shut off the fuel supply.

Direct current to the modulator controls the amount of gas flow. The less the current flow, the higher the flame. A duct thermostat is connected in series with the solenoid valve. It is used as a safety device if the duct temperature becomes too high.

Space heating is obtained by combining the remote temperature selector and discharge air sensor. It is combined into a single wall-mounted unit, shown in **Figure 26-69.**

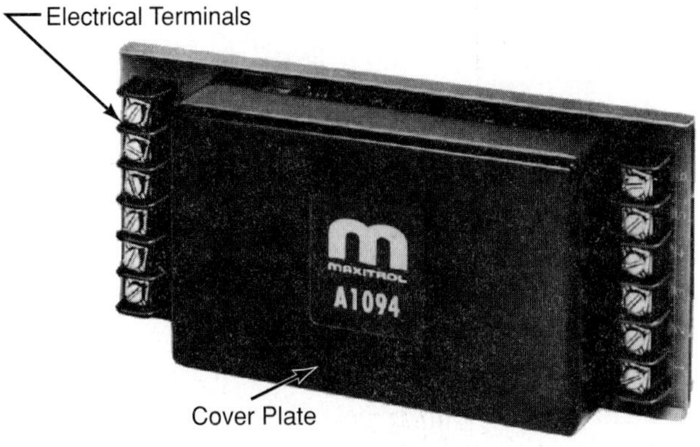

Figure 26-65. *Solid-state electronic amplifier may be placed at any convenient location. (Maxitrol Company)*

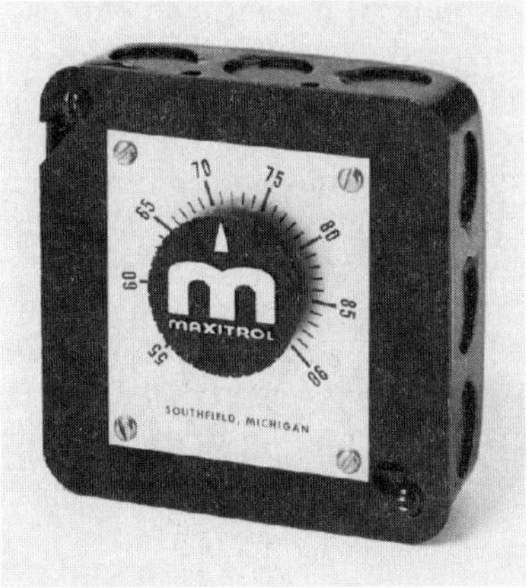

Figure 26-66. *This remote temperature selector is not temperature-sensitive and may be placed in any convenient location. (Maxitrol Company)*

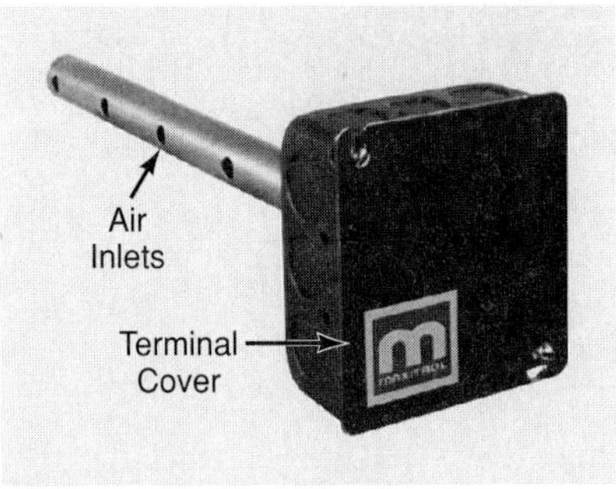

Figure 26-67. *Discharge air sensor. It sends signals to the amplifier of any temperature change from setpoint. (Maxitrol Company)*

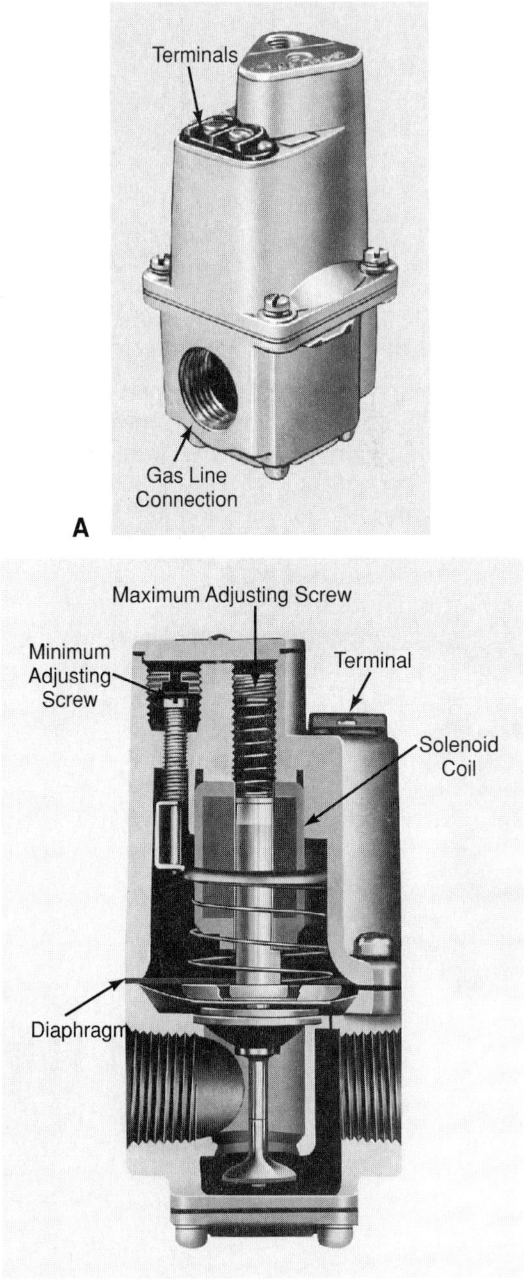

Figure 26-68. *Modulator/regulator valve performs both regulation and modulation to vary burner flame size (or burning rate). A—Outside view. Note gas line connection and electrical terminals. B—Cutaway of same modulator/regulator valve. Note adjusting screws used to vary amount of pressure required to operate valve. (Maxitrol Company)*

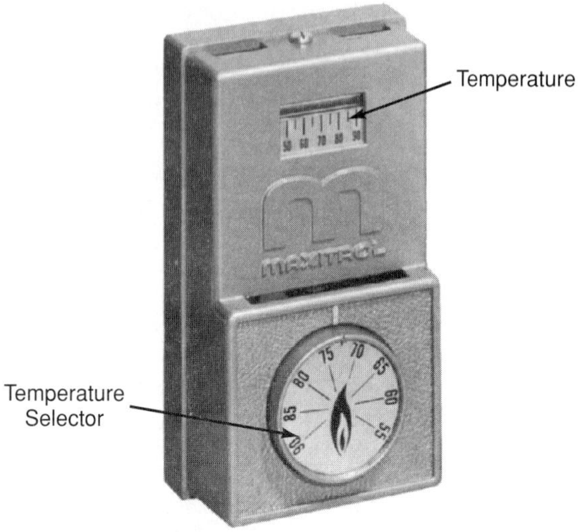

Figure 26-69. *Combination modulating makeup air thermostat and selector. (Maxitrol Company)*

26.18.2 Oil Furnace Controls

In the gun-type oil burner, a primary control usually starts the burner motor. The control can also stop the unit. This occurs if the flame goes out or if the flue temperature becomes too high.

Some primary units are mounted in the flue. When the thermostat signals for heat (closing of points), the primary control starts the gun-type oil burner motor. It turns on the ignition. This control is shown in **Figure 26-70.** It has a temperature sensing element. The element will shut down the unit unless the flue temperature rises in a few seconds. (This will indicate that the oil is burning.)

This same sensor will constantly check for flame temperatures. It will shut down the system if the flame goes out. It will also shut off the system if the thermostat or one of the limit controls opens the circuit.

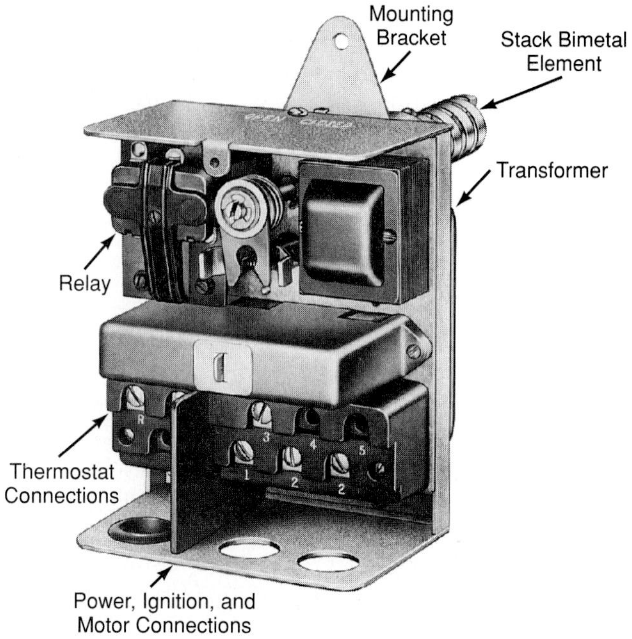

Figure 26-70. *Oil burner primary control. This control will cycle burner, operate electric ignition system, shut off unit if ignition fails, and scavenge unit after each cycle.*

Many commercial heating and process burners using gas or light oil fuels employ a modular burner management system. Flame monitoring is done by ultraviolet scanners or flame rod/photocell detectors and plug-in amplifiers and programmer modules. These are connected to a standard chassis and wiring base. Interchangeable programmer and amplifier modules determine the control methods. In the event of ignition failure (or following a safety shutdown), the unit locks out. This activates an alarm. Various programmer modules are available, depending upon the type of installation, **Figure 26-71.**

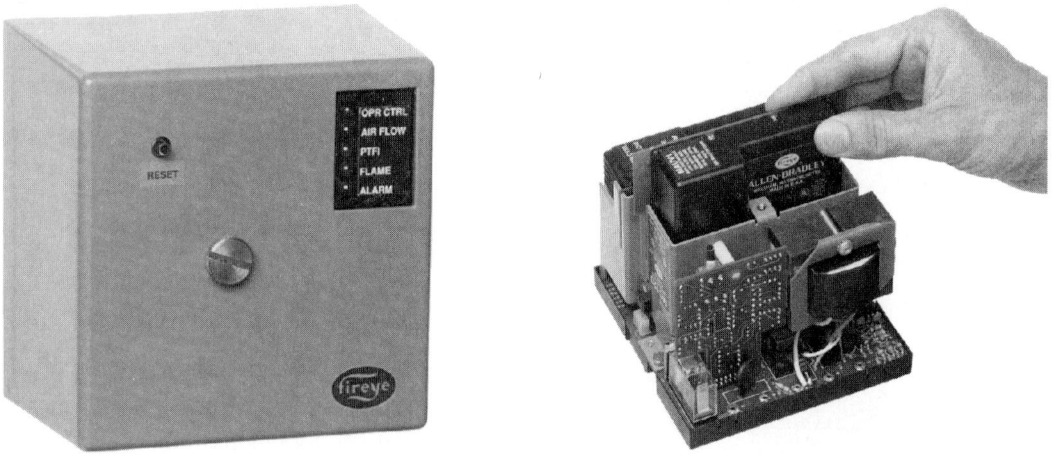

A B

Figure 26-71. *Flame-out safety control monitor used with commercial oil and gas furnaces. A—Exterior view of solid-state flame monitor. B—Monitor with cover removed. An amplifier module programmer is being inserted into the chassis. (Fireye)*

The unit's operation is based on an ultraviolet scanner. It is located within 18″ (46 cm) of the flame to be monitored, closer if possible. The scanner must not sight the ignition's spark directly. See **Figure 26-72.**

New models of oil burners use solid-state controls. Ignition (electrical) may be either continuous or intermittent. **Figure 26-73** shows a gun-type oil burner with a primary control and ignitor. When the thermostat calls for heat, the gun oil burner motor turns on. The oil pump starts, and the ignition is turned on. If fuel does not ignite in 15 seconds, the flame detector stops the burner motor. After the trial for the ignition period, the control provides a 5-second to 10-second ignition overrun time. If the fuel does not ignite within 30 seconds, the safety switch must be manually reset. The flame detector stops the burner motor and closes down the system. It then attempts to re-start. **Figure 26-74** shows the circuit.

A solid-state primary control is shown in **Figure 26-75.** It can be used as a replacement for older model primary controls. **You must be absolutely certain that the power is off before installing the unit.**

The control is wired as shown in **Figure 26-76.** The cadmium cell must be very carefully mounted. The cell must "see" the flame, as well as be heated to above 140°F (60°C). The correct mounting is shown in **Figure 26-77.** Another photocell flame detector mounting is shown in **Figure 26-78.** The flame detector must be lined up with an opening in the static pressure disk to be able to "see" the flame.

The transformers have a primary winding of 120 V, 240 V, or 208 V. The secondary winding usually provides 10,000 V. Some systems use 12,000 V.

When a 10,000 V transformer needs replacement, a 12,000 V unit should be used. This is especially advisable in cold air or cold oil situations. It is also advisable when line-voltage drops are known or suspected.

26.18.3 Electric Heat Controls

Baseboard units usually have individual thermostats. Primary controls are most often relays and limit controls. Units using blowers or fans have fan controls. Such controls may operate from a separate thermostat, or may be connected in parallel with the heating element.

Figure 26-73. *Gun-type oil burner. (Carlin Combustion Technology, Inc.)*

Central systems normally use sequence relays as primary controls. Blowers operate the same as on baseboard units. In addition, a safety control is usually provided. This shuts off the heating elements if the blower fails to operate. It will also shut them off if air fails to circulate.

26.18.5 Infrared Heat Controls

Electric infrared heat lamps may require as much as 32 kilowatts (kW). The heating unit may have as many as 16 lamps of 2000 W each (32,000 W). The electric circuit, in such cases, is controlled by a large-capacity relay (contactor). At 240 V ac, the line capacity will need to carry 175 A. The 175 A main circuit is usually divided

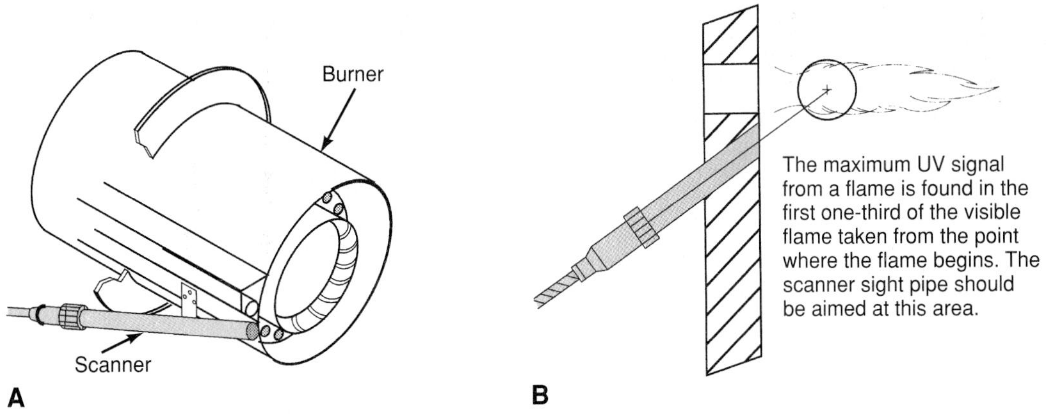

Burner
Scanner
A
The maximum UV signal from a flame is found in the first one-third of the visible flame taken from the point where the flame begins. The scanner sight pipe should be aimed at this area.
B

Figure 26-72. *Ultraviolet scanner installation. A—Typical scanner installation on burner. B—Scanner sight pipe aimed at first 1/3 of visible flame. This provides maximum UV signal. (Fireye)*

A

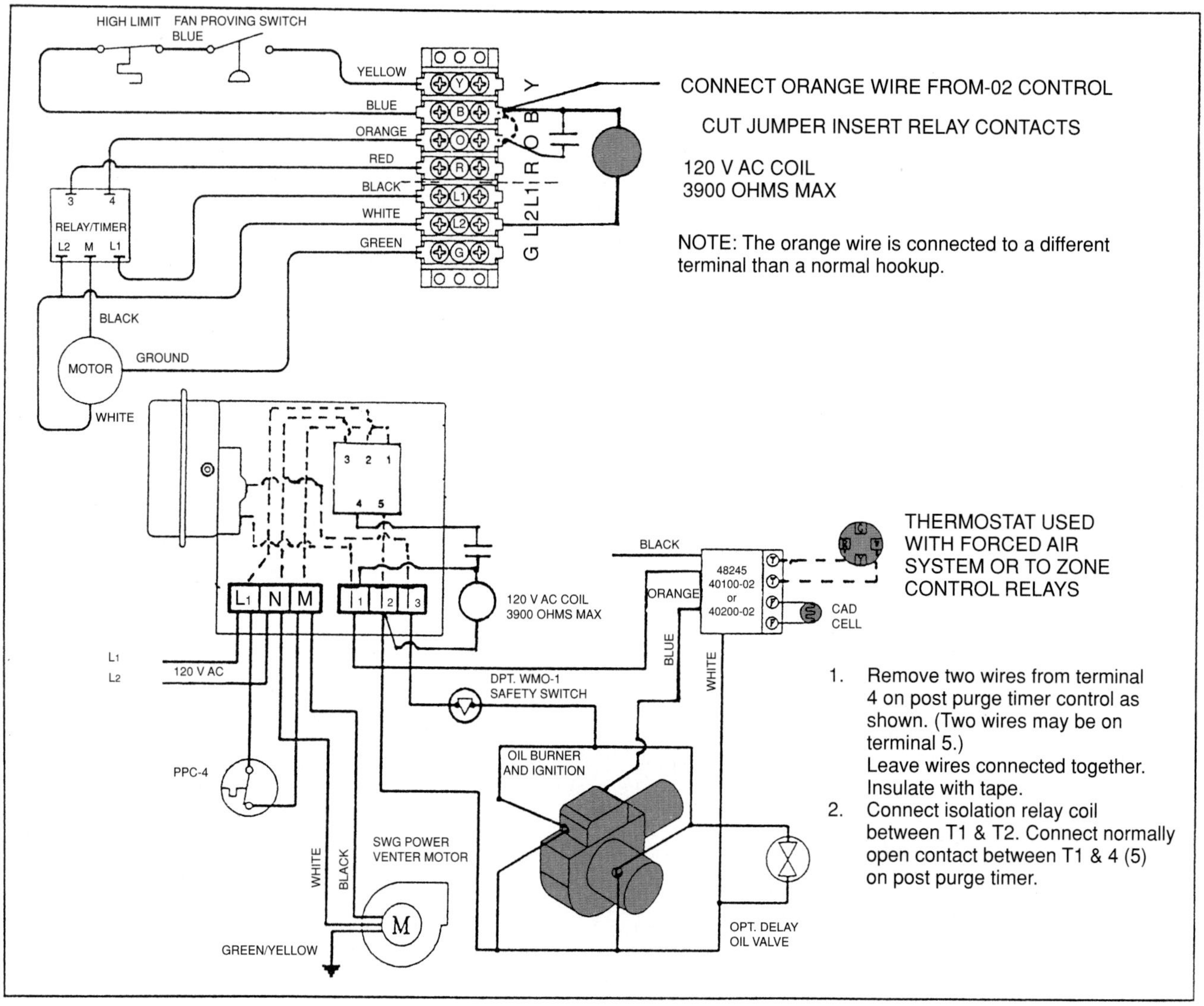

B

Figure 26-74. *Wiring diagrams for electronic oil burner primary control and ignitor. A—120 V ac coil with 3900 Ω resistance. B—Thermostat is connected to the low-voltage terminals. Note location of oil burner and ignitor in relation to the thermostat and cad (photoelectric) cell. (Carlin Combustion Technology, Inc.)*

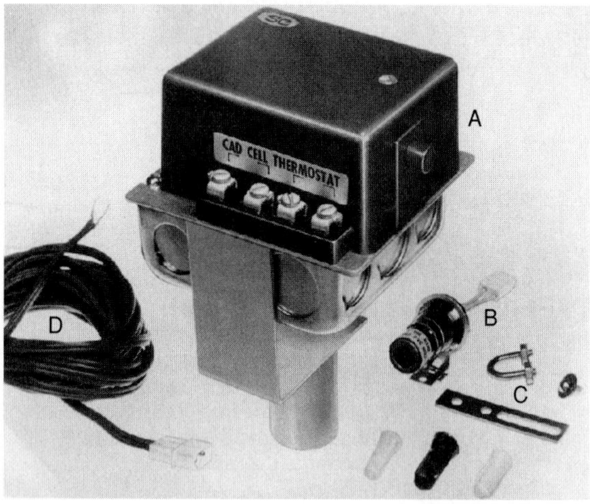

Figure 26-75. *Solid-state primary control for gun-type oil burner. A—Primary control. B—Cadmium cell flame detector. C—Cadmium cell mounting bracket and fittings. D—Connecting lead.*

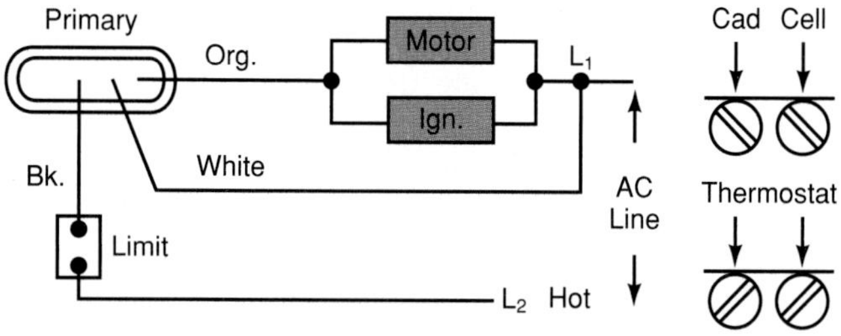

Figure 26-76. *Schematic wiring diagram for solid-state primary control for gun-type burner. Note that current is supplied to motor and ignitor at same time.*

into four or more separate circuits. Each circuit has a relay (contactor).

Either thermostat or solid-state sensors energize the operating coil of the contactors. The contactors are usually sequenced so that only one closes at a time. (Sequencing means that the contactors are set to close one

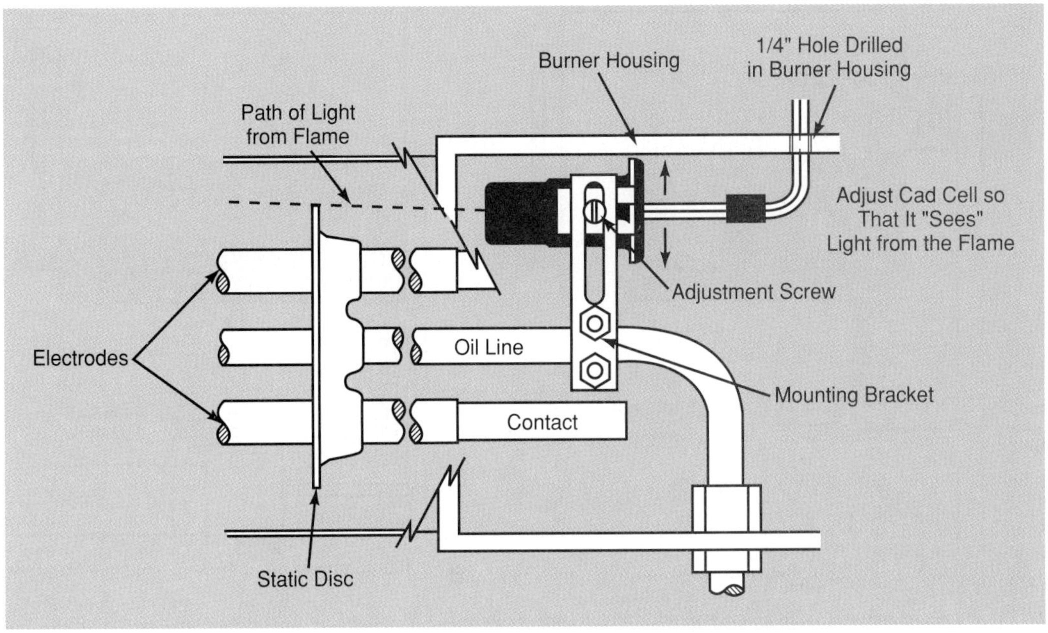

Figure 26-77. *Method of mounting photoelectric flame sensor in gun-type oil burner.*

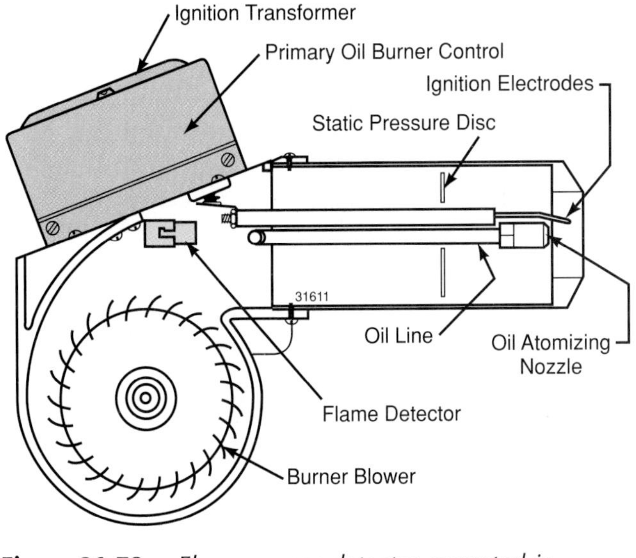

Figure 26-78. *Flame sensor detector mounted in gun-type oil burner. (White-Rodgers Division, Emerson Electric Co.)*

after the other, rather than all at once. This is usually done by a time-delay on each contactor switch which operates the coil of the next contactor.)

Solid-state temperature controls modulate the current flow with thermistors and triacs. They reduce part of each sine wave of the ac flow. This maintains a constant temperature. See Chapter 6.

26.18.6 Hydronic System Controls

Primary control features in water or hydronic heating systems are water level and temperature control. Water level controls are especially important on steam heating systems.

The control is generally a switch turned on and off by a float. If the water level drops near the danger point, the float will drop. It drops far enough to open the electrical circuit to the operating controls. This stops the boiler.

The switch generally operates in two stages. Lowering of the float will trip a switch. This will turn on the feed water pump or feed water solenoid valve. If the float drops still more, the system is shut off.

Some water level controls are of the probe type. This type immerses two electrodes in the water. A small current flowing in the water between the two will energize a holding relay. The system will be allowed to run. If the water level falls below the upper probe, the current flow will cease. The operating controls will shut down.

The operating principles of the two types of water level control are shown in **Figure 26-79.** Avoid air in the system, which could cause the controls to malfunction.

26.18.7 Comfort Cooling Controls

Controls for comfort cooling are of the same basic types as those in heating. There are operating controls, primary controls, and limit controls.

The operating controls are thermostats, pressurestats, and humidistats. Primary controls include motor starters and starting relays. Limit controls include overload circuit breakers, thermal overloads, internal motor overloads, refrigerant pressure limit controls, and oil pressure limit controls. Most of these controls are described in Chapters 8 and 13. The two most popular refrigerant controls are the thermostatic expansion valve (TEV) and the capillary tube. These controls are described in Chapters 5 and 13.

The schematic diagram in **Figure 26-80** is for a circuit used in a comfort cooling unit. This system uses a

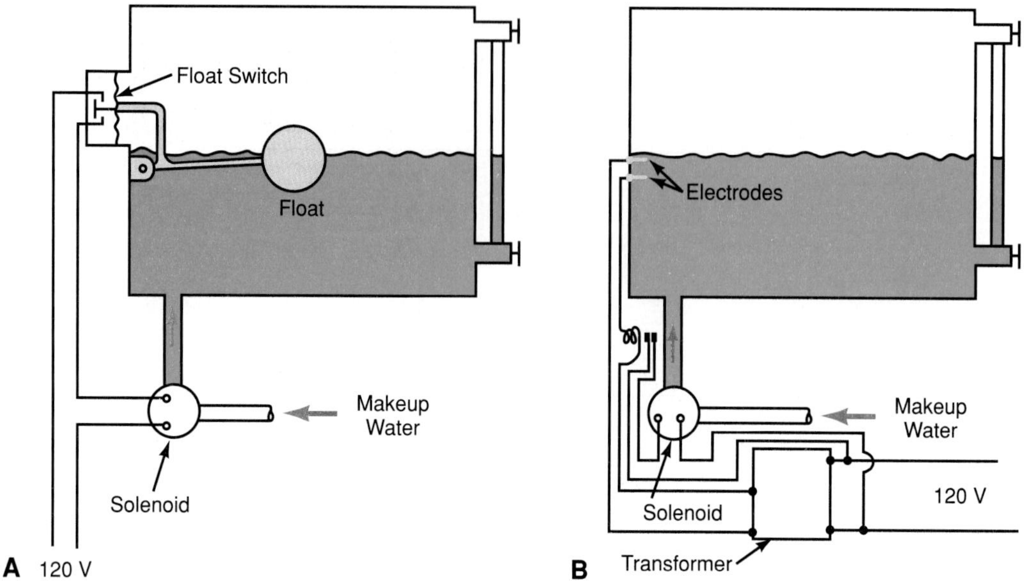

Figure 26-79. *Two types of water level controls. A—Float control. Float valve will open makeup water valve when water level drops. Some valves are connected to electric switches that will shut off the unit if water level gets too low. B—Probe-type water level control. Solenoid opens when current stops traveling across electrodes.*

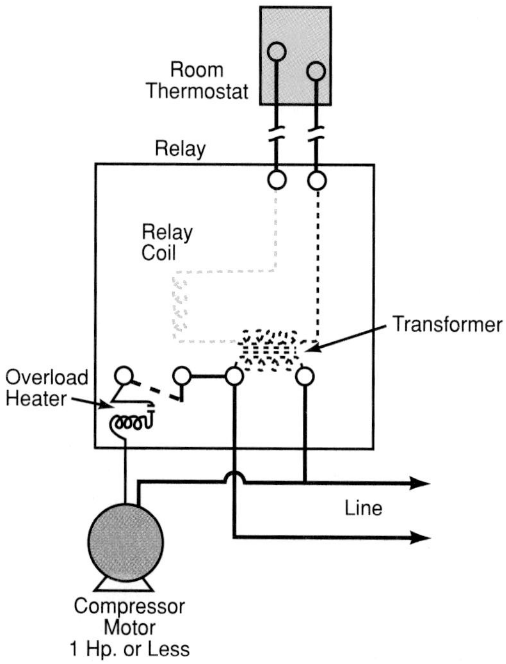

Figure 26-80. *Wiring diagram of thermostat relay combination for small comfort cooling units.*

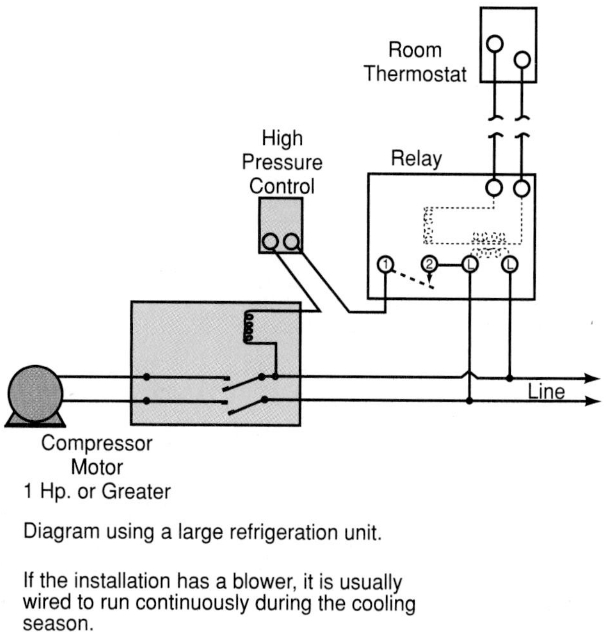

Figure 26-81. *Wiring diagram is for a comfort cooling unit that uses high-pressure safety cutout and motor starter. (White-Rodgers Division, Emerson Electric Co.)*

low-voltage, two-wire thermostat. The thermostat controls a relay that will close the motor circuit. If the motor cannot be connected directly to the line, and pressure safety devices are to be put in the system, the wiring will be somewhat like that shown in **Figure 26-81.** The high-pressure safety cutout is wired in series with the starter coil. It will open the circuit if pressures become too high.

Some systems cycle on command from a low-side pressure control. The thermostat operates a solenoid valve mounted in the liquid or suction line. When the thermostat temperature is satisfied, the solenoid valve will close. When the low-side pressure drops enough, the motor circuit will be opened. This is done by means of the pressure control connected to a magnetic starter. See **Figure 26-82.** The unit will then stop. A high-side switch is also provided in this control. This control will stop the compressor if the high-side pressure exceeds a preset limit.

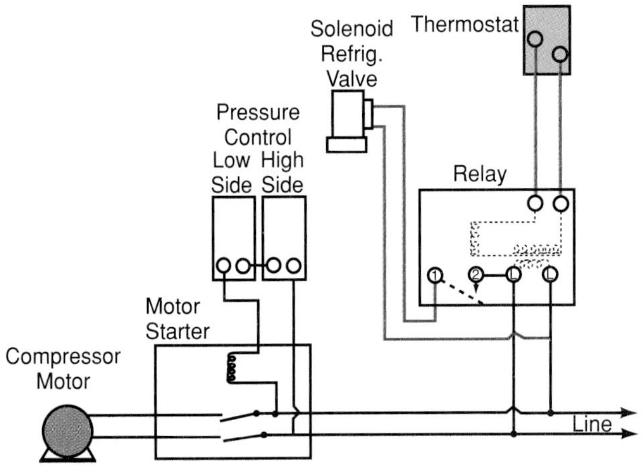

Figure 26-82. *Comfort cooling system wiring diagram. Note that thermostat operates the solenoid valve. System cycles as low-side pressures vary.*

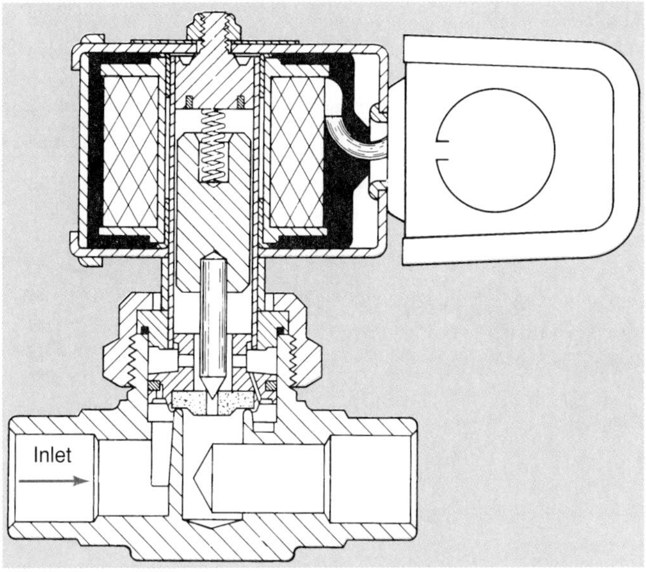

Figure 26-83. *Cross section of solenoid refrigerant control valve used in liquid lines on air conditioning systems. (Sporlan Valve Co.)*

The controls found in comfort cooling systems are:

- Thermostat (24 V or 120 V) or thermistor sensor (one-stage or two-stage).
- Evaporator icing control (freeze-up control), two-wire 24 V or 120 V to control dampers, valves, or compressors when evaporator temperature approaches 32°F (0°C).
- Multiple compressor sequence starting controls.
- Multicylinder compressor unloading sequencing controls.

Generally, large air conditioning evaporators use the thermostatic expansion valve for refrigerant control. Some installations use several such valves on one large evaporator to get maximum efficiency. Self-contained systems—especially those hermetically built—may use the capillary tube refrigerant control. These controls are described in Chapter 5.

In addition to the refrigerant control used on automatic systems, a solenoid valve is sometimes placed in the liquid line. This will automatically stop the flow of refrigerant to the evaporator the instant the condensing unit stops, or when the low-side pressure control opens. This is done so that the evaporator will not become flooded with refrigerant while the condensing unit is idle, and will be pumped down.

Figure 26-83 illustrates a solenoid refrigerant control valve. This valve uses 5 W at 120 V ac. It has an orifice 0.1″ (2.5 mm) in diameter. At a 5-lb. pressure drop across the orifice, the refrigerating capacity is 1.3 ton for R-12. It is 2.1 ton for R-22. For R-500, it is 1.62 ton.

In addition to the operating controls (thermostats), comfort cooling systems have primary controls and limit controls.

Larger refrigerating units are usually equipped with pressure controls. The low- and high-pressure controls are usually designed to lock the circuit open if any unusual pressures occur. The operator must then manually turn the system on. This allows a careful check for faults. The internal construction of a combination low-pressure and high-pressure control is shown in **Figure 26-84.**

Comfort cooling systems use several types of limit controls:

- Motor limit controls.
- Pressure limit controls.
- Temperature limit controls.
- Fluid flow limit controls.

Motor limit controls are described in Chapter 8. Such controls will stop the unit if current draw becomes too high. They will also stop the unit if the motor temperature rises to a dangerous level.

The anti-icing control is another type of temperature limit control. (This is in addition to the motor thermistor or bimetal protector.) It is located on the evaporator. Should ice accumulate there, this control will open and stop the system.

Fluid flow controls stop the system in the event chilled water flow ceases. They will also stop the system if conditioned airflow stops or slows to inefficient amounts. **Figure 26-85** shows an airflow signal switch or shutoff switch or both. A paddle is mounted on a pivoted arm. If airflow is too great, it will move back through a small arc. It will trip an electrical switch. **Figure 26-86** shows a similar switch for liquid flow, such as chilled water or condenser water.

26.18.8 Humidity Controls

On smaller units, humidity control systems are almost always electric. Pneumatic systems are usually used for large installations. Humidity is most often made higher by adding water vapor to the air. It is most often lowered by cooling air below its dew point temperature. This condenses moisture out of the air.

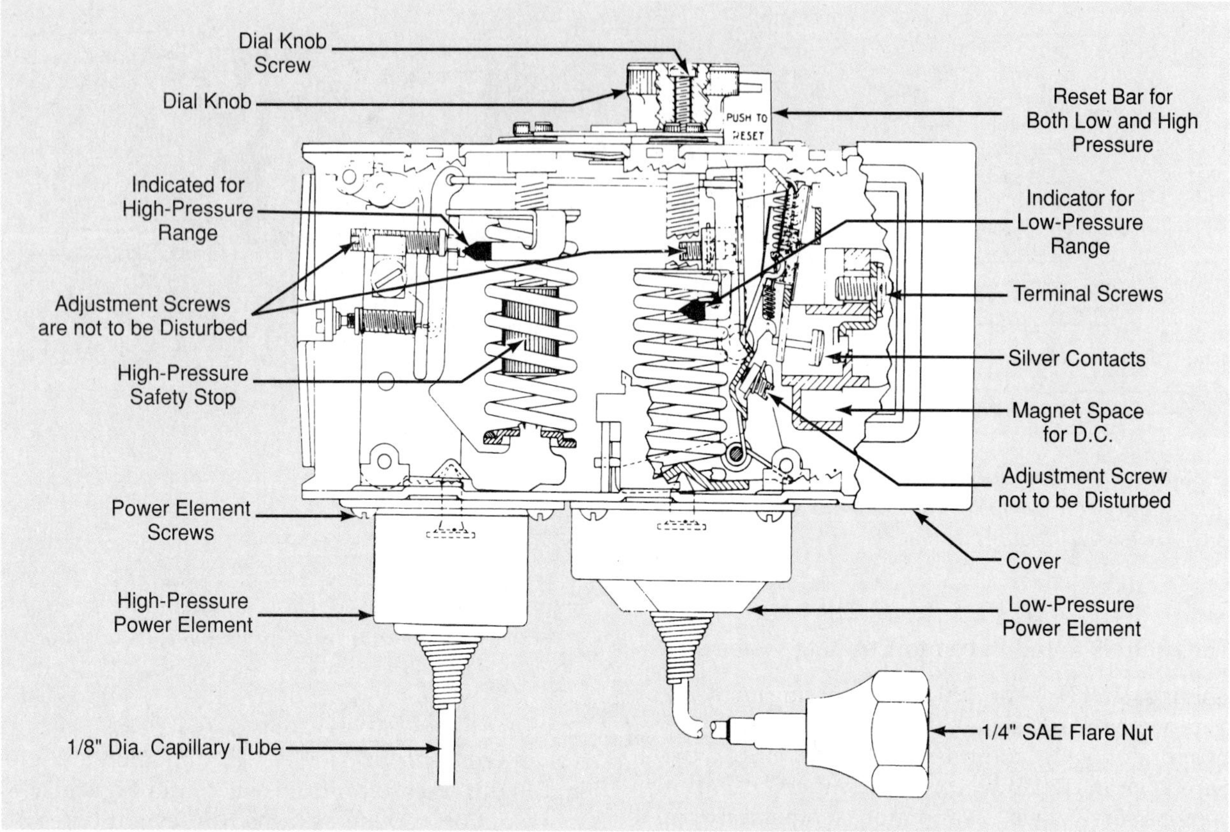

Figure 26-84. *Inner mechanism of air conditioning condensing unit pressure motor control. (Ranco North America)*

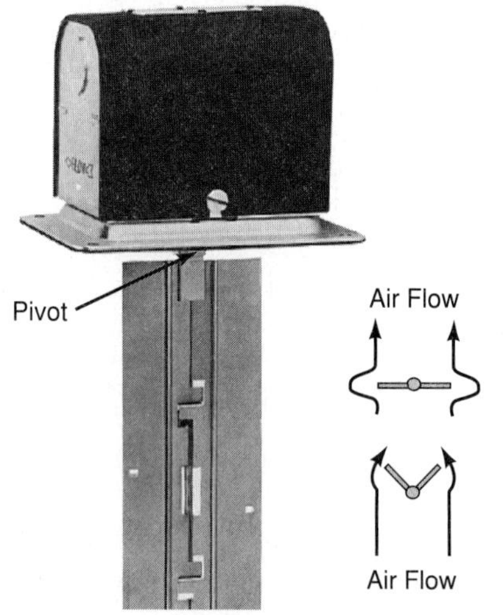

Figure 26-85. *Airflow signal switch prevents damage to system from too-high air velocities. If velocity gets too high, it will move the paddle, tripping the electrical switch. (ITT McDonnell & Miller)*

The electrical system uses a humidistat control and an electrical power source. Either a solenoid or a motor operates a valve or damper. The pneumatic system uses a humidistat control, piping, and a vacuum or pressure source. The controlled vacuum or pressure then acts upon a diaphragm-operated valve or a diaphragm-operated damper.

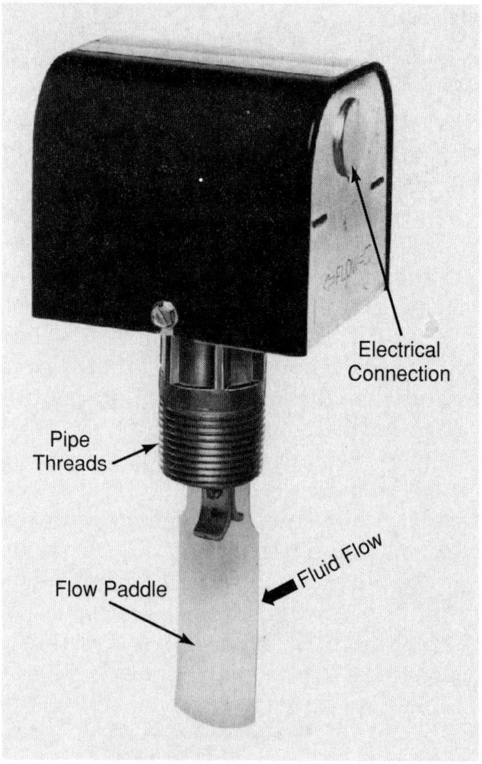

Figure 26-86. *Liquid flow switch. If chilled water or condenser water flow is not sufficient, this unit will close signal circuit, shut off the unit, or both. (ITT McDonnell & Miller)*

Humidistats

A ***humidistat*** is used as the control device in a humidity control system. It responds to changes in humidity; in doing so, it opens or closes a control system.

The sensing element of the humidistat is called a ***hygroscopic element.*** Such an element will stretch as the moisture content of the air increases. Commonly used hygroscopic elements are: human hair, wood, nylon ribbon, membranes.

Electronic solid-state sensors are also used. In electronic solid-state sensors, the sensor resistance changes as moisture content of the air changes. Some sensors used are:

- Hygroscopic salt (for example, lithium chloride).
- Carbon particles imbedded in a hygroscopic material. In these substances, the resistance decreases as the humidity increases.

The change in size, shape, or electrical resistance of the sensing element is used to operate a switch or a pneumatic system. **Figure 26-87** shows a humidistat that uses a multiple hair element. The control should be kept dust-free. The cover must permit free air circulation over the element.

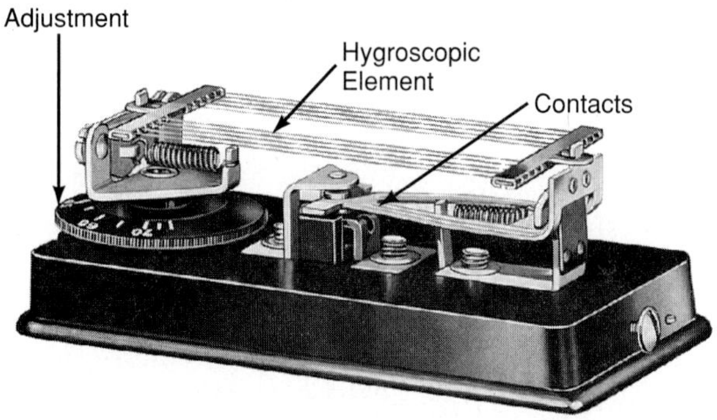

Figure 26-87. *A humidity control. Operating mechanism is shown.*

Figure 26-88 shows a typical electronic humidistat. Its output amperage varies, based on an internal resistance that changes with humidity. See Section 19.3.1 for more information on humidity measurement.

26.18.9 Electronic Cleaner Controls

Electronic cleaners have an electrical circuit that converts 120 V ac to about 9000 V dc. This circuit also has safety devices such as door interlocks, and service devices such as automatic washing and diagnostic circuits.

Figure 26-89 shows an electronic cleaner electrical circuit. Note the two stepup transformers, the door interlock, the full wave rectifiers, and the neon light circuit. The actual wiring diagram is shown in **Figure 26-90.**

Figure 26-88. *Electronic humidistat. Element varies resistance with changes in humidity. (Action Instruments, Inc.)*

The electronic cleaner is connected into the electrical circuit of a heating/cooling system. It uses a single speed fan. There is 120 V service to the cleaner and 240 V service to the fan motor. This is shown in **Figure 26-91.** The automatic water-cleaning system has a washing cycle with detergent, a rinsing cycle, and a drying cycle. See **Figure 26-92.**

Service to electronic cleaners includes checking rectifiers and high dc voltage. **High voltages are used in these electronic cleaners. They should be serviced only by someone with special training on the model being serviced. Figure 26-93** shows how the rectifiers are removed and the high voltage is checked. Note the heavy insulators on the meter leads. They protect against this very dangerous voltage.

26.19 Airflow Controls

Airflow controls constantly regulate air volume and temperature. If outside air is too cold, thermostats will cause power-operated devices to close the outside air damper. Thermostats also react if the recirculated air and outside air is out of balance. They will open one damper and close the other just enough to produce the correct mix. Damper motors are usually used. However, heated vapor element may also be used, as may air pressure or vacuum (pneumatic).

If air temperature is too high or too low (in the airflow to the room, in the recirculated air, in the fresh air, or in the exhaust air), the thermostat will react. It will properly adjust the dampers. Pneumatic motors are used to vary the air supply as outside temperature changes.

One motor for operating dampers uses power to open the damper. It works against spring pressure, and holding the dampers open until the thermostat points open. Spring pressure then closes the damper. The motor operates the damper through a gear reduction train. **Figure 26-94** shows such a damper-control motor and shaft.

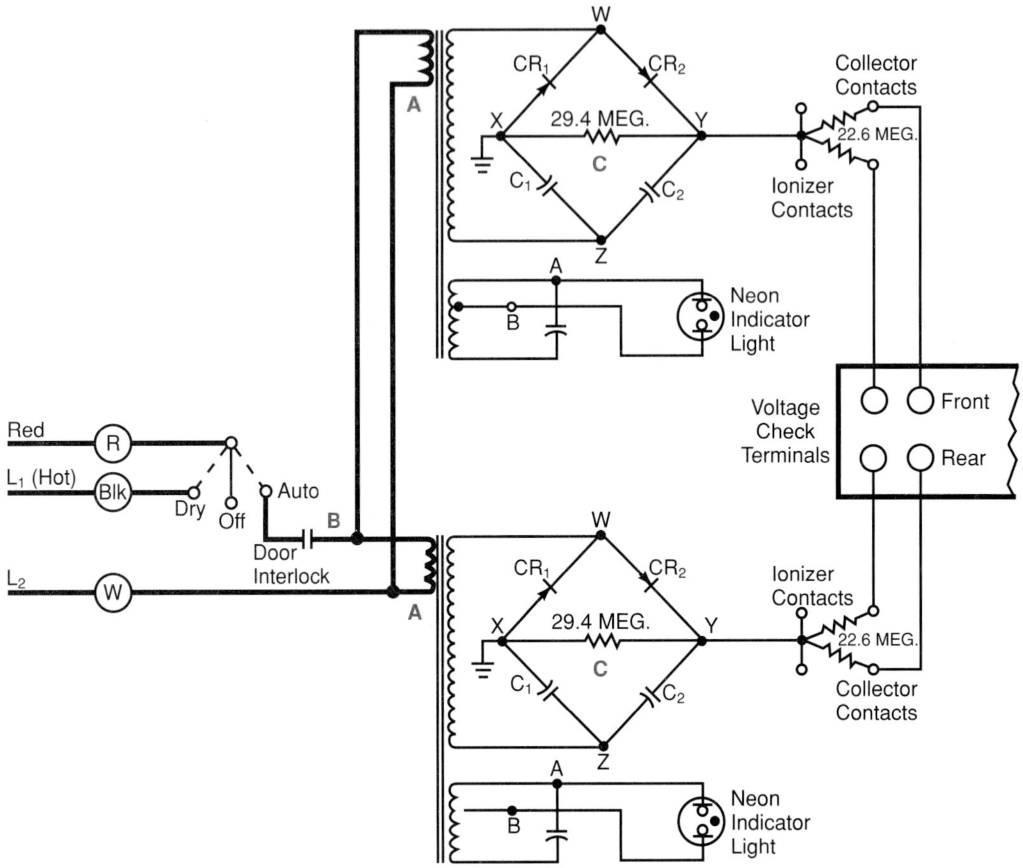

Figure 26-89. *Wiring circuit for an electronic air cleaner. A—Transformers. B—Door interlock (safety switch). C—Rectifiers. (Honeywell Inc.)*

Control devices that regulate air volume in a distribution system are called variable air volume (VAV) controllers. These devices use electronic components. They are usually part of a larger computer-controlled HVAC system.

26.19.1 Pneumatic Systems

Pneumatic (air) systems are often used to control air conditioning. Thermostats control a pressurized air line. Air in this line can operate pneumatic motors (a piston and cylinder or a diaphragm). Motors, in turn, operate dampers, valves, and switches.

The system's two main parts include sensing devices and pneumatic controllers. The operating pressures used are either 12 psig or 24 psig (17 psia or 39 psia [117 kPa or 269 kPa]). Some systems operate at a vacuum.

The 12 psig system actually uses an operating pressure of 3 psig to 15 psig (18 psia to 30 psia or 124 kPa to 207 kPa). The 24 psi system actually uses an operating pressure of 3 psig to 27 psig. (18 psia to 42 psia or 124 kPa to 290 kPa).

The regulator reduces the supply air pressure. The supply air pressure for these systems is as follows:

- 12 psig system = 18 psig to 20 psig (33 psia to 35 psia or 227 kPa to 241 kPa).
- 24 psig system = 30 psig to 35 psig (45 psia to 50 psia or 310 kPa to 344 kPa).

Air pressure controls are often used in large commercial and industrial systems. These control systems should be thoroughly checked each month.

26.20 Distribution Controls

The heating and air conditioning control systems reviewed so far must have a method for controlling the distribution of the air, water, or whatever medium is used. ***Distribution controls*** help to evenly and efficiently transfer the heating or cooling medium to the area where it is needed. These controls ensure that steam, water, or air is properly circulating in the system.

In steam systems, the zone control valves are operated upon a signal from a thermostat. In hot water (hydronic) systems, pumps and valves must work in proper sequence. They must operate in this way upon command from the operating controls.

In warm air systems, air movement may be controlled by using separate thermostats. They may control air movement in the complete system or in part of the duct system. These thermostats may operate the blower motor, duct controls, or the furnace primary controls.

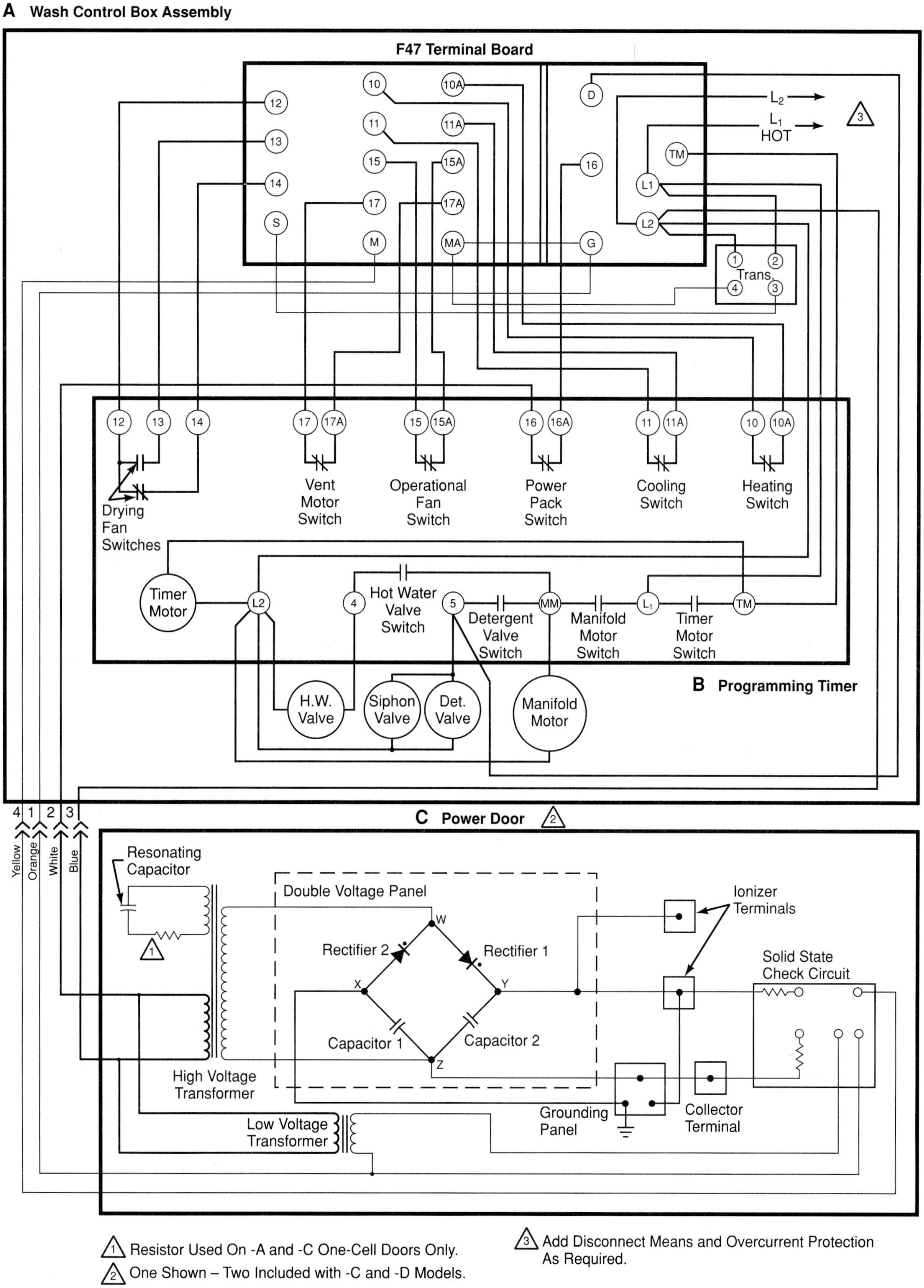

Figure 26-90. *Complete wiring diagram of an electronic air cleaner. The letters A, B, and C indicate the three main circuits. Note that the power door uses direct current in part of the circuit.*

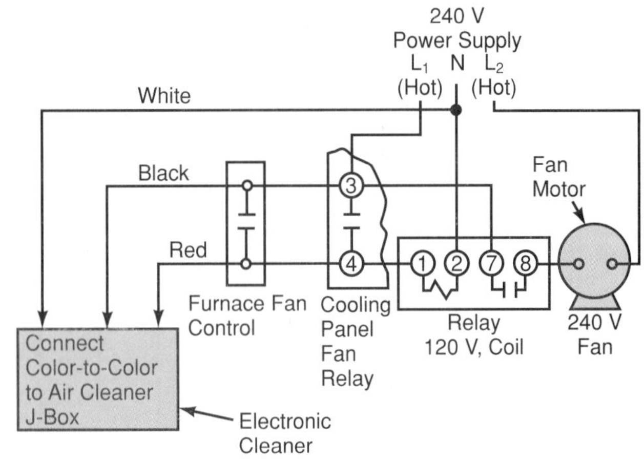

Figure 26-91. *A 120 V service electronic cleaner is connected to a 240 system. A 240 V fan motor is used.*

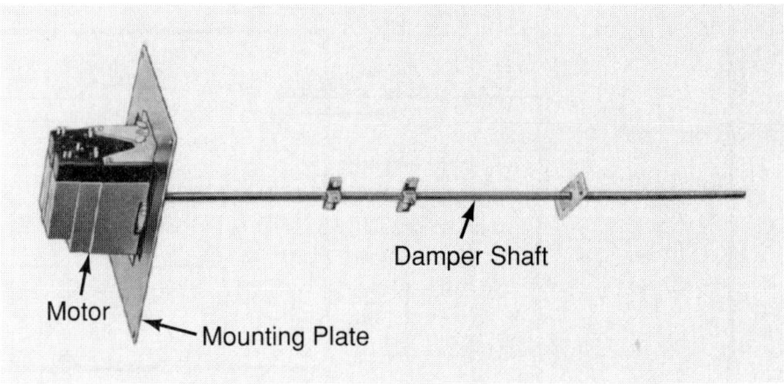

Figure 26-94. *An electric motor damper control. (White-Rodgers Division, Emerson Electric Co.)*

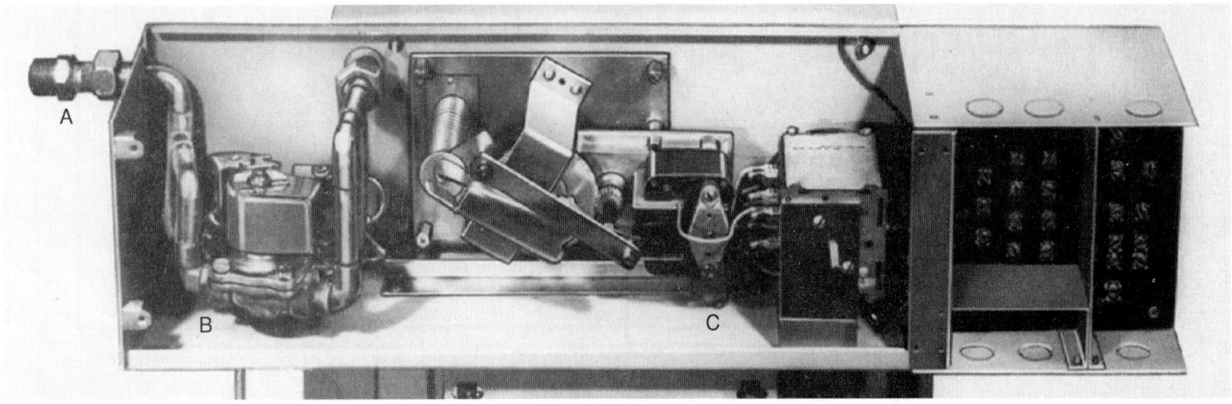

Figure 26-92. *Electronic cleaner washing controls. A—Warm water in. B—Solenoid. C—Timer.*

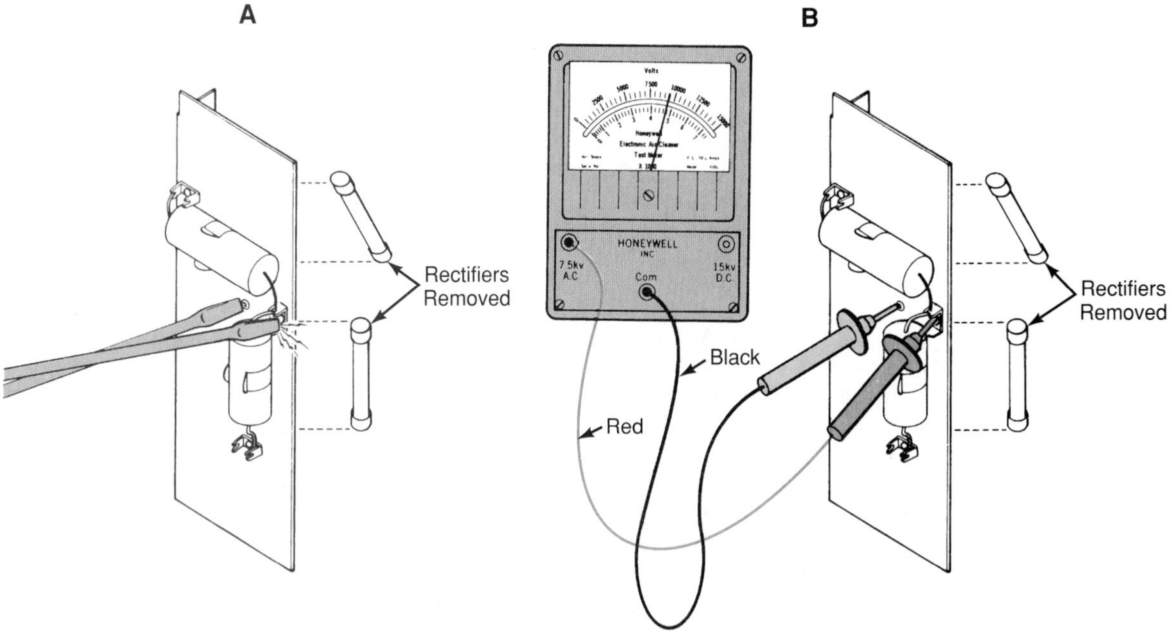

Figure 26-93. *Servicing an electronic air cleaner. A—Removing the rectifiers. B—Testing the high voltage. (Honeywell Inc.)*

26.20.1 Duct Controls

Safe distribution of conditioned air to the occupied spaces is important in warm air systems. This air must be distributed in the proper amounts to condition the space. It must not be too cold or too hot. It must have enough fresh air (for oxygen). The air must be flowing.

Many warm-air units are zone-controlled. These systems have dampers to regulate the flow of air to the zones. The dampers open and close upon command of a thermostat. Dampers may be powered by motorized valves, pneumatics, hydraulics, or solenoids. **Figure 26-95** is a sketch of a typical duct damper. It shows four ways to control damper action.

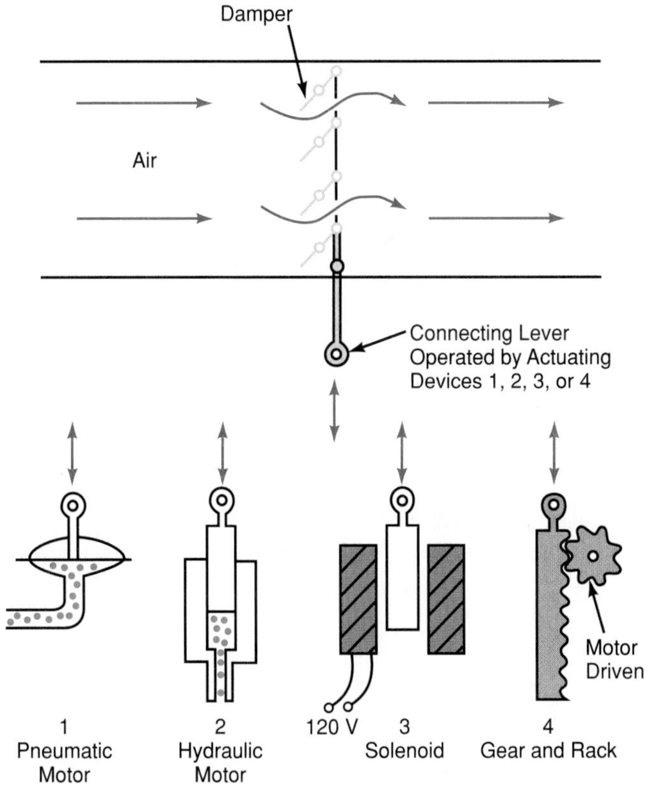

Figure 26-95. *Duct damper for controlling airflow and power devices used to operate it.*

26.20.2 Split-System Controls

Many systems provide either hot or cold water to the heat exchange unit in the conditioned space. Units are also designed to provide either hot or cold air to the space to be conditioned.

The three-pipe system, for example, carries both hot and cold water to a heat exchanger. The third pipe is a return line for either. **Figure 26-96** shows a thermostat connected to the two valves of a split system.

Many water heating or cooling systems bypass the heat exchange coil when no heating or cooling is needed. This action is obtained through a special solenoid valve. **Figure 26-97** shows such a valve.

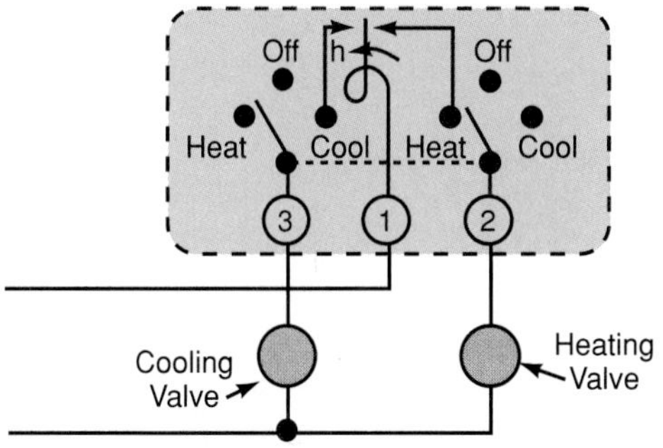

Figure 26-96. *Combination thermostat is connected to two valves of a split system. One is for cooling, the other is for heating. (Johnson Controls, Inc.)*

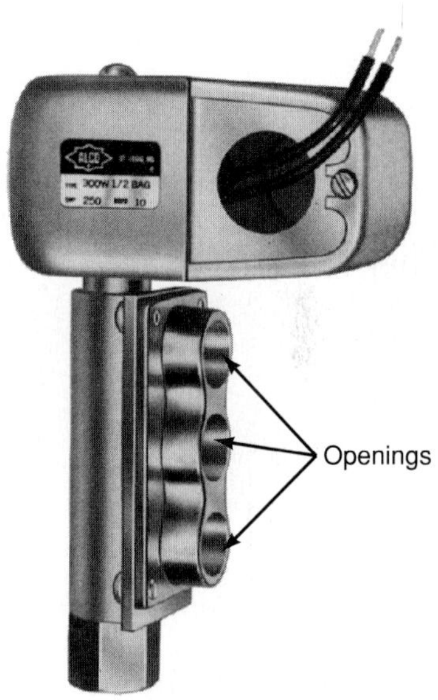

Figure 26-97. *Three-way valve. It is designed to bypass cold or hot water as necessary.*

26.20.3 Zone Controls

Many heating and cooling systems have ***zone controls.*** They maintain each zone (building area) at the desired temperature. For heating, the zone controls may activate an electrical heating circuit, or control the flow of conditioned air, hot water, or steam.

Zone cooling uses damper controls in ducts or automatic valves in the chilled-water circuit. These zone controls include:

- A thermostat or sensors.
- Proportional thermostats or sensors.
- Damper motors.
- A proportional damper positioner.

Electronic zone control devices automatically sense temperature in a number of zones. Then they regulate dampers within a distribution system. A device of this type is shown in **Figure 26-98.**

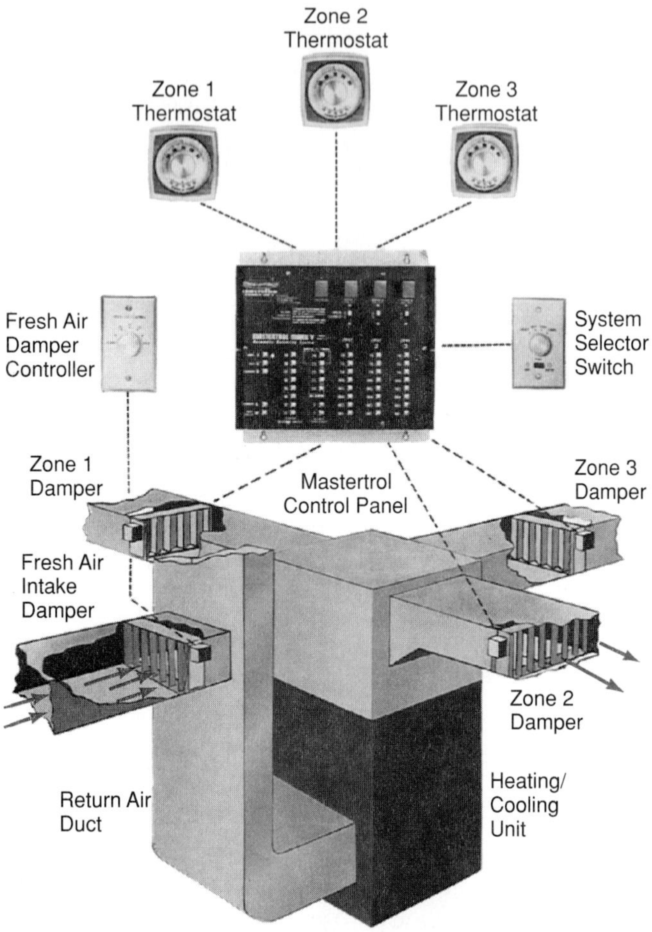

Figure 26-98. *Master control panel such as this is used to control heating, cooling, and air distribution in three different zones. (Trol-A-Temp, Division of Trolex Corporation)*

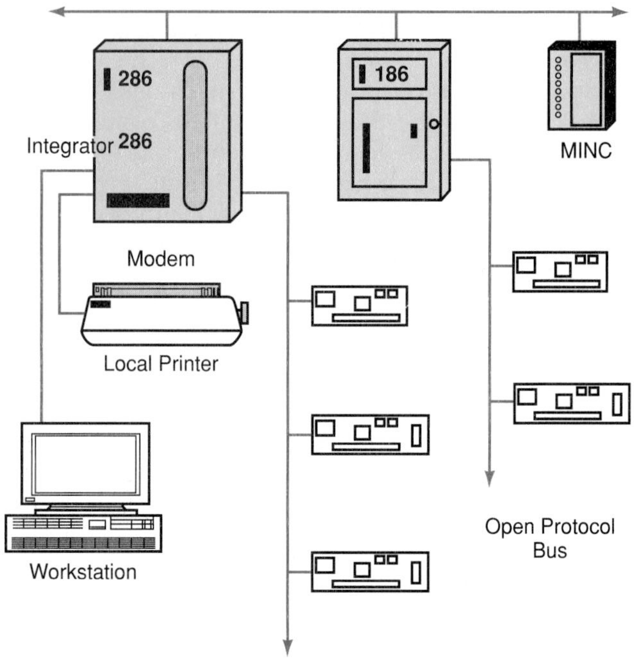

Figure 26-99. *Management control system. The major controller (labeled "Integrator, #286") can be connected with up to 128 internal input/output points. The unit adjacent to it, #186, is a stand-alone control of the type that can be located throughout the facility and network back to #286. The MINC (modular integrated network controller) is a smaller stand-alone controller connected back to the #286. (Teletrol Systems, Inc.)*

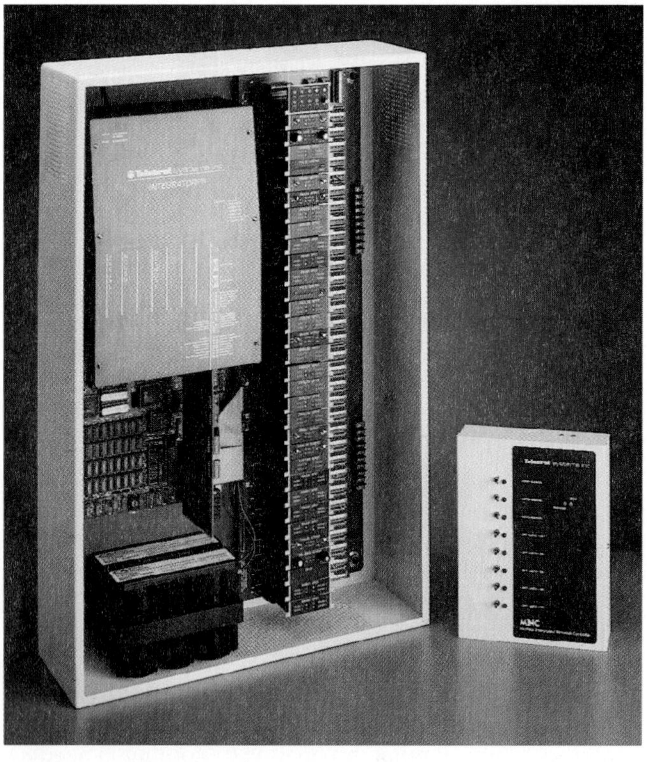

Figure 26-100. *A master control system used for commercial and industrial applications. (Teletrol Systems, Inc.)*

ENERGY MANAGEMENT MODULE

26.21 Total Energy Management Systems

Control systems and components reviewed thus far relate to heating, cooling, or HVAC systems within a building or residence. Most buildings in the United States were built when energy was relatively cheap and no one felt (or *understood*) the need to conserve. Energy sources, types, and cost have changed dramatically. Older buildings are now being modified and new buildings designed to operate using less energy. **Figure 26-99** shows an automation system with modular controllers for commercial and industrial control applications. The system has numerous small stand-alone controllers like those in **Figure 26-100.**

Total Energy Management (TEM) is also referred to as facility management, **Figure 26-101.** It is a new energy conservation concept. It recommends that each

Figure 26-101. *A facility management system like this one provides environmental control, energy management, lighting control, fire management, security functions, and overall facility monitoring. (Johnson Controls, Inc.)*

building should be viewed in terms of its total energy consumption. Previously, energy requirements of each separate system (heating, cooling, lighting, etc.) were the primary concern. Ideally, each system within the building would be ***optimized*** (changed to get maximum efficiency). However, each building represents many energy consumption systems interacting with each other. Compromises can be made in the total energy used. This has been referred to as "smart" building. A TEM system provides the flexibility to conserve energy and cut costs. At the same time it meets the needs of the building's users.

The following paragraphs will present background on system installation and usage, energy consumption, various types and functions of typical TEM systems, and a section on servicing.

26.21.1 System Determination and Usage

The first step in selecting a total energy management system is to determine the necessity. This is done through the use of an energy audit. The audit is needed to develop an ***Energy Utilization Index (EUI).***

An ***energy audit*** accurately determines the current energy consumption for a given area. Initially, the person doing the audit walks through the structure in question. All areas where waste and inefficiency are obvious are noted. Then an accurate recording of all energy consumed during the audit follows. (This usually takes a month or two.) An example of the audit form used in this process is shown in **Figure 26-102.** The energy utilization index is determined by dividing energy consumption in Btu by the square footage of conditioned space. This calculation is shown in the lower right corner of the energy audit form in **Figure 26-102.**

Additional information is then gathered as part of the total energy audit. This includes:

- Previous year's energy use data.
- Weather data (for accurate comparisons).
- Building data such as plans or accurate floor layouts.
- Equipment operation logs.
- Other information on any aspect of energy consumption within the structure.

The sum of all this information is then processed (computer programs for this purpose are available). The overall status of energy management for the structure is determined. Then, based on the cost, time, and efficiency, a total energy management system is chosen.

Once the system has been selected and installed, some computer programming will be required. This can vary from setting temperatures to implementing a complex computer program. These programs can be changed as needed. During system operation, the total energy management system should be verified. It must continue to perform as designed.

26.21.2 Energy Consumption

Energy use in a building is determined primarily by several factors. These include working environment, regional climates, and type of equipment required by the building occupant.

Energy Management Form

Building ______________________________

Month	Heating Deg. Days	Cooling Deg. Days	Electricity						Purchased		
			kWh	kWh/ Deg. Days	kW Demand		Cost		M (lbs.)	M (lbs.)/ Deg. Days	lbs/hr
					Actual	Billed	Total	Per Ut.			Actual
1	2	3	4	5	6	7	8	9	10	11	12
Jan.											
Feb.											
March											
1st Quarter											
April											
May											
June											
2nd Quarter											
July											
Aug.											
Sept.											
3rd Quarter											
Oct.											
Nov.											
Dec.											
4th Quarter											
Total Per Year											

Building Data

Gross Conditioned Area (ft)[2] ______________________________

Gen. Notes: ______________________________

____

____

Annual Energy Consumption In Btu

Quantity

1. ______________ kWh
Electricity
2. ______________ (M) lbs
Purchased Steam
3. ______________ MCF
Natural Gas
4. ______________ Gallons
Oil
5. ______________
Other Fuel

Figure 26-102. *This is a standard energy management form used in conducting energy audits. The energy utilization index (EUI) is calculated in the lower right-hand corner. (Form copyrighted by National Electrical Contractors Assoc.)*

________________ Year

Steam			Fuel							Total Energy Cost
			Oil			Check Gas☐ Coal☐ Other☐			Fuel/ Deg. Days	
Demand	Cost		Quant. (Gal.)	Cost		Quant.	Cost			
Billed	Total	Per Unit		Total	Per Unit		Total	Per Unit		
13	14	15	16	17	18	19	20	21	22	23

	Conversion Fac.		Btu/yr
×	3413	=	________
×	1,000,000.	=	________
×	1,030,000.	=	________
×	#2 - 138,700 #6 - 149,700	=	________
×	________	=	________
		6. Total	________

Energy Utilization Index

$$EUI = \frac{\text{Total Energy Consumption Btu/yr}}{\text{Gross Conditioned Area (ft)}^2}$$

$=$ ________________ Btu/ft^2/yr

Figure 26-102. *Continued.*

In a typical building, the level of energy consumption is determined by three main systems:

- ***Energized systems*** required for heating, cooling, lighting, machine operation, etc.
- ***Nonenergized systems,*** such as floors, walls, roof, and windows.
- ***Human systems,*** including personnel who work in the building. Also included are any other people who may use the building in a typical day.

Each of these systems can be modified to achieve significant savings of energy. Energized systems are the easiest to define in terms of direct energy usage. Therefore, the tendency is to concentrate conservation efforts in this area. This may result in total energy consumption greater than that which would be achieved if *all* systems are involved in the conservation effort. The areas of heating and cooling usually consume the largest single "block" of energy.

Heating and cooling should not be viewed as adding heat to or removing heat from room air to achieve a given temperature. Rather, heating and cooling are the processes of providing for a heat loss or heat gain in a building. Heat loss and gain occur at the same time. One is usually greater than the other, depending on the outside temperature.

The factors which influence heat loss or gain include:

- ***Infiltration.*** This involves passage of outside air into the building through doors, cracks around windows, and other openings.
- ***Transmission.*** This is heat loss or gain through ceilings, walls, floors, windows, and other building components.
- ***Ventilation.*** Forced airflow that takes place, by design, between the inside and outside of the building.
- ***Lighting.*** Heat is generated in direct proportion to the wattage involved.
- ***Solar.*** Heat generated as a result of the intensity and direction of the sun's rays.
- ***Equipment.*** Some equipment generates heat during operation.
- ***Occupants.*** People and their activities generate heat.

Heat loss and gain factors have been only briefly presented. However, they show that heating and cooling alone are complex problems in a TEM system. Complexity increases dramatically with the number of systems to be integrated. Chapter 27 covers these items in greater depth. There, actual heat-load calculations are discussed.

26.22 Energy Management System Types and Functions

There are three general types of energy management systems.

- Localized controllers.
- Remote controllers.
- Centralized computer control.

More information on these systems can be found in Sections 26.22.1 through 26.22.4.

Typical energy management functions can be classified in the same way as control system components. See Section 26.13. These included primary, sequential, and limit controls.

Primary control functions would include:

- Timed operation.
- Demand control.
- Temperature-compensated duty cycling.
- Air temperature cutoff.
- Duty cycling.

Sequential control functions are associated with increasing efficiency in a TEM system. They include:

- Economizer.
- Boiler temperature.
- Air distribution.

Limit control functions are operations associated with monitoring. Examples include:

- Boiler profile operation.
- Maintenance time scheduling.
- Safety alarms.
- Chiller profile operation.
- Efficiency scheduling.

Some miscellaneous additional functions within a system may include:

- Security.
- Fire detection.

26.22.1 Direct Digital Control

Total energy management systems require automatic control operations. ***Direct Digital Control (DDC)*** is the use of a digital computer to perform these operations. (See Section 6.6.14.) Each of the three types of TEM system controllers (localized, remote, and centralized computer) use direct digital control. The computer size varies from a small microprocessor to a minicomputer or mainframe computer. The size depends on the number and complexity of the operations required.

26.22.2 Localized Controllers

Localized controllers used in control systems provide independent control for specific systems or equipment. This control is provided at a relatively low cost. Each local controller is *independent*—it controls its specific system and has no interaction with any other controlling device.

Typical localized controllers include:

- Time clocks.
- Alarms.
- Local optimization devices (for example, those that control dampers based on zone temperatures).

Localized controllers are used for relatively simple HVAC system control. They are used where:

- Installation time allowed is minimal.
- Individual system efficiencies are more important than total system management.
- Minimum cost is desirable. See **Figure 26-103.**

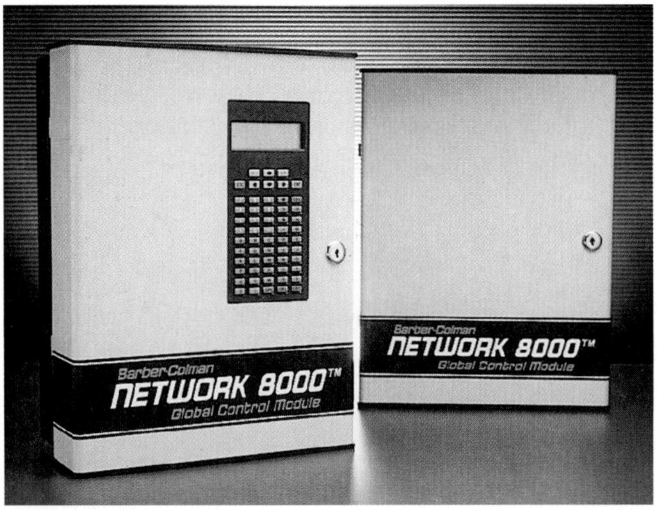

Figure 26-103. *This stand-alone energy management and temperature control system accepts direct digital input. The unit at right has a lock-on security cover to prevent unauthorized use.(Siebe Environmental Controls)*

26.22.3 Remote Controllers

A ***remote controller*** operates differently from a localized controller. More than one energy-consuming device can be controlled. It can also be located some distance from the devices it controls. These controllers, **Figure 26-104,** are used to minimize overall energy demand.

Figure 26-104. *A remote energy controller. Many energy-consuming devices can be controlled at the same time. (Control Systems International)*

The newer remote controllers are programmable through the use of microprocessors. See Sections 6.6.12 and 6.6.14. It is the most common type of total energy management system in use today. This is due to the number of units typically controlled and the lower cost of the microprocessor devices. An illustration of this type of controller is shown in **Figure 26-105.**

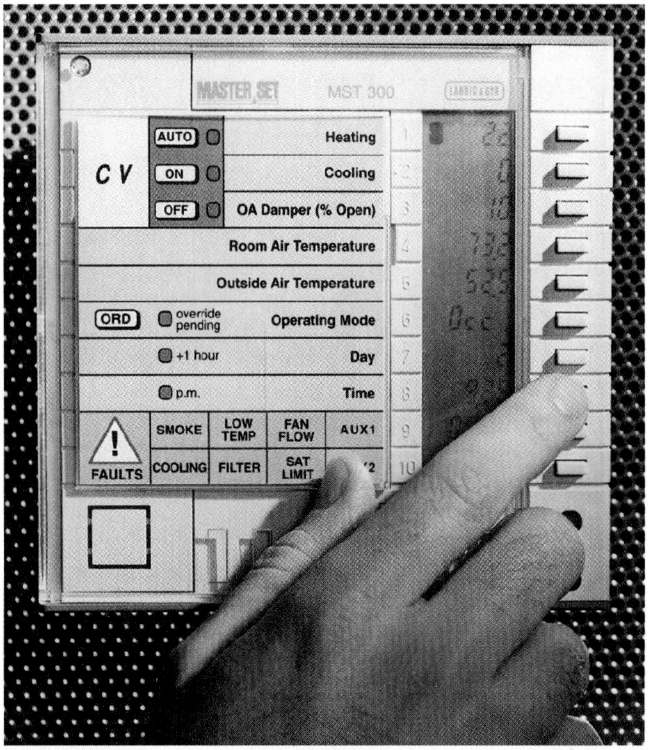

Figure 26-105. *This stand-alone energy management and temperature control system has direct digital control capabilities. (Landis & Gyr Powers, Inc.)*

Functions normally found in a remote controller include:

- ***Remote start/stop.*** The controller turns systems and devices on and off at certain times.
- ***Optimized start/stop.*** Devices are controlled based on some preprogrammed schedule to minimize energy use.
- ***Status monitoring.*** Used to indicate if the system is operating.
- ***Alarms.*** Devices that indicate if a system is operating incorrectly.
- ***Demand control.*** The overall demand for electrical power is monitored and modified for minimum energy consumption.
- ***Duty cycling.*** Turning systems on and off in cycles. This maintains minimum energy consumption with the least effect on the building's users.

A remote controller is used when multiple functions and systems (approximately 50 maximum) require control.

26.22.4 Centralized Computer Control

The centralized computer control system is the most elaborate of the Total Energy Management systems. One or more centralized computers make control decisions. These are based on operating data, programmed information, and data already stored in the computer memory. This is the most costly of the three types of TEM systems (localized, remote, and centralized computer control). However, it does offer the widest range of control functions. In large complex structures, this centralized system results in the best overall total energy consumption control. **Figure 26-106** shows a central control site for this type of Total Energy Management system.

The centralized computer controller is used in most newly constructed large building complexes. It also is being installed in many older structures that use large amounts of energy. School buildings and office complexes can use this type of control.

Functions of a centralized computer controller include all tasks basic to the remote and localized controllers. However, there are additional capabilities. Some of the systems available today will monitor utility usage and verify billing. They also schedule and notify personnel of all needed maintenance functions.

The centralized computer control TEM system is used where energy reduction and complex system control are required. Installation of such a system may require on-site availability of personnel knowledgeable in maintenance and programming. **Figure 26-107** shows a programming station for a computerized system.

The centralized computer control system is classified into two general categories, based on system construction:

- ***Packaged,*** in which the complete system is furnished by a manufacturer. The manufacturer also usually provides some level of service and follow-up after installation.
- ***Hybrid,*** in which a system is composed of components from several manufacturers. These systems may be designed by personnel within the company or through an outside firm. The same persons may then assume total system responsibility.

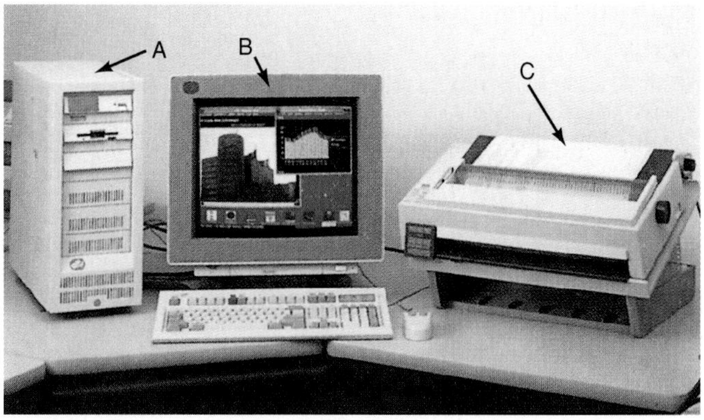

Figure 26-107. *A typical program station for a facilities management program. A—UltiVist™ system for control of building or buildings. B—Computer and status display screen. C—Printer. (Siebe Environmental Controls)*

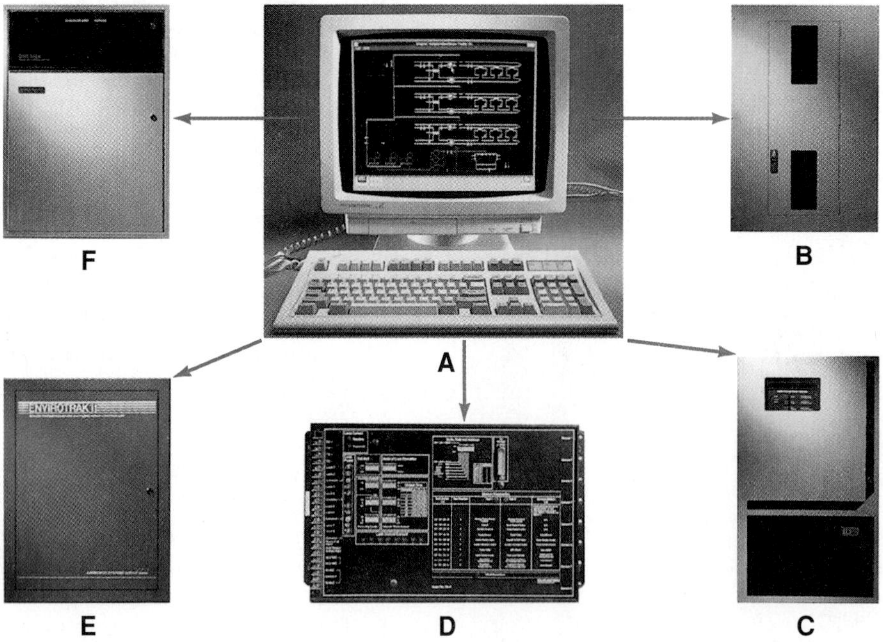

Figure 26-106. *UltiVist™ system for controlling environment. A—Color monitor with keyboard, IBM PS/2 computer. B—General Electric's lighting panel. C—Cerberus Pyrotronics' fire and smoke detector safety panel. D—Schlage Electronics' door access panel. E—Anemostat's Envirotrak panel (laboratory exhaust fume hood control). F—Siebe Environmental Controls' DMS 350A Digital Management System, used to control heating, ventilating, and air conditioning in facility. (Siebe Environmental Controls Co.)*

26.23 Control System Diagnostics and Repair

Diagnosis and repair of control system problems is the key to efficient system operation. Control systems of a simpler type are often more difficult to diagnose and repair than more complex systems. Simpler systems, such as controls for basic gas or oil furnaces, usually involve more mechanical and individual electrical components. The more advanced Total Energy Management systems involve the use of electronic circuit boards. They offer computer diagnostic advantages.

Manufacturers of controls at all levels typically provide a troubleshooting guide. Following is an example of troubleshooting instructions for an HVAC control system failure:

System fails to start:

1. Check fuses or circuit breakers and the power-in with a suitable test light or voltmeter.
2. Check the main switch and thermostat switch to make sure they are in closed position.
3. Check the thermostat. It must be on a setting that will close thermostat switch.
4. Check the contactor. It must be pulled in. If not, the contactor circuit is open:
   A. Contactor open but not buzzing. Contactor coil not powered. Check control circuit using voltmeter and ohmmeter.
   B. Contactor open and buzzing. Contactor coil is operating, but armature not pulled in. Either the armature is stuck or control coil voltage is too low.
   C. Contactor closed and motor hums. Motor is powered but will not start. Check capacitors, if used, for low voltage. Check for excessive pressures in system.
   D. Contactor closed and motor does not hum. Motor is not powered. Check for open circuit by continuity test.

Figure 26-108 shows part of a troubleshooting guide for a TEM system. Note that repair may include replacement of a given circuit board within the control device.

Troubleshooting Guide		
Problem	**Possible Fault**	**Procedure**
1. No LED Display on Display Panel.	STEP 1. Loss of 12 V ac input supply.	Take a voltage reading across terminals 12 V and 12 C on the terminal strip of the interface PC board. There should be a reading of 12 V ac. If not present, check for a tripped circuit breaker. Verify primary input (120 V ac) at the electrical outlet and 12 V ac output at the secondary of the transformer. If 12 V ac is verified, proceed to STEP 2.
	STEP 2. Blown 12 V ac fuse on interface board.	Check for improper wiring or shorts to ground and replace fuse if necessary. If the fuse has been replaced and there is still no display or if fuse was not blown, proceed to STEP 3.
	STEP 3. ON/OFF power switch is in the OFF position.	Slide switch to the UP position. If switch is in UP (ON) position, proceed to STEP 4.
	STEP 4. Loose 20-conductor ribbon cable connection.	Check the 20-conductor ribbon cable connection on the processor and interface PC boards to see that it is properly inserted. If it is, proceed to STEP 5.
	STEP 5. Faulty interface PC board.	Take a voltage reading across capacitor TB1 located on the lower left-hand corner of the interface PC board. There should be a reading of 12 V ac. If there is no voltage, replace interface PC board (see Service Section). If there is voltage, proceed to STEP 6.
	STEP 6. Faulty microprocessor PC board.	Take a voltage reading across the 12 V ac quick disconnects located on the upper left-hand corner of the microprocessor PC board. There should be a reading of 12 V ac. If there is no voltage, replace the 20-conductor ribbon cable. If there is voltage, replace the microprocessor PC board. (See Service Section).
2. LED Display frozen. Clock will not advance and program modes cannot be accessed.	STEP 1. Defective microprocessor PC board.	Replace microprocessor PC board. (See Service Section).

Figure 26-108. *This excerpt is from a typical energy management system troubleshooting guide. Note that complete replacement of a circuit board is sometimes recommended.*

26.24 Review of Safety

Electronic cleaner controls involve high voltage. They should be serviced only by a technician with special training in that particular model.

Safety controls should never be removed to keep a heating or cooling system operating. They also should not be bypassed for this purpose. Dangerous conditions or damage to the system may result.

Handling heating or cooling controls does not present any great shock hazard to you. The voltage and amperage levels are quite low. However, the instrument readings must be taken and interpreted properly. If they are not, the installation may be faulty. The equipment could then be seriously damaged. Persons within the structure may be placed in danger of serious injury.

Before working on any part of a system, always determine the pressure in that part. Carelessness could result in injury if the system is opened. Always measure the temperature of the system parts to be repaired, adjusted, or touched. This will avoid serious burns. Never guess at pressure or temperature.

Use instruments to check electrical circuits. Never assume that the power is off.

Remember that even a few milliamperes can injure or kill a person. Always use rubber gloves and rubber boots or shoes when working on electrical devices or circuits. Tools such as wrenches, pliers, and screwdrivers should have insulated handles.

26.25 Test Your Knowledge

Please do not write in this text. Place your answers on a separate sheet of paper.

CONTROL MECHANISMS MODULE

1. A pressurestat is a device _______.
   A. that increases pressure as needed
   B. used to shut down a component when the pressure exceeds set limits
   C. that can be used to measure pressures in a system
   D. Both A and C.
2. What is the approximate voltage of a low-voltage thermostat?
   A. Up to 10 V.
   B. Up to 20 V
   C. Up to 30 V.
   D. Up to 60 V.
3. Which type of thermostat must be carefully leveled?
   A. A digital thermostat.
   B. One with mercury tube contact points.
   C. A low-voltage thermostat.
   D. A Group 2 thermostat.
4. Which type of thermostats do millivolt systems have?
   A. Low-voltage.
   B. High-voltage.
   C. None.
   D. Group 2.
5. Line-voltage thermostats operate at _______ V.
   A. 30
   B. 120
   C. 240
   D. Both B and C.
6. How does a heat anticipator impact the thermostat?
   A. It resists dramatic fluctuations in temperature.
   B. It warms the thermostat a little more rapidly then the room is warmed.
   C. It eliminates over-run.
   D. Both B and C.
7. Which of the following is *not* a consideration in the placement of a thermostat?
   A. Location of living spaces.
   B. Hot or cold air from ducts.
   C. Concealed pipes.
   D. Chimney location.
8. A heat anticipator is wired _______.
   A. with a high-voltage thermostat
   B. in series with the thermostat's contacts
   C. in parallel with the thermostat
   D. None of the above.
9. Which type of thermostat contains a diaphragm that moves as its liquid expands and contracts?
   A. Bimetal thermostats.
   B. Electronic digital thermostats.
   C. Hydraulically operated thermostats.
   D. A thermostat equipped with a heat anticipator.
10. Thermostat location should be changed if affected by _______.
    A. radiant heat from the sun
    B. switches not operating properly
    C. room temperature
    D. Any of the above.

CONTROL SYSTEM COMPONENTS MODULE

11. Which refrigerant controls are generally used on air conditioning evaporators?
    A. AEV and capillary tube.
    B. Thermostatic expansion valve and capillary tube.
    C. Reversing valve and thermostatic expansion valve.
    D. Check valve and reversing valve.
12. A flame detector is used in an oil burner system to _______.
    A. shut off the system if the oil does not ignite
    B. shut off the system if the flame goes out
    C. monitor the temperature of the flame
    D. Both A and B.
13. The role of the solenoid refrigerant valve is to _______.
    A. stop refrigerant flow if condensing unit stops
    B. stop flow of refrigerant when low-side pressure control stops
    C. avoid flooding the evaporator when the condensing unit is idle
    D. All of the above.

14. The fan motor (blower) starts when the plenum chamber reaches approximately ______ °F and cuts out at approximately ______ °F to ______ °F.
    A. 140 (60°C), 80 to 90 (27°C to 32°C)
    B. 120 (49°C), 70 to 80 (21°C to 27°C)
    C. 155 (69°C), 90 to 100 (32°C to 38°C)
    D. None of the above.
15. What does a 24 V thermostat on a gun-type oil burner operate when the points close?
    A. A flame-out safety control.
    B. A low-voltage photosensitive flame detector.
    C. A primary control located on the furnace flue pipe.
    D. All of the above.
16. Which controls do steam heating systems often use?
    A. TEV.
    B. Float.
    C. Capillary tube.
    D. All of the above.
17. Humidity is most often lowered by ______.
    A. adding vapor to the air
    B. heating the air above its dew point temperature
    C. cooling air below its dew point temperature
    D. None of the above.
18. When servicing an electronic cleaner, you should ______.
    A. check rectifiers
    B. check high dc voltage
    C. remove the heavy insulators on the meter leads
    D. Both A and B.
19. Which of the following is true about VAV controllers?
    A. They regulate the volume of air.
    B. They use electronic components.
    C. They are usually part of a larger computer-controlled HVAC system.
    D. All of the above.
20. Where is a fan control thermostat mounted for a forced warm air furnace?
    A. Inside the plenum chamber.
    B. On the plenum chamber.
    C. Inside the fresh air intake damper.
    D. None of the above.

ENERGY MANAGEMENT MODULE

21. What is the basic premise of Total Energy Management?
    A. Only nonpolluting forms of energy are used.
    B. Each building is viewed in terms of its total energy consumption.
    C. One should view the energy requirement of each separate system.
    D. None of the above.
22. Which of the following determines energy use in a building?
    A. Climatic condition of the region.
    B. Working environment.
    C. Type of equipment required by those in the building.
    D. All of the above.
23. Which of the following is *not* a typical energy management function?
    A. Primary functions.
    B. Sequential functions.
    C. Direct digital controls.
    D. Limit functions.
24. Which of the following is *not* a typical localized controller?
    A. Remote start/stop.
    B. Time clocks.
    C. Alarms.
    D. Local optimization devices.
25. A remote controller is used when ______.
    A. independent control is needed
    B. low-cost control is needed
    C. multiple functions and systems are to be controlled
    D. All of the above.
26. Which of the following is *not* a benefit of a centralized computer control system?
    A. Widest range of control functions.
    B. Energy reduction.
    C. Control of complex systems.
    D. Little initial financial investment.
27. What is involved in an energy audit?
    A. A walk-through.
    B. Accurate recording of all energy consumed during the audit.
    C. Determination of an Energy Utilization Index (EUI).
    D. All of the above.
28. What will you most often use for diagnosis of control system problems?
    A. A technical textbook.
    B. The "guess and check" method.
    C. A manufacturer's troubleshooting guide.
    D. None of the above.
29. Why does service of an electronic cleaner require special training?
    A. They are very complicated.
    B. They involve high voltage.
    C. They are very "sensitive."
    D. None of the above.
30. Prior to working on any part of a system, you should ______.
    A. determine the pressure in that part of the system
    B. measure the temperature of parts to be repaired
    C. use instruments to check electrical circuits
    D. All of the above.

Direct Digital Control (DDC) is the use of a digital computer to perform automatic control operations. Some of the components of a DDC system are shown here. (Landis & Gyr Powers, Inc.)

Chapter 27

AIR CONDITIONING SYSTEMS—HEAT LOADS

Key Words:

degree-day
heat gains
heat lag
heat leakage
heat loads
heat losses
heat transfer coefficient
heat transfer rate
ponded roof
R-value
thermal conductivity
U-value

Learning Objectives:

After studying this chapter, you will be able to:

- Define heat load and identify its sources for both heating and cooling of space.
- Determine heat loads through the use of U- or R-values, square footage, and design temperature charts.
- **Practice approved safety procedures.**

Heating systems require an installation that will produce enough heat. The heat must keep the occupied space at a comfortable temperature and relative humidity. This requirement should hold true even at record low temperatures for the given locality. In much of the United States, typical conditions are as follows: 70°F (21°C) inside temperature when it is 0°F (−18°C) outside with a 15 mph wind blowing.

27.1 Heat Loads

An air conditioning/heating system must heat a space enough to make up for ***heat losses*** during heating. It must remove as much heat as the space accumulates during cooling.

Whenever a temperature difference exists, heat energy will flow from the higher to the lower temperature level. In heating, it is necessary to retard this heat flow as much as possible because the amount of heat lost must be replaced. In cooling, the system must remove the amount of heat gained.

The maximum heat load (loss or gain) is determined for a period of one hour. Charts (found in numerous manuals) are used to make the computation.

Heat load calculation manuals are available through various sources. The Air Conditioning Contractors of America (ACCA) publish the *Load Calculation for Residential Winter and Summer Air Conditioning,* known as Manual J. They also publish *Load Calculation for Commercial Summer and Winter Air Conditioning,* known as Manual N.

There are several major heat loads:

- Heat that is transmitted through walls, ceilings, and floors (conduction).
- Heat necessary to control moisture content in the air.
- Conditioning the air that enters the building by leakage and for ventilation.
- The sun produces heat in buildings, directly through the windows. It also produces heat by heating the surfaces it strikes (a cooling load).

- Energy devices—light fixtures, electric motors, electric or gas stoves, etc.—all produce heat. People, too, release a considerable amount of heat.

In all cases, the ***heat load*** can be described as either sensible heat load (temperature change) or latent heat load (moisture), evaporating or condensing.

27.1.1 Heat Loads for Heating

Heat loads for heating include all means by which heat will be lost from a building. They also include heat loss due to warming of cooler substances brought into the building. See **Figure 27-1.**

The major heat losses are:

- Conduction through walls, ceilings, and floors.
- Air leaking out of the building (exfiltration), and that which leaks into the building (infiltration).
- Combustion air leaving the flue from gas or oil furnaces, or from fireplaces.

Normally, all other heat losses are ignored. They are too small, relatively, to affect the size of the unit to be installed.

27.1.2 Heat Loads for Cooling

There are definite sources of heat gain in warm weather:

- Heat leakage into the building.
- Air leakage into the building.
- Ventilation air.
- Sun load.
- Heat from appliances and lights.
- Heat gain from occupants.

Heat gain refers to the heat added to a space that is being cooled. This heat must be removed to keep temperature and relative humidity at the values desired.

Heat gain in warm weather is produced by heat conduction. This takes place through the walls, ceilings, floors, windows, and doors of the enclosure. Also, heat enters the room by way of infiltrated air. People and animals in the room also give off heat.

Miscellaneous sources of heat are electrical devices (lights and motors), gas-burning devices, and steam tables. Another source of heat is the sun. The sun may be a considerable source of heat. See **Figure 27-2.**

Heat Leakage

Heat leakage is the heat that is conducted through the walls, ceilings, and floors. The total heat leakage is computed as follows: First, determine the area of each type of surface through which heat is leaking. Next find the U-value for each type of surface. The ***U-value*** represents the amount of heat that will pass from one side of a wall to the other. Total heat leakage is found by multiplying the heat leakage areas by their respective U-values. The U-value may also be referred to as the ***heat transfer coefficient.***

Heat leakage can also be computed using the thermal resistance of the structure. Thermal resistance is known as the ***R-value.*** It is the reciprocal of conductance (C) or the overall heat transfer (U).

Building materials have been tested to determine the amount of heat they will transfer. **Figure 27-3** shows temperature changes during the heating season for a typical wood-sided wall. Both noninsulated and insulated walls are shown.

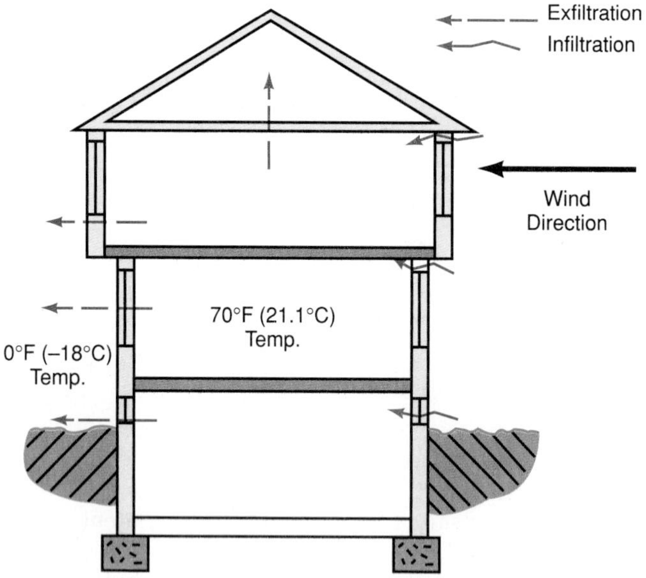

Figure 27-1. *Large heat losses occur during heating season due to cold air filtering into building and warm air filtering out. Heat is also lost through walls, floors, windows, and ceilings.*

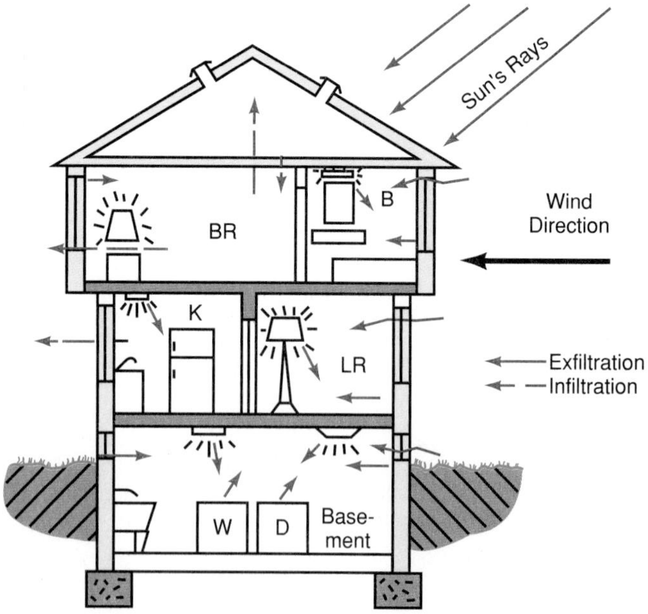

Figure 27-2. *Heat is gained in building during cooling season. Note heat leakage, air infiltration, sun load, lights, appliances, and moisture sources. B—Bathroom. BR—Bedroom. K—Kitchen. LR—Living room. W—Washer. D—Drier.*

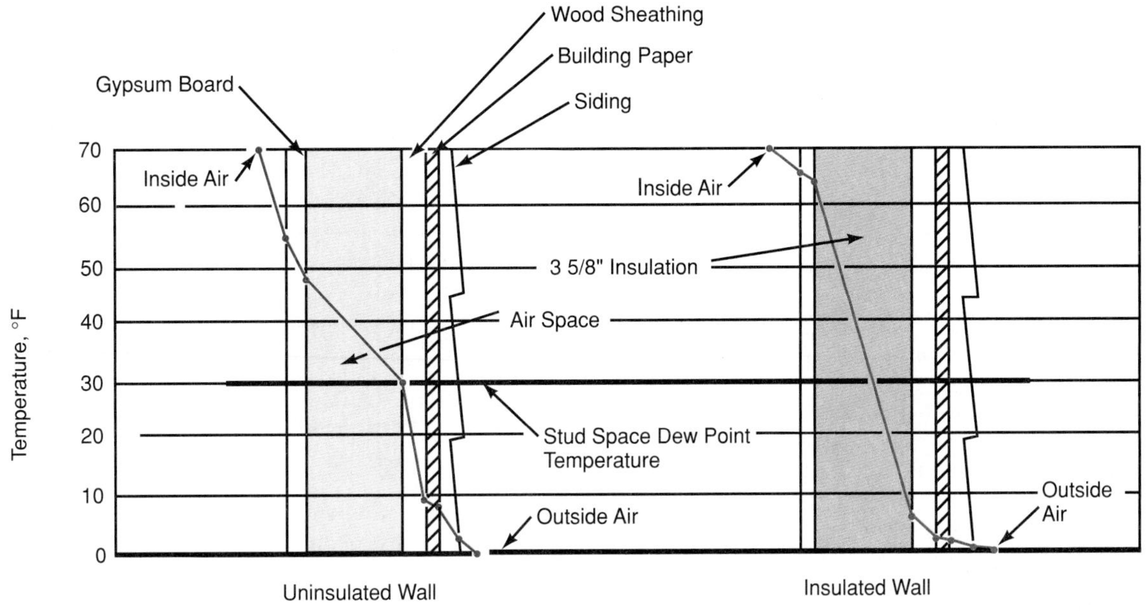

Figure 27-3. *Temperature change through uninsulated wall and insulated wall. Uninsulated wall allows temperature of inside wall surface to drop as low as 55°F (13°C). Insulated wall maintains temperature of inside wall surface at 67°F (19°C).*

The ***thermal conductivity*** (represented by the letter *K*) of a material is a measure of how quickly heat can travel through a material. The units of *K* are normally Btu in./ft^2/°F/hr., making it the amount of heat (in Btu) that travels through one square foot of material one inch thick when there is a temperature difference of one degree Fahrenheit.

The letter *C* is used to indicate heat transfer through a wall made of different substances.

$$\frac{1}{C} = \frac{X_1}{K_1} + \frac{X_2}{K_2} + \frac{X_3}{K_3}$$

X is the thickness of a material in inches.

$$C = \frac{1}{\frac{X_1}{K_1} + \frac{X_2}{K_2} + \frac{X_3}{K_3}}$$

Infiltration

Buildings are not airtight and air will leak through openings if there is any difference between inside and outside air pressure. Air may leak into or out of the building.

The air pressure difference is usually caused by wind. Parts of the building that the wind presses against are those through which air enters. The remaining areas are those through which the air escapes, **Figure 27-4.**

During the heating season, any cold air that filters in must be heated. Air that leaks out represents lost heat. During the cooling season, warm air that filters in must be cooled. Cooled air that filters out is a heat gain (some cooling effect is lost). If the building can be sealed, this infiltration and exfiltration can be minimized. However, care must be taken to provide enough fresh air for ventilation and combustion purposes.

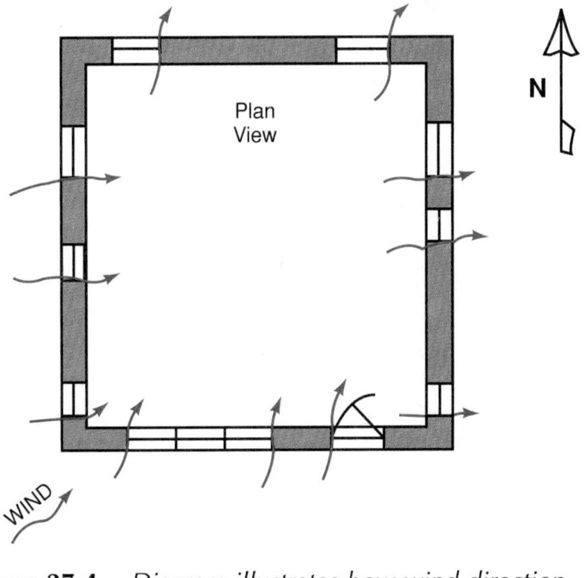

Figure 27-4. *Diagram illustrates how wind direction affects air leakage into and out of house.*

A way to prevent unwanted infiltration is to maintain a positive air pressure within the building. The pressurized air will filter out through cracks and openings in the building. With this practice, a special fresh air intake is needed. Incoming air can be conditioned before it is admitted to the rooms in the building.

Infiltration calculations can be based on the total volume of the building, or they can be based on the length and size of all the cracks in the building. **Figure 27-5** lists the air changes in buildings. A building with a volume of 10,000 ft^3, will have at least 10,000 ft^3 of fresh air infiltration per hour. If six people occupy this space,

Types of Space	No. of Air Changes/Hr.
1 Side Exposed	1
2 Sides Exposed	1 1/2
3 Sides Exposed	2
4 Sides Exposed	2
Entrances	2–3

Figure 27-5. *Approximate number of air changes desirable per hour for various room exposures.*

there is 10,000 ÷ 6 or 1667 ft^3 per hour for each person, or 1667 ÷ 60 = 27.8 cfm (0.79 m^3/min.). The air change will be reduced considerably if this building is constructed with vapor barriers. Fitting all doors and windows with weather-stripping will also reduce the air change. It may even be reduced to the point of unsafe ventilation. (There will be too little oxygen in the air.)

Heat Transfer Rate

The ***heat transfer rate*** is the amount of heat conducted through a structure for a given unit of time. It is usually represented by the letter *Q* and expressed in Btu/hr. The total heat transfer rate is found as follows: The heat transfer coefficient (U-value) is multiplied by the temperature difference and the area.

Heat Transfer Rate = Heat transfer coefficient × area × temperature difference

$$Q = (U \times \text{Area}) \times (T_O - T_I)$$

Where: T_O = Outside temperature
T_I = Inside temperature

U-value for Computing Heat Leakage

In computing heat transfer, the letter U has almost the same value as the letter C. The value of U, however, includes the additional insulating effect of an air film. This air film always exists on each side of the surface. See **Figure 27-6.**

$$U = \frac{1}{\frac{1}{F_i} + \frac{X_1}{K_1} + \frac{X_2}{K_2} + \frac{X_3}{K_3} + \frac{1}{F_o}}$$

F_i is the heat transfer through the inside air film. F_o is the heat transfer through the outside air film. U-value is a term indicating the amount of heat transferred through a structure (wall):

$$U = Btu/ft^2/°F/hour$$

U-values are based on a 15 mph wind on the outside and a 15 fpm (1/6 mph) draft on the inside wall surface.

The U-value for most types of construction can be obtained from reference books published by the American Society of Heating, Refrigerating and Air-Conditioning Engineers (ASHRAE). **Figure 27-7** is a condensed table for some of the more common constructions.

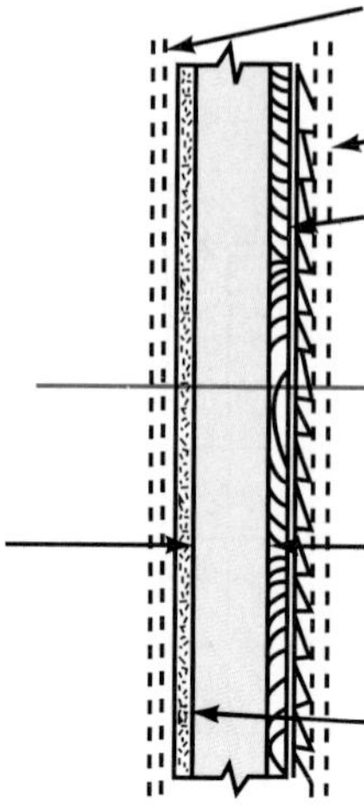

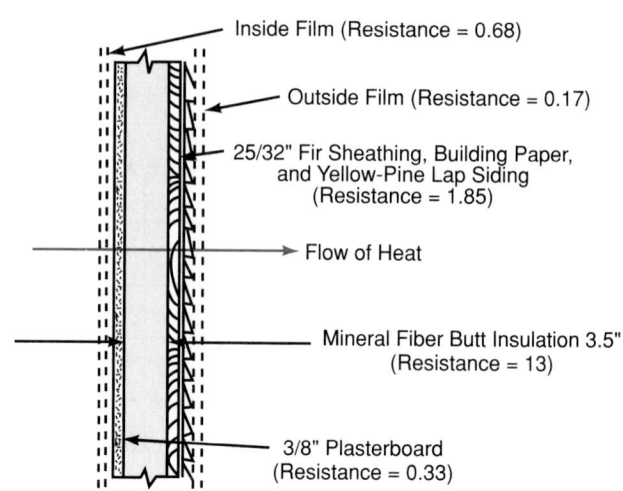

Figure 27-6. *Drawing of external wall section shows inside and outside air film. C values are reciprocal of the R-value of material, or C = 1/R. (Edison Electric Institute)*

Given the U-value, the design temperature conditions of 70°F (21°C) indoors and 0°F (−18°C) outdoors, and the area, calculate the heat load as follows:

Formula:

Heat flow = area × temperature difference × U-value

Total heat transfer (Q) = U × total surface × temperature difference

Example:

A structure has 400 ft^2 of surface. The temperature difference is 70°F. The structure has a brick veneer wall and no insulation. It has a U-value of 0.25.

Solution:

This U-value means that 0.25 Btu will transfer through each square foot of wall for each one degree F temperature difference in one hour.

Total heat transfer (Q) = 400 × (70°F − 0°F) × 0.25
= 400 × 70 × 0.25
= 28,000 × 0.25
= 7000 Btu/hr.

Example:

If total surface area is 1200 ft^2, heat transfer can be determined as follows:

Q = 1200 × (70°F − 0°F) × 0.25
= 1200 × 70 × 0.25
= 84,000 × 0.25
= 21,000 Btu/hr.

The same method of computing heat leakage is used when using the metric system. Only the units used are changed.

Figure 27-8 shows typical heat leakage listed in watts per square foot. To obtain the watt load per degree F, divide the "35 column" by 35. To convert watts to Btu/hr., multiply watts by 3.414368.

Heat Loads

Constants for Heat Transmission (U and R Values)

Expressed in Btu per hour per square foot per degree temperature difference, based on 15 mph wind velocity.

Masonry Construction

General Wall Classification	Masonry Thickness 8"	
	U	R
Brick–Plain		
1/2" Wallboard	.29	3.51
1" Polystyrene, 1/2" Wallboard	.13	7.54

Frame Construction

	U	R
Wood Siding on 1" Wood Sheathing, Studs, Gypsum Board 1/2"	.23	4.40
Wood Siding, Sheathing, Studs, Gypsum Board 1/2", 1" Polystyrene	.07	14.43
Wood Siding, Sheathing, Studs, 1/2" Flexible Insulation, Gypsum Board 1/2"	.15	6.7
Wood Siding, Sheathing, Studs, Rock Wool Fill, Lath and Plaster	.072	13.9
Note: Frame Walls with Shingle Exterior Finish Same as Walls with Wood Siding		
Stucco, Wood Siding, Studs, Gypsum Board	.30	3.32
Stucco on 25/32" Rigid Insulation, Studs, 1/2" Rigid Insulation and Plaster	.20	5.0
Stucco on 1/2" Rigid Insulation, Studs, Rock Wool Fill, Lath and Plaster	.074	13.5

Concrete Floors and Ceilings

	U	R
4" Thick Concrete, No Finish	.65	1.54
4" Concrete, Suspended Plaster Ceiling	.37	2.7
4" Concrete, Metal Lath and Plaster Ceiling, Hardwood Floor on Pine Sub-Flooring	.23	4.35
4" Concrete, Hardwood and Pine Floor, No Ceiling	.31	3.23

Pitched Roofs

	U	R
Asphalt Shingles on Wood Sheathing	.44	2.30
Asphalt Shingles, 1" Flexible Insulation	.136	7.93
Asphalt Shingles, Polystyrene 1"	.132	7.56

Windows and Skylights

	U	R
Single Glass	1.04	.96
Single Glass and Storm Window	.56	1.79
Double Glass, Intermediate Air Space 1/2"	.58	1.73
Hollow Glass Tile Wall 6" × 6" × 4" Blocks	.60	1.67

Brick Veneer on Frame Construction

	U	R
Brick Veneer, 1" Wood Siding, Studs, Gypsum Board 1/2"	.27	3.71
Rigid Insulation, Studs, Gypsum Board 1/2"	.25	4.00
Brick Veneer, 1" Wood Siding, Studs, 1" Polystyrene Insulation, Gypsum Board 1/2"	.07	14.43
Brick Veneer, 1" Wood Siding, Studs, Rock Wool Fill, Lath and Plaster	.074	13.5

Interior Walls

	U	R
Note: In general for cooling computations, base the calculations for heat gain from adjoining non-conditioned rooms on a differential equal to 1/2 the differential to outside.		
Gypsum Board 1/2" on Both Sides	.31	3.23
Gypsum Board 1/2" on Both Sides, Foam Insulation	.08	13.26

Flat Roofs with Built-up Roofing

Deck Material	No Ceiling		Metal Lath and Plaster Ceiling	
	U	R	U	R
Precast Cement Tile	.81	1.23	.43	2.33
Precast Cement Tile, 1" Insulation	.24	4.7	.19	5.26
4" Thick Concrete	.72	1.39	.40	2.50
4" Thick Concrete, 1" Insulation	.23	4.35	.18	5.55
2" Wood	.32	3.13	.24	4.17
Flat Metal Roofs	.95	1.05	.46	2.18
Flat Metal Roofs, 1" Insulation	.25	4.00	.19	5.26

Frame Floors and Ceilings

	U	R
Hardwood and Pine Flooring on Joists, Metal Lath and Plaster Ceiling	.23	4.37
Rough Pine Floor, Wood Lath and Plaster Ceiling	.28	3.57
No Floor, Lath and Plaster Ceiling	.62	1.61
No Floor, Metal Lath and Plaster Ceiling, 3 5/8" Rock Wool Fill	.079	12.65
No Floor, Lath and Plaster Ceiling, 1" Flexible Insulation	.17	5.9

Figure 27-7. *U- and R-values are given for walls, ceilings, floors, and partitions for various types of construction and for various thicknesses. (Reprinted by permission of the American Society of Heating, Refrigerating, and Air-Conditioning Engineers, Atlanta, Georgia)*

HEAT LOSS—Watts/ft^2

CONSTRUCTION DTD	35	45	55	65	75	85	95	105
EXTERIOR WALLS								
Masonry (8" concrete block)	3.60	4.63	5.66	6.68	7.71	8.74	9.77	10.80
Masonry (8" concrete block; 1/2" gypsum board) with R-7 (2 1/4") Fiberglas	1.05	1.35	1.65	1.95	2.25	2.55	2.85	3.15
Brick Veneer (1/2" gypsum sheathing, 1/2" gypsum board)								
with R-11 (3 1/2") Fiberglas	.75	.97	1.18	1.39	1.61	1.82	2.04	2.25
with R-13 (3 5/8") Fiberglas	.65	.83	1.02	1.20	1.39	1.57	1.76	1.94
Frame (Woodrock siding 1/2" gypsum sheathing, 1/2" gypsum board)								
with R-11 (3 1/2") Fiberglas	.85	1.09	1.33	1.58	1.82	2.06	2.30	2.54
with R-13 (3 5/8") Fiberglas	.75	.97	1.18	1.39	1.61	1.82	2.04	2.25
Frame (wood clapboard or shingles, plywood or wood fiber sheathing, 1/2" gypsum board)								
with R-11 (3 1/2") Fiberglas	.70	.90	1.10	1.30	1.50	1.70	1.90	2.10
with R-13 (3 5/8") Fiberglas	.60	.77	.94	1.11	1.28	1.45	1.62	1.79
WINDOWS								
Single Light	12.65	16.19	19.73	23.28	27.07	30.61	34.41	37.95
Double Glazed	6.80	8.70	10.61	12.51	14.55	16.46	18.50	20.40
Single with Storms	5.80	7.42	9.05	10.67	12.41	14.04	15.78	17.40
DOORS								
Door only	5.10	6.53	7.96	9.38	10.91	12.34	13.87	15.30
with Storm Door	3.40	4.35	5.30	6.26	7.28	8.23	9.24	10.20
FLOORS								
Concrete Slab No Insulation	8.30	10.67	13.04	15.41	17.78	20.15	22.52	24.89
with R-7 (1") Zer-O-Cel Urethane Perimeter (watts/lin. ft. exposed edge)	2.40	3.08	3.77	4.45	5.14	5.82	6.51	7.19
Wood (5/8" plywood) over vented space								
with R-13 (3 5/8") Fiberglas	.65	.83	1.02	1.20	1.39	1.57	1.76	1.94
with R-19 (6") Fiberglas	.50	.64	.78	.93	1.07	1.21	1.35	1.49
Over unheated basement								
with R-7 (2 1/4") Fiberglas	.95	1.22	1.49	1.76	2.03	2.30	2.57	2.84
with R-11 (3 1/2") Fiberglas	.70	.90	1.10	1.30	1.50	1.70	1.90	2.10
CEILINGS								
with R-19 (6") Fiberglas	.49	.63	.77	.91	1.05	1.19	1.33	1.47
with R-22 (6 1/2") Fiberglas	.45	.58	.70	.83	.96	1.09	1.22	1.34
FIREPLACES (Watts)								
tight damper	205	264	322	381	440	498	557	615
average damper	510	657	803	949	1095	1241	1387	1533
BASEMENT WALL, above grade* (U = .10, 70°F basement)	1.02	1.31	1.60	1.89	2.18	2.47	2.76	3.05

NOTE: The below grade portion of a heated basement must be calculated separately since heat loss relates to ground or ground water temperature rather than air temperature. The U-value of .10 represents typical concrete construction with insulation and a furred finish wall. Add above-grade figure to below-grade figure for total basement heat loss.

*Below Grade
HEAT LOSS, Watts/Ft^2 (U = .10, 70°F basement)

Ground Water† Temperature,°F	Floor	Wall
40	.879	1.758
50	.586	1.172
60	.293	.586

†Ground water temperature is available from the local weather bureau.

Figure 27-8. *Table of heat leakage values in watts per square foot. The DTD (Design Temperature Difference) is listed from 35°F (19°C) to 105°F (58°C). Note that from first column to last column, every number, including DTD, is multiplied by 3.00.*

Some common metric system heat transmission units are: joules/second; kilocalories/hour; and watts. These values are used with square meter or square centimeter areas.

Heat transmission using watts per square meter is the most popular method. The watt unit is both a U.S. conventional unit and an SI metric unit. It may be easily applied to heating and cooling unit capacities.

Multiply Btu/hr./ft^2/°F by 5.674 to obtain W/m^2/°C. To change metric units to U.S. conventional units, multiply W/m^2/°C by 0.1762 to obtain Btu/hr./ft^2/°F.

Multiply hr. ft^2/°F/Btu by 0.1762 to obtain m^2/°C/W. To change metric units to U.S. conventional units, multiply m^2/°C/W by 5.674 to obtain hr. ft^2/°F/Btu.

R-value for Heat Leakage

The ***R-value*** of a material is its ***thermal resistance.*** See **Figures 27-6** and **27-7** for R-values of some common construction materials. Heat transfer and heat leakage calculations may use R. It is also called "ru," which means resistance unit.

Thermal resistance is the reciprocal of the heat transmission coefficient (U):

$$R = \frac{1}{U}$$

For a composite wall (a typical building), the total R equals the sum of the individual reciprocals of the C values.

$$R_{Total} = \frac{1}{C_1} + \frac{1}{C_2} + \frac{1}{C_3} + \frac{1}{C_4} + \frac{1}{C_5}$$

or

$$R_{Total} = R_1 + R_2 + R_3 + R_4 + R_5$$

Individual R-values for a composite wall can be totaled. Then, the heat transfer coefficient will equal the reciprocal of the total resistance. The R-values can be found by taking reciprocals of the heat conductance (C values from tables). Tables showing the R-values also may be used. See **Figure 27-9.** Example: R-values for a typical brick veneer wall are as follows:

	R
Outside air film	0.17
Face brick veneer	0.39
Wood siding and building paper	0.86
Airspace	0.97
1/2" plaster (0.09) on gypsum lath (0.32)	0.41
Inside air film	0.68
Total R =	3.48

$$U = \frac{1}{R} = \frac{1}{3.48} = 0.287$$

The conservation of energy has become of great concern. Therefore, it is recommended that homes and apartments have thermal insulation. See **Figure 27-10.**

Construction	(R) Resistance Value
Surface (Still Air)	.68
Air Space	.97
Gypsum Wallboard 3/8"	.32
Outside Surface (15 mph Wind)	.17
Face Brick	.39
Concrete Block 4"	1.11
Urethane Insulation	9.1
Siding (Wood) 1/2" × 8"	.85
Building Paper	.06
Wood Sheathing	.98
Wood Floor 1"	.98
Linoleum or Tile	.05
Asphalt Shingles or Plywood	.95

Figure 27-9. *Table of typical thermal resistance (R) values for various parts of a building. (Reprinted by permission of the American Society of Heating, Refrigerating, and Air-Conditioning Engineers, Atlanta, Georgia)*

Thermal Insulation Values		
Area	U	R
Ceiling with Unheated Space above	.08	12
Exterior Wall	.07	14
Wall with Unheated Space on One Side	.10	10
Floor over Unheated Space	.07	14

Figure 27-10. *Some recommended thermal insulation values are listed for homes and apartments.*

A comparison of U.S. conventional and metric system values follows:

	U.S. Conventional	Metric
Specific heat at constant pressure	Btu/lb./°F	kj/kgK
Internal film coefficient	Btu/hr. ft^2/°F	W/m^{2}K
Total heat flow	Btu/hr. watts	kcal/hr.
R—Total resistance to heat flow	hr. ft^2/°F/Btu	m^{2}K/W
U—Overall heat transfer coefficient	Btu/hr. ft^2/°F	W/m^2 K
Velocity	ft./min.	m/s

Wall Heat Leakage Areas

In addition to finding the several U- or R-values for the building structure, the area of the walls will need to be calculated:

Wall heat leakage = U × wall area × temp. difference.

Areas to be measured are the outside dimensions of the building. These will result in slightly higher heat leakage loads than if inside dimensions are used. U-values based on outside dimensions are conservative.

To estimate the heat load, measure the entire building: walls, windows, ceilings, and floors.

To measure walls, take the outside length and width of the house and the inside ceiling height. To determine the total surface area, first measure the distance around the house. This will be the length plus the width, plus the length, plus the width. L + W + L + W = perimeter.

Total wall area is obtained by adding these values and multiplying by the wall height. For example, a house, as shown in **Figure 27-11,** is 24′ × 32′ (outside). It has an 8′ ceiling. The total area will be:

$$\begin{aligned} \text{Perimeter} &= L + W + L + W \\ &= 32' + 24' + 32' + 24' \\ &= 112' \\ \text{Area} &= \text{perimeter} \times \text{height} \\ &= 112' \times 8' \\ &= 896\ \text{ft}^2 \end{aligned}$$

This is the total wall area. Window and door areas must be subtracted.

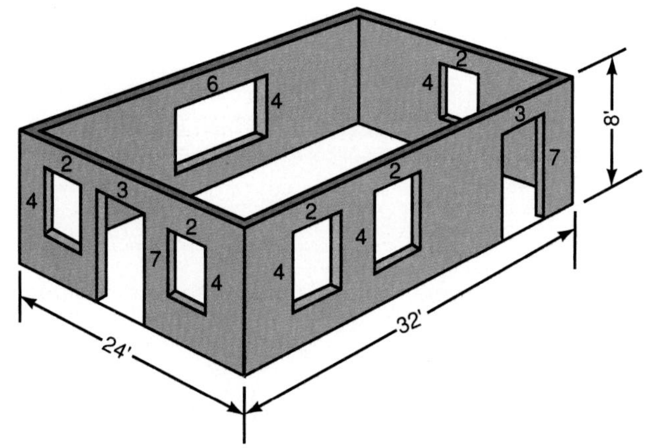

Figure 27-11. *Drawing of one-story home shows wall, window, and door areas. Building is 32′ (9.8 m) long by 24′ (7.3 m) wide. Room height is 8′ (2.44 m). Five windows are 4′ (1.22 m) high by 2′ (0.61 m) wide. Two doors are 7′ (2.13 m) high by 3′ (0.914 m) wide. Large window is 4′ (1.22 m) high by 6′ (1.83 m) wide.*

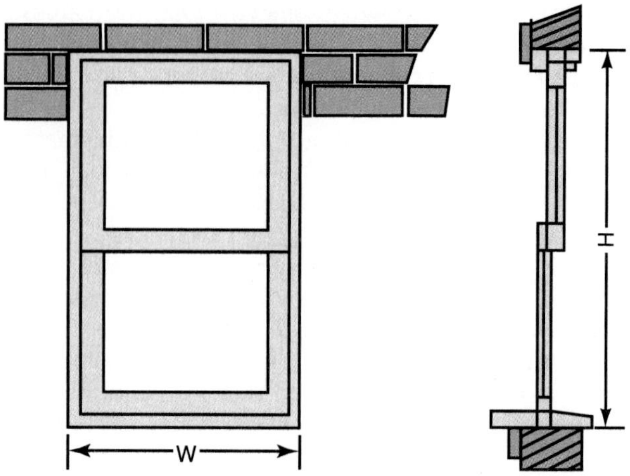

Figure 27-12. *Typical double-hung window showing width and height of window opening.*

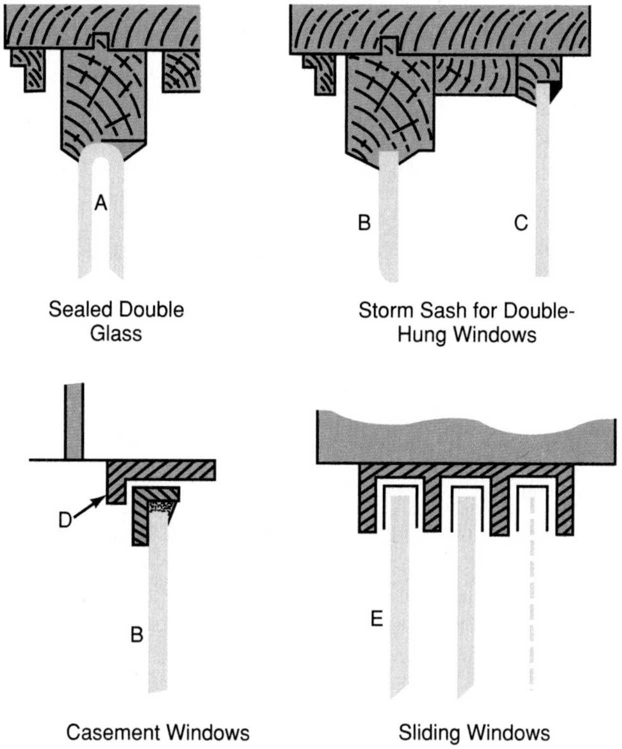

Figure 27-13. *Window construction. A—Dry air. B—Primary glass. C—Storm glass. D—Metal sash. E—Sliding window and storm with screen.*

Windows and Doors

Knowledge of the area of each window is needed in a heat leakage calculation. It is determined by measuring the opening in the wall. This would be the distance to the brick edges, as shown in **Figure 27-12.**

Window construction varies considerably. Windows may be single-pane, double-pane using a storm window, or permanent double-pane. Energy-efficient windows may even have three permanent layers of glass. See **Figure 27-13.**

The permanent window (single- or double-pane) is called the primary window. An additional framed pane of glass may be set into place to provide added insulation. This is called a storm window or sash.

The most efficient window construction is the permanent double- or triple-pane. Two or three panes of glass with sealed airspaces between panes provide excellent insulation. This airspace is dehydrated and evacuated. Then it is usually filled with nitrogen or some other dry gas to prevent sweating (condensation).

Windows are installed in a variety of ways. Some possibilities include:

- Fixed (picture windows).
- Single- or double-hung (where either one or both sashes move up and down).
- Sliding horizontal or casement (hinged on one side and open out with a crank mechanism).

The frame around windows may be made of wood or metal. Vinyl-clad aluminum frames are used to minimize frosting. The R- and U-values of window pane assemblies are shown in **Figure 27-14.**

R- and U-Values		
Type Window	U	R
Single-Glazed	1.04	.96
Double-Pane Insulating Glass	.49	2.0
Triple-Glazed	.32	3.1

Figure 27-14. *Thermal resistance (R) and heat transfer coefficient (U) of various window pane assemblies are specified. (Andersen Corp.)*

Warm air will condense on cold surfaces. This becomes a problem with windows and walls. Condensation occurs due to a combination of both temperature and relative humidity, (dew point). For example, air at 70°F (21°C) and 40% relative humidity, contacting a window at 45°F (7°C), causes condensation. This process is shown on the psychrometric charts in Chapter 19.

To prevent condensation, reduce the relative humidity or raise the temperature of the glass surface. Lowering the relative humidity in the home may not be practical or comfortable. Another solution would be to add a vapor barrier between the inside surface of the window and the room air. It will also serve as added insulation. Many companies make products to provide a temporary vapor barrier for the cold season.

The area used to calculate heat leakage for a door is the height times the width of the door opening.

Doors are constructed in a variety of designs. They may be made of solid wood or of wood veneers over foam cores. Recent energy efficient doors are constructed of a metal shell filled with insulation. Some doors have windows as part of their design. Large sliding glass doors are called patio doors. They are responsible for 20% of solar heating and heat leakage in some homes.

Proper door-to-wall sealing is very important to minimize heat leakage. The use of a rubber weather-stripping is common practice. On metal doors, a magnetic weather-stripping, similar to that used on refrigerators, can also be used.

When computing the wall heat leakage area, add the area of the doors in the outside walls to the area of the windows. Then, subtract this amount from the total wall area.

Example:

Back in **Figure 27-11,** there are five windows measuring 2′ × 4′, two doors measuring 3′ × 7′, and one window measuring 4′ × 6′.

$$
\begin{aligned}
2 \times 4 \times 5 = 8 \times 5 &= 40 \text{ ft}^2 \\
3 \times 7 \times 2 = 21 \times 2 &= 42 \text{ ft}^2 \\
4 \times 6 \times 1 &= \underline{24 \text{ ft}^2} \\
\text{Total opening area} &= 106 \text{ ft}^2
\end{aligned}
$$

The total wall area is 896 ft²

The net wall area is

$$896 \text{ ft}^2 - 106 \text{ ft}^2 = 790 \text{ ft}^2$$

The two values, 106 ft² of window area and 790 ft² of wall area, will be used later to find the building heat load.

Ceilings

Ceilings generally are made by fastening drywall to the joists. **Figure 27-15** shows several typical ceiling constructions. Heat leakage will be considerable if the joists do not have a floor over them. It will also be considerable if there is no insulation between the joists. See **Figure 27-7** for U-values for ceilings.

Using the sample house in **Figure 27-11,** the ceiling area is calculated as follows:

$$
\begin{aligned}
\text{Ceiling area} &= W \times L \\
&= 24' \times 32' \\
&= 768 \text{ ft}^2
\end{aligned}
$$

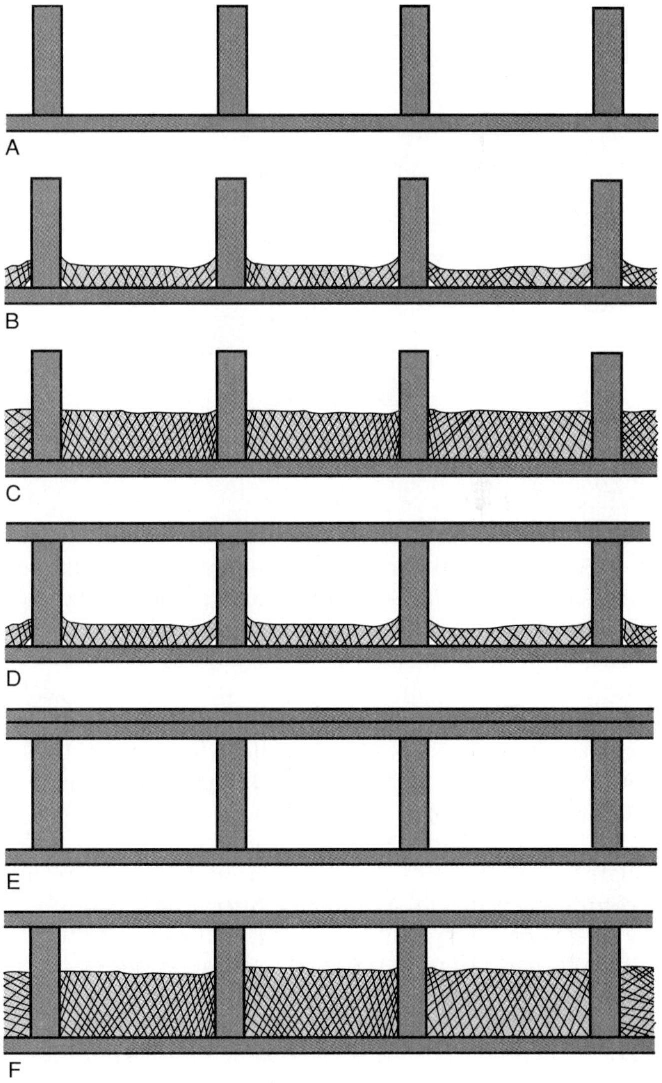

Figure 27-15. *Ceiling construction. A—No floor or insulation. B—No floor, 2″ insulation. C—No floor, 4″ insulation. D—Floor, 2″ insulation. E—Double floor, no insulation. F—Floor, 6″ insulation.*

Basement Heat Loss

Heat losses or gains for basements vary widely. **Figure 27-16** shows the heat loss for a basement built five feet into the ground. The deeper the basement, the less the heat loss. It is usually assumed that a basement is at 60°F (16°C). Leakage through the basement floor is usually not calculated. The heat leakage load is calculated through the first floor of the building (the basement ceiling).

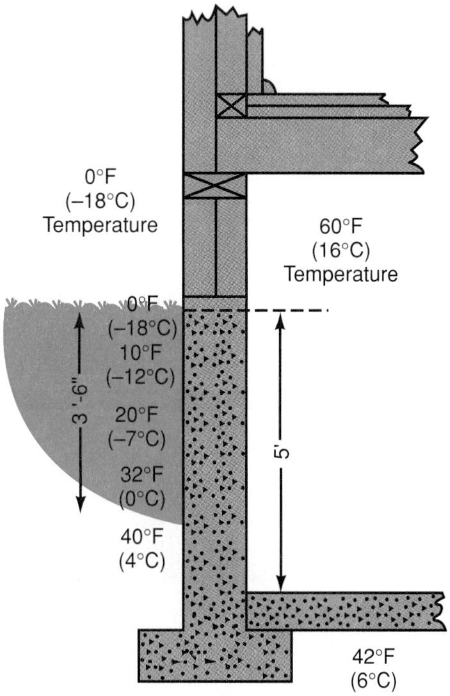

Figure 27-16. *Temperature conditions and construction of building with basement.*

Buildings built on a concrete slab have different heat losses than those built with a basement. See **Figure 27-17.** With the building on a slab, heat loss above ground is calculated in the typical manner. Ice and frost may form around the outside of a slab building. One method for minimizing this is to use a rigid urethane slab. The slab should be at least 2″ thick and installed 2′ to 4′ in the ground. **Figure 27-18** shows a typical installation.

Heat losses for the slab design are usually calculated as follows: The perimeter of the building is determined. The total length is multiplied by 18 Btu/hr. for each foot of length (0°F [−18°C] design temperature).

Another popular type of foundation leaves just enough space between the floor and the ground to allow access. This is referred to as a ***crawl space.*** The earth floor of a crawl space should have a vapor barrier on it. (This could be plastic sheeting, roofing paper, etc.) In addition, the floor can be insulated from underneath to provide maximum thermal protection. A crawl space must have sufficient venting to minimize moisture problems in the summer. The venting also minimizes the amount of cold air entering in the winter. Vents with dampers are installed to serve this dual purpose.

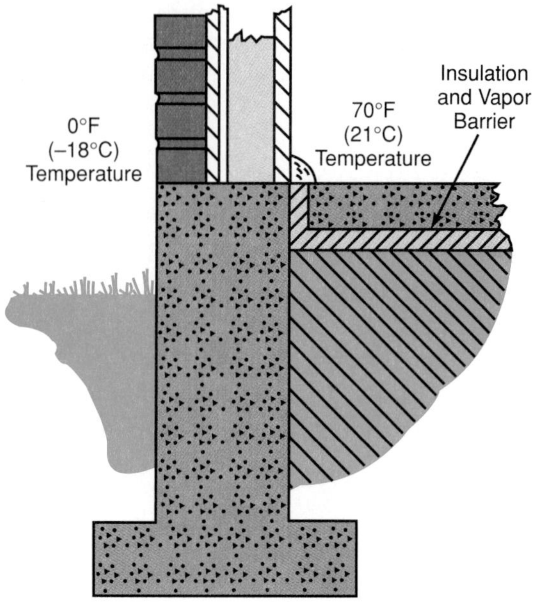

Figure 27-17. *Temperature conditions and construction of building built on concrete slab.*

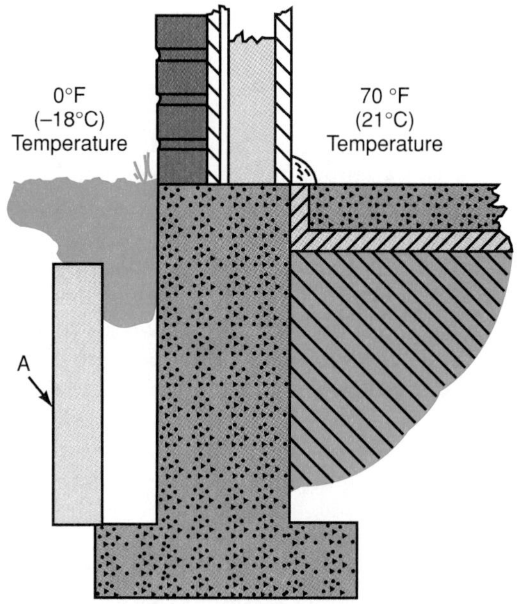

Figure 27-18. *A method is shown for preventing ice and frost from forming around perimeter of building built on slab. A—Rigid urethane insulation. Install as close to building as convenient.*

Recent construction practices use vapor barriers (plastic sheeting) between the basement walls and the surrounding ground. Refer to the American Society of Heating, Refrigerating and Air-Conditioning Engineers recommendations for additional data.

Sun Heat Load

Heat energy from the sun adds considerable heat load during the summer. The sun's rays in the northern hemisphere shine on the east, south, and west walls. They also shine on those roof sections that are exposed to the sun. Therefore, when computing total heat load, heat from the sun must be considered:

- On the east wall in the morning.
- On the south wall all day long.
- On the west wall in the afternoon.

This is shown in **Figure 27-19.**

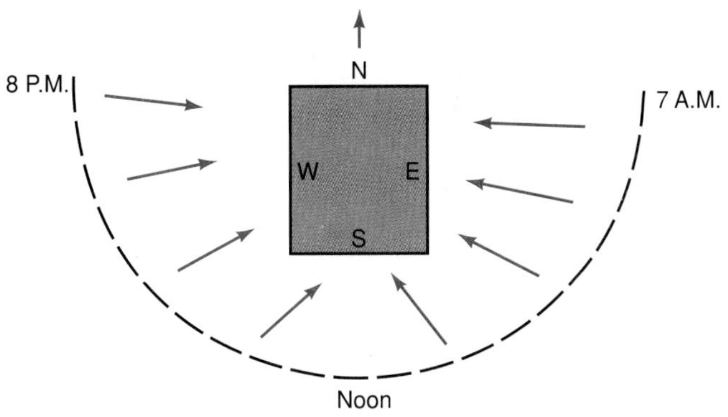

Figure 27-19. *Sun rays and their impact on the walls of a building during a 12-hr. period.*

The sun releases different amounts of heat to surfaces, depending upon the part of the world in which the building is located. The approximate maximum heat gain from the sun is 330 Btu per hr. per ft^2 (97 watts/ft^2 [1040 W/m^2]). This is for a black surface at right angles to the sun's rays near the equator (tropic). Any other color surface at an angle to the sun's rays will receive less heat.

At the 42nd parallel (a line going through New York City, Cleveland, and Salt Lake City), the maximum heat from the sun's rays is about 315 Btu per hr. per ft^2 (92 watts/ft^2 [993 W/m^2]).

Much of the heat from the sun is reflected back into the atmosphere. **Figure 27-20** indicates the window heat gain for windows facing different directions. This heat gain must be removed with air conditioning.

Effect of Sun on Windows	
Exposure	Heat Absorption Btu/hr./ft^2
Southwest	110
West	100
South	75
East	55
Single Skylights	110
Double Skylights	60
North	1
Northeast	2
Northwest	3

Figure 27-20. *Heat absorption when sun is shining on windows.*

Unless windows are protected with awnings, use a temperature of 15°F (8°C) higher than outside ambient temperature for correct results. Also add this amount to ambient temperature to take care of the effect of the sun shining on walls.

Approximate values obtained by using the 15°F (8°C) temperature correction generally are usable. However, there are many special cases that require careful study. Consider the changing position of the sun relative to the surfaces of the building. Also consider the time lag required for this heat to reach the building's interior.

Heat Lag

It takes time for the heat to travel through a substance that is heated on one side. ***Heat lag*** is the time needed for heat to travel through a substance that is heated on one side. The sun heats the outside wall of a building. However, several hours pass before this heat reaches the inner surfaces of the wall. In normal buildings, this time varies between three and four hours. With well-insulated or thick walls, the sun may be gone by the time the heat "soaks" through.

In the southwest, adobe walls are made quite thick. The sun heat moves into the wall while the sun shines. The wall is thick enough to prevent the heat from reaching the interior. During the night, when outdoor temperatures fall below those inside, the heat flow reverses itself and travels outward through the wall.

The south wall receives sunlight the entire day. The east and west walls receive sunlight only a small period of time. However, the south wall is not as strongly affected as the east and west walls. This is because the rays come from overhead. The length of daylight also changes throughout the season.

This heat lag causes the rooms to be heated even after the sun goes below the horizon and the outdoor temperature drops. See **Figure 27-21.**

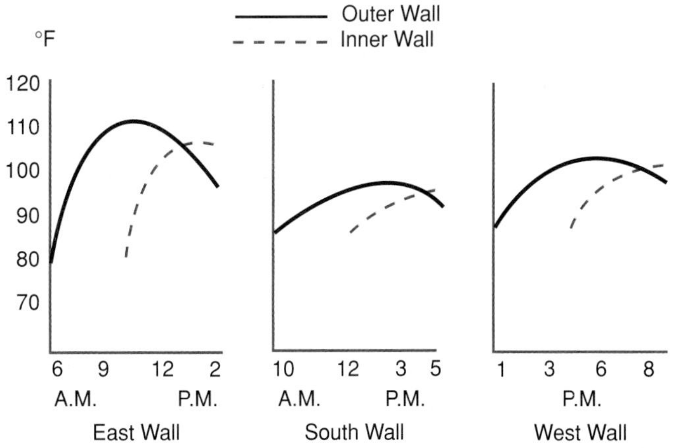

Figure 27-21. *Lag in interior wall temperature following exposure to sun.*

Heat Sources in Buildings

Heat sources may or may not be a benefit, depending on whether it is summer or winter. There are several sources of heat besides infiltration and sun load. All sources must be considered when figuring the comfort cooling load.

During the heating season, the heating system is aided by these other sources. Practically all energy expended in the building becomes heat. Yet, these sources are usually ignored when figuring heat loads of small buildings in the winter. They are small amounts compared to the total heat load in temperate zones.

However, when figuring the summer heat load (cooling heat load), all heat energy sources must be considered. (This includes heat released by human beings, stoves, lights, electric motors, etc.) **Figure 27-22** shows some of these heat sources. Notice that the two sources of heat—sensible heat gain and latent heat gain—are itemized. Latent heat raises the relative humidity.

During the cooling season, the heat released by persons must be taken into account. The heat released by one person weighing about 150 lb. (68 kg) is 74 watts (253 Btu/hr.) when at rest. It is about 440 watts (1,500 Btu/hr.) when that same person is working. About 25% to 45% of this heat is by moisture evaporation. This is a combination of moisture from the respiratory system and from the skin.

Window Heat Load for Cooling

There is considerable heat flow through ordinary window glass. It is approximately three times as great as flow through ordinary residential roofs and ceilings. Therefore, air conditioning areas containing a large amount of ordinary glass can become a problem. To reduce the heat conductivity through glass, a storm sash is used. To reduce the solar heat through glass, special types of glass with high heat-reflecting qualities may be used.

Special heat-absorbing glass can reduce the solar heat load by as much as 30%. Another method is to use glass tinted a bluish gray to reduce the solar glare and cooling load.

Roof extensions over a window will reduce the area exposed to the sun. Double-glazed windows exposed to sun rays reduce solar heat absorption by 15%. Awnings to shade glass windows exposed to the sun can reduce the heat load by 55%.

Humidifier Heat Load

During the heating season, water vapor must be added to the air for comfortable conditions. Heat to produce the water vapor may come from heated air, furnace heat, or electric heat.

The amount of heat needed is figured as follows: The number of volume changes per hour must be known. Generally, one change per hour is satisfactory for homes. The number of grains to be added per pound of air to obtain the required relative humidity must be known. See Chapter 19.

Formula:

Pounds of air per 24 hours
× increase in grains = grains/day
gr./day/7000 gr./lb. = lb. of water/day
lb. of water/day/8.34 lb./gal. = gal./day

To calculate:

$$\text{Volume} \times \text{changes/hr.} \times \frac{(gr_I - gr_O)}{33{,}000} = \text{gal./day}$$

Example:

A home has 12,000 ft^3. The grains to be added per pound of air to change the air from 35°F (2°C) and 90% RH to 72°F (22°C) and 40% are 20. Find the total gallons of water to be evaporated per day.

Solution:

$$\frac{12{,}000 \times 1 \times 20}{33{,}000} = \frac{240{,}000}{33{,}000} = 7.3 \text{ gal./day}$$

Heat Loads

Energy Source	Device	Hp	Heat: Sensible Btu/hr.	Heat: Sensible Watts	Heat: Latent Btu/hr.	Heat: Latent Watts
Electric	Lights/KW		3415	1020		
	Motors, Electric/Hp in Room	UP TO 1/2	4200	1230		
		UP TO 3	3700	1080		
		UP TO 20	2950	880		
	Motors, Electric/Hp Out of Room	UP TO 1/2	1700	500		
		UP TO 3	1150	340		
		UP TO 20	400	120		
	Stoves, Electric/KW		3415	1020		
Gas	Natural Gas/Ft^3		1100	320	300	88
	Artificial Gas/Ft^3		550	160	675	198
General	Heat From Meals/Meal		36	10		
	Steam Tables/Ft^2		400	120	800	230
Humans					140	41
		Sitting	370	110		
		Working	700–1500	200–440		
		Dancing	2000	590		

Figure 27-22. *Heat released by various energy sources within a building.*

Formula:

The amount of heat (in Btu/hr.) needed to evaporate the water is found as follows:

$$\text{Volume of house} \times \text{changes per hour}/13.55\ \text{ft}^3 \text{ of air/lb.} \times (\text{gr./lb. indoors} - \text{gr./lb. outdoors}) \times 970.3\ \text{Btu/lb.}/7000\ \text{gr./lb} = \text{Btu/hr.}$$

$$\text{Volume} \times \text{changes/hr.} \times \frac{gr_I - gr_O}{97.75} = \text{Btu/hr.}$$

Example:

Using the same numbers as before, the volume is 12,000 ft^3. The grains are 20. Find the required heat.

Solution:

$$\frac{12{,}000 \times 1 \times 20}{97.75} = \frac{240{,}000}{97.75} = 2455\ \text{Btu/hr.}$$

Air Conditioner Heat Load

Figure 27-23 is a form used to calculate the cooling heat load for a room. By multiplying the area of the floors, walls, and windows by multipliers, the amount of required energy can be determined. The multipliers are obtained by multiplying a typical U-value by the temperature difference. For example, the windows (in the shade) have a U-value of 1.25. If the temperature difference is about 12°F (7°C), the multiplier becomes 15.

$$12 \times 1.25 = 15$$

The following is a way to make a rough estimate. The chart identifies COP, which is the ratio of output divided by input. The output is the amount of heat absorbed by the system. Input is the amount of energy put into the system. Remember that, on the average, a medium-size room needs 5000 to 6000 Btu/hr. of cooling. The average window comfort cooling unit will adequately handle the cooling loads as follows:

0–6000 Btu/hr.	= 1/2 hp.,	COP = 4.71
6000–9000 Btu/hr.	= 3/4 hp.,	COP = 4.71
9000–11,000 Btu/hr.	= 1 hp.,	COP = 4.32

27.1.3 Total Heat Load

It is best to set up total heat load calculations in table form. **Figure 27-24** shows a typical heat load calculation for a 24′ × 32′ (7.3 m × 9.8 m) house. Refer to **Figure 27-7.** Note that the temperature difference for the ceiling is only 35°F (19°C). Also consider that the roof

Energy Required for Cooling				
Name: ______				
Address: ______				
Phone: ______				
Space used for: ______				
Interior Room Dimensions:				
Length: ______	Width: ______	Height: ______		
Windows:				
No.: ______	Facing: ______	Size: ______ × ______		
Window Loads				
Sun exposed (interior shades)				
West side:	______ sq. ft. ×	60 =	______ Btu/hr.	______ watts
South side:	______ sq. ft. ×	40 =	______ Btu/hr.	______ watts
Sun exposed (awnings)				
West/South side:	______ sq. ft. ×	35 =	______ Btu/hr.	______ watts
East, North, or shaded:	______ sq. ft. ×	15 =	______ Btu/hr.	______ watts
Wall Load				
South, West exposure:	______ sq. ft. ×	8 =	______ Btu/hr.	______ watts
East, North exposure:	sq. ft. ×	5 =	______ Btu/hr.	______ watts
Thin wall, all exposures:	______ sq. ft. ×	10 =	______ Btu/hr.	______ watts
Interior walls:	______ sq. ft. ×	4 =	______ Btu/hr.	______ watts
Interior glass partition:	______ sq. ft. ×	10 =	______ Btu/hr.	______ watts
Floor Load				
	______ sq. ft. ×	3 =	______ Btu/hr.	______ watts
Ceiling Load				
Occupied above:	______ sq. ft. ×	3 =	______ Btu/hr.	______ watts
Insulated roof:	______ sq. ft. ×	8 =	______ Btu/hr.	______ watts
Uninsulated roof:	______ sq. ft. ×	20 =	______ Btu/hr.	______ watts
Ventilation Load				
	______ sq. ft. ×	4 =	______ Btu/hr.	______ watts
Occupancy Load				
	______ sq. ft. ×	400 =	______ Btu/hr.	______ watts
Miscellaneous Loads				
Electrical watts:	______	× 3.4 =	______ Btu/hr.	______ watts
Other:		=	______ Btu/hr.	______ watts
		Total:	______ Btu/hr.	______ watts

Figure 27-23. *This table can be used to determine the cooling needed for a typical residence.*

Surface	Area	R	U	Temp. Diff.	Heating Leakage Btu/hr.
Wall, Gross	996	0.73	1.38		
Window	116	0.89	1.13	70	9176
Wall, Net	880	4.0	0.25	70	15400
Ceiling	768	1.61	0.62	35	16666
Floor	768	2.94	0.34	25	6528
				Total	47770

Figure 27-24. *Typical heat load calculation for 24′ × 32′ home having an 8′ ceiling height.*

serves as added insulation. It keeps the attic temperature higher than the outdoor temperature. The attic temperature can be accurately calculated by making the heat leaking into the attic in winter equal the heat leaking out.

$$\text{Ceiling area} \times (70°\text{F} - [\text{attic temp.}]) \times U_c = \text{Roof area} \times (\text{attic temp.} - 0°\text{F}) \times U_r.$$

Where: U_c = U-value of the ceiling
U_r = U-value of the roof

Note that most homes have an 8′ (2.4 m) ceiling, although many homes are now being built with 7′-6″ (2.3 m) ceilings. The lower ceiling helps prevent trapping hot air near the ceiling. The hot ceiling air can be used for heating.

Each heat leakage value is obtained by means of the following formula:

Heat leakage = area × U-value × temperature difference

A quick method used to estimate total heat loads is shown in **Figure 27-25.** Note that the heat load is based on room volume. The table also includes cooling. A method for estimating duct sizes is shown in **Figure 27-26.** This table is used as a companion to the heat load

Residential Forced Air System Design Guide (For Estimating Purposes Only)

Room Vol. ft³	A/C CFM	Winter Heat Losses Heating Btu @Supply °F.						Summer Heat Gains Cooling Btu @ Supply °F.		
		120°	130°	140°	150°	160°	170°	60°	55°	50°
200	14	750	910	1060	1210	1360	1510	300	375	450
300	20	1080	1300	1510	1730	1945	2160	430	540	650
400	27	1460	1750	2040	2330	2625	2915	580	730	875
500	34	1830	2200	2570	2940	3300	3670	735	920	1100
600	40	2160	2590	3025	3455	3890	4320	865	1080	1290
700	47	2540	3025	3550	4060	4565	5080	1015	1270	1530
800	55	2970	3565	4155	4750	5345	5940	1190	1485	1780
900	60	3240	3880	4535	5185	5830	6480	1295	1620	1940
1000	65	3500	4210	4915	5615	6320	7000	1400	1755	2100
1100	75	4050	4860	5670	6480	7290	8100	1620	2000	2430
1200	80	4320	5200	6040	6910	7780	8640	1730	2160	2600
1300	87	4700	5635	6570	7520	8455	9400	1880	2350	2820
1400	95	5130	6155	7180	8200	9235	10260	2050	2560	3080
1500	100	5400	6480	7560	8640	9720	10800	2160	2700	3240
1600	107	5775	6930	8090	9245	10400	11550	2310	2890	3460
1700	113	6100	7320	8540	9760	10950	12200	2440	3050	3660
1800	120	6480	7775	9070	10360	11665	12960	2590	3240	3880
1900	125	6750	8100	9450	10800	12150	13500	2700	3370	4050
2000	135	7300	8750	10200	11660	13120	14600	2920	3645	4370
3000	200	10800	12960	15120	17280	19440	21600	4320	5400	6480
4000	265	14300	17170	20035	22890	25750	28600	5720	7150	8580
5000	335	18100	21700	25325	28940	32560	36200	7240	9040	10850
6000	400	21600	25920	30240	34560	38880	43200	8640	10800	12960
7000	465	25100	30130	35150	40170	45200	50200	10040	12550	15060
8000	535	26900	34670	40450	46220	52000	53800	10760	14440	17330
10000	670	36180	43410	50650	57890	65125	72360	14470	18090	21700
12000	800	43200	51840	60480	69120	77760	86400	17280	21600	25920
14000	935	50500	60590	70680	80780	90880	101000	20200	25240	30290
16000	1065	57500	69000	80510	92010	103520	115000	23000	28750	34500
18000	1200	64800	77760	90720	103680	116640	129600	25920	32400	38780
20000	1335	72100	86500	100920	115340	129760	144200	28840	36040	43250
25000	1670	90200	108215	126250	144290	162320	180360	36070	45090	54100
		Based on 70°F Return Air						Based on 80°F Return Air		

Max. Air on One Outlet

Figure 27-25. *Approximate heat load chart for winter heating and summer cooling is based on volume of conditioned space. (Detroit Edison Co.)*

Residential Forced Air System Design Guide

Room Volume ft^3	Supply Duct (In.) R'nd. (Dia.)	Supply Duct (In.) Equiv.	Outlet (In.) Floor	Outlet (In.) Wall	Outlet (In.) Ceiling (Dia.)	Return Grille (In.)	Return Duct (In.) R'nd. (Dia.)	Return Duct (In.) Equiv.
	Air Distribution Sizes							
200	4	4 1/2 × 3			4	6 × 10	6	8 × 4
300	4	4 1/2 × 3			4	6 × 10	6	
400	4	4 1/2 × 3			4	6 × 10	6	
500	4	4 1/2 × 3			4	6 × 10	6	
600	5	10 × 2 1/4	2 1/4 × 10		4	6 × 10	6	
700	5	8 × 3 1/4	2 1/4 × 10	4 × 10	6	6 × 10	6	
800	5	5 × 4	2 1/4 × 10		6	6 × 10	6	
900	6	14 × 2 1/4	2 1/4 × 10		6	6 × 10	6	
1000	6	10 × 3 1/4	2 1/4 × 10		6	6 × 10	7	8 × 6
1100	6	8 × 4	2 1/4 × 12		6	6 × 10	7	
1200	6	6 × 5	2 1/4 × 12	10 × 6	6	6 × 10	7	
1300	6	6 × 5	2 1/4 × 12		6	6 × 10	7	
1400	7	14 × 3 1/4	2 1/4 × 14		6	6 × 14	8	8 × 7
1500	7	11 × 4	2 1/4 × 14	12 × 6	8	6 × 14	8	
1600	7	8 × 5	4 × 10	14 × 6	8	6 × 14	8	
1700	7	7 × 6	4 × 10	14 × 6	8	6 × 14	8	
1800	7		4 × 12	14 × 6	8	6 × 14	8	
1900	7		4 × 12	14 × 6	8	6 × 14	8	
2000	7		4 × 12	14 × 6	8	6 × 14	8	
3000	7 1/2	13 × 4				8 × 14	10	8 × 11
4000	9	8 × 8				6 × 24	11	8 × 13
5000	10	8 × 11				6 × 30	12	8 × 16
6000	11	8 × 13				8 × 30	13	8 × 18
7000	11 1/2	8 × 14				8 × 30	14	8 × 22
8000	12	8 × 16				18 × 18	15	8 × 24
10000	13	8 × 18				18 × 18	16	8 × 28
12000	14	8 × 22				18 × 24	18	8 × 36
14000	14 1/2	8 × 24				24 × 24	20	8 × 46
16000	15	8 × 26				24 × 24	20	
18000	16	8 × 30				24 × 30	20	
20000	17	8 × 34				24 × 30	22	8 × 60
25000	18	8 × 39				24 × 30	22	

Figure 27-26. *Duct sizes based on estimated values found in* ***Figure 27-25.*** *(Detroit Edison Co.)*

table. For a given room volume, a recommended supply and return duct size is given. Outlet and return grille areas are also shown. More information on the calculation of proper air distribution systems is found in Chapter 23. Standard worksheets are available for calculating total heat load. See **Figure 27-27.**

Heat gain calculations to determine the total building cooling load are similar to heat loss calculations. The temperature difference is based on the locality being considered. Indoor temperature is usually designed to be 75°F (24°C) at 50% relative humidity (RH). Therefore, if the summer design temperature is 100°F (38°C), the temperature difference is 25°F (14°C). This temperature difference is for load calculations only. In practice, a 10°F to 15°F (6°C to 8°C) difference is recommended.

Other heat sources must be considered. Sun load, electrical load, and occupants are large enough sources of heat to be included in the heat load calculations.

27.2 Design Temperatures

Contact the local weather bureau or local chapter of the American Society of Heating, Refrigerating, and Air-Conditioning Engineers (ASHRAE) for data on design temperatures.

Always choose the outdoor design temperatures (ODT) on the low side. Heating plants that are overworked cause excessive stack and chimney temperatures

Residential

Whole House Worksheet

Customer's Name__________________ Address__________________

City____________ State________ Zip________ Telephone Number____________

WINTER: Inside Design Temp________ °F–Outside Design Temp________ °F = Heating Temp Difference________ °F

SUMMER: Outside Design Temp________ °F–Inside Design Temp________ °F = Cooling Temp Difference________ °F

HEATING		COMMON DATA SECTION		COOLING	
BTUH LOSS	HEATING FACTOR	SUBJECT	SQ. FT.	COOLING FACTOR	BTUH GAIN
	FROM TABLE E	GROSS WALL		FROM TABLE E	
		DOORS & WINDOWS (Table A or B)			
		NET WALL			
		CEILING			
		FLOORS			
Infiltration Btu/hr =	Heating Table D	× 10 × 1.1/60 × Volume (Cu. Ft.)	Volume (Cu. Ft.) ×	1.1/60 × ΔT × Cooling Table D =	Infiltration Btu/hr
=		× 0.18333 ×	×	0.01833 × × =	
		SUB-TOTAL BTUH LOSS (per 10°F)			
×		ADJUSTMENT FACTOR (Table C)			
		TOTAL BTUH LOSS			
		PEOPLE ___ × 300 BTUH GAIN (Assume 2 persons per bedroom)			
		APPLIANCES BTUH			1200
		SUB-TOTAL BTUH GAIN (room sensible only)			
×		DUCT LOSS/GAIN FACTOR (Table F)			×
		SUB-TOTAL BTUH (Sensible Gain)			
		MOISTURE REMOVAL (sub total × 1.3)			× 1.3
		TOTAL BTUH LOSS/GAIN			

TABLE A–HEATING–DOORS & WOOD FRAME WINDOWS (PER 10°F)

For sliding glass doors - use factors for the same type window construction.

Window & Door Types	Frames: Wood	Frames: TIM	Frames: Metal	× Area	= Btuh Loss
Single Pane Clear	9.90	10.45	11.55		
With Storm	4.75	5.25	6.50		
Double Pane Clear	5.51	6.09	7.25		
With Storm	3.41	3.85	4.90		
Triple Pane Clear	3.80	4.39	5.46		
Jalousie Single	–	–	11.0		
Single w/ Storm	–	–	5.0		
Skylights Single	11.07	11.69	12.92		
Double	6.65	7.35	8.75		
Door Wood Only	4.60	–	–		
Wood w/ Storm	3.20	–	–		
Urethane Core (R-5)	–	–	1.90		
Urethane Core (R-5) w/Storm	–	–	1.70		
			TOTALS		

TABLE B–COOLING–DOORS & WINDOWS

Factors assume windows have inside shading by draperies or venetian blinds and sliding glass doors are treated as windows.

	SINGLE GLASS TEMP. DIFF.			DOUBLE GLASS TEMP. DIFF.			TRIPLE GLASS TEMP. DIFF.			× Area	= BTUH GAIN
Direction	15°	20°	25°	15°	20°	25°	15°	20°	25°		
N	18	22	26	14	16	18	11	12	13		
NE & NW	37	41	45	31	33	35	26	27	28		
E & W	52	56	60	44	46	48	38	39	40		
SE & SW	45	49	53	39	41	43	33	34	35		
S	28	32	36	23	25	27	19	20	21		
Skylights	164	168	172	141	143	145	132	136	140		
Wood ①	8.6	10.9	13.2	8.6	10.9	13.2	8.6	10.9	13.2		
Metal ②	3.5	4.5	5.4	3.5	4.5	5.4	3.5	4.5	5.4		
									TOTALS		

① For wood doors and polystyrene core metal doors

② For urethane core metal doors

TABLE D–INFILTRATION MULTIPLIERS

Winter Air Changes Per Hour

Floor Area	900 or Less	900-1500	1500-2100	Over 2100
Best	0.4	0.4	0.3	0.3
Average	1.2	1.0	0.8	0.7
Poor	2.2	1.6	1.2	1.0
For each fireplace add:		Best 0.1	Average 0.2	Poor 0.6

Summer Air Changes Per Hour

Floor Area	900 or Less	900-1500	1500-2100	Over 2100
Best	0.2	0.2	0.2	0.2
Average	0.5	0.5	0.4	0.4
Poor	0.8	0.7	0.6	0.5

TABLE C–ADJUSTMENT FACTORS–(HEATING)

°F. Temperature Diff.	30	40	50	60	70	80	90
Adjustment Factor	3	4	5	6	7	8	9

Figure 27-27. *Use this worksheet to calculate heat losses and gains. (The Trane Company) (Continued, next page.)*

HEAT LOSS & GAIN FACTORS

TABLE E
CONSTRUCTION FACTORS HEATING & COOLING

Heating Factor ①	TYPE OF CONSTRUCTION	Cooling Factor (°F. Temp. Diff.) 15°	20°	25°
	WALLS (Use Sq. Ft.)			
	Walls–wood frame w/ sheeting & siding, veneer or other finish			
2.71	A) No insulation, 1/2" Gypsum Board	5.0	6.4	7.8
0.90	B) R-11 Cavity insulation + 1/2" Gypsum Board	1.7	2.1	2.6
0.80	C) R-13 Cavity insulation + 1/2" Gypsum Board	1.5	1.9	2.3
0.70	D) R-13 Cavity insulation + 3/4" Bead Board (R-2.7)	1.3	1.7	2.0
0.60	E) R-19 Cavity insulation + 1/2" Gypsum Board	1.1	1.4	1.7
0.50	F) R-19 Cavity insulation + 3/4" Extruded Poly	0.9	1.2	1.4
	Masonry Walls			
5.10	A) Above grade No insulation	5.8	8.3	10.9
1.44	B) Above grade + R-5	1.6	2.3	3.1
0.77	C) Above grade + R-11	0.9	1.3	1.6
1.25	D) Below grade No insulation	0.0	0.0	0.0
0.74	E) Below grade + R-5	0.0	0.0	0.0
0.51	F) Below grade + R-11	0.0	0.0	0.0
	CEILINGS (Use Sq. Ft.)			
5.99	A) No insulation	17.0	19.2	21.4
1.20	B) 2"-2 1/2" insulation R-7	4.4	4.9	5.5
0.88	C) 3"-3 1/2" insulation R-11	3.2	3.7	4.1
0.53	D) 5 1/4"-6 1/2" insulation R-19	2.1	2.3	2.6
0.48	E) 6"-7" insulation R-22	1.9	2.1	2.4
0.33	F) 8"-9 1/2" insulation R-30	1.3	1.5	1.6
0.26	G) 10"-12" insulation R-38	1.0	1.1	1.3
0.23	H) 12"-13" insulation R-44	0.9	1.0	1.1
3.08	I) Cathedral type No insulation (roof/ceiling combination)	11.2	12.6	14.1
0.72	J) Cathedral type R-11 (roof/ceiling combination)	2.8	3.2	3.5
0.49	K) Cathedral type R-19 (roof/ceiling combination)	1.9	2.2	2.4
0.45	L) Cathedral type R-22 (roof/ceiling combination)	1.8	2.0	2.2
0.40	M) Cathedral type R-26 (roof/ceiling combination)	1.6	1.8	2.0
	FLOORS (Use Sq. Ft. OR Linear Ft.)			
	Floors over unconditioned space (use sq. ft.)			
1.56	A) Over basement or enclosed crawl space (not vented)	0.0	0.0	0.0
0.40	B) Same as "A" + R-11 insulation	0.0	0.0	0.0
0.26	C) Same as "A" + R-19 insulation	0.0	0.0	0.0
3.12	D) Over vented space or garage	3.9	5.8	7.7
0.80	E) Over vented space or garage + R-11 insulation	0.8	1.3	1.7
0.52	F) Over vented space or garage + R-19 insulation	0.5	0.8	1.1
0.24	**Basement Floors (use sq. ft.)**	0.0	0.0	0.0
	Concrete slab floor unheated (use linear ft.)			
8.10	A) No edge insulation	0.0	0.0	0.0
4.10	B) 1" edge insulation R-5	0.0	0.0	0.0
2.10	C) 2" edge insulation R-9	0.0	0.0	0.0
	Concrete slab floor duct in slab (use linear ft.)			
19.00	A) No edge insulation	0.0	0.0	0.0
11.40	B) 1" edge insulation R-5	0.0	0.0	0.0
9.30	C) 2" edge insulaton R-9	0.0	0.0	0.0

① Heating Factor for 10° Temperature Rise

TABLE F

Calculate only if duct is located in an unconditioned space.

DUCT LOSS MULTIPLIERS

Duct Location and Insulation Value	Duct Loss Multipliers: Winter Design Below 15°F	Winter Design Above 15°F
Case I - Supply Air Temperatures Below 120°F		
Exposed to Outdoor Ambient		
Attic, Garage, Exterior Wall, Open Crawl Space - None	1.30	1.25
Attic, Garage, Exterior Wall, Open Crawl Space - R2	1.20	1.15
Attic, Garage, Exterior Wall, Open Crawl Space - R4	1.15	1.10
Attic, Garage, Exterior Wall, Open Crawl Space - R6	1.10	1.05
Enclosed in Unheated Space		
Vented or Unvented Crawl Space or Basement - None	1.20	1.15
Vented or Unvented Crawl Space or Basement - R2	1.15	1.10
Vented or Unvented Crawl Space or Basement - R4	1.10	1.05
Vented or Unvented Crawl Space or Basement - R6	1.05	1.00
Duct Buried In or Under Concrete Slab		
No Edge Insulation	1.25	1.20
Edge Insulation R Value = 3 to 4	1.15	1.10
Edge Insulation R Value = 5 to 7	1.10	1.05
Edge Insulation R Value = 7 to 9	1.05	1.00
Case II - Supply Air Temperatures Above 120°F	Winter Design Below 15°F	Winter Design Above 15°F
Exposed to Outdoor Ambient		
Attic, Garage, Exterior Wall, Open Crawl Space - None	1.35	1.30
Attic, Garage, Exterior Wall, Open Crawl Space - R2	1.25	1.20
Attic, Garage, Exterior Wall, Open Crawl Space - R4	1.20	1.15
Attic, Garage, Exterior Wall, Open Crawl Space - R6	1.15	1.10
Enclosed in Unheated Space		
Vented or Unvented Crawl Space or Basement - None	1.25	1.20
Vented or Unvented Crawl Space or Basement - R2	1.20	1.15
Vented or Unvented Crawl Space or Basement - R4	1.15	1.10
Vented or Unvented Crawl Space or Basement - R6	1.10	1.05
Duct Buried In or Under Contrete Slab		
No Edge Insulation	1.30	1.25
Edge Insulation R Value = 3 to 4	1.20	1.15
Edge Insulation R Value = 5 to 7	1.15	1.10
Edge Insulation R Value = 7 to 9	1.10	1.05

DUCT GAIN MULTIPLIERS

Duct Location and Insulation Value	Duct Gain Multiplier
Exposed to Outdoor Ambient	
Attic, Garage, Exterior Wall, Open Crawl Space - None	1.30
Attic, Garage, Exterior Wall, Open Crawl Space - R2	1.20
Attic, Garage, Exterior Wall, Open Crawl Space - R4	1.15
Attic, Garage, Exterior Wall, Open Crawl Space - R6	1.10
Enclosed in Unconditioned Space	
Vented or Unvented Crawl Space or Basement - None	1.15
Vented or Unvented Crawl Space or Basement - R2	1.10
Vented or Unvented Crawl Space or Basement - R4	1.05
Vented or Unvented Crawl Space or Basement - R6	1.00
Duct Buried In or Under Concrete Slab	
No Edge Insulation	1.10
Edge Insulation R Value = 3 to 4	1.05
Edge Insulation R Value = 5 to 7	1.00
Edge Insulation R Value = 7 to 9	1.00

ESTIMATED PROCEDURES

1. Fill in customer information.
2. Record inside and outside design temperatures; find temp difference.
3. Measure length of each outside wall, multiply each by ceiling height. Record the total sq. ft. of exposed wall under "gross wall".
4. Using Tables A and B, determine the total area for windows & doors & enter in common data section.
5. Determine Net Wall by subtracting windows and doors from gross.
6. Measure & record total ceiling area.
7. Measure and record total floor area for floors over crawl space or basement. Total floor edge *length* (perimeter) in floor is a slab.
8. Using Table E select construction type and use the corresponding heat and cool factors on the form.
9. Determine BTUH Loss & Gain in Tables A and B by multiplying the area of glass and doors by the multiplier under the specified temperature difference. Enter total BTUH Loss/Gain on worksheet.
10. On worksheet, multiply the areas × the factors and total as instructed.

Figure 27-27. *Continued.*

and may cause fires. Oversized units will be less efficient and waste energy.

The ODT is never as low as the lowest temperature recorded for the area. This is because these extreme lows are usually of short duration. Residual heat in the building usually enables the furnace based on the design temperature to handle the load.

ASHRAE has a method of listing outside design temperatures (ODT). In this method, the ODT varies with latitude and elevation. When working with design temperatures, use the current values. (These are available from the local ASHRAE chapter.)

ASHRAE charts give three different values for each locality. The lowest temperature is for small (domestic and office) uninsulated buildings. The 99% temperature means that the outdoor temperature is at or above this temperature 99% of the time. It can be used for well-constructed and well-insulated buildings having a standard number of windows.

A 97.5% value also given for each locality means that the outside temperature is above the value listed 97.5% of the time. It is used for large buildings with considerable thermal capacity and small total window area.

Example: Detroit once used −10°F (–23°C) as ODT for all buildings. Now it is recommended that 0°F (–18°C) be used for small, uninsulated buildings. For well constructed, insulated buildings with a standard window area, +4°F (2°C) (99% factor) is used. For large buildings with a standard number of windows, +8°F (4°C) (97.5% factor) is used.

If the Inside Design Temperature (IDT) is 72°F (22°C), an ODT of 0°F (–18°C) means a 72°F (40°C) temperature difference (TD). The 99% ODT (+4°F [+2°C]) means a 68°F (38°C) TD. The 97.5% means a 64°F (35°C) TD.

$$\frac{72 - 64}{72} \times 100 = \frac{8}{72} \times 100 = 11\%$$

This represents an 11% savings in equipment size on this installation.

In many buildings, certain areas such as closets, hallways, and attics are unheated. These spaces receive their heat from heat leakage through partitions, ceilings, and floors. These unheated areas are assumed to be halfway between the indoor design temperature and the outside design temperature.

27.2.1 Degree-Day Method

The degree-day method is used for determining fuel consumption and heating cost during a season. It determines, by previous weather records, the average temperature of each day during the season. It also keeps a record of daily temperature during the season under study.

The degree-day method uses a 65°F (18°C) indoor temperature as a standard. If the average temperature outside is 15°F (−9°C), the temperature difference would be 50°F (28°C). Therefore, that day would have 50 degree-days. Each degree-day requires a certain heat load to keep the inside temperature at 65°F (18°C).

Knowing the number of degree-days since a fuel oil tank was filled enables you to accurately calculate the amount of fuel left in the tank. If the degree-days for a season are known, the heating cost for that season can be calculated.

27.3 Insulation and Vapor Barriers

A large number of different insulating materials have been developed for buildings. It is necessary that the insulation reduce heat loss. In addition, vapor barriers should be included in the insulation or in the walls. The barriers reduce moisture travel through the wall.

It is very important that hygroscopic (moisture absorbent) insulation is hermetically sealed. Aluminum sheet or tarred paper are two popular sealing materials. Even insulation not affected by moisture should be vapor sealed. The insulation will lose much of its insulating value if it fills up with moisture. This is particularly important in cooling applications.

The insulation selected should have sufficient strength to support itself and not shrink or settle. It must not deteriorate in the presence of moisture. It must not have any unpleasant odor. The insulation should be vermin-proof and fire-resistant.

The type of insulation used depends, to some extent, on the method of application. Easy-flowing bulk insulation can be blown into the space between the studs of an existing building. For new buildings, rigid insulation such as a plaster base can be used. It can be a part of the building wall or a substitute for the sheeting.

Flexible insulation is easy to install. It conforms to any irregularities in the construction. Batts of fiberglass, rock wool, and blankets of pulverized wood are examples of flexible insulation. **Figure 27-28** shows batt insulation being installed between studs.

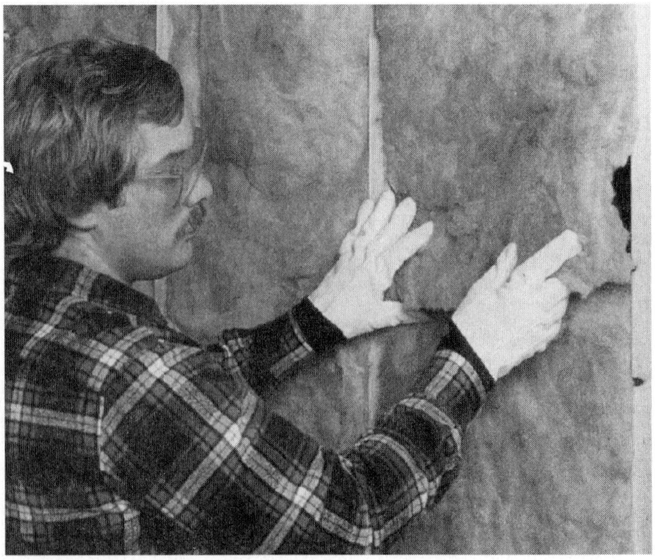

Figure 27-28. *Flexible batt insulation being installed between studs. (Owens-Corning Fiberglas Corp.)*

The Mineral Insulation Manufacturer's Association, Inc. (MIMA) recommends certain insulation for electrically heated and air conditioned homes:

Ceilings	R-19 through R-28
Walls	R-11 through R-19
Floors over unheated spaces	R-11 through R-19

The above R-values depend upon the location. Use the higher values for locations above the 42nd parallel. R-22 and R-13 values are recommended for special situations where extra insulation is necessary.

27.3.1 Ponded Roof

An ordinary roof may be heated by the sun to a temperature of 100°F to 150°F (38°C to 66°C). Ceilings under such roofs will become warm and radiate this heat through the space below.

Many buildings with flat roofs are provided with some summer comfort cooling by using ***ponded roofs.*** A 2″ to 3″ (5 cm to 8 cm) pond of water covers the roof surface. This type of cooling is especially well suited to one-story factory and market buildings. To be effective, the roof area should be as large as the floor area.

The cooling effect comes from the evaporation of water from the roof. Naturally, ponded systems are most effective in areas having a high temperature, low relative humidity, and bright sunshine. By ponding, roof temperature may be kept below that of the surrounding atmosphere.

Ponded roofs require a means of maintaining a constant level of water on the roof. Drains are needed to take away excess water due to rain. If the roof is large, wave breakers are needed to prevent waves from forming under high winds. The waves could cause a large quantity of water to be blown off the roof edge.

A ponded roof may reduce the required air conditioning capacity by as much as 30%. However, the added weight on the roof will result in higher construction costs. A ponded roof should never be added to an existing building unless the roof structure can support the additional weight of the water.

27.3.2 Building Insulation and Ventilation for Electric Heating

Electric heating is becoming more popular. Problems in electric heating usually involve the need for insulation or ventilation. Excess relative humidity also can pose a problem.

The cost of heating is another factor to be considered. Usually, it costs more to produce a unit of heat electrically than with most other types of fuels. Reducing heat loss through windows, walls, floors, and ceilings will reduce the cost of electric heating. Therefore, installing efficient insulating materials and double-glazed windows will help cut the heat load.

Electric heating requires a well-insulated and tight structure. Sometimes, however, this adds to the ventilation problem. If the building is occupied by many people, frequent air changes must be provided for ventilation purposes.

Spaces that are comfort cooled as well as electrically heated usually have the electric heating elements in the plenum chamber or the air duct system. Some systems also have electric elements located in each room. This allows individual control of the spaces being heated.

Relative humidity control in electrically heated buildings is usually different than from those using fuel-burning heating equipment. With fuel-burning equipment, a fairly large air quantity passes through the furnace and out of the stack. Usually, makeup air enters the building by leakage around the doors, windows, and cracks in the floor. With most electrically-heated structures, building construction permits little infiltration of this nature. Therefore, there is a possibility that the relative humidity is too high. Vapor formed in the space cannot easily escape from the occupied space. Dehumidifying equipment may be needed.

There are other factors, too, that help make the cost of heating with electricity favorable. There is no dirt or smoke generated in connection with the heating. Drapes, upholstery, and woodwork remain clean longer, and the cost of cleaning and redecorating may be reduced. In some areas, the electrical companies provide a lower price for electricity used for heating a structure, as an incentive.

27.4 Energy Conservation

When reviewing the calculations for determining heat loads, certain steps should be followed. Following these steps can result in significant energy savings. They include:

- Use increased insulation where possible (especially between roof and ceiling surfaces). The increase in the R-value that results is directly proportional to energy savings in terms of heat loss.
- Use the latest design temperatures when calculating required HVAC system capacity.
- Use proper inside design temperatures. These should be at the low end of a person's comfort level in the winter, and at the high end in the summer. Proper relative humidity and indoor design temperature (IDT) will maximize overall comfort.
- Eliminate unnecessary heat leakage around doors and windows with good sealing techniques.
- Use secondary heat sources (motors, machinery, etc.) as much as possible. This is especially true when calculating the required HVAC system capacity.
- Use the most efficient construction materials, such as shaded glass, Thermopanes, and metal foam-insulated doors.

27.5 Construction Types and Designs

Building construction will be a factor when designing or servicing an HVAC system. A building's design may affect:

1. The type of heating or air conditioning system.
2. The air or water distribution network.
3. The HVAC installation.

You should have a basic understanding of the more common construction practices. A typical residence can be classified as a single-floor type (ranch style), a two-floor type (colonial style), or a combination of both (split-level style). Each style may be built over a basement, a crawl space, or a cement slab. Each type of structure requires careful consideration when installing or servicing an HVAC system. The materials used in construction (brick, aluminum, insulation, etc.) must be taken into consideration. They all have significant effects on the heating and cooling of the structure.

You should be able to read construction prints or floor plans. These will provide information useful in making heating or cooling calculations. This would include dimensions, with locations of ductwork and electrical wires. A typical floor plan with some dimensional information is shown in **Figure 27-29.** Note the air conditioner at the center.

The following paragraphs include some specific construction differences and their effects on heat load calculations. Recent designs and new materials have reduced heat losses and gains.

27.5.1 Roof Design and Construction

Heat loss through the roof of a building is influenced by several factors. These include the type of roof construction, ventilation, and covering. The most common residential roof construction is the pitched roof using a prefabricated truss system. See **Figure 27-30.** The slope of a pitched roof is the vertical rise or height of the roof compared to the horizontal run. This is shown in **Figure 27-30.** Slope is given as "__ in 12." For example, a typical ranch style home will have a roof slope of 4 in 12. If you were in the attic and walked a distance of twelve feet in the direction of the roof pitch, the roof height would have increased by four feet. The pitch of a roof is the ratio of the vertical rise to the span (twice the run). It is given as a fraction. For example, if the total roof rise is 4′ and the total span is 24′, the pitch is 1/6 (4/24 = 1/6). The pitch of a roof is determined by the climate and the designer's interior space needs. The more snow that falls, the greater the pitch should be. This minimizes the snow weight at the center. A split-level is more likely to have a higher pitch to allow more headroom on the second story.

Recently, roof pitch has become an aspect of energy conservation as well. For example, the designer must consider such things as the space required for adequate insulation with a cathedral ceiling, or a roof pitch that would allow an efficient solar panel system.

Proper ventilation is also important in the roof enclosed area. Attics are used to vent household air through various ceiling and exhaust fans in the home.

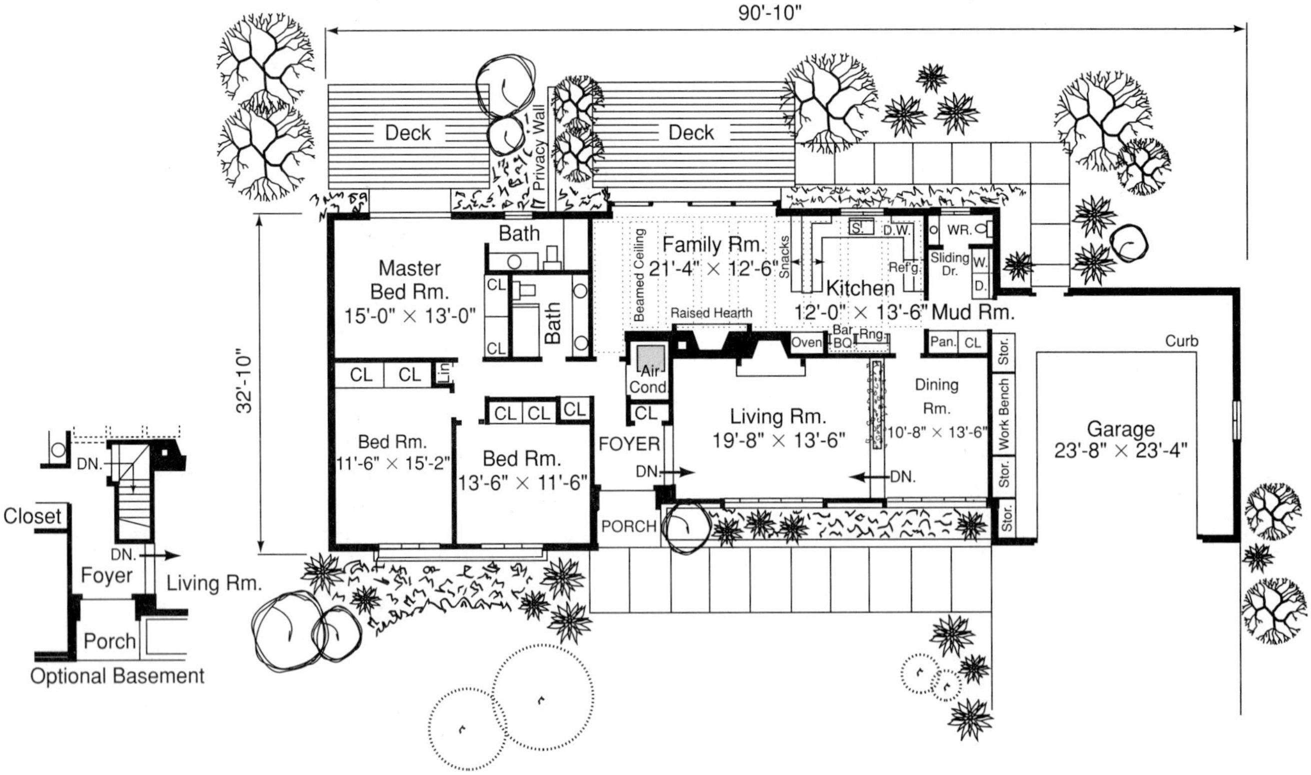

Figure 27-29. *Typical home floor plan shows dimensional information. Note the air conditioner at center. (©Home Planners, Inc.)*

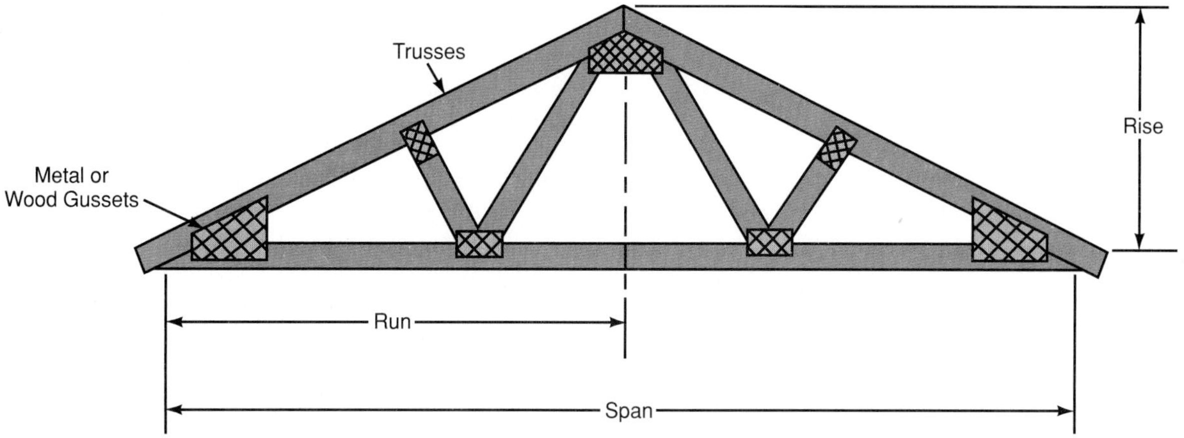

Figure 27-30. *Prefabricated roof truss. This unit is called a Fink truss.*

Most attic areas are vented so that, in summer, cool input air can come from the overhangs. This air picks up heat and is exhausted through roof vents or gable vents. See **Figure 27-31.**

Roof coverings are chosen based on the surrounding climate. Dark asphalt shingles help to absorb the sun's heating effect during winter months. The same feature, however, would work against cooling in the summer.

Roof coverings for pitched roofs are installed over sheathing. Insulation can be used beneath the sheathing. Roof coverings for flat roofs can consist of many layers, as in built-up roofing. Insulation can be a part of the roof covering.

The R-values for pitched roofs are higher than those for flat roofs. Refer to **Figure 27-7.** The design must take this into account, along with the color of the roof covering.

27.5.2 Wall Construction

Building wall construction has changed in the past few years. Heat leakage and moisture passage through the wall structure has been reduced. **Figure 27-32** shows typical brick veneer outside wall construction. During the heating season, the inside vapor barrier is necessary. In summer, the outside vapor barrier is required. If the dew point temperature in winter (based on the indoor relative humidity) is reached between the two vapor barriers in **Figure 27-32,** the location of the dew point must not be at the inside vapor barrier. If it is, condensation will take place on the room side of the inside barrier. If the dew point temperature in summer (based on the outdoor relative humidity) is reached between the two barriers, the dew point must not be at the outside vapor barrier. If it is, condensation will occur on the exterior side of the outside barrier. For all weather conditions,

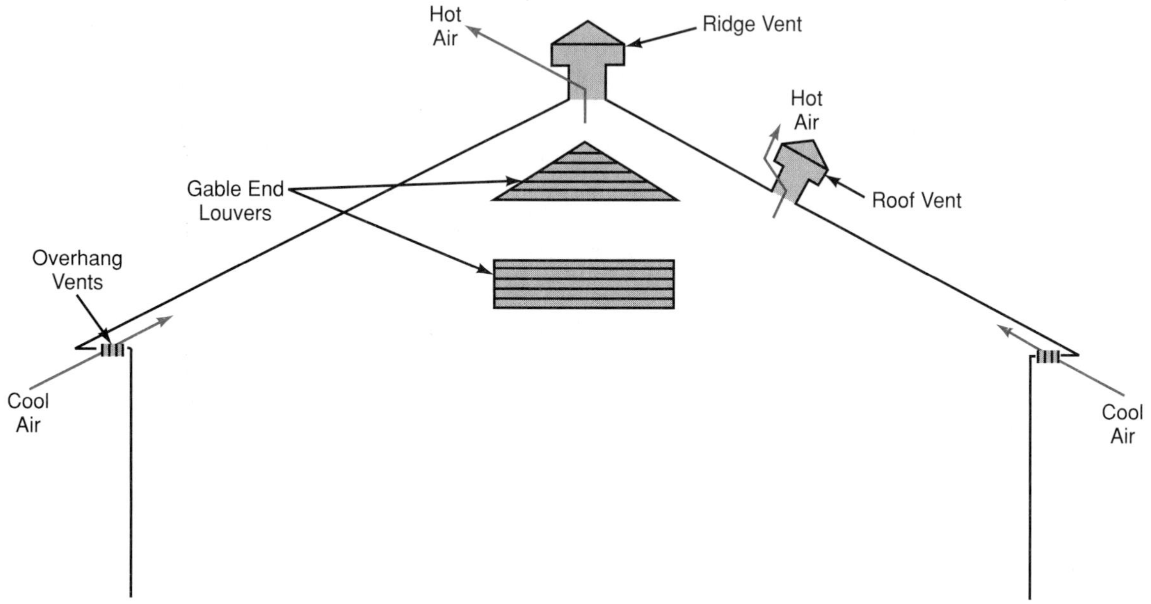

Figure 27-31. *Roof and attic ventilation. If the attic can be kept cool in the summer, less cooling will be needed inside the building. A second use of ventilation is to help keep snow from melting on top of the roof, running down, and freezing at the overhang. Ice dam on overhang can force water under shingles.*

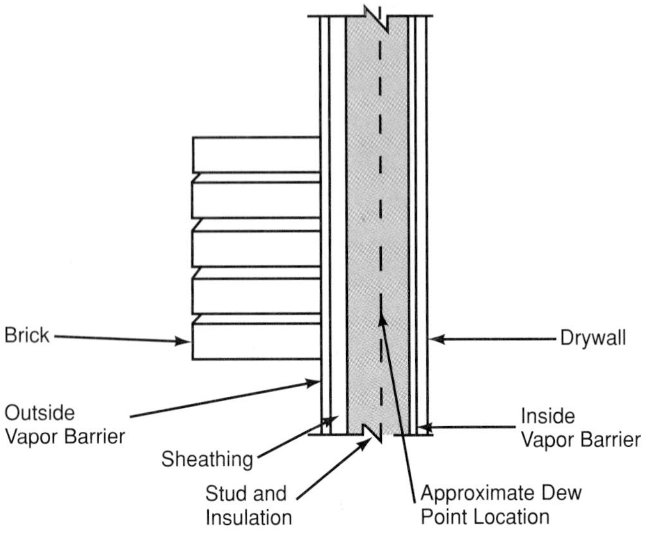

Figure 27-32. *Brick veneer wall construction.*

the dew point location should be halfway between the two barriers. This means the following should be avoided:

1. Trying to get the indoor relative humidity above 21% when the outdoor temperature is −20°F (−29°C) or lower.
2. Trying to get the indoor temperature below 67°F (19°C) when the outdoor conditions are higher than 98°F (37°C) and higher than 97% relative humidity.

In older homes, an approximate 1″ air space separates the brick veneer from the sheathing. Vapor barriers were not commonly used.

Where vapor barriers are used, they should be as tightly sealed as possible. This may even mean tarring the breaks in the seal. The barriers may be made of tarred paper, aluminum foil, or plastic film. Aluminum foil has a reflection value as well as being a vapor-tight seal. Ideally, a hermetically sealed (airtight) insulation should be used in areas where dew point temperatures occur during the heating or cooling season.

Another type of wall construction, more common in warm areas, is the stucco wall. Stucco is a thin coating of cement. It is spread over concrete blocks or a metal mesh and nailed to a wood stud wall. This type of finish can be applied by hand-troweling or machine spray.

A variety of all-wood wall constructions are also found. In most newer construction, all-wood walls are made of some type of hardwood siding. It is fit together over the sheathing. The remainder of the wall is similar to a brick construction. Some of the more decorative constructions include siding made from cedar and redwood.

Aluminum siding is nearly as common as the brick exterior. Aluminum siding is available in a variety of colors, textures, and styles. It is usually installed over the sheathing material. Newer aluminum siding has a layer of insulation bonded to the back side of the aluminum. The aluminum is installed in an overlapping pattern. Holes are provided on the bottom edges of the siding for moisture release.

27.5.3 Commercial Construction

Small commercial buildings use many of the same construction practices and materials as the typical residential buildings discussed in this chapter. Foundations, wall and window construction, and roof design may be similar. These types of buildings house small offices or retail shops.

In many instances, the low-rise shopping center has replaced the individually constructed shop. It is now one of the largest consumers of air conditioning and refrigeration products. These buildings are usually made of concrete block set on a concrete slab. Common walls between businesses may be made of concrete block (each with its own footing) or of a wood-stud construction. Both allow for floor plan changes to suit the tenant. The roof on the low-rise shopping center is usually flat with asphalt weather protection. The water drainage on a flat roof is critical. Care should be taken not to interfere with drainage when working on the roof. After a job is complete, the roof should be inspected. Any damages that may have occurred to the weather seal protection should be noted. Small industrial firms have similar construction.

High-rise buildings and large industrial buildings are much more complex in their construction. In these cases, you should become familiar with the general building design. Study the blueprints prior to doing any major HVAC work involving the building's structure.

27.6 Review of Safety

For safety, a heating system must have enough capacity to heat a structure without becoming taxed or overheated. The heating system should be slightly oversize, never undersize.

All designs of heating and cooling systems should be checked carefully. Be sure that enough fresh air enters the structure. It must provide adequate ventilation for the maximum number of occupants. It must provide for adequate combustion air during the heating season.

In some well-insulated homes, it may be necessary to open a basement window or provide some other air inlet. This enables the furnace and fireplace flues to draw properly—smoke and carbon monoxide will not be released into the building.

27.7 Test Your Knowledge

Please do not write in this text. Place your answers on a separate sheet of paper.

1. _______ causes the majority of heat loss.
   A. Conduction
   B. Exfiltration
   C. Infiltration
   D. All of the above.

2. Which of the following sources of heat are part of the cooling load for a building?
   A. Heat and air leakage.
   B. Ventilation air.
   C. Heat from appliances, lights, and occupants.
   D. All of the above.
3. When computing total heat leakage, what does "U" represent?
   A. Temperature difference.
   B. Overall heat transfer.
   C. Thermal resistance.
   D. Total heat leakage.
4. Outside design temperature (ODT) _______.
   A. varies with latitude and elevation
   B. provides four different values for each location
   C. does not consider the number of windows
   D. All of the above.
5. A humidifier that vaporizes one pound of water per hour will vaporize _______ gallons of water per day.
   A. 24
   B. 12
   C. 2.88
   D. 3.5
6. For which heating calculation(s) are heat gains from lights, motors, appliances, and people not taken into account?
   A. Commercial.
   B. Small buildings.
   C. Industrial.
   D. All of the above.
7. What is exfiltration?
   A. Air that leaks out of a building.
   B. An effect of negative air pressure.
   C. Air that leaks into a building.
   D. Both B and C.
8. _______ usually cause(s) air pressure difference.
   A. Fans
   B. Too many exhaust vents
   C. The wind
   D. All of the above.
9. The _______ the basement goes into the ground, the _______ the heat loss.
   A. deeper, more
   B. deeper, less
   C. less, less
   D. None of the above.
10. Which of the following will reduce condensation on a glass surface?
   A. Reducing the relative humidity.
   B. Raising the temperature of the glass surface.
   C. Adding a vapor barrier between the inside surface of the window and the room air.
   D. All of the above.
11. The _______ wall is not effected by the sun in the northern hemisphere.
   A. east
   B. west
   C. north
   D. south
12. How is the effect of the sun usually included in cooling load calculations?
   A. By adding 20°F (11°C) to the design temperature.
   B. By adding 5°F (3°C) to the design temperature.
   C. By adding 15°F (8°C) to the design temperature.
   D. It is automatically included. There is no need to add to the design temperature.
13. What is "heat lag"?
   A. The thickness of a wall and its insulation.
   B. The time it takes for heat to travel through a substance that is heated on one side.
   C. Adobe walls.
   D. It causes rooms to chill after the sun goes below the horizon.
14. What is air film?
   A. A thin layer of air and moisture on glass surfaces.
   B. A thin layer of still air that clings to all surfaces.
   C. A thin layer of still air that clings to warm surfaces.
   D. None of the above.
15. If the outside temperature is 20°F (−7°C) for 24 hours, how many degree-days accumulate?
   A. Thirty.
   B. Twenty-five.
   C. Forty-five.
   D. None of the above.
16. What is the design wind velocity used in computing heat loads?
   A. 5 mph.
   B. 10 mph.
   C. 20 mph.
   D. 15 mph.
17. Which dimensions are used when determining wall area?
   A. Outside dimensions.
   B. Inside dimensions.
   C. A combination of outside and inside dimensions.
   D. It depends upon the type of construction being evaluated.
18. Which has the greatest heat flow resistance, inside air film or outside air film?
   A. Inside.
   B. Outside.
   C. They are equal.
   D. It depends upon humidity and ambient temperature.
19. The degree-day method uses a _______ indoor temperature as a standard.
   A. 50°F (10°C)
   B. 65°F (18°C)
   C. 70°F (21°C)
   D. 72°F (22°C)
20. A typical ranch-type home will have a roof slope of _______ in _______.
   A. 2, 12
   B. 8, 16
   C. 12, 6
   D. 4, 12

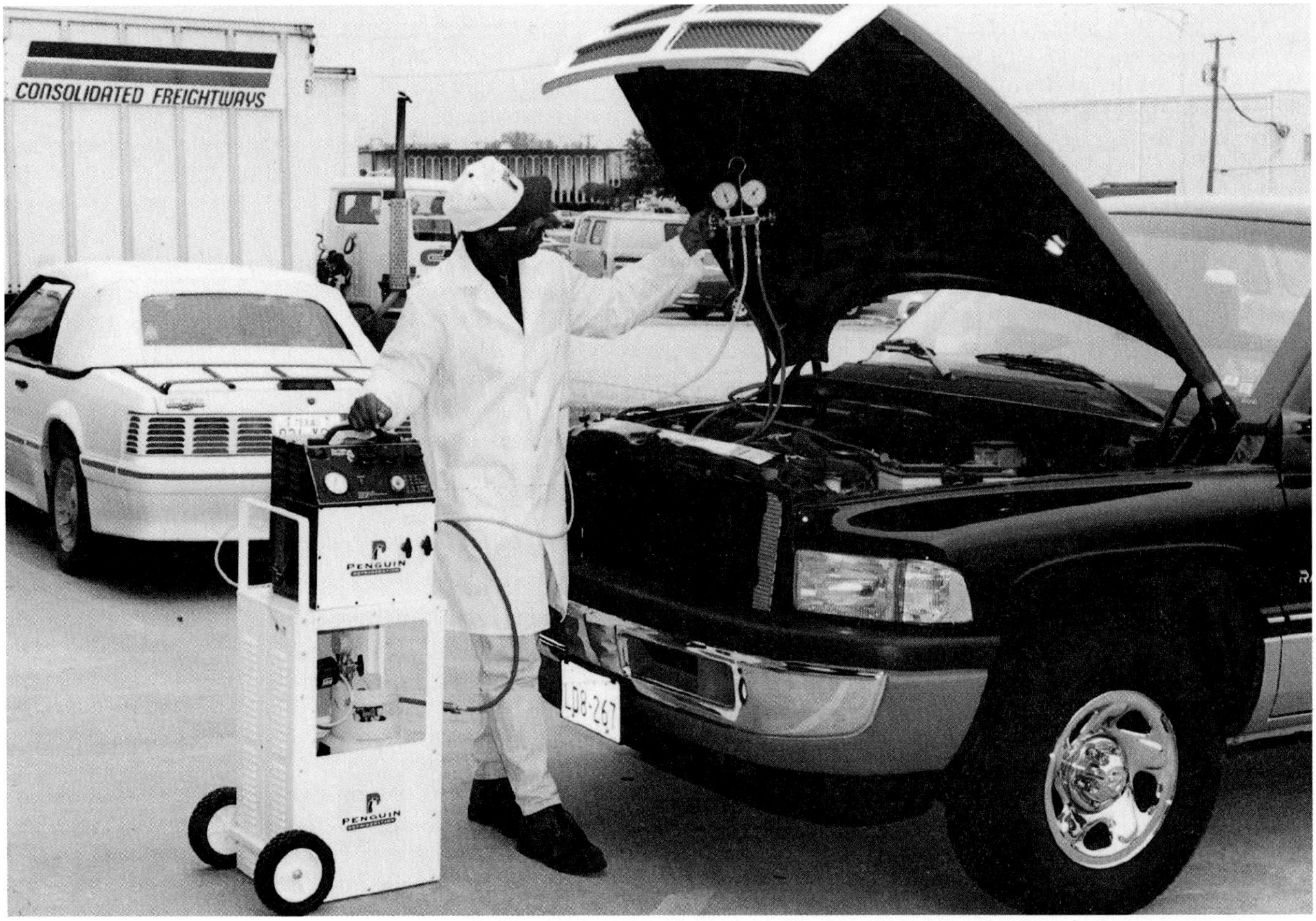

A technician servicing a late-model vehicle. (Penguin Refrigeration, Inc.)

Chapter 28

AUTOMOTIVE AIR CONDITIONING

Key Words:

accumulator
automatic climate control
HVAC controller
magnetic clutch
pressure cycling switch
servomotor
swash-plate compressor
thermostatic switch

Learning Objectives:

After studying this chapter, you will be able to:

- Explain how automotive air conditioning varies in design and application from stationary systems.
- List and describe the types of automotive compressors.
- Have a better understanding of service and repair of automotive air conditioning units.
- **Follow approved safety procedures.**

Many of the principles discussed in earlier chapters also apply to automotive air conditioning. There are many unique applications and slight variations in design that must be understood.

The automobile air conditioner uses a refrigerant system driven by the vehicle's engine to furnish cooling. Hot engine coolant usually provides heat to the passenger compartment when needed.

To fully understand this chapter, study Chapters 1, 2, 4, 5, 6, 8, 9, 15, 19, and 20 first. This will provide a solid foundation in fundamentals and ensure a better grasp of the principles of automotive air conditioning systems.

Conditioning the air in the interior of a vehicle involves heating, cooling, and dehumidification. Heat is usually provided by circulating hot coolant, from the vehicle engine. It is circulated through a ***heater core*** (small radiator-like device under the dashboard). The engine water pump forces hot coolant through heater hoses in the heater core.

When cooling is needed, the refrigerant system is brought into operation. An evaporator inside a plenum chamber cools the air circulated through the passenger compartment. (The plenum chamber is an enclosure under the dashboard.) Using a belt, the vehicle's engine drives the compressor to pump refrigerant through the system.

28.1 Purpose of Automotive Air Conditioning

An automobile passenger compartment is relatively small. Yet, a vehicle traveling at high speed on a hot day requires considerable refrigerating capacity in order to keep the interior at a comfortable temperature. Likewise, the same vehicle, traveling on a cold winter day, will require considerable heating capacity.

Operating the air conditioner also reduces the humidity of the air inside the vehicle. In addition, moisture (condensate) formed on evaporator surfaces collects dust and pollen in the air. These entrapped particles are carried away by the condensate. The condensate drains off the evaporator and outside of the vehicle. Most automatically controlled vehicle air conditioning systems operate the refrigerant system for all outside temperatures above 45°F (7°C). This provides low humidity air to aid in defrosting the vehicle's windows.

28.2 Automotive Air Conditioner Operation

The basic components of the automotive air conditioner are the same as other air conditioners. These include a compressor, condenser, refrigerant control and evaporator. The difference is in the operation and servicing of the system. **Figure 28-1** illustrates a technician servicing an automobile. The latest EPA regulations require certification for persons who repair or service automotive air conditioners. They must be able to properly use approved equipment.

A typical automotive air conditioning system for automobiles is shown in **Figure 28-2.** The belt-driven

Figure 28-1. *A technician servicing an automobile, using equipment approved by the EPA. (Robinair Division, SPX Corporation)*

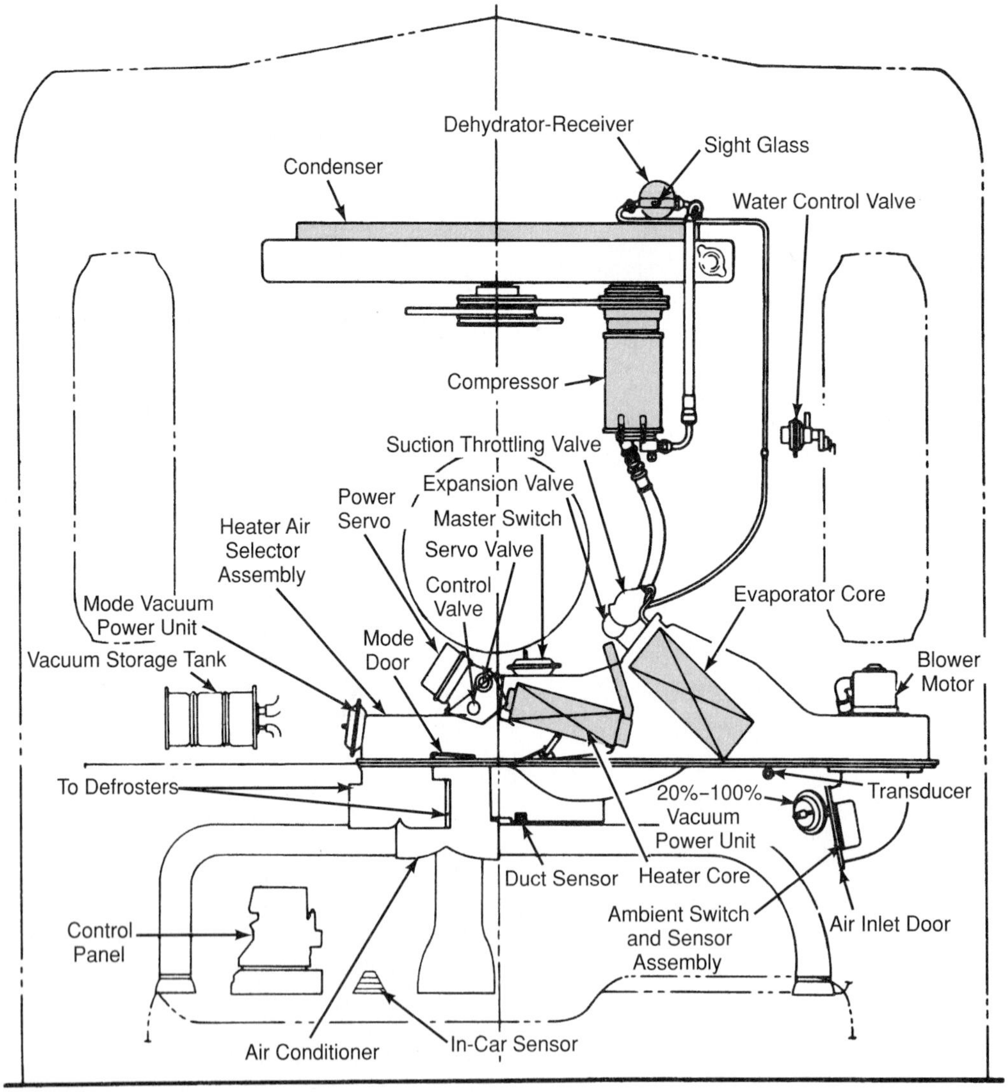

Figure 28-2. *Top view of year-round, all-season air conditioning system showing main components. (Cadillac Motor Car Div., General Motors Corp.)*

compressor is mounted on the engine, ahead of the vehicle radiator. This allows cool air to flow over the condenser. The evaporator is mounted inside the plenum chamber in the passenger compartment. All of these devices are connected together by lines and hoses.

Liquid refrigerant flows from the condenser to the liquid receiver. The refrigerant is dried and filtered. Then it flows through a control device and into the evaporator. In the evaporator, the refrigerant is vaporized and absorbs heat. The vaporized refrigerant finally flows back through the suction line to the compressor.

Low-pressure refrigerant vapor enters the compressor through the low side. The vapor is drawn into the cylinder and is compressed by the piston. It is then discharged through the high side back to the condenser. The heat of compression and the latent heat of vaporization absorbed by the refrigerant are given up to the air flowing over the condenser fins. At this point, the entire process repeats itself.

Meanwhile, a blower (fan) forces air through the evaporator. The resulting cool air is circulated to the vehicle's interior through air distribution ducts and grilles.

28.2.1 Cooling Capacity

Automobile air conditioning systems range in size and cooling capacity. Their output range is similar to the one- to four-ton residential or commercial units. A cooling capacity of 12,000 Btu/hr. (one ton) is minimum. Capacities up to 48,000 Btu/hr. are available for special applications—station wagons or vans, for example.

The capacity of the air conditioning system should match vehicle size. Undercapacity will result in inadequate cooling in hot weather. Overcapacity is uneconomical and causes frequent cycling of the system. Systems are usually designed to keep inside temperature 15°F to 20°F (8°C to 11°C) below outside (ambient) temperature with the vehicle traveling about 30 mph (48 km/hr.).

Figure 28-3 shows how horsepower required for air conditioning system operation varies as vehicle speed changes. As the vehicle speeds up, the capacity of the compressor will increase. As it slows down, the capacity will decrease. This variation in output is somewhat parallel to the changing heat load, except when the vehicle is parked or in slow moving traffic. One solution is a variable displacement compressor, which is covered in Section 28.3.2. At these critical times, compressor capacity may be below normal. A partial solution may be to idle the engine at a higher speed or use a lower transmission gear to obtain higher engine speeds.

Larger air conditioning systems can consume as much as 8 hp (6 kW) from the engine at high speeds. Capacity at this speed will be approximately 48,000 Btu/hr. This means that 2 hp (1.5 kW) is used for each ton of refrigeration. Compare this to 1 hp (0.75 kW) for each ton of refrigeration in a motor-driven, constant-speed compressor comparably built, with the evaporator and condenser more ideally located.

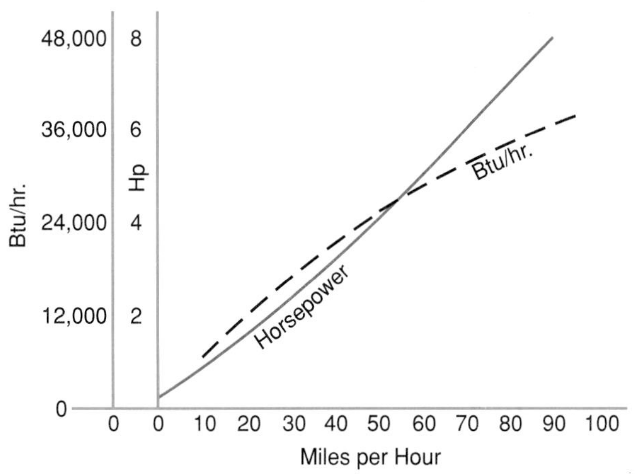

Figure 28-3. *Curves show relationship between car speed, heat load, and horsepower required to drive air conditioning system.*

28.3 Operating Conditions

The automobile air conditioner must provide comfort in vehicles in all weather. This control must be adequate during cold, mild, damp, and hot weather. It must provide heating, defogging, and de-icing and must also remove dust, smoke, and odor.

The compressor is belt-driven by the engine. The compressor speed will vary with engine speed. The system must have enough capacity to provide sufficient cooling. It must function at idling speed, in the sun, and in the wind. There must be considerable excess capacity for normal driving speeds.

Varying weather conditions can cause problems with the control of the temperature. Refrigerant flow problems may occur (both liquid and vapor) within the system. With the compressor operating and little or no refrigeration needed, the low-side pressure may drop too low.

Decreasing the low-side pressure lowers the evaporator temperature. The evaporator surface temperature should not be allowed to drop below 33°F (0.5°C). The evaporator surface may become covered with ice if it operates at 32°F (0°C) or lower for any length of time. This will stop air circulation through the evaporator. Cool air cannot enter the passenger compartment.

Also, operating the system with low-side pressure too low may cause oil pumping. This condition may damage the compressor valves and, if continued, may burn out the compressor.

A typical automotive air conditioning system will cool an automobile from 110°F (43°C) to 85°F (29°C) in about 10 minutes. The vehicle's interior may reach 150°F (66°C) when parked in the sun with the windows closed. The greatest heat load or heat gain is the sun load and heat conducted through car windows. See **Figure 28-4.**

Automotive air conditioning systems use anywhere from no fresh air (all recirculated) to 100% fresh air. When outside, or "fresh air," ducts are closed off, the

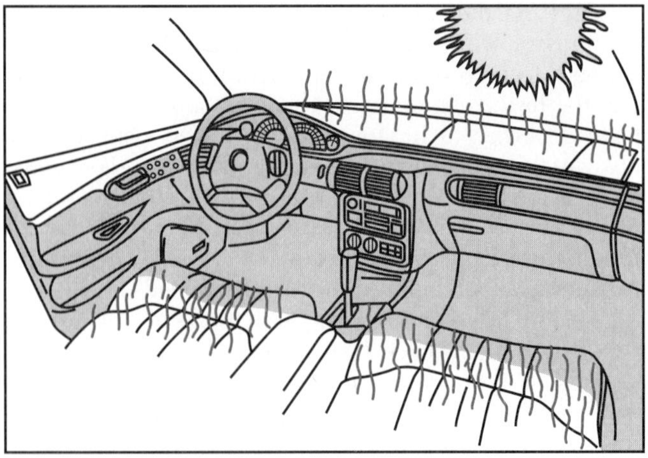

Figure 28-4. *Heat from the sun enters the car. (Chrysler Corporation)*

vehicle's inside air is circulated back through the evaporator. The same cool air is continually brought back through the evaporator plenum. This is known as using ***recirculated air.*** The recirculated air has previously been cooled by the evaporator. Therefore, it is possible to obtain colder interior temperatures than when using outside air. Remember, fresh air ducts must be closed for maximum cooling. The fan blower uses approximately 200 watts or 15 amps of power. It delivers from 250 cfm to 325 cfm of air.

Operation of an air conditioning system may reduce fuel economy by as much as 10%. This is primarily due to the power needed to turn the compressor shaft.

28.3.1 Magnetic Clutch

The ***magnetic clutch*** permits the engine to run without running the compressor. This will allow disengagement of the compressor when the system is not being used and when overcapacity could cause icing of the evaporator. The clutch engages or disengages the compressor belt drive pulley and the compressor shaft. The clutch is operated by forcing a clutch disk, mounted to the compressor shaft, against the belt pulley, using electromagnetism. **Figure 28-5** shows the construction of a magnetic clutch.

Two basic designs of magnetic clutches are used:

- A revolving magnetic coil that turns when the compressor revolves. It has two carbon brushes in contact with two copper rings mounted on the coil.
- The more common type, a stationary magnetic coil mounted on the compressor body. It has two electrical leads. **Figure 28-6** shows the design of one type of stationary coil electromagnetic clutch.

Disengaging the magnetic field will cause the belt pulley to "free-wheel" on its bearings. The compressor will stop.

There are several makes of electromagnetic clutches on the market. The electric current needed to magnetize the various clutches is 2 amps to 3 amps at normal voltage.

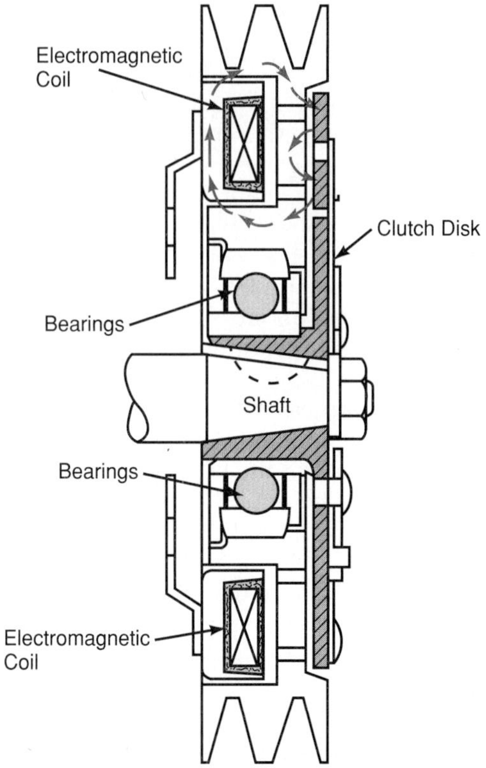

Figure 28-5. *Compressor pulley with stationary electromagnet, showing magnetic field.*

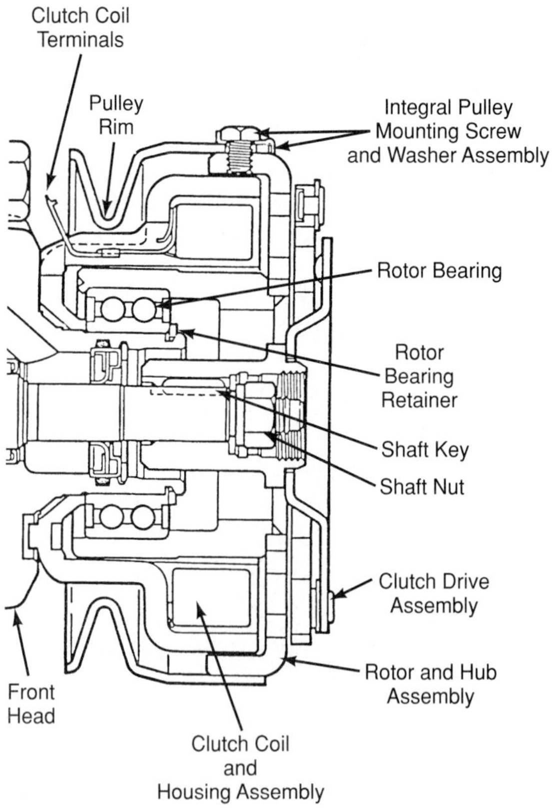

Figure 28-6. *Cutaway view of magnetic clutch. When stationary coil is energized, armature revolves with pulley and turns compressor shaft. (Pontiac Motor Div., General Motors Corp.)*

28.3.2 Compressors

The operation of the automobile air conditioning compressor is similar to those in nonautomotive applications. Its function is to compress low-pressure, low-temperature refrigerant vapor into a high-pressure, high-temperature vapor.

There are four basic types of compressors in general use in automotive air conditioning:

- Two-cylinder reciprocating compressors.
- Swash-plate compressors.
- Scotch yoke compressors.
- Scroll compressors.

Most automotive compressors are semihermetic. This type can be internally serviced.

Two-Cylinder Reciprocating Compressor

Conventional reciprocating compressors used for add-on units in automotive air conditioning usually contain two pistons. They are in a parallel V-type configuration.

The two-cylinder compressor is usually constructed of die cast aluminum. See **Figure 28-7.** The pistons are attached to a connecting rod that is driven by the crankshaft. The crankshaft is connected to the compressor clutch assembly driven by an engine belt.

The two-cylinder compressor was widely used in past automotive applications. However, it has given way to more efficient types of compressors.

Most conventional compressors contain high- and low-side service fittings. These fittings have a Schrader valve for attaching gauges and servicing the system. See Section 4.5.2 and Section 4.6.1.

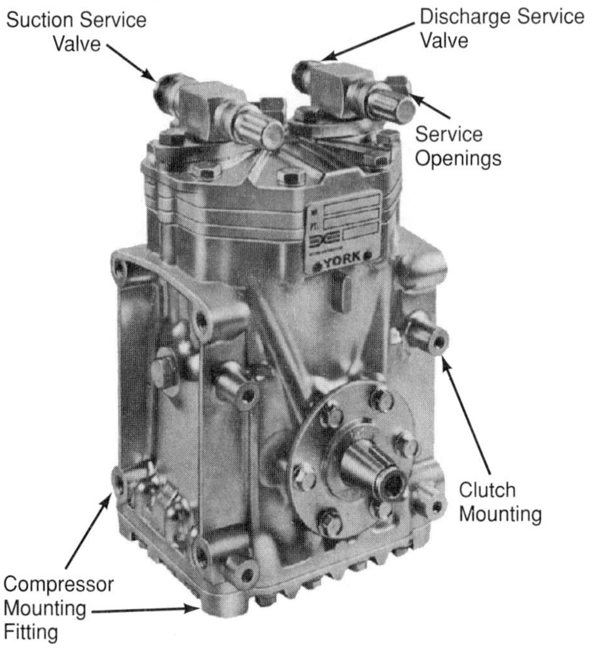

Figure 28-7. *A two-cylinder reciprocating compressor. The compressor body has a variety of mounting holes to permit vertical, horizontal, or inclined mounting. Service valves are flange-mounted, using two cap screws.*

Swash-Plate Compressor

In the ***swash-plate compressor*** (also known as wobble plate), piston motion is parallel to the crankshaft. The pistons are connected to an angled swash plate using ball joints.

There are five models of swash-plate compressors:

- Five-cylinder.
- Six-cylinder.
- Ten-cylinder.
- Five-cylinder variable displacement.
- Seven-cylinder variable displacement.

Each cylinder has a set of reed valves for intake and exhaust. The cylinders utilize connecting passages to form one common high-side port and one low-side port. These ports are located at the rear of the compressor. Also located at the rear of the compressor is a high-side pressure relief valve which prevents possible rupture of other system components.

The rotary movement in the swash-plate compressor is changed to a reciprocating movement. Therefore, the movement of the piston is accomplished by the swash plate and shaft. See **Figure 28-8.** A number of paired pistons are set in an interval of 72°. When one side of a piston is in the compression stroke, the other is in a suction stroke.

A six-cylinder swash-plate compressor is shown in **Figure 28-9.** It consists of three sets of opposing cylinders and three double-acting horizontal pistons. An axial plate, pressed to the shaft, drives the pistons. **Figure 28-10** shows the parts of a six-cylinder fixed displacement compressor.

The ten-cylinder fixed displacement swash-plate compressors are lighter and smaller than the five- or six-cylinder models. **Figure 28-11** illustrates the construction of the 10-cylinder swash-plate compressor.

A five-cylinder variable displacement compressor is shown in **Figure 28-12.** The variable displacement

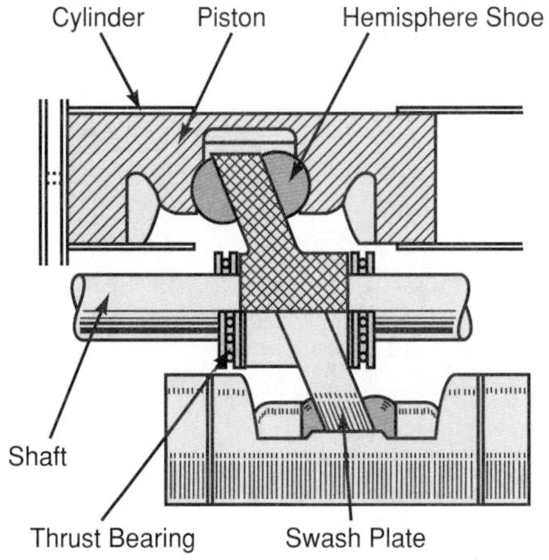

Figure 28-8. *A swash-plate compressor. The cylinder and piston are parallel to the shaft. (Toyota Motor Corporation)*

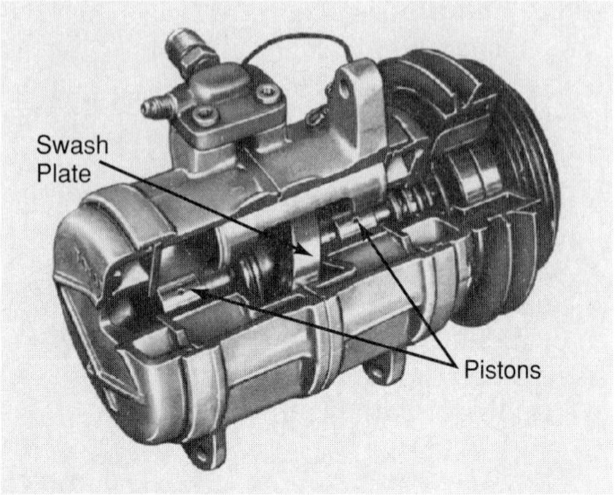

Figure 28-9. *Six-cylinder, swash-plate compressor. (Chrysler Corporation)*

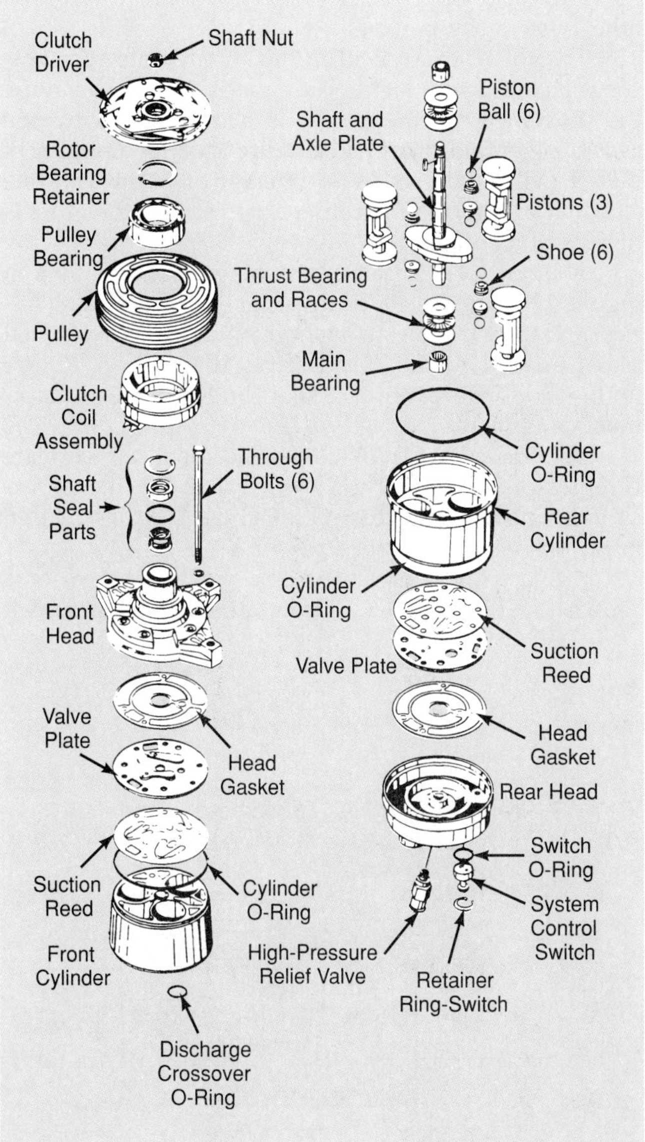

Figure 28-10. *Internal parts of swash-plate compressor. Unit has six cylinders. (Cadillac Motor Car Div., General Motors Corp.)*

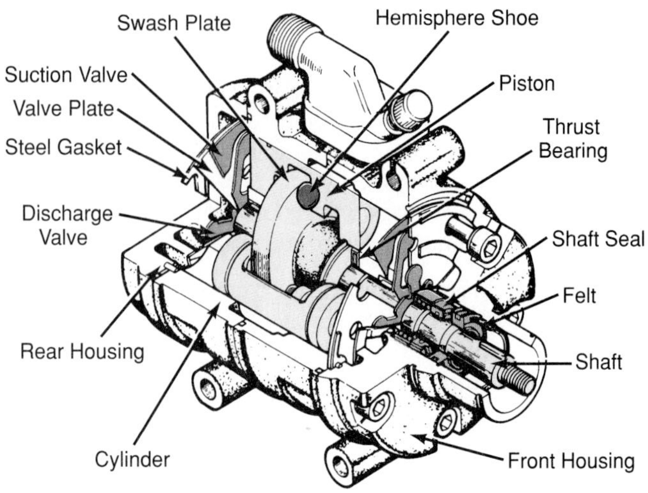

Figure 28-11. *Ten-piston swash-plate compressor. (Toyota Motor Corporation)*

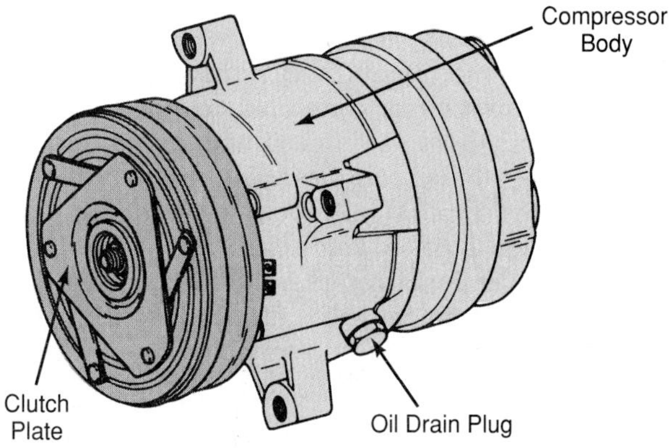

Figure 28-12. *Five-cylinder, variable displacement swash-plate compressor. Note location of oil drain plug and clutch. (Cadillac Motor Car Div., General Motors Corp.)*

swash-plate compressor differs from other swash-plate compressors in that it uses a plate connected to a hinge pin. The hinge pin allows the compressor to vary the swash plate angle. The swash plate angle is controlled by a bellows valve that senses suction pressure. This valve is known as the "control valve." During high-load conditions, the swash plate angle is large. The compressor displacement is maximum. During low-load conditions, the swash plate angle is smaller and compressor displacement is reduced.

The five-cylinder variable displacement compressor has a low-pressure safety cutoff switch at the rear of the compressor. This switch is designed to disengage the clutch during low refrigerant conditions. The five-cylinder variable displacement compressor is designed to compensate for engine power and air temperature variations. The efficiency is thereby improved over that of a fixed displacement compressor. **Figure 28-13** shows the parts of the compressor.

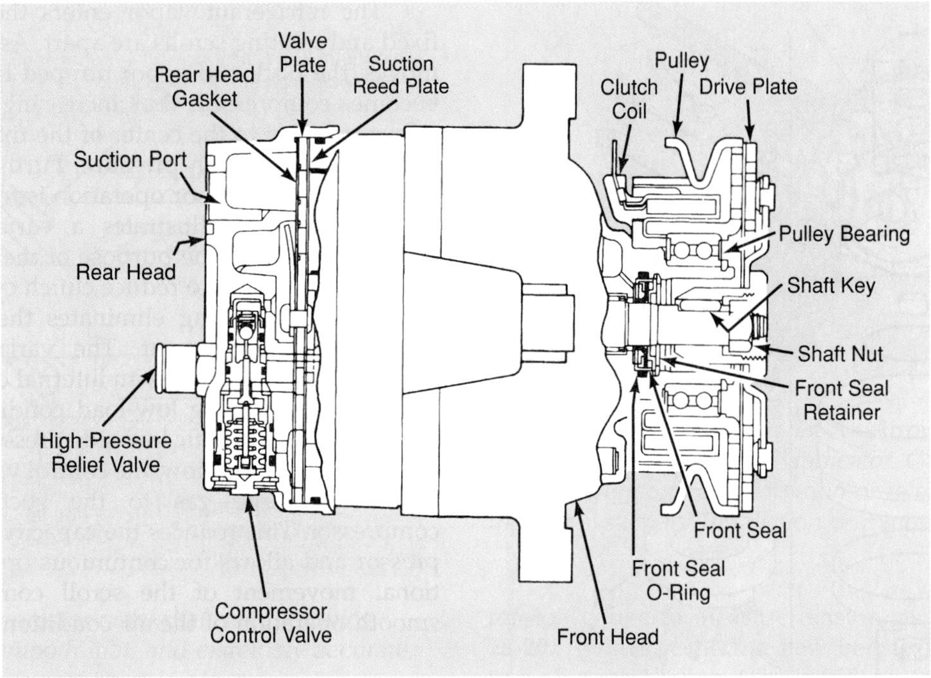

Figure 28-13. *Internal parts of five-cylinder, variable displacement swash-plate compressor. (Cadillac Motor Car Div., General Motors Corp.)*

Scotch Yoke Compressor

A four-cylinder radial compressor of a modified scotch yoke design changes rotary motion into reciprocating motion. The radial compressor is shorter in length and larger in diameter than most other compressors. The basic mechanism of a modified scotch yoke assembly contains four pistons. They are mounted 90° from each other. This is shown in **Figure 28-14.**

Opposed pistons are pressed into a yoke that rides on a slide block located on the shaft eccentric. Rotation of the shaft provides reciprocating motion with no connecting rods, **Figure 28-15.**

Refrigerant flows into the crankcase from the rear. It is drained through the reeds attached to the piston top during the suction stroke. During compression, refrigerant is discharged through the valve plate. It flows out of the connector block at the rear. **Figure 28-16** illustrates the internal part of a typical radial compressor.

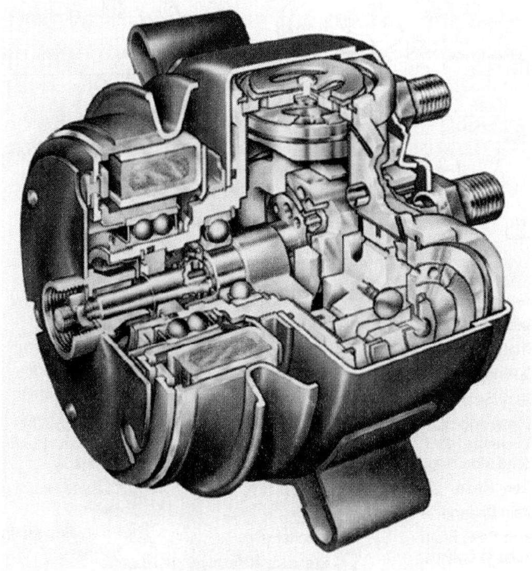

Figure 28-14. *Four-cylinder, Scotch yoke radial compressor can be mounted in any position around the crankshaft.*

Scroll Compressor

One type of compressor which has recently come into use in automotive air conditioning is the scroll com-

Figure 28-15. *Four-cylinder, Scotch yoke mechanism. Note rotation of shaft upon slide block to produce compression of refrigerant.*

On a system with an automatic tensioner, the tensioner will usually have some type of indicator marking. This will show if the belt has worn or stretched to the point that proper tension can no longer be maintained.

The belt may also power several other components, or all belt-driven components on serpentine belt applications. These could include the water pump, fan, and alternator.

28.3.5 Condensers

The condenser is mounted in front of the engine radiator. It is usually constructed of copper or aluminum with multiple rows of finned tubes. See **Figure 28-21.** Heat moves from the condenser into the airflow set up by the moving vehicle or by an electric or engine-driven fan. This heat removal causes vaporized refrigerant to change to liquid as it passes through the condenser.

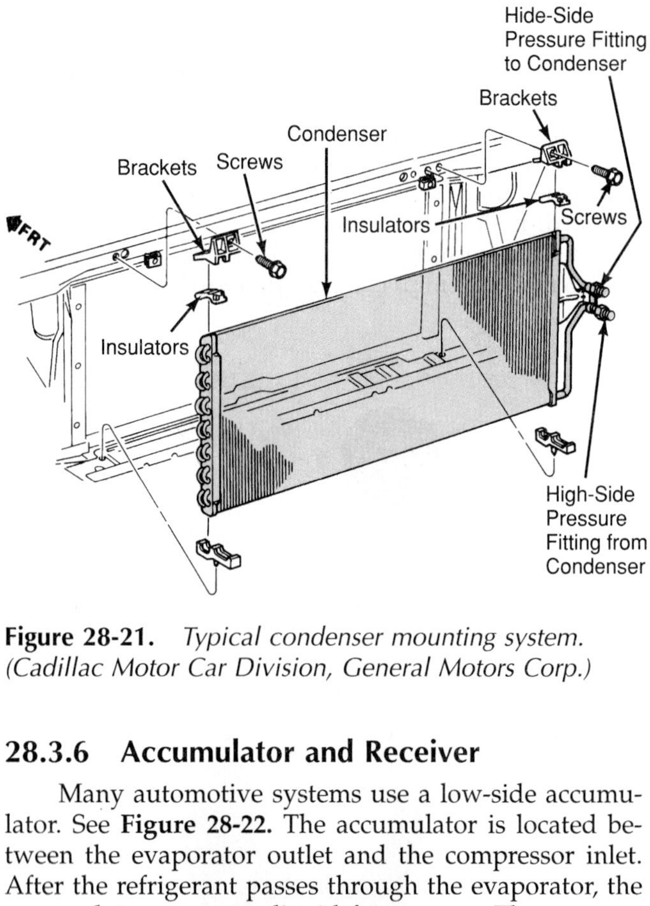

Figure 28-21. *Typical condenser mounting system. (Cadillac Motor Car Division, General Motors Corp.)*

28.3.6 Accumulator and Receiver

Many automotive systems use a low-side accumulator. See **Figure 28-22.** The accumulator is located between the evaporator outlet and the compressor inlet. After the refrigerant passes through the evaporator, the ***accumulator*** separates liquid from vapor. The accumulator retains the liquid and releases mostly vapor to the compressor. This ensures that liquid refrigerant will not enter and damage the compressor.

The accumulator is usually constructed of aluminum. It includes a desiccant (moisture-absorbing material) to remove any contaminants from the system. Accumulators may also have a service valve for manifold gauge hook-up. Most accumulators are sealed and

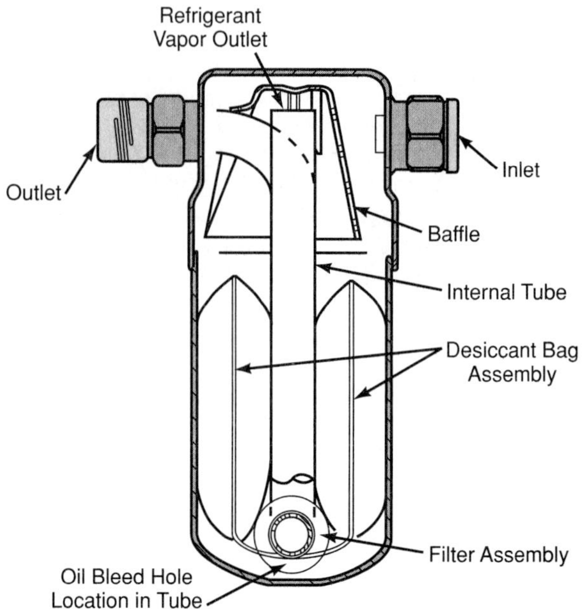

Figure 28-22. *Low-side accumulator. Note inlet and outlet. (Cadillac Motor Car Div., General Motors Corp.)*

must be replaced if defective. They must also be replaced if exposed to excessive contaminants or moisture. Many companies also utilize a suction accumulator drier.

Some automotive air conditioning systems have a receiver or receiver-drier located between the condenser and evaporator. The receiver also allows for some changes in refrigerant charge and liquid volume. (These changes are caused by expansion and contraction of refrigerant as temperature changes.) The receiver serves as a storage container for liquid refrigerant that enters from the condenser. It also helps to filter out contaminants in the system. Some receivers may contain a liquid indicator or sight glass. This helps in determining correct refrigerant charge. A receiver-drier is shown in **Figure 28-23.**

28.3.7 Evaporator and Heater Core

The evaporator is often mounted next to the heater core assembly. Both are enclosed in a plenum chamber. This allows forced air from the blower or fan motor to be directed through them.

Evaporators are used to both cool and dehumidify the passenger compartment air. They are usually constructed of aluminum with several rows of tubes and attached fins. Construction is similar to a condenser but much thicker to provide a more compact design. **Figure 28-24** shows an evaporator, blower, and duct installation mounted under the hood and dashboard.

In the evaporator, low-pressure liquid changes to low-pressure vapor. This change of state absorbs heat and cools the passenger compartment. Many evaporator assemblies include an accumulator for storage of liquid refrigerant.

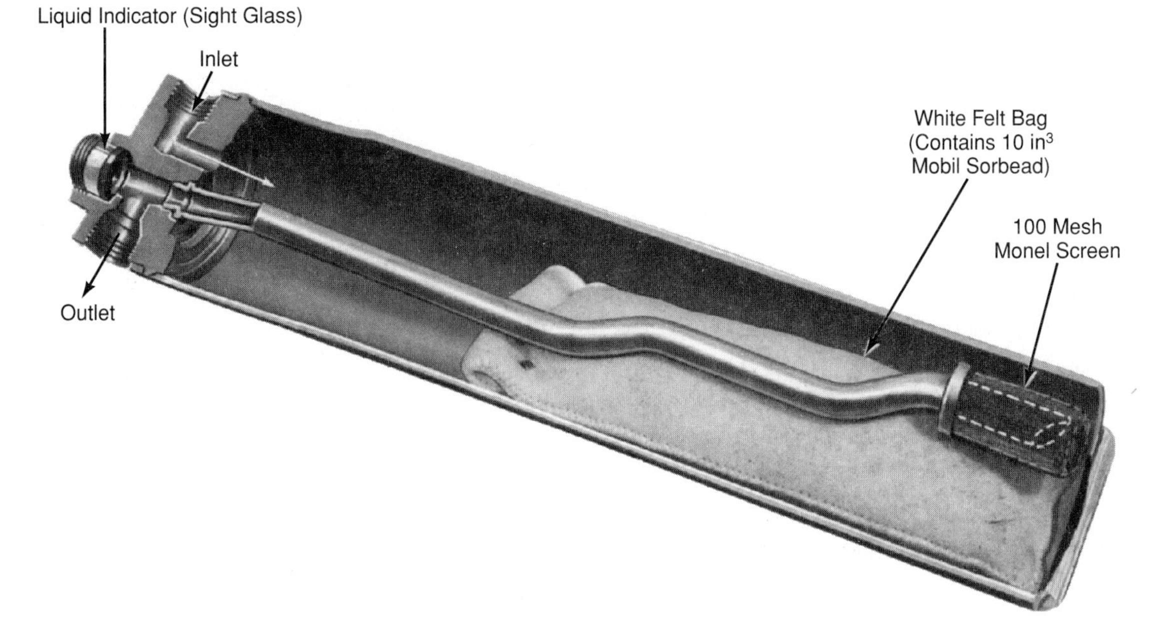

Figure 28-23. *Cross section of combination liquid-receiver, filter, drier, and sight glass. Drier material is in felt bag.*

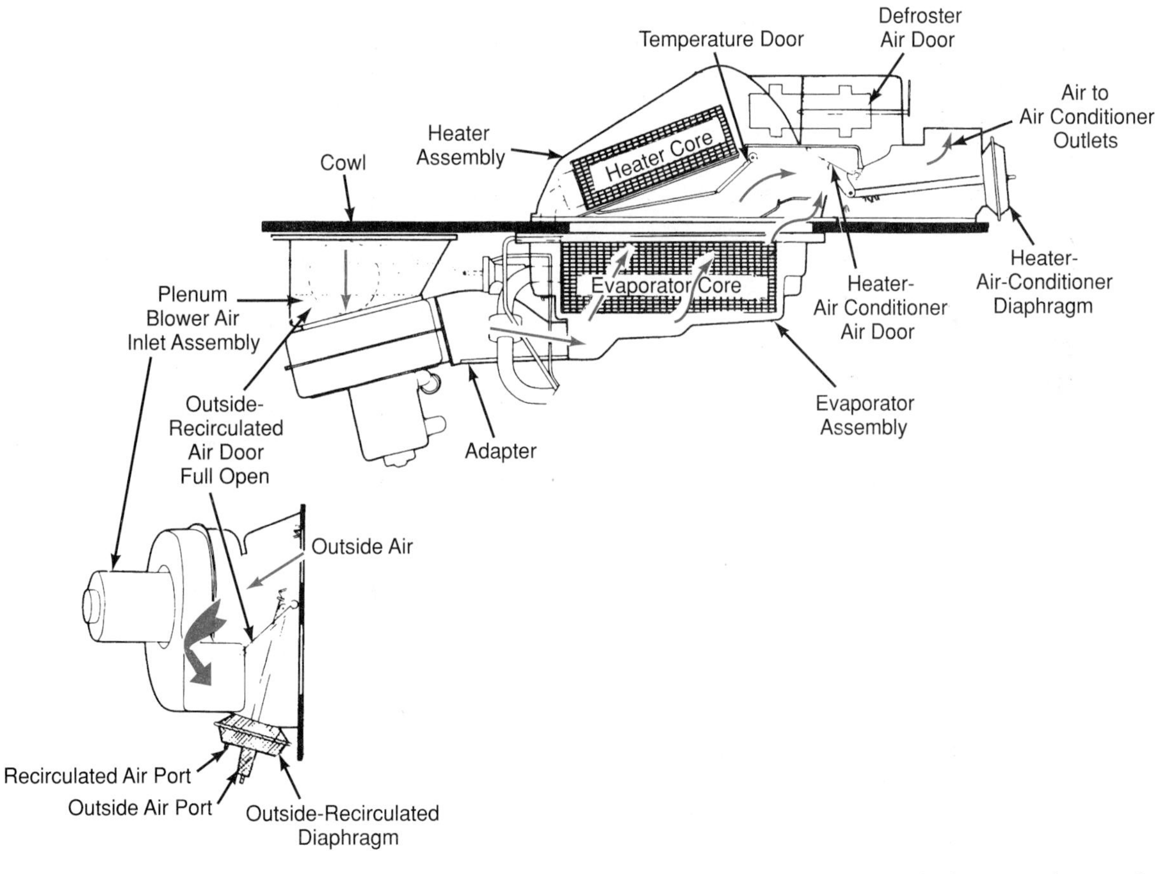

Figure 28-24. *Typical installation showing arrangement of blower, evaporator, heater coil, ducts, and controls. Heater core is under dashboard. Evaporator is under hood.*

The heater core is usually constructed of copper or aluminum. It is similar to a small engine radiator. It is mounted inside the blower housing. The heater core is attached to the engine through inlet and outlet hoses (heater hoses). An ethylene-glycol and water mixture is circulated by the engine water pump to the heater core. It warms the passenger compartment. The same blower fan is used for both evaporator and heater core.

28.3.8 Refrigerant Lines

Special flexible refrigerant lines are used in automotive air conditioning applications. They are used for several purposes:

- To carry refrigerant in its various states to different components throughout the system.
- To allow flex points from moving components to fixed components.
- To help dampen noise and vibration transfer throughout the system.

Flexible refrigerant lines are also called hoses. They are commonly covered with a foam or convoluted plastic sleeve. This protects them from abrasion due to engine vibration. These hoses are designed to be flexible and vibration proof.

Refrigerant lines are made of steel, copper, and aluminum. Flexible lines are generally made of a braided rubber hose that is compatible with the refrigerant being used. Refrigerant lines should be carefully routed to prevent them from rubbing against any part of the vehicle. Wear and corrosion would quickly cause leaks at the points of contact.

Refrigerant lines use fittings to allow for servicing. They are fastened to the system parts in various ways:

- Flared fitting.
- Single or dual O-ring fitting.
- Hose clamp fitting.
- Spring lock fitting.

The first three fittings are shown in **Figure 28-25.** A spring-lock coupling is shown in **Figure 28-26.** It is held together by a garter spring inside a circular cage. When the coupling is connected, the flared end of the female fitting slips behind the garter spring inside the cage of the male fitting. The garter spring and cage prevent the flared end of the female fitting from pulling out of the cage.

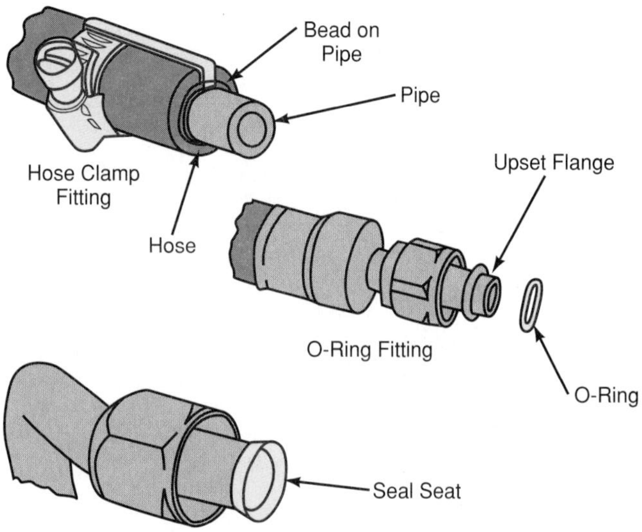

Figure 28-25. *Typical refrigerant line connection fittings are hose clamp, O-ring, and flare. (Harrison Radiator Div., General Motors Corp.*

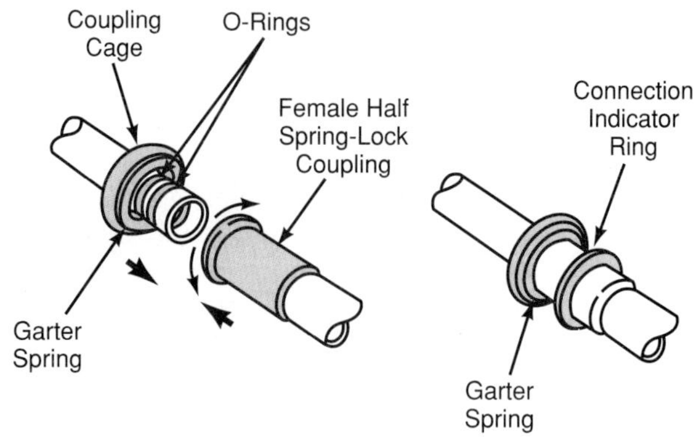

Figure 28-26. *Spring-lock coupling connection.*

The hoses vary in size, depending on the unit capacity and the state of the refrigerant. Vapor-carrying lines are larger than liquid lines. The size of the lines must match the fittings supplied by the manufacturer so that the system capacity will not be reduced.

Refrigerant lines should have large bends to prevent restriction of flow. They should also be supported, grommeted, and clamped. This will prevent wear from chafing and will prevent them from touching hot engine parts. Enough slack in the hose should be allowed at the compressor end to provide for movement of the engine on its motor (vibration absorber) mounts.

New lines come equipped with caps and plugs to keep the inside of the lines clean and dry. These caps and plugs should not be removed until just before installation to prevent contamination. Most refrigerant joints require lubrication prior to sealing. Refer to the vehicle manufacturer's specification before using a joint lubricant. Many manufacturers recommend the use of mineral oil for R-12 system joint lubrication. Only approved PAG (Polyalkylene Glycol) oil is recommended for R-134a systems. This lubrication will help provide for a leakproof connection.

28.3.9 Metering Devices

Automotive air conditioning metering devices are similar in operation to other refrigeration and air conditioning metering devices. They must allow high-pressure, high-temperature liquid refrigerant to pass through an orifice and then rapidly expand in volume. This expansion results in lower refrigerant pressure. The refrigerant begins to boil, thus absorbing heat.

There are three types of common metering devices used in automotive air conditioning:

- Thermostatic expansion valve (TEV).
- H-type expansion valve.
- Orifice (expansion tube).

Thermostatic expansion valves are used to throttle the amount of refrigerant entering the evaporator. TEVs operate on the temperature of the evaporator outlet and the low-side pressure. Chapter 5 explains the principle, installation, and service of this type of valve.

Many automotive TEVs have a pressure equalizer connection. There are two types of TEV pressure equalizers—externally-equalized and internally-equalized. The only difference is that the external line is connected to an evaporator pressure control valve or to the evaporator outlet line. This provides a means of sensing evaporator outlet pressure. The internal equalizer senses evaporator pressure internally through the equalizer pressure passage. See **Figure 28-27.**

The sensing bulb should be tightly-clamped at the outlet of the evaporator. It must be well insulated from engine heat.

H-type expansion valve operation is similar to that of the thermostatic expansion valve. Many H-type expansion valves contain a low-pressure cutoff switch in the base of the unit. Care must be taken not to scratch gasket mating surfaces when servicing the H-valve. **Figure 28-28** illustrates an H-type valve, a cycling clutch switch, and a low-pressure cutoff switch.

The expansion tube orifice can be located anywhere between the condenser outlet and the evaporator inlet. It is used in place of the thermostatic expansion valve or H-valve. The orifice tube is the dividing line between the system's high- and low-pressure sections. However, its operation does not depend on comparison of evaporator pressure and temperature. It is a fixed orifice. Its flow rate is determined by the pressure difference across the orifice. Cycling the compressor on and off, and varying the compressor displacement, controls flow.

Figure 28-29 illustrates a system using an orifice expansion tube. It is a common type of system because of its simplicity.

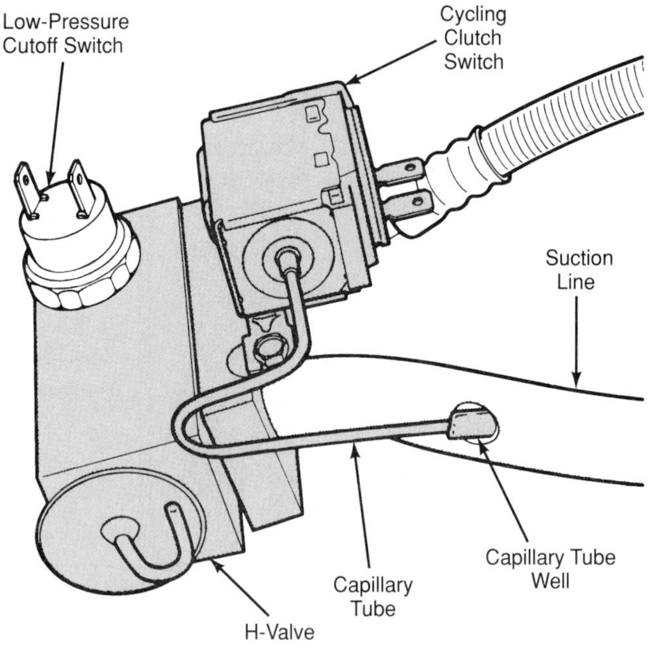

Figure 28-28. *This H-type expansion valve is mounted with cycling clutch switch. (Chrysler Corporation)*

Filter screens are located at both the inlet and outlet of the orifice tube. They catch contaminants that may have entered the system. An orifice tube assembly is usually nonserviceable and must be replaced when defective.

28.3.10 Refrigerant

The refrigerant that has been traditionally used by automobile manufacturers was R-12. Environmental

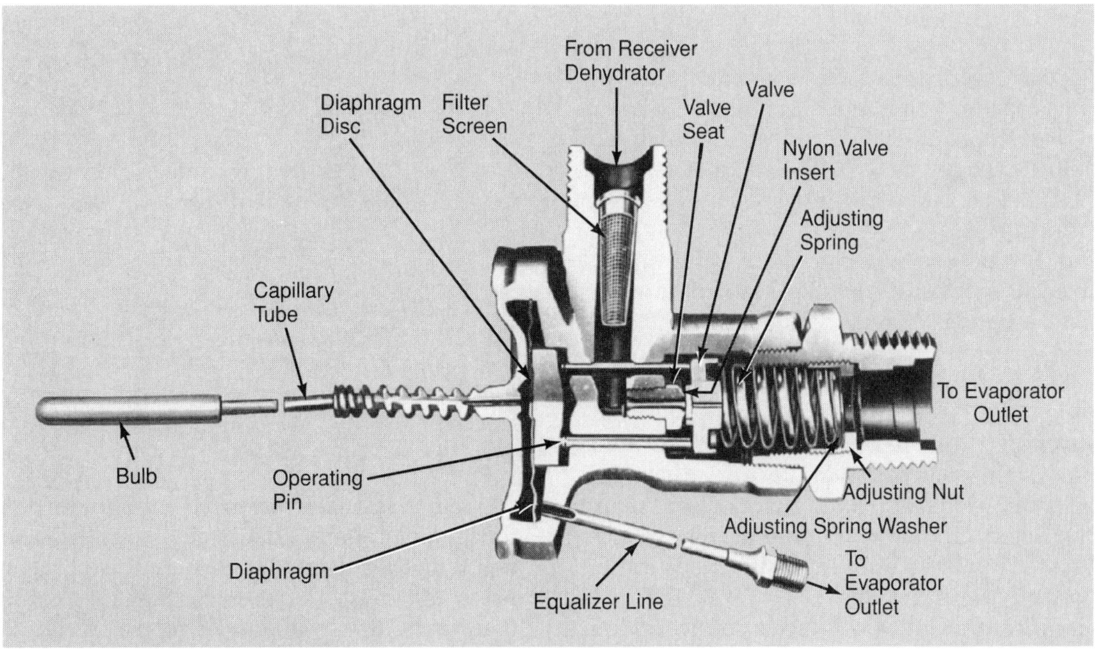

Figure 28-27. *Nonadjustable thermostatic expansion valve. Note inlet connection and outlet connections.*

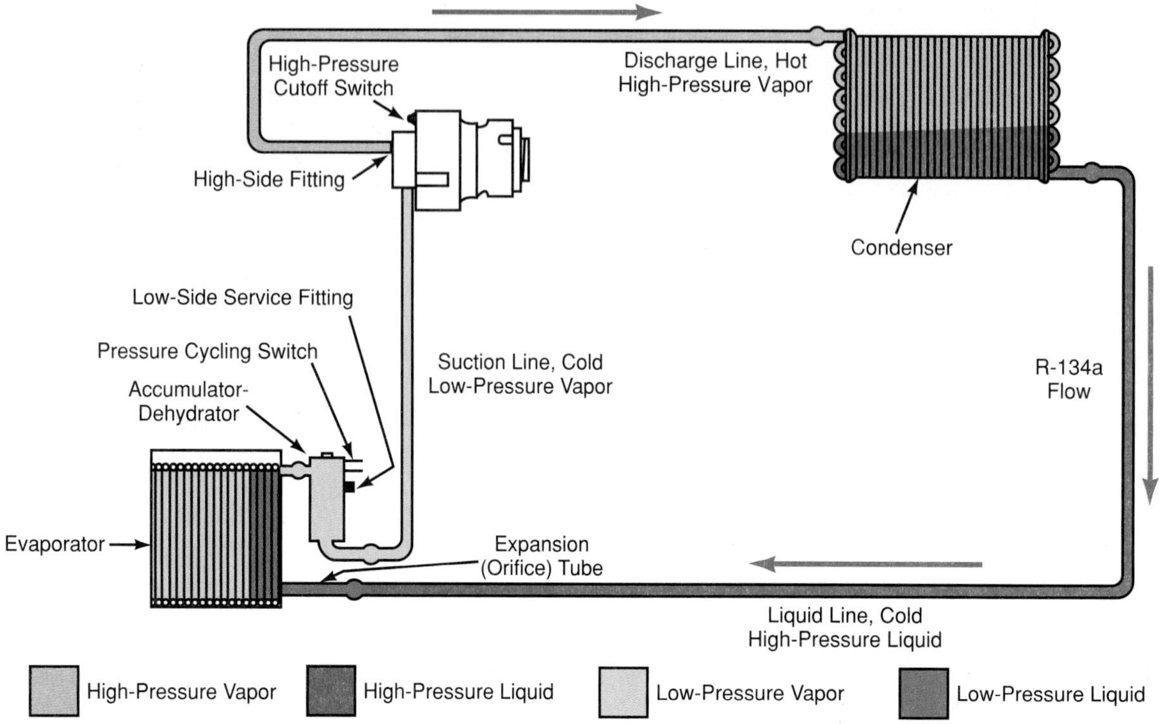

Figure 28-29. *Basic refrigeration cycle of a typical air conditioning system. Note refrigerant flow. (Buick Motor Div., General Motors Corp.)*

concerns have been raised by scientists concerning the use of R-12. Chlorofluorocarbons (CFCs) contained in R-12 may cause deterioration of the earth's ozone layer. This ozone layer, located in the upper atmosphere, helps limit the amount of the sun's ultraviolet rays that pass into our atmosphere.

These findings have caused rulings by the United Nations and U.S. Environmental Protection Agency. Venting of R-12 during service is no longer allowed. Additional rulings have lead to a total phaseout of R-12 production.

R-134a is currently being used as a replacement for R-12. R-134a has a slightly higher boiling point. This will require more heat transfer at the condenser to maintain the same cooling capacity as R-12. R-134a does not contain any CFCs. It is not known to cause any effects to the earth's atmosphere.

R-12 and R-134a are not compatible and should never be mixed in a system. Also, R-134a should never be used in a system as a replacement for R-12. The exception is in a system that has been converted to accept R-134a. The system manufacturer's directions must be used.

The conversion from R-12 to R-134a can be minimal on some systems, as most manufacturers have anticipated the change. They have incorporated many R-134a compatible components into the newer systems. Some older systems may require extensive modification to operate correctly with R-134a.

Some significant changes will face a service technician in repairing an R-134a system. The diameter of the low- and high-pressure valves have been enlarged to prevent accidental use of R-12 refrigerant. The service valves also have been designed with a quick disconnect joint, instead of the previous screw-in type. See **Figure 28-30.**

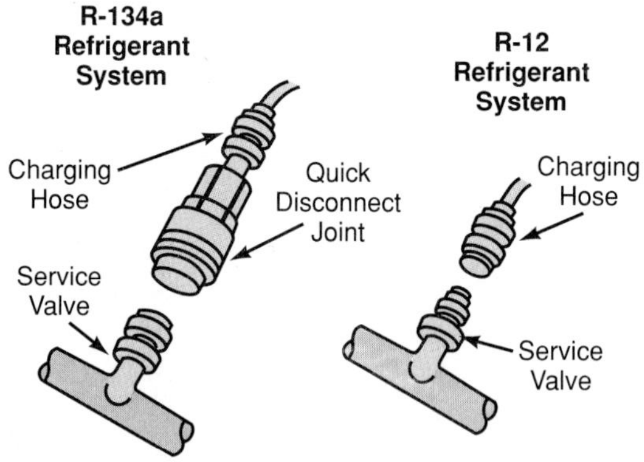

Figure 28-30. *Comparison of components required for the R-12 and R-134a refrigerant systems. (Chrysler Corporation)*

28.3.11 Oil

Only specially prepared oil should be used in the refrigerant system. This oil circulates throughout the system with the refrigerant. Most of the oil stays in the compressor. It must lubricate whether very cold or very hot. It must be dry, pure, and wax-free. Even a very small amount of moisture could freeze at the refrigerant metering device, or form a sludge.

Chapters 2 and 9 both describe refrigerant oil properties. Before adding oil to a system, the vehicle manufacturer's specifications must be consulted. Different compressors may use different oil viscosity. R-12 systems usually use 525 viscosity oil. R-134a systems usually use polyalkylene glycol or PAG synthetic oil.

As with refrigerants, the correct oil should be used for the system being serviced. If an incompatible oil is used, it will not be carried through the system. Thus, damage or failure of components may result.

28.3.12 Service Valves

A variety of service valves are used on automotive air conditioning systems. Included are the standard front- and back-seat (one-way and two-way) service valves (see Chapters 2 and 15). Also included are access valves (see Chapter 12).

All systems have some means by which high-pressure and low-pressure gauges can be connected to the system. This allows for pressure testing and permits charging or adding oil to the system.

R-12 and R-134a use different types of service valves to prevent cross-contamination of refrigerants. **Figure 28-31** shows a compressor fitting with an access valve connection for an R-12 system. **Figure 28-32** shows the two types of service valves used on R-134a systems. Previous R-12 systems required special adapters for different vehicle manufacturer's service valves. All R-134a refrigerant systems use the same low-side and high-side outside dimensions for service valves.

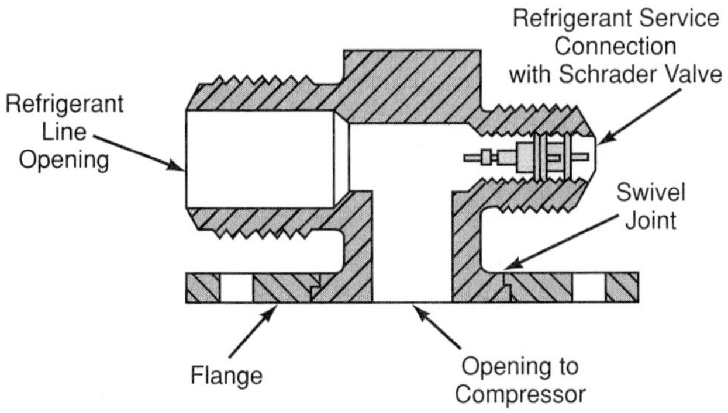

Figure 28-31. *A compressor mount equipped with Schrader valve attachment.*

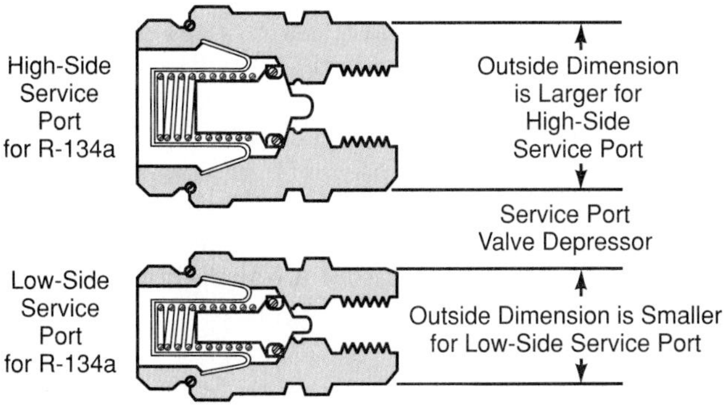

Figure 28-32. *High-side and low-side service ports. The difference in size prevents the technician from crossing the service lines from one side to the other. (Chrysler Corporation)*

The Society of Automotive Engineers (SAE) has developed specifications for service valves. All manufacturers must follow these specifications. This ensures that a standard R-134a gauge manifold can be used on any R-134a vehicle, regardless of the vehicle manufacturer.

28.4 Electrical Systems

Automotive electrical systems are powered by the vehicle's dc battery and alternator (generator). Most systems are 12 V (actually 12.6 V). They increase to 13 V to 15 V when the vehicle is running.

Electrical current is supplied to various components through wires. These wires are generally wrapped in convoluted tubing, with pin connector terminals on their ends.

28.5 Air Distribution

Air distribution is required for heating, ventilating, air conditioning, and defrost. It is accomplished through the use of ducts, vent doors, and louvers. Air is drawn into the vehicle's plenum through vents by a blower. These may be at the base of the windshield, or in the vehicle's passenger compartment. See **Figure 28-33.** The air is then passed through the heater core or the evaporator.

A control panel setting directs the air to the heater or the air conditioning ducts. This is accomplished by doors inside the ducts, controlled electrically, mechanically, or by vacuum.

28.5.1 Blowers

Radial flow blowers (squirrel cage or centrifugal) are used to circulate air in automobiles. The blowers are driven by dc motors, usually 12 V. Generally, the blowers are flexibly mounted to reduce noise. They may be single-speed, multi-speed, or variable-speed.

The blower motors are either single-shaft or double-shaft. They have sealed bearings that usually do not require oiling.

The motor's electrical load can be checked with an ammeter. The ammeter is inserted in the wiring to the blower motor. If the reading is higher than the service manual specifications, and if the motor is overheated, the windings may be shorted. Remove the motor. Check it again, and replace it if faulty. If the reading is lower than specifications, there may be a loose or dirty connection. Use a voltmeter to check for voltage drops. Most

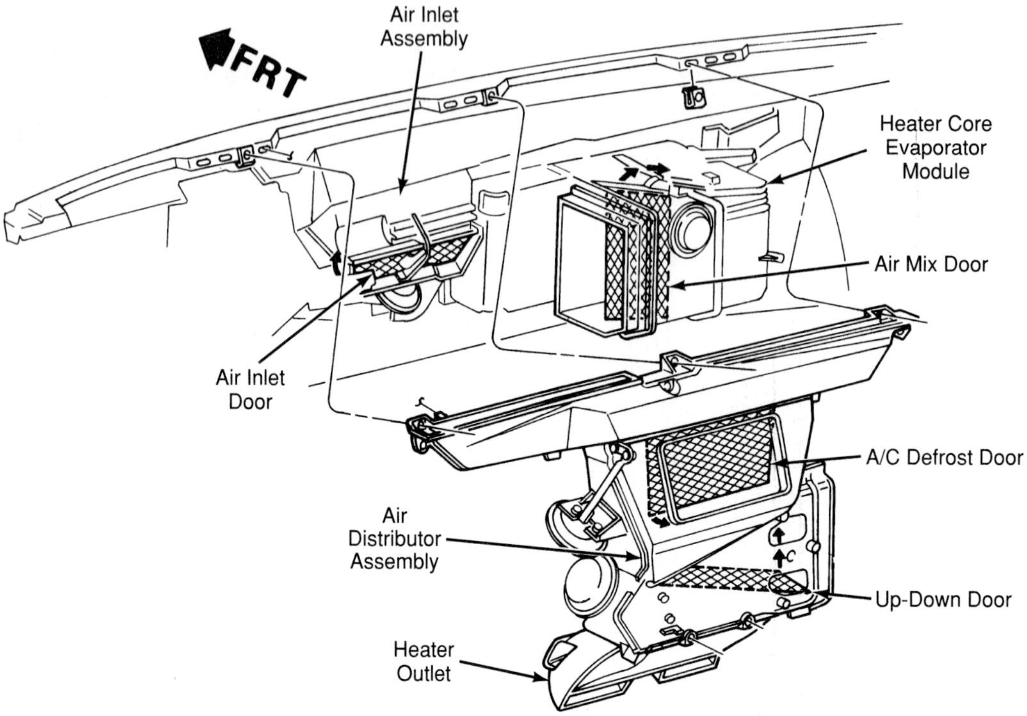

Figure 28-33. *Evaporator and plenum assembly with combined heater core and evaporator module. Note its location under the dashboard. (Cadillac Motor Car Div., General Motors Corp.)*

units, however, are sealed and require replacement if found to be defective.

28.5.2 Ducts

Air conditioning ducts are usually mounted on the face of the dashboard. They often have adjustable louvers. Defrost vents are located on the top of the dashboard. They direct air up and over the windshield. Heater ducts usually empty below the dashboard, near the floor of the passenger compartment. They sometimes continue into the rear heating area.

Heating and cooling ducts are made of metal or plastic. Fairly high air velocities are used. (The noise of air movement is not as critical as it is in an office or residence.)

The duct system includes the following:

- Fresh air inlet.
- Return air inlet.
- Evaporator housing.
- Drain pan and drain connection.
- Plenum chamber.
- Conditioned air outlets (heating or cooling):
  - Defrost or de-ice (windows).
  - Vents.
    - A. Instrument panel.
    - B. Doors.
    - C. Rear of vehicle.
- Dampers to change airflow:
  - Manual.
  - Vacuum-powered.
  - Electric-powered.
- Grilles or louvers.

The heater core and evaporator are usually in series (regarding airflow). The same duct system is used for both systems.

Air dampers may be controlled by cables, electric motors, or vacuum-powered diaphragms, also referred to as ***servomotors.*** The typical path of conditioned air is shown in **Figure 28-34.**

28.6 Insulation

Most automotive vehicle bodies are insulated. Fiberglass, glass wool, and various low K-value, nonsettling, flexible insulations are used in automobiles. (See Section 27.1.2 for a discussion of K-value.)

The large amount of window area in an automobile allows considerable heat leakage. A high sun load exists. However, the use of tinted glass reduces radiant heat load considerably. Light colored cars will absorb considerably less radiant heat than dark-colored vehicles.

Insulation is usually placed over the back of the rear seat in the vehicle. This prevents air conditioning the trunk space. The body of the vehicle must be tightly sealed at all joints. Door gaskets must be in good condition. The vehicle body is tested for air tightness, as well as water tightness.

28.7 Types of Control Systems

The control systems used in automotive air conditioning systems tend to vary widely. An automobile manufacturer may use a similar refrigeration system as

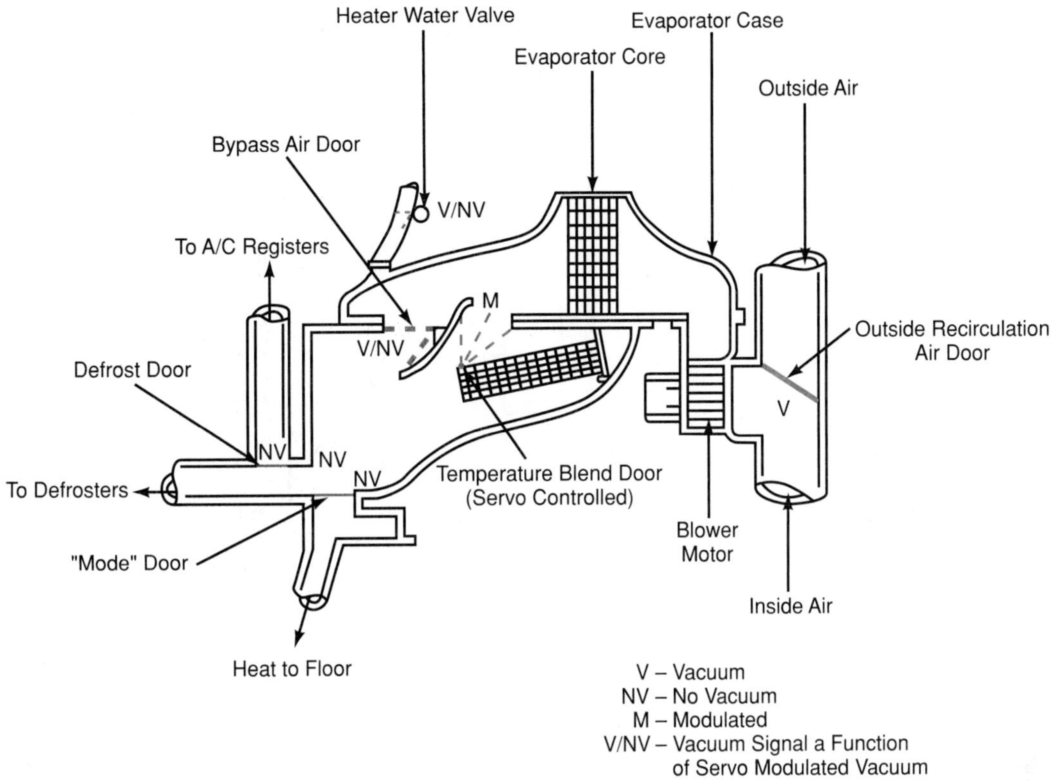

Figure 28-34. *A duct system and plenum chamber uses several doors to control the direction of airflow in ducts. (Ford Customer Service Div.)*

another manufacturer. However, their controls systems for compressor and air delivery can be very different. Therefore, problems may be difficult to diagnose and service.

Two types of control systems are widely used: manual and automatic. Both systems use the basic refrigeration and air distribution components. The difference in the two systems is in how these devices are controlled.

28.7.1 Manual Control Systems

Manual climate control systems require the driver to operate devices on the control panel to obtain the desired changes in the vehicle's inside temperature. See **Figure 28-35.** If the inside of the vehicle is too warm, the driver must change the blower speed or adjust the temperature control to a cooler setting.

The manual control system regulates the in-car temperature by way of several different designs. One design uses switches and cables connected to the control panel. Cables operate temperature and air distribution dampers. The switches operate the blower and compressor.

Another design also uses switches to operate compressor and blower operation. Instead of cables, however, this system makes use of vacuum actuators or servomotors. Vacuum, supplied by the vehicle's engine or pump, is directed through a switching device. This allows several combinations of air delivery and temperature control. **Figure 28-36** shows a typical vacuum control system.

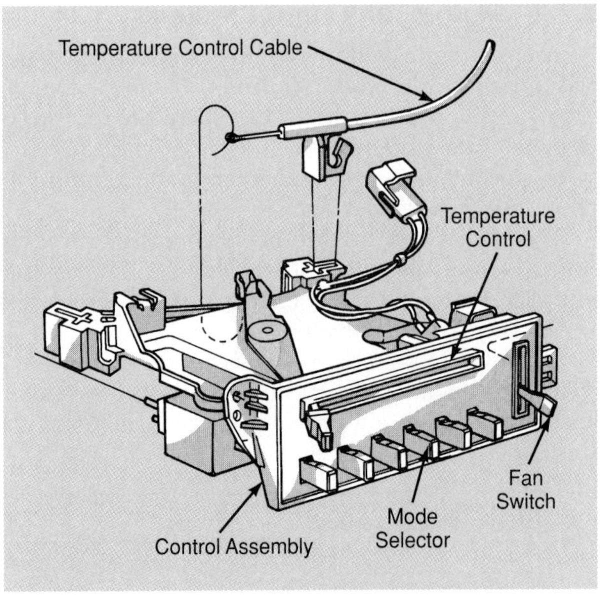

Figure 28-35. *A manual climate control panel. Note manual temperature control slide switch. (Chrysler Corporation)*

A third way to control a manual system is by the use of electronics. Driver control panel levers or buttons are connected in series with a microprocessor or computer. When the driver selects a desired setting, the computer selects the correct damper by means of an electric

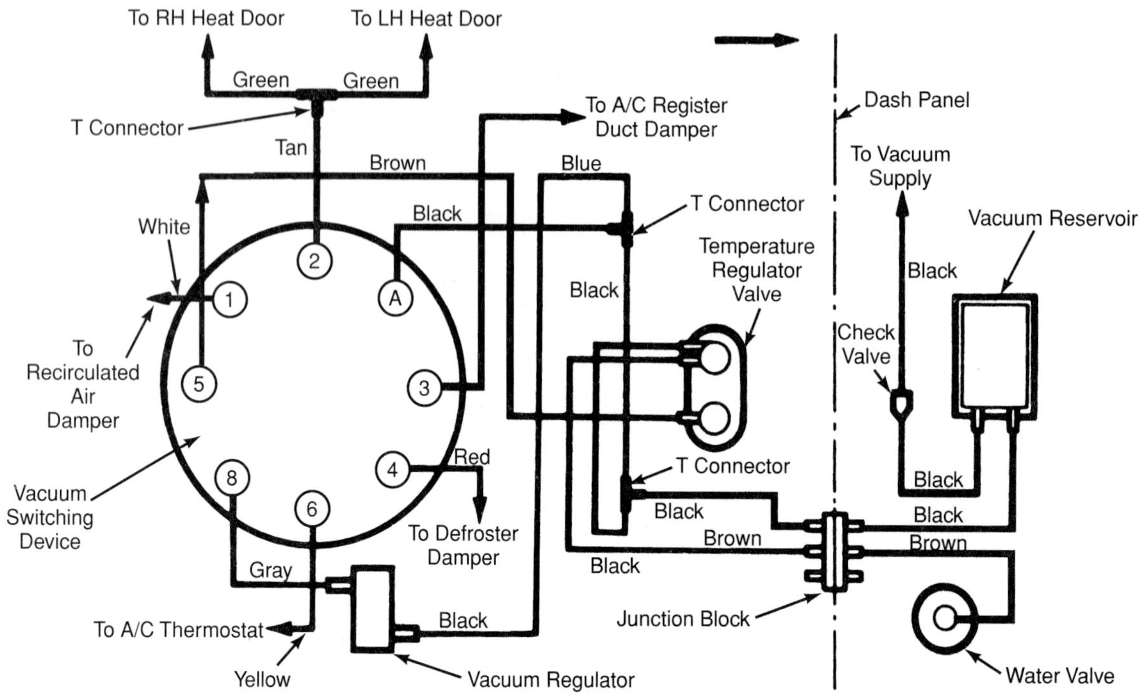

Figure 28-36. *Vacuum piping and vacuum actuators used in an automotive air conditioning system. Vacuum is supplied by engine. (Ford Motor Co.)*

motor. This controls blower speed or compressor running time through the use of a switching relay.

28.7.2 Automatic Control Systems

An automatic control system is sometimes referred to as ***automatic climate control.*** In this system, the driver simply inputs a desired temperature on the control panel. The system automatically adjusts numerous devices to obtain the desired interior temperature and airflow. See **Figure 28-37.**

The primary control for this system is an in-car temperature sensor, usually a thermistor. As the passenger compartment temperature changes, the internal resistance of the sensor changes.

One type of system operates through the use of an HVAC controller. It is the primary part of the system. It uses temperature input provided by sensors to calculate whether more heating or cooling is needed. Changes in damper position, blower speed, and compressor control are made by the controller. It determines all actions of the system components.

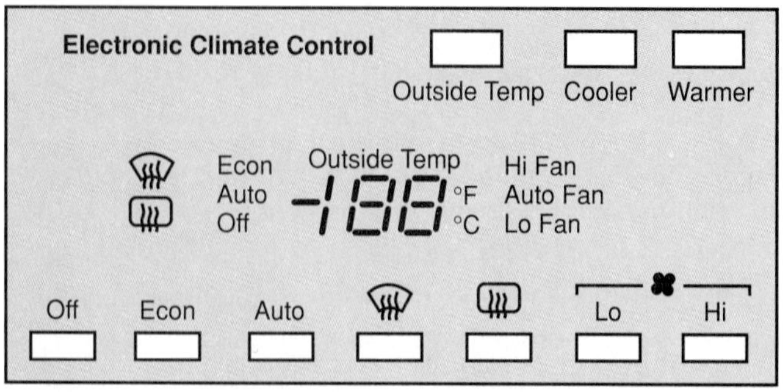

Figure 28-37. *An electronic control panel. Note various temperature selectors. (Cadillac Motor Car Div., General Motors Corp.)*

Another type of system operates with a more manual approach. The inside sensor is connected in series with dampers and blower controls in an HVAC controller. As the resistance of the sensor changes, it signals the unit. The signal indicates changes needed in damper position and blower speed. This may be made possible through the use of a vacuum or electric servos, or both. **Figure 28-38** illustrates one such system.

HVAC Controller

The HVAC controller may be part of the electronic control panel, or it may be mounted separately. It can also be incorporated into other electronic components that control many of the vehicle's systems. The ***HVAC controller*** receives inputs from several of the vehicle sensors and the climate control panel. It then processes the data to control the system. The HVAC controller is usually a sealed unit with multiple pin wire connections. It may not be internally serviceable and may need to be replaced if defected.

The primary information received by the controller is a result of thermistors. Some common sensing devices are:

- In-car temperature sensor.
- Outside temperature sensor.
- Sun-load sensor (light/photo sensor).
- Engine coolant temperature sensor.
- High-side temperature sensor.
- Low-side temperature sensor.

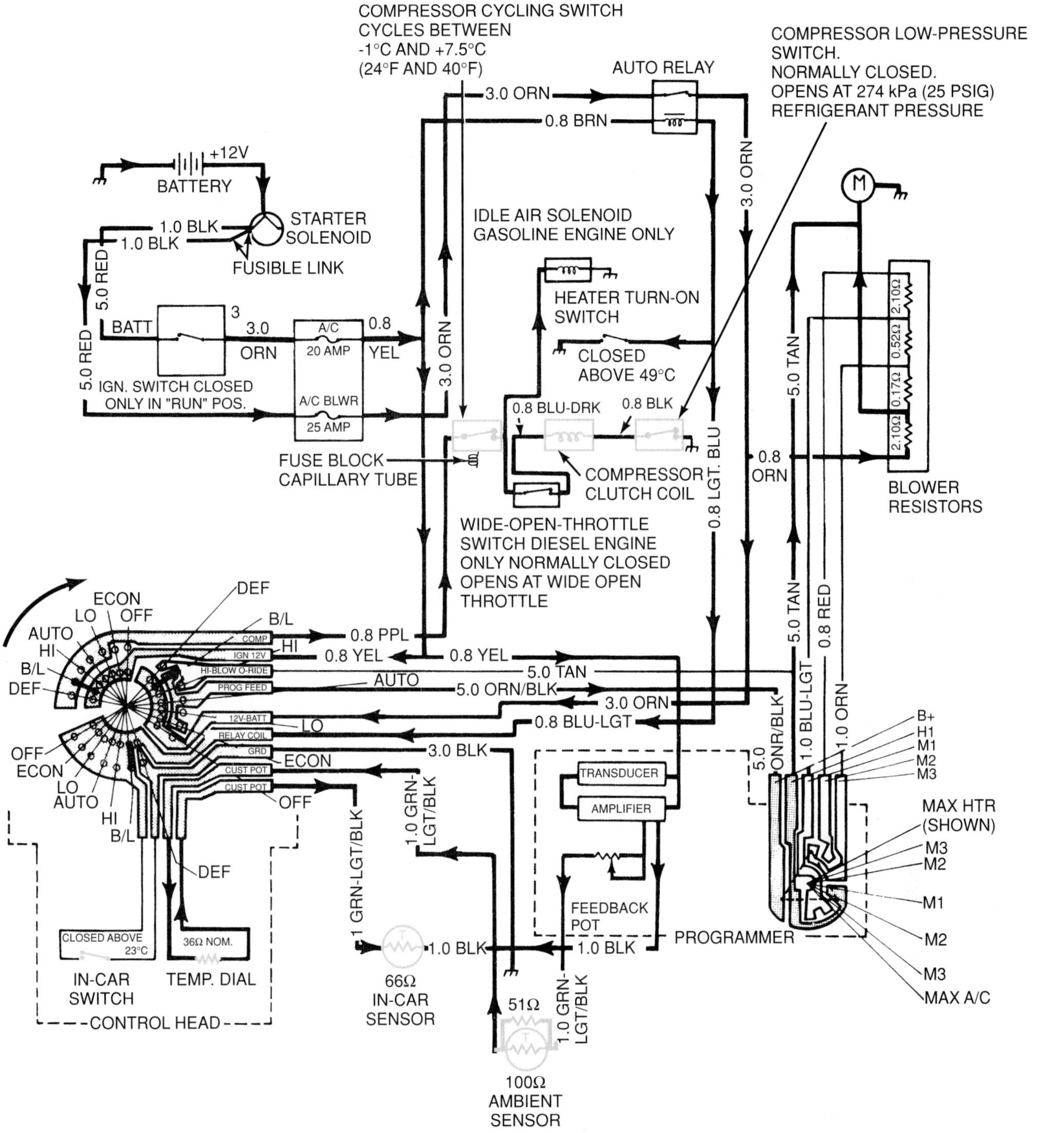

Figure 28-38. *Typical automobile automatic temperature control system. Note how in-car switch, temperature dial, in-car sensor, and ambient temperature sensor are connected in circuit. Also note compressor cycling switch (orange) with capillary tube and compressor low-pressure switch. These switches control the compressor clutch coil. (Buick Motors Division, General Motors Corp.)*

The in-car temperature sensor measures the actual temperature of the vehicle's passenger compartment. This information is used to regulate blower speed, air mix damper position, and compressor operation. It is also used to regulate coolant flow in the heater core. The in-car temperature sensor is usually mounted behind the instrument panel.

The outside temperature sensor indicates the outside or ambient air temperature. It is usually mounted near the radiator grille. This allows airflow over the sensor as the vehicle travels. This sensor can be used for controlling air mix in the vehicle. It can be used to prevent compressor operation during cold weather. It may also supply the driver with a display of the outside temperature.

The sun-load sensor is usually located at the base of the windshield in the instrument control pad. Its purpose is to sense the amount of incoming sun rays. On a bright day, this input may be used to allow additional cooling to compensate for sun load.

The coolant temperature sensor may be located in the coolant hose near the intake, or it may be threaded into the water jacket of the engine. This sensor may be used to control heater operation. It may also be used to disable the compressor if the engine overheats.

The high-side temperature sensor is located near the condenser, before the metering device. The low-side temperature sensor is located between the metering device and the evaporator. Both of these inputs may be used to monitor refrigerant pressure conditions. (These include conditions such as excessive high-side pressure or low refrigerant.) These are only a few of the inputs that may be used by the HVAC controller to control the climate control system.

Some air conditioning systems are connected to the engine controller. Under certain engine load conditions, the engine controller will disable the compressor. This is done to allow faster acceleration or to prevent stalling.

Some systems may contain several controllers. They are usually connected together by some type of data transfer line. In this way, they can communicate with each other, **Figure 28-39.** An example would be when the driver requests air conditioning operation through the climate control panel. The HVAC controller would request that the engine controller activate the compressor. If conditions were correct, the engine would activate the compressor. It might also activate the condenser cooling fan if the vehicle speed were low. The HVAC controller may also command a blower control module to operate the evaporator blower. This all takes place in a fraction of a second.

Electronic Control Diagnostics

Many electronic climate control systems have a self-diagnostic or a self-test capability. This feature helps you locate trouble and faulty components. In a self-diagnostic system, the computer can sense faults. It can alert you and/or the driver in various ways.

One of the basic ways is to flash a display on the climate control panel. For example: A temperature actuator error is shown by flashing a temperature bar display on the climate control display. This would alert you to perform diagnostics on that circuit of the system.

Some systems can be put into diagnostic mode by touching a combination of control panel buttons, or a jumper wire across two test terminals can be used. The system may display a numbered code or flash an indicator lamp. A chart in the service manual will explain what each number or code means. The chart explains the proper diagnostic steps to be taken.

A separate hand-held diagnostic unit may also be used to access diagnostic codes and information on the vehicle. See **Figure 28-40.**

The climate control panel or hand-held diagnostic unit may also serve as an information center. This may allow you to display a variety of information: blower voltage, damper door positions, high- and low-side temperatures, and other component status.

You can refer to the control panel information and the information in the manufacturer's service manual. The problem area can be quickly isolated. Steps can then be taken to correct the problem. Refer to **Figure 28-41.**

Special care should be observed when handling and testing electronic components. Many electronic components operate on lower voltages than other vehicle systems. Improper test procedures and static electricity can damage these components. They may cause failure or a reduced service life of the components. You should

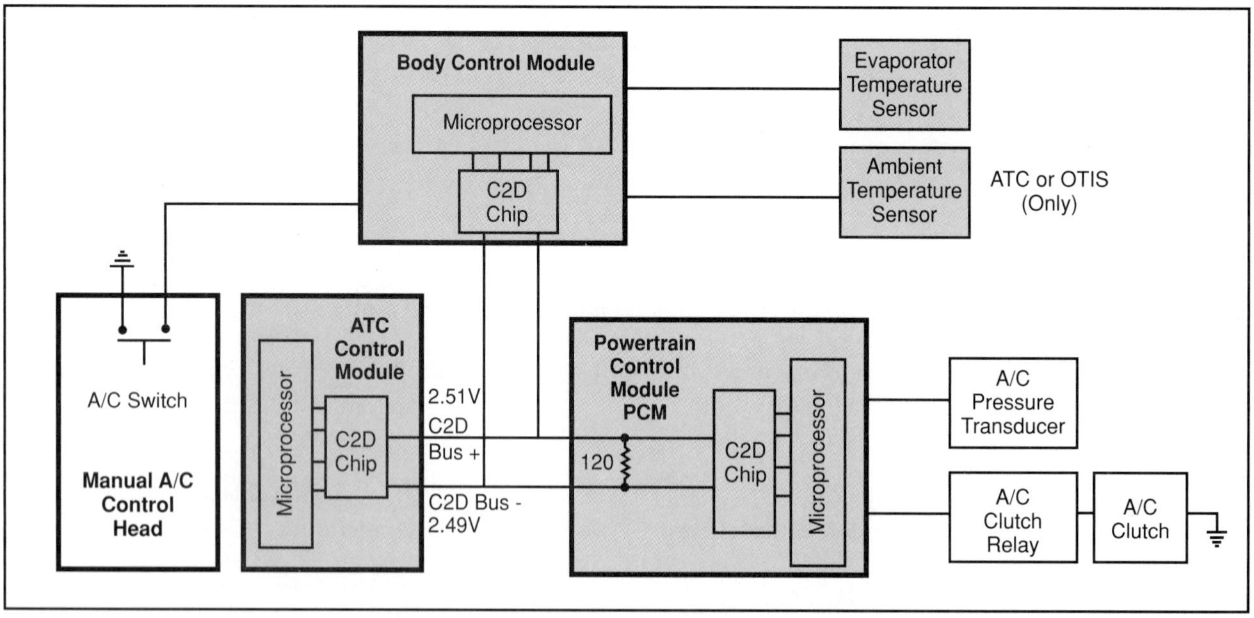

Figure 28-39. *Electrical control circuit for an automotive air conditioning system. Note the various sensors used to indicate temperature in given areas. Also note the modules with transfer line, enabling them to communicate.*

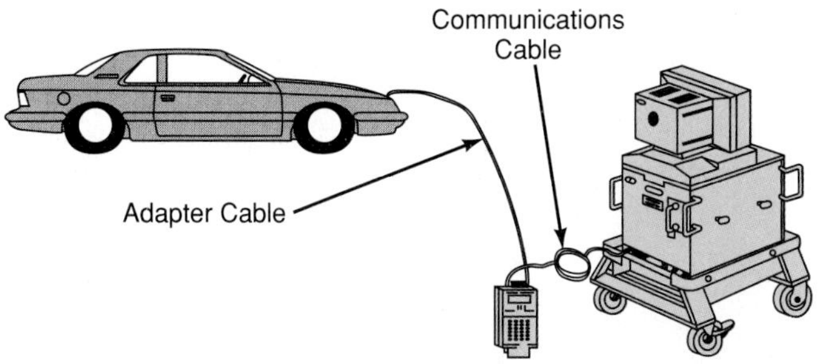

Figure 28-40. *Diagnostic equipment used for analyzing an automotive air conditioning system. The arrangement allows the technician to go through procedures with on-screen information. There is less need for diagnostic flow charts. (Chrysler Corporation)*

always follow proper service procedures as recommended by the manufacturer.

Compressor Protection and Control Switches

Several devices are used to protect the compressor from extreme pressures. These switches are usually wired in series with the compressor clutch circuit. However, they could be monitored by one of the vehicle's computers if so equipped.

The high-side pressure cutoff switch protects the compressor from operating at extremely high pressures. It is usually mounted on the compressor housing. When the high-side pressure exceeds a predetermined point, the switch opens and cuts off current flow to the compressor clutch. When the pressure drops, the switch closes and the compressor resumes operation.

The low-pressure cutoff switch is normally mounted on the accumulator. It prevents compressor damage by disengaging operation when the refrigerant charge is too low in the system. When low-side pressure falls below a predetermined point, the low-pressure cutoff switch contacts open, preventing current flow to the compressor clutch.

Some air conditioning systems also use a wide open throttle or throttle position sensor. Both types of switches are designed to disengage the compressor clutch during periods of rapid acceleration. This switch is mounted on the vehicle's throttle control or is a function controlled by the engine computer. Its purpose is to reduce engine load or power loss when needed. Examples of this would be when passing other vehicles, climbing a steep grade, or entering a highway. This normally closed switch is opened when the throttle is fully depressed, causing compressor clutch disengagement.

Cycling and Thermostatic Switches

There are two types of cycling switches used to control compressor operation—pressure cycling switches and thermostatic switches. They are used to maintain the

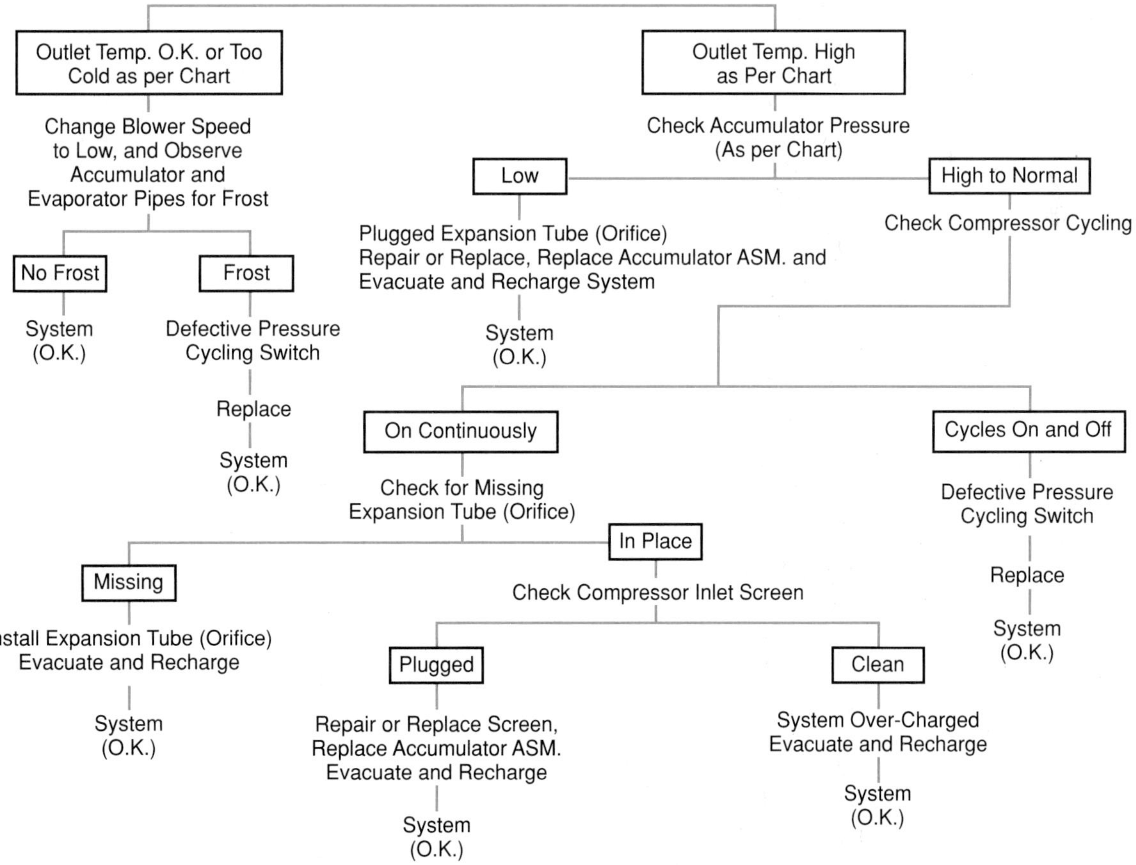

Figure 28-41. *A typical manufacturer's troubleshooting chart. Technicians use it to locate problems and service automobile air conditioners. (Buick Motor Div., General Motors Corp.)*

correct evaporator temperature. The switches provide maximum cooling without allowing the evaporator to drop below freezing. They also prevent compressor operation in cold weather conditions. These switches may be wired in series with the compressor control circuit, or they may be monitored by a computer.

A ***pressure cycling switch*** is usually mounted on the evaporator inlet line, after the metering device. When the low-side pressure drops below a predetermined point, the contacts in the switch open. As the low-pressure then rises, contacts in the switch stay open until a predetermined value. The closing pressure is usually 15 psi to 20 psi higher than the opening pressure. This prevents rapid cycling of the compressor.

Thermostatic switches are usually mounted on the evaporator case. A refrigerant-charged sensing bulb is clamped to the inlet side of the evaporator. The operation of this switch is similar to that of the pressure cycling switch. The only exception is that temperature—rather than pressure—controls the operation of the switch.

28.8 Truck and Bus Air Conditioning

The cabs of many trucks, buses, farm tractors, and earth movers are air conditioned. Some truck systems use a remote condenser mounted on the roof of the cab. This installation removes the condenser from in front of the radiator. The radiator can then operate at full efficiency. This is especially important during long pulls in low gear when the engine could overheat.

Figure 28-42 illustrates a one-piece unit that is utilized for small trucks. This system is charged with R-134a. It is designed to protect temperature-sensitive cargoes from 55°F (13°C) down to 0°F (−18°C) on truck bodies up to 16′ (4.9 meters) in length. The unit is run by a swash-plate 6-cylinder compressor and is belt-driven. It is controlled by a thermostatically-operated magnetic clutch. The condenser and evaporator fan motors are all enclosed in a singular housing. They are operated by 12-volt dc permanent-magnet motors. The unit has an automatic hot-gas defrost system.

Some buses have a combination roof-mounted air conditioning and heating system. An electronic panel in the evaporator case controls ventilation, heating, and air conditioning. The temperature control is mounted on the dash near the driver.

These systems are similar to an automobile air conditioner. They are installed and serviced in the same general manner.

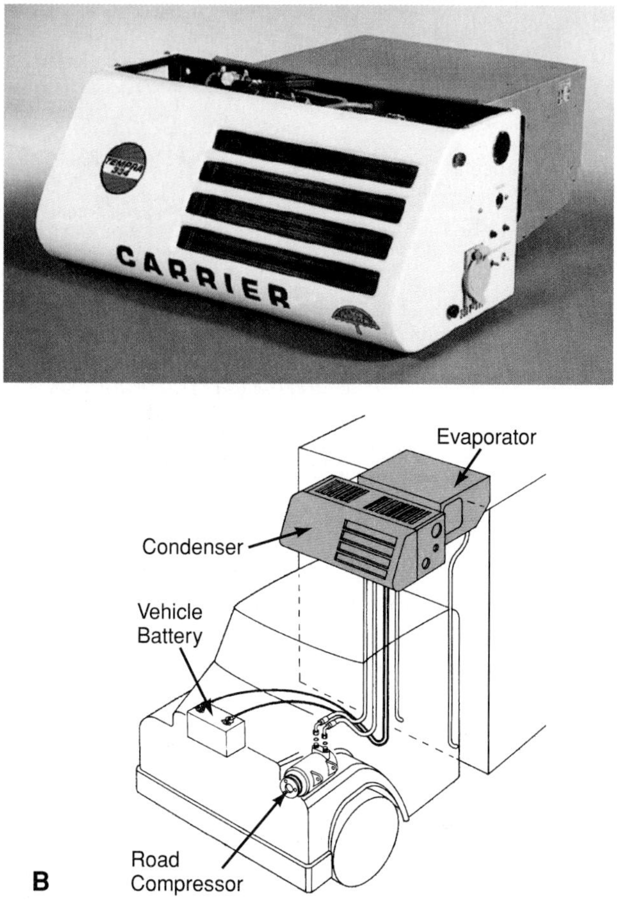

Figure 28-42. *Small truck refrigeration system. A—One-piece unit condenser and evaporator section mounted to the truck body. B—Note road compressor and line arrangement to evaporator and condenser. (Carrier Transicold Division, Carrier Corporation)*

28.9 Servicing Automobile Air Conditioners

Servicing automotive air conditioners is similar to servicing standard air conditioning systems. Study Chapters 12 and 15 carefully before trying to work on an automotive refrigerating system.

Servicing usually begins with a customer's complaint or occurs during an annual check of the system. Common owners' complaints are:

- No cooling or poor cooling.
- Noise.
- Intermittent cooling.
- Vibration.

There may be several causes for each complaint. Check the system thoroughly to find the cause.

The method of installing the gauge manifold is similar to the procedure described in Chapters 12 and 15. Shut off the engine when installing gauges to avoid injuries. Always clean the connections before removing any caps or plugs. **Figure 28-43** illustrates a technician using EPA-approved refrigerant recovery-recycle equipment.

Always attach a gauge manifold to the air conditioning system before service. Never use a manifold set that has been open to the air until after the manifold and lines have been purged (cleaned) and dried.

Figure 28-43. *Technician servicing an automobile. Note the refrigerant recovery-recycle equipment. (Penguin Refrigeration, Inc.)*

Service valves can be situated in many different locations. They can be found on compressor suction or displacement openings, accumulators, or lines. They may be part of a metering device. Some service valves must be back-seated to seal the gauge openings.

Before servicing an automotive air conditioning system, know what performance to expect from the system. Customer expectations of the cooling capacity could vary from actual system capacity in hot weather. **Figure 28-44** is a table of one manufacturer's operating conditions of an R-134a system at various temperatures.

28.9.1 Servicing Belts

An engine may be equipped with several belt-driven accessories. It may have several different belts to drive the components. Belt tension may be adjusted by use of an automatic belt tensioner. It may also be adjusted by moving the accessory components, or an idler pulley.

Belts stretch in use. Therefore, they should be periodically checked for tightness and condition. A loose belt will soon fail. Pulleys can also fail due to wear caused by belt slippage. Remove excessive oil from the pulleys. A shrill squeal when engine speed is increased usually indicates loose belts or glazed belt surfaces. A few drops of belt dressing on a glazed V-belt may stop the squeal temporarily. However, it is best to replace the belt.

Compressor belts with signs of oil contamination, cracks, or frayed edges should be replaced.

Always remember that a short belt life or a broken belt may be the result of an unusual overload (excessive pressures). Other causes may be pulley misalignment, oil leak, or incorrect tension.

Always loosen a belt before attempting removal. Forcing a belt over a pulley may damage and weaken the belt. This could also bend the pulleys.

28.9.2 Testing for Leaks

Checking for refrigerant leaks can be done by using the following methods:

- Trace chemicals.
- Halide torch.
- Ultraviolet fluorescent leak detector.
- Electronic leak detector.
- Foam leak detector (soap bubbles).
- Pressure rise method.

These leak-testing techniques are described in Chapters 12 and 15.

C.C.O.T. A/C System Diagnostic Procedure
"Insufficient Cooling"

Performance Pressure – Temperature Data
Verify Refrigerant Charge

Temperature of Air Entering Condenser		70°F (21°C)	80°F (27°C)	90°F (32°C)	100°F (38°C)	110°F (43°C)
*Compressor Out Pressure (Before Orifice)	A Series	110-150 psi 758-1034 kPa	130-170 psi 896-1172 kPa	160-200 psi 1103-1379 kPa	195-235 psi 1344-1620 kPa	230-250 psi 1586-1724 kPa
	B Series	120-160 psi 827-1103 kPa	150-190 psi 1034-1310 kPa	185-225 psi 1276-1551 kPa	215-255 psi 1482-1758 kPa	270-310 psi 1861-2137 kPa
*Evaporator Pressure (At Accumulator)**	A Series	24-30 psi 165-207 kPa	24-30 psi 165-207 kPa	24-30 psi 165-207 kPa	24-30 psi 165-207 kPa	26-30 psi 179-207 kPa
	B Series	25-30 psi 173-207 kPa	25-30 psi 173-207 kPa	25-30 psi 173-207 kPa	26-32 psi 180-220 kPa	26-32 psi 180-220 kPa
*Discharge Air Temperature – Left Center Outlet	A Series	34-38°F 1-3°C	34-38°F 1-3°C	35-39°F 2-4°C	38-42°F 3-6°C	40-44°F 4-7°C
	B Series	38-42°F 3-6°C	38-42°F 3-6°C	38-42°F 3-6°C	38-42°F 3-6°C	42-48°F 6-9°C

*Just before compressor disengages
** At sea level

Figure 28-44. *Performance chart for a system operating at various temperatures. Both conventional and metric pressures are shown. (Buick Motor Div., General Motors Corp.)*

Some technicians will put a refrigerant colored with reddish dye into the system. Then, red discoloration on the metal surfaces will reveal the source of the leak.

Use of a halide torch on R-12 systems can locate a leak that amounts to 1 lb. (0.45 kg) in about fourteen years. When using a halide torch, an exploring tube end sniffer is placed near the joint being checked. If there is a leak, some escaping refrigerant is drawn up the tube. It passes over a propane or acetylene heated copper element. If there is refrigerant vapor in the air sample, the flame will turn green.

Danger! When R-12 is burned, very poisonous phosgene gas is produced. Avoid breathing fumes when leak testing an air conditioning system with a torch type tester.

The ultraviolet fluorescent leak detection system is used on residential, commercial, and automotive systems. It uses a high-intensity ultraviolet lamp, **Figure 28-45,** and a mist infuser tool. Specially formulated fluorescent additives are used to find the smallest possible leaks in the system. It is effective on leaks as small as 1/4 ounce per year. It can be used on any type of refrigerant. A technician inserts a premeasured fluorescent additive into the refrigerant system with the mist infuser. Then the lamp is used. It pinpoints leaks in the fittings, tubing, coils, or compressor. The additive remains in the system, allowing future leak inspection.

Figure 28-45. *Ultraviolet lamp and fluorescent tracer dye locates an automotive leak. (Spectronics Corporation)*

Frequently, leaks are found by the use of an electronic leak detector. This is a hand-held electronic device with a pump that is calibrated for detecting the refrigerant used. The most sensitive models can detect leaks that would amount to 1 lb. in forty years. A technician moves a hand-held probe across the area being tested. An air sample is drawn into the device by a small pump built into the detector. If refrigerant is detected, the leak detector alerts you with an audible tone. The tone increases in intensity with the amount of refrigerant being detected. Most models detect both R-12 and R-134a.

Leaks can be detected when a vacuum is being drawn. With the vacuum pump running, shut off the vacuum valve on the manifold. If the vacuum gauge needle starts to creep back toward zero (atmospheric pressure), there is a leak in the system. The leak must be corrected before completing the vacuum operation for drying out the system. This test is a good practice when evacuating a system. It can help conserve refrigerant and needless venting into the atmosphere.

28.9.3 Oil in the System

To operate properly, the compressor must have the correct amount and type of oil. The oil must be clean and dry. Too much oil will cause oil pumping, reducing the efficiency of the system and possibly causing damage to the compressor valves. Too little oil will cause rapid wear of the compressor bearings, pistons, rings, and valves. It will also cause scoring of the shaft seal. Therefore, it is important to maintain the correct amount of oil in the system.

Most refrigerant reclaiming stations for R-12 or R-134a contain provisions for measuring oil removed from the system. The amount of oil removed during evacuation or reclaiming of refrigerant should be carefully measured. It should be replaced during charging. Always check the manufacturer's service manual for the correct oil to be used.

When system components are replaced, they may need some additional oil. Therefore, you should check the service manual for the correct amount of oil in a system. Additional oil may also be needed if the system has been repeatedly discharged. This occurs when a vehicle is involved in a collision.

Some compressors allow checking of the oil. Some must be removed from the vehicle to check the oil level, while others require a dipstick. The wire dipstick is inserted through a bolt hole in the compressor crankcase to check the oil level. Check service manual procedures if in doubt.

28.9.4 Testing the Compressor

The compressor must pump efficiently. If the compressor capacity decreases, maximum cooling will not be obtained.

A compressor should pump vacuum to 15″ Hg (381 mm Hg) in a short time against normal head pressure. If this cannot be done, the pistons, rings, or intake valves are leaking (worn).

Refrigerant is used for testing for compressor leaks. A careful examination of gauge pressures is used to evaluate compressor pumping capacity and valve condition.

Note! Never run a compressor unless it has the correct amount of clean refrigerant oil.

Some compressors have a screen installed in the compressor body under the suction line mounting. This screen removes foreign particles such as dirt, sand, and metal chips. This prevents damage to the compressor. The screen should be inspected each time the refrigerant is removed from the system. It should be cleaned if necessary. If the screen is blocked (clogged), or almost blocked, the system will not refrigerate. In addition, the compressor crankcase would be under a vacuum. Low-side pressure would be above normal, and high-side pressure would be below normal. Little or no refrigerant would circulate.

28.9.5 Charging the System

Charging and testing of both R-12 and R-134a systems are very similar. When opening a system for servicing or refrigerant disposal, certified refrigerant recovery equipment must be used. See Chapter 10.

Some automotive systems have a sight glass to aid in charging the system. See **Figure 28-46.** No bubbles should appear in the sight glass after the system has operated for a few minutes. If the system is short of refrigerant, bubbles will appear regularly in the sight glass. A system with little or no refrigerant will not have enough liquid to form bubbles.

An overcharged system may be detected by excessive head pressure. This condition will not show in the sight glass. High system head pressure will be shown on the high-side gauge. Check to see if sufficient airflow

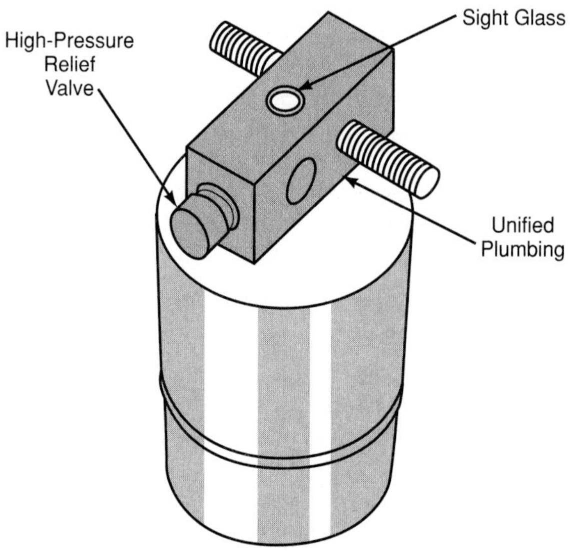

Figure 28-46. *Filter-drier with sight glass for use as a refrigerant level indicator. Presence of bubbles always indicates a problem. (Chrysler Corp.)*

is being passed through the condenser. If so, recover the excessive refrigerant.

The system is charged by means of the service manifold. Connect the manifold to the system. Then connect the charging cylinder to the manifold and purge the lines.

Every vehicle contains a unique refrigerant charge designed for optimum air conditioning performance. It is very important to charge the system to the manufacturer's recommended refrigerant weight. The most effective way to do this is as follows:

1. Fully reclaim all of the refrigerant in the system.
2. Evacuate the system to remove moisture.
3. Add the recommended refrigerant charge.

If only a small amount of refrigerant is needed, charge through the low side with the cylinder upright. If a complete charge is to be put into the system, and the system is under vacuum, charge the system through the high side in liquid form. Do this by inverting the cylinder and opening the service valves. Chapter 12 gives full instructions on the operation of the gauge manifold.

The manifold gauges are the best indicators of system performance. You can use gauge readings to determine:

- State of charge.
- Performance of the compressor.
- Operation of metering device.
- Efficiency of heat transfer.
- Location of possible restrictions.

A careful inspection of gauge readings is essential for proper diagnosis. Gauge readings should be used whenever you service a system for a refrigeration complaint. Note that the gauges in **Figure 28-47** are identified for use with R-134a.

Figure 28-48 illustrates a portable charging station. Note the gauge manifold, charging cylinder, and vacuum pump. The gauge manifold has three gauges. The two gauges back of the manifold handwheels are the suction gauge and the high-pressure gauge. The gauge on the right of the manifold is used during vacuum operations.

Figure 28-49 shows a technician using a charging station and a recovery-recycling machine. The portable cart-style charger evacuates and charges automatically. Refrigerant is added to the system in 0.2 lb. increments for partial charging or leak checking. Next to the service technician is a recovery and recycling unit. The EPA requires refrigerant to be removed from the system without allowing it to enter the atmosphere. The unit also allows for recycling of the recovered refrigerant.

The units have separate controls for recovery and recycling procedures. To recover refrigerant, a manifold gauge set is attached to the air conditioning system. The exhaust hose is connected to the recovery inlet. The manifold valves are opened, and the recovery start button is pressed. The unit will run until recovery is complete, and will then automatically shut off. When a storage tank on the recovery unit is filled, the unit can

Figure 28-47. *A two-valve manifold gauge set with R-134a dials. (Uniweld Products, Inc.)*

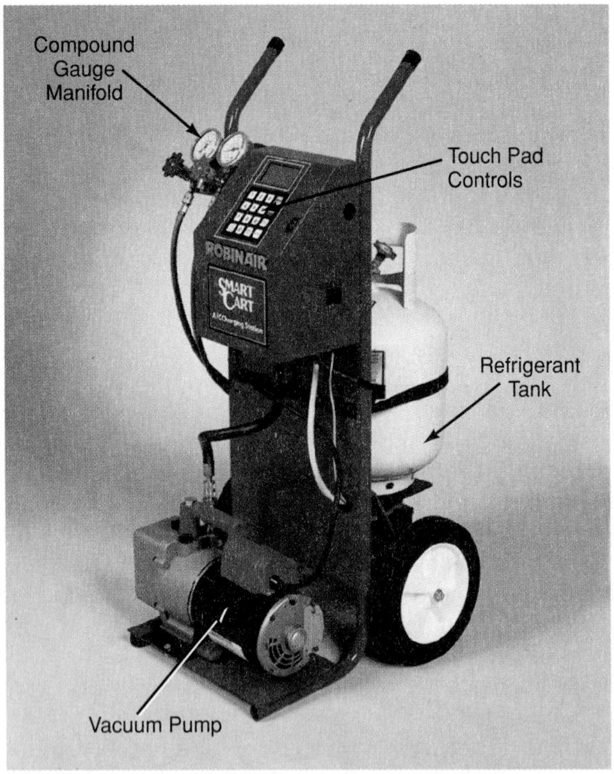

Figure 28-48. *Portable charging and discharging unit. Programming of unit for charging and evacuation is achieved by entering time needed for evacuation and amount of refrigerant to be charged. (Robinair Division, SPX Corporation)*

Figure 28-49. *Technician servicing an automobile air conditioning system. A—Vacuum-charging station. B—Refrigerant recovery and recycling system. (Robinair Division, SPX Corporation)*

then be used as a recycling system. The unit is operated until any dirt or moisture has been removed. When this is done, an indicator light will turn green. The refrigerant can then be reused.

Charging and recovery equipment come in many different styles. They can also be incorporated into one unit which may charge, recover, and recycle. R-12 equipment and R-134a equipment is not interchangeable. It should only be used for the type of system for which it is intended.

28.9.6 Periodic Maintenance

Tell the vehicle owner to periodically operate the air conditioning. It should be operated for a few minutes each month in fall, winter, and spring. This keeps compressor parts (especially the shaft seal) lubricated.

The owner should check the system each spring and fall as follows:

- Condenser (clean fins and tubes of leaves, lint, and insects).
- Refrigerant lines (check for signs of chafing or wear).
- Belts (check for belt deterioration and proper adjustment).

You should check the system each spring and fall or each 10,000 miles (16 100 km). See **Figure 28-50.**

The following should be done:

- Clean all parts externally, including the condenser.
- Straighten fins on the condenser.
- Check the refrigerant charge:
  - Sight glass.
  - Pressure in the system.
- Check oil level in the compressor (if applicable).
- Check for leaks.
- Make sure belts are in good condition and adjusted correctly.

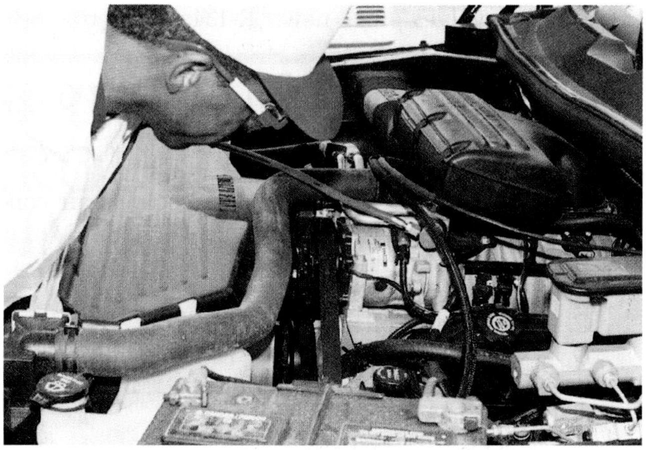

Figure 28-50. *Service technician checking automotive air conditioning system. (Penguin Refrigeration, Inc.)*

The air conditioner should be started very carefully on a vehicle that has been stored for a long period. Sometimes the compressor binds during storage. It is best to raise the hood and watch the compressor and belt while turning on the system. If the belt or compressor starts to slip, stop the engine at once. This indicates that the compressor is turning with difficulty or is "frozen" (stuck). Try to "free" the compressor by slow and careful turning. If it will not turn, remove it at once for reconditioning.

28.10 Review of Safety

Several very important safety precautions should be observed when working on automotive air conditioning systems.

The engine must be running to provide power for the air conditioner and air for the condenser. Therefore, connect the car's tail pipe to an air exhaust system when in an enclosed area. Do not touch the exhaust manifold or serious burns may result. Avoid putting tools, hands, or clothing in contact with the revolving fan. Moving belts are also dangerous. Loose clothing presents a hazard around any moving parts.

It is best not to wear rings, a bracelet, or a wristwatch. These metal parts may short an electric circuit and burn the wearer, or they may catch in a moving part and cause injury. One solution is to put a temporary shield over the fan. Usually, these are made of plastic and fasten to the radiator of the car.

Be careful about putting hands or tools near the spark plugs or plug wires. The electrical shock is not harmful itself. However, the shock may cause one to jump against something, fall down, jerk into moving engine parts. If possible, stop the engine before performing work on air conditioner parts under the hood. Always block the wheels of the car when running the engine.

Be sure to wear goggles and gloves when charging the system. They should be worn also when working on parts that contain refrigerant. R-12 boils at −21.7°F (−29°C). It will cause freeze injury if it contacts the skin or eyes.

Keep a protective cap on the refrigerant cylinder when it is not in use. When using the cylinder, fasten it to a part of the vehicle or to some sturdy stand. Otherwise, it may fall and break the connections of the valve.

If it is necessary to heat a refrigerant cylinder, use warm water only. Never use a torch, electric heat, steam heat, stove, or radiator. This may cause the cylinder to explode.

Avoid welding, brazing, or steam cleaning near the system unless refrigerant has been removed. Otherwise, the excessive pressures may damage the system and injure people in the area.

Breathing quantities of any refrigerant is harmful to a human or an animal. Ventilate the area to keep the vapor concentration to a minimum.

When discharging a system near an open flame, the refrigerant tends to break down. It forms toxic gases. These gases also tarnish metal and plated surfaces.

Most engine cooling systems are pressurized. If a pressure cap is removed when the engine is hot, hot water will erupt from the radiator. This can cause severe burns.

Blower fans can cause painful injuries to the hands. The sharp fins on the coils of the condenser and evaporator can cause deep cuts.

Manifold service lines must be kept clear of pulleys, belts, and fans.

28.11 Test Your Knowledge

Please do not write in this text. Place your answers on a separate sheet of paper.

1. The reciprocating compressor is commonly made of ________.
   A. die cast aluminum
   B. steel
   C. cast iron
   D. None of the above.
2. Wobble plate compressors may contain ________ pistons.
   A. 5
   B. 6
   C. 10
   D. Any of the above.
3. If the operating low-side pressure is too low, what can occur?
   A. Ice formation on the evaporator.
   B. Less circulation of air.
   C. Warm conditions in the vehicle.
   D. Any of the above.
4. The advanced automatic control system funct by the use of ________.
   A. a controller
   B. vacuum actuators
   C. buttons and microprocessor
   D. None of the above.

29.2 Troubleshooting Procedure

One of the key requirements for a service technician is the ability to follow a ***standard procedure.*** An example of a standard procedure follows:

1. Obtain a description of the problem from the owner.
2. From the problem description, determine the possible cause.
3. Identify a specific remedy for the problem.

Using such a standard procedure will save time, money, and frustration. By following the same sequence of activities, you will become more efficient through repetitive use of your skills.

Whenever possible, obtain a service manual or troubleshooting chart. The manual or chart should be written by the company whose equipment is being serviced. It may be available from the building owner or from your employer's resources. There are many different types of manuals in use. However, they use the same basic concept.

Most charts have three basic columns. See **Figure 29-2.** Normally, these column headings are:

- Problem *(Trouble, Complaint . . .).*
- Possible Cause *(Probable Cause, Have You Checked . . .).*
- Remedy *(Repair, You May Need To . . .).*

It is important that you follow the troubleshooting chart on this step-by-step basis.

Upon arrival at the customer's site, you should become familiar with the system in question. The system should be visually inspected. All components and wiring should be examined for any evidence of malfunction. If such evidence is found, you should then examine the system's electrical wiring and component diagram. It will detail all of the system's components.

Always approach the problem in a logical and systematic sequence. Never attempt to make a quick decision that may only temporarily fix the problem. Since it may not uncover the root or cause of the problem, such a decision will often result in a ***callback*** for the same reason as the original service call. An example of this would be a complaint of "inadequate cooling." After determining that the system is low on refrigerant, you could simply add refrigerant without locating the leak that caused the problem. This would provide a temporary solution and an incorrect remedy. You are likely to be called back to the same location very soon for the same problem.

29.2.1 Owner's Description of Problem

The first column of a troubleshooting chart normally lists problems. This column would be the complaint given to you by the owner. Usually the complaint is described in general terms.

An experienced technician begins troubleshooting by *carefully listening* to the owner's complaint. The owner often is not familiar with the principles of operation of the system. Frequently, he or she will use terms that are not the same as those used in the field. Therefore, you must listen carefully.

When analyzing the problem presented by the owner, obtain as much information as possible. This includes how the system is operating now and how it operated before any malfunction. Then, obtain from others in the building any additional information as to how the system has been functioning.

1

Trouble	Probable Cause	Suggested Remedy

2

Complaint	Cause	Repair

3

Problem	Have You Checked	You May Need To

4

PROBLEM	POSSIBLE CAUSE	REMEDY
F. Unit operates long or continuously.	1. Shortage of refrigerant.	1. Fix leak, add charge.
	2. Control contacts stuck or frozen closed.	2. Clean contacts or replace control.
	3. Refrigerated space has excessive load or poor insulation.	3. Determine fault and correct.
	4. System inadequate to handle load.	4. Replace with larger system.
	5. Evaporator coil iced.	5. Defrost.
	6. Restriction in refrigeration system.	6. Determine location and remove.
	7. Dirty condenser.	7. Clean condenser.
	8. Filter dirty.	8. Clean or replace.

Figure 29-2. *Four common types of troubleshooting chart headings. Chart 4 shows the problem, possible cause, and remedy.*

29.2.2 Checking Possible Cause

The next step would be to check the possible cause column of the troubleshooting chart. This listing should be analyzed in terms of the major components of the system. Each problem/complaint related to a malfunction in the system has a specific possible cause.

Figure 29-2 lists possible causes of a problem in a self-contained commercial food storage unit. As shown, the owner indicated the major problem is long or continuous unit operation. Possible causes could be a shortage of refrigerant, control contacts stuck or frozen closed, or any of the eight items identified. More extensive charts are shown later in this chapter.

The possible cause column should be investigated thoroughly. Once you identify the part of the system listed as the possible cause of the problem, you should be able to determine a specific cause or malfunction. The specific faulty part can then be identified.

29.2.3 Suggested Remedy

The final column on the troubleshooting chart may be labeled "Remedy" or may use a similar term. This is the third step when using a troubleshooting chart. You will perform the appropriate task from this third column.

There are many steps to follow in repairing refrigeration or air conditioning equipment. Each part is checked in a step-by-step manner. The actual procedure will vary. It will depend on the specific remedy selected, the type of part or device being checked, and the specific system. The sequence of procedures for checking and repairing or replacing a part will vary. Procedures for an electrical device will be different from that of a mechanical device.

Always follow basic service and safety procedures as you repair a system. Proper tools, gauges, electrical analyzing devices, and other necessary equipment must be used. See Figure 29-3.

29.3 Troubleshooting Charts

The use of a troubleshooting chart is relatively simple. You must understand that it is a helpful map which leads from Step 1 (Problem) to Step 2 (Possible Cause) to Step 3 (Remedy). Troubleshooting charts vary, depending upon the purpose of the equipment and the particular manufacturer. Be very careful to use the specific troubleshooting chart from the equipment's manufacturer. Any components that have been added to the system must also be taken into consideration.

Five different troubleshooting charts are illustrated here for study. They cover domestic refrigeration, commercial food storage, industrial refrigeration, ice machines, and gas-fired furnaces.

A *Hermetic System Diagnosis, Troubleshooting, and Service Chart* is shown in **Figure 29-4.** This chart is designed for use with hermetic refrigeration systems. Note the general introductory procedures and the system electrical check used if the compressor will not run. The troubleshooting chart is broken down into three basic columns: "Complaint," "Possible Cause," and "Repair."

Figure 29-3. *Technician using a combination volt-ammeter to check the electrical system of a combination heating and cooling system. (Southeast Oakland Vocational Education Center)*

Figure 29-5 shows a *Domestic and Light Commercial Systems Troubleshooting and Service Chart.* The introductory section lists some of the common complaints, common causes, and repairs. This is a written analysis of four common complaints (A, B, C, and D) pertaining to compressor malfunctions. These are shown in column form in the troubleshooting chart along with other frequent complaints.

At the bottom of each troubleshooting chart, this company emphasizes caution regarding electrical problems: "WARNING: ELECTRICAL POWER MUST BE DISCONNECTED WHEN TERMINAL PROTECTIVE COVER IS NOT IN PLACE TO PROTECT AGAINST ELECTROCUTION OR VENTED TERMINAL."

The *Industrial Refrigeration Troubleshooting and Service Chart,* **Figure 29-6,** is for use with industrial-type refrigeration equipment. It is assumed that only experienced, qualified persons would use this chart, and that the refrigeration equipment has been operational. It is written only for industrial refrigeration systems using external drive reciprocating compressors. **Safe operating practices should be followed before and during any troubleshooting or service.**

Figure 29-7 is an *Ice Flaker Machine Troubleshooting and Service Chart.* This chart first provides a sequence of operation and a schematic drawing of the unit. It is essential to know how the machine operates before attempting to use the troubleshooting chart and servicing the system. The troubleshooting chart is broken down into three basic columns: "Trouble," "Cause," and "Remedy."

The *Gas-Fired Forced-Air Furnace Servicing and Heating Analysis Guide* in **Figure 29-8** uses a format different from the standard three-column chart. However, the three-step concept can still be used. The first step is "Complaint," which is divided into two categories, "No Heat" or "Unsatisfactory Heat." The trouble is further identified by three conditions under No Heat and six under Unsatisfactory Heat. The complaint in the highlighted example was Unsatisfactory Heat. It was further indicated by the owner that there was too much heat. Following down the chart to the dots, you can identify the second step, "Possible Cause." The cause could be any of the four items indicated. The third step, "Test Method/Remedy," indicates that you should see Service Procedure S-17. You would then perform the service procedures as indicated. If all steps check out okay, you would proceed to check the other possible causes.

29.4 Customer Relations

The role of the refrigeration and air conditioning service technician has changed tremendously in recent years. As a service technician, you are no longer simply required to be knowledgeable in refrigeration and air conditioning. You must also meet the customer's needs and understand business operations and contractual agreements.

One of the keys to a successful business operation is good customer relations. ***Customer relations*** are based on evaluations by the consumer for whom service is being provided. The consumer will make these evaluations based on your job performance and attitude. As a technician, it is your responsibility to instill in the customer a sense of trust, value, and satisfaction with the work performed. This is accomplished mostly through your verbal communication with the customer and your general attitude and appearance. These factors combine to give a positive impression to the customer.

It is essential that you are always courteous when dealing with customers, even when things are not going well. This is especially true when the customer feels that a problem is not being handled properly. It is important that you treat the customer's needs as an emergency situation. In the eyes and mind of the customer, it *is* an emergency.

HERMETIC REFRIGERATION SYSTEM DIAGNOSIS

1. GENERAL
   Each complaint is followed by probable causes and suggested repairs. To isolate the possible cause, proceed in a systematic manner to determine the faulty component. This guide does not cover all possible troubles and deficiencies that may occur under conditions of operation.
2. ELECTRICAL CHECK:
   A—COMPRESSOR DOES NOT RUN—
   1. Check power at outlet receptacle. Your compressor is designed to operate (see serial no. and data plate) on 115-60-1 with a voltage range of 126.5-103.5. Your 240/220-50-1 will operate within a range of 264-198.
   2. Check thermostat for proper setting and continuity. Make sure control setting is not in an "Off" position. Continuity may be verified by following instructions on the individual compressor motor circuits.
   3. Look for obvious loose or broken wiring.
   4. Following systematically the instructions listed on the compressor motor circuitry, check the relay, overload, and, if employed, the start capacitor for continuity. Replace any components found faulty with the recommended service parts.

3. **TROUBLESHOOTING AND SERVICE CHART**

COMPLAINT	POSSIBLE CAUSE	REPAIR
A. Compressor will not start—no hum.	1. Line disconnect switch open. 2. Fuse removed or blown. 3. Overload protector tripped. 4. Control stuck in open position. 5. Control off due to cold location. 6. Wiring improper or loose.	1. Close start or disconnect switch. 2. Replace fuse. 3. Refer to electrical diagram. 4. Repair or replace control. 5. Relocate control. 6. Check wiring against diagram.
B. Compressor will not start—hums.	1. Improperly wired. 2. Low voltage to unit. 3. Starting capacitor defective. 4. Relay failing to close. 5. Compressor motor has winding open or shorted. 6. Internal mechanical trouble in compressor.	1. Check wiring against diagram. 2. Determine reason and correct. 3. Determine reason and replace. 4. Determine reason and correct, replace if necessary. 5. Replace compressor. 6. Replace compressor.

Figure 29-4. *Hermetic refrigeration system diagnosis for a self-contained commercial food storage unit. (Silver King Division of Stevens Lee Company)*

TROUBLESHOOTING AND SERVICE CHART		
COMPLAINT	POSSIBLE CAUSE	REPAIR
C. Compressor will not start—hums but trips on overload protector.	1. Improperly wired. 2. Low voltage to unit. 3. Relay failing to open. 4. Run capacitor defective. 5. Excessively high discharge pressure. 6. Compressor motor has a winding open or shorted. 7. Internal mechanical trouble in compressor (tight).	1. Check wiring against diagram. 2. Determine reason and correct. 3. Determine reason and correct, replace if necessary. 4. Determine reason and replace. 5. Check discharge shutoff, possible overcharge or insufficient cooling on condenser. 6. Replace compressor. 7. Replace compressor.
D. Compressor starts and runs, but short cycles on overload protector.	1. Additional current through overload protector. 2. Low voltage to unit. 3. Overload protector defective. 4. Run capacitor defective. 5. Excessive discharge pressure. 6. Suction pressure too high. 7. Compressor too hot—return gas hot. 8. Compressor motor has a winding shorted.	1. Check wiring diagram, check for added fan motors, pumps, etc., connected to wrong side of protector. 2. Determine reason and correct. 3. Check current, replace protector. 4. Determine reason and replace. 5. Check ventilation, restrictions in cooling medium, restrictions in refrigeration system. 6. Check for possibility of misapplication. Use stronger unit. 7. Check refrigerant charge (fix leak), add if necessary. 8. Replace compressor.
E. Unit runs ok, but short cycles.	1. Overload protector. 2. Thermostat. 3. High pressure cut-out due to: a. Insufficient air. b. Overcharge. c. Air in system. 4. Low pressure cut-out due to: a. Undercharge. b. Restriction in expansion device.	1. See D above. 2. Differential set too close—widen. 3. a. Check air to condenser—correct. b. Reduce refrigerant charge. c. Purge. 4. a. Fix leak, add refrigerant. b. Replace device.
F. Unit operates long or continuously.	1. Shortage of refrigerant. 2. Control contacts stuck or frozen closed. 3. Refrigerated space has excessive load or poor insulation. 4. System inadequate to handle load. 5. Evaporator coil iced. 6. Restriction in refrigeration system. 7. Dirty condenser. 8. Filter dirty.	1. Fix leak, add charge. 2. Clean contacts or replace control. 3. Determine fault and correct. 4. Replace with larger system. 5. Defrost. 6. Determine location and remove. 7. Clean condenser. 8. Clean or replace.
G. Start capacitor open, shorted, or blown.	1. Relay contacts not operating properly. 2. Prolonged operation on start cycle due to: a. Low voltage to unit. b. Improper relay. c. Starting load too high. 3. Excessive short cycling. 4. Improper capacitor.	1. Check and replace. 2. a. Determine reason and correct. b. Replace. c. Correct by using pump down arrangement if necessary. 3. Determine reason for short cycling (see E above) and correct . 4. Determine correct size and replace.
H. Relay defective or burned out.	1. Incorrect relay. 2. Incorrect mounting angle. 3. Line voltage too high or too low. 4. Excessive short cycling. 5. Relay being influenced by loose vibrating mounting. 6. Incorrect run capacitor.	1. Check and replace. 2. Remount relay in correct position. 3. Determine reason and correct. 4. Determine reason (see E) and correct. 5. Remount rigidly. 6. Replace with proper capacitor.
I. Space temperature too high.	1. Control setting too high. 2. Inadequate air circulation.	1. Reset control. 2. Improve air movement.
J. Suction line frosted or sweating.	1. Evaporator fan not running. 2. Overcharge of refrigerant.	1. Determine reason and correct. 2. Correct charge.
K. Liquid line frosted or sweating.	1. Restriction in dehydrator or strainer.	1. Replace part.
L. Unit noisy.	1. Loose parts or mounting. 2. Tubing rattle. 3. Bent fan blade causing vibration. 4. Fan motor bearings worn.	1. Find and tighten. 2. Reform to be free of contact. 3. Replace blade. 4. Replace motor.

Figure 29-4. *Continued.*

DOMESTIC AND LIGHT COMMERCIAL SYSTEMS

The "Troubleshooting and Service Chart" is quite self-explanatory; however, a discussion of some of the complaints, possible causes, and repair solutions may be of some additional assistance.

COMPLAINT "A" is Compressor will not start—no hum.

Possible causes are:

1. *Switch open.* Rather obvious, but maybe it would be wise to determine why it is open or who opened it.
2. *Fuse removed or blown.* Again, was there a reason?
3. *Overload protector tripped.* Here, too, it is not a case of waiting until the overload resets, but rather to determine why.
4. *Control stuck in open position.* Faulty contactors may be a cause, although every effort is made to provide the best quality contactors. And heed this warning: Don't use the insulated end of a screwdriver to hold the contactor in. In doing so you run the risk of burning out a good compressor.

COMPLAINT "B" is Compressor will not start—hums and trips on overload.

1. & 2. have been discussed (see Complaint "A").

3. *Starting capacitor defective.* It says "determine reason." Possibly a start capacitor was installed which had too low a voltage rating.
4. *Relay failing to close.* Is the correct one being used? There seems to be a considerable tendency to substitute something other than the one specified—if it works leave it. We say, please don't do it.
5. *Compressor has a winding open or shorted.* The repair specified says simply "replace compressor." This means if the cause indicated has been proved—without doubt—conclusively—no question about it—only then replace the compressor. Remember that replacing the compressor is generally the most costly repair bill an owner can get. So be sure—first.
6. *Internal mechanical trouble in compressor.* If the serviceperson has proved without question that none of the other possible causes are the reason for the condition, then and then only can it be mechanical trouble.

A "Troubleshooting Chart" of this kind is not the entire answer. There are probably a number of other reasons for the cause of each "complaint" listed, so keep in mind that application of knowledge gained through experience and common sense are as much a part of troubleshooting as the use of any chart.

COMPLAINT "C" is Compressor starts but stays on run winding.

How do you know this condition is occurring? If the amps stay higher than normal. Or if you don't hear the changeover.

1. through 3. Covered previously (see Complaints "A" and "B").

4. *Run capacitor defective.* If the run capacitor is failed closed, there will be a period of time when the current and running sounds will seem to indicate the relay has not switched. In a relatively short period of time, the start winding will burn, so time is of the essence in this case.
5. *High head pressure.* Be sure to check all the things listed in the "repair" column.

COMPLAINT "D" is Compressor starts and runs, but short cycles on overload protector.

1. Mentioned before (see Complaint "A").
2. *Low voltage to unit (or unbalanced if three-phase).* In the matter of three-phase unbalance—this is an instance in which it is probably wise to call in the power company, or check with the building owner to determine what other equipment is on the source of power to cause the unbalance.
3. *Overload protector defective.* Sometimes difficult to determine. One good clue is how it looks—does it show to have been overheated?

4. and 5. Answered previously (see Complaint "C").

6. *Suction pressure too high.* This will occur more often on refrigeration than air conditioning, especially on low temperature equipment.
7. *Compressor hot—insufficient gas cooling.* Usually a result of a low charge.

Most of the other complaints, causes, and repair suggestions are straightforward, and the best suggestion is to follow the chart.

WARNING: ELECTRICAL POWER MUST BE DISCONNECTED WHEN TERMINAL PROTECTIVE COVER IS NOT IN PLACE TO PROTECT AGAINST ELECTROCUTION OR VENTED TERMINAL.

TROUBLESHOOTING AND SERVICE CHART

COMPLAINT	POSSIBLE CAUSE	REPAIR
A Compressor will not start—no hum.	1. Line disconnect switch open. 2. Fuse removed or blown. 3. Overload protector tripped. 4. Control stuck in open position. 5. Control off due to cold location. 6. Wiring improper or loose.	1. Close start or disconnect switch. 2. Replace fuse. 3. Refer to electrical section. 4. Repair or replace control. 5. Relocate control. 6. Check wiring against diagram.
B Compressor will not start—hums and trips on overload protector.	1. Improperly wired. 2. Low voltage to unit. 3. Starting capacitor defective. 4. Relay failing to close. 5. Compressor motor has a winding open or shorted. 6. Internal mechanical trouble in compressor. 7. Liquid refrigerant in compressor.	1. Check wiring against diagram. 2. Determine reason and correct. 3. Determine reason and replace. 4. Determine reason and correct, replace if necessary. 5. Replace compressor. 6. Replace compressor. 7. Add crankcase heater and/or accumulator.
C Compressor starts, but does not switch off of start winding.	1. Improperly wired. 2. Low voltage to unit. 3. Relay failing to open. 4. Run capacitor defective. 5. Excessively high discharge pressure. 6. Compressor motor has a winding open or shorted. 7. Internal mechanical trouble in compressor (tight).	1. Check wiring against diagram. 2. Determine reason and correct. 3. Determine reason and correct, replace if necessary. 4. Determine reason and replace. 5. Check discharge shut-off valve, possible over-charge, or insufficient cooling on condenser. 6. Replace compressor. 7. Replace compressor.

Figure 29-5. *Domestic and light commercial systems troubleshooting and service chart. (Tecumseh Products Company)*

TROUBLESHOOTING AND SERVICE CHART		
COMPLAINT	POSSIBLE CAUSE	REPAIR
D Compressor starts and runs, but short cycles on overload protector.	1. Additional current passing through overload protector.	1. Check wiring diagram. Check for added fan motors, pumps, etc., connected to wrong side of protector.
	2. Low voltage to unit (or unbalanced if three phase).	2. Determine reason and correct.
	3. Overload protector defec tive.	3. Check current, replace protector.
	4. Run capacitor defective.	4. Determine reason and replace.
	5. Excessive discharge pressure.	5. Check ventilation, restrictions in cooling medium, restrictions in refrigeration system.
	6. Suction pressure too high.	6. Check for possibility of misapplication. Use stronger unit.
	7. Compressor too hot—return gas hot.	7. Check refrigerant charge (fix leak), add if necessary.
	8. Compressor motor has a winding shorted.	8. Replace compressor.
E Unit runs OK, but short cycles on.	1. Overload protector.	1. See D above.
	2. Thermostat.	2. Differential set too close—widen.
	3. High-pressure cut-out due to:	
	a. Insufficient air or water supply.	3a. Check air or water supply to condenser—correct.
	b. Overcharge.	b. Reduce refrigerant charge.
	c. Air in system.	c. Purge.
	4. Low-pressure cut-out due to:	
	a. Liquid line solenoid leaking.	4a. Replace.
	b. Compressor valve leak.	b. Replace.
	c. Undercharge.	c. Fix leak, add refrigerant.
	d. Restriction in expansion device.	d. Replace device.
F Unit operates long or continuously.	1. Shortage of refrigerant.	1. Fix leak, add charge.
	2. Control contacts stuck or frozen closed.	2. Clean contacts or replace control.
	3. Refrigerated or air conditioned space has excessive load or poor insulation.	3. Determine fault and correct.
	4. System inadequate to handle load.	4. Replace with larger system.
	5. Evaporator coil iced.	5. Defrost.
	6. Restriction in refrigeration system.	6. Determine location and remove.
	7. Dirty condenser.	7. Clean condenser.
	8. Filter dirty.	8. Clean or replace.
G Start capacitor open, shorted, or blown.	1. Relay contacts not operating properly.	1. Clean contacts or replace relay if necessary.
	2. Prolonged operation on start cycle due to:	
	a. Low voltage to unit.	2a. Determine reason and correct.
	b. Improper relay.	b. Replace.
	c. Starting load too high.	c. Correct by using pump down arrangement if necessary.
	3. Excessive short cycling.	3. Determine reason for short cycling (see E above) and correct.
	4. Improper capacitor.	4. Determine correct size and replace.
H Run capacitor open, shorted, or blown.	1. Improper capacitor.	1. Determine correct size and replace.
	2. Excessively high line voltage (110% of rated-max.).	2. Determine reason and correct.
I Relay defective or burned out.	1. Incorrect relay.	1. Check and replace.
	2. Incorrect mounting angle.	2. Remount relay in correct position.
	3. Line voltage too high or too low.	3. Determine reason and correct.
	4. Excessive short cycling.	4. Determine reason (see E above) and correct.
	5. Relay being influenced by loose vibrating mounting.	5. Remount rigidly.
	6. Incorrect run capacitor.	6. Replace with proper capacitor.
J Space temperature too high.	1. Control setting too high.	1. Reset control.
	2. Expansion valve too small.	2. Use larger valve.
	3. Cooling coils too small.	3. Add surface or replace.
	4. Inadequate air circulation.	4. Improve air movement.
K Suction line frosted or sweating.	1. Expansion valve passing excess refrigerant or is oversized.	1. Readjust valve or replace with smaller valve.
	2. Expansion valve stuck open.	2. Clean valve of foreign particles, replace if necessary.
	3. Evaporator fan not running.	3. Determine reason and correct.
	4. Overcharge of refrigerant.	4. Correct charge.
L Liquid line frosted or sweating.	1. Restriction in dehydrator or strainer.	1. Replace part.
	2. Liquid shut-off (king valve) partially closed.	2. Open valve fully.
M Unit noisy.	1. Loose parts or mountings.	1. Find and tighten.
	2. Tubing rattle.	2. Reform to be free of contact.
	3. Bent fan blade causing vibration.	3. Replace blade.
	4. Fan motor bearings worn.	4. Replace motor.

Figure 29-5. *Continued.*

PROBLEM	POSSIBLE CAUSE	REMEDY OR COMMENT
I. COMPRESSOR WILL NOT START.	A. No power to motor.	1. Check power to and from fuses; replace fuses if necessary. 2. Check starter contacts, connections, overloads, and timer (if part winding start). Reset or repair as necessary. 3. Check power at motor terminals. 4. Repair wiring if damaged.
	B. Control circuit is open.	1. Safety switches are holding circuit open. Check high pressure, oil failure, and low pressure switches. Also check oil filter pressure differential switch if supplied. 2. Thermostat is satisfied. 3. Check control circuit fuses if blown; replace. 4. Check wiring for open circuit.
II. MOTOR "HUMS" BUT DOES NOT START.	A. Low voltage to motor.	1. Check incoming power for correct voltage. Call power company or inspect/repair power wiring.
	B. Motor shorted.	1. Check at motor terminals. Repair or replace as necessary.
	C. Single phase failure in the three phase power supply.	1. Check power wiring circuit for component or fuse failure.
	D. Compressor is seized due to damage or liquid.	1. Remove belts or coupling. Manually turn crankshaft to check compressor.
	E. Compressor is not unloaded.	1. Check unloader system.
III. COMPRESSOR STARTS, BUT MOTOR CYCLES OFF ON OVERLOADS.	A. Compressor has liquid or oil in cylinders.	1. Check compressor crankcase temperature. 2. Throttle suction stop valve on compressor to clear cylinders and act to prevent recurrence of liquid accumulation.
	B. Suction pressure is too high.	1. Unload compressor when starting. Use internal unloaders if present. 2. Install external by-pass unloader.
	C. Motor control.	1. Motor control located in hot ambient. 2. Low power voltage. 3. Motor overloads may be defective or weak. 4. Check motor control relay. 5. Adjust circuit breaker setting to full load amps.
	D. Bearings are "tight."	1. Check motor and compressor bearings for temperature. Lubricate motor bearings.
	E. Motor is running on single phase power.	1. Check power lines, fuses, starter, motor, etc., to determine where open circuit has occurred.
IV. COMPRESSOR STARTS BUT SHORT CYCLES AUTOMATICALLY.	A. Low refrigerant charge.	1. Check and add if necessary.
	B. Driers plugged or saturated with moisture.	1. Replace cores.
	C. Refrigerant feed control is defective.	1. Repair or replace.
	D. No load.	1. To prevent short cycling, if objectionable, install pump-down circuit, anti-recycle timer or false load system.
	E. Unit is too large for load.	1. Reduce compressor speed. 2. Install false load system.
	F. Suction strainer blocked or restricted.	1. Check and clean or replace as necessary.
V. MOTOR IS NOISY OR ERRATIC.	A. Motor bearing failure or winding failure.	1. Check and repair as needed.
	B. If electronic starter, check calibration on control elements.	1. Adjust as necessary.
VI. COMPRESSOR RUNS CONTINUOUSLY BUT DOES NOT KEEP UP WITH THE LOAD.	A. Load is too high.	1. Speed up compressor or add compressor capacity. 2. Reduce load.
	B. Refrigerant metering device is underfeeding, causing the compressor to run at too low a suction pressure.	1. Check and repair liquid feed problems. 2. Check discharge pressure and increase if low.
	C. Faulty control circuit, may be low pressure control or capacity controls.	1. Check and repair.
	D. Compressor may have broken valve plates.	1. Check compressor for condition of parts. This condition can usually be detected by checking compressor discharge temperature.
	E. Thermostat control is defective and keeps unit running.	1. Check temperatures of product or space and compare with thermostat control. Replace or readjust thermostat.

Figure 29-6. *Troubleshooting chart for use with industrial refrigeration systems employing an external drive compressor. (Vilter Manufacturing Corporation)*

PROBLEM	POSSIBLE CAUSE	REMEDY OR COMMENT
	F. Defrost system on evaporator not working properly.	1. Check and repair as needed.
	G. Suction bags in strainers are dirty and restrict gas flow.	1. Clean or remove.
	H. Hot gas bypass or false load valve stuck.	1. Check and repair or replace.
VII. COMPRESSOR LOSES EXCESSIVE AMOUNT OF OIL.	A. High suction superheat causes oil to vaporize.	1. Insulate suction lines. 2. Adjust expansion valves to proper superheat. 3. Install liquid injection (suction line desuperheating).
	B. Too low of an operating level in chiller will keep oil in vessel.	1. Raise liquid level in flooded evaporator (R-12 systems only).
	C. Oil not returning from separator.	1. Make sure all valves are open. 2. Check float mechanism and clean orifice. 3. Check and clean return line.
	D. Oil separator is too small.	1. Check selection.
	E. Broken valves cause excessive heat in compressor and vaporization of oil.	1. Repair compressor.
	F. "Slugging" of compressor with liquid refrigerant that causes excessive foam in the crankcase.	1. "Dry up" suction gas to compressor by repairing evaporator. 2. Refrigerant feed controls are overfeeding. 3. Check suction trap level controls. 4. Install a refrigerant liquid transfer system to return liquid to high side.
VIII. NOISY COMPRESSOR OPERATION.	A. Loose flywheel or coupling.	1. Tighten.
	B. Coupling not properly aligned.	1. Check and align if required.
	C. Loose belts.	1. Align and tighten per specs. 2. Check sheeve grooves.
	D. Poor foundation or mounting.	1. Tighten mounting bolts, grout base, or install heavier foundation.
	E. Check compressor with stethoscope if noise is internal.	1. Open, inspect, and repair as necessary.
	F. Check for liquid or oil slugging.	1. Eliminate liquid from suction mains. 2. Check crankcase oil level.
IX. LOW EVAPORATOR CAPACITY.	A. Inadequate refrigerant feed to evaporators.	1. Clean strainers and driers. 2. Check expansion valve superheat setting. 3. Check for excessive pressure drop due to change in elevation, too small of lines (suction and liquid lines). A heat exchanger may correct this. 4. Check expansion valve size.
	B. Expansion valve bulb in a trap.	1. Change piping or bulb location to correct.
	C. Oil in evaporator.	1. Warm the evaporator, drain oil, and install an oil trap to collect oil.
	D. Evaporator surface fouled.	1. Clean.
	E. Air or product velocity is too low.	1. Increase to rated velocity. 2. Coils not properly defrosting. 3. Check defrost time. 4. Check method of defrost.
	F. Brine flow through evaporator may be restricted.	1. Chiller may be fouled or plugged. 2. Check circulating pumps. 3. Check process piping for restriction.
X. DISCHARGE PRESSURE TOO HIGH.	A. Air in condenser.	1. Purge noncondensibles.
	B. Condenser tubes fouled.	1. Clean.
	C. Water flow is inadequate.	1. Check water supply and pump. 2. Check control valve. 3. Check water temperature.
	D. Air flow is restricted.	1. Check and clean: a. Coils b. Eliminators c. Dampers
	E. Liquid refrigerant backed up in condenser.	1. Find source of restriction and clear. 2. If system is overcharged, remove refrigerant as required. 3. Check to make sure equalizer (vent) line is properly installed and sized.
	F. Spray nozzles on evaporator condensers plugged.	1. Clean.

Figure 29-6. *Continued.*

PROBLEM	POSSIBLE CAUSE	REMEDY OR COMMENT
XI. DISCHARGE PRESSURE TOO LOW.	A. Ambient air is too cold.	1. Install a fan cycling control system.
	B. Water quantity not being regulated properly through condenser.	1. Install or repair water regulating valve.
	C. Refrigerant level low.	1. Check for liquid seal, add refrigerant if necessary.
	D. Evaporator condenser fan and water switches are improperly set.	1. Reset condenser controls.
XII. SUCTION PRESSURE TOO LOW.	A. Light load condition.	1. Shut off some compressors. 2. Unload compressors. 3. Slow down RPM of compressor. 4. Check process flows.
	B. Short of refrigerant.	1. Add if necessary.
	C. Evaporators not getting enough refrigerant.	1. Discharge pressure too low. Increase to maintain adequate refrigerant flow. 2. Check liquid feed lines for adequate refrigerant supply. 3. Check liquid line driers.
	D. Refrigerant metering controls are too small.	1. Check superheat or liquid level and correct as indicated.
XIII. SUCTION PRESSURE TOO HIGH.	A. Low compressor capacity.	1. Check compressors for possible internal damage. 2. Check system load. 3. Add more compressor capacity.

Figure 29-6. *Continued.*

29.4.1 Technician Appearance and Conduct

Your appearance and conduct contribute to the company image. A neat personal appearance will help to create the sense of confidence that is necessary in dealing with the customer. It will affect the customer's attitude toward the service performed.

Arriving on time and displaying good work habits will create a desirable impression. Accurate and efficient work and respect for the customer's property will build the customer's trust. The customer is then more likely to be satisfied with the service. As a service technician, you should be aware that positive remarks concerning the company contribute to good customer relations. It is also important to have respect for company vehicles and equipment. This will reflect a concern for the total operation of the company.

29.4.2 Arriving on the Job

When arriving on the job, you should identify yourself as well as the company. The reason for the call should be stated. The customer should then be asked some specific questions: what has occurred, when it was first noticed, and how many times it has occurred. Any additional inquiries applicable to the situation should be made. A polite attitude when asking these questions will help you obtain the information needed to determine the problem and make the repair.

You should write on the service contract or work order any information that the customer volunteers concerning previous problems with the equipment. Also, any interest shown by the customer in add-on equipment or new contractual agreements should be noted.

When servicing is completed, you should present the proper billing forms for the customer to sign. See **Figure 29-9.** If applicable, indicate to the customer what can be done in the future so that the problem will not occur again.

29.4.3 Maintenance Service Contracts

Consumers have become very familiar with the concept of the ***maintenance service contract.*** All new cars today, for example, are sold with an extended maintenance contract available to the owners. By this means, the purchaser can extend the service contract on the vehicle. Maintenance agreements are used for many items: telephones, household appliances, television sets, VCRs, and other products. This wide application has created a public awareness of the benefits of maintenance contracts.

In the air conditioning field, many companies offer this service to their customers. Service contracts vary, depending on the particular company. The service agreement may cover all repairs and replacements that are needed in a given period of time. The agreement shown in **Figure 29-10** indicates specifically what the "twice-a-year" tune-up and cleaning includes. It identifies the basic tasks to be performed, such as cleaning the evaporator coil, replacing filters, and checking fan blades for tightness. The agreement also identifies what benefits the owner will receive from this contract. Included in the benefits are:

- Cooling and heating costs reduced by $200 per year.
- Efficiency of units improved by 21%.
- 24-hour emergency service, 365 days a year.

The agreement then identifies specific equipment make, warranty date, model number, and serial number. Note the no-risk three-month money-back guarantee. Such a guarantee encourages the homeowner to sign the contract.

ICE FLAKER MACHINE
TROUBLESHOOTING AND SERVICE CHART

SEQUENCE OF OPERATION

1. At start-up all thermostats are warm, electric power flows from fuse thru Bin Thermostat directly to the compressor (through High- and Low-Pressure Cut-Outs as applicable) and through the warm side of the Gear Thermostat to the Gearmotor (and in parallel to the Green light).
   1a. Original pull down of Flaker may be slow, 5 to 10 minutes before ice is produced. During the last half of pull down, the suction line at the compressor will be frosted, and the sight glass will show bubbles during pull down. After ice starts to be produced, frost will leave the compressor suction line and within 5 minutes the sight glass will show clear or only a small bubble in the top, which will disappear slowly.
2. As ice begins to form, the suction line from the evaporator (location of Gear Thermostat cap tube) gets cold enough (below 32°) to trip Gear Thermostat to the cold side bypassing the Bin Thermostat (resetting for later Ice Unloading period) so power flows directly through the Gear Thermostat to the Gearmotor (and Green light); no change is apparent.
   2a. [On rare occasions with some particle-free waters, the water will subcool substantially below freezing, then suddenly crystallize and stall the Gearmotor (designed to withstand a stall). The evaporator will continue to get colder until the bottom (location of the Defrost Thermostat cap tube) gets below freezing, tripping the Defrost Thermostat cold, sending power to the Defrost Valve (and Red light) which sends hot refrigerant through the evaporator, releasing the Auger and Gearmotor until the bottom of the evaporator reaches about 45°, tripping the Defrost Thermostat open, and normal refrigeration restarts. Since subcooling with crystallization occurs only 2 or 3 times out of 100, it is unlikely to happen and very unlikely to repeat.]
3. During ice formation, ice is drawn to the top of the evaporator, lowering the water level, opening the float valve. When water-flow is seen, it indicates ice formation even before ice appears in the bin coming from the ice tube. Water flows through the bottom of the auger and from a hole near the bottom flight into the ice chamber, cools to 32°, then freezes in thin layers on the wall of the evaporator cylinder where the auger flights remove it and spiral it upward. The ice thickens until it is a fairly dense mass as it rotates under the helical path of the deflector until it is forced into the discharge spout by the deflector. The ice then flows by gravity through the ice tube to give it velocity which spreads it in the bin.
4. When ice piles over about half the length of the 1/4" stainless tube which holds the Bin Thermostat cap tube, the thermostat opens, shutting off power to the compressor. (Turning on the Yellow light which is connected in parallel with the thermostat). The Green light stays on as the Gear Thermostat is still cold supplying power from the fuse to the Gearmotor (and Green light) to Unload Ice from the evaporator until the freezing surface is above 32°. The Gearmotor must remain on a couple minutes to complete unloading and usually stays on 5 to 15 minutes until the suction line out of the evaporator warms to about 45°. Then the Gear Thermostat trips warm connecting the Gearmotor to the output side of the bin thermostat (resetting for the next start-up) shutting off the Gearmotor and Green light.
5. When the bin is full, only the Yellow light is illuminated indicating power is on and fuse okay. Only a trickle of power is used through Yellow neon-type light during flaker-off periods. Flaker may start up while a small amount of ice is still on the 1/4" stainless steel tube. This is okay as the deflector can push ice through the ice tube creating some spread around the top of the bin until the stainless steel tube is fairly well covered again.

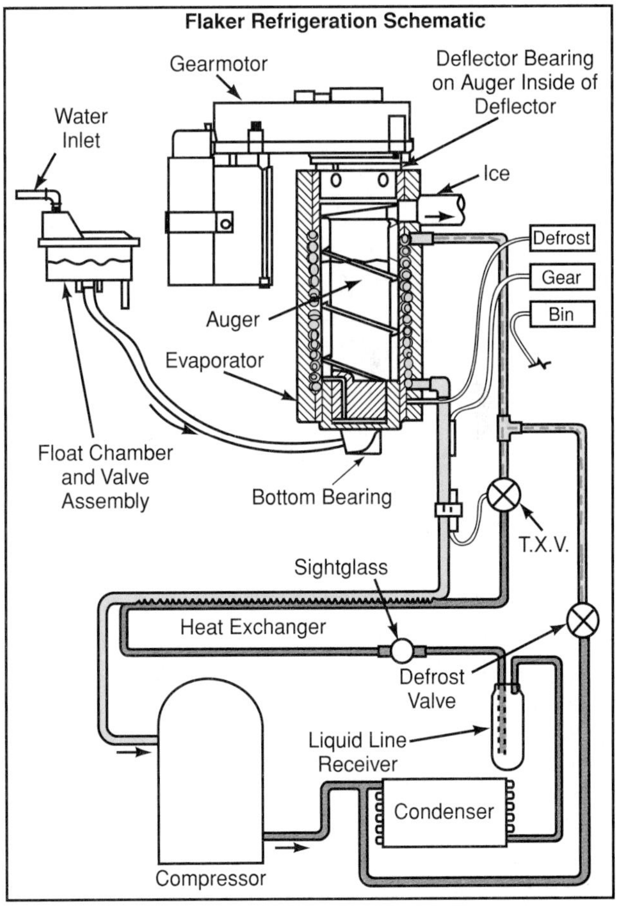

Figure 29-7. *Ice flaker machine service procedures and troubleshooting charts. (Kold-Draft Division, Uniflow Manufacturing Company)*

TROUBLE, CAUSE, AND REMEDY		
TROUBLE	CAUSE	REMEDY
1. Flaker will not operate. No lights on in control box.	1. Line fuse blown.	1. Check circuits for short or ground. Replace fuse.
	2. Loose connection in control box or in power supply line.	2. Check for power supply at controls in control box. Check connections to bin thermostat.
1a. Flaker will not operate. Yellow light on, but no ice on bin thermostat.	1a. Bin control set too warm in a cold room between 45° and 55°.	1a. Set bin thermostat colder (cw) but recheck with ice to be sure it will shut off.
	2a. Room below 45°.	2a. Add heat to the room.
	3a. Bin control has lost charge.	3a. Replace bin control.
2. Condensing fan and gearmotor operate (green light on) but not the compressor.	1. Inoperative capacitors or relay.	1. Replace capacitors or relay.
	2. Overload switch defective.	2. Replace overload switch.
	3. Loose connections or defective compressor.	3. Check for power at compressor C-R terminals, C-S terminals. With power off, remove "C connection", check ohms between C and R, also C and S.
2a. Water-cooled flaker: gearmotor operates (green light on) but not the condensing fan or compressor.	1a. High-pressure cut-out open; inadequate water supply.	1a. Check water supply and condenser water valve.
	2a. Water supply okay, high pressure cut-out won't close with condenser cool.	2a. Replace defective high-pressure cut-out.
3. Compressor operating but fan off.	1. Circuit not complete.	1. Check circuit.
	2. Fan motor burned out.	2. Replace motor.
4. Condenser fan operating but compressor unit operating intermittently.	1. Dirty condenser coil.	1. Clean coil.
	2. High or low voltage.	2. Correct to proper voltage within 10% of nameplate.
	3. Excessive refrigerant.	3. Remove some refrigerant, check sight glass.
5. Intermittent defrost, red light cycling on and off. Water level normal in float tank.	1. Water line elbow in bottom bearing in too far.	1. Back out elbow 1 turn.
	2. Defrost thermostat misadjusted, very cold supply water.	2. See service manual for defrost thermostat setting. Also see Abnormal Defrost Themostat Setting in the manual.
	3. TXV too far open.	3. Close TXV 1/4 turn, see service manual for proper setting.
	4. Deflector partially closed.	4. See service manual for correct deflector setting.
	5. Gearmotor not running.	5. Check power to gearmotor receptacle in bottom of control box.
	6. Gearmotor stalled with power on, bottom evaporator mounting screw too tight.	6. Back off evaporator mounting screw 1 turn to see if gearmotor will operate after it cools down sufficiently for overload to cut back in. Also see #7.
	7. Gearmotor stalled.	7. Remove gearmotor and auger assembly and check for operation on workbench (take care not to damage auger flights). Remove bottom bearing, check for condition and check for snug but not a tight fit on bottom of auger. Check clearance between top of auger and bottom of deflector (see Flaker Mechanism Assem. Instr. #5).

Figure 29-7. *Continued.*

TROUBLE, CAUSE, AND REMEDY		
TROUBLE	CAUSE	REMEDY
6. Wet ice.	1. High water level. 2. Undercharge, bubbles going through sight glass. 3. Misadjusted TXV.	1. Lower water level. 2. Check for leaks, add R-12. 3. Adjust TXV (see "Expansion Valve Adjustment").
7. Ice too hard.	1. Low water level. 2. Deflector incorrectly adjusted. 3. TXV closed much too far. 4. Moisture in system and TXV partially frozen shut.	1. Raise water level. 2. SEE DEFLECTOR ADJUSTMENTS PROCEDURE. 3. See service manual for proper suction pressure and suction line temperature. 4. Dehydrate and recharge system.
8. No ice with gearmotor, compressor and condenser fan operating. (Red light not on, no power to defrost valve.)	1. Very low refrigerant. 2. Stuck defrost valve, defrost line warm, suction pressure above 20 pounds.	1. Repair leak, see service manual for proper charge. 2. Repair or replace defrost valve.
9. Flaker does not turn off.	1. Misadjusted bin thermostat. 2. Bin thermostat will not open when set warmest with ice on thermostat cap tube. 3. Mislocated bin thermostat cap tube.	1. Adjust bin thermostat (ccw), check with ice on coil. 2. Replace bin thermostat. 3. Check location of 1/4" stainless cap tube holder parallel and below ice path coming from ice tube. Check thermostat cap tube located in the 1/4" stainless tube.
10. Flaker cycles off and on.	1. Ice falling on bin thermostat capillary tube. 2. Ice tube not on outlet spout.	1. Relocate tube. 2. Remount ice tube and check for restriction.
11. Low production.	1. High head pressure. 2. Inadequate water supply. 3. TXV misadjusted. 4. High ambient. 5. Deflector closed. 6. Low head pressure.	1a. Clean condenser. Improve water supply to water-cooled condenser. Improve ventilation. 1b. Replace head pressure control valve on remote condenser models. 2. Check and clean filters. Raise water level. 3. Adjust TXV. 4. Decrease ambient to 90°F. Max. 5. SEE DEFLECTOR ADJUSTMENTS PROCEDURE. 6a. Adjust water regulating valve. 6b. Add R-12 to remote condenser models. 6c. Refer to Remote Condenser section of Service Manual.
12. Flaker spouts coming off.	1. I.D. of discharge tubing too small. 2. Roll pins breaking on deflector. 3. Hose kinking.	1. Change to braided nylon tubing. 2. Readjust defrost thermostat. 3. Change to braided nylon tubing, reroute presently used tubing.

Figure 29-7. *Continued.*

SERVICE PROCEDURE S-17

CHECKING DUCT STATIC S-17

The maximum and minimum allowable external static pressures are found in the specification section. These tables also show the amount of air being delivered at a given static by a given motor speed or pulley adjustment.

The furnace motor cannot deliver proper air quantities (CFM) against statics other than those listed.

Too great of an external static pressure will result in insufficient air that can cause excessive temperature rise, resulting in limit tripping, etc. Whereas not enough static may result in motor overloading.

To determine proper air movement, proceed as follows:

1. With clean filters in the furnace, use a draft gauge (inclined manometer) to measure the static pressure of the return duct at the inlet of the furnace. (Negative Pressure.)
2. Measure the static pressure of the supply duct. (Positive Pressure.)
3. Add the two readings together for total external static pressure.

NOTE: Both readings may be taken simultaneously and read directly on the manometer if so desired. If an air conditioning coil or Electronic Air Cleaner is used in conjunction with the furnace, the readings must also include these components.

4. Consult proper tables for the quantity of air.

If the total external static pressure exceeds the minimum or maximum allowable statics, check for closed dampers, registers, undersized and/or oversized poorly laid out duct work.

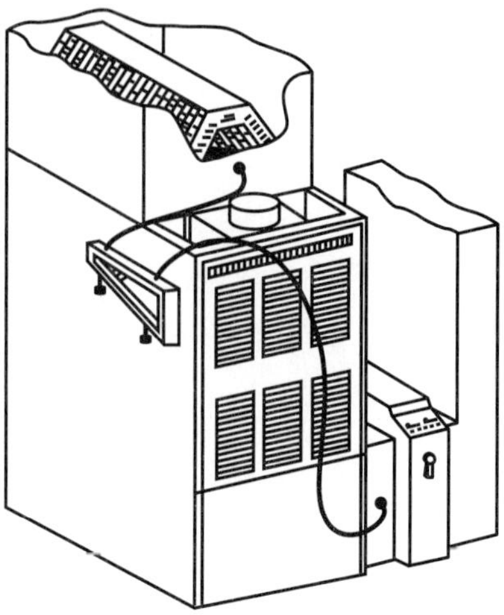

Checking Static Pressure

SERVICING
HEATING ANALYSIS GUIDE

POSSIBLE CAUSE	COMPLAINT									TEST METHOD–REMEDY	SEE SERVICE PROCEDURE
	NO HEAT			UNSATISFACTORY HEAT							
	SYSTEM WILL NOT START	BURNER WON'T IGNITE	BURNER IGNITES-LOCKS OUT	MAIN BURNER SHUTS OFF PRIOR TO T-STAT BEING SATISFIED	SHORT CYCLES	LONG CYCLES	SOOT AND/OR FUMES	TOO MUCH HEAT	NOT ENOUGH HEAT		
NO MAIN POWER	●									TEST VOLTAGE	S-1
FAULTY TRANSFORMER	●									TEST TRANSFORMER	S-4
FAULTY THERMOSTAT	●									TEST THERMOSTAT	S-3
FAULTY LIMIT SWITCH		●		●						TEST FAN AND LIMIT CONTROL	S-6
FAULTY FLAME SENSOR			●							TEST FLAME SENSOR	S-22
FAULTY IGNITION CONTROL MODULE		●	●							TEST IGNITION CONTROL MODULE	S-21
GAS SUPPLY VALVES IN OFF POSITION OR GAS VALVE OFF		●								TURN VALVES TO "ON" POSITION	
FAULTY INDUCED DRAFT BLOWER MOTOR	●									TEST MOTOR	S-8, S-9
OPEN DOOR SWITCH	●									CHECK DOOR CLOSED-TEST SWITCH	
FAULTY WIRING HARNESS		●								TEST WIRING	S-2
BROKEN OR SHORTED IGNITER		●								TEST IGNITER	S-20
FAULTY COMBUSTION RELAY	●			●						TEST RELAY	S-5
SENSOR NOT IN FLAME, LOW MICRO-AMPS			●							TEST FLAME SENSOR	S-22
FAULTY GAS VALVE OPR.		●								TEST GAS VALVE MAIN OPERATOR	S-10
OPEN AUX. LIMIT		●		●						PUSH MANUAL RESET	S-7
IMPROPER HEAT ANTICIPATOR SETTING				●	●	●				CHECK HEAT ANTICIPATOR SETTING	S-3B
IMPROPER AIR FLOW OR DISTRIBUTION				●	●			●	●	CHECK DUCT STATIC	S-17
CYCLING ON LIMIT					●				●	CHECK CONTROLS & TEMP. RISE	S-6, S-7, S-18
IMPROPER THERMOSTAT LOCATION					●	●		●	●	RELOCATE THERMOSTAT	
DELAYED IGNITION							●			TEST FOR DELAYED IGNITION	S-15
FLASHBACK							●			TEST FOR FLASHBACK	S-16
ORIFICE SIZE							●			CHECK ORIFICES	S-13
GAS PRESSURE		●					●		●	CHECK GAS PRESSURE	S-14
BURNED OUT HEAT EXCHANGER, CHECK TEMPERATURE RISE							●			CHECK TEMP. RISE	S-18
THERMOSTAT OUT OF CALIBRATION								●	●	RECALIBRATE OR REPLACE	
STUCK GAS VALVE								●		REPLACE GAS VALVE	
FURNACE UNDERSIZED									●	REPLACE FURNACE WITH PROPER SIZE	
NO LOW VOLTAGE	●									CHECK WIRING	S-2
FAULTY PRESSURE SWITCH		●		●						TEST PRESSURE SW.	S-19
BLOCKED OR RESTRICTED FLUE		●		●						CHECK FLUE	S-19
HIGH RESISTANCE GROUND			●							TEST GROUND	S-21
VENT SAFETY SWITCH OPEN	●									TEST CONTINUITY	S-23

Figure 29-8. *Heating analysis guide for gas-fired furnace. The following has been selected as an example. Step 1—The complaint was unsatisfactory heat (too much heat). Step 2—The possible causes are: improper air flow or distribution, improper thermostat location, thermostat out of calibration, or stuck gas valve. Step 3—The possible remedies are: check duct static (this information would be obtained from the Service Procedure S-17), relocate thermostat, recalibrate or replace thermostat, or replace gas valve. (Amana Refrigeration, Inc.)*

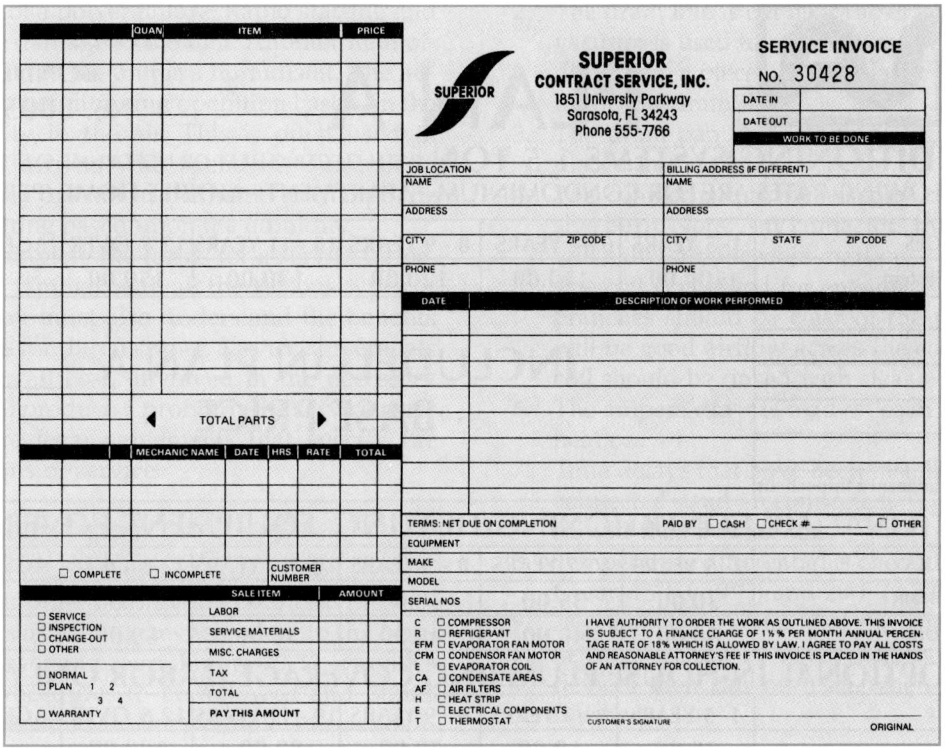

SERVICE INVOICE
NO. 30428

SUPERIOR
CONTRACT SERVICE, INC.
1851 University Parkway
Sarasota, FL 34243
Phone 555-7766

DATE IN
DATE OUT
WORK TO BE DONE

QUAN	ITEM	PRICE
	TOTAL PARTS	

MECHANIC NAME	DATE	HRS.	RATE	TOTAL

☐ COMPLETE ☐ INCOMPLETE CUSTOMER NUMBER

	SALE ITEM	AMOUNT
☐ SERVICE ☐ INSPECTION ☐ CHANGE-OUT ☐ OTHER	LABOR	
	SERVICE MATERIALS	
	MISC. CHARGES	
☐ NORMAL ☐ PLAN 1 2 3 4	TAX	
	TOTAL	
☐ WARRANTY	PAY THIS AMOUNT	

JOB LOCATION — NAME, ADDRESS, CITY, ZIP CODE, PHONE
BILLING ADDRESS (IF DIFFERENT) — NAME, ADDRESS, CITY, STATE, ZIP CODE, PHONE

DATE	DESCRIPTION OF WORK PERFORMED

TERMS: NET DUE ON COMPLETION PAID BY ☐ CASH ☐ CHECK # ☐ OTHER
EQUIPMENT
MAKE
MODEL
SERIAL NOS

C ☐ COMPRESSOR
R ☐ REFRIGERANT
EFM ☐ EVAPORATOR FAN MOTOR
CFM ☐ CONDENSOR FAN MOTOR
E ☐ EVAPORATOR COIL
CA ☐ CONDENSATE AREAS
AF ☐ AIR FILTERS
H ☐ HEAT STRIP
E ☐ ELECTRICAL COMPONENTS
T ☐ THERMOSTAT

I HAVE AUTHORITY TO ORDER THE WORK AS OUTLINED ABOVE. THIS INVOICE IS SUBJECT TO A FINANCE CHARGE OF 1½% PER MONTH ANNUAL PERCENTAGE RATE OF 18% WHICH IS ALLOWED BY LAW. I AGREE TO PAY ALL COSTS AND REASONABLE ATTORNEY'S FEE IF THIS INVOICE IS PLACED IN THE HANDS OF AN ATTORNEY FOR COLLECTION.

CUSTOMER'S SIGNATURE
ORIGINAL

Figure 29-9. *Service invoice. Note the various categories: Item, Description of Work Performed, Equipment, Make, Model, Terms, etc. (Superior Contract Services, Inc.)*

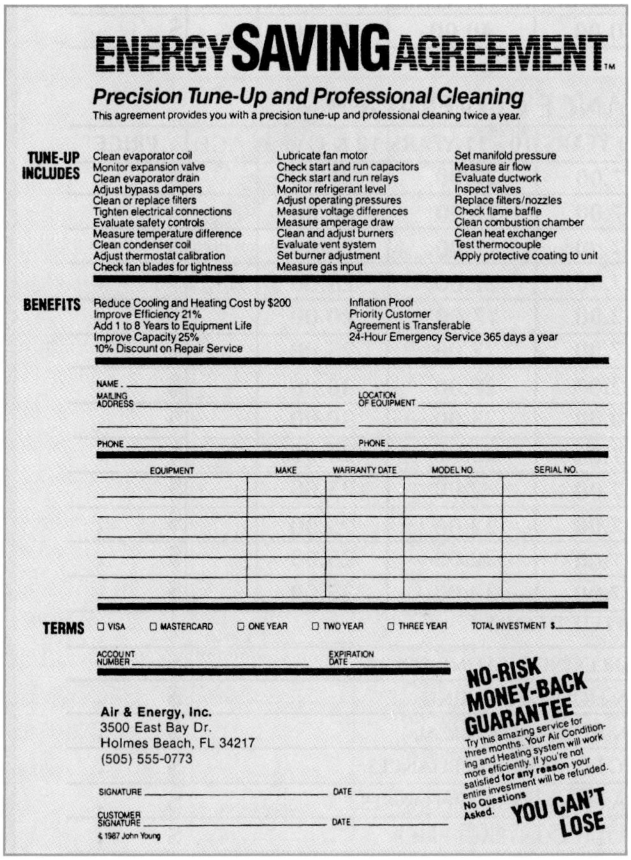

ENERGY SAVING AGREEMENT™

Precision Tune-Up and Professional Cleaning

This agreement provides you with a precision tune-up and professional cleaning twice a year.

TUNE-UP INCLUDES
Clean evaporator coil
Monitor expansion valve
Clean evaporator drain
Adjust bypass dampers
Clean or replace filters
Tighten electrical connections
Evaluate safety controls
Measure temperature difference
Clean condenser coil
Adjust thermostat calibration
Check fan blades for tightness
Lubricate fan motor
Check start and run capacitors
Check start and run relays
Monitor refrigerant level
Adjust operating pressures
Measure voltage differences
Measure amperage draw
Clean and adjust burners
Evaluate vent system
Set burner adjustment
Measure gas input
Set manifold pressure
Measure air flow
Evaluate ductwork
Inspect valves
Replace filters/nozzles
Check flame baffle
Clean combustion chamber
Clean heat exchanger
Test thermocouple
Apply protective coating to unit

BENEFITS
Reduce Cooling and Heating Cost by $200
Improve Efficiency 21%
Add 1 to 8 Years to Equipment Life
Improve Capacity 25%
10% Discount on Repair Service
Inflation Proof
Priority Customer
Agreement is Transferable
24-Hour Emergency Service 365 days a year

NAME
MAILING ADDRESS
LOCATION OF EQUIPMENT
PHONE
PHONE

EQUIPMENT	MAKE	WARRANTY DATE	MODEL NO.	SERIAL NO.

TERMS ☐ VISA ☐ MASTERCARD ☐ ONE YEAR ☐ TWO YEAR ☐ THREE YEAR TOTAL INVESTMENT $
ACCOUNT NUMBER
EXPIRATION DATE

Air & Energy, Inc.
3500 East Bay Dr.
Holmes Beach, FL 34217
(505) 555-0773

SIGNATURE DATE
CUSTOMER SIGNATURE DATE
© 1987 John Young

NO-RISK MONEY-BACK GUARANTEE
Try this amazing service for three months. Your Air Conditioning and Heating system will work more efficiently. If you're not satisfied **for any reason** your entire investment will be refunded. No Questions Asked.
YOU CAN'T LOSE

Figure 29-10. *A twice-a-year cleaning and tune-up agreement. Note the numerous items that the technician performs. (Air & Energy, Inc.)*

Another type of service agreement asks whether the customer can: inspect the air conditioning system, replace filters, clean coils, check motor and compressor, lubricate motor, calibrate thermostat, and other items. It then indicates that, if the owner cannot perform these tasks, the company will do so. For a given cost, the company will also furnish all parts, labor, and material necessary.

29.4.4 Contractual Agreement

Contractual agreements often are purchased as a result of an initial service call. The form is filled out by the technician after speaking with the owner. These contracts provide repair services as well as periodic maintenance.

A typical contractual agreement is shown in **Figure 29-11.** Note the technician is to indicate the equipment ages for air conditioning and heating units. The ages will determine the actual cost of the maintenance and service contract. For a system that is five years old, the annual contract charge is $100. The charge is $130 for one that is ten or more years old.

Contractual agreements vary, depending upon the particular company. Some contracts cover domestic refrigerators, automatic ice makers, and other major appliances. Others cover commercial appliances. The contractual agreement gives the company an opportunity to sell additional services. These will help reduce the owner's cost of operating the unit. An example is having a time delay installed in the system. The time delay prevents the unit from starting immediately after it has

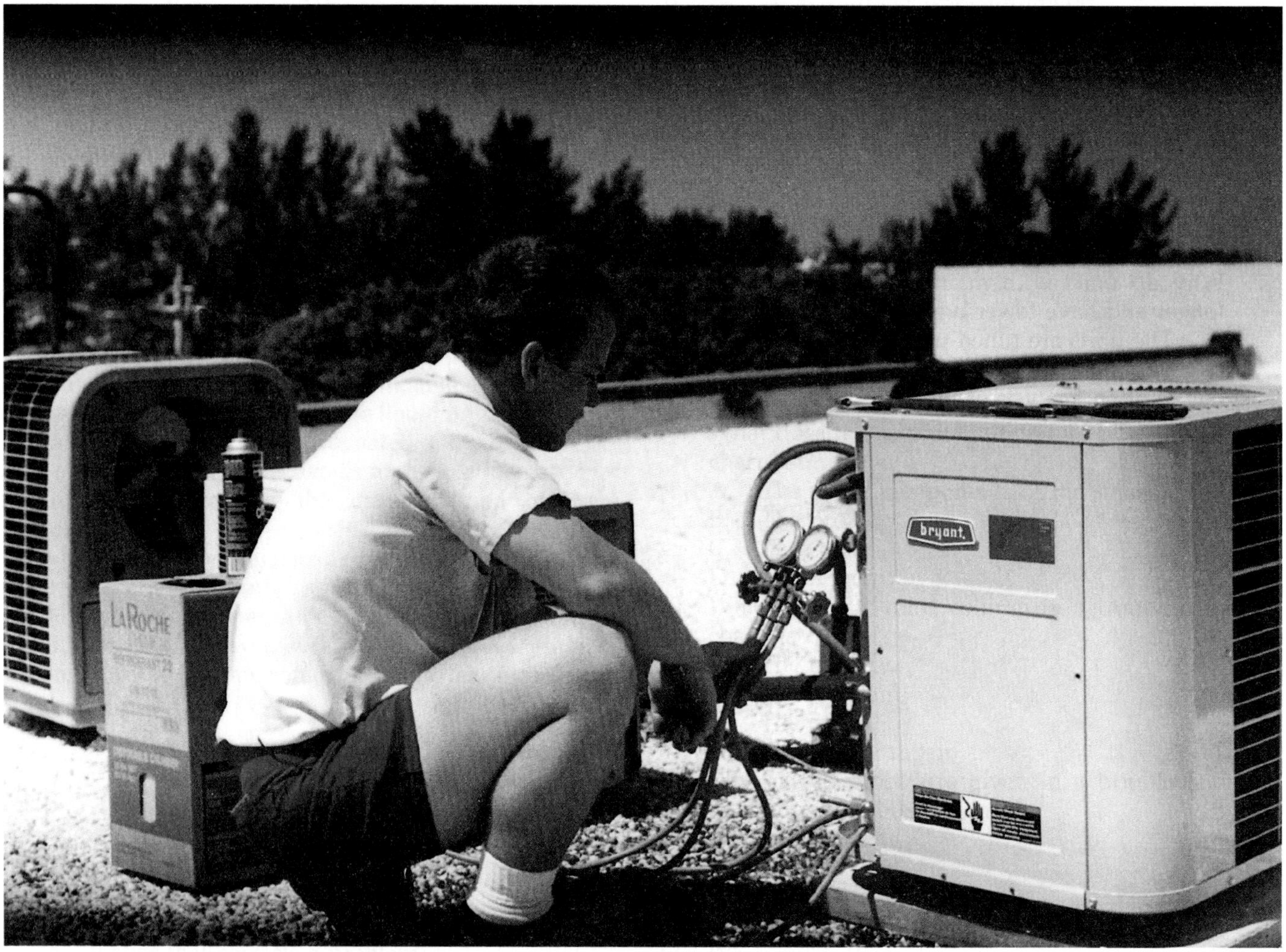

Servicing a rooftop compressor. (Superior Contract Services, Inc.)

Chapter 30

PASSING EPA EXAMS

Modules:

Key Words:

Clean Air Act—Section 608
core questions
exam preparation
test format
Type I: Small appliances
Type II: High-pressure or very high-pressure appliances
Type III: Low-pressure appliances
Type IV: "Universal Certification"

Learning Objectives:

After studying this chapter, you will be able to:

- Properly prepare for taking EPA, state, local, and organizational exams.
- Understand the EPA regulations as they relate to air conditioning and refrigeration.
- Be knowledgeable of the four EPA classifications—Type I, Type II, Type III, and Type IV.
- Develop proper examination techniques for taking the tests.
- Become knowledgeable of the areas that may be covered on the EPA exams.

PREPARING FOR EXAMS MODULE

30.1 EPA Technician Certification Exams

The purpose of this chapter is to provide basic information needed for taking EPA technician certification exams for refrigeration and air conditioning. It will also help in developing procedures for passing State, local, and organizational exams. See **Figure 30-1.** The first module of this chapter provides suggestions for preparing for the exam. The second module presents some typical questions in a format similar to that used on EPA exams.

The EPA certification exam for automotive air conditioning is a separate exam. State and local licensing may vary in each locality. Therefore, you will have to contact the State and local authorities for information. Becoming familiar with local or national air conditioning organizations may help you. It is important to observe all local regulations and licensing.

30.1.1 EPA Requirements

The depletion of the ozone has been an international concern. The threat of ozone molecules being destroyed by chlorine molecules was a major reason for the Environmental Protection Agency (EPA) to require technician certification.

The EPA currently requires technicians who service air conditioning and refrigeration equipment that uses CFC or HCFC refrigerant to be certified. This requirement also includes anyone who installs, maintains, or repairs equipment and may reasonably have the opportunity to release CFC or HCFC to the atmosphere. In addition, anyone who disposes of refrigerant or air conditioning equipment must be certified.

Since November 14, 1994, only technicians certified by an EPA certifying agency are permitted to purchase refrigerant. Certification requires taking and passing a

EPA
Environmental Protection Agency

CLASSIFICATION FOR TECHNICIAN CERTIFICATION:

TYPE I: Small Appliances.
TYPE II: High-pressure or very high-pressure appliances.
TYPE III: Low-pressure appliances.
TYPE IV: "Universal Certification."

MICHIGAN DEPARTMENT OF LABOR

APPLICATION FOR MECHANICAL CONTRACTOR LICENSE EXAM
MICHIGAN DEPARTMENT OF LABOR
BUREAU OF CONSTRUCTION CODES/BOARD OF MECHANICAL RULES
7150 HARRIS DR., P.O. BOX 30255
LANSING, MICHIGAN 48909

1. Hydronic heating and cooling and process piping.
2. HVAC equipment.
3. Ductwork.
4. Refrigeration.
5. Limited heating service.
6. Unlimited heating service.
7. Limited refrigeration and air conditioning service.
8. Unlimited refrigeration and air conditioning service.
9. Fire suppression.
10. Speciality license.

Texas Department of Licensing and Regulation
P.O. Box 12157 Austin, Texas 78711

TEXAS AIR CONDITIONING AND REFRIGERATION CONTRACTORS
EXAMINATION AND LICENSE APPLICATION

___ CLASS A AIR CONDITIONING
___ CLASS A COMMERCIAL REFRIGERATION
___ CLASS B AIR CONDITIONING
___ CLASS B COMMERCIAL REFRIGERATION

Figure 30-1. *The EPA exam has four categories. The first through third cover specific areas of servicing. Type IV covers all three areas. In addition to the EPA exams that are required nationally, each state may have its own requirements. For example, Michigan has ten types of licenses; Texas has four.*

test. The test questions and the organization giving the test must be EPA approved. The Clean Air Act, Section 608, charges the EPA with establishing and enforcing regulations pertaining to the refrigeration industry.

The EPA regulations cover a wide scope of topics. A summary of the final regulation follows:

- Requires service practices that maximize recycling of ozone-depleting compounds, both CFCs and HCFCs, during service and disposal of air conditioning and refrigeration equipment.
- Sets certification requirements for recycling and recovery equipment, technicians, and reclaimers.
- Restricts the sale of refrigerant to certified technicians.
- Requires persons servicing or disposing of air conditioning equipment to certify to the EPA that they have acquired recycling or recovery equipment and are complying with the requirements of the rule.
- Requires the repair of substantial leaks in air conditioning and refrigeration equipment with a charge of greater than 50 lb.
- Establishes safe disposal requirements to ensure removal of refrigerants from goods that enter the waste stream with the charge intact (e.g., motor vehicle air conditioners, home refrigerators, room air conditioners, etc.)

30.1.2 The Clean Air Act—Section 608

In order to understand EPA certification testing, one must have a better understanding of the Clean Air Act—Section 608. In 1990 the Clean Air Act became a law. Its regulations became effective on May 14, 1993. The goal of the Act was to increase environmental awareness, increase conservation, and reduce pollution. This section defines the procedures to reduce the release of chemicals into the atmosphere. In broad terms, the Clean Air Act—Section 608 does the following:

- Controls refrigerant purchase.
- Requires repair of large refrigeration leaks.
- Increases cooperation among countries in an effort to reduce emissions.
- Reduces industrial use of refrigerants.
- Discontinues the production of certain refrigerants.

The Clean Air Act—Section 608 delegates to the EPA the power to establish and enforce these regulations regarding the refrigeration / air conditioning industry.

30.2 Certification Types

Technicians must be certified in the equipment category that they will be servicing or installing. Each of the four certification levels directly correlates to the type of service provided and the type of equipment serviced. The following are the EPA classifications for technician certification:

- Type I: Small appliances.
- Type II: High-pressure or very high-pressure appliances.
- Type III: Low-pressure appliances.
- Type IV: "Universal Certification."

A further description of the types of certification follows. It is important to remember that the type of certification indicates the type of equipment on which you can work. Being certified in Type I does not mean you can perform service on Type II, III or IV equipment.

30.2.1 Type I Certification: Small Appliances

These are appliances with a factory refrigerant charge of five pounds or less. These may include the following:

- Freezers and refrigerators designed for home use.
- Window air conditioners and packaged terminal air conditioner (PTAC) units.
- Packaged terminal heat pumps.
- Humidifiers.
- Counter-type ice makers.
- Vending machines.
- Drinking water coolers.

30.2.2 Type II Certification: High- or Very-High Pressure Appliances

This includes applications using more than five pounds of charge of any refrigerant other than R-11, R-113, and R-123. Examples of refrigerants included are R-12, R-114, R-500, and R-502. The boiling point of the refrigerant used is the key to defining high and very high pressure. Boiling points between −58°F (−50°C) and 50°F (10°C) at atmospheric pressure are high-pressure. Refrigerants with a boiling point below −58°F (−50°C) at atmospheric pressure are considered very high-pressure. This type of certification includes the following types of equipment:

- Residential air conditioning and heat pumps.
- Commercial air conditioning equipment.
- Chillers (other than low-pressure).
- Commercial refrigeration.
- Transport refrigeration.

30.2.3 Type III Certification: Low-Pressure Appliances

These appliances are mostly centrifugal compressors using R-11 refrigerant. Refrigerants with a boiling point above 50°F (10°C) at atmospheric pressure are considered low-pressure. The equipment includes:

- Chillers.
- Equipment using R-11, R-113, R-123, and others.

30.2.4 Type IV Certification: Universal Certification

This type of certification is defined as covering all types of appliances. Technicians holding Type IV Certification may work on appliances covered under Types I, II, and III Certification.

30.3 Recovery Procedures

The Clean Air Act requires a technician to evacuate air conditioning and refrigeration units to specific levels. These levels ensure that discharge of refrigerant into the atmosphere will not occur. Two levels of vacuum are legally acceptable. One level is for recovery equipment manufactured prior to November 15, 1993. The other level is for recovery equipment manufactured after November 15, 1993, which has been tested and certified by an EPA organization. See **Figure 30-2** for the required levels.

30.3.1 Recovery Procedures, Type I Certification

The recovery procedures depend on the size and type of equipment being serviced. With Type I Certification, recovery devices may be divided into two types—self-contained and system-dependent recovery equipment.

Appliance Type	Inches Hg. Vacuum* Using Eqt. Manufactured:	
	Before Nov. 15, 1993	*After* Nov. 15, 1993
R-22 appliance** containing less than 200 lb. of refrigerant	0	0
R-22 appliance** containing 200 lb. or more of refrigerant	4	10
Other high-pressure appliance** containing less than 200 lb. of refrigerant (R-12, -500, -502, -114)	4	10
Other high-pressure appliance** containing 200 lb. or more of refrigerant (R-12, -500, -502, -114)	4	15
Very high pressure appliance (R-13, -503)	0	0
Low-pressure appliance (R-11, R-123)	25	29 (25 mm Hg. Absolute)

*Relative to standard atmospheric pressure of 29.9″ Hg.
**Or isolated component of such appliance

Figure 30-2. *Required evacuation vacuum levels for various appliances.*

A self-contained recovery unit has its own compressor to remove refrigerant from the refrigerating system. A system-dependent recovery unit depends on the compressor in the appliance or pressure of the refrigerant in the appliance.

For small appliances (Type I), recovery equipment must recover 90% of the refrigerant when the compressor is operating. If the compressor is not operating, 80% of the refrigerant must be recovered.

30.3.2 Recovery Procedures, Type II Certification

Type II Certification divides recovery by the refrigerant used and the amount of refrigerant in the equipment. EPA evacuation levels for Type II Certification are as listed:

- R-22 appliances with less than 200 lb. of charge, when using recovery equipment manufactured after November 15, 1993—evacuation level is 0″ Hg.
- R-22 appliances with more than 200 lb. of charge, when using recovery equipment manufactured after November 15, 1993—evacuation level is 10″ Hg.
- Other high-pressure appliances with less than 200 lb. of charge, when using recovery equipment manufactured after November 15, 1993—evacuation level is 10″ Hg.
- Other high-pressure appliances with more than 200 lb. of charge, when using recovery equipment manufactured after November 15, 1993—evacuation level is 15″ Hg.
- Very high pressure equipment, when using recovery equipment manufactured after November 15, 1993—evacuation level is 0″ Hg.

30.3.3 Recovery Procedures, Type III Certification

In this type of certification, chillers are initially recovered by using liquid recovery. The process is completed through vapor recovery. Chiller systems usually contain a large quantity of liquid refrigerant. Therefore, a liquid recovery will be faster and will recover most of the refrigerant. Specific liquid pumps are designed for recovery of low-pressure refrigerants. When liquid recovery is performed, it is important to take precautions to minimize the movement of contaminants to the recovery container.

Recovery for Type III Certification is divided by refrigerant used and the amount of charge in the equipment. The recovery requirement for low-pressure equipment, when using equipment manufactured after November 15, 1993, is an evacuation level of 25 mm Hg. absolute (25 mm Hg. absolute = 2500 microns = 29″ Hg.).

30.4 Leak Repairs

An annual leak rate of 35% or more in commercial or industrial equipment with a refrigerant charge of 50 lb. or more must be repaired. It is the duty of the technician to notify the equipment owner of the leak. The owner is required to repair the leak within thirty days.

Comfort cooling equipment with a charge of over 50 lb. is granted an annual leak rate of 15% of charge per year. (This excludes industrial process and commercial equipment.) Technicians who service such appliances must notify the equipment owner of the leak. The owner is required to repair the leak within thirty days.

30.5 Exam Preparation

First, you must determine the dates and times at which an EPA approved testing organization will be presenting the certification exam. These organizations normally provide a one- or two-day course, which assists in preparing for the exam. There is usually an additional fee for this program.

The second step is to determine what to study in preparing for taking the exam. Sample test questions are provided in Section 30.9. (Keep in mind that these questions are not identical to those on the exam.) After identifying weak areas, check the index to find the proper page for possible help. A review of Chapters 9 and 10 is recommended. It is important that the material used for exam preparation be current.

The third step is to begin review. Always keep a positive attitude throughout the studying process and during the exam. Additional information can be obtained from the EPA and local refrigeration organizations.

Studying for the tests can be divided into two areas. The first area includes those questions requiring what is called rote "specific" memorization. An example would be, "Who is the governor of your State?" Questions of this type are best studied shortly before the exam. This is because they are easily forgotten since they only require simple memorization.

The second type of question is more complex. It requires the evaluation of facts and situations. It is based on a situational understanding of many items. Such questions are slower and more difficult to learn than the rote "specific" memorization questions.

A useful technique to increase retention of materials is to be quizzed by another person. Simply use the sample questions. Such review will assist in determining those areas on which to focus.

30.6 Test Format

With most certification exams, the questions are divided into two groups. In Group One are questions concerning the laws which will affect technicians, ozone depletion, and industry practices. In Group Two, questions concerning safety, leak detection, shipping and disposal, and recovery techniques are addressed.

With most tests, you must pass a core-type (Group One) portion of the exam in order to be certified in any of the types of certification. With most approved agencies, the EPA certification tests have many similarities in contents and style. All tests are closed-book and taken in the presence of a proctor.

Tests for Types I, II, and III Certification include 25 core questions, and 25 additional questions for each type of certification. Universal certification exams consist of 100 questions—25 core questions and 75 additional questions (25 each from Types I, II, and III).

A passing score for the exam is 78%. Once certified, it is unnecessary to be recertified. You may choose to take additional tests at a later date to achieve further certifications.

30.7 Taking the Test

The certification exam will consist of multiple-choice questions. Each question will offer four suggested answers, only one of which is correct. Answers will be placed on a second document which will be electronically scored. Since a machine, rather than a person, will be scoring the exam, it is important to be precise. Do not doodle or make stray marks on the answer sheet. Complete the answer sheet carefully, being certain to fill in an answer for each question. Read each question and its suggested answers carefully. If a question is difficult to answer quickly, skip it and go on to the next one. Make certain to leave that number blank on the answer sheet and return to it at a later time. Skipping these questions initially will assist in effective time management during the test. After the last question, return to those questions skipped initially. If still uncertain which answer is correct, eliminate as many incorrect answers as possible. Then, make an educated guess using those remaining choices.

If time remains, recheck all work for accuracy. It is important to remember that there is no penalty for incorrect answers. Therefore, there is no difference between leaving an answer blank or getting it wrong. Answer all questions even if it is necessary to guess at an answer.

Remember, pacing oneself is important. Remain aware of progress throughout the exam. Bring a watch in case there is no clock in the room.

30.8 Areas for Research

The following are areas which may be covered on the EPA exam. Allow time to research and study these topics:

- Definition of ozone depletion potential.
- Method of transporting cylinders.
- Effects of recovering refrigerant at low ambient temperatures.
- Date when alternates for CFCs and HCFCs will be illegal to vent.
- Some violations of the Clean Air Act.
- Origin of most of the chlorine in the stratosphere.
- Names of some CFC refrigerants.
- Where superheated vapor is found.
- Definition of recycling.
- Definition of ternary blend refrigerant.
- Method of charging blended refrigerants.
- Definition of Type I, II, and III technician.
- Date for small appliance servicing to be certified.
- Date to certify the recovery equipment of small appliances.
- Function of low-loss fittings.
- Definition of small appliances.
- Definition of service aperture.
- When packaged terminal air conditioning units can be serviced by Type I technicians.
- Methods for transporting used refrigerants.
- When pressurized dry nitrogen is used.
- When system-dependent recovery equipment cannot be used.
- How low a vacuum pump should be capable of pumping down.
- Where oil foaming would occur.
- Why a regulator should be used on a nitrogen tank.
- Definition of an A1 refrigerant.
- Why a sensor and alarm is required for A1 refrigerants.
- The function of a hydrostatic tube test kit.
- Definition of a water box.
- The maximum nitrogen pressure for leak testing.
- For commercial refrigeration system leaking, the leak rate per year at which the leak must be repaired.

- When liquid removal should be used over vapor removal.
- With R-11, whether a liquid or vapor removal should be used to start, and why.
- Why a heater is used on recovery vessels.
- With a low-pressure chiller, what should be done after recovering the liquid refrigerant.
- Why water must be circulated through a chiller during the evacuation process.
- When charging a chiller, why vapor should be charged into the system before liquid.
- When nitrogen can be used to pressurize a system.
- The level to which recovery equipment should evacuate after November 15, 1993.
- The function of a purge unit on an R-11 chiller.
- Where a ruptured disc is connected on a centrifugal chiller.
- Why a low- and high-side access valve should be installed on small appliances when recovering refrigerant.
- Definition of a system-dependent recovery process.
- When the system-dependent recovery process can be used.
- Some safety practices for recovering refrigerant.
- Definition of phosgene gas.
- Requirements of Department of Transportation when shipping hazard-class refrigerant.
- Why R-12, -22, and -500 would tend to displace oxygen.
- Steps for inspecting a leaking system.
- With appliances containing more than 50 lb. of refrigerant, the leak rate per year at which the leak must be repaired.
- Maintenance check that should be performed on recovery equipment.
- Effects of ice on recovery vessels.
- What must be done if R-22 must be recovered from a system, but the recovery equipment has R-502 in it.
- Why quick couplers and self-sealing hoses are used during service.
- Whether a refrigerant that has been removed from system needing repair can be used again after the repair.
- What precautions should be taken when transferring refrigerant into an empty cylinder.
- Procedures used to recover refrigerant when replacing a compressor.
- What should be done after the required recovery vacuum is reached.
- In evacuating systems containing more than 200 lb. of R-12, the level to which the system must be evacuated.
- Evacuation levels required for small and large appliances before and after November 15, 1993.
- When CFCs can be evacuated to the atmosphere pressure.
- What would be described as a major repair on a system.

TYPICAL QUESTIONS MODULE

30.9 Types of Questions

The following questions are typical of the types of questions used in Types I, II, and III Certification exams. These questions are not part of the certification test, but merely representations of questions, and are designed to familiarize you with a variety of test questions.

30.9.1 General Core—Used in Types I, II, and III Certification Exams

1. Ozone may be defined as ______.
   A. a harmful gas
   B. similar to radon gas
   C. a protective gas that shields the earth
2. In what decade did the concern of ozone depletion start?
   A. 1950.
   B. 1960.
   C. 1970.
   D. 1980.
   E. 1990.
3. The type of refrigerant that is most harmful to the ozone layer is ______.
   A. HFC
   B. ammonia
   C. HCFC
   D. CFC
4. The refrigerant entering the compressor is ______.
   A. a liquid
   B. superheated vapor
   C. subcooled vapor
   D. superheated liquid
5. Which is included in the list of good practices for conserving refrigerant?
   A. Fix repairs.
   B. Tighten all system connections.
   C. Recover-recycle refrigerant.
   D. All of the above.

30.9.2 Type I Certification

1. When using passive recovery on a system with a non-operating compressor, the recovery must be made through ______.
   A. low side only
   B. high side only
   C. both low and high sides
2. Define a small appliance.
   A. An evaporative cooler.
   B. A packaged terminal air conditioner with 5 lb. of charge.
   C. All of the above.

3. Which definition best describes the term small appliance recovery category?
   A. Appliances containing more than 5 lb. of refrigerant.
   B. Includes refrigerators and freezer only.
   C. Products charged with 5 lb. or less of refrigerant.
   D. Products charged with 2 lb. only of refrigerant.
4. Packaged terminal air conditioning can be serviced by a Type I Technician if the unit ______.
   A. is charged with R-12
   B. is charged with R-22
   C. contains less than 5 lb. of refrigerant
   D. contains 15 lb. or more of refrigerant
5. When a tank of mixed refrigerant is received at a reclamation facility, they may ______.
   A. refuse the refrigerant for reclamation
   B. destroy the refrigerant
   C. resell the refrigerant as is
   D. report and fine the sender by EPA law

30.9.3 Type II Certification

1. Why would nitrogen be used when brazing?
   A. It eliminates the use of flux.
   B. It eliminates copper oxidation.
   C. It makes better, stronger joints.
   D. All of the above.
2. When a system has a burnout, the recovery or recycle equipment should be protected to ensure high-efficiency recovery. The recovery/recycle equipment should contain a filter drier in ______.
   A. the suction line of the system
   B. the liquid line of system
   C. cylinder inlet
   D. machine inlet
3. Refrigerant must be removed from the condenser outlet when ______.
   A. the evaporator is leaking
   B. the compressor is not operating
   C. the condenser is overhead
   D. the condenser is below the receiver
4. Once liquid refrigerant has been recovered, any vapor which is left is ______.
   A. bled to the atmosphere
   B. pumped into the high side
   C. pumped into a recovery unit
   D. condensed by the recovery unit
5. Refrigerant can be removed from the condenser when ______.
   A. the compressor is not working
   B. the condenser is mounted above ceiling height
   C. no receiver is present
   D. no discharge valves are present

30.9.4 Type III Certification

1. Why is a purge cycle required on low-pressure chillers?
   A. Operating pressure may be below atmospheric pressure.
   B. Because of the refrigerant's low boiling point.
   C. Non-condensables in the system.
   D. Chiller may be opened for maintenance.
2. One method of recovering refrigerant from a chiller that is cost effective and meets EPA requirement is to ______.
   A. vapor recover
   B. use a liquid pump
   C. recover liquid first, then vapor
   D. recover vapor first, then liquid
3. When removing R-11 or R-123 from a system, removal starts with ______.
   A. vapor removal
   B. liquid removal
   C. liquid/vapor removal
   D. oil separation
4. The maximum pressure of nitrogen when leak testing centrifugal systems is ______.
   A. 50 psig
   B. 10 psig
   C. 0 psig
   D. 25 psig
5. Recovery and recycling equipment manufactured before November 15, 1993, used for evacuation of low-pressure systems must evacuate the system to ______ before making a major repair.
   A. 5 inches of mercury
   B. 25 inches of mercury
   C. 29 inches of mercury
   D. 15 inches of mercury

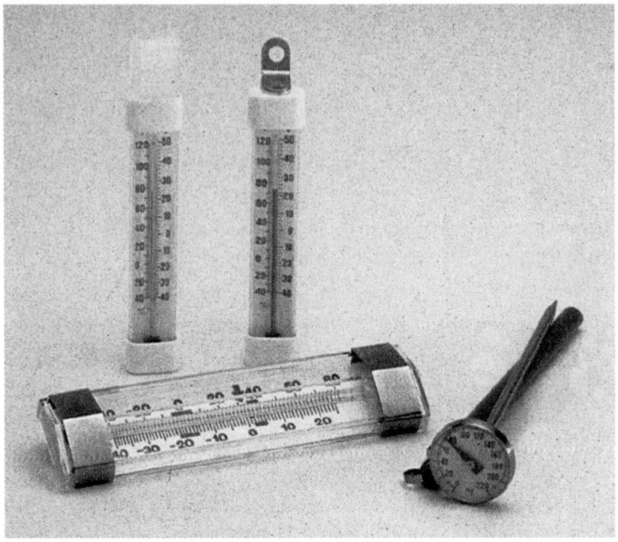
A

B

C

Various tools and pieces of equipment for servicing refrigeration and air conditioning systems. A—Thermometers. (Uniweld Products, Inc.) B—Time-sharing recorder, which can record two temperatures at one time. (Amprobe Instrument) C—Refrigerant leak detector. (Leybold Inficon, Inc.)

Chapter 31

TECHNICAL CHARACTERISTICS

Contents

31.1 Kata Thermometer

A Kata thermometer is used to measure air currents in open spaces. It is an alcohol thermometer with a Fahrenheit scale etched in the glass. Two scales are available. One reads from 95°F to 100°F (35°C to 38°C). The other reads from 125°F to 130°F (52°C to 54°C).

To use the thermometer, heat it to the higher °F value in a water bath. Thoroughly dry the thermometer and suspend it in the air current. Then record the time it takes, in seconds, to cool to the lower °F scale reading. Using this information, determine air movement in feet per minute from a Kata thermometer table.

31.2 Infrared Thermometer

Temperatures may be measured with an infrared thermometer, **Figure 31-1.** This optical-electronic instrument gives almost instantaneous readings. It is aimed at the surface for which the temperature is to be determined. The temperature is read directly from the temperature scale.

Different materials and types of surfaces give off heat at different rates. Therefore, it is necessary to set the instrument according to the nature of the surface. The adjustment for this is called the emissivity adjustment. Some emissivity adjustments are given in **Figure 31-2.**

Figure 31-1. *An infrared thermometer. This instrument focuses automatically for spot temperature measurements at distances from 0.35" to 40' (9 mm to 12 m). The measurement temperature range is from −50°F to 1800°F (−46°C to 982°C). (Wahl Instruments, Inc.)*

Infrared Thermometer Adjustment Values

Material		Emissivity
Aluminum	Bright	0.09
	Anodized	0.55
	Oxidized	0.2 to 0.3
Brass	Bright	0.03
	Oxidized	0.61
Chromium	Polished	0.08
Copper	Bright	0.05
	Oxidized	0.78
Iron and Steel	Polished	0.55
	Oxidized	0.85
Nickel	Polished	0.05
	Oxidized	0.95
Zinc	Bright	0.23
	Oxidized	0.23
Brick	Building	0.45
Paints	White	0.9
	Black	0.86
	Oil Paints (all colors)	0.92
Roofing Paper		0.91
Rubber		0.94
Silica		0.42 to 0.62
Water		0.92

Figure 31-2. *Emissivity of various surfaces.*

31.3 Weights and Specific Heats of Substances

Material	Weight lb./ft^3	Specific Heat Btu/lb.
Gases		
Air (normal temp.)	0.075	0.24
Metals		
Aluminum	166.5	0.214
Copper	552	0.094
Iron	480	0.118
Lead	710	0.030
Mercury	847	0.033
Steel	492	0.117
Zinc	446	0.096
Liquids		
Alcohol	49.6	0.60
Glycerin	83.6	0.576
Oil	57.5	0.400
Water	62.4	1.000
Others		
Concrete	147	0.19
Cork	15	0.48
Glass	164	0.199
Ice	57.5	0.504
Masonry	112	0.200
Paper	58	0.324
Rubber	59	0.48
Sand	100	0.195
Stone	138-200	0.20
Tar	75	0.35
Wood, Oak	48	0.57
Wood, Pine	38	0.47

31.4 Energy

There are two kinds of energy—potential and kinetic. ***Potential energy*** is like a body of water controlled by a dam. The water has a great *potential* for doing work. However, until the water flows through a water wheel, no energy is produced. A mass at height H has PE = MgH.

Kinetic energy is the energy of a moving body. The formula is KE = 1/2 MV^2 = Foot Pounds. M = Weight, divided by 32 (the acceleration due to gravity). V = Velocity in feet per second (fps). In metric, KE = 1/2 MV^2 = Joules. M = Mass in kilograms. V = Velocity in meters per second (m/sec).

31.5 Energy Equivalents (U.S. Conventional)

1 Btu	= 778 ft.-lb. = 252 gram-calories (g/cal) = 1054.8 joules (J)
1 horsepower	= 33,000 ft.-lb./min. = 550 ft.-lb./sec. = 746 W = 2545.6 Btu/h = 42.42 Btu/min. = 1.014 hp (metric)
1 horsepower hour	= 1 hp for 1 hr. = 1,980,000 ft.-lb. = 746 W/hr. = 0.746 kWh = 2545.6 Btu
1 watt (W)	= 3.414 368 Btu/h
1 kilowatt (kW)	= 1000 W = 1.34 hp
1 kilowatt hour (kWh)	= 1 kW for 1 hr. = 1000 W/hr.

31.5.1 Ice Melting Effect (IME)

1 ton of refrigeration	= 288,000 Btu/day = 12,000 Btu/h = 200 Btu/min. = 83.3 lb. of ice/hr. = 3.515 kW

31.6 Energy Equivalents (SI Metric)

Energy	
1 dyne cm	= 1 erg = 0.001 g·cm = 7.38×10^{-8} ft.-lb.
1 g·cm	= 980.6 ergs = 7.233×10^{-5} ft.-lb.
1 ft.-lb.	= 13,557,300 ergs = 13,825.5 g·cm
1 therm	= 100,000 Btu
Rate of Energy	
1 erg/sec.	= 1 dyne cm/sec. = 7.38×10^{-8} ft.-lb./sec.
1 g·cm/sec	= 980.6 ergs/sec. = 7.24×10^{-5} ft.-lb./sec.
1 ft.-lb./sec.	= 13,557,300 ergs/sec. = 13,800 g·cm/sec.

SI metric unit for energy is the joule. The erg and the dyne are part of the CGS metric system. (CGS stands for centimeter-gram-second. This is a metric system widely used in all branches of science throughout the world before adoption of SI units.)

31.7 Linear Measurement Equivalents (U.S. Conventional—SI Metric)

1 inch		= 2.54 cm = 25.4 mm = 25 400 μm
1 foot	= 12″	= 0.304 m = 30.48 cm
1 yard	= 3′	= 0.914 m = 91.44 cm
1 micron		= 0.000 039 4″
1 millimeter	= 1000 μm	= 0.0394″
1 centimeter	= 10 mm	= 0.3937″
1 meter	= 100 cm	= 39.37″
1 kilometer	= 1000 m	= 0.62137 miles

31.8 Fractional Inch Equivalents (Decimals and Millimeters)

Fraction	Decimal	mm	Fraction	Decimal	mm
1/64	0.0156	0.3967	33/64	0.5162	13.0968
1/32	0.0312	0.7937	17/32	0.5312	13.4937
3/64	0.0468	1.1906	35/64	0.5468	13.8906
1/16	0.0625	1.5875	9/16	0.5625	14.2875
5/64	0.0781	1.9843	37/64	0.5781	14.6843
3/32	0.0937	2.3812	19/32	0.5937	15.0812
7/64	0.1093	2.7781	39/64	0.6093	15.4781
1/8	0.125	3.175	5/8	0.625	15.875
9/64	0.1406	3.5718	41/64	0.6406	16.2718
5/32	0.1562	3.9687	21/32	0.6562	16.6687
11/64	0.1718	4.3656	43/64	0.6718	17.0656
3/16	0.1875	4.7625	11/16	0.6875	17.4625
13/64	0.2031	5.1593	45/64	0.7031	17.8593
7/32	0.2187	5.5562	23/32	0.7187	18.2562
15/64	0.2343	5.9531	47/64	0.7343	18.6531
1/4	0.25	6.5	3/4	0.75	19.05
17/64	0.2656	6.7468	49/64	0.7656	19.4468
9/32	0.2812	7.1437	25/32	0.7812	19.8437
19/64	0.2968	7.5406	51/64	0.7968	20.2406
5/16	0.3125	7.9375	13/16	0.8125	20.6375
21/64	0.3281	8.3343	53/64	0.8281	21.0343
11/32	0.3437	8.7312	27/32	0.8437	21.4312
23/64	0.3593	9.1281	55/64	0.8593	21.8281
3/8	0.375	9.525	7/8	0.875	22.225
25/64	0.3906	9.9218	57/64	0.8906	22.6218
13/32	0.4062	10.3187	29/32	0.9062	23.0187
27/64	0.4218	10.7156	59/64	0.9218	23.4156
7/16	0.4375	11.1125	15/16	0.9375	23.8125
29/64	0.4531	11.5093	61/64	0.9531	24.2093
15/32	0.4687	11.9062	31/32	0.9687	24.6062
31/64	0.4843	12.3031	63/64	0.9843	25.0031
1/2	0.50	12.7	1	1.000	25.4

31.9 Area Equivalents

1 in^2		= 0.0065 m^2
1 ft^2	= 144 in^2	= 0.093 m^2
1 yd^2	= 9 ft^2	= 0.836 m^2
1 yd^2	= 1296 in^2	

31.10 Volume Equivalents

1 in^3		= 0.016 L = 16.39 cm^3
1 ft^3	= 1728 in^3 = 7.481 gal.	= 28.317 L = 0.0283 m^3 = 28 317.00 cm^3
1 yd^3	= 27 ft^3 = 46,656 in^3	
1 gal.	= 0.1337 ft^3 = 231 in^3	= 3.79 L = 3785 cm^3
1 cm^3	= 0.0610237 in^3	
1 L	= 61.03 in^3 = 0.2642 gal.	= 1000 cm^3

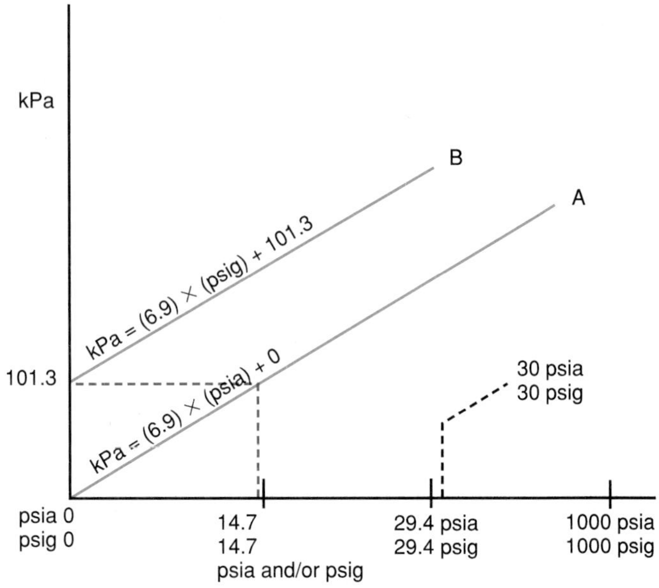

Figure 31-3. *The graph shows the following relationships. A—Between psia and kPa. B—Between psig and kPa. Most technicians prefer to use the lower graph line (A), which is simpler because it begins at zero. It is best to convert all psig readings to psia before trying to determine the kPa reading.*

31.11 Pressure Equivalents

Figure 31-3 shows how psia units are related to kPa units. It also shows how psig units are related to kPa units.

1 psia	= 0.068 atmosphere = 144 lb./ft^2 = 2.036″ Hg = 2.307′ water = 27.7″ water	 = 0.703 m water = 70.3 cm water = 51.7 mm Hg = 6.9 kPa
1 psig 0 psig	= 15.7 psia = 14.7 psia	= 108 kPa = 101.3 kPa
1 oz./in^2	= 0.128″ Hg = 1.73″ water	
1″ Hg	= 0.0334 atmosphere = 0.491 psi = 1.13′ water = 13.6″ water = 70.73 lb./ft^2	= 3.386 kPa = 25.4 mm Hg = 0.3453 m water
1′ water	= 0.0295 atmosphere = 0.434 psi = 62.43 lb./ft^2 = 0.03 atmosphere = 0.883″ Hg	= 2.985 kPa = 22.42 mm Hg = 0.305 m water

1 atmosphere	= 29.92″ Hg = 33.94′ water = 14.696 psi = 2116.35 lb./ft^2	= 101.28 kPa = 760 mm Hg = 10.33 m water
1 lb./ft^2	= 0.007 psi = 4.725 × 10^{-4} atmosphere = 0.01414″ Hg = 0.016′ water	= 0.048 kPa = 0.359 mm Hg = 0.0049 m water
1 kg/cm^2	= 14.22 psi = 2048.17 lb./ft^2 = 0.967 atmosphere = 28.96″ Hg = 32.8083′ water	= 10 m water = 97.98 kPa
1 m water	= 1.42 psi = 204.8 lb./ft^2 = 0.097 atmosphere = 2.896″ Hg = 3.28′ water	= 73.55 mm Hg = 9.78 kPa
1 mm Hg	= 0.019 psi = 2.78 lb./ft^2 = 0.001316 atmosphere = 0.039″ Hg = 0.0446′ water	= 0.133 kPa = 0.0136 m water

31.12 Velocity Equivalents

1 mi./hr.	= 1.47 ft./sec.	= 1.61 km/hr.
	= 0.87 knot	= 0.45 m/sec.
1 ft./sec.	= 0.68 mi./hr.	= 1.1 km/hr.
	= 60 ft./min.	= 0.305 m/sec.
	= 0.59 knot	
1 m/sec.	= 3.28 ft./sec.	= 3.6 km/hr.
	= 2.24 mi./hr.	
	= 1.94 knot	
1 km/hr.	= 0.91 ft./sec.	= 0.28 m/sec.
	= 0.62 mi./hr.	
	= 0.54 knot	

31.13 Liquid Measure Equivalents

Liquid Measure	U.S.	Metric
1 pt.	= 16 oz.	= 0.473 L
1 qt.	= 2 pt.	= 0.946 L
1 qt.	= 32 oz.	
1 gal.	= 4 qt.	= 3.785 L
1 gal.	= 8 pt.	
1 gal.	= 231 in^2	
1 ft^3	= 7.48 gal.	
1 gal.	= 8.34 lb. water	
1.136 qt.		= 1 L

31.14 Weight Equivalents

Avoirdupois		
1 oz.	= 437 grains (gr)	= 28.35 g
		= 0.028 kg
1 lb.	= 7000 gr	= 0.4536 kg
		= 453.6 g
1 lb.	= 16 oz.	= 453.6 g
1 gr	= 0.000143 lb.	= 0.064 80 g
1 ton	= 2000 lb.	= 907.2 kg
1 g	= 15.43 gr	= .001 kg
	= 0.03527 oz.	
	= 0.002205 lb.	
1 kg	= 2.2 lb.	
Specific Weights (Density)		
1 lb./in^3	= 1728 lb./ft^3	= 27.68 g/cm^3
		= 2.768×10^7 g/m^3
1 lb./ft^3	= 5.787×10^{-4} lb./in^3	= 0.016 g/cm^3
1 gm/cm^3	= 62.43 lb./ft^3	
1 kg/m^3	= 0.06243 lb./ft^3	

31.15 Flow Equivalents

1 ft^3/min.	= 7.481 gal./min.	= 28 317 cm^3/min.
	= 449 gal./hr.	= 28.32 L/min.
		= 1700 L/hr.
1 ft^3/hr.	= 0.0167 ft^3/min.	= 0.472 L/min.
	= 0.1247 gal./min.	= 28.317 L/hr.
	= 7.481 gal./hr.	= 472 cm^3/min.
1 gal./min.	= 0.1337 ft^3/min.	= 3.79 L/min.
	= 8.022 ft^3/hr.	= 3785 cm^3/min.
1 L/min.	= 0.0353 ft^3/min.	= 1000 cm^3/min.
	= 2.118 ft^3/hr.	
	= 0.2642 gal./min.	
	= 15.852 gal./hr.	

31.16 Heat Equivalents

1 Btu	= 252 calories (cal)
1 Btu	= 1054.4 J
1 kcal	= 1000 cal
1 kcal	= 4.1840 kJ
1 kcal/kg	= 4.1840 kJ/kg
1 Btu/lb.	= 0.5556 kcal/kg
1 Btu/lb.	= (4.1840 kJ/kcal)/(1.8°F/°C)
	= 2.3244 kJ/kg
1 kcal/kg	= 1.8 Btu/lb.
1 Btu/h	= 0.2931W

The most correct definition of the kilocalorie is 1 kcal = 4.1840 kJ. This defines the unit called the ***thermochemical calorie.*** The thermochemical calorie is slightly more accurate than the unit used earlier in this book (1 kcal = 4.187 kJ).

Unit of heat per pound Celsius or Centrigrade heat unit is the heat required to raise one pound of water 1°C. It equals 1.8 Btu.

1 Btu/ft^2/hr./°F = 4.88 kgcal/m^2/hr./°C
1 kgcal/m^2/hr./°C = 0.205 Btu/ft^2/hr./°F

31.17 Pumping Ratio

The ***pumping ratio*** is the ratio of the suction pressure to the condensing or head pressure expressed in absolute pressures. ***Absolute pressure*** equals gauge pressure plus 15.

Example:

If the low-side pressure of a system is 15 psig:

15 + 15 = 30 psia
and the high-side pressure is 150 psig:
150 + 15 = 165 psia,
the pumping ratio equals 30/165 = 1/5.5 = 5.5:1

The pumping ratio should not exceed a certain value for each particular refrigerant. If the pumping ratio is too high, the temperature of the high-pressure vapor going through the exhaust valve would overheat the mechanism. This would cause some of the oil in the exhaust pocket to become carbonized.

31.18 Compression Ratio

Compression ratio is the ratio of the total volume of the cylinder to the clearance space in a compressor, or that space which remains at the end of the compression cycle.

Example:

A compressor has a 2″ bore and a 2″ stroke and a clearance space of 0.010″. The volume in the valve ports equals 0.05 in^3. The piston displacement is:

$\pi r^2 \times S$

$\pi = 3.1416$

$r = 1$

$S = 2$

The volume equals $3.1416 \times 1 \times 2 = 6.2832$ in^3

The volume of the clearance space equals

$\pi r^2 \times r \times S$

$\pi = 3.1416$

$r = 1$

$S = 0.010$

The volume equals

$3.1416 \times 1 \times .010 = .031416 = 0.03$ in^3

Total clearance space: $0.05 + 0.03 = 0.08$ in^3

The total volume of the cylinder is:

$6.2832 + 0.08 = 6.3632 = 6.36$

The compression ratio is $6.36 \div 0.08 = 79.5{:}1$

31.19 Temperature Conversion Table

To use the following table, find the temperature to be converted in the center column. If converting to Celsius, read to the left. If converting to Fahrenheit, read to the right.

°C		°F	°C		°F
−273.15	**−459.67**		−212	**−350**	
−268	**−450**		−207	**−340**	
−262	**−440**		−201	**−330**	
−257	**−430**		−196	**−320**	
−251	**−420**		−190	**−310**	
−246	**−410**		−184	**−300**	
−240	**−400**		−179	**−290**	
−234	**−390**		−173	**−280**	
−229	**−380**		−169	**−273**	−459.4
−223	**−370**		−168	**−270**	−454
−218	**−360**		−162	**−260**	−436

°C		°F	°C		°F
−157	**−250**	−418	2.2	**36**	96.8
−151	**−240**	−400	2.8	**37**	98.6
−146	**−230**	−382	3.3	**38**	100.4
−140	**−220**	−364	3.9	**39**	102.2
−134	**−210**	−346	4.4	**40**	104.0
−129	**−200**	−328	5.0	**41**	105.8
−123	**−190**	−310	5.6	**42**	107.6
−118	**−180**	−292	6.1	**43**	109.4
−112	**−170**	−274	6.7	**44**	111.2
−107	**−160**	−256	7.2	**45**	113.0
−101	**−150**	−238	7.8	**46**	114.8
−95.6	**−140**	−220	8.3	**47**	116.6
−90.0	**−130**	−202	8.9	**48**	118.4
−84.4	**−120**	−184	9.4	**49**	120.2
−78.9	**−110**	−166	10.0	**50**	122.0
−73.3	**−100**	−148	10.6	**51**	123.8
−67.8	**−90**	−130	11.1	**52**	125.6
−62.2	**−80**	−112	11.7	**53**	127.4
−56.7	**−70**	−94	12.2	**54**	129.2
−51.1	**−60**	−76	12.8	**55**	131.0
−45.6	**−50**	−58	13.3	**56**	132.8
−40.0	**−40**	−40	13.9	**57**	134.6
−34.4	**−30**	−22	14.4	**58**	136.4
−28.9	**−20**	−4	15.0	**59**	138.2
−23.3	**−10**	14	15.6	**60**	140.0
−17.8	**0**	32	16.1	**61**	141.8
−17.2	**1**	33.8	16.7	**62**	143.6
−16.7	**2**	35.6	17.2	**63**	145.4
−16.1	**3**	37.4	17.8	**64**	147.2
−15.6	**4**	39.2	18.3	**65**	149.0
−15.0	**5**	41.0	18.9	**66**	150.8
−14.4	**6**	42.8	19.4	**67**	152.6
−13.9	**7**	44.6	20.0	**68**	154.4
−13.3	**8**	46.4	20.6	**69**	156.2
−12.8	**9**	48.2	21.1	**70**	158.0
−12.2	**10**	50.0	21.7	**71**	159.8
−11.7	**11**	51.8	22.2	**72**	161.6
−11.1	**12**	53.6	22.8	**73**	163.4
−10.6	**13**	55.4	23.3	**74**	165.2
−10.0	**14**	57.2	23.9	**75**	167.0
−9.4	**15**	59.0	24.4	**76**	168.8
−8.9	**16**	60.8	25.0	**77**	170.6
−8.3	**17**	62.6	25.6	**78**	172.4
−7.8	**18**	64.4	26.1	**79**	174.2
−7.2	**19**	66.2	26.7	**80**	176.0
−6.7	**20**	68.0	27.2	**81**	177.8
−6.1	**21**	69.8	27.8	**82**	179.6
−5.6	**22**	71.6	28.3	**83**	181.4
−5.0	**23**	73.4	28.9	**84**	183.2
−4.4	**24**	75.2	29.4	**85**	185.0
−3.9	**25**	77.0	30.0	**86**	186.8
−3.3	**26**	78.8	30.6	**87**	188.6
−2.8	**27**	80.6	31.1	**88**	190.4
−2.2	**28**	82.4	31.7	**89**	192.2
−1.7	**29**	84.2	32.2	**90**	194.0
−1.1	**30**	86.0	32.8	**91**	195.8
−0.6	**31**	87.8	33.3	**92**	197.6
0	**32**	89.6	33.9	**93**	199.4
0.6	**33**	91.4	34.4	**94**	201.2
1.1	**34**	93.2	35.0	**95**	203.0
1.7	**35**	95.0	35.6	**96**	204.8

°C		°F	°C		°F
36.1	**97**	206.6	349	**660**	1220
36.7	**98**	208.4	354	**670**	1238
37.2	**99**	210.2	360	**680**	1256
37.8	**100**	212.0	366	**690**	1274
43	**110**	230	371	**700**	1292
49	**120**	248	377	**710**	1310
54	**130**	266	382	**720**	1328
60	**140**	284	388	**730**	1346
66	**150**	302	393	**740**	1364
71	**160**	320	399	**750**	1382
77	**170**	338	404	**760**	1400
82	**180**	356	410	**770**	1418
88	**190**	374	416	**780**	1436
93	**200**	392	421	**790**	1454
99	**210**	410	427	**800**	1472
100	**212**	413	432	**810**	1490
104	**220**	428	438	**820**	1508
110	**230**	446	443	**830**	1526
116	**240**	464	449	**840**	1544
121	**250**	482	454	**850**	1562
127	**260**	500	460	**860**	1580
132	**270**	518	466	**870**	1598
138	**280**	536	471	**880**	1616
143	**290**	554	477	**890**	1634
149	**300**	572	482	**900**	1652
154	**310**	590	488	**910**	1670
160	**320**	608	493	**920**	1688
166	**330**	626	499	**930**	1703
171	**340**	644	504	**940**	1724
177	**350**	662	510	**950**	1742
182	**360**	680	516	**960**	1760
188	**370**	698	521	**970**	1778
193	**380**	716	527	**980**	1796
199	**390**	734	532	**990**	1814
204	**400**	752	538	**1000**	1832
210	**410**	770	543	**1010**	1850
216	**420**	788	549	**1020**	1868
221	**430**	806	554	**1030**	1886
227	**440**	824	560	**1040**	1904
232	**450**	842	566	**1050**	1922
238	**460**	860	571	**1060**	1940
243	**470**	878	577	**1070**	1958
249	**480**	896	582	**1080**	1976
254	**490**	914	588	**1090**	1994
260	**500**	932	593	**1100**	2012
266	**510**	950	599	**1110**	2030
271	**520**	968	604	**1120**	2048
277	**530**	986	610	**1130**	2066
282	**540**	1004	616	**1140**	2084
288	**550**	1022	621	**1150**	2102
293	**560**	1040	627	**1160**	2120
299	**570**	1058	632	**1170**	2138
304	**580**	1076	638	**1180**	2156
310	**590**	1094	643	**1190**	2174
316	**600**	1112	649	**1200**	2192
321	**610**	1130	704	**1300**	2372
327	**620**	1148	760	**1400**	2552
332	**630**	1166	816	**1500**	2732
338	**640**	1184	871	**1600**	2912
343	**650**	1202	927	**1700**	3092
982	**1800**	3272	1371	**2500**	4532

°C		°F	°C		°F
1038	**1900**	3452	1427	**2600**	4712
1093	**2000**	3632	1482	**2700**	4892
1149	**2100**	3812	1538	**2800**	5072
1204	**2200**	3992	1593	**2900**	5252
1260	**2300**	4172	1649	**3000**	5432
1316	**2400**	4352			

Interpolation Values

°C		°F	°C		°F
0.56	**1**	1.8	3.33	**6**	10.8
1.11	**2**	3.6	3.89	**7**	12.6
1.67	**3**	5.4	4.44	**8**	14.4
2.22	**4**	7.2	5.00	**9**	16.2
2.78	**5**	9.0	5.56	**10**	18.0

Courtesy of Thermatron Corporation

31.20 Seasonal Energy Efficiency Rating (SEER)

The EER (energy efficiency rating) and the COP (coefficient of performance) are combined into the SEER (seasonal energy efficiency rating). See Section 16.16. An average is obtained with the following formula:

SEER = 1.2 winter days/summer days
+ winter days EER + 0.8 summer days/summer
days + winter days (COP + 1)·3.414368

The EER is found in the usual ways as follows:

EER = Btu/h of heating/electrical power in watts

There is a better formula which uses all numbers given in Btu/h. Use the following formula:

$SEER_Q$ = 1.2 WD/SD + WD EER_Q
+ 0.8 SD/SD + WD (COP + 1)

The EER_Q is found as follows:

EER_Q
= Btu/h of heating/electrical power in watts 0.293

For all formulas, the COP is as follows:

COP
= Btu/h of cooling/electrical power in watts 0.293

31.21 Standard Temperature

The weight of a certain volume of air depends on its temperature. The standard temperature used for determining the weight of air is 32°F (0°C). This is the temperature of melting ice.

31.22 Standard Pressure

The weight of a certain volume of air depends on its pressure. The standard pressure for determining the weight of air is 29.92″ Hg (760 mm Hg.). This is the pressure of one atmosphere.

31.23 Standard Temperature and Pressure Sample

A sample of air at standard temperature and under standard pressure is known as a ***sample at STP conditions.***

31.24 Standard Air

Standard air and ***standard conditions*** are the same. Both conditions include pressure, temperature, and air density of 0.075 lb./ft^3 (1.2007 kg/m^3). These values are:

Condition	U.S. Conventional System	Metric System
Pressure	29.92″ Hg =	760 mm Hg
	14.696 psia = 0 psi	101.28 kPa
Temperature	69.8°F	21°C
Specific Volume	13.33 ft^3/lb.	0.833 m^3/kg

31.25 Heating Value of Fuels

Fuels commonly used for heating purposes are coal, oil, gas, and wood. Burning these fuels in atmospheric air produces the heat needed. The heating value of coal is given in **Figure 31-4.**

The commercial grades of heating oil have been established by the American Society for Testing and Materials (ASTM). Grade numbers are 1 through 6, although No. 3 and No. 4 are used very little. The grade of heating oil may be determined with a special heating oil hydrometer. Grade No. 1 has a gravity of between 45 and 38, as shown in **Figure 31-5.** The heating value of oil is also shown in this figure.

The heating values of common fuel gases are shown in **Figure 31-6.** (The heating value of natural gas varies greatly, depending upon its composition.)

The heating value of wood is approximately 6200 Btu/lb.

Fuel	Heat Released—Btu/lb.
Coal:	
Bituminous	12,000 to 15,000
Anthracite	13,000 to 14,000

Figure 31-4. *Heating value of coal. As noted, the heating value of coal varies considerably depending upon the moisture, sulfur content, and other impurities.*

Commercial Grade No.	Btu/gal.	A.P.I.** Gravity Range ASTM†	Average Weight per Gallon
1	137,000	45–38	6.8
2	140,000	40–30	7.1
3	140,000*		
4	141,000*	32–12	7.7
5	148,000	20–8	8.1
6	152,000	18–6	8.2

*Not in common use
**American Petroleum Institute rating
†American Society for Testing and Materials

Figure 31-5. *Table of heating values of fuel oils. Numbers 5 and 6 are high viscosity oils and need preheating before use.*

Fuel	Heat Released—Btu/ft^3
Gas:	
Natural	1000 to 1100*
Manufactured	500 to 600
LP (Liquid Petroleum)	2500 to 3200

*Check with the local gas company

Figure 31-6. *Heating value of fuel gases.*

31.26 Single-Source Energy

In some buildings, only a single source of energy—such as natural gas—is supplied. All other needed energy is generated in the building from this source. For example, gas-burning engines may be used to drive electric generators. The waste heat from the cooling of the engines and the exhaust may be used for other purposes. It may be used for heating water, space heating, or as heat for industrial processes. Also, the heat from lighting may be used in heating water or air.

31.27 Thermodynamic Laws

The ***First Law of Thermodynamics*** is a formula for converting heat into work or work into heat. The formula is: 778 ft.-lb. of work is equivalent to the heat energy of 1 Btu.

The ***Second Law of Thermodynamics*** states that heat will only flow from a body at a given temperature to another body which is at a lower temperature.

31.28 Heat Conductivity

Heat Conductivity of Miscellaneous Substances		
Material	**k***	**R****
Air	0.175	5.714
Concrete wall	8.00	0.125
Glass	5.0	0.2
Lead	243.0	0.004
Vacuum, high	0.004	250.0

Heat Conductivity of Miscellaneous Insulating Materials			
Material	**Density lb./ft³**	**k* Conductivity**	**R****
Cork, granulated	8.1	0.34	2.941
Cork, granulated impregnated with pitch	17.79	0.428	2.336
Balsa	7.05	0.32	3.125
Felt	16.9	0.25	4.0
Glass wool (curled pyrex)	4.0	0.29	3.448
Kapok	0.87	0.24	4.167
Mineral (slag) wool, loose packed	12.0	0.26	3.846
Polyurethane	1.5	0.16	6.25
Rock wool (fibrous rock, also felted)	6.0	0.26	3.846
Rubber, cellular	5.0	0.37	2.703
Sawdust, pine	18.76	0.57	1.754
Straw fibers, pressed	8.67	0.32	3.125
Wood fibers (kingia australis)	8.4	0.33	3.03
Wool, pure	4.99	0.26	3.846

Heat Conductivity of Proprietary Materials			
Trade Name	**Density lb./ft³**	**k* Conductivity**	**R****
Armstrong corkboard	7.3	0.285	3.509
Celotex	13.2	0.31	3.226
Dry-Zero	1.0	0.24	4.167
Nu Wood	15.0	0.32	3.125
United 100% pure corkboard	9.0	0.27	3.704
U.S. mineral wool	12.0	0.26	3.846
Ferro-Therm metal sheet (4 sheets)	4 oz./ft²/sheet	0.226	4.425

Note: k* = Btu·in./hr·ft²·°F
R** = Reciprocal of k

(Reprinted by permission of the American Society of Heating, Refrigerating, and Air-Conditioning Engineers, Atlanta, Georgia.)

31.29 Entropy

Entropy is the heat available measured in Btu per pound degree change for a substance. Entropy calculations are made from generally accepted temperature bases. For heating and steam power using water as the medium, the accepted base is 32°F (0°C). For domestic and most commercial refrigeration calculations, the base is −40°F (−40°C). For research and very low-temperature work, a base of a lower temperature may be selected.

Entropy is generally used only in engineering calculations. Entropy tables have been worked out and are found in most engineering handbooks.

31.30 Theory of Matter

To understand electricity and electronics, you must know the properties of ***matter.*** All matter (material) is made up of molecules. A molecule is the smallest part of a substance that has all the properties of that substance. All molecules are made up of atoms. A molecule may contain only one atom or many hundreds of atoms.

Atoms are made mainly of electrons, neutrons, and protons. All the material in the universe is made from only about 110 different atoms (the chemical ***elements,*** such as copper, or silica, or hydrogen). The atoms of each element are identical, but are different from the atoms of any other element. Differences in the atoms depend on the number of electrons, protons, and neutrons in their composition.

All substances have mass. ***Mass*** is determined by the amount of matter contained in a substance. The mass of a body, then, is determined by dividing the weight of the body by the acceleration due to gravity, 32.2 ft./sec/sec.

The total of mass and energy is constant. Mass can be turned into energy. Energy can be turned into mass. The relationship of mass and energy is described by Albert Einstein in the equation, $E = MC^2$. (E = energy. M = mass. C = speed of light.)

31.31 Electron Theory

Most principles of construction and operation of electrical equipment are based on theories, such as the electron theory described in this section. These theories have come to be accepted principles of operation.

As described in Section 31.30, all substances are made up of molecules. A molecule is the smallest part of a substance that has all the properties of that substance. Molecules, in turn, are made up of atoms. Atoms consist of electrons, protons, and neutrons. Electrons have negative charges of electricity. Protons have positive charges of electricity. Neutrons have no charge—they are neither negative nor positive.

In the lighter elements, the atom contains one proton for each neutron. In the heavier elements, the atom contains more neutrons than protons. For example:

- Helium = 2 protons, 2 neutrons
- Boron = 5 protons, 5 neutrons
- Mercury = 80 protons, 120 neutrons

A proton and a neutron are equal in mass. Either one is about 1845 times as great in mass as the electron mass. For example, imagine that the outer electrons of a typical atom structure are in an orbit 200 yards in diameter. The nucleus in the center would be 1/2″ in diameter.

31.32 The Atom

An elementary drawing of an atom, **Figure 31-7,** looks like the solar system with a sun and planets. The electrons revolve around the atom's ***nucleus*** (center). The nucleus is made of protons and neutrons.

The revolving electrons are of two types:

- Free electrons.
- Bound electrons.

If an atom has free electrons, the atom can transfer or conduct electrical energy. See **Figure 31-8.** If an atom collects an electron charge for an instant, it is negatively charged. If the atom loses an electron, it becomes positively charged. The electrons travel from atoms that have extra electrons to those with a lack of electrons.

When the outer electrons of an atom interact with electrons of another atom, they combine to form a ***compound.*** In some compounds, each atom will retain all its electrons. In others, one or more outer atoms will be gained or lost as a result of the bonding. The nucleus does not enter into chemical or electrical processes. Disrupting the nucleus of an atom requires a vast amount of energy, such as in atomic bomb explosion.

An atom entering a chemical change does so in the form of a charged particle called an ***ion.*** Ions are of two types, *negative* and *positive.* A ***negative ion*** is an atom that is negatively charged (has an extra electron). A ***positive ion*** is an atom that is positively charged (lacks an electron). Like-charged ions repel. Unlike-charged ions attract.

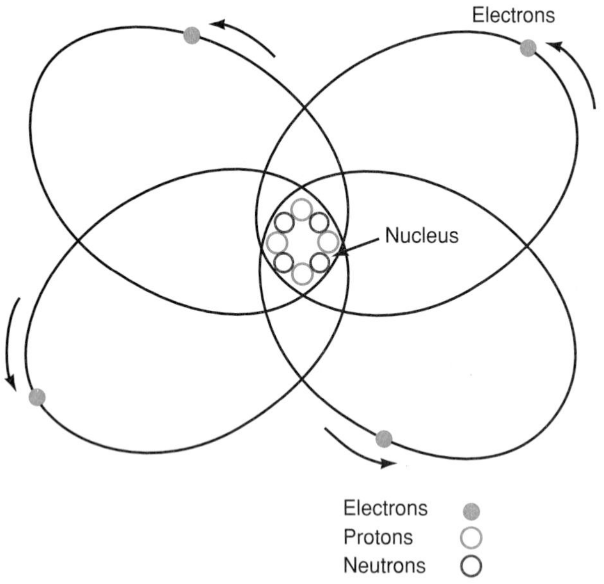

Figure 31-7. *Makeup of an atom, showing electrons revolving around a nucleus made up of protons and neutrons.*

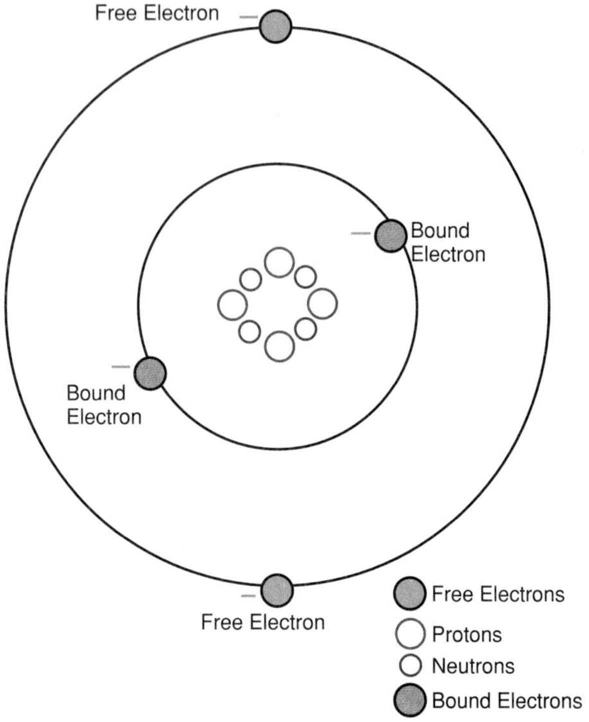

Figure 31-8. *An atom with two free electrons and two bound electrons.*

31.33 The Electron

In most atoms, the number of electrons equals the number of protons. See **Figure 31-9.** The negative charges equal the proton charges. The resulting *(net)* charge of the atom is zero if no chemical action or electrical flow is occurring.

An external force applied to an atom may remove one or more of its outermost electrons. These removable electrons are commonly called "free" electrons.

Two forces act on revolving electrons:

- Centrifugal force.
- Centripetal force.

Centrifugal force tends to make electrons move away from protons. ***Centripetal force*** tends to make electrons move toward protons. The electron orbit is a balance of these two forces.

The revolving electrons are attracted to the nucleus. Yet their travel speed keeps them a small distance from the nucleus by centrifugal force. The force trying to pull the electron into the center is called the ***negative electrical charge.*** (Negative sign is −.) The nucleus attraction is called the ***positive electrical charge.*** (Positive sign is +.) Based on this attraction, electricity is movement of the electrons (−) toward the nucleus (+).

As noted earlier, electrons are negative charges of electricity. **Figure 31-10** shows an atom including its nucleus. Protons are positive charges of electricity. If an

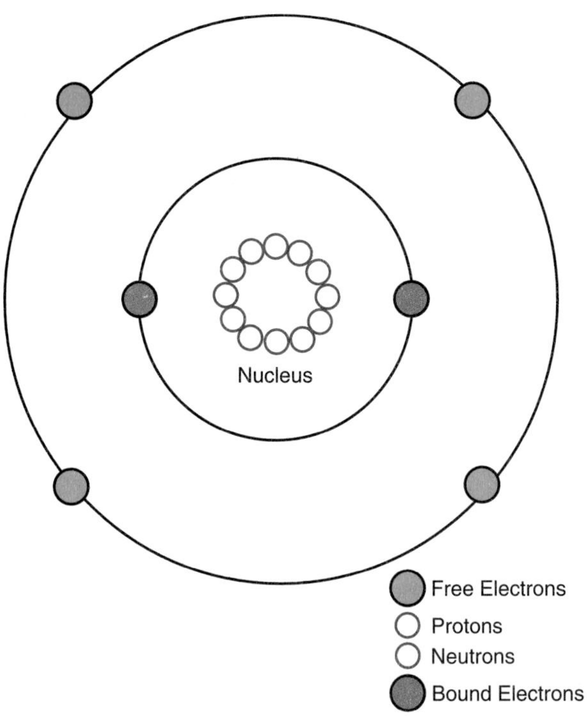

Figure 31-9. *A carbon atom with four free electrons, two bound electrons and six protons. Since the negative charges (electrons) and positive charges (protons) are equal, the net charge of the atom is neutral. The six neutrons have no electrical charge.*

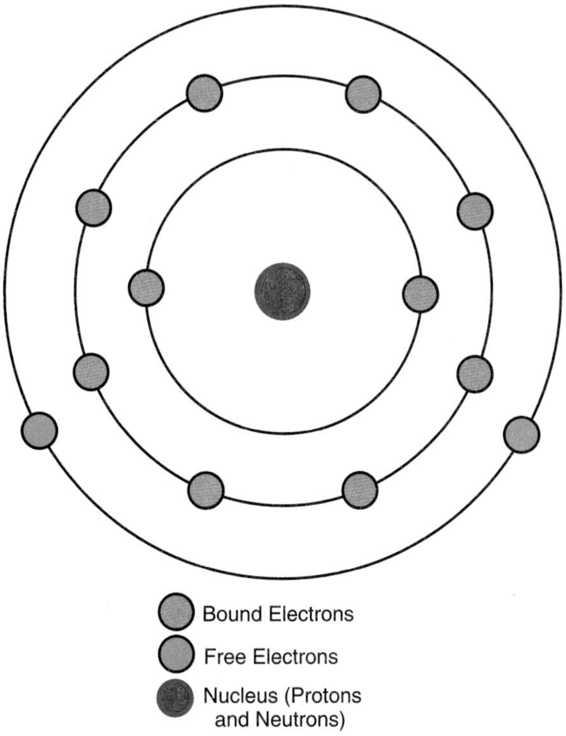

Figure 31-10. *An atom with a net positive charge. It has lost one of its free electrons, so that it has only 12 negative charges, while there are 13 positively charged protons in its nucleus.*

atom collects an extra free electron for a moment, it is negatively charged. If an atom loses a free electron, it is positively charged. The electrons travel from atoms with extra electrons to atoms with a lack of electrons.

The number of electrons that flow is extremely high. The term ***coulomb*** is used to represent the flow of 6.24×10^{18} or 6,240,000,000,000,000,000 electrons. One coulomb flowing for one second equals one ampere. The ampere is known as the quantity of electricity.

If there are too many protons, and electrons move to make up the difference, this movement or pressure to move is called ***electromotive force.*** Electric current consists of electrons traveling in a conductor. A conductor may be a solid, a liquid, or a gas.

31.34 Electrical Units and Symbols

Electrical Engineering Units and Constants (As adopted by NBS)			
Symbols and Units			
Quantity	**Symbol**	**Unit**	**Unit Abbreviation**
charge	*Q*	coulomb	C
current	*I*	ampere	A
voltage, potential difference	*V*	volt	V
electromotive force	$\mathscr{E}$	volt	V
resistance	*R*	ohm	Ω
conductance	*G*	mho (siemens)	A/V, or mho (S)
reactance	*X*	ohm	Ω
susceptance	*B*	mho	A/V, or mho
impedance	*Z*	ohm	Ω
admittance	*Y*	mho	A/V, or mho
capacitance	*C*	farad	F
inductance	*L*	henry	H
energy, work	*W*	joule	J
power	*P*	watt	W
resistivity	ρ	ohm-meter	Ωm
conductivity	σ	mho per meter	mho/m
electric displacement	*D*	coulomb per sq. meter	C/m^2
electric field strength	*E*	volt per meter	V/m
permissivity (absolute)	$\in$	farad per meter	F/m
relative permissivity	$\in_r$	(numeric)	
magnetic flux	ϕ	weber	Wb
magnetomotive force	$\mathscr{F}$	ampere (ampere turn)	A
reluctance	$\mathscr{R}$	ampere per weber	A/Wb
permeance	$\mathscr{P}$	weber per ampere	Wb/A
magnetic flux density	*B*	tesla	T
magnetic field strength	*H*	ampere per meter	A/m
permeability (absolute)	μ	henry per meter	H/m

Continued.

Electrical Engineering Units and Constants (As adopted by NBS)			
Symbols and Units			
Quantity	**Symbol**	**Unit**	**Unit Abbreviation**
relative permeability	μ_r	(numeric)	
length	L	meter	m
mass	m	kilogram	kg
time	t	second	s
frequency	f	hertz	Hz
angular frequency	ω	radian per second	rad/s
force	F	newton	N
pressure	p	newton per square meter	N/m^2
temperature (absolute)	T	Kelvin	K
temperature (international)	t	degree Celsius	°C

Recommended Unit Prefixes			
Multiples and submultiples	**Prefixes**	**Symbols**	**Pronunciation**
10^{12}	tera	T	ter'a
10^{9}	giga	G	ji'ga
10^{6}	mega	M	meg'a
10^{3}	kilo	k	kil'o
10^{2}	hecto	h	hek'to
10	deka	da	dek'a
10^{-1}	deci	d	des'i
10^{-2}	centi	c	sen'ti
10^{-3}	milli	m	mil'i
10^{-6}	micro	μ	mi'kro
10^{-9}	nano	n	nan'o
10^{-12}	pico	p	pe'ko
10^{-15}	femto	f	fem'to
10^{-18}	atto	a	at'o

Defined Values and Conversion Factors	
Meter	1 650 763.73 wavelengths of the transition $2p_{10} - 5d_s$ in ^{86}Kr
Kilogram	mass of the international kilogram
Second	1/31 556 925.974 7 of the tropical year 1900
Kelvin	In the thermodynamic scale, 273.16 K = triple point of water (fp, 273.15 K = °C)
Unified atomic mass unit. u	1/12 the mass of an atom of the ^{12}C nuclide
Standard acceleration of free fall	9.806.65 m s^{-2}, 980.665 cm s^{-2}
Normal atmosphere	101.325 N·m^{-2}, 1 013 250 dyne·cm^{-2}
Thermochemical calorie	4.1840 J. 4.1840 × 10^7 erg
Int. Steam Table calorie	4.1868 J. 4.1868 × 10^7 erg
Liter	0.001 000 028 m^3, 100.028 cm^3
Inch	0.0254 m, 2.54 cm
Pound (avdp)	0.453 592 37 kg, 453.592 37 g

31.35 Motor Size Calculations

Horsepower of a motor required to drive a refrigeration compressor can be calculated. The usual method uses the ***mean effective pressure*** (MEP) in the cylinder. This is the pressure that is bearing down on the piston head and which must be overcome by the electric motor when driving the compressor. The MEP is determined by a formula. Basic variables of the formula are the low-side pressure, the high-side pressure, and the ratio of the specific heat at constant pressure to the specific heat at constant volume Cp/Cv = K (for the kind of refrigerant). The formula follows:

$$\text{MEP} = P_1 \times \frac{K}{K-1}\left[\left(\frac{P_2}{P_1}\right)^{\frac{K-1}{K}} - 1\right]$$

P_1 = suction pressure, psia
P_2 = condenser pressure, psia
$K = C_p/C_v$

This value is multiplied by the area of the piston, by the length of the stroke, and by the rpm. The result will give the ***foot-pounds*** per minute needed to drive the compressor.

$$\text{Ft.-lb. per min.} = \text{MEP} \times \frac{\pi D^2}{4} \times S \times N \times R$$

D = piston diameter
S = stroke in inches
N = number of cylinders
R = rpm

Convert the foot-pound per min. into hp by dividing by 33,000:

$$\text{hp} = \text{ft.-lb./min./33,000}$$

The MEP may be determined from the above. It may also be determined by obtaining the indicator card of the compressor being studied. Engineers' handbooks set forth the methods of obtaining and using indicator cards.

Example:

The indicator card shows an MEP of 30 lb./in^2. The indicator hp necessary to drive a one cylinder compressor with a 2″ bore and a 2″ stroke, running at 200 rpm, would be as follows:

$$\text{hp} = 30 \text{ psi} \times \text{area} \times \text{stroke} \times \text{rpm}/33{,}000$$

The dimensions must be in foot-pounds = pressure × area × stroke. The resistance is the length of the compression stroke × rpm. The horsepower is:

$$\frac{30 \text{ psi} \times \frac{\pi D^2}{4} \times \frac{2}{12} \times 200}{33{,}000}$$

$$= \frac{30 \times \pi D^2 \times 2 \times 200}{4 \times 33{,}000 \times 12} = \frac{30 \times \pi 4 \times 2 \times 200}{4 \times 33{,}000 \times 12}$$

$$= \frac{\pi \times 200}{6600} = \frac{\pi}{33}$$

$= .0952$ or approximately 0.10 hp.

This is the theoretical hp. It neglects friction, oil pumping, starting load, and drive losses. Up to a 1-ton machine, this should be doubled. For example, 1/4 hp, calculated, will need a 1/2 hp motor. Up to 5 tons, this ratio gradually tapers off until adding 30% at 5 tons' capacity will take into consideration the above losses. The reason for the decrease is that some of the losses remain constant, while others do not increase as rapidly in proportion to the increase in the size of the unit.

31.36 Lowering Voltage Saves Power

During power shortages, some utilities reduce the voltage supplied to the customer. This may result in some power savings. The largest savings will come from noninductive loads, such as lighting, resistance heating, and electric cooking. In these appliances, a reduction in voltage applied causes some drop in current flow rate. Electrical power is equal to $V \times C - E \times I = W$. Therefore, if both the voltage and current are reduced, the $E \times I$ will result in the reduction in the amount of power used.

With inductive loads, there will not be as great savings. If the voltage is dropped, the current required will automatically increase. Therefore, power consumed ($E \times I = W$) will be affected very little. If "E" goes down and "I" goes up, the power required will be affected very little. This means there will be very little savings accrued in the operation of such appliances as refrigerators, air conditioners, and ventilating equipment during a power shortage. If motor speed is reduced, there will be some savings.

31.37 Resistances of Conductors, Semiconductors, and Nonconductors

The resistance value of conductors, semiconductors, and nonconductors (insulators), measured in ohms, varies from very large to very small numbers. This would require working with long numbers and long decimal numbers. Therefore, the usual practice is to use the significant number multiplied by a power of 10.

10^3 means $10 \times 10 \times 10 = 1000$
10^2 means $10 \times 10 = 100$
10^1 means 10
10^0 means 1

For decimals, a negative power of 10 (-10) is used:

10^{-1} means 0.1
10^{-2} means $0.1 \times 0.1 = 0.01$
10^{-3} means $0.1 \times 0.1 \times 0.1 = 0.001$

The following table lists the approximate resistance of conductors, semiconductors, and nonconductors. The resistance is in ohms per centimeter of length.

For conductors, the resistance in ohms usually varies from:

$0.000001 = 10^{-6}$
$0.00001 = 10^{-5}$
$0.0001 = 10^{-4}$
$0.001 = 10^{-3}$

For semiconductors, the resistance in ohms usually varies from:

$1 = 10^0$
$10 = 10^1$
$100 = 10^2$
$1000 = 10^3$
$10{,}000 = 10^4$
$100{,}000 = 10^5$
$1{,}000{,}000 = 10^6$

For nonconductors, the resistance in ohms usually varies within the range of 10^9 to 10^{18}.

31.38 Color Code for Resistors

The color bands on a resistor, as shown in **Figure 31-11**, indicate its resistance value. The first band (A) is the first digit of the resistance value. The second band

Color Code for Resistors

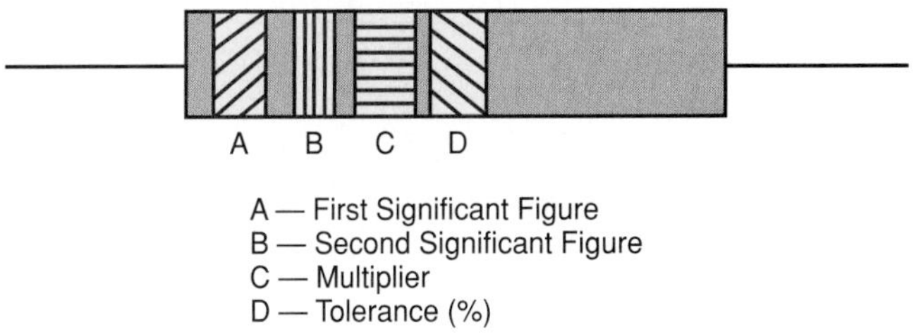

Color	Band A & B	Band C	Band D
Black	0	1	
Brown	1	10	
Red	2	100	
Orange	3	1,000	
Yellow	4	10,000	
Green	5	100,000	
Blue	6	1,000,000	
Violet	7		
Gray	8		
White	9		
Silver		0.01	±10
Gold		0.1	±5
No color			±20

Figure 31-11. *Color coding indicates a specific resistor value.*

(B) is the second digit of the resistance value. The third band (C) is the multiplier, and the fourth band (D) determines the tolerance.

A resistor that is color-coded orange, blue, red, and gold would have a resistance value of 36 × 100 = 3600 ohms. Since the fourth band is gold, the tolerance would be 5%. A 5% tolerance means the actual resistance can be between 3420 and 3780 ohms.

31.39 Resistance in Series and Parallel Circuits

The total resistance of resistances in *series* is the sum of the separate resistances. Total resistance of a parallel circuit is always less than any individual resistor or branch resistance. It can be computed by adding the reciprocals of each resistance. The reciprocal of that sum is equal to the total resistance. Refer to **Figure 31-12.**

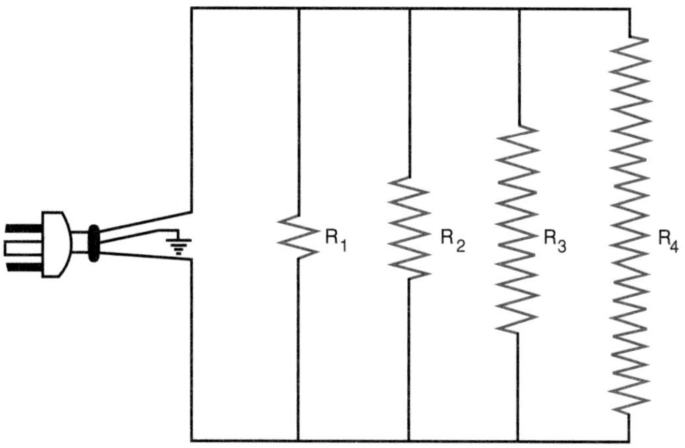

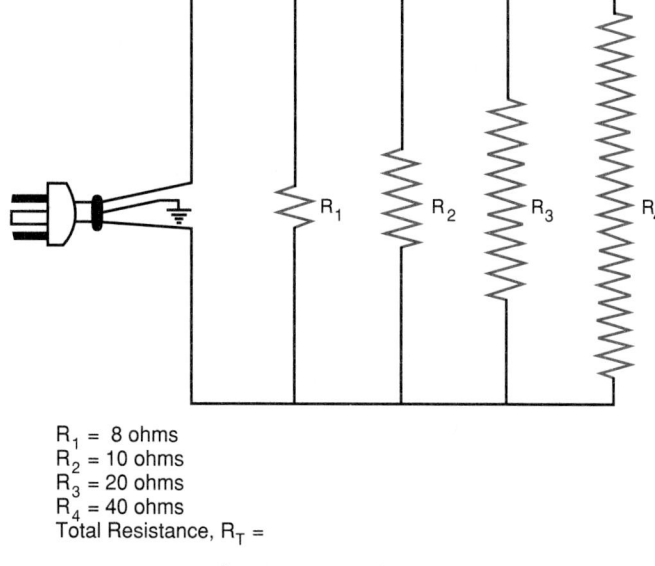

R_1 = 8 ohms
R_2 = 10 ohms
R_3 = 20 ohms
R_4 = 40 ohms
Total Resistance, R_T =

$$R_T = \frac{1}{\frac{1}{R_1} + \frac{1}{R_2} + \frac{1}{R_3} + \frac{1}{R_4}} =$$

$$\frac{1}{1/8 + 1/10 + 1/20 + 1/40} =$$

$$\frac{1}{5/40 + 4/40 + 2/40 + 1/40} = \frac{1}{12/40} = \frac{1}{.3} = 3.33 \text{ ohms}$$

R_T = 3.33 ohms

Figure 31-12. *Resistances in parallel and how their values are computed.*

31.40 Electrical Codes

The National Electrical Code (United States) and the Canadian Electrical Code each deal with specifications related to electrical circuits used in refrigerating and air conditioning. In addition, most local communities have codes dealing with these circuits.

A service technician should have available both national and local codes. In most cases, refrigeration service technicians do not install electrical circuits. You should, however, be familiar with the codes. This enables you to know whether the supply circuit is properly installed and safe to use.

31.41 Root Mean Square (rms) Values

The ***root mean square (rms)*** values of voltage and current are used to calculate power in a circuit when the current and voltage are not in phase. In this calculation, a mean voltage and mean current are used.

To find the root mean square value of a quantity that varies over a cycle, use the following steps:

1. Find the instantaneous values of the current and voltage and *square* them.
2. Find the *mean* over one cycle of the square of the instantaneous values.
3. Find the square *root* of this value.

This gives the rms (root mean square) value.

The calculation can be done two ways: with the currents in phase, or with the currents not in phase.

If currents are *in phase* and vary as a sine curve **(Figure 31-13)**, then the rms current (I_{rms}) is:

$$I_{rms} = I_{max} \times \left(\frac{\sqrt{2}}{2}\right) = I_{max} \times 0.707$$

If the voltage varies as a sine curve, then the rms voltage is:

$$V_{rms} = V_{max} \times 0.707$$

The average power is then:

$$P = I_{rms} \times V_{rms}$$
$$= (I_{max} \times 0.707) \times (V_{max} \times 0.707)$$
$$P = 0.5 \times I_{max} V_{max}$$

If currents are *not* in phase, but vary as sine curves, much of the previous calculation is still useful.

$$I_{rms} = I_{max} \times 0.707$$
$$V_{rms} = V_{max} \times 0.707$$

The average power is:

$$P = I_{rms} \times V_{rms} \times \cos \theta$$

The term "cos θ" is the cosine of the phase angle between the current and voltage. The phase angle, θ, can be determined with an oscilloscope.

The dual-trace oscilloscope or oscillograph is best to use. However, the phase angle can be determined on a single-trace oscilloscope. With a dual-trace oscilloscope, just note the amount of shift from right to left of the two traces. The shift can be from 0° to 90°. (Ignore negative shifts like "−70°".) Two wave traces on top of each other give a 0° shift. Two waves with a zero of one above a maximum of the other give 90°. Other amounts are in between as counted on the grid lines of the oscilloscope.

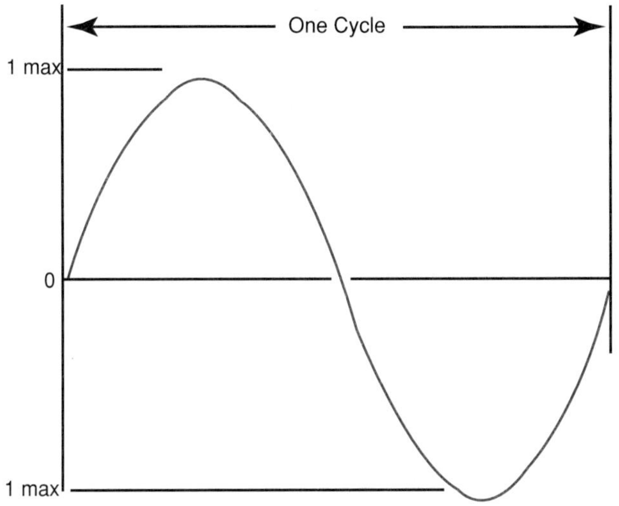

Figure 31-13. *The sine curve represents the current in a circuit.*

31.42 Galvanic Action Sequence

Certain materials produce electricity by chemical action. This is known as ***galvanic action.*** Some materials are much more active than others. Magnesium is one of the most active and platinum is one of the least active.

The corrosion of materials (usually metals) is often due to galvanic action. There is rapid deterioration at the place where galvanic action is occurring. Joints where two unlike galvanic materials touch should be electrically insulated from one another. This will minimize galvanic action.

Figure 31-14 lists the galvanic action sequence for a series of common materials. The items listed at the top are the most active and most likely to corrode. In galvanic action, these materials become an ***anode-positive.*** The items at the end of the list are least likely to corrode. In galvanic action, these become a ***cathode-negative.***

Most Active Metals ↑	Magnesium
	Magnesium Alloys
	Zinc
	Aluminum 2S
	Cadmium
	Aluminum 17ST
	Steel or Iron
	Cast Iron
	18-8 Stainless Steel
	Lead-Tin Solders
	Lead
	Tin
	Nickel
	Brass
	Copper
	Bronze
	Silver Solder
	Silver
	Graphite
	Gold
Least Active Metals ↓	Platinum

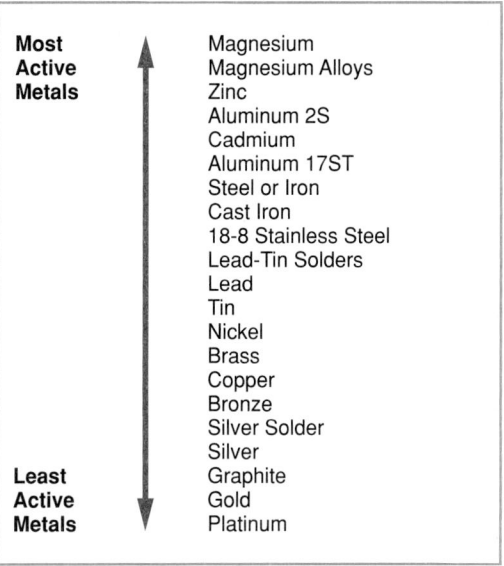

Figure 31-14. *Galvanic action sequence for commonly used metals.*

31.43 Electrolysis

Electrolysis is the movement through a substance of dc (direct current) electricity that causes chemical change in the substance or its container. Electrolysis can be used for producing useful results such as storage battery operation and electroplating.

However, it can also cause severe damage. Whenever there are two electrical conductors and moisture, dc electricity will flow if the conductors are of different activity levels. As the dc electricity flow continues (electrons from − to +), one of the conductors will become coated with a new chemical. The other will have material removed (leaving pits and holes). The pitted and "eaten away" areas are called corrosion areas. To avoid electrolytic corrosion, remove the moisture, seal off (insulate) the conductors, or use conductors of the same activity level.

31.44 Moisture-Holding Properties of Air

The moisture-holding capacity of air depends upon its temperature. The warmer the temperature, the greater the amount of moisture it will hold. **Figure 31-15** shows a table of moisture-holding properties of air. **Figure 31-16** shows the saturation pressure and the heat in the liquid. It also shows the total heat after vaporization for mixtures of air and water vapor.

Moisture and Air Relationships									
ASHRAE has adopted pounds of moisture per pound of dry air as standard nomenclature. Relation of other units are expressed below at various dew point temperatures.									
Equiv. Dew Pt. °F	Lb H_2O/ lb. dry air	Parts per million	Grains/ lb. dry air[a]	Percent Moisture %[b]	Equiv. Dew Pt. °F	Lb H_2O/ lb. dry air	Parts per million	Grains/ lb. dry air[a]	Percent Moisture %[b]
−100	0.000001	1	0.0007	—	0	0.0008	787	5.51	5.0
−90	0.000002	2	0.0016	—	10	0.0013	1 315	9.20	8.3
−80	0.000005	5	0.0035	—	20	0.0022	2 152	15.1	13.6
−70	0.00001	10	0.073	0.06	30	0.0032	3 154	24.2	21.8
−60	0.00002	21	0.148	0.13	40	0.0052	5 213	36.5	33.0
−50	0.00004	42	0.291	0.26	50	0.0077	7 658	53.6	48.4
−40	0.00008	79	0.555	0.5	60	0.0111	11 080	77.6	70.2
−30	0.00015	146	1.02	0.9	70	0.0158	15 820	110.7	100.0
−20	0.00026	263	1.84	1.7	80	0.0223	22 330	156.3	—
−10	0.00046	461	3.22	2.90.0	90	0.0312	31 180	218.3	—
					100	0.0432	43 190	302.3	—

a 7000 grains = 1 lb
b Compared to 70°F saturated

Normally the sensible heat factor determines the cfm required to accept a load. In some industrial applications, the latent heat factor may control the air circulation rate.

$$\text{Thus cfm} = \frac{\text{Latent heat, Btu/h}}{(W_1 - W_2) \times 4840}$$

Figure 31-15. *Moisture-holding properties of air. (Numbers, 1985, Altadena, CA, by Bill Holladay and Cy Otterholm)*

Ice, Water, Water Vapor					Dry Air		
Temp. F.	Sat. Press. in Hg.	Heat in Liquid Btu/lb.	Total Heat after Vaporization Btu/lb.	Vol. of Water Vapor cu.ft./lb.	Volume cu.ft./lb.	Specific Heat Btu/lb.	Amount of Water Vapor Saturate Grains
−40	3.790×10^{-3}	−177.1	1043.4	1.343×10^{5}	10.567	−9.61	.5508
−30	7.503×10^{-3}	−172.7	1047.8	7.441×10^{4}	10.820	−7.21	1.018
−20	1.259×10^{-2}	−168.2	1052.3	4.237×10^{4}	11.073	−4.81	1.830
−10	2.203×10^{-2}	−163.6	1056.7	2.475×10^{4}	11.326	−2.40	3.206
0	3.764×10^{-2}	−159.0	1061.1	1.481×10^{4}	11.579	0.00	5.480
10	6.286×10^{-2}	−154.2	1065.5	9060	11.832	2.40	9.161
20	.1027	−149.4	1069.9	5662	12.085	4.81	14.99
30	.1645	−144.4	1074.3	3608	12.338	7.21	24.07
32	.1803	−143.4	1075.2	3305	12.389	7.69	26.40
35	.2034	0.0	1076.5	2948	12.464	8.41	29.80
40	.2477	8.0	1078.7	2445	12.591	9.61	36.34
45	.3002	13.1	1080.9	2037	12.717	10.82	44.14
50	.3624	18.1	1083.1	1704	12.844	12.02	53.40
55	.4356	23.1	1085.2	1431	12.970	13.22	64.36
60	.5216	28.1	1087.4	1207	13.096	14.42	77.29
65	.6221	33.1	1089.6	1022	13.223	15.62	92.51
70	.7392	38.1	1091.8	868.0	13.349	16.83	110.4
72	.7911	40.1	1092.6	814.0	13.399	17.31	118.4
75	.8751	43.1	1093.9	740.0	13.475	18.03	131.3
80	1.0323	48.1	1096.1	633.0	13.602	19.23	155.8
85	1.2136	53.1	1098.3	543.3	13.738	20.43	184.4
90	1.4219	58.0	1100.4	467.9	13.854	21.64	217.6
95	1.6607	63.0	1102.6	404.2	13.981	22.84	256.4
100	1.9334	68.0	1104.7	350.2	14.107	24.04	301.5
120	3.4477	88.0	1113.3	203.2	14.612	28.85	569.0
140	5.8842	108.0	1121.7	123.0	15.117	33.67	1071.
160	9.6556	128.0	1129.9	77.27	15.622	38.48	2090.
180	15.295	148.0	1137.9	50.22	16.128	43.30	4598.
200	23.468	168.1	1145.8	33.64	16.632	48.12	16052.
212	29.921	180.1	1150.4	26.80	16.900	50.00	---

Figure 31-16. *Air-water vapor values. Properties of a mixture of air and water vapor for various temperatures from −40°F to 212°F (−40°C to 100°C) are shown. Note large volume occupied by 1 lb. of water vapor at the lower temperature.*

31.45 Evaporation Sources

Operation	Pounds of Moisture per Day
Bathing:	0.1 to 5
Clothes:	
Drying—average family—unvented	26.0
Washing—average family	4.0
Cooking:	
Breakfast	0.9
Lunch	1.2
Dinner	2.7
Dishwashing:	
Breakfast and lunch	0.2
Dinner	0.7
Humans:	
Average	0.4
Mopping:	
Per 100 ft^2	3.0

31.46 Desiccants

Desiccants are used in driers installed in refrigerator liquid lines and/or suction lines. They collect and remove moisture (water) from the system. Some common desiccants are:

- Activated alumina.
- Alumina gel.
- Calcium sulfate.
- Molecular sieve.
- Silica gel.

Most driers will remove moisture, sediment, and acids from the circulating refrigerant. They are usually placed in the liquid line close to the refrigerant control valve. Driers should always be located in a cool place in the line. This is done because heat may cause moisture to leave the desiccant and continue to circulate. Liquid line driers are usually installed permanently. They are replaced only when they lose their effectiveness.

Following a burnout, a large-capacity drier is usually placed in the suction line. Suction line driers should be removed as soon as the cleanup is completed. Driers that have been used in a refrigeration system should not be reactivated by heating.

31.47 Psychrometric Chart Formula

Psychrometric charts do not give relative humidity for dry bulb temperatures below about 20°F (−7°C). A formula can be used. A formula also is useful if a chart is unavailable. Use the following formula:

$$RH = 100e^{\left(\frac{T_W - T_D}{28.116}\right)} - 17.4935(T_D - T_W)e^{-\frac{T_D}{28.116}}$$

RH is the relative humidity in percent. T_D is the dry bulb temperature in °F. T_W is the wet bulb temperature in °F. The symbol e refers to the exponential function. The exponential function is taken to the power represented by the number above the symbol. (The function e is the usual one used where 2.71828 is the value of e.)

The formula for grains of water per pound of dry air is as follows:

$$\frac{\text{grain}}{\text{lb.}} = 0.09065RHe^{\frac{T_D}{28.116}}$$

31.48 Boyle's Law

Boyle's Law expresses a very interesting relationship between the pressure and volume of a gas. It is stated as follows:

"The volume of a gas varies inversely as the pressure, provided the temperature remains constant."

This means that if a quantity of gas has its pressure doubled, the volume will become half of what it originally was. Or, if the volume is doubled, the gas pressure will be reduced by half. If a perfect gas is considered, Boyle's Law may be expressed as a formula:

Pressure × Volume = A constant number.

Assuming this to be true, when either the pressure or the volume is changed, the corresponding pressure or volume will be changed in the opposite direction. Therefore:

Old Pressure × Old Volume = New Pressure × New Volume

Expressed in letter form:

Po × Vo = Pn × Vn
Po = Old Absolute Pressure
Pn = New Absolute Pressure
Vo = Old Volume
Vn = New Volume

Note: This formula will be true only if the pressures are expressed as absolute (psia).

Example:

5 ft^3 of gas at 20 psi is compressed to 60 psi. The temperature remains constant. What will be the new volume?

For calculation purposes, atmospheric pressure = 15 psi.

Po × Vo = Pn × Vn
Po = 20 psi + atmospheric pressure
= (20 + 15) = 35 psia (absolute pressure)
Pn = 60 psi + atmospheric pressure =
(60 + 15) = 75 psia

Po × Vo = Pn × Vn, substituting the preceding values in this formula:

35 × 5 = 75 × Vn or 35 × 5/75 = Vn
Vn = 35/15
Vn = 2.33 ft^3 at 60 psi

In the metric system, volume is meters cubed (m^3).

Pressure is in newtons per meter squared = N/m^2 = 1 pascal = 1 Pa.
Atmospheric pressure = 101 kPa
One bar = atmospheric pressure
1 bar = 100 000 Pa = 10^5 Pa

Be careful to use the same volume units and pressure units in the same problem.

31.49 Charles' Law

Gases behave consistently with temperature changes. This is stated in ***Charles' Law:***

"At a constant pressure the volume of a confined gas varies directly as the absolute temperature. At a constant volume, the pressure varies directly as the absolute temperature."

Absolute pressures and absolute temperatures must always be used in the equations. In the equation form:

At Constant Volume

Old Absolute Pressure (Po)
× New Absolute Temperature (Tn)
= New Absolute Pressure (Pn)
× Old Absolute Temperature (To).
Po × Tn = Pn × To

Example:

What is the pressure of a quantity of confined gas when raised to 60°F (16°C) if its original pressure was 35 psig and temperature 40°F (4°C), at a constant volume?

Solution:

(Note that absolute temperatures and pressure must be used.)

Po × Tn = Pn × To
Po = (35 + 15) = 50
Tn = (60 + 460) = 520
To = (40 + 460) = 500

$$Pn = Po \times \frac{Tn}{To}$$

Pn = 50 × 520/500 = 52 psia
Pn = 52 − 15 = 37 psig

At Constant Pressure

Old Volume (Vo)
× New Absolute Temperature (Tn)
= New Volume (Vn)
× Old Absolute Temperature (To).
Vo × Tn = Vn × To

Example:

5 ft^3 of gas at 37°F (3°C) is raised to 90°F (32°C) at constant pressure. What is the new volume?

Solution:

Vo × Tn = Vn × To
5 × (90 + 460) = Vn × (37 + 460)
5 × 550 = Vn × 497
5 × 550/497 = 2750/497 = Vn
Vn = 5.53 ft^3

In the metric system:

- The volume will be in meters cubed (m^3).
- The pressures will be in newtons per square meter (N/m^2), or pascals (Pa).

1 N/m^2 = 1 Pa

31.50 Gas Law

Boyle's Law and Charles' Law may be combined to solve true gas problems. The formula is:

Po × Vo/To = Pn × Vn/Tn
To = Old Absolute Temperature
Tn = New Absolute Temperature
Po, Vo, and To represent original conditions
Pn, Vn, and Tn represent new conditions

Absolute values for temperature and pressure must be used in this equation.

Example:

2 ft^3 of gas at 15 psi and 140°F is stored in a 4 ft^3 container at 30 psi. What is the new temperature?

Formula:

Po × Vo/To = Pn × Vn/Tn
Po = (15 + 15) = 30
Pn = (30 + 15) = 45
Vo = 2
Vn = 4
To = (140 + 460) = 600
Tn = Pn × Vn × To/Po × Vo
Tn = 45 × 4 × 600/30 × 2 = 108,000/60 =
1800 R. (F_A.)
= 1800 − 460 = 1340°F

In the metric system, the same formula is used but the units are different:

- The volume is in cubic meters (m^3).
- The pressure is in newtons per square meter (N/m^2) or pascals (Pa).

31.51 Adiabatic Expansion and Contraction of a Gas

When the term "adiabatic" is used, it refers to a natural process in a gas. The ***adiabatic process*** or property allows the gas to be expanded or contracted (compressed) without absorbing heat from outside or without transferring heat out of the gas.

Adiabatic expansion and compression of gases would occur if the gas were placed in a perfectly insulated cylinder with a frictionless piston. Heat could not enter the gas during expansion nor escape during compression.

During adiabatic compression and expansion of ideal gases, the work performed (compression and expansion) is obtained *from the gas.* Heat is generated when work is done on a gas as it is adiabatically compressed. The heat generated is not lost. It increases the temperature and therefore the pressure of the confined gas. During expansion, the pressure and temperature of the ideal gas decreases.

31.52 Isothermal Expansion and Contraction of a Gas

An ***isothermal condition*** exists when the expansion or contraction of a gas occurs without a temperature change. (See Boyle's Law, Section 31.48.) This condition can occur either during the expansion or the compression of a gas.

During expansion, the gas is cooled. Heat needed to keep the gas at a constant temperature must come from an outside source. Heat so obtained must exactly equal that given up by the gas during its expansion. This is necessary to keep the temperature constant.

A similar condition must exist during the isothermal compression of a gas. Heat must be removed in an amount equal to the heat energy of compression. This is necessary in order to maintain a constant temperature. Whether the compressor is air-cooled or water-cooled, heat removed must equal the heat input from the work of compressing the gas.

Gas compression and expansion operations cannot occur without a change in the confined gas temperature. If the action results in raising the confined gas temperature, it is called ***exothermal.*** If the action results in lowering the confined gas temperature, it is called ***endothermal.***

31.53 Gas and Vapor

A ***true gas*** exists as a gas at standard temperatures and pressures. ***Vapor*** is the gaseous state of a substance that is a liquid at standard atmospheric pressures and temperatures (such as water). However, vapor is also the term applied to refrigerants in the gaseous state inside the refrigerating system.

31.54 Refrigerant Properties

Several refrigerant tables and pressure-heat enthalpy diagrams are given in this section. Most of the physical properties of refrigerants may be found in these tables and charts.

Refrigerants are assigned a refrigerant ("R") number designation by the American Society of Heating, Refrigerating, and Air-Conditioning Engineers (ASHRAE). The number designation is based on the composition of the refrigerant. See **Figure 31-17** for a summary of this numbering system.

Series	Refrigerant Classification
−000	Methane-based compounds
−100	Ethane-based compounds
−200	Propane-based compounds
−300	Cyclic organic compounds
−400	Zeotropes
−500	Azeotropes
−600	Organic compounds
−700	Inorganic compounds
−1000	Unsaturated organic compounds

Figure 31-17. *Series numbering system used for determining refrigerant identification numbers. (American Society of Heating, Refrigerating, and Air-Conditioning Engineers, ASHRAE Standard 34–1992)*

31.55 Characteristics of Little-Used Refrigerants

Chapter 9 describes the characteristics and uses of the more popular refrigerants. Some refrigerants which were in common use at one time, but are now encountered much less frequently, are listed below. Various properties of these and other less-common refrigerants are given in **Figure 31-18.** Their physical properties and uses as refrigerants also are explained in the ASHRAE Handbook of Fundamentals.

Number	Chemical Name	Formula
R-13	Monochlorotrifluoromethane	$CClF_3$
R-13 B1	Monobromotrifluoromethane	$CBrF_3$
R-21	Dichloromonofluoromethane	$CHCl_2F$
R-30	Methylene Chloride	CH_2Cl_2
R-40	Methyl Chloride	CH_3Cl
R-113	Trichlorotrifluoroethane	CCl_2FCClF_2
R-114	Dichlorotetrafluoroethane	$CClF_2CClF_2$
R-160	Ethyl Chloride	C_2H_5Cl
R-290	Propane	C_3H_8
R-600	Butane	C_4H_{10}
R-744	Carbon Dioxide	CO_2
R-764	Sulfur Dioxide	SO_2

Refrigerant Number	Name	Chemical Symbol	Trade Name	Molecular Weight	Odor	Toxicity	Flammability	Pressure psia at 5°F	Pressure psia at 86°F	Latent Heat at 5°F	Sp. Heat of Liquid at 5°F	Critical Temperature of	Critical Pressure psia	Sp. Volume of Gas at 5°F	Density of Liquid in a 5° F.#/cu. ft.	CP/CV Ratio	Sp. Heat of Vapor at 86°F
R-764	Sulfur Dioxide	SO_2		64.06	Pungent	High	Non	11.81	66.45	172.3	.34	314.8	1141.5	6.421	92	1.256*	.34
R-40	Methyl Choride	CH_3Cl		50.489	Sweet	Med.	Slight	20.89	95.53	180.6	.45	289.6	969.2	4.530	61	1.20	.4
R-160	Ethyl Chloride	C_2H_5Cl	Alcozol	64.51	Etheral	Med.	Yes	4.65	27.10	177	.47	369	764	17.55	59.00	1.13	.42
R-13	R-13	$CClF_3$		104.46	Sweet	Low	Non	.....	.....	63.85	.247	84	561	.431	77	1.172	.....
										(-115°F)	(-22°F)			(-115°F)	(0°F)		
R-744	Carbon Dioxide	CO_2		44.005	Non	Low	Non	334.4	1039.0	116	.5	87.8	1066.2	.2673	61.22	1.30**	1.95
R-611	Methyl Formate	$C_2H_3O_2$		60.04	Slight	...	Slight	1.96	13.69	236	.515	417	870	46.7	.....	.....	.515
R-30	Methylene Chloride	CH_2Cl_2	Carrene No.1	84.9	Sweet	...	Yes	1.17	10.6	162.1	.34	421	670	50.58	.....	.....	.34
R-21	R-21	$CHCl_2F$	Thermon	102.92	Sweet	Low	Non	5.5	30.5	105.5	.26	353.3	750.0	8.83	90.1	.....	.26
R-114	R-114	$CClF_2CClF_2$		170.93	Sweet	Low	Non	.....	.....	58.9	.238	294	474	.488	73.1	1.088	.160
R-113	R-113	$C_2Cl_3F_3$		187.4	Sweet	Low	Non	.98	7.86	70.62	.199	417.4	495	27.04	103	.....	.26
R-290	Propane	C_3H_8		44.06	Sweet	Low	Yes	41.9	155	170.2	.56	302	661.5	2.48	34.33	.....	.55
R-170	Ethane	C_2H_6		30.04	Sweet	Low	Yes	236.0	675.0	150.5	.66	90.1	730	.533	26.96	.....	.83
R-600	Butane	C_4H_{10}		58.12	Sweet	Low	Yes	8.2	41.6	170.7	.51	308	529	9.98	38.41	.....	.51
R-13B1	Kulene 131	CF_3Br		.....	Etheral	Low	Non	77.93	261.8	44.88	.182	1535	587	.3854	112	.....	.10
R-115	Monochloropenta-fluoroethane	$CClF_2CF_3$		154.48	Sweet	Low	Non	38	148.9	.....	.....	175.9	.....	.82	.....	3.76	.....

*at 70°F **at 32°F #at 70°F

Figure 31-18. *Table of properties of little-used refrigerants.*

31.56 Metric Pressure-Heat Diagrams

With increased use of the SI Metric system, metric pressure-heat diagrams for refrigerants are being used more extensively.

The metric system pressure-heat diagram for R-134a is shown in **Figure 31-19. Figure 31-20** shows the metric system pressure-heat diagram for R-123.

Note that the enthalpy is given in kJ/kg (kilojoules per kilogram). The pressure is given in MPa (megapascals).

For the U.S. conventional unit pressure-heat diagrams for R-134a and R-123, see Chapter 9.

31.57 Moisture in Liquid Refrigerants

Moisture is soluble in most liquid refrigerants. More moisture may be safely present in some refrigerants than in others.

The amount of the allowable moisture varies with the temperature. **Figure 31-21** shows the solubility level in water for some of the commonly used and alternative refrigerants. In low-temperature applications, a very small amount of moisture may cause trouble.

31.58 Dryness of Refrigerants

Any gas that is holding water in the vapor state will, when cooled, reach a temperature at which free moisture (liquid water) will appear. The greater the concentration of water vapor in the gas, the higher the temperature at which free water will condense out.

This freeing of moisture is readily shown by frost on cold lines. It is shown as beads of water on glasses of cold liquids. In the refrigeration system, it is shown by ice in the expansion valve orifice. Free water must be present in a refrigerating system before ice can be formed.

The highest temperature at which free water is liberated, on cooling, is called the ***dew point.*** There is only one dew point temperature for a given moisture concentration in a gas such as air or a refrigerant vapor. The dew point determines the lowest temperature at which the refrigerant control will operate satisfactorily.

The dew point should be 10°F (6°C) lower than the low-side refrigerant temperatures. This will prevent the formation of sludges and plating. A refrigerant dew point of 5°F (3°C) lower than the low-side refrigerant temperature will prevent refrigerant control freezeup.

31.59 Refrigerant Oils

Requirements for a satisfactory refrigerant oil are rather severe. It must provide good lubricating qualities. The viscosity must be correct for the refrigerant and the machine in which it is used. The oil must be free of moisture.

The viscosity of a refrigerant oil is measured using the Saybolt Universal Viscosity Test. A Saybolt viscometer should be used. In this instrument, oil at 100°F (38°C) is allowed to flow through the standard Saybolt orifice. The time required in seconds for 60 cm^3 of the oil to flow through this orifice is recorded.

The amount of moisture in refrigerant oil may be measured by its resistance to the flow of a current of electricity through it. This is known as its ***dielectric property.*** A good refrigerant oil should have a minimum dielectric value of 25,000 V.

Another test that may be given to refrigerant oil is known as the ***floc test.*** This test applies to oils that are used with completely ***miscible*** (mixable) refrigerants. (Examples of these refrigerants are R-11, R-12, and R-22.) The test is conducted by mixing 10% refrigerant with 90% oil. The mixture is sealed in a glass tube. Then it is cooled slowly until a ***flocculent*** (cloudy) precipitate of wax appears. The maximum temperature at which this occurs is recorded as the ***floc point.***

The viscosity of oil changes with the temperature. Oils at very low temperatures may not pour and may become a plastic solid. The temperature at which oil will just flow is called ***the pour point.*** This temperature is recorded either in Fahrenheit or Celsius.

The flash point of an oil may be determined as follows. A quantity of the oil is heated slowly in an open container. A thermometer is used to determine the temperature of the oil. A lighted candle or other flame is brought to the oil's surface as it is heated. The temperature at which the vapors from the oil surface burn or flash is the oil's ***flash point.***

Refrigerant oils sometimes include a very small amount of antifoam inhibitor to reduce foaming. Compressor parts are sometimes given a phosphating treatment to improve lubrication. Tricresyl phosphate has also been added to refrigerant oils to improve lubrication.

Some of the newer replacement refrigerants require a different type of oil. Mineral oil-based lubricants cannot be used. Lubricants designed for these refrigerants include polyol ester, alkyl benzene, and polyalkylene glycol.

Testing procedures for the new lubricants are the same as for the mineral oil. The floc point test as described above can easily be used. The synthetic refrigerant compression lubricating oils are soluble with the refrigerants. They are compatible with all varieties of materials, including metals, wire coatings, etc.

31.60 Eutectic (Phase-Change) Materials

Proprietary compounds are available that provide an active thermal energy source to maintain a specified temperature range. (***Proprietary*** means made and sold

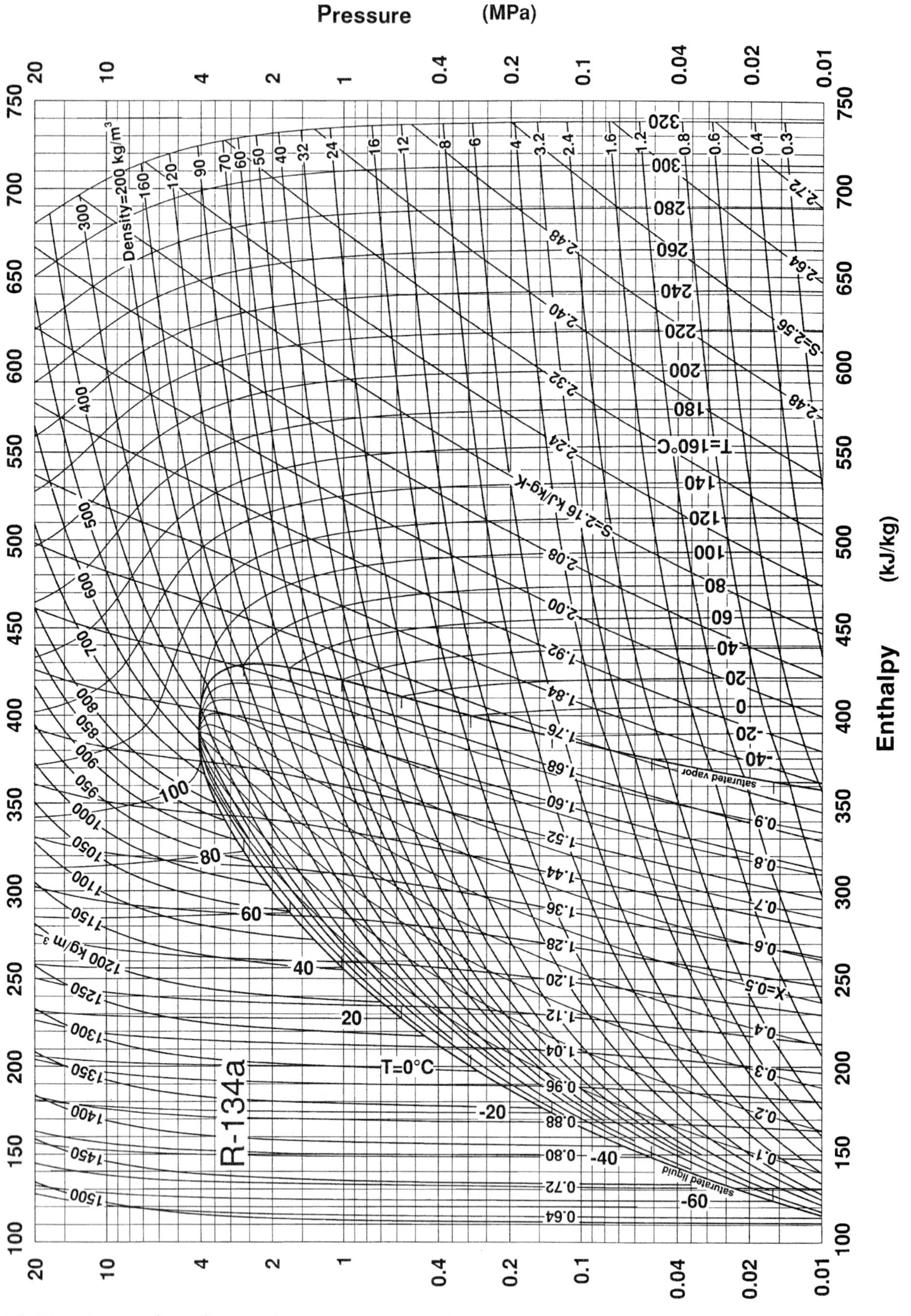

Figure 31-19. *Pressure-heat diagram for R-134a, expressed in metric units. (Prepared by Center for Applied Thermodynamic Studies, University of Idaho. Copyright 1992 American Society of Heating, Refrigerating, and Air-Conditioning Engineers)*

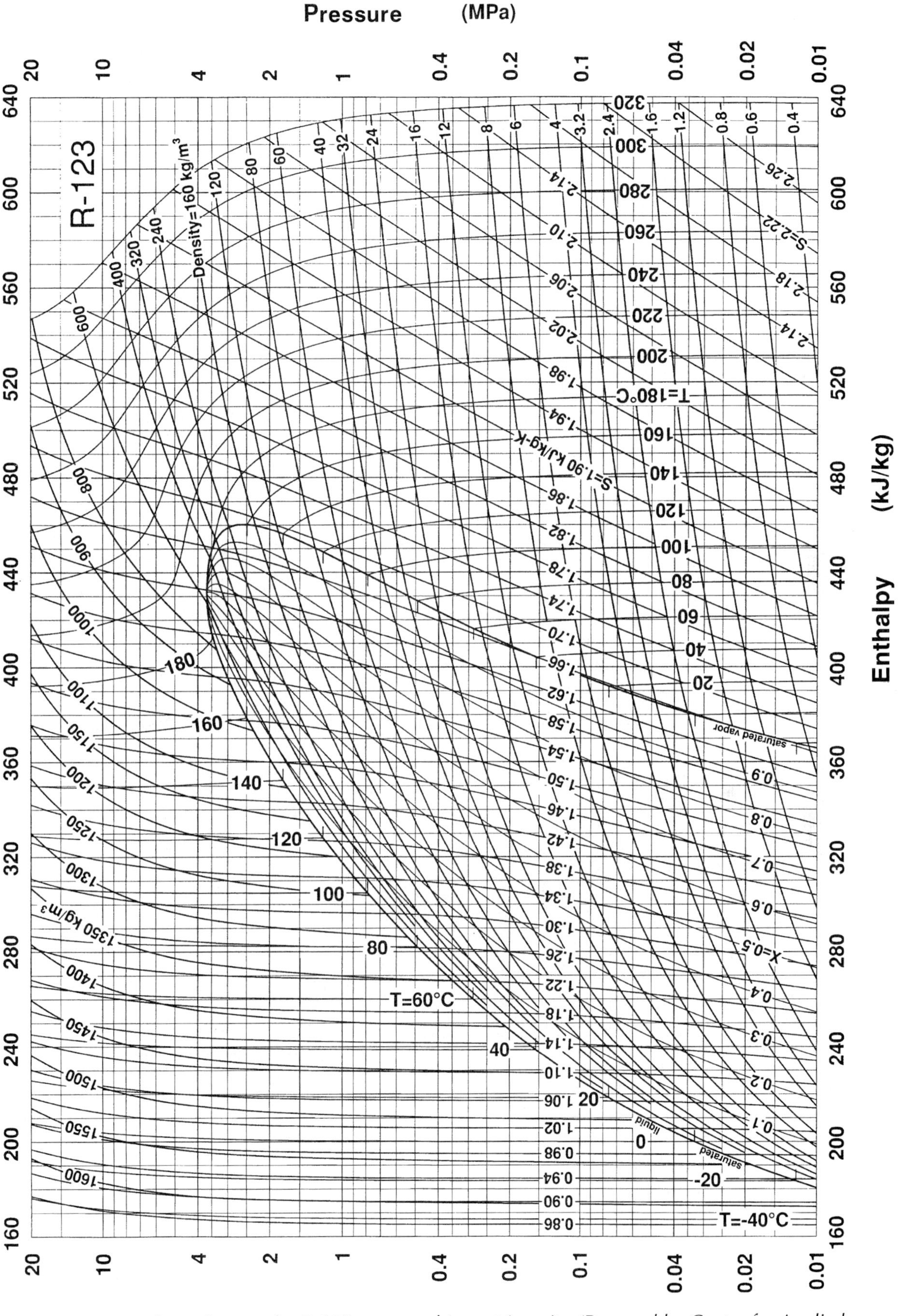

Figure 31-20. *Pressure-heat diagram for R-123 expressed in metric units. (Prepared by Center for Applied Thermodynamic Studies, University of Idaho. Copyright 1992 American Society of Heating, Refrigerating, and Air-Conditioning Engineers)*

Temp., °F	Solubility, ppm by weight								
	R-11	R-12	R-13	R-22	R-113	R-114	R-123	R-134a	R-502
160	460	700	—	4100	460	450	2600	4200	1780
150	400	560	—	3600	400	380	2300	3600	1580
140	340	440	—	3150	344	320	2000	3200	1400
130	290	350	—	2750	290	270	1800	2800	1220
120	240	270	—	2400	240	220	1600	2400	1080
110	200	210	—	2100	200	180	1400	2000	930
100	168	165	—	1800	168	148	1200	1800	810
90	140	128	—	1580	140	120	1000	1500	690
80	113	98	—	1350	113	95	900	1300	580
70	90	76	—	1140	90	74	770	1100	490
60	70	58	44	970	70	57	660	880	400
50	55	44	—	830	55	44	560	730	335
40	44	32	26	690	44	33	470	600	278
30	34	23.3	—	573	34	25	400	490	225
20	26	16.6	14	472	26	18	330	390	180
10	20	11.8	—	384	20	13	270	320	146
0	15	8.3	7	308	15	10	220	250	115
−10	11	5.7	—	244	11	7	180	200	90
−20	8	3.8	3	195	8	5	140	150	69
−30	6	2.5	—	152	6	3	110	120	53
−40	4	1.7	1	120	—	2	90	89	40
−50	3	1.1	—	91	—	1.5	70	66	30
−60	2	0.7	—	68	—	1	53	49	22
−70	1	0.4	—	50	—	0.6	40	35	16
−80	0.8	0.3	—	37	—	0.4	30	25	11
−90	0.5	0.1	—	27	—	0.2	22	18	8
−100	0.3	0.1	—	19	—	0.1	16	12	5

Data on R-134a adapted from Thrasher *et al.* (1993) and Allied-Signal Corporation.
Data on R-123 adapted from Thrasher *et al.* (1993) and DuPont Company.
Remaining data adapted from DuPont Company and Allied-Signal Corporation. Used by permission.

Figure 31-21. *Solubility level in water in commonly used and alternative refrigerants. (Reprinted by permission of the American Society of Heating, Refrigerating, and Air-Conditioning Engineers, Atlanta, GA, from the 1994 ASHRAE Handbook—Refrigeration)*

by only one firm.) This process is called ***thermal energy storage (TES).*** It is created by the heat of fusion or heat of crystallization.

These compounds absorb or release heat with little or no temperature change while changing physical state. The melting of ice is an example. In the process of cooling water, 1 Btu of heat is removed to lower the temperature of 1 pound of water 1 degree. This is true down to the freezing temperature. At the freezing temperature, 144 Btu/lb. must be removed to freeze 1 pound of water. The temperature remains constant until the water is frozen.

A separate compound or mixture is needed for different temperature ranges. **Figure 31-22** lists some substances that may be used to maintain various temperatures.

Heat is absorbed or released by the compound as it changes state. For low temperatures, the heat of fusion is used. For high temperatures, the heat of crystallization is used.

Characteristics of Thermal Energy Storage Compounds			
TES Material	Melting Temperature F	Melting Temperature C	Heat of Fusion Btu/lb.
Trans-Temp 12*	12	−11	115
N-Dodecane	10	−12	47
Ethylene Glycol	9	−13	63
Trans-Temp 27*	27	−3	142
H_2O	32	0	144
Trans-Temp 40*	40	4	75
N-Tetradecane	42	6	98
Trans-Temp 65*	65	18	80
N-Hexadecane	62	17	102
Acetic Acid	62	17	80
Trans-Temp 130*	130	54	72
Paraffin Wax	130	54	63
Tristearin	133	56	82
Dry Ice	−109	−78	241

*Proprietary material

Figure 31-22. *Characteristics of some eutectic thermal energy storage (TES) materials.*

Eutectic compounds used in eutectic plates, Chapter 13, contain substances to prevent expansion on freezing. Freezing might otherwise rupture the containers. They also contain substances to reduce oxidation of the container materials. Useful life of the containers is thus extended.

31.61 Cryogenic Temperatures

The cryogenic temperature range extends from −250.0°F to absolute zero (−459.69°F). The cryogenic ranges may also be expressed as follows:

−156.6°C to −273.16°C
116.5 K to 0 K
209.69°F to 0°R
F = Fahrenheit
C = Celsius
K = Kelvin
R = Rankine

Low-carbon steel should not be used for piping or pressure parts in cryogenic applications. The metal becomes very brittle when cold. Some nickel and/or some chromium improves the properties of steel at very low temperature.

31.62 Brine Freezing Temperatures

A ***brine*** is water mixed with a substance that will go into solution with the water. A brine will provide a fluid which can readily flow at temperatures below 32°F (0°C). See **Figure 31-23.**

Brines are of several types:

- ***Alcohol brines*** are usually made from ethyl alcohol.

Brine Specific Gravity	Freezing Temperature °F and °C						
	20 (–7°C)	10 (–12°C)	0 (–18°C)	–10 (–23°C)	–20 (–29°C)	–30 (–34°C)	–40 (–40°C)
Alcohol (Formula No. 1)							
Specific Gravity at 60°F		.9691	.9592	.9486	.9345		
Calcium Chloride							
Specific Gravity at 60°F	1.090	1.140	1.175	1.201	1.227	1.254	1.265
Percent of Chemical	10	17	20.5	23	25	27	28
Sodium Chloride							
Usable Only Down to 0°F							
Specific Gravity at 60°F	1.072	1.118	1.158				
Percent of Chemical	10	16	21				
Ethylene Glycol							
Specific Gravity at 60°F	1.05	1.07	1.075	1.08	1.09	1.096	1.105
Percent of Chemical	32	40	43	45	50	53	57

Figure 31-23. *Table of properties for various brines.*

- ***Salt brines*** are usually made with sodium chloride and/or calcium chloride. The eutectic point for a sodium chloride (common salt) solution is −6°F (−21°C). For calcium chloride, it is 60°F (−51°C).
- ***Glycol brines*** with noncorrosive properties are usually made from glycerin, ethylene glycol, and/or propylene glycol.

A hydrometer can measure the density of a brine solution. The freezing temperature may be worked out from the hydrometer reading.

This is the same as measuring to determine the freezing temperature of cooling solution in automobile radiators. If the hydrometer reading indicates the solution freezing temperature is too high, it may be lowered. This is done by adding more of the antifreeze compound to the solution. Note that with alcohol brine, the density decreases with the lowering of the freezing temperature. This is because alcohol is lighter than water. A hydrometer with an alcohol testing scale must be used.

Decimal Sizes of Numbered Drills

No.	Size of Drill In Inches	No.	Size of Drill In Inches	No.	Size of Drill In Inches	No.	Size of Drill In Inches
1	.2280	21	.1590	41	.0960	61	.0390
2	.2210	22	.1570	42	.0935	62	.0380
3	.2130	23	.1540	43	.0890	63	.0370
4	.2090	24	.1520	44	.0860	64	.0360
5	.2055	25	.1495	45	.0820	65	.0350
6	.2040	26	.1470	46	.0810	66	.0330
7	.2010	27	.1440	47	.0785	67	.0320
8	.1990	28	.1405	48	.0760	68	.0310
9	.1960	29	.1360	49	.0730	69	.0292
10	.1935	30	.1285	50	.0700	70	.0280
11	.1910	31	.1200	51	.0670	71	.0260
12	.1890	32	.1160	52	.0635	72	.0250
13	.1850	33	.1130	53	.0595	73	.0240
14	.1820	34	.1110	54	.0550	74	.0225
15	.1800	35	.1100	55	.0520	75	.0210
16	.1770	36	.1065	56	.0465	76	.0200
17	.1730	37	.1040	57	.0430	77	.0180
18	.1695	38	.1015	58	.0420	78	.0160
19	.1660	39	.0995	59	.0410	79	.0145
20	.1610	40	.0980	60	.0400	80	.0135

Figure 31-24. *Number-size twist drills.*

31.63 Twist Drill Sizes

Sizes of twist drills may be given in fractions of an inch, decimals, numbers, or letters. Metric drill sizes are indicated in millimeters and decimals of a millimeter. You will use mostly fractional, number, and letter drills. If much tapping and threading is necessary, drills will be used from each of these categories.

Fractional-drill sizes start at 1/16″. They increase by 1/64″ through the required range. Decimal-size drills are mostly used in manufacturing, hardly ever by a service technician. Number-size drills are marked from 1 through 80. The range extends from .0135″ up to less than 1/4″. A table of numbered drills is shown in **Figure 31-24.**

Letter-drill sizes extend from A through Z (roughly from 1/4″ through .413″). Letter-drill sizes are shown in **Figure 31-25.** Tap drill sizes are shown in **Figure 2-66.**

Decimal Sizes of Lettered Drills

Letter	Dia. In.	Letter	Dia. In.	Letter	Dia. In.
A	0.234	J	0.277	S	0.348
B	0.238	K	0.281	T	0.358
C	0.242	L	0.290	U	0.368
D	0.246	M	0.295	V	0.377
E	0.250	N	0.302	W	0.386
F	0.257	O	0.316	X	0.397
G	0.261	P	0.323	Y	0.404
H	0.266	Q	0.332	Z	0.413
I	0.272	R	0.339		

Figure 31-25. *Letter-size twist drills.*

Tube Size O.D.	Wrench Size Across Flats	
	Old	New
1/4 in.	3/4 in.	5/8 in.
3/8 in.	7/8 in.	13/16 in.
1/2 in.	1 in.	15/16 in.

Figure 31-26. *A table of flare nut wrench sizes.*

31.64 Piping Color Codes

Use	Color
Fire protection equipment	Red
Safe material	Green (or, if needed, white, gray or aluminum)
Protective material	Bright blue
Extra-valuable material	Deep purple
Dangerous material	Yellow or orange

31.65 Solders and Brazing Metals

The following table lists some common solder alloys used in refrigeration work:

Solder	Melting Point (°F)	Flow Point (°F)	Shear Strength (psi)
50-50 Tin-Lead	358	414	83.4
95-5 Tin-Antimony	450	465	327.0
Silver Solder			
45 Ag, 15 Cu, 24 Cd, 16 Zn	1120	1145	8340
Phosphorous-Copper	1310	1650	8340

31.66 Flare Nut Wrench Sizes

A table of flare nut wrench sizes is shown in **Figure 31-26.** Old and new wrench sizes across the flats of the flare nut are given. Tubing size is based on outside diameter (OD).

31.67 Solvents and Cleaning

Parts may become coated with lubricating oils or grease as a result of use or repair operations. They may also become coated with oxides, dirt, metallic particles, or abrasives. Many methods may be used successfully to clean these parts. Lubricants or greases made from animal or vegetable oils or fats—such as tallow, lard oil, palm oil, or olive oil—can usually be removed by saponification. (***Saponification*** is the process of making a soap of the oil or fat.) Parts are soaked in an alkaline solution. The oils react with the alkali to form water-soluble soap compounds.

Mineral oils which cannot be made into a soap solution—such as kerosene, machine oil, cylinder oil, and general lubricating oils—are usually cleaned by an emulsification process. Soaps, wetting agents, and dispersing agents are used. (***Emulsification*** means to disperse a liquid, such as oil, in another liquid. The oil does not dissolve but is suspended in the liquid and makes it thick.)

Dirt, abrasives, metal dust, and inert materials are also generally removed by one or both of these processes.

Solvent cleaning is used for removing most of the oils from coated pieces. They are immersed (dipped) in a solvent, such as mineral spirits. Solvent tanks should have safety lids and should be hooded and vented.

Vapor cleaning is also used to remove oils. Parts are held in a container where solvent vapors can condense on them. The condensed solvent washes away the oily coating, leaving surfaces dry and nearly clean.

Production degreasing machines have two or three compartments. The work is immersed in the first compartment containing a boiling solution of the solvent. It is then dipped into the second section, which contains clean cold solvent. Finally, it is hung in the third section. There, only clean solvent vapors condense on and wash over the work. The degreasing unit is self-purifying. Oils and waste gather at the bottom of the third section.

Job shop cleaning uses the third section only. **Solvents should be those approved under the Occupational Safety and Health Act (OSHA). Venting is extremely important for safety to the operator.**

Alkaline cleaning baths are used primarily to remove oils, greases, solid dirt particles, and metal particles. The parts are immersed in hot alkaline solutions. The chemicals saponify vegetable and animal oils and fats. They emulsify mineral oils and greases, and suspend the solid material.

The combination of heat, active chemicals, and agitation (shaking) are important. Soap is either added or is formed by the saponification of vegetable or animal fats. Caustic soda, soda ash, and causticized soda are the cheapest and most direct methods of producing alkalinity in the bath. Such materials, however, have less surface activity (as a general rule) than more complex materials. Sodium metasilicate, trisodium phosphate, and similar salts are often used to obtain alkalinity in a solution.

A number of proprietary preparations are used for ***emulsion cleaning.*** They consist of an emulsification agent which disperses organic solvents in water solutions. Emulsifiable cleaners are miscible (able to mix) with oils and can be washed off with water. However, a film of oil may remain on the work. This makes subsequent alkali cleaning necessary and results in added expense.

Alkaline materials are used in ***electrolytic cleaning.*** The bath is maintained at a temperature as near boiling as possible without causing heavy tarnishing. The gas that is evolved lifts off the soil, making a clean surface for subsequent operations. The work to be cleaned is usually made the cathode. There are many formulas available for this type of cleaning. The one used depends upon the materials to be cleaned and the degree of tarnish permissible. In many cases, unusual results can be obtained by switching polarity several times during cleaning. This is particularly true for carbon steel or cast iron.

31.68 Clean Rooms

Absolute control of temperature, humidity, air cleanliness, and air chemistry is necessary in many laboratories. Such rooms are called ***clean rooms.*** The aerospace industry and computer chip manufacturers are primary users of clean room technology.

Some assembly operations must also be conducted in clean rooms. Dust particles must not be allowed to enter mechanisms during assembly. Even those particles too small to be seen (in the 20-50 microns range) may interfere with the operation of very complex devices. See Chapter 23 for information on air cleaning.

31.69 Review of Safety

A safety code for mechanical refrigeration is sponsored by the American Society of Heating, Refrigerating, and Air-Conditioning Engineers (ASHRAE). This code has been developed according to rules and regulations of the American National Standards Institute. Copies are available from ASHRAE.

Many safety regulations and controls are provided by the Occupational Safety and Health Act. It includes regulations for the installation, operation, and service of refrigerating and air conditioning mechanisms.

Threshold Limit Values have been determined for certain substances. These are substances that may be toxic under certain conditions and lengths of exposure. These Threshold Limit Values were published by the American Conference of Governmental Industrial Hygienists. The threshold values for some of these substances for an eight-hour exposure time are shown in the following table.

Substance	ppm
Acetone	1000
Ammonia	50
Carbon dioxide	5000
Carbon monoxide	100
Carbon tetrachloride-Skin	10
Chlorine	1
Dichlorodifluoromethane	1000
Dichloromonofluoromethane	1000
Ethyl ether	400
Fluorine	0.1
Gasoline	500
Methyl acetylene	1000
Methyl alcohol (methanol)	200
Methyl chloride	100
Ozone	0.1
Sulfur dioxide	5
Xylene (xylol)	200

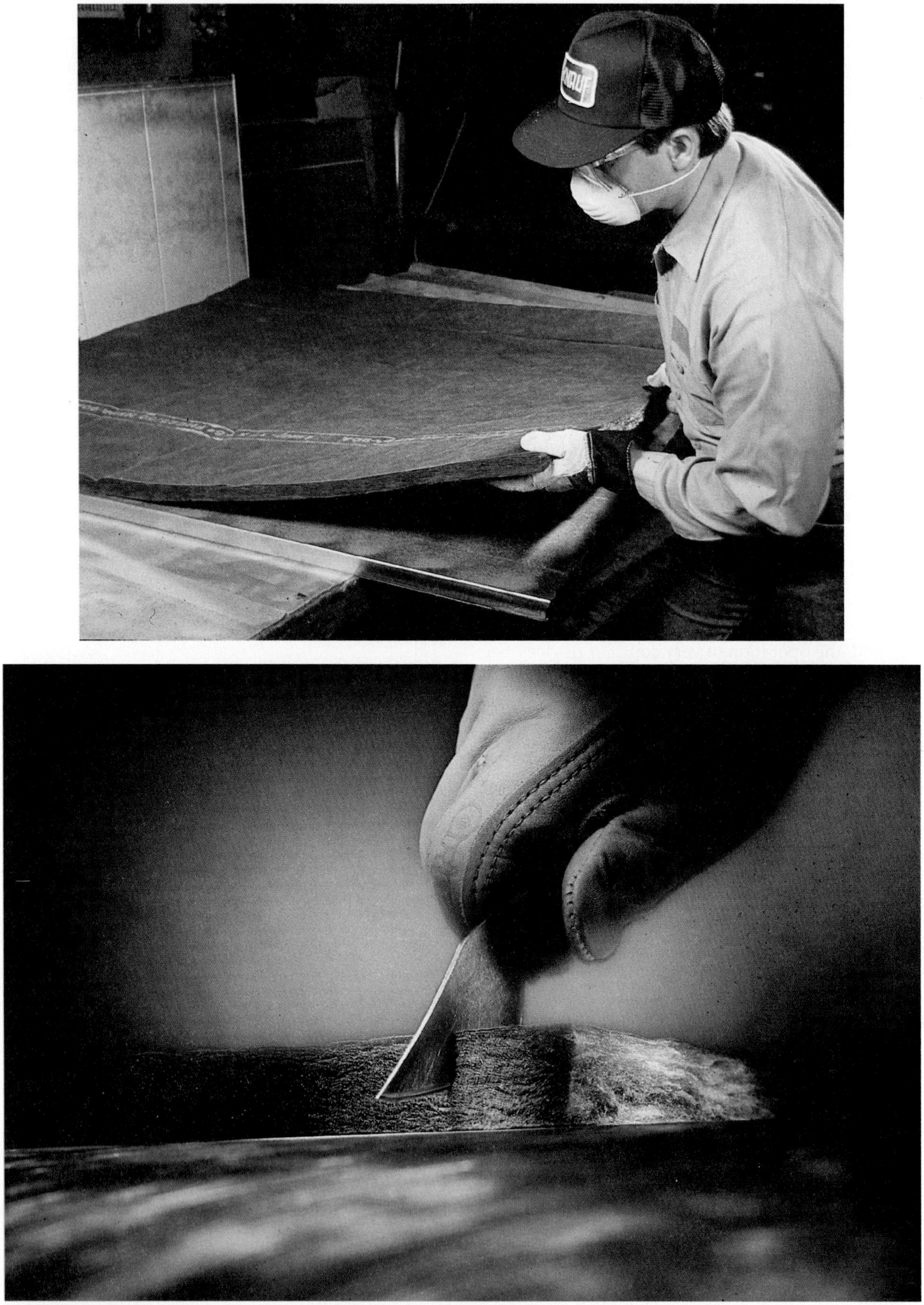

Duct liner, which comes in flexible rolls or rigid boards, provides better acoustical insulating performance than duct wrap. Most liners have a coated or mat-faced airstream surface. (Knauf Fiber Glass GmbH)

DICTIONARY OF TECHNICAL TERMS

The following pages comprise a complete dictionary of the most frequently used technical terms in the heating, ventilating, and air conditioning industry. These brief explanations will help your understanding of many terms used throughout the textbook.

A

Absolute humidity: Amount of moisture in the air, indicated in grains per ft^3.

Absolute pressure: Gauge pressure plus atmospheric pressure (14.7 psi).

Absolute temperature: Temperature measured from absolute zero.

Absolute zero temperature: Temperature at which all molecular motion ceases (−460°F and −273°C).

Absorbent: The ability to soak up another substance.

Absorber: A solution or surface that is capable of soaking up (taking in) another substance or energy form.

Absorption: The process of taking or soaking up into a substance.

Absorption chiller: A chiller that uses a brine solution and water to provide refrigeration without the aid of a compressor.

Absorption refrigerator: Refrigerator that creates low temperatures by using the cooling effect formed when a refrigerant is absorbed by chemical substance.

Accessible hermetic: Assembly of motor and compressor inside a single bolted housing unit.

Accumulator: Storage tank that receives liquid refrigerant from the evaporator and prevents it from flowing into the suction line before vaporizing.

Acid condition in system: Condition in which refrigerant or oil in system is mixed with acidic fluids.

ACR tubing: Tubing used in air conditioning and refrigeration. Ends are sealed to keep tubing clean and dry.

Activated alumina: A chemical that is a form of aluminum oxide. It is used as a drier or desiccant.

Activated carbon: Specially processed carbon used as a filter-drier; commonly used to clean air.

Active solar heating system: A system in which solar energy is absorbed in a collector, stored, and distributed by an auxiliary circulating system.

Actuator: That portion of a regulating valve that converts mechanical fluid, thermal energy, or electrical energy into mechanical motion to open or close valve seats.

Adiabatic compression: Compressing refrigerant gas without removing or adding heat.

Adsorbent: Substance with the property to hold molecules of fluids without causing a chemical or physical change.

Adsorption: The adhesion of a thin layer of molecules of a gas or liquid to a solid object.

Aeration: Act of combining a substance with air.

Agitator: Device used to cause motion in confined fluid.

Air: An invisible, odorless, and tasteless mixture of gases that surrounds the earth.

Air break: An inverted opening placed in the chimney of a gas furnace to prevent back pressure from outside wind from reaching the furnace flame or pilot.

Air cleaner: Device used for removal of airborne impurities.

Air coil: Coil on some types of heat pumps used either as an evaporator or a condenser.

Air conditioner: Device used to control temperature, humidity, cleanliness, and movement of air in conditioned space.

Air conditioning: Control of the temperature, humidity, air movement, and cleaning of air in a confined space.

Air-cooled condenser: Heat of compression is transferred from condensing coils to surrounding air. This may be done either by convection or by a fan or blower.

Air cooler: Mechanism designed to lower temperature of air passing through it.

Air core solenoid: A solenoid that has a hollow core instead of a solid core.

Air curtain: A system in which a blower is activated when a door is opened to blow across the open area, preventing the transfer of air between outdoors and indoors.

Air defrosting: Evaporator defrosting that occurs as evaporator warms when the compressor is not running.

Air diffuser: Air distribution outlet or grille designed to direct airflow into desired patterns.

Air gap: The space between magnetic poles or between rotating and stationary assemblies in a motor or generator.

Air handler: Fan-blower, heat transfer coil, filter, and housing parts of a system.

Airtight: Sealed to prevent the passage of gas.

Air-to-air heat pump: A heat pump that uses outdoor air, as opposed to a *geothermal heat pump.*

Air vent: Valve used to remove air from the highest point of a coil or piping assembly.

Air washer: Device used to clean air while changing its humidity.

Alcohol brine: Water and alcohol solution that remains a liquid below 32°F (0°).

Algae: Low form of plant life, found floating free in water.

Allen-type screw: Screw with recessed, hex-shaped head.

Allen wrench: Hexagonal (6-point) tip used to fit socket head screws or setscrews.

Alternating current (ac): Electric current in which direction of flow alternates (reverses). In 60-cycle (Hertz) current, direction of flow reverses every 1/120th of a second.

Altitude: The height at a point above a reference level, sea level, or the earth's surface.

Ambient compensator: An electronic device that provides a small amount of heat to the refrigeration compartment to ensure that the machinery continues to cycle when ambient temperatures are low.

Ambient temperature: Temperature of a fluid (usually air) that surrounds an object.

American Standard Pipe Thread: Type of screw thread commonly used on pipe and fittings to ensure a tight seal.

Ammeter: Electric meter used to measure cur rent.

Ammonia: Chemical combination of nitrogen and hydrogen (NH_3). Ammonia refrigerant is identified as R-117.

Amperage: Electron or current flow past a given point in a circuit.

Ampere: Unit of electric current equivalent to flow of one coulomb per second.

Ampere-turns: Term used to measure magnetic force. Represents product of amperes times number of turns in coil of electromagnet.

Amplifier: Electrical device that increases electron flow in a circuit.

Anemometer: Instrument for measuring the rate of airflow.

Angle valve: Type of globe valve design, having pipe openings at right angles to each other. Usually, one opening is in the horizontal plane and one is in the vertical plane.

Annealing: Cooling a metal slowly from a high temperature to make the metal soft.

Annual Fuel Utilization Efficiency (AFUE) rating: A rating system for furnaces that compares energy input and energy output.

Anode: Positive terminal of electrolytic cell.

Anticipator: A device used with a start-stop control to reduce the control differential.

Arcing: Band of sparks formed when an electrical discharge from a conductor jumps to another conductor.

ARI: Air-Conditioning and Refrigeration Institute.

Armature: Part of an electric motor, generator, or other device moved by magnetism.

Articulated connection rods: Short connecting rods in a compressor.

ASA: Formerly, abbreviation for American Standards Association. Now known as American National Standards Institute (ANSI).

Asbestos: Strong, fire-resistant, cancer-causing silicate.

ASME Boiler Code: Standard specifications issued by the American Society of Mechanical Engineers for the construction of boilers and pressure vessels.

Aspect ratio: Ratio of length to width of a rectangular air grille or duct.

Aspirating psychrometer: Device that draws a sample of air through it to measure the humidity.

Aspiration: Movement produced by suction.

ASTM Standards: Standards issued by the American Society for Testing and Materials.

Atmospheric Dust Spot Efficiency: Measurement of a device's ability to remove atmospheric air from test air.

Atmospheric pressure: Pressure that gases in air exert upon the earth (14.7 psi).

Atom: Smallest unit of an element that can exist alone or in combination.

Atomize: Process of changing a liquid to minute particles or a fine spray.

Auger: Device with a helical shaft that, when rotated, can be used to move material.

Automatic control: Valve action reached through self-operated or self-actuated means, not requiring manual adjustment.

Automatic defrost: System of removing ice and frost from evaporators automatically.

Automatic expansion valve (AEV): Pressure-controlled valve that reduces high-pressure liquid refrigerant to low-pressure liquid refrigerant.

Automatic ice cube maker: Refrigerating mechanism designed to automatically produce ice cubes in quantity.

Autotransformer: Transformer in which both primary and secondary coils have turns in common. Step-up or step-down of voltage is accomplished by taps on common winding.

Auxiliary evaporator: Small evaporator consisting of coils of tinned tubing below the shelves in a display case.

Azeotropic mixture: A liquid mixture having constant maximum and minimum boiling points. Refrigerants comprising the azeotropic mixture do not combine chemically, yet the mixture provides constant characteristics.

B

Back pressure: Pressure in low side of refrigerating systems; also called suction pressure or low-side pressure.

Back seating: Fluid opening/closing such as a gauge opening; to seat the joint where the valve stem goes through the valve body.

Bacteria: A form of unicellular microorganisms.

Baffle: Plate or vane used to direct movement of a fluid within a confined area.

Balance point: The point at which the heating capacity of a heat pump is equal to the heat losses of the structure it is heating.

Ball valve: A check valve that uses a ball to permit flow in one direction only.

Bar: Unit of pressure. One bar equals .9869 atmosphere (approximately one atmosphere, 14.51 psi).

Barometer: Instrument for measuring atmospheric pressure.

Bath: Liquid solution used for cleaning, plating, or maintaining a specified temperature.

Battery: Electricity-producing cells that use interaction of metals and chemicals to create electrical current flow.

Baudelot cooler: Heat exchanger in which water flows by gravity over the outside of tubes or plates.

Bearing: Low-friction device for supporting and aligning a moving part.

Bellows: Corrugated cylindrical container that moves as pressures change, or provides a seal during movement of parts.

Bellows seal: Method of sealing the valve stem. The ends of the sealing material are fastened to the bonnet and to the stem. Seal expands and contracts with the stem level.

Belt: A rubber-like, continuous loop placed between two or more pulleys to transfer rotary motion.

Bending spring: Coil spring that is placed on inside or outside of tubing to keep it from collapsing while bending.

Bernoulli's Theorem: In stream of liquid, the sum of elevation head, pressure head, and velocity remains constant along any line of flow, provided no work is done by or upon liquid on course of its flow; decreases in proportion to energy lost in flow.

Bimetal strip: Temperature regulating or indicating device that works on the principle that two dissimilar metals with unequal expansion rates, welded together, will bend as temperatures change.

Bioaerosals: Airborne microorganisms derived from viruses, bacteria, fungi, protozoa, mites, and pollen.

Blast freezer: Low-temperature evaporator that uses a fan to force air over the evaporator surface.

Bleeding: Slowly reducing the pressure of liquid or gas from a system by opening a valve slightly.

Bleed valve: Valve with small opening inside that permits a minimum fluid flow when valve is closed.

Blend: A mixture of various refrigerants.

Blown: With respect to fuses, a fuse that has been melted, breaking the electric circuit and preventing overload.

Boiler: Closed container in which a liquid may be heated and vaporized.

Boiler, high-pressure: A boiler operating with water temperature and water pressure above low-pressure boiler ratings.

Boiler horsepower: Seldom-used term equivalent to a heating capacity of 33,475 Btu/hr. (9804 watts).

Boiler, low-pressure: A boiler operating with up to 250°F (121°C) water temperature and 160 psi water pressure or less.

Boiling point: The temperature of a liquid at which it changes to a gas under a pressure of 14.7 psia (101.3 kPa).

Boiling temperature: Temperature at which a fluid changes from a liquid to a gas.

Bonnet: In a furnace, the sheet metal chamber where heat collects before being distributed.

Booster: Common term applied to the use of a compressor as the first stage in a cascade refrigerating system.

Bore: Diameter of a hole. Inside diameter of a cylinder.

Bourdon tube: Thin-walled tube of elastic metal flattened and bent into a circular shape that tends to straighten as inside pressure is increased. Used in pressure gauges.

Boyle's Law: Law of Physics: The volume and pressure of a gas vary inversely if the temperature remains the same. Example: If the pressure is doubled on a quantity of gas, its volume is reduced one-half. If the volume is doubled, gas has its pressure reduced by one-half.

Brazing: Method of joining metals with nonferrous (without iron) filler using heat between 800°F (427°C) and the melting point of base metals.

Breaker strip: Strip of wood or plastic used to cover the joint between the outside case and inside liner of the refrigerator.

Breeching: Space in hot water or steam boilers between the end of the tubing and the jacket.
Brine: Water saturated with a chemical such as salt.
British thermal unit (Btu): Quantity of heat required to raise temperature of one pound of water one degree Fahrenheit.
Building Related Illness (BRI): An illness caused by an airborne virus in a building.
Built-up terminal: Electrical terminal attached to a compressor dome.
Bulb, sensitive: Part of sealed fluid device that reacts to temperature. Used to measure temperature or to control a mechanism.
Bunker: Space where ice or cooling element is placed in commercial installations.
Burner: Device in which burning of fuel takes place.
Butane: Liquid hydrocarbon (C_4H_{10}) commonly used as fuel for heating purposes.
Bypass: Passage around a regular passage.
Bypass cycle: A cycle using a bypass line with either hot gas or liquid used to defrost an evaporator or for low pressure control.

C

Cabinet: The housing of a refrigerator.
Cabinet volume: The volume of the interior cabinet dimensions.
Calcium sulfate: Chemical compound ($CaSO_4$) that is used as a drying agent or desiccant in liquid line driers.
Calibrate: Position indicators to determine accurate measurements.
Callback: A service call to repair a problem that had been improperly repaired.
Calorie: Two different calorie units are used by scientists. The calorie used by medical science is a small heat unit. It equals the heat required to raise the temperature of one gram of water one degree Celsius (C). The calorie used by engineering science is a larger heat unit (see ***Kilocalories***).
Calorimeter: Device used to measure quantities of heat or determine specific heats.
Cam: Oblong mechanical component that produces a reciprocating motion when rotated.
Capacitance (C): Property of a nonconductor (condenser or capacitor) that permits storage of electrical energy in an electrostatic field.
Capacitive reactance: The opposition, or resistance, to an alternating current as a result of capacitance; expressed in ohms.
Capacitor: Electrical storage device used to start and run circuits on many electric motors.
Capacitor-start motor: A motor with a capacitor in the starting circuit.
Capacity: Refrigeration rating system. Usually measured in Btu per hour or watts.
Capillary tube system: A refrigerant control system in which pressure difference is maintained through the use of a thin capillary tube.
Carbon dioxide (CO_2): Compound of carbon and oxygen that is sometimes used as a refrigerant. Refrigerant number is R-744.
Carbon dioxide indicator: Instrument used to indicate the percentage of carbon dioxide in stack gases.
Carbon filter: Air filter using activated carbon as an air cleansing agent.
Carbon monoxide (CO): Colorless, odorless, and poisonous gas produced when carbon fuels are burned with too little air.
Carbon tetrachloride (CCl_4): Colorless, nonflammable, and toxic liquid used as a solvent.
Carrene: Refrigerant in Group A1 (R-11). Chemical combination of carbon, chlorine, and fluorine.
Cascade systems: Arrangement in which two or more refrigerating systems are used in series; uses evaporator of one machine to cool condenser of other machine. Produces ultra-low temperatures.
Cathode: Negative terminal of an electrical device. Electrons leave at this terminal.
Cavitation: Localized gaseous condition within a liquid stream.
Celsius temperature scale: Temperature scale used in metric system. Freezing point of water is 0°C, boiling point is 100°C.
Centigrade temperature scale: See ***Celsius temperature scale.***
Centimeter: Metric unit of linear measurement, equal to 0.3937″.
Central air conditioning: A system capable of providing heating, cooling, humidifying, and dehumidifying.
Centralized computer control: Energy control device, centrally located, that makes control decisions based on operating data, programmed information, and stored data. Can be used to optimize energy consumption of many devices throughout a building.
Central station: Central location of condensing unit with either wet or air-cooled condenser. Evaporator located as needed and connected to the central condensing unit.
Centrifugal compressor: Pump that compresses gaseous refrigerants by centrifugal force.
Centrifugal force: Force that pushes a rotating object away from the center of its rotation.
Centrifugal switch: An electrical switch that is opened and closed by centrifugal force.
Ceramic ignitor: Electric ignition system used in a water glycol solution, forced-air furnace. Electrically heated to create ignition of the gas-air mixture in the combustion chamber.

Change of state: Condition in which a substance changes from one state (solid, liquid, or gas) to another.
Charge: Amount of refrigerant placed in a refrigerating unit.
Charging board: Specially designed panel or cabinet fitted with gauges, valves, and refrigerant cylinders used for charging refrigerant and oil into refrigerating mechanisms.
Charles' Law: A law stating, that at a constant pressure, mass and temperature of a gas are inversely proportional.
Check valve: Device which permits fluid flow in only one direction.
Chemical refrigeration: System of cooling using a disposable refrigerant. Also called *expendable refrigerant system.*
Chiller: Air conditioning system that circulates chilled water to various cooling coils in an installation.
Chill factor: Calculated number, based on temperature and wind velocity, that indicates chill effect.
Chimney: Vertical shaft enclosing one or more flues for carrying flue gases to the outside atmosphere.
Chimney connector: Conduit (pipe) connecting the furnace to the vertical flue.
Chimney effect: Tendency of gas to rise when heated.
Chimney flue: Flue gas passageway in a chimney.
Chlorofluorocarbons (CFCs): Refrigerants that are composed of chlorine, fluorine, and a hydrocarbon (methane). CFCs deplete the ozone layer.
Choke tube: Throttling device used to maintain correct pressure difference between high-side and low-side in refrigerating mechanism. Capillary tubes are sometimes called choke tubes.
Circuit: Tubing, piping, or electrical wire installation that permits flow to and from an energy source.
Circuit breaker: Safety device that automatically opens an electrical circuit if overloaded.
Circuit, parallel: Arrangement of electrical devices in which the current divides and travels through two or more paths and then returns through a common path.
Circuit, pilot: Secondary circuit used to control a main circuit or a device in the main circuit.
Circuit, series: Electrical path (circuit) in which electricity to operate a second device must pass through first; current flow travels, in turn, through all devices connected together.
Clean room: A room in which special efforts are made to eliminate dust and other contaminants.
Clearance space: Small space in a cylinder from which compressed gas is not completely expelled. For effective operation, compressors are designed to have as small a clearance space as possible.
Climate: The average weather conditions for a region.
Climate control: Devices used to maintain an ideal climate in a space.
Closed circuit: Electrical circuit in which electrons are flowing.
Clutch, magnetic: Clutch built into automobile compressor flywheel, operated magnetically, which allows a pulley to revolve without driving the compressor when refrigeration is not required.
Code installation: Refrigeration or air conditioning installation that conforms to the applicable codes.
Coefficient of conductivity: Measure of the relative rate at which different materials conduct heat. A good conductor of heat has a high coefficient of conductivity.
Coefficient of expansion: A measure of the change in size of a material as the temperature changes.
Coefficient of performance (COP): Ratio of the work performed to the energy used.
Cogeneration: Using waste energy as a primary heat source. Example: The use of waste heat from an electrical energy generation system to heat a building.
Cold: The absence of heat; a temperature considerably below normal.
Cold ban: A plastic trim piece used to reduce heat flow between the outer and inner shell of a refrigerator door.
Cold junction: That part of a thermoelectric system which absorbs heat as the system operates.
Cold wall: Refrigerator construction that has the inner lining of the refrigerator serving as the cooling surface.
Collector: Semiconductor section of a transistor, connected to the same polarity as the base.
Colloids: Miniature cells peculiar to meats, fish, and poultry which, if disrupted, cause food to become rancid. Low temperatures minimize this action.
Combined Annual Efficiency (CAE) ratio: Rating system used for combined heating systems, which heat both air and water.
Combustible liquids: Liquid having a flash point above 140°F (60°C); known as Class 3 liquids.
Combustion: The process of igniting and burning.
Comfort chart: Chart used in air conditioning to show the dry bulb temperature, humidity, and air movement for human comfort.
Comfort cooler: System used to reduce the temperature in the living space in homes. These systems are not complete air conditioners as they do not provide complete control of heating, humidifying, dehumidification, and air circulation.
Comfort zone: Area on psychrometric chart that shows conditions of temperature, humidity, and sometimes air movement in which most people are comfortable.
Commercial system: A refrigeration or air conditioning unit that is used in commercial buildings.
Commutator: The part of the rotor in an electric motor which conveys electric current to the rotor windings.
Complaint: Statement of dissatisfaction with regards to a service.

Compound gauge: Instrument for measuring pressure.
Compound pump: A rotary pump that has two rotors in series.
Compound refrigerating systems: A system that has several compressors or compressor cylinders in series. The system is used to pump low-pressure vapors to condensing pressures.
Compound wound: Winding used in motors that run on dc current.
Compression: Term used to denote increase of pressure on a fluid by using mechanical energy.
Compression chiller: A chiller that achieves the required pressure difference through the use of a compressor.
Compression gauge: Instrument used to measure positive pressures (pressures above atmospheric pressures) only. Gauge dial usually runs from 0 to 300 psig (101.3–2200 kPa).
Compression ratio: Ratio of the volume of the clearance space to the total volume of the cylinder. In refrigeration it is also used as the ratio of the absolute low-side pressure to the absolute high-side pressure.
Compression ring: Upper piston ring.
Compressor: Pump of a refrigerating mechanism that draws a low pressure on cooling side of refrigerant cycle and squeezes or compresses the gas into the high-pressure or condensing side of the cycle.
Compressor, external drive: See ***Compressor, open type.***
Compressor, hermetic: A compressor in which the driving motor is sealed in the same dome or housing as the compressor.
Compressor, multiple-stage: A compressor having two or more compressive steps. Discharge from each step is the intake pressure of the next in series.
Compressor, open: A compressor in which the crankshaft extends through the crankcase and is driven by an outside motor. Commonly called an *external drive compressor.*
Compressor, reciprocating: A compressor that uses a piston and cylinder mechanism to provide pumping action.
Compressor, rotary: Compressor that uses vanes, eccentric mechanisms, or other rotating devices to provide pumping action.
Compressor seal: Leakproof seal between crankshaft and compressor body in open compressors.
Compressor, single-stage: Compressor having only one compressive step between low-side pressure and high-side pressure.
Computer: Series of electrical components which accepts inputs from an operator and controls outputs.
Computer languages: Specific wording or codes, such as BASIC, FORTRAN, COBOL, and C, which direct a computer to accept and store information and control outputs.
Condensate: A fluid formed when a gas is cooled to its liquid state.
Condensate pump: Device to remove water condensate that collects beneath an evaporator.
Condensation: Liquid or droplets that form when a gas or vapor is cooled below its dew point.
Condense: Action of changing a gas or vapor to a liquid.
Condenser: The part of the refrigeration mechanism which receives hot, high-pressure refrigerant gas from compressor and cools gaseous refrigerant until it returns to its liquid state.
Condenser, air-cooled: Heat exchanger that transfers heat to surrounding air.
Condenser comb: Comb-like device, metal or plastic, used to straighten the metal fins on condensers and evaporators.
Condenser fan: Forced-air device used to move air through air-cooled condenser.
Condenser, water-cooled: Heat exchanger designed to transfer heat from hot gaseous refrigerant to water.
Condensing furnace: High-efficiency, gas forced-air furnace that extracts the latent heat lost in conventional gas forced-air furnaces.
Condensing pressure: Pressure inside a condenser at which refrigerant vapor gives up its latent heat of vaporization and becomes a liquid. This varies with the temperature.
Condensing temperature: Temperature inside a condenser at which refrigerant vapor gives up its latent heat of vaporization and becomes a liquid. This varies with the pressure.
Condensing unit: The part of a refrigerating mechanism that pumps vaporized refrigerant from the evaporator, compresses it, liquefies it in the condenser, and returns it to the refrigerant control.
Condensing unit service valves: Shutoff valves mounted on the condensing unit to enable service technicians to install and service the unit.
Conduction: The flow of heat between substances by molecular vibration.
Conductivity: Ability of a substance to transmit heat or electricity.
Conductor: Substance or body capable of transmitting electricity or heat.
Console: A total unit or system of controls located in one area and enclosed. A window air conditioner is a console air conditioner.
Constant: Remains the same; unchanging.
Constrictor: Tube or orifice used to restrict the flow of a gas or a liquid.
Contaminant: Substance such as dirt, moisture, or other matter foreign to refrigerant or refrigerant oil in system.

Continuous absorption system: System that has a continuous flow of energy input.
Continuous operation: In constant use.
Contractual agreement: A written arrangement, enforceable by law, that is entered into between two parties.
Control: Automatic or manual device used to stop, start, or regulate the flow of gas, liquid, or electricity.
Control, compressor: See ***Motor control.***
Control, defrosting: Device used to automatically defrost the evaporator. It may operate by means of a clock, door cycling mechanism, or during the Off cycle.
Control, low-pressure: Cycling device connected to the low-pressure side of system.
Control module: An electrical component used in automotive air conditioning systems to receive sensor input and regulate climate control functions. Also referred to as a microcomputer.
Control, motor: Temperature or pressure-operated device used to control running of motor.
Control point: The condition being maintained by a proportional control.
Control, pressure motor: High- or low-pressure control connected into the electrical circuit and used to start and stop motor. It is activated by demand for refrigeration or for safety.
Control, refrigerant: Device used to regulate flow of liquid refrigerant into evaporator. Can be a capillary tube, expansion valve, or high-side and low-side float valves.
Control, temperature: Temperature-operated thermostatic device that automatically opens or closes a circuit.
Control system: All of the components required for the automatic control of a process variable.
Control valve: Valve that regulates the flow or pressure of a medium that affects a controlled process. Control valves are operated by remote signals from independent devices using any of a number of control media such as pneumatic, electric, or electrohydraulic.
Controller: A group of controls and circuits used to accurately and automatically operate a device.
Convection: Transfer of heat by means of movement or flow of a fluid or gas.
Convection, forced: Transfer of heat resulting from forced movement of liquid or gas by means of a fan or pump.
Convection, natural: Circulation of a gas or liquid due to difference in density resulting from temperature differences.
Cooler: Heat exchanger that removes heat from a substance.
Cooling coil: Coils cooled by a fluid that does not evaporate (such as brine). The evaporator is sometimes incorrectly referred to as a *cooling coil.*
Cooling tower: Device that cools by water evaporation in air. Water is cooled to wet bulb temperature of air.
Copper plating: Abnormal condition developing in some units in which copper is electrolytically deposited on compressor surfaces.
Core, air: Coil of wire not having a metal core.
Core, magnetic: Magnetic center of a magnetic field.
Core valves: Shrader valve used to gain access to a hermetic unit.
Corrosion: Deterioration of materials from chemical action.
Coulomb: The quantity of electricity transferred by an electric current of one ampere in one second.
Counter emf: Tendency for reverse electrical flow as magnetic field changes in an induction coil.
Counterflow: Flow in opposite direction.
Couplings: Mechanical device joining refrigerant lines.
"Cracking" a valve: Opening a valve a small amount.
Crankthrow: Distance between centerline of main bearing journal and centerline of the crankpin or eccentric.
Crankshaft seal: Leakproof joint between crankshaft and compressor body.
Crisper: Drawer or compartment in refrigerator designed to provide high humidity along with low temperature to keep vegetables—especially leafy vegetables—cold and crisp.
Critical pressure: Pressure at which vapor and liquid have same properties.
Critical temperature: Temperature at which vapor and liquid have same properties.
Cross-charged: Sealed container of two fluids that, together, create a desired pressure-temperature curve.
Cryogenic Food Freezing: See ***Fast food freezing.***
Cryogenic fluid: A substance that exists as a liquid or gas at temperatures of −250°F (−157°C) or lower.
Cryogenics: Refrigeration that deals with producing temperatures of −250°F (−157°C) and lower.
Current: Transfer of electrical energy in a conductor by means of electrons changing position.
Current-limiting fuse: A fuse that protects an electrical circuit by limiting the amount of current that flows through it, but does not "blow."
Current relay: Device that opens or closes a circuit. It is made to act by a change of current flow in that circuit.
Customer relations: The evaluation of the technician by the customer as a result of the technician's job performance and attitudes.
Cut-in: The temperature or pressure at which the control circuit closes.
Cut-out: The temperature or pressure at which the control circuit opens.
Cycle: A series of events or operations that repeat.
Cylinder: 1—Device that converts fluid power into linear mechanical force and motion. This usually consists

of movable elements such as a piston and piston rod, plunger or ram, operating within a cylindrical bore. 2—Closed container for fluids.

Cylinder, refrigerant: Cylinder in which refrigerant is stored and dispensed. Color code painted on cylinder indicates kind of refrigerant.

Cylinder head: Plate or cap that encloses compression end of compressor cylinder.

Cylindrical commutator: Commutator with contact surfaces parallel to the rotor shaft.

D

Dalton's Law: Vapor pressure created in a container by a mixture of gases is equal to sum of individual vapor pressures of the gases contained in mixture.

Damper: Device for controlling airflow.

Dasher: Stirring mechanism in a dispensing freezer.

Deaeration: Act of separating air from a substance.

Decibel (dB): Unit used for measuring relative loudness of sounds.

Deck (coil deck): Insulated horizontal partition between refrigerated space and evaporator space.

Defrost cycle: Refrigerating cycle in which evaporator frost and ice accumulation is melted.

Defrost timer: Device, connected into electrical circuit, that shuts unit off long enough to permit ice and frost accumulation on evaporator to melt.

Defrosting: Process of removing frost accumulation from evaporators.

Defrosting evaporator: Evaporator operating at such temperatures that ice and frost on surface melts during the Off cycle.

Degreasing: Removal of oil or grease from refrigerator parts with a solution or solvent.

Degree-day: Unit that represents one degree of difference from inside temperature and the average outdoor temperature; often used in estimating fuel requirements for a building.

Dehumidifier: Device used to remove moisture from air.

Dehydrated oil: Lubricant that has had most of its water content removed (dry oil).

Dehydrator: See *Drier.*

Dehydrator-receiver: Small tank that serves as liquid refrigerant reservoir and also contains a desiccant to remove moisture. Used on most automobile air conditioning installations.

De-ice control: Device for operating a refrigerating system in such a way as to provide melting of the accumulated ice and frost.

Delta transformer: Three-phase electrical transformer that has ends of each of three windings electrically connected to form a triangle.

Demand meter: Instrument that measures the kilowatt-hour usage of a circuit or group of circuits.

Density: Closeness of particles within a given substance. The weight per unit volume.

Deodorizer: Device that absorbs or adsorbs various odors, usually by principle of absorption. Activated charcoal is commonly used.

Department of Transportation (DOT): A governmental unit that regulates the transportation of refrigerants from one location to another.

Desert bag: A bag used to keep water cool in the desert. The fabric is not waterproof, so water leaks through and evaporates, cooling the water inside the bag.

Desiccant: Substance used to collect and hold moisture. A drying agent. Common desiccants are activated alumina and silica gel.

Design pressure: Highest pressure expected during operation. Sometimes calculated as operating pressure plus a safety allowance.

Detector, leak: Device used to detect and locate refrigerant leaks.

Dew: Condensed atmospheric moisture deposited in small drops on cool surfaces.

Dew point: Temperature at which vapor (at 100% humidity) begins to condense and deposit as a liquid.

Diac: A two-lead alternating current semiconductor that allows current to flow in both directions at a preset voltage.

Diagnostics: The process of identifying or determining the nature and circumstances of an existing condition.

Diaphragm: Flexible material usually made of thin metal, rubber, or plastic.

Dichlorodifluoromethane: Refrigerant commonly known as R-12.

Die casting: Process of molding low-melting-temperature metals in accurately shaped metal molds.

Dielectric fluid: Fluid with high electrical resistance.

Differential: The difference between cut-in and cut-out temperature or pressure of a control.

Diffuser: Attachments for duct openings that distribute the air in wide flow patterns.

Diode: Two-element electron tube that will allow more electron flow in one direction in a circuit than in the other direction; tube that serves as a rectifier.

Direct current (dc): Electron flow that moves continuously in one direction in a circuit.

Direct Digital Control (DDC): Use of digital computer to perform required automatic control operations in a total energy management system.

Direct expansion evaporator: Evaporator using either an automatic expansion valve (AEV) or a thermostatic expansion valve (TEV) refrigerant control.

Direct-spark ignition: A furnace control in which a spark is used to ignite the gas-air mixture. There is no constantly-burning pilot light.

Dispensing freezers: A freezer with built-in dispensing

equipment, used for serving ice cream and frozen drinks.

Displacement: Volume obtained by multiplying the area of the cylinder bore by the length of the piston stroke.

Distilling apparatus: Fluid-reclaiming device used to reclaim used refrigerants. Reclaiming is usually done by vaporizing and then condensing refrigerant.

Distribution controls: Systems that help evenly and efficiently transfer the heating or cooling medium to the area where it is needed.

District heating and cooling: Use of a central utility system designed to provide heating and cooling to large residential and industrial areas.

Dome-hat: Sealed metal container for the motor compressor of a refrigerating unit.

Door heater: A heater located around the door opening of a freezer, used to prevent ice buildup from freezing the door closed.

Double-duty case: Commercial refrigerator in which a part of space is for refrigerated storage and part is equipped with glass windows for display purposes.

Double-thickness flare: Copper, aluminum, or steel tubing end that has been formed into two-wall thickness, 37° to 45° bell mouth or flare.

Dowel pin: Accurately dimensioned pin pressed into one assembly part and slipped into another assembly part to ensure accurate alignment.

Downflow furnace: A furnace in which return air enters through the top and is pulled down through the heat exchanger. Also called *counterflow furnace.*

Draft gauge: Instrument used to measure air movement by measuring air pressure differences.

Draft indicator: Instrument used to indicate or measure chimney draft or combustion gas movement. Draft is measured in units of 1″ of water column.

Draft regulator: Device that maintains a desired draft in a combustion-heated appliance by automatically controlling the chimney draft to the desired value.

Drier: Substance or device used to remove moisture from a refrigeration system.

Drip pan: Pan-shaped panel or trough used to collect condensate from evaporator.

Dry bulb: An instrument with a sensitive element to measure ambient air temperature.

Dry bulb temperature: Air temperature as indicated by an ordinary thermometer.

Dry capacitor: See ***Electrolytic capacitor.***

Dry cell battery: Electrical device used to provide dc electricity, having no liquid in the cells.

Dry ice: Refrigerating substance made of solid carbon dioxide, which changes directly from a solid to a gas (sublimates). Its subliming temperature is −109°F (−78°C).

Dry system: Refrigeration system that has the evaporator liquid refrigerant mainly in the atomized or droplet condition.

Dual-pressure regulator: A combination of a high-pressure and a low-pressure regulator.

Duct: Tube or channel through which air is conveyed or moved.

Duct sweeper: A tool used to remove dirt and debris from ducts.

Dynamometer: Device for measuring power output or power input of a mechanism.

E

EPA: See ***Environmental Protection Agency.***

Eccentric: Circle or disk mounted off center on a shaft.

Economizer: A mechanism that removes flash gas from the evaporator.

Eddy currents: Induced currents flowing in a core.

EER: See ***Energy Efficiency Ratio.***

Effective area: Actual flow area of an air inlet or outlet. Gross area minus area of vanes or grille bars.

Effective latent heat: The amount of heat absorbed from the cabinet and evaporator.

Effectiveness (absorption systems): Method of evaluating absorption cooling systems, in which the cooling effect is divided by the work equivalent to the heat supplied to the absorber.

Effective temperature: Overall effect of air temperature, humidity, and air movement on human comfort.

Efficiency: Output of a device, system, or activity, divided by the input necessary to create the output. In a compressor, the efficiency would be the work output, as measured by pressure change, divided by the energy input (usually electrical).

Ejector: A device that uses high fluid velocity, such as a venturi, to create low pressure or vacuum at its throat to draw in fluid from another source.

Electric defrosting: Use of electric resistance heating coils to melt ice and frost off evaporators during defrosting.

Electrical circuits: The electrical wiring that permits flow from the energy source, through the circuit, and back to the energy source.

Electrical resistance: A resistance to (working against) the movement of electrons (flow of electricity).

Electric heating: System in which heat from electrical resistance units is used to heat a building.

Electricity: Electric current or power.

Electric water valve: Solenoid (electrically operated) valve used to turn water flow on and off.

Electrodeposition: Process in which metallic particles are applied to another metal surface through the use of an electric current.

Electrolysis: A chemical change in a substance caused by movement of electricity.

Electrolytic condenser-capacitor: Plate or surface capable of storing small electrical charges.
Electromagnet: Coil of wire wound around a soft iron core. When electric current flows through wire, the assembly becomes magnetized.
Electromagnetic energy: Energy that has both electrical and magnetic characteristics. Solar energy is electromagnetic.
Electromotive force (emf) voltage: Electrical force that causes current (free electrons) to flow or move in an electrical circuit. Unit of measurement is the volt.
Electron: Elementary particle or portion of an atom that carries a negative charge.
Electronic control diagnostics: Trouble codes that may be referenced on an automatic climate control system to diagnose problems.
Electronic leak detector: Electronic instrument that measures electronic flow across a gas gap. Electronic flow changes indicate presence of refrigerant gas molecules.
Electronic relay: Electronic switch, such as a triac, that controls a power consuming device.
Electronics: Field of science dealing with electron devices and their uses.
Electronic sight glass: Device that sends an audible signal when system is low in refrigerant.
Electrostatic air filter: A filter that gives dust particles an electric charge. This causes particles to be attracted to a plate so they can be removed from air.
Embrittlement: To become easily broken.
End bell: End structure of the plate of an electric motor, which usually holds the motor bearings.
Endothermal: Chemical reaction in which heat is absorbed.
End play: Slight movement of shaft along its center line.
Energized: Having current flow.
Energy: Actual or potential ability to do work.
Energy audit: Process of accurately determining the current energy consumption for a given area.
Energy Efficiency Ratio (EER): The ratio of the rated cooling capacity divided by the amount of electrical power used.
Energy management control system: Controllers used in a system which optimizes total energy usage in a building or residence.
Energy Utilization Index (EUI): A number used to compare energy usage for different areas. It is calculated by dividing the energy consumption by the square footage of the conditioned area.
Enthalpy: Total amount of heat in one pound of a substance calculated from accepted temperature base. Temperature of 32°F (0°C) is accepted base for water vapor calculation. For refrigerator calculations, accepted base is −40°F (−40°C).
Entropy: Engineering calculation used to determine heat available. Measured in Btu per pound degree change for a substance.
Environment: The surrounding conditions.
Environmental Protection Agency (EPA): A governmental agency empowered by the government to protect the environment.
Enzyme: Complex organic substance, originating from living cells, that speeds up chemical changes in foods. Enzyme action is slowed by cooling.
Epoxy: Synthetic plastic adhesive.
Equalizer: A device that is used to balance pressure in a system or balance liquid levels between two containers.
Equivalent length: Length of piping plus pressure losses due to bends, fixtures, etc.
Ethane (R-170): Refrigerant sometimes added to other refrigerants to improve oil circulation.
EUI: See *Energy Utilization Index.*
Eutectic: That unique mixture of two substances providing the lowest possible melting temperature.
Eutectic point: Freezing temperature for eutectic solutions.
Evacuation: Removal of air (gas) and moisture from a refrigeration or air conditioning system.
Evaporation: Term applied to the changing of a liquid to a gas. Heat is absorbed in this process.
Evaporative condenser: A device that uses an open spray or spill water to cool a condenser. Evaporation of some of the water cools the condenser water and reduces water consumption.
Evaporative cooling: A cooling method practical in hot, dry climates in which hot air is blown over hot water. As some water evaporates, the remaining water is cooled, and then used to cool air.
Evaporator: Part of a refrigerating mechanism in which the refrigerant vaporizes and absorbs heat.
Evaporator, dry: Evaporator in which the refrigerant is in droplet form.
Evaporator, flooded: Evaporator containing liquid refrigerant at all times.
Evaporator fan: Fan that increases airflow over the heat exchange surface of evaporators.
Excelsior: Fine curled wood shavings.
Exfiltration: Flow of air from a building to the outdoors.
Exhaust valve: A movable port that provides an outlet for the cylinder gases in a compressor or engine.
Exothermal: Chemical reaction in which heat is released.
Expansion joint: Device in piping designed to allow movement of the pipe caused by thermal expansion and contraction.
Expansion tank: A tank used to allow water to expand and contract with temperature changes.
Expansion valve: Device in refrigerating system that reduces the pressure from the high side to the low side.

Expendable refrigerant system: A system that discards the refrigerant after it has evaporated.

External drive: Term used to indicate a compressor driven directly from the shaft or by a belt using an external motor. Compressor and motor are serviceable separately.

External equalizer: Tube connected to the low-pressure side of a thermostatic expansion valve diaphragm and to the exit end of the evaporator.

Extrinsic semiconductor: An intrinsic semiconductor with impurities added; very sensitive to electrical forces.

F

Fahrenheit scale: Temperature scale with, under standard atmospheric pressure, a water boiling point of 212° and water freezing point of 32°.

Fail-safe control: A device that opens a circuit when a sensing element loses its pressure.

Fan: Radial or axial flow device used for moving or producing flow of gases.

Farad: Unit of electrical capacity. Capacity of a condenser which, when charged with one coulomb of electricity, gives a difference of potential of one volt.

Faraday experiment: Silver chloride absorbs ammonia when cool and releases it when heated. This is basis on which some absorption refrigerators operate.

Fast-acting fuse: A fuse that blows immediately when the rated load is reached.

Fast food freezing: Method that uses liquid nitrogen or carbon dioxide to turn fresh food into long lasting frozen food. It is often referred to as ***Cryogenic Food Freezing.***

Featheredging: Sanding a finish in a way that a level surface recesses evenly to a depressed point.

Feedback: Information on current operation of a system or device used by the control system to modify future operation.

Feedback control system: Control system that is constantly correcting the condition. Also called a "closed loop system."

Female thread: The internal thread on fittings, valves, machine bodies, etc.

Fiberglass: A composition of material consisting of glass fibers in resin.

Field pole: Part of the motor stator that concentrates the magnetic field of field winding.

Fill (cooling tower): Material in a cooling tower over which water flows.

Filter: Device for removing small foreign particles from a fluid.

Firepot: Refractory-lined combustion chamber.

Flammability: Tendency to ignite.

Flammable liquids: Liquids having a flash point below 140°F (60°C) and a vapor pressure not exceeding 40 psia (276 kPa) at 100°F (38°C).

Flapper valve: Thin metal valve used in refrigeration compressors that allows gaseous refrigerants to flow in only one direction.

Flare: An enlargement at the end of a piece of flexible tubing by which the tubing is connected to a fitting or another piece of tubing. This enlargement is made at about a 45° angle. Fittings grip it firmly to make the joint leakproof and strong.

Flare nut: Fitting used to clamp tubing flare against another fitting.

Flash gas: Instantaneous evaporation of some liquid refrigerant in evaporator, which cools the remaining liquid refrigerant to the desired evaporation temperature.

Flash point: Temperature at which flammable liquid will give off sufficient vapor to support a flash flame but will not support continuous combustion.

Flash weld: Resistance weld in which mating parts are brought together under considerable pressure while a heavy electrical current is passed through the joint to be welded.

Flexible duct: A duct that can be routed around obstacles by bending it gradually.

Flexible lines: Lines that provide for easy bending.

Float valve: Type of valve that is operated by a sphere or pan that floats on a liquid surface.

Flooded system: Type of refrigerating system in which the liquid refrigerant fills most of the evaporator at all times.

Flooded system, high-side float: Refrigeration system that has a float operated by the level of the high-side liquid refrigerant.

Flooded system, low-side float: Refrigerating system that has a low-side float refrigerant control.

Flow check piston: Piston assembly, with an orifice in the center, that can operate as an expansion valve.

Flow meter: Instrument used to measure velocity or volume of fluid movement.

Flue: Gas or air passage that usually depends on natural convection to cause the combustion gases to flow through it. Forced convection may sometimes be used.

Fluid: Substance in either a liquid or gaseous state.

Fluorescent: Capable of exhibiting fluorescence, the emission of electromagnetic radiation of visible light.

Flush: Operation to remove any material or fluids from refrigeration system parts by purging them to the atmosphere using refrigerant or other fluids.

Flux: Substance applied to surfaces to be joined by brazing or soldering to keep oxides from forming.

Foam leak detector: System of soap bubbles or special foaming liquids brushed over joints and connections to locate leaks.

Foaming: Formation of a foam in an oil-refrigerant mixture due to the rapid evaporation of refrigerant dissolved in the oil. This is most likely to occur when the compressor starts and the pressure is suddenly reduced.

Foot-pound: Unit of work. A foot-pound is the amount of work done in lifting one pound one foot.

Force: Accumulated pressure multiplied by an area.

Forced air: An air conditioning or heating system in which airflow is caused by a fan.

Forced-air heating: A heating system that uses a fan to circulate the heated air.

Forced-circulation evaporators: An evaporator that uses a fan to circulate air.

Forced convection: Movement of fluid by mechanical force such as fans or pumps.

Force-feed oiling: Lubrication system which uses a pump to force oil to surfaces of moving parts.

Frame: See ***Stator.***

Free wheeling: Continued rotation of a magnetic clutch on an automotive compressor when the clutch is disengaged.

Freeze drying: Process of food preservation wherein food is frozen and ice content changed rapidly into a vapor, which is then absorbed on an evaporator.

Freezer burn: Condition applied to food that has not been properly wrapped in a freezer and has become hard, dry, and discolored.

Freeze-up: 1—Formation of ice in the refrigerant control device, which may stop the flow of refrigerant into the evaporator. 2—Frost formation on an evaporator, which may stop the airflow through the evaporator.

Freezing: Change of state from liquid to solid.

Freezing point: Temperature at which a liquid will solidify upon removal of heat. The freezing temperature for water is 32°F (0°C) at atmospheric pressure.

Freezing point depression: Temperature at which ice will form in solution of water and salt.

Friction: Force of resistance produced when two surfaces rub together.

Frost back: Condition in which liquid refrigerant flows from the evaporator into the suction line; usually indicated by sweating or frosting of the suction line.

Frost control, automatic: Control that automatically cycles refrigerating system to remove frost on evaporator.

Frost control, manual: Manual control used to change operation of refrigerating system to produce defrosting conditions.

Frost control, semiautomatic: Control that starts defrost part of a cycle manually and then returns system to normal operation automatically.

Frost-free refrigerator: Refrigerated cabinet that operates with an automatic defrost during each cycle.

Frosting evaporator: Refrigerating system that maintains the evaporator at frosting temperatures during all phases of cycle.

Frozen: 1—Water in its solid state. 2—Preserved by freezing. 3—Seized (as in machine parts) due to lack of lubrication.

Fuel oil: Kerosene or any hydrocarbon oil as specified by U.S. Department of Commerce Commercial Standard CS12 or ASTM D296, or the Canadian Government Specification Board, 3-GP-28, and having a flash point not less than 100°F (38°C).

Fuel oil distillate: Fuel oil produced through distillation.

Furnace: Self-contained appliance designed to supply heated air through ducts to spaces remote from the appliance location.

Fuse: Electrical safety device consisting of strip of fusible metal in circuit, which melts when the circuit is overloaded.

Fusible plug: Plug or fitting made with a metal of a known low-melting temperature. Used as safety device to release pressures in case of fire.

G

Galvanic action: Corrosion of two unlike metals due to electrical current passing between them. The action is increased in the presence of moisture.

Gas: Vapor phase of a substance.

Gas, noncondensable: Gas that will not form into a liquid under the operating pressure-temperature conditions.

Gas valve: Device in a pipeline for starting, stopping, or regulating the flow of gas.

Gasket: Resilient (spongy) or flexible material used between mating surfaces of parts to give a leakproof seal.

Gauge, compound: See ***Compound gauge.***

Gauge, high-pressure: Instrument for measuring pressures in range of 0 psig to 500 psig (101.3 kPa to 3 600 kPa).

Gauge, low-pressure: Instrument for measuring pressures in range of 0 psia to 50 psia (0 kPa to 350 kPa).

Gauge manifold: Chamber device constructed to hold both compound and high-pressure gauges. Valves control the flow of fluids through it.

Gauge port: Opening or connection provided for installing a gauge.

Gauge vacuum: Instrument used to measure pressures below atmospheric pressure.

Generator: In an absorption system, the component in which the ammonia-water mixture is heated.

Geothermal: An underground or underwater temperature source used for the operation of a heating and cooling system (heat pump).

Geothermal heat pump: A heat pump that uses the constant underground or underwater temperatures as supply.

Glycol water solution forced-air furnace: Furnace with 50% glycol and 50% distilled water solution, which passes through a tube-and-fin heat exchanger to distribute heat through the furnace duct system.

Grain: Unit of weight equal to 1/7000 lb., used to indicate the amount of moisture in the air.

Gravity air: Air that naturally rises when heated and flows through warm air ducts. When it cools, it becomes denser (heavier) and flows down.

Gravity flow: The tendency of liquids to flow downward and rest at the lowest possible point.

Gravity heating: Heating system in which heated air is distributed by natural rising (no fans are used for circulation).

Grille: Ornamental or louvered opening placed in a room at the end of an air passageway.

Grommet: Plastic, metal, or rubber doughnut-shaped protectors, which line holes where wires or tubing pass through panels.

Ground, short circuit: Fault in an electrical circuit allowing electricity to flow into the metal parts of a mechanism.

Ground coil: Heat exchanger buried in the ground. May be used either as an evaporator or as a condenser.

Ground wire: Electrical wire that will safely conduct electricity from a structure into the ground.

Gun burner: Furnace burner that atomizes oil by pushing it through an orifice, into the combustion chamber.

H

Halide refrigerants: Family of refrigerants containing halogen chemicals.

Halide torch: Type of torch used to safely detect halogen refrigerant leaks in system.

Halogens: Substance containing fluorine, chlorine, bromine, or iodine.

Head: Pressure, usually expressed in feet of water, inches of mercury, or millimeters of mercury.

Head, static: Pressure of fluid expressed in terms of height of column of the fluid, such as water or mercury.

Head, total static: Static head from the surface of the supply source to the free discharge surface.

Head friction: Head required to overcome friction of the interior surface of a conductor and between fluid particles in motion.

Head pressure: Pressure that exists in condensing side of refrigerating system.

Head pressure control: Pressure-operated control that opens electrical circuit if high-side pressure becomes too high.

Head pressure safety cutout: Motor protection device wired in series with motor; will shut off the motor when excessive head pressures occur.

Head velocity: Height of fluid equivalent to its velocity pressure in flowing fluid.

Header: Length of pipe or vessel, to which two or more pipe lines are joined, that carries fluid from a common source to various points of use.

Heat: Form of energy that acts on substances to raise their temperature; energy associated with random motion of molecules.

Heat absorber: The low-pressure side of a refrigeration system. The evaporator absorbs heat.

Heat anticipators: A thermostatic anticipator.

Heat dissipator: The high-pressure side of a refrigeration system. The condenser dissipates heat.

Heat energy: Kinetic energy of molecules in motion.

Heat exchanger: Device used to transfer heat from a warm or hot surface to a cold or cooler surface. (Evaporators and condensers are heat exchangers.)

Heat gains: Heat added to a space being cooled.

Heat input method: Method of sizing motor in which the required energy from the motor is the amount of heat added to the vapor in the compressor.

Heat insulators: Poor conductors of heat.

Heat lag: The time it takes for heat to travel through a substance heated on one side.

Heat leakage: Flow of heat through a substance.

Heat leakage load: Total amount of heat that leaks from a structure.

Heat load: Amount of heat removed during a period of 24 hours.

Heat loss: Loss of warm air, resulting in a lower temperature.

Heat of compression: Additional temperature produced by increased pressure.

Heat of fusion: Heat released from a substance to change it from a liquid state to a solid state. The heat of fusion of ice is 144 Btu per pound (335 kJ/kg).

Heat pipe: High efficiency gas furnace that uses vertical liquid filled pipes. The pipes are heated by a burner at their base, and the liquid boils and vaporizes within the pipe. The furnace blower circulates air over the pipes for heating.

Heat pump: Compression cycle system used to supply heat to a temperature-controlled space. The same system can also remove heat from the same space.

Heat recovery system: Produces and stores hot water by transferring heat from condenser to cooler water.

Heat sink: Relatively cold surface capable of absorbing heat.

Heat transfer: Movement of heat from one body or substance to another. Heat may be transferred by radiation, conduction, convection, or a combination of these three methods.

Heat transfer coefficient: (U-value) A measure of the amount of heat that a material or combination of materials will allow through.

Heat transfer module: Primary system of heat transfer in a glycol water solution forced-air furnace. The heat transfer module contains the ignitor, burner, and primary solution circulating coil.

Heat transfer rate (Q): The amount of heat transfer through a given material per unit time.

Heating coil: Heat transfer device consisting of a coil of piping, that releases heat.

Heating control: Device that controls temperature of a heat transfer unit.

Heating value: Amount of heat that may be obtained by burning a fuel. The heating value is usually expressed in Btu per lb., Btu per gal., or kJ/kg.

Hermetic compressor: Compressor that has the driving motor sealed inside the compressor housing. The motor operates in an atmosphere of the refrigerant.

Hermetic motor: Compressor drive motor sealed within same casing which contains compressor.

Hermetic unit: Refrigeration system which has a compressor driven by a motor contained in compressor dome or housing.

Hertz (Hz): Correct terminology for cycles per second.

Hg. (mercury): Heavy silver-white metallic element; only metal that is liquid at ordinary room temperature.

High-efficiency gas furnace: Furnace that uses recycling of combustion gases or pulse combustion to obtain operating efficiencies from 85% to 95%.

High-limit control: Control that stops the flow of gas when the bonnet on a furnace is too hot. Also called a *safety stat.*

High-pressure cut-out: Electrical control switch, operated by the high-side pressure, that automatically opens electrical circuit if pressure is too high.

High side: The parts of a refrigerating system subject to the condenser pressure.

High-side float: Refrigerant control mechanism that controls the level of the liquid refrigerant in the high-pressure side of mechanism.

Horizontal furnace: A furnace in which air blows horizontally through the heat exchanger.

Horsepower: Unit of power equal to 33,000 ft.-lb. of work per minute. One electrical horsepower equals 746 W.

Hot gas: High temperature gas taken from the compressor used to defrost the evaporator.

Hot gas bypass: Piping system in refrigerating unit that moves hot refrigerant gas from condenser into low-pressure side.

Hot gas defrost: Defrosting system in which hot refrigerant gas from the high side is directed through the evaporator for short period of time at predetermined intervals in order to remove frost.

Hot junction: The part of thermoelectric circuit that releases heat.

Hot-surface ignition system: Furnace ignition system in which a silicon carbide element is heated in order to light the main burner. No pilot light is needed.

Hot water heating system: System in which water is circulated through heating coils.

Hot wire: 1—Resistance wire in an electrical relay which expands when heated and contracts when cooled. 2—Electrical lead which has a voltage difference between it and the ground.

Humidifier: Device used to add to the humidity.

Humidistat: Electrical control that is operated by changing humidity.

Humidity: Moisture. Dampness of air.

Hunting: The cycling above and below the set point.

Hydraulics: Having to do with the mechanical properties of water and other liquids in motion.

Hydrocarbons: Organic compounds containing only hydrogen and carbon atoms in various combinations.

Hydrochlorofluorocarbons (HCFCs): Type of refrigerant that is damaging to the ozone layer, but to a lesser degree than CFCs.

Hydrofluorocarbons (HFCs): A type of refrigerant that will not damage the ozone layer.

Hydrogen: A light gas that makes up a small part of the atmosphere. It is present in most fuels.

Hydrometer: Floating instrument used to measure the specific gravity of a liquid.

Hydronic: Heating system that circulates a heated fluid, usually water, through baseboard coils by means of a circulating pump controlled by a thermostat.

Hygrometer: Instrument used to measure amount of moisture in the air.

Hygroscopic: Ability of a substance to absorb and release moisture and change physical dimensions as its moisture content changes.

I

Ice bank: Refrigerating systems that form a bank of ice around the evaporator to provide reserve cooling capacity.

Ice melting effect: Amount of heat absorbed by melting ice at 32°F (0°C) is 144 Btu per pound of ice or 288,000 Btu per ton.

Identification plate: Provides information such as manufacturer, part number, and specifications. Frequently mounted on the outside housing of motors and compressors.

Idler: Pulley used on some belt drives to provide proper belt tension and to eliminate belt vibration.

Ignition system: Method of lighting a furnace burner.

Ignition transformer: Transformer designed to provide a high-voltage current. Used in many heating systems to ignite fuel.

Immersion freezing: Freezing of articles by dipping them into liquid refrigerant.

Impedance: Opposition in an electrical circuit to the flow of an alternating current that is similar to the electrical resistance to a direct current.

Impeller: Rotating part of a pump.

Incandescent: Glowing due to heat.

Incomplete combustion: Combustion with insufficient oxygen.

Indoor Air Quality (IAQ): The status of indoor air as measured by numerous factors: temperature, humidity, airflow, pollutants, occupants, etc.

Induced magnetism: Ability of a magnetic field to produce magnetism in a metal.

Inductance: Inducing voltage in a coil due to the change in the rate of flow of current in the coil.

Induction motor: An ac motor that operates on the principle of a rotating magnetic field. The rotor has no electrical connection, but receives electrical energy by transformer action from field windings.

Inductive reactance: Electromagnetic induction in a circuit creates a counter or reverse emf as the original current changes. It opposes the flow of alternating current.

Infiltration: Passage of outside air into building through doors, cracks, windows, and other openings.

Infrared: Invisible rays just beyond red in the visible spectrum.

Inhibitor: Substance that prevents a chemical reaction.

Inspection: To examine and compare with established guidelines.

Instrument: Used broadly to denote a device that has measuring, recording, indicating, or controlling abilities.

Insulation, electric: Substance that has almost no free electrons.

Insulation, thermal: Material that is a poor conductor of heat; used to retard flow of heat through wall.

Integrated circuit: A circuit that incorporates multiple transistors and other semiconductors to a single circuit, sometimes called a "chip."

Integrated circuit board: Electronic circuit made from transistors, resistors, etc., all placed into a package referred to as a "chip," since all circuits are on one base of semiconductor material.

Interlocked: Controlled by a switch that does not allow a component to operate when a hazardous condition exists.

Intermittent absorption system: Refrigeration system using a kerosene burner and ammonia, normally used in situations where gas and electricity is not available.

Intermittent cycle: Cycle which repeats itself at varying time intervals.

Interstate Commerce Commission (ICC): Government body that controls the design and construction of pressure containers.

Intrinsic semiconductor: Material that is neither conductor nor insulator, such as silicon and germanium, often used in temperature sensing devices.

Ion: Group of atoms or an atom which is electrically charged.

IR drop: Electrical term indicating the loss in a circuit expressed in amperes times resistance (I × R) or voltage drop.

Isothermal: Changes of volume or pressure under conditions of constant temperature.

J

Jet cooling system: Jet pump is used to produce a vacuum so water or refrigerant may evaporate at relatively low temperatures. These systems usually require a large condenser and have a low efficiency.

Jet pump: A centrifugal pump combined with an ejector, which can replace the compressor in some refrigeration systems.

Joint: Connecting point as between two pipes.

Joule: Metric unit of heat.

Joule-Thomson Effect: The change in the temperature of a gas on its expansion through a porous plug from a higher pressure to a lower pressure.

Journal, crankshaft: Part of shaft that contacts the bearing on the large end of the piston rod.

Junction box: Box or container housing group of electrical terminals.

K

Kata thermometer: Large-bulb alcohol thermometer used to measure air speed or atmospheric conditions by means of cooling effect.

Kelvin scale (K): Thermometer scale on which unit of measurement equals the Celsius degree and according to which absolute zero is 0°, the equivalent of −273.16°C. Water freezes at 273.16°K and boils at 373.16°K.

Kilocalorie: Great calorie (1000 calories) used in engineering science.

Kilometer (km): Metric unit of linear measurement equal to 1000 meters.

Kilopascal (kPa): Metric unit of pressure equal to 1000 Pascals.

Kilovolt ampere (kVA): Unit of electrical flow equal to volts multiplied by amperes and divided by one thousand.

Kilowatt: Unit of electrical power, equal to 1000 watts.

Kinetic energy: Energy of motion.

King valve: Liquid receiver service valve.

L

Lacquer: Protective coating or finish that dries to form a film by evaporation of a volatile constituent.

Ladder diagram: Electrical diagram that indicates order of electrical devices in a specific electrical circuit.

Lag: Delay in response.

Lamp, steri: Lamp that has a high-intensity ultraviolet ray used to kill bacteria. Also used in food storage cabinets and in air ducts.

Lapping: Smoothing a metal surface to high degree of refinement or accuracy using a fine abrasive.

Latent heat: Heat energy absorbed in process of changing form of substance (melting, vaporization, fusion) without change in temperature or pressure.

Latent heat of condensation: Amount of heat released (lost) to change from a vapor (gas) to a liquid.

Latent heat of vaporization: Amount of heat required to change from a liquid to a vapor (gas).

Leak detector: Device or instrument, such as a halide torch, an electronic sniffer, or soap solution, used to detect leaks.

Legionnaire's Disease Bacterium (LDB): Is thought to be transmitted by airborne routes, possibly by open air cooling towers or evaporative condensers in commercial systems. Disease is named after an outbreak of illness at an American Legion convention in July, 1976.

Limit control: Control used to open or close electrical circuits as temperature or pressure limits are reached.

Liquefied gases: A gas below a certain temperature and above a certain pressure, that becomes liquid.

Liquid: Substance whose molecules move freely among themselves, but do not tend to separate like those of gases.

Liquid absorbent: Chemical in liquid form that has the property to "take on" or absorb other fluids.

Liquid desuperheater: Valve that permits small flow of refrigerant to enter low side of systems to cool suction gas.

Liquid floodback: A surge of liquid returning to the compressor.

Liquid indicator: Device located in liquid line that provides a glass window through which liquid flow may be watched.

Liquid line: Tube that carries liquid refrigerant from the condenser or liquid receiver to the refrigerant control mechanism.

Liquid nitrogen: Nitrogen in liquid form used as a low-temperature refrigerant in expendable or chemical refrigerating systems.

Liquid receiver: Cylinder (container) connected to condenser outlet for storage of liquid refrigerant in a system.

Liquid receiver service valve: Two- or three-way manual valve located at the outlet of the receiver and used for installation and service purposes. It is sometimes called the king valve.

Liquid transfer method: Method of liquid refrigerant recovery in which the air conditioning unit is pressurized and refrigerant is removed by the created pressure difference.

Liquid-vapor valve refrigerant cylinder: Dual hand valve on refrigerant cylinders that is used to release either gas or liquid refrigerant from the cylinder.

Listing: To tilt to one side due to an unbalanced load.

Liter: Metric unit of volume, equal to 61.03 in^3.

Lithium bromide: Chemical commonly used as the absorbent in absorption cooling system. Water would then be the refrigerant.

Localized controllers: Independent energy control device located near the system it is controlling.

Locked rotor amperage: Initial current when initially closing a circuit, two to four times as high as the running current.

Locker plant: Manufacturing plant designed to prepare, freeze, and store food products.

Lockout relay: A device that shuts down a circuit whenever a safety control device is open.

Low pressure safety cutout: Motor protection device that senses low-side pressure. Control is wired in series with the motor and will shut off during periods of excessively low suction pressure. Also called *low-side pressure indicator.*

Low side: The parts of a refrigeration system subject to the evaporator pressure.

Low-side float valve: Refrigerant control valve operated by level of liquid refrigerant in low-pressure side of system.

Low-side pressure: Pressure in cooling side of refrigerating cycle.

Low-side pressure control: Device used to keep low-side evaporating pressure from dropping below certain pressure.

Low-side pressure limiter: See ***Low-pressure safety cutout.***

LP fuel: Liquefied petroleum used as a fuel gas.

M

Machine room: Area where commercial and industrial refrigeration machinery—except evaporators—is located.

Machine screws: Fine threaded fasteners manufactured to narrow tolerances.

Magnetic clutch: Device operated by magnetism to connect or disconnect a power drive.

Magnetic field: Space in which magnetic lines of force exist.

Magnetic flux: Lines of force of a magnet.

Magnetic gasket: Door-sealing material that keeps door tightly closed with small magnets inserted in gasket.

Magnetic permeability: Property of a material that determines its flux density under a magnetic field.

Magnetism: A field of force which causes a magnet to attract materials made of iron, nickel-cobalt, or other ferrous material.

Make-up air units: An air unit used to create a slight positive pressure in homes, reducing infiltration.

Male thread: External thread on pipe, fittings, and valves for making connections that screw together.

Manifold, service: Chamber equipped with gauges and manual valves, used by service technicians to service refrigerating systems.

Manometer: Instrument for measuring pressure of gases and vapors. Gas pressure is balanced against a column of liquid, such as mercury, in a U-shaped tube.

Mass: Quantity of matter held together so as to form one body.

Matched: In refrigeration systems, the correct balancing of the following items: heat load, condensing unit capacity, evaporator capacity, and total system capacity.

MBH: Thousands of British thermal units (1 MBH = 1000 Btu).

McLeod gauge: Instrument used to measure high vacuums.

Mean Effective Pressure (MEP): Average pressure on a surface when a changing pressure condition exists.

Mechanical cycle: Cycle that is a repetitive series of mechanical events.

Mechanism: Machinery.

Melting point: Temperature at which a substance will melt at atmospheric pressure.

Micro: Prefix denoting one-millionth; for example, a microliter is one-millionth of a liter.

Microcomputer: A computer containing a microprocessor.

Micrometer: Precision measuring instrument used for making measurements accurate to .001 to .0001″.

Micron: Unit of length in metric system equal to one-millionth of a meter.

Micron gauge: Instrument for measuring vacuums very close to a perfect vacuum.

Microorganism: A plant or animal of microscopic size.

Microprocessor: Electrical component consisting of integrated circuits that may accept information, store it, and control an output device.

Milli: Prefix denoting one thousandth (1/1000); for example, millivolt means one thousandth of a volt.

Minerals: A substance with a definite chemical composition and characteristics. In this text, the term refers to substances found in water (carbonate, sulfate, lime, iron, etc.) that produce scale formation inside tubing.

Minimum Stable Signal (MSS): Correct setting for an expansion valve where it is utilizing the evaporator efficiently but remains free from hunting.

Miscibility: Capable of being mixed.

Modulate: Control or adjust.

Modulating controls: A control capable of gradual adjustments, rather than simple on-off control.

Modulating refrigeration cycle: Refrigerating system of variable capacity.

Modules: Thermoelectric cooling units.

Moisture indicator: Instrument used to measure moisture content of a refrigerant.

Molecule: Smallest portion of an element or compound that retains chemical identity with the substance.

Mollier's diagram: Graph of refrigerant pressure, heat, and temperature properties.

Monochlorodifluoromethane: Refrigerant better known as R-22. Chemical formula is $CHClF_2$. Cylinder color code is green.

Motor: Rotating machine that transforms fluid or electric energy into a mechanical motion.

Motor, capacitor: Single-phase induction motor with an auxiliary starting winding connected in series with a condenser (capacitor) for better starting characteristics.

Motor burnout: Condition in which the insulation of an electric motor has deteriorated due to overheating.

Motor control: Device to start and stop a motor at a certain temperature or pressure.

Motor starter: High-capacity electric switches, usually operated by electromagnets.

Muffler: Sound absorber chamber in refrigeration system. Used to reduce sound of gas pulsations.

Mullion heater: Electrical heating element mounted in the mullion. Used to keep mullion from sweating or frosting.

Multiple-pass recycling machine: A refrigerant recycling machine that cycles the refrigerant through a filter-drier several times in order to separate and remove oil.

Multiple system: Refrigerating mechanism in which several evaporators are connected to one condensing unit.

Multipurpose fuse: A fuse that does not blow under small overload conditions of a short duration, but blows immediately when a large overload occurs.

Multistage system: A system used to produce very low temperatures.

N

Natural convection: See ***Convection, natural.***

Natural gas: A mixture of methane and other hydrocarbons found in the earth's crust, used as fuel.

Needle point valve: Valve having a needle point plug and a small seat orifice for low-flow metering.

Negative Temperature Coefficient Thermistor (NTC): Electronic thermistor that decreases in resistance as temperature increases.

Neoprene: Synthetic rubber that is resistant to hydrocarbon oil and gas.

Net capacity: The interior volume of a refrigeration cabinet.

Neutralizer: Substance used to counteract acids in a refrigeration system.

Neutron: That part of an atom core which has no electrical potential; electrically neutral.

Newton: Force required to accelerate an object that has a mass of 1 kilogram to 1 m/sec.^{2}.

Nitrogen: A gaseous element comprising three-fourths of the earth's atmosphere by weight.

Nitrogen dioxide (NO_2): Mildly poisonous gas often found in smog or automobile exhaust fumes.

No-frost freezer: Low-temperature refrigerator cabinet in which no frost or ice collects on freezer surfaces or materials stored in cabinet.

Noise dosimeter: Instrument used to measure sound.

Nominal size tubing: Tubing measurement that has an inside diameter the same as iron pipe.

Noncode installation: Functional refrigerating system installed where there are no local, state, or national refrigeration codes in effect.

Noncondensable gas: Gas which does not change into a liquid at operating temperatures and pressures.

Nonferrous: Group of metals and metal alloys that contain no iron.

Nonfrosting evaporator: Evaporator that never collects frost or ice on its surface.

Noninductive load: An electrical load consisting of resistance, which does not affect the power factor.

Normal charge: Thermal element charge that is part liquid and part gas under all operating conditions.

North Pole, magnetic: End of magnet out of which magnetic lines of force flow.

NTC: See *Negative temperature coefficient thermistor.*

O

Octave: Frequency difference between harmonic vibrations; the doubling of the frequency of sound.

Octyl alcohol-ethyl hexanol: Additive in absorption machines that reduces surface tension in absorber.

Odor: The property of air contaminants that affect the sense of smell.

Off cycle: Segment of refrigeration cycle when system is not operating.

Offset: In a proportional control system, the deviation between the set point and the control point. Also called *error.*

Ohm: Unit of measurement of electrical resistance. One ohm exists when one volt causes a flow of one ampere.

Ohmmeter: Instrument for measuring electrical resistance in ohms.

Ohm's Law: Mathematical relationship between voltage, current, and resistance in an electric circuit; discovered by George Simon Ohm. It is stated as follows: voltage (E) equals amperes (I) times ohms (R); or $E = I \times R$.

Oil, refrigeration: Specially prepared oil used in refrigerator mechanism which circulates, to some extent, with refrigerant.

Oil binding: Condition in which oil layer on top of refrigerant liquid may prevent it from evaporating at normal pressure and temperature.

Oil burner: A device for burning vaporized oil (gas) to produce heat.

Oilless bushing: Assembly in which a shaft passes through a sintering bushing (which is impregnated with oil). Bushing is permanently lubricated.

Oil level regulator: A device that controls the oil level in the compressor.

Oil pressure safety cutout: Motor protection device that senses oil pressure in the compressor. It is wired in series with the compressor and will shut it off during periods of low oil pressure.

Oil reservoir: Container that stores the compressors oil supply.

Oil ring: Lower piston ring.

Oil separator: Device used to remove oil from gaseous refrigerant.

Oil slugging: Oil being pumped out of the compressor.

One-way valve: A valve with only one opening that can be either opened or closed.

Open circuit: Interrupted electrical circuit, which stops flow of electricity.

Open compressor: Term used to indicate an external drive compressor (not hermetic).

Open-cycle refrigeration: Refrigeration system in which the refrigerant is released to the atmosphere after evaporation.

Open display case: Commercial refrigerator designed to maintain its contents at refrigerating temperatures even though the contents are in an open case.

Open system: Refrigerating system which uses a belt-driven or a coupling-driven compressor.

Operating controls: Devices used to cause a refrigeration system to maintain desired conditions.

Operating differential: The actual temperature or pressure difference in the conditioned area.

Operating pressure: Actual pressure at which the system works under normal conditions. This pressure may be positive or negative (vacuum).

Organic: Pertaining to or derived from living organisms.

Orifice: Accurate size opening for controlling fluid flow.

Orifice tube: Metering device consisting of a restricting tube with inlet and outlet screens.

O-rings: Sealing devices used between parts where there may be some motion.

Oscilloscope: Fluorescent-coated tube that visually shows an electrical wave.

Overload: Load greater than that for which the system was intended.

Overload protector: Device, either temperature, pressure, or current operated, that will stop operation of unit if dangerous conditions arise.

Oxidation: The chemical combining of oxygen with a specific material, resulting in deterioration of that material.

Oxygen: An elemental gas that comprises approximately 25% of the atmosphere. It is required for combustion.

Ozone: A form of oxygen, O_3, having three atoms to the molecule, usually produced by discharge of electricity through the air. The ozone layer is the outermost layer of the earth's atmosphere, that absorbs ultraviolet light from the sun and shields the lower layers and the earth from harmful rays.

P

Packaged terminal air conditioning: A combination heating and cooling unit designed for a single room or zone.

Package units: Complete refrigerating system including compressor, condenser, and evaporator located in refrigerated space.

Packing: Sealing device consisting of soft material or one or more mating soft elements. Reshaped by manually adjustable compression to obtain or maintain a leakproof seal.

Partial pressures: Condition where two or more gases occupy a space and each one creates part of the total pressure.

Parts per million (ppm): Unit of concentration of one element in another.

Pascal (Pa): Unit of pressure in the metric system.

Pascal's Law: Pressure imposed upon a fluid is transmitted equally in all directions.

Passive solar heating system: A solar energy system that is dependent upon the radiation striking directly on the surface to be heated.

PCB: See *Polychlorinated Biphenyl.*

Peltier Effect: When direct current is passed through two adjacent metals, one junction will become cooler and the other will become warmer. This principle is the basis of thermoelectric refrigeration.

Perimeter drier: An electrical resistance heat wire located in a freezer door to prevent condensation on the exterior of the cabinet and around the freezer door.

Perimeter hot gas tube system: System that has a tube located on the surface of the outer portion of the cabinet to prevent condensation from forming.

Permanent magnet: Material that has its molecules aligned and has its own magnetic field; bar of metal that has been permanently magnetized.

Permanent split capacitor motor: A motor with no relay, in which current flows through both the starting and running winding, making the motor sensitive to line voltage and resulting in low starting torque.

Permeable: Having openings that allow the passage of liquid or gas.

pH: Measurement of the free hydrogen ion concentration in an aqueous solution. A pH of 7 is neutral.

Phase: Distinct functional operation during a cycle.

Phase loss monitor: Motor protection device for polyphase motors that measures current flow to detect phase loss.

Phial: Term sometimes used to denote the sensing element on a thermostatic expansion valve.

Photoelectricity: Physical action wherein an electrical flow is generated by light waves.

Photon: Particle of electromagnetic energy found in solar radiation.

Photostatically: Method by which the molecular formation of an element changes due to light.

Photovoltaic cell: See *Solar cell.*

Piercing valve: A type of service valve used on hermetic units.

Piezoelectric: Property of quartz crystal that causes it to vibrate when a high frequency (500 kHz or higher) voltage is applied. Concept is used to atomize water in a humidifier.

Piston: Close-fitting part or plug that moves up and down in a cylinder.

Pitot tube: Tube used to measure air velocities.

Planck's constant: Constant value (6.626×10^{-34} J·s) which, when multiplied by the frequency of radiation, determines the amount of energy in a photon.

Plenum chamber: Chamber or container for moving air or other gas under a slight positive pressure.

Pneumatic system: An air conditioning system in which pneumatic motors are operated by pressurized air lines.

Pollen count: A measure of the amount of pollen in the air.

Polychlorinated Biphenyl (PCB): Dielectric fluid used in capacitors and transformers that is very toxic. Use of PCB in transformers and capacitors is strictly regulated by the Environmental Protection Agency.

Polyphase motor: Electrical motor designed to be used with a three- or four-phase electrical circuit.

Polystyrene: Plastic used as an insulation in some refrigerated structures.

Polyurethane: Any synthetic rubber polymers produced from the polymerization of an HO and NCO group from two different compounds. Often used in insulation and molded products.

Ponded roof: Flat roof designed to hold a quantity of water, which acts as a cooling device.
Porcelain: Ceramic coating applied to steel surfaces.
Portable service cylinder: Container used to store refrigerant. Two most common types are disposable and refillable.
Positive pressure: A pressure greater than atmospheric.
Positive temperature coefficient thermistor (PTC): Electronic thermistor that increases in resistance as temperature increases.
Potassium permanganate: Chemical used in carbon filters to help reduce odors.
Potential, electrical: Electrical force that moves, or attempts to move, electrons along a conductor or resistance.
Potential energy: Energy related to an object's position.
Potential relay: Electrical switch that opens on high voltage and closes on low voltage.
Potentiometer: Instrument for measuring or controlling by sensing small changes in electrical resistance.
Pound-force: Force applied to a 1-lb. mass to give it an acceleration of 32.173 ft./s^2 (gravitational acceleration).
Pour point: Lowest temperature at which a liquid will pour or flow.
Power: 1—Time rate at which work is done or energy emitted. 2—Source or means of supplying energy.
Power burner: A burner that has air blown into it by a blower.
Power element: Sensitive element of a temperature-operated control.
Power factor: Correction coefficient for the changing current and voltage values of ac power.
Power saver switch: A switch that disconnects heaters in a refrigeration cabinet.
ppm: See ***Parts per million.***
Precooler condenser: Used to cool the refrigerant prior to entering the main condenser.
Pressure: Energy impact on a unit area; force or thrust on a surface.
Pressure, absolute: See ***Absolute pressure.***
Pressure, atmospheric: See ***Atmospheric pressure.***
Pressure, back: See ***Back pressure.***
Pressure cycling switch: Pressure-controlled switch located on the inlet line of the evaporator to prevent rapid cycling of the compressor.
Pressure drop: Pressure difference at two ends of a circuit, or part of a circuit.
Pressure gauge: Instrument for measuring the pressure exerted by the contents on its container.
Pressure, gauge: Pressure above atmospheric pressure.
Pressure, head: Force caused by the weight of a column or body of fluids.
Pressure-heat diagram: Graph of refrigerant pressure, heat, and temperature properties. (Mollier's diagram).
Pressure limiter: Device that remains closed until a certain pressure is reached, then opens and releases fluid to another part of system or breaks an electric circuit.
Pressure motor control: Device that opens and closes an electrical circuit as pressures change.
Pressure-Operated Altitude (POA) valve: Device that maintains a constant low-side pressure, independent of altitude of operation.
Pressure, operating: Pressure at which a system is operating.
Pressure regulator, evaporator: Automatic pressure regulating valve mounted in the suction line between the evaporator outlet and the compressor inlet. Its purpose is to maintain a predetermined pressure and temperature in the evaporator.
Pressure, suction: Pressure in low-pressure side of a refrigerating system.
Pressure switch: Switch operated by a change in pressure.
Pressure water valve: Device used to control water flow. It is responsive to head pressure of refrigerating system.
Primary air: In a combustion system, the air mixed with fuel prior to ignition.
Primary coil: A tube-and-fin circular coil that contains a water-glycol solution, which surrounds the ignitor and burner. This coil is used in a water-glycol gas forced-air furnace.
Primary control: Device that directly controls operation of heating system.
Process tube: Length of tubing fastened to hermetic unit dome, used for servicing unit.
Product heat load: Sum of specific, latent, and respiration heat loads.
Products of combustion: The material produced when a substance is burned.
Propane: Volatile hydrocarbon used as a fuel or as a refrigerant.
Proportional: Being in the proper relative quantity or balance.
Protector, circuit: Electrical device that will open an electrical circuit if excessive electrical conditions occur.
Proton: Particle of an atom with a positive charge.
psi: Pounds per square inch.
psia: Pounds per square inch absolute. Absolute pressure equals gauge pressure plus atmospheric pressure.
psig: Pounds per square inch gauge.
Psychrometer: Instrument for measuring the relative humidity of atmospheric air. Also called *Wet Bulb Hygrometer.*
Psychrometric chart: Chart that shows relationship between the temperature, pressure, and moisture content of the air.
PTC: See ***Positive temperature coefficient thermistor.***
Puffback: The ignition of vaporized oil in the firepot.

Pulley: Flat wheel with a "V" groove. When attached to a drive and drive members, the pulley provides a means for driving the compressor.
Pulse: Term referring to one cycle of ignition and combustion of a gas-air mixture in a pulse combustion furnace.
Pulse combustion process: Repeated ignition of a gas and air mixture in a high efficiency gas furnace.
Pulse furnace: Furnace that has a "tuned" (resonant) combustion chamber. Part of the energy normally lost through the flue is returned to start next "pulse" of combustion.
Pump: Any one of various machines that force gas or liquid into–or draw it out of–something as by suction or pressure.
Pump, centrifugal: Pump that produces fluid velocity and converts it to pressure head.
Pump, fixed displacement: A pump in which the displacement per cycle cannot be varied.
Pump, reciprocating single-piston: A pump having a single reciprocating (moving up and down or back and forth) piston.
Pump, screw: Pump having two interlocking screws rotating in a housing.
Pump down: The act of using a compressor or a pump to reduce the pressure in a container or a system.
Purging: Releasing compressed gas to the atmosphere for the purpose of removing contaminants.
Pyrometer: Instrument for measuring high temperatures.

Q

Quenching: Submerging a hot object in cooling fluid.
Quick-connect coupling: A device that permits easy and fast connecting of two fluid lines.

R

R-value: The thermal resistance of a given material.
R-11, Trichloromonofluoromethane: Low-pressure, synthetic chemical refrigerant that is also used as a cleaning fluid.
R-12, Dichlorodifluoromethane: Popular refrigerant known as Freon 12.
R-22, Monochlorodifluoromethane: Low temperature refrigerant with boiling point of −41°F (−40.5°C) at atmospheric pressure.
R- 113, Trichlorotrifluoroethane: Synthetic chemical refrigerant which is nontoxic and nonflammable.
R-160, Ethyl chloride: Toxic refrigerant now seldom used.
R-170, Ethane: Low-temperature refrigerant.
R-290, Propane: Low-temperature refrigerant.
R-500: Refrigerant that is an azeotropic mixture of R-12 and R-152a.
R-502: Refrigerant that is an azeotropic mixture of R-22 and R-115.
R-503: Refrigerant that is an azeotropic mixture of R-23 and R-13.
R-504: Refrigerant that is an azeotropic mixture of R-32 and R-115.
R-600, Butane: Low-temperature refrigerant; also used as a fuel.
R-611, Methyl formate: Low-pressure refrigerant.
R-717, Ammonia: Popular refrigerant for industrial refrigerating systems; also a popular absorption system refrigerant.
Radial commutator: Electrical contact surface on a rotor, perpendicular to the shaft centerline.
Radiant heating: Heating system in which warm or hot surfaces are used to radiate heat into the space to be conditioned.
Radiation: Transfer of heat by heat rays.
Range: Pressure or temperature settings of a control; change within limits.
Rankine scale: Name given the absolute (Fahrenheit) scale. Zero (0°R) on this scale is −460°F.
Reactance: That part of the impedance of an alternating current circuit due to capacitance or inductance or both.
Receiver-drier: Cylinder (container) in a refrigerating system for storing liquid refrigerant and desiccant.
Receiver heating element: Electrical resistance heater mounted in or around liquid receiver. It is used to maintain head pressures when ambient temperature is low.
Reciprocal: Inverse.
Reciprocating: Back and forth motion in a straight line.
Reciprocating compressor: A compression driven by piston (positive displacement).
Reclaiming: Taking refrigerant that has been removed from a system and processing it in accordance with EPA rules.
Recording ammeter: Electrical instrument that uses a pen to record the amount of current flow on a moving paper chart.
Recording thermometer: Temperature measuring instrument that has a pen marking a moving chart.
Recovery: Removal of refrigerant from a system.
Rectifier, electric: Electrical device for converting ac to dc.
Recuperative coil: Secondary coil in glycol water forced-air furnace that extracts latent heat from combustion gases.
Recycling: Passing of flue gases from combustion in a furnace to a secondary heat exchanger to remove latent heat.
Reed valve: Compressor valve consisting of a thin, flat, high-carbon alloy steel.
Refractory: A material with a high melting point.

system and indicates presence of gas bubbles in liquid line.

Silica gel: Absorbent chemical compound used as a drier. When heated, moisture is released and compound may be reused.

Silicon-controlled rectifier (SCR): Electronic semiconductor that contains silicon. Controls current by timing pulses.

Silver brazing: Brazing process in which brazing alloy contains some silver.

Sine wave: Wave form of single frequency alternating current.

Single-pass recycling machine: A recycling machine in which the refrigerant is passed through a filter-drier once.

Single-phase motor: Electric motor that operates on single-phase alternating current.

Single-pipe system: System of steam heating in which a single pipe carries steam to radiator and is also used as a condensate return.

Single-pole, double-throw switch, (SPDT): Electric switch with one blade and two contact points.

Single-pole, single-throw switch, (SPST): Electric switch with one blade and one contact point.

Single-stage compressor: Compressor having only one compressive step between inlet and outlet.

Skin condenser: Condenser using the outer surface of the cabinet as the heat radiating medium.

Sleeve covers: The top opening cover on an ice cream cabinet.

Sling psychrometer: Measuring device with wet and dry bulb thermometers. Moved rapidly in air, it measures relative humidity.

Slip ring lubricating method: A lubricating method in which a brass ring lubricates the bearing.

Slug: 1—Unit of mass equal to the weight of object (in pounds) divided by 32.2 (acceleration due to the force of gravity). 2—Detached mass of liquid or oil which causes an impact or hammer in a circulating system.

Slugging: Condition in which a mass of liquid enters the compressor, causing hammering.

Smoke test: Test made to determine completeness of combustion.

Snow making: The process of producing artificial snow by means of a water spray into which compressed air is added, creating a fine mist that freezes rapidly.

Solar cell: Device that converts solar radiation directly to electricity. Also known as a *Photovoltaic cell.*

Solar collector: Device used to trap solar radiation, usually using an insulated black surface.

Solar energy: Energy contained in sunlight.

Solar energy systems: Systems used to collect, convert, and distribute solar energy in forms useful within a business or residence. A passive system uses no additional energy from other sources for the distribution of the solar generated heat. An active system may use blowers, supplementary coils, etc.

Solar heat: Heat created by energy waves from the sun.

Soldering: Joining two metals by adhesion of a metal with a melting temperature of less than 800°F (427°C).

Solenoid valve: Electromagnet with a moving core. It serves as a valve or operates a valve.

Solid fuel heating: The use of solid natural resources such as wood or coal to provide heat.

Solid-state electronic relays: See *Electronic relays.*

Solubility: The tendency of a substance to dissolve in another substance.

Solution: A liquid that has another liquid or solid completely dissolved in it. A lithium bromide water solution, commonly used in absorption systems, is water with lithium bromide dissolved in it. "Strong" and "weak" solutions are those with respectively high and low concentrations of another liquid or solid.

Sone: Sound loudness rating.

Sound tracer: Instrument that helps locate sources of sound.

South Pole, magnetic: That part of magnet into which magnetic flux lines flow.

Specific gravity: Weight of a liquid relative to water.

Specific heat: Ratio of quantity of heat required to raise temperature of a body 1° to that required to raise temperature of equal mass of water 1°.

Specific heat capacity: The amount of heat required to change a given mass of a substance from one temperature to another.

Specific volume: Volume per unit mass of a substance.

Splash system, oiling: Method of lubricating moving parts by agitating or splashing oil in the crankcase.

Split-phase motor: Motor with two stator windings. Both windings are in use while starting. One is disconnected by centrifugal switch after motor attains speed. Motor then operates on other winding only.

Split system: Refrigeration or air conditioning installation that places condensing unit outside or away from evaporator. Also applicable to heat pump installations.

Spray cooling: Method of refrigerating by spraying expendable refrigerant or by spraying refrigerated water.

Squirrel cage: Fan that has blades parallel to fan axis and moves air at right angles or perpendicular to fan axis.

Standard air: Air having a mass density of 0.075 lb./ft^3 (1.204 kg/m^3), a temperature of 70°F (21°C), and a pressure of 30″ Hg (760 mm Hg).

Standard atmosphere: Condition when air is at 14.7 psia pressure, at 68°F (20°C) temperature and a relative humidity of 36%.

Standard conditions: Used as a basis for air conditioning calculations: temperature of 68°F (20°C), pressure of 29.92″ of mercury (Hg), and relative humidity of 30%.

Standing pilot: Old system of furnace burner ignition, in which a pilot light is constantly burning.
Start capacitor: A capacitor used in the starting winding only.
Starve: Condition in which there is not enough refrigerant reaching the evaporator.
Starting relay: Electrical device that connects and disconnects starting winding of electric motor.
Starting winding: Winding in electric motor used only briefly, while motor is starting.
Static electricity: Electricity at rest.
Static head: Vertical piping run.
Static system: A system in which air is circulated through the condenser by natural convection.
Stationary blade compressor: Rotary pump that uses a nonrotating blade inside pump to separate intake chamber from exhaust chamber.
Stator, motor: Stationary part of electric motor.
Steam: Water in vapor state.
Steam heating: Heating system in which steam from a boiler is piped to radiators.
Steam jet refrigeration: Refrigerating system which uses a steam venturi to create high vacuum (low pressure) on a water container causing water to evaporate at low temperature.
Steam trap: Automatic valve that allows condensate to pass while preventing passage of steam.
Sterling cycle: A refrigeration cycle that can produce temperatures down to −450°F (−268°C).
Stethoscope: Instrument used to detect sounds and locate their origin.
Strainer: Device such as a screen or filter used to retain solid particles while liquid passes through.
Stratification: Condition in which there is little or no air movement in room; air lies in temperature layers.
Stroke: The distance traveled by a piston.
Subcooling: Cooling of liquid refrigerant below its condensing temperature.
Sublimation: Condition where a substance changes from a solid to a gas without becoming a liquid.
Substance: Any form of matter or material.
Suction line: Tube or pipe used to carry refrigerant gas from evaporator to compressor.
Suction pressure control valve: Device located in the suction line that maintains constant pressure in evaporator during running portion of cycle.
Suction service valve: Two-way, manually operated valve located at the inlet to compressor. It controls suction gas flow and is used to service unit.
Suction side: Low-pressure side of the system extending from the refrigerant control through the evaporator to the inlet valve of the compressor.
Superheat: 1—Temperature of vapor above its boiling temperature as a liquid at that pressure. 2—The difference between the temperature at the evaporator outlet and the lower temperature of the refrigerant evaporating in the evaporator.
Superheater: Heat exchanger arranged to take heat from liquid going to evaporator and use it to superheat vapor leaving the evaporator.
Surge: Regulating action of temperature or pressure before it reaches its final value or setting.
Surge tank: Container connected to the low-pressure side of a refrigerating system which increases gas volume and reduces rate of pressure change.
Swaging: Enlarging one tube end so the end of another tube of the same size will fit within it.
Swamp cooler: Evaporative type cooler in which air is drawn through porous mats soaked with water.
Swash plate: Device used to change rotary motion to reciprocating motion. Used in some refrigeration compressors.
Sweating: 1—Condensation of moisture from air on cold surface. 2—Method of soldering in which the parts to be joined are first coated with a thin layer of solder.
Sweet water: Term sometimes used to describe tap water.
Synchronous speed: A speed equal to that of a rotating magnetic field.
Synthetic: Produce through synthesis, not natural occurring.
Synthetic dust weight arrestance: Measurement of filter's ability to remove synthetic dust from test air.

T

Tail pipe: Outlet pipe from the evaporator.
Tandem: A term used to identify a system that connects two motor compressors at the motor end.
Tank, supply: Separate tank connected directly or by a pump to the oil-burning appliance.
Tap (screw thread): Tool used to cut internal threads.
Temperature: 1—Degree of hotness or coldness as measured by a thermometer. 2—Measurement of speed of motion of molecules.
Temperature-humidity index: Actual temperature and humidity of air sample compared to air at standard conditions.
Temperature sensing bulb: Bulb containing a volatile fluid and bellows or diaphragm. Temperature increase on the bulb causes the bellows or diaphragm to expand.
Test light: Light provided with test leads. Used to test or probe electrical circuits to determine if they are working properly.
Therm: Quantity of heat equal to 100,000 Btu.
Thermal conductivity: The ability of a material to transfer heat.
Thermal precipitation: The collection of dirt around warm air grilles.

Thermal relay (hot wire relay): Heat-operated electrical control used to open or close a refrigeration system electrical circuit. This system uses a resistance wire to convert electrical energy into heat energy.

Thermal resistance: See ***R-value.***

Thermistor: A semiconductor with electrical resistance that varies with temperature.

Thermocouple: Device that generates electricity, using the principle that if two unlike metals are welded together and the junction is heated, a voltage will develop across the open ends.

Thermodisc defrost control: Electrical switch with bimetal disc controlled by temperature changes.

Thermodynamics: Part of science that deals with the relationships between heat and mechanical action.

Thermoelectric refrigeration: Refrigerator mechanism that depends on Peltier effect. Direct current flowing through an electrical junction between unlike metals provides heating or cooling effect, depending on direction of current flow.

Thermometer: Device for measuring temperatures.

Thermomodule: Number of thermocouples used in parallel to achieve low temperatures.

Thermopile: Number of thermocouples used in series to create a higher voltage.

Thermostat: Device which senses ambient temperature conditions and, in turn, acts to control a circuit.

Thermostat droop: Added heat in line-voltage thermostat produced by the thermostat itself.

Thermostatic control: Device which operates system or part of system based on temperature change.

Thermostatic expansion valve (TEV): Control valve operated by temperature and pressure within evaporator. It controls flow of refrigerant. Control bulb is attached to outlet of evaporator.

Thermostatic motor control: Device used to control cycling of unit through use of control bulb. Bulb reacts to temperature changes.

Thermostatic switch: A switch controlled by temperature changes.

Thermostatic valve: Valve controlled by temperature change response elements.

Thermostatic water valve: Valve used to control flow of water, actuated (made to work) by temperature difference. Used in units such as water-cooled compressors and condensers.

Three-phase: Operating by means of combination of three alternating current circuits that differ in phase by one-third of a cycle.

Three-way valve: Flow control valve with three fluid flow openings.

Throttling: Expansion of gas through orifice or controlled opening without gas performing any work as it expands.

Throw: The distance air travels from a grille before slowing to 50 ft./minute.

Time-delay fuse: A fuse that does not blow until the overload has persisted for a set duration of time (normally about 10 seconds).

Timed On-Off control: Control needed when the existing differential is too great.

Timers: Clock-operated mechanism used to control opening and closing of an electrical circuit.

Timer-thermostat: Thermostat control which includes a clock mechanism. Unit automatically controls room temperature and changes temperature range depending on time of day.

Ton of refrigeration: Refrigerating effect equal to the melting of 1 ton of ice in 24 hours. This may be expressed as follows: 288,000 Btu/24 hr., 12,000 Btu/1 hr., 200 Btu/min.

Ton refrigeration unit: Unit that removes same amount of heat in 24 hours as melting of one ton of ice.

Torque: Turning or twisting force.

Torque, full load: Maximum torque delivered without overheating.

Torque, stall: Torque developed when starting.

Torque, starting: Amount of torque available to start and accelerate the load.

Torque wrenches: Wrench which may be used to measure torque or pressure applied to a nut or bolt.

Torr: A unit of pressure equal to 1/760 of an atmosphere (1 mm Hg), normally used for measuring vacuum pressure.

Total air balance (TAB): In an air circulation system, adjusting the system so that all rooms receive the proper amount of air.

Total Energy Management (TEM): Conservation concept where a building is looked at in terms of its total energy usage, rather than analyzing the requirements of separate systems.

Total heat: Sum of both the sensible and latent heat.

Toxicity: A measure of the amount of poison in a substance or the amount of harm it can cause.

Transducer: Device turned on by a change of power from one source for the purpose of supplying power in another form to a second system.

Transformer: Electromagnetic device that transfers electrical energy from the primary circuit into variations of voltage in a secondary circuit.

Transformer-rectifier: Combination transformer and rectifier in which input ac current may be varied and then rectified into dc current.

Transistor: Electronic device commonly used for amplification. Similar in use to electron tube. Depends on conducting properties of semiconductors in which electrons moving in one direction are considered as leaving holes that serve as carriers of positive electricity in opposite direction.

Transmission: Heat loss or gain from a building through exterior components such as windows, walls, floors, etc.

Triac: Three-lead semiconductor that allows current flow in two directions when a preset voltage is applied at one of the leads.

Trichlorotrifluoroethane: Complete name of R-113. Group A1 refrigerant in common use. Chemical compounds that make up this refrigerant are chlorine, fluorine, and ethane.

Triple point: Pressure-temperature condition in which a substance is in equilibrium (balance) in solid, liquid, and vapor states.

Troposphere: Part of the atmosphere immediately above the earth's surface in which most weather changes occur.

Troubleshooting: The analysis of a problem.

Truck, refrigerated: Commercial vehicle equipped to maintain cool temperatures.

Tube-within-a-tube condenser: Water-cooled condensing unit in which a small tube is placed inside a large unit. Refrigerant passes through the outer tube, water passes through the inner tube.

Tubing: Fluid-carrying thin-walled pipe.

Twin parallel: Two or more units installed in line by piping.

Two-pipe system: A heating system in which one pipe delivers steam to radiators and a second pipe is used to return condensate.

Two-position On-Off control: A control system in which the control device can only start and stop the equipment.

Two-stage vacuum pump: A vacuum pump used to remove vapor and moisture from a system.

Two-temperature valve: Pressure-opened valve used in suction line on multiple refrigerator installations that maintains evaporators in system at different temperatures.

Two-way valve: Valve with one inlet port and one outlet port.

U

U-value: Represents the heat leakage from one side of a wall to the other.

Ultrasonic: Having a frequency above the range of human hearing.

Ultraviolet: Invisible radiation waves with frequencies shorter than wave lengths of visible light and longer than X rays.

Unitary system: A factory assembled heating/cooling system in one package and usually designed for conditioning one space or room.

Universal motor: Electric motor that will operate on either ac or dc.

Unloader: A device that allows for easier compressor start-up by temporarily reducing high-side pressure at the cylinder head.

Upflow furnace: A furnace in which air is forced up through the furnace from the bottom.

Urethane foam: Type of foam insulation that is placed between the inner and outer walls of a container.

V

Vacuum: Pressure lower than atmospheric pressure.

Vacuum activators: Dampers and control valves used in automotive air conditioning system; controlled by the vacuum created by engine intake manifold vacuum.

Vacuum control system: Intake manifold vacuum is used to operate dampers and controls in some automobile systems.

Vacuum pump: Device used for creating vacuums for testing or drying purposes.

Valve: Device used for controlling fluid flow.

Valve, expansion: Type of refrigerant control that maintains constant pressure in the low side of refrigerating mechanism. Valve is caused to operate by pressure in low or suction side. Often referred to as an automatic expansion valve, or AEV.

Valve, service: Device used to check pressures and charge refrigerating systems.

Valve, solenoid: Valve made to work by magnetic action through an electrically energized coil.

Valve, suction: Valve in refrigeration compressor that allows vaporized refrigerant to enter cylinder from suction line and prevents its return.

Valve, water: In most water cooling units, a valve that provides a flow of water to cool the system while it is running.

Valve plate: Part of the compressor located between the top of compressor body and the head. It contains compressor valves and ports.

Vapor: A gas that is often found in its liquid state while in use. The preferred name for vaporized refrigerant.

Vapor, saturated: Vapor condition that will result in condensation into droplets of liquid if temperature is reduced.

Vapor barrier: Thin plastic or metal foil sheet used to prevent water vapor from penetrating insulating material.

Vaporization: Change of liquid into a gaseous state.

Vapor lock: Condition where liquid is trapped in a line because of a bend or improper installation. Such vapor prevents liquid flow.

Vapor pressure: Pressure imposed by a vapor.

Vapor pressure curve: Graphic presentation of various pressures produced by refrigerant under various temperatures.

Vapor recovery method: A refrigerant recovery system using relatively small equipment to remove refrigerant from residential, automobile, and light commercial units.

Vapor velocity: Speed or rate at which a gas moves.

Variable Air Volume (VAV) controller: Device having electronic components used to regulate the volume of air in a distribution system.

Variable control: A control that can make gradual adjustments to equipment, as opposed to a simple on-off control.

Variable pitch pulley: Pulley that can be adjusted to provide different pulley drive ratios.

VAV: See ***Variable air volume.***

V-belt: Belt commonly used in refrigeration work with a contact surface and pulley in a V-shape.

Velocimeter: Instrument that measures air speeds.

Velocity: Quickness of motion; swiftness; speed. Change in position with respect to time.

Ventilation: Airflow from one area to another.

Vibration absorbers: Soft or flexible substance or device that will reduce the transmission of a vibration.

Viscosity: Resistance to flow.

V_{max}: Maximum (peak) voltage in alternating current cycle.

Volatile: Easily vaporized; a liquid that changes to the gaseous state readily.

Voltage: 1—Term used to indicate the electrical potential or electromotive force in an electrical circuit. 2—Voltage or electrical pressure which causes current to flow. 3—Electromotive force.

Voltmeter: Instrument for measuring voltage in electrical circuit.

Volumetric efficiency: Term used to express the relationship between the actual performance of a compressor or of a vacuum pump and theoretical performance of the pump based on its displacement.

Vortex tube: Mechanism for cooling or refrigerating that accomplishes cooling effect by releasing compressed air through a specially designed tube.

V_{rms}: Root mean square voltage; average voltage equal to the maximum voltage multiplied by a constant.

W

Walk-in cooler: Larger, commercially refrigerated space. Often found in supermarkets or wholesale meat distribution centers.

Water-cooled condenser: See ***Condenser, water-cooled.***

Water coil: A coil submerged in water to take advantage of the water's temperature.

Water defrosting: Use of water to melt ice and frost from an evaporator.

Water hammer: Noise generated by back pressure of water when a valve is closed.

Water spray: A filtering device in which air is blown through a spray of water, removing solid contaminants, liquid contaminants, and water-soluble gas contaminants.

Watt: Unit of electrical power.

Wax: An ingredient in many lubricating oils that may separate from the oil.

Wet bulb hygrometer: See ***Psychrometer.***

Wet bulb temperature: Measure of the degree of moisture. It is the temperature of evaporation for an air sample.

Wet cell battery: Cell or connected group of cells that converts chemical energy into electrical energy by reversible chemical reactions.

Wet roof cooling: Method of heat reduction in a building in which a pond of water is kept on the roof at all times. Heat from the sun is dissipated through the evaporation of water, so the roof surface remains cool.

Wheatstone bridge: Electronic circuit consisting of resistors and a thermistor. A temperature change causes the bridge to become unbalanced, which sends a signal to the output device.

Wick system: Lubricating method used in external-drive motors.

Wind chill index: A measure of how cold it feels, based on dry bulb temperature and wind velocity.

Window unit: Air conditioner that is placed in a window.

Wobble plate: See ***Swash plate.***

Work: A force applied over a distance.

Work hardening: Increasing the strength of a material (normally a metal) by bending and deforming it.

Z

Zeotropic: A refrigerant blend, comprised of various refrigerants, that changes in volumetric composition and saturation temperature when used.

Zero ice: Trade name for dry ice. See ***Dry ice.***

Zone controls: Controls used to maintain each specific area or zone within a building at a desired condition. This is a type of distribution control often used in hydronic heating system.

ACKNOWLEDGMENTS

The production of a book of this nature would not be possible without the cooperation of the refrigeration and air conditioning industry along with other associations, groups, and individuals. In preparing the manuscript for **Modern Refrigeration and Air Conditioning,** industry and related groups have been most cooperative. The authors and publisher acknowledge the cooperation of these people and associations with great appreciation:

A-1 Compressors, Inc.
ABB Stal Refrigeration Corporation
Abbeon Cal, Inc.
Acme Electric Corp.
Action Instruments, Inc.
AC & R Components, Inc.
Addison Products Company
Aeroquip Corp.
Air Balance Inc.
Air Conditioning Contractors of America (ACCA)
Air-Conditioning and Refrigeration Institute (ARI)
Airserco Mfg. Co.
Aitken Products, Inc.
Alco Controls Division, Emerson Electric Co.
Allied Signal, Inc.
Alnor Instrument Co.
Amana Refrigeration, Inc.
American Metal Products Co., A Masco Company
American Panel Corporation
American Saw & Mfg. Company
American Society of Heating, Refrigerating, and Air-Conditioning Engineers (ASHRAE)
American Ultraviolet Co.
Americold Compressor Division of White Consolidated Industries, Inc.
Amprobe Instrument
ANAMET Industrial, Inc., An ANAMET Company
Andersen Corp.
Anemostat Products Div.
Armstrong Air Conditioning, Inc.
Auto Flo
Bacharach, Inc.
Bally Engineered Structures, Inc.
Baltimore Aircoil Company
Barber-Colman Co.
Beckett Corporation
R. W. Beckett Corp.
BOHN Refrigeration Products, a Unit of Heatcraft, Inc.
BOTSBALL, Howard Mfg. & Consulting, Inc.
Bristol Babcock, Inc.
Bristol Compressors
Bryant Air Conditioning/Heating
Buchbinder, Chicago, IL
Buick Motor Div., General Motors Corp.
Burnham Corporation
Cadillac Motor Car Div., General Motors Corp.
Carlin Combustion Technology, Inc.
Carlyle Compressor Company, Division of Carrier Corp.
Carrier Corp.
Carrier Corp., Replacement Components Division
Carrier Corp., Subsidiary of United Technologies Corporation
Carrier Transicold Division, Carrier Corp.
Chatleff Controls, Inc.
Chicago Valve Plate & Steel Co.
Chrysler Corporation
David Clark Co., Inc.
Cleveland Twist Drill Co.
The Coleman Co., Inc.
Comfortmaker, Inter-City Products Corp. (USA)
Compressed Air magazine
Conservation Energy Systems
Consolidated Industries Corporation
Control Systems International
Cooper Instrument Corporation
Cooper Tools, Nicholson
Copeland Corporation
CPS Products, Inc.
Dairy Equipment Co.
Danfoss Automatic Controls, Division of Danfoss, Inc.
DEC International, Inc.
Des Champs Laboratories, Inc.
DESA International

Detroit Edison Co.
Detroit Public Schools
Dodge Engineering
Dongan Electric Mfg. Co.
Dunham-Bush, Inc.
DuPont Company
DuPont Energy Management Co., Inc.
Duro/Indestro, Duro Metal Products Co.
Dwyer Instruments, Inc.
Eaton Corp.
Eaton Corp., Controls Division
Ebco Manufacturing Company
Edison Electric Institute
Edwards Engineering Corp.
Eklind Tool Co.
Electro-air/White-Rodgers Division,
Emerson Electric Co.
Electrolux AB
Elkay Mfg. Co.
Emerson Electric Co.
Enerstat Corporation
The ESAB Group, Inc.
Evcon Industries, Inc.
Farr Company
Fasco Motors Group
Fedders North America
Fireye
Flex-L International, Inc.
Florida Heat Pump Division, a Harrow Company
Fluidex Division, Parker-Hannifin Corp.
Fluke Corporation
Ford Customer Service Div.
Ford Motor Co.
Frick Co.
Frigidaire Company
Fusite Division, Emerson Electric Co.
The Gates Rubber Co.
Gem Products, Inc.
General Electric Co.
General Filters, Inc.
General Motors Corporation
Goodman Manufacturing Corporation
Goodway Technologies Corp.
Goss Inc.
Grainger, Division of W.W. Grainger, Inc.
Grasso, Inc.
Grundfos Pumps Corporation
Hackney Brothers Body, Wilson, NC
Hanson Technologies Corporation
Harrison Radiator Div., General Motors Corp.
Hart & Cooley, Inc.
Harvey-R.J. Instrument Corp.
Heat Controller, Inc.
Heatcraft, Inc.
Heatcraft, Inc., Refrigeration Products Division
Heil Heating and Cooling Products, Inter-City
Products Corp. (USA)
Henry Valve Co.
Hobart Corp.
Home Planners, Inc.
Honeywell Inc.
Hoshizaki America, Inc.
Hotpoint Div. General Electric Co.
Howe Corporation
Edward Hulyk Studio
Hussmann Corporation
HyCal Unit of General Signal
Hypertherm, Incorporated
Ice-O-Matic
ICI Americas, Inc.
Imperial Eastman, Imperial Division
Inter-City Products Corporation (USA)
ITT Fluid Handling Sales
ITT McDonnell & Miller
ITW Vortec Corporation
Jarrow Products, Inc.
Johnson Controls, Inc.
Johnson Controls, Inc., Systems & Services Div.
Klein Tools, Inc.
Knauf Fiber Glass GmbH
Kold-Draft Division, Uniflow Manufacturing Company
Koolatron Industries
Kramer Trenton Co.
Kysor/Warren Division of Kysor Industrial Corp.
Landis & Gyr Powers, Inc.
Larchmont Engineering
Lennox International, Inc.
Leybold Inficon, Inc.
The Lincoln Electric Company
Lucas-Milhaupt, Inc., A Handy & Harman Company
Lutron Electronics Co., Inc.
MagneTek
Maid-O'-Mist Div.
Maneurop Inc.
The Marley Cooling Tower Company
Marley Electric Heating, A United Dominion Company
Marsh Instrument Co.
Mast Development Co.
Maurey Mfg. Corp.
Maxitrol Company
McQuay International
McQuay, SnyderGeneral Corp.
Metrosonics, Inc.
Wm. W. Meyer & Sons, Inc.
Michigan Farmer
Micro Switch, Div. of Honeywell, Inc.
Midco International, Inc.
Mitsubishi Electronics America, Inc.
Mitsubishi Heavy Industries, Ltd.
Monarch Nozzle Company
Paul Mueller Company
Mueller Refrigeration Products Co., Division of
Mueller Industries, Inc.
Murray Corp.

Mycom Corp.
National Electrical Contractors Association
National Electrical Manufacturers Association (NEMA)
National Institute for Automotive Service Excellence (ASE)
National Refrigerants, Inc.
National Refrigeration Products
NIBCO, Inc.
Nor-Lake, Incorporated
Numbers, 1985, Altadena, CA, by Holladay and Otterholm
Owens-Corning Fiberglas Corp.
Packless Industries
Paragon Electric Co., Inc.
Parker-Hannifin Corporation, Seal Group
Parker-Hannifin Refrigerant Components Group
Peerless of America, Inc.
Penquin Refrigeration, Inc.
Pontiac Motor Division, General Motors Corp.
PSG Industries, Inc.
QMark, A Division of Marley Electric Heating
Qualimetrics, Inc.
Ranco North America
Recycling Specialists International
Reed Manufacturing Co.
Refrigerant Management Systems, Inc.
Refrigerating Specialties Div., Parker-Hannifin Corp.
Refrigeration & Air Conditioning Division, Parker-Hannifin Corp.
Refrigeration Research, Inc.
Refrigeration Service Engineers Society (RSES)
Refrigeration Technologies
Revcor, Inc.
Rheem Mfg. Co., Air Conditioning Div.
Rheem Mfg. Co., Scientific Products Div.
Ridge Tool Company
Ritchie Engineering Company, Inc.
Robertshaw Controls Co.
Robinair Division, SPX Corporation
Robur Corporation
Rosemount Analytical, Inc.
Rubatex Corp.
S-T Industries, Inc.
Sani-Serv
Scale Free Systems, Inc.
Schaefer Brush Mfg. Co., Inc.
Scientific Systems Corporation
Scotsman Ice Systems
Seal-Tite
Sealed Unit Parts Co., Inc.
Sensidyne, Inc.
SIBIR
Siebe Environmental Controls
Sierra Instruments, Inc.
Silver King Division of Stevens Lee Company
Simpson Electric Company
Skuttle Mfg. Co.
A.O. Smith Electrical Products Co.
Snap-on Tools Corp.
Solar Components Corporation
Solarex Corporation
Southeast Oakland Vocational Education Center
Spectronics Corporation
A.W. Sperry Instruments, Inc.
Sporlan Valve Company
Sprinkool Systems, Inc.
Square D Company
Standard Refrigeration Co.
Suburban Manufacturing Company
Suntec Industries, Inc.
Superior Contract Services, Inc.
Superior Valve Co., Division of AMCAST Industrial Corporation
Sweden-Alco Dispensing Systems, A Div. of Alco Foodservice Equipment Co.
Taylor Company
Taylor Environmental Instruments
Tecogen, A Division of Thermo Power Corporation
Tectronic Products Company, Inc.
Tecumseh Products Co.
Telaire Systems, Inc.
Teletrol Systems, Inc.
Temprite Div. of Elkay Mfg. Co.
Tempstar Heating and Cooling Products, Inter-City Products Corp. (USA)
Testo, Inc.
Texas Instruments, Inc.
Thermal Engineering Company, Division of Seakay Co., Inc.
Thermatron Corporation
Thermo Products, Inc.
Therm-O-Disc, Inc.
TIF Instruments, Inc.
Tjernlund Products, Inc.
Torrington Research
Toyota Motor Corporation
The Trane Co.
The Trane Co., Unitary Products Group
Tranter, Inc.
Tranter, Inc., Edgefield Division
Tranter, Inc., Texas Division
H.O. Trerice Co.
Trion, Inc.
Trol-A-Temp, Division of Trolex Corporation
TRW Greenfield Tap & Die Div.
TSI Incorporated
Tutco, Inc.
Tyler Refrigeration Corp.
Unique Air, Inc.
United States Environmental Protection Agency, U.S. Dept. of Health and Human Services
Uniweld Products, Inc.
Utica Boilers, Inc.
Vaisala, Inc.

F

G

H

I

M

N

O

P

Q

R

S

T

U

V